D1104011

HANDBOOK OF CHLORINATION AND ALTERNATIVE DISINFECTANTS

HANDBOOK OF CHLORINATION AND ALTERNATIVE DISINFECTANTS

Fourth Edition

Geo. Clifford White
Consulting Engineer

A Wiley-Interscience Publication

JOHN WILEY & SONS, INC.

New York / Chichester / Weinheim / Brisbane / Singapore / Toronto

This book is printed on acid-free paper. ♾

Published simultaneously in Canada.

Library of Congress Cataloging-in-Publication Data:

White, George Clifford.
Handbook of chlorination and alternative disinfectants / Geo. Clifford White.—4th ed.
p. cm.
Includes bibliographical references.
ISBN 0-471-29207-9 (hardcover)
1. Water—Purification—Chlorination. 2. Sewage—Purification—Chlorination. 3. Water—Purification—Disinfection. I. Title.
TD462.W47 1999 97-30189
628.1′662—dc21

Printed in the United States of America.

10 9 8 7 6 5 4 3 2 1

For my wife Augusta
and
all of my colleagues in this business
since April 1, 1937

CONTENTS

PREFACE TO THE FOURTH EDITION [1998]

Only a few years have passed since the publication of the third edition of this book. The catalyst for publishing this fourth edition was that the publisher, VNR, since acquired by John Wiley & Sons, Inc., received a letter in late 1995 from Ashland Chemical Co., the people who depend upon chlorine dioxide for their activities, asking when the fourth edition was to be published.

When Nancy Olsen of VNR contacted me for a decision, I thought of all kinds of reasons to put out a fourth edition. First, my consulting activities in the United States had dropped off considerably due to the overspending by the utilities that was required by more stringent federal regulations. So I could keep myself busy doing something useful. Little did I know how complicated this project would be.

During late 1993 and quite a bit of 1994, I spent a lot of time investigating the work of Dr. Bruce Ames' cancer research group at the University of California in Berkeley, California. In a few words Dr. Ames proved conclusively to me that THMs from chlorinated water were not and never would be a public health risk. This conclusion tells me that the biggest mistake the EPA made was believing the outmoded Harris Report (circa 1975). This report concluded that slightly less than lethal chlorine residuals caused tumors in rats! In view of this I wrote a report for my "overseas" clients titled "Degradation of Water Quality by Federal Mandate," in which I made the point that THMs are just trivia, which they are. See the Special Preface for more information.

In addition to the EPA–THM challenge there was so much new and more modern equipment, many more active companies, and a considerable number of company mergers. In addition to all this, the entire planet is going into a problem of water shortage due to overpopulation. This was one of my major priorities in the fourth edition because nature is now producing powerful mutated species of pathogenic organisms that penetrate our water supplies and can cause lethal results for the young and the elderly, or for anyone with immune deficiency. The planet, according to United Nations research, will be deficient in drinking water around the year 2025, and according to local researchers, California will be out of water by the year 2015.

Putting this fourth edition together has required an enormous amount of work, but I have enjoyed every minute of it. I take this opportunity to thank all of you who have contributed to this fourth edition in one way or another. I am most grateful for your help.

San Francisco G.C.W.

SPECIAL PREFACE TO THE FOURTH EDITION

WATER: EARTH'S MOST IMPORTANT NATURAL RESOURCE

In the last decade, we have seen the effect of a continuous increase in the population of the United States, as well as in Canada. This is partially due to an influx of immigrants from throughout the world searching for new opportunities. The nonimmigrant population appears to be growing as well.

There is also a steady increase in population throughout the world, mostly in the Third World countries. In addition to the CDC in Atlanta, Georgia, agencies such as the Pan American Health Organization, NATO, the World Health Organization, and others have been carefully following this population increase, pointing out that by about 2025 the projected world population of 8.9 billion people will be all that the planet's water supply can take care of, simply because there is no more water on Planet Earth now than there was during the days of Queen Cleopatra. White's own experience over many years of work around the world has demonstrated that the procedures and formulas used for estimating future populations are amazingly accurate.

One disturbing factor in analyzing future population figures is that California is growing faster than the rest of the world. The California State Water Resources Board has predicted that the State will run out of water in about the year 2015. This is one reason why there are several water reclamation plants under design now (1997).

Of all the water on earth, about 97.2 percent is in the oceans. This is the prime source of water used by humans, after its evaporation by the sun. Another 2.2 percent is locked in ice caps and glaciers. This leaves only 0.6 percent of all the earth's water as fresh water for human use via cities, industry, and agriculture.[1, 2]

The earth's moisture is in continuous circulation. This process is usually described as the water cycle, or the hydrologic cycle. It has been estimated that about 80,000 cubic miles of water from the oceans on the earth and 15,000 cubic miles from lakes and land surfaces evaporate annually. The total evaporation is equaled by the total precipitation, of which about 24,000 cubic miles falls on land surfaces. Microorganisms of various kinds are present at different stages in this cyclic process, which occurs in atmospheric, surface, and underground waters.

The Prelude to Population Control

The current trend in the United States is to be concerned with the condition of the environment and endangered species, instead of evaluating the importance of public health. One reason for this is lack of proper information on the status of Planet Earth at the end of the twentieth century. The word "environment" has, for all practical purposes, pushed aside the thought of public health issues, with people assuming all the time that the public health situation is well taken care of—except maybe for AIDS.

The Current Situation. Werner Fornos,[3] the president of the Washington-based Population Institute reported that the world's population is growing by almost 90 million annually, a much slower rate than the 100 million annual growth of recent years. But he also pointed out that the current 90 million rate is far too high for the planet to support.

As of January 1, 1997, the world's population was nearing 5.9 billion, and it was expected to be more than 6 billion by the year 2000. Fornos said it is possible to stabilize the population at 8 billion by 2025 (see Figure below) if people keep doing what works. He also indicated that he was encouraged by the present slowing of the birth rate, which he attributed in part to better education and job opportunities for girls and women.

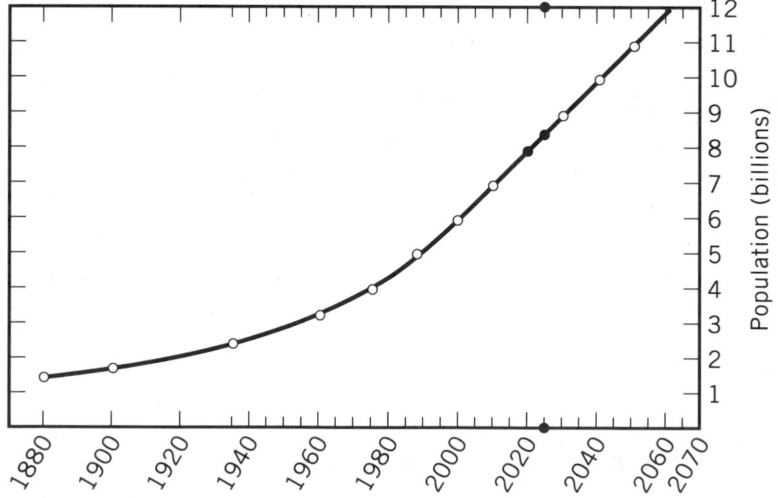

Planet Earth population curve (by White, 1993).

Fornos also pointed out that the 1994 world population and development conference, in Cairo, Egypt, had led to genuine progress in making voluntary family planning available for more couples. This report also found that some nations, including Thailand, Turkey, and Brazil, had sharply reduced their growth rates, but others, such as Nigeria, Ethiopia, and several Central American nations, were on course to double their population within 30 years.

Fornos also pointed out that other studies have found that the richer countries use a disproportionate share of the world's resources. For example, the United States has 4 percent of the world's population and uses 30 percent of the world's resources. Overall, rich countries have a fifth of the world's population, but they use four-fifths of the resources.

Nature Fights Back

To assist the planet in achieving a population compatible with its limited water supply, nature is providing more and more lethal organisms that cause waterborne diseases. It seems that these more lethal weapons appear in the news almost on a monthly basis (1997). As we create more antibiotics to protect ourselves against these diseases, nature allows these organisms to mutate. This increases the lethality of these microbiological organisms, causing more and more deaths. In addition, nature has created some powerful organisms that attack the immune systems of the very young and the elderly—another population control device.

What can we do? For the waterborne diseases we can protect our water supplies better by limiting the use of chloramination to its appropriate application as a distribution-system residual disinfectant, by applying powerful alternative disinfectants such as ozone and chlorine dioxide, and by better balancing the real need for microbial protection with concerns about the environmental and health effects of disinfection by-products. Earlier super-prechlorination methods, where a significant chlorine residual is maintained throughout any given water treatment plant from raw water intake to the effluent leaving the plant, need to be replaced.

A recent report concerning a series of experiments on water supplies in Third World countries found that the removal of *Cryptosporidium* oocysts by filtration was considerably higher when the raw water was prechlorinated to a significant free chlorine residual[2] (free chlorine has to contain 85 percent HOCl).

Important Features of Planet Earth[1, 2]

Whether life as we know it is unique to our planet is a question that has never been resolved. However, it is quite clear that the earth provides an excellent environment in which a rich diversity of living organisms has evolved. Of all the living organisms, none are more versatile than the microorganisms. Wherever higher organisms are present, microorganisms are never absent;

and in many environments that are devoid of, or are hostile to, higher organisms, the microorganisms exist and even flourish.

Owing to the fact that microorganisms cannot be seen with the naked eye, their existence in the environment is often unsuspected. Yet these unseen organisms carry out a number of functions vital to the life of higher organisms. Without microbes, the higher organisms would quickly vanish from the earth.

In their description of the Planet Earth, geologists divide it into three zones, the lithosphere, the hydrosphere, and the atmosphere. To these zones, we can add the biosphere. The lithosphere is the solid portion of the earth, composed of solid and molten rock and soil. The hydrosphere represents the aqueous environments such as the oceans, lakes, rivers, and the wetlands. Microorganisms abound in these habitats, from the tropics to both polar regions. The atmosphere is the gaseous region that surrounds the earth; it is relatively dense near the earth's surface, but it thins to nothing in the upper reaches (>100,000 ft). Although microbes are carried around the world on winds and other air currents (jet streams), they do not actually reproduce in the atmosphere.

The biosphere represents the mass of living organisms, found in a thin belt at the surface of the earth. These organisms have had, and continue to have, a most profound effect upon the earth itself. They are responsible for almost all of the oxygen found in the atmosphere, as well as for the enormous deposits of coal, iron, and sulfur found underground. Of almost equal importance, higher organisms and their corpses provide excellent microbial environments. Therefore, we find large populations of microbes associated with higher plants and animals.

In spite of the fact that the earth provides suitable environments for microbial growth, the same organisms are not found everywhere. Actually every environment, no matter how slightly it may differ from another, will have its own peculiar collection of microbes that differ from those in other environments. As microbes are indeed very small, so too are their environments. A single handful of soil will contain many microenvironments, with each providing conditions suitable for the growth of a restricted range of microorganisms. When thinking of these situations, we have to "think small."

New environments for microorganisms are continually being created. Many of them are the result of an unheard-of underground volcano somewhere in one of the vast oceans, the creation of a new lake after an enormous landslide, an earthquake, or the ravages of a forest fire caused by an unusual combination of a drought and a lightning storm. However, most of the new environments are artificial, produced by pollution of surface waters and lakes, the planting of exotic crops, the clearing of timber forests, the introduction of new pesticides and fertilizers, or the development of new water supplies for the growing population. Sometimes these new conditions make possible a further evolutionary step by the development of a new set of organisms. These new organisms are a part of the natural evolution of the planet and are in large part the result of nature's basic approach—described as its ability to establish a resistance to antibiotics developed by humans, as well as to other changes in the environment caused by humankind.

The great diversity of microbial life on our planet should not surprise us. It simply reflects the great diversity of habitats within which the microbes can grow and evolve. It must be remembered that these microorganisms are not passive inhabitants—they are very alert and reactive. Their activities affect their habitats and environments in a great many ways. Many cause disease in both humans and animals. The deleterious effects of these organisms are far worse than we would have predicted from their tiny sizes, including also food spoilage (molds), souring of milk, deterioration of clothing, and corrosion of drinking water piping systems. Foul tastes and off-color in potable water supplies are primarily due to microorganisms.

At the same time, microorganisms perform a great many important beneficial roles in nature. They are responsible for most of the decomposition of dead animal and plant bodies, thus returning important plant nutrients to the soil. Many of these organisms that live in the intestinal tracts of animals synthesize certain vitamins, thus freeing their hosts of the need to obtain these vitamins from their diets. Without microorganisms, animals such as cows, sheep, and goats would be unable to digest the cellulose of grass and hay, and they could not survive.

The Beginning of the Water Quality Scare

About 1970 the U.S. Public Health Service became concerned over an increase in waterborne disease outbreaks and possible surface water contamination by herbicides, pesticides, and other petrochemical products. In 1970, a community water supply survey discovered that the levels of dissolved organics in many water supplies exceeded the Public Health Service's recommended limit of 0.02 mg/L.[4] About the same time other government agencies began to examine various parts of the environment, making that a magic word. A few years later the Harris reports were released, which caused the transformation of the Public Health Service from an agency dedicated to the protection of human health to one emphasizing the protection of the environment. The focus changed from ensuring potable water quality to protect consumers' health to protection of the environment from pollution of all kinds, beginning with the protection of aquatic life by regulations issued by the Fish and Game Commission, another federal bureau.

In 1974, articles began to appear in the various technical journals blaming all types of human cancer on pollution of the environment.[5] Since then cancer researchers have found these conclusions to be flawed because a simplistic casual relationship was drawn between the development of tumors in rats and mice and the introduction of nearly lethal dosages of carcinogens.[6] One group of scientists tried very hard to prove that Americans were facing several natural disasters due to pollution of the environment. This group, from Oak Ridge Tennessee, was involved in a variety of projects relating to nuclear reactors, air and water pollution, and the possible environmental health effects of countless synthetic chemicals found in a variety of source waters; and it was headed by a great organizer and capable Ph.D. scientist, Robert Jolley.[7] These scientists focused on the detrimental effects of disinfection of potable water and wastewater by the use of chlorine, in spite of the fact that chlorine is the planet's near-universal water disinfectant. They were a great help to the EPA, as they wrote a series of books that were widely circulated in the United States and Europe. They concentrated their efforts on the examination of chloro-organics that entered surface waters from industrial discharges, sewage treatment plant effluents, agricultural area runoffs, and urban area runoffs.

In a 1977 publication,[7] Jolley's group emphasized that: "surface water pollutants identified thus far had ppb concentrations of highly chlorinated pesticides and hydrocarbons that represented a more serious problem with respect to water treatment, than does the presence of monochlorinated reaction products from sewage and cooling water treatment. Chlorinated pesticides and hydrocarbons have been found in some municipal drinking waters." This kind of information powerfully affected the people who were in the production of safe, high quality drinking water during several years of the group's activity, plus several editions of its book.

In the early 1970s, a portion of the prestigious U.S. Public Health Service was being transformed into the U.S. Environmental Protection Agency (EPA). Establishment of the EPA required a couple of years, involving several name changes. It did not take much time for the federal government, in Washington D.C., to set up the regulatory division of the EPA.

In 1972, the EPA reported that 46 organic chemicals were present in trace amounts in both the raw and finished water supplies at three locations along the Lower Mississippi,[8] and a 1974 EPA study identified 66 organic compounds in the New Orleans drinking water.[9] These were known as the Harris reports, and were most attractive to the EPA regulators, who focused on the dangers of the cancer-causing carcinogens in the chlorinated Mississippi River water serving the people in New Orleans and the surrounding area. These cancer data were based upon a primitive method of dosing rats and mice with trihalomethanes (THMs) in concentrations a mere fraction less than lethal to the rats and mice. This approach to cancer research is now outdated by at least 20 years. The aforementioned doses did cause tumors in the rats and mice, which caused the EPA to conclude that disinfection by chlorine caused cancer in humans. In this writer's opinion, the EPA could not have been "wronger."[5]

On the same day that the above report on the New Orleans was issued (Friday, November 8, 1974), Russell E. Train, Administrator of the EPA, announced that he was ordering an immediate nationwide survey to determine the concentration and potential health effects on certain organic chemicals in drinking water. Then, on December 18, 1974, Administrator Train named the 80 cities to be included in the National Organics Reconnaissance Survey.

The issue of chlorinated organics formed in the treatment of drinking water became the U.S. EPA's top priority in January of 1975. In 1974, the work of

Rook[10, 11] in the Netherlands and Bellar et al.[12] in the United States, demonstrated that chlorine used in some waters reacted with organic precursors to produce some trihalomethanes (mostly chloroform) in the finished water, which were not found in the respective raw water at the locations of study. This was thought to be a troublesome observation. To assess the general situation across the United States, a National Organics Reconnaissance Survey was conducted in 1975 by Professor James Symons and 10 coauthors.[13, 14] (See Table next page.)

This survey revealed trace amounts of four trihalomethanes (THMs): chloroform (trichloromethane, $CHCl_3$), bromodichloromethane ($CHCl_2Br$), dibromochloromethane ($CHClBr_2$), and bromoform (tribromomethane, $CHBr_3$) in the drinking water of 27 large cities in the United States. The average amounts in mg/L were as follows: $CHCl_3 = 0.39$, $CHCl_2Br = 0.0377$, $CHBr_3 = 0.0030$. Carbon tetrachloride, CCl_4, was not identified in 27 city water supplies; one had 0.002 mg/L and two had less than 0.002 mg/L CCl_4.

Regarding the NORS report, it is interesting to note that (1) only three metropolitan water systems exceeded the EPA interim maximum contaminate level of chloroform, (2) only two contained carbon tetrachloride, (3) ten water systems had 0.1 mg/L or less of chloroform, and (4) only 10 water systems exceeded the EPA limit of 4.0 mg/L of nonvolatile organic carbon. It is very difficult to think that the above figures started the water utilities' frenzy over the occurrence of DBPs. Furthermore this approach engendered the belief that human cancer was caused by pollution of the environment.[5]

We now know, owing to new and more reliable information, gathered in the last decade by experienced and reliable epidemiologists,[15–18] that chlorinated water was not and never could be a carcinogen.

The current (1997) EPA regulations are based upon outmoded information, created some two decades ago.[8, 19–22] A far better procedure would be for the EPA to forget the THMs and establish an acceptable chlorine demand for the finished water. In addition to this, owing to the increasing variety of pathogenic organisms, it would be more protective to choose a "consensus organism" instead of total coliforms for proof of disinfection. The chlorine dose and contact time needed to kill this more resistant organism would be controlled by a proven *Ct*. This would eliminate having to deal with THMs.

AWWA Water Quality Division Disinfection Committee Report

Survey of Water Utility Disinfection Practices.[23] This report was prepared by a 24-member committee headed by Chairman Charles N. Haas. The following is the committee's conclusions of their survey:

> Since the previous disinfection survey, there has been a steady shift in practices at many utilities. These changes have resulted from attempts to

Total THM Concentration in Treated Water from a Selected List of Metropolitan Areas from NORS Report[13]

Water Supply and Raw Water Source	*TTHM, mg/l*	*Water Supply and Raw Water Source*	*TTHM, mg/l*
Boston, MA		Concord, CA	
Norumbego Treatment Sta.	0.0048	Contra Costa Co. Water Dist.	
New York City		Calif. Water Proj. & San Joaquin Riv.	
Croton Reservoir	0.0299	Bollman Plant	0.0558
Little Falls, NJ		Atlanta, GA	
Passaic Valley Water Co.		Chattahoochee River Plant	0.0480
Passaic River	0.0710	Chattanooga, TN	
Philadelphia Water Dept.		Chattanooga River	0.0400
Torresdale Plant		Nashville, TN	
Delaware River	0.1060	Lawrence Plant	
Washington, DC		Cumberland River	0.0213
Dalecarlia Plant		Cincinnati, OH	
Potomac River	0.0512	Ohio River	0.0623
Baltimore, MD		Chicago, IL	
Loch Haven Reservoir		So. District Filter Plant	0.0293
Montebello Plant No. 1	0.0450	Indianapolis, IN	
Fairfax Co. Water Authority		White Riv. Plant & Wells	
Anandale, VA		White River	0.0420
New Lorton Plant		Detroit, MI	
Occoquan River Impoundment	0.0735	Waterworks Park Plant	
Miami, FL		Detroit River at Belle Isle	0.0244
Preston Plant		Milwaukee, WI	
Groundwater	0.4270	Howard Ave. Plant	
Ottumwa, IA		Lake Michigan	0.0191
Des Moines River	0.0008	Dallas, TX	
St. Louis, MO		Bachman Plant	
Central Plant		Trinity River, Elm Fork	0.0230
Missouri River	0.0722	Los Angeles, CA	
Denver, CO		Owens River Aqueduct	0.0510
Marston Plant		San Diego, CA	
Marston Lake	0.0270	Miramar Plant	
Cape Girardeau, MO		Colorado River Aqueduct	0.1040
Mississippi River	0.1413	San Francisco, CA	
Salt Lake City, UT		O'Shaugnessy Lake	
Mountain Dell Reservoir	0.0420	Yosemite Calif.	
Phoenix, AZ		San Andreas Filter Plant	0.0606
Verde Plant		Seattle, WA	
Salt & Verde Rivers	0.0440	Cedar River System	0.0159
Calif. Water Project		Cleveland, OH	
California Aqueduct		Division St. Filter Plant	0.0310
Sacramento & San Joaquin Rivers	0.0500		

balance maintenance of adequate disinfection with minimization of DBP concentrations in distributed water. Clearly, the diversity of utilities and source waters, has created a multitude of responses to these new challenges. On the basis of the current survey, the following major conclusions have been drawn:

- Utilities are making concerted efforts toward meeting THM regulations.
- To implement reductions in THMs, the most significant change by utilities was to alter the point of application of chlorine and the dosage of chlorine used.
- Many utilities have instituted changes in coagulation to minimize THM production.
- A sizable fraction (20 percent) of utilities have reported using ammonia addition as part of their THM minimization program.
- Some utilities have applied ClO_2 as a pre-oxidant or final disinfectant in order to reduce THM formation.
- Changes to reduce THM formation that involve major infrastructure investments, such as GAC contactors or ozone, have not been widely adopted.
- A sizable fraction of utilities that incorporated process modifications for THM reduction, reported operational problems as a result of the modifications.

In the years ahead, with changes both in regulation and in the state of knowledge, the committee anticipates further changes in practice. These changes will be reported in subsequent reports and surveys.

Disinfection and Alternative Disinfectants

With the onslaught of *Cryptosporidium* and the newer more virulent *Giardia,* people need to have available the most efficient disinfectant. However, when the EPA decided to make an enormous problem out of disinfection by-products (DBPs), it began a gradual degradation of the nation's water quality. Regulation of DBPs to parts per billion caused a great many utilities to switch to chloramines from the well-established free chlorine process that has worked extremely well. It takes about one hour or more of contact time for chloramines to achieve the same disinfection efficiency as free chlorine at 15 min. of contact time, when the chlorine-to-ammonia N weight ratio is 3 or 4 to 1. The worst part about chloramines comes after the chlorine residual disappears or declines so low that it has no stabilizing effect on the biological life in the distribution system. When this occurs, the remaining ammonia N acts as "caviar" for organisms that upsets the stability of the appropriate microbial growth.

Originally chloramines were used (1920s) to combat chlorinous tastes caused by industrial pollution of surface waters. In 1926 the terms "superchlorination" and "dechlorination" were used by Howard and Thompson[24] of Toronto, Canada. Superchlorination, when followed by dechlorination, removed the terrible taste and odors that occurred from normal chlorination procedures.

At about the same time, other water utilities were experimenting with using ammonia followed by chlorine. This did eliminate a lot of objectionable chlorinous tastes in some waters. After many years of frantic experimenting with ammonia and chlorine as well as chlorine dioxide, the breakpoint process was discovered in 1939. To perform the breakpoint process, a dosage of a 9 or 10 to 1 weight ratio of chlorine to ammonia N present in the water resulted in a final product that contained at least 85 percent free chlorine. The remaining 15 percent has been labeled as nuisance residuals. Almost all the ammonia N content is removed by this process. The higher the free chlorine percentage, the better the final product is. This, plus the appearance of the amperometric titrator several years later, provided a complete change in water disinfection practices as this device could accurately identify all the chlorine species in the total chlorine residual. However, some utilities never changed from a 3:1 chlorine-to-ammonia ratio, even though immediately following U.S. entry into World War II ammonia N was not available because of wartime requirements.

One of the problems with the chloramine process is getting complete mixing of the ammonia N. Best results were always obtained when it was added ahead of the chlorine. Under EPA regulations a water utility is allowed to use the free chlorine residual process if chlorine is added first, with only a few minutes of contact time, before addition of the ammonia N. This tends to bungle the chloramine process. Previously the preferred way to chlorinate at a treatment plant was to carry a chlorine residual throughout the plant and then dechlorinate the finished water to a distribution system residual of 0.5 to 0.75 mg/L free chlorine. This takes care of all the other unwanted biological problems that affect the other treatment plant processes. The result is a truly palatable water that meets all the California requirements.

Alternative disinfectants include chlorine dioxide (which was until recently banned in California owing to the chlorate and chlorite ions), ozone, bromine, iodine, and UV. Ozone is good for killing some viruses, but it is not a reliable disinfectant for two reasons: it is so volatile that its contact time is short, only a few minutes; and, most important, no one has been able to devise a reliable and acceptable automatic control system. Also ozone does such a good job of breaking up the organic matter that, in most cases, chlorine has to be added to the distribution system to maintain a proper biological balance. Bromine cannot qualify because it causes unacceptable-tasting water that is not palatable. Iodine could never be acceptable because there is only one major source for it. UV is extremely doubtful because it cannot kill anything that can be seen with the naked eye. To draw a sample of UV-treated water to be tested for coliforms, one should always use a pump. The pump will break up any protective solids that might have hidden the coliforms from the UV rays. The result will show the total disinfection efficiency.

REFERENCES

1. Brock, T. D., *Biology of Microorganisms,* 3rd ed., Prentice Hall, Englewood Cliffs, NJ, 1979
2. Pelczar, M. J., and Reid, R. D., *Microbiology,* 2nd ed., McGraw-Hill, New York, 1965.

3. Fornos, W., "Population Growth Slowing," Population Institute, Washington, DC, Dec. 28, 1996.
4. "Community Water Supply Study: Analysis of National Survey Finding," Bur. of Water Hygiene, Envir. Health Service USPHS, Dept. of HEW, Washington, DC, July 1970.
5. Epstein, S. S., "Environmental Determinants of Human Cancer," *Cancer Research,* **34,** 2425–2435 (Oct. 1974).
6. Ames, B. N., "Understanding the Cause of Aging and Cancer," Division of Biochemistry and Molecular Biology, Univ. of California, Berkeley, CA 94720, Feb. 22, 1994.
7. Jolley, R. L., and Pitt, W. W., Jr., "Chloro-organics in Surface Water Sources, for Potable Water," Oak Ridge National Laboratory, Oak Ridge, TN 37830, presented at the May 8–13, 1977, AWWA Disinfection Symposium at Anaheim, CA.
8. Anon., "Industrial Pollution of the Lower Mississippi River in Louisiana," U.S. Environmental Protection Agency, Cincinnati, OH, Apr. 1972.
9. Anon., "New Orleans Area Water Supply Study," U.S. Environmental Protection Agency, Cincinnati, OH, draft report released on Nov. 8, 1974.
10. Rook, J. J., "Formation of Haloforms during Chlorination of Natural Waters," *Water Treatment and Examination,* **23** (Part 2), 234 (1974).
11. Rook, J. J., "Haloforms in Drinking Water," *J. AWWA,* **68,** 168 (Mar. 1976).
12. Bellar, T. A., Lichtenberg, J. J., and Kroner, R. C., "The Occurrence of Organohalides in Chlorinated Drinking Waters," *J. AWWA,* **66,** 703 (Dec. 1974).
13. Symons, J. M., Bellar, T. A., Carswell, J. K., DeMarco, J., Kropp, K. L., Robeck, G. G., Seeger, D. R., Slocum, C. V., Smith, B. L., and Stevens, A. A., "National Organics Reconnaissance Survey for Halogenated Organics," *J. AWWA,* **67,** 634 (Nov. 1975).
14. Symons, J. M., et al., "Ozone, Chlorine Dioxide, and Chloramines as Alternatives to Chlorine for Disinfection of Drinking Water," U.S. EPA, Cincinnati, OH, Nov. 1977.
15. Ames, B. N., Gold, L. S., and Willett, W. C., "The Causes and Prevention of Cancer," *J. Am. Med. Assoc.,* Special Issue on Cancer, 1995.
16. Ames, B. N., and Gold, L. S., "Chemical Carcinogens: Too Many Rodent Carcinogens," *Proc. Natl. Acad. Sci. USA,* **87,** 7772–7776 (Oct. 1990).
17. Ames, B. N., Profet, M., and Gold, L .S., "Dietary Pesticides (99.99 percent All Natural)," *Proc. Natl. Acad. Sci. USA,* **87,** 7777–7781 (Oct. 1990).
18. Ames, B. N., Profet, M., and Gold, L. S., "Nature's Chemicals and Synthetic Chemicals: Comparative Toxicology," *Proc. Natl. Acad. Sci USA,* **87,** 7782–7786 (Oct. 1990).
19. Norman, T. S., Harms, L. L., and Looyenga, R. W., "Use of Chloramines to Prevent THM Formation at Huron, S.D., *J. AWWA,* **72,** 176 (Mar. 1980).
20. Barrett, R. H., and Trussell, A. R., "Controlling Organics: The Casitas Municipal Water District Experience," *J. AWWA,* **70,** 660 (Nov. 1978).
21. Love, Jr., O. T., Carswell, J. K., Miltner, R. J., and Symons, J. M., "Treatment for the Prevention or Removal of Trihalomethanes in Drinking Water," Appendix 3 to Treatment Guide for the Control of Chloroform and Other Trihalomethanes, U.S. EPA, Cincinnati, OH, 1976.
22. Anderson, M. C., Butler, R. C., Holdren, F. J., and Kornegay, B. H., "Controlling Trihalomethanes with Powdered Activated Carbons," *J. AWWA,* **73,** 432 (Aug. 1981).
23. AWWA Water Quality Division, Disinfection Committee, "Survey of Water Utility Disinfection Practices," *Journal AWWA,* pp 121–128 (Sept. 1992).
24. Howard, N. J., and Thompson, R. E., "Chlorine Studies and Some Observations on Taste Producing Substances in Water, and the Factors Involved in Treatment by the Super- and De-chlorination Method," *J. NEWWA,* **40,** 276 (1926).

1
CHLORINE: HISTORY, MANUFACTURE, PROPERTIES, HAZARDS, AND USES

HISTORICAL BACKGROUND

Elemental Chlorine

Chlorine is an element of the halogen family, but it is never found uncombined in nature. It is estimated to account for 0.15 percent of the earth's crust in the form of soluble chlorides, such as common salt (NaCl), carnallite ($KMgCl_3 \cdot 6H_2O$), and sylvite (KCl). In nature, therefore, it exists only as the negative chloride ion with a valence of -1.

Chlorine is a most unusual and versatile chemical, as its properties differ so widely in the gaseous, liquid, and aqueous states. For this reason, each phase will be treated separately.

Chlorine Gas

Chlorine was discovered in its gaeous state in 1774 by Karl W. Scheele, a Swedish chemist, when he heated a black oxide of manganese with hydrochloric acid:[1]

$$MnO_2 + 4HCl \xrightarrow{\text{heat}} MnCl_2 + Cl_2 + 2H_2O \tag{1-1}$$

The chlorine thus liberated is a strong-smelling, greenish-yellow gas with a pungent odor; it is extremely irritating to mucous membranes.

This element certainly must have been known to the medieval Arab chemist Geber (ca. 720–810), as he was the discoverer of aqua regia (3 parts HCl, 1 part HNO_3), used to dissolve the noble metals. When this mixture is heated, it gives off fumes similar to "Scheele's gas," about which there was a great controversy when it was studied by the great chemists Berthollet, Lavoisier, Gay-Lussac, Berzelius, Therrard, and Davy.[1]

Scheele called the gas he discovered "dephlogisticated muriatic acid" on the theory that manganese had displaced "phlogiston" (which hydrogen was then called) from the muriatic acid (HCl). Scheele also observed that the gas was soluble in water, that it had a permanent bleaching effect on paper, vegetables, and flowers, and that it acted on metals and oxides of metals.

During the decade following Scheele's discovery, Lavoisier successfully attacked and, after a memorable struggle, completely upset the phlogiston theory of Scheele. Lavoisier was of the opinion that all acids contained oxygen. Berthollet found that a solution of Scheele's gas in water, when exposed to sunlight, gives off oxygen and leaves a muriatic acid behind. Considering this residue proof of Lavoisier's theory, Berthollet called it oxygenated muriatic acid.[2] However, Humphry Davy (1778–1829) was unable to decompose Scheele's gas. On July 12, 1810, before the Royal Society of London, he declared the gas to be an element, which in muriatic acid is combined with hydrogen. Therefore, Lavoisier's theory that all acids contain oxygen had to be discarded. Davy proposed the name "chlorine" from the Greek *chloros,* variously translated "green," "greenish yellow," or "yellowish green," in allusion to the color of the gas.[1]

Pelletier in 1785 and Karsten in 1786 succeeded in forming yellow crystals of chlorine hydrate by cooling Scheele's gas in the presence of moisture. From this they inferred that it was not an element. In 1810, Davy proved that these crystals could not be formed by cooling the gas even to −40°F in the absence of moisture. It is now known that these crystals are in fact chlorine hydrate ($Cl_2 \cdot 8H_2O$) and will form under standard conditions with chlorine gas in the presence of moisture beginning at 49.3°F.

Chlorine Liquid

In 1805, Thomas Northmore liquefied Scheele's gas by compression. He noted that it became a yellowish amber liquid under pressure, and upon release of the pressure it volatilized rapidly and violently into a green gas. He further noted its pungent odor and severe damage to machinery.

Michael Faraday (1791–1867) also observed liquid chlorine. On March 5, 1823, he was visited in his laboratory (he worked as an assistant to Davy) by J. A. Paris while he was working with chlorine hydrate in a sealed tube. Paris noted a yellowish, oily-appearing substance in the tube and chided Faraday for working with dirty apparatus. When Faraday tried to open the tube, it shattered and the oily substance vanished. After studying the accident, Faraday wrote Paris: "Dear Sir: The oil you saw in the sealed tube yesterday turned out to be liquid chlorine. Yours faithfully, Michael Faraday."[1]

MANUFACTURE OF CHLORINE

History

From 1805 to 1888, Scheele's gas remained a laboratory curiosity—and a dangerous one—until it was produced on a commercial scale by cooling and compression in a suitable apparatus. This was made possible by Kneitsch's discovery in 1888 that dry chlorine did not attack iron or steel, thus making it possible to package chlorine as a liquid under pressure.

Until this time, chlorine was used as a bleaching agent in the form of a solution. In 1785, Berthollet prepared this solution by dissolving Scheele's gas in water and adding it to a solution of caustic potash. This was done at a chemical plant in Javel, then a small French town, now a part of Paris. Hence, the solution is known as Javelle water. James Watt, the inventor, obtained from Berthollet a license for the manufacture of Javelle water and brought it to Scotland for Charles Tennant, founder of the English chemical company. In 1789, Tennant, through extensive experimentation, produced another liquid bleaching compound, a chlorinated milk of lime. A year later, he improved it greatly by making it a dry compound, which has been known ever since as bleaching powder. The chlorine for the manufacture of bleaching powder was obtained by Berthollet's method of heating sodium chloride, manganese, and sulfuric acid in lead stills. During this time, chlorine was also being made on a limited scale by the Weldon process, which employed the basic reaction of Scheele's discovery, hydrochloric acid and manganese dioxide. This method was given considerable impetus when in 1836 Gossage invented his coke towers for the absorption of waste hydrochloric acid, resulting from the LeBlanc soda process. This invention resulted in cheap hydrochloric acid for the Weldon process.[1] However, in 1868, Scheele's method by the Weldon process became obsolete.* Henry Deacon and Ferdinand Hurter patented a process for producing chlorine by decomposing hydrochloric acid with atmospheric oxygen in the presence of a catalyst.[2] A mixture of hydrochloric acid and air is heated. About 70 percent of the hydrogen chloride can be converted to chlorine as it mixes with the air and steam. This mixture is condensed, and the steam absorbs the hydrogen chloride, forming a very strong muriatic acid mixed with hydrogen chloride gas. This mixture is passed through a superheater, raised to about 430°C, and then passed through a decomposer consisting of a brick- or pumice-lined chamber impregnated with cupric chloride, the catalyst. After this step, the mixture is washed first with water and then with sulfuric acid. The remaining mixture consists of nitrogen and oxygen containing 10 percent chlorine gas, which can be utilized without any difficulty in the manufacture of bleach liquids and powders. The original HCl is recycled so that the only additional materials are heat and air. Considering the amount of chlorine produced, the plant is extremely bulky. The reaction is:

$$4HCl + O_2 \xrightarrow{\text{heat}} + \xrightarrow{\text{catalyst}} 2Cl_2 + 2H_2O \qquad (1\text{-}2)$$

This reaction is reversible and incomplete. The rate of reaction is made satisfactory by the addition of heat and the catalyst, cupric copper.

*The Weldon process or Scheele's method is still used to advantage for making hypochlorites in remote areas (see Chapter 2).

The hydrogen chloride in this process is largely a by-product of the LeBlanc soda process. With the advent of the Solvay sodium ammonia process in 1870, the LeBlanc process fell into a sharp decline, causing the abandonment of the Deacon process in favor of electrolytic methods, which were then emerging from the experimental state.

Electrolytic Processes

History. In 1883, after years of research, Faraday postulated the laws that govern the action of passing an electric current through an aqueous salt solution. He coined the word "electrolysis" to describe the resulting phenomenon. These fundamental laws are among the most exact in chemistry and are as follows:

- The weight of a given element liberated at an electrode is directly proportional to the quantity of electricity passed through the solution. (The unit of electrical quantity is the coulomb.)
- The weights of different elements liberated by the same quantity of electricity are proportional to the equivalent weights of the elements.

Charles Watt obtained an English patent for the electrolytic manufacture of chlorine in 1851. However, at that time electric current generators of sufficient size were not available, and so the patent was only of academic interest. When this equipment became available, interest in electrochemistry was greatly stimulated. In 1890 the first commercial production of chlorine by the electrolytic method was introduced by the Elektron Company (now Fabwerkie-Hoechst A.G.) of Griesheim, Germany.[1] The first electrolytic plant to go into production in America was at Rumford Falls, Maine, in 1892 for the Oxford Paper Company. In 1894, Mathieson Chemical Company acquired the rights to the Castner mercury cell and began the first commercial production of bleaching powder at a demonstration plant in Saltville, Virginia. In 1897, this operation was moved to Niagara Falls, New York. These original Castner rocking cells operated successfully until they were shut down in 1960.[3,4]

At first, the electrolytic process for the commercial production of chlorine was primarily for the manufacture of caustic. Chlorine was a by-product. At the Niagara Falls plant, a small amount of chlorine was used for bleach and for the making of hydrochloric acid, the rest being discharged into the Niagara River. Not until 1909 was liquid chlorine manufactured on a commercial basis. It was first packaged in 100-lb-capacity steel cylinders made in Germany. The business grew slowly but steadily, most consumers using it for bleaching textiles, pulp, and paper. The first American tank car, with a capacity of 15 tons, was manufactured in 1909. The next year, 150-lb cylinders came into use, and in 1917 ton containers were made for the chemical warfare service.[1]

Current Practice. Most of the chlorine today (1990s) is manufactured by three types of electrolytic cells: diaphragm, mercury, and membrane. There are other methods of production, which are designed to fit the raw material containing the chlorine ion. These methods include the electrolysis of hydrochloric acid, the salt process, and the HCl oxidation process. Chlorine is frequently produced as a by-product of heavy metal recovery such as in the tungsten sponge process or the extraction of magnesium from magnesium chloride ore. Although the methods of electrolytic production of chlorine have not changed, the equipment use has changed dramatically.

ELECTROLYTIC CELL DEVELOPMENT

The Ideal Electrochemical Cell

This would be a cell composed of an anode, a cathode, and a separator, which isolates the liquids contained in the anode chamber and the cathode chamber. The function of the separator is to isolate the two chambers while allowing the migration of selected ions from the anode chamber to the cathode chamber. Brine composed of sodium chloride and water is introduced into the anode chamber, where oxidation of the sodium chloride takes place. Chlorine gas is released at the anode. The sodium ions are attracted to the negatively charged cathode and transported through the separator. If the separator is doing a perfect job, which in an ideal cell it would, all of the chloride would be contained on the anode side of the cell.

Water is reduced at the cathode and hydrogen gas is evolved. The remaining hydroxide ion combines with the sodium ion to form sodium hydroxide solution (caustic), which exits the cathode chamber. The ideal separator (membrane) would keep all the OH^- ions on the cathode side of the cell. See Fig. 1-1.

Recent Developments

In practice electrolytic cells are plagued with a variety of problems such as corrosion, erosion of electrodes, and plugging of the separator. All of these factors generate high maintenance costs. There is also great concern over the cost and availability of energy required to drive the oxidation reaction of the brine solution.

In recent years the industry has made important advances on all of these fronts. Energy consumption has been reduced significantly by improved cell design and improved electrical equipment required to convert AC to DC. Motor generator sets, mercury arc rectifiers, and mechanical rectifiers operate at efficiencies on the order of 97 percent. This compares with efficiencies of 90 percent experienced with mercury arc rectifiers at lower cell circuit voltages 40 years ago.

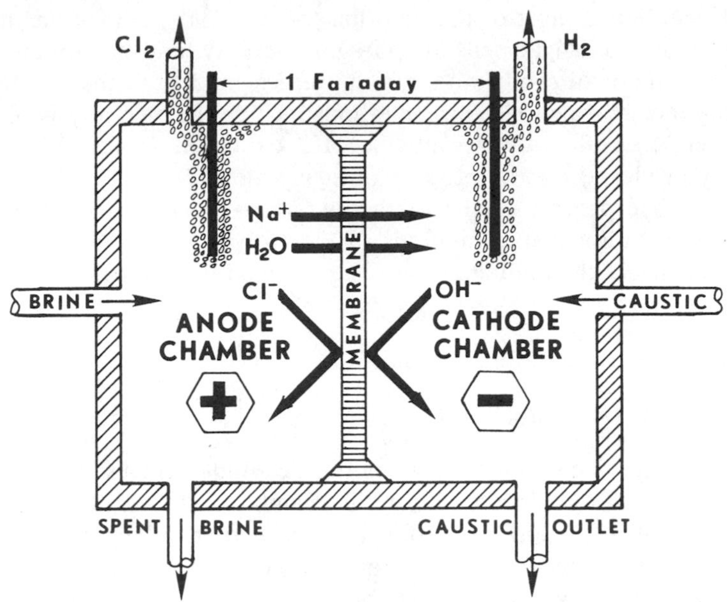

Fig. 1-1. The ideal membrane cell.

Another example is the case of centrifugal chlorine compressors. They are now available with unit capacities in excess of 600 tons/day chlorine and the ability to operate at 10 atm. discharge pressure. These units have superseded the 30–50 ton/day compressors with discharge pressure capability of only 3–4 atm.

The major advances in the industry have been in electrolytic cell design, as new materials of construction have revolutionized cell design. Such materials as valve metal (titanium or tantalum), various noble metal oxides, plasticized diaphragms, and special cladding processes that eliminate or retard the corrosion of cell components have been introduced. Two major achievements have been the development of dimensionally stable anodes and the membrane cell.

Dimensionally Stable Anodes

The chlor-alkali industry has sought for years a stable metal anode for chlorine production. Aerospace industry developments in the 1960s provided the material at a reasonable cost. In 1968 Diamond Shamrock Corporation, a world leader in the production of chlorine and caustic, announced the development and commercialization of new types of metallic anodes, which they named dimensionally stable anodes (DSA®).[5, 6, 50, 51]

These anodes have reduced capital and operating costs. As their name implies, they retain their size and shape during use and have a life longer than 10 years in diaphragm cells, compared to about 180 days for the graphite anode. Moreover, replacement of graphite anodes removes the major source of hydrocarbon contamination of chlorine, which poses a health problem in potable water chlorination (see "Impurities in the Manufacture of Chlorine," this chapter).

These anodes can be reactivated by redeposition of the metal coating. In addition to the saving in downtime and labor for anode replacement, the saving in graphite consumption alone is said to run as high as $4.00 per ton of chlorine. The DSA®s have played an important role in the future of the chlor-alkali industry. They are an important factor in the improvement and expansion of existing facilities.

Approximately 70 plants in North America are using DSA®s, producing 7.3 million tons of chlorine per year. The conversion to DSA® from other anode systems has resulted in a power saving of 16 billion kWh per year.[6, 7]

Anodes used before the introduction of the DSA®s were made of graphite or some form of impregnated carbon. They belong to the soluble type of anodes. Their life was short, about 5–6 months. They eroded in an irregular pattern, which resulted in variable and increased voltage requirements. This phenomenon reduced significantly the efficiency of the cell.

DSA® are based on inventions using mixed metal oxide coatings. The metal to which these oxide coatings are applied is called valve metal and is either titanium or tantalum. Tantalum is typically not used because of higher cost and toxicity. The type of coating to be used is generally determined by the parameters of the process in which it is to be used. For chlorine electrolysis the coatings are usually ruthenium-titanium oxides.[5] However, several types of coatings and a wide variety of substrate geometries, thicknesses, and shapes are available. Therefore DSA® can be custom-made to fit the application, whether it be cell modification or a process requirement.

Mixed oxide coatings of the DSA® type exhibit stable operation for long periods of time at low operating voltages. In the production of chlorine this results in a significant reduction in labor required to maintain cell operation, a reduction in downtime for cell cleaning and adjusting, plus an energy savings of 20 percent over the formerly used soluble graphite anodes. Although the operating experience of DSA® anodes in membrane cells is relatively short, the coating life could approach the coating life in diaphragm cell service.

Anode lifetime depends upon the current density at which the cell operates. At relatively low current densities of about 1.5–2.6 kA/m^2, which is typical for diaphragm cells, the DSA®s have operated well in excess of ten years. At 10 kA/m^2, which is common for mercury cells, these anodes operate for about two years before the structure must be removed from the cell and a new coating applied.

The mixed oxide coatings do not dissolve in mercury, and the coating surface is not easily wetted by mercury. These unique features permit operation of mercury cells with extremely small gaps between the anode and the cathode, commonly as low as 3 mm. (See Fig. 1-14). Anode–cathode gaps of 12–

15 mm are common in diaphragm cells. With the use of newly developed polymer-modified asbestos diaphragms, this gap can be reduced to 3–5 mm. In the development of the membrane cell, research is directed to the design of a zero gap cell. The smaller the gap, the greater the power savings.

Membrane Cell

Description. In the membrane process, the anolyte and the catholyte are separated by a cation exchange membrane that selectively transmits sodium ions but suppresses the migration of hydroxyl ions from the catholyte to the anolyte. This produces a catholyte effluent with a strong caustic soda solution containing a very low sodium chloride content. The advantages of the membrane process are its energy efficiency and its ability to produce, without any harmful effect upon the environment, a strong high-quality solution of caustic soda (NaOH).

History. Hooker Chemicals and Plastics and Diamond Shamrock Corp. both began membrane cell development programs in the 1950s.[8, 9] The largest producer of chlorine in the United Kingdom, the ICI corporation, began membrane cell development 20 years later.[10] Catalytic Inc. of Philadelphia is licensed in the United States to sell ICI membrane cells. The membrane separator concept was made feasible by the availability of selective ion exchange membranes. These membranes became available as a result of DuPont's research on fuel cells. The first of this family of membranes was Nafion, which consisted of a 2–10-mil-thick film of perfluorosulfonic acid resin.[11] This is a copolymer of tetrafluoroethylene and another monomer to which negative sulfuric acid groups are attached.[8, 11]

The world's first commercial membrane cell chlor-alkali plant rated to produce 40,000 metric tons/year caustic soda began successful operation in April 1975 for Asahi Chemical Co. at Nobeoka, Japan on Kyushu Island.[21] This plant produced caustic soda with concentrations varying from 18 to 21 percent at a current density of 5 kA/m^2, 4.2 V, and a current efficiency of 80.5 percent. The power consumption was 3496 kWh/metric ton NaOH and 0.89 metric ton chlorine.

It was discovered that the Nafion 315 membrane could not attain high current efficiency when producing high concentration caustic soda because the hydrophilic sulfonic acid groups lead to counter migration of hydroxide ions from the cathode to the anode compartments. In view of this, research began on other membranes. Asahi Chemical Co. developed a perfluorocarboxylic acid membrane that demonstrated a 90+ percent current efficiency at high concentrations of caustic soda in the electrolyzer. This was announced in 1976.[9] DuPont followed with a series of improved membranes. Their third generation membrane and presumably their best one so far for the chlor-alkali industry is the Nafion series, which has high current efficiency for a 35 percent caustic concentration in the electrolyzer.

In the meantime, the Asahi Glass Co. developed its third generation membrane, Flemion 753, in 1981. This is the membrane used by Asahi Chemical Industry Co., Ltd., ICI, and Uhde in their chlor-alkali electrolyzers.[9,12,13] Both DuPont and Asahi use a combination of both sulfonic and carboxylic groups to minimize the water hydration, which allows the back-migration of the hydroxyl ions. Dow Chemical Co., one of the largest chlor-alkali producers, developed its own membranes, which it expected to result in a membrane structure that would eliminate the back-hydration problem and furthermore would allow the construction of a "zero gap" electrolyzer.[14]

Theory of Operation. Referring to Fig. 1-2, the membrane cell is a simulation of the ideal cell. The operation of any electrolytic cell is limited by the ability of the components to perform as required. The necessity for research and development is to find the proper materials for the electrodes and the separator (membrane) to provide the greatest yield and purity of product with the least amount of energy and maintenance.

Salt, water, and electric current are the raw materials. Solid salt is dissolved in a saturator with fresh water and depleted brine. The saturated brine is chemically treated to precipitate impurities, which are removed through clarification and filtration. The brine is then further treated by an ion exchange process specific for brine. This additional purification is necessary in order to limit the calcium ion content to *less than* 0.05 mg/L. The membranes are greatly affected by calcium deposits.[45] If this limit is not achieved, cell performance will suffer, and maintenance costs will escalate.

Hydrochloric acid is added to the brine to neutralize part of the back-migrating hydroxide ions. These ions reduce the formation of objectionable by-products before the purified brine is fed to the anode compartments of the cells. The salt is electrically decomposed to form chlorine gas, which is evolved at the anode. The resulting sodium ions remain in solution and are transported through the membrane to the cathode, where they combine with hydroxide ions formed in the cathode chambers of the cells. The depleted brine in the anode chamber is treated to separate out any remaining chlorine. After this reprocessing, it is recycled to the brine saturator. Caustic solution that has formed in the cathode compartment flows to a caustic surge tank. Most of the caustic entering this tank is cooled and recycled to control the caustic concentration and temperature in the cells. Softened water is added to maintain the desired product concentration. Hydrogen, which is also produced in the cathode chambers, passes from the cells and may be vented or recovered for use in the plant. The feed brine contains about 320 g/L NaCl and 30 g/L max. of chlorate ion. Depleted brine has a pH range of 2–5 and contains about 170 g/L NaCl. Chlorine is 97–99.5 percent pure. It contains 0.5–3.0 percent oxygen. The caustic product, which is 30 percent by weight NaOH, contains 40–50 ppm NaCl and 5–15 ppm ClO_3^-. Hydrogen is 99.9 percent pure.

Tables 1-1 and 1-1a give typical operating data of membrane cells now in use.

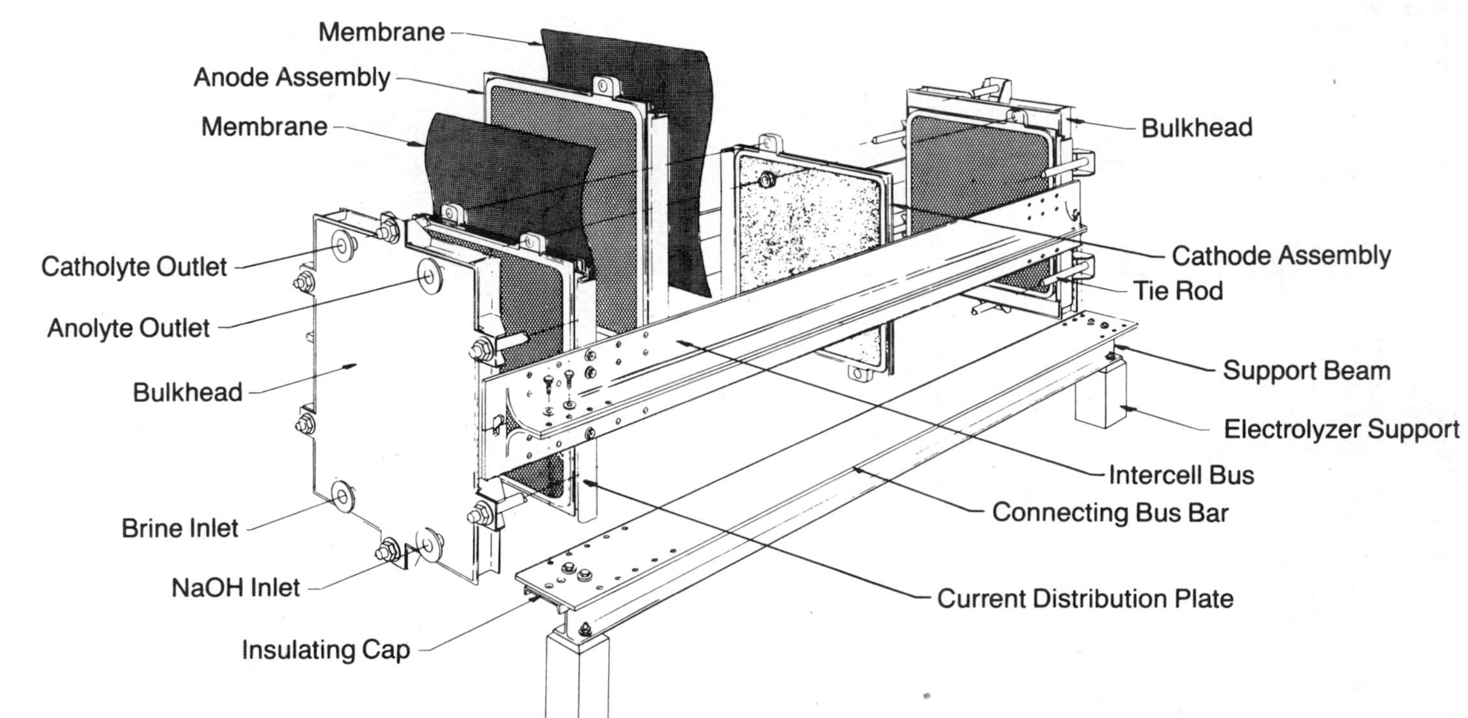

Fig. 1-2. The membrane cell (courtesy ELTECH Systems Corporation).

Table 1-1 Operating Data—Membrane Cells*

	Manufacturer and Model No. of Cell				
	Asahi	*Catalytic FM-21*	*Olin OM-PAC*	*PPG BIZEC*	*Uhde HUMM*
Cell current (kA)	2030	220	300	15.3	70
Current density (kA/m^2)		3.93	5.0	4.0	3
Current efficiency (%)	93–95	94–96	95–97	95	93–95
Voltage/cell	3.11	3.3	3.32	3.25	3.23
Chlorine production (ton/day)	69.4	6.05	9.9	23.0	15
Caustic production (ton/day)	75.13	6.82	11.1	26.0	2.5
Power consumption (kWh/ton Cl_2)	2100	2070	2080	2034	2570
Caustic concentration (%)	20–40	25–35	32	36	33–35
H_2 production (lb/day/cell)	40.26	1.64	62	29	940
Catholyte temp (°C)	90	90–95	95	90	85–90
Anode life (yrs.)	>5	>5	>4	>5	2–5
Membrane life (yrs.)	>2	>2	>2	>2	2
Membrane brand name	perfluoro-carboxylic	Flemion 723	varies	permionic	Nafion
No. of cells	100	24		50	100

Typical analyses, ASAHI electrolyzer:
Chlorine 99.4%, 0.5% O_2 (vol.)
Caustic soda 50% (wt), 50 ppm (wt) NaCl
Hydrogen 99.9% (vol.)
Brine feed 305–315 g/l NaCl 60°C
*See Table 1-1a for Eltech MGC membrane electrolyzer

Table 1-1a

Eltech MGC Electrolyzer Design Range	
Cells per electrolyzer	2–30
Current density (kA/m^2)	2–5
Circuit current (kA):	6–202
Production per electrolyzer:	
Chlorine (MTPD)	0.2–6.3
NaOH (MTPD)	0.2–7.0
Caustic concentration (wt. %)	30–35
Expected MGC Electrolyzer Operating Characteristics	
Basis: 3.1 kA/m^2	
32 wt. % NaOH from Cells	
Electrolyzer voltage	3.1 volts
Power consumption	2165 DCkWh/MT NaOH
Catholyte temperture	90°C
Anode coating life	>5 years
Membrane life	>2 years
Membranes available	DuPont and Asahi Glass

Monopolar vs. Bipolar Configuration. The term "electrolyzer" has an important meaning in the electrolytic process of making chlorine.[16] It refers to a cluster of individual cells. The arrangement of these cells to form an electrolyzer varies from one manufacturer to another, such as the potential production requirements.

Monopolar. When the cells of an electrolyzer are arranged electrically in parallel, it is a monopolar design. All of the anodes are electrically joined together, and so are all the cathodes. The electrical energy flows into all of the anodes simultaneously. It then passes through the membranes to the cathodes, and then exits the electrolyzer. The amperage supplied to the electrolyzer is not limited by the electrode area of one cell but by the electrode area of the entire electrolyzer.

This configuration yields a high amperage electrolyzer that exhibits low voltage. Also all the cells within the electrolyzer will operate at the same voltage regardless of variable resistances within the individual cells. Therefore it is not necessary to electrically isolate each cell from the remaining cells in the electrolyzer because they are all at the same potential.

The monopolar arrangement allows simplification of feed and discharge piping. Electrical short circuits, which cause corrosion, are greatly reduced. It also provides a safer environment for the operators. Electrical arcing between the anode and the cathode does not readily occur, so the membrane remains intact. This prevents the formation of an explosive hydrogen–chlorine mixture.

A more important advantage of the monopolar electrolyzer is that it minimizes the danger caused by a loss of brine flow in the cell. When this occurs, the resistance to the flow of electricity increases, but the current flow redistributes itself throughout the electrolyzer. This maintains equal but higher voltages in the cell, but not enough to cause the arcing described above. The operator, recognizing an abnormal voltage on the electrolyzer, has the opportunity to seek out and correct the brine flow problem before a hazardous condition results.

Bipolar. When the cells of an electrolyzer are arranged electrically in series, it is a bipolar design. The electric current enters one end of the electrolyzer and passes through each membrane cell in series until it exits the electrolyzer. The current enters into an anode, passes through the membrane to a cathode, through electrical connections into an anode, through another membrane into a cathode, and so on throughout the entire electrolyzer. Because the electrical resistances within the cells are additive, this configuration yields a high voltage–low amperage electrolyzer. The total amperage is limited by the electrode area of a single cell. For this design the voltage drop between cells is low because the cathode of one cell is connected directly to the anode of the next cell.

The rectifier for the bipolar electrical system is slightly less expensive than for the monopolar design.

The companies that favor the monopolar design cite the following disadvantages of the bipolar configuration:

1. Each cell must be electrically isolated.
2. Design of feed and discharge piping is complex.
3. Extra care must be exercised to prevent the failure of seals or welds and to prevent hydrogen penetration of bimetallic bonds.
4. If the brine feed to one cell is interrupted and not corrected quickly, the electrical resistance in the system becomes so high that the electrical energy arcs from the anode to the cathode, passing through both the membrane and newly formed vapor space in the cell. This arcing will burn holes in the membrane, permitting hydrogen gas to mix with chlorine gas in the anode chamber, forming an explosive mixture.

This is one of the most dangerous situations that can occur in a chlorine plant. For additional information on bipolar design, see Ref. 22.

Choice of Design. It is fair to say that both monopolar and bipolar electrolyzers have been thoroughly field-tested by the various manufacturers. In spite of the dismal picture painted by the proponents of the monopolar design itemized above, one of the most successful membrane cell designs is bipolar. However, each manufacturer has a good reason for a particular choice. In some cases local conditions might favor one over the other. As of 1997, the favorite choice seems to be the monopolar design.

Advantages of the Membrane Cell. The most important feature of the membrane cell is the savings in energy. Figure 1-3 illustrates this by comparing it to the mercury cell and the diaphragm cell.

The membrane cell is easy to build, easy to operate, and easy to maintain.

The synthetic membrane separator and the DSA® in combination with coated metal cathodes provide an electrolytic cell with characteristics that approach those of the ideal cell. About the only feature that is lacking is the ability to produce a 50 percent caustic solution as can be produced by amalgam (merc.) cell. To date the most concentrated caustic solution produced by a membrane cell is 30–40 percent. Note that in Fig. 1-3 the power consumption has been adjusted to the amount required to concentrate the caustic in the membrane and diaphragm cells up to 50 percent.

Current Developments

Membrane Gap Cell. Most membrane cells in operation in 1997 are defined as finite gap cells. This means that there is an established distance or space between the membrane and each of the electrodes. When the electrodes are

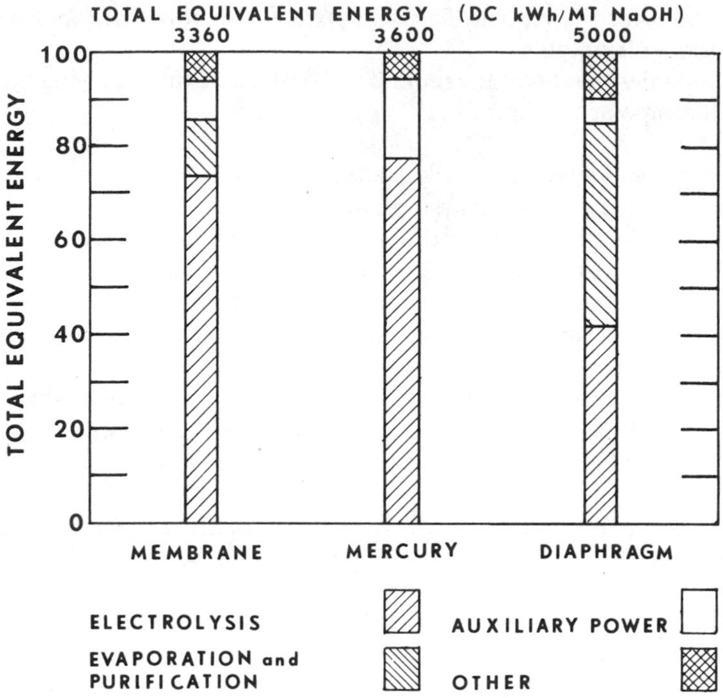

Fig. 1-3. Energy consumption comparison between membrane, mercury, and diaphragm cell processes.

separated from each other by only the thickness of the membrane, as is achieved, for example, in ELTECH's MGC electrolyzer, significant savings in power results. It is estimated that cell voltage could probably be reduced by 0.6 V at 3.1 kA/m^2.[15] Eventually cell operating potentials of about 2.9 V may be achieved by elimination of interelectrode gaps and membrane improvement.

Air Cathodes. Research has indicated that cell performance would be greatly improved by substituting the air cathode for the hydrogen cathode.[15] It is further predicted that electrolyzers equipped with the air cathode should be able to operate in the 2.1 V region at 3.1 kA/m^2. To accomplish this goal, a totally new electrode in a new cell configuration will be required.

In a hydrogen cathode system the water in the catholyte only must reach the electrode surface to be reduced to hydrogen gas and the hydroxide ion. An air cathode, by comparison, must receive the reactants from two different phases. Oxygen from air supplied to the cathode must penetrate the porous electrode. Likewise water from the catholyte must also penetrate the cathode from the electrolyte side. Oxygen is reduced to the hydroxide ion at the electrode on the presence of suitable catalysts. The air in the cathode is now depleted in oxygen and must diffuse back out of the electrode into the air chamber. The hydroxide ion must then move back into the catholyte.

The electrode used to accomplish all of this has a three-layer structure. The porous carbon layer is the region in which the reaction occurs. The second layer, which is also porous and is located between the active layer and the air chamber, prevents the penetration of electrolyte through the electrode. This prevents flooding of the air chamber by the catholyte. A third layer is the conductive grid located on the electrolyte side. It is used to distribute current to the active layer immediately adjacent to it.

Figure 1-4 illustrates schematically the operation of an air cathode system in a chlor-alkali electrolytic cell.[15]

There are three major differences in the air cathode cell compared to a conventional cell. First, an air chamber is included behind the electrode to provide the air for the cathodic reaction. Second, because hydrogen is not produced by the reduction of oxygen, hydrogen handling equipment is not required. Third, the cell operates at approximately 1.0 V lower potential than a conventional hydrogen-evolving cell.

Recent Developments. These membrane cells are being continuously updated by the industry along with additional chlor-alkali manufacturers entering

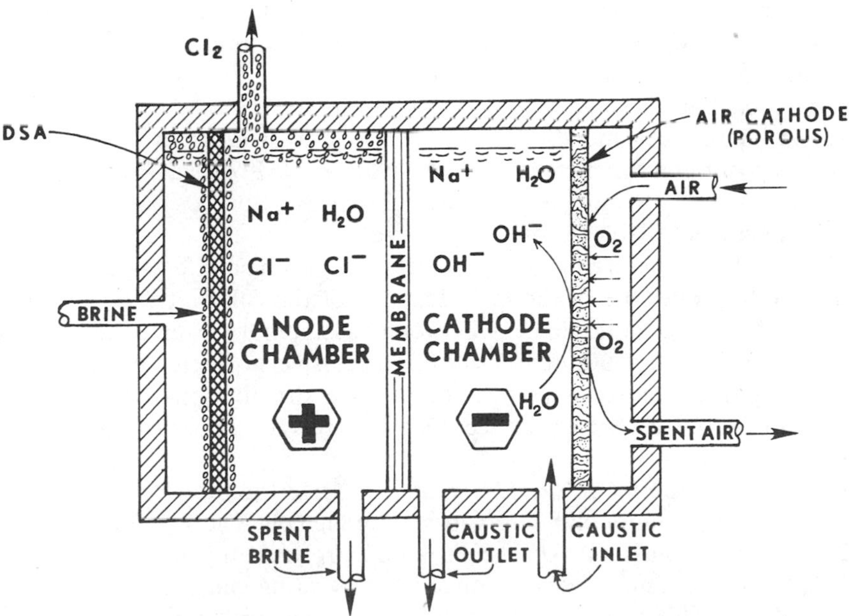

Fig. 1-4. Air cathode cell (courtesy ELTECH Systems Corporation).

the field. Table 1-2 shows operating characteristics of membrane cells as of 1991.[84]

The following membrane cells are the most recent in the chlor-alkali industry:

1. Tokuyama Soda TSE-270. This is a bipolar cell that comes in three different sizes. The unit active areas are 2.7, 1.25, and 0.85 m^2. The TSE-270 is the 2.7 m^2 size. There can be 30–120 cell units incorporated in one electrolyzer.
2. Hoechst-Uhde. This is a jointly developed bipolar cell to supplement their monopolar cell. The effective area of a unit cell is 1.2–3.0 m^2, and up to 100 unit cells can be assembled into one electrolyzer.
3. Krebskosmo M2B. This is a bipolar type and has internal channels for the distribution of the electrolyte and the collection of the gases. Two special features of the cell are:
   (a) The annular-shaped passages between the electrolyte compartments and distribution/collection channels allow a strong electrolyte flow with minimal leakage of electric current.
   (b) The partition wall between the anode and cathode compartment, which is made of a polytetrafluoroethylene (PTFE) foil, is supported by an expanded metal sheet that reduces the cell weight. This results in a much lower cell cost than do those designs made up of a rigid multilayer partition wall.

   The cell elements are built with a membrane size of 2.5 m^2 with 72 cell elements arranged in four stacks of 18 elements connected in parallel. One such cell block can produce 20 tons/day of chlorine.
4. Asahi Glass AZEC. This is one of their latest developments. It is a monopolar type, composed of cell units (see Fig. 1-5) 0.26 m wide and 1.45 m long = 0.38 m^2. It is a zero-gap unit featuring:
   (a) A Flemion-Dx membrane with increased hydrophilicity.
   (b) A proprietary, activated cathode, with a low hydrogen overpotential and stability against reverse current.
   (c) A simple, reliable zero-gap structure.
5. De Nora K40. This is a monopolar electrolyzer with an effective membrane area of 0.64 m^2. One of the main features of the cell unit is the use of a resilient compressible element between the membrane and the cathode assembly. It is capable of transmitting excessive pressures on individual contact points to adjacent points. This provides distribution of the pressures over the entire electrode surface. One electrolyzer consists of 20–60 cell units.
6. Oxy Tech MGC Electrolyzer. Formerly known as the Eltech MGC electrolyzer, this is a monopolar zero-gap membrane cell. The membrane size is 2.23 m^2 with a maximum of 30 cells per electrolyzer. It is illustrated in Fig. 1-2. Table 1-2 contains design range and operating characteristics.
7. ICI FM-21. This too is a monopolar electrolyzer incorporating a simple, pressed electrode structure of relatively small size. The anode is composed

Table 1-2 Characteristics of Membrane Cells (Courtesy of VCH Publishers, Inc., New York)

	Bipolar Type						*Monopolar Type*			
	Asahi Chemical Industry		*PPG*	*Tokuyama Soda*	*Hoechst Uhde*	*Krebskosmo*	*Asahi Glass*	*De Nora*	*OxyTech*	*ICI*
	Standard	*Super*	*BIZEC*	*TSE-270*	*BM*	*MZB*	*AZEC*	*K-40*	*MGC*	*FM-21*
Effective membrane area, m^2	2.7	5.08	3.83	2.7	1.2–3	2.5	0.2	0.64	1.5	0.21
Cells per electrolyzer	80–110	80–110	20–50	30–120	up to 100	4 × 18	30–540	20–60	2–30	1–120
Current load, kA	10.8	20.3	15.3	8.1–10.8	3–15		18–340	40–150	6–225	1–100
Current density, kA/m^2	4	4	4	3–4	2–5		3–4	3–4	2–5	1.5–4.1

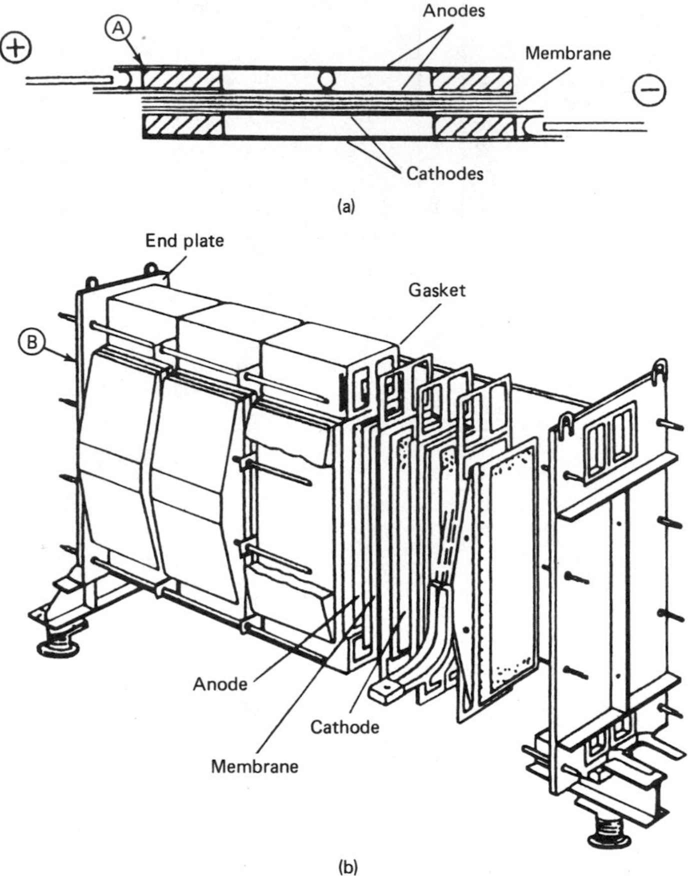

Fig. 1-5. Asahi Glass AZEC monopolar electrolyzer. (A) AZEC cell; (B) AZEC electrolyzer (courtesy of VCH Publishers, Inc., New York).

of a 1-mm-thick titanium panel between compression-molded joints of ethylene-propylene-diene monomer (EPDM) copolymer. This is a synthetic cross-linked elastomer. The cathode assembly is composed of a 1-mm-thick nickel panel between compression joints of molded EPDM. This series is repeated until the number of electrodes required for the desired cell has been assembled. Then the rear plate is attached, sandwiching the assembly into a complete cell.

Internal porting is used to eliminate external piping to the individual cell compartments.

This electrolyzer features coated titanium anodes. The cathodes are pure nickel and are available with a coating that reduces the hydrogen-overpotential. The electrodes are pressed from integral sheets of pure metal. This makes the necessary recoating process simple and allows rapid turnaround.

8. Chlorine Engineers MBC. This is known as the membrane bag cell. When diaphragm cells are retrofitted to become membrane cells, the ion-exchange membrane is installed in the shape of a bag. This bag encloses the anodes, and is made to enclose one or three anodes. The power supply rod passes through a hole in the bottom of the membrane bag for the connection to the baseplate, as in a diaphragm cell. The open end of the bag, facing upward, is fixed to the partition plate by a sealing plug.

   The use of a baglike ion-exchange membrane on the anode side rather than on the cathode side simplifies the setting of the membrane and decreases the electrolyte IR drop. This simplifies the cost of retrofitting a diaphragm cell.

Diaphragm Cells

History. The bulk of the chlorine produced in the United States and North America is made by the diaphragm cell. Approximately 70 percent of the total U.S. production is by these cells.[16] Although several hundred types of diaphragm cells have been designed, those most widely used are listed below in Table 1-3, which tabulates typical operating data of the modified cells.

Table 1-3 Operating Data; Diaphragm Cells

	Manufacturer and Cell			
Characteristics	*Eltech MDC-55*	*Hooker H-4*	*PPG V-1144*	*Uhde HU-60*
Current (kA)	135	150	72	80–120
Current efficiency (%)	95.8	96.6	95–96	
Voltage/cell	3.46	3.47	3.5	3.41
Power (kWh/ton Cl_2)	2475	2492	2500	2700
Cl_2 production (lb/day/cell)	9053	10024	4900	4780–7280
NaOH production (lb/day/cell)	10323	11309	5418	5380–8200
S/C ratio (lb NaCl/lb NaOH)	1.3	1.25		
NaOH in cell liquor (%)	11	11.5	11–12	
NaOH in cell liquor (g/l)	135	140	135–145	
Catholyte temp. (°C)	95	97	95	
H_2 production (lb/day/cell)	257.4	287	142.6	149.6–226.6
H_2 cu.ft/day/cell (760 mm O°C)	45866	51118	25403	
Anode life (mos.)	>120	132	84–132	
Diaphragm life (days)	300–600	300–1000	356–730	

Typical analyses—Eltech (MDC-55)

Chlorine Gas, percent by vol:		*Caustic Cell Liquor, grams/liter*		*Hydrogen, percent by vol.*	
Cl_2	96.5–98.0	NaOH	130–150	H_2	99.8
H_2	0.1–0.3	NaCl	175–210	N_2	0.07
CO_2	0.1–0.3	$NaClO_3$	0.05–0.30	O_2	0.10
O_2	1.0–3.0	Fe	<0.00005	CO_2	0.04
		Na_2SO_4	0.00–7.00		

These cells have undergone a transformation related to the introduction of DSA®s and polymer-based improvements of the diaphragms. These innovations occurred in the 1970s. The replacement of graphite anodes with DSA®s was a most significant improvement in the art of chlor-alkali production. Next in importance has been improvement in the original asbestos slurry diaphragms. The new modified asbestos diaphragms incorporate polymer application technology, which results in lower power consumption, a more rugged diaphragm, dimensional stability of the diaphragm which produces a longer diaphragm life, and more reproducible diaphragm operating characteristics.

Table 1-4 (below) illustrates the improvement in diaphragm cell operating characteristics from 1920 to 1992. The improvements are due largely to coated metal anodes and better diaphragm material.

Recent Developments. Diaphragm cells have been under attack since the discovery of the human health hazards of asbestos fibers. Although the actual risks are overblown by cancer researchers, the industry has been unrelenting in efforts to find an acceptable alternative diaphragm material. The deposited asbestos diaphragm developed by Hooker Chemical in 1928 was the diaphragm of choice until 1971 when Eltech developed the "Modified Diaphragm," a mixture of asbestos and a fibrous fluorocarbon polymer. The polymer was used to stabilize the asbestos, which lowers the required cell voltage. It also allowed the use of expandable DSA® anodes. In its formulation, the Modified Diaphragm is the most common diaphragm in use today, but it still contains 75 percent asbestos.[83]

In 1983 OxyTech Systems set out to develop a diaphragm that did not contain any asbestos,[85] owing to future EPA regulations that would ban the use of asbestos entirely. This effort led to the development of the Polyramix (PMX) diaphragm, a fibrid composite of fluorocarbon and chemically resistant metal oxide particles with a relative density twice that of asbestos. These particles can be vacuum-deposited onto a diaphragm by conventional methods. PMX has left the laboratory and has been under commercial evaluation for several years.[85]

Table 1-4 Examples of Changes in Diaphragm Cell Characteristics

	Producer and Cell				
Characteristic	*Allen-Moore*	*Hooker S*	*Hooker S-38*	*Hooker S4*	*Eltech MDC-55*
Year	1920	1938	1952	1965	1981
Anode material	G	G	G	G	DSA®
Current (kA)	1.3	6	27	55	135
Current density (kA/m^2)	0.4	0.6	1.3	1.3	2.5
Voltage/cell	3.6	3.3	3.9	4.0	3.46
Ton Cl_2/cell/day	0.04	0.20	0.90	1.83	4.53

G = Graphite

1. OxyTech MDC Cells. Table 1-5 lists the operating capabilities and characteristics of this cell, which is manufactured and licensed by OxyTech. It features woven steel-wire cathode screen tubes that are open at both ends and are welded into thick steel-tube sheets at each end. The tubes, tube sheets, and the outer steel cathode shell form the catholyte chamber of the cell (see Fig. 1-6) Copper is bonded rather than welded to the rectangular cathode shell on the two long sides parallel to the tube sheets.

Table 1-5 OxyTech Systems MDC Cells: Operating Capacities and Characteristics (Courtesy of VCH Publishers, Inc., New York)

	Model Number and Operating Range, kA					
	MDC-29			*MDC-55*		
Item	*35*	*to*	*80*	*75*	*to*	*150*
Chlorine capacity,						
metric ton/day	1.05		2.41	2.33		4.53
short ton/day	1.16		2.66	2.48		5.00
Caustic capacity,						
metric ton/day	1.21		2.76	2.59		5.18
short ton/day	1.33		3.04	2.85		5.70
Hydrogen capacity,						
m^3/day	335		765	720		1435
cubic feet/day	11830		27010	25420		50670
Current density,						
kA/m^2	1.21		2.76	1.37		2.74
$A/in.^2$	0.78		1.78	0.88		1.76
Cell voltage, V[a]						
steel cathode	2.90		3.62	3.00		3.62
activated cathode	2.80		3.51	2.90		3.51
Power consumed (d.c., steel cathode)[b],						
kW h/t	2310		2876	2390		2870
kW h/short ton	2100		2610	2175		2610
Power consumed (d.c., activated cathode)[b],						
kW h/t	2230		2786	2310		2780
kW h/short ton	2025		2530	2100		2530
Diaphragm life, years	1–2		0.5–1.0	1–2		0.5–1.0
Anode life, years	8–10		5–8	8–10		5–8
Cathode life, years	10–15		10–15	10–15		10–15
Distance between cells[c],						
m		1.60			2.13	
inches		63			84	

[a]Cell voltage includes loss in intercell bus.
[b]Power consumed per ton (metric or short) or chlorine produced.
[c]Distance centerline-to-centerline and side-by-side with bus connecting.

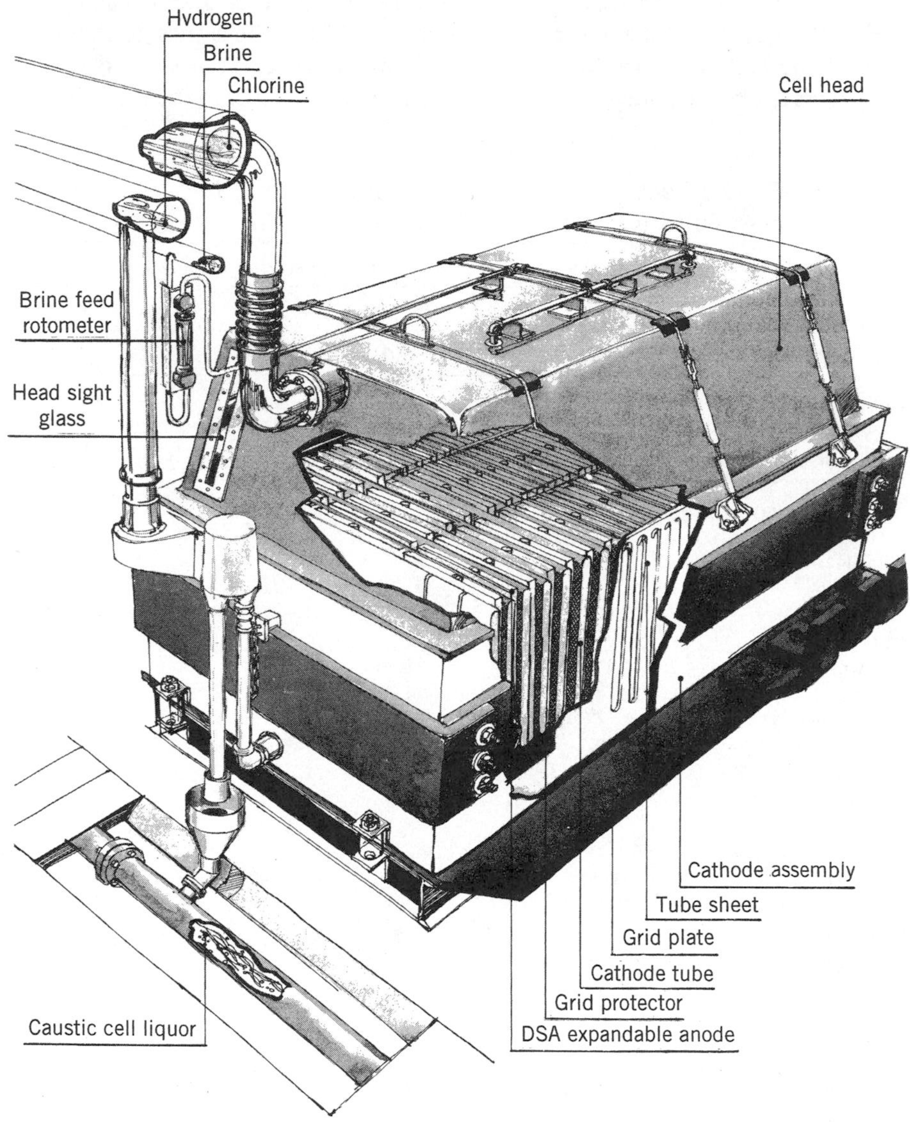

Fig. 1-6. ELTECH MDC-55 diaphragm cell (courtesy of ELTECH Systems Corporation).

Copper connectors attached at the ends of the bonded copper side plates complete the enclosure of the cathode with copper. The anodes are connected to a patented copper cell base that is protected from the anolyte by a rubber cover or a titanium base cover (TIBAC). Orientation of the cathode tubes is parallel to the cell circuit. This is the opposite of a Hooker-type cell. This arrangement accommodates thermal expansion of the cell and circuit without changing the anode-to-cathode alignment.[84] The combination of the Modified Diaphragm and expandable DSA® anodes reduces power consumption 10–15 percent from that of the outmoded asbestos diaphragms with fixed DSA® anodes.

2. Dow Cell. The Dow Chemical Company is the largest chlor-alkali producer in the United States, accounting for one-third of the U.S. production and one-fifth of the world capacity. Its production capacity is concentrated in a few sites. This caused Dow's cell development program to follow a different path from that of other technology developers. Dow uses its own cell design of the filter press bipolar type, and has operated this type of cell for more than 80 years,[84] so its cell development occurred in several stages. The current cell uses vertical anodes of graphite or DSA® coated titanium anodes, verticle cathodes of woven wire mesh bolted to a perforated steel backplate, and a vacuum-deposited asbestos diaphragm. A single bipolar element may have 100 m^2 of both anode and cathode area. This cell operates at lower current densities than others in the industry. These cells are usually operated with 50 or more cells in one unit or series. For its diaphragms, Dow uses a mixture of chrysotile asbestos and crocidolite asbestos, whereas the rest of the industry uses only chrysotile asbestos. Operating data on these cells have not been published. The Dow cell is illustrated in Fig. 1-7.
3. Glano Electrolyzer. This is a bipolar type of cell developed jointly by PPG Industries and Oronzio De Nora Impiana Electrochemici SpA. The electrolyzer consists of several bipolar cells clamped between two end electrode assemblies by means of tie rods. This forms a filter press type of electrolyzer. It is equipped with DSA® titanium anodes normally consisting of 11 cells, and was designed primarily for the large chlor-alkali plants. The V-1144 electrolyzer (Fig. 1-8) was the first commercial unit and is being used in eight plants. The second generation is the V-1161 electrolyzer, which uses the OxyTech Modified Diaphragm, narrower electrode gaps, lower current density, and DSA® anodes, all of which contribute to much lower power consumption than that of the V-1144 electrolyzer. The operating characteristics of the Glanor electrolyzer are shown in Table 1-6.
4. OxyTech-Hooker Cells. The first chlor-alkali industry cell with a deposited asbestos diaphragm was the Hooker type S-1 monopolar cell, introduced in 1929 in the United States. The design featured vertical graphite anode plates connected to a copper bus bar. The cathode was made of woven steel wire cloth or perforated steel fingers between the anodes. The

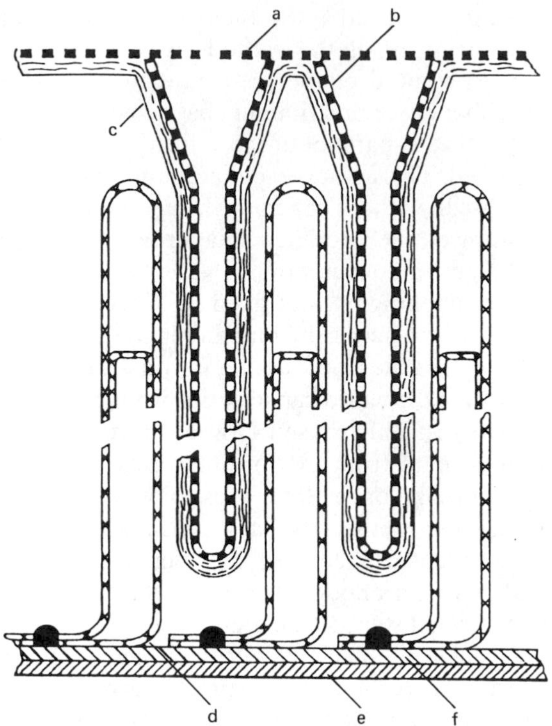

Fig. 1-7. Dow diaphragm cell, section view. (a) Perforated steel backplate; (b) cathode pocket; (c) asbestos diaphragm; (d) DSA anode; (e) copper backplate; (f) titanium backplate (courtesy of VCH Publishers, Inc., New York).

cathodes held the vacuum-deposited asbestos fiber diaphragms, which separated the anode and the cathode compartments. The cathode fingers did not extend completely across the cell but left a central circulation space. In the following 40 years, a family of Series S cells with similar characteristics evolved. More than 12,000 of these cells have been licensed for use in chlor-alkali plants around the world.

In 1973, a new H series of monopolar cells was introduced by Eltech (now OxyTech Systems) and Hooker Chemical using DSA® anodes. These cells have shown significant voltage savings over the S series. This allows increases in cell capacity without any increases in rectification capacity. By changing from the graphite anodes to the open DSA® anodes, and with the incorporation of open-ended cathode tubes extending across the cell, the required circulation space was satisfied. Table 1-7 contains the operating characteristics of the OxyTech Systems–Hooker H-series diaphragm cells.

5. HU Monopolar Cells. These cells were a joint development of Hooker (now OxyTech Systems) and Uhde. Therefore, these cells use the Oxy-

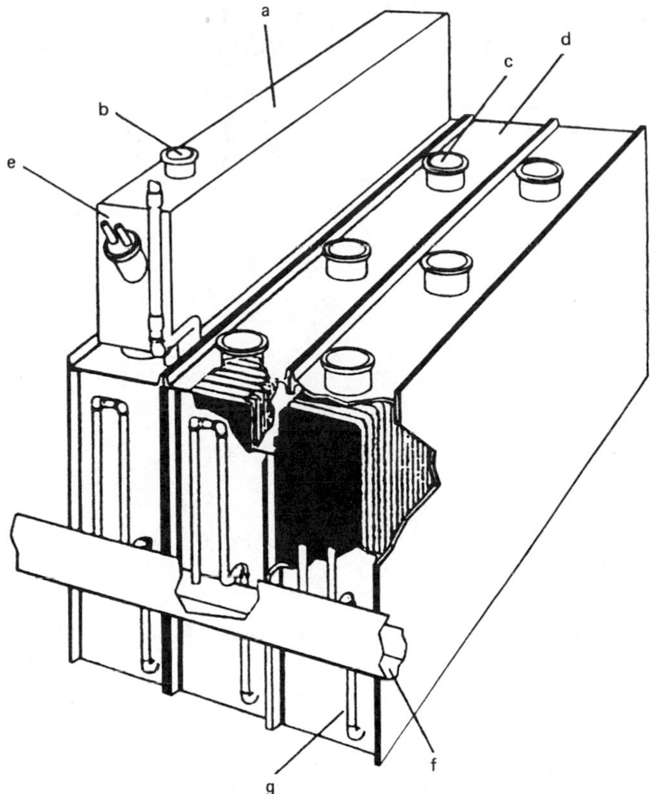

Fig. 1-8. Glanor bipolar electrolyzer. (a) Disengaging tank; (b) chlorine outlet; (c) hydrogen outlet; (d) bipolar element; (e) brine inlet; (f) cell liquor trough; (g) cell liquor outlet (courtesy of VCH Publishers, Inc., New York).

Tech Modified Diaphragm. The HU-type electrolyzer is rectangular and not cubic. It is narrow in the direction of current flow because the anodes are arranged in a single row. The cathode is long and narrow, so the current density is lower through the cathode shell. This allows closer anode-to-cathode fabrication tolerances and spacing. This particular configuration provides shorter electrolyzer current pathways through the cell room, thus saving piping and its associated components. This cell is illustrated by Fig. 1-9.

The electrolyzer design incorporates the creation of a second operating floor by raising the cells from the floor. This allows the interconnecting bus bars to be distributed over the entire length of the cell; each HU cell provides a bus bar for each individual anode. This arrangement allows easy access for connecting monitoring facilities for the electric current flowing through each anode. Because the HU bypass switch is installed underneath instead of adjacent to the circuit of cells, its connec-

Table 1-6 Glanor Bipolar Diaphragm Electrolyzers: Design and Operating Characteristics (Courtesy of VCH Publishers, Inc., New York)

Item	*Model V-1144*	*Model V-1161*
Cells per electrolyzer	11	11
Active anode area per cell, m_2	35	49
Electrode gap, mm	11	6
Current load, kA	72	72
Current density ϱ at 72 kA, kA/m^2	2.05	1.47
Cell voltage, V	3.50	3.08
Current efficiency, %	95–96	95–96
Power consumption (d.c.), kW h/t*	2500	2200
Anode gas composition (alkaline brine)		
Cl_2, %	97.3–98.0	97.0–98.0
O_2, %	1.5–2.2	1.5–2.2
H_2, %	<0.1	<0.1
CO_2, %	0.4	0.4
Cell liquor		
NaOH, g/l	135–145	135–145
$NaClO_3$, %	0.03–0.15	0.03–0.15
Production per electrolyzer		
Chlorine, t/d**	26.7	26.7
NaOH, t/d	29.8	29.8

*Per short ton of chlorine.
**Short tons.

Table 1-7 OxyTech Systems Hooker H-Series Diaphragm Cells: Design and Operating Characteristics (Courtesy of VCH Publishers, Inc., New York)

	H-2A		*H-4*	
Operating current, A	80 000		150 000	
Anode area, m^2	36.16		64.52	
$in.^2$	56 050		100 000	
Current density, A/m^2	2212		2325	
$A/in.^2$	1.43		1.50	
Cell voltage, V	3.44		3.44	
Approximate cell dimensions, m	1.87 × 2.66		2.58 × 3.11	
Diaphragm life, days	300–500		300–500	
Anode life, years	5–7	5–7		
Operating NaOH concentration, g/l	140	160	140	160
%	11.35	12.89	11.33	12.87
Current efficiency, %	96.4	94.6	96.6	94.9
Chlorine output,				
metric ton/day	2.45	2.41	4.60	4.52
short ton/day	2.70	2.65	5.07	4.98
Caustic soda output,				
metric ton/day	2.76	2.71	5.19	5.10
short ton/day	3.05	2.99	5.72	5.62

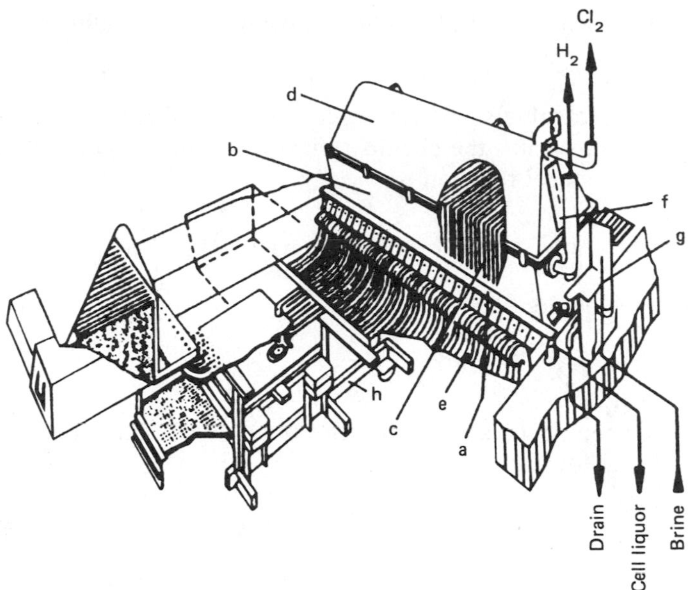

Fig. 1-9. Oxy/Uhde HU-type cells. (a) Cell bottom; (b) cathode; (c) anode; (d) cell cover; (e) bus bars; (f) brine level gauge; (g) brine flow meter; (h) bypass switch (courtesy of VCH Publishers, Inc., New York).

tion is made for each individual anode, and no additional contact bus bars are needed.

The HU type cells are able to cover 30–150 kA. The only difference between the cell models is the number of elements and consequently the length of the cell. The cell voltage and power consumption per ton of chlorine is identical for all cell types, as shown in Table 1-8. The design and operating characteristics for the various HU-Modified Diaphragm

Table 1-8 HU Series Diaphragm Cells: Specific Load, Cell Voltage, and Power Consumption (Courtesy of VCH Publishers, Inc., New York)

	Specific Load, kA/m²	
	1.5	*2.3*
Cell voltage, V	3.12	3.41
Power consumption (d.c., average), kW h/t*	2500	2700

*Per ton of chlorine.

cells are shown in Table 1-9. The cell configuration is illustrated by Fig. 1-10.

Chemistry of Electrolysis. A typical diaphragm cell is illustrated in Fig. 1-6, and Fig. 1-11 illustrates the chemistry involved, which is described below.

The overall chemical reaction is:

$$NaCl + H_2O + \text{electric current} \rightarrow NaOH + \tfrac{1}{2}Cl_2 + \tfrac{1}{2}H_2 \qquad (1\text{-}3)$$

The principal anode reaction is:

$$2Cl^- \rightarrow Cl_2 + 2e^- \qquad (1\text{-}4)$$

Chlorine formed at the anode saturates the anolyte, and an equilibrium is established, as follows:

$$Cl_2 + (OH)^- \rightarrow Cl^- + HOCl \qquad (1\text{-}5)$$

$$HOCl \rightarrow H^+ + OCl^- \qquad (1\text{-}6)$$

The principal cathode reaction is:

$$2H^+ + 2OH^- + 2e^- \rightarrow H_2 + 2OH^- \qquad (1\text{-}7)$$

Therefore the H^+ ion present as H_2O in the catholyte evolves at the cathode as hydrogen gas, leaving the hydroxyl ion (OH^-) behind in the catholyte. Since chlorine has evolved at the anode, the sodium ion is free to join the hydroxyl ion as it migrates from the anolyte chamber to the catholyte chamber by means of differential hydraulic head. The porous diaphragm is used to inhibit the migration of the OH^- ions from the cathode to the anode; otherwise the mixing of these solutions would result in the formation of hypochlorite instead of the evolution of chlorine.

By maintaining the pH of the anolyte solution between 3 and 4 and a differential head of 5 inches between the anolyte and catholyte solution levels, the operation of the cell can be kept in chemical equilibrium to produce a 97 percent chlorine gas at the anode, almost 100 percent hydrogen gas at the cathode and an 11–12 percent NaOH content in the catholyte effluent, which is replaced by purified brine at 60–70°C at the inlet.[17] The diaphragm is usually made by introducing an asbestos slurry into the brine and drawing a vacuum on the inside of the cathode compartment. This deposits the asbestos slurry on the perforated steel screens that serve as the cathode. In time this diaphragm material becomes clogged with impurities and must be removed and replaced.

Figure 1-12 illustrates the flow sheet of a typical modern chlor-alkali plant producing chlorine by electrolysis in a diaphragm cell.

The ingredients required are brine, water, and electric power. Brine is made available in one of two ways: (1) rock salt is delivered to the plant and dissolved

Table 1-9 HU Series Diaphragm Cells: Design and Operating Characteristics (Courtesy of VCH Publishers, Inc., New York)

Item	Cell Type						
	HU 24	*HU 30*	*HU 36*	*HU 42*	*HU 48*	*HU 54*	*HU 60*
Number of anodes	24	30	36	42	48	54	60
Anode surface area, m^2	20.6	25.8	31.0	36.1	41.3	46.4	51.6
Load, kA	30–45	40–60	50–70	55–85	60–95	70–105	80–120
Cl_2 production, t/d	0.90–1.36	1.19–1.82	1.49–2.12	1.64–2.58	1.79–2.88	2.09–3.18	2.39–3.64
NaOH (100%) production, t/d	1.01–1.54	1.35–2.05	1.68–2.39	1.85–2.91	2.02–3.25	2.36–3.59	2.69–4.10
H_2 production, kg/d	25–39	34–52	42–60	47–73	51–82	59–91	68–103
Cell length, m	2.1	2.6	3.0	3.5	3.9	4.4	4.8
Distance, cell-to-cell, m	1.5	1.5	1.5	1.5	1.5	1.5	1.5

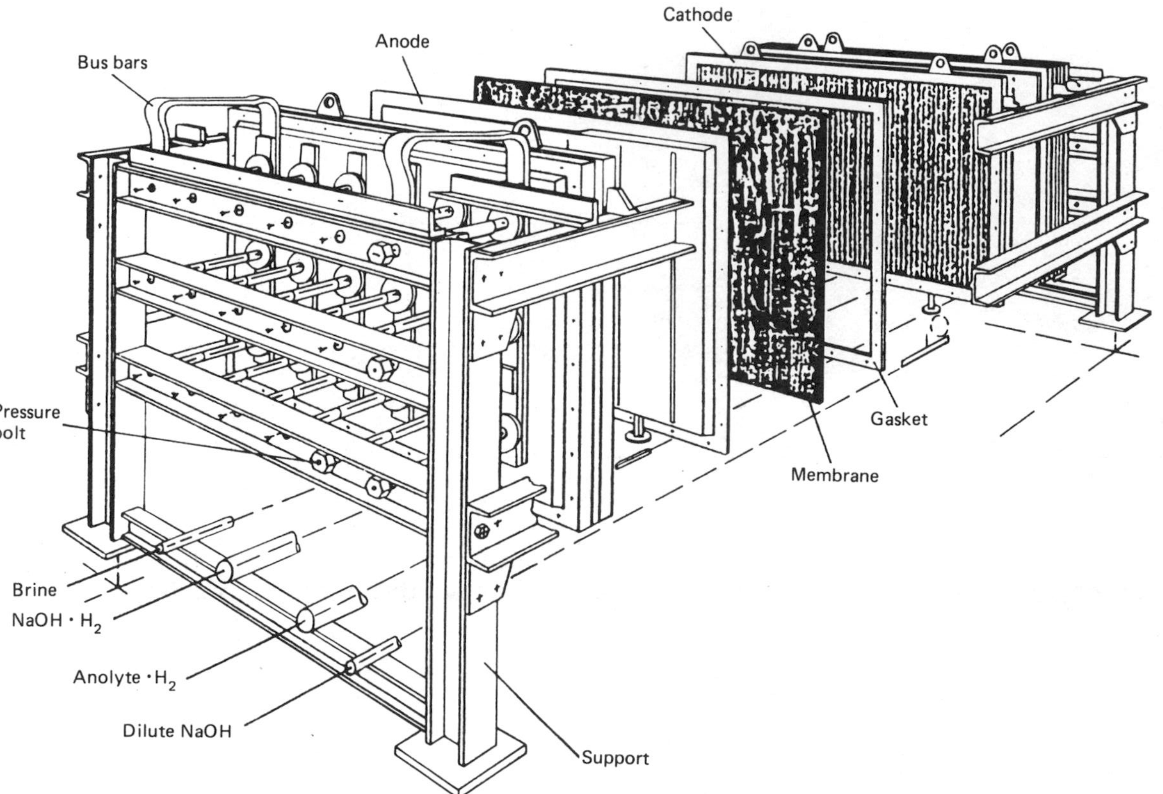

Fig. 1-10. Hoechst Uhde cell (courtesy of VCH Publishers, Inc., New York).

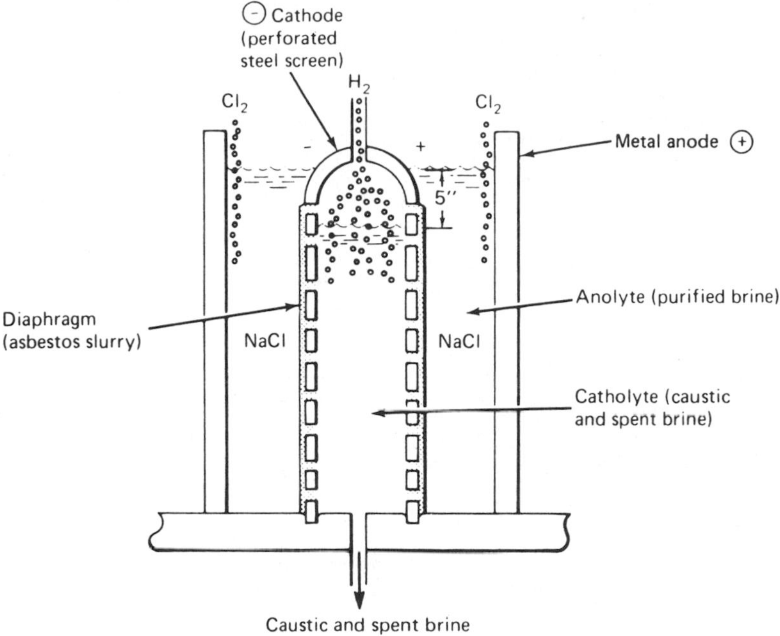

Fig. 1-11. Schematic diaphragm cell.

in water, or (2) the plant may be located adjacent to underground deposits of salt from which the brine is produced by water injection into a well. Impurities such as the calcium and magnesium ions of the sulfates and chlorides must be removed from the brine solution. This is accomplished by the addition of sodium carbonate and sodium hydroxide in a carefully controlled treatment process followed by sedimentation and filtration. The salt content of this brine must be increased to the optimum level of approximately 26.6 percent (322 g/L) of NaCl at 70°C. This is accomplished by heating the brine and saturating it with purified salt. The latter is obtained from the evaporation step in the processing of the cell liquor.

The water supply used for dissolving the salt should be completely softened for the removal of Ca, Mg, and Fe ions to minimize blockage in the cell diaphragm. This water should also be free of any ammonia ion to minimize the nitrogen content. This reduces the hazard of forming potentially explosive nitrogen trichloride in the production of chlorine.

The power requirement is approximately 3000 kWh per short ton of chlorine. This is usually a high-voltage AC source, which must be put through various pieces of switch gear to step down the voltage and rectify it to DC. Power to the cells is low voltage (4–5 V DC), and, depending upon the size and particular design of the cell, the current consumption ranges between 10,000 and 50,000 amperes for modern installations.[17]

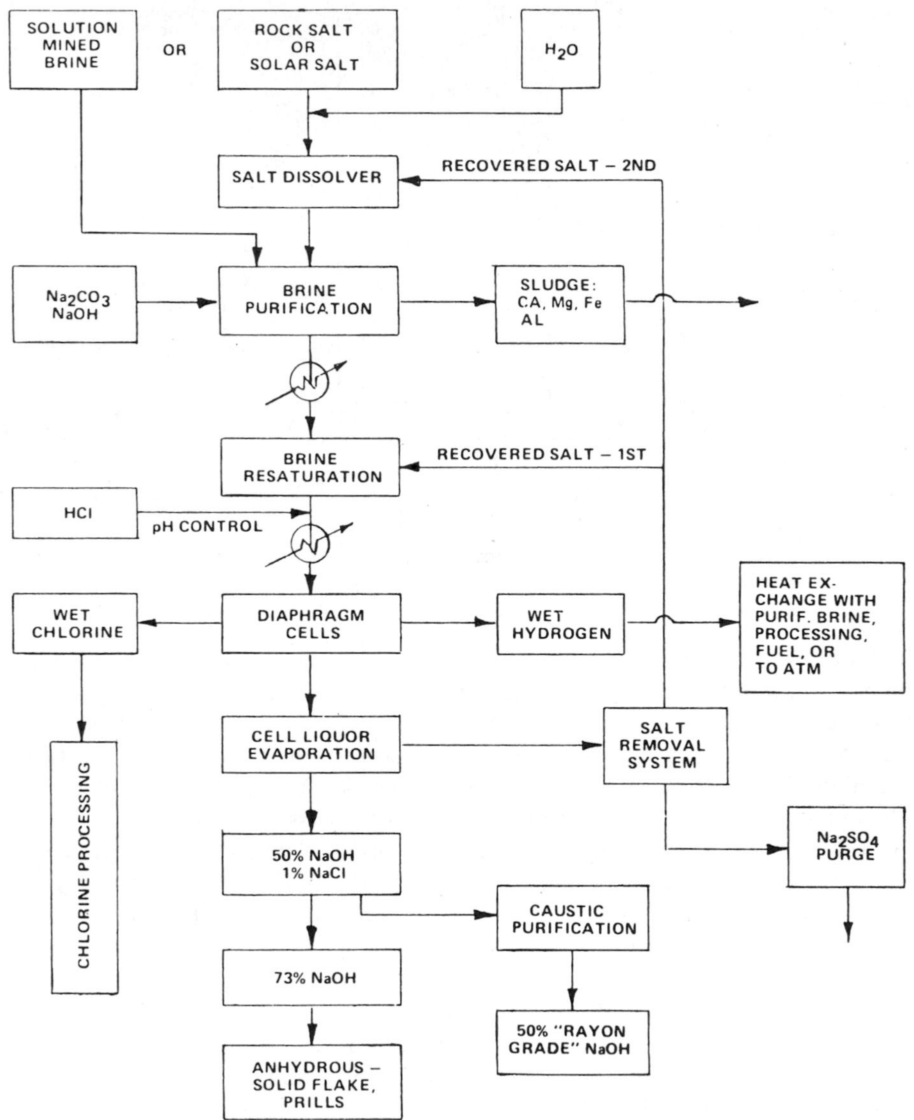

Fig. 1-12. Chlorine-caustic soda plant by the Hooker diaphragm cell process (courtesy of Occidental Chemical Corporation).

The products of this electrolysis process are chlorine gas, hydrogen gas (both of which are saturated with water vapor), and the spent cell liquor (caustic).

The gas leaving the cathode is about 99.8 percent hydrogen. It is scrubbed with water both to cool it and to remove any traces of salt or caustic, and it is then compressed for supplying various processes or is used as fuel.

The gas leaving the anode is about 97.5 percent chlorine. The remainder usually consists of a mixture of water vapor, oxygen, nitrogen, and carbon dioxide. At this point the gas is hot (210°F), moist, and extremely corrosive. It must be cooled and dried. Oddly enough, it is usually cooled by direct contact with water in a packed tower and then dried by scrubbing with sulfuric acid. The dried chlorine is then compressed to about 60 psi. After compression, it is fed into a special fractionating tower to remove impurities such as chloroform and chlorinated hydrocarbons. This innovation was introduced about 1930 in an effort to clean the chlorine so that it could be more easily handled by conventional chlorination equipment used for potable water and wastewater treatment. The impurities are removed at the bottom of the tower as a solution containing little or no chlorine. Usually these impurities are less than 0.2 percent by weight in untreated liquid chlorine.

After leaving the fractionating tower, the chlorine gas is then liquefied by refrigeration and pumped to storage tanks, from which it is then pumped into tank cars and ton containers. The trend today is for the manufacturer to ship liquid chlorine in tank cars to packagers, who use the cars as storage from which to fill 150-lb and ton cylinders, which are then shipped directly to the consumer. Each distributor has a liquid bleach manufacturing operation in order to utilize the "snift gas" that would otherwise be wasted in the cylinder-filling operation.

Manufacturing plants also have the problem of recovering the snift or blow gas as well as the chlorine lost in the water used for cooling the gas. Recovery of chlorine from these two waste sources is usually accomplished by some patented process. The Hooker process uses water to absorb the chlorine in the snift gas. This water is then used in the cooler. Upon leaving the cooler, it is heated with steam and then acidified, thus stripping the water of chlorine, which is then put back into the packaging cycle. The Diamond–Alkali process uses carbon tetrachloride to absorb chlorine; the carbon tetrachloride is then heated and stripped of chlorine.[17]

The spent liquor from the cells usually contains about 11.5 percent NaOH and 16 percent NaCl. Before this can be marketed, the salt must be removed, and the concentration of caustic must be raised to the optimum 50 percent. To raise it to this concentration, the liquid is first passed through a double- or triple-effect evaporator. As it proceeds through the evaporation cycle, the NaOH concentration increases, and the salt crystallizes and is separated from the caustic liquor by decantation and filtration. The salt is then washed free of caustic, dissolved, and recycled through the brine system. More salt is removed from the 50 percent caustic by cooling and settling.

According to Faraday's laws, one coulomb of electricity deposits exactly 0.00111801 g of silver; or, said another way, a current of one amp deposits 0.00111801 g of silver in one second. According to the second law, the quantity of electricity that liberates one gram equivalent weight of an element is the same for all elements. Because the equivalent weight of silver is 107.880, this quantity must be:

$$\frac{107.880}{0.00111801} = 96{,}493 \text{ coulomb} \tag{1-8}$$

Therefore, from Faraday's law we know that 96,493 coulomb (1 faraday) will liberate 1.0080 g of hydrogen and 35.457 g of chlorine in the electrolysis of salt. Converting this to amperes required per pound of chlorine per day, we get:

$$\frac{\text{amp} \times 86{,}400 \text{ sec/day}}{96{,}493 \text{ coulomb}} \times \frac{35.457 \text{ g } Cl_2}{454 \text{ g/lb}} = 0.07$$

$$\text{amp} \times 0.07 = \text{lb/day } Cl_2 \tag{1-9}$$

Referring now to Table 1-3, we see that 135,000 amperes (Eltech) with a current efficiency of 95.8 percent will produce 9053 lb chlorine per day. Check: 135,000 × 0.07 × 95.8 = 9053

Mercury Cells

History. The method of producing chlorine and caustic utilizing a mercury cathode was discovered simultaneously by two men on different continents. Each discoverer was unaware of the other's efforts. One was an American, Hamilton Y. Castner; the other an Austrian, Karl Kellner. Both applied for patents in 1892.

The first Castner cell installation (a demonstration plant for Mathieson Chemical Company, at Saltville, Virginia, in 1897, designed for 550-ampere operation), was later moved to Niagara Falls, New York, where it was operated successfully until 1960. At that time it was replaced by the E.11, the 1961 version of the Olin-Mathieson mercury cells, with a capacity of 100,000 amperes.[17, 20]

In the 1970s mercury cell operations in North America were found to be discharging effluents containing mercury in excess of safe limits established by environmental agencies. Some installations were shut down, but those that were not shut down improved the process to a point where the mercury loss was well below the maximum allowed contamination levels.

Although 70 percent of the chlorine produced today in the United States is by the diaphragm cell, the mercury cell is still a formidable competitor because of its ability to produce a 50 percent caustic solution without any further concentration procedures. This is one of the major advantages of the mercury cell.

The mercury cell has been improved considerably. Replacement of the "soluble" graphite anodes with DSA®s has increased the power efficiency of the cell as well as other overall operating features such as longer anode life, a smaller gap between the anode and the amalgam cathode, and less maintenance. Cleaning up the mercury loss has also contributed to the increased efficiency and lower operating cost of the cell.

It is interesting to note that Uhde of Germany, one of the leading manufacturers of mercury cells, lists 91 installations worldwide producing 10,886 tons of chlorine/day. Moreover 8 of these installations were commissioned in 1983 with a production of 1075 tons of chlorine/day.

Recent Developments. A parallel development with the changeover to DSA®s was the addition of computer monitoring and control. This led to improved short-circuit protection and a reduction of the energy consumption by computer-controlled anode adjustments. In view of the drastic increase in power costs during the late 1970s, this improvement was of great significance. The other important task, which was relentlessly pursued by the industry, was the installation of devices to reduce the incidents of mercury emissions. Furthermore, computer programming promotes the optimization of cell size, number of cells, optimum current density as a function of power cost, and capital cost.

Manufacturers. The following is a description of the various mercury cells available for use by the chlor-alkali industry.[84]

1. The Uhde Cell. These cells are available with cathode surface areas from 4 to 30 m^2 for chlorine production rates from 10 to 1000 tons/day for the complete cell installation. The anodes are suspended in groups by carrying frames supported near the cells on transverse girders equipped with lifting gear. The automatic equipment for protection and adjustment of the anodes depends upon the shunt measurement of the currents, which is controlled by a central main frame computer. This arrangement achieves the selection of the optimum K-factor for each cell. The vertical decomposers are provided with hydrogen coolers and are situated at the end of each cell. The amalgam flows into the decomposer by gravity. See Table 1-10 for the characteristics of this cell. Figure 1-13 illustrates the Uhde mercury cell.
2. De Nora Cell. The size of these cells varies from 4.5 to 36 m^2. This corresponds to electric currents from 4.5 to 400 kA. The cells are covered by a flexible, multilayer sheet of elastomer spread over the cell trough, and this cover is supported by the anode rods while providing a seal for them. The DSA anodes are held rigidly in strong carrying frames that are automatically adjusted by electric motors. Individual anode adjustment is

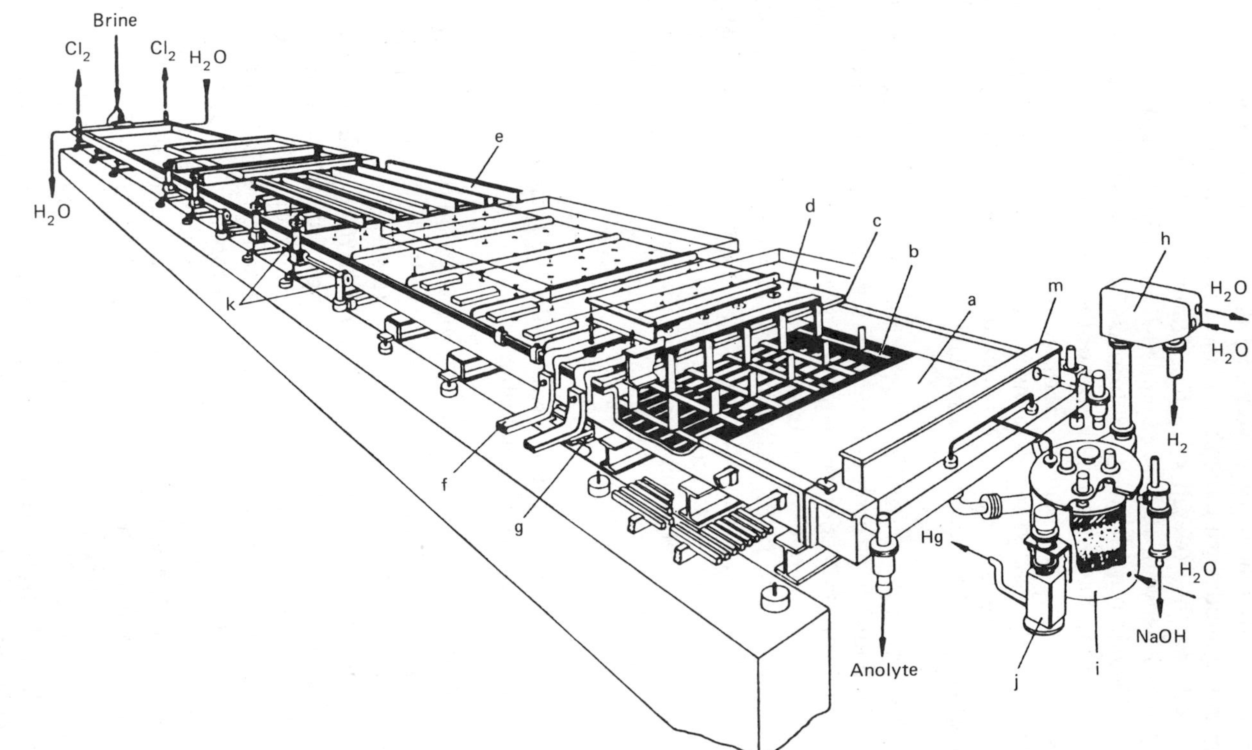

Fig. 1-13. Uhde mercury cell. (a) Cell base; (b) anode; (c) cover seal; (d) cell cover; (e) group adjusting gear; (f) intercell bus bar; (g) short-circuit switch; (h) hydrogen cooler; (i) vertical decomposer; (j) mercury pump; (k) anode adjusting gear; (l) inlet box; (m) end box (courtesy of VCH Publishers, Inc., New York).

not provided. Improved circulation of brine and gas removal within the cells reduced energy consumption. Consequently the brine concentration was increased from the usual 35–40 g/L to 60–70 g/L. This allowed about a 40 percent reduction in the brine recirculation rate. The graphite catalyst in the vertical decomposer is activated with molybdenum. The characteristics of these cells are shown in Table 1-10.

3. Krebskosmo Cell. The brine flows through a flowmeter on the inlet box and is distributed evenly over the width of the cell. The depleted brine is withdrawn at the end of the box. The anodes are fixed by copper rods to the anode carrier, where they are sealed to the rubber-lined steel cover by polytetrafluoroethylene (PTFE) bellows. The anodes are individually adjustable or as a group, and can be adjusted either manually or by electric motors.[84] The lifting gear is supported by the cell cover. Pressure sensors in the mercury circulation system will shut down the cell immediately if the mercury pump fails. The short-circuit switches are situated under the cells. See Table 1-10 for this cell's operating characteristics.
4. The Olin Mathiesen Cell. The special feature of this cell lies in the system of mounting and adjusting the anodes. Above each row of anode rods, a U-shaped copper or aluminum bus bar serves also as a support for the anode lifting gear; the anode rods are bolted to the U-shaped bus bar. The anodes are adjusted as a group, either manually or by a remote computer. This unit is identified as the remote computerized anode adjuster (RCAA) system. The electric currents are measured independently of the cell potentials by means of reed contacts. See Table 1-10 for operating characteristics of this cell.
5. The Solvay Cell. The bus bars on this cell are made primarily of aluminum. Above the cells is a cover that also serves as a convenient walkway that provides access to the anode rods. The titanium anodes are specially coated and are automatically adjusted by a computer. The vertical decomposers are located under the cells. The MAT type cell characteristics are shown in Table 1-10.

Occupational Health. Anyone working in the cell area must undergo regular health checks. The Western Chlorine Manufacturers in the Bureau International Technique Chlor (BITC) have prepared a "Mercury Code of Practice," in which rules are recommended for dealing with mercury.[84] These rules include protective measures and medical tests. The USEPA has established 18 rules relating to the cleanliness of cell rooms. Adherence to these rules eliminates any danger to the health of personnel caused by mercury. The maximum allowable concentration or threshold limit value (TLV) of mercury in the atmosphere is between 0.025 and 0.100 mg/m^3.[84]

Chemistry of Electrolysis. The mercury cell has two essential parts: (1) the electrolyzer and (2) the amalgam decomposer. In the electrolyzer, a salt solution is electrolyzed, making use of a DSA® anode and a flowing mercury

Table 1-10 Characteristics of Modern Mercury Cells (Courtesy of VCH Publishers, Inc., New York)

	Manufacturer					
Characteristic	*Uhde*	*De Nora*	*Krebskosmo*	*Olin Mathiesen*	*Solvay*	*Krebs Paris*
Cell type	300–100	24M2	232–70	E 812	MAT 17	15 KFM
Cathode area, m^2	30.74	26.4	23.2	28.8	17	15.4
Cathode dimensions, $l \times b$, m^2	14.6 × 2.1	12.6 × 2.1	14.4 × 1.61	14.8 × 1.94	12.6 × 1.8	9.6 × 1.6
Slope of cell base, %	1.5	2.0	1.8	1.5	1.7	
Rated current, kA	350	270	300	288	170	160
Max. current density, kA/m^2	12.5	13	13	10	10	10.4
Cell voltage at 10 kA/m^2, V	4.25	3.95	4.25	4.24	4.10	4.30
Number of anodes	54	48	36	96	96	24
Stems per anode	4	4	4	2	1	4
Number of intercell bus bars	36	32	18	24	24	12
Quantity of mercury per cell, kg	5000	4550	2750	3800		1650
Energy requirement per tonne of Cl_2, kW h d.c.	3300	3080	3300	3300	3200	3400

cathode. Chlorine gas is liberated at the anode, and sodium is deposited at the surface of the flowing mercury cathode, in which it dissolves to form a liquid amalgam. This amalgam flows into the decomposer, where it is decomposed with water to form sodium hydroxide and hydrogen gas. (See Fig. 1-14.)

The principal reactions are as follows:

1. Electrolyzer

a. At the anode:

$$Cl^- = \tfrac{1}{2}Cl_2 + e^- \qquad (1\text{-}10)$$

b. At the cathode:

$$Na^+ + (Hg)^+ + e^- = Na(Hg) \qquad (1\text{-}11)$$

Overall reaction:

$$NaCl + (Hg) \xrightarrow{\text{1 faraday}} Na(Hg) + \tfrac{1}{2}Cl_2 \qquad (1\text{-}12)$$

2. Decomposer

a. At the anode:

$$Na(Hg) = Na^+ + (Hg) + e^- \qquad (1\text{-}13)$$

b. At the cathode:

$$H_2O + e^- = OH^- + \tfrac{1}{2}H_2 \qquad (1\text{-}14)$$

Overall reaction:

$$Na(Hg) + H_2O \xrightarrow{\text{1 faraday}} NaOH + \tfrac{1}{2}H_2 + (Hg) \qquad (1\text{-}15)$$

The fundamental differences between this and the diaphragm cell are as follows: The spent brine or anolyte is withdrawn separately from the mercury amalgam. The caustic is produced in the decomposer as a by-product resulting from the preparation of the amalgam to be returned as the mercury cathode. The amalgam is the catholyte and does not mix with the anolyte. Both processes use about 1.7 tons of salt per ton of chlorine produced.

The net result is essentially the same as with diaphragm cells. The ingredients are the same, except for the inventory of mercury required.

Electrolyzer. The purified, saturated (305 g/L) alkaline brine solution is fed to the electrolyzer portion of the cells when the pH is adjusted to a range of 2.5–5 with HCl. This is somewhat dependent upon the amount of $CaSO_4$ that

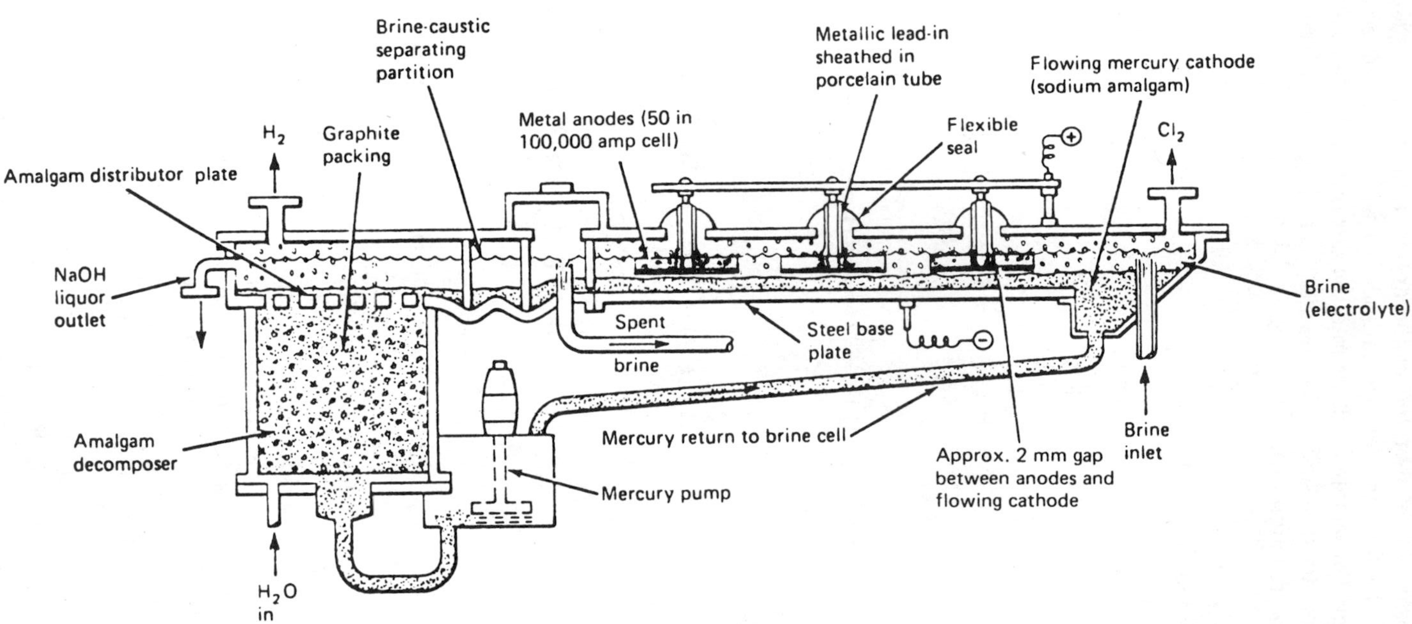

Fig. 1-14. Olin E-510 mercury cell with metal anodes (courtesy Olin Corporation).

can be tolerated in the brine. The pressure on the anode side is kept at atmospheric ±15 mm Hg. The direct current in the specified amount of the cell rating (Olin-Mathiesen E-11 is 100,000 amperes) is applied at a voltage of 4–4.5 V between the metal anode and the mercury cathode, with the chlorine being liberated at the anode. The spent brine is monitored so that it contains 260–280 g/L of NaCl as it leaves the cell and so that the temperature does not exceed 85°C. This is accomplished by regulation of brine feed to the cells. The spent brine is dechlorinated by stripping, concentrated by contact with solid salt, treated with NaOH to pH 10, settled, filtered, and recycled to the cells.

Decomposer. The decomposer is a vertical tower packed with lumps of broken graphite. A distributor plate spreads the amalgam over the top of the packing. Purified and softened water enters the bottom of this vertical tower below the packing and overflows as 50 percent sodium hydroxide above the packing. Sodium amalgam is decomposed in contact with the graphite packing and water, to form sodium hydroxide and hydrogen. Hydrogen is collected from the top of the decomposer. It is wet and contains some mercury vapor and entrained caustic spray. These contaminants are removed mainly by cooling in a scrubber or a condenser.

The caustic solution produced by this method is 50 percent (diaphragm cells produce 11–12 percent caustic and membrane cells 30–40 percent). It is customary to filter the 50 percent caustic solution for the removal of graphite particles picked up in the decomposer. These particles also contain some mercury, which is recovered from the sludge of the filtering operation.

The chlorine produced from this cell is treated in the same way as that produced from the diaphragm cells.

Operating Characteristics. Table 1-11 is a tabulation of mercury cell operating data. These data were furnished by the electrolyzer manufacturers shown in this table.

OTHER PROCESSES

Salt Process

The salt process for the production of chlorine is based upon the reaction between sodium chloride and nitric acid. It has been known for a long time that chlorides are changed to nitrates. Aqua regia was made by the alchemists in the eighth century from a mixture of niter, salt, and sulfuric acid. Evolution of chlorine in this reaction was simply a by-product.

Variations in fertilizer industry requirements and other lesser factors prompted investigation into the production of nitrate and chlorine from salt. This process was put on a commercial basis by Allied Chemical Company at Hopewell, Virginia, in 1936. The overall reaction of this process is:

Table 1-11 Operating Data: Mercury Cells

	Manufacturer and Model			
	Krebs ZTE-100-10M	*Uhde**	*deNora 30M2*	*Olin E-812*
Operating current (kA)	100	300–350	330	300
Cell to cell voltage (metal anodes)	4.15	4–4.25	3.9	4.24
Cathode current density (kA/m^2)	10	10–12	10	10
Power (kWh/ton Cl_2)	3000	3100–3300	3070	3310
Current efficiency (% NaCl)	96	96–98	96	97
Anode material	activated titanium	coated titanium	DSA®	metal
Anode life (yrs.)	3–5	5	2	2
Mercury inventory (lb/cell)	2900	9130	9500	7304
Caustic strength (percent)	50	50	50	50
Mercury loss (lb/ton Cl_2)	0.01 0.02	<.003	<.001	0.23
Chlorine production (ton/day)	3.3		11.0	9.9
Caustic production (ton/day)	3.75		12.39	11.11

*Typical analyses—Uhde cell with metal anodes

Chlorine Gas (by vol.)		*Caustic Soda Solution*		*Hydrogen*	
Cl_2	99.5%	NaOH	50%	H_2	99/9%
CO_2	0.2%	NaCl	30 ppm	Hg	1–10 μg/m^3
H_2	0.1%	Na_2CO_3	200 ppm		
Air	0.2%	Fe	2 ppm		
		Hg	0.05–3 mg/l		

$$3NaCl + 4HNO_3 \rightarrow 3NaNO_3 + Cl_2 + NOCl + 2H_2O \quad (1\text{-}16)$$

Refinements of this process include oxidation of the nitrosyl chloride for further recovery of chlorine:

$$2NOCl + O_2 \rightarrow N_2O_4 + Cl_2 \quad (1\text{-}17)$$

The nitrogen tetroxide can be taken as a product or recycled for acid for sodium nitrate manufacture.

In this process, dilute nitric acid (55 percent or less) is first concentrated by evaporation to 63–66 percent, mixed with sodium chloride, and heated with steam, where the reaction takes place. Nitrosyl chloride, chlorine, and sodium nitrate are produced in equal molar quantities. The solution is stripped of nitrosyl chloride and chlorine, which are then scrubbed, dried, and liquefied with refrigerated brine. The nitrosyl chloride–chlorine mixture is then put through a separating column; the chlorine leaves as a gas at the top, and the nitrosyl chloride leaves at the bottom as a liquid. The latter is sent to a recovery operation, and the chlorine is reliquefied and sent to storage.[17]

HCl Oxidation Processes

In recent years the market for chlorine has increased at a much greater rate than that for caustic, while the market for hydrochloric acid has declined. This situation has created a real demand for the production of chlorine from by-product hydrochloric acid. This demand has revived the Deacon process, which is attractive because of its simplicity. It has a mildly exothermic reaction, with low electrical power and thermal needs. The reaction—HCl oxidation—takes place in the vapor phase over a copper base catalyst as follows:[17]

$$4HCl + O_2 \xrightarrow{450\text{-}650^\circ C} 2Cl_2 + 2H_2O \tag{1-18}$$

There are no side reactions or competing reactions: the principal problem is to develop operating conditions that best balance the increased rate of reaction at higher temperatures against higher yields at lower temperatures. The Air Reduction Company has developed an improved Deacon process that makes it practical to produce chlorine at the one ton per day level from by-product HCl at about 27 percent chlorine by volume with air and about 90 percent chlorine with 95 percent oxygen. For economical commercial production, the size of the plant is limited to about 15 tons per day.[3]

The utilization of metal chlorides formed by hydrochloric acid with the metal oxide is followed by decomposition of the metallic chlorides by oxygen and heat.

The Grosvenor Miller process, with a fixed-bed system, utilizes the following reactions:[17]

$$Fe_2O + 6HCl \xrightarrow{250\text{-}300^\circ C} 2FeCl_3 + 3H_2O \tag{1-19}$$

$$2FeCl_3 + 1\tfrac{1}{2}O_2 \xrightarrow{475\text{-}500^\circ C} Fe_2O_3 + 3Cl_2 \tag{1-20}$$

A product gas composition approaching a maximum of 70 percent chlorine can be obtained in this process using a three- to five-bed continuous system. The procedure is briefly described as follows: Reactor I is laden with dry ferric chloride at 250–300°C and is fed oxygen.[3] The ferric chloride is converted to chlorine and ferric oxide. The gas contains approximately 30 percent chlorine, 70 percent unreacted oxygen, and some air. This mixture passes into Reactor II, which is maintained as a mixed bed of ferric oxide and ferric chloride at 500°C to serve as both chlorinator and oxidizer. Both hydrochloric acid vapor and oxygen (with excess HCl) are passed into Reactor II so that the HCl reacts with Fe_2O_3. The chlorine gas does not react, and so it and the excess HCl pass through Reactor II. The effluent gas from this reactor, containing chlorine gas, HCl vapor, steam, and excess oxygen, is passed into Reactor III (maintained at 250–400°C). This reactor was previously Reactor I, and thus is oxygen-laden. Reactor III strips the HCl vapor from the gas. Chlorine does not react, and so it passes through, carrying with it steam and excess oxygen until this reactor becomes saturated with chlorides. At this point, the functions of Reactor III are transposed to those of Reactor I, and gas flow is III to II to I. There are other variations of this process, such as the Dow Moving Bed process and others using molten metallic chlorides as catalysts.

The Kel-Chlor process[23] is a modification of the Deacon process, which has always been plagued with chemical equilibrium problems that resulted in low chlorine yields.

The Kel-Chlor process solves the equilibrium problems by combining a very active catalyst (nitrogenous compounds) with a powerful dehydrating agent (sulfuric acid), which reduces the activity of the steam to a negligible value and thus allows the reaction to proceed to completion. The result is a better HCl-to-chlorine conversion ratio. Plant experience has also demonstrated that corrosion problems have been solved as well. The Kellog Co. has claimed that chlorine from waste hydrochloric acid can be produced for $20/ton. Chlorine made by this process can be refined to a higher purity than ordinary commercial chlorine prepared by electrolysis.[23]

Electrolysis of Hydrochloric Acid Solutions

The Hoechst–Uhde Process.[24] Considerable quantities of aqueous hydrochloric acid or hydrogen chloride gas result each year as by-products from a variety of chemical manufacturing activities. These by-products are a form of industrial waste and are difficult to dispose of; therefore, a process utilizing them is of special interest. The I. G. Farben Industries began to develop a process in 1938 at the Bitterfield, Germany plant using bipolar diaphragm cells. The early cells had a limited production capacity, usually less than 50 tons of chlorine per day. A system producing one ton of chlorine per day will yield 10,000 cu ft of hydrogen; it requires 2060 lb of 100 percent muriatic acid, 1900 kWh AC, and 10,000 gal of cooling water at 15°C for Cl_2 and H_2. This process has had many fabricating and operating problems. A joint experimental effort by Farbwerke Hoechst AG and Friedrich Uhde GmbH resulted in a successful design, which was first constructed in 1963.

The first electrolysis unit in the United States was built in 1971–72 at the Mobay Chemical Corp. at Baytown, Texas.[24] This electrolysis plant was built by the Hoechst-Uhde Corp. and has a capacity of 180 metric tons per day chlorine (198 short tons). The Mobay operation uses waste hydrogen chloride gas from the manufacture of isocyanate. In addition to the manufacture of isocyanates, hydrochloric acid is a waste product in the manufacture of chlorine-bearing solvents, production of raw materials for detergents, production of chlorination products, and production of frigens and silicones.

Today's Hoechst-Uhde electrolyzers consist of 30 to 36 single elements working at current densities from 4 to 5 kA/m^2. The chlorine capacity varies from 90 to 180 metric tons of chlorine per day. A typical 30-element unit occupies a space of only 12 × 16 ft. Several of these plants are operating in Germany, France, Japan, and the United States.

IMPURITIES IN THE MANUFACTURE OF CHLORINE

Historical Background

Special attention has to be given to the production of chlorine when it is to be used in the treatment of potable water, wastewater, and water reclamation plants. The total amount of chlorine used in these plants is only about 6 percent of all the chlorine produced in the United States; the industry thinks in terms of thousands of pounds per hour, whereas treatment plant users think in terms of hundreds of pounds per day. Moreover, for all chlorine use with conventional chlorination equipment, the equipment meters and controls the chlorine in a vacuum system of about 18–20 inches of Hg, which is a major safety feature in its handling. This is one of the major reasons why a catastrophic leak is unlikely ever to occur at these installations. Catastrophic leaks are usually labeled as a one-in-a-million possibility. In view of this, the chlorine manufacturers have done a commendable job in providing a clean and moisture-free product for the users of chlorination and dechlorination equipment.

In the early days of chlorination (1920s), metering equipment was continually plagued by fouling from what was commonly called "gunk" or "taffy." It was difficult to prove the origin of this gunk until Wallace and Tiernan developed the bell-jar chlorinator (1922). This chlorinator employed a self-cleaning type of chlorine pressure-reducing valve (visible through a glass dome), which operated under a vacuum. When the gunk plugged this valve, stopping the flow of chlorine, the valve would automatically shift from a throttled position to a wide-open position by raising the water level, which in turn raised a float-operated mechanism. In most cases, the valve in this wide-open position would purge itself of the gunk, thus allowing the inrush of gas to spew it all over the bell jar. This would force the water level down, and the chlorinator would automatically resume operation. In the meantime, the customer was able to back up his complaint with a sample of gunk. After many of these complaints, the chlorine manufacturers introduced into their manufacturing process a fractionating tower[18] in an attempt to clean up their product and eliminate the gunk (ca. 1935).

In July 1977 occasional concentrations of CCl_4 above the detectable level (1 mg/L) were found in the finished water at Belmont and Queen Lane,[52] two Philadelphia plants receiving waters from the Schuylkill River. A subsequent investigation traced the source of these high concentrations to the chlorine supplied to these plants. When the chlorine supply was changed to another manufacturer, the problem disappeared. It was revealed that the manufacturer of the chlorine in question was using a carbon tetrachloride scrubbing system in the chlorine–hydrogen air separation system. The problem was solved by segregating the chlorine to be used for potable water sales from the chlorine extracted from the separation system. This episode was thoroughly examined on a national level by the Chlorine Institute and representative chlorine manufacturers in order to establish an interim maximum level for CCl_4 in chlorine used in potable water treatment. This interim level was established at 100 mg/L by EPA and agreed to by the manufacturers' association pending an assessment of what the chlorine industry is currently producing and what it is capable of producing with available technology.[52]

The current AWWA chlorine purity standard is contained in the ANSI/AWWA report of June 7, 1981.[53] This report limits the maximum concentration level of carbon tetrachloride to not more than 150 mg/L (0.015 percent). This report states that the chlorine supplied under this standard shall be "99.5 percent pure" by volume as obtained by vaporizing the chlorine as determined by ASTM Standard E 412-70 Assay of Liquid Chlorine (Zinc Amalgam Method). Other limitations listed are: moisture not to exceed 150 ppm (0.015 percent) by weight; heavy metals not to exceed 30 ppm (0.003 percent) expressed as Pb; mercury not to exceed 1 ppm (0.0001 percent); and nonvolatile total residue not to exceed 50 ppm (0.005 percent) by weight in chlorine tank cars and tank trucks and 150 ppm (0.015 percent) by weight as packaged in ton containers and cylinders.

Sources of Impurities

The major sources of impurities found in chlorine are:

1. Moisture entrapment during packaging.
2. Organic impurities in the salt.
3. Hydrocarbon sources introduced from valve lubricant, pump seals, and various packings used throughout the manufacturing process.
4. Recovery systems used to separate chlorine produced from entrained hydrogen and air.

These impurities are classified as follows:[25]

Gases	*Volatile Liquids*	*Volatile Solids*	*Nonvolatile Solids*
CO_2	bromine	hexachlorbenzene	$FeCl_3 \cdot 6H_2O$
H_2	carbon tetrachloride	hexachlorethane	
O_2	carbonyl chloride (phosgene)		$Fe_2(SO_4)_3 \cdot 9H_2O$
N_2	chloroform		nitrosyl chloride
NCl_3	HCl		nitrogen tetroxide
	methylene chloride		H_2SO_4
	moisture		

Consequences of Impurities

Public Health. During the days when chlorine manufacturers used graphite or carbon anodes, there was some fear that these anodes would form carbontetrachloride and chloroform in the production of chlorine. It was thought that these compounds would be a public health risk. However no evidence ever did confirm this fear. Now, however, in the 1990s, there are no known manufacturers in North America that use graphite or carbon anodes. There may be some such use in Third World countries.

Effect upon Chlorine Control Equipment. Of all the impurities listed above, moisture in the chlorine is the worst offender because it makes the chlorine highly corrosive, leading to the formation of ferric chloride. This causes gross fouling in the metering equipment (chlorinators).

The next most offensive impurities are hexachloroethane and hexachlorobenzene. These compounds form a taffy-like substance commonly called "gunk." They also cause serious equipment fouling. They can originate from valve lubricating compounds, packings, gaskets, and so on, used in the various manufacturing components.

Most impurities in chlorine are soluble in liquid chlorine. Ferric chloride seems to plate out in thin layers on the contact surface between the liquid chlorine and metal piping, and so forth. Deposition is most notable in areas of chronic flashing, usually where there is a restriction or a turbulent flow regime. Precautions can be taken to minimize the difficulty from ferric chloride. Ferric chloride appears to have the ability to spread from the liquid phase to the vapor phase so that the contamination carries through the chlorination equipment. The largest deposits occur at points of pressure reduction and areas of reliquefaction.

Both hexachlorethane and hexachlorobenzene tend to sublime at room temperature, and are usually found deposited at points of pressure reduction in the metering and control equipment.

For the benefit of consumers using chlorine control equipment in the sanitary field, chlorine producers have long made it a practice to limit the total impurities in liquid chlorine to about 200 mg/L and the nonvolatile impurities to 20–30 mg/L.[25]

Nitrogen Trichloride in Liquid Chlorine

Occurrence, Formation, and Significance. Nitrogen trichloride was first obtained in 1811 from the action of chlorine on a solution of ammonium chloride by Dulong, who lost an eye and three fingers as the result of an explosion. When generated in the laboratory, this compound appears as a yellow oil with a pungent odor resembling somewhat the odor of chlorine. It is practically insoluble in water, but easily soluble in most organic solvents such as benzene, carbon tetrachloride, ether, and so on. It has been reported that a drop of the oil explodes violently when touched with a feather dipped in turpentine.[54, 59]

Nitrogen trichloride is not listed as an impurity in chlorine produced in the United States or Canada, where producers have long been aware of the dangers of NCl_3 in liquid chlorine. Its occurrence in the manufacture of liquid chlorine in the United States and Canada has been virtually nil since about 1930.

Formation of nitrogen trichloride in the production of chlorine occurs if ammonia nitrogen is present in the brine fed to the electrolytic cells. It is soluble to the extent of 7.3 mg/L in liquid chlorine, whereas it is not soluble to any extent in either water or concentrated sulfuric acid. The latter is used in the production of chlorine to remove the moisture content. Therefore, once NCl_3 forms in the electrolytic cells, it will pass with the chlorine gas through the coolers, scrubbers (H_2SO_4), and acid-sealed pumps, and be condensed with the liquid chlorine. The greatest danger of explosion occurs at the point when the liquid chlorine vessel (ton container, evaporator, etc.) becomes exhausted of liquid chlorine, and only the chlorine vapor remains. This is the result of the solubility of NCl_3 in liquid chlorine. It concentrates itself in the layer of liquid chlorine next to the vapor phase. As the liquid is gradually used up, the concentration of NCl_3 keeps increasing in the small area of the vapor–liquid interface.

Uses of Nitrogen Chloride. In spite of all the dangers that existed from the presence of nitrogen chloride, it is interesting to note that it was used to bleach flour all across the United States* from about the middle or late 1920s until the entrance of the United States into World War II in 1941, as there was no more ammonia N available to the public. This was accomplished by a generating system designed and marketed by Wallace & Tiernan Co. It was produced by mixing an ammonium chloride solution with a solution of HOCl produced by a conventional gas chlorinator. The ratio of chlorine to ammonia N was sufficient to produce nitrogen chloride when the mixed solution was aerated in a packed tower. This was all done without any adverse effects or apparent hazards. When ammonia N was no longer available, Wallace & Tiernan had to convert the system to the generation of chlorine dioxide as the bleaching agent.[87]

Explosions Caused by Nitrogen Trichloride. The last reported chlorine explosions from NCl_3 in liquid chlorine in the United States and Canada were the two ton container explosions at chlorination stations operated by the New York City Water Department ca. 1929. These explosions occurred when the cylinders had been emptied to zero gauge pressure. The cause of these explosions was traced to the presence of nitrogen trichloride in the liquid chlorine. The ammonia nitrogen concentration in the prepared brine solution for the electrolytic cells was found to be about 15 mg/L, and the NCl_3 in the liquid chlorine was 500 mg/L (0.5 g/L).[54] Theoretically each mg/L of ammonia nitro-

*White's first job with W&T in 1937 was inspecting and monitoring the operation of all the flour-bleaching equipment in the seven western states.

gen in the brine will generate 50–60 mg/L of NCl_3 in the liquid chlorine. Owing to the fact that it is difficult to predict the "safe" allowable concentration of NCl_3 in chlorine, the manufacturers' consensus is a 5 mg/L limit.[54]

There are many chlorine producers outside the United States and Canada who are unaware of the dangers of NCl_3, its occurrence, and its prevention. An evaporator explosion was reported in India in 1965. In 1981 White investigated several evaporator explosions in South America. These were all the result of ammonia N in the chlorine cell water. The NCl_3 was present in concentrations between 50 and 300 mg/L. The ammonia nitrogen in the brine makeup water varied on a weekly basis from about 1.5 to 6 mg/L. The sudden appearance of ammonia nitrogen in the plant water supply was the result of a new sewage collection system, which discharged untreated domestic wastewater into the river supply upstream from the plant.

Prevention of Nitrogen Trichloride Formation in Liquid Chlorine. The most effective method of dealing with the NCl_3 problem is to remove all the ammonia nitrogen from the brine solution before it reaches the cells. This takes care of those situations where the brine solution is prepared from rock salt that contains ammonia nitrogen.

Two methods can be used to solve the NCl_3 problem. One is to remove the ammonia nitrogen from the brine water by breakpoint chlorination; the other is to subject the chlorine gas exiting the cells to ultraviolet light within the spectrum of 3600–4400 angstrom wavelength. The UV application has to be done before the chlorine enters the scrubbers. As it is difficult to monitor the effectiveness of the NCl_3 removal by UV, it is more practical and reliable to remove the ammonia nitrogen from the brine water by breakpoint chlorination, followed by aeration to remove as much as possible of the combined chlorine residual resulting from the breakpoint procedure. This method has proved very effective and reliable in North America. The UV method has been used successfully where small quantities of NCl_3 are involved. One such plant is the PPG Industries plant at Natrium, West Virginia. The decomposition of NCl_3 by UV is only about 50 percent complete.[54]

Silica Contamination

This is a rare situation but worth mentioning, as its occurrence caused considerable aggravation to one consumer.

The end result of chlorine contamination by silica is the formation of white crystals (SiO_2) at the entrance of the chlorine gas into the injector assembly. The injector inlet port plugs up rapidly and renders the chlorinator completely inoperable.

The source of silica is thought to be silica-contaminated brine water, or silicone grease used in valves by the chlorine packager. Silica-contaminated brine water is most likely to occur when the production of chlorine is a by-product of a metal refining process such as the extraction of magnesium from a magnesium chloride ore.

If the brine is contaminated with silica, electrolysis will convert the silica present to silicon tetrachloride. When the liquid chlorine is vaporized, the chlorine vapor carries the $SiCl_4$ through the chlorinator and into the injector. When chlorine comes in contact with the water in the injector, the $SiCl_4$ is immediately transformed into SiO_2 crystals, which eventually plug the injector inlet.

PHYSICAL AND CHEMICAL PROPERTIES

General

Chlorine has an atomic number of 17 (number of excess positive charges on the atomic nucleus) and an atomic weight of 35.457. Molecular chlorine, Cl_2, has a weight of 70.914. Two isotopes of chlorine, Cl^{35} and Cl^{37} occur naturally, and at least five other isotopes have been artificially produced.[1] Ordinary atomic chlorine consists of a mixture of about 75.4 percent Cl^{35} and 24.6 percent Cl^{37}. Chlorine usually forms univalent compounds, but it can also combine with a valence of 3, 4, 5, and 7. (See Chapter 4.)

In its elemental form, chlorine is a greenish yellow gas that can be readily compressed into a clear, amber-colored liquid that solidifies at atmospheric pressure at about −150°F. Chlorine gas forms into a soft ice upon contact with moisture at 49.3°F and at atmospheric pressure. (This is chlorine hydrate, $Cl_2 \cdot 8H_2O$.)

In commerce, chlorine is always packaged as a liquefied gas under pressure in steel containers. The liquid is about one and one-half times as heavy as water (denser), and the gas is about two and one-half times as heavy as air. The liquid vaporizes readily at normal atmospheric temperature and pressure. It has an unmistakable irritating, penetrating, and pungent odor. The properties of chlorine gas and liquid are listed in Tables 1-12 and 1-13, respectively, which are supplemented by Figs. 1 through 8 in the appendix.

Some of these properties merit comment.

Critical Properties

The *critical temperature* is that above which chlorine exists only as a gas (291.2°F) despite the pressure. The *critical pressure* is the vapor pressure of liquid chlorine at this critical temperature. The *critical density* is the mass of a unit volume of chlorine at the critical pressure and temperature.

Compressibility Coefficient

The compressibility coefficient of liquid chlorine is greater than that of any other liquid element. It represents the percent decrease in volume correspond-

Table 1-12 Properties of Chlorine Gas

	Ref.
Symbol: Cl_2	
Atomic wt.: 35.457	
Atomic number: 17	
Isotopes: 33, 34, 35, 36, 37, 38, 39	60
Density (see Appendix) at 34°F and 1 atm: 0.2006 lb/ft^3	61
Specific gravity at 32°F and 1 atm: 2.482 (air = 1)	62
Liquefying point at 1 atm: −30.1°F (−34.5°C)	67
Viscosity (see Appendix) at 68°F: 0.01325 centipoise (approximately the same as saturated steam between 1 and 10 atm)	63, 74, 76
Specific heat at constant pressure of 1 atm and 59°F: 0.115 Btu/lb/°F	64
Specific heat at constant volume at 1 atm pressure and 59°F: 0.085 Btu/lb/°F	64
Thermal conductivity at 32°F: 0.0042 Btu/hr/ft^2/ft	65
Heat of reaction with NaOH: 626 Btu/lb Cl_2 gas	
Solubility in water at 68°F and 1 atm: 7.29 g/L.	66

Combining Quantities

1 lb chlorine gas combines with

1.10 lb commercial hydrated lime (95% $Ca(OH)_2$

$$2Ca(OH)_2 + 2\,Cl_2 = Ca(OCl)_2 + CaCl_2 + 2H_2O$$

0.83 lb commercial quicklime (95% CaO)

$$2CaO + 2H_2O + 2Cl_2 = Ca(OCl)_2 + CaCl_2 + 2H_2O$$

1.13 lb caustic soda (100% NaOH)

$$2NaOH + Cl_2 = NaOCl + NaCl + H_2O$$

2.99 lb soda ash

$$2\,Na_2CO_3 + Cl_2 + H_2O = NaOCL + NaCl + 2NaHCO_3$$

Table 1-13 Properties of Liquid Chlorine

		Ref.
Critical temperature	144°C; 291.2°F	67
Critical pressure	1118.36 psia	67
Critical density	573 g/1; 35.77 lb/ft^3	67
Compressibility	0.0118% per unit vol per atm increase at 68°F	68
Density (see Appendix) at 32°F	91.67 lbs/ft^3	62
Specific gravity at 68°F	1.41 (water = 1)	69
Boiling point (liquefaction point) 1 atm	−34.5°C; −30.1°F	67
Freezing point	−100.98°C; −149.76°F	70
Viscosity (see Appendix) at 68°F	.345 centipose (approx. 0.35 × water at 68°F)	71, 75
1 volume liquid at 32°F and 1 atm	457.6 vol gas	62
1 lb liquid at 32°F and 1 atm	4.98 ft^3 gas	72
Specific heat	0.226 Btu/lb/°F	73
Latent heat of vaporization	123.8 Btu/lb at −29.3°F	70
Heat of fusion	41.2 Btu/lb at −150.7°F	73

ing to a unit increase in pressure when the liquid is held at constant temperature. This physical characteristic is the reason why the volume–temperature relationship of chlorine is very important. This is described below.

Volume–Temperature Relationship

The volume of liquid chlorine increases rapidly as its temperature increases. Because of this characteristic, coupled with noncompressibility, extreme care must be taken to prevent the possibility of hydrostatic rupture of containers or pipelines by expanding liquid chlorine due to a rise in temperature. All containers are filled to their prescribed weight of chlorine at 60°F so that 15 percent of the container volume is vapor space. Actually the *Chlorine Institute Manual* (4th ed., 1969) says on p. 5, paragraph 2.1.5b, "The maximum permitted filling density is defined by the D.O.T. [173.300(g)] as '. . . the percent ratio of the weight of gas (sic) in a container to the weight of water that the container will hold at 60°F. (One pound of water equals 27.737 cubic inches at 60°F.)' " The practical approach to an overly complicated D.O.T. definition is to know the density of water at 60°F (62.366 lb./ft^3) and the water volume of the container.

Example: A 55-ton car holds 10,564 gal of water. At 7.48 gal/ft^3 the car volume is 1412.30 ft^3 at 60°F. This weight of water amounts to 88,080 lb. Multiplying water weight by the D.O.T. factor of 1.25 gives 110,100 lb. This equals 55.05 tons of liquid chlorine (not gas, as described in the D.O.T. proclamation).

The vapor space provided by the above requirement is designed to accommodate a temperature rise sufficient to melt the fusible plug in the 150-lb cylinders and ton containers. The fusible plug metal is designed to melt at about 165°F, thus relieving pressure and preventing rupture of the container in case of fire or other exposure to high temperature.* An inspection of Fig. 8 in the appendix illustrates the volume–temperature relationship in a container loaded to its authorized limit. From this curve the container will be "skin-full" when the liquid chlorine temperature reaches 153.64°F. At this temperature (Fig. 6 in Appendix I) the vapor pressure is 290 psi, and at the upper melting temperature of the fusible metal (160°F) the vapor pressure would be about 310 psi. This appears to demonstrate that the D.O.T. definition for filling a container does not allow for enough space to match the fusible metal melting temperature of 165°F. However, in chlorine cylinders and ton containers, there fortunately exists an elastic volumetric expansion of the metal. When the containers are hydrostatically tested at 500 psig, it is not uncommon for a 3 percent expansion to be observed.[56] This would easily permit a temperature higher than 160°F without fear of rupture. Ton containers

*This assumption about 165°F as the melting point of the fusible plug is very misleading because no one has ever actually determined the melting point of a fusible plug. It is thought to be between 160–165°F. Chlorine packaged outside the United States are not known to use fusible plugs.

have an added expansion factor in the dished heads that have been described above. They can reverse from the concave posture to the convex position before rupturing. This has been observed in several cases of overpressuring ton containers,[57] due to nitrogen trichloride explosions.* This assumption is very misleading because no one has ever confirmed or actually measured the melting temperature of a fusible plug. It is thought to be about 160°F. There is no known case of a fusible plug in a ton cylinder having melted in a water or wastewater treatment plant.† The case of tank cars and stationary tanks is quite different from that of vessels with fusible plugs. Tank cars and stationary storage tanks used for potable water treatment or wastewater treatment are usually fitted with a spring-loaded Chlorine Institute safety valve combined with a breaking pin assembly that breaks at 225 psig. On cars used in pulp and paper chlorination, the safety valve relieves at 375 psig because it is air-padded at a higher pressure.

As of 1997 there are available a variety of automatic shut-off valves, as well as protective hoods on the manway covers to protect operators of tank cars or storage tanks.

Density of Chlorine Vapor

The density of chlorine vapor varies widely over changes in pressure and modestly over changes in temperature. This is a most important variable in calculating gas flow pressure drop in both vacuum and pressure systems. The relationships of vapor density for various pressures and temperatures are shown in Figs. 1 and 2 in Appendix I.

Density of Liquid Chlorine

The density of liquid chlorine varies only slightly with temperature. At 40°F it is 90.85 lb/ft^3, and at 140°F it is 79.65 lb/ft^3. (See Fig. 4 in Appendix I.)

Viscosity of Chlorine

This is the measure of internal molecular friction when a substance is in motion. It is necessary to know this property for both liquid and gaseous chlorine, as it is a variable in calculating the Reynolds number for the determination of friction losses in pipelines. The temperature–viscosity relationship for both chlorine liquid and the gas is shown in Fig. 3 in Appendix I.

---

*White has had the opportunity to investigate several NCl_3 explosions in liquid chlorine systems. All of these were where ton cylinders of chlorine were being used. In every case the dished heads of the cylinders blew outward from concave to convex. In each case the discharge piping on the evaporators ruptured.

†The worst aspect of fusible plugs is that they have been the reason for a great many serious leaks at potable water and wastewater chlorination installations. The reason is that they are made of brass, which corrodes easily due to the inherent presence of moisture in all chlorine supplies. These plugs should be replaced every four to five years.

Latent Heat of Vaporization

This is the heat required to change one mass of liquid to vapor without a change of temperature. If the liquid chlorine is at 70°F, it requires about 100 Btu to vaporize one pound of liquid chlorine. (See Fig. 5 in Appendix I.)

Vapor Pressure

This is the pressure of chlorine gas above liquid chlorine when the vapor and the liquid are in equilibrium. This pressure varies widely with temperature. It is necessary to know this relationship, particularly when the consumer is transferring chlorine from tank cars to vaporizing equipment. (See Fig. 6 in Appendix I.)

Specific Heat

This is the amount of heat required to raise the temperature of a unit weight of chlorine vapor one degree F. At atmospheric pressure and 59°F, 0.085 Btu/lb is required.

Solubility of Chlorine Gas in Water

Chlorine has a limited solubility in water. At atmospheric pressure and 68°F its solubility in water is 7.29 g/L. This does not represent the conditions for the application of chlorine in this text. Operation of chlorination equipment that produces chlorine solution is always at partial pressures (vacuum). At the vacuum levels currently being used, the maximum solubility is about 5000 mg/L. The upper limit of solubility recommended by all chlorinator manufacturers is 3500 mg/L. This arbitrary figure has been successful in preventing solution discharge systems from being adversely affected by gas pockets in the solution piping and off-gassing at the point of application.

Solubility of Liquid Chlorine in Water

This is a controversial subject. Knowledgeable people say that as soon as liquid chlorine is discharged into water, it flashes off into vapor, and during the flash-off the water temperature drops to 49°F or lower. At this temperature the chlorine vapor and the water combine to form a solid hydrate, $Cl_2 \times 8\,H_2O$, known as "chlorine ice." However, the interesting part of this phenomenon is that this chlorine ice is highly soluble in water.

Now let us examine Fig. 8 in Appendix I. This illustration shows the solubility of chlorine vapor at various water temperatures and pressures. To clarify and/or explain this property of solubility, the following discussion of an operating installation should prove helpful.

In 1940 the Standard Oil Co. in Richmond, California was using liquid chlorine to treat a cooling water system for its new wax plant.[77] The cooling water was seawater from San Francisco Bay. The chlorine was applied to the suction of a 50,000 gpm pump. The seawater temperature was about 50°F. The chlorination system was bizarre, to say the least. It consisted of six inverted 150-lb cylinders, all manifolded to a common control valve in the liquid discharge line to the pump suction. The point of liquid chlorine release was about 15 ft below the water surface adjacent to the pump suction. Figure 8 in Appendix I indicates the solubility would be approx. 12 lb/100 gal under the conditions described above. Chlorine was applied intermittently at a rate sufficient to produce a 4–5 mg/L residual at the wax plant condensers, some 5 min. after the point of chlorination.

The 5-min. chlorine demand of the seawater is usually about 1.5 mg/L. Therefore, the dosage was always approx. 6.5 mg/L. This means that the chlorine feed rate was on the order of 3900 lb/day. This is slightly less than 3 lb/min. If the solubility were 12 lb/100 gal, the 50,000 gpm pump could dissolve 6000 lb/min., which is far in excess of anything that might have been required of this installation. The chlorine was applied for a total of about 60 min./day. Therefore daily chlorine usage was approximately 180 lb/day.

The remarkable thing about this installation was the lack of any operating problems. There were no incidences of flash-off, off-gassing, or pump corrosion—all of which proves that under proper conditions liquid chlorine is in fact soluble in water.

Many years later White and Tracy[78] tried an experiment of dumping liquid chlorine in an open reservoir to see if there was any validity in using such a reservoir for the disposal of a leaking container in the case of an emergency. This took place at the Crystal Springs Reservoir, City of San Francisco Water Department. The test consisted of inverting a 150-lb cylinder and discharging the liquid chlorine at a point approximately 10 ft below the surface, where the water was quite deep. The water temperature was about 55°F.

The results were quite interesting. The liquid rising from the end of the discharge pipe came to the surface like an inverted ice cream cone. There was no flash-off or any indication of turbulence. The color of the cone was a pale amber. Within about 6 inches or less from the surface water in the reservoir the color changed to a pale green, indicating the formation of chlorine vapor. Oddly, there was no indication of any chlorine hydrate formation. Some ammonia solution was squirted onto the surface where it was thought that off-gassing might occur. Surprisingly, only a small white puff appeared on the surface just above the area where the water color had changed to a pale green. A slight chlorine odor was detected by the participants, who were some 10–12 ft from the chlorine "cone." This odor disappeared quickly after the cylinder was shut down.

To remove all doubts about the solubility of liquid chlorine in water, the reader is referred to an operating installation that happens to be unique in water and wastewater practices, the Sanitation Districts of Los Angeles County chlorination facility at the JWPCP on So. Figueroa Street in Carson, California. Over 16 years the capacity of the system rose from 30,000 lb/day to 100,000 lb/day, all this with no need to change any of the structures housing the equipment. This installation manufactures a 15,000 mg/L calcium hypochlorite solution with liquid chlorine, lime slurry, and plant effluent. The liquid chlorine is dissolved in a mixture of lime slurry and plant effluent that amounts to about 350 gpm at a chlorine feed rate of 45 tons/day = 90,000 lb/day. The water pressure at the chlorine injection point is approx. 30 psi. According to Fig. 8 in the appendix, the solubility of liquid chlorine is going to be at least 10.5 lb/100 gal of effluent plus the slurry. This calculates to approximately 53,000 lb/day of chlorine. As the system is capable of 100,000 lb/day without any problems due to "flashing" or off-gassing at the point of injection, this proves beyond a doubt that under the proper conditions liquid chlorine is highly soluble in water, whereas chlorine gas is not.

This installation is fully described in Chapter 2.

The above example establishes the fact that water can be used to absorb significant amounts of liquid chlorine—if used properly. This feature makes it incumbent upon the designers of chlorine storage areas to provide dramatic slopes to the floors under both ton cylinders and storage tanks so that chlorine spills can find their way quickly to a narrow but deep collecting slot in the floor. This delays considerably the flash-off phenomenon, to a point where the liquid can be discharged to a suitable body of water by either a pump or an eductor.

Chemical Reactions

Liquid chlorine in the absence of moisture will not attack ferrous metals; hence the use of steel containers. Since there is no such thing as absolutely "dry" liquid chlorine, extra wall thickness is provided to offset corrosion.

Liquid chlorine will attack and very quickly destroy PVC materials and rubber, hard or soft.

Dry chlorine gas will not attack ferrous metals, copper, or ferrous alloys. Dry chlorine gas will support combustion of carbon steel at 483°F.[27] Chlorine exists only as a gas above 291.2°F regardless of pressure (critical temperature).

Moist chlorine gas will destroy all ferrous metals including stainless steel, copper, and ferrous alloys. Gold, platinum, and tantalum are the only metals that are totally inert to moist chlorine gas. Silver is widely used with moist chlorine gas because the silver chloride formed upon contact with the moist gas is inert.

Aqueous solutions of chlorine are extremely corrosive. For this reason, PVC, fiberglass, Kynar, polyethylene, certain types of rubber, Saran, Kel-F, Viton, and Teflon are commonly used where both moist gas and chlorine solutions are encountered.

Chlorine reacts with ethyl alcohol and ether in trace amounts to form solid, waxy hexachloroethane. It also reacts with grease and oils to form a voluminous frothy substance. Solid complex hydrocarbons are formed by the reaction of chlorine and the various petroleum distillates.[28] At normal temperatures there are no reactions between chlorine gas and the methane derivatives, chloroform, wood alcohol, and carbon tetrachloride.

The chemical reactions of chlorine gas and chlorine solutions in the various phases of potable water and wastewater treatment are discussed in considerable detail in other chapters.

HAZARDS FROM CHLORINE VAPOR AND LIQUID

Toxic Effects

Liquid chlorine is a skin irritant and can cause severe damage, resembling a burn, to body tissues. As the liquid vaporizes rapidly to gas at atmospheric temperature and pressure, it is difficult to attribute this damage to the liquid or the gas phase. The gas in low concentrations is an irritant to mucous membranes and the respiratory system. The amount of gas inhaled determines the severity of the impairment of the respiratory system. Although it has not been recognized in the literature, there are two types of gassing by chlorine. The one usually referred to is from the fumes of chlorine gas in the dry state. The other type of gassing, and probably the more dangerous of the two, is by chlorine fumes from an aqueous solution. This occurs in the case of a *chlorine solution* leak or exposure in a confined area if the concentration of the solution is in excess of approximately 750 ppm titrable chlorine. Because the fumes are laden with moisture, they seem more tolerable to the respiratory tract, and the victim will unwittingly inhale excessive amounts of molecular chlorine. This results in a slow but significant production of pulmonary edema that could cause death by "drowning" while the victim is asleep. In this type of gassing, the victim should place his or her head downward as far as possible to drain the edema before retiring.

Gassing from dry gas is more common and much more disagreeable. As soon as the gas enters the throat area, the victim will sense a sudden stricture in this area—nature's way of signaling to prevent passage of the gas to the lungs. At this point, the victim must attempt to do two things: to get out of the area of the leak, proceeding upwind, and to take only very short breaths through the mouth. Normal breathing will cause coughing, which must be prevented if possible.

First Aid

In severe cases of inhalation, the first thing to do is to call the local fire department to administer a mixture of CO_2 and oxygen under an exhalation pressure not to exceed 4 cm water. While awaiting fire department arrival, the patient should be kept in the open, because in cases of severe exposure the victim's clothing will have absorbed a considerable amount of chlorine, which would be further inhaled if the victim were confined to a room. If blankets are available, the clothing should be removed and the patient kept warm and quiet. Never let a rescue squad use a Pulmotor on a gassed victim.

In an extremely severe case, the victim may stop breathing and may also turn blue (cyanosis). Obviously a physician should be summoned and artificial respiration started immediately. The Chlorine Institute recommends the Nielson armlift–back pressure method.

A physician can take steps to reduce the formation of pulmonary edema, arrest falling blood pressure, and supervise administration of oxygen. Other than that, there is little to be done in severe cases. The Chemical Warfare Service of the U.S. Army[28] has done considerable research on phosgene gas ($COCl_2$) poisoning, and has found that adrenalin, Pituitrin, and morphine are injurious and should never be used because they lower blood pressure, which is already falling fast in cases of severe exposure.

After the severely gassed victim has recovered from anoxemia and acute pulmonary edema, he or she will need careful watching and nursing to prevent development of pneumonia.

For additional details on the medical care of severely gassed patients, see the references for the article by Joyner and Durel.[29]

Treatment of mild cases occurring from time to time around chlorination equipment is considerably different. The first step is of course to get clear of the fumes, breathe lightly, move slowly without exertion, remain quiet, keep warm, and resist the impulse to cough.

The victim will be at first seized with fear and may become panicky because of the sensation of the closing of the throat and a feeling of suffocation. In a mild case, immediate relief can be achieved by sipping a teaspoon of either Anti-Chlor[30] vodka or whiskey.

An equivalent of one tablespoon every 15 min. until relief is obtained, or for one hour, should be administered. In preparing the mixture, add oil of peppermint to the alcohol and then spirits of chloroform, spirits of ammonia, and lavender, in the order given. Stir after each addition. Next, add this mixture to water, to which sugar has been added. To make spirits of chloroform, add 60 cc of the chloroform to 940 cc of grain alcohol.

Because alcohol is the principal ingredient in the formula, the same results can be obtained by sipping a similar amount of straight whisky. The alcohol counteracts the shock and panic, relaxes the muscles in the throat, eliminates the desire to cough, and restores normal breathing.

So far as is now known, there are no residual symptoms attributable to severe gassing by chlorine for most patients.[31] There are, however, some anxiety reactions that linger for many months. Victims with a history of asthma have had lingering effects of distress; mild cases have been known to cure bronchial ailments in young persons. Further clinical research is necessary to determine the long-range effects of gassing by chlorine.

Physiological Response

The United States Bureau of Mines gives the following physiological responses to various concentrations of chlorine gas:

Effect	*Parts of Chlorine per Million Parts of Air by Volume*
Least amount required to produce slight symptoms after several hours of exposure	1
Least detectable odor	3.5
Maximum amount that can be inhaled for one hour without serious disturbances	4
Noxiousness, difficulty in breathing, several minutes	5
Least amount required to cause irritation of the throat	15.1
Least amount required to cause coughing	30.2
Amount dangerous in 30 minutes to one hour	40–60
Kill most animals in a very short time	1000

CHLORINE LEAKS

Definitions

These leaks are usually referred to as emissions or releases. They are simply inadvertent discharges of either chlorine liquid or vapor (gas) into the surrounding atmosphere. In potable water and wastewater chlorination systems that meter chlorine under a vacuum, the leaks are usually minor ones. When liquid chlorine is involved, the leak is usually considered as a major leak. This is not necessarily the case because liquid chlorine is highly soluble in water, much more so than chlorine gas.

Leaks are characterized as minor, major, or catastrophic. A catastrophic leak at a potable water or wastewater treatment plant is a one-in-a-million type of event. Such a leak would be from a container rupture, a guillotine rupture of a chlorine supply line under container pressure, or a "blowout" of a fusible plug. The remoteness of such an occurrence is due mainly to the inherent operation under a vacuum at water and wastewater treatment plants, both municipal and industrial, but also to the special relationship between the users of chlorine and the producers and the packagers of the chlorine. Since only 5–6 percent of all the chlorine produced in the United States is used at these plants, the quality of the chlorine is restricted as to the moisture content, as well as restrictions for other items that might cause "gunk" to fall out in the chlorination system. This is very important because of the necessity to measure accurately the relatively small dosages of chlorine that are used in many of these treatment plants, plus the fact that the chlorine is metered under a vacuum of about 18–20 inches of Hg. All of this method of operation ensures a major safety condition at all times in these plants, which has yet to be recognized in U.S. fire codes.

Minor leaks usually occur at the start-up of a new installation or right after maintenance or inspection procedures. They usually are the result of a gasket failure, valve packing adjustment, or equipment malfunction.

Major leaks include such things as a guillotine break in a pipeline under pressure, a broken flexible connection, a fusible plug failure, and/or repair work while a system is under chlorine supply pressure. Not to be overlooked is the real possibility of spontaneous combustion of steel and chlorine if for some reason the temperature of the chlorine in the container vessel or the chlorine evaporator vessel reaches 483°F. This could occur in a fire or from some other overheating condition such as the improper use of an acetylene torch or the failure of the automatic heat control to a chlorine evaporator due to improper or faulty maintenance. (See "A 1600-lb Vapor Leak" and "Evaporator Blowout" in this chapter.)

The Uniform Fire Code: Realities and Options[89]

In the United States there are three major organizations producing model fire codes. The International Fire Code Institute (IFCI), headquartered in Whittier, California, publishes the Uniform Fire Code (UFC), which is used predominantly in the western United States. The Building Officials and Code Administrators, International (BOCA) is headquartered in the Midwest and the northeastern United States. The Standard Fire Prevention Code (SFPC), from Birmingham, Alabama, is published by the Southern Building Code Congress International, and covers the south-central and southeastern states. An additional organization, the National Fire Protection Association (NFPA), in Quincy, Massachusetts, does not produce a model code, but it does develop and publish numerous standards and technical documents. NFPA materials cover a number of subjects related to fire and building code issues, including hazardous materials and emergency response. NFPA standards are frequently referenced as code documents by governmental groups and the three model codes.

The use of chlorine has been seriously affected by fire and building codes. The significant changes relating to hazardous materials have occurred since 1988, when the UFC revised its hazardous materials section, Article 80. This Article in the 1985 UFC covered only five and one-half pages of the code text. The next edition of the UFC, issued in 1988, expanded Article 80 to 51 pages. This increase in the volume of text has resulted in numerous new and specific code requirements for handling all hazardous materials, including chlorine. The other two major model codes, BOCA and the Standard Code, followed the lead of the UFC and greatly expanded their own hazardous materials regulations.

When Article 80 was written, the chemical industry was not adequately involved in the fire code process. Many items in the 1988 UFC, which greatly affects the use of chemicals, particularly gases, were written with very little information and insufficient consideration for their impact on both industry and users of these materials. As is usually the case, it has proved much more difficult to change existing code language than to be part of the process in the beginning. In contrast to the UFC, the chemical industry (particularly the chlorine, compressed gas, and swimming pool chemical industries) was more actively involved in the initial writing of the BOCA Code and the SFPC Code hazardous material sections.

Chlorine, Rural Water Agencies, and the UFC

The membership of the California Rural Water Association is composed of small utilities that tend to use chlorine in 100- and 150-lb cylinders. Railroad tank cars and tank trailer shipments are not a concern in most cases. Therefore, Article 80 of the UFC is of particular interest to this agency.

The Uniform Fire Code is the predominant model fire code used in California. The UFC classifies chlorine as both a toxic gas and a corrosive gas, so the requirements for both classifications must be met. However, the code has a minimum quantity for each hazard classification that must be exceeded before it applies. This quantity is per "control area," a term that will be defined shortly. Any hazardous material defined by the code that is used or stored indoors is regulated if the quantity per control area exceeds the exempt amount listed in Tables 13A or 13B in Article 80 of the UFC. For corrosive and toxic gases (chlorine), this quantity is 810 cu ft, which equates to exactly one 150-lb cylinder of chlorine. This means that if you have only one 150-lb maximum cylinder of chlorine per control area, you are not regulated under the UFC. A control area is defined in section 8001.8 of the 1994 UFC as "a space bounded by not less than a one-hour fire resistant occupancy separation within which the exempted amounts of hazardous materials may be stored, dispensed, handled, or used." Except for stores, most buildings can contain up to four control areas.

One way that users of 100- and 150-lb chlorine cylinders can be exempted from the code is to utilize this feature of the code, dealing with exempt quantities per control area. The Uniform Building Code (UBC) gives specific details on what constitutes a one-hour fire-resistant wall. Such walls can be constructed of various substances, and do require elaborate plans or materials for construction. A water authority can install up to four control areas per building, including well-head structures. Thus, as many as four 150-lb chlorine cylinders per building are exempt from the UFC. If the building or control areas are sprinklered, the exempt amounts can be doubled.*

*No mention is ever made that sprinklered rooms where chlorine containers are being used are constantly releasing small amounts of chlorine in the storage rooms, so that after some period of time the sprinklers fail, owing to attack by the chlorine.

This means you can now have two 150-lb chlorine cylinders per control area, or up to eight per building, and not be under the UFC requirements. Because the quantity of chlorine in a single ton container exceeds the 810 cu ft quantity, ton containers always fall under the UFC.

A June 1996 development occurred that has affected the use of 150-lb and ton cylinders in the arid western states. This is the requirement of sealed total containment vessels for these chlorine cylinder containers instead of airtight rooms with chlorine scrubbers. The makers of these patented containers, TGO Technologies, Santa Rosa, California, have already sold 20 units. For further details on these containment vessels, see Chapter 9.

The most burdensome requirement for toxic and corrosive gases in the UFC is the need for a treatment system to mitigate a chlorine release. Treatment systems are required by the UFC if the amount of chlorine exceeds the exempt amount per control area. The code states that: "Treatment systems shall be capable of diluting, adsorbing, absorbing, containing, neutralizing, burning or otherwise processing the entire contents of the largest single tank or cylinder of gas stored or used. When total containment is utilized, the system shall be designed to handle the maximum anticipated pressure of release to the system when it reaches equilibrium." For chlorine this generally means a scrubber although some fire chiefs have approved dilution systems. The code requires that treatment systems must be designed to discharge a gas at a maximum concentration of one-half of the IDLH level. This level is currently (1997) 10 ppm for chlorine. It should be noted that the SFPC, the BOCA Code, and the NFPA Standard No. 55, all allow the use of chlorine emergency kits or cylinder containment vessels to serve as substitutes for the treatment system requirement.

If a facility must operate with quantities above the exempt level of the UFC, there is a second option. Section 103.1.2 of the UFC allows the chief to approve alternative materials or methods that comply with the intent of the code. If the chief is satisfied that the community can be adequately protected by alternative measures, such as trained responders using chlorine emergency kits, the requirement can be waived. In some areas covered by the UFC, local chiefs have stated that they believe the treatment systems are not the only answer, so they entertain this option. A case should be made to the chief and the local officials about specific situations. The Chlorine Institute takes the position that the use of trained responders, along with the appropriate chlorine emergency kit, can protect the community in most situations. Each site should be evaluated independently to determine if additional measures such as treatment systems are needed.*

One important new item in the UFC needs mentioning. Significant changes have been made to the code concerning piping systems for toxic and highly

*The chiefs should be made aware of the pressure control systems available to operators of treatment plant chlorination installations. See Chapter 9.

toxic gases. The 1994 edition of the UFC has less than one page, Section 800.4.3, devoted to piping systems for all hazardous materials. (The Chlorine Institute alone has a 57-page pamphlet on piping systems for dry chlorine.) The 1994 code did not allow the use of threaded or flanged piping for toxic or highly toxic gas service unless an exhausted enclosure around the gas piping was provided. Through the efforts of several industry groups, and the support of many UFC members, the UFC now allows the use of threaded piping up to 1.5 inches in diameter, and the use of flanged piping for toxic gas systems. These changes follow the basic recommendations in Chlorine Institute Pamphlet No. 6, entitled "Piping Systems for Dry Chlorine." The changes were approved at the August 1995 code hearings in Las Vegas, for publication in the 1997 edition of the UFC. Any questions concerning these changes can be addressed to the staff of the UFC in Whittier, California.

One additional, important step is to determine which code and which code year are legally binding in your location(s). Each facility should contact the local governmental jurisdiction and determine which fire codes are in effect. The year of the adopted code is also critical. Only legally binding codes can be enforced by the fire service or building officials. The codes themselves are not legally binding documents until an appropriate governmental agency votes to adopt them. For example, if the local government has only adopted the 1985 UFC, then the provisions of the more recent yearly editions are not enforceable. Also, a local jurisdiction may adopt only part of a model code, and it may add or delete specific sections, making the code unique to that area.

Steps Water Utilities Can Take

The UFC can affect every water treatment facility that uses chemicals, regardless of size. There is clearly a disagreement between many in the chemical and water treatment industries over the need for some of the requirements listed in the UFC. The voting members of the UFC are trying to protect their communities in the best manner possible, based upon the sources of information available to them. Most have limited knowledge about the properties of the thousands of chemicals they encounter as well as the standard industrial practices for managing these chemicals. As a critical first step, a dialogue should be established with local fire officials. You should show them how hazardous materials are handled, advise them of your training end emergency response plans, and invite them to drills. Work with them, and express, in writing, any code concerns you may have, and the reasons behind your concerns. The local government should also be contacted as code requirements can have a serious economic impact on a community.

A second, and equally important, step is to become as active as possible in the UFC code process. These code meetings occur twice each year at various locations throughout the United States. Also, most states have a state association of fire chiefs that meets within the state to discuss code proposals. California has two such groups: a northern and a southern group. Although it is time-consuming, this is the only way to let the chiefs know about the effects that future changes will have on your industry, before they become part of the next UFC.

Also, have the trade associations that represent you get involved at the two annual UFC code hearings. They can support or oppose pending code changes, as well as introduce code change proposals for the industry. The many members of the water and wastewater industry must make a united effort and become active in the code processes. They must work with both the code organizations and the local and state governments adopting the codes to make certain that the code requirements are justified and that they enhance safety in a cost-effective way.

Characteristics of a Major Liquid Chlorine Release

Brian Shera's Bucket. It is important to be aware of what is likely to take place in a catastrophic leak. The following is what occurred during an actual demonstration at a major chlor-alkali plant. The person who performed this demonstration was Brian Shera of Pennwalt Corporation.[79]

He first dug a hole in an open area of sandy soil with a posthole digger. The hole was about 6 inches in diameter and 10–12 inches deep. All observers were wearing an air pack breathing apparatus and were placed in an upwind position relative to the hole. The ambient temperature was about 60° F. Mr. Shera withdrew a bucketful of liquid chlorine from an adjacent storage system, walked a few steps, and poured it into the dug hole.

There was an immediate flash-off of chlorine vapor when the bucket was being filled so that the bucket was not full when he poured it into the hole. Mr. Shera calculated that the amount poured into the hole was close to 28 lb. There was a visible film of thin ice at the surface of the bucket after the original flash-off and just prior to the pouring of the liquid chlorine into the hole. When he poured the chlorine into the hole, there was another flash-off that ended quickly with another appearance of the ice film. If 25 percent flashed off when it was poured into the hole (7 lb), then there was approximately 20 lb remaining to be vaporized by the ambient air, whose temperature was about 65°F. The exposed area of the hole was about 0.2 sq ft. We all stood around watching the intermittent flash-off phenomenon for over an hour. When we left, there was a substantial amount of liquid chlorine left in the hole.

This is in accordance with Howerton's Report (see Chlorine Institute Report No. 71, Fig. 6), which shows that the vaporization will be 6.8 lb/sq ft/hr. The surface area of the hole was calculated to be approx. 0.20 sq ft (D = 6 in.); so the evaporation rate calculates to only 1.36 lb/hr.

The most significant characteristics of a major leak demonstrated by this performance were the long drawn-out vaporization cycle and the rapidity of the freezing cycle. It also verified that the flash-off phenomenon can only occur when the liquid is spilled into the atmosphere. In other words, a chlorine container of any kind cannot undergo the flash-off unless a rupture occurs that exposes the liquid surface to the atmosphere—such as a dished head blown off. However, the liquid that spills will flash off, and the remaining liquid will vaporize at a rate of approximately 7 lb/sq ft/hr.

IMPORTANT CHARACTERISTICS OF CHLORINE CONTAINERS

Introduction

The handling of any chlorine container always revolves around the variable temperature–pressure relationship of liquid chlorine and its continuity of balance with the ever-present chlorine vapor. When the vapor is withdrawn from a container in excess of the liquid's ability to absorb sufficient Btu's from the surrounding atmosphere to vaporize the liquid, then the liquid will begin to cool. This cooling process lowers the vapor pressure.

At one of its chlor-alkali plants, Dow Chemical Co. uses the temperature–pressure property of liquid chlorine to provide the ultimate in the safe handling of chlorine. Dow recirculates the liquid chlorine in its main supply storage vessel through a refrigeration system that cools the liquid to atmospheric pressure. The storage vessel is located in a containment structure with a properly sloping floor so that if a leak does occur, it can be quickly blanketed with an insulating foam. The entrapped liquid can then be harmlessly transferred to a safe disposal system.

Owing to its heavy layer of insulating material, a railcar chlorine container does not respond the same way to the temperature–pressure relationship during vapor withdrawal as does a noninsulated container. The former reaction is much slower because the rail car is insulated to avoid temperature increases due to storage in open areas.

Emptying a Tank Car

A lot can be learned about the characteristics of liquid chlorine from operational anecdotes. The following is one of them. The sequence of events that occur during the time when the liquid chlorine supply begins to wane reveals some interesting warning signs. This is the scenario as related by the chief operator of a large water treatment plant when he was asked the question, "Since you do not have any weighing scales, how can you tell when the car is becoming empty?" He promptly advised that the first sign of a depleted supply from a tank car during liquid withdrawal is a rise in the gas temperature exiting the evaporator. This, he explained, is the result of a decrease in the chlorine feedrate due to the onset of liquid exhaustion. When this occurs, the operator proceeds to the loading platform on the car to see if the liquid discharge flexible connection is getting cold.

If it feels cold, it means that the liquid supply is rapidly approaching empty. Then when this sign is followed by a loud rattling noise in the discharge line, the operator knows that is the last of the liquid chlorine. In the meantime the car has maintained its vapor pressure because even after all the liquid has gone, there is still about 1900 lb of chlorine gas remaining in the 55-ton car—if the vapor pressure is approx. 85 psi.

Pressure Reduction Test

The following demonstrates the use of the vacuum control system available in chlorination equipment for the purpose of determining the rate of a pressure reduction of chlorine in a 55-ton tank car at a conventional wastewater treatment plant. This plant had three 6000 lb/day chlorinators and evaporators. The chlorine supply pressure was about 85 psi. By withdrawing chlorine gas from the tank car to the evaporators at a rate of about 18,000 lb/day, the chlorine pressure in the car was reduced to 35 psi in 45 min. If this had been a noninsulated storage tank, the pressure reduction would have been faster and at a lower pressure. However, it has to be noted that withdrawing chlorine gas at a higher than normal amount from a container vessel will include a "mist" of liquid chlorine that can only be handled properly by having the gas go directly to an evaporator. Such a piping arrangement is illustrated in Chapter 9.

Visual Determination of the Chlorine Storage Tank Content

Another operating anecdote is used here to demonstrate one of the basic properties of liquid chlorine. The following problem occurred at a wastewater treatment plant equipped with two 50-ton bulk storage tanks.[80] These tanks were installed with electronic level indicators as the only means for monitoring the chlorine supply. Unfortunately but predictably, these liquid level monitors failed in less than six months. Fortunately these tanks were piped up to discharge either chlorine liquid or vapor to the chlorination system. When asked how the operating personnel were able to determine the status of the chlorine supply, the chief operator advised that by switching from liquid to vapor withdrawal at 3000 lb/day, a frostline would appear on the tank in about 20 minutes. This meant that the surface of the liquid had reached 32°F. This is an important tactic, which showed the operating personnel that it could be used to provide a pressure differential between the storage tank and the tank car that would allow them to load the tank without the use of the usually troublesome air padding system. It also demonstrates that the only procedure other than refrigeration that can lower the temperature of the liquid in a container is vapor withdrawal at a rate that exceeds the ability of the ambient air to transfer enough heat to vaporize the liquid chlorine.

Importance of Vacuum Withdrawal

This brings up an important point about the almost complete lack of knowledge of this tactic in the handling of chlorine. The situation is self-explanatory. The use of liquid chlorine in the treatment of potable water and wastewater is about the only situation where metering and control of the applied chlorine are always under a vacuum. This only accounts for about 6 percent of all chlorine produced in the United States. Practically all of the other 94 percent of operators handle the chlorine under pressure; therefore, this tactic is not available to them.

In Chapter 9, there is a detailed description of a pressure control system that illustrates how chlorine container pressures can be controlled to prevent overpressure leaks or other types of leaks.

CALCULATING CHLORINE LEAK RATES

Leak Formula

All of the following calculations are based upon the Chlorine Institute formula[86] shown in its Pamphlet No. 74 (1982) under 2.4.1, "Release Rate Formulas."

Liquid Release.

$$Q = 77A\sqrt{(P_1 - P_2) \times \rho} = \text{lb/sec} \tag{1-21}$$

where:

A = area of opening to atmosphere, ft^2
P_1 = upstream pressure, psi
P_2 = downstream pressure, psi
ρ = density of liquid chlorine upstream from the opening to atmosphere, lb/ft^3

Vapor Release.

$$Q = 36.64A\sqrt{\frac{P}{V}} \tag{1-22}$$

where:

A = area of opening to the atmosphere, sq ft
P = upstream pressure, psi

V = upstream specific volume, cu ft/lb

$$V = \frac{1}{\text{density of chlorine vapor at opening lb/cu ft}}$$

Tanker Truck Leak while Unloading

The following is a valid method for estimating the leak rate from a road tanker while transferring its contents to the consumer's storage tank. The leak is based upon the assumption that a separation in the loading lines has occurred.

The usual capacity of these tankers is 17 tons. They can empty their contents in 2.75 hours with a 30 psi pressure differential. Therefore, this flow rate is 34,000 lb/165 min. = 206 lb/min. = 3.43 lb/sec. Then, using the equation $Q = 77A\sqrt{(P_1 - P_2) \times \rho}$ and substituting 3.43 lb/sec for Q and 30 psi for $P_1 - P_2$ and 88 lb/cu ft for ρ, A is calculated as follows:

$$3.43 = 77A\sqrt{30 \times 88}$$

$$A = \frac{3.43}{77 \times 51.38} = 0.00087$$

Use $77A = 0.0667$ for convenience.

Now the tanker truck leak rate can be calculated for any differential pressure. The value of P_1 would be based upon the pressure in the road tanker at the time of the leak. Assuming that the tanker pressure is 90 psi, the leak rate would be:

$$Q = 0.0667\sqrt{(90 \times 88)}$$

$$Q = 0.0667 \times 88.99 = 5.94 \text{ lb/sec}$$

$$Q = 356.44 \text{ lb/min.}$$

This is obviously a worst-case scenario because the excess flow check valves would not only severely limit this flow rate but would more than likely stop the spill entirely. The latter has been the case in many railcar accidents when a derailed car had the valves on the car dome completely wiped out. These road tankers are fitted with 14,000 lb/hr (233 lb/min.) excess flow valves in order to reduce the transfer time as much as possible.

Guillotine Break in a Pipeline: Ton Cylinder Supply

The maximum size of a chlorine header system under pressure in any potable water or wastewater treatment plant should never be larger than one inch.

If the installation involves liquid withdrawal from ton cylinders, then evaporators will be an integral part of the chlorine supply system. Therefore, the worst-case scenario would be a rupture in the liquid header between the cylinders and the evaporators. To simplify the concept, let the calculations be confined to one cylinder, one evaporator, and 100 ft of one-inch header pipe.

The liquid exiting the cylinder must pass through a 3/8-inch tubing in the dished head—then through the cylinder shutoff valve, then through the auxiliary cylinder shutoff valve, and finally through the header valve. All of these components are flow restrictors compared to a one-inch pipe. So how can these restrictions be accounted for in calculating the chlorine leak rate?

Circa 1950, the operating personnel needed to know the maximum possible liquid withdrawal rate from a single ton cylinder at the East Bay Municipal Utility District wastewater treatment plant, Oakland, California. Their chlorinator capacity was 18,000 lb/day. Their test, which they performed several times, indicated that the maximum rate was only about 10,200 lb/day.[81] The pressure drop between the cylinder and the entrance to the chlorinator was on the average about 85 − 40 = 45 psi because there was a pressure-reducing valve between the evaporator and the chlorinator. The flow at this pressure drop has to be recalculated to reflect a zero pressure at the leak. To apply a worst-case situation, let us assume a cylinder pressure of 120 psi.

Using the Chlorine Institute formula:

$$Q = 77A\sqrt{(P_2 - P_2) \times \rho} = \text{lb/sec} \tag{1-21}$$

where:

$$Q = 10{,}200 \text{ lb/day} = 0.1181 \text{ lb/sec}$$

$$\rho = 88 \text{ lb/ft}^3$$

Substituting in the above formula, the value of the unknown, $77A$, can be found:

$$Q = 77A\sqrt{(45 \times 88)}$$

$$Q = 0.1181 \text{ lb/sec}$$

$$0.1181 = 77A \times 62.93$$

$$77A = \frac{0.1181}{62.93} = 0.00188$$

Assuming a cylinder pressure of 120 psi, chlorine density at 88 lb/ft^3, and substituting $77A = 0.00188$ in Eq. (1-21), the leak rate Q will be:

$$Q = 0.00188\sqrt{(120 \times 88)}$$

$$Q = 0.1899 \text{ lb/sec} \times 60 = 11.4 \text{ lb/min.}$$

This then is the worst-case leak rate from a single ton cylinder "on line," when there is a guillotine break in the liquid chlorine header piping. It is obvious that if ton cylinders are being used for liquid withdrawal, an evaporator is part of the system. Therefore, when there is a guillotine break in the liquid header, the contents of the evaporator become part of the leak.

Powell[82] found by actual test that the maximum liquid withdrawal from an inverted 150-lb cylinder was 20 lb/min. at 90 psi cylinder pressure. The restrictions in this instance were the cylinder valve and 3–4 ft of ⅜-inch tubing.

All chlorine evaporators are designed to vaporize chlorine at a temperature varying between 160 and 180°F, regardless of the chlorine feed rate. This means that the level of liquid chlorine in the evaporator remains fairly constant. It is safe to conclude that the evaporator content is never more than 100 lb. At 20 lb/min., the evaporator will empty in about 5 min. because of the chlorine header rupture. Therefore, the probable maximum chlorine release rate will be 11.4 + 20 lb/min. for the first 5 min. and then 11.4 lb/min. after that interval. This is for each cylinder "on-line" and each evaporator.

Such a leak will cool the room so quickly that vaporization is temporarily stopped. During this time, if the container floor area has been designed properly, the liquid chlorine will flow through the collecting slots in the floor and be hustled off to the scrubber system. This reduces enormously the amount of time required to clean up a major spill. This maneuver capitalizes on one of the little-known properties of liquid chlorine—its solubility in water. Under a slight pressure, such as the discharge from an eductor or a pump, its solubility is three to four times that of chlorine vapor.

A Major Leak from PVC Header Failure

This was a totally inexcusable chlorine leak from a ton cylinder in a small wastewater treatment plant in Alaska. A conventional chlorine system by Wallace & Tiernan Inc. had been operating there for several years without any particular problems. When the plant was being upgraded in 1991–92, the chief operator complained to the plumbing contractor that he wanted the rusty old steel chlorine header piping between the ton cylinder room and the chlorinator room to be replaced with PVC piping to eliminate the rusting "problem." This was done without anyone's questioning this vital change, not even anyone working for the plumbing contractor. It was also surprising that the well-experienced consulting engineers were never consulted because part of their upgrading was the installation of a sulfur dioxide dechlorination system. On top of all this, the operator should have made a phone call to Wallace & Tiernan in Seattle.

It is well known by those of us in this business that chlorine gas under pressure must not exceed 2–3 psi pressure in using PVC pipe because higher pressures will damage the PVC. The damage is gradual; as the pressure increases, the PVC begins to heat up. The greater the pressure, the more quickly the heat rises. At approx. 150°F the PVC melts. Figure 1-15 illustrates the reducing bushing adjacent to the cadmium-plated header valve that melted, causing the leak. This was the thinnest piece in the entire PVC piping header. Other piping that did not melt showed external blistering, indicating that melting of the rest of the piping was on the way. The plant operator was on duty from 7:30 A.M. to 4:00 P.M. on January 29, 1992, the day prior to the leak. A phone call came in on the Fire Department 911 circuit at 1:30 A.M. on January 30, describing a strong offensive chlorine odor. Furthermore a large cloud was sighted in the atmosphere near the WWTP at 1:48 A.M., January 30, 1992. This was identified as the chlorine leak. It is fair to say that the leak began close to 1:00 A.M. At about 2:00 A.M. the Fire Department took charge of the situation and called the operator, who arrived at the plant with an assistant to turn off the chlorine cylinder at about 2:30 A.M. His testimony of what he found upon entering the chlorine cylinder room identified the crisis (both men were wearing protective breathing apparatus). He said that on account of the density of the fog in the room, he had a lot of trouble getting to the cylinder shut-off valve. Moreover, he said it was so hot in the room they got out as fast as they could. The room temperature was estimated at about 100°F. The leak lasted a good 2.5 hours.

Because there were no gauges in the cylinder room, the chlorine pressure was unknown at leak time; and because there were no cylinder scales, the total amount of the leak was unknown. However, it is possible to make a fairly accurate estimate of the leak, using Chlorine Institute formulas, as follows:

At the first instant after the melt took place, at 150° F at about 1:15 A.M., it is fair to assume that room temperature was enough to produce 120 psi ton cylinder chlorine pressure. Therefore, with the melt hole size measured at 0.0002 ft^2, the leak rate would be about 5.88 lb/min or about 360 lb/hr. This amount of leak rate might reduce the container pressure from 120 psi to

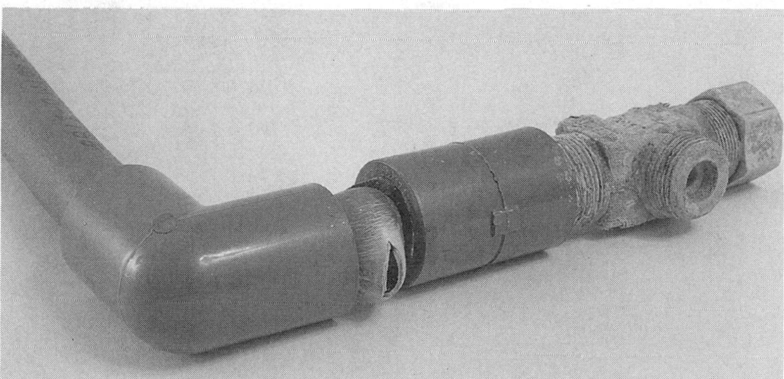

Fig. 1-15. Chlorine gas piping (PVC) under pressure (80–100 psi) immediately downstream from the ton cylinder connection, illustrating the hole in the PVC bushing that caused the enormous chlorine gas leak.

possibly 40 psi in about one hour, assuming that the room temperature declined accordingly. So at about 2:30 A.M., if the room temperature had declined, then when the cylinder was shut down, the leak rate would be calculated to be about 2.13 lb/min, or 128 lb/hr. However, it is known from the operator's testimony that during the "meltdown" leak, the room temperature was close to 100°F. This temperature would have kept the cylinder pressure to at least 70 psi. At this pressure the leak rate would be 3.52 lb/min, or 212 lb/hr.

One of the most important factors concerning this leak is the formation and travel of the leak plume. Immediately after the ton cylinder was shut down, the Fire Department opened the two large doors of the cylinder room, which opened into the WWTP yard. This allowed 2200 cubic feet of 100°F air containing some 500,000 ppm Cl_2, to form a plume in the local atmosphere, which at the time was 5°F and 90 percent humidity. Considering the fact that the leak lasted 2½ hours, the total amount of chlorine in the 2200 cu ft of air in the room was about 500 lb. It must be remembered that this leak went unnoticed for at least 2½ hours.

Before the arrival of the operators and the Fire Department, the chlorine was seeping out of a small ventilator grate (9 × 12 in.). The leak was detected by a resident who lived about 600 ft downstream from the WWTP. The chlorine odor that was detected was from the ventilator grate, probably after it had reached its maximum strength.

Owing to the weather conditions and the extreme heat of the chlorine-laden air from the cylinder room, the plume rose quickly and headed for the adjacent Kenai River (see Fig. 1-16). This is why only those residents who lived within 600–700 ft of the plant were affected by the leak. This group amounted to about seven or eight nearby residents.

Fig. 1-16. Portion of aerial photo showing the area affected by the enormous chlorine gas leak at the wastewater treatment plant.

The importance of this leak is in illustrating the big difference that heat makes at the location of the leak. In this case, the entire plume from the cylinder room rose quickly (chimney effect) and drifted northward with a 5 mph wind current. Witnesses said that the plume followed the Kenai River and rose so fast that only a small portion of the area was affected. Although the plume concentration leaving the cylinder room was estimated at 500,000 ppm, it rose so quickly so that when it passed over the affected area it was about 300 ppm.

Ton Cylinder Flexible Connection Failure

Assuming a worst-case situation, the flexible connection breaks at the auxiliary cylinder valve. Logic dictates that for a 120-psi cylinder pressure without the restriction of a header valve plus a 4 ft flexible connection, the release rate will exceed the rate from a header pipe rupture. A reasonable estimate would be a 20 percent increase: $(11.4 \times 0.2 = 2.28) + 11.4 = 13.68$ lb/min.

Fusible Plug Failure from Corrosion

Description. This is the most common problem of fusible plug failure. A ¾-inch plug contains a lead core about 3⁄16 inch in diameter in a brass body. The inherent moisture in "dry" chlorine begins an immediate attack on the vulnerable brass body. Therefore, the hole generated by the corrosion is shaped like a cone with the base of the cone on the inside of the ton cylinder. The end result of this corrosive attack is a pinpoint hole between the brass body and the threaded steel of the dished cylinder head.

Field observations by White indicate that this hole is never larger than 0.1 inch in diameter.

Liquid Release. For a worst-case situation the following calculations will be based upon a hole diameter of 0.15 inch, with the fusible plug located below the chlorine liquid level in the ton cylinder. Here again the cylinder pressure is assumed to be 120 psi.

Therefore:

$$Q = 77A\sqrt{(120 - 0) \times \rho} = \text{lb/sec}, \ \rho = 81 \text{ lb/ft}^3$$

$$A = \frac{\pi \times D^2}{4} = \frac{\pi \times (0.15)^2}{4} = 0.018 \text{ in.}$$

$$A - 0.000125 \text{ ft}^2$$

$$Q = 77 \times 0.000125\sqrt{120 \times 81} = \text{lb/sec}$$

$$Q = 77 \times 0.000125 \times 98.59 = 0.949 \text{ lb/sec}$$

$$Q = 56.94 \text{ lb/min.}$$

Vapor Release. This is an important comparison because knowing the huge difference in the release rate, the safety crew should attempt to rotate the leaking plug to the vapor area. If this is done, the escaping vapor will cool the liquid to 40°F in 3–4 min. This has to be taken into account when using the Chlorine Institute formula:

$$Q = 36.64A\sqrt{\frac{P}{V}} = \text{lb/sec} \tag{1-22}$$

The cylinder pressure will have been reduced enormously because the escaping vapor is at zero gauge pressure. There is little doubt that the cylinder pressure will be as low as 40 psi. Then V will be chosen for the density of chlorine vapor at 40°F, which is 0.77 lb/ft^3. Then:

$$V = \frac{1}{0.77} = 1.3 \text{ ft}^3\text{/lb}$$

Therefore:

$$Q = 36.64 \times 0.000125\sqrt{\frac{40}{1.3}} = 0.0254 \text{ lb/sec}$$

$$Q = 1.52 \text{ lb/min.}$$

Fusible Plug Blowout

There is no such occurrence on record, but it is a frequently used "scare tactic." Such an occurrence would be almost equivalent to a cylinder rupture. It is assumed that the total discharge will be liquid chlorine.

As it has already been shown how these plugs have failed to do what they were supposed to do, this is all the more reason why they should be eliminated in the making of ton cylinders.

If there is such an occurrence, it is assumed that the total discharge will be liquid chlorine. There will not be any "flash-off" unless the ambient atmosphere reaches the inside surface of the liquid chlorine in the container. For the sake of a rational calculation, it will be assumed that the cylinder pressure drops to 30 psi. This is equivalent to a liquid temperature of 20°F; therefore, the density of the liquid chlorine is 93 lb/ft^3.

So the leak rate is calculated as follows:

$$Q = 77A\sqrt{(P_1 - P_2)\,\rho} = \text{lb/sec.}$$

$$\rho = \text{density at 20°F} = 93 \text{ lb/ft}^3 \tag{1-21}$$

$$A = \frac{\pi \times (0.75 \text{ in})^2}{4} = 0.44 \text{ in.}^3 \times 0.00694 = 0.003 \text{ ft}^2$$

$$Q = 77 \times 0.003\sqrt{(30 - 0) \times 93} = 12.2 \text{ lb/sec}$$

$$Q = 732.09 \text{ lb/min.}$$

It is quite obvious that the calculations indicate an impossibility. The contents of the container could never be discharged at that rate; otherwise, the cylinder would be empty in less than 30 minutes! The scenario that is closer to what will happen is a sudden cooling of the liquid chlorine that brings the cylinder pressure to atmospheric, whereby the liquid chlorine will go into a freezing and thawing cycle that may take hours for the chlorine to escape (see above, "Brian Shera's Bucket"). Such a fusible plug blowout is considered a one-in-a-million occurrence. White's search of the records, as well as his six decades in the field of chlorination, has never turned up such a failure. Corrosion of fusible plugs that suffer from moist chlorine in the containers has produced several of these leaks. That is why packagers should replace these plugs after every five years of service in a container.

SUMMARY OF MAJOR LEAK EVENTS

1. Whenever there is a major leak, the flash-off phenomenon will always prevent a positive pressure condition in a containment structure. The sudden vaporization due to flash-off cools the closed area so fast and so much that a negative pressure in the room is the result.

2. There will always be a significant amount of liquid chlorine that must be dealt with as soon as possible. Because it is much more soluble in water than chlorine vapor, it can be disposed of quite easily by a water-operated eductor or a liquid chlorine pump.

3. The only way that liquid chlorine can be cooled by a leak is to withdraw vapor from it. Liquid flowing out of a container because of a major leak will not cool the cylinder or reduce the vapor pressure unless the leak is a large hole in the container such as a fusible plug blowout. When this type of leak occurs, the flash-off phenomenon begins as soon as the liquid chlorine is exposed to the open room. This will always cool the room so quickly that it will produce a negative head in the room.

4. Overtemperature situations have caused two very unlikely major leaks. One was a case at a wastewater treatment plant in a remote area, where the operator was upset by the rusty Sch 80 steel header piping in the ton container room. During some upgrading work at the plant, he had the plumbing contractor replace the steel pipe with PVC; and after 6–7 hours of operation, the PVC pipe under cylinder pressure melted a reducing bushing and caused a 500 lb chlorine leak. Owing to the high humidity of the local weather (95 percent), the leaking hot chlorine created a dense fog in the room. When

the Fire Department arrived and opened the double doors of the ton container room, the chlorine fog in the room formed a plume that rose in the atmosphere, as in a chimney effect. This characteristic of a heat-generated leak considerably reduces the effect on people at ground level.

Another major leak due to overheating caused a chlorine evaporator to be destroyed by spontaneous combustion at a temperature of about 425°F. Such an occurrence is very rare, but it does illustrate the phenomenon of heat. The overheating in this case was due to poor maintenance and lack of operating experience. Here the operator should have shut down the entire system not only when he saw steam vapor escaping from the evaporator, but when it was so hot that he could not enter the evaporator room. The cause of this accident was the failure of the temperature control switch to shut down the power supply to the evaporator, as the switch was in a jammed-open position due to a collection of debris in the electrical switch case. This was a tremendous leak, but fortunately it occurred at a lakeside potable water pumping plant, in a sparsely populated area. Again the heated chlorine-laden plume rose in the atmosphere quite rapidly so that no one in the surrounding area was conscious of the leak, which was estimated at about 14,000 lb. The chlorine supply was in a 20-ton storage tank.

CALCULATING THE AREA AFFECTED BY CHLORINE RELEASES

Whenever there is a major liquid chlorine spill, a vapor cloud is certain to form if the vapor is released to the atmosphere. This may not occur if the leak is from the gas phase of the system because of the initial dilution of the vapor with air at its source. However, if the leak is due to heat, such as those that have occurred (e.g., the rupture of an evaporator due to spontaneous combustion of steel by chlorine when the temperature in the evaporator casing reached 483°F) because of overheating due to the failure of an electrical switch designed to shut down the heating elements in the evaporator[90], an overheated plume is generated, which will rise in the atmosphere, producing a chimney effect.

Chlorine vapor is readily amenable to following air currents, whether they be ventilation air or atmospheric air, largely owing to the available moisture in the atmosphere. The higher the humidity, the greater is the attraction of chlorine. This behavior is synonymous with the suck-back phenomenon.

A great many researchers have investigated the phenomenon of major releases of hazardous chemicals. The equation used by most investigators is the Gaussian plume model equation, which predicts the length and the shape of the cloud formed from the initial release, provided that weather conditions are known. The cloud that emerges from this model is shaped like a cone sliced in half with the flat part at ground level and the apex at the source of the release. The value of the mathematical model is to assist authorities to set reasonable boundaries for evacuation after a release has occurred. The

Gaussian equation takes into account release rate, the standard deviation of the crosswind plume, width and height of the plume, height of the initial source, the downwind, crosswind, and vertical distances, and chlorine concentration, as follows:

$$C_{xyz} = \left[\frac{Q}{\pi \cdot U \cdot \Sigma y \cdot \Sigma z} \right] e - \left(\frac{h^2}{2\Sigma z^2} + \frac{y^2}{2\Sigma y^2} \right) \quad (1\text{-}23)$$

when:

C = concentration units/m^3
Q = release rate, units/sec
Σ_y, Σ_z = thc standard deviation of the crosswind plume (width and height in meters)
U = mean wind speed velocity (m/sec) at h.
h = release source height (meters)
x,y,z = downwind, crosswind, and vertical distances (meters).

There are three factors not accounted for in the above plume model: ambient temperature, relative humidity, and local terrain. These factors contributed significantly to the fatalities in the Youngstown accident (see below). A release in a fog-shrouded area is probably the worst case. Air movement in a low-lying fog-shrouded area is usually nil. The relative humidity is at the saturation point, which allows the moisture-seeking chlorine gas to saturate the fog shroud. Clothing on people in the release area will absorb the chlorine-laden moist air, thus multiplying the inhalation of chlorine. In such cases it is not sufficient merely to evacuate thc area quickly; exposed persons must remove their clothing as soon as possible. This adds another dimension of risk bccause a fog usually occurs in an area where the ambicnt tcmperature is quite low.

Although fire is to be avoided at all costs where chlorine containers are stored, a brisk fire adjacent to a chlorine release can be a big help. This was demonstrated in a recent derailment when a tank car was ruptured by a following propane car, which exploded and caught fire. The hcat from the burning propane produced a chimney effect, and the entire contents of the 90-ton tank car escaped without anyone's suffering from chlorine inhalation.

CHLORINE ACCIDENTS

History

A chlorine accident occurs when fumes of the gas or the liquid are inadvertently released into the atmosphere. However, a great many so-called chlorine accidents that receive wide coverage by the news media are those where no chlorine emission occurs—the chlorine tank cars involved in the freight train derailments or the sinking of chlorine barges in rivers. It would seem that every string of derailed freight cars contains at least one chlorine tank car. This of course is not the case. Approximately 30 chlorine tank cars are involved in wrecks each year. Compared with over 9000 tank cars in service in a recent year, each making from 12 to 16 trips per year, the number of wrecks is small. In 1981, for example, there were 24 tank cars involved in rail incidents in the United States, mostly from derailments. Of these incidents, six were reported to have resulted in chlorine emissions.[33]

A major cause of large releases of chlorine in the early years was the cracking of the tank at the anchor, which is the attachment of the tank body to the railroad car frame. These early designs used forged anchors. The problem was corrected by using welded anchors. There has been no report of an anchor failure since 1963.

Usually the most serious damage that is inflicted on a tank car is to the protective dome assembly that houses the outlet valves and safety relief valve. On a few occasions in the past 30 years, one or more valves on the dome containing two liquid chlorine and two gaseous chlorine outlets have been sheared off. This has resulted in only minor spills. Such a situation automatically triggers into operation the excess flow valves inside the tank. These valves act as automatic checks, thereby containing the liquid in the car.

The most serious accidents occur when the tank car is punctured in a derailment by the coupler of an adjoining car. To avoid puncture damage, all new 90-ton cars are built with thicker steel shells and head ends, and are provided with *shelf couplers.* These couplers prevent disengagement in the vertical plane, which usually occurs in a derailment. Existing 55-ton and 90-ton cars are being provided with shelf couplers. As of 1993 this retrofitting has eliminated tank ruptures due to railroad accidents that result in derailments of the tank cars.

The first incident of tank car rupture occurred in 1961 and caused the death of an 11-month-old infant.[29] This was the result of a serious spill from a 30-ton car in a rural area of Louisiana. The chlorine cloud resulting from this spill covered an area of about 6 square miles. It was about 80–90 feet high, 2½ miles wide, and extended 3½ miles downwind. Measurements of chlorine concentration were made that revealed levels exceeding 400 ppm 7 hours after the accident. There were 451 animal fatalities, including cows, horses, mules, hogs, and other domestic animals. The infant child died after about 15 minutes exposure 150 ft from the tank car. The mother, who had 11 other children, said that during the panic to get all her family away from the wreck she became confused and lost count. About 100 people were treated for vapor burns and respiratory irritation. All of the 15 hospitalized patients were discharged on the 16th day.

Fear of chlorine gas in large cities has led to bizarre situations worthy of comment. For example, the city of Chicago will not allow a chlorine tank car to pass through the Loop in order to get to the water treatment plants on the shores of Lake Michigan, but will allow tank cars carrying 15-ton containers to pass through it. Each container has six fusible plugs and two valves, each a source of leakage. The hazard is obviously greater from a load of such containers than from a tank car. New York City has restrictions against liquid chlorine that produce absurd situations for designers. Most have been a result of a single accident that should have gone unnoticed but actually developed into a serious situation nearly resulting in tragedy. In June 1944, a delivery truck carrying chlorine cylinders was traveling through the streets of Brooklyn.[31] When a leak in one of the 150-lb cylinders was discovered, the driver pulled up at the side of the street only a few inches from a series of gratings covering ventilating shafts leading to an adjacent subway station. The truck was parked in this position for approximately 20 minutes. Some 400 people were overexposed to chlorine. All but two were in the subway station. Two hundred eight persons of the 400 examined were admitted to eight different hospitals. Thirty-three persons showed evidence of moderate or severe poisoning and required hospital care for one or two weeks. All the rest were released after two or three days. If the driver had kept his truck moving, probably little or no overexposure would have resulted. This statement is substantiated by known cases of chlorine leaks occurring in tank cars in transit. One occurred in Indiana in 1935 when a tank car developed a leak en route and lost 60,000 pounds of chlorine without property damage or personal injury.[34] In this case the tank car kept moving.

Transport Accidents

Mexico. Some of the worst chlorine spills have occurred during the movement of chlorine in tank cars and tank trucks. One of the worst spills ever occurred August 1, 1981 at Estación Montana, San Luis Potosí, Mexico, population 400. A 38-car freight train hauled by two diesels and with a caboose left the tracks about 300 yards from the station. This train was hauling thirty two 55-ton chlorine tank cars. The official report listed 17 people dead from this accident. Four of the train crew died as a result of the derailment, and 13 people died as a result of chlorine vapor inhalation. The accident was caused by excessive train speed in a terrain with a descending slope of 4 percent, plus the failure of the braking system. Seven tank cars lost all of their contents. Two of these lost their contents almost immediately. One car had a 20-inch hole in one of the heads and lost all of its contents. Four of the cars lost their valves and their protective housing. The cars leaked for several days and eventually lost all of their contents. Vegetation and some corn crops were damaged by the chlorine. Trees and shrubs turned yellow; however, several weeks after the accident, it was noticed that the vegetation was growing again. The Mexican government no longer allows this many chlorine cars in a freight train.[35]

Youngstown, Florida (2/26/78). At 1:55 A.M., February 26, 1978, a freight train traveling at 45 mph with a string of 140 cars hauled by five diesel locomotives derailed near Youngstown, Florida. The train was 1½ miles long and had no radio communication. All of the locomotives and the seven lead cars derailed as did the 9th through the 44th cars. The latter group contained ten tank cars of hazardous materials, including two of chlorine. One of the chlorine cars was punctured by a car originally located seven cars behind. The puncture was about 0.75 sq ft in size, with several long radial cracks extending as much as 2.5 ft from the puncture. The other derailed chlorine car was badly damaged but did not leak. Both were 90-ton cars. The chlorine that leaked out of the ruptured car was the direct cause of 8 fatalities and 138 injuries. It is estimated that about 50 tons of chlorine was lost in the first few minutes in both the liquid and gas phase. About 12 hours after the wreck, visual observation indicated that 30 to 40 tons of liquid chlorine remained in the tank car. This gassed off slowly.

Seven hours after the derailment, the chlorine cloud was 3 miles wide, 4 miles long, and 1000 ft high. Wind velocity varied from 2 to 3 knots. The temperature was between 50 and 55°F. A systematic evacuation of residents within a 10-mile radius was made. The cloud was so dense that it obscured the highway 300 ft away. Before it could be blocked off, a number of motorists entered the area of the gas cloud. Their automobiles kept stalling due to the heavy concentration of chlorine in the air. Seven motorists, mostly teenagers, abandoned their cars and fled into the adjacent swamp, but died in vain attempts to flee the gas. It was later believed that it was this group who were responsible for the derailment, which was most likely due to sabotage. An eighth motorist drove successfully through the cloud, only to succumb to the effects of the gas a short distance down the road. Others suffered extreme nausea and respiratory distress. One hundred twelve persons were treated and released from five hospitals. Twenty-two were so severely injured that they remained hospitalized. The injured included motorists, train crew members, and law enforcement personnel. Witnesses reported that at night the chlorine cloud appeared as fog and conveyed no indication of danger. The only warning conveyed by the cloud was its pungent odor. By then it was too late. The punctured car was not equipped with shelf couplers or head shields.[36]

Crestview, Florida (4/8/79). At 8:00 A.M., April 8, 1979, 29 cars, including 26 placarded tank cars of hazardous materials, derailed while traveling at 40 mph on an S curve near Crestview, Florida. This was a 116-car Louisville and Nashville freight train containing cars loaded with chlorine, ammonia, acetone, sulfur, methanol, carbolic acid, and carbon tetrachloride. Two tank cars of anyhdrous ammonia ruptured and rocketed. Twelve other cars containing five kinds of hazardous material ruptured, including a car of chlorine that had an 18-inch puncture hole in the side of the tank. Chlorine and ammonia vapor dispersed, forming a cloud that grew until it threatened an area of 300 square miles downwind. More than 4500 persons were evacuated. The cloud posed a threat to the health of the population, and to wreck-clearing and emergency personnel, for nine days. There were no fatalities, and only 14 persons were slightly injured by exposure to the variety of hazardous chemicals. Three were hospitalized. At the time of the accident, weather conditions were daylight, with 7-mile visibility and temperature of 57°F, sky overcast and misting rain, and winds at 5 mph, gusting to 20 mph. Throughout the nine days of the emergency, the gases reacted with the moisture of mucous membranes of the victims' lungs, eyes, and noses, producing various acids that became trapped in the affected organs. Ten wreck-clearing workers were overcome despite the use of self-contained breathing apparatus and short work shifts. It is believed that failure of the breathing apparatus might have been due to beards on some of the work crew.

The cloud movement was observed by an Air Force aircraft, and entry into the air space was prohibited. The vapor cloud reached a 5000-ft altitude and extended 28 miles in four hours. The next day, the cloud height was 1000 feet, and chlorine was estimated to be leaking at a rate of 18,000 lb/hr. Emergency response and post-emergency investigation of this accident involved local, regional, state, and private industry forces and at least six federal agencies. The emergency activities reportedly were so uncoordinated that the first team of chlorine emergency specialists was actually turned back en route to the scene.[37]

As is usually the case, poor road-bed conditions were blamed for the accident. There were other factors that could have been more significant than the road-bed conditions. The train was probably going too fast (40 mph) to negotiate the S curve safely. The train was probably too long for the curve, and the power thrust of the engines might have been improperly applied. It is worth noting that none of the derailed cars were equipped with shelf couplers or head shields.

Mississauga, Ontario, Canada (11/12/82). This town is a suburb of Toronto. Late Saturday night, November 10, 1979, a 106-car freight train carrying tank cars of chlorine, caustic soda, and flammables derailed in Mississauga, Ontario, 20 miles west of Toronto, Canada in 20°F weather. A 90-ton car of chlorine was upended adjacent to eight burning cars of propane. Heat from the fire and the presence of chlorine vapor in the top of the car caused the steel in the chlorine tank car to burn spontaneously. This allowed the release of about 70 tons of chlorine. In this case the chimney effect caused by the burning propane drove the escaping chlorine upward into the atmosphere, where it was carried away by natural air movement. Evacuation boundaries were extended through Sunday, until at 8:30 A.M. (20 hours after the accident) police announced that virtually the entire 50 square mile area encompassing Mississauga was to be evacuated. That order affected 250,000 residents and was made because of a wind shift. It was reportedly the biggest evacuation in North America. The wire services reported that anyone breathing 3 ppm of chlorine for 15 min. would receive medical treatment.

Taking advantage of the low ambient temperature (20°F), water was poured into the remaining 20 tons of liquid chlorine in the ruptured car. This formed a thin sheet of chlorine hydrate ice on top of the liquid chlorine. This inhibited vaporization and thus expedited and simplified the transfer of this chlorine to other containers. Six days later all of the liquid was removed from the car, and the evacuation was terminated. There were no reports of injuries due to chlorine exposure or other hazardous materials involved in the wreck. The low ambient temperature was an important factor that helped control the chlorine emissions. Subsequent newspaper reports claimed that no one, except those at the site, ever smelled any chlorine. This was due to the fact that the chlorine "fog" went straight up in the atmosphere like from a tall chimney.[38]

Tanker Trucks

Egypt. The only chlorine tank truck accident that has been reported to date is the one that occurred in Alexandria, Egypt in December 1965. Five people died, including the truck driver and two would-be rescuers.[34]

A chlorine tank truck, loaded with seven tons of liquid chlorine, swerved to avoid hitting a passenger car, overturned, and sheared off a gas valve. Civil defense procedure was inadequate, the truck had neither a protective dome over the valves nor excess flow valves, and knowledge of emergency procedure was probably nonexistent. The most seriously exposed were those who tried to rescue the unconscious truck driver injured by the crash. Approximately 2 tons of chlorine leaked out before the gas valve had been sealed off. By this time some 500 persons had been exposed to the fumes.

Notable Consumer Accidents

General. There have been a variety of lost-time accidents from chlorine leaks at water and wastewater treatment plants and industrial plants. In California alone there are about 200 reported cases of lost-time accidents annually due to chlorine and its compounds and about 40 due to sulfur dioxide, according to the California Department of Industrial Relations. The bulk of these accidents occur in industry. This is not surprising, as the chlorine used at water and wastewater treatment plants and at power stations accounts for a mere 5 percent of the annual production of chlorine.

Only one employee fatality has been reported for a water or wastewater treatment plant operation. A utility man employed by a Municipal Sanitation District in California died from inhalation of chlorine. His vocal chords were burned, and he died two weeks later from pulmonary edema. Details of this accident have never been made available.

Some of the most serious consumer accidents are detailed below.

A Fatal One-Ton Cylinder Liquid Leak. This accident resulted in the most serious consequences of any chlorine leak in North America associated with the handling of chlorine containers at the consumer level. It caused the untimely and quite unnecessary deaths of two people close to but not associated with the handling of the chlorine. This particular accident occurred in May 1959. At the time, the leak was quickly blamed on a "faulty gasket." A team of newspaper reporters made a comprehensive investigation of this accident and arrived at the following conclusions:

1. The accident probably would not have occurred if there had not been a power failure during a brief rain squall.
2. The accident probably would not have occurred if the city had not removed the auxiliary municipal light plant service to the filtration plant one month before the accident.
3. If working gas masks were conveniently available, the leak could have been halted immediately, and no injuries would have resulted.
4. Residents adjacent to the plant would not have been affected if the chlorine container room had been farther than 65 ft away from their homes.

An analysis of this accident and the newspaper reporter's conclusions leads to the following observations.

The power failure plunged the chlorine container room into practically total darkness (a rain squall at 4 P.M. shut out most of the light). This situation caused a great delay in the operator's response to all of the duties required during the power failure. It is easy to imagine the panic that struck him at smelling a severe chlorine leak in the dark! He could not find the gas masks quickly; so he left the area immediately (as he should have) to go for help, as he was the sole operator on duty.

With the power out, the operator was unable to relieve the chlorine piping system of pressure, even though the container valve at the leaking connection might have been closed. During the time of a power failure, all of the liquid chlorine in the piping system between the containers and the metering equipment can exit through any leaking joint unless there are appropriate isolating valves. When power is available, the metering equipment is able to withdraw the liquid chlorine from the supply piping system very quickly, thereby averting a massive chlorine leak.

Therefore, auxiliary lighting to show the operator the way and auxiliary power to operate the chlorine withdrawal system are essential for safety in chlorine handling.

Easy access to chlorine gas masks is of top priority. Just as soon as the fire department arrived with the necessary assistance to provide gas masks and emergency lights, the operator was able to locate the leaking cylinder connection. By this time power had been restored, and the operator was able to relieve the system of pressure and repair the leak.

In situations where operators face overwhelming odds, they need all the isolating valves available to them to reduce the length of piping under pressure adjacent to the leak. This is why every container should be connected with an auxiliary cylinder valve attached to the cylinder outlet valve. The outlet of the auxiliary valve is then connected to the inlet of the flexible connection and the outlet to the stationary header valve. Some operators prefer an auxiliary header valve at the outlet of the flexible connection so that for liquid withdrawal the flexible connection can be shut off at each end and removed during each cylinder change without the fear of discharging liquid chlorine. The best way to operate is to use duplicate header systems so that one header can be completely emptied of liquid chlorine during a cylinder change.

The deaths that occurred in this accident could have been avoided had the residents been told to get out of their houses. The children of the deceased did run away from the house, an act that undoubtedly saved their lives. By staying in a house through which chlorine gas has passed, any occupant is subject to continuous chlorine exposure because the gas becomes absorbed immediately by upholstered furniture, drapes, carpets, and clothing. Body heat and perspiration enhance chlorine absorption by the clothing. This exposes the victim to continuous chlorine inhalation from both the clothing and house furnishings.[39]

A Four-Ton Liquid Leak. This leak allowed escape of the liquid chlorine contents of four one-ton chlorine cylinders which were "on line" at the time of the leak. Fortunately, there was not loss of life, but many residents nearby were "treated" for various degrees of chlorine inhalation.

This massive leak was the result of the following factors, in this order of importance:

1. The operator attempted to stop a leak at the stem of a chlorine header valve while the system was under full pressure from the four cylinders.
2. This leak was caused by the structural failure of a bushing (¾″ × 1″) in the 1-inch chlorine header into which the leaking header valve was threaded.
3. The failure of the bushing was the result of pernicious corrosion over a long period of time.

The lessons to be learned from this accident are as follows:

1. Never use bushings in the chlorine supply system.
2. Operators should be warned never to attempt repair of a chlorine leak while the system is under supply pressure.
3. The first step in repairing a leak is to relieve the system of the chlorine cylinder pressure. Necessary gauges must be included in the design to provide the operator with this vital information.
4. *Duplicate header systems should be provided* so that these piping systems can be replaced in their entirety on a regular basis (i.e., every 5 years for systems passing 2 tons or more per day and every 10 years for all others passing less than 2 tons per day).

A 1600-lb Gas Leak. This case was a freak accident compounded by mistakes resulting from lack of experience in coping with a massive chlorine leak.

A workman for a natural gas utility company was busy with a cutting torch working at several isolated locations in an industrial area. He was cutting into sections of empty, unused natural gas pipelines preparatory to sealing them off. By mistake he cut into a 6-inch underground chlorine pipeline about 7000 ft long. This pipeline was used to transport chlorine gas from a chemical plant to a nearby plastics plant. The pipe was under tank pressure for about 85 psi at the time of the accident.

This calculates to about 1600 lb of chlorine gas that escaped into the atmosphere. Fortunately, the line was equipped with automatic shut-off valves at both ends. When the pressure dropped as a result of the hole made by the cutter's torch, these valves automatically closed.

Concidentally with the action of the pipe cutter, which occurred at about 7:45 A.M., some 300–400 people were waiting to go to work at a nearby refinery. These people were all gathered in a small cluster waiting for the gates to open at 8:00 A.M. They were a short distance downwind from the leak. The morning air was cool, and the humidity was high—about 70 percent. The general location is only a few miles from the ocean surf, and the wind was blowing in their direction at 10–15 mph. Even though this is considered a massive leak, only 30–40 of these people were taken to the hospital. It is probable that the high humidity coupled with a strong wind was responsible for rapid dilution and dispersal of the chlorine into the atmosphere. However, the humidity had an adverse effect. The combination of body moisture and high humidity conspired to absorb chlorine into the workers' clothing. The people taken to the hospital were herded into a small room and were immediately exposed to more chlorine coming from their clothing. A hospital attendant realized what was happening, and they were ushered outside and then taken in one by one and had their clothing removed. They remained in the hospital long enough for a thorough observation and time to get a change of clothes. None stayed more than one night in the hospital.[39]

The lessons to be learned from this accident are as follows:

1. Pipelines carrying chlorine liquid or gas should never be buried in the ground. The preferred method is to place them in a grate-covered concrete channel at grade level or in overhead support systems.
2. Whenever a person is exposed to chlorine, that person's clothing should be removed as soon as possible and the body showered thoroughly with warm water if available to avoid further shock.
3. Never allow persons subjected to chlorine exposure to remain in a confined area, particularly a room with rugs, carpets, drapes, or upholstered furniture. Get them into the open air as soon as possible.
4. It is most important to remember that the effects of inhaling chlorine gas versus that from the "off-gassing" from a strong chlorine solution are totally different for the human pulmonary system. Chlorine gas pro-

duces a sharp throttling effect that simulates strangling. The off-gas from a chlorine solution does not "throttle" the throat muscles as gas does, but creates a mild-to-strong irritation in the person's breathing that allows the chlorine actually to reach the lungs. This causes a watery mucus to be generated in the lungs, depending on how long the person has been in the area of "off-gassing," or of a foglike cloud of chlorine vapor. Some cases have been known to be so serious that some people have died from an overdose because they went to sleep shortly after their exposure. Death then was caused by an excess of watery mucus in the lungs, the cause of drowning.

A 14,000-lb Liquid Leak

Introduction. This is the largest known leak involving a chlorination installation for a potable water treatment system. It is a leak that never should have occurred. It involved a chlorination system located at a remote pumping station that supplied the water treatment plant with water from an adjacent lake. Fortunately, this leak occurred in a sparsely occupied area of large farmlands.

The nearest residences were some 3000 to 4000 ft away toward the northwest. A major wastewater treatment plant was located about 6000 ft southeast of the pumping station. The pumping station where the chlorination equipment was installed is about 3 miles from the water treatment plant, which treats the source water from the lake. Because the heat that caused the leak was very great, it produced a chimney effect that formed a narrow hot plume that carried the vaporized liquid chlorine straight up into the surrounding atmosphere. Such heat-involved leaks are rarely if ever noticed by people at ground level. The chlorination installation was used solely on an intermittent basis to keep the water channel to the treatment plant free from slime and other biological growths. This in itself was part of the problem because the installation was not subject to any organized routine maintenance program. As a matter of fact, the personnel who visited the system at the time of this accident had never read any of the instruction books for either the evaporators or the chlorinators.

The chlorinator layout was arranged by the consulting engineers who designed the system, who decided for unknown reasons to put the chlorine diffusers into the discharge piping of the pumping plant pumps instead of into the suction of these pumps, although that would have solved only the mixing of the chlorine solution into the large volume of the pumped water.

Instead, their design required injector booster pumps, plus diffusers followed by mixing devices for the chlorine solution. This booster system necessitated handling chlorine solution at substantial pressures, which is totally undesirable. A leak in such a system becomes a difficult operator problem.

Chlorination System. This consisted of two chlorinators and two evaporators, with electrically heated evaporator water bath containers. These evaporators were installed with all the protective devices for both low and high temperature situations, plus an automatic electrical supply switch wired to shut down the evaporators in the event of a water bath temperature exceeding 200°F. The main power supply for the electric heater elements in the evaporators was controlled by an external safety switch designed to shut down the electrical power to both the evaporators when the water bath temperature reached 200°F.

Cause of the Accident. Because of the lack of routine maintenance, the aforementioned contractor switch became fouled with packed debris in the switch, so that it was stuck with the heaters energized. This debris was apparently caused by insects living in that area. Insect infestation of electrical devices and equipment is not uncommon in that part of the United States. The people assigned to make the inspection of this installation demonstrated fully that they had no knowledge of this part of the evaporator–chlorinator system. Therefore, with the emergency contactor switch stuck in a closed position, the heat in the water bath area rose to 485°F, causing the liquid chlorine temperature in the evaporator to ignite the steel of the liquid chlorine container, which melted immediately, without being even touched by the evaporator heater rods. It was simply the temperature of the environment. Another factor that allowed this leak to be so enormous was the abandonment of ton cylinders in favor of a 25-ton storage tank. This was done after several years of using ton cylinders. If they had still been using ton cylinders, the leak might have been as small as 8000 lb.

The piping and the accessories from the tank to the evaporators were all wrong. The only delivery pipe to the evaporators was the liquid chlorine line without any gauges on the manway cover. This prevented the operators from being able to withdraw gas from the tank or evaluate the chlorine supply system. This is a serious installation error. There was no way an operator could feed chlorine gas to the chlorinators or the evaporators. The operator should always be able to feed either chlorine gas or liquid chlorine. Because the storage tank was not covered, feeding chlorine gas could in many cases reduce the chlorine pressure on the storage tank. Furthermore, because the tank was not covered, it could easily absorb enough heat to operate without heat from the evaporators at 3200 lb/day for several hours during weather that did not drop much below 68°F.

Comments. The other problem that confronted this installation was that all this occurred after the system had been shut down for a considerable length of time. That is, the clorination facility had been reactivated recently and placed in service at a total combined feed rate of 3120 lb/day. The intermittent use of the facility was certainly a factor in poor operator awareness of chlorine system operation and emergency procedures. The operators did not know that the evaporator must be considered as a part of the storage tank. In shutting an evaporator system down, the first and foremost consideration is the handling of the liquid chlorine in the evaporator. The first action should be to shut off the discharge valve of the evaporator. Then if rising heat is a problem, the liquid chlorine will boil back into the storage tank. Then if the pressure in the storage tank reaches its upper limit, the safety relief valve on the tank manway cover will release gas to the atmosphere. This would be a comparatively small leak that would tell an operator where the problem was: excess temperature in the evaporator water bath. This would give the operator plenty of time to correct the situation. This of course did not occur. The operator shut off the inlet valve to the evaporator!

The primary question remains: why did the operator on duty fail to call for expert help when he saw steam vapor coming out of the evaporator outer cabinet?

Another error in the installation was piping the evaporators so that one evaporator could serve two chlorinators. This should never be done. Each evaporator must be dedicated to only one chlorinator. If standby equipment is required, then engineers should install another evaporator and its dedicated chlorinator.

Magnitude of the Leak. In this case, chlorinator #1 was feeding 2360 lb/day and chlorinator #2 was feeding 760 lb/day, for a total of 3120 lb/day, which equals a total of 130 lb/hr. It was estimated that the leak lasted a good 17 hours. As the leak was from the evaporator assembly, the chlorine from the chlorinators amounted to $17 \times 130 = 2210$ lb, if the chlorinators were somehow getting chlorine from the disabled evaporator. This is extremely doubtful. The best source of information was from the chlorine tank scale, which showed a loss of 14,000 lb. Obviously 130 lb/hr was going through the chlorinators for a few hours before the evaporator failed. But that is why it is called a 14,000-lb leak.

The Good News. Since this was a leak caused by a 485°F fire, the leak was not only confined to the chlorination equipment room, but the temperature that was involved created a "smokestack" effect for the escaping chlorine gas plume. This was the first thing one of the arriving witnesses saw. So anyone outside of the plant probably never even smelled the chlorine leak.

Help in Chlorine Emergencies

This includes all hazardous chemicals:

PHONE—TOLL FREE—DAY OR NIGHT

CHEMTREC/MCA

800–424–9300

For Alaska, Washington, D.C., and Hawaii, phone 202–483–7616. For Canada call collect to 613–996–7616

Effect of Chlorine Accidents on the Community

That locally stored chlorine is a potential hazard is undeniable. However, the risk of chlorine storage for use at potable water plants and wastewater plants must be put in perspective. There have been many more serious accidental emissions from ton containers than from tank cars while stored at these plants. A tank car is not nearly so vulnerable as a ton container. In spite of this, there is much more chlorine to worry about in a tank car or stationary storage tank than in a battery of ton containers.

The City of San Francisco has three wastewater treatment plants within the city limits. Two of these used tank cars for more than 25 years. The third plant used a battery of 12–15 ton containers from 1938 to 1981. Moreover, there were 50 ton stationary chlorine storage tanks located at the plants using tank cars. The safety record for chlorine handling at these plants was exemplary. There were a few minor accidental emissions, mostly at the plant using ton containers, where there was one known lost-time accident.

In 1978 a newspaper reporter wrote a story about the hazards of transporting chlorine tank cars across San Francisco Bay and storing these cars at the wastewater plants. The evening paper ran the story under the front page headline "S. F. SITS ON A GAS DISASTER" (*San Francisco Examiner,* Sept. 5, 1978). Although the reporter focused largely on the way the cars were transported, the San Francisco County Board of Supervisors voted to stop using chlorine gas as soon as possible and to switch to sodium hypochlorite. This was accomplished in March 1980. Sulfur dioxide was also discontinued in favor of sodium bisulfite at the same time. The city had to pay out of its own funds about 3 million dollars for this project. Federal grant money was denied the city on the argument that its existing chlorination–dechlorination facilities were safe and reliable.

Knowing that the hypochlorite was going to be significantly more expensive than chlorine gas, the City of San Francisco attempted to secure other large municipal users of chlorine to enter into a joint venture of making hypochlorite from liquid chlorine in tank cars and caustic solution. They all declined. Several of these users who were using tank cars made separate investigations of the safety aspects and the hazard potential of their use of tank cars. All of these investigations resulted in the decision to stay with the tank cars. The two overriding factors turned out to be the high cost of hypochlorite (40¢/lb vs. chlorine gas at 6–7¢/lb) and the fact that of the total amount of chlorine transported in and around these treatment plants, only 10 percent was being used at the treatment plants.

On November 16, 1979, a corroded bolt on the flange of a tank car *flexible connection failed,* spewing liquid chlorine at a sodium hypochlorite manufacturing plant across the tracks from one of the above wastewater plants that used tank cars. This hypochlorite manufacturing plant is located adjacent to the freeway that is the interchange between the San Francisco–Oakland Bay Bridge and the East Bay cities. This main thoroughfare had to be closed during the morning rush hour until the leak was secured. Although the leak was of fairly short duration, 10 people were admitted to the hospital and released after a short time. The incident did not provoke any negative community reaction.

Frequency and Magnitude of Chlorine Leaks

A study of leaks over the past 50 years indicates by frequency and magnitude of chlorine emission the following list of causes in order of potential hazard (with item 1 representing the greatest hazard):

1. Fire.
2. Flexible connection failure.
3. Fusible plug corrosion.
4. Freak accidents caused by carelessness and ignorance.
5. Valve packing failure.
6. Gasket failure.
7. Piping failure.
8. Equipment failure.
9. Collision accidents causing physical damage to containers.
10. Container failure.
11. Chlorine pressure gauge failure.

The following paragraphs discuss these causes.

1. Fire was listed by the Chlorine Institute as the most serious hazard primarily because companies that stored swimming pool chemicals were often wiped out when stored granular hypochlorite suffered a spontaneous explosion. If 150-lb cylinders of chlorine were in the area, these too would explode because the fusible plug does not melt when the cylinder becomes skin-full. This might cause a blowout of the cylinder valve before the plug does finally melt. Ton cylinders are different. When they become skin-full, the dished heads blow outward without rupturing the container. This demonstrates the strength and the integrity of all chlorine cylinders. It also demonstrates the uselessness of fusible plugs. This maneuver increases the volume of the cylinder, which is sufficient to allow the necessary expansion of the chlorine during the next 4°F rise required for the fusible plug to melt. *Fortunately, such fires as described above do not occur in the storage areas of the chlorine cylinders at water or wastewater treatment plants.* Moreover, these treatment plants use modern chlorination or sulfonation equipment to handle these vapors under a vacuum so that excessive pressures, for whatever reason, rarely if ever occur.

In the early years of ton cylinder usage, the industry suffered container explosions due to the spontaneous combustion of nitrogen trichloride. When chlorine is made from electrolytic cell water containing ammonia N in sufficient quantities, the chlorine produced will contain NCl_3, which is soluble in liquid chlorine. When the liquid is withdrawn, the NCl_3 evolves as a vapor and will explode. This first occurred at a New York City water treatment plant in the 1920s. It became a simple matter to remove the ammonia N from the cell water, but Americans did not do a very good job of exporting this information. In 1981, White was called to Bogota, Colombia to investigate such explosions. During his inspections of the local chlor-alkali plant, he counted more than a dozen U.S.-made ton cylinders with their dished heads still intact although they had been blown from a concave to a convex position.

There never has been any report of a container failure of 150-lb cylinders, one-ton cylinders, stationary storage tanks, or railcars at any potable water, wastewater, or industrial cooling water chlorination or sulfonation installation in the United States or Canada. Fires in these locations are practically unheard of. If a fire were to occur that caused a major liquid chlorine release, the chlorine would be vaporized immediately and would disappear in the chimney effect of the fire. This actually occurred in the Mississauga freight train derailment, where a 90-ton chlorine railcar was upended adjacent to a propane tank car that exploded. The heat from the propane fire caused the chlorine vapor in the tank car to ignite the steel in the 90-ton car, which released 70 tons of chlorine before being smothered by water, which iced over the remaining liquid. There were no reports of anyone ingesting any chlorine. (See above.)

2. Probably the most frequent cause of chlorine emissions resulting in overexposure to personnel is the failure of connecting lines between the container and the metering and control equipment. These connecting lines are traditionally made of annealed copper, 2000 psi strength, and cadmium-planted. Copper is used because it is flexible and has the proper structural strength; however, at each container change, the chlorine remaining inside the tubing will absorb moisture from the atmosphere, and a cycle of corrosion will begin. Therefore, any flexible connection has a life dependent upon how many times it is disconnected and left open to the atmosphere. To check the worthiness of a flexible connection, one should bend it carefully; if it screeches slightly, it is due for replacement. The noise produced by the bending is indicative of the magnitude of corrosion products on the inside of the tube.

3. Fusible plug failure without any evidence of elevated temperature caused by fire or direct sunlight is next in order of hazard magnitude. A fusible plug that is supposed to but does not melt at 158°F may leak from corrosion of a poor bond between the lead alloy and the plug retainer. There is only one fusible plug on a 150-lb container, at the base of the outlet valve. There are three in each of the dished heads of a ton container. Gas emissions through faulty fusible plugs have been numerous enough for engineers to question their safety value. The AWWA Committee report[28] *recommends elimination of fusible plugs.*

4. Carelessness is high on the list of causes of chlorine accidents, sometimes involving unusual situations. One case involved a 6-inch buried chlorine gas

line[40] originally entirely within the property of two chemical plants. Adjoining property became subdivided 20 years later, and some of the underground piping, including that for chlorine gas, became part of the pavement area of new streets. This chlorine gas line was first cut into by mistake by a welder who was supposedly inactivating obsolete natural gas lines. The heat from the torch burned a small hole in the pipe, so that the chlorine could support combustion of the carbon steel pipe. Within seconds, the hole was approximately 8 inches in diameter, resulting in an almost immediate discharge of the contents of 7000 feet of 6-inch pipe. A few months later, on successive days, this same line was broken twice again by a backhoe excavating for additional underground lines. After these experiences, the decision was made to lower the pipe from its former depth of approximately 2 ft to approximately 5 ft.

5. Valve packing failures have never caused any serious problems. If the leak is minor, it can often be corrected by tightening the packing nut. Serious leaks are taken care of by the application of a proper container safety kit.

6. Gasket failures are serious only when they occur on the seal between the dome and the tank of a chlorine tank car. Other gaskets are located so that the supply can be secured, the system emptied of gas, and the gasket replaced in a routine fashion. Gasket failures on tank cars have occurred, but these are rare, and the emission of chlorine is minor.

7. Piping failures have been rare, and are sometimes the results of using improper materials. Lines carrying liquid chlorine can be a potential hazard. One of the few fatalities attributed to chlorine in recent years was caused by the failure of a liquid chlorine line. The worker's death was the result of his constricted breathing caused by the amount of chlorine fumes, which made it impossible for him to climb out of the confined location of the leak.

8. Equipment failure usually refers to vaporizers used between the containers and the chlorine metering and control equipment. These failures are due primarily to corrosion by the chlorine, and are a function of the amount of wall thickness versus the amount of chlorine passing through the vaporizer. The frequency of this type of accident is low.

9. Chlorine emission accidents caused by collisions involving containers are now considered rare. This is a direct result of the shelf coupler requirements on all freight cars. All the major chlorine releases caused by derailments were due to couplers from adjacent cars being disconnected by vertical movement. This allowed the disconnected coupler to plunge into the chlorine tank car. This always caused a sizable rupture in the tank car. When couplers were not involved and the railcar simply tipped over, the worst result was having the tank car dome completely wiped off. In these cases the excess flow valves always seemed to shut tight so that the vapor emission through the two vapor valves was always easy to plug on-site.

10. Container failures, except those caused by fire, are extremely rare. The chlorine packagers throughout the United States and Canada are keenly aware of the potential hazards of handling chlorine containers. This has resulted in

a strict monitoring of container condition. Perusal of the Chlorine Institute accident reports indicates two container failures over a period of 15 years. Considering that shipments in 150-lb and ton containers amount to nearly 100,000 per year, this sort of accident is a rarity and is usually a result of using an overage cylinder. One of these accidents occurred at a military base. A ton container was being loaded at a dock. It slipped out of the sling and fell on its end. The tank split at the "chime," which is the joint between the cylinder and the dished head. Investigation of this accident revealed that this cylinder had been kept in service by the military for 45 years.

Chlorine tank car container failures are practically nonexistent for a very good reason: these containers are readily accessible for interior inspection on a programmed basis.

11. This type of gauge is vulnerable to leaks because of the silver or Hastelloy diaphragm that protects the brass bourdon tube. The bourdon tube flexes to indicate pressure changes. Constant flexing of the diaphragm causes metal fatigue that results in diaphragm failure. When this occurs, the brass bourdon tube corrodes quickly from the inherent moisture in the chlorine. This results in a severe leak. These gauges should be replaced about once every five years. To protect personnel as well as to provide the best possible method of minimizing the danger of a leak from a diaphragm failure, chlorine pressure gauges and pressure switches should be installed as shown in Chapter 9, Fig. 9-14.

Safety Precautions: Gauges, Valves, Fusible Plugs

Handling chlorine need not be a serious hazard if the personnel working with it are properly educated and trained in its handling.

The following are some guidelines for assuring the safe handling of chlorine:

1. Install leak detector sensor at appropriate locations.
2. Provide proper instruction and supervision to workers charged with responsibility for chlorination equipment.
3. Provide proper and approved self-contained breathing apparatus for persons working where there is a possibility of exposure to chlorine gas fumes. Locate the breathing apparatus close to the potential leak area far enough away so that it is accessible in case of a major leak.
4. Survey the areas of most likely chlorine emissions in an attempt to predict the downwind travel in case of accidental release. Use a wind sock to determine air movement so as to establish upwind areas.

A convenient aid in formulating this prediction is the use of "Downwind Vapor Hazard Nomographs." These are based on equations developed by Sutton of the United Kingdom and modified by Calder and Milly of the U.S. Army Chemical Corps.[41] Additional information on these nomographs may be obtained from the office of the Chief Chemical Officer, Department of the Army, Washington, DC.

A paper of more recent vintage often used by engineers to predict the spread of toxic fumes is the one published by the Chlorine Institute titled "Estimating Area Affected by a Chlorine Release" by A. E. Howerton,[32] in March 1969. This publication contains other pertinent references.

5. Never store combustible or inflammable materials in or near chlorine containers.

6. Never apply direct heat to a chlorine container. Never attempt any welding operation on an "empty" chlorine gas line without having purged it with air.

7. Always keep available and close to the chlorine containers a water supply that can be used to keep the containers cool in case of fire or by personnel if they accidentally come into bodily contact with chlorine gas or liquid.

8. In case of a leak, determine if the rupture is in a faulty container or in the control apparatus or connecting pipe.

9. If the leak is in a container, an appropriate emergency kit should be brought into action.

10. If the leak is in the control apparatus or connecting pipe, at least two persons should don breathing apparatus, find the leak by means of ammonia fumes, and secure the valves at the containers. The operation of the control equipment will drain the system of chlorine pressure. When the system is down to atmospheric pressure (zero gauge), steps can be taken to make the necessary repairs.

11. After workers have been exposed to a chlorine leak of sufficient magnitude while working with self-contained breathing apparatus, their clothes should be removed and their bodies showered. The clothes should be aired adequately. The danger here is that the normal perspiration absorbed by clothing retains a tremendous amount of chlorine gas, which will be released continuously after the exposure. When the workers have left the leak area, they may think they are still being exposed to a leak because of the chlorine given off by the clothing. Therefore, always remove breathing apparatus in an open area—never in a room or a confined location.

12. Spraying water on leaking containers may make the leak worse as a result of corrosion. Water in sufficient quantity and velocity can be used to confine or limit the spreading of moderate leaks.[42]

13. Never try to disperse chlorine gas directly from a container to an open body of water. Chlorine gas is only very slightly soluble in water at atmospheric pressure. White[43] has demonstrated that it is far more beneficial to attempt the dispersal of liquid chlorine in a body of water if the conditions are favorable. Liquid chlorine goes into solution much more rapidly than gaseous chlorine does at atmospheric pressure. In the case of a leaking container, it is far more desirable to dump the container and its contents in a body of water if it is at least 10–12 ft deep. The body of water will cool the container, which will slow down the leak rate, while the depth of water will allow the chlorine to go into solution so that there will be no off-gassing. Also the Fish and Game people will be happy to know that the fish and other wild life in

the area will automatically remove themselves from the "chlorine zone." This is their planet too.

14. Chlorine leaks must be given prompt attention.

15. When entering a chlorine equipment area, always be on the alert. Take shallow breaths when entering and until it is ascertained that no leaks are in progress.

16. Keep upwind of chlorine leaks. Although chlorine gas is two and a half times as heavy as air, it will always follow air currents, as it has an affinity for moisture in the air. It is therefore a fallacy that chlorine always settle to the ground or to the floor.

17. Be aware of proper first aid procedures. Never give anything by mouth to an unconscious person.

18. If a container develops a leak in transit, keep the vehicle moving. Conversely, if a stationary container develops a leak, try to transport it quickly to a predetermined disposal site until the emergency response team arrives.

19. It is advisable to rely upon chlorine control and metering equipment for direct disposal into a natural stream or treatment plant facilities.

20. Do not attempt to rely upon direct disposal methods of the contents of chlorine containers, evaporators, or liquid chlorine piping unless these systems are connected to a predesigned chlorine absorption tank. (See Chapter 9.)

PRODUCTION AND USES OF CHLORINE

Annual Production

The amount of chlorine produced in the United States is reported by the Chlorine Institute in two categories: gaseous and liquid. Gaseous production generally refers to captive generation by industries that use the gas in the manufacture of their products. This has a tendency to distort somewhat the chlorine usage figures. The most productive year to date (1997) was 1979. In that year 12.3 million tons of gas was produced as compared to 7.3 million tons of liquid chlorine. The latter was packaged and delivered to consumers.[16] This compares with the 1950 production of 2 million tons of gas and one million tons of liquid chlorine.[16]

Energy Consumption in the Production of Chlorine

Table 1-14 shows the energy consumed to produce chlorine. The figures shown are for both English and metric units.[58]

End Uses of Chlorine

General. The best estimate of end uses that are significant is based upon a 1981 first quarter study by the Diamond-Shamrock Corporation[44] shown in

Table 1-14 Energy Consumption in Chlorine Production

Energy Input	*English Units*	*Metric (SI) Units*
Steam amount	10,500 lb/ton	5,250 kg/kkg[a]
Steam equivalent fuel[b]	14.1×10^6 Btu/ton	16.4 GJ/kkg[c]
Electricity amount	3.043 kWh/ton[d]	12.1 GJ/kkg
Electricity equivalent fuel	26.2×10^6 Btu/ton[e]	30.5 GJ/kkg
Total	40.3×10^6 Btu/ton	46.9 GJ/kkg

[a]1 kkg = 1000 kg = 2,204.6 lb
[b]Based upon 1340 Btu/lb steam
[c]1 GJ = Giga Joule = 10^6 Joules; 1 Btu = 1,054.8 J
[d]1 kWh = 3.6×10^6 Joules
[e]Based upon 50 percent self-generated electricity; overall: 8,610 Btu/kWh

Table 1-15. A 1981 tabulation of chlorine end use by the Pennwalt Corporation[45] is shown in Table 1-16. This data is still applicable for 1997.

Water and Wastewater Treatment. It should be of interest to the environmentalists who are forever concerned about the potential hazards to the environment caused by the chlorination of drinking water and wastewater that this use accounts for only 5–6 percent of the market, whereas the chemical industry accounts for about 50 percent of the market. The chlorinated hydrocarbons used in pesticides that have direct access to the environment in rainfall runoff are estimated at about 15 percent of the market.

The Power Industry. The electric power industry uses a substantial amount of chlorine to prevent biofouling in heat exchangers. This biofouling seriously affects the heat exchange efficiency of the condensers. A 1978 survey found the industry using approximately 10,800 lb of chlorine and hypochlorite.[46] This amounts to 1.8 percent of the figure shown in Tables 1-15 and 1-16 for water treatment. This makes the industrial percentage in Table 1-15 look

Table 1-15 End Uses of Chlorine

Market	*Chlorine Use (1000 S.T.)*	*Percent Distribution*
PVC-VCM	2067	18.5
Pulp and paper	1160	10.5
Organic chemicals	1758	16.0
Chlorinated ethanes	1327	12.0
Propylene oxide	814	7.5
Inorganic chemicals	664	6.0
Water and wastewater treatment	600	5.5
Fluorocarbons	828	7.5
Chlorinated methanes	442	4.0
Miscellaneous	1418	12.5
Total	11078	100.0

Table 1-16 End Uses of Chlorine

Market	*Percent*
PVC production	20
Slovent manufacture	24
Organic chemicals	20
Inorganic chemicals	12
Pulp and paper	12
Water and wastewater (municipal, 2%; industrial, 3%)	5
Miscellaneous	7
Total	100

credible because there are many industrial users of chlorine for biofouling control in condensers other than the power industry. There are also many industrial users of chlorine in process water treatment as well as in wastewater treatment.

The Chemical Industry. Chlorine is a most unusual and versatile substance, and has more diverse uses than any other chemical known—from rocket fuels to the manufacture of food products. One of its most unusual products is monochloroacetic acid, used to make carboxylmethyl cellulose, which, in turn, is used as a detergent builder, as a filler in ice cream,[17] and in the manufacture of thioglycollic acid (used in home permanents). Since 1950, the largest single use for chlorine has been in the manufacture of ethylene oxide and glycol, used to make antifreeze fluids and synthetic fibers.

Before the tremendous increase in the chemical industry, most of the chlorine was used in the textile industry for bleaching purposes. In 1965, some 30,000 tons were used in textile bleaching.[47] This was less than 1 percent of the total U.S. production in 1965.

Most chlorine manufacturing plants are situated where inexpensive sources of either power or salt, or both, are available. It is also necessary to plan for geographical demand, as chlorine must be sold on an FOB freight equalized basis. Although location is an important consideration, it is estimated that between 70 and 75 percent of chlorine production is captive. Because chlorine and its co-products must be carefully balanced, it sometimes becomes advantageous to utilize methods other than the electrolysis of brine for production. For example, the HCl oxidation processes are advantageous when there is an overabundance of hydrochloric acid. All these factors play an important part in location and manufacturing methods.

As the chemical industry accounts for the largest share of the total U.S. chlorine production, it is of interest to list the end uses of chlorine compounds. There are 10 inorganic and 25 organic compounds derived from chlorine which find more than 50 end uses.[48] The following is a list of chlorine compounds and some of their end uses:

Aluminum chloride	aircraft, automotive cutting oils, paints, metal processing, pigments, plastics
Allyl chloride	adhesives, aircraft, anesthetics, drugs, electronics, explosives, food additives, insecticides, paints, soft drinks, solvents, water repellants
Chlorinated benzenes	adhesives, aircraft, bactericides, detergents, drugs, fumigants, fungicides, household chemicals, protective coatings, solvents
Chlorinated paraffins	automotive, cutting oils, metal processing, plastic sanitizing
Chloral	aerosols, anesthetics, dry cleaning, insecticides, mothproofing
Chlorine trifluoride	rocket fuel, missiles
Chloroform	aerosols, anesthetics, drugs, dyestuffs, fumigants, perfumes, plastics, refrigeration
Dichlorobenzene	insecticides, organic solvents
Epichlorhydrin	detergents, dyestuffs, electronics, explosives, missiles, nuclear reactors, pigments, synthetic rubber
Ethyl chloride	anesthetics, automotive, drugs, oil processing, plastics, refrigeration
Ethylene chlorhydrin	antifoaming agents, bactericides, cutting oils, dry cleaning, explosives, food additives, germicides, paints, pulp and paper, polyfoams, rocket fuels, soft drinks, solvents, synthetic fibers, textiles, tobacco
Ethylene dichloride	adhesives, aircraft, automotive, electronics, plastics
Ferric chloride	water treatment, wastewater treatment
Fluorocarbon	aerosols, aircraft, anesthetics, drugs, fire extinguishers, food additives, germicides, paints, plastics, polyfoams, synthetic fibers
Glycerine	synthetic resins, tobacco, explosives, electronics, food processing
Hexachlorocyclopentadiene	fire retardants, household chemicals, insecticides, plastics
Hexamethylene diamine	aircraft, automotive, household chemicals, missiles, textiles
Hydrazine	aircraft, drugs, electronics, metal processing, refrigeration
Hydrochloric acid	aircraft, electronics, food processing, drugs, gasoline, household chemicals, missiles, pulp and paper, pigments, plastics, solvents, water treatment
Methyl chloride	adhesive, aerosols, aircraft, anti-foaming agents, bactericides, drugs, food additives, oil processing, plastics, refrigeration, silicones
Methylene chloride	aerosols, drugs, plastics, paint remover, solvents
Monochloroacetic acid	herbicides, detergents, pharmaceuticals, food additives, textiles

Perchloroethylene	dry cleaning, metal processing, missiles, nuclear reactors, solvents
Phosgene	adhesives, aircraft, automotive, electronics, paints, protective coatings, synthetic fibers
Phosphorous trichloride	aircraft, automotive, disinfectants, drugs, electronics, fire retardants, fungicides, gasoline, household chemicals, hydraulic fluids, missiles, paints, perfumes, pigments, plastics, protective coatings
Propylene glycol	polyester resins, cellophane
Propylene oxide	aircraft, automotive, disinfectants, drugs, electronic, food additives, fungicides, paints, plastics, protective coatings, solvents
Titanium tetrachloride	aircraft, automotive, dyestuffs, electronics, fire retardants, gasoline, missiles, pigments, plastics, titanium
Trichlorocyanurate	bactericides, bleaches, detergents, drugs, household chemicals, swimming pools
Trichloroethylene	dry cleaning, food processing, electronics, missiles, oil processing, solvents
Trichlorethane	aerosols, aircraft, dry cleaning, food processing, solvents
Vinyl chloride	aircraft, electronics, household chemicals, paints, pulp and paper, plastics, synthetic fibers, textiles
Zirconium tetrachloride	aircraft, disinfectants, drugs, dyestuffs, fire retardants, gasoline, pigments, plastics, synthetic fibers

REFERENCES

1. Baldwin, R. T., "History of the Chlorine Industry," *J. Chem. Ed.,* 313 (1927).
2. Mond, L., "History of the Manufacture of Chlorine," *J. Soc. Chem. Ind.* (London), 713 (1896).
3. Sommers, H. A., "The Chlor-Alkali Industry," *Chem. Engr. Prog.* **61,** 94 (1965).
4. Gardiner, W. C., "Castner, A Pioneer Inventor in Alkali-Chlorine," *Proc. Chlorine Bicentennial Symposium,* Electrochem. Soc., Princeton, NJ, 1974.
5. Holden, H. S., and Kolb, J. M., "Metal Anodes," *Kirk-Othmer Encyclopedia of Chemical Technology,* 3rd ed., Vol. 15, p. 172, John Wiley & Sons, Inc., New York, 1981.
6. Anon., "Dimensionally Stable Anodes," ES-EC-1 Eltech (Diamond Shamrock Corp.), Chardon, OH, 1979.
7. Thomas, V. H., and Rudd, E. J., "Energy Saving Advances in the Chlor-Alkali Industry," paper presented at the Chlorine Conference, Soc. of Chem. Ind., London, UK, June 1982.
8. Dahl, S. A., "Chlor-alkali Cell Features New Ion-Exchange Membrane," *Chem. Engineering,* 60 (Aug. 18, 1975).
9. Anon., "Membrane Cell Technology," Eltech, Chardon, OH, Feb. 1983.
10. O'Brien, T. F., and Gilliatt, B. S., "The FM 21, A Novel Approach to Membrane Cell Design," paper presented to the Chlorine Plant Operations Seminar, Atlanta, GA, Feb. 17, 1982.
11. Hora, C. J., and Maloney, D. E., "Nafion Membranes Structured for High Efficiency Chlor-Alkali Cells," paper presented at 152nd National Mtg. Electrochemical Soc., Atlanta, GA, Oct. 10–14, 1977.
12. Isfort, H., "Uhde's Membrane Cell Technology," paper presented at the Chlorine Conference, Soc. of Chem., Ind., London, UK, June 1982.
13. Udagawa, H., private communication, Asahi Chemical Co., Feb. 1983.

14. Anon., "A Revolution in Chlor-Alkali Membranes," *Chem. Week,* 35 (Nov. 17, 1982).
15. Gestaut, L. J., Thomas, V. H., and Moomaw, J. A., "Air Cathodes for the Chlor/Alkali Industry," Eltech Systems Corp., Chardon, OH, 1982.
16. Anon., "North American Chlor-Alkali Industry Plants and Production Data Book," Chlorine Institute Pamphlet 10, New York, Jan. 1982.
17. Sconce, J. S., *Chlorine: Its Manufacture, Properties and Uses,* Reinhold, New York, 1962.
18. Penfield, W., and Cushing, R. E., "Bathing the Green Goddess," *Ind. Eng. Chem.* **31,** 377 (1939).
19. Sisler, H. H., *College Chemistry,* p. 201, Macmillan, New York, 1935.
20. Schultze, H. W., "The Chlorine Industry; Past, Present and Future," *Proc. Chlorine Bicentennial Symposium,* Electrochem. Soc., Princeton, NJ, 1974.
21. Iammartino, N. R., "New Ion-Exchange Membrane Stars in Chlor-Alkali Plant," *Chem. Engineering,* 86 (June 21, 1976).
22. Kienholz, P. J., "Bipolar Chlorine Cell Development," *Chlorine Bicentennial Symposium,* San Francisco, CA, p. 198, 1974.
23. Anon., Kel-Chlor, M. W. Kellog Co., Houston, TX, company bulletin, 1980.
24. Anon., "Recovering Chlorine from Waste HCl," *Env. Sci. Technol.,* **9,** 16 (Jan. 9, 1975).
25. Laubusch, E. J., "Standards of Purity for Liquid Chlorine," *J. AWWA,* **51,** 742 (1959).
26. Adams, F. W., and Edmonds, R. G., "Absorption of Chlorine by Water in a Packed Tower," *Ind. Eng. Chem.,* **29,** 447 (1937).
27. Heinemann, G., Garrison, F. G., and Haber, P. A., "Corrosion of Steel by Gaseous Chlorine," *Ind. Eng. Chem.,* **38,** 497 (1946).
28. Committee Report, "Chemical Hazards in Waterworks Plants," *J. AWWA,* **27,** 1225 (1935).
29. Joyner, R. E., and Durel E. G., "Accidental Liquid Chlorine Spill in a Rural Community," *J. Occup. Med.,* **4,** 3 (Mar. 1962).
30. Hedgepeth, L. L., "Handling Chlorine to Avoid Trouble," *J. AWWA,* **26,** 1602 (1934).
31. Chasis, H., Zapp, J. A., Bannon, L. H., Whittesberger, J. L., Helm, J., Doheny, J. J., and McLeod, C. M., "Chlorine Accident in Brooklyn," *Occup. Med.,* **4,** 152 (Aug. 1947).
32. Howerton, A. E., "Estimated Area Affected by a Chlorine Release," Chlorine Institute Report 71, March 1969.
33. Anon., unpublished accident reports, 1981, Chlorine Institute, New York, 1983.
34. Anon., unpublished accident reports, 1930–82, Chlorine Institute, New York, 1983.
35. Perez, F. J., Plant Mgr., Celulosa y Derivados, S. A. Monterrey, Mexico, Letter to Chlorine Institute, Oct. 22, 1981.
36. Anon., National Transportation Safety Board Accident Report No. NTSB-RAR-78–7, Washington, DC.
37. Anon., National Transporation Safety Board Accident Report No. NTSB-RAR-79–11, Washington, DC.
38. Anon., unpublished accident reports, 1982, Chlorine Institute, New York, 1982.
39. Unpublished chlorine accident correspondence, G. C. White, San Francisco, CA, 1950–83.
40. *Los Angeles Times,* Sept. 13, 1966.
41. Anon., "The Handling and Storage of Liquid Propellants," Office of the Director of Defense Research and Engineering, Washington, DC, 1961.
42. Hopkins, E. W., and Faber, H. A., "Chlorine Dispersion with Fog Nozzles," *Water Sew Wks.,* **98,** 350 (Aug. 1951).
43. White, G. C., unpublished data on discharging liquid chlorine from 150-lb cylinder.
44. Krupp, R. C., Diamond Shamrock Corp., private communication, Dec. 11, 1981.
45. Trojak, G. F., Pennwalt Corp., private communication, Nov. 25, 1981.
46. Chow, W., Electric Power Research Inst., private communication, Jan. 8, 1982.
47. Sconce, J. S., private communication, 1968.
48. Anon., "Chlorine Facts," Chlorine Institute, New York, 1968.
49. Molnar, C. J., and Dorio, M. M., "Effects of Brine Purity on Chlor-Alkali Membrane Cell Performance," paper presented at the 152nd National Mtg. Electrochemical Soc., Atlanta, GA, Oct. 10–14, 1977.

50. U.S. Patent No. 3,632,498, "Electrode and Coating Therefor," H. B. Beer, filed Feb. 2, 1968.
51. Horacek, J., and Puschaver, S., "A Comparison of Economic Factors Involved in the Choice of Chlorine–Caustic Cell Anode System," paper presented at Chlorine Institute annual conf., 1972.
52. Cairo, P. R., Lee, R. G., Aptowicz, B. S., and Blankenship, W. M., "Is Your Chlorine Safe to Drink?" *J. AWWA,* **71,** 450 (Aug. 1979).
53. Anon., Am. Nat. Std. for Liquid Chlorine, ANSI/AWWA B301-81, June 7, 1981.
54. Unpublished Chlorine Institute file on nitrogen trichloride (ca. 1935).
55. White, G. C., unpublished investigation of chlorinator injector fouling, Orange Co. Water District, Fountain Valley, CA, June 16, 1982.
56. Doyle, J. H., private communication, Chlorine Institute, New York, Feb. 14, 1983.
57. White, G. C., unpublished report on overpressure in ton containers causing gross deformation without rupture. Bogotá, Colombia, 1981.
58. Meyers, J. G., "Energy Consumption in Manufacturing," The Conference Board, Energy Information Center; The Ford Foundation, Energy Policy Project, NSF-RANN 1973.
59. DuLong, P. L., *Schweigger's J. Chem. Pharm.,* **8,** 32 (1812).
60. Sconce, J. S., *Chlorine, Its Manufacture, Properties, and Uses,* Amer. Chemical Society Monograph Series No. 154, Reinhold, New York, 1962.
61. National Research Council, *International Critical Tables,* Vol. 1, McGraw-Hill, New York, 1926.
62. Kapoor, R., and Martin, J., "Thermodynamic Properties of Chlorine," Engr. Res. Inst. Univ. of Mich., 1957.
63. Wobser, R., and Muller, F., *Kolloid-Beihefte,* **53,** 162 (1941).
64. Lange, N. A., *Handbook of Chemistry,* Handbook of Publishers Inc., Sandusky, OH, 10th ed., 1961.
65. Eucken, A., and Hoffman, *Z. Physik, Chem.* **B5,** 422 (1929).
66. Adams, F. W., and Edmonds, R. G., "Absorption of Chlorine by Water in a Packed Tower," *Ind. Eng. Chem.,* **29,** 447 (1937).
67. Pellaton, M., "Constantes Physiques du Chlorine," *Jour. de Chemie Physique,* **13,** 426 (1915).
68. National Research Council, *International Critical Tables,* Vol. 3, McGraw-Hill, New York, 1928.
69. Lacey, J. I. (Hooker Electrochemical Co.), *Anal. Chem.,* **20,** 379 (1949).
70. Giaque, W. F., and Powell, T. M., "Chlorine, The Heat Capacity, Vapor Pressure, Heats of Fusion and Vaporization, and Entropy," *J. Am. Chem. Soc.,* **61,** 1970 (1948).
71. National Research Council, *International Critical Tables,* Vol. 7, McGraw-Hill, New York, 1930.
72. Lange, N. A., "Euber einige Eigenschaften des Verflussigten Chlor," *Z. Angewandte Chemie,* **13,** 683 (1900).
73. National Research Council, *International Critical Tables,* Vol. 1, McGraw-Hill, New York, 1926.
74. National Research Council, *International Critical Tables,* Vol. 5, McGraw-Hill, New York, 1929.
75. Steacie, E. W. R., and Johnson, F. M. G., "The Viscosities of the Liquid Halogens," *J. Am. Chem. Soc.,* **47,** 756 (1926).
76. Trautz, V. M., and Ruf, F., "Die Reibung, Warmeleitung und Diffusion in Gasmischungen," *Annalen Physik,* **20,** 127 (1934).
77. White, G. C., private field inspection of installation and discussion with operating personnel, Standard Oil Refinery, Richmond, CA (ca. 1940).
78. White, G. C., and Tracy, H., field demonstration of liquid chlorine discharging into an open body of water, Crystal Springs Reservoir, San Mateo County, CA (ca. 1970).
79. Shera, Brian, demonstration of a 30–50-lb liquid chlorine leak, Pennsylvania Salt Co., Tacoma, WA, 1965.
80. White, G. C., personal inspection with Chief Operator, North Point WWTP, City of San Francisco (ca. 1950).

81. White, G. C., personal investigation, East Bay Municipal Utility District WWTP, Oakland, CA, 1950.
82. Powell, D., videotape of 600-lb liquid chlorine leak, with commentary, Powell Fabrication and Mfg. Co., St. Louis, MO, 1985.
83. Curlin, L. Calvert, B., Tilak, V., and Hanssen, C. B., "Alkali and Chlorine Products, Chlorine and Sodium Hydroxide," *Kirk-Othmer Encyclopedia of Chemical Technology,* John Wiley & Sons, Inc., New York, 1990.
84. "Chlorine," Vol. A6 in *Uhlman's Encyclopedia of Industrial Chemistry,* pp. 399–481, VCH Verlagsgesellschaft MbH, D-6940 Weinheim, Germany, 1986.
85. Curlin, L. C., and Romine, R. L., "Polyramix Moves Into Commercial Use," paper presented at the Eltech Seminar, Cleveland, OH, Oct. 3, 1990.
86. Anon., "Calculating the Area Affected by Chlorine Releases," Chlorine Institute Pamphlet 74, ed. #1, rev. #1, June 1982.
87. White, G. C. "Monitoring, Inspection, and Repair of Flour Bleaching Installations with NCl_3" unpublished report to Wallace & Tiernan Co. by White, 1937.
88. "Implications of Cancer Causing Substances in Mississippi River Water," Environmental Defense Fund, 1525 18th St. N.W., Washington, DC 20036.
89. Trojak, Gary F., "The Uniform Fire Code: Realities and Options," Chlorine Institute, Washington, DC, 1996.
90. White, G. C., "Spontaneous Combustion of an Evaporator Chlorine Container Vessel," unpublished report, 1995.
91. "Sulfur Compounds," *Kirk-Othmer Encyclopedia of Chemical Technology,* 3rd ed., Vol. 22, p. 107, John Wiley & Sons, Inc., New York, 1983.

2
HYPOCHLORINATION

HISTORICAL BACKGROUND

Early Uses

One of the first known uses of chlorine for disinfection was in the form of hypochlorite, known as chloride of lime. Snow used it in 1850 in an attempt to disinfect the Broad Street Pump water supply in London after an outbreak of cholera caused by sewage contamination. Sims Woodhead used "bleach solution" as a temporary measure to sterilize potable water distribution mains at Maidstone, Kent (England) following a typhoid outbreak in 1897.[1]

Subsequent to Scheele's discovery of the element chlorine in 1774, the compounds made from this element were used expressly for bleaching (particularly textiles) and not for sanitation. Prior to the discovery of chlorine as a bleaching agent, textiles were bleached by the cumbersome method of using the sun's rays known as crofting. The textiles to be bleached were spread on large grass fields.

In 1785, Berthollet prepared a bleaching agent by dissolving "Scheele's gas" in water. In 1789, he improved this bleaching liquid by mixing it with a solution of caustic potash (KOH). This work was carried out at a French chemical plant in Javel, a former town in France, now a part of Paris. This solution was called Javelle water and still goes by that name today.[2]

A short while later, Labarraque replaced the expensive potassium hydroxide with caustic soda obtained from soda ash. This development resulted in what was probably the first use of sodium hypochlorite as a bleach. Because of common usage, it too became known as Javelle water. It completely replaced Berthollet's expensive potassium hypochlorite, and soon crofting went out of existence. About 1810, the introduction of the Leblanc process for the manufacture of soda ash made the sodium hypochlorite process even more secure by making available more soda ash at a greatly reduced price. Other methods for producing sodium hypochlorite were developed, but the chlorination of caustic soda remains the most popular method.

In 1978, Tennant[3] patented a successful liquid bleach for the Union Alkali Company by passing chlorine through a milk of lime solution. This is known as "bleach liquor." Tennant continued his experimentation and developed a bleaching powder by passing chlorine gas over slaked lime. This process was patented in 1799 as Tennant's bleach, and had a tremendous impact upon society. Here was a solid form of chlorine bleach that could be easily transported and needed only to be dissolved in water to be available for use. It was a boon to the textile industry, and became a cornerstone of the early chemical industry. Much of our chemical engineering heritage dates to early efforts to make soda ash, caustic soda, chlorine, and bleaching powder for use in bleaching cloth.

Freshly prepared bleaching powder contains about 36 percent available chlorine.[1] Its storage life is short, especially in warm climates. This degradation at higher temperatures demonstrated a need for a more stable compound. The addition of quicklime to bleaching powder produced what is known as "tropical bleach," which is fairly stable at tropical temperatures and contains 25 to 30 percent available chlorine. This product, developed about 1920, is still used in significant amounts in the underdeveloped areas of the world.

The first U.S.-produced dry calcium hypochlorite appeared on the market in 1928. This bleaching compound contains about 70 percent available chlorine, and is sold under the trade names of HTH, Perchloron, Pittchlor, and others. This product has largely replaced Tennant's bleaching powder in the United States.

Liquid bleach (sodium hypochlorite) came into widespread use in about 1930 for laundry, household, and general disinfecting uses. Today it is the most widely used of all the chlorinated bleaches. More than 150 tons/day is now used in the United States alone. Its preparation is a modification of Labarraque's method, using lower residual alkali, which simplifies purification and sedimentation while maintaining a pH of about 11 for stability.

CHEMISTRY OF HYPOCHLORITES

Potable Water

The application of either sodium or calcium hypochlorite in potable water and waste treatment achieves the same result as does that of chlorine gas:

Sodium hypochlorite:

$$NaOCl + H_2O \rightarrow HOCl + Na^+(OH)^- \qquad (2\text{-}1)$$

Calcium hypochlorite:

$$Ca(OCl)_2 + 2H_2O \rightarrow 2HOCl + Ca^{++}(OH^-)_2 \qquad (2\text{-}2)$$

The active ingredient is the hypochlorite ion (OCl^-), which hydrolyzes to form hypochlorous acid. The only difference between the reactions of the hypochlorites and chlorine gas is the side reaction of the end products. The reaction with the hypochlorites increases the hydroxyl ions by the formation of sodium hydroxide (Eq. 2-1) or calcium hydroxide (Eq. 2-2); the reaction with chlorine gas and water increases the H^+ ion concentration (see Chapter 4) by the formation of hydrochloric acid. There is reason to speculate that a chlorine gas solution at pH 2 to 3 will always be somewhat more effective than a solution of hypochlorite at pH 11 to 12 at the immediate area of the point of application, simply because there is more of the active ingredient HOCl and possibly of some extremely active molecular chlorine on account of the low pH of the chlorine gas solution. It is a well-known fact that at pH 11 or 12 the HOCl is almost completely dissociated to the ineffective hypochlorite ion, as follows:

$$HOCl \leftrightharpoons H^+ + OCl^- \tag{2-3}$$

This high pH condition will exist only momentarily at the interfaces of the hypochlorite solution and the water or waste to be treated.

Wastewater

There is reason to speculate that in wastewater chrorination, where much higher dosages are used than in potable water, better germicidal efficiency would result from chlorine gas, particularly in poorly buffered wastewater. In these cases chlorine gas solution tends to lower the pH of the treated wastewater, increasing the effectiveness of the HOCl.

However, Sawyer[4] reported in 1957 greater efficiency with hypochlorite solutions in the disinfection of wastewater than with chlorine gas solutions. He theorized there may be undesirable side reactions occurring in the lower pH environment of the chlorine gas solutions, robbing it of some of its disinfecting powers. These side reactions may be in the nature of the formation of organic chloramines, with little or no germicidal efficiency, and yet will appear as part of the total chlorine residual. Sawyer's laboratory work indicated that 1 lb available chlorine in the form of hypochlorite was equivalent in disinfecting power to about 1.5 lb of aqueous chlorine gas solution. This fact was not substantiated in treatment plant application. However, the most impressive findings of Sawyer's work showed significantly higher amounts of residual chlorine—as measured by amperometric titration—after 15 min contact time in the case of hypochlorite-treated samples for all dosages.

Example: At a chlorine dosage of 10 mg/L (Providence sewage) the amperometric residual for aqueous chlorine gas solution at the end of 15 min contact was 0.80 mg/L versus 1.40 mg/L for sodium hypochlorite solution. Similarly, sewage from Worcester, Massachusetts, showed, for the same dosage and contact time as above, 1.75 mg/L for aqueous chlorine gas solution versus 2.75 mg/L for sodium hypochlorite solution.

Unfortunately these observations when applied to field conditions were not of the same magnitude or consistency, but they indicate that any given wastewater is likely to demonstrate a higher chlorine demand when aqueous chlorine gas solution is used than when sodium hypochlorite is used. This becomes an economic consideration for wastewaters of poor quality and high chlorine demand. These situations call for highly efficient mixing to eliminate the possibility of undesirable side reactions. Such reactions are more prone to occur in waters of high chlorine demand in the low pH area surrounding the point of application of aqueous chlorine gas solutions. This theory has yet to be proved but should be investigated.

From 1993 to 1996, White investigated the use of the Stranco High Resolution Redox Control System and found that it eliminated any possibilities of errors caused by the presence of nitrites and organic N in the measuring of the true oxidative power of the chlorine residual measured by the residual analyzer. This then is the method to use whenever there is any doubt about the accuracy of the measured residual chlorine. This does not mean the elimination of the use of chlorine residual analyzers because the plant operator needs the analyzer to prove the proper operation of the chlorination equipment.

The next consideration is the total effect on the pH of the affluent due to chlorination. This depends upon three factors: (1) the buffering capacity of the wastewater; (2) the chlorine dosage; and (3) whether chlorine is in the form of hypochlorite or aqueous gas solution. Sawyer concluded that the variation in pH of the final sewage mixture is not a significant factor. However, as the pH drops below 7.5, the ratio of monochloramine to dichloramine shifts toward the formation of more dichloramine; therefore, the disinfection efficiency should improve because dichloramine has at least twice the germicidal efficiency of monochloramine.[5] The pH of the effluent in the 36 plants investigated by White[6] was usually between 6.9 and 7.3. Only a few plants noted any appreciable downward shift because of chlorination. One plant using a very high dose of chlorine (25–30 mg/L), which lowers the pH to about 6.3 from 7.0, reported considerable improvement in disinfection efficiency. Such increased efficiency has been reported by others.[7] In another case (primary effluent), experiments were carried out to determine the difference between the efficiency of hypochlorite and aqueous gas solution. A hypochlorite dose of 15 mg/L raised the pH from 7.1 to 7.4, but it was lowered to 6.85 with the same dose of an aqueous chlorine gas solution. Likewise, another primary effluent in the same city, but with a high chlorine demand and a pH of 7.4 prior to chlorination, reached a pH of approximately 8.0 after a hypochlorite dose of 40 mg/L. After a similar dose of aqueous chlorine gas solution, it reached a final pH of approximately 6.5. In both of these instances the disinfection efficiency was significantly increased as the pH was lowered owing to the use of the aqueous gas solution.[8]

However, of the 36 plants White observed in 1972 and an additional 20 from 1973 to 1976, the final pH was not in itself a major factor contributing to disinfection efficiency with the two exceptions mentioned above.[6]

An additional consideration must be noted in comparing hypochlorite with aqueous chlorine gas solution, and that is the concentration of available chlorine in the solution. By design the chlorine concentration in the injector discharge of a conventional chlorinator is limited to 3500 mg/L, as this is the upper limit of containment of the molecular chlorine (off-gassing) inherent in these solutions. However, a hypochlorite solution of 15 percent available chlorine has an available chlorine concentration of 150,000 mg/L or about 43 times the concentration of a conventional chlorinator discharge. This may be of great significance in the practical application of chlorine because it relates to the segregation phenomenon.[9] This phenomenon, briefly stated, is that to mix two liquids, one of them a chemical to be intimately mixed with the process flow, the mixing is most efficient when the chemical added is in the smallest possible amount as compared to the process flow. In other words, using a chemical solution of the highest possible concentration produces the best mixing. To date this segregation phenomenon has not been investigated for disinfection efficiency relative to the concentration of the chlorine solution applied.

Sodium Hypochlorite

Characteristics. Sodium hypochlorite is often referred to as liquid bleach or soda bleach liquor, and is the most widely used of the hypochlorites for potable water and waste treatment purposes. Although it requires much more storage space than does the high-test calcium hypochlorite and is more costly to transport over long distances, it is more easily handled and gives the least maintenance problems with pumping and metering equipment.

The preparation of sodium hypochlorite is a relatively simple procedure, involving the reaction of chlorine with caustic soda. This reaction proceeds as follows:

$$\underset{\text{caustic soda}}{2\,NaOH} + \underset{\text{chlorine}}{Cl_2} \rightarrow \underset{\text{sodium hypochlorite}}{NaOCl} + \underset{\text{sodium chloride}}{NaCl} + H_2O + \text{Heat} \tag{2-4}$$

On the basis of molecular weight, 1 lb chlorine reacts with 1.128 lb caustic soda to produce 1.05 lb sodium hypochlorite and 0.83 lb sodium chloride. In practice, an excess of caustic soda is used as a stabilizer. Table 2-1 shows the quantities of chlorine and caustic soda required to make 1000 gallons of sodium hypochlorite of various strengths.[10, 11]

The strength of a soda bleach solution is commonly expressed in terms of its available chlorine content as "trade percent" or "percent by volume." A more accurate expression is the actual weight percent of the available chlorine or sodium hypochlorite. The relationship between these values is:

$$\text{trade percent (percent by volume)} = \frac{\text{g/L available } Cl_2}{10} \tag{2-5}$$

$$\text{weight percent available } Cl_2 = \frac{\text{trade percent}}{\text{specific gravity of solution}} \tag{2-6}$$

Table 2-1 Quantities of Chlorine and Caustic Required to Make 1000 Gallons of Sodium Hypochlorite of Various Strengths[10]

Available Chlorine			*As NaOCl*			*Excess*		*Caustic Soda Required*		*Water Required for Diluting Caustic (gal.)*	
Gm/l	*Trade %*	*Wt. %*	*Gm/l*	*Wt %*	*S.G. 70°F*	*NaOH Gm/l*	*Cl_2 Req lbs.*	*lbs Solid*	*Gals 50% liq*	*Solid*	*50% liq*
10	1.0	.98	10.5	1.0	1.014	1.0	83	104	16	991	979
25	2.5	2.4	26.2	2.5	1.035	2.5	209	261	40	979	949
50	5.0	4.7	52.5	4.9	1.070	5.0	417	523	80	958	897
100	10.0	8.8	105.0	9.2	1.140	10.0	834	1045	160	915	794
120	12.0	10.3	126.0	10.8	1.168	12.0	1001	1254	192	898	753
150	15.0	12.4	157.5	13.0	1.210	15.0	1251	1568	240	873	691

$$\text{weight percent sodium hypochlorite} = \frac{\text{g/L sodium hypochlorite}}{\text{specific gravity of solution} \times 10} \quad (2\text{-}7)$$

$$\text{weight percent sodium hypochlorite} = \frac{\text{g/L available Cl}_2 \times 1.05}{\text{specific gravity of solution} \times 10} \quad (2\text{-}8)$$

The weight in terms of the lb/gal for a particular strength is not a constant. It varies, depending on the amount of excess sodum hydroxide the manufacturer uses to promote stability.

Table 2-2 gives approximate values of weights suitable for calculating dosages of sodium hypochlorite. In Table 2-2, to find the available chlorine in pounds per gallon, multiply the trade strength by 0.08345.

Stability of Solutions. Sodium hypochlorite solutions are vulnerable to a significant loss of available chlorine in a few days. This is a major problem with this type of chlorination system. The user must dedicate laboratory time to monitoring the decay rate in available chlorine. This serves two purposes: (1) it establishes an understanding with the supplier to arrive at the optimum cost for a given trade strength of solution; and (2) it establishes the most cost-effective quantity per delivery and frequency of delivery to minimize loss of chlorine in the stored hypochlorite solution. Of several users questioned about this surveillance, none was monitoring the available chlorine decay rate.

The stability of hypochlorite solutions is greatly affected by heat, light, pH, and the presence of heavy metal cations. These solutions will deteriorate at various rates, depending upon the specific factors:

1. The higher the concentration, the more rapid the deterioration.
2. The higher the temperature, the faster the rate of deterioration.
3. The presence of iron, copper, nickel, or cobalt catalyzes the deterioration of hypochlorite.

Iron is the worst offender. In minute quantities it causes rapid deterioration of these solutions.[12] The source of iron is usually the caustic used in preparing

Table 2-2 Approximate Values of Weights Suitable for Calculating Dosages of Sodium Hypochlorite

Trade % Avail. Cl_2 gm/l 10	*Approx wt. of One Gallon (lb)*	*Available Cl_2 (lb/gal.)*	*Gpd req'd to Dose 1 mgd to 1.0 ppm*	*Gpm req'd to Dose 1 mgd to 1.0 ppm*
1.00	8.45	0.0083	100	0.0694
5.00	8.92	0.42	20	0.0138
5.25	8.95	0.44	19	0.0132
10.0	9.50	0.83	10	0.0069
15.0	10.1	1.25	6.6	0.0046

these solutions. Iron in quantities as low as 0.5 mg/L will cause rapid deterioration of a 15 percent solution in a few days.

The copper content should be kept as low as possible, not to exceed 1 mg/L in the finished solution. It is generally present because of the copper flexible connections and brass body chlorine line valves used in the chlorine supply system. Great care must be taken *by the producer* to prevent, insofar as is possible, active corrosion of these parts. This can be done by keeping them internally free of moisture. This is a difficult task.

The most stable solutions are those of low hypochlorite concentration (10 percent), with a pH of 11 and iron, copper, and nickel content less than 0.5 mg/L, stored in the dark at a temperature of about 70°F.

Figure 2-1 illustrates the chlorine strength decay rate over a period of 160 days for three different sources of hypochlorite under controlled conditions as indicated. These data reflect a "best case" situation.

In 1942 the Transportation corps of the U.S. Army made a thorough investigation of sodium hypochlorite solution in an effort to determine the most effective strength over a period of 30 days. This hypochlorite was to be used with the hypochlorination systems on all troop transports used in the Pacific Theatre of World War II. The Army investigation determined that a 10 percent available chlorine solution was the strongest solution with the least decay for a 30-day period for the temperatures encountered.

Caustic (NaOH) is used in the making of these solutions purely as a stabilizing factor. A large excess of alkalinity, however, does not stabilize a hypochlorite solution any more than the slight excess given in Table 2-1. However, *if the pH drops below 11,* decomposition becomes more rapid.

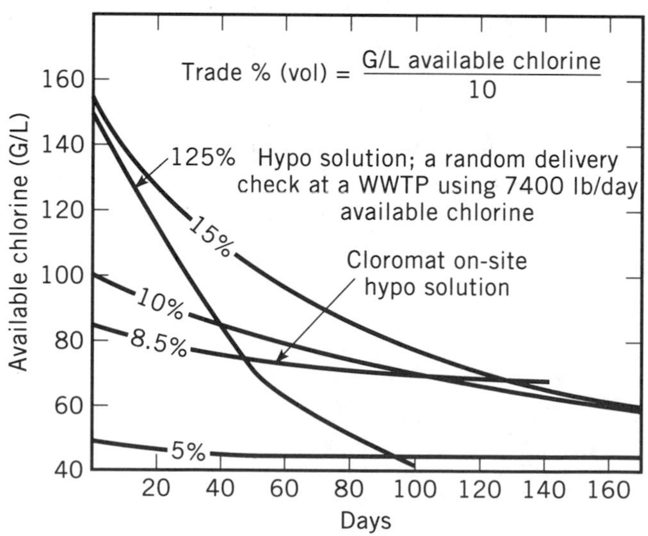

Fig. 2-1. Decay rate sodium hypochlorite solutions.

Another way of defining the decay of various concentrations of sodium hypochlorite solutions at different temperatures is by the half-life method[13] shown in Table 2-3. When the rate of decay is nonlinear, as in hypochlorite solutions, this method is the best way of expressing "shelf life."

Hypochlorite does not lose its strength at a constant rate per day, but at a decreasing rate as it loses strength. It is estimated that the rate of decomposition of 10 and 15 percent solutions nearly doubles with every 10°F temperature rise.

It is important to note that a 167 g/L solution (16.7 trade percent) stored at 80°F will decay in strength 10 percent in 10 days, 20 percent in 25 days, and 30 percent in 43 days.[14]

The influence of light on a solution of sodium hypochlorite is easy to demonstrate by putting one portion in a clear container and the other in an amber container and exposing both to sunlight. The half-life of a 10–15 percent available chlorine solution will be reduced about three or four times by sunlight. For stronger solutions up to 20 percent, the result is a reduction of half-life of about six times.[13]

These solutions will freeze, but at temperatures considerably lower than the freezing point of water. (See Fig. 2-2).

Sodium hypochlorite is commercially available in strengths of approximately 5–15 percent. If stronger solutions are attempted, solid sodium chloride is formed, presenting an additional problem of purifying; most makers limit the strength to 15 percent to avoid this problem.

These solutions are packed in various ways: in 5-gal carboys, in rubber-lined 55-gal steel drums, and in some localities in tank trucks. However, the most popular method is a corrugated-type carton with four one-gallon plastic jugs (usually colored red), all of which are returnable for refill.

There is no fire hazard connected with the storage of this chemical, as there is with some of the powdered hypochlorite. It should, however, be kept away from any equipment that could be damaged by corrosion caused by spillage in the normal handling process. Corrosion from these solutions is far more devastating than that caused by an aqueous solution of chlorine made from chlorine gas. Handling of sodium hypochlorite solutions is at best a messy situation. The handling and storage area should have a concrete deck that

Table 2-3 Half-life Method

Percent Available Chlorine	*Half-life, Days*			
	212°F	*140°F*	*77°F*	*59°F*
10.0	0.079	3.5	220	800
5.0	0.25	13.0	790	5000
2.5	0.63	28.	1800	
0.5	2.5	100.	6000	

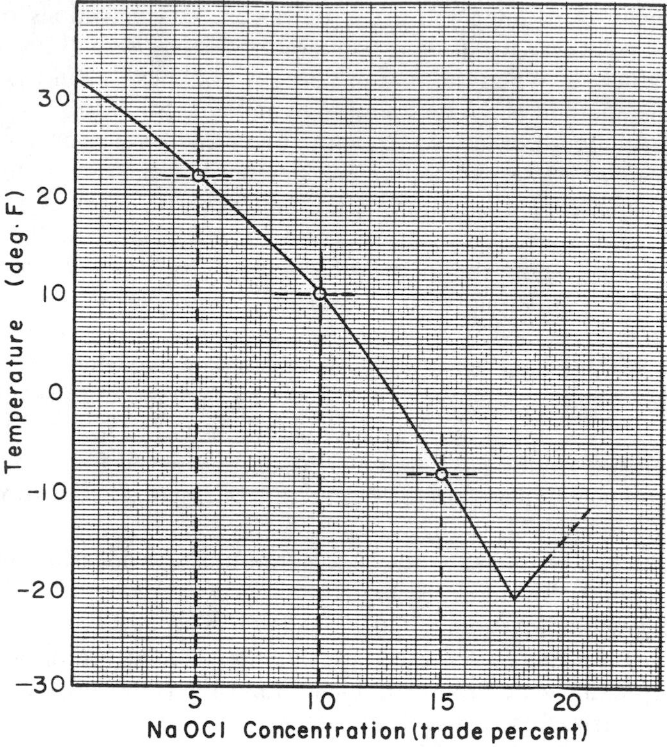

Fig. 2-2. Freezing temperatures of hypochlorite solutions (courtesy Dow Chemical U.S.A.).

can be flushed with copious amounts of water. Otherwise, the installation becomes impossible to maintain properly.

Suggested Specifications. All bulk sodium hypochlorite shipments for large installations should be purchased on the basis of specifications delineating the available chlorine, iron, and copper content, and excess caustic. Copper should be limited to a maximum of 0.5 mg/L and iron to 1.0 mg/L. Excess caustic should be limited by a pH not to exceed 11.2. If these limitations cannot be met by the suppliers, then compromises with price and chlorine strength will have to be made. Furthermore, it should be specified that the material be free of sediment and suspended solids. All shipments of bulk sodium hypochlorite should be analyzed upon receipt for available chlorine, iron, copper, and excess caustic.

Sodium Hypochlorite: Major Operating Problems

These problems were unknown until users of chlorine gas in large potable water and wastewater treatment plants switched to hypochlorite to comply with the highly questionable Uniform Fire Code that dwelled on the danger of chlorine gas leaks from ruptured 150-lb cylinders, ton cylinders that fell apart, and 90-ton tank cars that got shot up. The fire marshals ignored the largest chlorine gas leak that ever occurred at a water treatment plant, which was due to the use of sodium hypochlorite. This occurred at a large treatment plant in the East, when the tanker truck driver of a load of ferric chloride, dumped his load of pH 4 into the hypo tank by mistake. This lowered the hypo pH from about 12 to 5 almost instantly, releasing the chlorine all in a great mass, about 12,000 lb. This is the worst kind of a gas leak because the chlorine is in the form of a very moist vapor, and this penetrates the lungs with hardly any pain. This is very damaging to the human lungs.

This is only part of the problem. The following[58] is based upon difficulties related to me by operating personnel who never had such problems after many years (20) of operation with chlorine gas in ton cylinders and 55- and 90-ton tank cars. In all this time they never needed any scrubbers that trap operating personnel in airtight enclosures.

The effect of the following continuous problems has been not only to increase interest in using on-site generation of chlorine (see Chapter 3), which eliminates concern about the UFC, but to increase interest in on-site manufacture of calcium hypochlorite, similar to that in the plant built by the Los Angeles Sanitary Districts at their South Figueroa Street plant in Carson, California (described later in this chapter). As for hypo, the problems are many. Moreover, hypo costs three times as much as chlorine gas. The most serious worker safety problems have been related to the hypo off-gassing. Hypo off-gasses a great deal of vapor during its natural and continuous decomposition. (See Fig. 2-1, the 12.5 percent hypo solution at a WWTP using 7400 lb/day chlorine; that is a lot of off-gassing in 30 days.)

In one case where the hypo suction piping is about 100 ft of 2-inch pipe, the vendor estimated that this pipe generates approximately 1 cu ft of gas per day. Both plants have experienced the explosion of ball valves and 4-inch plastic strainers. This happened in each case when the hypo was left trapped between two closed valves. In the case of the exploding valve, it was caused by the hypo trapped inside the valve. In addition to this, there have been several different types of valves that cracked. To correct this problem, operators have had to retrofit all of their hypo valves by drilling a hole in one side of the PVC ball. This allows the gas to leak off and not pressurize in the valve.

Gas that vents off in enclosed spaces is another important safety issue. The hypo does release Cl_2 gas even under static conditions. Enclosed areas such as a pump house or a filter plant tunnel are a real hazard, requiring the same safety gear as that required for Cl_2 gas. Moreover, Cl_2 gas never appears in these places.

The biggest potential problem is the incompatibility of sodium hypochlorite and any acid. Quite often hypo is stored near either hydrofluorosilicic acid or sulfuric acid in a treatment plant tank farm. Any mixing of these acids with the hypo will cause a major Cl_2 release.

Such an off-gassing situation causes many operational problems. Most common is the loss of suction pumps that run intermittently. The gas accumulates in the suction side of the pump and displaces the hypo. Check valves used as foot valves have had limited success. Usually the pump has to be left running continuously at a low feedrate, to keep the suction side flooded.

There also have been problems with air accumulating in the downstream piping. This air often accumulates in sections of pipe that go up a wall, then run horizontally, and then run down again. The high spots in the pipe act as pneumatic chambers and fill with gas before the gas pushes through. Operating personnel tried to solve this problem with back-pressure valves, but ultimately ended up having to eliminate all these raised portions of pipe.

When the Cl_2 feedrate was flow-paced, operators would see a steady drop in residual while the air (containing Cl_2 gas) was accumulating. During this time period, the chlorine feedrate would go up. Then after the air had passed, there would be a sharp rise in the chlorine residual.

Employee exposure is another current problem. There is not a single staff member who does not have bleached-out clothes, and usually these clothes are filled with holes. They try to wear aprons and rain gear, but it is very difficult to avoid leaks from casual contacts.

There have also been problems with material compatibility. First, operators invested in cross-linked polyethylene tanks that were only giving about 4 years of life. In the future they are going to try lined fiberglass, or lined steel for any tanks that are not being used in small disposable applications. The piping systems have been very difficult to cope with because the hypo leaks through most of the mechanical fittings, and most plastic systems have to be glued together.

A survey of these systems revealed that even glued joints leaked after several years of service! Operators have had a hard time finding hypochlorite pumps for their particular needs for accuracy, capacity, and dependability. One operator recently had a pressure reducing valve, downstream from a hypo feed pump, that broke. It sprayed hypo all around the work area, which eventually corroded some electronic equipment and caused the hypo feed station to come off-line.

Another operational problem that has been experienced by users is the calcification of fittings, especially injection points and solution diffusers.

California Department of Health Services: Policy Memo 96-003

This memo recommends the following procedures for the handling of sodium hypochlorite to minimize the formation of chlorites and chlorates that occurs during the on-site storage time plus the delivery time.

1. Store hypo at 70°F, and indoors. This could require air-conditioned tank farms, or extensive heat exchange or cooling coils in storage.

2. Dilute stock to 10 percent. (This recommendation dates back to World War II, when the U.S. Army Transportation Corps was asked by White to evaluate the most stable percent strength for hypo on a troop ship making a round trip to Guadalcanal.) This will require additional bulk storage.
3. Minimize storage time of stock. This would be next to impossible for operators trying to minimize the generation of chlorates and chlorites. Regarding these two compounds, see the latest status of requirements in Chapter 12, Chlorine Dioxide.

Granular Calcium Hypochlorite

Manufacturing Techniques. This, the prevailing form of dry bleach in North America, is marketed at 70* percent available chlorine content. The first calcium hypochlorite marketed was German Perchloron, which consisted of hemibasic calcium hypochlorite and impurities. The available chlorine content was about 65 percent.

Mathieson Chemical Company developed a more stable calcium hypochlorite by mixing equivalent amounts of sodium hypochlorite and calcium chloride. In this process, a slurry of lime and caustic soda is chlorinated and then cooled to −10°F. At this temperature, the crystals that form are centrifuged to remove the mother liquor and the insoluble impurities. These crystals are then added to a slurry of chlorinated lime that contains calcium chloride in an amount equivalent to the sodium hypochlorite content of the crystals. When this mixture is warmed, a calcium hypochlorite dehydrate precipitate is formed, and the sodium chloride remains in solution. The precipitate is then filtered, and the resulting cake is granulated, sized, and dried. The result is a product containing over 70 percent available chlorine and less than 3 percent lime.[1]

The Pennwalt Corp. (formerly the Pennsylvania Salt Co.) developed the American version of the Perchloron process, an improvement over the German process by the same name. This process forms calcium hypochlorite dihydrate crystals and also retains a sufficient amount of the large hemibasic crystals to permit filtration of the material. This product has an available chlorine content in excess of 70 percent. This process reduces the lime content of the German method from 12–16 percent to 3–5 percent.

PPG Industries (formerly Columbia Southern) utilizes an entirely different approach. First, hypochlorous acid is made by the addition of chlorine monoxide to water. This acid solution is then neutralized with a lime slurry to produce a solution of calcium hypochlorite. The clear liquor is first spray-dried and then granulated and vacuum-dried to yield a product containing 70 percent available chlorine and 4 to 6 percent lime. All of these products contain

*Since February 1979 there has been an industry reduction to 65 percent to reduce fire hazards.

some calcium carbonate and other insolubles, which cause precipitates when dissolved in tap water. The amount of precipitates formed depends on the dilution factor and the total hardness of the diluting water.

Storage of Granular Calcium Hypochlorite. The storage of granular calcium hypochlorite, whether the 65 percent or the 35 percent available chlorine, is a major safety consideration. It should never be stored where it is subject to heating or allowed to contact any organic material of an easily oxidized nature. The decomposition of calcium hypochlorite is exothermic, and will proceed rapidly if any part of the material is heated to 350°F. Many fires of spontaneous origin have been caused by improperly stored calcium hypochlorite. The decomposition releases oxygen and chlorine monoxide.

Under normal conditions, it will lose 3 to 5 percent available chlorine per year. Contact with water or moisture from the atmosphere induces the decomposition. Oxygen release with contact with a corrugated cardboard carton can result in spontaneous combustion of the carton, depending on its proximity to the oxygen being released by the decomposing hypochlorite.

Lithium Hypochlorite

Lithium, an excellent granular hypochlorite, is prepared by mixing a strong solution of lithium chloride with a strong solution of sodium hypochlorite. A large portion of sodium chloride precipitates, and the 30 to 35 percent solution of lithium hypochlorite is evaporated and drum-dried. The finished product is white, free-flowing, granular, and dust-free.

Lithium Corporation of America, manufacturers of this product, lists the following ingredients:

	Percent by Wt.
Available chlorine	35
$LiOCl$	30
$NaCl$	34
Na_2SO_4 & K_2SO_4	20
$LiCl$	3
$LiClO_3$	3
$LiOH$	1
Li_2CO_3	2
H_2O	7

Lithium hypochlorite is readily soluble in water, and has the distinct advantage over the calcium hypochlorite solutions of being clear when prepared, thereby eliminating the need for the settling and discarding of sludge. When stored properly in closed containers at 75°F and 75 percent relative humidity for four months, it retains more than nine-tenths of its available chlorine.[15]

It does not affect the alkalinity or pH as much as do the other hypochlorites. The pH of a 100 ppm solution of available chlorine at 25°C is only 10.0.

This product was introduced in 1964 as a laundry bleach and dishwashing compound because of its excellent solubility characteristics. Then it was introduced to the swimming pool industry in competition with all the other disinfectants.[16] Although the cost of this material is somewhat greater than that of the other hypochlorites, on the basis of available chlorine, it is gaining in popularity because of the ease with which it can be handled. It has the same disinfecting characteristics as any other chlorine compound that produces HOCl as a hydrolysis product.

The same safety precautions must be followed for the storage of granular lithium hypochlorite as those for granular calcium hypochlorite.

Hypochlorite Tablets

PPG Industries Inc.[59] was the first industry in the manufacturing of chlorine gas to introduce a tablet for continuous chlorination of small water supplies. It has combined its technology with Hammond Services, using the PPG 3-inch (65 percent available chlorine) calcium hypochlorite tablets and patented chlorinators designed by Hammond Services along with PPG expertise to offer a total chlorination system to meet user requirements. The Hammonds tablet chlorination system is illustrated in Fig. 2-3.

Users are warned that these systems are designed to be used with only PPD 3-inch calcium hypochlorite tablets. Use with any other tablet will result in loss of chlorination control and voids all warranties associated with the unit.

Liquid Calcium Hypochlorite*

Calcium hypochlorite liquor can be made directly from the reaction of chlorine and hydrated lime in solution. However, the maximum strength of these liquors is limited to about 35 percent available chlorine. These solutions are made by passing chlorine into a milk of lime suspension according to the equation:

$$\underset{148.192}{2Ca(OH)_2} + \underset{148.828}{2Cl_2} \longrightarrow \underset{142.994}{Ca(OCl)_2} + \underset{110.994}{CaCl_2} + \underset{36.032}{2H_2O} \tag{2-9}$$

Theoretically, 1.043 lb actual $Ca(OH)_2$ or 0.791 lb actual CaO will react with 1 lb chlorine to produce 1.008 lb calcium hypochlorite and 0.782 lb calcium chloride. In actual practice, it is usually assumed that the hydrated lime contains 95 percent $Ca(OH)_2$ and that the quicklime contains 95 percent CaO.

*See discussion of on-site manufacturing of calcium hypochlorite at the Los Angeles Sanitation District Figuero St. Plant in Chapter 3.

Fig. 2-3. Hammonds model 3550 hypochlorite tablet chlorination system (courtesy PPG Industries, Inc.).

On this basis, the equivalents of the chemical reaction, which do not include the excess lime required for stability, are as follows: 1.0 lb Cl_2 + 1.10 lb commercial hydrated lime or 1.0 lb Cl_2 + 0.833 lb commercial quicklime, which will produce 1.008 lb calcium hypochlorite and 0.78 lb calcium chloride.

In practice, an excess of from 10 to 20 lb $Ca(OH)_2$ (hydrated lime) or 7 to 14 lb CaO (quicklime) should be used per 1000 gal of calcium hypochlorite to ensure proper stability by keeping the pH at 11.2 or above.

The following formulae apply to making 1000 gal of calcium hypochlorite of various strengths:[11]

$$\text{lb } Cl_2 = \text{g/L available chlorine} \times 8.34 \qquad (2\text{-}10)$$

$$\text{lb } Ca(OH)_2 = \text{lb } Cl_2 \times \frac{1.043 \times 100}{\%\ Ca(OH)_2 \text{ in lime}} + 10 \text{ to } 20 \text{ lb} \qquad (2\text{-}11)$$

$$\text{lb CaO} = \text{lb } Cl_2 \times \frac{0.791 \times 100}{\%\ \text{CaO in quicklime}} + 7 \text{ to } 14 \text{ lb} \qquad (2\text{-}12)$$

Table 2-4 shows the theoretical minimum and maximum amounts of lime required to make liquid calcium hypochlorite.

Table 2-4 Pounds Required per 1000 Gallons $Ca(OCl)_2$

Avail. Cl_2 gm/l	lb Cl_2	Hydrated Lime $Ca(OH)_2$ 95%			Quicklime CaO 95%		
		Theor.	*Min.*	*Max.*	*Theor.*	*Min.*	*Max.*
10	83.4	91.7	101.0	110.3	69.4	76.4	83.5
20	166.8	183.5	192.7	202.0	138.8	145.9	152.9
30	250.2	275.2	284.5	193.7	208.2	215.3	222.3
40	333.6	367.0	376.2	385.5	277.7	284.7	291.7
50	417.0	458.7	468.0	477.2	347.1	354.1	361.1
60	500.4	550.4	559.7	568.9	416.5	423.5	430.5
70	583.8	642.2	651.4	660.7	485.9	492.9	499.9

The preparation of liquid calcium hypochlorite is not an easy process. First, the quality of lime available must be considered. Lime with a low magnesium or aluminum oxide content is desirable, as these impurities cause excessive sludge and poor settling properties. The problem of getting the hydrated lime to settle out of solution is the most difficult to overcome, and depends upon both the lime and the arrangement of equipment.

CHOICE OF EQUIPMENT

Introduction

There is no engineering limitation to the quantity of hypochlorite that can be metered in potable water and wastewater treatment. The hypochlorites are universally applied as aqueous solutions although there is no reason why the granular material cannot be metered through gravimetric or volumetric dry chemical feeders. The latter method does pose monumental problems of corrosion if the material is not kept completely free of moisture. This also raises the question of possible spontaneous combustion of granular hypochlorites. At the moment the recommendations are exclusively for metering and controlling the *solutions* of sodium, calcium, or lithium hypochlorite, rather than the granular forms of calcium or lithium hypochlorites.

Feeding equipment for the hypochlorites is a direct competition with chlorine gas feeding equipment. The latter is always the chemical equipment of choice because it is compact, flexible, and easy to maintain and operate, and it adapts well to automation. Furthermore, chlorine gas is three times less expensive than hypochlorites. This is one good reason why the use of hypochlorites has been limited to small supplies. Recently however, since the so-called adoption of the unrealistic Uniform Fire Code, some large installations have had to switch to hypochlorite, much to their dismay because of the mess of handling of large quantities of it. The fire marshals who published the UFC claim that handling chlorine in tank car quantities is much too hazardous in populated areas. This is not necessarily so. Prior to the issuance of the UFC the use of hypochlorite for potable water and wastewater treatment was always confined to very small water supplies or small wastewater systems. (To realize higher safety factors, see on-site generation in Chapter 3, from very small to very large installations.)

Small Water Supplies

The first question to be decided in selecting chlorination equipment for small water supplies is whether to use chlorine gas in cylinders or hypochlorite. Gas chlorinators do have practical limitations as to minimum feedrates. Chlorinators that utilize rotameters for indicating chlorine feed rate can be obtained with ranges as low as 0.1–1.2 lb/24 hr. However, the smallest metering orifice consistent with good accuracy and a minimum amount of maintenance is from 0.5 to 10.0 lb/24 hr. The one exception to this is a chlorinator that uses a bubbling type meter still manufactured by Wallace & Tiernan, which has a practical operating range as low as 0.1 lb/24 hr.*

A good rule of thumb in selecting hypochlorite or gas feeding equipment is based upon the daily usage of the chemical. This is related to the economics of chemical cost, storage space, and maintenance of equipment. A hypochlorinator installation usually requires more attention than gas feeding equipment, owing to the necessity of refilling solution containers with chemical. Generally the dividing point is close to 3 lb/day.

If the usage is below this figure, hypochlorite equipment should be seriously considered. At current prices, one pound of chlorine gas in 150-lb cylinders costs approximately 15 cents; the equivalent in hypochlorite purchased in 4-gal lots is between 50 and 75 cents per pound of chlorine.

If 3 lb of available chlorine consumption is accepted as the upper limit for hypochlorite feeding equipment, let us relate this to water and wastewater to be treated.

If a dose of 1 mg/L for water treatment is considered adequate, then 3 lb/day will treat approximately 360,000 gpd (1 mg/L = 8.33 lb/mg; 3.0/8.33 = 0.36 mg).

However, there is a growing trend to size chlorination equipment for application bordering on superchlorination, primarily to be certain that possible virus contamination may be dealt with. Therefore, a design criterion of 2 mg/L might be more realistic. This would decrease the amount of water treated by 3 lb/day chlorine from 360,000 gal to only 180,000 gpd or 125 gpm. On this basis, it is conceivable to use hypochlorite feeding equipment on supplies utilizing an intermittent pumping arrangement.

Hammonds Tablet Chlorination System.[59] Using the above figures signifies all of their sizes of chlorination units fit well into those calculations (see Fig. 2-3).

*This unit, the MSP, was discontinued by Wallace & Tiernan in the 1960s.

There are four different tablet capacities varying from 15 lb to 550 lb. This calculates to four different equipment sizes from 12 lb/day of chlorine to 720 lb/day. Figure 2-3 is an illustration of Hammonds' largest Model 3550 system. This unit uses a multistage centrifugal pump that has a flow range of 20–100 gpm, a line pressure of 0–120 psi, and 4.5–30 lb/hr of chlorine that is provided by 550 lb of PPG 3-inch tablets containing 65 percent of available chlorine.

The other three sizes have tablet containers containing 15, 75, and 150 lb of PPG tablets. These three sizes have the following chlorine production capacities: 0.5, 2.0, and 12.0 lb/hr. Gravity feed systems are also available.

These systems have several obvious advantages. They are simple hook-up/plug-in, self-contained packaged units that eliminate the need for risk management plans, as well as the need for secondary containment. These units provide accurate and consistent chlorine application, which is largely due to the tablets, owing to their long shelf life.

Pumped Supplies. For example, suppose that such a system utilized a 250 gpm pump but operated only 12 hr/day. This would be a candidate for a hypochlorinator, but it would also be in competition with a small gas chlorinator. If it were a deep well, and the chlorine could not be applied down the well (for construction reasons), then the hypochlorinator would have the advantage because the gas chlorinator would require a booster pump (to inject chlorine solution against the pressure in the main) as an additional piece of equipment to maintain and operate. In general then, it can be considered that up to about 200 gpm for a pumped supply is the practical limit for hypochlorinators.

Figure 2-4 shows schematically the installation of a hypochlorinator on a pumped surface supply. Situations like this raise a question about the point of application. If the pump suction is accessible, always apply the chlorine in the open body of water adjacent to the foot valve on the pump suction. This is preferable to the discharge side of the pump. The latter point of application, because of the static head involved, results in more wear and tear on the reciprocating parts of the hypochlorinator.

Never tap the pump suction line for the application of chemical. This becomes a source for an air leak, which will impair the efficiency of the water pump.

If the hypochlorite solution discharge line is much longer than 15 ft to the point of application at the pump suction, it is well to consider the advisability of installing a back pressure valve at the point of application. This keeps the solution line from emptying each time that the pump shuts down.

Figure 2-5 shows the installation of a hypochlorinator with a well pump. For simplicity, the point of application is made on the discharge side of the pump. In some cases where submersible pumps are used, this may not be possible, and so it becomes necessary to run the solution line down the well. When this alternative point of application is used, always determine as accurately as possible the location of the pump bowls and then terminate the solution line 10 ft below this point.

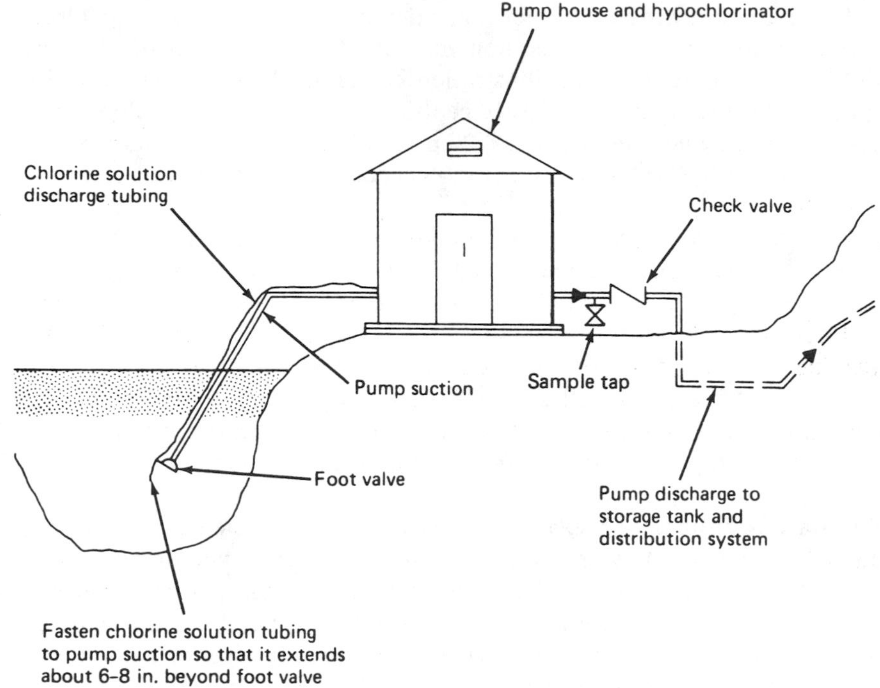

Fig. 2-4. Hypochlorinator installation for a pumped supply. Hypochlorinator is an electric-drive diaphragm pump with manual dosage control arranged to start and stop automatically both water pumps.

Electrically driven hypochlorinators can be arranged to provide continuous automatic proportional control, as shown in Fig. 2-6. In this case, there are three pumps, each of a different capacity, but discharging into a common line. A propeller-type flow meter can be utilized to transmit a milliamp signal in proportion to the flow. This signal is then used to operate an SCR unit, which drives a DC motor at variable speeds in accordance with the milliamp signal. Although this is presented here to illustrate the possibilities, such a situation probably would not occur in actual practice. There are very few small water systems that find it necessary to resort to multiple pumping systems.

Gravity Systems.*

These systems usually present the designer with the most difficult problems in selecting a proper chlorination system. There is usually no electric power

*This describes the use of a flowmeter used to operate and control an automatic hypochlorinator made by Wallace & Tiernan that was used for a variety of systems particularly where electric power was not available. This unit is no longer made, but here its service on troop ships during World War II is kept for historical reasons.

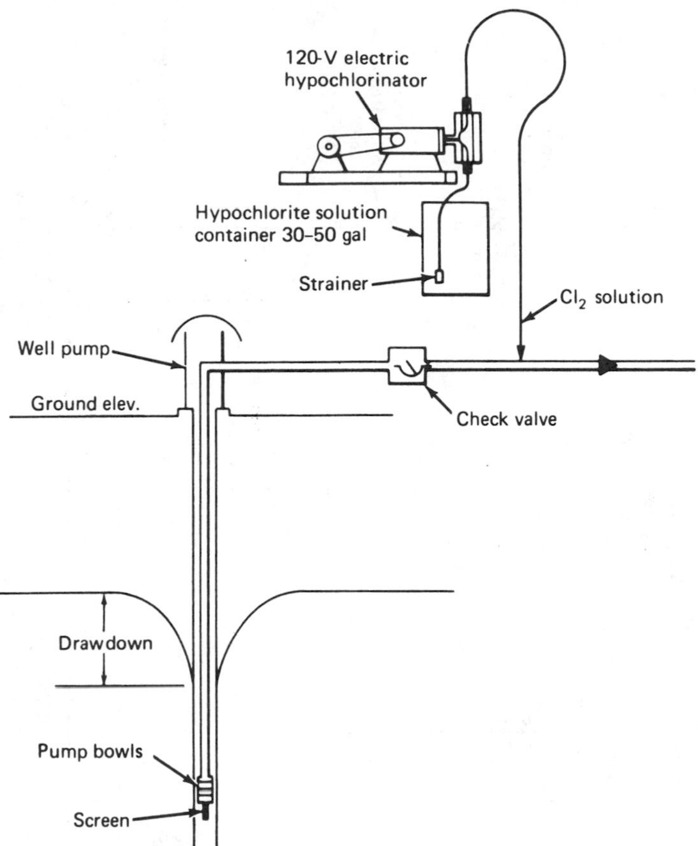

Fig. 2-5. Hypochlorinator installation with a well pump (hypochlorinator is arranged to start and stop with well pump).

available. The size of the distribution and transmission piping is usually designed for fire flows that are far in excess of the normal domestic consumption. For these gravity systems, particularly those without electric power, the equipment of choice is a water-operated (hydraulic ram) hypochlorinator that can be paced by any of several types of water meters. This is the same unit that is used for ships, whether they be aboard a troop ship or a cruise ship. They are illustrated in Fig. 2-6, Fig. 2-7, and Fig. 2-8.

Unfortunately Wallace & Tiernan discontinued this particular model sometime in the early 1960s. However, White is leaving Fig. 2-7 in because of the part these hypochlorinators played, by the installation of water treatment systems on all the troop ships that left San Francisco Harbor, in World War II.

Gravity Systems with Power Available. Whenever power is available, it is often possible to use a system that treats a constant flow either steadily or

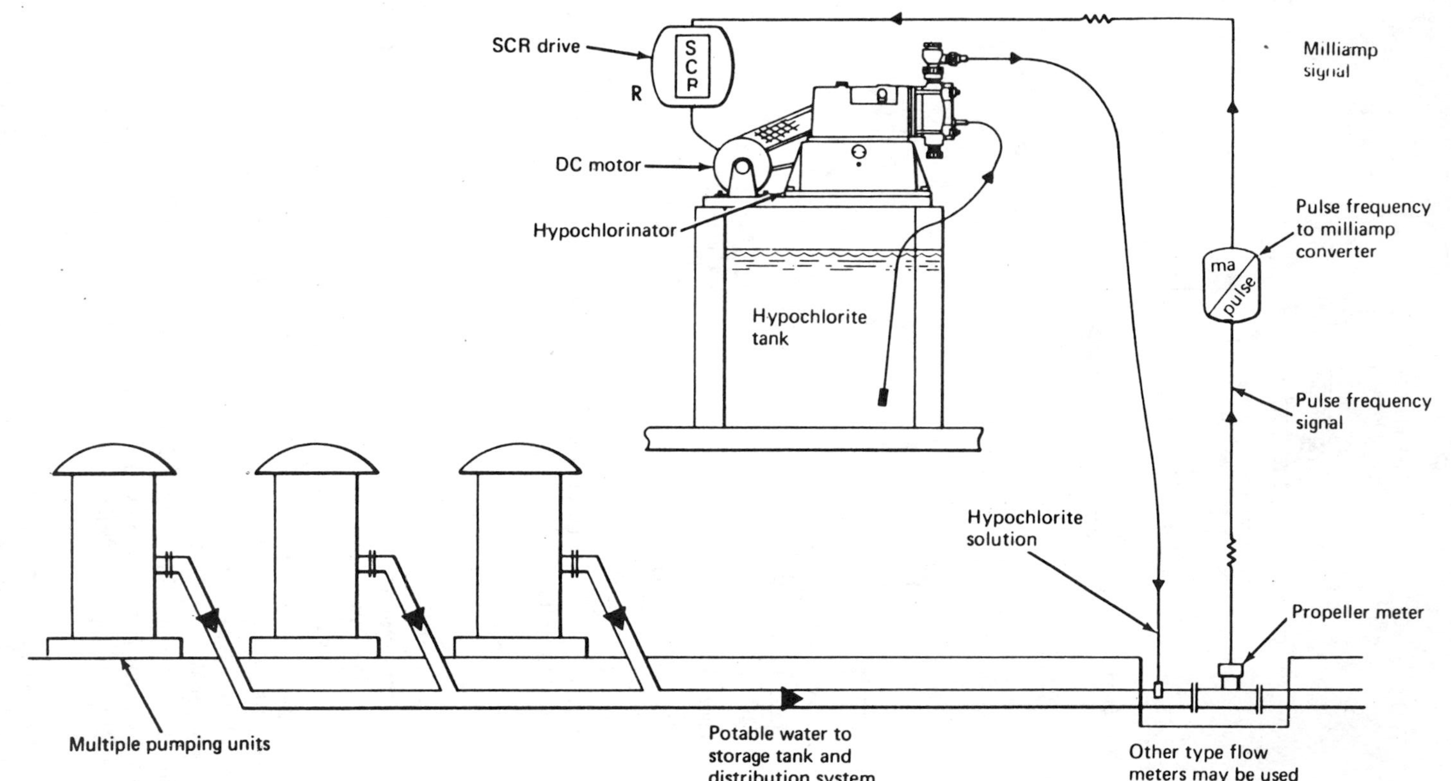

Fig. 2-6. Electric-drive hypochlorinator arranged for automatic flow-proportional control. (The DC motor provides hypochlorinator with variable speed drive from SCR unit.)

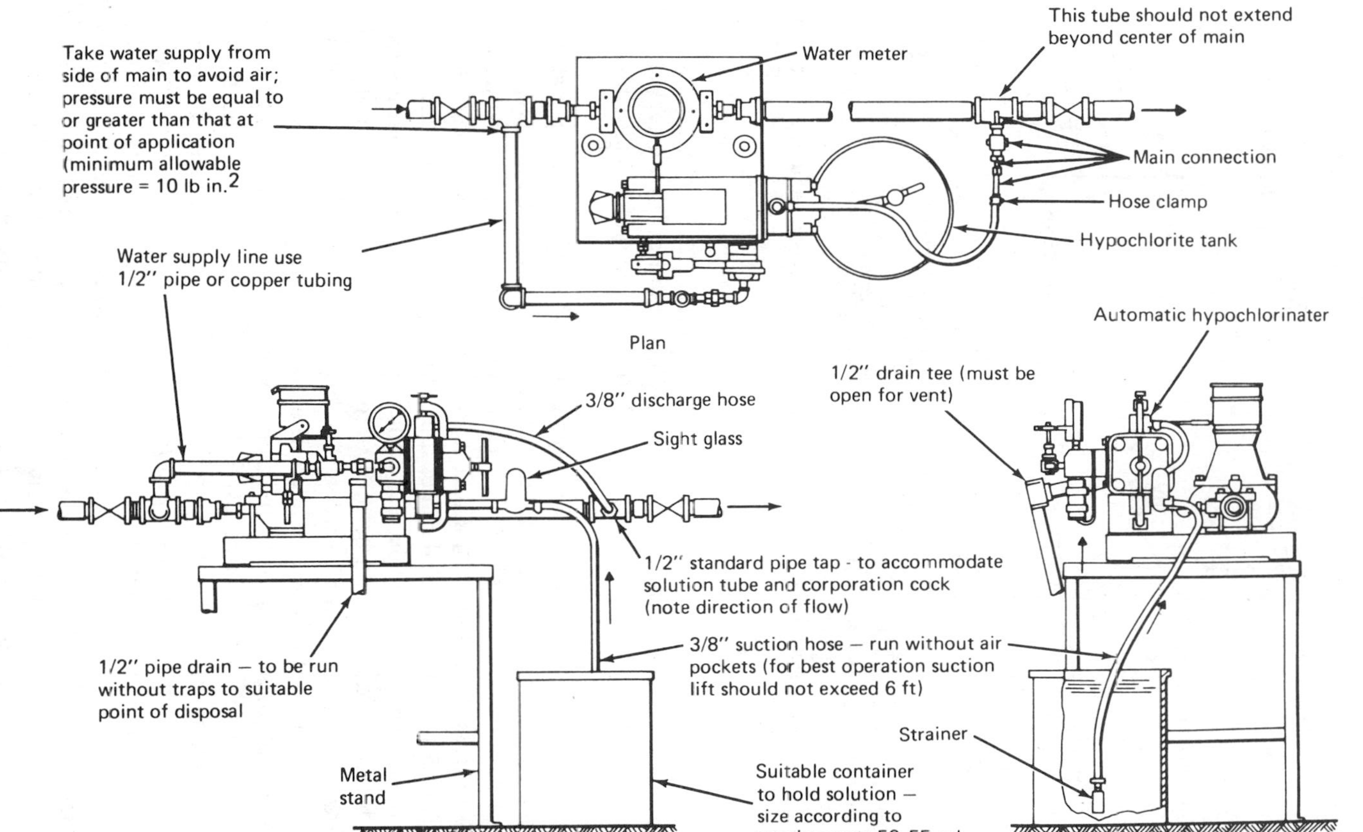

Fig. 2-7. Installation of an automatic hypochlorinator paced by a main-line meter (courtesy Wallace & Tiernan).

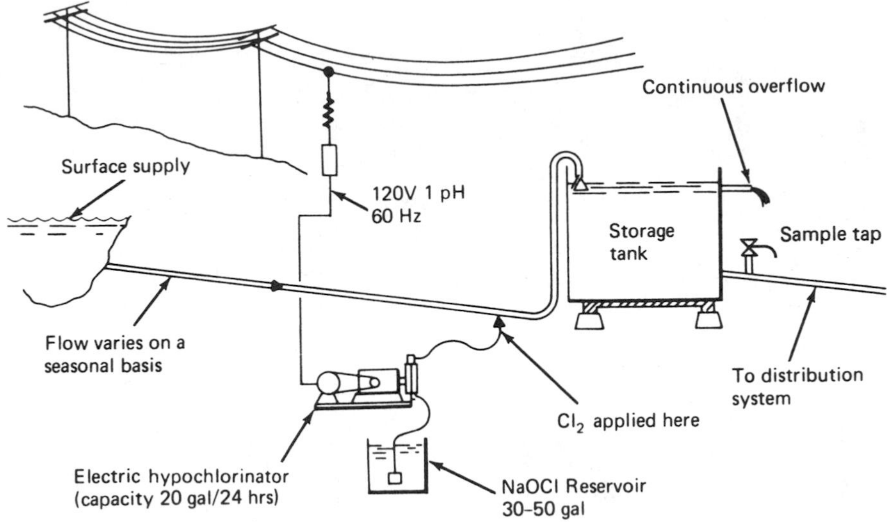

Fig. 2-8. Electric hypochlorinator in gravity supply. (Entire flow, including tank overflow, is chlorinated).

intermittently. The difference here is that an electric motor-driven hypochlorinator may be utilized, eliminating the need for a pacing meter. This simplifies the installation.

Figure 2-9 illustrates a typical gravity supply that originates from a stream or spring and discharges to a holding tank. The flow through this tank is constant on a daily or weekly basis, varying only seasonally. The entire flow should be treated, as the chemical cost to treat the overflow is negligible—about two dollars or three dollars per million gallons. The hypochlorinator is simply plugged into 120 V single-phase power and allowed to run continuously. The chlorine residual should be measured at the tap shown and not at the overflow. The water in the overflow line does not have the benefit of any contact time, since it is short-circuited directly across the tank from the inlet.

If the client does not want to treat any overflow, an arrangement such as that shown in Fig. 2-10 can be used. Here the water entering the tank is controlled by an inlet valve that operates from a fully closed position to a fully open position without any throttling action in between. This valve (Cla-Val Company) can be arranged with an electrical contact so that when it opens it will energize a relay that starts the electric drive on the hypochlorinator.

Assuming that the tanks are open, the residual at the sampling point should be ample to allow for any after contamination in the open tank. If the chlorine residual loss due to sunlight on the open tank is appreciable, then the tank should be covered. Most health authorities recommend closed tanks to avoid contamination by birds, rodents, and other wildlife.

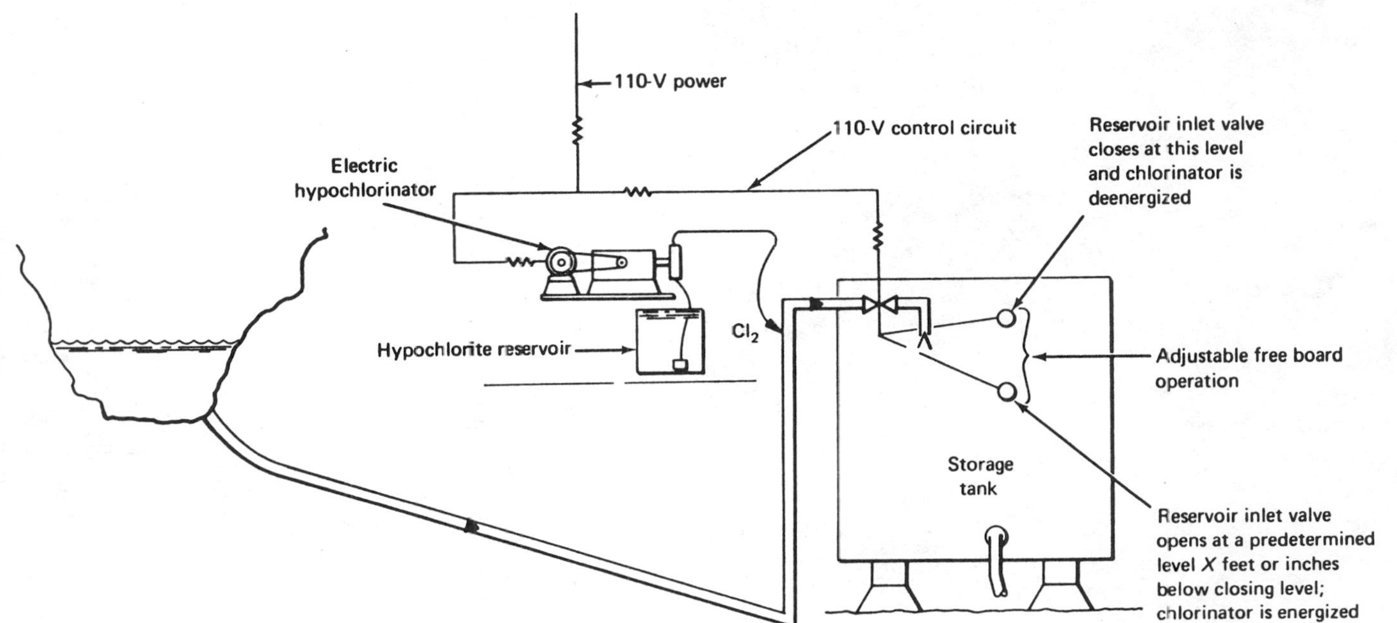

Fig. 2-9. Electric hypochlorinator in gravity supply. This system is unique in that the chlorinator unit is in operation only when the reservoir inlet valve is open. This valve opening is always full open, so the chlorinator always treats the same flow at each opening of the reservoir valve.

Fig. 2-10. Installation of automatic hypochlorinator aboard ship.

In the general area of water treatment, other chemicals are often required, such as hexametaphosphate for sequestering iron or manganese deposition, alum for filters, and soda ash for pH and alkalinity correction. These can also be metered by the same type of equipment used to meter hypochlorite.

Gravity Filter Systems. In those cases where the filtration system rides on the line with no clear well facilities, the chlorination should precede the filtration to keep the filter free from slime and organic deposits. The hypochlorinator in this case would have to be the water-operated type paced by a meter, or, if power is available, the electric type paced electrically from the meter through a variable speed drive SCR unit, which requires a 10 to 50 mA control signal from the water meter. Pacing an electric drive hypochlorinator from a water meter on a pulse frequency start–stop basis is not recommended, as it does not provide continuous chlorination proportional to flow. The SCR drive unit does provide proper chlorination for these conditions.

If the filtration system utilizes clear well storage facilities, then it is possible to arrange the clearwell to fill and draw on an intermittent basis, so that by means of an altitude-type valve with an electrical contact, the electric hypochlorinator may be started and stopped automatically.

This type of system creates a pressure break by reason of the operation of the clear well, and so it is limited to those cases where a loss of head due to the pressure break is of no significance.

SAMPLE CALCULATIONS. Use of any hypochlorinator installation necessitates calculating the amount of hypochlorite to be added to the solution container. Usually the capacity of these containers is 30 or 55 gal. The amount of hypochlorite to be added is only approximate. Final adjustment of dosage is done on the hypochlorinator.

Let us assume that a start-and-stop hypochlorinator operates with a 50 gpm well pump. Also assume that a 1 ppm dosage is adequate, that the capacity of the solution container is 30 gal, and that the hypochlorinator can pump 20 gal solution per day.

$$\text{step 1: convert 50 gpm into mgd} = 0.072 \text{ mgd}$$

$$1 \text{ ppm} = 8.33 \text{ lb chlorine per mg} = 8.33 \times 0.072 = 0.599 \text{ lb/day } \textit{rate}$$

Let us further assume that the hypochlorinator will operate at 50 percent capacity, or 10 gpd, if running continuously. Then there must be approximately 0.6 lb chlorine in every 10 gal of solution. So to each full 30-gal container, $0.6 \times 3 = 1.8$ lb available chlorine must be added. If 10 percent sodium hypochlorite is used, this will yield approximately 0.83 lb chlorine per gallon. (See Table 2-2). So adding 2 gal (1.66 lb available chlorine) would be close enough. The hypochlorinator could then be adjusted to give the desired residual.

Shipboard Installations*

Historical Background. During World War II, it was the consensus that the quality of the water in the Southwest Pacific was questionable at all times. Therefore it soon became mandatory that all troop ships be equipped with chlorination facilities for any water used by the crew and troops. As amoebic dysentery was a potential hazard in this area, it was decided that all installations must be capable of dosing 10 ppm chlorine followed by 45 minutes' detention, after which it was dechlorinated by means of an activated carbon filter.[17] Part of the chlorinated water was bypassed, so that water going to the ship's system carried about 0.3–0.5 O-T residual chlorine.

In the light of S. L. Chang's work at Harvard in 1941 and 1944,[18, 19] it developed that these requirements were adequate for the destruction of cysts, provided that the original chlorine demand of the raw water was not excessive.

*This part, illustrating the meter-controlled hypochlorinator, is being kept because of its value to the potable water systems that were installed on board all the troop ships leaving San Francisco harbor during World War II.

This method of treatment provided not only a safe water but one that was extremely palatable. Passage through the activated carbon filter enhanced the taste of the treated water considerably.

Figure 2-7 illustrates a typical shipboard chlorinator installation. Note that there are dual hypochlorite solution tanks with a capacity of approximately 55 gal. These are rubber-lined, with a compression fit inspection hatch with hand-hold and an integral mounting stand for the hypochlorinator. The tanks are fitted with special supports that are secured to the deck. The cover is bolted to the tank so that there is no possibility of any spillage. The hypochlorinator is usually located between the retention tank and carbon filter. The retention tank is constructed with special end-around baffles to prevent any short-circuiting. The hypochlorinators were paced by either 1½″ or 2″ meters, depending upon the size of the troop ship (Class C1 or C2). These installations were always main-line systems in which all the water treated passed through the meter. The maximum pressure drop through the system was by design 10 psi, which limited the maximum amount of water that could be chlorinated by a 1½″ meter unit to 85 gpm and that of a 2″ meter to 175 gpm.

In the early part of the Pacific war, shipboard chlorinator installations were confined to troop and supply ships. It was assumed that the fighting ships (battle cruisers, aircraft carriers, destroyers) had no water supply problems because they were able to fulfill all their water requirements by means of the shipboard evaporators. This turned out to be an erroneous assumption. During the invasion of Tarawa, most of the Marine assault forces were suffering from dysentery when they hit the beach. Investigation into the cause of this epidemic indicated that it was probably contaminated drinking water. Just prior to the invasion (which witnessed the largest amassing of ships in one place up to that time—approximately six hundred vessels), it was concluded that the seawater in the vicinity of the vessels became grossly polluted from their own waste discharges. Even though the shipboard evaporators were functioning properly, it was discovered that at the end of each evaporation cycle a small amount of seawater siphoned over into the distilled portion, just enough to cause gross contamination of the entire potable water supply.

Following this discovery, drastic changes were made in supplying these ships with water. Several fuel tankers were immediately converted into water supply ships, each with a 300 lb/day gas chlorinator installation. These water tankers were able to chlorinate all the water as it came aboard or as it was being transferred to another vessel. Provision was made to recirculate and rechlorinate en route if necessary. The filling and loading rates were so high (400 to 1000 gpm) that gas chlorinators were mandatory. This system prevailed until the end of the war.

Shortly after World War II, the U.S. Navy Bureau of Ships began to equip all fighting ships with shipboard chlorination systems to treat the entire water system. Details of these requirements are covered in the *U.S. Navy Manual of Preventive Medicine.*[20] All water, evaporated or otherwise, must show a

0.2 mg/L free chlorine residual* at the end of 30 minutes of contact time. The equipment used is functionally identical with that illustrated in Fig. 2-7.

These units are specially made to meet the U.S. Navy shock-test requirements, and must be practically nonmagnetic. These installations do not utilize the special detention tank or carbon purifier used on board the troop ships, as it is necessary to chlorinate the evaporator effluent and the water in the holding tanks only as it is being transferred from one tank to another. This water is low in organic matter and other pollutants, and so elaborate treatment is not necessary.

In recent years the U.S. Navy and the Coast Guard have stepped up their potable water disinfection program so that all of their vessels from the largest aircraft carrier down to the smallest patrol boat are equipped with disinfection systems. A great many of these ships are equipped with brominators that utilize a polybromide impregnated cartridge. (See Chapter 16 for further details.)

In 1946 the U.S. Public Health Service decreed that all present-day ocean going vessels carrying passengers and utilizing skin tanks for potable water storage must chlorinate this water as it is pumped to the ships' water systems. A skin tank is defined as one that utilizes the ship's hull as a part of the tank. The theory here is that an opening in the hull or the "skin" of the ship would cause instant contamination. In installations on the American President Lines, an electrically operated, constant-feedrate hypochlorinator can be used, as it is a usual practice to chlorinate only while the water stored in the skin tanks is being transferred to the double-bottom or other tanks. The proper contact time is achieved by the amount of storage available in these transfer tanks.

Shipboard installations must be carefully analyzed to determine whether or not a meter-paced unit or a constant-feed unit is required. A much better choice is always the water-powered meter-paced unit, which eliminates the dependence on the ship's power system. During times of electrical aberrations, the chlorination system is certain to be neglected, and there is no easy way to compensate for water not chlorinated during the electrical outages.

The one disadvantage of the water-powered unit is the necessity for providing a suitable drain for the power water, which must go to waste at atmospheric pressure of about 1 to 1.5 gpm, equivalent to about 70 tons of water per day. Obviously the hypochlorinator must be located at a point in the ship where this wastewater can flow back into the ship's storage system, usually to the double-bottom tanks.

Other special considerations for a shipboard installation are related to the possible high ambient temperatures in the region of the installation and the elevated temperature of the water passing through the system. Ambient temperatures of 120°F have been reported aboard ship, and water temperatures have been known to go as high as 90°F. This requires that the water meter be equipped with a special hot-water disk. The suction and discharge valves

*To achieve a free chlorine residual, 85 percent of the total residual must be HOCl.

of the hypochlorinator pumping head must be made of material that not only will resist corrosion but also will not expand owing to this temperature rise. Certain ceramics are suitable for the ball and seats of these valves.

Suction and discharge tubing of the high pressure type, with natural or reclaimed rubber lining, is the most suitable. Neoprene is not at all suitable for this service. Certain types of plastics that "flow" or otherwise deteriorate at these temperatures must not be used in any of the hypochlorinator components.

For the operator's assistance, a liquid level staff that measures the amount of solution remaining in the hypochlorite tanks should be provided, together with a chart that shows the quantity of hypochlorite per inch of tank height required to provide the prescribed solution strength and a 10- to 16-ounce measuring cylinder, so that the proper amount of sodium hypochlorite can be added for any given amount of solution required to fill the tank.

Recent Waterborne Outbreaks in Cruise Ships. One of the more popular vacation adventures today is a trip on board a passenger cruise vessel. In 1966 about 300,000 passengers sailed on these vessels. By 1973 the number had reached about 750,000, 75 percent of whom sailed from New York City and two Florida ports—Miami and Port Everglades.[46] This number continues to grow.

The first large outbreak reported in the period 1970–75 aboard a passenger cruise vessel occurred in January 1970, where 10 passengers and 36 crew members, aboard the British ocean liner *Oronsay,* contacted typhoid fever; one passenger died.[44] An investigation revealed that the most likely source of the outbreak was the ship's water supply. Later that year the Centers for Disease Control in Atlanta, Georgia initiated a Vessel Sanitation Program in an effort to prevent waterborne and foodborne outbreaks aboard cruise ships.

This inspection program is under the direction of the U.S. Public Health Service. The standards used resemble those adopted by the World Health Organization for seagoing vessels. The vessel inspection consists of a 100-point checkist that covers water, refrigeration, food preparation, and the personal cleanliness of the food handlers. John Yashuk, chief sanitation officer for the Public Health Service Miami-based quarantine division, has advised that the inspection program is a voluntary one between the U.S. Public Health Service and the cruise industry. The Health Service has no regulatory powers; it can only recommend. However, when a ship fails the inspection and the Health Service advises against its sailing, the vessel usually complies.

This of course is not the case where U.S. flagships are concerned. These ships must comply with all Health Service requirements.

A 1972 survey by the CDC found that food handling practices and water systems aboard selected ships demonstrated a significant potential for the transmission of foodborne and waterborne disease.[45] Outbreaks of illness on board passenger cruise vessels during 1970–75 were caused by Shigella flexneri, Salmonella, and Vibrio parahaemolyticus. Vehicles for the etiological agents were water, multiple foods, seafood cocktail, and shrimp and lobster. *Staphylococcus aureus* caused two outbreaks of foodborne illness on board aircraft during the same period. The poison vehicles then were custard and ham.

The most notorious outbreak was on board the TV "Love Boat," otherwise known to unsuspecting passengers as the *M/S Sun Viking*.[47] The ship was commissioned in 1972 and is one of three 18,500-ton Finnish-built ships of Norwegian registry that comprise the fleet of the Miami-based Royal Caribbean Cruise Line. The ship has a passenger capacity of 800 and a total staff of 324. The ship's potable water was found to contain an unacceptable concentration of coliform organisms. An extensive study of the drinking water supply revealed that this was the source of the gastrointestinal illness. The CDC investigated four outbreaks of gastrointestinal illness aboard the *M/S Sun Viking* in 26 months.[47]

The water purification system of course was suspect. Here is a synopsis of the system: Water was bunkered through the fresh water filling line into three fresh water skin tanks and the laundry tank. The water delivered to the three fresh water tanks was batch-chlorinated to a level of 2 ppm. (The fresh water supply is from the salt water evaporators.) After the so-called batch chlorination system, the water is pumped without filtration through two UV disinfecting units and then stored in three potable water tanks.

The illness suffered by the passengers resembled acute gastroenteritis. The water system was suspect as the common source of the four outbreaks because at the time of the last outbreak the ship's UV disinfection system was not operating properly and the ship's drinking water was found to be contaminated with coliform bacteria. Moreover the UV system had no verifiable backup system that could be related to a chlorine residual measurement system.

An illustration of how elusive the causes of a diarrheal illness outbreak might be is the case of the *T.S.S. Fairsea*.[48] This is a 25,000-ton vessel of Liberian registry staffed by an Italian crew and based in Los Angeles, California. The ship carries 950 passengers and a crew of 426. On five consecutive cruises in the period April 23 to June 18, 1977, outbreaks of diarrheal illness occurred. On the five different cruises the illness attack rate among the passengers varied from 20 to 60 percent, which is a shocking situation.[49] Unfortunately all of the investigations were unable to uncover the etiology of these outbreaks. To this day the *Fairsea* problem remains a mystery.

Three days after she failed the U.S. Public Health inspection for the sixth time in 1977, the luxury Dutch cruise ship *Statendam* sailed out of Miami on December 2 for a 10-day Caribbean cruise.[50] Aboard were 671 passengers and a crew of 372. By the time the vessel docked again in Miami, at least 82 passengers and 13 crew members had been stricken with gastrointestinal afflictions. Investigators found that there was a serious malfunction in the chlorination system of the potable water supply. This was directly related to the chlorine residual analysis system used to monitor the chlorine residual. Apparently the low residual alarm system was inoperative.

Many other violations of the Public Health Service code for cruise ships occur on a regular basis. The question of "how safe is it to sail on a cruise liner that fails to meet the sanitary standards of the U.S. Public Health Service" continues to be asked because of the following examples.

A January 30, 1979 inspection of the luxury liner *Queen Elizabeth 2* made by the Los Angeles Quarantine Station cited the ship for poor food handling and preparation practices.[51] At the same time the U.S. Public Health Service in Miami advised that the Swedish American Line's *Linblad Explorer* flunked the test because of problems with her potable water distribution system.[51]

In February 1979, more than 120 vacationers on a week-long cruise through the Caribbean aboard the Miami-based cruise ship *Festivale* came down with stomach cramps, nausea, and diarrhea.[53] There were so many things wrong with the vessel's method of food preparation and storage that the investigators never did pinpoint the exact origin of the *Salmonella* poisoning, which caused the Carnival Cruise Lines to cancel the next cruise at a loss of $1 million. However, it is encouraging to note that the current chief of all these ship inspections, John Yashuk, believes that now 80 percent of all cruise ships that call at U.S. ports regularly meet the standards, although in July 1975 when the new codes went into effect, not a single ship could meet the standard. In spite of these encouraging statistics, the *Festivale* failed all seven of the service's sanitary inspections between July 1975 and July 1979.*

When White and his wife embarked (about 1980) on cruise ship vacations, he did not suffer from the air-conditioning system that was used by a box-type freight line that carried mostly fruit and vegetables from San Francisco down the Pacific coast through the Panama Canal, down the east coast of South America, through the Strait of Magellan, and back up the Pacific coast to San Francisco via the west coasts of South America and the United States—a 50-day trip. However, when that operation ceased in about 1985, the couple switched to the Royal Viking Line that had just moved into San Francisco. For reasons yet undetermined, White suffered terribly from the recirculated air-conditioning system, as did other passengers. It was so bad that there was no point in going to the movies; one could not hear the dialogue because of the enormous amount of sneezing and coughing. White went to great lengths with the Captain and crew to get the problem corrected. He suspected that it had to be due to recirculation of the same air without any disinfection by UV or chlorine. These were ships made in Scandinavia. Why did they not use the great amount of fresh air out in the ocean? White corrected the problem for himself by having the crew close off the recirculated air duct to his cabin. In a subsequent trip on an Italian-made ship, the recirculated air was fresh air, and White never had a problem. The bad news was that he could never get the health services systems to get involved, particularly because there were lots of disease outbreaks during that time—the middle and late 1980s.

*It is interesting to note that the EPA was never involved in any of these public health problems.

Choice of Chemical

Three factors govern the choice of hypochlorite to be used: (1) cost; (2) type of metering equipment to be used; and (3) availability. Other factors being equal, the chemical of choice would always be sodium hypochlorite solution. There are fewer maintenance problems with metering equipment, and there is no fire hazard connected with its storage. It is costly to transport over long distances and takes greater storage space than calcium hypochlorites. Per pound of available chlorine, it is more expensive than the calcium hypochlorites purchased in moderate quantities.

In remote areas where transportation cost is significant, the granular hypochlorites [$Ca(OCl)_2$] are the chemicals of choice. However, the potential insolubles of these products are sufficient to cause maintenance problems with metering equipment. If the hardness of the water available for dissolving this granular material is in excess of 75 to 100 ppm, the water should be softened before any attempt to make a hypochlorite solution. This can be done by adding equal amounts of soda ash with the calcium hypochlorite, allowing it to settle for 24 hours, and then decanting the clear liquor; or the makeup water can be softened by a conventional ion exchange unit. Another possibility is to sequester the potential insolubles (calcium ions) by the addition of 2 to 5 ppm of a sodium hexametaphosphate (sodium polyphosphate) to the makeup water before addition of the calcium hypochlorite. This will yield a sodium hypochlorite solution with very little insoluble material to cause equipment fouling. There are commercial laundry preparations of granular calcium hypochlorite and sufficient sodium polyphosphate to yield a nearly complete conversion to sodium hypochlorite when they are dissolved for use. But all these chemicals are expensive. It is therefore worthwhile to investigate the cost of lithium hypochlorite, which does not have these potential insolubles that plate out and plug up the metering equipment.

If for one reason or another (usually transportation and storage expense) a granular hypochlorite is the chemical of choice, lithium hypochlorite (LiOCl) may be the most economical, considering the time and expense required to remove or sequester the insolubles of $Ca(OCl)_2$. No matter what precautions are taken with the preparation of calcium hypochlorite solutions, the operator of the metering equipment should be prepared to recirculate muriatic acid from the point of entry to the point of discharge of such equipment to remove the calcium that will eventually deposit to some extent throughout the travel of the hypochlorite solution. The intervals of purging with acid can be more widely spaced if more precautions are taken to remove the insolubles first.

Maintenance

Maintenance of these and similar systems using hypochlorite requires the periodic flushing of the hypochlorinator with muriatic acid to remove the white, scaly deposit resulting from the interaction of hypochlorite with the hardness in the makeup water as well as the deposition of the inherent impurities in the hypochlorite. These deposits interfere with the efficiency of the internal valve action, causing errors in pumping rates and clouding the suction sight glass. When the sight glass becomes so coated with a deposit that its operation is obscured, then it is time to flush the apparatus with muriatic acid.

LARGE HYPOCHLORITE SYSTEMS

Current Practices

In the years the stress on safety and the fear of a chlorine accident caused large metropolitan areas to consider the use of hypochlorite rather than chlorine gas systems where large amounts of the liquid–gas chlorine is stored in either stationary tanks, railcars, or ton cylinders. This has occurred in spite of the good safety record of such installations. Since 1908, when chlorine gas was first used in the United States, there has been only one fatality from a chlorine accident at a water or wastewater installation in the United States. There have been nine transportation-related fatalities resulting from massive derailment of tank cars. Between 1908 and 1961 there was only one fatality that was transport-related. Since shelf couplers have been installed on all rail tankers, derailment accidents causing chlorine leaks have ceased.

Despite this record and the considerable additional cost of hypochlorite over chlorine gas (two to four times) and its inherent unwieldy and cumbersome handling problems, the city of New York changed from gas to hypochlorite at certain of their wastewater treatment plants in 1969.*

The first to experiment with the use of large amounts of hypochlorite was the power industry. Electric generator stations use considerable quantities of chlorine for slime control of the condenser cooling water. In one such case, a hurricane along the northeast coast of the United States ripped loose a ton cylinder of chlorine at one of these power stations and carried it a considerable distance. No damage was done, but the management decided it was time to eliminate any possible hazard from chlorine gas.

Sometime in 1955, the Narragansett Electric Company of Providence, Rhode Island, effected the switchover to hypochlorite.[22] Other stations in Chicago and New York followed suit.

A notable experiment in the application of hypochlorite for wastewater disinfection has been done by the Metropolitan Water Reclamation District of Greater Chicago (MWRDGC). Installation of disinfection facilities using sodium hypochlorite at 15 percent "trade" strength was begun in 1969 at the 330 mgd North Side Plant, the 900 mgd West-Southwest Plant, and the 220 mgd Calumet Plant.[21, 30, 39, 40, 41] Since July 1972, the District has been

*Since the development in the early 1990s of on-site generation of 0.8 percent hypochlorite, this nonhazardous chemical has revived considerably the application of hypochlorite, rather than chlorine gas.

continuously chlorinating all the effluents from these plants.[42] The newest facility of MWRDGC is the John E. Egan plant at Schaumburg, Illinois, which was first used in 1976.

Currently the city of Houston, Texas uses sodium hypochlorite at its largest plants and liquid-gaseous chlorine at the smallest plants.[41]

In 1980, the city of San Francisco switched from chlorine tank cars to sodium hypochlorite. This move prompted three of the largest wastewater treatment plants in the San Francisco Bay Area to evaluate the situation of chlorine gas versus hypochlorite. After separate and independent investigations these plants decided not to change to hypochlorite for the following reasons: (1) the reliability and safety procedures of the chlorine storage system were satisfactory; (2) the amount of chlorine delivered to these plants was less than 10 percent of the total amount of chlorine moving into the area; and (3) the cost of hypochlorite was too great compared to chlorine.

However, a few years later San Francisco decided to switch to hypochlorite at its three WWTPs, primarily because a large apartment building was being built right alongside the area where 55-ton tank cars for the North Point WWTP were parked. This and a big scare caused by a newspaper article describing the dangers of shipping these tank cars on barges across the bay caused the city to switch to hypochlorite. They hired White as their special consultant for this switch to hypo at their two plants using tank cars and one that was using ton cylinders.

Some of the large users have switched back to chlorine gas. After several years of trial some power plants have given up hypochlorite because of the inherent difficulties in handling it in large amounts. Others have done so to save money. In 1975 Cleveland decided to make the change from hypochlorite to gas at the Easterly Wastewater Treatment Plant. This plant was originally designed for hypo.

The interest shown in the use of hypochlorites created a great deal of interest in the use of on-site chlorine generating apparatus. Presently, in 1997, the Chemical Services Co. of Campbell, CA, has installed several of their hypochlorite solution generators. This unit is described in Chapter 3.

Hypochlorite Quantities Required

To get an idea of quantities involved, let us examine the chlorine requirement for disinfection of a secondary treated effluent discharging into a receiving water. Proper disinfection to maintain the receiving waters safe for water contact sports is usually about 100–125 lb chlorine per million gallons of treated effluent.

Using a high-strength sodium hypochlorite of 10 percent by weight chlorine would require the following amount of sodium hypochlorite:

$$\text{percent available } Cl_2 \text{ by wt.} = \frac{10\%}{\text{s.g.} = 1.14} = 8.8\%$$

Each gal NaOCl contains 9.5 × 8.8% = 0.84 lb Cl_2. If the dosage is 100 lb/mg, then 100/0.84 = 119 gal 10 percent NaOCl/mg. Therefore a 200 mgd plant would require 200 × 119 = 23,809 gal 10 percent NaOCl/24 hr. Assuming peak rate of 2½ times average, 500 mgd × 119 = 59,500 gpd or 59,500/1,440 = 41 gpm. So the metering equipment should be sized to handle a maximum of 50 gpm of 10 percent hypochlorite solution.

Comparing the half-lives of various strengths of hypochlorite, it appears that 10 percent strength is the most stable strength, but not the most economical. Large installations are probably suited for a maximum storage period of one week. There would be very little deterioration in the strength of a 10 percent solution in this length of time. Manufacturers of sodium hypochlorite prefer to provide a strength of 15 percent.

The choice of one over the other is primarily a matter of economics. The 10 percent solution has a greater stability than the higher strengths, and so, other things being equal, it should be favored. However, storage facilities are such a large cost factor in the overall installation that the economy of the 15 percent solution must be considered as well as the deterioration due to age.

To calculate dosages and storage facilities, the reader is referred to Table 2-2.

HYPOCHLORITE FACILITY DESIGN

Pumped Systems

Introduction. Feeding and control systems should be categorized by the quantity of hypochlorite solution to be applied up to rates of approximately 380 gal/hr.

These pumps can be automatically controlled by either a variable speed drive by SCR (selenium control rectifier) control or by a stroke length positioner (electric or pneumatic) or both. Therefore, a compound loop control system is practical. In practice the flow signal is sent to the SCR controller, which provides a 20:2 metering range and the chlorine residual signal is sent to the stroke positioner, which has a 10:1 range. This is more than ample for any wastewater disinfection application.

Wallace & Tiernan Encore 700 44-Series Pump[23, 24]

Introduction. This pump made its debut in early 1996. It is a completely re-engineered venerable 44-Series pump. The result is a mechanical diaphragm pump that is probably unrivaled for robustness, economy, simplicity and serviceability, regardless of the application. It can pump. It can pump up to 360 gph against back pressures up to 175 psi. It has many operating benefits, along with capacity control via stroke length and stroke frequency. This unit is illustrated by Fig. 2-11.

Fig. 2-11. Wallacc & Ticrnan Encore™ 700-44 series diaphragm metering pump.

Operating Benefits. This unit was designed to ensurc rcliable metering performance. The clear PVC cartridge valves provide built-in visual indication of operation along with numerous single and double ball valve designs, which are able to handle mild solutions, aggressive chemicals, high viscosity polymers, and various types of slurries. This unit is based upon a robust long-life design. It combines a steel and modular iron non-loss-motion mechanical assembly, with an epoxy painted cast iron gear-box, 316 SS fasteners, tapered roller bearings, and very special robust gears.

Then there is a ten-turn micrometer stroke length adjuster that reads feedratc scttings in increments of 25 percent from zero to 100 percent. Operating turndown on the pulley drive is 40 to 1 with a standard induction motor. With an optional variable speed DC motor and SCR Control Unit, the total turndown is 800 to 1.

There are also numerous process control options that are detailed in W&T publication No. TI 440.400UA.

Wallace & Tiernan Premia 75 Solenoid Metering Pumps[60]

Introduction. The Premia 75 represents a complete line of solenoid pumps from single mode control to microprocessor control with reliability, economy, and true metering performance. Whether the application calls for a simple manual operation or a complex process control, there is an economical Premia 75 available for the project.

There are six models available to handle capacities up to 500 gal of process water per day with back pressures up to 300 psi. These units are also available with a wide range of accessories and options, including high viscosity and automatic degasifying arrangements, to handle the most demanding applications. (See Fig. 2-12.)

Anyone interested in these solenoid diaphragm metering pumps should obtain Wallace & Tiernan publication No. TI 460.150-3UA, as there are six categories of these pumps that are available in different sizes. For instance, the Micro and Mega groups each have 20 sizes to handle 20.83 gph, the Mini DC group has four sizes to handle 1.83 gph, the Mini group has eight sizes to handle 1.83 gph, the Econo group has eight sizes to handle capacities to 1.25 gph, and the Mono group has four sizes to handle 1.25 gph.

Features. The following details cover the most important parts of the solenoid metering pumps:

1. Five Function Valve. This device provides relief from any excess situation that might damage the unit. It also allows metering into the atmosphere,

Fig. 2-12. Wallace & Tiernan Premia™ 75 solenoid metering pumps.

as well as creating a discharge restriction for repeatable output. It prevents siphoning through the pump when the point of injection is below the pump, or into the suction line of another pump.

The pumphead has an air bleed that is used as a priming aid to manually remove air from the pumphead, while the discharge drain depressurizes and drains the discharge line without loosening tubing or connections, or exposing the operator to injurious chemicals.

2. Cartridge Valves. These high precision guided ball and seat valves provide a superior seal, for reproducible metering accuracy along with fast and foolproof service. Single and double ball versions are available to handle mild solutions, aggressive chemicals, and viscosities up to 10,000 cps.
3. Liquid End Materials. There is a wide array of these materials, including SAN for integral sight flow indication, PVC, polyvinylidene fluoride, 316 SS, and glass-filled polypropylene. For fluids that entrain air, select the PVC automatic degassifying liquid end, available in capacities up to 48 gpd.
4. The Straight Through Flow Path. This precision-engineered flow path provides the utmost in efficient fluid metering. The standard liquid end arrangements can handle viscosities up to 3000 cps. With spring-loaded valves, the Prima 75 solenoid pumps can handle viscosities up to 10,000 cps.
5. Premium Composite Diaphragm. This part is made to stringent specifications to ensure long life under the most demanding applications. The design incorporates Teflon-facing, for the highest degree of chemical resistance, and nylon reinforcements, all bonded to a preformed elastomeric support. They have also added convolutions for unconstrained rolling action, as well as an integral O-ring for complete sealing, plus a brass insert to assure volumetric accuracy, even at varying discharge pressures.
6. Hot Rated Pump. To protect the solenoid from operating heat, it is separately encapsulated in a *fin-cooled* thermoconductive enclosure that effectively dissipates the heat.
7. State-of-the-Art Electronics. These devices are used on the Premia 75 circuit boards to ensure performance and dependability. The timing circuit is virtually unaffected by temperature, or EMI and any other electrical disturbances. A transient voltage suppressor protects the electronics. Components are surface-mounted to reduce wired connections and improve reliability.

Control Panel. This is mounted directly on the pump itself as shown in Figure 2-13. It has a 16-character dot matrix backlit display that provides the operator with all the necessary operating conditions and alarms. Besides English, Spanish and French versions are available.

Fig. 2-13. Pump control panel for Wallace & Tiernan Premia™ 75 solenoid metering pumps.

The following is a detailed description of this panel:

1. A Locking Stroke Length Adjuster. This is used to maintain precise feedrate settings. A special two-piece shaft keeps the dial snug to the graduations throughout the scale. The feedrate is infinitely adjustable from zero to 100 percent.
2. A Stop LED. This illuminates when an external switch closure stops the pump.
3. External Circuit Breaker. To resume pumping after overload conditions have been corrected, simply push this button on the Mega and Micro control panel.
4. Two Relay Outputs. These are available on the Micro for enhan. A power relay is optional, rated at 250 AC, 0.5 A. Contact closures occur for the alarm or a condition that is chosen via the six-station membrane touch pad. Selections include stop, loss of mA input signal, full count, memory register overflow, and circuit failure.
5. Stop Input Control. This is standard on the Micro and optional on the Mega. In this mode, a nonvoltage contact closure, such as feedback from a low liquid level gauge, stops the pump in either manual or external control, and illuminates the red indicator LED. The pump resumes normal operation when the contact opens.

6. A Six-Station Membrane Touch Pad. This simplifies stroke frequency and function selection. The stroke rate is adjustable from 1 to 100 percent. In addition to simple manual operation, stroke counting and timed interval operation are standard with the Micro.
7. A Power-On/Stroke LED. This illuminates when the pump is switched on and strobes with each stroke.
8. External Control. These modes are standard on the Micro, but optional on the Mono, Econo, and Mega. With the Micro, the choice is from pulse input, 4–20 mA, or 20–4 mA input, ratio control, and memory register to store pulses exceeding 125 contacts per minute. Pulse input is available on the Mono, Econo, and Mega. A 4–20 mA input is also available on the Mega. Chemical metering systems and their accessories along with preventive maintenance kits are described in W&T publication No. TI 460.150-4UA. The Premia 75 is fully described in W&T publication No. SB 460.150UA.

Other Pumped Systems

Pulsafeeder of Rochester, New York,[25] also makes a suitable line of diaphragm pumps with automatic feedrate control for hypochlorite solutions. The Pulsafeeder control approach is to resolve the entire variation of solution discharge as a function of stroke length, maintaining the stroke speed constant at all times. This means that a system using the Pulsafeeder would have to have a multiplier to combine the flow and residual signals into a single signal. Therefore, the concept of compound loop control is not yet available with a Pulsafeeder system.

Diaphragm metering pumps suitable for this service can tolerate back pressures as high as 150 psi. This is far in excess of any need that could occur in a wastewater system.

The City of Chicago prefers diaphragm metering pumps for its smaller plants. The capacities of these small plants vary, with average daily flows from 10 to 50 mgd. The O'Hare WRP is a 30–55 mgd plant using chlorine dosages of 1.0–3.5 mg/L, the egan WRP is a 23–32 mgd plant using chlorine dosages of 1.5–3.5 mg/L, and the Hanover WRP is a 10–15 mgd plant using chlorine dosages of 3.0–5.0 mg/L. The larger plants are all in excess of 100 mgd. Chlorination at these plants is required only seasonally.

Systems requiring pumping rates in excess of the capacities available in diaphragm pumps can use a centrifugal pump. Feedrate control in these situations is by the use of a modulating rate valve downstream from the pump discharge.

Two classes of centrifugal pumps are available for hypochlorite service. One is the precious metal alloy, and the other is the fiberglass type. The Duriron Co., of Dayton, Ohio, makes a titanium pump that is excellent for this service. This pump has a long and satisfactory record of pumping chlorinator injector discharge solutions into transmission lines where pressures are too high for the usual injector booster pump upstream from the injector. This pump can also be used as a transfer pump or a low-lift pump for any hypochlorite installation. The "Durco Titanium" line of pumps comes in a variety of sizes for either low- or high-head purposes.

The Fybroc Division of Metpro Corporation, Hatfield, Pennsylvania, makes a line of fiberglass pumps that are also suitable for corrosive solutions. They are ideal for low-head service such as at a wastewater treatment plant.

Pumps other than those mentioned above could be carefully investigated for their suitability for hypochlorite service. A chlorine manufacturer put out a bulletin on the advantages of hypochlorite that recommended graphite pumps as suitable for use with high-strength (15%) hypochlorite solutions. Upon investigation, the manufacturer of such a pump (Union Carbide Corp.) advised that its Karbate line of graphite pumps is *not recommended* for this purpose.[26]

Plunger- or gear-type pumps are not recommended owing to their high maintenance costs and poor reliability. There is insufficient operating experience with these types of pumps.

Gravity Systems. The preferred system is to meter the hypochlorite through a modulated rate valve and allow the solution to flow by gravity to the point of application. If the hydraulic gradient from the storage tanks to the point of application cannot provide gravity flow, the hypochlorite could be pumped to an intermediate head tank. This can be done with a sonic-type level control switch allowing automatic intermittent operation of the supply pumps, which would always pump at a constant rate. The modulating rate valve would be somewhere on the downstream side from the intermediate head tank.

The Metropolitan Water Reclamation District of Greater Chicago has used all three systems of hypochlorite delivery, and has found the gravity system to be the most reliable and the one that requires the least amount of maintenance.

A 1991 update from the city of Chicago[54] advised that it preferred the gravity feed using control from a rate valve but only at plants in excess of 100 mgd. Those plants are as follows: Stickney, Calumet, Northside, and Lemont. They are no longer required to chlorinate the effluents from these plants.

Eductor System. Eductors can be used effectively up to about 1015 psi total back pressure. They are normally powered by pumps using the plant effluent. The principle of operation is similar to that of a chlorinator injector system. Of the different systems described, this one is the least reliable for controlling flows. Another disadvantage is the loss of available chlorine caused by the presence of ammonia nitrogen in the plant effluent. The breakpoint reaction when using a mixture of hypochlorite solution of about pH 10 takes place almost instantaneously. Therefore, in the eductor system there will be a loss of chlorine to complete this reaction by a factor of 10 parts (by wt) chlorine for each part of ammonia nitrogen. As an example, the eductor system at the Calumet treatment plant of the MWRDGC requires about 65 gpm of water to provide the necessary minor flow of sodium hypochlorite (about 2–4 gpm). If the treated effluent is used for powering the eductor system and if the effluent is not nitrified, the ammonia nitrogen content may be as high as 20 mg/L. Therefore, the chlorine consumption in the eductor water would be as follows:

$$65 \text{ gpm} \times 1440 = 0.0936 \text{ mgd}$$
$$20 \text{ mg/L} \times 8.34 = 167 \text{ lb/mg}$$
$$167 \text{ lb/mg} \times 0.094 \text{ mgd} \times 10 \text{ mg/L } Cl_2 = 157 \text{ lb/}Cl_2\text{/day}$$

So regardless of the chlorine dosage, 157 lb of available chlorine will be used in the reaction with the effluent eductor water. Assuming a 15 percent trade strength hypochlorite solution this would amount to about 126 gal of hypochlorite.*

Monitoring the System. All hypochlorite systems should be equipped to indicate and total the flow of hypochlorite to each point of application. Frequent checks of the solution strength should be made so that the operator can easily verify the chlorine dosage. It is imperative that the operator be able to verify the chlorine dosage from a visual indicator. This is one of the most important tools an operator can have. Therefore, the selection of a hypochlorite flow measuring device should be made with great care and consideration. Two types of flow meters are available. One is the magmeter, and the other is the rotameter. When a rotameter is used, it should be equipped with both an indicator and a transmitter.

It goes without saying, but the reader is reminded, that monitoring with a chlorine residual analyzer is necessary for the operator to confirm that the chlorination system is operating properly. When diaphragm pumps are used, pulsation dampeners should be installed on both the suction and the discharge lines adjacent to the pumps. These devices are necessary to minimize the pulsation effect, which interferes with the operation of the flow meter.

Materials of Construction

Storage Tanks. The first large storage tanks built for the Chicago projects were of the filament-wound fiberglass type. These tanks proved unsatisfactory, having a life of only about four years.[38] About 1975 these tanks were converted to hand lay-up fabricated fiberglass tanks utilizing a vinyl resin binder. This change was made based upon positive results obtained from another operating installation. The performance of these new tanks was acceptable.[28]

*One gallon of 15 percent hypochlorite contains 1.25 lb available chlorine.

The experience of the Metropolitan Water Reclamation District of Greater Chicago with underground concrete tanks has indicated that fiberglass lining of concrete is undesirable. Laminate failures have caused clogging of valves, pumps, and diffusers. The use of these tanks has been discontinued. For the MWRDGC at the John E. Egan wastewater plant, which was put into operation in 1976, the hypochlorite storage tanks were put underground because of their size (12 ft in diameter and 35 ft long, approximately 30,000 gal). These are plastic-lined, continuous-weld, full-weight carbon steel tanks. The lining, which was applied on-site, consisted of two coats of fiberglass-reinforced polyester material applied at a nominal thickness of 35 mils.

Rubber-lined steel tanks would be just as satisfactory as any type of PVC lining. This method has been used in various aspects of chlorination for many years.

Storage tanks should be equipped with a level gauge and transmitter for continuous readout of its contents. They should also have vents and some method of manually sampling both the incoming delivery and the contents of the tank.

The fill piping system should discharge through the top of the tank. A pressure relief and overflow pipe should also be provided. The fill pipe connection to the delivery vehicle should be a Hastelloy C or a titanium nipple securely braced to the tank. The fill pipe itself can be PVC, R.L. steel, Saran-lined steel, or Resistoflex (Kynar). A high capacity drain or sewer with a proper drainage gradient adjacent to the storage tanks should be provided in the event of a storage tank rupture. Underground tanks should be equipped with equivalent facilities consisting of a completely corrosion-resistant sump pump.

A 1991 update on the experiences at Chicago indicated that the operators experienced storage difficulties. The original fiberglass tanks developed hairline cracks that resulted in leakage of the contents. They corrected this problem by specifying that fiberglass tanks have an interior corrosion-resistant barrier fabricated with isophthalic resin, reinforced with a 20–30-mil glass veil, backed with a 100-mil chopped strand fiberglass laminate. The details of this tank construction were included in the specifications by several tank manufacturers in the Chicago area.

Piping. Aboveground piping can be Sch 80 PVC, Kynar, hard-rubber-lined steel, or Saran-lined steel. The steel-lined pipe presents some installation difficulties. The preferred material is Kynar for both pipe and fittings. This pipe is also known as Fluoroflex-K. The latter is a trademark of Resistoflex,[29] and Kynar is a Pennwalt Corp. trademark.

Underground piping should be some type of lined steel pipe. Chicago favors either a PVC or a polypropylene lining. Hard rubber or Saran lining is also acceptable and is preferred by others.

After 10 years of service, Chicago[54] also found hairline cracks in the PVC piping that transports the hypochlorite. When it reaches this condition, it is simply replaced. However, for piping that is exposed to sunlight, CPVC is used exclusively because it resists UV radiation better than PVC. They also experienced pipe joint leaks, as others have too. These occur mostly at screw-type joints. They have had much better results with solvent weld joints. Age and local points of vibration hasten the pipe failures, which are a result of brittleness.

Valves. Plug valves are preferable to ball valves. Ball valves are subject to stem breakage. This limits a ball valve operating life to about six months.[28] Plug valves made of steel and lined with PVC or polypropylene are preferable.

Diffusers. The distribution of the hypochlorite solution at the point of application needs special attention because of the very high concentration of the solution (15 percent = 150,000 mg/L). This compares to the chlorine solution concentration of 3500 mg/L (maximum) in a conventional chlorinator discharge. Although the segregation phenomenon favors better mixing with a higher concentration of the process chemical, efficient dispersion is the goal for a hypochlorite diffuser. The diffusers should be designed with perforations for across-the-channel installations using about a 25–30 ft/sec velocity at the perforations.

The diffusers should be made of either PVC, Kynar, or rubber-lined and covered steel pipe.

Flow Meters. These can be either Teflon-lined magnetic flow meters manufactured by Fischer and Porter or PVC Straight-Through Vareameters with transmitter (electric or pneumatic) manufactured by Wallace & Tiernan.

Rate Control Valves. The most common control valve in use is the Saunders-type using a rubber diaphragm with teflon facing. The valve body can be all PVC or a steel body lined with either PVC, Kynar, hard rubber, or Saran. These valves can be equipped with electric or pneumatic actuators. Plug-type valves are also available for this service. PVC ball-type construction is not recommended for this service. Maintenance cost is too high and reliability is low.

For flows up to 5 gpm, Fischer and Porter uses its Chloromatic PVC and Teflon valve; above 5 gpm a Saunders-type diaphragm valve is used.

Eductors. These must be all-PVC construction because of the corrosivity of the hypochlorite solution. The eductor used at the MWRDGC Calumet wastewater treatment plant is a Penberthy Model No. 168P (PVC). This unit operates at about 35 psi and uses approximately 70 gpm wastewater for a maximum back pressure of 10 ft.[30] It can deliver about 3 gpm of 15 percent hypochlorite.

Hazards of Hypochlorite

The use of hypochlorite as an alternative to liquid (gaseous) chlorine in reasonably large quantities is primarily for safety reasons. However the hazards due to the presence of hypochlorite must not be overlooked. These hazards derive from storage accidents.

One such accident occurred in Knoxville, Tennessee on April 8, 1983.[53] A lethal cloud of chlorine gas escaped from sodium hypochlorite tanks used in the disinfection system for the wastewater treatment plant. The system also uses ferric chloride as a coagulant, which is normally shipped in railcars. When railcars are not available, ferric chloride is shipped by tank trucks. The hypochlorite is always delivered by tank trucks similar to those used for shipping ferric chloride.

In this instance the ferric chloride was shipped by truck, and because the truck connections were compatible, the driver, who had never made a delivery to the plant before, made the connection, pressurized the truck, and unloaded approximately 600 gal of ferric chloride, which mixed with approximately 3000 gal of 10–12 percent sodium hypochlorite.

Owing to the low pH of the $FeCl_3$ and the concentration of the reactants, molecular chlorine was released instantaneously from the hypochlorite. A cloud of Cl_2 began rolling out of the hypochlorite tank vent. Fortunately an emergency response plan that had been worked out by the city was implemented as soon as the cloud was sighted. This included local evacuation and rerouting of all mobile traffic near the area.

Precautions must be taken to make certain that only hypochlorite can be put into a hypochlorite storage tank. Any acidic chemical will generate the release of molecular chlorine from a sodium hypochlorite solution.

When the City and County of San Francisco switched to hypochlorite at all three of their watewater treatment plants, they also switched from SO_2 to bisulfite solution. Precautions were taken in the design of these systems to prevent the possibility of unloading bisulfite into a hypochlorite tank or vice versa. This mixture produces a heat of reaction sufficient to cause disintegration of a fiberglass tank. The heat generated is so great that an explosive force would surely be produced.

In spite of all the various precautions devised to prevent these accidents caused by tanker truck drivers, they continue to occur when an inexperienced driver is involved.

Another hazardous situation could occur at installations where both chlorine and ammonia are used for treating potable water. Adding ammonium hydroxide to hypochlorite will generate lethal quantities of nitrogen trichloride. Moreover, this mixture might produce a violent explosion if the right concentrations were reached.

Still another possibility is the generation of chlorine dioxide on-site. The ingredients for generating chlorine dioxide are commonly chlorine, acid (HCl), and sodium chlorite solution ($NaClO_2$). If an acid delivery gets into the sodium chlorite, molecular chlorine will be released instantaneously. If hypochlorite is used as the chlorine source, the acid–hypochlorite mixture will be an additional hazard. Mixing the sodium chlorite (pH 12) with the hypochlorite (pH 12) may not create a lethal combination, but precautions should prevent such an accidental mixture.

Operating Costs

The operating cost of any imported hypochlorite system will depend entirely upon the amount of chemical to be delivered at one time and the total amount to be consumed over a contract period. The following are examples of hypochlorite and chlorine liquid–gas prices in various metropolitan areas.

CHICAGO, ILLINOIS. The Metropolitan Sanitary District of Greater Chicago (MSDGC) purchases 15 percent (trade) sodium hypochlorite under contract agreements with two suppliers: K. A. Steele and Barton Chemical.

As of 1996, part of Chicago's supply, the hypo used at the Egan WWTP came from a company that makes steel products where the chlorine they use is a waste product. This keeps the price of this supply at 37–40 cents/lb versus 48–52 cents/lb from the rest of the suppliers. The unit price paid for sodium hypochlorite is usually adjusted each month, based upon quantity requirements of a facility, published prices of certain raw materials, transportation, fixed costs, and the Consumer Price Index. Utilizing this rather involved calculation, one of the potential suppliers in the Chicago area supplied 15 percent hypochlorite on the basis of one tank truck (4000 gal) minimum delivery at a cost of 38 cents/gal for March 1993. The other supplier delivered hypochlorite to the smaller plants at a cost of 60 cents/gal. The 38 cents/gal figure calculates to 31 cents/lb available chlorine. The 60 cents/gal figure calculates to 48 cents/lb available chlorine.[31]

Chlorine gas in the Chicago area is grossly dependent upon the size of the containers, because of a peculiar safety ruling of long standing. The various water treatment facilities for the city of Chicago are compelled to receive their chlorine supply by the truckload (14-ton containers). By law, chlorine tank cars are not allowed to have access to the Lake Michigan water treatment plants. Therefore, the cost per ton of chlorine gas at these plants was \$140/ton (7 cents/lb) in 1982, down from \$204/ton (10 cents/lb) in 1976. If 90-ton tank cars were allowed access to these plants, the chlorine would have cost \$120/ton (6 cents/lb) in 1982, down from \$140/ton in 1976.[32]

HOUSTON, TEXAS. Sodium hypochlorite at 15 percent chlorine is available from local suppliers at 50 cents/lb in tank truck quantities.

Local suppliers quoted liquid chlorine in ton containers at \$240/ton (12 cents/lb) and 90-ton cars at about \$150 ton (8 cents/lb) (1990 prices).[33]

SAN FRANCISCO BAY AREA. Sodium hypochlorite as of 1996 cost 85 cents/gal. Each gallon contains about one pound of available chlorine. This is being used at all the San Francisco potable water and wastewater treatment plants. It is delivered in 5000-gal tanker trucks. This compared to the 1995 price of chlorine in ton containers of 25 cents/lb.

Jones Chemical will deliver 12.5 percent hypochlorite anywhere in the Bay Area (1996) for 80–90 cents/lb. Chlorine in ton cylinders is 28 cents/lb, and in tank cars it is 16 cents/lb.

Imperial West Chemical in Pittsburg, California also sells 13–14 percent hypo for 47 cents/lb of chlorine.

All Pure Chemical Co. in Tracy, California will deliver ton cylinders of chlorine for 25 cents/lb, also delivering chlorine in tank cars for 13 cents/lb and 12.5 percent hypo for 80 cents/lb.

Continental Chemical Co. will deliver ton containers for $176/ton (9 cents/lb) at 5 tons per week. PPG and Hooker Chemical delivered chlorine in 90-ton tank cars at $65/ton in a glut market (1982). This is only 8 cents/lb.

Whenever there is an upsurge in the sale of caustic, the price of chlorine drops dramatically because it is a waste product in the making of caustic (NaOH). Many times in the 1980s and 1990s, 90-ton tank cars of chlorine made in the United States were shipped to treatment plants in Southern California for free, while some users only had to pay the freight!

As can be seen from the above discussion, the cost of chlorine gas and hypochlorite varies considerably, depending upon the locality, demand, and availability of both caustic and chlorine.

The price spread between hypochlorite and chlorine gas increases significantly as the distance from the source of chlorine gas manufacture and the user increases.

The most optimistic estimate is that imported hypochlorite will cost at least three, or, more likely, four times the cost of liquid–gas chlorine.

Cost comparisons of the chlorination facility between liquid chlorine and hypochlorite should include storage and supply facilities, metering equipment, instrumentation, and monitoring equipment.

Generally, the metering and feeding equipment for chlorine gas is more expensive than that for hypochlorite, but the expense of storage facilities for hypochlorite is far greater and more than offsets the equipment difference. Maintenance of a hypochlorite system requires more worker-hours than the gas system.

Reliability

For maximum reliability the hypochlorite flow control system should consist of two separate and independent systems that accurately control the flow of hypochlorite to provide a predetermined chlorine residual in the plant effluent.

If pumps are required to deliver the hypochlorite or operate eductors, both standby equipment and standby power should be available to prevent interruption of hypochlorite flow to the point of application.

ON-SITE MANUFACTURE OF HYPOCHLORITE

Introduction

The on-site manufacture of hypochlorite is normally based upon the interaction of chlorine gas and either sodium hydroxide or lime. This usually depends upon the quantity of hypochlorite to be produced. Two systems are described below. The first example is a small plant application using less than 4000 lb/day chlorine. The other system involves massive amounts of chlorine: 50,000 lb/day or more.

Potable Water System (4000 lb/day)

Sometimes in certain extraordinary cases in the application of chlorine for water treatment, such as in bleaching organic color, the amount of chlorine required depresses the pH, thereby interfering with the bleaching action of chlorine. In one such instance,[35] calcium hypochlorite was made in situ. Figure 2-14 shows a schematic diagram of how this was accomplished. Here it is interesting to see how a conventional installation of a lime feeder and chlorinator was arranged to produce, in an efficient manner, a calcium hypochlorite solution so that the chlorination process of bleaching could be carried out

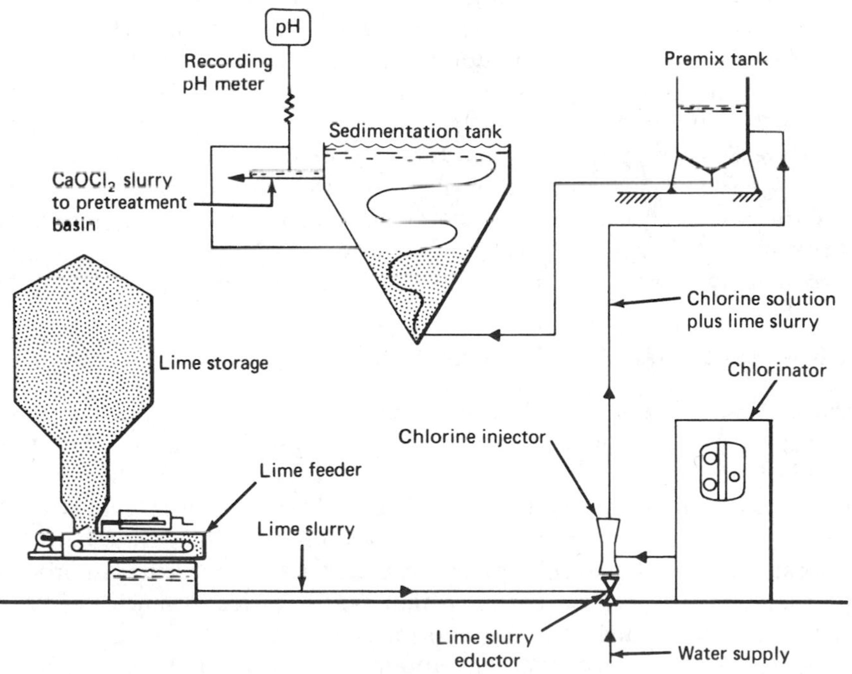

Fig. 2-14. On-site manufacture of hypochlorite.

without depressing the pH of the treated water. Here too, the most difficult problem to solve was to properly settle out of solution the hydrated lime and its associated impurities. This flash mixing process of making calcium hypochlorite consists of three basic elements: (1) a premix tank, (2) a calcium hypochlorite sedimentation tank, and (3) a pH detection system.

In Fig. 2-14, chlorine solution discharging at approximately 15 gpm from a chlorinator is made to flow through the premix tank to the sedimentation tank and thence to raw water in one continuous operation. At a point located well upstream from the premix tank, lime slurry is fed into the chlorine solution at approximately 15 gpm, the concentration of which is equivalent to the quantity of free chlorine present. The purpose of the premix tank is to remove the turbulence caused by the lime injector and the chlorine injector prior to entrance into the sedimentation tank. The gradually sloping sides of the sedimentation tank cause a decrease in velocity of the solution. This gradual slowing down of the upward movement of the solution toward the overflow weir allows the settleable material to form a sort of mat, allowing clear liquor above to enter the raw water. This lime suspension hydraulically grades itself with the large particles composing the lower section of the suspension and the fine particles forming the top or upper layer. The effluent from the sedimentation tank contains less than 0.1 percent total solids, owing to the small velocity at the crest of the tank. The solution flows by gravity to the point of application. After the desired feed rate is set on the chlorinator, the lime feeder is set in accordance with the figures shown in Table 2-4. The final determining factor in setting the lime feeder is the ratio of the pH of samples taken from within the sedimentation tank and the pH of the effluent of the sedimentation tank. When these two values of pH are in equilibrium, the lime feeder is set correctly, indicating that the quantity of lime collecting in the "lime layer" is not in excess of that required by the reaction. Underfeeding of lime is noted by both a drop in pH and a free chlorine odor coming off the surface of the two tanks. The average dosage by this system amounts to approximately 13 mg/L calcium hypochlorite to 26 mgd water treated, or approximately 2800 lb chlorine and 2900 lb lime per day.

Wastewater System (25,000–100,000 lb/day Cl_2)

Introduction. The system described here is unique because it can make either calcium or sodium hypochlorite on-site using liquid chlorine. The choice between lime or caustic is based largely upon economics. Caustic is always more expensive than lime but is easier to handle. This system concept is totally different from the vacuum principle. It does not use either chlorinators or evaporators. It uses liquid chlorine that is mixed with a lime slurry to produce a hypochlorite solution concentration that can be varied as much as 5000 to 15,000 mg/L, depending upon the chlorine requirement.

It is important to realize that only about 6–7 percent of all liquid chlorine manufactured in North America is used for drinking water and wastewater treatment. The amount required for wastewater is some 5–7 times that required for water treatment. The chemical industry and the pulp and paper industry account for the other 94–95 percent. Large-capacity chlorination systems for wastewater treatment plants such as this one present many difficult problems with respect to safety. These installations all use the vacuum principle to feed chlorine vapor into the process stream. This requires vaporizers to convert the liquid chlorine to gas and special vacuum-regulating equipment for feedrate control. However, the JWPCP operates on an entirely different principle; it uses only liquid chlorine absorbed into a lime slurry. The end product is calcium hypochlorite.

The calcium hypochlorite is made from liquid chlorine withdrawn from 90-ton railcars. The lime slurry and liquid chlorine are both injected into the hypochlorite makeup water, which is plant effluent. These two chemicals enter the solution piping at opposing flanges of a 12-inch cross.

This particular installation compelled the designers to consider a constant-strength solution because the point of application of chlorine is some 4000 feet from the point of generation. This is much too far to consider remote injectors.

This system's desirability is related in part to the incompressibility characteristic of water. Therefore, regardless of chlorine solution flow variation, the impact of this change occurs instantaneously at the point of application some 4000 ft away. This is accomplished by the production of a constant-strength chlorine solution at the chlorine station.

The Sanitation Districts of Los Angeles County, California-JWPCP.[55–57] This plant is located on South Figueroa Street, Carson, California. The chlorination system has gone through several stages of development and refinement since the Sanitation Districts engineers pioneered this project in 1957. White followed these developments closely because he believed this type of chlorine application had many merits. In 1990 and 1996 he was asked to review the system and report on its safety and reliability.

The chlorination facility is far removed from the wastewater treatment plant that discharges about 400 mgd, containing 60 percent secondary effluent and 40 percent primary effluent. This effluent discharges into a tunnel under the San Pedro Hills that goes into an ocean outfall. This outfall terminates in a sophisticated diffuser dispersion system about two miles offshore in the Pacific Ocean. Components of this system are illustrated by Figures 2-15 through 2-19.

Owing to its size—100,000 lb/day plus—the JWPCP system using liquid chlorine greatly reduces the size of the facility. Its success is based upon one of the basic characteristics of liquid chlorine, which is often misunderstood—its relatively high solubility in water when injected into water under pressure. At a water pressure of 30 psi and 70°F, its solubility is 10.5 lb/100 gal of water. So a 300-gpm stream of process water can absorb 31.5 lb of liquid chlorine per minute. The pressure in the hypochlorite pipeline at the point of chlorine injection is 30 psi at this installation. However, the secret to the ultimate success of this system was the untold man-hours of on-site experimentation

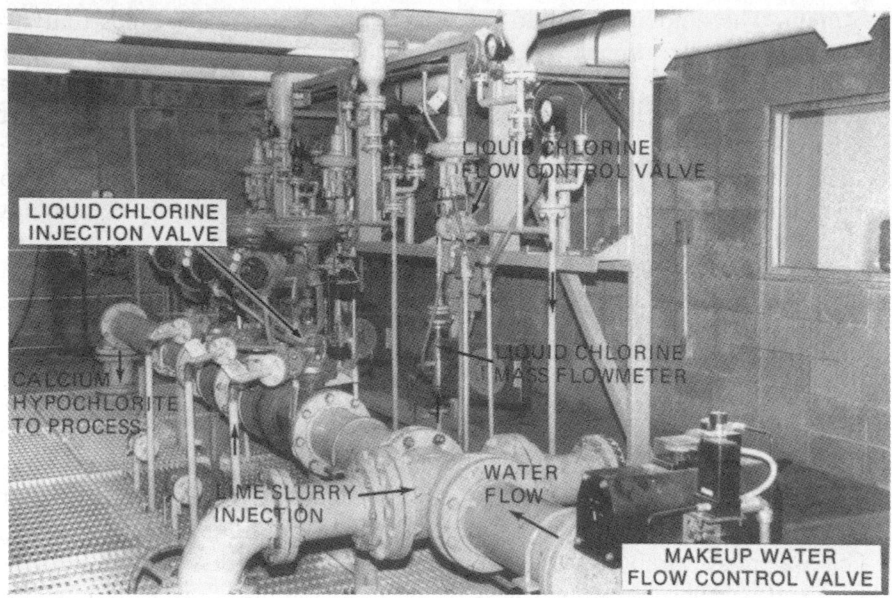

Fig. 2-15. On-site calcium hypochlorite production system showing the liquid chlorine and lime slurry injection system (courtesy County Sanitation Districts of Los Angeles County, California).

Fig. 2-16. Liquid chlorine injection valves (courtesy County Sanitation Districts of Los Angeles County, California).

Fig. 2-17. Chlorination leak scrubbers for the injection room illustrated in Fig. 2-15 (courtesy County Sanitation Districts of Los Angeles County, California).

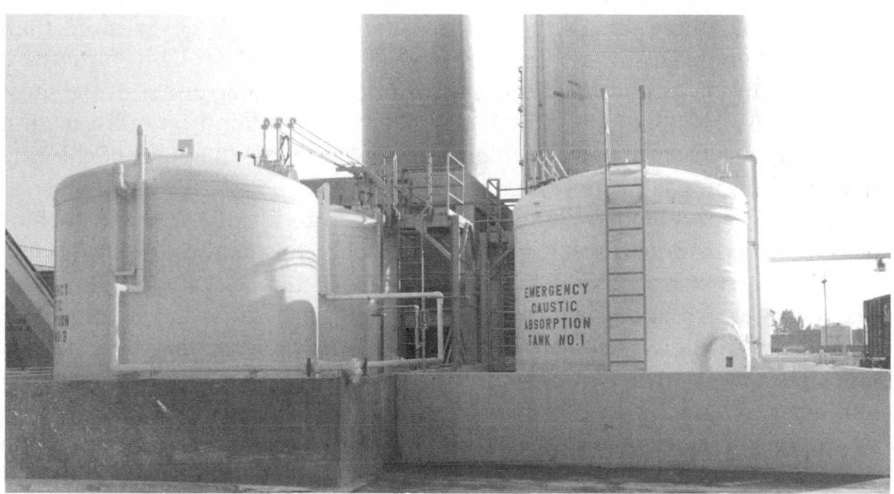

Fig. 2-18. Emergency caustic absorption storage tanks for railcar depressurization (courtesy County Sanitation Districts of Los Angeles County, California).

Fig. 2-19. Chlorine unloading area, showing the arrangement for two rows of three 90-ton railcars (courtesy County Sanitation Districts of Los Angeles County, California).

to discover the proper liquid chlorine pressure upstream from the control valves and at the injection point—plus having the right components to do the job.

This is the only known chlorination system for either drinking water or wastewater treatment in the United States (or anywhere) that applies liquid chlorine directly to the process stream. This statement highlights one of the many safety features of this type of installation. There is never more than 320 lb of chlorine in the system that is outside the railcar. This is the case when either of the two most northerly cars is on line. When either of the most southerly cars is on line, the amount is 195 lb. Another important observation is the complete absence of any evidence of off-gassing at the point of chlorine injection. This in itself proves the high solubility of liquid chlorine.

The final product in this system is a 7500 mg/L solution of calcium hypochlorite. A constant-strength solution is maintained by flow control of both liquid chlorine and makeup water. Lime slurry is added for pH control. The plant effluent is used as makeup water for the hypochlorite solution and is controlled by an automatic flow-rate valve downstream from the pumping system. This eliminates any control loop lag time in spite of the fact that the point of chlorine application is 4000 ft from the chlorination facility.

The slime slurry that makes up the hypochlorite solution is from pebble lime that is processed by lime slakers and stored in two 19,450-gal underground tanks. The slurry is pumped to the injection room by centrifugal pumps, and is controlled with eccentric ball valves.

One of the obscure features of this type of installation is the ease with which its capacity can be increased by simply increasing the strength of the hypochlorite solution. This requires a modest increase in both the liquid chlorine flow and the lime slurry. Since this facility was first put into operation, the capacity has been increased from 30,000 to 100,000 lb/day of chlorine as calcium hypochlorite. The only changes that were required were three new, more technically advanced liquid chlorine flow meters and an additional lime slurry pump.

Safety Features

Chlorine Scrubber System. Any chlorine leaks in the chlorine and lime slurry injection room are neutralized by a venturi scrubber that recirculates the air in the room and uses caustic as the neutralizing agent. Normal atmospheric ventilation in this room is shut off, and the scrubber system is then automatically started upon detection of chlorine by a conventional leak detector. The air is withdrawn at floor level and returned at the ceiling level. This unit is designed to bring the room air down to 1 ppm of chlorine, in the event of a chlorine release, in 2 hours.

Railcar Scrubber System. In 1995 a railcar leak containment structure, with a once-through caustic scrubber, was constructed that provided enough capacity to neutralize the contents of one 90-ton chlorine railcar. The system consists of a 10,000-gal caustic scrubber, a venturi, and blowers for assisted gas flow, and a 100,000-gal caustic tank, plus a containment that will operate at a negative pressure while the system is in operation. The system was designed and built by EST of Quakertown, Pennsylvania. It is capable of neutralizing a 15,000 lb/hr (4.17 lb/sec) chlorine leak, while not discharging more than 5 ppm chlorine in its exhaust.* (See Chapter 1, section on "Chlorine Leaks.")

Additional Safety Features. A vital part of the system is a dry air facility, which is used primarily to pad the railcars to the necessary operating pressure of 175 psi. Pressure switches are used on the liquid chlorine lines to open and close automatic block valves in the air-pad lines. The setpoints are chosen to always maintain a lower railcar pressure than the air-pad discharge pressure of the compressor, so that the chlorine cannot migrate up the air-pad lines. Dry air is also made available for purging the entire system. It is neatly arranged for operator convenience and on-site monitoring.

Expansion tanks in the chlorine liquid piping are appropriately located with both on-site and remote monitoring of system chlorine pressure. All the necessary sensors and alarm devices are strategically located at sensitive parts of the facility. The one-inch carbon steel piping that transports liquid chlorine from the railcar containment building to the injection room is double contained piping. Any piping leaks will migrate inside the outer jacket back to the

*No mention is made of what kind of a leak the system was designed for, or how the figure of 15,000 lb/hr was determined. There are many factors that control the size and length of a chlorine leak that must be considered.

containment building. The sensing and monitoring equipment consists of the most technically advanced instruments that are currently available.

Any municipality or other utility that might be seriously interested in the use of this unique system should consider it an imperative to make an on-site study of this facility. The operating personnel can provide information on the variety of problems that had to be corrected to achieve the accuracy and the reliability that this system currently enjoys. As for safety, it is definitely a role model for future installations of this kind.

Outline of System Components

Introduction. The following discussion is based upon the design created for the Sacramento, California Regional WWTP in 1974. This was designed as a Breakpoint Chlorination system to achieve nitrogen removal of a secondary effluent to supplement the biological nitrification–denitrification process. At the time, it was considered necessary only during cannery season when the biological treatment could not handle the cannery load. In the meantime, the EPA lowered its requirements so the breakpoint system was never built. This was a mistake because the disinfection chlorination system capacity requirements escalated so much in the following years that the on-site hypochlorite generating system would have been ideal, instead of ten 10,000 lb/day chlorinators and evaporators!

System Description. The on-site manufacturing process is a simple one. It merely involves the simultaneous injection of liquid chlorine and lime slurry or caustic solution into a transmission line carrying the makeup water. The hypochlorite solution produced is about 8000 mg/L in strength. It is normally poised at a pH of 7.5 to 8.5. Although this is an alkaline solution, it is preferable to transport the solution in corrosion-resistant piping such as PVC, KYNAR, or rubber-lined or Saran-lined steel pipe. The secret of a successful installation of this type is the ability to achieve complete and immediate mixing of liquid chlorine to prevent evolution of any chlorine gas.

The complete system includes storage facilities for chlorine tank cars, a 175 psi air-pad system with air driers, liquid chlorine flow control system, liquid chlorine injectors, makeup water supply system, pH control system, hypochlorite solution lines, and lime or caustic facilities.

Chlorine Facilities

Chlorine Supply System. The layout of railroad siding and the design of chlorine headers, loading platforms, air padding, air drying, expansion tanks, and so on, is the same as for liquid–gas systems using conventional chlorination equipment described in Chapter 9 with one exception: the air pad on the tank car must be continuous at 175 psi for reasons described below.

Reserve Tank. The same applies here to the use of the reserve tank concept described in Chapter 9. This ensures a continuous uninterrupted supply of liquid chlorine. Moreover, it eliminates the need for weighing devices.

It is also practical to use a resettable reverse totalizer on the readout signal provided by the liquid chlorine flow meter signal. With an accurate signal from the liquid chlorine rotameter to the flow meter totalizer, the reverse counter on the totalizer can be set to alarm at 500–700 lb left in the car. Then the operators should check with the totalizers on the flow meter to verify the amount left in the car. The alarm on the reserve tank serves as the final verification that the car is empty of liquid chlorine.

Chlorine Control System. It has been stated elsewhere in this text that it is impractical to measure liquid chlorine with conventional metering devices. The pressure drop across the meter, although very small, causes the liquid to flash to gas, destroying the meter's accuracy. This is true at the pressures usually encountered in conventional practice. The flashing phenomenon can be eliminated if the liquid supply is pressurized to 175 psi. At this pressure, Los Angeles County operators have confirmed that the liquid chlorine flow meter reading is as accurate as a similar meter would be if measuring Cl_2 gas or water flow. (See Fig. 2-15).

The chlorine control system can be designed to respond to either a plant flow signal or a combination of flow and chlorine residual. This is the same as a conventional system. The major difference is that the hypochlorite makeup water flow rate, rather than the chlorine feed rate, is the primary response to changes in plant flow and/or chlorine residual. The chlorine feed rate follows as a secondary response.

The flow rates of liquid chlorine and lime slurry (or caustic) are varied automatically to comply with changes in the makeup water flow, so as to maintain a constant concentration of hypochlorite solution going to the point of application.

Chlor-Alkali Mixing. This is the key to a successful operating system. In order to achieve a stable hypochlorite solution without the evolution of molecular chlorine, a constant back pressure of 35–40 psi must be maintained at the point of liquid chlorine injection. The alkali solution (CaO slurry or NaOH) is injected diametrically opposite the chlorine injection.

The high head loss from the artificially padded liquid chlorine line upon discharge from the special chlorine injection valve, permits the chlorine to vaporize immediately inside the makeup water pipe. If the pressure drop across the injection valve and the back pressure in the hypochlorite solution line are not maintained, a complete and stable solution will not be produced. The necessary back pressure can be sustained by the installation of an automatic pressure regulating valve in the solution line downstream from the mixer. If the hypochlorite solution is discharged through more than one diffuser, each diffuser header must be equipped with a pressure regulator. A single regulation in the mainline is not sufficient to keep the proper back pressure and provide uniform flow when more than one diffuser is involved.

Solution Line and Diffusers. These appurtenances should be constructed of the same materials and designed for the same hydraulic conditions as for a conventional chlorination system. (See Chapter 8.)

Although the alkaline hypochlorite solution is mildly corrosive, it will be necessary at prescribed intervals to purge the calcium scale deposit in the solution lines caused by the use of lime slurry. Removal of this scale deposit is easily accomplished by lowering the pH of the hypochlorite solution (temporarily) to about 4 or 5. Obviously these procedures require corrosion-resistant piping, valves, and fittings.

Makeup Water System. The water to make the hypochlorite solution should be chlorinated plant effluent, even if the ammonia nitrogen (NH_3-N) content is high. A small portion of chlorine injected into this water will be consumed immediately by the ammonia nitrogen in the breakpoint reaction. This reaction between Cl_2 and NH_3-N is complete in a few seconds at the pH and chlorine concentrations used in this method. For example, 350 gpm of makeup water containing 20 mg/L NH_3-N will consume 834 lb/day chlorine in the breakpoint reaction. For an 8000 mg/L hypochlorite solution, this represents 3 percent chlorine loss with 20 mg/L NH_3-N present in the makeup water.

A variable speed booster pump is essential for the makeup water system. Instantaneous response of chlorine dosage rates is made possible by holding a constant hypochlorite concentration in the solution line. Therefore, the rate of flow in the solution line is varied according to plant flow and residual changes. Regardless of solution line length, chlorine dosage changes at the chlorination station are reflected immediately at the point of application.

If the system is designed to produce a hypochlorite solution with a chlorine concentration of 7000–8500 mg/L, the makeup water flow would have to be about 240–290 gpm/1000 lb/hr chlorine feed rate.

Essential instrumentation for this system requires that a flow meter be installed in the common discharge of the makeup water line. The flow signal from this meter is combined through a ratio station with the chlorine flow signal to provide a signal to the hypochlorite concentration flow recorder.

Alkali System. The chemical of choice is pebble lime; however, some systems designed for nitrogen removal where peaks cannot be handled by biological processes would be better suited to a caustic installation, particularly if the intermittent operation is 70–75 days or less per year. The choice is mainly one of economics, but from a chemical standpoint lime may be preferable because the calcium ion present provides a protective coating on the hypochlorite solution line.

Burned pebble lime costs about $40/ton as compared to caustic soda at about $170/ton. Even though the storage and handling system for lime is more expensive than that for caustic, the cost-effectiveness is heavily in favor of lime. In the quantities required for this type of operation, quicklime is preferable to hydrated lime; however, source of supply and chemical cost would still govern the choice. Less mechanical equipment is required for the hydrated lime system, and the least mechanical equipment is required for the caustic soda system. However, the overall costs invariably favor the quicklime system.

Quicklime is readily available by truck or in railroad carload lots. It can be conveyed pneumatically to storage silos with ease and without dust or other air pollution problems. Proper quicklime (pebble) handling equipment includes the following:[36]

1. Air compressors for vacuum and pressure requirements.
2. Transfer piping.
3. Pressure–vacuum separation device (lime is removed from delivery vehicle by vacuum and delivered to storage by pressure).
4. Bulk storage facilities.
5. Filters on air intakes and discharges.
6. Monitoring, alarm, and control devices.

The pneumatic conveyor system for unloading quicklime should have interlocking safety devices to shut down the system if excessive vacuum or pressure develops. It should be further interlocked with level measuring devices in the lime storage tank to cause the conveyor to shut down when a full tank is reached. Low-level warning devices should also be installed in the storage tanks.

Feeding of bulk lime to the slaker is normally done by gravity. The slaker package consists of a slaker, lime feeder, control panel, and accessory control valves. The lime feeder should be equipped with a variable speed drive motor so that it can be manually adjusted to run at any of the numerous lime feed rates. Automatic operation of the slaker–feeder combination can be achieved, by a level probe system in the lime slurry storage tank, to activate or deactivate the combination as low or high levels are attained. Manual adjustment of the lime feeder to the slaker is sufficient to supply slurry of the proper concentration. The slaker equipment normally includes grit removal components, and the overall system must include belt or other conveying equipment to move this material to storage facilities for disposal. The bulk storage and slaking facilities can be eliminated by purchasing liquid slurry, but at greatly increased costs.

Lime slurry from the slakers flows by gravity to lime slurry storage tanks where the lime is kept in suspension by turbine mixers. Lime slurry from the storage tanks is pumped by a variable speed progressive cavity positive displacement pump into the hypochlorite pipeline for combination with chlorine. A piston or diaphragm positive displacement pump is not satisfactory for this application. The lime slurry pump pumps lime as demanded by a pH control system, which samples the contents of the hypochlorite pipeline and signals the lime pump to pump as required to maintain a preset pH level required for satisfactory hypochlorite production. To maintain a slightly alkaline pH, approximately 0.80 ton of CaO is required per ton of Cl_2.

Reliability. The low intial cost of all the mechanical equipment and piping lends itself to complete duplication of the entire system. This provides a system reliability comparable to or better than that of most conventional systems. Power failures affect this system with equal results to those in a conventional system, whether it be on-site generation or imported hypochlorite.

Cost Comparison. The disinfection facility was designed for 110 mgd. The nitrogen removal process by breakpoint chlorination was estimated to require 70 days of operation per year.

The comparison plant was designed for 60 mgd to produce a commonly nitrified secondary effluent using on-site generated chlorine for nitrogen removal.

Scaling the above two plants to a common size, the following is an economic evaluation of the on-site hypochlorite manufacturing (Los Angeles System) versus the on-site generated system. This comparison is of considerable interest owing to the enormous dollar savings in capital cost of chlorination equipment, particularly where chlorine usage approaches tank car quantities.

The comparative costs are as follows:[37]

		Annual
	Cap. Cost $	Oper. Cost $
Nitrification–denitrification	5,640,000	1,250,000
Breakpoint chlorination	32,000	1,295,000

Amortized capital costs (25 years at 7 percent) plus operating* and maintenance costs based on 1974 prices—$/mg (annual average):

On-site hypochlorite manufacturing: $37.00.
On-site hypochlorite generation: $107.00.

For a similar plant that would reduce the ammonia nitrogen (biologically) continuously by nitrification followed by denitrification, the following are the relative costs based upon the Sacramento analyses. If nitrogen removal by chlorine is a requirement for a given effluent, or if the chlorine dosage require-

*Costs are based on chlorine @ $144/ton and caustic soda ca. $168/ton. A system using lime for caustic buffer would be even less costly.

ments for disinfection approaches a maximum of 10,000–12,000 lb/day, the *on-site hypochlorite manufacturing* system is of considerable interest. All of the cost figures for the Sacramento design (which was for only 70 days operation per year) were based upon the most expensive alkali system, that is, caustic soda instead of lime. It should further be noted that this system of hypochlorite manufacture must be as described because it is not feasible without the addition of alkali.

Practical Considerations. Because only a single operating system falls into this descriptive category, one may hesitate regarding its feasibility. The answer to this question is straightforward and unequivocal. The first such unit was put into operation about 1958. This was a much smaller system than the prototype, but all of the instrumental anomalies were sorted out in the first system so that the current system enjoys either continuous or intermittent operation, whichever is required, with the same flexibility that is enjoyed by conventional chlorination systems.

Manufacture in Underdeveloped Areas

Many underdeveloped areas of the world suffer great losses in human life from such waterborne diseases as cholera, typhoid, dysentery, and virus infections. In these areas, the cost of transporting a disinfectant over long distances makes the use of chlorine a luxury. Therefore, an alternative method for producing hypochlorite with materials available locally is presented here to show how it can be done without electric power and without chlorine gas as a raw material. The following discussion is based upon an operating installation described by Stone[38] in 1950. The materials available were common salt and manganese dioxide, mined nearby. Figure 2-20 illustrates this installation.

About 85 miles from where the hypochlorite was to be made was a sulfuric acid plant. There was also available locally a supply of low-grade slaked lime. With these materials it is possible to make a bleaching powder with about 35 percent available chlorine. Here is how it is done:

The manganese dioxide and common salt are mixed and placed in a reaction tank, which is suspended above an open type water boiler. Above the chemical reaction tank is a sulfuric acid tank with a control valve that allows regulation of the flow of acid to the reaction tank (Fig. 2-20). The rate of chlorine gas generation is regulated by the flow of sulfuric acid and the temperature of the water. The heated water, in turn, heats the chemicals in the reaction tank, accelerating the following reaction:

$$MnO_2 + 2NaCl + 2H_2SO_4 \rightarrow Cl_2 \uparrow + MnSO_4 + Na_2SO_4 + 2H_2O \quad (2\text{-}13)$$

The proportion of these chemicals required is directly related to their molecular weights.

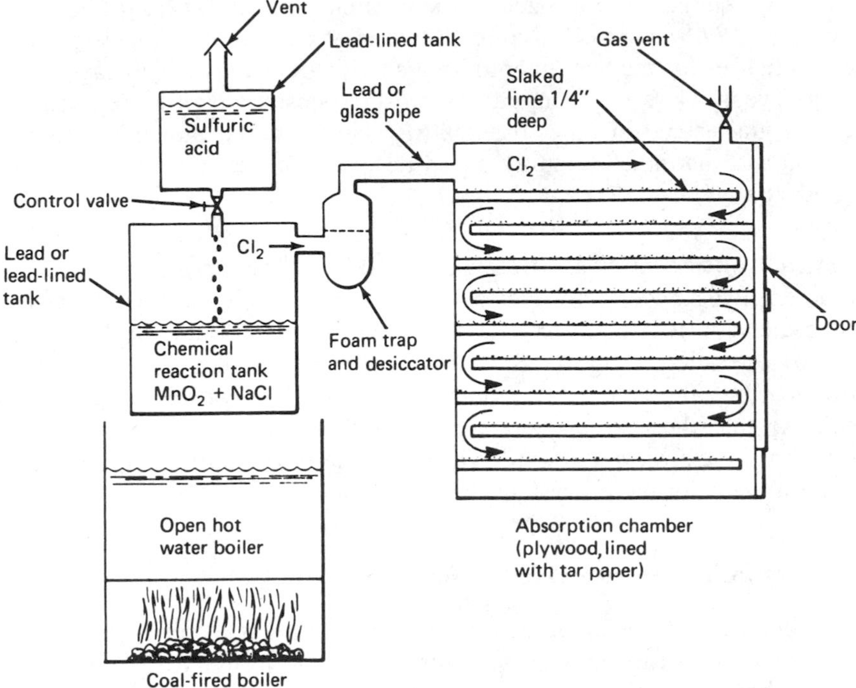

Fig. 2-20. Manufacture of hypochlorite in underdeveloped areas.[38]

The chlorine gas generated is passed through a foam trap and desiccator, where it is dried before going to the absorption chamber. The absorption chamber consists of several trays containing a quarter-inch layer of carefully slaked lime. The chlorine enters at the top of the chamber and, being heavier than air, passes downward over the trays, reacting to form bleaching powder:

$$Cl_2 + Ca(OH)_2 \rightarrow CaOCl_2 + H_2O \qquad (2\text{-}14)$$

The resulting product contains about 70 percent calcium oxychloride; therefore when this powder is dissolved in water, it hydrolyzes to form calcium hypochlorite and calcium chloride:

$$\underset{\text{calcium oxychloride}}{2\,CaOCl_2} \rightarrow \underset{\text{calcium hypochlorite}}{Ca(OCl)_2} + \underset{\text{calcium chloride}}{CaCl_2} \qquad (2\text{-}15)$$

An outlet or a vent valve in the top of the absorption chamber is left open until the chlorine gas has displaced the air. It is closed when a chlorine gas odor can be detected at this outlet. The generation of gas is allowed to proceed until the chemical reactions are completed. It was found that this method, with all its primitiveness, would produce about 30 lb of bleaching powder with 35 percent available chlorine in 12 hours of operation. This particular installation, limited only by its physical dimensions, could produce enough chlorine to treat a little more than 1 mgd at 1.0 ppm chlorine dose.

REFERENCES

1. Sconce, J. S., *Chlorine: Its Manufacture, Properties and Uses,* Reinhold, New York, 1962.
2. Mond, L., "History of Manufacture of Chlorine," *J. Soc. Chem. Ind.,* 713 (1896).
3. Baldwin, R. T., "History of the Chlorine Industry," *J. Chem. Ed.,* 313 (1927).
4. Sawyer, C. N., "Hypochlorination of Sewage," *Sewage Ind. Waste,* **29,** 978 (1957).
5. Fair, G. M., Morris, J. C., and Chang, S. L., "The Dynamics of Water Chlorination," *J. NEWWA,* **61,** 285 (1947).
6. White, G. C., "Disinfection Practices in the San Francisco Bay Area," *J. WPCF,* **46,** 89 (Jan. 1974).
7. Krusé, C. W., Olivieri, V. P., and Kawata, K., "The Enhancement of Viral Inactivation by Halogens," *Water and Sewage Works,* **118,** 187 (June 1971).
8. White, G. C., private communication, K. Fraschina and A. Bagot, City and County of San Francisco, 1971.
9. Rietema, K., "Segregation in Liquid–Liquid Dispersions and Its Effect on Chemical Reactions," *Chem. Eng. Sci.,* **8,** 103 (1958).
10. Anon., "Chlorine Bleach Solutions," Solvay Div. Allied Chem. Co. Bull. 14, New York, 2nd Ed., 1960.
11. Anon., "Chlorine," Columbia Southern Chemical Corp., Pittsburgh (now PPG Ind.), 1952.
12. Baker, R. J., "Characteristics of Chlorine Compounds," *JWPCF,* **41,** 482 (1969).
13. "Chlorine Bleach Solutions," Solvay Div. Chem. Co. Bull. 14, New York, 2nd Ed., 1960.
14. "Chlorination of Sewage With Hypochlorites," Dow Chem. Co. Form #125–1086–68, 1968.
15. Anon., Lithium Corp. of America, Product. Bull., New York, 1966.
16. Kirk, B. A., and Lindeke, W. A., "A New Pool Sanitizer Discussed by Maker," *Sw. Pool Age,* 36 (Jan. 1968).
17. White, G. C., unpublished data, shipboard chlorination systems, 1942.
18. Chang, S. L., and Fair, G. M., "Destruction of *Entamoeba histolytica,*" *J. AWWA,* **33,** 1705 (1941).
19. Chang, S. L., "Destruction of Microorganisms," *J. AWWA,* **36,** 1192 (1944).
20. Anon., "*Manual of Naval Preventive Medicine,*" Dept. of Navy, Bur. of Med. and Surgery, Chap. 6, part IV, Washington, DC, Dec. 1962.
21. Bacon, V. W., "Chicago MSD Progress Report on Chlorination," *Water and Sew. Wks.,* **114,** 350 (Sept. 1967).
22. Springs, J. D., "Hypochlorination for Slime Control," *Power,* 102 (June 1957).
23. "Wallace and Tiernan 44 Series Solution Metering Pumps," Cat. File 440, 100 Rev. 5–76.
24. Peterson, R., BIF, private communication, Walnut Creek, CA, 1977.
25. Pulsafeeder Catalog No. 7120 "Pulsa 7120," Rochester, NY, 1976.
26. Straight, W. S., Union Carbide Co., personal communication, Oct. 12, 1976.
27. "Feedrator II Liquid Chemical Feed Dispenser," Fischer and Porter Co. Warminster, PA, Cat. 70 FR 1000, 1977.
28. Barbolini, R. R., Metropolitan Sanitary District of Greater Chicago, personal communication, Nov. 4, 1976.
29. Resistoflex Corp., "Flexible and Rigid Piping Accessories," Bull. Sk-5 Roseland, NJ, 1975.
30. Barbolini, R. R., private communication, Aug. 4, 1970.
31. K. A. Steel, Chemicals Inc., Melrose Park, IL., private communication, Feb. 15, 1983.

32. Halter, C., Deputy commissioner of Water Operations, City of Chicago, private communication, Dec. 23, 1982.
33. Knox, D., Wallace & Tiernan Div. Pennwalt Corp., private communication, Houston, TX, Dec. 1982.
34. Jones Chemical Co, Milpitas CA, private communication, 1983.
35. Murray, B. W., "A Calcium Hypochlorite Manufacturing Process for Water Treatment Plant Use," *Water and Sew. Wks.,* **100,** 318 (Sept. 1963).
36. Nagel, C., private communication, Sanitation Districts of Los Angeles County, 1972.
37. In-house analysis by Brown & Caldwell Engineers and G. C. White acting for Sacramento Area Consultants, County of Sacramento Regional Plant, 1975.
38. Stone, R., "The Small Scale Manufacture of Bleaching Powder in Backward Areas," *J. AWWA,* **42,** 283 (1950).
39. Bacon, V. W., "How Chicago Saved $2.5 Million," *American City,* 16 (Oct. 1967).
40. Dorolek, R. J., "Wastewater Plant Effluent Chlorination Made Easy and Inexpensive," *Water and Wastes Eng.,* **48** (Oct. 1968).
41. Tech. Practice Committee, Water Pollution Control Federation MOP No. 4, "Chlorination of Wastewater," Washington, DC, 1976.
42. Lue-Hing, C., Lynam, B. T., and Zenz, D. R., "Wastewater Disinfection: A Case Against Chlorination," paper presented at Forum on Disinfection With Ozone, Chicago, IL, June 2–4, 1976.
43. Chinn, R. Supt. North Point Treatment Plant, San Francisco, private communication, Nov. 1982.
44. Davies, J. W., Cox, K. G., and Symon, W. R., "Typhoid at Sea: Epidemic aboard an Ocean Liner," *Can. Med. Assoc. J.,* **106,** 877 (1972).
45. Merson, M. H., MD, Hughes, J. M., MD, Wood, B. T., Yashuk, J. C., and Wells, J. G., "Gastrointestinal Illness on Passenger Cruise Ships," *J. Am. Med. Assoc.,* **231,** 723 (Feb. 17, 1975).
46. Merson, M. H., Hughes, J. M., Lawrence, D. N., Wells, J. G., D'Agnese, and Yashuk, J. C., "Food and Waterborne Disease Outbreaks on Passenger Cruise Vessels and Aircraft," *J. Milk Food Technology,* **39,** 285 (Apr. 1976).
47. Center for Disease Control, "Diarrheal Illness among Passengers on the Cruise Ship *M/S Sun Viking,*" Public Health Service, CDC, Atlanta, GA, Mar. 8, 1977.
48. Center for Disease Control, "Gastrointestinal Illness, *TSS Fairsea,*" Public Health Service, CDC, Atlanta, GA, Nov. 7, 1977.
49. Davis, W. A., "Many Luxury Liners Fail U.S. Sanitation Test," *Chicago Tribune,* Apr. 16, 1978.
50. Blumenthal, R., "Many Cruise Ships Fail Health Tests," *New York Times,* Dec. 19, 1977.
51. Borcover, A., "Ship Violations Listed," *Chicago Tribune,* Jan. 1979.
52. Clary, M., "U.S. Inspections Help to Improve Ship Sanitation," *Chicago Tribune,* Sept. 30, 1979.
53. Brower, G. R., "A Chlorine Gas Cloud From Sodium Hypochlorite?," *AWWA Op Flow,* **10,** 3 (Mar. 1984).
54. Barbolini, R., Private communication, Metropolitan Water Reclamation District of Greater Chicago, Apr. 1991.
55. White, G. C., personal inspection, 1991.
56. Friese, P., and La Roche, J., private communication, County Sanitation Districts of Los Angeles County, Mar. and Apr. 1996.
57. White, G. C., personal inspection, 1996.
58. Mazza, P., Operations Division San Francisco Water Department, personal communication, 1996.
59. Clifford, K. L., *Hammonds PPG Tablet Chlorination System,* Hammonds Technical Services, 15760 Hardy Road, Suite# 400, Houston, TX 77060, 1996. Customer Service Ph. 800-582-4224. Or PPG Industries, One PPG Place, Pittsburgh, PA 15272.
60. Anon., *Premia 75 Solenoid Metering Pumps;* Wallace & Tiernan Co. Inc.., 1901 W. Garden Road, Vineland, NJ 08360. Customer Service Ph. 609-507-9000.

3
ON-SITE GENERATION OF CHLORINE

HISTORICAL BACKGROUND

The Beginning

From time to time interest is revived in the possibility of producing chlorine electrolytically at the point of use, thereby eliminating the potential hazard of chlorine stored on-site in containers. One of the first installations of this kind was at Brewster, New York in 1893. The installation was known as the Woolf process. This process is particularly efficient where saline water is present. Therefore, the Central Electricity Generating Board of London experimented with such systems as an alternative method of chlorinating steam power plant condenser cooling water at electric generating stations. They found that the method was inefficient and beset by a great many operating problems, and the experimentation was terminated.

A little-known device that has since been very helpful to a lot of us is the instrument that Mr. Wallace, of Wallace & Tiernan, made for doctors to treat patients with a cold. And who hasn't had a cold? The device was created about 1920, but he decided to take it off the market when doctors began using it for a lot of things that were not the common cold.

White did not find out about this instrument until the 1950s. He went right to Mr. Wallace to get all the details, so that he could compare his own cold remedy chlorination system with Wallace's unit. (It must be okay because White has not had a cold in many years.)

On-site generation utilizing the electrolytic process dates back to the 1930s, when Wallace & Tiernan made electrolytic chlorinators for YMCA swimming pools. Success in this field has always been marginal because the cost of the electrolytic equipment is always much greater than that of conventional equipment using bottled chlorine gas.

Electrodes were always a major source of trouble in the electrolytic systems. The most common of these have been platinum-coated; others have been made of carbon and iron, graphite with a lead shield inside a stainless steel sheath, and titanium electrodes coated with rare metals. The process is inefficient at best but it does have interesting possibilities, in particular the safety factor and the fact that raw materials might all be at the point of application, thereby eliminating storage requirements. However, one of the current difficulties is the ever increasing cost of the electric power required for electrolytic production.

Early Experience in the United States

On-site generation of hypochlorite in the United States was largely inspired by the use of hypochlorite solution during World War I (1914–1918). This solution became known as the Carrell–Dakin solution.[1] Its success in the antiseptic treatment of open wounds led to the on-site generation of this solution in hospitals. One of the first electrolytic cells for this purpose was developed by Van Peursem et al.[2] This cell was designed to produce the equivalent of the Carrell–Dakin solution.[3]

Wallace & Tiernan first made electrolytic chlorinators to provide a safe means of swimming pool chlorination for those installations where pools were located in buildings where people slept. As early as 1939 Wallace & Tiernan established a policy that chlorine gas equipment should not be installed in such buildings. For this purpose they developed an electrolytic chlorinator.[4]

This unit aroused the interest of Pan American Airways, which, at the time (1936), was developing refueling sites on its San Francisco-to-Sydney and Orient flights. Use of the electrolytic chlorinator for water supplies at these way stations was ideal. World War II changed all of these plans.

Current Interest

After World War II, the enthusiasm for on-site generation disappeared until the hazard potential of chlorine gas stored in containers was evaluated, owing to the proliferation of chlorine gas installations at wastewater and potable water treatment plants. At about the same time small electrolytic generators began appearing on the market (1950s) for use in backyard swimming pools. The cost of these units and the manufacturer's inability to provide satisfactory service discouraged the use of this equipment.

In the 1970s the popularity of on-site generation began to rise once again, largely because of the potential hazards of liquid–gas systems using chlorine stored in containers and the availability of federal funds for the necessary research and development of reliable equipment.

Now in the 1990s, after the advent of the unrealistic *Uniform Fire Code,* there has been a great surge in the interest of on-site generation. There is very little danger, if any, that an on-site generator could produce any situation that could be classified as a *major leak.* Moreover, current on-site generation systems produce chlorine solutions containing only 0.8 percent chlorine, and storage of this concentration of chemical is not classified as a hazardous. So these systems start off with a *clean bill of health.* Operating personnel are very favorably disposed toward these systems because the operation of this equipment does *not* require any special training in the use of *hazardous material.*

IMPORTANCE OF RAW MATERIAL

Seawater Systems

All of the on-site generators described in this text are designed primarily to use seawater, but they can also use brine. There is one exception: the Ionics, Inc. Cloromat unit, which uses a conventional membrane cell that produces both chlorine and caustic. This results in a conventional sodium hypochlorite solution. The seawater units can use a brine solution instead of seawater. The brine solution concentration is limited to 30,000 mg/L, and whatever goes into these units is discharged into the process stream that is to be treated. Therefore whether it is seawater or brine, the TDS concentration in the generator discharge is going to be about 30,000 mg/L. This limits the application of these "seawater" units to disinfection of wastewater, cooling water chlorination, off-shore well drilling installations using chlorine to control marine growth, and other such marine installations. Only under special circumstances can they be used for potable water treatment and water reuse applications (see the "Summary" of this chapter). They are not applicable to hydrogen sulfide control in sewage collection systems because seawater exacerbates the formation of H_2S in sewage.

These units are designed to be able to use saline waters without any pretreatment except screening or microstraining. However, in some instances filtration has proved to be necessary.

Seawater varies in composition from ocean to ocean. The TDS ranges from 30,000 to 36,000 mg/L, and 19,000 mg/L chlorides is about average. Table 3-1 shows the composition of seawater according to the Hydrographic Laboratory of Copenhagen.

Brine Systems

The quality of the raw material is of great concern in the operation of any electrolytic process. Chlorine manufacturers have long realized that successful cell operation is dependent upon the use of a pure brine. The best product,

Table 3-1 Composition of Standard Seawater

Cations	*mg/L*	*Anions*	*mg/L*
Na^+	11,035	Cl^-	19,841
Mg^{++}	1,330	$SO_4^=$	2,769
Ca^{++}	418	HCO_3^-	146
K^+	397	Br^-	68
SR^+	14	F^-	1.4

Total salinity, 36047 mg/l
Total alkalinity, 119.8 mg/l

which needs very little pretreatment except for the brine makeup water, is food-grade salt, which is mostly refined solar salt extracted from seawater by evaporation. Solar salt that is not refined must be treated by the on-site system. This is known as "stack" salt.

Then there is mined salt, or brines naturally occurring in the earth's crust. This material, unless it is of exceptional quality, also has to be treated at the site. Any underground brines should be checked for ammonia nitrogen (NH_3-N) content, as this is most undesirable impurity in the electrolytic production of chlorine.

Salt or brine impurities seriously affect the operation and maintenance of any type of membrane cell. The allowable calcium ion content in the brine must be less than 0.05 mg/L according to Eltech. (See Chapter 1.)

EQUIPMENT DEVELOPMENT

Membrane Cell

Circa 1968–70, the U.S. Environmental Protection Agency became actively concerned about the environmental impact of storm water overflows (from large combined sewer systems) discharging into confined receiving waters such as the Charles River, San Francisco Bay, Chesapeake Bay, and so on. These massive overflows have long been considered hazardous to health (in water contact sports and shellfish growing areas) in the absence of disinfection. Because the application of chlorine would have to be at the point of storm water overflow, the chlorine containers would surely have to be transported through congested areas, thus creating the potential hazard of a chlorine spill.

The EPA, considering these factors, funded a study for on-site generation of chlorine carried out by Ionics, Inc., of Watertown, Massachusetts.[5, 6] This project resulted in the development of an extremely efficient electrolytic cell similar to, but more advanced than, some of the older designs currently used by manufacturers of chlorine gas. Details covering the development and features of this cell are well documented.[5]

The two-compartment membrane cell with expanded electrodes as developed by Ionics, Inc. is illustrated in Fig. 3-1. The most important feature of this cell is the membrane, which separates the anode compartment from the cathode compartment. This membrane separator concept is not new. Hooker Chemicals and Plastics Corp. began a program in 1950 that later culminated in the introduction of the MX chlor-alkali cell for commercial production of liquid–gas chlorine.[7] The concept reached economic feasibility with the availability from DuPont of Nafion membrane material. The Nafion-family membranes consist of a 2–10-mil-thick film of perfluoro-sulfonic acid resin; a copolymer of tetrafluoroethylene; and another monomer to which negative sulfonic groups are attached. (For more detailed information on the development of membrane cells, see Chapter 1.)

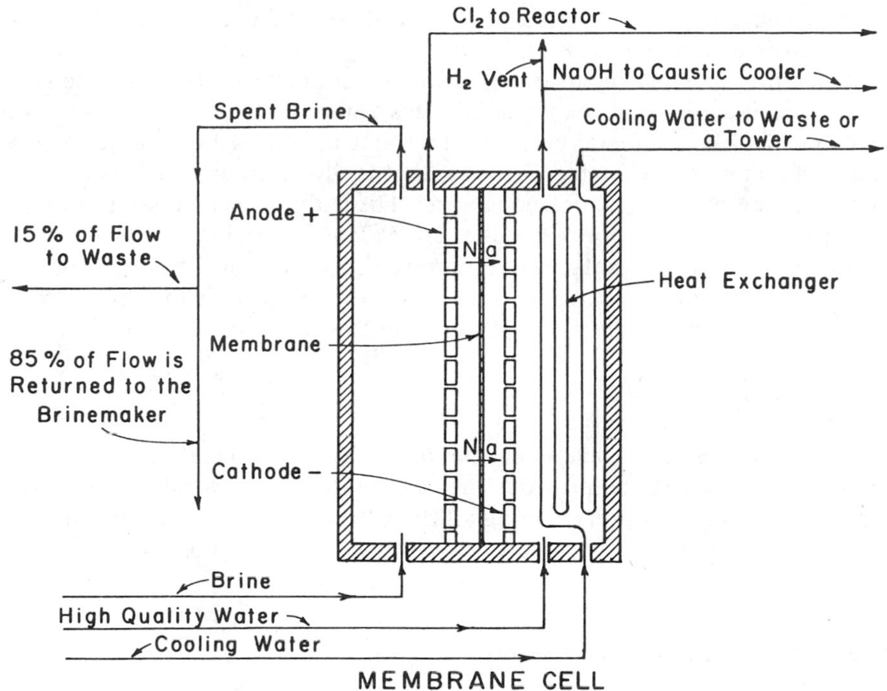

Fig. 3-1. Cloromat membrane cell with expanded electrodes.

The world's first *commercial membrane cell* chlor-alkali plant has been operating successfully since April 1975 for Asahi Chemical Co. at Nobeoka, Japan using Nafion 315 membrane cells.[8] However, a perfluorocarboxylic acid membrane developed by Asahi Chemical has been reported to give a higher current efficiency and is being used to replace the Nafion membranes.[8]

In a membrane cell (Fig. 3-1) the anode and the cathode compartments are separated by the cation-exchange membrane. This membrane inhibits negative ions (anions) from moving through the membrane but allows the positive ions (Na^+, K^+, H_3O^+ cations) to move freely. This effect is known as the Donnan exclusion.

There is no direct hydraulic flow from the anode compartment to the cathode compartment. The only water that passes through the membrane is endosmotic water, which is associated with the ions being transferred. The anolyte liquor (spent brine) can be sent to waste or partially recycled. Chlorine from the anode chamber is sent to a water-cooled reactor where it is mixed with the caustic solution.

Brine is fed to the anode compartment. Water is fed to the cathode compartment to sweep out the sodium hydroxide (caustic) that is produced. The cathode compartment is cooled by water flowing through a heat exchanger.

The caustic and the hydrogen produced in the cathode compartment discharge from a common port. The hydrogen is vented to the atmosphere.

The cathode compartment cooling water cannot be scale-forming. The electrodes are described as "expanded." This refers to their shape. They are rectangular pieces of metal perforated so that they look like a grating. The anode is dimensionally stable (DSA®); that is, it is nonsacrificial. It is made of pure titanium with a platinized coating. The cathode has the same configuration as the anode except that it is sacrificial, as it is made of iron.

One of the major advantages of a membrane cell is that the anode and the cathode can be placed close to the membrane. This increases the current efficiency and reduces the space required in stacking the cells side by side.

Ionics, Inc. Guide[27]

This is a brief announcement of the *Ionics Environmental Water Quality Monitoring Guide,* which provides the latest information available from the Ionics Instrument Division covering all the methods of water quality analysis. A brief description of this guide can be found in Chapter 6. The guide first appeared in 1993.

The Ionics, Inc. Cloromat System

Introduction. Something of great importance to users of all on-site generation systems was discovered after two of the Chloromat installations in the San Francisco Bay Area had been in operation for about two years in the 1950s: the need to separate the personnel responsible for generating the chlorine from those operating the treatment plant, who were responsible for the application of chlorine. The plant operator has enough to do without having the responsibility for seeing if the chlorine supply were sufficient, and in proper operation (Ionics, Inc. uses the membrane cell *described above*) for chlorine generation. The cells are assembled in modular form as in a filter press. The cells within a module are connected in series electrically, but connected in parallel hydraulically. Manifolds molded into the cell frames provide the connections for parallel fluid flow to and from the cells. The modulc and individual cells are designed for easy dismantling and reassembly required for annual maintenance. To increase the production of a unit, cells are added to the cell module in increments of about 165 lb/day chlorine per cell. A 2500-lb unit consists of 15 cells. The system is designed to produce an 8 percent hypochlorite solution. For this strength solution a 2500 lb/day unit requires about 2 gpm of high-quality water to sweep out the caustic from the catholyte compartment of the 15 cells involved. The flow diagram of a 2500 lb/day unit is shown in Fig. 3-2. The brine maker requires about 1 gpm so that the total amount of water to be produced by the water treatment unit is on the order of 3–4 gpm.

A water supply of 25 gpm is required for cooling the cells, the caustic cooler, and the chlorine–caustic reactor (see Fig. 3-2). This must be a water supply completely free of any scale forming substance.

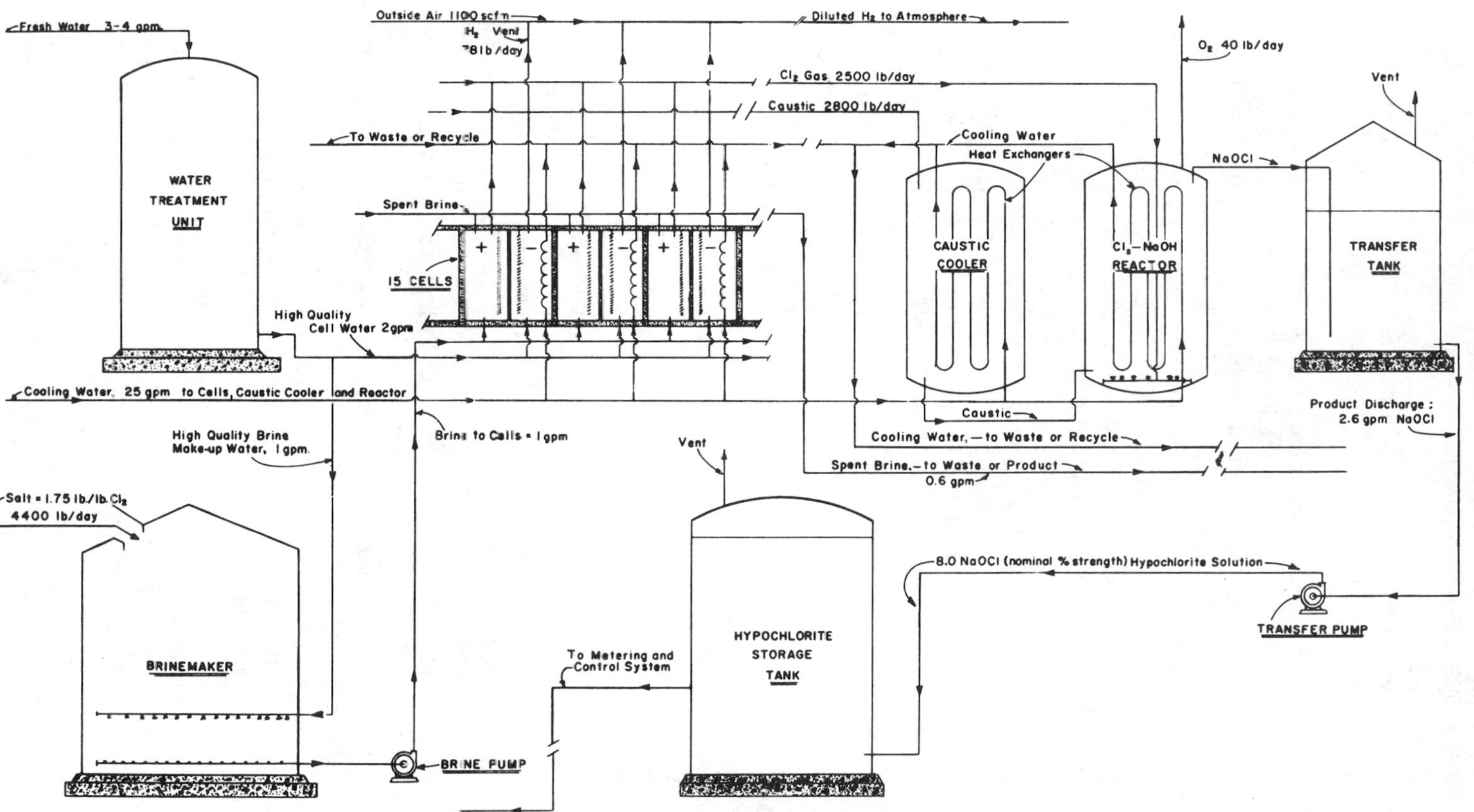

Fig. 3-2. Ionics, Inc. Cloromat hypochlorite generating system 2500 lb/day chlorine capacity.

The brine solution pumped from the brinemaker to the cells is about 1 gpm containing 3 lb/gal NaCl (4400 lb salt/day). This calculates to 1.75 lb salt/lb chlorine.

The spent brine should be discharged to waste unless recycling is planned as part of the system. Recycled spent brine must be subjected to special treatment before it is allowed to return to the system. The spent brine flow is about 0.6 gpm for a 2500 lb/day unit (see Fig. 3-2).

All of the mass balance figures shown in Fig. 3-2 are based upon food-grade salt. This represents 1.75 lb NaCl/lb chlorine, at 2.0 kWh/lb of chlorine. In the San Francisco area electrical energy costs about 8.5 cents/kWh.

When considering this system the designer must be careful to specify the grade of salt to be supplied. According to Ionics, Inc.,[9] food-grade salt, which is refined, is so free from the usual impurities normally associated with sodium chloride that the hypochlorite solution thus produced is characterized as "Rayon Grade Bleach." This means that this hypochlorite does not contain the usual impurities (heavy metal ions) that contribute to the instability of the hypochlorite solution during average storage conditions. For example, the usual 10 percent commercial bleach (NaOCl) deteriorates from 10 to 8.5 percent in 40 days as compared to the Cloromat 8.43 percent solution, which deteriorates to 7.5 percent in the same length of time. However, it becomes a question of economics whether or not food-grade or stack-grade (unrefined) solar salt should be used.

Stack-grade salt requires a pretreatment system, which, in terms of power, is estimated at 0.32 kWh/lb chlorine.

Food-grade salt in the San Francisco Bay Area at the harvester costs (1977) 2.0 cents/lb, stack grade is 1.0 cents/lb.

The Cloromat system is a proprietary manufacturing process. When it is compared to other on-site processes, there will always be claims and counterclaims for the various processes.* The designer is therefore warned to consider in detail all of the suggested pretreatment requirements of raw materials presented and recommended by the various manufacturers. These will include, but will not be limited to, the treatment of cell water, cooling water (if any), brine, and recycled brine. Other factors are the variations required in the materials of construction. These can be dependent upon the degree of pretreatment and the use of reclaimed raw material.

For example, there will always be a small amount of molecular chlorine in recycled brine; so the brinemaker tank should be made of filament wound fiberglass if brine recycling is involved.

Space Requirements. Any comparison between alternative methods of chlorination should include the space required to house both the equipment and chemical storage facilities. For example, a 5000 lb/day conventional chlori-

*One of the most important considerations of any on-site generation system is the strength of hypochlorite solution produced. This involves size of storage tanks, pumps, and piping systems.

nator facility using ton containers requires about 1000 sq ft for all the equipment (chlorinators, evaporators, analyzers, etc.) plus space for 16 containers.

Based upon two Cloromat installations (ca. 1980), a facility capable of producing 5000 lb/day including standby equipment requires about 5100 sq ft. This includes space for the electrolyzer cells, metering pumps, analyzers, chemical storage, and a brine purification system. This does not include much-needed laboratory space to carry out monitoring of the chlorine generation process.

Operating Costs. The following costs are based upon a user analysis report covering a 12-month span of continuous operation.[19] All costs analyzed cover

Cost Summary of the Chloromat Systems for the Generation of Chlorine

Raw Materials			
Salt consumption	1.75 lb/lb Cl	@ $.02/lb =	$.035/lb Cl
Feed water to cells	9 gal/lb Cl	@ $.0007/gal =	$.006/lb Cl
Sodium hydroxide	0.01 lb/lb Cl	@ $.16/lb =	$.002/lb Cl
Soda ash	0.02 lb/lb Cl	@ $.14/lb =	$.003/lb Cl
Water softeners salt beads	0.02 lb/lb Cl	@ $.05/lb =	$.001/lb Cl
		Total	$.047/lb Cl
Electrical Energy			
Electrolytic cells	2 kWh/lb Cl	@$.085/kWh =	$.170/lb Cl
Brine purification system	.32 kWh/lb Cl	@$.085/kWh =	$.027/lb Cl
		Total	$.197/lb Cl
Anode and Membrane Replacement			
Calculated cost from 12 months experience			$ .01/lb Cl
Mechanical and Electrical Labor Cost			
Routine maintenance			$.005/lb Cl
Periodic maintenance			$.009/lb Cl
Operating Labor Cost (6 hrs per shift)			
Routine daily inspection and calibration plus electrolyzer and brinemaker maintenance:			$.054/lb Cl

Summary of Chlorine Production Cost

Raw materials	$.047
Electrical energy	.197
Anode and membrane replacement	.010
Routine maintenance	.005
Periodic maintenance	.009
Operating labor	.054
Parts and materials	.006
Total chlorine cost per lb	$.328

only those pertaining to the production of chlorine. These costs are separated into the appropriate categories.

Comparison with Purchased Hypochlorite. At the same time that the user analysis report was made, two local suppliers were quoting 14 percent trade strength sodium hypochlorite at $.47/gal delivered. To calculate the amount of "available chlorine" in one gallon, multiply trade strength by 0.08345. Therefore, local hypochlorite at 1.17 lb chlorine per gal = $.40/lb Cl. However, in actuality the cost of chlorine will be more than this because of the high decay rate of these solutions. Purchases of hypochlorite in quantity should be negotiated to arrive at a fair price on the basis of a two-week decay rate (see Chapter 2).

Electrocatalytic—Formerly Engelhard Industries

Introduction. Electrocatalytic is a British-based company that has been dealing primarily with the use of cathodic protection aboard seagoing vessels since about 1957. This was largely the result of Charles Engelhard's Hydrogen Detector patent, issued in 1952. It was adopted by the U.S. Navy as the standard monitor for monitoring hydrogen in submarine battery rooms. The cathodic protection activity for ships quickly spawned interest in shipboard chlorination to control marine growths of all kinds. Therefore, the proximity of seaboard use of seawater chlorination and cathodic protection resulted in the acquisition of Engelhard Industries' Chloropac by Electrocatalytic Inc. in 1990. The Chloropac systems remain the same with the exception of a few refinements. The Chloropac line has been supplemented by the Electrocatalytic Pacpuri System 3 on-site sodium hypochlorite generator. These systems are described below.

In 1972, the U.S. Department of the Interior Office of Saline Water awarded a contract to construct, install, and operate a seawater hypochlorite generator at its desalting test facility, Wrightsville Beach, North Carolina. The contract for this project was awarded to Engelhard Industries following an 18-month test of a prototype unit at the OSW facility.[10]

The unit developed by Engelhard for desalting plants is also applicable to other seawater situations, the most important being the control of marine fouling organisms in seawater-piping systems. Originally the Engelhard Chloropac system was designed specifically for shipboard use to combat slime and marine growths of all kinds, which normally thrive in the seachests and the ship's seawater systems. Protection from the proliferation of these growths is needed wherever seawater is used (i.e., condenser cooling, general engine room services, circulating water in the ship's air-conditioning system, fire system, and other seawater piping throughout the vessel). Moreover, chlorination is now mandatory for the potable-water supply aboard ship even though the water is produced by the ship's distillation process.

One other important application, which gives validity to this system and thereby enhances its reliability, is the widespread use of the Engelhard Chloropac on seawater drilling platforms and seawater supertanker storage platforms. These storage systems require reliable prevention of marine growth in the pump passages and storage-system piping.

Projecting this concept of making hypochlorite from saline water will, therefore, surely include its use for wastewater disinfection whenever a reasonable supply of seawater is available.

Raw Material. All of the Electrocatalytic systems are designed to handle an electrolyte equivalent in strength to a normal seawater (i.e., 19,000 mg/L chlorides and 30,000 mg/L TDS). This holds true for all of the systems whether it be recycled seawater or brine. The brine is always diluted to the optimum seawater chloride content before it is sent to the cells.

This presents a significant parameter, namely, the amount of raw product required to provide the chlorine-generating capacities. This amounts to a minimum of 20 gpm of electrolyte (seawater) for a module containing a series grouping of 2 to 10 cells. Each 10-cell module can produce up to 240 lb/day chlorine as hypochlorite. The modules are connected hydraulically in parallel so that a 480 lb/day unit consisting of two 10-cell modules requires 40 gpm seawater, and so on. The maximum concentration of chlorine in the hypochlorite solution generated in the once-through seawater system is about 1200 mg/L, and the TDS will be on the order of 30,000 mg/L. If refined brine is used instead of seawater, the TDS will be about 20,000 mg/L.

Cell Configuration. The Electrocatalytic cell is a flow-through or "open-type" cell where the saline water is subjected to electrolytic decomposition on an incremental basis as the salt water flows through a series of these cells. The cell is designed so that the salt water flows through an annular opening at a velocity of 5–7 ft/sec. This velocity is supposed to continuously flush out precipitates of insoluble anions normally found in seawater or other brackish waters. The patented Chloropac® cell is illustrated in Fig. 3-3. It is constructed of three titanium cylinders. Two cylinders are placed axially in line and connected by their flanges with an insulating cylindrical spacer to form a smooth-bore pipe. The third cylinder is small in diameter and longer than the first two pipes. This third cylinder is sealed at each end and placed inside the pipe formed by the first two (see Fig. 3-3.). The salt water electrolyte flows in the annular space between the inside of the outer cylinders and the outside of the inner cylinder.

The inside surface of the outer cylinder on the left is coated with a proprietary platinum alloy that allows it to respond as though the entire cylinder were a solid platinum alloy. This cylinder connected to the positive terminal generates molecular chlorine on its inner surface. The outside of the inner cylinder adjacent (to the right) receives the electric current as a cathode and releases the cathode products (i.e., sodium hydroxide and hydrogen). The

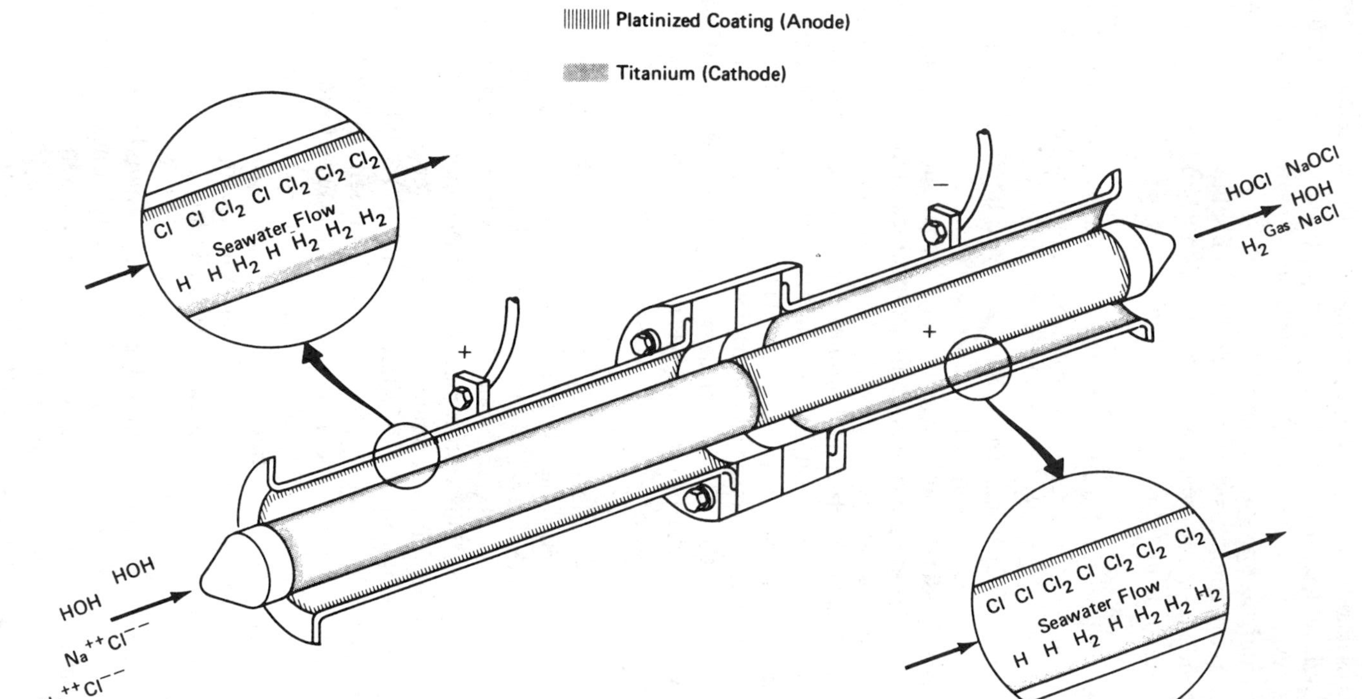

Fig. 3-3. Electrocatalytic seawater cell anode reaction 2Cl→ + 2e, cathode reaction 2H +2OH→H_2↑ + 2OH + 2e, overall reaction NaCl + H_2O + D.C.→NaOCl + H_2 gas (courtesy Electrocatalytic, Systems Department).

right-hand half of the cell operates in the same fashion except that the roles of the anode and cathode surfaces are reversed. Here the outside of the inner titanium surface is coated with platinum alloy, and the current passes from the outside of the inner cylinder to the inside of the second outer cylinder. The flowing stream of electrolyte mixes the products produced at the anodes and cathodes, which produces a weak sodium hypochlorite solution and minute bubbles of hydrogen gas. This chemical reaction is summarized by the equation shown in the caption of Fig. 3-3.

It should be noted that both the inner and the outer pipes of this cell are made of titanium, whether platinized or not. Titanium is resistant to salt water corrosion. Moreover, titanium possesses the unique chemical characteristic of being able to form a protective oxide coating so that it will receive but not emit a current in the 8–12 V DC range. The platinized anodes are consumed at the rate of 6 mg/year. This calculates to an expected anode life of about 6 years.

As long as the system stays within the power design specified by the manufacturers, the titanium cathode will not be consumed in the electrolytic process. It is therefore labeled as an "infinite life" electrode by the manufacturer.

Chloropac Generator. The Electrocatalytic system is usually arranged for 10–12 cells in series. This is identified as a module, and the arrangement of these modules is further identified with a model number as to equivalent chlorine capacity in lb/day. Figure 3-4 illustrates a 960 lb/day generator. This

Fig. 3-4. Electrocatalytic Chloropac seawater cells system of 40 cells in series; capacity 960 lb/day available chlorine (courtesy Electrocatalytic, Systems Department).

unit consists of 40 cells arranged so that there are two groups of 20 cells in series. However, each of these two groups is connected in parallel hydraulically.

The individual cells are arranged in pairs so that the first and last cells of any group are properly grounded to eliminate the possibility of stray current corrosion. Each cell is then electrically connected in series to the center pair. These anodes are connected to the positive power source. Since the cells are connected in series hydraulically, the strength of the hypochlorite solution produced increases from one cell to another.

Assuming that a seawater with a total salt content of 32,000 mg/L will contain about 19,000 mg/L of chloride ions, each cell will generate about 100 mg/L of hypochlorite. Therefore, each group of 10–12 cells produces a 1000–1200 mg/L hypochlorite solution. At this concentration there is no need for product cooling or hydrogen gas venting (H_2 venting is required if a hypochlorite storage tank is used).

The electrolyte flow through any given cell is limited to 20 gpm. Therefore, assuming a seawater content of 19,000 mg/L chloride ions, any saline water of this concentration will produce 1.0 lb Cl_2/hr/cell at the 20 gpm flow of electrolyte. Therefore, the Chloropac generator system is arranged so that the modules of hypochlorite production are hydraulically connected in parallel consisting of series-connected units of 10 lb/hr capacity each. A 20 lb/hr system would consist of two groups of 10 cells in series arranged in parallel.

Chloropac Systems

Once-Through Seawater System. This is the simplest system used for an unlimited supply of saline water containing from 1.5 to 4.0 percent salt and where a weak sodium hypochlorite solution is acceptable. Such a salt supply will result in hypochlorite solution strengths varying from 100 to 1000 mg/L available chlorine. Typical users of this salt water source include: seaboard utilities; industrial condenser cooling water systems directly dependent upon a salt water source; offshore oil production facilities; desalting facilities; and seagoing vessels.

Once-Through Brine System. This system is attractive if there is an abundant supply of either salt or brine, and the cost is lower than that for the more efficient recycling system. Brine solution is prepared in a salt dissolver and then mixed with feedwater until it reaches the approximate salinity of seawater, 3–3.5 percent salt, which is the optimum strength for electrolytic decomposition. This solution is fed to the electrolytic generator and converted to a weak NaOCl solution as in the seawater system above. The typical user might be an inland industrial plant or waterflood facility located in an area of natural salt beds or strong brackish groundwater.

Seawater Recycle System. In this system, the initial weak hypochlorite solution is discharged from the generator to a holding tank and then recycled to join the incoming brine flow. Recycling on a continuous basis gradually raises the strength and temperature of the solution. The maximum strength of solution is about 4000 mg/L available chlorine. The rise in temperature of this process tends to speed up the decay of the hypochlorite solution; so a minimum storage time of the final product should be considered.

Seawater or Brine Recycle with Cooling of Product. Whenever recycling is involved, maximum extraction of chlorine is dependent upon the optimum temperature of the product. This requires a cooling system.

The system is designed to provide maximum economy of chlorine extraction from the brine. The brine is first diluted to nominal seawater salinity (about 19,000 mg/L chlorides), and then it is recycled through the Chloropac generator until the hypochlorite generated reaches about 1.0 percent concentration. The amount shown in the Electrocatalytic literature is 7520 mg/L.[11] Figure 3-5 illustrates the operation of the brine recycle system. The entire process of brine dilution, product recycling, and product cooling is automatically controlled.

Automatic Dosage Control. The Electrocatalytic systems are responsive to automatic control. The concentration of the hypochlorite solution can be controlled by a saturable reactor which responds to a 0–5 mV signal. This controller, which is a transformer within a transformer, can utilize a cascade system whereby a flow signal is combined through an electronic multiplier with a chlorine residual analyzer signal to produce a final control signal. This signal changes the power input, which in turn changes the number of Faraday units applied to the constant flow of brine through the cells. This changes the amount of chlorine generated in accordance with the control signal.

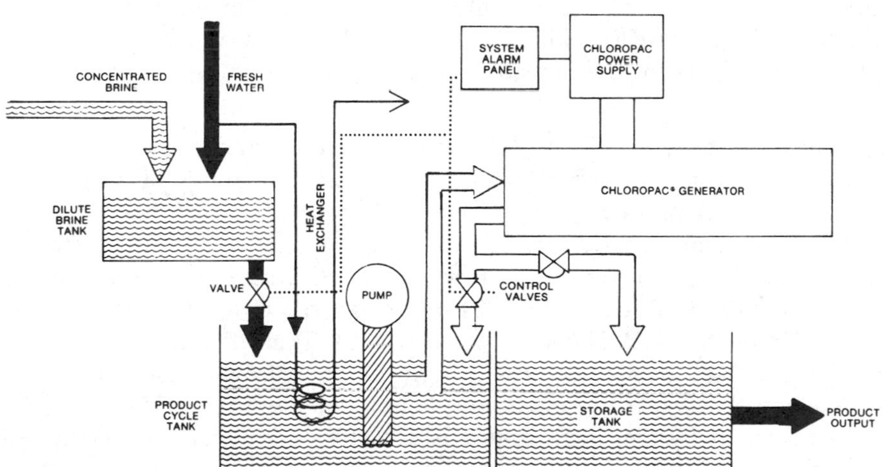

Fig. 3-5. Electrocatalytic Brine System (courtesy Electrocatalytic, Systems Department).

Cost Comparisons. Electrocatalytic claims 2.3 kWh/lb of chlorine at 70°F and 2.6 kWh/lb at 50°F, assuming a nominal seawater concentration of 19,000 mg/L of chloride ion. The estimated salt requirement for the Electrocatalytic brine system is 3 lb salt/lb of chlorine.

The Houston analysis by Matson and Coneway[12] compared on-site generation of a system producing a 0.8 percent hypochlorite solution and one producing an 8.0 percent solution. Both of these systems were evaluated on the basis of a concentrated brine system. Cost comparisons for individual systems are burdened with many unknown pitfalls; however, it is pertinent to note that the 8 percent hypochlorite system was by far the best buy for the dollars invested. This system demonstrated a 30 percent advantage over the 0.8 percent system for a 5-ton/day facility. This is not meant to be an indictment against the seawater systems but a warning to the designers of the problems involving system selection.

SANILEC[13] claims that using 80 percent strength seawater at 25°C will require 2 kWh to produce 1 lb of chlorine at full load production.

Much more operating experience is needed with both the seawater and the brine systems to determine the chlorine production costs of these systems.

Electrocatalytic Pacpuri System 3

System Description. This modular electro-chlorination system was introduced by Electrocatalytic Ltd. in 1986. It was designed primarily for potable water treatment. This sodium hypochlorite generating system is the result of an intensive development program and proven process on-site performance. The hypochlorite solution is automatically fed from the storage tank by PID compound loop control, from both chlorine residual and flow rate analysis. The intensive program of design and development carried out by Electrocatalytic in Great Britain was due to a widespread government ban on the use of liquid chlorine containers for use in water supplies.

System Components. The system may be described as follows: (1) A salt saturator forms the brine, which is mixed with softened water in the generator; (2) the generator produces the specified sodium chloride solution strength; and (3) this solution passes through a flow meter fitted with a flow detector switch into (4) the Pacpuri cell, where a low DC voltage current from the rectifier unit passes through the cell from (5) anodes to (6) cathodes. This is where part of the NaCl is converted to NaOCl. Then (7) this hypochlorite solution goes into the storage tank, where (8) the hydrogen gas byproduct is diluted and (9) is vented to the atmosphere by high efficiency fans.

Each Pacpuri System 3 generator is a factory-assembled unit complete with brine-water injectors, pressure gauges, manual isolating and pressure-regulating valves, and cell flow indicators, plus interlocking flow, level, and temperature switches.

Automatic Control. The hypochlorite solution is then injected into the water to be treated, by dosing pumps under PID control in a compound-loop circuit. This automatically compensates for any fluctuations in chlorine residual or water flow rate by continuous signals from the chlorine residual analyzer and the water flow transmitter.

Empicon. The entire process is controlled from the Electrocatalytic Empicon unit. It consists of a microprocessor controller, with an optional chart recorder, an ASII printer, and a remote control duplicate Empicon via modern telemetry. The sophisticated software in this information station controller is dedicated to the Pacpuri System 3. It is configured to a wide range of selectable components and available options. This unit controls the entire process from softening, through hypochlorite generation, to dosing and water quality monitoring.

Capacity Ranges. The Pacpuri System 3 is available in a wide range of modular sizes. The largest is 8000 lb/day of equivalent chlorine as 0.8 percent NaOCl.

ON-SITE GENERATION OF SODIUM HYPOCHLORITE

SANILEC Systems I[13]

Use of these systems is probably the most practical and interesting development in the use of chlorine for the disinfection of public water supplies, making it simple to stabilize the microbial life in water distribution systems—otherwise known as *dead-end* problems.

Clor-Tec Standard Process System[23]*

Since the publication of the Uniform Fire Code and the recent activity by the EPA concerning hazardous chemicals, an enterprising company has been formed to provide the availability of the SANILEC Brine system to any and all water utilities in need of something simple to avoid the handling of a hazardous chemical, such as 12 to 15 percent hypochlorite solution or chlorine in 150-lb and one-ton containers.

This is being accomplished by the Chemical Services Company, which manufactures the Clor-Tec On-Site Sodium Hypochlorite Generation Systems (see Fig. 3-6). The company is located in San Jose, California and Clearwater, Florida, and has had extensive experience in designing, installing, and servicing electro-chlorination systems, beginning in Hawaii in 1988. It now has well over 300 installations in mainland United States, operating at capacities from

*This is their version of the SANILEC system.

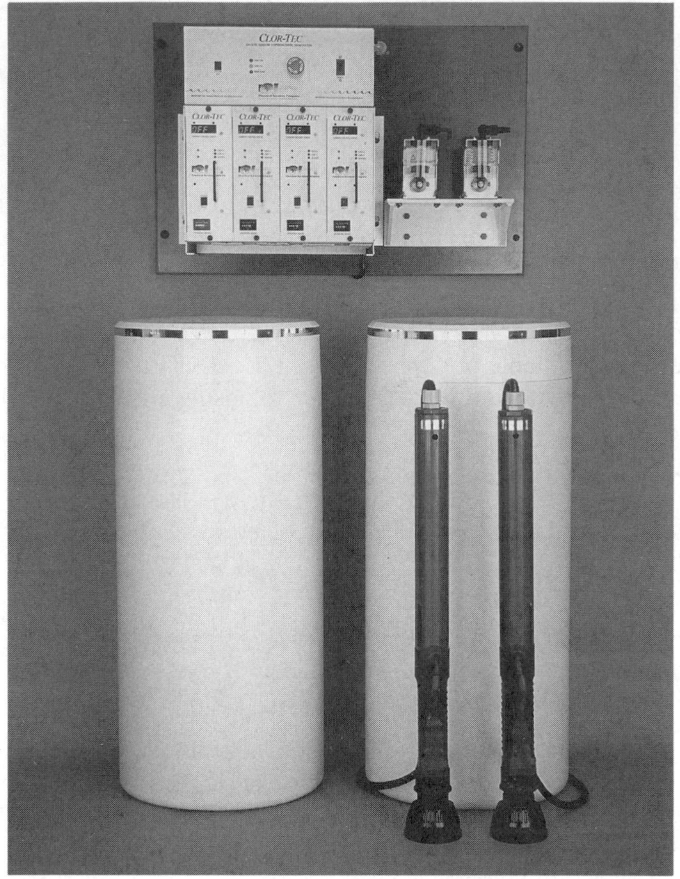

Fig. 3-6. Clor-Tec on-site sodium hypochlorite generation system.

6 to 500 lb/day of chlorine. It also has units available with capacities of 1000 and 2000 lb/day of chlorine.

The Clor-Tec standard process system is illustrated in Figs. 3-7 and 3-8. All units use 3.5 lb of salt, 2.3 kWh electricity (AC), and 15 gal of water = 1 lb chlorine.

The Clor-Tec system is controlled by the sodium hypochlorite level in the storage tank. It operates until the tank is full, and then it shuts off. When the level drops, it restarts automatically. It is also self-regulating and will shut off if a low brine concentration is detected.

Clor-Tec is compatible with SCADA and process control systems, allowing remote monitoring of the entire process.

Flow Diagram. The design and operation of the Clor-Tec system is simple and straightforward:

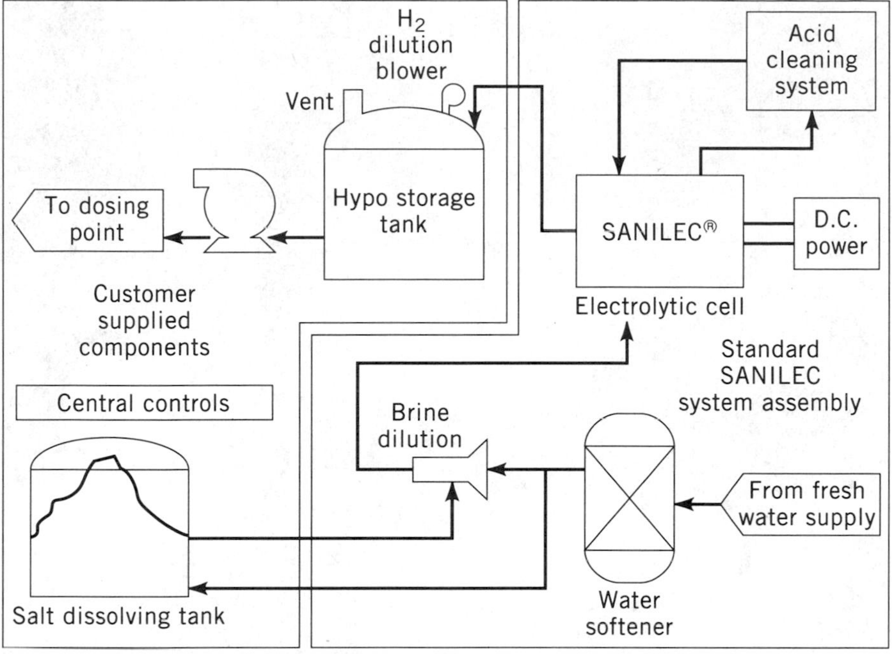

Fig. 3-7. Brine flow diagram.

1. Salt is dissolved with water to form a concentrated brine solution.
2. The brine proportioner dilutes and supplies a 3 percent solution to the reactor tank.
3. In the reactor tank the cell pack(s) electrolyzes the diluted brine into an 0.8 percent sodium hypochlorite solution as shown in the following equation:

$$\underset{\text{salt}}{NaCl} + \underset{\text{water}}{H_2O} + 2e = \underset{\text{hypo}}{NaOCl} + \underset{\text{hydrogen}}{H_2}$$

4. The hypochlorite solution flows into the storage tank via gravity.
5. Hydrogen is safely vented to the atmosphere.
6. A metering pump delivers the solution to an injection point in the treatment system.

Applications. Clor-Tec can be used for potable water or wastewater treatment in any application that requires the addition of chlorine for disinfection or oxidation, plus the following:

- Air scrubber odor control.
- Sewage odor control.

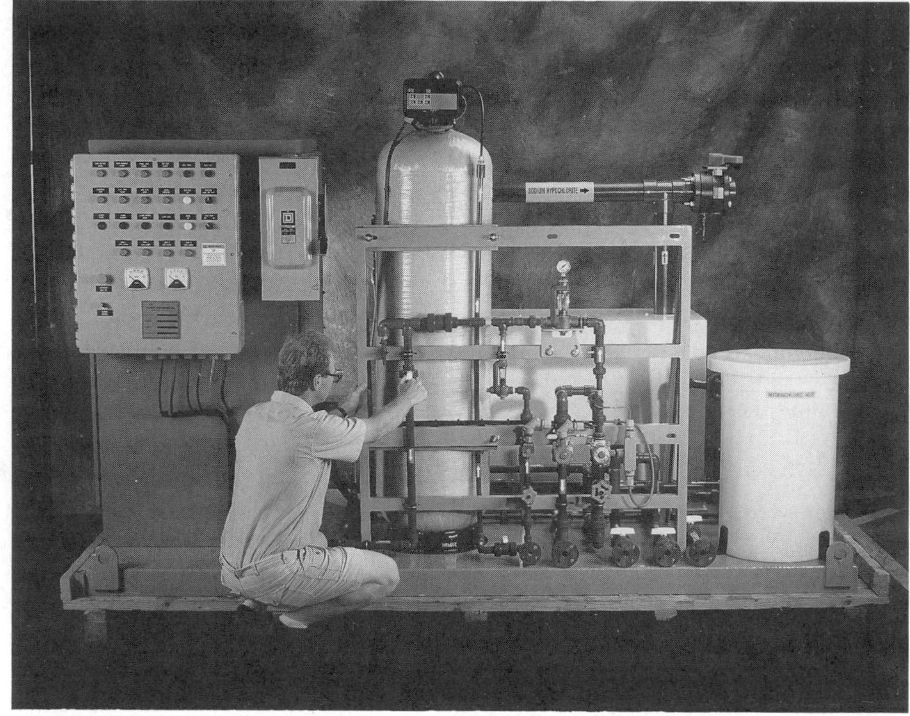

Fig. 3-8. The SANILEC FB500 system without tanks, salt dissolver, and dosing system. This unit produces 500 lbs equivalent chlorine per day. (Courtesy Exceltec International Corporation.)

- Industrial cooling towers.
- Food processing.
- Laundry bleaching.
- Swimming pools.
- Irrigation.
- Chemical destruction.

Further Activity. Details of the activities of Chemical Services Company and its SANILEC Brine Systems are given in Chapter 6.

SANILEC Systems II

Introduction. On-site hypochlorite generating systems by this trade name have an interesting origin. The parent company of SANILEC is ELTECH Systems, one of the leading producers of equipment for the chlor-alkali industry over the past 30 years. The parent company must be credited with one of the most significant improvements in the chlor-alkali industry for many years, the "Dimensionally Stable Anodes," DSA®. Through its subsidiary, Electrode Corporation, ELTECH Systems has licensed over 95 percent of the North American chlorine capacity to use the energy-saving DSA technology. The continuing development of chlorine production efficiency led to the development of cells for the production of chlorine from seawater.

With this background and demonstrated expertise, ELTECH International Corporation (EIC), a wholly owned subsidiary of ELTECH Systems, launched the SANILEC system for on-site generation of hypochlorite to capture whatever market was available on the basis of expediency, safety, and/or economics. Table 3-2 shows a summary of SANILEC installations by application and represents over one million lb/day of chlorine equivalent. These systems are installed worldwide, with the smallest plant in operation producing 20 lb/day of available chlorine and the largest plant producing 57,600 lb/day of available chlorine. These systems are installed in 39 countries worldwide.

SANILEC offers two types of systems for on-site generation. One is for the use of saline waters similar to seawater situations or various dilutions thereof, depending upon local conditions. The other is the use of a concentrated brine as raw material. These systems have been described in detail by Bennett and Cinke.[14] These authors emphasize that many important facts distinguish electrolytic cells which use seawater from those which use prepared brine solutions. Therefore, SANILEC set about to design two distinctly different types of cells: one for seawater and one for pure brine.

Table 3-2 SANILEC®: Summary of Installation Applications (Date: 1/91)

Application	*Qty.*	*%*	*Capacity (Lb./Day)*	*%*
Cooling water—fossil power plants	34	12.3	357,528	48.1
Cooling water—nuclear power plants	18	6.5	184,520	24.8
Cooling water—industrial plants	29	10.5	115,799	15.6
Wastewater treatment	26	9.4	8,145	1.1
Potable water treatment	44	15.9	47,500	6.4
Swimming pool disinfection	33	11.9	1,880	0.3
Oil recovery and waterflood projects	2	0.7	12,100	1.6
Other applications, including:	91	32.9	15,339	2.1
Textile manufacturers	9	3.2	410	
Industrial bleaching	1	0.4	100	
Beverage water treatment	3	1.1	220	
Cyanide destruction	1	0.4	50	
Odor control	17	6.1	4,190	
Oceanariums	8	2.9	2,415	
Air conditioning cooling systems	22	7.9	5,454	
Pharmaceutical water	1	0.4	50	
Sugar mills	16	5.8	1,010	
Misc.	13	4.7	1,440	
TOTALS:	277	100.0	742,811	100.0
		kg/day	337,029	

Seawater System. These systems suffer from impurities inherent in seawater; therefore, the cell configuration must be designed accordingly. These impurities cause bulky deposits that interfere with the electrolyte flow. These deposits are a result of natural seawater hardness, which is about 1800 mg/L, caused by calcium and magnesium in seawater containing 19,000 mg/L chloride. The precipitates of these ions not only interfere with the hydraulics of the cell system but also act to insulate the cathode, causing a reduction in process efficiency. By using a turbulent flow regime through the cell, SANILEC can operate for 1 to 2 months continuously before cell cleaning is required. Cleaning consists of recirculating a 10 percent muriatic or sulfamic acid solution through the cells for 1–2 hours. This system therefore requires the availability of standby cells to accomplish routine maintenance. The SANILEC seawater electrolyzer is shown in Fig. 3-9. The electrodes of the seawater cell are classed as dimensionally stable. The anode is the expanded metal type, whereas the cathode is a thin solid sheet of metal, which is a proprietary nickel alloy.

The "seawater system" is shown in Fig. 3-10. This system requires straining of the seawater to remove particles larger than 800 microns before it flows to the cells. Each cell module receives a constant flow of seawater through cells connected in series, both hydraulically and electrically. Power for these electrolyzers is supplied by a rectifier, which converts AC to DC. Hypochlorite production is controlled by DC current variation from the rectifier, either manually or automatically, by plant flow pacing plus residual control, to trim the chlorine concentration as shown in Fig. 3-10. The concentration of the hypochlorite solution produced from seawater is limited to a range of 150–3000 mg/L available chlorine (0.015–0.300 percent).

The hypochlorite solution leaving the electrolytic cells goes to a gas release tank where the hydrogen produced in the electrolytic process is diluted with air to less than 2 percent and vented to the atmosphere.

For an alternate method to the disposal of hydrogen by air dilution, ELTECH has developed a patented hydrocyclone, shown in Fig. 3-11, which separates 95 percent of the undissolved hydrogen from the seawater/hydrogen mixture leaving the electrolyzer. The discharged hydrogen is vented to atmosphere through a water seal tank. This system can provide hypochlorite solutions with up to 15 psi back pressure without having to pump the hypochlorite solution. This feature eliminates the need for gas release tanks or hydrogen dilution blowers.

The proper amount of seawater must be delivered to the cells at a minimum of 40 psi. Water flows required vary with the size of the system. The 3200 lb/day unit produces a 1775 mg/L available chlorine solution using eight 400 lb/day cells and requires a seawater flow rate of 150 gpm. A 60 lb/day system produces a 160 mg/L available chlorine solution using one 60 lb/day cell and 40 gpm seawater flow.

Power consumption at full production depends upon the temperature and the salinity of the seawater as well as the concentration of hypochlorite pro-

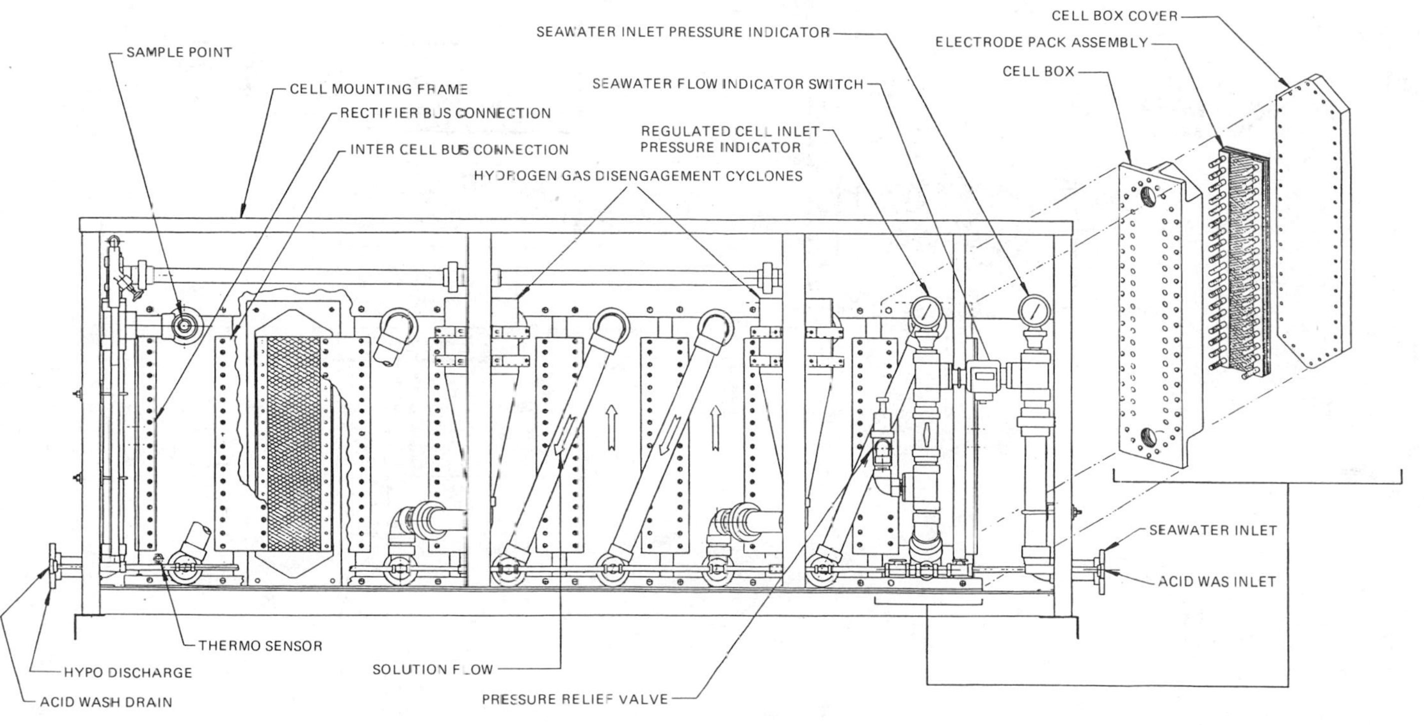

Fig. 3-9. Seawater Electrolyzer (courtesy ELTECH International Corporation).

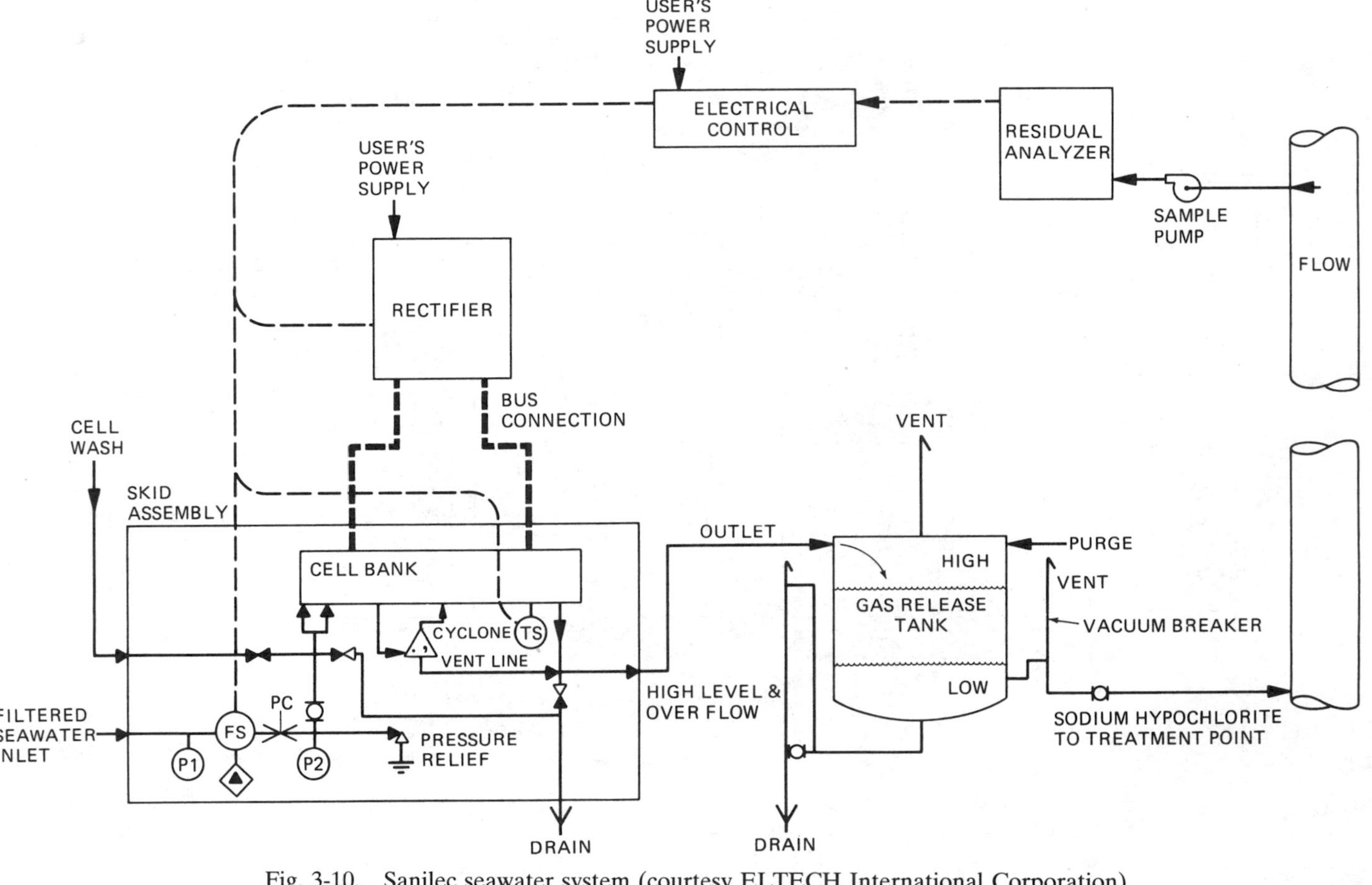

Fig. 3-10. Sanilec seawater system (courtesy ELTECH International Corporation).

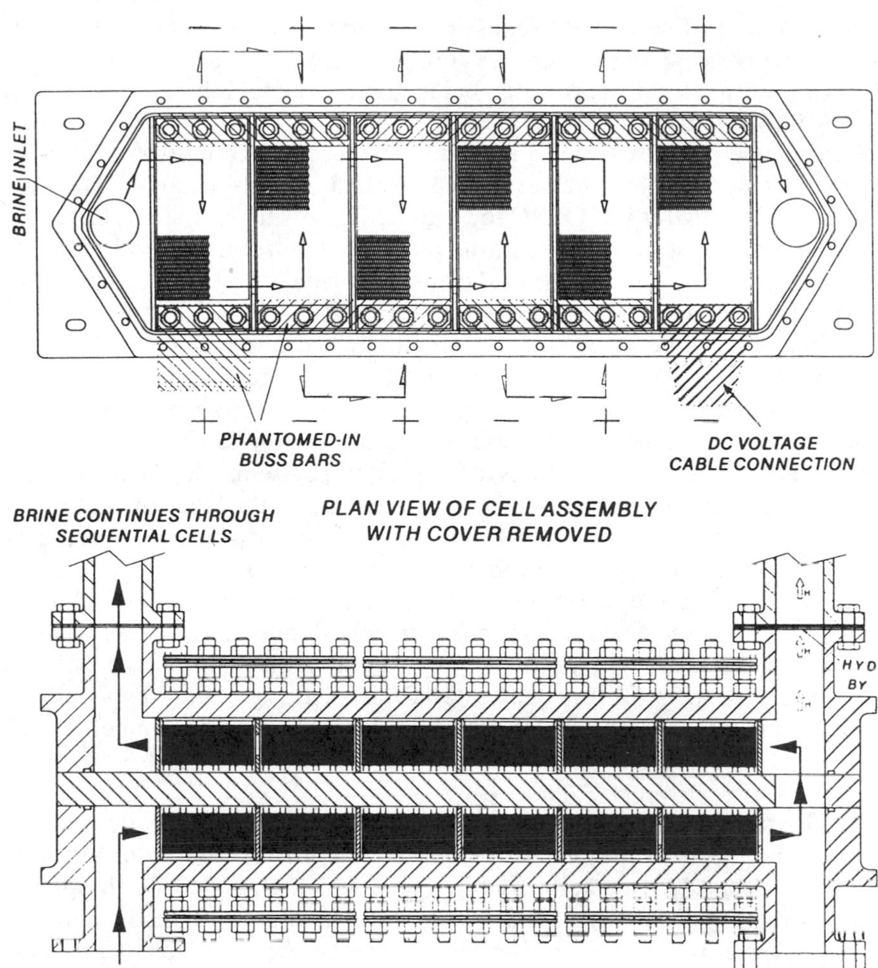

Fig. 3-11. Sanilec brine cell (courtesy ELTECH International Corporation).

duced by the electrolyzer. For example, the 3200 lb/day system described above will require 1.7 kWh DC per pound of available chlorine at a seawater temperature of 25°C and a chloride content of 19,000 mg/L.

Seawater salinity of 100 percent is generally defined as 18,900 mg/L of chloride ion. In practice, seawater strength can be and is significantly diluted, especially in harbors and waterways with large inflows of surface water runoff. The SANILEC system is not typically designed to operate on seawater with less than about 9500 mg/L of chloride ion. The use of seawater for electrolytic cells contributes to a certain amount of chlorides and dissolved solids in the system effluent, which may be a significant factor in the application of the seawater hypochlorite. Additionally, electrolysis of seawater will produce a certain amount of suspended solids in the form of magnesium hydroxides, calcium hydroxides, and carbonates. For example, a 10 mg/L dose of chlorine will add to the process water being treated some 189 mg/L chlorides, 355 mg/L TDS, and 2 mg/L SS.

Present operating experience indicates that one of the most crucial parts of any seawater system, whether it be SANILEC, Electrocatalytic, DeNora, W&T OSEC, W&T Mini OSEC, or others, is the supply system. First, the seawater must be strained; second, the piping system from the intake suction to the electrolytic cells must be designed to provide a flushing velocity of 4–5 ft/sec. Moreover, all of this piping should be PVC, CPVC, Kynar, or steel pipe lined with rubber or Saran. Special attention has to be paid to the types of joints used with plastic pipe transporting hypochlorite. Random leaks from threaded joints lined with Teflon tape have been reported by many users. Success has been attained in the use of fusion-type joints.

Recycling is not used in the SANILEC seawater system. Although recycling can be used to lower cell voltage, this advantage cannot balance the accompanying inherent loss in current efficiency and pumping power. Therefore, recycling seawater systems is not recommended by SANILEC.

SANILEC points out one important factor in the production of hypochlorite from seawater: these solutions generated from seawater are inherently unstable. They deteriorate significantly within 48 hours because the seawater brine contains a vast array of ions that catalytically decompose the hypochlorite solution. Therefore, recycling of seawater systems is not recommended by SANILEC.

Brine Systems. Figures 3-12 and 3-13 show seawater hypochlorite generating systems. Electrolytic systems of chlorine and/or hypochlorite production, utilizing prepared brine solutions rather than seawater, are easier to operate than seawater systems. Selected grades of salt (food-grade is recommended) allow the use of cation resin exchangers to remove hardness from the process water. This in turn allows much slower flow rates (6 gpm) and hypochlorite concentrations exceeding 8000 mg/L. The brine cells all require in the range of 2.8–3.5 lb of salt, 2.2 kWh of power, and 15 gal of water per pound equivalent chlorine.[14]

All SANILEC brine systems require food-grade salt, as do all competitive systems. The brine system is simple. Salt is dissolved with softened water to form a concentrated brine solution, which is diluted to 28 g/L NaCl (optimum) in an automatic brine dilution panel and passed through the cell. Hydrogen is disentrained from the solution—either in the cell in smaller (100 lb/day chlorine equivalent or less) systems, and vented to the atmosphere, or in larger systems carried with the solution to a hydrogen disentrainment system for direct discharge or blower dilution before discharge.

Designs. SANILEC brine cells have three basic designs. A small portable cell has been developed that provides a low volume of Cl_2 production for

Fig. 3-12. Sanilec seawater hypochlorite generating plant, Saudi Arabia, capacity 57,600 lb/day equivalent chlorine (courtesy ELTECH International Corporation).

Fig. 3-13. Sanilec seawater hypochlorite generating system at a desalination plant in Saudi Arabia, capacity 36,000 lb/day equivalent chlorine (courtesy ELTECH International Corporation).

general water treatment. This monopolar cell may be powered by household current, photovoltaic cells, or auxiliary power devices for use in remote areas. Designed as a batch process, the system provides ideal simplicity for Third World countries. For example, the cell is designed for easy cleaning with household vinegar, and hundreds of these systems are being distributed to remote areas under the auspices of the World Health Organization.

A slightly larger system (50 and 100 lb/day chlorine equivalent) is based upon a bipolar cell and is designed for automatic dosing applications. These systems are ideally suited for chlorine replacement well heads in densely populated locations where there is light industrial usage. Owing to the design of the cell, no system cooling is required.

Finally, the largest systems (standard modular sizes up to 2000 lb/day) are based upon the monopolar seawater cell operated horizontally rather than vertically. These generators are skid-mounted complete with DC rectifiers, acid cleaning systems, and water softening equipment (see Fig. 3-10). Despite the salt requirements of these systems, they have proved to be cost-effective for municipal and industrial water treatment in the United States as well as internationally. In some cases, several of the largest skids (2000 lb) have been used together, generating thousands of pounds per day of chlorine equivalent.

WALLACE & TIERNAN OSEC, AT CHERTSY; BRA.367.1

On-site Electrolytic Chlorination System[25]

Introduction. This model unit is the very latest design from the W&T Ltd., Priory Works in Tonbridge, Kent, England and its associated manufacturing companies in the United States, Germany, Canada, Australia, Mexico, Brazil, and agents throughout the world.

A leader in chlorination technology for over 80 years, Wallace & Tiernan introduced two innovative systems, in 1980 and 1993, for the electrolytic generation of hypochlorite.[15] Standard systems are available in capacities from 4 to 2500 lb/day. See Fig. 3-14. Larger systems use multiple module arrangements. Each system is custom-designed and engineered for each installation. The operating characteristics of each of these systems are based upon experience with both brine and seawater. Of all the suppliers of on-site chlorine generating technology, Wallace & Tiernan has the longest record.[4] White inspected and operated one of the first of its locally made units to be used at Pan American Airways refueling stations on the San Francisco-to-Sydney and Orient run, ca. 1939. World War II changed all of this rather quickly in 1941.

The Electrolyzer. The electrolyzer (Fig. 3-15) is a once-through flow unit designed for generating efficiency and simplicity of field maintenance. The anodes, which are the only parts susceptible to passivation (wear through

Fig. 3-14. W&T series 85.510 OSEC seawater system capacity 2500 lb/day available chlorine. (Courtesy Wallace & Tiernan.)

electron transfer), are made of oxide-coated titanium (valve metal substrate). A minimum service period of two years can be expected of these anodes. The cathodes are made of Hastelloy C. The electrode chassis is lightweight and is arranged so that it can be easily removed from the electrolyzer casing for servicing without disturbing the plumbing.

The electrolyzer consists of one, two, four, or eight or more casings, depending upon the chlorine production capacity. Each casing is divided into four cells, each with a set of electrodes (see Fig. 3-16). The cells are arranged in a bipolar configuration. Brine enters the casings and floods the cells. A DC current impressed upon the electrolyzer converts the sodium chloride to molecular chlorine and sodium hypochlorite. There also remains in solution unreacted brine and hydrogen gas, which is a product of electrolysis. This is shown in the following diagram:

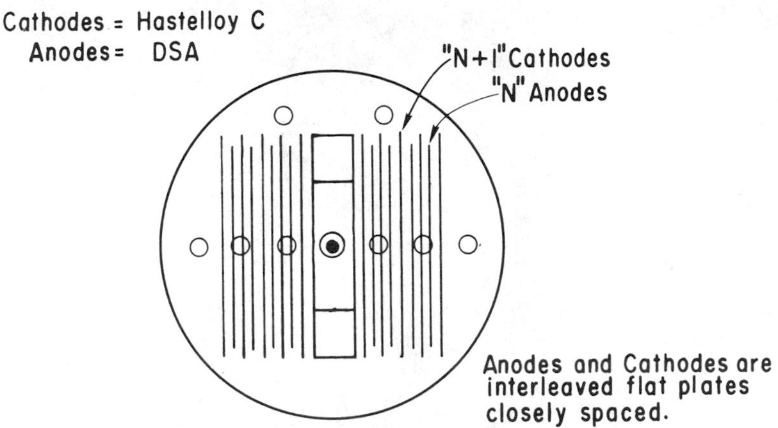

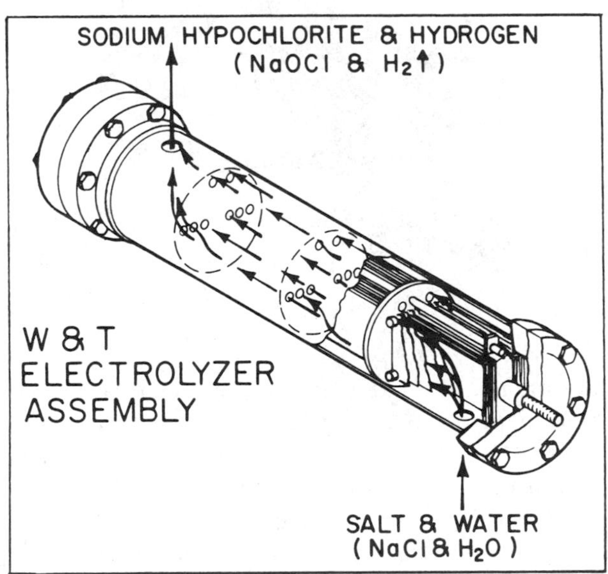

Fig. 3-15. OSEC electrolyzer (courtesy Wallace & Tiernan).

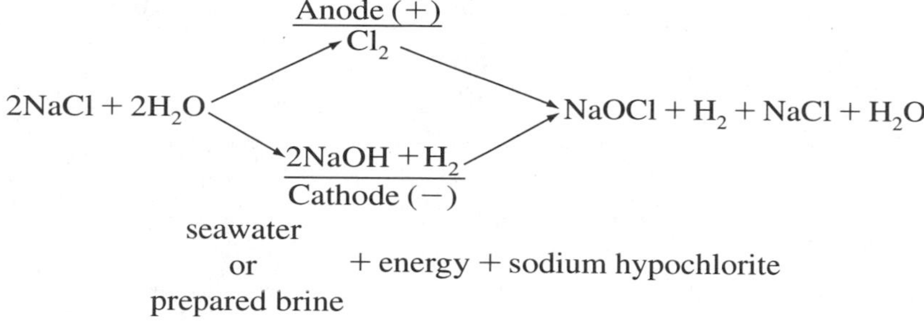

The design configuration of the electrolyzer accelerates the removal of hydrogen gas from the generation zone by thermal convection. The separated hydrogen gas passes through gas ports in the compartment partitions, traveling laterally to a discharge connection. The brine, electrolyte, and hypochlorite pass from one cell to the next through ports located below the solution level. In multiple casing arrangements, the electrolyte and the hypochlorite pass through an outlet connection in the first cell of the next casing. In the final casing the electrolyte and the hypochlorite are discharged together with the hydrogen gas to a storage tank, which is designed to scavenge the hydrogen from the final product.

As brine progresses through the successive electrolyzer cells, its degree of conversion to hypochlorite increases. Concentration of the final product is about 8000 mg/L available chlorine when generated from prepared brine; and about 1800 mg/L from seawater.

Most salts and practically all waters contain some degree of hardness in the form of calcium and/or magnesium. These impurities are prone to form insulating precipitates between the electrodes, which inhibit the electrolytic process. This precipitation reduces the conversion efficiency and increases maintenance costs. Wallace & Tiernan claim that their design minimizes this ubiquitous problem if properly maintained.

Figure 3-16 illustrates all of the components of a typical on-site electrolytic chlorination system, Wallace & Tiernan version. This illustration shows both the brine and seawater applications.

Differences Between Seawater Model and Brine Model. This is an important consideration in the choice of these models, regardless of manufacturer.

The Wallace & Tiernan brine system produces about 1 lb of chlorine for every 3.5 lb of salt; that is about 8000 mg/L available chlorine solution using a 28,000 mg/L TDS brine. With a brine system the NaCl is efficiently used at the expense of a little less efficient power utilization.

A seawater system provides an almost "free" source of salt; however, power efficiency is paramount. The Wallace & Tiernan seawater system will generate an 1800 mg/L available chlorine solution using seawater at 34,000 mg/L TDS or about 1 lb chlorine for every 19 lb salt. In this model the water velocity is

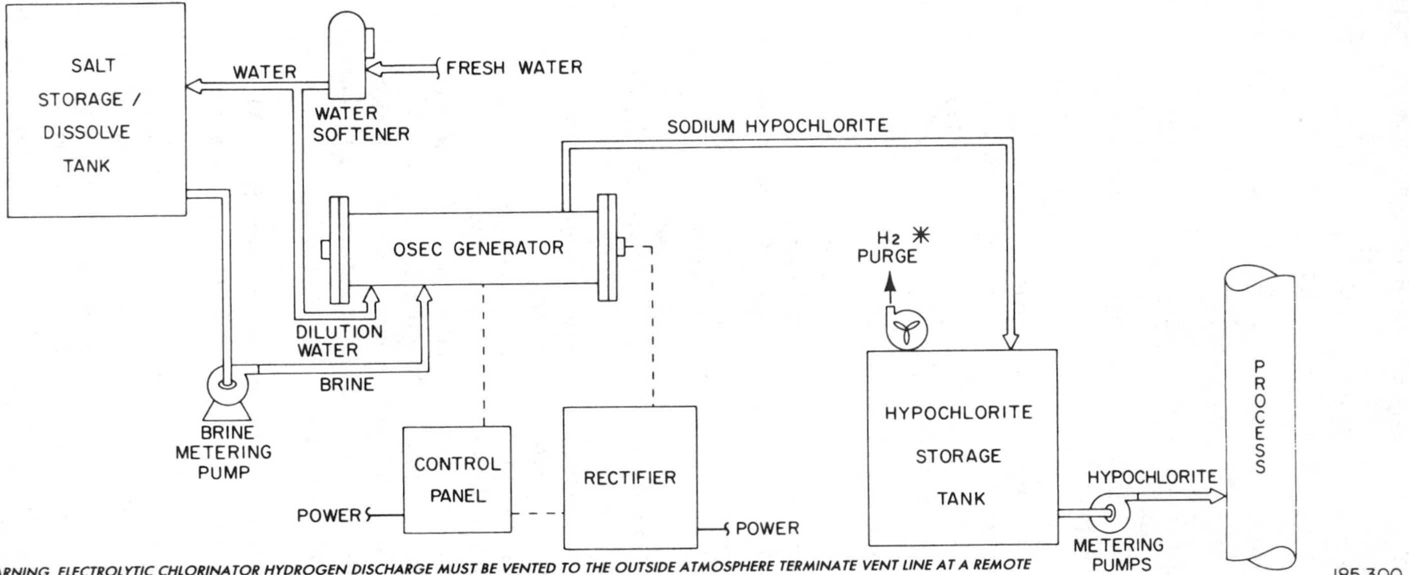

Note
* WARNING ELECTROLYTIC CHLORINATOR HYDROGEN DISCHARGE MUST BE VENTED TO THE OUTSIDE ATMOSPHERE TERMINATE VENT LINE AT A REMOTE LOCATION TO INSURE DISPERSAL OF HYDROGEN AND ELMINATE EXPOSURE TO ANY SOURCE OF IGNITION

Fig. 3-16. Wallace & Tiernan OSEC brine system (courtesy Wallace & Tiernan).

kept high and the electrode spacing is increased in order to minimize problems with scale formation inherent in seawater use. The scale formation is a formidable problem owing to the high concentration of calcium and magnesium ions.

Manufacturer's Recommendations

Seawater Systems. Seawater temperature must be min. 40°F; salinity, min. 17 g/L. Seawater must be filtered to remove all particles larger than 0.045 inch. These systems must be acid-flushed with 2 percent hydrochloric acid. This removes the inherent buildup of calcium and magnesium precipitates on the electrode surfaces. The length and the frequency of flushing are empirical.

Capacities are available in two arrangements: (1) low capacity systems are 25, 50, 75, and 100 lb/day units; (2) high capacity systems are 575, 1200, and 2400 lb/day units. From the lowest to the highest capacity unit the seawater requirement is from 7.5 to 120 gpm. Power consumption will vary, depending upon temperature and salinity of the seawater. This variation is probably on the order of 1.7–3.0 kWh/lb chlorine.

Brine Systems. To minimize the formation of calcium and magnesium precipitates within the electrolyzer, the salt should be as free from hardness as possible. Evaporated salt, Southern rock salt, stack and/or soda salts, and foodgrade salt are all satisfactory materials for making the cell brine. The supply water for brine preparation should not exceed 25 mg/L as $CaCO_3$ hardness. The temperature of this water should lie between 35 and 80°F.

Power consumption will be about 2.5 kWh/lb chlorine and salt consumption about 3.5 lb/lb chlorine.

Operating Experience. Several important features were either confirmed or corrected by the use of pilot plant test installations in an attempt to fulfill the design expectations. One was the determination of the flushing period duration and frequency. The optimum time, based upon seawater on opposite coasts of the United States, appeared to be a two-hour recycled flushing of the electrolyzer cells with 2 percent hydrochloric acid every two weeks. This calls for backup equipment or storage of hypochlorite to span the flushing period.

During these tests the manufacturers claim that by using 3 percent saline seawater at 50°F, the power requirement of 2.5 kWh/lb chlorine was verified.

Also during these tests design changes were made that resulted in successful operation in seawater at 40°F on the shores of the United Kingdom.

Upon observation of product strength decay at an ambient temperature of 65°F and an initial product strength of 700 mg/L, 9 percent decay occurred after 4 hours of storage, 24 percent after 8 hours, and 61 percent after 48 hours.[20]

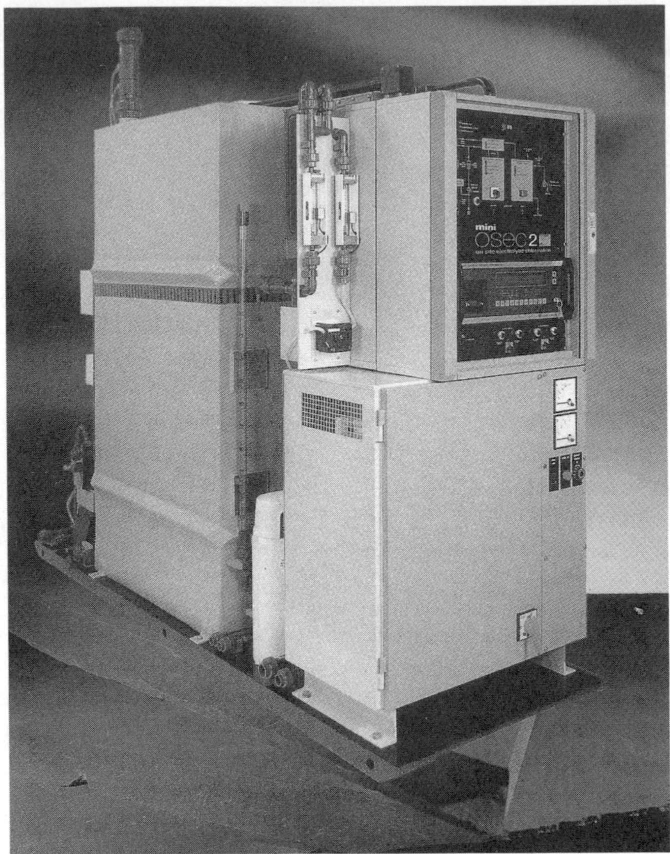

Fig. 3-17. Wallace & Tiernan mini-OSEC-2 on-site electrolytic chlorination system.

Wallace & Tiernan Mini-OSEC-2, On-site Electrolytic Chlorination System[21]

Introduction. This unit was developed in about 1993, from the already successful 2000 lb/day OSEC unit that is described above. This unit is definitely an important alternative for the chlorination of small water supplies. Its maximum capacity of 37 lb/day is equivalent to the operating capacity of a 150-lb cylinder of chlorine.* See Fig. 3-17.

These on-site generating units are available just in time to overcome the annoyances of the Uniform Fire Code—which has destroyed the reputation of our best disinfecting chemical application (the century-old use of chlorine

*When using a 150-lb cylinder of chlorine, it is standard practice never to exceed 40 lb/day withdrawal rate of chlorine at room temperature (68°F). Withdrawal rates in excess of this rate will gradually lower the chlorine pressure in the cylinder, thereby decreasing the rate of withdrawal.

gas)—by producing a chlorine solution that is not listed as a hazardous chemical. Therefore the user does not have to worry about HazMat. Furthermore this unit makes a 0.8 percent sodium hypochlorite solution that is completely stable and therefore harmless to operating personnel, and it does not undergo any degradation in the 265-gal storage tank. Sodium hypochlorite suppliers, on the other hand, deal in hypochlorite solutions that are from 12.5 to 15 percent chlorine. The terrible problems that operating personnel have encountered where water treatment plants using chlorine gas have elected to switch to hypochlorite to meet the UFC requirements are described in Chapter 2.

This concentration of chlorine, at pH 11 to 12 NaOCl, off-gases during storage because of natural decay with time. These are definitely not stable solutions. This instability causes lots of operator problems, plus the generation of chlorate and chlorite ions. Also, these ions cause severe damage to the condition of the blood of dialysis patients (see Chapter 2).

System Description. This is a skid-mounted system, as shown in Fig. 3-17, that can generate up to 37 lb/day of chlorine gas. This is the usual limit of a single 150-lb chlorine gas cylinder. The one item that is not skid-mounted is the salt saturation tank.

Design of the Mini OSEC 2 unit is based upon the generation of sodium hypochlorite solution from salt, water, and electricity, not chlorine gas as in many on-site systems. A controlled flow of brine (salted water) is fed to the system electrolyzer (generating unit). There, an impressed DC current electrolyzes the brine, which transforms the salted water into a solution of sodium hypochlorite, plus a small amount of hydrogen gas. See Fig. 3-15 for the chemistry.

As the brine progresses through the successive electrolyzer cells, its degree of conversion to hypochlorite gradually increases in concentration to the design limit of 0.08 percent HOCl concentration. This is one of the design secrets of this system. Finally the solution flows into a 265-gal skid-mounted sodium hypochlorite storage tank.

Technical Requirements.[22] The following are the important requirements necessary to achieve a satisfactory operating installation

- Salt must be high quality, preferably food-grade. This is necessary to avoid formation of calcium and magnesium deposits in the electrolyzer. Between 2.8 and 3.5 lb of salt will be required to produce 1.0 lb of chlorine, depending upon water quality and site conditions.
- Brine flow is controlled by a variable speed peristaltic pump equipped with flow meters and a low flow alarm switch.
- Supply water requirements are: minimum to maximum water pressures 30 to 75 psi. Minimum water temperature at entry to the skid is 45°F, and maximum at entry to the skid is 77°F.
- Supply water consumption is 0.6–0.9 gph/lb of chlorine.

- A water softener is required for waters with a hardness factor above 17 mg/L. The softener should be the regenerative type, with a twin tank designed for automatic switchover. Regeneration of the exhausted resin bed is accomplished by backflushing with brine. A suitable sanitary drain must be provided for the backwash discharge.
- Power requirement is 440 V 3 phase, 50/60 Hz.
- Power consumption is 22.5A (6.0 KVA) at maximum capacity of 37 lb/day. Actual running power consumption will be considerably less. Nominal DC power consumption will be more like 1.8–2.2 kW DC/lb of equivalent chlorine.
- The electrolyzer consists of four cells containing electrodes that are all connected in series. The anodes are ruthenium-coated titanium. The cathodes are made of hastelloy. A temperature sensor is mounted in the electrolyzer housing. A plate-type heat exchanger is mounted in the discharge line from the electrolyzer.
- As there is a moderate bit of hydrogen off-gas in this process, there is a centrifugal fan that supplies dilution air to the skid-mounted unit. It purges the electrolyzer chamber and the storage tank prior to discharge outside the building. Everything in this part of the system is designed to ensure that whatever amount of hydrogen is being vented, the concentration will always be below the explosion level.
- The hypochlorite solution storage is a 265-gal tank on a skid. It is equipped with float-type level switches for low level start and a high level switch for pump cutoff. Supplementary storage tanks do not require venting.

For further details, read Wallace & Tiernan Publication TI 85.020-1UA.[22]

Oxi-Co Waste Treatment Systems

Introduction. In 1988, during an International Water Quality Conference in Cali, Colombia, an on-site chlorine generating unit was introduced as an MOGGOD System. The name of the supplier was not identified, but it was similar to the Oxi-Co unit. Several operating installations in Mexico and South America were discussed.

The patented Oxi-Co system is based upon an electrolytic cell that uses ion-selective membrane technology.[24] This is supposed to produce a mixed oxidant gas from the food-grade salt (NaCl) solution. Its manufacturer claims that this gas has proved extremely effective in many types of water treatment, yet the cell water never passes through the Oxi cell itself. Instead, it is said that the *mixed oxidant gas* is aspirated as needed into the process water to be treated, and that disinfection occurs more rapidly than in traditional systems. Fig. 3-18 illustrates the Oxi-Co 6A unit that has a capacity of 13.2 lb/day of "mixed oxidants."

Proof of Oxi Method Advantages.[24] Laboratory and field tests are reported to show that this method is simple and reliable. Applications claimed are:

Fig. 3-18. The Oxi 6A unit, 13.2 lb/day.

1. Killing *E. coli* equally as well as chlorine, and *Legionella* and *Giardia* better than chlorine.
2. Destroying cooling tower algae at residual chlorine levels of 0.5 ppm and ORP levels below 500 at a pH of 8, which is impossible with chlorine alone.
3. Swimming pool chlorination without the offensive odor, taste, or eye irritation found with straight chlorination methods.
4. Treating drinking water to reduce iron and manganese to acceptable levels, where previous treatment with hypochlorite was unsuccessful.
5. Rapid water purification. Tests in Costa Rican drinking water distribution pipes showed that acceptable total and fecal coliform kills occurred within 33 ft of an 0.8 lb/day mixed oxidant injection point in a 2-inch-diameter pipe with a 3.1 ft/sec flow rate, which is equivalent to a 1.5 mg/L dosage rate.
6. Achieving acceptable chlorine residual levels at the end of a drinking water distribution line where previous disinfection with chlorine had produced no residual, even at elevated dosage rates at the same injection point.

Why Oxi Units Are Supposed to be More Effective than Traditional Systems.[24] A mixture of oxidants is often more effective than a single oxidant against a broad spectrum of microorganisms. Part of the reason is that different oxidants have different ranges of physical and chemical conditions within which they operate effectively as disinfectants.

For example, they vary as to pH and temperature within which they work well as disinfectants. Combinations of oxidants can act synergistically to disinfect in a way that individual oxidants cannot. A mixture of oxidants is generated in the Oxi system. Aspirated directly into the water, this mixed oxidant is far superior as a disinfectant to chlorine alone. Oxidants identified in mixed-oxidant-treated water include chlorine, hypochlorous acid, ozone, hydrogen peroxide, short-lived oxygen, hydroxyl radicals, and others.[24] Users should be sure to confirm these claims by using an ORP system to compare the mixed oxidant procedure with ordinary chlorination systems.

MIOX Corporation Automated Systems[25]

Innovative Technology. MIOX products are based upon a propietary membraneless electrolytic cell that produces a liquid stream of mixed oxidants, including ozone, chlorine dioxide, and hypochlorous acid, which are highly effective in disinfecting water.* The electrolytic cell generates an unusually strong oxidizing solution using a wide variety of saltwater feedstocks, and requires very low power. The MIOX technology is both elegant and robust, allowing a careful matching of a diversified product line with customer requirements and preferences. Units can be located on-site anywhere, and can be tailored to meet customer needs with respect to the concentration of oxidants necessary, and the amount of disinfection required. Power can be provided from any international source, as well as solar panels or batteries.

Effective Results. Mixed oxidants offer a major advantage over chlorine alone, because they work synergistically to kill microbes. The resulting anode products, which are a blend of primary disinfectants in use today, provide substantially more disinfection power than does any single element alone.

System Capacity. The manufacturer's information on this subject is hopeless because its literature says that *the dosage required and amounts of water treated are determined by raw water quality.* The only thing we know with any certainty is that the largest unit can treat about 0.5 mgd, but MIOX does not classify the water quality limitations.

This certainly leaves a lot of doubt in the mind of any future customer who might be attracted by the chlorine dioxide and ozone lure.

Operation Safety Factor. The mixed oxidant solution is extremely dilute, so that it is safe to use and nontoxic. There is no chemical off-gassing, and the equipment can be put into a pump house or treatment facility with no

*As of 1997 no one has been able to explain or confirm the chemical formation of ozone, chlorine dioxide, or hypochlorous acid (HOCl) as the "mixed oxidants," or how hydrogen peroxide (H_2O_2) joins this family of "mixed oxidants." Users are strongly urged to measure the resulting oxidant level (ORP) using the Strantrol HRR Control System. See Chapter 9 for details.

special requirements. In other words, these installations do not fit the list of hazardous chemicals (See Fig. 3-19).

Pure Water Products, Inc.[28]

Chlorine Factory On-site Chlorine Generating System. This company, which was established in 1986, has devoted most of its efforts to residential pool owners, with more than 5000 units installed. Three patents have been issued on this chlorine-generating system. Recently the company has noticed a rise in interest among small water well system owners.

Therefore, anyone who might be interested in the chlorination of a small water supply should take the time to read about these units. This is a commercially expandable on-site chlorine generating system.

Each unit is equipped with a wall-mounted power supply for every pair of electronic generating cells. This unit can be expanded to 12 of these cells, fed

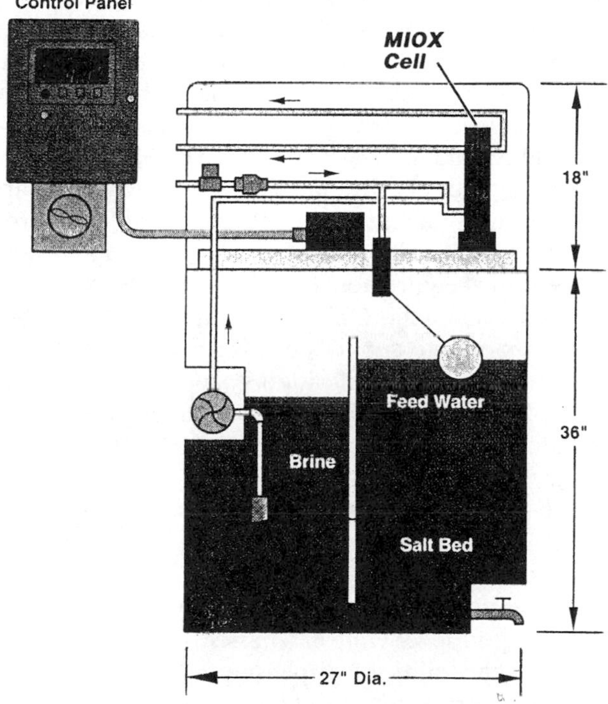

Fig. 3-19. Miox Automated System SAL-020. This is an on-site mixed oxidant generator that produces up to 20 gal/hr of mixed oxidant solution. The system is designed for easy salt loading, and can operate up to two weeks on a single loading of salt, while treating up to 20,000 gal of water per hour. The unit automatically adjusts to operating conditions, is self-diagnosing, and can be equipped for remote monitoring and control signals. It eliminates the need for handling hazardous materials (high concentrations of chlorine vapor). This eliminates the problem of system corrosion.

by a common brine tank with 200 lb of salt. Because these units can produce 1.2 + lb/day chlorine per cell, when they are expanded to 12 cells the chlorine capacity becomes 1.2 × 12 = 14.4 lb/day.

Model C40-CES28. This unit has the following features:

1. A brine tank constructed of PVC, with a removable lid to allow easy access to the salt compartment. This tank is capable of providing brine for up to 12 cells. It has a visual salt level indicator to alert the operator when the salt level is low. Unused brine water is recycled through the brine tank.
2. A power supply that is wall-mounted, weather-resistant, and capable of providing up to 250 watts per hour. This much power is sufficient for two cells. All the electronic parts are cooled by forced air. Each transformer is controlled by an adjustable switch that adjusts the chlorine production from about 8 to 30 amps per cell.
3. The chlorine generating system. This is the fundamental part of the Chlorine Factory; it consists of electronic cells that are encompassed in an injection-molded housing. However, on units larger than two cells, they are manifolded together and are not submerged in the process solution. These cells are easily accessible for service and repair. The cathode is made of titanium, and the anode is also titanium but is coated with a patented platinum coating. These two electrodes are separated by an ion-selective membrane. Each cell is able to produce about 1.2 lb/day chlorine. Additional cells and power supplies can be added for larger chlorine requirements.
4. A reverse osmosis system. Only fresh water is used in the generation of chlorine. In order to supply pure fresh water for this process, the reverse osmosis filtering purification system must be used. It also removes about 90 percent of the hardness minerals commonly found in domestic water. This pretreated water is used for the brine tank makeup water and the pH tank.
5. The chlorine gas that is generated is converted to chlorine solution before leaving the system. However, the system can be plumbed so that the chlorine gas can be introduced directly into the process water.
6. A pH tank. The system contains a tank with a single manual valve and an automatic liquid level control to assist in pH control. This control system has the capability of dispensing a portion of the dilute base product to waste, or to the process water being treated.
7. Optional chlorine storage tank. Diluted chlorine solution (HOCl) can be accumulated in a storage tank with two outlets, one for chlorine solution and the other for any chlorine off-gas. The chlorine solution can be delivered to separate destinations by chlorine feed pumps. The storage tank is designed with an automatic level control system. The

tank system is also arranged so that the strength of the chlorine solution can be changed by operating personnel.

8. The chlorine delivery system can be wired into an automatic controller so that the chlorine feed system and the pH-adjusting device can be controlled automatically.
9. The venturi device can be used to independently withdraw any accumulation of chlorine off-gassing from both the brine tank and the chlorine solution storage tank.
10. A warranty contract is included with the cost of the system. Pure Water Products, Inc. provides a five-year limited warranty for the Chlorine Factory unit, plus all equipment manufactured by Pure Water Products, Inc.

De Nora Seaclor® Systems

Introduction. The name of Oronzio De Nora, Milan, Italy, is one of the oldest names in the technology of electrolytic production of chlorine. This company pioneered the development and use of the mercury cell, which still accounts for a major proportion of worldwide chlorine production. It has recently developed a seawater electrolyzer that is being used extensively in the Mediterranean area and Middle East countries. These systems have been supplied in capacities up to 5000 lb/day available chlorine. The electrolyzers are arranged in a vertical position (see Fig. 3-20) in contrast to the Wallace & Tiernan and Electrocatalytic units, which operate in a horizontal position.

Fig. 3-20. DeNora Seaclor© electrolyzers (courtesy Oronzio de Nora Impianti Elettrochimici, Milan, Italy).

Electrolyzer. The electrolyzers are of modular construction formed by electrolytic cells arranged electrically and hydraulically in series and firmly bound together to constitute an electrode assembly, which is placed in a closed container of corrosion-resistant materials. The cells are arranged in the bipolar configuration. The anodes are dimensionally stable DSA® coated metal, which are widely used in all electrolytic chlorine production processes. The cathodes are made of valve metal (titanium).

Standard electrolyzers are available in unit capacities from 40 to 5800 lb/day of equivalent chlorine. This allows the possibility of standard systems with capacities up to 3400 lb/day of equivalent chlorine without a multiplicity of small cells.[16, 17]

Seawater Systems. Seawater must be screened to prevent particles in excess of 0.04 inch in diameter (1 mm) from entering the electrolyzer. The seawater should contain a minimum of 10 g/L NaCl concentration; 25 g/L is preferred. The temperature of the seawater should be not less than 40°F or more than 86°F.

The maximum hypochlorite concentration that can be generated from seawater containing 19,000 mg/L chloride ion and 36,000 mg/L total salinity is 2500 mg/L available chlorine. This is without a recycle.

De Nora claims a power consumption of 1.7–2.0 kWh/lb equivalent chlorine from a standard seawater at 68°F.

Hydrogen gas is scavenged at the hypochlorite exit of each electrolyzer.

Prepared Brine Systems. The De Nora electrolyzers are equally adaptable to prepared brine solutions and to seawater. When a prepared brine solution is used, the hypochlorite solution produced can achieve a concentration of 8000 mg/L of available chlorine. This is similar to other seawater electrolyzers. To minimize maintenance problems, the brine dilution water hardness should not exceed 25 mg/L as $CaCO_3$, and must be at a temperature not less than 40°F. The raw salt should similarly be as free of calcium and magnesium hardness as is possible. Any significant concentration of these salts will result in precipitation of the salts in the electrolytic components. The precipitates formed inhibit the electrolytic process and thereby significantly reduce the efficiency of the system. Additionally, the maintenance time required to overcome process efficiency reduction due to precipitates is directly proportional to the concentration of calcium and magnesium salts.

Power consumption in these systems will be about 2.25 kWh/lb available chlorine and salt consumption about 3.5 lb/lb chlorine. This is consistent with other manufacturers of similar equipment for this purpose.

SUMMARY

On-site Capabilities

All of the leading manufacturers of this equipment claim nearly equivalent chlorine production per kWh depending upon seawater salinity and temperature—about 2.5 kWh/lb available chlorine when the seawater salinity is 3 percent and the temperature about 68°F.

The hypochlorite solutions produced from seawater are usually limited to about 1800 mg/L available chlorine, and those produced from brine about 8000 mg/L. These strengths vary from the once-through systems to the recycled systems and from manufacturer to manufacturer.

Choice of Systems

General Considerations. The primary consideration for an on-site generating system might be the availability of raw material such as salt and/or electric power. Or it could be the superior safety inherent in the use of on-site generated hypochlorite. Evaluating these factors will require a choice between seawater and prepared brine systems, presuming that both raw materials are available.

Hypochlorite generated from a prepared brine solution is generally quite stable and can be stored at 8000 mg/L concentration for significantly longer periods than high concentration commercial hypochlorite.[15]

Chlorine solutions generated from seawater are highly unstable because of the heavy metal ions present in seawater. Normal decay of solution strength is about 2 to 3 percent per hour. Therefore storage of solution is not practical.

Health Considerations. The use of either brine or seawater for chlorine generation will add sodium to the process stream. Health officials are generally against any process that increases the sodium content of potable water. The seawater system will add 13 parts sodium for each part chlorine. The brine system adds only 2.3 parts sodium per part chlorine.

In the Middle East where fresh water is scarce, the only reliable source of potable water is desalted seawater. Therefore the use of seawater chlorine generating systems is widespread in this area.[18] They are preferred over brine systems because they eliminate the necessity of obtaining solar salt by evaporation. Moreover, chlorine is not readily available in these countries.

In areas where desalted seawater is the only potable water source, disinfection is imperative in order to protect the quality of the finished water from possible aberrations that usually occur in a desalting process. Seawater is a potential health hazard, particularly when the source is adjacent to a community.

Wastewater Treatment. The broadest spectrum of use of these systems in North America is probably for the disinfection of wastewater, where the potential hazard of storing chlorine gas is the overriding factor.

The next consideration in making a decision between a brine system and a seawater system is the resultant effect of TDS in the treated wastewater. The dilution factor in using a seawater system will be about 4–5 to 1, meaning that there would be 4–5 times more TDS using seawater instead of brine.

This would lead us to the conclusion that a seawater system should not be used for any water reuse situation.

It is also questionable whether or not a seawater system should be considered for chlorine production when used for odor control in sewage collection systems. The large amount of seawater required, because of the low chlorine concentration in the hypochlorite, would undoubtedly cause an additional load on the sewage flow, which in turn would promote the generation of hydrogen sulfide, thereby causing more and more odor. Force mains can generate hydrogen sulfide in concentrations sufficient to require 20–30 mg/L of chlorine for proper control. Seawater hypochlorite generators would be largely self-defeating in these cases.

Typical Applications. In addition to the applications described above, seawater systems have found particular favor for the following uses:

- Offshore oil platforms.
- Ships' water systems.
- Electric utility sites.
- Slime control in seawater cooling systems.
- Remote unattended locations.
- Food processing.

REFERENCES

1. Sweeney, O. R., and Baker, J. E., "An Electrolytic Apparatus for the Production of Antiseptic Sodium Hypochlorite Solution," Iowa State College Bulletin 111, Ames, IA, Jan. 1933.
2. Van Peursem, R. M., Pospishu, B. K., and Harris, W. D., "Antiseptic Hypochlorite by Electrolysis," *Iowa State College J. Sci.,* **4,** (1929).
3. Griffith, I., "The Dakin or Carrel-Dakin Solution," *Am. J. Pharm.,* **89,** 497 (1917).
4. "W & T Electrolytic Chlorinator Type EVC-M," Wallace & Tiernan Company, Inc., Tech. Pub. No. 201 Newark, NJ, 1941.
5. Michalek, S. A., and Leitz, F. B., "On-site Generation of Hypochlorite," *J. WPCF,* **44,** 1697 (Sept. 1972).
6. Leitz, F. B., "On-site Hypochlorite Generator for Treatment of Combined Sewer Overflows," Report No. 11023 DAA 03/72, EPA, Washington, DC, 1972.
7. Dahl, S. A., "Chlor-Alkali Cell Features New Ion-Exchange Membrane," *Chem. Eng.,* 60 (Aug. 18, 1975).
8. Iammartino, N. R., "New Ion-Exhange Membrane Stars in Chlor-Alkali Plant," *Chem. Eng.,* 86 (June 21, 1976).
9. D'Elia, R. A., Mfg. Rep., Ionics, Inc., private communication, San Mateo, CA, 1977.
10. Baur, F., "Contract Granted for Sodium Hypochlorite Generator," *Water and Sew. Wks.,* **119,** 76 (June 1972).
11. "Engelhard Brine Chloropac®," Catalog 75.001, Union, NJ, 1977.
12. Matson, J. V., and Coneway, C. R., "Economics of Disinfection," paper presented at the IOI Forum on Ozone Disinfection, Chicago, IL, June 2–4, 1976.
13. Seawater Data Systems Tech. Information Bulletin E-SC-21, Diamond Shamrock-Sanilec Systems, Electrode Corp., Chardon, OH, 1976.
14. Bennett, J. E., and Cinke, J. E., "On-site Hypochlorite Generation for Water and Wastewater Disinfection," Electrode Corp., Chardon, OH, 1975.

15. Anon., "Wallace and Tiernan On-site Electrolytic Chlorination Systems (OSEC)," Cat. File No. 85.500, Jan. 1980.
16. Anon., "Sealcor® Systems for On-site Generation of Hypochlorite Solution from Seawater," Oronzio De Nora Impianti Elettrochimici Spa, Milan, Italy, 1983.
17. Spinelli, F., private communication, Mar. 11, 1983.
18. Kott, Y., private communication, Technion University, Haifa, Israel, Sept. 1982.
19. Staff Report, "Hypochlorite Generator System Comparative Cost Analysis," Union Sanitary District In-House Report, Fremont, CA, 1982.
20. Bryant, J. L., private communication, Wallace & Tiernan, Seattle, WA, July 1983.
21. *Wallace & Tiernan Mini OSEC 2 On-site Electrolytic Chlorination System,* publication No. SB 85.020 UA, July 1994.
22. *Wallace & Tiernan Mini OSEC 2, On-site Electrolytic Chlorination,* Publication TI 85.020-1UA, Aug. 1994.
23. Chemical Services Co., *Chlor-Tec Standard Process,* 2528 Seaboard Avenue, San Jose, CA, 95131, 1996.
24. Oxi-Co, Inc. Water Treatment Systems, *Oxi-Co. Mixed Oxidant Gas Generated On-site for Disinfection (MOGGOD),* 218 Southgate Avenue, Virginia Beach, VA 23462, 1996.
25. OSEC at Chertsy; BRA.367.1, *Wallace & Tiernan OSEC,* 1996.
26. MIOX Corp., *Automated Disinfection Systems,* 5500 Midway Park Place NE, Albuquerque, NM, 87109, 1996.
27. Weeks, C., *Ionics Environmental Water Quality Monitoring Guide,* Ionics, Inc., 65 Grove Street, Watertown, MA 02172-2882, 1993. Customer Service Ph. 800-346-1730.
28. Tucker, D. M. *Chlorine Factory On-site Generation System,* Pure Water Products Inc., 1054 Shary Circle, Suite C, Concord, CA 94518, 1996. Customer Service Ph. 510-827-0291.

4 CHEMISTRY OF CHLORINATION

FUNDAMENTALS OF CHLORINE CHEMISTRY

Introduction

This chapter describes the chemistry of chlorine gas molecules and their reactions when dissolved in aqueous solutions. The purpose of this presentation is to show all of the fundamental reactions so that the practical application of chlorine to potable water, industrial process water, and wastewater can be better understood and analyzed.

Our knowledge of the fundamental chemistry of chlorination has been enlarged considerably in the past 50 years. This has contributed to significant advancement in the field. However, the more we learn, the more we realize how fortuitous it is that chlorine, applied in its simplest form (Cl_2), can be a very potent disinfectant. The phenomenon of chemical simplicity must surely be an important contributing factor to its germicidal efficiency. As a disinfectant, it is without equal, despite its shortcomings.

It is well known that the amount and the complexity of pollutants reaching our potable water supplies are increasing at an alarming rate. This has a direct effect on the chemical reactions of chlorine in aqueous solutions. In general it can be said that the following compounds are of significance in their reactions with chlorine, insofar as water and waste treatment are concerned:

1. Ammonia
2. Amino acids
3. Proteins
4. Total organic carbon (TOC)
5. Nitrites
6. Iron
7. Manganese
8. Hydrogen sulfide
9. Cyanides
10. Organic nitrogen

Before discussing the reactions of these compounds with chlorine, it is desirable to become acquainted with how the chlorine molecule is handled as a disinfectant. Chlorine gas (Cl_2) is dissolved either directly in water by a solution-feed chlorinator to form hypochlorous acid, or by a specially controlled process in a solution containing caustic to yield a hypochlorite bleach solution. The exception to these systems occur when a direct gas feed chlorinator or the Water Champ* is used. In these cases, chlorine gas is dispersed directly into the process stream. When the latter solution is used as a disinfectant, it is diluted with water to form hypochlorous acid, as in the first method. The first part of this discussion will concern the fundamental reaction of chlorine and water and the formation of the oxidizing agent HOCl (hypochlorous acid).

Hydrolysis of Chlorine Gas

When chlorine gas is dissolved in water, it hydrolyzes rapidly according to the following equation:

$$Cl_2 + H_2O \rightarrow HOCl + H^+ + Cl^- \tag{4-1}$$

The rapidity of this reaction has been studied by many investigators. Complete hydrolysis occurs in a few tenths of a second at 64°F; and at 32°F, only a few seconds are needed.[1] This unusually rapid rate of reaction is best explained if the mechanism is a reaction of the chlorine molecule with the hydroxyl ion rather than with the water molecule. This can be represented as follows:

$$Cl_2 + OH^- \leftrightharpoons HOCl + Cl^- \tag{4-2}$$

The rate constant for this reaction is about 5×10^{14}, indicating that the reaction occurs at almost every collision of ions.[2] This reaction is of great practical importance because it relates to the chemistry of aqueous chlorine solutions discharging from conventional chlorination equipment. The resulting solution in a chlorinator discharge is limited by design to 3500 mg/L. At this concentration the most highly buffered injector water would result in a pH no higher than 3. At this pH the amount of molecular chlorine in equilibrium with HOCl is substantial. Concentrations higher than 3500 mg/L cause excessive chlorine gas release at the point of application, which is extremely undersirable. Likewise if negative pressures exist in the chlorine solution piping, this contributes to the release of molecular chlorine at the point of application. In addition to the degassing effect, the release of gas in the solution piping has been known to adversely affect the hydraulic gradient between the injector and the point of application. Injector systems are usually designed to maintain at least

*The Water Champ is unique because it is the only device on the market that pulls the chlorine gas directly from the chlorinator and into the process stream without the necessity for using an auxiliary injector at the point of application. (For details see Chapter 9.)

2 psig at the injector discharge. At this pressure and a temperature of 68°F, the solubility of chlorine in water is only about 7.5 g/L.[3]

To demonstrate the relationship of the molecular chlorine–hypochlorous acid equilibrium for both buffered and unbuffered water, Tables 4-1 and 4-2 have been compiled from a computer printout provided by the Bioengineering Research and Development Lab, U.S. Army, Fort Detrick, Maryland.[4] The results are based upon the Cl_2–HOCl equilibrium; the C_3^- ion formation from Cl_2 and the chloride ion; a mass balance for all chlorine species; and an ion balance on Cl^-. Thus the mole percent for HOCl in the tables is based upon a lengthy and complex cubic equation,* which is best described as follows:

$$\text{Percent HOCl} = \frac{100 \times (\text{HOCl})}{[(\text{HOCl}) + (\text{Cl}_2) + (\text{OCl}^-) + (\text{Cl}_3^-)]} \tag{4-3}$$

Table 4-1 illustrates what happens in the chlorine solution discharge from a chlorinator ranging in feed rates to produce concentrations varying from 500 to 3500 mg/L, at 60°F. It also demonstrates the necessity for maintaining a constant high concentration of chlorine (e.g., 1500–2000 mg/L at a low pH in the generation of chlorine dioxide). The molecular chlorine present in the solution coming in contact with the sodium chlorite provides the impetus for a fast and complete reaction.

Table 4-2 demonstrates the stability of a chlorine water solution buffered with either sodium hydroxide or calcium hydroxide. These figures are of interest for on-site manufacture as well as on-site generation of hypochlorite.†

In any hypochlorite solution the active ingredient is always hypochlorous acid:

$$\text{NaOCl} + \text{H}_2\text{O} \rightarrow \text{HOCl} + \text{Na}^+ + \text{OH}^- \tag{4-4}$$

$$\text{Ca(OCl)}_2 + 2\ \text{H}_2\text{O} \rightarrow 2\ \text{HOCl} + \text{Ca}^{++} + (\text{OH})^= \tag{4-5}$$

When a chlorine solution, such as the solution discharge of a conventional chlorinator (unbuffered), is subjected to negative pressure conditions, the solubility is reduced, a phenomenon that usually results in the release of molecular chlorine at the point of application, provided the diffuser is in an open body of water such as an open channel.

For example, at atmospheric pressure and 68°F, the maximum solubility of chlorine is about 7395 mg/L. However, if the solution is subjected to a negative head of 9 inches of Hg, the solubility is reduced to about 5560 mg/L.[4] Therefore, all systems that are not closed should be designed to avoid negative pressure

---

*This equation is shown in Appendix I.

†Tables 4-1 and 4-2 are of no practical value to the plant operator because the injector-operated chlorinator is designed to produce a chlorine solution limited to 3500 mg/L, and the pH of this solution will vary from 3 to 5.

Table 4-1 Percent Molecular Chlorine and Hypochlorous Acid in a Water Solution Buffered from pH 1-6 at 15°C at Atmospheric Pressure

	Solution Concentration (mg/liter)									
	500		*1000*		*1500*		*2000*		*3500*	
pH	Cl_2	*HOCl*	Cl_2	*HOCl*	Cl_2	*HOCl*	Cl_2	*HOCl*	Cl_2	*HOCl*
1	54.30	45.65	64.67	35.25	69.94	29.95	73.29	26.57	78.91	20.89
2	17.66	82.31	27.41	72.52	33.95	65.93	38.78	61.05	49.70	49.97
3	2.48	97.51	4.73	95.25	6.79	93.17	8.68	91.26	13.57	86.28
4	0.26	99.72	0.52	99.46	0.77	99.20	1.02	98.45	1.76	98.19
5	0.026	99.74	0.05	99.71	0.078	99.68	0.104	99.66	0.181	99.58
6	0.000	97.68	0.005	97.67	0.008	97.67	0.010	99.67	0.018	97.66

Table 4-2 Percent Molecular Chlorine, Hypochlorous Acid, and OC^- Ion in a Water Solution Buffered from pH 6-9 at 20°C

	Solution Concentration (mg/liter)								
	5000			*7000*			*10000*		
pH	Cl_2	*HOCl*	OCl^-	Cl_2	*HOCl*	OCl^-	Cl_2	*HOCl*	OCl^-
6.5	.0063	92.28	7.71	.0088	92.28	7.71	.0126	92.28	7.71
7.0	.0017	79.10	20.89	.0024	79.10	20.89	.0034	79.10	20.89
7.5	.0004	54.84	45.51	.0005	54.49	49.51	.0007	54.49	45.51
8.0	.0001	27.46	72.54	.0001	27.46	72.54	.0001	27.46	72.54
8.5	.0000	10.69	89.31	.0000	10.69	89.30	.0000	10.69	89.30
9.0	.0000	3.65	96.35	.0000	3.65	96.35	.0000	3.65	96.35

conditions in the chlorinator solution lines. This prevents off-gassing at the point of application where a diffuser assembly is installed; otherwise, serious corrosion in the surrounding area could occur, as well as offensive chlorine odors.

Chemistry of Hypochlorous Acid

Effect of pH. The next most important reaction in the chlorination of an aqueous solution is the formation of hypochlorous acid. This species of chlorine is the most germicidal of all chlorine compounds with the possible exception of chlorine dioxide.

Hypochlorous acid is a "weak" acid which means that it tends to undergo partial dissociation as follows:

$$HOCl \rightleftharpoons H^+ + OCl^- \tag{4-6}$$

to produce a hydrogen ion and a hypochlorite ion. In waters of pH between 6.5 and 8.5 the reaction is incomplete, and both species are present to some degree. The extent of this reaction can be calculated from the following equation:

$$K_i = \frac{(H^+)(OCl^-)}{(HOCl)} \tag{4-7}$$

Ki, the ionization constant, varies in magnitude with temperature. The values of this constant shown in Table 4-3 have been computed from the acid dissociation constant, *pKa,* based on J. C. Morris's's[5] best fit formula, developed in 1996 as follows:

$$pK_a = \frac{3000.0}{T} - 10.0686 + 0.0253T \tag{4-8}$$

where T = 273 + degrees centigrade.

Table 4-4 shows the precent undissociated HOCl species for the various temperatures and pH values from 4 to 11.7. The percent OCl^- ion is the difference between these numbers and 100.

The percent distribution of the OCl^- ion (hypochlorite ion) and undissociated hypochlorous acid can be calculated for various pH values as follows:

Table 4-3 HOCl Ionization Constant

Temperature (°C)	*0*	*5*	*10*	*15*	*20*	*25*	*30*
$K_i \times 10^{10-8}$ (moles/liter)	1.488	1.753	2.032	2.320	2.621	2.898	3.175

Table 4-4[a]

pH	Percent HOCl 0°C	5°C	10°C	15°C	20°C	25°C	30°C
5.0	99.85	99.82	99.80	99.79	99.74	99.71	99.68
5.5	99.53	99.45	99.36	99.27	99.18	99.09	99.00
6.0	98.53	98.28	98.00	97.73	97.45	97.18	96.92
6.1	98.16	97.84	97.50	97.16	96.82	96.48	96.15
6.2	97.69	97.29	96.88	96.45	96.02	95.60	95.20
6.3	97.11	96.62	96.10	95.57	95.05	94.53	94.04
6.4	96.39	95.78	95.14	94.49	93.84	93.21	92.61
6.5	95.50	94.75	93.96	93.16	92.37	91.60	90.87
6.6	94.40	93.47	92.51	91.54	90.58	89.65	88.78
6.7	93.05	91.92	90.75	89.58	88.43	87.32	86.27
6.8	91.41	90.03	88.63	87.23	85.85	84.54	83.31
6.9	89.42	87.77	86.10	84.43	82.82	81.29	79.86
7.0	87.04	85.08	83.10	81.16	79.29	77.53	75.90
7.1	84.22	81.92	79.63	77.39	75.26	73.27	71.44
7.2	80.91	78.25	75.64	73.11	70.73	68.52	66.52
7.3	77.10	74.08	71.15	68.35	65.75	63.36	61.22
7.4	72.78	69.42	66.20	63.18	60.39	57.87	55.63
7.5	67.99	64.33	60.88	57.68	54.77	52.18	49.90
7.6	62.79	58.89	55.27	51.98	49.03	46.43	44.17
7.7	57.27	53.23	49.54	46.23	43.32	40.77	38.59
7.8	51.57	47.48	43.81	40.58	37.77	35.35	33.30
7.9	45.82	41.79	38.25	35.17	32.53	30.28	28.39
8.0	40.18	36.32	32.98	30.12	27.69	25.65	23.95
8.1	34.79	31.18	28.10	25.50	23.32	21.51	20.01
8.2	29.77	26.46	23.69	21.38	19.46	17.88	16.58
8.3	25.19	22.23	19.78	17.76	16.10	14.74	13.63
8.4	21.10	18.50	16.38	14.64	13.23	12.07	11.14
8.5	17.52	15.28	13.46	11.99	10.80	9.84	9.06
8.6	14.44	12.53	11.00	9.77	8.77	7.97	7.33
8.7	11.82	10.22	8.94	7.92	7.10	6.44	5.91
8.8	9.62	8.29	7.23	6.39	5.72	5.18	4.75
8.9	7.80	6.70	5.83	5.15	4.60	4.16	3.81
9.0	6.29	5.39	4.69	4.13	3.69	3.33	3.05
9.5	2.08	1.77	1.53	1.34	1.19	1.08	0.98
10.0	0.67	0.57	0.49	0.43	0.38	0.34	0.31
10.5	0.21	0.18	0.15	0.14	0.12	0.11	0.10
11.0	0.07	0.06	0.05	0.04	0.04	0.03	0.03
11.5	0.02	0.02	0.015	0.013	0.012	0.01	0.01
11.7	0.01	0.01	0.01	0.01	0.007	0.007	0.006

[a]Computer printout courtesy D. S. Cherry, N.C. State Univ., Raleigh, N.C.

$$\frac{(\text{HOCl})}{(\text{HOCl}) + (\text{OCl}^-)} = \frac{1}{1 + \dfrac{(\text{OCl}^-)}{(\text{HOCl})}} = \frac{1}{1 + \dfrac{K_i}{(\text{H}^+)}} \tag{4-9}$$

Example: At 68°F and pH 8, the percent distribution of HOCl is:

$$100 \times \left[1 + \frac{2.61 \times 10^{-8}}{10^{-8}}\right]^{-1} = \frac{100}{3.61} = 27.5\%^*$$

Effect of Ionic Strength (*I*)

The concentration of positive and negative ions in solution affects the ability of molecules to dissociate into their respective ions. In the case of HOCl, it is the H^+ ion and the OCl^- ion. There is a powerful mutual attraction of oppositely charged ions; therefore, the more ions (the more TDS), the stronger the forces are the dissociation of the molecules. The more ions (the more TDS), the greater the ionic strength ±. Figure 14 in the Appendix illustrates the HOCl dissociation curve for four ionic strengths with water at 77°F. Examination of this curve shows 50 percent HOCl at pH 7.5 for $I = 0.001$. The TDS at this ionic strength is 40 mg/L. At the same pH but in a water with an ionic strength of $I = 0.01$ (TDS = 400 mg/L, the HOCl concentration drops to 45 percent. Waters with a TDS concentration of 40 mg/L would be those obtained from melting snow packs. Those with very high TDS concentrations are to be found in many groundwaters and some surface waters. Those with the highest TDS concentration (i.e., greater than 400 mg/L) would probably be highly polished reclaimed wastewater effluents.

The ionic strength of any solution can be calculated from a known TDS concentration by use of the following equation:

$$K_i = \frac{[f_H{}^+)(\text{H}^+)][f_{OCl}{}^-)(\text{OCl}^-)]}{(\text{HOCl})} \tag{4-9a}$$

where f is the activity coefficient of the ionic species, H^+ and OCl^-, and K_i is the ionization constant. For very dilute solutions (TDS 10 mg/L) $f = 1.0$. So as I approaches zero, f approaches unity.

To determine the ionic strength of a particular water, the activity coefficients have to be determined. These can be approximated by the Debye—Hückel equation. This equation and the appropriate constant needed to solve this equation are to be found in the Appendix.

---

*It must be noted that in lime-softened water, or other high pH waters, the percent of free chlorine (HOCl) is only 0.04 at pH 11 and 4.13 at pH 9 at 68°F.

Hypochlorite Solutions

The exact same chemical reaction occurs when hypochlorite solutions are used instead of aqueous chlorine solutions. If, for example, common bleach (sodium hypochlorite) is used, it disperses in water to form hypochlorous acid:

$$NaOCl + H_2O \rightarrow HOCl + NaOH \qquad (4\text{-}10)$$

The hypochlorous acid formed by this reaction proceeds to dissociate as in Eq. (4-6).

Familiarity with these factors is essential to an understanding of the behavior of dilute chlorine solutions because HOCl and OCl^- ion have very different germicidal efficiencies, as will be discussed later in this chapter. To properly understand the chemistry of the basic reaction of chlorine in an aqueous solution, Eq. (4-1), it is helpful to understand the structure of the chlorine atom.

Chlorine has the periodic number 17, which indicates that there are 17 positively charged protons in the nucleus of the chlorine atom. These protons in the nucleus are neutralized by 17 negatively charged electrons in a series of outer shells. Two of the outer shells contain an irrevocable number of electrons, two in the first and eight in the second. These electrons are not available for reaction. This, then, leaves seven electrons of a possible eight in the last outer shell. Since this shell lacks one electron, it is not closed. This means that there are seven electrons available for reaction. If the chlorine atom were to lose these seven electrons in a chemical reaction, the atom would then have an excess of seven positive protons, giving the chlorine atoms a valence of +7. Furthermore, the nature of the chlorine atom is such that this outer open shell can take on a maximum of only eight electrons in its outer shell to give one negative electron in excess over the protons; this results in a valence of −1 on the chlorine atom. When this situation of −1 valence occurs, the outer shell of the chlorine atom is filled to its maximum of eight electrons, thus giving a stable chlorine radical, Cl^{-1}, which is recognized as the chloride radical.

Valence of Chlorine

Example of the various valences of the chlorine atom in different compounds are as follows:

Sodium perchlorate	$Na^{+1}Cl^{+7}O_4^{-8}$
Sodium chlorite	$Na^{+1}Cl^{+5}O_2^{-6}$
Chlorine dioxide	$Cl^{+4}O_2^{-4}$
Sodium hypochlorite	$Na^{+1}O_2^{-2}Cl^{+1}$
Hypochlorous acid	$H^{+1}O^{-2}Cl^{+1}$
Hydrochloric acid	$H^{+1}Cl^{-1}$
Monochloramine	$N^{-3}H^{+1}Cl_2^{+2}$

Dichloramine	$N^{-3}H^{+1}Cl_2^{+2}$
Nitrogen trichloride	$N^{-3}Cl_3^{+3}$

Therefore, the chlorine atom can have a valence ranging from +7 to −1.

Next, we see that all molecules and compounds must have a sum valence of zero, which means that all of the electrons are balanced by the protons. The chlorine molecule can therefore be considered as two chlorine atoms, $Cl^{+1}Cl^{-1}$, each having a positive and a negative valence balancing to zero. When the chlorine molecule becomes completely reduced to two Cl^{-1} ions, each with a valence of −1, only the positive chlorine atom of the molecule has undergone a change. It changes from valence +1 to valence −1, gaining two electrons for the chlorine molecule. Thus, chlorine is by definition an oxidizing agent. The rule is as follows: Any radical that loses electrons is being oxidized and is therefore a reducing agent. Conversely, any radical that gains electrons is being reduced and is an oxidizing agent.

When chlorine reacts with water, a special type of oxidation–reduction reaction takes place. The chlorine molecule with the sum valence of zero enters into what is known as a disproportionation reaction with water to form HOCl with a Cl^{+1} radical and HCL with a CL^{-1} radical.[6] The sum valence of the reaction

$$Cl_2 + H_2O \rightarrow H^{+1}O^{-2}CL^{+1} + H^{+1}Cl^{-1} \tag{4-11}$$

is still zero, indicating that no electrons have been gained or lost, thus signifying that no oxidation or reduction has occurred and that no available chlorine has been lost in forming HOCl. Now, when oxidation of a substance by HOCl takes place, the Cl^{+1} radical in the hypochlorous acid will steal two electrons from the substance being oxidized and will then become a chloride radical with a valence of −1. This gain of two electrons by definition shows that the oxidizing capacity of the HOCl is equal to two equivalents of chlorine or one mole of Cl_2.

$$H^+ + Cl^- + HOCl + 2e^- \rightarrow H_2O + 2\ Cl^- \tag{4-12}$$

Available Chlorine

The term "available chlorine" has no place in the field of water and waste treatment. This term was established as the basis for comparing the potential bleaching or disinfecting power of chlorine compounds. To compare one bleaching compound with another, it was necessary to be able to establish the oxidizing power of any such compound. This had to be accomplished by a quantitative analysis of the chlorine that was available for oxidizing, or, as seen above, how much chlorine with a valence of greater than −1 was present in the compound. At that time, the only quantitative method in use was the starch–iodide test, now known as the iodometric method.[7] When potassium iodide is added to a solution containing chlorine available for an oxidizing reaction, it will liberate iodine quantitatively from the potassium iodide, as follows:

Hypochlorous acid:

$$HOCl^{+1} + 2\ KI^{-2} + HAc^{*} = I_2{}^0 + KCl^{-1} + KAc + H_2O \qquad (4\text{-}13)$$

Sodium hypochlorite:

$$NaOCl^{+1} + 2\ KI^{-2} + 2\ HAc = I_2{}^0 + NaCl^{-1} + 2\ KAc + 2\ H_2O \qquad (4\text{-}14)$$

Monochloramine:

$$NH_2Cl^{+1} + 2\ KI^{-2} + 2\ HAc = I_2{}^0 + KCl^{-1} + KAc + NH_4Ac \qquad (4\text{-}15)$$

Dichloramine:

$$NHCl_2{}^{+2} + 4\ KI^{-4} + 3\ HAc = 2\ I_2{}^0 + 2\ KCl^{-1} + 2\ KAc + NH_4Ac \qquad (4\text{-}16)$$

From the foregoing basic chemistry, two things are evident: (1) each chlorine radical with a valence of +1 will liberate elemental iodine I_2 on a quantitative basis; and (2) because all of the chlorine compounds (those that contain at least one chlorine radical with ± 1 valence) are made from elemental chlorine Cl_2, it really takes one molecule of Cl_2 to liberate one molecule of I_2.

To show how the available chlorine percentage is determined, let us examine the case of sodium hypochlorite (NaOCl):

Na	O	Cl	mol wt = 74.5
23	16	35.5	= 74.5

$$\text{available chlorine (by wt)} = \frac{35.5}{74.5 \times 100} = 47.7\%$$

In Table 4-5, multiply the percent by weight of chlorine actually present by 2, to give 95.4 percent available chlorine. This shows that the term "available chlorine" is a misnomer, as it is the calculated weight of elemental chlorine (Cl_2) that is required to liberate the same amount of elemental iodine. As only half of the elemental chlorine molecule is of positive valence, when in solution, the available chlorine content of any chlorine compound that has a Cl^+ radical will always be twice the amount of this radical present.

*This equation is shown in Appendix I.

Table 4-5 Available Chlorine

Compound	Mol wt.	Mol of Equivalent* Chlorine	Percent of Chlorine by Weight	
			Actually Present	Available
Cl_2	71	1	100	100
HOCl	52.5	1	67.7	135.4
NaOCl	74.5	1	47.7	95.4
$Ca(OCl)_2$	143	2	49.6	99.2
NH_2Cl	51.5	1	69.0	138.0
$NHCl_2$	86	2	82.5	165.0
NCl_3	120.5	3	88.5	265.5

*Number of mol of chlorine having an oxidizing capacity equivalent to one mol of the compound.

Further proof of this lies in the manufacture of hypochlorite. For example, 10 percent by wt of active chlorine in a sodium hypochlorite solution contains 0.834 lb of active chlorine per gallon. This means that the manufacturer had to use 0.834 lb of chlorine gas (elemental chlorine, Cl_2) to make every gallon of the solution. Furthermore, if 10 gal of this solution were required to give a 0.5 ppm chlorine residual to one million gallons of water treated, it would simply require 8.34 lb of chlorine gas dissolved in one million gallons of water to produce the same residual.

In summary, the term "available chlorine" is a misnomer. Actually, it refers to the oxidizing power of the compound tested, and is always twice the value of the active chlorine by weight present in that compound. The term "available chlorine" is analogous to alkalinity as "calcium carbonate." There could even be a case of a compound being tested that does not contain any chlorine at all; for example, it is proper to use the term "available chlorine" when comparing the oxidizing power of ozone. The reader is cautioned that this term is not to be confused with the term "free available chlorine," which is the concentration of hypochlorous acid and hypochlorite ions existing in chlorinated water. To qualify as a *free chlorine* residual, HOCl must be 85 percent of the total chlorine residual measured.

Chlorine and Nitrogenous Compounds

The most important and undoubtedly the most complex chemistry of water and wastewater chlorination is its reaction with various forms of nitrogen naturally occurring in water.

On this account the concentrations of ammonia N and organic N should be one of the first measurements that a lab technician makes.*

*To quantify the amount of organic N present in a water or wastewater sample, subtract the ammonia N concentration for the Tkn as described in *Standard Methods*.

If the water to be treated did not contain nitrogenous compounds, the chlorination of water would be much simpler. (One of the great attractions of chlorine dioxide is that, unlike chlorine, it does not react with ammonia. See Chapter 12.) The total residual would always be free available chlorine. There would be no problem with quantitative determination of residuals. The disinfecting efficiency of chlorine could be predicted and controlled within a negligible margin of error. Problems of taste and odor from chlorination would probably be nonexistent.

However, this is not the case. Nitrogen appears in most natural waters and in varying amounts as either organic or inorganic nitrogen. These compounds of nitrogen and their relationship to chlorination will be considered in the general grouping as follows:

Inorganic Nitrogen	*Organic Nitrogen*
Ammonia	Amino acids
Nitrites	Proteins
Nitrates	

The chemical state of any nitrogen compound found in nature is a function of time in the overall life processes of all plants and animals. The amounts of these various forms of nitrogen relate directly to the sanitary quality of the water to be treated. These compounds fit very definitely in time on the nitrogen cycle of nature's own processes of purification. A look at the nitrogen cycle (Fig. 4-1) will show the relationship that exists between the various nitrogen compounds and the changes that are likely to occur in nature.[8] Consider the nitrogen cycle as a clock. Organic nitrogen from plant protein occurs at, say, 6 A.M., while organic nitrogen from animal protein occurs at 8 A.M. The first appearance of inorganic nitrogen in the form of ammonia occurs at noon. The completion of the stabilization of the nitrogen compounds occurs at about 4 P.M. in the form of nitrates. These are then available to start the process all over again. This is obviously an oversimplification of the process. It is the relationship in time that is important. Further, it should be remembered that nitrogen and its compounds are in a continuous state of flux throughout this cycle. The various reactions are continuously competing with one another. Nature has conveniently set up a stored supply of nitrogen in the atmosphere, which can reach the earth's surface in two ways. One is by electrical discharge during a storm. In this case, the nitrogen is converted to N_2O_5, which hydrolyzes on contact with water to HNO_3, and falls to earth in solution with the rain. The second way is by nitrogen-fixing bacteria and algae, which have the ability to extract nitrogen directly from the atmosphere.

Certain conclusions can be drawn about the degree of pollution of water if the chemical composition of the nitrogen compound is known. For example, it is safe to say that water containing only nitrates is rather remote in time from any pollution, that water containing nitrites will be highly suspicious, and that waters containing mostly organic nitrogen and ammonia will have

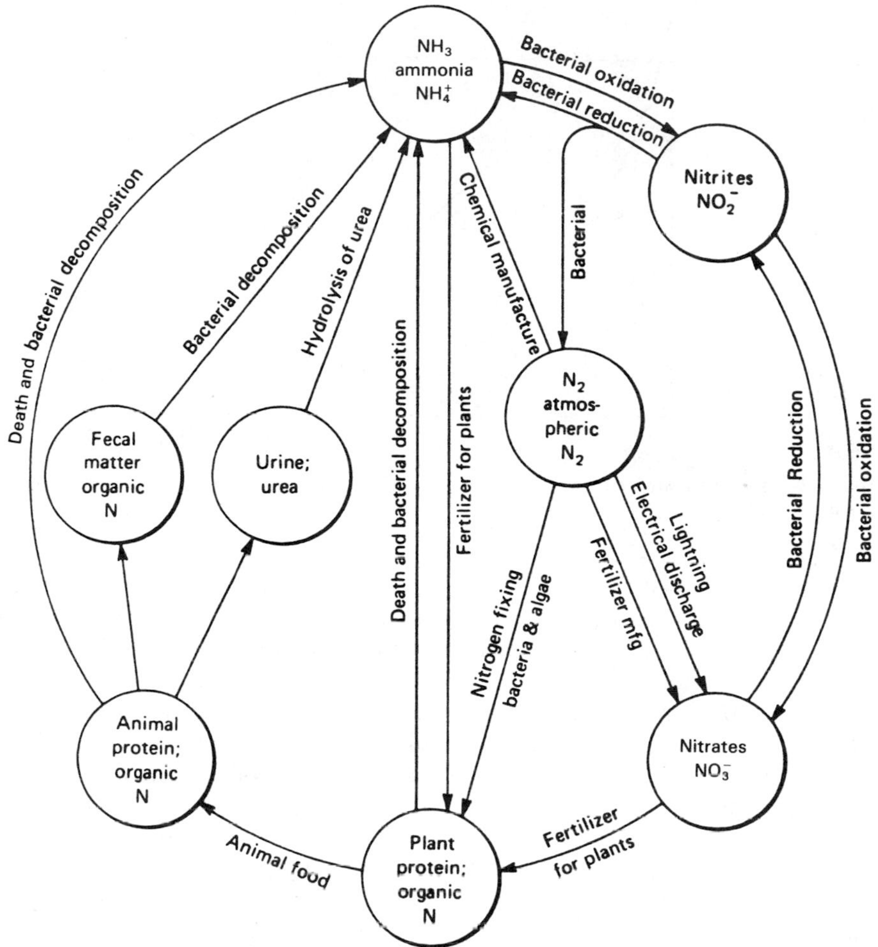

Fig. 4-1. The nitrogen cycle. After Sawyer.[8]

been subjected to recent pollution. However, the nitrogen method of pollution evaluation is not sensitive enough to predict whether a polluted body of water constitutes a public health hazard. The general relationhip of these compounds occurring in polluted water undergoing purification by aerobic processes is best illustrated in Fig. 4-2.

The reaction of chlorine with any compound containing the nitrogen atom with one or more hydrogen atoms attached will form a compound broadly classified as an N-chloro compound, or, more commonly, as chloramine. There are two distinct classes of chloramines—organic and inorganic. The inorganic chloramines are formed by the reaction of chlorine in an aqueous solution with ammonia N naturally occurring in the potable water or wastewater being treated. When organic N is present in a drinking (potable) water system, it will

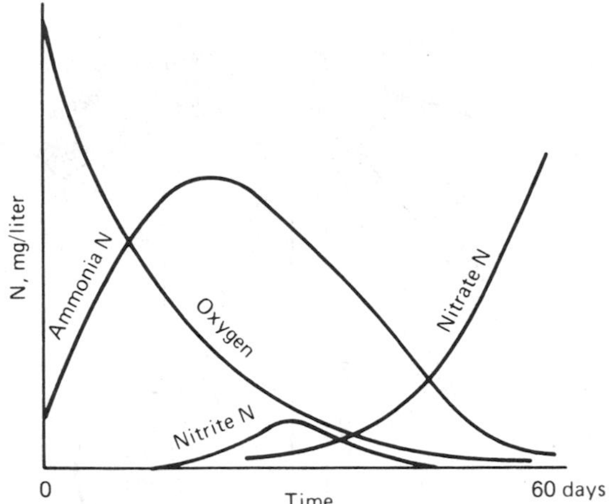

Fig. 4-2. Relationship of nitrogen compounds occurring in polluted water under aerobic conditions. After Sawyer.[8]

react instantaneously with the free chlorine (HOCl), to form a nongermicidal organochloramine. This compound will titrate as dichloramine in the forward titration procedure of both the amperometric method and the DPD-FAS titrimetric procedure.

Other methods, such as the colorimetric procedures, cannot detect this interfering compound. It is important to know that in wastewater chlorination the monochloramine produced hydrolyzes over a period of time (30–40 min) with the organochloramine, to form a nongermicidal chloramine. This is why the germicidal efficiency of chlorinated wastewater effluents will decrease over time, as this hydrolysis reaction will continue for hours. The concentration of organic N is the amount of difference between the Tkn and the ammonia N concentration as shown in *Standard Methods,* 19th Edition.

Organic N: A New Measuring Instrument

Whenever analytical measurements have to be made for either potable water, wastewater, or reclaimed water, two of the most important compounds that have to be quantified are organic N and ammonia N. Now (1997) there is available a new instrument, the DN-1900, Organic N analyzer by Tekmar-Dohrmann Co. (See Fig. 4-3) This instrument can measure the total N content within 2–3 minutes and the total organic N content within 5 minutes. The manufacturer claims that the accuracy is more than equivalent to the *Standard Methods* laborious procedure for the Tkn–ammonia N procedure.

It is the first analyzer to offer two analysis channels. One for the determination of total nitrogen (TN), and a second for nitrogen oxides, nitrates/nitrites (NN). This instrument automatically compares the difference between these determinations and reports the difference as organic N(ON), which is the equivalent of Tkn.

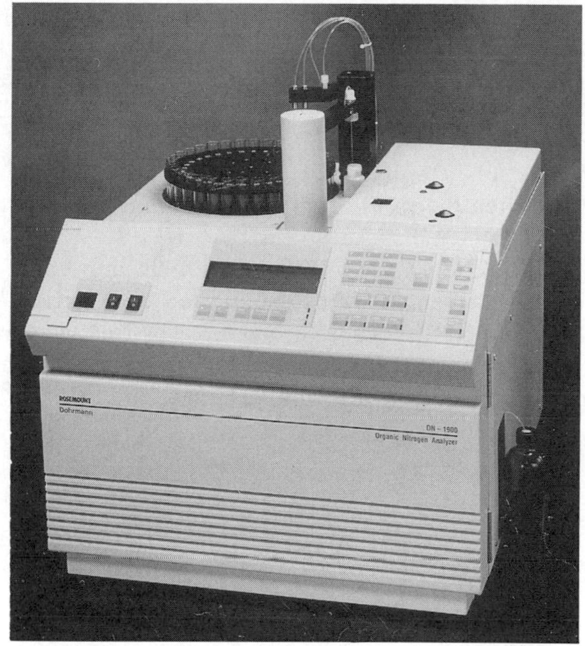

Fig. 4-3. DN-1900 Nitrogen analyzer.

For more detailed information, call Tekmar-Dohrmann in Cincinnati, Ohio at 800-543-4461, or U.S. sales at 800-538-7708.

THE BREAKPOINT PHENOMENON

Introduction

The chemistry of this phenomenon is based upon the inorganic reaction of chlorine with ammonia nitrogen. In dilute aqueous solutions (1–50 mg/L) the reaction between ammonia nitrogen and chlorine forms three types of chloramines, in the following competing reactions:

$$HOCl + NH_3 \rightarrow NH_2Cl \text{ (monochloramine)} + H_2O \tag{4-17}$$

$$NH_2Cl + HOCl \rightarrow NHCl_2 \text{ (dichloramine)} + H_2O \tag{4-18}$$

$$NHCl_2 + HOCl \rightarrow NCl_3 \text{ (trichloramine*)} + H_2O \tag{4-19}$$

These reactions are in general by steps, so that they all compete with each other. A series of complex reactions with all of these substances involves the

*More commonly called nitrogen trichloride.

chlorine substitution of each of the hydrogen atoms in the ammonia molecule. These competing reactions are heavily dependent upon pH, temperature, contact time, initial chlorine-to-ammonia ratio, and most of all the initial concentrations of chlorine and ammonia nitrogen. Note that in all three equations the chlorine atom is positively charged

The reaction of Eq. (4-17) will convert all of the free chlorine to monochloramine at ph 7 to 8 when the ratio of chlorine to ammonia is equimolar (5:1 by wt) or less—that is, 4:1, 3:1, as so on.

The rate of this reaction Eq. (4-17) is extremely important, as it is pH-sensitive. According to reaction rates established by Morris,[9, 10] the fastest conversion of HOCl to NH_2Cl occurs at pH 8.3. The following are calculated reaction rates for 99 percent conversion of free chlorine to monochloramine at 25°C with a molar ratio of 0.2×10^{-3} mol/L HOCL and 1.0×10^{-3} mol/L NH_3:

pH	*seconds*
2	421
4	147
7	0.2
8.3	0.069
12	33.2

The reaction slows appreciably as the temperature drops. At 0°C, it requires nearly five minutes for 90 percent conversion at pH 7.

The pH dependence of this reaction is described accurately on the basis of the HOCl—OCl^- equilibrium and the NH_3—NH_4^- equilibrium.

The reaction of Eq (4-18) will form dichloramine between pH 7 and 8 if the ratio of chlorine to ammonia is 2 mol chlorine to 1 mol ammonia nitrogen (10:1 by wt). The rate of this reaction is much slower than that of Eq. (4-17). It may take as long as one hour for 90 percent conversion[9] and up to five hours at pH 8.5 and above when ammonia nitrogen concentrations are very low. As the pH approaches 5, the reaction speeds up appreciably. This reaction is dependent upon pH, initial ammonia nitrogen, and temperature. The reaction time of Eq. (4-18) is known to be minutes when the initial nitrogen concentration is in excess of 1 mg/L and the pH is favorable.

The reaction of Eq. (4-19) will form some nitrogen trichloride when the pH is beween 7 and 8 if the chlorine-to-ammonia nitrogen ratio is 3 mol of chlorine to 1 mol of ammonia nitrogen (15:1 by wt). At present very little is known about the kinetics of this reaction, particularly in concentrations of less than 10 ppm (10^{-4} *M*). Nitrogen trichloride does form, even at equimolar ratios of chlorine to ammonia nitrogen, if the pH is depressed to 5 or less. At one time it was thought that it would not form above pH 5. It is known to exist in water treatment plants when the pH is as high as 9.[11] This occurs at very high chlorine-to-ammonia nitrogen ratios (25:1 by wt).

In waterworks practice, if the chemistry of Eq. (4–17) is practiced, it is known as either the chlorine–ammonia process, the chloramine process, or chloramination. Equations (4-18) and (4-19) are related to the "breakpoint" phenomenon. It was Griffin's work that led to the discovery of this phenomenon in 1939.[12, 13] Griffin was attempting to explain the sudden loss of chlorine residuals and the simultaneous disappearance of ammonia nitrogen at treatment plants that were experimenting with higher than usual chlorine residuals (2–15 mg/L).

These high residuals were used in an attempt to destroy obnoxious taste and odors. Griffin was startled to find that increasing the chlorine dose in certain waters not only did not increase the residual but reduced it significantly. He called this point of maximum reduction of residual the "breakpoint." Figure 4-4 illustrates this phenomenon, with point *A* designated as the breakpoint.

The importance of the breakpoint phenomenon is its relationship to the control of taste and odors either naturally present in the water or caused by the addition of chlorine. Griffin found that tastes and odors occurring in the region on the left of *A* disappeared on the right of *A* in Fig. 4-4. Other significant benefits were experienced in plant practice, the most important being the increased germicidal efficiency. The killing power of chlorine on the right of *A* is 25-fold more than that on the left of *A*.

Attempts to explain Griffin's findings[13, 14] led to serious scientific investigations between 1946 and 1950. The most extensive laboratory investigation was accomplished at Harvard University. The work of Fair, Morris, Chang, Weil,

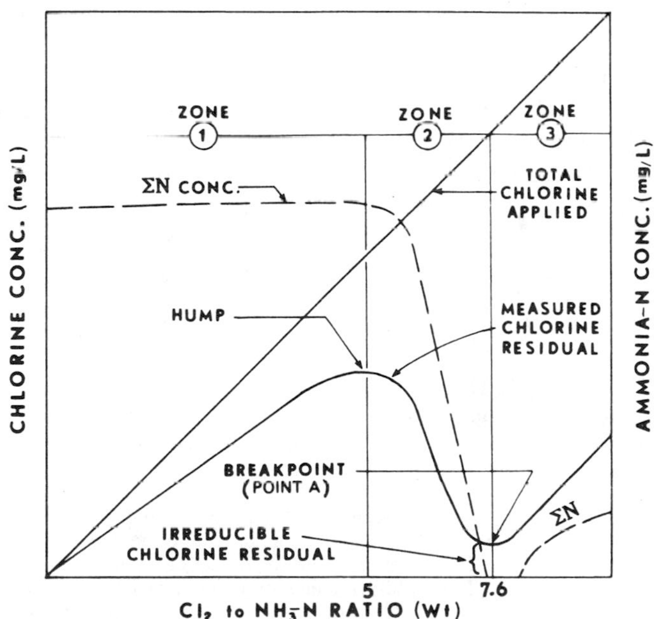

Fig. 4-4. Theoretical breakpoint curve.

Burden, Culver, and Granstrom is largely responsible for our present knowledge of the physical chemistry of chlorination.[9, 15–22]

Palin's* work in England during the same period is also noteworthy.[23, 24] Williams's work on a plant scale substantiates the findings of both Palin and the Harvard group.[25–27] Williams's work is of significant importance because it documents a period of 20 years and because it demonstrates the practical value of this knowledge. It is of historical interest to note that all of these investigations substantiated that of Holwerda done almost 20 years earlier.[28] The work at Harvard was an effort directed primarily toward the explanation of the mechanism that triggers the breakpoint. These investigators used the most sophisticated and accurate method then available for the identification of the various chlorine residual fractions—namely, the light absorbance principle of the spectrophotometer.

The Breakpoint Curve

The breakpoint curve is a graphic representation of chemical relationships that exist as varying amounts of chlorine are added to waters containing small amounts of ammonia nitrogen. The theoretical breakpoint curve is shown in Fig. 4-4. It was originally developed as a result of Griffin's work in 1941–44.[13, 14] This curve has several characteristic features. The principal reaction in Zone 1 is the reaction between chlorine and the ammonium ion indicated in Eq. (4-17). This results in a chlorine residual containing only monochloramine all the way to the hump in the curve. The hump occurs, theoretically, at a chlorine-to-ammonia nitrogen weight ratio of 5 : 1 (molar ratio 1 : 1). This ratio indicates the point where the reacting chlorine and ammonia nitrogen molecules are present in solution in equal numbers. As the molar ratio begins to exceed 1 : 1, some of the monochloramine starts a disproportionation reaction† to form dichloramine in accordance with Eq. (4-18.)[22]

To the right of the breakpoint, Zone 3, chemical equilibria require the buildup of free chlorine residual (HOCl). In practical applications of breakpoint chlorination, reactions occur that result in the formation of nitrogen gas, nitrate, nitrogen trichloride, and other end products. These reactions

*Palin developed the DPD-FAS titrimetric method in England in 1950 long after the amperometric method was developed in the United States by H. C. Marks and J. R. Glass in 1942. In about 1955, Palin got the amperometric method banned in the United Kingdom. This forced Wallace & Tiernan of Great Britain to develop the Residometer for measuring the DPD colorimetric methods (see Chapter 5).

†A disproportionation reaction is one that transforms a substance into two dissimilar compounds by a process involving simultaneous oxidation and reduction. Therefore, the chemical equilibrium in Zone 2 favors the formation of dichloramine and the oxidation of ammonium ion according to Eqs. (4-17), (4-18), and (4-19). This results in a marked reduction of ammonia nitrogen caused by the oxidation of chlorine. The above reactions proceed in competition with each other to produce a breakpoint at a theoretical Cl_2 to NH_4^+ weight ratio of 7.6 to 1. At the breakpoint (Point *A*) ammonia nitrogen either is at a minimum or disappears entirely.

consume chlorine and cause the $Cl_2:NH_4^+$ ratio to exceed the stoichiometric value of 7.6:1 and affect the shape of the breakpoint curve.*

As the chlorine-to-ammonia nitrogen ratio increase beyond about 12–15:1, the reaction of Eq. (4-19) sets in. Under these conditions, the formation of nitrogen trichloride will occur even at pH values as high as 9. As the chlorine dose is increased beyond point *A* in Zone 3, the free available chlorine residual will increase in an amount equal to the increase in the dosage. Therefore, the breakpoint curve in Zone 3 should plot at a 45 degree angle.

It should be emphasized that the shape of the breakpoint curve is affected by contact time, temperature, concentration of chlorine and ammonia, and pH. High concentrations increase the speed of the reactions. As the pH decreases below 8.3, the reactions are retarded. The higher the temperature, the faster the reactions. The shape of the curve is different for different contact times.

In potable water practice this is known as free residual chlorination rather than the breakpoint process. In potable water treatment the practical significance of the curve is briefly as follows:

- *Zone 1.* The residuals in this zone up to the hump are all monochloramine. The residuals in this zone do not form trihalomethanes, nor do they usually contribute to tastes and odors
- *Zone 2.* As the hump is passed, the monochloramine plus the addition of more free chlorine begins to form dichloramine, which is about twice as germicidal as monochloramine. However, this may not be the best part of the curve for the production of a palatable water. A pure dichloramine residual has a noticeable disagreeable taste and odor, but monochloramine does not. It is generally considered best to avoid this part of the curve in order to avoid taste and odor problems.
- *Zone 3.* At the dip of the curve (point *A*) and beyond, free chlorine residual will appear. The total residual will be made up of the nuisance residuals plus free chlorine. If nitrogen trichloride is formed, it will appear in this zone. In practice it has been found that a ratio of free chlorine to total residual of 85 percent or greater will result in the most palatable water.

The Breakpoint Reaction

The mechanism by which the breakpoint phenomenon occurs was first proposed by Rossum in 1943.[30] He calculated the distribution of mono- and

*The following values have been observed at the breakpoint; Griffin and Chamberlin observed 10–12.5:1 at pH values between 6 and 9;[13] Palin observed 9.5:1 at pH 6, 8.25:1 at pH 7, and 8.4:1 at pH 9;[24] Moore et al. observed an average value of 9.0:1 at pH 6–9;[29] Metropolitan Water, District of Southern California has observed values of 9:1 at pH 7.9 and 11:1 at pH 7.5 in their two different supplies.[35] This is also the point at which there exists an irreducible minimum of chlorine residual, commonly described as "nuisance residuals." These residuals titrate as predominantly dichloramine with a trace of free chlorine and monochloramine. Griffin also discovered that in Zone 3 there will be a return of ammonia nitrogen.

dichloramine by means of mass action equations. Although he recognized the possible occurrence of nitrogen trichloride, he considered the breakpoint reaction as one of the relationships of mono- and dichloramine. He confirmed Griffin's finding that ammonia does reappear to the right of the dip in the curve, and was one of the first to observe the destruction of free chlorine (HOCl) by sunlight.

The equilibrium condition between mono- and dichloramine in equimolar concentrations was established by the Harvard group, using spectrophotometric methods.[31] This work demonstrated that the equilibrium condition between mono- and dichloramine can be expressed as follows:

$$\frac{(NH_4^+)(NHCl_2)}{(H^+)(NH_2Cl)} = 6.7 \times 10^5 = K \qquad (4\text{-}20)$$

Curve *A* in Fig. 4-5 shows this equilibrium relationship for mono- and dichloramine at different pH levels based on Eq. (4-20).

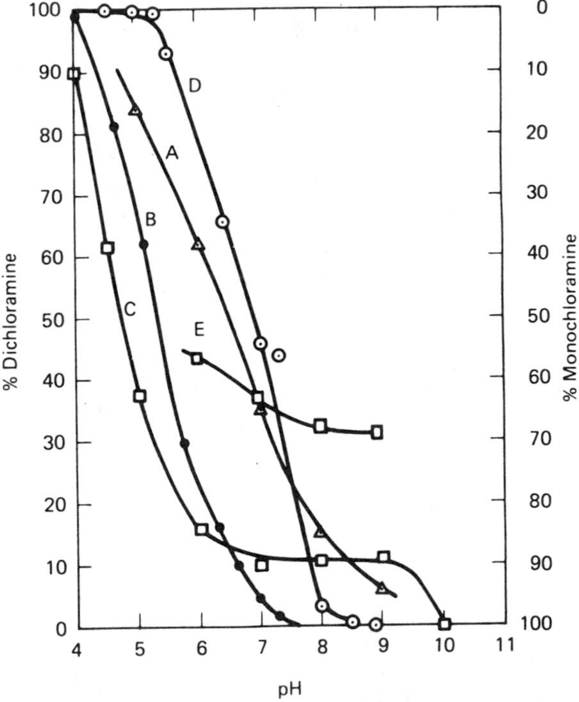

Fig. 4-5. Distribution of monochloramine and dichloramine with relation to pH. Curve A = Fair et al.[31] Cl—NH_3 wt ratio 5:1; curve B = Palin[24] Cl—NH_3 5:1, 2-hr contact; curve C = Baker[33] Cl—NH_3 4:1 2-hr contact; curve D = Chapin[32] (excess NH_3); curve E = Kelly and Sanderson[34] Cl—NH_3 2:1 (curve A computed from Eq. 4-20, $K_{eq} = 6.7 \times 10^5$).

The work of other investigators is also shown. It is interesting to note that Chapin's work[32] in 1929, using a variation of the starch–iodide technique for the residual fractions, compares more closely to the expression in Eq. (4-20) than do the others. However, Palin's[24] and Baker's[33] work tend to agree with each other. Palin separated the residual fractions by neutral O-T FAS titration. Baker used the amperometric titration method, as did Kelly and Sanderson.[34] For unknown reasons, the work of Kelly and Sanderson does not compare favorably with thc rest. Williams's work,[36, 37] which is on a plant operation, more nearly substantiates the Harvard investigation.

From all of this information, Morris[9] suggested that the key to the final breakpoint reaction was the formation and decomposition of dichloramine. This work, in turn, inspired the investigation of the disproportionation reaction of monochloramine in the breakpoint phenomenon.[21] This investigation has led to the following information about the breakpoint phenomenon:

As the chlorine-to-ammonia nitrogen ratio proceeds from 5:1 to 10:1 and greater, two important reactions take place that tend to shift the relative concentrations of mono- and dichloramine:

1. The spontaneous conversion of monochloramine forms dichloramine and ammonia. This is known as the disproportionation of monochloramine.
2. The decomposition of dichloramine decreases its concentration.

These two reactions are thought to occur only in the presence of excess free chlorine.

The conversion of mono- to dichloramine takes place by means of hydrolysis of a monochloramine molecule to form HOCl, which then reacts with another monochloramine molecule to form dichloramine as follows:

$$\begin{array}{l} NH_2Cl + H_2O \rightarrow HOCl + NH_3 \\ \qquad\qquad\qquad\quad \downarrow \\ \qquad\qquad\qquad\quad NH_2Cl \rightarrow NHCL_2 + H_2O \end{array} \tag{4-21}$$

$$HOCl + NH_2Cl \rightarrow NHCl_2 + H_2O \tag{4-22}$$

This is considered to be a first-order reaction, which does not seem to be affected by the pH or the buffering of the solution.

This is also a second-order reaction, which proceeds in parallel with the previous one, but is pH- and buffer-dependent. It is thought to proceed by means of acid catalysis. This reaction probably occurs between a monochloramine acid complex catalyst, which then reacts with another monochloramine molecule to form dichloramine, as follows:

$$\begin{array}{c} NH_2Cl + \text{Acid} \leftrightharpoons [NH_2Cl \cdot \text{Acid}] \\ [NH_2Cl \cdot \text{Acid}] + NH_2Cl \leftrightharpoons NHCl_2 + NH_3 + \text{Acid} \end{array} \tag{4-23}$$

(In any catalytic reaction, the catalyst is never lost.)

These two reactions then make up the spontaneous conversion of mono- to dichloramine, known as the disproportionation of monochloramine.

Although the rate of formation of monochloramine is dependent on the concentration of ammonia and free chlorine, it is dependent also on pH (because of the equilibria $H^+ + OCl^- \leftrightarrows HOCl$ and $NH_4^+ \leftrightarrows H^+ + NH_3^-$) but is not affected by the buffering of the solution, whereas the rate of formation of dichloramine is acid-catalyzed—in other words, it is dependent upon the buffering of the solution.

An excess of ammonia suppresses disproportionation reaction of monochloramine. These must be an excess of HOCl for Eqs. (4-22) and (4-23) to proceed.

The next step in the breakpoint reaction is the decomposition of the dichloramine, which is dependent on the hydroxyl ion activity. Morris[9] has proposed the following mechanism for the decomposition of dichloramine:

First, the dichloramine ionizes as a weak acid:

$$NHCl_2 \rightarrow H^+ + NCl_2^- \tag{4-24}$$

This proceeds to react with the hydroxyl ion (OH^-) in either of two possible ways:

$$NCl_2^- \xrightarrow{(OH)} N - Cl + Cl^- \text{ (slow)} \tag{4-25}$$

$$N - Cl + OH^- \rightarrow NOH + Cl^- \text{ (fast)} \tag{4-26}$$

or:

$$(NCl_2)^- + (OH)^- \rightarrow NCl(OH)^- + Cl^- \tag{4-27}$$

$$NCl(OH)^- \rightarrow NOH + Cl^- \tag{4-28}$$

Equation (4-28) shows the formation of an intermediate reaction product, the nitroxyl radical NOH. This can proceed to decompose into the end products of the reaction in three ways. Each way is valid because the end products depend on many variables, such as molar ratio of chlorine to ammonia, initial ammonia nitrogen concentration, pH, temperature, and unknown side reactions of chlorine and organic matter.

The first way assumes that the predominant end point of the breakpoint reaction might be N_2O, as first postulated by Chapin.[32] This would require the formation of hyponitrous acid from two nitroxyl radicals through dimerization:

$$2\ NOH \rightarrow H_2N_2O_2 \tag{4-29}$$

The hyponitrous acid slowly decomposes to give dinitrogen monoxide:

$$H_2N_2O_2 \rightarrow N_2O + H_2O \tag{4-30}$$

This mechanism calls for two moles of chlorine for every atom of nitrogen that is oxidized. Morris[39] points out that Palin found that it required less than two moles of chlorine to oxidize each atom of nitrogen, which suggests the possibility that nitrogen gas (N_2) is the predominant end product. In view of this statement, it is interesting to note that the later work (1970) on nitrogen removal from wastewater by Pressley et al.[40] concludes that it requires less than two moles of chlorine to oxidize one atom of nitrogen and that the end products of the reaction were found to be N_2, NO_3^-, and NCl_3.

The second way the breakpoint reaction might proceed would be to assume that the main end product is nitrogen gas (N_2). The decomposition of dichloramine to produce the intermediate reactive product NOH would be as follows:[39]

$$NHCl_2 + H_2O \rightarrow NOH + 2\ H^+ + 2\ Cl^- \tag{4-31}$$

This would be followed by three competing reactions involving the intermediate, NOH, as follows:

$$NOH + NH_2Cl \rightarrow N_2 + H_2O + H^+ + Cl^- \tag{4-32}$$

$$NOH + NHCl_2 \rightarrow N_2 + HOCl + H^+ + Cl^- \tag{4-33}$$

$$NOH + 2\ HOCl \rightarrow NO_3^- + 3\ H^+ + 2\ Cl^- \tag{4-34}$$

These hypothetical reactions are compatible with the end products of the breakpoint reaction.

Equation (4-19) is a part of this reaction, as it accounts for the formation of nitrogen trichloride NCl_3) when an excess of chlorine is present. The reverse of this reaction:

$$NCl_3 + H_2O \rightarrow NHCl_2 + HOCl \tag{4-35}$$

is required to demonstrate the well-known fact that NCl_3 is quite stable under conditions of water chlorination only when an excess of chlorine is present, and to account for the slow decomposition of NCl_3. It is, however, easily aerated because of its low solubility in water.

The third breakpoint reaction could occur in the presence of a large excess of free chlorine:

$$H_2N_2O_2 + HOCl \rightarrow 2\ NO + H_2O + H^+ + Cl^- \tag{4-36}$$

This reaction would require 2.5 mol of chlorine for every nitrogen atom oxidized. This is what Griffin[13] found for large excesses of free chlorine in the pH range 6–9. (To convert ratios in moles to weight ratios multiply by 5; i.e., CL:N = 71:14 = 5.)

It can be concluded then that the breakpoint phenomenon will occur when the ratio of chlorine to ammonia nitrogen by weight is from 9 to 1 or greater, provided that the pH of the environment is favorable. Reaction times are a function of initial ammonia nitrogen concentration but will range from minutes to hours for given pH and temperature conditions. The optimum pH for the fastest reaction is between 7 and 8. Above and below this range the reaction slows appreciably. Lowering the temperature retards the reaction.

Other conclusions of academic interest may also be drawn. Free available chlorine is converted from its positive valence state to its negative valence, or chloride, state as a result of oxidizing the nitrogen atom. The end products are most probably released as gases of nitrogen compounds. If there is any carbonaceous material, chlorine will react with these compounds, with carbon exerting a chlorine demand and being released as carbon dioxide. This release of gases as a result of the breakpoint reaction is readily observed in rotameters installed in the chlorine solution lines. These rotameters are downstream from the injector and will carry a solution of HOCl at a concentration in the range of 500 to 3500 ppm. The pH is usually about 2. Copious quantities of gas bubbles are usually observed. Most of these bubbles will consist of CO_2 and be related to the alkalinity of the injector operating water. The nitrogen gases will be related to the amount of ammonia nitrogen present in the water. In treated wastewater, the organic nitrogen probably will not be released at this point, since the reaction time is not sufficient. A scientific investigation of this reaction might enhance our knowledge of chlorination chemistry. It might also serve to establish some design criteria because the gas released in this phenomenon is sufficient to cause abnormal friction losses in solution lines, and because in some waters the bubbles are of such magnitude that rotameters will not function.

Nitrites have been suggested as a possible product of decomposition. This is unlikely, since nitrites readily react with free chlorine and are oxidized extremely rapidly to nitrates:[38, 92]

$$NO_2^- + HOCl \rightarrow NO_3^- + H^+ + Cl^- \qquad (4\text{-}37)$$

Nitrites will not react with chloramines in the pH range of 6 to 9. However, at pH 4 there is some indication of a reaction with dichloramine.

Unidentified Monochloramine Decomposition Product

In spite of the intensive as well as comprehensive study of this compound by Valentine et al.[90, 91] using all the current technologies, its significance and practical consequences are unknown at this time. It is clearly very polar or ionic because it could not be extracted into several common organic solvents. The species, whatever it is, does not appear to be a cation. Although its existence as an anion is most consistent with its ion chromatographic behavior, this is not conclusive owing to the potential sorption to the resins and the formation of an anionic reaction product that appears as an unidentified peak.

Presumably it can act as both an oxidant and a reductant. This is demonstrated by its disappearance when sulfite and chlorine are added. The very slow reaction kinetics with free chlorine above pH 6 and the rapid kinetics near pH 3 suggest that both OCl^- and HOCl are not reactive species. Therefore, this suggests the presence of some other species that dominates at low pH, such as aqueous chlorine gas or chlorine monoxide.

In any case, the unidentified product could be expected to accumulate in the breakpoint region under typical water treatment conditions, if it turns out that it forms under those conditions. Its reactivity with activated carbon and two common amine-based resins indicates that it may be reactive toward organic compounds found in drinking water. In spite of the fact that this unidentified product could account for only several percent of the initial oxidant concentration, it should not be ignored until it is identified and fully characterized. Its reaction kinetics should be evaluated under more realistic conditions. Whether or not it might be a health concern is not so important as understanding chlorine chemistry, which is undeniably complex.

Hand et al.[97] found by using spectrophotometry that there was one peak and one unknown. The species associated with this unknown peak appears to form at pH values between 4 and 10. Nothing more is known about it.

During all the various work projects on the subject of chlorine chemistry, a significant discrepancy has been found between the results of the DPD-FAS titrimetric method and the UV spectrophotometric procedure.[90] The results should match, but they do not. At the start of the reaction time, the two methods match. As the reaction time progresses, the discrepancy becomes larger. The UV spectrophotometer method always indicated more total oxidant (mono + di) than the DPD-FAS method. It was concluded that there was an interference using the UV spectrophotometric method caused by an unknown or unknowns.

Significance of Trichloramine in the Breakpoint Reaction

The mechanism of breakpoint has never been precisely defined because it is not well understood. This lack of understanding is due largely to the many competing reactions that can take place with the highly reactive chlorine species. In the past there has not been sufficient information about the individual reaction steps that lead to N_2 formation. However, in 1990 Yiin and Margerum[94] addressed this problem by isolation of several of the reactions where NCl_3 is a reactant. They have proposed a redox reaction model that is initiated by trichloramine as follows:

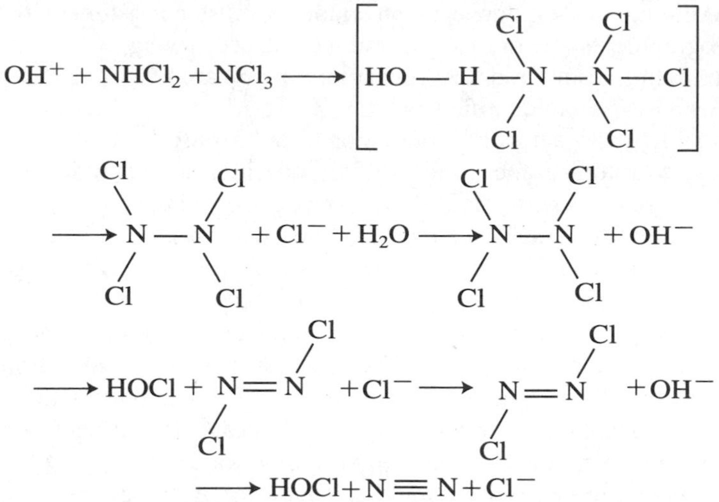

$$OH^+ + NHCl_2 + NCl_3 \longrightarrow \left[HO-H-N(Cl)_2-N(Cl)_2-Cl \right]$$

$$\longrightarrow Cl_2N-NCl_2 + Cl^- + H_2O \longrightarrow Cl_2N-NCl_2 + OH^-$$

$$\longrightarrow HOCl + ClN=NCl + Cl^- \longrightarrow ClN=NCl + OH^-$$

$$\longrightarrow HOCl + N\equiv N + Cl^-$$

This is described as the base-assisted reaction of $NHCl_2$ and NCl_3. The base in this case is OH^-, which helps to remove the proton from dichloramine as it reacts with trichloramine to displace Cl^-. Yiin and Margerum have proposed that the first product is tetrachlorhydrazine (N_2Cl_4), as shown above. This compound decomposes rapidly by OH^- attack to remove Cl^+ and establish the formation of an N—N double bond with Cl^- elimination to give $Cl-N=N-Cl^-$. This reacts by a similar mechanism to give another HOCl molecule, N_2, and Cl^-.

The initial base-assisted reaction between $NHCl_2$ and NCl_3 is much more favorable than the corresponding reactions of $OH^- + NCl_3 + NCl_2$ or $OH^- + NHCl_2 + NHCl_2$. This indicates that OH^- and $NHCl_2$ are needed to generate a strong nucleophile (NCl_2^-) that can react with a strong electrophile (NCl_3) to form an N—N bond and to eliminate Cl^-. The N—Cl bond strength of NCl_3 is weaker than that of $NHCl_2$, so it is easier to eliminate Cl^- from NCl_3. The speed of the base-assisted reaction between $NHCl_2$ and NCl_3 is remarkable. In 0.1 *M* OH^-, the rate constant for the $NHCl_2 + NCl_3$ reaction is more than 10^{10} times greater than the rate constant for hydrazine formation from $NH_3 + NHCl_2$.[99]

There does not seem to be any information about the proposed N_2Cl_4 and N_2Cl_2 intermediates, but the driving force to form N_2 should be very favorable, and they would be expected to be short-lived intermediates in base reactions.[100]

Selleck[95] points out that the above reaction scheme parallels the one proposed in 1973 by Jander and Engelhardt[96] for a reaction between monochloramine and dichloramine as follows:

$$NHCl_2 + NH_2Cl \rightarrow NClN-NClH \xrightarrow{2[OH^-]} N_2$$

In retrospect it is most interesting to find that Chapin's work[93] in 1931, long before the discovery of breakpoint, concluded that the redox reaction was dependent upon the formation of trichloramine (NCl_3)! Moreover, the fact that all the analyses had to be performed by wet chemistry indicates the enormity of Chapin's undertaking for his investigation. As it turned out, he was one of the first investigators of chlorine chemistry to use typical water treatment concentrations of chlorine dosages.

Selleck–Saunier Breakpoint Chemistry

In an attempt to fully understand the chemistry of the breakpoint phenomenon, Professor Selleck and graduate student Saunier decided to review this reaction using both potable water and wastewater. At the time there were disagreements in the literature concerning the chemical pathways to the end products. This work by Saunier[41, 42] under the guidance of Professor Robert E. Selleck, University of California, Berkeley, resulted in the creation of a computer model based upon more than 80 experimental runs, within a pH range of 6–9; ammonia nitrogen from 1 to 20 mg/L; chlorine-to-nitrogen molar dose ratios of 1.6:3.5; and temperatures between 12°C and 21°C. The model as developed by this research can be used to compute the production of the various species of chlorine compounds formed during the breakpoint reaction; the pH of the water immediately and chlorine addition; and the amount of chemical required for buffering the reaction to the optimum pH for the greatest speed of the reaction, as well as the decrease in pH as the reaction proceeds. Based on Saunier's[41] research, the following set of reactions appears to be the most reasonable:[43]

$$NH_4^+ + HOCl \rightarrow NH_2Cl + H_2O + H^+ \qquad (4\text{-}38)$$

$$NH_2Cl + HOCl \rightarrow NHCl_2 + H_2O \qquad (4\text{-}39)$$

$$0.5\ NHCl_2 + 0.5\ H_2O \rightarrow 0.5\ NOH + H^+ + Cl^- \qquad (4\text{-}40)$$

$$0.5\ NHCl_2 + 0.5\ NOH \rightarrow 0.5\ N_2 + 0.5\ HOCl + 0.5\ H^+ + 0.5\ Cl^- \qquad (4\text{-}41)$$

and finally:

$$NH_4^+ + 1.5\ HOCl \rightarrow 0.5\ N_2 + 1.5\ H_2O + 2.5\ H^+ + 1.5\ Cl^- \qquad (4\text{-}41a)$$

The above reactions are based on the formation of NOH, apparently a catalytic intermediary compound, which Saunier and Selleck believed is a result of the formation of hydroxylamine (NH_2OH) as an intermediate reaction.[41, 42] These reactions are as follows:

$$NHCl_2 + 2\ H_{2O} \rightarrow NH_2OH + HCl + HOCl \qquad (4\text{-}42)$$

$$NH_2OH + HOCl \rightarrow NOH + HCl + H_2O \qquad (4\text{-}43)$$

$$NOH + NHCl_2 \rightarrow N_2 + HOCl + HCl \tag{4-44}$$

The formation of the hydroxylamine and the resulting formation of the catalytic compound NOH greatly increase the speed of the breakpoint reaction, as the ammonia nitrogen concentration increases with increased NOH production.

The breakpoint occurs through the sequential formation of monochloramine and dichloramine with the subsequent catalytic decomposition of dichloramine to produce an end product of nitrogen gas, with a partial return of free chlorine residual (HOCl) to the solution. These reactions confirm that 1.5 mol (gram molecular weight) of chlorine are required to oxidize 1.0 mol of ammonia to nitrogen gas.

Stoichiometrically, the breakpoint reaction requires a weight ratio of chlorine to ammonia nitrogen ($Cl_2:NH_4^+{-}N$) at the breakpoint of 7.6:1 as shown below:

$$\begin{aligned} \text{Molecular wt HOCl} &= 70.9 \text{ (as } Cl_2\text{)} \\ \text{Moles HOCl required} &= 1.5 \\ \text{Molecular wt } NH_4^+ &= 14 \text{ (as N)} \\ \text{Moles } NH_4^+ \text{ required} &= 1.0 \end{aligned}$$

Therefore, $Cl_2:NH_4^+{-}N = (1.5)(70.9):(1.0)(14.0) = 7.6:1$; so for each milligram per liter of $NH_4^+{-}N$, 7.6 mg per liter of chlorine is required to reach the breakpoint. In actual wastewater-treatment practice, as was demonstrated by the Rancho Cordova project, 10 mg/L chlorine was required for each 1.0 mg/L ammonia nitrogen present in the process influent.[41] Approximately 70 percent of this breakpoint dosage was consumed to produce nitrogen gas (N_2) from the ammonium ion (NH_4^+) at pH set points between pH 7 and pH 8. The oxidation of NH_4^+ to NO_3^- consumed 8 to 19 percent of the total chlorine dosed to the system. Overall, about 96 percent of the total chlorine dosage was accounted for in reactions between chlorine and nitrogeneous species in specific chemical pathways and free chlorine residual remaining in solution following breakpoint.

Free Hydroxyl Radical

Owing to recent research and pilot plant studies (see Chapter 14) that have evaluated the potency of the "free" hydroxyl radical [OH], particularly where T & O control is important, it has occurred to White that perhaps the breakpoint chemistry described above should be rewritten to show that the hydroxyl ion might be the free hydroxyl radical. This thought was based upon the numerous instances in the 1940s and 1950s where waters with chlorophenol tastes and odors were completely eliminated by the induced breakpoint method. This method involves the addition of ammonia N at the rate of 1 lb/mg, followed by chlorine at the rate of 20 lb/mg. To verify this assumption, White referred the question to the leading authority on free hydroxyl radical chemistry, Professor Wm. H. Glaze at the University of North Carolina, Chapel Hill (Professor Glaze is, at this writing, at Virginia Tech, Blacksburg, Virginia). White included pages 169 to 173 from *The Handbook of Chlorination,* 2nd Ed., 1986, along with his query. Here is Glaze's answer: "The examples you draw from on the pages you attached are not appropriate examples of hydroxyl radical chemistry. The confusion is due to the fact that chemists use terms such as radical to indicate pieces of molecules (such as OH within NOH) as well as 'free' radicals. Also, the term 'hydroxyl' is used to indicate OH in several types of chemical environments. The hydroxyl radical that we are concerned with nowadays, is the 'free' uncharged radical, OH, not OH that appears in molecules such as NOH or methanol (CH_3OH), and not the charged species OH^- which is the thing that causes solutions to have basic pH."

ORP Control Systems

Introduction. The answer given above by Professor Glaze is absolutely correct; there is no doubt about that. However, in the years 1992 to 1997 White worked as a special consultant to Stranco for the evaluation of its High Resolution Redox Control System, for use in the disinfection of potable water and wastewater.

White's investigation of Stranco's ORP systems convinced him that using its ORP measurement was consistent with the evaluation of water and wastewater disinfection because the errors inflicted upon chlorine residual measurements, by organic compounds in these process waters, were eliminated by ORP measurements.

All of this led to a search by White to find the potential strength of residuals, such as free chlorine, combined chlorine, chlorine dioxide and ozone, followed much later by ozone plus hydrogen peroxide as well as ozone plus UV for measure of the oxidation level.

Based upon Eq. (4-31) from laboratory studies by Professor Morris[9] at Harvard University, White opted to believe that the NOH in this equation became the key to the final "breakpoint reaction" by the formation and decomposition of dichloramine, as follows:[106]

$$\text{Dichloramine, basic: } NHCl_2 + 2H_2O + 4e^- = 2Cl^- + NH_3 + 2OH^-$$

This means that the hydroxyl ion in the breakpoint reaction ($2OH^-$) is probably 0.40 mV ORP.

As the variety of equations describing this reaction in this text contain NOH, White has reached the conclusion that the breakpoint reaction is dependent upon the OH ion. If that is so, then the oxidation power of the hydroxyl radical (not ion) wherever it occurs, is most important to disinfection chemistry.

The Hyperzone Reaction. This is the reaction that occurs when ozonated solutions are treated with hydrogen peroxide. This reaction literally makes the ozone "superpowerful." The reaction is as follows:

$$H_2O_2 + O_3 = 2\ OH\cdot + 3\ O_2$$

Note: 2 OH· is the powerful hydroxyl radical.

This unpaired electron makes the 2 OH· molecule extremely active, as can be seen from the table of standard oxidation potentials (below), where ozone is 2.1 and the hydroxyl radical is 2.8. As two of these radicals are formed in the Hyperzone process, that makes this process arithmetically 2.7 times more powerful than ozone alone.

It has been known for a long time that the addition of hydrogen peroxide to an ozonated water increases the oxidation of the ozone—but it took ORP to explain the difference. Actually the ORP difference could be more than the arithmetical difference when hydrogen peroxide-to-ozone ratios and contact times are properly investigated.

The best and only way to accomplish this task is by using the Stranco High Resolution Redox Control System for the Hyperzone Process, as it is the only means by which the maximum ozone concentration can be measured accurately at the maximum ozone contact time. Because every ozonated water usually contains multiple oxidants, ORP analysis is the only way the maximum ozone concentration can be determined immediately prior to the addition of the hydrogen peroxide.

Comparative Oxidation Potentials—mVs

1. Fluorine: 3.0
2. Hydroxyl radical: 2.8
3. Ozone: 2.1
4. Hydrogen peroxide: 1.8
5. Potassium permanganate: 1.7
6. Hypochlorous acid: 1.5
7. Chlorine dioxide: 1.5
8. Chlorine: 1.4
9. Oxygen: 1.2
10. Hydroxyl ion: 0.4

ORP Analysis of the Breakpoint Reaction

Another good reason for the use of ORP is for a quick and accurate method of analyzing the efficiency of the oxidation process that occurs in the B-P process. There will always be "nuisance" residuals that may or may not have any germicidal effect. That is why a "free residual" must always contain at least 85 percent HOCl.

ORP measurement using the Stranco High Resolution Redox Control System has proved to be the method of choice for making this oxidation efficiency measurement. See Chapters 6, 8, 9 and 10 for details.

Side Reactions of Breakpoint Chlorination

The following are some of the chemical reactions other than the direct oxidation of ammonia to nitrogen gas. The reaction products and chlorine consumption for such reactions are governed by factors such as the type and degree of pretreatment, initial $Cl_2 : NH_4^+—N$ ratio, pH, and alkalinity. These reactions are as follows:

Description	*Reaction Stoichiometry*	
Breakpoint reaction	$NH_4^+ + 1.5\ HOCl \rightarrow 0.5\ N_2 + 1.5\ H_2O + 2.5\ H^+\ 1.5\ Cl^-$	(4-45)
NCl_3 formation	$NH_4^+ + 3\ HOCl \rightarrow NCl_3 + 3\ H_2O + H^+$	(4-46)
Nitrate formation		
1-From ammonia	$NH_4^+ + 4\ HOCl \rightarrow NO_3^- + H_2O + 6\ H^+ + 4\ Cl^-$	(4-47)
2-From nitrite	$NO_2^- + HOCl \rightarrow NO_3^- + H^+ + Cl^-$	(4-48)

pH and Alkalinity Considerations

The nature and the concentration of the breakpoint chlorination end products, the chlorine dosage required to reach breakpoint, and the rate of the breakpoint reaction are all affected by the initial pH (following chemical addition) and the pH change that occurs as the breakpoint reaction proceeds. The initial pH in the reaction zone and the pH change through breakpoint depend upon the pH and the alkalinity of the process influent stream, the ammonia concentration, the chlorine dosage, and the amount of alkalinity supplementation.

Acidity is generated in breakpoint chlorination applications (nitrogen removal) from both the hydrolysis and the dissociation of chlorine gas (when Cl_2 gas solutions are used), and the oxidation of ammonia nitrogen.

When the acidity generated is from the hydrolysis and the dissociation of chlorine gas, the following reaction occurs:

$$1.5\ Cl_2 + 1.5\ H_2O \rightarrow 1.50\ OCl^- + 3\ H^+ + 1.5\ Cl^- \qquad (4\text{-}49)$$

If acidity is from the oxidation of ammonia, the following two reactions prevail:

$$NH_4^+ + 1.5\ OCl^- \rightarrow 0.5\ N_2 + 1.5\ H_2O + H^+ + 1.5\ Cl^- \qquad (4\text{-}50)$$

and

$$NH_4^+ + 1.5\ Cl_2 \rightarrow 0.5\ N_2 + 4\ H^+ + 3\ Cl^- \tag{4-51}$$

To counteract the acidity generated in the above reactions, either lime or caustic can be used as follows:

$$\textit{Lime:}\ 2\ CaO + 2\ H_2O \rightarrow 2\ Ca^{++} + 4\ OH^- \tag{4-52}$$

$$\textit{Caustic:}\ 4\ NaOH \rightarrow 4\ Na^+ + 4\ OH^- \tag{4-53}$$

Stoichiometrically, three moles of hydrogen ions are liberated in the hydrolysis and dissociation of chlorine gas to provide sufficient chlorine for the oxidation of one mole of ammonia nitrogen; assuming the initial pH in the reaction zone is alkaline. One mole of hydrogen ion is liberated in the oxidation of ammonia to nitrogen gas.

Saunier's Research

Saunier studied the kinetics of the breakpoint reaction in both tap water and tertiary effluent.[41] Facets of this research will be discussed here only as they may relate to tertiary effluents.*

One of the most important findings by Saunier was the confirmation that there is a considerable time difference required to complete the breakpoint reaction as related to the ammonia nitrogen concentration, other factors being equal.

Figure 4-6 illustrates the distribution of the chlorine residual species (as predicted by Saunier's model) when the NH_3—N is 0.5 mg/L and the contact

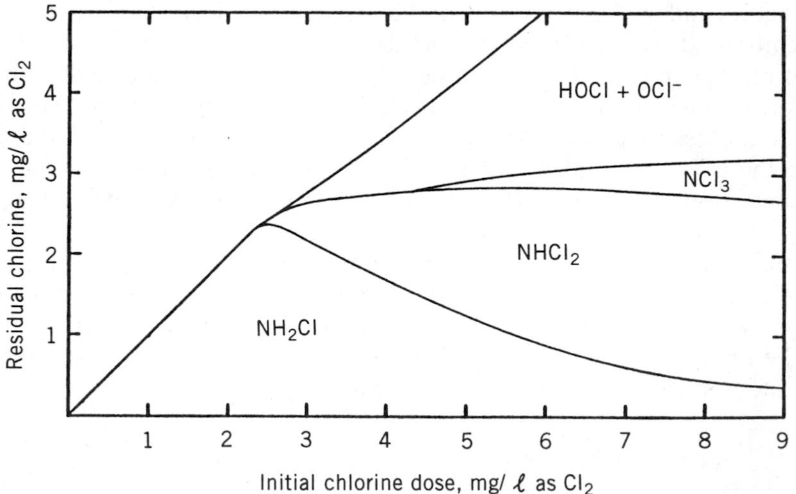

Fig. 4-6. Chlorine dose residual curves predicted by the model after 2.5 min. contact time (pH = 7.4, NH_3—N = 0.5 mg/L, temp = 15°C).

*This includes water for reuse.

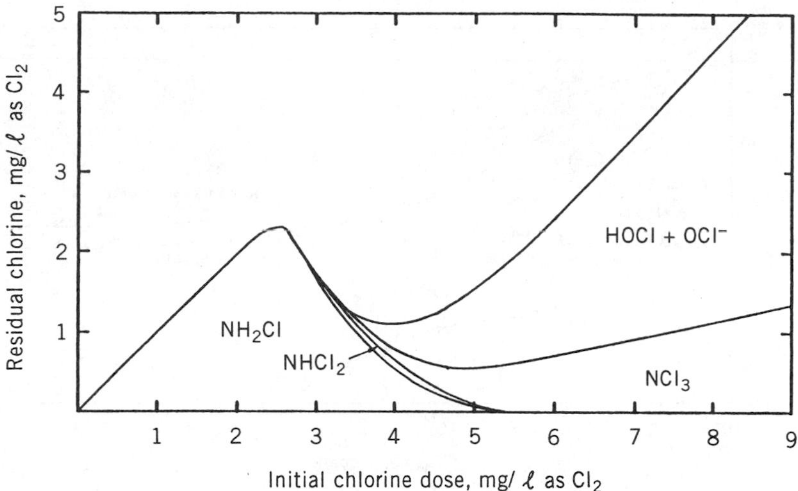

Fig. 4-7. Chlorine dose residual curves predicted by the model after 20 min. contact time (pH = 7.4, NH_3—N = 0.5 mg/L, temp = 15°C).

time is 2.5 min.; Fig. 4-7 is the same situation except that the contact time is 20 min. Figure 4-8 is the predicted breakpoint kinetics but with an ammonia nitrogen concentration of 2.5 mg/L.

Compare these curves with Fig. 4-9, which is a tertiary effluent with an NH_3—N content of 18 mg/L and molar ratio of Cl:N of 1.77. The breakpoint reaction in Fig. 4-9 occurs in less than 3 min. (at the disappearance of NH_2Cl),

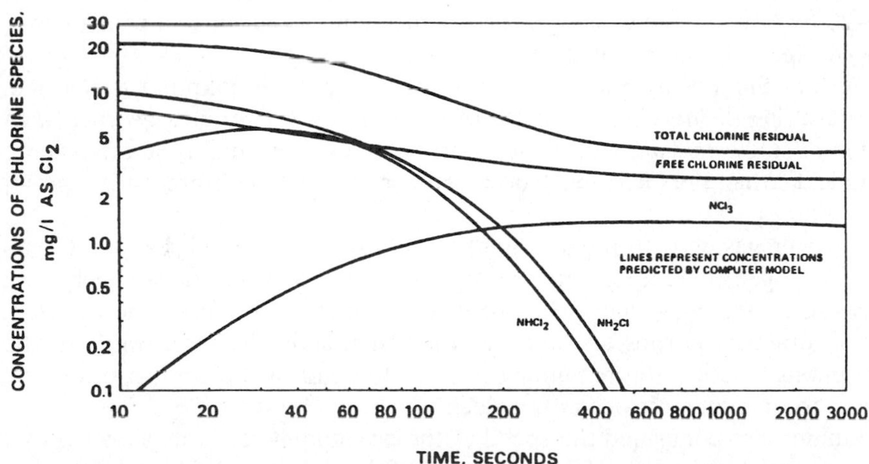

Fig. 4-8. Predicted breakpoint kinetics in a plug flow reactor (pH = 7.5, NH_3—N = 2.5 mg/L, temp = 15°C and Cl_2/N = 9.0 wt). The lines represent concentrations predicted by the model.

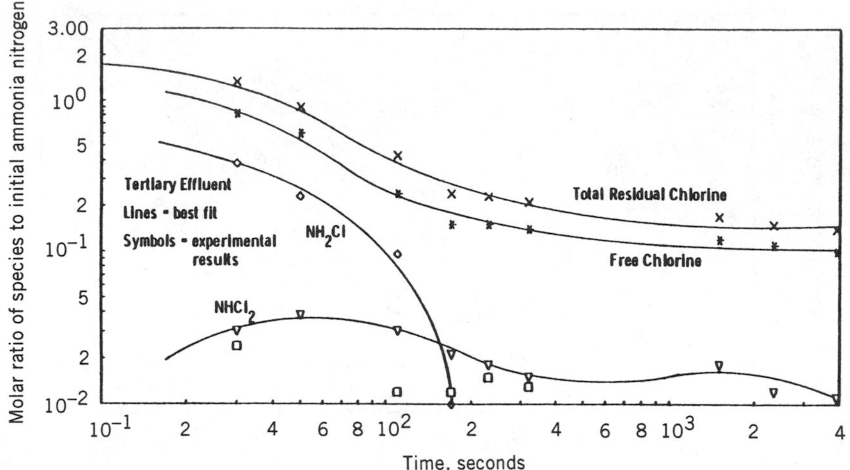

Fig. 4-9. Breakpoint chlorination kinetics of a tertiary effluent in a plug flow reactor (pilot plant experiment). pH = 7.5 after 150 sec, temp = 18.8°C, N = 18.03 mg/L and Cl_2/N = 1.77 molar ratio. Lines = best fit; symbols = experimental results.

whereas it takes about 20 min. for water containing less than 1.0 mg/L NH_3—N and about 8–10 min. when the NH_3—N concentration is 2.5 mg/L, other factors such as pH and temperature being equal. The model prediction of Fig. 4-9 is in close agreement with the Rancho Cordova project findings,[43] where molar ratios of Cl_2—NH_3—N of 2:1 produced the breakpoint reaction in 60–90 sec. at NH_3—N initial concentrations of 15 to 20 mg/L. So for tertiary effluents nitrified to NH_3—N concentrations of less than 2 mg/L, it will require 15–20 min. to complete the breakpoint reaction and produce a controllable, stable, free chlorine residual.

Further findings by Saunier showed that after the breakpoint is reached, the resulting *dichloramine fraction* of the *residual is nearly as germicidal as the free HOCl residual,* and furthermore the HOCl residual is 30 times more germicidal than the OCl^- ion. These findings are indeed different from previous findings.

Nitrogen gas and nitrate are the final end products of the breakpoint reaction. Nitrogen trichloride* formed during the reaction decreases slowly with time. Although NCl_3 does not appear to be an end product of the reaction, for all practical purposes it is an end product, as in the time frame of the process in practice, the remaining NCl_3 at the end of the contact time) will revert to ammonia nitrogen after dechlorination.

Saunier also confirmed the speed of the breakpoint reaction, showing that it occurs most rapidly at pH 7.5. He also confirmed that the NCl_3 concentration

*Nitrogen trichloride is a most effective germicide.

increased with increasing Cl-to-N ratios at all levels of pH. He showed also that the organochloramines interfered mainly with the dichloramine and indirectly with the NCl_3 DPD-FAS analysis, primarily at pH values above 7.5. Traces of organic nitrogen (0.02 mg/L) formed organochloramines in sufficient amount to give false DPD-FAS dichloramine readings at pH values above pH 7.5. The oxidation of the organic nitrogen proceeded faster at pH values of less than 7.5.

In summarizing Saunier's findings, particularly as they relate to tertiary effluents, it can be concluded that if it is desired to achieve a free chlorine residual for bacteria and virus destruction, when the ammonia nitrogen concentration is less than 2.0 mg/L, the process should be carried out at a control pH of 7.0–7.2 with a minimum contact period of 30 min. Moreover, Saunier's work shows some instability of the free chlorine residual in the usual control sampling time of 1 to 3 min. after the point of application. Therefore, residual control in situations where the initial ammonia nitrogen concentration is less than 2.0 mg/L should be based on the total chlorine residual measurement.

At this level of chlorination with most wastewater effluents, pH adjustment due to the addition of chlorine would probably not be required. Most effluents have an alkalinity of 150 to 200 mg/L which is easily capable of maintaining the pH at status quo with chlorine dosages as high as 25 mg/L.

Significance of Organic Nitrogen

Chlorination of waters containing organic nitrogen presents a variety of problems not encountered in waters containing ammonia nitrogen. Organic nitrogen looms as a formidable stumbling block in the production of an acceptable water because of the complexity of the compounds that contain organic nitrogen.

A little background on the source and nature of these compounds will be helpful to us in understanding the problem.

Standard Methods[7] gives analytical procedures for the determination of two nitrogen classifications that derive from other than ammonia nitrogen:

1. Albuminoid nitrogen, which gives only an approximation of the proteinaceous matter present in the water. The nitrogen in this determination is thought to be related to the nitrogen content of the various amino acids.
2. Total organic nitrogen, which is contributed in various degrees by proteins, polypeptides, and amino acids. In other words, the total organic nitrogen determination measures the nitrogen associated with proteins and all the hydrolytic products thereof.

The proteins are the most difficult entities to deal with by chlorination. Proteins are a constituent of plant and animal life normal to the aquatic life and are found in both human and animal waste. All proteins are made up of nitrogen, carbon, oxygen, and hydrogen, and some contain sulfur. Of these elements, nitrogen, carbon, and sulfur exert a chlorine demand. They are complex high molecular weight organic compounds found in various stages of hydrolysis. The hydrolysis is accomplished by hydrolytical enzymes, and the products of degradation are as follows:

$$\text{protein} \longrightarrow \text{proteoses} \longrightarrow \text{peptones} \longrightarrow \text{polypeptides}$$

$$\downarrow$$

$$\alpha \text{ amino acids} \longleftarrow \text{dipeptides} \longleftarrow$$

The α amino acids are then deaminized by enzymic action, and free fatty acids and other acids result; the free acids serve as food for the microorganisms, which, in turn, are converted to carbon dioxide and water.[8]

From the products of hydrolysis, we see that the proteins represent one end of the spectrum and the amino acids the opposite end. Therefore, the chlorination of organic nitrogen compounds can be classified into these two categories. Likewise, the determination of loss of total organic nitrogen will represent the measure of oxidation of proteins by chlorine, whereas the loss of albuminoid nitrogen will measure the oxidation or decomposition of the amino acids involved.

Taras[44] found that the total organic nitrogen consumed after chlorination follows a well-defined general pattern. The ammonia nitrogen is lost completely within one hour of contact time; the simple and unsubstituted amino nitrogen of many common amino acids is consumed more slowly over an extended period of time (many hours); protein nitrogen shows only a negligible loss, even after many days. Taras also demonstrated that this total nitrogen consumption is an exponential function of time.

By way of contrast and clarification, it should be noted that the albuminoid nitrogen content of many simple amino compounds is reduced by as much as 75 percent within the first hour of contact with chlorine, but from proteinaceous matter the reduction is only very slight.

Chlorination of this group of compounds is further complicated by the many varieties encountered. The particular protein involved will have a variable chlorine reaction, depending on the number of amino groups available for reaction with chlorine, which is a function of the peptide linkage in the protein. Therefore, the chemistry of chlorination of organic nitrogen compounds is at best extremely complex because of the various hydrolysis products that may be encountered. Fortunately, most of the proteins are colloids that are partially removed in normal water treatment processes.

The significant difference between the reaction of chlorine with organic nitrogen compounds and that of ammonia is one of time, the comparative loss of nitrogen, and the formation of complex organic chloramines.

Both Griffin[13, 14] and Williams[11] found that waters containing a mixture of both ammonia and organic nitrogen did not display the classic dip of the breakpoint reaction but, rather, a plateau effect. Figure 4-10 illustrates what

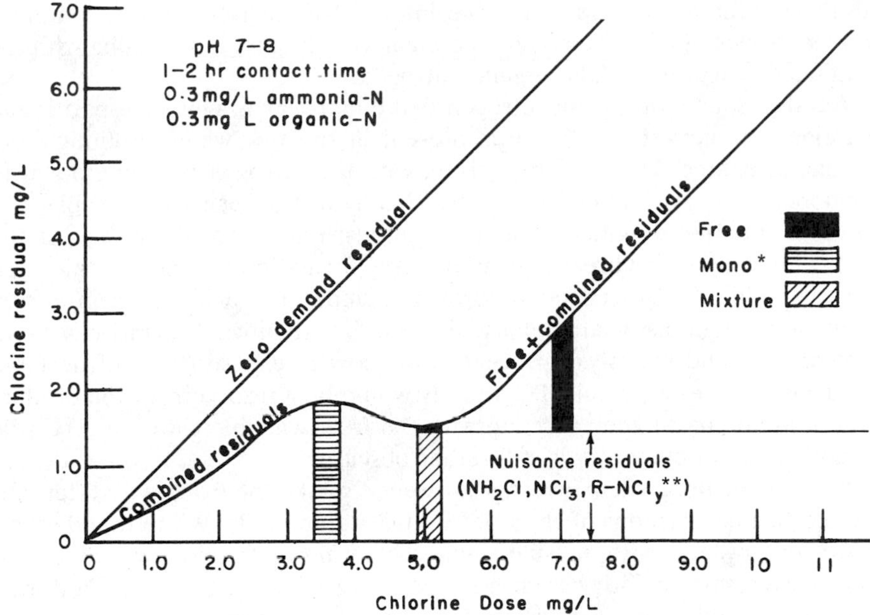

Fig. 4-10. Relationship between ammonia nitrogen and organic nitrogen and chlorine. *Predominately monochloramine; R—NCl_y titrates as dichloramine. **R—NCl_y = organochloramines.

may be expected with a water containing 0.3 ppm ammonia nitrogen and 0.3 ppm organic nitrogen, with about 50 percent of the latter attributable to the simple amino acids and the rest proteinaceous matter. The results shown here are based largely on observations by White and by Williams[11, 27] and Palin.[24] Additional scientific evidence is needed to better predict the free available residual curve with waters characterized by significant amounts of organic nitrogen. Figure 4-10 indicates very little drop in chlorine residual beyond the hump. This signifies two things: (1) continuing and competing reactions between mono- and dichloramine, and (2) very little loss of nitrogen—hence the plateau effect. Beyond the plateau, free available residual begins to appear; but note that the irreducible minimum residual is considerably greater than in the reaction with ammonia nitrogen (Fig. 4-6). Further, it is to be noted that the combined available residual can be made up of equal amounts of mono- and dichloramine, and as the ratio of chlorine to nitrogen increases, significant quantities of nitrogen trichloride can begin to form. This usually occurs when the Cl : NH_3–N ratio exceeds 14 : 1. Therefore these waters have the ability to produce: proportionately greater nuisance residuals; dichloramine, which contributes to taste problems; and nitrogen trichloride, which has an obnoxious odor. In some cases, the ratio of di- to monochloramine has been known to be 2 or 3 : 1 in the presence of excess free chlorine. The residuals in these systems are unstable, inasmuch as the free chlorine continues to react to decompose the mono- and dichloramine that are forming with the continued reaction with the organic nitrogen.

Another source of organic nitrogen that might play a significant part in the chlorination chemistry is the urea present in the raw water resulting from sewage discharge. Urea will hydrolyze, with the nitrogen breaking down to ammonia as one of the end products. This hydrolysis usually takes place in three to four hours under normal conditions; but if there is a lack of the urease enzyme, which breaks down urea, the formation of ammonia nitrogen is greatly inhibited. Urea does not form urea-chloramine at low concentrations, owing to its high dissociation constant,[45] but if a significant quantity of urea-N is present and hydrolysis proceeds at a slow rate, a situation of unstable residuals could easily result. The urea-N would be a reservoir for the production of ammonia to keep reacting with the free available chlorine. The role of urea in water chemistry is still rather obscure.

In Great Britain, where investigators seem to take the trouble to differentiate the various fractions of chlorine residual compounds, White found swimming pool systems with unusual stable dichloramine fractions, regardless of time or excess free chlorine, such as he found in 1982 at the San Jose, California WWTP. The following is a description of the swimming pool problems.

In 1959, Lusher[103] reported that with a total chlorine residual of approximately 6.0 mg/L and free chlorine of approximately 3.0 mg/L, the dichloramine fraction was as high as 2.4 mg/L; so the remaining monochloramine residual was 0.6 mg/L. This condition persisted for days. Increasing the free chlorine residual to 16.0 mg/L did not reduce the dichloramine fraction, but raised it to between 3.0 and 4.0 mg/L The bathing load in this instance was rather high—70 to 80 bathers per hour. Since that time there has been considerable evidence that this phenomenon of high dichloramine values with free residual chlorination is common in British swimming pools where bathing loads are unbelievably high. It is not unusual for 200,000 gal swimming pools to have bathing loads of 2000 per day.

In 1968 Malpas,[104] of the British Wallace & Tiernan Co., reported the following in a large pool: at a pH of 8.65, the free chlorine was zero, the monochloramine 0.4 mg/L, and the di- 2.6 mg/L. The absence of free chlorine was not surprising because the sample took three days to get to the laboratory for analysis. The chlorine fractions were all analyzed by the amperometric method. At that time the British strongly suspected that the stability of the dichloramine fraction was due to the presence of organic N, which they thought retarded the residual process, or that it might be from the formation of an N-chloro compound exhibiting the characteristics of dichloramine. Their suspicions were correct, but little did they know that this compound showing up in the dichloramine fraction is totally nongermicidal.

In 1967, Lomas[105] demonstrated the probability that the anomalous results of chlorination giving rise to stable dichloramine systems are due to the presence of creatinine, a nitrogenous constituent of urine. It was concluded that creatinine forms a mono-chlor-substitution product with chlorine (chlorcreatinine), which is stable in the presence of free chlorine residual, and exhibits characteristics of dichloramine upon quantitative analysis. It was also further concluded that urea will not form a stable dichloramine at pH 8. This was not confirmed by what Malpas discovered.

Lomas further declared at that time that the toxic and bactericidal properties of such compounds as chlorcreatinine were unknown. Consequently, the practical significancc of thc mono- and dichloramine fractions in swimming pool water is unknown, particularly in the presence of free chlorine residual and organic N. As of 1982, we in the United States had found out that the organochloramine that showed up in the dichloramine fraction of a forward titration procedure was indeed nongermicidal when found in a domestic wastewater effluent. We are quite sure this is the same for potable surface waters.

Another significant factor is the chlorine demand exerted by the carbon content of some of the simple amino acids, such as cysteine and glycine.[9] For glycine, Palin[24] noted that there were similarities to the chlorine–ammonia breakpoint phenomenon. Morris's[9] work on the kinetics of chlorine and glycine reveals that up to a 1:1 molar ratio there is only a small loss of oxidizing chlorine. Beyond this ratio, the loss of chlorine increases rapidly until it is complete at a molar ratio of 1.5 chlorine to 1 glycine. This is a lower ratio than the 2:1 for the chlorine–ammonia breakpoint system (10:1 by weight). At the higher chlorine concentrations, there is a distinct departure from the chlorine–ammonia pattern. Significant amounts of oxidizing chlorine are not found in the solutions again until a molar ratio of 4:1 is reached. Upon analyzing the solutions, Morris[9] found that it was the carbon, not the nitrogen, that was being oxidized when the molar ratios were 1.5:1 and 2:1. The original nitrogen was still present either as ammonia or as organic nitrogen. The oxidized carbon is released as carbon dioxide.

The reaction between chlorine and organic nitrogen, if given sufficient time, is certain to form organic chloramines of some kind. These chloramines display characteristics of dichloramine because they always appear in the dichloramine fraction when subjected to present amperometric and DPD-ferrous titrimetric methods of analysis. Recent experience with completely nitrified and highly polished reclaimed wastewater reveals that a small fraction of these organic chloramines will intrude into the monochloramine reading.

Another reaction involving organic nitrogen and chlorine is the relationship between the organic chloramines and the monochloramine formed when ammonia N is present. White et al.[70] found that there is a definite shift of the monochloramine fraction to the dichloramine (organic chloramine) fraction with time. This is probably caused by the disproportionation reaction of monochloramine, whereby HOCl is produced by the monochloramine, which then reacts with more organic N to form more organic chloramines. This causes a simultaneous reduction of monochloramines.

Some organic chloramines are known to hydrolyze to produce HOCl as one end product. The net result of organic chloramines is an overall reduction in the effectiveness of the total chlorine residual. In the free residual process they become a major portion of the nuisance residuals and are known to cause taste and odor problems

Not all species of organic N react with chlorine. In a study of the El Dorado, Arkansas municipal water system, White discovered that a concentration of 0.5 mg/L of organic N in the groundwater supply did not react in several breakpoint curves that were run. This water supply was in an area of a lot of oil well activity.

Organic N: Practical Considerations

In spite of the complexities of organic N compounds, their effects on the chlorination of water are easily determined but are rarely predictable. Their chemical reactions are site-specific. In some groundwaters, such as those found in areas where there are active oil fields, the organic N present in the water may not react at all with chlorine.

The appearance of organic N in surface water is usually an indication of long-term pollution. In groundwaters it may have nothing to do with pollution. By comparison, the presence of ammonia N represents recent pollution.

In chlorination of either potable water or wastewater in the presence of organic N, both free chlorine and monochloramine react to form organochloramines. Free chlorine reacts almost instantaneously, whereas monochloramine takes several minutes. The longer the contact time, the greater the shift will be from monochloramine to organochloramine. With very few exceptions, all organochloramines are nongermicidal and nontoxic to acquatic life. This characteristic adds another element of complexity to the situation because the organochloramines appear in the total residual measurement. Therefore, the operator will never know the negative effect of the organic N present unless forward titrations are carried out to identify and quantify all the chlorine residual species.

If it were not for the strict coliform discharge requirements that affect all of California's dischargers, this phenomenon might have gone completely unnoticed. When the San Jose, California municipal WWTP added nitrification to its treatment train, the efficiency of its disinfection process put it in violation on a daily basis in spite of large increases in chlorine dosage. The nitrification process used at the San Jose WWTP achieved total removal of ammonia N. This meant that there should be a significant amount of free chlorine residual in the effluent. This caused lots of trouble; however, the chief chemist at the plant (Virginia Alford) solved the mystery when she decided to analyze the chlorine residuals for all the chlorine species by amperometric forward titration. She discovered that a 9 mg/L total residual was made up of approximately 50 percent free chlorine and 50 percent organochloramine, which titrated in the dichloramine fraction. This occurred in 1982, and Virginia Alford was the first person to discover this phenomenon.[101, 102] As there was an organic N concentration in the effluent that was between 4 and 6 mg/L on a daily basis, it was obvious that the organic N was depleting the disinfection power of the total chlorine residual. In the case of potable water the usual organic N concentration is on the order of 0.3–0.6 mg/L.

The lesson to be learned from the above scenario is that operating personnel must measure all the chlorine species in all residual measurements and use the total residual measurement as a type of verification. Additionally it is helpful to the operator to be constantly aware of both the ammonia N and organic N concentration in the process water, whether it be potable water or wastewater. There are only two acceptable methods for these determinations: (1) amperometric titration and (2) DPD-FAS titrimetric analysis. The UV spectrophotometer method does not agree with these methods (see Ref. 90).

Chlorine Demand Research by Feben and Taras

General Discussion. To help recognize these systems as well as the general overall relationship between chlorine and nitrogenous compounds, Feben and Taras[46, 47] in 1950 made a notable contribution in detailed study of the chlorine demand of Detroit water.

Taras pursued this work[44, 48] during the next few years, and revealed a definite and significant relationship between chlorine demand and the character of the nitrogen compounds present. These findings are of such practical importance that they are presented here to demonstrate not only how the chemist can identify these troublesome compounds but also how the problem of organic nitrogen can be controlled.

Taras's findings are as follows:

1. There is a definite mathematical relationship between the amount of chlorine consumed in one contact period and the amount consumed in any other period.
2. A study of the constants in this mathematical relationship of chlorine consumed provides clues as to the nature of the nitrogen compounds reacting with the chlorine.
3. This mathematical relationship makes it possible to maintain specific chlorine residuals in the finished water.
4. The source of the chlorine-consuming nitrogenous compounds can be categorized as ammonia, amino acids, and proteins.
5. The total nitrogen consumed follows a well-defined general pattern.

To find the mathematical relationship for chlorine consumed as a function of time (chlorine demand), a series of chlorine determinations are made on a sample of water with varying contact times from 30 min. to 36 hours. These results are plotted on logarithmic paper, as seen in Fig. 4-11. In order to achieve such a plot, it is necessary for the chlorine dose to be sufficient always to provide a measurable residual at the end of the contact time.

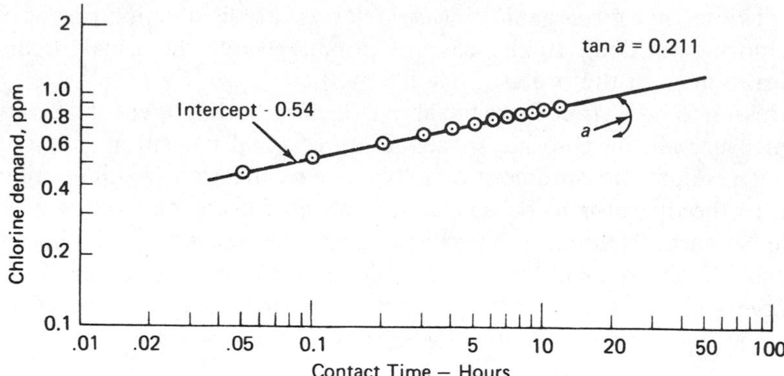

Fig. 4-11. Relationship between chlorine consumed and contact time.

Mathematical Relationship. It is preferable to use the amperometric method for determining chlorine residual. Two curves can be plotted: one for free chlorine residual; the other for total chlorine residual. The results should plot in a straight line, which is of the general form:

$$D = kt^n \tag{4-54}$$

where

D = chlorine demand, ppm (chlorine dose − chlorine residual)
t = chlorine contact time in hours
k = chlorine demand after one hour, ppm
n = slope of curve (tan a)

For the example illustrated in Fig. 4-9, the equation then becomes:

$$D = 0.54t^{0.21} \tag{4-55}$$

This equation can be put into a more practical form if the 30-minute and 1-hour chlorine demands are known. Equation (4-54) can be rewritten to read:

$$Dt = D_1 \, t^n \tag{4-56}$$

where D_1 is the chlorine demand after one hour contact, and Dt is the chlorine demand after t hours of contact. Then, when t = 0.5 hour, Eq. (4-56) can be written:

$$D_{0.5} = D_1 \times 0.5^n \tag{4-57}$$

and expressed logarithmically as:

$$\log D_{0.5} = \log D_1 + n \log 0.5 \tag{4-58}$$

then:

$$n = \frac{(\log D_1 - \log D_{0.5})}{-\log 0.5} \tag{4-59}$$

or:

$$n = \frac{(\log D_1 - \log D_{0.5})}{0.30103} \tag{4-60}$$

or:

$$n = 3.322 (\log D_1 - \log D_{0.5}) \tag{4-61}$$

Substituting Eq. (4-61) in Eq. (4-56), it becomes:

$$D_t = D_1 t^{3.322(\log D_1 - \log D_{0.5})} \tag{4-62}$$

Equation (4-62) may then be expressed in the logarithmic form:

$$\log D_t = \log D_1 + 3.322 \log t (\log D_1 - \log D_{0.5}) \tag{4-63}$$

or:

$$D_t = D_1 \left(\frac{D_1}{D_{0.5}}\right)^{3.322 \log t} \tag{4-64}$$

where

D_t = chlorine consumed in t hours (chlorine dose minus chlorine residual)
D_1 = chlorine consumed in one hour
$D_{0.5}$ = chlorine consumed in 30 min.

With this equation, the chlorine demand at the end of any contact and t can be determined, provided that the chlorine demands at one hour and 30 min. are known.

Practical Significance. The usefulness of the basic equation derived from measuring chlorine residuals as plotted in Fig. 4-11 is the variation in the exponent value n. This reveals the speed of the reaction and gives a clue to the nature of the organic material involved in the reaction with chlorine. The inorganic ions (NH_3^-, Fe^{++}, $S^=$, etc.) all react almost instantaneously, causing rapid initial chlorine demand. This decreases the slope of the curve, so that the exponent n approaches zero.

The higher the value of the exponent, the more complicated the organic material. For example, well waters generally show lower exponents (0.01 to 0.05) than do treated surface waters (0.10 to 0.20). This is so because the nitrogen content of well water is mainly in the form of the inorganic ammonium ion, which reacts rapidly with chlorine, whereas surface waters generally suffer from sewage pollution and therefore contain the more complex compounds of amino acids and proteins.

In amino acids, the 15-min. chlorine demand is usually directly related to the amount of albuminoid nitrogen present. The exponential reaction constant as a function of time is dependent upon the individual structure of the amino acid. An increase in the structural complexity results in higher values of the reaction constant n, Eq. (4-56), and will therefore exhibit prolonged chlorine demand.

Because the proteins are among the most complex compounds found in nature, they are difficult to decompose by chlorination. They have a high molecular weight and consist generally of 16 percent nitrogen, 50 percent carbon, 22 percent oxygen, and 7 percent hydrogen. Some contain as much as 2 percent sulfur; others, phosphorus and small amounts of iron. The presence of significant amounts of proteinaceous material in chlorinated water is indicated by a relatively high value of the reaction constant n. Some proteinaceous substances have yielded values of n as high as 0.90.[47]

Therefore, if we plot on logarithmic paper the chlorine demand of various dosages versus contact time, determination of the value of n will serve as a clue to the nature of the nitrogen content present. A significant rise in the value of the exponent would indicate a rise in the organic nitrogen present and, further, a deterioration in the raw water quality.

The practical significance of organic nitrogen increase, and its relation to chlorine demand, becomes a major parameter of raw water quality as follows:

1. Waters containing inorganic nitrogen (usually NH_3) are the easiest to handle. Stable chlorine residuals can be achieved after one hour or less of contact time, depending on the nitrogen concentration.
2. Waters containing organic nitrogen will not produce stable residuals except after many hours of contact. This fact presents difficulties because it means that the residuals leaving the confines of the plant are most likely unstable, and that special treatment procedures are required. The stability of these residuals depends upon the complexity of the proteinaceous compounds as well as upon their concentration. The amino acid group is identified by the albuminoid nitrogen, which is available for fairly rapid destruction by chlorination. This can usually be entirely destroyed by free residual chlorination in one to two hours as can ammonia nitrogen in 30 min. to one hour. The proteins are identified by the total organic nitrogen, and are the most resistant substances to decompo-

sition by chlorine. The residuals produced are the least stable of the three, inasmuch as the oxidation reaction of chlorine continues for days.

These reactions are characterized by the loss of nitrogen when subjected to the appropriate analyses. For example, when the total nitrogen is in the form of the ammonium ion, the reaction with chlorine will produce a loss of almost 90 percent of the nitrogen. When the total nitrogen is in the form of both ammonia and albuminoid nitrogen, the loss may be only 75 percent or less, depending on the relative concentration of each. In sharp contrast to this is the almost negligible loss of nitrogen when it is present in the form of proteins.[44, 49] White found a groundwater in El Dorado, Arkansas containing some 0.3–0.5 mg/L of organic N that did not react to free chlorine, in the mid-1970s.

CHLORINE AND OTHER INORGANICS

Alkalinity

Since chlorine solutions are highly corrosive, the application of chlorine to a process stream will often raise the question of chlorine corrosion. Corrosion by chlorine is related directly to pH, which is dependent upon alkalinity. Therefore it is appropriate to know the effect of chlorine on the alkalinity of a water.

If a water is dosed with chlorine to the extent of the chlorine demand of the water, all of the chlorine applied will end up as chloride ion (Cl^-) as follows:

$$Cl_2 + H_2O \rightarrow HOCl + HCl \tag{4-65}$$

$$HOCl + Cl\ demand + HCl \rightarrow 2\ HCl \tag{4-66}$$

This calculates as 1.4 parts alkalinity for each part chlorine:

$$2\ HCl + H_2O + CaCO_3 \rightarrow CaCl_2 + CO_2 + 2\ H_2O \tag{4-67}$$

$$\text{Alkalinity as } CaCO_3 = \frac{100}{Cl_2} = \frac{100}{71} = 1.4$$

The reaction in Eq. (4-67) occurs at pH 4.3, which is the endpoint of the alkalinity titration, and where CO_2 exists.

When HOCl is not reduced, only one Cl goes to HCl, and the reaction consumes just half of the alkalinity, or 0.7. As can be seen from the above, the subject of alkalinity destruction by chlorine is not a simple one.

Take the case where water has sufficient alkalinity to maintain a 7.0 pH, and none of the chlorine applied is consumed by chlorine demand. In this case 50 percent of the chlorine applied will go to HCl. Eighty percent of the remaining 50 percent will be undissociated HOCl and unreactive in the alkalinity reactions (see Table 4-4). However, the other 20 percent of the remaining 50 percent at pH 7 is H^+ and OCl^-.

The alkalinity reduction is calculated as follows:

$$50\% + (0.20 \times 50\%) = 60\% = 0.60$$

$0.60 \times 1.4 = 0.84$ parts alkalinity reduction by chlorine at pH 7.0.

The rule of thumb for alkalinity correction by the use of caustic to maintain pH is: a pound of caustic to a pound of chlorine.[82] One part of caustic produces 1.25 parts of alkalinity. If a water at pH 7 is dosed with chlorine, and the demand is 6 mg/L, then the alkalinity reduction is 1.3 parts of alkalinity for each part of chlorine. This rule is reliable.

Carbon

Chlorine reacts with inorganic carbon in much the same manner as it reacts with organic carbon. The reaction is rapid, and can be expressed as:

$$C + 2\ Cl_2 + 2\ H_2O \rightarrow 4\ HCl + CO_2 \qquad (4\text{-}68)$$

This is the reaction that takes place in the dechlorination process using granular carbon filter beds. The production of HCl consumes approximately 2.1 parts of alkalinity as $CaCO_3$ for each part of chlorine removed. The weight ratio of chlorine to carbon is 1.0:0.0845.

Cyanide

In an alkaline solution of pH 8.5 or higher, chlorine reacts with cyanide to form cyanate. This is a primary step in treating plating wastes for the destruction of cyanides, and may be described as follows:

$$2\ Cl_2 + \underset{\text{(caustic)}}{4\ NaOH} + \underset{\text{(cyanide)}}{2\ NaCN} \rightarrow \underset{\text{(cyanate)}}{2\ NaCNO} + 4\ NaCl + 2\ H_2O \qquad (4\text{-}69)$$

This reaction requires 2.73 parts of chlorine and 3.07 parts of caustic for each part of cyanide to convert to cyanate.

The complete destruction of cyanide by chlorine is usually carried out at a pH of 8.5 to 9.5, and follows this equation:

$$5\ Cl_2 + 10\ NaOH + \underset{\text{(cyanide)}}{2\ NaCN} \rightarrow \qquad (4\text{-}70)$$

$$2\ NaHCO_3 + 10\ NaCl + N_2 \uparrow + 4\ H_2O$$

The cyanide decomposes, liberating nitrogen as a gas; the carbon atom joins to form sodium bicarbonate. The other decomposition product is water. The weight ratio is 6.82 parts of chlorine and 7.69 parts of caustic to decompose one part of cyanide (CN^-).

Hydrogen Sulfide

Hydrogen sulfide gas is frequently found dissolved in underground waters (rarely in surface supplies). It is also a constituent of septic sewage and is characterized by its obnoxious odor, best described as that of rotten eggs. It reacts instantaneously with chlorine either to precipitate free sulfur or to form dilute sulfuric acid. The determining factor in the formation of these end products is the pH.

Table 4-6, by Nordell,[50] shows how chlorine and hydrogen sulfide react at different pH levels.

The complete oxidation of hydrogen sulfide to the sulfate form is as follows:

$$H_2S + 4\ Cl_2 + 4\ H_2O \rightarrow H_2SO_4 + 8\ HCl \qquad (4\text{-}71)$$

Therefore 8.32 mg/L of chlorine are required to oxidize 1 mg/L of hydrogen sulfide to the sulfate form. However, the chlorine required to oxidize hydrogen sulfide to sulfur and water is only 2.08 mg/L chlorine to 1 mg/L hydrogen sulfide:

$$H_2S + Cl_2 \rightarrow S\downarrow + 2\ HCl \qquad (4\text{-}72)$$

In this instance the sulfur precipitates as finely divided white particles, which are sometimes colloidal in nature.

Table 4-6 Oxidation of Sulfides by Chlorine at Different pH Levels. (Contact time 10 minutes)

Final pH	*Sulfides as ppm H_2S*	*Cl_2 Added mg/l*	*Cl_2 Residual mg/l*	*Cl_2 Consumed mg/l*
3.2	5	50	6	44
5.0	5	50	7	43
6.2	5	50	7	43
6.4	5	50	9	41
6.8	5	50	11	39
7.1	5	50	18	32
7.6	5	50	18	32
9.0	5	50	25	25
10.1	5	50	25	25

(Amount of chlorine required to oxidize 5 ppm sulfides expressed as H_2S is $8.32 \times 5 = 41.6$ mg/l.)

In Table 4-6 it can be seen that the sulfides were all oxidized to sulfates at pH values below 6.4. At pH around 7.0, about 70 percent was oxidized to sulfate and 30 percent to water and sulfur;* at pH values of 9 and 10, approximately 50 percent was oxidized to sulfate and 50 percent to sulfur and water.

It should be noted that for complete oxidation both sulfuric acid and hydrochloric acid are end products that will consume alkalinity. Each part of H_2S oxidized in this reaction will consume 10 mg/L alkalinity as $CaCO_3$. For the reaction that precipitates sulfur, only 2.6 mg/L alkalinity as $CaCO_3$ is consumed.

In actual practice, the destruction of hydrogen sulfide in potable water is not so simple as portrayed by the stoichiometric reactions of Eqs. (4-71) and (4-72)—there are side reactions that interfere with the production of the end products. (See Chapter 6.)

Iron

The reaction of chlorine with iron is a very convenient one, as it can be used for two different purposes: to remove iron from water and to produce a coagulant for both water and sewage treatment.

In water supplies, iron is usually associated with underground waters and is generally in the form of ferrous bicarbonate, which is slightly soluble in water. Chlorine reacts with the ferrous ion and converts it to the ferric form. Depending upon the hydroxyl ion activity, the ferric chloride formed will quickly hydrolyze to ferric hydroxide. The latter precipitates as a reddish fluffy mass, depending on the concentration of ferric ion. Omitting the intermediate reaction of the formation of ferric chloride, the reaction is as follows:

$$2\ Fe(HCO_3)_2 + Cl_2 + Ca(HCO)_3)_2 \rightarrow 2\ Fe(OH)_3 + CaCl_2 + 6\ CO_2 \quad (4\text{-}73)$$

This reaction produces a rapid release of carbon dioxide, which causes a significant rise in pH. The reaction can proceed over a wide pH range (4–10), but the optimum pH is above 7. Each part of iron to be removed requires 0.64 part of chlorine. This reaction consumes 0.9 part of alkalinity as $CaCO_3$ for each part of ferrous ion oxidized to ferric ion. Either free or combined chlorine will produce this reaction, which is instantaneous for inorganic iron in solution. If the iron is in the form of an organic compound, the reaction is considerably inhibited.

It is often desirable to use the ferric ion as a coagulant in the water treatment process. This is most conveniently accomplished by converting the easy-to-handle ferrous ion to ferric by oxidation with chlorine. For example, pickle

*When chlorine is used to control H_2S formation, the treated water will likely appear milky because of the formation of free sulfur. This is known as the Tyndall effect.

liquor, a by-product of steel mills, is the liquid form, and copperas is the granular form of ferrous sulfate commonly used with chlorine to make a coagulant, as follows:

$$6\ FeSO_4 \cdot 7\ H_2O + 3\ Cl_2 \rightarrow 2\ FeCl_3 + 2\ Fe_2(SO_4)_3 + 42\ H_2O \quad (4\text{-}74)$$

Each of the ferric salts will hydrolyze further to ferric hydroxide ($Fe(OH)_3$), utilizing the hydroxyl ion of the water and thus forming a brown gelatinous floc. To complete this reaction, 7.8 parts of ferrous sulfate are required for each part of chlorine. The resulting solution will contain about 29.9 percent ferric chloride and 70.1 percent ferric sulfate. Each part of chlorinated ferrous sulfate consumes 0.54 part of alkalinity as $CaCO_3$. This reaction proceeds over a wide pH range. With pickle liquor, adequate floc can be produced at as low as pH 4; however, the optimum pH should be kept at 7 or higher, since the resulting ferric hydroxide hydrolyzes more rapidly in this range.

Manganese

Free available chlorine reacts to oxidize soluble manganese compounds, but chloramines or combined chlorine residuals have little effect. The oxidized manganese drops out as a precipitate in the form of manganese dioxide, as follows:

$$MnSO_4 + Cl_2 + 4\ NaOH \rightarrow MnO_2 + 2\ NaCl + Na_2SO_4 + 2\ H_2O \quad (4\text{-}75)$$

This reaction requires 1.3 parts of free chlorine for each part of manganese oxidized, and 3.4 mg/L of alkalinity as $CaCO_3$ will be consumed. The reaction will proceed in a pH range of 7 to 10, with the optimum close to 10. It usually takes from two to four hours to reach completion. If the manganese is in the form of an organic complex, the reaction time will be even longer and unpredictable.

Methane

Methane gas is sufficiently soluble in water to be both explosive and a fire hazard. Waters in this category are known to exist, and one of their characteristics is a high chlorine demand (10–25 mg/L). It was thought that chlorine reacted directly with methane to form carbon tetrachloride, as follows:

$$CH_4 + 4\ Cl_2 \rightarrow CCl_4 + 4\ HCl \quad (4\text{-}76)$$

This reaction between molecular chlorine gas and methane gas is possible at temperatures of 300°C and higher. However, there is no reaction between methane and hypochlorous acid. The high chlorine demand of these waters is attributed to the presence of high amounts of oxidizable organics.

Nitrites

Chemistry of Formation. The effect of nitrites on the chlorination process is often overlooked and misunderstood. Nitrites appear as transitory compounds when for a variety of reasons ammonia nitrogen is being oxidized to nitrates. The following equations illustrate the chemical pathways of this reaction.

$$2\ NH_4^+ + 3\ O_2 \rightarrow 2\ NO_2^- + 4\ H^+ + 2\ H_2O \tag{4-77}$$

Generally bacteria of the *Nitrosomas* genus are involved in conversion of ammonium to nitrite under aerobic conditions. This is the reaction that often occurs within the biological slime on filter media.

The nitrites can be oxidized to nitrate generally by *Nitrobacter* according to the following reaction:

$$NO_2^- + 0.5\ O_2 \xrightarrow{\text{bacteria}} NO_3^- \tag{4-78}$$

To oxidize 1 mg/L of ammonia nitrogen requires about 4.6 mg/L of oxygen (excluding synthesis of nitrifiers).

Reaction with Chlorine. Nitrites exert a significant chlorine demand in the presence of free chlorine as follows:

$$HOCl + NO_2^- \rightarrow NO_3^- + HCl \tag{4-79}$$

Since it takes two atoms of chlorine to make one molecule of HOCl, it takes five parts of chlorine to oxidize one part of nitrite (as nitrogen). Therefore each mg/L of nitrite represents 5 mg/L chlorine demand in the presence of free chlorine. *However, a combined (chloramine) chlorine residual will not oxidize or react with nitrites.*

Nitrites are a deceptive factor in the chlorination process because they interfere with chlorine residual measurements. Apparently nitrites can oxidize KI to I_2. This is a reaction similar to that of free chlorine and chloramines. Therefore, because there is no reaction between nitrites and chloramines, any residual measurement using the acid-iodide ($pH < 4$) procedure will show higher residuals when nitrites are present with chloramines.

When the phenylarsine oxide–iodine procedure is used (back titration) for measuring chloramine residuals the nitrites will be oxidized by the iodine solution used for back-titrating the PAO. Therefore, the apparent chlorine residual will be higher than the true value (see Chapter 5).

Seawater Chemistry

Effect of Bromides. The chemistry of seawater is unique when chlorination is involved owing to the presence of bromides in the seawater. These bromide compounds result in a bromide ion content of 50–70 mg/L, which will always be well in excess of any chlorine applied for chlorination purposes. A good example is the chlorination of seawater used for cooling water in an electric power generating station. Chlorination is used to control the marine growths in the steam condenser tubes because these growths interfere with the heat transfer efficiency of the condensers.

When the chlorine solution from a conventional chlorinator mixes with the seawater (pH 8.3), the chlorine reacts immediately with the bromide ions to form hypobromous acid:[60, 83]

$$HOCl + Br^- \rightarrow HOBr + Cl^- \qquad (4\text{-}80)$$

The rate constant (K_0) for this reaction is 2.95×10^3 liter per mol per sec. The hypobromous acid will dissociate, as does hypochlorous acid, depending upon the pH, temperature, and ionic strength, as follows:

$$K = \frac{(H^+)(OBr^-)}{(HOBr)} \qquad (4\text{-}81)$$

where

$$K = 2 \times 10^{-9} \text{ at } 20°C$$

The reaction rate between chlorine and the bromide ion, Eq. (4-80), is rapid. Using documented rate constants, Selleck et al.[84] calculated the time required to convert 99 percent of the chlorine to bromine at a temperature of 25°C and a pH of 8.3* for various percentages of seawater. Figure 4-12, taken from Selleck's calculations, shows that the reaction of Eq. (4-80) proceeds to 99 percent completion within 10 sec in highly saline waters (i.e., 19,000 mg/L chlorides).

Effect of pH. The importance of pH in the reaction of chlorine species and the bromides in seawater must be emphasized. Free bromine (Br_2) is extracted from seawater and from the Dead Sea by chlorination:

$$Cl_2 + 2\,Br^- + 2\,H^+ \rightarrow Br_2 + 2\,HCl \qquad (4\text{-}82)$$

However this reaction will not proceed to completion until the seawater is acidified to pH 4. (See Chapter 16.)

Chloramines will not oxidize bromides in either the neutral pH zone of 7–9 or in the acidic zone of 4–6. This is why the chlorine–ammonia process does not increase the TTHM potential when bromides are present in potable water.

---

*Seawater is always at pH 8.3 in the salinity range of 12,000–19,000 mg/L chlorides.

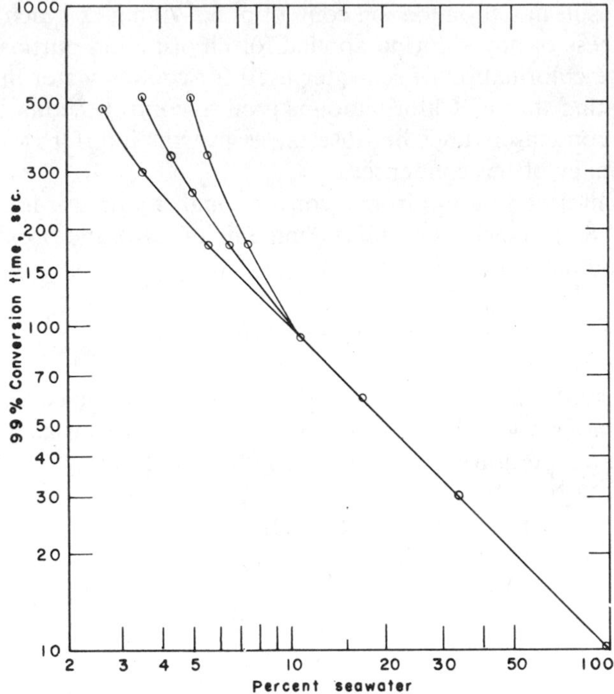

Fig. 4-12. Time required for 99 percent conversion of free chlorine to HOBr at 25°C and pH 8.3[86] (courtesy *Journal WPCF*).

At pH 4, chloramines will not oxidize the bromide ion to free bromine, but HOCl will. This bit of fundamental chemistry is the basis for the invention of a free chlorine residual analyzer that is not affected by the presence of chloramines. (See Chapter 9.)

Dissociation of Hypobromous Acid. This is a salient factor in the chlorination of seawater because HOBr is an effective biocide. This must be assumed because all seawater chlorination studies have reported residuals as chlorine, when in reality the residuals are most likely to be a mixture of bromine species. Seawater chlorination has been practiced by the electric power industry for eight decades, and the success of this practice of reporting residuals as chlorine for the control of marine growths is sufficient evidence for us to conclude that the resulting bromine species are potent biocides. The dissociation of HOBr plays an important role. Some comparative experiments have been made on coliform destruction using chlorine generated from seawater versus chlorine solution with fresh water and conventional chlorination equipment. These experiments have tended to show a better coliform kill with on-site generated hypochlorite from seawater.

The explanation might be found in the dissociation of HOCl versus HOBr. From Eq. (4-81) it can be calculated that at pH 8, the undissociated HOBr is 83 percent versus HOCl at only 28 percent.

Effect of Ionic Strength. Since seawater chlorination produces species of bromine residuals owing to excess bromide ions present in seawater, the dissociation values for HOCl based upon ionic strength of seawater cannot be used. From the dissociation value for HOBr described above, the ionic strength of seawater ($I = 1.0$) will have little effect on the potency of the undissociated HOBr.

The Role of Ammonia Nitrogen. The presence of ammonia nitrogen in seawater complicates the chemical reactions between the chlorine species and the bromides. When this situation occurs, there are two competing reactions operating simultaneously: (1) chlorine reacts with the ammonia N practically instantaneously at pH 8.3 to form chloramines, and (2) the chlorine will react at about the same speed with the bromide ion. However, since the bromide ion concentration is on the order of 50–70 mg/L and the ammonia-N concentration not more than 2–3 mg/L, the chlorine–bromide reaction might be the faster. The result is probably the immediate formation of both monochloramine and hypobromous acid. These two compounds start another round of reactions as follows:

$$NH_2Cl + Br^- + 2\ H_2O \rightarrow HOBr + NH_4OH + Cl^- \tag{4-83}$$

$$HOBr + NH_4OH \rightarrow NH_2Br + 2\ H_2O \tag{4-84}$$

$$NH_2BR + HOBr \rightarrow NHBr_2 + H_2O \tag{4-85}$$

$$NHBr_2 + H_2O \rightarrow NOH + 2\ H^+ + 2\ Br^- \tag{4-86}$$

The intermediary NOH formed in Eq. (4-86) is the catalyst, which allows the dibromamine to trigger the breakpoint as follows:

$$NHBr_2 + NOH \rightarrow N_2 + HOBr + H^+ + Br^- \tag{4-87}$$

The reactions described by Eqs. (4-83) through (4-87) are sequential and rapid. Very little monobromamine is formed. Moreover, there is no evidence of tribromamine formation. The most significant aspect of the above reaction is the rapid formation of dibromamine, which triggers the breakpoint. Selleck points out that owing to the presence of excess bromide ion, the removal of ammonia nitrogen from seawater by these reactions is independent of the molar ratio of HOBr to NH_3—N.[85] Therefore the presence of ammonia N in seawater does not present any problems for the chlorination process.

The above reactions take place only in the range of seawater pH, 8.3.

GERMICIDAL SIGNIFICANCE OF CHLORINE RESIDUALS

Introduction

Studies have been made on the nature of the killing mechanism of chlorine on bacteria,[20, 51–54] cysts,[20, 55, 56] and spores.[20, 57] The inactivation mechanism of viruses by chlorine and other oxidants has never been resolved. An infective virus particle or virion consists of a core composed of or containing one kind of nucleic acid, which is encased in a protein shell or capsid. The capsid is built from a number of morphological subunits identified as capsomeres. It is thought that the oxidant penetrates the capsid by chemical transformation and then attacks the nucleic acid.

Exactly how chlorine kills bacteria, cysts, and spores is still an academic puzzle. During the period of its early use (ca. 1910) as a disinfectant, it was believed that the germicidal power of chlorine was due solely to the liberation of nascent oxygen from the hypochlorous acid formed by the reaction of chlorine and water ($HOCl \rightarrow HCl + O$). This theory led to considerable confusion on the subject and thwarted many serious investigations. It was not until 1944 that Chang[56] dispelled this belief by showing that hydrogen peroxide and potassium permanganate liberated considerable quantities of nascent oxygen but showed little germicidal efficiency. He further proved that there was no liberation of oxygen involved in the chlorine reaction, and that the disinfecting agent is hypochlorous acid. He favored the formation of the toxic substance theory.

In 1946, Green and Stumpf[51] concluded that chlorine reacted irreversibly with the enzymatic system of bacteria, thereby killing the bacteria. They were able to show that a bacterial suspension became sterile when bacteria lost the power to oxidize glucose. They also found that when the enzymes that contain sulfhydril groups were oxidized by chlorine, the oxidation was irreversible, thus abolishing enzyme action and resulting in the destruction of the bacteria, and that the most significant of this group is the enzyme triosephosphate dehydrogenase. In 1953, Ingols et al.[58] concurred with this theory; but, using monochloramine, they found that even though they restored the sulfhydril groups, the bacteria were not restored. This led them to believe that although the sulfhydril group may be the most vulnerable to a strong oxidant such as hypochlorous acid or chlorine dioxide, there were changes in other groups caused by monochloramine and dichloramine that may be important in bringing about death to bacteria.

Later (1962), Wyss[54] pointed out that probably the effective mechanism of death to microorganisms is the phenomenon of unbalanced growth. That is, destruction of part of the enzyme system will so throw the cell out of balance that by progress of its own metabolism the cell dies before the necessary repairs are made.

Whatever the chemical action, it is generally agreed that the relative efficiency of various disinfecting compounds is a function of the rate of diffusion of the active agent through the cell wall. Figure 4-13 shows the essential elements of a bacterial cell. It is assumed that after penetration of the cell wall is accomplished, the disinfecting compound has the ability to attack the enzyme group, whose destruction results in death to the organism. Factors that affect the efficiency of destruction are:

1. Nature of disinfectant (kind of chlorine residual fraction).
2. Concentration of disinfectant.
3. Length of contact time with disinfectant.
4. Temperature.
5. Type and concentration of organisms.
6. pH.

Hypochlorous Acid

HOCl is the most effective of all the chlorine residual fractions. This fraction is known officially in the industry as free available chlorine residual.[59, 60] Hypochlorous acid is similar in structure to water; hence, the formula HOCl is preferred to HClO. The germicidal efficiency of HOCl is due to the relative ease with which it can penetrate cell walls. This penetration is comparable to that of water, and can be attributed to both its modest size (low molecular weight) and its electrical neutrality (absence of an electrical charge).

Other things being equal, the germicidal efficiency of a free available chlorine residual is a function of the pH, which establishes the amount of dissociation of HOCl to H^+ and OCl^- ions. Table 4-4 shows the percentage of undissociated HOCl in a chlorine solution for various pH values and temperatures. Lowering the temperature of the reacting solution suppresses the dissociation; conversely, raising the temperature increases the amount of dissociation.

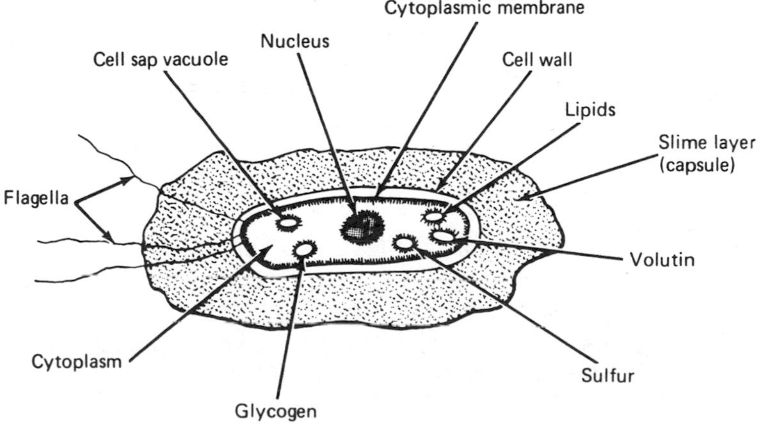

Fig. 4-13. Schematic diagram of a bacterium.

The rate of dissociation of HOCl is so rapid that equilibrium between HOCl and the OCl^- ion is maintained, even though the HOCl is being continuously used. For example, if water containing 1 mg/L of titrable free available chlorine residual has been dosed with a reducing agent that consumes 50 percent of the hypochlorous acid, the remaining residual will redistribute itself between HOCl and OCl^- ion according to the values shown in Table 4-4. This is commonly referred to as the "reservoir" effect.

Hypochlorite Ion

The OCl^- ion, which is a result of the dissociation phenomenon, is a relatively poor disinfectant because of its inability to diffuse through the cell wall of microorganisms. The obstacle to this passage is the negative electrical charge, as substantiated to some extent by the fact that the activation energy for disinfection by HOCl is in the range for diffusion ($E = 7000$ calories), whereas that of the OCl^- ion is more characteristic of a chemical reaction ($E = 15{,}000$ calories). The relative efficiencies of HOCl and the OCl^- ion were well documented by Professor Fair and his colleagues at Harvard.[20, 37] These findings took into consideration the work done by Butterfield et al.[61] in 1943 and by Butterfield and Wattie[62] in 1944. This was a concerted effort to relate disinfection to chemistry.

It is well known that the disinfecting efficiency of free available chlorine residual decreases significantly as the pH rises. At a pH above 9 there is little disinfecting power. At this pH level and at 20°C, 96 percent of the titrable free available chlorine will consist of the OCl^- ion. This is an indication of the low germicidal efficiency of the OCl^- ion. The Harvard group developed an equation devised to calculate the total free available chlorine residual required to kill a given percentage of a specified organism. This equation requires the use of a constant, which is the ratio of germicidal efficiency of the OCl^- ion to HOCl:

$$R = A \frac{1 + [K_i/(\mathrm{H}^+)]}{1 + B[K_i/(\mathrm{H}^+)]} \tag{4-88}$$

where

R = the required total titrable available chlorine residual
A = the amount of HOCl alone (expressed as titrable chlorine) required to destroy the bacteria
K_i = the dissociation constant of HOCl for a given temperature
B = the ratio of the efficiency of OCl^- to HOCl.

Fair et al.[31] observed concentrations of titrable free available chlorine required to kill 99 percent *Escherichia coli* in 30 minutes at 2–5°C at different pH values:

$$R = 0.005 \frac{1 + [2.2 \times 10^{-8}/(H^+)]}{1 + 0.012[2.2 \times 10^{-8}/(H^+)]} \tag{4-89}$$

The value of B used (0.012) indicates that the OCl^- ion is about 1/80 as efficient as HOCl for the conditions stated. This is not in very close agreement with investigations by Selleck at the University of California, Berkeley. His work demonstrates that the OCl^- ion is only 20 times less effective than HOCl rather than 80 times.[63]

A similar investigation was made on the destruction of cysts, and further demonstrated the greater efficiency of HOCl over the OCl^- ion. Experimental cysticidal residuals were obtained for a 100 percent destruction of 30 cysts per ml of *Entamoeba histolytica* over a temperature range of 3–23°C and with a contact time of 30 minutes. The relative efficiencies of the OCl^- ion to HOCl for inactivation of cysts are summarized as follows:

Temperature, °C	*OCl^- to HOCL Relative Effective Ratio*
3	1/150
10	1/200
18	1/250
23	1/300

Similarly, Eq. (4-88) can be used to calculate the amount of R required to destroy this organism under specific conditions of temperature contact time and pH. Because cysts are the most difficult microorganisms to destroy with a chlorine residual, it is reasonable to expect that other organisms, such as spores and viruses, would show a relative effective ratio of the OCl^- ion to HOCl somewhere between the extremes of *Esch. coli* at 1/20 and *E. histolytica* at 1/300. It is evident that because of this effective ratio the low germicidal efficiency of OCl^- must be taken into account, and that the effectiveness of free available chlorine residual is seriously impaired when the pH exceeds 9.0. The reason for this is obvious. At pH 9 and a temperature of 15°C the percentage of HOCl in a given solution is 4.13; at pH 7.5 it is 57.68. (See Table 4-4.) This prevails in spite of the HOCl "reservoir" effect.

Chloramines

Background Discussion. That chloramines are slower to kill microorganisms than free available chlorine has been common knowledge for some time. Probably the first person to become aware of this was Alexander Houston[64] in 1925. The first scientific laboratory work on the germicidal efficiency of chloramine was done by K. Holwerda[28] at the Laboratory for the Purification of Water at Manggarai in the Dutch East Indies. His work investigated chloramines at pH 4.5, 6.8, and 8.5. At pH 4.5, the chlorine compound would most likely be 100 percent dichloramine (see Fig. 4-5); at pH 8.5, it would probably be 100 percent monochloramine. Holwerda's findings indicated a time factor as high as 80:1 required for monochloramines over free available chlorine, and as low as 20:1 for the same germicidal efficiency. This compares with Butterfield and Watties's findings[62, 65, 66] that chloramine required approximately a hundredfold increase in contact time over free available chlorine at pH 9.5. Kabler[67] also concluded that, to obtain the same kill with the same amounts of free available and combined available chlorine residual, the latter required 100 times more exposure time.

The above historical data have categorized chloramines as a poor disinfectant compared to free chlorine. However, recent studies indicate that if chloramines are allowed contact times of 45–60 minutes, they are able to match the efficiency of free chlorine based upon the destruction of coliform organisms. (See Chapter 8.) This is based upon studies of coliform distribution systems.

Monochloramine. From the work by Butterfield and Wattie mentioned above, later confirmed by Kabler, it might be concluded that for the same conditions of contact time and temperature, and a pH range of about 6–8, it will take at least 25 times more combined available chlorine than free available chlorine to produce the same germicidal efficiency. Furthermore, it can be assumed that if the chlorine-to-ammonia nitrogen ratio is less than 5:1 and if the pH is 7.5 and higher, the combined residual will probably be 100 percent monochloramine. This difference in potency of monochloramine and HOCl might be explained by the difference in their oxidation potentials, assuming that the action of chloramine is of an electrochemical nature rather than one of diffusion, as seems to be the case with HOCl. (See below, Fig. 4-17.) The work of Ingols et al.[58] might also explain the difference between free available and combined available chlorine residuals. They indicated that by using monochloramine only, its reaction with vulnerable enzyme groups (in bacteria) may be a reversible reaction. Another explanation of the low potency of monochloramine is that it may be a function of hydrolysis. The hydrolysis constants of chloramine in general can be expressed as follows:

$$K_n = \frac{(\mathrm{HOCl})(\mathrm{RR'NH})}{(\mathrm{RR'NCl})} \tag{4-90}$$

From this equation it can be seen that one of the products of hydrolysis is free available chlorine (HOCl), which tends to be low in strong solutions, but which increases in concentration in weak solutions.[16] The hydrolysis constant of monochloramine, as determined by Corbett et al.,[68] was found to be $K_n = 2.8 \times 10^{-10}$ at 15°C in accordance with:

$$NH_2Cl + H_2O \leftrightharpoons NH_3 + HOCl \tag{4-91}$$

This is in fair agreement with later work by Morris,[39] who found that $K_n = 1.0 \times 10^{-11}$ (approximately). These two investigations were carried out under quite different conditions: Corbett et al. performed their study at pH 14 and 10^{-2} *M*; Morris, at pH 5 and 10^{-4} *M*. Morris calculated that a pure solution of NH_2Cl at a concentration of 2.0 mg/L as chlorine (2.82×10^{-5} *M*) will be 0.58 percent hydrolyzed at pH 7 and 25°C.

However, the work done at the San Jose/Santa Clara Water Pollution Control Plant[70] over a two-year period (1980–82) indicated that in an effluent containing organic N, monochloramine has a much greater germicidal efficiency than a nitrified effluent using the free chlorine approach. Ammonia N has to be added to these nitrified effluents. This investigation demonstrated clearly that in a completely nitrified effluent, monochloramine residuals, based upon a chlorine-to-ammonia N ratio of 6:1, can achieve far better coliform kills than free chlorine residuals of the same concentration and the same contact time (i.e., 50 min.) because the organic N reacts immediately with the free chlorine (HOCl) to form the nongermicidal organochloramines. This imposter chloramine compound appears in the dichloramine fraction of the amperometric titration procedure. However, at the time (1981) it was not realized that a nitrified effluent containing less than 0.1 mg/L of ammonia N could not contain any monochloramine. It was not until later in 1982 that the lab chief, Virginia Alford, decided to measure the chlorine residuals by the forward amperometric titration process. This revealed that 9.5 mg/L total residual contained only 4.5 mg/L of free chlorine, about 0.5 mg/L monochloramine, and about 4.5 mg/L dichloramine. By this time we all saw the mistake that had been made, which was not to realize that monochloramine apparently could not exist to any meaningful extent in a nitrified effluent. The amount shown was a little "spillover" of dichloramine. Therefore, because an organochloramine is nongermicidal, half of the total residual was used up to no avail by the organic N immediate reaction with free chlorine. Other experiences quoted elsewhere indicate that monochloramine can be equivalent to free chlorine, provided that the contact time is at least 45–60 minutes and the Cl to N ratio is 6:1. The potency of monochloramine in the San Jose incident is due to its slow reaction time with the organic N content in the effluent. The other problem at San Jose was the immediate loss of 15–17 mg/L of free chlorine residual due to its bleaching action on the color of the wastewater effluent.

What is currently realized (1997) is that where a wastewater is nitrified, ammonia N must be added at a chlorine-to-ammonia N weight ratio of 6 to 1 for maximum disinfection potency because the reaction between chloramines and organic N is very slow, 40 to 50 min. This delays the formation of the nongermicidal organochloramines. This is also why the disinfection efficiency of monochloramines begins to deteriorate over time, such as longer than

50 min. Remember, practically all wastewater effluents carry about 3–8 mg/L of organic N.

Some investigators monitoring the potency of monochloramine in water distribution systems believe that it is a function of its slow rate of diffusion through the bacteria cell wall. During this time some hydrolysis product (HOCl) is forming, which increases the potency of the monochloramine residual because the organic N, if any, has already been accounted for.

True Dichloramine. Although Holwerda was probably not aware of it, his work definitely demonstrated that dichloramine is a more potent germicide than monochloramine. At pH 4.5 and with 0.5 mg/L chloramine (probably 100 percent dichloramine) disinfection was achieved in 20 min., whereas at pH 8.6 (probably 100 percent monochloramine) it required 60 min. This was based upon the observance of gas formers. On another series using plate counts, it took 10 min. at pH 4.5 and 60 min. at pH 8.6 to achieve a one plate count with the control showing 388.

Fair et al.[20] found that at pH 4.5, which they observed to be 100 percent dichloramine (see Fig. 4.5), a residual of 2.0 mg/L gives a 100 percent kill of cysts (*E. histolytica*), whereas at pH 10.0, when only monochloramine is present, the required cysticidal residual is approximately 7.5 mg/L. They used a concentration of 30 cysts/ml at a temperature of 23°C and a contact time of 30 min. Their overall results with cysts show that dichloramine is about 60 percent and monochloramine about 22 percent as efficient as HOCl. Because of the formation of the relatively ineffective OCl^- ion in HOCl solutions at high pH values, the chloramines are more efficient cysticidal agents above a pH of about 9. On the limited number of experiments performed with spores of *B. anthracis,* dichloramine showed an efficiency of 15 percent that of HOCl for a 30-min. contact period, whereas the efficiency of monochloramine was too low to be measured.[20]

Kelly and Sanderson[34] attributed the inactivation of poliomyelitis and coxsackie viruses to dichloramine rather than to monochloramine. They were able to achieve 99.7 percent inactivation with a combined available chlorine residual in three hours at pH 6, whereas at pH 10 six to eight hours was required for polio virus. They determined that dichloramine was 43 percent of the total at pH 6, and at pH 10 it was 32 percent. (See Fig. 4-5.) Although these fractions were determined by the amperometric method, the results vary widely from the curve by Fair et al., which was determined by a more sophisticated method—light absorbence as measured by a spectrophotometer.

It can probably be concluded from all these data that dichloramine is twice as potent as monochloramine as a germicidal agent. However, additional scientific data are necessary to establish this as a fact. This would be a welcome contribution to the field of wastewater chlorination. However, in water treatment it is only of academic interest, as dichloramine is to be avoided because it contributes to taste and odor problems. Moreover, it is doubtful that there ever would be a condition whereby a pure dichloramine residual could be isolated and measured as such. Those residuals that exhibit themselves as dichloramine via all of the well-known analytical procedures are probably a mixture of organochloramines and not a specific specimen of dichloramine. This is the weakness and the failure of the analytical procedures. A pure dichloramine made in the laboratory is unstable in the presence of HOCl. It is quite conceivable that in waters lacking organic nitrogen but containing ammonia N, pure dichloramine will form in the free residual process. However, it will disappear when the competing reactions pass beyond the breakpoint zone.

To date there are no analytical methods that can identify pure dichloramine as differentiated from the organochloramines that appear in the dichloramine procedure.

Nitrogen Trichloride. This chlorine compound is known to be an effective oxidizing agent. It has been used for over 40 years in the bleaching of flour, as a fungicide, and for control of insect pests on fruit in storage rooms and cars during shipment. Its oxidation potential and its bactericidal efficiency have never been evaluated, but this compound may possibly contribute substantially to the oxidation of organic matter and the destruction of microorganisms in water and wastewater chlorination in those instances when it does appear. Nitrogen trichloride imparts an objectionable taste and a foul odor when present in potable water. Fortunately it is unstable in sunlight and highly insoluble in water, and it aerates easily. It is characterized by its pungent chlorinelike odor. It causes severe eye tearing in low concentrations. It is difficult to capture by any analytical procedure—it is too volatile. Nitrogen trichloride cannot exist without the presence of HOCl. It has been known to form in the distribution system long after leaving the treatment plant. This situation can be corrected by converting the free chlorine residual to chloramines by post-ammoniation. However, the preferred method is complete dechlorination of the finished water followed by rechlorination.

It is never necessary to determine the concentration of NCl_3 present. The operator needs only to know if it is there. This is easily determined by human responses, either by the eyes tearing or by its odor. It is the only chlorine species that causes tearing of the eyes.

Organic Chloramines

General Discussion. This discussion excludes the manufactured organic chloramines used for sanitizing purposes and swimming pool treatment (i.e., the chloroisocyanurics, chlorinated hydantoins, etc.). These compounds have hydrolysis constants as high as 10^{-4}. They will yield HOCl concentrations equivalent to hypochlorites of the same available chlorine content.[16] This is achieved by hydrolysis at a slow but steady rate.

The organic chloramines of interest here are those that are formed during the chlorination of potable water or wastewater. These organochloramines are directly related to the organic nitrogen content in potable waters, which varies from a trace up to a maximum of 3 mg/L. Any concentration greater than 0.25 mg/L is certain to cause taste and odor problems.

Well oxidized secondary effluents might contain from 6 to 8 mg/L organic N, whereas well-polished filtered and nitrified effluents will usually contain from 2 to 4 mg/L organic N. These chloramines can be measured by forward titration using either the amperometric or the DPD-FAS method. They will appear in the dichloramine fraction. These chloramines are nongermicidal, hence the term "nuisance residuals," and they are practically always found in wastewater effluents.

Studies of Germicidal Efficiency. As early as 1966 Feng[87] discovered a great disparity in the germicidal efficiency between ammonia chloramines and those found in an environment of pure organic nitrogen compounds. He reported that methionine, which is one of the indispensable amino acids for biological growth and is expected to be present in wastewater, forms a measurable chlorine residual with no germicidal power. Feng also investigated the lethal activities of the glycine, taurine, and gelatin chloramines. His work shows that taurine chloramines are as lethally active as ammonium chloramines at pH 9.5, but that their germicidal efficiency falls off as the pH decreases. The glycine chloramines are as germicidally active as monochloramine at pH 4 but are totally inert at pH 7, and the gelatin chloramines are active at pH 9.5 but are inert at pH 7 and 4. There are certain to be other such organic nitrogenous compounds, which contribute to the total chlorine residual, that have little or no germicidal effect.

Sung[88] made a controlled laboratory study of 15 organic compounds representing seven groups to evaluate their individual and combined effect upon the chlorination process. Nine of the 15 compounds were found to interfere with the germicidal efficiency of the chlorination process. Of these nine compounds, five were organic nitrogen compounds. Cystine and uric acid were the severest inhibitors of the nitrogen group. When five of the interfering compounds were mixed together, their combined effect was found to be more pronounced than any of their individual effects, but did not equal the sum of their individual effects. Sung compared the germicidal efficiency of a simulated wastewater with and without the interfering organic compounds. He found that the germicidal efficiency of wastewater containing the interfering compounds by themselves and the resulting chlorine residuals had little or no germicidal effect. The greatest interference was observed to be caused by cystine, tannic acid, humic acids, uric acids, and arginine.

Cystine is an amino acid connected by two sulfur groups that is known to react with chlorine. Tannic, humic, and uric acids are capable of exerting a significant chlorine demand when present in water or wastewater. Arginine is a basic amino acid. The reaction between chlorine and arginine is almost instantaneous.

The organic compounds that had little or no interfering effects on the chlorination process were: acetic acid, cellubiose, dextrose, glutamic acid, uracil, and lauric acid.

The significance of the above findings by Sung[88] confirms the theory of interference in the chlorination process by the presence of organic nitrogen. It further points to the fact that present analytical techniques do not provide for separating the chlorine residual fractions into those of equal germicidal efficiency.

It is also interesting to note that Esvelt et al.[89] found that the toxicity of combined chlorine residuals diminished with time. This finding, together with those of Sung,[88] demonstrates quite convincingly that there are a significant number of organic compounds in wastewater that will react with chlorine to form organic chloramines of little or no germicidal potential, and, moreover, that these compounds appear to increase in concentration with time. This apparent increase of this chlorine residual fraction with time would also explain the loss of germicidal efficiency of combined chlorine residuals in wastewater with time. The supposition is that the predominantly monochloramine residual present hydrolyzes to form free chlorine (HOCl), and both the monochloramine and HOCl react with the organic nitrogen compounds to form the ineffective organic chloramines. (See discussion of the San Jose experience, Chapter 8.)

Nuisance Residuals. These are the residuals, other than free chlorine (HOCl), that appear beyond the breakpoint. This is Zone 3 in Fig. 4-4. They usually consist of chloramines that titrate in the dichloramine fraction. As described before, they are probably made up of mostly organic chloramines.

At the breakpoint there are at least three competing reactions, which tend to produce free chlorine, monochloramine, dichloramine, and nitrogen trichloride. Pure dichloramine is unstable in the presence of free chlorine; nitrogen trichloride cannot form without free chlorine; and monochloramine is almost always present when nitrogen trichloride is formed. Therefore, a reasonably good guess as to the composition of nuisance residuals in actual practice might be: (1) when NCl_3 is formed, 85–90 percent organic chloramines, 8–12 percent monochloramine, and 2–3 percent NCl_3; (2) when NCl_3 is absent, 90 percent organic chloramines and 10 percent monochloramine.*

Investigators' Consensus of Germicidal Efficiency

In 1964 Clarke et al.[71] prepared graphics (see Fig. 4-14) showing the relative germidical efficiency of HOCl, OCl^-, and monochloramine. These curves represent a composite of available data adjusted to yield a common base for

*Organic chloramines that appear in the dichloramine fraction of the forward amperometric titration procedure are known to intrude into the monochloramine fraction. This distorts the monochloramine reading.

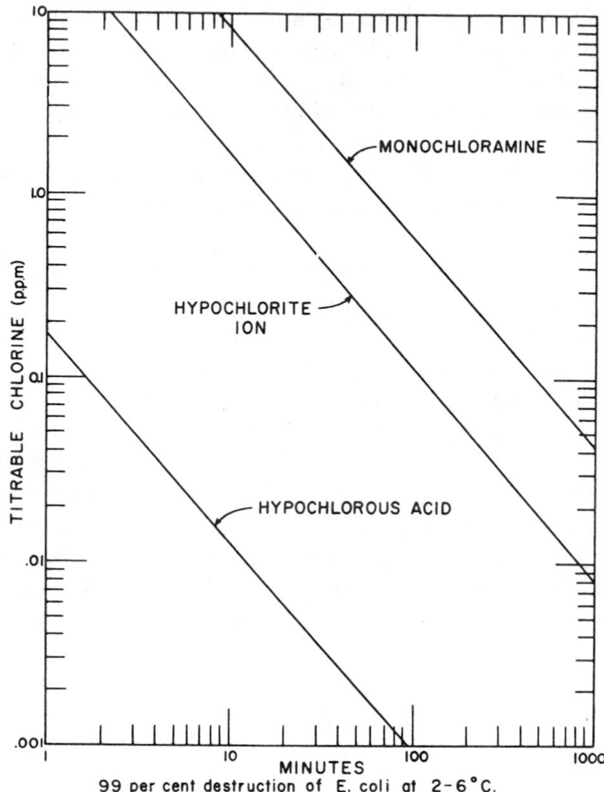

Fig. 4-14. Comparison of germicidal efficiency of hypochlorous acid, hypochlorite ion, and monochloramine.

comparison, and are presented here to summarize the preceding discussion of the germicidal efficiency of the different fractions of available chlorine residual. Scientific data are lacking to properly locate on this curve the germicidal efficiency of the dichloramine and nitrogen trichloride fractions.

It should be noted that these curves represent chlorine dosages applied to chlorine demand-free solutions sufficient to achieve only a 2 log reduction at very low temperatures and unreported contact times. The lower end of the curves may not be completely reliable. Full-scale plant operation on water reuse situations have demonstrated that for contact times of 45–60 min. and 4–5 log reductions of coliforms, monochloramine residuals are almost equal to free chlorine residuals and in some cases superior to free chlorine. (This anomaly is described in Chapter 8.)

Morris's Lethality Coefficient

In 1967 Morris[72] presented a tabulation of germicidal concentrations giving 99 percent (2 logs) inactivation within 10 min. of contact time. From this Morris derived a lethality coefficient:

Table 4-7 Values of λ at 5°C*

Species	*Enteric Bacteria*	*Amoebic Cysts*	*Viruses*	*Spores*
HOCl as Cl_2	20	0.05	1.0 up	0.05
OCl^- as Cl_2	0.2	0.0005	<0.02	<0.0005
NH_2Cl as Cl_2	0.1	0.02	0.005	0.001

*λ in $(mg/liter)^{-1}$ $(min)^{-1}$

$$\lambda = 0.46/C_{99:10}$$

where C is the concentration of the chlorine compound in mg/L. The values of λ computed from this tabulation are shown in Table 4-7.

This tabulation illustrates the superiority of free chlorine over both monochloramine and the hypochlorite ion. This is an interesting and useful comparison, but here again the organism destruction is based upon a short contact time (10 minutes) and only 2 logs removal. From plant operation experience White believes that longer contact times and greater total coliform removals (4–5 logs) close the gap between the germicidal efficiency of free chlorine and monochloramine. This has been demonstrated by Professor R. E. Selleck, University of California,[73] the Sanitation Districts of Los Angeles County,[74] and the San Jose/Santa Clara, California Water Pollution Control Plant.[70] (See Chapter 8.)

It must also be observed that in comparing free chlorine (85 percent HOCl) and true chloramines on a contact time basis where the wastewater contains 3 to 4 mg/L of organic N, the free chlorine residual will be depleted by exactly the amount of organic N, but it will take at least 40 min. for the organic N to start depleting the chloramine residual. (Also, see Chapter 10, which describes the reason why ORP measurements are superior to other methods for evaluating the germicidal efficiency of chlorine residuals.)

ELECTROCHEMICAL PROPERTIES OF CHLORINE RESIDUALS

Oxidation Reduction Potential—ORP*

Historical Background. General dissatisfaction with the orthotolidine method of measuring chlorine residuals led a number of investigators to try the oxidation–reduction potential concept for evaluating the germicidal efficiency of the various species of chlorine residuals. This concept is commonly referred to as "ORP" or "redox systems." In its simplest terms it is an electrode potential reading.

*See Chapter 9 for the important applications of ORP in both water and wastewater treatment facilities.

The originators of this method as it related to the chlorination of water were Rideal and Evans,[75] who suggested it as early as 1913. They believed that the germicidally active chlorine could best be measured by the potential created across an electrode system in the solution. Most investigators studying the disinfecting power of chlorine found that its efficiency varies widely under different conditions. The chief factor influencing the efficiency of free chlorine (HOCl) is the pH of the solution.

In 1933 Schmelkes[76] demonstrated that the level of potential of a chlorine residual does in fact affect its germicidal efficiency and that the potential is related to pH, chlorine concentration, and the ratio of chlorine to ammonia.

Figure 4-15 shows that free available chlorine has a much higher potential than do chloramines, and it varies with pH.

Figure 4-16 shows the relationship between the chlorine–ammonia ratio and potential level, indicating again the lower potential of the chloramines. During these early investigations it was generally believed that the germicidal efficiency of chloramines was less than that of free chlorine. Studies had suggested that the electrochemical properties of chlorine might be directly related to its germicidal efficiency.

In 1939, Schmelkes et al.[77] pursued this concept by investigating the "differential" potential of various types of free and combined chlorine residuals. Their work, which confirmed the previous investigations, is shown in Fig. 4-17. However, they considered an additional variable, described as the "base

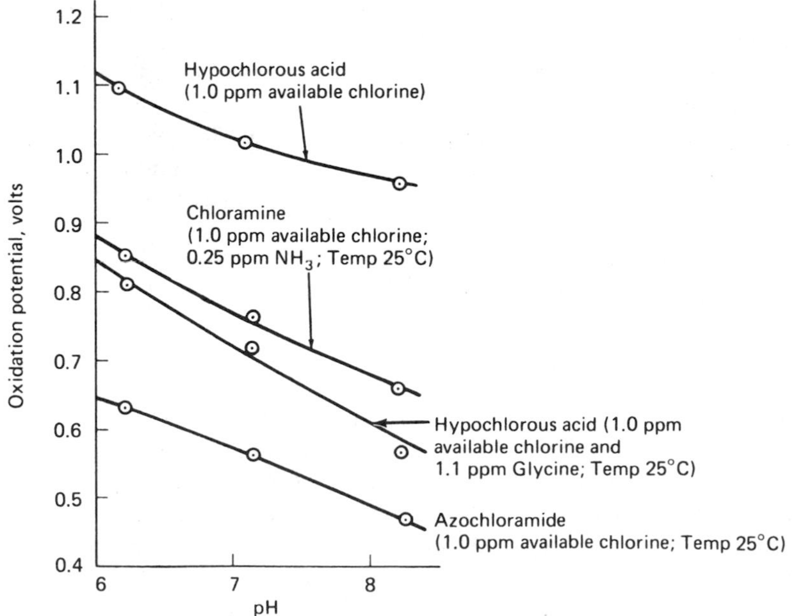

Fig. 4-15. Oxidation potentials of chlorine compounds.

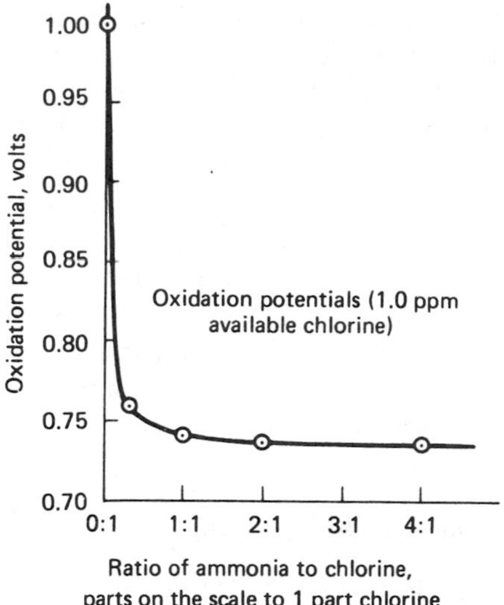

Fig. 4-16. Relationship of chlorine to ammonia ratio and oxidation potential.

potential" of different waters. They found that raising the poise from 170 mV base potential to 190 mV might not produce the same germicidal efficiency as raising the poise in another water from 190 mV to 210 mV. Therefore, the degree to which a water was poised had a significant effect in evaluating the potential as related to germicidal efficiency. Although this work left some doubt as to the reliability of the concept, it provided hope for future investigators.

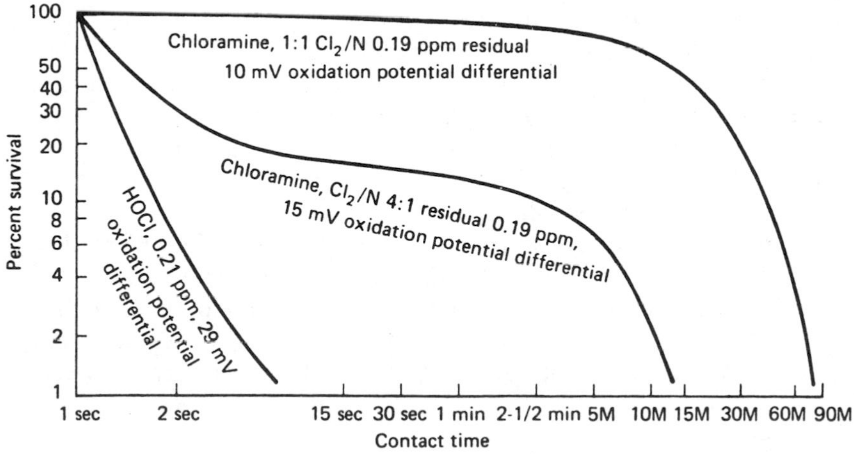

Fig. 4-17. Relationship of germicidal efficiency and oxidation potential differential.

In the next few years, S. L. Chang of Harvard University did some notable work on the destruction of microorganisms by chlorine.[55, 56, 78] Part of this work was an investigation of the oxidation potential concept of germicidally active chlorine.[79] Chang concluded that because of interfering substances, the uncertain amount of dissolved oxygen in raw waters, and the uncertainty of the reversibility of these systems, it was impossible to evaluate the potential measurements of residual chlorine. He did find that free chlorine, inorganic chloramines, and organic chloramines all had their own characteristic oxidation potentials, but that the potential for one compound at cysticidal concentrations gives no information on the cysticidal efficiency of any other compound. This in itself renders this concept useless for evaluating the germicidal efficiency of different chlorine compounds.

Practical Application of the ORP Concept

Thanks to Frank Strand, the use of ORP to quantify the oxidative power of the various species of chlorine residuals in any given situation has long since been accomplished, by Stranco Corp. of Bradley, Illinois. This was done by developing ORP probes that are able to withstand a high oxidative environment, such as a 15 mg/L chlorine residual. Strand's first success was in 1970[80, 81] for control of the chlorination and pH systems in large outdoor and indoor swimming pools. These two installations were so successful that ORP control has been used in this category of pools for over 25 years. The Stranco system is called the "High Resolution Redox Control System." Stranco's next success was an automatic chlorination control system for cooling tower recirculating water, and currently it has made a great success in the disinfection of potable water and wastewater* (see Chapter 9).

The swimming pool system is made up of two parts: ORP control of the chlorine addition and ORP control of the pH. A control panel starts and stops the chlorinator, and the pH control starts and stops a $NaCO_3$ chemical metering pump. The system maintains precise control of the pH, which assures a constant percentage of undissociated HOCl. The ORP control maintains a high level of free chlorine (1.5–2.5 mg/L), which helps account for its success. A survey of three of these operating installations by White verifies their success and practicality for swimming pool application (see below). In every case the Strantrol system, if it has any faults at all, errs on the side of safety. The free chlorine residual was always in the 1.5–2.5 mg/L bracket, and the pool water appeared brilliant and sparkling. At pH 7.5 this is an ideal situation.

In 1975 White investigated four swimming pools. Three of these pools were using the Strantrol system, and one was using pH control and conventional chlorine residual control. The following relationships were investigated:

*It is now (1997) standard practice for large commercial pools, both open and covered, to use the Stranco System of High Resolution Redox Control.

1. At a constant free chlorine residual of 2.1 mg/L, ORP measurements were made at pH levels from 6 to 10 (see Fig. 4-18). It is significant to note that the difference between the Gilroy and the Petaluma waters is on the order of 60 mV for identical pH levels. These curves also show that any given water with a constant chlorine residual will show an ORP shift of about 60 mV when there is a one-point shift in pH.
2. With the pH constant at 7.5, ORP measurements were made with free residuals ranging from 0.1 to 10 mg/L (see Fig. 4-19). The curves shown demonstrate the logarithmic nature of the chlorine residual versus ORP values. This increases the sensitivity of this method of chlorine residual measurement. It takes 37 mV difference in the ORP value for a change

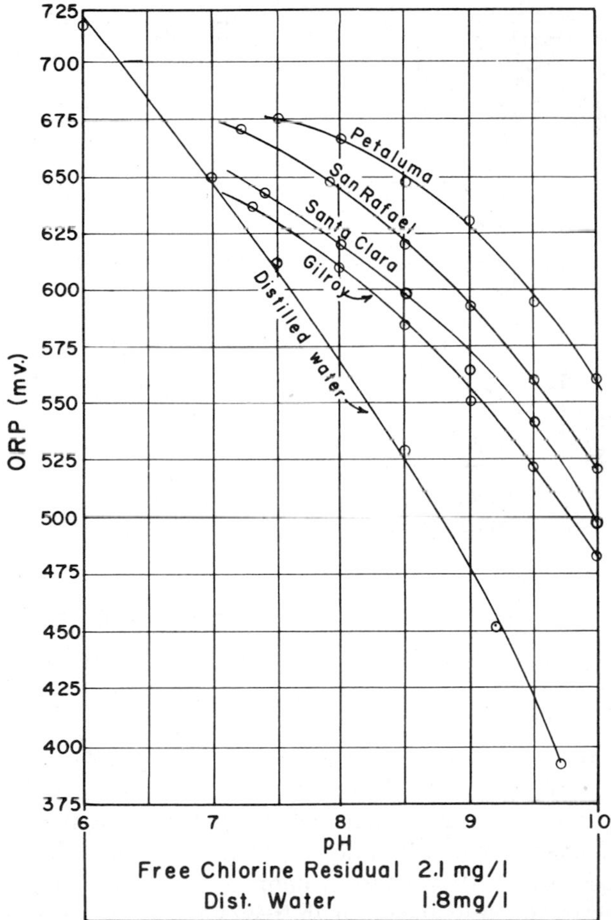

Fig. 4-18. Relationship between oxidation potential and pH at a fixed chlorine residual.

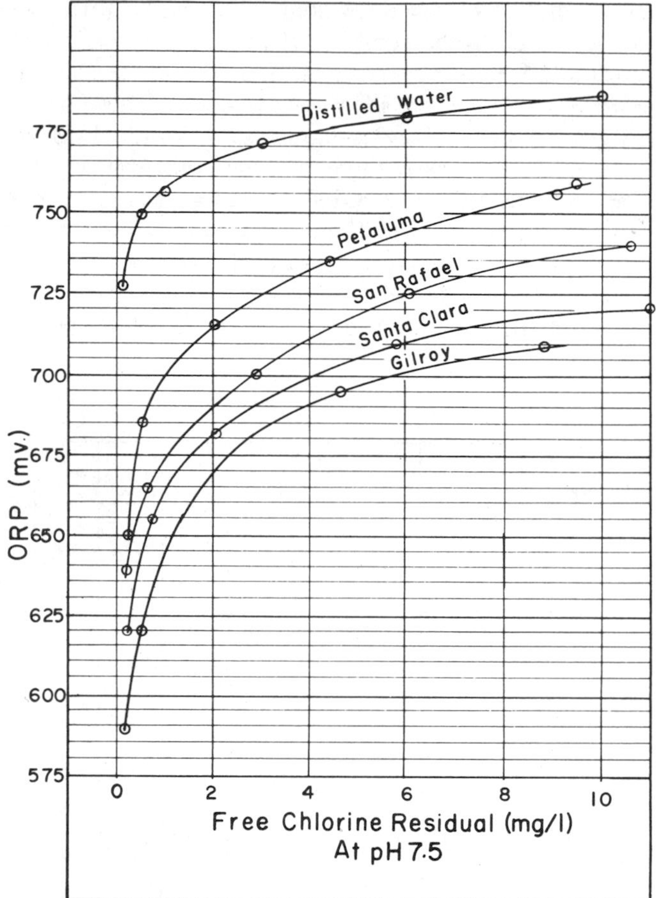

Fig. 4-19. Relationship between oxidation potential and free chlorine residual at a constant pH in five different waters.

of only 0.5 mg/L in the free chlorine residual. This kind of accuracy is what inspired a new ORP control system by Stranco for dechlorination. (See Chapter 10.)

As the above investigations illustrate that every water has a different ORP "poise," each installation has to be calibrated to that poise. This is simply achieved by an ORP measurement before chlorine is applied. If the water source or the character of the water is changed, a new calibration must be made. The success of this system for swimming pools is due to the close control of the pH and the maintenance of high free chlorine residuals. This prevents the typical formation of organic N compounds (due to low residual chlorine) caused by urea and perspiration of the swimmers. These compounds, if allowed to persist, will totally upset the water quality, and the pool will have to be shut down, drained, and flushed. The close control of the pH ensures the presence of a constant amount of a free chlorine (HOCl) residual, which is the most active chlorine residual species.

REFERENCES

1. Shilov, E. A., and Soludushenkov, S. N., "Hydrolysis of Chlorine," *Comptes Rendus Acad. Sci. l'URSS,* **3,** 17, No. 1 (1936), and *Acia Physiochim URSS,* **20,** 667, No. 5 (1945).
2. Morris, J. C., "The Mechanism of the Hydrolysis of Chlorine," *J. Am. Chem. Soc.,* **68,** 1692 (1946).
3. Adams, F. W., and Edmonds, R. G., "Absorption of Chlorine by Water in a Packed Tower," *Ind. Eng. Chem.,* **29,** 447 (1937).
4. Rosenblatt, D. H., and Small, M. J., private communication, U.S. Army Medical Bioengineering Research and Development Lab., Fort Detrick, MD, 1976.
5. Morris, J. C., "The Acid Ionization Constant of HOCl from 5 to 35°C," *J. Phys. Chem.,* **70,** 3798 (Dec. 1966).
6. Chamberlin, N. S., and Snyder, H. B., "Technology of Treating Plating Wastes," Tenth Annual Waste Conf., Purdue Univ., 1955.
7. Anon., *Standard Methods for the Examination of Water and Wastewater,* 12th ed., pp. 91, 185, 186, 208, 209, and 210, American Public Health Assoc., New York, 1965.
8. Sawyer, C. N., *Chemistry for Sanitary Engineers,* Chapter 25, McGraw-Hill, New York, 1960.
9. Morris, J. C., Weil, Ira, and Culver, R. H., "Kinetic Studies on the Break-Point with Ammonia and Glycine," unpublished copy from senior author, Harvard Univ., 1952.
10. Morris, J. C., "Kinetic Reactions between Aqueous Chlorine and Nitrogen Compounds," Fourth Rudolphs Res. Conf., Rutgers Univ., June 15–18, 1965.
11. Williams, D. B., private communication, Brantford, Ontario, Canada, 1967.
12. Griffin, A. E., "Effect of Applying Chlorine Before Ammonia and Vice Versa," private communication, 1938.
13. Griffin, A. E., and Chamberlin, N. S., "Some Chemical Aspects of Break-Point Chlorination," *J. NEWWA,***55,** 371 (1941).
14. Griffin, A. E., "Chlorine for Ammonia Removal," Fifth Annual Water Conf. Proc. Engrs. Soc. Western Pennsylvania, p. 27, 1944.
15. Morris, J. C., "The Chemistry of the pH Factor in Pools and Its Relation to Reactions with Nitrogenous Substances," paper presented at Water Chemistry Seminar, N.S.P.I. Convention, Chicago, IL, Jan., 1964.
16. Robson, H. L., *Encyclopedia of Chemical Technology,* Vol. 4, pp. 908–928, John Wiley & Sons, New York, 1964.
17. Weil, I., and Morris, J. C., "Kinetic Studies on Chloramines," *J. Am. Chem. Soc.,* **71,** 1664 (1949).
18. Weil, I., and Morris, J. C., "Equilibrium Studies on N-Chloro Compounds," *J. Am. Chem. Soc.,* **71,** 3123 (1949).
19. Morris, J. C., Salazar, J. A., and Weinman, M., "Equilibrium Studies on the N-Chloro Compounds," *J. Am. Chem. Soc.,* **70,** 2036 (1948).
20. Fair, G. M., Morris, J. C., and Chang, S. L., "The Dynamics of Water Chlorination," *J. NEWWA,* **61,** 285 (1947).
21. Granstrom, M. L., "The Disproportionation of Monochloramine," Ph.D. dissertation in Sanitary Engineering, Harvard Univ., 1954.
22. Morris, J. C., Weil, I., and Burden, R. P., "The Formation of Monochloramine and Dichloramine in Water Chlorination," paper 117th meeting, Am. Chem. Soc. Detroit, MI, Apr. 16–20, 1950.
23. Palin, A. T., "Chemical Aspects of Chlorine," *Inst. of Water Engrs. (England),* p. 565 (1950).

24. Palin, A. T., "A Study on the Chloro-Derivatives of Ammonia and Related Compounds with Special Reference to Their Formation in the Chlorination of Natural and Polluted Waters," *Water and Water Engineering* (*England*), p. 151 (Oct. 1950), p. 189 (Nov. 1950), p. 248 (Dec. 1950).
25. Williams, D. B., "How to Solve Odor Problems in Water Chlorination Practice," *Water and Sew. Wks.,* **99,** 358 (1952).
26. Williams, D. B., "Control of Free Residual Chlorine by Ammoniation," *J. AWWA,* **55,** 1195 (1963).
27. Williams, D. B., "Elimination of Nitrogen Trichloride in Dechlorination Practice," *J. AWWA,* **58,** 248 (1966).
28. Holwerda, K., "On the Control and Degree of Reliability of the Chlorinating Process of Purifying Drinking Water, Especially in the Relation to the Use of Chloramines for This Purpose," *Meddeelingen van den Dienst der Volksgezondheid in Nederlandsch-Indie,* **17,** 251 (1928) and **19,** 325 (1930).
29. Moore, W. A., Megregian, S., and Ruchhoft, C. C., "Some Chemical Aspects of the Ammonia–Chlorine Treatment of Water," *J. AWWA,* **35,** 1329 (1943).
30. Rossum, J. R., "A Proposed Mechanism for Break-Point Chlorination," *J. AWWA,* **35,** 1446 (1943).
31. Fair, G. M., Morris, J. C., Chang, S. L., Weil, I., and Burden, R. P., "The Behavior of Chlorine as a Water Disinfectant," *J. AWWA,* **40,** 1051 (1948).
32. Chapin, R. M., "Dichloramine," *J. Am. Chem. Soc.,* **51,** 2112 (1929).
33. Baker, R. J., "Types and Significance of Chlorine Residuals," *J. AWWA,* **51,** 1185 (1959).
34. Kelly, S. M., and Sanderson, W. W., "The Effect of Chlorine in Water on Enteric Viruses II., The Effect of Combined Chlorine on Poliomyelitis and Coxsackie Viruses," *Amer. J. Publ. Health,* **59,** 14 (1960).
35. Barrett, S., private communication, Metropolitan Water District of Southern California, June 1983.
36. Williams, D. B., "Monochloramine and Dichloramine Determinations in Water," *Water and Sew. Wks.,* **98,** 429 (1951).
37. Williams, D. B., "Free Chlorine, Monochloramine and Dichloramine in Water," *Water and Sew. Wks.,* **98,** 475 (1951).
38. Hulburt, R., "Chlorine and Orthotolidine Test in the Presence of Nitrites," *J. AWWA,* **26,** 1638 (1934).
39. Morris, J. C., and Wei, I., "Chlorine–Ammonia Breakpoint Reactions: Model Mechanisms and Computer Simulation," paper Annual Meet. Am. Chem. Soc., Minneapolis, MN, Apr. 15, 1969.
40. Pressley, T. A., Bishop, D. F., and Roan, S. G., "Nitrogen Removal by Breakpoint Chlorination," paper, Am. Chem. Soc., Chicago, Sept. 1970.
41. Saunier, B. M. "Kinetics of Breakpoint Chlorination and Disinfection." PhD. Thesis, Univ. of California, Berkeley (1976).
42. Saunier, B. M., and Selleck, R. E. "The Kinetics of Breakpoint Chlorination in Continuous Flow Systems," paper presented at the AWWA Ann. Conf., New Orleans, LA, June 22, 1976.
43. Stone, R. W. "Rancho Cordova Breakpoint Chlorination Demonstrations," report by Sacramento Area Consultants, Sept. 1976.
44. Taras, M. J., "Effect of Free Residual Chlorine on Nitrogen Compounds in Water," *J. AWWA,* **45,** 47 (1953).
45. Chang, S. L., private communication, 1968.
46. Feben, D., and Taras, M. J., "Chlorine Demand of Detroit Water," *J. AWWA,* **42,** 453 (1950).
47. Feben, D., and Taras, M. J., "Chlorine Demand Constants," *J. AWWA,* **43,** 922 (1951).
48. Taras, M. J., "Chlorine Demand Studies," *J. AWWA,* **42,** 462 (1950).
49. Williams, D. B., "The Organic Nitrogen Problem," *J. AWWA,* **43,** 837 (1951).
50. Nordell, E., *Water Treatment for Industrial and Other Uses,* 2nd ed., p. 213, Reinhold, New York, 1961.

51. Green, D. E., and Stumpf, P. K., "The Mode of Action of Chlorine," *J. AWWA,* **38,** 1301 (1946).
52. Knox, W. E., Stumpf, P. K., Green, D. E., and Auerbach, V. H., "The Inhibition of Sulfhydril Enzymes as the Basis of the Bactericidal Action of Chlorine," *J. Bact.,* **55,** 451 (1948).
53. Marks, H. C., and Strandkov, R. B., "Halogens and Their Mode of Action," *Annals of NY Academy of Sci.,* 163 (1950).
54. Wyss, O., "Disinfection by Chlorine, Theoretical Aspects," *Water and Sew. Wks.,* **109,** R155 (1962).
55. Chang, S. L., and Fair, G. M., "Viability and Destruction of the Cysts of *Endamoeba histolytica,*" *J. AWWA,* **33,** 1705 (1941).
56. Chang, S. L., "Destruction of Microorganisms," *J. AWWA,* **36,** 1192 (1944).
57. Marks, H. C., Wyss, O., and Strandkov, F. B., "Studies on the Mode of Action of Compounds Containing Available Chlorine," *J. Bact.,* **49,** 299 (1945).
58. Ingols, R. S., Wyckoff, H. A., Kethley, T. W., Hogden, H. W., Fincher, E. L., Hildebrand, J. C., and Mandel, J. E., "Bacterial Studies of Chlorine," *Ind. and Eng. Chem.,* **45,** 996 (1953).
59. Anon., *Water Quality and Treatment,* 2d ed., p. 206, American Water Works Assoc., New York, 1951.
60. Farkas, L., Lewin, M., and Bloch, R., "The Reaction between Hypochlorites and Bromides," *J. Amer. Chem. Soc.,* **71,** 1988 (1949).
61. Butterfield, C. T., Wattie, E., Megregian, S., and Chambers, C. W., "Influence of pH and Temperature on the Survival of Coliform and Enteric Pathogens When Exposed to Free Chlorine," *Pub. Health Rpts.,* **58,** 1837 (1943).
62. Butterfield, C. T., and Wattie, E., "Relative Resistance of *E. coli* and *E. typhosa* to Chlorine and Chloramines," *Pub. Health Rpts.,* **59,** 1661 (1944).
63. Selleck, R E., private communication, Univ. of California, Berkeley, July 1981.
64. Houston, Sir A. C., "19th and 20th Annual Reports of the Metropolitan Water Board, London, England," 1925 and 1926.
65. Butterfield, C. T., and Wattie, E., "Influence of pH and Temperature on the Survival of Coliforms and Enteric Pathogens When Exposed to Chloramine," *Pub. Health Rpts.,* **61,** 157 (1946).
66. Butterfield, C. T., "Comparing the Relative Bactericidal Efficiencies of Free and Combined Available Chlorine," *J. AWWA,* **40,** 1305 (1948).
67. Kabler, P. W., "Relative Resistance of Coliform Organisms and Enteric Pathogens in the Disinfection of Water with Chlorine," *J. AWWA,* **43,** 553 (1953).
68. Corbett, R. E., Metcalf, W. S., and Soper, F. G., "The Reaction Between Ammonia and Chlorine in Aqueous Solutions," *J. Chem. Soc.* (*London*), 1927, Part II (1953).
69. Faust, S. D., and Hunter, J. V., *Principles and Application of Water Chemistry,* p. 31, John Wiley & Sons, New York, 1969.
70. White, G. C., Beebe, R. D., Alford, V. F., and Sanders, H. A., "Problems of Disinfecting Nitrified Effluents," paper presented at Second National Symposium on Municipal Wastewater Disinfection, Orlando, FL, Jan. 26–28, 1982.
71. Clarke, N. A., Berg, G., Kabler, P. W., and Chang, S. L., "Human Enteric Viruses in Water: Source, Survival and Removability," International Conf. Water Pollution Research, Pergamon Press, London, Sept. 1964.
72. Morris, J. C. "Aspects of the Quantitative Assessment of Germicidal Efficiency," *Disinfection: Water and Wastewater,* J. D. Johnson (Editor), Ann Arbor Science, Ann Arbor, MI, p. 1, 1975.
73. Selleck, R. E., Saunier, B. M., and Collins, H. F., "Kinetics of Bacterial Deactivation with Chlorine," *J. Env. Eng. Div. ASCE,* p. 1197 (Dec. 1978).
74. Selna, M. W., Miele, R. P., and Baird, R. B. "Disinfection for Water Reuse," paper presented at the Disinfection Seminar at the AWWA Annual Conf. Anaheim, CA, May 8, 1977.
75. Rideal, S., and Evans, U. R., *J. Soc. Pub. Analysts and Other Analytical Chemists, London* (Aug. 1913).

76. Schmelkes, F. C., "The Oxidation–Reduction Potential Concept of Chlorination," *J. AWWA,* **25,** 695 (1933).
77. Schmelkes, F. C., Horning, E. W., and Campbell, G. A., "Electro-chemical Properties of Chlorinated Water," *J. AWWA,* **31,** 1524 (1939).
78. Chang, S. L., "Studies on *Endamoeba histolytica.* III, Destruction of Cysts of *Endamoeba histolytica* by a Hypochlorite Solution, Chloramines in Tap Water and Gaseous Chlorine in Tap Water of Varying Degrees of Pollution," *War Medicine,* **5,** 46 (1944).
79. Chang, S. L., "Applicability of the Oxidation Potential Measurements in Determining the Concentration of Germicidally Active Chlorine in Water," *J. NEWWA,* **59,** 79 (1945).
80. Hubbard, L. S., "Electronic Robot Controls Water Quality at Eisenhower Pool," *Park Maintenance* (March 1970).
81. Abbott, R., "Automation Solves Problems at Michigan Univ. Pool," *Swimming Pool Age* (May 3, 1971).
82. Baker, R. J., private communication, Mar. 1, 1983.
83. U.S. Patent No. 2,443,429, "Procedure for Disinfecting Aqueous Liquids," Marks and Strandkov, and Wallace & Tiernan Co., Inc. Belleville, NJ, June 15, 1948.
84. Selleck, R. E., et al., "Optimization of Chlorine Application Procedures and Evaluation of Chlorine Monitoring Techniques," Univ. of California Publication No. UCB-ENG-4180, Berkeley, CA, 1976.
85. Selleck, R. E., private communication, Univ. of California, Berkeley, Sept. 1983.
86. Hergott, S. J., Jenkins, David, and Thomas, J. F., "Power Plant Cooling Water Chlorination in Northern California," *J. WPCF,* **50,** 2590 (Nov. 1978).
87. Feng, T. H. "Behavior of Organic Chloramine in Disinfection." *J. WPCF,* **38,** 614 (1966).
88. Sung, R. D. "Effects of Organic Constituents in Wastewater on the Chlorination Process," Ph.D. Thesis, Univ. of California Davis, 1974.
89. Esvelt, L. A., Kaufman, W. J., and Selleck, R. E., "Toxicity Assessment of Treated Municipal Wastewater," paper presented at the 44th Ann. Conf. of WPCF, San Francisco, CA, Oct. 4–8, 1971.
90. Valentine, R. L., Brandt, K. I., and Jafvert, C. T., "A Spectrophotometric Study of the Formation of an Unidentified Monochloramine Decomposition Product," *Water Research,* **20,** 8, 1067–1074 (1986), printed in Great Britain by Pergamon Journals LTD.
91. Valentine, R. L., and Wilber, G. G., "Some Physical and Chemical Characteristics of an Unidentified Product of Inorganic Chloramine Decomposition," in *Water Chlorination; Chemistry, Environmental Impact and Health Effects,* Vol. 6, Chap. 63, pp. 819–852, 1987.
92. Diyandoglu, V., Marinas, Jr., B., and Selleck, R. E., "Stoichiometry and Nitrite Kinetics of the Reaction of Nitrite with Free Chlorine in Aqueous Solutions," *Envir. Sci. Tech.,* **24** (1990).
93. Chapin, R. M., "The Influence of pH upon the Formation and Decomposition of the Chloro Derivatives of Ammonia," *J. Am. Chem. Soc.,* **53,** 912–20 (1931).
94. Yiin, B. S., and Margerum, D. W., "Non-metal Redox Kinetics: Reactions of Trichloramine with Ammonia and with Dichloramine," *Inorg. Chem.,* **29,** 11, 2135–2141 (1990).
95. Selleck, R. E., "A Literature Review of Chlorine–Ammonia Reaction Chemistry," unpublished report, University of California Sanitary Engineering Research Lab., Richmond, CA, Sept. 25, 1990.
96. Jander, J., and Engelhardt, U., "Nitrogen Compounds of Chlorine, Bromine and Iodine," in *Developments in Inorganic Nitrogen Chemistry,* Vol. 2, C. B. Colburn (Editor), Elsevier Scientific Pub. Co., Amsterdam, pp. 70–228, 1973.
97. Hand, V. C., Margerum, D. W., and Huffman, R. P., "Kinetics and Mechanism of the Decomposition of Dichloramine in Aqueous Solutions," *Inorg. Chem.,* **22,** 10, 1449–1456 (1983).
98. Diyamandoglu, V., Marinas, B. J., and Selleck, R. E., "Stoichiometry and Kinetics of the Reaction of Nitrite with Free Chlorine in Aqueous Solutions," *Envir. Sci. Tech.* **24** (1990).
99. Yagil, G., and Anbar, M. J., *J. Am. Chem. Soc.,* **84,** 1797–1803 (1962).
100. Nagy, J. C., Kumar, K., and Margerum, D. W., *Inorg. Chem.,* **27,** 2773–2780 (1988).

101. White, G. C., Beebe, R. D., Alford, V. F., and Sanders, H. A., "Wastwater Treatment Plant Disinfection Efficiency as a Function of Chlorine and Ammonia Content," in *Water Chlorination: Environmental Impact and Health Effects,* Vol. 4, Book 2, R. L. Jolley et al. (Editors), Ann Arbor Science, Ann Arbor, MI, 1983.
102. White, G. C., Beebe, R. D., Alford, V. F., and Sanders, H. A., "Problems of Disinfecting Nitrified Effluents," *Municipal Wastewater Disinfection, Proc. of Second Ann. Nat. Symposium,* Orlando, FL, EPA report 600/9-83-009, July 1983.
103. Lusher, E. E., "Swimming Bath Water Treatment," Baths Service (England), p. 16 (1959).
104. Malpas, J. F., Wallace & Tiernan Ltd., London, private communication, 1968.
105. Lomas, P. D. R., "The Combined Residual Chlorine of Swimming Bath Water," *J. Assoc. Pub. Analysts* (England), **5,** 27 (1967).
106. Pontius, F. W., *Water Quality and Treatment,* 4th Ed., by American Water Works Association, published by McGraw-Hill, New York, 1990.
107. Anon., "Oxidation of Complex Organic Nitrogen by Hydrogen Peroxide," *Chemical Engineering,* p. EE-17, (Sept. 1994).

5

DETERMINATION OF CHLORINE RESIDUALS IN WATER AND WASTEWATER TREATMENT

DEVELOPMENT OF ANALYTICAL METHODS

Historical Background

When chlorine compounds were first introduced, about 1902, as a means of water disinfection, the only method available for testing residual chlorine was the starch–iodide one. Between 1902 and 1908, the proponents and investigators of the process were merely feeling their way and were not ready to accept such a concept as residual chlorine. As a matter of fact, considerable doubt existed as to the bactericidal efficiency of chlorine, so that results were evaluated entirely on the total bacteria and coliform kill. The latter was determined by the coliform presumptive test.

After the momentous Jersey City litigation in 1910, wherein the court ruled unequivocally in favor of chlorine as a successful disinfecting process for potable water, the workers in this field turned their attention to the proper control of chlorination. In 1909 E. B. Phelps[1] first proposed orthotolidine as a qualitative indicator of residual chlorine. Then in 1913 Ellms and Hauser[2] developed the quantitative test for residual chlorine using orthotolidine. In addition, they developed colorimetric standards for its use. This was a great contribution, since it paved the way for a scientific approach to the study of chlorination. Between 1917 and 1920 Wolman and Enslow[3] made studies on chlorine absorption in water and demonstrated the suitability and reliability of the orthotolidine test. Their work was the first scientific approach to the use of chlorine in water treatment. Until about 1930 the use of chlorine in water treatment was increasing at a tremendous rate. This inspired more and more studies of the various phenomena of controlling the application of chlorine. Refinements in the use of orthotolidine for measuring residual chlorine continued. The color standards formulated by Ellms and Hauser were modified, first by Meur and Hale[4] in 1925, then by Scott[5] in 1935, and again in 1939.[6] A final refinement of these standards was made in 1943 by Chamberlin and Glass.[7]

In 1940 the bombing of Britain focused attention on the need for an adequate field test for high chlorine residuals to be used in emergency sterilization of water mains. Contributions on the measurement of high residuals by Chamberlin[8] and Griffin[9] are the most notable. As a result of this work, it appeared necessary to overhaul the entire procedure of chlorine residual measurement by orthotolidine. This led in 1943 to the *AWWA Joint Committee Report*,[10] which was based upon the research work of Chamberlin and Glass.[7,11] This work is also the basis for both the procedure and color standards of the orthotolidine test appearing in the 19th edition of *Standard Methods*.[12]

In 1939 the discovery of the breakpoint phenomenon revealed that there was more than one kind of chlorine residual. The types were identified as free available and combined chlorine residuals. Further work indicated that the breakpoint phenomenon could not be properly evaluated or controlled without an adequate method of differentiation between these two types of residual. This led to more important contributions to the technique of chlorination. First came the orthotolidine flash test by Laux[13,14] in 1940, followed by the OTA (orthotolidine arsenite) test developed by Hallinan[15,16] in 1944. In the meantime, an even more important development in the method for measuring chlorine residuals was being explored. This was the amperometric method, first introduced for measuring chlorine residuals by Marks[17] in 1942.

Subsequent refinements resulted in changes in both the application of chlorine and its measurement. The orthotolidine test gave considerable ground to the OTA test, and the amperometric test was refined[18] to give the various fractions of free available and combined chlorine residuals. Marks[19,20] in 1951 presented an amperometric method capable of further differentiation of the total chlorine residual into free chlorine, monochloramine, and dichloramine by a series of titrations.

While Marks in the United States was investigating the amperometric method, Palin in England was exploring different colorimetric procedures to differentiate the various fractions of free available and combined available chlorine residuals. Palin[21] first reported on the use of *p*-aminodimethylaniline in 1945, as a method superior to orthotolidine. Although this indicator had been reported on by Moore[22] in 1943 in the United States, Palin's work was considerably more extensive than Moore's. Further work by Palin convinced him that he should discard this indicator in favor of a new approach. In 1949 Palin[23] announced a new and improved method, using ferrous ammonium sulfate as the titrating agent and neutral orthotolidine as the indicating reagent. This approach showed considerable promise in that it was shown to be valid for measuring free available chlorine, monochloramine, dichloramine, and even nitrogen trichloride. Further refinements were reported by Palin[24] in 1954, but owing to limitations of water temperature and errors due to interferences Palin explored other colorimetric indicating reagents. In 1957[25] he reported on his investigation of diethyl-*p*-phenylene diamine, soon to be known as DPD. This method, using the various reagents in tablet form, is the most widely accepted method in England.

In 1966 the Chester Beatty Research Institute at the Royal Cancer Hospital, London, England, issued a report that led to the abandonment in 1969 of orthotolidine as a residual chlorine test in the British Isles. The report, "Precautions for Laboratory Workers Who Handle Carcinogenic Aromatic Amines," gave a list of chemicals—including orthotolidine—regarded as potential causes of tumors in the urinary tract. Consequently, the only colorimetric procedure now used in the British Isles is Palin's DPD method.*

Other noteworthy contributions toward finding better methods for determining chlorine residuals have been made. The orthotolidine titration technique was introduced by Connell[26] in 1947 as an attempt to differentiate free available from combined available residual chlorine. This method was based on the development of a cherry red color when insufficient O-T is present, followed by the conversion of the red chlorine-substituted holoquinone to the yellow holoquinone by rapid addition of sufficient O-T. This method, of questionable practical value, has never been adequately evaluated. The method assumes no reaction with chloramines present with the O-T reagent in the time it takes to perform the titration. Because O-T is the reagent, the usual limitations apply: The temperature of the sample is a limiting factor affecting the accuracy of the free chlorine residual. Furthermore, other investigators do not agree that it is proper procedure to convert the red color of the chlorine-substituted holoquinone to the yellow holoquinone.[8, 9, 11]

Chamberlin[11] lists as one of the conditions for the accuracy of the O-T test that the sample should be added to the reagent and not vice versa, as in the above procedure.

Next came the investigation of Methyl Orange as a substitute for orthotolidine. Chang[36] used this method in 1944 for the measurement of free chlorine in his investigation of the relative cysticidal power of free and combined chlorine.

Taras[27, 28] investigated this method and described both the volumetric and colorimetric procedures in 1946 and 1947. As recently as 1965 the methods proposed by Taras were more fully evaluated by Sollo and Larsen[29] in their quest for a better and simpler field method of differentiating free and combined chlorine residual.

This method never gained popularity or widespread credibility. It is mentioned here as an historical item and also because the EPA "Analytical Reference Service Report No. 40," 1971, showed that the Methyl Orange procedure demonstrated the best accuracy for measuring both free and combined chlorine (separately). The other methods tested were: Leuco Crystal Violet, SNORT, DPD-Tritrimetric, DPD-Colorimetric, Amperometric, and OTA.†

The Methyl Orange method is based upon the almost instant decolorization of methyl orange by chlorine on a quantitative basis, whereas combined chlo-

*Owing to its carcinogenic properties, *Standard Methods* eliminated all orthotolidine procedures in the 15th edition. However O-T methods are still used in the United States.

†The EPA Report 40 cited is listed as a reference on page 300 of *Standard Methods,* 15th edition.

rine (chloramines) is much slower in its bleaching effect on methyl orange. Free chlorine bleaches methyl orange quantitatively on the basis of two molecules of chlorine to one of methyl orange. Therefore the weight ratio of one molecule of M.O. to two molecules of chlorine is 2.34 : 1. One milliliter of 0.005 percent M.O. is instantly decolorized by 21.9 mg of chlorine, and 1 ml of 0.005 percent M.O. solution contains 50.0 mg of M.O. This yields an M.O.-to-chlorine weight ratio of 2.28 : 1 as compared to the molecular weight ratio of 2.34 : 1. This is well within the margin of experimental error. Therefore the arithmetical relationship of this M.O. reaction with chlorine is:

$$\text{mg/L } Cl_2 = 0.217 \times \text{ml } 0.005\% \text{ M.O. solution} + 0.04 \qquad (5\text{-}1)$$

(1.0 ml M.O. is equivalent to 0.23 mg/L free chlorine.)

The analyst has the choice of the volumetric method,[26, 27] where a titration is carried to an end point using an exact amount of M.O. titrant; or the colorimetric method of Sollo and Larsen,[29] using an excess of M.O. and comparing to permanent color standards, or by determining the light absorbance of the color with a spectrophotometer properly calibrated for the test.

The reagent, methyl orange, is a primary standard, is stable, maintains its titer indefinitely, and reacts with chlorine over a wide range of temperatures. However, the useful range of this method is limited to 2 mg/L free chlorine. EPA Report No. 40 was based upon free chlorine concentrations of 0.44 and 0.98 mg/L and combined chlorine of 0.66 mg/L. This method has not been used except in a few laboratory projects.

In 1967, Black and Whittle[30, 52] reported on a new indicator solution identified as leuco crystal violet. It possesses excellent properties for the quantitative measurement of free chlorine, and was rated second to M.O. in EPA Report No. 40. This scheme of measuring chlorine residuals was developed as a result of investigating the efficiency of free iodine versus free chlorine in both potable water and swimming pools. This method has not been widely acclaimed in spite of the fact that it is superior to the O-T method for either free or combined residuals. It is described on page 4-70 *Standard Methods,* 19th edition.

In 1963, the U.S. Army Medical Research and Development Command began supporting research to develop an improved field method for measuring free available chlorine (FAC). In 1967 the Army adopted the modified orthotolidine–arsenite method (MOTA) for measuring FAC in water.[37] This method is an improvement over Hallinan's OTA method.[15] The MOTA method was an attempt to eliminate or minimize interfering substances; however, the performance was not successful enough for the Army. Combined chlorine, nitrites, and iron and manganese compounds continued to cause significant errors in the determination of FAC.

In 1965 the School of Public Health at the University of North Carolina began an investigation sponsored by the U.S. Army to find a method that could accurately measure FAC in the presence of combined chlorine under all conditions of military operations.[31–33] Moreover the method had to be a simple field operation, like the O-T method.

The method chosen to be developed involved the use of a stabilized neutral O-T solution. Known as the SNORT method, it is based upon the development of the blue meriquinone colors of O-T in the neutral range of sample–reagent mixture. This method was first used by Harrington[34] at Montreal and then by Caldwell[35] at Springfield without benefit of a stabilizer. The SNORT procedure developed for the Army, which included a test kit with disk like the Wallace & Tiernan and Hellige O-T kits, was presented at the 1976 fall meeting of the Armed Forces Epidemiological Board. After reviewing the test results, the board recommended against adoption of this method because of the false-positive FAC reading in the presence of combined chlorine. This factor erased any advantage of SNORT over the MOTA method.

In 1972 Bauer et al.[38] reported on the use of a treated strip of paper and a color chart that would offer the user a rapid means of measuring free chlorine. This test was based upon the reaction of free chlorine with the compound syringaldezine. The developer, Miles Laboratory, claimed no interference from combined chlorine. The U.S. Army Medical Environmental Research Unit was suitably impressed with the possibilities of this method. It engaged Guter and Cooper to evaluate the method.[39, 40] This investigation concluded that a syringaldazine solution was the most specific for FAC of the five hand-held portable field test kits that were evaluated. However, as might be expected, organically polluted water resulted in appreciable false-positive FAC readings for the majority of the kits.[40]

In the meantime, the U.S. Army Medical Bioengineering Research and Development Laboratory, Aberdeen Proving Ground, Maryland, continued its own investigation of tests for FAC. This they did in the presence of significant concentrations of combined chlorine compounds. These tests concluded that the syringaldazine (FACTS) method was the most specific for FAC.[41, 42] The Army pursued its search for an FAC test accurate under field conditions, having decided from World War II and Vietnam experiences that only a free chlorine residual of a given magnitude would provide the drinking water safety needed. In 1975 the Army investigative unit suggested that the DPD method might be modified to decrease the combined chlorine breakthrough into the FAC reading.[43] Palin followed through on this suggestion by proposing the addition of glycine immediately after mixing the DPD tablet with the sample. Glycine is supposed to suppress the combined chlorine reaction with DPD reagent, thereby inhibiting the combined chlorine breakthrough into the FAC reading. This modification was examined by Meier et al.[50] in 1978 for the Army. The Army was not satisfied with this modification as compared with the syringaldazine method. Palin made a further modification, which he called Steadifacs.[44, 45] This modified DPD procedure utilizes the addition of thioacetamide solution. When added in the correct amount, it will immediately dechlorinate completely a sample containing combined chlorine without affecting a previously developed DPD color representing the FAC in the same portion of the same sample.

Syringaldazine (FACTS) Method

See p. 4-47, *Standard Methods,* 19th edition.

Both the U.S. Army and the Drinking Water Research Center at the Florida International University, Miami, Florida carried out more testing and observations of various methods of FAC detection and measurement.[47, 48] The investigators concluded that the measurement of FAC by the FACTS method was acceptable in the presence of significant concentration of chloramines, and was superior to the DPD method, which suffers appreciably from chloramine intrusion. This method uses the Iodometric Electrode Technique. It measures FAC residuals over a 0.1 to 10 mg/L range. (FACTS is an acronym that stands for Free Available Chlorine Tests by Syringaldazine.)

In 1982 EPA approved the FACTS method for the determination of free chlorine residuals.[49] In 1983 Gibbs et al.[53] and Cooper et al.[55] summarized their work exploring for the Army the best method for measuring FAC in the presence of chloramine under military field conditions. This work concluded that the FACTS method was superior.

Finally, in 1983 Cooper et al.[54] reported the FAC residuals measured by FACTS and amperometric titrations to be equivalent.

Organic N

In 1982, White[77, 79–81] found exactly how and to what extent this compound affects the measurement of all chlorine residuals. (See Ref. 81.) Now, in 1997, there is an instrument that measures quickly and accurately both ammonia N and organic N in both potable water and wastewater. (See Chapters 4, 6, and 8.)

CURRENT STATUS OF ANALYTICAL METHODS

See pages 4-36 and 4-38, *Standard Methods,* 19th edition, the Introduction to the measurements of chlorine residuals.

Colorimetric

General Considerations. The most important factor to consider when using colorimetric methods is that the user will probably perform field measurements with a test kit that utilizes a rotating disk or a slide with permanent glass color standards. Therefore, the operator of these kits must depend upon visual acuity to determine the residual. However, the investigators who probed the accuracy of these methods used sophisticated laboratory equipment (spectrophotometer) that can identify residuals to the second decimal place. There are currently available field test kits that use spectrophotometers, but they are not yet widely used. One of the most popular and probably the most sophisticated of these test kits is described below.

Another factor is the shelf life of chemicals used to perform the tests. This must be determined in order for investigators to have confidence in the test results. Chemical solutions or tablets should always be stored in a cool, dark, and dry place.

Test equipment should be selected on the basis of the expected accuracy of the measurements and the range of residual concentrations.

It should be noted that *Standard Methods* calls for the use of either a spectrophotometer or a photometer for all colorimetric methods. Permanent color standards are an alternative and are acceptable if range and readability fit the user's requirements.

Orthotolidine

General Discussion. This method is obsolete, and its use is diminishing year by year. It may disappear in a few years because it is a health hazard to the people making the powder and the solutions. It is a potential carcinogen in humans—it affects the bladder and the urinary tract.

Owing to its use over so many years, it is discussed here because much of the investigative work in the past relied upon O-T as a method for measuring residuals. It is a simple and reliable method for measuring total residual chlorine in potable water. Only one reagent is involved. It covers a range of 0–10 mg/L and in the 0–5 mg/L range residuals can be estimated with reliability to the nearest 0.25 mg/L.

It is not an acceptable method for measuring chlorine residuals in wastewater because the sample reagent mixture is in such a low pH environment (<2). At this pH the chlorine residual species becomes highly active and is partially consumed before the O-T color develops. This results in residual measurements lower than the true residual. Investigations have shown that the discrepancy is 2–2.5 times lower for O-T than amperometric measurement in primary effluents.

The free chlorine species can be distinguished by the O-T method, provided that the sample reagent mixture is chilled to 1°C. This takes time, and thus the additional contact time can produce a slight error due to total chlorine residual decay.

One other factor to be aware of while examining investigations of germicidal efficiency, particularly those comparing chlorine and ozone, is to determine if the medium used contains ammonia nitrogen. Some bacteriological researchers inadvertently used bacterial cultures that contained ammonia nitrogen, which converted the free chlorine to chloramines, thus distorting the results.

Measuring Residuals Greater than 10 mg/L. The upper limit of residual measurement with O-T is 10 mg/L. However, there always exists the need to measure residuals in the 50 mg/L range. These residuals are required for water main sterilization, reservoirs, subdivisions, ships' tanks, hospital water systems, new office buildings, etc. This is accomplished by the following method.

The Drop Dilution Method. This method was developed about 1941 for determining residuals from 10 to 100 mg/L.[9] This method consists of the addition of one or more drops of the chlorinated water sample to a cell or a vessel of known volume containing orthotolidine and distilled water. This method is simple and rapid for approximating residuals between 10 and 100 mg/L. Using a standard chlorine residual comparator, calibrate the dropper furnished with the comparator. Use this dropper exclusively for measuring the chlorinated water sample. One drop from such a dropper usually equals 0.05 mL. The standard comparator utilizes cells that have a 15 mL capacity. With this information, the procedure is as follows:

1. Collect sample in a small glass container.
2. Add 0.5 ml O-T to the center tube and fill it to the scribed mark with distilled water. (The standard comparator cell holds 15 ml of sample to the scribed mark.)
3. Fill the other tube with distilled water. (This tube compensates for the natural color of the water sample.)
4. Add one drop of chlorinated sample to the center tube, mix it, and read the result immediately.

If no color appears, add additional drops of chlorinated sample, one at a time, until a reading within the range of color disk is obtained.

Computation of Residual. First, it shall be assumed that the dropper capacity is 0.05 mL per drop and that the cell capacity is 15 ml. Then we have:

$$\text{mg/L chlorine present} = \frac{\text{capacity of cell}}{\text{ml of chlorinated sample added}} \times \text{comparator reading} \tag{5-2}$$

Example: If two drops of sample produced a comparator reading of 0.25 mg/L Cl_2, then:

$$Cl_2 \text{ mg/L} = \frac{15}{(0.05 \times 2)} \times 0.25 = 37.5 \text{ mg/L} \tag{5-3}$$

Interfering Substances. The 1943 *Joint Committee Report on the Control of Chlorination*[10] declared that the color developed by the reaction of chlorine and O-T could be considered residual chlorine if the water sample contained no more than the following amounts of interfering substances: 0.3 mg/L iron,

0.01 mg/L manganese (manganic), and 0.10 mg/L nitrite. These limits were never changed. These substances interfere with other colorimetric methods.

The substance that is most likely to contribute significant interference is the nitrite ion, NO_2^-.[60] If the presence of nitrites is suspected, the test should be carried out so that the color development proceeds in total darkness. Then the test should be repeated in artificial light. In this way the color interference by nitrites will be nearly eliminated. Colors due to nitrites will develop even in the dark if nitrites are present in excess of 1.0 mg/L. Nitrites are the intermediary in the conversion of ammonia nitrogen to nitrate by biological action. The difficulty arises when there is sufficient nitrogen present to maintain all the chlorine residual as chloramine. The nitrites present causes false residual readings, which deceive the operator, and the chloramine residual is unable to oxidize the nitrites. The result is an algal bloom in a swimming pool, and in a water treatment plant it is a proliferation of bacteria. The reaction between nitrite and orthotolidine is not well understood, but it is certain that the reaction is one of oxidation by nitrites. This confirms previous reports that nitrites are an intermediary state in the conversion of ammonia to nitrates. This would indicate that nitrites can act as either an oxidizing or a reducing agent.

Leuco Crystal Violet Method

Chemistry of the Method. The details of this method appear on page 4-70, item 4100-I B, *Standard Methods,* 19th edition. This method measures separately the FAC and the TRC (total residual chlorine). The LCV reagent reacts instantaneously with free chlorine to form a bluish color. Interference from combined chlorine is avoided by completing the test for free chlorine (taking a color depth reading) within 5 min. after adding the reagent.

The total residual chlorine reading involves the reaction of free and combined chlorine with iodide ion to produce hypoiodus acid, which in turn reacts instantaneously with LCV to form the dye, crystal violet. The color of this dye is stable for days. This reaction requires the addition of pH 4 buffer and potassium iodide solutions to perform the release of I_2, which forms the hypoiodous acid.

Two separate sets of color standards are required for this method: one for free chlorine (bluish colors) and one for total residual chlorine (violet color). Commerically prepared standards are available in test kits for free chlorine determinations but not for TRC. Color values expressed as chlorine residual can be measured by a calibration curve established for either a filter photometer or a spectrophotometer.

The minimum detectable concentration is 10 μg FAC and 5 μg TRC. The practical range for this method if 0–10 mg/L for both FRC and TRC.

Interference with FRC occurs when the combined chlorine concentration is 5.0 mg/L or greater. The addition of 5.0 mL of sodium arsenite solution ($NaAsO_2$ at 5 g/L) will minimize this interference.

The major interference in the FRC measurement is from the *manganic ion,* which increases the apparent chlorine residual reading. When the manganic ion is known to be present, the photometric procedure is used to determine the absorbance due to the manganic ion separately. This measurement is subtracted from the total absorbance to yield that produced by FAC alone.

In the presence of monochloramine, nitrites will cause serious interference in the determination of free chlorine. Addition of sodium arsenite tends to minimize this interference.

Procedure. See pages 4-36 to 4-47, *Standard Methods,* 19th edition.

FACTS Method (Syringaldazine)

See page 4-47 *Standard Methods,* 19th edition.

General Discussion. A saturated solution of syringaldazine is used as the indicating reagent. It is stable when stored as a solid or as a solution. It is oxidized by free available chlorine on a 1:1 molar basis, yielding a colored product with an absorption maximum at 530 nm. The color product is only slightly soluble in water; therefore, at chlorine concentrations greater than 1 mg/L, the final reaction mixture must contain 2-propanol to prevent product precipitation and color fading. The pH of the reagent and sample mixture is critical. It must be held between 6.5 and 6.8. Otherwise color develops too slowly or too fast, which results in fading. A buffer reagent, which produces a pH of 6.7, must be used along with the indicator. Care must be taken when chlorinated samples derive from predominantly acid or alkaline water so that the amount of buffer added is the amount necessary to produce a reagent–sample mixture pH within the 6.5–6.8 limitation.

Interferences common to other methods for determining FAC do not affect the FACTS procedure. Monochloramine up to 18 mg/L, dichloramine up to 10 mg/L, and oxidized forms of manganese up to 1 mg/L do not interfere. Concentrations of ferric iron up to 10 mg/L do not interfere, nor do nitrite concentrations less than about 250 mg/L. Strong oxidizing agents such as iodine, bromine, and ozone will produce a color.

The minimum detectable FAC is 0.1 mg/L or less. The range of FAC measurement is 0.1–10.0 mg/L.

Temperature has a minimal effect on color reaction. The maximum error observed over a range of 5–35°C is ± 10 percent.

It is necessary to use either a filter photometer or a spectrophotometer to convert depth of color (light to dark pink) to mg/L FAC. *This method measures only FAC.*

Some commercial color kits are available with four permanent glass color standards of 0.5, 1.0, 1.5, and 2.0 mg/L FAC.

Procedure. See pages 4-45 and 4-46, *Standard Methods,* 19th edition.

DPD Colorimetric Method I

See pages 4-45 and 4-47, No. 4500 Cl G, *Standard Methods,* 19th edition.

General Discussion. This is the only colorimetric method with the capability of differentiating the various species of chlorine residuals. The indicator, which produces gradations of red color in the presence of chlorine residuals, is *N,N*-diethyl-*p*-phenylenediamine (DPD). The most significant interference is oxidized manganese. To compensate for this, a blank must be used. The minimum detectable concentration is about 10 μg/L as chlorine. The range of the test is 0–4 mg/L without sample dilution.

Palin Tablet Method. All of the reagents required are incorporated into four tablets which have to be completely dissolved before executing the residual measurements. These are as follows:

Tablet	*Contents*
DPD No. 1	DPD indicator with EDTA and buffer
DPD No. 2	Stabilized KI for monochloramine activation
DPD No. 3	Stabilized KI for dichloramine activation
DPD No. 4	All reagents in a single tablet

Depending upon the tests desired the appropriate procedure is selected:

FAC	Use Tablet No. 1
FAC + combined chlorine	Use Tablets No. 1 and 3
Monochloramine	Use Tablet No. 2
Dichloramine	Use Tablet No. 3
Total available chlorine	Use Tablet No. 4

The colorimetric procedures using the Palin tablet method require the use of a calibrated spectrophotometer to measure the color development of the indicator, or the use of permanent glass color standards available in a variety of commercial kits. These kits should be carefully chosen for range and residual interval spread.

Monochloramine Breakthrough. This phenonemon occurs when monochloramine is present in significant concentrations. Palin has suggested the use of thioacetamide solution to inhibit or suppress the monochloramine breakthrough into the FAC reading. When this reagent is used, Palin calls it the "DPD Steadifac" method.[44] Details of this alternative are given in the section "DPD Ferrous Titrimetric Method," below.

DPD Colorimetric Method II

The Residometer Method. This is the basic DPD method described above, except that it measures only free and total available chlorine residuals. It utilizes a calibrated spectrophotometer and solution reagents instead of tablets. The measuring instrument is the Wallace & Tiernan Residometer with a digital readout to the second decimal place (See Fig. 5-1). It was developed by Wallace & Tiernan in Great Britain, when Palin got the amperometric method banned in England. The Residometer is widely used in England because it is much faster and more accurate than Palin's methods. It is equipped with batteries for portable use and 110 V power for laboratory use. It is mounted in a carrying case the size of a conventional attaché case. One model is equipped with a pH electrode for pH measurements from 0 to 14. Three reagents are supplied for chlorine residual measurements and two for pH. The reagents have been found to have a shelf life of at least 18 months under ordinary conditions. The dropper bottles used to measure the reagent dosage have proved to be well within the accuracy of the DPD method. The solution reagents allow the user to perform the analytical tasks with greater speed and ease than when the tablets are used. The speed of color development in thc DPD measurements enhances the accuracy of the method.

This method is affected by the same interferences as those described in Method I, namely, oxidized manganese.

Procedure. The Residometer comes with complete instructions on how to use the solutions for both free and total chlorine residuals.

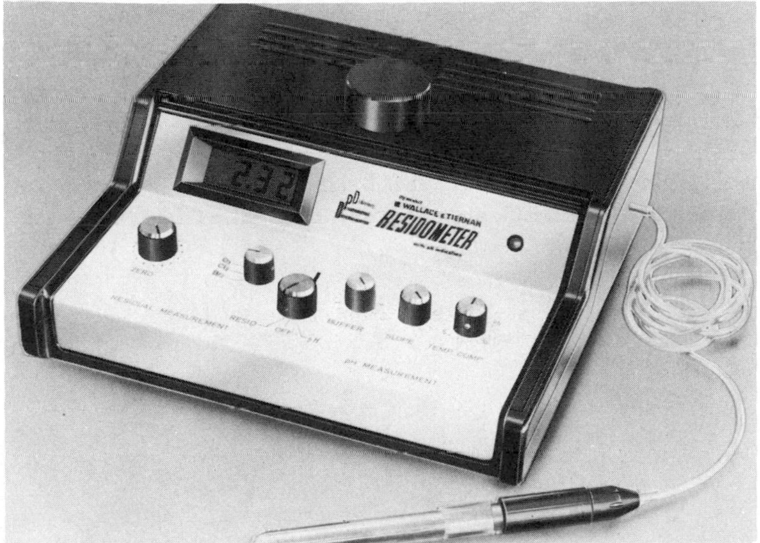

Fig. 5-1. Residometer DPD (courtesy Wallace & Tiernan).

Titrimetric Methods: Iodometric Method I

See pages 4-38 and 4-39, *Standard Methods,* 19th edition.

General Discussion. The principle of this method is based upon the phenomenon of the release of elemental iodine, which is quantitatively proportional to the chlorine residual present. The liberated iodine is titrated with a standard solution of sodium thiosulfate using starch as the indicator.[61, 62] However, there are certain conditions that limit and control this phenomenon of iodine release.

Procedure. The test is performed as follows:

1. The pH of the sample to be tested is adjusted to the desired range by a buffer soltuion.
2. The addition of potassium iodide is immediately followed by the release of elemental iodine, which will cast a brown color on the sample.
3. A reducing agent is added only until the brown cast disappears.
4. Then a starch solution is added, which turns the sample blue in the presence of elemental iodine.
5. The titration with the reducing agent is then continued only until the blue color disappears.

The following equations show the chemistry of the reaction:
Free chlorine residual:

$$HOCl + KI \rightarrow KCl + I_2 + I_2 + H_2O \tag{5-3}$$

Chloramine or combined residual:

$$NH_2Cl + KI \rightarrow KCl + I_2 + NH_4OH \tag{5-4}$$

(*Note:* The above equations are not balanced because the other products in the reaction, HCl and H_2O, are purposely omitted.) Titration by a reducing agent (thiosulfate) is as follows:

$$I_2 + 2\ \underset{\text{(thiosulfate)}}{Na_2S_2O_3} \rightarrow 2\ Na_2S_4O_6 + 2\ NaI \tag{5-5}$$

or

$$I_2 + 2\ S_2O_3^{=} \rightarrow S_4O_6^{=} + 2\ I^- \tag{5-6}$$

It is preferable to carry out the above reactions at pH 3 or 4 because at neutral pH the reaction is not stoichiometric owing to the oxidation of some thiosulfate to sulfate. This method can be carried out with phenylarsene oxide or thiosulfate. PAO is preferable because it reacts much faster than thiosulfate, which reacts in a stepwise fashion.

This starch–iodide iodometric titration procedure is the oldest method used to determine chlorine residuals. This method was used at one time to distinguish only those residuals (monochloramine) that released iodine from KI at neutral pH. This is no longer practiced; now the pH of the system must be reduced to 4 to measure all of the chlorine residual species. Acetic acid was found to be more desirable than either sulfuric or hydrochloric acid because it gave the best accuracy of pH control at pH 4. If ferric or manganic compounds are present, they will give false residual readings. To compensate for these interferences, the procedure should include a blank titration. Nitrites are a serious interference, but can be overcome by the sulfamic acid procedure described below.

This method has been popular for establishing temporary chlorine standards in spite of the fact that the minimum practical reading is 1.0 mg/L TRC. However, its popularity is limited by several drawbacks: the large sample required, low sensitivity, and the relative instability of the thiosulfate and starch solution reagents, together with the tricky end point, which is sensitive to temperature.

In spite of these difficulties, the method is still widely used for standardizing chlorine water used in laboratory studies of chlorine demand, tastes, and odor control, and coliform destruction. This is the most convenient method for measuring chlorine concentrations in the range of 50–2500 mg/L chlorine.

For the convenience of laboratory technicians, the procedure is outlined below.

Standardization of Chlorine Water

Reagents.

1. Glacial acetic acid (concentrated).
2. KI Crystals, USP.
3. Starch solution, 1 percent starch (water-soluble). Boil 10 min., and decant after it stands overnight.
4. Standard sodium thiosulfate 0.1 N. Dissolve 24.82 g $Na_2S_2O_3 \cdot 5H_2O$ in 1 liter freshly boiled distilled water.

Procedure.

1. Place a few crystals of KI in a wide-mouthed Erlenmeyer flask.
2. Add approximately 10 ml distilled water.
3. Add about 2 ml glacial acetic acid.
4. Add 10–50 ml chlorine water by means of a volumetric pipette. The tip of the pipette should almost touch the surface of the water in order to avoid surface loss.

5. Titrate with standard 0.1 N sodium thiosulfate, using starch as an indicator near the end of the titration.

Note: In step 4, the chlorine will liberate iodine (I_2) from the KI present, giving a brown cast to the sample. The brown cast will be discharged upon addition of the reducing agent, sodium thiosulfate, and at this point 1 ml of starch indicator is added. The starch will produce a blue color in the presence of free iodine, and the titration is continued until this color disappears. Read the amount of thiosulfate used at first disappearance of the blue color, and disregard reappearance of the color upon standing.

Calculation. The calculation is based upon the following: ml 0.1 *N* thiosulfate used × 3.545, divided by ml of sample tested × 1000 = mg/L chlorine concentration of chlorine water, or:

$$\frac{\text{ml } 0.1\ N \text{ thiosulfate} \times 3.545 \times 1000}{\text{ml sample}} = \text{mg/L Cl} \qquad (5\text{-}7)$$

This procedure will be applicable to waters generally in the range of 500 to 2000 mg/L chlorine.

Iodometric Method II—Measuring Residuals in Wastewater

See page 4-39, *Standard Methods,* 19th edition.

General Discussion. When determination of residuals in wastewater became an important consideration in the 1960s, the majority of the effluents examined had only primary treatment. In the usual procedure, such as Iodometric I, the iodine release is performed first, and the titrant is added to reduce the iodine to iodide, which becomes the measure of the chlorine residual. It was found that during the time required to neutralize the iodine by the titrant, some of the iodine was being consumed by the organic matter in the sewage. To overcome this occurrence, the back-titration method was devised. It is applicable to both iodometric and amperometric procedures, as follows.

Back-Titration Method. See Iodometric Method II, pages 4-39 and 4-40, *Standard Methods,* 19th Edition. This is the preferred method for primary effluents and highly polluted or poorly treated effluents. See Chapter 8 for other procedures for well-oxidized effluents.

Procedure. The amount of sample to be taken for titration is governed by the concentration of chlorine in the sample. For residual chlorine of 10 mg/L or less, 200 ml sample should be titrated. For greater concentrations, proportionately less of the sample should be used.

Titration. Place 5.0 ml 0.00564 *N* phenylarseneoxide (PAO) solution in a flask or a white casserole. Add excess KI (approximately 1 g) and 4 ml pH 4 acetate buffer solution, or sufficient buffer to reduce the pH to between 3.5 and 4.2. Pour in the sample and mix with a stirring rod. Just prior to titration with 0.0282 *N* iodine, add 1 ml starch solution for each 200 ml sample. Titrate to the first appearance of blue color that persists after mixing. As 1 ml 0.00564 *N* PAO consumed by a 200 ml sample represents 1 mg/L available chlorine, 5 ml PAO solution is sufficient for residual chlorine concentrations up to 5 mg/L. For residual chlorine concentrations of 5–10 mg/L, 10 ml PAO solution is required.

Calculation:

$$\text{mg/L chlorinc} = \frac{(A - 5B) \times 200}{C}$$

where

A = ml 0.00564 *N* PAO solution
B = ml 0.0282 *N* iodine
C = ml of sample

This method is least desirable when there may be interference from color or turbidity in the wastewater. Otherwise it gives results comparable to the amperometric method.

Control of Nitrite Interference. Serious interference from nitrites was reported by the California State Department of Health, Burcau of Sanitary Engineering in 1972.[57] They found residuals as high as 16 mg/L with *no chlorine added.* This interference occurred with the idiometric method. Based upon the reaction in Eq. (5-4), it would appear that nitrites (NO_2^-) must act to release I_2 from KI.

Control of this interference is relatively simple; it depends upon the use of sulfamic acid, which will convert the nitrites to nitrates.

To a 200 ml sample of wastewater add 1 ml of a 5 percent solution of sulfamic acid (NH_2SO_3H), which will convert the nitrites to nitrates. Mix the solution and let it stand for 10 min. Then add excess standard phenylarsene oxide (0.00564 *N*) and excess KI (1 g) and titrate with standard iodine (0.00282 *N*), using starch as the indicator.

This is thc samc as the procedure outlined in Method I except that sulfamic acid is used instead of phosphoric or acetic acids to depress the pH of the sample mixture. Sulfamic acid will depress the pH to the level necessary to perform the quantitative release of I_2 from KI in proportion to the total chlorine residual. The final pH will be about 4.3.

When using this procedure it is desirable to check the pH of the sample–sulfamic acid mixture; if the pH is above 4, add some pH 4 buffer (acetic acid).

Standardizing 0.0282 *N* Iodine Solution with PAO. This solution will remain stable for months if it is stored in the dark, in a brown bottle, and at moderate temperature conditions. It is only necessary to check the normality weekly or biweekly. This can be done with PAO solution as described below in the amperometric titration section.

DPD Ferrous Titrimetric Method

See pages 4-43 to 4-45, *Standard Methods,* 19th edition.

General Discussion. This method, developed by Palin,[25, 63] is the only one of the three he developed that has survived. (The other two were the *p*-amino-dimethylanaline method[21] and the neutral O-T–ferrous ammonium sulfate titration procedure.[24] The method was first introduced in 1956. It utilizes an indicator solution of diethyl-*p*-phenylenediamine. When used as a colorimetric test, it is much preferred over either the acid or the neutral O-T methods or the OTA method, for its sharper free chlorine–chloramine differentiation. This method requires five reagents and has a range up to 4 mg/L without dilution. The minimum detectable concentration is about 18 μg/L as chlorine.

Principle. The titrimetric procedure requires the use of a standardized ferrous ammonium sulfate as the titrant and *N,N*-diethyl-*p*-phenylenediamine as the indicator (DPD). In the absence of iodide ion, FAC (free available chlorine) reacts instantly with DPD indicator to produce a red color. Subsequent addition of a small amount of iodide ion acts catalytically to cause monochloramine to produce color. Further addition of potassium iodide to an excess evokes a rapid response from dichloramine.* The color at each stage is titrated to a colorless end point.

Procedure. Reagents include:

1. Standard ferrous ammonium sulfate (FAS) solution, 1 ml = 0.100 mg available chlorine.
2. DPD No. 1 powder.
3. Potassium iodide crystals.

The above DPD No. 1 powder is a combined buffer–indicator reagent. If liquid reagents are used, the buffer solution and the DPD solution must be kept separate.

*Monochloramine is always present whenever NCl_3 is formed in a potable water or wastewater chlorination reaction.

Procedure for Free Available Chlorine. To 100 ml sample add approx. 0.5 g DPD No. 1 powder. Mix the solution rapidly to dissolve the powder, and titrate immediately with FAS solution (Reading *A*).

Procedure for Combined Available Chlorine Compounds. Add to the above solution one very small crystal of potassium iodide. Mix it, and continue titration immediately (Reading *B*). Add several crystals (approx. 0.5 g) of potassium iodide. Mix to dissolve the KI, and after the solution stands about 2 min. continue the titration (Reading *C*). If this fraction is fairly high, use double the quantity of potassium iodide (i.e., approx. 1.0 g).

Procedure for Nitrogen Trichloride. To 100 ml sample add one small crystal of potassium iodide. Mix the solution and then add approx. 0.5 g DPD No. 1 powder. Titrate immediately with FAS solution (Reading *D*).

Calculations. For 100 ml sample, 1 ml FAS solution = 1 mg/L available chlorine:

A = free available chlorine
B = free + mono
C = free + mono + di + ½NCL_3
D = free + mono + ½NCl_3*
E = total available chlorine

Free chlorine	$= A$
Monochloramine	$= B - A$
Dichloramine	$= D - C$
Nitrogen trichloride	$= 2(D - B)$
Total available chlorine	$= E$

Total Available Chlorine. This may be obtained in one step by adding the DPD No. 1 powder and the full quantity of potassium iodide to the sample at the start and letting it stand for about 2 min. A combined DPD and potassium iodide reagent known as DPD No. 4 powder is available, thus providing a single reagent for total available chlorine. For 100 ml sample, use about 0.5 g of this powder, or, if high concentrations of available chlorine are present, use about 1.0 g. After the sample stands the required 2 min. or so, titrate with the FAS solution to obtain the total available chlorine (Reading *E*).

Monochloramine Breakthrough. In the event of a substantial amount of monochloramine present with free chlorine, an alternative procedure is recommended to prevent monochloramine breakthrough into the free residual reading. This breakthrough can be as high as 2 percent/min. of monochloramine

*Monochloramine is always present whenever NCl_3 is formed in a potable water or a wastewater chlorination reaction.

present. To overcome this, add 0.5 ml of 0.25 percent solution of thioacetamide to the 100 ml sample immediately after mixing the DPD reagent. This stops further reaction with the combined chlorine in the free measurement. Continue immediately with FAS titration to obtain free chlorine. Then obtain total available chlorine from the above procedure (Reading *E*) without thioacetamide. When thioacetamide is used with the DPD procedure (either colorimetric or titrimetric), it is called the "Steadifac" method, by Palin.[44]

The reaction of the Steadifac solution occurs in a molar ratio of 1 : 1, from which it may be calculated that one drop (0.05 ml) of 0.25 percent solution of this acetamide added to a 10 ml sample can eliminate 11.8 mg/L of chloramine.

Interferences. The most significant interfering substance likely to be encountered in water is oxidized manganese. This can be corrected by the addition of 0.5 ml of sodium arsenite solution with 5 ml of buffer solution to the titration flask. Add 5 ml DPD indicator solution, mix the sample, and titrate with standard FAS titrant until the red color is discharged. Subtract this reading from Reading *A* (free chlorine), obtained by the normal forward-titration procedure (described above), or from the total available chlorine, Reading *E,* also described above.

If the combined reagent in powder form is used, first add KI and arsenite to the sample and mix the solution; then add combined buffer–indicator reagent.

Amperometric Titration Method

See page 4-41, *Standard Methods,* 19th edition. See also p. 4-43, for the Low Level Amperometric Titration Method. For Chlorine Dioxide see pp. 4-53 to 4-58.

General Discussion. This is one of the most widely used methods of measuring chlorine residuals in North America. The titrator is easy to use, accurate, and reliable so long as reasonable care is exercised in maintaining electrode sensitivity. This is accomplished simply by soaking the electrodes in a dilute chlorine solution for a few minutes before operating the titrator. If the titrator is to be used in the back-titration mode, then soak the electrodes in a weak (light brown) iodine solution.

Principle of Titrator Operation. The amperometric method is a special adaption of the polarographic principle. This is a class of electrometric titration whereby the current that passes through the titration cell between an indicator electrode and an appropriate depolarized reference electrode (at suitable applied EMF) is measured as a function of the volume of a suitable titrating solution.[64] In general, the end point of an amperometric titration can appear either as the cessation of current at the equivalence point (for complete reduction of the titrant), as a sudden increase in current (when the titrant is the oxidant and overcomes the reducing agent in the electrolyte), or as an abrupt change in current at the equivalence point.

In any cell consisting of two electrodes contacting an electrolyte, the impressed voltage will be opposed by a countervoltage, due to accumulation (polarization*) of the electrolysis products or exhaustion (depolarization) of the material being electrolyzed at the electrode surfaces.

The current that can flow in the electrolyte between the electrodes is said to be decreased by concentration polarization, which is sometimes described as concentration overpotential or activation potential. The flow of current in a cell will cause the accumulation of reducing agents at the cathode (negative electrode) and oxidizing agents at the anode (positive electrode). The addition of an oxidizing agent to the cathode solution or a reducing agent to the anode solution will decrease the accumulation at the respective electrodes. This phenomenon, known as depolarization of the electrodes, reduces the counter EMF, thereby allowing more current to flow.

Historical Development. The principle of amperometric titrations was proposed as early as 1897 by Salomon.[65] It was used in 1905 by Nernst and Merriam,[66] who showed that the diffusion current is proportional to the concentration of the electroreducible substance. Amperometric titrations are classified into two categories: (1) one-indicator electrode with reference electrode and (2) dual-indicator electrodes.

Marks first made this method available for measuring chlorine residual in 1942.[17] In his investigation and development of this method, he used the single-indicator electrode (gold) with a silver chloride reference electrode. The titrating agent used was sodium arsenite. He later found a more discriminating reducing agent (phenylarsene oxide), which allowed modification of the procedure to give better differentiation of the various chlorine residual fractions.[18–20] In 1949 the first amperometric titrator for residual chlorine developed by Marks was made available. A current version of this unit, utilizing a single-indicator electrode, is illustrated in Fig. 5-2 and is shown schematically in Fig. 5-3.

The dual-indicator electrode system was introduced by Foulk and Bawden in 1926 for the titration of iodine with thiosulfate.[67] Not until 1966 was the dual-electrode system proposed for residual chlorine determination.[68] This system is illustrated in Fig. 5-4.

The Single-Indicator Electrode. Figure 5-3 illustrates schematically the operation of this unit.[20] This titrator consists of a platinum (cathode) indicating

*Polarization is defined as an effect produced on the electrodes of a cell by the deposition on them of gases liberated by the current. It is caused chiefly by hydrogen, which increases the resistance and sets up a counter EMF, and can be visualized as numerous tiny bubbles of gas covering the surface of the electrode. These tiny bubbles insulate the electrode, thereby increasing the resistance to current flow through the electrolyte.

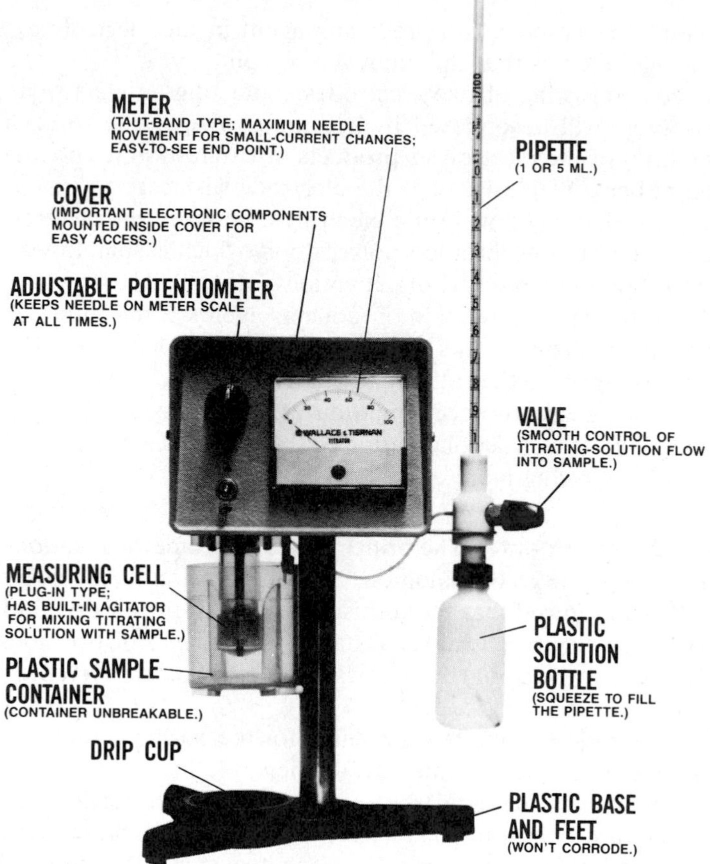

Fig. 5-2. Amperometric titrator (courtesy Wallace & Tiernan).

electrode (which is polarized to some extent) and a silver electrode immersed in a saturated salt solution. This is the reference electrode and is the nonpolarizablc anode. The potential of this silver electrode is such that it provides an internal voltage and makes it unnecessary to impress an external voltage. Because of the low concentrations of chlorine to be measured, the surface area of the indicating electrode is made large to increase the sensitivity of the cell. In this case the output of the cell is such that a 0.01 mg/L change in chlorine concentration will produce a current of about 2.5 microamps, which will move the pointer about seven divisions. Efficient agitation at the indicating electrode surface, which is necessary to obtain a higher and more uniform diffusion current, it achieved by a high-speed rotating Lucite sleeve, which fits over the cylinder containing the platinum electrode with minimum clearance. The electrical path through the water sample (electrolyte) is made short and of low electrical resistance by putting a salt bridge in the intervals in the narrow space between the turns of the platinum spiral electrode. To perform a titration, a sample that contains the oxidizing agent—either chlorine (HOCl)

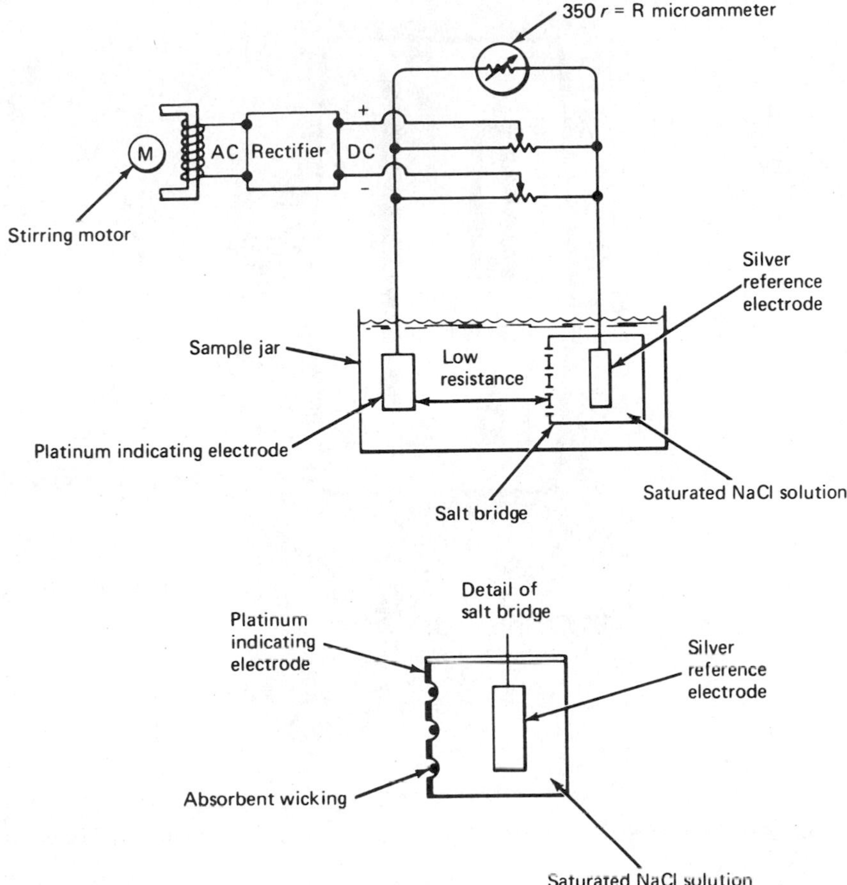

Fig. 5-3. Schematic amperometric titrator with single-indicating electrode (courtesy Wallace & Tiernan).

or iodine (I_2)—is placed in position, and the agitator is started. The microammeter will deflect to the right, upscale, depending upon the concentration of the oxidizing agent. The reducing agent (phenylarsene oxide) is then added, which decreases the concentration of the oxidizing agent and accordingly decreases the current through the cell. When the oxidizing agent has been completely reduced by the titrating agent (PAO), the end point is indicated by no change in the current upon further addition of reagent.

The reverse of this procedure, known as back-titration, is performed by using an oxidizing agent (iodine) as the titrating solution to overcome an excess of reducing agent previously added to the sample. In this case the end point is observed as the first appearance of a current caused by the exhaustion of the oxidizing agent by the reducing agent.

The Dual-Indicator Electrode. The Bailey–Fischer and Porter portable amperometric tritrator (17T2000) utilizes dual-indicator electrodes. (See Fig.

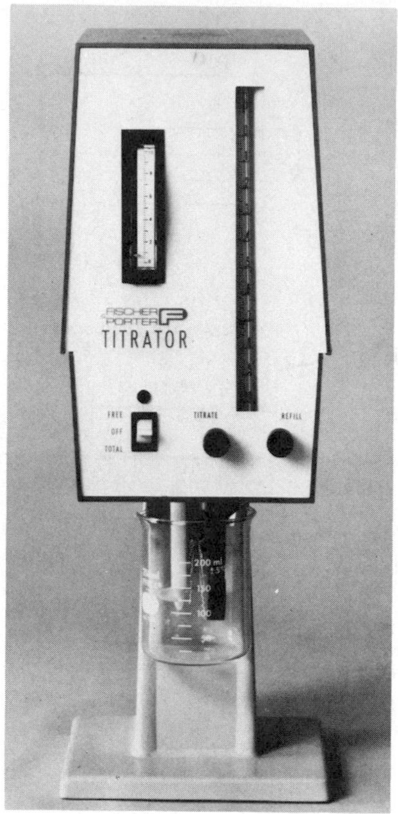

Fig. 5-4. Amperometric titrator with dual-indicating electrodes (courtesy Fisher and Porter Co.).

5-4.) Figure 5-5 illustrates the essential components of the titrator. It uses a platinum and copper electrode pair rather than the previously used dual platinum electrodes. This new arrangement incorporating the Pt–Cu pair provides a good current–voltage curve, yielding a usable plateau, wherein small voltage variations do not produce significant current changes. (See Fig. 5-6.) At the selected electrode potential (200 mV, TRC; 100 mV, FAC) the cell generates a stable current level, as a function of the chlorine concentration, which can be applied to several stages of amplification. This amplification provides the unit with a sensitivity of 0.005 mg/L. A Pt–Pt pair produces a current versus voltage curve, which is essentially a straight line for free chlorine (FAC). Therefore, a small shift in electrode potential generates a significant change in current, making amplifier input–output unpredictable and of limited use.

The final display output (meter reading) is made more readable in the critical end-point region by applying the cell output to a log-linear amplifier

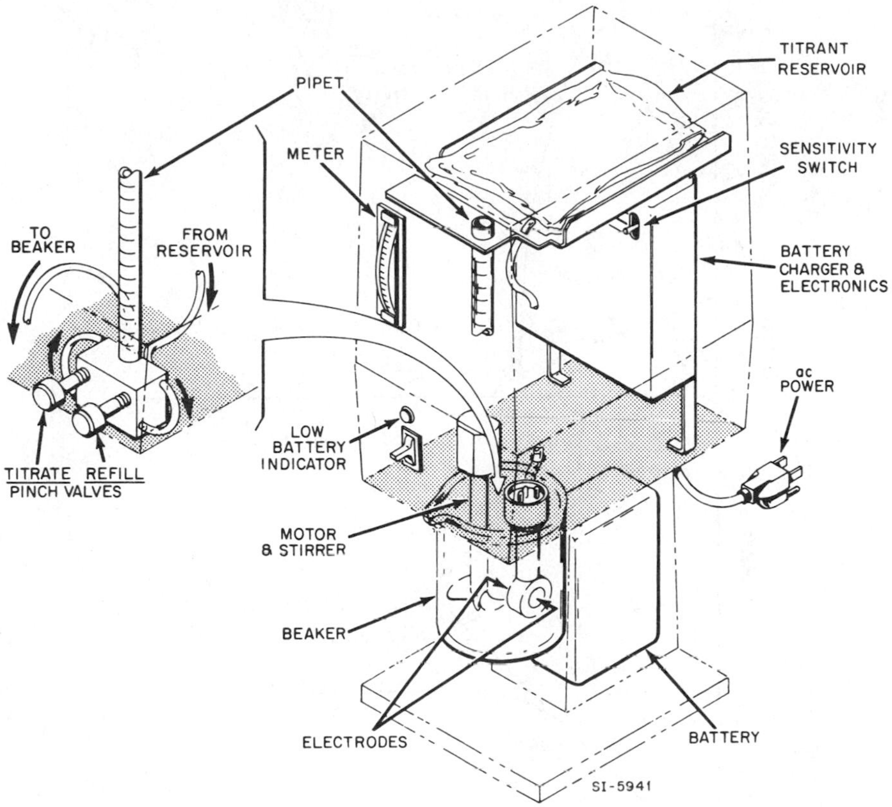

Fig. 5-5. Diagram showing essential components of F and P titrator (courtesy Fisher and Porter Co.).

circuit. The upper curve in Fig. 5-6 compares the log function to the conventional display. For any given chlorine concentration the meter response is minimal at the beginning of the titration, increasing drastically as the end point is approached.

This system is known as biamperometric titration. The procedure is the same as that used in the single-electrode system. When the oxidizing agent is being titrated by a reducing agent, both electrodes remain depolarized, and a current flows in proportion to oxidizing agent and is indicated on the output meter.

When the last traces of the oxidizing agent have been destroyed by the reducing agent, the cathode is without a depolarizer and the flow of current is immediately arrested, signifying the endpoint. This is indicated by the milliammeter. Conversely, if the titrant is an oxidizing agent, such as iodine, which is added to overcome a reducing agent, as in the back-titration procedure, the end point occurs at the first appearance of a current. This occurs as soon

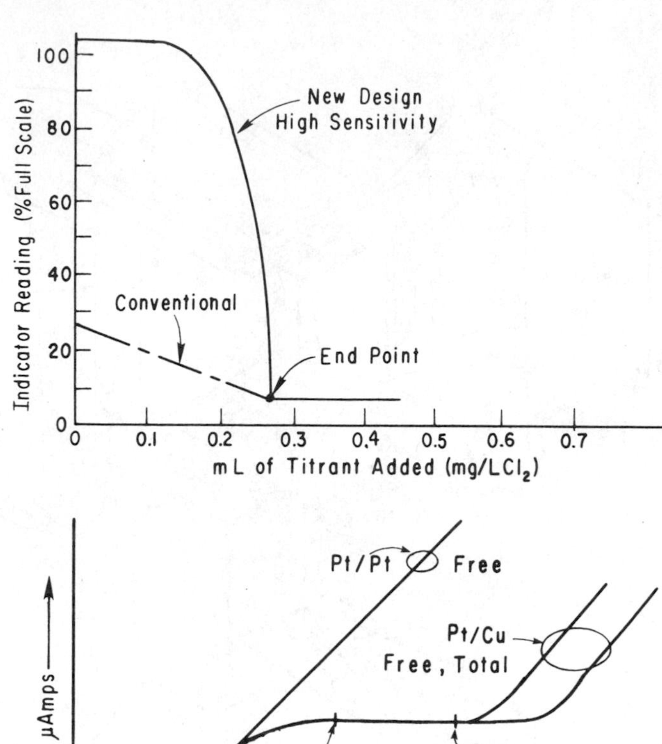

Fig. 5-6. Relationship between current and applied voltage across the Pt/Cu electrodes and log function readability versus conventional display (courtesy Fisher and Porter Co.).

as there is an excess of iodine, which depolarizes the cathode. The anode is already depolarized by the iodide ion.

Precision and Sensitivity. The precision of the single-indicator electrode described above is on the order of ±0.05 mg/L, and the sensitivity is 0.01 mg/L. These units can be made supersensitive to measure residuals down to 0.001 mg/L by the following modification, which was performed on a Wallace & Tiernan Model A-790 Titrator:[58]

1. Dilute the PAO reagent four times.
2. Encase the sample jar in aluminum foil and ground it.

3. Remove the ammeter and insert a Rochester converter that will put out a milliamp signal.
4. Connect the milliamp output to an electrician's type of voltmeter.
5. This will measure the electrical potential across the electrodes.
6. This will allow the user to read residuals down to 0.001 mg/L.

Operating Characteristics. The following statements appeared in *Standard Methods,* 15th edition: "The method is not as simple as the colorimetric methods and requires greater operator skill to obtain the best reliability. Loss of chlorine can occur because of rapid stirring in some commercial equipment. Electrode cleanliness and conditioning are necessary for sharp end points" (p. 286). "Amperometric titration requires a higher degree of skill and care than colorimetric methods. Chlorine residuals over 2 mg/L are measured best by means of smaller samples or by dilution with water that neither has residual chlorine nor a chlorine demand" (p. 278).

None of these statements can be supported by the present author's personal experience or by operators with whom he has worked over the past 40 years on both potable water and wastewater. Measuring residuals up to 20 mg/L without dilution is a routine task with an amperometric titrator. As for the skill required, one only needs to know how to handle a pipette and a burette and how to interpret an ammeter.

The one technique that is often overlooked is care of the electrodes. Each manufacturer has specific instructions for electrode care; however, the most important task to perform is to sensitize the electrodes. For forward-titrations, sensitize the electrodes with a dilute free chlorine residual (5 mg/L) by letting them soak in this solution in a sample jar. When back-titrations are made in wastewater situations, let the electrodes soak in a dilute iodine solution (light brown color in sample jar). When routine residual measurements are being made for *both* forward- and back-titrations, separate titrators should be used because they reduce the rinsing time required to remove the iodine in performing the free chlorine measurement.

The electrodes should be sensitized as described above for about 10 min. before use if the titrator is not being used on a daily basis. Otherwise, they should be sensitized at weekly intervals. Frequent use of the titrator is very rewarding because its repeatability builds the operator's confidence. Colorimetric methods always leave some doubt in the operator's mind because color depth acuity varies from individual to individual. Therefore, errors will often occur in colorimetric methods that do not occur in the amperometric method. Moreover, all colorimetric methods that do not use a spectrophotometer readout require that the operator estimate visually the color depth between the color standards, such as the difference between 1.0 and 1.5 mg/L, and so on. Quite often operators have a difficult time distinguishing between 1.0 and 1.5 mg/L.

Chemistry of the Amperometric Method. The success of this method is largely due to the characteristics of the reducing agent phenylarsene oxide

(C_6H_5AsO). Other reducing agents were tried and discarded for a variety of reasons. For example, thiosulfate, which is used in the starch–iodide method, does not sufficiently discriminate between free and combined available chlorine. Further, it acts in a reducing process by steps, causing a time lag that interferes with the end point. Sodium arsenite, which was used in some of the early investigations of this method, had possibilities because it would react with HOCl only in the absence of potassium iodide and with any iodine released by chloramines in the presence of iodide. However, it would not react quantitatively below pH 6, which is essential in order to pick up dichloramine, which liberates iodine only at a pH much below 6.

Phenylarsene oxide reacts only with free chlorine at pH 7 in the absence of potassium iodide (KI). PAO reacts with monochloramine in the presence of 50 mg/L KI at pH 7 and with dichloramine in the presence of 250 mg/L KI at pH 4. Actually PAO does not directly titrate either mono- or dichloramine. Iodine is liberated quantitatively by these fractions in the presence of the amounts of potassium iodide shown and at the pH levels indicated. The PAO then titrates the iodine liberated.

These two general reactions are described by the following equations:

$$C_6H_5AsO + HOCl + H_2O \rightarrow C_6H_5AsO(OH)_2 + HCl \qquad (5\text{-}8)$$

$$C_6H_5AsO + I_2 + 2H_2O \rightarrow C_6H_5AsO(OH)_2 + HI \qquad (5\text{-}9)$$

The strength of PAO was selected at 0.00564 *N* so that in using a 200 ml sample 1 ml of PAO is equivalent to 1 ppm of chlorine. Therefore, when the end point of the titration is reached, the volume of PAO used represents the chlorine concentration in milligrams per liter. One mole PAO reacts with two equivalents of halogen.

In determining free chlorine, the pH must not be greater than 7.5 because of sluggish reactions at higher pH values, or less than 6.5 because at lower pH values some combined chlorine may react even in the absence of iodide. In determining combined available chlorine, the pH must not be less than 3.5 because substances such as oxidized manganese interfere at pH values lower than 3.5, or greater than 4.5 because the reaction of combined residual chlorine is not quantitative at higher pH values.

The tendency of monochloramine to react more readily with iodide than does dichloramine provides a means of further differentiation. The addition of a small amount of potassium iodide in the neutral pH range enables the estimation of monochloramine content. Lowering the pH into the acid range and increasing the KI concentration allows the separate determination of dichloramine.

Preparation for Titration

Apparatus. An amperometric titrator, the end-point detection instrument, consists of a two-electrode cell connected to a microammeter with an adjustable potentiometer. Accessories should include a 200 ml sample jar with displacement cup for accurately measuring 200 ml, a 1.0 ml buret with 0.01 ml graduations, and a finger pump for charging the buret with the reducing reagent.

To prepare the titrator for operation, the operator should check the following:

1. If the unit utilizes a silver–silver chloride electrode (Fig. 5-4), make certain that there are sufficient salt tablets in this cell (⅔ full) and that enough distilled water has been added to cover the tablets.
2. Make certain that the electrical contacts between cell and microammeter are clean and making proper contacts.
3. Be sure that the platinum electrode surface is free of any deposits. If it is dirty, clean it by lightly rubbing the platinum surface with scouring powder, using only the fingers. Be sure to avoid disturbing the porous wicking that lies between the turns of the platinum ribbon.
4. If the titrator has been used routinely for free chlorine measurements only, and it is desired to measure combined or total chlorine residual, the platinum electrode must be sensitized to iodine. This will be noticed during a titration when the potassium iodide is added. When KI is added and the needle deflects downscale momentarily and remains there, the cell is said to have lost its sensitivity to iodine. The sensitivity is easily restored by adding enough free iodine to the water in a sample jar to create a yellowish color, agitating sample for 2 or 3 min., and then allowing it to stand in this solution for 10 or 15 min. After this treatment, the cell unit should be rinsed thoroughly to remove all traces of free iodine.
5. All glassware used in this procedure should be chlorine-demand-free. The sample jar should be treated with water containing at least 10 mg/L of chlorine for three hours or more before use and then rinsed with chlorine-demand-free chlorine water.
6. To prevent errors in titration from contaminants, it is good laboratory practice to fill the pipette to the top and to run the contents to waste before beginning a series of titrations.

Reagents:

1. Standard phenylarsene oxide titrant (a 0.00564 *N* solution of PAO, C_6H_5AsO).
2. pH 4 acetate buffer solution.
3. pH 7 phosphate buffer solution.
4. Potassium iodide solution (5 parts CP grade potassium iodide in 95 parts distilled water).
5. Electrolyte tablets (made from USP sodium chloride).

Procedure:

I. Free Available Chlorine:

1. Connect the titrator to a source of 115 V single-phase power.
2. Fill the pipette with phenylarsene oxide solution. Remove all air from the pipette and tubing and then discharge the PAO. Refill the pipette to the top calibration mark.
3. Measure a 200 ml sample of the water to be tested in the sample container, and place it in position on the titrator.
4. Add 1 ml of phosphate buffer solution pH 7 to the water sample. If it is known that the natural pH of the sample lies between 6.0 and 7.5, it is not necessary to use the pH 7 buffer solution.

   Note: Droppers are usually furnished for these reagents and are scribed to the 1 ml mark; therefore a dropperful of solution should be used whenever 1 ml of solution is called for.
5. Start the agitator by turning the switch to "on."
6. Adjust the potentiometer so that the microammeter needle reads near maximum on the scale. If the needle is above maximum when the adjusting knob is rotated completely counterclockwise, then the titration must be started with the knob in this position.
7. Begin adding the phenylarsene oxide solution just under the surface of the sample. If there is free chlorine present as the titrant is being added, the cell current will decrease, resulting in a downscale movement of the microammeter needle. If the needle is above maximum at the beginning of the titration, the needle will remain above maximum until enough phenylarsene oxide solution has been added to reduce the free chlorine residual (and hence the current) to a point where the needle will read less than maximum. As the titrant is added, it will be necessary to adjust the potentiometer from time to time to bring the needle back on scale. As the end point is approached, the response of the microammeter to each increment becomes more sluggish, and smaller increments of titrant should be added. The end point is just passed when a single drop of PAO no longer causes a downscale (decrease of current) movement of the needle.* The buret is then read, and the last increment (one drop) of titrant is subtracted from the reading. The result is the free available chlorine concentration in mg/L.

---

*This end point is true for nearly all waters. There are some cases when the needle will deflect downscale after the end point has been reached. In such cases the true end point occurs when for equal amounts of PAO added the amount of deflection changes from large to small. At this point the last increment of titrant is subtracted from the total reading as before.

II. Monochloramine. This follows immediately and in the same sample as in *I* above.

1. Note the buret reading at the endpoint in *I*.
2. Add 0.2 ml of potassium iodide to the original sample. The furnished dropper usually delivers 20 drops/ml. Therefore, 4 drops correspond to 0.2 ml. If monochloramine is present, the needle will deflect to the right (probably off scale) immediately upon addition of the potassium iodide.
3. Proceed with the addition of PAO to the endpoint as in *I*. Note the buret reading. The difference between the buret readings at the end points of *I* and *II* is the amount of monochloramine present.

Note that this titration is performed at pH 7.

III. Dichloramine. This determination follows the procedure in *II* and with the same sample.

1. Decrease the pH of the sample to 4 (3.5 to 4.5) by adding 1 ml of buffer solution pH 4.
2. Add 1 ml of potassium iodide solution. If dichloramine is present, the needle will deflect to the right at this point.
3. Proceed with addition of PAO until the end point is reached as before.

Note the reading on the buret. The difference between this reading and the previous reading at the end of II is the amount of dichloramine present in mg/L.

IV. Total Available Chlorine Residual. This procedure bypasses all the separate fractions described previously and lumps all these into one total reading.

Starting with a fresh 200 ml sample, proceed as follows:

1. Add 1 ml of buffer solution pH 4.
2. Add 1 ml of potassium iodide solution. At this point the needle first will deflect to the left and then go upscale (probably off scale). Any chlorine residual, free or combined, will release iodine from the potassium iodide quantitatively.
3. Proceed with the addition of PAO to the end point as before.

Note the reading of the buret. This represents the total of all fractions of available chlorine residual present in the sample.

The indicating electrodes must be sensitized to iodine as explained previously.

V. Free and Combined Available Chlorine Residual. It is often desirable to make separate determinations of free and combined available chlorine residual. This can be done with only one sample.

1. Measure free chlorine as in *I*.
2. Add 1 ml of buffer solution pH 4.
3. Add 1 ml of potassium iodide.

Items 2 and 3 should be added simultaneously. At this point all of the mono- and dichloramine liberates iodine from potassium iodide quantitatively. If combined chlorine residual is present, the needle will deflect first to the left and then sharply to the right upon the addition of potassium iodide.

4. Add PAO until the end point is reached as before.

Note the reading on the buret. This reading represents the total available chlorine, and the difference between this reading and the reading in *I* is the combined available chlorine residual—all in mg/L.

Note: If free available chlorine residuals are made following procedures using potassium iodide, the cell unit must be rinsed thoroughly to remove traces of potassium iodide and pH 4 buffer solution.

VI. Nitrogen Trichloride. It is a well-established fact that nitrogen trichloride can exist simultaneously with free chlorine at a pH as high as 8.5 or even 9[25, 72, 73] if free ammonia is available to react with the chlorine and if chlorine is added in sufficient quantity to produce the breakpoint phenomenon. In the amperometric method it is believed that any nitrogen trichloride titrated will appear partly in the first fraction as free chlorine and partly in the third fraction as dichloramine.[20] Dowell and Bray[69] found that the starch–iodide titration includes only 80 percent of the total nitrogen trichloride present, while the acid–orthotolidine method includes only about 60 percent of the total. Palin's methods have been shown tentatively to account for all the nitrogen trichloride present. However, sufficient reliable data are lacking to establish the ability of the amperometric method to differentiate the total nitrogen trichloride from the other fractions.

Williams[71] uses the following method:

1. A carefully collected sample, subjected to a minimum amount of agitation to prevent aeration of nitrogen trichloride, is placed in the titrator. Before the agitator is started, an excess of pH 4 buffer and an excess of potassium iodide are added, to "fix" all the chlorine fractions, including NCl_3.
2. The above is followed by the procedure for determining total available chlorine residual. This is reading *A*.
3. Next, the titrator cell, agitator, and sample jar are thoroughly rinsed. Another sample is placed in the titrator jar; with the sample jar held in

the operator's hand, to achieve the maximum amount of turbulence, the agitator is turned on to aerate out all the nitrogen trichloride.

4. The sample jar is quickly returned to the normal position. The rest of the procedure is the same as that for HOCl, NH_2Cl, and $NHCl_2$.

The sum of these fractions will be reading *B;* the difference $(A - B)$ will be the nitrogen trichloride present, assuming that all the NCl_3 will appear in the total available chlorine determination (step 1).

If so desired, the nitrogen trichloride can be extracted by the use of carbon tetrachloride rather than by aeration. This method, however, is more cumbersome than the one described.

The amperometric method can also be used to determine total available iodine and bromine. The procedure for measuring chlorine dioxide is fully described in Chapter 12.

FAC Residuals at Short Contact Times. In some special instances it is desirable and necessary to be able to determine the free available chlorine residual fraction when the contact time is only a matter of seconds. This method is based on the ability of phenylarsene oxide to react with free chlorine at pH 7 but not with combined chlorine.

By adding an excess of PAO solution to a properly buffered sample, the amount of free chlorine residual in the sample at the time the PAO is added can be determined by titrating the remainder of the excess PAO with a standardized chlorine solution.

The apparatus and the reagents are the same except for the standardized chlorine solution (0.02 percent HOCl). This solution should be standardized with PAO solution so that 1 ml chlorine solution is equivalent to 1 ml of PAO solution.

Procedure:

1. To the empty but clean titrator jar add 1 ml pH 7 buffer solution and 5 ml PAO.
2. Add precisely 200 ml of sample to the jar, and mix the solution as thoroughly and rapidly as possible, probably not in the vicinity of the titrator.
3. Place the sample jar in the titrator and proceed with titration using the standardized chlorine solution (0.02 percent HOCl) as the titrant.
4. At each addition of the titrant the microammeter needle will make a small momentary deflection to the right, returning each time. The end point is reached when the needle deflects discernibly to the right, indicating that the excess PAO has all been oxidized by the HOCl solution and that the last addition of titrant has produced a current. This last addition should be subtracted from the buret reading. This is reading *A*.

Calculation. The free chlorine residual mg/L = ml PAO − *A*.

Note: If reading *A* is greater than 4.0, repeat the procedure using less PAO solution; if reading *A* is less than 0.2, repeat the procedure using more PAO solution.

The purpose of this procedure is to determine the free chlorine residual after a contact time of only seconds. The contact time between the chlorine and the sample is measured from the time the chlorine is applied to the water being treated until the time the sample reacts with the PAO in the sample jar. The time it takes to transport the sample jar to the titrator and to run the titration is not a part of the contact time.

Determination of Residual Chlorine in Wastewater

Primary Effluents. In 1948 when it became commonplace to chlorinate primary effluents to enhance their acceptability, Wallace & Tiernan suggested a new approach to amperometric titration of chlorine residuals.[59] Primary effluents presented a problem because they exerted an immediate iodine demand by reducing agents that were unaffected by the chloramine residual present. To avoid this loss of iodine the titration procedure was reversed as follows: An excess of titrant (PAO) is added to the sample before addition of the iodide. Then when the sample pH is reduced to 4 with buffer and the iodide added, the iodine released is immediately consumed by the PAO. The remaining PAO is back-titrated with a standardized iodine solution.

Back-Titration Procedure:*

1. Sensitize electrodes with a dilute solution of iodine—a light brown color is satisfactory. This will produce sharp end points.
2. Allow the electrodes to soak in this solution in the sample jar for 15–20 min.
3. Thoroughly rinse out the sample jar after sensitizing is completed.
4. Place a 200 ml sample of wastewater in the titrator.
5. Start the agitator.
6. Add 5 ml phenylarsene oxide (PAO) solution to sample and mix it. If total residuals are known to be greater than 5 mg/L, add 10 ml PAO.
7. Add 4 ml pH 4.0 buffer solution (or sufficient to ensure a sample pH between 3.5 and 4.2) to the sample and mix it.
8. Finally add 1 ml KI solution and mix it.
9. Adjust the microammeter pointer so that it reads about 20 on the scale.
10. Add 0.0282 *N* iodine solution in small increments from a 1 ml pipette or a 1 ml buret.

*Since the discovery of the deleterious effects of organic N on the disinfection process, this procedure should be used only to confirm the forward-titration procedure. See below, section on "Organic Chloramine Interference."

11. As iodine is added to the sample, the pointer remains practically stationary until the end point is approached. Just before the true end point, each increment of iodine solution causes a temporary deflection of the ammeter to the right, but the pointer drops back to about its original position. The true end point is reached when a small addition of iodine solution gives a definite and permanent pointer deflection to the right (upscale).
12. Note the volume of iodine solution used to reach the end point, and calculate the total residual chlorine as follows:

$$\text{Total chlorine residual mg/L} = \text{ml phenylarseneoxide} - 5 \times \text{ml iodine solution} \tag{5-10}$$

This calculation assumes a 200 ml sample and a PAO solution that is 0.00564 *N* and an iodine solution that is 0.0282 *N*.

Monitoring the Normality of Iodine Solution. In the past the 0.0282 *N* iodine solution was reported to be so unstable that it should be made up fresh daily. Experience spanning many years discloses that this was a gross misrepresentation. If stored in the dark (in a brown bottle) at moderate temperatures, it will remain stable for months.

However, it is prudent to check the normality on a weekly or biweekly basis. The following is a simple field procedure that can be accomplished quickly with an amperometric titrator:

1. Sensitize the titrator electrodes to iodine by placing the sample jar in place containing a weak iodine solution—one that gives a light brown cast. Leave the electrodes immersed for 15 min.
2. Rinse the electrodes thoroughly with tap water.
3. Add 5 ml PAO to 200 ml distilled water and place the solution in the titrator.
4. Titrate with 0.0282 *N* iodine solution.
5. The end point is reached when a small addition of iodine produces an ammeter needle deflection to the right (upscale) that holds for 15–20 sec. At this point all of the PAO has been oxidized.
6. Now read the amount of iodine used from the buret (or pipette). If 1.00 ml iodine solution neutralizes the 5 ml PAO solution, the iodine solution is 0.0282 *N*. If the iodine solution has deteriorated, the volume of iodine solution reaching the end point will be somewhat greater than 1.00.

Sample Calculation. Upon standardizing the iodine solution, it is found that 1.2 ml 0.0282 *N* I_2 is required to neutralize 5.0 ml PAO in a 200 ml sample, and that the PAO is 0.00564 *N*.

The chlorine residual is calculated from the following equation:

$$\text{chlorine residual mg/L} = \text{ml PAO} - (5 \times 0.0282\ N\ I_2) \qquad (5\text{-}11)$$

As 1.2 ml I_2 is required to neutralize 5 ml PAO, 1.0 ml of PAO will only neutralize:

$$\frac{5.0}{1.2} = 4.17 \text{ ml PAO}$$

Therefore the calculation for residual, using Eq. (5-11), becomes:

$$Cl_2 \text{ residual (mg/L)} = \text{ml PAO} - (4.17 \times \text{ml } I_2)$$

Nitrogen Trichloride. Nitrogen trichloride is produced in two ways outside of the laboratory in potable water and wastewater treatment systems: (1) in the manufacture of chlorine when the electrolytic cell water contains ammonia nitrogen (see Chapter 1); (2) in water systems containing ammonia nitrogen when the chlorine-to-ammonia N ratio is about 12:1 or greater at pH up to 9 (see Chapter 4). Its occurrence is rare in potable water treatment. At the Rancho Cordova 3 mgd nitrogen removal study using chlorine, nitrogen trichloride generation was not a problem because of its rapid decay.[51] Palin[23, 73] and Cooper and Gibbs[48] have spent an enormous amount of time investigating the detection and measurement of NCl_3 in potable water. It is a tantalizing exercise but a seemingly futile one because NCl_3 solutions are made up in the laboratory under controlled conditions. Nitrogen trichloride does not occur in plant operation in the same way that the solutions are made in the laboratory; so there is little relevance for detection under laboratory conditions compared to field conditions.

Two factors relating to nitrogen trichloride formation should be kept in mind: (1) HOCl must always be present for NCl_3 to exist, and (2) monochloramine is always found in the presence of NCl_3. This is the result of the simultaneous competing breakpoint reactions described in Chapter 4.

Nitrogen trichloride is a gas. It is *practically* insoluble in water, but it is soluble in liquid chlorine and in most solvents such as carbon tetrachloride, and so on. The strongest solution of NCl_3 achieved by Cooper et al.[47] was 13.8 mg/L. The solutions were stored at 0°C or used immediately. These solutions and those by Palin were made under carefully controlled conditions and reacting concentrations substantially different from any field conditions.

From a practical viewpoint the operator is more interested in whether or not NCl_3 formation is occurring in the treatment process, rather than how much is formed. Moreover, the operator can detect the presence of NCl_3 more quickly than a laboratory technician can. The olfactory nerve can detect concentrations as low as 0.02 ppm,[71, 72] but the eyes can detect it before the nose can. At the slightest concentration the eyes will smart, and prolonged exposure will cause severe tearing. As the concentration increases, the effect is the same as that of tear gas. Unlike the case of chlorine, the respiratory system is only slightly affected, even at high concentrations.[74]

The only documentation covering the measurement of NCl_3 generated under field conditions at a treatment plant was by Williams.[71,72] His method of NCl_3 measurement by amperometric titration is described in this chapter.

Chemistry of Nitrite Interference. In the early 1970s the Bureau of Sanitary Engineering, California State Department of Health, carried out numerous investigations related to disinfection of wastewater effluents by chlorination. At that time the starch–iodide forward-titration was a popular method. The procedure called for acidifying the sample to a pH between 1 and 2 and performing the I_2 release by the combined chlorine residual. In several instances false residuals were reported; that is, residuals were being measured when the chlorinators had been shut down.[57] An investigation discovered that the false residuals were due to the presence of nitrites. It has since been confirmed that at a pH below 3.0, nitrite will oxidize KI to I_2. The rate depends upon acidity and KI concentration.[78]

For a very long time the starch–iodide titration was carried out at a pH of 1 or 2 to eliminate the iodine organic matter reactions.[78]

Nitrite interference in both the starch–iodide and the amperometric end-point procedures is eliminated by the use of pH 4 buffer (acetic acid).

The question of nitrite interference in the back-titration procedure using PAO has been raised. In this method excess PAO is added to the sample containing chlorine residual. The pH is depressed to 4 after the addition of KI. The iodine released by the KI–chlorine residual reaction is immediately consumed by the PAO. Iodine will react with nitrite to form nitrate, but this reaction is so slow at pH 4 that nitrite interference is nonexistent.

The short-term solution to the nitrite interference described above, by the California State Department of Health, was to destroy the nitrites present in the samples with sulfamic acid. However, sulfamic acid cannot be used for a long-term solution because it hydrolyzes to form HOCl. This adds another dimension to the interference.

Baker's Alternative Procedure.[75] About 1974 chlorine residuals in the receiving waters were discovered to be toxic to aquatic life. Since that time the measurement of these residuals has come under close scrutiny. Wastewater and cooling water discharges became the prime targets of this scrutiny. The basis of Baker's procedure is the assumption that chlorine species capable of any significant performance in the chlorination of potable water, wastewater, cooling water, and swimming pools have an ORP high enough to oxidize iodide to iodine at pH 7.[76] Driving this reaction to completion at pH 4 does not represent reality. One of the implications of this hypothesis concerns dechlorination. Current practice may require dechlorination of nontoxic residuals appearing in the "dichloramine" fraction (see Chapter 6).

This procedure demonstrates the hypothesis described above most dramatically with nitrified wastewater effluents where only a trace of ammonia N is present and the organic N content is 3.0 mg/L or greater.

Procedure:

1. Titrate for free chlorine with PAO.
2. Titrate for mono at pH 7 in the usual way with PAO.
3. Take another sample, add excess KI (1 ml) and PAO, back-titrate at pH 7 with standardized chlorine solution. One cannot back-titrate with I_2 because it will be consumed by organic matter at pH 7—but not at pH 4.
4. Then take sample 3 and drop the pH to 4. Whatever titrates in this procedure will be the "dichloramine" species.

In wastewaters with organic N concentrations greater than 1.0 mg/L, the "dichloramine" species is most likely to be impotent organochloramines (see below, "Organic Choramine Interference," for deleterious effects of organic N on disinfection of wastewaters).

SUMMARY AND RECOMMENDATIONS

General Considerations

The selection of an analytical method involves several factors based upon local conditions and requirements. It involves a variety of situations, such as potable water or wastewater; field or laboratory use; field monitoring or analyzer calibration; private, semiprivate, or public water supplies. There are methods to cover all of these situations. They are described below with situation recommendations based upon many years of experience.

Potable Water Treatment

Small Supplies: Private and Semipublic. These supplies serve summer camps, resorts, public campgrounds, motels, highway restaurants, private subdivisions, and so on.

Monitoring chlorine residuals for this type of water system is similar to working under U.S. Army "field conditions." Therefore, it would follow that the method selected by the Army, after intensive investigation, would be the method of choice. This is the FACTS method, using a testing kit with permanent glass color standards (0.2, 0.4, 0.7, 1.0, 2.0, 5.0, and 10.0), made by the Ames Division of Miles Laboratory. Two solutions are required: buffer and syringaldazine. This method measures only free chlorine. This is important for the small nonregulated water supply because a free chlorine residual is imperative for the short contact times inherent in these water systems.

Municipal Supplies

Chlorination Stations. Many water utilities do not have a central treatment plant but have chlorination stations scattered throughout the system. A good example would be a city that has a multiple-well field with several chlorinator stations. The field test choice in a case such as this could be either a portable amperometric titrator (Fig. 5-2 or Fig. 5-4) or a portable DPD colorimetric test kit with a spectrophotomer such as the Residometer (Fig. 5-1). Operating requirements for these supplies would probably depend on either free or total chlorine residual monitoring.

Treatment Plants. Practically all treatment plants have laboratory facilities. Therefore, operators have an additional choice: the DPD–ferrous titrimetric method. The amperometric titrator and the Residometer would remain as options. All of these methods measure either free or total chlorine residual.

In the case where the plant needs to monitor all the chlorine species, the choice is either the DPD titrimetric or the amperometric method.

Continuous Chlorine Residual Analyzers. These instruments provide intelligence only on a secondary level. Their information output is not absolute. The primary source of intelligence is derived from the information provided by the analytical method used for calibrating the analyzer. Therefore, the calibration method must be quantitatively precise and qualitatively definitive. Operating personnel have three choices: amperometric titration, DPD–FAS titration, or the DPD colorimetric method based upon spectrophotometric analysis. Other methods lack the precision and quality requirements for accurate determinations of residual levels.

Chloramines. All chloramine installations should have testing equipment capable of monitoring free chlorine monochloramine and dichloramine residuals. This is for process control of THMs and taste and odor control. The ability to measure these species provide helpful information relating to the Cl-to-N ratio location on the breakpoint curve. The appearance of dichloramine will affect the taste and the odor of the treated water. Therefore, to measure all chlorine residual species, use either an amperometric titrator or the DPD–FAS titrimetric method.

Wastewater Treatment

Primary Treatment Plants. The quality of these effluents is such that the alternative procedure, identified as the back-titration method, should always be used. Two methods are available, the amperometric titrator or the idometric (starch–iodide–iodate) method. The latter is acceptable when total residuals are 1 mg/L or greater. Colorimetric methods have never proved useful in measuring chlorine residuals for primary effluents.

Secondary Treatment or Well-Oxidized Effluents. The back-titration method, using the amperometric endpoint, is probably the most popular today. Next in popularity is the long-standing iodometric (starch–iodide) II method using the back-titration procedure. Many users of the starch–iodide (iodometric II) route have found the DPD–ferrous titrimetric method preferable.

For treatment plants with moderate to severe disinfection requirements (2.2 to 240/100 ml MPN), a forward-titration procedure, either amperometric or DPD–FAS titrimetric method, should be used for the following reasons: White et al.[77] found that greater efficiency of mixing, at the point of application of chlorine, resulted in a greater proportion of mono- to dichloramine in the final residual, resulting in greater disinfection efficiency. Therefore, the forward-titration procedure has an important place in monitoring the disinfection process. It is further recommended that use of the forward-titration procedure to determine mono- and dichloramine be supplemented by running total chlorine residuals by the back-titration procedure. The mono- to dichloramine relationship and its effect upon disinfection efficiency are described in Chapter 8.

Tertiary (Filtered) Effluent

Nitrified. These effluents will show both free and combined chlorine residuals. To properly monitor the disinfection process it is of significant benefit to run forward-titrations supplemented by total chlorine residuals if dechlorination is required. If the effluent is completely nitrified, monochloramine will be absent; so the second step can be omitted. However, the third step is important because the ratio of free to dichloramine residual becomes important to disinfection efficiency. The two methods of choice are the amperometric and the DPD–FAS titrimetric methods.

Nonnitrified. Residual measurements for these effluents should be by the forward-titration procedures to quantify mono- and dichloramine fractions, followed by a separate total chlorine residual by either back-titration or forward-titration. The latter serves as a check on the forward-titration for the mono and dichloramine fractions. The methods of choice are the amperometric titration procedure and DPD–FAS titrimetric method.

Organic Chloramine Interference

Now, in the 1990s, it is most important that the organic N compound that occurs in surface water supplies, and in the effluents of domestic wastewater treatment plants, causes significant errors in the measurement of total chlorine residuals as measured by conventional chlorine residual analyzers. This is so because the organochloramines that are formed are nongermicidal. The work by Wajon and Morris,[46] published in 1978, is misleading, as it claimed that the FAC (free available chlorine–HOCl) in these waters could not always be defined as FAC unless the types of organic N compounds were known. They both doubted that when FAC was identified in a wastewater effluent, it was really not FAC. White[77, 79–81] has never found this to be the case in his many investigations of wastewater disinfection systems during the last 15 years. When the phenomenon of organochloramines was first discovered at the San Jose, California, WWTP after it began practicing nitrification, which produced a free chlorine residual, 10–15 mgL of chlorine dosage was lost due to "bleaching" of the effluent color. This proved without a doubt that even when the organic N was in the effluent, but without any ammonia N, free chlorine would always be there.

However, identifying the amount of interference caused by organochloramines requires that a forward-titration procedure be performed, using either the amperometric method or the DPD–FAS titrimetric method, because they titrate as dichloramines, and a true dichloramine cannot be formed in a nitrified effluent.

This phenomenon was first discovered by Lomas, who was investigating England's swimming pool chlorination practices in 1967 (see p. 481, *Handbook of Chlorination,* 1st Ed., 1972). When free chlorine is absent and monochloramine is the disinfecting residual, and if organic N is present, the monochloramine in time will react with the organic N to form the nongermicidal dichloramine species. In order for true dichloramine to form in the presence of monochloramine, the chlorine-to-ammonia N ratio has to be at least 6:1. by weight.

Owing to the above phenomenon, it is imperative that all wastewater treatment plant operators use the forward-titration procedure because both free chlorine and monochloramine react with organic N to form these nongermicidal chloramines. When free chlorine is present, this reaction is almost instantaneous, whereas with monochloramine it takes from 35 to 40 min.

REFERENCES

1. Phelps, E. B., Testimony in re: In Chancery of New Jersey: Jersey City vs. Jersey City Water Supply Co., Vol. 12, pp. 6921 et seq.; decision rendered May, 1910.
2. Ellms, J. W., and Hauser, S. J., "Orthotolidine as a Reagent for the Colorimetric Estimation of Small Quantities of Free Chlorine," *J. Ind. and Eng. Chem.,* **5,** 915 and 1030 (1913).
3. Wolman, A., and Enslow, L. H., "Chlorine Absorption and Chlorination of Water," *J. Ind. and Eng. Chem.,* **11,** 209 (1919).
4. Meur, H. F., and Hale, F. E., "Present Orthotolidine Standards for Chlorine," *J. AWWA,* **13,** 50 (1925).
5. Scott, R. D., "Improved Standards for the Residual Chlorine Tests," *Water Wks. and Sew.,* **82,** 399 (1935).
6. Scott, R. D., "Refinement in the Preparation of Permanent Chlorine Standards," *Ohio Conf. Water Purif.,* **18,** 39 (1939).
7. Chamberlin, N. S., and Glass, J. R., "Colorimetric Determination of Chlorine Residuals Up to 10 ppm, Part II, Modified Scott Permanent Chlorine Standards," *J. AWWA,* **35,** 1205 (1943).
8. Chamberlin, N. S., "The Determination of High Chlorine Residuals," *Water Wks. and Sew.,* **89,** 496 (Nov. 1942).
9. Griffin, A. E., and Chamberlin, N. S., "Estimation of High Chlorine Residuals," *J. Awwa,* **35,** 571 (1943).
10. Anon. Committee Report, "Control of Chlorination," *J. AWWA,* **35,** 1315 (1943).

11. Chamberlin, N. S., and Glass, J. R., "Colorimetric Determination of Chlorine Residuals up to 10 ppm, Part I, Production of Orthotolidine–Chlorine Colors," *J. AWWA,* **35,** 1065 (1943).
12. Anon., "Standard Methods for the Examination of Water and Wastewater," 19th ed., American Public Health Assoc., Washington D.C. (1995).
13. Laux, P. C., "Break-Point Chlorination at Anderson, Indiana," *J. AWWA,* **32,** 1027 (1940).
14. Laux, P. C., and Nickel, J. B., "A Modification of the Flash Color Test to Control Break-Point Chlorination," *J. AWWA,* **34,** 1785 (1942).
15. Hallinan, F. J., "The OTA Test of Residual Chlorine," *J. AWWA,* **36,** 296 (1944).
16. Hallinan, F. J., and Gilcreas, F. W., "Use of OTA Test for Residual Chlorine," *J. AWWA,* **36,** 1343 (1944).
17. Marks, H. C., and Glass, J. R., "A New Method for Determining Residual Chlorine," *J. AWWA,* **34,** 1227 (1942).
18. Marks, H. C., Bannister, G. L., Glass, J. R., and Herrigel, E., "Amperometric Methods in the Control of Water Chlorination," *Ind. and Eng. Chem. Anal. Ed.,* **19,** 200 (1947).
19. Marks, H. C., Williams, D. B., and Glasgow, G. U., "Determination of Residual Chlorine Compounds," *J. AWWA,* **43,** 201 (1951).
20. Marks, H. C., "Residual Chlorine by Amperometric Titration," *J. NEWWA,* **66,** 1 (Mar. 1952).
21. Palin, A. T., "The Determination of Free Chlorine and of Chloramine in Water Using *p*-Aminodimethylaniline," *Analyst* (England), **70,** 203 (1945).
22. Moore, W. A., "The Use of *p*-Aminodimethylaniline for the Determination of Chlorine Residuals," *J. AWWA,* **35,** 427 (1943).
23. Palin, A. T., "The Estimation of Free Chlorine and Chloramine in Water," *Inst. of Water Engrs.* (England), **3,** 100 (1949).
24. Palin, A. T., "Determination of Chlorine Residuals in Water by the Neutral Orthotolidine Method," *Water Wks. and Sew.,* **101,** 74 (Feb. 1954).
25. Palin, A. T., "Determination of Free Chlorine and Combined Chlorine in Water by the Use of Diethyl-*p*-phenylene Diamine," *J. AWWA,* **49,** 873 (1953).
26. Connell, C. H., "Orthotolidine Titration Procedure for Measuring Chlorine Residuals," *J. AWWA,* **39,** 209 (1947).
27. Taras, M. J., "The Micro-titration of Free Chlorine with Methyl Orange," *J. AWWA,* **38,** 1146 (1946).
28. Taras, M. J., "Colorimetric Determination of Free Chlorine with Methyl Orange," *Ind. and Engr. Chem. Anal. Ed.,* **19,** 342 (1947).
29. Sollo, F. W., Jr. and Larsen, T. E., "Determination of Free Chlorine by Methyl Orange," *J. AWWA,* **57,** 1575 (1965).
30. Black, A. P., and Whittle, G. P., "New Methods for the Colorimetric Determination of Halogen Residuals," *J. AWWA,* **59,** 471 (1967).
31. Johnson, J. D., Overby, R., and Okun, D. A., "Analysis of Chlorine, Monochloramine, and Dichloramine with Stabilized Neutral Orthotolidine," paper presented at 85th Ann. Conf. Amer. Water Works Assoc., Portland, OR, June 28, 1965.
32. Johnson, J. D., and Overby, R., "Stabilized Neutral Orthotolidine, (SNORT) Colorimetric Method for Chlorine," paper presented Ann. Conf. Am. Chem. Soc., Detroit, MI, Apr. 1965. Publication #163, School of Public Health, Univ. of North Carolina, Chapel Hill, NC.
33. Johnson, J. D., and Overby, R., "Stabilized Neutral Orthotolidine Method for Free Available Chlorine," private communication, 1967.
34. Harrington, J. H., "Photo-cell Control of Chlorination," *J. AWWA,* **32,** 859 (1940).
35. Caldwell, D. H., "Automatic Residual Chlorine Indicator and Recorder," *J. AWWA,* **36,** 771 (1944).
36. Chang, S. L., "Destruction of Microorganisms," *J. AWWA,* **36,** 1192 (1944).
37. Anon., "Evaluation of the Stabilized Neutral Orthotolidine Test (SNORT) for Determining Free Chlorine Residuals in Aqueous Solutions," Sanitary Engineering Special Study No. 24-3-68/70 USA EHA, Edgewood Arsenal, MD, 1970.
38. Bauer, R., Phillips, B. F., and Rupe, C. O., "A Simple Test for Estimating Free Chlorine," *J. AWWA,* **64,** 787 (Nov. 1972).

39. Guter, K. J., and Cooper, W. J., "The Evaluation of Existing Field Test Kits for Determining Free Chlorine Residuals in Aqueous Solutions," U.S. Army Medical Environmental Engineering Research Unit, Edgewood Arsenal, MD, Report No. 73-03, Oct. 1972.
40. Guter, K. J., Cooper, W. J., and Sorber, C. A., "Evaluation of Existing Field Test Kits for Determining Free Chlorine Residuals in Aqueous Solutions," *J. AWWA,* **66,** 38 (Jan. 1974).
41. Cooper, W. J., Meier, E. P., Highfill, J. W., and Sorber, C. A., "The Evaluation of Existing Field Kits for Determining Free Chlorine Residuals in Aqueous Solutions, Final Report," U.S. Army Medical Bioengineering Research and Development Laboratory, Technical Report 7402, Aberdeen, MD, Apr. 1974.
42. Meier, E. P., Cooper, W. J., and Sorber, C. A., "Development of a Rapid Specific Free Available Chlorine Test with Syringaldazine (FACTS)," U.S. Army Medical Bioengineering Research and Development Laboratory, Aberdeen Proving Ground, MD 21010, May 1974.
43. Cooper, W. J., Sorber, C. A., and Meier, E. P., "A Rapid Specific Free Available Chlorine Test with Syringaldazine (FACTS)," *J. AWWA,* **67,** 34 (Jan. 1975).
44. Palin, A. T., "A New DPD-Steadifac Method for the Specific Determination of Free Available Chlorine," *J. Inst. Water Engrs. and Scientists,* **32,** 327 (1978).
45. Palin, A. T., "A New DPD-Steadifac Method for Specific Determination of Free Available Chlorine in the Presence of High Monochloramine," *J. AWWA,* **72,** 12 (1980).
46. Wajon, J. E., and Morris, J. C., "The Analysis of Free Chlorine in the Presence of Nitrogenous Organic Compounds," paper presented at the Division of Environmental Chemistry, ACS, Anheim, CA, Mar. 13–17, 1978.
47. Cooper, W. J., Roscher, N. M., and Slifker, R. A., "Determining Free Available Chlorine by DPD Colorimetric, DPD-Steadifac (Colorimetric) and FACTS Procedures," *J. AWWA,* **74,** 362 (July 1982).
48. Cooper, W. J., and Gibbs, R. P., "Equivalency Testing of the Free Available Chlorine Test with Syringaldazine (FACTS)," U.S. Army Medical Bioengineering Research and Development Laboratory, Fort Detrick, Frederick, MD 21701, July 1, 1982.
49. Kimm, V. J., Director, Office of Drinking Water (WH-550), Directive, Sept. 7, 1982.
50. Meier, E. P., Cooper, W. J., and Highfill, J. W., "Evaluation of the Specificity of the DPD-Glycine and FACTS Test Procedures for Determining Free Available Chlorine (FAC)," U.S. Army Medical Bioengineering Research and Development Laboratory, Fort Detrick, Frederick, MD, 1978.
51. Stone, R. W., "Rancho Cordova Breakpoint Chlorination Demonstration," report prepared by the Sacramento Area Consultants, Sept. 1976.
52. Whittle, G. P., and Lapteff, Jr., A., "New Analytical Techniques for the Study of Water Disinfection," paper presented at National ACS meeting, Dallas, TX, Apr. 1973.
53. Gibbs, P. H., Cooper, W. J., and Ott, E. M., "A Generalized Statistical Experimental Design for Comparison Testing of Analytical Procedures," report in process to *J. AWWA,* Apr. 1983.
54. Cooper, W. J., Gibbs, P. H., Ott, E. M., and Patel, P., "Equivalency Testing of Test Procedures for Free Available Chlorine: Amperometric Titration, DPD and Facts," *J. AWWA,* 1983.
55. Cooper, W. J., Mehran, M. F., Slifker, R. A., Smith, D. A., and Villate, J. T., "Comparison of Several Methods for the Determination of Chlorine Residuals in Drinking Water," U.S. Army Medical Bioengineering Research and Development Laboratory, Fort Detrick, Frederick, MD 21701, Apr. 1982.
56. Carns, K. E., private communication, East Bay Municipal Utility District, Mgr. Water Quality, Jan. 13, 1983.
57. Jopling, W., "Measurement of Chlorine Residuals," *Calif. Water Poll. Control Bulletin,* p. 25 (Jan. 1972).
58. Brooks, A. S., private communication, Univ. Wisconsin, Milwaukee, WI, Oak Ridge Conference, Oct. 23, 1975.
59. Marks, H. C., Joiner, R. R., and Strandskov, F. B., "Amperometric Titration of Residual Chlorine in Sewage," *Water and Sew. Wks.,* **95,** 175 (May 1948).
60. Hulbert, R., "Chlorine and the Orthotolidine Test in the Presence of Nitrite," *J. AWWA,* **26,** 1638 (1934).

61. Hallinan, F. J., "Thiosulfate Titration for Determining Chlorine Residuals," *J. Am. Chem. Soc.,* **61,** 265 (1939).
62. Wilson, V. A., "Determination of Available Chlorine in Hypochlorite Solution by Direct Titration with Sodium Thiosulphate," *Ind. and Engr. Chem. Anal. Ed.,* **7,** 44 (1935).
63. Palin, A. T., "A Study on the Chloro-Derivatives of Ammonia and Related Compounds With Special Reference to Their Formation in the Chlorination of Natural and Polluted Waters," *Water and Water Engineering* (England), p. 151 (Oct. 1950); p. 189 (Nov. 1950); p. 248 (Dec. 1950).
64. Kolthoff, I. M., and Lingane, J. J., *Polarography,* Vols. 1 and 2, 2nd ed., Chap. 47, Interscience, New York, 1952.
65. Salomon, E., "On a Galvanometric Titration Method," *Z. Physik. Chem.,* **24,** 55 (1897); **25,** 366 (1898); *Z. Elektrochem.,* **4,** 71 (1897).
66. Nernst, W., and Merriam, E. W., "Amperometric Titrations," *Z. Physik. Chem.,* **53,** 235 (1905).
67. Foulk, C. W., and Bawden, A. T., "A New Type of End Point in Electrometric Titration and Its Application to Iodimetry," *J. Am. Chem. Soc.,* **48,** 2045 (Aug. 1926).
68. Morrow, J. J., "Residual Chlorine Determination with Dual Polarizable Electrodes," *J. AWWA,* **58,** 363 (Mar. 1966).
69. Dowell, C. T., and Bray, W. C., "Experiments with Nitrogen Trichloride," *J. Amer. Chem. Soc.,* **39,** 896 (1917).
70. Stock, J. T., *Amperometric Titrations,* Interscience, New York, 1965.
71. Williams, D. B., private communication, 1968.
72. Williams, D. B., "Elimination of Nitrogen Trichloride in Dechlorination Practice," *J. AWWA,* **58,** 248 (1966).
73. Palin, A. T., "Determination of Nitrogen Trichloride in Water," *Proc. Soc. for Water Treatment and Examination,* **16,** 127 (1967).
74. White, G. C., unpublished reports on operation and maintenance of Wallace & Tiernan Agene Gas (NCl_3) systems used in flour mills for flour bleaching, 1937–41.
75. Baker, R. J., "Measurements of Chlorine Compounds," paper presented at Electric Power Research Institute and Maryland Power Plant Siting Program, Johns Hopkins Univ., Baltimore, MD, Jan. 16–17, 1975.
76. Baker, R. J., private communication, Wallace and Tiernan Div., Pennwalt Corp., April 1975.
77. White, G. C., Beebe, R. D., Alford, V. F., and Sanders, H. A., "Problems of Disinfecting Nitrified Effluents," *Municipal Wastewater Disinfection,* Proc. of Second Ann. Nat. Symposium, Orlando, FL, EPA Report 600/9-83-009, July 1983.
78. Baker, R. J., private communication, Wallace & Tiernan Div., Pennwalt Corp., Jan. 13, 1984.
79. White, G. C., "Wastewater Disinfection: Negative Effects of Organic Nitrogen," *CWPCA Bulletin,* p. 90 (Apr. 1987).
80. White, G. C., Discussion of: "The Toxicity of Chlorinated Wastewater: Instream and Laboratory Case Studies" by Gerald M. Szal et al., *Research Journal WPCF,* Vol. 63, Sept./Oct. 1992.
81. White, G. C., unpublished data describing the variability of the organic N problem as it affects disinfection by chlorine, 1982–96.

6
CHLORINATION OF POTABLE WATER

MICROBIAL FLORA OF NATURAL WATERS

Introduction. It was not until 1854 that scientists proved that a cholera epidemic was caused by a polluted drinking water supply. Since that time extensive bacteriological studies have been made to establish sources of these and other possible deadly organisms, and to find procedures and methods for detecting, identifying, and destroying them.

Although seawater is the largest natural environment inhabited by microorganisms, mostly on the surface layers, similar types of these organisms abound in surface waters such as rivers and lakes. These waters have been known to contain a variety of bacteria, algae, protozoans, and viruses.

In addition to research on deadly organisms, water microbiology is concerned with the natural microbial flora of oceans, lakes, rivers, and wetlands. These bodies of water harbor a multitude of microorganisms capable of producing a variety of chemical changes that are instrumental in maintaining the normal balance of marine life, as well as contributing to various geochemical processes.

Atmospheric Water. The moisture contained in clouds and precipitated as snow, sleet, hail, and rain constitutes atmospheric water. The microbial flora of this water is contributed by the air. In effect, the air is washed by the atmospheric water, which carries with it particles of dust to which the microorganisms are attached. Then, most of these organisms are removed from the air during the early stages of precipitation. The precipitation process also removes the miscellaneous debris that is discharged into the atmosphere by volcanic eruptions, natural and man-made forest fires, and natural and man-made oil and industrial fires, plus all the discharges from the super weapons of war. This is one of the many reasons why the enormous oil fires set by Saddam Hussain in the Gulf War never traveled very far from the source. It was originally predicted that they were going to devastate the planet, but nature intervened.

For each type of organism there is a cardinal temperature. Some have a cardinal temperature as low as 5–10°C, whereas some have one as high as 75–80°C. Oceans average 5°C, and deep waters average 1–2°C. Frozen environments are never sterile; there are always hidden pockets of liquid water.

Surface Water. Bodies of water such as lakes, streams, wetlands, and oceans represent surface water. These waters are susceptible to almost continuous contamination with microorganisms from precipitation, which arrives on-site in the form of surface runoff from soil and city drainage systems. These systems may be combined sewer overflows or just the drainage from separate storm water systems, and any other wastes that may be deliberately discharged directly into these bodies of water. Obviously the microbial populations of these waters vary greatly in kind and number owing to the great diversity in climates and human population around the globe. These populations are greatly dependent upon the nutrients available to support the microbial flora. In spite of the fact that atmospheric water eventually may reach the surface as springs or seepage, most of it finally finds its way to the sea. This water can be subterranean water if it occurs where all the pores in the soil or the rock-containing materials are saturated with it. Microbial life as well as suspended particles are removed from soil in varying degrees, depending upon the permeability characteristics of the soil and the depth to which the water penetrates.

Spring Waters. These waters consist of groundwater that reaches the surface through fissures or exposed porous soil. Wells are created by sinking a vertical pipe shaft, complete with a strainer assembly, and either a pump suction pipe or a submersible pump, depending upon the depth of the water table.

Well Waters. Wells less than 100 ft in depth are classified as shallow wells. Those over that are considered deep wells. From the bacteriological viewpoint, wells and springs that are properly located produce a very high quality water. However, a great many wells contain iron, manganese, and sulfur compounds that provide the sheathed organisms such as *Sphaerotilus, Crenothrix,* and *Leptothrix* with the necessary energy for life. Chemical treatment is always required for the removal of these compounds and associated organisms. Normally the microbial content of the wells is negligible when these compounds are absent. This is why special treatment with chlorine is usually required whenever groundwater and surface water are mixed in a distribution system.

PURPOSE OF WATER CHLORINATION

Just as water is close to being a universal solvent, so chlorine has nearly become in the the last century a universal water treatment chemical.

The primary objective of water supply chlorination is disinfection. Because of chlorine's oxidizing powers, it has been found to serve other useful purposes in water treatment, such as taste and odor control, prevention of algal growths, maintaining clean filter media, removal of iron and manganese, destruction of hydrogen sulfide, color removal by bleaching of certain organic colors, maintenance of distribution system water quality by controlling slime growths, restoration and preservation of pipeline capacity, restoration of well capacity, main sterilization, and improved coagulation by activated silica.

None of the alternatives to chlorine have been able to compete with its versatility and relative ease of use.

HISTORICAL BACKGROUND OF CHLORINE

Use as a Germicide

Chlorine was first introduced to water treatment as a disinfectant about the turn of the twentieth century. Since that time it has become by far the predominant method used for this purpose. This popularity is deserved because of its potency and range of effectiveness as a germicide. It is easy to apply, measure, and control; it persists reasonably well; and it is relatively inexpensive. Other agents may equal to or even excel aqueous chlorine in any one of these characteristics, but there is none that combines them in such an advantageous way. Ozone, bromine, iodine, chlorine dioxide, silver ions, ultraviolet, and ultrasonics have been investigated. Some of these have found important uses in special situations, but none of them so far has been a serious contender for replacing chlorine's near-universal use.

As a result, of all the municipal water supplies that are being chemically disinfected, at least 99 percent use chlorine.[1] Very few other processes enjoy this kind of monopoly. Even in countries where ozone is preferred, theoretically for physiological reasons, chlorination is almost universally employed in practice as an adjunct to ozone.

Probably the first known use of chlorine as a germicide was by Semmelweis, when in 1846 he introduced the scrubbing and cleansing of hands in chlorine water between contacts with each mother after he found that child bed (puerperal) fever was being transmitted from mother to mother in the maternity wards of the Vienna General Hospital. Then, in 1881 the German bacteriologist Koch demonstrated under controlled laboratory conditions that pure cultures of bacteria could be destroyed by hypochlorites. The earliest record of a suggestion to chlorinate water, even before water was known to be a carrier of disease germs, is a statement by Dr. Robley Dunlingsen in his *Human Health,* published in Philadelphia in 1835: "To make the water of the marshes potable, it has been proposed to add a small quantity of chlorine or one of the chlorides in small but sufficient amounts to destroy the foulness of the fluid."[2] Who originally made this proposal is not known. That water is a mode of transmission of disease has been known for only one and one-half centuries. Dr. John Snow first theorized in 1849 that water was the mode of communication of cholera.[3] He was later able to demonstrate his theory in the Broad Street pump episode in London in 1854 by removing the pump handle, thereby staying an epidemic of cholera that had already claimed some five hundred lives. Dr. Snow also found the source of the infection: a soldier recently returned from active service in India, residing in a rooming house adjacent to the broken sewer that contaminated the water drawn from the Broad Street pump, was a carrier of cholera.[4]

Early Uses in Water Treatment

The first use of chlorine as a continuous process in water treatment was probably in the small town of Middelkerke, Belgium, in 1902. This was known as the ferrochlor process. Ferric chloride for coagulation was mixed with calcium hypochlorite, which hydrolyzes to form a ferric hydroxide floc and hypochlorous acid as the disinfectant.[5] At Ostende, Belgium, in 1903, chlorine gas was generated by mixing potassium chlorate and oxalic acid. These installations, designed by Maurice Duyk, a chemist for the Belgian ministry of public works, are the earliest known continuous applications of chlorine for disinfection of potable water.

Chlorine in the form of hypochlorites (bleaching powder and chloride of lime) was first used as a temporary expedient to stay typhoid epidemics. The first recorded use was in 1896 when a typhoid epidemic occurred at the Austria-Hungary naval base of Pola on the Adriatic Sea.[2] In 1897 Sims Woodhead used a solution of bleaching powder for disinfecting the Maidstone, England, water supply during a typhoid epidemic.[6] The first known continuous use of chlorine in England was at Lincoln in 1905.[7,8] A serious typhoid fever epidemic had occurred at Lincoln late in 1904, and was traced to the water supply. The advice of A. C. Houston, bacteriologist, and G. McGowan, chemist to the Royal Commission on Sewage Disposal, was sought. The source of the water supply was the River Witham and two small tributaries. Houston and McGowan recommended that the water be treated with chloros (alkaline solution of sodium hypochlorite of about 10–15 percent available chlorine), the most easily available chlorine compound at the time.[8] Treatment began on February 11, 1905, under the direction of D. B. Byles, starting with a dose of 10 ppm and finally settling on one of 1 ppm. The amount of water treated was 1 to 1½ mgd. This treatment continued until 1911, when a new supply of water came into use.

First Uses in the United States

The introduction of chlorination in the United States occurred on May 22, 1888, when patents on electrolytic treatment of water were issued to Albert R. Leeds, professor of chemistry at Stevens Institute of Technology at Hoboken, New Jersey.[2] Webster's British patent of January 27, 1887, was the forerunner and possibly the model of later patents issued in Great Britain and the United States for the use of electrode-generated current to disinfect potable water and wastewater. Leeds's theory of purifying water was to remove organic impurities found in natural waters or wastewaters that cannot be removed by mechanical filtration or chemical precipitation by treating the water with the gases obtained by the decomposition of water containing an acid or a salt in solution, the decomposition being effected by use of an electric current. The acid employed was to be either hydrochloric, nitric, phosphoric, chromic, or sulfuric, or the salts of these acids, or a mixture of these acids or salts. Leeds thought that the best results would come from hydrochloric acid, or its salt sodium chloride. So far as is known, Leeds did not follow up his patent with any operating installations.

Also in 1888 a patent was granted to Omar H. Jewell and his son William M. Jewell, both of Chicago, for a combination of electrodes with a mechanical filter. The patent was assigned to the Jewell Pure Water Company. The apparatus consisted of electrodes mounted vertically in the dome of a pressure sand filter. Passage of electric current through these electrodes and a solution of carbonic acid was supposed to produce an insoluble blanket of sodium bicarbonate on top of the filter sand to remove material that the sand could not. Nothing came of this device, which could not be demonstrated to be of any value.

Some years later William Jewell designed another electrolytic process, which was used by George W. Fuller in January 1896 for experimental work at Louisville, Kentucky. Twelve of Jewell's electrolytic generators were installed. These units required ten pounds of salt to produce one pound of chlorine. This installation was capable of providing a chlorine dosage of 0.25 mg/L into the filtered effluent. This same device was used again in 1899 on an experimental basis at Adrian, Michigan. Jewell experimented briefly with chlorine gas by this process but abandoned it in favor of chloride of lime.

In 1893 Albert E. Woolf used an electrolytic process, similar to the one used by Leeds, to generate chlorine in the sewage effluent at Brewster, New York. Disinfection was required here because the discharge was into the east branch of the Croton River, a part of the New York City water supply. This chlorination system operated successfully for 18 years, until it was destroyed by fire in 1911.

The first notable and successful chlorination installation in the United States is credited to George A. Johnson at the Bubbly Creek Filter Plant in Chicago in 1908, to serve the Chicago stockyards.[8–10] The raw water here contained a large amount of sewage, and treatment consisted of filtration in conjunction with copper sulfate application. The large stock suppliers complained that animals drinking this water gained less weight than they did when city water was supplied to them.

Under pressure of a lawsuit brought by the City of Chicago against the Union Stockyard Company, the contractors who built the filter plant were enabled to fulfill their guarantee when Johnson substituted chloride of lime for the copper sulfate treatment. The rate of application was 45 lb of chloride of lime (approximately 30 percent available chlorine)/mg 7½ hours before filtration. The treated water showed better bacteriological results than the city water: Bubbly Creek water showed 0.34 percent cases of *E. coli* versus 12.8 percent for city water.

The Jersey City Case

Again in the same year, Johnson, who was a consulting engineer for the firm headed by George Fuller (Herring and Fuller), was hired by Dr. Leal to chlorinate the Jersey City water supply.[6, 11] This installation is significant because it followed litigation over the quality of the new water supply. Jersey City had entered into a contract with the builders, who later identified themselves as the Jersey City Water Supply Company, to provide a new water supply. This work consisted mainly of constructing a large reservoir to impound water from the Rockaway River, a tributary of the Passaic, and a pipeline to carry 40 mgd of water from the reservoir at Boonton to Jersey City, 23 miles away. The contractors fell into financial difficulties, and the project had to be completed by the bonding company five years after the contract was awarded in 1899. Prior to its completion in 1904, Jersey City started a suit against the Jersey City Water Company, claiming that the work was not being performed according to the specifications, which read in part:

> The water to be furnished must be pure and wholesome for drinking and domestic purposes; and the works shall be so constructed and maintained by the contractor that the water delivered therefore shall be pure and wholesome and free from pollution deleterious for drinking and domestic purposes.

The city's claim was based upon the pollution in the reservoir, which occurred two or three times a year during high water caused by storm runoff. The city claimed that the pollution could be corrected by a filtration plant. The defendants countered that if proper sewage disposal facilities were constructed by the city, this pollution would not occur.

The court rendered an opinion that the construction of a filter plant was not the responsibility of the defendants, and suggested, not as its own opinion but from evidence given by the city, that the water could be made to comply with the contract by the construction of sewers and sewage disposal works for various towns in the watershed.[6]

Dr. Leal was convinced that this would not entirely eliminate the pollution, and strongly advised that the company be allowed to suggest to the court its own method for the complete fulfillment of the contract requirements. He pointed out that the greater percentage of bacteria and *E. coli* found at the point of delivery in Jersey City was due primarily to the washing of soils, roads, streets, and manured fields rather than from any sewage contamination.

Leal's request was granted in a court decree filed June 4, 1908. Dr. Leal had in mind the electrolytic process of making chlorine, and on June 16, 1908, hired the firm of Herring and Fuller to build and operate a hypochlorite plant under the direction of G. A. Johnson. The plant itself was started on September 26, 1908. It was an enormous and cumbersome affair, as might be expected, consisting of three 10,500-gal concrete chloride of lime solution storage tanks as well as a large building at the Boonton reservoir outlet.[11] After several months of operation, the dosage was reduced from approximately 1.0 to 0.2 ppm, requiring a high of 0.35 ppm during times of high water. Bacterial reduction in the raw material was from 200 to 20,000 per cc to from 20 to 30 per cc in the delivered water.

After several months of practical operation, extended testimony was taken in the court of chancery by ex-Chancellor Magie from a score of the foremost experts of that time. In his opinion, handed down in May, 1910,[11] Magie said:

> From the proofs taken before me, of the constant observations of the effect of this device, I am of the opinion and find that it is an effective process which destroys in the water the germs, the presence of which is deemed to indicate danger, including the pathogenic germs, so that the water after this treatment attains a purity much beyond that attained in water supplies of other municipalities. The reduction and practical elimination of such germs from the water was shown to be substantially continuous.

In the search for witnesses in this case, a toxicologist was sought to declare the process of chlorination a poisonous one. No one could be found who had this opinion.

It is interesting to note that the expert testimony developed the following theory of chlorination chemistry: Hypochlorous acid formed in the reaction of chloride of lime and water liberates oxygen in a very active state and leaves hydrochloric acid. The latter drives off the weaker carbonic acid and unites with calcium to form calcium chloride. Hypochlorous acid will form whether or not the process is electrolytic production of molecular chlorine or hydrolysis of chloride of lime.

However, the nascent oxygen theory of the killing action of chlorine presented in testimony in this case prevailed as the accepted theory until 1944, when it was finally disproved by S. L. Chang. (See Chapter 4.) The nascent theory was also responsible for the popularity of ozone.[5] Another point of interest developed was the fear of producing "free chlorine." This term referred to molecular chlorine and not hypochlorous acid, which is present-day terminology for free chlorine. This fear apparently stemmed from the knowledge of the toxicity of molecular chlorine, particularly in its contact with the human respiratory tract.

The ruling by Magie in the Jersey City case gave chlorination a much needed boost but did not settle the argument as to the efficiency or desirability of chlorination as a water treatment process. There was a great deal of public sentiment as well as that of engineers against the idea of chlorination of potable water.[5, 12] Many prominent persons in the water works industry proposed outright abandonment of a water supply source if it had to be disinfected, regardless of the method. These same persons were also against any kind of pollution, for which they must be commended. In the meantime, they gained considerable support from the public, who took great pride in a pure and wholesome water supply. This attitude was largely responsible for the revival of the Jersey City case some 10 years after chlorination had been in successful operation.

In March, 1919, Jersey city officials, still irked by the two decisions against them in 1908 and 1910, appealed to the highest state court, the Court of Errors and Appeals, contending that chlorination does not afford the remedy claimed for it, asserting that in cold weather, "bacteria are stunned but not killed." Expert testimony and scientific evidence presented by the defendant clearly demonstrated to the court that chlorination was a reliable disinfectant that did in fact effectively destroy pathogenic organisms under all weather conditions.

Prior to the use of chlorine for disinfection and after the discovery of bacteria in water—thus making water a carrier of disese—filtration was investigated as a practical means of mechanically removing these bacteria. This was accomplished at first by allowing the water to slowly settle through formations of specially graded sand, which trapped some of the bacteria but not all. Filtration became greatly improved by chemical precipitation and sedimentation prior to sand filtration. The first known use of a municipal filtration system was the one built in London in 1829.[11] The apparent success of the filtration system in Hamburg, Germany, during the 1892 cholera epidemic convinced sanitationists that filtration should be seriously considered as a treatment process for polluted water. Sand filter plants were built as early as 1874 at Poughkeepsie, New York. George Johnson, one of the outstanding proponents of chlorination for disinfection, stressed the inadequacy of filtration as the sole means of treatment for polluted water, and was able to amply demonstrate his opinions by scientific data collected at the Bubbly Creek filter plant in Chicago[10, 11] (see above).

Subsequent statistical evidence made public health experts aware of the need to establish disinfection as a process, either by itself or in conjunction with filtration, or to abandon polluted supplies altogether. The statistics included data not only on death rates from typhoid fever but on the general community health.

Typhoid Fever and Waterborne Outbreaks

The death rate from typhoid fever became a yardstick in the United States for public health authorities. Just before the turn of the century it was firmly established that this disease could be waterborne.

Military campaigns often produced virulent outbreaks. In the American Civil War, typhoid casualties were estimated at more than 75,000 cases, with more than 27,000 deaths. In the Franco-Prussian War, typhoid accounted for more than 73,000 cases, with more than 8,000 deaths; this latter death toll was higher than that from combat. The British suffered severe outbreaks in the Boer War, totaling more than 57,000 cases, with some 8,000 deaths.[13] A fateful individual case was that which took the life of Albert, Prince Consort to Queen Victoria, in 1861.

The French and German armies adopted vigorous sanitation procedures as a result of their typhoid experience, so that on the eve of World War I the typhoid morbidity was down to eight cases per 100,000. It was in the Russo-Japanese War of 1904–5 that for the first time troops went to the front with units for boiling drinking water as a method of disinfection. By this time it had been established that boiling for a given length of time would kill the causative organism. In 1910 Whipple estimated that causes of typhoid fever could be grouped as: waterborne, 40 percent; food, 25 percent; ordinary contagion, including fly transmission, 30 percent; shellfish and all others, 5 percent.[8]

In 1900 the average death rate from typhoid in the United States was 36 per 100,000 population. Thus, more than 25,000 persons died of typhoid fever that year. The death rate dropped to 20 per 100,000 in 1910[13] and to 3 per 100,000 in 1935.[18] By 1960, fewer than 20 persons died from typhoid fever in the entire United States.[14]

Between 1920 and 1936, 470 waterborne outbreaks of typhoid fever in the United States and Canada caused nearly 1200 deaths and illness to approximately 125,000 persons. During this same period, there were reported more than 16,000 cases of waterborne typhoid fever, more than 100,000 of gastroenteritis, more than 200 of bacillary dysentery, 1400 of amoebic dysentery, and 28 of jaundice. More than 30 percent of the U.S. outbreaks originated in well water. The cases of amoebic dysentery and jaundice were unique in water supply and public health history.[15]

The principal causes of these outbreaks, in order of magnitude, were: surface pollution of shallow wells; cross-connection with a polluted supply; contamination of a spring or an infiltration gallery by pollution of a watershed; contamination of a brook or a stream by pollution on a watershed; use of polluted water from a river or an irrigation ditch without treatment; inadequate chlorination as the only treatment; inadequate control over the filtration and chlorination process; seepage of surface water or sewage into a gravity conduit.

One of the most notable examples of chlorination of the water supply for the safeguarding of public health is reflected in the efforts made by the Kansas State Health Department.[16] Between 1908 and 1917, at least 30 Kansas towns started using hypochlorite solutions for disinfection. In 1917, chlorine gas equipment was installed at Coffeyville and Valley Falls; by 1919, Atchison, Leavenworth, Herington, Independence, Iola, Kansas City, Ottawa, Sabetha, Topeka, and Wichita followed suit; by 1942, all 60 of the surface water supplies and 40 of the groundwater supplies in the state were so treated. Then in 1942, a disastrous waterborne bacillary dysentery epidemic of 3000 cases occurred in Newton, Kansas. Eventually, in 1956, the state health department ordered the chlorination of all public water supplies. By 1960, 462 of the 463 public supplies were being chlorinated.

Another factor that greatly favored safeguarding a water supply by chlorination was the tremendous cost to the cities in damages resulting from these waterborne epidemics. One example was the 1929 typhoid fever epidemic at Olean, New York, which cost the city $350,000.[15] Chlorination proved, among other things, to be cheap insurance.

The Mills–Reincke Phenomenon and the Hazen Theorem

As early as 1893, two public health researchers, Mills and Reincke, discovered after studying a great many communities that when a polluted water supply was replaced by a pure supply, the general overall health of the community was greatly improved—to a greater degree than could be accounted for by the reduced prevalence of typhoid fever and other recognized typical waterborne diseases. This discovery became known as the Mills–Reincke phenomenon. In 1903, Allen Hazen, a pioneer in the water works industry, discovered that when a community water supply was changed from bad to excellent by adequate treatment, for every person saved from death by typhoid three other persons would be saved from death by other causes, many of which were probably never thought to have any connection with, or to be especially affected or influenced by, the quality of the public water supply. This change in the death rate observation was known as the Hazen theorem. Therefore, disinfection of public water supplies goes further than just the control of waterborne diseases.

WATERBORNE DISEASES

The sole purpose of disinfection of potable water is to destroy pathogenic organisms and thereby eliminate and prevent waterborne diseases. The following diseases are known to be transmitted by water, but not necessarily by water alone. Food and personal contact are other means by which these diseases may spread.

Typhoid Fever

This disease was once a scourge that destroyed armies more effectively than weaponry and a pestilence that haunted towns because of its mystery. Typhoid masked itself in symptoms common to other ills and was often mistaken for typhus or was classified merely as one of many fevers. In North America it has been responsible for more sickness and death than any other waterborne disease. One frightening aspect is that the causative organism may be carried for a lifetime by an individual who has been infected but has recovered.* Not everyone who has contracted the disease and recovered becomes a carrier, but a great many do.

*America's best-known typhoid carrier was Mary Mallon, otherwise known as Typhoid Mary, who came to symbolize a fearsome drama in the public mind. She worked as a cook for several upper-class New York families before being tracked down in 1907 as a carrier responsible for several outbreaks of the disease. When her feces were examined bacteriologically, it was found that she had practically a pure culture of the typhoid bacterium, *Salmonella typhi.* She remained a carrier for many years, probably because her gall bladder was infected, and organisms were continuously being excreted from there into her intestine. She was imprisoned for three years when she refused to have her gall bladder removed. She was released from prison when health authorities accepted her pledge that she would not cook or handle food for others and that she would report to the health department every three months. She disappeared, changed her name, and cooked in hotels, restaurants, and sanitariums, leaving behind a wake of typhoid fever victims. After five years she was captured as a result of the investigation of an epidemic at a New York hospital. She was promptly arrested and imprisoned. She remained in prison for 23 years, where she died in 1938, 32 years after she was discovered to be a typhoid carrier.[11, 13]

The causative organism of typhoid fever, *Salmonella typhi,* was discovered by Karl J. Eberth in 1880 (in some textbooks, it is referred to as *Eberthella typhosa*). This durable organism resides and proliferates solely in the intestines of humans (not other animals). It is readily destroyed by heat or such disinfectants as chlorine, iodine, bromine, and ozone. Under natural conditions in soil, water, or feces, it can remain alive for weeks or months. Beard[19] confirmed that typhoid could survive up to five months in frozen debris and ice. When the ice was allowed to melt, the water became infected again.[19] Typhoid bacilli enter the body through the mouth, invade the mucosa of the small intestine, and traverse the intestinal lymphatic and mesenteric nodes to reach the blood. Subsequent proliferation occurs in the liver, gall bladder, biliary tract, and duodenum, with particular concentration in the lymphoid tissue of the small intestine mucosa. The bacillus is discharged in both the urine and feces, with stools becoming bacteriologically positive about two weeks after onset.

As the organisms can retain their virulence and remain viable for long periods of time, any water supply contaminated by sewage is potentially a carrier of the disease if a carrier of the causative organism contributes to the sewage. Gradually over the years as the spread of this organism has been controlled by chlorination and other means of disinfection (boiling, iodine, ozone), the number of carriers has been diminished. The incidence of typhoid in public water supplies is practically nil. In 1981 the CDC in Atlanta, Georgia did not report a single case. When a U.S. case is reported, it is usually a foreign traveler. Between 1946 and 1976, 610 cases were reported; 507 of these cases were traced to private water supply systems.[20] Continued diligence is necessary because the populace has not developed any natural resistance to *Salmonella typhi.*

Typhoid fever has put a strange blotch across the pages of history: it was a scourge that destroyed armies more effectively than weapons; this pestilence haunted towns because of its mystery; it masked itself in symptoms common to other illnesses; in the nineteenth century mortality rates rose as high as 300 per 100,000; and military campaigns have produced particularly virulent outbreaks, as discussed above.

Cholera

Cholera was originally known as Cholera Morbus in its death-dealing form. It has been known in India since about 400 B.C., and the name later became known the world over because of the magnitude of its attack rate, which resulted in a high mortality rate. It ravaged entire cities. The organism stayed in India until 1817, when it escaped for the first time.[21] It spread at the leisurely pace of human travel in those times to China, Russia, Europe, and Britain. It took some 20 years to cross the Atlantic into North America. It came to Britain in 1831 and killed 21,000 people. There have been several epidemics in London caused by returning British soldiers who became carriers after contracting the disease while on military duty in India.[21]

In the twentieth century it spread to the Philippines, Indonesia (Sulawesi province), the Middle Eastern countries, Africa, and Italy.

At one time cholera was common in Europe and North America, but the disease has been virtually eliminated in those areas by effective water treatment practices.

The causative organism was originally named *Vibrio comma* because of its shape. It is now labeled *Vibrio cholerae.* There is a new vicious strain of cholera organism, known as the El Tor *Vibrio,* which is slightly different from other strains. It is named after an Arab refugee camp in the Sinai Peninsula that was hit by the disease in the 1960s. The El Tor cholera pandemic that reached Naples, Italy, in 1973, began on the island of Sulawesi (Celebes) in 1960 and 1961. Then it turned up in the Philippines, and two years later in Korea, whence it spread to China.

In 1981 there were 913 culture-confirmed cases of cholera El Tor, serotype Ogawa, that occurred in Bahrain. The overall attack rate was 27 per 10,000.[21] This outbreak was caused by drinking water. The Naples outbreak was caused by shellfish contaminated by sewage containing the El Tor organism.

Houston found in the 1930s that the virulence of *V. cholerae* is reduced 99 percent by one week's storage in water, and that the organism is easily killed by chlorine.[23]

The countries affected by cholera could easily wipe out this haunting scourge simply by practicing better hygiene, improved sanitation, and more effective water treatment practices.

Amoebic Dysentery

Often referred to as amebiasis, this disease is a potential threat throughout the world, and is terrifying because it is almost impossible for a patient to experience complete recovery unless correct diagnosis and proper medical attention are immediately available. This disease was one of the most feared by U.S. armed forces serving in the Pacific during World War II. The causative organism (*Entamoeba hystolytica*) is a cyst, and can therefore lie dormant in the intestinal tract and go undetected by the carrier for long periods of time, only to initiate a recurring illness without warning. The best protection against this disease is removal of the organism from water supplies by filtration. *Entamoeba histolytica* is large enough to be trapped by sand filters; high doses of chlorine (10 mg/L) and a long contact time (one hour or more) are required to effect disinfection. One of the most notable of the several outbreaks of this disease in the United States was at the Congress Hotel in Chicago during the 1933 World's Fair, when the water system became contaminated with sewage through defective flushing valves on the toilet bowls. The most recent outbreak in the United States was in 1955.[25]

Bacterial Gastroenteritis

There are sporadic outbreaks of this intestinal disorder, which produces a wide variety of symptoms, such as nausea, vomiting, diarrhea, abdominal cramps, fever, and headache. The organisms usually responsible for such waterborne outbreaks belong mainly to the following genera of organisms. These generally including both bacteria and viruses, keep increasing in number and lethality with the passing of time.

Shigellosis. The genus *Shigella* include *S. dysenteriae, S. paratyphi* (both responsible for paratyphoid in humans), *S. shiga, S. sonnei,* and *S. flexneri.* The mode of transmission of these bacteria is by the fecal–oral route, which means they can be waterborne. In the period 1946–75 there were 49 outbreaks of shigellosis in the United States, which affected 10,869 people. The most common serotype isolated in these outbreaks was *S. sonnei.* The species *S. flexneri* was isolated from patients following an outbreak on a cruise ship. This too was traced to the ship's water supply.[26] This genus is also responsible for most foodborne outbreaks of gastroenteritis. In general these bacteria are not long-lived in the environment; so they are rarely isolated from bathing beaches or large bodies of water.

Salmonellosis. Salmonellosis is the generic name of the diseases caused by organisms of the genus *Salmonella.* These organisms are most usually pathogenic to humans and other warm-blooded creatures. The genus *Salmonella* contains over 1000 distinct types, which have different antigenic specificities in their O antigens.

Until the Riverside, California outbreak in 1965, it was thought that *Salmonella typhi* (typhoid fever, described above), *Salmonella paratyphi,* and *S. schottmuelleri* (paratyphoid) were the only significant organisms of the genus *Salmonella* capable of carrying waterborne diseases. This outbreak was an epidemic of acute gastroenteritis that affected 18,000 people. There were three deaths. After a lengthy investigation it was concluded that the causative organism was *Salmonella typhimurium,* which was transported through the city water system.[27–29] The water supply was underground well water. Until this outbreak occurred, it was generally believed that *Salmonella* species were only associated with food poisoning (except for *S. typhi*).

The dominant organism in food poisoning is *S. enteriditis.* There are so very many strains of *Salmonella* that it is difficult to identify the particular strain in a foodborne outbreak that might be associated with a contaminated water supply. The CDC listed 18 foodborne *Salmonella* outbreaks affecting 1573 people in 1975 and a similar number of outbreaks in 1974 affecting 5499 people. The common factors in these outbreaks indicate that food prepared away from home has a potential for causing gastroenteritis.

Toxigenic *E. coli.* This bacterial pathogen has been found to cause travelers' diarrhea in Mexico, commonly called "Montezuma's Revenge." The median duration of the illness is about five days, and it occurs about six days after the travelers arrive in Mexico. There are, of course, other contributing pathogens, such as *Shigella* and *Salmonella.* Enterotoxigenic *E. coli* has also been found to be responsible for diarrhea in travelers to Brazil and Kenya and in persons living in Brazil, Asia, Japan, and the United States.[30] The World Health Organization believes that 80 percent of all intestinal illnesses in Third World countries is due to lack of sanitation and proper hygiene. The highest disease rate and death rate is from diarrhea, whose pathogen is commonly waterborne.[31] The causative agent is most likely enterotoxigenic *E. coli.* In 1969 the diarrhea death rate in New York state was 2.4 per 100,000; in India it was 312 per 100,000.[31] Death from diarrhea is thought to be the result of dehydration. Other symptoms produced by this coliform pathogen include abdominal cramps, nausea, headache, vomiting, and fever.

In all cases investigated, the primary cause is either poor or no sanitation, absence of disinfection, malfunction of equipment, or improper design.

***Campylobacter* Enteritis.** Outbreaks of this disease have been reported as waterborne.[22, 32] One in Sweden during October 1980 that affected about 2000 people was found to be caused by the species *C. jejuni.* This organism was isolated in the stools of patients. Neither food nor drink was found to be a common factor. Epidemiologic evidence indicated that the infection was spread via the water mains, but the total coliform count gave no indication of fecal contamination.

The first outbreak of this disease occurred in 1978 in Bennington, Vermont. There were 3000 cases of diarrheal illness representing 19 percent of the population. Typical symptoms were abdominal cramps, diarrhea, and headache. The illnesses were associated with drinking water from the town water supply, which obtained unfiltered, but chlorinated, water from a nearby brook. *Campylobacter fetus* and *C. jejuni* were cultured from the feces of 42 percent of the cases, but not from 23 people who served as controls. Oddly enough, coliforms were not found in the water at the time of the outbreak. However, after the chlorination dosage was increased and a boil-water order was issued, no further cases were reported. Another outbreak occurred in Britain at a boarding school, affecting 234 students and 23 staff members. The causative agent was again *C. jejuni,* which was traced to an unchlorinated water supply.

Mycobacteria. Pelletier et al.[449] found that recovery of *Mycobacterium* from chloramine-treated water in the Boston area was resistant to persisting chloramine residuals. Their investigation of nontuberculosis bacteria revealed that these organisms were widely distributed in the natural environment. They have been increasingly responsible for serious infections in the immunocompromised population. Moreover, the investigators found that the incidence of *Mycobacterium avium* complex (MAC) isolation from the general patient population in Massachusetts had increased fivefold from 1972 to 1983. Since person-to-person transmission of nontuberculosis bacteria had not been generally observed, contaminated water was suspected as a source.

These organisms have been recovered frequently from environmental water sources,[450] including rivers, ponds, and lakes. Treated municipal water systems, bottled and distilled water, and water used for critical care procedures, such as hemodialysis, are known to have been contaminated by mycobacteria.[451, 452] Recovery of mycobacteria from chlorinated sources has been made in swimming pools, water taps, drinking fountains, aquaria, and whirlpools, which demonstrates that these microorganisms are resistant to accepted disinfection practices.[453–455] In earlier studies that examined the effect of free chlorine upon nontuberculosis mycobacteria, Pelletier et al. showed that *M. avium* will survive after 24 hours of exposure to 1.0 mg/L of free chlorine. Although this is a strong indication that nontuberculosis bacteria are resistant to chlorine, there had only been a limited investigation of their resistance to chloramines prior to the investigation by Pelletier et al.

The results of the investigation by Pelletier et al.[449] clearly demonstrate the ineffectiveness of chloramines as an appropriate disinfectant. At a 1.0 mg/L residual there was only a 15 percent reduction of a surface-water-derived strain of *Mycobacterium avium* after 24 hours of contact time. At a 3.0 mg/L residual there was an 80 percent survival at 8 hours of contact time, but no survival was detected after 24 hours of contact time. At a 6.5 mg/L residual there was no survival after 2 hours of contact time. Resistance of *M. intracellulare* was similar. After a 4-hour contact time in 1.2 mg/L chloramine residual, 90 percent of the original inoculum was viable. Raising the chloramine residual to 3.0 mg/L resulted in the complete inactivation of the organisms in 12 hours. All inactivation procedures used chlorine-demand-free water (CDFW).

***Yersina* Enteritis.** An outbreak of gastroenteritis that affected 87 persons was reported from the state of Washington in January 1982.[32] An investigation identified *Y. enterocolitica* as the causative agent. The water supply used for the packing of soybean curd (tofu) was found to contain the causative organism. A water purification system that included disinfection was required to eliminate this health problem.

Pseudomonas. *Pseudomonas aeruginosa,* which is considered to be an opportunistic pathogen, is commonly found in water systems. It is primarily a soil organism, and frequently grows in filter systems. It has been known to cause gastroenteritis in newborn babies.[33] This "nonpathogenic" bacterium is a particular problem in hospitals, where infections are acquired from catheterizations, tracheostomies, and intravenous intrusions. An epidemic of diarrhea in a hospital premature-baby nursery was traced to *Pseudomonas* bacteria in an aerating cap on a spigot used by ward personnel.[34, 35] This raises an important point about the role of disinfection in water supplies: the advent of antibiotics has broadened or increased the range of conditions in which gram-negative bacteria, such as *Pseudomonas,* may become pathogenic. Therefore the presence of these organisms in water supplies assumes greater importance than it did in the past because of the increased virulence of these organisms.

Schistosomiasis

This serious parasitic disease is produced by a blood fluke, such as *Schistosoma mansoni,* which lives in human abdominal veins and expels eggs through the urine or feces. Within a few hours after discharge into fresh or brackish water, these ova mature into free-swimming embryos, known as *Miracidia,* which escape the ova and then seek and penetrate a snail host. Within the tissues of the snail they grow and reproduce large numbers of larvae known as *Cercariae.* Literally thousands of these larvae emerge from a single snail each morning when the body of water in which the snail is living is warmed by the sun. The *Cercariae* are considerably larger than the *Miracidia,* can swim vigorously but not very far, but can be carried a long way by water currents. They tend to remain near water surfaces, where they can easily make contact with swimmers or waders. Upon contact with human skin the *Cercariae* attach themselves by means of suckers and then start a process of penetration. Within hours they are in the bloodstream and are carried to the liver, where they grow to maturity in a few weeks. They mate and move to small blood vessels in the walls of the intestine or bladder.

It is estimated that this disease may affect some 200 million persons who live in tropical climates throughout the world.[36] It is thought that the disease can be contracted by drinking water contaminated with the *Cercariae.* This is a serious threat to public health because of the possibility of its being widespread by immigration of persons from areas where the disease is endemic. Once infected with the blood flukes, a victim never gets rid of the disease but finally succumbs after the vital organs are incapacitated. This cycle may take many years of gradual debilitation of the patient, who can function adequately most of the time and who is always a carrier of the potential infection to others. Chlorination appears to provide the most effective treatment of drinking and bathing water, but its application to streams or other uncontrolled waters is not practical.[38] Control may be best accomplished by breaking the life cycle of the *Schistosoma* through environmental control, the principles of which include: preventing discharge of *Schistosoma* ova into water courses; applying molluscicides to rid streams and ditches of snail hosts; killing the *Cercariae;* and protecting populations from contact with infected waters.[36, 37]

Giardia

The flagellated protozoan *Giardia lamblia* is responsible for the disease giardiasis in humans. These parasitic flagellates are distributed worldwide and are the most commonly reported human intestinal parasites in the United States and Great Britain.[39, 40] This is a debilitating febrile disease. The fever is accompanied by severe cramps, headache, and diarrhea. Children are more vulnerable than adults.[41] This organism has eight flagella and a ventral sucker which it attaches to the intestinal mucosa. This sucker causes diarrhea in the human species. *Giardia* cysts are carried by dogs, cats, and wildlife in addition to man. The disease has been referred to as "beaver fever" because it is thought to be spread by beavers who make their homes in the watershed.

Because the primary source of *Giardia* infestation is human feces, giardiasis is found all over the world. This is why it has often been called the "travelers' disease." In addition to travelers, it is a common affliction of skiers, campers, hikers, hunters, fishermen, and vacationers who drink untreated surface waters from mountain streams or other assorted recreation areas that do not have properly treated water. It is believed that leaking septic tanks and poorly treated sewage, together with a proliferation of backpackers and campers, have been responsible for the widespread outbreaks of giardiasis from coast to coast in the United States. People also have passed it along by practices as diverse as having poor personal hygiene, participating in oral and anal sex, or drinking polluted water.[23]

Outbreaks of waterborne disease due to parasites and other etiologic agents have increased in the United States for the past quarter-century despite regulations and measures taken to ensure safe drinking and recreational water for all. Obviously these regulations are not strict enough, or the enforcement thereof is not being pursued where it is needed most.

One study found that more waterborne disease outbreaks were reported during 1971 to 1985 than during any 15-year period since 1920. During this interval, 502 outbreaks involving 111,228 cases of illness were reported to the Centers for Disease Control (CDC) and the U.S. Environmental Protection Agency (USEPA). Of these, 92 outbreaks, involving 24,365 individuals, were directly attributable to the protozoan parasite *Giardia lamblia,* making it the predominant cause of identifiable waterborne disease.[319]

Between 1965 and 1982 more than 60 outbreaks of waterborne giardiasis occurred in the United States, mainly in New England, the Rocky Mountains, and the Pacific states. Also in the 1980s larger communities suffered outbreaks—Wilkes-Barre, Pennsylvania and Reno, Nevada.[23] The first reported waterborne outbreak in the United States occurred in 1965 in Aspen, Colorado, an extremely popular ski resort. This is significant because it has been confirmed that cold temperatures indigenous to streams and lakes at the higher elevations provide an agreeable environment for *Giardia* cysts. This makes it imperative to have more rigorous disinfection and more reliable filtration for these waters.

Studies of giardiasis outbreaks have demonstrated that conventional water treatment problems have been the major contributory factors. Two-thirds of these outbreaks have been caused by inoperative chlorination equipment, inadequate chlorination, and deficiencies in both coagulation and filtration. Chlorination deficiencies have been a primary cause in a majority of the outbreaks.[320, 321]

Giardia cysts have been found in the San Francisco water supply at the source. One of the highest-quality waters in the United States, it derives from melted snow high in the Sierra Nevada mountains adjacent to Yosemite National Park 185 miles east of San Francisco. This discovery is a stern warning that the natural barriers against transmission of waterborne diseases have finally been broken. This means that the pristine mountain water supply is a thing of the past, and that disinfection of these supplies by current chlorination practices without coagulation, sedimentation, and filtration is no longer sufficient.

The *Giardia* organism was first discovered by Leeuwenhoek, who found the organism in his own stools. It was named for Giard, who studied the parasite, and for Lamblia, who first described it. *Giardia lamblia* was originally considered nonpathogenic because many of the individuals who harbored it did not get sick.[322]

Members of the genus *Giardia* are single-cell protozoans that are intestinal parasites. They have two stages in their life style: (1) a vegetative or reproductive stage, called a "trophozoite," and (2) a dormant stage, called a "cyst." The cyst stage is the one most likely to occur in drinking water supplies. The organisms seem to be activated or stimulated by stomach acid, which usually acts as a barrier to infection by many bacteria. This acid activation causes the release of trophozoites that attach to the upper small intestine where they reproduce by simply splitting in two. Periodically these trophozoites detach from the upper intestine walls, and they encyst as they pass through the bowel. This process takes from one to two weeks.[323] This results in acute symptoms of stomach disorders consisting of diarrhea, vomiting, fatigue, and severe cramps.

Death of humans from *Giardia* infections is quite rare. From 1971 to 1985 only four such deaths were reported in the United States. From 1965 to 1980 about 23,000 cases of giardiasis were reported, in spite of the fact that it was not a reportable disease during those years. In 1983 there were 300 giardiasis cases reported that were due to water from a poorly maintained reservoir at McKeesport, Pennsylvania. In addition to being part of the city water supply, this reservoir also served as a neighborhood swimming pool during warm weather. This outbreak occurred in spite of the fact that the water from this reservoir was treated by conventional water treatment with both pre- and postchlorination, coagulation, sedimentation, and filtration.[323]

As of the 1990s, *Giardia* is becoming a serious threat to not only health but also life in the United States, Giardiasis is extremely difficult to diagnose, even with stool specimens.[467] Therefore, many people with intestinal pains and diarrhea spend unnecessary hours and an excessive amount of money visiting doctors and undergoing agonizing tests without a proper diagnosis, and all they have is giardiasis.[467] Doctor Martin Wolfe,[467, 468] a tropical medicine specialist in Washington, D.C., has had immediate success using an antiprotozoal. Unfortunately, parasitology is not taught in many medical schools today. Moreover, up until the middle 1980s, *Giardia* was not generally considered something that could cause serious illness, but now it has spread to child day-care centers.

A recent CDC survey of day-care centers in Fulton County, Georgia showed that 25 percent of the children are infected with *Giardia,* while in New Haven, Connecticut, the rate has run as high as 50 percent.[468] Oddly, in 1984, just when *Giardia* cases were exploding and the CDC was undergoing a series of budget cutbacks, it stopped keeping track nationwide. The CDC did, however, account for a record 26,650 cases that year, nearly double the number of cases a few years earlier. Today the situation is far more serious. Pennsylvania leads the nation in waterborne disease outbreaks—almost all from *Giardia*—with 15,508 cases reported from 1979 through 1990. Over the past several years, some 250,000 Pennsylvanians have had to boil their water, largely because of *Giardia* contamination.

In New York State the number of reported cases of giardiasis jumped to 2,553 in 1990, up from 961 in 1986; Wisconsin had 1,911 cases in 1990, up from 118 in 1981; Washington State had 796 in 1990, more than ten times the 75 they first counted in 1974. In Vermont, giardiasis is now the number one reported disease.[468]

Contaminated water is only part of the picture. A noodle salad served at a Connecticut picnic in 1985 resulted in sickness to 16 people. The CDC tracked the infection to a woman who had mixed the cold salad with her hands. She did not have *Giardia* but one of her small children did—though without any symptoms. The woman had apparently assisted her child after a bowel movement and thought her hands were clean—according to current practices. But with *Giardia,* just ten or one hundred organisms can infect you. A hundred would barely make a smudge under your fingernail.*

During 1990, in a Minnesota nursing home for the elderly, an "adopted grandparent" program sent *Giardia* from the toddlers to the elderly, infecting 88 people in six weeks. All of this is reminiscent of "Typhoid Mary."

It is not easy to distinguish a *Giardia* infection from many other ailments. One basic sign that is helpful is that a fever does not accompany a *Giardia* infection. Fortunately, there are a few special signs to watch for. See Table 6-1.

Most intestinal problems do not last as long as giardiasis, which can keep one miserable for as long as two weeks with stomach cramps, nausea, constipation, diarrhea, and noxious gas. These symptoms may vanish spontaneously then come back repeatedly. Between these episodes the victim will feel constantly fatigued. After these recurring bouts, the victim may experience weight loss. *Giardia* organisms are believed to attach themselves to the walls of the upper intestine, crippling the body's ability to absorb nutrients.

Table 6-1 lists ten organisms that are likely to cause intestinal illness, together with their symptoms, sources, and treatment.[469] No matter which one is involved, the patient should drink eight to ten glasses of water per day. The major damage in most cases is dehydration. If the symptoms are severe or if you are running a fever, see your doctor!

As of 1997, *Giardia* can and does pass through outdated filters in water systems supplying 20 million Americans. Moreover, chlorination practices

*One soiled diaper can harbor millions of the *Giardia* parasites, and only 10 are required to produce an infection.

Table 6-1 Organisms Associated with Waterborne Diseases

Organism	*Symptoms and Source*	*Duration and Treatment*
Giardia lamblia	Bloating, gas, diarrhea, abdominal cramps, nausea, fatigue, weight loss, lactose intolerance, constipation. Symptoms usually appear 7 to 14 days after consuming contaminated food or water.	Recurrent, usually lasting 2 weeks; liquids and prescription drugs Quinacrine, Flagyl, or Furazolidine, taken for 5 to 7 days.
Campylobacter jejuni	Fever, watery and sometimes bloody diarrhea, abdominal pain. Symptoms usually appear 2 to 7 days after eating raw meat or poultry, or drinking unpasteurized milk.	1 to 2 weeks; liquids, antibiotics if symptoms persist.
Clostridium perfringens	Diarrhea, abdominal pain, nausea, fever, gas in mild cases; malabsorption of nutrients in severe cases. Symptoms usually appear 8 to 15 hours after eating food cooked or stored at temperatures too low to kill bacteria.	6 to 24 hours; liquids, severe cases may require penicillin.
Escherichia coli	Cramps, stomach rumblings, gas, nausea, and vomiting usually appear 5 to 48 hours after eating or drinking contaminated food, water, or unpasteurized milk.	3 days to 2 weeks; liquids, bland foods.
Listeria monocytogenes	Fever, headache, chills, nausea, and vomiting usually appear 48 to 72 hours after eating contaminated foods such as unpasteurized milk and cheese.	1 to 4 days; penicillin or other antibiotics may be necessary for infants, pregnant women, the elderly, cancer patients, or others with suppressed immune systems.

Norwalk viruses	Diarrhea, nausea, vomiting. Symptoms appear 1 to 2 days after eating food contaminated by feces.	1 to 2 days; liquids.
Salmonella	Nausea, crampy abdominal pain, followed by diarrhea, fever, and sometimes vomiting. Symptoms usually appear 12 to 48 hours after eating contaminated poultry, eggs, beef, pork, or unpasteurized milk.	1 to 4 days; avoid milk products, eat bland foods, drink liquids. Antibiotics may be necessary only for infants or the elderly.
Shigella	Intense abdominal pain and watery diarrhea that worsens with time; dehydration and weight loss in severe cases. Symptoms usually appear 1 to 4 days after contact with infected feces.	Mild cases 4 to 8 days, severe cases 3 to 6 weeks; liquids, hot water bottle for abdominal pain, antibiotics.
Staphylococcus aureus	Severe nausea and vomiting, sometimes with abdominal cramps and diarrhea, less often with headache and fever. Symptoms appear abruptly 2 to 8 hours after eating contaminated meat, fish, or dairy products.	3 to 6 hours; liquids, sometimes salt replenishment.
Vibrio vulnificus	Severe crampy abdominal pain, weakness, watery and sometimes bloody diarrhea, low-grade fever, chills. Symptoms appear abruptly 15 to 24 hours after eating contaminated raw or undercooked shellfish.	2 to 4 days; liquids.

used today (chloramination) to achieve the EPA MCL for THMs are worthless for protection against *Giardia.* The only successful treatment is superchlorination followed by partial dechlorination to achieve the necessary chlorine residual contact time "envelope," according to information developed by Hibler's research (see below in this section). *Giardia* cysts are extremely resistant to chlorine. Jarroll et al.[43] found that it required a chlorine concentration of 4 mg/L and a 60 min. contact time to kill all cysts in a chlorine-demand-free solution at pH levels of 6, 7, and 8 (at 5°C). This in effect means that in a system with chlorine demand the chlorine dosage would have to be sufficient to produce a 4 mg/L free residual at the end of 60 min. These findings compare favorably with those of Meyer and Jarroll.[40]

Giardia cysts are even more resistant to chloramines than to chlorine. This is to be expected, as they generally follow the disinfection resistance pattern of *Entamoeba* cysts. Bingham and Meyer[318] found that complete destruction of *Giardia* cysts required 4.5 hours of contact time with 2.6 mg/L chloramine residual at a water temperature of 50°F. Sample water pH was not an important factor. Cyst death was slightly more rapid at pH 7.0 than at pH 8.5.

In order to be as accurate as possible, it would be well to establish an ORP mV level at a 30-min. contact time that is required for 4 and preferably 5 logs removal for the inactivation of *Giardia.* This would eliminate any potential problems likely to occur in establishing a chlorine residual level by a chlorine residual analyzer. There is plenty of evidence that total chlorine residual readings on these instruments are inherently subject to producing errors caused by nitrites, organic N, and other natural organic matter.

Inactivation of *Giardia* is best achieved by the multiple barrier approach. This would include removal of cysts by conventional water treatment processes such as flocculation, coagulation, sedimentation, and filtration. Any remaining cysts can then be inactivated by terminal disinfection. It is strongly recommended that prechlorination be applied as far upstream in the treatment train as is possible so as to achieve the optimum contact time of 60 min. This is similar to the recommendation for inactivation of *Entamoeba histolytica* cysts.

Pilot plant tests of diatomaceous earth (DE) filtration resulted in almost 100 percent removals of *Giardia lamblia* cysts for both coarse and fine grades of DE over a wide range of conditions.[324] The cyst analysis procedure was performed by pathologist C. P. Hibler, a renowned expert in the field. Between 1979 and 1986 more than 6000 water samples from 301 municipal sites in 28 states had been analyzed by his laboratory. These samples were from springs, creeks, lakes, and wells.[325]

Inactivation by ultraviolet irradiation is not effective. Therefore UV is not an appropriate method of disinfection for *Giardia* cysts.[49] In an article[320] that appeared in the *J. AWWA* in January 1986, "Waterborne Giardiasis: Where and Why," it was concluded that the primary cause of most outbreaks was inadequate or interrupted treatment. Disinfection problems or failure to provide filtration was prevalent in a majority of the treatment-related outbreaks.

Where treatment was provided, inadequate coagulation and poor filtration were identified as the cause.

The AWWA Research Foundation Report by Hibler et al.[326] is the most comprehensive treatise on the inactivation of *Giardia* by chlorine to date (1990). This work allows the water purveyor to choose a *Ct* value using the free chlorine residual process that will produce a 3 log inactivation at pH of 6, 7, and 8 and water temperatures of 0.5°C, 2.5°, and 5.0°C. The EPA* has published a lengthy document with several tables[327] giving *Ct* values for various log removals, pH values, and temperatures. These EPA tables fully reflect a gross lack of public health concern. These tables were constructed by extrapolation of Hibler's data. After a discussion with Hibler[328] it was concluded that the data for pH 8 were so scattered as to be too unreliable to be subjected to extrapolation. Haas and Heller[329] came to the same conclusion in their 1989 investigation. In view of these opinions the most reliable *Ct* values are shown in the table below and are for a 2 mg/L free chlorine residual measured by either amperometric titration or the DPD-FAS titration procedure. Do not use any colorimetric method. These *Ct* values are for 3 log inactivation at the pH values and water temperatures shown, and are based upon the AWWA research project by Hibler.

	0.5°C	*2.5°C*	*5.0°C*
pH 6	200	180	150
pH 7	270	170	300

There need not be any *Ct* values for water temperatures warmer than 5°C, as the *Ct* values shown will simply provide an ample safety factor. The germicidal efficiency of free chlorine increases significantly as the water temperature rises.

Owing to diligence and dedication by researchers in this field, the difficult task of identifying and enumerating *Giardia* and *Cryptosporidium* is becoming simpler and more reliable. Recent work by Le Chevalier et al.[335] describes a technique that can detect approximately 12 times more of these organisms in surface water supplies than the previously used zinc sulfate flotation method. This newly developed method is identified as the IFA-Percoll† technique, which can simultaneously detect *Giardia* and *Cryptosporidium.* This technique can also be useful for evaluating filter plant performance.

A chlorination system for a high quality surface water that does not require filtration can be designed to inactivate *Giardia* cysts, as has been done in some instances for amoeba cyst inactivation. In either case partial dechlorination after terminal disinfection must be considered. Chlorine dioxide, alone

*These values for *Giardia,* furnished by the EPA, are not even close to the California requirements for 5-log removal of all pathogens.

†The IFA (immunofluorescence assay)-Percoll method requires the use of an epifluorescent miscroscope and the availability of monoclonal antibodies specific to *Giardia* and *Cryptosporidium.*

and applied sequentially with other disinfectants, has proved very effective for *Giardia* inactivation (see Chapter 12).

Cryptosporidium

This parasitic organism has been termed the new "superbug" of the water industry because it has many similarities to *Giardia.* The *Cryptosporidium* oocyst, which is the dormant form of the protozoan found in the environment, is only half the size of a *Giardia* cyst. *Cryptosporidium* is an extremely hardy organism. The oocyst is the form that occurs naturally in the environment, where it can survive for up to two years owing to its protective cyst wall. Therefore, because of its smaller size and tougher cyst wall, it is far more difficult to deal with than *Giardia.* Moreover, it is highly resistant to reasonable free chlorine residuals, so that chlorine is practically ruled out as an acceptable disinfectant. Fig. 6-1 illustrates the comparative dimensions of *Giardia* and *Cryptosporidium* cysts.

First described in 1895, it was given the name *Cryptosporidium parvum* when it was found in the domestic mouse. Currently there are a disputed number of distinct species of this organism because cross-transmission studies have revealed that an organism from one animal host can infect other animals, including humans.[330]

This organism joined the list of organisms responsible for waterborne outbreaks where there was an outbreak of gastroenteritis in 1984 in Braun Station,

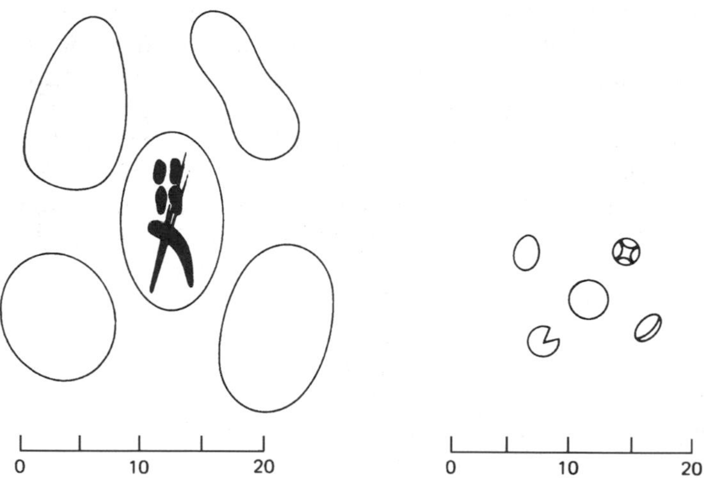

Fig. 6-1. Schematic sketches showing shape and dimensions of *Giardia* and *Cryptosporidium* cysts (courtesy Jerry E. Ongerth, University of Washington, Department of Environmental Health). Left, *Giardia* (*lamblia* or *muris*) cyst with classical internal structures (center) with common alternate characteristic shapes; right, *Cryptosporidium* parvum oocyst (center) with common alternate characteristic shapes.

a suburb of San Antonio, Texas. It was linked to sewage leaking into the well water supply for this community. The presence of oocysts was found in the stools of 47 of 79 residents.

In Sheffield, England during the summer of 1986, a peak of cryptosporidiosis was recorded among individuals whose only epidemiological source link was drinking water supplied from a common reservoir that was believed to have been contaminated by surface runoff following a heavy rain. The link here was that oocysts were identified in the second source water as well as in cattle that grazed on the land surrounding the reservoir. Also in the summer of 1986, there were 78 laboratory-confirmed cases of cryptosporidial illness that were linked to the consumption of untreated water in New Mexico.[319]

A 1987 outbreak of gastroenteritis affecting some 13,000 residents of Carroll County, Georgia sent an important warning to the water industry because the only significant risk factor associated with the outbreak was exposure to the public water supply, which was filtered and chlorinated. Furthermore, the treatment facility was operating within the EPA guidelines. During the outbreak, samples tested negative for coliforms, the turbidity range was 0.07–0.18 NTU, and chlorine residuals were on the order of 1.5 mg/L at the treatment plant.[319] *Cryptosporidium* oocysts were identified in 39 percent of the patients tested during the outbreak. Low numbers of oocysts were identified from samples collected at the treatment plant and in the dead ends of the system during the outbreak, which occurred in January and February 1987.

Investigation of the WTP operating procedures revealed that filters were frequently restarted without being subjected to backflushing. Turbidity measurements taken from the filter effluents within 3 hours after restarting showed abnormal values—as high as 3.2 NTU. It was thought that the lack of proper backflushing may have allowed the inadvertent discharge, into the finished water, of clumps containing infectious oocysts that had been trapped on the filters.[319]

Two additional *Cryptosporidium*-related outbreaks were reported in Ayrshire, Scotland and Oxfordshire-Swindon, England, in 1988 and 1989. Both were remarkably like the Sheffield incident. The one in the Oxford–Swindon area is noteworthy because a boil-water order was given to 600,000 people. Once again, *Cryptosporidium* oocysts were identified from affected people and the water supply.

The disease cryptosporidiosis can be contracted by humans from ingestion of as little as one cyst. On ingestion, the oocyst sheds its protective wall in the human gastrointestinal tract. In this new form a developmental and reproductive cycle begins. The resulting disease has symptoms similar to giardiasis, including diarrhea, stomach cramps, nausea, dehydration, and headaches. During the acute stage, up to 10 million trophozoites per gallon of feces may be found. This is the reproductive form of the organism.[330]

Since this waterborne disease affects humans, it threatens our water supplies. The disease, cryptosporidiosis, is a serious one because of the classes of individuals affected: those with underdeveloped or suppressed immune systems, including infants and the elderly, and those who have contracted AIDS, for whom it is life-threatening. Taken together, these groups are estimated to constitute close to 25 percent of the total population.

Cryptosporidiosis was first recognized as a disease in cattle in 1971. The first human case occurred in 1976, with very few others reported during the next five years. The AIDS epidemic of the 1980s and 1990s produced much greater numbers. More than 400 cases of cryptosporidiosis in AIDS patients were reported between 1981 and 1990. It was cited as the major cause of death in these individuals.

The disease is self-limiting in individuals with healthy immune systems. Currently there is no cure for cryptosporidiosis, and once infected, the host is a lifetime carrier subject to relapses of symptoms. The elderly are particularly vulnerable to this disease.

Inactivation of this organism is unbelievably difficult. The only methods that appear to be practical are ozone, which may require both pre- and post-ozonation, and chlorine dioxide (see Chapter 12). Korich et al.[331] were able to accomplish slightly greater than 90 percent removal by exposing the oocysts to 1.0 mg/L of ozone during a 5-min. contact time. Exposure to a 1.3 mg/L residual of chlorine dioxide required a one-hour contact time to achieve a 90 percent oocyst reduction. However, subsequent work has shown chlorine dioxide to be much more effective than this data would indicate. Both free chlorine and monochloramine required an 80 mg/L chlorine residual and a 1½-hour contact time. Because all the above disinfectants were used in "demand-free" situations, the residuals cited were the exact amount of disinfectant dosage.

In the late 1980s considerable research was carried out on the techniques of identification and inactivation of this oocyst. The researchers were aware of the impending *Ct* rule; so all the studies related residual, contact time, and oocyst survival—if any.

J. E. Peeters[332] seeded demineralized water with controlled numbers of *Cryptosporidium parvum* oocysts. These organisms were subjected to various doses of ozone and chlorine dioxide. In their preliminary trials the investigators found that 1000 oocysts per mouse produced 100 percent infection. In water containing 10,000/ml oocysts dosed with 1.1 mg/L of ozone, the infectivity of the oocysts was completely eliminated with a contact time of 6 min. for neonatal mice. Therefore, the *Ct* in this case is 6.66. Because the water used had no ozone demand, the dosage equaled the residual. When the level of oocysts was raised to 50,000/ml, an ozone dose of 2.27 mg/L and a contact time of 8 min. were required for 100 percent inactivation of the oocysts. The *Ct* in this case is 18.16. The application of chlorine dioxide did result in inactivation of the oocysts, though somewhat less than with ozone. These researchers jumped to the conclusion that European practices with ozone are sufficient to inactivate the crypto oocysts at concentration levels less than 10,000/ml because these installations use ozone dosages of 1.5–4 mg/L, resulting in residuals of about 0.4 mg/L at the end of 6 min. contact time.

Korich et al.[331] concluded that disinfection as commonly practiced in municipal surface waters *without filtration* should not be expected to destroy the oocyst stage of this parasite. For example, their investigation revealed viable sporozoites of *C. parvum* oocysts detected after exposure to 10 mg/L chlorine in excess of 4 hours of contact time. At pH 7 and 25°C, the estimated *Ct* value for free chlorine would have to be 7200 for a 2-log inactivation of these oocysts. This is 720 times greater than the *Ct* value for *Giardia lamblia,* and 248 times that reported for *Giardia muris.* Monochloramine was much less effective, which is to be expected. As would be expected, ozone turned in the best performance. At a 1 mg/L ozone residual, 1-log inactivation occurred in 3 min. In 5 min. it was 1–2 logs, and in 10 min. it was only 2–3 logs inactivation. This means that the *Ct* value for 2-log inactivation is between 5 and 10. This is about 30 times the *Ct* value for *Giardia lamblia* inactivation by ozone.

Investigations by Joret et al.[333] were reported at an IWEM seminar on "*Cryptosporidium* in Water," October 26, 1990, London, England. Their study produced the minimal and maximal *Ct* values for 2, 3, and 4 $\log_{10}$ reduction of infectivity at 7°C and 20°C by ozone as shown in the following table:

Inactivation (log_{10})	*7°C*	*20°C*
2	3.2–4.8	1.6–2.6
3	6.6–8.8	2.4–3.2
4	10.4–11.8	3.2–4.4

Langlais et al.[334] compared the inactivation levels of seven strains of free amoeba (trophozoites and cysts) by chlorine (HOCl) and ozone. It was found that both *Acanthamoeba* and *Naegleria* cysts show a high degree of resistance to free chlorine. A *Ct* value of 17 is required to achieve only 2 logs of inactivation, whereas ozone requires a *Ct* value of only 1.3 for the same level of inactivation. This is with an ozone residual of 0.4 mg/L and a contact time of 3 min. These studies involved bubbling ozone into a demand-free water to develop a 0.4 mg/L measured residual. Then 150 ml ozonated water samples were spiked with the oocysts and held for 4 min. *Ct* values of 3.22 or greater were found to achieve a 3–4 log reduction of oocyst infectivity.

All free living amoeba strains tested except *Acanthamoeba polyphaga* were sensitive to ozone disinfection. A 30-sec contact time with a 0.4 mg/L ozone residual was sufficient for 100 percent inactivation of these free living amoeba (*Ct* = 0.2). *Nagleria gruberi* echiroles and *Hartmanella vermiformis* were not included in this test.

For various other organisms subjected to ozonation, *Ct* values for 2-log inactivation ranged from 0.7 to 1.1. With *Ct* values of 17 to 34, free chlorine can inactivate 2 logs of *Nagleria* cysts. However, ozone can achieve the same inactivation with an ozone residual of 0.4 mg/L and 2 min. contact time. Ozone can also inactivate 2 logs of *Acanthamoeba* cysts with 0.4 mg/L residual in less than 3 min.

Chlorine can achieve a 2-log inactivation of trophozoites with *Ct* values of 0.7–1.5, whereas ozone can do the same with a 0.2 mg/L residual in only 30 sec.

Two of the critical problems in the evaluation of microorganisms classed as protozoans such as *Giardia* and *Cryptosporidium* are their accurate detection and proof of inactivation. This is why there is not necessarily close agreement on the conclusions of the various researchers. Le Chevalier et al.[335] made a comprehensive evaluation of this situation and concluded that the immunofluorescence (IFA) method was much preferred over the zinc sulfate flotation and Lugol's iodine method for the simultaneous detection of these two organisms. Although this method requires the use of an epifluorescence microscope, it uses monoclonal antibodies that are specific to both *Giardia* and *Cryptosporidium.* Commercial development of these antibodies has made them available to water utility laboratories. This parasite detection method is commonly known as the IFA-Percoll procedure. It not only detects both of these organisms but finds far more of them than do other procedures.

Current research by Le Chevalier et al.[336] is being focused on the morphology of the oocysts to determine the effects of disinfection. The primary objective is to determine whether or not the oocysts are potentially viable after disinfection.

Legionnaire's Disease

Since 1977 the Centers for Disease Control in Atlanta, Georgia, have investigated numerous outbreaks of this disease.[44] A clear mode of transmission has not been found. However, potable water as the basic source of the causative organism is strongly implicated in these and other investigations in the United States and abroad.

The name of the disease derives from a mysterious explosive outbreak of pneumonia following an American Legion Convention in Philadelphia, Pennsylvania in 1976.[45, 49] Pneumonia occurred primarily in persons attending the convention. Twenty-nine of the 182 cases were fatal. The causative pathogen was previously unknown in the annals of microbiology. The spread of the bacterium appeared to be airborne, but the source was not found. The epidemiologic analysis suggested that exposure may have occurred in the lobby of the headquarters hotel or in the area immediately surrounding the hotel.

Subsequent investigations of Legionnaire's disease at institutions have shown that multiple sites have yielded the causative agent, *Legionella pneumophila.* These sites include cooling towers, evaporative condensers, water heaters, faucets, and showerheads. Therefore, it is highly probable that the source in the American Legion outbreak was aerosols from the air-conditioning system.

The earliest documented outbreak of Legionnaire's disease occurred at St. Elizabeth's Hospital (Washington, D.C.) in July and August 1965.[46] It was thought that this outbreak was in some way related to the excavation of the hospital grounds for the installation of a lawn sprinkler system. It was found that patients whose beds were near windows in buildings close to the excavation and patients who had access to the excavation were the ones who became ill. No attempt was made to recover the disease agent from the soil, which would have been most difficult at that time. However, in 1977 stored serum-pairs were tested, and *L. pneumophila* was documented in 85 percent of the pairs. Since 1977, *L. pneumophila* has been discovered in mulch and soil.[46]

This disease is now known as legionellosis, which is the generic term for a group of diseases caused by the bacteria in the family Legionellaceae, genus *Legionella.* This family has one genus that contains at least six species: *L. pneumophila, L. dumoffii, L. bozemanii, L. gormanii, L. micdadei,* and *L. longbeachii.* Although the species are genetically and antigenetically distinct, they are morphologically and biochemically similar. *Legionella pneumophila* is the most commonly isolated species and has been found to have six serogroups.[46]

Legionellosis is currently recognized in two distinct forms: Legionnaire's disease, characterized by pneumonia with significant mortality rates; and Pontiac fever, which is a nonpneumonic, self-limiting febrile illness with a very high attack rate.[47] The incubation period for Legionnaire's disease is 2–20 days; for Pontiac fever it is only 5–66 hours.[48] Though the disease is respiratory in nature, person-to-person transmission has not been conclusively established. It is still not known why *L. pneumophila* causes two such different illnesses, or why the bacterium affects only 1 percent of the people exposed to it in the case of legionellosis, and as many as 95 percent in the case of Pontiac fever.[46]

It is paradoxical that an organism so fastidious in its nutritional requirements and so difficult to grow in the laboratory apparently persists and succeeds so well in an inanimate environment. The precise ecological niche and nutritional supply have not been defined.

Legionella pneumophila is now considered to be a common environmental contaminant. It may be ubiquitous. It has been isolated from many aquatic sites, including ponds, lakes, potable water, and equipment using potable water such as cooling water and evaporative condensers. *Legionella gormanii* has been isolated from a creek bank, and *L. micdadei* has been isolated from nebulizer water.

The identification of *Legionella* in 23 lakes in Georgia and 67 lakes (793 samples) in North Carolina, Florida, and Alabama[48] is cause for concern although legionellosis does not occur by ingestion of water. It apparently attacks via the airborne route, as a result of aerosolization, and most likely utilizes algae or amoebae as transport mechanisms. Investigations have also demonstrated that *L. penumophila* is able to colonize and grow in certain parts of water distribution systems. Warm water seems to enhance this ability. *Legionellae* have been found growing in certain species of free-living fresh-water amoebae and soil amoebae. These amoebic species have been isolated from humidifiers, and if they are inhabitants of slime in showerheads, this could explain the presence of large numbers of *Legionellae* in samples from such fixtures. It has been estimated that a person inhaling a single infected amoeba could receive a dose of up to 1000 *Legionella* cells.[48]

It can be reasonably concluded that potable water probably is no more than the transport vehicle of this pathogen. However, this can have serious implications because it tends to colonize and grow to high densities in components usually found in hotels and institutions. These components are storage tanks, cooling towers, evaporative condensers, showerheads, and water faucets.

Legionellae are susceptible to chlorination. Chlorination has been found to be effective at the 1–2 mg/L dosage level. Precise data for contact time and free residuals are not available. Institutions, hotels, and so on, should make it a practice to sterilize their water systems on a regular basis, similar to the recommended practice of main sterilization. This could be as frequently as once each six months.

Water suppliers whose systems have large holding reservoirs should be aware that *Legionellae* can work in such environments.

Worms

Since the disease schistosomiasis described above is carried by an organism sometimes referred to as a worm, it is pertinent to mention other worms found in water supplies. Free-living nematodes (NAIS), visible to the naked eye and resembling a tiny white worm, have been reported to occur in water supplies. In 1961 a survey of 22 cities indicated that these nematodes are fairly common in areas using surface waters. Although these nematodes are nonpathogenic, it is thought they might be able to ingest pathogenic organisms that could later transmit a waterborne disease.[50] Large numbers of one species, *Diplogaster medicapitus* of the Rhabitidae family, are present in the effluent of trickling filters and activated sludge process of sewage treatment and have been isolated in treated water supplies. They have been found to be capable of ingesting *Salmonella* and *Shigella* bacteria and Coxsackie virus. Of the ingested pathogens, 6–16 percent survived one day, and 0.1–1 percent survived two days. These nematodes resisted 2.5–3.0 mg/L chlorine residuals in contacts of up to two hours. It is unlikely that the nematodes could ingest pathogens at a sewage plant and be carried by the polluted water through the water system in one day; but even though the chance is remote, the threat exists.

Although the problem of nematodes is primarily aesthetic, they can impart a strong earthy or musty odor to a water supply if allowed to propagate within the treatment plant. When the finished water shows more than ten worms per gallon, it should be investigated.

The most practical approach to preventing a nematode infestation is to prechlorinate the raw water to a free residual of 0.5 mg/L and six hours of contact time. Although many of the nematodes will not be killed, they will be sufficiently affected so that they cannot survive and will therefore be settled out in the flocculation process.[50]

Tularemia

This disease of rodents, humans, and some domestic animals derives its name from Tulare County, California, where it was discovered. In humans it produces an irregular fever that continues for several weeks and results in moderate to severe debilitation.

The usual means of transmission is by direct contact although the causative organism *Pasteurella tularensis* has been isolated in certain streams in Montana that are sources of water supplies.[51] These organisms are easily destroyed by chlorination.

Acute Gastrointestinal Illnesses (AGI)

All reported outbreaks of unknown etiology are classified by the CDC and EPA as AGI. In the period 1971 to 1977 a total of 192 outbreaks of waterborne disease affecting 36,757 persons was reported in the United States. There was no etiology in 57 percent of the outbreaks and 58 percent of the illnesses.[68] These illnesses are therefore listed as AGI. Before the cooperative effort between the CDC and EPA to document, investigate, and report waterborne outbreaks in 1971, AGI outbreaks were not always documented. For example, in the 10-year period 1961–70, of 130 waterborne outbreaks affecting 46,374 people, 30 percent were classified as a gastrointestinal outbreak of unknown origin.[20, 69] In 1981, 44 percent of the outbreaks, affecting 43 percent of the cases, were classified as AGI in the CDC Annual Summary.

Recently Described Conditions with Conspicuous GI Symptoms

At a 1983 gastroenterology seminar held in London,[297] several important new microbial agents and infective conditions were discussed as to the clinical features, pathogenesis, epidemiology, occurrence, and infection site. These syndromes are listed in Table 6-2.

Viral Diseases

General Background. Not all viral diseases are waterborne. The human enteric viruses (poliomyelitis, Coxsackie, and echo) do not produce their associated diseases via the waterborne route.[52] It is also possible to have these viruses in the bloodstream without having the disease. These facts appear to contradict the nature of the diseases involved, which complicates our understanding of viral diseases, especially in potable water supplies.

Table 6-2 Recently Described Conditions with Prominent Gastrointestinal Symptoms

Syndrome	*Causative Organisms*	*Site of Infection*	*Main Gut Symptoms*
Toxic shock syndrome	*Staphylococcus aureus*—enterotoxin F-producing strains	Vaginal and other sites	Diarrhea
Legionnaire's disease	*Legionella pneumophila*	Lower respiratory tract	Diarrhea and vomiting: abdominal distension
Traveller's diarrhea	*E. coli*—enterotoxigenic strains; *Campylobacter* spp.	Gut	Diarrhea
Gay-bowel syndrome	*Giardia lamblia* *Entamoeba histolytica*	Gut	Diarrhea
Infant botulism	*Clostridium botulinum*	Gut (?)	Constipation

Adenovirus, which is not an enteric virus, has been reported to have caused a disease outbreak in a water supply.[53]

The viruses of known concern to potable water producers are: hepatitis A, Norwalk, and rotavirus. It has been said that the latter two are responsible for about 77 percent of acute waterborne gastroenteritis.[54] These three viruses can and do travel the waterborne route. It is claimed by some investigators that rotavirus can survive modern wastewater treatment processes. These viruses travel the fecal–oral route.

Coping with Viruses in Potable Water. It is the consensus of virologists that the raw water of all surface water supplies in the United States and Canada contains naturally occurring human enteroviruses. These investigators and public health agencies believe that it is necessary to adopt the multiple barrier approach in order to protect the public from sickness attributable to viruses.

A comprehensive study by O'Connor et al., published in 1982,[55] concluded that virus removal and inactivation are essentially complete when modern treatment practices are followed. This means the use of prechlorination, flocculation, coagulation, sedimentation, and filtration followed by terminal disinfection. Prechlorination will enhance the reliability of the treatment plant to remove and inactivate the viruses because of the extended contact time through the treatment processes. If the coagulation step is to be effective, the water must have a reasonable amount of turbidity.

In those instances where riverbank filtration is used to simulate slow-sand filtration, flooding the banks will nullify the purpose of this treatment step and cause a rapid increase in the virus concentration in the treatment plant.

Another situation that causes virus population increase is sludge blanket breakthrough in the sedimentation step.

The effect of chlorination system failures is well known and need not be discussed here.

Another promising method of virus removal is storage. For this method to be effective, the water to be stored must be biologically active. Virus removal by storage can easily achieve levels of one plaque-forming-unit.[53] Storage is reliable if there is no short-circuiting. The use of lagoons has been studied in Israel for many years. Three log reductions of viruses in wastewater after 28 days storage have been observed.[56]

Consensus Organism Indicator. It is appropriate here to consider how important it would be to have established the *Ct* factor of a "consensus" organism, which would be used so that pathogenic bacteria, cysts, viruses, protozoans and other waterborne disease producing organisms would automatically be inactivated, by the applied chlorine. Now (1997) with all the information that has been collected in the last seven years about waterborne diseases, it does not seem possible at this time to isolate an organism that would serve that purpose.

Moreover, the problem is compounded by the fact that there is no observed relation between the microbial indicators, such as total coliforms, fecal coliforms, or standard plate counts, and the presence of viruses or other organisms. Also there is no observed relation between the detection of *Salmonellae* and viruses and vice versa. Nor can the enteroviruses or coliphages be used as virus indicators with our present knowledge of the organisms.

For example, a potable water can be contaminated by a small source that introduces viruses but no fecal indicators such as total or fecal coliforms. Likewise, if there is fecal contamination from a populace that is not "shedding" a virus, there will be fecal indicators but no virus. All of this shows why an indicator organism might be useful.

White believes that in order to eliminate the public health risk as much as possible in our potable water supplies, large or small, the use of an indicator organism is warranted—eg, implementation of the *Ct* value of the Cocksackie, AZ virus as established by Clark and Kabler[180] and Bauman and Ludwig[175, 185] (see Figures 6-14, 6-15, and 6-16).

Adenoviruses. These viruses, originally isolated from tissue culture of human adenoids, are responsible for febrile pharyngitis, swimming pool conjunctivitis, and acute respiratory catarrh. They are not enteric viruses but apparently can be waterborne, as they have been detected in surface waters. There are some 28 serologic types of this group of viruses. Types 4 and 7 are found in most cases of acute respiratory disease. Types 1, 2, and 5 are frequently associated with febrile respiratory infections in young children. Types 6 and 10 cause conjunctivitis, which can be transmitted in swimming pools and possibly other surface waters. They are common in recruit camps where attack rates have been reported as high as 70 percent. They are also seen in children's institutions due to the crowding together of susceptible young hosts.[57] Although these viruses are resistant to antibiotics,[41] they are the most susceptible to chlorination of all the viruses normally detected in surface waters.

Enteroviruses. These viruses derive their name from the fact that the alimentary tract forms a reservoir where they may be found. They enter the body through the mouth. Infection is spread mainly by the fecal–oral route because of poor hygiene, poor sanitation, or a poor-quality water supply. These viruses cause the greatest concern in water supply practice because of the severity of the diseases they carry as well as their resistance to disinfection. The enteric viruses of least concern to the water producer are the poliomyelitis, coxsackie, and echoviruses, for reasons described above. These are not considered to be waterborne, in spite of the fact that they are excreted in the feces of infected patients.

Poliovirus. This virus is the cause of what used to be a common infectious disease of the central nervous system known as infantile paralysis, or poliomyelitis. The virus was thought at one time to have been transmitted by water supplies because it is readily detected in the feces of infected patients. However, a great many studies have failed to confirm this theory. The introduction of mass vaccination in the United States about 1960 has reduced the occurrence of poliomyelitis to something less than 5 cases per 100,000 per year. It appears that permanent immunity may be possible, provided that booster doses are taken at specified intervals. Therefore, it is practically impossible to prove whether or not the disease is waterborne without the occurrence of an epidemic.

Coxsackie. This virus is thus named because it was first isolated in the potable water supply at Coxsackie, New York. There are many types of this virus, which cause Bornholm disease (named after a Danish island in the Baltic Sea, where there was an epidemic of pleurodynia, an inflammation of fibrous tissue causing acute pain in the side), herpangina (a contagious disease of children characterized by fever, headache, and vesicular eruption in the throat), and epidemic pharyngitis. Some types of Coxsackie viruses are the most resistant of all viruses to chlorination.

Echoviruses (Enteric, Cytopathogenic, Human Orphan). These viruses are so named because on first isolation they appeared to be associated with no known human infection. They have since been shown to cause acute respiratory symptoms, enteritis, and meningitis.[58] There are more than 20 different types.

Infectious Hepatitis (Hepatitis A). The etiological agent is a virus whose characteristics are not yet completely known. Another type, serum hepatitis, is clinically indistinguishable from infectious hepatitis, although the virus is different. Infectious hepatitis is an acute viral inflammation of the liver characterized by fever, malaise, gastrointestinal symptoms (nausea and vomiting), and jaundice. The fact that infectious hepatitis can be waterborne is well documented. Its incidence has been rising to such an alarming extent since about 1950 that it is recognized as a disease of considerable environmental significance and one that is of great concern in water supply practice. The most explosive and devastating waterborne outbreak of infectious hepatitis occurred in New Delhi, India, in 1955–56, resulting in more than 50,000 cases.[59] The raw water was known to have been contaminated by domestic sewage, but was treated by conventional methods. The failure of the treatment process was traced to an inadequate chlorination facility.

Viral Gastroenteritis

AGI. This is the label for acute gastrointestinal illness of unknown etiology. CDC reported 4430 cases of waterborne illness from 32 outbreaks for the calendar year 1981. AGI accounted for 1893 cases from 14 outbreaks, with Rotavirus accounting for 1761 cases from only one outbreak. It is speculated that the majority of AGI cases are the result of Norwalk-like viruses.

Norwalk Virus. This virus has been associated with epidemics of acute gastrointestinal illness occurring in families, schools, and communities.[60–62] It is regarded as a common pathogen of older children and adults in the United States. This is an interesting observation, because the original outbreak caused by this virus occurred in an elementary school in Norwalk, Ohio, October 30–31, 1968. Its role in pediatric diarrhea has not been thoroughly investigated. The incubation period is 18–48 hours, and symptoms last from 24 to 48 hours. Usual symptoms are abdominal cramps, headache, malaise, low grade fever, nausea, and either vomiting or diarrhea or both. This sickness has often been characterized as 24-hour flu.[60, 62, 63]

Some 25 separate outbreaks of nonbacterial gastrointestinal illnesses were studied serologically for evidence of infection with the Norwalk virus and the rotaviruses. Eight of the 25 appeared to be related to the Norwalk virus. In one of the 25 there was evidence of Rotavirus infection. These observations suggest that the Norwalk virus or serologically related agents play an important role in epidemic nonbacterial gastroenteritis in adults and older children.[62, 63]

The outbreaks studied occurred in communities, cruise ships, schools, summer camps, and colleges. The 1968 Norwalk outbreak was studied by means of stool swabs. These swabs failed to detect *Salmonella, Shigella,* enteropathogenic *E. coli, Staphylococcus aureaus, Aeromonas,* noncholera *Vibrio,* or enterococci. The attack rate was 50 percent (116 of 232) of students and teachers. There were several schools in the town, and food for lunches was distributed to all schools. There was no illness at any of the other schools. It was concluded that the source was the water well at the affected school, which was the only school that had its own well.[64]

In August 1980 there was a potable-water-related outbreak of acute gastrointestinal illness in the community of Lindale, Georgia.[65] The attack rate was determined in 10 neighborhoods. This rate increased significantly near a textile plant where there was known to be a cross-connection between an industrial water supply system (which contained fecal coliforms) and the community potable water system. The infectious agent was thought to be waterborne. There is ever-increasing evidence that acute gastrointestinal illness is indeed waterborne and one of the more important infectious agents is Norwalk virus.[66]

Rotavirus. This virus is known to cause both sporadic and epidemic cases of gastroenteritis in infants and young children throughout the world. Approximately one-half the infants and young children hospitalized with diarrhea are infected with rotavirus. In temperate climates the peak prevalence is in the winter months. Infection is uncommon in the summer. Transmission is thought to be by the fecal–oral route. Diarrhea of 5–8 days' duration is the main feature of rotaviral disease. It is frequently accompanied by vomiting and fever. Death due to this virus is rare but has been reported. Rotaviruses have been observed in diarrheal feces from the young of numerous mammalian species.[60] Waterborne rotavirus infection has been reported in Europe.[57, 67] This virus is not believed to be so important as the Norwalk virus as the cause of acute gastrointestinal illness (AGI).

Summary of Waterborne Diseases

Examination of the statistics tends to indicate that waterborne outbreaks are on the upswing. This may be due to better reporting, or it may be that some of the barriers are being overstressed. Public health agencies are deeply committed to the theory of multiple barriers or multiple points of control between sewage discharges and water supply intake. These barriers or points of control include wastewater treatment, land confinement, dilution, time, distance, and potable water treatment. Any type of treatment is fallable; so reliance on natural barriers should be maintained as long as possible. However, it is obvious that disinfection of potable water is the last line of defense.

The Safe Drinking Water Act of December 1974 resulted in improved surveillance of drinking water supplies. The provisions of this act apply to all water systems serving 15 connections or at least 25 individuals for a minimum duration of 60 days. This act was followed by the National Interim Primary Drinking Water Regulations created by the EPA in 1975 and implemented in 1977. These regulations established maximum contaminant levels for specified microbiological and chemical contamination, as well as turbidity limits and monitoring frequency.

A perusal of the annual reports by Gunther F. Craun in the *J. WPCF* June issue (literature review), under the heading "Disease Outbreaks Caused by Drinking Water," 1971–80, reveals some interesting data about the supplies involved in these outbreaks.

The supplies that suffered the most outbreaks were the noncommunity supplies such as cruise ships, resorts, summer camps, boarding schools, and other small private supplies. In one year these supplies accounted for as much as 75 percent of the outbreaks. However, it is the community supplies that account for the greatest number of cases, about 85 percent of cases each year. The traveling public suffers greatly from waterborne illnesses, particularly in Mexico.

The causes of these outbreaks fall into the following categories: contaminated ground water, 65 percent of outbreaks and 63 percent of cases; inadequate or interrupted chlorination, 31 percent of outbreaks and 44 percent of cases. In some years contamination of the municipal distribution system causes as much as 70 percent of cases.

It remains to be seen whether or not waterborne diseases are on the upswing. It may be that the reporting has become more comprehensive.

DISINFECTION SYSTEM INVESTIGATION

Introduction

Ammonia N and Organic N. The concentration of these two items are the first to be analyzed. Their values will be of great help in deciding if they may be causing any interference in the analytical measurements of the chlorine residuals.

Chlorine Demand. This value is most important for the system operator to obtain. It is vital to the final determination of the problem, if any. First, use the following list of dosages; 1, 2, 3, 4, and 5 mg/L. Then analyze the residuals of these dosages at the end of 10, 20, 30, and 60 min. by either the complete forward amperometric titration method or the DPD-FAS method.

If there is 0.1 mg/L or less of ammonia N, 0.1 mg/L monochloramine in the residual, and any dichloramine residual, this indicates that the dichloramine residual is really a nongermicidal organochloramine, owing to the organic N content in the water sample.

All of the above serves to orient the operator in the proper way to handle the chlorination system.

The Chlorine demand × Contact Time = *Ct* Factor

This is the most important variable for determining or predicting the germicidal efficiency of any disinfectant.[128] It has been used most successfully by the wastewater dischargers in California since the 1970s. The chlorine demand (C) of the water or wastewater to be examined is the most important part of this factor. The lab technition should determine the C demand for a variety of contact (t) times to evaluate which will be the most suitable for the particular water being tested, as shown above.

Compared to potable water disinfection, the burden of proof for the wastewater dischargers is much easier because one group of organisms is required to be inactivated: total coliforms.

Potable water may be contaminated by a variety of pathogenic organisms. These include, but are not limited to, enteric viruses, protozoans, and cysts that could be responsible for waterborne disease outbreaks. Owing to this dilemma, it is extremely difficult to choose a "consensus" organism. Such an organism would have to be easily identifiable, and its inactivation rate would have to be similar or equal to that of the most resistant organism that causes a waterborne disease outbreak.

Ct Values by EPA for *Giardia* and Viruses

Currently the EPA has developed *Ct* values for *Giardia* cysts and enteric viruses. Given that a *Ct* value has been established for a series of organisms known to cause waterborne outbreaks, there are other factors that affect the *Ct* factor, as follows:

1. The chlorine species to be found in the residual *C* is important. There is a great deal of difference between the disinfecting power of free chlorine and that of monochloramine. Therefore, the *Ct* value has to be determined on the basis of a given species. If organic N is present in the raw water, then the nongermicidal organochloramine residual must be accounted for. It appears in the dichloramine fraction. If both monochloramine and dichloramine appear in the total residual, and if the water contains organic N, the monochloramine will slowly hydrolyze with the organic N to form organochloramines. This is a site-specific situation that is directly related to contact time.*

2. The contact time, *t,* is as important as *C.* The method of determining contact that provides the greatest factor of safety is to use a tracer. Apply the tracer at the proposed point of disinfectant application, and record the time required for the first appearance of the tracer at the end of the contact chamber. This is standard practice in California for wastewater dischargers. The tracer used is Rhodamine B dye.

Another common method is to provide a chlorine spike at the point of application, and time the appearance of the spike at the analyzer sampling the discharge from the contact chamber.

3. Water temperature has a significant effect on the inactivation of these organisms. The lower the temperature is, the lower the disinfection efficiency. When Hibler developed the *Ct* values for *Giardia* cysts at 0.5, 2.5, and 5.0°C, he recommended that the values for 5.0°C be used for warmer waters simply to provide a greater safety factor.[328]

4. pH is a consideration. When free chlorine is required, it is essential to maintain the water being disinfected at a reasonably steady pH between 7.0

*The types of chloramines were not identified.

and 7.5. In potable water treatment this range has been found to be the most effective for disinfection by free chlorine.

5. The effect of sunlight is a factor. When free chlorine is required, and if the contact chamber is exposed to sunlight, there will be a significant loss of chlorine residual from the sunlight (UV rays). This phenomenon varies with both latitude and altitude as well as cloud cover. Therefore, residual control is imperative. There also has to be a programmed dosage change to account for the lack of UV rays after sundown.

6. The contact chamber design is important. If the contact chamber is going to be an open channel type, it must be designed to achieve 85–90 percent plug flow conditions (see Chapter 8). This minimizes short-circuiting. The preferred design, particularly when free residual chlorination is being practiced, is a closed conduit with a slight surcharge. Closed pipes flowing full can easily achieve 97 percent plug flow conditions and, at the same time, eliminate loss of residual due to sunlight.

7. Adequate mixing at the point of chlorine application is as desirable as plug flow conditions in the contact chamber. (Chapter 8 discusses this subject in detail.)

Ionics Water Quality Monitoring Guide[504]

Introduction. In 1993 Ionics published a booklet that is a guide to the various on-line methods of monitoring water quality. This guidebook is most important for anyone dealing with water quality. Because Ionics makes a chlorine generating system (described in Chapter 3), this booklet should be of interest to readers of this chapter.

As there are numerous methods of measuring and/or monitoring contaminants in potable water, and as new methods—along with new analytical instruments—are appearing frequently, Ionics is making every effort to provide updated information, which is being added to the guidebook.[504]

Available Technologies. The following is a brief listing of the contents of the Ionics guide:

1. Monitoring specifications.
2. Twenty-four discharger points of Eastman Kodak Co.
3. ICI Digi-Chem Controls Reaction End Point.
4. Nuclear resin destruction.
5. Sample handling considerations.
6. High organic monitoring.
7. Analytical methods for carbon and oxygen demand.
8. Biochemical oxygen demand.
9. Reliable "Ionics Policeman."
10. Demand analysis for total carbon, total organic carbon, and total oxygen.
11. Chemical oxygen demand analysis.

12. Oxidizing efficiency using the dichromate method.
13. On-line chemical analyses.
14. Colorimetric methods.
15. Selective ion electrodes.
16. Analyses for dissolved metals.
17. Programmable stress sequences.
18. Customer support capabilities by the Ionics Instrument Division.
19. Shelters for analyzer systems.

EARLY DEVELOPMENT OF CHLORINATION EQUIPMENT

Historical Background

The next important development in the progress of chlorination was the invention of apparatus that could use chlorine gas directly from a cylinder. Until about 1910 the only successful practical method was using hypochlorite solution. The one successful electrolytic installation at Brewster, New York, which operated continuously for 18 years on sewage effluent, was replaced by a hypochlorite installation after the original one was destroyed by fire in 1911. At best the hypochlorite method was cumbersome, extremely awkward, and messy to handle and apply. In June 1910, C. R. Darnall, a major in the Medical Corps, U.S. Army, began to experiment with liquid molecular chlorine compressed in steel cylinders.[17] The cylinders, which were not then available in the United States, had to be imported from Germany. Major Darnall not only made the first practical chlorinator using chlorine gas,[18] but also published the first work on the bactericidal efficiency of elemental chlorine dissolved in water.[17, 70] Darnall's apparatus (U.S. Patent 1,007,647), illustrated in Fig. 6-2, consisted of a pressure-reducing mechanism, a metering device, and an

Fig. 6-2 (facing page). Major Darnall's chlorinator. When valve on chlorine cylinder is opened the chlorine gas passes through piping and control valve to the receiver 20 in the diaphragm-type chlorine pressure-reducing valve and simultaneously through the back-pressure assembly 34. Because of the pressure buildup in 34, the pressure in 20 rises and lifts the metallic diaphragm 21, which operates the lever assembly 30–32 so as to close the chlorine control valve 14. As the gas passes through chamber 34, the pressure in chamber 20 falls and the lever assembly 30–32 operates to open valve 14, thus permitting more gas to pass.

The quantity of chlorine may be increased or decreased by shifting weight 17 on lever 16. Once adjusted, the chlorine flow is presumed uniform. The water flow is maintained constant by means of the equalizing tank 13.

Supposedly this system was meant to be automatic start-and-stop based on opening and closing valve 4 as follows:

If valve 4 is closed, the flow of water out of the tank will stop and presumably raise the pressure in pipe 34 via pipe 36, thereby reacting on the chlorine diaphragm assembly 21, causing the flow of chlorine through valve 14 to shut off.

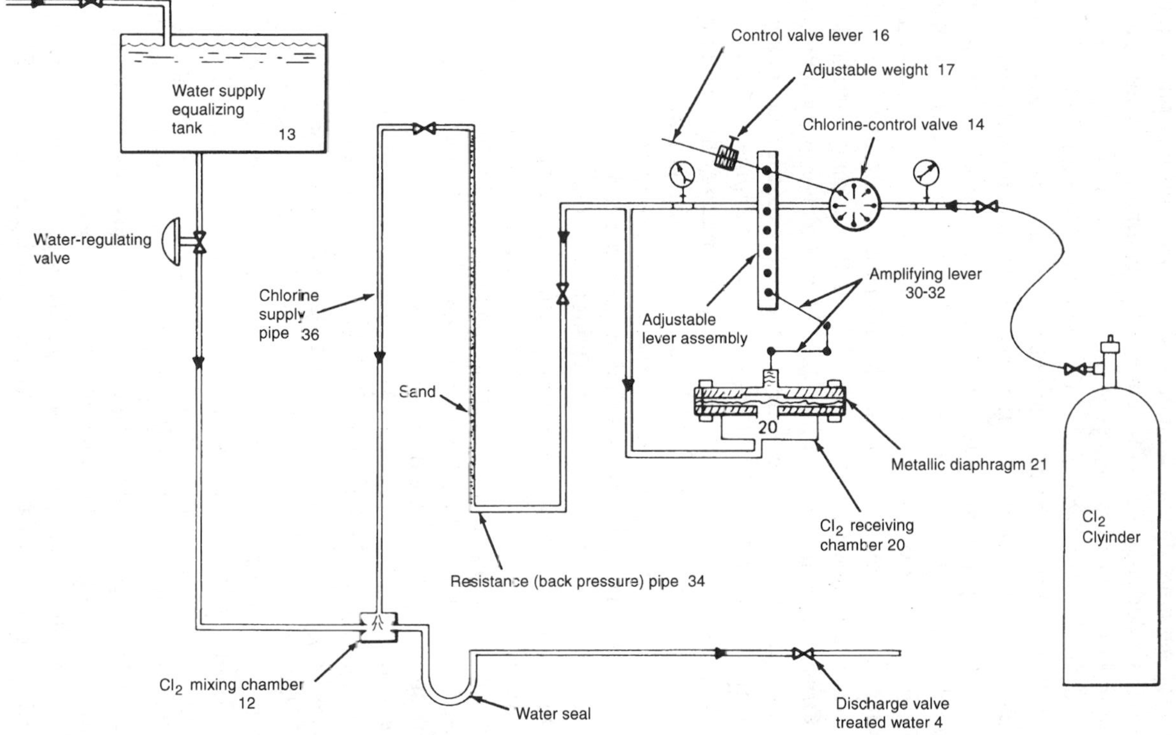

Fig. 6-2. Major Darnall's chlorinator (see page 370 for complete figure legend).

absorption chamber. Only one of these units was ever installed—at Youngstown, Ohio.

Although Darnall never entered into the equipment field, he intended to set up license agreements with communities to install his equipment for waste disinfection. Under such an arrangement, the community would pay him a royalty based on per capita, per year, or total amount of water treated. He approached the Electro Bleach Gas Company of Niagara Falls, New York, which in 1909 was one of the first commercial producers of liquid chlorine in the United States, and proposed that all the liquid chlorine used in his installations be supplied by the company, and that, in return, the company should agree to sell liquid chlorine for this purpose to no one else. E. D. Kingsley, president and founder of the company, agreed, provided that Darnall furnish the chlorine feeding device in working form within six months. When Darnall failed to do so, Kingsley assigned the job of developing such a device to George Ornstein, then a consulting chemist to the company, employed to develop applications for liquid chlorine in bleaching textiles and paper pulp.[38]

Meanwhile, in December 1912, John Kienle, chief engineer of the Wilmington, Delaware, water department, began experimenting with the application of chlorine gas from cylinders.[20] He fed the chlorine gas to an absorption tower, through which a counterflow of water absorbed the chlorine, which was then applied as an aqueous solution. A balanced, weighted pressure-reducing valve was connected to the cylinder for chlorine feed-rate control but became inoperative and so was discarded; only the valve on the cylinder was available for control. Kienle recommended to his water board that installation of such chlorination equipment be made, but only by a manufacturer that agreed to demonstrate efficiency of control for 30 days and guarantee the apparatus against deterioration for one year.

Following Kienle's experiments at Wilmington, Ornstein developed a chlorine control apparatus (U.S. Patent 1,142,361), which consisted of a high- and a low-pressure gauge, the latter being calibrated to indicate the flow of gas through a fixed orifice. The metered gas was then directed into a stoneware absorption tower, where it became dissolved in the water flowing by gravity out of the bottom of the tower as an aqueous chlorine solution. This unit was first tested on a plant scale in November 1912, at the Western New York Water Company plant at Niagara Falls. Under the direction of Harry F. Huy, the bacteriological results using this apparatus were documented and compared with the chloride of lime method.[38] The results exceeded expectations. This experimental apparatus was greatly improved, and on February 1, 1913, the Ornstein chlorinator, as manufactured by the Electro Bleach Gas Company, was put into operation at Kienle's Wilmington, Delaware, plant. This is considered the first commercial solution feed chlorinator installation in the United States.[18, 38] John Kienle shortly thereafter became associated with the Electro Bleach Gas Company to manage the sales of its chlorine feeding equipment. Under this guidance, installations of this equipment were made in both potable water and wastewater plants, as far west as Salt Lake City, Utah; Monterey, California; and Seattle, Washington.

About this time, Wallace & Tiernan entered the field. The firm set about making a gas chlorinator for an installation at Dover, Delaware, on the Rockaway River, a tributary that fed the Boonton, New Jersey, reservoir. This, its first direct-feed gas chlorinator, was installed February 22, 1913.[18] This installation included a gas pressure gauge, a temperature–pressure compensating device, and an ingenious inverted glass siphon for measuring low rates of chlorine gas flow, which is still in use today in modern equipment. The chlorine gas was applied directly to the water being treated, where it was dissolved by means of porous ceramic diffusers. This new chlorinator was simple and dependable in operation, and so it soon gained popular acceptance. Late in 1913, Wallace & Tiernan began developing solution-feed chlorinators, which proved more reliable than the direct-feed units. The direct-feed gas diffusers were plagued by cold weather. Within a year, Wallace & Tiernan began to dominate the field. By September 1914, it had 23 chlorinators installed in 18 different cities; in 1916 it installed nine direct-feed units in Croton, New York aqueducts. In 1917, Wallace & Tiernan became sole licensee of the Ornstein patent, and the Electro Bleach Gas Company discontinued marketing its liquid chlorine feeding equipment. By 1918, Wallace & Tiernan equipment was installed as far west as California; by 1920, the company was represented by sales and service facilities in major cities throughout the United States; and in 1925 the British company was established. After World War II the company was represented around the world; its impact on the water works industry was extraordinary. In the late 1920s, when it was confirmed that a combination of chlorine and ammonia could eliminate many tastes and odors due to chlorination, Wallace & Tiernan promptly supplied the appropriate metering and control equipment to handle anhydrous ammonia.

In the years that followed, the name, "Wallace & Tiernan" became synonymous with "chlorination." The company was responsible for all the significant developments in chlorination apparatus. Among these achievements were: the unique vacuum solution-feed chlorinator utilizing the aspirator-type injector; the development of the visible vacuum principle of equipment (ca. 1920), which resulted in an industrywide change in the manufacturing of chlorine gas to produce a cleaner, less troublesome product. Figure 6-3 is an illustration of the bell-jar chlorinator, invented by C. F. Wallace. This particular model (A-255) was the first automatic flow-paced solution-feed chlorinator (ca. 1925).

This was followed by the development of the amperometric titrator in 1942 by Henry Marks of Wallace & Tiernan. This invention provided the analytical techniques required for the control of the free residual process in accordance with the breakpoint phenomenon. Shortly thereafter the continuous amperometric chlorine residual analyzer was made available (ca. (1950), followed in 1953 by apparatus capable of automatic residual control. Other significant developments were: the first chlorine leak detector (ca. 1955) and the

Fig. 6-3. Flow paced automatic bell-jar chlorinator Wallace & Tiernan Series A-255 (courtesy Wallace & Tiernan).

compound-loop principle of automatic chlorine residual control (U.S. Patent 2,929,393), introduced in 1960.

In the early days of chlorination (1920s) Wallace & Tiernan devoted a considerable amount of time to residual control research. Systems were developed that could continuously control to a specific O-T residual. The first of these was installed at Little Falls, New Jersey, in 1929; the second at Rahway, New Jersey, in 1930.[71] A modification of these units was installed in 1932 and 1934 on the Los Angeles System.[72] The Lower San Fernando installation is shown in Fig. 6-4.

Advances in Equipment

During these early years the most successful competitors of Wallace & Tiernan were Pardee and Everson Manufacturing. Wallace & Tiernan controlled about

Fig. 6-4. Colorimetric chlorine residual controller lower San Fernando Reservoir Los Angeles Department of Water and Power.

90 percent of the chlorinator business until Fischer and Porter entered the field in early 1950s. At that time plastic pipe and fittings were beginning to replace rubber hose and rubber-lined pipe for chlorine solutions. The use of injection-molded plastic components transformed the manufacturing concepts of chlorinators overnight. Fischer and Porter capitalized on these developments. They displayed a reasonably comprehensive line of cabinet-style molded plastic chlorinators in the 1000–2000 lb/day capacity range. Wallace & Tiernan bell-jar chlorinators were dedicated to using expensive molded hard rubber parts along with other expensive parts made with silver and tantalum. Their only resource was to abandon the bell-jar line and follow the path of Fischer and Porter, which proved prudent and successful. Some years later one of the key chlorinator design engineers left Fischer and Porter to start his own chlorinator company, which is now known as Capital Controls. It was estimated that these three companies shared 97 percent of the chlorinator business in 1983.

The most significant advances in equipment have resulted in wider range of chlorine flow control, a greater variety and sophistication of automatic controls, and the development of reliable continuous chlorine residual measuring equipment. In 1970, Fischer and Porter developed an analyzer that can measure free chlorine alone in the presence of significant amounts of combined chlorine.[300] Analyzers that measure total chlorine residual have been available since 1950, and have been greatly improved for better reliability.

Advances in safety include remote vacuum chlorine supply systems. This allows the gas from the container to travel to the chlorination equipment under a vacuum, thereby minimizing opportunities for leaks. Development of the remote injector concept has greatly simplified hydraulic problems and piping configurations in chlorinator installations.

Installations are designed to be compatible with strict safety procedures, which include use of container safety kits, leak detectors, and air and nitrogen purge systems of chlorine headers. Safety drills with breathing apparatus and emergency kits are routine exercises at many water departments.

DEVELOPMENT OF CHLORINATION PRACTICES

Historical Background

It was found that in addition to destroying disease-producing organisms chlorine can also destroy the nuisance organisms that abound in the water supply: those that cause taste and odor, foul the filter media, and degrade the quality of water in the distribution system. Although one of the primary objections to chlorine was that it was enforced medication, a more obvious objection soon rallied many more enemies of chlorination—the chlorinous taste, a by-product of early disinfection practices. Initially chlorination for disinfection was administered on the basis of total dosage; little or no account was taken of water quality or the reactions from the added chlorine. A determination of the effectiveness of the chlorine added was not possible for at least 24 hours, when bacteriologic test results were available.

In 1919, Wolman and Enslow[73] made a most significant contribution to chlorination practice by demonstrating that chlorine absorption could vary widely from water to water. Thus began the first concept of chlorine demand.*

In the same year Alexander Houston reported to the Metropolitan Water Board of London that variable dosages of chlorine will destroy tastes and odors and, moreover, that, the more the chlorine dose is increased, the more certain will be the absence of tastes following dechlorination. These discoveries encouraged an examination of the effects of different chlorine dosages. Although there were those who either had no problems resulting from the use of chlorine as a disinfectant or chose to ignore them, there were others who found that they could produce a more palatable water with much higher dosages of chlorine.

In 1926 Howard and Thompson[74] reported on an exhaustive study begun in 1922 on taste and odor control of the Toronto water supply by means of what they termed superchlorination followed by dechlorination. Experiments at Cleveland[75] also tried heavy doses of chlorine to eliminate tastes resulting from industrial waste discharges into the receiving waters of the Cleveland

*Chlorine demand is defined as the difference between the amount of the chlorine added to the water and the amount of chlorine (free available or combined available) remaining at the end of a specified contact period.

water supply. The most objectionable of these tastes derived from phenols, the same compounds that have plagued most of the water supplies of Europe, particularly those from the Seine and the Marne in France and the Rhine and the Ruhr in Germany.

Cox[76] reported in 1926 that heavy doses of chlorine could control algae, tastes, and odors in the New York City water supply system. Cox called this method double chlorination, consisting of splitting the chlorine solution discharge from one chlorinator to two points of application instead of one. Use of the heavier dose of chlorine for taste and odor control was ahead of all other treatment. The second point of application was postchlorination for disinfection.

At about the same time that Cox and Braidech were experimenting with heavy doses of chlorine, McAmis introduced a new approach to chlorinous taste and odor control: the use of ammonia along with chlorine.[77] The success of the ammonia–chlorine process was so extensive that it practically replaced plain chlorination. This process enjoyed its greatest popularity from about 1930 until 1942. The reason for its decline was twofold: the discovery by Griffin in 1939 of the breakpoint phenomenon, which provided a means for more reliable chlorinous taste and odor control; and the shortage of ammonia nitrogen due to U.S. participation in World War II.

The discovery of the breakpoint phenomenon marked the beginning of the free residual chlorination process. It also sparked a great deal of interest in the kinetics of chlorination and the discovery of the various chlorine residual fractions and analytical methods for their differentiation. It also revived interest in the concept of superchlorination (to a free residual) followed by dechlorination. However, conventional water treatment plants continued to be designed with a variety of locations for the addition of chlorine. In addition to provisions for chlorinating the raw water and the finished water, there was usually a point of application ahead of the filters and another at the entrance of the sedimentation basins. Multiple chlorine application points usually contribute to an overcomplicated chlorine solution piping system, which requires elaborate and expensive flow splitting devices.

USERS OF CHLORINATION AND DECHLORINATION PROCESSES

Contemporary Practices

All editions of this book relate largely to water and wastewater systems that use more than 10 lb of chlorine per day. These systems use 5–6 percent of the total chlorine production in the United States.

There are hundreds of chlorination systems in North America and maybe thousands around the globe that use from 1 to 5 lb of chlorine per day for treating potable water. Obviously that amount of chlorine cannot disinfect wastewater. On the average, secondary wastewater effluents require at least five times more chlorine to achieve proper disinfection than the amount required for potable water.

On-site generation of chlorine is always gaining in popularity around the world for a variety of reasons. In the United States, more and more on-site equipment keeps appearing in the advertising publications. These units cover all the sizes from 1 to 100 lb/day. If more than 100 lb/day is required, the supplier simply adds more units (more electrolyte cells). See Figs. 3–13 and 3–14, Chapter 3.

The Chlorination–Dechlorination Concept

System Description. When reliable chlorine residual controllers became available in the 1960s, the superchlorination–dechlorination approach was an attractive method for controlling tastes, odors, and sedimentation tank and filter biofouling, as well as providing reliable high-quality water in the distribution system. The term "superchlorination," used in the early days, has been replaced by "free residual chlorination."

It is extremely important to point out that a total residual measurement must contain at least 85 percent HOCl in order to be identified as a "free residual." The EPA never mentions this when listing the chlorine residual species in its MCLs (maximum contaminant levels). Using the chlor–dechlor method can achieve the modern approach to the various objectives of chlorination described in the preceding text. The system must be able to apply enough chlorine at the influent of the treatment process to maintain an adequate free chlorine residual into the treated water storage. Then at this point the residual should be controlled automatically by dechlorination.[293] This is in lieu of manually adjusting the prechlorination dosage so that it does not interfere with the postchlorination dosage. Experience has shown that the chlorination process has suffered unnecessarily in a great many instances because the operators were unable, by reason of design, to achieve proper flexibility of application. First, if the plant experiences seasonal problems with tastes and odors, the operator is reluctant to increase the dosage to achieve control of these tastes and odors for fear of having too high a residual entering the distribution system. Second, the operator knows that when he or she has achieved a substantial free chlorine residual there will be considerable loss from the action of sunlight on the open basins. This causes a wide variation of residual chlorine leaving the plant, which cannot be tolerated. This phenomenon is complicated further by the weather. Changes in the cloud cover introduce another variable, which contributes to additional variation in the residual. Under these conditions it is impossible to calculate hours ahead what the prechlorination dosage must be to achieve the desired residual in the treated water effluent. The chlor–dechlor method eliminates all of these problems.

The Chlor–Dechlor Facility. The usual arrangement of this type of system is to use two flow-paced automatic chlorinators for prechlorination. One unit is for standby. One sulfonator is used to apply sulfur dioxide to the finished water. This unit can be operated as a compound-loop system or by direct residual control. The success of the system depends upon the proper maintenance of the chlorine residual analyzer, which is used to control the sulfonator. The analyzer can be either the amperometric type or the membrane type. Colorimetric analyzers are not recommended because their response time is much too long.

Since the advent of the Stranco Corp. High Resolution Redox System, and the W&T Deox 2000™ analyzer, there has been a totally different approach to this subject. (See Chapter 10).

The chlorination system used at this plant shown in Fig. 6-5 uses a single point of application, that applies enough chlorine so that the chlorine residual in the effluent is large enough to require dechlorination. This is much different from the system required in Chapter 10.

The arrangement of the equipment to control the residual leaving the plant is simple. There is no bother with postchlorination, as the postchlorinator becomes a sulfonator. This unit dechlorinates the treated water to a specified fixed residual. This residual is controlled by the chlorine residual analyzer equipped with a controller. The sulfonator is usually fitted with a flow-paced chlorine metering orifice that receives a flow signal from the plant effluent flow meter. Therefore, the sulfonator is arranged to receive two separate signals (flow and residual). This defines the compound loop system. If a flow meter in the treated water line is not available, it will be necessary to rely entirely on direct residual control. This method of control has been successful because it has good accuracy over a 7 to 1 range. Never attempt to utilize the plant influent flow meter signal to provide flow pacing for the sulfonator.

The degree of success of this method will depend largely upon the ability to provide the chlorine residual analyzer controller with a well-mixed sample. (Mixing is described in Chapters 8 and 9).

If there is too much loss of free residual in the sedimentation tanks and filters due to sunlight, and if it is not practical or cost-effective to cover these open structures with fiberglass, and so forth, it may be necessary to superimpose an automatic prechlorination dosage change between day and night operations. This has proved to be very successful.

If the primary control is the residual signal, and if the flow of water through the plant does not follow the usual smooth diurnal flow change pattern, as in the case of step rate flow due to pump sequencing, then the chlorine metering orifice should be controlled by an electric step rate controller. This step rate flow signal is imperative even if residual control is not involved.

An Operating Installation. The following description illustrates the salient features of a chlor–dechlor prototype installation that has been in operation for many years. Figure 6-5 shows the plant layout and the points of application of chlorine and sulfur dioxide.

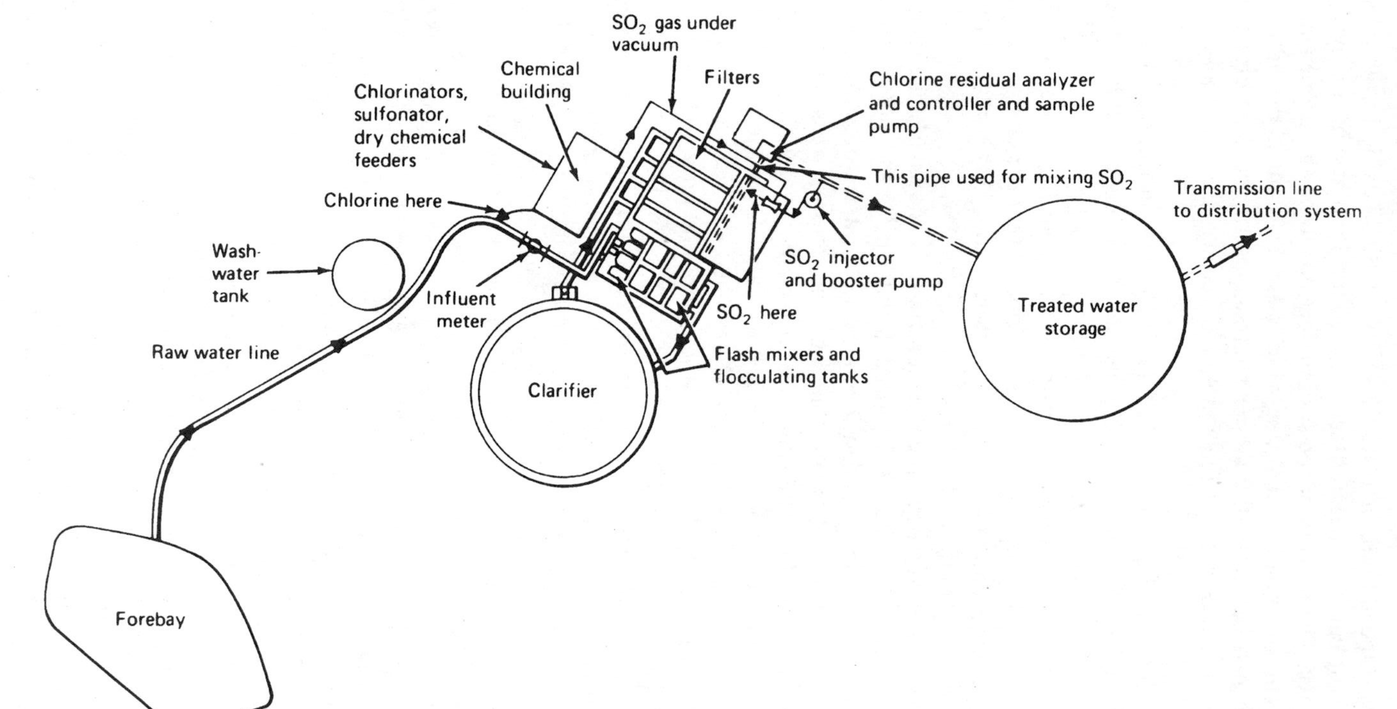

Fig. 6-5. The modern approach: Chlorination and dechlorination using residual control. Chlorinators are controlled by electric signal from influent meter plus automatic dosage control. A time clock is used to select two basic prechlorination dosages—one for daytime, and one for nighttime. The sulfonator trims the outgoing residual either up or down in accordance with the set point on the residual analyzer.

The importance of the continuous record of the chlorine residual is illustrated graphically in Figs. 6-6 and 6-7, which show two charts: one from the residual analyzer and one from a corresponding SO_2 flow record of the sulfonator. It can be seen that when the SO_2 feed rate has varied from a low of 20 lb/24 hr to a high of 85 lb/24 hr, the chlorine residual is practically constant at 0.9 mg/L. Note that at 8:00 A.M. there is an upset in the chlorine residual, due to the backwashing of some filters, with water having zero residual going by the analyzer. At this particular installation the SO_2 is mixed in approximately 25 ft of 24-inch line. The steady line on the analyzer chart reveals that it is a good mix. The plant flow rate averages 5 mgd. The prechlori-

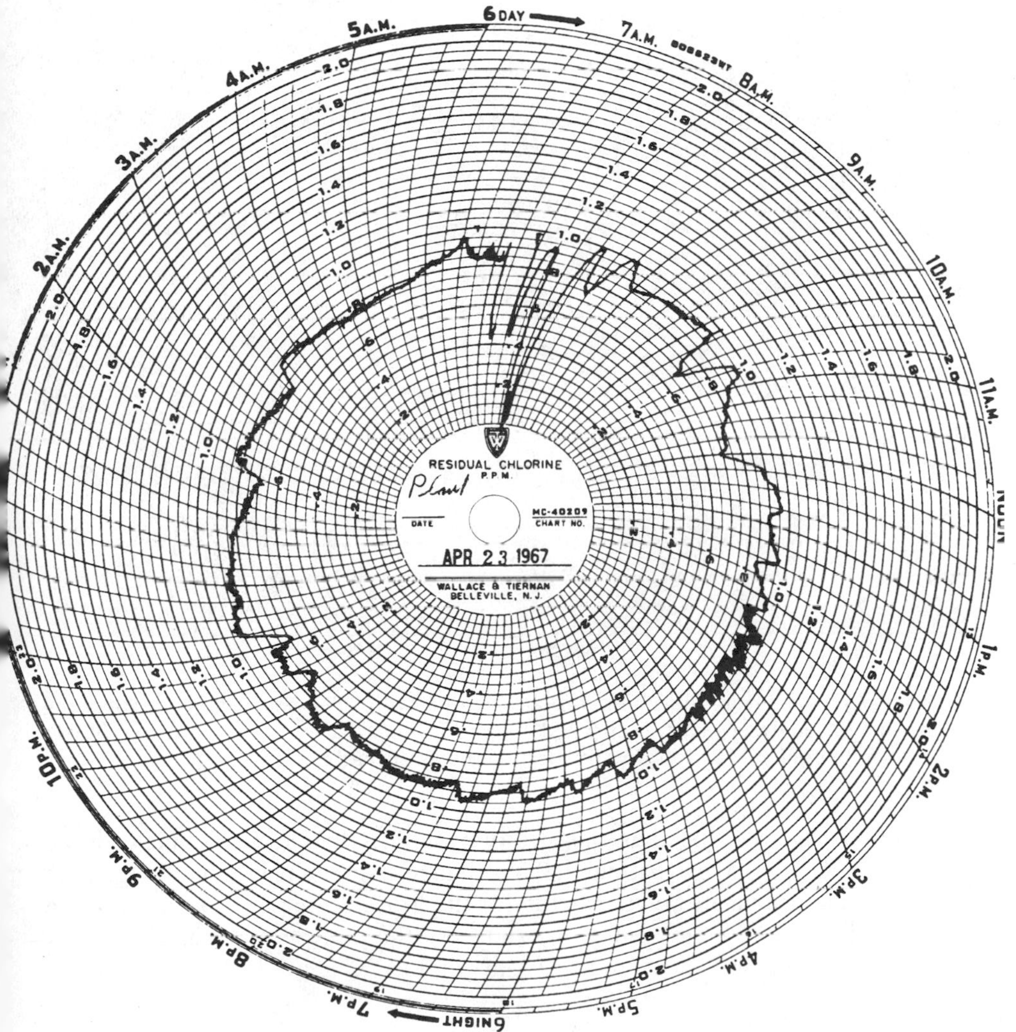

Fig. 6-6. A chart from the analyzer showing free chlorine residual.

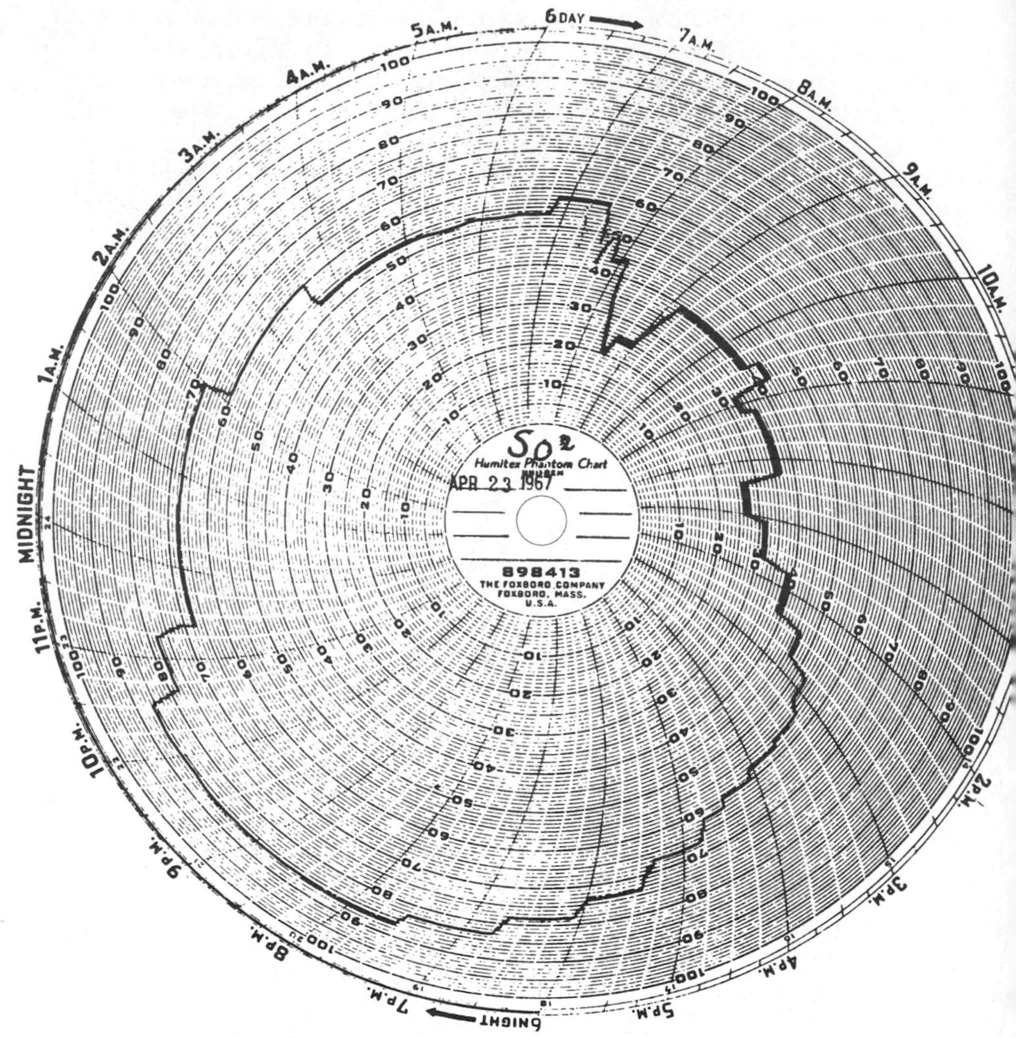

Fig. 6-7. Chart from the sulfonator showing record of SO_2 feed rate corresponding in time to analyzer chart. Reprinted from *Journal of the American Water Works Association* **60**(5) May 1968, permission of the Association.

nation dosage is changed automatically by a time clock from 6–7 mg/L in the daytime to 3.5–4.5 mg/L at night. This change in prechlorination dosage is to effect an approximate compensation for the free chlorine residual destruction by sunlight. On a bright sunny day (lat. 35°) the free residual chlorine decay due to sunlight was measured to be as high as 0.5 mg/L/hr.* The residence

*A comprehensive examination of this phenomenon in 1984 at 39.5° N lat. and 5000 ft elev. revealed that free chlorine destruction by sunlight on a bright sunny day can be as high as 2.5 mg/L/hr.

time in the open basins is long, about 4 hours at average flow. Cloudy days can be distinguished from sunny ones by mere inspection of these charts. The prechlorinator is paced by a flow signal from a metering unit at the influent to the plant. Because there is no metering equipment at the effluent of the plant, the sulfonator is controlled solely by the chlorine residual analyzer.

Plants using this system in the western part of the United States use from 2 to 7 mg/L prechlorination dosage to solve seasonal taste and odor problems, and are easily able to have enough free residual at the effluent to be trimmed by SO_2 to a final residual of 0.5–0.75 mg/L.

Complete Dechlorination Followed by Rechlorination. There are many waters that could benefit from this technique, in particular those that develop taste and odor problems caused by objectionable concentrations of the nuisance beyond the breakpoint. In this procedure dechlorination does not remove the ammonia N bound up in these residuals. It remains to form true monochloramine in the rechlorination step. The objectionable nuisance residuals are removed, which is the great advantage of this procedure. As rechlorination is designed to be in the range of 3–5 to 1 chlorine to the remaining ammonia nitrogen, there is no danger of any nuisance residuals forming. The end result is in fact the ammonia–chlorine process preceded by the free residual process.

This is an extension of common practice in many plants, which follow the rule: free chlorine through the plant and chloramine in the distribution system. There are of course exceptions. There are some waters that cannot be cleaned up with free residual chlorination because the raw water quality is so poor that it needs some effective pretreatment to lower the 20-min chlorine demand to a reasonable level, such as 1–1.5 mg/L.

CHLORAMINATION: THE AMMONIA–CHLORINE PROCESS

Historical Background

The use of ammonia with chlorine in water treatment has an unusual history. Both its discovery and its use were largely by accident. Fritz Raschig[80] in 1907, while working with aniline and hypochlorite, noticed that no color developed when these two compounds were reacted in the presence of ammonia. This made him want to experiment further with these compounds. By reacting two parts by weight of chlorine and one part of ammonia, he was able to produce a compound resembling a faint yellow oil. He termed this compound "chloramine" in accordance with the following equation:

$$NaOCl + NH_3 \rightarrow NH_2Cl + NaOH \quad (6\text{-}1)$$

Raschig's experimental work provided the theoretical basis for the later application of chlorine and ammonia in the water treatment field. Race[81–83] was probably the first to use the ammonia–chlorine process. He was in search of a method that would reduce the cost of chlorination, and was influenced by some observations by Rideal,[294] who noted that during the chlorination of sewage bactericidal action continued even after all the hypochlorite had "disappeared." Rideal attributed this continuing action to the influence of ammonia in the sewage. After performing successful laboratory experiments with ammonium hypochlorite, Race adopted this method in 1916 at the principal treatment plant at Ottawa, Ontario. Purely by an accidental error of some sort, his findings indicated that ammonium hypochlorite had three times the germicidal action that bleaching powder (calcium hypochlorite) did. Plant scale operations convinced Race that it was a more economical method than chlorine alone and, furthermore, that the finished water was free from objectionable tastes and odors, which he attributed primarily to the decrease in dosage.

In view of present-day knowledge, both of his conclusions were incorrect. Race also noted the elimination of aftergrowths following the prolonged use of ammonium hypochlorite. The first installation in the United States was made at Denver in 1917.[295] Alexander Houston[24] paved the way, in the development of preammoniation, to prevent taste formation caused by the reactions of chlorine and organic matter.

Although the first successful application of the ammonia–chlorine process for taste and odor control in the United States was by McAmis[77] in 1926, it was not until the work by Harold[86] and Adams[87] in England, and by Lawrence[88] and Braidech[89, 90] at Cleveland, Ohio, that the possibilities of the process were fully realized. After the intensive investigation (five months and nine thousand tests) of Lawrence and Braidech, this process for taste and odor control achieved nationwide prominence. This was due in part to a previous investigation by Ellms and Lawrence[91] in 1921 on the same water, in which they submitted that waters polluted by phenols did not respond to superchlorination followed by dechlorination with activated carbon.

The Cleveland investigation concluded that preammoniation prevents tastes and odors with phenol concentrations as high as 1.0 mg/L, and that combined residuals up to at least 0.6 mg/L can be maintained without producing chlorinous taste and odors. This was considered a high chlorine residual because at this time an acid O-T residual of 0.2–0.3 mg/L was considered adequate.

The installations at Cleveland began in November 1929 at the Division plant, followed two months later at the Baldwin Street plant. Other successful installations were reported at Springfield, Illinois, by Spauling in 1929,[92] at Lansing, Michigan, in 1929 by Harrison,[93] and at Lancaster, Pennsylvania, by Ruth in 1931.[94] These successes created a great demand for the process, which resulted in some miserable failures because of ignorance of its limitations.

This process enjoyed its greatest popularity between 1929 and 1939. Two years after the Cleveland installations, there were 190 other installations. In 1938, based on replies to a questionnaire from 2541 supplies in 36 states, 407 of these supplies used ammonia with chlorine.[98] In 1958 Griffin and committee sent out a questionnaire to 114 municipalities. Among the 85 answers, only seven reported preammonia; 26 reported postammonia; two, both pre- and postammonia. The surprising number of users of postammoniation was probably for treatment of high pH waters. The process fell into a rapid decline shortly after the discovery of the breakpoint phenomenon in 1939. This, coupled with an inability to purchase ammonia during the war years (1941–45), relegated its use to cases of special treatment.

General Discussion of the Ammonia–Chlorine Process

This process involves the addition of ammonia and chlorine compounds separately to a water processing system. These compounds are usually anhydrous ammonia and hypochlorous acid. The combination forms chloramines. This procedure can also be called chloramination or the chloramine process. In general, when the ammonia is applied first, it tends to prevent the formation of compounds that would otherwise produce chlorinous taste and odors. However, there are many instances where ammonia is added to a water containing a combination of free and combined residual for the sole purpose of converting all the free residual to combined residual. This has been well documented by Williams.[78, 79] The water that Williams was dealing with was polluted with a significant amount of treated wastewater discharges containing a significant amount of organic N, probably 4–6 mg/L. The source of these nitrogen compounds (chlorcreatine) were known to be from urine in the treated wastewater.

Now we know (1990) that when a sample of this wastewater is titrated for a chlorine residual by either the amperometric method or the DPD-FAS method, the entire dichloramine fraction will titrate as a nongermicidal organochloramine. This was discovered by the chief laboratory chemist (Virginia Alford) at the San Jose/Santa Clara WWTP after they had started denitrification in 1981. The loss of ammonia N produced chlorine residuals with about 4–6 mg/L free chlorine and about the same amount of dichloramine while the total residual was from 10–12 mg/L. However, it took several years before this valuable information reached wastewater operators around the United States. This technique of free residual conversion to chloramines has been called dechlorination by ammonia. This is a misnomer, because it does not remove any chlorine compounds. One disadvantage of this procedure is the inability of the ammonia addition to remove any of the nuisance residuals that may have formed in the free residual process. Prior to 1975 it was common practice to add ammonia to a water containing free residual to purposely provide chloramine residual in the distribution system.

This usually ends in failure, because it is impossible to achieve an accurate 3 to 1 or 4 to 1 chlorine to ammonia ratio unless a motorized mixer is used. This is due to the fact that the amount of ammonia N added is so small. This addition of ammonia N eventually degrades the water quality in dead ends of the distribution system after the chlorine residual disappears.

Conversion of free residual chlorine to chloramines is a special case in high pH waters, including waters that are softened by the excess lime process, because disinfection has already occurred before the conversion takes place. However it was recently discovered that if the high pH water is chlorinated with a Water Champ instead of a water-operated injector, the chlorine gas is more effective than HOCl. The preferred ratio of chlorine to ammonia N has been documented as 3 : 1. This ratio appears to produce the best-tasting water. The ratio for maximum disinfection efficiency is 6 : 1; therefore, each application should be ratio-tested in the laboratory to find acceptable taste close to maximum disinfection efficiency, and a ratio that produces the best-tasting water.

Basic Chemistry of Ammonia N Reactions with Chlorine.[508] The easiest way to understand these reactions between ammonia and chlorine is to write them in their molecular forms where N has a valence of minus 3. That means we are looking at the NH_3 parts of the following nitrogen compounds: $(NH_4))_2 SO_4$, NH_4Cl, NH_4OH, and others. We can then write:

1. Monochloramine: $NH_3 + Cl_2 = NH_2Cl + HCl$
2. Dichloramine $NH_3 + 2Cl_2 = NHCl_2 + 2HCl$
3. Trichloramine $NH_3 + 3Cl_2 = NCl_3 + 3HCl$

Trichloramine = nitrogen trichloride.

The Critical Part. Reaction 1 above occurs at a pH of 8.5 or higher. This produces a preponderance of monochloramine. Reaction 2 occurs at a pH between 4.4 and 5.0, with a preponderance of dichloramine at lower pH values. The distribution of monochloramine and dichloramine follows a curve dependent upon pH. At a pH in the 7 to 8 range it is mostly monochloramine. Reaction 3 does not occur in dilute medium pH conditions. Trichloramine is a very volatile chloramine. It certainly forms at pH below 5 or so, and absolutely below pH 4.0.

The Agene process was used by Wallace & Tiernan in the 1930s for bleaching flour with NCl_3, which was created by mixing a chlorine solution from a chlorinator with a solution of ammonium sulfate, with reaction in a packed tower. The NCl_3 was air-stripped from the tower and injected into the flour as a gas. This was a perfect low pH system, because the highly reactive nitrogen trichloride was not allowed to accumulate in the tower.

Preformed chloramines for chloramination in public water supplies are feasible, but the clue to success is pH control. If they could be prepared at pH 10, there would be very little problem of keeping them in chlorine solutions of 3000 mg/L or less.

Induced Breakpoint. The application of ammonia should be evaluated on the merits of each individual case. Its use to create, or induce, a breakpoint*

*If ammonia is used to induce a breakpoint, use a Cl to N ratio of 12 : 1 or greater.

has been excellent. Often a water utility receives chloraminated water from a water supplier that causes all sorts of problems in dead-ends, storage reservoirs (nitrites), and chlorine relay stations due to the ammonia N in the water. In these cases the user has no other choice but to remove the ammonia N by the breakpoint process (chlorinating the supply with a 10–12 to 1 ratio of Cl_2 to NH_3-N by weight). The induced breakpoint is a simple but effective procedure. Ammonia is added to an otherwise ammonia-free water, upstream from the chlorine application. Then chlorine is added at a ratio of about 12 parts to 1 part of ammonia N.

A most successful instance of this procedure occurred about 1965 in the Reno, Nevada water supply. Ammonia was applied to three separate raw water sources from the same watershed at the rate of 1 lb/mg, followed 10 min. later by 19 to 20 lb/mg chlorine. This produced a water free of taste and odor with a free chlorine residual between 1 and 2 mg/L. When chlorine was used alone on this water, a chlorophenol-like taste developed that did not respond to super doses of chlorine, but persisted for months. The source of the phenol-like compounds contributing to these chlorinous tastes was unknown.* Therefore, each water that develops obnoxious tastes and odors resulting from chlorination may benefit by the use of the ammonia–chlorine process for the prevention of such tastes. This procedure began a half-century ago and is still being used today (1997).

After the discovery of the breakpoint phenomenon in 1939, the ammonia–chlorine process fell into a sharp decline because the free chlorine residual produced by the breakpoint procedure proved to be equally effective in either controlling or eliminating taste and odors. Additional benefits that were not necessarily common to chloramine residuals were also discovered. The demise of the ammonia–chlorine process was nearly complete during World War II. It was practically impossible to get any ammonia nitrogen compounds during the war (1941–45). Its use all but disappeared then, and former users switched to the free residual process without any further problems.

A good example is that of the San Francisco Hetch-Hetchy Supply transport system, which had been chloraminated from the beginning of the installation in 1938 (by White). This installation was intended to eliminate enormous crenothrix growths in the coastal tunnel. These growths were caused by large cracks in the cement lining due to large earth movements during tunnel construction. When the ammonia supply was suddenly cut off by the U.S. Army, the Tesla Portal tunnel chloramination system continued as a "free chlorine" system that produced a 2 mg/L free residual at the end of the tunnel several miles downstream from the Tesla Portal, and chloramination was never resumed.

However, there were some practitioners of chloramination, who managed shortly after the war (1955) to get ammonia N, who believed in the use of chloramine residuals to maintain water quality in the distribution system.

---

*The source of these compounds was found to be the quaking aspen tree (*Poplar trimulus*) on the watershed.

Unfortunately the current (1990s) emphasis is on the use of chloramines as a secondary disinfectant. This can only degrade the efficiency of the disinfection process by chlorine. However, this is still a controversial subject among a lot of the expert practitioners.

Germicidal Efficiency

The EPA accepted chloramines as a secondary disinfectant in 1978,[99] and outlined certain conditions that would allow the process to be used. The most acceptable situation was to apply the chlorine first and then apply the ammonia after 10 min. of contact time with chlorine alone.

As of 1983 the ammonia–chlorine process was accepted by the EPA as a primary disinfectant, provided that contact time is adequate and that there is proof of disinfection—largely because the addition of ammonia before chlorine prevents the formation of THMs. There are some waters that cannot comply with EPA MCL limitations of THMs when the chlorine is applied before the ammonia. Therefore, these waters must rely upon the ammonia–chlorine process to meet EPA standards, unless they adopt alternative disinfectants, such as chlorine dioxide or ozone.

Recent research in the use of chloramines for water reuse situations reveals an entirely different concept of the germicidal efficiency of chloramines. The plant scale investigations by the Sanitation Districts of Los Angeles County[101] developed criteria for disinfection of wastewater to be used for bathing and reuse. Their work revealed that chloramines were almost as good as free chlorine as a bactericide. This efficiency was based upon good mixing and contact times of 50–60 min. with the same chlorine dose. Tihs success was probably due to the fact that the organic N in the wastewater effluents reacted with the free chlorine to produce a nongermicidal organochloramine that titrated in the dichloramine fraction of a forward-titration procedure. This phenomenon was accidentally discovered in about 1982, but was not published until about 1987. All the references from 100 to 108 in this chapter were published several years before the discovery of this organic N phenomenon at the San Jose, California WWTP in 1982. (See Chapter 8.)

A study by Selleck et al.[102] at the University of California Sanitary Engineering Research Laboratory, Richmond, California, revealed that the germicidal efficiency of chloramines, based upon total coliform kills, is almost equal to that of free chlorine, provided that the contact time is one hour or more, the chlorine-to-ammonia N ratio is 6:1, and the organic N is less than 4 mg/L.

The Metropolitan Water District of Southern California made a lengthy pilot plant study to evaluate THM formation and disinfection efficiency of chloramine using a chlorine-to-ammonia ratio of 3:1 by weight.[103] Taste and odor formation and other chlorine-to-ammonia ratios were not investigated. They studied three schemes of ammonia–chlorine application in addition to free chlorine. These schemes were:

1. Sequential chloramination: chlorine upstream from the rapid mixer, then ammonia 17 min. later downstream from flocculation and upstream from sedimentation.
2. Concurrent chloramination: chlorine upstream from rapid mixer, then ammonia 50 sec later in the rapid mix basin.
3. Preammoniation: ammonia added in the rapid mix basin followed by chlorine in the flocculation basin.

The lowest levels of THM formation were observed for the concurrent and preammoniation schemes.

The point of application of chlorine and ammonia did not affect the occurrence of coliform bacteria in the pilot plant effluent. They were uniformly absent. This finding agrees with others: given long contact times, free chlorine and chloramine residuals are equivalent disinfectants (based upon coliform kill).[101, 102, 170–172] The contact time from influent to effluent in the pilot plant was 3 hours 21 minutes. However, coliforms were observed on 13 occasions upstream from the sedimentation basins (17 minutes contact time) using the preammoniation scheme.

The use of the Millipore standard plate count procedure (m-SPC) bacteria indicated that preammoniation produced higher pilot plant effluent total plate counts than concurrent or sequential addition of ammonia or free chlorine. This disparity between the total plate counts of the different chloramination schemes was thought to be the result of preammoniation interference with the physical removal of particulate material (including bacteria). However, no statistical difference in effluent turbidity was found for the ammoniation schemes. Therefore, higher plate counts appear to be due to the reduced bactericidal efficiency and not compromised filtration efficiency.

One of the major difficulties resulting from the MWD chloramination system occurred when the MWD water entered the users' system and mixed with their "different source water." Storage reservoir supplies became contaminated with the ammonia N content of the MWD water, causing nitrite formation, excess algae growths, and other harmful microbiological growths.

One of the large users, City of Los Angeles Water Department, receives its MWD water just downstream from the MWD Jensen WTP. It solves all its potential problems from the ammonia N by treating this MWD plant effluent with the breakpoint process to remove the ammonia N applied by the MWD.

Another serious problem, which caused the MWD to discontinue the addition of ammonia N, was the shutdown of some 4000 kidney dialysis machines in Los Angeles County, only a few months after the chloramination system was put into operation. The trouble with the water going to the dialysis machines was due to the necessity of its being dechlorinated prior to its entry into these machines. When the free chlorine system was being used, the maintenance people were always aware when the activated carbon needed replacement, because free chlorine consumes the carbon. However, dechlorination of chloramines is much different; it simply consumes the ability of the carbon to dechlorinate, so the operator has no visual way of judging when the carbon needs replacing.

The only way to solve this problem is by the installation of a standby or duplicate carbon filter. This caused a temporary shutdown of the chloramination system for about 2 years until all dialysis water systems had two carbon filters. The operators could then tell when to switch filters simply by checking the chlorine residual in the filtered water.

Factors Affecting the Efficiency of the Process

These factors are most important; they are based upon several years of careful observation of this process at several water treatment plants.

The preferred sequence of chemical addition is usually ammonia first and then the chlorine. The ammonia must be well dispersed before the chlorine is added. This is a critical factor. When the chlorine is added, it must be mixed as quickly as possible (2 sec). This is necessary because the reaction between ammonia and chlorine solution at the pH range of 7–8.5 is practically instantaneous. If mixing is slow or poor, side reactions between the chlorine and organic matter can interfere with the formation of true chloramines. An example would be organic matter subject or prone to bleaching by chlorine solution.

White et al.[172] Found that in a high quality wastewater tertiary effluent, mixing at the point of application greatly affected the bactericidal efficiency of the chloramine process. In this particular case the ammonia was added to a completely nitrified and filtered effluent, which passed through a pumping cycle before the chlorine was added. The chlorine solution was mixed with turbine-type mechanical mixers. The total coliform count with the mixers was usually about 13–20/100 ml MPN. With no mixing the count was 1300/100 ml MPN or greater. Water temperature affects the reaction time between ammonia and chlorine. The reaction is significantly retarded at temperatures below 50°F. Treatment plant operators in cold areas should be cognizant of this because it can affect the location of the point of application of each chemical.

Summary. When adding ammonia N for the chloramination process, the only thing we have to know is that the chlorine must be added first in the range of three or four parts of chlorine to one part of ammonia N by weight. Rapid mixing at the point of application of both these chemicals is vital to the success of this process. The only other time ammonia N is used in potable water systems is to induce the B-P, to eliminate a foul taste that is usually caused by some unknown chemical compound, man-made or natural. Because nearly all drinking water sources contain less than 0.2 mg/L ammonia N, and rarely any organic N in excess of 0.3 mg/L, these compounds will not interfere with the chloramination process. However, it is imperative that their concentrations be checked on a routine basis. Those surface waters that freeze over in the winter must be checked for ammonia N because the freezing causes a buildup of ammonia N in the surface waters.

In wastewater treatment, the presence of ammonia N is the source of a great many operational problems, along with organic N. In these systems the ammonia N varies from nitrified effluents at 0.1 mg/L to 20 or 30 mg/L for secondary nonnitrified effluents, and organic N ranges from 3 to 9 mg/L. The deleterious effect of organic N is the production of nongermicidal organochloramines. This occurs over a period of time (30 or more min.) caused by the hydrolysis conversion of monochloramine residual to organochloramines. The amount of this conversion represents an error in the total chlorine residual measured by an on-line analyzer. The conversion increases with contact time.

Pretreatment Problems

In using the ammonia–chlorine process as a potable water pretreatment process, in which rapid sand filters are involved, certain precautions must be taken. An addition of ammonia in excess of chlorine can promote the growth of nitrifying bacteria in the filter beds. As chloramines will not react to destroy nitrites, and the excess of ammonia acts as a nutrient, the nitrifying bacteria will proliferate. Then it becomes necessary to destroy the bacteria and the nitrites by free chlorine. If a proper ratio of chlorine to ammonia is maintained, including the natural ammonia content of the water, and if an adequate chloramine residual (0.5–1.0 mg/L) is maintained at the discharge of the filters, such difficulties with nitrifying bacteria will not occur. To eliminate all possibility, the operator should not use colorimetric methods for residual determination when using a combination of chlorine and ammonia N, because if nitrites do develop, they cause a significant false chlorine residual analyzer readings. Use, instead, the amperometric forward titration procedure or the DPD-FAS titrimetric method to monitor the prechlorination process.

Certain Effects of Chloramination

General Discussion. The addition of ammonia N to drinking water may compromise the water quality at the consumer's tap. In the years since the discovery of the breakpoint phenomenon, free residual chlorine* (total residual must contain 85 percent HOCl) has proved to be the best disinfectant available to the water industry. Although it is true that chloramination† has been practiced for more than 60 years in the United States, its use is site-specific, whether it is for T & O control or biological stability control in the distribution system. The operators of chloramine systems must be aware of the fact that should the residual die in the distribution system, the ammonia

*Free residual chlorine is defined to mean that 85 percent of the total residual must be HOCl.
†Since the Bruce Ames cancer research group have found that the formation of THMs is pure trivia, chloramination is rarely necessary.

N will remain, and immediately the microbial life in the distribution system will begin to disintegrate. This can cause serious dead-end problems.

The water producers in Western Europe and Great Britain still operate under the European Economic Commission limitation of allowing no more than 0.2 mg/L ammonia N at the consumer's tap. This limitation by itself substantially eliminates the use of chloramination in any of the treated water. The Europeans have to spend a great deal more money on potable water treatment than do their U.S. counterparts. All over Germany and the Netherlands, groundwater recharging is a common pretreatment process.

Distribution Systems. When chloramines are used, the distribution system must be continuously monitored at strategic locations for mono- and dichloramine residuals and DO. In some cases identifying only the total chlorine residual is not enough, particularly if any organic N is present in the water. If it is present, monochloramine will, over a period of 30–40 minutes, hydrolyze to react with the organic N to form organochloramines that are nongermicidal. This species titrates quantitatively in the dichloramine fraction of the forward-titration procedure. Although this reaction contributes to biological instability, a more serious situation follows when the monochloramine residual disappears. When this happens, free ammonia N, which is a powerful biological nutrient, reappears. Therefore, this occurrence will cause biological instability in that portion of the distribution system. The result is usually foul tastes, odors, and dirty and off-colored water at the consumer's tap.

When attempting to clean up this portion of the distribution system, one always must ask which method to use: free chlorine or chloramines? White[337] always chose the free residual route because it was many times quicker than using chloramines. The consumers suffered more over a short span of a few days with the free chlorine method, whereas the chloramine route took weeks. Moreover, in every case the former method was superior to the chloramine procedure because cleanup progress could be monitored. Using dosages that produced some free residual and measuring both free and combined residual, it was found that as the cleanup progressed, the ratio of free to combined chlorine increased until it reached 85 percent free chlorine. At this point it was considered that the cleanup was complete. This was verified by the taste and color of the flushing water.

In spite of White's preference, and the European limit on ammonia N of 0.2 mg/L at the consumer's tap, the use of free chlorine versus chloramines in the maintenance of distribution systems remains a controversial subject in the United States. One of the reasons for the controversy is based upon the theory that free chlorine residuals are used up before they are able to penetrate the protective membrane that surrounds the bacteria involved. This is only a theory.

Reservoir Nitrification. The California experience resulting from the use of chloramination to control THMs has created serious and expensive water quality problems, due largely to the vast numbers of storage reservoirs that are all riding on the distribution systems. These reservoirs are required by the water producers to provide their systems with the necessary flexibility and reliability.

In a progressive step to provide safer water supplies, the Sanitary Engineering Division of the California Health Services has decreed that all open reservoirs must be covered. When this is not feasible or practical, then the discharge from the reservoir must undergo conventional water treatment that includes filtration and disinfection. If the reservoir is covered, and if it receives chloraminated water, partial nitrification probably will occur. This will bring on a proliferation of ammonia oxidizing bacteria (AOB) and nitrites. This occurrence causes biological instability in the reservoir that produces severe algal blooms accompanied by disagreeable tastes and odors.

The Metropolitan Water District of Southern California, serving 15 million consumers, had been using free residual chlorination for more than 40 years without incident. To meet regulatory THM limits, MWD switched to chloramines in 1983, and experienced problems. First there was the need to switch back to free chlorine while for nearly a year the carbon filters for the dialysis machines were rearranged to eliminate the chloramine residuals. The following summer, two of the large covered reservoirs suffered from partial nitrification. Because of the uncontrollable algal bloom from this nitrification event, the water in the reservoirs had to be taken out of service, and the water in the reservoir had to have the ammonia N removed by breakpoint chlorination. This biological phenomenon, generally thought to be caused by autotrophic ammonia-oxidizing bacteria (AOB),[338] is not uncommon. It has occurred frequently in secondary wastewater treatment plants and swimming pools. It has not been a frequent occurrence in potable water treatment plants because chloramination is not generally used where covered reservoirs are used in distribution systems.

Nitrification. This is a microbiological process in which ammonia (NH_3) is oxidized sequentially to nitrite (NO_2^-) and nitrate (NO_3^-). It is a two-step process that is carried out by two distinct chemolithotrophic bacteria. These organisms obtain their energy from the oxidation of reduced forms of inorganic nitrogen and derive most of their cell carbon by fixing carbon dioxide via the Calvin cycle.[339] Heterotrophic bacteria are also able to carry out nitrification, but at a much slower rate than by autotrophic nitrification.[340–342]

In the final step, ammonia is oxidized to nitrites as follows:

$$NH_4^+ + \tfrac{3}{2}SO_2 \rightarrow NO_2^- + H_2O + 2H^+ - 65\text{kcal/mol} \qquad (6\text{-}2)$$

Hydroxylamine is an intermediate in this energy-yielding reaction, and very small amounts of nitrous and nitric acids are produced.[343]

Although *Nitrosomonas* bacteria are the ones most frequently considered to be the operatives of this step, other genera have been known to oxidize ammonia to nitrite autotrophically.[344] These genera are: *Nitrosolobos, Nitrosococcus, Nitrosovibrio,* and *Nitrosospira.*

In the second step, nitrite is oxidized to nitrate without detectable intermediates:

$$NO_2^- + H_2O \rightarrow NO_3^- + 2H^+ - 20\ \text{kcal/mol} \quad (6\text{-}3)$$

In soil and fresh water, this step is carried out exclusively by members of the genus *Nitrobacter.* In nature, nitrite usually does not accumulate during nitrification unless a specific inhibitor of nitrite oxidation or a selective toxin for *Nitrobacter* is present.[345] Of significant importance are recent data showing that the growth of nitrifiers is inhibited in the presence of sunlight,[346] which indicate that a dark environment such as a covered reservoir will speed up their growth. However, White has observed enormous algal blooms in both swimming pools and secondary wastewater effluents exposed to sunlight during periods of unwanted episodes of partial nitrification. Researchers also have found that nitrifying bacteria secrete organic compounds that can stimulate the growth of heterotrophic bacteria.[344, 347] Where drinking water is the consideration, the removal of ammonia (nitrification) is a primary step in producing a water that will be biologically stable in the distribution system. Rittman and Snoeyink have documented planned biological removal of ammonia to achieve distribution system stability.[348] European utilities have demonstrated the successful use and benefits derived from the practice of biological removal of ammonia from drinking water.[349, 350] When ammonia is deleted from the water supply, not only is less chlorine required, but the chlorine that is applied will provide a free residual. This will quickly restore and maintain the biological stability of the system on a reliable basis.

When nitrites proliferate in this uncontrolled and unplanned nitrification event, they must be destroyed. This can only be done by free chlorine; chloramines cannot oxidize nitrites. Five parts of chlorine are required to oxidize one part of nitrite. Nitrites will proliferate in a covered reservoir if the water entering that reservoir has been chloraminated. At the onset of this proliferation, the chloramine residual will begin to decay rather sharply, and the HPC levels will escalate. The immediate partial solution for this problem at the two MWD covered reservoirs, after they were put back into service, was to add sufficient chlorine to the reservoir influent to bring the chlorine-to-ammonia ratio up to at least 4 : 1.

Occurrence of AOB (Ammonia-Oxidizing Bacteria). In order to fully understand how nitrification develops, MWD of Southern California conducted a study to examine the factors that influence the AOB in a chloraminated distribution system.[351] Samples were collected over an 18-month period from three sources: (1) raw project surface water, (2) two covered finished water reservoirs that had previously experienced nitrification episodes, and (3) conventional treatment plant effluent. Sediment and biofilm samples were collected from the interior wall surfaces of two finished water pipelines and one of the covered reservoirs. The AOB were enumerated by the MPN technique;

isolates were isolated and identified. The resistance of naturally occurring AOB in chloramines and free chlorine also was examined.

The monitoring program indicated that the levels of AOB identified as members of the genus *Nitrosomonas* were seasonally dependent in both raw water and finished water. The highest levels occurred in the warm summer months. The AOB concentrations in the two floating cover reservoirs had MPNs of <0.2 to >300/ml. This correlated significantly with temperature and concentration levels of heterotrophic plate count bacteria (HPC).

When the water temperature was below 60–65°F, AOB were not detected in the chloraminated reservoirs. The study indicated that nitrifiers occur throughout the chloraminated distribution system. Higher concentrations of AOB were found in the reservoir and pipe sediment materials than in the pipe biofilm samples. The AOB were found to be about 13 times more resistant to monochloramine than to free chlorine. After 33 min of exposure to 1.0 mg/L of monochloramine (pH 8.2, 74°F) only 99 percent (2 logs) of an AOB culture was inactivated. Chloramine residuals currently in use are 1.5 mg/L at a 3:1 chlorine-to-ammonia N ratio. Results of this study have indicated that this residual level may not be adequate to control the growth of AOB in the distribution system because it was found that the AOB not only could survive in the reservoir water columns in the presence of 1.3–1.5 mg/L chloramine residual, but also were capable of growing in the presence of these disinfectant levels. These organisms seem to grow best under conditions of mild alkalinity (pH 7.5–8.5), warm water temperature (77–82°F), darkness, extended detention time, and the presence of free ammonia. All these conditions were satisfied at the two MWD covered reservoirs.

The fact that AOB were recovered when the nitrite concentration was below the detectable level suggests that the measurement of AOB is a more sensitive indicator of nitrification than nitrite analysis, particularly when the nitrites had already been converted to nitrate. The AOB levels correlated highly with the R2AHPCs[351] in one of the reservoirs, suggesting that these heterotrophs may be good indicators of the presence of nitrifiers in some chloraminated systems. It would be most useful to have such a surrogate indicator, owing to the lengthy incubation period required for enumerating the AOB.[351]

Because the presence of free ammonia is the root of the problem, it becomes imperative that Cl-to-N ratios between 3:1 and 5.5:1 be investigated for T & O acceptability at the consumer's tap. The higher the ratio is, the lower the ammonia level.

Topudurti and Haas[447] found that there is a direct transfer of Cl^+ from monochloramine to phloroacetophenone (PAP), which generates the formation of chloroform ($CHCl_3$). This may be a significant deterrent to switching from free chlorine to chloramination. PAP is in a class of compounds known as acetogenins. These types of compounds contain groups that are common to natural pigments. PAP was previously used as a model compound by

Morris and Baum[448] in their study of the reactions between FAC and several compounds that increased the chlorine demand of the experimental water.

Effect upon Kidney Dialysis Patients. Kjellstrand et al.[300] reported that chloramine residuals in tap water pass through reverse osmosis membranes in dialysis machines quite easily. Furthermore, they directly induce oxidant damage to red blood cells, resulting in methemoglobin formation and damage to the HMPS with which red cells defend themselves against oxidant damage, which induces hemolysis and short red cell survival time. Moreover, they sensitize the patients to oxidant drugs such as primaquine, sulfonamides, and so on.

The effect upon the patient is apparently directly from the chloramines and not from any nitrogen compound associated with the chloramines. When Kjellstrand experimented with ascorbic acid as a dechlorinating agent there was no detectable methemoglobin formation. Ascorbic acid reduces chloramines to hydrochloric acid and ammonia.[300]

Activated carbon has been thoroughly investigated as an acceptable method for the dechlorination of both free chlorine (HOCl) and chloramines ($NHCl_2$, NH_2Cl, and NCl_3).[302–304] The most difficult species to dechlorinate on a predictable basis are the chloramines. Meyer and Klein[301] investigated the use of granular activated carbon (GAC) as a practical method to remove chloramines from dialysis water. They found that the dechlorination capacity of three different carbons differed as much as one order of magnitude. They further estimated that a 1.5 mg/L chloramine residual could be reliably removed during 156 five-hour dialysis with a GAC column.

Ascorbic acid is a more positive method and is appealing because of its simplicity. There is no danger of any chloramine breakthrough into the dechlorinated water, provided that the ascorbic acid dosage is adequate and the application is reliable. The GAC method requires careful monitoring to avoid chloramine breakthrough as the carbon approaches exhaustion.

Experience has shown that the use of GAC for dechlorinating chloramines is site-specific. Therefore, any water containing chloramines to be dechlorinated should be bench-tested to determine precisely the equipment required for a particular carbon source to be used.

Effect upon the Aquarium. Chloramines represent a two-pronged toxicity dilemma to aquatic life. Both chloramine residuals and un-ionized ammonia are toxic to fish in very low concentrations. The mechanism of chloramine toxicity is as follows: Chloramines pass readily through the permeable gill epithelium with an insignificant amount of cell damage. Once the chloramines have entered the bloodstream, they chemically bind to iron in hemoglobin in red blood cells. This cripples the ability of the cells to bind oxygen. This condition is known as methemoglobinemia. It is similar to the oxidation of hemoglobin caused by nitrite toxicity.[305, 306]

When monochloramine is dechlorinated by a reducing compound that produces sulfurous acid (H_2SO_3), ammonium chloride is formed as follows:

$$H_2O + NH_2Cl + H_2SO_3 \rightarrow NH_4Cl + H_2SO_4 \qquad (6\text{-}4a)$$

Ammonium chloride ionizes to form ionized NH_4^+ and un-ionized NH_3 nitrogen (ammonia). Therefore, all of the ammonia utilized to produce monochloramine is returned to the water as shown in Eq. (6-4a), $NH_4^+ + Cl^-$. This represents total ammonia. To avoid fish toxicity, un-ionized ammonia must be limited to 0.025 mg/L. Un-ionized ammonia concentration can be calculated from total ammonia. (See Chapter 7.) The maximum allowable total ammonia is about 0.4 mg/L at pH 8.3 and 70°F in a marine aquarium.

Therefore, ammonia removal is also a must for the aquarium when chloramines are dechlorinated chemically. In aquatic systems such as an aquarium, ammonia accumulates from nitrogenous wastes released directly from the fish and from deamination of protein in food and wastes by heterotrophic bacteria. However, nature provides a mechanism that serves to control ammonia buildup, nitrification. It is the primary means of ammonia removal in fish culture systems.[307] By this process, ammonia is oxidized to nitrate in two separate steps, each step dependent upon and controlled by a specific group of bacteria.

$$NH_4^+ + \tfrac{3}{2}SO_2 + \textit{Nitrosomonas}\ \text{Group} \rightarrow NO_2^- + 2H^+ + H_2O \qquad (6\text{-}4b)$$

$$NO_2^- + \tfrac{1}{2}SO_2 + \textit{Nitrobacter}\ \text{Group} \rightarrow NO_3^- \qquad (6\text{-}4c)$$

Simply placing fish into an aquarium with a recirculation filter will initiate the growth of these two groups of bacteria because they exist naturally in the environment. Unfortunately, these bacteria grow slowly; so they need a conditioning period to attain the necessary population to limit the production of the ammonia produced by the aquatic culture.

Removal of the ammonium ion (NH_4^+) is best accomplished by naturally occurring zeolites known also as clinoptilolites. Zeolites have a high selectivity for NH_4^+ removal, which decreases the toxic un-ionized ammonia concentration. They are relatively inexpensive and commercially available. They have been used for many years in industry as a freshwater demineralizer.[305, 307]

Finally it should be pointed out that the toxicity of ammonia decreases with pH. If the pH of the system is below 7, the ammonia produced will not be a problem in the aquarium. For example, in water at pH 6.9 and 75°F, 99.58 percent of the ammonia will be in the nontoxic ionic form; however, at the same temperature, but at a pH of 8, ammonia will be 9.49 percent ionic.[305]

Activated carbon has been used with moderate success to dechlorinate chloramines in potable water treatment processes, but its principal role is to adsorb organics in order to eliminate taste and odors. It is a popular filter material used in the aquarium industry for the filtration of fresh water and marine aquaria. It is a porous material that removes molecules from solution by binding them to the carbon's surfaces by a process known as *adsorption.* When carbon removes chloramines, ammonia is formed in the early stages of the adsorption process, followed by a decrease in ammonia as it is converted to nitrogen gas. This reaction occurs with predictable results if the carbon has been seasoned by exposure to residual chlorine. When chloramines come in contact with virgin carbon, they are destroyed, but some of them revert to ammonia as the surface of the carbon oxidizes, and chloride ions are released into solution.[305] If the carbon has been previously exposed to chlorine, the carbon surface is coated with oxides, which prevent the formation of ammonia by chloramine reversion. Therefore, the surface reaction of the chloramines and carbon results in a combination of adsorption and catalytic oxidation, releasing nitrogen gas and chloride ions and lowering the water pH.

General Summary. For dialysis sytems, ascorbic acid for dechlorination seems to have all the practicality and reliability assets. Activated carbon and some green sand zeolite are also capable of achieving complete dechlorination.

Aquaria require much more treatment than dialysis systems because both chlorine residuals and ammonia are toxic to aquatic life. Dechlorination plus ammonia removal can be accomplished in one step using activated carbon, provided that the system is carefully planned and properly designed. Dechlorination can be easily accomplished by chemicals such as the sulfite ion species or thiosulfate. The latter is not recommended, however, because it is too expensive and has a longer stepwise reaction. Following chemical dechlorination the ammonia can be removed by the use of zeolites. The zeolite removal process is similar to a water softening system.

The best way to remove ammonia N is by the simple B-P process.

American vs. European Practice: A Paradox. The paradox is the current European limitation on the allowable ammonia nitrogen concentration in finished water (0.2 mg/L) versus the allowable addition of ammonia nitrogen to U.S. drinking water to form chloramines, which prevent the formation of THMs*.

European Practice. The ammonia N–chlorine process was never practiced as such in European treatment schemes because most of the surface waters contained an abundance of ammonia nitrogen. The source of ammonia nitrogen was mainly untreated sewage and industrial wastes. Therefore, the European treatment schemes used prechlorination for nitrogen removal. This in turn allowed the practice of free residual chlorination. In some instances as much as 20 mg/L chlorine was required to achieve the breakpoint.

---

*This paradox is responsible for the Europeans being able to provide a higher quality water to the consumer.

The detection of trihalomethanes where free residual chlorination of polluted waters was practiced and the general undesirability of ammonia nitrogen in potable water resulted in some significant changes in European water treatment practices. There is no question that ammonia interferes considerably with the disinfection process, particularly when that process is free residual chlorination. This circumstance, coupled with the observation that trihalomethane concentrations increase as free chlorine residuals increase, was considered by the European community of water producers as grounds for elimination of ammonia nitrogen in potable water supplies. Although it is never mentioned as such, one of the dominant arguments against ammonia nitrogen, particularly in U.S. supplies, is the nutrient factor (see below), which eventually contributes to water quality degradation in reservoir systems, long transmission lines, and widespread distribution systems unless adequate chlorine residuals can be maintained.

European water purveyors were thus made subject to an agreement by the ministers of the European Economic Community on December 19, 1978, limiting the allowable ammonia nitrogen concentration in water delivered to the consumer to 0.05 mg/L. Moreover, the ministers decreed that the method of nitrogen removal to comply with the limitation must be biological nitrification. Realizing that such a drastic change in the treatment procedures would take time, the use of chlorine for nitrogen removal was allowed until installation of the recommended system was found practical.

As of 1979 the agreed-upon treatment train for nitrogen removal where applicable and possible included: raw water ozonation at a fixed dosage; groundwater storage followed by coagulation, sedimentation, and filtration followed by ozonation; and then biological nitrification followed by chlorine dioxide for effluent disinfection.

Actually, effluent disinfection by either chlorine or chlorine dioxide was required, for ozone was not generally accepted as the final disinfectant. Intermediate points of chlorination or chlorine dioxide are acceptable, provided that they do not interfere with biological nitrification. This is a two-step process of ammonia oxidation, which involves the formation of nitrites. The conversion of nitrites to nitrates is the final step in ammonia removal. This is accomplished by the *Nitrobacter* species of bacteria. Oxidation of nitrites and/or destruction of the *Nitrobacter* bacteria by chlorine or chlorine dioxide may either interrupt or destroy the nitrification process. Therefore the addition of either chlorine or chlorine dioxide at intermediate points must be carefully selected.

Rittman and Snoeyink[309] presented a comprehensive review (1984) of European practice, together with a microbiological theory demonstrating how biological processes within a water treatment plant can remove the organic and inorganic substrates that promote biological instability, the root cause of biofouling in a distribution system. A biologically stable water is one that does not support the growth of microorganisms to a significant extent, whereas an unstable water results in high numbers of microbes in the distribution system unless persisting chlorine residuals are maintained. (In contrast, chlorine dioxide's biocidal activity is substantially uncompromized by pH, and is far more effective at high pH than either ozone or chlorine; see Chapter 12.)

Rittman and Snoeyink[309] investigated both French and British practices of nitrification using biological filters, fluidized bed filters (also called biological sedimentation filters), rapid sand filters, and granular activated carbon beds. All of these processes are biofilm reactors that achieve nearly 100 percent ammonia removal even at low temperatures. Although the primary objective of these processes is to remove ammonia nitrogen, they simultaneously remove a significant proportion of the natural chlorine demand* of the raw water. This allows the Europeans to disinfect at lower chlorine dosages, which reduces THM levels. If chlorine dioxide is used, its demand is likewise lowered, thereby avoiding elevated concentrations of chlorate and chlorite ions.

U.S. Practice. In 1980 the EPA changed its stance on the use of chloramine as a disinfectant for potable water. It was allowed to become a primary instead of a secondary disinfectant. This provided an easy solution to those water producers who were plagued with unacceptable THM levels caused by free chlorine residuals. This revived the ammonia–chlorine process in the United States.

Many water producers in the United States have encountered chlorammation-associated problems (eg, nitrification) when they switched to chloramines to comply with THM regulations. It probably would have been preferable for water producers to clean up their raw water first to reduce chlorine demand. This would allow significantly lower doses of chlorine and/or chlorine dioxide and would provide biologically stable water to enter the distribution system.

For example, a limitation on chlorine demand could be a surrogate parameter for raw water quality. If the limit were exceeded, then the water might reasonably be required to undergo special pretreatment to reduce the 20 min. chlorine demand to an acceptable level.

Blending Chloraminated Water with Chlorinated Water. This is an extremely difficult operation to perform, once the water has traveled an hour or so in the distribution system.

The Metropolitan Water District of Southern California changed its primary disinfectant from free chlorine to chloramine in 1984 to meet the EPA requirements for THM formation. They preferred the activated carbon treatment, but found it too expensive; so their only choice was to go to choramines. Problems of taste and odor complaints were anticipated, so they designed a laboratory study to determine the effects of blending chloramine residuals with chlorine residuals by developing blend–residual curves to simulate the systems involved.[311] The potential problems of blending are related to a mixture that proceeds through the breakpoint curve, which allows the formation

*Chlorine demand is defined as the difference between the amount of the chlorine added to the water and the amount of chlorine (free available or combined available) remaining at the end of a specified contact period.

of dichloramine and possibly nitrogen trichloride. These two compounds are notorious for causing consumer complaints. MWD decided on a 3:1 Cl-to-N ratio (by wt) for chloramination. Using this ratio and accounting for the almost negligible concentration of naturally occurring ammonia N, they developed an induced breakpoint curve for each of their two supplies: (a) the Northern California water supply, known as state project water (SPW); and (b) the Colorado River water, labeled CRW. The induced breakpoint of the SPW at pH 7.5 occurred at a Cl-to-N ratio of 11:1, and for the CRW at pH 7.9 the ratio was 9.2:1. These breakpoints occurred after 4 hours in the dark at 25°C. They followed this with a series of breakpoint curves, due to blending. Two of these, shown in Figs. 6-8 and 6-9, illustrate the same two waters at two different contact times of 5 min. and 4 hours. These figures demonstrate that at short contact times all of the chlorine species may exist together in the breakpoint region.

From this experimental work, the MWD staff developed a computational model that allows the prediction of acceptable blends of chloraminated and chlorinated waters without experimentally determining each case. They anticipate the need of further analyses as a check on the distribution system to see if there may be occurrences of Cl-to-N ratios greater than 5:1. If such a situation exists, it becomes bacteriologically significant at ratios of about 6.5:1. From this point forward the chlorine residual decreases rapidly to nearly zero at the breakpoint, where disinfection ceases entirely. This is a more serious situation than a consumer taste and odor complaint. Such a situation develops into nitrification in covered reservoirs, which produces nitrites that cause severe algae blooms, as well as serious dead-end problems. All the occurrences

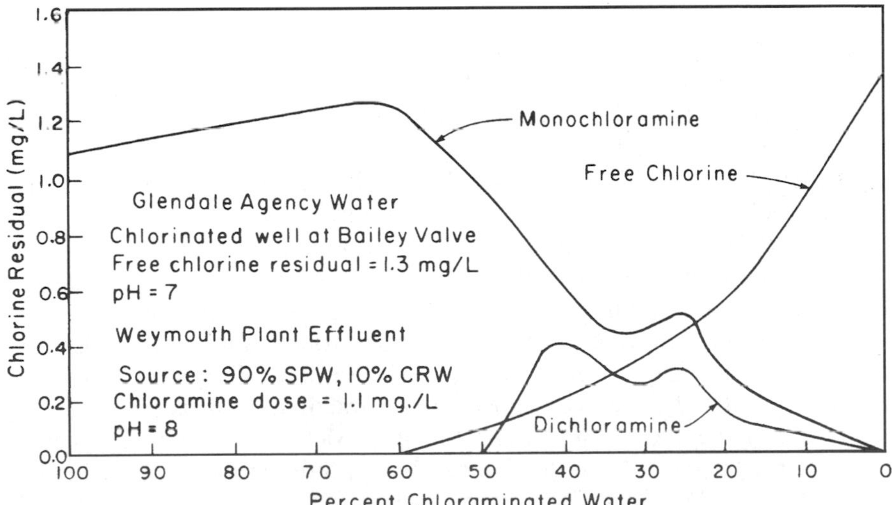

Fig. 6-8. Chlorine residual curves for Glendale California chlorinated well water blended with Weymouth treatment plant chloraminated effluent after 5-min. contact time[311] (courtesy Metropolitan Water District of Southern California).

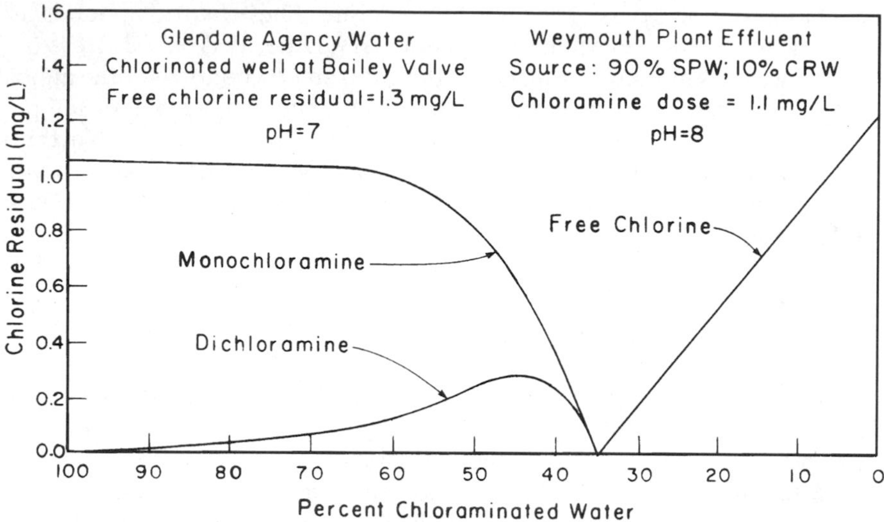

Fig. 6-9. Chlorine residual curves for Glendale California chlorinated well water blended with Weymouth treatment plant chloraminated effluent after 4-hr contact time[311] (courtesy of Metropolitan Water District of Southern California).

that MWD had to cope with make the cost of using the activated carbon method very desirable, as they never knew what was going to happen next. This is why the City of Los Angeles applies the breakpoint process to the chloraminated water that they receive from MWD.

Advances in Chlorination

Historical Research. Attempts to explain Griffin's findings[104, 105] led to serious scientific investigations during 1946–50. The work of Fair, Morris, Chang, Weil and Burden,[106–108] is largely responsible for our present knowledge of the kinetics of chlorination.

In 1950, Palin[109, 110] in England did some masterful work on the kinetics of the breakpoint reaction and the distribution of various chlorine residual fractions. In 1954, Granstrom[111] explained the breakpoint phenomenon. Then Williams of Canada, working with a most difficult water, proceeded to substantiate the laboratory findings of both Palin and the Harvard group.[112–114]

In the pursuit of his research work on chlorination, Palin directed his attention toward the analytical measurement of chlorine residuals. He wanted to be able to quantify each chlorine residual fraction, such as free chlorine, monochloramine, dichloramine, and nitrogen trichloride. His many years of work resulted in the DPD method of residual differentiation.[115, 116] In addition to the chlorine fractions, Palin developed procedures for chlorine dioxide, chlorite ion, bromine, and ozone.[117] The titrimetric procedure using ferrous ammonium sulfate as the titrant is a favorite with researchers.

However, the amperometric titrator remains the most popular method among operators who need to differentiate between free and combined chlorine. As the titrator is portable, it is well suited to field work for on-site determination of chlorine residuals. It is also the most popular method in wastewater and water reuse situations. Orthotolodine has been replaced by other colorimetric procedures, but largely replaced by the colorimetric DPD tablet method and the FACTS method. Orthotolidine lost its popularity when it was discovered that the orthotolidine powder was a carcinogen that affected the urinary tract of workers who prepared the indicator solution.

The laboratory research on the dynamics of chlorination and the breakpoint phenomenon was done by Wei and Morris[118] in 1972 at Harvard and by Saunier and Selleck[119] at the University of California, Berkeley in 1976. This work explained in much greater detail the kinetics of chlorination and provided a more precise explanation of what actually does trigger the breakpoint. Moreover computer models were developed that provide the practitioners with a tool for predicting what will occur in a given situation.

Although the laboratory research work taught the industry about the theoretical aspects of chlorine and ammonia reactions, it was the work by Williams[120] on a plant scale basis that showed the practical effects of chlorinating a water containing both ammonia nitrogen and organic nitrogen. His work demonstrated that chlorination in the presence of organic nitrogen formed an unusually stable chloramine compound which differentiates as the dichloramine fraction in either the amperometric or DPD-FAS procedures. This has since been confirmed by many others, but the amazing discovery about this "dichloramine" fraction, was made at a wastewater treatment plant in 1982, when the chief lab chemist (Virginia Alford) found that this fraction was really a nongermicidal organochloramine! Some of the most important advances in chlorination have been in improvement in the metering and control equipment. Automatic residual control was not available in North America until about 1960. Before that time Wallace & Tiernan had made some custom-designed residual controllers (using orthotolidine color as the control) for special situations. About 1950 Wallace & Tiernan of Great Britain began offering automatic residual control. Even though Wallace & Tiernan U.S.A. had already developed the amperometric titrator and the amperometric residual recorder by 1950, it was another decade before they believed that their new line of chlorination equipment was capable of combining flow pacing with residual control. Since 1960 a wide variety of control modes have been available from both Wallace & Tiernan and Fischer and Porter. Since about 1976 Capital Controls has been offering flow pacing combined with residual control. These developments have increased the reliability of the chlorination process.

The Free Residual Process

Role of Ammonia Nitrogen. This process is the implementation of the breakpoint phenomenon. The chemistry of this phenomenon is described in Chapter 4.

Free residual chlorination as a process was a direct result of the search for a better way to produce a palatable water than by the ammonia–chlorine process, also known as the chloramine process. This search was accomplished in the field and laboratory almost simultaneously on opposite ends of the United States by Griffin,[121, 122] and O'Connell[123] in 1939.* After literally thousands of tests on different waters, with different dosages and contact times it was found that the optimum residual for the most palatable water should contain 85 percent free chlorine.

This process is based upon the fact that a free residual (<5 mg/L HOCl) does not impart any off-flavor to the water. Moreover, this species of chlorine residual fraction is by coincidence the most germicidal.

In the free residual process, sufficient chlorine is added to destroy the ammonia nitrogen. This occurs when the ratio of chlorine to ammonia nitrogen (as N) is about 10:1 by weight. However, the stoichiometric ratio (theoretical) is 7.6:1. In practice it is known to vary from 8.5:1 to 11:1.

If ammonia nitrogen is present in the water, chlorine reacts rapidly with it to form mono- and dichloramine. These compounds will appear in the combined residual. Depending upon the factors of concentration and ratio of chlorine to ammonia nitrogen, NCl_3 (nitrogen trichloride) may form, which causes an offensive odor at the tap. Chlorine reacts slowly with most organic nitrogen compounds to form organic chloramines, which are stable even in the presence of high concentrations of free residual. Unfortunately these N-chloro compounds have little if any germicidal power. Moreover, most of the N-chloro compounds impart objectionable off-flavors to the treated water, depending upon their concentration. These compounds make up part of the combined residual fraction, the other fraction is the free chlorine.

Figure 6-10 illustrates the chlorine and ammonia reaction that produces the breakpoint phenomenon. At the breakpoint (bottom of the curve) there exists an irreducible minimum of total chlorine residual, the "nuisance residuals," which consist of free chlorine, mono- and dichloramine, and organochloramines. To the right of the dip there will be found increasing amounts of free residual together with the nuisance residuals and possibly some nitrogen trichloride. Nitrogen trichloride is not soluble in water; so it is extremely volatile. Therefore, water that generates nitrogen trichloride in the distribution system will produce an offensive odor at the instant of withdrawal from a tap. Reaction time between the chlorine and ammonia nitrogen is all-important. The only satisfactory way to provide proper control is to perform a bench scale study using different Cl-to-N ratios, different dosages, and different contact times. If the water tends to form NCl_3, try to find a contact time that occurs before the water leaves the plant; then the finished water can be

---

*Both Griffin and O'Connell were employees of Wallace & Tiernan. Griffin worked out of the W&T factory, while O'Connell worked as a special investigative chemist out of the San Francisco office. White was his assistant. Their project investigation was how to eliminate, by the use of chlorine, the objectionable taste in the Stanford University well water supply.

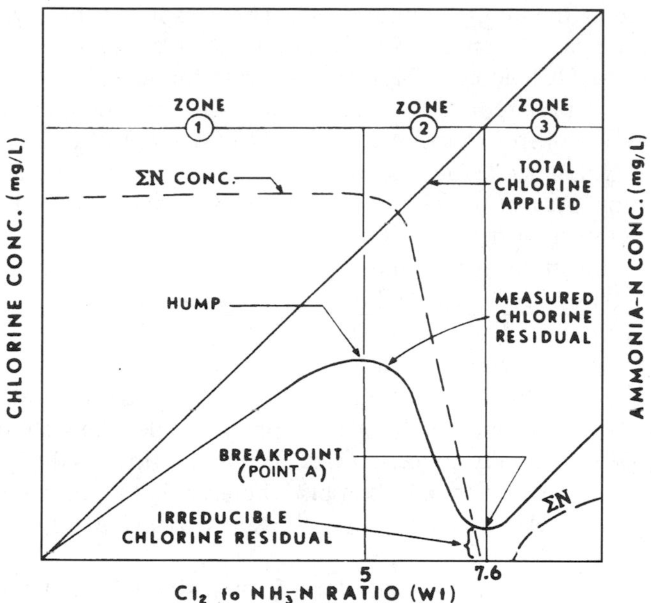

Fig. 6-10. Effect of chlorine and ammonia reaction illustrating the breakpoint phenomenon.

subjected to postaeration, which readily removes any NCl_3. It is also readily destroyed by sunlight.

The success of the free residual process depends upon maintaining a free residual that is always at least 85 percent of the total residual, because free residual is taste-free and is the most germicidal of all the chlorine residual species.

The speed of reaction between chlorine and ammonia nitrogen to produce a free residual varies from water to water because this reaction is not only temperature- and pH-dependent, but is grossly affected by the concentrations of both chlorine and ammonia N. If organic nitrogen is present in the untreated water, it compounds the problems associated with this process. This occurrence is described below.

Effect of pH. The free residual process is grossly affected by the pH of the water, because of the chemistry of hypochlorous acid (see Chapter 4). It dissociates in water as follows:

$$HOCl \rightarrow H^+ + OCl^- \tag{6-5}$$

All analytical techniques for measuring free residual include the sum of the undissociated hypochlorous acid (HOCl) and the hypochlorite ion (OCl^-). However, the efficiency of the free residual process relates directly to the concentration of the undissociated HOCl. The reason for this lies in the fact that the germicidal efficiency of HOCl is believed to be about 100 times more than that of the OCl^- ion.[100] The ratio of formation of each of these species is a function of pH (see Table 4-4, Chapter 4). For example, at pH 5 an aqueous solution of free chlorine contains 99.74 percent undissociated HOCl at 20°C. At pH 8 the undissociated HOCl is only 27.69 percent; the remainder is the OCl^- ion. As can be seen from Table 4-4, dissociation is also affected by the temperature of the water but to a lesser extent.

Figure 6-11 illustrates the amount of titrable free chlorine residual required to produce 1.0 mg/L HOCl. At pH 8, 5 mg/L titrable free residual is required to produce 1 mg/L undissociated HOCl. If the pH were 7.5, only 2 mg/L of titrable free chlorine residual would be required to yield 1 mg/L HOCl, etc. The effect of pH on the free residual process affects the selection of chlorine dose, as can be seen from the above example. Consider the example of virus destruction. It is generally conceded that a free chlorine residual is the most reliable for virus destruction. Clark et al.[124] believe that it takes about 0.3 mg/L of undissociated HOCl residual to inactivate Coxsackie A2 virus in about 20 min. at 5°C.

If the pH of the water to be treated is 8.5, then 10 mg/L of titrable chlorine is required to produce 1.0 mg/L undissociated HOCl (see Fig. 6-12). The free chlorine residual required at this pH level will be 10×0.3 or 3 mg/L. This then is the minimum dose that will be required to provide 0.3 mg/L of 100 percent undissociated HOCl at the end of 20 min.[124, 125]

Another consideration of the free residual process is the effect of sunlight on HOCl. While it is a relatively stable compound in that it resists loss due

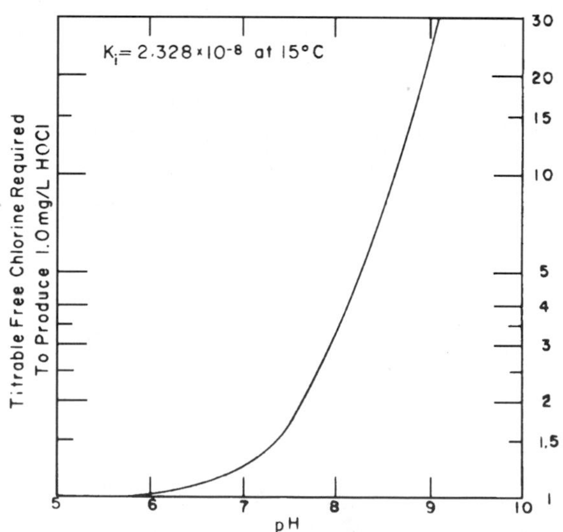

Fig. 6-11. Effect of pH on the formation of hypochlorous acid.

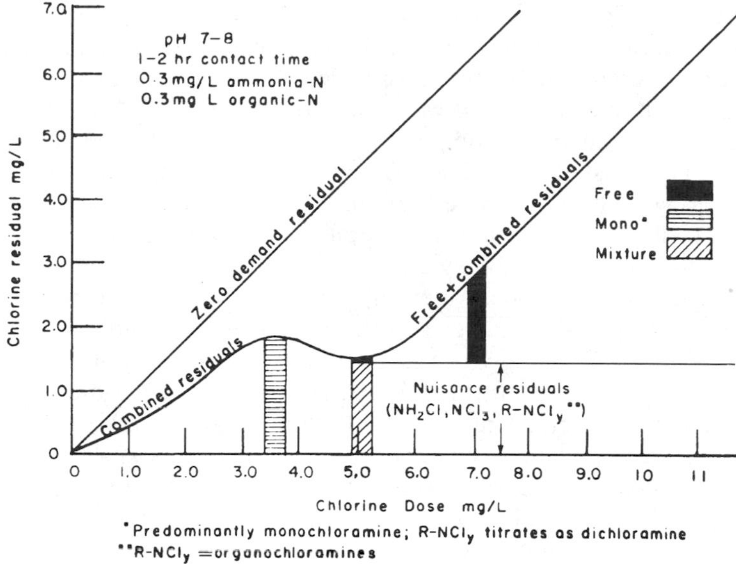

Fig. 6-12. Effect of chlorine–ammonia reaction in a water containing organic nitrogen.

to aeration, losses due to bright sunlight have been reported as high as 2 mg/L HOCl in 4 hours. Chloramines act in reverse fashion; they do not suffer appreciable loss due to sunlight but are subject to losses as high as 15 percent of the residual by aeration. If nitrogen trichloride is present, all of it is quickly lost because of either sunlight or aeration or both.

Reservoir Effect of OCl^- Ion. The notion is valid that at high pH values the distribution of the two species of titrable chlorine—HOCl and OCl —is such that dissociation of the OCl^- ion and the hydrogen ion provides a reservoir for the formation of HOCl. In accordance with Le Chevalier's principle, as soon as the HOCl in Eq. (6-5) is used up (i.e., reduced to HCl), more HOCl is immediately formed from the OCl^- ion and H^+ ion to maintain the chemical equilibrium in Eq. (6-5).

Role of Organic Nitrogen. Waters containing organic nitrogen in addition to ammonia nitrogen are quite another matter. Organic nitrogen in concentrations as low as 0.3 mg/L will seriously interfere with the chlorination process, and looms as a formidable obstacle in producing an acceptable water. Two areas that have been troubled with the organic nitrogen problem are Little Falls, New Jersey[126] and Brantford, Ontario. The latter is well documented.[114, 120] Griffin[104] found that waters high in albuminoid ammonia (organic nitrogen) displayed a plateau rather than the typical sharp dip associated with breakpoint curves. Figure 6-14 illustrates this characteristic. Palin[110] and others have confirmed these findings. In Chapter 4, Taras was able to clearly demonstrate how continuing chlorine demand tests could provide clues that can alert an operator to anticipate organic nitrogen problems derived from increasing pollution in the raw water supply. Researchers have found that there are a lot of chemical compounds that consist of "organic N." This is supposed to establish a significant difference between ammonia nitrogen and organic nitrogen, owing to the enormous varieties of organic nitrogen compounds that might occur in a water supply. In actual practice where surface waters are used in both potable water and wastewater, White found only one difference; this was a groundwater supply in El Dorado, Arkansas, where 3 mg/L of organic N did not react with the chlorine applied in establishing breakpoint curves. When any of these are present in excess of 0.3 mg/L this will complicate the control of the free residual process. The simple ammonia nitrogen compound (NH_3) reaction with chlorine rarely takes longer than 30 min. for the complete destruction of the ammonia and a 75–80 percent loss of nitrogen. Depending upon the temperature this would be for an ammonia nitrogen concentration of 0.3–0.5 mg/L. However the reaction between chlorine and organic nitrogen may go on for days before going to completion with the chlorine.[120] Depending upon the particular nitrogen compound, the reaction may never go to completion.[127] These last two studies were done with separate and different organic N compounds. Those that appear in surface waters do not seem to vary.

Organic nitrogen in potable water is a direct result of wastewater contamination. However, there is thought to be a significant amount entering surface supplies due to natural runoff that leaches and dissolves organic compounds from industrial waste dump sites. The source of most organic nitrogen compounds is proteinaceous matter and urine, both of which are present in copious quantities in all domestic wastewater discharges. Urine contains substances that react with chlorine to form extremely stable N-chloro compounds that titrate as dichloramine by either the amperometric method or the DPD method of residual analysis, but are nongermicidal. One of these constituents of urine is creatinine, which forms a chlor–creatinine compound and shows up on the curve of Fig. 6-14 as a nuisance residual. There are other N-chloro compounds (organic chloramines) in this same category that have little or no bactericidal value, as they appear to be unreactive.[127] Their significance is not clearly understood, but the consensus is they are not only undesirable, but nongermicidal.

In Fig. 6-14 it is to be noted that there is very little drop in the chlorine residual beyond the hump, signifying continuing and competing reactions between mono- and dichloramine, no perceptible loss of nitrogen in the reactions, and that the irreducible minimum residual is considerably larger than in water, containing only ammonia nitrogen. Here again, it is the ratio of the free chlorine beyond the plateau to the total chlorine residual that is of importance. The combined fraction will contain N-chloro compounds responding to analysis as mostly dichloramine, and always some monochloramine. As the ratio of chlorine to nitrogen approaches 20:1, obnoxious quantities of nitrogen trichloride are certain to develop. Williams[113, 114] has discovered that best results occur when enough chlorine is applied to produce a total of approximately 6.0 mg/L combined chlorine residual, resulting in about 5.0 mg/L free residual (HOCl), 0.4 mg/L monochloramine, and 0.6 mg/L dichloramine, as differentiated by the amperometric forward-titration method. To prevent formation of nitrogen trichloride in the distribution system, this water is subsequently treated with an excess of ammonia to convert the outgoing residual to monochloramine (80 percent) and dichloramine (20 percent). The water can be dechlorinated and rechlorinated, whichever makes the most palatable water at the least cost.

In reviewing the work of Williams and others in that era, it seems that their quality problems were due primarily to pollution by organic compounds, which must have diminished considerably in the last three or four decades. This change could be from a progressive reduction in untreated industrial waste discharges into surface water systems. Another important factor in the kinetics of chlorine and nitrogenous compounds is the relationship of monochloramine to organochloramines. In water reuse systems where contact times can be monitored and are designed for maximum effectiveness, it has been discovered that, after a certain point, as the contact time is lengthened, the germicidal efficiency starts to decrease. This phenomenon is noticeable when the contact time exceeds about 40 to 50 min. in most systems. Unfortunately, this organochloramine shows up as total chlorine residual on a chlorine analyzer. In these particular systems it was discovered that the combined residual showed a decrease in the monochloramine fraction and a simultaneous increase in the dichloramine fraction, which in 1982 was found to be a nongermicidal organochloramine. The amount of error involved from this phenomenon is determined by analyzing the chlorine residual by the complete forward titration procedure using only the amperometric procedure or the DPD-FAS titrimetric method.

Morris et al.[137] reported in 1980 that a large number of nitrogenous organic compounds occur naturally in water supplies and react readily with aqueous chlorine. This results in the formation of N-chloro compounds that exert a significant chlorine demand, and some can produce chloroform, particularly in the pH range of 8.5–10.5. The compounds identified by Morris et al. are adenine, 5-chlorouracil, cytosine, guanine, purine, thymine, and uracil.

This same investigation also revealed that summer blooms of blue-green algae can result in a significant increase in the organic N content of a potable water supply. This raises concern about the potability of a finished water during occurrences of summer blue-green algae. Aside from the taste and odors associated with such occurrences, high levels of organic N material released by these algae could result in increased THM formation as well as combined N-chloro species that demonstrate little or no germicidal value. Moreover most of these N-chloro compounds yield false-positive free chlorine residuals.

In potable water supplies, whenever organic nitrogen is present in concentrations higher than 0.3–0.5 mg/L, the operator can expect difficulties with the free residual chlorination process. This much organic nitrogen indicates a highly undesirable low-quality water that has probably suffered from recent contamination by wastewater effluent. Special pretreatment processes should be considered for these waters. White has suggested that a water quality parameter for any potential drinking water supply be limited to 0.3 mg/L organic N[128] and a 30 min. chlorine demand of no more than 1.5 mg/L.

Maximum THM Potential. The potential for the formation of THMs is directly related to the chlorine demand. Trihalomethanes are formed by the reactions between natural organic compounds (humic and fulvic acids) in the environment, naturally occurring bromides, and chlorine used in the water treatment processes. One of the techniques to determine the potential THM formation is a procedure long known as the "chlorine demand" of a water.

In the case of THM formation the recommended procedure is to add sufficient chlorine to the sample so that a measurable free chlorine residual will persist for at least seven days. The samples should be stored in screw-top or crimp-top bottles filled to exclude air space and placed in a 25°C water bath in the dark. Depending upon local circumstances the detention time may be more or less than seven days.[156]

The total trihalomethane (TTHM) concentration is measured at daily intervals. These concentrations are then plotted to generate a TTHM curve as shown in Fig. 6-13. When the curve reaches a plateau, that is the maximum THM potential (MTP).

This extended chlorine demand test can serve as a useful tool for the water producer. It is a significant water quality parameter that should be evaluated annually. To make use of Taras's chlorine demand concept described in Chapter 4, chlorine residuals should be measured after 30 min. and 60 min. of contact time and then at hourly intervals for the first 24 hours.

Each water supply will react differently to this test. Chlorine dosages to provide a measurable free chlorine residual to produce a plateau will vary from 2 to 20 mg/L, and the reaction time may be from 5 to 15 days. These variations will be a function of the water quality and the nature and the concentration of the precursors.

The Hackensack Experience[457]

The recent chloramine versus free chlorine experience of the 122-year-old Hackensack Water Company, serving 750,000 consumers, was inspired by the desire to control THM formation to 0.1 mg/L or less, in order to meet new regulations. Chlorination was used from 1940 to 1967, when the company switched to free chlorine. In 1976, it switched back to chloramination and immediately ran into some serious problems with dialysis patients as well as bad public relations from widespread fish kills in home aquaria. After 6 days of chloramination, it had to return to the B-P chlorination process. This required a dosage of 4–8 mg/L (depending upon the season) to produce a free residual through the filters of 0.3–0.5 mg/L. This single point of chlorine

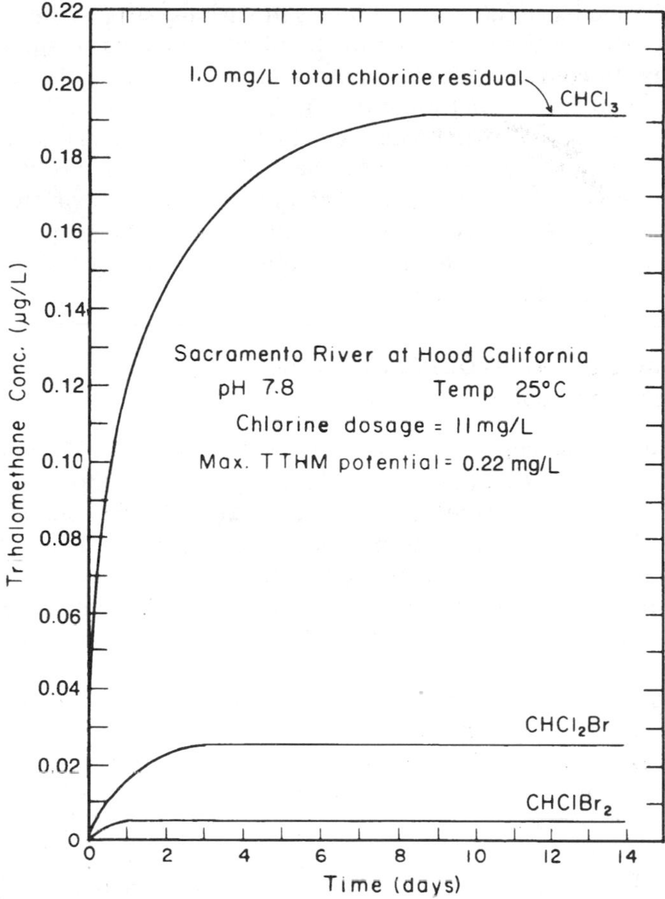

Fig. 6-13. Total trihalomethane formation potential due to the free residual process.[156] It is important to note the total THM potential is in micrograms per liter.

application usually resulted in THMs in excess of 0.1 mg/L. This is not surprising with such a high chlorine demand, caused no doubt by a high TOC demand.

To solve this problem, operators had to break down the prechlorination dose of 4–8 mg/L into four stages: three pre- and one post-. The prechlorination dose was split between the raw water headers, the basin inlet, and the prefilter point. The dosage was also reduced in progressive stages. The total prechlorination dose was virtually cut in half. The prefilter dose was made large enough to ensure a slight free chlorine residual through the filters to assist in the removal of the manganese in the raw water. This is surprising because both iron and manganese present should have been removed by the ozone. The postchlorination dose had to be sufficient to produce at least a 1.0 mg/L free chlorine residual in the plant effluent. This higher residual was necessary because of some bacteriological problems in various parts of the distribution system. After a short contact time in the plant, ammonia was added to the free residual to convert it to chloramines. This revision of the chlorination process reduced the THM formation to about 0.07 mg/L. This is a typical scenario for a water source that has a high chlorine demand. It must be remembered that in these cases where the THM formation has been reduced by either ozone or any other improvement in the pretreatment processes, the reduction of THMs is solely due to the reduction in the chlorine demand. It is also typical of such a water to react favorably to chloramines, another reason why these experiences are site-specific.

Total Organic Carbon (TOC) Removal by Anion Exchange Resins

Introduction. The current limitation on TOC is 4 mg/L. There are only a few city water supplies that ever exceed this amount. It is thought that humic and fulvic acids may be "Precursors" (more likely chlorine demand) for the formation of THMs and other disinfection by-products by their reaction with free chlorine (HOCl). These organic compounds are ubiquitous in the environment, and they account for 50–60 percent of the total organic carbon (TOC) present in potable water supplies. Therefore, the removal of TOC is a direct route to the elimination of DBPs (THMs and others) generated by the reaction of TOC and free chlorine, which results in the removal of about 0.1 mg/L of the total chlorine demand. Consequently, the primary objective in water treatment practices should be the removal of TOC instead of the more complex approach of using disinfectants other than HOCl.

The IX Method. The fulvic and humic acids that make up the majority of the TOC exist largely as negatively charged anions at pH above 6. Therefore, removal of these substances by anion exchange resins (IX method) should be a reliable possibility. Previous work showed that the Rohm and Haas acrylic anion exchange resin IRA 958 was effective for the removal of TOC.[356] Kim and Symons[355] proceeded to investigate the anion exchange (IX) method of TOC and THM precursor removal. They took into account that previous work done in 1979 and 1980 demonstrated that the THM production, upon chlorination, varied as a function of the molecular weight of the humic substances.[357, 358]

The several treatment techniques that will remove precursors of DBPs include coagulation, oxidation, adsorption, ion exchange, and membranes. Of these, membranes are the best but expensive. The most common approach for going beyond the 30–50 percent removal achieved by coagulation, settling, and filtration is adsorption on granular activated carbon. GAC performs well but is not long-lasting. It lacks acceptable reliability.

Performance. The University of Houston investigation by Kim and Symons[355] demonstrated that the IX method performed very well even though regeneration was necessary. This can be done in place with caustic and salt. They found that the IX bed could be reused eight times without interfering with its performance. In-place generation is not possible for GAC. Their best performance was to follow the IX bed with a GAC bed. Here, the IX bed acted as a "rougher," protecting the expensive GAC and permitting extended performance. After 4000 bed volumes (four weeks), when the test was stopped (the total time to exhaustion was not determined), the combined system was still removing 93 percent of the influent THMFP (THM precursor). The effluent water quality at that time was TOC = 0.2 mg/L and THMFP = 0.0065 mg/L. This compares with the TOC levels in the current finished drinking water from Lake Houston of 2–3 mg/L.

Chlorine Demand. There are sufficient data from more than a half-century of operating experience to declare that the 20-min. chlorine demand of a raw water can be used as a surrogate parameter for water quality. It is suggested that if the chlorine consumed in this 20-min. period exceeded the arbitrary allowable level, the water in question would have to undergo some pretreatment process to reduce the natural chlorine demand to the acceptable level.

To put the chlorine demand level in perspective, it is important to mention that the source of all the water on this planet—seawater—has a 20 min. chlorine demand that rarely exceeds 1.5 mg/L. White suggests that the maximum allowable source water chlorine demand should not exceed 2 mg/L.

A 1996 DBP survey shows a steady shift in the practices of many utilities in order to maintain a balance between adequate disinfection and minimizing DBP concentrations in accord with the Washington, D.C., EPA THM requirements.

Clearly, the diversity of utilities and source waters has created a multitude of responses in these new challenges to control the THM trivia. On the basis of the current survey, the following major conclusions have been reached:

- Utilities are making concerted efforts toward meeting THM regulations.
- To implement reduction in THMs, the most significant change by utilities was to alter the point of application of chlorine and the dosage of chlorine used.
- Many utilities have instituted changes in coagulation to minimize THM production.
- A sizeable fraction (20 percent) of utilities have reported using ammonia addition as part of their trivial THM minimization program.
- Some utilities have applied ClO_2 as a preoxidant or final disinfectant in order to reduce trivial THM formation.
- Changes to reduce THM formation that involve major infrastructure investments, such as GAC contactors or the addition of ozone, have not been widely adopted.
- A sizable fraction of utilities that incorporated process modifications for trivial THM reduction reported operational problems as a result of the modifications.

In the years ahead, with changes both in regulations and in the state of knowledge, the committee anticipates further changes in disinfection practices. The committee promises that these changes will be reported in subsequent reports.

City of Phoenix Municipal Water System

Introduction. Like many other municipalities, the City of Phoenix has to deal with an ever increasing population that demands more and more water. This water has to be of highest quality to meet the new EPA amendments to the 1986 SDWA. In view of these developments, the City organized a master plan that included a pilot study of many different treatment strategies to improve the water quality in order to meet the proposed new EPA standards. These strategies are described below.[354]

Water System Description. The City of Phoenix owns and operates a potable water supply and distribution system that serves a population of more than 1.1 million. The sources of this water supply are: the Salt River Project, the Central Arizona Project (CAP), and groundwater from the valley's aquifer. This water is treated at five major facilities, which include four water treatment plants for the Salt River Project with a total output of 430 mgd and the Central Arizona Project WTP at 80 mgd. Additionally, there are 55 wells that supply about 80 mgd of groundwater located in the Verde Well Field, which is adjacent to the Verde River. The distribution system includes over 5000 miles of water mains and more than 460 million gallons of water storage capacity. The raw water sources for the City's SRP water treatment plants are the Salt and Verde rivers.

Water Quality. The raw water quality parameters include TOC, ranging from 2.5 to 3.5 mg/L; pH, 7.8–8.6; hardness as $CaCO_3$, 150–200 mg/L; turbidity, 5–30 NTU; and seven-day THM formation potential, 0.20–0.25 mg/L. The finished water THM quarterly distribution system concentration is 0.055–0.070 mg/L. Alum dosages are 15–25 mg/L, and chlorine is 2.5–4.0 mg/L to produce a free residual in the settled water. Additional chlorine can be applied to either the settled or filtered water so that a free residual of 0.8–1.2 mg/L is carried into the distribution system.

Development of Treatment Strategies. A range of treatment strategies was developed to minimize the DBPs. The primary disinfectants chosen were chlorine and ozone. Chloramines were not considered because of the prohibitively high *Ct* values required to achieve primary disinfection. For secondary disinfection both chlorine and chloramines were selected, based upon their ability to maintain stable residuals in the distribution system. Ozone was not considered owing to its inability to maintain a distribution system residual.

GAC was included in these strategies as the basis of design for evaluating both adsorption and biodegradation. Membrane technology, such as nanofiltration for the removal of TOC, DOC, and "THM precursors," was also included in the treatment strategies. The strategies were divided into two categories: (1) those using chlorine as a secondary disinfectant (Nos. 1 through 11) and (2) those using chloramine as a secondary disinfectant (Nos. 12 through 21).

The possible microbiological effects of ozonation were considered because of the increase in AOC after ozonation. Consequently, the use of ozone requires a biologically active filter medium to stabilize the AOC in the ozonated water. Therefore, to stabilize the microbiological quality of the finished water, any strategy using ozone and chloramines includes a brief period of free residual chlorination following the biologically active filter.

For the purpose of screening the various strategies, they were grouped as follows:

a. Coagulation, oxidation, and disinfection.
b. Provision for adsorption.
c. Membrane technology.

THMs were evaluated as simulated distribution system THMs, identified by SDS-THM.

Of Strategy Nos. 1, 2, and 12, based upon previous experience of others, Strategy No. 2 was retained. This eliminates prechlorination by moving the point of primary disinfection downstream from coagulation, flocculation, and sedimentation. Similarly, Strategy No. 12 was retained, which uses chlorine as a primary disinfectant following sedimentation and chloramination for secondary disinfection. Fig. 6-14a illustrates Strategy Nos. 1, 2, and 12.

Strategy No. 3 was retained because bench and pilot testing indicated that lowering the pH during coagulation improved DOC removal. The use of chloramines as a secondary disinfectant along with lowering the pH reduced the SDS-THM concentration from a range of 0.04–0.06 mg/L to less than 0.025 mg/L. This made Strategy No. 13 eligible for review. Fig. 6-14b illustrates Strategy Nos. 3 and 13.

Strategy Nos. 7 and 9, which used ozone as the primary disinfectant and chlorine as the secondary disinfectant, were eliminated because of the increase in the formation of SDS-THMs along with lack of DOC removal by ozone.

Strategy No. 10 was retained because DOC removal can be achieved by adding adsorption to the strategy.

Strategy Nos. 17 and 21 were also retained. They used ozone as the primary disinfectant, which indicated a marked potential for reducing THMs by minimizing the contact time of free chlorine. Fig. 6-14c illustrates these two strategies.

Strategy No. 8 uses a combination of ozone and hydrogen peroxide as the

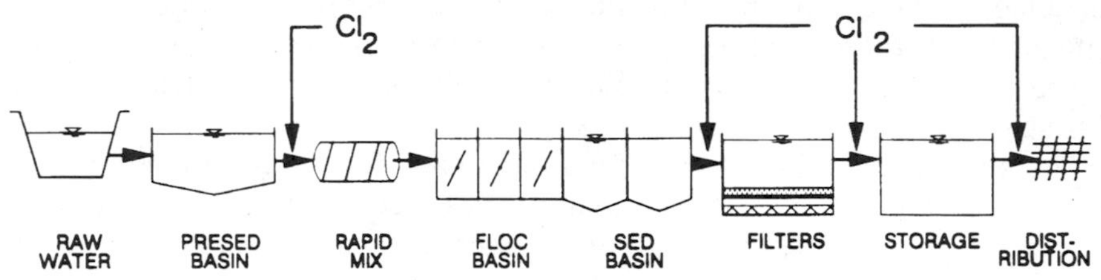

STRATEGY NO. 1

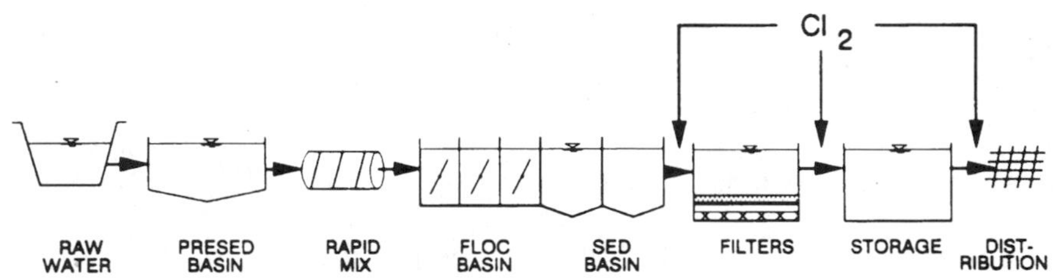

STRATEGY NO. 2

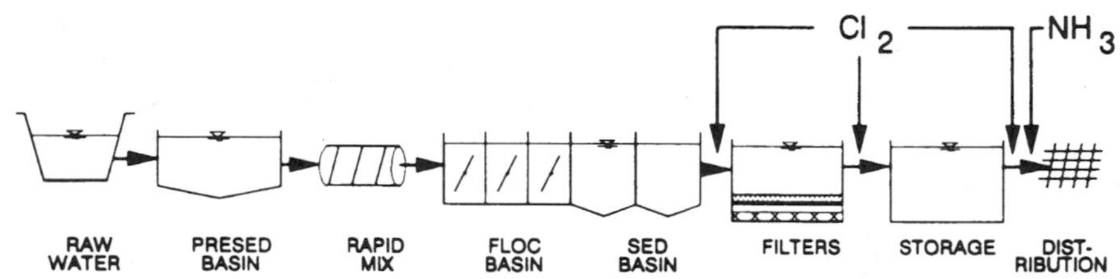

STRATEGY NO. 12

(a)

Fig. 6-14a. Strategies No. 1, 2 chlorine/chlorine, and 12 chlorine/chloramine (courtesy City of Phoenix, Arizona).

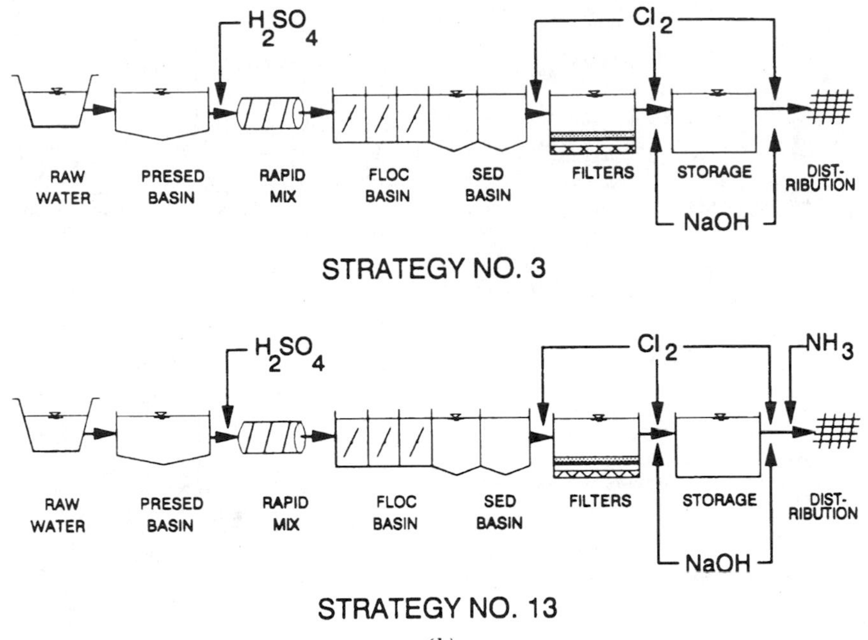

Fig. 6-14b. Strategies No. 3 chlorine/chlorine and 13 chlorine/chloramine (courtesy of City of Phoenix, Arizona).

primary disinfectant. This usage is commonly called the "peroxone process" and is listed as an Advanced Oxidation Process (AO_XP). This strategy, using chlorine as the secondary disinfectant, increased the THM potential compared to the present baseline treatment with lower pH. Therefore, this strategy was eliminated.

The results of Strategy No. 18 were similar to those for Strategy No. 17. Therefore, to simplify comparisons of strategies, No. 18 was eliminated, and No. 17 was used for cost estimates to represent peroxone as the primary disinfectant.

Strategy Nos. 4 and 14 were used to investigate the use of PAC adsorption for the removal of natural organic material (NOM) in the water supply. PAC dosages of 10–50 mg/L along with alum at pH 6.5 did not reduce the SDS-THMs or improve DOC removal. Therefore, these strategies were eliminated from the comparison of strategies.

Strategy No. 9 was not retained because the substitution of ozone for chlorine as a primary disinfectant does not reduce SDS-THMs when chlorine is used as the secondary disinfectant.

Strategy Nos. 5, 10, 15, and 19 were studied as a part of the pilot project to determine if GAC was effective in the reductions of DOC and THMs.

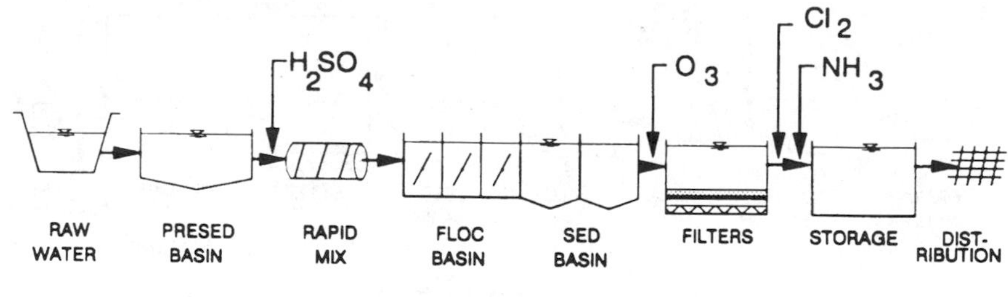

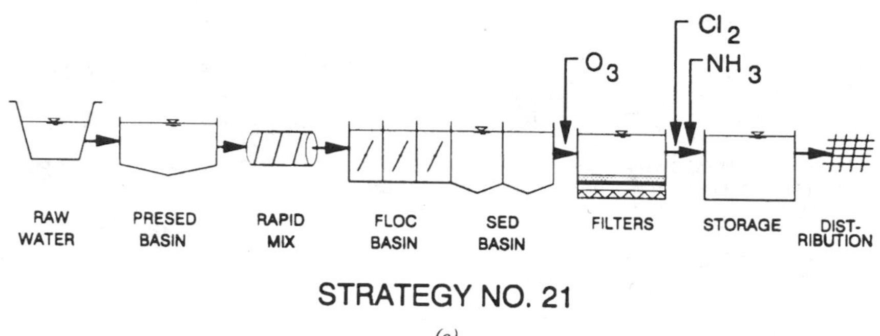

(c)

Fig. 6-14c. Strategies No. 17 and 21 ozone/chloramine (courtesy of City of Phoenix, Arizona).

These substances were reduced by 75 and 95 percent, respectively—strictly by adsorption. Therefore, these strategies were retained for comparison. They are illustrated in Fig. 6-14d and Fig. 6-14e.

Strategy Nos. 6, 11, 16, and 20 were all used for the evaluation of membrane technology. They were performed with coagulation at pH between 5.8 and 6.5 to decrease bicarbonate alkalinity, which reduces the fouling and improves recovery. A nanofilter softening membrane with pore sizes in the 150–250 molecular weight range was used as a polishing step following conventional treatment. The pilot study results indicated that 80 percent reduction of DOC and THMs could be achieved by these membranes. However, owing to the hardness of the water the membrane manufacturer advised that the water recovery rate would recede to a range of 80 to 85 percent over a period of time longer than that used for the pilot study. Therefore a 15–20 percent volume rejection could be undesirable, considering water volume availability in the Phoenix area.

Nevertheless, these strategies were retained for further evaluation because of the impressive removals and improvement of water quality over the baseline strategy. These strategies are illustrated in Fig. 6-14f and Fig. 6-14g.

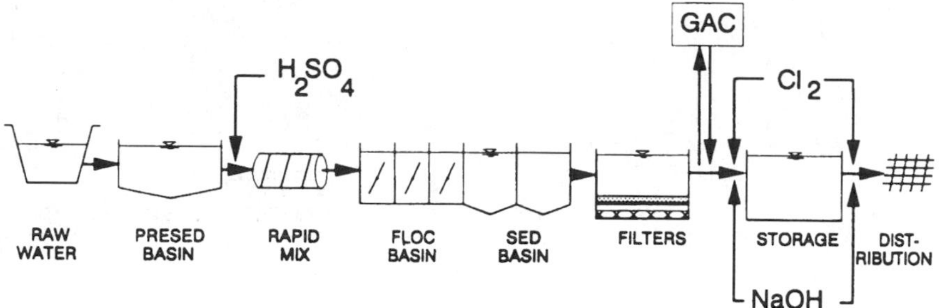

STRATEGY NO. 5

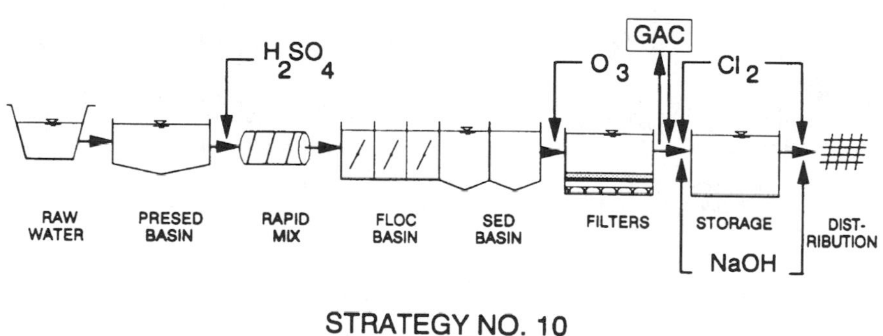

STRATEGY NO. 10

(d)

Fig. 6-14d. Strategies No. 5 and 10 chlorine/chlorine and ozone/chlorine (courtesy of City of Phoenix, Arizona).

Summary. The key component in the wide array of treatment options was the attempt to control (insignificant) DBPs because of anticipated future water quality regulations. Evaluation of these treatment options has resulted in the following conclusions:

1. If the MCL for THMs is reduced to 0.05 mg/L or less, Phoenix will have to replace free chlorine with chloramines unless supplemental NOM removal processes are added.
2. From a treatment standpoint the least costly approach to meet the lower THM MCL is the use of chloramine as a secondary disinfectant with either ozone, chlorine, or chlorine dioxide.

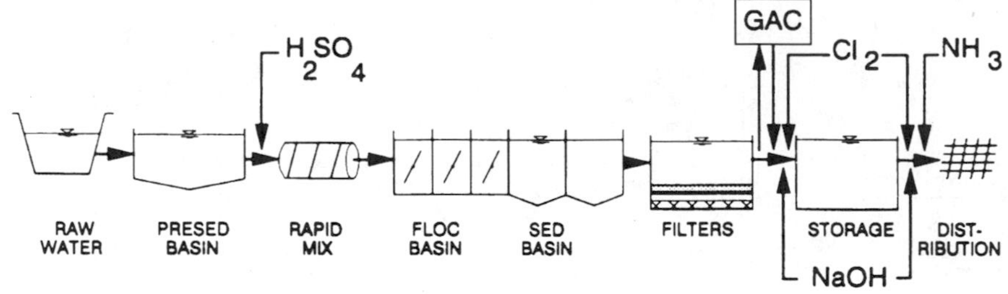

STRATEGY NO. 15

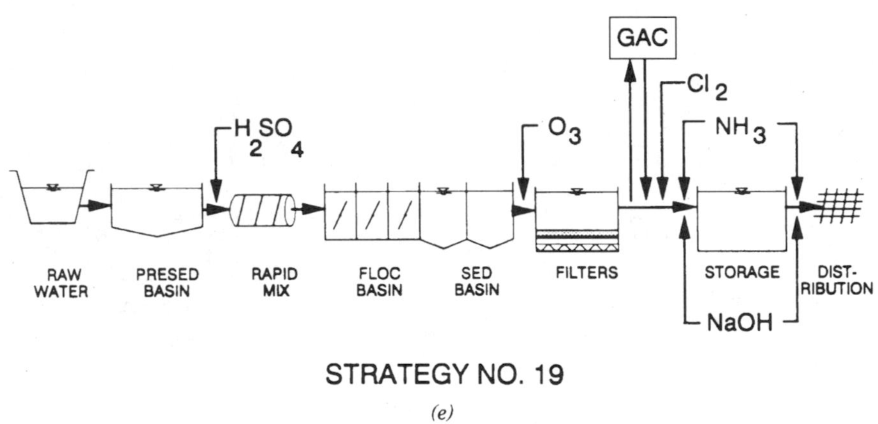

STRATEGY NO. 19

(e)

Fig. 6-14e. Strategies No. 15 and 19 chlorine/chloramine, and ozone/chloramine (courtesy City of Phoenix, Arizona).

POTABLE WATER REUSE PROJECT*

Denver, Colorado Pilot Study

Introduction. In 1985, the city of Denver Water Department initiated one of the most significant and ambitious projects in the history of water treatment: its recycled wastewater project, which was developed to determine if potable water of comparable quality to Denver's existing supply could be continuously and reliably produced from a secondary wastewater effluent. This comparison was chosen because the Denver water supply is derived from a relatively protected source, and there is no reason to believe that it will fail to satisfy any future health standards that may be developed. This ensures the margin of safety necessary to apply this technology many years hence. Surveys showed that if water were produced that would be comparable to Denver's existing supply, consumers would be more inclined to accept it as an alternate source.

*For current (1997) reuse projects now under design or construction, see Chapter 8.

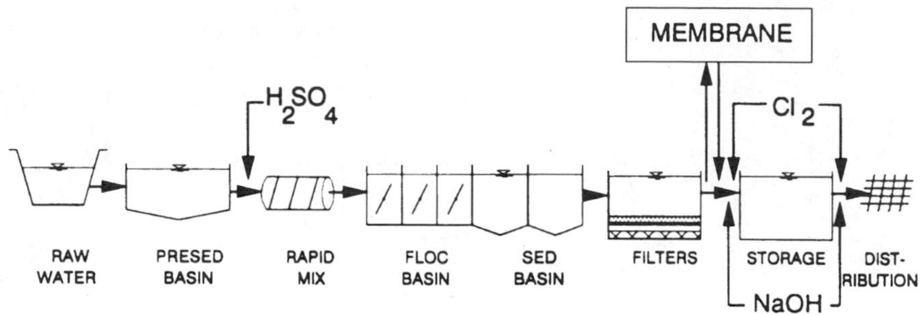

STRATEGY NO. 6

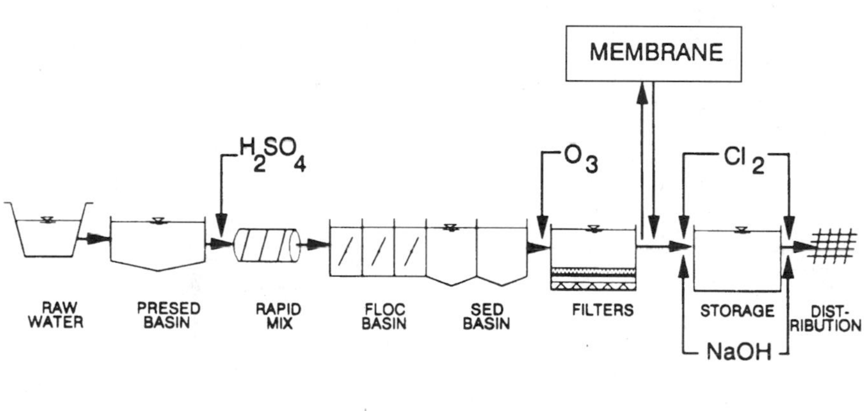

STRATEGY NO. 11

(f)

Fig. 6-14f. Strategies No. 6 and 11 ozone/chlorine (courtesy of City of Phoenix, Arizona).

This is a $30 million project to treat 1.0 mgd flow of secondary wastewater effluent. To avoid all possible DBPs, this effluent is not to be chlorinated prior to any treatment.[445, 446]

Project Objectives. Any proposal for direct potable reuse (without groundwater recharge) raises many questions regarding health effects, safety, reliability, public acceptance, and both technical and economic feasibility. Therefore, this full-scale pilot project has to determine product safety, demonstrate process dependability, generate public awareness, achieve regulatory agency acceptance, and provide data for large-scale systems.

Project Treatment Train. The multiple barrier approach was selected to duplicate containment removal capabilities by succeeding process steps. Also,

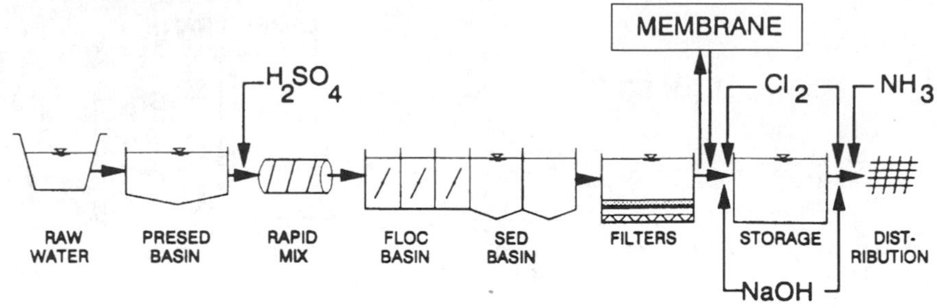

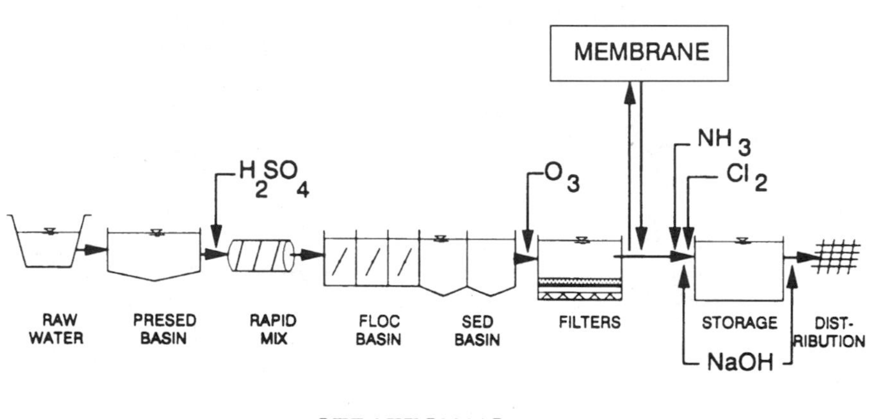

(g)

Fig. 6-14g. Strategies No. 16 and 20, chlorine/chloramine and ozone/chloramine (courtesy City of Phoenix, Arizona).

sufficient backup systems were included so that continuous operation would be possible with normal mechanical failures. Additional flexibility was provided by allowing the option to bypass any process step. This created alternate treatment combination possibilities. The treatment train is illustrated by Fig. 6-15. The process steps are as follows:

1. High-pH lime clarification.
2. Recarbonation—to lower pH.
3. Filtration.
4. UV irradiation.
5. Activated carbon adsorption.
6. Reverse osmosis or ultrafiltration (membrane technology).

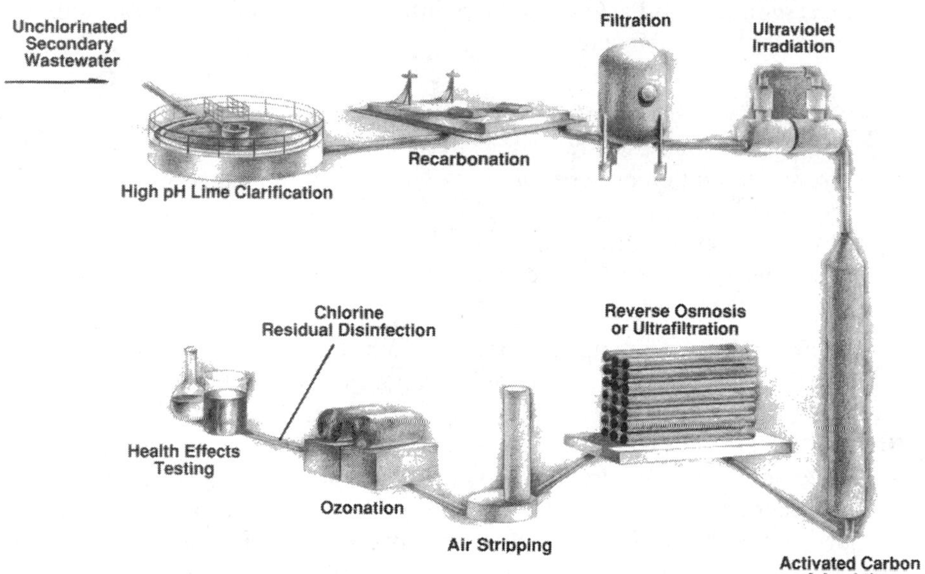

Fig. 6-15. Project treatment train (courtesy city of Denver, Colorado).

7. Air stripping.
8. Ozonation.
9. Postchlorination with a specified contact chamber.
10. Health effects testing.

Current Evaluation. From current available data the results are significantly optimistic. It is quite apparent that the finished water quality of this treatment train will produce a higher-quality water than many finished water effluents from existing surface supplies. One very favorable sign is the ability of the system to lower the influent TOC from peaks of 12 mg/L down to 4–5 (mg/L) in the finished water. One problem that remains unsolved is incomplete ammonia removal. The finished water at the point of disinfection contains 2–3 mg/L ammonia N. This means that final disinfection would be by chloramination.

CURRENT PRACTICES IN THE UNITED STATES

Review of Modern Practices

Chlorine Dosages. In March 1978 a questionnaire developed by the AWWA Disinfection Committee was mailed to numerous water utilities in an attempt to document current practices. There were some 350 respondents to this questionnaire.

The respondents indicated that the points of application could be a combination of any of the following:

- Pre–raw water storage.
- Precoagulation/post–raw water storage.
- Presedimentation/postcoagulation.
- Postsedimentation/prefiltration.
- Postfiltration (disinfection).
- Distribution system (disinfection/water quality).

The chlorine dosage range reported was a minimum of 0.2 mg/L to a maximum of 40 mg/L. The maximum dosage reported by 96 percent of the respondents was 15 mg/L. The all-time record is 120 mg/L reported at Ottumwa, Iowa in 1950.[157, 158] The raw water was from the Des Moines River. This range of dosages demonstrates dramatically the variation in water quality throughout North America. When the free residual process is practiced, a winter ice cover on surface supplies will almost automatically double the summertime chlorine demand. This is a result of the ammonia N increase due to the ice cover.

To combat pollution and taste and odor problems, various combinations and dosages of chemicals are in use today in both the United States and Canada. These include combinations of chlorine and chlorine dioxide, ammonia, potassium permanganate, and activated carbon, sometimes followed by dechlorination with either sulfur dioxide or activated carbon filtration.

Prechlorination of low quality water is most often the operator's salvation. It is one of the most important tools for maintaining the efficiency of a water treatment plant. In these situations, which are numerous, it would be virtually impossible to turn out an acceptable water if it were not for the unique ability of chlorine to maintain a persisting residual throughout the process. In these cases disinfection is just a side effect. However a lot of these water suppliers will have to reevaluate the chlorination process in order to meet mandatory TTHM regulations.

Two examples of low quality water that require chlorine dosages up to 16.0 mg/L are the Grand River at Brantford, Ontario, Canada[159] and the Passaic River at Little Falls, New Jersey.[160]*

In North America water has been treated for decades by a prechlorination dose sufficient to provide a substantial free residual in the flocculation or sedimentation basins. Many of these plants follow this by an intermediate dose of chlorine or chlorine dioxide sufficient to carry a residual through the filters. This practice will probably have to be modified in some instances. Switching from free residual chlorine to chloramines or chlorine dioxide might be the answer

*Note: The three water treatment plants using the Seine River as the domestic water supply for Paris all prechlorinate the water with 16–18 mg/L of chlorine dioxide. This is followed by ozonation and chlorine as it enters the distribution system.

in some cases (see Chapter 12). Other cases might need to change the entire pretreatment processes, which could include ozone. (See Chapter 13.)

In those cases where free residuals through the treatment plant eliminate objectionable tastes and odors in the finished water, the use of postammoniation to convert the free chlorine to chloramines can sometimes be beneficial. This has been practiced at Brantford, Ontario and other plants for decades.

Dechlorination by sulfur dioxide to trim the chlorine residual as it enters the distribution system is being practiced in many places.

The case of clean waters where chlorine is used solely for disinfection is remarkably different from the case of low quality waters. Two typical examples are the supplies for the City of San Francisco and the East Bay Communities in Alameda and Contra Costa counties in California. These waters are derived directly from melted snow in separate runoff areas high in the Sierra Nevada mountains. The San Francisco supply is transported by a 165-mile aqueduct, and the East Bay supply travels about 90 miles to various local storage reservoirs. The chlorine dosage required for disinfecting these waters is 0.8–1.2 mg/L. The San Francisco supply was plagued from the start by assorted difficulties. The most humiliating one occurred a few months after the system was put in operation. Part of the 165-mile aqueduct consists of a concrete-lined tunnel 25 miles long through the Coast Range mountains. Although the concrete tunnel lining is up to 12 inches thick, many cracks developed due to heavy ground caused by groundwater. The groundwater is laden with filamentous bacteria (*Crenothrix*), which infiltrate the tunnel and give rise to luxuriant growths on the tunnel lining. This created a debacle in the distribution system. Industries were forced to shut down because of this biofouling and the sloughing of the filamentous debris. To combat the problem, the City of San Francisco, after having investigated several treatment methods, concluded that a persistent chlorine residual was the only answer. Therefore, a chloramine station was built (1937) to inject chlorine and anhydrous ammonia at the entrance to the tunnel. A 1.5 mg/L dose produced a combined residual of about 1.0 mg/L at the end of the tunnel 25 miles away. When ammonia became difficult to get during World War II, the system was converted to the free residual process, which is still in use (1983). San Francisco water does not generate THMs above the EPA maximum contamination limits.

The superior quality of the San Francisco water supply is illustrated in Appendix II. This is the 1990 annual water quality report. Examination of this table shows why San Francisco water is a favorite among the beverage industry and bottled water suppliers.

There are many areas in the United States that are not so fortunate as the San Francisco Bay Area. Two of these are St. Louis and Kansas City, Missouri. St. Louis, using Mississippi River water, prechlorinates at an average 5 mg/L. This is supplemented by intermediate doses of chlorine to preserve the efficiency of the filter system. Kansas city, using Missouri River water, prechlorinates at about 12 mg/L. These chlorine applications are not for disinfection.

This is to keep the water treatment processes from deteriorating and for taste and odor control.[161]

The City of Chicago, using a cleaned-up Lake Michigan water, prechlorinates at only 1–1.2 mg/L. This is comparable to the melted snow waters of California. This prechlorination dose is designed to provide an absolute minimum of 0.25 mg/L free chlorine through the filters. It is supplemented by a postchlorination dose of 0.25–0.5 mg/L to give a 0.75 mg/L free chlorine residual entering the distribution system. To exemplify the effect of pollution on chlorination, whenever there is an upset with the Chicago Ship Canal, amounting to a reversal of flow which dumps canal water into Lake Michigan, the chlorine demand escalates to 10 mg/L, and even at this dosage there is so much ammonia N in the raw water that a free chlorine residual is not attainable.[162]

Surface waters in some areas of Pennsylvania require prechlorination doses of 7–8 mg/L, and the City of Philadelphia provides sufficient chlorination equipment capacity to dose as high as 30 mg/L at the raw water basin outlet (ice cover situation) and up to 4 mg/L chlorine dioxide for pretreatment.[161]

It is significant to note the quality of the water provided by the billion-dollar California Water Plan. This water is a mixture of several rivers, primarily the Sacramento and San Joaquin. It requires only a 2–4 mg/L prechlorination dose to produce a 0.7 mg/L free residual in the filter plant finished water without any intermediate chlorination. This is a tribute to the diligence and surveillance of both the California State Department of Health and the California Water Resources Quality Control Board for their programs designed to preserve the water quality of the receiving waters.

Chlorine Residuals. To this question there were 226 respondents to the AWWA Disinfection Committee Questionnaire.[163] The highest reported was 5 mg/L and the mean was 1.4 mg/L (total Cl). The figures for free residuals were about the same: maximum 4.75 mg/L and mean 1.0 mg/L.

Out of 332 respondents, 235 were using DPD, 165 amperometric titrator, 127 acid O-T, 11 neutral O-T, 1 FACTS, and 1 starch–iodide.

Contact Time. One of the questions asked in this questionnaire was designed to determine the contact time between the point of application of chlorine and the first consumer. Strange as it seems, 90 percent reported 10 hours, and 6 percent reported 2 min. The median was 60 min., and the mean was 237 min.

Chlorine Demand. Chlorine demand tests reportedly take place in about 40 percent of the respondents' facilities. Twenty of the utilities reported maximum demands between 5 and 10 mg/L chlorine, and another 11 reported demands between 10 and 22 mg/L. Of 123 respondents the median demand was 1.8 mg/L. However the maximum reported was 65 mg/L at a maximum contact time of 12 hours.

In 1993 White found what he believes to be a surface water with the lowest possible chlorine demand, when he was working on a reevaluation project for the City of Santiago, Chile. The water supply is from melted snow on the adjacent Andes Mountains. However, this water travels several miles over a ground cover, heavily laden with a brown claylike soil, before reaching the WTP. The chlorine demand of this naturally filtered water was only 0.2 mg/L after 20 min. of contact time!

Deficiencies in Current Practices

The Community Water System Survey. The most comprehensive information available on the production of potable water in the United States is the Community Water System Survey of 1970.[129] This survey points directly to the deficiencies in current practices. As is always the case, whether it be potable water or wastewater treatment, it is the small communities and the small producers that have the greatest deficiencies. The large producers are endowed with more sophisticated treatment techniques, adequate funding, and more qualified operating personnel than the small ones. This report reveals that in communities of less than 500 population, only 50 percent met the Drinking Water Standards of 1962 (USPHS). At the same time in communities of 100,000 or more, 73 percent met these standards. Of equal or greater significance, the survey revealed that 19 percent of surface and mixed-source water facilities not practicing disinfection were not properly protected at the source, and of those facilities practicing disinfection, 16 percent were inadequate and 7 percent had inadequate control of the disinfection process. Overall, this survey indicated that too many Americans were not being provided with potable water that met contemporary standards of good practice.

Too many unnecessary risks are being taken in the production and distribution of potable water, particularly in the numerous small water systems (e.g., ski and other resorts, roadside restaurants, bus stops, motorway rest areas, trailer camps, farms, suburban homes, isolated institutions, organized summer camps, and other similar small water systems). Many of these small supplies are pumped into a pressure tank and thence to a close-coupled distribution system with practically zero contact time. Obviously the only line of defense for these systems is the disinfection facility. Therefore, this facility must be designed with a proper factor of safety to allow for variation in water-quality parameters affecting disinfection. The same rationale would also hold true for large systems under similar circumstances.[165]

Viruses. One of the major causes for concern is that the frequency of viral diseases and acute gastrointestinal illnesses (AGI) has been on the rise in the past few decades, while typhoid has almost disappeared in the United States. Until 1971 gastroenteritis as such was not a reportable disease. Since 1971 the EPA and the CDC have entered into a cooperative reporting scheme classifying such illnesses along with other outbreaks of known etiology. This has gone a long way in solving the origin of these outbreaks. During the period 1961–70 a total of 26,546 cases of gastroenteritis were definitely attributed to water.[166] Of 52 waterborne outbreaks in the United States in 1971–72 there were 22 outbreaks of AGI amounting to 5615 from a total of 6817 cases of all waterborne illnesses.[167] The concern here is that these cases may all be of viral origin. See the section on "Viral Diseases" in this chapter. There are over 100 viruses excreted in human feces that have been reported to be in contaminated water. Any of these could cause a waterborne disease, and some have done so.

In order to cope with viruses, the disinfection process will have to address itself to longer minimum contact times and higher residuals. Surface waters that rely solely on disinfection will have to be reevaluated. In these cases it may be necessary to adopt filtration as an additional barrier.

Proof of Disinfection. To date we have been relying upon the coliform group of organisms as proof of disinfection. So far this indicator has served us well for bacterial infection. However, it cannot be relied upon as an indicator for viral infection. Microbiologists are in total agreement that there is no observed relation between total coliforms, fecal coliforms, or standard plate counts and the presence of viruses. Also there is no observed relation between the detection of *Salmonella* and viruses and vice versa. With our present knowledge of viruses, the coliphages or enteroviruses cannot be used as indicators. This means that proof of disinfection must derive from a properly treated water; this would involve filtration and the assurance of a substantial chlorine residual in the distribution system. It has been demonstrated that water treatment processes such as those used to treat Missouri River water downstream from Kansas City, where fecal coliform densities are as high as 16,000 per 100 ml, are a reliable barrier for viruses.[55] This investigation was unable to isolate any viruses in the treated water.

Sanitary biologists have expressed dissatisfaction with coliforms as indicator organisms for years. They insist that this group of organisms is not resistant enough to chlorine or other disinfectants to allow any factor of safety. There have been many cases of enteric organisms such as the genera *Klebsiella* and *Enterobacter* (*cloacae*) found responsible for positive samples occurring in the routine coliform sampling procedure. The discovery of such potentially hazardous organisms in a distribution system is unsettling, particularly when the water has been coagulated, settled, filtered, and chlorinated sufficiently to carry a residual (0.5 mg/L) throughout the distribution system. There have been cases where the residual was kept much higher in an attempt to destroy these organisms, but without complete success. It is thought that these encapsulated organisms, once they are in the distribution system, may be harbored in protective slime, scale, or sediment. The *Klebsiella* germs are encapsulated organisms and can cause severe enteritis in children. In adults they can cause pneumonia and upper respiratory tract infections, septicemia meningitis, peritonitis, and urinary tract infections. The presence of *E. cloacae* in a water system implies that there could be contamination of fecal origin, although *E. cloacae* is not known to have caused any waterborne disease. However, *Klebsiella* is a hazardous organism.[169]

In view of the above, the work of Engelbrecht et al. is of considerable importance.[168] Their investigation evaluates two promising groups of organisms, which include two acid-fast cultures, *Mycobacterium fortuitum* and *M. phlei,* and a yeast, *Candida parapsilosis.* The resistance of these two groups of organisms to chlorine is believed to be in the range necessary to inactivate both bacillary pathogens and waterborne viruses. Further research for a better indicator organism is recommended.

The precise reason why *Klebsiella* and *E. cloacae* can be isolated in distribution systems in the presence of a significant free chlorine residual after the water has been coagulated, settled, filtered, and disinfected remains a mystery. In this chapter, under "Distribution Systems and Transmission Lines," some probable causes are discussed.

In spite of the above anomalies it is still the consensus that the best method of assuring the microbiological safety of drinking water is to maintain good clarity, provide adequate disinfection (which includes maintenance of a disinfectant residual in the distribution system), and make frequent measurement of the total coliform density and general bacterial population (SPC) in the distributed water.[174] The emphasis is always on a persisting disinfectant residual in the distribution system.[173, 174] The Community Water Supply Study[164] undertaken in 1969 to investigate the status of the surveillance program in each of 969 water supply systems found that only 10 percent met the sampling criteria, whereas 90 percent either did not collect sufficient samples or collected samples that showed poor bacterial quality or both. In general the survey showed that the probability of finding coliform bacteria in a distribution system decreases as the residual chlorine increases. Overall the Community Survey specifically showed that in chlorinated water supplies a chlorine residual must be maintained throughout the distribution system in order for operators to have confidence that disinfection has been accomplished.

Robeck[174] has been able to demonstrate with data that, when properly used, the chlorine residual test satisfies the problems occurring with bacteriological sampling. The sampling problems in many utilities occur when there is too high a proportion of the sampling in the free flowing, high turnover area of a system, and too small a percentage in the troublesome areas: reservoirs, dead ends, and low flow areas.

It could be concluded then that the safest drinking water system is one that maintains an adequate chlorine residual throughout the distribution system, which is verified by frequent sampling or continuous residual monitoring. This notion presupposes that this water meets the turbidity standards as well.

OBJECTIVES OF CHLORINATION

Disinfection Guidelines

Free Residual Process. Disinfection practices should be governed by the chlorine residual–contact time envelope, based upon the destruction of a consensus organism.[128] This concept was first proposed by Baumann and

Ludwig in 1962.[175] The major factors affecting the germicidal efficiency of the free residual process are: chlorine residual concentration, contact time, pH, and water temperature. Increasing the chlorine residual, the contact time, or the water temperature increases the germicidal efficiency. Increasing the pH above 7.5 drastically decreases the germicidal efficiency of free chlorine. (See Chapter 4.)

A review of the literature shows that the resistance of pathogens to free residual chlorination varies over a wide spectrum of organisms. Most of the viral pathogens are considerably more resistant than the bacterial pathogens. Cysts in most instances are the most resistant to free chlorine. Varma and Baumann[176] compiled the chlorine residuals and contact times needed to kill vegetative bacteria, viruses, and amoebic cysts from a comprehensive review of the literature. From this review they plotted the available data, shown graphically in Fig. 6-16, for the chlorine concentration–contact time relationship to achieve the destruction of these four classes of organisms. Table 6-3 shows the sources of the data for the plots in Fig. 6-16, which were considered to be the most up-to-date and reliable. Figure 6-17 represents a plot of disinfection versus free chlorine residuals at 0 to 5°C as compared to Fig. 6-16, which is at a temperature range of 20 to 29°C. As there are many locales where the minimum water temperature for both well water supplies and surface waters

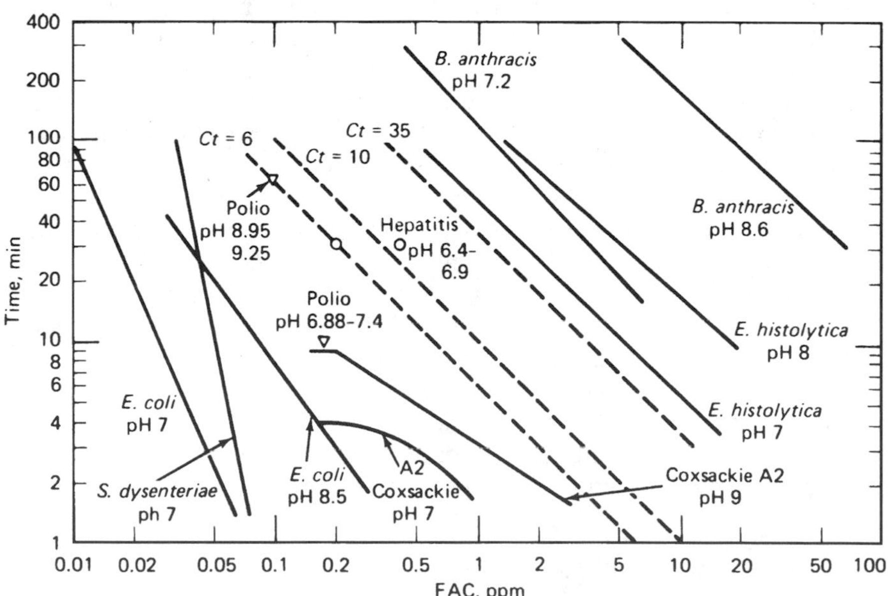

Fig. 6-16. Disinfection versus free available chlorine residuals. (Time scale is for 99.6 to 100 percent kill. Temperature was in the range of 20 to 29°C, with pH as indicated.) Reprinted from *Journal American Water Works Association* **54,** 1379, Nov. 1962, by permission of the Association.

Table 6-3 Sources of Data for Disinfection Time—Chlorine Concentration

Organism	*0°C Temp. Range*	*pH*	*Source*
E. coli	2–5	7, 8.5	Butterfield[177]
E. coli	20–25	7, 8.5	Butterfield[177]
Salm. typhosa	20–25	9.8	Butterfield[177]
S. dysenteriae	20–25	7.0	Butterfield[177]
Poliomyelitis	20–30	6.85–9.25	Lensen[178]
Poliomyelitis	0	7, 8.5	Weidenkopf[179]
Coxsackie A2	3–6	7–9	Clark and Kabler[180]
Coxsackie A2	27–29	7–9	Clark and Kabler[180]
E. histolytica	3	7, 8	Fair[181]
E. histolytica	20–25	7, 8	Snow[182]
B. anthracis	4	7.2, 8.6	Brazis[183]
B. anthracis	22	7.2, 8.6	Brazis[183]
P. tularensis	15.5–18.5	7.3	Foote[51]
Hepatitis	Room	6.4–6.9	Neefe[184]

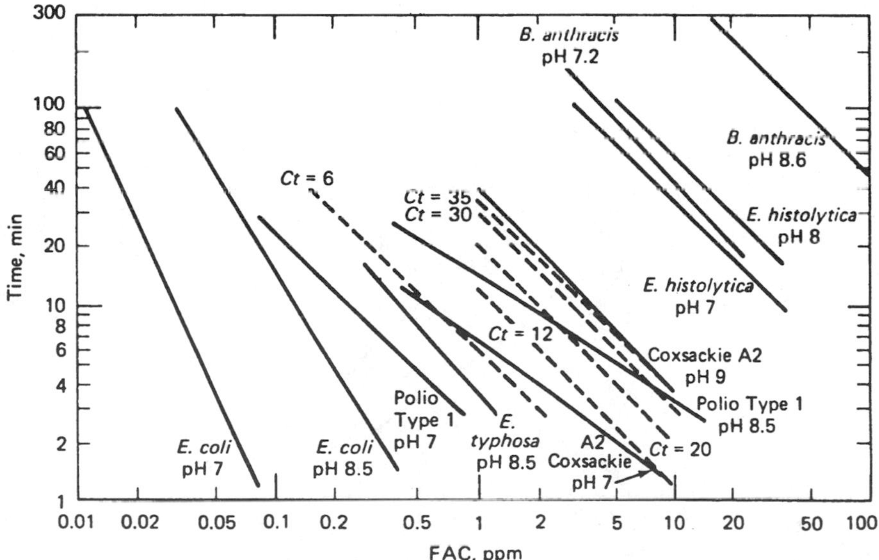

Fig. 6-17. Disinfection versus free available chlorine residuals. (Time scale is for 99.6 to 100 percent kill. Temperature was in the range of 0 to 5°C, with pH as indicated.) Reprinted from *Journal American Water Works Association* **54,** 1379, Nov. 1962, by permission of the Association.

may never fall as low as 0°C, Baumann and Ludwig[175] constructed Fig. 6-18, which represents a transposition of the plots on Fig. 6-18 at 20 to 29°C and Fig. 6-18 at 0 to 5°C to a plot representing a water temperature of 10°C. This transposition was accomplished by means of the Van't Hoff-Arrhenius reaction rate equation:

$$\log_e \frac{t_1}{t_2} = \frac{E(T_2 - T_1)}{RT_1T_2} \tag{6-6}$$

in which T_1, T_2 = upper and lower temperatures between which reaction rates are compared, and t_1, t_2 = times in minutes required for equal percentage of kill to be effected at temperatures T_1 and T_2 at a fixed concentration of disinfectant. E = activation energy (calories). R = gas constant, 1.99 cal/°K, 10°C = 238°K. The result of the application of this mathematical transformation procedure is Fig. 6-18.

The curves illustrated in Figs. 6-16, 6-17, and 6-18 are plotted on log–log paper to give straight lines. These lines follow the general equation $y = ax^b$, where b is the slope. As b is negative, the curve is hyperbolic. The curve shown in solid lines have slopes that approximate −1. If y is plotted as time t, and x as free chlorine concentration C, the equation may be written:

$$t = aC^b \tag{6-7}$$

where b is a positive number expressing the relationship between C and t, and a is a constant for a given organism, water pH, and water temperature.

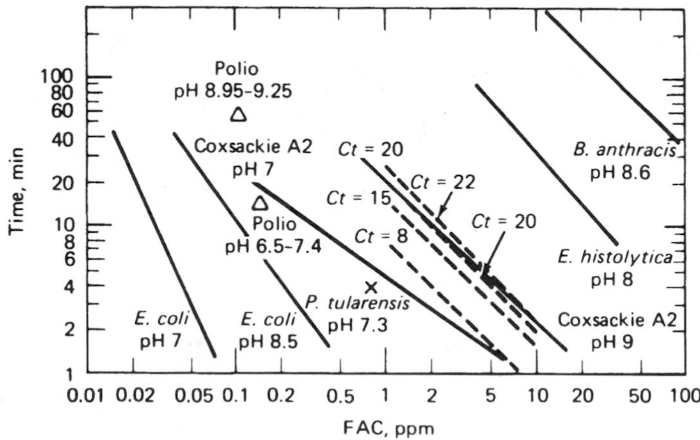

Fig. 6-18. Disinfection versus free available chlorine residuals. (Time scale is for 99.6 to 100 percent kill. Temperature was 10°C, with pH as indicated.) Reprinted from *Journal American Water Works Association* **54,** 1379, Nov. 1962, by permission of the Association.

The slope is expressed as $-b$; then if the slope approximates -1, the equation may be written:

$$t = aC^{-1} \tag{6-8}$$

or

$$a = Ct \tag{6-9}$$

The dotted lines in Figs. 6–16, 6–17, and 6–18 are constructed from the above equation to represent envelopes of disinfection time–chlorine concentration of various pH values. For example, in Fig. 6–17 the envelope $Ct = 35$ represents the envelope for the organism Coxsackie A2 at pH 9, and the envelope $Ct = 12$, for pH 7. Other envelopes for intermediate pH values are constructed by interpolation. This is based upon the HOCl concentration available from the total titrable chlorine at a given pH.

From all of the data shown in Figs. 6–16, 6–17, and 6–18 the question now becomes one of how much chlorine should be applied and how long the residence (contact) time should be. Baumann has suggested that the minimum dose be sufficient to inactivate Coxsackie A2 virus in the treatment of small water supplies. If this reasoning is applicable to small water supplies, it should be extended to all water supplies simply to encourage better disinfection practices and provide a wider margin of safety.

Baumann[175,185] selects Coxsackie A2 virus as the indicator organism; if it is destroyed by chlorination, then all other pathogenic bacteria and viruses would also be killed. This would not include *Entamoeba histolytica, B. anthracis, Giardia lamblia,* or possibly Hepatitis A. Table 6-4 gives the required values for the values of a in $Ct = a$. These are shown as the dotted lines in Figs. 6-16 and 6-17.

Application of this information is most useful for small supplies that suffer from the problem of short contact time.

Table 6-4 Disinfection Time–Chlorine Concentration Envelopes for 0–5°C and 10°C

	Value of a *in* Ct = a	
Water pH Range	*0–5°C*	*10°C*
7.0–7.5	12	8
7.5–8.0	20	15
8.0–8.5	30	20
8.5–9.0	35	22

EXAMPLE: A well supply with a pH of 8.2 and a possible minimum temperature of 10°C shows $Ct = 22$. The well pump discharges to a pressure tank and thence to a small distribution system, so that the estimated contact time when the pump is on is 6 min.

Therefore $C = 22/6 = 3.7$ mg/L free available chlorine residual (not dosage). This would call for dechlorination. For a small supply, such as bus stops, trailer courts, camp sites, or motels, the activated carbon pressure filter would be the preferred choice as a method of dechlorination. However, there is an alternative: A holding tank could be placed just downstream of the pump and pressure tank to provide a minimum of twenty minutes contact time. This then would require a free chlorine residual of only $22/20 = 1.1$ mg/L, which would not require dechlorination.

Larger supplies, such as municipally operated systems, usually have considerably greater contact times; however, the envelopes are just as valid for large municipal supplies as they are for small supplies. Part of the reasoning behind higher doses of chlorine for small supplies is that proper supervision and laboratory control are not usually available; the systems are usually designed without any alternative to long-enough contact time, and are not under strict public health supervision. Most treatment plants have from one and one-half to three hours of contact time from influent to treated water storage at average flows, so that the contact time for disinfection is usually easy to achieve.

Whether or not Coxsackie virus A2 could be considered the consensus organism is of course debatable. However, the idea is appealing, as it embodies all the elements of a positive and thorough approach to disinfection. In the free residual process it would be most helpful to know precisely the chlorine concentration–contact time envelope required to achieve the desired disinfection efficiency.

Chlorine–Ammonia Process. This is currently a controversial subject (1997), because many people in the potable water supply business believe that monochloramines have more disinfection power than free chlorine. Of course these believers, rarely if ever, define "free chlorine" as 85 percent HOCl. If organic N is present in the water, the free chlorine is converted immediately to a nongermicidal organochloramine that appears in the dichloramine fraction of a forward-titration procedure using either the amperometric or the DPD-FAS titrimetric procedure. Whatever the amount of organic N present, that is the exact amount of free chlorine that will be consumed immediately. When chloramines are present, this conversion requires a long contact time, usually up to at least 40 min.

Originally (1930s) the chloramine process was used solely for taste and odor control. Therefore, the process did not receive much attention after the discovery of the breakpoint phenomenon.

As investigators have always known that all pretreatment processes lower the chlorine demand of the source water, and also the formation of the THMs, it makes a lot more sense to choose a chlorine demand target for a given contact time to see if the source water is acceptable. All through the literature on this subject, describing the control of THMs, it is always called "removal of the precursors," which is nothing more than lowering the "chlorine demand."

This process needs a lot more study to see how powerful its supposed disinfection power really is. Any kinetic study should include the following variables: effect of pH, effect of water temperatures, stability in sunlight, and germicidal efficiency over a reasonably broad spectrum of conditions. The germicidal efficiency should be measured for a wide range of organisms including, but not limited to, enteric viruses, amoebic cysts, *Giardia,* hepatitis, and enteric bacteria. The objective of the germicidal efficiency investigation would be to provide a chlorine residual concentration–contact time envelope for several different organisms similar to the information provided by Baumann described above. Among the problems that continue to plague operating crews are the variety of major difficulties in dead-end problems that occur on widespread distribution systems where the chloraminated water supply is added to a groundwater system already in operation. This is the result of the remaining ammonia N after the chlorine residual disappears. The nitrogen is a basic nutrient to the microbiological life in the distribution system. This causes a wide variety of objectionable taste, color, and corrosion problems.

There are three other areas that need careful study: (1) evaluation of germicidal efficiency for a variety of chlorine to ammonia N ratios from 3 to 1 to 6 to 1; (2) comparison of tastes and odors for these different ratios; and (3) comparison of areas 1 and 2, applying ammonia first versus chlorine first allowing 10 min. of contact with chlorine before ammonia addition.*

Taste and Odor Control

Introduction. The term "taste and odor" should logically be replaced by the term "flavor" as applied to the palatability of water or, for that matter, anything that is ingested by humans, whether food or drink. Taste itself is a sensation caused by buds on the tongue and limited to acid, bitter, salt, and sweet. All other sensations of "taste" are either combinations or are modified by smell. An additional sensation that the tongue can detect is feel or touch—slick or oily as well as metallic, dry, or astringent. "Flavor" could be used as an all-inclusive term.

The origins of taste and odors in water supply are of two categories: natural and synthetic. Natural sources include: aquatic growth, such as algae and diatoms and other organic compounds from decaying vegetative matter; and inorganic compounds, such as hydrogen sulfide, sulfates, and other sulfurous compounds.

Synthetic sources of taste and odor in water supplies are industrial and domestic waste discharges. The worst are the discharges from manufacturing

*In the author's opinion, it is better to spend time on these investigations than on worrying about the control of THMs.

plants that contain phenols, cresols, certain amines, and mercaptans. Pulp and paper mills that discharge sulfites also cause serious taste and odor problems. With the possible exception of hydrogen sulfide, the worst offenders by far are the organic substances.

Substances causing odor are volatile and usually liquids. Most odorous substances are soluble in both water and organic liquids. Only a very few inorganic substances have odor, but they do provide the nutrients to promote the growth of odor-producing algae.

The first step in attacking a taste and odor problem is to review the literature. Several hundred articles have been published in various journals dealing with tastes and odors in the past four decades. To save time, the serious student, researcher, or operator should first read a 1977 review by S. D. Lin[186] and another by E. J. Middlebrooks in 1965.[187] These two reviews are replete with references. Another valuable reference is the AWWA handbook of specific experiences of taste and odor control in the United States and Canada.[188]

The next step is to learn about the Threshold Odor Test, detailed in *Standard Methods.* This reference and papers by Gerstein[189] and Sigworth[190] describe in detail some of the ways to construct continuous and batch-type odor monitoring devices.

There are three ways to deal with odorous contaminants: keep them out of the water, remove them from the water, or destroy them in the water. The tastes and odors produced by algae have been generally described as aromatic, fishy, grassy, earthy, musty, and septic. Threshold odors caused by algae may be in some areas as low as 1–14, but in other areas go up to 30–40 and occasionally as high as 90 or more. Algal odor is generally objectionable, even when the threshold number is low. For satisfactory results the threshold odor should be reduced to 5 or less.[191]

Identification and quantitative analysis of the offending organism are important facets of taste and odor control, as the number of standard areal units per milliliter varies with the particular species of algae. For example, for *Asterionella* it may be 3000; for *Synura,* 200;[191] and for *Dinobryon,* 30.[192]

Principal Odor-Producing Algae. *Synura* is a very potent odor producer. A comparatively few colonies per milliliter will cause a perceptible odor resembling ripe cucumber of muskmelon. It can also produce a bitter taste leaving a persistent dry, metallic sensation on the tongue. When present in large numbers, it may cause a fishy odor.

Dinobryon imparts a prominent fishy odor when the standard areal count reaches 30/mL. This organism develops in the southern end of Lake Michigan in June and July of almost every year.

Asterionella (500 units or more) imparts an aromatic geraniumlike odor that changes to fishy when present in large numbers.

Tabellaria has a similar effect.

Synedra produces an earthy to musty odor, and also inhibits proper floc formation.

Stephanodiscus contributes a vegetable oily taste but produces very little odor.

Ceratium is an armored flagellate that is abundant in California and is responsible for odors varying from fishy to septic. It is likely to proliferate rapidly during any season.

Anabena, Anacystis (formerly known as *Microcystic, Polycystis,* and *Clathrocystis*), and *Aphanizomenon* are blue-green algae well known for developing very foul "pigpen" odors in water. In small concentrations these three algae impart a grassy to moldy odor to the water. They may in large enough masses cause luxuriant blooms. The foul odor undoubtedly develops from products of decomposition as the algae begin to die off in large numbers.

Green algae are not often associated with tastes and odors in water. Their growth may help to somewhat inhibit or keep in check the blue-green algae and the diatoms, and thereby be helpful in the control of water quality.

Dictyosphaerium, one of the worst offenders among the green algae, will produce a grassy to nasturtium odor as well as a fishy odor in larger concentrations. Some of the swimming green algae, such as *Volvox,* may also produce fishy odors.

Jenkins et al.[193] have reported several odorous organic sulfur-containing compounds produced in decaying blue-green algae cultures and reservoir waters containing blue-green algal blooms. These compounds included methyl mercaptan, dimethyl sulfide, isobutyl mercaptan, and N-butyl mercaptan.

Actinomycetes is an order of filamentous, branching bacteria. It was formerly identified with the general grouping of blue-green algae now called cyanobacteria, and encompasses several families of bacteria.[26] This group of organisms has long been suspected as the source of earthy odors in water supplies, and therefore has been subjected to intensive investigation.[194–197] These earthy-musty T/O compounds are considered to be the major source of consumer complaints.

Appreciable progress has been made in evaluating the relationship between volatile products of actinomycetes and the musty-earthy odor problems affecting water supplies across the nation. Modern research techniques have led to isolating two major earthy-musty smelling compounds: geosmin and 2-methylisoborneol (MIB). They are metabolites of actinomycetes and blue-green algae (cyanobacteria) and show very low threshold odor concentrations.[198–200] Work done by Dougherty and Morris[201] in 1967 on the Cedar River identified the causative agent, which they called "mucidone." They classified this compound as a metabolite of actinomycetes; so it was probably geosmin. Their work using chlorine and activated carbon indicated what others have found, that oxidation by chlorine is relatively ineffective but that activated carbon adsorption (powdered), to be effective (lower the threshold number to an acceptable level), requires a 25 mg/L dose. They also concluded that, to be effective, an oxidant would have to be able to break the carbon–carbon double bond of the T/O compound produced by actinomycetes.

Others who have attempted to control or remove the taste and odor from actinomycetes have had moderate success with combinations of activated carbon and potassium permanganate.[188] Silvey et al.[194] reported in 1950 that activated carbon and chlorine dioxide were somewhat effective, but that ozone, bromine, and oxygen were not. The Metropolitan Water District of Southern California had no success in removing geosmin or MIB by air stripping.[197] The realization that both oxidation and adsorption are only marginally effective in controlling the T/O compounds of actinomycetes began to focus attention on preventing proliferation of actinomycetes. This, however, is a formidable task, as these organisms are ubiquitous in the environment. They are widely distributed in terrestrial, freshwater, and marine habitats. Moreover, the terrestrial organisms have easy access to surface waters during runoff conditions. The control of actinomycetes taste- and odor-producing compounds remains in doubt. The Metropolitan Water District of Southern California has had some success controlling the blooms of these organisms, which occur in the bottom of the reservoirs. They have designed and built a mobile chlorination system using a long snakelike chlorine solution line and diffuser that can reach into all areas of these reservoirs.

Synthetic Sources of Taste and Odor. In the following text, certain values will be given for various organic compounds as threshold taste and odor concentrations. This means that the values given are the minimum amounts that will produce a response to the taste and odor detection system of the observer—the consumer complaint level of response.

Table 6–5 gives the threshold concentration of various chemicals that produce objectionable tastes and odors in water supplies.[202] The tastes are variously described as phenolic, iodoform, medicinal, and so on.

The potentially worst offenders for taste and odor production are the discharges from the manufacture of chemicals, dyes, medicinal products, coke (quench water), ammonia recovery, wood oil, phenols, cresols, petroleum products, textiles, and paper products. Of all the various chemicals that are contributory to off flavor in potable water, the ones that have received the most attention are the phenols, because of the intensification of the natural tastes of such chemicals by marginal chlorination in the early years of its use. Table 6–5 illustrates how micro quantities of certain chemicals can produce objectionable tastes and odors in water supplies. Almost from the very beginning of chlorination practice, the supplies for Chicago, Cleveland, and Toronto, among others, have been plagued with off-flavor from these chemicals. These problems were studied, evaluated, and eventually corrected by the diligence and wisdom of men such as Baylis and Vaughn at Chicago, Ellms and Braidech at Cleveland, Howard and Thompson at Toronto, and Vaughn and Besozzi at Whiting and Hammond, Indiana.

It is beyond the scope of this text to include all the organic compounds that find their way into the nation's drinking water and are potential taste and odor producers, or those that are toxic when ingested. By 1983, more than 65,000 organic compounds had been manufactured in the United States since

Table 6-5 Chemicals that Produce Tastes and Odors

Chemical	*Average*	*Range*
Acetic acid	24.4	5.07–81.2
Acetophenone	0.17	0.0039–2.02
N-amyl acetate	0.08	0.0017–0.86
Aniline	70.1	2.0–128
Benzene	31.3	0.84–53.6
N-Butanol	2.5	0.012–25.3
P-Chlorophenol	1.24	0.02–20.4
o-Cresol	0.65	0.016–4.1
m-Cresol	0.68	0.016–4.0
Dichloroiso propylether	0.32	0.017–1.1
2,4-Dichlorophenol	0.21	0.02–1.35
Ethylacrylate	0.0067	0.0018–0.0141
2-Meracaptoethanol	0.64	0.07–1.1
Methylamine	3.33	0.65–5.23
Methyl ethyl pyridine	0.05	0.0017–0.225
β-Naphthol	1.29	0.01–11.4
Phenol	5.9	0.016–16.7
Pyridine	0.82	0.007–7.7
Quinoline	0.71	0.016–16.7
Trimethylamine	1.7	0.04–7.15
N-Butyl mercaptan	0.006	0.001–0.06

World War II. This number increases by 3000 per year. There is great concern that the proliferation of these chemicals could devastate the nation's water supply. The EPA is working to find ways to effectively treat the volatile organics in drinking water.[235]

Role of Chlorination in Taste and Odor Control

Algal Tastes and Odors. Early chlorination practice consisted of application of the absolute minimum amount to ensure maximum bacteria kill consistent with good public health practice. This in most cases consisted of dosages as low as 0.25 mg/L with 15 min. O-T residuals recorded as 0.05 mg/L. Compare this with current practice of dosages from 1.0 to 5.0 mg/L. Earlier it was also thought that the lower the dose, the lower the threshold taste from chlorination. During this time public opinion was very much against chlorination as forced medication; therefore, the most direct and obvious mode of attack by the public was to complain of the terrible taste it caused. As a result widespread public opinion made bad-tasting water synonymous with chlorination. This attitude inspired many able people in the waterworks profession to explore more vigorously the mechanics of chlorination. The first assault was directed at improving or eliminating tastes and odors. In the 1920s, Howard of Toronto[203–205] led the way in exploring doses of chlorine beyond the marginal bactericidal doses. He described this approach as superchlorination. At about the same time Bushnell[206] reported that raising marginal doses to 3 to 5 mg/L, described as overchlorination, eliminated foul odors thought to be caused by gnats breeding inside a covered reservoir. Griffin[121] pursued the exploration of superdoses of chlorine (up to 25–30 mg/L) in a wide variety of conditions and practically revolutionized chlorination practice. Not only did his work lead to the discovery of the breakpoint phenomenon, but it proved that the free chlorine residual fraction was not necessarily in itself a producer of off-flavor. In those cases where it was demonstrated that the residual in the finished water was too great or seemed offensive, the situation was usually corrected by dechlorination. The success of using heavy doses of chlorine in the raw water, for the correction and control of tastes and odors that presumably originated in algae growths and/or seasonal reservoir situations, is well documented.[207–211] It is evident from this information that offensive tastes from algae blooms caused by *Synura, Synedra, Dinobryon, Asterionella, Anabena, Ceratium,* and *Anacystis* can be controlled by proper prechlorination that will produce a sizable free chlorine residual—1 to 5 mg/L. Taste and odor problems are not necessarily caused entirely by either odor-producing algae or trade wastes but usually by a combination that manifests itself in a continuing degradation of raw water quality. Although it is desirable to control the plankton count in storage reservoirs before it reaches the treatment plant, Riddick[221] has pointed out that there are so many contributing factors of geology and terrain that prevent such control, that proper facilities at the treatment plant are the best assurance of taste and odor control. This would include a chlorination facility capable of carrying a free residual through the entire treatment process.

Industrial Wastes. Heavy doses of chlorine were first used by Howard[203] not to correct only off-flavors caused by algae but primarily those caused by trade wastes. It was apparent that the worst offenders were the phenolic compounds that had their source in the wastes from coke and natural gas manufacturing plants. The addition of low dosages of chlorine to a water bearing phenols will produce a chlorophenol compound that imparts an extremely objectionable medicinal taste to the water. The reaction between chlorine and phenolic compounds has been thoroughly investigated.[212–215]

Ettinger and Ruchhoft[213] demonstrated that the result of chlorinating phenols was a reaction by steps whereby the taste-producing intensity was enormously increased by partial chlorination until a maximum intensity was reached, after which the further addition of chlorine resulted in a progressive decrease of the taste intensity until the chlorophenol tastes disappeared completely. These phenolic materials include cresol, naphthol, and dichlorophenol. In a later investigation, using more sophisticated research tools, such as gas chromatography, Burtschell et al.[214] discovered that the phenolic compound that was always present when there was a chlorophenolic taste, was identified as 2,6-dichlorophenol.

It was further discovered in these investigations that, in order to eliminate tastes from these industrial wastes, the system would have to achieve a truly stable free chlorine residual and be allowed sufficient time for the reaction to go to completion. This means that time and pH environment are influencing factors. Ammonia in the raw water definitely slows down the reaction, but this can be overcome by increasing the chlorine dosage. As in all chemical reactions, water temperature is also a factor. The system behaves much differently in the winter months than it does in the summer months. The colder the water, the longer the contact time and the more chlorine required.

Others have demonstrated also that there is a definite ratio between chlorine required and phenoliclike substances present. However, each different water to be treated must be considered unique, and the problem has to be evaluated on an individual basis for dosage, time of contact, and so on. It has to be concluded, however, that the application of chlorine can definitely be used to destroy off-flavors caused by phenolic compounds because of the many successes reported in the literature.[205, 216–218]* However, the subject of taste and odor control has become so complex, owing to the general deterioration of potable water supplies throughout the United States and Canada, that there is no single solution to the problem. Heavy doses of chlorine with plenty of contact time are not a cure-all. Supplies that have a history of taste complaints are generally equipped to combine chlorine with some form of activated carbon treatment. For example, the City of Chicago, at its South Water filtration plant, is able to apply either chlorine or activated carbon to the raw water tunnel about 1100 ft ahead of coagulation. When trade wastes produce a threshold odor value of 10 or more, carbon is added at this point; however, when the *Dinobryon* count reaches 3000 or more per ml, chlorine is usually applied here. The reason for this flexibility is that sometimes it is more desirable to apply carbon first and sometimes chlorine. Depending on the trade waste, particularly if it is predominantly phenols, it may be more economical to let the carbon do the work of adsorbing the phenols to prevent chlorophenol tastes than it will be to destroy the phenols with heavy doses of chlorine and use the carbon for the adsorption of the off-flavors resulting from chlorination. It must be remembered that if phenols are only partially chlorinated to produce chlorophenols, very high doses of activated carbon are required to adsorb the resulting chlorophenols.

Several added benefits have been discovered in the application of heavy doses of chlorine to raw water. This type of treatment controls algae growth that tends to clog filter media and inhibit flocculation. Residual chlorine in the sedimentation basins prevents septicity in the sludge blanket. Therefore, the maintenance of a free chlorine residual in these areas has improved sedimentation and filtration. Another effect has been color reduction in some

---

*Also see Chapter 12 for the use of chlorine dioxide as a specific treatment for the destruction of phenols.

waters caused by chlorine bleaching the organic matter, giving the treated water a polished look.

There is no definite upper limit on the chlorine dosages required. Usually the range is from 5 to 10 mg/L with a few instances requiring as high as 25 to 30 mg/L; the record is 120 mg/L at Ottumwa, Iowa.[219, 220] It is interesting to note that the successful users of heavy prechlorination doses for taste and odor control had to rely on dechlorination facilities, and those who failed either did not have enough chlorination capacity to achieve high-enough doses or were not equipped to dechlorinate and therefore chose not to release to the consumers water with an abnormally high chlorine residual.

Use of Carbon with Chlorine. Another lesson learned in those cases where heavy prechlorination doses were used was that the consumption of powdered activated carbon was reduced significantly. Also, the application of carbon following prechlorination did not interfere with either process. Hence off-flavors caused by chloro products were eliminated by the carbon while the consumption of chlorine by the carbon amounted to only about 1 lb of chlorine for 20 lb of carbon applied. When phenolic wastes are known to be in the water, it takes heavier doses of carbon to adsorb the chlorophenols than it does to adsorb the raw phenol wastes. The use of carbon or chlorine first is a question of economics: should the phenols be destroyed by chlorine, or should they be adsorbed first by the carbon?

Seeking the Solution to a Taste and Odor Problem

Identify Water Quality. The way to success in dealing with taste and odor problems is anticipation, by considering local conditions. Some waters at certain times of the year cannot tolerate any combination of chlorine dosage and contact time without producing objectionable tastes and odors. Some of these are discussed in the chlorine–ammonia process. Knowledge of raw water quality—such as threshold odor, organic nitrogen, ammonia nitrogen, degree of pollution, concentrated chloroform extract, pH, and temperature—is of great importance. The threshold odor data are the most important, there being no substitute for this information.

The means provided by a chlorination facility to control taste and odors is: remove taste- and odor-causing substances by oxidizing them to odorless and tasteless compounds; and control or prevent the growth of odor-producing algae and microorganisms during the treatment process and in the distribution system.

The following course of action is recommended when the T/O problem occurs in the treated water before it enters the distribution system.

Laboratory Experimentation. Experiment first in the laboratory, because if something will not work in the lab, it will not work in the plant.

Chlorine Dosage. Run a series of chlorine demand curves for contact times of 5, 15, 30, 45, and 60 min. with a dose of chlorine normally used. If the curves plotted from the residuals at this time indicate a possible breakpoint, run some intermediate contact times to locate the breakpoint. Check the T/O for the original dose and compare to other dosages, higher and lower than the normal dose.

Ammonia. If the problem still persists, try adding ammonia (before chlorine) in dosages to correspond to Cl to N ratios of 3, 4, 5, 6, 7, 10, and 12 to 1. Pick two contact times: 30 and 60 min. Check the residuals at these times and compare the T/O with a control sample.

Activated Carbon and Potassium Permanganate. If the combination of free chlorine and chloramines fails to achieve the desired results, try some combinations of chlorine and activated carbon (before and after chlorination). Also try some combinations with potassium permanganate. In addition to these combinations it will be necessary to try a variety of dosages. The combination might seem awesome, so it is always prudent to get some advice from the chemical suppliers for carbon and potassium permanganate.

Chlorination System Requirements for Taste and Odor Control

In order to provide the maximum choice and flexibility of control by the operator, a chlorination facility must include the following: (1) chlorination equipment adequate to produce a sufficient free chlorine residual throughout the treatment process; (2) separate automated equipment for each point of application of prechlorination, to avoid the so-called split-feed of the chlorine solution from one chlorinator; (3) automated dechlorination facilities based on residual control; (4) provision for possible preammoniation application; (5) automated postchlorination and postammoniation if distribution system residuals are requied; (6) alternate points of application for activated carbon and chlorine—carbon first followed by a detention period of 20–45 min. and then chlorine for those special cases where hydrocarbon wastes are a predominate factor; (7) for taste and odor control, a minimum contact time of at least one or two hours if any ammonia is present, and longer if the organic nitrogen concentration is significant—0.3 to 0.5 mg/L; (8) provisions for adequate chlorine dosage and sufficient contact time prior to the raising of the pH for waters that undergo softening at high pH values (10–11); (9) provision for application of chlorine dioxide when indicated; (10) provision for the application of both activated carbon and potassium permanganate to the raw water if filtration is part of the process (potassium permanganate should not be used unless it is followed by filtration); (11) a continuous taste- and odor-monitoring system in a special "odor-free" room. Because taste and odor problems do not necessarily end at the treatment plant, just as much consideration must be given to the off-flavors that may develop in the distribution system. It is practically axiomatic that, in systems with a tendency to develop off-flavors, these unpalatable flavors will develop as soon as the available chlorine residual disappears in the distribution system. Tastes and odors from the application of chlorine are not likely to occur from the chlorine compounds themselves up to the limits listed below:

Free chlorine ($HOCl$)	20.0 mg/L
Monochloramine (NH_2Cl)	5.0 mg/L
Dichloramine ($NHCl_2$)	0.8 mg/L
Nitrogen trichloride (NCl_3)	0.02 mg/L

Nitrogen trichloride can cause a severe odor problem because its solubility in water is negligible. It is difficult to capture and measure in a sample, as it will aerate with the slightest bit of agitation. However, there is no mistaking the presence of NCl_3. In concentrations too low to get a response from the olfactory system it will cause the eyes to tear quite profusely. The odor of NCl_3 is entirely different from odors of free chlorine and the other chloramines. Not much energy is required to aerate large quantities of NCl_3. It can, however, become an air pollution problem.

Williams[114, 221, 222] demonstrated over a period of 20 years that if the ratio of monochloramine to dichloramine is kept at or greater than 2:1, objectionable tastes due to dichloramine will be at a minimum. This ratio of chloramines to achieve palatability is site-specific. It is not a hard and fast rule.

Studies by Ryckman and Grigoropolous[215] and Erdei[195] indicate the necessity for the removal of ammonia, organic nitrogen, phenolic compounds, and other organic extracts, all of which interfere with the chlorination process, with the result that it is becoming virtually impossible to provide palatable water in cases where these compounds are present in significant concentrations. These compounds are abundant in man-made wastes and should be prevented from reaching the potable water supplies. This is an impossible task; however, every effort should be made to minimize their entry into these supplies. The proliferation of toxic wastes threatening U.S. water supplies is awesome. Since World War II, many thousands of new organic compounds have been manufactured, and the number increases significantly each year. It is inevitable that these compounds will reach our water supplies, going either into the ground or into the rivers and streams. The consequences can be devastating.

Distribution Systems and Transmission Lines

General Discussion. The problem of delivering a palatable and safe water to the consumer does not end as the water leaves the treatment plant, pumping station, or well discharge. The difficulties of maintaining water quality in transmission conduits and distribution systems are legion. The problems of water quality control in distribution systems are of two kinds. Probably the most prevalent are problems of taste, odor, and dirty water. The others involve deterioration of bacteriological quality. The question of the proper solution to these problems is one of the most controversial subjects of modern water treatment practice.

There seem to be more proponents of free residual chlorine as the proper treatment than those favoring chloramine residuals. Then there are those who favor either ammonia-induced breakpoint or ammonia-controlled free residual chlorination, as well as those who favor chlorine dioxide. It should be emphasized that it is imperative in a chloramine application to maintain a residual throughout the system and to enforce a flushing program; for when the chlorine residual disappears, the ammonia nitrogen returns, and it is a nutrient that promotes bacterial proliferation.

Nature of the Problem. The deterioration of water quality in a distribution system is usually attributed to three phenomena: (1) biofouling, due to the proliferation of microorganisms, which cause tastes, odors, and dirty water, with loss of carrying capacity of the pipes and sometimes severe corrosion; (2) chemical and electrolytic corrosion, resulting in undesirable end products, such as metallic and brackish tastes as well as failure of hot water heaters and residence water piping; (3) appearance of coliform organisms in the distribution system, indicating recontamination of an otherwise safe water.

Each of these phenomena contributes to consumer complaints. The most difficult problem to handle is the proliferation of microorganisms.

Photographic evidence that a large and diverse microbe community can adhere to and colonize the interior walls of underground water distribution mains and the surfaces of suspended particulate matter in drinking water systems has been provided by Ridgway and Olson[236] and Ridgway et al.,[237] using a scanning electron microscope. Such microbial colonization can apparently take place in summer months in spite of intermittent low-level free chlorine residuals (0.1–0.2 mg/L). There is ample evidence that microorganisms attached to pipe walls or present in partially anaerobic sediments in the invert of the pipe in low-flow and dead-end portions of the distribution system are responsible for taste, odor, and color problems of potable water systems.

Proliferation of bacterial colonies on pipe surfaces and suspended particulate matter leads to a type of biofouling that can harbor and encapsulate organisms such as *Klebsiella* p. and *Enterobacter cloacae.* These organisms have been isolated in distribution systems in the presence of free chlorine residuals. This phenomenon is not yet understood, but it is believed to occur in older systems where slime layers of bacterial colonies are layered with particulate matter followed by an inorganic scale of silica and/or calcium, as shown in Fig. 6–19. This sequence keeps repeating itself until a sloughing process occurs that tends to keep the layering in equilibrium. It is thought that the reappearance of coliform organisms in these systems is caused by sloughing adjacent to the inlet of the sample line. This sloughing involves the release of a piece of the top layer of the scale, thereby exposing the slime layer which harbors the coliforms that are drawn into the sample line.

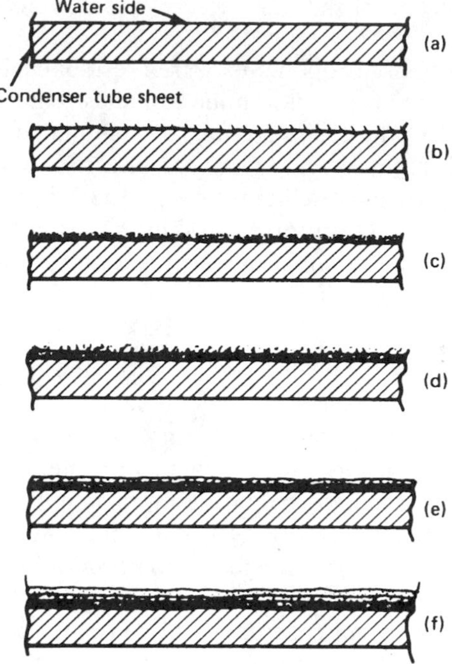

Fig. 6-19. Mechanics of biofouling: (a) The pipe is new, i.e.; smooth and clean so that nothing can attach or accumulate. (b) The pipe surface becomes abraded or rough, followed by attachment of gelatinous slime forming organisms. (c) The gelatinous film serves as a microstrainer and entraps sediment. Some bacteria can promote deposition of inorganic salts as well, which sometimes is mistaken for a scale deposit. (d) Successive layers of slime and particulate matter form. (e) As time passes and if bacteria are allowed to proliferate, the mass becomes more dense. (f) It is believed that microbial colonies establish themselves in these protective layers and in time the layers spall off, discharging organisms that are well protected by this debris.

This notion, if valid, would suggest that all sampling in any distribution system should provide a representative sample of the flowing stream. To determine the possibility of the type of contamination that reflects the reappearance of indicator organisms, sampling should be done at the consumer's tap. Sample connections to the distribution main should extend into the pipe about one-quarter of a pipe diameter. If this is not convenient, then use a fire hydrant.

The photographs taken with the electron microscope by Ridgway and Olson illustrate graphically many of the organisms discussed below. They also show clearly the scale-type layer, inorganic salts, bacteria attachment material, filaments coated with debris, actinomycetes adhering to the mineral layer (scale), *Gallionella* attached to the pipe surface, extracellular slime, and capsular material.

Dead-End Surveillance Program

This is the program offered by Le Chevalier. The typical water system measure chlorine residual (also temperature and pH) at sites where coliform and HPC samples are collected. Most water quality supervisors know where the problem areas are in the distribution system. When chlorine residuals start to drop from expected levels (these levels may vary from system to system, but are typically between 0.2 and 0.5 (mg/L), the first action taken will be to flush the dead-end area. For a system with a large number of problem areas, crews may operate on a regular basis (during summer months) to flush known problem areas (even without prior testing). The requirement to maintain a "detectable residual" (in many states, detectable equals 0.2 mg/L) has increased this practice. If an area has a chronic problem, the system may install a continuous bleed system to keep the water from stagnating. There are examples where new pipes have been installed simply to improve circulation in a particular zone for maintenance of chlorine residuals.

White has found the use of the Stranco Redox system to be more successful than chlorine residual measurements and easier to perform. Only a probe system with a millivolt readout device is required. These systems also require far less maintenance. (See Chapters 8 and 9.)

Bacteriology of Water Systems

Introduction. Bacteria are always present in water, even in the effluent from the most efficient and modern treatment plant. The kinds and numbers of these bacteria are dependent upon their environments. Many bacteria that may not be able to flourish in a particular environment are able to survive in a dormant state for indefinite periods of time and proliferate rapidly when the environment changes to a more favorable one. The most important factor for a favorable environment is the available food supply. Waters vary within wide limits as to the kind and quantities of food substances they carry; likewise, the types of bacteria that flourish in the pipes of these systems are equally variable.

Impounded surface waters that are allowed to produce abundant growths of algae and that are not filtered will contain great quantities of bacterial food in the bodies of algae that die in the pipes of the distribution system. Filtration effectively removes solid foods but not the soluble ones, such as proteinaceous matter, nitrates, phosphates, sulfur, ferrous and manganous compounds, and many others.

Underground supplies seem to be the worst offenders. They are low in dissolved oxygen, a situation that contributes to anaerobic conditions, and are more likely to contain significant quantities of soluble iron and sulfur compounds, which are energy for the two most offensive species of bacteria that infest water distribution systems. These are the so-called iron bacteria

(*Crenothrix, Leptothrix, Clonothrix, Cladothrix,* and *Gallionella*) and sulfur bacteria (*Beggiatoa, Thiobacillus thioxidans,* and *Sphaerotilus*).

Iron Bacteria. The iron bacteria obtain energy by oxidizing soluble ferrous compounds to insoluble ferric compounds. The amount of deposition of these insoluble compounds is large in comparison to the enclosed cells. These compounds may be deposited in a sheath that surrounds the organism or be secreted so as to form stalks or ribbons attached to a cell. The iron may be obtained from the pipe itself or from the water in the system.

There are also bacteria that do not oxidize ferrous iron but may indirectly cause it to be dissolved or deposited. In their growth they either liberate iron by utilizing organic radicals to which the iron is attached, or they alter environmental conditions to permit solution or deposition of iron. Less ferric hydrate may be produced, but taste, odor, and fouling may be promoted by these bacteria.

Iron bacteria are normal inhabitants of soils, and so they readily find their way into water supplies. They are never found as a single species in a pure culture. Analysis of slimes and tubercles consist of many bacteria intermingled with the crenoform filamentous iron bacteria.[223] Most of these iron bacteria can equally utilize the soluble manganese compounds, which cause black deposits as compared to the reddish brown deposits associated with iron deposition.[224, 225]

The filamentous types most commonly found are *Crenothrix, Cladothrix, Clonothrix, Leptothrix,* and *Sphaerotilus.* These "crenoform" organisms cause the most serious type of biofouling if they are allowed to proliferate. These long filamentous organisms have been known to grow "massive curtains" up to 2 and 3 ft in length in concrete tunnels.

Because iron is the predominant metallic deposit, they are classified as iron bacteria. These bacteria are considered a special group because their appearance closely resembles certain species of algae and fungi. There are two general types: filamentous and stalked.

These bacteria are sometimes lumped together under one general term of crenoform organisms, as they all cause equally serious biofouling.[226] This group of organisms can be differentiated by the way the branching of the filaments occurs (Fig. 6-20). They utilize the manganous salts better than they do the ferrous salts, and therefore may have manganic oxide deposits on their sheaths, giving rise to black, shiny deposits. They also oxidize organic matter for energy and thus do without the metallic salts.

Sphaerotilus belongs to a different cultural group than the other three, but performs in similar fashion. This organism occurs abundantly in heavily polluted waters rich in organic material such as wastewaters of sugar factories and paper mills. Figure 6-21 illustrates the sheath and cells, which occur in chains. The sheath may appear to branch, but this is recognized as false branching, as can be seen in Fig. 6-21. The cells emerging from an open end or a break in the sheath wall are called "swarm cells." The hollow sheath is

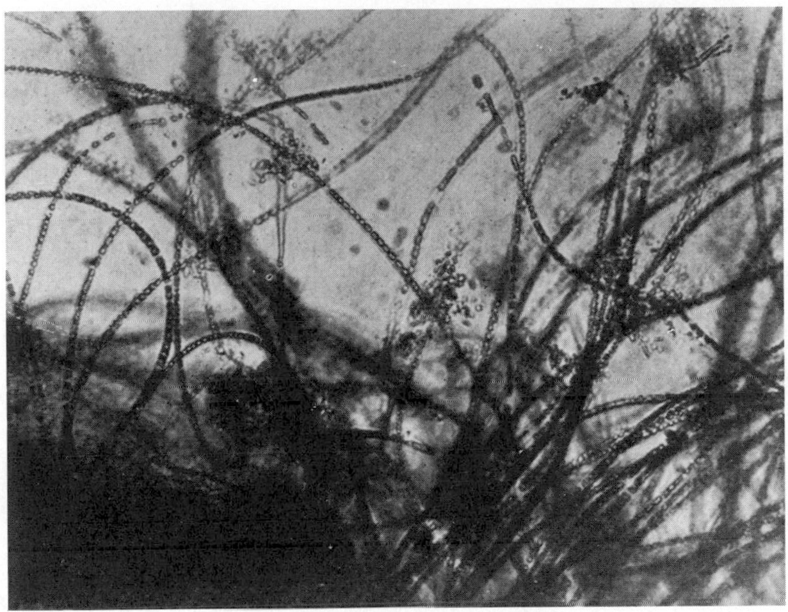

Fig. 6-20. Crenothrix (200 X).

of organic nature, but may become encrusted with an accumulation of ferric iron deposits.[227]

The stalked bacteria, the *Gallionella* (Fig. 6-22), were formerly called *Spirophyllum* because of their twisted-ribbon appearance. It was found that this ribbon is really a stalk of almost pure colloidal iron hydroxide excreted by the small oval bacterial cell at one end. These organisms cannot utilize manganese, and their growth is depressed by the presence of soluble organic matter.

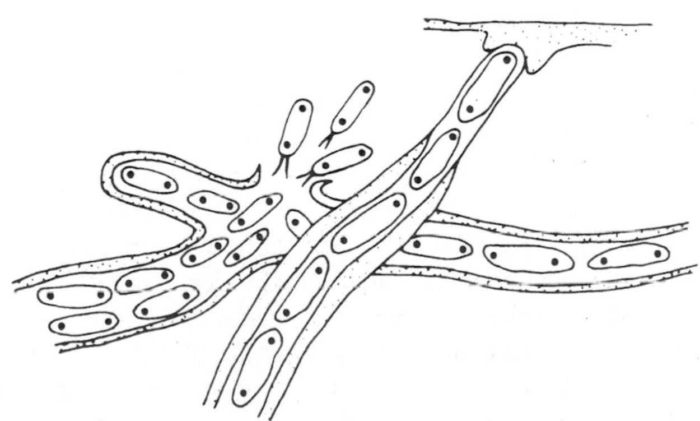

Fig. 6-21. Sphaerotilus.

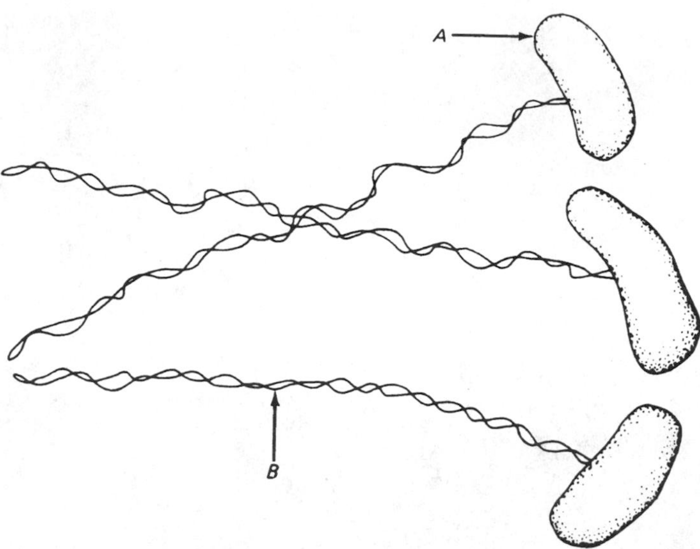

Fig. 6-22. Gallionella, showing bacterial cell (*A*) and colloidal ferric hydroxide (*B*) deposited from the concave side of the cell, which form flat bands or ribbons extending from the cell. The individual ribbons twist and become entangled with other ribbons.

They are therefore true iron bacteria in the strictest sense of the term. Another peculiarity of these organisms is that their optimum growth occurs at about 6–10°C; so they are often found in cold well waters or in the wintertime. In the summer months, they give way to *Crenothrix* and *Leptothrix.*

There are other members of the stalked bacteria family not usually thought of as iron bacteria, such as *Siderophacus* and *Nevskia.* The latter often contain globules of fat or sulfur, which tends to confuse the classification. The family of Siderocapsaceae includes 10 genera whose cells are usually surrounded by a thick mucilaginous capsule containing iron or manganese compounds. Only one species of this group of 10 genera, *Ferrobacillus ferrooxidans,* has been isolated in North America, but they are all widely distributed in European iron-bearing waters.[228]

Sulfur Bacteria. These bacteria, such as *Beggiatoa* and *Thiobacillus,* obtain the energy necessary for growth by oxidizing the sulfide ion to colloidal sulfur, which they store in their cells. By metabolic action this stored sulfur is eventually oxidized to sulfates. The most common of this group (*Beggiatoa*) is shown in Fig. 6-23. These organisms grow in long filaments in which can be seen granules of free sulfur and often a purple pigment. They grow in a scum on the surface of sulfide-bearing waters and are easily detected by microscopic examination. This filamentous group of organisms, if allowed to grow, causes serious taste and odor problems in a distribution system.

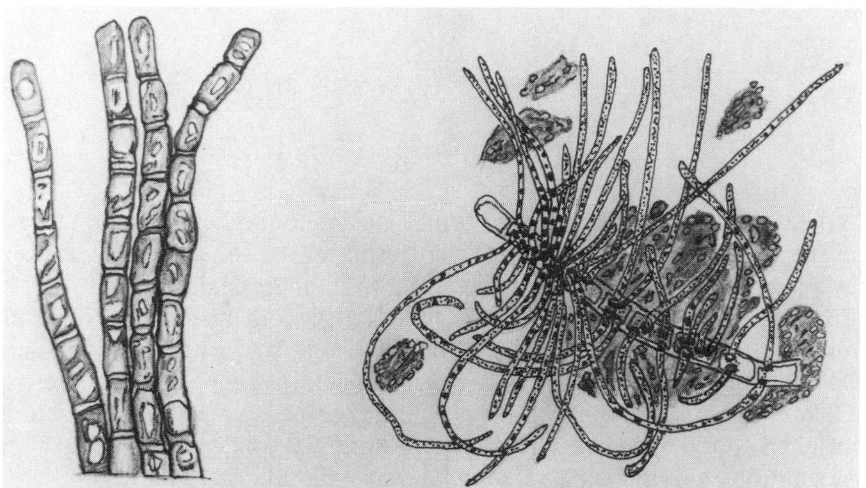

Fig. 6-23. Beggiatoa (2000 X) and trichomes attached to algal slime (400X).

Other sulfur bacteria that cause severe corrosion, taste, odor, and discolored water are the sulfate-splitting bacteria (*Desulfovibrio desulfuricans*). In anaerobic environments these bacteria convert sulfate and other sulfur compounds to hydrogen sulfide.

Most of the bacteria that inhabit water pipes are attached forms, which build colonies resembling drops of colored jelly, encrusting masses varying in size from one-quarter inch up to tubercles as large as 2 inches in diameter. Because they are attached forms, tap samples will not reveal the magnitude of growth inside the pipes. For this reason every water distribution system should maintain accessible for routine inspection pieces of pipe in different locations that can be easily removed so that the inside of the pipe may be examined for these attached growths.

It is well known that bacteria growing inside water pipes can cause severe corrosion.[223, 229–231] Thomas[232] described how water lines at a steel plant had a life of only a few months. An investigation revealed that this corrosion was due to the sulfate-splitting bacteria that flourished in the distribution system. This was corrected by chlorination. The action is explained as follows.[231]

The sulfate-reducing bacteria splits the $SO_4^=$ ion to form hydrogen sulfide. The hydrogen sulfide either reacts with iron to form ferric sulfide or escapes through the porous ferric hydroxide scale and is oxidized into sulfuric acid by the sulfur bacteria (*Beggiatoa*). This proceeds according to the following equations:

$$SO_4^= + 10H^+ \xrightarrow{\text{sulfate-reducing bacteria}} H_2S + 4H_2O + \text{energy} \qquad (6\text{-}10)$$

and

$$2H_2S + O_2 \xrightarrow{\text{sulfur bacteria}} 2S + 2H_2O \qquad (6\text{-}11)$$

$$2S + 3O_2 + 2H_2O \xrightarrow{\text{sulfur bacteria}} 2H_2SO_4 \qquad (6\text{-}12)$$

When a tubercle is formed, the sulfuric acid is found between the tubercle and the metallic surface of the pipe enclosed by the tubercle. This explains why all such tubercles show evidence of pitting underneath them.

Iron bacteria also form tubercles, and cause both anaerobic conditions within the tubercle and low pH conditions, so that the cycle of this environment plus the catalytic action caused by the iron bacteria keep putting more iron into solution under the tubercle. This cycle continues with the iron bacteria, which live on the outside of the tubercle, continuing to deposit the iron going into solution from the wall of the pipe, thus enlarging the tubercle.

Corrosion of iron pipe will not proceed when an equilibrium is established between the metal ion concentration in the water and the concentration of electrons in the metal. It is the bacteria that upset this equilibrium by putting more and more metal ion into solution.

Tubercle Formation. Figure 6-24 illustrates the formation of a tubercle. This is one of the most common phenomena of biofouling in a piping system. Tubercles seem to form in areas where pipeline velocities are low (<3 ft/sec), or when there are long periods of low velocity. All the evidence indicates that they are of biologic origin,[226, 232] apparently the outgrowth of a collection of fairly large colonies of crenoform organisms. These clusters of biologic deposits first start forming on the invert of the pipe. Long threads appear next in these jellylike clusters. The threads begin to acquire a sheath, which thickens by deposition of iron oxide, which the bacteria have taken out of solution from their water environment. Because the tubercles are hollow and spherical in shape, and all tubercles show evidence of pipe wall corrosion by pitting within the tubercle, this suggests the evolution of CO_2, which would assist in the expansion of the sheaths to a bubblelike form. This would also help to explain the cause of pitting due to a low pH environment. An alternative explanation is that the tubercles form oxygen concentration cells, which are also conducive to pitting. Sometimes these tubercles grow as large as two inches in diameter.

In an attempt to determine the nature of common tubercle deposits, the literature reports iron oxides varying from 25 percent to 90 percent, with several analyses showing appreciable silica content.

This type of biofouling not only promotes severe corrosion but also reduces the carrying capacity of the piping system. Often these tubercles are first discovered when the pipe perforates under the tubercle (see Fig. 6-24). In some cases the tubercles will give off a pigpen odor when broken while still moist. This is a positive indication of organisms undergoing putrefaction. This

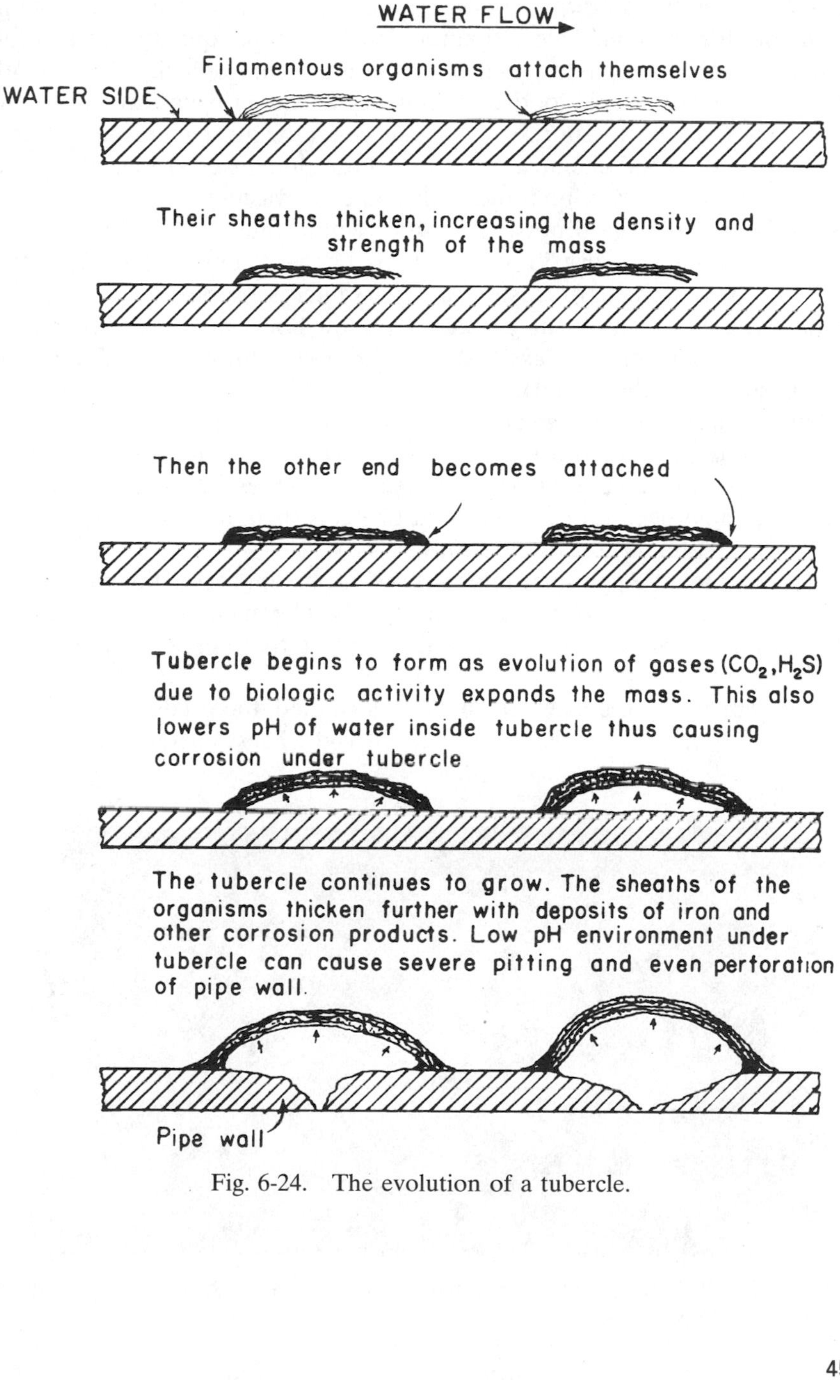

WATER FLOW →

Filamentous organisms attach themselves

WATER SIDE

Their sheaths thicken, increasing the density and strength of the mass

Then the other end becomes attached

Tubercle begins to form as evolution of gases (CO_2, H_2S) due to biologic activity expands the mass. This also lowers pH of water inside tubercle thus causing corrosion under tubercle

The tubercle continues to grow. The sheaths of the organisms thicken further with deposits of iron and other corrosion products. Low pH environment under tubercle can cause severe pitting and even perforation of pipe wall.

Pipe wall

Fig. 6-24. The evolution of a tubercle.

explains why this type of biofouling can contribute to significant taste and odor problems.

Other Types of Biofouling. Slime-forming organisms can lead to biofouling in a distribution system. This ultimately leads to water quality degradation. Some are fungi (thallophytes) that do not contain chlorophyll. Consequently they must live on organic food or on energy available from inorganic compounds. They are mostly spore-forming and resist the most adverse conditions of moisture, heat, and chemical poisons. These organisms are usually found in waters that are heavily polluted with domestic wastewater.

Bacteria are found in profuse quantities within the zoological slimes that lead to biofouling in a distribution system. They are unicellular plants that sometimes grow in chains or clusters. They are microscopic in size and proliferate rapidly. One organism can produce in a few hours millions of organisms in a colony. They can be classified into three structural types: the coccus, which is spherical; the bacterium or bacillus, which is rod-shaped; and the spirillum, which is curved. Some types are aerobic, growing only in the presence of dissolved oxygen; others are anaerobic, growing in the absence of dissolved oxygen. Some are capable of producing spores; others surround themselves with a gelatinous sheath or capsule that protects them against heat, dryness, or chemical treatment for long periods.

The predominant organism found in about 50 percent of the slimes examined in the laboratory are short, gram-negative rods. The majority of these belong to the genus *Aerobacter; Aerobacter aerogenes* is the most common species (see Fig. 6-25).

Of the remaining 50 percent of the slimes studied, the predominant organisms are aerobic gram-positive, spore-bearing rods. These bacteria, when prop-

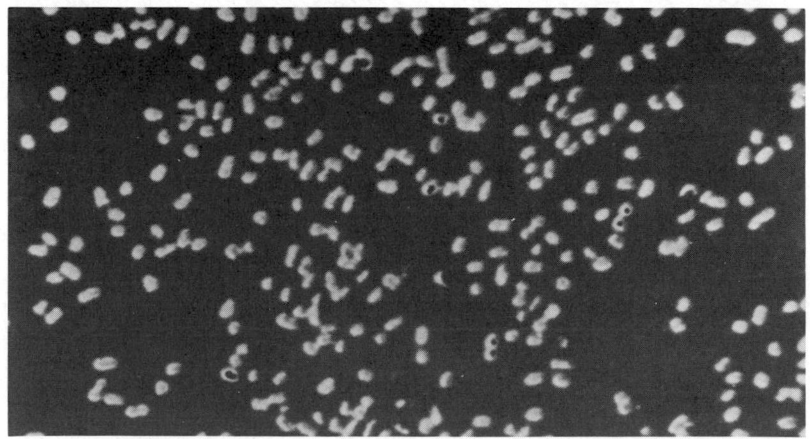

Fig. 6-25. Small slime-forming capsulated coccobacilli (approximately 600 X) (short gram-negative rods).

erly stained, exhibit a visible capsule. These are of the genus *Bacillus* (see Fig. 6-26). *Bacillus subtilis* and *B. megatherium* have been found in many samples studied. This group develops as mucoid colonies, which provide a gelatinous slime binder.

For a more detailed description of the nature, morphology, and structure of bacterial slime growths and the transformation of iron by bacteria in water, see Starkey[233] and Nason.[234]

Marine Biofouling Organisms

Introduction. Before the invasion of North America by the zebra mussels, the only troublesome organisms were found in seawater intake systems. The systems most affected were steam-powered electric generating stations, which are usually located adjacent to seawaters and their tributaries because of their unlimited supply of cooling water for the condensing phase of the steam turbine operation. The fouling of the seawater intake piping involved the destruction of hard-shell organisms which impeded the flow of water, and has nothing to do with the primary problem of biological slime in the condenser tubes which affects the heat transfer of the condenser.

In these cases, seawater presented an entirely different set of problems in their water circuits than did fresh water because of the higher forms of life that only existed at that time in seawater. These forms of life included the black mussels and barnacles that fouled the intake structures.

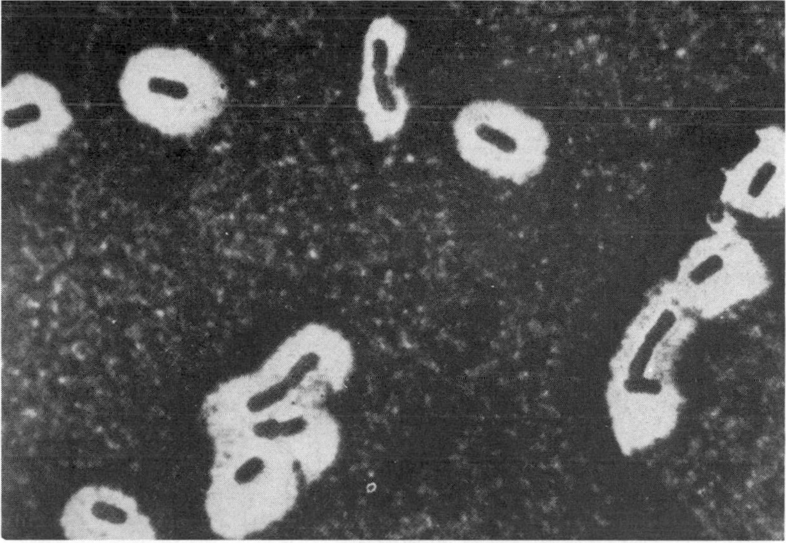

Fig. 6-26. Large capsulated bacteria (approximately 550 X) (gram-positive spore-forming rods).

Mollusca. Members of the family Mollusca are the most troublesome of the seawater fouling organisms.[462, 463] This family includes snails and other single-shelled Gastropoda and the clams, oysters, mussels and other bivalve (two-shelled) organisms. The most common of these is the edible black mussel *Mytilus edulis,* which is found throughout the world. It has been responsible for most of the fouling difficulties in the British Isles, the North Atlantic, and the western coasts of the United States.

The black mussel (Fig. 6-27) varies in size, up to a maximum length of four inches. Its life cycle is typical of many other of the fouling invertebrates. Within five hours after the female eggs have been fertilized, the embryo is completely free-swimming in seawater at 68°F. At the end of 48 hours, the organism has attached to some stable hard material—rock, concrete, glass, rubber, or similar support. At this point, the shell begins to develop. At the end of 69 hours, the fleshy portions of the organism are completely enclosed by the elementary single shell, and the organism is still capable of rapid swimming and is able to crawl on solid surfaces, vertical walls, and the tops of horizontal tunnels. Some time later, the true bivalve shell develops.

Contrary to popular belief, the organism, which is now attached by long, clear threads (known as byssus threads), is still capable of motion, and if conditions of poor feeding, inadequate oxygen supply, or mechanical or chemical irritation develop, the organism will break the byssus threads from their point of attachment and crawl by a "foot," which projects out near the hinge. This is similar to the locomotion of a snail. The mussel can also move by secreting new byssus threads and breaking off older ones.

Fig. 6-27. The black mussel.

The greatest growth rate occurs when the organism is completely submerged at all times. *Mytilus* grows three times faster in total darkness than in sunlight. Optimum growth occurs between 50 and 60°F and is seriously inhibited by heating for even a few minutes a day.

Under normal conditons, the adult mussel will leave the shell open about 15–20 degrees about three-quarters of the time. Water containing food is forced through the digestive tract, and any solid or suspended matter is deposited on the mucous surfaces. The digestive juices then digest the material usable as food and reject the rest. If any unfavorable conditions develop, or if anything disturbs the organism, the shell will close in a fraction of a second, and only after an extended period will the mussel open slightly and feed cautiously. This ability to close the shell so quickly and so tightly makes adults difficult to kill.

The American oyster is a common member of the Mollusca responsible for severe fouling of power-plant inlets on the South Atlantic and in the Gulf Coast states. It is easily recognized by its rough shell and white interior. Its life cycle is similar to that of the edible mussel.

Zebra Mussels.[460] The invasion of these organisms into North American fresh waters is a recent phenomenon. Their source is believed to be the Caspian sea, from which they have gradually spread over the last 150 years[461] to European fresh water harbors. Compared to our major mollusk, the black mussel, the zebra is tiny, often less than one inch in length. Its official name is, *Dreissina polymorpha.* The name "zebra" derives from the attractive striped patterns on the shells.

This mussel is regarded as the most potentially damaging natural intrusion into the U.S., water distribution system in years. In Europe, although the freshwater ports were sufficiently polluted to prevent the survival of the species during the early years, the success of the European cleanup program of the harbors created an environment where these organisms could thrive. Populations of 100,000 mussels/sq ft are not unusual. This far exceeds the density of growth demonstrated by the native black mussel in the United States.

It appears that sometime in 1985 or 1986, a ship left Europe carrying ballast water containing young zebra mussels, called "veligers." This ship delivered the young mussels when it discharged the ballast water somewhere between Detroit and Lake Huron. A freighter that is without cargo needs ballast water for stability in rough seas and to be certain that the propeller is always submerged.

The freshwater ports in the Great Lakes have provided the zebra mussels with a hospitable environment. The population is now established and expanding. Some experts believe these mussels could increase tenfold each year. They also expect this species to eventually inhabit about two-thirds of U.S. waters. In parts of Lake Erie, zebra mussels have been found in densities as great as 70,000/sq ft. Figure 6-28 illustrates the spread of these organisms into the Great Lakes. They entered the Ohio/Cumberland river systems in the

North American Range of the Zebra Mussel
as of 21 September 1991

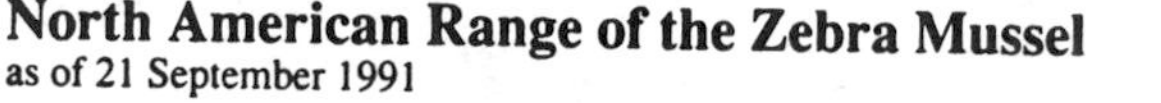

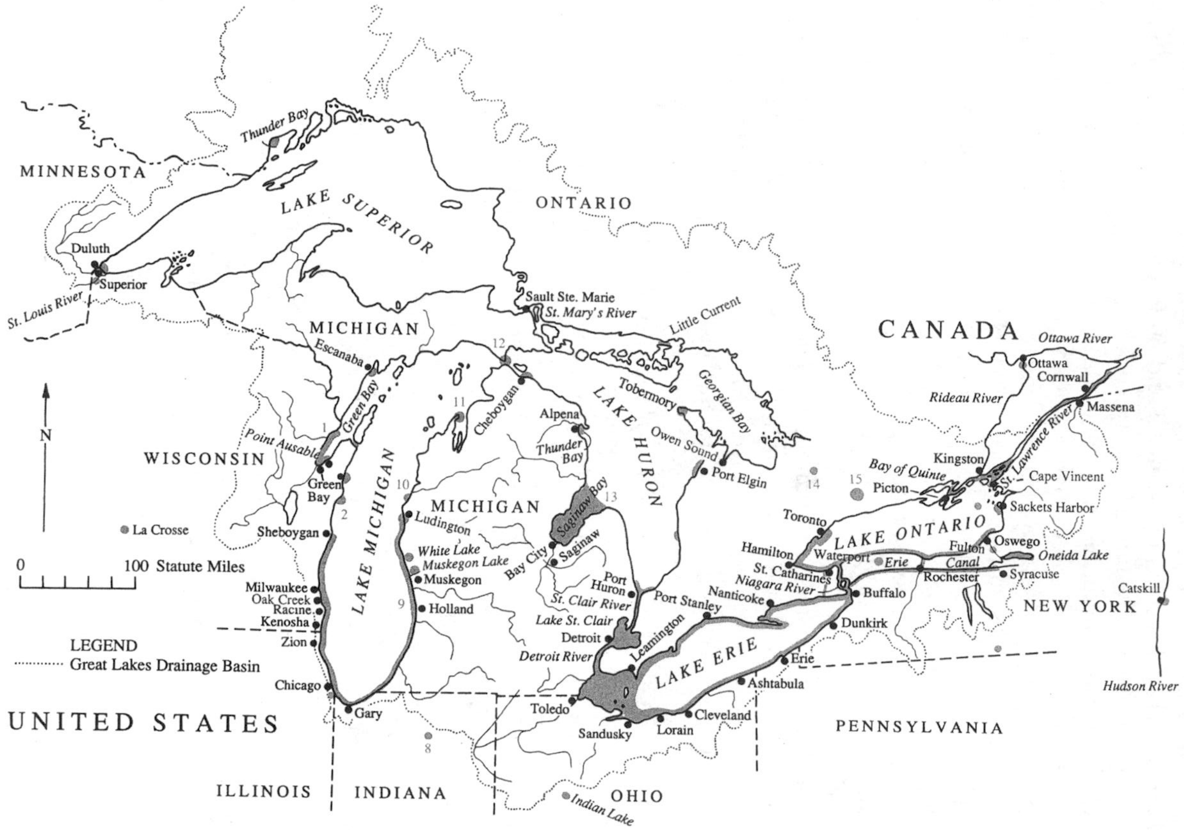

Fig. 6-28. Great Lakes, showing locale of zebra mussel infestation as of September 21, 1991 (New York Sea Grant).

mid-1990s. Experts project that these mussels will inhabit the entire Eastern United States within 15 years.

Damage to U.S. water systems by these organisms has already occurred. Their rate of regeneration is enormous, over 30,000 eggs a year by each female. This growth, coupled with their ability to attach to almost any clean hard surface and their tendency to form large clumps, means that the zebras can easily clog submerged pipe intake lines. Water intakes are popular spots for the mussels because of the continuous flow of water that contain algae, phytoplankton, and other organisms that represent the food chain for mussel survival.

Control of Marine Fouling

Historical. As disastrous as all this may be, these mussels can be effectively controlled by the use of the free chlorine process. The earliest known experimental work regarding control of marine growths was made for the Portobello Generating Station, Edinburgh, Scotland in 1919.[462] The first work in this field in the United States was by W. J. O'Connell, Jr., about 1926.[464] The control of marine growths was started in 1929 at the Northpoint station of the Long Island Lighting Company. Early plant scale work was done at the Lynn Gas and Electric Company plant at Lynn, Massachusetts, reported by Patten[465] in 1944. Patten's report was supplemented by a detailed discussion of the black mussel and a description of chlorination as a means of control from experiments conducted at Kure Beach, North Carolina, by Clapp.[466]

Chlorination Process. It has been found that the *hard-shelled* organisms are most effectively controlled by *continuous chlorination.* Maintaining a free residual of 0.5 to 1.0 mg/L throughout the intake pipe will usually destroy and prevent the growth of these organisms. It should be emphasized that chlorine dosages and contact time are site-specific. The water involved must be subjected to chlorine demand tests to determine the dosage requirements. Then the chlorine residuals must be continuously monitored by two different procedures. As the free chlorine process is mandatory, a free residual is one where the free available chlorine is 85 percent of the total residual. The difference between the two values is the combined residual. The combined residual, which is relatively ineffective in this application, is caused by the presence of ammonia nitrogen and organic nitrogen in the water being treated. The chlorine dosage requirement will vary seasonally for a variety of reasons; one major factor is water temperature. Chlorine dioxide has been found especially effective for control of both zebra mussel and religers (see Chapter 12).

A plan to control the zebra species should be in place even before they are detected. Once they are detected, a program to eliminate them from the system must be immediately initiated.

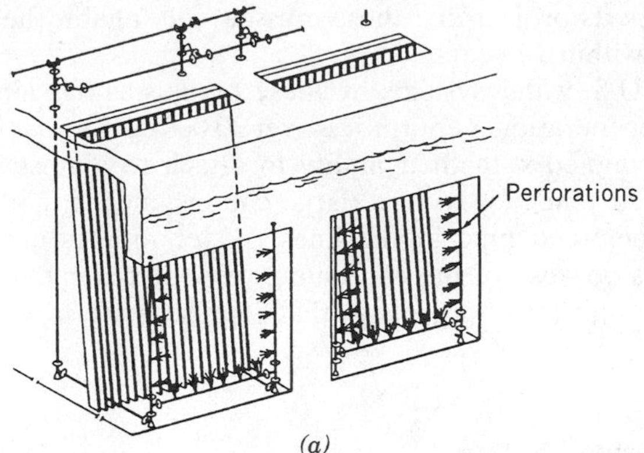

Fig. 6-29a. Perforated chlorine solution diffuser for installation ahead of bar racks.

Chlorine Diffusers

One of the most important components of the chlorination system is always the diffuser used to disperse the chlorine solution evenly across the pipe diameter. Its design is also based upon the situation at each different location. A variety of designs are available from many years of experience in the electric power industry. The theory and the practice of diffuser design are covered quite thoroughly in Chapters 8, 9, and the Appendix.

Model studies conducted at Worcester Polytechnic Institute found the diffusers shown in Fig. 6-29a and Fig. 6-29b to be the most satisfactory of several configurations that were tried.[463] Although these diffusers were studied in 1964, current experience with the tiny zebra mussels shows that they can and do plug the bar racks. Therefore, many installations will choose a diffuser that will apply chlorine ahead of the bar racks to prevent the zebras from clumping.

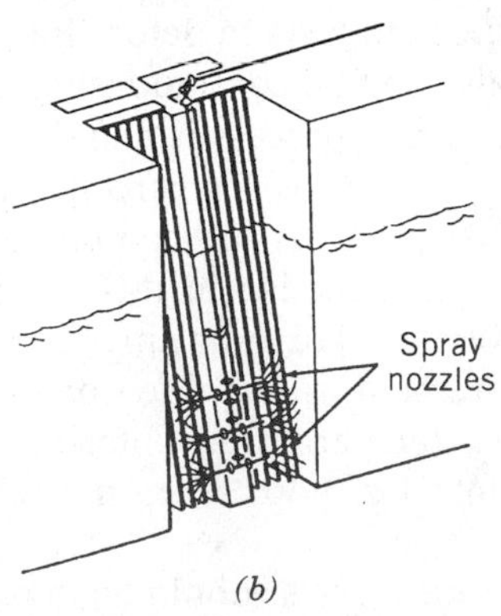

Fig. 6-29b. Chlorine solution diffuser with spray nozzles for installation ahead of bar racks.

Where hard-shelled organisms are not involved, the diffusers can be installed behind the bar racks. Figure 6-29c illustrates this arrangement.

Over the years of battling the black mussels, some power companies have installed duplicate intake pipes so that one can be taken out of service at any time if the control system fails for any reason, or for routine inspection.

Coliform Regrowth. The frequency and persistence of coliform "regrowth" in the distribution network has become a major concern to many water utilities. This continuing problem has occurred in what are considered biologically stable systems even though the water leaving the treatment plant is free of coliforms, and cross-connections are nonexistent.

The American Water Works Service Company, Haddon Heights, New Jersey, reported in 1984 that six of its Midwestern water utilities had been experiencing low levels of coliforms in their distribution systems during the past four years.[312] In all cases, traditional corrective measures had failed to reduce the levels of bacteria concentration until the chlorine residual was increased to 6 mg/L. These systems suffered from an unexplained appearance of coliforms in distribution systems that habitually maintained free chlorine residuals of 2–3 mg/L.[313]

In one of these systems a zinc phosphate compound was added to the treated water as it entered the distribution system. This treatment seems to have stopped the appearance of coliform organisms. Moser[313] offers the following explanation based upon his belief that the coliforms may be harbored in the gasket material at the pipe joint: the zinc compound added to the water deposits a metallic film at the pipe joints, thereby encapsulating the surviving organisms. This explanation derives from the fact that the distribution system piping involved in the coliform reappearance phenomenon had been polypigged prior to the zinc phosphate treatment. This treatment removes tubercles and other scaly deposits that are likely to harbor or encapsulate microorganisms.

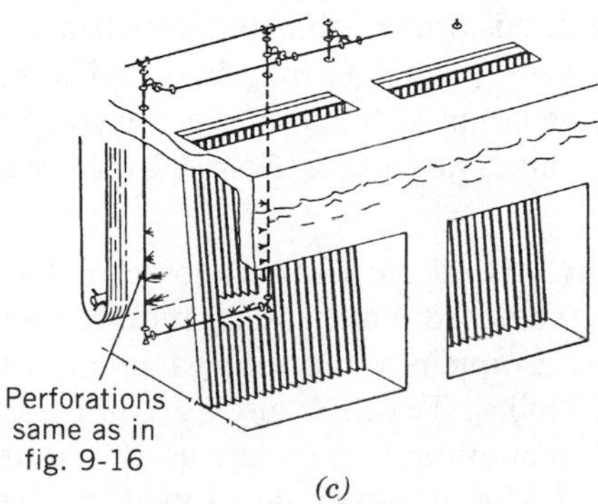

Fig. 6-29c. Perforated chlorine solution diffuser for installation behind bar racks.

Olivieri,[314] who investigated the regrowth problem that has plagued the above systems, believes the answer is the result of a microbiological process. This is the development of a biofilm by the microorganism that protects the organism from the chlorine residual. He further believes that the microorganism develops this film as a consequence of the hostile environment: free chlorine residual and a starvation level of nutrients. White,[315] who also investigated one of these systems, thinks that age of the pipe, joint material, scaling conditions, and a system free from biofouling are all contributing factors for the survival of a sturdy strain of coliforms. As one noted microbiologist said, "When nutrients in the system are restricted to the poverty level as in a clean system the opportunistic organisms that do survive become street tough."

In 1990 Schotts and Wooley made an in-depth study of bacterial–protozoan interaction in the presence of chlorination. It was found that when the bacteria were alone, they were killed by free chlorine residuals of 0.25–1.0 mg/L. However, when they were ingested by protozoans, these same bacteria survived in significant concentrations in the presence of free chlorine residuals of 1.0–10 mg/L up to 24 hours.[352] This is an interesting discovery because it provides a rational explanation for coliform outbursts in distribution systems where free chlorine residuals as high as 10 mg/L were unable to eliminate these organisms. The bacteria used in this investigation were potential human pathogens such as *Klebsiella pneumonia, K. oxytoca, Enterobacter cloacae, Ent. agglomerans, Citrobacter freundi, E. coli, Campylobacter jejuni, Legionella gormanii, Salmonella typhimurium, Yersinia enterocolitica,* and *Shigella sonneii.* The protozoans were *Tetrahymena pyriformis, Acanthamoeba castellanii,* and *Bodo edax.*

A study of *Legionella* contamination in the distribution system of a newly constructed water treatment facility found it to be extremely difficult to eliminate. Contamination was found in both hot and cold water systems. Repeated flushings after exposure to free chlorine residuals as high as 30 mg/L were not able to decontaminate the system.[353] Within two days after flushing of the system containing these high residuals, representative positive isolates were identified as *L. pneumophila* serogroup I. Because *Legionella* is not significantly chlorine-resistant, this proved that superchlorination is not always able to reach heavily colonized organisms in valves and their associated interior parts, either through stagnation or from physical protection caused by interior design features of the valves and other distribution system components.

Bacteria Enumeration. Until a few years ago the method for bacteria detection and enumeration (standard plate count) required a protracted length of time—sometimes days. A rapid new method was exhibited at the 1984 Annual AWWA Conference, Dallas, Texas. It uses a color-guided test instrument that can perform the enumeration of as few as 100 bacteria per ml within a 2–3 min. test period.[316] This test was developed by the Baylor College of Medicine, Department of Virology and Epidemiology.[317]

Procedure and other details of this test appear on p. 918 of *Standard Methods,* 16th ed. This rapid procedure greatly simplifies distribution system quality monitoring.

ORP Control

Rechlorination. The Safe Drinking Water Act (SDWA) and its amendments place a new emphasis on distribution system water quality. This in turn is generating increased pressure on many distribution system managers to identify and adopt new strategies for managing disinfectant residual concentration.[493]

Owing to this dilemma, managers of many water distribution systems are finding it difficult to maintain the disinfectant residual balance necessary to meet the new federal and state requirements. However, by adopting an unconventional strategy that involves identifying areas of chlorine residual deficiency and treating these areas to achieve and maintain specific Oxidation Reduction Potential (ORP) parameters, adequate chlorine residuals can be restored without chlorine overfeed.[496, 499]

Chlorination serves two basic purposes in public water systems: substantial reduction of pathogens in source water and maintenance of low bacterial counts in the distribution system. The SWDA stipulates that every public water supply now must disinfect, and a measurable disinfectant residual concentration must be maintained in the finished water and all the sample points in the distribution system.[494]

In Ohio, regulations took effect in June 1993 that require 95 percent of all samples at remote traps to contain zero total coliform and no less than 0.2 mg/L free chlorine (85 percent HOCl), or 1.0 mg/L of combined chlorine.

Residual chlorine essentially serves as a disinfection safety margin for preventing bacterial growth in the distribution system. But owing to factors such as turbidity masking and disinfection stressing, rapid deterioration of the chlorine residual can present problems, especially at the outer extremities of medium to large distribution systems.[494, 499]

Although chlorine is very reactive in water, residual fluctuations (chlorine demand) are common in the most rigorously managed program. As chlorine is added to water, many residual compounds are formed. Some are effective as disinfectants and some are not. One of the many advantages of ORP measurement (millivolts) is its ability to detect these variations in the oxidant profile, which is constantly changing. Chlorine residual measurement can distinguish free from total chlorine, but only ORP can measure the effect of the changes in the oxidant profile. This is so largely because of compounds that occur in surface waters that are measured as chlorine residuals but are nongermicidal.

Some of the early research on the relationship between ORP and microbiological kill rate was conducted by Ebba Lund at the University of Gothenburg in Sweden in the late 1950s and early 1960s. Her research demonstrated a direct link between disinfection power and the ORP millivolts achieved by adding various oxidizers. These experiments demonstrated that a given millivolt potential for a specific time (*Ct*) would assure the destruction of the polio virus.[497]

Later studies by Carlson and Haesselbarth in Berlin on the kill rate of *E. coli* by chlorine led to similar conclusions.[500] Their reaction kinetic studies led them to define the rate of sterilization as a function of ORP and pH, independent of the chlorine concentration.

Control of Biofouling in Distribution Systems

Free Chlorine vs. Combined Chlorine Residual. The application of chlorine is imperative to control biofouling in distribution systems, whether it be regrowth of coliform organisms or the growths that cause taste, odor, dirty water, or corrosion. The controversial aspect of this treatment is how to apply it. Some strongly believe that the best method is to push a free chlorine residual to the far reaches of the distribution system. Others believe just as strongly that a combined residual that is predominantly monochloramine should be used.*

The proponents of the combined available residual method cite the following:

1. The combined residual, although less potent, is more persistent and will eventually penetrate farther for a longer time.
2. If a free residual is pushed forward and the system is badly contaminated, this free residual in its travel will gradually be turned back toward the dip and then to the hump of the breakpoint curve and will eventually become all combined available residual. Further, during the process whereby the free residual is being converted to combined residual, tastes will result from the inevitable formation of nitrogen trichloride and dichloramines.
3. This eventual formation of combined residual chlorine is certain to occur in the consumer service pipes long after the distribution pipes have been cleaned.
4. The free residual method reacts much faster than the combined residual, which produces sloughing of the debris in the system caused by the biofouling process. This produces horribly dirty water and objectionable tastes and odors.

Case Histories. The proponents of the free residual are usually those who are confronted with a serious situation that needs drastic action, and are quick to admit serious consumer complaints while the system is being cleaned.

The ammonia chlorine process was used by Arnold[238] in 1936 to control the growth of filamentous bacteria by intrusion waters in the coast tunnel of the San Francisco water supply. The application of ammonia was discontinued

*This is a very controversial subject; many experienced operators prefer total chloramine residuals.

when it was difficult to obtain during the war years, and was never resumed. Ackerman[239] describes controlling tuberculation using chloramine treatment.

Harvill et al.[95, 96] used the unique method of utilizing ammonia to produce an induced breakpoint. By varying the ammonia dosage, they were able to control the magnitude of the residual at the dip. The ammonia was applied after the chlorine. This is a variation of the postammoniation technique recommended by proponents of chloramine residuals for distribution systems. This technique was used for about eight years until the system was cleaned, at which time the ammonia was discontinued.[97]

Alexander[243] reported in 1944 upon the successful use of free residual chlorination in cleaning and maintaining a distribution system free from biofouling originally caused by iron- and sulfur-related organisms. Gradually pushing through a 1.0 mg/L free residual over the entire system did not result in complaints of chlorinous tastes. However, the success described by Alexander came about after the system was thoroughly cleaned of massive tenacious films of organic slimes. This cleaning was done by isolating parts of the system and subjecting them to sterilization dosages of 50 to 100 mg/L chlorine plus hydraulic and air purging of the pipe lines.

Brown[244] reported the use of free chlorine residuals as high as 1.5 mg/L to clean the Santa Rosa, California, system, but there were numerous complaints about chlorinous odors, which were apparently caused by the formation of nitrogen trichloride. Dechlorination, to limit the residuals going into the system, was tried, after the system had been cleaned, in an attempt to reduce complaints about chlorinous tastes and odors, sometimes described as a "Clorox cocktail" by the newspapers. This dechlorination did reduce these complaints, but soon the black fluffy slime returned to the taps. The use of dechlorination was discontinued in favor of a free chlorine residual in the distribution system in order to restore water quality to the consumer.

Wilson[244] reported the successful use of free residual chlorine, with 0.5–0.8 mg/L in the distribution system, which took three years and resulted in many consumer complaints about chlorinous tastes. After the system was pronounced clean, the chlorine residual entering the system was reduced from 2 to 1 mg/L with a noticeable reduction in the number of complaints.

Pomeroy and Montgomery[244] experimented at Torrance, California, between 1940 and 1949, by using both chlorine and ammonia and free residual chlorination. They were unable to clean the system by using chloramine residuals. After the system was rid of the putrefactive condition, free residual chlorination produced a water of excellent quality. Although they attributed a large part of the success of cleaning the system to chlorine treatment without ammonia, an intensification of swampy and musty odors—causing a storm of complaints—was experienced when free residual chlorination was substituted for the ammonia–chlorine treatment.

Blair[245] reported in 1954 on the use of free chlorine residuals to clean red water troubles caused by *Crenothrix* that had been plaguing the Palo Alto, California, water system.

Implementing a Cleanup Program. This carefully planned program at Palo Alto attempted to carry a 1.0 mg/L free chlorine residual througout the entire distribution system. No objectionable tastes and odors were reported whenever the residual was as high as 0.4 mg/L, but when it dropped to 0.2 mg/L or less, complaints were prevalent. After the entire system was cleaned satisfactorily, further complaints were traced to consumer service lines. A phenomenon reported by Wilson[246] was also reported by Blair. In nearly all these instances, the objectionable taste and odor occurred in the morning after the water had stayed in the pipes overnight, and all the free chlorine had been consumed. After this water and its chloro products had been discharged from the pipes, the taste and odors disappeared, only to recur when consumer use dropped off, and the free chlorine residual again disappeared. The distribution system residual might often show 0.6 mg/L free chlorine residual while the consumer tap would not show even a trace of residual.

The foregoing demonstrated the effectiveness of chlorination in cleaning a distribution system suffering from biofouling organisms. The following is a summary of guidelines for carrying out a program to clean such a system:

1. These systems are usually supplied with underground waters that are more or less deficient in oxygen, so that the first step is to run a dissolved oxygen survey to identify the zones of degradation.
2. Implement a flushing program at the worst places, starting with the zones of least dissolved oxygen.
3. Try to push through the entire system a free chlorine residual of at least 1.0 mg/L.
4. Enlist the aid of the local newspaper in indicating to the consumers the reasons for the program and how they can help by flushing their own service pipes.
5. Keep a continuous record of complaints.
6. Repeat the dissolved oxygen survey as necessary.
7. Utilize portable chlorine residual recorders where possible, not only to record the residual variations but to identify the free residual chlorine.
8. Make spot checks of distribution system residuals versus adjacent consumer tap residuals. If these checks show a wide disparity, institute a special consumer service flushing program at now cost to the consumer for water used.

Chlorine Residuals to Maintain Bacterial Quality

Historical Background. The National Academy of Sciences and the National Research Council issued a joint report in 1955 and a clarification in 1959[247] advising that the establishment of a universal standard for maintaining residual chlorine in water in distribution systems was not desirable at that time, owing to the wide variations in circumstances encountered. This report, pertaining solely to military installations, acknowledges not only the desirability of chlorine residuals to maintain better bacteria control but that loss of the residual could be a warning of possile sabotage. Furthermore, this report recognizes that combined residuals can be maintained within a distribution system much more easily than can free chlorine residuals, but that the combined residuals are such relatively weak disinfectants that it is questionable whether the two types of residual (free and combined) should be considered analogous or even comparable unless the combined residuals used are from 10 to 20 times the usual value of free residuals.

When the U.S. Public Health Service standards on water quality were revised in 1942, emphasis was placed on samples of water collected from the distribution system. Water leaving the treatment plant or pumping station could be well within the permissible bacteriological standards but might show considerable deterioration in the distribution system. The result was an increased effort to maintain chlorine residuals in those distribution systems confronted with this deterioration of quality.

Baylis and Kuehn[247] reported that the city of Chicago maintains chlorine residuals throughout its system of approximately 0.4–0.5 mg/L, some free chlorine residuals and some chloramine.

Crabill[247] reported that:

1. High-level chlorine residuals—0.4—maintain a better quality of water in the distribution systems than do low-level residuals—0.05.
2. The quality of the distributed water is reduced when chlorine residuals are lost.
3. When chlorine residuals are lost, supplemental chlorination will improve water quality.
4. There is a definite relation between water temperature and the persistence of chlorine residuals.

Plowman and Rademacher[248] found that the loss of combined residual chlorine was low when the water was in the 44 to 60°F range, but that during the summer, when the range was 68 to 73°F, there was a rapid loss of residual.

These conclusions indicate a real value in maintaining chlorine residuals throughout the system.

Umbehauer[247] reports that the use of 2.0 mg/L residuals in the far-flung El Paso, Texas, distribution system has been highly successful from a bacteriological standpoint with fewer than the anticipated number of consumer complaints.

The public health aspect of maintaining chlorine residuals in a distribution system for added consumer protection is making it virtually a necessity. It is well known that a distribution system can become contaminated from cross-connections with nonpotable water, or a water main and sewer line break at crossover points, thus causing outbreaks of a waterborne disease. This emphasizes the necessity for maintaining a chlorine residual in the distribution system.

Eliassen and Cummings's analysis of waterborne disease outbreaks showed that the greatest number of cases among users of public water supplies resulted from contamination of the distribution system.[249] One of the largest outbreaks occurred at Newton, Kansas, with an epidemic of 3000 cases of bacillary dysentery in 1942. In 1952, the Kansas State Board of Health requested that cities using surface supplies maintain a minimum of 0.4 mg/L of free available residual chlorine at all consumer taps.[16] Later the same year, 30 cases of amoebic dysentery broke out among factory workers in South Bend, Indiana, caused by sewage contamination of the distribution system.[250] The 1952 request for distribution system residuals and the chlorination of all supplies was made into an order by the Kansas State Department of Health in 1956.

The investigation of the Cincinnati system reported by Buelow and Walton[240] concluded that the change from combined to free chlorine residual together with an increase in the residual concentration greatly reduced the average monthly coliform counts. Prior to the changeover from combined to free residual, the plant effluent was regulated to maintain 0.85 ± 0.20 mg/L total chlorine; to this 0.2 mg/L ammonia was added. When the changeover was made, the chlorine residual was raised to 2.0 mg/L total; about 80 percent was free chlorine. This study emphasized what the Community Water System Survey[129] showed—that a chlorine residual must be maintained throughout the system in order to establish proof that disinfection has been achieved. Also the recurring cases of coliform appearances in the distribution system samples when none were detected in the treatment plant effluent indicate the necessity for continuous disinfection throughout the system. The data collected in the Cincinnati investigation strongly support the need to take bacteriological samples in known problem areas such as reservoirs, dead-ends, and the periphery of the system. It has been suggested that chlorine residuals could in some cases be substituted for elaborate bacteriological sampling, provided of course that the residual is properly monitored.

In 1974 the committee on water quality in transmission and distribution systems outlined some basic requirements for a water entering a distribution network.[241] For example, if a water is of low quality with a high nutrient content, it should be carefully flocculated and filtered to produce a product low in aluminum and iron with a turbidity no higher than 0.2 JTU. This water should be able to hold a chlorine residual fairly well if the piping system is relatively clean. The goal is to maintain a 0.2–0.3 mg/L total residual in the far reaches of the system. If the residual decays in the system, rechlorination may be necessary.

In 1973 the City of Chicago converted both the Central and South Water Filtration Plants to provide free chlorine residuals throughout the city.[162] In view of the mass of evidence in the literature and case histories of water quality improvement in distribution systems, this change had been contemplated for some time. After one year of operation the records indicated that the switch-over was a complete success, in that there had been fewer consumer complaints and better bacteriological quality, and residuals could be maintained in remote dead-end areas with a modest amount of special flushing techniques.

Snead et al.[242] made a comprehensive report in 1980 for the EPA, demonstrating the benefits and the necessity of chlorine residual in distribution systems, with particular emphasis on free chlorine residuals.

The foregoing case histories reflect success with free chlorine residuals.

Relay Chlorination. The availability of modern equipment to monitor and record either free or combined chlorine residuals simplifies the task of maintaining effective residuals throughout a distribution system. The Hartford-type installation, applicable to most systems, is described next.[251]

The water is rechlorinated solely by automatic residual control. A chlorine analyzer provides the intelligence to control the chlorination equipment without the necessity of installing expensive flow-sensing devices in the distribution system. This installation provides the means of obtaining residuals in that part of the system where the bacteriological quality of the water is substandard. This also is the part of the system where it is impossible to obtain residuals with chlorination at the treatment plant alone. This single point of rechlorination provides the means of meeting the U.S. Public Health standards. The attempt is to keep 0.8–1.2 mg/L free residual chlorine; however, owing to flow reversals, the residual deteriorates to as low as 0.3 mg/L and climbs to as high as 2.0 mg/L for a few hours each day. This approach has proved successful in many instances. It is particularly useful where open balancing reservoirs deplete the residual in the distribution system.

ORP Control of Relay Chlorination

Because of ORP's ability to measure the ever-changing chlorine demand of the oxidant profile in the distribution system, the restoration of bacterial control and provision of adequate chlorine residuals where needed can be achieved effectively by treating the system to maintain specific ORP parameters.

Under the ORP system of measurement to control rechlorination in a distribution system, a redox probe reports the ORP levels in the system to a controller, which in turn modulates the rate of chlorine feed, based upon changing oxidant demand. By monitoring the total demand, the system matches changing chlorine feed requirements, to hold the chlorine residual at the value required to accurately control bacteria growth as well as meet disinfection requirements.[496, 499] (See Chapter 9 for design details.)

Practical Considerations. Careful planning of the distribution system and the proper location of balancing reservoirs can largely prevent situations that tend to cause wide fluctuations in distribution system residuals. As an example, more attention should be given to the prevention of dead-ends. A periphery type of grid system will be needed to improve the circulation of a distribution system. The literature abounds with dead-end water quality problems, but apparently the designers of such systems are totally unaware of the consequences of poor circulation.

Research is also needed to properly determine whether free residual or combined residual chlorine is preferable in a distribution system. As a broad general rule, it is better to have a predominantly monochloramine residual, rather than a free residual, entering a relatively clean system. The chance of developing nitrogen trichloride is nil if the residual entering the system is primarily monochloramine. However, if a system is dirty, then free residual chlorine must be used to do the cleaning as soon as possible. Such applications are bound to produce nitrogen trichloride and dichloramines in quantities sufficient for consumers to complain of tastes and odors. These obnoxious chlorine compounds will persist until the free chlorine residuals penetrate to all parts of the system, which is accomplished by a continuous flushing program with sufficient high free residuals to do the job.

Another area that needs more attention is that of the corrosion potential of the system residual. As an example, Williams[222] used a combined residual at Brantford, Ontario, to maintain proper water quality and prevent the formation of tastes and odors in the distribution system. Owing to the presence of significant concentrations of organic nitrogen (1–3 mg/L) in the raw water and the necessity of using the free residual chlorination process at the treatment plant, nitrogen trichloride was produced at the treatment plant. If free chlorine residuals were allowed to enter the distribution system, then nitrogen trichloride would develop in the distribution system, causing taste and odor complaints. To prevent the subsequent formation of nitrogen trichloride in the distribution system, postammoniation was practiced. Sufficient ammonia was added to convert all the free chlorine residual to combined residual, and enough in excess of this to produce a combined residual consisting of 80 percent mono- and 20 percent dichloramine going to the distribution system. If the organic nitrogen content were high, the dichloramine fraction would be higher, and vice versa. The total combined residual going to the distribution system was kept between 0.6 and 0.8 mg/L.

At one time this water passing through the distribution system developed a progressively deteriorating palatability, with the resulting taste described as "bitter irony." A thorough investigation revealed a pickup of zero to 0.8 mg/L ferrous iron and a shift in the chlorine residual to 40 percent mono- and 60 percent dichloramine, with the monochloramine deteriorating from about 0.6 mg/L to about 0.15 mg/L. The drop-off of chlorine residual in the system was proportional to the pickup of ferrous iron. There was also a drop-off in ammonia content as high as 75 percent in two hours as the water passed through the distribution system. However, samples stored in the laboratory showed complete stability of chlorine residual and ammonia content. The difficulty was finally discovered to be aggressive water. Analysis showed that the heavy prechlorination doses combined with the addition of aluminum sulfate for turbidity removal resulted in a sufficient loss in alkalinity and a lowering of the pH to make the treated water aggressive. This aggressive water contained enough free CO_2 to attack the iron pipe, causing ferrous bicarbonate to go into solution. This resulted in the dechlorination of the monochloramine fraction, producing microquantities of ferric hydroxide, which eventually flushed from the system without causing red water troubles. During these times (1950s), the stability of the dichloramine fraction of the residual was somewhat of an anomaly, but it was explained by Williams that when the organic N increases, so does the dichloramine proportion of the total combined residual. Owing to sewage pollution, the raw water contained a rather high amount of organic N. As sewage contains urine and other compounds containing significant quantities of creatinine, the stable dichloramine fraction is probably a chlorcreatinine compound, an organic N compound. We now know (1990s) that the chlorine residual fraction that appears in the forward-titration procedures of only the amperometric and DPD-FAS methods is a nongermicidal organochloramine. That explains its stability. This was not discovered until about 1982.

Williams confirmed the fact that there is no off-flavor resulting from the dichloramine fraction. The monochloramine fraction is considerably more susceptible to dechlorination by ferrous iron than is dichloramine, and so the gradual disappearance of monochloramine residual as the water passes through the distribution system is not unexpected.

The problem of the iron pickup and the resulting "bitter irony" taste was solved at Brantford by the addition of lime to the treated water, which restored the alkalinity and raised the pH to normal values in accordance with Langelier's equilibrium index. Now the water has the ability to lay down a slight protective film of calcium carbonate on the metal pipes.

Various experiences, extending over many years, some of which are described in the foregoing text, show that chlorine residuals in a distribution system do the following: control growths in the system; improve the palatability of the water; give added consumer protection against waterborne diseases; and protect against accidental and possibly intentional contamination of the system.

BACTERIAL REGROWTH IN WATER SYSTEMS

General Discussion

Introduction. With monitoring changes required by the EPA 1989 coliform rule, the problems associated with coliform regrowth in finished water are increasing rapidly, simply because the EPA is more worried about THMs than pathogenic organisms by restricting the level of total chlorine residuals in the distribution system—and insists on using the chloramine process, which allows microbiological life to thrive on the ammonia N that remains when the chlorine residual disappears! Moreover the EPA has yet to mention that to qualify as a "free chlorine" residual, the total residual must contain at least 85 percent HOCl.

The occurrence of coliform bacteria in otherwise high-quality drinking water has been plaguing the water industry for years.[359] In 1930, the AWWA Committee on Water Supply reported on the problems of "B. Coli" regrowth.[360] In the nearly 70 years since that time there has been little progress in solving the regrowth problem. The basic problem that must be faced is the potential that exists for serious waterborne disease outbreaks. If coliform organisms can grow in treated water supplies as compared to a host of disease-causing organisms, the coliform group is quite fragile and relatively easy to inactivate.

The efforts by the EPA in passage of the Safe Drinking Water Act in 1989 were definitely a step in the right direction.[470] This Act and its amendments placed a new emphasis on distribution system water quality, which requires all water systems to maintain a chlorine residual in the distribution system. Water system managers had to identify and adopt new strategies for managing disinfectant residual concentration, which itself is directly related to microbiological regrowth.

Explanation of Terms Involving Regrowth. Biofilm, breakthrough, and growth are key terms.

1. *Biofilm.* The most important factor affecting regrowth, biofilm is an organic or inorganic surface deposit consisting of microorganisms, microbial products, and detritus.[361–363] Biofilm may occur on pipe surfaces, sediments, inorganic tubercles, suspended particles, or virtually any substratum immersed in the aquatic environment. It may be evenly distributed or occur as sporadic, random patches on a surface.

Biofilm formation occurs with the transport and accumulation of microorganisms and nutrients at an interface, followed by metabolic growth, product formation, and, finally, detachment, erosion, or sloughing of the biofilm from its surface.[362, 363] The rate of biofilm formation depends upon the chemical-thermodynamic properties of the interface, the physical roughness of the surface, and the physiological characteristics of the attached microorganisms.[364] Sheer forces generated by fluid velocity and the effects of disinfectants on extracellular polymer substances (EPS) may be important in the release of biofilms from surfaces.

The attachment of bacteria to surfaces in flowing aquatic environments has important ecological consequences:[364]

- Macromolecules tend to accumulate at liquid–solid interfaces, creating a favorable environment in an otherwise nutrient-deficient situation.
- Low nutrient concentrations in the water plus high flow rates can result in the transport of tremendous quantities of nutrients to fixed microorganisms.
- EPS anchor attached bacteria and may be a factor in nutrient capture.

- Bacteria embedded in EPS matrixes are protected from disinfectants by a combination of physical and transport phenomena.

These observations and others have led microbiologists to conclude that most bacteria in aquatic environments exist at solid–liquid interfaces.

2. *Breakthrough.* This is considered to be an increase of bacteria in the distribution system resulting from excessive numbers of bacteria passing through or avoiding the treatment process. This phenomenon is associated with the traditional causes of waterborne illness.

3. *Growth.* This is an increase in the bacterial numbers in the distribution system resulting from cell reproduction. Significant growth always occurs at the expense of the organic or inorganic substrate. Therefore, an understanding of the nutritional condition within the pipe network is critical to control of growth within the system.[458, 459]

Benefits of Bioassays. These procedures provide a useful means for determining whether a given water can support microbial growth. However, the majority of this growth occurs in the biofilm formation associated with pipe surfaces and not in the bulk water. The bulk water bathes these surfaces and thus provides a constant flux of nutrients. The bioassays that are designed to determine the nutrient status of the bulk water can play a major role in evaluating biological stability.[458]

The studies performed by Herson et al.[459] underscore the need for the development of alternative techniques to provide bacteriological monitoring of surfaces. These researchers found that coliform and indigenous noncoliform organisms are able to accumulate on surfaces, resulting in dramatic differences in microbial numbers between the bulk water phase and the surfaces. However, in nutrient supplementation experiments, no apparent increase in microbial numbers was found in the bulk water phase after the addition of more nutrients although microscopic examination of particle surfaces revealed profuse aggregates. This is important because current standard procedures for determining water quality are based upon methods that were developed using unattached bacteria.

Types of Regrowth

Introduction. Although it is poorly defined, the term "regrowth" is well rooted in the vocabulary of the water utilities. In this text "regrowth" will be used exclusively to describe multiplication of microorganisms in the distribution system, either in the water or on pipe surfaces as biofilm. Growth of bacteria in drinking water may be the result of a variety of different situations. Likewise, the approach to solving these situations will vary.

Loss of Residual. Growth of HPC and coliform bacteria may occur in some sections of a distribution system that cannot maintain an effective disinfectant residual.[164, 365] Resolving the problem in this situation is relatively simple. Flush the system well, and then increase the disinfectant dosage to provide an effective residual in all parts of the system. Monitor the system for both disinfectant residual and DO. This may reveal the necessity for relay chlorination to handle remote parts of the system or other parts that may be prone to dead-end problems such as insufficient DO. The DO concentration is most important because lack of it brings on septicity, which causes biological instability. Also, do not overlook the necessity for removal of chlorine-demand-causing compounds through revised treatment practices, pipeline pigging, relining, or pipe replacements.

Monitoring by ORP. With the advent of the Safe Drinking Water Act in 1989,[493] every public surface water supply must disinfect, and must maintain a measurable disinfectant residual concentration in the finished water, and at all sample points in the distribution system. The most difficult task a water system operator has to deal with probably is monitoring the water quality in the distribution system, primarily because of the accuracy of the method used to identify the chlorine residual in various parts of the distribution system.

The most accurate and best available chlorine analysis method is either the complete forward-titration of the amperometric method or the DPD-FAS method.

If the water system contains significant concentrations of ammonia N, organic N, nitrites, or other organic compounds, there can be significant errors in the total residual measurement by either of the above methods. This in turn affects the oxidant profile (true chlorine residual), which is of primary importance in maintaining a distribution system with a stable microbiological life throughout the system.[495]

The ability of ORP to measure the ever-changing chlorine demand and oxidant profile in the system allows the restoration of bacterial control and the creation of an adequate chlorine residual in all areas of a water distribution system. Only the ORP method can accurately detect variations in the oxidant profile, which is constantly changing in a distribution system. It is clearly the best method available to accurately establish the oxidant profile.[496, 499] This is accomplished by maintaining specific ORP parameters with an ORP control system.* (See Chapter 9 for system details.)

Furthermore, unlike chlorine residual measurement systems, ORP can detect the effect of pH. As free chlorine (HOCl) dissociates according to pH, undissociated HOCl has approximately 60 to 100 times the oxidative power of its dissociated form.[471] Therefore, the same free residual chlorine will be twice as effective at pH of 7.2 as a 7.8. Therefore, ORP not only can control the chlorine feed rates; *it tracks the chlorine demand.*[494]

*In these situations, the best oxidant measuring procedure is to use ORP m V values that will eliminate the errors caused by the nongermicidal chlorine residuals, caused by the presence of organic N compounds in the potable water supply.

Breakthrough. This will be indicated by the appearance of coliform bacteria along with high levels of HPC bacteria in the finished water effluents. Breakthrough of coliform organisms in treatment plants may occur even when effluent samples apparently are of good microbiological quality. This may be due to the repair of the cellular lesion in badly injured coliform bacteria that have been resuscitated in the biofilm.[366] Several reports[367, 368] describe how the detection of injured coliform bacteria in plant effluents has helped the operator to detect and correct these microbiological problems.

Research has shown that increased resistance to disinfection results from attachment or colonization of microorganisms on various surfaces, including macroinvertebrates (such as Crustacea, Nematodes, Insecta, and Platyhelminthes),[369, 370] turbidity particles,[371–375] algae,[377] and carbon fines.[376, 377] Bacteria on these particles may survive disinfection and enter the system. Solving this type of problem requires intensive examination by operating personnel. The sources of contamination in the Grand Rapids, Michigan system[445] included a turbid discharge from one of the plant's filters, seepage of rainwater into the filter beds, cross-connections, and leaks in the clear well. Each one of these contamination events resulted in a prolonged occurrence of coliforms in the distribution system.

Breakthrough and regrowth are characterized by a large initial occurrence of coliform organisms, followed by a gradual decline in the bacterial levels over time, perhaps as long as several months. The only solution to breakthrough and regrowth problems is to locate and then eliminate the source of contamination.[377]

Coliform Occurrence

General Discussion. The regrowth phenomenon is best characterized by the continuous and persistent appearance of coliform organisms in high quality drinking water. Several factors distinguish coliform growth in distribution systems:

1. Coliforms are not detected, or counts are very low in treatment plant effluents even when sensitive methodologies are employed.
2. High densities of coliforms are routinely detected in the distribution system samples.
3. Coliforms persist in the distribution system samples despite the maintenance of a disinfectant residual.
4. The coliform episode lasts for a long period of time—several years.[388]

Most of the time it is difficult to distinguish between a true coliform event and an unexplained coliform occurrence. Relatively few studies have actually identified coliform organisms in distribution system biofilms. Procedures used to be formulated to help operators and regulators to distinguish regrowth events from episodes of unrecognized contamination.

Regrowth in Water Supplies

Microscopic Evidence. There is ample evidence that most pipe surfaces in distribution systems are colonized by microorganisms. Electron microscopy of systems encrustations has revealed several common characteristics:

1. A hard but porous surface.
2. A multitude of crystals beneath the surface veneer.
3. Microorganisms predominating near the surface layer.
4. Microcolonies of similar-shaped organisms, indicating growth at the biofilm surface.[378]

Inductively coupled plasma spectrometric analysis of tubercles scraped from a cement-lined ductile-iron pipe showed that the material was 98.7 percent iron.[388]

X-ray energy-dispersive microanalysis of distribution system pipe surfaces has shown that tubercles are composed predominantly of iron, calcium, silicon, phosphorus, aluminum, and sulfur.[379, 380]

Allen et al.[378] showed large populations of bacteria in main encrustations collected from seven water utilities throughout the United States, Diatoms, algae, and filamentous and rod-shaped bacteria were commonly encountered.

In spite of the differences in microscopic observations, the data clearly indicate that distribution systems are readily colonized by a variety of microorganisms. These observations suggest that bacterial growth routinely occurs in all distribution pipe networks. It is not clear at this time why coliform regrowth is an apparent problem only in certain systems. As the water industry develops a more comprehensive data base on bacterial nutrients, information should be available to better address this question.

Cultural Evidence: HPC Bacteria. Cultural examination of distribution system biofilms has demonstrated large variations in the number of HPC bacteria. Tuovinen and Hsu[381] found viable counts that ranged between 40 and 3.1×10^6 bacteria/gram in tubercles collected from water distribution system pipelines in Columbus, Ohio. The tubercles were found to contain several types of organisms, such as sulfate reducers, nitrate reducers, nitrite oxidizers, and various unidentified HPC microorganisms.

Bacteria genera associated with distribution system biofilm included *Artrobacter, Flavobacterium, Maraxella, Acinobacter, Bacillus, Pseudomonas, Alcaligenes,* and *Achromobacter.* Nagy and Olson[382] observed a correlation between the years in service of a pipeline and the density of bacteria. They estimated that HPC bacterial levels increased one log for every 10 years of service. They did not find any correlation between the maintenance of free chlorine residuals in the water and the HPC densities in the biofilm. In another study by Nagy et al.[383] of the biofilms in the Los Angeles Water & Power Mono Basin Aqueduct, HPC levels were observed as high as 1.9×10^4 bacteria/cm^2 in the presence of a 1–2 mg/L free chlorine residual.

Donlan and Pipes[384] inserted coupons into cast iron corporation cocks as sampling devices for various areas of the Philadelphia, Pennsylvania Suburban distribution system. When water temperatures were warm (20–25°C), bacteria colonized and grew on the coupons to 10^4–10^8 cells/cm^2 within 28–115 days. Under cold water conditions (5–9°C), bacterial densities dropped to 10^2–10^5 cells/cm^2 within 29–84 days. The investigators found that water velocity and chlorine residuals did not correlate with biofilm viable counts.

Cultural Evidence: Coliform Bacteria. Despite the strong evidence for biofilms in distribution systems, little information is available concerning the occurrence of coliform bacteria at potable water interfaces. Seidler et al.[385] recovered *Klebsiella* growing in slime layers on staves of redwood distribution storage tanks. These bacteria were responsible for coliform densities in private drinking water systems that exceeded the federal membrane filter guideline as much as 10–40-fold. Olson[386] reported recovering *Escherichia coli* from the organic surface layer of a mortar-lined pipe in the distribution system of the Metropolitan Water District of Southern California, and Victoreen[387] isolated *Enterobacter cloacae* and *Serrutin marcescens* from tubercles in the Wilmington, Delaware distribution system.

Investigators at a water utility in New Jersey showed that coliform bacteria found in the water of the distribution system originated in the system biofilm.[388] Monitoring results showed that coliform levels increased 20-fold as the water moved away from the treatment plant through the distribution system. Calculations revealed that the increased levels could not have been due to growth in the water but must have come from the biofilms in the transmission line near the treatment plant. Identification of coliform bacteria showed that species diversity increased as the water flowed through the study area. (See Fig. 6-30). This demonstrated that the distribution system was quite hospitable to coliform organisms, showing that it could support growth of a variety of coliform species.

Other studies[389] of bacteria in tubercles flushed from the New Haven, Connecticut distribution system yielded the following isolates: *Enterobacter cloacae, E. agglomerans, E. alvei, E. sakazakii, Citrobacter freuendii, Klebsiella pneumoniae,* and *K. oxytoca.*

Because coliform bacteria occur at specific, discrete locations within the distribution system,[388] a random sampling of pipe surfaces may not detect these organisms. It is necessary to understand the mechanism that allows bacteria to colonize and grow in drinking water supplies. Otherwise it would not be possible to control future coliform episodes.

Factors Influencing Microbial Growth

Introduction. The unique growth requirements for coliform bacteria in drinking water have not been studied. Recent investigations have related bacterial growth to the following:

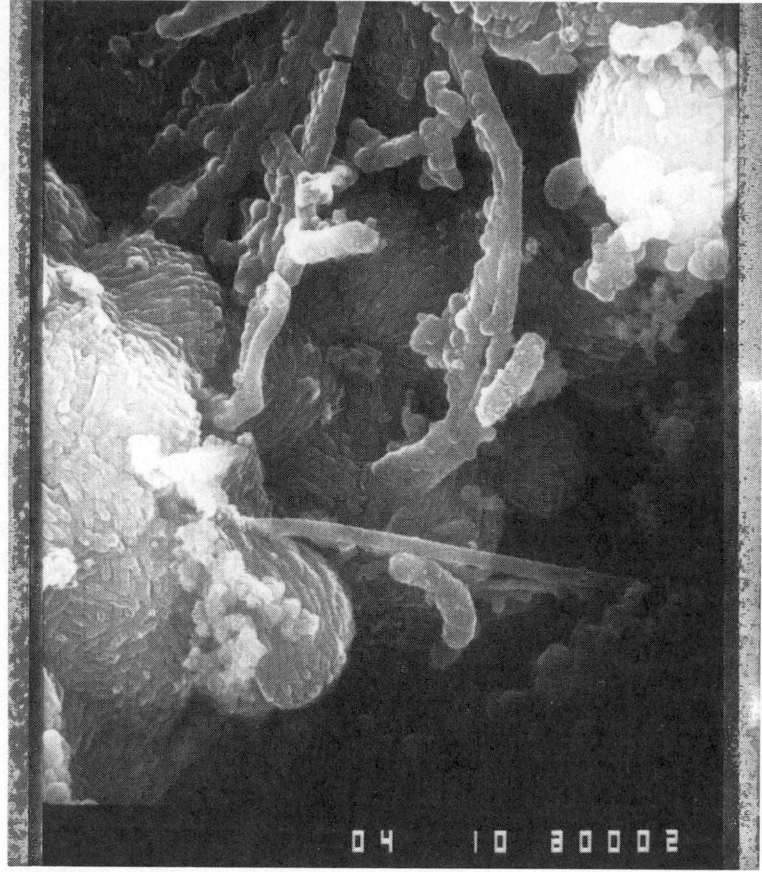

Fig. 6-30. Electromicrograph of a distribution system tubercle showing the association of bacillus-shaped organisms with filamentous bacteria. (Courtesy American Water Works Service Company Inc.)

1. Environmental factors, such as temperature and rainfall.
2. Availability of nutrients.
3. Impotence of disinfectant residuals.
4. Corrosion and sediment accumulation.
5. Hydraulic effects.

Environmental Factors: Temperature. Water temperature is perhaps the most important rate-controlling parameter in the microbial growth process. Temperature directly or indirectly affects all of the factors that govern microbial growth.

Temperature influences treatment plant efficiency, microbial growth rate, disinfection efficiency, decay of disinfectant residuals, corrosion rates, and distribution system hydraulics—for example, water velocity caused by consumer demand. However important it is, the water utilities can do very little to control water temperature.

Most investigators have observed significant microbial activity in water at 15°C or higher.[360] Fransolet et al.[390] found that water temperature influenced not only the growth rate but the lag phase and cell yield as well. The length of the lag phase was found to be quite important. For *Pseudomonas putida,* the lag in the growth phase was on the order of three days at 7.5°C but only 10 hours at 17.5°C. At low temperatures, cells would be washed out of the distribution system before significant growth could be achieved. Fransolet et al.[390] found that the growth of *E. coli* and *Enterobacter aerogenes* was very slow at water temperatures below 20°C.

Environmental Factors: Rainfall. Rainfall has been suggested by some investigators[391, 392] to be a catalyst for coliform growth. Lowther and Moser[392] found that TOC levels in raw water were at their highest when turbidity increased after rainfall. LeChevalier et al.[391] observed a seven-day lag between rainfall events and the occurrence of coliform bacteria in distribution system water samples. (See Figure 6-33). The authors speculated that the rainfall washed the available nutrients into the watershed; and then, after a transit period and growth lag, this resulted in an increase in bacterial densities.

Bacterial Nutrients. In order to grow, microorganisms must obtain all the substances they require for the synthesis of cell material and for the generation of energy from the environment. For coliforms and HPC bacteria, the principal nutrient sources are phosphorus, nitrogen, and carbon.

Phosphorus in the environment occurs almost exclusively as orthophosphate (PO_4^{3-}) with a valence state of +5. Because it is not consumed by microbial activity such as TOC, the turnover rate of phosphorus in aquatic habitats is more important than the level of orthophosphate in the distribution system. Determination of the concentration of phosphorus in drinking water is more involved than just measuring the phosphate levels in the water samples. The recycling of phosphate within the biofilm also must be determined.

Nitrogen occurs in water samples as organic N, ammonia N, nitrite, and nitrate. All of these compounds are extremely undesirable in drinking water.[393] Ammonia N is an electron donor for autotrophic bacteria and can promote bacterial growth in the system. Rittman and Snoeyink[394] found that ammonia concentrations in groundwater supplies frequently are high enough to cause biological instability. The proliferation of ammonia oxidizing in large covered finished water reservoirs in Southern California destroyed chlorine residuals, increased nitrite levels, and stimulated the growth of HPC bacteria.[396] Because autotrophic bacteria grow slowly, they require long retention times and warm water temperatures for these types of problems to occur. The exact role of nitrogen in the growth of coliform bacteria is unclear because some strains of *Klebsiella* can fix molecular nitrogen.[396]

Organic carbon is utilized by HPC bacteria to produce new cellular material (assimilation) and as an energy source (dissimilation). Most organic carbon in water supplies is natural in origin because it derives from living and decaying vegetation. These compounds may include humic and fulvic acids, polymeric carbohydrates, proteins, and carboxylic acids.

In the USEPA's National Organic Reconnaissance Survey, the nonpurgeable TOC concentration in finished drinking water at 80 different locations ranged from 0.05 to 12.2 mg/L, with a median concentration of 1.5 mg/L.[397] Because HPC bacteria require carbon, nitrogen, and phosphorus, in a ratio of approximately 100 : 10 : 1 (C : N : P), organic carbon is often a growth-limiting nutrient.

Assimilable organic carbon (AOC) is the portion of the TOC that can be readily digested by aquatic organisms that is used for growth. Often the AOC comprises only a fraction of the TOC, 0.1–9.0 percent.[398]

LeChevalier et al.[388, 391] showed that AOC levels declined in drinking water as it flowed through the distribution system. Removal of AOC occurred very quickly, within a short distance from the treatment plant. A reasonable relationship was found between AOC levels and growth of HPC bacteria. Growth of an *E. coli* isolate was inhibited by AOC levels less than 54 μg/L. Other studies[391] demonstrated that the occurrence of coliform bacteria could be associated with AOC levels greater than 50 μg/L. Van der Kooij et al.[398–400] found that HPC bacterial growth in distribution system water will not occur at AOC levels less than 10–15 μg of acetate carbon equivalents per liter (ac-C eq/L). AOC levels of 15–50 μg ac-C eq/L produced variable growth results. However, bacterial growths have always been observed when water contains AOC levels greater than 50 μg ac-C eq/L. Because water supplies in North America have been found to contain between 1 and 2000 μg ac-C eq/L, it is not surprising that many systems experience regrowth problems.

Advantages of ORP Monitoring

The primary advantage of using ORP on potable water systems is the complete elimination of nitrogen compound problems in maintaining adequate chlorine residuals in the distribution system.

All operators know how difficult it is to maintain a clean distribution piping system owing to faulty chlorine residuals caused by organochloramines that are nongermicidal, due to the presence of organic N in the water supply.

The Stranco High Resolution Redox Control System* provides an accurate and continuous oxidant level produced by the true chlorine residual. The system is not adversely affected by the presence of compounds that cause inaccuracies in chlorine residual analyzers.

This system provides operating personnel with a continuous chlorine oxidant level in a self-cleaning accurate installation that is easy to operate.

---

*The Stranco Group spent several years developing this highly accurate and dependable system before providing it for public use.

Reliability of Chlorine Residuals

Free Chlorine. Several investigators found that the maintenance of substantial free chlorine residuals did not always correlate with reduced bacterial counts in samples from the distribution system.[388, 401–404]

Nagy et al.[383] reported that a 1–2 mg/L chlorine residual reduced bacterial levels in the Los Angeles Aqueduct biofilm by 2 logs, but the remaining bacterial level was too high. It was necessary to lower the bacterial level by 3 logs, which required chlorine residuals of 3–5 mg/L. Ridgway et al.[405] found that 15–20 mg/L chlorine residuals were necessary to control biofouling of reverse osmosis membranes. Characklis et al.[406] reported that 12.5 mg/L free chlorine residual was required to reduce the thickness of experimental biofilm 29 percent in an annular fouling reactor. None of their experiments resulted in the complete removal of the biofilm.

During the persistent coliform episode at Muncie and Seymour, Indiana, chlorine residuals were boosted as high as 15 mg/L for days in an attempt to control and eliminate coliform bacteria in the respective distribution systems.* Coliform occurrence could not be reliably controlled with free chlorine residuals less than 6 mg/L.[392, 402] In the above cases Earnhardt reported recovering more than 50/100 mL coliforms in samples containing free chlorine residuals of 10–12 mg/L. Factors that promote this kind of bacterial survival in these chlorinated water supplies include attachment to surfaces, bacterial aggregation, age of biofilm, encapsulation, previous growth conditions, alteration of the bacterial cell wall, incrustation, corrosion, and choice of disinfectant.[407–413] These factors may act in a multiplicative manner, making disinfection of biofilm organisms extremely difficult.[412]

Positive Effects of Flushing. White believes from his experience, that flushing and monitoring inactive areas of the distribution piping is the only solution to regrowth problems. This includes proper pH levels, acceptable heterotrophic plate counts, at least 0.5 mg/L chlorine residual, preferably with 85 percent HOCl, and sufficient DO.

From his many experiences with on-site clean-up of many distribution systems, White has found that high velocity flushing to be very helpful. One thing he seemed to find with considerable consistency was this: as the free chlorine dosage (that he always used) proceeded through the distribution piping, the free chlorine residual changed to monochloramine where the system had yet to be cleaned. Therefore, White kept on cleaning the system until all samples were free chlorine residuals (85 percent HOCl).

Chloramines. LeChevalier et al.[412, 413] indicated that various disinfectants may interact differently at biofilm interfaces. In a comparison study,[414] investi-

*White[417] was witness to the Muncie, Indiana episodes along with R. H. Moser.

gators found that low levels (1.0 mg/L) of either free chlorine or monochloramine could reduce viable counts by greater than 2 logs for biofilms grown on copper, galvanized, or PVC pipe surfaces. However, when the organisms were grown on iron pipes, free chlorine residuals of 3–4 mg/L were ineffective for biofilm control. In this instance, only monochloramine residuals greater than 2.0 mg/L were successful for reducing biofilm viable counts. Haas et al.[415] modeled the interaction of free chlorine and monochloramine with biofilm surfaces and suggested that free chlorine, owing to its high reaction rate, was largely consumed before it penetrated the biofilm. Because monochloramine is more limited in the types of compounds with which it will react,[416, 502] it is better able than free chlorine to penetrate the biofilm layer and thereby inactivate the attached organisms.

The inability of a disinfectant to penetrate distribution system biofilms can account for the occurrence of coliform bacteria in highly chlorinated water. A better understanding of the interaction of disinfectants with distribution system interfaces is necessary before appropriate strategies for biofilm control can be formulated.

Corrosion and Sediment Accumulation. The observation that iron pipe surfaces protect attached bacteria from inactivation by a free chlorine residual suggests that corrosion interfaces with biofilm disinfection efficiency. Chlorine is well known for reacting with ferrous iron to produce insoluble ferric chloride,[417] as follows:

$$2\ Fe(HCO_3) + Cl_2 + Ca(HCO_3)_2 \rightarrow 2\ Fe(OH)_3 + CaCl_2 + 6CO_2$$

Each part of iron as Fe that is oxidized requires 0.64 mg/L chlorine. White[417] indicated that if iron is present in a complex form, free chlorine is more effective than combined chlorine in breaking up the complex iron compound so that ferrous oxidation can proceed. It is likely that biofilm organisms complex ferrous ions from metal surfaces within the glycocalyx layer. Free chlorine, therefore, not only reacts with the extracellular polysaccharides but liberates ferrous ions, which also consume the chlorine residual. It is interesting to note that in several systems[388, 389] high levels of coliform bacteria were associated with iron tubercles.

The accumulation of sediment and debris in distribution system lines can provide habitats for microbial growths plus protection from disinfectants. Dixon et al.[418] found that residual aluminum in filter effluent samples can form a hydrous floc deposit on pipe walls. This increases the concentration of organic compounds, which provide protection of bacteria from the disinfection process.

Flushing and mechanically scraping (pigging) distribution mains will remove loose sediments and tubercles that play host to bacterial activity. Whenever this is done, it is highly desirable to follow the mechanical action with a 50 mg/L dose of chlorine.[417, 419] In one study,[420] routine flushing of the distribution system to remove accumulated sediment resulted in a significant improvement in the bacteriological quality of the water.

Hydraulics. White[417] has observed the many ways hydraulics of a distribution system can affect the mirobiological growth on pipe surfaces. Increasing the velocity allows for greater flux of nutrients to the pipe surface, greater transfer of disinfectant, and greater shearing of biofilm from the pipe surface. Alternatively, stagnation in the system leads to a loss of the disinfectant residual, which allows subsequent microbial growth. This can also lead to an anaerobic condition that causes septicity because of insufficient DO in the water. This is the reason why dead-end areas show significant deterioration in microbial quality. Similarly, stagnation of water in service lines can promote high bacterial count at the consumer's tap.[42, 388] Donlan and Pipes[384] demonstrated that water velocity has an inverse relationship to biofilm counts.

Reversal of water flows can easily shear biofilms, and water hammer events can dislodge tubercles from pipe surfaces.[360, 417] Opheim et al.[389] found that bacterial levels in an experimental pipe system increased tenfold when flows were started and stopped. Larger releases were noted when the pipe system was exposed to physical and vibrational forces. Application of hydraulic models, modified to include AOC levels, bacterial growth rate, and disinfection kinetics, could be useful for a better understanding of microbial growth in a distribution system.[360]

Control of Regrowth

Treatment and Monitoring. Effective control of the bacterial occurrences in the distribution system of any water supply is largely dependent upon the quality of the finished water leaving the treatment plant. Selection of a suitable monitoring system is necessary to determine the exact cause of bacterial occurrences so that an approximate control strategy can be formulated. This approach is almost certain to require far more samples than are necessary for compliance purposes. It also may require different analytical methods.

Flushing and Cleaning. These procedures have been a tradition in the maintenance of drinking water distribution systems. Procedures for designing and conducting a flushing program have been outlined by AWWA.[422] However, once the problem of biological growth has become severe, flushing and mechanical cleaning may not be sufficient to control the regrowth problem.

Increased flushing of the New Haven, Connecticut distribution system actually increased the coliform levels in the water. Presumably this could be done if pieces of the biofilm were torn away from the pipe surfaces by changes in the shear or oxidative processes.[423] Three days after the flushing at Muncie, Indiana, 126/100 ml coliforms were recovered just a few blocks away from the treatment plant.[424] Flushing sections of the Seymour, Indiana distribution system did not eliminate coliform occurrences.[392] Flushing and mechanically cleaning a section of the New Jersey American Water Company system did not eliminate coliform bacteria because the organisms were found to originate in other parts of the system.

In practice it is difficult to perform these procedures on transmission mains and truck lines without extreme effort, high costs, and usually considerable disruption of service to the consumers.

Disinfection. Recent research has suggested that monochloramine may be more effective for biofilm control than free chlorine.[413, 414] It has been the experience of many in the water industry that secondary disinfection with a 100 percent combined chlorine residual can effectively control bacterial levels in the distribution system.[423–432] As reviewed by Kreft et al.,[429] more than 70 U.S. utilities effectively use chloramines for the disinfection of drinking water in the distribution system. MacLeod and Zimmerman[432] reported that before conversion to chloramines 56.1 percent of the water samples taken from the distribution were positive for coliform bacteria, and that after conversion only 18.2 percent of the samples contained coliform organisms. Although there may be many reasons for the reduced coliform counts, the system has remained coliform-free since February 1984.[433]

White suggests that when one is experimenting with chloramines in the distribution systems, some attention should be given to evaluating different chlorine-to-ammonia N ratios. He advises that in every system he has checked, everyone claims the ratio to be 3:1. Obviously this is the ratio that was universally used in the United States when chloramines were used to control chlorinous taste problems before the discovery of the breakpoint process. As it has been found in wastewater disinfection that a 6:1 ratio has the highest disinfection potential owing to the presence of some dichloramine, chlorinous tastes should not occur.

LeChevalier et al.[414] showed that there was a threshold level at which monochloramine was effective for biofilms on iron pipes: a 2.0 mg/L monochloramine residual. This number will certainly vary in accordance with site-specific conditions of water quality and pipe conditions.

The Hackensack, New Jersey Water Company converted to chloramine in its distribution system in 1982.[434] The system was given an initial chloramine dose of 2.0 mg/L, but because of sporadic presumptive coliform occurrences and evidence of nitrification in the distribution system, the company increased the chloramine dose to 3.0 mg/L in 1986. During that year only a few coliform bacteria were discovered in the summer months. In August 1986, chloramine doses were increased to 40 mg/L so that the average distribution system residual ranged from 2 to 3 mg/L for the remainder of the summer. Since November 1986, there have been no coliform bacteria or nitrification occurrences in the distribution system.

Utilities experiencing coliform regrowth problems have been relying upon high free chlorine residuals in the distribution system to control bacterial occurrences.[388, 401, 403, 404, 423, 424, 435, 436] In general, free chlorine residuals of

3–6 mg/L have been necessary to control coliform regrowth. However, Earnhardt[424] reported receiving 51/100 mL coliform bacteria in samples containing between 10 and 12 mg/L free chlorine. Recent studies suggest that biofilm control can be achieved using chlorine residual levels ranging from 2 to 4 mg/L. Further research is needed to confirm these results in a full-scale distribution system.

Corrosion Control. Application of corrosion inhibitors can improve the disinfection efficiency of free chlorine in its attack on biofilm bacteria in iron pipes. LeChevalier et al.[414] demonstrated that application of polyphosphate and zinc orthophosphate, plus pH and alkalinity adjustment, resulted in a 10–100-fold improvement in biofilm disinfection by free chlorine. Lowther and Moser[392] suggested that corrosion may have been related to the occurrence of coliform bacteria in the Seymour, Indiana distribution system. They reported that levels of coliform bacteria decreased within a few weeks following the application of zinc orthophosphate. This chemical has also been successfully used at other Indiana operations to control coliform occurrence.[437] Martin et al.[435] reported that adding lime to treated water supplies was an effective method of bacterial control by pH adjustment because high pH levels were bactericidal, in spite of the fact that at high pH levels free chlorine has been shown to lose its efficiency owing to the smaller fractin of HOCl.[417] It is possible that this may be the result of the "reservoir" effect of the HOCl ion relationship.

Hudson et al.[436] increased the pH to 10.2 and the free chlorine residual to 3–5 mg/L in the Springfield, Illinois distribution system network to control the coliform occurrence. In both of these situations, reduced corrosivity of the water could have resulted in improved free chlorine disinfection of the biofilm organisms.

Application of zinc orthophosphate was coupled with elevated pH levels at the American Water Works Service Company.[388] In this case, corrosion control did not affect coliform occurrence. It is not clear, however, how closely the corrosivity of the water was monitored. Fluctuations in water quality parameters (pH, chlorine residual, temperature), as observed in the New Jersey situation, should have made consistent corrosion control difficult.

One impact of the proposed lead regulations, which include corrosion control measures, may be improved disinfection of biofilm organisms. Obviously, additional research is necessary to determine the level at which corrosion interferes with free chlorine disinfection efficiency.

Biological Treatment. This type of drinking water treatment is the practice by which microbial activity is encouraged within the treatment process. Biologically stable water is produced when microbial activity has been sufficient to remove all nutrients that might support significant bacterial growth in the treated effluents. This type of treatment has been used for many years in Germany and more recently in Canada. The Canadians have had such remarkable success with the BAC system that they have been able to promote the appearance of a more efficient category of carbons. These carbons can be backwashed effectively without disturbing the biological colonization on the carbon. (See Chapter 13.)

The various benefits of biological treatment are: removal of TOC, reduction of concentrations of bacterial nutrients, removal of micropollutants, improved T & O control, reduction of chlorine demand which automatically reduces the level of DBPs, and a general improvement of all the processes in the treatment train.

Not all of these advantages are important to all water utilities, such as those with high-quality source water (e.g., San Francisco, Denver, and Portland, Oregon). Most systems are concerned with T & O programs, chlorine demand, DBPs, and bacterial regrowth. Owing to the proposed strict regulations now on the horizon, the implementation of biological treatment might be the answer to these problems. When this occurs, we in the United States will be able to impose a 0.2 mg/L limit on ammonia N at the consumer's tap. Then we can enjoy the luxury of the free chlorine residual process, as the people under the guidance of the European Economic Community have been doing for the last two decades.

Micropollutants. Application of biological treatment for the removal of micropollutants may be effective, depending upon the type of contaminant. Studies have shown that diphenalamines (products of microbial degradation of pesticides), naphthyl, styrene, chlorobenzene, 2-4D, and 1-naphthyl methylcarbamate are biodegradable at concentrations <100 μg/L.[438, 439] Wang et al.[440, 441] found that the BAC process was effective for the removal of phenol (80 percent), cyanide (65 percent), and various heavy metals, including Fe, Mn, and Cu (80–92 percent). However, other compounds such as chloroform, tetrachloroethylene, and trichloroethylene are degraded poorly under aerobic conditions.[442, 444] These compounds are far more susceptible to mineralization by certain advanced forms of UV radiation and other advanced oxidation processes. (See Chapter 16.)

Application of biological processes for the removal of micropollutants is highly site-specific. Other treatment processes should be considered in these cases, such as ozone, GAC, peroxone, hydrogen peroxide, and UV. (See Chapters 13–16.)

Chlorine Demand: Tastes. The reduction of chlorine demand in drinking water has been a prime objective in many foreign water treatment processes during the last four decades. This is why so much time, effort, and money are being spent on biological treatment processes. Bablon et al.[443] reported on how the use of BAC filtration combined with ozone in a French facility lowered the chlorine demand of the finished water, and this in turn eliminated both tastes and odors. It also resulted in a stable chlorine residual in the distribution system. At the same time, lowering the chlorine demand automatically lowered the concentration of DBPs.[443]

Summary

The water industry has been given important information on the phenomenon of coliform regrowth by the comprehensive and diverse investigations described above. The various environmental influences on coliform occurrences are, in general, as follows:

1. Temperature and rainfall.
2. Available nutrients (carbon-TOC, nitrogen, and phosphorus).
3. Ineffectivc disinfectant residuals.
4. Corrosion and sediment accumulation.
5. Effects of variable hydraulic conditions.
6. Lack of sensitive and accurate methods for enumerating and identifying bacterial growths.

The water industry now has a better understanding of the ecology of bacteria in the distribution system, of how treatment changes alter chemical and nutrient composition of water, and of cryptic bacterial breakthrough events. Morc intensive research is absolutely necessary, along with pilot-scale investigations of biological treatment processes that can produce higher-quality water. These processes should be oriented so that they remove TOC, limit AOC, and reduce chlorine demand. This will go a long way in solving distribution system problems.

White suggests that there is now a great need for upgrading all surface source waters that are used for drinking watcr supplies. The USEPA must set limits on industrial waste dischargcs and provide the surveillance required to see that all of the wastes are treated at conventional municipal WWTPs. These limits should be such that they would improve the quality of the source water rather than just prevent further degradation. The municipal WWTPs then would have the power to charge the industrial waste discharger an appropriate sum for gallonage treated. This would provide a significant incentive for the discharger to reduce the quantity being discharged. This type of program has worked very well in California for the past half-century.

Restoring Pipeline Capacity

The application of chlorine to a long transmission systems is somewhat akin to the maintenance of chlorine residuals in a distribution system. The principal difference is that there are seldom any consumers supplied by the transmission system, and therefore the public health aspect and the taste and odor control problems are usually not factors. These systems usually terminate at the treatment plant. If not, then the transmission line is simply part of the distribution system. It has been demonstrated that filamentous organisms can create a condition inside pipes to significantly increase the friction loss, thereby reducing the design capacity.

Rogers[253] reported a slime deposit ⅜ to ½ inch thick attached to the cement lining of a 42- and 48-inch transmission line 34 miles long. This deposit was 75 percent ferric oxide and contained *Crenothrix.* The capacity of the line dropped from 30 million to 20.4 million gal in three years. Continuous chlorination at 0.4–0.5 mg/L at the end of the line, prevented growth of the *Crenothrix* and thereby maintained the *C* factor in the line.

Griswold[254] used chloramine treatment on a 24-inch supply line 12 miles long. After 15 years of continuous treatment with a terminal residual of 1.0–1.5 mg/L combined chlorine, the line showed an average loss of 10 percent in capacity in seven years. The line is now cleaned every five or six years, whereas prior to chloramine treatment the line had to be cleaned every 20 months; but, in order to achieve the nominal carrying capacity of this line on a continuous basis provided by the chloramine treatment, the line would have to be cleaned every seven months if the treatment were discontinued.

The San Diego aqueduct, which travels some 70 miles from the west portal of the San Jacinto tunnel to the San Vicente reservoir in San Diego, originally had a capacity of 104 cfs. This dropped to 94 cfs, and slime growths were suspected. Streicher[255] reported on the success of intermittent chlorination of this line to maintain the original capacity of 104 cfs. After much experimentation it was found that a 2.2 mg/L chlorine dose for two hours twice a week was sufficient.

The experience at Little Rock, Arkansas, described by Jackson and Mayhan,[256] graphically illustrates how much more difficult it is to clean a transmission line after the onset of slime growths than it would be to prevent the growth from forming. A 39-inch transmission line 33 miles long was put into operation in 1938 with a *C* factor of 147 and a carrying capacity of 25.32 mgd. In one year this line was reduced in capacity by 20 percent.

The slime deposit of black jellylike organic material varied from a thin film to a mat ¼ inch thick and was heavily interspersed with iron oxide. The dominating organisms were found to be large gram-positive encapsulated bacteria and *Crenothrix.* The water supply originated at Lake Winona, a soft water of only 15 mg/L hardness and an iron content of approximately 0.4 mg/L. Chloramine treatment was begun in 1939 at 1.25 mg/L with a residual of 0.55 mg/L (starch–iodide) at the terminus of the line, which was the influent to the treatment plant. This treatment proved to be only moderately successful in restoring the line capacity. One year later free residual chlorine treatment was begun, and after nine years of a 5 mg/L dose resulting in a 1 mg/L residual at the plant the line was considered to be almost like new. The *C* factor had risen to 133 with a 22.75 mgd capacity in 1950. At this point, as the line was considered to be relatively clean, intermittent treatment with superdoses of chlorine was substituted for continuous treatment. Doses of 12 mg/L for about 2 or 3 hours per day resulted in residuals of 8 to 10 mg/L at the plant.

Dechlorination of these residuals was found to be unnecessary, as they dissipated rapidly in the large settling basins. The intermittent treatment, which consumed only about 250 lb/day chlorine, resulted in an annual savings of approximately fifty thousand dollars, thus demonstrating the exorbitant cost of the cure as compared to the cost of prevention.

The first step in any program designed to maintain a transmission line is to anticipate the problem of slime growths. When the line is first put into operation, the coefficient of friction should be determined while it is in the new, clean condition. This will establish an irrefutable reference point to determine when the line needs cleaning, and will provide a means for evaluating any needed chemical treatment. The friction factor should be checked each month of the first year of operation to get a slime growth profile.

The effect of corrosion and tuberculation on the friction coefficient of metal pipelines is well known. However, for many years it was thought that concrete pipe or cement-lined steel pipe would maintain their capacities indefinitely. This has proved to be an incorrect assumption, as it has been abundantly demonstrated that filamentous organisms, freshwater sponges, and other organisms are able to attach themselves in massive quantities to concrete and various types of cement surfaces. The carrying capacity of such lines drops off rapidly without any increase in roughness, simply because of the reduction in the effective diameter of the pipeline. Derby[257] cites the *C* factor of a 5-ft concrete pipe 50 miles long dropping from 140 to 95 because of massive growths of fresh-water sponges *Asteromyenia plumosa* and *Trochospongilla leidyi.*

Another unknown factor that has caused serious trouble in transmission systems is the possibility of groundwater intrusion into the concrete-lined tunnels carrying the water supply. These waters are often the source of iron and sulfur bacteria infestation. This was the cause of luxuriant growths of *Crenothrix* in the coast tunnel of the San Francisco water supply and the source of sulfur-bearing water in the Santa Barbara supply. The San Francisco coast tunnel capacity decreased 30 percent in three weeks.[238]

Control of such growths in transmission lines requires careful study to determine the proper course of action. This may be cleaning followed by chemical treatment with chlorine, chloramine, and possibly copper sulfate. It may not be possible to accomplish satisfactory cleaning. If chemical treatment is to be used without cleaning, it should be determined whether or not the line velocities are sufficient to flush out the organisms destroyed by the chemical treatment. For long conduits of varying cross sections, chemical treatment may be cheaper and more satisfactory than mechanical cleaning. Intermittent chemical treatment must also be considered where taste and odor problems are not involved. If the organisms cause taste and odor, some program of continuous treatment will probably be necessary to prevent these growths from forming. Open channels are subject to sunlight, which has the ability to destroy free chlorine residuals. Therefore these situations call for some kind of variation of the chemical treatment—possibly a combination of chloramine and copper sulfate, intermittent or continuous. If intermittent, the treatment should be confined to the hours of darkness if it is found that free residual chlorine is most effective.

Aid to Coagulation

Prechlorination to free residual is nearly always effective in improving coagulation or reducing the coagulant dose or both. This phenomenon has been known for some time.[258, 259] It is not clearly understood how chlorine acts as a coagulant aid although this is probably due to its oxidizing effect on organic matter.

Chlorine is also used in the preparation of the ferric ion as a coagulant for use in certain types of color and turbidity removal processes. In these instances chlorine solution from conventional chlorination equipment is added to either a solution of pickle liquor or copperas, both of which are ferrous sulfate. Chlorine converts the ferrous ion to ferric, which hydrolyzes to form ferric hydroxide, a fluffy gelatinous floc. Copperas is a granular free-flowing material easily handled in a dry chemical feeder. The chlorine is added downstream from the solution chamber outlet of the copperas feeder. These systems have special application where the optimum pH for coagulation is in the range of 8 to 9. It requires one part of chlorine to react with 7.8 parts copperas to convert all of the ferrous to ferric ion as follows:

$$6\ FeSO_4 \cdot 7\ H_2O + 3\ Cl_2 \rightarrow 2\ Fe_2(SO_4)_3 + 2\ FeCl_3 + 7\ H_2O \qquad (6\text{-}13)$$

Chlorinated copperas has been known to form a tough, rapidly settling floc with as little as 0.5 grain per gallon dosage.[260]

Chlorine is also used in the preparation of activated silica, a coagulation aid.

Aid to Filtration

At one time is was thought that the efficiency of filtration depended in part upon the establishment of a very thin film of organic material on the sand grains, and that a chlorine residual would destroy this organic film, thereby decreasing the efficiency of the filter.[261] This notion was soon discarded when it was discovered that the proliferation of microorganisms in the filter media would seriously impair the effectiveness of filtration. Baylis of Chicago insisted that *all* surface supplies should be chlorinated as soon as the water reached the treatment plant, as is recommended for polluted supplies.[262]

Baumann et al. reported in 1963 on a controlled study of prechlorination and its effect on filter efficiency.[263] They found that the prechlorinated filter operated considerably longer than the control filter, which was not prechlorinated. Other observations included: prechlorination reduces the depth of suspended solids into the filter sand; less sand was removed during the cleaning process; oxidizable tastes and odors were destroyed; oxidizable organic matter was confined to the surface cake at the sand surface; the bacterial quality of the filter effluent was considerably improved as compared to the unchlorinated control filter.

An adequate free residual chlorine is necessary to maintain filter runs and to prevent the formation of mud balls, which not only impair the filtration efficiency but interfere with proper backwashing. This chlorine residual prevents organic growths that contribute to slime buildup on the filter media. These growths not only reduce the mechanical efficiency of filtration but contribute to the degradation of the bacterial quality of the filtered water.

Certain types of algae are also well known for contributing to the clogging of filters. The most serious offenders are the diatoms, which are present during all seasons of the year—*Asterionella, Fragilaria, Tabellaria,* and *Synedra.*

In Chicago, when the water to be filtered contained approximately 700 organisms/mL, principally *Tabellaria* and *Fragilaria,* the filter runs were only 4.5 hours. Three days later, when the count was down to 100/mL the filter runs increased to 41 hours.[191] In Washington, D.C., filter runs were reduced from an average of 50 hours to less than one hour by the sudden influx of the diatom *Synedra,* which had a concentration in the raw water reaching 4800 cells/mL.[191]

If the filter-clogging blue-green algae—*Anacystis, Rivularia, Anabena,* and *Oscillatoria*—are allowed to grow, they form a loose slimy layer over the sand grains, thereby reducing the flow of water through the filter.

Organic growths on the sand grains or other media within the filter reduce the length of filter runs and cause the formation of mud balls, which seriously impair backwashing efficiency. These same organic growths provide an anaerobic environment, which contributes to the rise of the ammonia nitrogen content in the filter effluent. If this is not counteracted by free residual chlorination, nitrifying bacteria such as *Pseudomonas* may proliferate, causing the conversion of the available nitrogen to nitrites. At this point, with the development of nitrites, the environment becomes most suitable for the proliferation of other bacteria, thereby degrading the bacterial quality of the effluent. This situation can be prevented only by a free residual chlorine. Combined chlorine will not oxidize the nitrites to nitrates; only free chlorine (HOCl) can do this.

A filter that has become fouled by organic growths can often be restored by prechlorination. At first the only residual that will appear in the effluent will be all combined residual—meaning that it is predominantly monochloramine. This suggests the presence of significant amounts of ammonia developing within the filter media because of the organic growths. If the prechlorine dose is gradually increased, the organic material will become more and more oxidized. Within a short period of time—possibly one or two weeks—the total residual of filter effluent should begin to show a trace of free chlorine. When the free chlorine content is 80–85 percent of the total residual, it may be assumed that the filter is relatively free from organic growth. Backwashing will become more effective, filter runs will be considerably increased, and the bacterial quality will return to acceptable standards. The application of chlorine ahead of filters that have not been properly cared for by prechlorination will not necessarily achieve the desired results. Some filter media that have been neglected for a long time simply cannot be cleaned by adequate chlorination; these media should be discarded and replaced with new material.

As free residual chlorination is an aid to both coagulation and filtration, it is patently clear that the case for carrying a free chlorine residual throughout the entire treatment process becomes strong indeed. This also indicates the necessity of a special point of application of chlorine just ahead of the filters.

Hydrogen Sulfide Control and Removal

Hydrogen sulfide is probably the most obnoxious and troublesome compound to be dealt with in a potable water supply. It is an almost impossible task to produce a palatable water that is free of taste and odor at all times if hydrogen sulfide is present in the raw water in significant concentrations. Not only does the treatment process require continuous surveillance, but the distribution system and consumers' hot water sytems require monitoring.

Hydrogen sulfide occurs mainly in well waters. Occurrence in surface supplies is primarily by groundwater intrusion; however, with the rising pollution of natural waters by sewage and industrial wastes, surface waters may become contaminated with hydrogen sulfide.

Sulfides in well water are probably produced through chemical and bacterial changes under anaerobic conditions far underground. Sulfates may be reduced to sulfides by organic matter under anaerobic conditions, and the resultant metallic sulfide changed to hydrogen sulfide by the action of carbonic acid.

The sulfate-reducing bacteria (*Desulfovibrio desulfuricans*) are another source of hydrogen sulfide production. In anaerobic environments these bacteria convert sulfates and other sulfur compounds to H_2S. They have a growth range of pH 5.5 to 8.5 and are found to exist in temperatures of 0 to 100°C, with an optimum range of 24 to 42°C.[264]

Another group of bacteria also plays an important part in sulfur-bearing waters. These are the sulfide-oxidizing forms. The most prevalent are *Beggiatoa* and *Thiobacillus*. *Beggiatoa* are filamentous white sulfur bacteria that obtain the energy necessary for their growth by oxidizing the sulfide ion to colloidal sulfur, which is then stored in their cells.

Hydrogen sulfide is a flammable and extremely poisonous gas.

Brief exposures (30 min. or less) to H_2S concentrations as low as 0.1 percent by volume of air may be fatal. The gas is highly soluble in water to the extent of 4000 mg/L at 20°C and one atmosphere. The minimum detectable concentration by taste in water is given as 0.05 mg/L.[264] In aqueous solutions it hydrolyzes as follows:

$$H_2S \leftrightarrows HS^- + H^+ \tag{6-14}$$

The hydrosulfide ion (HS^-) further dissociates as follows:

$$HS^- \leftrightarrows S^= + H^+ \tag{6-15}$$

At 18°C the hydrolysis constant for Eq. (6–14) is:

$$K_h = 9.1 \times 10^{-8} = \frac{[H^+][HS^-]}{[H_2S]} \tag{6-16}$$

And for Eq. (6–15):

$$K_h = 1.2 \times 10^{-15} \frac{[H^+][S^-]}{[HS^-]} \tag{6-17}$$

Figure 6–31 illustrates the distribution of H_2S, HS^-, and $S^=$ for various pH levels. AT pH 7, hydrogen sulfide is approximately 50 percent of the total dissolved sulfides; at pH 5, it is practically 100 percent of the total; at pH 9, there is nearly all hydrosulfide ion. Therefore the existence of hydrogen sulfide in sulfur-bearing waters is pH-dependent. This scientific fact has been well documented.[265, 268, 269, 274]

Whenever the equilibrium between the hydrosulfide ion and the hydrogen sulfide in solution is upset, as when H_2S is removed by oxidation, the shift will be to form more H_2S from the remaining dissolved sulfides to reestablish the equilibrium. This stored sulfur gradually disappears by metabolic action, being itself oxidized to sulfate to yield more energy.

The primary objection to waters containing hydrogen sulfide is the offensive rotten-egg taste and odor. The secondary objection is the marked corrosiveness of these waters to both metals and concrete structures. Another serious prob-

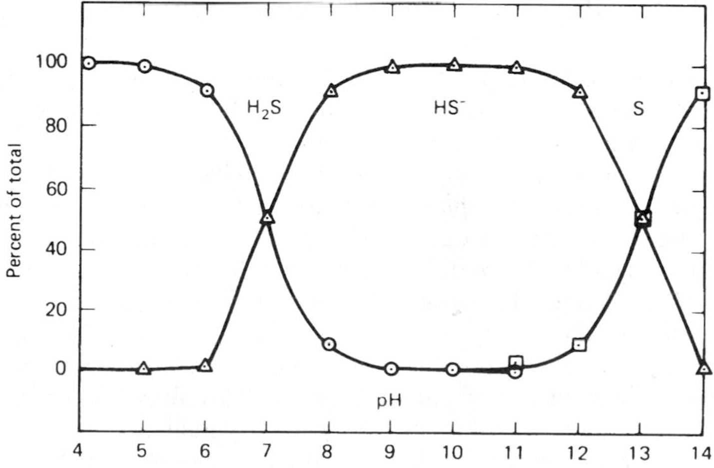

Fig. 6-31. Effect of pH on hydrogen sulfide–sulfide equilibrium.

lem is their ability to promote luxuriant blooms of the various types of filamentous sulfur bacteria that lead to a general degradation of water quality in the system.

The presence of H_2S will turn silverware black, discolor lead-base paint, make bathing in tubs or showers extremely unpleasant, and stain all plumbing fixtures. The sulfides in the water react with iron from the mains to form a suspension of iron sulfide that causes a discoloration of the water. This suspension makes laundering almost impossible.

The presence of hydrogen sulfide is noticeable to the extent of 0.5 mg/L even in cold water.[265] If the pH is high, the odor may be slight. Concentrations of H_2S as low as 0.2 mg/L will promote the growth of *Beggiatoa.*[264] The range of hydrogen sulfide concentration in usable sulfur waters is below 10 mg/L. Notable exceptions are the Wadsworth Plant of the Southern California Water Company[266] and the Santa Barbara supply,[267] where the H_2S content in the raw waters sometimes reaches 20 mg/L.

The chemistry of the oxidation and removal of hydrogen sulfide is extremely complex. Supposedly the oxidation proceeds to form either elemental sulfur, sulfate, or both, as follows:

Aeration or oxidation by dissolved oxygen in the water:

$$2S^{=} + 2O_2 \rightarrow SO_4^{=} + S\downarrow_0 \tag{6-18}$$

and chlorination:

$$H_2S + Cl_2 \rightarrow 2HCl + S\downarrow_0 \tag{6-19}$$

and:

$$H_2S + 4Cl_2 + 4H_2O \rightarrow 8HCl + H_2SO_4 \tag{6-20}$$

Theoretically Eq. (6-19) requires 2.1 (mg/L chlorine for each mg/L H_2S, and Eq. (6–20) requires 8.5 mg/L for each mg/L H_2S. This reaction was amply demonstrated in field experiments by Powell.[274]

The assumption is that if enough chlorine is added to satisfy the natural chlorine demand of the water plus enough to react with the H_2S present, then free sulfur will be formed in Eq. (6–19), and that if more chlorine is added to satisfy the stoichiometric requirements of Eq. (6–20), then all the H_2S will be converted to sulfates. Unfortunately, this is not the case.

When chlorine is added, regardless of the amount, to a sulfur-bearing water, colloidal free sulfur will be formed if enough chlorine is added to more than satisfy the natural chlorine demand of the water. This formation of free sulfur is readily evident as a milky blue turbidity (the Tyndall effect).

Choppin and Faulkenberry[270] state that the oxidation of the alkali sulfides is not simple and direct and may yield as end products polysulfides, sulfites, and thiosulfates in addition to elemental sulfur and sulfates. The relevancy of this statement is based upon practical evidence that most sulfur-bearing waters that have been treated for hydrogen sulfide removal will exhibit some magnitude of odor in hot water systems. Although the door is not that of hydrogen sulfide, it is a sulfurous odor. Monscvitz and Ainsworth[267] believe that this odor is the result of the formation of polysulfides (HS_n^-) in the oxidation reaction of hydrogen sulfide. Given time in days, these polysulfides will eventually oxidize to sulfates in the presence of dissolved oxygen in the water. If dissolved oxygen is not present, as happens in distribution systems, these polysulfides may be reduced back to hydrogen sulfide.

The oxidation of hydrogen sulfide will produce colloidal sulfur, which can be removed by filtration. Side reactions will produce polysulfides, which must be dealt with in order to eliminate threshold odors in the finished water. Monscvitz and Ainsworth suggest the conversion of colloidal sulfur and the remaining polysulfides to sulfates by first adding sulfite (which forms thiosulfate) and then converting the thiosulfate (S_2O_3) to sulfate by rechlorination as follows:

$$\begin{array}{l} S^0 + SO_3^= \leftrightarrows S_2O_3 \\ \downarrow \ \uparrow \\ HS_n^- \end{array} \tag{6-21}$$

$$S_2O_3 + HOCl \rightarrow S_4O_6^= + SO_4^= \tag{6-22}$$

The reaction of Eq. (6–21) is very rapid. The formation of tetrathionate ($S_4O_6^=$) is not significant, as it will be converted to sulfate within a short time, with no contribution to any threshold odor. The sulfite ion may be added as sulfur dioxide in an aqueous solution or as sodium metabisulfite. This reaction, taken to completion, is recognized as the typical dechlorination reaction.

Above pH 9.0, polysulfides do not appear to form. This is probably one reason why any lime-softened sulfur-bearing water will not produce threshold odors resulting from these compounds.

Probably the most important factor in hydrogen sulfide removal is that of contact time. Given time and an oxidizing agent, all dissolved sulfides can be converted to sulfates.

Aeration with contact times of up to three hours have shown hydrogen sulfide removals of 2–3 mg/L and 35–45 percent removal of dissolved sulfides.[271] Shorter contact times and raw water concentrations of 5 mg/L H_2S reduced to 1 mg/L by aeration, followed by coagulation and filtration, require free residual chlorine of up to 0.35 mg/L for the complete removal of H_2S.[272] Derby[273] reported in 1928 on the satisfactory removal of up to 7 mg/L by conventional lime-softening and chlorination. Concentrations as high as 32 mg/L have been successfully removed by pre- and postchlorination, coagulation, and filtration.[266]

The final concern of the water producer is to protect the quality of the product in the distribution system. This is difficult to accomplish because of the possible growth of sulfur bacteria. Although the oxidizing forms that grow in an aeration tower may be beneficial because of their ability to convert the sulfide ion to sulfates, they may find their way to the distribution system and in their life cycle slough off into dead-ends and contribute sulfur compounds that under anaerobic conditions will start the entire hydrogen sulfide production cycle over again. For this reason chlorination for the control of these organisms in the distribution system is of utmost importance.

Hydrogn sulfide can be successfully removed by aeration and chlorination, followed by the addition of sulfur dioxide or sulfite for the removal of polysulfides and colloidal sulfur, and then rechlorination to convert all the remaining sulfur compounds to sulfates.

The formation of hydrogen sulfide is pH-dependent and is encouraged by a lower pH.

Oxidation of sulfur-bearing waters will produce colloidal sulfur that may cause a milky blue turbidity (the Tyndall effect).

As the hydrogen sulfide is removed by oxidation, the equilibrium shifts to form more H_2S from the remaining dissolved sulfides; depending upon the amount of oxidant and contact time, this reaction can continue until all the dissolved sulfides are removed.

When sulfur-bearing waters are oxidized, a combination of sulfur compounds is formed, including colloidal sulfur, polysulfides sulfites, and sulfates.

Aeration, chlorination, coagulation, filtration, and softening have been used successfully for the removal of H_2S.

Chlorination is extremely important for providing not only an oxidant but also a germicide to prevent the growth of unwanted sulfur bacteria.

Complete removal of H_2S can best be accomplished by oxidation by chlorine, conversion of resulting colloidal sulfur and polysulfides by metabisulfite or sulfur dioxide to thiosulfates, and then conversion to sulfates by rechlorination.

The Iron and Manganese Problem

Nature and Occurrence. Both iron and manganese cause serious problems in potable and industrial water systems. Although it is beyond the scope of this text to cover the subject of iron and manganese removal, it is pertinent to discuss the occurrence of these elements in water supplies, their significance, and the role of chlorination as related to compounds of these substances.

These two elements are relatively abundant in the earth's crust. The lithosphere contains approximately 5 percent iron and 0.1 percent manganese.[276] Iron exists in soils and minerals mainly as insoluble ferric oxide, and manganese is present as manganese dioxide. Iron also occurs as ferrous carbonate (siderite), which is slightly soluble. As groundwaters usually contain substantial amounts of carbon dioxide (30–50 mg/L), appreciable amounts of ferrous carbonate may be dissolved to form soluble (150 mg/L) ferrous bicarbonate as follows:

$$FeCO_3 + CO_2 + H_2O \rightarrow Fe(HCO_3)_2 \tag{6-23}$$

This is the commonest form in which iron is found in water supplies in troublesome amounts. Iron in natural water supplies may also be present as ferric hydroxide, ferrous sulfate, and colloidal or organic iron.[265]

The insoluble ferric compounds will only go into solution under reducing conditions—in the absence of oxygen. The same holds true for the compounds of manganese.

Manganese most commonly occurs in water supplies as manganese bicarbonate, which is even more soluble than ferrous bicarbonate. Acid mine waters frequently contain manganous sulfate as well as ferrous sulfate; some shallow wells and surface waters contain colloidal or organic manganese.

The occurrence of iron and manganese in water supplies is usually limited to wells and impounded surface supplies. It is rare to find either in flowing streams of normal pH and alkalinity. In waters in the regions of acid mine drainage wastes, particularly along parts of the Allegheny and Monongahela rivers, where the pH is less than 5, there are significant quantities of both iron and manganese compounds in solution.

From all the evidence available it appears that biologic activity is a powerful factor in the dissolving of iron and manganese.[269, 277] It is apparent that anaerobic conditions must develop in order for appreciable amounts of iron and manganese to gain entrance to the water supply.[275] For example, groundwaters that contain appreciable amounts of iron and/or manganese are always devoid of dissolved oxygen and are high in carbon dioxide. High carbon dioxide content indicates that bacterial oxidation of organic matter has been extensive, and the absence of dissolved oxygen shows that anaerobic conditions were developed.

Waters that normally do not contain any iron or manganese while flowing in a stream are certain to contain these compounds after the water has been impounded. The amount that goes into solution depends upon the character of the soil and the amount of plant life. Decomposition of organic matter in the lower strata of the water (hypolimnion) in the reservoir results in the elimination of dissolved oxygen and the production of carbon dioxide, so that the iron and manganese compounds in the flooded soil and rocks are converted to soluble compounds. These soluble compounds rise to the surface in those areas where the fall overturn occurs. At this point they are oxidized and precipitated and then sink to the lower portion of the reservoirs, where resolution occurs in the absence of oxygen. Thus waters near the surface of reservoirs are most likely to be free from iron and manganese. Therefore, it is essential that multiple-port outlet structures be provided for impounded supplies to allow selection of waters somewhere between the surface water containing algae and the deeper water (devoid of oxygen) that may contain iron and/or manganese. The amount of iron and manganese that gains entrance to either impounded surface supplies or groundwater varies throughout the United States. The manganese concentration in the hypolimnion of reservoirs varies from 2.0 mg/L in New Jersey to as much as 20 mg/L or more in the lakes of the Tennessee Valley Authority.[277] The time required for the appearance of significant amounts of iron and/or manganese in man-made lakes also varies with the location. Some areas develop significant concentrations in one year, whereas others may require ten years.

Significance. Although there is no evidence that humans suffer from drinking water containing iron or manganese, these substances contribute to some of the most serious problems ever to confront the waterworks industry. Waters containing ferrous bicarbonate stain everything with which they come in contact a yellowish to reddish brown. Manganese-bearing waters free of iron will produce black stains. Waters containing ferrous bicarbonate usually contain some manganese, and this combination will produce stains varying from dark brown to black. Depending on the concentration of these compounds, consumers' complaints will start first with staining and streaking problems in the laundering process; next, there will be red water or dirty water; finally, there will be large visible chunks of material that have sloughed off the distribution piping system. Most industrial wet processes, such as those in the textile, pulp, and paper, and beverage industries, cannot tolerate this kind of water.

In addition to the physical phenomena of the presence of iron and manganese in the water, both contribute to and promote the growth of crenoform organisms in the distribution system. These filamentous organisms utilize both iron and manganese in their metabolism, and deposit within the pipelines to form heavy, gelatinous, stringy masses that slough off at intervals, causing a variety of problems for both domestic and industrial consumers. In addition to the unsightly mess that results, the growth of these organisms impairs the hydraulic carrying capacity of the entire system.

Because of the potential damage that these two elements can cause to a water supply system, the U.S. Public Health Service Drinking Water Standards of 1962 placed a limit of 0.3 mg/L for iron and 0.05 mg/L for manganese in potable water supplies. An AWWA task group[278] suggested limits of 0.05 mg/L for iron and 0.01 mg/L for manganese for an "ideal" quality water for public use. Traditionally manganese was reported as a combination with iron regardless of the ratio. Prior to the 1962 standards the limit was set at 0.3 mg/L for both iron and manganese. Although 0.3 mg/L of iron could be tolerated under certain conditions, this amount of manganese alone would cause severe difficulties. In view of the 1962 standards, iron and manganese should be reported separately. Manganese in excess of 0.02 mg/L will almost surely cause problems in the distribution system. A large segment of waterworks people believe that all public water supplies should be devoid of manganese. Although this is imperative for some industries, most domestic systems can tolerate up to 0.01 mg/L of manganese.

To emphasize the difficulty that may be encountered from these compounds it is pertinent to cite a classic example:

The Metropolitan Water District of Salt Lake City, Utah, has its source of supply water impounded at the Deer Creek reservoir on the Provo River. The pipeline from this reservoir to the distributing reservoir in Salt Lake City is some 40 miles long. The water is chlorinated at the outlet of the Deer Creek reservoir just as it enters the long pipeline. This treatment is for disinfection and prevention of slime growths in the long transmission line. A few years after this system had been placed in operation, it became coated with a black deposit of iron and manganese. When this coating is about 1/32 inch thick, it breaks off and colors the water. The manganese content is usually a few hundredths of a mg/L and sometimes goes as high as 0.1 mg/L but is continuous throughout the year. The outlet of the reservoir is at the bottom where the dissolved oxygen becomes depleted after spring turnover until it reaches zero. The iron and manganese content in this water is due to the reducing conditions in the bottom of the reservoir, where both iron and manganese from the soil and biological life readily go into solution. After spring turnover, the iron and the manganese rise to the top, where they are oxidized by the available dissolved oxygen in the water. This is followed by precipitation to the lower levels, where they again go into solution, only to become oxidized again by the chlorine and precipitate out in the pipeline. All types of chlorination programs were tried to eliminate this deposition. When chlorine was reduced, matters became worse because this resulted in a buildup of pipeline growths. Therefore, the district has had to resort to pipeline cleaning on a regular basis. The solution to this problem is an outlet structure with the ability to withdraw water at selected depths to avoid the areas of iron and manganese concentrations. Unfortunately the present outlet structure is irrevocable.

The Role of Chlorine

Iron Removal. Generally speaking, there are three methods for removing iron: simple oxidation, by aeration or chlorination, or a combination of both; precipitation with lime at a pH above 8; or ion exchange. The major problem in these methods is removal of the insoluble precipitate from each.

Chlorine, either free or combined, reacts to oxidize ferrous iron as follows:

$$2Fe(HCO_3)_2 + Cl_2 + Ca(HCO_3)_2 \rightarrow 2Fe(OH)_{3\downarrow} + CaCl_2 + 6CO_{2\downarrow} \quad (6\text{-}24)$$

The soluble ferrous bicarbonate is oxidized to the insoluble ferric hydroxide, which can be removed by sedimentation and/or filtration, depending on how heavy a floc is produced. Although this reaction will take place over a wide range of pH (4–10), the optimum pH is 7.0.[279] The colder the water, the slower these reactions are. This reaction takes a maximum of one hour, and is most rapid at pH 7.0. Each part of iron as Fe oxidized requires 0.64 mg/L chlorine.

This reaction consumes 0.9 mg/L alkalinity as calcium carbonate ($CaCO_3$) for each mg/L iron as Fe oxidized.

If the iron present is in the complex organic form, free residual chlorine is more effective than combined in breaking up the iron complex so that oxidation by chlorine can proceed.

It is common practice to preaerate the raw water, which will oxidize some of the iron but, most important, will reduce the carbon dioxide content, causing a rise in pH where further oxidation by chlorine is more effective. Although chlorine is usually applied after aeration, it is nearly always desirable to prechlorinate to control the crenoform organisms and to chlorinate again just after aeration for further oxidation of the iron.

The application of chlorine to iron-bearing waters is imperative, whether or not it is considered a part of the iron removal process, simply to prevent and control the growth of the crenoform organisms, which, if allowed to proliferate, can devastate the entire system and render the iron removal process useless. It should also be emphasized that when iron is present in small quantities (0.3 mg/L) where iron removal is not a factor, chlorine should be used to prevent the growth of the crenoform organisms, which have been known to proliferate in waters containing iron as low as 0.1 mg/L.

Manganese Removal. Chlorine will oxidize soluble manganous manganese to the insoluble manganic form as follows:

$$Cl_2 + MnSO_4 + 4NaOH \rightarrow MnO_2\downarrow + 2NaCl + Na_2SO_4 + 2H_2O\downarrow \quad (6\text{-}25)$$

The manganese dioxide produced may be removed by filtration as follows: Insoluble MnO_2 plates out on the sand grains. These deposits act as a catalyst to make possible complete extraction of manganese as MnO_2.

The optimum pH for this reaction is between 7 and 8. At pH 8 and with alkalinities on the order of 50 mg/L, the time required may be two to three hours. As the pH increases, the time requirement diminishes to the pH values in the softening zone, where oxidation appears to be complete within minutes.[277, 279] If the pH approaches 6.0, the time required may be as much as 12 hours. Temperature of the water does not appear to be a significant factor.

It should also be noted that manganese readily precipitates at pH 2 in heavy concentrations of chlorine. Proof of this is the troublesome deposit of manganese often found on the injector throats of chlorination equipment.

Chlorine in the form of free residual is imperative, and is in addition to that required to react with iron, ammonia, hydrogen sulfide, and so on. There should be 1.3 mg/L chlorine for each part of manganese as Mn.[279, 280] For each part of Mn oxidized, 3.4 mg/L alkalinity as $CaCO_3$ is consumed.

Difficulties for both iron and manganese are much more widespread than is commonly thought. Probably the most troublesome cases are those that are not readily categorized for physical removal of these compounds by an engineered treatment plant. Myers[281] reported a number of such case histories.

The end result is always the same: degradation of water quality in the distribution system; sloughing off of microbiological debris; and red or black water not fit to use. These cases occur throughout the United States, and are most common in underground supplies. The problem usually hits hardest those who can least afford to take proper corrective steps—the small rural community water supply or the individual well owner.

A typical case is a well that consistently contains less than 0.4–0.6 mg/L iron and probably 0.1 mg/L manganese. Most wells are capable of supporting troublesome growths of crenoform organisms, and so chlorine is applied to control these growths. Following chlorination there is likely to be a delayed reaction with chlorine and manganese. In such cases deposition of manganese in the remote or dead ends of the distribution system is common, indicating that 10 to 12 hours is required for the complete oxidation of manganese. This problem can be overcome by adding a sequestering agent along with the chlorine as follows: Apply the chlorine down the well, 10 ft below the pump bowls, as described in Chapter 9. Then add, at a rate of 2 mg/L, sodium hexametaphosphate down the well in the same line as the chlorine solution. This has been found in actual practice to be permissible, probably because there is not enough time for any reaction to set in between the additions of the chlorine and hexametaphosphate. The latter will prevent the deposition of the manganese in the distribution system.

The procedure to follow for chlorination is: If the water by proper chemical analysis shows more than 0.3 mg/L iron, it should be chlorinated to prevent the growth of crenoform organisms. If there is more than 0.03–0.05 mg/L manganese present, it will probably precipitate out in a delayed reaction. If this occurs, add the hexametaphosphate as described.

This procedure has been known to be successful on iron contents as high as 0.6–0.8 mg/L and higher in some rare cases. As chlorine is a must on any water over the iron limit, results will dictate whether or not the iron and manganese will have to be removed by a conventional treatment process.*

It should also be added that chlorination may be required in some instances where crenoform organisms proliferate when both iron and manganese are below the U.S. Public Health Service limits of 1962.

Color Removal

The practice of free residual chlorination has been known to bleach true color in waters when color removal by coagulation and filtration has failed. The bleaching effect of chlorine is well known in swimming pool operation. An adequate free chlorine residual produces a polished, sparkling look to the water because of its ability to bleach organic matter.

True color removal by chlorine is most effective in the acid pH zone, between pH 4.0 and pH 6.8. Usually highly colored, low-turbidity waters are

*See Chapter 12—the use of chlorine dioxide for iron and manganese removal.

either naturally acid or so lightly buffered that the application of chlorine may be sufficient to reduce the pH. In some cases the simple reduction of pH may cause the color to disappear. Therefore, in decolorizing with chlorine, care must be taken so that when the upward readjustment of pH is performed for corrosion control, it does not result in color return. This can be predicted by a simple laboratory procedure.

Color removal by free residual chlorine is usually instantaneous, and temperature does not seem to be a factor. There is no rule for predicting the optimum dosage. The amount of color removal will vary with local conditions.[279]

Where color is accompanied by turbidity, and coagulation is practiced, prechlorination to a free residual will aid materially in color removal by two means: through oxidation of part of the color, and through its action as a coagulant aid. Thus two benefits may be expected: better color reduction and decreased coagulant dosage.

In highly polished nitrified wastewater effluents White has observed that a free chlorine residual can produce a sparkling blue effluent, whereas the same effluent dosed with NH_4OH to produce a combined chlorine residual shows a slight greenish-straw color. This demonstrates the bleaching ability of free chlorine.* So far as is known, chloramine or other combined chlorine residuals do not have any bleaching power. If all else fails in an effort to prove that a chlorine residual is in fact a true free available residual and not some complex organic chloramine, the bleaching quality of a free residual is certain proof.

Chlorination of High pH Waters

Described below are two fairly new ways to overcome the difficulties involved in the chlorination of high pH waters. A great many public water supplies are lime-softened, resulting in pH values of 10 or more. One of the more notable sources of such supplies is the Missouri River. In the heavily populated areas along its course, the chemical and bacteriological characteristics of this water vary widely, putting a formidable burden on the treatment plant. It is common practice first to soften the water with lime, which is followed by two-stage flocculation, coagulation, and settling. Chlorine is usually applied just ahead of secondary settling, followed by filtration and postammoniation.

Considerable bacteria reduction occurs in the softening process. As the chlorine-consuming inorganic compounds of iron, manganese, and sulfur are largely removed by precipitation in the lime-softening process, the remaining chlorine demand is due primarily to nitrogenous compounds still in solution.

At pH 10 and higher, the reaction between free chlorine and the nitrogenous compounds is extremely slow. Temperature is also a factor. As the temperature drops, the chlorine ammonia reaction slows down. Figure 6-32 illustrates the reaction of chlorine and lime-softened Missouri River water.[282] This water contained 0.9 mg/L ammonia nitrogen, and the temperature was 65°F. Note

*When chlorine is used to bleach organic color, the required dose may be as high as 15–20 mg/L.

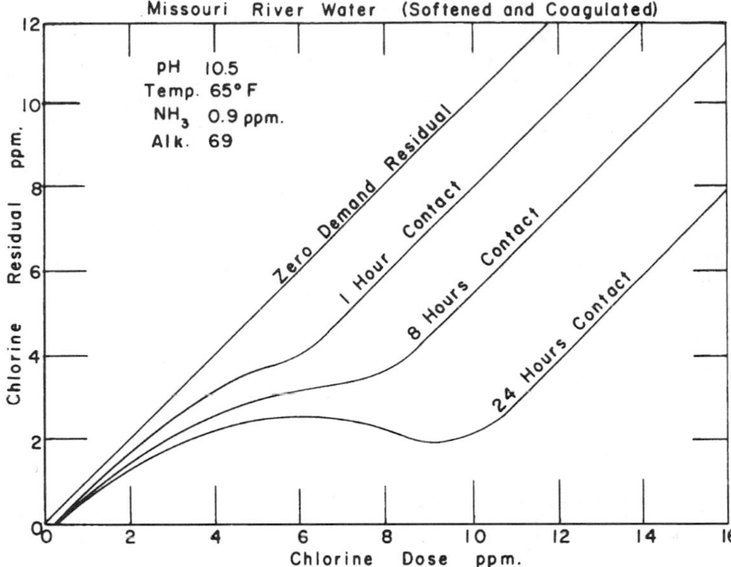

Fig. 6-32. Relationship of chlorine and nitrogenous compounds in high-pH waters.

that after one hour's contact very little chlorine was consumed, and only after 24 hours did a typical free residual chlorine curve appear. Figure 6-33 compares the chlorination of the same river water before and after softening when subjected to a 60-min. contact period.

The importance of such application of chlorine is in the time factor. Even though the reactions between the chlorine and ammonia do not have sufficient time to go to completion within the treatment plant, there is a considerable benefit in a high free chlorine residual during this period. Chlorine doses of 5 mg/L lower the alkalinity approximately 7 mg/L, which may result in some lowering of the pH and tend to increase the efficiency of the free residual. Postammoniation is desirable to stabilize the free chlorine residual for two reasons. At pH values above 9, the efficiency of monochloramine is known to be superior to that of free chlorine. (See Chapter 4.) If the free chlorine residual were allowed to go into the distribution system, the continuing slow action of chlorine with the nitrogenous compounds already in the water could eventually, if given the proper amount of time, develop nitrogen trichloride and dichloramine, which could produce objectionable tastes and odors at the consumers' taps. Postammoniation prevents the formation of these compounds. The key to successful chlorination of high pH waters is a long residence time with free chlorine residual followed by postammoniation.

Since the Water Champ introduces chlorine gas directly and totally into the process water, operators have been experiencing much better chlorination results by using it, at a saving of up to 25 percent on chemical costs. The

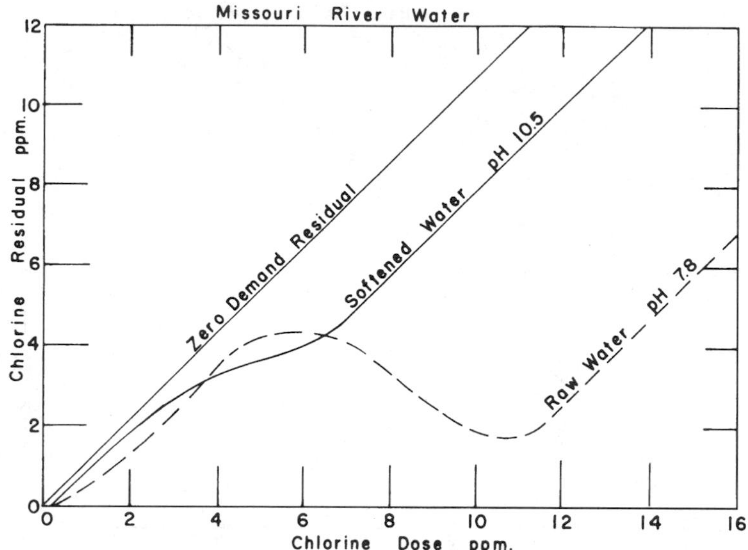

Fig. 6-33. Comparison of chlorination of raw water and softened water.

reason for this is simple chemistry. The oxidation potential of chlorine gas is practically the same as that of a 100 percent solution of HOCl. However, the percentage of HOCl in a pH 10 water is less than 5 percent.

Preference for ORP Control. When using either the W.C. or a water injector, instead of having to worry about the accuracy of a chlorine residual analyzer, one can use ORP control to identify exactly what the chlorination system is accomplishing. (See Chapter 9 for all the details.)

Miscellaneous Applications of Chlorine

Desalting Plants. Current chlorination practice for desalinization plants has gradually evolved so that design engineers are providing pretreatment for the control of marine growths and the destruction of hydrogen sulfide and postchlorination for disinfection.

Prechlorination. Whenever seawater is used for any purpose, it will support and promote the various marine growths: mussels, sponges, bryozoa, and soft forms. Chlorination to control these growths can be on an intermittent basis, as in cooling water treatment. This treatment maintains plant capacity and prevents damage and plugging of tubular heat transfer equipment, thus avoiding costly downtime for repairs and cleaning.

Anytime that chlorination is in or around seawater, the design engineer or the operations group should be advised that using the seawater for the chlorinator injector supply is much preferred over freshwater because the bromides in the seawater are converted to bromamines by the chlorine. As bromamines are somewhat more powerful than HOCl, the use of seawater simply increases the efficiency of the chlorination process.

An unexpected development has occurred at several desalinization plants—the appearance of hydrogen sulfide in the raw salt water. Its presence is attributed to decaying vegetation in tropical water supplies. If the supply is from deep wells, it is thought that the organic matter that gives rise to hydrogen sulfide has been trapped in the coral growth.

The presence of hydrogen sulfide in a desalting plant using an evaporative technique is very costly because it will cause severe corrosion to almost any metal. Sometimes the H_2S comes in slugs, making control difficult. It can be best eliminated by chlorination of the raw water unless the amounts present are excessive (>5.0 mg/L), when chlorination combined with aeration may be more economical.

Postchlorination. The freshwater effluent should be chlorinated as a safety precaution even though the U.S. Public Health Service recognizes the pasteurization effect of the evaporative techniques. At times in the operation of these plants when shutdowns occur, the freshwater system may become contaminated with the raw water. The treatment plant is by definition one huge cross-connection with a contaminated supply. For this reason most engineers include postchlorination in their plant design.

Reflecting Pools. It is common practice to include various kinds of water fountains to enhance the appearance of public recreation centers. These bodies of water require careful control to prevent the growth of any unsightly algae. At the same time the treatment must not give off any obnoxious chemical odors that would be offensive to the public.

These water systems usually contain cascades that provide a certain amount of aeration. It therefore is imperative that chlorination be maintained at a free residual level and that none of the chloramines be allowed to form. It is the chloramines, particularly dichloramine and nitrogen trichloride, that aerate out and cause offensive odors near the pool area.

The best way to treat these waters is to maintain a free chlorine residual of 0.75–1.25 mg/L continuously at a pH of 9.0–10.0 and never to use any algicide that contains nitrogen compounds. If proper free residual chlorination is practiced, algicides will not be required. The chlorination equipment, whether chlorine gas solution feed or hypochlorite, must be controlled by an amperometric residual analyzer calibrated for free chlorine residual. This will assist in preventing the formation of chloramines. The assurance of a *continuous* adequate free chlorine residual will eliminate the necessity of supplementing the chlorine treatment with an algicide.

All recirculated waters will require the continuous addition of caustic, sodium carbonate, or sodium bicarbonate to maintain the elevated pH when using chlorine gas, and most waters will require the addition of these chemicals even if hypochlorite is used. The latter may not add enough caustic to raise the pH to the required level except in waters softened by the lime soda process.

Restoring Wells. Most wells suffer from loss of efficiency with time. As the demand for additional water steadily increases, it becomes just as important to maintain the efficiency of existing wells as it is to develop new wells.

There are several causes for the decline in well production: lowering of the groundwater table; loss of pump efficiency, owing to worn, corroded, or encrusted parts; plugging of the aquifiers by microorganisms and/or scale deposits; biofouling of the well screen area by microorganisms; and fouling of the well screen area by deposits of scale and corrosion products or by mud, sand, and silt.

Natural phenomena bring about the deposition of scale and corrosion products and the proliferation of microorganisms that cause biofouling.

Water reaching the underground aquifers as rain picks up carbon dioxide. This is further supplemented by the evolution of carbon dioxide from decaying organic matter in the soil as the rain water penetrates into the earth. This addition of carbon dioxide greatly increases the solvent power of the water, so that the potential scale-forming compounds are readily dissolved. The simple mechanical action of pumping a well causes a drawdown of the static water level, which decreases the pressure in the vicinity of the well. This decrease in pressure plus the turbulence in the pump bowl area results in the release of carbon dioxide, which decreases the solubility of the water. Therefore, it is at this point that the scale-forming compounds of calcium, magnesium, iron, and silica are deposited.

The existence of biofouling organisms in underground waters is well known.[283] The prime offenders are the filamentous iron bacteria; *Crenothrix, Cladothrix, Leptothrix,* and *Gallionella.* Sulfur-bearing waters contribute to the filamentous organism *Begiattoa.* The area of the United States between Lake Michigan and Lake Superior to the north and to the Gulf of Mexico on the south has an overabundancc of iron bacteria growing in underground supplies. The area immediately adjacent to Green Bay, Wisconsin, has earned a reputation of harboring the most lush growth of iron bacteria of any place in the United States and perhaps in the world. It is also known that a great many areas in the western part of the United States have underground waters that support troublesome growths of biofouling organisms deep within the aquifers. These areas include, but are not limited to, certain areas of the Rocky Mountains, the alluvial plains of Utah, and the Sacramento and San Joaquin valleys and the Los Angeles basin in California.

Two separate factors are involved in the restoration of well capacity: (1) stimulation or redevelopment; and (2) cleaning, reconditioning, and replacement of worn parts in the pump and ancillary devices.

Chlorine is used primarily in conjunction with other chemicals and methods involved in the stimulation procedure as a supplement following acid or polyphosphate treatment. It should always be used as a final chemical treatment before putting a well back in service, simply to eliminate contamination that may have resulted from the mechanics of the stimulation procedure. Erickson[284] has described the various methods in considerable detail. The methods used are dependent upon the nature of the problem and the type of aquifer involved.

The successful restoration of well productivity by chlorine alone was reported by Brown[285] in 1942. A group of four wells, with a total initial production of 7090 gpm, had deteriorated in four years to 3350 gpm. One treatment with superchlorine doses restored the total capacity to 6480 gpm, and the second chlorine treatment, a short time later, to 7200 gpm. Houston[286] reported (1946) on the use of heavy doses of chlorine to restore two wells that had lost 50 percent of their productivity. Chlorine treatment of 40–50 lb of chlorine *down the well* was followed by several hundred gallons of muriatic acid to dissolve the scale and encrusted material.

Chlorination of well water with dry ice for stimulation was reported by Suter[287] in 1938. This application to a well suspected of declining production due to biofouling completely restored the well to its original capacity.

The use of dry ice results in the evolution of carbon dioxide. This creates turbulence, which mixes the chlorine and builds up a pressure that forces the heavily chlorinated water back into the aquifers. The use of dry ice, while inexpensive, is sometimes difficult to control. It has given way to other more modern methods using compressed air and vibratory explosions.[284]

The application of chemicals to effect a penetration is often done by the surging technique. This consists of pumping for 10 to 15 min. at a time. As the pump is shut down, the water in the riser column upstream of the check valve surges back down into the well. This is repeated several times, and then the chlorine is allowed to "soak" in the well for 12 to 24 hours before being flushed out.

When a well is suspected of loss of productivity due to biofouling, there is no better way to alleviate this condition than with chlorine. It is not always a simple matter to prove that biofouling is the cause. The organisms responsible for these deposits belong to a group not ordinarily isolated in a routine sanitary water analysis. Many of them will not grow on the ordinary culture media employed; some of them are very difficult to cultivate on any media; and they are seldom dispersed in the water in appreciable concentration.

For the destruction of the iron bacteria group (*Crenothrix, Leptothrix, Clonothrix,* and *Gallionella*), a 200 mg/L concentration in the well is effective.The destruction of sulfur bacteria may require as much as 500 mg/L. The chlorine should be applied down the well 10 ft or so below the pump bowls and, after the stimulation procedure, be allowed to soak for several hours before flushing. If scale and corrosion deposits are known to exist in the well screen and adjacent areas, the chlorine may be supplemented with acid after an acid inhibitor is applied first to protect against corrosion of the metal parts.

In using chlorine for the elimination of biofouling organisms, there is the danger of oxidizing ferrous compounds to ferric hydroxide, which might result in the formation of gelatinous flocs that would settle out in the well only to cause some plugging at a later date. If investigation reveals that this might be a factor, polyphosphates are used along with the chlorine to sequester the formation of ferric precipitates. However, care must be exercised in the use of polyphosphates. In wells with multiple screens in water-bearing strata separated by clay or shales, the action of polyphosphates on these materials will cause the disintegration of the clay and shales, which will infiltrate the sand and gravel strata and tend to reduce the productivity of the well.

The treatment of some biofouling problems occasionally calls for sterner and more costly methods of chemical treatment. Piatek[288] reported (1967) that attempts at Sayreville, New Jersey, to control such fouling in the screens, riser column, and transmission line required such long periods of high dosages of chlorine that a corrosion problem developed. The problem was solved by separating the cleaning procedure into two parts: one for the well and riser column, and the other for the transmission line. Chlorine dioxide was used instead of chlorine. In this case the chlorine dioxide was produced by combining hypochlorite with an aqueous solution of anthium dioxide. This mixture was added along with a polyphosphate. For an 8-inch well containing 40 ft of water and 20 ft of air space, a solution of 100 lb of Calgon in 50 gal of water was followed by 5 lb of 70 percent hypochlorite dissolved in 5 gal of water. This solution was surged once in the well to assure a good mix, and then 15 gal of anthium dioxide was added. This total mixture was surged for two hours by pumping 15 min. and shutting down for 15 min., and repeating. Then it was allowed to stand overnight. This was repeated every two months. Such experiences indicate the necessity of continuous treatment with both chlorine and polyphosphate added down the well at nominal dosages for the continuous destruction of the organisms that cause fouling in the system elsewhere than in the aquifers.

Injection Wells. Reclaimed or other waters are being used more and more to replenish the underground aquifers in those areas that are short of water supply. Injection wells are also used extensively for maintaining oil well production,[289] and for the prevention of seawater intrusion of freshwater aquifers, as well as to prevent land subsidence due to withdrawal of oil from underground deposits.

It is the consensus[289, 290] that any water used for injection into underground aquifers should be sterilized—free of bacteria. This will provide protection from the growth of filamentous organisms that might plug the aquifers, and will also inhibit the growth in the aquifers of bacteria native to the underground formation that might otherwise proliferate owing to the introduction of a water containing oxygen and nutrients to their growth.

It may be appropriate for users of injection water to establish a very high initial chlorine dosage to provide several mg/L free chlorine residual and then taper off to 1.5 mg/L.[290]

Disinfection of Water Mains and Storage Tanks

Chlorine is used extensively in the disinfection of both water mains and storage tanks. It can be applied as an aqueous solution from chlorine gas in cylinders, as a hypochlorite solution from either granular calcium hypochlorite or liquid sodium hypochlorite, or as calcium hypochlorite in the tablet form.

When using chlorine gas from cylinders, it is imperative that a conventional chlorinator with injector and booster pump be used. Otherwise, the handling of the chlorine gas can be awkward and hazardous.

Calcium hypochlorite contains 65 percent available chlorine by weight in the granular or tabular form. The tablets, six to eight to the ounce, are designed to dissolve slowly in water. Calcium hypochlorite is dissolved into water to make about a 1 percent (10,000 mg/L) solution, which is injected into the main.

Sodium hypochlorite is packaged in strengths from 5.25 to 16 percent available chlorine. It is available as a liquid in containers varying in size from one-quart bottles to 5-gal carboys, and may also be purchased in bulk for delivery by tank truck in some locations.

Water Mains. The hypochlorite solutions are usually injected into water mains by gasoline- or electric-powered chemical feed pumps or by hand pumps. Special pumps designed to handle only hypochlorite solutions are not mandatory. Ordinary cast iron pumps give many hours of satisfactory operation if they are properly flushed and cleaned after each use period. Hand pumps are also satisfactory.

The methods of applying chlorine to water mains are continuous-feed, slug-dosage, and tablet. These are described in detail in the AWWA committee 8360D report.[291] The AWWA standard for the continuous-feed method calls for chlorination to continue until the entire main contains chlorine solution. The recommended dosage is 50 mg/L, but the requirement is to have no less than a 10 mg/L residual throughout the entire length of the main at the end of 24 hours.

For large systems, the slug method is used. This consists of a slug or column of water containing a concentration of at least 300 mg/L to expose all the interior surfaces for a period of at least three hours.

The tablet method is limited to short extensions (up to 2500 ft) and smaller diameter mains (up to 12 inches). Because the preliminary flushing step must be eliminated, this method should be used only when scrupulous cleanliness has been maintained.

Table 6-6 gives the amounts of chlorine required for various size pipes. Table 6-7 shows the number of hypochlorite tablets reqiured.

The confirmation of adequate disinfection is by proper bacteriologic tests made after final flushing.

The recommended procedure for chlorine residual determination is the drop dilution method using acid O-T as the reagent. (See Chapter 5.)

Table 6-6 Chlorine Required to Produce 50 mg/L Concentration in 100 Feet of Pipe by Diameter

Pipe Size (in.)	*100 Percent Chlorine (lb)*	*1 Percent Chlorine Solution (gal.)*
4	0.027	0.33
6	0.061	0.73
8	0.108	1.30
10	0.170	2.04
12	0.240	2.88
18	0.483	5.80
24	0.875	10.10
36	2.220	26.50

Storage Tanks. The disinfection of elevated tanks, covered ground storage tanks, and ship tanks is achieved most conveniently by the use of portable chlorination equipment. The chlorine solution is added at the rate of 50 mg/L while the tank is being filled.

Portable chlorination equipment must be used carefully and under expert supervision. A typical arrangement, as reported by Tracy,[292] consists of limiting the solution strength to 500 mg/L using a 25 gpm pump with 100 ft of 1-inch chlorine solution hose terminating in a PVC spray nozzle. All workers must wear oxygen-type gas masks, and operators in the tank will require raincoats as well. This operation requires three people, changing nozzle operators every 20 min. Although it is an uncomfortable job, it is not particularly hazardous. The washing operation is carried out with all the valves closed. All of the chlorine solution will flow to the bottom, and any particular matter will be carried along or sink to the bottom in the strong chlorine solution. This is then flushed from the reservoir after the washing is completed. Washing reservoirs with hypochlorite solutions is awkward and cumbersome. The use

Table 6-7 Number of Hypochlorite Tablets of 5 g Required for Dose of 50 mg/L*

	Diameter of Pipe (in.)					
Length of Section (ft)	*2*	*4*	*6*	*8*	*10*	*12*
13 or less	1	1	2	2	3	5
18	1	1	2	3	5	6
20	1	1	2	3	5	7
30	1	2	3	5	7	10
40	1	2	4	6	9	14

*Based on 3¾ gm available chlorine per tablet.

of chlorine solution from portable equipment is not only less expensive but can accomplish the task in a much shorter time.

Clor-Tec Sodium Hypochlorite On-site Generation System[503]

Introduction. This is one of the more interesting development in the use of chlorine for the treatment of potable water supplies. It is designed to handle all the miscellaneous applications of chlorine with rechlorination problems in the distribution system, outlying reservoirs, injection wells, and restoration of wells, plus the disinfection of water mains and storage tanks. The technology of this system is described in Chapter 3.

In addition to applying chlorine for routine disinfection of potable water, Chemical Services Company recently developed a unique program for the stabilization of microbial growths in the distribution systems, by reestablishing the loss of proper chlorine residuals that occurs in many distribution systems. This is known as "dead-end" problems. They have also recently developed a reservoir circulation and chlorination system, which again maintains the proper chlorine residuals required in any such system to maintain a stable microbial life in both the reservoir and adjacent piping.

Moreover, this process eliminates any necessity for using 12–15 percent hypochlorite, which has caused many problems for operating personnel. Before the HazMat ruling that caused the switch from chlorine gas to hypochlorite, no one in his or her right mind would ever have made this switch.

System Developer. Chemical Services Company began with the application of on-site chlorine generation in 1988 in Hawaii. Since that time it has installed over 300 of these units in the mainland United States. These installations involve both large and small units. Units available range in capacity from 1 lb of chlorine/day to 2000 lb/day. The company maintains distribution centers in Campbell, California and Clearwater, Florida.

Clor-Tec SANILEC Brine Cell Process

Description. The operation of the SANILEC brine system is simple in principle. Solid salt is dissolved in softened water to form a brine solution having a 30 percent salt concentration.

This brine is further diluted with softened water to a concentration of 2.5 to 3.0 percent for feed to the electrolytic cell, by the following equation:

$$\underset{\text{Salt}}{NaCl} + \underset{\text{Water}}{H_2O} + \underset{\text{Electricity}}{2e^-} = \underset{\text{Hypo}}{NaOCl} + \underset{\text{Hydrogen}}{H_2}$$

The final product contains 0.8 percent, plus or minus 0.1 percent, sodium hypochlorite. The brine is also used to regenerate the water softening equipment. This system is illustrated by Fig. 6-34.

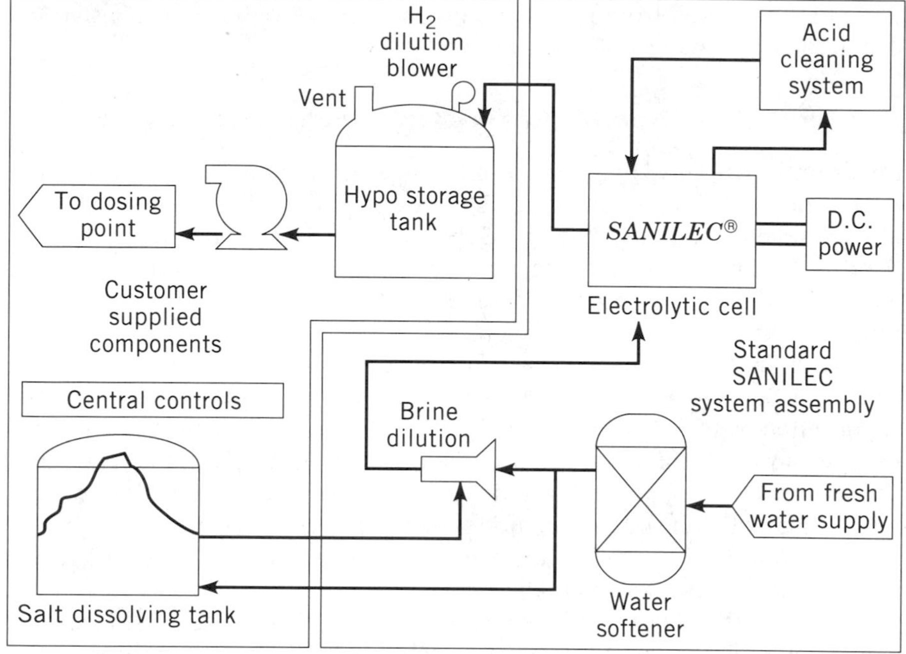

Fig. 6-34. Brine flow diagram.

The SANILEC brine system uses two or four cell assemblies stacked horizontally as shown in Fig. 6-35. Brine flows through the cells in a serpentine series, allowing the hypochlorite concentration to increase in strength while hydrogen is vented after each cell pass to the atmosphere, to eliminate electrode blanketing effects. The individual electrode packs are connected in an electrical series.

Conclusions. Considering the fact that this on-site generating system is not on the HazMat list of hazardous chemicals, and since the chemical cost is about equivalent to that of chlorine gas, there is no reason for anyone to switch from chlorine gas to 12–15 percent sodium hypochlorite solution, regardless of the size of the installation.

The troubles and the extra expense to those who have had to make this switch are almost beyond belief. The plant personnel are the ones suffering, while those responsible for the switch keep saying "we have to be in compliance."

Summary of Recommendations to Achieve the Objectives of Chlorination

Chlorine Demand. This is one of the most important parameters that defines the quality of potable water. Next and equally important are the compounds

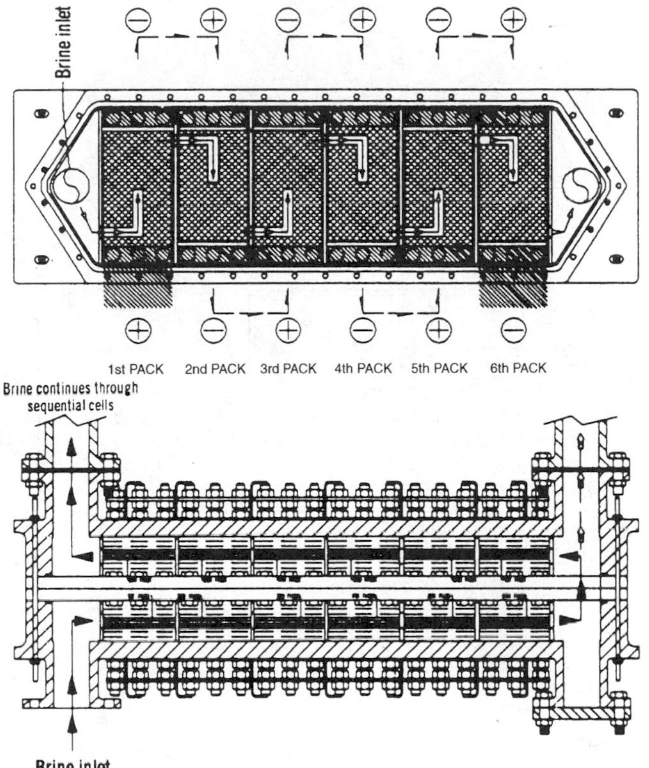

Fig. 6-35. Clor-Tec SANILEC brine cell process by Chemical Service Co.

of ammonia N and organic N, which are so important that there is now available (1997) a high quality instrument that measures these two compounds quickly and accurately (see Chapter 4).

The chlorine demand should be routinely measured as described by Feben and Taras in 1950 (see Chapter 4). It is interesting to note that their signal for running these chlorine demand tests came when the ammonia N content of the water reached 0.15 mg/L and the organic N 0.20 mg/L. From such studies, each water producer can establish chlorine demand constants, which will provide a rational control of the application of chlorine. Furthermore, these studies will provide a historical background for the raw water quality—all this instead of worrying about miniscule amounts of trivial THMs.

It is apparent from the above discussion that knowing the value of chlorine demand is very important. So White is retaining Fig. 6-36 in this edition—as before ORP was available to track the chlorine demand of a potable water or wastewater, this system was the only way that chlorine demand could be determined. It is extremely complicated compared to ORP!

ORP. Since about 1992 the Stranco Corp. has been studying the on-site control of chlorine application at many water treatment plants, using its High

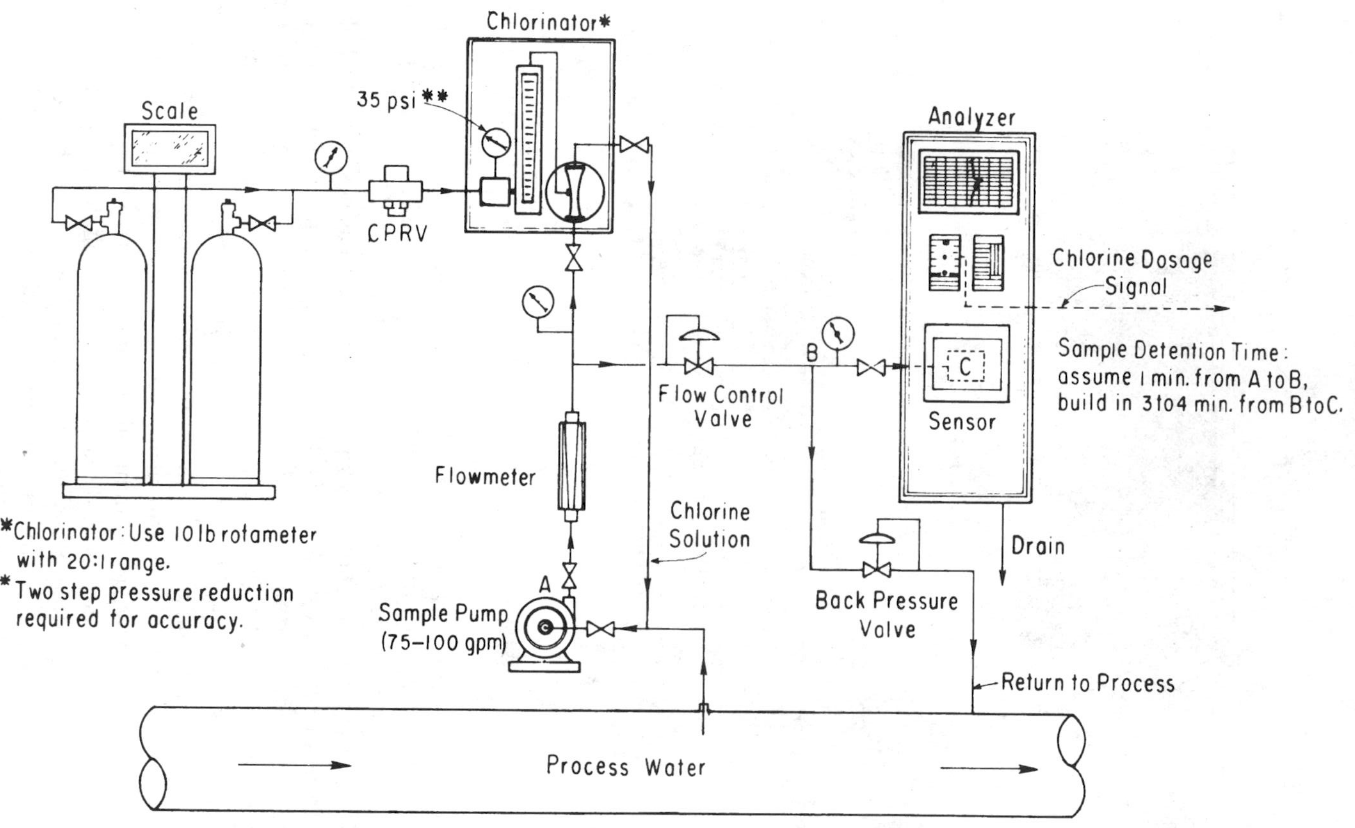

Fig. 6-36. Continuous chlorine demand monitoring system.

Resolution Redox Control System (see Chapter 9). This work has proved conclusively that ORP not only provides accurate control of the chlorine applied, but it tracks the chlorine demand.[496, 499, 500] All this is so simply because ORP measurements are not adversely affected by compounds normally found in water that continuously contribute to errors in chlorine residuals measured by chlorine residual analyzers. So if an operator uses ORP control, the system in Fig. 6-36 will become irrelevant.

Chlorine Residual Analysis. When using the free residual process it is customary to monitor and control on the basis of the free residual fraction. This is not enough. It is suggested that whenever the chlorine residual analyzer is calibrated, a forward-amperometric titration for all three fractions (free, mono-, and dichloramine) should be carried out. The best results from the free residual process are obtained from a total residual that contains *85 percent free available chlorine* (HOCl).

In spite of the emphasis given to the importance of ORP, chlorine residual analysis is indispensable. It is not only a backup to ORP, but it is necessary to assist operators in monitoring the operation of the chlorination equipment.

Nitrogen Trichloride Formation Spells Trouble. Nitrogen trichloride can be generated in copious quantities during the free residual process in some waters. Its presence is easily recognized because it burns the eyes long before its odor is noticed. Highly polluted waters are usually contaminated with organic nitrogen compounds derived usually from wastewater discharges. Typical conditions for NCl_3 formation at pH levels of 7 to 9 require a chlorine-to-ammonia ratio of about 12 : 1, ammonia N concentration 0.5 mg/L or greater, and organic N concentration 0.3 mg/L or greater. The higher the concentration of the nitrogen compounds, the greater the production of NCl_3. This compound produces a pungent geraniumlike odor at low concentrations (0.02 mg/L). However, it is practically insoluble in water; so it usually off-gases easily and quickly and is difficult to measure by available methods. Owing to its pervasive and unmistakable odor, and its characteristic eye-tearing quality, it is not necessary to attempt detection with analytical methods. Therefore, those waters subject to developing significant amounts of nuisance residuals (greater than 20 percent of the total chlorine residual) should be able to generate NCl_3, which is easily scavenged by postaeration. This will also reduce the dichloramine fraction well below the taste- and odor-producing level. Nitrogen trichloride aerates easily and decomposes rapidly in sunlight. This ability to decay rapidly in the atmosphere, including the hours of darkness, reduces significantly the possibility of its becoming an air pollution problem.

Points of Chlorine Application. It was found to be very practical to use only a single point of application of chlorine at the influent of the treatment process, in a sufficient amount to maintain a free chlorine residual throughout the entire treatment train. When this was not possible, a secondary point of application was provided, usually just ahead of the filters.

White has had great success with this single point of chlorine application, with the following observations. Often there was considerable loss of chlorine residual due to sunlight over the sedimentation and filter basins. This was largely corrected by automatically changing the dosage rates, one for nighttime and one for daytime. In some areas, cloudy weather required a small daytime change in dosage. Sometimes the operators overcame that problem by increasing the daytime dosage, followed by dechlorination of the effluent entering the distribution system to a level of about 0.5–0.75 mg/L.

Monitoring the System. The various locations of chlorine applications in a treatment plant do not need to be monitored if the plant is using the single point of chlorination procedure, as described above.

For a variety of reasons, operators have different viewpoints on this subject. In addition to the plant effluent, some operators feel the need to monitor chlorine residuals at the influent to the sedimentation basins and the inlet to the filters, all on a continuous basis. This requires a complicated installation of analyzers and sample pumps. Residuals should also be monitored in the distribution system. The preferred locations are at trouble spots where there may be degradation of water quality, insufficient chlorine residual, and/or consumer complaints.

Monitoring stations in the distribution system will help determine both the need for and the location of relay chlorination stations.

Samples of raw water should be monitored on a monthly basis for ammonia N and organic N. If either is present in amounts of 0.2–0.3 mg/L or greater, the samples of the finished water and the samples taken from the distribution system should be monitored for all the chlorine residual species by forward-titration, using either the amperometric method or the DPD-FAS titration procedure.

Now, in 1997, with the availability of qualified ORP instrumentation, the problem of continuous monitoring of the chlorination process in all parts of a potable water treatment system can be easily solved with the "operator-friendly" ORP method (see above and Chapter 9).

Controlling the Chloramine Process. This process carries with it the potential to supply the treated water with nutrients that promote the proliferation of bacteria and other microorganisms that conspire to degrade the quality of any potable water. These nutrients will become available when the chloramine residual is consumed. At this point the chloramine reverts to ammonia nitrogen. If chlorine is added when the residual is depleted, the ammonia nitrogen is immediately converted to chloramine. This is the reason for chlorine relay stations. Chloramination systems must be designed and operated so that measurable chloramine residuals persist continuously throughout the distribution network.

The most practical way to monitor areas of potential water quality degradation is by ORP. It is accurate and needs very little maintenance, and the installation involves only a pair of probes and a millivolt meter.

What the Consumer Can Do[515]

What Do I Do If My Drinking Water Tastes Funny? Four suggestions are:

1. Store some drinking water in a closed glass container in the refrigerator (warm drinking water has more taste than cold drinking water). Although some plastic bottles are OK for storing drinking water in the refrigerator, some types of plastic will cause a taste in the water. If you are having trouble, use a different type of plastic.
2. Use an electric mixer or blender to beat or blend the drinking water for five minutes. The mixing will remove some of the bad taste, but not all of it. Remember that to be smelled, the chemicals that cause the odor must leave the water, get into the air, and enter your nose. When you beat or blend the water, you hasten the chemicals leaving the water and get rid of some of the odor-causing chemicals prior to drinking the water. Then there are fewer chemicals to smell when you drink.
3. Some people object to the chlorine taste of their drinking water. Boiling tap water for five minutes should remove most of the disinfectant, if not all of it. Of course, some of the minerals in the water will be concentrated a little by the boiling, but this should not be a problem in most cases. After the water cools, refrigerate it. Remember that once the disinfectant is removed, the water must be treated like any other food. Keep it covered, and use it as quickly as possible.
4. Adding one or two teaspoons of lemon juice to refrigerated drinking water may result in a pleasant-tasting drink. If the problem is the rotten-egg odor, you may wish to consider a piece of home treatment equipment that will remove hydrogen sulfide, a nontoxic (in small amounts) but offensive chemical that causes this problem. If you have a water softener that is on both the hot and cold water, the chlorine will react with the softening materials inside the softener, and the chlorine will be removed. Thus, you may not have a chlorine taste, even though chlorine is added by the water supplier. **You should report any unusual taste and odor to your water supplier.**

REFERENCES*

1. Morris, J. C., "Future of Chlorination," *J. AWWA,* **58,** 1475 (Nov. 1966).
2. Baker, M. N., "The Quest for Pure Water," Amer. Works Assoc., New York (1930).
3. Cohen, M. D., "John Snow—Autumn Loiterer?" *Proc. Roy. Soc. Med.,* **62,** 99 (Jan. 1969).
4. "Manual of British Water Supply Practice," compiled by Inst. Of Water Engrs., Heffer & Son Ltd., Cambridge, England (1950).
5. Whipple, G. C., "Disinfection as a Means of Water Purification," *Proc. AWWA,* 266 (1906).
6. Leal, J. L., "Sterilization Plant of the Jersey City Water Supply," *Proc. AWWA,* 100 (1909).

7. Anon, "Watre Quality and Treatment," 2d ed., Amer. Water Works Assoc., New York (1951).
8. Houston, A. C., *Studies in Water Supply,* Macmillan & Co., Ltd., London (1913).
9. Hooker, A. B., "Chloride of Lime Sanitation" (1913).
10. Jennings, C. A., "Significance of the Bubbly Creek Experiment," *J. AWWA,* **40,** 1037 (1948).
11. Johnson, G. A., "Hypochlorite Treatment of Public Water Supplies," *Am. J. Pub. Hlth.,* 562 (1911).
12. Orchard, W. J., Wallace & Tiernan Company, private communication, 1959.
13. Anon, "Strange Fever," *M. D. Magazine,* 221 (Nov. 1969).
14. Laubusch, E. J., "Chlorination and Other Disinfection Processes," Chlorine Inst., New York (1964).
15. Gorman, A. E., and Wolman, A., "Significance of Water-borne Outbreaks," *J. AWWA,* **31** (1939).
16. Culp, R. L., "History and Present Status of Chlorination Practice in Kansas," *J. AWWA,* **52,** 888 (1960).
17. Darnall, C. R., "Purification of Water by Anhydrous Chlorine," *Am. J. Pub. Hlth.,* 783 (1911).
18. Tiernan, M. F., "Controlling the Green Goddess," *J. AWWA,* **40,** 1042 (Oct. 1948).
19. Beard, P. J., "The Survival of Typhoid in Nature," *J. AWWA,* **30,** 124 (1938).
20. Craun, G. F., and McCabe, L. J., "Review of the Causes of Waterborne Disease Outbreaks," *J. AWWA,* **45,** 74 (Jan. 1973).
21. Senevirtine, G., "The Wandering of a Vicious Cholera Strain," *San Francisco Chronicle* (Sept. 10, 1973).
22. Dufour, A. P., "Disease Outbreaks Caused by Drinking Water," *J. WPCF,* **54,** 980 (June 1982).
23. Morello, C., "Sinister Parasites Play Havoc with Pennsylvania Water Supply," *San Francisco Sunday Examiner and Chronicle* (Marc. 11, 1984).
24. Houston, A. C., "Water Purification," *Municipal Sanitation,* **3,** 4, 148 (Apr. 1932).
25. Laubusch, E. J., "How Safe Is Your Chlorine Residual?" *Pub. Works* (Mar. 1959).
26. Brock, T. D., *Biology of Microorganisms,* 3rd ed., Prentice-Hall, Englewood Cliffs, NJ, 1979.
27. Greenberg, A. E., and Ongerth, H. J., "Salmonellosis in Riverside, California," *J. AWWA,* **58,** 1145 (Sept. 1966).
28. Ross, E. C., Campbell, K. W., and Ongerth, H. J., "*Salmonella typhimurium* Contamination of Riverside, California Supply," *J. AWWA,* **58,** 165 (Feb. 1966).
29. Collaborative Report, "A Waterborne Epidemic of Salmonellosis in Riverside, California, 1965." *Am. J. Epdemiol.,* **93,** 33 (1971).
30. Craun, G. F., "Waterborne Outbreaks," *J. WPCF,* **49,** 1268 (June 1977).
31. Lloyd, B., and Morris, R., "Effluent and Water Treatment Before Disinfection," paper presented at the International Symposium: Viruses and Disinfection of Water and Wastewater, Univ. of Surrey, Guildford, U.K., Sept. 1–4, 1982.
32. Dufour, A. P., "Disease Outbreaks Caused by Drinking Water," *J. WPCF,* **55,** 905 (June 1983).
33. Hunter, C. A., and Ensign, P. R., "An Epidemic of Diarrhea in New-born Nursery Caused by *P. Aeruginosa,*" *Am. J. Pub. Hlth.,* **37,** 1166 (1947).
34. Roueche, B., "Three Sick Babies," *The New Yorker* (Oct. 5, 1968).
35. Culp, R. L., "Disease Due to Non-pathogenic Bacteria," *J. AWWA,* **60,** 157 (1968).
36. Scott, J. A., "Schistosomiasis Control in Water Supply Sources," *J. AWWA,* **61,** 352 (1969).
37. Herringer, E. J., "Schistosomiasis Control Is an Engineering Problem," *Pub. Works* (Jan. 1949).
38. Faber, H. A., "How Modern Chlorination Started," *Water and Sew. Wks.,* **99,** 455 (Nov. 1952).
39. Wolfe, M. S., "Giardiasis," *Pediatric Clinics of North America,* **26,** 295 (1979).
40. Meyer, E. A., and Jarroll, E. L., "Giardiasis," *Am. J. of Epidemiol.,* **111,** 1, (1980).
41. Pelczar, M. J., Jr. and Reid, R. D., *Microbiology,* 2nd ed., McGraw-Hill, New York, 1965.
42. Rice, E. W., Hoff, J. C., and Schaeffer, III, F. W., "Inactivation of *Giardia* Cysts by Chlorine," *App. and Env. Microbiol.,* **43,** 250 (Jan. 1982).

43. Jarrol, E. L., Bingham, A. K., and Meyer, E. A., "Effect of Chlorine on *Giardia lamblia* Cyst Viability," *App. and Env. Microbiol.,* **41,** 483 (Feb. 1981).
44. Garbe, P. L., private communication, Centers for Disease Control Atlanta, GA, July 5, 1983.
45. Fraser, D. W., et al., "Legionnaires Disease," *New Eng. J. of Medicine,* 297 (Dec. 1977).
46. Herwaldt, L. A., and Fraser, D. W., "Legionellosis: Legionnaires' Disease and Related Diseases," U.S. Dept. of Publ Hlth. reprint p. 45 (1981).
47. Broome, C. V., and Fraser, D. W., "Epidemiologic Aspects of Legionellosis," *Epidemiologic Reviews,* **1,** 1 (1979).
48. Dufour, A. P., and Jakubowski, W., "Drinking Water and Legionnaires Disease," *J. AWWA,* **74,** 631 (Dec. 1982).
49. Fraser, D. W., and McDade, J. E., "Legionellosis," *Scientific American,* **241,** 82 (Oct. 1979).
50. Chang, S. L., "Viruses, Amoebas, and Nematodes and Public Water Supplies," *J. AWWA,* **53,** 288 (1961).
51. Foote, H. B., Jellison, W. L., Stienhaus, E. E., and Kohls, G. M., "Effect of Chlorination of *Pasteurella tularensis* in Aqueous Suspension," *J. AWWA,* **35,** 7 (July 1943).
52. Olivieri, V., and Cabelli, V., personal communication, Sept. 2, 1982.
53. Morris, R., Finch, P., and Sharp, D. N., "Effect of the Kingsbury Lakes on the Microbiological Quality of the River Thames, U.K.," a paper presented at the International Symposium on Viruses and Disinfection of Water and Wastewater, University of Surrey, Guildford, U.K., Sept. 1–4, 1982.
54. Cabelli, V., "Waterborne Viral Infections," paper presented at the International Symposium on Viruses and Disinfection of Water and Wastewater, University of Surrey, Guildford, U.K., Sept. 1–4, 1982.
55. O'Conner, J. T., Hemphill, L., and Reach, C. D. Jr., "Removal of Virus from Public Water Supplies," EPA report 600752–82–024, Cincinnati, OH, Aug. 1982.
56. Kott, Y., "Effluent Usage and Disposal," paper presented at International Symposium: Viruses and Disinfection of Water and Wastewater, Univ. of Surrey, Guildford, U.K., Sept. 1–4, 1982.
57. Timbury, M. C., *Notes on Medical Virology,* 4th ed., Churchill Livingstone, Edinburgh and London, 1973.
58. *Van Nostrand's Scientific Encyclopedia,* 4th ed., Van Nostrand Reinhold, New York, 1968.
59. Dennis, J. M., "Infectious Hepatitis at New Delhi," *J. AWWA,* **51,** 1288 (1959).
60. Blacklow, N. I., and Cukor, G., "Viral Gastroenteritis Agents," Chap. 90 in E. H. Lennette, A. Balows, W. J. Hausler, Jr. and J. P. Truant (Eds.), *Manual of Clinical Microbiology,* 3rd ed., Am. Society of Microbiology, Washington, DC, 1980.
61. Wilson, R., et al., "Waterborne Gastroenteritis Due to Norwalk Agent: Clinical and Epidemiological Investigation," *Am. J. Pub. Hlth.,* **72,** 72 (1982).
62. Greenberg, H. B., et al., "Role of Norwalk Virus in Outbreaks of Non-Bacterial Gastroenteritis," *J. Inf. Dis.,* **139,** 564 (May 1979).
63. Kaplan, J. E., Gary, G. W., Baron, R. C., Singh, N., Schonberger, L. B., Feldman, R., and Greenberg, H. B., "Epidemiology of Norwalk Gastroenteritis and the Role of Norwalk Virus in Outbreaks of Acute Nonbacterial Gastroenteritis," *Annals of Internal Medicine,* **96,** 756–761 (1982).
64. Adler, J. L., and Zickl, R., "Winter Vomiting Disease," *J. Inf. Disease,* **119,** 668 (1969).
65. Kaplan, J. E., Goodman, R. A., Schonberger, L. B., Lippy, E. C., and Gary, G. W., "Gastroenteritis Due to Norwalk Virus: An Outbreak Associated With a Municipal Water System," *J. Inf. Dis.,* **146,** 190 (Aug. 1982).
66. Taylor, J. W., "Norwalk Related Viral Gastroenteritis Due to Contaminated Drinking Water," *Ann. J. Epidemiol.,* **114,** 584 (1981).
67. Zatotin, B. A., Libiyainen, L. T., Bortnik, F. L., Chernitskaya, E. P., et al., "Waterborne Group Infection of Rotavirus Etiology," *Microbiol. Epidemiol. Immunol.,* **11,** 99–101 (1981).
68. Craun, G. F., "Disease Outbreaks Caused by Drinking Water," *J. WPCF,* **51,** 1751 (June 1979).
69. Craun, G. F., "Disease Outbreaks Caused by Drinking Water," *J. WPCF,* **45,** 1566 (Jan. 1973).

70. Kienle, J. A., "Use of Liquid Chlorine for Sterilizing Water," *Proc. AWWA,* 267 (1913).
71. Cutler, J. W., and Green, F. W., "Operating Experience with a New Residual Recorder Controller," *J. AWWA,* **22,** 755 (1930).
72. Goudey, R. F., "Residual Chlorination on the Los Angeles System," *J. AWWA,* **28,** 1742 (1936).
73. Wolman, A., and Enslow, L. H., "Chlorine Absorption and the Characteristics of Water," *J. Ind. and Eng. Chem.,* **11,** 209 (1919).
74. Howard, N. J., and Thompson, R. E., "Chlorine Studies and Some Observations on Taste Producing Substances in Water, and the Factors Involved in Treatment by the Super- and De-chlorination Method," *J. NEWWA,* **40,** 276 (1926).
75. Watzl, E., "Superchlorination and Dechlorination over Carbon for a Municipal Water Supply," *Ind. and Eng. Chem.,* **21,** 156 (1929).
76. Cox, C. R., "Double Chlorination," *J. AWWA,* **16,** 55 (1926).
77. McAmis, J. W., "Prevention of Phenol Taste with Ammonia," *J. AWWA,* **17,** 3,341 (Mar. 1927).
78. Williams, D. B., "Control of Free Residual Chlorine by Ammoniation," *J. AWWA,* **55,** 1195 (Sept. 1963).
79. Williams, D. B., "Elimination of Nitrogen Trichloride in Dechlorination Practices," *J. AWWA,* **58,** 248 (Feb. 1966).
80. Raschig, F., "Chloramine," *Verh. Ges. Dent Naturforsch. Aertzl. (Germany),* **11,** 120 (1907).
81. Race, J., *Chlorination of Water,* John Wiley & Sons, New York, 1918.
82. Race, J., "Chlorination and Chloramines," *J. AWWA,* **3,** 63 (Mar. 1918).
83. Race, J., "Discussion of Pre-ammoniation of Filtered Water," *J. AWWA,* **23,** 411 (Mar. 1931).
84. Adams, B. A., "The Iodoform Taste Aquired by Chlorinated Water," *Med. Officer,* **869,** 33 (Dec. 1925).
85. Houston, Sir A. C., 19th Ann. Rept., Metro. Water Bd., London, "Chemical and Bacteriological Examination of London Water," 1925.
86. Harold, C. H. H., "Further Investigation into the Sterilization of Water by Chlorine and Some of Its Compounds," *J. Royal Army Corps,* **45,** 190, 251, 350, 429 (1925).
87. Adams, B. A., "The Chloramine Treatment of Pure Water," *Med. Officer,* **5,** 55 (1926).
88. Lawrence, W. C, "Studies in Water Purification Processes at Cleveland," *J. AWWA,* **23,** 6, 896 (June 1931).
89. Braidech, M. M., "The Ammonia–Chlorine Process as a Means for Taste Prevention and Effective Sterilization," *Ohio Conf. Water Purif.,* **9,** 67 (1930).
90. Braidech, M. M., "Practical Application of Ammonia–Chlorine Process in Sterilization of Cleveland Water Supply," *J. AWWA,* **22,** 1297 (Sept. 1930).
91. Ellms, J. W., and Lawrence, W. C., "Investigation of Tastes and Odors in the Cleveland Water Supply," *Eng. News-Rec.,* **86,** 1039 (1921).
92. Spaulding, C. H., "Pre-Ammoniation at Springfield, Illinois," *J. AWWA,* **21,** 1085 (Aug. 1929).
93. Harrison, L. B., "Chlorophenol Tastes in Water of High Organic Content," J. AWWA, **21,** 542 (Apr. 1929).
94. Committe Report, "Control of Tastes and Odors in Public Water Supplies," *J. AWWA,* **25,** 1490 (Nov. 1933).
95. Harvill, C. R., Morgan, J. H., and Mauzy, H. L., "Practical Application of Ammonia Induced Breakpoint Chlorination," *J. AWWA,* **34,** 275 (1942).
96. Harvill, C. R., Morgan, J. G., Hagar, M. C., and Todd, A. R., "Maintenance of Chlorine Residual in the Distribution System," *J. AWWA,* **34,** 1797 (1942).
97. Harvill, C. R., private communication, 1966.
98. Joint Committee Report, "Chlorine–Ammonia Treatment," *J. AWWA,* **33,** 2079 (Dec. 1941).
99. Anon., "Interim Primary Drinking Water Regulations; Control of Organic Chemical Contaminants in Drinking Water," Environmental Protection Agency, Washington, DC, *Federal Register,* Part II (Feb. 9, 1978).

100. Morris, J. C., "Aspects of Quantitative Assessment of Germicidal Efficiency," in J. D. Johnson (Ed.), *Disinfection: Water and Wastewater,* Ann Arbor Science, Ann Arbor, MI, 1975, p. 1.
101. Selna, M. W., Miele, R. P., and Baird, R. B., "Disinfection for Water Reuse," paper presented to the Disinfection Seminar at the Ann. Conf. AWWA, Anaheim, CA, May 8, 1977.
102. Selleck, R. E., Saunier, B. M., and Collins, H. F., "Kinetics of Bacterial Deactivation with Chlorine," *J. Env. Engr. Div. ASCE,* **104,** EE6, 1197 (Dec. 1978).
103. Means, E. G., McGuire, M. J., Otsuka, D. J., and Tanaka, T. S., "Impact of Chlorine and Ammonia Application Points on Bactericidal Efficiency of Free Chlorine and Chloramines in Pilot Plant Studies," paper presented at the Ann. Conf. AWWA, Las Vegas, June 9, 1983.
104. Griffin, A. E., and Chamberlin, N. S., "Some Chemical Aspects of Break-Point Chlorination," *J. NEWWA,* **55,** 371 (1941).
105. Griffin, A. E., "Chlorine for Ammonia Removal," Fifth Annual Water Conf. Proc. Engrs. Western Penn., p. 27, 1944.
106. Fair, G. M., Morris, J. C., Weill, Ira, and Burden, R. P., "The Behavior of Chlorine as a Water Disinfectant," *J. AWWA,* **40,** 1051 (Oct. 1948).
107. Fair, G. M., Morris, J. C., and Chang, S. L., "The Dynamics of Water Chlorination," *J. NEWWA,* **61,** 285 (1947).
108. Morris, J. C., Weil, Ira, and Burden, R. P., "The Formation of Monochloramine and Dichloramine in Water Chlorination," paper presented at 117th meeting, Am. Chem. Soc., Detroit, MI, April 16–20, 1950.
109. Palin, A. T., *Chemical Aspects of Chlorine,* Inst. of Water Engrs. (England), p. 565, 1950.
110. Palin, A. T., "A Study of the Chloro-Derivatives of Ammonia and Related Compounds with Special Reference to Their Formation in the Chlorination of Natural and Polluted Waters, " *Water and Water Engineering* (England), p. 151 (Oct. 1950), p. 189 (Nov. 1950), p. 248 (Dec. 1950).
111. Granstrom, M. L., "The Disproportionation of Monochloramine," Ph.D. dissertation in Sanitary Engineering, Harvard Univ., 1954.
112. Williams, D. B., "How to Solve Odor Problems in Water Chlorination Practice," *Water and Sew. Wks.,* **99,** 358 (1952).
113. Williams, D. B., "Control of Free Residual Chlorine by Ammoniation," *J. AWWA,* **55,** 1195 (Sept. 1963).
114. Williams, D. B., "Elimination of Nitrogen Trichloride in Dechlorination Practice," *J. AWWA,* **58,** 148 (Feb. 1966).
115. Palin, A. T., "Determination of Free Chlorine and Combined Chlorine in Water by the Use of Diethyl-*p*-Phenylene Diamine," *J. AWWA,* **49,** 873 (July 1957).
116. Palin, A. T., "Methods for the Determination, in Water, of Free and Combined Available Chlorine, Chlorine Dioxide and Chlorite, Bromine, Iodine, and Ozone, using Diethyl-*p*-Phenylene Diamine (DPD), "*J. Inst. Water Engrs.,* **21,** 537 (1967).
117. Palin, A. T., "Analytical Control of Water Disinfection with Special Reference to Differential DPD Methods for Chlorine Dioxide, Bromine, Iodine, and Ozone," *J. Inst. Water Engrs.,* **28,** 139 (1974).
118. Wei, I. W., and Morris, J. C., "Dynamics of Breakpoint Chlorination," Division of Engineering and Applied Physics, Harvard Univ., Cambridge, MA, May 1973.
119. Saunier, B. M., and Selleck, R. E., "The Kinetics of Breakpoint Chlorination in Continuous Flow Systems," paper presented at the AWWA Ann. Conf., New Orleans, LA, June 22, 1976.
120. Williams, D. B., "The Organic Nitrogen Problem," *J. AWWA,* **43,** 847 (Oct. 1951).
121. Griffin, A. E., "Reactions of Heavy Doses of Chlorine in Water," *J. AWWA,* **31,** 2121 (Dec. 1939).
122. Griffin, A. E., "Observations on Breakpoint Chlorination," *J. AWWA,* **32,** 1187 (July 1940).
123. O'Connell, Jr., W. J., unpublished report "Superchlorination and Ammonia–Chlorine Treatment, Governors Ave. Well Incident," Stanford Univ., Palo Alto, CA, 1939.
124. Clark, N. A., et al., "Human Enteric Viruses in Water: Source, Survival and Removability," *International Conf. Water Pollution Research,* Sept. 1962, Pergamon Press, London.

125. White, G. C., Bean, E. L., and Williams, D. B., "Chlorination and Dechlorination: A Scientific and Practical Approach," *J. AWWA,* **60,** 540 (May 1968).
126. White, G. C., unpublished notes of plant survey, 1959.
127. Sung, R. D., "Effects of Organic Constituents in Wastewater on the Chlorination Process," Ph.D. dissertation, Univ. of California, Davis, CA, 1974.
128. White, G. C., "Disinfection: The Last Line of Defense for Potable Water," *J. AWWA,* **67,** 410 (Aug. 1975).
129. "Community Water Supply Study: Analysis of National Survey Finding," Bur. of Water Hygiene, Envir. Health Service USPHS, Dept. of HEW, Washington, DC, July 1970.
130. Anon., "Industrial Pollution of the Lower Mississippi River in Louisiana," U.S. Environmental Protection Agency, Cincinnati, OH, Apr. 1972.
131. Anon., "New Orleans Area Water Supply Study," U.S. Environmental Protection Agency, Cincinnati, OH, draft report released on Nov. 8, 1974.
132. Rook, J. J., "Formation of Haloforms during Chlorination of Natural Waters," *Water Treatment and Examination,* **23** (Part 2), 234 (1974).
133. Bellar, T. A., Lichtenberg, J. J., and Kroner, R. C., "The Occurrence of Organohalides in Chlorinated Drinking Waters," *J. AWWA,* **66,** 703 (Dec. 1974).
134. Symons, J. M., Bellar, T. A., Carswell, J. K., DeMarco, J., Kropp, K. L., Robeck, G. G., Seeger, D. R., Slocum, C. V., Smith, B. L., and Stevens, A. A., "National Organics Reconnaissance Survey for Halogenated Organics," *J. AWWA,* **67,** 634 (Nov. 1975).
135. Trussell, R. R., and Umphres, M. D., "The Formation of Trihalomethanes," *J. AWWA,* **70,** 604 (Nov. 1978).
136. Lange, A. L., and Kawczynski, E., "Controlling Organics: The Contra Costa County Water District Experience," *J. AWWA,* **70,** 653 (Nov. 1978).
137. Morris, J. C., Ram, N., Baum, B., and Wajon, E., "Formation and Significance of N-Chloro Compounds in Water Supplies," EPA Report No. 600/2-80-031, Mun. Env. Res. Lab., Cincinnati, OH, July 1980.
138. Symons, J. M., et al., "Ozone, Chlorine Dioxide, and Chloramines as Alternatives to Chlorine for Disinfection of Drinking Water," U.S. EPA, Cincinnati, OH, Nov. 1977.
139. Norman, T. S., Harms, L. L., and Looyenga, R. W., "Use of Chloramines to Prevent THM Formation at Huron, S.D.," *J. AWWA,* **72,** 176 (Mar. 1980).
140. Barrett, R. H., and Trussell, A. R., "Controlling Organics: The Casitas Municipal Water District Experience," *J. AWWA,* **70,** 660 (Nov. 1978).
141. Rook, J. J., "Haloforms in Drinking Water," *J. AWWA,* **68,** 168 (Mar. 1976).
142. Selleck, R. E., private communication, Aug. 1978.
143. Love, Jr., O. T., Carswell, J. K., Miltner, R. J., and Symons, J. M., "Treatment for the Prevention or Removal of Trihalomethanes in Drinking Water," Appendix 3 to Treatment Guide for the Control of Chloroform and Other Trihalomethanes, U.S. EPA, Cincinnati, OH, 1976.
144. Anderson, M. C., Butler, R. C., Holdren, F. J., and Kornegay, B. H., "Controlling Trihalomethanes with Powdered Activated Carbons," *J. AWWA,* **73,** 432 (Aug. 1981).
145. Miller, R., and Hartman, D. J., "Feasibility Study of Granulated Activated Carbon Adsorption and On-site Regeneration," EPA Report, No. 600/52-82-087, Mun. Env. Res. Lab., Cincinnati, OH, Nov. 1982.
146. Oulman, C. S., Snoeyink, V. L., O'Connor, J. T., and Taras, M. J., "Removing Trade Organics from Drinking Water Using Activated Carbon and Polymeric Adsorbents," EPA Report No. 600/52-81-077-078-079, Mun. Env. Res. Lab., Cincinnati, OH, July 1981.
147. Quinn, J. E., and Snoeyink, V. L. "Removal of Total Organic Halogen by Granular Activated Carbon Adsorbers," *J. AWWA,* **72,** 483 (Aug. 1980).
148. Weber, Jr., W. J., and Pirbazari, M., "Effectiveness of Activated Carbon for the Removal of Toxic and/or Carcinogenic Compounds from Water Supplies," EPA Report No. 600/52-81-057, Muni. Env. Res. Lab., Cincinnati, OH, June 1981.
149. Symons, J. M., Stevens, A. A., Clark, R. M., Geldreich, E. E., Love, O. T., Jr., and DeMarco, J., "Removing Trihalomethanes from Water," *Water Engr. and Mgmt.,* p. 50 (July 1981).

150. Staff Report, "The Activated Carbon Dilemma," *Water and Sew. Wks.*, **125,** 34 (Dec. 1978).
151. J. M. Montgomery Engineers, "Alternative Disinfectants for Trihalomethane Control," Report to the Metropolitan Water District of Southern California, Oct. 1981.
152. Anon., "Mainstream," AWWA, Denver, CO, Mar. 26, 1982.
153. Kennedy/Jenks Engineers, "Trihalomethane Control Investigation," Report to the Alameda County Water District, Fremont, CA, Apr. 1982.
154. Harms, L. L., and Looyenga, R. W., "Preventing Haloform Formation in Drinking Water," EPA Report No. 600/2-80-091, Mun. Env. Res. Lab., Cincinnati, OH, Aug. 1980.
155. Rice, L. M., and Bolding, M. E., "An Alternative Solution to the THM Problem," *Water Engr. and Mgmt.,* p. 59 (May 1981).
156. McGuire, M. J., Shepherd, B. M., and Davis, M. K., "Surface Water Supply Trace Organics Survey: Maximum Trihalomethane Potential," Metropolitan Water District of Southern California, Water Quality Lab. Report, Feb. 1980.
157. Brown, H. A., "Superchlorination of Ottumwa, Iowa," *J. AWWA,* **32,** 1147 (July 1940).
158. Brown, H. A., "Triple Chlorination at Ottumwa, Iowa," *Water and Sew. Wks.,* **97,** 267 (July 1950).
159. Williams, D. B., Informal Presentation at AWWA Research Committee Meeting on Taste and Odor Control, Chicago, IL, Nov. 12 and 13, 1974.
160. Inhofer, W. R., and DeHooge, F. J., "Free Residual Chlorination of Passaic River Water at Little Falls, New Jersey," paper presented at the AWWA Ann. Conf., Boston, MA, June 16–21, 1974.
161. American Water Works Association, Special Taste and Odor Research Committee Meeting, Chicago, IL, Nov. 12 and 13, 1974.
162. Willey, B. J., Duke, C. M., and Rasho, J., "Chicago's Switch to Free Chlorine Residuals," *J. AWWA,* **68,** 441 (Aug. 1975).
163. Hoehn, R. C., and Johnson, J. D., "An Analysis of Disinfection, Water Quality Control, and Safety Practices in 1978 in the United States Water Industry," AWWA Water-Quality Division, Disinfection Committee, 1978.
164. McCabe, L. J., Symons, J. M., Lee, R. D., and Robeck, G. G., "Survey of Community Water Supply Systems," *J. AWWA,* **62,** 670 (Nov. 1970).
165. White, G. C., "Disinfection: The Last Line of Defense for Potable Water," **67,** 410 (Aug. 1975).
166. McDermott, J. H., "Virus Problems and Their Relation to Water Supply," presented at Virginia section AWWA Conference, Roanoke, VA, Oct. 25, 1973.
167. McCabe, L. J., "Significance of Virus Problem," paper presented at AWWA Water Quality Conference, Cincinnati, OH, Dec. 3 and 4, 1973.
168. Engelbrecht, R. S., Foster, D. H., Masarik, M. T., and Sai, S. H. "Detection of New Microbial Indicators of Chlorination Efficiency," paper presented at the AWWA Water Technology Conference, Dallas, TX, Dec. 1–3, 1974.
169. Ptak, D. V., Ginsburg, W., and Willey, B. F., "Identification and Incidence of *Klebsiella* in Chlorinated Water Supplies," *J. AWWA,* **65,** 604 (Sept. 1973).
170. Sorber, C. A., Williams, R. F., Moore, B. E., and Longley, K. E., "Alternative Disinfection Schemes for Reduced Trihalomethane Formation: Vol. 1. Prototype Studies," U.S. EPA Report #600/52-82-037, U.S. Environmental Protection Agency, Cincinnati, OH, Aug. 1982.
171. Ward, N. R., Means, E. G., Olson, B. H., and Wolfe, R. L., "The Inactivation of Total Count and Selective Gram-Negative Bacteria by Inorganic Monochloramines and Dichloramines," paper presented at AWWA Water Quality Technology Conf., Nashville, TN, 1982.
172. White, G. C., Beebe, R. D., Alford, V. F., and Sanders, H. A., "Wastewater Treatment Plant Disinfection Efficiency as a Function of Chlorine and Ammonia Content," in R. L. Jolley et al. (Eds.), *Water Chlorination: Environmental Impact and Health Effects,* Vol. 4, Book 2, p. 1115, Ann Arbor Science, Ann Arbor, MI, 1983.
173. Buelow, R. W., and Walton, G., "Bacteriological Quality versus Residual Chlorine," *J. AWWA,* **63,** 28 (Jan. 1971).

174. Robeck, G. G., "Substitution of Residual Chlorine Measurement for Distribution Bacteriological Sampling," paper presented at AWWA Ann. Conf., Minneapolis, MN, 1974.
175. Baumann, R. E., and Ludwig, D. D., "Free Available Chlorine Residuals for Small Non-Public Water Supplies," *J. AWWA,* **54,** 1379 (Nov. 1962).
176. Varma, M. M., and Baumann, E. R., "Superchlorination–Dechlorination of Small Water Supplies," State Proj. Rept. Project 353-S, Iowa State Univ. Engr. Exp. Sta., Ames, IA, 1959.
177. Butterfield, C. T., Wattie, E., Megregian, S., and Chambers, C. W., "Influence of pH and Temperature on the Survival of Coliform and Enteric Pathogens When Exposed to Free Chlorine," *Pub. Hlth. Reports,* **58,** 1837 (1943).
178. Lensen, S. G., Rhian, M., and Stebbins, M. R., "Inactivation of Partially Purified Poliomyelitis Virus in Water by Chlorination, Part II," *Am. J. Pub. Hlth.,* **37,** 869 (1947).
179. Weidenkopf, S. J., "Inactivation of Type I Poliomyelitis Virus with Chlorine," *Virology,* **5,** 56 (1958).
180. Clark, N. A., and Kabler, P. W., "Inactivation of Purified Coxsackie Virus in Water by Chlorine," *Am. J. Hyg.,* **59,** 1159 (1954).
181. Fair, G. M., Morris, J. C., and Chang, S. L., "The Dynamics of Chlorination," *J. NEWWA,* **61,** 285 (1947).
182. Snow, W. B., "Recommended Residuals for Military Water Supplies," *J. AWWA,* **48,** 1510 (Dec. 1956).
183. Brazis, R. A., et al., "Special Report to Department of the Navy, Bureau of Yards, and Docks: Sporicidal Action of Free Available Chlorine," R. A. Taft San. Engr. Center, Cincinnati, OH, 1957.
184. Neefe, J. R., et al., "Inactivation of the Virus of Infectious Hepatitis in Drinking Water," *Am. J. Pub. Hlth.,* **37,** 365 (1947).
185. Baumann, R. E., "Safe Disinfection for Household Water Systems," *Pub. Works* (May 1964).
186. Lin, S. D., "Tastes and Odors in Water Supplies: A Review," *Water and Sew. Wks., Ref. No.,* p. R-141 (1977).
187. Middlebrooks, E. J., "Taste and Odor Control," *Water and Sew. Wks., Ref. No.,* p. R-122 (1965).
188. Spitzer, E. F. (Ed.), *Handbook of Taste and Odor Control Experiences in the U.S. and Canada,* Amer. Water Works Assoc. Denver, CO, 1976.
189. Gerstein, H. H., "Odor Monitor and Threshold Tester," *J. AWWA,* **43,** 373 (1951).
190. Sigworth, E. A., "The Threshold Odor Test," *Water and Sew. Wks., Ref. No.,* p. R-92 (1964).
191. Palmer, C. M., "Algae in Water Supplies," U.S. Dept. of H.E.W. PHS #657, Washington, DC, 1962.
192. Vaughn, J. C., "Tastes and Odors in Water Supplies," *Env. Sci. and Tech.,* **9,** 703 (Sept. 1967).
193. Jenkins, D., Medsker, L. L., and Thomas, J. F., "Odorous Compounds in Natural Waters: Some Sulfur Compounds Associated with Blue-Green Algae," *Env. Sci. and Tech.,* **1,** 9, 731 (Sept. 1967).
194. Silvey, J. K. G., Russell, J. C., Redden, D. R., and McCormick, W. C., "Actinomycetes and Common Tastes and Odors," *J. AWWA,* **42,** 1018 (1950).
195. Erdei, J. F., "Control of Taste and Odor in Missouri River Water," *J. AWWA,* **55,** 1506 (1963).
196. Safferman, R. S., Rosen, A. A., Mashui, C. I., and Morris, M., "Earthy-Smelling Substance from a Blue-Green Alga," *Env. Sci. and Tech.,* **1,** 5, 429 (May 1967).
197. Lalezary, S., Pirbazari, M., McGuire, M. J., and Krassner, S. W., "Trace Taste and Odor Compounds from Water," paper presented at AWWA Ann. Conf., Las Vegas, NV, June 8, 1983.
198. Gerber, N. N., and Lechevalier, H. A., "Geosmin, an Earthy Smelling Substance Isolated from Actinomycetes," *Appl. Microbiol.,* **13,** 935 (1965).
199. Safferman, R. S., Rosen, A. A., Mashni, C. I., and Morris, M. E., "Earthy Smelling Substance from a Blue Green Alga," *Env. Sci. and Technol.,* **1,** 429 (1967).
200. Medsker, L. L., Jenkins, D., and Thomas, J. F., "Odorous Compounds in Natural Waters: An Earthy-Smelling Compound Associated with Blue Green Algae and Actinomycetes," *Env. Sci. and Technol.,* **2,** 461 (1968).

201. Dougherty, J. D., and Morris, R. L., "Studies on the Removal of Actinomycetes Musty Tastes & Odors in Water Supplies," *J. AWWA,* **59,** 1320 (Oct. 1967).
202. Baker, R. A., "Threshold Odors of Organic Chemicals," *J. AWWA,* **55,** 913 (1963).
203. Howard, N. J., "Removal of Taste & Odor," *J. AWWA,* **18,** 766 (1926).
204. Howard, N. J., and Thompson, R. E., "Chlorine Studies on Taste Producing Substances," *J. NEWWA,* **40,** 276 (1926).
205. Howard, N. J., and Thompson, R. E., "Progress in Superchlorination at Toronto," *J. AWWA,* **23,** 387 (1931).
206. Bushnell, W. B., "Over-Chlorination for Taste Control," *J. AWWA,* **17,** 653 (1925).
207. Lloyd, J. M., "Superchlorination at Tyler Texas," *J. AWWA,* **31,** 2130 (1939).
208. Lower, J. R., "Superchlorination at Upper Sandusky, Ohio," 19th Ann. Rept. Ohio Conf. Water Purif., p. 69 (1940).
209. Calvert, C. K., "Superchlorination," *J. AWWA,* **32,** 299 (1940).
210. Harlock, R., and Dowlin, R. "Use of Chlorine for Control of Odors caused by Algae," *J. AWWA,* **50,** 29 (1958).
211. Riddick, T. M., "Controlling Taste and Odor and Color with Free Residual Chlorination," *J. AWWA,* **43,** 545 (1951).
212. Adams, C. D., "Control of Tastes and Odors from Industrial Wastes," *J. AWWA,* **38,** 702 (1946).
213. Ettinger, M. B., and Ruchhoft, C. C., "Stepwise Chlorination on Taste and Odor Producing Intensity of Some Phenolic Compounds," *J. AWWA,* **43,** 561 (1951).
214. Burtschell, R. H., Rosen, A. A., Middleton, F. M., and Ettinger, M. B., "Chlorine Derivatives of Phenol Causing Taste and Odor," *J. AWWA,* **50,** 205 (1959).
215. Ryckman, D. W., and Grigoropolous, S. G., "Use of Carbon and Its Derivatives in Taste and Odor Removal," *J. AWWA,* **50,** 1268 (1959).
216. Harrison, L. B., "Super-Chlorination of Phenol Wastes," *J. AWWA,* **17,** 336 (1927).
217. Hale, F., "Successful Superchlorination and Dechlorination for Medicinal Taste of a Well Supply," *J. AWWA,* **23,** 373 (1931).
218. Baty, J. B., "Taste and Odor Control by Superchlorination," *Can. Engr.,* **78,** 19 (1940).
219. Brown, H. A., "Superchlorination at Ottumwa, Iowa," *J. AWWA,* **32,** 1147 (1940).
220. Brown, H. A., "Triple Chlorination at Ottumwa, Iowa," *Wat. and Sew. Wks.,* **97,** 267 (July 1950).
221. Williams, D. B., "A New Method for Odor Control," *J. AWWA,* **41,** 441 (May 1949).
222. Williams, D. B., "Dechlorination Linked to Corrosion," *Water and Sew. Wks.,* **100,** 104 (Mar. 1953).
223. Wilson, C., "Bacteriology of Water Pipes," *J. AWWA,* **37,** 52 (Jan. 1945).
224. Beger, H., "Iron Bacteria in Water Works and Their Practical Significance," *Gas-u.-Wasser,* **80,** 886 (Dec. 11, 1937).
225. Beger, H., "The Biology of Iron Bacteria," *Gas-U.-Wasser,* **86,** 779 (Oct. 23, 1937).
226. O'Connell, Jr., W. J., "Characteristics of Microbiological Deposits in Water Circuits," *Refining,* 66–83 (1941).
227. Peclzar, Jr., M. J., and Reid, R. D., *Microbiology,* 2nd ed., McGraw-Hill, New York, 1965.
228. *Standard Methods for the Examination of Water and Wastewater,* 12 ed., Am. Pub. Hlth. Assoc., New York, 1965.
229. Reddick, H. G., and Linderman, S. E., "Tuberculation of Mains as Affected by Bacteria," *J. NEWWA,* **46,** 146 (1932).
230. Von Wolzogen Kuhr, C. A. H., and Van Der Vlugt, L. S., "Aerobic and Anaerobic Iron Corrosion in Water Mains," *J. AWWA,* **45,** 33 (1953).
231. Weers, W. A., and Middlebrooks, E. J., "A Review of the Theory and Control of Corrosion," *Water and Sew. Wks.,* **114,** 156 (May 1967).
232. Thomas, A., "Role of Bacteria in Corrosion," *Water Wks. and Sew.,* **89,** 367 (1942).
233. Starkey, R. L., "Transformation of Iron by Bacteria in Water," *J. AWWA,* **37,** 963 (Oct. 1945).
234. Nason, H. K., "Chemical Methods in Slime and Algae Control," *J. AWWA,* **30,** 437 (Mar. 1938).

235. Love, Jr., O. T., Miltner, R. J., Eilers, R. G., and Fronk-Leist, C. A., "Treatment of Volatile Organic Compounds in Drinking Water," EPA-600/8-83-019, Cincinnati, OH, May 1983.
236. Ridgway, H. F., and Olson, B. H., "Scanning Electron Microscope Evidence for Bacterial Colonization of a Drinking-Water Distribution System," *Appl. and Environ. Microbiol.,* **41,** 274 (Jan. 1981).
237. Ridgway, H. F., Means, E. G., and Olson, B. H., "Iron Bacteria in Drinking-Water Distribution Systems: Elemental Analysis of *Gallionella* Stalks, X-Ray Energy-Dispersive Microanalysis," *Appl. and Environ. Microbiol.,* **41,** 288 (Jan. 1981).
238. Arnold, G. E., "Crenothrix Chokes Conduits," *Engr. News. Rec.,* **116,** 774 (May 1936) and *Water Wks. and Sew.,* **85,** 263 (Apr. 1938).
239. Ackerman, J. W., "Capacity of Cast Iron Main Sustained by Chloramine Treatment," *Water Wks. and Sew.,* **83,** 159 (May 1936).
240. Buelow, R. W., and Walton, G., "Bacteriological Quality vs. Residual Chlorine," *J. AWWA,* **63,** 28 (Jan. 1971).
241. Victoreen, H. T., "Control of Water Quality in Transmission and Distribution Mains," *J. AWWA,* **66,** 369 (June 1974).
242. Snead, M. C., Olivieri, V. P., Kruse, C. W., and Kawata, K., "Benefits of Maintaining a Chlorine Residual in Water Supply Systems," U.S. EPA Mun. Env. Res. Lab. Report No. EPA 600/2-80-010, June 1980.
243. Alexander, L. J., "Control of Iron and Sulfur Organisms by Super-Chlorination and Dechlorination," *J. AWWA,* **36,** 1349 (Dec. 1944).
244. Panel Discussion, "Control of Growths in California Distribution Systems," *J. AWWA,* **42,** 849 (Sept. 1950).
245. Blair, G. Y., "Combating Pipeline Growths by Maintaining Chlorine Residuals throughout a Distribution System," *J. AWWA,* **46,** 681 (July 1954).
246. Wilson, C., "Odor and Taste Control as Influenced by Consumer Pipes," *Water and Sew. Wks.,* **95,** 156 (Oct. 1948).
247. Panel Discussion, "Value and Limitation of Chlorine Residuals in Distribution Systems," *J. AWWA,* **51,** 215 (1959).
248. Plowman, H. L., and Rademacher, J. M., "Persistence of Combined Available Chlorine Residual in Gary-Hobart Distribution System," *J. AWWA,* **50,** 1250 (1958).
249. Eliassen, R., and Cummings, R. H., "Analysis of Water-borne Outbreaks 1938–45," *J. AWWA,* **40,** 1301 (1948).
250. Offutt, A. C., Poole, B. A., and Fassnacht, G. G., "A Water-borne Outbreak of Amebiasis," *J. Pub. Hlth.,* **45,** 486 (1955).
251. Minkus, A. J., "Rechlorination in the Hartford Distribution System," *J. NEWWA,* **72,** 251 (Sept. 1958).
252. Williams, D. B., Brantford, Ont., private communication, 1970.
253. Rogers, M. E., "Restoring Pipeline Capacity at Wichita, Kansas," *J. AWWA,* **37,** 713 (1945).
254. Griswold, L. J., "Maintaining Transmission Line Capacity with Chlorine and Ammonia," *Water and Sew. Wks.,* **96,** 472 (1949).
255. Streicher, L., "San Diego Aqueduct Capacity Restored by Chlorination," *Water and Sew. Wks.,* **100,** 333 (Sept. 1953).
256. Jackson, L. A., and Mayhan, W. A., "Chlorination Maintains Supply Line Capacity," *Water and Sew. Wks.,* **98,** 248 (June 1951).
257. Derby, R., "Control of Slime Growths in Transmission Lines," *J. AWWA,* **39,** 1107 (1947).
258. Weston, R. S., "The Use of Chlorine to Assist Coagulation," *J. AWWA,* **11,** 446 (1924).
259. Griffin, A. E., "Chlorination a Ten Year Review," *J. NEWWA,* **68,** 97 (1954).
260. Billing, L. C., "Experiences with Chlorinated Copperas as a Coagulant," *Water Wks. and Sew.,* **81,** 73 (1934).
261. Streeter, H. W., and Wright, C. T., "Prechlorination in Relation to the Efficiency of Water Filtration Processes," *J. AWWA,* **23,** 22 (1931).
262. Baylis, J. R., "Improving the Bacterial Quality of Water," *Water Wks. and Sew.,* **86,** 96 (Mar. 1939).

263. Baumann, R. E., Willrich, T. L., and Ludwig, D. D., "Prechlorination," *Agr. Engr.*, **44,** 138 (Mar. 1963).
264. McKee, J. E., and Wolf, H. W., "Water Quality Criteria," 2d. ed., Calif. State Water Quality Control Board, Sacramento (1963).
265. Nordell, E., *Water Treatment for Industrial and Other Uses,* Reinhold, New York (1951).
266. Foxworthy, J. E., and Gray, H. K., "Removal of Hydrogen Sulfide in High Concentrations from Water," *J. AWWA,* **50,** 872 (July 1958).
267. Monscvitz, J. T., and Ainsworth, L. D., "Hydrogen Polysulfide in Water Systems," Am. Chem. Soc. Mtg., Div. Water, Air, and Waste Chemistry, Minneapolis, (Apr. 1969).
268. Black, A. P., and Goodson, J. B., Jr., "The Oxidation of Sulfides by Chlorine in Dilute Aqueous Solutions," *J. AWWA,* **44,** 309 (Apr. 1952).
269. Sawyer, C. N., *Chemistry for Sanitary Engineers,* McGraw-Hill, New York (1960).
270. Choppin, A. R., and Faulkenberry, L. C., "The Oxidation of Aqueous Sulfide Solutions by Hypochlorite," *J. Am. Chem. Soc.,* **59,** 2203 (1937).
271. Wells, S. W., "Hydrogen Sulfide Problems in Small Water Systems," *J. AWWA,* **46,** 160 (Feb. 1954).
272. Sammon, L. L., "Removal of Hydrogen Sulfide from a Ground Water Supply," *J. AWWA,* **51,** 1275 (1959).
273. Derby, R. L., "Hydrogen Sulfide Removal and Water Softening at Beverly Hills, Calif.," *J. AWWA,* **20,** 813 (1928).
274. Powell, S. T., and Von Lossberg, L. G., "Hydrogen Sulfide Removal," *J. AWWA,* **40,** 1277 (1948).
275. Fair, G. M., Geyer, J. C., and Okun, D. A., *Water and Wastewater Engineering,* Vol. 2, John Wiley & Sons, New York (1968).
276. Robinson, L. R., and Dixon, R. I., "Iron and Manganese Precipitation in Low Alkalinity Ground Waters," *Water and Sew. Wks.,* **115,** 514 (Nov. 1968).
277. Griffin, A. E., "Significance and Removal of Manganese in Water Supplies," *J. AWWA,* **52,** 1326 (Oct. 1960).
278. Bean, E. L., "Progress Report on Water Quality Criteria," *J. AWWA,* **54,** 1313 (Nov. 1962).
279. Griffin, A. E., and Baker, R. J., "The Breakpoint Process for the Free Residual Chlorination," *J. NEWWA,* **73,** 250 (Sept. 1959).
280. Edwards, S. E., and McCall, G. B., "Manganese Removal by Breakpoint Chlorination," *Water and Sew. Wks.,* **93,** 303 (Aug. 1946).
281. Myers, H. C. "Manganese Deposits in Western Reservoirs and Distribution Systems," *J. AWWA,* **53,** 579 (May 1961).
282. Tuepker, J. L., "Chlorination of High pH Waters," Panel Discussion, Ann. A.W.W.A. Mtg., San Diego, CA, May 21, 1969.
283. Griffin, A. E., "Well Rehabilitation by Chlorination," *Water and Sew. Wks.,* **102,** 277 (June 1955).
284. Erickson, C. R., "Cleaning Methods for Deep Wells and Pumps," *J. AWWA,* **53,** 155 (Feb. 1961).
285. Brown, E. D., "Restoring Well Capacity with Chlorine," *J. AWWA,* **34,** 698 (1942).
286. Huston, W. E., "Restoring Well Capacity with Chlorine," *J. AWWA,* **38,** 761 (1946).
287. Suter, M., "Cleaning of Wells," *J. AWWA,* **30,** 1130 (1938).
288. Piatek, A., "Preventing Filamentous Scale in Well Water," *Water and Waste Engr.,* 55 (Dec. 1967).
289. Griffin, A. E., "Water Treatment for Water Flooding," *Producers Monthly* (1954).
290. A. W. W. A., Task Group Report, "Experience with Injection Wells for Artificial Ground Water Recharge," *J. AWWA,* **57,** 629 (1965).
291. A. W. W. A. Committee, 8360D Report, "Disinfecting Water Mains," *J. AWWA,* **60,** 1085 (1958).
292. Tracy, H., "Tank Disinfection," *J. AWWA,* **43,** 85 (1951).
293. White, G. C., "Chlorination and Dechlorination: A Scientific and Practical Approach," *J. AWWA,* **60,** 540 (May 1968).

294. Rideal, S., "The Influence of Ammonia and Organic Nitrogenous Compounds on Chlorine Disinfection," *J. Royal San. Inst.* (England), **31,** 33 (1910).
295. Deberard, H. I., "Chloramine at Denver, Colo., Solves Aftergrowth Problems," *Engr. News-Rec.,* **79,** 210 (1917).
296. Hulbert, R., "Chlorine–Ammonia Treatment Yields Nutrites in Effluent," *Engr. News-Rec.,* **109,** 315 (1933).
297. Shanson, D. C., "Infections and the Gut," Gastroenterology Seminar, St. Stephen's Hospital, Chelsea, London, U.K., Hospital Update, p. 756, June 1983.
298. Markell, E. K., Havens, R. F., and Kuritsubo, R., "Intestinal Parasitic Infections in Homosexual Men at a San Francisco Fair," *Western J. Medicine,* p. 177 (Aug. 1983).
299. Petit, C., "Castro District Survey," *San Francisco Chronicle* (Sept. 17, 1983).
300. Kjellstrand, C. M., Eaton, J. W., Yawata, Y., Swofford, H., Kolpin, C. F., Buselmeier, T. J., Von Hartitzsch, B., and Jacob, H. S., "Hemolysis in Dialized Patients caused by Chloramines," paper prepared by the Department of Medicine, Chemistry and Surgery, Univ. Minnesota, Minneapolis, MN, 1974, published in Switzerland in *Nephron,* **13,** 427 (1974).
301. Meyer, M. A., and Klein, E, "Granular Activated Carbon Usage in Chloramine Removal from Dialysis Water," School of Medicine, Nephrology Division, Univ. of Louisville, published in "Thoughts and Progress," *Artificial Organs* (Louisville, KY), **7,** 484 (Apr. 1983).
302. Bauer, R. C., and Snoeyink, V. L., "Reactions of Chloramines with Activated Carbon," *J. WPCF,* **45,** 2292 (1973).
303. Snoeyink, V. L., and Suidan, M. T., "Dechlorination by Activated Carbon and Other Reducing Agents," in J. D. Johnson (Ed.), *Disinfection of Water and Wastewater,* Ann Arbor Science Publishers, Inc., Ann Arbor, MI, 1975.
304. Kim, R. B., and Snoeyink, V. L., "The Monochloramine–GAC Reaction in Adsorption Systems," *J. WPCF,* **50,** 122 (Jan. 1978).
305. Blasiola, G. C., "Protecting Aquarium and Pond Fish from the Danger of Chloramines," *Freshwater and Marine Aquarium Magazine,* 1984.
306. Smith, C. E., and Russo, R. C., "Nitrite Induced Methemoglobinemia in Rainbow Trout," *Prog. Fish Cult.,* **37,** 150 (1975).
307. Gratzek, J. B., and Hayter, C., "An Experiment in Filtration for the Freshwater Aquarium," *Pets: Supplies and Marketing* (June 1979).
308. Atkins, P. F., Scherger, D. A., Barnes, R. A., and Evans, F. L., "Ammonia Removal by Physical-Chemical Treatment," *J. WPCF,* **45,** 2372 (Dec. 1973).
309. Rittman, B. E., and Snoeyink, V. L., "Achieving Biologically Stable Drinking Water," paper presented at the Ann. Conf. AWWA, Dallas, TX, June 14, 1984.
310. Hack, D. J., "Survey On the Use of Chloramine in Water Supplies of 50 States," paper presented at Chloramination Seminar, Ann Conf. AWWA, Dallas, TX, June 10, 1984.
311. Barrett, S. E., Davis, M. K., and McGuire, M. J., "Blending Chloraminated Water with Chlorinated Water: Considerations for a Large Water Wholesaler," presented at Ann. Conf. AWWA, Dallas, TX, June 14, 1984.
312. Anon., "Coliforms Resist Treatment in Six Midwestern Systems," *AWWA Mainstream* (Feb. 1984).
313. Moser, R. H., private communication, American Water Works Service Co. at Ann. Conf. AWWA, Dallas, TX, June 12, 1984.
314. Olivieri, V. P., private communication, Ann. Conf. AWWA, Dallas, TX, June 13, 1984.
315. White, G. C., "Investigation of Muncie Indiana Waterworks Disinfection System," confidential report to American Water Works Service in Muncie, Indiana, Aug. 1980.
316. Anon., Bacteria Detection Instrument Code No. 300-25, Rothmoore Analytical Houston, TX, June 1984.
317. Wallis, C., and Melnick, J. L., "An Instrument for the Immediate Quantification of Bacteria in Potable Waters," Baylor College of Medicine, Houston, TX, 1984.
318. Bingham, A. K., and Meyer, E. A., "Disinfection of *Giardia muris* Cysts in Chloraminated Water," Dept. of Microbiology and Immunology, Oregon Health Sciences University, Portland, OR, Nov. 12, 1981.

319. Sterling, C. R., Ph.D., "Waterborne Cryptospiridiosis," Chap. 3, Department of Veterinary Science, Univ. of Arizona, Tucson, AZ 85721.
320. Anon., "Waterborne Giardiasis: Where and Why, *J. AWWA,* p. 85 (Jan. 1986).
321. Braidech, T. E., and Karlin, R., "Causes of a Waterborne Giardiasis Outbreak," *J. AWWA,* p. 48 (Feb. 1985).
322. Lin, S-D., "*Giardia lamblia* and Water Supply," *J. AWWA,* p. 40 (Feb. 1985).
323. Jakubowski, W., "Waterborne *Giardia:* It's Enough to Make You Sick," "Roundtable," *J. AWWA,* p. 14 (Feb. 1985).
324. Lange, K. P., Bellamy, W. D., Hendricks, D. W., and Logsdon, G. S., "Diatomaceous Earth Filtration of *Giardia* Cysts and Other Substances," *J. AWWA,* p. 76 (Jan. 1986).
325. Hibler, C. P., "Analysis of Municipal Water Samples for Cysts of Giardia," *Advanced Giardia Research,* pp. 237–245, Univ. of Calgary Press, Calgary, Canada, 1988.
326. Hibler, C., P., Hancock, C. M., Perger, L. M., Wegrzyn, J. G., and Swabby, K. D., "Inactivation of *Giardia* Cysts with Chlorine at 0.5°C to 5.0°C," Department of Pathology, Colorado State University, Fort Collins, CO 80523, AWWA Research Report, 1987.
327. National Primary Drinking Water Regulations, Final Rule, 40 CFR Parts 141 & 142, Part II, EPA, *Federal Register,* June 29, 1990.
328. Hibler, C. P., private communication, Aug. 1990.
329. Haas, C. N., and Heller, B., "Kinetics of Inactivation of Giardia Lamblia by Free Chlorine," *Water Res.,* **24,** 2, 233–238 (1990). Printed in Great Britain.
330. Silverman, G. P., "Cryptospiridium, the Industry's New Superbug," *Opflow, AWWA* (Sept. 1988).
331. Korich, D. G., Mead, J. R., Madore, M. S., Sinclair, N. A., and Sterling, C. R., "Effects of Ozone, Chlorine Dioxide, Chlorine and Monochloramine on *Cryptosporidium parvum* Oocyst Viability," Department of Microbiology and Immunology and Department of Veterinary Science, Univ. of Arizona, Tucson, AZ 85721, Nov. 1989.
332. Peeters, J. E., "Effect of Disinfection of Drinking Water with Ozone and Chlorine Dioxide for Inactivation of *Cryptosporidium," Appl. Envir. Microbiol.,* **55,** 6, 1519–1522 (1989).
333. Joret, J. C., Langlais, B., Perrine, D., Bourbigot, M. M., and Valentis, G., "*Cryptosporidium* in Water," Centre de Recherche Compagnie General des Eaux, Chemin de la digue, BP 76, 78,600 Maisons Laffitte, France, Oct. 26, 1990.
334. Langlais, B., Perrine, D., Joret, J. C., and Chenu, J. P., "The *Ct* Value Concept for Evaluation of Disinfection Process Efficiency: The Particular Case of Ozonation for Inactivation of Some Protozoan Free Living Amoeba and *Cryptosporidium," Proc. 1990 Spring Conference, New Developments: Ozone in Water and Wastewater Treatment,* pp. 391–412, Intl. Ozone Assoc., Norwalk, CT 1990.
335. LeChevalier, M. W., Trok, T. M., Burns, M. O., and Lee, R. G., "Comparison of the Zinc Sulfate and Immunofluorescence Techniques for Detecting *Giardia* and *Cryptosporidium,*" *J. AWWA* (Sept. 1990).
336. LeChevalier, M. W., Private communication, Nov. 1990.
337. White, G. C., Free Residual Chlorination plus Main Flushing Operation at San Francisco Water Dept., Tracy, CA, Palo Alto, CA, Santa Rosa, CA, Eureka, CA, San Luis Obispo, CA, Reno, NV, and Salt Lake City, UT, 1938–58.
338. Wolfe, R. L., Means, E., G., III, Davis, M. K., and Barrett, S., "Biological Nitrification in Two Covered Finished Water Reservoirs Containing Chlorinated Water," *J. AWWA,* p. 109 (Sept. 1988).
339. Means, E. G., et al.,"Bacteriological Impact of a Changeover from Chlorine to Chloramination Disinfection in a Water Distribution System," *Proc.* AWWA Annual Conference, Denver, CO, June 22–26, 1986.
340. Verstraete, W., and Alexander, M., "Heterotrophic Nitrification," in J. I. Prosser (Ed.), *Nitrification,* IRL Press, Washington, DC, 1986.
341. Killham, K., "Heterotrophic Nitrification," in J. I. Prosser (Ed.), *Nitrification,* IRL Press, Washington, DC, 1986.

342. Beck, E., "Lithotrophic and Chemoorganotrophic Growth of Nitrifying Bacteria," in D. Schlessinger (Ed.), *Microbiology,* Am. Soc. of Microbiology, Washington, DC, 1978.
343. Wood, P. M., "Nitrification as a bacterial Energy Source," in J. I. Prosser (Ed.), *Nitrification,* IRL Press, Washington, DC, 1986.
344. Watson, S. W., Valois, F. W., and Waterbury, J. B., "The Family of Nitrobacteracae," in M. P. Starr et al. (Eds.), *The Prokaryotes,* Springer-Verlag, Berlin, Heidelberg, 1981.
345. "Nitrates: An Environmental Assessment," National Academy of Science, Washington, DC, 1978.
346. Alleman, J. E., Karimda, V., and Pantea-Kister, L., "Light Induced *Nitrosomonas* Inhibition," *Water Res.,* **21,** 499 (1987).
347. Jones, R. D., and Hood, M. A., "Interaction Between an Ammonium Oxidizer, *Nitrosomonas* sp. and Two Hetrotrophic Bacteria, *Nocardia Atlantica* and *Pseudomonas* sp.," *Microb. Ecol.,* **6,** 271 (1980).
348. Rittman, B. E., and Snoeyink, V. L., "Achieving Biologically Stable Drinking Water," *J. AWWA,* **76,** 10, 106 (Oct. 1984).
349. Goodall, J. B., "Biological Removal of Ammonia," in H. Sontheimer and W. Kuhn (Eds.), *Oxidation Techniques in Drinking Water Treatment,* EPA Report 57019-79-020, USEPA, Cincinnati, OH, 1979.
350. Richard, Y., "Biological Methods for the Treatment of Groundwater," in H. Sontheimer and W. Kuhn (Eds.), *Oxidation Techniques in Drinking Water Treatment,* EPA Report 57019-74-020, USEPA, Cincinnati, OH, 1979.
351. Wolfe, R. L., Lieu, N. I., Izagurre, G., and Means, E. G., "Ammonia-Oxidizing Bacteria in a Chloraminated Distribution System: Seasonal Occurrence, Distribution, and Disinfection Resistance," *Appl. Envir. Microbiol.,* pp. 451–462 (Feb. 1990).
352. Schotts, Jr., E. B., and Wooley, R. E., "Protozoan Sources of Spontaneous Coliform Occurrence in Chlorinated Drinking Water," EPA/600/52-89/019, Mar. 1990.
353. Voss, L., Button, K. S., Lorenz, R. C., and Tuovinen, O. H., "*Legionella* Contamination of a Preoperational Treatment Plant," *J. AWWA,* p. 7 (Jan. 1986).
354. Cline, G. C., and Russell, J. S., "An Evaluation of Treatment Strategies for the Control of Disinfection Byproducts—Water Quality vs Cost," presented at the AWWA Annual Conf., Cincinnati, OH, June 1990.
355. Kim, H-S., and Symons, J. M., "Use of Anion Exchange Resins for the Removal of Trihalomethane Precursors," presented at the AWWA Annual Conf., Cincinnati, OH, June 1990.
356. Fu, P. L. K., and Symons, J. M., "Removal of Natural Organic Matter Using Anion Exchange Resins," *J. AWWA,* **82,** 10, 70–77 (Oct. 1990).
357. Schnoor, J. L., Nitzschke, J. L., Lucas, R. D., and Veenstra, J. N., "Trihalomethane Yields as a Function of Precursor Molecular Weight," *J. Env. Sci. Tech.,* **13,** 1134–1138 (1979).
358. Oliver, B. G., and Visser, S. A., "Chloroform Production from the Chlorination of Aquatic Humic Material: The Effect of Molecular Weight, Environment and Season," *J. Water Res.,* **14,** 1137–1141 (1980).
359. LeChevalier, M. W., "Coliform Regrowth in Drinking Water: A Review," *J. AWWA,* pp. 74–86 (Nov. 1990).
360. Committee on Water Supply, "Bacterial Aftergrowths in Distribution Systems," *Am. J. Public Health,* **20,** 485 (1930).
361. Marshall, K. C., *Interfaces in Microbial Ecology,* Harvard Univ. Press, Cambridge, MA, and London, 1976.
362. Characklis, W. G., "Fouling Biofilm Development: A Process Analysis," *Biotechnol. Bioengrg.,* **23,** 1923 (1981).
363. Safe Drinking Water Committee, "Biological Quality of Water in the Distribution System," *Drinking Water and Health,* Vol. 4, Natl. Acad. Press, Washington, DC, 1982.
364. Fletcher, M., and Marshall, K. C., "Are Solid Surfaces of Ecological Significance to Aquatic Bacteria?" in K. C. Marshall (Ed.), *Advances in Microbial Ecology,* Vol. 6, Plenum Press, New York, 1982.

365. Geldreich, E. E., et al., "The Necessity of Controlling Bacterial Populations in Potable Waters: Community Water Supply," *J. AWWA,* **64,** 9, 596 (Sept. 1972).
366. Waters, S., McFeters, G. A., and LeChevalier, M. W., "Reactivation of Injured Bacteria," *Appl. Envir. Microbiol.,* **55,** 3226 (1989).
367. McFeters, G. A., Kippin, J. S., and LeChevalier, M. W., "Injured Coliforms in Drinking Water," *Appl. Envir. Microbiol.,* **51,** 1 (1986).
368. Clark, T. F., "New Culture Medium Detects Stressed Coliforms, *Opflow,* **11,** 3, 3 (1988).
369. Tracy, H. W., Camarena, V. M., and Wing, F., "Coliform Persistence in Highly Chlorinated Waters," *J. AWWA,* **58,** 9, 1151 (1966).
370. Levy, R. V., et al.,"Novel Method for Studying the Public Health Significance of Macroinvertebrates Occurring in Potable Water," *Appl. Envir. Microbiol.,* **47,** 889 (1984).
371. Hoff, J. C., "The Relationship of Turbidity to Disinfection of Potable Water," in C. W. Hendricks (Ed.), *Evaluation of the Microbiology Standards for Drinking Water,* EPA-570/9-78-00C, USEPA, Washington, DC, 1978.
372. Hejkal, T. W., et al., "Survival of Poliovirus within Organic Solids during Chlorination," *Appl. Envir. Microbiol.,* **38,** 114 (1979).
373. LeChevalier, M. W., Evans, T. M., and Seidler, R. J., "Effect of Turbidity on Chlorination Efficiency and Bacterial Persistence in Drinking Water," *Appl. Envir. Microbiol.,* **42,** 159 (1981).
374. Ridgway, H. F., and Olson, B. H., "Chlorine Resistance Patterns of Bacteria from Two Drinking Water Distribution Systems," *Appl. Envir. Microbiol.,* **44,** 972 (1982).
375. Herson, D. S., et al., "Attachment as a Factor in Protection of *Enterobacter cloacae* from Chlorination," *Appl. Envir. Microbiol.,* **53,** 1178 (1987).
376. LeChevalier, M. W., et al., "Disinfection of Bacteria Attached to Granular Activated Carbon," *Appl. Envir. Microbiol.,* **48,** 918 (1984).
377. Camper, A. K., et al., "Bacteria Associated with Granular Activated Carbon Particles in Drinking Water," *Appl. Envir. Microbiol.,* **52,** 434 (1986).
378. Allen, M. J., Geldreich, E. E., and Taylor, R. H., "The Occurrence of Microorganisms in Water Main Incrustations," *Proc. AWWA WQTC,* Philadelphia, PA, 1979.
379. Tuovinen, O. H., et al., "Bacterial, Chemical and Mineralogical Characteristics of Tubercles in Distribution Pipelines," *J. AWWA,* **72,** 11, 626 (Nov. 1980).
380. Ridgway, H. F., and Olson, B. H., "Scanning Electron Microscope Evidence for Bacterial Colonization of a Drinking Water Distribution System," *Appl. Envir. Microbiol.,* **41,** 274 (1981).
381. Tuovinen, O. H., and Hsu, J. C., "Aerobic and Anaerobic Microorganisms in Tubercles of the Columbus, Ohio, Water Distribution System," *Appl. Envir. Microbiol.,* **44,** 761 (1982).
382. Nagy, L. A., and Olson, B. H., "Occurrence and Significance of Bacteria, Fungi and Yeasts Associated with Distribution Pipe Surfaces," *Proc. AWWA WQTC,* Houston, TX, 1985.
383. Nagy, L. A., et al., "Biofilm Composition, Formation and Control in the Los Angeles Aqueduct System," *Proc. AWWA WQTC,* Nashville, TN, 1982.
384. Donlan, R. M., and Pipes, W. O., "Selected Drinking Water Characteristics and Attached Microbial Population Density," *J. AWWA,* **80,** 11, 70 (Nov. 1988).
385. Seidler, R. J., Morrow, J. E., and Baglet, S. T., "*Klebsiella* in Drinking Water Emanating from Redwood Tanks," *Appl. Envir. Microbiol.,* **33,** 893 (1977).
386. Olson, B. H., "Assessment and Implications of Bacterial Regrowth in Water Distribution Systems," EPA-600/52-82-072, 1982.
387. Victoreen, H. T., "Water Quality Deterioration in Pipelines," *Proc. AWWA WQTC,* Kansas City, MO, 1977.
388. LeChevalier, M. W., Babcock, T. M., and Lee, R. G., "Examination and Characterization of Distribution System Biofilms," *Appl. Envir. Microbiol.,* **53,** 2714 (1987).
389. Opheim, D., Growchowski, J., and Smith, D., "Isolation of Coliforms from Water Main Tubercles," *Abstracts, Ann. Mfg. Am. Soc. Microbiol.* (1988).
390. Fransolet, G., Villers, G., and Maschelin, W. J., "Influence of Temperature on Bacterial Development in Waters," *Ozone Sci. Engrg.,* **7,** 3, 205 (1985).

391. LeChevalier, M. W., Schultz, W., and Lee, R. G., "Bacterial Nutrients in Drinking Water," in M. W. LeChevalier, B. H. Olson, and G. A. McFeters (Eds.), *Assessing and Controlling Bacterial Regrowth in Distribution Systems,* AWWARF, Denver, CO, 1990.
392. Lowther, E. D., and Moser, R. H., "Detecting and Eliminating Coliform Regrowth," *Proc. 1984 AWWA WQTC,* Denver, CO, 1990.
393. Grant, W. D., and Long, P. E., *Environmental Microbiology,* John Wiley & Sons, New York, 1981.
394. Rittman, B. E., and Snoeyink, V. L., "Achieving Biologically Stable Drinking Water," *J. AWWA,* **7,** 10, 106 (Oct. 1984).
395. Wolfe, R. L., et al., "Biological Nitrification in Covered Reservoirs Containing Chloraminated Water," *J. AWWA,* **80,** 9, 109 (Sept. 1988).
396. Oskov, I., "*Klebsiella,*" in N. R. Krieg and J. G. Holt (Eds.), *Bergy's Manual of Systematic Bacteriology, Vol. 1, Bergy's Manual of Systematic Bacteriology,* Vol. 1, Williams & Wilkins, Baltimore, MD, 1984.
397. Symons, J. M., et al., "Natural Organics Reconaissance Survey for Halogenated Organics," *J. AWWA,* **67,** 11, 634 (1975).
398. Van Der Kooij, D., Visser, A., and Hunen, W. A. M., "Determining the Concentration of Easily Assimilable Organic Carbon in Drinking Water," *J. AWWA,* **74,** 10, 540 (1982).
399. Van Der Kooij, D., Visser, A., and Oranje, J. P., "Multiplication of Fluorescent *Pseudomonas* at Low Substrate Concentrations in Tap Water," *Antonie van Leeuenhock,* **48,** 229 (1982).
400. Van Der Kooij, D., and Hunen, W. A. M., "Measuring the Concentration of Easily Assimilable Organic Carbon (AOC) Treatment as a Tool for Limiting Regrowth of Bacteria in Distribution Systems," *Proc. AWWA WQTC,* Houston, TX, 1985.
401. Reilly, J. K., and Kippen, J. S., "Relationship of Bacterial Counts with Turbidity and Free Chlorine in Two Distribution Systems," *J. AWWA,* **75,** 6, 109 (June 1983).
402. Goshko, M. A., et al., "Relationship Between Standard Plate Counts and Other Parameters in Distribution Systems," *J. AWWA,* **75,** 11, 568 (Nov. 1983).
403. Olivieri, V. P., et al., "Recurrent Coliforms in Water Distribution Systems in the Presence of Free Residual Chlorine," in R. L. Jolley et al. (Eds.), *Water Chlorinatoin: Chemistry, Environmental Impact and Health Effects,* Lewis Publ., Chelsea, MI, 1985.
404. Ludwig, F., "The Occurrence of Coliforms in the Regional Water Authority Supply System," report submitted by the South Central Conn. Regl. Water Auth., New Haven, CT, 1985.
405. Ridgway, H. F., et al., "Biofilm Fouling of RO Membranes—Its Nature and Effect on Treatment of Water for Reuse," *J. AWWA,* **76,** 6, 94 (June 1984).
406. Characklis, W. G., et al., "Oxidation and Destruction of Microbial Films," in R. L. Jolley et al. (Eds), *Water Chlorination: Environmental Impacts and Health Effects,* Vol. 3, Ann Arbor Science Publ., Ann Arbor, MI, 1979.
407. Carson, L. A., et al., "Factors Affecting Comparative Resistance of Naturally Occurring and Subcultured *Psuedomonas aeruginosa* to Disinfectants," *Appl. Envir. Microbiol.,* **23,** 863 (1972).
408. Berg, J. D., Matin, A., and Roberts, P. V., "Growth of Disinfection-Resistant Bacteria and Simulation of Natural Aquatic Environments in the Chemostat," in R. L. Jolley et al. (Eds.), *Water Chlorination: Environmental Impact and Health Effects,* 1981.
409. Kutcha, J. M., et al., "Enhanced Chlorine Resistance of Tap Water—Adapted *Legionella pheumophila* as Compared with Agar-Medium-Passed Strains," Appl. Envir. Microbiol., **50,** 21 (1984).
410. Wolfe, R. L., Ward, N. R., and Olson, B. H., "Inactivation of Heterotrophic Bacterial Populations in Finished Drinking Water by Chlorine and Chloramines," *Water Res.,* **19,** 1393 (1985).
411. Stewart, M. S., and Olson, B. H., "Mechanisms of Bacterial Resistance to Inorganic Chloramines," *Proc. AWWA WQTC,* Portland, OR, 1986.
412. LeChevalier, M. W., Cawthon, C. D., and Lee, R. G., "Factors Promoting Survival of Bacteria in Chlorinated Water Supplies," *Appl. Envir. Microbiol.,* **54,** 649 (1988).
413. LeChevalier, M. W., Cawthon, C. D., and Lee, R. G., "Inactivation of Biofilm Bacteria," *Appl. Envir. Microbiol.,* **54,** 10, 2492 (1988).

414. LeChevalier, M. W., Cawthon, C. D., and Lee, R. G., "Disinfecting Biofilms in a Model Distribution System," *J. AWWA,* **82,** 7, 87 (July 1990).
415. Haas, C. N., LeChevalier, M. W., and Geoffry, M., "Modeling of Chlorine Inactivation of Drinking Water Biofilms" (unpubl.).
416. Jacangelo, J. G., Olivieri, V. P., and Kawata, K., "Mechanism of Inactivation of Microorganisms by Combined Chlorine," AWWARF, Denver, CO, 1987.
417. White, G. C., "Handbook of Chlorination (2nd. ed.), Van Nostrand Reinhold Co., New York, 1986.
418. Dixon, K. L., Lee, R. G., and Moser, R. H., "Residual Aluminum in Drinking Water," American Water Works Service Company, Voorhees, NJ, 1988.
419. Mackenthon, K. M., and Keup, L. E., "Biological Problems Encountered in Water Supplies," *J. AWWA,* **62,** 8, 250 (Aug. 1970).
420. Seidler, R. J., and Evans, T. M., "Persistence and Detection of Coliforms in Turbid Finished Drinking Water," EPA-600/52-82-054. USEPA, Cincinnati, OH, 1982.
421. Brazos, B. J., O'Conner, J. T., and Abcouwer, S., "Kinetics of Chlorine Depletion in Household Plumbing Systems," *Proc. AWWA WQTC,* Houston, TX, 1985.
422. AWWA, *Maintaining Distribution-System Water Quality,* Denver, CO, 1986.
423. Center for Disease Control, "Detection of Elevated Levels of Coliform Bacteria in a Public Water Supply," *Morbidity Mortality Weekly Rept.,* **34,** 142 (1985).
424. Earnhardt, Jr., K. B., "Chlorine Resistant Coliforms—The Muncie, Indiana Experience," *Proc. AWWA WQTC,* Miami Beach, FL, 1980.
425. Brodtman, Jr., N. V., and Russo, P. J., "The Use of Chloramines for Reduction of THMs and Disinfection of Drinking Water," *J. AWWA,* **71,** 1, 40 (Jan. 1979).
426. Norman, T. S., Harms, L. L., and Looyenga, R. W., "The Use of Chloramines to Prevent THM Formation," *J. AWWA,* **72,** 3, 176 (Mar. 1980).
427. Shull, K. E., "Experiences With Chloramines as Primary Disinfectants," *J. AWWA,* **73,** 2, 102 (Feb. 1981).
428. Mitcham, R. P., Shelley, M. W., and Wheadon, C. M., "Free Chlorine versus Ammonia-Chlorine: Disinfection, THM Formation and Zooplankton Removal," *J. AWWA,* **75,** 4, 196 (Apr. 1983).
429. Kreft, P., et al., "Converting from Chlorine to Chloramines: A Case Study," *J. AWWA,* **77,** 1, 38 (Jan. 1985).
430. Dice, J., "Denver's Seven Decades of Experience with Chloramination," *J. AWWA,* **77,** 1, 34 (Jan. 1985).
431. Means, E. G., et al., "Effects of Chlorine and Ammonia Points on Bactericidal Efficiency," *J. AWWA,* **78,** 1, 62 (Jan. 1986).
432. MacLeod, B. W., and Zimmerman, J. A., "Selected Effects on Distribution System Water Quality as a Result of Conversion to Chloramines," *AWWA WQTC,* Portland, OR, 1986.
433. MacLeod, B. W., personal communication, 1989.
434. Fung, L., personal communication, 1989.
435. Martin, R. W., et al., "Factors Affecting Coliform Bacteria Growth in Distribution Systems," *J. AWWA,* **74,** 1, 34 (Jan. 1982).
436. Hudson, L. D., Hankins, W. J., and Bataglia, M., "Coliforms in a Water Distribution System: A Remedial Approach," *J. AWWA,* **75,** 11, 564 (Nov. 1983).
437. Lowther, E. D., unpublished data, 1989.
438. Boethling, R. S., and Alexander, M., "Microbial Degradation of Organic Compounds at Trace Levels," *Envir. Sci. & Technol.,* **13,** 989 (1979).
439. Rittman, B. E., "Biological Processes and Organic Micropollutants in Treatment Processes," *Sci. Total Envir.,* **47,** 99 (1985).
440. Wang, B. Z., et al., "Purification of Polluted Source Water with Microflocculation/Direct Filtration–Biological Activated Carbon Process," *Aqua,* **6,** 321 (1986).
441. Wang, B. Z., Tian, J. Z., and Yin, J., "Purification of Polluted Source Water with Ozonation and Biological Activated Carbon," *Aqua,* **6,** 351 (1986).
442. Bouwer, E. J., and McCarty, P. L., "Transformations of 1- and 2-Carbon Halogenated Aliphatic Organic Compounds under Methanogenic Conditions," *Appl. Envir. Microbiol.,* **45,** 1286 (1983).

443. Bablon, G., et al., "Removal of Organic Matter by Means of Combined Ozonation/BAC Filtration, a Reality on an Industrial Scale at the Choisy-Le-Roi Treatment Plant," *Proc. AWWA Annual Conf.,* Denver, CO, 1986.
444. Baozhen, W., et al., "A Preliminary Study of the Efficiency and Mechanism of THM Removal in the Ozonation and BAC Process," *Ozone Sci. Engrg.,* **6,** 261 (1985).
445. Lauer, W. C., "Water Quality for Potable Water Reuse," *Tech.* (Kyoto), **23,** 2171–2180 (1991).
446. Lauer, W. C., "Denver's Potable Water Reuse Demonstration Project Process Selection for Potable Reuse Health Effects," Denver, Water Department, 1600 West 12th Ave., Denver, CO 80254, Mar. 1989.
447. Topudurti, K. V., and Haas, C. N., "THM Formation by the Transfer of Active Chlorine, from Monochloramine to Phloroacetophenone (PAP)," *J. AWWA,* pp. 62–66 (May 1991).
448. Morris, J. C., and Baum, B., "Precursors and Mechanisms of Haloform Formation in the Chlorination of Water Supplies," in R. L. Jolley et al. (Eds.), *Water Chlorination: Environmental Impacts and Health Effects,* Vol. 2, Ann Arbor Science, Ann Arbor, MI, 1978.
449. Pelletier, P. A., Carney, E. M., and du Moulin, G. C., "Comparative Resistance of *Mycobacterium avium* Complex and Other Nontuberculosis *Mycobacterium* to Chloramine," Joint Clinical Immunotherapy Program at Boston Univ. School of Medicine and the New England Baptist Hospital and Cellcor Therapies, Inc., Newton MA, presented at the Annual AWWA Conference, Philadelphia, PA, June 1991.
450. Collins, C. H., Grange, J. M., and Yates, M. D., "A Review: Mycobacteria in Water," *Appl. Bacteriol.,* **57,** 193–244 (1984).
451. Carson, L. A., Bland, L. A., Cusick, C. B., and Favero, M. S., "Prevalence of Nontuberculosis Mycobacteria in Water Supplies of Hemodialysis Centers," *Appl. Envir. Microbiol.,* **54,** 3122–3125 (1988).
452. Kazda, J., "The Importance of Water for the Distribution of the Potentially Pathogenic Mycobacteria, Growth of Mycobacteria in Water Models," *Zbl. Bakt.,* **158,** 170–176 (1973).
453. du Moulin, G. C., and Stottmeier, K. D., "Waterborne Mycobacteria: An Increasing Threat to Health," *ASM News,* **10,** 525–529 (1986).
454. Havelaar, A. H., Bervald, L. G., Groothuis, D. G., and Baas, J. G., "Mycobacteria in Semipublic Swimming Pools and Whirlpools," *Zbl. Bakt.,* **180,** 505–514 (1985).
455. Park, U. K., and Brewer, W. S., "The Recovery of *Mycobacterium marinum* from Swimming Pool Water and Its Resistance to Chlorine," *J. Envir. Health,* **38,** 390–392 (1976).
456. Pelletier, P. A., du Moulin, G. C., and Stottmeier, K. D., "Mycobacteria in Public Water Supplies: Comparative Resistance to Chlorine," *Microbiol. Sci.,* **5,** 147–148 (1988).
457. Schwartz, B. J., and Fung, L. C., "The Hackensack Water Company Experience: Free Chlorine vs. Chloramines," presented at the Annual AWWA Conference, Philadelphia, PA, June 1991.
458. Rice, E. W., Scarpino, P. V., Reasoner, D. J., Logsdon, G. S., and Wild, D. K., "Correlation of Coliform Growth Response with Other Water Quality Parameters," *J. AWWA,* **83,** 7, 98 (July 1991).
459. Herson, D. S., Marshall, D. R., Baker, K. H., and Victoreen, H. T., "Association of Microorganisms with Surfaces in Distribution Systems," *J. AWWA,* **83,** 7, 103 (July 1991).
460. Cobban, W., "Zebra Spells Trouble for Treatment Plant Operators," *Opflow,* Oct. 1991 (monthly publication of Amer. Water Works Assoc.).
461. Moore, S. G., private communication, New York Sea Grant (1990).
462. Dobson, J. G., "The Control of Fouling Organisms in Fresh Organisms in Fresh and Salt Water Circuits," *Trans.* ASME (Apr. 1946).
463. Cole, S. A., "Control of Bryazoa and Shellfish in Circulating Water Systems," Wallace & Tiernan Special Report (1964).
464. O'Connell, Jr., W. J., "Characteristics of Microbiological Deposits in Water Circuits," *Refining,* 66–83 (1944).
465. Patten, I. A., "Project Study for the Mitigation of Marine Fouling," paper presented at Semiann. Mtg., ASME, San Francisco (June 27–30, 1944).

466. Clapp, W. F., "Some Biological Fundamentals of Marine Fouling," paper presented at Semiann. Mtg., ASME, San Francisco (June 27–30 1944).
467. Wolfe, M., private communication, Washington, DC, Nov. 1, 1991.
468. Moser, P. W., "Danger in Diaperland," *Health,* p. 77, Sept./Oct. 1991).
469. Fahey, V., "How Do You Know which Bug Has You Down"?, *Health,* p. 79, Sept./Oct., 1991.
470. Brock, T. D., *Biology of Microorganisms,* 3rd ed., Prentice Hall, Englewood Cliffs, NJ, 1979
471. Reiff, F. M., Consulting Engineer for Pan-American WHO, personal communication, 1996.
472. Pelczar, M. J., and Reid, R. D., *Microbiology,* 2nd ed., McGraw-Hill, New York, 1965.
473. White, G. C. "Water Quality Degradation in the USA by Federal Government Mandate" (1994).
474. Ames, B. N., Gold, L. S., and Willett, W. C., "The Causes and Prevention of Cancer," *J. Am. Med. Assoc.,* Special Issue on Cancer, 1995.
475. Ames, B. N., and Gold, L. S., "Chemical Carcinogens: Too Many Rodent Carcinogens," *Proc. Natl. Acad. Sci. USA,* **87,** 7772–7776 (Oct. 1990).
476. Ames, B. N., Profet, M., and Gold, L. S., "Dietary Pesticides (99.99 percent All Natural)," *Proc. Natl. Acad. Sci. USA,* **87,** 7777–7781 (Oct. 1990).
477. Ames, B. N., Profet, M., and Gold, L. S., "Nature's Chemicals and Synthetic Chemicals: Comparative Toxicology," *Proc. Natl. Acad. Sci USA,* **87,** 7782–7786 (Oct. 1990).
478. Jolley, R. L., and Pitt, W. W., Jr., "Chloro-organics in Surface Water Sources, for Potable Water," Oak Ridge National Laboratory, Oak Ridge, TN 37830, presented at the May 8–13, 1977, AWWA Disinfection Symposium at Anaheim, CA.
479. Ames, B. N., Shigenaga, M. K., and Hagen, T. M., "Oxidants, Anti-oxidants, and the Degenerative Diseases of Aging," *Proc. Natl. Acad. Sci. USA,* **90,** 7915–7922 (1993).
480. Block, G., Patterson, B., and Subar, A., "Fruit, Vegetables, and Cancer Prevention: A Review of the Epidemiological Evidence," *Nutr. Cancer,* **18,** 1–29 (1992).
481. Gold, L. S., Slone, T. H., Stern, B. R., Manley, N. B., and Ames, B. N., "Rodent Carcinogens: Setting Priorities," *Science,* **258,** 261–265, 1992.
482. Epstein, S. S., "Environmental Determinants of Human Cancer," *Cancer Research,* **34,** 2425–2435 (Oct. 1974).
483. Ames, B., N., "Understanding the Cause of Aging and Cancer," Division of Biochemistry and Molecular Biology, Univ. of California, Berkeley, CA 94720, Feb. 22, 1994.
484. Gold, L. S., Slone, Thomas, H., and Ames, B. N., "Prioritization of Possible Carcinogenic Hazards in Foods," Life Science Division, Lawrence Laboratory, and Division of Biochemistry and Molecular Biology, Barker Hall, Univ. of California, Berkeley, CA 94720, 1996.
485. National Cancer Institute, "Carcinogens and Anti-carcinogens in the Human Diet: A Comparison of Naturally Occurring and Synthetic Substances," available from the National Academy Press, ph. 1-888-624-6242, 417 pp., ISBN)-309-05391-9, 1996.
486. Ames, B. N., "The Topic of Cancer," *Sierra Magazine* vs. Ames, Bruce, "Science and the Environment," *American Spectator,* June 1993.
487. Ames, B. N., and Gold, L. S., "The Rodent High Dose Cancer Test," Life Sciences Division, Lawrence Berkeley Laboratory, Berkeley, CA 94720, Jan. 16, 1996, http://potency.berkeley.-edu/cpdb.html.
488. Brody, J. E., "Strong Views on Origins of Cancer—Scientist at Work—Bruce N. Ames," *New York Times,* Science Tuesday, July 5, 1994.
489. Ames, B. N., "The Causes of Aging and Cancer: The Misinterpretation of Animal Cancer Tests," *Human and Ecological Risk Assessment,* **2,** 6–9 (1996).
490. Ames, B. N., "Does Current Cancer 'Risk Assessment' Harm Health?" The American Society for Cell Biology, Congressional Biochemical Research Caucus, Washington, DC, Nov. 5, 1993.
491. Ames, B. N., "Carcinogens and Public Policy," presentation for the Commonwealth Club of San Francisco, CA, July 13, 1990.
492. Cohen, B. L., "Test of the Linear–No Threshold Theory of Radiation Carcinogens, for Inhaled Radon Decay Projects," Aug. 1994, Univ. of Pittsburgh, PA 15260; Copyright 1995, Health Physics Society.
493. "Surface Water Treatment Rule," *Federal Register,* **54,** 124 (June 29, 1989).

494. American Society of Civil Engineers, AWWA, "Water Treatment Design," 2nd ed., p. 199, 1990.
495. Disinfection Committee, "Survey of Water Utilities Disinfection Practices," *J AWWA,* 121–128 (Sept. 1992).
496. Kim, Y. H., "Evaluation of Redox Potential and Chlorine Residual as a Measure of Water Disinfection," International Water Conf. 54th Ann. Mtg., Pittsburgh, PA, Oct. 11–13, 1993. Customer Service-808-882-6466
497. Lund, E., "Inactivation of Poliomyelitus Virus by Chlorination at Different Oxidation Potential Levels," Archly fur die Gesamte Virusforschung, Sonderabdsuch aus Band XIII, Heft 4, Vienna, 1–18, 1963.
498. U.S. EPA, "Control of Biofilm Growth in Drinking Water Distribution Systems," EPA/625/R-92-001, 1992
499. Strand, R. L., and Kim, Y. H., "ORP as a Measure of Evaluating and Controlling Disinfection of Potable Water," paper presented at AWWA Water Quality Conf., Miami, FL, Nov. 7–11, 1993. Customer Service-800-882-6966
500. Carlson, S., and Hasselbarth, U., "Fundamentals of Water Disinfection," *J. Water SRT—Aqua,* **40,** 6, 346–356 (1991).
501. Lawler, D. F., and Singer, P., "Analyzing Disinfection Kinetics and Reactor Design: A Conceptual Approach versus the SWTR."
502. Neden, D. G., Jones, R. J., Smith, J. R., Kirmeyer, G. J., and Foust, G. W., "Comparing Chlorination and Chloramination for Controlling Bacteria Regrowth," Research and Technology, *J. AWWA,* p. 80 (July 1992).
503. Chemical Services Co., "Clor-Tec On-site Sodium Hypochlorite Generating Systems," 2528 Seaboard Ave San Jose, CA 95131, June 1996.
504. Weeks, C., "Ionics Environmental Monitoring Guide for Water Quality" by Ionics Inc., 65 Grove Street, Watertown, MA 02172-2882, 1993. Customer Service Ph. 800-348-1730.
505. Fornos, W., "Population Growth Slowing," Population Institute, Washington, DC, Dec. 28, 1996.
506. AWWA Water Quality Division, Disinfection Committee, "Survey of Water Utility Disinfection Practices," *Journal AWWA,* pp 121–128 (Sept. 1992).
507. Perlman, D., "AIDS Continues Deadly March around the World," *San Francisco Chronicle,* Nov. 28, 1996.
508. Baker, R. J., "Basic Chemistry of Ammonia N Reactions with Chlorine in Solution," private communication, Nov. 19, 1994.
508. Adams, W. S., "Early Days at Mount Wilson," *Publications of the Astronomical Society of the Pacific,* **59,** 213 (1947).
509. Adams, W. S., "The Founding of the Mount Wilson Obsevatory," *Publications of the Astronomical Society of the Pacific,* **66,** 269 (1954).
510. Howard, R., "Eight Decades of Solar Research at Mount Wilson," *Solar Physics,* **99,** 171 (1985).
511. Wright, H., *Explorer of the Universe: a Biography of George Ellery Hale,* E. P. Dutton & Co., New York, Inc., 1966.
512. Gilman, P. "History of 150 ft. Solar Tower" Caltech World Wide Web, http: caltech.edu/history.html.
513. "Signals from the Sun" in *Signals from the Stars* Chapt. 3, pp. 62–98, Scribner Publishing, New York, 1931.
514. "The Spectrohelioscope and Its Work, Part III: Solar Eruptions and Their Apparent Terrestrial Effects," *Astrophysical Journal,* **73,** (1931).

*Some references listed herein may be found in the Special Preface to the 4th Edition. A few references, while not cited in either the text of Chapter 6 or in the Special Preface, contain important information and were used as sources.

7 CHLORINATION OF WASTEWATER

INTRODUCTION

Uses of Chlorine

The use of chlorine in wastewater treatment falls into the following categories: (1) odor control in wastewater and foul air, (2) prevention of septicity, (3) control of activated sludge bulking, (4) cyanide destruction, and (5) disinfection. Each of these categories will be discussed in detail.

Historical Background

The use of chlorine has generally been associated with the search for a means to control disease in humans. Until the late nineteenth century the idea persisted that disease was spread by odors, and that therefore the control of odors would stop the spread of disease. Hence it is not surprising that chlorine was used as a deodorant long before its value as a germicide was recognized. Although bacteria was discovered about 1680, it was not until about 1880 that investigations revealed that certain bacteria—now described as pathogens—caused specific diseases.

It was subsequently discovered that small concentrations of various chlorine compounds could destroy these organisms. The earliest recorded practice of sewage chlorination on a large scale was in 1854 when the Royal Sewage Commission used chloride of lime to deodorize London sewage. During the next several decades many British patents were issued to control the use of chlorine and chlorine compounds for treating sewage. In 1859 an investigation for the Metropolitan Board of Works, London,[1] by Hofman and Frankland showed that a dosage of 400 lb of chlorinated lime per million gallons (15 mg/L) could delay putrefaction of the raw sewage for four days.

The first known application of chlorine for disinfection was described by William Soper of England in 1879, when he reported the use of chlorinated lime to treat the feces of typhoid patients before disposal into a sewer. The first use of chlorine on a plant scale for disinfection of sewage was made at Hamburg, Germany in 1893. This application was the result of a disastrous waterborne typhoid epidemic there. The first recorded use for this purpose in the United States was in 1894 at Brewster, New York,[2] a small village in the Croton drainage area of the New York City water supply. The chlorine was produced by the Woolf process. This installation was unique in that it made chlorine on the spot by electrolytic decomposition of a brine solution and discharged the resultant hypochlorite solution into the sewage. This plant operated successfully until it was destroyed by fire in 1911.

Meanwhile, the results reported by British and German investigators using chlorinated lime were confirmed by studies made in the United States by Phelps and Carpenter in 1906–7 at the Massachusetts Institute of Technology.[3] Plant-scale studies by Phelps and others at Red Bank, New Jersey, Baltimore, and Boston in 1907–8 marked the beginning of effective chlorination practice in the United States. Clark and Gage[4,5] established many of the basic principles of chlorination at the Lawrence Experiment Station prior to 1911. The practical value of chlorinating raw, septic, or treated sewage with chlorinated lime was soon confirmed by studies at installations in Philadelphia, Chicago, Providence, and elsewhere. By 1911, eight sewage treatment plants in New Jersey were reported to be using chlorinated lime to protect water supplies, shellfish areas, and bathing beaches.

Until the development of a suitable gas chlorinator in 1913, the adoption of chlorination for disinfection and odor control of sewage was very slow because the chlorinated lime was found to deteriorate in storage and was relatively expensive, messy to handle, and awkward to apply.

Liquid chlorine became available as a commercial product in 1909 at Niagara Falls. The first chlorinator for metering and applying chlorine gas was developed by George Ornstein of the Electro Bleach Gas Company in 1912. In 1913, Wallace & Tiernan marketed the first successful gas chlorinator based upon the Ornstein patent, which revived interest in sewage chlorination. Among the first municipalities to employ liquid chlorine for the disinfection of sewage effluents were Altoona, Pennsylvania, and Milwaukee, Wisconsin, in 1914, followed by El Dorado, Kansas, in 1915 and Philadelphia in 1916.

The use of chlorination in wastewater processes has grown tremendously over the years since the development of suitable equipment. In 1958, it was reported that over 2200 plants serving a population of almost 38 million were equipped with chlorination facilities.[6] This represented about 30 percent of all treatment plants in the United States and about 50 percent of the population served by treatment facilities.

Contemporary Chlorination Practices

Chlorination is now established as an integral part of wastewater treatment practice in the United States and Canada. Several hundred technical articles have been written on the subject. It is the consensus that the primary use of chlorine is for disinfection. As early as 1945, Enslow and Symons[7] stated that by definition there are only three processes of sewage treatment: primary treatment, secondary treatment, and disinfection. In recent years tertiary treatment has become an additional process. Sludge disposal is relegated to being a by-product of the process.

Active interest in wastewater disinfection began in the United States about 1945. Up to that time the primary use of chlorine in sewage disposal systems was for odor control, hydrogen sulfide destruction, and prevention of septicity. Most of the sewage treatment plants practicing disinfection during that time belonged to the U.S. Armed Forces. It was military policy during World War II that sewage effluents at all Army bases in the United States had to be chlorinated. Today, as a result of the 1970 Federal Water Pollution Control Act, almost all wastewater treatment plants are subjected to some disinfection requirement.

In 1961 the first chlorine residual controlled disinfection system for wastewater was installed at Napa, California. In 1975 chlorine was first used on a plant scale for nitrogen removal from wastewater. Both of these applications proved successful.

The uses of chlorine in wastewater treatment practice may be summarized as follows: disinfecting; controlling odor and preventing septicity; improving grease and scum removal; preventing filter ponding; controlling flies; controlling activated sludge bulking; controlling odor in the sludge-thickening process; controlling waste-activated sludge disposal; controlling foaming; destroying cyanides; destroying phenols; and foul air scrubbing.

Chemistry of Chlorine and Wastewater

The chemistry of the chlorination of potable water, wastewater, and industrial waste is fundamentally the same for each process. The reactions differ only because of the differences in species and amounts of interfering substances, both organic and inorganic. The interfering substances are those that either contribute to excessive consumption of chlorine or impair the bactericidal efficiency of the chlorine residual.

The most important of these interfering substances found in wastewater discharges are: ammonia nitrogen, organic nitrogen, hydrogen sulfide, tannins, cystine, uric acid, humic acid, pickle liquor, cyanides, and phenol.

All of these compounds have been at one time or another isolated and studied for their effect upon the application of chlorine.

Ammonia Nitrogen. With the exception of highly nitrified effluents there is usually an appreciable amount of ammonia nitrogen in all wastewater effluents. The range is on the order of 10 to 40 mg/L. The ammonium ion exists in equilibrium with undissociated ammonia and hydrogen ion, and the distribution is dependent upon pH and temperature. The relative distribution can be defined as follows:

$$NH_4^+ \leftrightharpoons NH_3 + H^+, \text{ where } K = 5 \times 10^{-10} \text{ at } 20^\circ\text{C} \tag{7-1}$$

where

NH_3 = undissociated ammonia
NH_4^+ = ammonium ion
H^+ = hydrogen ion
K = dissociation constant

According to the dissociation constant, the pH value at which undissociated ammonia and ammonium ion are present in equal proportions (pK) is about pH 9.3 at 20°C. Above pH 9.3 undissociated ammonia (NH_3) predominates; below pH 9.3 ammonium ion (NH_4^+) predominates.

At usual wastewater pH levels the predominant chlorine reaction proceeds as follows:

$$HOCl + NH_4^+ \leftrightarrows NH_2Cl + H_2O + H^+ \tag{7-2}$$

when the chlorine-to-ammonia nitrogen weight ratio is less than 5:1. If the pH drops below 7, dichloramine ($NHCl_2$) will begin to form, and at a much lower pH nitrogen trichloride will form. However, as soon as the 5:1 weight ratio of chlorine to ammonia nitrogen is exceeded, a new set of reactions takes place, as described in Chapter 4.

The ammonia–chlorine reactions in wastewater follow the same pathways as in potable water. This is thoroughly discussed in Chapter 4. It is abundantly clear from the vast numbers of observations of chlorine in wastewater that the speed of reaction is fastest with inorganic compounds. These reactions seem to take precedence over the penetration of bacteria.

The speed of the chlorine–ammonia reaction in wastewater is grossly pH-dependent. For example, at 25°C and with a molar ratio of 0.2×10^{-3} mol/L HOCl and 1.0×10^{-3} mol/L NH_3–N, there will be a 99 percent conversion to monochloramine in 0.2 sec at pH 7 and 421 sec at pH 2. The most rapid reaction occurs at pH 8.3 (0.069 sec). This is a significant factor in the location of the injectors with respect to the diffusers when using plant effluent for injector operating water. If the travel time of the chlorine solution is more than 3 min., then about 10 percent of the chlorine feed rate will be consumed in the breakpoint reaction during the travel time to the diffusers. Therefore, the injectors should always be as close to the diffusers as possible.

However, because the chlorination process operates in the neutral pH range, the speed of the chlorine–ammonia reaction to monochloramine is finite but almost instantaneous and takes precedence over bacteria penetration, provided there is proper mixing at the point of application.[8] It can be computed that, in a wastewater containing 30 mg/L ammonia, the molecules of ammonia will outnumber the organisms by a factor of about 10^{13}. This may be a significant factor in the kinetics of wastewater disinfection.

Organic Nitrogen. Although the reactions between chlorine and ammonia nitrogen can be predicted with some certainty, the reactions between chlorine and compounds containing organic nitrogen were relatively obscure until the predictable reaction was discovered when San Jose/Santa Clara, California, started nitrifying its filtered secondary effluent in about 1981. This phenomenon is explained in detail in Chapter 8. Organic nitrogen compounds are present in all wastewaters containing domestic sewage. These treated effluents discharge significant amounts of proteinacious matter, amino acids, and the various nitrogenous compounds of urine. Studies of some of these compounds have revealed that they are stable over a long period of time, even in the presence of free chlorine.[9] These chlorinated organic nitrogen compounds are now classified as nongermicidal organochloramines. In the procedures for chlorine residual determination, these compounds show up in the dichloramine fraction of the complete forward-titration procedure of either the amperometric procedure or the DPD-FAS method. They cannot be identified by a chlorine residual analyzer measuring total residual. One investigator identified creatinine, a compound of urea, as responsible for the long-lasting ineffective "dichloramine" residuals in swimming pools,[10] and Williams confirmed this at the water treatment plant at Brantford, Ontario. Other investigations have shown that cystine and tannins have the most inhibiting effect upon wastewater disinfection, closely followed by uric acid and humic acid[11] and some of the amino acids.[12] So far as is known, the organochloramines do not enter into any reactions with inorganic compounds present in the wastewater.

Tannins. These compounds of organic nitrogen are used in leather-processing works. They exert an extremely high chlorine demand. In one instance, slugs of tanning wastes raised the chlorine demand of a typical primary-treated domestic waste from 12 mg/L to over 60 mg/L. The situation was largely corrected by distributing the waste from the tanning operation over a longer period of time, thus taking advantage of a dilution factor. Tannins should be treated at the source before entry into the collection system.

Sulfites. Inorganic chemicals containing the sulfite ion are used in leather-processing and wine-making. Sulfites will promote the generation of hydrogen sulfide in any sewage collection system, particularly if the wastewater temperature rises above 19°C. Sulfites are easily neutralized by chlorine. This is the reverse of dechlorination. Sulfites should be neutralized at the source.

Pickle Liquor. Pickle liquor from steel-mill operations in significant amounts will definitely affect treatment plant chlorination procedures. This substance ($FeSO_4 \cdot 7H_2O$) is the result of pickling steel products to remove the scale. Ferrous sulfate combines with chlorine to form ferric chloride. Each 7.8 mg/L of full-strength pickle liquor will consume approximately 1.0 mg/L of chlorine, depending upon the strength of the former. This reaction, if enough ferric chloride is formed, will have a decided beneficial reaction at any plant practicing prechlorination. The ferric chloride hydrolyzes to ferric hydroxide, a fine coagulant at the pH range found in most wastewater processes. However, if only effluent chlorination is practiced, the floc formed in the chlorine contact chamber or outfall line makes the effluent turbid and of a reddish brown color.

Phenols. Phenols are present in many chemical plant effluents, and will exert a definite additional chlorine demand in wastewater processing. Care must be taken to destroy the phenols before they reach the receiving water because they contribute to extremely obnoxious taste and odor problems in water treatment processes. Theoretically, 10 mg/L chlorine is required to destroy each mg/L of phenol. However, owing to the presence of other compounds usually associated with phenols, this figure can soar to 20 mg/L of chlorine.

Chlorine Demand. The only other specific parameter to cite in connection with the application of chlorine is its relation to the strength of the waste to be treated. It can be said with assurance that the chlorine consumption will increase as the BOD of the waste increases, and will vary as the BOD varies. This is the reason why most domestic wastes have a diurnal variation of chlorine demand from as low as 2:1 up to a high of 5:1. The latter is a result of the effect of infiltration in the collection system, so that the treatment process may be handling practically all runoff water between 2 A.M. and 6 A.M.

A rule-of-thumb estimate of 15-min. chlorine demand for various wastewaters is as follows:

Raw fresh domestic waste	12–15 mg/L
Raw septic domestic waste	15–40 mg/L
Primary effluent	12–16 mg/L
Biofilter effluent (secondary)	4–8 mg/L
Trickling filter effluent	4–10 mg/L
Well oxidized secondary effluent	3–8 mg/L
Multimedia filter effluent	3–6 mg/L
Slow sand filter effluent	2–4 mg/L
Nitrified filtered effluent	2–10 mg/L
Septic tank effluent	30–45 mg/L

These figures can be used to calculate chlorinator capacity for disinfection. As these are 15 min. contact time demands, make the following adjustment to arrive at chlorine dosage: Assume total contact time equal to 45 min. Add 1.5–2.0 mg/L die-away in the contact chamber, and then add the required residual calculated from the Selleck–Collins model.

The figures as shown for raw sewage are adequate to size chlorination equipment for odor control or septicity control.

THE ODOR PROBLEM

Significance

If the combined efforts of both the designers and the operators of a wastewater collection and treatment system result in obnoxious odors, the public will consider the project a complete failure. There is probably no other type of process in which odor is used as quickly by the general public to measure success or failure. It does not matter if the process produces an effluent that meets the most stringent requirements of the local regulatory body. It must not emit any objectionable odors.

Sources of Odors

Odors are the result of putrefaction of the solids in wastewaters. The amount of these solids is on the order of 0.08 percent by weight, and the suspended solids are present either in true colloidal suspension or in solution. Removal of the suspended matter in the waste is of vital importance to the operator. Proteins make up a large percentage of the organic suspended matter, having complex structures containing four elements in varying proportions: carbon, nitrogen, oxygen, and hydrogen. One variety contains a small amount of phosphorus. Proteins may be classed as colloids, and thus contribute to the colloidal nature of domestic wastes. When the oxygen supply in the wastewater is depleted, the microorganisms, enzymes, and inorganic compounds present cause putrefaction to start. Putrefaction can produce a variety of odoriferous compounds—indol, skatole, leucine, tyrosine—all derivatives of amino acids, which are products of the decomposition of proteins.[13]

Many distinctive and unpleasant odors are caused by bacteria and enzymes acting on proteinaceous matter. Ammonia, hydrogen sulfide, volatile fatty acids, and mercaptans are also products of putrefaction. Fats, another class of compounds that give off offensive odors, contain but three elements: carbon, hydrogen, and oxygen. They putrefy in two ways: by oxidation, to give a tallowy odor; or hydrolytically, so that the fats break down to fatty acids of a volatile nature. These acids are usually referred to as the "goat acids," the name suggesting the type of odor to expect.

Odor-producing substances, such as amino acids, mercaptans, and goat acids, usually persist only within the confines of the treatment plant. These odors can be greatly reduced by the liberal use of landscaping with shrubs and flowers, which have a great capability of both masking and absorbing odors. However, experience has shown that when hydrogen sulfide is controlled, other obnoxious odors are also controlled. The other compounds are difficult to detect and analyze, and so it is assumed that the detection and measurement of hydrogen sulfide will indicate when putrefaction may be occurring and consequently when conditions in the wastewater could produce foul odors.

Identifying the Problem

The easiest odor-producing compound to identify and control is hydrogen sulfide. It is generally assumed that when hydrogen sulfide odors are eliminated, the odors from other compounds (cadaverine, indole, mercaptans, and skatole) will also be eliminated or greatly suppressed.

Therefore, evaluation of H_2S emmissions will identify the problem. This can be done by two methods.

Lead Acetate Impregnated Tiles. This method is generally preferred for the confines of the treatment plant. These tiles are placed to form a network around the plant or treatment units as described by Chanin et al.[14] The tiles are rated on a daily basis and evaluated weekly as shown in Table 7-1.

Air-Sampling H_2S Detectors. A variety of these devices are available, which are generally of three different types: (1) A lead acetate impregnated tape type draws a continuous air stream to the tape. The tape changes color with H_2S concentration. The light transmissibility from the color spot on the tape is automatically recorded as H_2S concentrations. Examples of these units are those made by Arizona Instrument Co., Research Appliance Corp., and Houston Atlas, Inc. (2) There is a wet chemistry automatic coulometric type made by ITT Barton. (3) Solid state electrolytic cell sensors are offered by Texas Analytical Controls and Bio Marine. Many other devices can be found in the publications *Water & Waste Digest* and *Pollution Control News.*

The selection of these units for monitoring and/or control of a treatment process should be based primarily upon the effective detection range and the probable maintenance time required for proper operation. For odor control and monitoring purposes in wastewater treatment the H_2S air sampling unit should be capable of detecting concentrations as low as 1 mg/L. The lead acetate tape units have proved successful with respect to detection requirements and maintenance time.

One additional piece of instrumentation necessary to solve odor monitoring is a wind direction and velocity recording unit. This unit provides the information necessary to establish the validity of odor complaints.

Hydrogen Sulfide Characteristics

Hydrogen sulfide is a lethal and obnoxious gas generated in wastewater systems and caused by putrefaction. It is the main cause of sewage odors and is detectable in very low concentrations (<1.0 mg/L). It has these characteristics:

1. It is a deadly poisonous gas, and has resulted in many deaths to operating personnel. Many sewer maintenance workers have lost their lives because

Table 7-1 Lead Acetate Hydrogen Sulfide Detector Tiles—Numerical Rating System

	Observed Color								
	White or Yellow	*Very Light Brown*		*Light Brown*		*Brown*		*Dark Brown or Black*	
Starting Color	*24 or 48 hr*	*24 hr*	*48 hr*	*24 hr*	*48 hr*	*24 hr*	*48 hr*	*24 hr*	*48 hr*
White or yellow	0	1	½	2	1	3	2	4	3
Very light brown	—	½	¼	1	½	2½	1½	3½	2½
Light brown	—	—	—	¼	¼	2	1	3	2
Brown	—	—	—	—	—	—	—	2	1

Plot on daily basis number corresponding to color change at each station.

Also: Change Per 7 Days $= \frac{7}{\text{Total Days}} \times \Sigma$ (Color Change Numbers For Period)

A weekly number of greater than about 3.5 would indicate an odor problem.

of this gas. Death can occur within a few minutes' exposure to concentrations as low as 2000 mg/L by volume in the atmosphere.*

2. It gives off a "rotten egg" odor that is intolerable.
3. It combines with moisture in the humid air above the water line to form sulfuric acid, which is devastating to concrete structures.[16]
4. It is explosive at a concentration of 4.3 percent by volume in the atmosphere. H_2S is entirely different from other odor-producing compounds. It will not diffuse in air to any appreciable extent, and so it persists for great distances and tends to flow in a laminar configuration. Its odors have been known to be obnoxious at distances up to six miles from the point of origin, and in a band no wider than 100–200 ft, with a concentration of dissolved sulfides at the point of origin no greater than 10–12 mg/L.[17]
5. It will, over a short period of time, paralyze the olfactory nerve to the extent that an operator working in the area of a nontoxic concentration loses the ability to detect H_2S by odor. This is one of the reasons it is so dangerous in a poorly ventilated area.

At 5 mg/L concentration in air (by volume) the odor of H_2S is moderate and easily detectable. Concentrations of 2 ppm are considered inoffensive and are not hazardous. This is the optimum goal for the control of hydrogen sulfide in wastewater practice. Other effects are as follows:

10 mg/L: eye irritation starts.
30 mg/L: strong unpleasant odor of rotten eggs.
1100 mg/L: coughing; loss of smell in 2–15 min.
200–300 mg/L: red eyes; rapid loss of smell; breathing irritation.
500–700 mg/L: unconsciousness and possibly death in 30–60 min.
700–1,000 mg/L: rapid unconsciousness; breathing stops.
1,000–2,000 mg/L: instant unconsciousness; death in a few minutes.
43,000 mg/L: lower explosive limit.

Source of Hydrogen Sulfide in Wastewater

Of all the compounds present in wastewater, those containing sulfur have the greatest potential for generating hydrogen sulfide. The putrefaction of organic matter causes the generation of hydrogen sulfide.

The sulfate ion ($SO_4^=$) is the principal sulfur compound in wastewater. When organic matter is present and oxygen is absent, bacteria of the species Desulfovibrio desulfuricans will reduce sulfate to sulfide, using oxygen to

*A specific antidote for the toxicity of H_2S was reported by the Esso Medical Center, Fawley, Southhampton, U.K. An ampoule of amyl nitrite is broken into a cloth and the casualty breathes the vapor. When medical aid is available, 10 ml of 3 percent sodium nitrite is injected intravenously over 2–3 min. If symptoms recur, a further 5 ml may be given.[15]

oxidize organic matter. Letting C represent organic matter, the reaction may be written as follows:

$$SO_4^{=} + 2C + 2H_2O \xrightarrow{\text{bacteria}} 2HCO_3^{-} + H_2S \qquad (7\text{-}3)$$

In this reaction 96 grams of sulfate will make available 64 grams of oxygen, leaving 32 grams of sulfide. Using an equation showing an average formula for the reacting organic matter would probably indicate that about 42 grams of organic matter would be oxidized. Sulfite, thiosulfate, free sulfur, and other inorganic sulfur compounds sometimes found in wastewater can be similarly reduced to sulfide. The sulfur-bearing organic compounds of the mercaptan group contribute significantly to the odor problem.

Effect of Sulfate Concentration. Most of the sulfate in sewage comes from the water supply. Sulfur in proteins is metabolized by the body and excreted as sulfate. This amounts to only 10–20 mg/L of sulfate added by the sewage. The rate of sulfide generation is affected very little by sulfate concentration if some is present. Variations of sulfate concentrations in the water supply of a given area have no significant effect. Pomeroy reported that in one sewer line the sulfate concentration was reduced to 1 mg/L before there was any significant slowing of sulfide generation.[18] The total amount of sulfide produced in a sewer is not greatly affected by sulfate concentrations as long as there is a substantial surplus.

Other Sulfide Compounds in Wastewater. There are several forms of sulfides found in wastewaters that contribute to the generation of hydrogen sulfide. This mixture is made up of insoluble metallic sulfides, dissolved sulfides, and the secondary sulfide ion, $S^{=}$. (There are also organic sulfur compounds in which the sulfur is arbitrarily assigned a valence of minus two; these are called "organic sulfides." They do not respond to analytical tests used to measure inorganic sulfide, and they do not have the same significance.)

There are also volatile organic sulfur compounds that are very important odor components in wastewater. There are three principal types: mercaptans (thiols), thioethers (sulfides), and disulfides.

There are also nonvolatile sulfur compounds that do not cause the problems associated with the volatile compounds unless they are broken down by biological action to yield inorganic sulfide. These naturally occurring compounds in wastewater are principally the albuminoid proteins.

Hydrogen Sulfide Chemistry

General Discussion. This gas is highly soluble in water: 3618 mg/L at 20°C and one atmosphere. In aqueous solutions it hydrolyzes as follows:

$$H_2S \leftrightharpoons HS^- + H^+ \tag{7-4}$$

The hydrosulfide ion (HS^-) further dissociates as follows:*

$$HS^- \leftrightharpoons S^= + H^+ \tag{7-5}$$

AT 18°C the hydrolysis constant for Eq. (7-4) is:

$$K_h = 9.1 \times 10^{-8} = \frac{([^+][HS^-]}{[H_2S]} \tag{7-6}$$

Figure 7-1 illustrates the distribution of H_2S, HS^-, and $S^=$ for various pH levels. At pH 7 hydrogen sulfide is approximately 50 percent of the total dissolved sulfides; at pH 5 it is practically 100 percent of the total; at pH 9 dissolved sulfides are nearly all sulfide ion. Therefore, the existence of hydrogen sulfide in wastewater is pH-dependent.

Whenever the equilibrium between the hydrogen sulfide ion and the hydrogen sulfide in solution is upset, as when H_2S is removed by oxidation (Cl_2) or air stripping, the shift will be to form more H_2S from the remaining dissolved sulfides. This shift reestablishes the equilibrium of Eq. (7-4). This stored sulfur gradually disappears by metabolic action, being itself oxidized to sulfate to yield more energy.[19, 20]

The chemistry of the oxidation and removal of hydrogen sulfide is extremely complex. Supposedly the oxidation proceeds to form either elemental sulfur, sulfate, or both. These reactions are described in the section "Chlorine for Odor Control" in this chapter.

Categories of Sulfides Important to H_2S Generation. These categories are as follows:[21]

1. *Total sulfides* include *dissolved* H_2S and HS^- plus the *acid-soluble* metallic sulfides present in the suspended matter. The secondary sulfide ion $S^=$ mentioned above is negligible. Coopper and silver sulfides are so insoluble they can be ignored.
2. *Dissolved sulfides* is that portion remaining after the suspended solids have been removed by flocculation and settling.
3. *Un-ionized hydrogen sulfide* may be calculated from the concentration of dissolved sulfide, the pH of the sample, and the practical ionization constant of hydrogen sulfide. (See *Standard Methods,* 10th Edition, p 4–129, 4500–5^2)

*and for Eq. (7-5):

$$K_n = 1.2 \times 10^{-15} \frac{[H^+][S^=]}{[HS^-]} \tag{7-7}$$

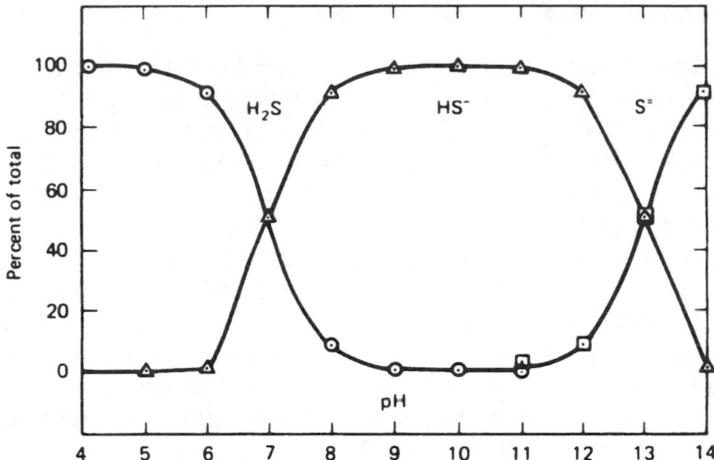

Fig. 7-1. Effect of pH on hydrogen sulfide–sulfide equilibrium.

There are several methods outlined in *Standard Methods* for finding the concentration of total and dissolved sulfides. It is important to remember the distinction between these two categories of sulfides in order to understand the problem of odor control. As a general rule, those who live with the problem of H_2S generation in a sewage system run total sulfide determinations first because the procedure is faster. If this concentration is 0.4 mg/L or greater, then a dissolved sulfide test is performed because 0.3 mg/L dissolved sulfides is considered the tolerable limit. Concentrations higher than this will cause odor problems. Below this limit the evolution of H_2S from the wastewater will not cause an odor problem. Experience has shown that in areas of high fluid turbulence the concentration of H_2S in the surrounding atmosphere will be less than 2 mg/L. At this concetration there will not be an odor problem.

Hydrogen Sulfide Generation

Causes and Occurrence. In general, it can be said that hydrogen sulfide generation is a result of septic conditions in the wastewater. The onset of this septicity occurs when the oxygen in the sewage or in the enveloping atmosphere is depleted.[22–24]

Factors contributing to the generation of H_2S are as follows: high BOD in the wastewater; high sulfates in the waste (sulfate content of the contributing water supply is not a factor[23]); high sewage temperatures; slugglish and stagnant flow conditions; lack of air cover in force mains; sludge deposits.

Hydrogen sulfide is generated in both free-flowing sewers and force mains. It always forms in a force main, provided that the detention is in excess of 20 or 30 min., but it does not necessarily form in free-flowing sewers.

Free-Flowing Sewers

General Discussion. Pomeroy and Bowlus[23] summarize the fundamental concept of sulfide generation in sewers as follows: "In free-flowing sewers, sulfides are produced only by slimes on the submerged surface of the sewer and by deposited sludge. In the flowing body of the sewage, sulfides are not generated, but on the contrary, are destroyed by oxygen which is continually being absorbed from the surface."

Sewage flowing in a conduit showing a rapid increase in sulfide concentrations will not necessarily show the same increase when a sample is enclosed in a bottle, because the H_2S generation is produced by sludge deposits in the conduit. Figure 7-2 illustrates the processes occurring in free-flowing sewers under sulfide buildup conditions. To prevent sulfide access to the flowing stream, the oxygen concentration in the stream is critical. The required range

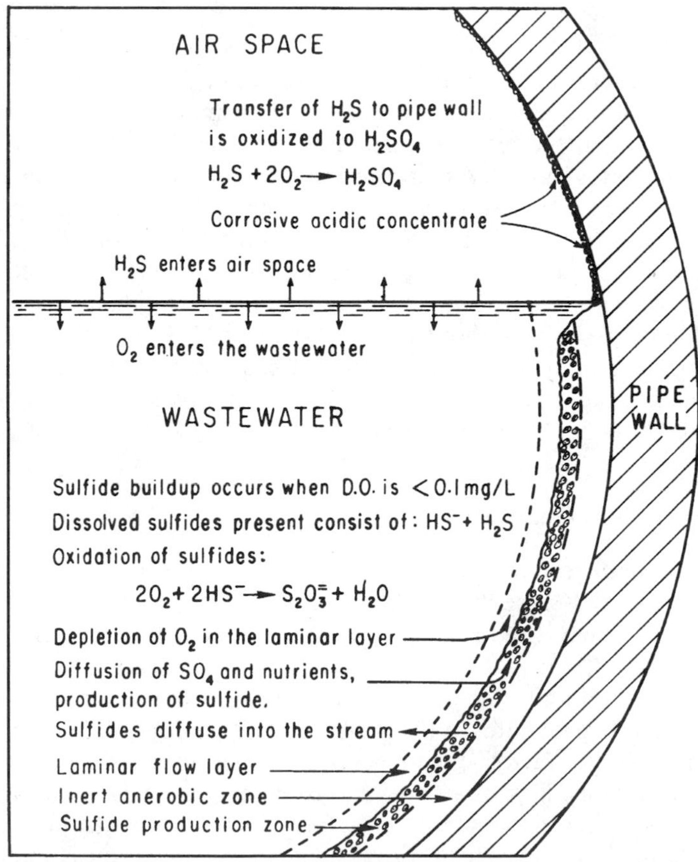

Fig. 7-2. Processes occurring in free-flowing sewers under sulfide buildup conditions.

is between 0.1 and 1.0 mg/L oxygen.[21] Before all the sulfides produced can pass into the stream, conditions must be completely anaerobic.

The rate of sulfide generation is roughly proportional to the BOD, which is a measure of the oxygen utilized by microorganisms in oxidizing the sewage to stable end products (CO_2, H_2O, NO_3^-). As this is a biological phenomenon, it is markedly affected by temperature. Pomeroy combines these two factors into a single factor called "effective BOD." Using 20°C as a standard temperature, and with the biological activity increasing 7 percent for each degree C rise (geometrically), the effective BOD may be expressed as follows:

$$\text{Effective BOD } (E_{\text{BOD}}) = \text{Standard BOD} \times (1.07)^{t-20} \qquad (7\text{-}8)$$

Pomeroy also makes the observation that, for a specified temperature and flow condition, there is a limiting sewage strength below which a buildup of sulfide will not occur. In free-flowing sewers, the BOD would have to be greater than 50 mg/L to generate sulfide. Below this figure, it is doubtful that sulfides would be generated.

As sulfide-producing microorganisms can function over a wide range of pH, the latter is not generally a factor. However, raising the pH significantly by adding caustic or lime is known to greatly diminish the generation of sulfides.

The generation of sulfide in the slime layer of Fig. 7-2 will also be dependent upon the ratio of sulfates to organic matter at a ratio of about 2.3 to 1. One will be used up before the other, and the relatively scarce one will then be the constituent that limits the sulfide production rate. If there is an abundant supply of organic matter, but sulfate is scarce, then sulfide generation will be proportional to sulfate concentration and independent of the amount of organic matter.

It was once thought, because of long detention times in the collection system, that the age of the sewage was responsible for sulfide generation. What matters is not the age but how rapidly the sewage flows. If the velocity is adequate, a considerable degree of purification takes place in long lines, thus preventing a sulfide buildup. At the same time, undue turbulence is to be avoided, as it releases the H_2S and may give rise to odor complaints at manhole locations. On the other hand, high velocities (5 ft/sec and higher) scour the slimes, preventing sulfide buildup. This scouring action is not noticeable until the velocity exceeds 3 ft/sec.[23] Pomeroy and Bowlus state: "For any particular temperature and sewage strength combination, there is a limiting flow velocity above which sulfide buildup will not occur." (This applies only to free-flowing sewers.)

Forecasting Sulfide Buildup. It is the consensus that as little as 0.2 mg/L DO will prevent H_2S buildup—the slime layer is the critical factor. It is desirable to keep this in mind when sulfide buildup is a factor.

The Sacramento Regional County Sanitation District consolidated 12 existing wastewater treatment plants into one major facility. This consolidation resulted in 27 miles of collecting sewers with travel times up to 24 hours before reaching the regional plant. This required a long and laborious study of the system in order to provide rational solutions to prevent hydrogen sulfide buildup.[25, 26]

In 1946, Pomeroy and Bowlus[23] developed the first equations relating to conditions necessary for H_2S generation in gravity sewers. In 1950, Davey[27] reported on the effect of velocity on H_2S generation. This work was later modified by Pomeroy into what is known as the *Z* formula:

$$Z = \left[\frac{E_{\mathrm{BOD}}}{S^{0.50}Q^{0.33}}\right]\frac{P}{b} \tag{7-9}$$

where

Z = function defined by the equation
$E_{\mathrm{BOD}} = (\mathrm{BOD})_5 \times (1.07)^{T-20}$ mg/l
T = temperature, °C
S = slope, ft/ft
Q = discharge, cfs
P = wetted perimeter, ft
b = surface width, ft

The value of *Z* obtained is interpreted as follows:

$Z < 5000$	sulfide is rarely produced
$5000 < Z < 10000$	conditions marginal for buildup
$Z > 10000$	sulfide buildup is common

This formula has been used successfully in predicting generation of sulfide in sewers but there is some uncertainty as to its reliability.[28]

Equations for Partially Filled Pipes. In 1977 Pomeroy and Parkhurst[29] presented a quantitative method for sulfide forecasting. The work by the Sacramento Area Consultants[15, 26, 30] in their study of the regional collection system concluded that the best predictive equations available were those developed by Pomeroy and Parkhurst in 1977.[29] The final report confirmed the accuracy of these equations,[30] which are as follows:

$$\frac{d[\mathrm{S}]}{dt} = \frac{3.28M'(E_{\mathrm{BOD}})}{r} - \frac{2.10m[\mathrm{S}](\mathrm{SV})^{0.375}}{d_m} \tag{7-10}$$

where

$\frac{d[\mathrm{S}]}{dt}$ = rate of change of total sulfide, mg/L-hr
M' = sulfide flux coefficient

M is empirically determined to suit conditions. It is usually about 0.4×10^{-3} m/hr when dissolved oxygen is low (<0.5 mg/L) and approaches zero as DO concentrations increase. Pomeroy suggests 0.32×10^{-3}.

E_{BOD} = 5 day BOD $\times$ $(1.07)^{T-20}$ mg/L
T = °C
r = hydraulic radius, ft
m (or N) = empirical coefficient for sulfide losses

Pomeroy indicates that m ranges from a conservative value of 0.64 to a less conservative value of 0.96. The Sacramento Calif. County study found the best fit was for m = 0.64. This was based upon field study results.[30]

[S] = total sulfide concentration, mg/L
S = slope, ft/ft
V = velocity, ft/sec
d_m = mean hydraulic depth, cross-section divided by surface width

The negative term of Eq. (7-10) is proportional to the sulfide concentration. When the sulfide losses equal the sulfides generated and the buildup is zero, the total sulfide concentration approaches a limit:

$$\frac{3.23M'(E_{BOD})}{r} = \frac{2.10m(SV)^{0.375}}{d_m} \times [S]_{lim} \tag{7-11}$$

Substituting P/b for d_m/r, where P is the wetted perimeter and b is the surface width of the stream, and using the conservative value of 0.64 for m, Eq. (7-11) becomes:

$$[S]_{lim} = 0.78 \times 10^{-3} \frac{E_{BOD}}{(SV)^{0.375}} \times \frac{P}{b} \tag{7-12}$$

To calculate the sulfide generation in a series of pipe sections with changing hydraulic conditions, the following relationship occurs:

$$t_2 - t_1 = \left(\frac{1.10d_m}{m(SV)_{0.375}}\right)\left(\log\frac{[S]_{lim} - [S]_1}{[S]_{lim} - [S]_2}\right) \tag{7-13}$$

where

$t_2 - t_1 = \Delta t$ = flow time (hours) in a given pipe reach with constant slope s, diameter, and flow.
1.10 = meters to feet conversion factor
$[S]_1$ = sulfide concentration at start of reach, mg/L
$[S]_2$ = sulfide concentration at end of reach, mg/L

$[S]_{lim}$ = limiting sulfide concentration from Eq. (7-12)
m = empirical coefficient (either 0.64 or 0.96).

By rearranging Eq. (7-13) we get the following expression for $[S]_2$:

$$[S]_2 = [S]_{lim} - \frac{[S]_{lim} - [S]_1}{\log^{-1}\left(\dfrac{m(SV)^{0.375}\,\Delta t}{1.10\,d_m}\right)} \tag{7-14}$$

and the value of $[S]_2$ for the first reach becomes $[S]_1$ for the second reach, and so on. By calculating $[S]_{lim}$ and t and knowing the hydraulic conditions for each reach of the system, the sulfide buildup can be calculated for a series of reaches.

Force Mains

General Discussion. It is well known that in very large pipes where the sewage has time to become thoroughly anaerobic, such as in long force mains, a substantial amount of sulfide can be generated in the stream as compared to the amount on the walls. If all the generation were in the stream, then the sulfide concentration discharging from the pipe would be proportional to the detention time but independent of the pipe diameter. This is by no means the case. Pomeroy found[23, 31] that a small pipe produces more sulfide in the same time than does a large pipe. A pipe 34″ in diameter produced 14 mg/L of sulfide per hour; a 96″ and 144″ complex produced 0.75 mg/L per hour. Pomeroy estimates that in a 12″ pipe about 11 percent of the sulfide is produced in the stream, whereas in a 48″ pipe about 54 percent is produced in the stream. These figures are only an approximation. Pomeroy developed a mathematical relationship so that the total sulfide production in force mains can be calculated.

Forecasting Sulfide Buildup. The calculation of S_5, which is the weight of sulfide produced in the sewage system, assuming that the amount is proportional to the volume of the pipe and the effective BOD, C_{EBOD}, can be written:[31]

$$S_5 = N\frac{\pi D^2 L}{4} C_{EBOD} \tag{7-15}$$

where

N = a coefficient, lb/day/cu ft/mg/L
D = pipe diameter in ft
L = length of filled pipe in ft

The calculation of S_w, the weight of sulfide produced on the wall, lb/day, may be written[31] for circular pipes:

$$S_w = M\pi DLC_{\text{EBOD}} \tag{7-16}$$

where

M = coefficient, lb/day/sq ft/mg/L

Adding Eqs. (7-15) and (7-16), the total sulfide production S_T may be expressed:

$$S_T = C_{\text{EBOD}}\pi DL\left(M + \frac{ND}{4}\right) \tag{7-17}$$

Substituting P for $N/4M$, Eq. (7-17) becomes

$$S_T = MC_{\text{EBOD}}\pi DL\ (1 + PD) \tag{7-18}$$

For calculating the increase in concentration of sulfide in sewage going through a force main, Pomeroy uses the following equation:

$$\Delta\ C_5 = \frac{44.6\ MtC_{\text{EBOD}}}{D}\ (1 + PD) \tag{7-19}$$

Converting pipe diameters to inches, Eq. (7-19) becomes:

$$\Delta\ C_5 = \frac{535\ MtC_{\text{EBOD}}}{d}\left(1 + \frac{Pd}{12}\right) \tag{7-20}$$

where C_5 is the concentration of sulfide in mg/L, and t is the time of passage in minutes. Based on data collected by Pomeroy, it appears that the value of P should be about 0.12. This corresponds to production of about 0.4 mg/L of sulfide per hour in the body of the stream. Substituting this value of P and using K in place of 535 M, Eq. (7-20) becomes:

$$\Delta\ C_5 = KtC_{\text{EBOD}}\ \frac{(1 + 0.01\ d)}{d} \tag{7-21}$$

From available field data, Pomeroy concluded that up to a time of 10 min. little sulfide is likely to be produced, and that K would be on the order of 0.0010 in Eq. (7-21). Between 10 and 60 min., K would be on the order of 0.0020. For detention times of one to five hours, K values would be above 0.0020. Pomeroy suggests $K = 0.0026$, which gives a value of M as 4.85×10^{-6} or say 5×10^{-6}. This he believes to be a conservative figure which will predict more sulfide than will probably be generated.

Chlorine for Odor Control

Historical Background. The widespread utilization of chlorine for odor control has been well documented.[13,22,24,32–39] The task has been accomplished both by up-sewer cholorination and by prechlorination. Probably the most extensive work on up-sewer chlorination has been done by the sanitation departments of Los Angeles and Orange counties in California, where it began in 1930 and 1931.[23]

These two areas, based upon their experiences of over 60 years, consider it far better to design a collection system that will avoid undue buildup of sulfides in the sewage rather than resort to chemical treatment.

Up-sewer chlorination is usually practical only at pumping stations where space can be provided for a safe facility. But these pumping stations are usually adjacent to residential areas, thereby presenting a hazard. Considerably less hazardous is the use of copperas for controlling sulfides. Copperas, however, will not control sulfides to less than about 1 mg/L, whereas complete destruction by chlorine is possible. Two possibilities are: purging the sewer walls of slimes by raising the pH with slugs of caustic, and injecting air into force mains. If the force main is not designed to handle air injection, this method may create prohibitive pressures from air accumulation. The disadvantages of chlorine is that although it will destroy sulfides, converting them to colloidal sulfur and sulfates, reasonable dosages may not prevent re-formation of sulfides before the sewage reaches the treatment works.[38] These instances indicate the necessity of relay chlorination, such as that practiced at various times by Los Angeles County.

Chemistry. In dilute aqueous solutions, chlorine reacts instantaneously with sulfides to form colloidal sulfur, Eq. (7-22), or sulfates, Eq. (7-23), depending upon the pH, temperature, and ratio of chlorine to sulfides present.[39, 42]

$$Cl_2 + H_2O \rightarrow HOCl + H_2S \rightarrow S^\circ \downarrow + HCl + H_2O \qquad (7\text{-}22)$$

This reaction requires 2.2 mg/L chlorine per mg/L of sulfide as hydrogen sulfide. A high pH is required for this reaction to go to completion. At pH 10, the amount oxidized to free sulfur is only slighty more than 50 percent.[42]

The addition of 8.87 mg/L chlorine to each part of sulfide theoretically will oxidize the sulfides to sulfates:

$$4H_2O + S^{=} + 4Cl_2 \rightarrow SO_4^{=} + 8HCl \qquad (7\text{-}23)$$

Nagano[43] was able to show by a series of tests made in 1950 that the chlorine–sulfide reaction most consistent with experimental evidence is as shown in Eq. (7-23)—that is, chlorine oxidizes sulfur to sulfates in the ratio of 8.87 : 1. Actually the chemistry of this reaction is so complex that, depending upon the pH and the temperature of the sewage, some free sulfur and polysulfides are probably formed along with the sulfates. Given sufficient time in an oxidizing environment, the polysulfides will become sulfates, but in a reducing environment they can easily revert to hydrogen sulfide.

Chlorine Requirement. If the wastewater to be treated for hydrogen sulfide control is available, the practical way to determine the optimum chlorine dosage is the chlorine demand test. Use the procedure given in *Standard Methods* and determine the chlorine consumed during a 5-min. contact period. The correct dosage will yield a measurable residual at the end of 5 min., which is sufficient time for all the chlorine–sulfide reactions to go to completion.

Prechlorination Facility. Every wastewater treatment plant should have provisions for prechlorination. This provision can be used as a standby system for disinfection and/or as a supplement for RAS (return activated sludge) chlorination. In the absence of chlorine demand tests, the equipment capacity should be at least 15 mg/L for peak dry weather flow, assuming a fresh domestic sewage. If the wastewater is to be a mixture of industrial wastes and septic domestic sewage, the optimum chlorine requirement might escalate to 30 or 40 mg/L.

It is not practical to control the chlorine dosage by the chlorine residual method. A chlorine residual is not necessary to control hydrogen sulfide generation. Moreover a chlorine residual analyzer cannot perform on raw sewage.

The ORP method (oxidation–reduction potential) has been tried in the past at many different plants with little success. However, the Stranco High Resolution Redox Control System was introduced after the publication of the third edition of this book, and it should be investigated carefully for the control of problems with the RAS, as this control system has been found to be ideal for both odor control and disinfection of wastewater. (See Chapter 8.) There are a great many WWTPs using the Stranco system.

Currently the preferred method with a good record is based upon the control of chlorination dosage, which will prevent generation of H_2S in an aerated sample of chlorinated wastewater. This is obviously a place for the use of ORP. The H_2S that breaks out of an aerated sample of sewage is directly proportional to the dissolved sulfides in the sewage. Proper chlorination control using this air sampling technique by ORP can be readily accomplished, provided that the sample is aerated properly. One example of adequate aeration is found in a typical sewage sampling device. The wastewater is pumped at a rate of 2–3 gpm to the sampling device, where it discharges to waste over a weir. The air suction cone of the H_2S detector is suspended over the weir so that any H_2S being aerated out of the sample flow will be trapped by the detector. One of the most successful installations of this kind utilized an automatic proportional sampler of the raw sewage being prechlorinated for odor control.[44] Chlorination control was maintained so that the *dissolved sulfides* were always slightly less than 0.3 mg/L, which has been established as the critical limit. The air entering the cone of the H_2S detector suction line is pumped by the detector through the analyzer, where it comes into contact with a lead acetate saturated tape. The H_2S in the air sample produces a black spot with a density proportional to the amount of H_2S in the atmosphere. The density of the spot is converted into an electric signal fed into a controller. The controller operates the chlorine metering orifice in accordance with a field-observed empirical setpoint. This system allows the operator to adjust the setpoint to meter the chlorine in accordance with some acceptable level of H_2S concentration in the atmosphere adjacent to the wastewater surface.

The chlorine control system must be equipped with a bias feature, which limits the minimum dosage of chlorine to approximately 8 mg/L. If this were not provided, there could be times when H_2S was not being generated, and the control system would eventually reduce the chlorine feed rate to near zero. Then with a sudden appearance of H_2S, the control system would be unable to react in time to prevent H_2S breakout in the treatment plant area. There are several ways to achieve this minimum dosage control.

It is wise to use two H_2S detectors: one to control the chlorine dosage and the other to monitor plant area odors. The latter analyzer should be a portable type. There are a variety of H_2S analyzer-detectors available for this type of system, such as those of the Research Appliance Corp. (RAC), Houston-Atlas Co., Texas Analytical Instruments, and others. Wet chemistry types of analyzers are not recommended.

The control analyzer should be fitted with a 0–3 mg/L H_2S range that can be converted to 0–10 mg/L in the field. The portable monitor analyzer should have the same capability. Both units should have built-in H_2S concentration recorders. This historical record is of vital importance to the operator.

Local conditions will determine the necessity of a flow pacing signal in addition to the H_2S signal. A flow signal is highly desirable and should be considered. It is well worth the expense even if the flow meter range is limited to 3 to 1. The flow must be measured near the plant influent. Practically all modern influent pumping stations are designed to use variable speed pumps, which follow the diurnal flow change of the collection system. However, if a step-rate pump system controls the influent flow, then a step-rate control system is an imperative for the chlorine feed rate control—in addition to the H_2S signal.

Automatic chlorine dosing for odor control using an H_2S analyzer to detect the odor potential of the sewage is the surest way to achieve successful control. The analyzer eliminates the need to rely upon human response as a detection device. It is well known that continuous exposure to H_2S in low concentrations will in a brief span of time paralyze the olfactory nerve, completely destroying the human response.

OTHER USES OF CHLORINE IN WASTEWATER

Preventing Septicity

General Discussion. One of the generally overlooked advantages of prechlorination is the ability of chlorine to prevent septic action by maintaining the waste in a fresh condition, in which state the various types of biologic treatment processes are most efficient. One important result of maintaining a fresh waste is improved clarification. The application of chlorine ahead of the primary clarifier not only improves the settling rate of normal sewage but eliminates septicity in settled sludge, thereby preventing it from rising in the clarifier. This has been reported by many plant operators.[41] One of the most dramatic cases of improved settling occurred when a small domestic waste treatment plant was nearly put out of operation by cannery waste,[45] which was about 40 percent of the total flow, consisting of peaches, pears, and tomatoes. The mixture of the cannery waste and domestic sewage became septic during the time it took to pass through the primary clarifier. Laboratory tests showed that either waste alone would not become septic.

The result of this septicity was a thick, floating scum on the clarifier, pigpen odors, and suspended solids removal of only 36 percent. It was found that the addition of chlorine prevented the septic action, which floated the solids to the surface of the clarifier. It was further found that the magnitude of chlorine dosage controlled the length of time that the solids would remain settled. By laboratory experiment with Imhoff cones, a settling period of 2 hours was considered adequate. This amounted to a chlorine dose of about 80 percent of the immediate chlorine demand. In practice, this proved sufficient. After the application of chlorine, the thick scum disappeared overnight, the pigpen odor was eliminated, and the suspended solids removal rose to 64 percent.

The comment has often been heard that prechlorination interferes with digester performance, but there is no scientific evidence to support this. From a chemical standpoint, it is highly unlikely because the chlorine would be converted to chloride long before it got to the digester, and the resulting chloride concentration would be so low as to be negligible.

The use of chlorine to prevent and/or control the septicity of sewage has applications other than the extreme case described above. Precisely how much chlorine is required to prevent septicity over a given time span can be determined by using the methylene blue stability test.[46] This can be used as described above for less severe cases of septicity in sedimentation tanks as well as long periphery trunk sewers carrying treated effluents from several plants to a common discharge point.

The problem of septicity is closely allied to the generation of hydrogen sulfide. However, septic conditions can occur in situations where hydrogen sulfide formation is not a factor. In the case where solids in a sedimentation basin float to the surface, this is caused by the formation of methane gas in the bottom of the sedimentation tank due to anaerobic conditions. Chlorine addition to relieve this type of condition raises the question of how much is sufficient.

Methylene blue becomes decolorized in the presence of anaerobic conditions caused by reducing bacteria, loss of oxygen, or the formation of hydrogen sulfide. Thus, so long as the blue color persists in a sample containing methylene blue, no septicity or sulfide odor will exist.

Methylene Blue Stability Test. The first step is to determine the 15-min. chlorine demand of the sewage to be examined for septicity. This will be a guide to finding the optimum dosage for a given situation. If we assume this chlorine consumption in 15 min. is 14 mg/L, then it would be appropriate to dose four samples at 3, 5, 10, and 15 mg/L, respectively, provided that preventing septicity for at least 3 hours is required.

Collect sewage samples in 250-ml glass-stoppered bottles so that no air space is left between the liquid and the stopper. Using the same chlorine solution as used in the chlorine demand test, dose the four 250-mL samples as described above. After a 5-sec vigorous mix of the chlorine solution, apply the stopper as directed above and store the samples in the dark at 25–30°C if possible. As a clean bottle does not resemble the biological conditions in a sewage system, this elevated temperature is an attempt to approach actual conditions.

At the end of one hour add 0.5 ml of a 0.10 percent solution of methylene blue to each 250-mL sample. Be sure to have an equivalent control sample without chlorine added. Observe these samples over a 6–8-hour period and record the status of the blue color in each of the samples as compared to the control. Do this at hourly intervals. If septic conditions develop, the MB color will fade. The optimum chlorine dose will be somewhere between the faded and the nonfaded sample. The time involved in these occurrences is the probable length of time that the "optimum" chlorine dose will keep the sewage fresh and thereby prevent septicity. When the MB color fades, that point is the onset of septicity. Chlorine can be applied to prevent septicity in raw sewage up to at least 24 hours.[47]

Sludge Bulking

Nature of the Problem. The consequences of sludge bulking in the activated sludge process can be disastrous. This phenomenon is responsible for many of the upsets that occur in the activated sludge process and can seriously impair the overall efficiency of the entire plant. Effluents discharged during serious bulking conditions have been known to cause pollution in the receiving waters worse than would have occurred if the plant had been completely bypassed. Therefore sludge bulking must be prevented at all costs.

Chlorine has been used for at least 40 years for control of sludge bulking problems.[41, 48] It is a reliable and economical method.

Activated sludge contains large populations for several types of microorganisms. There are two important morphological groups: (1) the floc-formers and (2) the filamentous sludge bulkers. The floc-forming group of bacteria include the following genera: *Zoogloea, Pseudomonas, Arthrobacter,* and *Alcaligenes.* The filamentous group (the causative organism of sludge bulking) includes members of the genera *Sphaerotilus, Thiothrix, Microthrix, Parvicella, Beggiatoa,* and others.*

Satisfactory operation of the activated sludge process depends upon controlling the level of the filamentous organism population.

---

*For illustrations of some of these filamentous organisms, see Chapter 6.

Identifying the Problem.[69, 93] There are two tests to perform in order to identify sludge bulking: (1) measurement of the sludge volume index (SVI) and (2) daily microscopic observations of the RAS organisms.

The most common causes of sludge bulking (poor settling) are: (1) low dissolved oxygen, (2) low food-to-microorganism ratio (F/M), and (3) nutrient deficiency (nitrogen or phosphorus or both). However, bulking is a physical phenomenon caused by filamentous bacteria.

Sludge Volume Index.[35] This is the volume in milliliters occupied by one gram of activated sludge after the aerated liquor has settled for 30 min. (See page 2–79 of *Standard Methods,* 19th edition.) The operating personnel should monitor both the DO in the aeration tanks and the depth of sludge in these tanks, and then calculate the sludge volume index (SVI) of the mixed liquor. When the SVI reaches 150, sludge bulking may begin. Sludges with good settling characteristics in diffused-air aeration systems operating with mixed-liquor suspended-solids concentrations of 800–3500 mg/L will show an SVI range from 35 to 150. Therefore with a fixed and limited capacity return-sludge pump (RAS recirculating pump), the sludge concentration that can be maintained in the mixed liquor (without escaping in the effluent) is reduced as the SVI increases. However, keeping the mixed liquor suspended solids low at high SVI is not important. What is important is the *clarifier flux,* that is, the total pounds per day of suspended solids per square foot of settling area in the clarifier. For example, doubling the MLTSS from 1100 to 2200 with an SVI of 200 would give the same performance if the surface area of the clarifier were also doubled. For acceptable clarifier operation the maximum solids flux goes down as the SVI goes up. Therefore, a rising value of SVI is indicative of trouble ahead, and prompt action should be taken to bring it under control. The first step in the control program is to set a target SVI. There is general agreement that it should be from 60 to 110, provided that the wastewater flow compares to the plant design flow.

Microscopic Observation. Satisfactory operation of the activated sludge process depends upon the population control of the filamentous organisms in the sludge. These organisms interfere with the growth and survival of the floc-formers upon which the process depends. Fortunately the filamentous organisms are more sensitive to chlorine than are the floc-formers and are therefore destroyed first. The floc-formers grow inside the floc, where they are protected from the chlorine by a diffusion barrier. Moreover, chemical reactions between chlorine and compounds inside the floc restrict the penetration of chlorine to the outside shell of the floc.

Figure 7-3 shows a typical healthy bloom of *Thiothrix* in the activated sludge floc. The important element of the photograph is the bridging of the filaments between the floc particles. This is without chlorination showing the onset of bulking.

Figure 7-4 shows the same activated sludge sample during chlorination. There are fewer filaments, and the bridging has disappeared. The empty sheaths and the missing cells are the effects of chlorination.

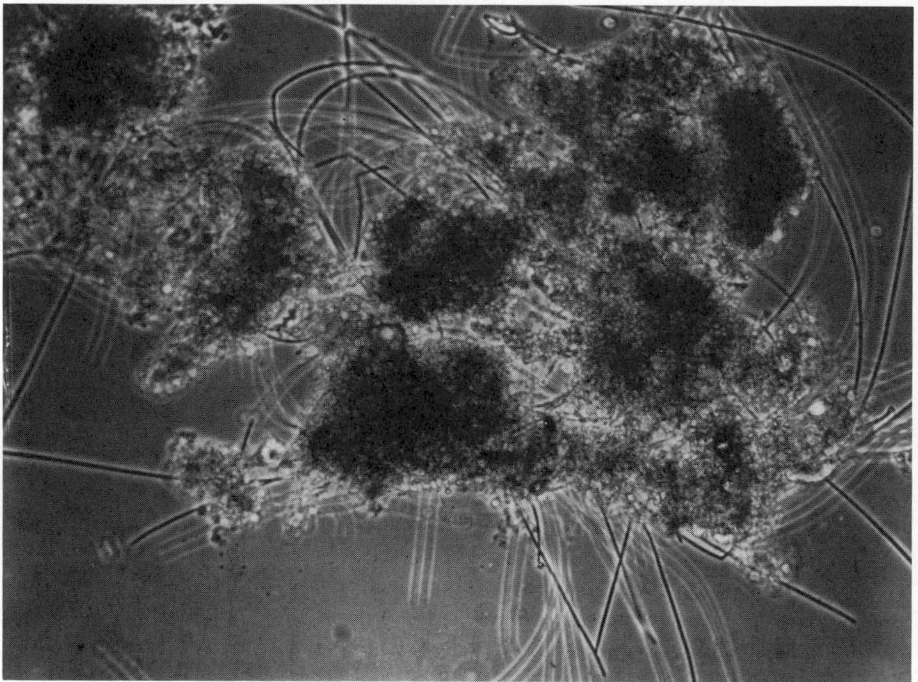

Fig. 7-3. Activated sludge during bulking conditions resulting from healthy filamentous *Thiothrix* (100 × mag.).

Figure 7-5 illustrates the final phase of bulking control by chlorination. At this point chlorination should be discontinued.

Research on this subject at the San Jose/Santa Clara Water Pollution Control Plant, San Jose, California, revealed that in addition to using a target SVI to regulate the chlorine dose to the RAS, observations of the number and condition of filamentous organisms extending from the activated sludge floc were most helpful.[69] The microbiology laboratory staff at the above plant used a simple but effective filament counting technique, as follows:[69, 94]

1. Transfer 50 microliters mixed liquor sample to a glass slide.
2. Cover the sample completely with a 22- by 30-mm cover slip.
3. Using 100× total magnification and starting at the edge of the cover slip, observe consecutive fields across the entire length of the cover slip. (At San Jose, this is 17 fields).
4. The eyepiece is fitted with a single hairline. Count the number of times that any filamentous organism intersects with the hairline.
5. Sum the number of intersections for all fields examined. This is the filament count.

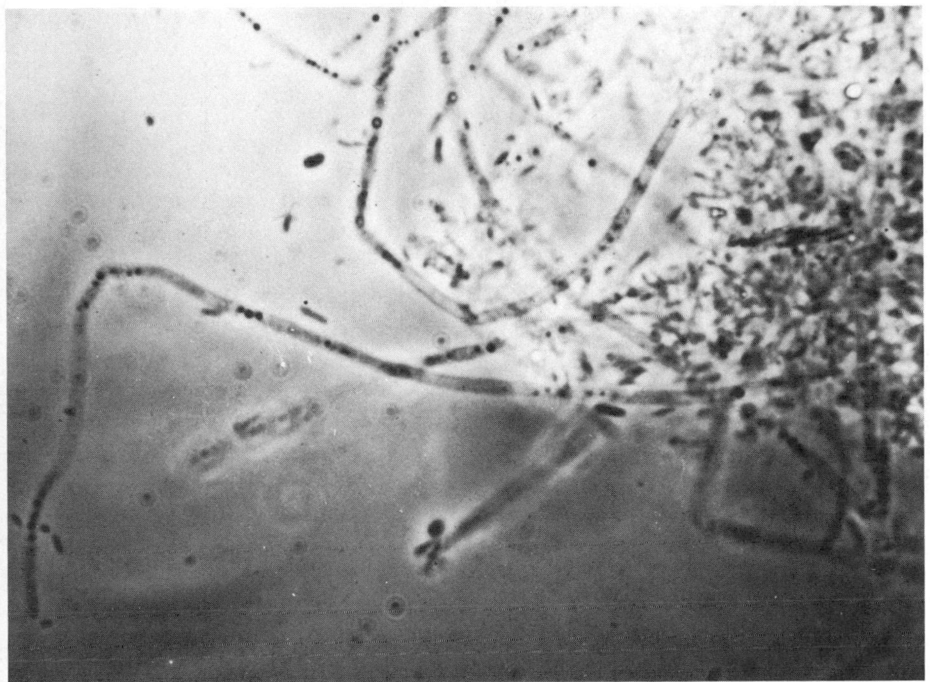

Fig. 7-4. Activated sludge bulking during chlorination. The *Thiothrix* are devoid of sulfur and the cells are deformed (100 × mag.).

Daily observations of the organisms in the activated sludge serve as a basis for checking the level of the filamentous organisms. This provides the information required to adjust the chlorine dosage or frequency of RAS passes by the point of chlorination.

Increases in extended filament length do not necessarily precede increases in SVI, nor does a decrease in filament length precede a decrease in SVI. Therefore, microscopic enumeration of filamentous bacteria does not allow prediction of changes in the settling characteristics of the sludge. However, microscopic observation of the chlorinated RAS does show the onset of the effectiveness of chlorination, which in turn establishes when to reduce the chlorine dose. The progressive effects of chlorine cause cells inside the filament sheaths to become deformed, to pull away from the sheath, and to lyse and create empty spaces in the sheaths. Finally, the sheaths become empty and broken. When the filamentous organism in *Thiothrix,* the early effect of chlorine is the disappearance of sulfur inclusions in the sheaths. In dealing with a sheathed bacterium such as *Thiothrix,* sludge settleability may be influenced by the empty sheaths even after the organism is no longer viable. Therefore, the reduction in SVI may continue for two or three days after chlorination has ceased because the sheaths continue to break down and are wasted from the sludge inventory.

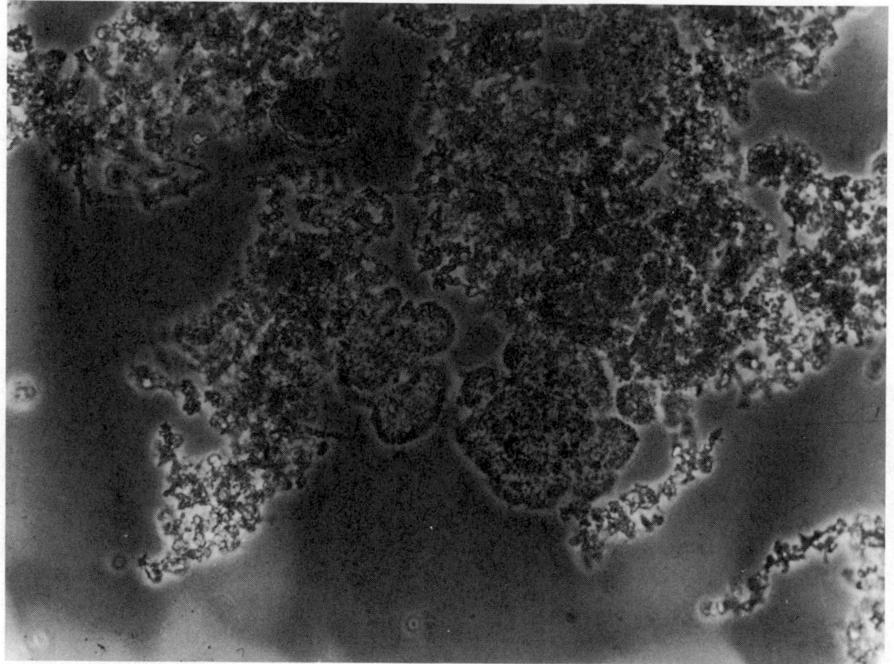

Fig. 7-5. Normal non-bulking activated sludge, showing floc-formers and absence of filamentous organisms (100 × mag.).

Nitrogen Deficiency. There are times when some plants are subjected to heavy seasonal cannery loads, during which time ammonia N is completely consumed by cell synthesis, and none remains to produce chloramines. During such an event nitrites are likely to appear in concentrations from 1 to 3 mg/L. Since there is no ammonia N to form chloramines, all of the chlorine will exist as HOCl (free chlorine). Free chlorine reacts immediately with the nitrites to oxidize them to nitrates. Chlorine is consumed in this reaction at a rate of 5 parts chlorine to 1 part NO_2^-–N (nitrite ion). Therefore, if RAS chlorination proves ineffective in the presence of good initial mixing, it would be prudent to check for both ammonia N and nitrite in the RAS. (Chloramines do not react with nitrites).

Another situation may occur during these periods of nitrogen deficiency. It is possible that both NH_3–N and NO_2^-–N may be absent, and RAS chlorination is ineffective. This would most likely be due to the greater reactivity of free chlorine than chloramines. The result is higher chlorine consumption, which must be compensated for by increasing the dose. Therefore, during these periods of nitrogen deficiency, it may be necessary to add ammonia N to the RAS system. Either aqueous ammonia (NH_4OH) or anhydrous ammonia (NH_3) may be used.

Control of Sludge Bulking by Chlorination

Historical Background. In 1945 Tapleshay[49, 50] began investigating the use of chlorine for the prevention and control of sludge bulking. He theorized that chlorine restores the balanced load between the organic matter and the biological life in the sludge. This investigation of the sludge bulking phenomenon encompassed 50 treatment plants in the United States. He found that the best method of control was possible when the chlorine dose was correlated to the sludge index (SI = amt. of dry solids in the return sludge) expressed as follows:

$$\text{lb/day } Cl_2 + \text{SI} \times F \times W \times 0.0000834 \tag{7-24}$$

where

SI = sludge index (Mohlman)
F = return sludge rate in mgd
W = suspended solids in return sludge in mg/L

Chamberlin[41] found in his studies that the usual chlorine dose required to control bulking averaged about 5 mg/L based upon the return sludge flow. This application of chlorine usually resulted in a turbid effluent for the first 12 hours as the reaction began to stabilize. This situation caused some who had tried chlorination to discontinue it and call it a failure. If chlorination had been continued in these cases, the operator would have found the turbidity gradually decreased along with the sludge volume index. Usually a decided improvement was noted in three or four days. In these early years some relied solely on intermittent chlorination, whereas others practiced continuous chlorination. However, current knowledge of the problem has provided a more reliable program than those for control.

Chlorination System

Operating Plant Example. The following calculations are based upon field data operating in the conventional mode (see Fig. 7-6):

Average daily flow	= 10 mgd
Aeration detention time	= V/Q = 6 hours
Hydraulic detention time, where R = RAS return rate, mgd	= $\dfrac{V}{Q + R}$ = 4.4 hours
Aeration volume (including mixed liquor channel)	= 2.5 million gal
Clarifier area	= 12,500 ft^2
Clarifier volume	= 1.3 million gal
Clarifier overflow rate gpd/ft^2	= 800 av.
	= 1050 peak

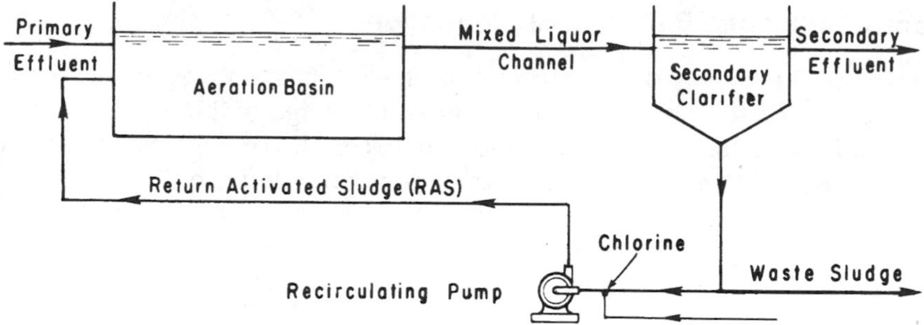

Fig. 7-6. Return activated sludge (RAS) system showing point of chlorination and relevant structures.

Mixed liquor total suspended solids (MLTSS)	= 1375 mg/L
Mixed liquor volatile suspended solids (MLVSS) (MLVSS is about 75–80 percent of MLTSS)	= 1030 mg/L
Average suspended solids in clarifier (TSS)	= 2000 mg/L
Volatile suspended solids in clarifier (VSS)	= 1500 mg/L
Return activated sludge suspended solids (RASTSS)	= 5300 mg/L
RAS return rate = 35 percent of total plant flow	= 3.5 mgd

Note: During normal conditions the clarifier is the repository of about 20 percent of the total suspended solids in the activated sludge system. However, during bulking conditions this percentage can escalate to 50 percent.

The food-to-microorganism ratio is expressed as F/M = 0.30. The food is the lb BOD applied (primary effluent = 135 mg/L), and the microorganisms are the total system volatile suspended solids (TSVSS), including aerators, clarifiers, and mixed-liquor channel.

The mean cell residence time (MCRT) is 6.3 days, and the sludge yield is 0.7 lb VSS per lb BOD removed. Therefore, the MCRT is related to

$$\text{F/M} = \frac{\text{lb BOD applied}}{\text{total system volatile inventory}}$$

The proper RAS rate can be estimated from the following relationship:

$$\frac{\text{RAS flow}}{\text{Primary eff. flow}} = \frac{\text{MLTSS}}{\text{RASTSS} - \text{MLTSS}} \tag{7-25}$$

It is best to use a target value for the RASTSS in the above equation; otherwise, as the RAS thins out, Eq. (7-25) will call for a higher RAS flow. The objective is to thicken the RAS—more organisms per unit volume. Reducing the RAS flow increases the detention time in the aeration basins, and as the RAS flow is reduced, the sludge concentration tends to increase owing to greater compaction and more settling time in the clarifier.

Substituting the figures from the example above into Eq. (7-25) we get:

$$\frac{\text{RAS flow}}{\text{Primary eff. flow}} = \frac{1375}{5300 - 1375} = 0.35$$

Therefore, RAS flow = 0.35 × 10 = 3.5 mgd.

Note: Clarifier suspended solids should be sampled—a "sludge judge" tube may be used—because chlorination is normally practiced under bulking conditions when an abnormally large amount of sludge inventory may be in the clarifiers due to poor settling; that is, do not assume it is the same as the mixed liquor concentration.

Initial Chlorine Dose. It is recommended that operators start at a low dose of 4 lb chlorine/day/1000 lb VSS. The VSS is total inventory in the entire system (i.e., aeration basins, clarifiers, and mixed-liquor channel). If more chlorine appears to be needed, it is prudent to make conservative increases—about 10–20 percent per day.

A sample calculation for the chlorine feed rate is as follows:

VSS′ in aeration basins:

$$\text{MLVSS mg/L} \times 8.34 \times \text{volume (million gallons)} = \text{MLVSS lb}$$
$$1030 \times 8.34 \times 2.5 = 21{,}475$$

MLVSS in clarifiers:

$$1500 \times 8.34 \times 1.3 = \underline{16{,}260}$$
$$\text{Total inventory} = 37{,}735 \text{ lb}$$

So the daily feed rate is:

$$\frac{37{,}735}{1000} \times 4 \text{ lb } Cl_2 = 150 \text{ lb/day}$$

The next step is to check the number of exposures of the RAS to the chlorine applied. The minimum number is considered to be 3.

$$\text{total sludge returned per day} = \text{RAS flow} \times 8.34 \times \text{RASTSS} \times \text{Volatility} \quad (7\text{-}26)$$

Substituting:

$$\text{total sludge lb/day} = 3.5 \text{ mg} \times 8.34 \text{ lb/mg} \times 5300 \text{ mg/L} \times 0.75$$
$$\text{RASVV} = 116{,}000 \text{ lb/day}$$

The total inventory is 37,735 lb (see above); so the exposure frequency is:

$$\frac{116{,}030}{37{,}735} = 3.05 \text{ times per day}$$

The above calculation can be done using TSS if they are used for the RAS and inventory calculations; however, the volatile inventory has already been calculated.

As the exposure frequency is marginal (3 times/day), it is suggested that if the effects of chlorination are not observed within 48 hours in this example, and provided that other parameters are in order, the sludge return rate should be increased. This can be done if the clarifier solids flux loading will allow it.

Clarifier solids flux check: RAS rate (lb/day) must be RASTSS, so:

$$\text{RASTSS} = \frac{\text{RASVSS}}{\text{percent volatile}} = \frac{116{,}000}{0.75} = 154{,}700 \text{ lb}$$

The total clarifier area is 12,500 ft^2. Then:

$$\text{clarifier flux} = \frac{154{,}700}{12{,}500} = 12.4$$

A safe flux range is 20–25 lb/ft^2/day. Therefore, the RAS flow in the example above could be safely increased 40–45 percent. In this event the RASTSS concentration will probably decrease slightly. The flux should be rechecked when this reduction in RAS concentration is known.

Final chlorine dosage check at point of injection:

$$\frac{150 \text{ lb/day}}{Q_{\text{RAS}} \times 8.34} = \frac{150}{3.5 \times 8.34} = 5.1 \text{ mg/L}$$

This dosage can and does require 6–7 mg/L.

The chlorine must be applied to the sludge at a point where there is effective initial mixing. The suction side of a centrifugal RAS pump is ideal.* This will provide the necessary initial mixing, which is most important. It assures that the filaments in the total sludge mass will be inhibited, instead of small amounts of the sludge being fully oxidized in passing clumps.

Chlorine should be applied continuously until the target SVI is achieved. Intermittent chlorination allows filaments to pass through undamaged and be able to proliferate in those parts of the aeration basin where they have an environmental advantage.[69, 93]

Warning of Initial Side Effects. When chlorination of the RAS is first initiated to correct a bulking problem, temporary deterioration in the effluent quality will occur. Therefore, the operators of single-stage systems can expect the turbidity to double, the suspended solids to increase by 20 percent, and the BOD to increase by no more than 10–15 percent.

If, however, after RAS chlorination has been initiated, a marked increase in BOD occurs in the effluent (40 percent or more), this is a warning signal that chlorine is inhibiting the biological processes, and the dose must be reduced.

Two-Stage Sludge Systems. In these systems, for example, that of the San Jose/Santa Clara plant,[69] using high chlorine doses in the first-stage activated sludge system led to a deterioration in the first-stage effluent quality. In this example, it allowed the system to operate at high loadings and to achieve adequate removals of the major portion of the organic load to the biological treatment system. Thus the best use of the first stage was to achieve adequate overall removals of organics at the price of some deterioration in effluent quality, and the second stage functioned to polish the effluent before it entered the tertiary filters.

Sludge Transport

Since 1976, the U.S. Environmental Protection Agency permits have required phase-out programs for ocean disposal of activated sludge. All ocean dumping of sludge had to stop by 1981. Nearly all of the cities established timetables to cease dumping sludge into rivers, lakes, and ocean beds to meet the deadline. The responses to these actions caused a variety of changes in the handling of sludge. Of particular interest is the movement of sludge from point of origin to a terminus for further treatment or disposal.

If during the transport operation the sludge becomes septic, hydrogen sulfide will be generated in amounts sufficient to cause intolerable odor and corrosion

*The pump should not be constructed with brass or bronze parts due to possible chlorine corrosion. Fiberglass or all iron pumps are preferred. Moreover, chlorine application should be arranged so that it cannot be applied while a pump is out of service.

problems. Pumping sludge in a force main will cause serious problems unless high concentrations of chlorine are applied to prevent septicity. The dose of chlorine will be high enough to cause corrosion of the sludge pumps. Dosages to prevent septicity are usually on the order of 50–125 mg/L.

Work performed by the Sanitation Districts of Los Angeles suggested a different approach: to transport the sludge as a mixture of sludge and raw or treated sewage—whichever is more convenient.[51]

To better understand the phenomenon of sulfide generation in the case of sludge transport, the importance of the Pomeroy *i* factor must be explained. This important variable concerns the rate of H_2S buildup in a sewage collection system:

$$i = \frac{\Delta S_{OX}}{R_R + R_{SO}} \tag{7-27}$$

where

ΔS_{ox} = oxidation rate of sulfides (sulfides removed),mg/hr/L
R_R = rate of DO reaction with impurities in process stream, or oxygen removed if no sulfides are present,mg/hr/L
R_{so} = oxygen uptake rate, mg/hr/L

Determination of the *i* factor or oxidation capacity of the sewage stream is as follows: Agitate a sample of the sewage in question so that the DO reaches about 6 mg/L. To this sample, add sulfides to determine the sulfide oxidation capacity. Assume that the total of sulfides removed is 4 mg/hr/L, and the total loss of DO is 12 mg/hr/L. The *i* factor is $4/12 = 0.33$. During this test the DO and dissolved sulfides are maintained above 0.3 mg/L.

When the Los Angeles County sulfide lab ran tests to determine the oxidation capacity of the raw sewage versus raw sewage plus raw sludge, it found the *i* factor went from 0.26 in the raw sewage to 0.43 in the sewage–sludge mixture. This signifies that the oxidation capacity of a sewage stream is increased nearly twofold when raw sludge is added. Therefore, whenever sludge is to be transported, the use of a sewage flow to carry the sludge should be investigated.

To determine the feasibility of any sludge transport system the probable condition of the sludge at the terminus must be investigated. Septic conditions and hydrogen sulfide generation must be avoided if at all possible. This can be done conveniently with reasonable chlorine doses. The sludge or sludge and wastewater mixture should be subjected to the methylene blue stability test. (This test is described under the heading "Preventing Septicity" in this chapter.) Using this test the operator can determine the chlorine dose necessary to prevent septicity in the flowing system for a given length of time. The time involved should be the length of time required for the flow to travel from the point of origin to the terminus. It is important for the operator to understand that *it is not necessary to carry a chlorine residual to the end of the force main.* The methylene blue test is based upon chlorine dose, not residual.

It is good practice to purge the sludge transport pipeline with caustic on a periodic basis. This will remove biological slimes, which might otherwise reduce the oxidation capacity of the system. It will also act to prevent an increase in hydraulic friction head. The recommended dosage is 400 lb 50 percent NaOH per 450 gpm, applied over a 30-min. period, about once every 4 to 6 weeks.

Grease Removal

A great many waste treatment works are troubled by grease. Preaeration, vacuum flotation, and aerochlorination are common methods of removing grease. Mahlie[53] lists the adverse effects of grease as follows: "It blinds screens, accumulates on walls, destroys the paint on steel structures, gives rise to odors, interferes with secondary treatment processes by clogging trickling filters, spray nozzles and sand beds, and interferes with sludge digestion. It also may have an unsightly effect on the receiving waters."

The addition of chlorine for improved grease removal was first reported in 1937 at the Woonsocket, Rhode Island, plant,[54] where 2 mg/L of chlorine gas was added to the air used for preaeration for grease removal. With a contact time of approximately 6 min., increased grease removal over using air alone ranged from 189 to 442 percent. About 1938, Keefer and Cromwell[55] tried chlorine dosages from 1 to 10 mg/L with contact times of 5, 10, and 15 min. at the Baltimore, Maryland, plant. Chlorine gas was introduced into the diffused air lines. Best results were obtained with a chlorine dose of 5 mg/L and a contact time of 5 min. Increased removals compared to air alone varied from 148 to 800 percent.

These results were so phenomenal that aerochlorination was tried at the two activated sludge plants of Lancaster, Pennsylvania. This treatment is of particular interest because the chlorine was applied as a solution immediately ahead of the air diffusers in the preaeration channel. Chlorine dosages averaged about 2 mg/L with 2–3 min. of contact. With this treatment, 70–80 percent of the grease was removed, as compared to only 50 percent with air alone. Although the data on the use of aerochlorination for grease removal are limited, they do indicate a definite increase in removal over air alone. This is further substantiated by the use of chlorine for grease removal in animal-rendering works. Chlorine appears to break up the emulsions, allowing the grease to float and thus facilitating its collection. Operators have noticed this phenomenon for years. Moreover, the amount of grease recovered makes this a profitable venture, which more than pays for the chlorination system in these rendering works.

Conventional wastewater prechlorination practices have been found to increase the amount of scum removed from the surface of settling tanks. The scum appears to accumulate in a more dense or compact mass, requiring less skimming and resulting in an improved general appearance of the settling basins and clarifiers, compared to systems not using prechlorination.

Chlorine for grease removal may be applied as a solution ahead of the primary clarifiers (prechlorination) or ahead of the aeration tanks. It is not necessary to satisfy the chlorine demand of the wastewater. Often as little as 2–5 mg/L will suffice to bring about an improvement in the grease removal.

Aerochlorination seems to be more effective than conventional procedures.[54] In this case, the dry chlorine gas is mixed with the diffused air going to the aeration tanks. Conventional vacuum feed solution type chlorination equipment is used; however, air, instead of water, is forced through the injector to operate the equipment. With this method, 2–10 mg/L chlorine as chlorine gas and 0.02—0.20 cu ft of air per gallon of wastewater with a 3–20-min. aeration period is the customary practice.

Trickling Filters and High-Rate Recirculating Biofilters

Chlorine has proved beneficial in the operation of both trickling filters and high-rate recirculating biofilters. Trickling filters invariably require chlorine for odor control. The worst offenders are those that operate on an intermittent cycle. During the downtime the wastewater is allowed to get septic, giving rise to the formation of odorous gases, which are aerated into the atmosphere through the spray nozzles during the filtering cycle. About 1928 Morris Cohn[40] at Schenectady was one of the first plant superintendents in the United States to experiment with the use of chlorine for controlling odors at trickling filters. The chlorine is applied to the dosing tank of the trickling filter unit. Sometimes effective trickling filter odor control can be achieved by using the prechlorination point of application, but this should not be depended upon. Proper control is usually obtained by satisfying only a fraction of the immediate chlorine demand (approximately 25 to 30 percent). The chlorine varies from 2 to 6 mg/L. Chlorination to less than the chlorine demand is somewhat difficult to control. Effectiveness depends on the number of complaints received; so, rather than risk any complaints, some operators make occasional dissolved oxygen tests at the influent to the dosing tank. The chlorine applied is sufficient if there is some DO at this point. If not, the chlorine dose is increased, and the secondary effluent is recirculated through the filter until DO appears again in the dosing tank.

High-rate recirculating biofilters are generally odor-free, and so chlorine for this purpose is not a consideration. This is a result of their continuous operation, which produces a more stable biological environment. However, recirculating filters do suffer from ponding and filter fly nuisance, as does the trickling filter. Chlorine, if used properly, can be effective in controlling these two problems.

Ponding is due to the clogging of the interstices by filamentous growths or excessive accumulation of solids. If the filter is overloaded and the clogging material has penetrated deeply, it may be too late for the chlorine to be of any help. Chlorine is most successful when used in the spring to induce unloading of material accumulated in the filter during the winter. The chlorine attacks the organisms in the zoogleal mass, causing them to loosen so that they can be flushed out of the bed. Removal of this material requires heavy doses of chlorine.

Sufficient chlorine should be applied at the filter influent to produce an amperometric residual of between 4 and 8 mg/L at the nozzles.[55]

Contrary to popular belief, filter ponding on high-rate filters can be prevented and controlled by continuous chlorination to a residual of about 0.5 mg/L.[56] This small amount of chlorine continuously applied seems to create a stable environment that will not allow excessive growths and does not interfere with normal biological processes within the filter.

Some filters never experience ponding, whereas others experience it as a regular seasonal occurrence in varying degrees. For this reason, operators must find by experience the best way to apply the chlorine. Some find the answers in programming chlorination for a few hours at night one or two days a week. The most difficult to control are the intermittent trickling filters.

The trickling filter and the recirculating filter present ideal places for the propagation of the filter fly, *Psychoda alternata,* which develops rapidly during the warm summer months. Eggs are laid on the surfaces of the zoogleal film in irregular masses containing from 30 to 100 individuals. The eggs hatch into larvae, which penetrate the film, leaving a breathing tube projecting. In this position they feed and are transformed into pupae. When pupation is completed, the shell bursts, and the fly emerges. The life cycle may be completed in as little as 12 days. The larvae and pupae are most abundant between 3 and 12 inches below the surface of the filter, although they may be distributed throughout the bed.

There is no entirely effective method for controlling this nuisance. Chlorination has had some success either by limiting the thickness of the zoological film or by removing it entirely in the top layer of the filter. The number of larvae and pupae seems to be related to the thickness of the zoogleal film on the filter rock. Best control is achieved in the larvae and pupae stage. Chlorine has no effect on the adult fly. Therefore, application of chlorine for the control of *Psychoda* must result in partial removal of the biological growth in the upper layers of filter rock, which is the breeding ground. This should be done only in the fly season. Chlorination is usually programmed in a fashion similar to that of correcting ponding—namely, 2-10 ppm O-T residuals at the nozzles to produce some sloughing of the filter rock film.

Chlorination for trickling filter odor control and filter fly control and high rate recirculating filters for ponding control are good operating practices. In many installations the quality of the filter effluent has been improved by this intermediate point of chlorination.

BOD Reduction

The ability of chlorination to reduce the BOD permanently is often overlooked in the perception of wastewater treatment. As early as 1859, chlorination of wastewater in the form of chloride of lime was practiced in England for the purpose of delaying putrefaction of the wastewater long enough for the receiving waters to be carried to the open sea. This in effect was chlorination for BOD reduction. It was this application and other applications in the United Kingdom that led to the origin of the five-day BOD parameter. This length of time was chosen because five days happens to be the longest time it takes any major receiving water in the British Isles to flush itself to the nearest open sea.

During periods of critical assimilation capacity* of the receiving waters, effluent chlorination has been used successfully to alleviate this situation by its ability to reduce the effluent BOD. This BOD reduction has allowed the receiving water to maintain the desired DO levels.

Susag[57] showed by example how effluent chlorination is more economical than additional treatment facilities to relieve the assimilation capacity of the Mississippi River during critical periods. The Minneapolis–St. Paul Sanitary District treatment plant utilizes the high-rate activated sludge process, which has a predicted BOD removal efficiency of 75 percent. This treatment level was considered adequate for all but about 7 percent of the time (26 days in the year 1980). From this study, the cost of effluent chlorination for BOD reduction ranged from 8 to 40 percent of that for additional treatment. Moreover, considering the toxicity limits on chlorine residuals, which require dechlorination, there is little or no incentive to use chlorine for BOD removal. There are no data on the effect of dechlorination on BOD reduction by chlorination.

BOD reduction by chlorination is by no means confined to effluent chlorination. Chlorination of wastewater for any purpose reduces the BOD. This effect is shown most dramatically in prechlorination of a strong sewage (high BOD). Rawn[59] reported a 25 percent BOD reduction as a result of a comprehensive up-sewer chlorination procedure for the Los Angeles County Sanitation District. The full value of this treatment was, however, not only BOD reduction. The sewage arriving at the treatment plant was in a fresh, rather than a septic, state, which had the ultimate effect of trebling the capacity of the plant.

A number of studies of BOD reduction by effluent chlorination have been carried out in the United States.[40, 52, 57, 60, 61] The consensus on the effect of chlorination is as follows:

1. For each mg/L of chlorine absorbed there will be a 2 mg/L reduction of BOD.[40, 62]

*The assimilation capacity of a receiving water is a function of its dilution capacity, the DO content, and the oxygen-consuming organic load of the receiving water itself.

2. The reduction appears to be permanent.
3. The reduction increases with increasing chlorine dose.
4. The unit BOD reduction per mg/L chlorine absorbed is greatest at the dosage producing the least residual, and decreases with increasing chlorine dosage. This is irrespective of total initial BOD.
5. The greatest BOD reduction per pound of chlorine applied up to the first appearance of a chlorine residual occurs in the strongest sewage (highest BOD).
6. The greatest BOD reduction in percent up to the first appearance of a chlorine residual occurs in the most highly treated sewage.
7. Up to approximately 80 percent removals can be expected by chlorination of highly purified sewage (i.e., activated sludge effluents).[52]

The work by Susag indicates that chlorine does not act solely as an oxidant of oxidizable organic matter; if it did so, the percent reduction would remain constant for all incubation periods, which it does not. It is evident that chlorine reacts to reduce and retard BOD, partly by oxidizing organic matter and partly by substitution and addition to unsaturated and saturated compounds to produce compounds that are either inert or resistant to bacterial action.

In summary, the use of chlorine to reduce BOD is not a practical premise for the design of a wastewater treatment system. However, the knowledge that it will reduce BOD is helpful in certain situations that could, for instance, make up-sewer chlorination preferable to some other method of treatment.

Nitrogen Removal

Purpose. Nitrogen in its various forms can deplete dissolved oxygen levels in receiving waters, stimulate aquatic growth, or, above a certain concentration, demonstrate lethal toxicity to aquatic life. Ammonia nitrogen depletes oxygen and produces toxic effects. Toxicity is of special concern and can be eliminated by converting the ammonia N to N_2 to nitrate.

In 1970 the European Inland Advisory Commission studied the effect of ammonia nitrogen on freshwater fish. A great many European and U.K. surface waters carried high levels of ammonia N (2–5 mg/L). This study indicated that the allowable limit for un-ionized ammonia (NH_3–N) (rather than $NH_3 + NH_4^+$—total ammonia) be limited to 0.025 mg/L because this fraction is far more toxic than the ionized fraction (NH_4^+). Unfortunately, it is impossible to directly measure the un-ionized fraction in the relevant concentration range. Ammonia N probes can only detect as low as 0.03 mg/L. However, un-ionized ammonia can be *calculated* from easily measured quantities: total ammonia, pH, TDS, or salinity, and temperature. The following expression[58] is used to calculate the un-ionized ammonia:

$$NH_3\text{–}N = \frac{\text{total ammonia}}{10^x + 1} \tag{7-28}$$

where x = pK_a – pH.

Example:

$$\begin{aligned} \text{pH} &= 8.3 \\ \text{Total ammonia} &= 3 \text{ mg/L} \\ \text{Temperature} &= 18°\text{C} \\ \text{TDS or salinity} &= 15 \text{ g/kg} \end{aligned}$$

and from tables,[63] pK_a = 9.605, we have:

$$NH_3\text{–}N = \frac{3}{10^{9.605-8.3} + 1}$$

$$= \frac{3}{14.33 + 1} = 0.20$$

The above approach has been incorporated into the discharge requirements by the California Water Quality board with the NH_3–N limit as calculated by Eq. (7-28). Using the limitation of 0.025 mg/L for the receiving water, Eq. (7-28) allows the discharger to calculate the maximum amount of un-ionized ammonia that could possibly exist in the receiving waters. Using this method, there will be no chance of ammonia buildup in the receiving waters.

Current Practices. The EPA "Process Design Manual for Nitrogen Control" (Oct. 1975) described several methods of nitrogen removal. Air stripping using cooling towers has been tried in many places. The results have always been disappointing. The most reliable method appears to be biological: either fixed growth or suspended growth reactors. These methods produce effluents that range in concentration from an undetectable trace of NH_3–N to about 4 or 5 mg/L, depending upon overloading conditions.

Chlorine is used in those cases where the effluent concentration must be limited to about 0.5 mg/L NH_3–N. Occasionally chlorine is used seasonably to assist the biological process.

Nitrogen Removal by Chlorine

Introduction. Removal of ammonia nitrogen by chlorine is an extension of the breakpoint reaction. The use of chlorine to remove ammonia nitrogen has been studied on two levels: (1) where chlorine was the sole process, and (2) where chlorine followed a nitrification process. In the first instance ammonia nitrogen concentrations were in the range of 15 to 25 mg/L, and in the second instance the range was from 0.5 to 2.5 mg/L. The importance of these two ranges of ammonia nitrogen concentration is the contact time requirement. The speed of reaction to complete the breakpoint reaction is a function of the concentration of the reactants (i.e., chlorine and ammonia). The higher the ammonia concentration, the faster the reaction, and conversely, the lower the concentration, the slower the reaction. Therefore, larger contact chambers are required for the removal of lower concentrations of ammonia N. These situations are described below.

Rancho Cordova Project. A full-scale demonstration of nitrogen removal by breakpoint chlorination was carried out at the Rancho Cordova secondary treatment plant. Sacramento County, California, between December 1975 and March 1976.[64, 65] During this time the automatic chlorination facility was operated 24 hours a day 5 days a week. The process flow rates varied from about 0.1 to 1.2 mgd. Influent ammonia nitrogen concentrations were ordinarily in the range of 15 to 25 mg/L. The breakpoint process succeeded in a constant removal of about 97 percent of ammonia nitrogen.

A number of specific observations and conclusions made as a result of the Rancho Cordova breakpoint chlorination demonstration program are enumerated below:

1. The dosage of chlorine at Rancho Cordova required to reach breakpoint and maintain a controllable free residual in the process stream averaged 10 mg/L for each 1.0 mg/L ammonia nitrogen present in the process influent, which is the plant effluent.

2. Approximately 70 percent of the breakpoint chlorine dosage was consumed to produce nitrogen gas (N_2) from ammonia (NH_4^+) at pH setpoints between pH 7 and 8. The oxidation of NH_4^+ to NO_3^- consumed 8–19 percent of the total chlorine dosed to the system. Overall, about 96 percent of the total chlorine dosage was accounted for in reactions between chlorine and nitrogenous species in specific chemical pathways and free chlorine residual remaining in solution following breakpoint. (See Fig. 7-7).

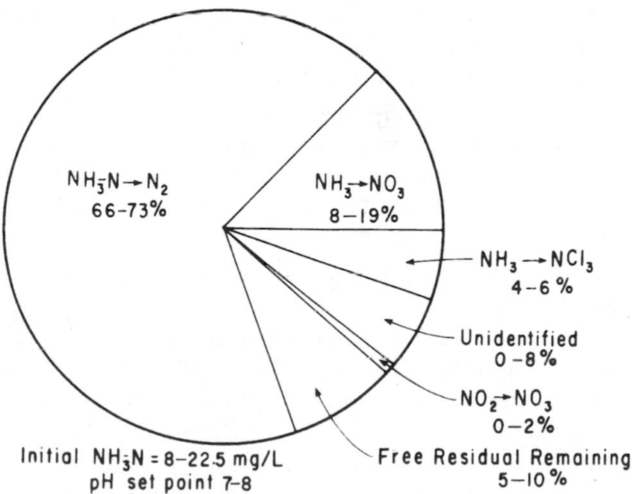

Fig. 7-7. Chlorine consumption during nitrogen removal by breakpoint chlorination.

3. Nitrate (NO_3^-) production in breakpoint chlorination was not found to be pH-sensitive, with about 1.0 mg/L NO_3^- (as N) produced from NH_4^+ across a final system pH range of pH 6.5 to pH 8.5. The production of NO_3^- from NO_2^- was wholly dependent upon the influent NO_2^- concentration.

4. Nitrogen trichloride (NCl_3) production was observed to be fairly insensitive to pH across a range of final system pH values from pH to pH 8. The median value for NCl_3 production was about 0.4 mg/L (as N) when breakpoint effluent was used as the source of chlorine injector water. While the amount of chlorine consumed in the formation of NCl_3 was relatively small (4–6 percent of total dosed), NCl_3 generation affects the minimum ammonia concentration that can be achieved in breakpoint, as its concentration decays slowly in dilute solution, and it is converted to ammonia upon dechlorination with sulfite ion ($SO_3^=$).

5. If the breakpoint process influent is used as injector water, reactions between chlorine and ammonia do occur in the injector water and can consume chlorine in undesirable side reactions. Therefore, injector water should always come from the reacted process effluent. At Rancho Cordova when secondary effluent (process influent) was used as the injector water source, the NCl_3 formed in the injector discharge increased the NCl_3 in the process effluent by about 0.2 mg/L.

6. The concentration of organic nitrogen compounds was not affected by breakpoint chlorination. *Note:* This cannot be true. In nitrified effluents organic N compounds react instantaneously with free chlorine (HOCl) to form nongermicidal organochloramines that titrate as partly (0.2 mg/L) monochloramine and the rest dichloramine, found only by the completely forward-titration procedure of the amperometric method or the DPD-FAS titrimetric method. See the San Jose/Santa Clara WWTP experience with nitrification in Chapter 8. This is why Fig. 7-7 shows that 8 percent of the compounds are unidentified. This is undoubtedly due to the presence of organic N in the wastewater, since we now know that free chlorine reacts instantaneously with organic N to form a nongermicidal organochloramine that titrates as pure dichloramine in the forward amperometric and DPD-FAS titrimetric procedures.

7. The rate of reaction for breakpoint chlorination was found to vary depending upon the pH control point (final system pH), with fastest rates observed at a setpoint of pH 7.0. The time to completion was found to be between 60 sec and 90 sec at pH 7.0. The reaction rate slowed considerably at pH setpoint 6.5, with gradual reductions in rate observed as pH increased from pH 7.3 to pH 8.5.

8. Variations in the amount of mechanical mixing intensity in the zone of breakpoint chemical application had no effect upon overall system chemical consumption and effluent quality. Mechanical mixing, to facilitate a rapid and thorough blending of process chemicals and the influent stream, was important in damping free residual oscillations for control purposes.

9. Sodium hydroxide (NaOH) was used throughout the study as an alkalinity supplement. The amount of NaOH required to neutralize all breakpoint

acidity (1.53 lb NaOH/lb Cl_2) was essentially identical to that predicted from chemical stoichiometry.

10. The small amount of NCl_3 formed did not present any odor problem, partly because of the closed pipe reactor and partly because of the reliability of the sophisticated control system.

11. Although it was built into the control system, the continuous NH_3–N process influent analyzer signal to the chlorination system was not necessary for the successful operation of the process.

12. The key control parameters, in addition to the plant flow signal, were the chlorine dosage trim signal from the free residual analyzer and the ability of the sodium hydroxide feed system to maintain a setpoint of the process pH within ±0.2 pH unit.

13. The DPD-FAS titrimetric method for the determination of the chlorine residual species proved most reliable and a time-saving method of analysis.

14. The reaction time is so rapid at high NH_3–N concentrations that the nuisance residuals have little chance to form.

Saunier's Research. Saunier made a comprehensive study of chlorine–ammonia nitrogen chemistry from which he constructed a computer model that predicts reaction times for various concentrations of ammonia nitrogen. This work is discussed in detail in Chapter 4, but repeated here to cover situations where ammonia N concentrations to be removed are 2.5–3.0 mg/L or less.

These effluents may contain ammonia nitrogen concentrations in the range of 0.5–3.0 mg/L. The chlorine-to-ammonia nitrogen ratio will be on the order of 10:1 by weight as found for the Rancho Cordova project.

Based upon Saunier's model, to remove 0.5 mg/L ammonia N will require at least 20 min. of contact time. If the ammonia N concentration falls below 0.5 mg/L, the contact time will have to be longer. However, for an ammonia N concentration of 2.5 mg/L, the contact time required is on the order of 10 min.

The key control parameter in these cases of low ammonia N concentrations will be the signal generated by the free chlorine analyzer. This analyzer must have the ability to detect free chlorine in the presence of combined chlorine over a range of 1 to 15 mg/L free chlorine in the presence of about 25 percent combined chlorine. The chlorine dose correction called for by the analyzer must have a correction frequency related to the contact time required to complete the breakpoint reaction. As long as the Cl:N ratio is optimum (10:1), the formation of nitrogen trichloride will not be significant enough to cause a nuisance.

Foul Air Scrubbing[96]

General Discussion. Scrubbing towers are in common usage at wastewater treatment plants. The principal objective is to control odors and thereby prevent an air pollution problem. Equally important is their use to prevent foul air from reaching the plant ventilation system. These towers are imperative when conditions do not allow control of odors at the source. Ventilation air must be free from lethal H_2S gas at all times. Many deaths occur in sewer-related accidents because of the presence of H_2S gas.

Scrubbing towers are also widely used at pumping plants and other collection system structures where uncontrolled sewage odors occur.

There are two general types of scrubbing towers: chemical and biological. The biological tower is dependent upon availability of a treated effluent. These are packed towers that utilize the buildup of a zoological slime on the packing material to oxidize the hydrogen sulfide and other sewage-related gases. These towers recirculate the effluent in a fashion similar to that of a high-rate recirculating filter. This method is usually reserved for large treatment plants. Complete H_2S removal is not always attainable; if it is needed, this method requires the use of a backup system of either chlorine or activated carbon.

Chemical tower systems require the use of either chlorine, caustic potassium permanganate, or activated carbon or some combination of these chemicals. For severe cases in odor-sensitive areas, the wet scrubber is followed by an activated carbon tower impregnated with either potassium or sodium hydroxide.[92] The most successful type of wet scrubber is the packed tower shown in Fig. 7-8. The foul air enters at the bottom, and the scrubbing water enters the top. This is called a countercurrent tower. The packing rings in the tower are designed to provide intimate contact of the recirculated water with the foul air in order to ensure maximum efficiency of odor removal.

Current practice consists of using secondary effluent when the scrubber location is at a treatment plant, and freshwater for collection system installations. All of these scrubbers operate as a recirculated system, using a continuous blowdown with a constant makeup water supply. A proper chemical balance must be maintained. A chlorine scrubbing tower is the most practical type because it is predictable, reliable, and flexible.

A Chlorine Wet Scrubber

General Discussion. The wet scrubber system (see Figs. 7-8 and 7-9) consists of a tower packed with plastic saddles. These saddles provide the necessary intimate contact between the foul air and the scrubbing water. The scrubbing water is recirculated on a continuous basis by a centrifugal pump with the suction connected to the bottom of the chemical mixing tank. The pump discharges into the top of the tower as shown in Fig. 7-8. The foul air is blown into the bottom of the tower with a centrifugal fan. Chlorine and caustic are added to the mixing tank. These are the only makeup fluids used in the system. They are usually sufficient to eliminate the need for continuous makeup water. Visual inspection supplemented by a total sulfide concentration determination will indicate the need to blow down some or all of the recirculation water.

There are several advantages to a chlorine scrubber as compared to a caustic scrubber. A chlorine scrubber is operated at neutral pH so there is

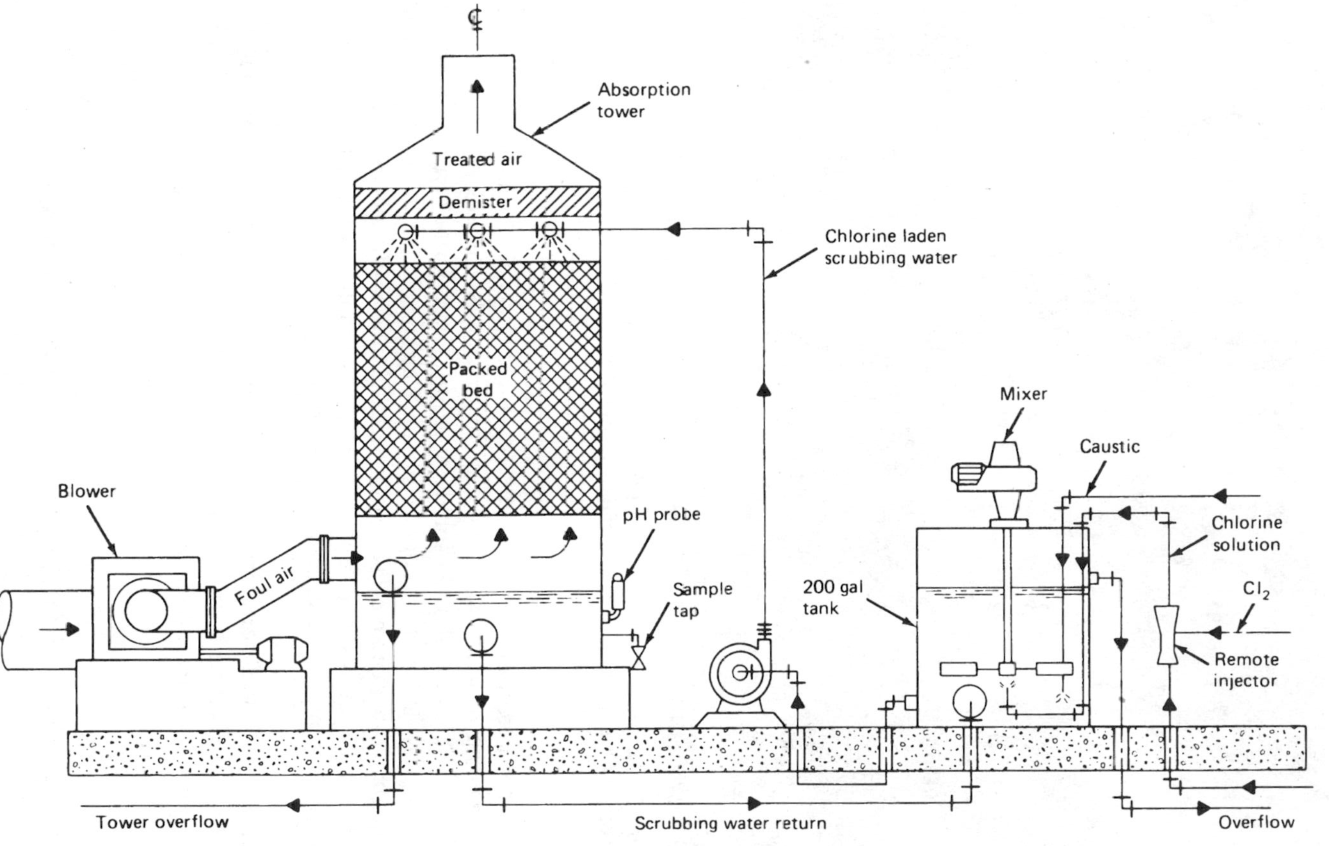

Fig. 7-8. Wet scrubber system for H_2S removal from foul air.

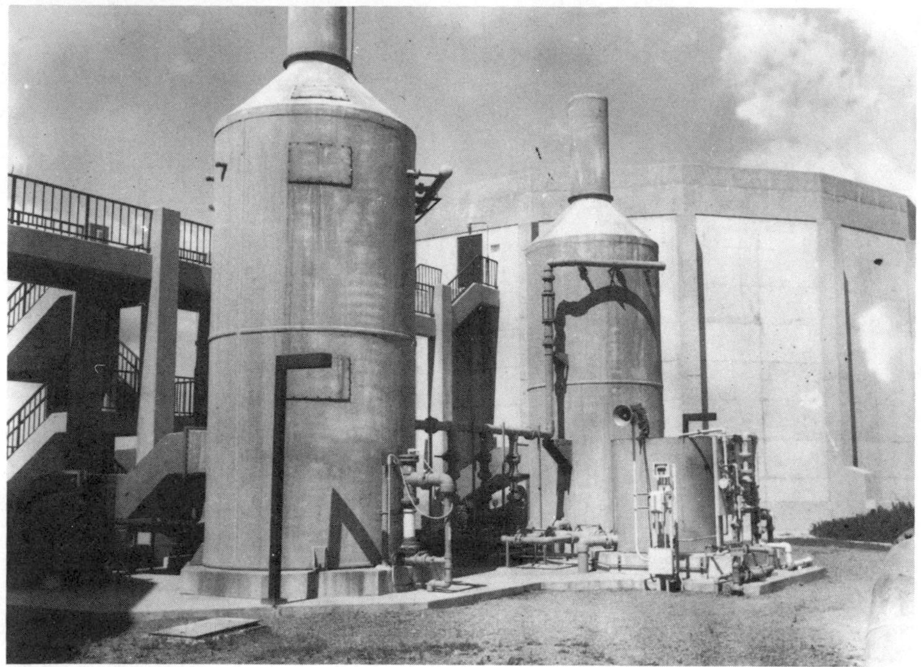

Fig. 7-9. Wet chlorine scrubber tower for H_2S removal from ventilation air (courtesy Envirotech Operating Services Fairfield-Suisun WWTP).

no precipitation of calcium or magnesium ions, which occurs in caustic towers. There is no corrosion from the chlorine because caustic is added to replace the alkalinity lost in the chlorine–sulfide reaction. The chlorine scrubber provides instantaneous and certain odor removal. Chlorine is far more reliable and flexible than other chemical scrubbers because it is easily controlled by simple monitoring of the residual. The presence of a small residual (0.5–1.0 mg/L) ensures the complete destruction of sulfides.

Chlorine–Sulfide Chemistry. Chlorine does not actually remove H_2S; it converts the H_2S and hydrosulfide ion (HS^-) to sulfates. In aqueous solutions hydrogen sulfide hydrolyzes as follows:

$$H_2S \leftrightarrows HS^- + H^+ \tag{7-29}$$

The hydrosulfide ion (HS^-)* further dissociates as follows:

$$HS^- \leftrightarrows S^= + H^+ \tag{7-30}$$

*This is commonly referred to as the sulfide ion.

See Fig. 7-1 for the distribution of H_2S, HS^-, and $S^=$ at various pH levels. At pH 7, hydrogen sulfide is approximately 50 percent of the total dissolved sulfides. Whenever the equilibrium between the sulfide ion (HS−) and the hydrogen sulfide in solution is upset, as when H_2S is removed by oxidation, the shift will be to form more H_2S from the remaining dissolved sulfides to reestablish the equilibrium. Assuming wastewater effluent as the scrubbing water at approximately 22°C (72°F), the solubility of H_2S in water at a pressure of one atmosphere is 3.432 mg/L. See Fig. 7-10a and Fig. 7-10c, which illustrate the chlorine odor control system at the Irvine Ranch Water District Reclamation Plant. Figure 7-10b shows the general arrangement of the scrubbing loose in Fig. 7-10a.

There arc two important reactions between chlorine and hydrogen sulfide: The first and most significant is the reaction that converts the H_2S to sulfates. This requires 8.32 parts of chlorine to each part of hydrogen sulfide:

$$H_2S + 4Cl_2 + 4H_2O \rightarrow 8\ HCl + H_2SO_4 \tag{7-31}$$

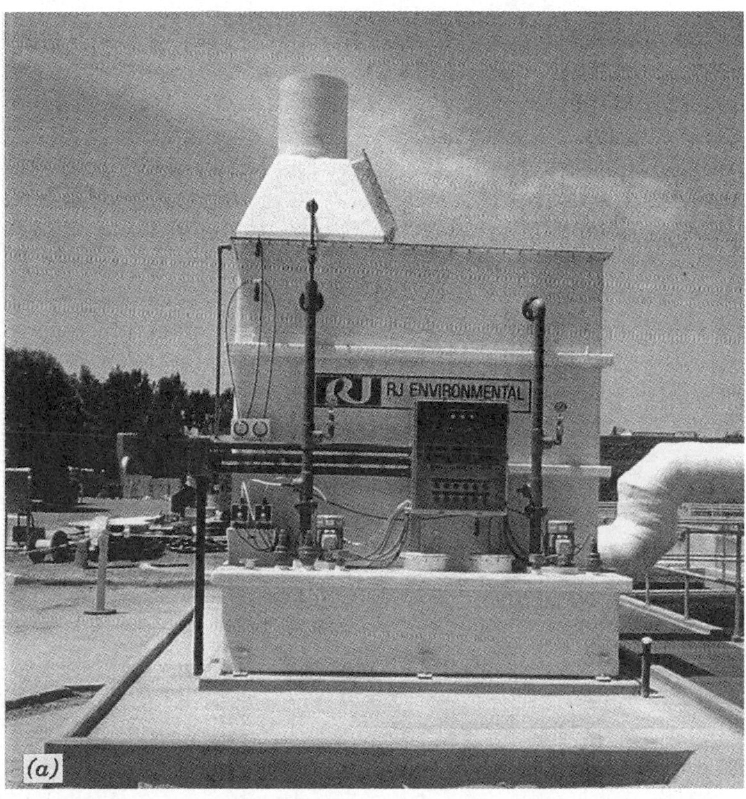

Fig. 7-10a. Odor control system Michelson wastewater reclamation plant, Irvine, California (courtesy R-J Environmental, Inc.).

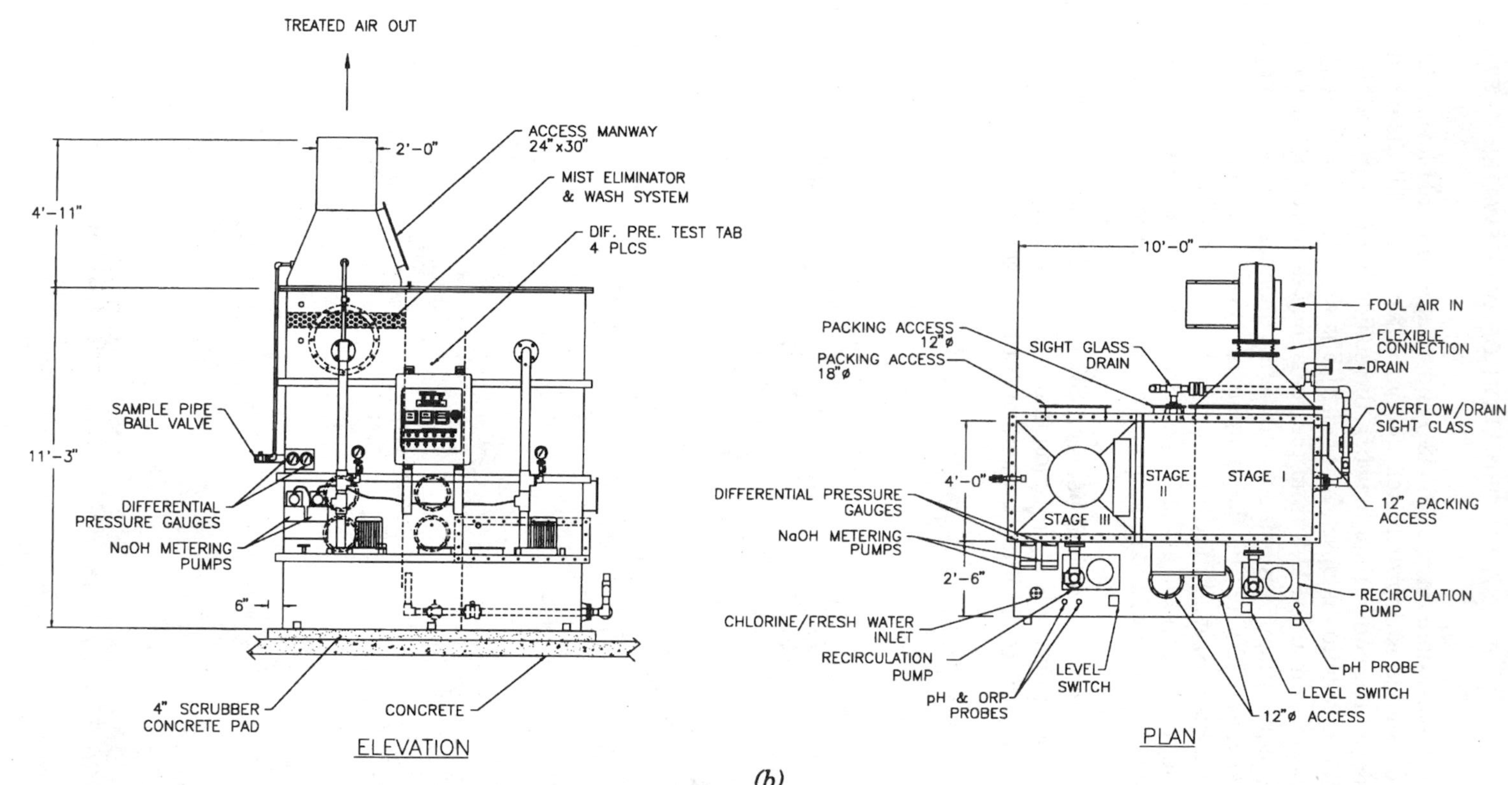

(b)

Fig. 7-10b. Irvine ranch water district odor control system, general arrangement (courtesy R-J Environmental, Inc.). Notes: 1) Resin: Hetron 922—surface veil Nexus 1012. Liner: 2 ply 1.5 oz. chopped strand glass (E grade). Total minimum liner thickness: 100 mL. 2) Total wall thickness: per design calculations. 3) Fabrication shall be in accordance with ASTM-4097 and PS-15-69. 4) Quality assurance shall be in accordance with ASTM D2563. 5) *True orientation is in plan view only.* 6) All bold holes are to straddle the tank's natural centerlines. 7) All bolt holes are to be backfaced for SAE washers. 8) Tank exterior to be surface coated, color per customer.

(c)

Fig. 7-10c. R-J 6000 ft^3/min. three stage odor control system, Michelson wastewater reclamation plant, Irvine, California (courtesy R-J Environmental, Inc.).

in the pH range 6.0 to 9.0. The *optimum pH for this reaction is 6.0. For each part of H_2S removed, 10 parts of alkalinity (as $CaCO_3$) will be consumed.*

The other important reaction is an intermediate one, which precipitates free sulfur:

$$H_2S + Cl_2 \rightarrow 2HCl + S \downarrow \tag{7-32}$$

This reaction requires only 2.10 parts chlorine for each part H_2S. For each part of H_2S removed, 2.6 parts alkalinity (as CaCO3) will be consumed. The pH range is 5.0 to 9.0. *The optimum pH for this reaction is 9.0.*

In a scrubbing operation the intent is to convert as much of the H_2S as possible to sulfates. But owing to the broad pH range for both of the above reactions it is desirable to maintain the pH in the scrubber system as near to 7.0 as practical. The blowdown from the system will always appear milky owing to the presence of colloidal sulfur formed by the intermediate reaction of Eq. (7-32). Maintaining the pH at 7.0 in the system will prevent long-term corrosion.

Loss of alkalinity in the system due to the addition of chlorine must be prevented by the addition of caustic.

In addition to absorbing the hydrogen sulfide and other odor-producing compounds, the scrubbing water will also absorb some carbon dioxide that may be in the foul air. This will tend to lower the pH and can be accounted for by continuous pH monitoring.

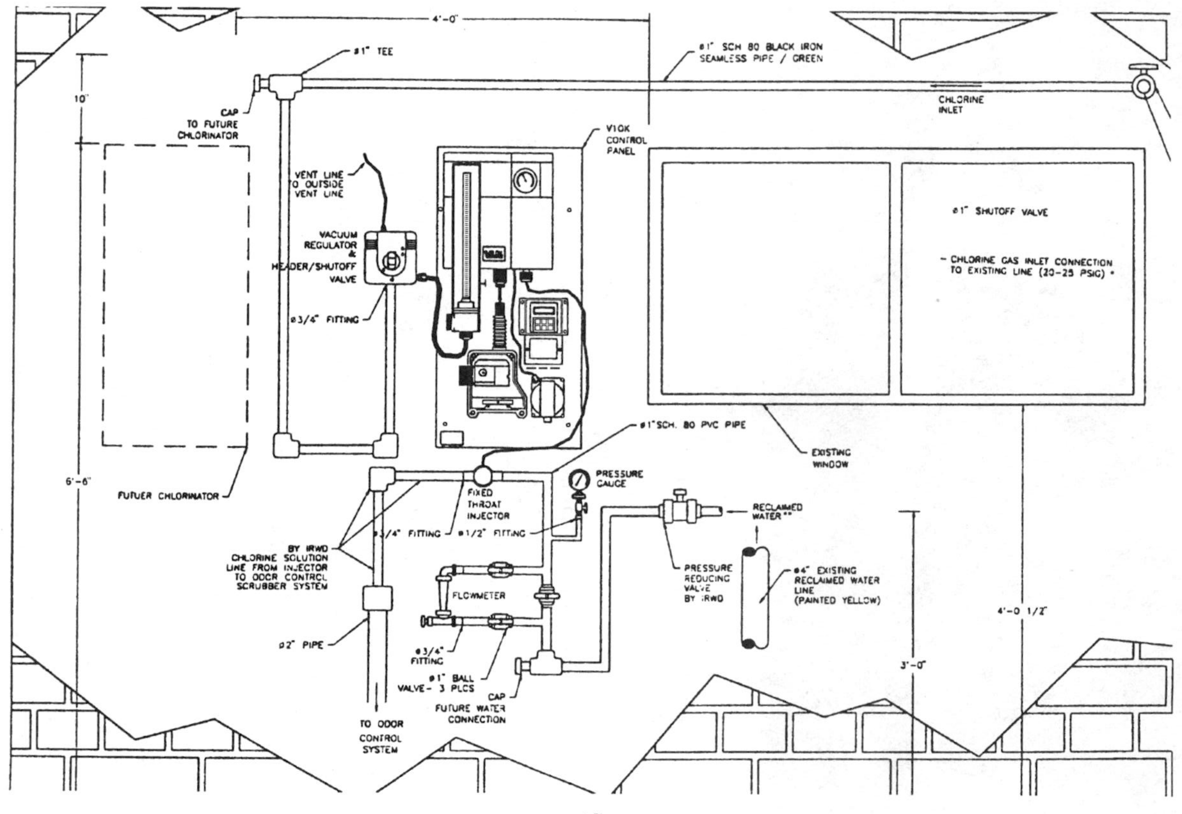

(d)

Fig. 7-10d. Irvine ranch water district odor control system, chlorination equipment layout.

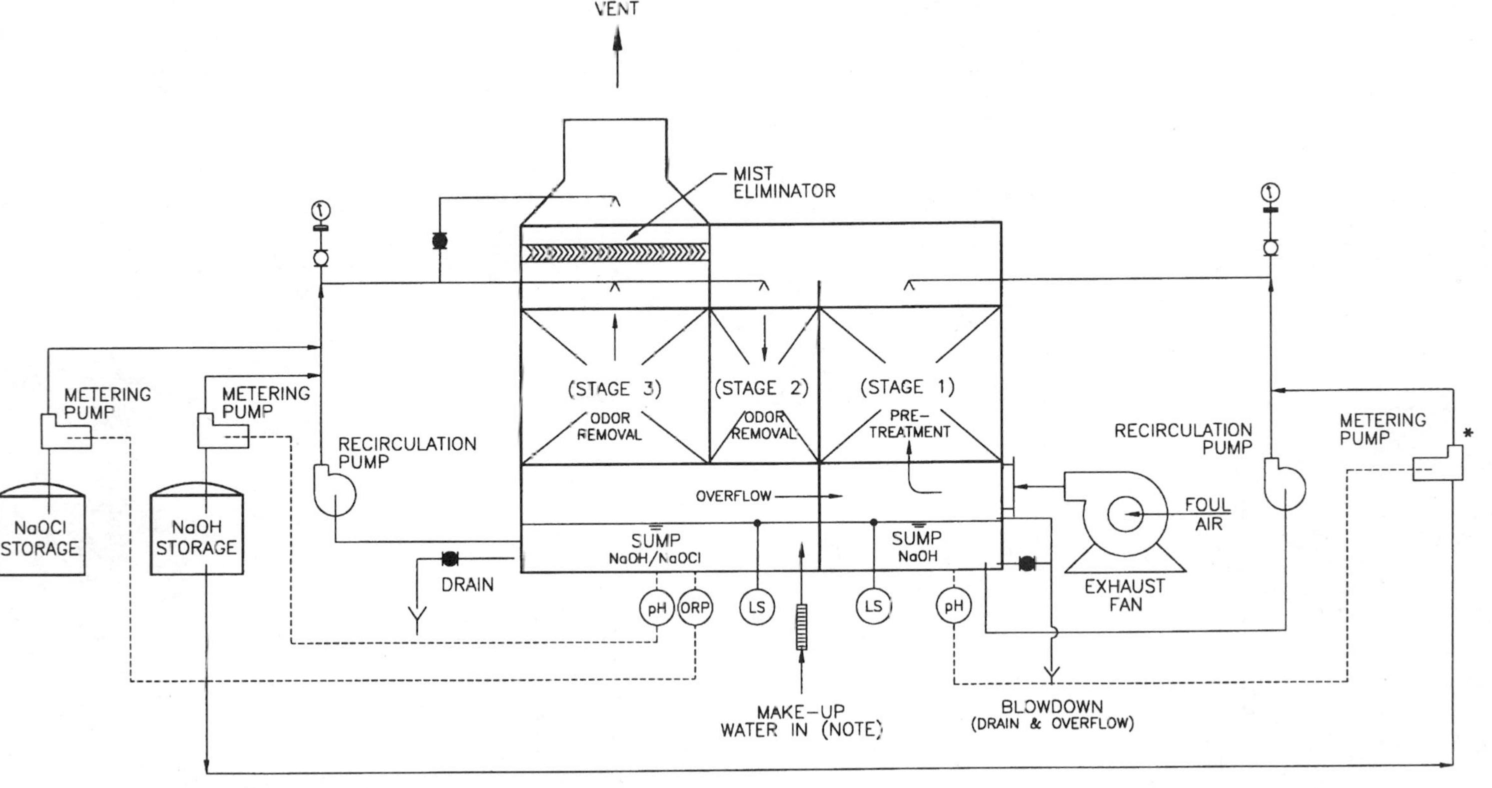

Fig. 7-10e. R-J odor control process diagram, Michelson odor control system (courtesy R-J Environmental, Inc.). Note: RJ Environmental strongly recommends that water hardness is less than 100 ppm $CaCO_3$ for makeup water to reduce maintenance. *Optional.

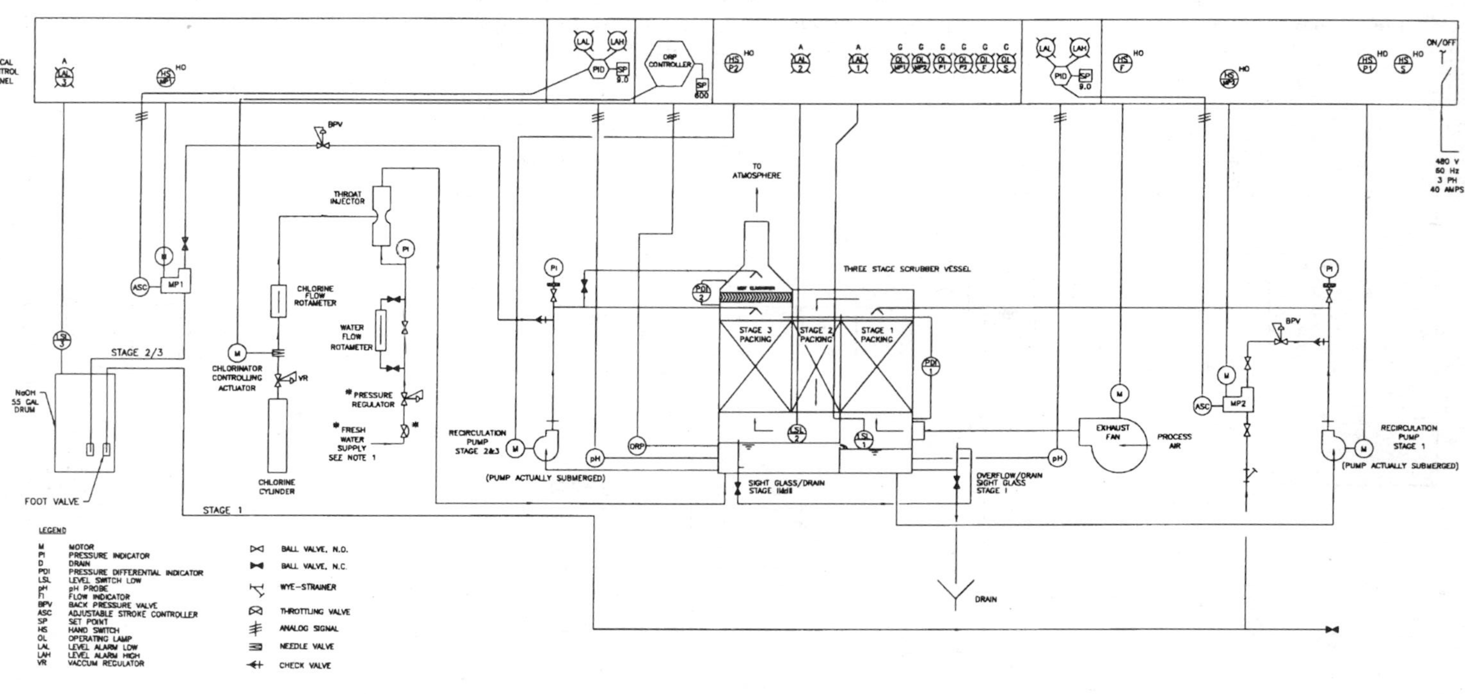

(f)

Fig. 7-10f. Irvine ranch odor control system process and instrumentation diagram (courtesy R-J Environmental, Inc.). Note: RJ Environmental strongly recommends that water hardness is less than 100 ppm $CaCO_3$ for makeup water to reduce maintenance. Minimum 40 psi required. *To be provided by others.

One other facet of chemistry important to the application of chlorine is the presence of ammonia nitrogen in the scrubbing water. If nonnitrified plant effluent is used, the NH_3–N concentration will be in the range of 12 to 22 mg/L. The chlorine dosage of the scrubbing water must be kept at a ratio of not more than six parts chlorine to one part NH_3–N. Otherwise, chlorine will begin to be consumed by the breakpoint reaction.

Sample Tower Calculation. Assume ventilation air at 20,000 SCFM to scrub a possible maximum H_2S concentration of 30 ppm. Fig. 7-9 and Fig. 7-10a illustrate this size tower.

$$30 \text{ ppm} \times 20{,}000 \text{ ft}^3/\text{min} = \frac{600{,}000}{10^6} = 0.6 \text{ ft}^3/\text{min}.$$

Weight of air = 0.075 lb/ft^3
S.G. of H_2S = 1.1895
$0.60 \times 0.075 \times 1.1895 = 0.0535$ lb/min = 77.08 lb/day
Chlorine requirement = 8.32×77.08 lb/day = 642 lb/day

Note: Each part H_2S requires 8.32 parts chlorine.

Next is the calculation of scrubbing water requirement. Tower company manufacturers recommend a liquid to gas ratio of 1.5–2 : 1:

$$\text{wt. of ventilation air} = 20{,}000 \times 0.075 = 1500 \text{ lb/min}.$$

Use a ratio of 1.75; then:

$$1500 \times 1.75 = 2625 \text{ lb/min}$$

$$2625/8.34 = 314.75 \text{ gpm}$$

This is the amount of recirculated water required.

Next, calculate chlorine dosage to see how near it will be to the breakpoint.

$$315 \text{ gpm} = 0.454 \text{ mgd}$$

$$Cl_2 \text{ dose} = \frac{77.08}{0.454} = 169.8 \text{ lb/mg} = \frac{169.8}{8.34} = 20.36 \text{ mg/L}$$

Therefore the dose, 20.36 mg/L is far from the breakpoint. Assume NH_3–N concentration in recirculating water is as low as 10 mg/L. The chlorine dose would have to be well in excess of $6 \times 10 = 60$ mg/L before chlorine starts being consumed by NH_3–N.

Next, calculate the amount of 50 percent caustic required to replace the alkalinity consumed by the chlorine: One gallon of 50 percent sodium hydroxide solution contains 6.38 lb of NaOH and has an S.G. of 1.52. Each part of H_2S removed by chlorine consumes 10 parts of alkalinity. Or, said another way, each part of chlorine added consumes about 1.25 mg/L alkalinity in the neutral pH range. (See Chapter 4.) One pound of caustic produces 1.25 mg/L alkalinity.

Therefore, replace the alkalinity consumed by chlorine on a pound for pound basis. This amounts to 642 lb/day. As there will be some carbon dioxide absorbed in the recirculating process, a daily caustic addition should be about 700 lb/day or 700/6.38 = 109.72 or about 110 gal 50 percent NaOH per day.

Other recommendations: Add chlorine to produce a total chlorine residual of about 1–2 mg/L. Because 700 lb/day chlorine will require a 2-inch injector, adjust the injector throat stem to deliver 30 gpm. This requires a 2-inch disk meter in the injector water line.

Add enough caustic solution to maintain a pH between 7.0 and 7.5.

Monitor pH and chlorine daily, and make a visual inspection of recirculating water overflow to determine its "milkiness." This will establish the frequency of system blowdown. The milkiness results from the inherent formation of colloidal sulfur caused by the reaction in Eq. (7-32).

Monitoring the H_2S in the ventilation air can be done in several ways. The most reliable method is to use a detection device similar to the Jerome or Interscan portable H_2S detector. Other methods include lead acetate–impregnated tiles that turn from white to black when exposed to H_2S. The Sanitation Districts of Los Angeles County use Union Carbide "Gastec" tubes. The tubes are filled with lead acetate–impregnated material. Air is injected at one end of the tube, and the material discolors as the H_2S is trapped. Concentrations can be read directly on a calibrated scale, depending upon the length of discoloration in the tube. Tubes are available in different concentration ranges. Los Angeles uses the 0–120 ppm tubes.

Caustic Towers. Another H_2S scrubbing method is the recirculation of scrubbing water dosed with 50 percent NaOH. The principle of this system is to raise the scrubbing water pH to about 12 so that the H_2S is converted to the HS^- ion. This does not remove the potential of H_2S formation in the scrubbing water. The overflow from this system is discharged into the plant effluent or into the adjacent collection system where it is "removed" by dilution.

The main problem with this system is the water "softening" action of the caustic. Raising the pH precipitates the calcium hardness down to a concentration of 13 mg/L. This requires shutdown and removal by flushing with 5 percent hydrochloric acid. Sanitation districts of Los Angeles County use these towers for scrubbing foul air from ventilation systems in and around the sewer collection system. They use the following design parameters:

Air flow velocity	350 ft/min
Recirculation rate scrubbing water	25 gpm/1000 cfm
Depth of packing media	8 ft
Removal efficiency at up to 150 mg/L H_2S	90–95 percent

Multiple Use Systems. Also available in the marketplace are multi-stage systems that utilize both an oxidant stage and a caustic stage for scrubbing. This hybrid type of system exploits the benefits of the Cl_2 and NaOH (or NaOCl) oxidant system but also reduces oxidant requirements by incorporating a first-stage caustic scrubber (NaOH).

Figure 7-10d illustrates the chlorination equipment used in the Irvine Ranch odor control system at the reclamation plant.

The system is a single-pass three-stage absorption system, consisting of a gas conditioning/pretreatment stage, followed by two vertical co-current/countercurrent packed bed absorption sections. The design of each stage provides a minimum overall performance of 99.0 percent. RJ Environmental provides such a system in a convenient package arrangement. An example of this packaged multiple use system utilizing Cl_2 and NaOH can be seen at the Michelson Water Reclamation Facility at Irvine Ranch, California. The various parts of the multiple system are shown in Figs. 7-10a through 7-10f.

The object of the first stage is to knock down a percentage of the H_2S with NaOH per the following reaction:

$$H_2S + 2NaOH \rightarrow 2Na_2S + 2H_2O$$

An air exhaust fan pulls foul odorous air from the sources via a network of FRP ducting/volume control dampers and passes it through the scrubber system. The gases pass through three efficient packed bed sections. The foul air first enters the preconditioning section, in a countercurrent flow, and contacts a recirculation stream including the chemical blowdown from stages 2 and 3. Most of the chemical content of this stream is utilized by recirculation in the pretreatment stage before blowing down. A NaOH pump and pH controller are included in this stage to reduce oxidant requirements in stages 2 and 3.

In the second stage, the gas flows co-currently, and it is contacted with fresh $NaOH/Cl_2$ or NaOH/NaOCl solution, so that much of the odorous material is oxidized. See the chemistry previously shown in this section. A final "polishing" takes place in the third stage, where gas and liquid are contacted countercurrently. This arrangement of gas absorption assures (1) complete and guaranteed odor removal with efficiencies in excess of 99.0 percent, and (2) complete chemical utilization prior to discharge from the system.

The composition of the chemical sump is maintained at a preset pH and ORP by injecting NaOH and Cl_2 or NaOCl, respectively. Chemical injections are controlled with pH and ORP controllers. Water losses due to evaporation and blowdown are compensated by adding water via a manual rotameter if required. An overflow above the liquid level ensures that the chemical sump can never be overfilled. A low level alarm, set at below the designed sump level, provides system warning.

The chemical sump and absorption stages are all housed in a single FRP chamber with access ports for easy and quick access to any part of the system. The spray nozzles in each section are easily removable.

Gas preconditioning followed by two stages of high efficiency scrubbing guarantees a minimum overall H_2S removal efficiency of 99.0, while minimizing chemical requirements.

RJ environmental has developed a configuration that incorporates a vertical recirculation pump of fiberglass-reinforced plastic. This has resulted in tremendous advantages over hirozontal pump design. With the pump submerged in the recirculating fluid, there is no mechanical seal or seal water required. With no mechanical seal, the installer will not need to coordinate seal flush piping, a reliable water source, drainage, or possibly associated heat tracing and insulation. With the pump submerged, there is no need for relying on the installer's connection of the pump suction piping; so the risk of leakage in the years of service ahead is reduced. In cold weather applications, the entire liquid storage area may be insulated and heated as a single application, without any concern for exterior liquid-bearing recirculation piping. By utilizing the space above the liquid storage sump, this new arrangement allows application in smaller areas.

The complete system is shipped to the job site as single-piece construction, greatly reducing installation time. All system components are easily accessible from the outside. The pumps are located on the deck and can be removed from service in a very short period of time. The complete system (scrubber and fan) is made of premium grade vinyl ester resin. The pumps are made of FRP or CPVC, and piping is schedule 80 PVC.

Carbon Columns. These are used in sensitive areas where 100 percent H_2S removal is required when caustic towers are used as the primary H_2S remover. Sometimes they are used to polish the discharge from biological towers. Biological towers cannot be depended upon to achieve 100 percent removal. Therefore supplementary treatment with chlorine or virgin activated carbon is required to ensure complete removal. Los Angeles County uses carbon columns to polish off the discharge from caustic scrubbers and seems satisfied with the results.

OTHER METHODS OF H_2S GENERATION CONTROL

Caustic Addition

The Sanitation Districts of Los Angeles County have had many years of experience with the intermittent use of caustic to destroy the slime layer in the far reaches of free-flowing sewers that cause sulfide problems. This treatment is most effective. When the sewage temperature reaches about 20°C in the winter, the treatment is done once every two weeks to once a month; and in the summertime, when the sewage temperature reaches 25°C, the treatment is done once every two weeks.

Treatment consists of a shock dose of 50 percent NaOH at the rate of 620 lb/mg applied over a 30-min. period. This is a shock dose of 620 lb

0.1 N caustic to 20,880 gal of sewage. Application of more caustic than the above dose will not improve the effect.

Air-Oxygen Injection

General Discussion.[67] These two methods are based upon the oxygen uptake of the sewage. To prevent H_2S generation it is necessary to oxidize the sulfides:

$$2HS^- + 2O_2 \rightarrow S_2O_3^= + H_2O \tag{7-33}$$

This is the predominant reaction in sulfide oxidation. It forms thiosulfate. Approximately one part of oxygen is required to oxidize one part of sulfide, as shown by the above equation.

Sewage increases its capacity to oxidize sulfides with age or with the diameter of the pipe. Conversely, the DO diminishes as the size of the line increases.

Typical oxidation rates of sulfides in free-flowing sewers can be as high as 5.0 mg/L sulfides oxidized per hour per liter, provided the DO can be maintained at 0.3 mg/L or greater. Of this total oxidation, only about 0.5 mg/L is from direct chemical action by the DO. The rest is from biological oxidation. Biological oxidation rates vary widely in different sewage. Fresh sewages display lower biological oxidation rates than do older sewages. In large lines the DO is usually less than 0.3 mg/L, and as a result the oxidation rate will be considerably less than 0.5 mg/hr/L.

The oxygen uptake rate is:

$$O_2 \text{ uptake} = \frac{[3600 \times 560 \times 10^{-6} \times (SV)^{0.375}]O^D}{d_m} \tag{7-34}$$

Simplifying:

$$O_2 \text{ uptake} = \frac{[2.016\,(SV)^{0.375}] \times O^D}{d_m} \tag{7-35}$$

where

O_2 uptake = mg/hr/L
S = slope, ft/ft
V = velocity, ft/sec
O^D = DO deficit, mg/L
d_m = mean hydraulic depth, ft

This is the sole means of measuring the sulfide destruction ability of a given sewage. Equation (7-34) was developed by the Hydrogen Sulfide Laboratory, Los Angeles County Sanitation District.[67]

Oxidation Capacity of Raw Sewage. This factor is known as the *i* factor, expressed as follows:

$$i = \frac{\Delta S_{\text{ox}}}{R_R + R_{\text{SO}}} \tag{7-36}$$

where

ΔS_{ox} = oxidation rate of sulfides (sulfides removed), mg/hr/L
R_R = oxygen consumed if no sulfides are present, mg/hr/L
R_{so} = oxygen uptake rate, mg/hr/L

(Determination of the *i* factor is described in "Sludge Transport," this chapter.) The *i* factor is a function of the oxygen uptake rate divided by the total oxygen demand when the DO is greater than 0.3.

EXAMPLE: Assume S_{ox} = 4 mg/hr/L and $R_{\text{R}} = R_{\text{SO}}$ = 12 mg/hr/L. Then $i = 4/12 = 1/3$. If the available DO in the sewage is only 3.0, then only 1 mg sulfides can be oxidized per liter per hour. This is the *i* factor. Fresh sewage may have an *i* factor of 0.2, whereas fresh sewage that is aged may have an *i* factor of 0.6. The *i* factor varies from sewage to sewage.

It only requires a small amount of DO (0.2) to prevent instream generation of H_2S. This is true regardless of the solids content of the sewage. The slime layer is the critical factor.

The Biotrickling Filter[95]

Odor Control in Sewage Collection Systems. The Los Angeles County Sanitation Districts of Southern California have been diligently researching methods of controlling odors due to unavoidable release of hydrogen sulfide. Odor control at many wastewater facilities is getting more difficult to achieve. This problem is due in large part to the fact that the neighbors at these locations are getting less tolerant of these low-level releases.

The Joint Water Pollution Control Plant (JWPCP) for Los Angeles County must control odor emissions that are not detectable at the plant's boundaries. Therefore, this plant is responsible for the operation of 22 odor control units that treat 74,000 cu ft/min. of process off-gases. The major odorous substance in these off-gases is hydrogen sulfide. See Figures 7-11 and 7-12.

The Joint WPCP treats 325 mgd of wastewater and has more than 9000 miles of sewer collection piping. As can be seen from the flat terrain of the

Fig. 7-11. Aerial photo of flat terrain portion of Los Angeles County wastewater collection system.

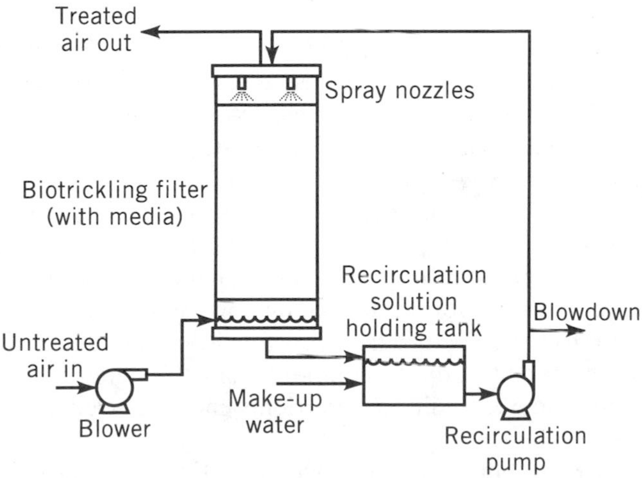

Fig. 7-12. Biotrickling filter pilot plant, Los Angeles County Sanitation District for odor control.

major portion of Los Angeles County (Fig. 7-11) it is not surprising that a large portion of the plant's inlet waters contain high levels of dissolved sulfides. Therefore, the plant managers follow the standard practice of covering and ventilating almost all wastewater treatment processes at the facility to prevent hydrogen sulfide and other odorous emissions.[97]

A variety of chemicals and air scrubbing methods have been tested over nearly two decades of implementing odor control at the plant. As of (1996) the plant has used a two-stage process, which is the most cost-effective system to date. The first stage uses liquid sodium hydroxide in countercurrent-flow packed towers to remove most of the hydrogen sulfide. In the second stage, dry, nonimpregnated activated carbon removes any remaining hydrogen sulfide and many of the odorous volatile organic compounds (VOCs) in the off-gases.

The annual operations and maintenance cost for the 22 odor control units (liquid caustic and activated carbon scrubbers) is approximately \$1.4 million, including chemical and labor costs.

Recognizing the high costs of the odor control systems, the Districts team has been pursuing a research program on low-cost biological odor control methods.

The likelihood of occurrence of the generation of hydrogen sulfide odors is directly related to the terrain in which the sewage collection system lies. As can be seen in Fig. 7-11, the terrain is flat as far as the eye can see. Any area that has a flat collection system, like many parts of Texas, is going to be plagued with all the hydrogen sulfide odor problems known to humankind for the duration. So it has to be dealt with, as the Sanitation Districts of Los Angeles County are doing.[97]

Filter Details

Fig. 7-12 illustrates the type of the filters used by the Sanitation Districts of Los Angeles County during their extensive research on odor control. These filters are similar to biofilters, but contain conventional, high-porosity, wet scrubber packing materials instead of compost, peat moss, wood chips and other organic materials. A recirculating liquid flows over the biotrickling packing material, either co-current or countercurrent to the flow of air being treated, and absorbs contaminants, such as volatile sulfur compounds. Bacteria, which grow as thin films on the packing material and are present in the recirculating solution, convert these contaminants into other compounds such as sulfuric acid. Reaction by-products can be disposed of an environmentally safe manner. Depending upon the compound or compounds that the biotrickling filter must remove, additional nutrients may be needed in the scrubbing solution.

These biotrickling filters are intended to optimize the growth of sulfur-oxidizing bacteria, principally thiobacilli. These bacteria use hydrogen sulfide as their energy source, converting hydrogen sulfide to sulfuric acid as follows:

$$H_2S + 2O_2 \rightarrow H_2SO_4$$

Because the by-product of biological hydrogen sulfide oxidation is sulfuric acid, the recirculation solution must have adequate blowdown rates to ensure that the acid concentration does not increase to a point that it limits the solubility of hydrogen sulfide in the water. Although the sulfur-oxidizing bacteria can survive a pH well below 1.0, an optimum balance for both the growth of the bacteria and the solubility of hydrogen sulfide is achieved in the range of pH 2.0 to pH 3.0. The low pH blowdown water can be mixed with inlet wastewater to neutralize the acid.

Compared with other H_2S control technologies, a biotrickling filter's operational costs can be very low because microbial oxidation occurs at ambient temperatures, no chemicals are consumed, and a minimal amount of labor is required for maintenance.

Probable Costs. Although the capital costs of constructing a biotrickling filter might be high because of its size, the savings in chemicals and labor would quickly made up the difference, in less than a couple of years. Operating and maintaining a sodium hydroxide chemical scrubber costs about $17 per million cubic feet of air treated.

By converting the scrubber system to a biotrickling filter odor control system, the costs would be decreased by about 80 percent. Operating a 1000 cu ft/min. biotrickling filter unit would save $7,000 per year over a traditional chemical scrubber, and would pay for its construction in two years. Moreover, if all of JWPCP's caustic scrubbers were converted to biotrickling filters, the yearly operational savings would be close to $400,000.

U-Tube Aeration. This concept of wastewater aeration has been tried on a limited scale.[25, 26] Figure 7-13 illustrates a typical U-tube installation. Air or oxygen is forced (pumped) or aspirated into the top of the downdrop, and the air/wastewater mixture then passes into an enlarged section. In this section, increasing head losses will occur, which will be dependent upon the air-to-water input ratio (by volume). Downleg velocities should be on the order of 1.5 ft/sec to provide enough detention time for optimum oxygen transfer. The upleg section is decreased in size to provide velocities of 4 ft/sec. This will prevent solids from depositing in the bottom of the U-tube. The key variable in U-tube design and operation is the air-to-water ratio. This ratio should be between 0.05 and 0.10. The critical value is about 0.1, where complete dispersion of air in the wastewater no longer occurs, and the head loss increases sharply. Oxygen transfer using pure oxygen instead of air is much higher, and head losses are less.

According to Pomeroy, the biological oxidation of sulfide in a typical sewage is about 10–15 mg/hr/L after the sewage has been flowing for a few hours. If the sulfide and DO concentrations are not less than 1 mg/L, this rate is

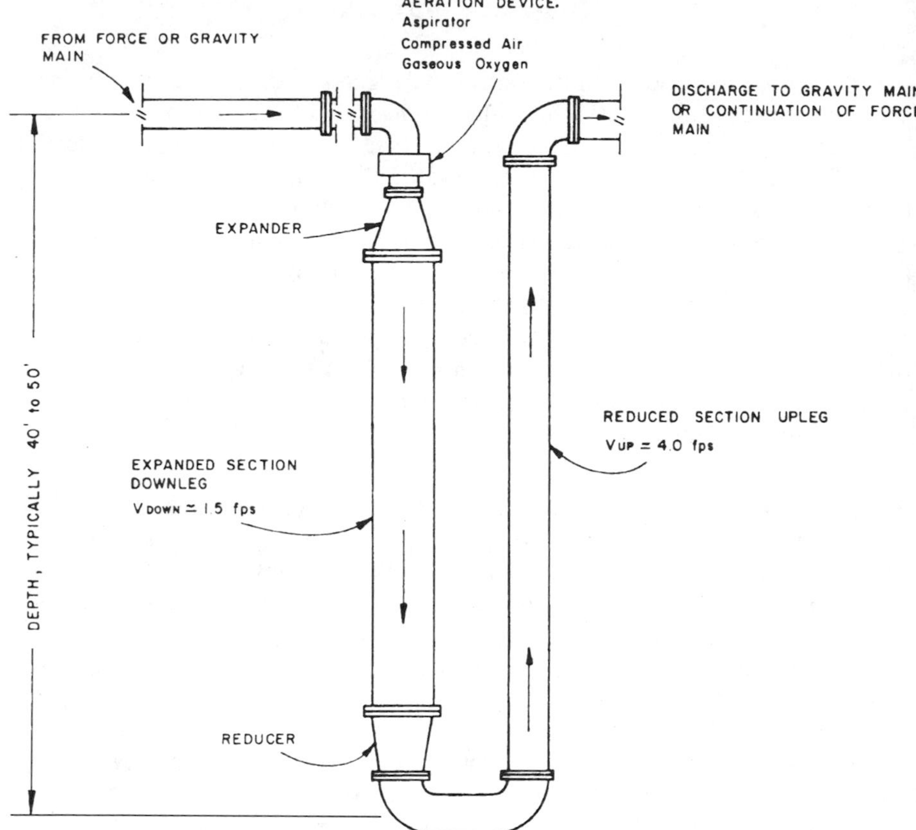

Fig. 7-13. Typical U-tube aeration installation.

independent of these concentrations.[28] Hoag et al.[25] confirmed this rate in their field studies.

Air and/or oxygen injection does not usually result in complete removal of the sulfides. However, this method can be used to reduce the sulfides to slightly below the critical level of dissolved sulfides, which is considered to be 0.3 mg/L.

Location of the U-tube is a difficult decision to make. Once the location is established, it cannot be moved and still remain cost-effective. Some U-tubes have been located near the end of a force main as a means to oxidize the sulfides before the sewage in the force main is released to atmospheric pressure. In these cases the dissolved sulfides must be kept below the critical level at the end of the force main.

Pure oxygen flow rates used in the Sacramento studies were 5–7 SCFM while air was 15–30 SCFM.

Force Mains

Compressed air injection into a force main has been used with varied success. Studies by the Sanitation Districts of Los Angeles County resulted in dividing force mains into two categories, each with its own formula for calculating the quantity of air required for sulfide control.[68]

Category I. This category includes force mains 14 inches in diameter or larger. In this category the oxidation rate within the pipe exceeds the sulfude generation. Therefore, only small amounts of air are required to completely remove any sulfide being generated within the force main piping. Moreover, the DO demand of the slime layer attached to the pipe wall is near zero at the air level, which controls the sulfide buildup. The mathematical relationship for calculating the quantity of air is as follows:

$$(\text{SCFM})_{\text{control}} = \frac{4.0 \times 10^{-3} \times D \times L}{1\text{n } P_{ps}} \tag{7-37}$$

where

D = pipe diameter, in.
L = length of force main, ft
ln = natural logarithm
P_{ps} = abs. pressure where air is injected, atmospheres

$$= \frac{34 + \text{Static head} + \text{Friction loss}}{34}$$

1 atmosphere = 34 ft*

Category II. This category is for force mains smaller than 14 inches in diameter. As the ratio of sulfide oxidation to sulfide generation is proportional to diameter, this ratio will always be lower for smaller pipes than for larger pipes. In this category the capacity of the sewage to oxidize sulfide is less than the capacity of the slime layer to produce sulfides. Therefore, the only way to successfully eliminate sulfide buildup in this category is to partially suppress generation within the slime layer by oxygen addition sufficient to create an aerobic zone well within the anaerobic zone of the slime layer. The mathematical expression for determining the quantity of air for force mains smaller than 14 inches in diameter is as follows:

$$(\text{SCFM})_{\text{control}} = \frac{5.35 \times 10^{-3} \times D \times L}{1\text{n } P_{ps}} \tag{7-38}$$

*At sea level.

In calculating air requirements for this category, it is suggested that a 20 percent factor of safety be used if there are any flat slopes that amount to 30–50 percent of the total force main length.

Oxygen transfer in a force main is a function of pressure and turbulence. The amount of turbulence and interfacial area determines the rate of oxygen transfer. Around each bubble of air there is a layer of water. The faster the bubble travels (turbulence), the thinner the layer of water, and therefore the greater the O_2 transfer efficiency. This phenomenon explains the effect of the physical slope of a force main. At the lower end of the force main the pressure is higher, and the O_2 transfer efficiency will be higher so there will be "overcontrol" at the lower end. Therefore, in these cases operators should strive for control in the last 25 percent of the length of this section.

Temperature is an important factor. Sewage temperatures at 20°C and above are certain to produce sulfide generation in a force main.

The addition of air does not add to the friction loss in a rising force main, provided that there are no negative slopes. The friction effect of air as an additional volume for the pipe to carry is canceled by the "lift" effect of the air. Force mains with negative slopes are usually not amenable to the compressed air treatment. The hydraulics of air addition generally cause difficulties that negate any success derived for sulfide control. It is best to avoid the use of air injection for these situations.

Another important factor relating to sulfide generation in force mains is the pumping cycle, which affects the turnover in the piping system. If there is sufficient DO in the wet well to oxidize the sulfides generated in the force main during the downtime, and if complete turnover occurs within one pumping cycle, no sulfides will form.

The most important factor in a compressed air system is the air compressor. It should be sized so that it only operates 50 percent of the time; otherwise maintenance becomes a chronic problem. The compressor capacity should be twice that needed for continuous operation. One should arrange to have an adjustable timer to provide an off time up to 15 min. Then adjust the timer to operate the compressor for 7.5 min. on and 7.5 min. off. If this provides too much oxygen transfer, change the timer to 5 min. on and 10 min. off. The compressor should never be off for more than 15 min..

The following is an example of a compressor selection in an existing force main (Category I), which is shown in Fig. 7-14. The pumps are variable speed. The velocity in the force main will vary from 0.5 to 2.5 ft/sec. Therefore the detention time will vary from 1 hour 40 min. to about 8 hours.

From Eq. (7-36) we get:

$$\begin{aligned} \text{SCFM} &= \frac{4 \times 10^{-3} \times D \times L}{\ln P_{ps}} \\ &= \frac{4 \times 10^{-3} \times 1.17 \times 15 \times 10^{3}}{\ln 4.82 = 1.6} \\ &= 43.88 \end{aligned}$$

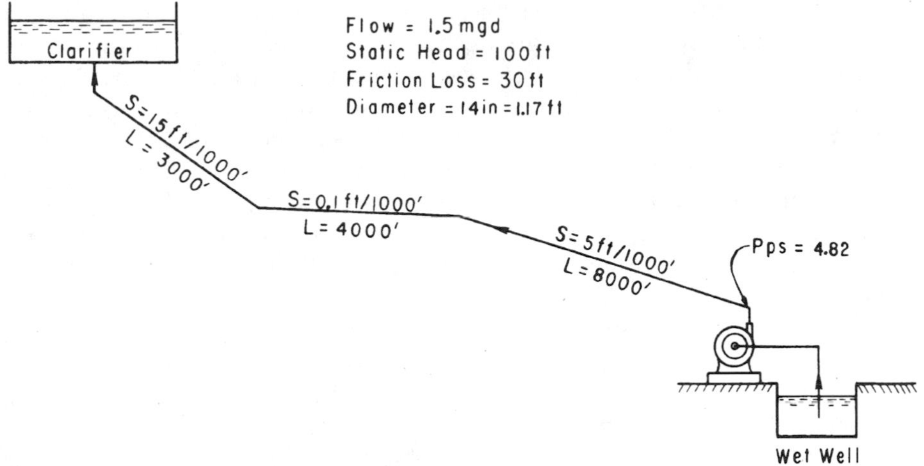

Fig. 7-14. Force main aeration calculation diagram.

Therefore, the compressor should have a nominal capacity of 100 SCFM (three-stage). Operate with the timer for 15 min. on and 15 min. off.

Hydrogen Peroxide. Los Angeles County Sanitation Districts studied the use of hydrogen peroxide for sulfide control and concluded that it was an expensive way to put DO in sewage. The Sacramento studies tried H_2O_2 as an alternative and found that the ratio of hydrogen peroxide to dissolved sulfides would have to be at least 3 to 1 to reduce the sulfides to a level of 0.5 mg/L. Therefore, the dosage to reach the critical level of dissolved sulfides (0.3 mg/L) might be as high as 4 or 5 to 1. The 1984 cost of 50 percent H_2O_2 in 4000 gallon lots was approximately 37 cents/lb. In estimating the overall cost of H_2O_2 half-life decay due to temperature and impurities within the system should be added to the effective dosage rate.

One other significant factor discovered in the use of H_2O_2 is the lack of effect it has on the BOD, COD, or DO concentrations at the end of force mains except when dosages are 60 mg/L.[25]

INDUSTRIAL WASTES

Cyanide Wastes

Significance. Cyanides are toxic to aquatic life, interfere with normal biological processes of natural purification in streams, present a hazard to agricultural uses of water, and are a menace to public water supplies and bathing. Cyanides exert a toxic action on living organisms, animals, and human beings by reducing or eliminating the utilization of oxygen, in a manner similar to asphyxiation. The action at the toxic level is both rapid and fatal.

If the microorganisms in a surface water responsible for proper oxygen balance lose their efficiency or are destroyed, oxygen depletion of the stream ensues. Fish in such a stream stand the chance of being killed either directly by the cyanide content or indirectly by the destruction of the organisms upon which they feed. The toxic threshold level for some species of fish is reported to be 0.05–0.10 mg/L cyanide as (CN) radical.[71, 72] This toxicity increases as the temperature increases and as the dissolved oxygen decreases. The toxic action is apparently related to photosynthesis, for it has been demonstrated that fish in the absence of sunlight can thrive in water containing 1000 mg/L cyanide.[40]

Livestock and other animals are likewise endangered when using a stream polluted with cyanides. Lethal doses to animals and humans amount to only 4 mg/L/lb of body weight.[40] A stream used for bathing or a domestic water supply should not contain even a trace of cyanide.

Cyanide wastes definitely impair the biological processes of waste treatment plants although these processes do reduce the cyanide concentration. The amount of reduction of cyanide concentration in these various processes has never been determined, and therefore it is not possible to establish a maximum allowable limit in the raw sewage entering the plant.

Owing to federal legislation implemented by the Environmental Protection Agency, all cyanide wastes discharging into any wastewater treatment collection system must be reduced to cyanates. All cyanide wastes discharging into the environment, whether it be in land disposal or in surface water, must be reduced completely to carbon and nitrogen. The latter situation has been standard practice for many years.

In addition to the cyanides present in metal-finishing wastes, there may be copper and zinc ions, which are also toxic to fish and other aquatic life. The toxic threshold level for some species of fish is reported to be 0.10–0.20 mg/L copper and zinc.

Occurrences. Although cyanide compounds are widely used in industry, these five processes are responsible for most of the cyanide wastes causing stream pollution and presenting problems in waste treatment plant operation: metal plating, case hardening of steel, neutralizing of acid "pickle scum," refining of gold and silver ores, and scrubbing of stack gases from blast and producer gas furnaces.

Metal Plating. The greatest source of cyanide-bearing waste is found in the metal-finishing industry, whose electroplating plants are distributed throughout the United States. Most of the cyanide wastes are rinse waters, spillages, and drippings from plating solutions of cadmium, copper, silver, gold, and zinc. These plating solutions will vary in concentrations of alkali and "free" cyanide (sodium or potassium), but the various metal baths are similar in concentration. The following shows the contents of a solution for a brass-plating operation:

Copper cyanide	4 oz/gal	30,000 mg/L
Zinc cyanide	¼ oz/gal	9200 mg/L
Sodium cyanide	7½ oz/gal	57,000 mg/L
Sodium carbonate	4 oz/gal	30,000 mg/L
	Total CN = 39,000 mg/L	

Case Hardening of Steel (Nitriding). The process of heat treating steel by cyanide consists of immersing the steel part at a predetermined temperature in a molten bath of a mixture of sodium cyanide, barium chloride, sodium chloride, potassium chloride, and strontium carbonate. When the part has been held for the proper time in this molten bath, it is withdrawn and quenched in a continuous flow of water. The discharge from such a bath may be expected to run as high as 50–100 mg/L of free cyanide.[71]

Neutralizing of Acid Pickle Scum. In the pickling of steel sheet for the removal of mill scale, the metal is usually dipped in dilute sulfuric and hydrochloric acid. As the steel leaves this pickling bath, a coating called "pickle smut," consisting of materials not soluble in the acid—iron carbide, silicon, silicon carbide, copper, and copper sulfides—remains on the surface. In addition, ferrous sulfate and ferrous chloride are formed and cling to the surface. If the sheet is then air-dried or neutralized with caustic, the ferrous iron is converted to ferric iron, which covers the pickle smut and forms a coating difficult to remove and destructive to the finish of the sheet. It can be removed by agitation in a solution of sodium cyanide (0.2 oz/gal) and sodium hydroxide (0.2 oz/gal). The resulting waste from this solution contains free cyanide in concentrations between 200 and 1000 mg/L.[71]

Refining Gold and Silver Ores. The present method of gold recovery from ore is the McArthur-Forrest process, first patented in 1887, which is based upon the solubility of gold in a cyanide solution. The gold is ultimately precipitated from this solution by zinc dust. The chemical reactions are as follows:

$$4Au + 8KCN + O_2 + 2H_2O \rightarrow 4KAu(CN)_2 + 4KOH \qquad (7\text{-}39)$$

The precipitation of gold by zinc dust is represented by:

$$2KAu(CN)_2 + Zn \rightarrow 2Au \downarrow + K_2Zn(CN)_2 \qquad (7\text{-}40)$$

The potassium–zinc–cyanide solution remaining after the gold is precipitated in Eq. (7-40) is the waste to be treated.

Alternatively, gold can be dissolved in a sodium cyanide solution:

$$4Au + 8NaCN + O_2 + 2H_2O \rightarrow 4NaAu(CN)_2 + 4NaOH \qquad (7\text{-}41)$$

The gold is then precipitated with zinc dust as follows:

$$4NaAu(CN)_2 + 2Zn \rightarrow 4Au \downarrow + 2Na_2Zn(CN)_4 \quad (7\text{-}42)$$

The recovery of silver from its ore is similar. The reaction is:

$$2Ag + 4NaCN + O + H_2O \rightarrow 2NaAg(CN)_2 + 2NaOH \quad (7\text{-}43)$$

The silver is then precipitated by zinc dust or aluminum powder. The remaining cyanide solution must then be treated for the destruction of cyanides before disposal.

Scrubbing or Stack Gases. Cyanide is a by-product of the manufacture of carbides. The cyanide gets into the flue gas and appears in the wastewater from the flue gas scrubber. This wastewater must then be treated for the destruction of the cyanides.

All of the wastes described above are so toxic to humans and animals alike that even in remote areas such as Nevada, where there are many gold and silver processing plants, the state department of health requires that the cyanide be completely destroyed, even when these wastes are stored in protected ponds.

Chemistry of the Treatment

There are two types of cyanide wastes: free cyanide (NaCN or KCN) and combined or complex cyanide ($Na_2Au(CN)_2$).

Destruction of Free Cyanide to Cyanate. When chlorine in its various forms (represented as Cl_2 below) is applied to a free cyanide such as sodium cyanide, oxidation takes place, producing cyanogen chloride:

$$NaCN + Cl_2 \rightarrow CNCl + NaCl \quad (7\text{-}44)$$

This reaction is practically instantaneous, and, is independent of pH.[73] Cyanogen chloride is also a toxic waste, and, as it volatilizes readily (similar to NCl_3), its formation is to be avoided. Cyanogen chloride can be converted to a more stable compound with proper pH control. In the presence of alkali, represented by NaOH, it decomposes to cyanate at pH 8.5 to pH 9.0 as follows:

$$CNCl + 2NaOH \rightarrow NaCNO + NaCl + H_2O \quad (7\text{-}45)$$

At a controlled pH of 8.5–9.0, it takes 10–30 min. for 100 percent completion of this reaction—that is, the conversion of free cyanide to cyanate with chlorine. As the pH increases, the reaction time diminishes. At pH 10–11, the time is on the order of 5–7 min. If the pH drops to as low as 8.0, cyanogen chloride will begin to form. This is to be avoided.[73]

The theoretical requirements for destruction of free cyanides to cyanates are as follows: 2.73 parts of chlorine for each part of free cyanides as CN;

1.125 parts of caustic as NaOH or 1.225 parts of hydrated lime ($Ca(OH)_2$) per part of chlorine applied. In actual practice, it requires more chlorine because other chlorine-consuming compounds are usually present.

Usually the waste is of rather high alkalinity, and so the alkali requirements average out less than one part—usually 0.6–0.8 part of alkali per one part of chlorine. If ammonia is present in the waste, causing formation of chloramines, it is not economically desirable to chlorinate beyond the breakpoint to a free chlorine residual, as chloramines will react to destroy the free cyanide. It simply takes a little longer time. For chloramine treatment, allow an additional 15 min. The pH control is critical and is to be measured after chlorination.

Either the flow-through or the batch system of treatment is acceptable when it is necessary only to destroy cyanides to cyanates. The flow-through system can be more economical than the batch system because it is possible to control the reaction between chlorine and cyanide to form cyanate before the reaction between chlorine and cyanate can begin. This economic advantage may not be a primary consideration when it is imperative always to destroy the free cyanide completely.

Destruction of Cyanates to Carbon and Nitrogen. Cyanates are several hundred times less toxic than are the cyanides, but in some instances it is mandatory to destroy them. The cyanates present as the end product in the destruction of the cyanides at pH 8.5 to pH 9.0 are not readily decomposed by water or the excess alkali present in the treated waste unless free chlorine is present. The cyanates in the presence of chlorine slowly hydrolyze to form ammonium carbonate and sodium carbonate:[73]

$$3Cl_2 + 4H_2O + 2NaCNO \rightarrow 3Cl_2 + (NH_4)_2CO_3 + Na_2CO_3 \quad (7\text{-}46)$$

Chlorine does not take part chemically in this reaction, but does aid in completing the reaction within one to one and one-half hours. It is thought that chlorine acts as a catalyst in Eq. (7-46). The chlorine in Eq. (7-46) must be free available HOCl to complete this reaction. After the hydrolysis of Eq. (7-46) takes place, the chlorine and caustic rapidly oxidize the ammonium carbonate to nitrogen gas, and the carbonates are converted to bicarbonates as follows:

$$\begin{gathered} 3Cl_2 + 6NaOH + (NH_4)_2CO_3 + Na_2CO_3 \rightarrow \\ 2NaHCO_3 + N_2\uparrow + 6NaCl + 6H_2O \end{gathered} \quad (7\text{-}47)$$

As part of this reaction, but not shown, small amounts of inert nitrous oxide (N_2O) and volatile nitrogen trichloride (NCl_3) are also formed.

The complete destruction of cyanides and cyanates to carbonates, nitrogen, and nitrous oxide theoretically requires: 6.82 parts of chlorine per part of free cyanide as CN; 1.125 parts of caustic (NaOH) or 1.225 parts of hydrated lime ($Ca(OH)_2$) per part of chlorine applied.

In practice, it requires slightly more chlorine to deal with other chlorine-consuming compounds that may be present. Any ammonia initially present will have to be destroyed by the breakpoint phenomenon in order to ensure the formation of free chlorine. This requires approximately 10.0 mg/L chlorine per mg/L of ammonia nitrogen.

The alkali requirements are usually less than one part (0.6–0.8) of alkali per part of chlorine, owing to the frequently high initial alkalinity of these wastes.

Either a flow-through or a batch system can be employed for the complete destruction of cyanides. In either case, the minimum holding time for all reactions to go to completion is one and one-half hours. In a flow-through system, short-circuiting becomes critical; therefore, it is necessary to monitor the waste at several points, utilizing ORP measurements.

Removal of Metallic Ions. In the electroplating industry, the plating baths and rinse waters contain an excess of cyanide above that required to form the soluble metal cyanide complexes. The excess cyanide is known as free cyanide; in the complex metallic form, it is referred to as combined cyanide. The latter forms can be those of cadmium: $Cd(CN)_2 \cdot 2NaCN$; of copper: $Cu(CN) \cdot 2NaCN$; of silver: $AgCN \cdot NaCN$; and of zinc: $Zn(CN)_2 \cdot 2NaCN$.

The chlorine and alkali requirement for the destruction of cyanides and cyanates in metal plating is based on the total cyanide present = free plus combined.

Chlorination of metallic cyanide wastes differs from that of free cyanides as follows:

1. The metals are more or less converted to insoluble oxides, hydroxides, or carbonates.
2. These insoluble compounds require additional facilities or sedimentation and sludge disposal.
3. Additional chlorine is required to oxidize certain metals to a higher form, specifically copper and nickel.
4. When nickel is present, it interferes somewhat with the destruction of cyanides to cyanates.
5. When iron is present as ferrocyanide, it is readily oxidized to ferricyanide, which resists destruction by the alkaline chlorine process.

The chlorination of such plating wastes to the soluble alkali cyanates and the insoluble metallic precipitate is shown by the following reaction, which forms insoluble zinc carbonate:

$$Zn(CN)_2 \cdot 2NaCN + NaCN + Na_2CO_3 + 10NaOH + 5Cl_2 \rightarrow ZnCO_3 \downarrow + 5NaCNO + 10NaCl + 5H_2O \quad (7\text{-}48)$$

or by the reaction forming insoluble zinc hydroxide:

$$Zn(CN)_2 \cdot 2\,NaCN + NaCN = 12NaOH + 5Cl_2 \rightarrow \qquad (7\text{-}49)$$
$$Zn(OH)_2 \downarrow + 5NaCNO + 10NaCl + 5H_2O$$

The above reactions are rapid in the presence of a free chlorine residual. If chloramines occur, the reaction will not go to completion, and insoluble metallic cyanides may be found in the sludge. This can be avoided by monitoring to a free chlorine residual or to the proper ORP level. The volume of sludge will be on the order of 1 percent of the initial waste when the total cyanide concentration is on the order of 50–75 mg/L.

When copper or nickel plating rinse waters (which are cyanide-free) are mixed with other cyanide wastes and treated, additional chlorine is consumed in converting the cuprous copper and nickelous nickel to higher forms as follows: 0.56 part of chlorine per part of cuprous copper; 2.25 parts of chlorine per part of nickelous nickel; 1.125 parts of caustic (NaOH) or 1.225 parts of hydrated lime ($Ca(OH)_2$) per part of chlorine applied.

Interference by Nickel. Nickelous nickel forms a cyanide complex that interferes with the destruction of cyanides to cyanates. It is reported that this complex cyanide cannot be completely destroyed to cyanate in a relatively short time, even in the presence of a free chlorine residual. However, it creates no problem when the cyanides are to be completely destroyed to carbon dioxide, nitrogen, and nitrous oxide.

In a flow-through plant in which the cyanides are to be converted to the cyanates, the presence of nickel may be solved as follows:

1. Separate nickelous waste from the cyanide waste.
2. Apply nickelous waste to inlet of settling basin along with the cyanide waste after chlorination (nickelous waste does not contain cyanide).
3. Allow the nickelous oxide or hydroxide formed by the reaction with the alkali to settle out. Any excess free chlorine in the chlorinated cyanide waste rapidly converts the nickelous nickel to the black nickelic oxide, which hydrolyzes further to the insoluble hydroxide.

Interference by Iron. Photographic process wastes contain potassium and sodium ferricyanides among other significant chemicals deleterious to the quality of receiving waters. Both ferrocyanide and ferricyanide are toxic to some fish in the presence of sunlight at levels as low as 2 mg/L and therefore should not be discharged without treatment to bodies of water used for beneficial purposes.

While the ferrocyanide is readily oxidized to ferricyanide by chlorine, it does not appear that it can be destroyed directly by alkaline chlorination. It is therefore necessary to decomplex the iron cyanide so that it can be destroyed by chlorine. The addition of mercuric chloride to the waste, to decomplex the iron cyanide, has been suggested.[74] This would form mercuric cyanide, which can be destroyed by the alkaline chlorination process, but the mercury would then have to be recovered from the waste.

Process Control. The monitoring and the control of cyanide destruction to either cyanates or to carbon dioxide and nitrogen require sophisticated laboratory expertise.

For the flow-through or continuous process treatment, the only satisfactory monitoring technique is the use of ORP control. This technique is thoroughly described in Chapters 9 and 10. The ORP measuring cell consits of a measuring electrode and a reference electrode (Fig. 7-15). There are two conveniently spaced ORP plateaus in the destruction of cyanides by the alkaline chlorination process. Figure 7-16 shows the first plateau, where cyanide is converted to cyanates at approximately +400 mV. The complete destruction of the cyanates occurs at the second plateau, some +200 mV higher.

These values are relative, depending on the poise of the waste. For example, the first plateau on some wastes might be as low as +325 mV, and the second at +500 mV. Therefore, the poise of the waste must be determined in order to establish the setpoint of the ORP controller.

If the cyanides are to be converted to cyanates only, then it is necessary to run quantitative tests for cyanides and cyanates to determine the optimum chlorine dosage. It is from these tests that the first plateau of the oxidation reduction potential is determined.

The following procedures are suggested:

- The modified Liebig titrimetric method, for the determination of total cyanide in raw waste and the total residual cyanide in the treated waste,[75]

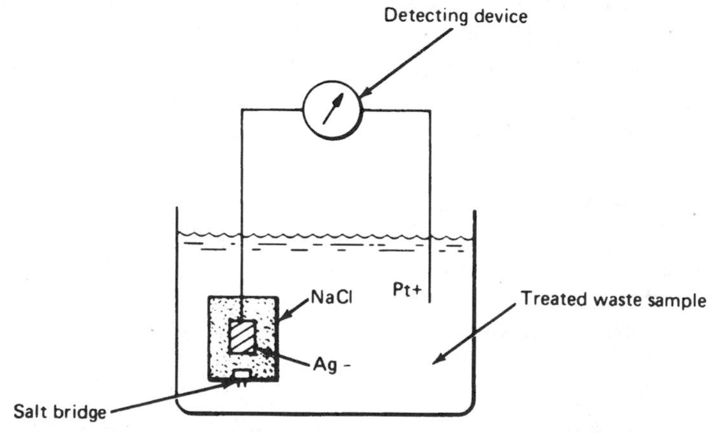

Fig. 7-15. ORP cell for cyanide waste treatment.

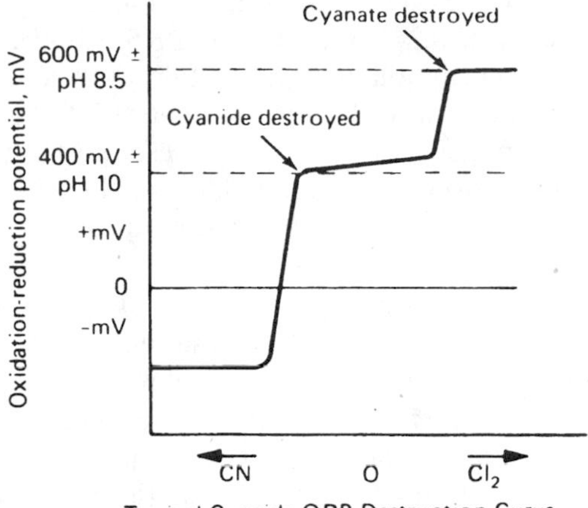

Fig. 7-16. Typical ORP versus cyanide destruction curve.

is applicable where great accuracy is not required and where it is not necessary to quantitatively measure the presence of minute quantities of total residual cyanide.

- The pyridine–benzidine method[76] and the pyridine–pyrazolone method[77] can be used for the microanalysis of any residual total cyanide left in the waste after treatment.
- A modification of the Herting method[78] is used for determining any residual cyanate left in the waste after treatment.

For the complete destruction of cyanides in a flow-through system, it is necessary to carry a free chlorine residual of 2–5 mg/L at the end of the required contact time. Measuring the ORP at this residual level will determine the poise of the particular waste and thereby establish the setpoint of the ORP controller.

It is necessary at all times to be able to determine quantitatively the chlorine residual at the selected control points in the treatment system. This is done with the amperometric titrator.[79] For distinguishing between free and combined chlorine, only a properly modified amperometric titration procedure is satisfactory. Heavy metal ions, such as cuprous and silver, as well as high concentrations of cupric ion interfere with the amperometric titration. In the complete destruction of cyanide wastes, dichloramine is always present, and the total combined chlorine may amount to as much as 25 times that of the free chlorine. It is possible to titrate free chlorine in the presence of up to 25 mg/L of monochloramine, but the dichloramine must be less than 5 mg/L. The remedy is to apply opposing voltage to make the proper change in the zero point of the ammeter scale on the titrator. This is easily done by use of a 1.5 Vt dry cell.[79]

For batch process control, the following is suggested:

- For the destruction of cyanides to cyanates, control to a free chlorine or chloramine residual of 0.2–0.5 mg/L at pH 8.5–9.0 after chlorination.
- For the complete destruction of cyanides to carbon dioxide and nitrogen, continued chlorination at a pH not less than 8.5–9.0 is required until a free chlorine residual (0.5–1.0 mg/L) is present after not less than one and one-half to two hours' retention. This would be at the suction side of the circulating pump and ahead of the point of chlorine application.

When metallic ions are present, the destruction of cyanides to cyanates also requires the appearance of a free chlorine residual.

Proper laboratory analysis will require dechlorination of the sample and usually distillation as well. These procedures are necessary to prepare the sample for the quantitative estimation of total cyanides and cyanates.

In using chlorinated samples for a total cyanide or cyanate analysis, it is essential (1) to remove any remaining cyanogen chloride before dechlorination by increasing the pH of the sample to 10–11 by addition of alkali, preferably NaOH, and (2) to remove all residual chlorine by a reducing agent such as sodium sulfite.

Distillation is necessary to obtain the total cyanide in the raw or treated waste when metallic cyanide complexes are present and to separate the cyanide from objectionable turbidity, color, and other substances that interfere chemically with specific tests for cyanides.

Hypochlorite. If the quantities of cyanide to be destroyed are small enough, it is perfectly acceptable to use hypochlorite. The same amount of available chlorine is required. The only difference is that caustic is not required when hypochlorite is used. The following equations demonstrate why equal amounts of available chlorine are required, whether it be hypochlorous acid made from chlorine gas and water:

$$Cl_2 + H_2O \rightarrow HOCl + HCl \tag{7-50}$$

or sodium hypochlorite, which is made from chlorine gas, water, and caustic:

$$HOCl + NaOH \rightarrow NaOCl + H_2O \tag{7-51}$$

Therefore, using chlorine water:

$$\begin{aligned} &5HOCl + 5HCl + 10NaOH + 2NaCN \rightarrow \\ &\quad 2NaHCO_3 + N_2\uparrow + 10NaCl + 9H_2O \end{aligned} \tag{7-52}$$

or using hypochlorite:

$$5NaOCl + H_2O + 2NaCN \rightarrow 2NaHCO_3 + N_2 + 5NaCl \tag{7-53}$$

Chlorination Facility

The first chlorination plant for the destruction of cyanides in the United States was built in 1942.[71] Since that time the alkaline chlorination system has been widely accepted. Some of these installations have been well documented.[80–83]

Many plant installations operate mainly on the batch method but utilize automatic pH control and ORP measurement to control the destruction to cyanates. In this form, the individual waste treatment system is allowed to discharge the cyanates directly to the municipal or other waste collection system. In some localities, wastes from individual plants are collected in a tank truck and taken to a central batch treating plant. Other plants, for example, those in areas of automobile manufacturing, munitions manufacturing, and ore processing, may be the sophisticated completely automated flow-through type.

It is customary to recirculate the cyanide waste through the chlorinator injector wherever possible. If a batch system is indicated, smaller, less expensive equipment may be used, whereas in a flow-through system more units of larger capacity may be required because the chlorinator "sees" the waste only once in the flow-through system. In both systems, the water to power the operation of the injector is always that of the waste itself. Figure 7-17 illustrates a typical batch system.

Let us assume that the cyanides are to be converted only to cyanates. The total cyanide content of the waste is 1500 mg/L, and the capacity of the batching system is 5000 gal/day. Conversion to cyanates requires a chlorinator capable of delivering 4 mg/L Cl_2 per mg/L cyanide = 6000 mg/L total capacity in 5000 gal.

$$6000 \times 8.33 = 49{,}980 \text{ lb/mg} \qquad \text{mg} = \text{million gallons}$$
$$5000 \text{ gal} = 0.005 \text{ mg}$$
$$0.005 \times 49{,}980 = 249.9 \text{ lb chlorine required}$$

A 2000 lb/day chlorinator would be the unit of choice. This unit operating at capacity could treat the entire batch in three hours by recirculating all the waste through the injector, as shown in Fig. 7-14. The caustic feeder should be able to match the 2000 lb rate of the chlorinator; this would amount to a rate of approximately 313 gal/day, so that the caustic feeder should have a capacity of approximately 350 to 400 gpd.

Theoretically it requires 1.125 parts of NaOH to neutralize each part of chlorine. In practice, because these wastes have such high alkalinities and high pH, it usually requires about 0.9–1.0 part caustic per part chlorine. Assume 1.0 part caustic to each part chlorine.

Caustic is usually handled at 50 percent strength. The specific gravity is 1.52, and thus each gallon of 50 percent caustic = 6.38 lb NaOH.

$$\frac{2000 \text{ lb } Cl_2 \text{ per day}}{6.38 \text{ lb/gal}} = 313 \text{ gal/day}$$

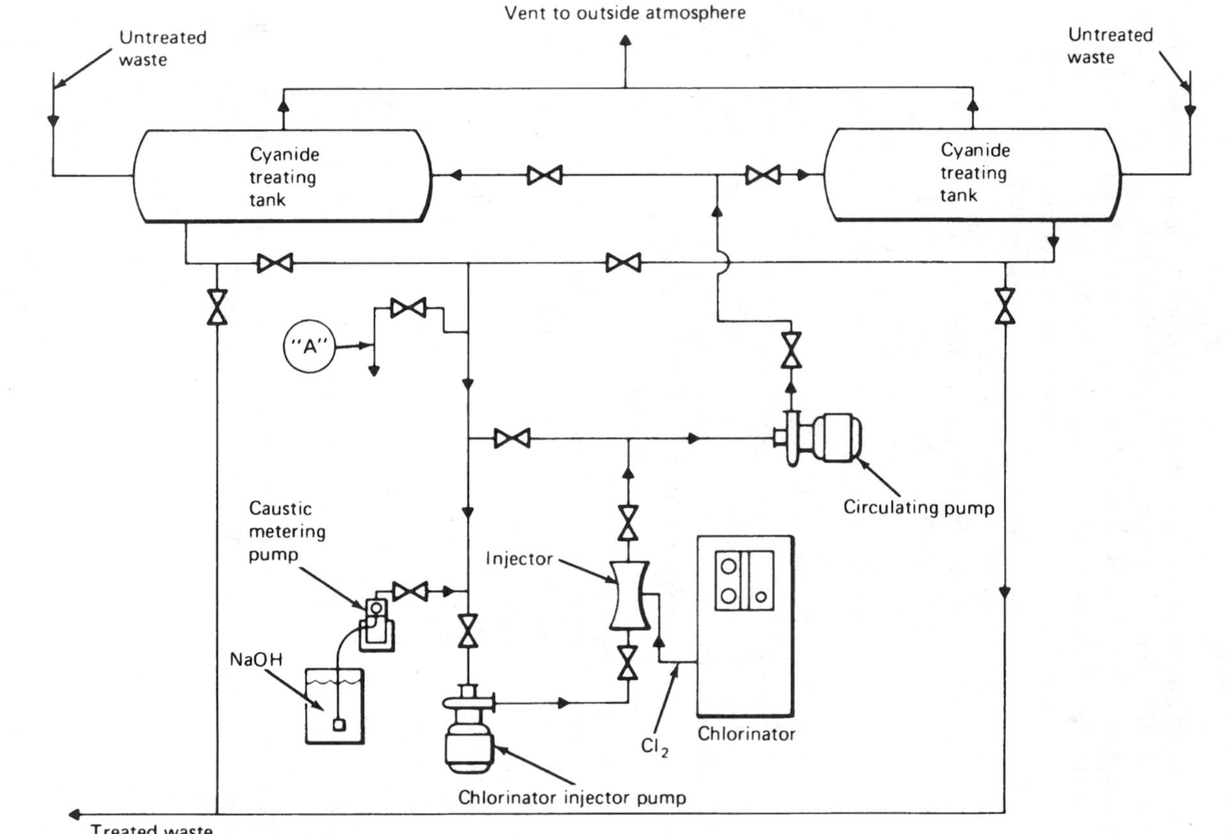

Fig. 7-17. Typical schematic diagram of batch-type chlorination system for the destruction of cyanides. (Free chlorine residual at point "A" denotes complete destruction of cyanides to carbon dioxide, nitrogen, and nitrous oxide.)

The system should have an indicating pH meter. Automatic operation or a time clock system is optional. Confirmation of cyanide to cyanate can be achieved by an indicating ORP unit.

Figure 7-18 illustrates a completely automated flow-through system. Here the pH is automatically controlled by using a ratio system to provide the proper amount of caustic to the chlorine being fed. In this case, it has been decided that the complete destruction of cyanide to nitrogen and carbonates is required. The success of this system is based on the ORP control and monitoring system. Once the setpoint of the ORP controller has been determined by laboratory procedure, described previously, the ORP controller takes over the control and monitoring of the chlorination system. This, in conjunction with automatic pH control, integrates a completely automatic flow-through system.

Frequent laboratory checks should be made to verify the ORP controller setpoint as well as the validity of the pH electrodes.

Controlling the Oxidation of Cyanides by the Chlorination Process

The Potential Curve or Characteristic for the Oxidation of Cyanides. When chlorine in increasing amounts is applied to a cyanide-bearing stream, the pH of which has previously been adjusted to 8.5 to 9.0, the potential of the treated waste increases along a characteristic curve similar to that illustrated by Fig. 7-19. The potential levels plotted vary somewhat with the potential cell; therefore, the potentials shown in the figure are only examples of the different levels recorded by a representative cell.

Furthermore, it is important to remember that a change in pH affects the potential. A drop in pH results in an increase in potential, whereas a rise in pH causes a decrease in potential. It is therefore essential for best control that the pH of the raw waste be maintained within reasonably close limits prior to chlorination.

The potential system, for purpose of further discussion, is divided into four potential levels: (1) the reduced or cyanide potential level; (2) the cyanate–chloramine potential level; (3) the free chlorine potential level; and (4) the controlling potential level.

The Reduced or Cyanide Potential Level. The reduced potential level is the potential level of the raw waste and the raw waste chlorinated with an insufficient amount of chlorine (less than 2.73 parts of chlorine per part of CN). The constituent most affecting the potential is the cyanide. This reduced potential level can therefore also be called the "cyanide potential level." In the potential system involved with the chlorination of the waste, it is the lowest potential level.

In Fig. 7-19, the cyanide potential level (−300 mV) is maintained with increasing applications of chlorine until practically all of the cyanide has been

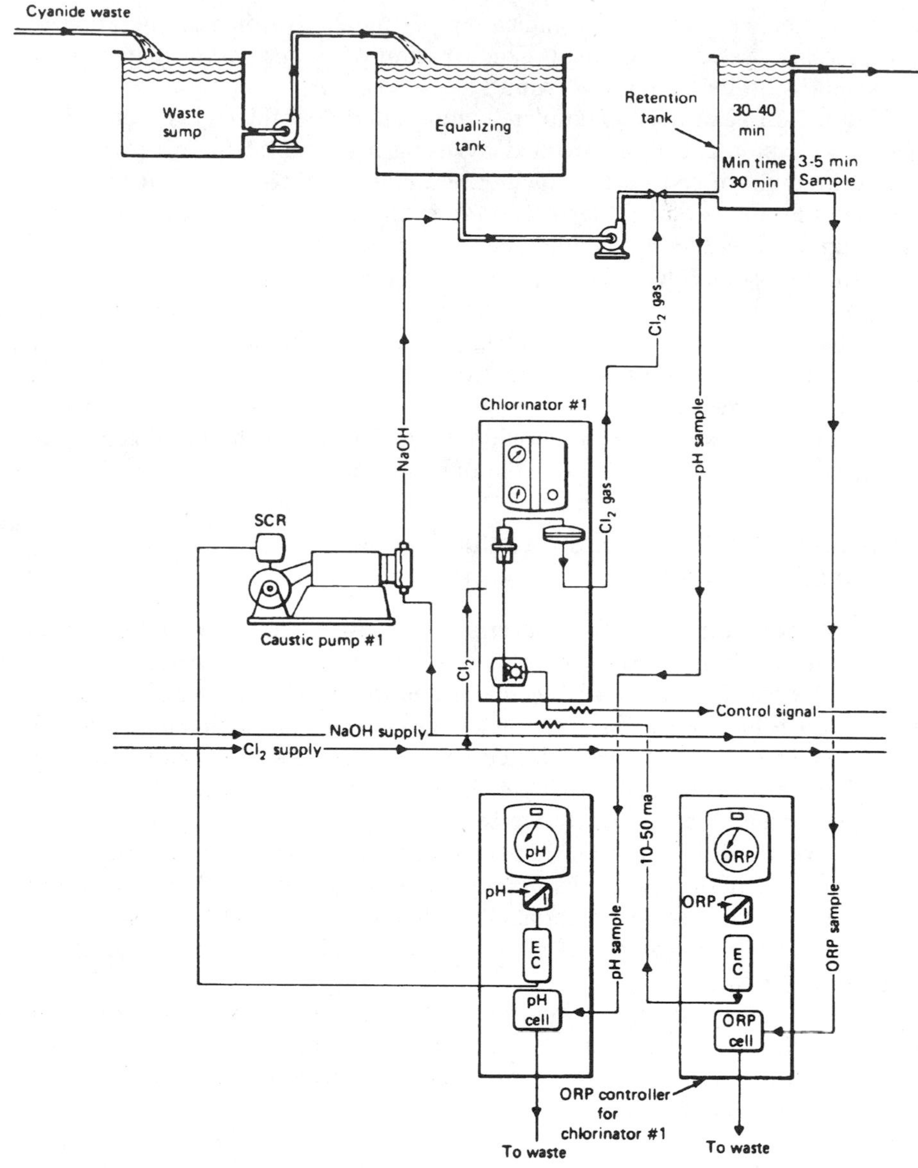

Fig. 7-18. Fully automatic alkaline-chlorine waste destruction system for continuous flow-through treatment. Chlorinator #1 controlled by ORP controller; this is first-stage destruction to cyanates. Caustic pump #1 controlled by pH controller. Chlorinator #2 controlled by chlorinator #1; this is second-stage destruction to C and N. Caustic pump #2 controlled by chlorinator #2. Chlorinator #3 and caustic pump #3 are backup units operated by alarms on ORP monitoring system.

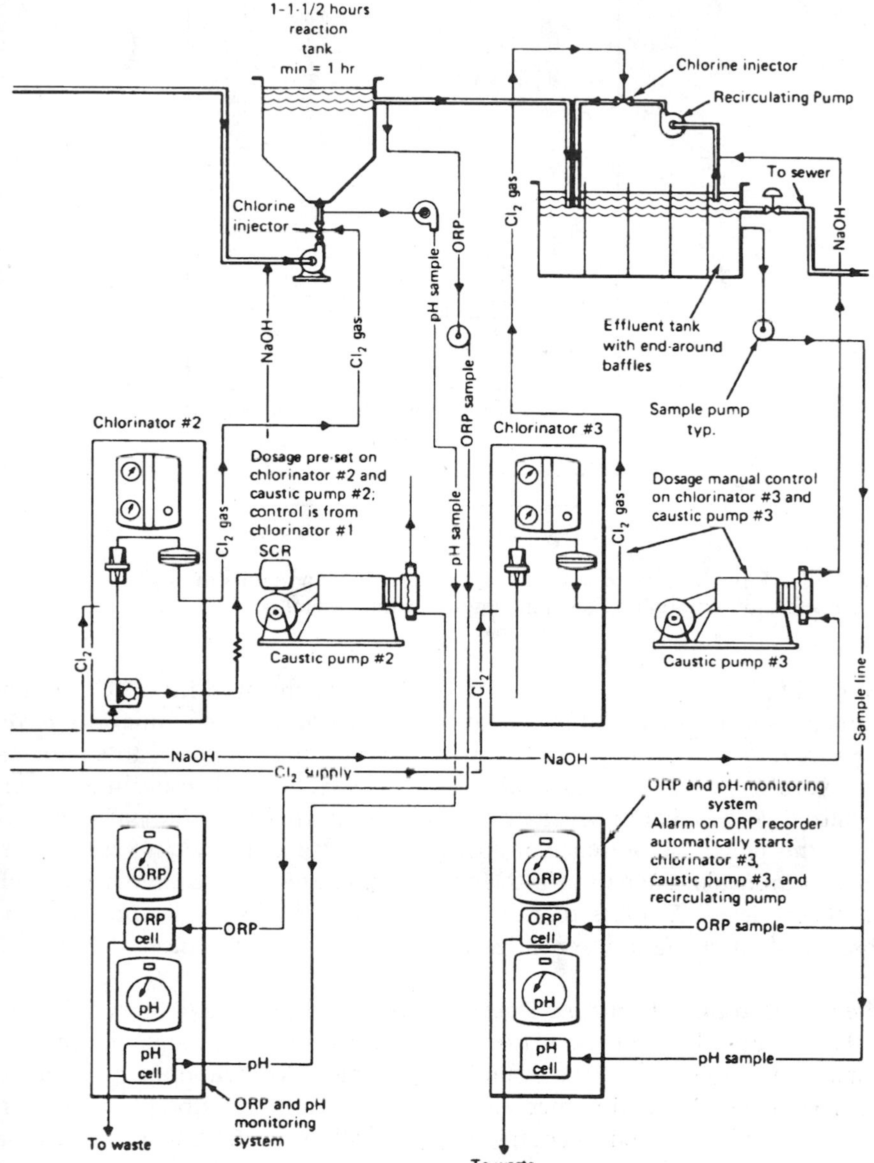

Fig. 7-18. (*Continued*).

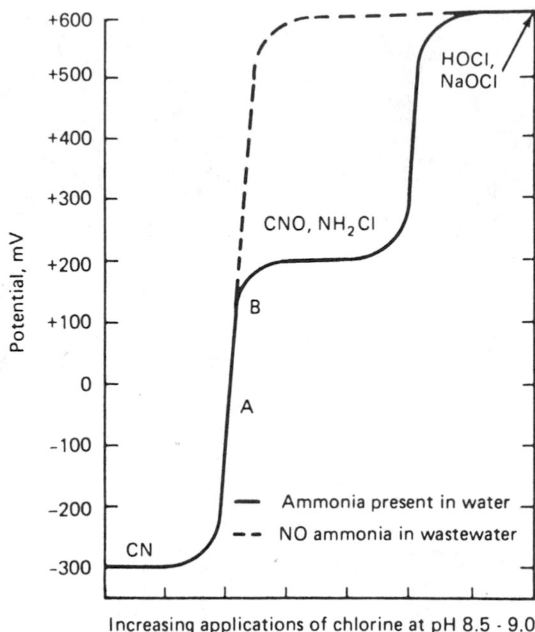

Fig. 7-19. Relationship of potential versus increasing amounts of chlorine presence and absence of ammonia in a cyanide destruction system.

oxidized to cyanogen chloride, and more than 90 percent of the CNCl, within 3 min., has been converted to cyanates, at which time the potential will have risen to zero or +50 mV. Additional retention time, sufficient to complete the hydrolysis of the cyanogen chlorides to cyanates, will result in a higher stabilized potential at the cyanate potential level with no additional chlorine. Up to this point, the presence or absence of ammonia in the raw waste make little difference in the end result. However, as chlorination continues beyond the theoretical 2.73 parts of chlorine to one part of CN, the presence or absence of ammonia in the raw waste is of extreme importance.

The Cyanate or Chloramine Potential Level. The cyanate potential level is the potential level of the waste chlorinated with just enough chlorine to completely convert all of the cyanide to cyanates. As mentioned previously, this level is not reached immediately, since a retention period of 30 min. or more is usually required to complete the hydrolysis of the last small fraction of cyanogen chloride.

As chlorination proceeds beyond the application necessary to reach the cyanate potential level, which in practice requires a somewhat greater ratio of chlorine to cyanide than the theoretical 2.73 : 1 ratio, two situations may occur: (1) that in which ammonia is present in the raw waste, and (2) that in which ammonia is not present in the raw waste.

In the presence of ammonia in the raw waste, any chlorine applied in excess of that required just to convert the cyanide to cyanates appears as chloramine residual. The cyanate and chloramine potential levels are identical, and this level is shown on Fig. 7-19 at about +200 mV. Usually, when ammonia is present in the waste, the cyanate potential level is the highest potential level attained, as it is seldom feasible to convert all of the ammonia to chloramines. However, if enough chlorine is added to react completely with all of the ammonia in addition to the cyanide initially present in the waste, further increases in chlorine application appear as a free chlorine residual, in which case the potential rises to a new maximum at about +600 mV.

When no ammonia is present in the raw waste, any chlorine applied in excess of that required just to convert the cyanide to cyanates appears as a free chlorine residual, and the potential rises immediately to a value of about +600 mV, as shown on Fig. 7-19 by the broken curve.

The Free Chlorine Potential Level. It has been mentioned that the free chlorine potential level may be attained in two different ways, determined by the presence or the absence of ammonia in the waste. The solid curve in Fig. 7-19 illustrates the potential characteristic with ammonia present, and the broken curve shows the characteristic when no ammonia is present in the raw waste. Both curves are identical while the first reaction is being completed—namely, the oxidation of cyanide to cyanogen chloride and the subsequent hydrolysis of this cyanogen chloride to cyanates.

As oxidation of the cyanates by additional chlorination requires not only the presence of a free chlorine residual but also a contact period of perhaps 45 min., the free chlorine potential level may be reached momentarily when no ammonia is present by applying more than enough chlorine to complete the cyanate reaction but not enough to completely convert all of this cyanate to nitrogen and carbon dioxide. In this case, the potential will remain at the free chlorine level as the oxidation of the cyanates proceeds until all of the excess free chlorine is used up, at which time the potential will fall to the cyanate level again.

The slowness of this cyanate–free chlorine reaction also accounts for the plateau on the solid curve at +200 mV in the presence of ammonia, which represents the cyanate–chloramine potential level, as any excess chlorine applied after the cyanates have formed reacts quickly with the ammonia to form chloramines. Because chloramines will not oxidize the cyanates to nitrogen and carbon dioxide, it is necessary to apply enough excess chlorine to (1) combine with all the ammonia, and (2) react with all the cyanates before the free chlorine potential level can be reached and maintained. Thus the length of the cyanate–chloramine plateau at +200 mV is directly associated with the amount of ammonia present in the waste.

The Controlling Potential Level. For purposes of control, it is important to sample the chlorinated waste as soon after the application of chlorine as is possible to obtain a sample that has significance or meaning with respect to the process. In practice, this has been found to be 3 min. (the control potential retention period). Point A in Fig. 7-19 represents the midpoint between the cyanide potential level and the cyanate–chloramine potential level—in this case, −50 mV.

Below this potential, cyanides are likely to be present in increasing small amounts with decreasing small applications of chlorine. Above this potential, a chlorine residual, free or chloramine as the case may be, will be present in increasing small amounts with increasing small applications of chlorine.

With the process under automatic potential control, the potential will at times fluctuate above and below the setpoint. This raises the possibility of having present small amounts of cyanides in the treated waste at times if the control potential point is set at the no cyanide–no chlorine residual level of −50 mV.

In view of the axiom that cyanides and a chlorine residual cannot be present in solution at the same time, it appears safer to have any possible fluctuation in the control potential occur in the zone in which chlorine residuals in small amounts will always be present after a 3 min. retention period. In the representative cell (Fig. 7-19), such a potential control or setpoint has been set at *B*, +100 mV.

This controlling potential may be computed as follows:

Formula. Controlling potential level in mV = calibrated cyanide potential in mV plus 80 percent of the potential difference in mV between the calibrated cyanide and cyanate potentials.

Example:

Calibrated cyanide potential = −300 mV
Calibrated cyanate potential = +200 mV
Potential difference = 500 mV
80% of potential difference = 400 mV
Controlling potential level = −300 mV + 400mV = +100 mV

In some cases, it may be possible to calculate this controlling potential level on the basis of a smaller percentage figure for the potential difference. This applies particularly to plants converting only to cyanates and where the 80 percent figure may produce a continuing choice residual that is considered excessive.

In terms of cells other than the representative cell mentioned in this discussion, knowledge of the cyanide and cyanate potential levels for a particular cell can be seen to be of utmost importance. Only from this knowledge is it possible to determine the proper control and alarm points for the individual cells associated with the control system. Normally the calibrated potential levels for individual cells should agree within +50 mV.

If the potential for any cell disagrees beyond this limit, it is suggested that the gold measuring electrode be removed from the cell and placed in a beaker of distilled water (200 ml) containing about 2 g of sodium cyanide for about 3 min.

Calibration of Potential Cells. The techniques for the calibration at both potential levels are similar, and for each level in potential the procedure is as follows:

For the cyanide or reduced potential level, use a cyanide solution buffered to pH 8.4, made as follows: To 10–12 quarts of tap water* in a clean porcelain pail, add 100 g or 0.25 lb sodium bicarbonate ($NaHCO_3$) and 2 g sodium cyanide (NaCN); then dissolve these substances by stirring.

Drain the sample pump and potential cell and recirculate this solution through the cell assembly. Depending upon the recorder and the chart range, it may be necessary to reverse the electrical connections at the cell in order to get this potential on scale. Record the potential until it becomes constant. This potential is the *calibrated cyanide potential* and should be so noted, with due regard for any reversal of cell connections.

For the cyanate potential level, use a hydrogen peroxide test solution buffered to pH 8.4, as it has the same potential as cyanates. It is made as follows: To 10–12 quarts of tap water in a clean porcelain pail, add 100 g or 0.25 lb sodium bicarbonate ($NaHCO_3$) and sufficient hydrogen peroxide to equal 2 g H_2O_2. Proceed as above, except that it will not be necessary to reverse cell leads. The recorded potential is the *calibrated cyanate or peroxide potential.*

Treatment and Destruction of Phenols

Phenolic compounds are toxic to aquatic life, and in minute quantities are deleterious to water supplies for their taste- and odor-producing capabilities. Wastes from many chemical plants, coke, gas-manufacturing, rubber-making plants, and places where sheep dips are used and livestock are disinfected contain these objectionable substances. Phenols are the hydroxyl derivatives of benzene, including various phenols, cresols, and related compounds of higher order.

Phenols can be completely destroyed through oxidation by chlorine, chlorine dioxide, and ozone. Economic considerations eliminate both chlorine dioxide and ozone as practical methods for the destruction of phenols in wastewaters.

Biological oxidation of phenols can be accomplished under rigorously controlled treatment conditions. However, slugs of phenolic compounds discharged to a conventional waste treatment plant (activated sludge or high-rate filters) will not have time to biologically adjust, so that there can be

*If the tap water contains appreciable amounts of iron or manganese, it is advisable to calibrate the cells with test solutions made up with distilled water.

oxidation and destruction of the phenols. In these instances, phenols will appear in the effluent, which cannot be tolerated. Furthermore, if the effluent from such a plant is chlorinated for disinfection purposes, the level of chlorination will be such that chlorophenols will be formed. This is the worst possible situation, since one part in 10,000 million parts[40] of chlorophenols will impart a detectable off-flavor in a water supply. In addition, chlorophenols are more toxic to aquatic life than are the straight phenols.

Destruction of phenols by chlorooxidation is the preferred chemical method, and is economically practical if the phenolic content of the waste is not in excess of 200 mg/L. The chlorooxidation process must be carried out at pH levels between 7 and 10 and with a minimum of 100 mg/L free chlorine residual.

Although the ratio of chlorine to phenol to accomplish destruction has been estimated at between 6:1 and 10:1,[83] most work indicates that large excesses of chlorine are required. Three-minute residuals should be on the order of 1000–3000 mg/L, as a result of the enormously high chlorine demand of phenolic wastes.

Chamberlin and Griffin[84] found that a coke plant still waste required 5000 mg/L of chlorine to destroy 100 mg/L of phenols. If the theoretical ratio is valid (10:1), then there were other compounds in this waste that had about a 4000 mg/L chlorine demand. It should also be recognized that the contact time for the reaction to take place is largely dependent upon the amount of excess chlorine available. The minimum chlorine residual in the destruction process should be about 100 mg/L. Although this seems unusually high, extremely obnoxious odors will occur during the treatment process if it is not maintained, indicating incomplete oxidation by chlorine.

Some free chlorine residual must be present to avoid these odors. Temperature also has an effect on the speed of the reaction. The rate of phenol destruction increases with increased temperature.[85]

Depending upon the amount of excess chlorine residual, the contact time may vary from minutes (5–10), with residuals of 1000–3000 mg/L, to 2–5 hours, with residuals of 50–100 mg/L. There is an apparent stepwise reaction with chlorine that first forms trichlorophenols. These compounds are subsequently cleaved by chlorine to yield aliphatic acids as the end product of the phenol destruction.[85] This high residual–contact time relationship brings up the problem of the treated effluent, which, having been treated for phenol destruction, is now a toxic waste because of too high a chlorine residual. This situation makes a holding or contact tank imperative so that the chlorine residual–contact time relationship can be resolved without the necessity of providing dechlorination facilities.

When the phenol content of a waste is in excess of 200 mg/L, consideration of a phenol recovery process is recommended. There are several methods available. Depending on the circumstances, the destruction of phenols may be done by a combination of biological treatment and chlorooxidation. Phenol-tolerant microorganisms can completely metabolize phenolic compounds. The removal of these compounds in a biological treatment plant would therefore depend upon establishing such microorganisms in the biological processes. When this has been done, these processes have been known to be able to continuously remove up to 50 mg/L of phenols.

Chlorination Facility. The chlorination facility for phenol destruction is similar to that for cyanide destruction. The waste to be treated is pumped from a holding tank through the chlorinator injector at a constant rate to a chlorine contact or holding tank.

Since phenol destruction will take place only in the pH 7 to 10 range, the pH of the solution discharged from the chlorine injector must be maintained accordingly. Therefore, caustic must be added at the approximate rate of 1.125 parts NaOH to each part chlorine, to replace the alkalinity destroyed by the chlorine. Depending on the buffering action of the waste, this rate may vary from 0.8 to 1.5 parts NaOH for each part chlorine. Therefore, the caustic feeder should be capable of feeding approximately 0.235 gal 50 percent NaOH per pound of chlorine.

The control of the caustic feeder should be done by an automatic pH recorder-controller. The control of the chlorinator feed rate can be predetermined by laboratory analysis for the optimum dosage for complete destruction. Since the chlorinator recirculates a constant amount of waste to be treated, the chlorinator control is done by manual dosage setting.

Textile Wastes

Wastes resulting from the processing of cotton and wool have a high BOD content (500–1500 mg/L) and are usually highly colored. These wastes are usually subjected to both chemical precipitation and biological types of treatment.[86] Chlorine plays a dual role in the treatment of these wastes: (1) it is used to make a coagulant (ferric chloride), and (2) it bleaches the color out of the treated effluent. One of the most popular coagulants in chemical precipitation of such wastes is the ferric ion.

Since chlorine is required as a bleaching agent, it might well be made available also as an oxidizer to convert the ferrous ion in the form of copperas (ferrous sulfate) to ferric sulfate as follows:

$$6\ FeSO_4 \cdot 7\ H_2O + 3\ Cl_2 \rightarrow 2Fe_2(SO_4)_3 + 2FeCl_3 + 42H_2O \quad (7\text{-}54)$$

The ferric sulfate hydrolyzes to ferric hydroxide, which forms a brown, heavy, gelatinous floc as follows:

$$Fe_2(SO_4)_3 + 6\ H_2O \rightarrow 2\ Fe(OH)_3 + 3\ H_2SO_4) \quad (7\text{-}55)$$

and similarly:

$$FeCl_3 + 3\ H_2O \rightarrow Fe(OH)_3\downarrow + 3\ HCl \quad (7\text{-}56)$$

For such a facility, the relatively easy-to-feed copperas ($FeSO_4$), which is a granular material, is fed through a volumetric feeder into a solution mixing chamber. This solution is pumped or ejected into the discharge of the chlorinator injector solution line, where the ferric sulfate is formed. The solution is then delivered to the chemical mixing chamber to provide the floc for the treatment process.

An additional point of application, proportioned by rotameters, is split off from the injector discharge ahead of the copperas addition, to be used as chlorine for bleaching color from the treatment plant effluent or for a disinfectant if combined wastes are involved.

Other Applications of Chlorine in Waste Treatment

Faber[87] reported in 1947 on the recovery of wool grease in the wool scouring process by the use of hypochlorite. Before raw wool can be processed by the textile manufacturer, it must be cleaned to remove adhering dirt, vegetable matter, perspiration, and grease. The liquor that results from this cleaning process is an emulsion of grease, organic matter, and fine solids in suspension. It is a brown and thickly turbid liquor, usually covered with a greasy scum, always alkaline and highly putrescible. Each pound of wool scoured will usually produce from 1 to 10 gal of waste. For this reason the wool waste varies considerably in strength but usually contains approximately 5000 to 10,000 mg/L suspended solids and 5000 to 15,000 mg/L grease, whereas the BOD range is between 5000 and 6500 mg/L.

The hypochlorite process reacts with the alkalinity present and the soluble soaps to produce a precipitate, so that the wool scouring waste is rapidly separated into its solid and liquid phases as a result of the oxidation of organic matter by the hypochlorite. This process is reported to be able to remove about 98 percent of the grease, of which 70 to 75 percent can be recovered and sold.

Because of this process, practically all of the suspended solids and grease can be removed, together with 80–90 percent of the BOD. This represents a genuine accomplishment in the treatment of an industrial waste.

Chlorine is also widely used in rendering plants, to convert inedible poultry, fish, and animal offal and waste products into animal feed and recovered grease. This is accomplished by cooking the inedibles at high temperatures for a period of several hours. The cooked material is pressed to remove the grease, and the pressings are ground. The cooking process gives off odorous vapors, which are captured by condensing them in large quantities of water and by burning of the noncondensible vapors. About 150 gpm are required for a five-ton cooker. The water that scrubs the stack gases is chlorinated to about 6–8 mg/L, which eliminates the odors and provides a substantial increase in grease removal.[88]

Strong liquid wastes result from the washing of trucks, barrels, and floors. Granstrom[89, 90] reported in the early 1950s that alum and chlorine are the chemicals of choice to deal properly with these wastes. The alum reacts with the bicarbonate alkalinity to form a floc, and the produced CO_2 reduces the pH. The chlorine oxidizes the reduced material and coagulates the protein. The protein coagulation is most effective at pH 4.0–4.5. The optimum dose found by Granstrom was 80 mg/L alum and 260–350 mg/L chlorine. The BOD reduction by chlorine was 35–90 percent for 260–350 mg/L dosages. The reader is cautioned about the wide variation in the strength of these wastes, depending upon the individual operating conditions.

Coagulation and precipitation of proteins by chlorine and a coagulant were reported in 1931 by Halvorson et al.[91] on fish-packing-house waste. This report also found that the best removal of protein by chlorination was at pH 4. The chlorine dosage, depending upon the protein content of the waste, varied from 100 to 1000 mg/L.

The Scott–Darcey Process

Although this process is no longer used, it should be preserved for historical reasons by a brief description here.

L. H. Scott was superintendent of water and sewage treatment at Oklahoma City, and H. J. Darcey was chief engineer of the Oklahoma state department of health. About 1930, they experimented with scrap iron and chlorine to provide a process whereby all the capabilities of chlorine and iron could be used by a single integrated installation. This patented process was utilized by Wallace and Tiernan, who developed equipment to produce ferrous chloride for odor control, ferric chloride for coagulation, and chlorine for disinfection. The Scott–Darcey process was based on the fundamental reactions of chlorine and iron in an aqueous solution:

$$HOCl + 2HCl + Fe \rightarrow FeCl_2 + HCl + H_2O \tag{7-57}$$

$$FeCl_2 + Cl_2 + H_2O \rightarrow FeCl_3 + H^+ + Cl^- + OH^- \tag{7-58}$$

This process was able to provide odor control, chemical precipitation, and disinfection with one integrated piece of apparatus utilizing such relatively inexpensive raw materials as scrap iron and chlorine. The installation consisted of a reaction tank that was charged with scrap iron. To this tank was added chlorine solution (HOCl). The reaction of the water, chlorine, and iron would stabilize in about 24 hours, so that with the proper chlorine/water ratio the solution could be considered predominantly ferrous chloride.

The success of this part of the process depended upon a centrifugal injector, which simultaneously recirculated the solution in the reaction tank and added chlorine gas to the solution under a vacuum. This ferrous chloride solution was used to react directly with H_2S for odor control, producing ferrous sulfide. A coagulant solution was made available by the use of another centrifugal injector, which added the required amount of chlorine to the odor control solution, converting it to ferric chloride and simultaneously discharging it to the point of application.

Chlorine for disinfection was available as a part of the system. One of the valuable side reactions of this process was that ferrous and ferric ion contributed beneficially to conditioning the sludge prior to digestion.

In the years of 1939 to 1942, White sold, and supervised the installation and operation of, several Scott–Darcy installations in central and coastal California. The results of this process were impressive, but the ability to obtain scrap iron after the United States entered World War II, in December 1941, completely disappeared overnight. Then, with the tremendous and sudden effort of building Army and Navy bases all over California, no attempts were made to improve the operation of civilian wastewater treatment plants. So this process lost its popularity then, and it has never been revived. Maybe in the future it will be rediscovered because of the great advantages of using the ferrous and ferric ion in waste treatment processes, and particularly for its long-lasting effect on the settleable solids, which eventually become part of the sludge digestion process. This facet of waste treatment has been overlooked in recent years because of more attractive results with flocculation, using chemicals that are specific for producing a floc but are relatively poor in enhancing the characteristics of disposal of the sludge formed by these flocculants, together with settleable organic matter.

REFERENCES

1. Hofman, A. W., and Frankland, E., "Report on the Deodorization of Sewage," report to the Metropolitan Board of Works, London, Aug. 12, 1859.
2. Baker, M. N., "Sewage Purification in America," *Engr. News,* 41 (July 13, 1893).
3. Phelps, E. B., and Carpenter, W. T., "The Sterilization of Sewage Filter Effluent," *Technology Quarterly,* **19,** 382 (1906).
4. Clark, H. W., and Gage, S. D., "Experiments upon the Disinfection of Sewage and the Effluent of Sewage Filters," 43rd Ann. Rept. Massachusetts State Board of Health, p. 339, 1911.
5. Clark, H. W., and Gage, S. D., "A Review of 21 Years Experiments upon the Purification of Sewage at the Lawrence Experiment Station," 40th Ann. Rept. Massachusetts State Dept. of Health, p. 251, 1908.
6. Thoman, J. R., and Jenkins, K. H., "Statistical Summary of Sewage Chlorination Practice in the U.S.," *Sew. and Ind. Wastes,* **30,** 1461 (1958).
7. Enslow, L. H., and Symons, G. E., "Sewage Treatment Processes," *Sew. Wks. J.,* **17,** 5, 984 (Sept. 1945).
8. White, G. C., "Chlorination—How It Works," paper presented at the 5th Annual Symp., Bureau of Eng., California State Dept. of Health, May 1970.
9. Lusher, E. E., "Swimming Bath Water Treatment," *Baths Service* (England), p. 16 (1959).
10. Lomas, P. D. R., "The Combined Residual Chlorine of Swimming Bath Water," *J. Assoc. Pub. Analysts* (England), **5,** 27 (1967).
11. Sung, R. D., "Effects of Organic Constituents in Wastewater on the Chlorination Process," Ph.D. Dissertation, Univ. of California, Davis, CA, 1974.
12. Feng, T. H., "Behavior of Organic Chloramine in Disinfection," *J. WPCF,* **38,** 614 (1966).
13. Ryan, W. A., "Control of Odors," *Sew. Wks. J.,* **5,** 89 (1933).
14. Chanin, G., Elwood, J. R., and Chow, E. H., "Atmospheric Hydrogen Sulphide Studies," *Sew. and Ind. Wastes,* **26,** 1217 (1954); "Correction," ibid, **26,** 1449 (1954).

15. Anon., "Highlights: Wastewater Wisdom Talk," Water Poll. Control Fed., Feb. 1980.
16. Swab, B. H., "Effects of Hydrogen Sulfide on Concrete Structures," *J. San. Engr. Div. ASCE,* **1** (Sept. 1961).
17. White, G. C., unpublished field studies, Modesto, CA, 1946.
18. Pomeroy, R. D., private communication, Nov. 7, 1983.
19. McKee, J. E., and Wolf, H. W., "Water Quality Criteria," 2nd ed., California State Water Quality Control Board, Sacramento, CA, 1963.
20. Nordell, E., *Water Treatment for Industrial and Other Uses,* Reinhold, New York, 1951.
21. Anon., "Process Design Manual for Sulfide Control in Sanitary Sewerage Systems," U.S. Environmental Protection Agency, Oct. 1974.
22. Heukelekian, H., "Sewage Chlorination for Odor Control," *Water Wks. and Sew.,* **89,** 302 (1941).
23. Pomeroy, R., and Bowlus, F. D., "Progress Report on Sulfide Control Research," *Sew. Wks. J.,* **18,** 597 (July 1946).
24. Heukelekian, H., "Utilization of Chlorine during Septicization of Sewage," *Water and Sew. Wks.,* **95,** 179 (May 1948).
25. Hoag, L. N., McLaren, F. R., and Hall, G. H., "Control of Odors and Corrosion in the Sacramento Regional Wastewater Conveyance System," a report by Sacramento Area Consultants, Sacramento, CA, Sept. 1976.
26. Meyer, W. J., and Hall, G. H., "Prediction of Sulfide Generation and Corrosion in Sewers," a report by J. B. Gilbert and Associates, Sacramento, CA, April 1979.
27. Davey, W. J., "Influence of Velocity on Sulfide Generation in Sewers," *Sew. and Ind. Wastes,* **22,** 1132 (1950).
28. Pomeroy, R. D., private communication, Nov. 1977.
29. Pomeroy, R. D., and Parkhurst, J. D, "The Forecasting of Sulfide Build-up Rates in Sewers," *Prog. Wat. Tech.,* **9,** 621–628 (1977).
30. Hall, G. H., McLaren, F., and Hyde, W. S., "Analysis and Design for Sulfide and Corrosion Control in the Sacramento Regional Wastewater Collection System," *Calif. Water Poll. Control Bulletin,* 40 (Jan. 1981).
31. Pomeroy, R., "Generation and Control of Hydrogen Sulfide in Filled Pipes," *Sew. and Ind. Wastes,* **31,** 1082 (Sept. 1959).
32. Hommon, C. C., "Control of Odorous and Destructive Gases in Sewers and Treatment Plants," *Water Wks. and Sew.,* **89,** 277 (1942).
33. Bowlus, F. D., and Banta, P. A., "Control of Sewage Condition by Chlorination," *Water Wks. and Sew.,* **89,** 277 (1932).
34. Anon., "California Operators Symposium—Odor Control," *Sew. Wks. J.,* **14,** 883 (1942).
35. Wisely, W. H., "Experiences in Odor Control," *Sew. Wks. J.,* **13,** 956 (1941).
36. Wisely, W. H., "Experiences in Odor Control," *Sew. Wks. J.,* **13,** 1230 (1941).
37. Nelson, M. K., "Sulfide Odor Control," *J. WPCF,* **35,** 1285 (Oct. 1963).
38. Backmeyer, D. P., and Drautz, K. E., "Miami Tries to Control Sulfides in Force Main," *Wastes Engrg.,* 290 (June 1963).
39. Black, A. P., and Goodson, J. B., Jr., "The Oxidation of Sulfides by Chlorine in Dilute Aqueous Solutions," *J. AWWA,* **44,** 309 (1952).
40. Anon., "Chlorination of Sewage and Industrial Wastes," Manual of Practice No. 4, WPCF, Washington, DC, 1951.
41. Chamberlin, N. S., "Chlorination of Sewage," *Sew. Wks. J.,* **20,** 304 (1948).
42. Nordell E., *Water Treatment of Industrial and Other Uses,* Reinhold, New York, 1961.
43. Nagano, J., "Oxidation of Sulfides during Sewage Chlorination," *Sew. and Ind. Wastes,* **22,** 884 (July 1950).
44. Chanin, G., "Solving Odor Problems by Prechlorination of Flow, Paced by Automatic Continuous Monitoring of H_2S content of Atmosphere," *Water Wks, and Wastes Engrg.,* 42 (Jan. 1964).
45. Castro, A. J., "The Treatment of Cannery Wastes at Santa Clara, California," *Calif. Sew. Wks. J.,* **14,** 2, 22 (1942).

46. Griffin, A. E., "The Regulation of Chlorine Application in Sewage Odor Control Work," *Water and Sew. Wks.,* **80,** 218 (June 1933).
47. White, G. C., unpublished report to Union Sanitary District, 1978.
48. Smith, E. E., "Control of Activated Sludge Bulking by Chlorination," *Water Wks. and Sew.,* **90,** 209 (June 1943).
49. Tapleshay, J. A., "Control of Sludge Index by Chlorination of Return Sludge," *Sew. Wks. J.,* **17,** 1210 (1945).
50. Tapleshay, J. A., "Sludge Density Control," *Water Wks. and Sew.,* **93,** 116 (1946).
51. Livingston, J., and Harland, R., private communication, Hydrogen Sulfide Lab., L. A. County Sanitation District, Los Angeles, CA, Nov. 1972.
52. Griffin, A. E., and Chamberlin, N. S., "Exploring the Effect of Heavy Doses of Chlorine in Sewage," *Sew. Wks. J.,* **17,** 730 (1945).
53. Mahlie, W. S., "Oil and Grease in Sewage," *Sew. Wks. J.,* **12,** 727 (July, 1940).
54. Faber, H. A., "Chlorinated Air Proves Aid in Grease Removal," *Water Wks. and Sew.,* **84,** 171 (1937).
55. Keefer, G. E., and Cromwell, E. C., "Grease Separation Enhanced by Aero-chlorination," *Water Wks. and Sew.,* **85,** 97 (1938).
56. White, G. C., operating report to H. N. Jenks, Consulting Engineer, Palo Alto, CA, 1943.
57. Susag, R. H., "B. O. D. Reduction by Chlorination," *J. WPCF,* **40,** 434 (1968).
58. Anon., "Water Quality Control Plan San Francisco Bay Basin (2)," Calif. Water Quality Board, San Francisco Bay Region, 1982.
59. Rawn, A. M., "Chlorination in Sewage Treatment," *Water Wks. and Sew.,* **81,** 97 (1934).
60. Baity, H. G., and Bell, F. M., "Reduction of the Biochemical Oxygen Demand of Sewage by Chlorination," *N. Carolina Water and Sew. Wks. Assoc. J.,* **6,** 151 (1928).
61. Enslow, L. H., "Recent Developments in Sewage Chlorination 1926," Proc. 9th Texas Water Wks. Short School, p. 317, 1927.
62. Babbitt, H. E., and Baumann, E. R., *Sewerage and Sewage Treatment,* 8th ed., John Wiley & Sons, New York, 1965.
63. Skarheim, H. P., "Tables of the Fraction of Ammonia in the Undissociated Form for pH 6 to 9, Temperature 0–30°C, TDS 0–3000 mg/l, and Salinity 5–35 g/kg," San. Engr. Res. Lab., Sch. of Pub. Hlth, Univ. of California, Berkeley, CA, SERL Report No. 73–5, June 1973.
64. Stone, R. W., "Rancho Cordova Breakpoint Chlorination Demonstration," report prepared by Sacramento Area Consultants, Sept. 1976.
65. Stone, R. W., Saunier, B. M., Selleck, R. F., and White, G. C., "Pilot Plant and Full-Scale Ammonia Removal Investigations Using Breakpoint Chlorination," paper presented at the 49th Ann. Conf. WPCF, Minneapolis, MN, Oct. 5, 1976.
66. Saunier, B. M., "Kinetics of Breakpoint Chlorination and Disinfection," Ph.D. Dissertation, Univ. of California, Berkeley, CA, 1976.
67. Livingston, J., private meeting, Los Angeles County Sanitation District, Hydrogen Sulfide Lab., Compton, CA, Mar. 20, 1972.
68. Livingston, J., and Harland, R., "Control of Sulfides in Force Mains Using Compressed Air," in-house report to F. D. Dryden, Sanitation Districts of Los Angeles, CA, Nov. 24, 1971.
69. Beebe, R. D., Jenkins, D., and Daigger, G. T., "Activated Sludge Bulking Control at the San Jose/Santa Clara California Water Pollution Control Plant," paper presented at 55th Ann. Conf. WPCF, St. Louis, MO, Oct. 1982.
70. Easley, R. S., and Vidal, B., private communication Sanitation Districts of Los Angeles County, Whittier, CA, Feb. 16, 1984.
71. Dobson, J. G., "The Treatment of Cyanide Wastes by Chlorination," *Sew. Wks. J.,* **19,** 1007 (Nov. 1947).
72. McKee, J. E., and Wolf, H. W., "Water Quality Criteria," Sacramento, Calif. St. Water Qual. Control Bd., 2nd ed., 1963.
73. Chamberlin, N. S., and Snyder, H. B., Jr., "Technology of Treating Plating Wastes," Tenth Ann. Wastes Conf., Purdue Univ., May 9–11, 1955.

74. Zehnpfennig, R. G., "Possible Toxic Effects of Photographic Laboratory Wastes Discharged to Surface Water," *Water and Sew. Wks.*, **115,** 136 (1968).
75. Ryan, J. A., and Culshaw, G. W., "Use of *p*-dimethylaminobenzylidene Rhodamine as an Indicator for the Volumetric Estimation of Cyanide," *Analyst* (England), **69,** 370 (1944).
76. Aldridge, W. H., "New Method for the Estimation of Microquantities of Cyanide and Thiocyanate," *Analyst* (England), **69,** 262 (1944).
77. Epstein, J., "Estimation of Microquantities of Cyanide," *Anal. Chem.*, **19,** 272 (1947).
78. Herting, O., "Analysis of Commercial Cyanides. Estimation of Cyanic Acid," *Zeit. Angew. Chem.*, **14,** 585 (1901).
79. Marks, H. C., and Chamberlin, N. S., "Determination of Residual Chlorine in Metal Finishing Wastes," *Anal. Chem.*, **24,** 1885 (Dec. 1952).
80. Carmichael, D. C., "Cyanide Wastes at Dupont," *Cons. Engr.* (Dec. 1952).
81. Weyermuller, G., and Morris, H. E., "Monsanto Controls Chemical Waste Disposal," *Chem. Processing* (Oct. 1955).
82. Lowder, L. R., "Modifications Improve Treatment of Plating Room Wastes," *Water and Sew. Wks.*, **115,** 580 (1968).
83. Hill, E. A., and Neff, F. J., "Cyanide Waste Oxidized in the Plating Room," *Plating* (Aug. 1957).
84. Chamberlin, N. S., and Griffin, A. E., "Chemical Oxidation of Phenolic Wastes with Chlorine," *Sew. and Ind. Wastes,* **24,** 750 (June 1952).
85. Eisenhauer, H. R., "Oxidation of Phenolic Wastes," *J. WPCF,* **36,** 1116 (1964).
86. Chamberlin, N. S., "Application of Chlorine and Treatment of Textile Wastes," *Am. Dyestuff Reporter* (June 21, 1954).
87. Faber, H. A., "The Hypochlorite Process for Treatment of Wool Scouring Wastes and for Recovery of Wool Grease," *Sew. Wks. J.,* **19,** 248 (1947).
88. White, G. C., unpublished chlorination results at Reno Rendering Wks., Reno, Nevada, 1956, 1968, 1969.
89. Granstrom, M. L., "Rendering Plant Waste Treatment Studies," *Sew. Wks. J.,* **23,** 1012 (1951).
90. Granstrom, M. L., "Rendering Plant Wastes Treatment Studies #II Pilot Plant Investigation," *Sew. Wks. J.,* **24,** 1478 (1952).
91. Halvorson, H. O., Cade, A. R., and Fullen, W. J., "Recovery of Proteins from Packinghouse Waste by Superchlorination," *Sew. Wks. J.,* **3,** 488 (1931).
92. Bizzari, R. E., Popeck, J. R., Pickard, D. W., and Drapp, J. E., "H_2S Odor Control on Tampa's Major Sewage Systems," paper presented at 55th Ann. Conf. WPCF, St. Louis, MO, Oct. 5, 1982.
93. Beebe, R. D., and Jenkins, D., "Control of Filamentous Bulking at the San Jose/Santa Clara Water Pollution Control Plant," paper presented at 53rd Ann. Conf. Calif. Water Poll. Control Assoc., Long Beach, CA, Apr. 29, 1981.
94. Adams, A., private communication, Envirotech Operating Services, San Mateo, CA, 1980.
95. Morton, R. and Caballero, R., "The Biotrickling Story" *Water Environment and Technology,* 39 (June 1996).
96. Jain, R., and Scanlan, M., "R-J Environmental, Low Profile Packaged Odor Control System," R-J Environmental Inc., 6197 Cornerstone Ct. East, Suite #108, San Diego, CA 92121.
97. Redner, J., project engineer, Los Angeles County Sanitation District Laboratory, 920 So. Alameda Street. Compton, CA 90221, personal communication, 1996.

8 DISINFECTION OF WASTEWATER

CONCEPTS OF DISINFECTION

Importance of a Chlorination Disinfection Facility

One of the most useful features of a wastewater disinfection facility is its ability to serve as a monitor for the combined wastewater treatment processes. If all the unit processes are not performing properly, the chlorination system will not be able to deliver an effluent that meets the NPDES discharge requirement. This important attribute of disinfection is often overlooked. This is true in California where the requirements are the strictest in the nation because it is mostly an arid state with mainly surface discharges into coastal bathing areas. Therefore, the consensus organism to eliminate and control is total coliforms rather than fecal coliforms, which is allowed by the EPA in many states. To compare these two discharge requirements, in California, a discharge requirement might be 200 MPN/100 ml total coliforms, whereas a fecal coliform requirement would be the equivalent of 1000 MPN/ml (see below, under "Rationale for Coliform Concentration Requirements").

Importance of Disinfection

Today the emphasis is on the disinfection of all wastewater effluents in the United States. Sewage disinfection is defined as the process of destroying pathogenic microorganisms in the wastewater stream by physical or chemical means. This is best accomplished by the use of chemical agents such as aqueous solutions of chlorine, chlorine dioxide, hypochlorite, bromine, bromine chloride, ozone, or combinations of these chemicals. Other means that have been used in special conditions and with varying degrees of success are ultraviolet radiation, gamma radiation, sonics, heat, and silver ions. Present disinfection practices depend almost exclusively on chlorine compounds.

From the viewpoint of health, the disinfection process is the most important stage of wastewater treatment. The objectives of wastewater disinfection are: to prevent the spread of disease and to protect potable water supplies, bathing beaches, receiving waters used for boating and water contact sports, and shellfish growing areas.

Public Health Agency Perspective. The public health agency is deeply committed to the theory of multiple barriers or multiple points of control between a sewage discharge and a water supply intake. These barriers or points of control include wastewater treatment, land confinement, dilution, time, distance, and potable water treatment. Any type of treatment is fallible; so reliance on natural barriers should be maintained as long as possible. Where the natural barriers are eroding, increased emphasis and reliability requirements must be placed on the artificial barriers of treatment processes. The factors that operate against and diminish the effectiveness of natural barriers are: increased population, increased mobility of the population, increased recreation, more leisure time, increased sewage discharge, and increased water use. Inequities of rainfall distribution combined with increased water consumption produce increased recycling of wastewater. This leads to an overall decrease in dilution, time, and distance factors between sewage discharges and potable water intakes.

All of these factors support the efforts of regulatory agencies to improve the quality of wastewater effluents prior to their discharge into the environment. For many areas of use this is the only protection available. *Therefore, disinfection is the last remaining barrier against the transmission of waterborne diseases.*[1]

Legal Concepts. English law has used the doctrine of public nuisance to abate public health hazards. An examination of centuries-old common-law precedents emphasizes that doctrines of common-law nuisance were directed toward abatement of risk rather than abatement of disease. There was no need to show that disease had actually occurred or was likely to occur. The law operates on commonsense public health recognition that conditions conducive to disease should be eliminated.[113] There are many public health hazards today that are the proper subject of judicial scrutiny and judicial action despite conflicting or contradictory evidence on the other side. *It is health risk rather than health harm that is the subject of the action.*

Lawyers who practice in the field of environment and public health protection are usually significantly more conservative than the EPA or most state public health agencies. These lawyers advocate maximum control of pathogenic organisms. This philosophy derives from their numerous cross-examinations of scientific experts. Such intensive examination over the years seems to have led lawyers to the conclusion that the scientific community knows little about the capability of the environment to assimilate the wide variety of pollutants discharged daily.

The legal profession is critical of the EPA position that seeks to justify "cost-effective" analysis, because health risks from the discharge of waste into receiving waters are impossible to quantify. Moreover, lawyers cannot accept putting costs and benefits on these risks because there are always costs and other variables that cannot be quantified.

Contemporary Practices in the United States. Over the last few decades the various regulatory agencies—local, state, and federal—have been seeking a set of guidelines for proof of disinfection in various receiving water situations.

The most persistent and aggressive pursuit of wastewater disinfection requirements has been by the California State Department of Public Health, United States. The adequacy of disinfection is evaluated in compliance with a prescribed MPN (most probable number) of coliform organisms as determined by *Standard Methods*[2] for the "confirmed" test procedure. For example, in ocean and saline bay waters used for recreation, 80 percent of the receiving water samples must fall within a coliform MPN of 1000/100 ml. This is approximately equivalent to a median MPN of 230/100 ml.* For other situations, more restrictive median or average coliform concentrations may govern. The discharger is required to disinfect to the degree necessary to maintain a suitable quality in the receiving waters. Often the discharger is given the option of demonstrating compliance by meeting the receiving water quality in the effluent itself. This eliminates the costly process of monitoring several sampling stations in the receiving waters. In most situations the discharger usually is required to apply the disinfection requirements only to the effluent quality. This practice has evolved to a point in the State of California where practically all cases are based on the MPN total coliform in the plant effluent.

The evolution of these requirements is of particular interest because it relates largely to geographical considerations.

Other states for various reasons have not been as concerned with the protection of receiving waters as has California. The predominant condition that led to the adoption of these coliform requirements was the extensive development and use of the California coastline and coastal waters for recreational and shellfish growing purposes. Some 720 miles in length, the California coastline has some of the most beautiful bathing and water sports beaches in the world.

Evolution of Disinfection Requirements

It was the fouling of one of the most beautiful expanses of resort beach areas in California that captured the attention and energies of the California State Department of Public Health (see Fig. 8-1). Beginning about 1920, the Bureau of Sanitary Engineering, a division of this agency under the able direction of C. G. Gillespie, entered into a 20-year battle with the City of Los Angeles to clean up its Hyperion outfall discharge, which was solely responsible for the fouling of a 10-mile stretch of this magnificent beach. Obviously Los Angeles was a huge stumbling block in the overall California antipollution program envisioned by Mr. Gillespie and his colleagues.

*The MPN of 230/100 ml is used when the statistical procedure utilizes a five tube dilution. For a three tube dilution the MPN would be 240/100 ml.

Fig. 8-1. Los Angeles County Beach, California, 1974 (Los Angeles Times photo).

In order to convince the City of Los Angeles that the Hyperion outfall discharge (170 mgd) was polluting this beach area (which currently attracts about one million people on summer weekends), the Bureau of Sanitary Engineering, California State Department of Health, made a year-long study of 10 miles of beach in Santa Monica Bay in 1941–42.[3]

The report of this investigation, dated June 26, 1943, led to the immediate quarantine of this 10-mile stretch of beach, on the grounds that both the beach and the surf waters were polluted with sewage and were therefore dangerous to human health. After a length of time considered sufficient to correct this hazardous condition of gross pollution, the California State Board of Public Health took the City of Los Angeles to court on a suit based on pollution as determined by the "coliform count" in the bathing waters. Other factors were considered, but it was the coliform count that became the most persuasive piece of evidence.

In their lawsuit against the city, the State Board of Health maintained that 1000 *Escherichia coli*/100 ml as a limiting standard in the surf waters ensured the safety of the bathers. The State of California won the suit because the judge hearing the case held that the coliform standards used by the Board of Public Health were reasonable, after it was demonstrated that there was clear evidence of pollution and physical nuisance in areas where these bacterial limits were exceeded.

This historic case established a statistical coliform concentration baseline that defines the difference between polluted and pollution-free recreational contact waters in the open surf. This has been accepted by the sanitary engineering profession in the United States as a landmark achievement.

Rationale for Coliform Concentration Requirements

The very comprehensive study of inland recreational waters by Stevenson[4] in 1950–53 concluded that an MPN coliform concentration of 2300/100 ml may be a threshold quality associated with an increase in the incidence of disease. For those who might have an interest in a comparison between fecal coliform and total coliform, with the use of data developed some years later on the fecal coliform content of the same waters and by the use of ratios, a geometric mean fecal coliform content of 400/100 ml was determined to be equivalent to the 2300/100 ml total coliform number.

In theory two different standards could be set for a potable water and for a wastewater discharge. One would be a standard of assured safety where there is no health concern, whereas the other standard would be the threshold of unsafe water (2300/100 ml) and might be used as an indication that quarantine action or abatement action should be undertaken. The California State Department of Health has recommended discharge requirements that, based on its experience and judgment, will result in a water quality where there is no health concern. In other words, its quality requirements reflect an inherent factor of public health safety, whether for potable water, wastewater discharging to variable-use receiving waters, or water reuse situations. These concepts are described as follows:

1. The limit for surf waters is about 500 times the pollution allowed by the U.S. Public Health Service standards for potable water (2.2 MPN/100 ml). The 2.2 standard has been universally accepted by water and health experts for many years. Comparing relative ingestion of potable water versus seawater in the course of surf bathing (2–3 ml per swim), the figures seem compatible.
2. There is no scientific evidence to indicate that water within this standard causes ill health.
3. The level of indicator organisms is seldom reached or exceeded in saline waters where the cause is not obviously recent waste contamination.
4. A less severe standard might show "approved" areas to lie within visible areas of grease and detritus of waste origin and would therefore seem to be lacking in common sense and decency.

The application of this standard for disinfection first appeared as a required chlorine residual after a specified contact time that would produce the desired quality in the receiving waters. At the time it was thought that disinfection would meet these standards if a 0.5–0.75 mg/L orthotolidine residual at the end of a 30 min. contact time was accomplished. Contact chambers were built to give a theoretical 30 min. detention time at average flow based on the volume of the chamber. Effective mixing and contact chamber short-circuiting was never considered. During this period a great many chlorination facilities went into operation (1947–57), and it became obvious that a residual–contact period requirement often produced effluents of quite different bacterial quality at different plants. After many years of testing and surveillance of these installations, the Bureau of Sanitary Engineering of the California State Department of Health concluded that it was a practical, feasible, and superior way to prescribe a coliform count directly to the plant effluent, rather than a chlorine residual value as evidence of disinfection.

Current Coliform Requirement in California

The numbers presently in effect are: 80 percent of samples less than 1000/100 ml for coastal bathing waters (equivalent to a median of 230/100 ml), a median of 70/100 ml for shellfish growing areas, and a median of 23/100 ml for confined waters used for bathing or other water contact sports, assuming that the dilution is at least 100 to 1. The requirement for discharge into ephemeral streams or other areass where there is public exposure to effluents receiving little dilution is for an essentially coliform-free effluent, that is, a median MPN not greater than 2.2/100 ml. There is a subtle implication of the necessity for good operation and adequate treatment to achieve the

23/100 ml requirement. This has been found to be a meetable standard, but it requires a properly operated treatment plant with an effective disinfection system to consistently achieve it. The severe affluent standard of 2.2/100 ml implies the necessity for some type of advanced treatment (i.e., filtration) prior to disinfection to reliably meet this level of disinfection effectiveness, and thereby suggests some virus-removal capability for the system beyond that which normally occurs. These are not alternative requirements, only implications of what might be needed to meet the requirement.

There may be cases where the quality of the plant effluent bears no relation to the receiving water quality. For example, a receiving water might have a consistent coliform concentration as high as 2300/100 ml or even greater. If the State Department of Health had decided that disinfection of a wastewater discharging to such a receiving water was necessary, the discharger would not be allowed simply to meet the water quality of the receiving water (2300/100 ml or greater) because this would not be evidence of effluent disinfection.

Conclusions. The California coliform index imposed upon every wastewater discharger carries with it strong implications of the degree of treatment required to achieve this index. Each and every wastewater discharger is evaluated separately for the individual circumstances of geography and the receiving water situation, and there are no exceptions.

The California coliform index system is based upon discharge requirements that will result in a water quality where there is no health concern. This is a function of the expertise, experience, and judgment of those involved.

A review of the California surface water coliform concentration requirements of all the major utilities using surface water supplies to produce potable water demonstrates that a long-range program of strict wastewater discharge requirements can result in the preservation of receiving water quality.

Considering chlorine demand as a surface water quality parameter, the surface waters of California, an arid state, pass with flying colors. This performance translates into good wastewater treatment programs as well as a strict measure of control for industrial and food packagers' wastes going into the wastewater collection system. These discharges in practically all cases requires the dischargers to treat their waste with an acceptable biological treatment system before dumping the waste into the sewage collection system.

The long-term philosophy of the Bureau of Sanitary Engineering supplemented by the able assistance of the various Regional Water Quality Control Boards is illustrated by the median MPN of 18 watersheds supplying surface water, which averaged 138 MPN/100 ml, with a maximum of 450 MPN/100 ml and a minimum of 6 MPN/100 ml. White believes this is an accomplishment to be envied by other agencies.

All waters processed for reuse must not only meet the 2.2/100 ml coliform MPN and 2 JTU turbidity standard, but should also be subjected to coagulation and filtration in order to be assured of significant viral destruction. This recommendation is based upon the many years of operation of the Santee water reclamation project in San Diego County, California, under the supervision of the California State Board of Health.[129]

All wastewater discharges into saline or fresh waters should be disinfected to 240/100 ml coliform MPN if there is to be any water-contact recreation in these waters. Similarly, all wastewater discharges should be disinfected to 70/100 ml coliform MPN in shellfish growing areas.

Operators have emphasized that in addition to disinfection, the chlorination system operation has proved to be an important continuous monitoring device for the entire wastewater treatment process.

The Effect of 1997 Conditions

While the total coliform index still remains the most profound single indicator of the distinction between a water without a waterborne disease potential and one that might be suspect, the situation has become extremely urgent since the arrival of a much more rugged *Giardia* cyst, along with an even more devastating organism, the *Cryptosporidium* oocycst. All of this makes using fecal coliforms as an acceptable wastewater quality indicator totally worthless, as it constitutes a degradation in water quality standards. It should be remembered that 200 MPN/100 ml of fecal coliforms is equal to 1000 MPN/100 ml of total coliforms. This gets back to what White has been emphasizing for many years—a "consensus organism." If this concept were adopted, a more resistant organism would be used to operate the chlorine disinfection system, for example, using the free chlorine residual (85 percent HOCl) or total chloramine residual and contact time required to inactivate a more resistant organism such as the polio virus. The use of millivolts of ORP to confirm the chlorine residual measurement is also recommended. This backup has been shown to be very effective and most pleasing to the operators.

Administration of Requirements

Administratively, this is how the control system presently operates in California:

1. The discharger applies to the appropriate Regional Water Quality Control Board for permission to discharge wastewater at a given location.
2. The Board then notifies all interested agencies for recommendations.
3. The State Department of Public Health submits its recommendation for the disinfection requirement to the Regional Board.
4. The Board then holds a public hearing to discuss and establish the requirements with the discharger.

The Regional Water Quality Control Board can issue a cease-and-desist action on a discharger for violation of any portion of the requirements including the requirements on disinfection. Further, the Board can place and has placed a ban on further connections to the discharger's collection system if the requirements are not met, and can impose a heavy fine ($10,000) for each day of violation.

Total Coliforms vs. Fecal Coliforms as a Standard

The fecal coliform determination is the latest in a long history of selective tests to separate the strains of coliform bacteia found in wastewater. It has been largely adopted by the U.S. Environmental Protection Agency for various disinfection requirements: shellfish areas, recreational waters, and so on.

Coliforms from the intestines of dogs and cats are mainly *E. coli.* The fecal coliforms from humans and livestock account for about 97 percent of the total. Fish do not have permanent coliform flora in their intestines. The presence of coliforms in fish is evidence of pollution in the water of their habitat. Fecal coliforms are not abundant in soil (as *E. coli*), as they die off rapidly when deposited in the soil.[5]

Currently, there are no means for distinguishing between fecal coliforms of humans and those of other warm blooded animals.[6] Their presence in significant numbers is, however, indicative of fresh pollution. All fecal coliforms in a stream may be accepted as being of fecal origin, whereas an unknown portion of the total coliform bacteia may be of other origins. Determination of fecal coliforms provides a superior indicator of fecal pollution. There is no argument on this point. The question remains: Is the fecal coliform concept a valid parameter for disinfection? White has been sure that fecal coliform counts are useless ever since he found high fecal coliform counts in an industrial effluent discharging into the Ohio River. His and other investigations could not locate the source of these fecal coliforms.

However, because the discharge requirements were based upon a fecal coliform limit (the counts were so high and the chlorine residual had to be so low), the only solution was to build an additional chlorine contact chamber to provide a one-hour detention time at peak flow. There was no room for a holding pond for the effluent. The differential fecal coliform test demonstrates the ability to enumerate coliform bacteria originating from fecal sources while suppressing those of soil origin. Consequently, it is a very useful tool in the Sanitary Survey of surface waters where numerous differing sources can contribute to the total coliform content.

The California Department of Health has used both the total coliform and fecal coliform enumeration in its surface water studies since 1960, being of the opinion that if a single indicator test for bacteriological quality were to be used, the fecal coliform test would be the best for *freshwater* areas because of its selective ability.[7]

Although the use of a fecal coliform number as a river water quality objective is appropriate, the use of a fecal coliform standard for a measurement of disinfection would not be appropriate. All available data indicate that the fecal coliform strains are more fragile than the total coliform group and can be more easily destroyed or inactivated by disinfection or natural purification processes. It has been amply demonstrated that on the basis of chlorine demand tests, fecal coliforms can be completely destroyed while significant numbers of total coliforms remain.[5]

There is only sketchy information on the relative resistance of pathogenic agents to chlorination. The data suggest that bacterial agents may be as hardy as coliforms, whereas most viruses are more resistant. The total coliform group is a more conservative indicator of effective disinfection, and total coliforms are more numerous in wastewater than the fecal coliform group. It is conservatively estimated that the ratio of fecal coliforms to total coliforms in saline waters is 1:70. In San Francisco Bay it has been observed that when the ratio of fecal coliform to total coliforms is 1:70, that area is generally out of the known pollution areas.[8]

However, there is a considerable difference in the fecal-to-total coliform ratio from saline to freshwater areas. Instead of the 1:70 ratio observed above, it is estimated that in wastewater effluents the total coliforms contain about 30 percent fecal coliforms.[9]* This comparison is most significant when disinfection is related to a final coliform count rather than a log reduction in the total coliforms present before disinfection.

Disinfection Efficiency: Bacteria Survival vs. Percent Kill

The effectiveness of a disinfectant dose for a given contact time is usually expressed as a ratio of logs reduction of initial to final bacteria count, or as a percent destruction of the initial bacteria count. In order for investigators to keep the proper perspective when evaluating reports of disinfecting procedures, the minimum acceptable surviving number of total coliforms should not be in excess of 1000/100 ml MPN. For comparing disinfectants of secondary effluents, the final MPN should be no more than 230/100 ml, total coliform concentration. Let us see what this means when looking at studies reporting log reduction versus percent destruction.

Assuming a well-oxidized secondary effluent, the total coliform concentration before disinfection is probably on the order of 1×10^6 (1,000,000), so a 4-log reduction would produce a final MPN of 100/100 ml, a 99 percent reduction would yield a final count of 10,000/100 ml, and a 99.99 percent reduction would yield 100/100 ml. So for this magnitude of initial count, a 99.99 percent kill is the same as a 4-log reduction.

A well-oxidized and filtered tertiary effluent would probably have an effluent coliform count of 50,000/100 ml before disinfection. The coliform requirement for a disinfected tertiary effluent is usually 2.2/100 ml. Therefore, a log reduction greater than 4 is required, and the percent kill must be greater than

*During the decade 1975–85, the consensus changed to 25 percent.

99.99. The point to be made is that disinfection efficiencies reported as 99 or 99.9 percent are meaningless. Actually, disinfection efficiency studies based upon coliform destruction should specify the range of initial coliform as MPN/100 ml in addition to the log reduction. Then the mathematical model developed by the lab of Selleck (Collins et al.) in 1970[10]* can be used for verification. This model is as follows:

$$y/y_0 = [1 + 0.23\ Ct]^{-3} \tag{8-1}$$

or

$$Ct = \frac{\sqrt[3]{y_0/y}}{0.23} - 1 \tag{8-1a}$$

where

y_0 = initial coliform MPN/100 ml
y = final coliform MPN/100 ml
c = chlorine residual, mg/L, at the *end* of contact time t
t = contact time in minutes

This model has been verified by several practitioners and researchers since it was first published.

Bacterial Indicator Concepts

The parameter of major importance that would provide proof of disinfection is the resistance of the indicator organism to the disinfectant. Researchers over the years have expressed dissatisfaction with the coliform group as an indicator organism because it is not resistant enough to chlorine to allow any safety factor. For a group of organisms to be an ideal indicator, the following conditions should demonstrate disinfection efficiency:

1. The indicator organism must be more resistant to disinfection than the pathogenic organisms.
2. The indicator must be present in the sample whenever pathogenic organisms are present.
3. The indicator must occur in greater numbers than the pathogens.
4. A simple, rapid, and unambiguous procedure must be capable of enumerating the indicator organisms.
5. The indicator should not regrow or otherwise increase in numbers in the aquatic environment after disinfection.

*In the past 25 years, White has found the use if this formula to be most successful in determining the important *Ct* factor for any wastewater effluent.

6. The indicator organism must be randomly distributed in the influent stream.
7. The presence of other organisms must not inhibit the growth of the indicator organism.
8. The indicator organism should be nonpathogenic to humans.

The lack of disinfection resistance of the coliform group was clearly demonstrated when the organism *Klebsiella* was found in the City of Chicago's water distribution system.[11] The members of the *Klebsiella* genus can cause severe enteritis in children, and pneumonia and upper respiratory tract infection, septicemia, meningitis, peritonitis, and urinary tract infection in adults. Discovering such a hazardous organism in a distribution system, carrying water that had been coagulated, filtered, and chlorinated sufficiently to carry a small residual in the system, was indeed unsettling. In this instance it was concluded that *Klebsiella* was the organism responsible for most of the positive samples occurring in the routine coliform sampling procedures. The *Klebsiella* group are encapsulated organisms, and once in the distribution system, may be harbored in protective slime and sediment. For these reasons they are more resistant to disinfection than the coliforms. In 1974, Engelbrecht et al.[12] evaluated two promising groups of organisms believed to be resistant to chlorine in the range necessary to inactivate both bacillary pathogens and waterborne viruses. These groups are the acid-fast cultures *Mycobacterium fortuitum* and *M. phlei* and a yeast, *Candida parapsilosis.*

However, it must be recognized that the concept of proof of disinfection for wastewater discharges and water reuse situations is entirely different from that of water to be used for potable purposes. With the situation as it exists in 1997 with respect to the organics in wastewater, it is unlikely that it will be possible to pursue the reclamation of wastewater for potable use. Therefore, proof of disinfection for wastewater and water reuse should always be on the basis of destruction of the indicator organisms. This could conceivably be on the basis of a chlorine residual contact time envelope, provided that this combination of criteria could be developed for a "consensus" organism.[13,14]

WASTEWATER REUSE*

Historical Background

Wastewater reuse, in the United States, has been practiced since about 1920. Historically, California water shortages are known to occur at regular cyclic intervals. Back in the 1920s and 1930s when the state's population was growing at a rapid rate, the city of Berkeley was almost devastated by a big fire (in 1923), owing to the lack of enough water at a sufficient pressure. One of the

*See Chapter 6 for a description of the Denver, Colorado 1 mgd recycling pilot plant, built in 1985. This plant recycles secondary wastewater effluent that produces potable water.

first systems to put its total discharge to beneficial use was the activated sludge sewage treatment plant in the Golden Gate Park of San Francisco. This operation began about 1930. In 1935, there were 62 communities using treated wastewater for crop irrigation, and regulations for this use had been in existence for about 15 years in California. During these times there were three very forceful individuals who realized that the future of California was burdened by the lack of a dependable and sufficient water supply. Mulholland, of Los Angeles, obtained the rights to the source water of the Mono River some 350 miles north of Los Angeles as early as 1910; then O'Shaugnessy of San Francisco manuevered the federal government into allowing the City of San Francisco to build the Hetch–Hetchy Dam in Yosemite National Park over the intense objections of the earliest "environmentalist" back in 1915. This dam provides the surface water source for San Francisco and the Peninsula. Governor Pardee helped get the Pardee dam built, which now supplies water to the municipalities in the East Bay. These two sources, the Hetch–Hetchy Dam and the Pardee Dam, started supplying water on a regular basis in the late 1920s and early 1930s.

In about 1940, the Shasta Dam in Northern California was built, and many years later (about 1960) Governor Pat Brown organized a questionable project that took the Shasta Dam water out of the Sacramento River near Stockton and had it pumped over the Tehachipi Mountains (3000 ft) to supply the consumers of the Metropolitan Water Distinct (MWD) of Southern California.

To underscore how serious the water problem has always been in California with the continuing increase in population, White remembers his first trip to the polls, when he became of age (21) in 1931, when he voted for the bond issue for the Colorado River Aqueduct, the original supply for the MWD of Southern California. There were several serious droughts that occurred during the construction of these projects. During those times the people in most of Southern California had to retrieve all the rainfall possibly by placing containers under the roof-runoff drains. This water was used for bathtub and dishwater.

By 1970, there were about 600 reclamation projects operating on a continuous basis in the United States, about half of them in the State of California. The first use, of course, was for crop irrigation. These crops were fiber, fodder, and seed crops, and there was little opportunity for public contact. Use of reclaimed water for landscape irrigation (i.e., irrigation of parks, playgrounds, golf courses, freeways, rights-of-way, and so forth) also has a fairly long history, but only in recent years has there been a sudden increase in the number of such installations. Where crop irrigation installations increased 40 percent, landscape irrigation installations multiplied ninefold. There is a trend, then, and a general swing toward uses of reclaimed water where the public may have more exposure and more contact with it.

Public attention has been drawn to many outstanding reclamation systems in California such as the Santee project, which includes swimming in addition to boating and fishing, and the Contra Costa project, for supplying industry cooling water needs from reclaimed wastewater.

Moreover, it is a matter of fiscal responsibility for any agency producing water and treating wastewater to have plans for treating wastewater effluent to produce potable water. Otherwise such an agency would not qualify for federal funds.

Proposed and Operating Water Reuse Projects (1997)

Northern California Projects

Cities of San Jose and Santa Clara.[132] This joint project is known as the South Bay Water Recycling (SBWR) project. It is important to know that the California Department of Health Services (DOHS) requires thorough disinfection of wastewater treatment plant effluent provided for application in areas available to the general public. As a result, chlorination plays a key role in most recycled projects, such as the SBWR now under construction.

This project is designed to recycle up to 20 mgd of tertiary treated effluent which would otherwise be discharged into the south end of San Francisco Bay. This recycled water will initially be delivered to more than 200 customers, primarily for landscape irrigation with some industrial reuse. The 20 mgd will be applied mostly during the six-month summer "dry season," and will annually replace approximately 10,000 acre ft of potable water, enough for roughly 40,000 residents. Prior to recycling, the water is treated at the San Jose/Santa Clara Water Pollution Control Plant to meet the DOHS Title 22 standards. This plant serves a population of 1.2 million people in the high-technology manufacturing center of Northern California, commonly called "Silicon Valley," which delivers an average daily flow of about 130 mgd. The SJ/SC tertiary plant is currently designed to treat up to 167 mgd.

Treatment includes primary sedimentation, secondary air activated sludge treatment, and advanced or tertiary treatment, consisting of a separate stage activated sludge nitrification process and mixed media filtration (sand, garnet, and anthracite coal). The effluent is consistently measured to have a total suspended solids (TSS) concentration less than 1 mg/L, and a five-day biochemical oxygen demand (BOD5) of 3 mg/L or less.

Disinfection by chlorine gas is accomplished by the use of the Water Champ System of in-stream injection of about 10,000 lb/day* as a super-mixing device. (See Fig. 8-2.)

The chlorine gas is transported in liquid form by 90-ton railcars, and is gassified by evaporators before being drawn into the effluent flow by the

*This method introduces chlorine gas at 60 ft/sec that has an ORP, at any effluent solution pH, of 1.36, which is 40 percent higher than that of the monochloramine solution delivered when the chlorinators are using the conventional water injector system.

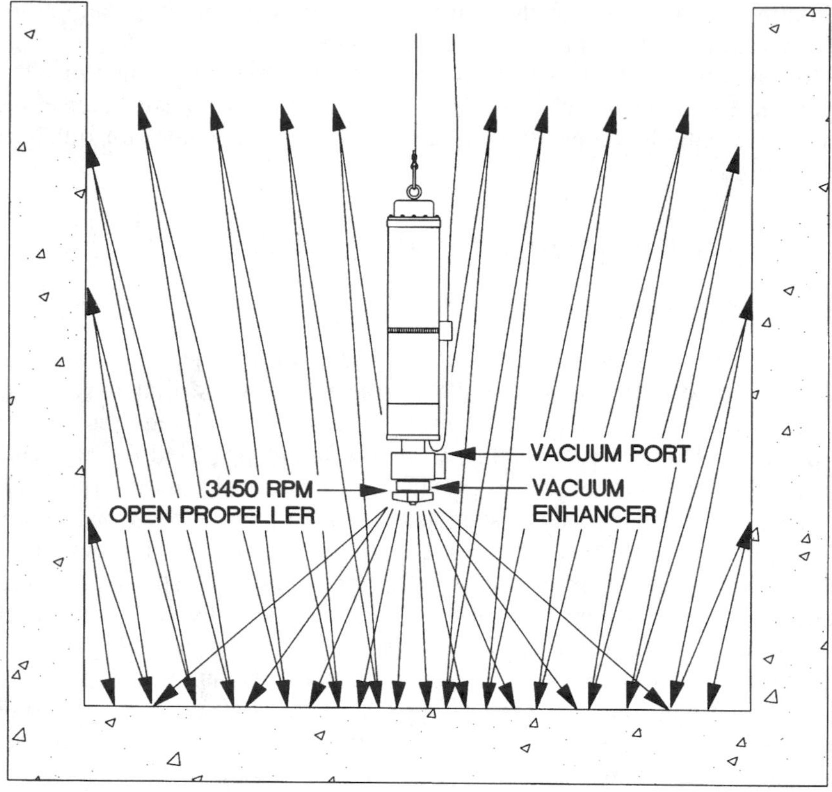

Fig. 8-2. This illustrates the unusual mixing activity available from the Water Champ (courtesy Gardiner Equipment Company, Houston, Texas).

Water Champ units. The initial dosage is approximately 9 mg/L. This maintains a chloramine residual of about 3 mg/L. After 50 min. of residence time in the contact basin of serpentine design, the final effluent is routinely measured to have a total coliform count of less than 2 MPN/100 ml. Prior to its reaching the contact chamber where the chlorine is applied, a solution of ammonia hydroxide is added to the nitrified effluent in a ratio of about 4 or 5 to 1 of chlorine in order to have a chloramine solution. This eliminates all the devastating problems caused by the high concentrations of organic N in the nitrified effluent.

Current Title 22 regulations for unrestricted use of recycled water require maintenance of a disinfection residual of 5 mg/L after 90 min. of contact time, or a total *Ct* value of 450. To meet this standard, SBWR (South Bay Water Reuse) facilities have been designed to provide a second stage of disinfection by chlorine injection into the 108-inch-diameter pipeline, which connects the plant to the transmission pumping station. The 4000-ft-long pipeline provides an additional 90 min. of detention at a recycled water flow of 20 mgd. Boosting the chlorine dosage after the first stage of disinfection to approximately 7 mg/L will provide a final residual of at least 5 mg/L, with a *Ct* value of 450 or more.

Although the effectiveness of chlorine disinfection is currently measured by the survival of coliform organisms, discussions in the industry now suggest that other indicators (e.g., bacteriophages) may provide a more accurate indication of pathogenicity. SBWR researches are currently evaluating the use of the MS2 bacteriophage as a reliable indicator of disinfection effectiveness. Depending upon the results of these investigations, the chlorine dosage may be modified as required to ensure the appropriateness of all applications of recycled water.

White encourages this sort of research, because some 30 years ago there were water pathogen outbreaks caused by hither to unknown viruses that attacked consumers in several well maintained potable water systems. He suggested at the time (1962) that a "consensus" organism such as the Coxsackie A2 virus be used (see Chapter 6, under "Disinfection Guidelines"). This could still be appropriate, but it could easily increase the Ct to 700 because in all wastewaters we are dealing with chloramine residuals and not free chlorine as in potable water supplies.

The Dublin San Ramon Services District (DSRSD)

Clean Water Revival. This district is developing a project that would take highly treated wastewater that can be used for landscape irrigation and further purify it through microfiltration and/or reverse osmosis (following microfiltration if both are used.) After this treatment this water would then be injected into the Tri-Valley groundwater basin.

This project aims to increase disposal capacity, improve groundwater quality, and increase local water resources.

There are two basic alternatives being evaluated for this project. The first is called the "Main Basin Alternative." It would result in injecting about 2.54 mgd of microfiltered and reverse-osmosis-treated water into the main groundwater basin. Locations in Pleasanton and the unincorporated areas of Alameda County are being studied. The injection wells would be placed a prescribed distance from wells used to draw drinking water from the basins.

The second process is called the "Fringe Basin Alternative." This would result in injecting highly treated wastewater that has been processed by microfiltration into the fringe basins in the Dublin area, as there are no municipal wells in these basins that pump drinking water. The alternative includes extracting water at a later date for landscape irrigation and other nonpotable uses.

"Clean Water Revival" is one of two projects in the Livermore–Amador Valley now (1996) under review by Zone & Water Agency, with respect to its long-term master plan for water recycling and groundwater quality improvement. The City of Livermore also has produced a similar well-injection project.

Current plans (1997) are to present the final Environmental Impact Report and projects to the DSRSD Board of Directors for a decision on whether or not to proceed. As unusually heavy December 1996 rain storms filled the reservoirs, White is worried that a decision by the Board will be delayed. California is still going to run out of water about 2015, however.

The Dublin San Ramon Services District (DSRSD) and the East Bay Municipal Utility District (EBMUD) in Northern California, a joint recycled water authority, has been formed in the realization that a water shortage crisis is certain to occur in California in about 2015. In preparation for this crisis, these two district operations created a joint authority to deliver recycled water to the San Ramon Valley. The DSRSD will sell its recycled water to the joint agency, which will, in turn, sell it back to each agency for delivery to the respective customers.

The EBMUD provides potable water and delivery services to customers in the northern part of the San Ramon Valley, and the DSRSD has water and wastewater customers plus a wastewater treatment plant in the southern end of San Ramon Valley.

Large landscaped areas, such as golf courses, parks, and common greenbelts, are potential customers for recycled water. The San Ramon Valley, including Dublin, has approximately 200 potential customers located near the DSRSD wastewater treatment plant. This combination of supply and demand makes the San Ramon valley a good prospect for a successful recycled water project. Therefore, both agencies will save some costs, to the benefit of their respective customers.

The project cost was estimated at $55 million in 1996 dollars for a 5860 acre-ft/yr (1.9 billion gallons) project, to be financed through revenue bonds. Repayment of the capital costs will come from sales revenue of recycled water over the next 20 years.

Monterey County Water Reuse Projects

Salinas Valley Reclamation and Irrigation Projects.[124] As agricultural activity and urban development in the Salinas Valley have increased in the past 60 years, water has been pumped from the groundwater aquifers faster than it can be replaced by rainfall runoff. This has caused the groundwater levels to drop below the sea levels. This allows the seawater from Monterey Bay to intrude the two groundwater aquifers located at 180 and 400 ft below the surface.

This intrusion has extended nearly six miles inland on the 180 ft aquifer and two miles on the 400 ft aquifers. This intrusion has rendered the groundwater totally unfit for either domestic or agricultural use. This affects nearly 16,000 acres of agricultural land.

Replacement of groundwater occurs primarily from percolation of surface water from the Salinas River and its tributaries. Increasing this percolation would keep the river water flowing all year; to achieve this goal, two dams were built, the Nacimiento in 1957 and the San Antonio in 1965. However, these reservoirs are not now enough to maintain the water required for use in all the agricultural areas.

The water shortage problem is basically one of distribution. Therefore, two projects are required. This also means that two agencies are involved: the Monterey County Water Resources Agency (MCWRA) and the Monterey Water Pollution Control Agency (MCWPCA). These two agencies have developed two separate projects for solving the water distribution problem: the Castroville Seawater Intrusion Project (CSIP)[125] and the Salinas Valley Reclamation Project (SVRP).[124] The CSIP consists of a pipeline distribution system to Castroville area farms, including a number of supplementary groundwater wells. The SVRP project consists of building a tertiary wastewater reclamation plant, adjacent to the existing Salinas Valley WWTP, with a transmission pipeline to the CSIP. This existing plant, also known as the Regional Treatment Plant in Marina, is a secondary facility that treats asbout 20 mgd using physical and biological processes. The upgrade of this facility will add tertiary treatment and disinfection facilities.[8]

Disinfection. MRWPCA selected chlorine gas for disinfection[128] at the wastewater reclamation. This decision was based primarily upon cost. The cost using UV was estimated at $57,900,000 versus $54,200,000 for chlorine.

Capacity. These two projects will provide 19,500 acre-ft (6.35 billion gallons) of reclaimed water for irrigation to more than 12,000 acres of cropland, thus reducing groundwater pumping by two-thirds. These projects will reduce the number of groundwater wells from 130 to 22.

Costs. The total cost of these two projects was estimated at $78 million in 1995 dollars. These projects were scheduled to be completed sometime in late 1997.

Safety. Using highly treated wastewater to irrigate vegetable crops is not a new idea; it has been studied and done for years. No adverse health effects have ever been found. In fact, crops irrigated with reclaimed water has been known to produce higher yields, with better quality and a better appearance, than those grown from well waters. (It is also widely believed that pesticide spraying has no ill effects on the vegetable or fruit crops because all of these growing products contain a lot more carcinogens than those in the pesticides. As these growing plants are not equipped with claws or beaks, nature has provided them with chemical protection. See Chapter 6.)

Southern California Projects

Goleta Sanitary and Water Districts of California

Water Recycling Plant and Distribution System. This area is located on the coast of California about 10 miles north of Santa Barbara. It suffers from chronic water shortages as well as budgetary constraints. This situation prompted the sanitary and water districts of Goleta to collaborate on a project to reclaim the effluent from their secondary WWTP, and distribute it in a pipeline system for landscape irrigation and agricultural reuse.

In addition to having water for landscape irrigation available on a continuous basis, it was decided that substituting the reclaimed wastewater would not only provide immediate drought relief, but would reduce the need for developing expensive new water supply sources. There would be no problem in maintaining the beauty of the coastal environment and its revenue-producing recreational areas.

The water reuse plant will have a 3 mgd capacity, which according to the water districts will have a potential potable water saving of over one billion gallons per year. The distribution system will contain nearly six miles of 8-inch- to 18-inch-diameter piping serving 49 potential landscape irrigation users in an 842-acre service area. This project will produce reclaimed wastewater that will meet the California Department of Health Services, Title 22, Division 4, standards for unrestricted landscape uses of reclaimed water.

Los Angeles and Orange County Joint Projects. This involves the Orange County Water District (OCWD) together with the County Sanitation Districts of Orange County (CSDOC). They are planning a water reuse project that will produce up to 90 mgd of extremely well-treated wastewater with a quality equivalent to that of potable water. This water will be used primarily for groundwater recharge. Here again, this is a result of water supplies in California becoming increasingly unreliable in the extremely arid region of the southern part of the state.

The OCWD is a pioneer in the use of treated wastewater for augmenting local supplies. The two agencies mentioned above, OCWD and CSDOC, have been successfully working together on relcaimed water projects since 1975, beginning with Water Factory 21 and more recently in the Green Acres Project, which was completed in 1991.

According to the Metropolitan Water District (MWD) of Southern California, serving nearly 16 million consumers claim they will need to develop approximately 360 billion gallons of new water supplies by the year 2000 to sustain the region during prolonged drought conditions. This is going to put an enormous burden on their supplies: the State water from Northern California and the Colorado River water.

Water Factory 21 is an important part of OCWD's approach to reducing dependency upon imported water. This project treats secondary effluents by using lime recalcination, multimedia filtration, carbon adsorption, disinfection, and reverse osmosis. The product water is blended with deep well water and injected into the groundwater basin to maintain a hydraulic barrier against seawater intrusion. The quality of this injected water has consistently met or exceeded regulatory requirements.

The Green Acres project currently delivers about 108 million gallons of water per year to 28 end-user sites. This water is treated to meet Title 22 requirements for agricultural and landscape irrigation plus some industrial uses. OCWD plans to extend the service reach of this project and increase the output to about 16.3 billion gallons per year.

It is interesting to note that as imported water prices have increased, the advancements in treatment technologies have made wastewater reuse more cost-effective.

The OCR Project.[122] The OCWD and the CSDOC are exploring the feasibility of an innovative water reuse project that would treat secondary wastewater effluents at a cost competitive with other alternative water sources. They plan to use various advanced technologies, such as microfiltration (MF), reverse osmosis (RO), and ultrafiltration (UF).[123] These processes would increase the removal of bacteria and viruses as well as the removal of total dissolved solids (TDS) and total organic carbon (TOC).

The proposed project would transfer the highly treated water some 13 miles along the Santa Ana River through a 6-ft-diameter pipeline that would distribute the water in OCWD's 1600 acres of existing recharge facilities in Anaheim, California.

In addition, the 6-ft-diameter pipeline would travel through the center of the industrial center of Orange County. Then the OCR project water could be arranged to distribute water for landscape irrigation for use by water-reliant industries along the corridor.

This OCR project would be developed in three phases. As envisioned, the project would produce about 45 mgd in the year 2000, and would increase in increments to 68 mgd and then 89 mgd by the year 2020. The reclaimed water would provide a supply to OCWD's existing recharge basins during most of the months of the year, increasing water reliability in the region.

Water Quality. In addition to improved water reliability and efficiency, the OCR project also offers water quality benefits. The quality of this reclaimed wastewater would meet all health standards required by the State Department of Health Services, and would be even better than the quality of the Santa Ana River water that is currently received from upstream discharges for groundwater recharge.

Costs. Based upon the preferred treatment train, Phase I of this project would cost from \$152 to \$176 million in 1995 dollars. Phase II is estimated to require \$53 to \$66 million for capital costs.

The Irvine Ranch Water District[134]

Michelson Water Reclamation Plant. This plant is located in Irvine, California at 15600 Sand Canyon Avenue, in San Diego County. It is owned and operated by the Irvine Ranch Water District (IRWD). This water district is one of the youngest districts and probably the most progressive water agency in the whole of California.

Recently, in the 1990s, this water reclamation plant has been supplemented by an odor control plant that is shown in Fig. 8-3. This system is fully described in Chapter 7.

The district was formed in 1961 to provide both irrigation and domestic water for a developing community. In 1963, a decision by the Board of Directors of this agency resulted in the expansion of the system to provide both sewage collection and treatment that produced reclaimed water. This decision resulted in the building of the Michelson Water Reclamation Plant, which went into service about 1967. The IRWD has been a leader in the design and development of rules and regulations for the use of reclaimed water. They developed a dual distribution system, with one set of pipes for potable water and another for reclaimed water.

Stringent water quality criteria established by the California Department of Health Services are met and exceeded by the MWRP. The high quality "polished" water that leaves this plant has earned the IRWD the first "unrestricted use permit" to be issued by the State of California. Consequently, the MWRP reclaimed water can be used for almost everything except drinking.

The reclaimed water is currently being used for crops, golf courses, parks, school grounds, greenbelts, street medians, and freeway landscaping. Recently, the IRWD pioneered another use for this reclaimed water: providing high-

Fig. 8-3. 6000 CFM packaged three-stage odor control system, Michelson ERP, Irvine, California.—same as Fig. 7-106

rise office buildings with the reclaimed water for flushing toilets. Additional office towers and other buildings are scheduled for this use in the future.

The MWRP is a 15 mgd activated slude plant, where the actual daily production of reclaimed water varies according to seasonal and operational demands. The reclaimed water distribution system is made up of about 150 miles of piping from 54 inches to 2 inches in diameter. The system includes pump stations and three storage reservoirs to meet the irrigation demands.

The IRWD maintains a state-certified laboratory to monitor every step of the reclamation process. Samples are taken every day for analysis. The results of these tests help plant personnel to monitor and control the overall treatment process.

The results of these analytical tests are reported regularly to the California State Health Department and the other regulatory agencies as required.

The drinking water supplied to the consumers of this district is also monitored and analyzed on a regular schedule at this California state-certified laboratory to assure proper water quality and safety.

Features of the Michelson Water Reclamation Plant. The operations center of this plant is located at 3512 Michelson Drive, Irvine, California 92715.

Headworks. This is where the raw sewage enters the plant. The sewage flow velocity is substantially reduced at this point of entry, where the bar screens trap the heavy particles of sticks and stones along with grit, sand, and dirt, and they are ground up before the wastewater moves on to the next step.

Primary Clarifiers. This is where the process (wastewater) flow leaves the headworks. It resides here undisturbed for about two hours in this step. This allows most of the solids (sludge) to settle out of the wastewater. The sludge is pumped out of the clarifiers and into the Orange County Sanitation District plant for treatment and disposal.

Flow Equalization Basins. These basins receive the local sewage flows in peaks, in response to water use patterns by the customers. The peak flow times are usually around 8:30 A.M. and 10:00 P.M. These basins hold the primary effluent and control its introduction into the system. This feature allows for efficient and economical plant operation.

The primary effluent then flows from the flow equalization basins on to the aeration tanks.

Aeration Tanks. These tanks have an abundance of microbiological life. These organisms have a lifestyle that causes decomposition of the various kinds of organic matter in the wastewater flow. This water remains in these tanks for about five hours. Air is pumped into these tanks in order to keep the microbiological life style aerobic instead of anaerobic.

Secondary Clarifiers. This is the next step in the treatment process. When the microbes have completed their job of consuming the wastewater material in the previous step, the process flow enters these clarifiers to remove more settleable solids. Therefore, the sewage that flows out of these secondary clarifiers is very clean, with about 90 percent of all the contamination removed. This effluent flows over weirs, leaving solids behind. The process flow remains in these secondary clarifier tanks for about two hours. The effluent from these secondary clarifiers is then pumped to the dual-media filters.

Dual-Media Filters. The secondary filter effluent is treated here with alum and polymer to assist in the coagulation of any solid particles. At this point the process flow is filtered through layers of coal and sand—the dual-media. When the process flow leaves the filters, it is estimated that more than 99 percent (2 logs) of the contaminants have been removed, and the water remaining in sparkling clear.

Chlorine Residual Contact Tanks. This is the final step in the overall process. The chlorine dosage applied here plus the maximum contact time of two hours will definitely inactivate any pathogenic bacteria and viruses that might be harmful to people or the environment. The total reclamation process takes less than 12 hours. White is very eager to learn how efficient this system would be in inactivating the oocysts of *Cryptosporidium.*

Final Step. These are the storage reservoirs. The reclaimed water leaving the treatment process is suitable for immediate application to landscapes and agricultural use; or it can be pumped to other storage reservoirs until it is needed.

Honors. The Michelson Water Reclamation Plant and its dual-distribution system was honored for two consecutive years by the U.S. EPA. This plant has also been recognized as the best large nondischarging treatment plant. It has also received first and second place awards in the EPA's National Wastewater Management Excellence Awards Program.

San Diego, California: North City Water Reclamation Plant

Introduction. This plant, illustrated in Fig. 8-4, is nearing completion and expects to be in full operation in 1998. It has been designed to produce a tertiary effluent, which means that the effluent will be filtered.

After the sewage has been collected it will pass through a bar screen and then a grit chamber, followed by primary clarification, where the influent is subjected to aeration in tanks where oxygen is applied. At this point the effluent is treated to secondary clarifiers and anthracite coal tertiary filters, followed by the application of chlorine. Immediately after chlorination the

Fig. 8-4. North City water reclamation plant, San Diego, California.

effluent then must pass through a specially designed contact chamber that will provide a one-hour detention time at peak daily flow.

Distribution System. This pipeline system will convey the effluent from this reclamation plant throughout the northern region of the city service area. This distribution system, the first of its kind in the City of San Diego, will provide customers with an average of 7 million gallons of reclaimed water per day, thereby conserving potable water that would otherwise be used for these needs.

Potential customers of this reclaimed water include science centers, nurseries, business parks, hotels, shopping malls, golf courses, highways, and homeowners' associations. Initially the first users will be those that use large quantities of water for irrigation purposes.

Repurification Project. This system is well into the design stage. It will furnish potable water from the above described water reclamation plant. The result of this project will be to furnish drinking water as clean as or cleaner than the current potable supply. In other words, this repurified water would meet standards that are far in excess of current standards. This will occur largely by the use of reverse osmosis and some new ozone combination.

Henderson, Nevada Water Reuse Program[126]

Introduction. This city, in the arid southwest United States, uses a wastewater treatment program that provides an effluent of three different qualities, one each for irrigation, groundwater aquifer recharge, and replenishment to Lake Mead, which is the major source of water in the Las Vegas Valley. Henderson is a suburb of Las Vegas that has been growing at a phenomenal rate for the past decade. It now has 110,000 residents (1997). This population increase places enormous pressure on southern Nevada's wastewater treatment capacity and limited water supplies, making water resource management crucial for its municipalities and water utilities.

Treatment Facility. This is a $30 million dollar, 10 mgd plant. Since July 1994, this facility has provided secondary and tertiary treatment, using filtration, phosphorus removal, and nitrification. This plant is different from most other reuse systems because it has the flexibility to use a combination of wastewater treatment processes by its design.

This facility operates like a water treatment plant attached to the end of a wastewater treatment plant, with the effluent being removed at the appropriate quality level for each of the reuse options. The facility is designed to meet any one of the three water quality standards.

It is designed to treat the entire flow to the highest quality standards, but systems and controls divert portions of the effluent from further treatment as necessitated by the disposal or reuse alternative. The effluent is extracted from the process train when the necessary level of treatment for the selected option is achieved. This avoids the expense of treating the wastewater to unnecessarily high water quality standards.

The design of this reclamation system is such that it can be easily expanded to 20 mgd, which at the current rate of growth for this city will be necessary by about the year 2002.

For example, the discharge into Lake Mead has the most stringent nutrient load requirements, with an effluent phosphorus concentration requirement of 0.25 mg/L or less. To meet this regulation, the effluent is treated with aluminum sulfate and other chemicals to precipitate out the phosphorus. Effluent for the irrigation application does not require phosphorus removal; therefore, it bypasses that treatment. Likewise, effluent for groundwater recharge is removed prior to filtration.

Quality Parameters. Effluent for irrigation has the strictest requirements: BOD and TSS, 30.0 mg/L; total coliforms, 2.2/100 ml; and turbidity, 2.0 NTU. For groundwater recharge, the only requirements are BOD and TSS, both 30 mg/L. The discharge requirements into both Lake Mead and Las Vegas Wash are BOD and TSS, 30 mg/L; fecal coliforms, 200/100 ml (1000 total coliforms); total ammonia N (April–September), 0.7 mg/L; total phosphorus (March–October), 0.25 mg/L; pH, 6.0–9.0; total residual chlorine, 0.30 mg/L.

Costs. This 10 mgd facility $30 million dollars in 1994. Owing to Henderson's reclamation program, the State of Nevada receives flow credit from the federal government that allows them to pump additional water from Lake Mead for the Las Vegas Valley drinking water supply.

The Delhi, New York City WWTP[127]

Discussion. This plant was engaged in a four-month pilot study conducted by the New York City Department of Environmental Protection, which demonstrated that a dual-stage, continuously back-washing, deep bed sand filter was able to remove 3 logs (99.9 percent) of *Giardia* cysts and *Cryptosporidium* oocysts from water within the New York watershed.

Now, since the demonstration of this four-month study, New York City officials believe that this system can perform the removal of these particular vicious organisms as well as or better than the most expensive microfiltration techniques.

Water Reuse Projects Summary—1997

There is no doubt that reclaimed water and water for reuse are being rapidly developed in the southwestern United States. Specifically, Arizona and California have several plants in operation; and Arizona, in particular, has over 20 new projects or expansions currently in some state of design.

Treatment of reclaimed water typically involves secondary treatment plus additional filtration and disinfection. The governing standard for reclaimed water is the California Title 22—Effluent Reuse Regulation. Arizona also has the Aquifer Protection Permitting Requirement, which has very stringent requirements for nutrient levels in the reclaimed water that often necessitate a nitrate removal process. In many cases, an NPDES permit is not required, since there is 100 percent reuse and no discharge.

Current disinfection for reclaimed water in California requires a long (90 min.) retention time in a chlorine contact basin, whereas in Arizona it does not. For reclaimed water, Arizona has chosen UV disinfection as the method of choice, without any chlorine residual control required.

Reclaimed water from wastewater treatment plants is currently used for nonpotable applications, including decorative lakes and ponds and the irrigation of golf courses, parks, and greenery along roads and highways. Reclaimed water is also being injected as a barrier protection in coastal applications to prevent salt water intrusion into existing freshwater aquifers.

Arizona has developed "scalping" water reclamation plants, which are designed to remove reclaimed water without doing any solids handling. They scalp reclaimed water from the overall wastewater flow, and return all the waste products back to the pipe for treatment at the downstream wastewater treatment facility. This procedure allows reclaimed water to augment potable water requirements. New water reclamation plants are currently being funded by local developers to supply the water required for new communities featuring lakes and golf courses.

Additionally, there are plants being designed to allow possible future direct return of highly treated reclaimed water into either reservoirs or aquifers.

Two current design projects include the Advanced Water Treatment Plant in San Diego, California, and the Scottsdale Water Campus in Scottsdale, Arizona. The contract to finalize the design of the San Diego Advanced Water Treatment Plant has been let to the engineering consulting firm of Malcolm Pirnie, Inc. at its Carlsbad, California office, under the direction of Mr. Rick Kennedy. He reports that the preliminary design criteria are based upon further treatment of reclaimed water from the new San Diego North City Water Reclamation Facility, as follows:

This San Diego treatment plant process design calls for micro- or ultrafiltration followed by reverse osmosis, then a degassifier followed by ion exchange, and then ozonation followed by final disinfection by residual chlorination. Pilot plant equipment trials are being run in the San Diego area to test various pieces of equipment and their performance. The San Diego Plant estimated date for plant installation and startup is in the year 2001.

The final effluent of this plant will meet drinking water standards, and satisfy all health department concerns for potable water. The overall goal is to be able to use this water to help recharge the San Vicente reservoir, to augment the drinking water supply for the City of San Diego.

California Requirements as of 1996

Since the third edition of this book was published in 1992, in the knowledge that the planet is going to run out of water in about 2025 owing to overpopulation, the realization has grown that wastewater reuse is going to be a necessity in many parts of the world. In the United States the hardest-hit areas are going to be in the states of California, Nevada, Arizona, and New Mexico. California is expected to run out of potable water in 2015.

As of 1997, there are several wastewater reclamation projects under design in California. Some of these plants are going to be designed to produce potable water; so, the California requirements are going to be based upon the production of potable water. These are set forth here as a reminder.

California Wastewater Reclamation Criteria (Title 22, Division 4, Chapter 3 of the California Code of Regulations) for uses where an oxidized, coagulated, filtered, disinfected effluent is required are as follows:

1. Filtered effluent turbidity equal to or less than 2 NTU.
2. Total coliform count equal to or less than 2.2/100 ml.
3. Virus inactivation efficiency, 4 logs removal (99.99 percent reduction) based upon polio virus.

California Requirements as of 1992

Rationale. In 1968, the California State Board of Public Health adopted standards for the quality of reclaimed water used for crop irrigation, landscape, and recreational impoundments. It is important to realize that scientists from all fields acknowledged that the establishment of specific quality limits for reclaimed water in terms of BOD and suspended solids, either soluble materials or other commonly used criteria, was not practical. The monitoring cost would be too great for smaller reclamation operations. The group decided to use, as the keystone of the standards, the coliform bacteria concentration in the reclaimed water. The coliform bacteria concentrations that are allowable for different reclamation uses are supplemented in the standards by terms such as "oxidized wastewater," "filtered wastewater," and other descriptive terms that broadly identify the type of reclaimed water required without specifically identifying numerous quality limits. Briefly these adopted standards are as follows:[15]

Primary Effluent. It can be used for surface irrigation of processed food crops, orchards, and vineyards; and for irrigation of fodder, fiber, and seed crops. Virtually no public contact or ingestion is possible.

Oxidized Effluent (Secondary). With a median coliform MPN of 23/100 ml, it can be used for landscape irrigation, spray irrigation of processed food crops, landscape impoundments, and milk-cow pastures. Public contact is possible, but ingestion is very unlikely.

Oxidized Effluent (Secondary). With a median coliform MPN of 2.2/100 ml, it can be used for surface irrigation of produce (makes this operation impractical) and restricted recreational impoundments. Public contact and minor ingestion are possible.

Filtered Effluent (Tertiary). With a median coliform MPN of 2.2/100 ml, it can be used for spray irrigation of produce and unrestricted recreational impoundments. Public contact and minor ingestion are likely.

Other Applications. At present there are many cooling water applications using secondary effluent meeting the 2.2/100 ml MPN coliform requirements. Water reuse is here to stay, at least in California. Now, with a severe water shortage predicted for as soon as 2015, there are several water reuse projects either under design or under construction. These projects are going to have to meet requirements for producing potable water.

Currently this water is used extensively to recharge groundwater supplies, which accomplishes two other major objectives: (1) provides a salt water barrier; and (2) prevents land subsidence due to withdrawal of natural gas and oil reserves.

Importance of Disinfection. Disinfection is the most important link in this chain of treatment. Chlorine must be relied upon to do the bulk of the disinfecting and to provide a persisting residual. Ozone may be called on to provide the assurance of viral inactivation.

The Environmental Health Laboratory at the Hebrew University in Jerusalem has studied the health risks resulting from spray irrigation of nondisinfected wastewater. Its studies were undertaken to obtain data about the number and types of enteric bacteria dispersed into the air as a result of spray irrigation. Katzenelson and Teltch[47] discoverd that coliform bacteria were found in the air at a distance of 400 yards downwind from the irrigation line, and in one case a *Salmonella* bacterium was isolated 65 yards from the source of irrigation. These findings inspired an epidemiological survey of the incidence of enteric communicable diseases in 77 agricultural communal settlements practicing spray irrigation with nondisinfected partially treated oxidation pond effluent, as compared with those in 130 similar settlements not practicing any form of wastewater irrigation. Katzenelson et al.[148] reported in 1976 that the incidence of shigellosis, salmonellosis, typhoid fever, and infectious hepatitis was two to four times higher in communities practicing wastewater irrigation. Moreover, it was found that there were no differences in the incidence of enteric diseases between these two communities during the winter nonirrigation season. *These studies have brought about recommendations for strong wastewater-treatment measures, including effective bacterial and viricidal inactivation by disinfection to prevent the spread of enteric diseases due to airborne contamination of communities adjacent to spray irrigation projects.*

VIRUSES

The Virus Hazard

Introduction. It caused some concern that the frequency of infectious hepatitis remained at a static level of 50,000–60,000 cases per year in the United States, while the incidence of typhoid fever dropped from approximately 2000 cases in 1955 to only 300 cases in 1968.[13] This indicated that although waterborne bacterial infections have been all but eliminated, water utility people may have a more severe task when dealing with waterborne viral infections. The problem of viruses in water supplies has received a great deal of attention over the years and is well documented.[16–21] In evaluating the virus hazard, two factors should concern the protectors of the water supplies, whether for wastewater disinfection or potable water treatment.

These factors are the origin of infectious hepatitis and the origin of the significant increase in gastroenteritis. Gastroenteritis is not a reportable disease, as distinguished from relatively well-defined illnesses (e.g., infectious hepatitis, shigellosis, salmonellosis, and typhoid). Yet it can be estimated that the number of gastroenteritis cases occurring per year is hundreds of thousands and possibly millions. (See Chapter 6.)

During the period 1961–70 a total of 26,546 cases of gastroenteritis were definitely attributed to contaminated water.[19] Of 52 waterborne-disease outbreaks in the United States in 1971–72, there were 22 outbreaks of gastroenteritis, amounting to 5615 cases from a total of 6817 cases of waterborne illnesses.[20] The concern here was that these cases might have been the result of some unidentified viruses. There are more than 100 viruses excreted in human feces that have been reported to be in contaminated water. Any of these could cause a waterborne disease.

Other aspects of viral infections that are a great cause for concern are included in the evidence examined by McDermott, who pointed out that poliomyelitis virus had hurdled the technical barriers of water treatment of the Paris, France, water supply, which consists of coagulation, filtration, and disinfection by both chlorine and ozone.[5]

In the same discussion, McDermott referred to the work by Plotkin and Katz showing that the minimum infective dose by a virus is 1 pfu.[18] Although this statistic may be open to question, the contemplation of this situation and the possibility of 100 or more viruses capable of contributing to a waterborne disease should be of extreme concern to water producers.

All of these concerns were confirmed and substantiated by the 1970 report of the Committee on Environmental Quality Management, ASCE Sanitary Engineering Division.[16] Some of the important conclusions of this report are as follows:

1. There is no doubt that the virus of infectious hepatitis can be transmitted by drinking water, and epidemiological opinion uniformly supports this conclusion.
2. Although evidence is scanty, it should also be assumed that the enteric viruses and other possible causative agents of viral gastroenteritis can be transmitted by drinking water.
3. There is no doubt that a positive coliform index means that virus may be present; however, absence of coliform may not mean that virus is absent. The coliform index, therefore, although a good laboratory tool, is not a relaible index for viruses. Greater assurance of the absence of virus would be a turbidity of less than 0.1 Jackson Unit and an HOCl residual of 1 mg/L after a contact period of 30 min.
4. The evidence available indicates that a risk of hepatitis infection results from the consumption of raw or steamed underpurated shellfish taken from sewage polluted waters and that the Public Health Service Coliform Standard (70 coliforms/100 ml) has been shown by experience to be a reliable indication of risk-free shellfish waters. The Committee of Environmental Quality Management believes that a high level of protection would be provided by activated sludge treatment and chlorination of the effluent to a level producing an amperometric chlorine residual of 5+ mg/L after 30 min. contact.*
5. Virus multiplication in polluted water appears not to be a significant possibility, and from the control point of view it can be disregarded.

*Amperometric residual is stipulated to be the total chlorine residual with no reference made to any free chlorine residual fraction.

6. Viruses are present in certain river waters, and failure to isolate them results presumably from their low concentrations and the relatively ineffective sampling and concentration procedures employed.
7. Enteroviruses and the virus of infectious hepatitis can survive for prolonged periods under conditions prevailing in drinking-water reservoirs. Long detention times therefore cannot be considered as a safety factor.
8. Enteric viruses differ in resistance to free chlorine. Adenovirus 3 is less resistant than *E. coli*, whereas poliovirus 1 and Coxsackie virus A2 and A9 appear to be more resistant than any of the other enteroviruses studied.

Viruses in Sewage Contaminated Portable Water Supplies. In the United States the enteric virus concentration in raw sewage probably ranges from 2 or 3 to more than 1000 infectious virus units/100 ml with peak levels occurring in late summer and early fall.[17] If the coliform bacteria concentration of raw sewage is estimated at 10^7–10^8/100 ml organisms, then the enteric virus concentration is perhaps 5–7 orders of magnitude lower.[22,23]

Although available evidence indicates that enteric virus concentrations in drinking water are likely to be very low, it is important to be aware of the fact that as little as one virus infectious unit is probably capable of producing an infection in humans.[18] Most enteric virus isolations have been made from heavily polluted surface waters, but Berg and co-workers detected enteric viruses in Missouri River water having fecal coliform concentrations as low as 60/100 ml.[24] Although virtually nothing is known about enteric virus levels in U.S. potable water supplies, monitoring of the potable water supplies of Paris, France, in the 1960s revealed that about 18 percent of the 200 samples examined contained enteric viruses; and the average virus concentration was estimated at one infectious unit/300 L.[25] The heavily polluted Seine River is one of the major sources of the Paris water supply. In the 1950s, White made a personal investigation of the three large privatized water treatment plants serving the Paris area. He was accompanied by the design engineers as well as the operators. He found that prechlorination of the river water was as high as 16 mg/L, followed by 5 mg/L of ozone, plus postchlorination for distribution system stability.

Nupen (1974) and co-workers reported finding enteric viruses in 10-L samples of drinking water in South Africa.[26]

Although there is little quantitative information available on enteric virus levels in sewage-contaminated surface and ground waters, there is plenty of evidence that wastewaters are a primary source of enteric virus contamination of the human environment. With the possible exception of a few poliomyelitis outbreaks, there is no evidence of waterborne outbreaks in the United States caused by other specific viruses. The most prevalent waterborne disease in the United States continues to be gastroenteritis of unknown etiology. Therefore, for reasons described above, it is imperative that wastewater-treatment processes and wastewater-reuse systems address themselves to this virus hazard. (See Chapter 6.)

Virus Inactivation

General Discussion. To date, the most comprehensive study of virus inactivation on a pilot-plant scale has been the work by the County Sanitation Districts of Los Angeles County. The results of this two-year study are contained in the "Pomona Virus Study—Final Report" prepared for the California State Water Resources Control Board and the U.S. Environmental Protection Agency in February 1977. This work was summarized in a paper by Selna et al.[27]

The Sanitation Districts of Los Angeles County have been active participants in various water reuse programs since the mid-1950s. Various discharges of disinfected secondary effluent have been spread in percolation basins for the purpose of replenishing groundwater used for domestic supplies. Conveyance of this water to the percolation sites is through open flood control channels, and in transit it is unintentionally used for recreational activities including body contact. About 80 mgd of chlorinated secondary effluent is subject to this unplanned recreational use in flood control channels. These channels are classified as "unrestricted recreational impoundments" by the California State Department of Health; therefore, wastewater discharged to such channels must comply with Title 22 of the California Administrative Code. This document contains the effluent quality and treatment system requirements for recreational reuse as decreed by the California State Department of Health. In order to qualify for such use, secondary effluent must be coagulated, settled, filtered, and disinfected to achieve a median total coliform MPN of 2.2/100 ml or less, and to provide an effluent that will protect swimmers against viral illnesses. Because this required treatment is expensive, from both a capital and an operational standpoint, the prime objective of the Pomona study was to investigate alternate methods of tertiary treatment that might be more cost-effective than the required Title 22 System and still produce an effluent with the required degree of public health protection.

Treatment System Studies. This work investigated the following four systems:

1. The Title 22 treatment called for by the California State Department of Health was a 40 gpm secondary effluent (NH_3–N, 20 mg/L) treated with alum and ionic polymer followed by flash mixing, flocculation, sedimentation, dual media filtration, and disinfection.
2. A 25 gpm secondary effluent, similar in quality to system 1, was treated with alum and ionic polymer followed by flash mixing, dual media filtration, and disinfection.

3. A 100 gpm secondary effluent, also similar in quality to systems 1 and 2, was treated by using two-stage carbon absorption followed by disinfection.
4. A 25 gpm nitrified secondary effluent (NH_3–N, 0.1 mg/L), similar in quality to the others, was treated with alum and anionic polymer followed by flash mixing, dual filtration, and disinfection.

Disinfection Process

Chlorine. Systems 1, 2, and 3 were capable of disinfection by either ozone or chlorine. System 4 was disinfected with chlorine only, since nitrification is not considered to enhance the performance of ozone.

Chlorine was applied in systems 1, 2, and 3 to produce two levels of chlorine residual at the end of the contact chamber (i.e., 5 and 10 mg/L). These residuals required chlorine dosages of about 10 to 15 mg/L, respectively.

In system 4 the free chlorine residual at the outlet of the contact chamber was maintained at 4 mg/L. This required a chlorine dosage of 10 mg/L.

All chlorine residuals were measured by the DPD-FAS titrimetric method.

Careful attention was given to the degree of mixing at the point of application. A ⅓ hp flash mixer was installed in a 32 gal confined mixing chamber. This calculates to a G factor of 450. A conventional perforated-type chlorine diffuser was used discharging about 3 inches from the mixer impeller.

The chlorine contact chambers were designed to have plug flow characteristics. After an optimization study to achieve a coliform MPN of 2.2/1000 ml, a detention time of 120 min. was selected. A tracer study of the prototype contact chamber revealed a modal time of 98 min. and a minimum time of 58 min. So for the sake of comparison with laboratory studies it might be stated that the bacterial kills and virus inactivation were accomplished with a one-hour contact time and not a two-hour time.

Ozone. The ozone system was arranged to provide a *dosage* level of 10 mg/L in system 1, 10–50 mg/L in system 2, and 6 mg/L in system 3. In each case the ozone contact time was 18 min. The ozone contactors consisted of six 14-inch diameter PVC columns 18 ft high.

Predisinfection Effluent Quality. The quality of the wastewater at the point of disinfection is of considerable interest. The nonnitrified effluent contained 20 mg/L NH_3–N, with suspended solids on the order of 1.5 mg/L and turbudities of 1–1.5 FTU.* The pH was 7.5, and TDS was 580 mg/L. At this pH the undissociated HOCl in system 4 was about 50 percent.

Results. The virus inactivation results from the Pomona study are illustrated in Figs. 8-5, 8-6, and 8-7. The effectiveness of combined chlorine residual is

*FTU = turbidity measurement using Formazin polymer.

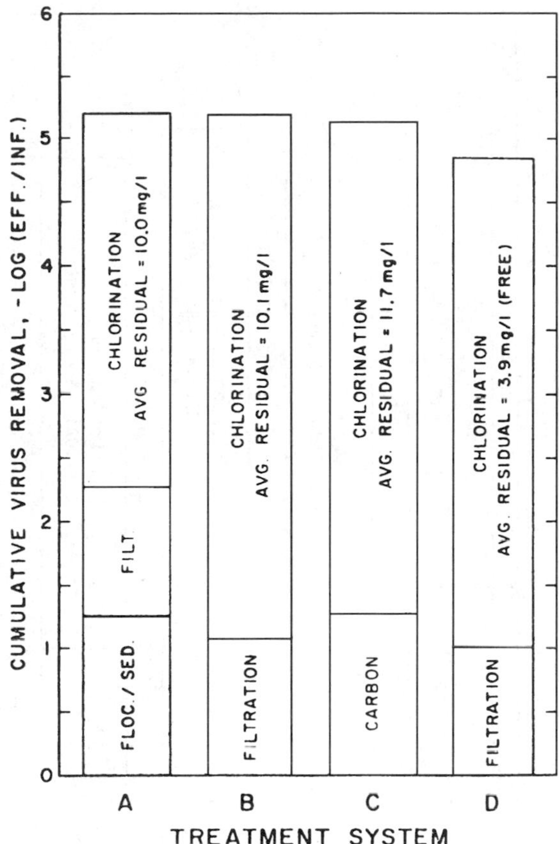

Fig. 8-5. Virus removal at high chlorine residuals.

a real surprise, which advances a totally new concept: that chloramines do in fact have viricidal efficiency potentially equal to that of free chlorine. Until the revelation of the Pomona study it was believed that the only viricidal chlorine compound was free chlorine (HOCl). This is not seen to be the case for tertiary effluents.

It must be pointed out that none of the samples used in these tests in the mid-1970s were not analyzed for organic N because it was not until early 1982 that organic N content was found to seriously affect free chlorine dosages, as it reacts instantaneously with free chlorine (HOCl) to form a nongermicidal organochloramine that becomes part of the total chlorine residual. (See the San Jose, California Water Pollution Control Plant case later in this chapter.)

Moreover, it was found that in system 4 using free chlorine increased the predisinfection effluent concentration of chloroform from 0.06 to 1.98 mg/L, whereas in the other systems using all combined chlorine increased the chloroform content on the order of 0.01 to 0.06 mg/L (approximately). This

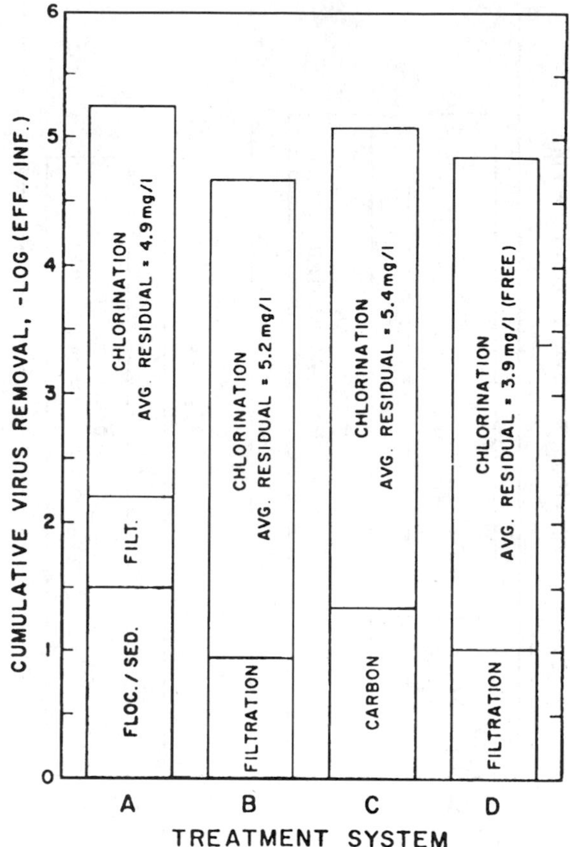

Fig. 8-6. Virus removal at low chlorine residuals.

observation plus the demonstrated effectiveness of chloramines as a viricide leads to the conclusion *that there is no benefit from free residual chlorination of tertiary effluents, which are required to* meet the Title 22 requirements of the California State Department of Health *where indirect potable reuse is called for.*

A dominant feature of Fig. 8-5 is that the majority of virus removal occurs in the disinfection step. The data presented in this figure indicate that reliance must be placed on the disinfection step rather than the filtration or carbon absorption steps prior to disinfection.

Other Conclusions. Other important conclusions drawn from the Pomona study are:

1. Virus inactivation in tertiary treatment systems employing combined chlorine residuals of 5–10 mg/L ranged from 4.7 to 5.2 logs. These results were obtained in poliovirus seeding experiments. The additional

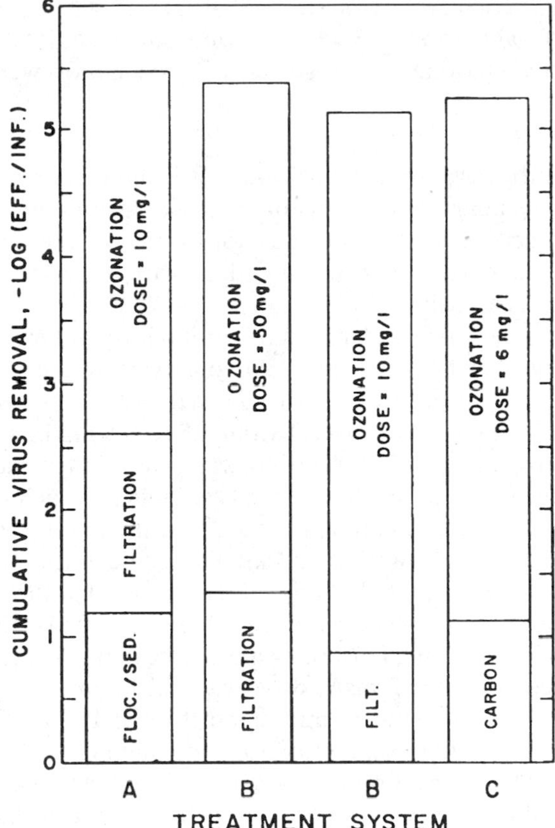

Fig. 8-7. Virus removal at ozonation experiments.

virus removal at the 10 mg/L residual over the 5 mg/L residual was minimal but the coliform kill was consistently better at the higher residual.

2. The system employing free residual chlorination produced about 4.9 logs virus removal when the average free chlorine residual was 4 mg/L.
3. All of the chlorination studies indicated consistent capability for attainment of the 2.2/100 ml MPN coliform requirement by all of the systems.
4. In the experiments using ozonation, virus removal ranged from 5.1 to 5.5 logs (see Fig. 8-7); however, attainment of the 2.2/100 ml MPN coliform standard was hampered by water quality variations.
5. Based upon the results of the virus experiments, it was concluded that system 2 (direct filtration) or system 3 (carbon absorption) tertiary treatment systems are more cost-effective than system 1 (the one required by the California State Department of Health).
6. The direct filtration system (2) is the lowest cost with chlorination, and was estimated at 13.7¢/1000 gal for total capital and operating

cost. This compares to system 1 at 21.5¢/1000 gal with chlorine, 17.2¢/100 gal for system 3 with chlorine, and 19.9¢/1000 gal for system 4 with free chlorine. The systems using ozonation were more costly than this.

Comparison with Other Investigations. It is difficult at best to compare virus inactivation studies owing to evaluation of detection limits, and seeding and analytical procedures; however, it is worthwhile to compare the Pomona study, which is an outstanding investigation, with other noteworthy virus inactivation studies as follows:

Ludovici et al.[28] performed a large number of pilot-plant experiments using the tertiary effluent from the Tucson, Arizona wastewater treatment plant. This effluent ranged in pH from 7.3 to 7.6; NH_3–N from 2.8 to 5.5 mg/L; organic N from 1.4 to 5.6 mg/L; BOD from 1.5 to 8.5 mg/L; and COD from 15 to 33 mg/L. No turbidity information was given, but obviously this is a high quality effluent except for the 5.6 mg/L organic N, which will have some inhibiting effect on the chlorination process. The effluent was seeded variously with polio 1, Coxsackie B1, and Coxsackie B2. The chlorine dosages were 2 and 4 mg/L, which means that all the residuals reported at the end of 30 min. contact time had to be combined chlorine. These were 0.31–1.2 mg/L for the 2 mg/L dose and 0.99–2.76 mg/L for the 4 mg/L dose in the polio 1 experiments. The mean reduction for the 2 mg/L dose was 96.19 percent, with a mean of 99.3 percent for 4 mg/L—all at 30 min. contact time. The Coxsackie B1 and B2 viruses were much more susceptible to the combined chlorine residuals, where a mean of 99.8 percent reduction was achieved with a 4 mg/L dose for Coxsackie B1 and a 100 percent reduction in the Coxsackie B2 experiments with the same chlorine dose.

Concurrently with these experiments Ludovici et al. investigated destruction of total coliforms. With a maximum Y_0 coliform MPN of 19,100/100 ml,* the maximum final coliform MPN during the Coxsackie B2 seeding experiment was 1.1/100 ml with a 4 mg/L chlorine dose, and a 30 min. contact time. The chlorine residual range was 1.5–2.94 mg/L. The same results occurred in the polio 1 seeding experiments, demonstrating the great enhancement of the chlorination process by a high-quality effluent having a low initial coliform count.

These virus studies by Ludovici et al.[28] are noteworthy and deserve some comment. For virus inactivation, reporting destruction as a pecent of the original number of organisms is a good comparative method, but it may not be compatible with what a regulatory agency is likely to concede as appropriate disinfection. We need to have more discussion on a standard of disinfection where there is a public health threat from viruses in certain wastewaters.

---

*Low initial coliform count (Y_0 in the Collins model) is anything less than 50,000/100 ml. By comparison, some raw potable water supplies exceed 3000/100 ml at the intake.

Other virus inactivation studies include the Louisville, Kentucky, experiments by Pavoni and Tittlebaum.[29] They applied ozone to a 40,000 gpd activated sludge plant effluent seeded with F_2 virus (bacteriophage) at a concentration of 10^{11} plaque units per ml and a rate of 1 ml to 1 L of sewage. They reported "virtually 100 percent efficiency (inactivation) after a contact time of 5 minutes with a total ozone dosage of approximately 15 mg/L and a residual of 0.15 mg/L."

Nupen et al.[26] treated a high-quality effluent for water reuse and found that a free chlorine residual beyond the breakpoint, which produced a pH of about 6.0 and a residual of not less than 0.6 mg/L, inactivated polio virus in 35 min.

One of the major difficulties in the inactivation of viruses is their variable sensitivity to disinfectants. Liu and co-workers tested 16 types of human enteric viruses for resistance to free chlorine in treated Potomac River water. The criterion used was time in minutes required for a 99.99 percent inactivation by a 0.5 mg/L free chlorine residual at pH 7.8 and 2°C.[21] The results are listed below in descending order:

polio type II (36.5 min.)
Coxsackie B5 (34.5 min.)
E. coli type 29 (18.2 min.)
E. coli type 12 (16.7 min.)
polio type III (16.6 min.)
Coxsackie B3 (15.7 min.)
adenovirus 7a (12.5 min.)
polio type I (12.0 min.)
Coxsackie B1 (8.5 min.)
adenovirus 12 (8.1 min.)
Coxsackie A9 (7.0 min.)
E. coli 7 (6.8 min.)
adenovirus 3 (4.3 min.)
reovirus 2 (4.2 min.)
reovirus 3 (4.0 min.)
reovirus 1 (2.7 min.)

So the picture developed to date by virus inactivation studies is a bit murky.

Future Considerations of Virus Destruction

The results of the experiments described above clearly illustrate the importance of predisinfection processes in treating wastewater effluents. Raw water for potable water supplies is hopefully of much higher quality than the well-oxidized and filtered effluent of a tertiary wastewater plant. Therefore, we have to recognize the existence of two different standards of disinfection related to virus inactivation: one for potable water and the other for wastewater. For the latter, it seems hopeless to expect significant virus destruction unless the effluent is of tertiary quality. As for raw potable water, it is the consensus that a 1.0 mg/L free chlorine residual at the end of 30 min. contact time at a pH not to exceed 8.0 will probably destroy all pathogenic viruses.

For tertiary effluents, based upon the Pomona study it appears that a *combined* chlorine residual between 5 and 10 mg/L with a two-hour contact is as good or better than a 4 mg/L *free* chlorine residual for the same contact time. This is a new concept. Other disinfectants (e.g., chlorine dioxide) might prove to be better viricides than either chlorine or ozone. There is no question of the efficacy of ozone as a viricide, but on a cost-effective basis it seems to be very little better than chlorine with the disadvantage that it may not produce an effluent to meet the coliform limit. All of this gives impetus to the future investigation of chlorine dioxide as a combination viricide and bactericide.

One other item of significance in considering the treatment of wastewaters for virus inactivation is the report by Boardman and Sproul[30] that none of the particulate systems investigated—clay, hydrated aluminum oxide, and calcium carbonate—protected the T7 phage from inactivation by chlorine. Therefore, viral absorption on an exposed surface provides negligible protection from disinfection, which raises the question of how much turbidity might be tolerated in a high quality effluent. Total encapsulation of the virus would appear to be the major mechanism by which a particle may be afforded protection due to adsorption.

OTHER DISINFECTION CONSIDERATIONS

The Regrowth Phenomenon

Significance of Regrowth. Numerous studies have demonstrated the regrowth phenomenon of coliform and fecal coliform organisms after disinfection. This may be due to the destruction of bacterial predators and may depend upon the presence of certain nutrients in the wastewater of the receiving waters.[7] It has been observed both in wastewaters and in receiving waters downstream from a disinfected sewage discharge. It is the judgment of authorities such as Geldreich that pathogenic bacteria such as *Salmonella* and *Shigella*, of the same family as the coliform group, also regrow. If the disinfection process has an effectiveness against such pathogens comparable to that with coliforms, a disinfection requirement resulting in a reduction of the numerous coliform bacteria down to a level of 23 or 230/100 ml would virtually assure the absence of the pathogenic bacteria present in sewage in smaller numbers. This would eliminate the pathogen regrowth potential, whereas a more liberal criterion of 200 or 400 fecal coliforms would not. Therefore, the phenomenon of regrowth after disinfection, whether a chlorine compound or ozone be used as the disinfectant, is not of any adverse public health significance.

The Clumping Phenomenon. It has been suggested over the years that "clumping" of organisms in the suspended solids of wastewater effluent contributes to irrational data on the efficiency of disinfection after the breakup of these clumps. Investigations by White[31] confirm an extremely small margin of disinfection reliability for a primary effluent. The extension of this observation to storm water overflows leads White to believe that it is not possible to adequately disinfect primary effluents or storm water overflows. This is exemplified by the fact that a primary effluent could be chlorinated with reasonable dosages of about 200 lb/mg to achieve a 230 MPN/100 ml in the effluent at a one-hour-plus contact time, but if the sample for this accomplishment were taken in the same place but on the discharge side of the sample pump, the MPN would rise to 30,000/100 ml. This clearly demonstrates the futility of attempting disinfection of primary effluents. The cause of this dramatic increase in MPN is the clumping phenomenon. The sample pump breaks up the clumps, thereby releasing great quantities of organisms protected from the disinfectant by the clumps. All of these data lend credence to state regulatory agencies asking for better and more effective predisinfection unit processes such as secondary treatment.

Toxicity of Chlorine Residuals

General Discussion. Chlorine residuals of low concentration, whether free or combined, are toxic to most fish and other aquatic life, depending upon the species and the time of exposure. Most oxidants such as chlorine and ozone are known to be irritants to both freshwater and saltwater fish.

The resistance of fish to toxic substances varies according to the size of the fish. Fingerlings are very susceptible, whereas carp and other large fish are highly resistant to toxic agents. The resistance is also proportional to the size of the fish scales: the larger the scales, the greater the resistance. This is why goldfish are hardier than trout.

A review of the literature reveals a general agreement by investigators as to what constitutes a lethal chlorine residual.

In 1968, Tsai[32] reported that the effects of chlorinated domestic sewage effluents on a fish population may stem not only from indirect degradation of wather quality and alteration of the stream bottom, but also from direct action on the fish similar to that of industrial discharges. When wastewater is chlorinated, other toxic compounds may also be formed. For example, if thiocyanate is present, this can be converted to the highly toxic cyanogen chloride by the chlorination process. It is also well known that under certain conditions an array of chloro-organic compounds are formed during the disinfection of wastewater by chlorination, and some of these compounds may also be toxic to aquatic life.[33] Very little is known about how these complex reactions affect aquatic life.

In 1971, Esvelt et al.[34] reported in a comprehensive study that the average daily emission of toxicity to the San Francisco Bay system was about 56 percent from municipal sources and 44 percent from direct releases by industry. The studies were performed long before it was thought necessary to be concerned about the concentration of the chlorine residual in the effluent, given the enormous dilution factor of the receiving waters. The total chlorine residuals encountered (as measured by the back-titration procedure with an emperometric endpoint) ranged from 1 to 8 mg/L. The test fish used throughout the project were golden shiners. Using continuous-flow on-line bioassays they reported a 96-hour TL_{50}* in municipal wastewaters to be 0.2 mg/L total chlorine residual. Even though the fish bioassay procedure is not necessarily a good measure of toxicity,[35] this investigation revealed the following important conclusions:

1. Toxicity removal by biological treatment was 75 percent, whereas toxicity removed by lime precipitation was 40 percent. Both ion exchange removal of ammonia and sorption of organics when coupled to lime precipitation provided an overall 65 percent removal.
2. MBAS† and NH_3–N represent significant toxicants in municipal wastewaters.
3. Chlorination increased the toxicity of treated municipal wastewasters in all instances.
4. A chlorinated and dechlorinated effluent (with a slight excess of sulfite ion) was less toxic than either the unchlorinated or the chlorinated effluent.
5. Dechlorination completely removed the chlorine-induced toxicity.

Also in 1971, the Michigan Department of Natural Resources conducted four separate studies at different wastewater-treatment plants and reported that for rainbow trout the 96-hour TL_{50} concentration below two plants was 0.023 mg/L, and that for fathead minnows concentrations less than 0.1 mg/L were toxic in the plant effuents.[36]

Another investigation demonstrated how the use of the orthotolidine method of measuring and control of wastewater chlorine residuals resulted in gross overchlorination beyond disinfection requirements.[37] This practice resulted in a major fish kill in the lower James River (Virginia).

Another case involving fish kills occurred in the Sacramento River (California) below a wastewater-treatment plant discharge. Chlorine-induced toxicity in the wastewater was immediately suspected. Consequently the regulatory agencies conducted bioassays to determine the source of the problem.[38] River water collected above, at, and below the discharge site was used in the static bioassays. King salmon fry were used as the test fish. The test fish were held in the receiving waters 150 ft upstream from the waste plume and in the waste plume 100, 200, and 300 ft downstream from the discharge where very little dilution of the waste discharge with the receiving water occurred. All of the

*This means that 50 percent of the fish subjected for 96 hours to the specified residual will die. TL = tolerance limit.

†MBAS = Methylene blue active substances.

test fish below the discharge died within a captive 14-hour period of exposure,* whereas all the fish upstream survived. The total chlorine residuals measured by the back-titration amperometric endpoint procedure ranged from 0.2 to 0.3 mg/L during the test period. Therefore, it could only be concluded that the fish kills were caused by the toxicity in the wastewater—presumably the chlorine residual.

Probably the most comprehensive study of the toxicity of chlorine residuals to aquatic life in the receiving waters is the 1975 report by Arthur et al.[39] The disinfection system under investigation with both chlorine and ozone was able to produce an affluent with coliform levels less than 1000/100 ml. All of the studies were in freshwater systems. Both fish and invertebrates were included in the study. The chlorinated effluent was more lethal than the ozonated effluent or the chlorinated–dechlorinated effluent, which confirms the observations and conclusions of previous investigators. The fish were more sensitive than the invertebrates to the chlorinated effluent in the 94-hour tests. The respective 94-hour TL_{50} values of total residual chlorine to fish and invertebrates ranged from 0.08 to 0.26 mg/L and from 0.21 to greater than 0.81 mg/L respectively.

A discussion of the effect of chlorine residuals would not be complete without a consideration of the literature review and analysis of the aquatic life criteria as related to treatment of wastewaters provided by Brungs in his 1976 report.[40]

It should also be of interest to realize the very different way the French handle the problem of industrial discharges that go into the River Seine before reaching the water treatment plants. They use large cages for measuring the change in the speed of the fish swimming upstream. When the fish slow down, the plant operators are advised that a potentially offensive discharge has been made upstream from the plant intake.

Need for Dechlorination†

It was shortly after the Esvelt, et al. report published in 1971[41] that the Water Quality Control Board, State of California, in cooperation with the Department of Fish and Game, began issuing orders for certain wastewater-treatment plants to supplement the chlorination system with dechlorination facilities. The first wastewater-treatment plant to add dechlorination (by sulfur dioxide) in California was the city of Burlingame, ca. 1972. The orders specified a chlorine residual not to exceed 0.1 mg/L. For technical reasons explained in Chapter 10 this meant total dechlorination with a slight excess of SO_3-ion.

*It is no wonder that these fish died; they were trapped in a small tank with a substantial chlorine residual and with no dilution water available.

†This idea for the use of dechlorination of wastewater was the best idea the investigative group had. This group consisted mainly of Esvelt, Kaufman, Selleck, Collins, and White, who out-tricked the Fish and Game Commission, which was way out of line by using fish tanks to hold the trapped fish until they died from the existing chlorine residual.

In the meantime the Sanitary Engineering Research Laboratory, University of California at Richmond, under the direction of Dr. Warren Kaufman, investigated the toxicity of the sulfite ion (a product of over-dechlorination with sulfur dioxide) and found none up to more than 10 mg/L.[42]

As of 1976, there were about 30 chlorination–dechlorination systems operating in California wastewater treatment plants and many more in the design-purchase stage. The California Water Resources Control Board and the Fish and Game Department are in agreement, by following the conclusion from the bay studies on toxicity, that a chlorinated and dechlorinated sewage effluent is less toxic to aquatic life than either a chlorinated or an unchlorinated effluent (which has not been treated by a chlorination process).

Problems leading up to the chlorine toxicity dilemma have been gross overchlorination and poor control systems. It has been argued in some quarters that optimizing the chlorination system and controlling the dechlorination of the effluent to a comfortable 0.5 mg/L chlorine residual would result in a sewage plume that would lose this small residual quickly in the receiving waters by the effects of chlorine demand and dilution.*

A study by Stone[43] revealed some interesting facts about the die-away of chlorine residuals in San Francisco Bay. When a chlorinated secondary effluent was diluted with seawater, there was no uptake of chlorine residual. The residual depletion was strictly by dilution: the seawater exerted no apparent chlorine demand on the combined chlorine residual, which varied from 0.9 to 7.6 mg/L at the end of 24 min. before dilution with seawater.

Freshwater receiving streams are known to exert a chlorine demand, which in most cases could quickly absorb a 0.5 mg/L chlorine residual. This should be investigated on a case-by-case basis because dechlorination to 0.5 mg/L has many more advantages than the use of excess sulfite ion. As of 1997, this has never been investigated, which demonstrates continued negligence by the EPA.

AVAILABLE METHODS OF DISINFECTION

Introduction

The scope of possible methods of wastewater effluent disinfection is great and includes natural processes (predation and normal die-away), environmental factors (salinity, solar radiation), and methods having certain industrial applications (ultrasonics, heat). Only those methods that appear to have possible general application for wastewater and water reuse disinfection will be explored in detail in this text. In this section general features or characteristics that have influenced the overall use of the various disinfection methods are

*All of these investigations that used the Fish and Game procedure of trapping the fish in tanks until they die from high chlorine dosages are, to say the least, totally misleading in trying to evaluate the toxicity of disinfection by chlorine.

identified and included in the design coverage except for gamma radiation and iodine, for reasons described below.

Chlorine

Chlorine has been and will probably continue to be the dominant disinfectant of wastewaters. It is available in different forms, and the characteristics of these forms greatly influence the system design. Because of the importance of chlorine in wastewater treatment the specific features of gaseous chlorine and chlorine compounds are discussed in detail elsewhere in this text (see table of contents).

Liquid–Gas Chlorine. This is the basic chlorine compound. It is used in large volumes by the chemical industry, where derivatives of chlorine find use as pesticides for agriculture, plastics, food preservatives, and pharmaceuticals. It is used in large quantities as a bleaching agent for paper and textiles. Only about 4–5 percent of the total annual production in North America is used for sanitary purposes: household bleaches, restaurant sanitizers, potable water treatment, wastewater treatment, swimming pools, cooling waters, and other industrial process water treatment.

Liquid–gas chlorine is manufactured commercially by the electrolysis of a saturated salt solution. The gas collected in the process is moist and must be dried by passing it through a concentrated sulfuric acid solution to remove the moisture. It is then liquefied by a combination of compression and cooling and is stored in steel containers from 150 lb cylinders to 90-ton tank cars. Liquid–gas chlorine is the principal form of chlorine used in wastewater disinfection. It is also used in wastewater treatment for odor control, destruction of hydrogen sulfide, prevention of septicity, control of activated sludge bulking, and so on.

Hypochlorite. "Available chlorine" can be provided in the form of either sodium or calcium hypochlorite. The most popular form is the sodium hypochlorite. Calcium hypochlorite is much too difficult to manage, owing to excessive maintenance problems resulting from the deposition of the calcium ion throughout the system. Therefore, calcium hypochlorite could only be considered an emergency alternative.

Sodium hypochlorite is a clear liquid available in concentrations of 5, 10, and 15 percent by weight trade strength (available chlorine).

Calcium hypochlorite is available either as a dry granular white powder or in tablet form in strengths of either 35 or 65 percent chlorine by weight.

Imported sodium hypochlorite is being used in some large wastewater treatment plants as a measure to avoid the potential hazard of liquid–gas chlorine delivered and stored in containers under vapor pressures of 80–110 psi.

On-site Hypochlorite Generation. Complete systems are available for the on-site manufacture of hypochlorite solutions by electrolysis, which also avoids the potential hazard of handling the liquid–gas chlorine in pressurized containers. Electrolytic cell systems are available for use with either seawater, brackish water, or concentrated salt brines. The hypochlorite is produced in much the same way that the liquid–gas chlorine is manufactured, except that there is no need to separate the chlorine gas and the sodium hydroxide, which are the products of the electrolysis. This also eliminates the necessity of the sulfuric acid drying step required in the manufacture of liquid–gas chlorine. In addition to the formation of sodium hypochlorite, hydrogen gas is a product of the electrolytic action. It is diluted with air and vented to the atmosphere in concentrations well below the combustible capability of hydrogen. Equipment for this on-site production includes the electrolytic cells, rectifiers, electric switchgear, brinemaker, brine-treatment unit (where required), water-treatment system for cellwater, cooling equipment, and storage tanks for brine and hypochlorite. For economy the process is operated at a constant rate, and excess hypochlorite solution is stored for high-demand periods. Experience with these systems at wastewater treatment plants is relatively limited.

Another method of on-site manufacture of hypochlorite of considerable merit is the use of tank car quantities of chlorine gas supplemented by either calcium hydroxide or sodium hydroxide solution to produce an 8000–9000 mg/L hypochlorite solution. This system is economically appealing from an equipment cost consideration, when the average daily chlorine feed rate exceeds 5–6 tons/day (see Chapter 2).

Chlorine Dioxide

Chlorine dioxide is a highly selective oxidant that is more similar to ozone than it is to chlorine. It is unstable as a compressed gas, and must be generated at the point of use. It cannot be stored in steel containers as can chlorine. Historically, chlorine dioxide has been generated on-site as an aqueous ClO_2 solution by reacting a solution of sodium chlorite with the aqueous solution of a conventional chlorinator injector discharge. New technology, that reacts chlorine gas with specially processed solid sodium chlorite, is the present (1998) state-of-the-art (CDG Technology, Inc., New York, NY), and has substantially resolved all of the problems historically associated with chlorine dioxide generation, making chlorine dioxide more of a wastewater treatment candidate, especially for tertiary treatment of water intended for reuse. Chlorine dioxide does not combine with the ammonia nitrogen normally present. Therefore, in a nitrogen-laden wastewater it is reputed to have a disinfection efficiency for both bacterial and viral destruction comparable to that of free chlorine. Experience with chlorine dioxide on wastewaters is limited. It is significantly more expensive than chlorine (see Chapter 12).

Bromine, Bromine Chloride, and Iodine

Bromine, bromine chloride, and iodine have been used in various ways as an alternative to chlorine. Bromine and bromine chloride are relatively soluble in water, more so than chlorine; however, bromine is much too hazardous a chemical to handle in the treatment of wastewater. Bromine chloride is much easier to handle than bromine because it has a vapor pressure of about 30 psi at room temperature. The materials required to meter bromine chloride are considerably different from those of chlorine; so a separate species of equipment is required for feeding and metering this gas (see Chapter 14).

Bromine compounds have an advantage over the various chlorine compounds concerning the toxicity of residuals to aquatic life in the receiving waters: bromine residuals usually die away rapidly compared to chlorine residuals.* However, this characteristic makes bromine or bromine chloride residual control virtually impossible. Moreover, the bromine compounds are more expensive than those of chlorine.

Iodine is a gray-brown crystalline solid that is only slightly soluble in water. It is derived from kelp or oil field brines, and is mined from deposits in South America. It has been used as an effective method of water treatment on an emergency basis. There are so many unknown factors about iodine as a wastewater disinfectant that this lack of information, coupled with its high cost and uncertain availability, conspires to eliminate it from consideration as a practical wastewater disinfectant. The higher molecular weight of both bromine and iodine puts them at a distinct competitive disadvantage with respect to chlorine.

Ozone

Ozone is an unstable gas that must be produced at the point of use. It is made commercially by the reaction of an oxygen-containing gas (air or pure oxygen) in an electric discharge. It is a powerful oxidant and has been used since the early 1990s for odor and color removal as well as disinfection of potable-water supplies in Western Europe and Canada. It has been investigated recently for use in a process for polishing tertiary effluents, for both color removal and disinfection. From these investigations it appears that ozone in combination with either chlorine or chlorine dioxide could solve the disinfection problem of both bacterial and viral contamination in tertiary wastewater effluents. This is particularly significant where there is consideration of wastewater reuse (see Chapter 13).

---

*Bromine residuals in wastewater effluents containing organic nitrogen are remarkably stable. This stability is directly related to the formation of organobromamines, which are similar to organochloramines and are probably nongermicidal.

Ultraviolet Radiation

Ultraviolet radiation (UV) from the sun has a sterilizing effect on microorganisms, but most of the radiation from this natural source is screened out by the atmosphere before reaching the earth's surface. Ultraviolet radiation can be produced by special lamps (mercury vapor) and is presently used in special situations for high quality potable water disinfection. The disinfection reaction occurs on the thin film surfaces of water where microorganisms can be readily exposed to the radiation. With wastewater, this lethal action cannot be exerted through more than a few centimeters. Beyond this limiting distance, the high absorption of the rays by the water and the suspended solids dissipates the ultraviolet energy. Consequently, the problems of providing effective exposure of sewage effluents containing varying amounts of interfering suspended solids and ordinary turbidity to the UV rays are such that the practical application must be to a very thin sheet of wastewater flow of nearly uniform thickness. Monitoring requirements for proof of disinfection of a UV system are nonexistent.

The UV disinfection concept is applicable when the water or the wastewater is of a high quality. The process must be carefully monitored and supplemented by terminal chlorination for drinking water. The use of UV in wastewater has possibilities where NPDES requirements are not stricter than 200/100 ml fecal coliforms. (See Chapter 17.)

However, in sampling a UV-treated effluent for coliform counts, the sample must be pumped similarly to a sample going to a chlorine analyzer. This demonstrates the effect of the clumping phenomenon, if any. This is why the total suspended solids and turbidity must be monitored to ensure proper disinfection by UV. It must be remembered that UV cannot kill anything that can be seen with the human eye. For further information see Chapter 17.

Gamma Radiation

In addition to UV radiation, gamma radiation has been investigated recently as a method of wastewater treatment and disinfection. In sufficient dosages, gamma radiation is an effective sterilant and is used as a method of sterilizing surgical instruments. Unlike UV radiation, gamma rays are capable of great penetration. Gamma radiation has the ability to alter organic and inorganic molecules, and this effect may benefit tertiary treatment processes. The most convenient source of energy for this irradiation is cobalt 60, which is available in virtually unlimited quantity. The cost of radiation energy is high, and gamma radiation as a disinfection process for wastewater is not economically or otherwise competitive with other methods.

The proponents of gamma radiation point out that cesium 137 is the major component of nuclear waste material, which is a by-product of nuclear power plants and is thus available as a source of gamma rays for water purification, and its use diminishes the amount of chlorine required for disinfection.

Woodbridge and Cooper[44] claim that six and seven orders of magnitude reduction in the concentration of microorganisms have been obtained by irradiation with less than 5 min. exposure. The reported advantages of this method include reliability, beneficial side effects and no residual effects. Disadvantages are principally associated with safety needs, excessive cost, and virtually no operating experience with this method. The application requires considerably more engineering design information than is now available. It should not be ruled out, however, as an adjunct to present methods. Because of lack of information about it as a wastewater disinfectant, nothing more will be said in this text on the subject.

Comparative Costs of Different Methods of Disinfection

Cost comparisons are unfair at best because all things are not equal between the various methods. One has to weigh the advantages and disadvantages of each method; cost is only one of many factors in evaluating the various methods. The method that will predominate: is one that gets the job done easily; is known to have a minimum health and safety risk; is easy to apply, measure, and control; and is one for which the handling equipment is reliable and easy to operate. The chlorination–dechlorination method fits this description and will certainly remain popular for a very long time.

In 1976 the EPA published a report on the various methods of wastewater disinfection.[45] Table 8-1 shows the comparative capital and process cost of the various methods studied.

Missing from the above tabulation is chlorine dioxide. The capital cost increase over chlorine for a chlorine dioxide installation is small: probably 15 percent. The process chemical cost for chlorine dioxide, assuming chlorine at 15¢/lb and sodium chlorite at 70¢/lb as well as a 2 mg/L dose and 30 min. contact time,[46] is shown in Table 8-2.

GENERAL CONCEPTS OF WASTEWATER DISINFECTION

Many modern wastewater disinfection practices in the United States had their origin in California. In the early 1920s, health officials were alarmed at the possible long-term deleterious effects of raw or poorly treated sewage discharging into the surf waters along the Pacific coast and freshwater streams throughout the state. A comprehensive investigation of the consequences of sewage discharges as related to public health indices resulted in a lawsuit against the City of Los Angeles. The presiding judge upheld the State Health Department contention that wherever samples taken from the surf exceeded a 1000 MPN/100 ml statistical coliform concentration, it constituted sewage contamination injurious to public health, and that the offending dischargers must be made to provide proper sewage treatment.

Over the years, the California State Department of Health has formulated a set of requirements for all receiving waters, including confined saline waters,

Table 8-1 Cost Summary

Plant Size (mgd)	1	10	100
Capital Cost	$K	$K	$K
Process			
Chlorine	60	190	840
Chlorine/SO_2	70	220	930
Chlorine/SO_2/aeration[b]	120	360	1,580
Chlorine/carbon	640	2,800	8,400
Ozone/air[a]	190	1,070	6,880
Ozone/oxygen[a]	160	700	4,210
Ultraviolet[a]	70	360	1,780
Bromine chloride	50	130	410
Activated Sludge	1,450	5,790	39,800
Disinfection Cost	¢/Kgal.	¢/Kgal.	¢/Kgal.
Process			
Chlorine	3.49	1.42	0.70
Chlorine/SO_2	4.37	1.75	0.89
Chlorine/SO_2/aeration[b]	7.66	2.39	1.19
Chlorine/carbon	19.00	8.60	3.28
Ozone/air	7.31	4.02	2.84
Ozone/oxygen[a]	7.15	3.49	2.36
Ultraviolet[a]	4.19	2.70	2.27
Bromine chloride	4.52	3.04	2.65
Activated Sludge	55.90	20.20	14.00

[a] Tertiary treatment stage is not included in these costs.
[b] Aeration is not required following dechlorination by SO_2 because a properly designed system will not remove any DO in the effluent. (*author's note*)

estuaries, surface waters, ephemeral streams, and shellfish growing areas. An additional constraint is considered whenever the receiving water is used for bathing or water contact sports. The numbers applied for these situations vary from 230 MPN/100 ml for confined saline waters down to 2.2/100 ml for sewage discharging into ephemeral streams or negative estuaries.

This concept of receiving water quality based upon coliform concentration carries with it the implication that certain degrees of treatment are imperative to achieve the various numbers specified. In other words, the designer should not attempt to depend wholly upon a disinfection system to achieve the desired coliform count in the plant effluent. For example, disinfection of a sewage discharging into a confined saline water or estuary cannot consistently accomplish the required 230/100 ml MPN coliform without secondary treatment. Attempts to "disinfect" raw sewage or stormwater overflows are a waste of chemicals, time, and energy. Similarly, tertiary treatment is almost always a

Table 8-2 Comparative Cost of Chlorine Dioxide

Design capacity (mgd)	1	10	100	150
Chlorine dioxide (¢/Kgal.)	4	2	1	1

necessity to meet a 2.2/100 ml MPN standard. When such a severe requirement is placed upon an effluent, there is a further implication of virus destruction. Virus destruction can only be accomplished on high quality effluents, regardless of the disinfectant used.

Nitrification* of an effluent is no longer considered a necessity when a low coliform count is required. It was once thought that if an effluent was nitrified, then the practice of free residual chlorination could be assured, which would result in a more reliable and efficient disinfection system. In recent years, it has been found that nitrification to produce a free chlorine residual is not necessarily worth the effort.

Disinfection studies should represent the degree of disinfection based upon the total coliform concentration in the plant effluent after disinfection.

A mathematical model has been developed that relates the coliform concentration before and after disinfection with chlorine residual at the end of a specified contact time. This model clearly demonstrates that a higher quality of effluent results in lower numbers of coliforms to be destroyed. This translates to higher efficiency of disinfection.

To date there does not seem to be a better indicator organism for proof of disinfection than the total coliform group.

Wastewater reuse requires a different approach. In these cases the sewage discharge is being used directly for land irrigation, spraying of crops, industrial cooling water, other makeup water requirements, groundwater recharge, prevention of salt water intrusion, and prevention of land subsidence due to underground withdrawals. All of these applications depend heavily upon the unit process of disinfection. It becomes the most important link in this chain of treatment for these applications.

Whenever wastewater is used in situations where there is the possibility of human contact or ingestion, the problem of widespread virus infection becomes the most serious concern of public health officials everywhere.

A study[135] by the Los Angeles County Sanitation Districts revealed that, contrary to previous beliefs, combined chlorine residuals can be nearly as effective as comparative free chlorine residuals in the destruction of viruses. It is highly probable that this reaction is due to the interference of organic N that almost always appears in wastewater effluents from 4 to 6 mg/L. The reaction with free chlorine (in a nitrified effluent) is instantaneous and forms a nongermicidal organochloramine that appears as a dichloramine species in the total chlorine residual, when the residual analysis is done by a complete forward-titration procedure. This problem was not discovered until 1982. (See discussion of "Organic Nitrogen," below.)

All of the data currently available demonstrate conclusively what the Selleck–Collins mathematical model tells us: that disinfection efficiency is related directly to the quality of the effluent, and that for any quality of effluent the

*The only reason for nitrification is to limit the concentration of un-ionized ammonia in the receiving waters. Ammonia is highly toxic to aquatic life.

degree of disinfection is directly related to the total chlorine residual and contact time, provided that mixing is rapid, and that the contact chamber demonstrates plug flow conditions.

The phenomenon is regrowth, which occurs temporarily in some cases downstream from the point of disinfection, is not considered significant to public health. The public health practitioners recognize this phenomenon, and they firmly believe that all pathogenic organisms are destroyed in the disinfection process.

A source of constant worry to any practitioner of disinfection is the clumping phenomenon. It is theorized that clumps of suspended or colloid-like particles, such as are present in raw sewage, primary effluents, and poorly treated secondary effluents, may be able to pass through the disinfection system only to break up downstream and spew into the effluent gross amounts of coliforms and possibly pathogens that were sheltered from the disinfectant.

This concept is of grave concern to public health practitioners and is a major reason for assuming that raw sewage and/or primary effluents should not be considered as candidates for the disinfection process.

The toxicity of chlorine residuals to aquatic life is well documented and has given rise to the addition of the dechlorination step to complete the disinfection process, whenever chlorine (or chlorine dioxide) is the disinfectant. A chlorinated–dechlorinated effluent has been proved to be less toxic than either a chlorinated or a nonchlorinated effluent.

Available methods of disinfection of wastewaters include all of the halogens (Cl_2, I_2, Br_2, BrCl, ClO_2), ozone, ultraviolet radiation, gamma radiation, and possibly some combinations with sonics. Cost-effective analyses of these various methods always place the chlorination–dechlorination method in the most favorable position.

From evaluations of the art of disinfection it is clear the process will not provide the desired results unless the other unit processes of the wastewater-treatment system are performing properly. Therefore, a disinfection system is a protective device for public health as well as a sensitive monitor of the entire wastewater treatment process.

CHEMISTRY AND KINETICS OF DISINFECTION BY CHLORINE

Reactions with Wastewater

General Discussion. There are numerous constituents present in wastewater that react immediately with the HOCl from the chlorinator injector discharge or the hypochlorite solution. Consequently, free chlorine (HOCl + OCl^-) is probably consumed or converted to some form of chloramine in a matter of seconds after mixing with the wastewater stream, owing to the presence of ammonia nitrogen. Very little is known about this specific reaction. Simultaneous with the Cl_2–NH_3 reaction are the other inorganic reactions of chlorine with reduced substances such as $S^=$, HS^-, $SO_3^=$, NO_2^-, Fe^{++}, Mn^{++}, and so on. These substances react with both the free chlorine and combined chlorine (NH_2Cl, $NHCl_2$) to reduce these compounds so that the active chlorine compound is eventually reduced to the stable chloride ion, which is nonbactericidal. At this point, there is no measurable residual. The chlorine consumption in the first minute of reaction is probably due to the reactions with inorganic substances. The reactions occurring in the following 2 or 3 min. are probably due to organic chlorine demand. These are much slower reactions. The 10-min. chlorine demand of a fresh domestic sewage may be as low as 5 mg/L to produce a measurable residual; however, this figure may escalate to 40 mg/L if the same sewage becomes septic. Ammonia N does not begin to consume chlorine until the Cl–N dosage ratio exceeds 5:1 (i.e., a point beyond the hump on the breakpoint curve).

The most significant chemical reactions between chlorine and the various chemical constituents in wastewater effluents are those with the various nitrogenous compounds, either inorganic (NH_3–N, NO_2) or organic (proteins and their degradation products). These reactions are described in Chapter 4.

Ammonia Nitrogen. With the exception of highly nitrified effluents, there is usually an appreciable amount of ammonia nitrogen in all wastewater effluents. The range is on the order of 10–40 mg/L. The ammonium ion (NH_4^+) exists in equilibrium with ammonia nitrogen and hydrogen. The distribution is dependent upon pH and temperature. The relative distribution can be defined as follows:

$$NH_4^+ \leftrightharpoons NH_3 + H^+ \text{ with } K = 5 \times 10^{-10} \text{ at } 20°C \quad (8\text{-}2)$$

where
NH_3 = ammonia molecule
H^+ = hydrogen ion
K = dissociation constant

According to the dissociation constant, the pH value at which NH_3 and NH_4^+ are present in equal proportions (pK) is about pH 9.3 at 20°C. Above pH 9.3, NH_3 predominates; below pH 9.3, NH_4^+ predominates.

At usual wastewater pH levels the predominant chlorine reaction proceeds as follows:

$$HOCl + NH_4^+ \leftrightharpoons NH_2Cl + H_2O + H^+ \quad (8\text{-}3)$$

when the chlorine-to-ammonia nitrogen weight ratio is 5:1 or less. If the pH drops below 7, dichloramine ($NHCl_2$) will begin to form; and at a much lower pH, nitrogen trichloride will form. However, as soon as the 5:1 weight ratio of chlorine to ammonia nitrogen is exceeded, a new set of reactions takes place. These reactions are described fully in Chapter 4.

The disinfecting power of chlorine in wastewater is greatly enhanced by good mixing at the point of application. White et al.[49,50] found a significant reason for this relationship: good mixing ensures the maximum formation of monochloramine in Eq. (8-3). If mixing is poor, the chlorine species tend to split between monochloramines and organochloramines (organochloramines titrate as dichloramine). However, pure dichloramine does not form when the pH is in the neutral range and the chlorine-to-ammonia nitrogen weight ratio is 6 : 1 or less. This is an important consideration because the species described above as organochloramines have no germicidal efficiency. The organochloramines derive from the organic nitrogen, which is always present in substantial amounts (3–15 mg/L). In highly nitrified and filtered effluents, the organic nitrogen is present in amounts from 0.75 to 3.0 mg/L. The lower the organic nitrogen content, the more germicidally efficient the chlorination process will become. The significance of organic nitrogen as it affects the efficiency of disinfection is described below.

Organic Nitrogen. In raw untreated municipal wastewaters organic nitrogen compounds are present as both soluble and particulate. Most occur as insoluble compounds. The soluble compounds are mainly in the form of urea and amino acids. Secondary biological treatment reduces the soluble organic N compounds to a range of approx. 10–14 mg/L. Filtration will remove the remainder, leaving about 3 mg/L in the filtered effluent. Filtration preceded by flocculation can reduce this residual to about 0.75 mg/L.

Chlorine is known to react and combine with urea, amino acids, and proteinaceous organic nitrogen compounds to form organochloramines of dubious germicidal efficiency. Very little is known about the kinetics of these reactions, their reversibility, or their germicidal power. Extensive investigations surrounding the presence and identification of this species of chlorine compounds in swimming pools and potable water have demonstrated that they titrate as dichloramines in the forward-titration procedures using either amperometric or DPD-FAS methods. How much these compounds interfere with the monochloramine determinations is unknown.[51,52]

White et al.[50] have been able to prove conclusively that the chloramines appearing in the dichloro fraction of a combined residual (organochloramines) have a significantly lower germicidal efficiency than those appearing in the monochloramine fraction. Now in 1997 we know they are totally nongermicidal.

Observations by White, Selleck, and Collins have indicated that the germicidal efficiency of a combined chlorine residual has a tendency to decrease with time. This decrease has been noticed in some secondary effluents. It is most noticeable in primary effluents when the contact time exceeds 45–60 min. It is thought that this is related to the presence of organic nitrogen. In the cases observed there was a noticeable shift in the chloramine species. Although the total residual remained relatively stable between 45 and 60 min., the monochloramine concentration decreased, and the "dichloramine" increased. Therefore, it is theorized that in effluents containing 0.5 mg/L or more organic nitrogen, the monochloramine formed by the presence of ammonia nitrogen in the wastewater slowly hydrolyzes with time to react with organic nitrogen compounds present to form organochloramines, thereby decreasing the overall germicidal efficiency of the remaining residual.

In 1966 Feng[53] discovered a great disparity in the germicidal efficiency between ammonia chloramines and those found in an environment of pure organic nitrogen compounds. He reported that methionine, an indispensable amino acid for biological growth present in wastewater, forms a measurable chlorine residual with no germicidal power. Feng also investigated the lethal activities of the glycine, taurine, and gelatin chloramines. His work shows that taurine chloramines are as lethally active as ammonium chloramines at pH 9.5, but that their germicidal efficiency falls off as the pH decreases. The glycine chloramines are as germicidally active as monochloramine at pH 4 but are totally inert at pH 7, and the gelatin chloramines are active at pH 9.5 but are inert at pH 7 and pH 4. There are certain to be other such organic-nitrogenous compounds, which contribute to the total chlorine residual, that have little or no germicidal effect.

Sung[54] made a controlled laboratory study of 15 organic compounds representing seven groups to evaluate their individual and combined effect upon the chlorination process. Nine of the 15 compounds were found to interfere with the germicidal efficiency of the chlorination process. Of these nine compounds, five were organic nitrogen compounds. Cystine and uric acid were the severest inhibitors of the nitrogen group. When five of the interfering compounds were mixed together, their combined effect was found to be more pronounced than any of their individual effects, but it did not equal the sum of their individual effects. Sung compared the germicidal efficiency of a simulated wastewater with and without the interfering organic compounds. He found that the wastewater containing the interfering compounds by themselves and the resulting chlorine residuals had little or no germicidal effect. The greatest interference was observed to be caused by cystine, tannic acid, humic acids, uric acids, and arginine.

Cystine is an amino acid connected by two sulfur groups that is known to react with chlorine. Tannic, humic, and uric acids are capable of exerting a significant chlorine demand when present in water or wastewater. Arginine is a basic amino acid. The reaction between chlorine and arginine is almost instantaneous.

The organic compounds that had little or no interfering effects on the chlorination process were: acetic acid, cellubiose, dextrose, glutamic acid, uracil, and lauric acid.

The above findings by Sung[54] confirm the theory of interference in the chlorination process by the presence of organic nitrogen. They further point to the fact that present analytical techniques do not provide for separating the chlorine residual fractions into those of equal germicidal efficiency.

It is also interesting to note that Esvelt et al.[55] found that the toxicity of combined chlorine residuals diminishes with time. This finding, together with those of Sung,[54] demonstrates quite convincingly that there are a significant number of organic compounds in wastewater that will react with chlorine to form organic chloramines of little or no germicidal potential, and, moreover, that these compounds appear to increase in concentration with time. The apparent increase of this chlorine residual fraction with time would also explain the loss of germicidal efficiency of combined chlorine residuals in wastewater with the passage of time, as described elsewhere in this text.* The following information should be a big help. Now in 1997 we have known for several years that these organochloramines are totally nongermicidal and consistently interfere with the accuracy of chlorine residuals measured by current chlorine residual analyzers. Their effects are easy to calculate, if the sample is from a nitrified wastewater effluent. In this situation the organochloramine shows up in all of the dichloramine fraction, and about 10 percent in the mono fraction.

If the sample is from a non-nitrified wastewater effluent, the free chlorine will be totally absent, and the monochloramine fraction will probably titrate 100 percent in the above forward titration if the sample is measured within 5 min of the dosing. In order to allow the monochloramine to react with the organic N, it is suggested that another forward titration be made after 40–50 min contact time. This time the organochloramine should show up in the dichloramine fraction. These two steps should be very helpful to plant operators.

Effluent Quality as Related to Dosage

Importance. The single most important effluent quality parameter as it effects the efficiency of disinfection is the indicator organism concentration in the treated effluent before the application of chlorine. This holds true for nonnitrified effluents based upon observations by White over a period of many years (1970–90).

The Selleck–Collins Model. The basic concept of adequate wastewater disinfection is expressed in the mathematical model developed by Selleck et al.[56] in 1970. This work resulted in the formulation of an equation based upon a comprehensive pilot plant study of a primary effluent. The message portrayed by the equation has been substantiated many times since its publication: if there is good mixing at the point of chlorine application, and if there are plug flow conditions in the contact chamber (no short-circuiting), one can expect a definitive coliform reduction with a given chlorine residual at the end of a

*The loss of germicidal efficiency with time is caused by the reaction of monochloramine with the organochloramines formed because of the presence of organic N. The monochloramine gradually hydrolyzes to form organochloramine. When this occurs, the disinfection efficiency is reduced proportionally.

specified contact time. This is the Ct relationship commonly referred to elsewhere as the chlorine concentration–contact time envelope.

The original model was subsequently finetuned by Selleck and Collins, based upon plant-scale studies.[57] This work reinforces the practical aspects of the original model, which is represented by the following equation:

$$y = y_0[1 + 0.23\ Ct]^{-3} \tag{8-4}$$

where

y = MPN in chlorinated wastewater at end of time t
y_0 = MPN in effluent prior to chlorination*
C = total chlorine residual, mg/L, at the end of contact time
t = contact time, minutes

The mathematical model of Eq. (8-4) was developed from a pilot plant system that had excellent mixing in a highly turbulent regime and an ideal plug flow contact chamber. Based upon many plant observations, good mixing occurs when the velocity gradient G is approximately 500, and the contact time t_i is not less than 30 min. Contact times longer than one hour should be avoided in effluents containing organic N in concentrations higher than 5 mg/L. Long contact times in these effluents allow the monochloramine fraction of the chlorine residual to hydrolyze and become converted to organochloramines. These organochloramines have low germicidal qualities; therefore, the potency of the total chlorine residuals in these effluents shows a marked decrease with time.

This model is a valuable tool for sizing chlorination equipment for a new plant, and for evaluating an existing plant. Some examples follow.

Primary Effluent. The value of y_0 is usually about 38×10^6/100 ml. Assume discharge into a surf water. In California the y requirement is 1000/100 ml MPN total coliforms. Substituting in Eq. (8-4):

$$\frac{1000}{38 \times 10^6} = [1 + 0.23\ Ct]^{-3} \tag{8-5}$$

See key stroke sequence for an HP-21 calculator in the appendix and solve for Ct. Assume t = 30 min., and solve for C:

$$Ct = 142 \qquad \text{and} \qquad C = 4.73$$

To allow for the probable immediate (3–5 min.) chlorine demand of 6–8 mg/L plus the die-away in the contact chamber (25+ min.) of about 1 mg/L, the required chlorine dosage will be about 5 + 8 + 1 = 14 mg/L.

*This is y_0 in Eq. (8-4).

Secondary Effluent. The median coliform concentration in a well-oxidized secondary effluent will be about $2 \times 10^6 = y_0$. Assuming a discharge into a confined body of water, the total coliform requirement might be 23/100 ml MPN. In this case *Ct* calculates to 188. This magnitude calls for a contact time longer than 30 min. So *C* for 45 min. is 4.2. To estimate the chlorine dosage, assume a 5 mg/L initial chlorine demand and 1.5 mg/L residual decay in the contact chamber. Then the required dosage will be 4.2 + 1.5 + 5 = 10.7, say 12 mg/L.

Filtered Effluent. The ease with which these effluents can be treated depends a great deal upon whether or not the filtered effluent has been preceded by coagulation and sedimentation. A conventional water reuse situation would consist of filtration of secondary effluent preceded by coagulation and sedimentation. The y_0 of such an effluent would probably range between 3000 and 10,000 coliforms per 100 ml. In California, whenever tertiary effluent is required, the coliform requirement is usually the same as that for potable water, namely, 2.2/100 ml. Assuming $y_0 = 10{,}000$ and $y = 2.2$, then *C* calculates to 2.25 mg/L for $t = 30$ min. A tertiary effluent with the predisinfection processes described above would require chlorine dosages on the order of 5–7 mg/L.

A filtered effluent with chemical coagulation but without sedimentation produces an effluent with coliform concentrations considerably higher, on the order of 50,000/100 ml. In this case using $y = 2.2/100$ ml, *C* calculates to 3.96 mg/L for $t = 30$ min. This is almost twice the residual required in the preceding case.

These examples clearly demonstrate the effect of effluent quality as it relates to the y_0 coliform concentration. The easiest effluent to disinfect that White has investigated to date is a secondary effluent followed by 100-day ponds. Here the y_0 rarely exceeds 4000/100 ml, and the final coliform is usually less than 3/100 ml using chlorine dosages on the order of 3.5–4.0 mg/L and 15 min. contact time. The chlorine residuals at the end of 15 min. are usually on the order of 2 mg/L. This is what an optimized system can do when y_0 is a low figure.

In order to provide a touch of conservatism, White always uses the chlorine residual *C* in the Selleck–Collins equation as the residual measured at the end of the contact chamber and the contact time t_i as the first appearance of the dye at the end of the contact chamber.

The Selleck–Collins Model.[58,61] This model is a refinement of the Selleck–Collins model shown in Eq. (8-4). It is described by the following equation:

$$y/y_0 = (RT/b)^{-n} \tag{8-6}$$

where:

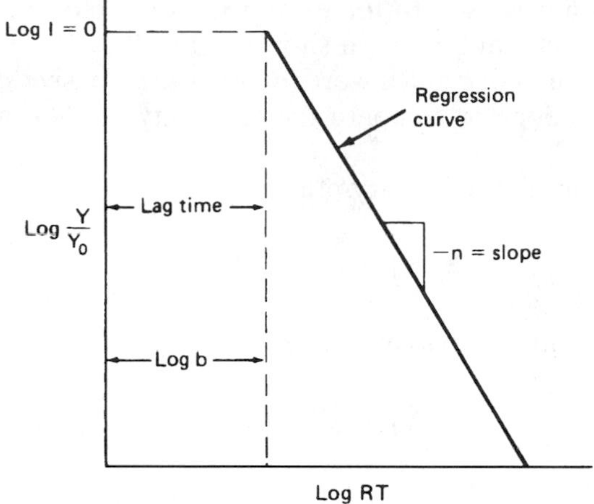

Fig. 8-8. Arithmetic plot of $Y/Y_o = \left(\frac{RT}{b}\right)^{-n}$.

y_0 = initial bacterial concentration before chlorination
y = bacterial concentration at end of contact chamber or at time T in minutes
R = chlorine residual at the end of time T, mg/L
T = contact time in minutes
b = the x intercept when $y/y_0 = 1$ or $\log(y/y_0) = 0$ (see Fig. 8-8)*
n = slope of the curve

An easy way to use this equation is to plot the log values on arithmetic paper: log y/y_0 on the y-axis and log RT on the x-axis. Examination of the equation shows that when $y = y_0$, there is no kill; so y must be less than y_0 to have any kill. When there is no kill, $y/y_0 = 1$ and $\log 1 = 0$. Therefore, the equation begins with 0 on the y-axis.

Now when $RT = b$, then $RT/b = 1$ and $y/y_0 = (RT/b)^{-n} = (1)^{-n} = 1.0$. Therefore b is determined when the regression plot intercepts the x-axis. If this equation is plotted on log–log paper, the intercept is 1.0; but on arithmetic paper the intercept is 0 because $\log 1 = 0$. The intercept is the point where $\log RT = \log b$. Every point on the regression curve to the right of the zero intercept represents $RT > b$.

When data are not available to plot a bacteria kill regression curve, the suggested value for b is 4 in working with total coliforms and 3 for fecal coliforms. When these values are used for b and for $n = -3$, the equation $y/y_0 = (RT/b)^{-n}$ is practically identical with the Selleck–Collins 1970 model, which is $y/y_0 = (1 + 0.23\,Ct)^{-3}$ where $c = R$ and $t = T$. The insertion of 1 in

*b is sometimes called the lag-time of bacterial kill because kill does not occur until $RT > b$. See Fig. 8-8.

the latter equation was used to force the regression plot into a straight line at low values of bacterial reduction shown in Fig. 8-9.

The following values of *b* and *n* were found in an extensive study by Roberts et al.,[59] which was done to compare the feasibility of ClO_2 as a disinfectant to chlorine.

Palo Alto, California, secondary effluent:

$$y/y_0 = (RT/3.95)^{-2.79} \tag{8-7}$$

San Jose, California, secondary effluent:

$$y/y_0 = (RT/4.06)^{-2.82} \tag{8-8}$$

Nitrified Effluents

Inadvertent Loss of NH_3–N. The chemistry of wastewater chlorination takes on a different aspect when a secondary effluent becomes nitrified. This situation occurs inadvertently in some plants when there is an overload due to cannery wastes. This was a common occurrence in a conventional secondary activated sludge plant that experienced heavy seasonal loads of cannery wastes. During these periods of overloading, the ammonia nitrogen in the effluent would disappear completely. When this occurred, the organic nitrogen increased from about 6 to 11 mg/L, and simultaneously the disinfection efficiency of the chlorination system fell off suddenly and dramatically.[60] In such a situation one would expect the appearance of the free chlorine species and a subsequent improvement in the disinfection efficiency. This never happened,

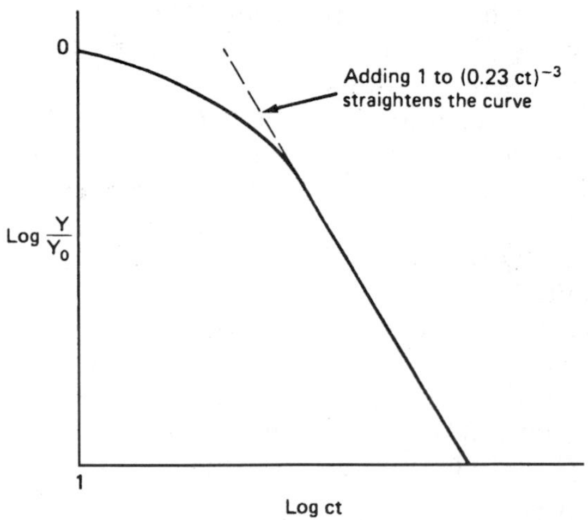

Fig. 8-9. Selleck–Collins 1970 model: explanation of unity factory.

to the amazement of all involved in the project. Subsequently it was determined that the chlorine applied reacted totally with the organic N present (13 mg/L) to form organochloramines.

In an attempt to correct for the dramatic decrease in disinfection efficiency, operating personnel increased the chlorine dosage to the maximum capability of the system. Although this increase in chlorine dosage represented a twofold increase (20 mg/L), it was not enough to achieve the NPDES requirement of 230/100 ml MPN total coliform concentration. However when the ammonia N concentrations exceeded about 1 mg/L, there was never any problem in achieving this disinfection requirement with chlorine dosages below the breakpoint (i.e., less than 10 mg/L).

To overcome this disappearance of ammonia nitrogen during the canning season, the plant influent was seeded with the secondary digester supernatant, which contained approximately 550 mg/L ammonia nitrogen. There was sufficient quantity of this liquor to dose the plant influent with 10 mg/L NH_3–N. This remedial action restored the disinfection efficiency at a chlorine dose of 10 mg/L. This plant was later converted to produce a nitrified and filtered effluent.

Filtered Effluents. Most nitrified effluents are filtered before disinfection. Filtration may in some instances be preceded by sedimentation; if not, the addition of a coagulant is usually provided in lieu of sedimentation. The NPDES requirements for these effluents are usually severe enough that virus removal becomes desirable—hence the need for filtration. Owing to the difficulty in achieving disinfection in the absence of ammonia nitrogen described above, White et al. investigated several plants with nitrified tertiary effluents.[49,50]

All of the plants investigated were required to meet an NPDES total coliform concentration of 2.2/100 ml MPN. All effluents were filtered except one. Some plants had coagulant addition capability ahead of filtration. Some used coagulants, and some did not. There was no set pattern, and the effluent quality with and without coagulants was undetermined.

The most disturbing element of this investigation revealed that in spite of higher quality effluents (y_0 = 30,000–60,000), none of the plants conformed to the Selleck–Collins[56] mathematical model. This model has been used with reliable success for predicting required combined chlorine residuals for given contact times to achieve a particular coliform concentration requirement. In general, the chlorine dosages required to achieve a 2.2/100 ml coliforms standard were often twice those needed in plants with a similar treatment process that *did not nitrify*. This was difficult to believe because the residuals in the nitrified effluents were usually 50–60 percent free chlorine as measured by amperometric titration and DPD-FAS titration.

The San Jose, California water pollution control plant was thoroughly investigated over a two- to three-year period in search of answers to this chlorination anomaly. The chlorine dosage required to achieve a 2.2/100 ml coliform count

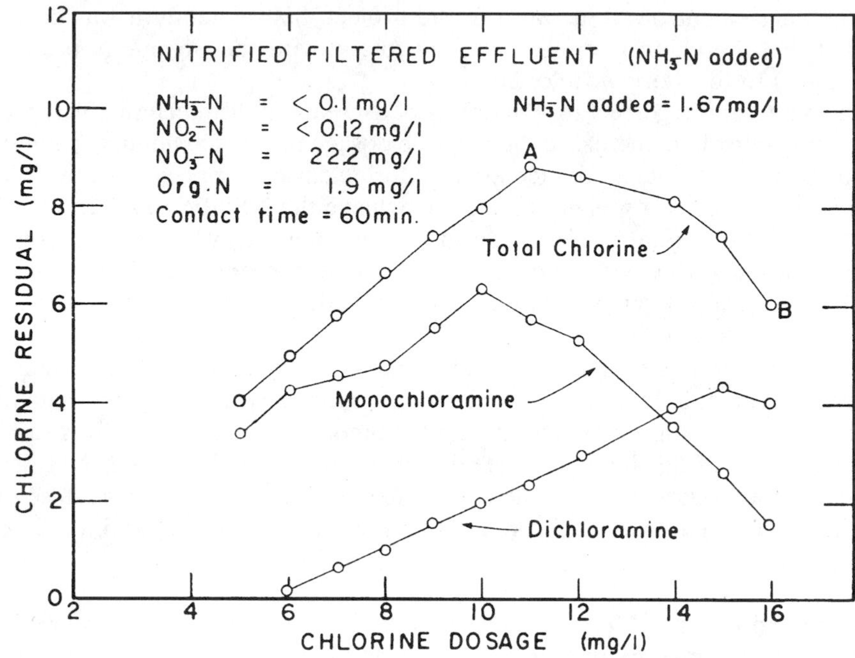

Fig. 8-10. Breakpoint curve: nitrified effluent with ammonia-N added.

in the nitrified effluent varied from a low of 17 mg/L to a high of 22 mg/L. The *total chlorine residual* at the end of 49 min. of contact time required to meet this goal consistently was about 9 mg/L. This residual contained about 50–60 percent free chlorine. About 90–95 percent of the remainder titrated as dichloramine, the rest as monochloramine.*

The Selleck–Collins model predicts a total chlorine residual of 2 mg/L at 49 min. contact time to achieve a 2.2/100 ml coliform level.

A laboratory study using the addition of ammonia N was made to find the chlorine-to-ammonia ratio with the best germicidal efficiency. (See Fig. 8-10.) This turned out to be 6:1 chlorine to ammonia N by weight. This confirmed the finding of Selleck et al.[62] They showed that the most effective germicidal chloramine residual occurs when the chlorine–ammonia N ratio is on the "breaking" side of the breakpoint curve, as shown in Fig. 8-10 between points *A* and *B*. It was further found that a chlorine dosage of 12 mg/L would produce

*The nongermicidal organochloramines appeared in the dichloramine fraction. This is why this phenomenon was unknown for so long. Who would think of doing a complete forward amperometric titration on a wastewater effluent? The organochloramines derive from the presence of organic N in the nitrified effluent. Since there was no trace of ammonia N in these samples, there could not be any monochloramine formed, or any other species of chloramine. Therefore, the appearance of monochloramine is the result of interference from organochloramines in the forward-titration procedure.

a combined residual of 7 mg/L at the end of 49 min. of contact time, which was sufficient to achieve a 2.2/100 ml coliform count in the effluent. Plant operation with the above chlorine and ammonia N dosages proved to be consistent with laboratory findings. This treatment consistently produced an effluent meeting the coliform requirement without exceeding the limit of 0.025 mg/L un-ionized ammonia N in the receiving waters. The chlorine dose and residual requirement to achieve the disinfection standards have been consistently reduced by 5 and 2 mg/L, respectively.

Presence of Free Chlorine Residual. One interesting observation has been made at the San Jose plant. The color of the final effluent is perceptibly different after the addition of ammonia. When ammonia is not added and free chlorine is present in the contact chambers, the effluent is bright and sparkling with a tinge of blue, similar to the water in a well-operated swimming pool. When ammonia is added, the effluent retains its "clean" look, but the color is a very light beige that can be characterized as a straw color with a tinge of very light green. This difference in color explains in part why so much more chlorine is consumed when ammonia nitrogen is removed, and free chlorine residuals are produced. The additional chlorine is consumed in bleaching the organic color in the nitrified effluent. This is typically consistent with potable water treatment practice where free chlorine residuals are used to bleach colored water. **This also can be taken as proof positive that all of the titrations identifying the free chlorine species were and are in fact free chlorine residuals.***

There were so many skeptical responses to the findings described above about the apparent impotence of the free chlorine residuals that one additional step was taken to prove the existence of the free chlorine residuals. This was the use of the Olivieri Biofac method.[63] This method is based upon the fact that a bacterial virus f_2 of the *E. coli* K-13 strain is inactivated in a few seconds by free available chlorine but is affected very little by combined chlorine. The San Jose effluent was subjected to this comparison with the following results: a 10 mg/L chlorine dose was applied to each of two aliquot samples. Sample *A* was nitrified with only a trace of NH_3–N remaining. Sample *B* was the same effluent as *A* except that it was spiked with 1.67 mg/L NH_3–N. Each sample was spiked with the f_2 virus. At the end of 5 min. of contact time, sample *A* measured a total chlorine residual of 7.4 mg/L. Of this total residual 4.5 mg/L was free available chlorine. Sample *B* measured a total chlorine residual of 8.7 mg/L; there was no free chlorine residual. This sample titrated 7.2 mg/L monochloramine and 1.2 mg/L dichloramine. The control sample

*This is put in bold type because when Professor J. C. Morris heard of this event, he insisted there was no way that free chlorine could exist under these circumstances. The clincher was the loss of 15–17 mg/L of chlorine due to this bleaching. The San Jose plant managers always required tours of the plant to take place at specified times. This gave the plant operators sufficient time to switch from combined chlorine to free chlorine operation, which showed a beautiful blue effluent!

without chlorine measured 780 f_2 units. The sample with free chlorine measured zero units, and the sample with combined chlorine residual measured 470 f_2 units.

Examination of other plants did provide some clues as to why so much additional chlorine is required to produce a disinfecting residual, and why disinfection can be achieved with a lower dose with sufficient ammonia present to form chloramine residuals.

Negative Effects of Organic N. When chlorine is added to a completely nitrified effluent where the ammonia N is only a trace, there will usually be an organic N concentration of about 3 mg/L. This is an important factor because only two species of chlorine residual will be produced, free available chlorine and "dichloramine," which is not true dichloramine but organic chloramines resulting from the presence of organic N. A forward-titration reveals some interference by the organochloramines in the monochloramine fraction. As there is no information available on the kinetics of organochloramine formation in the presence of free chlorine, it is difficult to predict with any accuracy the disinfection efficiency of the chlorine–contact time envelope by the Selleck–Collins model. Organic nitrogen compounds have a wide range of reaction times with chlorine to form the dichloramine fraction. Furthermore, the germicidal efficiency of these organic chloramines is most likely to be nil. Given this background, it is easy to see that the 9 mg/L chlorine residual at San Jose at the end of 49 min. contact time is not so impressive when half of it is composed of organic chloramines.* When ammonia nitrogen is *absent,* there is no formation of a true monochloramine.

To anticipate problems of ammonia N concentration variations plus the possible interference of organic N, the operator should determine chlorine residuals for all the species by forward-titration, amperometric or DPD-FAS titrimetric, followed by a titration for total chlorine residual to confirm the forward titration results. (Also see below, "Advice to Operators.")

One plant was found that proved to be the exception. This was the Fairfield-Suisin WWTP at Fairfield, California. This plant treats a daily wastewater flow that ranges from 7 to 14 mgd. The secondary effluent is filtered and nitrified before chlorination. Chlorine dosage ranges from 7.5 to 9.25 mg/L. The NPDES disinfection requirement is 2.2/100 ml MPN total coliforms. An investigation revealed the control residual (2–3 min. contact time) was 6.0 mg/L TRC. This consisted of 5.4 mg/L free residual (FRC) and 0.60 mg/L "dichloramine" residuals, by forward amperometric titration. The TRC was measured by back-titration with a separate titrator.

The residual at the end of the contact chamber was 2.8 mg/L, TRC. This was made up of 2.2 mg/L FRC and 0.60 mg/L "dichloramine." The samples used to measure the residuals at the end of the contact chamber were found to contain 0.04 mg/L NH_3–N and 0.60 mg/L organic N. Of all the nitrified

*The other half is free chlorine.

effluents investigated, this plant and one other* used the smallest chlorine dosage to achieve a 2.2/100 ml MPN coliform concentration. An equally important observation was that the Fairfield plant exhibited the lowest concentration of organic N. Organic N concentration in the other plant effluents (except Las Virgenes) were on the order of 3–5 mg/L. These observations seem to substantiate White's conviction that the organic N content in a nitrified effluent represents a major interference with the germicidal efficiency of free chlorine. It reacts swiftly to form impotent organochloramines, thus "dechlorinating" the free available chlorine.

Some nitrification plants are able to control the ammonia N concentration in the effluent to a level of 2–3 mg/L, which is sufficient to produce a combined residual containing about 80–85 percent monochloramine in the presence of 3–5 mg/L organic N. This solves the problem of excessive chlorine consumption and provides more reliable disinfection results. One such plant found that the saving in chlorine (due to allowing 2–3 mg/L NH_3–N to remain) amounted to $37,000 per year or 4.1 percent of the total plant budget.[64,65] The successes of these plants and the San Jose plant have been duplicated by others. The problem that usually arises is the inability to control the ammonia N level to the 2–3 mg/L range. Those plants are well advised to stay with complete nitrification and add either ammonium hydroxide (38 percent solution) or anhydrous ammonia to eliminate the possibility for generating free chlorine.

Nitrification Summary

General Discussion When the ammonia nitrogen is removed completely and chlorine is applied for disinfection, the free chlorine with its higher oxidizing power is subjected to excessive consumption by the various reducing compounds found in municipal wastewaters. As an example, organic color is bleached by free residual chlorine and is known to consume as much as 10–15 mg/L FRC in a few minutes.

The presence of organic N acts as an equally formidable competitor for the free chlorine because the resulting reaction forms organic chloramines well known to possess little if any germicidal efficiency.

The addition of ammonia N to these effluents has been shown to result in a big reduction in chlorine consumption together with an increase in disinfection efficiency, provided that the chlorine is mixed well with the NH_3–N spiked effluent. *Combined chlorine will not be consumed by organic color.*

Plant-scale investigations have determined that 6:1 Cl to N ratio is the most germicidal. This should be confirmed in the laboratory for each case.

---

*The Tapia water reclamation plant of the Las Virgenes Water District, Calabasas, California, in 1973–74 was using approximately 8 mg/L chlorine dosage and 2 mg/L free chlorine residual at the end of 49 min. contact time. Organic N was between 0.9 and 1.4 mg/L. Ammonia N was not detectable. The nitrified effluent was not filtered.[114]

These investigations have also substantiated the findings of Collins et al.,[66] Selleck et al.,[62] and Selna et al.,[67] that if given enough time (about one hour) combined chlorine will eventually catch up to free chlorine in bacterial efficiency. These findings, which corroborate one another, may be due to the simple phenomenon that chloramines take longer to diffuse through the cell walls of the bacteria than does free chlorine.

One other effect that is present when free chlorine is the primary disinfectant is the likelihood of bacterial encapsulation by calcium ions present in the wastewater. Collins et al.[66] reported a grossly poorer bacterial kill with free chlorine in a plant using the lime precipitation process, and it was believed that encapsulation could have been the cause. This phenomenon was first reported by Ingols[68] while studying the effect of calcium ion on the disinfection efficiency of swimming pool treatment methods.

Deleterious Effects of Organic N. Since the discovery of the breakpoint phenomenon in 1939, it has been known that concentrations of organic N as low as 0.3 mg/L can adversely affect potable water chlorination. However, in wastewater treatment these adverse effects were neither known nor seriously considered until the advent of nitrified effluents.

Organic N compounds are common to all wastewater effluents. The concentration levels of these compounds vary with the effluent quality. Activated sludge effluents usually contain between 3 and 6 mg/L, while filtered secondary effluents contain between 0.75 and 1.5 mg/L of organic N. These compounds react with both free chlorine and chloramines to produce nongermicidal organochloramines. In a nitrified effluent where ammonia N is absent, there cannot be any monochloramine. Therefore, the residual in this case will contain only free chlorine and organochloramine. The problem here is that the organochloramine fraction can only be identified by forward-titration procedures, either amperometric or DPD–FAS titrimetric. It titrates as dichloramine, which it is not. As this fraction is not germicidal, it has to be subtracted to evaluate the effectiveness of any such residual.

The reaction between free chlorine and organic N is much faster than the same reaction between monochloramine and organic N. Therefore, the absence of ammonia in a nitrified effluent results in a greater loss of free chlorine to the organic N reaction than occurs when ammonia is present, owing to the slow reaction between organic N and monochloramine (see above discussion of the San Jose, California, experience).

When a treatment plant is in a nitrification mode so that the ammonia N either is not detectable or has a content less than 0.5 mg/L, there is sure to be a free chlorine residual in the effluent. If organic color exists in the effluent of any of these cases, a significant amount of free chlorine will be consumed by the bleaching effect of free chlorine. The chlorine lost by bleaching varies from 10 to 15 mg/L. As nearly all wastewater effluents contain organic color, the method of choice to eliminate this problem is to add ammonia N at a ratio of one part to six parts by weight of chlorine. This ratio produces a chloramine residual of maximum germicidal efficiency. This residual should be coupled with a minimum detention time of 50 min. for maximum disinfection efficiency.

When the chlorine-to-ammonia N ratio is at the proper 6 to 1, the chlorine residual is on the downslope of the breakpoint curve; that is, the residual is decreasing in magnitude. As this ratio increases with declining ammonia N concentration because of nitrification, the operator has lost the battle to meet the effluent coliform discharge requirements, owing to insufficient chlorine residual. This is a common occurrence in the smaller plants (1–6 mgd) across the United States that experience ammonia N concentration excursions from 1 to 10 mg/L. This results in extremes of biological instability throughout the plant processes. The inability to control these ammonia N levels is the crux of the problem. Many of these plants go in and out of the breakpoint region, which occurs at a chlorine-to-ammonia N ratio of 10 to 1. This becomes an operator's nightmare!

The answer to this problem is quite simple, but the application is a nuisance. Ammonia N has to be added in the proper amount in an attempt to maintain a ratio of chlorine to ammonia of no greater than 6 to 1. If this is too difficult to achieve, use 5 to 1 or 4 to 1. At the latter ratios the chlorine dosage might have to be increased 5–10 percent because the germicidal efficiency of chloramines decreases somewhat at the lower ratios.

Anticipating the Problem. To minimize this potential operating problem, a historical record should be kept of organic N concentration in the effluent prior to chlorination and the excursions of the dichloramine fraction in the chlorine residuals. Sampling for oganic N content should be done on an occasional basis throughout a year to get a pattern that reflects seasonal variations as well as high and low flow conditions.

Chlorine residuals should be in duplicate, particularly when used for analyzer calibration. Colorimetric methods should not be used. The only reliable methods are amperometric titration and the DPD-FAS titrimetric procedure. If the amperometric titrator is used, two units are preferred, one for measuring total residual by the back-titration procedure and the other for the forward-titration procedure to determine all the chlorine species—particularly the dichloramine fraction. The latter will indicate the relative interference due to the presence of organic N.

Plants that do not nitrify should try never to let the ammonia N level drop below 3 mg/L.

Other Effects from Partial Nitrification. This phenomenon will cause the operator more operating problems than the loss of disinfection efficiency. The instability of the ammonia N concentration promotes the formation of nitrites, which is the first step in the nitrification process.[120,121]

Nitrification is a microbiological process in which ammonia is oxidized sequentially to nitrite and nitrates. It is a two-step process that is carried out by two distinct groups of chemolithotrophic bacteria. In the first step, ammonia is oxidized to nitrite by the energy and activity of *Nitrosonomas* bacteria, as follows:

$$H_4^+ + 3/2 O_2 \rightarrow NO_2^- + H_2O + 2H^+, \qquad -65 \text{ kcal/mol} \qquad (8\text{-}9)$$

In the second step, nitrite is oxidized to nitrate without detectable intermediates:

$$N_2O^- + H_2O \rightarrow NO_3^- + 2H^+, \qquad -20 \text{ kcal/mol} \qquad (8\text{-}10)$$

This step is carried out exclusively by the genus *Nitrobacter*. When this step is completed, there will be no operating problems. When the nitrites are not oxidized by step two, Eq. (8-10) above, all kinds of luxuriant biological growths will appear overnight.

The only solution to this situation is to oxidize the nitrites. This can only be achieved with a free chlorine residual; chloramines cannot oxidize nitrites. Usually the operator is stymied because the only residual that can be produced is all combined chlorine. In these cases when the Water Champ was being used instead of the conventional injector, the milliseconds of molecular chlorine contact time with the wastewater were sufficient to oxidize the nitrites. Only the Water Champ injection system can provide molecular chlorine contact with the process flow.

Therefore, it is necessary in these situations for the operating personnel to have their lab staff keep a record of nitrite levels. The next step is to oxidize the nitrites with free chlorine, provided that the chlorination equipment is of sufficient capacity to achieve breakpoint chlorination. If they are not oxidized, the nitrites will surely generate a biomass growth and make the sewage look like pea soup in a very short time. If free chlorine can be provided, it takes five parts of chlorine to oxidize one part of nitrite (by weight). If the ammonia N content of the wastewater is 4 mg/L at a nitrite level of 2 mg/L, it will take approximately 40 parts of chlorine to destroy the ammonia N and reach the breakpoint, plus another 10 mg/L chlorine to oxidize the nitrites. It is obvious that all of the above-suggested records should be carefully gathered to prevent this partial nitrification phenomenon. Oddly, it seems to plague the smaller plants.

Advice to Operators. Many of the smaller plants (i.e., less than 20 mgd) suffer from wide variations in ammonia N concentration in the effluent. This not only affects the disinfection efficiency, but it also seriously affects the overall effluent quality. To maintain control of this problem the operators must take the following steps, which should become a daily routine:

1. Chlorine residual measurements should be made with either the amperometric procedure or the DPD-FAS titrimetric method. Do not use any colorimetric method.
2. Two separate procedures should be used as follows: (a) forward-titration that measures free chlorine, monochloramine, and dichloramine; (b) confirmation of the separate species measurement above with a single total chlorine residual. If the amperometric method is used, then two titrators should be used, one for forward-titrations and one for total chlorine residuals.
3. Carefully monitor the chlorine-to-ammonia N ratio. This is extremely important. Here is why: A ratio of 6 to 1 is the most germicidal ratio because a small amount of pure dichloramine is formed. Pure dichloramine is half again more germicidal than monochloramine. Any ratio greater than 6 to 1 is the beginning of the breakpoint process. During this process, the chlorine residual begins to decay because the chlorine is in the process of destroying the ammonia N. At ratios of 9 or 10 to 1, the chlorine residual will be at its lowest level. Increasing the chlorine dosage at this point will initiate the development of a free chlorine residual. The free residual will quickly hydrolyze to form nongermicidal organochloramines that appear in the dichloramine fraction of the forward-titration. At the same time, the ammonia N has been removed; and dichloramine cannot exist without ammonia N. Obviously, during this period, beginning with an increasing chlorine-to-ammonia N ratio beyond 6 to 1, the coliform kill will be diminishing rapidly, and there goes the disinfection process. Ratios less than 6 to 1 will not cause any of the difficulties described here.
4. To properly relate the effects of the above scenario it is imperative for the plant operators to know the concentration of the y_0 coliforms. This is the level of coliforms just prior to chlorination. The importance of this value is described above under "Effluent Quality as Related to Dosage."

The above routine procedures will tell the operator when it is advisable to add ammonia N to the effluent prior to chlorination and how great a deleterious effect the organic N might be having. The latter is directly related to the level of organochloramines caused by the presence of organic N. This effect is measured by the concentration of dichloramine as determined by the forward-titration procedure. The answer to "Why do we have to worry about the organochloramines formed by the presence of organic N?" is quite simple. When sufficient ammonia N is present, it takes a much longer time for monochloramine to hydrolyze and be converted to an organochloramine; whereas when free chlorine is present, it hydrolyzes quickly to form organochloramine. Therefore, the deleterious effect of the presence of organic N can be easily monitored by the forward-titration procedure, instead of by running quantitative analyses of organc N. The concentration of dichloramine in wastewater effluents tells it all.

FORMATION OF ORGANOCHLORINE COMPOUNDS

Foreword

In the 3rd edition there was a footnote advising that the discussion of organochlorine compounds addressed only DBPs, and not as these compounds affect the efficiency of the disinfection process. Readers of the 4th edition should note that the following material is based upon Refs. 69 through 75, all of which were published in the mid-1970s.

Now, more than 20 years later, there is an entirely different story. In the passing years, cancer researchers have found without any doubt that the "rodent carcinogen" kill method used for evaluating the health risks of potable water consumers at that time was flawed. In effect, current researchers have found that the haloform compounds (THMs) are trivia, and the miniscule amounts in the NORS report are not a health risk. Actually, there are so many carcinogens in the fruits and vegetables we eat, that a dinner vegetable salad has more carcinogens in it than 200 gal of chlorinated drinking water. Therefore, all the information on the haloform compounds (DBPs) described in the following discussion and treatment plant investigations must be viewed in a new light.

General Discussion

These compounds enter surface waters from many sources, both point and nonpoint. Little is known about the chemical stability, biological degradation, distribution, and ecosystem behavior of the majority of these compounds. Most of the surface water pollutants identified thus far are present in microgram per liter concentrations or less. At this concentration level, the highly chlorinated pesticides and hydrocarbons probably represent a more serious problem with respect to potable water treatment than the chloro-organics formed during wastewater disinfection. These organochlorine species of compounds are formed by the chloramine reaction in chlorinated wastewater effluents that discharge into surface waters.[69]

The formation of a wide variety of organochlorine compounds as a result of wastewater chlorination practices has been well documented by Jolley[69,70] and others.

The importance of the formation of these halogenated-organic compounds is their possible public health risk when they appear in potable water. The carcinogenic nature of several of these compounds has been demonstrated in the laboratory, by using high concentrations and laboratory test animals. However, a direct cause-and-effect link with cancer in humans has not yet been established, and probably never will be.

The EPA has made a concerted effort to accumulate as many data as possible on the public health risk of these compounds. Their primary concern is the formation of the trihalomethanes (chloroform and bromoform) as a result of chlorination practices.[71] These compounds are known carcinogens; so it is prudent to pursue a course of practice that would reduce or eliminate the formation of these compounds.

It is interesting and important to observe that there is no evidence that chlorination practices using chloramines will form any trihalomethanes; therefore, sewage discharges not treated beyond the breakpoint will not form trihalomethanes. *This makes the practice of free residual chlorination of wastewater effluents highly speculative,* if one believes that these microscopic amounts of THMs can be harmful to one's health.

The Rancho Cordova Investigation

A recent comprehensive study by Stone[72] covered halomethane formation in the following situations: activated sludge effluent containing 20 mg/L ammonia nitrogen chlorinated to the breakpoint with dosages of 200 mg/L or more and resulting in substantial free chlorine residuals (7–12 mg/L); same effluent but with chlorination for disinfection (3–4 mg/L residual at 30 min. contact); same effluent except nitrified and filtered and chlorinated; a primary effluent dosed with 11 mg/L chlorine; and an oxidation pond (nitrifying) effluent downstream from the activated sludge effluent.

The results of this study reveal a rapid formation of chloroform (CCl_4) resulting from breakpoint chlorination, as was expected. At a pH of 7, 0.88 mg/L formed in 2 min. and 1.23 mg/L in 5 min. (Now, in 1997, medical researchers have found that chloroform in these amounts is beneficial to human health.)

The activated sludge effluent chlorinated for disinfection purposes formed only 4 μg/L of chloroform in four hours.

The highly polished activated sludge effluent dosed with chlorine at 11 mg/L led to the formation of three times more halomethane compounds than formed by disinfection chlorination of the activated sludge effluent. This is undoubtedly a result of the formation of free chlorine residuals in the highly polished effluent. In this environment of free residuals there is a steady growth of halomethane formation.

Chlorination of the primary effluent to 14 mg/L caused no change in chloroform concentration, a slight decrease in dichloroethane, and an appreciable increase in dibromochloromethane. Overall, a steady increase in total halomethane concentration was noted to a level of approximately 0.0021 μg/L at four hours contact time. This is almost the same as that measured for the polished activated sludge effluent at the same contact time, even though differences in the form of chlorine residuals exist. The residual in the primary effluent is all combined chlorine.

Chlorination of the oxidation pond effluent to a dosage of 6.7 mg/L yielded no change in THM concentration from unchlorinated levels through a four-hour chlorine contact period even though chlorine residuals of 2.5–3.5 mg/L were maintained throughout that time. Carbon tetrachloride and dibromochloromethane concentrations in the pond effluent were unaffected by chlorination. The low ammonia nitrogen concentration of the nitrifying pond effluent ensured that the chlorine residual existing throughout the four-hour contact period was a mixture of both free and combined residuals.

A sample of the chlorine injector water was analyzed for THM concentration to determine if the high chlorine concentrations (1000–3000 mg/L) present in the injector water could cause THM formation prior to the injection of the chlorine solution into the process liquid. The injector water source was activated sludge effluent that had not been subjected to any previous application of chlorine. At the time of this particular sampling, the injector water chlorine concentration was on the order of 1800 mg/L at a pH of 2.5. The residence time in the injector water system from the time of chlorine addition to the point of application was about 20 sec. Approximately 0.0023 μg/L of THM was observed to be formed in the injector water under the above conditions. However, in the overall system, dilution of the injector water into the process stream results normally in a 200 or 300 to 1 dilution so that the net impact of THM formation due to the chlorination system itself is negligible. It is likely that the low pH value and the short contact time in the injector water system may be responsible for keeping THM concentrations below that observed for breakpoint chlorination of the process stream.

The above observations of THM formation in sewage treatment processes were also compared with data on municipal tap water and raw surface water. Analysis of potable water taken from a municipal water system (surface water) in northern California showed THM levels almost identical to those reported as a median value of the NORS report,[73] with measured values of 0.025 mg/L chloroform and 0.0043 mg/L bromodichloromethane. The total halomethane concentration was observed to be 0.235 μm/L for the northern California potable water sample as compared to 0.218 for the median value of the NORS report. Analysis of samples collected above and below the wastewater discharge of the Rancho Cordova plant into the American River demonstrated that with the dilution factor of 250 : 1, current analytical techniques would be unable to detect the contribution of THM concentrations from this discharge to background THM levels in the American River.

Obviously the above levels of halomethanes are of no consequence, as was found in the NORS report in the 1970s. This is another confirmation that THMs are trivia because they are totally inconsequential. Why the EPA cannot or will not see this is pathetic for the various utilities involved.

The Occoquan Watershed Survey

A yearlong study of watershed runoff in this water service area located in northern Virginia near Washington, D.C. was reported by Hoehn et al. in

1976.[74] This report involved a system owned and operated by the Fairfax County Water Authority, providing potable water to more than 600,000 inhabitants in this rapidly urbanizing area. The Occoquan watershed collection system received both agricultural and urban runoff as well as treated sewage from 11 plants in the area. The purpose of the report was to determine the effects of various factors in a given runoff area that might contribute to the formation of trihalomethane in the potable water supply at the downstream end of this runoff system. The findings of this comprehensive study support other evidence that chloroform concentrations appearing in surface waters do not necessarily have their origin in chlorinated wastewater discharges.

For example, the upstream control at Catharpin on Bull Run had a mean concentration of 0.0022 mg/L chloroform in all samples; whereas at Bull Run 2.3 miles below the last of 11 chlorinated sewage treatment plant discharges the samples showed a mean concentration of 0.0032 mg/L of chloroform. However, although the intake of the water treatment plant downstream from the Bull Run sampling point showed only 0.0030 mg/L chloroform (mean) the treated potable water showed a mean of 0.2336 mg/L chloroform. *The latter must be attributed to the practice of free residual chlorination at the treatment plant.*

Pomona, California Virus Study

This comprehensive investigation of various tertiary effluent treatment processes has been described above. The chloro-organic compounds investigated confirm other findings concerning the formation of chloroform due to chlorination procedures. It was found that free residual chlorination of a tertiary effluent increased the average chloroform concentration 330-fold (0.0006–0.1980 mg/L) compared to a sixfold increase during combined chlorine residual chlorination. The tertiary effluent before chlorination contained chloroform concentrations on the order of 0.006–0.0015 mg/L.

CHLORINATION FACILITY DESIGN

Factors Unique to a Wastewater Chlorination System

General Discussion. There are several design considerations unique to a wastewater facility, which will be described here separately. Those features common to both potable water and wastewater are described in Chapter 9.

1. The chlorinator capacity is usually 10 times that of a potable water facility of the same hydraulic capacity.
2. The injector system is supplied with treated effluent and therefore must be pumped. This requires standby power for reliability. Moreover the chlorine–ammonia reaction time must be considered in the design of the solution lines to prevent unnecessary chlorine consumption.

3. Special considerations are necessary for the design of diffusers to provide maximum dispersion and prevent off-gassing.
4. The chlorine residual analyzers are different from potable water analyzers because they must handle a lower-quality sample and are always arranged to measure total chlorine residual.
5. The control scheme is committed to a specific contact time and a prescribed NPDES discharge requirement.
6. Most chlorination facilities are followed by a dechlorination system. This affects the chlorination facility design.

Factors Affecting System Efficiency

Introduction. The strict requirements set forth by the California regulatory agencies for wastewater disinfection focused on the performance of every wastewater chlorination system. Therefore, having been involved in every aspect of a great many wastewater chlorination systems, White decided in 1972 it was time to evaluate the performance of these systems and to determine the elements of an optimum system.[76] Over a six-month period in 1972, White made a personal investigation requiring at least three and sometimes four visits to 34 plants. From 1972 to 1976 inclusive, another 12 plants were investigated to see if the findings of the 1972 survey could be corroborated. The plant effluents investigated in the 1972 survey included 11 primary, 15 activated sludge, 4 secondary using high-rate recirculating biofilters, and 4 secondary followed by oxidation ponds. The second group included both secondary and tertiary effluents. From 1976 to 1982, White spent many days investigating the perceived anomalies of the disinfection process when applied to nitrified effluents.

White was hired in 1993 as a special consultant to Stranco Corp. to study the efficiency and reliability of its specially designed ORP system for the control of the chlorine disinfection process for all types of wastewater treatment plants, and he now believes that a new approach is required. White's three-year study of the Strantrol High Resolution Redox Control System demonstrated without a doubt that this ORP control system is the best available disinfection control system. It not only automatically follows the plant flow changes; it also tracks the changes in the chlorine demand of the wastewater. Moreover, it is not affected by any of the chemical compounds found in all wastewater effluents, which routinely cause errors in chlorine residual analyzers.

The effectiveness of this system has been demonstrated at a variety of currently operating installations. For more information, see Chapters 9 and 10.

Elements of an Optimum ORP System. White's investigations revealed that regardless of the treatment process, the most effective chlorination systems included the following elements:

1. A Stranco Corp. High Resolution Redox Control System (see Chapter 9), which is based upon ORP millivolt control.

2. A continuous chlorine residual recording analyzer to confirm the proper operation of the chlorination equipment.
3. Adequate initial mixing at the point of application of the chlorine.
4. Sufficient contact time (not less than 40 min. at peak dry weather flow) in a contact basin that has a minimum of short circuiting (i.e., 80–90 percent plug flow).
5. Competent and dedicated personnel.
6. Laboratory facilities sufficient to provide proper support to operating personnel.
7. Reliability.

These factors are examined below. The chlorine demand, pH, temperature, and initial coliform concentration (before chlorination) of the wastewater are beyond the operator's control but must be reckoned with. This becomes the duty of the operating personnel, but they have to have an optimum facility for accomplishing the job.

Optimized Mobile Pilot Plant Study[77–79]

Description of Plant and Study. In 1979, the California Department of Health Services completed a comprehensive design optimization study[79] on eight secondary and tertiary wastewater treatment plants. The major part of the project consisted of concurrent studies on a mobile optimized chlorination pilot plant and the existing full-scale system at each of the eight study sites. The trailer-mounted pilot plant contained optimum design features, including rapid initial mixing, reliable automatic chlorine residual control, and plug-flow-type contact tanks providing up to two hours of detention time. Disinfection efficiency was measured by total coliform bacteria and iodometric chlorine residual tests. Concurrent fish bioassays on the same effluents were performed by the Department of Fish and Game. In addition to the above concurrent studies, special pilot studies were made of initial mixing, residence time distribution, effects of chlorine residual and contact time, and dechlorination.

Results and Conclusions of Study:

1. At all locations the optimized pilot plant used significantly less chlorine than the full scale plant. This amounted to an overall average savings in chlorine of 46.7 percent or 8.9 mg/L. The higher chlorine dosages required by the full scale plants were due variously to inadequate contact time, unreliable chlorine control system, and/or inadequate operation and maintenance. The reduction in chlorine dosages was accomplished by an overall average reduction in effluent chlorine residual of 3.3 mg/L, which implies significant savings in sulfur dioxide used for dechlorination.
2. All plants studied except one met their disinfection requirements.

3. Overall, the pilot plant effluents were 43 percent less toxic to test fish than the full scale plants.

4. The chlorine dosages and residuals required to disinfect the various effluents to the same total coliform level differed fivefold in pilot plant studies. The dosages required to achieve an effluent MPN of 2.2/100 ml total coliforms with 60 min. contact time varied from 4 to 22 mg/L in different effluents. The corresponding control residuals ranged from 2.8 to 11.5 mg/L. Poorer-quality effluents and those containing industrial wastes required higher chlorine dosages and residuals than good-quality domestic effluents.

5. Nitrified effluents required higher chlorine dosages for disinfection than some of the nonnitrified effluents. This phenomenon is explained elsewhere in this chapter.

6. The Selleck–Collins mathematical model $N/N_0 = (RT/b)|\uparrow$ in good-quality effluents adequately described the relationship between the chlorine residual–contact time envelope (RT) and total coliform survival.

7. At contact times longer than 60 min. a noticeable decrease in the bacterial kill rate was observed. This phenomenon is partially explained under the subheading "Contact Time" in this chapter.

8. Tracer tests were necessary to assess adequately the performance of chlorine contact tanks. The use of the length-to-width ratio alone was found not to be sufficient.

9. Dechlorination by sulfur dioxide removed all chlorine-induced toxicity from the effluents.

10. The use of sulfur dioxide as a dechlorinating agent requires automatic dosage control based upon flow pacing, continuous chlorine residual measurement, and rapid initial mixing. No additional contact tanks are necessary.

11. Operator attendance is mandatory to keep the chlorine control system in proper working order. This includes daily cleaning and calibration of the chlorine residual analyzers. Therefore, improved operator training is necessary to maintain and properly operate the disinfection system.

12. The pilot plant chlorine control system performed significantly better than the full scale systems because of the following factors: (a) very short loop time, 60 sec; (b) adequate initial mixing, $G = 500$; (c) constant flow rate; (d) instrument compatibility; and (e) good operation and maintenance.

The pilot plant control system was able to maintain the control residual within the desired ±0.5 mg/L range. It was observed that poorly designed and/or operated chlorine control systems were responsible for a major portion of the excessive chlorine dosage used at most of the full scale plants studied. The pilot chlorine contact tank also performed better than most of the full scale tanks due to better plug-flow characteristics and longer minimum contact time.

Sizing the Chlorinator

Influence of Plant Flow Meter. The first step in the design of a chlorination facility is the determination of the maximum chlorine needed for the given situation, including prechlorination, intermediate, and postchlorination applications. Although this text is specific for disinfection (postchlorination), total capacity must include the requirements for the other purposes stated above. Chlorinators should be divided into two groups, one group for prechlorination and intermediate points of application and the other group for postchlorination (disinfection). Equipment should be arranged so that the first group can provide standby service for the disinfection equipment. A third group may be desirable if both pre- and postchlorination will be continuous. The third group should be arranged for both intermediate chlorination and standby for either pre- or postchlorination.

Chlorinator capacity is a function of the flow signal and the chlorine demand of the wastewater. Attention is called here to the term "flow signal." Since it is imperative that the chlorine feed rate be controlled in proportion to the flow, the chlorinator capacity must be sized to the capacity of the primary flow meter. Although it is true that chlorinators have overriding dosage control, this feature must be reserved for variations in the chlorine demand of the wastewater. For example, at a proposed plant, the primary flow meter to be used has a range of 0–10 mgd, which is the ultimate design capacity of the facility. Furthermore, the chlorine demand is to be a maximum of 150 lb/million gallons. The chlorinator capacity for that plant must be $10 \times 150 = 1500$ lb/day regardless of the initial low flows that may be expected for the first several years. This capacity is required because the chlorinator needs the full range of the primary meter flow signal so that it can also provide a dosage range to meet the diurnal chlorine demand variation.

This concept applies to both pre- and postchlorination, but not to intermediate chlorination because the latter chlorine feed rate is usually on a manual control basis.

In the above example the chlorinator would be ordered as a 2000 lb unit fitted with a 1500 lb rotameter and a 2000 lb spare rotameter.

Effect of Chlorine Demand. The required chlorine dosage for prechlorination is difficult to predict with any degree of accuracy in the absence of some historical evidence. If the sewage is moderately fresh, the dosage may be as low as 10–12 mg/L. If sulfite wastes such as from a tannery or a winery are a factor, the chlorine demand might be as high as 40–50 mg/L. If the sewage is septic upon arrival, the demand will most likely be from 30 to 40 mg/L.

Intermediate points of application of chlorine require approximately the following equipment capacities:*

1. Secondary sedimentation tanks (10 mg/L)
2. Return activated sludge-control of bulking (5 mg/L)

---

*All dosages are based upon treatment plant flows (PDWF), whereas sludge thickener dosage is for this specific flow rate.

3. Ahead of biofilters or trickling filters (10–15 mg/L)
4. Sludge thickener line—odor control (50 mg/L)

The use of chlorine at these points of application is rarely on a continuous basis; so the prechlorination equipment can be used for these purposes. If prechlorination is to be used continuously for odor control, additional equipment should be supplied for the intermediate points of application. This group of equipment should then be sized so as to provide standby service for either pre- or postchlorination.

Industrial wastes have a profound effect on the disinfection requirements of domestic wastes. This is further complicated by the seasonal nature of some industrial wastes and the extent to which these wastes are pretreated before being discharged into the domestic collection system.

For domestic wastes with not more than 1–2 percent industrial waste, the following are minimum capacity guidelines for the designer:*

Primary effluent	150–200 lb/mg
Secondary effluent	50–75 lb/mg
Secondary plus ponds	50 lb/mg
Tertiary (not nitrified)	30–50 lb/mg

Effluents that require a free chlorine residual require at least 10 parts of chlorine for each part of ammonia nitrogen.

If industrial wastes are present to the extent of 10–25 percent of the total wastewater flow, it may be necessary to increase the chlorine requirement for primary effluents by a factor of two and perhaps more, depending on the nature of the wastes. This may apply to secondary effluents as well, depending upon how effectively the treatment process can cope with the industrial waste. The more uncertain the factor of industrial waste, the more uncertain becomes the required quantity of chlorine for disinfection. In this case, laboratory determination of chlorine demand will be necessary.

CHLORINE CONTROL SCHEMES†

Flow Pacing

Chlorinators for both pre- and postchlorination should be provided with flow proportional control. The prechlorinator should be controlled from the influent flowmeter and the postchlorinator from an effluent flowmeter. Never attempt to control the postchlorinator from the influent meter or vice versa; it will not work because of the lag time between the two measuring points.

---

*These dosages are based upon NPDES requirements of 200/100 ml MPN fecal coliforms or 1000/100 ml MPN total coliforms.

†For more details on control strategy illustrated by instrumentation schematics, see Chapter 9.

As discussed previously, the flow proportional signal should be used to control the chlorine metering orifice in the chlorinator. This can be done either pneumatically or electrically. If it is to be done pneumatically, the signal must be linear (3–15 psi). If it is to be done electrically, there are a variety of linear signals. The one most preferred is the analog milliamp signal, with outputs of 1–5, 4–20, and 10–50 mA. The most common is 4–20 mA.

Although there is local dosage control provided on the chlorinator cabinet for individual chlorinator control, a ratio relay (0.4–4.0 range) should be installed in the flow meter electrical signal circuit. This instrument will provide the operator with the means to make precise dosage adjustments without interfering with the range and the accuracy of the flow pacing signal.

ORP Millivolt Control

Introduction. As of 1997, this method of control has been found to be superior to all the methods of chlorine residual control. These methods suffer from the formation of organic N compounds that are present in all treated wastewater effluents because they form nongermicidal organochloramines, which register as effective chlorine residual on chlorine residual analyzers' readout.

Stranco High Resolution Redox (ORP) Control System. This system was developed in the early 1990s, and has since proved to be completely reliable when operated according to the manufacturer's instructions at the time of installation. (See Chapter 9 for more details.)

The technique of this redox system is totally different from the efforts produced by chlorine residual analyzers, in several different ways:

1. ORP measures the actual oxidant and reductant activity, whereas chlorine residuals suffer quantitative errors from organic N compounds that are a part of all wastewater effluents.
2. Redox can measure potential (millivolts) of both sides of the chlorine and sulfite couple. This means it can measure both the chlorine and sulfite activity. This allows the operator to lock in a slight residual of either chlorine or sulfite.
3. Redox responses are logarithmic, meaning more precise dechlorination control, where zero residual is the discharge requirement. This not only perfects the dechlorination process, but it saves the users a sizable amount in chemical costs, up to 25 to 50 percent per annum.
4. Properly designed redox electrodes are very forgiving of process contamination, because they do not measure current. The electrodes can continue to operate properly even after being partially coated with source water contaminants.

Biological Nutrient Removal (BNR). A study by the operating staff at the Beloit, Wisconsin Water Pollution Control Facility indicated that the use of ORP monitoring of anoxic and anaerobic conditions is a valuable method for achieving consistent biological nutrient removal (BNR).[130]

Phosphorus Removal. The Beloit staff began investigating the plant's phosphorus removal capability in 1992 after the Wisconsin Department of Natural Resources announced a new effluent requirement of 1.0 mg/L for phoshporus on a monthly average for this 11 mgd WWTP. At this point it was fortunate that the staff members working on this project were familiar with the success of the ORP control and monitoring system, used in the application of chlorine at their industrial pretreatment facility to achieve odor control.

Owing to the fact that ORP directly measures the oxygen conditions, in either the positive (aerobic) state or the negative (anerobic) state, it provided the most logical means of monitoring the oxygen status in the selector.

Therefore, the staff was not surprised when dramatic changes in phosphorus removal efficiency were reflected in dramatic changes in the ORP readings.

Batch Reactor Wastewater Treatment. In the meantime, a group from both Hong Kong and Canada researched the use of ORP control of anoxic sequences in batch sewage treatment systems.[131] Although their work was not quite the same as that of the Beloit staff, it did confirm the Beloit findings. Their goal was a bit different from that of the Beloit group. Their main research objective was to demonstrate the use of ORP for automated control of a sequencing batch reactor in wastewater treatment systems. The control depended upon the unique nitrate breakpoint phenomenon as shown in Fig. 8-11. Its presence signifies the termination of anoxic conditions (the presence of nitrates, the absence of oxygen) and the beginning of true anaerobic conditions (the absence of both nitrates and oxygen).

The wastewater treatments studied included aerobic-anoxic sludge digestion and biological phoshporus removal. For both processes, the operating strategies consisted of a control or fixed-time reactor, and an experimental or real-time reactor. Thus another important use was developed for ORP, which will achieve better control of compounds that cause pollution.

Chlorine Residual Control

General Discussion. In addition to the flow proportioning control, residual control for disinfection is necessary because of the diurnal variations in chlorine demand and the necessity for close regulation of the disinfection process to meet NPDES discharge requirements.

The key factor for a successful residual control system is the ability to obtain a homogeneous sample downstream from the chlorine diffuser within 30 sec after the application of chlorine at average dry weather daily flow. At peak flow this time would be reduced to about 15 sec. These time values are based

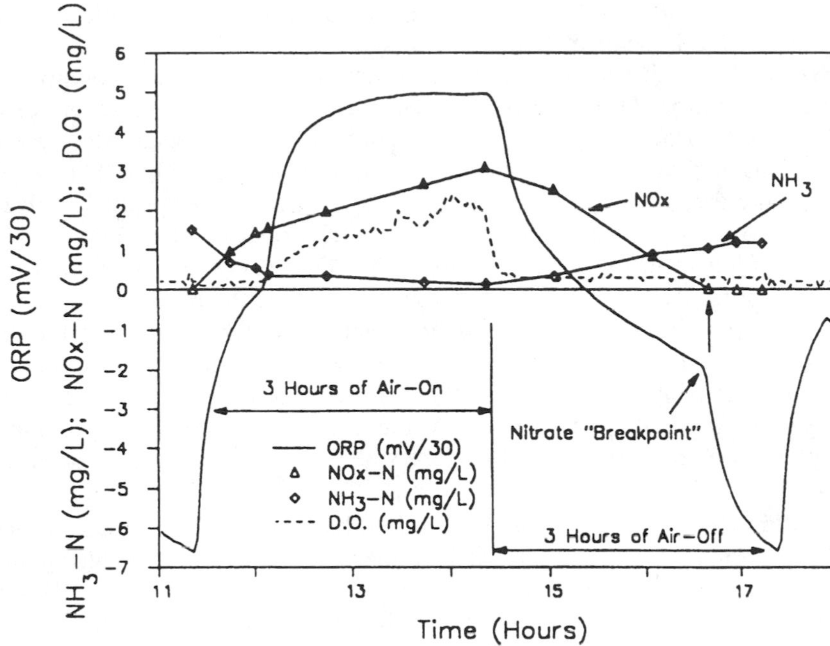

Fig. 8-11. Nitrate breakpoint occurring in ORP-time profile.[131]

upon a mixing time of not more than 3–5 sec. It will then take the sample another 45–60 sec to reach the cell in the analyzer, as it has to pass through a filter and the internal piping within the analyzer; therefore, the earliest the sample can reach the measuring cell is about 1½–2* min. after the chlorine has been mixed with the wastewater. If the chlorine has been well mixed, the control residual will lie on the flat part of the residual die-away curve shown in Fig. 8-11. Another mandatory requirement of a well-designed system is to utilize the remote injector concept by installing the injector as close to the point of application as is physically possible. The average loop time should be on the order of 2 min., with a maximum not to exceed 5 min.

Residual control is synonymous with dosage control. The analyzer transmits a signal that calls for either more or less chlorine. The control strategy can be either the floating type or a separate PID loop. There are a variety of arrangements, which are described below.

Compound-Loop Control. This has been the predominant method of residual control for wastewater treatment since it was first introduced in 1961. This system was unique to Wallace & Tiernan equipment when it was first

*At average dry weather flow.

introduced. In 1984, Wallace & Tiernan announced the elimination of this control method in favor of a dual signal system.

This method utilizes two signals that are sent to different components in the chlorinator. The chlorinator combines these two separate pieces of intelligence to produce a given chlorine flow rate. The wastewater flow meter sends a pacing signal to the motorized chlorine orifice valve. This signal changes the area of the valve opening. The signal from the residual analyzer is sent to the vacuum regulating valve in the chlorinator. The change in this valve position changes the differential "head" across the above-described metering orifice. The flow through an orifice varies with area and differential pressure across the orifice ($Q = A\sqrt{2gh}$). This method has many advantages, which are described in Chapter 9. The principal advantage is its uncommonly wide-range capability of at least 100 to 1. One other important feature is the ability of the residual signal to control when the flow signal becomes unreliable at low flows. See Chapter 9 and Fig. 9-85.

Multiplied Signal Control.* This method has been in widespread use as a process instrumentation technique for many years. When used for chlorination and dechlorination control, the wastewater flow signal from the treatment plant flow meter and the chlorine residual signal are sent to a multiplier. The output of the multiplier controls the chlorine metering orifice positioner. This method is limited to the practical range of the chlorine metering valve and rotameter, which is significantly less than the range of a compound-loop system with equivalent accuracy, described above and in Chapter 9.

Dual Signal Control. This is a variation of the multiplied signal. It was introduced by Wallace & Tiernan in 1978. The Wallace & Tiernan electric positioner is equipped with a manually controlled dosage potentiometer. The use of the positioner potentiometer is the basis for automatic residual control. The residual analyzer is fitted with a potentiometer. The potentiometer signal is integrated by the circuitry in the positioner so that the positioner responds to two separate signals—one from the flow meter and one from the potentiometer in the analyzer. An override switch on the front of the chlorinator permits manual dosage adjustment.[81]

Direct Residual Control. This method operates without benefit of a process flow signal. This has proved to be a successful and reliable method.[82] The success of this system depends upon three factors: (1) the residual signal must be transmitted to the chlorine orifice positioner (via a PID controller); (2) the positioner travel time from zero to maximum must be on the order of 15 sec; (3) the loop time should be limited to approximately 45 sec at peak flow conditions. See Chapter 9, Fig. 9-84 for a typical potable water system schematic.

*This is commonly called "compound-loop," which is erroneous.

INJECTOR SYSTEMS

General Description

The injector system is the heart of the entire chlorination facility. If this system is inoperative, no other part of the system can function. The various parts of this system include: (1) the operating water supply to the injector; (2) the injector; (3) the injector vacuum line from the chlorinator; (4) the injector discharge system, described as the chlorine solution line; (5) the diffuser at the point of application; and (6) the necessary gages required to monitor and troubleshoot the system. Design details are fully presented in Chapter 9. The following text serves to cover situations that are peculiar to wastewater installations, which may be different from potable water systems.

Water Supply

Every injector system is designed to create sufficient vacuum (16–22 in. Hg) to move the chlorine from the supply containers into and through the chlorinator to the injector suction port, where it is dissolved into the injector water supply. The injector must then be able to deliver the chlorine solution to the chlorine diffuser. The injector water supply pressure must be sufficient to create the required vacuum and overcome the friction loss in the solution piping and diffuser. Additionally this pressure must overcome any positive static head between the injector centerline and the hydraulic gradient at the diffuser. The amount of water required must be sufficient to limit the chlorine solution strength to 3500 mg/L. A broad rule of thumb is 40 gal of water/day/lb of chlorine, more or less, depending upon the local conditions.

Plant effluent is universally used for the injector water supply. Nonnitrified effluents contain 15–25 mg/L ammonia nitrogen. If the breakpoint reaction between ammonia N and chlorine is allowed to proceed in the chlorine solution line, 10 parts of chlorine will be consumed for each part ammonia N before the chlorine solution reaches the point of application. If proper precautions are not taken, the chlorine wasted in this reaction could amount to about 10 percent of the total chlorine used. This reaction is both time- and pH-dependent. At pH 3 the reaction will go to 90 percent completion in about 4 min. The usual pH of chlorine solution derived from wastewater is between 2 and 3, depending upon the chlorination rate. Therefore, chlorine solution lines should be designed to limit the solution travel time to not more than 4 min. If this is not practical, the injector should be moved nearer to the point of application.

Hydraulic Design Considerations

Positive Head. Static pressure at the discharge of the injector should be limited to no more than 4–5 psi, for the following reasons: (1) Injector operating pressures and water quantity escalate rapidly as the back-pressure on the injector rises beyond 5 psi, particularly when the chlorine feed rate is in excess of 500 lb/day.* (2) Head loss through the diffuser is used as a chlorine mixing device. Allow 8 to 10 ft of loss through the diffuser for this function. To meet this requirement, injectors should be located as close to the diffuser as possible and at an elevation above the hydraulic gradient at the diffuser location. Injectors may be installed in any direction—upright, horizontal, or upside-down.

When these conditions cannot be met and the static head at the injector is 10–15 psi or more, a chlorine solution pump should be installed adjacent to the injector discharge as shown in Chapter 9, Fig. 9-108. For wastewater installations, the Fybrock Division of Robt. D. Norton Co., Fairfield, New Jersey, can furnish Fiberglas pumps with a good experience record for pumping chlorine solution. The installation must be arranged so there is always a positive pressure (approx. 2–3 psi) on the discharge of the injector.

Negative Head. An open-channel diffuser can be a great nuisance and even a hazard if it is not properly designed. Anytime a negative head exists in the chlorine solution line, molecular chlorine will break out of the chlorine solution and cause serious chlorine gas emission at the diffuser. Therefore, if the sewage level at the diffuser is below the injector throat, the hydraulic gradient from the injector to the diffuser must be calculated to provide a reasonable injector back-pressure. Assume this back-pressure is 5 ft (approx. 2 psi). Then the friction loss through the diffuser holes plus the line losses minus the difference in elevation between sewage level and injector throat must equal approximately 5 ft of head. This is one of the reasons for having a compound gage on the solution line at the discharge of each injector. The minimum recommended size for diffuser holes is ⅜ inch.† The recommended velocity through the holes to provide initial mixing is 22–26 ft/sec. All open-channel diffusers should be constructed for easy removal. They may become plugged, or the hole configuration might have to be changed. This requires a flanged connection so that the entire diffuser can be lifted bodily from the channel. Diffusers are available with projecting pins that fit into slotted wall brackets for easy removal.

Figure 8-12 illustrates a problem in diffuser design for an open channel. Assume a 200 gpm flow through a 3-inch injector and 8000 lb/day chlorine so as to limit the chlorine solution strength to 3500 mg/L chlorine. Use 4-inch chlorine solution piping to the diffuser. Applying Bernoulli's energy equation, we get:

$$P_1 + E_1 + \frac{V_2^2}{2_g} - \text{losses} = P_2 + E_2 + \frac{V_2^2}{2_g} \tag{8-11}$$

*For chlorinators 500 lb/day and smaller, this back-pressure limitation is not a factor.
†This is for wastewater system diffusers.

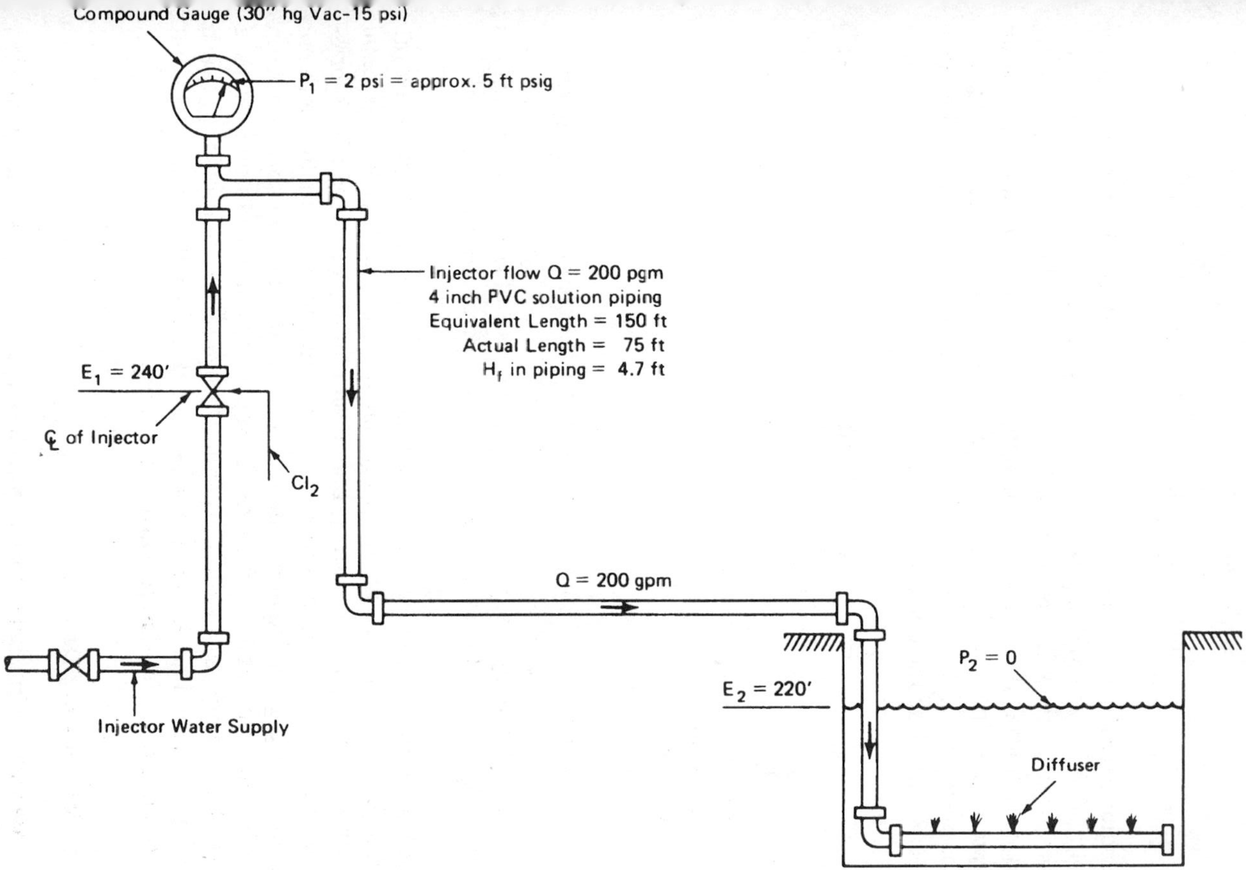

Fig. 8-12. Example of diffuser design for open channel application under negative head conditions.

Ignoring velocity head differences from Fig. 8-12, we get the following:

$$5 + 240 - \text{losses} = 0 + 220$$

$$-\text{losses} = -25 \text{ ft}$$

The equivalent length of the chlorine line is assumed to be 150 ft; this is made up of a 75-ft linear pipe length plus five elbows at 15 ft each. Therefore, the friction losses in the solution piping will be about 5 ft. So, to prevent a negative head condition at the diffuser by maintaining 5 ft of head at the injector discharge, the diffuser will have to consume 25 − 5 = 20 ft of head loss.

The next step is to choose a diffuser hole size. The minimum acceptable size is ⅜ inch. Use the equation

$$Q = CA\sqrt{2gh} \tag{8-12}$$

where

C = orifice coefficient = 0.75
A = area of orifice ft^2
h = head loss, ft
Q = discharge, cfs

Using A = 0.008 and h = 20, Q calculates to 0.0215 cfs = 9.7 gpm. Therefore, 200/9.7 = 20.6; so use 20 to 21 holes.

Diffusers should always be designed on the basis of one diffuser for each injector, unless more than one injector will be operated simultaneously.

Remote Injectors. Figure 8-13 shows the installation of a 2-inch remote injector. Large injectors (3-in. and 4-in.) should be installed in the horizontal position when possible. This low profile position reduces injector back-pressure and allows for easy disassembly of the injector. The vacuum line between the chlorinator and the injector should be sized so that the total pressure drop in this line does not exceed 1.5 inches of Hg at maximum chlorine flow. (See Chapter 9 for sample calculation.) A shutoff valve should be provided at each end of the vacuum line. A vacuum gage should be installed adjacent to, but downstream from, the shutoff valve at the injector end.

Injector Monitoring Devices

Flow Meters. Adjustable throat injectors (i.e., 2, 3, or 4 in.) should be provided with a flow meter of reasonable accuracy over a 3:1 range. A direct-reading rotameter utilizing a bypass orifice type installation is satisfactory. However, an in-line propeller meter is a more expensive device but makes for a neater piping arrangement.

Fig. 8-13. Remote injector utilizing injector water to provide sample to chlorine residual analyzer.

Flow meters are necessary for the operator to achieve optimum injector adjustment for the local conditions to ensure maximum efficiency, proper chlorine solution strength, and adequate back-pressure.

Gauges and Alarms. Gauges should be provided to show the operating water pressure for each injector in the system. Each injector should be provided with a chlorine solution pressure gauge mounted immediately downstream from the injector discharge. These must be compound gauges reading to 30 inches of Hg vacuum and to 15 psi pressure and equipped with a silver diaphragm protector. These gauges are necessary for the proper analysis of the injector operation. They provide the operator with the back-pressure readings for various quantities of water passing through the injector. In some situations this is critical information that cannot be obtained in any other way. It also constitutes proof of proper or improper diffuser design and/or chlorine solution line design.

Every installation should be provided with a loss of water pressure alarm to indicate a failure in the water supply. Overpressure switches are not necessary for the injector system.

Chlorine Solution Lines

The piping downstream from the injector is the chlorine solution line. It is permissible to manifold the injector discharge from two or more chlorinators into one point of application,* but a solution line to the point of disinfection should not be manifolded to any other point of application. The most desirable arrangement is for each injector (in a multiple injector system) to have its own solution line and diffuser. For pipe materials, see "Materials of Construction." Chlorine solution flow proportioning systems are not worth the expense, and most end in failure. Some waters, when carrying a chlorine solution, release a tremendous amount of carbon dioxide and other gases. The bubbles of gas in the solution passing through the rotameter can cause sufficient vibration to severely limit the accuracy of the reading. Glass tube rotameters should not be used because this vibration can cause the rotameter float to shatter the glass tube. The only rotameters satisfactory for chlorine solution lines are the straight-through metal (Hastelloy C) or PVC tube type with dial indication.

Diffusers

There are three categories of diffusers in wastewater treatment: prechlorination, intermediate chlorination, and disinfection. These diffusers are required to perform different tasks; so they must be discussed separately.

Prechlorination. The choice of chlorine solution diffuser for the best results at this point of application is a challenge to the designer. Usually the diffuser will be in a closed conduit flowing partially full, thereby conforming to open-channel flow. There is practically no mixing capability in an open channel, and, owing to the large amount of debris in the raw wastewater influent, mechanical mixers are totally impractical. Therefore, the designer must select the diffuser with the best mixing characteristics. Figure 8-14 shows one of the most satisfactory types of prechlorination diffuser.† This type can be made of either PVC or hard-rubber-lined and rubber-covered steel pipe. The curved and perforated sections are heat-treated to conform to the radius of the conduit and are laid in a recess in the conduit. The pipe is perforated, so that chlorine solution is discharged mostly below the minimum water surface level. This diffuser eliminates the problem of the accumulation of debris, which is always prone to collect on any object projecting into the flowing sewage. The diffuser should be designed to be easily removed. Removal may be necessary to clean the slot of silt and other debris as well as to inspect for plugging of perforations. Experience indicates that over a period of three or four years chlorine diffusers

*Provided that the solution line discharges into a multiple diffuser system so that there is one diffuser per injector.

†One such diffuser has been in operation without removal or maintenance problems for more than 40 years.

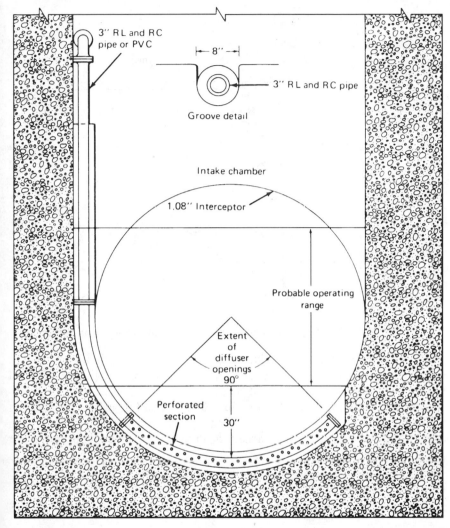

Fig. 8-14. Perforated diffuser shaped to fit into recess of circular conduit especially for prechlorination.

do plug up in wastewater plants. The deposits that cause the trouble appear to be the result of highly chlorinated organic compounds containing a large portion of grease, which provides a waxlike binder. The above type of diffuser must be incorporated into the original design of the influent pipeline.

Figure 8-15 shows the perforated type of diffuser laid on the invert of the pipe and running axially with the conduit. This type is used for existing systems when there is no other way to apply the chlorine solution.

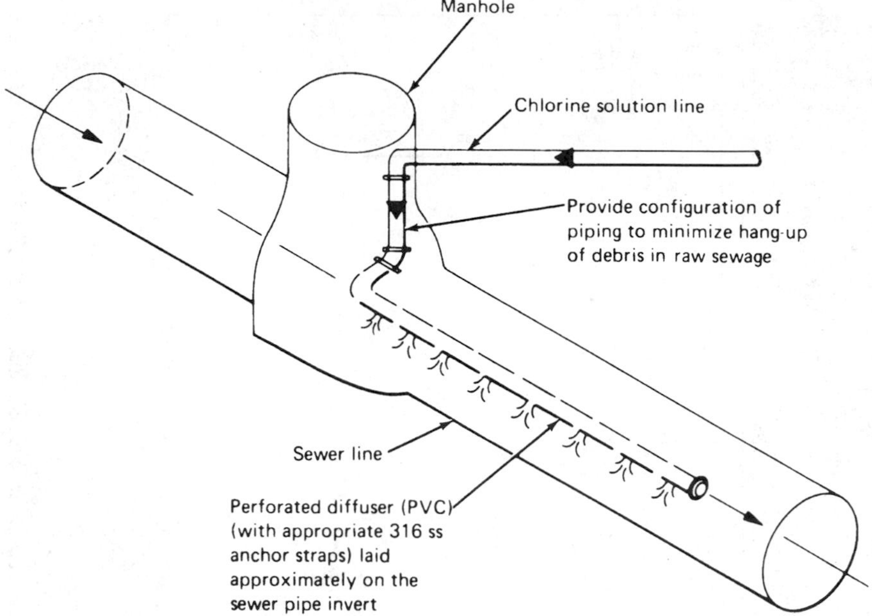

Fig. 8-15. Chlorine diffuser laid on pipe invert (prechlorination only).

Hanging spray nozzle diffusers have been used for prechlorination diffusers. Some have been successful, but again experience has shown that the type of diffuser illustrated in Fig. 8-14 is to be preferred.

Figure 8-16 illustrates the diffuser in a closed conduit, when the influent line to the plant is a force main. Although it is well known that the best mixing occurs when the chlorine is discharged into the center of the pipe, this is not

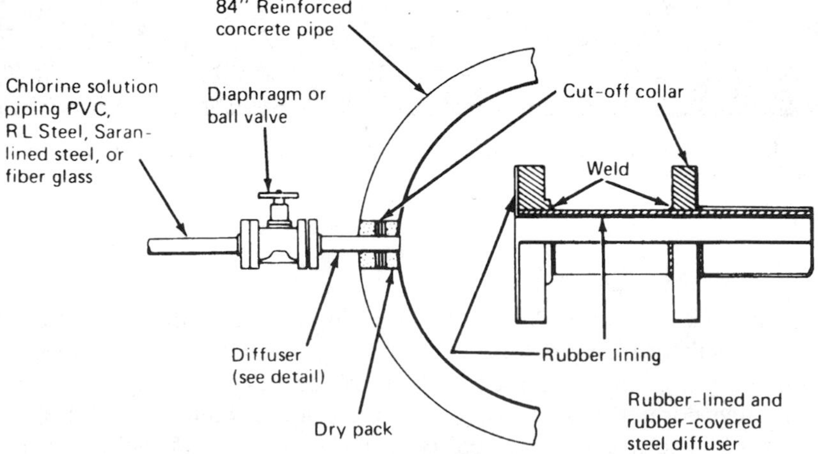

Fig. 8-16. Prechlorination diffuser flush with pipe wall to prevent debris fouling (orifice velocity 20–25 ft/sec). (Courtesy Chlorine Specialties Inc.)

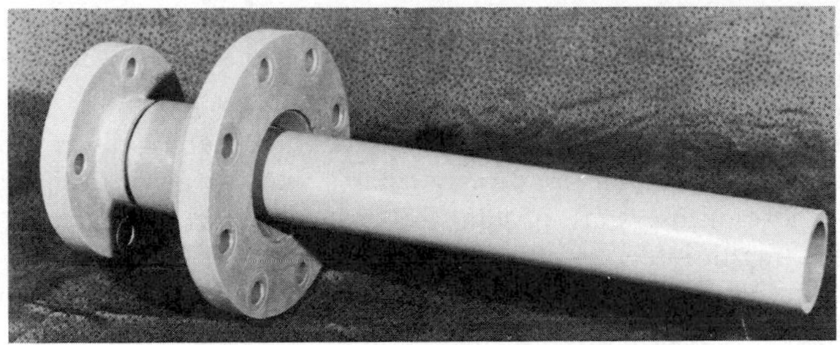

Fig. 8-17. Chlorine solution diffuser (courtesy Chlorine Specialties Inc.).

practical because of the debris that will accumulate on the projected diffuser. Therefore, the diffuser must end flush with the inside wall of the conduit, but it should be designed for a velocity of 22–26 ft/sec in order to project the chlorine solution toward the center of the conduit. Successful results have been obtained by following these guidelines. This velocity across the diffuser orifice will result in an 8–10-ft pressure drop. Therefore, the chlorine solution line and injector water supply will have to be designed to accommodate this additional injector back-pressure.

Intermediate Points of Application. The mixing and the dispersion of chlorine solution into return activated sludge, supernatant liquor, secondary sedimentation, and recirculating biofilters do not present the difficulties associated with pre- and postchlorination.

For example, return activated sludge, supernatant liquor, and other points of intermediate chlorination are usually into a closed pipe. For these situations a pipeline diffuser is used (see Figs. 8-17 and 8-18). This type of diffuser is designed to deliver the chlorine solution to the center of the pipe. This provides a complete mix in 10 diameters of the pipe. Return activated sludge chlorine should be applied in the recirculation pump suction.

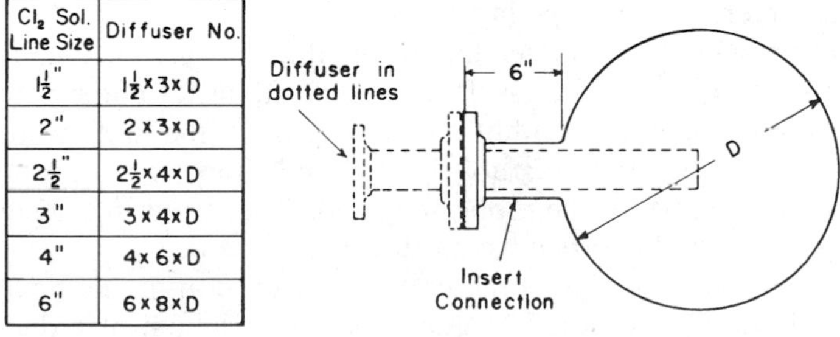

Cl_2 Sol. Line Size	Diffuser No.
$1\frac{1}{2}$"	$1\frac{1}{2} \times 3 \times D$
2"	$2 \times 3 \times D$
$2\frac{1}{2}$"	$2\frac{1}{2} \times 4 \times D$
3"	$3 \times 4 \times D$
4"	$4 \times 6 \times D$
6"	$6 \times 8 \times D$

Fig. 8-18. Dimension of pipeline diffuser.

Secondary sedimentation tanks are usually accommodated by the perforated type of horizontal diffuser in an open distributing box ahead of such structures.

Disinfection (Postchlorination). At this location there is never a problem of debris accumulation; so the designer has a wide array of design choices. However, these diffusers are of vital importance to one of the most critical factors affecting disinfection efficiency—initial mixing. Therefore the design and configuration of these diffusers is described below under "Chlorine Solution Mixing."

Materials of Construction

The materials used for the injector system are those employed in good water works practice for the water system up to the inlet of the injector assembly. From this point forward, a corrosive chlorine solution will be encountered, which requires special materials. The chlorine solution lines can be either Sch 80 PVC, Fluoroflex-K (Kynar), rubber-lined steel, Saran-lined steel, Kynar-lined steel, or certain types of fiber cast pipe.

Valves on solution lines can be either diaphragm or ball type. Diaphragm types are usually flanged, rubber-lined, or PVC-lined cast iron Saunders-type valves. The ball-type PVC valve is preferable up to the 2½-inch size, although the ball-type valve is available in much larger sizes. The diaphragm type should be considered for sizes 3-inch and larger.

The injector vacuum line between the chlorinator and the injector is the most controversial of all the piping used for the chlorination facility as to material selection. This line carries moist chlorine gas under a vacuum. It is preferable to use Sch 80 PVC or Kynar pipe, although some types of fiber cast pipe are also suitable. Ball-type PVC valves should be used instead of diaphragm-type valves on this line. There is no known corrosion-resistant metal pipe available for this use. Saran and Saran-lined steel pipe have been used for this purpose; glass pipe has been found to be impractical. The only time that this particular chlorine-carrying line becomes a consideration is in the case of remote injector installations. It should be noted that when the injector is located in the chlorinator room, the manufacturer supplies Sch 80 PVC for the interconnecting piping.

The diffusers and the piping leading to the diffusers are customarily made up of Sch 80 PVC pipe and fittings. If the specifications for underwater piping require steel construction for additional strength, all the underwater piping and diffusers must be made of rubber-lined and rubber-covered steel pipe. The diffuser holes must also be rubber-covered. This results in extremely high construction costs, and is rarely worth the expense.

Nuts and bolts for assembly of the underwater portion should be 316 stainless steel. All other bolts should be galvanized or cadmium-plated steel. Pins and slotted wall brackets for diffusers are available in PVC.

As described previously, all gauges on the solution and vacuum lines should be silver-diaphragm-protected. This is a standard item with the chlorinator manufacturers. Vacuum line gauges are for vacuum only; solution line gauges must be compound gauges as described in the section on gauges.

Chlorine Solution Mixing

General Discussion. The question of optimum mixing of chlorine solution into an ammonia-laden wastewater has been a formidable puzzle for designers. The principal competing reactions in a chlorine–wastewater scheme are HOCl with ammonia N versus HOCl and the microorganisms.

However, how long HOCl persists during the initial mix remains a controversial question among many investigators. Under normal conditions of wastewater disinfection the reaction between HOCl and NH_3–N is so rapid it is probably not measurable. However, it can be estimated. Assume the following condition: pH = 7.0, temperature = 25°C, NH_3 in wastewater = 14 mg/L as N, and chlorine dosage = 10.5 mg/L. Using Professor Morris's rate constants, the HOCl and NH_3–N reaction is 99 percent complete in 0.2 sec. At pH 4 it takes 147 secs. The peak rate occurs at pH 8.3. Therefore, it is the opinion of both White and Collins that under normal conditions there is no measurable effect of HOCl. Moreover, since the molecules of NH_3 outnumber organisms by a factor approximating 10^{13}, it is practically certain that all of the disinfection occurring in the initial mixing phase is by monochloramine.*

There is one other competing reaction that has a profound effect upon disinfection efficiency—the chlorine–organic N reaction. This was observed in a completely nutrified effluent[50] which was converted from free chlorine residual to a combined residual by adding Nh_4OH in a Cl to N ratio of 6:1. With little or no mixing, the final residual consisted of 45 percent monochloramine and 55 percent organochloramine. When there was thorough rapid mixing, the total residual was composed of 85 percent monochloramine and 15 percent organochloramine. The total coliform concentration at the end of a 50 min. contact time was 1300/100 ml for little or no mixing and 30/100 ml for thorough mixing. This phenomenon has yet to be explained. It is believed that where poor mixing is involved, a hydrolysis reaction takes place whereby the monochloramine reacts with the organic N to form organochloramines.

The difficulty with initial mixing is the lack of information that would allow a valid description of what to expect from different "levels" of mixing.

Historical Background. The earliest known mention of the importance of initial mixing in the chlorination of wastewater was by Rudolfs and Gehm.[84] They reported in 1936 on a comprehensive study of wastewater disinfection

*An exception to this is the Pentech and Fischer and Porter Systems addition system, where chlorine gas instead of chlorine solution is used. In this case, molecular chlorine has a definite effect in the initial mixing phase.

by chlorination. As a result of this study, they advised that "maximum bacterial kill can be obtained . . . provided good mixing is allowed. . . ." However, they made no attempt to define "good mixing."

The first known documentation of chlorine dispersion in a wastewater effluent occurred at the City and County of San Francisco, Richmond–Sunset WWTP in 1940.[83] The chlorine diffuser consisted of a series of hanging nozzles across a 4-ft-wide channel. The nozzles were slotted, and produced a thin layer of chlorine solution in the shape of a 90-degree fan. The nozzles were positioned about 18 inches below the wastewater surface in a 3-ft-deep channel. Residuals were measured by starch–iodide titration. The diffusers were located upstream from a Palmer–Bowlus flume. The residuals were measured about 50 ft downstream from the diffuser. The results demonstrated quite conclusively that there was gross stratification of the chlorine solution due to laminar flow and the absence of turbulence. Ten feet beyond the point of residual measurement the effluent discharged into an outlet structure where there was a high degree of turbulence. The chlorine residuals became uniform through this structure as a result of this turbulence.

In 1948 Eliassen, Heller, and Krieger et al.[88] presented a statistical approach to sewage chlorination. This study examined the effect of initial mixing upon bacterial destruction. They compared three levels of mixing intensity: (1) no mixing; (2) 10 swirls with glass rod at approx. 100 rpm; (3) 15 sec. in a Hamilton-Beach egg-beater. They found a significant difference in bacterial destruction by chlorine between the unmixed and the mixed samples. However, they found no improvement in the samples mixed by the egg-beater. These findings tend to support more recent findings, which are discussed later in this text.

Huekelekian and Day[90] reported in 1951 on a study of five sewage treatment plants using chlorine for disinfection. They found that in those plants not provided with special initial mixing compartments, gross segregation of chlorine concentration will occur. This is the result of low-velocity (0.5–0.75 ft/sec) laminar flow. Figure 8-19 illustrates the data collected by Heukelekian and Day demonstrating the gross differences in the homogeneity of chlorine residuals with and without mixing in a chlorine contact chamber.

About 1968 Harvey Collins, a graduate student under the direction of Professor Robert Selleck, began a pilot plant study at the Sanitary Laboratory of the University of California at Berkeley, in an attempt to delineate the kinetics of wastewater disinfection with chlorine.[56,87] This work demonstrated conclusively that initial mixing had a profound effect upon the kinetics of the disinfection reaction.

Other research work in the 1970s occurred at Johns Hopkins University by Krusé et al.[89] and at the University of California at Berkeley by Stenquist and Kaufman,[93] Collins and Selleck[58] and Selleck, Saunier, and Collins,[62] plus field studies by Longley.[91,92] Longley's field studies were supplemented by a design concept of a disinfection system in 1982.[86] In 1973, Deaner[102] reported on the use of fluorescent dyes for establishing the dispersion performance of chlorine solution mixing devices. All presented evidence that initial mixing

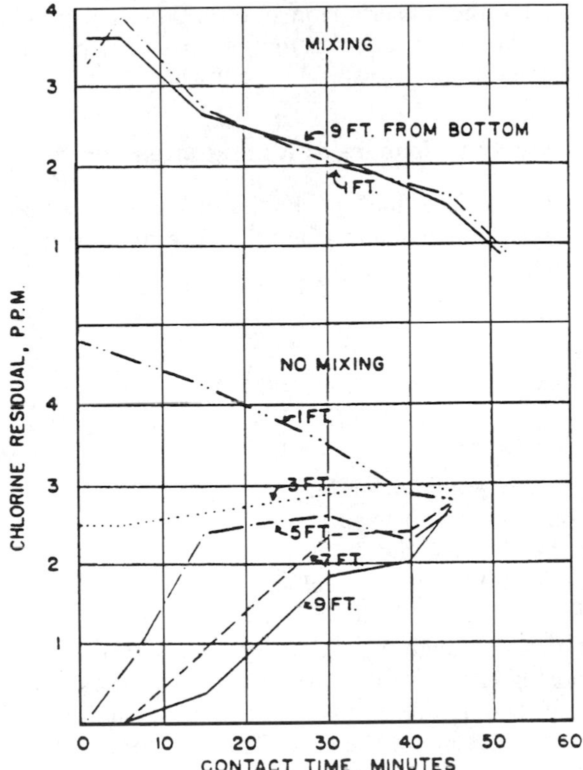

Fig. 8-19. Segregation in a chlorine contact tank: mixing versus no mixing.

has a singular effect on the efficiency of disinfection. These findings were further confirmed by the California State Department of Health mobile pilot plant study reported by Sepp and Bao in 1980.[79] A variety of field investigations between 1970 and 1980, covering some 40 treatment plants, convinced White that initial mixing—its quality and intensity—is an important factor in wastewater disinfection; however, the intensity and the quality of initial mixing are difficult to define.

Calmer et al.[94] reported in 1984 the results of a combined literature search and field study of the initial mixing phenomenon. They concluded that there was little or no benefit to coliform destruction where the mixing intensity was greater than $G = 500$. This indicates a wide disparity in beliefs about what the optimum level of mixing intensity should be. Additional research needs to be performed on operating systems, preferably at full scale, to establish the appropriate level of mixing intensity.

The work by Vrale and Jordan[61] in 1971 is often overlooked. It has an important but almost hidden message. Their work indicated that, although the G factor (velocity gradient) is a useful measure of mixing intensity, it should not be used as the sole criterion, and that the type of mixing may have a profound effect. The implication is that jet mixing is probably superior to other types of mixing using two streams of flowing water.

Velocity Gradient: The Measure of Mixing Intensity

General Discussion. The fluid mechanics of mixing has been represented by the velocity gradient for many years.[85] It is a dimensionless number expressed as follows:

$$G = \sqrt{\frac{P}{\mu V}}$$

where

G = mean velocity gradient, sec^{-1}
P = power requirement, ft lb/sec
V = mixing chamber volume, ft^3
μ = absolute fluid viscosity, lb-sec/ft^2

When water flows through a pipe or along any solid surface or through another fluid moving at a lesser velocity or in an opposite direction *the motion is resisted by drag*. The nature of this drag is fluid shear, and the cause of this phenomenon is the viscosity of the fluid. The work of shear results in *energy dissipation* as heat and is described as "head loss" or friction loss. The velocity gradient is related directly to the total shear per unit volume per unit of time. The G number provides a somewhat better understanding of turbulence and problems involving internal energy dissipation *as it relates to head loss, which in turn relates to mixing*.

In chlorination of wastewater, intensity of mixing is a key factor in disinfection efficiency. Defining the role of the G number as a parameter for disinfection efficiency as it is related to the intensity of mixing is important because experience has demonstrated that it is not the sole factor. The type of mixing is also important (e.g., a jet system or a turbine mixer, etc.).

Current Practices. Various consulting engineering firms in California use G numbers varying from 500 to 1000. The Sanitary Districts of Los Angeles County use a different approach. They rely on turbine mixers rated at 1 hp/1000 gals contained in the mixing compartment. White used to believe that the G number should be approximately 1000 for superior mixing, and that the intensity of mixing required was partly dependent upon the magnitude of coliform destruction required.

Later operating experience (1978–82) indicated that acceptable values of G can be significantly less than 1000 regardless of disinfection requirements.

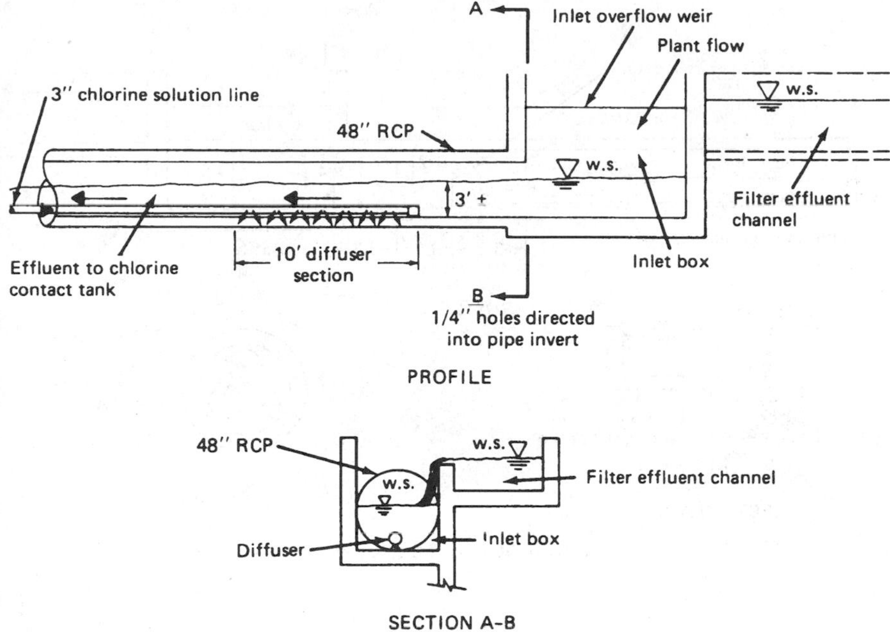

Fig. 8-20. Original Los Angeles–Glendale diffuser.

At the San Jose/Santa Clara WWTP in San Jose, California, turbine mixers provide a G number 475 for a consistent total coliform concentration in the filtered effluent of 2.2/100 ml at 60 min. t_i contact time and 12 mg/L chlorine dose (2 mg/L NH_3–N is added before chlorination to the filtered–nitrified effluent). Another treatment plant discovered there was enough mixing at the point of application due to hydraulic turbulence and jet action of the diffuser that the action of the mechanical mixers was unnecessary; so they were turned off.

Recent Developments. The most compelling example was the performance of a jet-type chlorine diffuser devised by Egan[95,96] in 1978 at the Los Angeles–Glendale water reclamation plant (filtered secondary effluent), Los Angeles, California. Egan retrofitted a perforated horizontal diffuser, shown in Fig. 8-20, to a jet diffuser, shown in Fig. 8-21. The jet system alone contributes a G number of 189. Between baffles B and C, $G = 205$, and the countercurrent effect produces $G = 232$, for a total $G = 626$.* This is known as the Egan Effect. The calculations are based upon reasonable hydraulic assumptions. The countercurrent value of G is based upon the energy dissipation caused by the opposing flows of chlorine solution and wastewater. The total G value

*See the appendix for calculations.

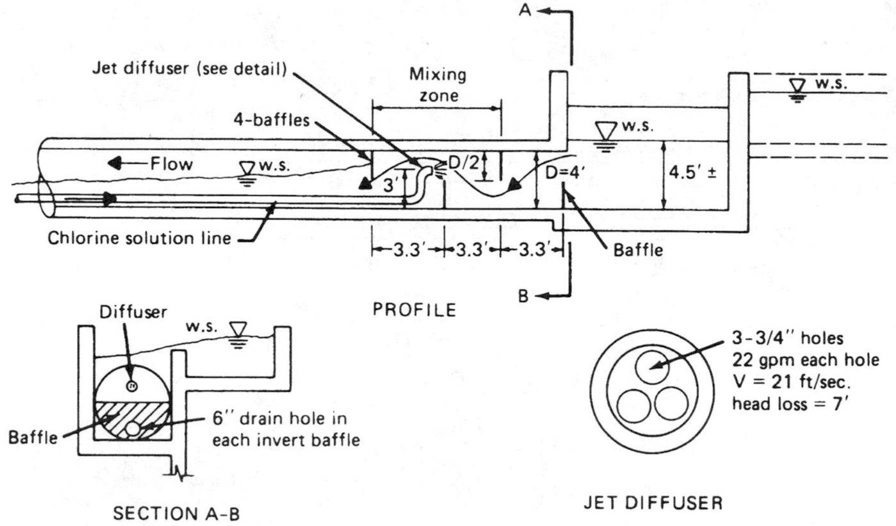

Fig. 8-21. Egan retrofit of Los Angeles–Glendale diffuser.[95,96]

is not surprising because the efficiency of the Egan mixing scheme resulted in consistent compliance with the NPDES requirement of 2.2 MPN/100 ml total coliforms at a chlorine dosage of only 4.8 mg/L (reduced from 10–12 mg/L), which resulted in contact chamber effluent residuals of 0.4–0.8 mg/L after 3 hours of detention. The original diffuser (Fig. 8-20) did not produce consistent compliance with the coliform requirement at elevated chlorine dosages of 10–12 mg/L.[112] The disinfection efficiency of the jet diffuser and baffles corresponds well to the Selleck–Collins model.

A dye study was performed, which demonstrated the ability of the mixing system to produce a homogeneous mixture of chlorine solution and wastewater. Therefore, it follows that intensity of initial mixing may not need to be as high as $G = 1000$, regardless of disinfection requirements.

The subject of mixing intensity as it relates to disinfection efficiency needs more research and laboratory study of the fluid mechanics of mixing chlorine with wastewater. The objective would be to quantify intensity versus homogeneity of the mixture in a given time frame.

Diffusers as Mixers

General Discussion. The notion that a chlorine solution diffuser could be designed to provide adequate initial mixing using the velocity gradient as the principal parameter was originated by Egan at the Los Angeles–Glendale Reclamation Plant ca. 1976.[96] The validity of such a notion is supported by the results outlined above.

The jet diffuser shown in Fig. 8-21 has a three-port nozzle through 67 gpm chlorine solution flows. At 22 gpm per ¾-inch port, the calculated head loss is 4.03 ft. This calculates to $G = 189$, which in itself is not impressive.

The most prevalent situation for diffuser location is in an open-channel-type structure. There are some situations where the diffuser can be located in a surcharged conduit. However, designing a diffuser for both of these situations requires the use of the same principles of fluid mechanics plus the phenomenon of the Egan effect.

Designing a Grid-Type Open-Channel Diffuser. First, choose a structure location upstream but adjacent to the contact chamber where the average daily flow is on the order of 1.0–1.25 ft/sec. Second, select a target G number, such as 500. Third, assume a head loss due to the jet action of the diffuser of from 7–10 ft. Fourth assume that the injector is adjacent to the mixing chamber.

A typical grid diffuser is shown in Fig. 8-22. The following are the calculations for this hypothetical diffuser:

EXAMPLE: Assume wastewater flow 10 mgd, and flow in channel to contact chamber = 0.75 ft/sec. Also assume mixing occurs in 3 sec. At 0.75 ft/sec this horizontal area of mixing is $L_m = 2.25$ ft.

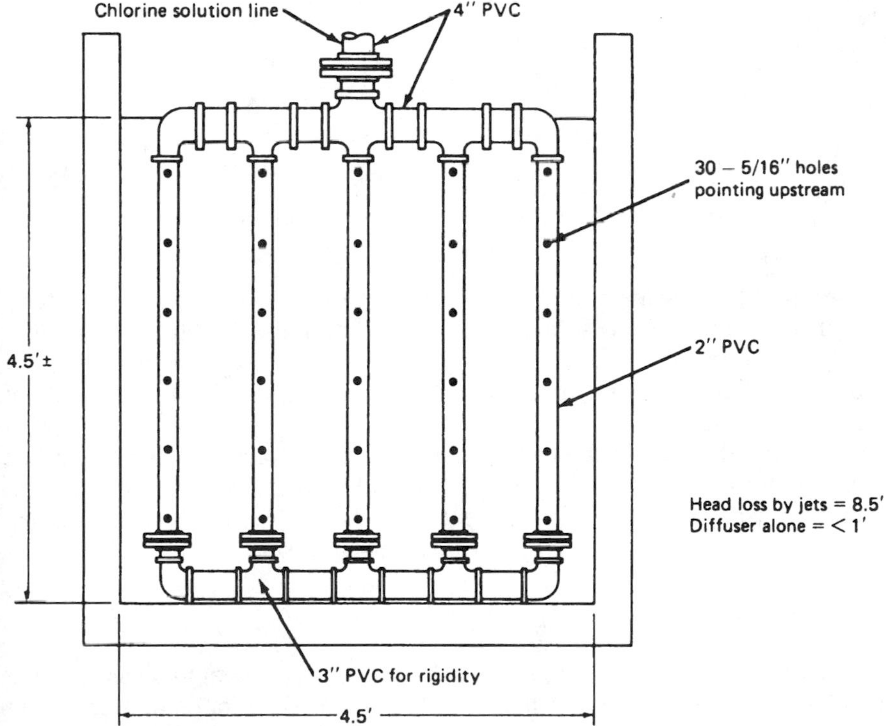

Fig. 8-22. Open-channel grid diffuser based upon the Egan concept (courtesy of Chlorine Specialties Inc.).

From the velocity gradient formula:

$$G = \sqrt{\frac{P}{2.5 \times 10^{-5} \times V}} \tag{8-13}$$

$$P = \frac{Q \times h}{*3960 \times \text{eff.}}$$

where

Q = total flow through diffuser, gpm
h = loss in diffuser orifice, ft
eff. = 1.0

Assume that plant flow is 10 mgd and chlorine dosage is 15 mg/L. This requires about 1250 lb/day chlorine; therefore, a 2-inch adjustable injector could be used. However, a mixer-diffuser needs to use as much water as is practical; so a 3-inch injector is preferable. Use a 5⁄16-inch-diameter hole and 4 gpm per hole. This requires an injector operating water supply of 120 gpm. This size hole produces a head loss h = 8.5 ft. Therefore, the mixing power generated by the diffuser in Fig. 8-22 is:

$$P = \frac{120 \times 8.5}{3960 \times 1.0} = 0.26$$

$$V = 3 \times 3 \times 2.25 = 20.25$$

$$G = \sqrt{\frac{550 \times 0.26}{2.5 \times 10^{-5} \times 20.25}} = 548.18$$

Therefore, a chlorine solution diffuser that exerts no more than 10 ft of friction loss in the chlorine solution line is well within the limits of acceptability for injector operation. The major difference between the previous concept of diffusers and the one shown in Fig. 8-22 is having double the amount of injector water flow.

Restricting the flow in the channel by the diffuser plus the countercurrent effect (perforations directed upstream) will add about 200 to the G number. Therefore it appears that achievement of G = 500+ is practical and much less energy-intensive than the use of mechanical mixers.

Mechanical Mixers

Foreword. When it comes to mixing chlorine in solution or as a gas, nothing can compete with the Water Champ, unless it would be in a situation where

*One hp is the amount of work necessary to lift 3960 gal of water one foot in one minute.

an extremely large quantity of a hypochlorite solution was being used to chlorinate a wastewater effluent.

Discussion. Mechanical mixers have been widely used in potable water treatment and more recently for mixing chlorine solution in wastewater. Ideally the mixing device should be able to homogenize the chlorine solution and the wastewater in a fraction of a second. The Mixing Equipment Company of Rochester, New York claims that it is practical to provide complete mixing within 1–3 sec (regardless of flow), depending upon how much power input is allowable.[98] Such a mixer is illustrated in Fig. 8-23.

This arbitrary mixing time is for either closed conduit or open channel flow using propeller mixers. Based upon a comparison of existing installations with various types of mixing, a complete mix in 3 sec would be rated as adequate. Figure 8-24 illustrates the use of a mechanical mixer in open channel flow. Sizing procedures for these mechanical mixers are proprietary secrets, but for general guidelines the Mixing Equipment Company advises that it sizes mixers for dispersing chlorine solution in a flowing stream on the basis of 0.3–0.6 hp/mgd flow. The mixing basins are sized for theoretical retention times of 5–15 sec. This information is helpful for the design of an optimum velocity gradient.

Let us take a condition of 5 mgd flow with 10 sec theoretical detention time in the mixing chamber, assuming a 3 hp mixer as the design parameter, which is the upper limit of design quoted by the manufacturers. Then G, the mean velocity gradient, is:

Fig. 8-23. Propeller mixer (courtesy Link Belt Co.).

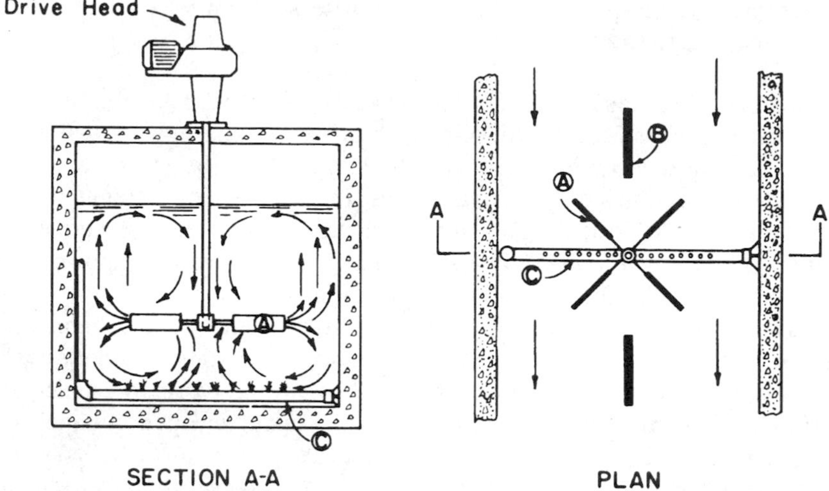

Fig. 8-24. Propeller mixer in open channel flow showing location of baffles. A = radial flow impeller, B = baffles, and C = chlorine diffuser (courtesy Mixing Equipment Co., Rochester, New York).

$$G = \sqrt{\frac{3 \times 550}{2.735 \times 10^{-5} \times 77.37}} = 948$$

Hydraulic Mixers

Closed Conduits. Mechanical mixers for closed conduits running full are not necessary, provided that certain design criteria are followed. White found by hydraulic laboratory dye studies and a plant-scale field demonstration that if the chlorine were injected at the center of the flow, the chlorine solution would be completely mixed in 10 diameters of pipe length, provided that the Reynolds number was 2000 or greater.[98] This rule of thumb is valid for pipe sizes up to 30 inches. The dosing and sampling arrangement is illustrated in Fig. 8-25. If the pipe is larger than 30 inches in diameter, use the across-the-pipe diffuser illustrated in Chapter 9. Perforate the diffuser across the middle one-third of the diameter. Make a similar device for the analyzer sampling probe. Whenever possible, use the chlorinator injector pump to provide the analyzer sample flow. This serves to shorten appreciably the loop time for residual control systems. This arrangement can be used when it is possible to utilize a short run of surcharged pipe immediately upstream from the contact chamber. The designers should always be alert and innovative to seize the opportunity to capitalize on hydraulic turbulence for mixing chlorine and wastewater.

Static Mixers. These can be used to advantage where 10 diameters of pipe length are not available. Figure 8-26 illustrates the concept of a static mixer.

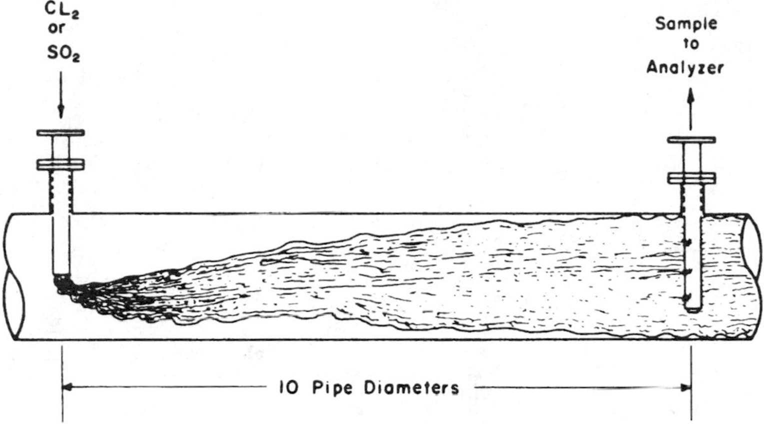

Fig. 8-25. Pipeline as a mixing device.

The mixer elements are made to fit inside pipe in lengths equal to 1.5D. The manufacturer's rule of thumb for the number of elements required is based upon local conditions of velocity. For flows 2 ft/sec or greater two elements are recommended. For flows that go as low as 0.5 ft/sec, four elements are required to produce proper mixing. The G factor can be calculated for each situation but it is not the principal criterion.

Hydraulic Jump. This device was first used as a chemical mixing device in a potable water treatment plant in 1927.[99] Figure 8-27 illustrates a typical hydraulic jump. The diffuser should be located as shown. This is the preferred location even though it may cause some chlorine odors in the turbulent zone. Placing the diffuser in the turbulent zone does not provide an equivalent mix. Moreover, since the turbulent zone moves laterally with flow changes, it is almost impossible to choose an optimum location. The hydraulic jump should

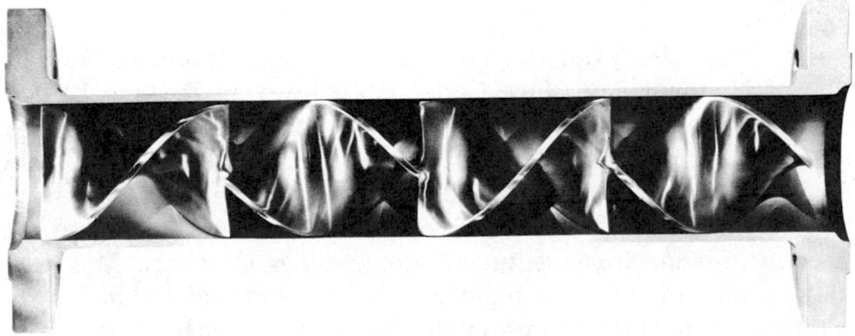

Fig. 8-26. Static mixer (courtesy Kenics Corp., Danvers, Massachusetts).

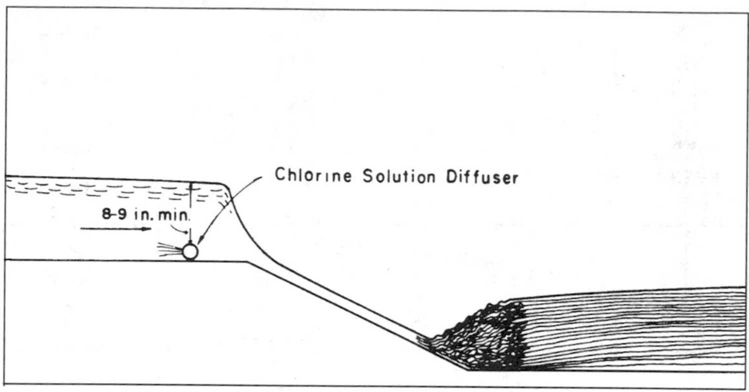

Fig. 8-27. The hydraulic jump as a mixing device.

be designed to provide at least 9 inches of cover over the diffuser. The diffuser perforations should be pointing upstream, and the velocity through these holes should be approximately 22 ft/sec. If properly designed, a hydraulic jump can be an excellent mixing device, but it requires about a 2 ft fall as shown to provide the necessary turbulence.

The Parshall flume and the Palmer–Bowlus flume should not be considered as mixing devices. They do not generate sufficient turbulence.

A partially closed sluice gate or other type of gate valve placed immediately downstream from the chlorine diffuser can develop a turbulent regime sufficient to achieve adequate mixing.

This type of mixing device has performed as well as or better than mechanical mixers in documented cases.[94] The chlorine diffuser should be placed immediately upstream from the sluice gate. If a 36-inch sluice gate is throttled to provide a 6-inch head loss, this will produce a G value of 614, assuming a 10 mgd flow and a 3-sec. mixing time. If the diffuser is designed for $G = 200$, the total G will be more than adequate for good mixing.

The Pentech System

This system was made available for wastewater chlorination by the Pentech Division of Houdaille Ind. Inc., Cedar Falls, Iowa, in the mid-1970s. The system is illustrated in Fig. 8-28. It utilizes either the direct injection of chlorine gas under a vacuum or chlorine solution from the discharge of a conventional chlorinator. When chlorine gas is injected directly, the pump shown becomes the "injector pump," and the jet nozzle assembly takes the place of the injector.

The entire effluent stream is forced through the reactor tube. This provides plug flow in a highly turbulent regime. This system provides almost instantaneous dispersion of chlorine throughout the mass of effluent. Pentech claims this mixing system will generate a G number as high as 10,000, and that total

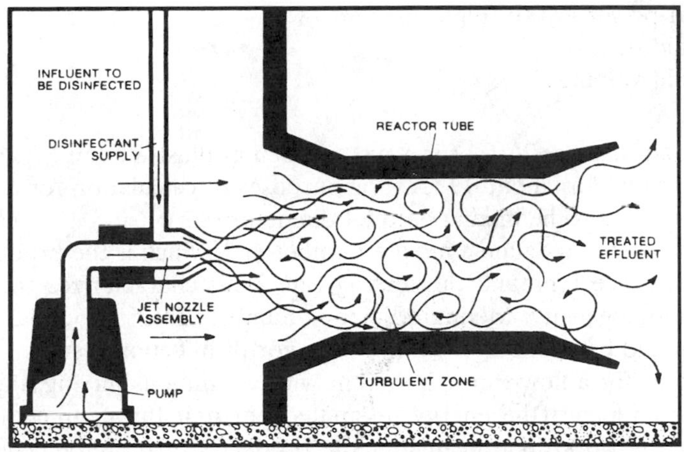

Fig. 8-28. Pentech injector mixer (courtesy Pentch-Hondaille).

mixing occurs in about 0.2 sec with the ability to cope with variable flow rates and to provide a system hydraulic gradient that does not introduce a significant head loss.[100]

Pentech recommends a minimum energy dissipation rate to be maintained so that the turbulent mixing zone will have a mixing rate t^{-1} ($\sec^{-1}$) of somewhere between 10 $\sec^{-1}$ and 5 $\sec^{-1}$, depending upon the size of the system, corresponding to mixing residence times of 1.0 sec or less. These limits require mixing in times far less than those described above for other systems. In accordance with fluid dynamic principles, the mixing rate t^{-1} is directly related to the specific turbulent energy dissipation rate in the turbulent mixing zone and is inversely related to the square of the scalar macroscale, L_s, of the turbulence structure of the mixing zone as follows:

$$t^{-1} = K(e/L_s^2)^{1/3} \tag{8-14}$$

where

e = specific turbulent energy dissipation rate
L_s = scalar macroscale
K = constant, which is 0.489 for the cgs system of measurement

The specific turbulent energy dissipation rate e is further defined as:

$$e = \frac{P}{pV} \tag{8-15}$$

where

P = net power lost to fluid
p = fluid density
V = fluid volume

The scalar macroscale L_s, for a system such as illustrated in Fig. 8-28, may be approximated as about $0.131D$ for purposes of calculation for equipment design, where D is the mixing parallel diameter (cm).

It is also useful to define a mixing number θt^{-1}, which is the product of the mixing residence time and the mixing rate. This characterizes the product stream inhomogeneity. Mixing efficiency numbers from about 1.5 to 15 or greater should be applied to achieve superior disinfection results.

Moreover, for a flow-through system with continuous mixing, the specific energy requirement (the energy dissipated per unit throughput of product stream or the work done in mixing the product stream) should be at least about 0.2 hp/mgd of treated effluent. For a given level of mixing, the energy requirement will increase with increasing values of L_s, but will generally be in the range of from about 0.2 hp/mgd to 3 hp/mgd of effluent to be treated.

The average residence time θ for the mixer illustrated in Fig. 8-28 may be readily determined from the volume of the turbulent mixing cone. The effluent and the disinfectant may generally be assumed to be mixed to within acceptable disinfectant concentration gradient limits upon reaching a point adjacent to the base of the mixing cone at its intersection with the mixing parallel. The volume V of the mixing cone thus defined may be calculated as follows:

$$V = \frac{D^3}{24 \tan(\alpha/2)} \tag{8-16}$$

where

D = diameter (cm) of the intercepting conduit at the point of intersection with the mixing cone
α = included angle of the mixing cone

By way of example, a 5 mgd flow is to be treated utilizing 5 hp input and the jet mixer shown in Fig. 8-28. The mixer parallel diameter is 20 in. × 2.54 = 50.8 cm and $\alpha = 28°$. Therefore, the mixing rate t^1, the residence time θ, and the energy requirement are calculated as follows:

$$V = \frac{D^3}{24 \tan(\alpha/2)} = \frac{(50.8)^3}{24 \tan 14°} = 68{,}827 \text{ cm}^3 = 2.43 \text{ ft}^3$$

$$\theta = \frac{2.43 \text{ ft}^3}{5 \text{ mgd} \times 1.55 \text{ cfs/mgd}} = 0.31 \text{ sec}$$

$$e = \frac{P}{pV} = \frac{(5)(550)(30.48)(454.5)(980)}{(1)(68,827)} = 542,436 \text{ cm}^2/\text{cm}^3$$

$$L_s = (0.131)(20)(2.54) = 6.65 \text{ cm}$$

$$t^1 = 0.489 \left[\frac{542,436 \text{ cm}^2}{44.22 \text{ cm}^2 \text{ sec}^3}\right]^{1/3} = 11.28 \text{ sec}^{-1}$$

$$e = \frac{5 \text{ hp}}{5 \text{ mgd}} = 1 \text{ hp/mgd}$$

Unfortunately this system has not been rigorously evaluated, so no conclusions can be drawn about its practicality or cost-effectiveness.

Fischer and Porter System[101]

This is what might be termed a modified Pentech system that requires a specially designed mixing chamber. The complete system is illustrated in Fig. 8-29. In operation a small portion of the wastewater to be chlorinated is

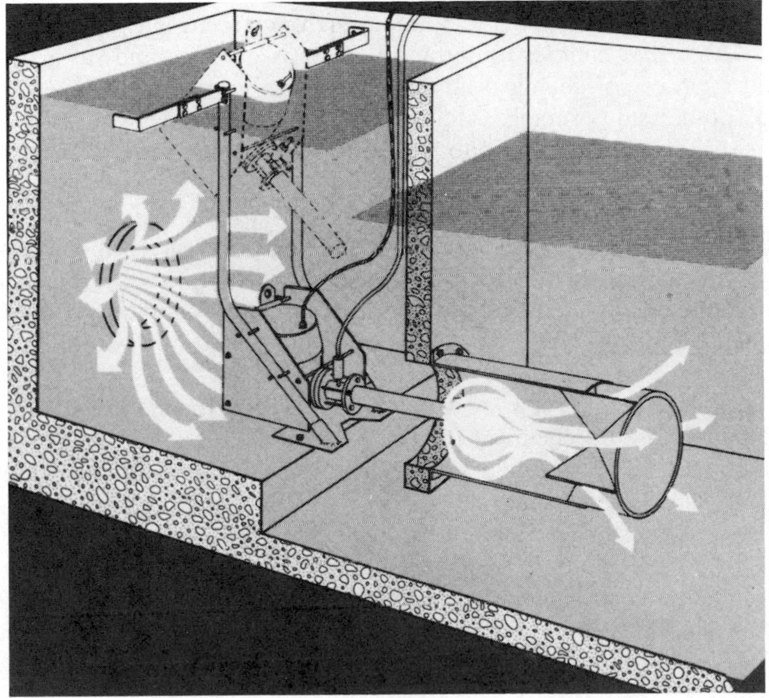

Fig. 8-29. Molecular chlorine–wastewater mixing device (courtesy Bailey-Fischer and Porter Co.)

pumped to the ejector-diffuser assembly by a submersible pump. This flow creates a vacuum that pulls chlorine gas into the diffuser portion of the assembly, where it is discharged under pressure into the turbulent mixing zone in the fiberglass reactor tube. This tube is mounted in a baffle wall in the mixing chamber, which seals the channel and directs all the remaining wastewater flow through the tube, where it is mixed with the chlorine. Rapid kills are claimed because the chlorine is in its molecular form. There is no current documentation to substantiate a claim for better kills by this method than for the chlorine solution method.

Although a constant-volume pump is used to provide the power to maintain the chlorine gas flow, the wastewater flow through the reactor tube can vary within the usual diurnal ranges encountered in most wastewater plants. The chlorination equipment can be controlled by the usual conventional methods of flow spacing and residual control.

These systems are designed to produce a turbulence of G = approx. 1650. They are most popular for treatment plants in the flow range of 0.5 to 3 mgd, but can be designed for much larger treatment plants.

Capitol Controls Co.: Chlor-A-Vac™ Series 1420 Chemical Induction Unit

Introduction. This is a self-contained, high-powered, and highly efficient mixing device that handles both vapors, such as chlorine and sulfur dioxide gas, and a variety of chemical liquids, such as sodium hypochlorite, sodium bisulfite, potassium permanganate, and so on. This unit has worldwide coverage, largely because Capitol Controls Co. is owned by the Severn-Trent Water Co. of Great Britain.

The Induction Unit. The Chlor-A-Vac is ideal for use in all-vacuum gas feed systems. It replaces the commonly used water or wastewater operated chlorinator injector systems. It also eliminates the necessity to have a diffuser or mechanical mixer. The series 1420 unit is illustrated in Figs. 8-30a, 8-30b, and 8-30c.

It is available in eight separate capacities to meet all the major requirements of gaseous and liquid chemical feed rates, plus a Chlor-A-Vac Mosquito®, designed to meet the specific needs of small plants. The Mosquito® unit can feed up to 100 lb/day of chlorine gas or 4 gpm of a liquid chemical.

The seven series 1420 Chlor-A-Vac units, cover a range of 750 to 10,000 lb/day of chlorine, and 27 to 65 gpm of liquid chemicals. Literature describing these units and a variety of installation diagrams are available from Capital Controls, which can also provide operating information from existing installations.

Chemical Reactions. As shown in the flow diagram (Fig. 8-30c) a small part of the process water (water or wastewater) flows through an injector

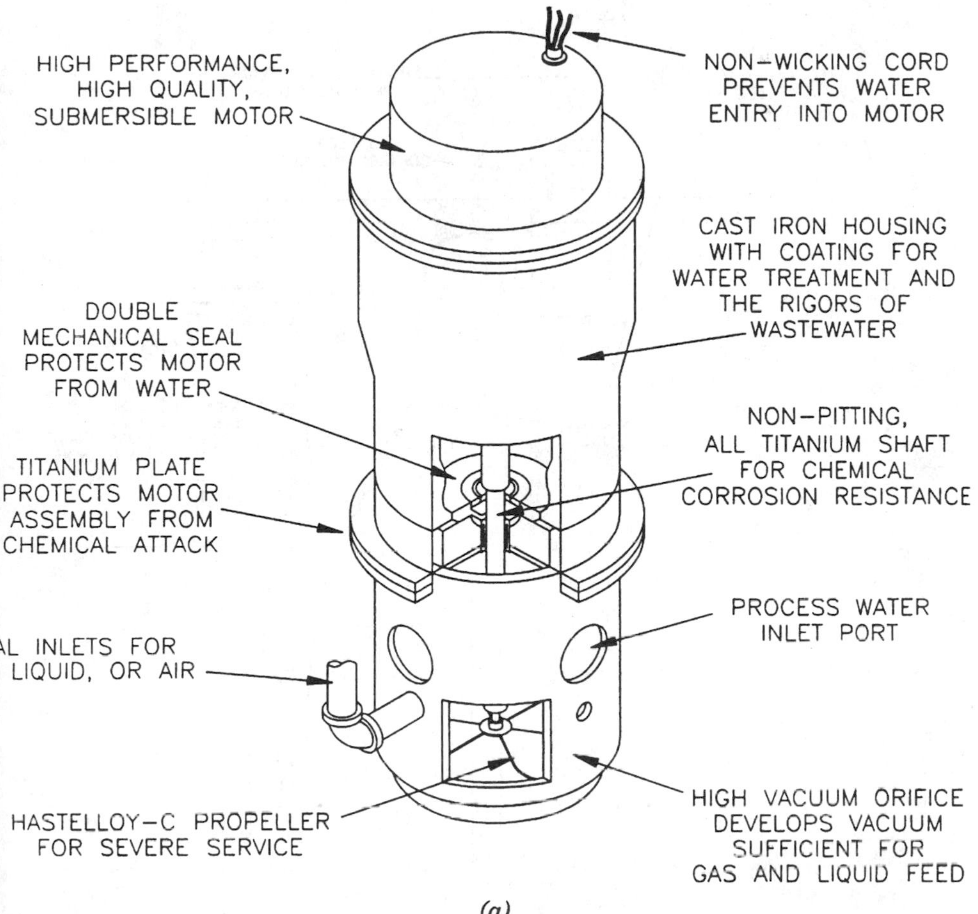

Fig. 8-30a. Chlor-A-Vac series 1420 induction unit (courtesy Capital Control Co.).

which creates the vacuum that pulls the chlorine, or liquid chemical, into the process water. Therefore, where the addition of chlorine is used for disinfection, the reaction with the process water is primarily the formation of free chlorine, which is HOCl, not Cl_2. This then makes the oxidative power of the Chlor-A-Vac totally dependent upon the ORP of HOCl, which is dependent upon pH, whereas the Water Champ has a higher oxidative power because it draws the chlorine gas directly into all the process water.

The ORP of 100 percent HOCl is 1.49 V, whereas that of Cl_2 is 1.36 V, and that of HOCl at pH 7.5 is 0.89 V. This gives the Water Champ a much higher overall ORP than the Chlor-A-Vac. In spite of all this, the Chlor-A-Vac is an excellent mixing device because Capital Controls has improved its overall efficiency. This is a new device that out-performs everything in the mixer market.

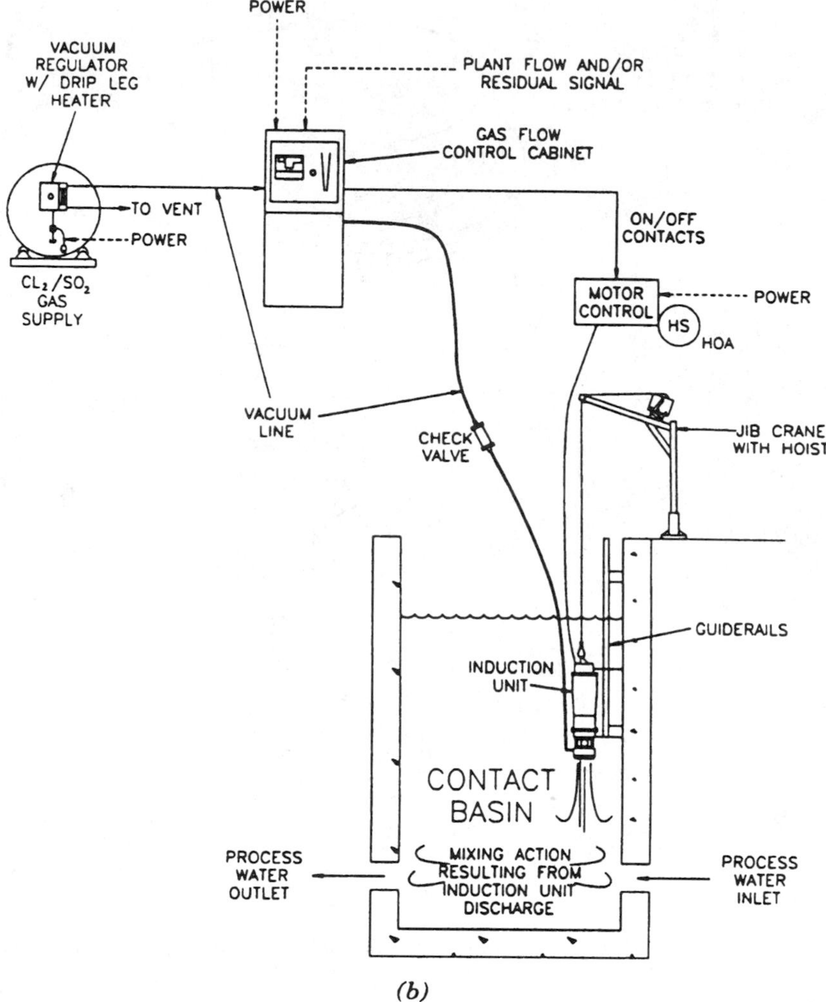

Fig. 8-30b. Chlor-A-Vac series 1420 typical chlorination system installation (courtesy Capital Control Co.).

The Water Champ

Historical Background. This device was invented because the inventor, Jack Gardiner[115] of Houston, Texas, was involved in attempting to eliminate the necessity of using potable water to operate all the prechlorinators at Texas wastewater primary treatment plants. This was happening in the late 1970s and early 1980s. The only chlorination at these plants was prechlorination for odor control, which in itself was a real problem in Texas, owing to long flat sewer lines that caused an enormous buildup of hydrogen sulfide. As these were only primary treatment plants, it was impossible to use the effluent for

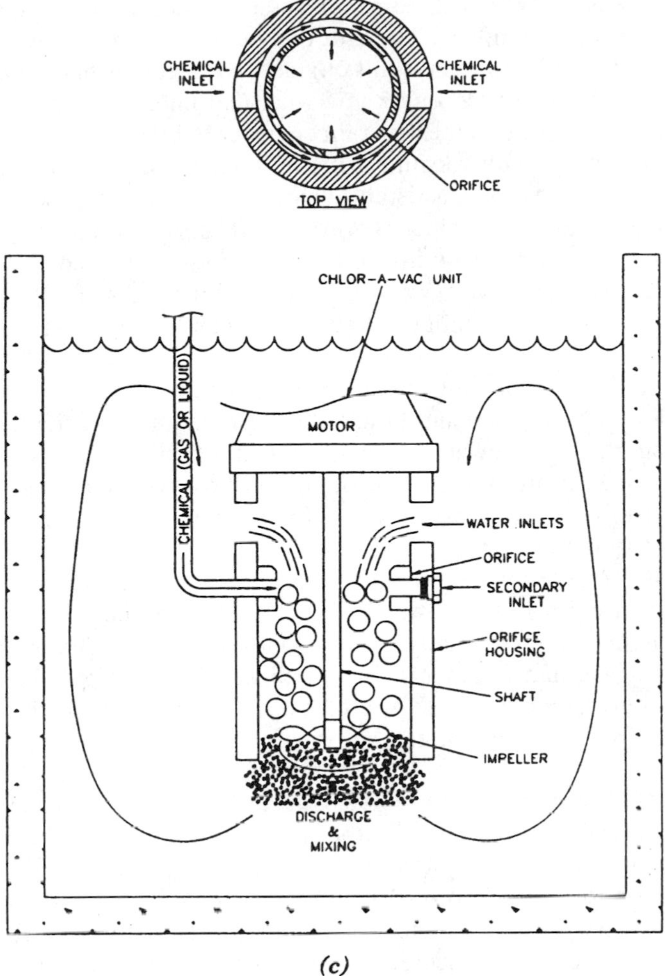

Fig. 8-30c. Chlor-A-Vac series 1420 flow diagram of induction unit (courtesy Capital Control Co.).

the chlorinator injector's water supply. Gardiner solved that problem, and that is why he named the device the Water Champ.

As of 1997, the W.C. remains unique, as no one has been able to copy it—and White found out why, after Gardiner hired him as a consultant in 1986. White could not figure out exactly how the W.C. could perform, even though he had a complete set of full-scale drawings. So he took these drawings to a well-known university professor who specialized in hydraulics. After the professor had studied the drawings thoroughly, he advised that for this unit to work, part of the wastewater to be treated would have to pass through an injector, in order to generate a proper vacuum level to withdraw the chlorine gas from the chlorinator.

During this period of time, Gardiner's assistant engineer quit the company and started his own company. The unit that he built was the only competition for the W.C. However, this assistant engineer to Gardiner had no more information than White had; no one knew what Gardiner knew because he told no one. Therefore, the unit the assistant engineer built had an auxiliary injector that maintained a steady vacuum using an injector that was being operated by an external injector. The assistant engineer's company was sold after a few years, and Gardiner's ex-engineer severed his relations with the new company, which is operating as a competitor to the W.C. but without the ability to apply the much higher oxidative power of all the chlorine gas (Cl_2) to the entire flow of plant effluent. This unit is now being sold under the name "Chlorovac."

Features. This unique and revolutionary injection system is now used for the injection of both gases and liquid chemicals required in the treatment of potable water and wastewater. The savings that occurred in the elimination of the use of freshwater for the chlorinator injector system are notable; it will be shown later that the greatest saving that accumulates from this device is *energy,* due to the higher oxidative power of Cl_2.

This unit is revolutionary because it eliminates not only the injector water supply system but also the diffuser assembly and the need for a mechanical mixer. (The injector system also was revolutionary when Charles F. Wallace of Wallace & Tiernan Company introduced vacuum-operated chlorinators in 1920.[116]) Another important feature of this device is that it can be retrofitted to any conventional vacuum feed chlorinator.

Principle of Operation. The Water Champ is part of the dynamic triad of any vacuum feed chlorinator, the other two parts of the triad (delta system) being the chlorine supply system and the metering and control equipment. The device consists of a motor-driven open impeller that creates a vacuum in a precisely structured chamber surrounding the impeller. The way the vacuum is created is a trade secret. This vacuum is transmitted to the chlorinator in a PVC pipe similar to that used in all conventional remote injector systems.

The internal mechanism that creates the vacuum by the Water Champ is fully illustrated in Fig. 8-31a.

The chlorine gas, or other gas or liquid, is injected directly into the process stream without any dilution water. This particular feature is what makes this device unique when compared to other competitive injection systems. In the absence of dilution water it has been found that, in the case of chlorine gas, the resulting reaction with the process stream is more complete, and therefore the results are chemically superior. This has been observed at many installations. For equal chlorine dosages, more chlorine is consumed in the plant effluent, with significantly better disinfection efficiency. This reduces the amount of chlorine required for disinfection as well as that for the SO_2 required for dechlorination.

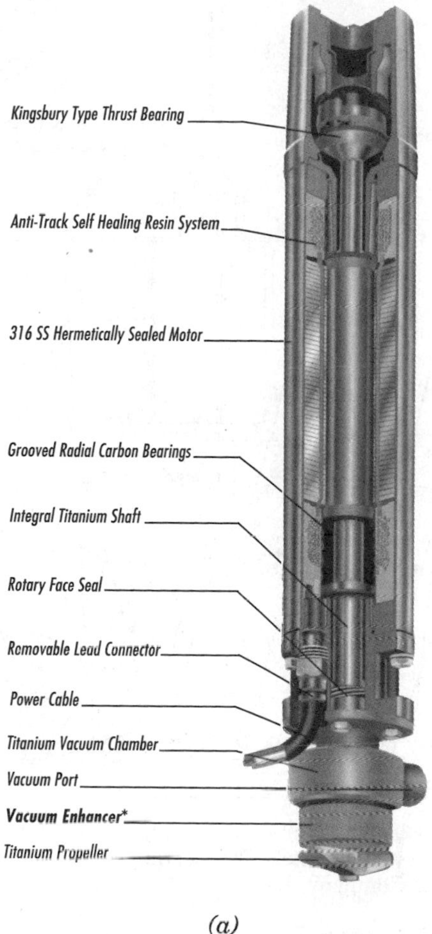

Fig. 8-31a. Water Champ vacuum induction system (courtesy Gardiner Equipment Co., Inc.). *Patented vacuum induction system.

The rotating speed of the impeller is 3450 rpm, which blasts the chemical into the process stream at 60 ft/sec. This velocity component is downward toward the bottom of the mixing chamber. This downward velocity component is sufficient to create a major vertical velocity component after striking the bottom of the chamber. (See Fig. 8-31.) This energy release provides a thorough mix in seconds over the entire width and depth of the channel. Figure 8-32 illustrates how the process flow pattern converts part of the upward velocity component into a horizontal component. This produces a scouring action in the bottom of the process channel. This is important in wastewater disinfection because it reduces significantly the deposition of the suspended solids that always appear in chlorine contact chambers. Referring again to

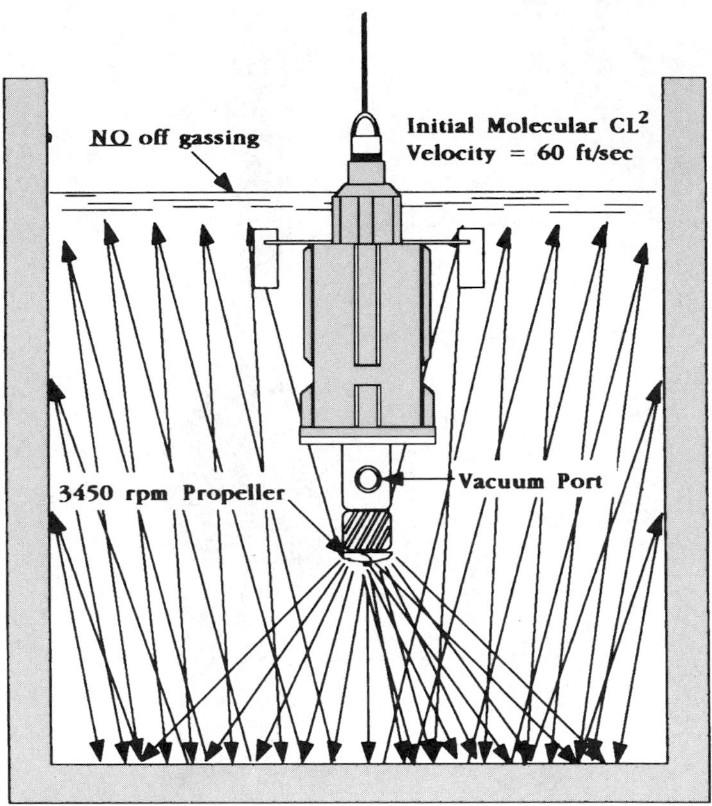

Fig. 8-31. The Water Champ. A submersible vacuum-type molecular chlorine vapor induction unit showing open channel cross section of diffusion and mixing (courtesy of Gardiner Equipment Company, Inc.).

Figs. 8-31 and 8-32, this device is only furnished with submersible motors. It is also to be noted that the mixing is so rapid and thorough that there is no off-gassing at the point of application.

Chemical Reactions. The most important effect of the rapid and thorough mix provided by the Water Champ is the greatly improved reaction between chlorine and the wastewater. It has revealed three different significant reactions, all of which are explained by the oxidation reduction potential (ORP) of both Cl_2 and HOCl.

The ORP of Cl_2 is 1.36 V, regardless of pH and that for 100 percent of HOCl is 1.49 V. At pH 7.5, the ORP of HOCl is only 60 percent of 1.49 = 0.894 V at 41°F. This illustrates precisely why the Water Champ can perform as it does.

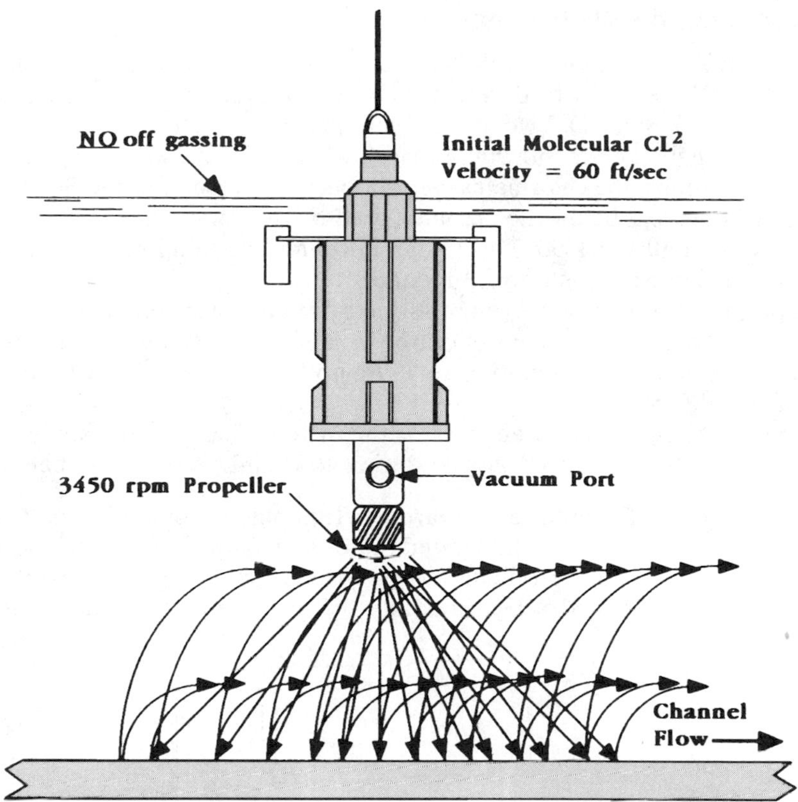

Fig. 8-32. The Water Champ. A submersible vacuum-type molecular chlorine vapor induction unit showing open channel profile of diffusion and mixing (courtesy of Gardiner Equipment Co., Inc.).

Specific Advantages of the Water Champ

1. The equivalent dosages of both the Water Champ and a conventional chlorine solution injection system will result in an equivalent bacteria kill but with a 1 to 1.5 mg/L lower residual for the Water Champ. This translates into equivalent chemical savings in the dechlorination process. This fact alone would verify that the reaction by molecular chlorine is more complete than that of either HOCl or the chloramines.

2. Nitrites can be destroyed if and when they occur in the presence of ammonia levels high enough to prevent the presence of free chlorine residuals. Nitrites cannot be oxidized by chloramines. Only free chlorine (HOCl) at a 5 to 1 chlorine-to-nitrogen ratio or an instantaneous contact with molecular chlorine (Cl_2) can oxidize nitrites to nitrates. This phenomenon indicates that with the Water Champ the molecular chlorine survives long enough to eliminate the nitrites before the molecular chlorine is converted to chloramines.

Superiority of the Water Champ

The South Bend, Indiana Installation. This is a 48 mgd activated sludge WWTP that is required to disinfect its effluent seven months of each year (April 1 to November 1). Two W.C.'s were installed, one for Cl_2 and one for SO_2. In the first seven-month run, in 1994, when the W.C.'s had replaced the competitors' units, the chlorine usage dropped 32 percent and the SO_2 usage dropped 24 percent below the amounts used by the competitors' units in the same seven months of 1993. This saving in chemical cost allowed the City to pay for this equipment in only 13 months.

These sort of results have been reported by many other users under a large variety of situations. The plant operator in this case reminded us that in the entire seven months of operation in 1994, both units operated without one mechanical difficulty.

The most comprehensive analysis of the Water Champ is the study done by the City of Sunnyvale, California during 1987 and 1988.[119] (See below.)

New Installation Techniques. Gardiner Equipment Company has sold so many of these units, both in the United States and abroad, that they have had to solve many installation problems. Three different situations are illustrated in Figures 8-33a, 8-33b, and 8-33c. Probably the most impressive one is Fig. 8-

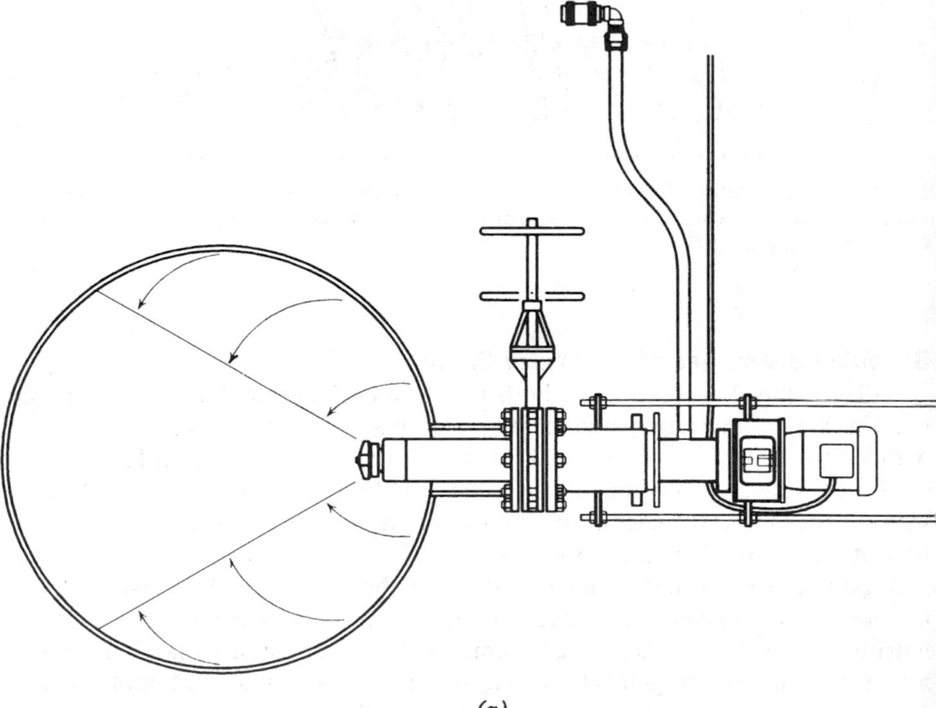

(a)

Fig. 8-33a. Water Champ installation full-flowing transmission pipeline (courtesy Gardiner Equipment Co.).

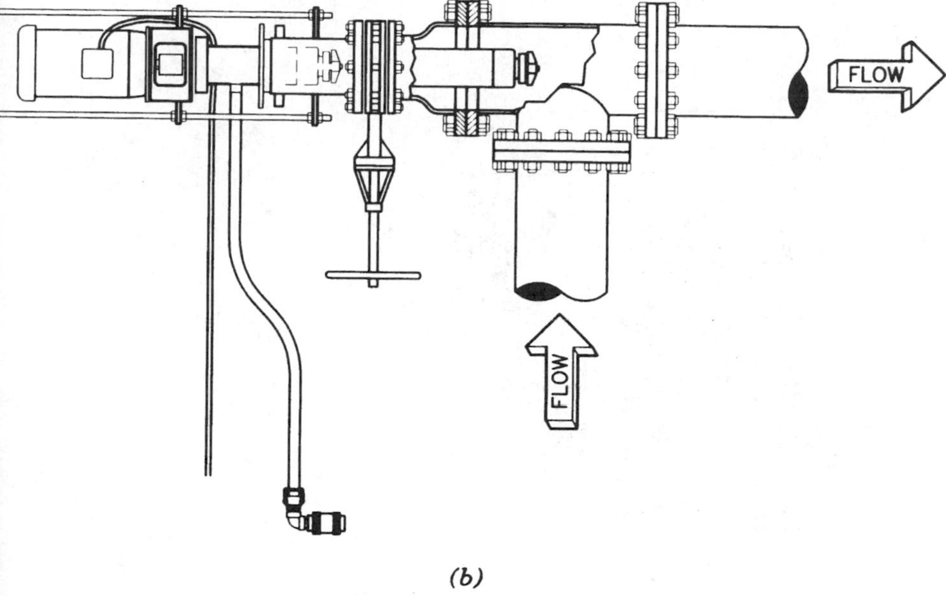

(b)

Fig. 8-33b. Water Champ installation full-flowing pipe in distribution system (courtesy Gardiner Equipment Co.).

33a, which shows how to install a unit in a large pipe flowing full. One of the first of these was for the Thames Water Company in London, England. This solved a persistent residual control problem in a large distribution system pipeline. The energy produced by the W.C. is sufficient to result in a perfect mix of the chlorine applied.

Segregation Phenomenon

Definition. This phenomenon[117] is an extremely complex chemical relationship that exists during the mixing of two fluids, with the chemical interaction being the dominant factor. In view of the importance of optimum mixing between a disinfectant and a wastewater stream to provide maximum disinfection efficiency, Selleck[118] raised the question of the optimum concentration of the disinfectant stream to be applied to the wastewater flow. Simply stated, the segregation phenomenon implies that mixing becomes most efficient between two solutions (a binary mixture) when the solution (chlorine or other chemical) to be mixed with a process flow (wastewater) is of the highest practical concentration. This would imply that a 15 percent trade strength hypochlorite solution (150,000 mg/L) would mix more efficiently than a chlorine solution discharge from a chlorinator (3500 mg/L, max.). Furthermore, it would follow that the mixing efficiency of the Water Champ would be made even better. The segregation factors of each of the above methods of chlorine addition are analyzed below.

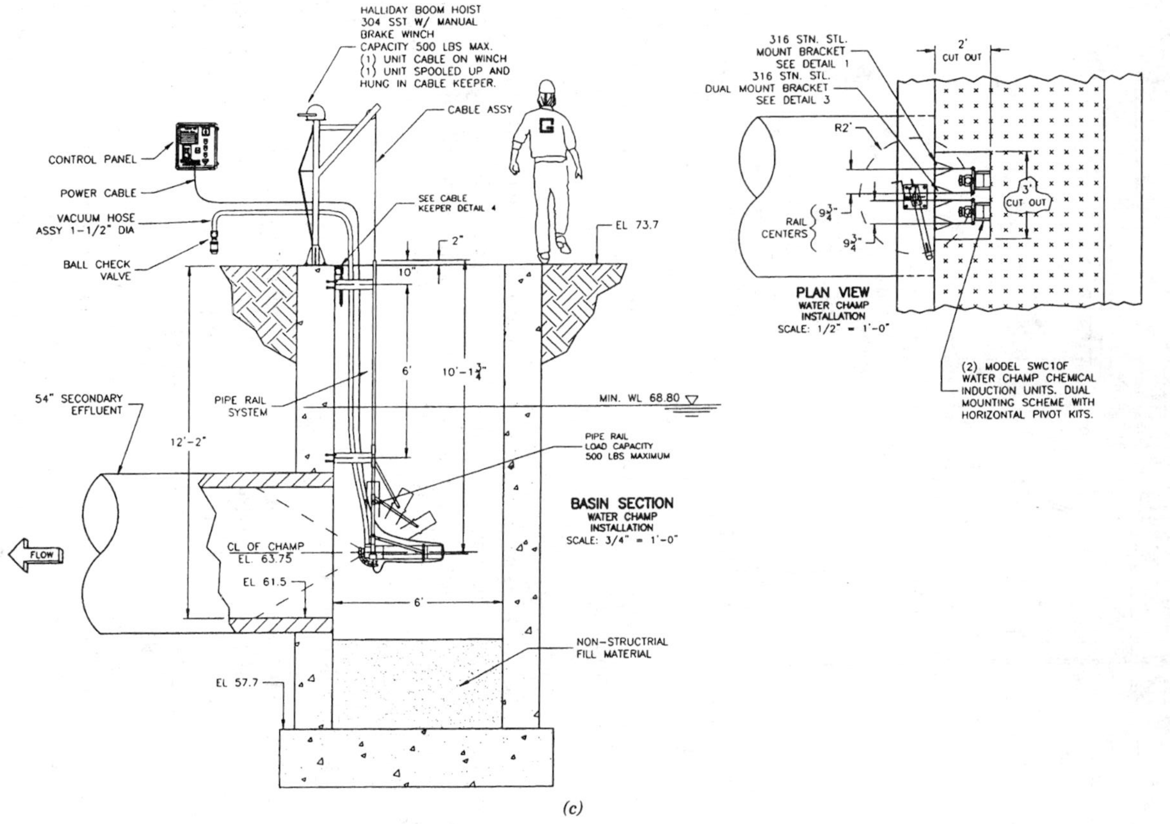
HALLIDAY BOOM HOIST
304 SST W/ MANUAL
BRAKE WINCH
CAPACITY 500 LBS MAX.
(1) UNIT CABLE ON WINCH
(1) UNIT SPOOLED UP AND
HUNG IN CABLE KEEPER.
CABLE ASSY
CONTROL PANEL
POWER CABLE
VACUUM HOSE
ASSY 1-1/2" DIA
BALL CHECK
VALVE
SEE CABLE
KEEPER DETAIL 4
2"
10"
EL 73.7
6'
10'-1 3/4"
54" SECONDARY
EFFLUENT
PIPE RAIL
SYSTEM
MIN. WL 68.80
12'-2"
PIPE RAIL
LOAD CAPACITY
500 LBS MAXIMUM
BASIN SECTION
WATER CHAMP
INSTALLATION
SCALE: 3/4" = 1'-0"
FLOW
CL OF CHAMP
EL. 63.75
EL 61.5
6'
NON-STRUCTRIAL
FILL MATERIAL
EL 57.7
316 STN. STL.
MOUNT BRACKET
SEE DETAIL 1
316 STN. STL.
DUAL MOUNT BRACKET
SEE DETAIL 3
2'
CUT OUT
R2'
3'
CUT OUT
RAIL
CENTERS
9 3/4"
9 3/4"
PLAN VIEW
WATER CHAMP
INSTALLATION
SCALE: 1/2" = 1'-0"
(2) MODEL SWC10F
WATER CHAMP CHEMICAL
INDUCTION UNITS. DUAL
MOUNTING SCHEME WITH
HORIZONTAL PIVOT KITS.

(c)

Segregation Factors. The segregation factor varies from zero to 1.0. Therefore, the closer this factor approaches zero, the greater the mixing efficiency. To demonstrate this phenomenon, let us compare the segregation factors, assuming a 100 mgd wastewater flow and a chlorine dosage of 10,000 lb/day, as follows:

1. The maximum strength of the chlorine solution discharge from a vacuum-operated chlorinator, which equals 3500 mg/L.
2. A 12 percent sodium hypochlorite solution.
3. A molecular chlorine vapor under 5 inches of Hg vacuum.

Chlorine Solution. At this dosage and 3500 mg/L strength, an injector water flow of 240 gpm is required. Assume that with a jet-type diffuser adequate mixing can be achieved in 15 sec. The segregation factor then is a weight ratio of chlorine solution to the wastewater flow in 15 sec:

$$\text{240 gpm/4} = 60.0$$

$$\text{100 mgd/1440} \times 4 = 17{,}361.11$$

$$\text{Segregation factor} = 60/17{,}361 = 0.00346$$

Hypochlorite. This solution contains 120,000 mg/L of chlorine per gallon, and each gallon contains 1.0 lb of chlorine. Therefore this process will require 10,000 gal of 12 percent hypochlorite solution per day, or 6.94 gpm, and the segregation factor is calculated as follows:

$$\text{6.94 gpm/4} = \text{1.74 gal/15 sec}$$

$$\text{100 mgd/1440} \times 4 = \text{17,361 gal/15 sec}$$

$$\text{Segregation factor} = 1.74/17{,}361 = 0.00010$$

Water Champ. This device pulls the molecular chlorine under 5 inches of Hg vacuum and blasts it into the wastewater stream at 60 ft/sec. Therefore:

Fig. 8-33c. Water Champ installations at Vacaville, California wastewater treatment plant (courtesy Gardiner Equipment Co., Inc.). Notes: Contractor is responsible for the following: 1) Assuring the rail system is plum and that the motor traverses the rail system free from binding. 2) Locating, mounting, and hookup of the control panel. See the controls section of the submittal or the O&M for additional information. 3) Cutting or moving handrails as necessary to install lifting hoist. 4) Water Champ must remain submerged at all times while running.

$$10{,}000 \text{ lb/day } Cl_2/1440 \times 4 = 1.736 \text{ lb/15 sec}$$

$$100 \text{ mgd/day} = 17{,}361.11 \text{ gal/15 sec} \times 8.34$$

$$= 144{,}791.67 \text{ lb/15 sec}$$

$$\text{Segregation factor} = 0.00001$$

This verifies that the Water Champ provides mixing efficiency that approaches perfection.

Other Chemicals. There are many other chemicals besides chlorine and sulfur dioxide currently in use at water and wastewater treatment plants. The Water Champ is an obvious selection for the application of liquid flocculants such as alum, ferric chloride, and polymers because the need for flash mixers would be eliminated.

It can also be used to great advantage for sodium hypochlorite at any trade strength and for sodium bisulfite solution for dechlorination. These installations will depend upon the proper selection of a flow-rate control valve between the chemical supply and the Water Champ.

It cannot be used to feed chlorine dioxide solution. This solution must be transported under pressure to the point of application in order to prevent the formation of ClO_2 vapor pockets in the solution line. These pockets of vapor are subject to puffs of spontaneous explosions that disrupt the entire ClO_2 generating process.

Cost-Effectiveness. The City of Sunnyvale, California made an economic study of the Water Champ based upon a 5 hp, 4000 lb/day unit, which was on trial at their tertiary WWTP during 1987 and 1988.[119] This report led to the City's purchase of three 5 hp submersible units after developing the following operating costs:

1. *Water savings:* The conventional injector system used a total of 0.40 mgd of tertiary effluent. The cost to treat this water is \$224.48/mg.

$$\text{Annual savings} = (0.40 \text{ mgd})(365 \text{ days})(\$224.48/\text{mg}) = \$32{,}774$$

2. *Power savings:* The following power costs were eliminated: two injector water pumps = 40 kW, one chlorine mechanical mixer = 6 kW, and one sulfur dioxide mixer = 6 kW, for a total of 52 kW.

$$\text{Annual savings} = (52 \text{ kW})(8760 \text{ hr/yr})(\$0.08/\text{kWh}) = \$36{,}442$$

3. *Sulfur dioxide savings:* These savings are the result of the average lower residual due to a more complete chemical reaction of chlorine by the Water Champ, which is 1.5 mg/L at the end of the contact chamber in a wastewater

treatment plant. Sunnyvale found it necessary always to dose SO_2 at twice the chlorine residual.*

$$\text{Sulfur dioxide savings} = (1.5 \text{ mg/l})(2)(17 \text{ mgd})(8.34)(365 \text{ days/yr})$$

$$= 155{,}250 \text{ lb/yr}$$

$$\text{Annual savings} = 155{,}250 \text{ lb/yr} = 77{,}625 \text{ tons} \times \$350/\text{ton}$$

$$= \$27{,}170/\text{yr}$$

4. *Water Champ operating costs:*

$$\text{Three units @ 5 hp each} = \text{approx. } 10 \text{ kW}$$

$$\text{Annual cost} = (10 \text{ kW})(8760 \text{ hr/yr})(\$0.08/\text{kWh}) = \$7{,}008$$

$$\text{Total annual savings} = \$32{,}774 + \$36{,}442 + \$27{,}170 - \$7008$$

$$= \$89{,}378$$

Cost Summary. Based upon a unit cost of $15,000 for each 5 hp submersible Water Champ and assuming the installation cost not to exceed $5,000, the simple payback would be $50,000/$89,378 = 0.559 year or 6.7 months. Using this calculation the City purchased three units.

Currently, in 1997, additional savings in operating cost, plus a greater mixing efficiency in mixing, are being experienced by many users in both potable water and wastewater systems.

CONTACT CHAMBERS

Function

The chlorine contact chamber must be designed to provide the optimum distribution of residence time for contact between the disinfectant and the microorganisms to be destroyed. There are many considerations required to develop this efficiency. Therefore, the chlorine contact facility should be considered as an integral part of the overall wastewater-treatment process, which means it becomes a unit process such as sedimentation, aeration, sludge digestion, and so on.

A chlorine contact chamber is not a mixing chamber, and should not be used for that purpose. The applied chlorine should be thoroughly mixed with the wastewater prior to entry into the contact chamber.

*This situation is not uncommon in plants that use flocculants. Some of these chemicals can consume the sulfite ion. The exact nature of this problem is being studied.

Distribution of Residence Time

General Theory. The distribution of residence time may differ appreciably in chambers of different geometrical configuration, although chamber volumes and flow rates are identical. The ideal reactor is one that can provide equal residence time for all the molecules in the chamber. This is described as 100 percent plug flow.

Figure 8-34 illustrates the dye tracer response in an ideal plug-flow chamber. This ideal situation compares to three other rectangular chambers less than ideal. Curve No. 1 illustrates the results of proper longitudinal baffling, with a modal time of 0.9 and approximately 95 percent plug-flow characteristics. Curve No. 2 is typical for what may be expected of circular tanks and rectangular tanks with poor baffling arrangements. Curve No. 3 is an actual dye response curve representing a seemingly well-baffled tank that exhibits excessive short-circuiting (back-mixing). This tank is considered properly baffled since the peak dye concentration occurs at the theoretical detention time (V/Q). However, the mere achievement of unity modal time does not satisfy all the requirements of the flow pattern in a contact chamber. In this case the tank was cross-baffled, which is to be avoided if possible, as this configuration promotes short-circuiting, which in turn significantly diminishes the plug-flow characteristics.

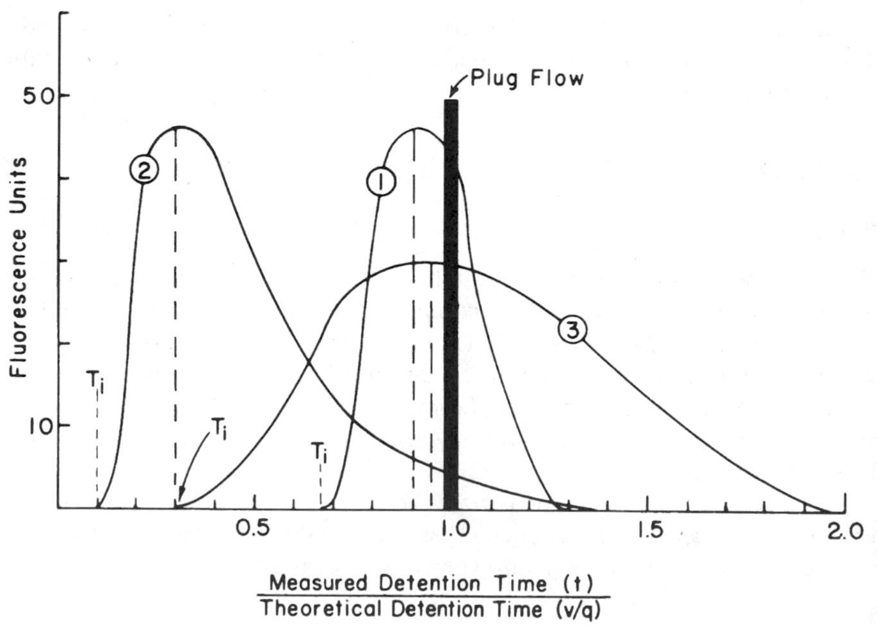

Fig. 8-34. Dye tracer studies of distribution residence times for various chlorine contact chambers configurations.

Effects of Short-Circuiting. Ignoring initial mixing effects, the batch-type reactor (beaker) produces results equivalent to any reactor that obtains 100 percent plug-flow characteristics. The effects of the residence time distribution on disinfection efficiency are illustrated in Fig. 8-35, which compares the destruction of coliform bacteria in a batch reactor with that of a continuous flow system afflicted with short-circuiting characteristics. The chlorine residuals are approximately equal in the two reactors. Figure 8-35 clearly demonstrates the gross effect of residence time distribution upon disinfection effi-

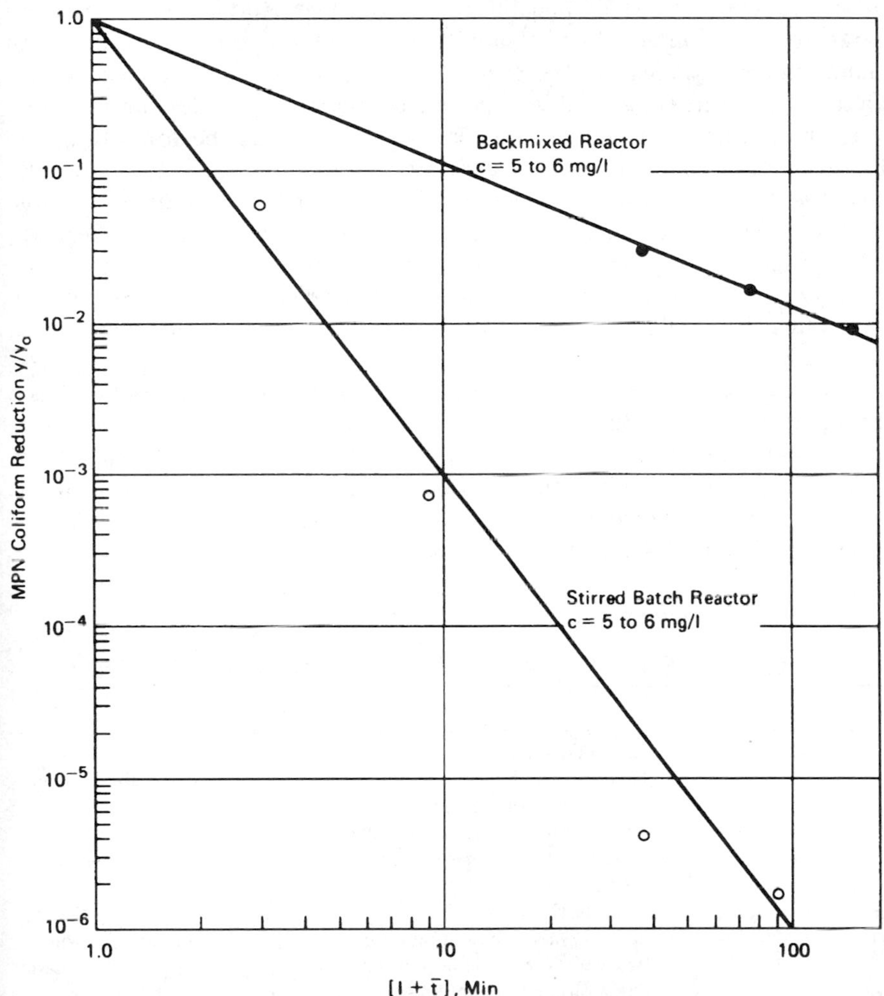

Fig. 8-35. Comparison of reactor (contact chamber) performance. t = contact time; c = amperometric chlorine residual (total) at the end of the contact time t, mg/L; Y = final coliform bacteria MPN/100 mL; Y_o = initial coliform bacteria MPN/100 mL.

ciency between the ideal plug-flow chamber (batch reactor) and the chamber with gross short-circuiting characteristics. It also demonstrates that the design criterion for residence time based upon chamber volume divided by flow rate (V/Q) is meaningless. For example, there exists a difference of four orders of magnitude in the coliform kill in the two reactors for a contact time t of 37 min.

Analysis of Residence Time. The conventional method of analyzing residence time uses the dye tracer dispersion technique. Since chlorine residuals normally carried in wastewater contact chambers do not affect the fluorescent quality of the dye, it is permissible to conduct a dye study during chlorination. The procedure consists of injecting a dye such as Rhodamine WT at the entrance to the contact chamber and measuring the dye concentration with a fluorometer at the exit of the chamber. The best way to introduce the dye is into the chlorine solution discharge line or the injector water supply. However, this requires shutting off the flow of chlorine, as the low pH of the chlorine solution plus the high concentration of the chlorine will significantly affect the fluorescent quality of the dye. This point of application is desirable because it provides a reliable indicator of diffuser efficiency as a mixer.

The shape of the dye concentration–time curve provides the necessary data to analyze the hydraulic flow pattern of any contact chamber. The conventional parameters used to describe the performance of a contact basin are presented in Table 8-3.

Each parameter shown in Table 8-3 can be used to predict contact basin performance. Figure 8-36 illustrates a typical dye dispersion curve. This curve represents a long, narrow, and shallow gunite-lined channel with an L to W ratio of 21 : 1, an unbaffled pipe entrance, and a sharp-crested weir at the outlet.[103] The importance of the shape of the dye dispersion curve is discussed below (see "Contact Chamber Evaluation").

Table 8-3 Contact Chamber Performance Characteristics

Parameter	*Definition*	
T	Theoretical detention time	
t_i	Time interval for the initial indication of the tracer in the effluent	
t_p	Time to reach peak concentration	
t_g	Time to reach centroid of effluent curve	
t_{10}, t_{50}, t_{90}	Time for 10, 50 and 90 percent of the tracer to pass at the effluent end	
t_i/T	Index of short-circuiting	Ideally, all these parameters will approach 1.0 under perfect plug-flow conditions
t_p/T	Index of modal detention time	
t_g/T	Index of average detention time	
t_{50}/T	Index of mean detention time	
T_{90}/t_{10}	Morril Dispersion Index; indicates degree of mixing; as t_{90}/t_{10} increases, the degree of mixing increases	

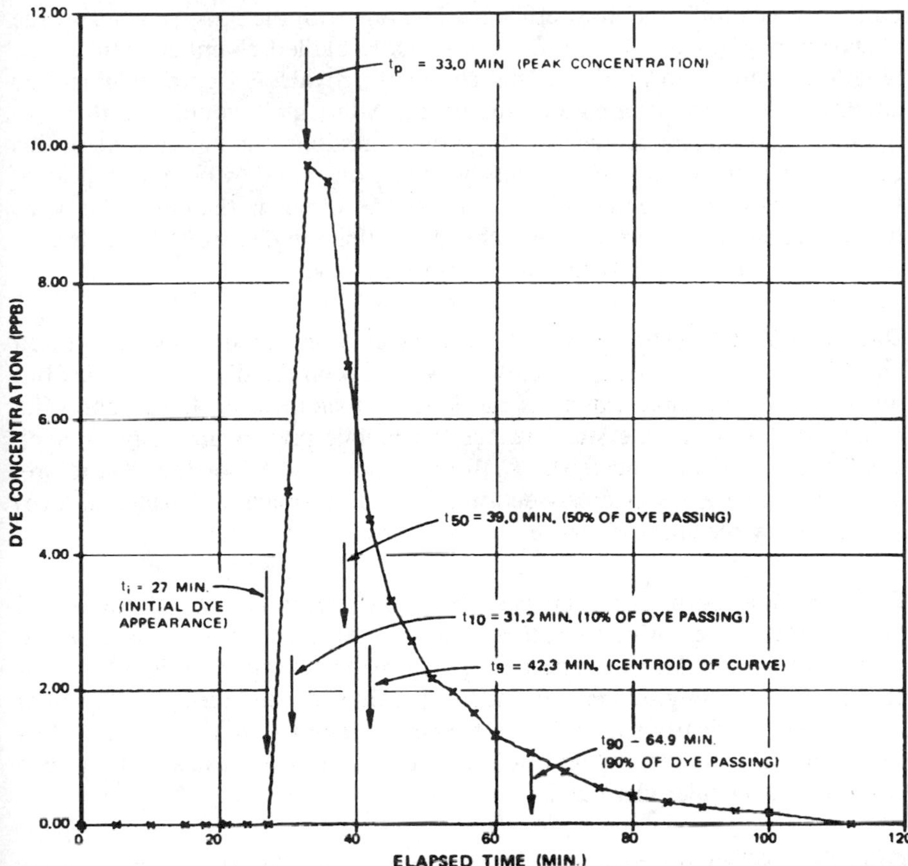

Fig. 8-36. Typical dye dispersion curve. Theoretical detention time (V/Q) = 50 min.; modal time = 0.7–0.8; t_i/T = 0.36–0.54 (depending upon wind direction); Morrill index = 2.1; chemical engineering disperson index = 0.08; percent plug flow = 60–70 (depending upon wind direction). (From Marske and Boyle.[103])

Effect of Chamber Configuration

Length to Width Ratio (*L/W*). The field study by Sepp and Bao[79] indicated that the L/W ratio is not in itself an accurate descriptor of plug-flow characteristics. Their pilot tank had a much higher L/W ratio than the plant-scale tanks but did not have a correspondingly smaller dispersion number. Studies by Marske and Boyle, and Trussell and Chao, have indicated the dispersion number usually decreases with increasing L/W ratio, but the correlation is poor. Other factors must play a role such as the depth-to-width ratio (H/W), extent of dead space, and eddy currents caused by poor plug-flow characteristics at the baffle turning points. The field investigation by Marske and Boyle[103] evaluated seven different chamber configurations. The ones with longitudinal baffles proved to be the most efficient. The one with the most practical value is shown in Fig. 8-37. This is a longitudinally baffled chamber with a flow length-to-width ratio of 72:1. This chamber provides 95 percent plug-flow conditions and exhibits a modal time of 0.7. Although it would be difficult to improve upon the percentage of plug-flow conditions, the modal time of this tank could be increased to 0.9, as has been demonstrated by others, with some slight changes such as elimination of the square corners in the tank. This work by Marske and Boyle substantiates the claims that long, narrow channels and/or conduits make the best chlorine contact chambers.

Depth to Width Ratio (*H*/*W*). The contact chamber analysis by Trussell and Chao[104] shows that depth can have an effect on the dispersion index but not nearly to the same extent as the length-to-width ratio. Based-upon the results of the plant-scale study versus the mobile pilot-plant study by Sepp and Bao,[79] the data indicate the *H*/*W* ratio should be 1.0 or less. Therefore, a compromise is a square cross-section at peak flow (maximum water surface) and a slightly rectangular section at lower flows.[105]

Circular Chambers. Circular chambers, least of all circular clarifiers, are not acceptable as chlorine contact chambers unless they are specially designed with an outer annular ring. The poor performance of a conventional circular clarifier versus a longitudinally baffled serpentine flow and the special annular ring clarifier is clearly shown in the Marske and Boyle investigation.[103] Any other circular chambers would be expected to perform just as poorly as a conventional circular clarifier.

Outfalls. When available, these structures provide almost perfect plug-flow conditions when flowing full or partly full. Figure 8-38 illustrates the plug-flow characteristics of a 4000 ft outfall line discharging into San Francisco Bay. This represents a dye-tracer study by Kennedy Engineers for the City

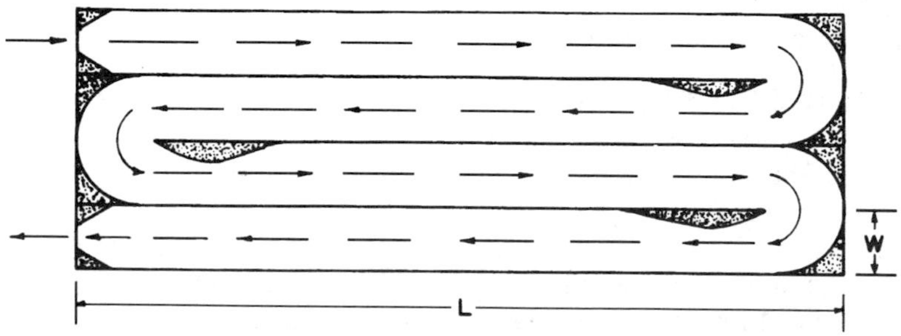

Fig. 8-37. Chlorine contact chamber with longitudinal baffling and optimum plug flow characteristics. Flow length: width = 72:1; L:W = 18:1; modal time = 0.70; percent plug flow = 95. (From Marske and Boyle.[103])

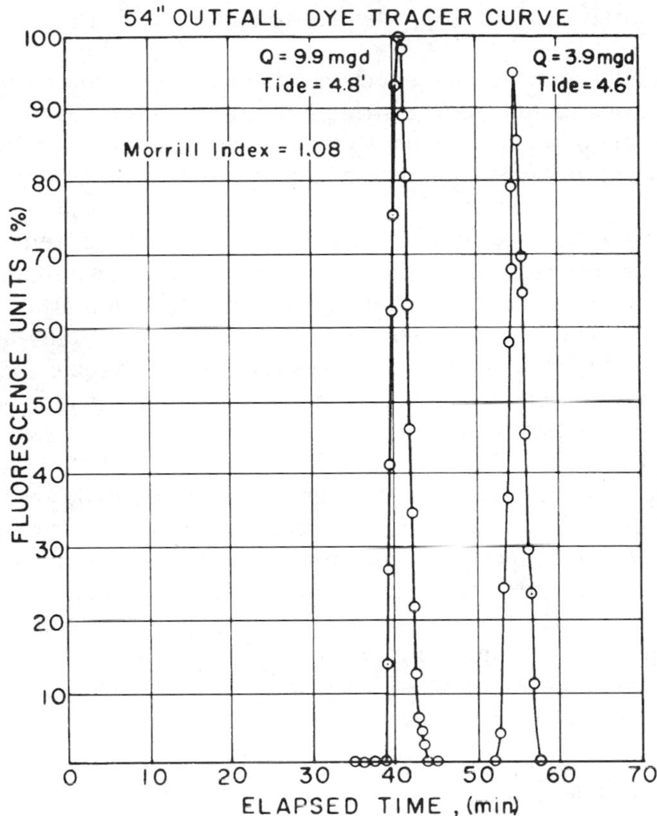

Fig. 8-38. A free-flowing outfall dye dispersion curve.

of Richmond, California. In order for an outfall to qualify as a contact chamber, an effluent sampling site must be available for chlorine residual monitoring and dechlorination control.

Other Physical Characteristics Affecting Residence Time

Marske and Boyle[103] also studied the following physical characteristics used in the design of a contact basin: (1) the depth of the basin as it relates to the effect of surface wind; (2) the effect of outlet weir configuration; and (3) the effect of turbulence. The effects of these physical characteristics are summarized below.

Wind. Two tracer tests conducted on a very long, narrow contact channel approximately 3 ft deep indicated that wind may cause surface currents resulting in some short-circuiting. In the tests, a downstream wind resulted in a relative time ratio, t_i/T, of 0.36, whereas an upstream wind resulted in a t_i/T of 0.54.*

Consequently, owing to wind effects, a unidirectional shallow basin or trench may not provide the plug-flow distribution of residence times that would be expected from the geometrical configuration. In this case, the designer would be well advised to replace this type of chamber with a closed conduit.

Weirs. Tracer tests, performed on a 3.5-ft-wide Cipolleti weir and on an 18-ft-wide sharp-crested weir, indicated that all hydraulic performance parameters were improved when the sharp-crested weir was employed. The value of t_i/T increased from 0.19 to 0.27, and the percentage of plug flow in the basin increased from 38 to 58 percent. Consequently, it is recommended that contact chamber overflow weirs extend across the entire width of the final channel of the contact chamber.

Turbulence. There is no necessity to provide turbulence such as is required in the mixing phase. Turbulence in a contact chamber will not improve coliform destruction; it may in fact be detrimental to the disinfection process. Turbulence during the contact period may reduce the monochloramine residual by aeration, and turbulence may cause back-mixing. The monochloramine residual is the most potent disinfecting compound of any combined residual occurring in wastewater discharges; so it should not be subjected to any loss by aeration in the contact chamber.

Effects of Baffles

General. Longitudinal baffles are superior to horizontal baffles. The latter cause many times more back-mixing in the chamber than do longitudinal baffles.

Special Structural Design Situations. Sometimes it is impossible to design a contact chamber with longitudinal baffling because of seismic design considerations. In these cases the horizontal baffling is usually integrated in the design as structural members. In these instances all of the corners should be eliminated as shown in Fig. 8-39 to minimize dead spaces and short-circuiting. Figure 8-40 is a photograph of Fig. 8-39 after construction. This illustrates the special smooth concrete surface, which is especially amenable to the cleanliness of this chamber.

Whenever the effluent MPN coliform requirement is 2.2/100 ml, the specifications for the concrete in the contact basins should call for a porcelain-like smooth finish free from any kind of pits or recesses, as these are breeding

*The term t_i is the time interval between the injection of the tracer and its first appearance in the effluent, and T is the contact time, calculated from V/Q, where V is the basin volume and Q is the flow rate.

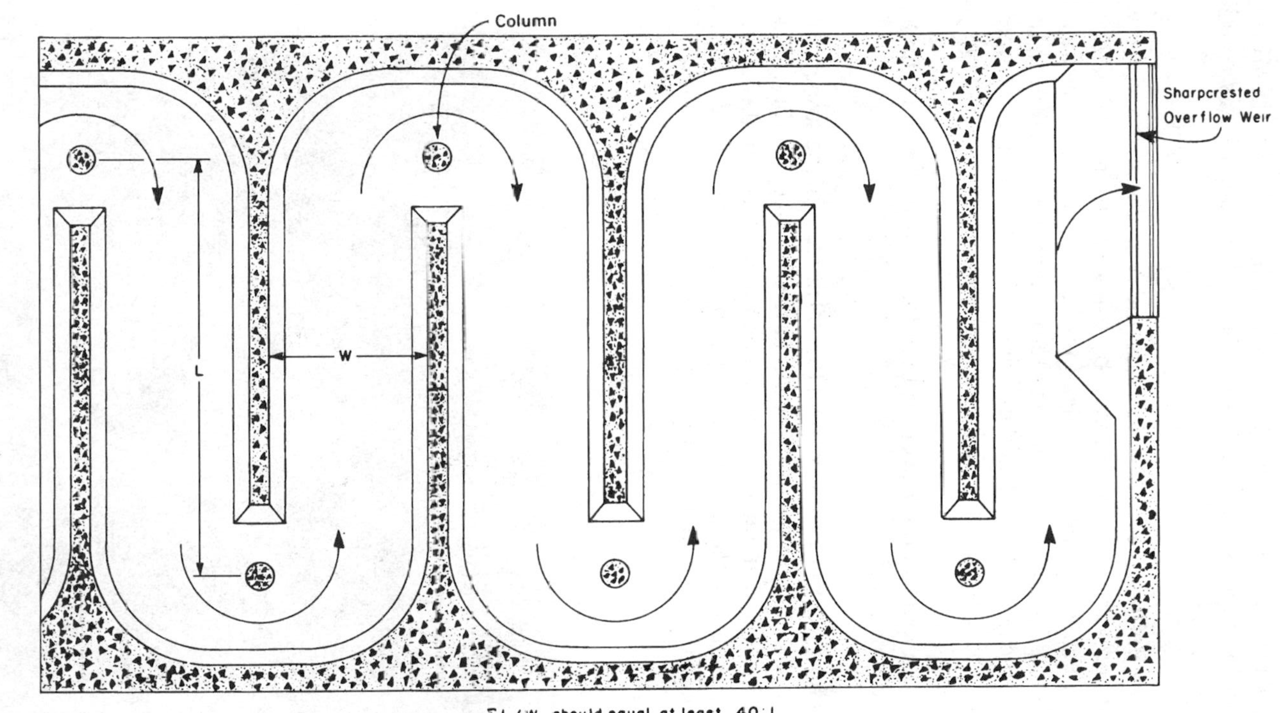

Fig. 8-39. A well-baffled contact chamber with horizontal baffles from contact drawings (courtesy of San Jose/Santa Clara, WWTP).

areas for the bacteria. Moreover, such situations call for special attention to frequent cleaning. Actually contact chambers designed with either longitudinal or horizontal baffling for effluents that are to achieve a coliform MPN of 2.2/100 ml must be kept as free from slime and algae deposits as a well-kept swimming pool.

Modification Baffles. Hart[106] and Hart and Vogiatzis[107] have shown that significant improvements can be made by using diffusion baffles. These are illustrated in Fig. 8-41. These modifications have been able to change the t_i/T from 0.08 to 0.50 and the dispersion index from 0.056 to 0.004. In terms of disinfection efficiency, the modified unit showed a 24 percent improvement,[107] and chlorine consumption was reduced by 10 percent for equivalent performance.

One type of modification proven to be effective is a V-notch launder-type weir at the chamber exit.

Provisions for Cleaning. It is known that 50 percent of the suspended solids remaining in the effluent will precipitate in the contact chamber. This is the result of velocities that range from a minimum of about 0.06 ft/sec to a maximum of about 0.4 ft/sec. Therefore, the contact chamber must be provided with a means for easy cleaning on a regular basis.

Fig. 8-40. Contact chamber shown in Fig. 8-39 after construction at the San Jose/Santa Clara WWTP.

Contact Chamber Evaluation

General. All contact chambers should be evaluated in situ as soon after being put into operation as is possible. The information developed must be based on the statistical analysis of the dye dispersion curve of the basin at various flow rates. This describes the hydraulic performance of the contact chamber as related to disinfection and considers the shape of the entire curve rather than the central tendency values.

The purpose of evaluating a contact chamber is to determine the capability of a given chamber to provide disinfection sufficient to meet the NPDES requirements. The most important factor in evaluating a chamber is the time it takes for the initial appearance of the dye at the chamber exit. This time (t_i) must not be less than 30 min. and should be closer to 60 min. The next most important index is the model time, represented by t_p in Table 8-3. This figure should be close to 90 percent of the theoretical contact time T = Q/V.

Finally, the overall performance can be characterized by either the dispersion index d or by the Morrill index. Both of these methods are described below.

Dispersion Index. The chemical engineering dispersion index d was introduced by Thirumurthi.[108] This index is calculated from the variance of the

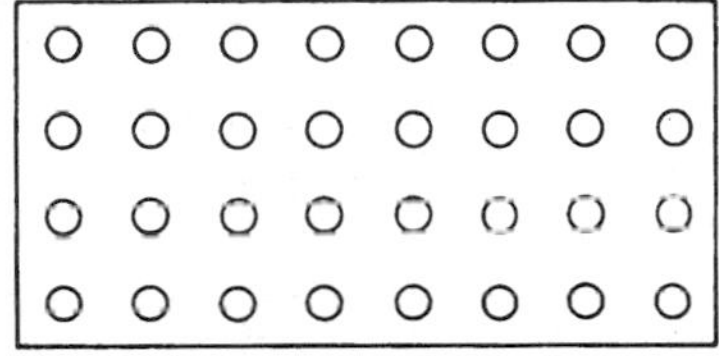

Typical Baffle

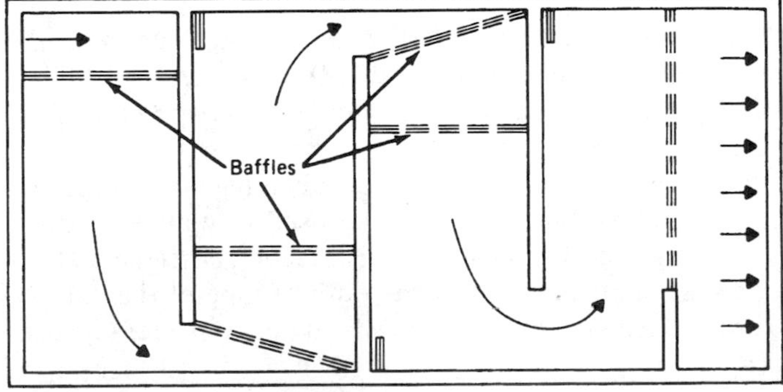

Contact Chamber with Diffusion Baffles

Fig. 8-41. Use of diffusion baffles to improve dispersion index in existing chlorine contact chambers [106] (courtesy *Journal WPCF*).

dye dispersion curve, which relates dye concentration to time. As conditions approach ideal plug flow, the value of d approaches zero. The algebraic calculations are delineated in Refs. 103 and 109. The dispersion index d calculated by this method includes all points on the dye dispersion curve illustrated in Fig. 8-34. The index described by this mathematical approach demonstrates a strong statistical probability of correctly describing the contact chamber efficiency, which is related to the dispersion index. The dispersion index is:

$$d = \frac{D}{uL} \tag{8-17}$$

where

D = longitudinal dispersion coefficient, ft^2/sec
u = L/T, average chamber velocity, ft/sec
L = reactor length, ft
σ^2 = variance of the tracer curve, which is a measure of the spread of the curve.* It is the square of the standard deviation.[109]

In computing the dispersion index, boundary conditions must be observed (i.e., whether the chamber is a closed or an open vessel). Outfalls would be characterized as open vessels. Most chlorine contact chambers fall into the closed vessel category where the velocity changes abruptly at both the inlet and the exit of the chamber. It is well known that the open vessel category has the characteristics nearest to those of plug flow; so computation of the dispersion index for these vessels will be ignored.

For closed vessels the dispersion index is computed from the following expression:

$$\sigma^2 = 2d - 2d^2(1 - \varepsilon^{-1}/d) \tag{8-18}$$

When there is a small amount of reduced variance such that d calculates to less than 0.05, the second term in Eq. (8-18) can be neglected. The complete solution of a typical dye tracer study is shown in the appendix, including the key sequence for an HP 21 calculator.

Use of the dispersion index has a major advantage over other parameters for contact chamber evaluation. It can be used in mathematical models to predict both hydraulic performance and disinfection efficiency. The dispersion number represents an expression of the entire shape of the curve. This is in contrast to other dispersion parameters, which only consider one or two points of the traces.

*Tracer curves often exhibit a long tail caused by recycling of the tracer from dead spaces and backmixing.

Morrill Index. Another method of analysis, which is considered practically equal in statistical confidence to the dispersion index, is the Morrill index.[110] This is a mathematical representation of dispersion such that ideal or 100 percent plug flow produces a Morrill index of 1.00. This analysis utilizes only two points in the dye dispersion curve shown in Fig. 8-38. The Morrill index is characterized by the expression t_{90}/t_{10}. This means that the time in minutes required for the passing of 90 percent of the dye (65 min.) divided by the time in minutes required for the passing of 10 percent of the dye (31 min.) equals the Morrill dispersion index, or 65/31 = 2.10. This compares with the longitudinally baffled chamber with the ideal length-to-width ratio of 72:1 and a Morrill index of 1.48.

Therefore, it would be safe to declare that any Morrill index from minimum to maximum flow rates on the order of 1.5 to 2.5, respectively, could be considered acceptable for disinfection purposes. A sample calculation of the Morrill index is shown in the appendix.

Dispersion in Long Narrow Structures

Closed Conduits and Round Pipes. Under the heading "Outfalls," above, these structures flowing full or partly full are considered to have nearly ideal plug-flow characteristics. Therefore very few actual data have been documented on this type of contact chamber. It is believed that for a long, straight outfall the dye curve shown in Fig. 8-38 would be representative. Trussell and Chao[104] and Trussell and Pollock[80] have discussed this case and offer the following empirical equation as suitable where turbulent flow intensity is N_R = 10,000 or greater:*

$$d = 895000 f^{3.6} (D/L)^{0.859} \tag{8-19}$$

where

D = pipe diameter, ft
L = length of pipe, ft
f = Darcy–Weisbach friction coefficient

Experience has demonstrated that a 30–45 min residence time at peak dry weather flow at a velocity of 1.5–2.0 ft/sec will produce nearly ideal plug flow.

Sample Calculation of Dispersion Index

1. Given:

*Empirical equation is from F. Sjenitzer, "How Much Do Products Mix in a Pipeline?" *The Pipeline Engineer*, D-31 (Dec. 1958).

$$Q = 5.2 \text{ mgd} = 8.06 \text{ cfs} = 484 \text{ cfm}$$

$$L = 0.6 \text{ mile}$$

$$t = \text{contact time} = 30 \text{ min.}$$

2. Calculate diameter for 30 min. contact time:

$$t = \frac{(\text{area})(\text{length})}{\text{flow rate}} = \frac{\frac{\pi D^2}{4} L}{Q}$$

$$D = \sqrt{\frac{4tQ}{\pi L}} = \sqrt{\frac{4 \times 30 \times 484}{\pi(3170)}} \tag{8-19a}$$

$$D = 2.41 \text{ ft, use standard size: } 30 \text{ in.} = 2.5 \text{ ft}$$

3. Solve for Reynolds number:

$$N_R = VD/\nu \tag{8-19b}$$

where
V = pipeline velocity, ft/sec
D = pipe diameter, ft
ν = kinematic viscosity, ft^2/sec at 70°F

$$V = Q/A = \frac{8.06 \text{ cfs}}{\frac{3.14 \times 6.25}{4}} = 1.64 \text{ ft/sec}$$

$$N_R = \frac{1.64 \times 2.5}{1.06 \times 10^{-5}} = 386{,}792$$

4. Select friction factor as follows: From Fig. 11 in the appendix select e as 0.003 for average concrete pipe, 2.5 ft in diameter. This intersects the es/d ordinate at approx. 0.0015 for $\varepsilon = 0.003$. Then from Moody's diagram (Fig. 10, appendix) select f from ε/D ordinate = 0.0015, where it intersects Reynolds number 3.9×10^5. This is $f = 0.0225$.

5. Calculate dispersion number from Eq. (8-19):

$$\begin{aligned} d &= 89{,}500\ (f)^{3.6}(D/L)^{0.859} \\ &= (89{,}500)(0.0225)^{3.6} \times (2/3170)^{0.859} \\ &= 0.0002 \end{aligned}$$

6. *Conclusion*: Perfect plug flow.

Open Channels. Trussell and Chao[104] have shown that the dispersion index can be estimated by the following expression:

$$d = \frac{24.4n(R_H)^{5/6}}{L} \tag{8-20}$$

where

n = Manning's coefficient
R_H = hydraulic radius, ft
L = length of channel, ft

By manipulation such as inserting coefficients of channel geometry and making simplications according to good practice, Trussell has presented the following as an abbreviated form of Eq. (8-20):

$$d = 0.14/\beta, \text{ where } \beta = \text{channel } L/W \tag{8-21}$$

This expression should be used with some caution because it is for long, straight channels without structures causing eddy currents or other hydraulic disturbances. Trussell has plotted field data collected from three sources[70,80,104] shown in Fig. 8-42. This suggests the use of a nonideality coefficient as a

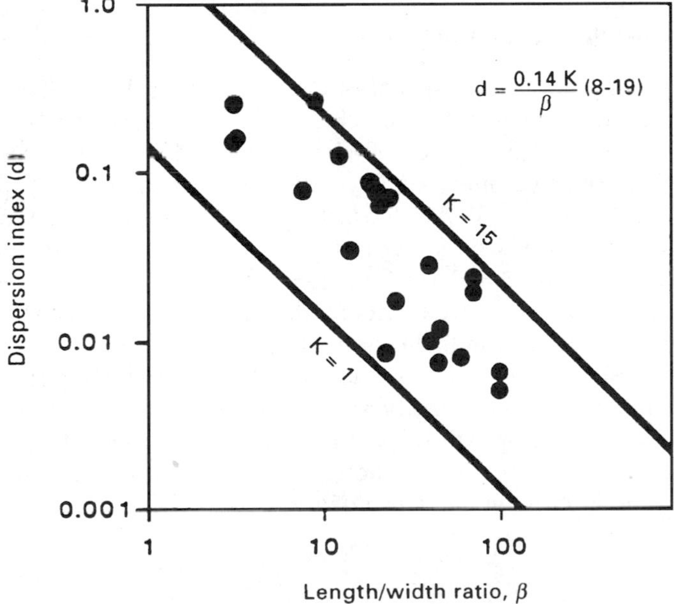

Fig. 8-42. Dispersion index versus length to width ratio of long channels. (From Trussell and Pollock.[80])

measure of effectiveness of a design that would approach the predicted performance of a straight channel:

$$k = d\beta/0.14 \tag{8-22}$$

The data in Fig. 8-42 show coefficients of nonideality ranging from 1.33 to 15 and dispersion numbers ranging from 0.0054 to 0.25. Ten of the 22 plants studied by the three sources had a nonideality coefficient of approx. 3–4. This suggests that such performance is a reasonable expectation for a rational design. Seven of the 22 plants had dispersion numbers near 0.01 or less; therefore, low dispersion conditions are attainable.

Sample Calculation Using Eq. (8-20). Channel flow is 20 cfs and velocity is 1.5 ft/sec. Assume that Manning's coefficient is $n = 0.015$, the channel is 3.65 ft wide, and the water depth is also 3.65 ft; then $R_H = 10.95$ ft, $L =$ 800 ft, and:

$$d = \frac{24.4 \times 0.015 \times (10.95)^{5/6}}{L = 800 \text{ ft}}$$

$$= 0.00334$$

This is equivalent to ideal plug-flow characteristics.

Optimum Design Considerations

Sanitation Districts of Los Angeles County, Example.[111] The optimum design of a contact chamber can be achieved if sufficient information is available. The factors that govern the design are as follows: (1) initial coliform concentration prior to disinfection (y_0); (2) final coliform concentration in the disinfected effluent (y); (3) the 2–3 min. chlorine demand, or the chlorine dosage required to produce the desired residual; and (4) the residual–contact time envelope (Ct) to achieve the final coliform concentration (y). When this information is known for a given effluent, the contact chamber can be designed for minimum annual cost and chlorine contact time.[19]

A differential equation based on the annual cost (capital amortization plus operation and maintenance) can be developed. To find the minimal cost this equation is differentiated with respect to t, and the result is set equal to zero. The answer represents the point where annual cost with respect to a change in contact time is a minimum. This equation takes the form:

$$\frac{\partial AC}{\partial t} = a - b\frac{K}{t^2} = 0 \tag{8-23}$$

Therefore:

$$t = \sqrt{\frac{b}{a}}\, K \tag{8-24}$$

where

AC = annual cost
$K = Ct$
c = chlorine residual, mg/L at time t
t = contact time, minutes

The AC (annual cost) consists of capital amortization plus operation and maintenance. This cost is made up of the following factors:

1. CRF (capital recovery factory), which consists of the cost of the chlorine contact chamber, the dechlorination station, the chlorine, the sulfur dioxide, and the labor. This is usually taken at i (interest) = 7 percent and n = 20 years.
2. Chlorine contact tanks = \$$y$/gal × z gpm × t (minutes).
3. Cost of dechlorination station.
4. Cost of chlorine.
5. Cost of sulfur dioxide.

From this a differential equation can be formulated, which will be as follows: AC = CRF (dollars for chlorine contact tank cost) + dechlorination station cost + annual chlorine cost (based on dosage required) + annual sulfur dioxide cost (based on maximum residual required).

When the above term AC is factored out, it will declare numerically the relation between contact time and chlorine residual at the end of this contact time, which is related to the values of a and b. These are a function of items 1–5 described above. The solution to this equation provides the minimum contact time for a given flow of wastewater to produce the coliform concentration limit (y) for a given situation.

Referring to Eq. (8-24), $K = Ct$; this is the chlorine residual contact time envelope. For a given situation the upper and lower limits for this factor should be selected. From this, contact time t should be tabulated, together with chlorine residual. Let us assume that the lower limit is 500 at a given peak flow to accomplish the desired disinfection; then the upper limit based on an average flow would be

$$\frac{Q_1 = \text{peak flow}}{Q_2 = \text{ave flow}} \times 500 = K$$

for contact chamber design conditions. Therefore, C and t for this value of K can be computed, and these will be the design parameters for contact time and chlorine residual.

Kennedy–Jenks Engineers Example.[105] This method requires knowledge of the wastewater quality, the same as in the above example.

1. The *Ct* envelope versus the log kill of coliform organism (y/y_0) is drawn. If there are known seasonal variations of y_0, an upper and a lower *Ct* envelope may be constructed, one for the lower limit of y_0 and one for the upper limit. The term y is the coliform concentration at the end of time t and is the disinfection requirement prescribed for the effluent.
2. The next step is to conduct some laboratory chlorine demand studies in order to establish the chlorine dose required to produce the desired residual at several different contact times.
3. The information developed in steps 1 and 2 will provide the necessary information to find the cost of chlorine and sulfur dioxide.
4. The next step is to determine the contact chamber construction cost for various selected contact times.
5. All of the above information is then tabulated against contact times. Then the cost of each function for a given time t is totaled and converted to a "present worth figure."
6. The last step is to draw a curve or a series of curves using the present worth figure as the ordinate and contact time as the abscissa. The series of curves may be one each from the two *Ct* envelopes, one dry weather flow and one wet weather flow, and another for the present worth figure of the chemicals. The lowest point on the curve to be used will represent the optimum contact time for a given known wastewater quality and disinfection requirement.

Optimum Contact Time.[79] The following approach to produce an adequately disinfected effluent at minimum cost determines the minimum contact time. Using Eq. (8-25) and simple economic analysis, a differential equation for partial cost of chlorination can be written, from which the most cost-effective contact time can be calculated:

$$N_1/N_0 = \left(\frac{RT}{b}\right)^n \qquad (8\text{-}25)$$

where

N_0 = average (median or geometric mean) concentration of coliform bacteria immediately upstream from chlorine application
N_1 = NPDES total coliform concentration in effluent
R = chlorine residual (mg/L) after contact time T
T = contact time, minutes
n = slope of the curve
b = the x-intercept when $N_1/N_0 = 1.0$ mg/L, minutes

The symbols used in the following analysis are:

CRF = capital recovery factor (amortization)
S = N_1/N_0 required bacterial destruction
T = mean residence time in contact chamber, minutes
C_c = unit cost of chlorine, \$/lb
C_s = unit cost of sulfur dioxide, \$/lb
C_t = unit cost of contact chamber, \$/gal
D = total chlorine demand, mg/L
CE = DE = cost of chlorination equipment or dechlorination equipment
Q = mean effluent flow, 10^6 gal/day

Annual cost of chlorine is:

$$365Q(R + D)8.34 \times C_c = 3044Q(R + D)C_c \tag{8-26}$$

Annual cost of sulfur dioxide is:

$$365Q(R)8.34 \times C_s = 3044QRC_s \tag{8-27}$$

Total cost of chlorine contact chamber is:

$$\frac{10^6}{24 \times 60} QTC_t = 694QTC_t \tag{8-28}$$

Total partial annual cost is:

$$\mathrm{CRF}(694QTC_t + \mathrm{CE} + \mathrm{DE}) + 3044Q[(R + D)C_c + RC_s] \tag{8-29}$$

From $S = (RT/b)^n$

$$R = bS^{1/n}/T \tag{8-30}$$

Substituting this into Eq. (8-29), we get:

$$\begin{aligned}\text{Total cost} = {} & \mathrm{CRF}(694QTC_c + \mathrm{CE} + \mathrm{DE}) \\ & + 3044[(bS^{1/n}T^{-1} + D)C_c + bS^{1/n}T^{-1}C_s]\end{aligned} \tag{8-31}$$

Differentiate Eq. (8-31) with respect to T, to optimize detention time, and then equate to zero:

$$\frac{d(\text{total cost})}{dT} = \mathrm{CRF}694QC_t - 3044Qbs^{1/n}T^{-2}(C_c + C_s) = 0 \tag{8-32}$$

Therefore, optimum residence time is:

$$T = 2.1 \frac{[bS^{1/n}(C_c + C_s)]^{1/2}}{\text{CRF} \times C_t} \tag{8-33}$$

where $n = -3$.

Sample calculation using Eq. (8-33): Assume CRF = 0.095 at 8 percent interest for a 20-year period; CE = \$60,000 for a 10 mgd plant; cost of chlorine, \$0.10/lb; cost of sulfur dioxide, \$0.15/lb; cost of contact chamber, \$0.35/gal. Assume also that:

$N_0 = 2.4 \times 10^6$ MPN/100 ml;
$N_1 = 240$ MPN/100 ml; D = 5.0 mg/L; $S = 1 \times 10^{-4}$
$b = 6$ and $n = -3$

Then for an optimized system:

$$T = 2.1 \left[\frac{6.0(10^{-4})^{-1/3}(0.10 + 0.15)}{0.095 \times 0.35}\right]^{1/2}$$

$$= 2.1 \left[\frac{6.0(10)^{4/3}(0.10 + 0.15)}{0.095 \times 0.35}\right]^{1/2}$$

$$= 2.1\sqrt{964.49}$$

$$= 65.22 \text{ min.}$$

Summary of Initial Mixing and Contact Chamber Requirements

Initial Mixing. This event should occur in the most turbulent zone of the effluent entrance to the contact chamber where the velocity is on the order of 2–3 ft/sec. If mixing is to be achieved by a grid-type diffuser, the diffuser should be located in this turbulent zone.

The mixing device should be able to generate a G number of about 500. The objective is to provide complete mixing of the chlorine solution with the wastewater flow in not more than 3 sec. A variety of solutions to the problem of initial mixing are available to the designer. The use of a closed conduit for the mixing zone should not be overlooked.

Contact Chambers. The most important parameter in the evaluation of these structures is t_i. This is the time it takes for the first appearance of dye at the exit of the chamber. This should never be less than 30 min.

The design objective is to achieve a dispersion index no larger than 0.02, and preferably 0.01. This can be done using longitudinal baffles where the L/W ratio is greater than 40, provided that the turning zones are designed to eliminate eddy currents, and the *H/W* ratio is 1.0 or less. Where possible and convenient, longitudinal baffles with an *L/W* ratio of 70 are preferred.

The exit structure of the contact chamber should be provided with either a diffuser baffle or a launder-type weir. These embellishments must be able to cope with the change in *H* due to variations in hydraulic flow.

In general the design of a contact chamber should provide a distribution of residence times approaching ideal plug-flow characteristics. This cannot be overemphasized. It is apparent, therefore, that a long, straight pipeline will fulfill these conditions.

Existing contact chambers can be retrofitted to achieve significant improvement in disinfection efficiency.

RELIABILITY PROVISIONS FOR CHLORINATION FACILITY

The need for continuous and dependable disinfection has been stressed. Chlorination system failure can be due to a number of causes, so the design of the system must include provisions to either prevent failure or allow immediate corrective action to be taken. Although assured reliability is essential, design provisions for this are often slighted.

Chlorine Supply

As with any chemical feed process, one of the most frequent interruptions in treatment is caused by the exhaustion of the chlorine supply. Five features are essential in order to maintain continuous chlorine feed: (1) an adequate reserve supply of chlorine sufficient to meet normal needs and bridge delivery delays and other possible contingencies; (2) chlorine container scales; (3) a manifolded chlorine header system; (4) an automatic device for switching to a full chlorine container when the one in use becomes empty; and (5) an alarm system to alert operating personnel to imminent loss of chlorine supply. These five features are discussed elsewhere in this text. Without them it is not possible to assure uninterrupted chlorine feed even with full-time operator attendance and no equipment breakdowns.

The chlorine header system is needed for two reasons: (1) to provide a connected on-line chlorine supply that is adequate to assure uninterrupted flow of feed for whatever period the system may be unattended; and (2) to allow switchover to a full cylinder without interruption of feed.

Power Failure

Power outage usually results in water supply failure, which in turn automatically shuts down the chlorination system. A range of special provisions can be employed to assure reliability of the power and water supply, depending upon the particular situation. A standby power source and duplicate pumps are the measures most often taken here.

Standby Equipment

The design of the chlorine feed system should provide for continued operation in cases of equipment failure. Where both pre- and postchlorination are to be practiced, separate chlorination systems should be provided for each plus a standby system. If prechlorination is not continuous, it may be possible to use the prechlorination system as the standby for disinfection. The units, piping, and accessories should be designed with this application in mind. If prechlorination must be carried out continuously, or if no prechlorination is to be done, a standby system capable of replacing the postchlorination system during repairs, maintenance, or emergencies should be provided. Standby equipment of sufficient capacity should be available to replace the largest unit during shutdowns. This includes standby pumps for the injector water supply.

Spare Parts

In addition to standby equipment, the equipment manufacturer should be consulted regarding vulnerable components. These components should be a part of the plant's inventory of spare parts.

Water Supply

As mentioned above, during a power failure the injector water system will be shut down unless there is an alternative supply that does not require power, such as an elevated tank. Standby equipment to provide injector water in the event of a power failure would consist of an engine-driven injector supply pump. Every injector water supply system should have such a standby pumping unit. There is no way to operate the chlorination system without an adequate water supply.

Chlorine Residual Analyzers

Now (1997), in the era of reliable ORP control of the disinfection systems for wastewater effluents, chlorine residual analyzers must be used to monitor the chlorination equipment. In addition to this, the chlorination system can be used as a backup to the ORP system if for any reason such a need were to occur.

Provisions should also be made for standby sample pumps. An adequate supply of all necessary spare parts should be on hand at all times.

Operator Documentation of ORP Control System

Determination of Treated Effluent "Poise." This is most important, as it records the chlorine demand on a continuous basis. This information is extremely valuable to operating personnel because a change in chlorine demand is notification of a change in the quality of the "raw" water or some problem in the treatment system, or both.

All that is necessary is continuous measurement just upstream from the point of final chlorination. Multiple points may be required, particularly if there are additional points of chlorination upstream from the effluent point of chlorination.

ORP Operating Information. Records of the following at the plant effluent should be kept on a twice-daily basis, one at peak flow and one at low flow, if convenient for operating personnel:

1. ORP, mV.
2. Chlorine feed rate, lb/day.
3. Plant treated water effluent, mgd.
4. Chlorine residual, mg/L.

Records and Reports. Reliable, continuing records should be kept to establish proof of performance and effectiveness of the control system, and to justify decisions, including expenditures, and various recommendations. These records serve as a source of information for plant operation, modification, and maintenance. Adequate records of disinfection are also important from the standpoint of regulatory agencies. Records will facilitate public health surveillance and enable the regulatory agencies to assess compliance with state regulations. Two classes of records should be maintained at a wastewater treatment plant using chlorination: (1) descriptive, planning, and inventory records related to the physical plant; and (2) performance records.

Physical Facilities. The following records, referring to the chlorination-dechlorination unit, should be available for reference at the plant:

1. Design engineer's report, including basis of design, equipment capacities, population served, design flow, reliability features, and other data.
2. Contract and "as-built" plans and specifications.
3. Shop drawings and operating instructions for all equipment.
4. Cost of each equipment item.
5. Detailed plans of all piping and electrical wiring.
6. A complete record of each piece of equipment, including name of manufacturer, identifying number, rated capacity, and dates of purchase and installation.
7. Supply of chlorine and dechlorinating agent, including reserves and availability estimate.

Records of Operation. Daily records should be kept of the following:

1. Continuous ORP mV readings from a chart recorder or a data logger.
2. Results of bacteriological analyses: coliform count exiting contact chamber and other required tests.
3. If possible, chart of ORP mV versus coliform count.
4. On dechlor applications, daily grab samples verifying the ORP mV setpoint.
5. Chlorine: daily quantities used, including dosage, for both pre- and postchlorination including method used, and continuous recording of chlorine residuals.
6. Wastewater flow: preferably continuous recording; total daily flow treated; maximum, minimum, and average daily flow.
7. Daily inspection: operational problems, equipment breakdowns, periods of chlorinator outage, diversions to emergency disposal, and all corrective and preventive action taken.

Records of Equipment Inspection, Maintenance, and Repair. In addition, the following important records and plans should be kept at each facility:

1. Routine equipment maintenance schedule and record.
2. Annual equipment inspection and maintenance record.
3. Plan for prearranged repair service.
4. Emergency plan for chlorinator failure.
5. Emergency plan for accidental chlorine or sulfur dioxide release.
6. Remove ORP probes from process to inspect and clean weekly.
7. Monitor consistency of ORP mV readings on a month to month basis to predict the end of probe life expectancy of 3 years.
8. Maintain fluid levels in automatic probe cleaning system.
9. Recheck periodically, the calibration of the instrument output (4–20 mA signal) to chlorinator or sulfonator.

Reports. Monthly operating reports should be prepared, containing information on chemical usage, wastewater flows, laboratory analyses, ORP response, and significant operational problems.

REFERENCES

1. Collins, H. F., Chief, Environmental Health Division, California Health Services, private communication, June 1983.
2. *Standard Methods for the Examination of Water and Wastewater,* 19th ed., American Public Health Assoc., 1996.
3. "Report of a Pollution Survey of Santa Monica Bay Beaches in 1942," California State Board of Health, June 26, 1943.
4. Stevenson, A. H., "Studies of Bathing Water Quality and Health." *Am. J. Public Health,* **43,** 529 (1953).

5. California State Department of Health Symposium, "Fecal Coliform Bacteria in Water and Wastewater," Berkeley, CA, May 21, 1968.
6. Unz, R. F., "Fecal Coliforms and Fecal Streptococci in the Bacteriology of Water Quality," *Water and Sew. Wks.*, **115,** 238 (1968).
7. Jopling, W., "Statement in Support of California Recommended Disinfection Requirements," paper presented by the California State Department of Public Health at a joint meeting of Arizona, California, Nevada, and EPA regarding waste discharge requirements and water quality objectives for the Colorado River, Las Vegas, NV, Oct. 28, 1975.
8. "National Symposium on Estuarine Pollution," Stanford Univ., Palo Alto, CA, 1967.
9. Jopling, W., and Young, C., private correspondence, California State Dept. of Health, Berkely, CA, 1976.
10. Collins, H. F., Selleck, R. E., and White, G. C., "Problems in Obtaining Adequate Sewage Disinfection," *ASCE J. San. Eng. Div.*, **97,** 549 (Oct. 1971).
11. Ptak, D. V., Ginsburg, W., and Willey, B. F., "Identification and Incidence of *Klebsiella* in Chlorinated Water Supplies," *J. AWWA,* **65,** 604 (Sept. 1973).
12. Englebrecht, R. S., Foster, D. H., Masarik, M. T., and Sai, S. H., "Detection of New Microbial Indicators of Chlorination Efficiency," paper presented at the AWWA Water Technology Conference, Dallas, TX (Dec. 1–3, 1974).
13. White, G. C., "Disinfection: The Last Line of Defense for Potable Water," *J. AWWA,* **67,** 410 (Aug. 1975).
14. White, G. C., "Disinfection Committee Report," paper presented at the AWWA Annual Conference, Minneapolis, MN, June 1975.
15. Jopling, W. F., "Water Re-use Standards for the State of California," paper presented at the annual WPCF Conference, Anaheim, CA, May 9, 1969.
16. Committee on Environmental Quality Management, "Engineering Evaluation of Virus Hazard in Water," *ASCE J. San. Engr. Div.,* **96,** 111 (Feb. 1970).
17. Sobsey, M. D., "Enteric Viruses and Drinking Water Supplies," *J. AWWA,* **67,** 414 (Aug. 1975).
18. Katz, M., and Plotkin, S. A., "Minimal Infective Dose of Attenuated Polio-Virus for Man," *Am. J. Public Health,* **57,** 1837 (1967).
19. McDermott, J. H., "Virus Problems and Their Relation to Water Supply," paper presented at Virginia Sect. Meeting, AWWA, Roanoke, VA, Oct. 25, 1973.
20. McCabe, L. J., "Significance of Virus Problems," paper presented at AWWA Water Qual. Conf., Cincinnati, OH, Dec. 3 and 4, 1973.
21. "Effect of Chlorination on Human Enteric Viruses in Partially Treated Water from Potomac Estuary," Proc. Congr. Hearings, Proc. Serv. No. 92–94. Washington, DC, 1973.
22. Clarke, N. A., and Kabler, P. W., "Human Enteric Viruses in Sewage," *Health Lab Sci.,* **1,** 44 (1964).
23. Geldereich, E. E., and Clarke, N. A., "The Coliform Test: A Criterion for the Viral Safety of Water," in V. Snoeyink and V. Griffin (Eds.), *Proc. 13th Water Qual. Conf.,* Univ. of Illinois, Urbana, IL, 1971.
24. Berg, G., "Reassessment of the Virus Problem in Sewage and in Surface and Renovated Waters," *Prog. Water Technol.,* **3,** 87–94 (1973).
25. Coin, L., et al., "Modern Microbiological Virological Aspects of Water Pollution," *Ad. Water Pollution Research,* Proc. 2nd International Conf., Pergamon Press, New York, 1966, pp. 1–10.
26. Nupen, E. M., Bateman, B. W., and McKenny, N. C., "The Reduction of Virus by the Various Unit Processes Used in the Reclamation of Sewage in Potable Waters," paper presented at the Virus Symposium, Austin, TX, Apr. 1974.
27. Selna, M. W., Miele, R. P., and Baird, R. B., "Disinfection for Water Reuse," paper presented for the Disinfection Seminar at the Ann. Conf. AWWA, Anaheim, CA, May 8, 1977.
28. Ludovici, P. P., Philips, R. A., and Veter, W. S., "Comparative Inactivation of Bacteria and Viruses in Tertiary Treated Wastewater by Chlorination," in J. D. Johnson (Ed.), *Disinfection: Water and Wastewater,* Ann Arbor Science, Ann Arbor, MI, 1975, p. 359.

29. Pavoni, V. L., and Tittlebaum, M. D., "Virus Inactivation in Secondary Wastewater Treatment Plant Effluent Using Ozone," Water Resources Symp. No. 7, in J. F. Malina, Jr. and B. P. Sagik (Eds.), *Viruses in Water and Wastewater Systems,* Univ. of Texas, Austin, TX, Apr. 1974.
30. Boardman, G. D., and Sproul, O. V., "Protection of Viruses During Disinfection by Absorption to Particulate Matter," paper presented at the 48th Annual Conf. WPCF, Miami, FL, Oct. 1975.
31. White, G. C., "Disinfection Facility Evaluation City of San Francisco," unpublished report, 1976.
32. Tsai, C. F., "Effects of Chlorinated Sewage Effluents on Fishes in Upper Patuxent River, Maryland," *Chesapeake Sci.,* **9,** 83 (June 1968).
33. Jolley, R. J., "Chlorine-Containing Organic Constituents in Chlorinated Effluents," *J. WPCF,* **47,** 601 (Mar. 1975).
34. Esvelt, L. A., Kaufman, W. J., and Selleck, R. E., "Toxicity Assessment of Treated Municipal Wastewaters," paper presented at the 44th Annual Conf. of the WPCF, San Francisco, CA, Oct. 4–8, 1971.
35. Esvelt, L. A., private correspondence, Oct. 1971.
36. "Chlorinated Municipal Waste Toxicities to Rainbow Trout and Fathead Minnows," Water Pollution Control Research Series No. 18050 GZZ 10/71, Bur. of Water Management, Michigan Dept. of Nat. Resources for the EPA, Oct. 1971.
37. Bellanca, M. A., and Bailey, D. S., "A Case History of Some Effects of Chlorinated Effluents on the Aquatic Ecosystem in the Lower James River in Virginia," paper presented at the 48th Ann. Conf. WPCF, Miami Beach, FL, Oct. 5–10, 1975.
38. Collins, H. F., and Deaner, D. G., "Sewage Chlorination versus Toxicity—A Dilemma?" *ASCE J. Environ. Eng. Div.,* 761 (Dec. 1973).
39. Arthur, J. W., Andrew, R. W., Mattson, V. R., Olson, D. T., Glass, G. E., Halligan, B. J., and Walbridge, C. T., "Comparative Toxicity of Sewage Effluent Disinfection to Freshwater Aquatic Life," EPA Report 600/3-75-012, Research Lab., Duluth, MN, Nov. 1975.
40. Brungs, W. A. "Effects of Wastewater and Cooling Water Chlorination on Aquatic Life," EPA Report 600/3-76-098, Research Lab., Duluth MN, Aug. 1976.
41. Esvelt, L. A., Kaufman, W. J., and Selleck, R. E., "Toxicity Removal from Municipal Wastewaters," SERL No. 71-8, Sanitary Engineering Research Lab., Univ. of California, Berkeley, CA, Oct. 1971.
42. Kaufman, W. J., private correspondence, 1972.
43. Stone, R. W., Kaufman, W. J., and Horne, A. J., "Long-Term Effects of Toxicity and Biostimulants on the Waters of Central San Francisco Bay," SERL Report 73-1, Univ. of California, Richmond, CA, 1973.
44. Woodbridge, D. D., and Cooper, P. C., "Reduction of Chlorination by Irradiation," paper presented at the AWWA Disinfection Seminar, Anaheim, CA, May 8, 1977.
45. A Task Force Report, "Disinfection of Wastewater." U.S. Environmental Protection Agency, No. 430/9-75-012, Washington, DC, Mar. 1976.
46. Love, O. T., Jr., Carswell, N. K., and Symons, J. M., "Comparison of Practical Alternative Treatment Schemes for Reduction of Trihalomethanes in Drinking Water," paper presented at the IOI Workshop on Ozone and Chlorine Dioxide, Cincinnati, OH, Nov. 17–19, 1976.
47. Katzenelson, E., and Teltch, B., "Dispersion of Enteric Bacteria by Spray Irrigation," *J. WPCF,* **48,** 710 (Apr. 1976).
48. Katzenelson, E., Buium, I., and Shuval, H. I., "Risk of Communicable Disease Infection Associated with Wastewater Irrigation in Agricultural Settlements," *Science,* **194,** 944 (Nov. 26, 1976).
49. White, G. C., Beebe, R. D., Alford, V. F., and Sanders, H. A., "Wastewater Treatment Plant Disinfection Efficiency as a Function of Chlorine and Ammonia Content," in R. L. Jolley et al. (Eds.), *Water Chlorination: Environmental Impact and Health Effects,* Vol. 4, Book 2, Ann Arbor Science, Ann Arbor, MI, 1983.

50. White, G. C., Beebe, R. D., Alford, V. F., and Sanders, H. A., "Problems of Disinfecting Nitrified Effluents," Municipal Wastewater Disinfection, Proc. of Second Ann. Nat. Symposium, Orlando, FL, EPA Report 600/9-83-009, July 1983.
51. Lomas, P. D. R., "The Combined Residual Chlorine of Swimming Bath Water," *J. Assoc. Pub. Analysts* (England), **5,** 27 (1967).
52. Fair, G. M., Morris, J. C., Chang, S. L., Weil, Ira, and Burden, R. P., "The Behavior of Chlorine as a Water Disinfectant," *J. AWWA,* **40,** 1051 (1948).
53. Feng, T. H., "Behavior of Organic Chloramine in Disinfection," *J. WPCF,* **38,** 614 (1966).
54. Sung, R. D., "Effects of Organic Constituents in Wastewater on the Chlorination Process," Ph.D. Dissertation, Univ. of California, Davis, CA, 1974.
55. Esvelt, L. A., Kaufman, W. J., and Selleck, R. E., "Toxicity Assessment of Treated Municipal Wastewaters," paper presented at the 44th Ann. Conf. of WPCF, San Francisco, CA, Oct. 4–8, 1971.
56. Collins, H. F., Selleck, R. F., and White, G. C., "Problems in Obtaining Adequate Sewage Disinfection," *ASCE J. San. Engr. Div.,* **97,** SA 5, Proc. #8430 (Oct. 1971).
57. Collins, H. F., White, G. C., and Sepp, E., "Interim Manual for Wastewater Chlorination and Dechlorination Practices," California State Dep. of Health, Feb. 1974.
58. Collins, H. F., and Selleck, R. E., "Process Kinetics of Wastewater Chlorination," SERL Report No. 72-5, Sanitary Engineering Research Laboratory, Univ. of California, Berkeley, CA, Nov. 1972.
59. Roberts, P. V., Aieta, E. M., Berg, J. D., and Chow, B. M., "Chlorine Dioxide for Wastewater Disinfection: A Feasibility Evaluation," Tech. Report No. 251, Civil Engr. Dept., Stanford Univ., Palo Alto, CA, Oct. 1980.
60. White, G. C., unpublished field studies, San Jose–Santa Clara Water Pollution Control Plant, San Jose, CA, 1973–76.
61. Vrale, L., and Jordan, R. M., "Rapid Mixing in Water Treatment," *J. AWWA,* **63,** 52 (Jan. 1971).
62. Selleck, R. E., Saunier, B. M., and Collins, H. F., "Kinetics of Bacterial Deactivation with Chlorine," *J. Env. Eng. Div. ASCE,* p. 1197 (Dec. 1978).
63. Snead, M. C., Olivieri, V. P., and Dennis, W. H., "Biological Evaluation of Methods for the Determination of Free Available Chlorine," in W. J. Cooper (Ed.), *Chemistry in Water Reuse,* Vol. 1, Chap. 19, Ann Arbor Science, Ann Arbor, MI, 1981.
64. Bhupinder, S. D., and Baker, R. A., "Controlling Nitrification to Reduce Energy Usage and Treatment Costs," report to Central Contra Costa San. Dist., Walnut Creek, CA, 1981.
65. Bhupinder, S. D., and Baker, R. A., "Role of Ammonia-N in Secondary Effluent Chlorination," *J. WPCF,* **55,** 454 (May 1983).
66. Collins, H. F., Selleck, R. E., and Saunier, B. M., "Optimization of the Wastewater Chlorination Process," paper presented at the National Conference on Environmental and Res., sponsored by EE Div. ASCE, Seattle, WA, July 12, 1976.
67. Selna, M. W., Miele, R. P., and Baird, R. B., "Disinfection for Water Reuse," paper presented in the Disinfection Seminar at the AWWA Annual Conf. Anaheim, CA, May 8, 1977.
68. Ingols, R. S., private communication, Georgia Inst. of Tech., Atlanta, GA, July 1977.
69. Jolley, R. L., and Pitt, W. W., Jr., "Chloro-organics in Surface Water Sources for Potable Water," paper presented for the Disinfection Symposium at the Ann. AWWA Conf., Anaheim, CA, May 8, 1977.
70. Jolley, R. L., "Chlorine Containing Organic Constituents in Chlorinated Effluents," *J. WPCF,* **47,** 601 (1975).
71. Symons, J. M., "Interim Treatment Guide for the Control of Chloroform and Other Trihalomethanes," report by the EPA Env. Res. Lab., Cincinnati, OH, June 1976.
72. Stone, R. W., "The Formation of Halogenated Organic Compounds in Wastewater Chlorination," unpublished report by Brown and Caldwell Cons. Engrs., Walnut Creek, CA, 1977.
73. Symons, J. M., Bellar, T. A., Carswell, J. K., Demarco, J., Kropp, K. L., Robeck, G. G., Seeger, D. R., Slocum, C. J., Smith, B. L., and Stevens, A. A., "National Organics Reconnaissance Survey for Halogenated Organics in Drinking Water," *J. AWWA,* **67,** 634 (Nov. 1975).

74. Hoehn, R. C., Randall, C. W., Bell, F. A., Jr., and Shaffer, P. T. B., "Trihalomethanes and Viruses in a Water Supply," paper presented at the ASCE National Conf. on Env. Eng. Res. Dev. and Design, Seattle, WA, July 12–14, 1976.
75. Breidenbach, A. W., "Interim Report on Montgomery Simulation: Study of Formation and Removal of Volatile Chlorinated Organics," EPA MERL, Cincinnati, OH, July 8, 1975.
76. White, G. C., "Disinfection Practices in the San Francisco Bay Area," *J. WPCF,* **46,** 84 (Jan. 1974).
77. Sepp, E., and White, G. C., "Manual for Wastewater Chlorination and Dechlorination Practices," State of California, Dept. of Health Services, Sanitary Engineering Section, Mar., 1981.
78. Sepp, Endel, "Optimization of Chlorine Disinfection Efficiency," *J. Environ. Engrng. Div. ASCE,* **107,** EE1 (Feb. 1981).
79. Sepp, E., and Bao, P., "Design Optimization of the Chlorination Process, Vol. I, Comparison of Optimized Pilot System With Existing Full-Scale Systems," Wastewater Research Div., Mun. Env. Res. Lab., Cincinnati, OH, and California Dept. of Health Services, Berkeley, CA, 1980.
80. Trussell, R. R., and Pollock, T., "Design of Chlorination Facilities for Wastewater Disinfection," paper presented at Pollock, T., "Design of Chlorination Facilities for Wastewater Disinfection," paper presented at Preconference Workshop, Ann. Conf. WPCF, Atlanta, GA, Oct. 2, 1983.
81. Anon., "Wallace and Tiernan V-Notch Chlorinators," Wallace & Tiernan–Pennwalt, Belleville, NJ, Cat. No. 25.052, Rev. 8-79.
82. Finger, R., private communication, Renton WWTP, Renton, WA, Apr. 1984.
83. Fraschina, K., "Chlorine Residual Distribution in Open Channel Flow," Richmond–Sunset Wastewater Treatment Plant, San Francisco, CA, 1940.
84. Rudolfs, W., and Gehm, H., "Sewage Chlorination Studies," Bulletin No. 601, New Jersey Agric., Exper. Sta., New Brunswick, NJ, Mar. 1936.
85. Camp, T. R., and Stein, P. C., "Velocity Gradients and Internal Work in Fluid Motion," in *Civil Engineering Classics: Outstanding Papers by Thomas R. Camp,* ASCE, New York, 1973, p. 203.
86. Longley, K. E., "Engineering Design of a Disinfection System," paper presented at National Conf. Environmental Engineering ASCE, Minneapolis, MN, July 14–16, 1982.
87. Collins, H. F., "Process Kinetics of Wastewater Chlorination," Ph.D. Dissertation in Engineering, Univ. of California, Berkeley, CA, 1971.
88. Eliassen, R., Heller, A. N., and Krieger, H. L., "A Statistical Approach to Chlorination," *Sew. Wks. J.,* **20,** 1008 (1948).
89. Krusé, C., Kawata, K., Olivieri, V., and Longley, K., "Improvement in Terminal Disinfection of Sewage Effluents," *Water and Sew. Wks.,* **120,** 57 (June 1973).
90. Heukelekian, H., and Day, R. V., "Disinfection of Sewage With Chlorine; III Factors Affecting Coliforms Remaining and Correlation of Orthotolidine and Amperometric Chlorine Residuals," *Sew. and Ind. Wastes,* **23,** 155 (Feb. 1951).
91. Longley, K. E., "Mixing and Chlorine Disinfection," paper presented at the Ann. WPCF Conf., Minneapolis, MN, Oct. 4, 1976.
92. Longley, K. E., "Turbulence Factors in Chlorine Disinfection of Wastewater," *Water Research,* **12,** 813 (Nov. 1978).
93. Stenquist, R. J., and Kaufman, W. J., "Initial Mixing in Coagulation Processes," Univ. of California, Berkeley, CA, SERL Report 72-2, Feb. 1972.
94. Calmer, J. C., Sanchez-Adams, R. M., and Tobin, L. D., "Chlorine Mixing Energy Requirements for Coliform Disinfection of Non-nitrified Secondary Effluents," paper presented at Ann. Conf. WPCF, New Orleans, LA, Oct. 1984.
95. Egan, J. T., "Chlorine Solution Diffusers and Mixers in Effluent Channels," *Calif. Water Poll. Control Assoc. Bulletin,* p. 38 (Apr. 1978).
96. Egan, J. T., private communication, City of Colorado Springs, CO, Nov. 1983.
97. Mixing Equipment Co., private communication, Rochester, NY, 1973.

98. White, G. C., Bean, E. L., and Williams, D. B., "Chlorination and Dechlorination, A Scientific and Practical Approach," *J. AWWA,* **58,** 540 (1968).
99. Levy, A. G., and Ellms, J. W., "Hydraulic Jump as a Mixing Device," *J. AWWA,* **17,** 1 (Jan. 1927).
100. Mandt, M. G., private communication, Pentech Div., Houdaille Ind. Inc., Cedar Rapids, IA, Mar. 1977.
101. Anon., Enhanced Disinfection System Fischer and Porter Specification 71 ED 1000, June 1981.
102. Deaner, D. G., "Effect of Chlorine on Fluorescent Dyes," *J. WPCF,* **45,** 507 (Mar. 1973).
103. Marske, D. M., and Boyle, V. D., "Chlorine Contact Chamber Design—A Field Evaluation," *Water and Sew. Wks.,* **120,** 70 (Jan. 1973).
104. Trussell, R. R., and Chao, J. L., "Rational Design of Chlorine Contact Facilities," *J. WPCF,* **49,** 659 (1977).
105. Calmer, J. C., and Adams, R. M., "Design Guide Chlorination–Dechlorination Contact Facilities," in-house report, Kennedy–Jenks Engineers, San Francisco, CA, July 1977.
106. Hart, F. L., "Improved Hydraulic Performance of Chlorine Contact Chambers," *J. WPCF,* **51,** 2868 (Dec. 1979).
107. Hart, F. L., and Vogiatzis, Z., "Performance of Modified Chlorine Contact Chamber," *J. Env. Engr. Div. ASCE,* **108,** EE3, 549 (June 1982).
108. Thirumurthi, D., "A Breakthrough in the Tracer Studies of Sedimentation Tanks," *J. WPCF,* **41,** Part 2, R405 (Nov. 1969).
109. Levenspiel, O., *Chemical Reaction Engineering,* John Wiley & Sons, New York, 1962, p. 250; 2nd ed., 1972, p. 260.
110. Morrill, A. B., "Sedimentation Basin Research and Design," *J. AWWA,* **24,** 1442 (1932).
111. Stahl, J. F., "Chlorine Contact Chamber Design, Pomona Water Renovation Plant," in-house report for Sanitary Districts of Los Angeles County, CA, Whittier, CA, ca. 1972.
112. Egan, J. T., private communication, City of Colorado Springs, CO, May 4, 1984.
113. Karaganis, J. V., "Sewage Treatment: The Present Situation Won't Do," *Water and Sew. Wks.,* **125,** Editorial (Nov. 1978).
114. Hedenland, L. D., private communication, Las Virgenes Municipal Water District, Calabasas CA, Dec. 10, 1974.
115. Gardiner, Jack, "Water Champ," Gardiner Equipment Co., Houston TX, 1986.
116. Wallace & Tiernan Inc., Belleville, NJ, "Gas Flow Control Apparatus," U.S. Patent No. 2,929,393, Mar. 22, 1960.
117. Levenspiel, O., *Chemical Reaction Engineering,* 2nd ed., John Wiley & Sons, New York, 1972, Chap. 10, p. 326.
118. Selleck, R. E., Professor, Univ. of California, Berkeley, CA, private communication, 1976.
119. Weir, Charles, "Water Champ—Final Report," City of Sunnyvale, CA, WWTP, May 16, 1988.
120. Wolfe, Roy L., Means, Edward G., III, Davis, Marshall K., and Barrett, Sylvia, "Biological Nitrification in Two Covered Finished Water Reservoirs Containing Chlorinated Water," *J. AWWA,* p. 109 (Sept. 1988).
121. Wolfe, Roy, L., Lieu, Nancy I., Izagurre, George, and Means Edward G., "Ammonia-Oxidizing Bacteria in a Chloraminated Distribution System: Seasonal Occurrence, Distribution, and Disinfection Resistance," *Appl. Envir. Microbiol.,* pp. 451–462 (Feb. 1990).
122. Orange County Water District and County Sanitation Districts of Orange County, Orange County Regional Water Reclamation Project Feasibility Study Report, 1995.
123. Mills, William R., Jr., and Van Haun, James A., "Proposed Wastewater Reclamation Plant equipped with Microfiltration, Reverse Osmosis, Final Disinfection for Groundwater Recharge," Orange County Water District, Fountain Valley, CA, May 1996.
124. Monterey County Water Resources Agency, "Salinas Valley Reclamation Projects," Salinas, CA, 1991.
125. Monterey County Water Pollution Control Agency, "The Castroville Seawater Intrusion Project," Monterey, CA, 1991.

126. Mayers, Robert J., and Segler, Kurt R., "Thirsty in the Desert," *Water Environment & Technology,* p. 42 (Oct. 1995).
127. Rider, David, and Suozzo, James, "Dual-Stage Filter Removes Protozoan Cysts from Effluent," *Water Environment & Technology,* p. 15 (June 1996).
128. Noesen, Mathew, and Jaques, Robert, "UV Pilot Testing at Monterey Regional Water Pollution Control Agency," *California Water Pollution Control J. Bulletin,* p. 19 (Winter 1995).
129. White, George C., "Effect of California Coliform Standards upon Wastewater Dischargers, Water Reclamation Projects, Recreation and Surface Water Supplies," paper presented at the Disinfection Workshop, Ann. WPCF Conf., Houston, TX, Oct. 7, 1979.
130. Hetzler, Jens T., and Spielman, John, "ORP, A Key to Nutrient Removal," *Operations Forum Magazine,* **12,** No. 2 (Feb. 1, 1995).
131. Wareham, David G., Hall, Kenneth J., and Mavinic, Donald S., "Real-Time Control of Wastewater Treatment Systems Using ORP," *Water Science Technology,* **28,** No. 11–12, 273–282 (1993).
132. Rosenblum, Eric, "South Bay Water Recycling Project," Research and Design Engineer, WWTP, 700 Los Esteros Road, San Jose, CA 95134. Nov. 8, 1996.
133. Spielman, John, personal communication, "ORP Operation Parameters," Scientific Utilization, Inc., 201 Electronics Blvd, SW, P. O. Box 6787, Huntsville, AL 35824-0787.
134. Young, Ronald, "Michelson Water Reclamation Plant," Irvine Ranch Water District, 15600 Sand Canyon Ave., Irvine, CA 92718, Customer Service Ph 1-714-453-5850.
135. Redner, John, project engineer, Los Angeles County Sanitation District Laboratory, 920 So. Alameda Street. Compton, CA 90221, personal communication, 1996.

9
CHLORINE FACILITIES DESIGN

PREFACE

The conventional chlorination facility for use in potable water and wastewater treatment consists of three principal parts: chlorine supply, metering system, and injector system. In addition there is ancillary equipment: safety equipment, metering and control instrumentation, and chlorine residual analyzers.

CHLORINE STORAGE AND SUPPLY SYSTEMS

Chlorine is packaged in special steel containers of various sizes, as follows:

1. 100- and 150-lb cylinders.
2. Ton containers.
3. Single-unit tank cars.
4. Multiple-unit tank cars (TMU) containing 15 one-ton cylinders.
5. Tanker trucks of 15–20 tons capacity.
6. Stationary storage tanks.

The selection of the various sizes is discussed in this chapter.

100- and 150-lb Cylinders

The pertinent dimensions and tare weight of these containers are shown in Fig. 9-1. Minor variations in these dimensions depend upon the cylinder age and the manufacturer. The 150-lb cylinder is so popular that the 100-lb size may be considered obsolete. Some packagers have available a 35-lb cylinder, which is suitable for laboratory or test work on a small scale. (See Fig. 9-2 and Table 9-1.)

The packager fills these cylinders with liquid chlorine to approximately 85 percent of total volume; the remaining 15 percent is occupied by the chlorine gas. There must be strict adherence to these figures in order to prevent hydrostatic rupture of the cylinder in the event of abnormally high ambient temperatures. As the temperature rises, the liquid chlorine expands. Theoretically, the cylinder could get hot enough for the liquid chlorine to completely fill the remaining 15 percent occupied by gas and thereby rupture the cylinder.

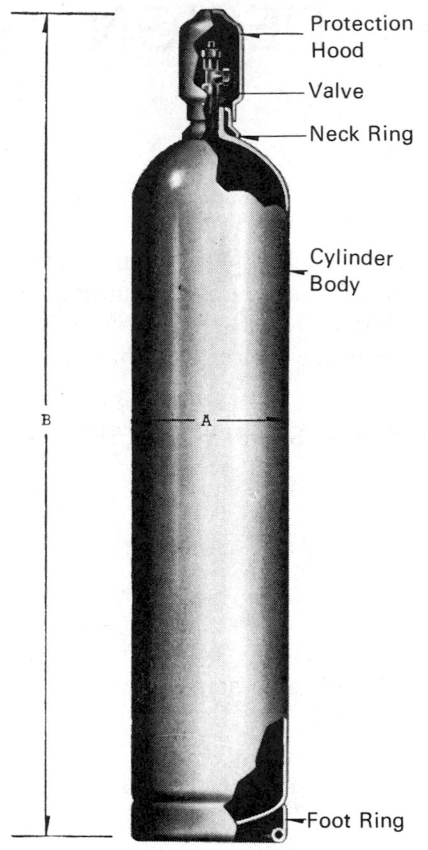

Net Cylinder Contents	Approx. Tare, Lbs.*	Dimensions, Inches	
		A	B
100 Lbs.	73	8 1/4	54 1/2
150 Lbs.	92	10 1/4	54 1/2

*Stamped tare weight on cylinder shoulder does not include valve protection hood.

Fig. 9-1. Chlorine cylinder dimensions (courtesy of PPG Industries, Chemical Division).

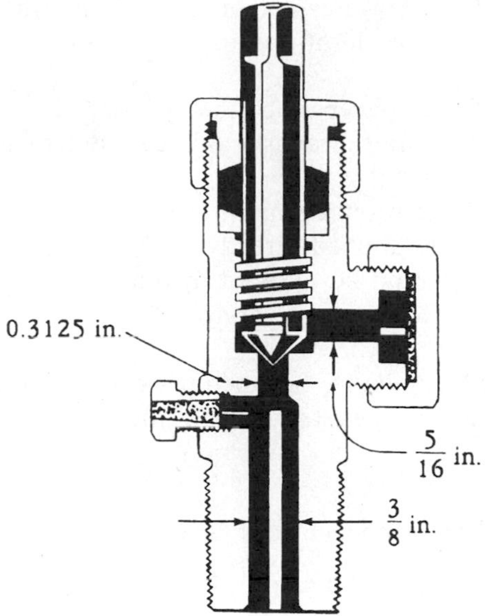

Fig. 9-2. Dimensions of 150-lb chlorine cylinder valve with fusible plug.

However, the outlet valve on these cylinders is fitted with a fusible plug, the core of which will melt at approximately 158°F, thus preventing rupture of the cylinder during instances of abnormally high temperatures. When the plug melts, the liquid chlorine discharged through the core opening (⅛-inch diameter) cools so rapidly that it freezes, momentarily halting the flow of liquid chlorine. By this time the danger of cylinder rupture is over, and a trained operator wearing air or oxygen breathing equipment can apply the chlorine safety kit, stop the leak, and remove the cylinder from the area.

The gross weight of a full 150-lb cylinder varies from 250 to 285 lb. Therefore these cylinders are best handled by a two-wheel cylinder handtruck. *Never use slings, or try to pick up cylinders with hoisting equipment attached to the*

Table 9-1 Various Sizes of Chlorine Container Vessels (in Gallons and Cubic Feet)

	Gallons	*Cubic Feet*
150-Lb cylinder	14.4	1.93
Ton cylinder	192	25.67
30-Ton tank car	5762	770.32
55-Ton tank car	10,564	1412.30
90-Ton tank car	17,298	2312.57
Water = 62.366 lb/ft^3 at 60°		

protective cap or valve. The water volume of a 150-lb cylinder is 14.4 gal. The most important design considerations are as follows:

1. Direct sunlight must never reach the cylinder.
2. The maximum withdrawal rate should be limited to 40 lb per day per cylinder.
3. Minimum allowable room temperature is 50°F.
4. Heat must never be applied directly to the cylinder.
5. Sufficient space should be allowed in the supply area for at least one spare cylinder for each one in service.

Provisions should be made so that the operator can determine the amount of chlorine left in the supply system. This is most effectively done by weighing scales.

Capital Controls Electronic Cylinder Scale[130]

System Description. Figure 9-3 illustrates a new concept in scales for weighing 100- or 150-lb chlorine cylinders. This scale utilizes an electronic support instead of a center post and uses low-profile "no lift" platforms for weighing the cylinders. The unique weighing base design permits off-center cylinder placement with no loss of accuracy.

This is the Capital Controls Co. ADVANCE™ Electronic Dual-Platform Cylinder Scale—Series 1361; it is ISO-certified. This electronic dual platform scale can be used for either chlorine or sulfur dioxide 150-lb cylinders. It consists of a dual platform weighing base and a wall-mounted weight indicator. Each platform has a gross weight capacity of 350 lb and a tare capacity of −180 lb. It has an accuracy of plus or minus 0.5 percent of the gross weight capacity, even if the cylinders are placed off-center on the platform. The electronic weight indicator has a separate LCD display for each scale platform. The electronic display features ½-inch-high characters for easy visibility. Jumpers allow easy field selection of displayed weight in pounds or kilograms.

The cylinder platforms are low-profile with rounded edges for easy-on/easy-off movement of the gas cylinders with no lifting. Built-in stops are provided to prevent overload damage.

The scale system is fitted with separate 3½ digit LED display for each platform, readable from a distance up to 10 ft. The weight indicators can be in either pounds or kilograms. The weight indicator electronics are housed in a NEMA 4X (IP65) enclosure to prevent corrosion from the small amount of chlorine that always escapes in changing cylinders.

The system operates on 120 V AC, 240 V AC, and 50/60 Hz single phase power source. Ambient temperature operation limits are from 35°F to 135°F.

Automatic switchover devices are readily adapted to this scale.

Capital Controls has five scales in this series for use in the market, each offering a slightly different feature/benefit for the user. The bases are all the

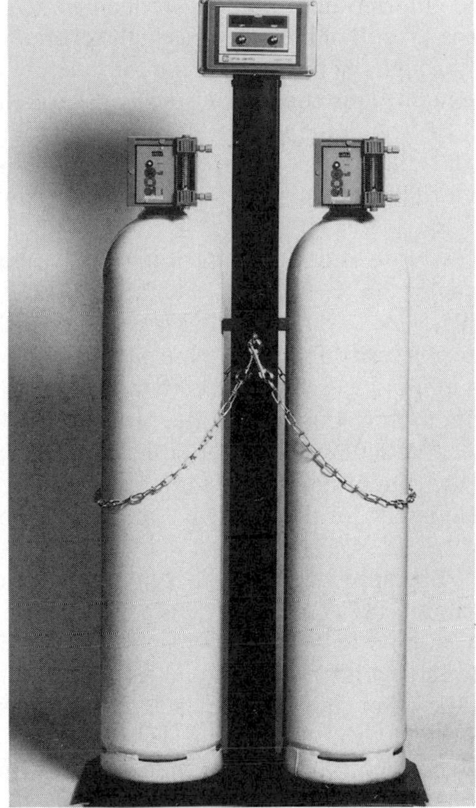

Fig. 9-3. Two-cylinder platform scal, Model 1361 B.

same, but the display (weight indication) differs slightly. Model 1363B has a remote indicator and no center post. It can be used when operating personnel want to have a weight indication outside the room. The other three scales offer variations to provide for a different location or type of weight indicator.

There is one additional scale, Model 1364B, specifically designed to handle the wider ammonia cylinders. Physically it looks the same, but its base dimensions are different from the others. The other scales are 1360B and 1363B.

Various Operating Situations. Ideally the chlorine supply system should always be at a lower temperature than the chlorinator when withdrawal is from the gas phase. This reduces the possibility of reliquefaction at the chlorinator.

The distance between the chlorine supply and the chlorinator should always be as short as conveniently possible.

The greatest difficulty in operating such installations is caused by reliquefaction of the chlorine gas. This occurs mostly at the first point of pressure reduction; when it occurs, impurities in the chlorine gas are deposited at this point, which is the chlorine inlet pressure-reducing valve of the chlorinator. This phenomenon is a result of the warm gas in the cylinder passing very slowly (inches per hour) through the piping between the cylinder and chlorinator and cooling during the night. This cooling causes the gas to reliquefy. The amount and the frequency of trouble depend on how much cooling takes place. This is a function of the difference in ambient temperatures between day and night, the volume of gas between cylinder and chlorinator, and the velocity of gas flow.

The lower the feed rate and the greater the distance between cylinder and chlorinator, the greater the chance of reliquefaction, other things being equal. Reliquefaction will not occur if the cylinder is kept cooler than the chlorinator. This brings up the necessity for some type of insulation, discussed later in this chapter. The minimum allowable temperature for the chlorine storage area is about 50°F. Below this temperature the flow of chlorine becomes sluggish and erratic, particularly for smaller installations, from 1 to 20 lb/day. Therefore all 150-lb cylinder installations utilizing one or two cylinders should be housed in adequately insulated areas with provision for heating to at least 65°F.

Heat should never be applied directly to a chlorine cylinder. Steel will ignite spontaneously at about 483°F in the presence of chlorine. For example, it is possible to "burn" a hole in a chlorine cylinder by directing the rays of an infrared lamp onto the cylinder at the liquid level. With heat on one side, the liquid chlorine on the other side acts as a catalyst and will support the burning of the steel* until the chlorine has become exhausted.

Sufficient space should be allowed for the scale and storage for one spare cylinder for each one connected in service. The distance between the cylinders should be sufficient so that standard four-foot flexible connections can be used. Each cylinder hookup should consist of an auxiliary cylinder valve and flexible connection.†

Filters and traps ahead of all chlorinator control apparatus are highly desirable, to prevent the impurities inherent in chlorine from reaching the chlorinator control mechanisms. Most small chlorinators have a built-in strainer of some type, which should be preceded by a convenient inlet trap.

If the storage area for in-service cylinders is properly designed, external chlorine pressure-reducing valves are not required. If the location of these cylinders is remote, an external reducing valve should be installed as close as possible to the cylinders.** If the chlorine supply line is longer than 10 or 15 ft and is subject to much variation in temperature due to poor insulation from ambient temperature changes, an external chlorine pressure-reducing

*See the cause of the 14,000-lb leak in Chapter 1.

†Cylinder-mounted chlorinators do not require flex connections or auxiliary cylinder valves. Nevertheless they are still subject to reliquefaction.

**When remote vacuum systems are used, external pressure reducing valves may be required to provide a two-step pressure reduction owing to ambient conditions.

valve should be installed close to the cylinders to avoid reliquefaction. If the vapor pressure in the chlorine cylinder is 100 psi, the gas between the cylinder and the chlorinator will reliquefy if the temperature drops below 80°F. If a reducing valve is utilized to reduce the pressure from 100 to 40 psi, the temperature would have to drop to below 32°F before reliquefaction will occur. (See Chapter 1.)

Automatic Switchover. The ability to switch from an empty cylinder to a full cylinder automatically increases the reliability of any chlorination system. Automatic switchover devices can perform in pressure systems or vacuum systems.

For systems up to 500 lb/day the vacuum system is ideal. A typical system is illustrated in Fig. 9-4. The vacuum regulator check units shown adjacent to the cylinder may be located remotely from the chlorinator (e.g., for separate housing location of cylinders).

The system functions as follows: Gas under pressure enters the vacuum regulator check unit. As the gas flows through the two valves, the cylinder pressure is reduced to the vacuum in the system created by the chlorinator injector. The valves will not open unless the minimum operating vacuum is produced (approx. 10 in. Hg). Flow of gas will be indicated by a depressed red indicator button. If the first valve allows gas to flow when a vacuum is not present, the second valve will remain closed and contain the pressure in the vacuum regulator check unit, which is designed to contain full container pressure. These vacuum regulator check units are fitted with mechanical latches called "detents." One valve remains open until the cylinder or cylinders connected to this valve are exhausted. This situation allows the operating vacuum to rise significantly higher than normal (3–5 in. Hg). This rise in vacuum level provides sufficient force to unlatch the second unit, which then takes over the gas supply function.

For safety reasons a **standby pressure relief valve** must be installed as shown in Fig. 9.4. This valve must be vented to atmosphere well above ground level.

Positive pressure systems are described later in this chapter.

Effect of Uniform Fire Code

TGO Technologies, Inc.[115]

100- and 150-lb Cylinders. These cylinders can now be installed and operated in and from a total containment vessel to comply with the latest revision of the California Fire Code, Title 24, Part 9, Article 80, adopted in January 1996. This Code requires that every chlorination system using chlorine supply cylinders must have a treatment system for mitigating accidental releases of chlorine gas or liquid chlorine.

These systems are required to reduce to 5 ppm the atmospheric chlorine concentration in the area surrounding the accidental leak before diverting the release to the external atmosphere.

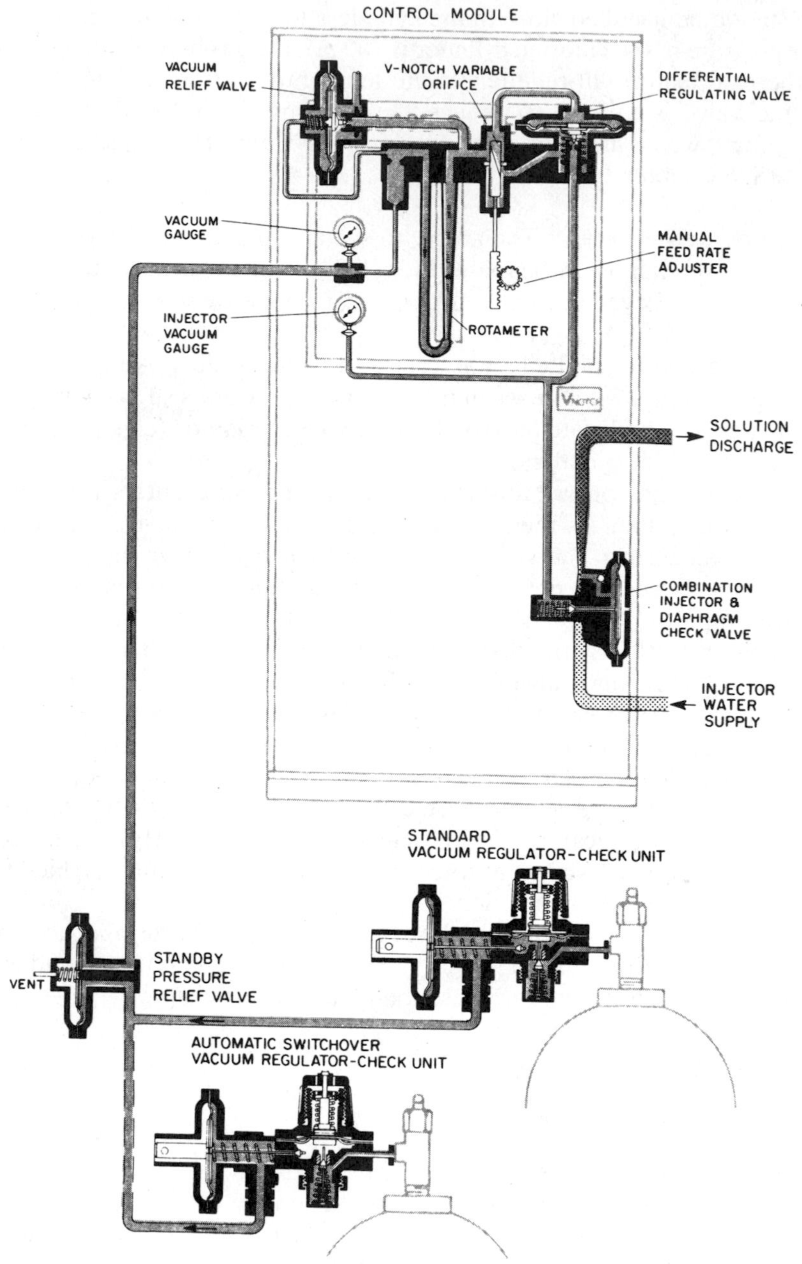

Fig. 9-4. Automatic cylinder switchover system, vacuum differential type (courtesy Wallace & Tiernan).

The TGO Technologies containment vessels eliminate the need for costly scrubber systems. Since accidental chlorine emission from either a 100- or a 150-lb container is extremely rare, this option can be expected to be very popular, given that the containment vessel and its accessories generally cost no more than about $25,000 per installation.

The containment vessel installation is illustrated in Fig. 9-5 for both 100- and 150-lb chlorine or sulfur dioxide cylinders. These installations should be made so that the containment vessel is never subjected to sunlight or rain.

General Discussion. TGO Technologies is a practicing packager of chlorine; its work is based upon many years of practical experience. The benefits of using an operational, total containment vessel are as follows:

1. It prevents operating personnel from being trapped inside a confined space while handling chlorine gas. Owing to the expansive nature of chlorine gas, a minor leak in a confined space can easily create a lethal environment. Although a room with a scrubber may protect the environment, it may do so at the risk of the operator. Experienced packagers and processors of chlorine do not allow their operators to package this product in enclosed rooms or confined spaces.

2. It eliminates construction of a special room and requires the installation of a caustic scrubber. The containment vessels need only to be sheltered from sunlight and rain. This allows the operator to be in an open air or outdoor environment while containing the cylinder in the process. A release, should one occur, is contained within the vessel. Installation of vessels for compliance typically costs a fraction of the cost of a sealed room plus a scrubber. The release inside the containment vessel is easily put back into the process at a controlled rate through the vacuum chlorinator system.

3. These total containment vessels are simply set in place with seismic restraint and are easy to install. Very little engineering or construction is

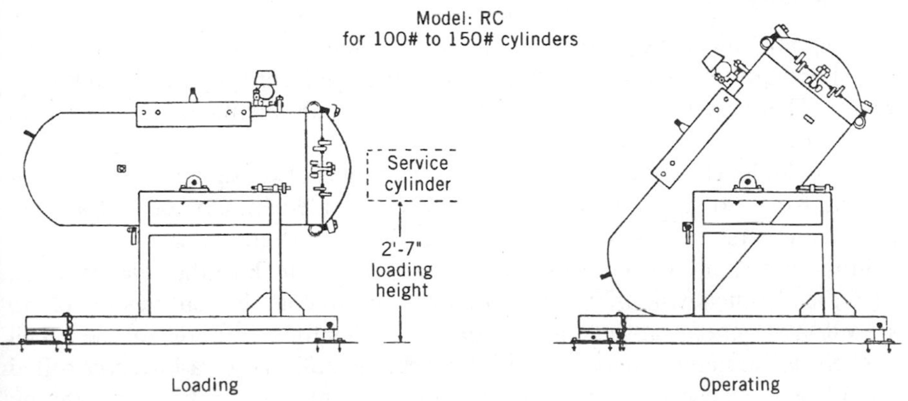

Fig. 9-5. Chlorine cylinder containment vessels, 100 lb and 150 lb (courtesy of TGO Technologies).

required. The chlorination vacuum system is simply attached to the fail-safe nitrogen-operated shut-off valve that is mounted on the outlet port of the total containment vessel, eliminating the need for a scrubber.

4. There is dependable operation with minimal maintenance. Additional backup emergency power is not necessary for the total containment vessel, as it will contain a release in the event of a power loss. Owing to the simplicity of design, no additional pumps, scrubbers, tanks, or caustic solution replacements are required. This eliminates the maintenance associated with unnecessary equipment.

5. The use of a simple seismic detection device that is connected to the fail-safe valve will shut down the system during an earthquake. When the quake is over, the system will reset automatically. If a seismic event occurred in an enclosed room with a scrubber, it could render the scrubber inoperable.

6. Chlorine stored or processed within rooms or confined spaces are classified as hazardous material and must be handled in accordance with the Hazmat group. Therefore, it is always preferable if possible to avoid any Hazmat occupancy environments because OSHA requires a six-member trained response team to respond to the release of any toxic gas. Two response team members, who enter the space, must be in self-contained breathing apparatus and chemical suits with harnesses. Two team members must remain outside the room with tethers to extract the first two from the room if necessary, one team member stands by for decontamination, and the team supervisor coordinates the response. Since many facilities do not have a six-member team available to address a release, Hazmat occupancy environments are not practical.

For all the above reasons, the use of total containment vessels for both 150-lb and one-ton cylinders should be considered the safest and most cost-effective way to satisfy the requirements of Article 80 of the Uniform Fire Code.

Ton Cylinders

This type of chlorine cylinder containment arrangement is illustrated by Fig. 9-6, showing the truck off-loading requirement adjacent to the containment vessel. This system must also meet all the requirements of the Uniform Fire Code.

This cylinder containment facility requires a specially designed truck unloading platform, so that the ton cylinder can be rolled into place in the high pressure containment vessel that encloses the ton cylinder.

In the event that the chlorine cylinder develops a leak (either gas or liquid Cl_2), the chemical will all be contained within the containment vessel. There are automatic chemical-sensing shut-off valve assemblies mounted at the vessel port on the containment vessel. This valve system allows for a low, controlled rate of leak evacuation flow from the containment vessel. This prevents the escape of any toxic gas or liquid from the chlorine cylinder into the environment.

Fig. 9-6. Ton container vessel (courtesy TGO Technologies).

As mentioned previously for 150-lb cylinders, the controlled-leak-rate evacuation from the containment vessel allow for elimination or reduction of the large scrubber systems for disposal of such a leak, which is normally needed to handle a high-volume leak rate.

Further details are available from TGO Technologies. (See Ref. 115. Customer Service phone is 800-544-6604.)

Ton Container Features

General Discussion. Unlike the 150-lb cylinders, either gas or liquid may be withdrawn from ton containers; consequently each container has two outlet valves. Also, unlike the small cylinders, the ton containers have six fusible plugs—three in each of the dished heads. Figure 9-7 illustrates the ton container. (See also Figs. 9-8 and 9-9.) They are transported either by truck (Fig. 9-10) or by a multiple-unit tank cars (TMU). A truck can carry a maximum of 14 containers; a TMU, 15, as shown in Fig. 9-11. *The water volume of a ton container is 192 gals.*

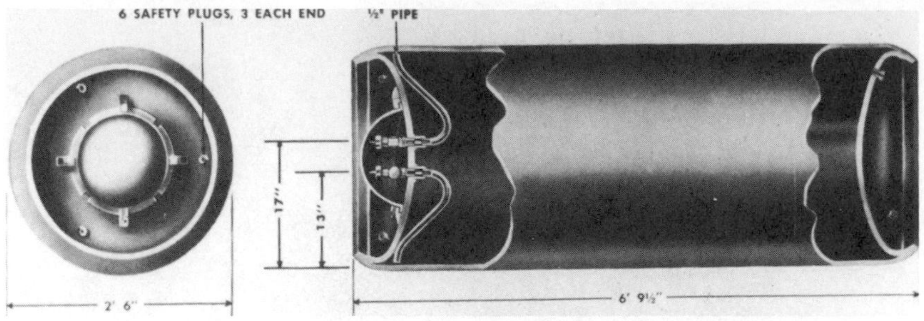

Fig. 9-7. Ton container for chlorine (courtesy Chlorine Institute).

The gross weight of these containers (3500 lb) dictates that proper handling equipment must be used. The container is designed for use in the horizontal position. Each cylinder must be positioned so that the outlet valves line up in the vertical orientation before being connected to the supply system. In other words, the container must be positioned so that one eductor tube is in the gas position and the other in the liquid when the container is full.

The worst part of the ton cylinder design is the use of fusible plugs, which serve no apparent purpose. The fusible plug does not melt until the cylinder is subjected to about 163°F, whereas the dished heads on the ton cylinder will blow from concave to convex at about 157°F. The only thing that White has ever seen them do is to cause serious leaks that are unstoppable. They suffer from corrosion due to the small amount of moisture in the liquid chlorine container. Some have been known to produce leaks after only three years of service.

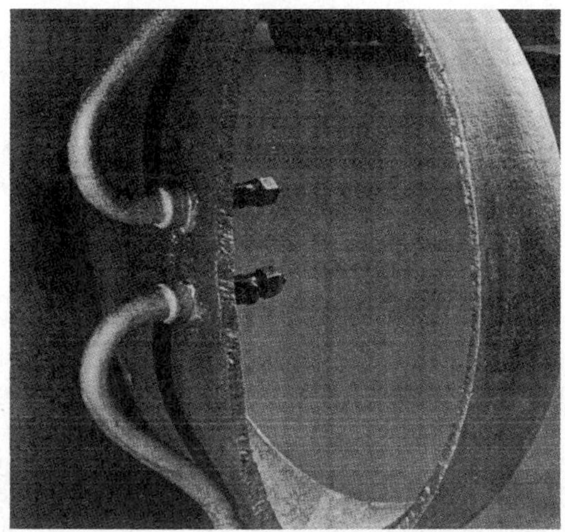

Fig. 9-8. Cutaway of 1-ton cylinder showing ½ in. diameter withdrawal tubes, 22 in. long, and outlet valves.

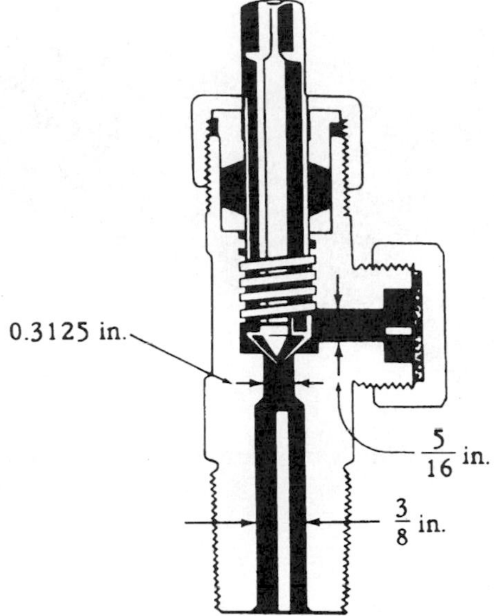

Fig. 9-9. Dimensions of ton cylinder valve.

Handling Equipment. Proper handling equipment includes the following:

1. Two-ton capacity electric hoist
2. Lifting bar
3. Cylinder trunnions

Fig. 9-10. Ton container truck and trailer rig with boom (courtesy PPG Industries, Chemical Division).

Fig. 9-11. TMU car (courtesy Chlorine Institute).

4. Monorail for hoist
5. Cinch straps*

Good examples of proper handling equipment are illustrated by Figs. 9-12, 9-13, and 9-14.

Not only must the cylinders be moved from the transport to the supply position; it is most important that each cylinder be connected so as to be easily rotated in order to align the outlet valves vertically. This is accomplished by a pair of trunnions (Fig. 9-13), which serve not only for positioning the outlet valves but also for the spacing and support for each cylinder. Further, if the trunnion is properly designed, it should function to contain the cylinder in the event of a collision with an incoming cylinder on the traveling hoist, and to prevent an empty cylinder from rotating without an external force of at least 15 ft lb.†

One of the critical design dimensions for ton container installations is the distance from the bottom of the monorail to the floor of the container room. The monorail must be high enough to pick up a cylinder off the truck and also high enough to lift one cylinder over another that is on the floor. Usually the governing distance is the height of the truck bed above the container floor. Figure 9-14 illustrates how to arrive at the minimum distance of the monorail above the floor.

*Where seismic forces are of concern, plastic cinch straps should be used to prevent the cylinder from becoming dislodged from the trunnions.[70]

†A recent earthquake with the epicenter at Livermore, CA and a 5.8–6.2 Richter force failed to move banks of ton containers at two wastewater treatment plants nearby. Plant container alignments were 90 degrees different from each other. Similarly, the October 1989 San Francisco quake, known as the Loma Prieta 7.6 quake, did not affect ton cylinders on trunnions without cinch straps.

Fig. 9-12. Typical chlorine storage with trunnions for rotation and spacing of ton containers (courtesy of Chlorine Specialties, Inc.).

Gas Withdrawal. Space requirements depend upon whether the cylinders are for liquid or gas withdrawal, the rate of withdrawal, the quantity price break for the number of cylinders delivered at one time, and the length of time containers can be used without incurring demurrage charges.

At room temperature the maximum gas withdrawal rate from a ton cylinder is approximately 400 lb/day. If the maximum chlorinator capacity is 400 lb/day, then one cylinder in service will suffice; for a 500 lb/day unit, two ton cylinders must be in service simultaneously. Theoretically, gas withdrawal can be used up to any capacity if enough cylinders are connected to the supply header. Switching from gas withdrawal to liquid withdrawal utilizing an evaporator is necessary when a continuous rate of 1500 lb/day is reached. The one exception to this rule of thumb is in intermittent-operating installations, as

Fig. 9-13. Trunnion for ton container (courtesy of Chlorine Specialties, Inc.).

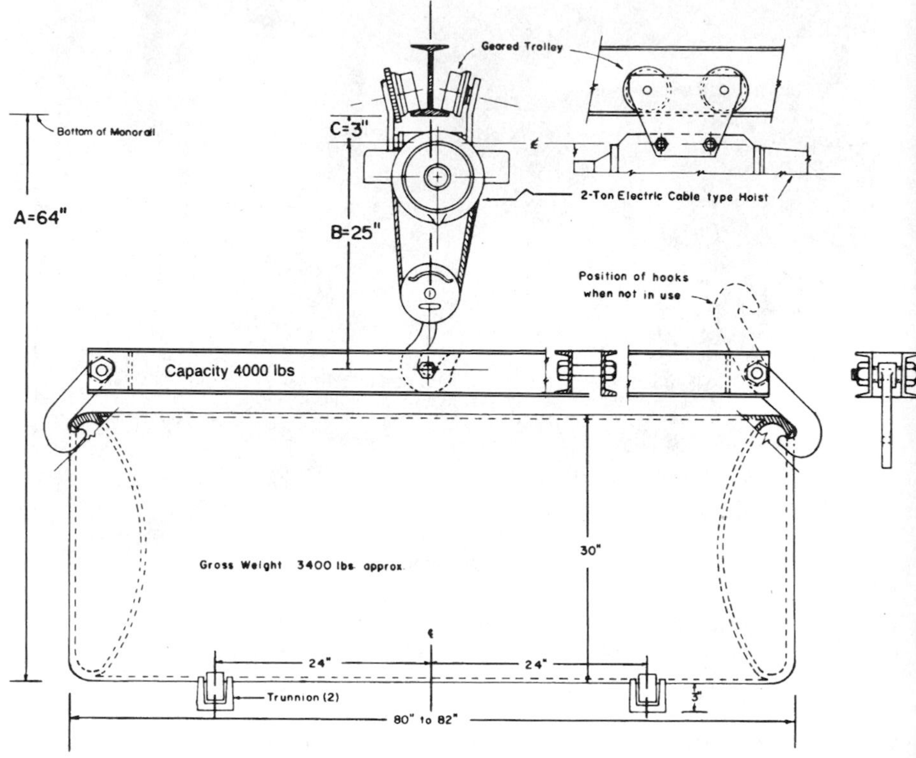

Fig. 9-14. Monorail height location. To determine monorail height above grade add truck bed height to distance "A" as illustrated. Check dimensions "B" and "C" with hoist manufacturer (courtesy Chlorine Specialties, Inc.).

employed on cooling water circuits. At such installations, it is customary practice to withdraw rates up to 1000 lb/day for 30 min. if the temperature of the storage area never goes below 50°F. In warmer areas, 1500 lb/day gas withdrawal is safe up to one hour from a single ton cylinder. This assumes a temperature–pressure restoration period of no gas withdrawal that is at least twice as long as the withdrawal period.

For up to 1500 lb/day maximum continuous withdrawal, space should be provided for twelve cylinders in simultaneous service, four standby cylinders, and four empty spaces for the next delivery. Beyond this rate, an evaporator should be installed. However, there is one exception, as follows:

In hot climates where the "in shade" summer temperatures exceed 95–100°F on a consistent basis, it is desirable to consider the use of an evaporator. Normally those climates experiencing summer temperatures of 100°F usually experience winter temperatures of 15°F. Therefore, the evaporator concept takes care of both extremes of climatic temperature. Liquid withdrawal of chlorine to an evaporator from a supply system is least affected by ambient temperature. The only precaution is to prevent direct sunlight on the cylinders. For winter operation no special precautions are required (such as artificial heating) because the gas temperature at the outlet of the evaporator will usually *exceed* 100°F regardless of room temperature, and the temperature of the gas in the vacuum line to the injector will never fall below the critical temperature (35°F) where chlorine hydrate occurs.

Evaporators are available in capacities of 4000 lb/day, 6000 lb/day, and 8000 lb/day. In a pinch, one cylinder can discharge liquid to an evaporator at a rate as high as 12,000 lb/day. This means that an evaporator can be used to conserve space for cylinder storage if necessary. The optimum storage requirements should be based on the quantity discount price break offered by the local chlorine supplier. This usually occurs at a quantity of five, thus dictating space for fifteen cylinders: five in service, five empties, and five vacant spaces for the incoming cylinders.

If the gas phase is used, the same consideration must be given to the design of ton cylinder storage space as for 150-lb cylinders. All ton cylinder installations using gas withdrawal should be equipped with a special filter installed as close as possible to the last ton container (Fig. 9-15). If the header between the last cylinder and the chlorinator is subject to temperature variations of 20°F or more over a 24-hour period, then it is desirable to install an external pressure-reducing valve. This valve prevents liquefaction of the chlorine in the header system and the chlorinator mechanism caused by wide variations of the ambient temperature. This valve should be installed immedi-

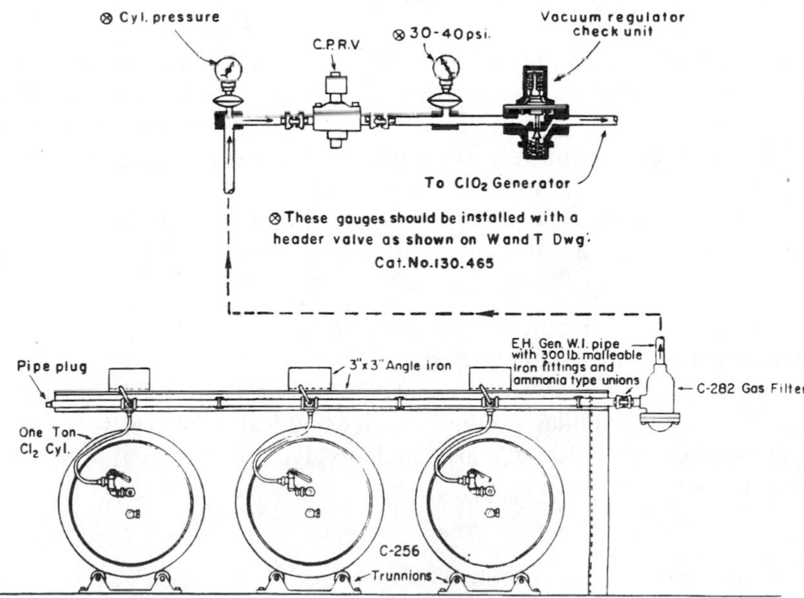

Fig. 9-15. Gas withdrawal system for ton containers.

ately downstream from the filter shown in Fig. 9-15. This filter is a combination sedimentation trap and filter. The filter medium is spun fiberglass, which is held in place by a stainless steel insert and screen assembly. This filter was designed specifically by chlorine specialties to protect all the chlorination equipment downstream from the CPRV.

Liquid Withdrawal. The chlorine header system for liquid withdrawal is somewhat different from that for gas withdrawal. The piping and the support system are the same, except that the flexible connections to the auxiliary header valve are connected to the bottom cylinder outlet valve (the top valve is for gas withdrawal). Figure 9-16 illustrates the proper method for connecting ton cylinders for liquid withdrawal. Valves at both ends of the flexible connection protect the operator when changing cylinders. Although this method makes it possible to trap liquid chlorine in the flexible connection under adverse conditions such as a fire, operators generally prefer this method because it eliminates their exposure to chlorine when changing cylinders. Therefore, they adjust their habits to follow all the required steps for maximum safety and minimum risk.

Figure 9-17 shows a chlorine gas filter. Because of the evaporator in the system, the filter is located immediately downstream from the evaporator outlet. The filter must always be installed in the gas phase. It is not possible to filter out chlorine impurities in the liquid phase, because the impurities are in solution. This places the filter just upsteam from the chlorine pressure-reducing valve, which now becomes an automatic shut-off valve as well.*

A complete liquid withdrawal system is illustrated in Fig. 9-18, which also shows proper spacing of the evaporator–chlorinator system.

It is to be noted that some engineers prefer to use duplicate header systems from the containers to the evaporators. In this way one header can be taken out of service for the required periodic cleaning without interruption of the entire facility. Most systems rely on a single header; however, duplicate headers have advantages from an operator's viewpoint.

In contemplating the use of liquid withdrawal versus gas withdrawal, the following advantages of a liquid withdrawal system should be considered:

1. The danger of reliquefaction of chlorine between the containers and the chlorinator is all but eliminated.
2. Fewer cylinders need to be connected at one time. Liquid withdrawal rates of a ton cylinder can be as high as 10,000 lb/24 hours.
3. The evaporators, although insulated, do give off some heat in the equipment room.

*This pressure-reducing and shut-off valve is always a part of the evaporator system. It is electrically interlocked with the evaporator water bath temperature, and automatically shuts off the chlorine supply in the event the water bath temperature falls below 150°F.

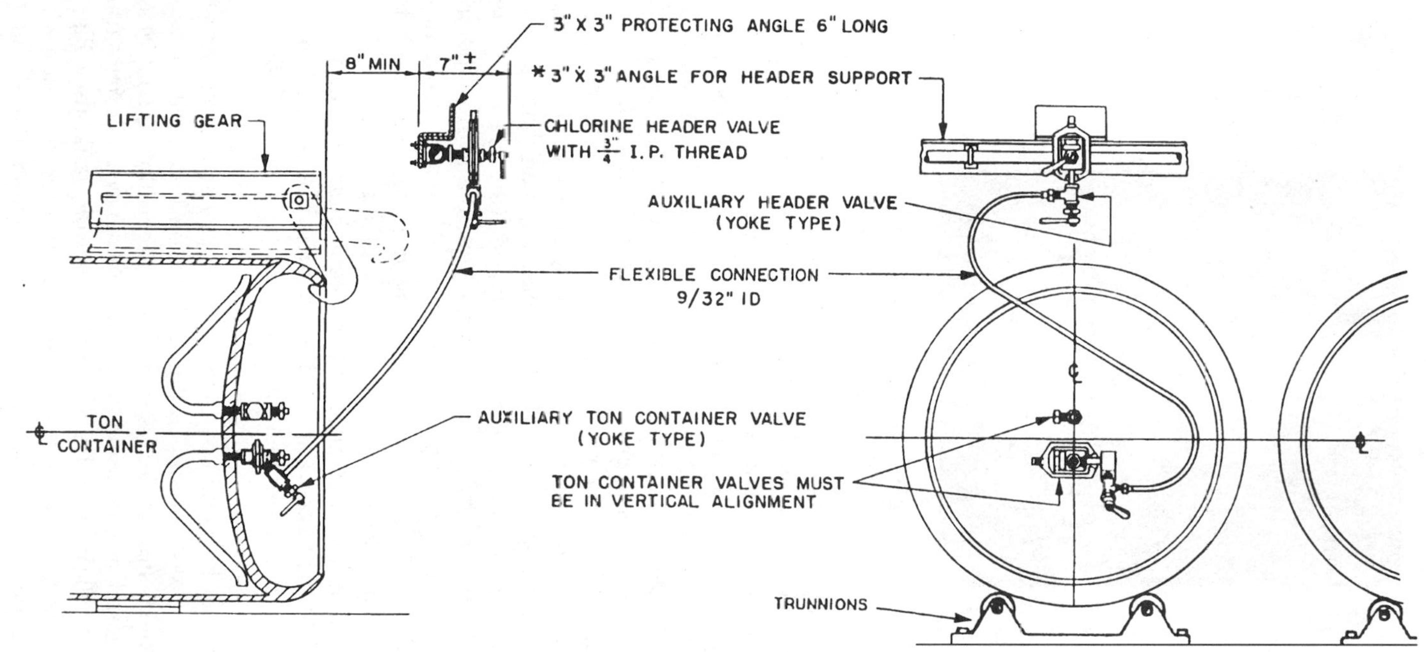

Fig. 9-16. Illustration of the proper method for connecting ton cylinders for liquid chlorine withdrawal.

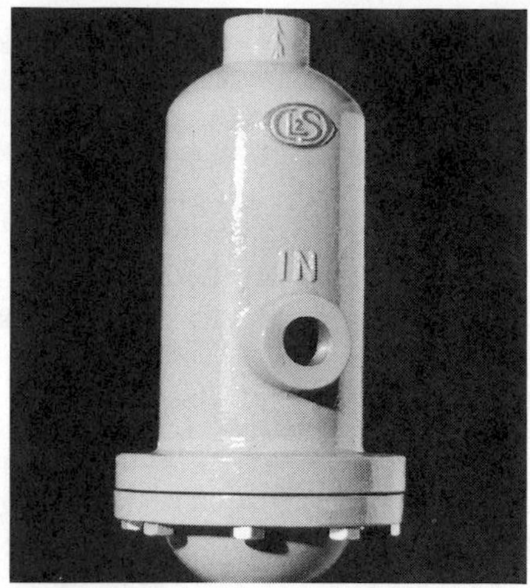

Fig. 9-17. Chlorine gas filter (courtesy Chlorine Specialties, Inc.).

Liquid withdrawal systems do not have the same critical design problems with regard to temperature considerations as gas withdrawal systems, except for inadvertent trapping of liquid in the header system, which constitutes a temperature–pressure hazard. If liquid chlorine is trapped between two shut-off valves and the ambient temperature rises a few degrees, the liquid chlorine will try to expand. Since it cannot expand, it will exert a pressure in accordance with the vapor pressure–temperature curve shown in the *Chlorine Institute Manual.*[1] For example, if a container connection full of liquid chlorine is trapped by yoke shut-off valves, and at each end a ton container or a tank car (at an ambient temperature of 80°F) is removed, then the vapor pressure would be 100 psi; if the connection full of liquid were allowed to reach a temperature of 90°F (e.g., by placing it in the sun) the hydrostatic pressure of the contained liquid would probably exceed the tensile strength of the already-fatigued flexible connection, resulting in a rupture of the flexible connection. (See valve designs in Figs. 9-19 and 9-20.)

A note of caution about manifolding ton containers withdrawing liquid: always be sure that the temperature of the cylinders is about the same; never connect "hot" cylinders to the manifold simultaneously with cylinders already in use.

The cylinders can be placed in carport-type open structures with only a sunshield when the system is operating entirely on liquid withdrawal (Fig. 9-12). If the cylinder storage area is remote from the chlorine control system, it is desirable to install an expansion tank in the header system using a frangible disk and pressure alarm in series with the expansion tank. (See Fig. 9-21.) This device is necessary to protect the system against the condition where an operator might close both the outlet valve on the header and the inlet valve to the evaporator, thereby trapping liquid in the header. Any subsequent

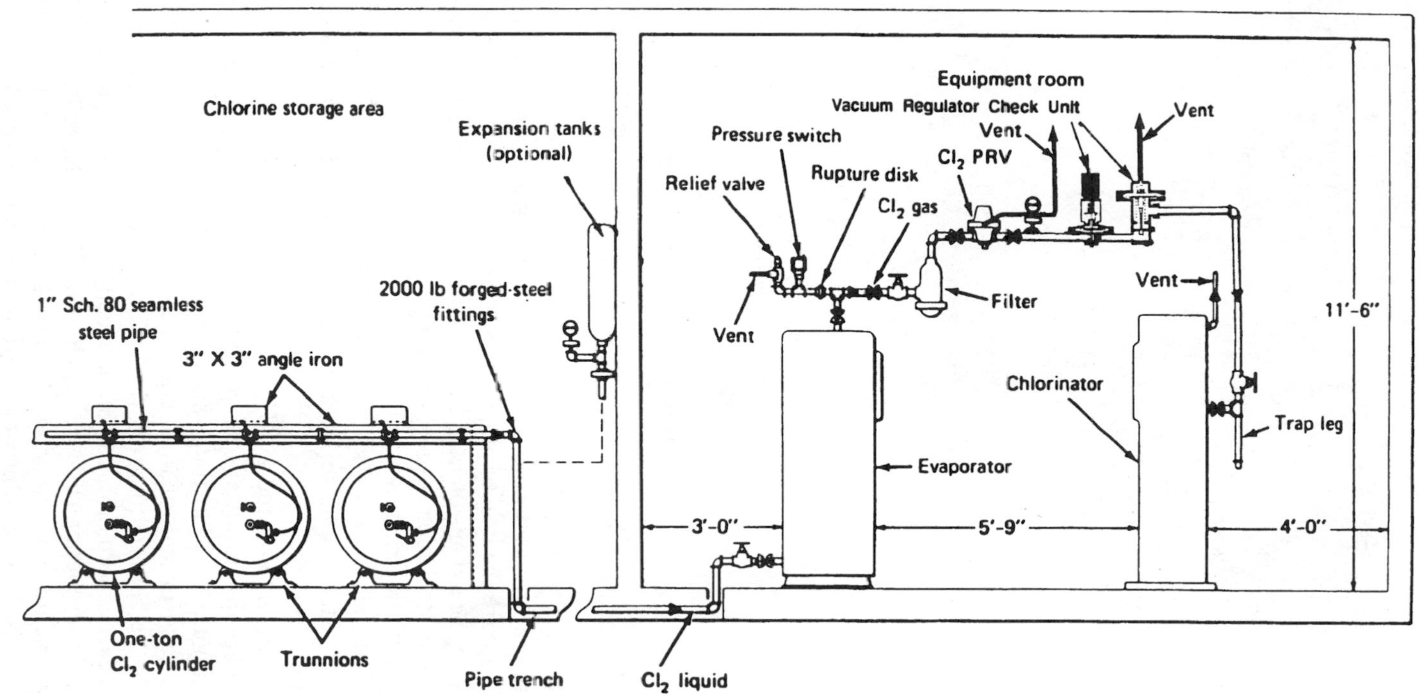

Fig. 9-18. Liquid withdrawal system showing chlorinator and evaporator spacing and ancillary equipment.

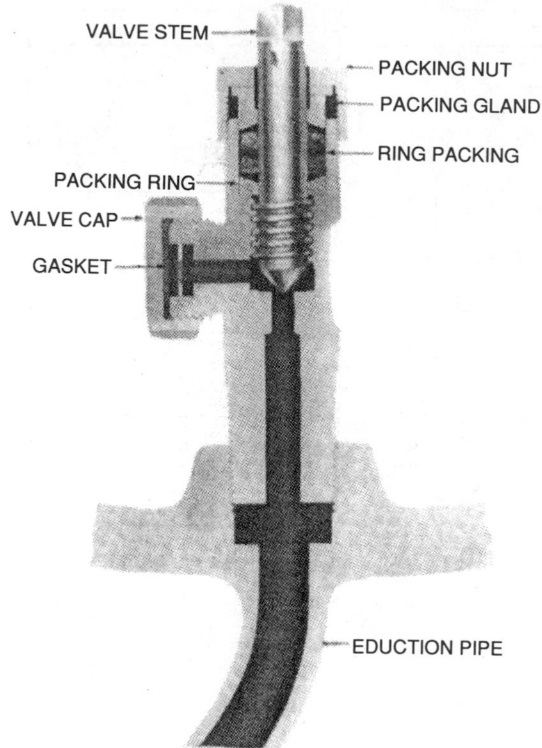

Fig. 9-19. Cross section of valve assembly for ton cylinders.

ambient temperature rise would result in a pressure rise in the liquid chlorine sufficient to rupture the header piping. This is a direct result of the hydrostatic pressure due to the expansion of the liquid chlorine (or SO_2), which is greater than that of the steel pipe or copper flexible connection as a function of a rise in ambient temperature.

Because chlorine headers must be cleaned occasionally, it is desirable in most cases to install duplicate headers between the cylinder area and the chlorination equipment.

In either liquid or gas withdrawal systems using ton containers or tank cars, a sediment trap should always be provided where the gas line enters the chlorinator. Figure 9-18 illustrates such a trap. This minor item saves much maintenance time on chlorination equipment. The chlorine gas line nearly always enters the chlorinator from the vicinity of the ceiling. The necessary downdrop to each chlorinator is a potential collection place for chlorine sludge caused by the inherent impurities in the chlorine. These deposits are the result of the reliquefaction phenomenon. A trap 12 inches long and capped will catch most of this debris and prevent it from entering the control mechanism of the chlorinator.

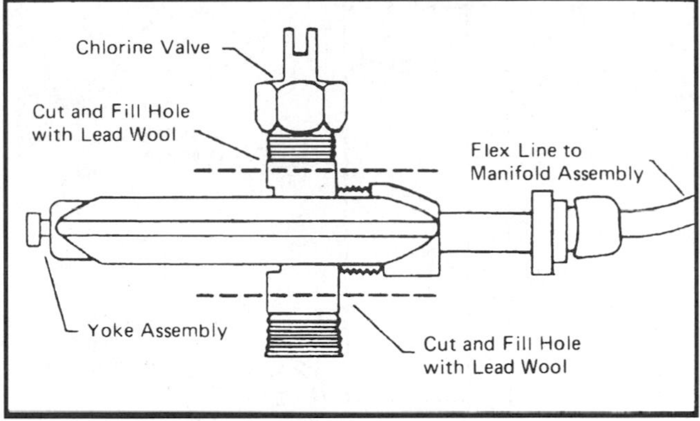

A modified chlorine valve can be used to seal off flex lines.

Below: Cutaway side view of flex line "cork."

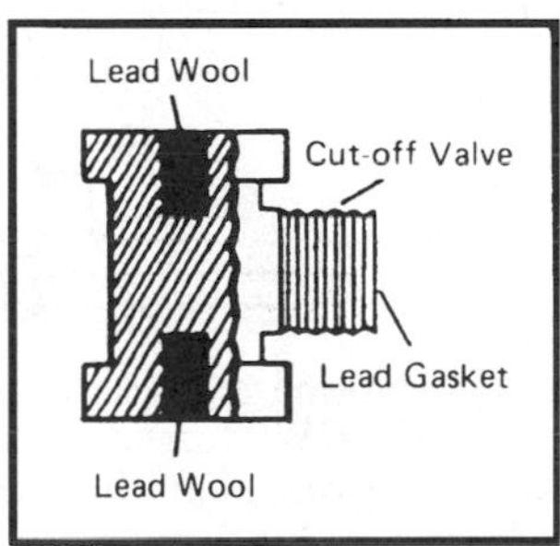

When chlorine cylinders are taken out of service, the flex line to the manifold assembly should be sealed so that moisture cannot enter and cause corrosion. An effective device for "corking" the flex line can be prepared by cutting off the top and filling the cut holes in the valve with lead wool. An out-of-service flex line can then be sealed by connecting it to the modified valve with a standard yoke assembly.

Fig. 9-20. Illustration of how a modified chlorine header valve can be used to prevent moisture entry by corking the flexible connection when not in use.[96] Reprinted from *Opflow* **8**(2) (Feb. 1982), by permission. Copyright © 1982, American Water Works Association.

Platform Scales. These scales are available for weighing one or more cylinders, up to a maximum of five. They are available in dial indicating and/or dial indicating plus recording models.

Weighing Devices Using a Hydrostatic Load Cell with Remote Dial Readout. These scales are popular for installations using ton cylinders. Figure 9-22 illustrates a load cell unit for weighing one ton cylinders. This unit features a remote digital readout at a much lower cost than beam- or pipe-lever-type scales. This type of scale is also available with electronic-type load cells. These systems are capable of providing a variety of chlorine inventory information.

External Chlorine Pressure Reducing Valve (CPRV). For all chlorine supply headers, whether they be gas withdrawal or liquid withdrawal, CPRVs

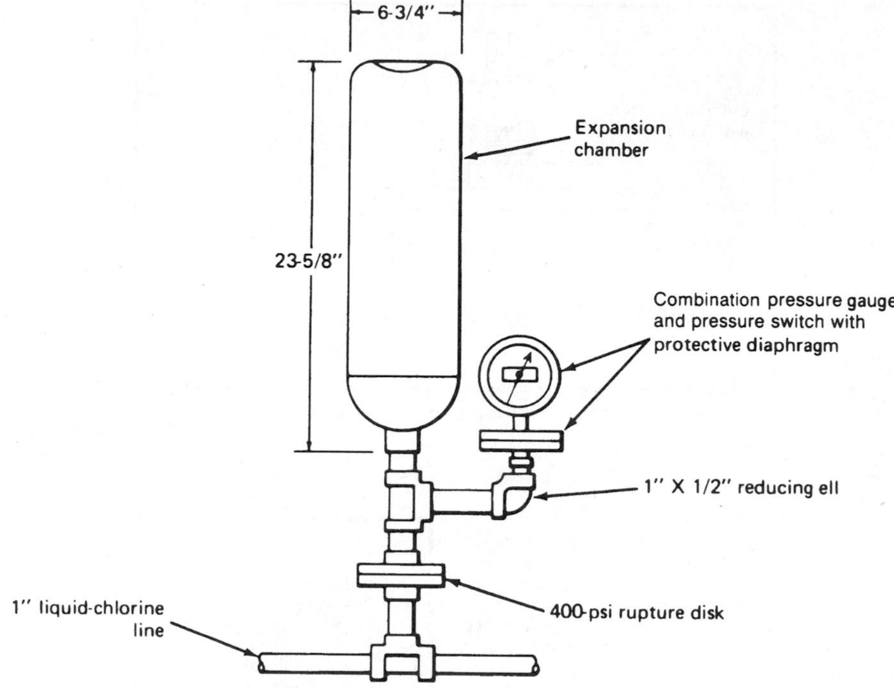

Fig. 9-21. Liquid chlorine expansion chamber. Connection is to liquid phase only. All fittings are 2000 lb CWP forged steel. All piping is seamless carbon steel Sch 80. Teflon tape should be used at all threaded joints.

should be installed immediately downstream from the gas filter, to prevent any possibility of reliquefaction in the gas header. Reliquefaction is highly undesirable because it causes immediate fallout of the inherent impurities in chlorine at any point where a pressure drop occurs. The larger the pressure drop, the greater the fallout of the impurities. The CPRV reduces the fallout potential because it provides a two-step pressure drop to the overall vacuum control system of any chlorinator installation. This virtually eliminates the maintenance on the remote vacuum pressure regulator that is designed to convert the chlorine supply system pressure (75–120 psi) to a vacuum. The external CPRV will require the maintenance of impurities removal on the valve stem and seat assembly. This is a relatively simple procedure. There is another important advantage to the two-step pressure drop system: it has been found in the warmer climes that the vacuum regulator check units will shut down and stop functioning when the supply pressure exceeds 120 psi, and the two-drop system prevents this occurrence. These valves are available in capacities of 500, 2000, 10,000, and 32,000 lb/day.

Chlorine Gas Filter. Every installation using ton containers and/or evaporators should use an adequate gas filter. Currently, the best available chlorine

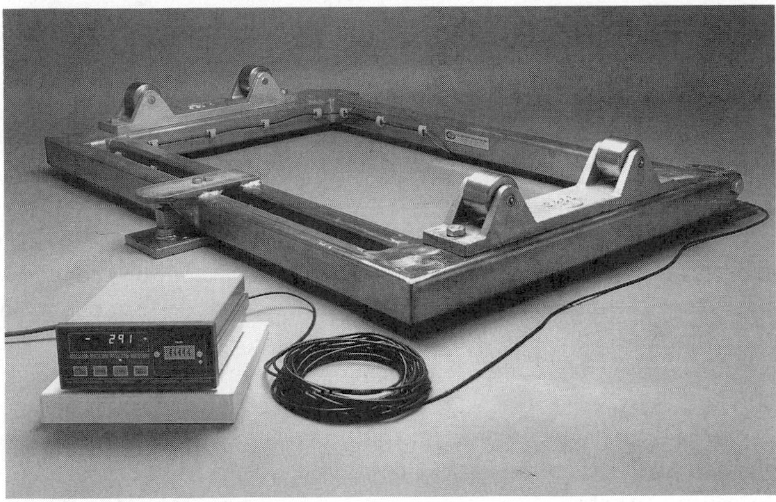

Fig. 9-22. Ton cylinder load cell type scale with digital readout (courtesy Chlorine Specialties Inc.).

gas filter is the Chlorine Specialties Model No. C-282, shown in Fig. 9-17. The filter medium is made of spun glass that is specially impregnated, contained in a removable stainless steel cylinder which is easily removed. The filter has two chambers. The lower portion acts as a trap for reliquefied gas and/or foam droplets from an evaporator. It is designed for a 560 psi working pressure and a chlorine gas flow in excess of 32,000 lb/day.

For ton container vapor withdrawal systems the filter should be located as close as possible to the last cylinder (see Fig. 9-15). For liquid withdrawal systems where evaporators are used, the filter should be installed immediately downstream from the evaporator shutoff valve on the vapor discharge line (see Fig. 9-18).

Expansion Tanks. These tanks are necessary only when there is danger of liquid chlorine becoming trapped in the supply line. Then, if there were a significant ambient temperature rise, hydrostatic pressure from the liquid chlorine trying to expand would rupture the pipe. One thing is certain, however: spring-loaded relief valves discharging to atmosphere should never be used on liquid chlorine lines. The chance of such a valve reseating properly is remote.

Figure 9-21 shows an expansion tank, originally recommended by the Chlorine Institute, with only a pressure switch. Every expansion tank on any given installation should be equipped with a pressure indicator in addition to the pressure switch. Otherwise, operating personnel would not be able to locate the overpressure situation. There are gauges available that are a combination of a pressure switch and a pressure indicator. For example, the U.S. Gauge Co.

Model 3050 utilizes a frangible disk that will rupture at some desired pressure between 300 and 400 psi, thereby allowing the liquid trapped in the system to enter the expansion tank.* This immediately produces some vapor pressure in the expansion tank, which actuates the pressure switch and sounds the alarm.

Figure 9-23 illustrates a typical ton cylinder handling facility for a chlorination facility with a maximum capacity of 24,000 lb/day. It indicates space requirements and the arrangement of handling facilities.

Systems employing long liquid lines and/or chlorine storage tanks usually have electrically operated valves at both ends of the lines. When there is a power failure or a pressure drop, these valves close. An expansion tank is required on such lines—equal in volume to at least 20 percent of the line volume between the automatic valves. The chlorine pressure-reducing and shut-off valve on the discharge side of the evaporator will also close on power failure, trapping the liquid in the evaporator. This will cause an immediate pressure rise in the evaporator, as the liquid chlorine in the evaporator will tend to reach the temperature of the water in the surrounding water bath. The vapor pressure in the chlorine chamber of the evaporator will rise until there is equilibrium in the temperature gradient between the liquid chlorine and its surrounding water bath. This would be close to 160°F, which is equivalent to a vapor pressure of 315 psi.

This condition is exaggerated when two or more evaporators are manifolded together and operating at maximum capacity at the time of power failure. In such cases, the vapor pressure could reach 425 psi.

To alleviate this condition, one or more expansion tanks must be connected into the system on the liquid inlet line to each evaporator just downstream from the automatic electric shut-off valve on the liquid supply header. Figure 9-18 illustrates the proper arrangement of an expansion tank to take care of any liquid that might get trapped in the evaporator; it also illustrates the conventional location of an expansion tank to handle any liquid that might get trapped in the supply header system.†

For installations where the supply system is close-coupled to the evaporators, there is no appreciable danger from closing valves at the outlet of the supply system simultaneously with the closing of the inlet to the evaporator; therefore, expansion tanks are not required. It is also deemed necessary to provide pressure relief of the evaporator between the inlet shut-off valve and the downstream shut-off valve in some distances.

Operators should be thoroughly and properly instructed never to close valves in this system so as to trap the pressure in any part of the liquid portion. The basic rule is never to shut off the outlet of the supply container unless

*If for any reason the header piping is subjected to a vacuum, the rupture disk will be damaged. Therefore, these disks must be specified for vacuum operation on the pressure side.

†Figure 9-18 also illustrates the 1975 mandatory evaporator relief system (see Fig. 9-63). The expansion chamber concept is the safest because it prevents a chlorine emission.

the system is being secured. In these instances, the pressure on the entire system is then allowed to go to zero gauge pressure.

Chapter 11 outlines operating procedures for securing a liquid supply system so as to eliminate the necessity for relying on expansion tanks for those systems that do not use automatic electric shut-off valves on the liquid lines.

Gauges. If the system is utilizing gas withdrawal and it is close-coupled, gauges in the chlorine gas header can be omitted because the gauge on the chlorinator will suffice. However, if this gauge fails, it does not leave the operator with any means to determine whether or not the system is "live." An extra gauge is always useful.

If the system utilizes an external pressure-reducing valve, there should be one gauge upstream from this unit and one downstream.

For a liquid withdrawal system, the evaporator has a gauge showing the system pressure. It is useful to have another gauge on the liquid line between the containers and the inlet shut-off valve to the evaporator. A gauge downstream from the pressure-reducing and shut-off valve should be considered, unless the system is a one-chlorinator installation that is close-coupled. In the latter case, the chlorinator gauge could be sufficient. However, when chlorinators and evaporators and manifolded together, it is always desirable to have gauges downstream from each reducing valve. Similary, when these systems are interconnected, it is also desirable to have gauges upstream from each reducing valve. When more than one evaporator is involved, there should be a gauge on the liquid chlorine inlet line to each evaporator upstream from the evaporator inlet shut-off valve and downstream from the evaporator isolating valve. These gauges are extremely helpful to the operator, for safety reasons as well as for troubleshooting.

All gauges that measure liquid chlorine and vapor under pressure are a source of chlorine leaks. These leaks occur when the protective diaphragm develops a pinhole from a flaw or rupture from metal fatigue. The chlorine attacks the brass bourdon tube, which will spring a leak from corrosion. Brass corrodes quickly from the inherent moisture in packaged chlorine. Unless these gauges can be isolated safely, in the event of a leak they represent a serious risk for a chlorine leak. All of these gauges should be installed as shown by Fig. 9-24. Moreover, these gauges should be replaced about every five years.

Alarms. Pressure switches that sound an alarm are helpful, and should be considered on most installations. Two kinds of pressure alarm devices are available. One is an adjustable mercoid-type switch with adjustable contacts that close for either rising or falling pressure—the two functions are not now available in the same housing. The other is a combination indicating gauge and pressure switch. The U.S. Gauge Co. Model 3050 is adjustable to sound an alarm at both high and low pressures. Its contacts are adjustable over the entire range of the gauge.

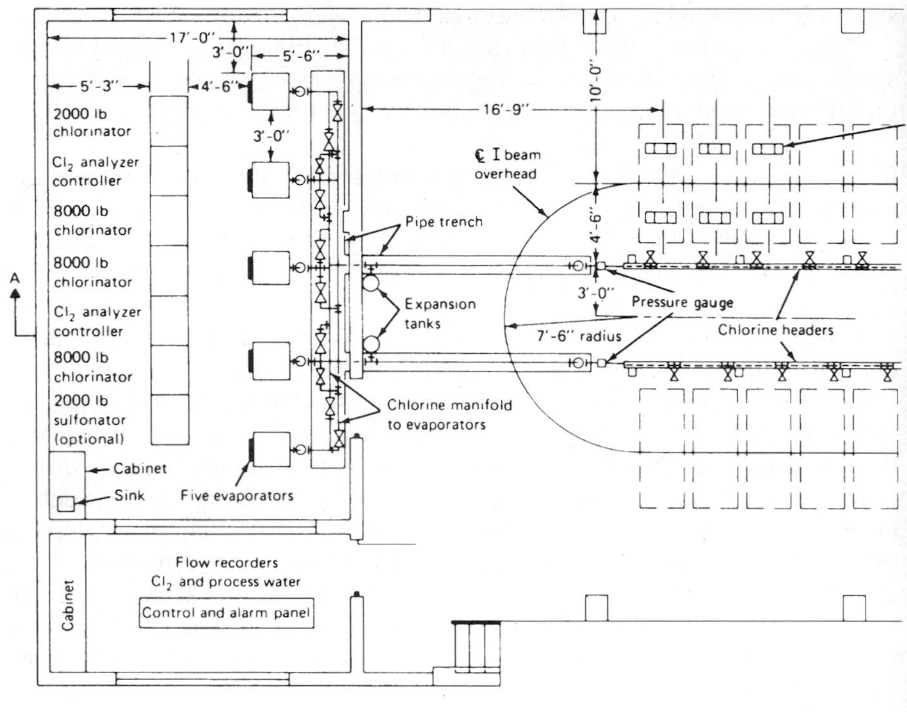

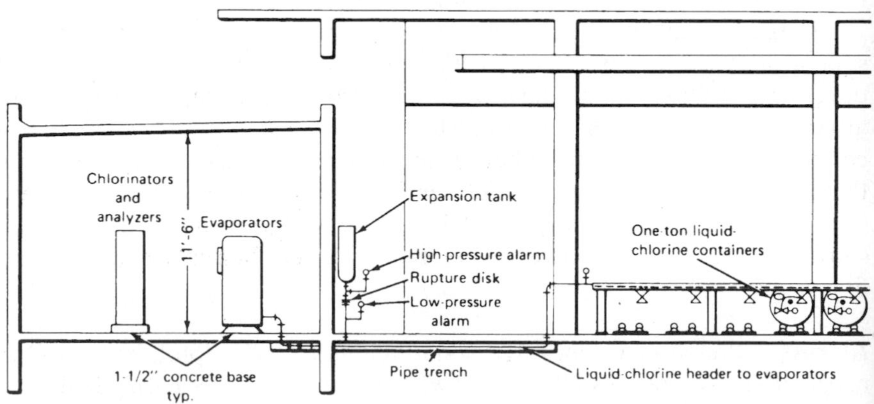

Fig. 9-23. Chlorine storage and handling facilities, equipment and space requirements for 24,000 lb/day capacity.

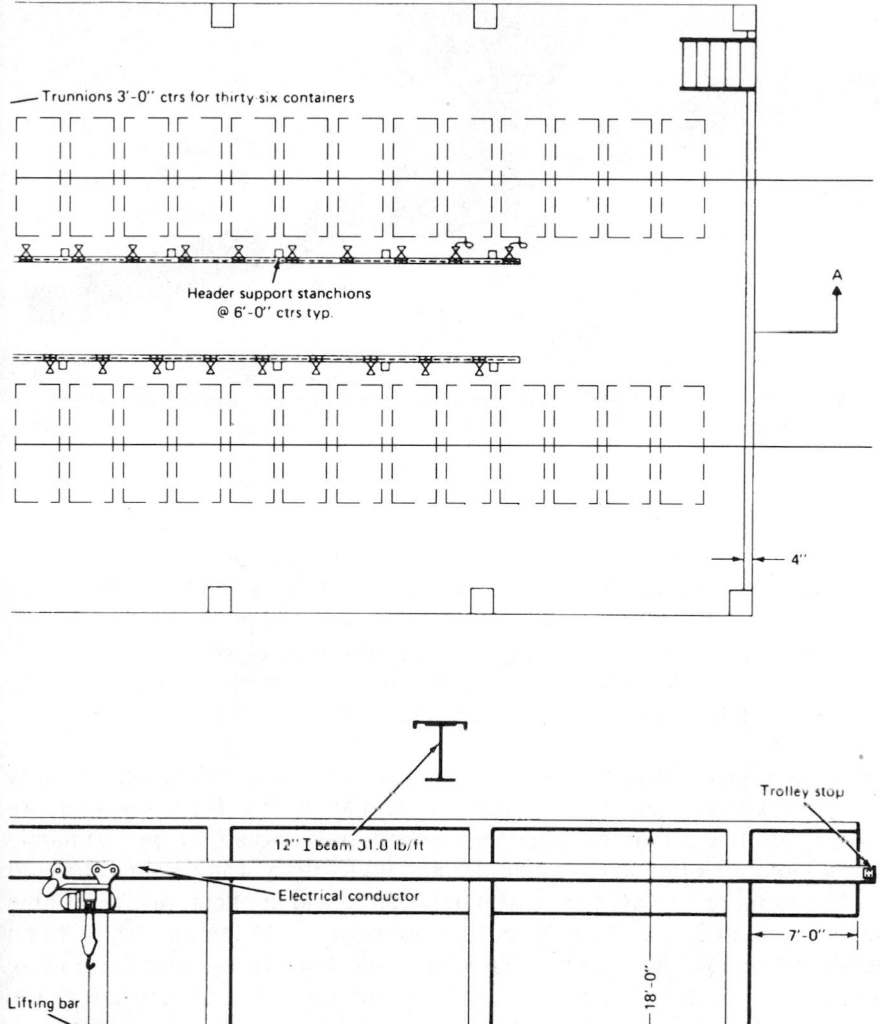

Fig. 9-23. (*Continued*).

Fig. 9-24. Proper safety method for installing all chlorine pressure gauges for both liquid and vapor.

All liquid withdrawal systems should have a high-pressure alarm, located on the liquid supply header between the chlorine containers and the evaporators, which will be activated when the pressure reaches 150 psi.

Gas withdrawal systems do not need high-pressure alarms.

A low-pressure alarm is highly desirable. It warns the operator of imminent exhaustion of the chlorine supply. In some cases it can be used in lieu of weighing devices.

For example, consider that five-ton containers are in service simultaneously in a chlorination system with a usage rate of 4000 lb/day. If the low-pressure alarm is set at 30 psi on the liquid supply line, there would be approximately 30 to 40 lb of chlorine in each cylinder and 30 to 40 lb in the evaporator and piping system, or a total of approximately 150 to 200 lb of chlorine remaining in the system at 30 psi. At 4000 lb/day, the usage is 167 lb/hour. Thus, when the alarm sounds, the operator has about one hour to prepare for putting another bank of cylinders in service. The situation is the same whether the withdrawal is liquid or gas.

Ton Cylinders: Reliability and Safety Considerations

Design Strength. The enormous strength of ton cylinders can be best described by the fact that when they are hydrostatically tested at 500 psig, it isnot uncommon for a 3 percent physical expansion of the container to occur without any damage whatsoever.[92] Figure 6 in the appendix shows that this would easily permit a temperature of 160°F in a "skin full" condition without any fear of rupturing. These cylinders also have an additional expansion factor in the dished heads, which will reverse to a convex posture before rupturing.

During an investigation of nitrogen trichloride explosions at a chlor-alkali plant near Bogota, Colombia, White[93] examined more than a dozen ton cylinders with the dished heads reversed. None of these cylinders exhibited any damage from the explosions. The damage occurred in the gas header immediately downstream from the evaporator discharge piping.[93]

Metal Fatigue. Many times in the past few decades metal fatigue in chlorine cylinders has been mentioned as a hazard to the life of ton cylinders. To date, there has only been reported failure by container rupture, which occurred when a cylinder owned by the U.S. Army split at the seam between the dished head and the cylinder body. The cylinder had been in use about 45 years. This brings up the question of whether or not cylinders should be scrapped after a specified length of service, as now a great number of these cylinders, made during World War II, are more than 50 years old. Many such cylinders are still in use. That they have survived this long is quite remarkable. Compared to a bulk storage tank, one of these cylinders undergoes countless cycles of pressurization and depressurization.

Fusible Plugs. The use of fusible plugs on ton cylinders is a questionable practice because of the hazard quotient. Fusible-plug failure is the leading cause of major leaks when ton cylinders are used. It is for this reason that the American Water Works Association passed a resolution in 1935 asking that these plugs be banned from ton cylinders.[87] There are three fusible plugs on each of the concave cylinder heads. The three plugs on the head where the cylinder outlet valves are located are a major cause for concern. These valves and their associated flexible connectors prevent easy access to a leaking plug with the emergency kit. These plugs were originally incorporated in the cylinder design as a safety pressure relief device in case of a fire in the cylinder area. However, flammable material or combustibles have long since been banned from the storage area of any compressed gases, so the possible need for fusible plugs has passed.

After a further look into the logic of fusible plug performance, the following interesting anomaly was discovered:

1. When properly filled, a ton cylinder becomes "skin full" at 155.1°F.[88]
2. These plugs are made with a brass body and a core that melts somewhere between 158 and 165°F.[89]
3. To bridge the gap of 3 to 10°F, the dished heads in the container are supposed to pop out, going from concave to convex![90] Presumably the cylinders were purposely designed to account for the above-described temperature gap.

In spite of this anomaly, the design might have worked to the user's advantage, as the cylinder designers had to take this bizarre set of circumstances into account by making the cylinders much stronger than would have been required otherwise. When NCl_3 is formed during the production of liquid chlorine, it remains soluble only in the liquid form of chlorine. Therefore, at the moment of liquid exhaustion in the cylinder or any other container, the NCl_3 is released as a vapor and explodes. In this particular case the presence of NCl_3 was due to ammonia in the electrolytic cell water.*

Fusible plugs are a liability because their brass body disintegrates rapidly if there is the slightest amount of moisture in the chlorine above the allowable limit. The resulting corrosion finds its way easily through the steel of the cylinder because of the inability of brass to resist corrosion by moist chlorine. The resulting leak is the primary cause of major chlorine releases when ton cylinders are used. Next in severity and frequency is the failure of the flexible connectors.

If fusible plugs are inevitable because of an industry attitude, they should be limited only to the head that does not hold the outlet valves. Moreover, these plugs should be used only in the area where chlorine vapor is present when properly aligned. This would greatly reduce the leak rate. If plugs must be used on both heads, then consideration should be given to using the 150-lb cylinder outlet valves that are fitted with a fusible plug.

Fire Protection. If the local fire marshal insists upon additional protection in case of a fire, there is good reason why sprinklers cannot be used to keep the cylinders cool and the pressure down during a fire. It must be remembered that the air in any closed room housing multiple cylinders of chlorine suffers from general corrosion, no matter how adequate the ventilation. Off-gassing is inevitable when cylinders are being changed. Cumulatively, these small amounts of chlorine emissions will quickly damage any modern sprinkler used for fire protection. In other words, if sprinklers were installed, they would become inoperative in a short time and be unable to release water at the time of a fire.

Spraying water on a cylinder during a fire to keep the pressure from escalating is a perfectly safe procedure. Failure of metallic components due to chlorine corrosion always occurs from within the container when an external leak is not in progress; for example, consider the "sweating" of any cylinder subjected to vapor withdrawal in excess of its allowable rate. Also see below, discussion of the "Arizona Blanket," under the heading "Chlorine Storage Pressure Control System."

Manifolding Cylinders. The Chlorine Institute recommends that for liquid chlorine withdrawal, the only acceptable way is to manifold the cylinders in accordance with their drawing (Fig. 9-25). This arrangement eliminates the potential hazard of connecting a full cylinder with a higher vapor pressure than that of the cylinders already "on-line." Such a situation could lead to

*This was the cause of the Bogota explosions.

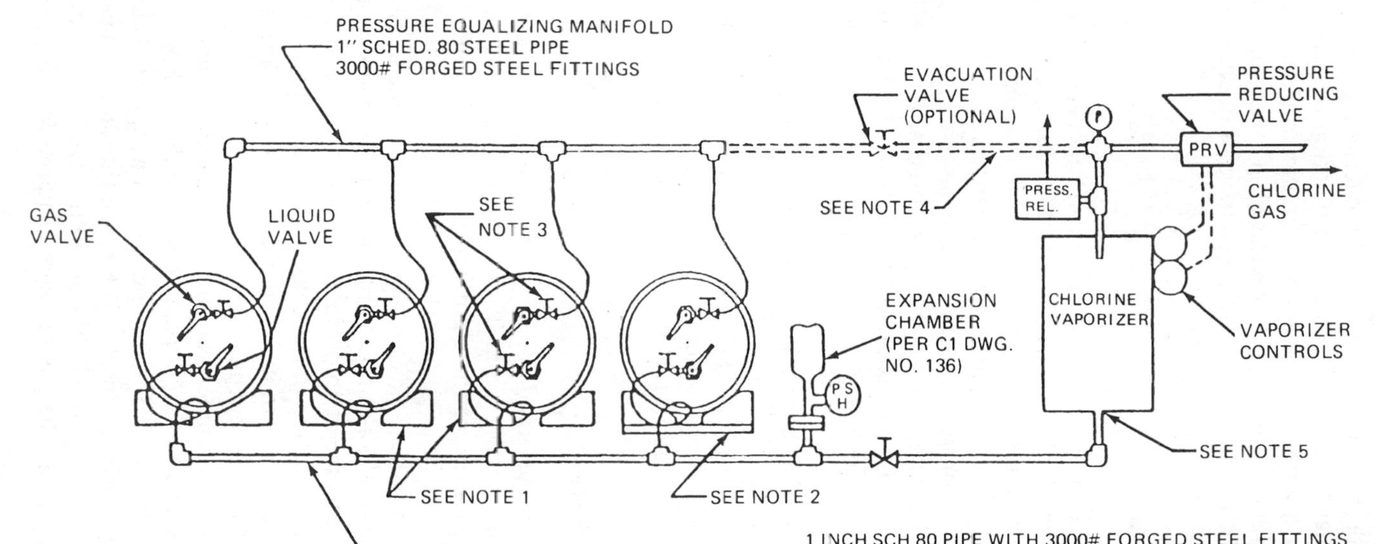

Fig. 9-25. Suggested concept of piping arrangement to accomodate manifolding ton cylinders for liquid chlorine withdrawal (courtesy Chlorine Institute). *Operating instructions*: Open all valves first; all liquid valves must be closed before closing any gas valves; during normal operation, keep the evacuation valve closed. *Note*: 1) All containers must be at the same elevation. 2) Trunnions should be used to support the cylinders. 3) Cylinders must be connected properly. 4) If an evacuation valve is installed, the vaporizer outlet tube must be at a higher elevation than the tops of the containers. 5) All cylinders must be at the same temperature before withdrawing chlorine.

overfilling of those cylinders already connected. This overfilling scenario never has been reported, nor has anyone ever seen an installation that conforms to the CI drawing at either a water or a wastewater treatment plant.

In manifolding ton cylinders for vapor withdrawal, it seems obvious that the overfilling potential is nonexistent because cylinders of unequal pressure will automatically come to the same pressure.

TGO Technologies Inc.[115]

Ton Cylinders. These chlorine and sulfur dioxide cylinders are also subject to compliance with the January 1996 issue of the California Fire Code Article 80, Title 24, Part 9, which requires both chlorination and sulfur dioxide facilities to have a treatment system for mitigating any accidental release of chlorine or sulfur dioxide, either gas or liquid. These treatment systems are required to reduce the concentration of either gas in the enclosed area of the leak to 5 ppm before it is released to the external atmosphere. Article 80 of the California Code allows the option of using total containment vessels for ton cylinders.

Several of these units are already in place (1997), and many more are being installed. As TGO Technologies is a Santa Rosa, California firm, it is not surprising that one of the first installations was in November 1996 at the Solano Irrigation District Chalk-Hill water treatment plant. These ton cylinder total containment vessels are illustrated in Fig. 9-6.

Container Withdrawal Rates[2]

General Discussion. The sustained maximum chlorine withdrawal rate will vary from plant to plant, depending primarily upon the temperature surrounding the container, characteristics of the chlorinator installed, and the size of the container.

Chlorinators require a minimum gas pressure for proper operation, again depending upon the type of chlorinator. Most vacuum type units require at least 10 psi but are more reliable at 15 psi inlet pressure. The gas pressure at the point of withdrawal is a function of the chlorine liquid temperature in the container. This relationship is shown in Table 9-2. The table is derived from graphs that can be found in the appendix.

For a given chlorinator, the temperature at which the minimum required gas pressure is reached is termed the "threshold temperature." For example, Table 9-3 shows the pressure needed for a typical vacuum-operated chlorinator occurs at about 0°F (14 psi). Therefore, 0°F is the threshold temperature for that type of chlorinator. The liquid chlorine in the cylinder or container must be kept at 0°F to maintain the gas pressure required for proper operation.

When gas is withdrawn from the container, the liquid in the container is chilled. To maintain the liquid at the threshold temperature, heat must pass

Table 9-2 Vapor Pressure vs. Container Liquid Temperature

Temperature, °F	*Chlorine*	*Sulfur Dioxide, psi*	*Ammonia*
0	14		16
10	21		24
20	28	2	34
30	37	7	45
40	47	12	59
50	59	18	75
60	71	26	93
70	86	34	114
80	102	45	138
90			
100			

from the air surrounding the cylinder through the walls of the container and into the liquid. Therefore the maximum sustained rate of withdrawal is directly dependent upon the rate at which this heat transfer takes place. Several factors influence the heat transfer rate, including ambient temperature, air circulation, humidity of the air, amount of liquid remaining in the container, and size and type of container. In general the important factors are: (1) ambient air temperature and (2) size and type of cylinder. Ambient air temperature is simply the air temperature surrounding the container.

Container Withdrawal Factor. The size and the construction of the cylinder can be related to the ambient temperature with a number called a "withdrawal factor." Table 9-3 lists several withdrawal factors.

Sample Calculation. The equation used to calculate the maximum withdrawal rate from a container is as follows:

Table 9-3 Withdrawal Factors

Gas	*Withdrawal Factor*
Chlorine	
150-lb cylinder	1.0
1-ton container	8.0
Sulfur dioxide	
150-lb cylinder	0.75
1-ton container	6.0
Ammonia	
150-lb cylinder	0.4
250-lb cylinder	0.8 estimated
800-lb container	3.2 estimated

$$(\text{Temp. of room} - \text{Threshold temp.}) \times \text{Withdrawal factor} = \text{Max. withdrawal rate, lb/day}$$

Assume room temperature or that ambient temperature surrounding the container is 70°F; the withdrawal temperature is 10°F. The withdrawal factor is 8 for a ton container. Then we have $(70 - 10) \times 8 = 480$ lb/day. The rule of thumb for acceptable vapor withdrawal from a one-ton cylinder is 300 to 350 lb/day at room temperature. Higher rates of withdrawal will cool the liquid in the container to the point that the vapor being withdrawn will contain a mist of liquid chlorine. For a variety of reasons, this will require more system maintenance, and the liquid chlorine carryover will severely attack the plastic parts in the chlorination equipment.

When the operator wants to implement the pressure control system described in this chapter, the withdrawal factor (8) of the ton cylinder can be used to calculate the withdrawal factor for any size chlorine storage container including railcars, provided that the surface area of the storage container is known. It is a simple ratio and proportion exercise in arithmetic. The surface area of a ton cylinder is approximately 63.5 sq ft. Assuming that a chlorine storage tank has 450 sq ft, the problem is solved as follows. Let Y equal the storage tank withdrawal factor:

$$Y/450 = 8/63.5$$

Then:

$$Y = 450 \times 8/63.5 = 57$$

Now the threshold temperature must be selected. If a value is too low, then the chlorine liquid misting problem will be certain to begin. This should be avoided. In this hypothetical case a threshold temperature of 40°F is suggested. Then the withdrawal rate is:

$$(75° - 40°) \times 57 = 35 \times 57 = 1984 \text{ lb/day}$$

This will probably allow vapor withdrawal without the liquid chlorine misting phenomenon. However, the chlorine piping between the storage tank and the chlorinator must be in accordance with that shown in Fig. 9-61.

Automatic Switchover, Ton Containers

General Discussion. Unattended chlorination stations must have special provisions to preclude the possibility of chlorine supply failure. Traditionally chlorine supply systems have been provided with scales for determining the amount of chlorine remaining in the connected cylinders. If this is the sole means of accounting for the chlorine supply, it requires frequent checking and operator judgment to determine when the cylinder will become empty. This diminishes facility reliability. The operator may be faced with repeated interruptions in chlorination.

To solve the problem of chlorine supply reliability a great many stations have adopted the automatic switchover concept in lieu of scales. There are two types: (1) the pressure system and (2) the vacuum system.

The Pressure System. This system has been in use since about 1966. It is illustrated in Fig. 9-26. This arrangement of pressure-reducing valves, gauge, valves, and piping usually requires less capital outlay than is required for weighing devices. Use of this system greatly improves the overall reliability of the chlorination facility.

The system consists of two independent manifolds, each equipped with a manually adjusted chlorine pressure-reducing valve (CPRV) and the necessary gauges. CPRV-1 is set 10–20 lb higher than CPRV-2. Therefore, only gas will flow through regulator 1, and only from manifold *A*. When all the liquid chlorine in the cylinders connected to manifold *A* vaporizes, and the gas pressure drops to the set limits, CPRV-2 will open, allowing the full cylinders on manifold *B* to supply the chlorine. With the higher pressure setting on CPRV-1, gas from manifold *B* cannot flow back into the cylinders connected to manifold *A*. The lower pressure reading on each gauge is evidence to the operator that chlorine is now coming from manifold *B*. After the empty cylinders connected to manifold *A* have been replaced by full ones, the manual valves ahead of the regulators are reversed so that the gas from manifold *B* now

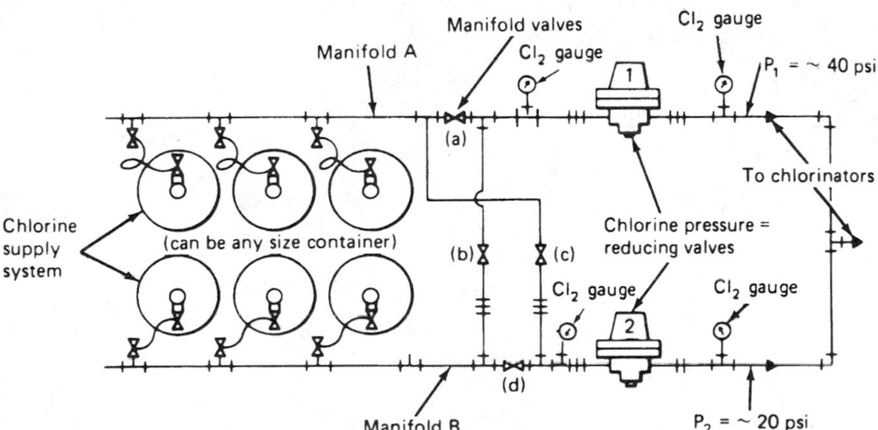

Fig. 9-26. Automatic chlorine cylinder switchover (gas phase only). By reversing manifold valves a, b, c, and d, chlorine can be routed from manifold A to CPRV #2 and manifold B to CPRV #1.

passes through CPRV-1 (higher pressure), which will again cause automatic switchover when the cylinders are empty.

The Vacuum System. The Wallace & Tiernan version of this system utilizes the dual arrangement of the vacuum-regulator check unit used in the remote vacuum arrangement for ton containers.[3] The switchover system is illustrated in Fig. 9-27. For automatic switchover capability two vacuum regulators are fitted with mechanical detents. One regulator feeds gas until the container to which it is connected is nearly exhausted. The resulting rise of vacuum to higher than normal provides sufficient force to unlatch the detent in the second regulator which then takes over the gas supply function. The original supply continues to feed with the new supply, insuring exhaustion of gas from the original supply container. Referring to Fig. 9-27, it is imperative that the vacuum-regulator check unit be vented to outside atmosphere well above ground level where chlorine fumes cannot cause injury to personnel or damage

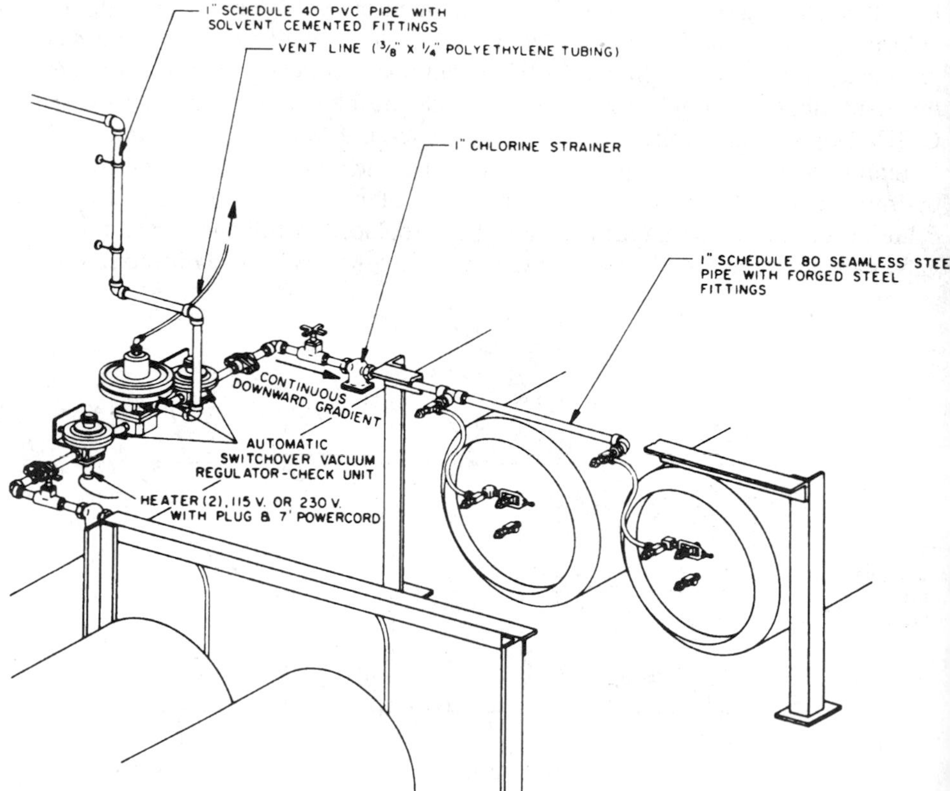

Fig. 9-27. Automatic chlorine cylinder switchover: Vacuum type, gas phase (courtesy Wallace & Tiernan).

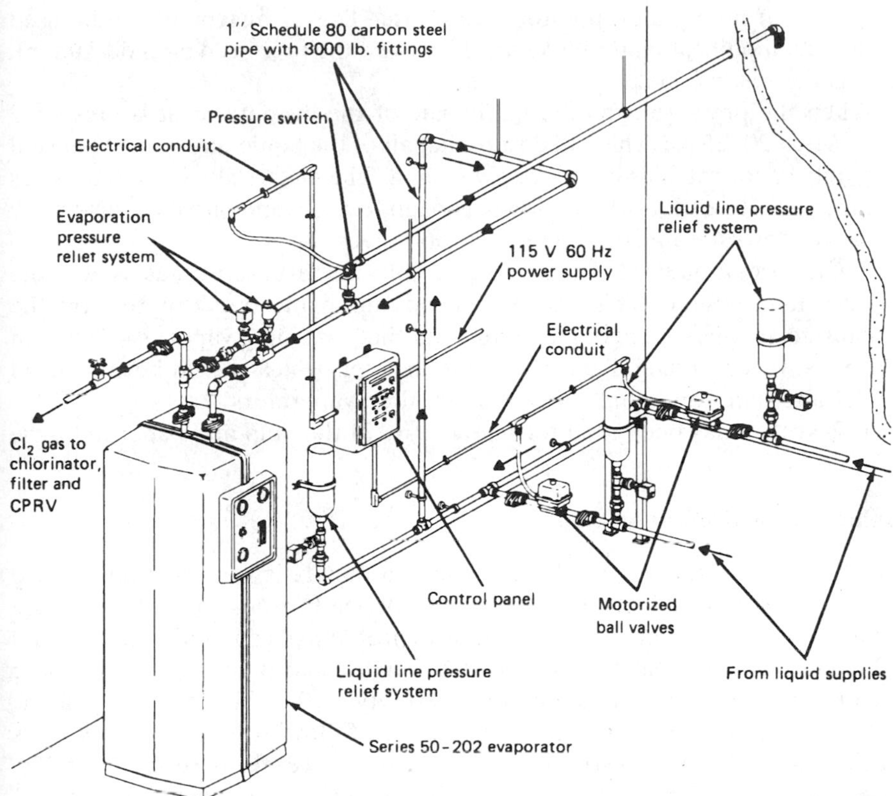

Fig. 9-28. Automatic chlorine supply switchover system for liquid withdrawal (courtesy Wallace & Tiernan).

to foliage.* Avoid areas routinely used by personnel, near windows, and near recirculated or ventilation air ducts.

This automatic switchover unit remains in the closed position and will not open until a vacuum is produced on the downstream side by the chlorinator injector. The entire unit is designed to withstand full container pressure. If the first valve passes gas without a vacuum, the second valve will remain closed and contain the cylinder pressure within the unit. In the extremely unlikely event that the second valve passes gas (without a vacuum), the built-in pressure relief valve will allow this gas to pass out the vent to atmosphere.

Liquid Chlorine. Automatic switchover for ton containers operating from the liquid phase is shown in Fig. 9-28. The pressure switch is in the common

*When multiple chlorinators are involved, serious consideration should be given to the use of a mini-absorption tank for these and other similar vents.

liquid line if multiple evaporators are in use. This arrangement can be used with some modification for tank cars. However, the preferred method for tank cars is the reserve tank.

When the pressure is on the liquid side of the evaporator, it is presumed to be set at 20–25 psi. This will ensure that all of the liquid has been exhausted and only vapor remains in the supply system. The electrically operated valves in the supply line should open slowly enough to accommodate the near-empty situation downstream from these shut-off valves.

Another method used is a sonic liquid chlorine level sensor, as used in the reserve tank system. This system depends upon the operator to strip the exhausted supply system of any liquid and the remaining vapor. As shown in Fig. 9-29, this stripping line is a separate header, which should be connected to the alternate top liquid connection on the evaporators.

This system has been used for many years in the pulp and paper industry.

Selection of Container Size

The first step in the design of a chlorination facility is to decide on the size of the chlorine containers to be used. This depends primarily on the average daily consumption of chlorine. It becomes a problem of economics and logistics although a few mitigating circumstances might take precedence over these factors, such as length of haul and accessibility to the chlorine storage area.

For the purpose of economic evaluation, the following are the *relative* costs of chlorine in the various size containers. Actual prices fluctuate considerably, depending upon the area, but will usually remain in the same ratios.

Size Container	*Cents/lb* Cl_2
100- to 150-lb cylinder	15
Ton containers	7½
55 to 90 ton tank car	3¾
Storage tank	3½

Another cost factor to be considered is the demurrage charge on the containers. Depending upon the chlorine packager, this may be a flat charge for any container kept longer than 30, 60, or 90 days, or it may be reflected in a price break for ordering multiple quantities at one time plus a demurrage charge on each container kept longer than 90 days.

Other things being equal, and considering the practical aspects of an installation, 150-lb cylinders are the containers of choice when the average daily consumption of chlorine does not exceed 50 lb. These cylinders are readily handled by one person using a special two-wheel hand truck.

When the daily consumption is in excess of 50 lb, ton containers are to be considered. More space must be allotted and special handling equipment must be provided for these containers.

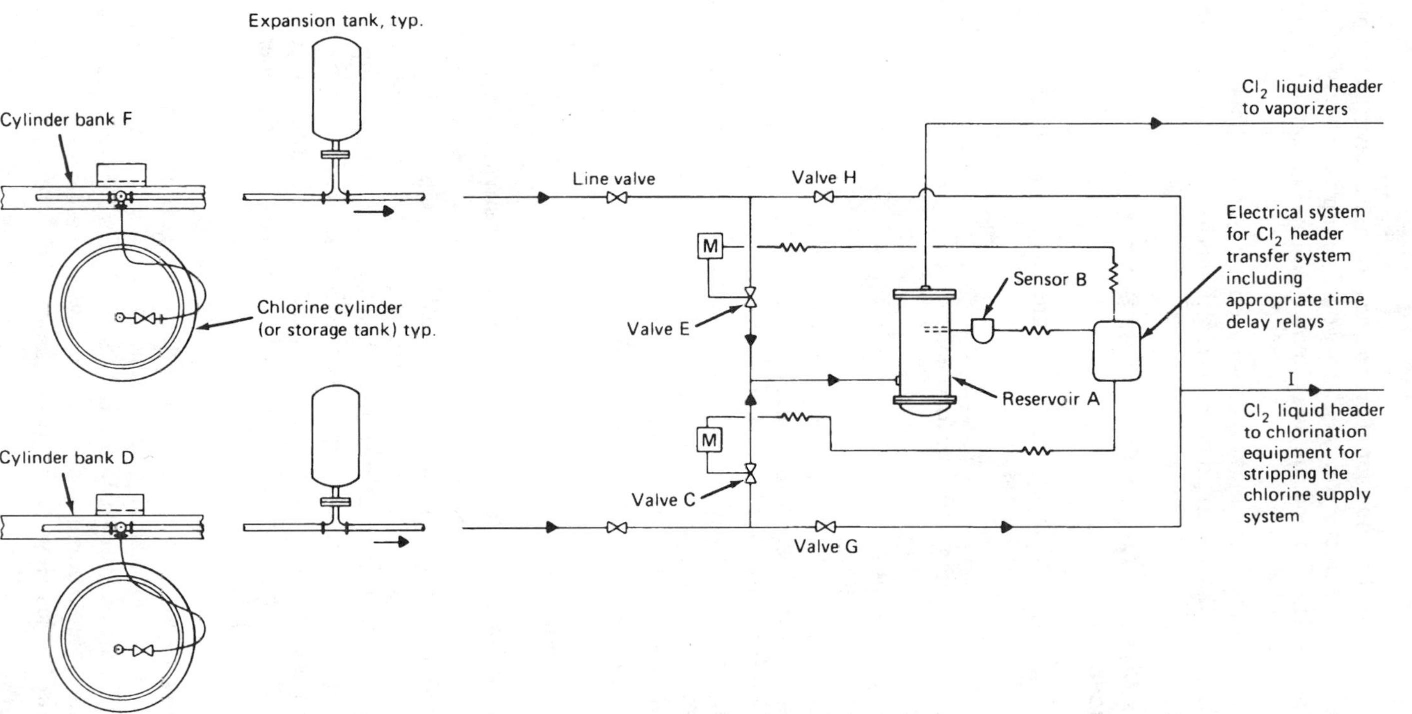

Fig. 9-29. Automatic liquid chlorine supply switchover system using a sonic level sensor.

Tank cars should be considered when the average daily consumption of chlorine reaches two tons per day, which would require a weekly delivery of 14 ton cylinders. This is about the maximum quantity that can be handled by truck rigs for one load.

The use of a storage tank may be considered for any installation using tank cars, particularly if the average daily consumption is four tons per day or more.

Designer's Checklist for Chlorine Supply System: Ton Cylinders

The following is a guide to remind the designer to check the chlorine supply layout for all necessary accessories:

1. Cylinder weighing scales or load cells.
2. Trunnions for ton containers.
3. Ton cylinder lifting bar.
4. Chlorine gas filter.
5. External chlorine pressure-reducing valve.
6. Liquid chlorine expansion tank.
7. Appropriate gas and liquid supply pressure gauges.
8. High- and low-pressure indicating switches for alarms (high-pressure alarm is used only on liquid systems).
9. Condensate traps at inlet to chlorinators.
10. Appropriate header valves and shut-off valves.

Single Unit Tank Cars

General Description. Chlorine is available in five sizes of single-unit cars: 16, 30,* 55, 85, and 90 tons. As of 1997 there are very few railcars available besides the 90-ton cars, all of which fit the following description.

Each car is equipped with an outlet dome that is made identical for all tank cars approved by the Chlorine Institute. This dome contains two liquid outlet valves that are in line with the longitudinal axis of the car and two gas outlet valves on an axis at right angles to the liquid valves. In the very center of the dome is a safety relief valve that will expel gas to atmosphere under overpressure conditions.

Excess Flow Valve. In each liquid outlet line, there is installed a "safety check valve," described as an "excess flow valve." This is an important item that is a vital part of every chlorine rail car, tanker truck, or storage tank. It is part of the "tank dome." The situation with these valves, has changed somewhat in the last few years. There now (1997) are more road tankers

---

*The 16- and 30-ton cars are being phased out of service in most areas.

and more storage tanks, plus the 90-ton rail cars which as the only size rail car available.

Under each liquid chlorine valve in the tank dome, there is an excess flow valve. This valve consists of a rising ball (see Fig. 9-34) that closes the valve when the rate of liquid chlorine flow exiting the tank exceeds a predetermined value. It does not respond to pressure in the car. The valve is designed primarily to close automatically, stopping the flow of liquid chlorine out of the tank car, in the event that the tank car dome angle valve is sheared off in an accident, such as a multiple- or single-car derailment. It may also close if a catastrophic leak occurs downstream from it, so that there is only atmospheric pressure on the downstream side of the valve. This has happened more than once in transporting railcars with liquid chlorine.

These valves are not designed to act as an emergency shut-off device during the tank car or tank truck unloading process.[89] Currently these valves that are associated with the Chlorine Institute regulations have maximum operating flow rates of 7,000, 11,000, or, 15,000 lb/hr.

It is important for operating personnel to realize that they must be very careful when opening the liquid valve is a new system that has not yet had any chlorine, either liquid or gas. If the liquid valve is not opened slowly, the excess flow valve will close with a bang, causing it be difficult to open. This is one of two reasons why every new chlorine system should be initiated by filling the piping, etc., with gas.[112] If there is a leak somewhere, which there usually is, a gas leak is far easier to deal with a liquid leak. Moreover, when the liquid valve is opened, it can be opened as quickly as the operating personnel desire, without any fear of the excess flow valve jamming shut. The slightest pressure above atmospheric in the downstream piping will always prevent that.

Dimensions and Construction. Tank car dimensions are shown in Table 9-4.

Table 9-4 Dimensions of Tank Cars

	*Length Over Strikers**	*Overall Height†*	*Height to Valve Outlet†*	*Extreme Width‡*
TMU	42′ 4″–47′ 0″	6′ 8″–7′ 6″	—	9′ 6″–10′ 1″
16-Ton	32′ 2″–33′ 3″	10′ 5″–12′ 0″	9′ 3¼″–10′ 0″	9′ 2″–9′ 6½″
30-Ton	33′ 10″–35′ 11½″	12′ 4½″–13′ 7″	11′ 3″–11′ 9″	9′ 3″–9′ 10″
55-Ton	29′ 9″–43′ 0″	14′ 3″–15′ 1″	12′ 6″–13′ 4″	9′ 3″–10′ 7½″
85-Ton	43′ 7″–50′ 0″	14′ 11″–15′ 1″	13′ 2″–13′ 4″	10′ 5½″–10′ 6½″
90-Ton	45′ 8″–47′ 2″	14′ 11″–15′ 1″	13′ 2″–13′ 4″	10′ 5½″–10′ 6½″

*Add 2′ 6″ for length over center line of coupler knuckles.
†Heights are for empty cars, and are measured from top of rail. Heights for loaded cars may be 4″ less.
‡Width over grab irons.
Note: Height to manway platform is 6 to 10″ less than height to center line of valve.

Figures 9-30 and 9-31 illustrate the pertinent features of tank car construction. These cars are provided with a cork or a combination fiberglass and foam insulation from 4 to 6 inches thick, covered with an outer steel jacket, to minimize vaporization and pressure buildup in transit.

The only opening in the tank, at the center on top of the car, is closed by the dome plate, which is sealed by a lead gasket and secured with a ring of bolts.

The mechanism for chlorine withdrawl from a tank car is shown in Fig. 9-32. Four forged steel angle valves are mounted on the cover plate which is

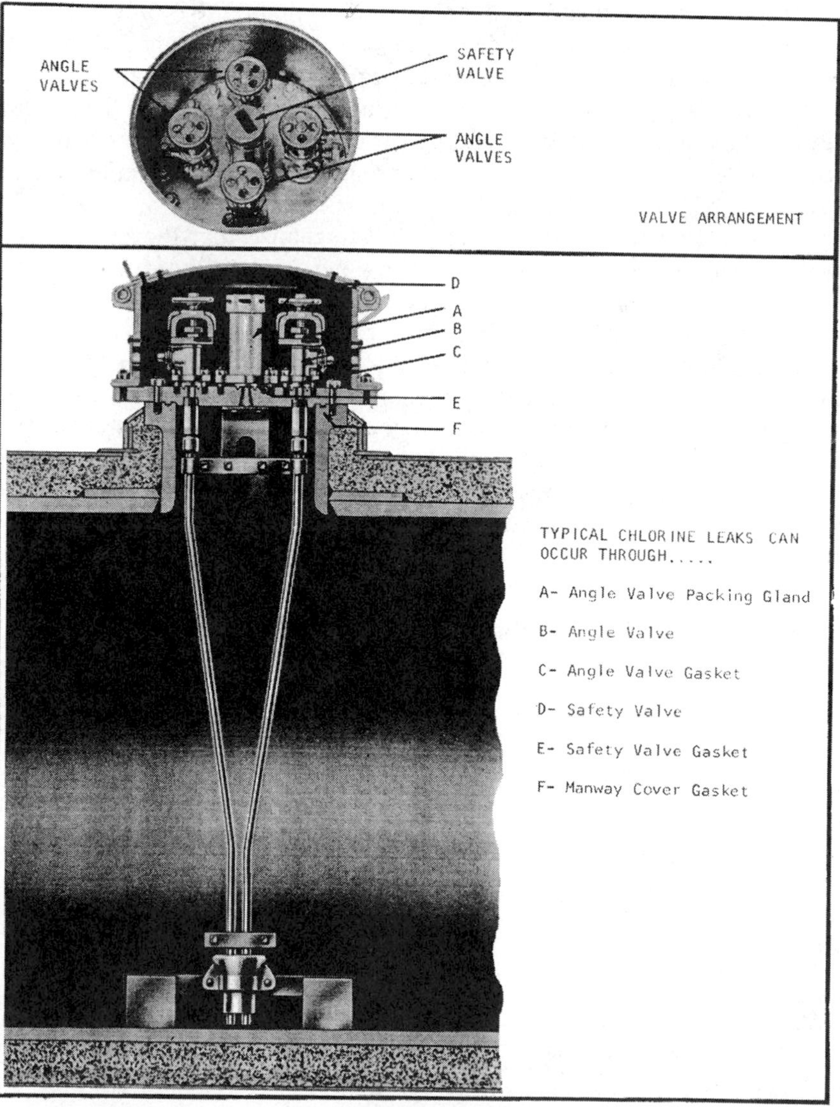

Fig. 9-30. Cross-section of a tank car manway and one-inch-diameter liquid withdrawal tubes.

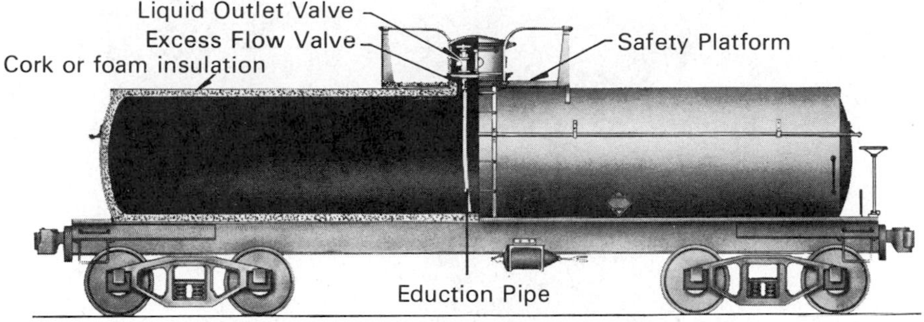

Fig. 9-31. Single unit chlorine tank car (courtesy PPG Industries, Chemical Division).

secured to the dome; a fifth is a safety relief valve. These valves have Monel seats and stems, and are protected by a false dome with a hinged cover. Openings around the false dome, protected by close-fitting covers, permit access for connection to the valves. The two angle valves (Fig. 9-33) on the longitudinal axis of the car are for liquid withdrawal. An eduction pipe extends from each liquid discharge valve to the bottom of the car. Just beneath each of these valves in the eduction pipe are excess flow valves, as shown in Fig. 9-34. These protective devices are designed to close automatically in case of a major break in the angle valve or in the chlorine unloading pipeline. These valves seat only when the chlorine flow reaches approximately 7000 lb/hour. This flow rate shutoff varies with car size.

The Unloading Site. For the layout on a single-unit car unloading site, the designer should consult the *Chlorine Institute Manual,*[1] which quotes from the ICC regulations for chlorine unloading.

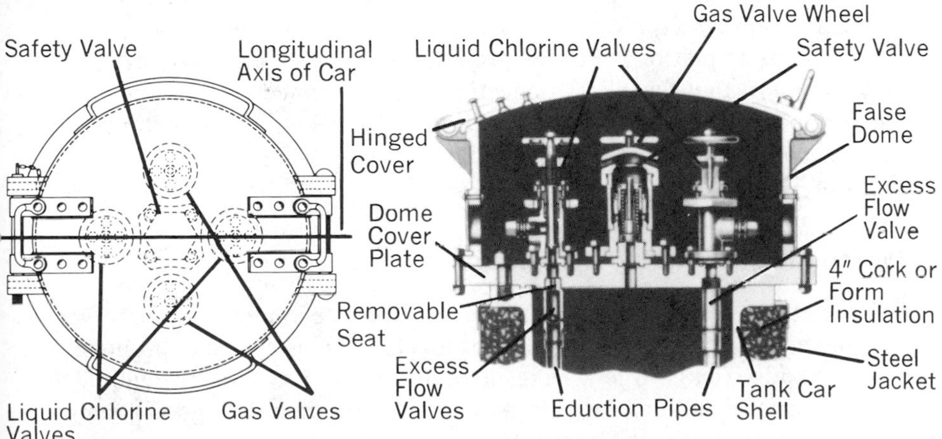

Fig. 9-32. Chlorine tank car dome (courtesy PPG Industries, Chemical Division).

Fig. 9-33. One-inch angle valves for corrosive products.

A dead-end siding should be provided for chlorine unloading. Tracks should be level. The car should be protected by a locked derail at least one carlength away from the end of a car hooked up for unloading. If the car must be on an open siding, both ends should be protected. If a switch is involved, it should have a lock; the keys for the switch and the derail should only be in the hands of the person responsible for unloading. (See Figs. 9-35 and 9-36.)

An operating platform must be provided at the unloading point to provide easy access to the dome for connecting and disconnecting the loading lines and for operation of the valves. (See Figs. 9-37 and 9-38.) This platform should be of such height that it is suitable for working on cars of any of the sizes listed in Table 9-4.

Midland Manufacturing Corporation[114]

Introduction. This company has recently been producing specialty valves and valve actuators for all corrosive products, such as chlorine, bromine, and sulfur dioxide. This is largely the result of the many users who are building their own chlorine or sulfur dioxide transportation equipment. A good example is the Metropolitan Water District of Southern California.

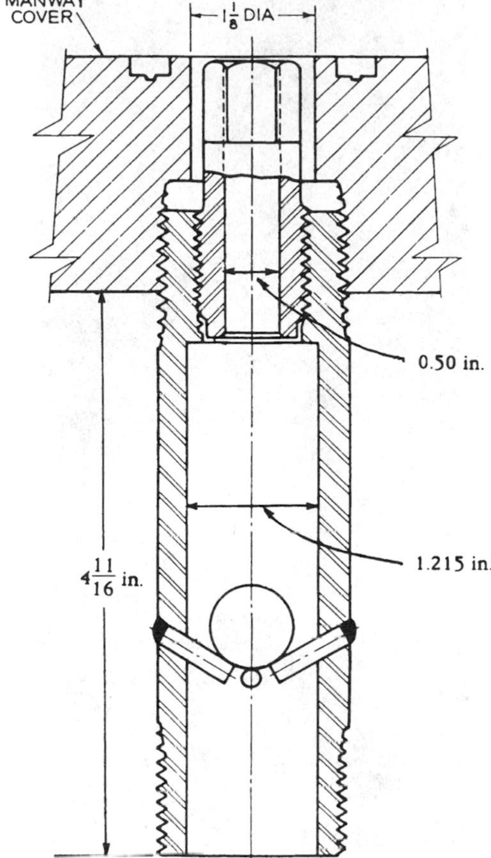

Fig. 9-34. Excess flow valve, 7000 lb/hr capacity.

Depending upon the particular corrosive chemical involved, the user has the choice of the following materials:

- Either 300 or 316 stainless steel
- Monel
- Nickel
- Hastelloy B
- Hastelloy C

One-Inch Angle Valves. The outstanding features of these valves, which can be used in the tank dome of chlorine storage tanks or road tankers, are:

1. Three seals in series at every potential leak-site. Also the redundancy of elastomeric O-rings and gaskets made of Teflon, or other inert materials, ensures leaktight operation.

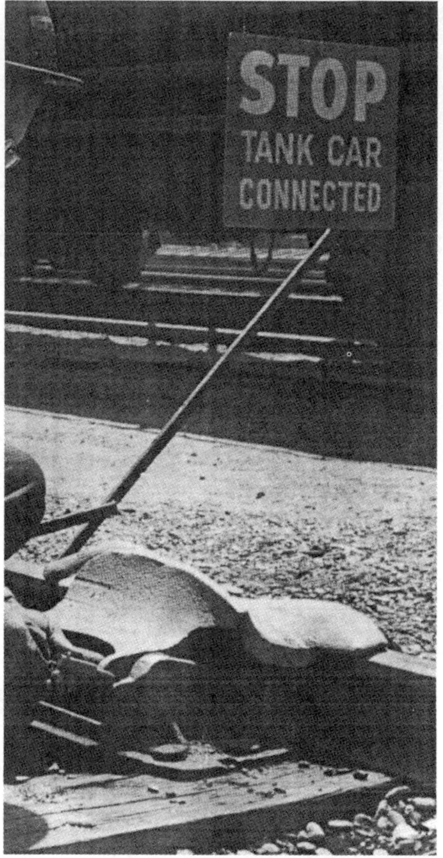

Fig. 9-35. Railcar derail system.

2. Replaceable side port, which can be made of hardened material. A flanged discharge connection is also available.
3. Elevated seat, which permits self-draining without trapping product.
4. Spring-loaded packing to compensate for wear and variable expansion rates of metallic parts and PTFE packing.
5. "Soft" PFTE seat seal, which provides long-term pressuretight sealing. The seal retainer is self-centering and does not rotate with the stem.
6. No remachining required in maintenance. Only periodic replacement of O-rings and PTFE seals is needed.

See Fig. 9-33 for a Midline valve.

1¼-Inch Excess Flow Check Valve. This valve is illustrated by Figs. 9-39 and 9-40. The outstanding features of the valve are:

Fig. 9-36. Railcar wheel lock.

1. It can be furnished in a variety of flow rates from 2000 lb/hr to 37,000 lb/hr of liquid chlorine. The highest rate was the design of a particular situation where the user wanted to be able to transfer from one container to another in the shortest possible time during hazardous conditions. The lower rates are those desirable for users who are not dependent upon these valves under hazardous conditions, such as those who are using road tankers or storage tanks. Everyone involved must remember that these valves will not go into operation when the pressure downstream from the valve is greater than atmospheric pressure.
2. It is very sensitive to flow rate changes, as reflected by the closing pressure differential.
3. Rapid reset: owing to a pressure bypass, the check valves, after closing, reset quickly after the angle valves are closed.
4. Nongalling assembly: Straight threads are used for the plug assembly into the body instead of tapered pipe threads. This eliminates galling that is common to nickel-bearing material. Monel is 65 to 70 percent nickel, which will readily gall when wedged together.
5. Durable construction: instead of four small pins, a ½-inch-deep bar spans across the bottom of the valve body, keeping the float from dropping out of the bottom of the valve.

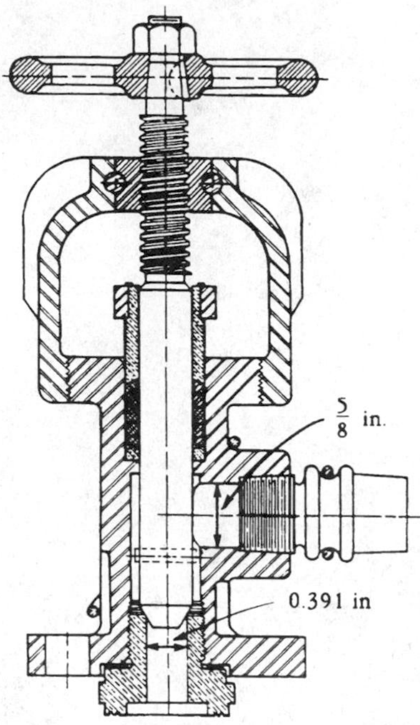

Fig. 9-37. Dimension of tank car shutoff valve.

Chlorine Valve Actuators. These are for the automatic control of the chlorine tank dome angle valves. They are made in one- and two-inch sizes. The exterior housings are made of stainless steel and Monel. The interior parts of the air motor are sealed from atmospheric contaminants. This actuator is illustrated by Fig. 9-41.

The outstanding features of these actuators are:

1. They are intrinsically safe because of the air-driven motors. Elimination of electricity in a potentially hazardous area is always an important safety step.
2. They are easy to use. The air hose is permanently attached to the air motor; so by pulling down from a reel or spring-loaded overhead tool retractor, the air motor can be readily positioned on the valve. No air hookup or tools are required to position and operate the actuator on the valve.
3. They are portable and lightweight. The air motor assembly weighs about 15 lb; so it is light enough to be easily handled, like a portable hand power tool.
4. Control panels can provide remote operation of these valves in case of an emergency that requires either opening or closing the valve, or simply jogging the valve.

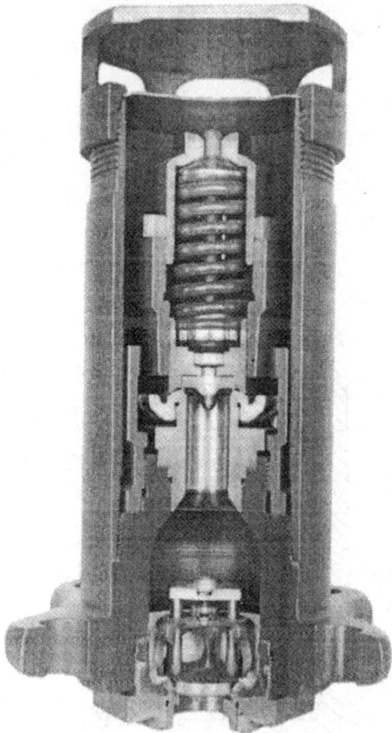

Fig. 9-38. Cross section of a standard tank car pressure relief valve.

Safety Relief Valve. Here again is a safety relief valve designed primarily for chlorine road tankers and storage tanks, for installation on the tank dome. This valve is illustrated by Fig. 9-42. These valves are designed to handle many other corrosive products besides chlorine, and a variety of acids.

The outstanding features of these valves are:

Fig. 9-39. Midland excess flow check valve (1¼ inch).

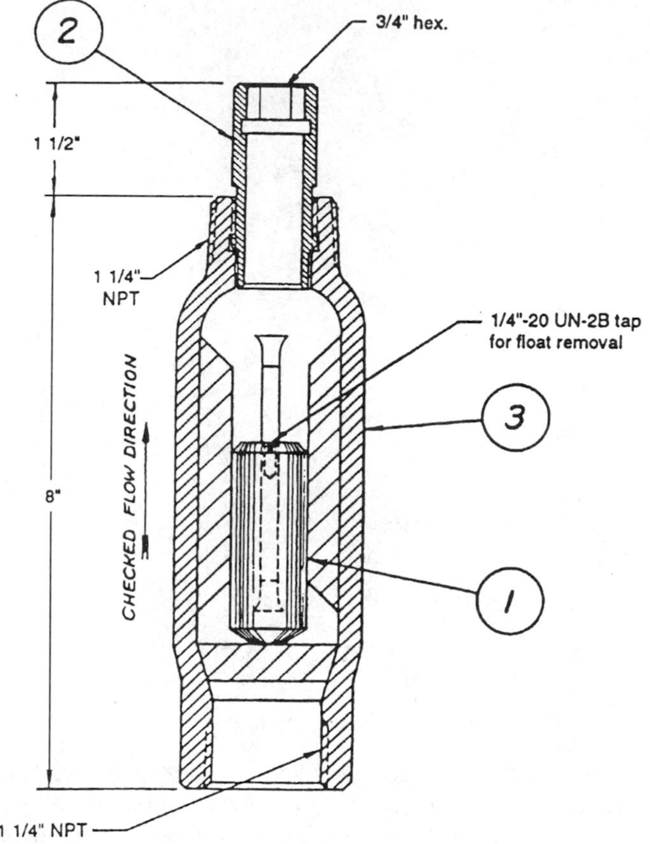

Fig. 9-40. Cross section of the Midland excess flow check valve A-120 (1¼ inch). Maximum flow rate is 32,000 pounds of water per hour. (1) The float, made of steel in both the A-120-CS and A-121-CS; (2) the seat, made of stainless in both the A-120-SS and A-121-SS; (3) the body, made of Monel in both the A-120-ML and A-121-ML. A-120 is for a threaded installation, A-121 for a welded installation.

1. They have a full range of pressure settings up to 450 psig. This requires only making a spring change for the start to discharge pressure.
2. The overall height is less than 12 inches, and they will fit inside the chlorine emergency kit. When mounted horizontally, they are only 6 inches high.
3. The sealed chamber contains the critical operating parts, which are the guides and the spring. Even when these valves discharge chlorine, nothing can get inside to corrode these parts.
4. Bubble-tight performance is assured by a metal-to-metal seat and resilient seals.

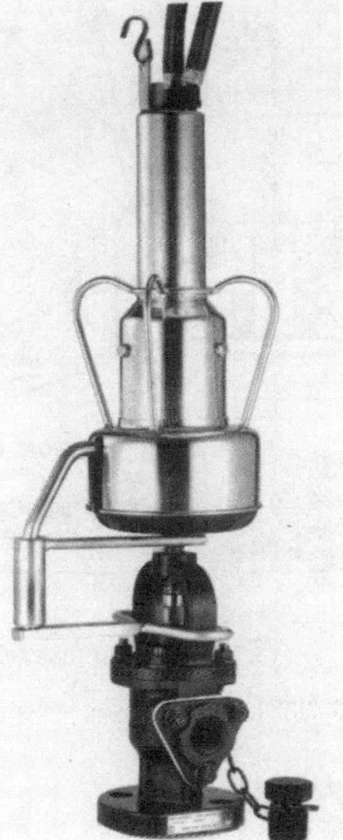

Fig. 9-41. Midland valve actuators, 1 inch and 2 inch.

5. They are attractively priced for items made of corrosion-resistant materials.
6. They are also available in combination with rupture disks.

Flexible Connections. Two types of flexible connections for use between the car dome and the piping system are available: the annealed copper loop and the flexible reinforced metal hose. The connections are fitted with two-bolt ammonia-type unions with a lead gasket joint. It is customary and desirable to connect a pressure gauge to one of the gas valves on the car dome. An L-shaped pipe assembly is used to permit mounting the gauge where it can be seen from the unloading platform.

One flexible connection is used to join one of the liquid withdrawal valves to the liquid header piping, and the other is used to connect one of the gas withdrawal valves to the air padding system. This gas connection can also be used to connect the gas withdrawal valve to the gas header piping.

Flexible connections should be plugged at both ends with corks immediately upon removal from the tank car, and be stored in a heated compartment. This minimizes corrosion and extends the active life of the flexible connections.

Fig. 9-42. Midland safety relief valve.

Header System. Two completely separate headers are required between the tank car (or storage tank) and the evaporators: one for liquid and one for gas. It is at times necessary to withdraw gas from the supply system. In the event of a leak in the tank car discharge system or header piping, the ability to withdraw gas can be used to reduce the tank car pressure. Accordingly, a field test was made with a 55-ton car approximately half full at the beginning of the test.[4] Three 8000-lb chlorinators operating simultaneously were used for this test. Withdrawal of approximately 2030 lb of chlorine reduced the car pressure from 98 psi to 37 psi in 3¾ hours. Ambient temperature was 55°F. The pressure drop caused a liquid chlorine cooling effect of 47°F. This represents 43 lb liquid chlorine per degree F drop. However, from 37 psi to 20 psi required only 30 min. and 208 lb of chlorine to produce an additional cooling of 24°F, or 9 lb chlorine per degree F.

The presence of two header systems is also desirable for standby service. Tank car consumers using a duplicate header system have reported frequent use of the second or standby header for routine maintenance. Each of these headers should be equipped with a pressure gauge immediately downstream from the shut-off valve on the unloading platform.

Air Padding System. This is required for unloading the tank car and for pressurizing cars in cold climates. Chlorine is usually shipped cold. Assuming a median temperature of 40°F, the vapor pressure will be 45 psig. Use only

Table 9-5 Tank Car Capacity

	55-ton Car	*90-ton Car*
Water capacity, gal	10564	17298
Total car volume, ft^3	1410	2335
Liquid Cl_2 volume 60°F, ft^3	1240	2035
Gas volume space, ft^3	170	300

clean dry air free of oil and foreign matter. Air must be dried to at least a −40°F dewpoint measured at atmospheric pressure; −50°F is preferable.

The air padding system should incorporate the following elements:

1. Oil-free air compressor with after-cooler and receiver
2. Air dryer
3. Air filter (oil absorber type)
4. Dewpoint indicator
5. Dewpoint alarm
6. Vapor pressure alarm system

Table 9-5 itemizes the volume characteristics of a 55-ton and a 90-ton car.

Table 9-6 gives the amount of padding air required to pressurize a 55-ton and a 90-ton car to various pressures.

The air flow required to unload a tank car at a given rate can be calculated by the following formula:

$$\text{SCFM Air} = \frac{\text{lb/day Chlorine} \times (P_{car} - 25)}{1.8 \times 10^6} \tag{9-1}$$

Selection of an oil-free air compressor is the objective of a reliable air padding system. An example is the Corken Pump Co. of Oklahama City, Oklahoma. Any selected compressor should be supplemented with a Deltech filter for compressed air. The compressor should be equipped with an ASME code receiver of appropriate size with an air-cooled after-cooler mounted on the receiver.

Table 9-6 Dry Air Required to Pressurize Tank Car*

	Volume of Free Air (ft^3) Required to Achieve Car Pressure Shown (psi)					
Car Size	*60*	*75*	*90*	*105*	*120*	*130*
55-ton car	170	340	510	680	850	100
90-ton car	300	600	900	1200	1500	180

*Assume median Liquid Cl_2 temp. = 40°F, and car pressure 45 psig.

Suitable air dryers are available from Deltech Engineering Co., Century Park, Newcastle, Delaware 19720 and Lectrodryer, Box 4599, Pittsburgh, Pennsylvania 15205. The air dryer can be the heat reactivated type capable of operating up to 150 psi. The air flow capacity of the dryer should be 1.5 times the compressor capacity discharging at a maximum dewpoint of −40°F. This part of the system should be equipped with a dewpoint indicator and a high humidity alarm.

A critical part of the air padding system is the check valve,* which prevents chlorine vapor from entering the system. Installations using ordinary ball check valves have invariably suffered from chlorine vapor getting into the air pad system. Spring-loaded check valves have proved to be somewhat more reliable, but they too have caused corrosion problems.[5]

The Chlorine Institute Pamphlet No. 4, "Consumer Air Padding of Chlorine Single Unit Tank Cars," illustrates a power operated shut-off valve instead of a check valve. This is the preferred method. Figure 9-43 illustrates a power-operated shutoff valve system used for many years by the Pennwalt Corporation, Tacoma, Washington.[6]

Each air padding system should have its own separate and independent air supply. When both chlorine and sulfur dioxide are present, the air padding systems must never be used interchangeably. Each liquid–gas supply must have its own air padding system.

Whenever air padding is deemed unnecessary, then nitrogen purging facilities should be considered mandatory. This system consists of a sufficient number of nitrogen cylinders containing nitrogen gas at 2000 psi and a common pressure regulator with an outlet pressure regulated at 150 psi. It should be arranged to allow the operator to purge the entire header system back into the containers. This system is most advantageous in the handling of liquid chlorine.

Absorption Tanks. In addition to purging and air padding systems, tank-car systems should be provided with a liquid chlorine absorption tank. This tank should be able to absorb all the liquid chlorine in the header piping and evaporators. The absorption tank should be made of reinforced Fiberglas or rubber-lined steel. It should have a vent, a special connection for caustic, and one for makeup water. There should also be a sampling tap to determine the effective absorption capacity of the solution. The piping of the liquid chlorine to the tank must have a barometric loop (see reserve tank system). The downdrop of this loop and the sparger (diffuser) should be made of Kynar pipe. The size of the absorption tank is based upon the stoichiometric combination of caustic (NaOH) and chlorine, which is 1.13 lb of caustic for 1.0 lb chlorine.

*This is the only instance where a check valve is optional in the chlorine supply system. Check valves at other locations (flex connections, header piping, etc.) should never be used. In a chlorine atmosphere they are unreliable because of corrosion products. Moreover, they would be hazardous to operating personnel.

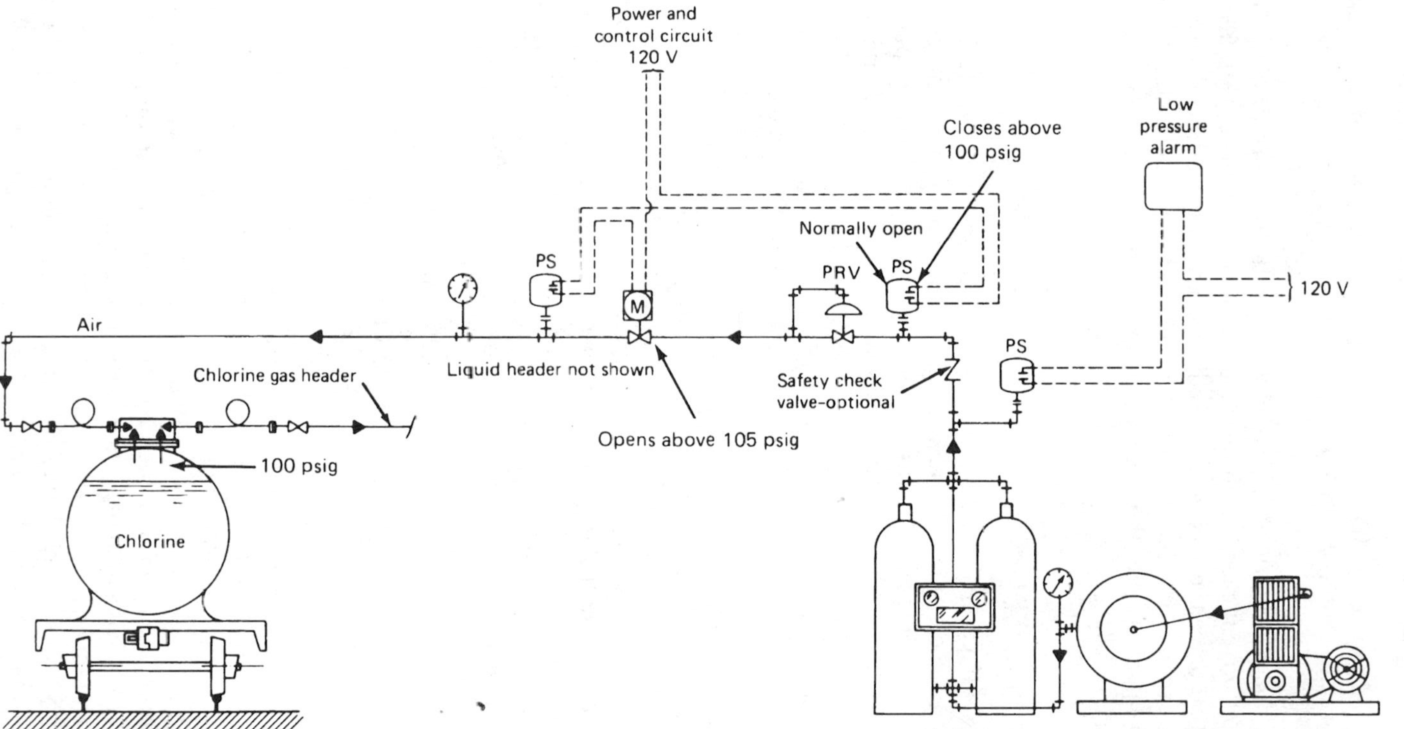

Fig. 9-43. Power-operated shutoff valve, air padding system. PS = pressure switch; PRV = pressure reducing valve; M = motorized ball valve. Operation: The shut-off valve M will be open only when dry air pressure exceeds 105 psig and when tank car pressure is less than 100 psig. If either of these conditions is not met valve M will be closed.

The absorption tank must have a minimum depth of 8 ft, and the caustic solution is usually kept at 0.5 to 1.0 lb/gal.

EXAMPLE. 1000 lb chlorine requires 1130 lb caustic for neutralization. One gallon of 50 percent caustic contains 6.38 lb caustic. Therefore, 1130/6.38 = 177 gal of 50 percent caustic is required for each charge of the absorption tank to neutralize 1000 lb of liquid chlorine. So to provide a tank having about 1.0 lb/gal caustic, the absorption tank should have a capacity of approximately 1200 gal.

Chlorine Leaks Compared to Ammonia and Sulfur Dioxide

Discussion. Of particular interest is the case of an enormous leak that occured when a local ammonia tank truck had an accident that caused the truck to turn over and spill its entire contents on a road in a heavy populated area. Owing to the chemical properties of ammonia, there was no harm from this enormous leak according to the local emergency response team. Because ammonia is lighter than air, the gas rose quickly into the atmosphere; and as it is highly soluble in water, the emergency response team flooded the ammonia that spilled on the ground, so it went into the local sewer system as an innocous solution. The chief of the emergency response team was thereafter convinced that ammonia in any form was not much of a threat during a leak.

The same is true for sulfur dioxide but for a totally different reason. It must be remembered that the cylinder pressure at room temperature is low, about 35 psi. In most locations the ton cylinders have to be covered with an artifically heated blanket to prevent the SO_2 from liquefying because of the low pressure.

We must work diligently to prevent chlorine leaks. Practitioners in this business know that chlorine leaks can be disastrous if we do not prepare properly for them.

A PLAN TO DEAL WITH A MAJOR CHLORINE LEAK

Definition of a Major Leak

Trying to deal successfully with a major leak is a formidable task. The two most discussed possibilities are a direct hit by an aircraft and a planned act of sabotage. The latter is usually dismissed on the basis that proper security measures can provide the necessary deterrence. The air crash scenario is usually dismissed as an improbability. However, an air crash accompanied by exploding fuel would result in the following: The aircraft impact or subsequent explosion would probably rupture the chlorine container(s). The ensuing fire would instantly vaporize the liquid chlorine, and the chlorine vapor would rise quickly with the heat of the fire. This sequence of events would serve to greatly diminish or eliminate chlorine exposure in the surrounding area. This was clearly demonstrated when a freight train derailment severely damaged a 90-ton chlorine tank car. The car was ruptured by the couplers from an adjoining butane tanker. All 90 tons of chlorine were released. The butane tank car explored, and all of its contents were consumed by fire. The heat from this fire vaporized the liquid chlorine, which disappeared into the upper atmosphere because of the rising hot air from the butane fire. A subsequent investigation revealed that no one in the surrounding area or at the scene of the accident was found who had experienced any exposure to chlorine (see Chapter 1).

The consensus definition of the most probable major leak is a guillotine break in the liquid chlorine header between the chlorine supply system and the chlorine evaporators.*

Important Aspects of a Major Leak. If a leak is to be considered a major one, there has to be a liquid spill. Any major gas leak can be dealt with quickly by the use of container kits and the proper use of chlorinator injectors to evacuate the vapor that is leaking. When a liquid spill is involved, the designer must make provisions for collecting the liquid in a confined sump and be able to hustle it off to either a scrubber or an absorption tank.

A major leak will never create a high atmospheric pressure condition in the room where the leak has occurred. Because of the enormous cooling effect in the leak area due to the liquid chlorine attempting to vaporize, the room pressure will be negative. This situation assures the flow of fresh outside air into the leak area. Therefore, a containment room for chlorine storage should always be designed for proper continuous outside ventilation. The fresh air from the outside should enter the storage room at ground level and exit at rooftop level. However, during a major leak event, the fresh air should enter at ceiling level and discharge to the scrubber at floor level, as shown in Fig. 9-44. The scrubber system should be a one-pass system; never use a recirculating system, as it would cause untold corrosion damage in the room where the leak occurred.

Liquid Chlorine Collection System

This part of the design focuses on the storage room floor configuration. The floor should have a dramatic slope (2½ in./10 ft) to a common point terminating in the liquid collection sump. The collecting slots should be narrow (2 in. max) and deep (5–6 in.) to shield the liquid from room temperature. This will significantly diminish the liquid evaporation rate. When liquid chlorine spills on a flat surface, "flash off" as chlorine vapor (liquid evaporation). This will occur intermittently until the room reaches temperature equilibrium. This causes a thin sheet of chlorine hydrate ice to form on the remaining liquid, which limits further evaporation until the ambient temperature melts the icy

*See Chapter 1 for the rate calculation for this type of chlorine leak.

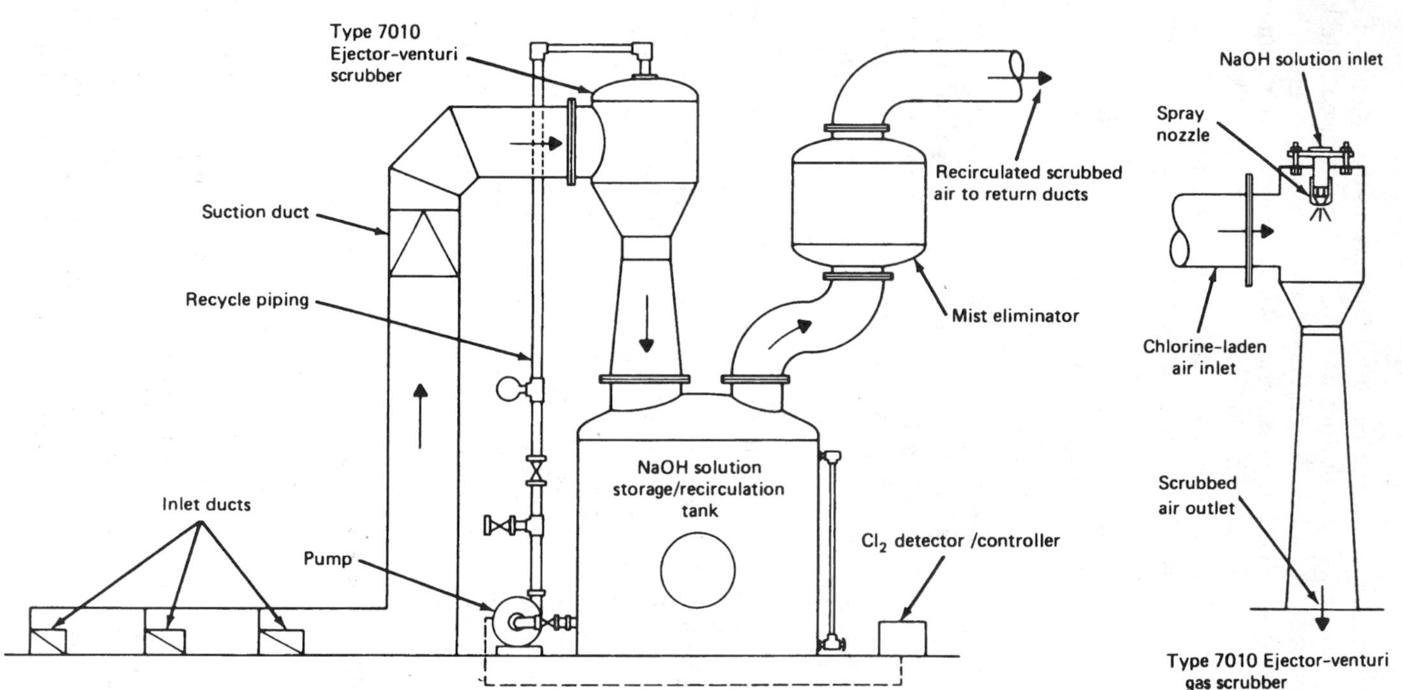

Fig. 9-44. Chlorine scrubber system (courtesy Ametek-Schutte and Koerting Div., Durham, North Carolina).

film. During this freezing and thawing cycle the vaporization rate of the remaining liquid is typically 8 lb of chlorine per square foot per hour.

The collection slots in the floor should terminate in the lowest part of the sloping floor. At this point the floor should be constructed to accept either a pump or an eductor.* Dealing with the liquid chlorine spill in the design of a neutralizing system is a number one priority because it reduces by a factor of ten the time required for a scrubber system to complete its objective.

Fundamentals of Estimating Leak Rates

One of the major errors usually committed when calculating leak rates from one-ton cylinders, railcars, and/or bulk storage tanks is trying to estimate the relevance of the physical dimensions of a given leak. When a guillotine break in a one-inch chlorine header is mentioned, this dimension is used as the area of the chlorine leak. In reality the chlorine liquid has to pass through a long series of restrictions to the flow: the 22 inches of ½-inch tubing inside the ton cylinder, the ton cylinder shutoff valve, the auxiliary cylinder valve, 4 ft of 9⁄32-inch inside diameter flexible copper tubing (maybe an auxiliary header valve), and a header valve. These restrictions are shown in Figs. 9-2, 9-6, 9-9, and 9-19. There is no method of quantifying these restrictions in terms of making it possible to calculate the chlorine leak rate from a broken or ruptured one-inch-diameter pipe. The only possible way that this problem can be solved is by simulating such a leak. Such a simulation was performed at the EBMUD, Oakland, California WWTP in the early 1950s—but for an entirely different reason. Operators needed to know the maximum liquid chlorine withdrawal rate from a single one-ton cylinder in order to verify the necessity to go to bulk storage. The amount of liquid chlorine that the one-ton cylinder could deliver to three 6000 lb/day chlorinators was only 10,200 lb/day with a 45-lb pressure drop between the cylinder and the chlorinator. Converting this flow rate to the pressure drop due to a header rupture, assuming a worst case of 120 psi pressure drop, was only 11.4 lb/min or 16,416 lb/day (see Chapter 1). The same approach has to be made for the case of noninsulated bulk storage tanks and insulated railcars. All of these situations have been calculated by White, and they appear in Chapter 1.

Since publication of the 1988 edition of the Uniform Fire Code (UFC), and subsequent revisions in 1991 and 1994, many fire marshals have made a concerted effort to use the capacity of the excess flow valves in railcars and bulk storage tanks as a basis for the capacity of a neutralizing system when there is a guillotine break in the piping system downstream from the excess flow valve. Examination of Fig. 9-34 shows that there are significant flow restrictions in this check valve, not to mention other components that are also

*Pumps are available from both the Duriron Co. and Powell Fabrication and Manufacturing, Inc. These pumps routinely handle liquid chlorine.

illustrated. If there is a sudden rapid change in the liquid chlorine flow rate due to a major leak, the liquid tends to vaporize and thus gets "gas-bound." This behavior slows the flow appreciably, preventing the excess flow valve from closing. The point of this discussion is that the rating of an excess flow valve is not relevant to the design of a neutralizing system when one is considering a major leak in the piping system downstream from the liquid chlorine discharge valve in a railcar or a storage tank. In some instances such a leak did occur, and the check valve did not close. An excess flow valve will close in an operating situation only when the system is started for the first time, or after the system has been shut down for routine maintenance and all the chlorine has been emptied from the piping system. Then if the operator opens the discharge valve too quickly, the valve will jam shut. The only other times when these valves have been known to close have been in a railroad accident where the car flipped over in such a way that the discharge valves on the car dome were sheared off. This is the result of zero pressure on the downstream side of the valve.

Illustrations that exhibit chlorine flow restrictions are Figs. 9-2, 9-9, 9-19, and 9-34. These restrictions limit the ability to calculate major leak patterns. The only way to arrive at reasonable leak flow rates is to simulate a leak on-site and use scales for actual weight loss.

NEUTRALIZING A MAJOR CHLORINE LEAK

Systems Description

Historical Background. The oldest and most popular method of dealing with a major chlorine leak is the use of an absorption tank. This was developed by the pulp and paper industry, where chlorine is used in enormous quantities for bleaching and control of biological growth during pulp preparation. The absorption tank was filled with enough caustic to neutralize all the chlorine contained in the piping system and its components downstream from the chlorine supply shut-off valve. This was deemed the only logical way to deal with a leak where all of the chlorine being used was under supply tank vapor pressure. This is not necessarily the case for the 5–7 percent of all the chlorine manufactured in the United States that is used in the treatment of drinking water and wastewater. In these applications, chlorine is metered and controlled under a 12–18 inches of Hg vacuum. This vacuum is created by the power of a venturi device, called an "injector," and the power of the venturi is obtained from a supply of water, usually at 50–60 psi pressure. Each chlorinator is fitted with an injector capable of feeding its maximum capacity. This makes each chlorinator a primary safety device because the injector can dispose of the chlorine in the piping system in a few minutes. In addition to this feat, the chlorinators (if piped properly) can reduce and control pressure in the supply system (i.e., the chlorine cylinders or storage tanks).

The chlor-alkali plants have made significant advancements in developing safer ways to store their chlorine production. The latest innovation involves the recirculation of liquid chlorine in an enormous spherical vessel through a refrigeration system that keeps the liquid at atmospheric pressure. All of this is supplemented by a containment structure with a sloping floor to confine a liquid leak in as small a space as possible. During a leak, an insulating foam is sprayed on top of the spill to prevent vaporization while the liquid is pumped to a neutralizing tank.

The Uniform Fire Code of 1988 changed the industry approach to safety precautions associated with the handling of a major leak. The major obstacle of this code is the requirement that the neutralizing system be able to handle the full contents of the largest single storage container. The code has led to a lot of confusion because the people who generated the code do not understand the basic characteristics of either liquid or gaseous chlorine.

Fume Scrubber

The system illustrated by Fig. 9-44 was the first system to be considered for major chlorine leaks. Typically this scrubber is designed to recirculate the containment room air until all of the chlorine spill has been neutralized. The system usually is designed to provide one complete room air turnover every 10 min. The scrubber system depends upon a chlorine detector to close the normal ventilation system and to activate the scrubber recirculating pump. This delivers caustic to the inlet of the venturi. Simultaneously, room air is drawn into the suction throat of the venturi, where it mixes with the caustic. This is similar to a chlorinator injector operation. The standard venturis on this type of system are only 85–90 percent efficient in the chlorine reaction with the caustic. The remaining 10–15 percent of the chlorine and all the inert gases along with the caustic descend into the top of the tank from the venturi outlet. These gases then are forced up the vent stack and out the mist eliminator. The scrubbed air is returned to the chlorine-contaminated containment room. The caustic tank is designed for any given expected major chlorine spill. The stoichiometric ratio of caustic (NaOH) and chlorine is 1.13 lb caustic per pound of chlorine. The scrubber operates until the chlorine concentration in the room air is reduced to 1 ppm.

During a leak episode it is necessary to monitor the room air for chlorine concentration and the capacity of the caustic to absorb the remaining chlorine. This is accomplished by titration procedures.

Spent Caustic Disposal. The spent caustic will be a sodium hypochlorite solution with sodium chloride and residual sodium hydroxide. This is easily disposed of at a water or wastewater plant provided it can be metered in small quantities over a given period of time. The hypochlorite can be easily destroyed by catalytic decomposition using nickel and iron as catalysts.* The use of

*If available, seawater will destroy the hypochlorite solution in a few hours owing to the presence of heavy metal ions.

sulfites to dechlorinate the hypochlorite solution is not recommended. The heat of reaction between the sulfite ion and hypochlorite is far too great at these concentrations.

Conclusions. This type of design has been all but abandoned, for a variety of reasons. This system has to shut off all ventilation, so no fresh air can be used for dilution. Because the recirculating scrubber is only 85–90 percent effective, the chlorine vapor will be increasing in volume during the initial phase of the leak. This causes positive pressure in the containment room. This does not comply with the UFC guidelines. In addition to the room pressure increase, the recycled chlorine vapor will contain both NaOH and NaOCl mist particles, which will not be removed by the mist eliminator. Furthermore, this mist will be created whenever the scrubber system is activated for testing or neutralizing a leak. This not only is a health and safety issue but also causes severe erosion to electrical equipment and other metal components in the room.

Another serious flaw in the recycle system is the lack of fresh air dilution. This can affect the emergency response team's decision to enter the room to proceed with its efforts to stop the leak. These Chemtrec teams are scattered all over the United States, and their entrance limitations for repair of a leak (only a container leak) vary with the local authority.

The Single-Pass Absorption System

One type of single pass absorption system, a 500 cfm dual venturi scrubber system, was tested by a field test, done by Powell Fabrication & Manufacturing Co. of St. Louis, Missouri, in July 1985. This system depends entirely upon the venturi units to supply the power to evacuate the contaminated room air. Absorption was not the function of the venturis. The absorption was stated to occur by virtue of the hydraulic and chemical kinetics in the proprietary reactors mounted adjacent to but downstream from the venturis.

The Single-Pass Three-Stage Absorption System

This scrubber system was designed by RJ Environmental, Inc. of San Diego, California specifically to meet the 1988 Uniform Fire Code (revised in 1991 and 1994), Section 80.303 of Article 80, as it pertains to indoor storage of compressed gases. It was designed to meet the UFC maximum allowable discharge concentration of the chlorine vapor, to one-half of the IDLH (Immediate Danger to Life and Health) at the point of discharge to the outside atmosphere. For chlorine, the IDLH was 30 ppm, which in June 1994 was reduced to 10 ppm. Therefore, the maximum allowable discharge concentration in the scrubber vent stack is currently 5 ppm, as stated in the UFC. The RJ scrubber, though, is designed to treat a release rate much higher than the UFC requirement. A full scale test with a chlorine release rate at about 100 lb/min. resulted in vent stack chlorine concentrations of less than 2 ppm. The entire unit is a skid-mounted package measuring 13 ft long by 7 ft wide and 8 ½ ft high. (See Figs. 9-45 and 9-46.)

Fig. 9-45. One ton emergency chlorine leak scrubber system (courtesy RJ Environmental, Inc.).

Scrubber. The RJE Vapor Scrubber is a three-stage chemical absorption system consisting of a horizontal crossflow spray system followed by two horizontal crossflow packed bed sections. An induced draft fan pulls vapors through the scrubber, where intimate contact with a recirculating caustic solution results in the complete absorption and removal of chlorine or sulfur dioxide vapors. A high efficiency mist eliminator is located in the gas stream, prior to exhaust, to remove any residual caustic solution.

A chlorine detector or manual remote start switch activates the system in two steps. The caustic pump is activated first to permit proper wetting of packing in the scrubber stages before starting the exhaust fan, with a 0 to 3 sec adjustable time delay. The time delay is typically set for 3 to 5 sec. This feature allows the scrubber to be ready prior to passing any chlorine-laden gases through it. Fig. 9-47 illustrates the scrubber process flow system, and Fig. 9-48 illustrates the process and instrumentation diagram.

The exhaust fan is placed downstream of the scrubber. This feature allows the complete system to be under negative pressure until the gases are completely scrubbed.

Major System Components:

1. FRP three-stage scrubber absorber*
2. Integral caustic storage tank

*The absorber is placed on top of a caustic storage tank, which is an integral part of the system.

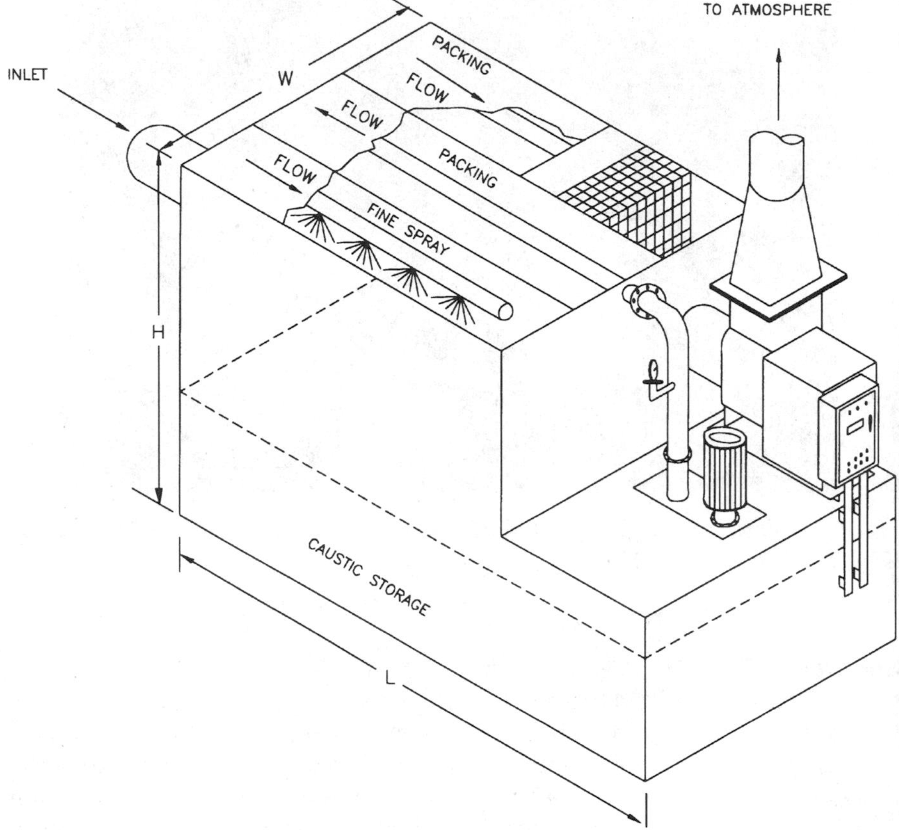

Fig. 9-46. Isometric view of one ton scrubber system (courtesy RJ Environmental, Inc.).

3. FRP air exhaust fan
4. Caustic recirculating pump
5. Electrical control panel
6. Piping, skid, and single unitary construction

Major System Components Sizing. The overall scrubber system package consists of the major system components described below. The overall size of the system is presented below, including the scrubber and caustic storage, pump, fan, and controls.

Scrubber and Caustic Storage. The complete scrubber system is made of FRP and consists of three stages for chlorine absorption. The caustic storage tank is an integral part of the scrubber system. The caustic storage tank is designed to hold about 20 percent caustic solution. The caustic storage tank is integrally molded into the scrubber housing and is made of premium grade

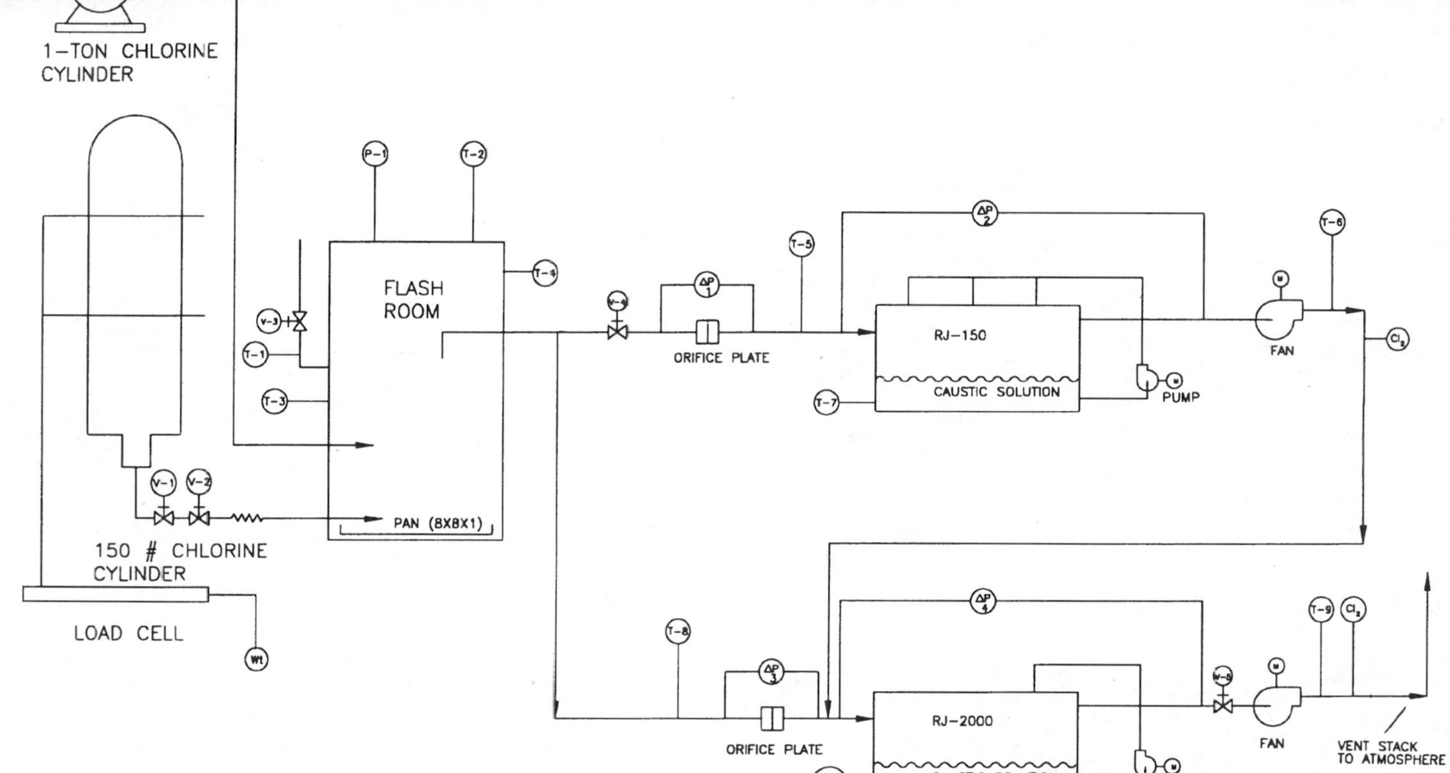

Fig. 9-47. Chlorine scrubber test process flow and instrumentation diagram (courtesy RJ Environmental, Inc.). T-1, intake air to flash room; T-2, flash room; T-3, flash room; T-4, flash room; T-5, RJ-150 air inlet; T-6, RJ-150 air outlet; T-7, RJ-150 sump; T-8, RJ-2000 air inlet; T-9, RJ-2000 air outlet; T-10, RJ-2000 sump; T-11, ambient (room); T-12, open; P-1 flash room pressure; Wt, load cell; Cl_2, exhaust stack chlorine level; ΔP-1, RJ-150 orifice plate 0–2"; ΔP-2, RJ-150 scrubber dP (0–5"); ΔP-3, RJ-2000 orifice plate (0–2"); ΔP-4, RJ-2000 scrubber dP (0–5").

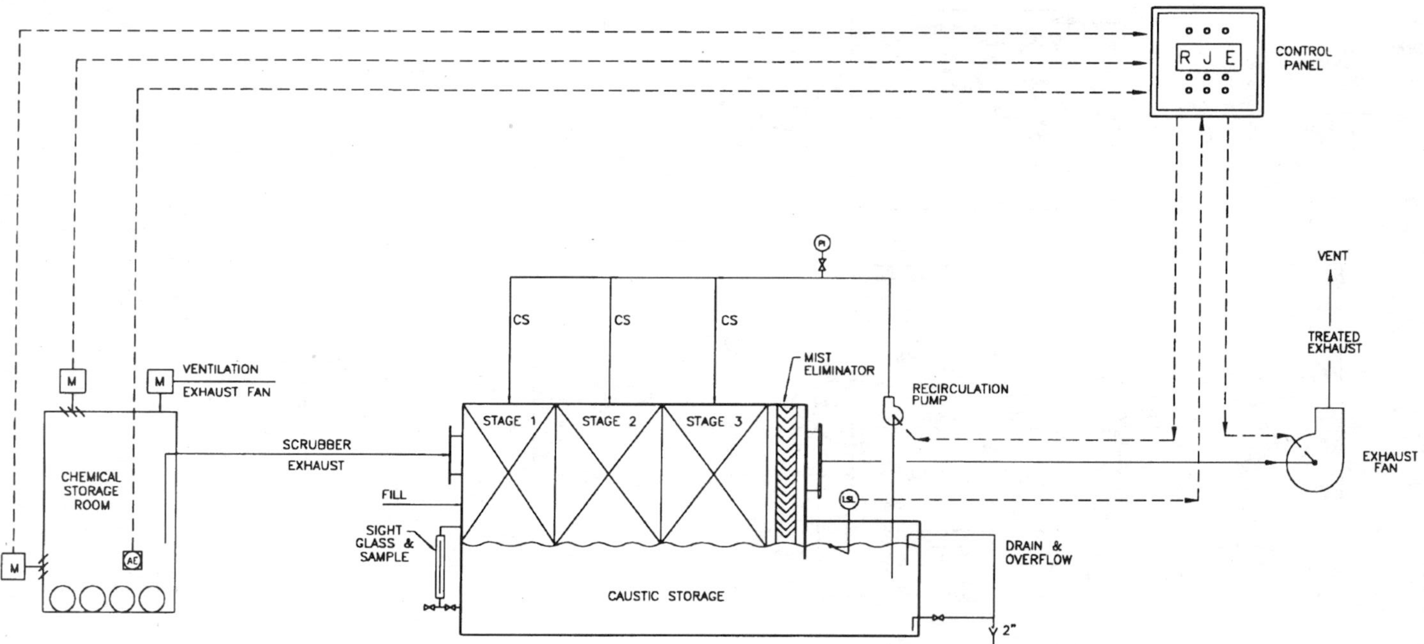

Fig. 9-48. Emergency chlorine vapor scrubber process diagram (courtesy RJ Environmental, Inc.). AE, vapor leak; CS, chemical solution; PI, pressure indicator; LSL, level switch low; ⋈, ball valve.

vinyl ester resin. The overall footprint of the scrubber and storage tank is as follows:

Length, ft: 13.0
Width, ft: 7.0
Height, ft: 8.5
Caustic storage capacity (20 percent), gal: 2100

Since the caustic storage tank and scrubber section are one part, there is sufficient free board height available for any expansion of liquid due to water formation.

Caustic Recirculation Pump (the pump does not require seal-water):

Recirculation Rate, gpm: 550
TDH, ft wc: 55
Brake HP: 14.1
Motor HP: 20
Construction: FRP
Type: Seal-less vertical centrifugal

Exhaust Fan:

Air flow rate, acfm: 3000
Duct pressure losses, in. WC: 2.0
Scrubber/stack press. losses, in. WC: 3.5
Total static pressure, in. WC: 5.5
Fan motor BHP: 4.2
Motor HP: 5.0

Electrical Control Panel. The scrubber system include a prewired electrical control panel that includes the following:

- Motor starter for exhaust fan and recirculation pump
- Timer delay for exhaust fan
- Control voltage transformer (480V, three phase to 110V, single phase)
- System HAND-OFF-AUTO switch and system ready light
- Pump HAND-OFF-AUTO switch and light
- Fan HAND-AUTO switch and light
- Low level switch, audible alarm, reset
- Alarm light for fan fail
- Alarm light for pump fail

A single 480V/three-phase power source will be required for the control panel.

Utilities Requirement:

Electrical Power Requirement:

- Air exhaust fan: 5 HP
- Caustic recirculation pump: 20 HP
- 480V, 3 PH, 50 AMP

RJE Major System Features/Advantages

Proven Design. Full scale tests were conducted on two RJE vapor scrubbing systems in April 1992 at a nationally recognized testing laboratory accredited by the International Conference of Building Officials (ICBO).

Testing was performed under rigidly controlled procedures with continuous on-line data recording and videotaping of each of the tests. Liquid chlorine was released directly from the cylinders in a specially designed flash room (13′ × 12′ × 12′). The systems were evaluated with chlorine release rates from 30 lb/min to 100 lb/min.

The Uniform Fire Code requires a maximum concentration of chlorine in the scrubber exhaust of 5 ppm. The chlorine concentrations in the RJE scrubber exhausts were 2 ppm or less during all tests!

In addition to proven system designs, the scrubber system offers many advantages that are not available with conventional system.

Vertical "Seal-less" Centrifugal Pump. To provide emergency vapor scrubber systems as maintenance-free and mechanically reliable as possible, RJ Environmental has developed a configuration that incorporates a vertical recirculation pump of fiberglass reinforced plastic. This has resulted in tremendous advantages over horizontal pump design; for example:

1. Elimination of mechanical seal: With the pump submerged in the recirculating fluid, there is no seal water required.
2. Elimination of seal flush piping: With no mechanical seal, the installer will not need to coordinate seal flush piping, a reliable water source, drainage, or possibly associated heat tracing and insulation.
3. Reduction of submerged exterior connections: With the pump submerged, there is no need for relying on the installer's connection of the pump suction piping; thus the risk of leakage is reduced in the years of service ahead.
4. Simpler heating and insulation: In cold weather applications, the entire liquid storage area may be insulated and heated as a single application, without concern of exterior liquid-bearing recirculation piping.
5. Smaller footprint and low profile: By utilizing the space above the liquid storage sump, this new arrangement allows application in smaller areas.

6. Single-piece (unitary) construction: The complete system is shipped to the job site as single-piece construction, greatly reducing installation time

Low Horsepower. The full scale system only requires about one-half to two-thirds the horsepower of eductor-venturi scrubbers because of low pressure recirculation of chemicals. The table below shows the horsepower requirements.

Air Flow Rate, cfm	*RJE H.P.*	*Eductor Type H.P.*
3000	<25	40
4000	<25	60
5000	<25	60+

Induced Draft Fan. This feature provides "negative pressure" thoughout the system, including the room, ducting, and scrubber.

Low Pressure Recirculation. 30 psig (vs. 70 psig) further enhances the safety of the system.

Scrubber Reaction Chemistry and Relevant Calculations for Chlorine Concentration vs. Time (Typical Calculations). We will assume an initial concentration in the room equivalent to the maximum scrubber inlet concentration. The actual concentration will be far less than this because of the mixing of room air. This assumption will provide the longest time to "clear" the room of any vapor.

By performing a material balance, the following differential equation is obtained for the decrease of chlorine vapor in the room as a function of time:

$$-V\frac{dc}{dt} = Q_G c$$

where

V = room volume, ft^3
c = Cl_2 vapor concentration in room, ppm
Q_G = room ventilation rate, ft^3/min.
t = time, min.

Upon integration, the following equation is obtained:

$$\frac{c_o}{c_i} = \exp - \frac{Q_G t}{V}$$

where

c_i = initial chlorine vapor concentration, ppm
c_o = chlorine vapor concentration at any instant, ppm

As an illustration, the total room volume is assumed to be 10,000 ft^3, the ventilation rate is 3000 cfm, and there is an initial chlorine vapor concentration of 180,787 ppm. These values can be used, as shown below, to determine the time needed to reduce the concentration to 1 ppm.

Calculations. The equation derived above relates chlorine concentration in the room to time. Rearranging the equation, the time required to reduce the concentration to 1 ppm is:

$$t = \frac{10{,}000}{3{,}000} \ln \frac{1}{180{,}787}$$

$$t = 40 \text{ min.}$$

Caustic Requirement. Sodium hydroxide solution is used to neutralize the chlorine. The reaction between sodium hydroxide and chlorine is:

$$2\ NaOH + Cl_2 \rightleftharpoons NaCl + NaOCl + H_2O + 44{,}600 \text{ Btu/mole}$$

Two moles of sodium hydroxide (80 lb) is required to neutralize each mole of Cl_2 (70.9 lb). To neutralize all the Cl_2 leaked from the cylinder, the minimum weight of sodium hydroxide is:

$$\frac{2350}{70.9} \times 80 = 2652 \text{ lb}$$

The caustic solution in the scrubber sump is 2100 gal of 20 percent by weight of sodium hydroxide. Each gal of 20 percent solution contains 2.04 lb of NaOH. Thus, NaOH available = 2.04 × 2100 = 4284 lb.

Therefore, the percentage of excess chemical = 62 percent.

Final Solution Composition. Based on the above reaction, neutralization of Cl_2 and caustic is as follows:

For each lb of chlorine:
NaOH required: 1.13 lb
NaCl produced: 0.825 lb
NaOCl produced: 1.05 lb
Water produced: 0.25 lb

Table 9-7 Scrubber Performance Predictions vs. Actual Results

	Inlet Conc. ppm	*Outlet Conc.* ppm*
Predictions	909,000	4.7
Actual	750,000–850,000	1.5

*After 3rd Stage

Scrubber Recirculation Rate. Tables 9–7 through 9–10 present calculations for the scrubber recirculation rate and available chemical at the start and the end of the release event.

Sump Temperature Rise. Neutralization of chlorine with sodium hydroxide is an exothermic reaction. The heat of reaction is 44,600 Btu/mole of Cl_2.

Reaction is in the liquid phase. If there is no heat loss to the air and other components of the scrubber system, the temperature rise of sodium hydroxide solution can be calculated by using the following terms: $H = mC_p \Delta T$; $H =$ total heat released by reaction, Btu; $m =$ total weight of caustic solution, lb; $C_p =$ heat capacity of caustic solution, Btu/lb; and $\Delta T =$ temperature rise, °F.

Heat capacity for 20 percent weight caustic solution is 0.9 Btu/lb. One gallon of 20 percent NaOH is equal to 10.21 lb. Therefore, the available sump solution temperature rise, if there were no heat transfer out of the system, would be:

$$\Delta T = \frac{(2350/70.9) \times 44{,}600}{2100 \times 10.21 \times 0.9} = 77°\text{F}$$

Table 9-8 150-lb Chlorine Release

Time (min.)	*Release Rate (lb/min.)*	*Exhaust Chlorine (ppm)*
1	28.5	0.0
2	25.8	0.0
3	25.8	0.0
4	23.2	0.0
5	23.2	0.0
6	19.6	0.0
7	6.2	0.4
8*	0.3	0.7

*Tank empty.

Table 9-9 1-Ton Chlorine Release

Time (min.)	*Release Rate (lb/min.)*	*Exhaust Chlorine (ppm)*	*Flash Room Temp.*	*Scrubber Inlet Temp.*	*Scrubber Outlet Temp.*	*Scrubber Sump Temp.*
1	52	0.0	48	47	84	81
2	42	0.0	41	44	84	82
3	40	0.0	37	45	85	82
4	39	0.0	30	47	85	82
5	37	0.0	27	45	85	82
10	37	0.0	22	46	88	86
15	35	0.0	22	45	92	90
20	35	0.0	21	47	97	95
25	33	0.0	28	55	107	105
30	34	0.0	27	49	104	102
45	29	0.0	72	55	111	110
60*	—	0.0	72	63	114	113
120	—	0.0	69	61	109	108
180	—	0.0	71	65	106	105
240	—	0.0	67	62	106	104
300	—	0.0	71	70	105	103
360	—	0.0	74	74	101	98

*Chlorine Release stopped at 51 minutes after start.

Most of the heat, however, will be lost because of air moving through the system, which can be analyzed using the following terms: m_1 = mass of NaOH, 20 percent, lb; m_2 = mass of air, lb; m_3 = mass of chlorine reacted, lb; C_p = specific heat; and dH = change of enthalpy, btu.

The heat generated by the chlorine absorption is calculated as follows:

$$dH = \frac{m_3 \times 44{,}600 \text{ Btu/lb} \cdot \text{mol}}{70.9 \text{ lb/lb} \cdot \text{mol}}$$

By conservation of energy, the total change above is equal to the change of energy in the liquid sump, dH_1, added to the change of energy in the air traveling through the system, dH_2:

$$dH = dH_1 + dH_2$$

Due to the initimate contact in the scrubber vessel between the airstream and the recirculating liquid, the two temperatures approximate one another, as can be seen in the performance test data.

Assuming the same initial temperatures of the sump and ambient air (to be conservative, as released chlorine vapor will be very cold):

$$dT_1 = dT_2 = \frac{dH_1}{m_1 C_{p_1}} = \frac{dH_2}{m_2 C_{p_2}}$$

Table 9-10 550-lb Chlorine Release

Time (min.)	*Release Rate (lb/min.)*	*Exhaust Chlorine (ppm)*	*Flash Room Temp.*	*Scrubber Inlet Temp.*	*Scrubber Outlet Temp.*	*Scrubber Sump Temp.*
1	99	0.1	31	33	91	91
2	91	0.2	19	25	89	87
3	84	0.3	12	21	90	88
4	77	0.2	7	18	90	88
5	71	0.3	6	17	91	89
6	65	0.2	7	16	91	89
7	22	0.5	31	33	92	90
8*	10	0.6	49	49	93	91
9	—	0.2	56	52	95	92
10	—	0.4	57	53	96	94
15	—	0.3	57	56	100	99
20	—	0.3	58	58	102	101
30	—	0.5	60	59	102	101
45**	—	1.7	71	78	102	100
60	—	2.2	78	82	102	101
75	—	1.1	84	93	102	101
90	—	1.2	90	99	103	101

*Release stopped.
**Propane burner started at about 30 minutes into test.

Solving these simultaneous equations, the temperature rise is 62°F. All equipment is rated to service at 150°F.

Chemical Utilization During Equipment Checkouts. Carbon dioxide will be absorbed by the caustic solution during weekly testing of the scrubber system. The reaction between sodium hydroxide and carbon dioxide is:

$$CO_2 + 2NaOH \rightarrow Na_2CO_3 + H_2O$$

Two moles of sodium hyroxide (80 lb) is required to react with each mole of carbon dioxide (44 lb).

It is recommended that the scrubber system be exercised 15 min. every two weeks. Based on these data, the following amount of caustic will be consumed:

CO_2 conc. in ambient air, percent by volume: 0.033 percent
Air flow rate, cfm: 3000
CO_2 flow rate, cfm: 0.99
CO_2 flow rate, lb-mole/min.: 2.57×10^{-3}
CO_2 flow rate, lb/min.: 0.11
NaOH required, lb NaOH/lb CO_2: 1.82
NaOH consumed, lb/min.: 0.20
NaOH consumed per cycle of 15 min.: 3.0

No. of cycles/yr: 26
NaOH consumed/yr, lb: 78
or
NaOH consumed, gal, 50 percent: 12

About 15 gal. of 50 percent caustic should be added every year.

Plant-Scale Field Test

Introduction. The RJ Environment scrubber system has been thoroughly tested at the Southwest Research Institute in San Antonio Texas. These tests were made for three different magnitudes of liquid chlorine spills. These were from a 150-lb cylinder, a one-ton cylinder, and a custom designed 550-lb leak. The results and documentation of these leaks are shown on Tables 9-7 to 9-10.The scrubber used at the field test is shown in Fig. 9-49, and the 550 lb leak is shown in Fig. 9-50. This illustration was taken through a special viewing window in the leak containment structure.

The leaks described above were repeated several times under controlled conditions with continuous data collection. This produced the following major findings: (1) room temperatures, (2) room pressures, (3) scrubber inlet temperature, (4) scrubber exhaust temperature, (5) chlorine leak release rate, and (6) overall scrubber performance.

Installation Details. Each of the described leaks was discharged into a containment room described as the "flash" room. This room contained a metal

Fig. 9-49. RJ Environmental's "RJ 2000" 1-ton emergency chlorine/sulfur dioxide scrubber system (courtesy RJ Environmental, Inc.).

pan 8 ft × 8 ft and 1 ft deep to contain the liquid chlorine spill. The room itself measured 12 ft × 13 ft with an 11 ft ceiling. Sealed windows in the flash room allowed complete visible access to the liquid leak discharge and the vaporization of the chlorine at all times. Figure 9-50 shows the 550-lb leak discharging from two short pieces of ¾-inch black steel pipe that terminates in a 45-degree elbow.

The flash room was coated and caulked with appropriate material to prevent both leakage and chlorine absorption into the wall surface. The leakage discharged into the flash room was drawn into the scrubber system suction line by a 3000 cfm exhaust fan, where it entered the first horizontal spray scrubber.

Witness. As this demonstration, on April 27, 1992 at the Southwest Research Institute in San Antonio, Texas, was most assuredly a once-in-a-lifetime experience of unusual significance, the witnesses are named:

1. Paula R. Diepolder, P. E., Brown and Caldwell Consultants, Houston, TX
2. Michael R. Gonzalez, Senior Engineer, Southwest Research Institute, Austin, TX
3. Welsey W. Hoffmaster, Senior Engineer, Los Angeles Co. San. District, CA
4. Roop C. Jain, P. E., RJ Environmental, Inc., San Diego, CA
5. Hernane V. Mayang, P. E., James M. Montgomery Engineers, Pasadena, CA
6. William T. Murray, CH2M Hill Engineers, Austin, TX

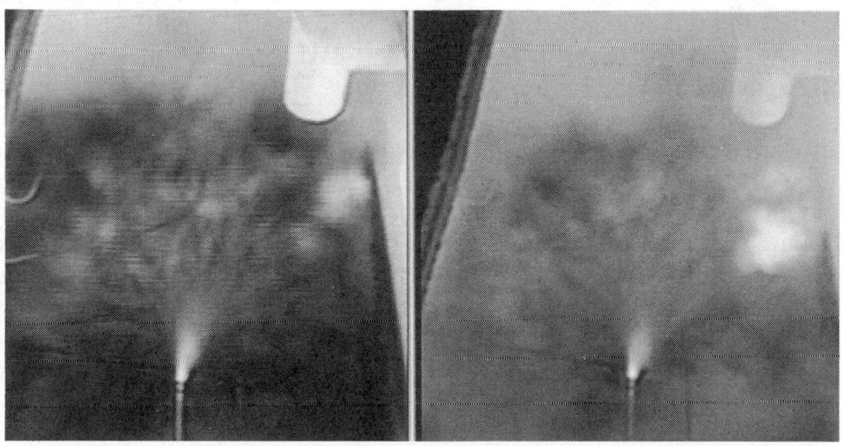

Fig. 9-50. Full-scale test at a nationally accredited ICBO (International Conference of Building Officials). Photo shows 550 lbs of liquid chlorine releasing at 99 lbs per minute (one second intervals). Simulation of a catastrophic rupture of a 1-ton cylinder (courtesy of RJ Environmental).

7. Steen (Amos) Robinson, Mgr., RJ Environmental, Inc., San Diego, CA
8. Geo. Clifford White, C. E., M. E., Consultant, San Francisco, CA

Removal of Liquid Chlorine. A major chlorine leak means that liquid chlorine has been released to atmosphere on the storage room floor. The floor should be sloped 2 in./10 ft to a sump that is designed to act as a suction well for a liquid chlorine eductor. The liquid can then be hustled off quickly, before waiting for it to vaporize, to the caustic tank for quick neutralization, as is done with an absorption tank. If this part of the system is designed properly, the time required for neutralizing a liquid leak will probably be reduced by a factor of 10–20.

It is important to understand the hydraulic kinetics when an eductor (or injector) is used for transporting liquid chlorine. When the liquid chlorine is collected in a sump or a slot in the floor, atmospheric pressure is the only motive force available until the water or caustic starts flowing through the eductor. Then the eductor provides the vacuum needed to create a pressure differential in the flow system. This differential will cause the liquid chlorine to flow under a small negative pressure (10–15 in. Hg) to the eductor throat without any off-gassing that might cause a gas binding situation.

Evaporator Pressure Relief Lines. These lines can be manifolded together into a common header pipe that terminates in a chlorine sparger in the caustic tank. The piping to the caustic tanks must contain a barometric loop; this prevents "suck back" of the caustic in the tanks. Each of the evaporator relieve valves must be fitted with a 25–30 psi rupture disk on the discharge side of the valve to protect it from corrosion in the event of a leak from one of the other valves.

Another method would be to use an air purge system of 1 std ft^3/min. This could prevent the migration of caustic to the relief valve seats. The use of a 400 psi rupture disk in the reverse position is standard practice because in the reverse position the rupture disk will burst at less than 25 psi.

Foaming Prevention. After the system has been in operation during a leak episode, there is a good chance that the recycled caustic will develop foam on the surface of the caustic in the tank if the scrubber is not properly designed. This foam may eventually be released in the discharge stack. Powell (a major manufacturer of scrubbers) advises that 25 sq ft of tank surface per venturi reactor is sufficient to eliminate any possibility of foaming. When sizing the caustic tank, always include a 10 percent excess of caustic for any proposed leak episode.

Materials of Construction. The preferred materials for a long-life project with very few major leak episodes will always be rubber-lined steel tanks, halar-lined pipe and fittings including venturi/reactor bodies, Teflon-lined plug valves, and titanium pumps. The only reason for considering plastics would be the cost differential. Lined metal tanks and fiberglass reinforced plastic have been used successfully to build these systems. In general, it is critical that any materials considered be chosen to handle the products of release and neutralization, including chlorine, sodium hypochlorite, and sodium chloride, as well as the sodium hydroxide neutralizing agent. For linked tanks, any lining chosen must not deteriorate with chlorine-containing compound. For fiberglass, a vinyl ester resin is chosen because of its excellent corrosion characteristics. Careful consideration must be given to the possibility of damage to the plastic components from seismic forces, reaction temperatures, and sunlight.

Summary—Design of Components for a One-Ton Cl_2 Scrubber:

1. The scrubber will be rated at 3000 cfm at 550 gpm of caustic.
2. The caustic tank will hold 2100 gal sodium hydroxide.
3. Caustic strength will be 20 percent.
4. Caustic volume will be 2100 gal.
5. Pump data are: vertical seal-less vinyl ester FRP pump, driven by a 3500 rpm, 20 hp motor at 550 gpm and 55 ft head.
6. Chlorine vapor capture rate will be 3000 cfm.
7. Chlorine neutralizing capacity will be 2350–2400 lb in 30 min., which equals 112,800 lb/day.

CHLORINE MONITORING IN CONTAINERS

Chlorine Tank Car Supply Monitoring Systems

General Discussion. Single-unit tank cars pose a problem in determining when exhaustion of the liquid chlorine is imminent. Reliance on pressure drop is not the total answer, and where air padding is used on a continuous basis (cold climate), temperature drop monitoring is not reliable. Track scales are the only direct means, but these constitute an exorbitant expense, about $50,000 to $75,000. Scales for storage tanks are much less expensive (about $5,000).

A convenient method for solving the supply monitoring problem would be to meter the flow of liquid chlorine exiting the car. Unless the car is padded to a pressure of 175–200 psi, it is not possible to measure with any accuracy the flow of liquid chlorine because of its physical characteristics. As the liquid flows past a point of slight pressure drop, as in a flow meter, the liquid flashes to vapor causing a large change in density. This makes calibration of a flow meter practically impossible. *A magnetic flow meter will not respond to liquid chlorine.*

Tank cars are well insulated so that they maintain vapor pressure equilibrium until all of the liquid has been discharged.

EXAMPLE. A 90-ton car displays a vapor pressure of 70 psi. When all the liquid is gone, there will remain in the car approx. 2500 lb of chlorine gas, provided that the car has not been air-padded during use. This is a significant amount of chlorine that should be used by the consumer.

Chlorine Pressure Switch. A chlorine pressure switch installed in the liquid header immediately upstream from the evaporators can be used for monitoring or automatic switchover service. A pressure switch set to alarm at 20 psi will advise the operator that approximately 1050 lb of chlorine is left in the car. This may be sufficient time to switch to a new car. The other alternative would be to set the pressure switch to 20 psi and use the alarm circuit to activate the motorized valves in an automatic switchover system. The time between the occurrences described above is totally dependent upon withdrawal rate and to some extent upon ambient temperature. Therefore, a reliable and accurate pressure switch is essential for liquid chlorine supply monitoring. The US Gauge Co. Model 3050 is a combination indicating gauge and pressure switch with adjustable contacts for both high and low pressures over the entire range of the gauge, 0–160 psi.

Three Detection Methods for Liquid Chlorine Exhaustion. The liquid chlorine exhaustion point has been detected in three ways: (1) A cylindrical vessel has been used that is mounted vertically and equipped with a liquid level sensor, which alarms when the level falls some predetermined distance in the cylinder. (2) All of the chlorinators have been equipped with chlorine flow recorders and a totalizer that will alarm at 500 or 1000 lb chlorine remaining. (3) The liquid chlorine temperature has been continuously monitored adjacent to the point where the flexible connection to the tank car joins the liquid header. As the liquid chlorine approaches the point of exhaustion, it attempts to vaporize as it exists the tank car. This premature vaporization causes a sudden drop in the temperature of the liquid as it enters the header piping at the unloading platform. A temperature sensor equipped with an alarm will notify the operator of imminent liquid exhaustion. The reliability of these methods is dependent upon the withdrawal rates of liquid chlorine at the moment of exhaustion.

Reserve Tank for Liquid Chlorine Supply

A system used for more than half a century in the pulp and paper industry is the reseve tank concept. This system was developed by the late Brian Shera of Pennwalt Corp., formerly Pennsylvania Salt Co., Tacoma, Washington.

This is a liquid chlorine flow-through tank, as shown in Fig. 9-51, which provides an active reserve when the car has been emptied of liquid. The tank may also be installed vertically if preferred. These tanks have been made in sizes from 1000 to 16,000 lb of chlorine. The capacity should provide approximately 1–3 hours reserve of chlorine at peak demand. Liquid chlorine enters and leaves the tank through pipes extending to the bottom of the tank.

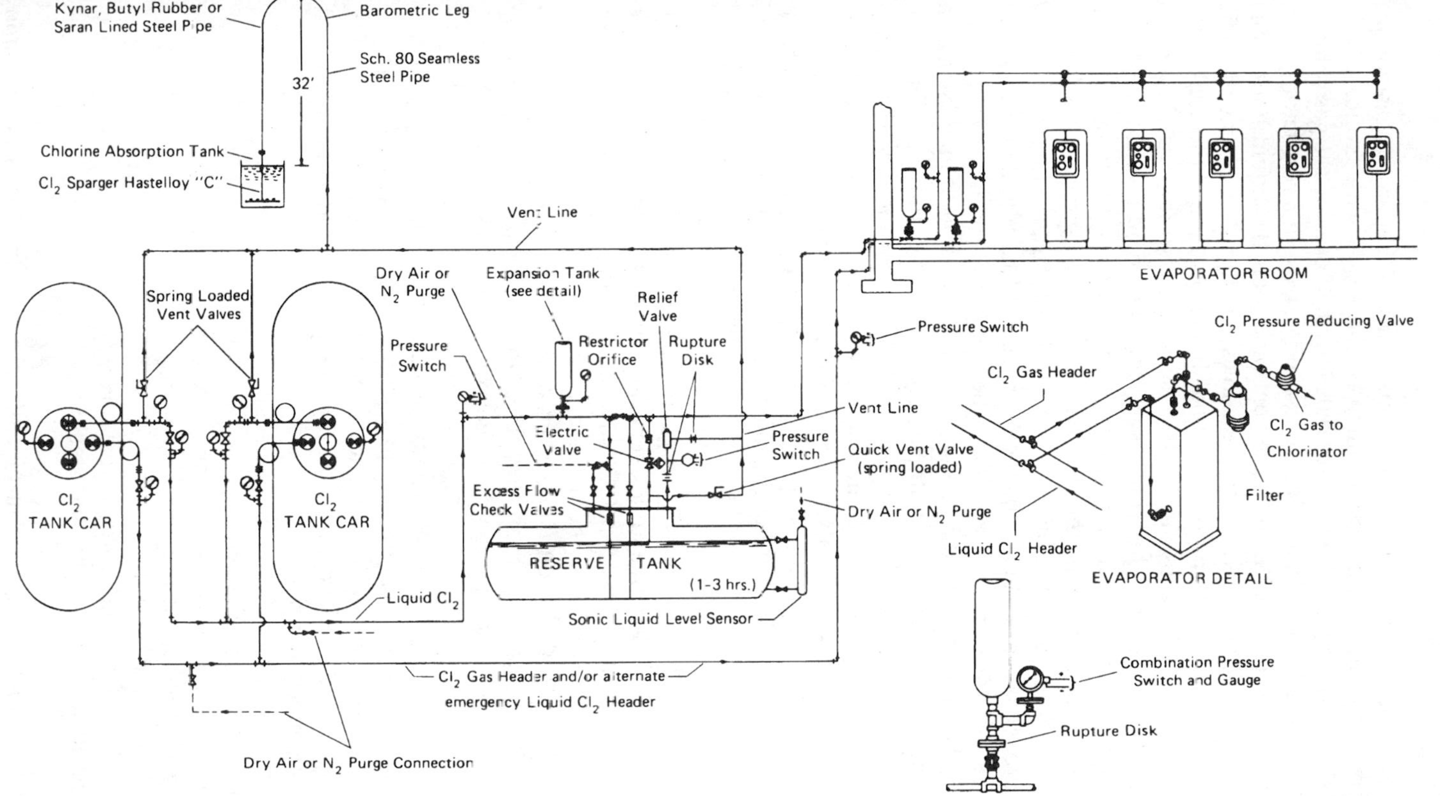

Fig. 9-51. Reserve tank system for chlorine supply system: Note chlorine absorption tank in upper left part of the system piping. Note: This arrangement is for a system that does not require an air pad on the tank car. Dry air and N_2 connections shown are only for purging Cl_2 out of the piping.

Venting of gas is necessary to provide a minimum of 20 percent gas volume as a safety factor against excessive filling of the vessel with liquid. This is accomplished by a dip pipe of an appropriate length to provide the gas volume. The vent piping is fitted with an electric shut-off valve and a restrictor orifice. Upon leaving the restrictor orifice, the vent pipe joins the liquid lines going to the vaporizers. This restrictor orifice allows a minor flow of gas that will flash over from the small amount of liquid being bled from the tank, thus keeping the liquid level constant in the reserve tank so long as chlorine is flowing from the tank car. This vent also allows the liquid level to rise to its normal height when the reserve tank is being filled. When the liquid chlorine flow from the tank terminates, the chlorine level in the reserve tank will drop, and this drop will be detected by the ultrasonic level sensor. The sensor sends a signal through a control box, which will engage an alarm and also close the electric shuf-off valve in the vent line. The operator now has two choices: shut off the liquid flow from the reserve tank and bleed the remaining gas in the tank car to the process; or immediately switch to a full car. When the time comes to switch to a new car, the reset button on the electric valve is actuated so as to open the vent line and allow refilling of the reserve tank. The reserve-tank piping and valving accessories are equipped with both automatic and manual pressure relieving devices, all discharging to the absorption tank. This prevents any overpressuring of the reserve tank. The pressure relief line has a rupture disk followed by a pressure switch alarm. This is followed by a spring-loaded relief valve protected from moisture entry by a rupture disk which is installed backwards for ruptures at low pressure. In addition, there is a spring-loaded manual quick vent valve to the absorption tank. Figure 9-51 illustrates all of the appropriate expansion tanks and pressure switches required to give the operator the necessary operating information.

The following are the recommendations of the Chlorine Institute for chlorine storage tanks, as set forth in Pamphlet No. 5, "Facilities and Operating Procedures for Chlorine Storage," 2nd edition, January 18, 1962:

- Minimum design volume: 25.6 ft^3 for each 2000 lb of chlorine to be stored.
- Minimum working pressure: 225 psig plus ⅛-inch corrosion allowance.
- Design and fabrication: compliance with the ASME-UPV Code, 300°F, 70 percent weld efficiency, spot X-ray or 100 percent X-ray of longitudinal and circumferential seams.
- Nozzle necks: Sch 160, seamless steel pipe.
- Piping: Sch 80, seamless steel.
- Flanges: 300 lb. weld-neck, slip-on, or screwed.
- Fittings: buttweld, extra heavy.
- Fittings: screwed, 3000 lb.
- Valves: 300 lb, flanged, forged steel body.
- Valves: 300 lb, screwed, forged steel body.

The purge connections shown in Fig. 9-51 are essential for inspection and repair of the system. They also prevent the emission of chlorine to the atmosphere and the entrance of moisture to the piping and tank system, which would accelerate corrosion and deteriorate the entire system. This is the only way to provide a completely closed system, which prevents penetration of atmospheric moisture.

Fully Automatic Tank Car Valve Shut-off System (CVA)*

Historical Background. The Clorox Company of California spent several years and large sums of money developing this valve shut-off system—only for tank cars.[85] The Clorox Company has a nationwide operation for producing sodium hypochlorite from chlorine in tank cars. There are a great many of these installations in the United States. In 1979 the company experienced a serious leak at its Oakland, California plant. The leak occurred after a full tank car had been connected. A small leak developed in a two-bolt flange near the car valve. The operator tried to stop the leak by tightening the bolts without shutting the car valve. As he tightened one bolt, the nut was stripped off the other bolt, partially opening the flange. This caused a significant leak of liquid chlorine. However, the size of the leak was not sufficient to activate the excess flow valve in the tank car. The leak required evacuation of people working in the immediate area and the rerouting of traffic on a heavily traveled freeway. This incident motivated the Clorox Company to seek a permanent solution to this type of problem; hence the development of this unique emergency tank car security system.[85] It is probably the most significant contribution toward safety in chlorine handling in the last half-century. The system incorporates the latest technology in mechanical, electrical, and instrumentation engineering. The reliability of the system has been established by rigorous and lengthy field testing and operation.

System Description. Three separate components make up the entire system. They are: a DC motor-operated valve actuator, an instrument control module, and leak detector units.

The motor-operated valve actuator is the heart of the system. It is a unified piece of machinery that is shaped to fit inside the circular tank car dome. The actuator fits over the tank car outlet valve so that there is no need to make any modifications or adjustment of the tank car valve. The gear train operated by the DC motor is designed to exert a 50 ft-lb torque on the valve stem. The torque is designed to be self-limiting, thus preventing the actuator from shearing the valve stem or damaging the valve in any other manner.

In addition to the valve actuator, the Clorox Company developed a Chlorine Detector Logic Panel (CDLP) to increase the sensitivity and the reliability of the chlorine leak detector. The CDLP electronic package incorporates the

*Car valve actuator.

logic to *close the valve only.* Closing time is 18 sec. The tank car valve must be opened manually by using a handwheel attachment on the top of the CVA.

The CVA will close under the following conditions:

1. Chlorine detection by Fischer & Porter Detectachlor.
2. Manual chlorine alarm.
3. Manual or automatic fire alarm.
4. Low air pressure.
5. 120 V power failure.
6. Low battery power.
7. Disconnected service cord to the CVA.
8. Manual trip at the Chlorine Detector Logic Panel or at the CVA.

Power failure will not impair the system owing to the separate self-contained DC power source which is under continuous recharge. Unfortunately this valve shut-off system is not available commercially at this time.

Alternative Devices. A wide variety of automatic shut-off valves are available for reliable use with liquid chlorine. They can be operated by either an electric solenoid, air pressure, or water pressure. Such a valve should be installed outside the manway enclosure between the liquid discharge valve and the flex connection. This valve is an important safety device for a railcar or bulk storage system in the event of a leak in the downstream chlorination system.

Halogen Automatic Actuators for Chlorine Valves[132]

Preface. This company has developed emergency automatic shut-off actuators that fit all sizes of chlorine container valves. The most important one has to be the actuator that takes the place of the Clorox actuator system that is not available to the public. White remembers when the Clorox actuator first entered service. Many users of tank cars sought to obtain permission to use the Clorox design for their installations. The attorneys for Clorox stopped that possibility, owing to the potential liability for Clorox. In the long run, that was probably a prudent choice.

Now that the Halogen Valve Systems has developed emergency valve actuators for all types of chlorine containers we have reached a new level of safety and dependability for all chlorine gas dispensing facilities. This should certainly impress the Fire Chiefs who are responsible for the requirements of the Uniform Fire Code.

Cylinder Valves. Halogen Valve Systems employs several techniques to mate their actuators with typical chlorine dispensing apparatus. The Halogen Model CC-1-R1 designed for mounting in conjunction with a vacuum regulator, on a 150-lb cylinder is illustrated in Fig. 9-52. The same actuator, adapted for mounting on a ton container valve, is illustrated in Fig. 9-53.

These actuators are designed with special nickel alloys and materials to provide resistance to chlorine vapor. The electrical components are housed in sealed enclosures. The actuator mounts directly to the existing hardware, and does not interfere with dispensing connections or manual valve operation. A battery that is constantly charged supplies power to the actuator. Local power interruptions have no effect on the system unless the outage lasts more than several days. In that case, the system shuts down automatically, before the battery is depleted. A solar powered option is also available. This provides total independence from local electrical power.

The system is designed to only power the valve closed. The valve is always opened manually. Personnel can open, or close, the valve with the actuator in place. However, the actuator defaults to the automatic mode after manual operation. Unattended stations may be automatically shut-off by leak detec-

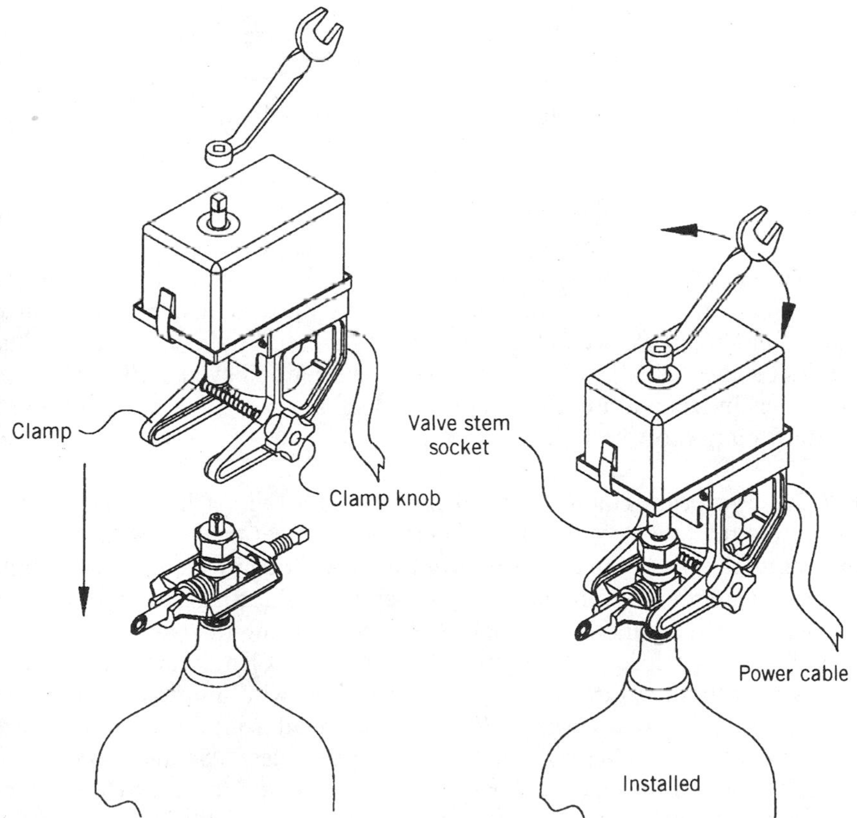

Fig. 9-52. Halogen emergency chlorine shutoff valve for 150-lb cylinders.

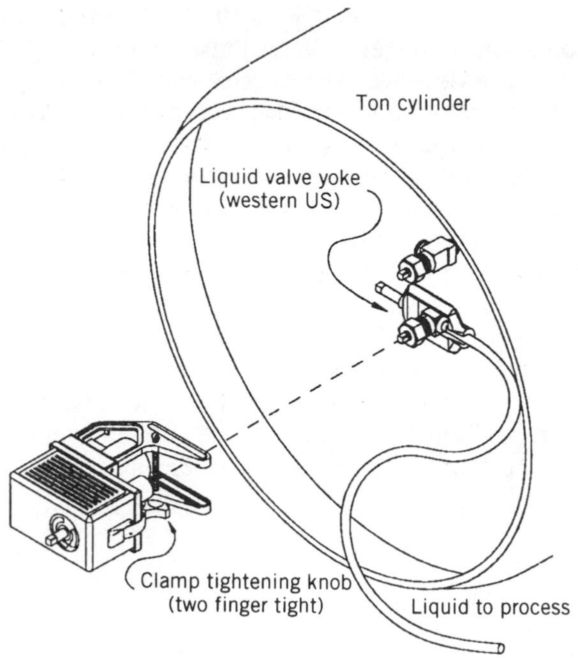

Fig. 9-53. Halogen automatic emergency shutoff valve for ton cylinders.

tors, by fire, or by seismic sensors. Shut-off may be selected from remote command and control centers or from just outside the gas room.

On April 9, 1997, White was given a complete demonstration of the Halogen valve actuator on a 150 lb cylinder. The results were astonishing! The valve closed in less than two seconds after it was triggered by the leak alarm. This provides the greatest safety factor that has ever been achieved in the operation of 150 lb cylinders and ton containers. This should provide a major relief to the interpretation of the UFC.

Tank Car System. Halogen has also designed a rail car hood system that replaces the rail car barns that have been used to meet the "exhausted enclosure" requirements of the Uniform Fire Code. Although with automatic shut-off of the rail car valves, the necessity for additional tertiary safety devices is an expensive redundancy. The hood is retracted during rail car changes. It is then reinstalled before commencing the unloading of a fresh car. A single hood and scrubber system is adequate for sidings with multiple spots when only one rail car is active at a given time. An overhead trolley system transports the containment hood between rail car spots. See Figs. 9-54 and 9-55.

Since the hood is brought under substantial negative pressure by the scrubber fan, leaks in the vicinity of the valve dome are quickly directed to the absorption system. For leaks in the piping to the plant, the actuators can

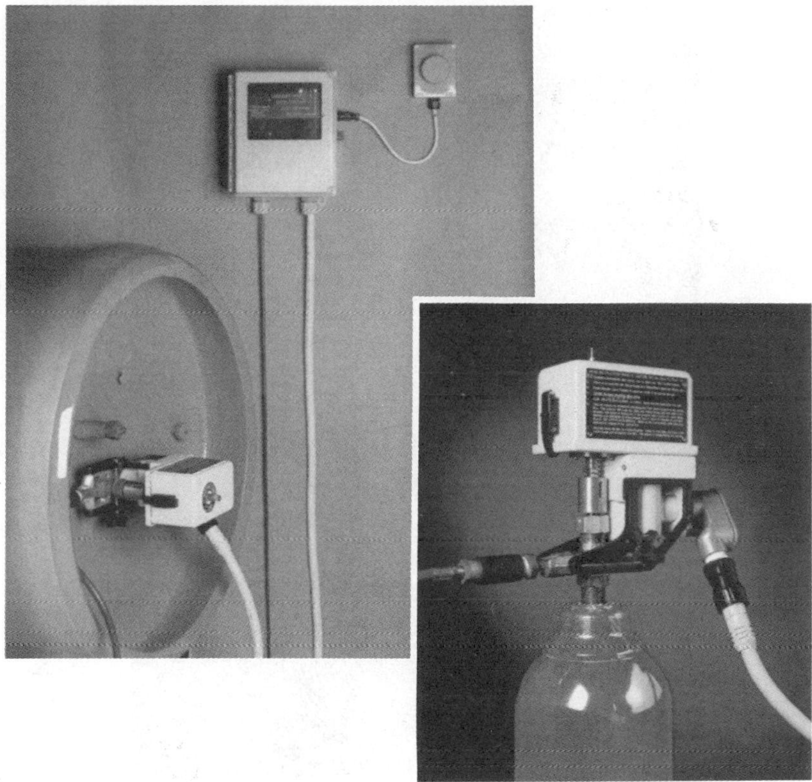

Fig. 9-54. Halogen automatic emergency chlorine shutoff valves. Illustration of model CC-1-R1 for both ton containers and 150-lb cylinder. Halogen valve systems model CC-1 actuator installed on the liquid valve of a chlorine ton cylinder. Control panel and emergency shut-off button on wall.

quickly isolate the chlorine source at the rail car. The connecting piping can then be rapidly evacuated by an eductor system circulating caustic through an absorption tank.

The proprietary ductwork within the hood is designed to pull all of the chlorine downward and below the rail-car platform. The internal vacuum pick-ups are flexibly mounted to accommodate variations in rail-car dimensions. Fresh air is admitted through a blow-in door located on the top of the hood. An open blow-in door is an indication of negative pressure and that scrubbing is underway. The downward flow of gas is predictable and takes advantage of the dense properties of chlorine gas.

At a volume of about 700 cubic feet, the atmosphere within the hood is replaced several times per minute by a typical 3,000 cubic feet per minute scrubber. This is particularly favorable when compared to a barn enclosure that can easily exceed 60,000 cubic feet in volume and require a good deal of time to scrub.

Fig. 9-55. Protective hood for operating personnel using Halogen emergency chlorine shut-off valve (Courtesy Halogen Co. Inc.).

Should Hazmat personnel be required to enter the hood to terminate an event, the generous fresh air make-up would be advantageous. This is far superior to putting railcar systems in a closed building. It is not only several times cheaper, but also much safer for operating personnel.

Stationary Storage

General Discussion. Up until the publication of the Uniform Fire Code in 1988, the use of stationary chlorine storage at both water and wastewater treatment plants was fairly popular owing to the increased availability of bulk chlorine deliveries by truck. Now, however, it is too expensive for a water or wastewater treatment plant to have an on-site chlorine storage tank and be required to have it caged in a building that meets the UFC requirements. Moreover, such a system is very unpopular with operating personnel.

There is one big exception, which is illustrated in Fig. 9-56. This is a 17-ton semi-trailer chlorine tank that is a part of the chlorine tanker fleet operated

Fig. 9-56. Chlorine semi-trailer tanker (courtesy Metropolitan Water District of Southern California).

by the Metropolitan Water District of Southern California, Los Angeles. Newer models are 19-ton units. These trailers are now being loaded with chlorine at their own loading facility, which is only 60 miles away from the various treatment plants in the Los Angeles Basin Area.

Originally, in the early 1960s, this MWD chlorine trailer fleet had to travel to the nearest chlor-alkali plant, which was located at Henderson, Nevada (near Las Vegas) some 300 miles from the various treatment plants. This facility allows MWD operators to use their trailers instead of storage tanks. This is a lot more economical because their new loading station has rail access; and because chlorine is a waste product in the production of caustic, quite often they can get a 90-ton tank car of chlorine for free, except for freight charges. These chlorine tank trailers are available from: Anderson Columbiana, Gainesville, Texas; Mississippi Tank, Hattiesburg, Mississippi; Mid Nebraska Tank, Grand Island, Nebraska; J&J Truck Bodies, Somerset, Pennsylvania; GATX, Chicago, Illinois; and ACF, New York, New York.

Unloading System. When the decision has been made to utilize a storage tank to be loaded from either a tank car or a tank truck, a special unloading system has to be designed. A typical storage tank and tank-car unloading system is shown in Figs. 9-57 and 9-58. The essential features are:

1. Unloading platform
2. Storage tank and sun shield
3. Weighing device
4. Air padding system

Fig. 9-57. Typical stationary chlorine storage system (courtesy of East Bay Municipal Utility District).

5. Eductor
6. Chlorine gas and liquid headers
7. Gauges
8. Pressure switches and alarms
9. Expansion tanks
10. Flexible connections

The storage tank capacity for truck tankers usually is 25 tons, but the storage tank capacity for cars should be commensurate with the size of tank cars likely to be delivered. Cars usually have a capacity of 55 or 90 tons of liquid chlorine. The volume of liquid chlorine increases considerably with increasing temperature. Therefore, to provide adequate room for this expansion, the Chlorine Institute recommends that the weight of chlorine in a tank never exceed 125 percent of the weight of water at 60°F that the tank will hold. On this basis, each ton of chlorine will require 192.2 U.S. gal of tank capacity.

The storage tank should be designed in accordance with the Chlorine Institute recommendations, as set forth in their pamphlet No. 5, "Facilities and Operating Procedures for Chlorine Storage." Tanks should be designed for 120 percent of the maximum expected working pressure, or not less than 225 psi. To allow for corrosion, the wall must be ⅛ inch thicker than required by the design formula code.

The tank should be designed so that it can accept a tank car dome assembly. This consists of the four outlet valves, safety valve, eductor tubes with excess flow valves, and cover assembly. It is best to purchase this entire assembly

Fig. 9-58. Typical stationary chlorine storage system (courtesy of East Bay Municipal Utility District).

from one of the chlorine tank car manufacturers, such as American Car and Foundry, Union Tank Car Company, or General American Transportation Company.

Weighing Device. This is imperative. Weighing can best be accomplished by the use of either a lever scale system or load cells. These systems cost about the same, but the load cell is the more popular, as it lends itself to remote readout and/or recording using a 4–20 mA output signal.

Air Padding System. This requirement is mandatory for stationary tanks. It is used for both tank car unloading (transfer) and process. It is necessary for purging the tank to allow inspection. For details see "Single Unit Tank Cars" in this chapter.

Ejector System. Some kind of system must be provided to evacuate the tank for both inspection and repairs. This can best be accomplished by using a standard 2-inch chlorinator injector. The inlet of this aspirator type of injector is connected to a supply of water at about 50 psi. The throat of the injector (suction side) is connected to the gas withdrawal line on the tank.

Piping and Header System. Figure 9-59 illustrates a typical piping system for a chlorine storage system. It is important to point out the necessity of two header systems: one for gas withdrawal and one for liquid withdrawal. Both of these enter the chlorination system at the liquid inlet side of the evaporator. Gauges should be provided as shown.

The flexible connections from the tank car to the tank may be of either annealed copper tubing or flexible metal hose.

Copper tubing, with a 2-ft-diameter expansion loop and silver-soldered copper nipples on each end, has been the most widely used. On each end an ammonia-type union is threaded onto the copper nipple.

In recent years flexible metal hose has been available and seems well suited to this application. It should be made of corrugated Monel and Monel wire braid and Monel nipples that are helium arc welded to the hose. The ammonia unions are threaded to each end. The total length of these flexible connections need be only 10 ft.

Expansion tanks, equivalent to 20 percent of the header volume, should be used between the storage tank and the inlet to the evaporator.

As an added precaution to prevent overfilling of the tank in the event of a failure in the weighing system, dip tubes can be threaded into the dome section just under each gas outlet valve. Figure 9-59 shows how these tubes come into operation. The storage tank is being filled from the tank car via line *L*. Gas is being withdrawn to the chlorination system via line *G*, which joins the chlorine liquid line at the evaporator. Whenever the tank is filled "too full," "liquid will be discharged through line *G* to the evaporators. The operator will immediately notice the cooling in line *G*, signifying that the liquid transferred from the car has reached the bottom of the dip tubes. The operator should then stop the unloading process.

The length of the tubes can be arbitrarily calculated on the basis of the maximum ambient temperature to be expected in the vicinity of the tank. For example, a 100°F the liquid chlorine of a "full" tank would occupy 92 pecent of the volume. This corresponds to 85 percent at 70°F, which is the recommendation of the Chlorine Institute. The depth of liquid to fill 92 percent of the tank can be calculated. The distance from the liquid to the inlet of the gas withdrawal valve determines the length of the dip tubes.

In some areas of the United States and Canada, chlorine is available in tank trucks (Fig. 9-60).* It is usually imperative that the user transfer the contents of the truck to a storage tank rather than tie up the truck for the length of time of consumption of the contents. Storage tanks for tank truck operations follow the same design requirements as those loaded from tank cars, except for size. The capacity of most tank trucks ranges from 15 to 20 tons. Thus, the storage tank capacity need not usually exceed 25 tons. Depending on

*Provided that the storage tanks and trucks are fitted with a Chlorine Institute tank car dome assembly.

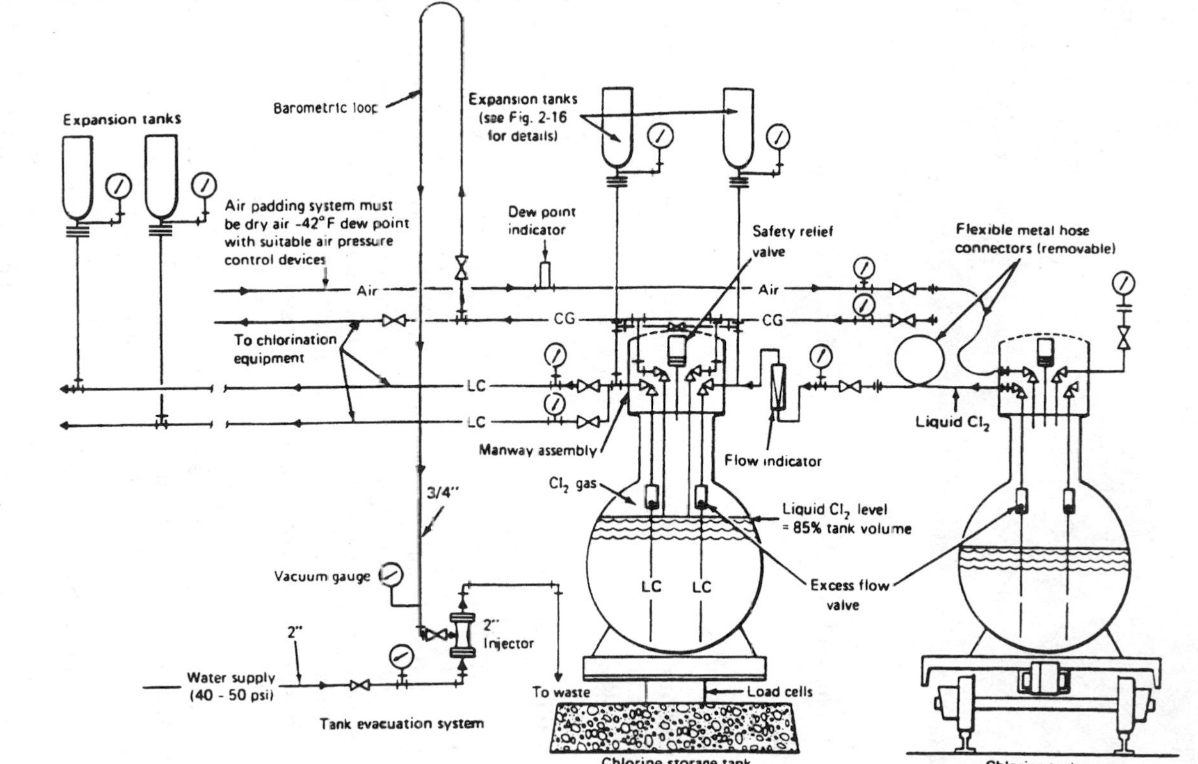

Fig. 9-59. Bulk chlorine storage and tank car unloading system. LC refers to liquid chlorine; CG refers to chlorine gas. The liquid chlorine header is in duplicate. A separate gas header is provided so that any car that arrives with too high a pressure (>125 psi) may be relieved to a lower pressure. (Dip tubes must extend far enough into the tank so that liquid contents cannot exceed 85% of the total tank volume.)

Fig. 9-60. Bulk chlorine tank truck (courtesy Evans Tank Co. Lubbock, Texas).

the length of haul and the chlorine consumption, the user should consider duplicate tanks.

Materials of Construction

Chemical Reactions. Liquid chlorine is always packaged in steel containers. Liquid chlorine that is absolutely free of moisture will not react with iron or steel that is similarly free of moisture. From a practical standpoint, it is not possible to package chlorine free of moisture; therefore, a very small amount of ferric chloride develops in each container of chlorine. This moisture is unavoidable because there is always some moisture in the atmosphere, and because containers are cleaned by being flushed with water. Although packagers are meticulous about drying the containers after flushing, first with steam and then with hot, dry air, enough moisture still remains to form some ferric chloride, a corrosion product of chlorine and iron or steel. This corrosion process is very slow. The design of containers and pipe lines should provide extra wall thickness to take care of this factor.

The two most significant chemical reactions of liquid chlorine and the materials of construction are:

1. Liquid chlorine* will spontaneously ignite and support combustion of carbon steel at 483°F. (See Chapter 1.)
2. It will attack and dissolve PVC at ambient temperatures.

The lessons to be learned here are:

1. Never apply heat directly to a chlorine container. An infrared lamp applied to the liquid portion of the cylinder can cause the burning of a hole through a steel cylinder.

*Chlorine cannot be in the liquid phase at 483°F.

2. Never use PVC or similar plastic materials anywhere in the liquid chlorine system or anywhere where chlorine gas is under pressure that is related to the vapor phase of the liquid–gas supply system. This usually means between the chlorine containers and the inlet pressure-reducing valve of the chlorinator.

Supply System. The chlorine supply system should consist of steel and cast iron products. The *supply system* is defined as that part of the system that begins at the chlorine containers and terminates at the inlet to the chlorinator.* From the chlorinator and beyond, the materials of construction are entirely different and are discussed elsewhere.

The supply system piping must be Sch 80 black seamless steel, and fittings must be 2000-lb forged steel. *Do not use bushings* (they cannot meet the 2000-lb criterion). Use reducing fittings instead.

All unions should be ammonia-type with a lead gasket joint. *Never use a ground joint union.* Filter bodies and reducing valve bodies are usually cast iron. Expansion tanks should be of welded steel construction, but can be a standard 100- or 150-lb chlorine cylinder. Valves for the chlorine supply system should be Chlorine Institute-approved. Two types of valves are used, one for main line shutoff purposes and the other for isolating cylinders (header valves). Header valves are identical to the outlet valves of ton containers: bronze bodies with Monel seat and stem. Main line valves can be either the ball type or the rising stem type. The ball-type is more popular because it utilizes a lever, which not only indicates at a glance the position of the valve but also makes it easier to operate the valve.

All gauges on the supply system must be equipped with a protector diaphragm. The diaphragm should be of silver, and the diaphragm housing can be either Hastelloy C or silver-cladded steel. Shut-off valves should not be used ahead of gauges. Gauges require a minimum of maintenance; so when replacement is needed, the entire supply line should be drained of pressure before replacing or removing the gauge. The value of a shut-off valve for this purpose is lost because a valve in a chlorine supply system loses its reliability if it is not operated on a frequent basis.

In assembling the piping system, either welded or threaded construction can be used; welded is preferable. If threaded construction is used, the contractor must be cautioned to use sharp dies, and all threaded pipe must be cleaned with solvent before assembly. Pipe dope should not be allowed; instead use Teflon tape for the thread lubricant.

Valves: Liquid and Gas Service. Valves on the chlorine supply side of the installation should be approved by the Chlorine Institute. The auxiliary header

*There is an exception: when the remote vacuum system is used, PVC pipe and fittings may be used downstream from the vacuum regulator.

and container valves are identical to cylinder valves—but without fusible plugs—with bronze bodies with Monel stems and seats. Line valves are of the ball type or the rising stem type. The ball type is more popular, as it uses a handle indicating immediately whether the valve is in the open or the closed position. The valves have a steel body with Teflon seat and Monel ball.

Construction. Extreme care must be taken during the fabrication of the chlorine supply system. Sharp dies must be used to cut the threads, and all joints must be cleaned of oil, dirt, and debris before assembly. The most satisfactory thread compound is Teflon tape; litharge and glycerine are usually detrimental to the quality of the pipe work.

For that portion of the system using PVC pipe, socket weld joints should be installed where possible. Every thread cut on a piece of PVC represents a potential weakness in this spot, as apparently adverse internal stresses are created during the thread-cutting process. Threaded joints will crack at the threads if subjected to even mild vibration or shock. The older the joint, the weaker it gets.

Chlorine Storage Tanks: Safety Features

Description. These tanks that are used at water and wastewater treatment plants are designed to conform to the same pressure and corrosion requirements of railcars and road tankers. They are also equipped with precisely the same dome assembly. These dome assemblies are available from several manufacturers that specialize in these lines. The only difference in design is that railcars are insulated, but the other are not.

In installing these fixed base storage tanks, it is standard practice to mount them on a hinged-type foundation so that the chlorine content can be weighted by an electronic load cell or other suitable pressure-responsive device.

Safety Considerations. There has never been a structural failure of any kind in stationary tanks or road tankers due to metal fatigue or a welded joint failure. With the availability of the scientifically advanced techniques of ultrasonics, users of these tanks no longer have to depend upon visual internal inspection of the tanks; the physical condition of the inside of these tanks can be determined by ultrasonic devices. However, hydraulic pressure testing must be done about once every five years. Tanks should never be opened to the atmosphere except for a pressure test because exposure to atmospheric humidity contributes to internal corrosion. After a pressure test, the tanks have to be dried with steam and hot air.

Stationary tanks should be equipped with indicators that measure the temperature and the pressure of both liquid and vapor discharges. This information is imperative for the operator to have when attempting to control or stop a major chlorine release downstream from the tank. This is explained below in the discussion of pressure control.

The most important safety feature of a storage tank is the elimination of the extremely vulnerable fusible plugs and the multitude of ⅜-inch flexible copper connections associated with the use of ton cylinders.

Stationary chlorine storage tanks have been in use for nearly half a century without a reportable tank leak.

Ventilation. Because of the attitude of the Uniform Fire Code (UFC), installations of these tanks may have to be in an enclosed environment. For these installations the operator should have the option of turning on a forced draft ventilation system that draws fresh air into the closed area at floor level and discharges it through the roof. Furthermore a chemistry-lab-type confined "hood" should be located in the area immediately above the tank dome assembly. Thus operators can be treated to fresh air blowing by their faces and out through the roof during the process of routine maintenance and inspection, and while connecting for a supply of chlorine or disconnecting after a filling operation.

Pressure Control. This is another important feature of stationary tanks; they react quickly and predictably to a pressure control system, provided that the adjacent piping is correctly installed. The basic principle of a pressure control system is the ability to withdraw chlorine vapor at a rate that exceeds the vaporization rate of the given ambient temperature. This cools the liquid chlorine in the tank, in turn lowering the chlorine vapor pressure. The amount of pressure reduction is dependent upon the chlorinator system capacity. This procedure can also be used to achieve the 30 psi pressure differential required to empty a road tanker. Not all road tankers are equipped with a compressor for unloading purposes. In very cold climes or other circumstances, this practice might not be feasible. In any case, this pressure control maneuver is always applicable if there is a major leak downstream from the tank. In these cases, the temperature and pressure gauges described above are very important for monitoring the pressure control system (PCS). (See Figs. 9-61 and 9-62, and the following text.)

Special emphasis is given in the following section to the Arizona Blanket, used to keep the liquid chlorine cool at all times. If it is installed properly, it can maintain the vapor pressure between 65 and 70 psi, 24 hr/day, even in the warmest of climes. It provides many benefits in the operation of a chlorination system, in many cases eliminating the need to maintain an air padding system for loading the storage tank. When evaporators are in use, as they are most likely to be where a bulk tank is involved, the evaporator efficiency increases as the vapor pressure decreases.

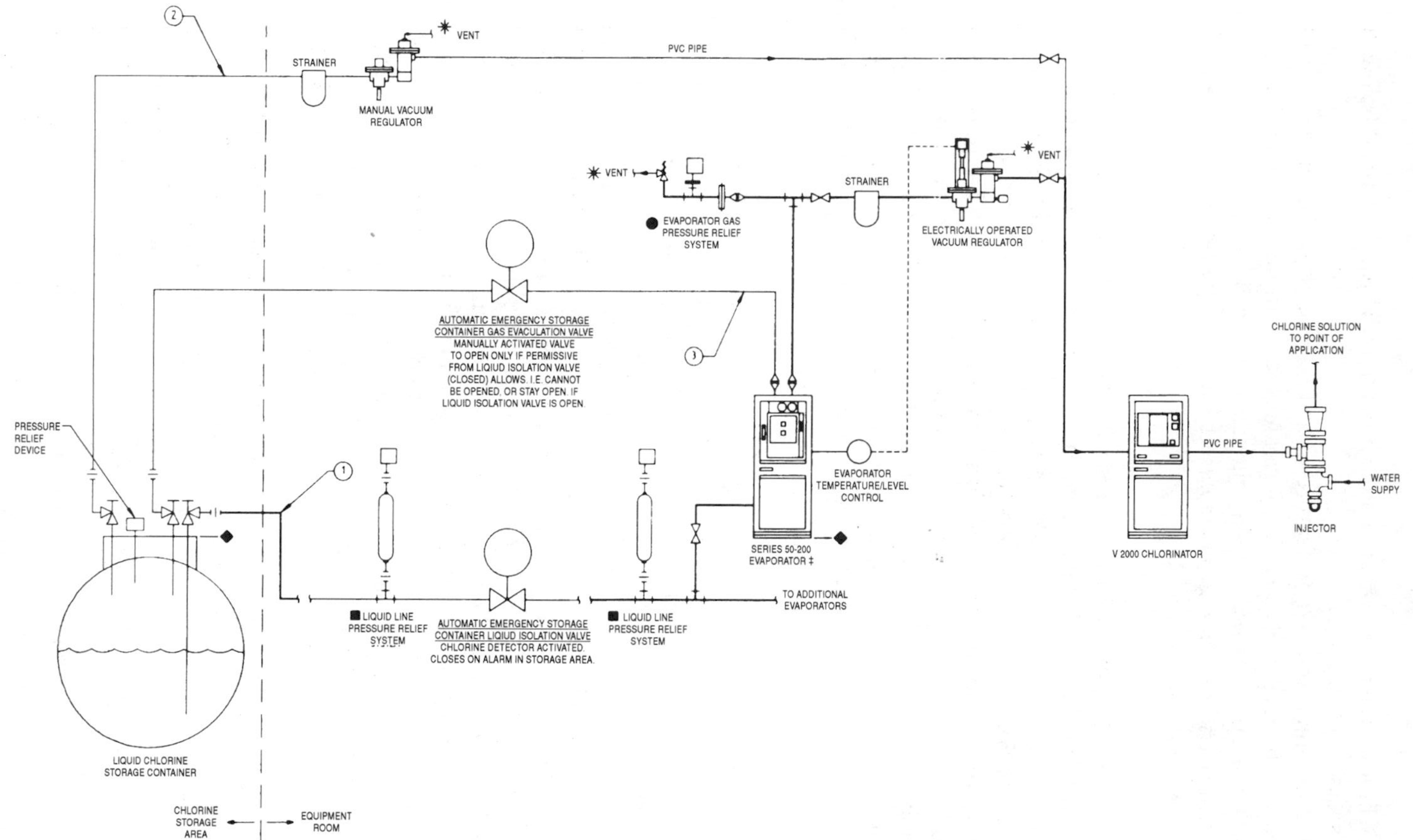

Fig. 9-61. Piping arrangement for emergency pressure reduction in the chlorine storage container by gas evacuation with liquid chlorine isolation valves (courtesy of Wallace & Tiernan).

NOTE:

— (heavy line) TYPICAL PIPING FOR LIQUID WITHDRAWAL THROUGH EVAPORATOR

— (light line) OPTIONAL PIPING PER DESIGN REQUIREMENTS

✳ THE VENT LINE MUST TERMINATE IN AN AREA WHERE GAS FUMES CANNOT CAUSE DAMAGE OR INJURY TO PERSONNEL. DO NOT TERMINATE THE VENT LINE AT A LOCATION ROUTINELY USED BY PERSONNEL SUCH AS WORK AREAS OR PATHWAYS NOR NEAR WINDOWS OR VENTILATION SYSTEMS INTAKES. VENT LINES TO BE INSTALLED ON CONTINUOUS DOWN GRADIENT WITHOUT TRAPS

● THE SYSTEM MUST BE INSTALLED WITH THE RELIEF VALVE IN A VERTICAL POSITION.

■ THE SYSTEM MUST BE INSTALLED WITH THE CLOSED END OF THE EXPANSION TANK FACING UP

⫲ SPECIAL EVAPORATOR ARRANGEMENT WITH GAS <u>INLET</u> CONNECTION.

◆ BOTTOM OF EVAPORATOR MUST BE NO LOWER THAN TOP OF LIQUID CHLORINE STORAGE CONTAINER, TO PREVENT HYDROSTATIC LIQUID CHLORINE DISCHARGE FROM EVAPORATOR IF AUTOMATIC EMERGENCY STORAGE CONTAINER VALVES MALFUNCTION.

LINE 1 LIQUID CHLORINE WITHDRAWAL LINE FOR NORMAL FEED OF CHLORINE THROUGH EVAPORATOR AT RATED CAPACITY

LINE 2 <u>OPTIONAL</u> TEMPORARY GAS WITHDRAWAL DIRECTLY FROM STORAGE CONTAINER. FEED RATE NOT TO EXCEED MAXIMUM SUSTAINED GAS WITHDRAWAL RATE FROM CONTAINER OR CAPACITY OF VACUUM REGULATOR

LINE 3 <u>OPTIONAL</u> EMERGENCY CHLORINE STORAGE CONTAINER PRESSURE REDUCTION LINE. FLOW CAPACITY OF ANY EXCESS FLOW DEVICE IN THIS LINE MUST BE ADEQUATE TO ALLOW CHLORINATOR(S) TO OPERATE AT MAXIMUM CAPACITY. THOROUGH INSPECTION AND POSSIBLE REPLACEMENT OF PARTS REQUIRED, IN THIS LINE, AS WELL AS EVAPORATOR DISCHARGE LINE THROUGH THE CHLORINATOR, AFTER UTILIZATION OF THIS SYSTEM IN AN EMERGENCY.

A ALL GAS LINES TO BE PITCHED BACK TOWARD SUPPLY CONTAINER TO ALLOW ANY RE-LIQUIFIED CHLORINE TO DRAIN BACK TO CONTAINER

B TEMPERATURE OF EQUIPMENT ROOM SHOULD BE HIGHER THAN TEMPERATURE OF STORAGE AREA TO PREVENT RE-LIQUIFACTION IN PRESSURIZED GAS LINES DURING GAS WITHDRAWAL

C REGARDING THE EMERGENCY STORAGE CONTAINER <u>EVACUATION</u> AND <u>ISOLATION VALVES</u>, THESE VALVES MUST OPERATE WHEN CALLED FOR IN AN EMERGENCY SITUATION. ANY VALVE USED IN CHLORINE SERVICE IS SUBJECT TO CONTAMINATION FROM IMPURITIES IN THE CHLORINE. TO ASSURE PROPER OPERATION OF THESE VALVES, IT IS RECOMMENDED THAT THEY BE EXERCISED (OPERATED TO ASSURE VALVE MOVEMENT) REGULARLY

D ALL PIPE AND FITTINGS FROM STORAGE CONTAINER TO VACUUM REGULATOR TO BE SCHEDULE 80 SEAMLESS CARBON STEEL PIPE WITH 3000 LB FORGED STEEL FITTINGS

E THIS DRAWING IS INTENDED TO DEPICT TYPICAL WITHDRAWAL PIPING FROM LIQUID CHLORINE STORAGE CONTAINERS IN A GENERAL MANNER. THIS DRAWING DOES NOT SHOW ALL PIPING, VALVES AND FITTINGS WHICH MAY BE NECESSARY FOR A PARTICULAR INSTALLATION. CHLORINE INSTITUTE RECOMMENDATIONS SHOULD BE FOLLOWED FOR ADDITIONAL REQUIREMENTS

Fig. 9-61. (*Continued*).

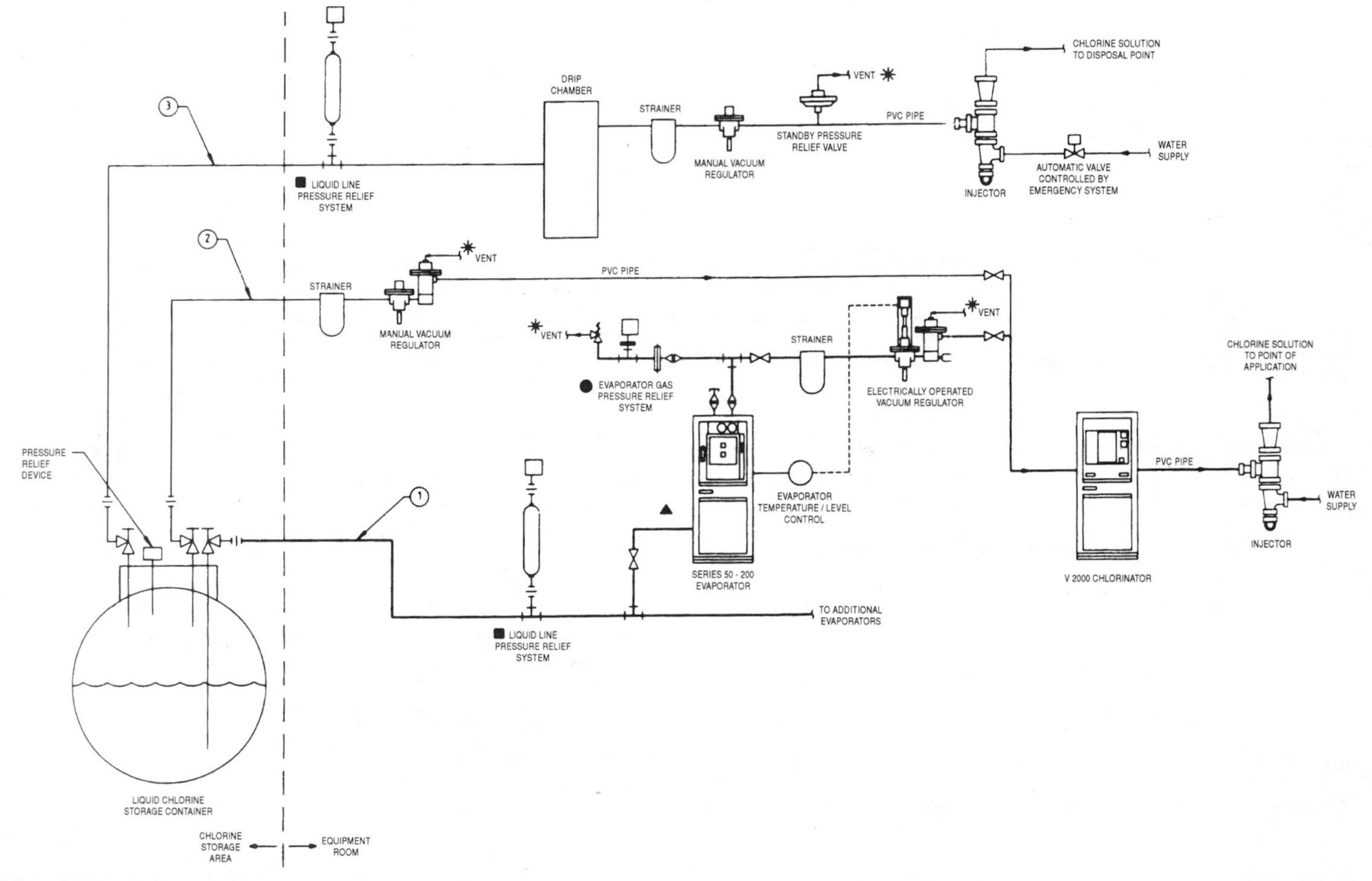

Fig. 9.62 Piping arrangement for emergency pressure reduction in the chlorine storage container with a separate pressure reduction

NOTE:

━ TYPICAL PIPING FOR LIQUID WITHDRAWAL THROUGH EVAPORATOR

— OPTIONAL PIPING PER DESIGN REQUIREMENTS.

✳ THE VENT LINE MUST TERMINATE IN AN AREA WHERE GAS FUMES CANNOT CAUSE DAMAGE OR INJURY TO PERSONNEL. DO NOT TERMINATE THE VENT LINE AT A LOCATION ROUTINELY USED BY PERSONNEL SUCH AS WORK AREAS OR PATHWAYS NOR NEAR WINDOWS OR VENTILATION SYSTEM INTAKES. VENT LINES TO BE INSTALLED ON CONTINUOUS DOWN GRADIENT WITHOUT TRAPS.

▲ USE BOTTOM CONNECTION WHEN 2 OR MORE EVAPORATORS ARE MANIFOLDED. FOR 1 EVAPORATOR THIS CONNECTION MAY BE AN ALTERNATE TO TOP INLET.

● THE SYSTEM MUST BE INSTALLED WITH THE RELIEF VALVE IN A VERTICAL POSITION

■ THE SYSTEM MUST BE INSTALLED WITH THE CLOSED END OF THE EXPANSION TANK FACING UP

LINE 1 LIQUID CHLORINE WITHDRAWAL LINE FOR NORMAL FEED OF CHLORINE THROUGH EVAPORATOR AT RATED CAPACITY.

LINE 2 OPTIONAL TEMPORARY GAS WITHDRAWAL DIRECTLY FROM STORAGE CONTAINER. FEED RATE NOT TO EXCEED MAXIMUM SUSTAINED GAS WITHDRAWAL RATE FROM CONTAINER OR CAPACITY OF VACUUM REGULATOR.

LINE 3 OPTIONAL EMERGENCY CHLORINE STORAGE CONTAINER PRESSURE REDUCTION LINE. MAXIMUM INITIAL WITHDRAWAL RATE LIMITED TO 12,000 LBS/DAY WITH 10,000 LB/DAY VACUUM REGULATOR. ADDITIONAL REDUCTION LINES WITH DRIP CHAMBERS, STRAINERS, VACUUM REGULATORS AND INJECTORS REQUIRED FOR HIGHER INITIAL WITHDRAWAL RATE. THOROUGH INSPECTION AND REPLACEMENT OF PARTS REQUIRED AFTER UTILIZATION OF THIS SYSTEM IN AN EMERGENCY. FLOW CAPACITY OF ANY EXCESS FLOW DEVICE IN THIS LINE MUST BE ADEQUATE TO ALLOW INJECTOR(S) TO ERATE AT MAXIMUM CAPACITY.

A. ALL GAS LINES TO BE PITCHED BACK TOWARD SUPPLY CONTAINER TO ALLOW ANY RE-LIQUIFIED CHLORINE TO DRAIN BACK TO CONTAINER

B. TEMPERATURE OF EQUIPMENT ROOM SHOULD BE HIGHER THAN TEMPERATURE OF STORAGE AREA TO PREVENT RELIQUIFACTION IN PRESSURIZED GAS LINES DURING GAS WITHDRAWAL

C. UNDER NORMAL OPERATING CONDITIONS, EITHER LIQUID (LINE 1) OR GAS (LINE 2) CAN BE USED FOR WITHDRAWAL. DO NOT ATTEMPT BOTH GAS AND LIQUID WITHDRAWAL SIMULTANEOUSLY.

D. ALL PIPE AND FITTINGS FROM STORAGE CONTAINER TO VACUUM REGULATOR TO BE SCHEDULE 80 SEAMLESS CARBON STEEL PIPE WITH 3000 LB FORGED STEEL FITTINGS

E. THIS DRAWING IS INTENDED TO DEPICT TYPICAL WITHDRAWAL PIPING FROM LIQUID CHLORINE STORAGE CONTAINERS IN A GENERAL MANNER. THIS DRAWING DOES NOT SHOW ALL PIPING, VALVES AND FITTINGS WHICH MAY BE NECESSARY FOR A PARTICULAR INSTALLATION. CHLORINE INSTITUTE RECOMMENDATIONS SHOULD BE FOLLOWED FOR ADDITIONAL REQUIREMENTS.

Fig. 9-62. (*Continued*).

CHLORINE STORAGE PRESSURE CONTROL SYSTEM

Attitude of Fire Departments

Based on the author's experience of many years, most fire departments would prefer to never have to enter a chlorine storage building to cope with a container leak.

Normally, this never should be necessary because the local Chemtrec Emergency Response unit is responsible for handling container leaks. The plant operator should be instructed to call them first, not the fire department; which is generally less familiar with the procedures to follow in the event of either chlorine or sulfur dioxide leaks.

The following design for chlorine storage tank systems that can quickly reduce chlorine vapor pressure will reduce the leak rate of any major release. This arrangement should be acceptable to any fire marshal in interpreting the Uniform Fire Code.

Many years ago, White made some field tests using vapor withdrawal from 50-ton storage tanks at one of the San Francisco wastewater treatment plants. At a withdrawal rate of only 3000 lb/day from a one-half-full tank, a frost line was created in about 30 min. This was the only means available to determine the amount of chlorine remaining in the tank; their electronic liquid level sensor had failed after a few months of operation. It was obvious that this procedure, vapor withdrawal, could slow a catastrophic leak dramatically. Therefore, an emergency response team would have no qualms about entering the storage area to repair a container leak.

Pressure Control: Storage Supply

In addition to chlorinator injectors, each storage tank must have a vapor evacuation injector to purge the tank before entry for visual inspection. The use of these injectors can reduce the storage tank pressure to 30 psi in about 20–30 min, depending upon ambient temperature and withdrawal rate. (See Figs. 9-61 and 9-62.)

Based upon recent field experimentation with chlorine gas, withdrawal from a 90-ton insulated tank car required the removal of 43 lb of chlorine to reduce the liquid temperature per degree F from 98 psi to 37 psi in 3.75 hours and only 9 lb per degree F from 37 psi to 20 psi. The chlorine gas withdrawal rate was approximately 13,000 lb/day. A noninsulated storage tank would require a longer time to achieve the same cooling rate.

A system with multiple chlorinator injectors plus two or three 4-inch vapor evacuation injectors at the tank would be capable of reducing the pressure in a 35-ton storage tank to near atmospheric pressure in a few minutes, assuming a chlorine withdrawal rate of 2500 lb/hr.

Although it has never been publicized, Dow Chemical Co. in Pittsburg, California stores liquid chlorine in an enormous sphere housed in a containment structure with a sloping concrete floor to contain the leaking liquid. The liquid chlorine is continuously recirculated through a refrigeration system that maintains the vapor at atmospheric pressure. When a leak occurs, a detector activates a foam system that insulates the leaking liquid from ambient air, thereby preventing any vaporization.

A significant added attraction of this pressure control system is its ability to quickly obtain the necessary differential pressure required to unload a tanker truck without the need for an expensive dry air padding system or a truck equipped with a chlorine vapor compressor.

Storage Tank: Vapor Evacuation Injector

Somc may wondcr about thc practicc of applying chlorine vapor under pressure to the suction side of the injector. It was discovered that a plant in Las Vegas, Nevada uses this method exclusively when filling storage tanks from a tanker truck. The only complaint offered by the chief operator was mild off-gassing at the point of injector solution discharge.

White's design is a bit more sophisticated. It plans to provide sufficient 3–5 percent caustic (or lime) solution to convert all of the molecular chlorine to $NaOCl$ or $CaOCl$ in the injector solution discharge line. This would eliminate the off-gassing mentioned above. Additionally there would be a barometric leg in the vapor vacuum line to prevent injector water suck-back into the storage tank. This was done at the San Francisco chlorine storage tanks (ca. 1948) with complete success.

This pressure control system approach reduces significantly the chlorine scrubber requirements. It also makes the absorption tank approach more interesting as an alternative to a venturi scrubber.

Arizona Blanket

This unique idea is presented to illustrate how it improves the safety of the pressure control system. Lowering the liquid chlorine temperature increases the heat transfer efficiency of an evaporator, and vice versa. For a storage tank, then, the Arizona Blanket is the perfect solution, particularly for warmer climes.

A blanket of absorption material is placed completely around the tank—except for the ends. A small pipe perforated with small holes is placed along the crown of the tank. This pipe is then connected to a water supply using a Dole flow control valve, which would allow a flow of approximately 0.5 gpm. All that is required is to keep the blanket moist. This allows intermittent water flow onto the blanket, depending upon the ambient temperature. This blanket can easily keep the tank vapor pressure at 65 psi around the clock in the hottest weather; it has a long and successful record. It is called the Arizona Blanket because Abe Frick of the W&T Los Angeles office (ca. 1936) worked this out with a prominent consulting engineer, John Carollo, of Phoenix,

Arizona. They used it exclusively on ton containers with great success. White also has used it on ton containers and a 25-ton storage tank.

The floor system under the storage tank should slope dramatically to the longitudinal centerline of the tank for quick positive drainage of either water from the blanket or liquid from a chlorine leak. At this center point there should be a 2-inch-wide, 6-inch-deep trough to collect the drainage. Disposal of this drainage will depend upon local conditions. In the case of ton containers, floor slots also should be provided so that the liquid portion of a spill can be quickly hustled off to the scrubber system.

Now we have the Susco Co. heated blanket for raising operating pressure on SO_2 cylinders. (See Chapter 10.)

Evaporators

The evaporators are crucial to the successful operation of the pressure control system. It is imperative that the vapor from the storage system pass through the evaporators before going to the chlorination system. The evaporators will eliminate any liquid chlorine "mist" that occurs when the vapor withdrawal rate exceeds the ability of the ambient air temperature to completely vaporize the chlorine. This liquid chlorine "mist" attacks and destroys the plastic components of the chlorination system.

Therefore, it is incumbent upon the evaporator manufacturer to modify its evaporators so that the liquid inlet to the evaporator enters at the bottom and the vapor inlet enters at the top. This is as it should be, as this is the only version acceptable to the Chlorine Institute for manifolding ton containers that also requires both liquid and vapor headers to all connected containers for pressure balance.*

Moreover, when there is more than one evaporator involved, the bottom liquid inlet is always preferable.

General Discussion. When the rate of chlorine withdrawal exceeds 1500 lb/day, an evaporator should be installed. This changes the supply system to liquid withdrawal, which has different characteristics from those of gas-withdrawal systems.

Evaporators are available in capacities from 400 lb/day to 10,000 lb/day. In a pinch, one cylinder can discharge liquid to an evaporator at a rate as high as 10,000 lb/day. This means that an evaporator can be used to conserve space for cylinder storage if necessary. The optimum storage requirements should be based upon the quantity-discount price break that is offered by the local chlorine supplier. This usually occurs at a quantity of five, thus dictating space

*The Metropolitan Water District of Southern California uses one of these evaporators as part of its 30,000 lb/day mobile chlorination system.

for five in service, five empties, and a vacant space for the incoming five, or a total space for 15 cylinders.

Electric Heater Type. The most widely used evaporator is the electric heater type, available from Capital Controls, Bailey–Fischer and Porter, and Wallace & Tiernan. These units are equipped with G.E. Calrod heating elements of various sizes depending upon the vaporization requirement. It takes approximately 65,000 Btu/hr to vaporize 8000 lb of chlorine. However, the evaporator must have a wide margin of safety to allow for the partial filling of the chlorine vessel with impurities inherent in the manufacture of chlorine. Therefore, to provide a sufficient safety factor, the chlorine gas that is vaporized must contain at least 20°F of superheat to prevent misting or liquid chlorine fallout in the gas discharge piping.

Misting occurs when the evaporator is pushed beyond its capacity. This is detrimental to the chlorinator because the little globules of mist contain the various impurities inherent in the production of chlorine. These impurities will plate out at the various stages of pressure reduction. This is another reason why a chlorine gas filter should always be installed immediately downstream from the evaporator. The important details to examine when comparing evaporators are as follows: volume of chlorine container vessel (extra volume is required for inherent sludge deposits); surface area in contact with water bath; volume of vapor space as determined by depth of penetration of gas discharge pipe into the container vessel; magnitude of hydrostatic pressure test; and allowable working pressure.

Beatty's Mixture. For more than 20 years the Metropolitan Water District of Southern California has been using a liquid mixture of water and corrosion inhibitor instead of a continuous flow of freshwater for the evaporator water bath. The mixture consists of 1 pint of closed-system corrosion inhibitor No. B-239 to 10 gal of demineralized water. Each 8000 lb/day evaporator installation is supplied with 5 gal of this solution. Operators add the solutions to the evaporator water bath on a biweekly basis from the 5-gal containers. In 1982, MWD reported[7] that use of this fluid for evaporator heating had eliminated scaling problems with immersion heaters, and had eliminated the need for cathodic protection of water bath and chlorine container vessels.

Hot Water Type. Evaporators are also available for use with recirculated hot water. One type utilizes the intermittent recirculating flow of hot water pumped in a circuit between a heat exchanger and the evaporator water bath. A temperature probe actuates the recirculating pump to maintain the water bath between 170 and 180°F. In another hot water arrangement, a treatment plant utilizes a closed-loop hot water system, in which water at 200°F is intermittently pumped through a coil in the evaporator hot water bath at approximately 10 gpm. This arrangement requires an independent water bath makeup system.

Steam Type. Vaporizers are available that use live steam instead of recirculated hot water. These units are available from Whitlock Mfg. Co. in capacities as high as 5000 lb/hr of chlorine. Custom-made evaporators are also available from specialty manufacturers in Seattle, Washington and Toronto, Canada. These vaporizers can be furnished for either steam or water heated by a special propane-fueled heater.

Chlorine Pressure Reducing and Automatic Shutoff Valve.* Integral with the evaporator is the electrically interlocked chlorine pressure shut-off valve on the discharge line of the evaporator. The circuit that operates this valve, whether an air solenoid for a pneumatically operated valve or an electric motor operator, is connected to the low-temperature alarm circuit. The alarm circuit sounds and deenergizes the CPRV circuit when the water bath temperature drops to 150°F. This protects the chlorinator against possibly receiving severely damaging liquid chlorine from the evaporator.

Remote Vacuum. This is a recent innovation by chlorinator manufacturers. The primary objective of this concept was to increase the rate of withdrawal from 150-lb containers. This idea launched the tank-mounted chlorinator. This arrangement allowed the rate of withdrawal from one cylinder to increase from about 40 lb/day to 100 lb/day. This was a boon to the small chlorinator market. Another important feature was its ability to transport the chlorine under a vacuum from the storage area to the metering and control equipment. The use of this concept was not intended to involve the large-capacity systems, such as those using evaporators.

The remote vacuum concept allows the user of vapor withdrawal to have a simple but reliable automatic cylinder switchover system.

Operating a vapor withdrawal system that reduces the supply pressure from 85–100 psi down to a vacuum of 20–25 inches of water will create additional maintenance problems at the point of pressure reduction. This is where the impurities inherent in chlorine will precipitate out. The device in question is the vacuum regulator check assembly (VRC). However, with liquid withdrawal systems, there is no need to increase the vapor withdrawal rate because that is all handled by the evaporator. Therefore, it is highly desirable to precede the VRC unit with a manually operated chlorine pressure-reducing valve (CPRV). It will greatly improve the reliability of the system, particularly in the warmer climes. In areas of hot weather the CPRV is imperative because these VRCs will automatically shut down when the supply pressure reaches 120–130 psi. (See Fig. 9–15.)

One of the best features of the remote vacuum system is its ability to lock down the entire piping system under a vacuum by simply shutting off the

*This valve is an imperative for remote vacuum systems.

chlorine supply. This feature allows the operator to purge the entire system of chlorine vapor.

Cathodic Protection and Insulation. All evaporators using a water bath should be equipped with a cathodic protection system to protect both the water bath tank and the outside of the chlorine container from aggressive water corrosion. This system is provided with an indicating ammeter on the evaporator instrument panel to verify cathodic protection.

The outside of the water bath should be insulated with a ½-inch covering of urethane foam.

Accessories. After a recent (1991) 14,000 lb leak (see Chapter 1) caused by the failure of the emergency electrical power shut-off switch,* the following so-called optional accessories are now considered mandatory: an automatic water-bath control system, including water-bath and chlorine gas temperature gauges, along with high temperature alarms for both the water bath and chlorine gas.

Electrical Requirements; Electric Heater Type:

1. A three-wire 240 or 480 V circuit is used for the heater elements in the evaporator water bath. The load requirement is 12 kW for 6000 lb/day and 18 kW for 8000 lb/day
2. A two-wire 120 V circuit is needed for the following functions:
   (a) Air solenoid or electric operator on chlorine pressure reducing and shut-off valve downstream from the evaporator, interlocked with low-temperature alarm.
   (b) Low-temperature alarm.
   (c) High temperature alarm—mandatory.
   (d) Solenoid valve to makeup line to water bath.
   (e) Water level pressure switch.
3. Alarms for each evaporator should include low temperature of the water bath and low water bath level.

Evaporator Pressure Relief System

Historical Background. Until as recently as 1974 there was no mandatory requirement to provide any pressure relief device for the liquid chlorine vessel in an evaporator. However, the manufacturers of chlorine evaporators have always been conscious of the possibility of liquid vessel ruptures. The prevaling belief was and still is that any pressure relief system creates more of a hazard than it might prevent. Therefore, evaporators have always been designed

*This enormous leak was caused by an evaporator water-bath temperature (in excess of 485°F), not by an NCl_3 impurity explosion.

with enough strength to hold any vapor pressure that could conceivably be encountered. Furthermore, the overall system design mitigates against any possibility that might allow the liquid vessel to get "skin full" of liquid chlorine. As a final precaution, the classic design of an evaporator is to have the connections to the liquid vessel made with lead gasketed "ammonia-type" unions. These unions act like relief valves under extremely high pressures. However, because the liquid chlorine vessel is fabricated according to the ASME Boiler and Pressure Vessel Code, it is subject to rigid inspection regardless of who the manufacturer may be and must therefore be certified accordingly.

As of 1997, White has never heard of the liquid chlorine vessel of an evaporator suffering from a rupture, not even after an NCl_3 explosion. The one mentioned above in the 14,000 lb leak suffered from a meltdown that would happen to any metal vessel, weak or strong. As of December 1975,[8] all pressure vessels manufactured in accordance with Division 1 Section VIII of the ASME code must be protected from overpressuring by means of a safety device. This safety device need not be provided by the vessel manufacturer, but must be provided prior to placing the vessel in service. This code defines the general requirements of pressure relief devices. The relief valve must be able to relieve the pressure in the liquid chlorine vessel when this pressure exceeds 110 percent of the rated working pressure of the vessel.

The Chlorine Institute specifies in Pamphlet No. 9 that chlorine vaporizing equipment must have a pressure relief device.[9] This can be either a rupture disk or a spring-loaded relief valve, or both, preferably discharging to an adsorption system. When both are used, the section between should be equipped with a vent or a pressure alarm.

The State of California safety orders for "Unfired Pressure Vessels" indicate that in addition to compliance with the ASME code, the following control is also required:

> 467. Controls: (a) Any pressure vessel not specifically covered or exempted elsewhere in these orders shall be protected by one or more safety valves or rupture disks set to open at not more than the allowable working pressure of the vessel* and by such other controlling and indicating devices as are necessary to insure safe operation.

Current Practice. Both Bailey–Fischer and Porter[10] and Wallace & Tiernan,[11] provide as optional equipment a relief valve system for all their various types of evaporators. Both illustrate the location of the relief valve on the gas and not the liquid phase of the evaporator connections. This current arrangement is shown in Fig. 9-63. It should be noted that all rupture disks

*Both Bailey-Fischer and Porter and Wallace & Tiernan use a working pressure design of approximately 500 psi at 212°F and a hydrostatic pressure test of 1450 psi at 125°F.[12,13]

can be damaged if subjected to a vacuum. They must be specified to withstand 25 inches of Hg vacuum on the pressure side.

Relief Valves. The Bailey–Fischer and Porter relief valve Model 71P1412 has been manufactured for over 20 years. This valve opens at about 275 psig and seats tightly at about 200 psig. Bailey–Fischer and Porter report a high confidence level for this valve.

The Wallace & Tiernan valve carries Part No. U25470. It opens at 560 psig and closes at 550 psig. This valve is purchased from Crosby Valve Co. and other suppliers such as Dresser Industries and Ferris Valve Co.[12]

Safety Considerations. The discharge of the relief valve system (vent) brings up serious questions about the hazards of chlorine leaking to the atmosphere at high pressures. Whenever either of these valves opens to relieve pressure, it becomes subject to atmospheric corrosion. The maintenance routine should therefore require that the valve be overhauled after each opening or closing cycle. Moreover, the valve must be protected at all times from chlorine vapor by installing a rupture disk as recommended by the manufacturer (upstream from the valve). This keeps the valve clean and dry during the periods of nonuse.

Owing to the potential hazard of a chlorine leak, it is always advisable but not mandatory to discharge the vent from this system into a chlorine absorption tank. Therefore, all of these systems should be designed so that at a later date a barometric loop and absorption tank can be conveniently added to the relief valve vent. (See section describing "Reserve Tank for Chlorine Supply.")

Absorption Tank for Relief Valve Vent System. In addition to a suitable size absorption tank (containing NaOH), the discharge piping between the relief valve discharge and the absorption tank mush contain a barometric loop. (See Fig. 9-51.) The barometric loop is mandatory with an absorption tank because it prevents the almost certain intrusion of moisture into the chlorine gas supply piping.

The absorption tank should be capable of neutralizing a maximum of 150 lb liquid chlorine from each evaporator during an overpressure crisis situation. This is about the amount of liquid chlorine that would be contained in each evaporator if the liquid vessel were full of liquid chlorine. Although this is an improbable situation, it provides a generous safety factor to the size of the absorption system. To this amount of chlorine should be added the amount in the liquid piping and expansion tanks between the supply tanks and the evaporators.*

*It is reasonable to expect that the rupture disks on the expansion tanks might also be overpressured at the same time.

Monitoring Relief System. As shown in Fig. 9-63, there will be a pressure switch to monitor a critical overpressure situation sufficient to rupture the frangible disk. However, it is recommended that if this relief system is vented to the atmosphere and not to an absorption chamber, the vent should terminate in an airtight compartment that is monitored by a leak detector. The detector

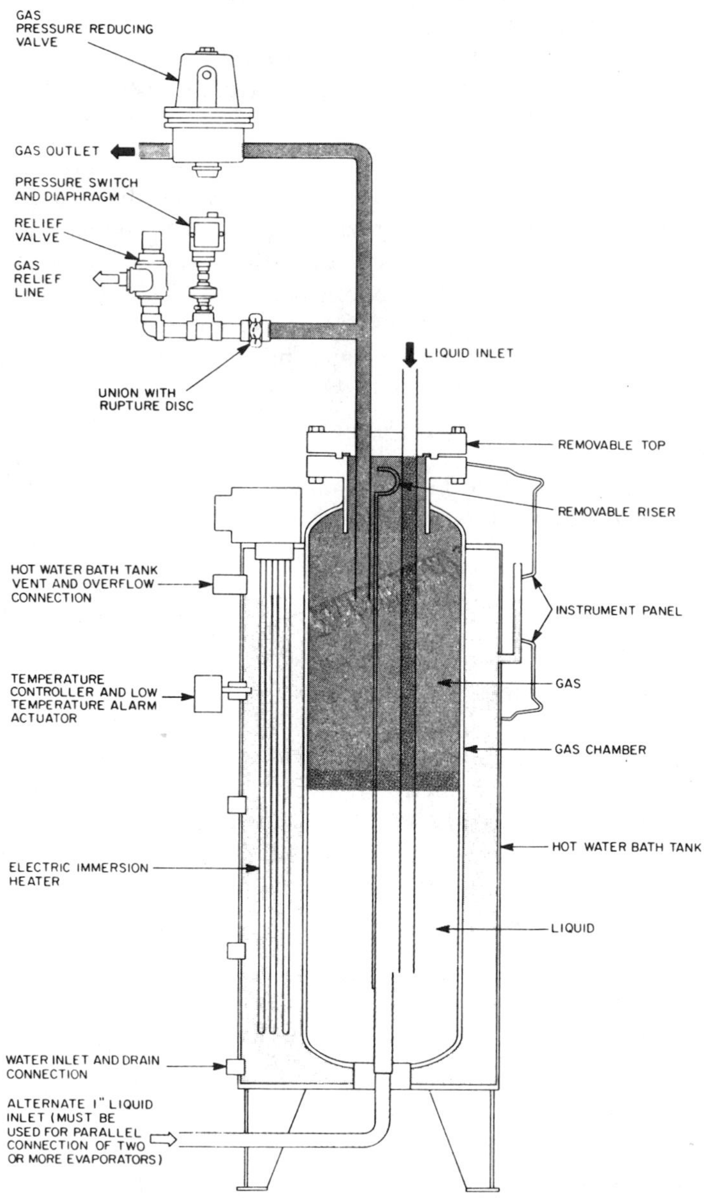

Fig. 9-63. Evaporator showing relief system (courtesy Wallace & Tiernan).

can be used to activate a small eductor, such as a chlorinator injector, which will discharge the leaking chlorine as a chlorine solution to an acceptable point of disposal, rather than discharging the vapor directly to the outside air.

Chlorinator Vents

All chlorinators and sulfonators have vents that release to atmosphere. In view of the Uniform Fire Code, they must be included as a safety hazard in the event of a vapor release. These vents are described as follows:

1. *Chlorine pressure reducing valve:* This item is frequently used to provide a two-step reduction to the operating vacuum level, rather than burden the chlorinator pressure to vacuum reduction system with one step. These valves have a vent to allow a leak in the valve-stem packing gland to escape to atmosphere. This vent must be able to "breathe;" it cannot ever be subjected to pressures above atmospheric.
2. *Vacuum regulating check unit:* These units are used on all remote vacuum chlorinator installations to relieve overpressure situations. Typically the vent from these units cannot be subjected to more than 2 psi pressure. This limit should be confirmed by the chlorinator manufacturer.
3. *Vacuum relief system:* There are many installations other than the remote vacuum units that use this kind of a system. These vents fall into the same category as the CPRV vent. They are designed to allow air into the chlorinator to relieve the vacuum upon shutdown and release chlorine to atmosphere during an unwanted overpressure situation.

All of these vents can be handled safely by terminating them in a common boxlike compartment outside the chlorination system structure. This box should have the ability to breathe the outside air through some ten to twelve ⅛-inch holes. A chlorine leak detector sensor should be installed to detect any chlorine leak into the box.

Evacuation of any chlorine release can be easily handled by connecting an injector vacuum line to the box. Typically the injector should be a standard adjustable 2-inch chlorinator injector connected to a 40–50 psi water or wastewater supply. Operation of the injector is done by opening a solenoid valve located on the injector inlet whenever the leak detector senses a chlorine leak.

The injector discharge containing the chlorine leak, now in a water solution, can be disposed of in the plant influent or in a proper wastewater collection system where there is minimal possibility of any off-gassing problems.

Safety Equipment and Accessories

Breathing Apparatus. There are two types of breathing apparatus, the canister-type gas mask and the oxygen or air-type breathing unit. The canister-type gas mask is limited in effectiveness; it is appropriate for changing chlorine cylinders or normal maintenance work. It is not satisfactory for use in repairing a leak. Therefore, either of the following types of equipment should be furnished: the air-type breathing unit (with 30-min. air supply, as manufactured by Mine Safety Appliance Company or Scott Aviation Company) or the oxygen breathing apparatus (as manufactured by MSA). The latter is similar to a canister type. When the seal on the unit is broken, the unit manufactures its own oxygen, which lasts for 45 min. These canisters must be discarded after the seal is broken.

Chlorine Container Emergency Kits. Every chlorination station should have at least one chlorine container emergency kit. These kits are available for 150-lb cylinders, ton containers, tank cars, tanker trucks, and stationary storage tanks. These kits are available from Chlorine Specialties Inc., San Francisco, California and Indian Springs Mfg. Co., Baldwinsville, New York. The kits are designed to seal off a leaking fusible plug, a leaking outlet valve, a tank car relief valve, or a moderate-size rupture in the container shell. Emergency Kit C for tank cars, tanker trucks, and storage tanks is illustrated in Fig. 9-64. Kit A is for 150-lb cylinders and Kit B is for ton containers.

Leak Detectors

General Discussion. There are two categories of leak detectors: (1) continuous monitors of the working environment ambient air and (2) hand-held

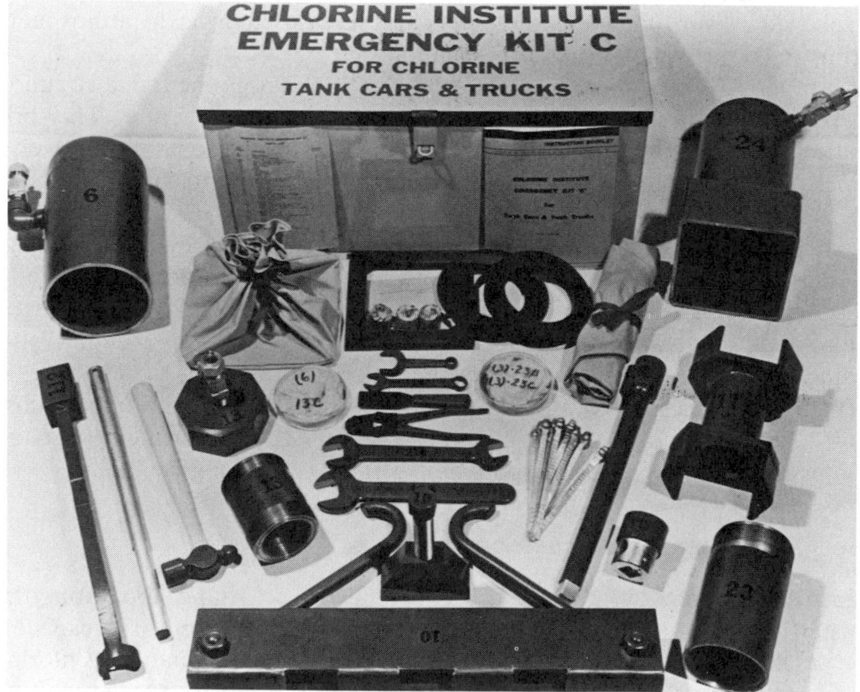

Fig. 9-64. Tank car emergency kit (courtesy Chlorine Institute NY and Chlorine Specialties San Francisco).

personnel leak locators. There should be at least one continuous detector for every chlorination station to prevent hazardous situations for both personnel and the surrounding population. Furthermore, these leak monitors are necessary to meet OSHA requirements for maximum contamination levels of chlorine (1 ppm) in the working area.

The second category of detectors is for searching or checking for leaks in the fabricated components of the chlorine supply system, which is under chlorine equilibrium vapor pressure.

As the continuous monitors are for area leaks, more than one detector per installation may be required. For example, one each should be provided for the tank car and/or storage tanks area, the ton cylinder room, and the evaporator–chlorinator room, and at the discharge of chlorine vent lines. Remote injector locations do not warrant a detector. Most leak detectors are designed as single-point samplers; therefore, multiple sampling point units should be used for detection at multiple sources. Figure 9-65 illustrates the single-point system,[14] and Fig. 9-66 shows the central system module capable of monitoring multiple sensors for a variety of sampling points.

Capital Controls Company, Inc. The ADVANCE™ Single Point Gas Detector—Series 1610B[14] provides continuous detection of chlorine or sulfur dioxide gas in a normally clean environment.[14] These gas detectors are ideal for the protection of personnel and property, wherever the chlorine or sulfur dioxide may be unloaded, stored, or used. (See Fig. 9-65.)

These units are highly sensitive, so they monitor gas levels far below OSHA current requirements for either gas. The sensor responds immediately to the presence of gas, and quickly recovers after the gas has cleared from the leak area. The sensor is also designed to eliminate false alarms caused by interference gases and other unusual environmental conditions.

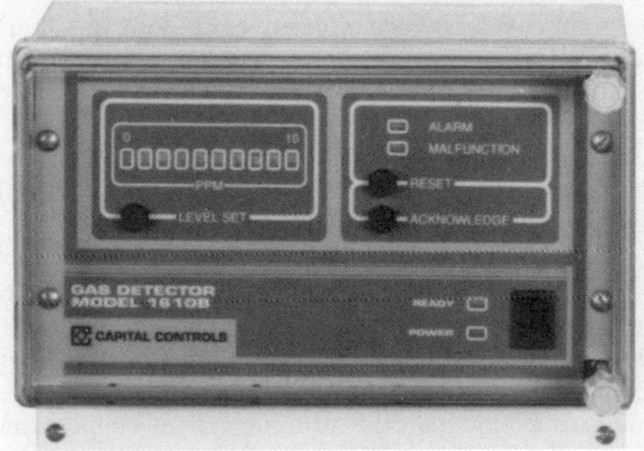

Fig. 9-65. Capital Control's single portal gas detector, series 1610B.

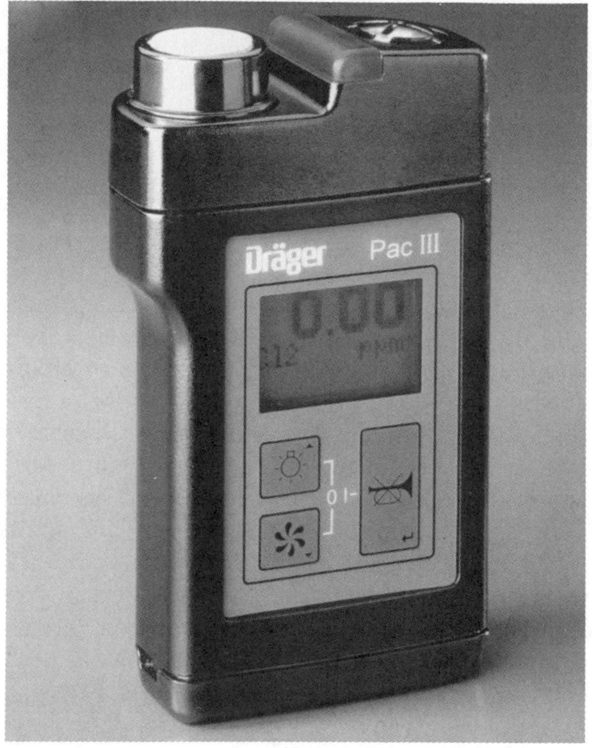

Fig. 9-66. Draeger Pac III (courtesy Draeger International and ICI, U.K.).

The Series 1610B design includes protection against radio-frequency/electromagnetic interferences (RFI/EMI) typically present at industrial and municipal plants. The modular design for the gas detector provides for easy installation of the receiver and sensor module.

This single-point detection system for either chlorine or sulfur dioxide gas consists of an electrochemical type sensor that does not require any chemical additions. It is housed in a corrosion-resistant, NEMA 4X enclosure suitable for wall mounting. In the presence of a gas, a current flow will develop, which will be transmitted through electrical wiring to the receiver. The maximum separation between the receiver and sensor should be no more than 1000 ft.

The receiver shall contain the following components:

1. A power switch and LED indicators for power when the system is ready.
2. A setpoint alarm level that is field-set via a pushbutton on the front of the receiver and indicated on the color bar graph by a flashing bar segment.
3. An LED alarm indicator and annunciator with corresponding contact.
4. An LED malfunction indicator with jumper-selectable manual or automatic reset relay contact.

5. A RESET button for clearing the alarm and malfunction circuits.
6. A timer for sensor stabilization, and an LED bar graph display with a range from 0 to 10 ppm.
7. A 4–20 mA DC output signal provided to transmit scanned levels.

The minimum detectable concentration of chlorine or sulfur dioxide gas is 0.5 ppm by volume. The response time is 30 sec for 80 percent of the instrument range up to 10 ppm of either gas at 20°C, after stabilization. The instrument is protected against radio frequency/electromagnetic interference that is typically present at industrial and municipal treatment plants. These units operate from 117 to 235 V AC, plus or minus 10 percent at 50/60 Hz power supply. Battery terminals are always provided for use in the event of a power failure.

An optional power backup system (Model 1640) can be provided, with an internal 18 V DC battery. This will automatically provide power in the event of a power failure. No manual switching will be required. The unit will automatically and continuously recharge to supply maximum support to the gas detector and/or the remote indicator.

An optional remote indicator (Model 1630) can be provided to remotely indicate and alarm when a preset gas level is exceeded. A cumulative bar-graph indicator will provide level indication. An annunciator provides audible indication of an alarm condition and can be turned off, but the indicator ALARM light will flash until the condition is cleared.

Capital controls warrants its gas detectors for 18 months from the date of invoice, or 12 months from the date of installation. Capital Controls is also ISO 9001 (International Standard Organization)–certified to provide quality and precision, and specializes in disinfection technologies, water quality monitors, and instrumentation for potable water and wastewater.

More than 35 years of industrial and municipal application experience in potable water and wastewater systems is incorporated into the equipment design to provide highest-quality solutions for the global market.

Another capital controls detector, the ADVANCE™ Multipoint Gas Detector—Series 1620B[14a] provides continuous detection of chlorine or sulfur dioxide gas in a normally clean air environment for up to eight different locations. This highly sensitive detector monitors gas vapor levels below the OSHA requirements, for either chlorine or sulfur dioxide gas. (See Fig. 9-67.)

The minimum detectable level for either gas is 0.5 ppm, and maximum separation between receiver and sensor is 1000 ft.

The sensors in these detectors respond immediately to the presence of these gases, and recover quickly after the gas has been cleared. The sensors are also designed to eliminate false alarms that may be caused by interference gases or other appropriate environmental conditions.

The Series 1620B design also includes protection against radio-frequency electromagnetic interferences (RFI/EMI) typically present at industrial and municipal chlorination/dechlorination installations. This unit operates from

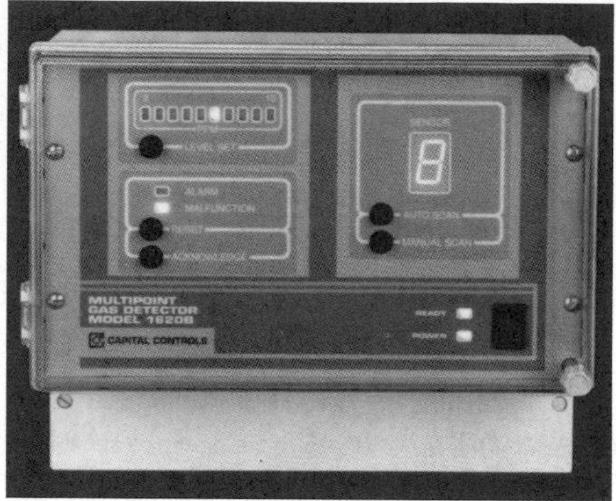

Fig. 9-67. Capital Control's ADVANCE™ multiport gas detector, series 1620B.

117 to 235 V AC, plus or minus 10 percent at 50/60 Hz power supply. Battery terminals are provided for use in the event of a power failure.

This multipoint gas detection system for either chlorine or sulfur dioxide gas consists of sealed electrochemical-type sensors that are housed in NEMA 4X corrosion-proof enclosures. When there is a leaking gas flow, an electric current will be developed that is transmitted immediately to the receiver. The receiver processes and displays these incoming signals from the sensor modules that are housed in the NEMA 4X enclosure.

The receiver shall contain the following components:

1. A power switch and LED indicators for power when the system is ready.
2. A setpoint alarm level that is set via a pushbutton on the front of the receiver and indicated by a flashing bar segment on the bar graph.
3. An LED alarm indicator and annunciator with corresponding contact.
4. An LED malfunction indicator with selectable manual or automatic reset relay contacts.
5. An acknowledge button for silencing the audible alarm.
6. A reset button for cleaning the alarm and the malfunction circuits.
7. An LED bar graph display in the range of 0 to 10 ppm.
8. An LED numerical sensor indicator, for display of the sensor being scanned and with pushbuttons for selecting either automatic or manual scan.
9. A 4–20 mA DC output signal provided to transmit scanned levels.

An optional power back-up is available that automatically provides power in the event of a power failure. This unit provides the necessary recharge to supply the maximum support to the gas detectors and or the remote indicators. Another option available is a remote indicator and alarm, when a preset gas level is exceeded. A cumulative bar-graph indicator is used to provide the level of indication. An annunciator provides audible indication of an alarm condition. Also an ALARM light will flash until the alarm condition is cleared.

The Series 1620B is also provided under the same warranty as the Series 1610B as described above, and carries the same ISO 9001 certification.

Draeger Polytron Sensing Systems[120]

Chlorine (Cl_2 L) Sensing Heads. These systems are based upon patented electrochemical sensing technology that provides fast, accurate, specific, and reliable responses to either chlorine or bromine gas. (See Fig. 9-66).

A special feature of the Polytron Cl_2 L Sensing Heads is the system self-test function, which not only checks the electrical circuit but, more important, the integrity of the sensing element, thus eliminating the need to perform routine field calibrations with chlorine gas. There is a two-year sensor warranty, and the expected sensor life is in excess of five years.

These Polytron sensing heads are used in safety, industrial hygiene, and engineering applications in the chemical, pulp and paper bleaching, potable water treatment, wastewater treatment, water reuse treatment, and other industries where Cl_2/Br_2 monitoring is required.

These sensing heads can interface with Draeger Polytron five- and twelve-channel system controllers. They are also compatible with data acquisition systems that accept a 4–20 mA signal.

Remote mounting up to 10,000 ft from the control system is the allowed limit.

Technical Specifications:

1. *Measuring ranges:* 0–5, 0–10, or 0–50 ppm Cl_2/Br_2.
2. *Response time:* less than 60 sec to 90 percent of the exposed concentration (at 25°C).
3. *Electrical class:* intrinsically safe when used with approved zaner barriers. Approved for: CENELEC EEx ia IIC T4. Designed for: Class 1, Division 1, Groups A, B, C, and D.
4. *Power requirements:* 8–30 V DC (20 mA maximum).
5. *Signal output:* 4–20 mA to a maximum loop impedance of 500 ohms.
6. *Wire requirements:* two-wire, stranded, twisted pair; should be shielded.
7. *Enclosure:* unplasticized PVC, resistant to the corrosive effects of chlorine and bromine gas (IP-53 rated). Mounting brackets are included.
8. *Environmental:* temperature: −5–110°F continuous operation. Humidity: 5–95 percent relative humidity continuous operation. Atmospheric pressure: 13.0–16.0 psi.
9. *Size:* 4.33 inches in diameter × 8.15 inches in height. Weight: 1.76 lb.

Contact National Draeger or an authorized representative for additional technical and ordering information on the Polytron Chlorine (Cl_2 L) Sensing Heads or the Polytron System.[120]

EIT

This company markets a line of toxic gas sensors.* The chlorine gas detector is its Chlor-Guard Model No. 5152.[16] It incorporates a gas diffusion sensor that does not require any maintenance. This sensor generates a current directly proportional to the concentration of chlorine. Response time is 90 percent in 15 sec, and the sensor may be located 500 ft from the alarm module.

The alarm module contains alarm lights, horn, and alarm relays. When a low level of chlorine gas is sensed, the yellow warning light will flash, and where a higher level of chlorine concentration is detected, the red lamp will glow steadily. The concentrations of chlorine gas at which these alarms occur are adjustable on-site from 1 to 10 ppm.

The Enterra toxic gas sensors are classified as voltammetric, diffusion-limited electrochemical devices. Each sensor contains a noble metal sensing electrode and a reference electrode immersed in a supporting acidic electrolyte. A diffusion membrane isolates the sensing electrode from the ambient air. In operation a small voltage is applied across the two electrodes, the voltage level being a function of the gas of interest. When the sensor is exposed to chlorine in the ambient air, diffusion begins. This allows the gas into the sensor through the diffusion membrane. When the gas comes in contact with the polarized sensing electrode, an oxidation reaction occurs at the counter (reference) electrode. This reaction generates a current in the system that is proportional to the concentration of chlorine gas at the diffusion membrane–sensor interface. The electrode reactions for chlorine are:[17]

Sensing electrode:

$$Cl_2 + 2e^- \rightarrow 2Cl^- \tag{9-2}$$

Counter electrode:

$$Pb + H_2SO_4 \rightarrow PbSO_4 + 2H^+ + 2e^- \tag{9-3}$$

ATI Series A15 Gas Sens—Modular Gas Detector[119]

Introduction. This is a new system (1994) that is equipped with self-checking gas sensors along with flexible components that provide a variety of options to meet individual gas detection and alarm requirements.

*See schematic Fig. 10-2, sulfur dioxide detector.

The Gas Sens detection systems consist of individual modules that can be located anywhere in a chlorination system where a gas leak might occur. The sensor transmitters are available in either NEMA 4X or the explosion-proof version, together with ATI's exclusive Auto-Test automatic sensor testing system, greatly reducing operator testing requirements.

Operating Ranges. For chlorine the range is from 0 to 10 ppm, and for chlorine dioxide it is 0 to 2 ppm.

Receiver Modules. These supply the electronic brains for the detection and alarm system. Each module includes a digital display of gas concentration, isolated analog output, and four relay outputs. Receivers may be located up to 1000 ft from the sensor/transmitter for remote indication. Or they can provide local control functions, such as valve shut-off, while transmitting a 4–20 mA signal to remote displays or data loggers.

Universal power supply modules provide DC power to receivers. The power supply is housed in a compact module similar to the receiver, and will accept inputs from 85 to 255 V AC or DC, without adjustment. The power supply unit also provides a power failure relay and charging for an optional battery backup unit.

Module Features. Receiver modules provide an interface between the detection system and external alarming and data logging requirements. One module is used with each sensor/transmitter and includes a variety of features. Some of these features are as follows:

1. *LED display:* This indicates leaking gas concentration directly in ppm or ppb. The display may be operated in high intensity mode for outdoor use or in the normal mode for indoor applications.
2. *Analog output:* An isolated 4–20 mA output is standard. The output will drive loads up to 1000 ohms for use in recording, data logging, or computer input.
3. *Two-alarm setpoints:* Alarm setpoints are factory-adjusted to standard values, but may be set to any value from 5 to 100 percent of range. The front panel LEDs, marked WARNING and ALARM, indicate the status of each alarm setpoint. A standard alarm time delay of 2 sec, or a longer delay of 10 sec, may be selected. In addition alarms may be switch-programmed to activate above or below the setpoint.
4. *Three-alarm relays:* Output relays are SPDT with unpowered contacts for use in activating external signaling devices or control elements, or for input to telemetry or annunciator systems. Each relay may be assigned to either alarm setpoint for application flexibility. The relays are factory-set to energize on alarm, but may be switch-programmed for fail-safe operation. They may also be set for either latching or nonlatching operation.

5. *Trouble alarm and relay:* If the sensor/transmitter input is lost, a trouble light (LED) on the front panel will flash, and an associated relay will be activated. For those systems equipped with the sensor Auto-Test feature, this alarm will also activate if the sensor does not respond to the automated gas test.
6. *Front panel reset switch:* A single front panel switch marked A/R (Acknowledge/Reset) serves a number of functions. When an alarm occurs, the switch will silence an audible horn wired to the module and will change the alarm lights from flash to steady on. After the alarm condition has cleared, the switch may be used to reset any latching alarms. The switch will also activate an electronic module test, inhibit alarm contacts, and activate the sensor Auto-Test.
7. *Remote reset input:* Terminals are provided for connection of a remote reset switch so that alarms can be acknowledged from a remote location or through a telemetry system.
8. *Pluggable terminal blocks:* External electrical connections are made to plug-in terminal blocks. Should module service ever be needed, modules can be replaced in minutes.

Sensor Auto-Test. A major expense in the operation of gas detector systems is the cost of regular testing to ensure that the sensors are responding correctly to the ambient environment. This requires a technician to inspect the sensors weekly, and manually to apply a small amount of gas to check the response. To solve this problem, ATI developed a unique system to reduce the maintenance time.

This system consists of an electrochemical chlorine gas generator closely coupled to the sensor. Every 24 hours, the receiver automatically activates this generator, producing a small amount of gas that diffuses into the sensor, just as it would if a gas leak had occurred.

The microcomputer in the receiver analyzes the output of the transmitter to determine if the sensor is responding normally. When proper sensor response is detected, the generator is turned off, and the system goes back to normal operation. However, if there is no sensor response detected, the TROUBLE light on the receiver will flash, and the trouble relay will activate. During testing, alarm relays are inhibited so that external alarms are not activated.

This Auto-Test feature ensures that each sensor is regularly tested with gas. Premature sensor failure or blockage of the sensor membrane is quickly detected. Moreover, self-testing will alert maintenance personnel when a sensor has reached the end of its useful life. As sensors normally last anywhere from 12 months to over three years, this feature allows users to determine accurately when sensor replacement is required.

In addition to all of the above, battery backup units and horn/strobe options are available for these ATI detectors. (For further information, call Customer Service, 800-959-0299.)

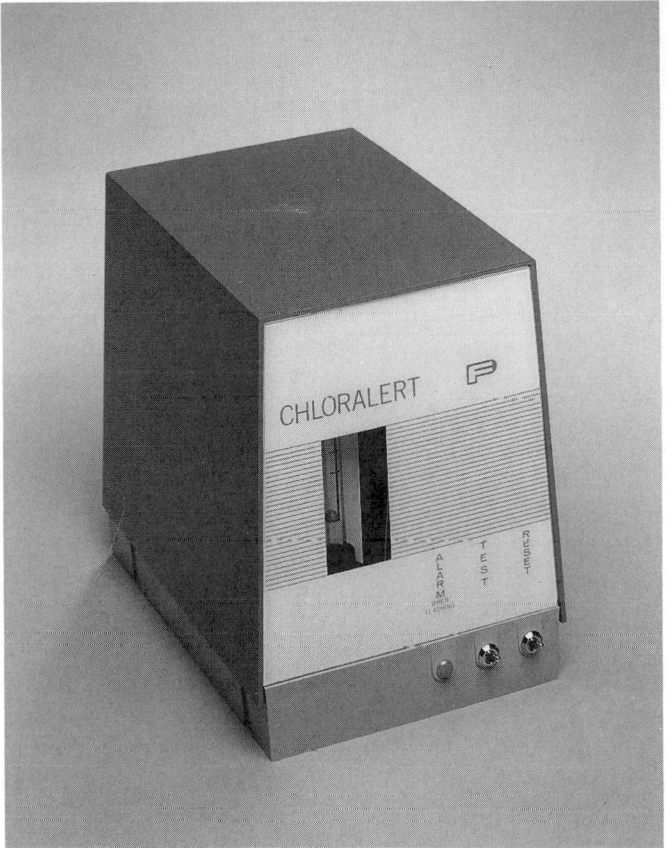

Fig. 9-68. Chloralert™ chlorine detector (courtesy Bailey–Fisher and Porter Co.)

Bailey–Fischer and Porter Chloralert

This chlorine leak detector, Series 17CA 1000 Chloralert, is a low-concentration ambient-air chlorine detector.[18] The detection level is fixed at 1 ppm chlorine by volume (3 mg/m^3). This instrument will sense and alarm at chlorine levels below the threshold of olfactory detection to conform to the OSHA maximum tolerable limit. This unit is suitable for use in any location where chlorine is manufactured, used, or stored, provided that the Chloralert is protected from rain, snow, sleet, and subzero temperatures. (See Figs. 9-68 and 9-69.)

The Chloralert uses a glycerine-based electrolyte containing a small amount of potassium bromide. Dual platinum electrodes are partially immersed in the KBr* electrolyte. A low voltage is applied across the electrodes, which intensifies the electrochemical response.

*When chlorine comes in contact with this electrolyte, Br_2 is released by the oxidation action of chlorine. The free bromine triggers the alarm circuit.

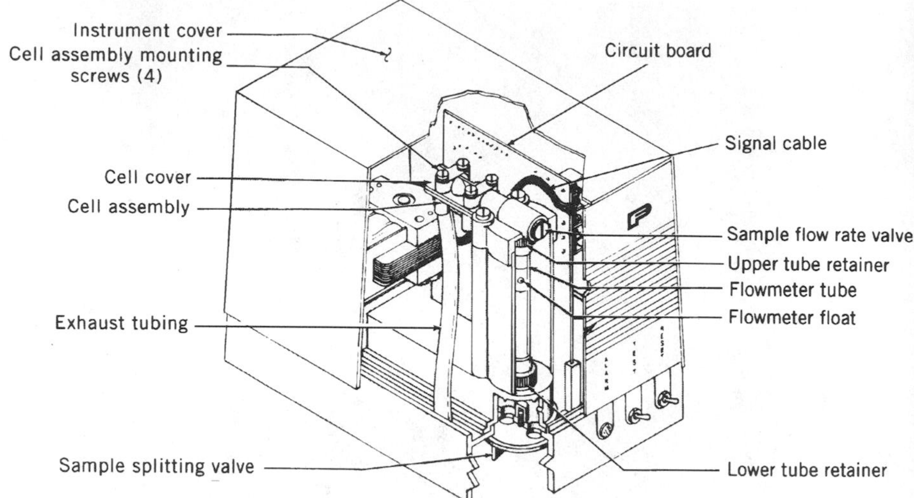

Fig. 9-69. Bailey–Fisher and Porter Chloralert chlorine leak detector, series 176A 1000, functional illustration.

An internal blower draws an air sample into the unit, where it passes through a flow indicator to permit adjustment of the sampling rate. From the flowmeter the air sample flows to the electrode sensing cell, where an electric current is generated by the presence of minute concentrations of chlorine. This current is sensed and monitored by a self-contained electronic circuit and is used to trigger the alarm system. The instrument response time is a matter of 2–3 sec, and it desensitizes equally fast, once the leak disappears.

Once the Chloralert is actuated, the circuitry is specifically designed to demand that the alarm be acknowledged with the spring-loaded reset switch. If the air sample still contains chlorine above the 1 ppm level, the alarm contacts will transfer to their normal (nonalarm) position whilc the reset switch is depressed, and will immediately return to the alarm position when the reset switch is released. The test switch, when depressed together with the reset switch, substitutes a simulated sensing cell current, thus allowing a sample check to be made of the circuit performance.

This instrument is designed for single sample point application and is limited to 25 ft from the source of air sampling. The sensing cell is designed for electrolyte replacement once a year.

Design Features. The instrument is

1. Simple to operate: A self-contained test switch permits a functional check of the circuitry.

Fig. 9-70. Wallace & Tiernan Acutek 35 chlorine gas detector system[124].

2. Low-maintenance: Under normal operation, the unit needs cleaning once a year.* There are no reagent bottles to refill. There is a plug-in alarm relay.
3. Sensitive: It will sense and alarm at chlorine levels below the threshold of smell to conform to the OSHA Short-Term Exposure Limit of 1.0 ppm.
4. Adaptable: The small, lightweight unit is easily wall mounted where space is available.
5. Convenient: The self-contained sampling system permits unit to be mounted remotely, even out of doors when it is protected from inclement or subzero weather, with the sample pipe run to the chlorine area.
6. Design for a cell response time, with instantaneous electrochemical reaction at the electrodes.

Factory Mutual has approved the Chloralert as a low-concentration chlorine detector.

Wallace & Tiernan Acutek 35 Gas Detector System[124]

This 1996 gas leak detector replaces the W & T Model 50.135 Rev 8–90. It utilizes the latest on-line gas monitoring technology, which is able to provide a flexible and reliable self-testing detection system that is in a compact modular design, with sensors that can detect the presence of chlorine, sulfur dioxide, and ammonia in ambient air.

Its modular component design (see Fig. 9-70) allows for a wide range of configurations that can be neatly and simply installed, using standard enclosures.

*Eight months if continuously operated at maximum temperature of 122°F.

The receiver is available in a choice of either a detector or a monitor version that has an LED display and an mA output signal. Both provide two selectable levels of alarm. A low level warning alarm may simply indicate transient leak conditions, whereas a second alarm can initiate a full alert in the event of more serious leaks.

For real peace of mind, there is an optional auto-test unit, which incorporates an integral gas generator that automatically tests the sensors daily, sounding alarms in the event of sensor failure. This added safety feature eliminates the costs incurred when manual testing procedures are used.

The flexibility of the modular system allows for the simple installation of either a single or a multipoint gas detection system. Each detection point requires the installation of one sensor/transmitter and one receiver module. The sensor/transmitter is installed in the area to be monitored, but the receiver module can be located up to 1000 ft away.

A power supply module sufficient for use with up to two receivers is housed with the receivers in a NEMA 4X corrosion-resistant enclosure.

Included within the enclosure is a piezo-electric horn to provide an audible local warning. Also available is a battery-backup system.

Standard ranges of gases that can be detected are as follows:

1. Chlorine: 0–10 ppm.
2. Sulfur dioxide: 0–20 ppm.
3. Ammonia: 0–100 ppm.

For further details see the Wallace & Tiernan publication TI 50.130 UA.

Ecometrics Series 1710 Gas Detector[129]

This is a fast-acting, fail-safe detector for a chlorine or sulfur dioxide gas leak. It accurately covers the range of 30 ppm of gas in the ambient air. It is suitable for both industrial and municipal applications. See Fig. 9-71.

This is a compact system that includes am electrochemical gas sensor and a microprocessor-based alarm indicator unit, both designed for wall mounting.

Gas Sensor. The Series 1710 Sensor is housed in a waterproof enclosure with an LED to indicate that power is being supplied to the sensor. The sensor is stable and practically maintenance-free, owing to the design of the cell. However, the cell is easy to replace whenever it is necessary.

This sensor is a high-resolution transducer that reacts quickly to the changing levels of ambient gas, transmitting a 4–20 mA signal, proportional to the level of the specific gas to the alarm indicating unit.

Alarm Indicating Unit. This unit has an LCD display with two lines of 16 alphanumeric characters in each line. The first line displays the gas level in

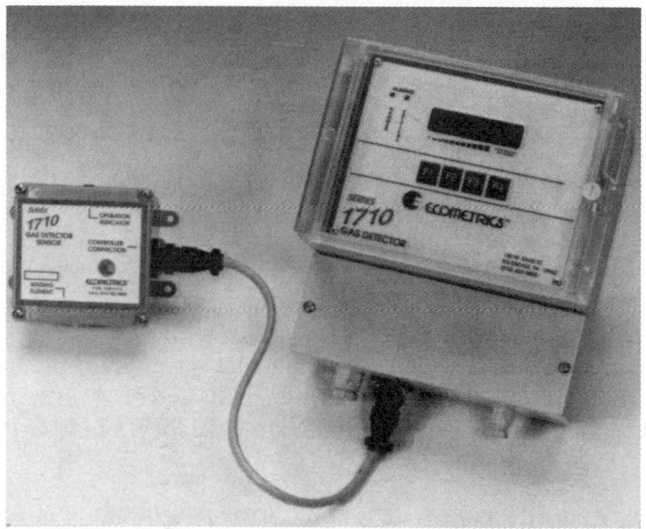

Fig. 9-71. Ecometric series 1710 chlorine gas detector[129].

ppm as well as the alarm status. The second line indicates the gas level as a percentage of the danger setpoint on a bar-graph display.

Any alarm condition is signaled by an audible alarm and is indicated on the LCD display. In addition, there are two red LEDs corresponding to DANGER and CRITICAL gas levels.

An optional 12 V battery, built into the alarm indicator, provides backup power in the event of AC power failure.

Alarm Settings. There are four alarm conditions:

1. Danger.
2. Critical.
3. Failure.
4. Power failure with optional battery backup).

Conditions 1 and 2 are individually selectable between 0.1 and 29.9 ppm.

The alarm set-up is achieved simply by a set-up screen on the alarm indicator unit's keyboard and stored in EPROM memory. The memory is secure until overwritten regardless of the main power or battery backup power status. (For Customer Service, see Ref. 129.)

Foxcroft Equipment & Service Co. Inc.[128]

Guardian FX–1500/C12. This is a chlorine gas detector monitor. Models are available for sulfur dioxide, ammonia, and hydrogen sulfide, in addition to chlorine. Its early warning system allows personnel to respond quickly, giving them time to take corrective action.

The three-electrode sensor combines fuel cell technology with a diffuser limited barrier, resulting in a maintenance-free sensor that is stable for long periods. This sensor is relatively unaffected by temperature and humidity variations. It simply will not "go to sleep."

The remote unit provides three alarm levels, with each level indicating a specific concentration of chlorine off-gas. These levels are fixed and cannot be adjusted. A system test switch allows receiving electronics to be tested and assures proper operations. Auxiliary relays for each alarm level are provided for external alarming

When there is a chlorine leak, the audible alarm sounds at the highest level to warn personnel that chlorine gas has reached danger zone status, and the area should be evacuted. The LED display shows chlorine concentration levels in the ambient air in ppm.

Remote Unit Features

1. Three alarm levels.
2. System test.
3. Audible alarm.
4. Corrosion-resistant NEMA 4X enclosure.
5. Visual alarms.
6. Relays for external alarming.
7. Cell failure indicator.
8. LED display.

Guardian II FX–1502 Dual Channel Gas Detector. This unit accepts two 4-20 mA signals from two of the same or two different toxic gas sensors. Each channel will alarm independently in the case of a toxic gas leak. Otherwise it has exactly the same features as the single-channel unit described above.[128]

Leak Locators

American Gas and Chemical Co. This Northvale, New Jersey company markets two products that assist the operator in locating chlorine, ammonia, or sulfur dioxide leaks. These products are identified by number for each specific gas.

For chlorine, CDP-100 is a chlorine-sensitive spray that reacts chemically with minute chlorine leaks, causing a visible color change from white to yellow at the point of leak. It is easily removed with a damp cloth.

For personnel protection and leak searching, there is a rechargeable, hand-held detector model, CGT-701. This instrument can locate chlorine leaks as low as 1 ppm. It uses a solid-state sensor that requires no maintenance and cannot be poisoned.

Interscan Corporation[122]

Introduction. This company makes a wide variety of leak detectors for chlorine, chlorine dioxide, sulfur dioxide, and hydrogen sulfide. However, they choose to identify these units as either gas analyzers or monitors.

System Configurations. Several configurations are available, which cover every requirement from personal protection to complete on-line multipoint analysis and data acquisition. These are briefly as follows:

1. *Personal alarm monitors:* These units miniaturize the detection and alarm capability, and dosimeters add data acquisition to provide a personal time history as well as personal alarm protection.
2. *The LD series continuous monitoring systems:* These units feature ultra-tough construction and a wide variety of alarm options. They are internationally acknowledged to be the most rugged and reliable single-point monitoring system available. This is the system recommended as the typical "leak detector" for a chlorination system installation.
3. *The standard portable and compact portable analyzers:* These are designed primarily for survey work. They operate off integral rechargeable batteries, and can be used for occasional longer-term studies. Intrinsically safe versions of these portable analyzers are available.
4. *The data logging compact portable analyzer:* This unit has powerful data acquisition capability right inside the analyzer. These units are ideal for time-history project studies.
5. *The multipoint systems:* These systems are used for monitoring several different locations within a local area of the overall installation of the operating equipment. These systems are custom-designed to meet the exact needs of a given project. They offer a cost-effective and completely organized system to protect any given area. Full data-acquisition capability is available with these multipoint systems, and with all the other systems.
6. *Rack-mount configured analyzers:* These units are intended for those applications in which line power operation is desired, but the industrial type packaging of the LD series is not required. These units can be installed in a 19-inch rack, or they can be used on a bench. They are also readily incorporated into larger instrument systems.

Principle of Operation. The Interscan voltammetric sensor (U.S. Patent No. 4,017,373) is an electrochemical gas detector operating under diffusion-controlled conditions.

Gas molecules from the sample are adsorbed on an electrolytic sensing electrode after passing through a diffuser medium, and are electrochemically reacted at an appropriate sensing electrode potential.

This reaction generates an electric current directly proportional to the gas concentration. This current is converted to a voltage for meter or recorder readout.

The diffusion-limited current, i_{lim}, is directly proportional to the gas concentration according to the following equation:

$$i_{lim} = nFADC/d$$

where F is the Faraday constant (96,500 coulombs), A is the reaction interfacial area in cm^2, n is the number of electrons per mole of reactant, d is the diffusion pathlength, C is the gas concentration ($moles/cm^3$), and D is the gas diffusion constant, representing the product of the permeability and solubility coefficients of the gas in the diffusion medium.

An external bias maintains a constant potential on the sensing electrode, relative to a nonpolarizable reference counter electrode in the two-electrode Interscan sensor. "Nonpolarizable" means that the counter electrode can sustain a current flow without suffering a change in potential. Therefore, the counter electrode acts also as a reference electrode, eliminating the need for a third electrode, plus a feedback circuit, as would be required for sensors using a polarizable air counter electrode.

Specifications for all Analyzer Models:

1. Accuracy, analog and digital units: plus or minus 2.0 percent of full scale.
2. Repeatability: plus or minus 0.5 percent of full scale
3. Minimum detectability: 1.0 percent of full scale.
4. Linearity: plus or minus 1.0 percent of full scale.
5. Zero drift: plus or minus 1.0 percent of full scale.
6. Span drift: less than 2.0 percent of full scale (24 hours)
7. Lag time: less than one second.
8. Rise time: 20 sec to 90 percent of final value (or better)
9. Fall time: 20 sec to 10 percent of original value (or better)
10. Calibration: Use Interscan's electronic calibration service.

To obtain further information, see Customer Service in Ref. 122.

FLOW OF CHLORINE IN PIPES UNDER PRESSURE

The fluid mechanics of liquid and gaseous chlorine are entirely different, and so are discussed separately.

Liquid Chlorine

The flow of liquid chlorine is restricted to the piping between the supply containers and the evaporators. This flow of liquid will not be similar to that of other liquids. For this reason, the friction loss of flowing liquid chlorine cannot be accurately predicted, as has been confirmed by field tests with rates of liquid chlorine flow up to 18,000 lb/day in approximately 1000 ft equivalent length of one-inch Sch. 80 pipe.[20] This is so because of the phenomenon of the liquid chlorine "flashing" to vapor whenever there is a sudden change of the flow rate in the system. Most chemical processes function at some fairly precise optimum capacity, but a chlorinator facility is designed to have a wide range of flow capability—at least 20 to one.

Therefore, when there is a sudden change in the flow rate, the liquid in the evaporators suddenly vaporizes rapidly, causing an abrupt pressure drop in the system, due to the change in demand. The vaporization process will extend back into the pipeline leading from the storage containers and create pockets of gas impeding the flow of the liquid. This vapor flashing occurs first at points of highest friction loss, such as entrances and exits of valves and fittings.

Finally, after the vaporization becomes stabilized in the evaporators, the pressure will rise in the evaporators, causing the flashing to cease in the pipeline. Then the system performs normally until there is another abrupt change in the liquid chlorine flow rate. The longer the pipeline, the longer the system takes to stabilize. In a 500 ft line, the pressure will begin to restore itself in about 5 to 10 min., depending on the rate of flow change, and will take two or more hours for complete restoration of pressure. This phenomenon is also related to temperature. If the chlorine pressure in the storage tank were at 80 psig, then the vapor temperature would be only 68°F. A long liquid line on a warm day or one exposed to the sun might be considerably warmer than 68°F, which would tend to warm the chlorine above its vapor pressure and to cause flashing in the line. The line would, however, be cooled by the flow of liquid chlorine, and the flashing would cease when it was sufficiently cooled below the vapor pressure. Therefore, depending upon the ambient temperature, the pressure in the chlorine containers, and the rate of change of liquid flow in the system, many combinations of pressure fluctuations will be noted. Sudden pressure drops of 10 to 15 psi in a one-inch line 500 to 1000 ft long with a maximum flow of 24,000 lb/day would not be unusual. Flows of 24,000 lb/day and less are considered "low" by chlorine manufacturers. This is only 1000 lb/hr, compared with the excess flow rate valve restricting withdrawal rates from tank cars at 7000 lb/hr.

Figure 9-72 shows the estimated pressure drop in various-size pipes carrying liquid chlorine. On the basis of sudden changes in flow under various kinds of climatic conditions and container pressures, it is recommended that the pipe size be limited to a maximum pressure drop of 0.25 psi per 100 ft for lines 500 ft and longer and 0.5 psi per 100 ft for shorter lines. Thus, the following size lines should be limited to the capacities shown in Fig. 9-72:

Line Size	Maximum Liquid Cl_2 Flow lb/Day Length of Line: Less than 500 ft	500–1500 ft
¾″	24,000	17,000
1″	48,000	33,600
1¼″	100,000	72,000
1½″	168,000	115,000

Actually, ¾ pipe should not be used for lines longer than about 100 ft. Furthermore, most manufacturers have standardized on one-inch inlets and outlets for the various chlorine piping accessories. Although the flows shown in Fig. 9-72 may seem enormous for such small lines, it should be noted that the velocity of liquid chlorine in a one-inch pipe at 48,000 lb/day is only slightly more than 1 ft/sec.

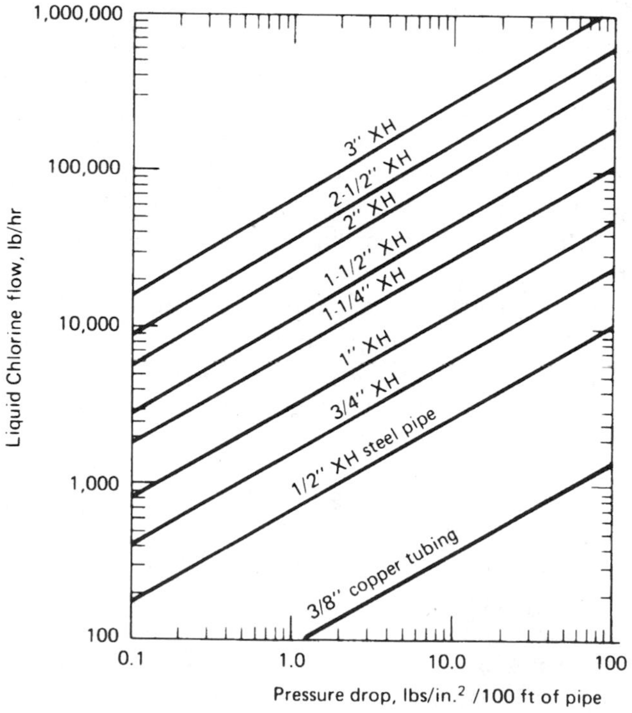

Fig. 9-72. Friction loss in liquid chlorine piping (clean new pipe; vaporization in the pipes will cause much higher drops) (courtesy Hooker Chemical Co.).

Example:

$$V \text{ ft/sec} = \frac{Q \text{ cu ft/sec}}{A \text{ sq ft}} \tag{9-4}$$

The inside diameter of a one-inch Sch. 80 pipe is 0.957 inches; so:

$$A \text{ sq ft} = \left(\frac{0.957}{12}\right)^2 \cdot \frac{\pi}{4} = 0.00499 \text{ sq ft}$$

Q = 48,000 lb chlorine per day; so:

$$\frac{48{,}000}{1440 \text{ min.} \times 60 \text{ sec}} = 0.5556 \text{ lb/sec}$$

Liquid chlorine at 68°F weighs 87.8 lb/cu ft; so:

$$Q = \frac{0.5556}{87.8} = 0.00632 \text{ cu ft/sec}$$

Therefore:

$$V = \frac{0.00632}{0.00499} = 1.27 \text{ ft/sec}$$

Therefore, the figures given in Fig. 9-72 are conservative for liquid chlorine pipe systems. The designer is cautioned to be meticulous about arriving at the proper equivalent length of pipe, accounting for elbows, sudden enlargements, ammonia unions, tees for pressure gauges, switches, and line valves. Most of these values are given in the appendix to this book.

The flashing phenomenon of liquid chlorine is one reason the flow cannot be properly measured. The usual differential pressure method for measuring the flow of liquids, including the variable area meters (rotameters), cannot be properly calibrated, since the basic principle of operation depends on a pressure drop that initiates the flashing phenomenon. At this point, there are both gas and liquid, with entirely different flow characteristics, passing through the measuring device simultaneously. Under proper conditions, a rotameter will give at best a rough approximation. The conductivity factor of liquid chlorine is so low that magnetic meters cannot be used either.

With the advent of reliable mass flowmeters (1990), such as Micro Motion, accurate measurement of liquid chlorine flow rates became a reality. These devices are currently being used by the Sanitation Districts of Los Angeles County, California.

Gaseous Chlorine

As the flow of chlorine gas follows the laws of fluid dynamics, the friction loss in such a system can be predicted with reasonable accuracy.

In pipelines carrying gas, the designer should have two special concerns: the piping system from the evaporators to the chlorinators, and the line carrying the gas from the chlorinator to the injector under a vacuum. When remote injectors are used, this calculation of line size is critical because the head loss tolerance is low—1½ to 2 inches of Hg vacuum. The chlorinator is relying upon the injector for its operating energy, so that little should be wasted in friction loss. However, on the pressure side between the evaporators and the chlorinator, a 10 psi pressure drop can be tolerated. A critical situation develops when the velocity of the gas under pressure is so great that it undergoes sufficient cooling to cause condensate or ice formation on the outside of the pipe.* Therefore, if the velocity is kept below 35 to 40 ft/sec, this phenomenon will probably not occur, and the pressure drop will be easily tolerable.

Most manufacturers arrange for one evaporator to serve each chlorinator independently. The openings out of the evaporator are one-inch; the chlorine filters and regulating valves are also made for one-inch pipe. Therefore, the pipe size for any 8000 lb/day chlorinator and evaporator should always be one inch. Let us investigate the velocity and possible head losses that will occur at maximum output of the chlorinator.

Gas leaving the evaporator will be at approximately 100°F. Let us assume a gauge pressure of 85 psi, indicating that the gas has a certain amount of superheat, which is normal for a properly operating evaporator.

The density (ρ) of the gas = 127 lb/100 cu ft (see Fig. 1 in Appendix I), and:

$$Q = 8000 \text{ lb/day} = 0.0926 \text{ lb/sec}$$

Converting to cu ft/sec:

$$\frac{0.0926 \text{ lb/sec}}{1.27 \text{ lb/cu ft}} = 0.0729 \text{ cu ft/sec}$$

Now:

$$V = \frac{Q}{A}, \qquad A \text{ for a one-inch Sch. 80 pipe} = 0.00499 \text{ sq ft}$$

so:

*As the gas cools, it becomes denser and the friction loss increases.

$$V = \frac{0.0729 \text{ ft}^3/\text{sec}}{0.00499 \text{ ft}^2} = 14.6 \text{ ft/sec}$$

This tolerable velocity exists between the evaporator and the external chlorine pressure-reducing and shut-off valve. (See Fig. 9-73.) Downstream from the CPRV, conditions change abruptly. Usually the CPRV is adjusted to give a downstream pressure of 40 psi. Let us now see what the velocity is at the reduced pressure of 40 psi. The temperature of the gas will drop about 25°F owing to this pressure reduction, so that the density of the chlorine gas will be $\rho = 0.7$ lb/cu ft. Therefore:

$$Q = 8000 \text{ lb/day} = 0.0926 \text{ lb/sec} = \frac{0.0926}{0.7} = 0.1325 \text{ ft}^3/\text{sec}$$

and:

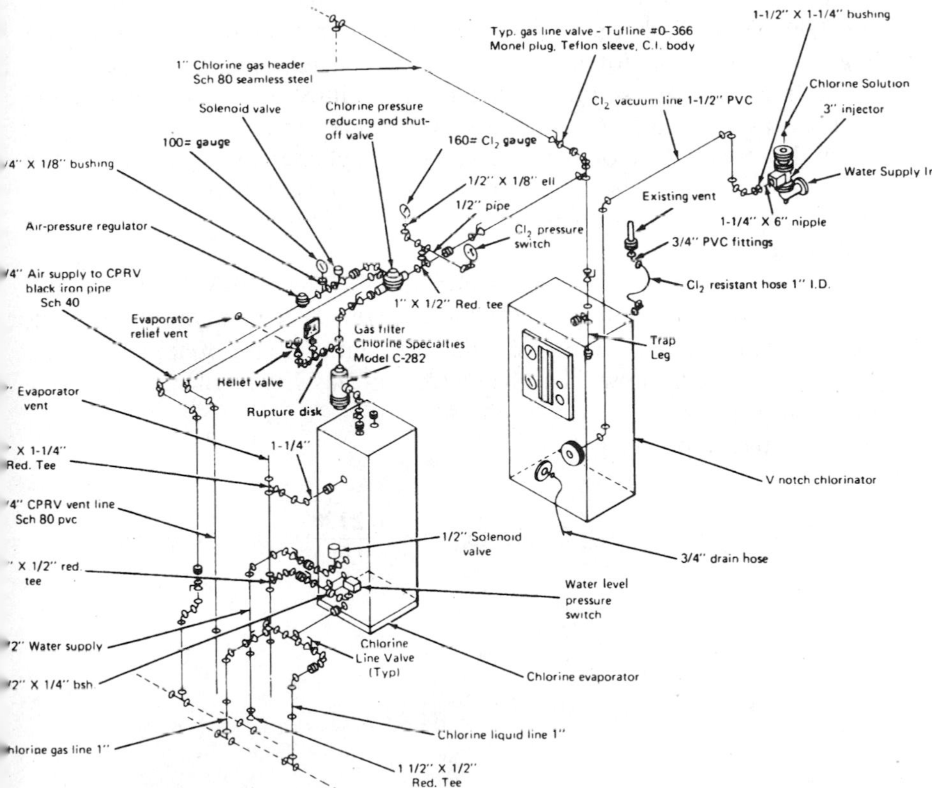

Fig. 9-73. Piping schematic evaporator, filter, CPRV, and chlorinator. Chlorine gas and liquid line fittings are 2000 lb forged steel. All unions on chlorine gas and liquid piping are ammonia type with lead gasket.

$$V = \frac{0.1325}{0.00499} = 26.51 \text{ ft/sec}$$

As the velocity increases, the pressure drop due to friction losses increases exponentially. This further reduction in pressure reduces the density of the gas, thereby initiating further increase in the velocity, which further compounds the friction loss problem. The above velocity, 26.5 ft/sec, will not cause excessive friction losses in a conventional layout, but does approach the upper limit of maximum allowable velocity.

The rule of thumb for designing chlorine gas supply systems is to limit the velocity at maximum flow to 50 ft/sec and preferably not more than 35 ft/sec.

Whenever evaporators are manifolded together, the designer must provide piping large enough to stay below this velocity limit. In these instances the rotameter and chlorine pressure-reducing valve will have one-inch pipe size connections, as will the chlorinator, but the piping required may have to be two inch.

As an example, consider the case of manifolding three evaporators, capacity 8000 lb/day, discharging to a common header through a CPRV, thence through a transmitting rotameter, for recording and totaling the chlorine flow, and thence to each of three chlorinators, as illustrated in Fig. 9-74. Note that the CPRV has a one-inch inlet and outlet, but has a nominal capacity of 32,000 lb/day. Similarly, a one-inch straight-through metal tube rotameter can easily handle 24,000 lb/day of chlorine. It is permissible to allow the high velocity of chlorine gas through these two devices, but the rest of the piping must be at least 1½ inches.

The gas will cool from the inlet temperature at the CPRV of 90 to 100°F down to an average of about 65 to 70°F between the CPRV and the chlorinator. The CPRV will be adjusted to give a downstream pressure of approximately 35 psi at a 24,000 lb/day flow. This gives a chlorine gas density of approximately 0.65 lb/cu ft.

Using 1½ inch pipe, we get:

$$Q = 24{,}000 \text{ lb/day} = 0.2778 \text{ lb/sec} = \frac{0.2778}{0.65} = 0.4265 \text{ ft}^3\text{/sec}$$

and:

$$V = \frac{0.4265}{0.01225} = 34.82 \text{ ft/sec}$$

This is close to the upper limit of velocity, which starts a vicious cycle: as the velocity increases, the gas becomes cooler; as the gas cools, it becomes denser, thus increasing the friction loss. As the friction loss increases, it be-

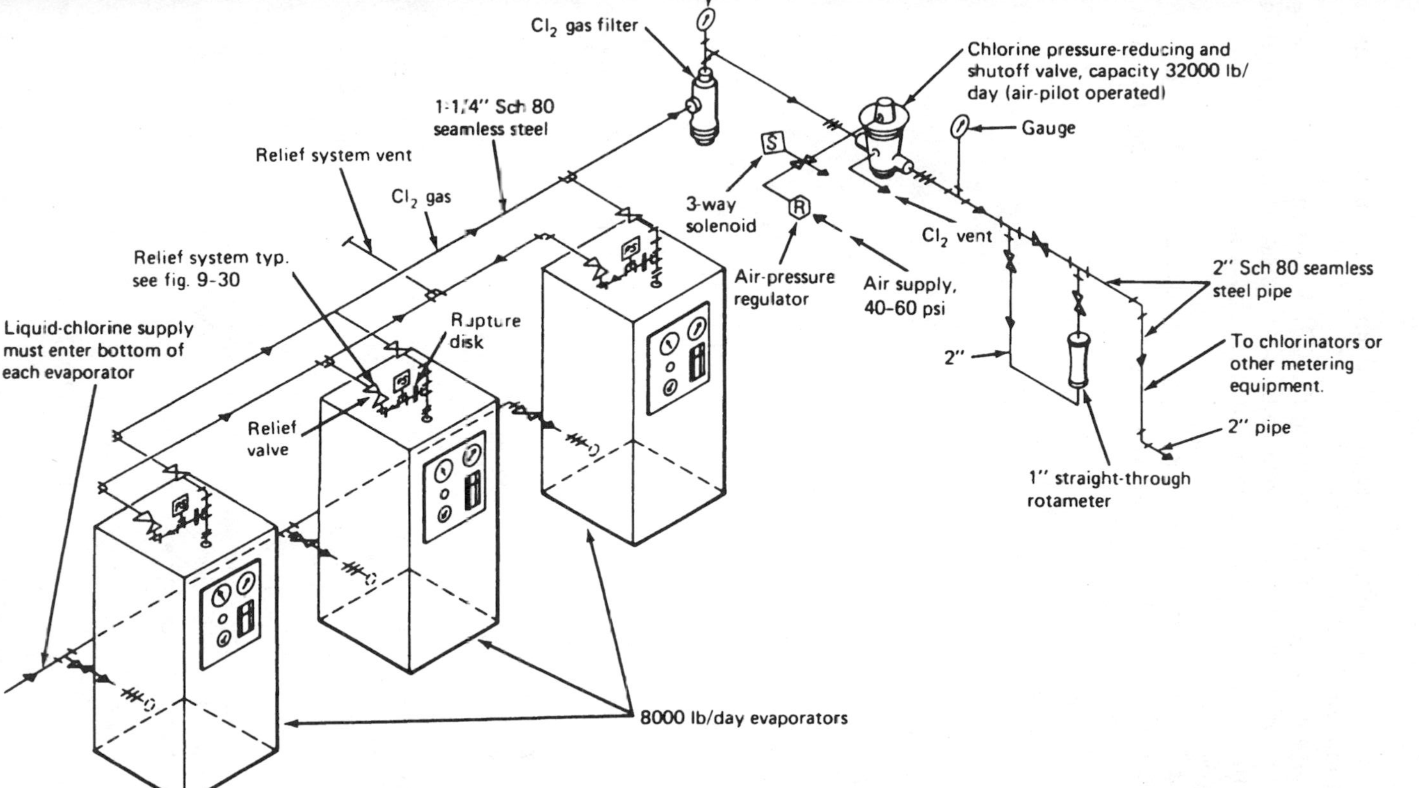

Fig. 9-74. Typical arrangement of manifolding three 10,000 lb/day evaporators supplying three 10,000 lb/day chlorinators. Note: The chlorination equipment such as the evaporators, gas filter, CPRV, and rotameter are all equipped with 1" inlet and outlet pipe fittings but pipe to handle gas flow is considerably larger. This is to reduce friction loss and prevent super-cooling of the gas due to high velocities.

comes more and more difficult to regulate the density through the rotameter in order to maintain its calibration.

The equivalent length of pipe between the CPRV and chlorinator may be as high as 60 ft, even though the two units are no more than 10 to 15 ft in pipe length apart, depending upon the number and kinds of fittings in the piping layout.

Another method of analysis is to use the Reynolds number (N_r), a dimensionless quantity that relates velocity, viscosity, and size of pipe. It can be expressed in the following ways:[21]

$$N_r = \frac{pVD}{u'} = \frac{6.32\ W}{ud} \tag{9-5}$$

where

V = velocity, ft/hr
D = diameter, ft
W = mass flow, lb/hr
d = diameter, in.
p = density, lb/cu ft
u = viscosity, cp
u' = viscosity, lb/hr-ft
$u' = 2.42u$

Taking the case of 24,000 lb/day flow in a one-inch pipe, the Reynolds number is calculated as follows:

$$N_r = \frac{6.32 \times 1000}{0.0127 \times 0.957}$$

The temperature of the gas at this velocity in a one-inch pipe will drop to about 40°F; hence $u = 0.0127$ cp, and therefore:

$$N_r = 654{,}600$$

Now let us make the same calculation for the case where the maximum recommended allowable velocity in a one-inch pipe is two evaporators manifolded together = 16,000 lb/day flow; then:

$$N_r = \frac{6.32 \times 666.7}{0.0129 \times 0.957}$$

The temperature of the gas at this velocity in a one-inch pipe will be about 50°F; hence $u = 0.0129$ cp, and therefore:

$$N_r = 364{,}175$$

Now for the 24,000 lb flow in a 1½ pipe, which shows a velocity under the 50 ft/sec maximum limit, the Reynolds number will be:

$$N_r = \frac{6.32 \times 1000}{0.0131 \times 1.500} = 321{,}630$$

The temperature of the gas at this velocity is estimated to be approximately 60°F.

It would appear that the size of the pipe must be such that the Reynolds number is kept below 365,000 for any piping system between the CPRV and the chlorinator.

CHLORINATORS

Historical Background

Direct Gas Feed. The first gas chlorinators developed ca. 1906–10 were of the direct gas feed type. No matter how ingenious or competent the design, these chlorinators suffered disastrously from flooding by water at the point of application. This occurred in spite of a negative head at the point of application. Maintenance due to corrosion was excessive, to say the least.

The Suck-back Phenomenon. This occurrence is related to the affinity chlorine has for water. If the chlorine feed is shut off or if the feed rate is low enough (<0.25 lb/day), chlorine gas in the feed line adjacent to the diffuser will gradually be absorbed by the water. If this condition persists, a vacuum begins to build up in the feed line, which will eventually pull the water from the point of application all the way back to the chlorine cylinder—in spite of the vapor pressure in the cylinder! This is the suck-back phenomenon, which plagued all direct-feed chlorinators until special corrosion-resistant back-pressure valves were designed to prevent its occurrence.

Development of a Corrosion-Resistant Direct-Feed Chlorinator

Introduction. All of the problems that plagued the previous direct-feed chlorinators, described above, have been practically eliminated by a new model, developed by Wallace & Tiernan, its Series 20-057, marketed in 1989. Figure 9-75 illustrates this unit. It is intended to be used during emergency situations (power failure) or as a standby unit when other units fail or are taken out of service for routine maintenance. These units can also be used to supplement the original systems to meet unusual operating demands.

There has been a long-standing practice of fluxing molten metal by chlorine gas. This application covers many metal fabrication products that require

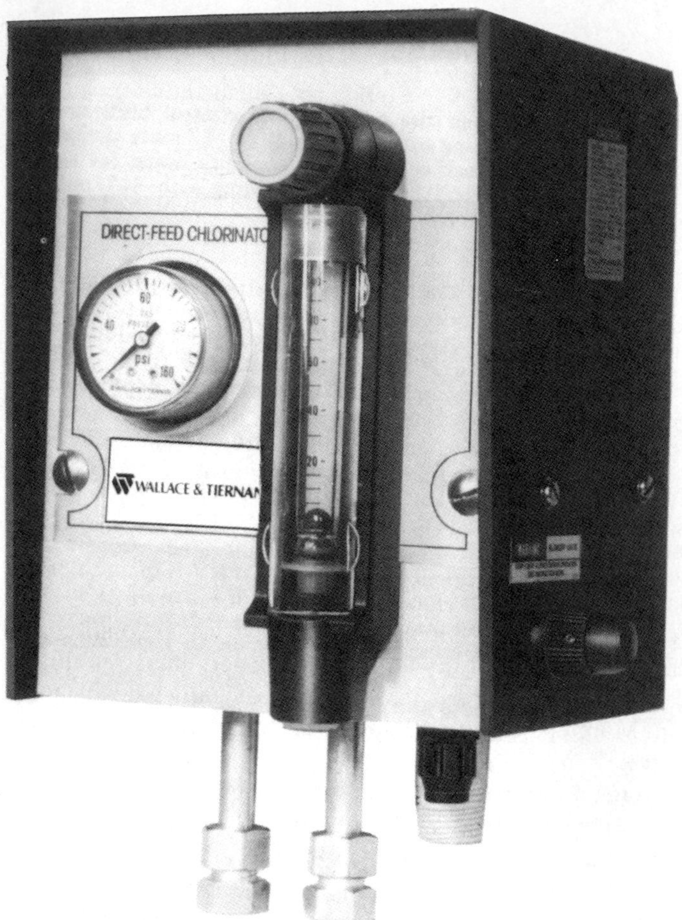

Fig. 9-75. Corrosion-resistant direct-feed chlorinator, Series 20-057 (courtesy Wallace & Tiernan).

highly refined metals such as aluminum, iron, magnesium, manganese, titanium, and so on. Aluminum fluxing is described below. It should be noted that W&T specifically prohibits the use of this chlorinator for swimming pools, for obvious reasons.

System Description. The W&T Series 25-057 is a direct-mounted cylinder unit that reduces the supply pressure to about 35–40 psi. Optional cylinder units automatically switch to a full cylinder when the one on-line runs out. The capacity of the unit is 300 lb/day. The control unit is designed to withstand supply pressures up to 150 psi and is fitted with brackets for wall mounting. This control unit includes a two-tank manifold, pressure relief valve, rotameter, control valve, and pressure gauge. The preferred method of installation is to choose the two-cylinder, scale-mounted automatic switchover system, as shown in Fig. 9-76.

There are two backcheck valves that protect the control unit against flooding. Components of the gas handling system are made of chlorine-resistant alloys or such halogenated polymers as Kynar and TFE. The flow schematic is illustrated by Fig. 9-76.

Points of Application. These units can be used in any pipeline, open tank, well, or channel. The allowable pressure at the point of application is 15 psi

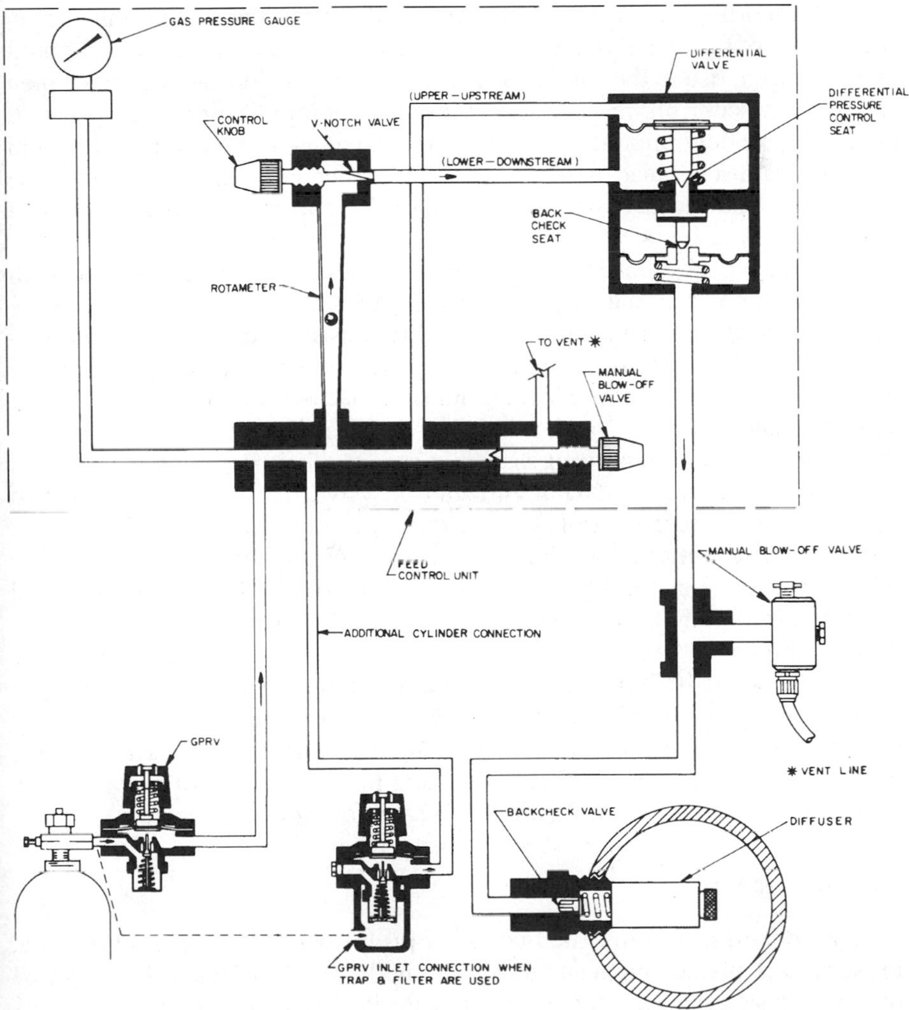

Fig. 9-76. Schematic diagram, corrosion-resistant direct-feed chlorinator, series 20-057 (courtesy Wallace & Tiernan).

using the single cylinder unit and 10 psi with the automatic switchover unit. The chlorine is applied through a porous Kynar diffuser. This diffuser can be cut on-site to meet specific requirements for feed rat and point of application configuration. The Kynar diffuser can be inserted into a standard main connection.

Special Application

Aluminum Fluxing. The use of chlorine in the production of aluminum is common enough to be given mention.

Dry chlorine gas is bubbled into the molten aluminum that is contained in the furnace. Chlorine reacts to oxidize other metallic impurities that are inherent in the process of making aluminum. These metallic oxides, which are composed mostly of iron, rise to the surface as a brown scum. The furnace operator then scrapes this scum off the surface to waste. The application varies, and is entirely dependent upon the operator's judgment as to the color of the scum when each furnace load has received the proper amount of chlorine. The amount of chlorine used varies from 1 to 7 lb of chlorine per 1000 lb of molten aluminum.

When the capacity of the above-described Series 20-057 direct-feed chlorinator is insufficient, the following arrangement of standard chlorination components may be used: If the chlorination facility consists of conventional vaporizing equipment, as described in this chapter, one or more evaporators are manifolded together, with chlorine entering the bottom of each evaporator as a liquid and leaving as a gas at the top. After the chlorine joins in a common discharge, a gas filter and trap assembly are installed immediately, followed by a chlorine pressure-reducing and shut-off valve. This valve is interlocked with the temperature controls on each evaporator, and will shut down the system in the event of power failure or low water temperature. Immediately following this is a metal tube rotameter to monitor the flow of chlorine, and downstream of this is the flow-regulating valve. The chlorine enters each furnace through a bank of "stingers"—the chlorine diffusers. Each stinger can handle about 10 lb of chlorine per hour. There may be up to six or eight stingers on each bank. They have flexible metal hose connections so that they can be raised and lowered into the molten aluminum. On large systems, it is desirable to have a standby nitrogen system to purge the chlorine out of the system at the conclusion of fluxing each batch of molten aluminum.

The Bell-Jar Era

In an all-out effort to solve the two major problems of chlorine gas metering, the suck-back phenomenon and impurities in the gas, C. F. Wallace developed the vacuum solution-feed bell-jar line of chlorinators. This line of chlorinators enjoyed great success from ca. 1922 to ca. 1960. Figure 6-4 illustrates a flow-paced automatic unit—maximum capacity 500 lb/day. The operation of the chlorine pressure-reducing valve was visible inside the bell jar. The vacuum created by the injector was transmitted to the bell jar, where chlorine impurities present were spewed into the glass jar and could be easily retrieved for analysis. The evidence provided by operators of these chlorinators resulted in the production of a special "clean" grade of chlorine for water supplies, wastewater, and cooling waters. The suck-back phenomenon was resolved by the incorporation of the vacuum relief system, which automatically allowed air to enter the chlorinator, thereby breaking any vacuum that might result from suck-back. All solution-feed chlorinators today have vacuum relief systems. The two salient points that were responsible for the success of bell-jar chlorinators were: (1) they were made entirely of corrosion-resistant materials, and (2) the operation was completely visible. Their major disadvantage was feed-rate capacity limitation. For feed rates in excess of 500 lb/day the design became complicated, which resulted in excessive production cost.

Chlorination Equipment Development as of 1997

Wallace & Tiernan V10K System[123]

Preface. This is the latest design of a 500 ppd completely automatic solution feed chlorinator. This chlorinator represents the culmination of the collective experience from Wallace & Tiernan centers around the world. The result is an efficient low-capacity gas feeder with a standardized flexible design that provides the exact configuration for the user's application. See Fig. 9-77.

The V10K System. It consists of a vacuum regulator mounted at the gas supply, a wall-mounted gas control unit with a rotameter for indication of feed rate, and a water-operated injector that provides the vacuum source that drives the entire system. Using automatic switchover regulators, the V10K provides an uninterrupted supply of gas to maintain continuous disinfection treatment. This is accomplished by the use of automatic positioners for the V-notch plug control device.

These units are available in two capacities; the 200 ppd model has its own size feed-gas regulator in combination with a ¾-inch injector, and the 500 ppd model has its own size feed-gas regulator in combination with a 1-inch injector

System Features. The following are the special features of this new chlorination unit:

1. Versatility: It handles all water treatment gases—chlorine, ammonia, sulfur dioxide, and carbon dioxide. For CO_2, owing to its high cylinder pressure, a pressure reducing valve must be used at the exit of the cylinder before it is connected to the V10K unit.
2. Proven accuracy and reliability of the V-notch flow control technology.

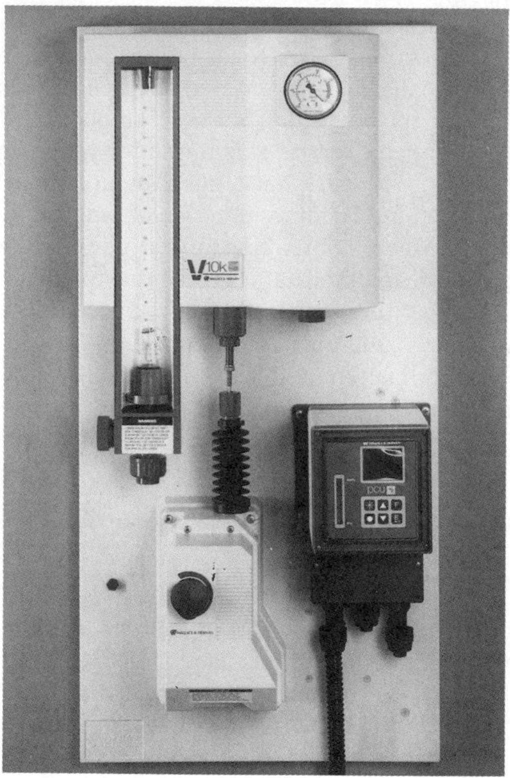

Fig. 9-77. Wallace & Tiernan V10K solution feed chlorinator 500 lb/day.

3. Premium construction: It features a one-piece molded headblock, for reliability and endurance
4. Large 5-inch and 10-inch rotameters available in 13 capacities up to 500 ppd of chlorine to provide the highest degree of readability.
5. Serviceability: Components are easily accessable for servicing without tools
6. Flexible control modes: manual to fully automatic control schemes.
7. Differential-type regulation: This permits lower vacuum levels with efficient and economical injector operation.

For further information please see the W&T Bulletin No. SB 25.100 UA, or call Customer Service, 800-507-9000.

Development of the Spring-Loaded Diaphragm Concept

General Discussion. The famous bell-jar line of chlorinators pioneered by Wallace and Tiernan in the 1920s was replaced virtually overnight by the introduction of PVC injection molding. Once again the original spring diaphragm concept was revived, and its success was made possible by research into plastics and corrosion-resistant diaphragms, particularly by DuPont. The very first series of chlorinators marketed by Wallace & Tiernan was based upon the spring-loaded diaphragm principle (ca. 1913). Owing to severe corrosion problems, this concept was dropped as soon as the visible vacuum models by C. F. Wallace were introduced (ca. 1920). When PVC injection molding and Teflon diaphragms became available, the chlorinator design approach reverted to the original spring-loaded diaphragm principle, utilizing the technology of the newly arrived plastics industry (ca. 1955).

This design concept, pioneered by Fischer and Porter, was a much needed stimulus for the chlorinator industry. This was a small specialty manufacturing industry at that time, but once the ingredient of quality competition was introduced, the results for the consumer were dramatic. To counter the impact of the Fischer and Porter "all plastic" line, Wallace & Tiernan introduced its line of "plastic" chlorinators but with a revolutionary idea of chlorine feed-rate control. Wallace & Tiernan introduced the V-notch orifice concept, which it had been keeping under wraps for at least 10 years. This concept of chlorine metering control proved to be unique. From this point forward Wallace & Tiernan developed unbelievably accurate and flexible control systems. Its success spurred its competitors to greater achievements, some of which are described below.

Theory of Operation. These chlorinators consisted of the following components:

1. Inlet chlorine pressure-reducing valve
2. Indicating meter (rotameter)
3. Chlorine metering orifice
4. Manual feed-rate adjuster
5. Vacuum differential-regulating valve
6. Pressure-vacuum relief valve
7. Injector

Figure 9-78 is a flow diagram of a Wallace & Tiernan V-800 chlorinator illustrating a "pressure" type system* as opposed to the remote vacuum arrangement. The chlorine gas enters the system through the pressure-vacuum regulating valve, at which point the inlet chlorine pressure from the supply system is reduced to some constant level of negative (vacuum) pressure (the level of vacuum varies with each manufacturer). The gas then passes through

*There are a great many of this type of unit in operation. It was replaced in 1989 by the V-2000 Series.

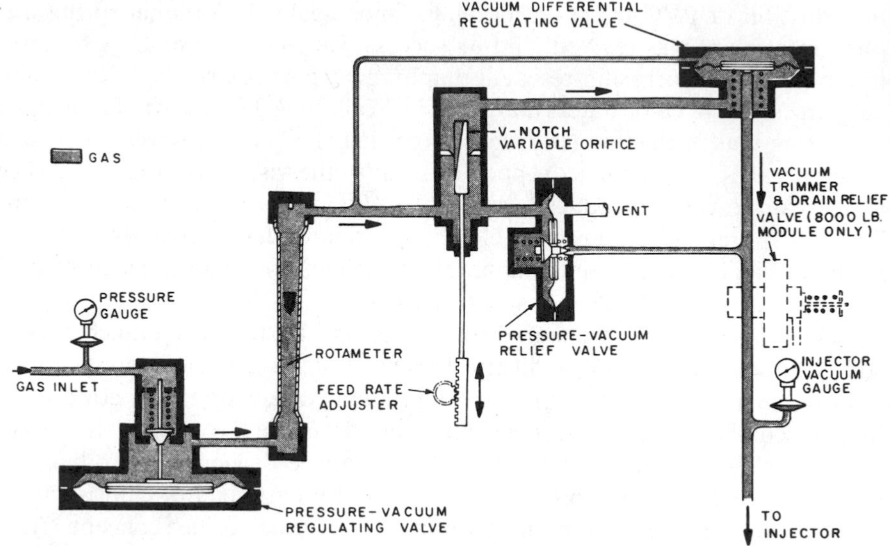

Fig. 9-78. Flow diagram V-800 series vacuum solution feed chlorinator (courtesy Wallace & Tiernan)

the metering orifice (V-notch) to the differential vacuum regulator and then to the injector vacuum line.

The chlorine flow through the V-notch orifice is based upon the classic flow formula $Q = AV$, where A is the area of the orifice opening (position of V-notch orifice positioner), and V is the velocity of the gas through the orifice. This gas velocity is best expressed in terms of the differential pressure across the orifice to produce a given velocity. This is equal to $C\sqrt{2gh}$, where h equals the differential vacuum across the differential vacuum regulator, and C is the velocity coefficient of the V-notch orifice. The pressure-vacuum regulating valve illustrated in Fig. 9-78 is designed to maintain a constant pressure (usually a slight vacuum) upstream from the metering orifice and control device. The vacuum differential regulating valve is designed to maintain a constant pressure drop (h) across the metering orifice (V-notch orifice).

Sonic Flow Concept

The most recent development in chlorinator design that affects accuracy and control modes is the concept of sonic flow. Previously described concepts assumed that the gas flow through the metering orifice (feed-rate valve) was a function of the differential pressure across that valve (h). This is true for a wide range of differential pressures; however, if the velocity through the valve is increased to the speed of sound in the gas flow at that point, a different set of conditions is encountered. Once the sonic velocity is reached, the flow through the valve is no longer a function of the pressure drop (h) across the valve. Under these conditions, gas flow is directly proportional to the area of the opening in the control valve and is entirely independent of the downstream pressure, which is a function of the injector vacuum. Therefore, when sonic flow conditions are attained, the differential vacuum regulator is no longer a necessary component of chlorinator design. Figure 9-79 illustrates a typical sonic flow design by Capital Controls. Sonic flow chlorinators are limited to 2000 lb/day capacity owing to the additional energy required by the injector systems to achieve sonic velocities across the chlorine feed-rate valve.

Wallace & Tiernan S10k Sonic Chlorinator[127]

Introduction. The S10k chlorinator is an all-vacuum-operated chlorinator that is sonically regulated. Direct cylinder mounting puts the vacuum-regulating valve right at the source, reducing the cylinder pressure to a vacuum as soon as the gas leaves the cylinder. It provides for economic low-capacity gas feed applications for municipal and industrial water and wastewater treatment for disinfection, plus treatment of swimming pools and special industrial process waters. (See Fig. 9-80.)

Its ability to handle all the gases used in the treatment of potable water and the disinfection of wastewater, as well as its flexible mounting configurations for cylinders, manifolds, or ton containers, provides versatility for all installations. Two basic arrangements are available in capacities of 200 and 500 ppd of chlorine gas. With its fewer internal parts, one can be confident that the S10k will provide reliable and dependable service.

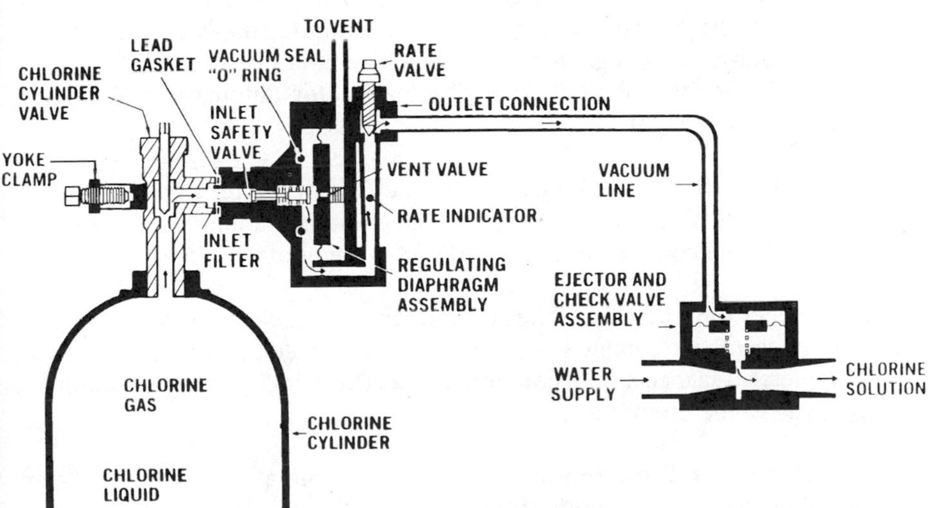

Fig. 9-79. Flow diagram sonic flow type chlorinator (courtesy Capital Controls Co.).

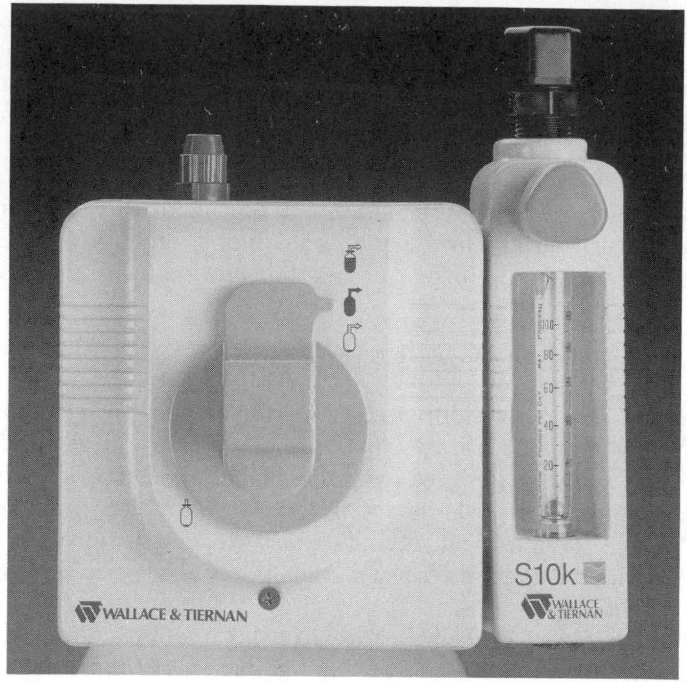

Fig. 9-80. Wallace & Tiernan S10K sonic chlorinator[127] solution feed cylinder, mounted.

Features

Positive Indication of Operating Status. It is equipped with easy-to-read icons that tell the operator the status of the chlorine supply in the cylinder (is it on standby, operating, empty, or off?) The operator can tell at a glance whether the container is in the standby status (for automatic switchover), operating, empty, or off.

Positive Shutoff. An OFF position on the face of the regulator allows for positive shut-off. Containers can be changed without admitting air, dirt, or moisture into the control unit, and without shutting off the injector.

Unique Secondary Check. The 500 ppd regulator includes a special secondary check designed to confine gas under pressure, should the primary valve seat not completely seal because of contamination. This minimizes the possibility of venting gas to the atmosphere.

Built-in Automatic Switchover. This feature eliminates the need for external switching devices. The nonisolating feature allows cylinders to be emptied thoroughly, for complete gas consumption.

Handling of All Water Treatment Gases. The S10k chlorinator can handle all typical potable water and wastewater gases: chlorine, sulfur dioxide, carbon dioxide, and ammonia.

Captive Yoke Mounting. A unique self-aligning captive yoke clamp makes it easy to line up and connect the chlorinator to the gasketed outlet of a container valve. The rugged yoke is designed to the Chlorine Institute recommendation per drawing 189. A similat captive yoke design is used for ammonia cylinders. A ton container kit includes a drip leg, to trap initial spurts of liquid, plus a heater for evaporation and a replaceable filter.

Detachable Rotameters. Two sizes of rotameters, 3-inch and 5-inch, are available in 15 capacities, between 1.2 and 500 ppd chlorine, with comparable capacities for other gases. These rotameters can be integral to the unit or mounted remotely for installation flexibility. Rotameters can be ganged together for multiple points of application.

Easy to Install, Operate, and Maintain. The S10k chlorinator is simple in design, compact, and easy to manage. The injector installation only requires a connection to a water supply and a PVC pipe to carry the chlorine solution to the point of application, whether it be a main or an open channel. A knob adjustment on the rotameter changes the gas feed rate, which is indicated on a high resolution 3-inch or 5-inch scale.

Technical Data

Accuracy. The gas feed is plus or minus 4 percent of the indicated flow.

Operating Range. The manual control operating range is 20 to 1 for any rotameter. The automatic control range is 10 to 1.

Control Modes. Modes are: manual control, start and stop, or program, flow proportional, direct residual, or multiple rate control, plus multiple-point operation.

Distance, Supply to the Control Unit. For flexibility, it is not necessary to install the vacuum regulating valve close to the control unit. The devices can be a few feet to several hundred feet apart, depending upon maximum feed rate, the size of the vacuum piping, and system performance requirements.

Injector Operating Water. This must be clean water without any debris. The fixed-throat differential-type injectors operate on inlet water pressures up to 300 psi at 100°F, and 150 psi to a maximum of 130°F.

Pressure at Application Point. This should never be more than 75 psi when using hose or polyethylene tubing. However, the system can handle pressures up to 160 psi when the injector discharge line is made of PVC piping that will handle that pressure.

Automatic Control. The S10k can handle automatic feed-rate control from simple to complex schemes. The control system consists of an actuator that can be either the SCU (Signal Conditioning Unit) or the PCU (Process Control Unit). Both of these systems are described later in this chapter.

For further information see W&T Publication TJ 25.200UA, or call Customer Service at 609-507-9000.

Remote Vacuum Design

The original concept of this design was to control the chlorine feed-rate system under a vacuum from the supply cylinder to the injector inlet. This approach made it possible to increase the withdrawal rate of a 150-lb chlorine cylinder from a maximum of 40 lb/day at 68°F to 100 lb/day. The design proved so successful that it was eventually applied to ton cylinders. This caused a lot of safety problems because the operators never knew the vapor pressure level in the container vessel. Furthermore, the greater the cylinder pressure was, the more intensive the maintenance required, owing to the greater pressure drop across the vacuum regulator unit. This combination causes a greater dropout of chlorine impurities on the stem and seat of the vacuum regulator unit. The so-called tank-mounted units should never be used on ton cylinders because the risk of a major chlorine vapor release is too great. Some operators have been seriously injured from these units because of an overpressure situation due to fouling of the vacuum regulator check unit.

When properly used, the remote vacuum system offers two advantages. In shutting down an installation using the W&T 10,000 lb model, the entire piping system from the shutoff valve on the supply system to the injector will remain under a vacuum. Therefore, where chlorine tanker trucks are used, these trucks can be drained of any remaining chlorine vapor. This allows the tankers to be immediately refilled without having to purge and dispose of remaining chlorine at the loading station. The other advantage is the ability to adapt to automatic switchover of the chlorine supply system. This is limited to the 500-lb and 2000-lb arrangements.

All of the remote vacuum units, which include the 500-lb, 2000-lb, and 10,000-lb arrangements, should include an external CPRV downstream from the chlorine filter and upstream from the vacuum regulator check unit. This improves system reliability, decreases maintenance, and prevents automatic shutdowns from overpressure events. The vacuum regulator check units of some species will automatically shut down when the supply pressure reaches about 120–125°F. This occurs frequently in warmer climes.

It is to be noted that the W&T 10,000-lb arrangement utilizes a combination vacuum trimmer and drain-relief valve with an antiflood vacuum-breaker, instead of a conventional vacuum relief valve, as shown in Fig. 9-81.[22] The 500-lb and 2000-lb W&T units use a conventional vacuum relief valve with a remote vacuum arrangement, as shown in Fig. 9-82.[22]

It should also be noted that although automatic switchover capability is limited to the 500-lb and 2000-lb arrangements, pressure sensing for automatic switchover is available for any capacity.

Venting Requirements. All chlorinators and sulfonators that operate under a vacuum have vents designed to discharge chlorine to the outside air during an overpressure situation or some other malfunction. Typically there are four types of vents, as follows:

1. The vacuum relief type is used by nearly all chlorinators with injector systems but without remote vacuum. These units have always been vented to the outside air at roof level without any moisture traps.

2. The vacuum regulator check units, such as those used with the remote vacuum, must be vented. They can vent chlorine up to a maximum pressure

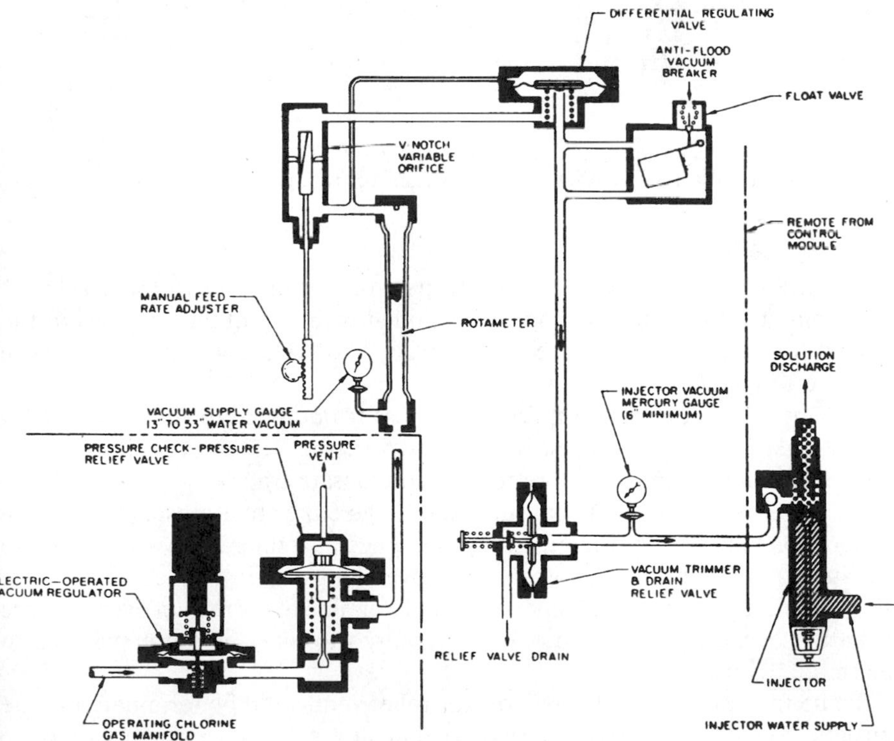

Fig. 9-81. Flow diagram of the remote vacuum 10,000 lb arrangement of the V-2000 series V-notch chlorinator (courtesy Wallace & Tiernan).

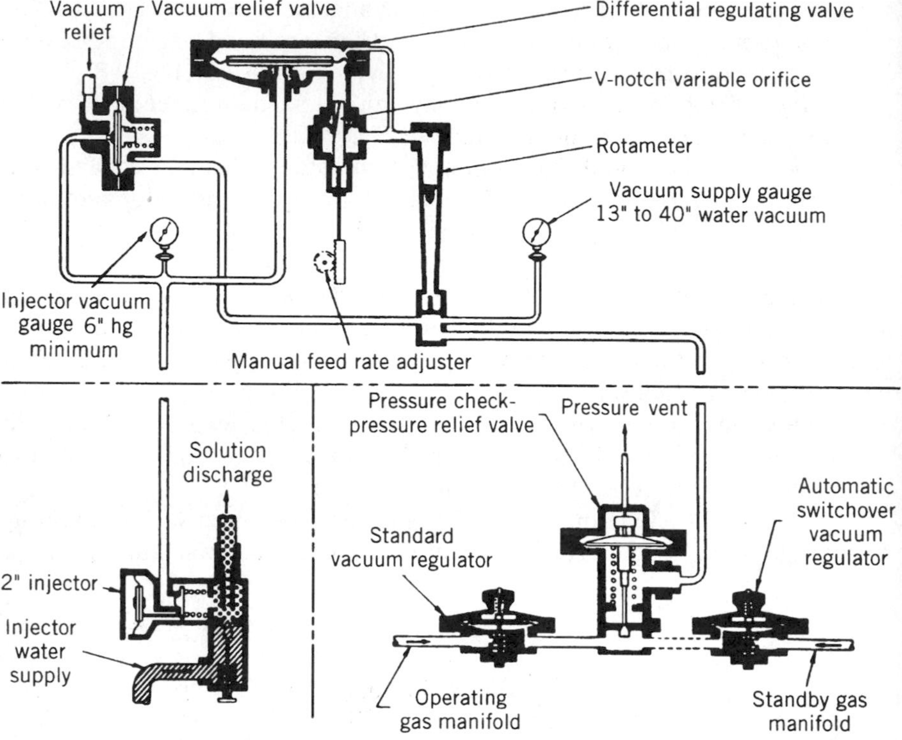

Fig. 9-82. Flow diagram of the remote vacuum 2000 lb arrangement of the V-2000 series V-notch chlorinator with automatic switchover (courtesy of Wallace & Tiernan).

of 2 psi. Such an event is due to some type of malfunction within the VRC. The vents for these units cannot be included with any other vent where the common discharge pressure could exceed 2 psi. This would cause severe damage to the VRC.

3. Most chlorine pressure-reducing valves are equipped to vent any packing gland leakage. This vent must be able to "breathe." Therefore, it must discharge to atmosphere with appropriate condensate traps.

4. The evaporator relief vent must discharge either to atmosphere or to an absorption tank via a barometric loop. Leakage past the relief valve is remote because they are designed to relieve above 500 psi. Before that occurrence, the operating personnel will have had plenty of time to correct an overpressure situation because all installations are required to have pressure switches to alarm at 125 psi.

To be fully protected, the evaporator relief vent could be terminated in the airtight container described above so that any chlorine leakage would be disposed of by the eductor. An equally effective method is to discharge the leaking chlorine into an absorption tank via a barometric loop. There may be multiple discharges into the absorption tank; so the downstream exit of the relief valve has to be protected from corrosion due to leaks from other sources being discharged into the absorption tank. This is accomplished by installing a 400-lb rupture disk in the inverted position on the outlet of the relief valve. In this position the rupture disk fails at about 25 psi.

The approach that uses the airtight container to terminate the vent lines and the eductor (injector) to discharge the escaping gas is sometimes referred to as the "Renton System" because of where the system originated—the Seattle Metro WWTP at Renton, Washington.

Control Strategies*

General Discussion. All chlorinators are arranged for basic manual control. This control is accomplished by opening and closing a valve characterized for a nominal range of 10 : 1. Most manufacturers provide rotameters calibrated over a 20 : 1 range. Chlorinator models are usually separated into capacity categories of 100, 500, 2000, and 8000 lb/day. Some manufacturers stretch the capacity of the 8000 lb/day size to 10,000 lb/day, depending upon certain site conditions.

For each of the models described above there is a variety of metering tubes and other internal accessories available to cover the entire capacity range of each category. For example, for a 2000-lb chlorinator, the available rotameter tube sizes are: 50, 75, 100, 150, 250, 500, 1000, 1500, and 2000 lb/day.

Flow Pacing. This is accomplished by the use of an analog signal transmitted from the process flow meter to the motor drive on the shaft that positions the chlorine metering orifice. This valve is designed and characterized so that the flow of chlorine varies *directly* as the process flow changes. Therefore, the signal transmitted by the process flowmeter must be linear with process flow. This signal can be either electric (4–20 mA) or pneumatic (3–15 psi). In the Wallace & Tiernan chlorinators the chlorine orifice valve is commonly referred to as the V-notch plug. The Fischer and Porter valve is known as the Chloromatic valve. Depending upon local conditions, an optional item for flow-paced control is the use of a ratio relay. This is desirable when the primary meter is oversized for future conditions.

> EXAMPLE: A treatment plant is designed for 50 mgd in the year 2000. Current peak daily demand is only 20 mgd. Install a ratio relay with a range of 0.4–4.0 in order to get the full range of the chlorinator.

When one is considering flow pacing by itself, there is a tradeoff between range and accuracy. A chlorine dosage that is flow-paced will only be as

*See the appendix for control function definitions.

accurate as the transmitted flow signal. The accuracy of this signal depends upon the primary metering element. Designers should specify the accuracy they want over a specified range (i.e., 3 : 1, 5 : 1, 10 : 1). Many makers of primary elements are reluctant to claim ±2 percent accuracy over a range that exceeds 4 : 1. This is a crucial factor in precise chlorine dosage control. It is discussed in further detail below.

Direct Residual Control. For new systems where retrofitting is not a problem, this method has considerable appeal. The key elements that make up a successful operating system are: (1) selection of the proper injector water pump, (2) careful design of the chlorine diffuser and residual sample tap, and (3) a short loop time.

The appropriate way to describe the direct residual control concept is by an example. A 48-inch pipe carries an average daily flow of 12 mgd and summer peaks of 24 mgd. Low flow during the early morning hours is 2 mgd. This represents a 12 : 1 flow range and a velocity range of 0.31–2.96 ft/sec. The first step is the diffuser design, to provide proper initial mixing at the point of chlorine injection. Assume that the chlorine dosage will be a maximum of 2.5 mg/L. This amounts to 500 lb/day at 24 mgd flow. This is the upper limit for a fixed-throat injector; so choose a 2-inch adjustable throat injector capable of passing at least 70 gpm. This is for mixing reasons, as will be shown. Use an across-the-pipe diffuser as shown in Fig. 9-83 with two cluster jets of three holes, each pointing upstream. Using the Egan concept of upstream jet energy for mixing, design chlorine diffuser perforations for 25 ft/sec velocity. This requires six ½-inch holes at 12 gpm per hold for a total amount of injector water of 72 gpm. This diffuser will develop a head loss of 11.75 ft.

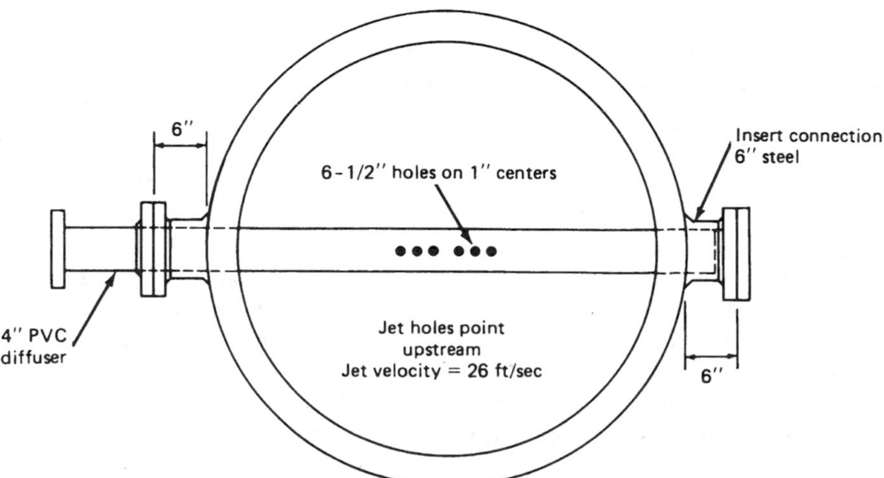

Fig. 9-83. Egan type cluster-jet counterflow diffuser.

The next step is to calculate the G factor:

$$G = \sqrt{\frac{550 \times P}{\mu V}} \tag{9-6}$$

$$P = \frac{Qh}{3960} = \frac{72 \times 11.75}{3960} = 0.21 \tag{9-7}$$

In order to calculate G, the volume of the initial mixing area affected by the energy of the jet will have to be decided. Experience in wastewater chlorination indicates that mixing will occur in 2–3 sec. The longer the time selected for mixing, the lower the G value will be. Assume 3 sec at average flow, 12 mgd, where $V = 1.48$ ft/sec. This means the mixing volume is 55.77 ft^3. Then:

$$G = \sqrt{\frac{550 \times 0.21}{2.35 \times 10^{-5} \times 55.77}} = 296.86$$

The next step is to calculate G for the dissipation in energy caused by the counterflow—this is called the Egan effect. Referring to Fig. 9-83, it is assumed that the obstruction of the 3-inch diffuser with the two cluster jets and the impinging effect of the cluster jets will dissipate some of the energy in the 12 mgd counterflow. The diffuser restriction increases the velocity of the mainstream from 1.48 ft/sec to 1.65 ft/sec and $V^2/2g = 0.04$ ft. Assume a head loss of 0.04 ft in the mainstream at this point. Then:

$$P = \frac{8343 \text{ gpm} \times 8.34 \times 0.04}{60 \times 550} = 0.08 \text{ hp}$$

and:

$$G = \sqrt{r\,\frac{550 \times 0.08}{2.35 \times 10^{-5} \times 55.77}} = 183.23$$

Therefore it is reasonable to believe that the total energy dissipated can be translated into a G factor equal to the jets, 297, plus the counterflow, 183, for a total $G = 480$. This amount of turbulence is sufficient for adequate mixing.

Sampling Tap. The downstream sampling tap, which is also the injector pump suction, should be a duplicate of the chlorine diffusers with the holes pointing upstream, except that the number of holes must be increased to reduce head loss to one foot. This arrangement will ensure that the sample received by the analyzer will be consistently homogeneous. Moreover, using the injector pump in this fashion will tend to short-circuit the main line flow in such a way as to prevent abnormally low velocities during low flow periods. (See Fig. 9-84.)

Loop Time. Estimating the loop time (system response) is the next step. Referring to Fig. 9-84, the loop-time circuit is defined as *A-B-C-D-E*. The travel time through all of these segments will be constant by design except for *A-B* and *C-D*. These segments of the loop will vary with the velocity of chlorine in the vacuum line and that of water in the 48-inch pipe from low flow to peak flow conditions.

From Fig. 9-84 use the following distances:

$A\text{-}B = 300$ ft
$B\text{-}C = 30$ ft
$C\text{-}D = 10$ diam. $= 40$ ft
$D\text{-}D_1 = 10$ ft
$D_1\text{-}E = 15$ ft

Injector pump flow by pump selection and control valve will be 72–75 gpm. Sample flow to the analyzer pump suction will be 1 gpm, and pressure in 48-inch pipe is estimated at 40 psi.

The next step is to calculate the flow in each of the segments:

1. *A-B*, chlorine gas flow at 500 lb/day max. Try a 1-inch PVC vacuum line:

$$W = 500 \text{ lb/day} = 0.0058 \text{ lb/sec}$$

Density of Cl_2 gas at injector vacuum of 20 inches of Hg = 0.06 lb/ft^3. So:

$$Q = \frac{0.0058}{0.06} = 0.0967 \text{ cfs}$$

and:

$$V = \frac{0.0967}{0.005} = 19.34 \text{ ft/sec}$$

Now calculate for friction loss to see if 1-inch pipe is large enough:

$$\Delta P = \frac{11.89 \times L \times f \times (W)^2}{10^9 \times \rho \times d^5} = \text{in. Hg}$$

where

ΔP = friction loss in vacuum line, in. Hg
W = lb/day Cl_2 or SO_2
L = equiv. length of vacuum line, ft

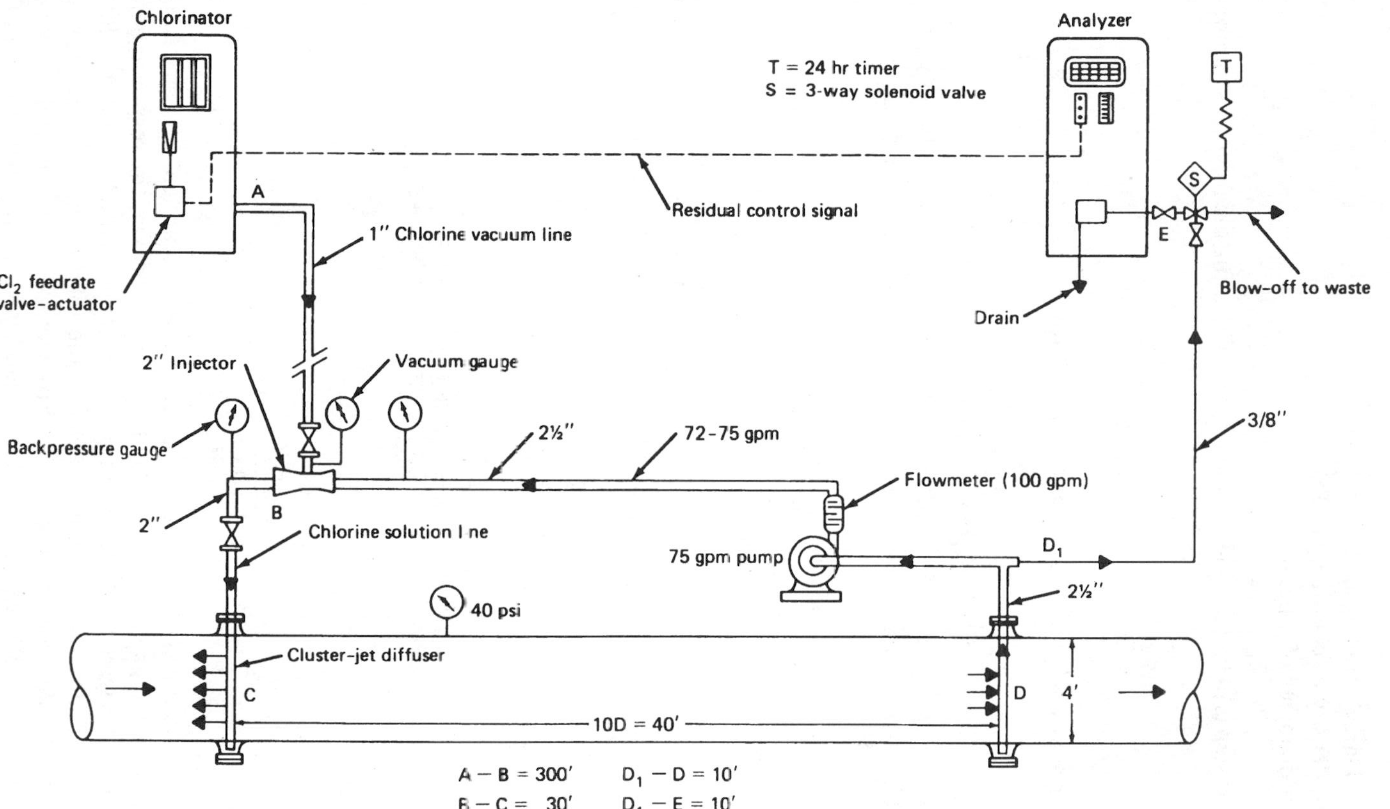

Fig. 9-84. Direct residual control system using injector water pump to provide analyzer sample, which minimizes control loop dead time.

f = Darcy's friction factor
ρ = density of chlorine gas, lb/ft^3
d = diameter of pipe, in.

The next step is to find the friction factor f by calculating the Reynolds number:

$$N_R = \frac{6.32 \times W}{u \times d \times 24}$$

where u = viscosity of chlorine gas = 0.0133; so:

$$N_R = \frac{6.32 \times 500}{0.0133 \times 0.957 \times 24} = 10345$$

From Fig. 9 in the appendix, the friction factor $f = 0.07$. Therefore:

$$\Delta P = \frac{11.89 \times 300 \times 0.07 \times (500)^2}{10^9 \times 0.06 \times (0.957)^5}$$

$$= 1.3 \text{ in Hg.}$$

Therefore, a 1-inch line is big enough. Thus at 500 lb/day the gas velocity will be about 19 ft/sec, and at low flow, 40 lb/day, the vacuum will be higher and the gas less dense:

$$Q = 40 \text{ lb/day} = 0.00046 \text{ lb/sec}$$

and:

$$Q = \frac{0.00046}{\rho} = \frac{0.00046 \text{ lb/sec}}{0.03 \text{ lb/ft}^3} = 0.01543 \text{ cfs}$$

So:

$$V = \frac{0.01543}{0.005} = 3.09 \text{ ft/sec}$$

Therefore, the travel time for the chlorine in the vacuum line will vary from 100 sec at a 40 lb/day rate to 8 sec at 500 lb/day. However, the response time following a dosage change occurs at a rate of 50 percent change in the first 25 percent of the time required to achieve the total dosage change of, say, 100 sec at 40 lb/day chlorine feed rate. For example, if the controller called for a 5 lb/day rate change at the 40 lb/day feed rate, the response time in the vacuum line to change the feed rate to 42.5 lb/day would be about 25 percent of 100 sec or 25 sec. This demonstrates the necessity to keep the chlorinator

Table 9-11 Loop Time

	Loop Segment	*Low Flow (2 mgd, seconds)*	*Peak Flow (24 mgd, seconds)*
A-B	Cl_2 Vacuum line	100	8
B-C	Cl_2 Solution line	1	1
C-D	48 In. pipeline	160	14
D-D_1	Sample line, pump suction	3	3
D_1-E	Sample line to analyzer	6	6
Total loop time		270	32*

*Response time for the sample within the analyzer is about 15–20 seconds.

module as close to the injector as possible. *It also demonstrates that remote vacuum and automatic switchover systems should be designed so that the vacuum line between the chlorine supply system and the chlorinator will be kept to a minimum,* (i.e., not more than ± 15 ft).

2. *B-C.* This is the chlorine solution line and diffuser segment. The injector should be as close as physically possible to the diffuser inlet. Assume a distance of 10 ft. At 72–75 gpm the velocity in a 2-inch pipe will be about 8 ft/sec and will cause a head loss of about 1.5 ft. The response time for this segment will be negligible—about 1 sec.

3. *C-D.* This is the 40-ft length of 48-inch pipe between the chlorine diffuser and the pump suction, which is also the sampling tap. At peak flow rate (24 mgd) V = 2.96 ft/sec and travel time to sample tap = 14 sec; at low flow (2 mgd) V = 0.25 ft/sec, and travel time will be 160 sec to the sample tap.

4. D-D_1. This is the segment where the sample is carried in the pump suction which is a 3-inch pipe. Assume this is 10 ft. At 75 gpm, V = 3.4 ft/sec; so travel time is a constant 3 sec.

5. D_1-E. This segment is the sample line to the analyzer from the injector pump suction. The flow requirement to the analyzer is only 1 gpm maximum. Therefore, use a ⅜-inch PVC pipe (brass or copper not allowed). The velocity in this pipe will be 1.67 ft/sec at 1.0 gpm. Assume this pipe is 10 ft long. Travel time will be a constant 6 sec.

Table 9-11 summarizes the response time elements. This represents a 9:1 change in response rate for a 12:1 flow treatment range. It is obvious that a great improvement could be made if the chlorinator could be moved to shorten the vacuum line to a reasonable distance of 10–20 ft. However, this wide range of response can be handled by a proper control system. It is recommended that the upper limit of response time for low flow conditions never exceed 5 min.

An embellishment would be to use an alternative sample point for flows less than 5 mg/day, located 20 ft downstream from the diffuser. This is usually unsatisfactory because the sample tends to be stratified with various levels of chlorine residual. Moreover, because this is an exercise in direct residual control, the above concept requires a flow meter to switch the valves on the two pump suction points.

The control system shown in Fig. 9-84 requires the electric valve operator for the chlorinator feed rate to be able to change the feed rate from zero to maximum in at least 15 sec. The analyzer is required to have a 4–20 mA transmitter, which will provide a linear signal from zero to full scale residual over a specified range of 0–1, 2, 5, 10, or 20 mg/L chlorine residual. This signal is transmitted to an adjustable residual setpoint controller with proportional, integral (reset), and derivative control functions. This controller drives the chlorine feed-rate valve. Potable water chlorination systems have been known to operate well on a 0.1 mg/L dead band where the residual span is 0–2 mg/L. One wastewater plant has operated successfully with a dead band of 0.2 mg/L where the residual span is 0–5 mg/L.[23]

A most important requirement in the operation of continuous residual analyzers is the necessity to purge the sample line at frequent intervals in order to scour away the biological slimes that are prone to build up in the sample lines. Referring to Fig. 9-84, a sample blow-off line controlled by a timer is shown. The purging cycle should be once every 24 hours for 2–3 min. The orifice on the downstream side of the ⅜-inch solenoid valve should be a ⅜ × ¼ reducer ending in about 6 inches of ¼-inch pipe.

The injector water pump for the system shown will be required to have a TDH of approximately 60 psi to provide 100 psi inlet pressure to the injector.

Compound Loop Control. This control strategy was first introduced in 1960 by Wallace & Tiernan (U.S. Patent No. 2,929,393). The control scheme is based upon two separate but independent signals, so that each can change the chlorine feed rate separately. One signal represents process flow changes, and the other registers chlorine residual changes. The first chlorine residual control for a wastewater plant utilized this system.[79] The installation was at the Napa Sanitary District treatment plant, Napa, California in 1961. Since that time there have been several hundred such installations. However, in 1984 Wallace & Tiernan elected to drop this system in favor of the remote vacuum method. As the compound loop system is being used extensively, and it can be made available on special order, the following discussion is deemed worthwhile.

The difference between this method and other systems using two signals is as follows: each signal controls a different component within the chlorinator, and each performs an independent chlorine feed-rate change by a different function of gas flow mechanics. Other systems using dual signals integrate the two signals by external instrumentation which transmits a single multiplied signal to the chlorine feed-rate valve. (*Exception:* The Fischer and Porter Chloromatic feed-rate valve contains built-in multiplier instrumentation.)

As illustrated by Fig. 9-85, the process flow signal is transmitted to the motor-operated chlorine feed-rate valve, and the chlorine residual signal goes to the vacuum differential regulating valve. This concept is based upon the fundamental law of fluid mechanics: $Q = AV$, where Q = chlorine flow, A = area of the chlorine orifice opening, and V = velocity through the orifice. The velocity through the orifice follows the following law of fluid mechanics: $V = \sqrt{2gh}$, where h is the head loss required to initiate a gas flow change. In this method h is the vacuum differential across the chlorine orifice which is controlled by the vacuum differential regulating valve.

Chlorine Feed-Rate Valve. This component is described by Wallace & Tiernan as the V-notch orifice positioner. The motor drive is capable of changing the feed rate from zero to maximum in 15 sec. There is an internal feedback loop that controls and verifies valve travel for any signal from 4 to 20 mA. This valve is controlled by the signal from the primary flow meter.

Vacuum Differential Regulating Valve. This valve changes the chlorine feed rate by changing the vacuum differential across the chlorine rate valve orifice. On Wallace & Tiernan equipment the vacuum range is from 8 to 88 inches of water, zero to maximum feed rate. This vacuum level can be controlled by a vacuum transmitter capable of changing the vacuum over the complete range in about 5 sec. This vacuum level can also be controlled by a motorized vacuum valve, which has a much slower vacuum rate change, approximately 15 min. from 8 to 88 inches of water. This provides floating control by a cycle and duration timer.

Table 9-12 illustrates the mathematics of variable vacuum control as a function of chlorine feed rate. This is informative because it is recommended procedure to adjust the vacuum level of the vacuum regulator in the midrange or 50 percent of valve travel. This is shown as 28 inches of water vacuum.

Theory of Operation. Figure 9-85 illustrates a potable water chlorination system that was retrofitted from flow pacing control to compound loop control using a programmable controller to perform the intelligence functions. The programmable controller for each chlorinator installation requires the control strategy to be designed for specific field conditions. This particular system described by Fig. 9-85 had the following characteristics:

1. The chlorinator was located at the outlet of a reservoir where reverse flow occurred on a daily basis.
2. The waterflow meter in a 60-inch concrete pipe had a range far exceeding any practical use, 0–200 cfs. The actual flow range in the high demand season was from 10 to 120 cfs—a 12:1 range.
3. The piping configuration was such that upstream supplies entering this pipeline occasionally contributed some measurable chlorine residual.

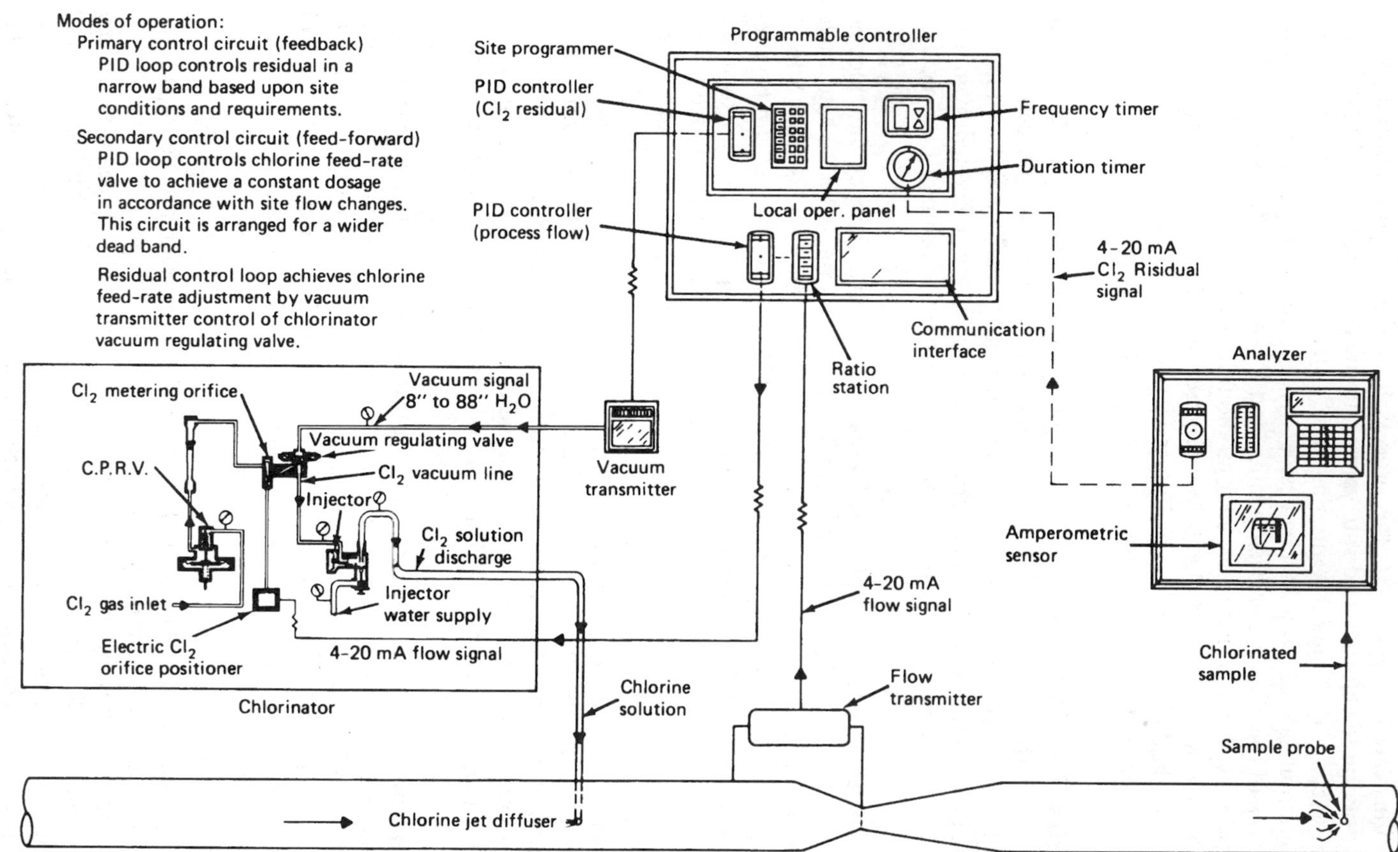

Modes of operation:
Primary control circuit (feedback)
PID loop controls residual in a narrow band based upon site conditions and requirements.
Secondary control circuit (feed-forward)
PID loop controls chlorine feed-rate valve to achieve a constant dosage in accordance with site flow changes. This circuit is arranged for a wider dead band.
Residual control loop achieves chlorine feed-rate adjustment by vacuum transmitter control of chlorinator vacuum regulating valve.
Programmable controller
Site programmer
PID controller
(Cl_2 residual)
Frequency timer
Duration timer
Local oper. panel
PID controller
(process flow)
4-20 mA
Cl_2 Risidual
signal
Communication
interface
Ratio
station
Analyzer
Cl_2 metering orifice
Vacuum signal
8″ to 88″ H_2O
Vacuum regulating valve
C.P.R.V.
Cl_2 vacuum line
Injector
Vacuum
transmitter
Cl_2 solution
discharge
Amperometric
sensor
Injector
water supply
4-20 mA
flow signal
Cl_2 gas inlet
Electric Cl_2
orifice positioner
4-20 mA flow signal
Chlorinated
sample
Flow
transmitter
Chlorine
solution
Chlorinator
Sample probe
Chlorine jet diffuser

Fig. 9-85. Compound-loop residual control system using a programmable controller.

Table 9-12 Arithmetic of Variable Vacuum Control Wallace & Tiernan Equipment

Chlorine Feed Rate (%)	*Chlorine Flow as a Factor*	*Vacuum Differential* $= Cl_2 \times$ *Flow*	*Chlorine Flow as a Factor* $= Cl_2$ *Flow*^{2}	*Vacuum Signal to Chlorinator (in.)* H_2O *(Variable Vacuum + Fixed Vacuum = Total)* *Flow Factor × Range*	*Fixed Vacuum Zero Signal*	*Vacuum Gauge in.* H_2O
100	1.00	100	1.00	80	8	88
90	0.9	81	0.81	64.8	8	72.8
80	0.8	64	0.64	51.2	8	59.2
70	0.7	49	0.49	39.2	8	47.2
60	0.6	36	0.36	28.8	8	36.8
50	0.5	25	0.25	20.0	8	28.0
40	0.4	16	0.16	12.8	8	20.8
30	0.3	9	0.09	7.2	8	15.2
20	0.2	4	0.04	3.2	8	11.2
14.3	0.143	2	0.02	1.6	8	9.6
*						
10	0.1	1	0.01	0.8	8	8.8
0	0	0	0	0.0	8	8.0

*Practical limit of usable range = 7:1.

Owing to the above characteristics it was decided to use the PC in the following manner:

1. Primary control was a PID feedback loop between the chlorine residual analyzer and the vacuum transmitter. This loop controlled the chlorine flow through the vacuum differential regulator. This control loop was designed with a narrow dead band.

2. Secondary control was achieved by a second feed-forward PID loop, which controlled the chlorine feed-rate valve to provide a constant dosage (lb Cl/cfs) all in accordance with water flow information from the flow meter. This loop was designed to operate on a wider dead band. This loop was only for abrupt flow change corrections. If an alarm condition occurred for X seconds or minutes, the feed-rate valve motor would open or close for N seconds.

3. If the chlorine residual level enters the PID alarm zones for X seconds, the mathematical constant that converts the water flow rate to chlorine dosage (lb Cl/cfs) units is changed, causing a greater or lesser dosage when the feed-forward loop operates. This dosage computation change protects the system from wide variations of chlorine demand due in part to chlorine residuals in the system upstream from the chlorine diffuser.

4. The reverse flow sensor system shuts down the chlorinator for a specific length of time to allow for flow oscillation to cease. The chlorinator is restarted when steady outflow occurs. Reverse flow detection can be performed by the

use of an auxiliary analyzer or a thermal probe sensing system. See "Reservoir Outlets" in this chapter.

5. The loop time for the chlorination system was determined in the field. It was 2 min. 25 sec at 47 cfs. Deducting the travel time from *C* to *D* at 47 cfs, the loop time constant calculated to 1.62 min. From this loop time characteristic the correction cycle frequency exhibited a variation from 2.0 min. at 120 cfs to 5.5 min. at 10 cfs and up to 11 min. at flow reversal conditions.

6. The next step was to determine the required length of correction duration for an arbitrary dead-band setting of 0.25 mg/L on the residual control PID loop. Using a 1000 lb chlorine rotameter and allowing each correction to be one-half the dead band or 0.125 mg/L, the chlorine feed-rate change required for each correction from 5 to 120 cfs flow is shown in Table 9-13. Since the chlorine feed-rate valve response is a constant at about 50 lb/sec, it was decided to use a wider dead band for flow pacing control so that at about 30–40 cfs flow the chlorine feed-rate adjustment would be taken over by the residual control loop, where the feed-rate change can be as low as 10 lb per correction. Table 9-13 itemizes the chlorine feed-rate change per correction based upon one-half the dead band change in dosage = 0.125 mg/L.

Using the loop time information developed on-site, the PC can be programmed with mathematical constants so that the chlorine feed-rate change will respond quickly and accurately to wide flow fluctuations and chlorine demand changes. The use of dual PID loops provides greater flexibility over the wider range of conditions. Assuming that the PID flow pacing loop has a ±3 percent accuracy over a 4:1 flow range and the residual control loop has at least a 7:1 range, a worst-case situation is provided with at least a 28:1 range—far in excess of the normal necessity. Other features of this type of

Table 9-13 Chlorine Feed-Rate Change per Correction

Water Flow			
cfs	*mgd*	*Cycle (min)*	*Cl_2 Flow Change (lb/day)*
120	77.5	2.0	80.60
110	71.1	2.0	73.94
100	64.6	2.0	67.18
90	58.2	2.5	60.53
80	51.7	2.5	53.77
70	45.2	2.5	47.01
60	38.8	2.5	40.35
50	32.3	2.5	33.54
40	25.8	3.0	26.83
30	19.4	3.0	20.18
20	12.9	3.5	13.42
10	6.5	5.5	6.76
5	3.2	9.5	3.33

dual signal control include the ability to isolate and operate manually each of the loops, either locally or remotely in case of analyzer failure.

The most important feature is the automatic takeover by the residual control loop when the process flow drops below the accurate range of the flowmeter. More than 35 years of operating experience has demonstrated that this method of dual signal feedback control is a favorite with operating personnel because it is easy to adjust, calibrate, maintain, and understand.

When restarting the system or when recalibrating, the operator manually positions the vacuum valve to read 28 inches H_2O vacuum. This is followed by adjusting the chlorine feed-rate valve to give the appropriate chlorine dosage. The PC allows site-specific corrections and adjustments to account for loop time and historical flow patterns. Dual adjustable dead bands provide more flexibility and greater reliability of the control system.

Dual Signal Control Systems

Introduction. This type of control strategy employs the use of two separate signals multiplied together to produce a single output that controls the chlorine feed-rate valve within the chlorinator. This is the method used by the three most active chlorinator manufacturers. This strategy is described in various ways, including compound loop control. It is not, however, the same as the compound loop control system described above.

In order to compare the systems offered by the different manufacturers *a common chlorination problem in potable water treatment was presented to them.* The strategies described for each manufacturer were based upon the following field conditions:

A 60-inch reinforced concrete pipe, served by a balancing reservoir, has an existing chlorine diffuser and a residual sample tap already in place. A venturi meter with a dp transmitter that generates a 4–20 mA output signal is located nearby. The flowmeter range is 0–200 cfs; however, historical records show that maximum flow has never exceeded 120 cfs. Low flows are on the order of 5 cfs owing to almost flow reversal conditions.

The loop time at 47 cfs has been established as 2 min. 25 sec. The distance from the chlorine diffuser to the residual sample point is 116 ft. Chlorine dosage is approximately 0.75 mg/L. The flow signal is not considered reliable under 50 cfs. Assuming one correction per loop time, frequency of correction will vary from 2 min. up to 7 min.

Chlorine Residual Analyzers

ATI—Series A15[121]

Introduction. This monitor is based upon a polarographic membrane sensor that measures residual chlorine directly. In most cases this is accomplished without the need for sample treatment of any kind. The result is an operating

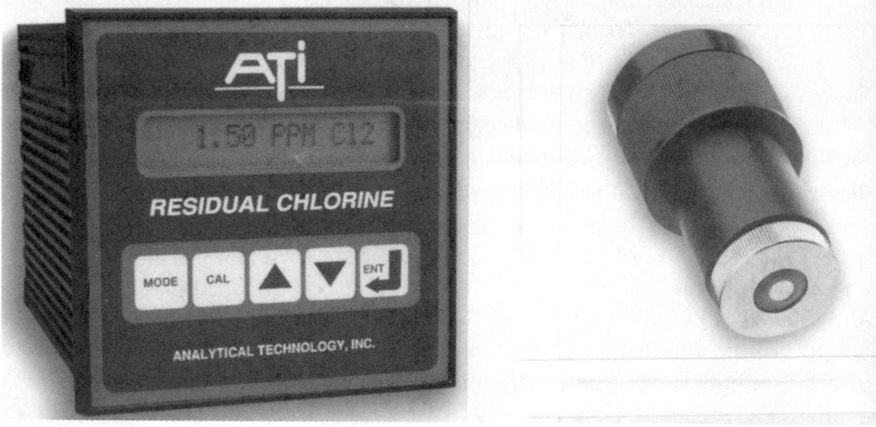

Fig. 9-86. ATI Residual Cl_2 Monitor Series A15[121] (courtesy Analytical Technology Inc.).

system that provides dependable measurements continuously in potable water systems. This system is illustrated by Figs. 9-86 and 9-87.

System Description:

Electronic Monitor. This is an electronic package that provides a real time display of chlorine concentration, control outputs, an isolated analog output,

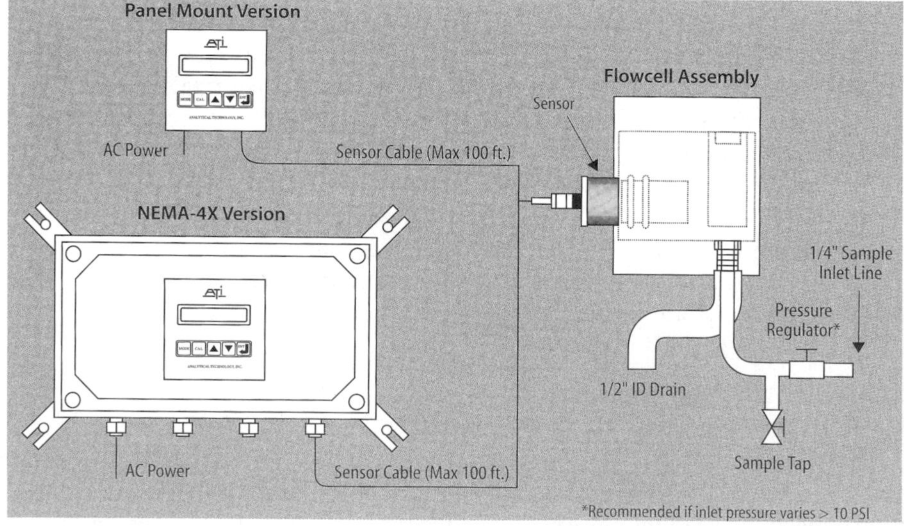

Fig. 9-87. Schematic for ATI residual chlorine monitor.

and an external alarm. The monitor is housed in a compact ¼ DIN panel mounted in an aluminum enclosure. It is also available in a corrosion-resistant NEMA 4X Fiberglass enclosure.

There are five switches located on the front panel that provide access for operators to monitor programming functions. This panel also allows the operators to view such information as sample temperature, alarm setpoints, and analog output value. An access code number is required in order to change any of the calibration or set-up parameters, thus protecting the system from unauthorized tampering.

Series A15 monitors provide two selectable display ranges: 0 to 2 mg/L or 0 to 20 mg/L. An isolated 4–20 mA output is provided for external recording or data logging. This output may be programmed for any required span, using the front switches. It may also be changed to 0 to 20 mA and used with a shunt resistor to provide zero-based voltage outputs, if required.

Contact outputs provided in the monitor include two programmable control relays with variable dead band and variable time delay functions. Control relays may be programmed on/off, pulse frequency, or pulse duration modes of operation for chemical feed control. An additional alarm relay is provided that will actuate on either low or high chlorine condition, or on the failure of the control system to maintain a proper chlorine residual concentration. All variables such as alarm or control setpoints, deadband, and time delay are programmable from the front panel keys.

Residual Chlorine Sensor. Residual chlorine measurement has traditionally been accomplished by injecting chemicals into the sample and measuring the products of a chemical reaction. This often involved measuring the amount of iodine released from the potassium iodide added to the sample. ATI's chlorine sensor eliminates the need for chemical addition by directly measuring the chlorine in the process solution.

The sensor consists of a pair of electrodes immersed in a conductive electrolyte and isolated from the sample by a chlorine-permeable membrane. The chlorine residual migrates through the membrane in a process called "diffusion." This reduces the chlorine to chloride on the surface of the working electrode. This process causes a flow of electrons through an external measuring circuit, with the current flow being linearly proportional to the chlorine concentration. The absolute response of the sensor is also temperature-dependent, and an RTD in the sensor provides a temperature input to the electronics assembly to allow for automatic temperature compensation.

Chlorine residual sensors have been developed for two different measurements. One version is optimized for the measurement of free chlorine (85 percent of HOCl). This sensor is best suited for free chlorine measurements in potable water, cooling towers, and industrial wash-waters. It can be used for monitoring chlorine concentrations from 0 to 200 mg/L.

The second version of the sensor is used for the measurement of choramine residuals. Therefore, this version is well suited for chloraminated potable water supplies and high quality (secondary) wastewater effluents that require disinfection.

Free chlorine sensors should always be used with a flowcell assembly and may require CO_2 buffering if the pH is above 8. Chloramine sensors may be used either submerged or in a flowcell. Submerged sensors require flow velocities of at least 0.4 ft/sec for proper operation.

Flowcell Assembly. Residual chlorine sensors provide the best sensitivity and stability when the sample flow is controlled. Therefore, a constant-head overflow assembly is provided to control both flow rate and water pressure in the area of the membrane.

The sample inlet flow and pressure from the system being monitored can vary widely without any effect upon the measurement. The only requirement is that the inlet flow rate be kept above the minimum flow of about 7 gal/hr.

To allow total observation of the sensor condition so that any fouling can be easily seen, it is made of clear acrylic material.

Special Features:

1. Display: 16 character alphanumeric liquid crystal display with backlight.
2. Sensitivity: 0.001 mg/L above 0.020 mg/L.
3. Response time: 90 percent in less than 60 sec.
4. Repeatability: plus or minus 0.05 mg/L.
5. Control relays: two SPDT, 5A, 239 V AC resistive, with programmable setpoints.
6. Control relay function: programmable on/off, pulse with modulation (PWM), or pulse frequency modulation (PFM).
7. Power: 120/230 V AC, 50/60 Hz, 5 V A max.
8. Operating temperature: electronics −20°C to +50°C.
9. Sensor materials: noryl and stainless steel.

For Customer Service, phone 800-959-0299.

Capital Controls Co.[117]

Introduction. In recent years this company has advanced its activities worldwide with a comprehensive list of disinfectant feed and generating equipment. Its new Captrol® Series 1450 and 1451 controllers are a series of microprocessor controllers designed specifically to handle the feed of chemicals in potable water and wastewater applications.

Model 1450. This is a special-purpose controller designed for use with a Capital Controls automatic gas feeder valve. The controller can be located remotely from the feed equipment, or attractively mounted in floor or wall cabinets. Standard features include:

1. Three chlorination and two dechlorination control modes.
2. Built-in multiplier for feed-forward dechlorination control modes.
3. Automatic transfer from compound to single loop control, when residual or flow signal is lost.
4. Bumpless transfer between manual and automatic control.
5. Digital displays.
6. Alarm indicators.
7. Alarm contacts.
8. Control switch inputs.
9. Gas flow output signal.

Please see Fig. 9-88.

The controller automatically positions the control valve to maintain the desired gas feed rate, or the chlorine residual, which is based upon electrical input signals from a flow meter or a chlorine residual analyzer.

A 4–20 mA/DC output signal that represents gas flow can be used as input to computer control systems for ratio control of chloramination and dechlorination systems, or for remote indication, recording, or totalizing gas flow.

Also the controller is field-adjustable for chlorination, such as: flow proportioning, residual, or compound loop control for dechlorination; or flow propor-

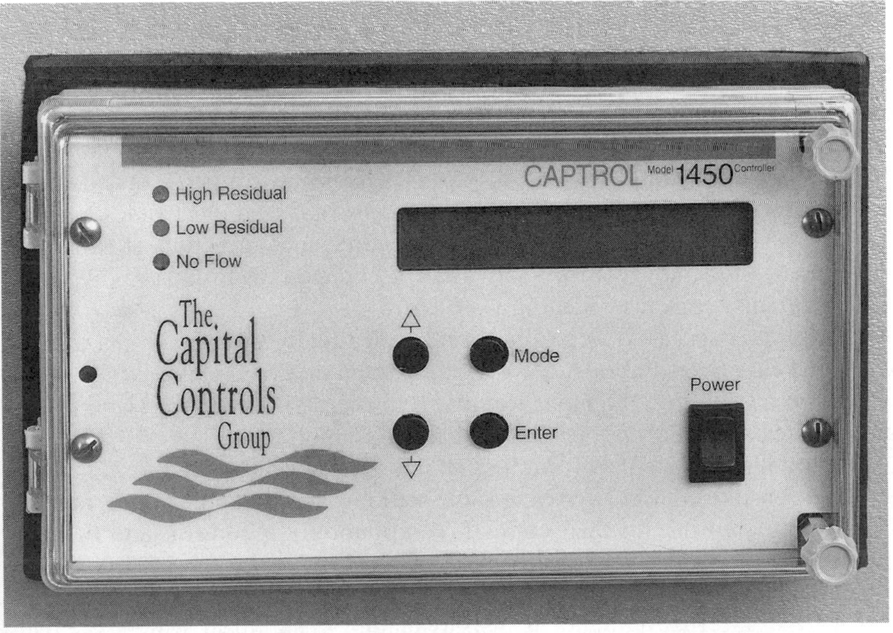

Fig. 9-88. Capital Control's micropressor controller model #1450.

tioning and feed-forward control for dechlorination using the built-in multiplier.

Residual control is accomplished by a single-mode integral control loop with adjustable process lag time. All setup, tuning, and control adjustment is done from the front panel where the input and output signals have been factory-calibrated. This microprocessor design simplifies the controller tuning to a maximum of just three adjustments: dosage, process lag time, and integral. Digital displays support rapid setup and tuning, and provide all information needed for precise and timely control of the process.

Front panel alarm indicators and internal contacts are provided for high and low residual setpoint deviation, and low process water flow. Setup, tuning, and control are accomplished with just four pushbutton switches, which are conveniently located on the front panel.

Switch inputs are available for connection to the following:

1. High/low vacuum relay contacts.
2. Duty/standby operation.
3. Connection to a residual inhibit switch that causes the controller to ignore the residual input.

A nickel–cadmium battery is a vital part of the controller. It provides full data protection for a maximum of four days.

For enhanced personnel safety, the enclosure is designed with separate access to the controller electronics and wiring terminals. Additional details are available in Bulletin A1.11450.2.

Model 1451. This microprocessor-based controller is similar to Model 1450, except that it is designed specifically to use two inputs, such as monitoring and control of disinfection, oxidation, and potable water quality.

The standard output provides a 4–20 mA signal to control a broad range of chemical feed applications. The microprocessor-based electronics and the digital alphanumeric displays simplify the setup and operation of this unit. Fully field-configurable control modes can be changed easily and rapidly, using the front panel controls. (See Figure 9-89.)

Microprocessor-based electronics automatically adjust either a gas control valve or a chemical solution metering pump to maintain the desired feed rates, which are based upon the input signals from the plant flow meter and/or the chlorine residual analyzer, or the water quality monitor.

User-configurable control modes, which are specific to automating both chlorination and ammoniation of potable water treatment applications, include flow-pacing chlorine residual control, compound-loop control, and cascade control.

For dechlorination control with sulfur dioxide, dual-input feed-forward control with an internal amplifier is also available. Pushbutton switches conveniently located on the front panel provide for rapid setup and simplified

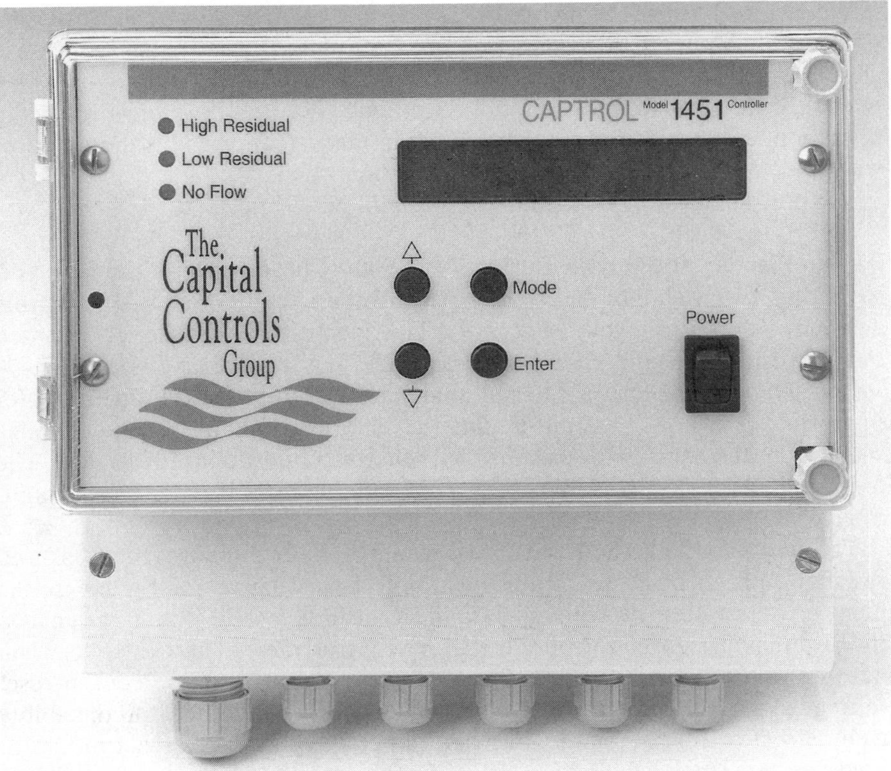

Fig. 9-89. Capital Control's micropressor controller model #1451.

operation, including the manual positioning of either the gas control valve or the solution metering pump.

Field conversion to any of the control modes is accomplished by simple pushbutton mode commands selected by the operating personnel. An internal battery protects the setup parameters for up to four days in the event of a power outage.

The control range adjustment of the dosage (loop gain) on the flow-input signal extends upward to a ratio of 4 to 1 or down to zero, thus permitting the accommodation of an oversized flow meter, control valve, or metering pump.

This Model 1451 unit makes it possible to consistently and precisely control around a setpoint in the compound loop system by automatically adjusting the lag time for changes in the process flow rate. It also provides uninterrupted operation in the event that one of the compound loop signals is lost, by automatically transferring to a single loop system.

LED alarm indicators, mounted on the front panel for easy viewing, will alert the operator to conditions of low or no plant water flow, as well as high or low deviation of residuals beyond limits set by the operator. Each alarm includes relay contacts for control of the remote warning and control devices.

When a gas system is being used, safe operation is enhanced by the provision of a discrete input for the connection of high/low vacuum switches, which will close the gas control valve. This cuts off the gas feed, after a 30 sec time-out, in the event of abnormal vacuum conditions. For plants requiring backup operation, a duty/standby discrete switch input system provides the operator with selection of the active and backup system.

Further information is available in bulletin No. 315.0005.0.[117]

Bailey–Fischer and Porter Series 71RC5000 Chlorine Residual Controller. This is a wall-mounted residual controller for automatic chlorinators and sulfonators as shown in Fig. 9-90. This electronic controller is mounted in a corrosion-resistant NEMA 4X (IEC 529, IP65) protective enclosure.

It is arranged to receive external analog signals from the treatment (water or wastewater) flow transmitter and the chlorine residual analyzer, and a signal from the chlorinator or sulfonator that is proportional to gas flow. The following spans are available: 0.25, 0.5, 1.0, 2.0, 5.0, 10.0, and 20.0 mg/L of chlorine.

The controller is microprocessor-based and digitally compares the measured residual with an operator-created setpoint. This value is multiplied by the plant process water flow rate signal, and transmits a 4–20 mA DC signal to the automatic (chlorinator or sulfonator) control valve. The controller shall have an adjustable, proportional band setting of 2 to 1000 percent, a reset setting from 0.02 min., a derivative setting from 0.01 to 80 min., and the ability to be direct or reverse acting, and shall operate in adaptive reset mode for longer system lag times. These units are factory-configured and have selectable automatic or manual output.

They are capable of alarming under the following conditions:

1. First-stage high or low residual.
2. High or low deviation between residual and setpoint.
3. Low treatment plant water flow rate.

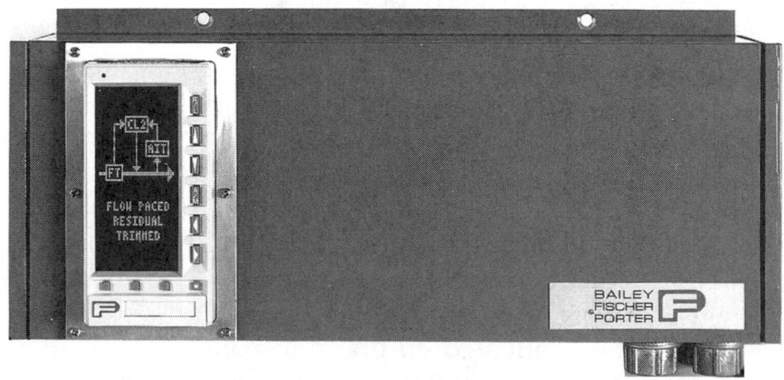

Fig. 9-90. Bailey–Fisher and Porter residual controller for automatic chlorination and sulfonators, series 71RC 5000.

Current output limiters are built into the program, which will limit the controller output should the second stage alarm levels be reached, thereby preventing the chlorine residual from rising above or falling below the established limits.

External alarm contacts rated at 3 A, 120 or 240 V, will be provided for high or low deviation between residual and setpoint, as well as for high or low first-stage alarms and low water flow.

Each of these units incorporate a dot matrix display on multiple screens that includes the following:

1. System schematics for compound-loop chlorination and dechlorination.
2. Flow pacing and residual for chlorination only.
3. Feed-forward with flow pacing for dechlorination.
4. Digital and bar-graph displays of residuals.
5. Setpoint and output to the chlorinator or sulfonator.
6. Adjustable time trends between 1 and 40 min. for recording residual.
7. Chlorine or sulfur dioxide gas flow and water flow.
8. Totalization of gas and water flows.
9. Controller tuning parameters.
10. Adjustable alarms for first- and second-stage high and low deviation between setpoint and residual.
11. Low water flow.
12. Adjustable high and low output signal limiters.

All alarms will flash when in the alarm condition. It is also important to note that when one of these instruments is being used where the maximum chlorine residual that is permitted is low or even zero, the unit to be used should be F&P specification 17SD4000 for the Z-CHLOR® Center Zero Dechlorination Control System.

For literature on chlorinators and sulfonators, refer to F&P specification numbers 70C1760/80, 70C4400, 70C5500, 70C6600, and 70C7700.

Wallace & Tiernan Process Control Systems[125]

Preface. Since the third edition of this book was published, W&T has made several changes in control systems. The following text describes both a process control unit and a signal conditioning unit. The illustrations show details of how they perform.

PCU—Process Control Unit.[125] This is a microprocessor-based control unit that is specifically designed for automatic residual or compound loop control of disinfection and dechlorination plus a variety of chemical feed systems in other water conditioniing applications. See Fig. 9-91.

It is used typically as a setpoint controller. The PCU provides accurate control of all gas feed equipment as well as chemical metering pump stroke

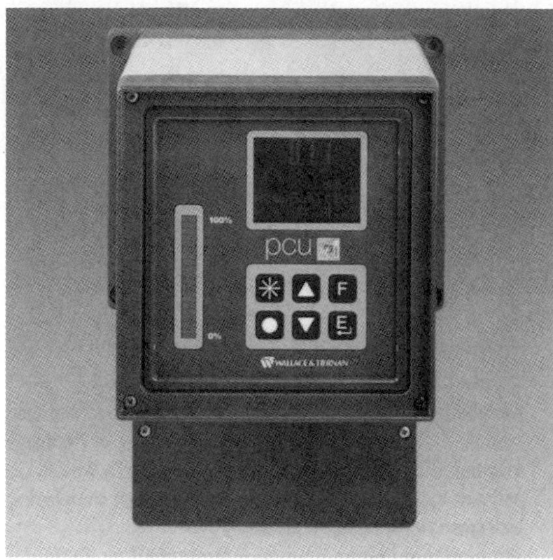

Fig. 9-91. Process control unit (courtesy Wallace & Tiernan).

length or variable speed drive applications. Easy operation, setup, and calibration using menu-structured electronics make sophisticated control functions simple with the PCU.

Operation. The PCU is a full feature P and I controller. It provides automatic process control of chemical feed equipment in response to two process inputs, typically process flow rate and chlorine residual. The PCU processes the input signals and can control a motor-driven actuator, which positions the V-notch plug on the chlorinator or the stroke length on a metering pump.

The controller can be programmed to perform four different types of control methods: flow proportional, feed-forward, chlorine residual feedback, and compound loop (dual signal feedback). In addition, the controller can be configured for "center-zero" feedback control, typically used in dechlorination control applications.

In the residual feedback mode, the controller compares a user-entered setpoint to the input signal. Proportional and integral control techniques are then used to adjust the position of the actuator. In compound loop, the setpoint control output is multiplied by the flow signal in order to scale the actuator position to plant flow. With the loss of the residual signal, the controller will automatically default to the flow proportional mode with manual dosage control. See Fig. 9-91 for inputs and outputs.

The PCU controller also has an isolated 4–20 mA output signal available for retransmission, remote indication, or the residual control of a metering pump through an SCR variable speed or variable frequency drive. Two programmable digital inputs (operating mode initialized from a remote contact) are also available.

Displays. Included here are: backlit LCD displays (contrast-selectable); four-digit numeric and five-character alphanumeric displays of measured residual; two 12-character alphanumeric message displays of operating and setup parameters; output bar graph, 20-segment in 5 percent increments; keypad selectable for flow rate or actuator position.

Features:

1. Continuous feedback control.
2. Four control modes.
3. Selectable alphanumeric LCD display of parameters as well as output bar graph.
4. Menu-driven electronics for clear operator instructions.
5. Isolated 4–20 mA output.
6. RS-485 serial interface.
7. Four alarm relays—each user-configurable for any one of 16 different conditions.
8. Nema 4X corrosion-resistant enclosure.
9. Automatic self-test and diagnostics menu.
10. Bumpless transfer when changing control modes.

For further detailed information, obtain a copy of W&T Publication No. TI 40.200UA at 1901 W. Garden Road Vineland, New Jersey 08360, phone 201-759-8000.

Wallace & Tiernan Signal Conditioning Unit[126]

SCU—Signal Conditioning Unit. This microprocessor-based SCU provides automatic flow-paced control of the gas feed rate for both gas feed systems and chemical metering pumps. This controller is specifically designed and developed for chlorination and dechlorination using either gas chlorinators or metering pumps for either hypochlorite or bisulfite. (See Fig. 9-92.)

It is also ideal for remote manual control of a chlorinator or metering pump, or it can be used to dose the flow input on a metering pump's SCR variable speed drive.

Operation. The SCU controller provides a proportional output for a 4–20 mA input, typically a flow signal. The output can control a motor-operated actuator, which positions the chlorinator V-notch plug, controls the chlorine feed rate, or the stroke length on a metering pump.

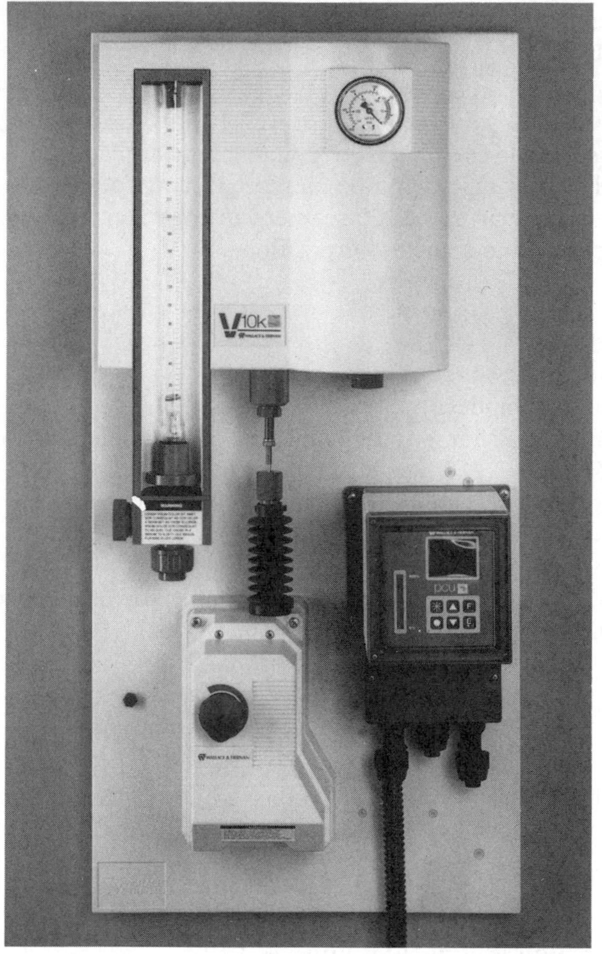

Fig. 9-92. Signal conditioning unit (SCU) by Wallace & Tiernan. SCU is mounted on a V10K chlorinator.

The control output can be scaled from 10 to 400 percent by a dosage factor selected at the SCU. The flow input can also be scaled from 10 to 400 percent by the flow factor, independent of dosage. An isolated 4–20 mA output signal (separate from the control output) is available for retransmission, remote indication, or pacing an SCR variable speed or variable frequency drive on a metering pump.

The SCU utilizes a menu structure to organize user operation and setup functions. Operating parameters, dosage, and other settings are entered via keypad with menu guidance. There are five top-level menus logically arranged to prompt the operator for easy access to operational information.

Features:

1. Large alphanumeric backlit LCD display of output (actuator position) and input flow pacing signal.
2. Nema 4X corrosion-resistant enclosure.
3. Simple operation, setup, and calibration.
4. Dosage and flow scaling.
5. Available 4–20 mA output signal.
6. Electronic and mechanical manual override.
7. Menu driven electronics.
8. On-line diagnostics.

REDOX CONTROL (ORP)

Historical Background

This method of chlorine control is commonly referred to as ORP (oxidation reduction potential). It is simply the reading of an mV (millivolt) potential across a pair of electrodes. Many years ago, a long-time colleague of mine, Richard Pomeroy, described the ORP phenomenon quite succinctly: "Any substance capable of absorbing electrons may act as an oxidizing agent, and any substance that can yield electrons may act as a reducing agent."[116] Therefore, some oxidizers show a very strong demand for electrons, so they are called "strong oxidizing agents," and others take electrons only when they are easy to get, and are the "weak oxidizing agents." In the appendix, the reader will find a table that lists all the important oxidizing agents discussed in this edition, along with their ORP value.

Many years ago Wallace & Tiernan developed an ORP process that was used successfully for the destruction of cyanide wastes, and chlorine manufacturers were using it successfully to make various strengths of hypochlorite. During this same era (early 1940s) White was trying to solve hydrogen sulfide odors in raw sewage entering the East Bay Municipal Utility District WWTP located in Oakland, California, adjacent to the entrance to the San Francisco Bay Bridge. He gave up after several attempts with different measuring probes. In the meantime Frank Strand, in Bradley, Illinois, was developing a successful ORP control system for the chlorination of swimming pools (see Chapter 4). In 1975 White made a personal field investigation of four pools using the Stranco system in the San Francisco Peninsula (see Chapter 4). Following their success in swimming pools, Strand made a success of ORP for the control of intermittent chlorination of cooling tower water to prevent slime formation (biofouling) in the plant's heat exchangers. This was followed in the early 1990s by on-site testing of their process for wastewater disinfection.[99–101] The work of Lund[109] in 1963 and of Carlson[111] in 1968, who both proved conclusively that using ORP mV readings was far superior to the use of chlorine species residual measurements, indeed provided a pathway for Frank Strand to pursue an ORP system for wastewater disinfection. This he accomplished in the early 1990s.

By 1997, this process has become very successful because ORP can not only control the chlorine feed rate due to flow rate and chlorine demand changes, but it can also track the chlorine demand.[105] The latter was a prime objective of Wallace & Tiernan when, in the early 1950s, they developed a very complicated "automatic chlorine demand control" system. It failed because it was too complicated.

Now we finally have an accurate, simple, low maintenance system that plant operators love, for controlling both chlor and dechlor systems.[101]

Some Basic Chemistry

There is a conceptual analogy between acid–base and reduction–oxidation reactions, similar to the way that acids and bases have been interpreted as proton donors and electron acceptors.[102] Owing to the fact that there are no free electrons, every oxidation reaction is accompanied by a reduction reaction, and vice versa. Or expressed another way, an oxidant is a substance that causes oxidation to occur while being reduced itself.

The oxidation state (millivolt number) represents a hypothetical charge that an atom would have if the ion or molecule would have if the ion or molecule were to dissociate:

$$O_2 + 4H^+ + 4e^- = 2H_2O\text{; reduction}$$

$$4Fe^{2+} = 4Fe^{3+} + 4e^-\text{; oxidation}$$

$$O_2 + 4Fe^{2+} + 4H^+ = 4Fe^{3+} + 4e^- + 2H_2O\text{; redox reaction}$$

The oxidation state, or the oxidation number (mV) represents a hypothetical charge that an atom would have if the ion or molecule were to dissociate.[103]

Any chemical in solution that is capable of entering into an oxidation or reduction reaction causes a potential difference between a standard half cell (reference electrode) and a measuring electrode. The measuring electrode (also known as the sensing electrode) for potable water, wastewater or biosolids has to be a pure noble metal such as 99.999 percent platinum.[104]

The electrode circuitry for pH is the same as described above except that the measuring electrode is made of glass, which is a specific ion electrode (hydrogen ion).[106]

Fundamentals of Redox Potential Chemistry[107,110]

The redox potential measures the net potential from an aqueous solution composed of oxidants (chlorine species) or reductants (sulfites). This gives redox potential the unique ability to detect whether the chlorine species present at any given time is sufficient to meet the demand, or sulfite addition is sufficient to neutralize the chlorine to achieve complete dechlorination.

As redox potential responses are logarithmic, as shown in Fig. 9-93a, this makes them most sensitive at extremely low levels of chlorine or sulfite content. Figure 9-93b illustrates the redox potential as a function of free chlorine (85 percent HOCl), chloramines, and sulfite concentrations.

This clearly shows the dramatic change on the redox potential when the residual is near zero, which makes detection and control of both chlorine and sulfite at extremely low levels (0.10 mg/L) a practical matter.

Under redox measurement and control of sulfite feed, a redox probe reports the ORP levels in the source water to the controller, shown in Fig. 9-92. The controller modulates the rate of sulfite feed based upon the changing chlorine and sulfite activity. The redox technique is totally different from the efforts produced by chlorine residual analyzers in several different ways, as follows:

1. It measures the actual oxidant and reductant activity, instead of residuals. Residual measurements suffer from compounds that regularly appear in the source water that cause serious quantitative errors in the measurement of chlorine residuals.
2. Redox can measure the potential (millivolts) of both sides of the chlorine and sulfite couple. This means that it can measure both the chlorine and the sulfite activity. This allows the operator to lock in a slight residual of either chlorine or sulfite.
3. Redox responses are logarithmic, meaning more precise dechlorination control, where zero residual is the discharge requirement. This not only

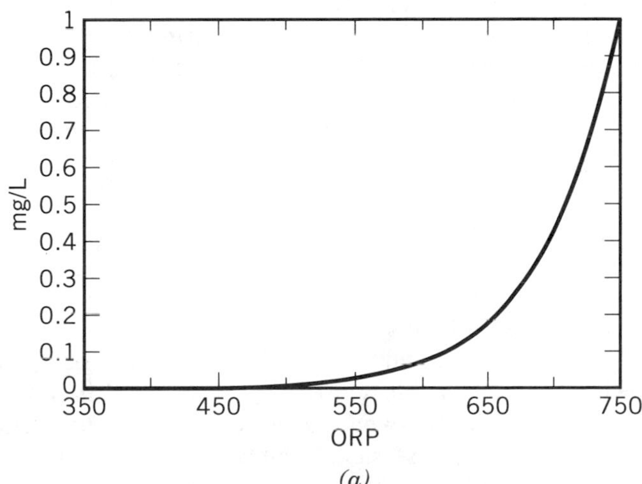

(a)

Fig. 9-93a. ORP mV versus chlorine residual (Cl_2 residual 0.0–1.0 mg/L).

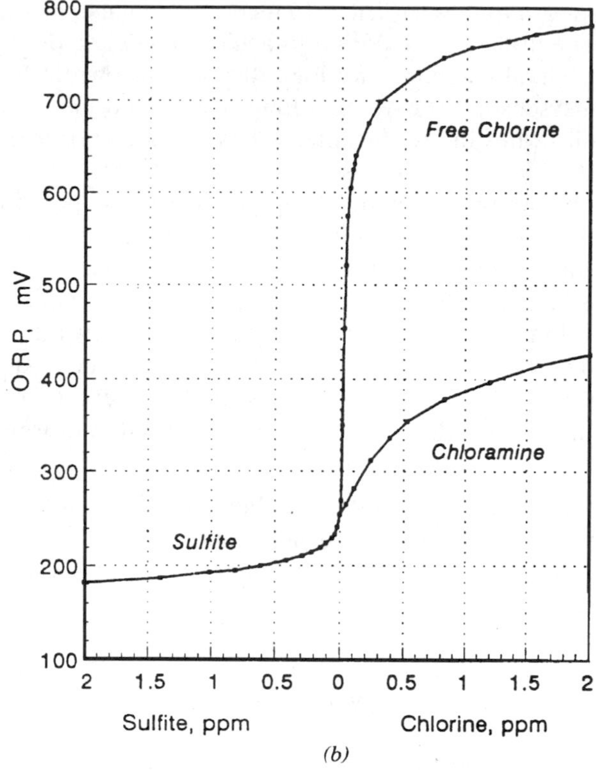

(b)

Fig. 9-93b. ORP versus chlorine and sulfite concentrations (tap water, 21°C, pH 7.5).

perfects the dechlorination process, but it saves the user a sizable amount in chemical costs, up to 25–50 percent per annum.

4. Properly designed redox electrodes are very forgiving of process contamination because they do not measure current. The electrodes can continue to operate properly even after being partially coated with source water contaminants.

The Illustrations Describing the Stranco HRR System

Figure 9-94 shows the extreme difference that there is in the oxidation power of the various chlorine residual species.[108]

Figure 9-95 illustrates the data logger for the Strantrol Chlor–Dechlor System, showing the submersible sensor that is dropped in a channel or an open tank. It sends the ORP signal to the controller. The controller then converts this signal to a 4–20 mA signal that actuates chemical feeding equipment. It also indicates the process flow. The system also provides an automatic probe cleaning device plus a data logger and a voice modem.

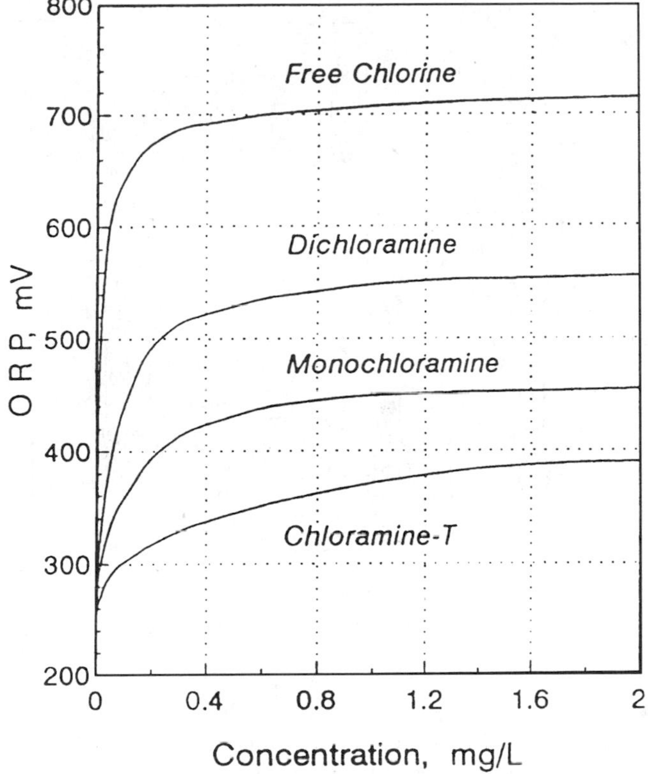

Fig. 9-94. ORP of various chlorine compounds (Victorin *et al.*)

Figure 9-96 depicts a typical application of Strantrol controllers for a complete chlor–dechlor system for a wastewater treatment plant. One HRR probe (sensor) is located in the contact chamber about 5 min. downstream from the location of the chlorine application diffusers or the Water Champ.* A second HRR probe is suspended about 20 sec downstream from the point of application of sulfur dioxide.

Figure 9-97 illustrates a compact, low-cost, easy-to-install rechlorination station. These units are often required in potable water distribution systems to maintain the stability of an aerobic microbiological system in the far reaches of the distribution system. They also assure compliance with regulations in cases where unattended pump stations are involved.

Figure 9-98 shows one of the latest advancements, in the features of the Strantrol 900 System. It operates and adjusts feed rates up to three processes

*This is a super mixing device, better than diffusers. See description in this chapter.

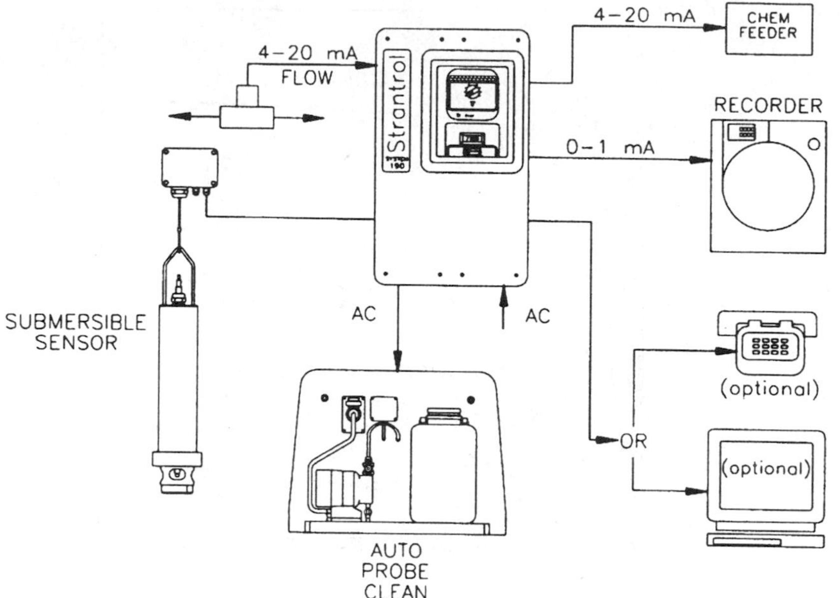

Fig. 9-95. Automatic control of Chlor-decklor systems (courtesy of Stranco Co.).

by using the Sensor Interface, a sensor management unit that converts the probe signal to a usable signal and transmits it to the 900 System.

Figure 9-99 illustrates the typical microorganism inactivation rate developed by Ebba Lund[109] from her extensive work on inactivation of the polio virus. It shows that redox measurement accurately predicts the rate of virus inactivation, regardless of the oxidizers used in the experiment.

Lund found that once a minimum redox potential has been exceeded, the inactivation rate progresses proportionally with respect to increasing redox potential. Various oxidants, such as chlorine, $KMnO_4$, and cystine, were employed in the study, and all those cases illustrated how unreliable the residual measurement is to predict the disinfection efficiency.

Why the Stranco Redox System is Unique. Probably the most important discovery made by Stranco to eliminate electrode measuring errors was the absolute necessity of having the reference electrode made of 99.999 percent pure platinum, plus having the silver–silver chloride measuring electrode immersed in a gel solution. This resulted in such things as eliminating the hydrogen overvoltage phenomenon, previously reported to cause sensing errors owing to the electrolysis of water (separation into H_2 and O_2) commonly encountered at potentials used for chlorinated waters. These gases enter into redox reactions at the electrodes, causing system instability.

The purpose of the measuring electrode is to act as a pool for the oxidizing species present in the water. The potential difference between the oxidation–

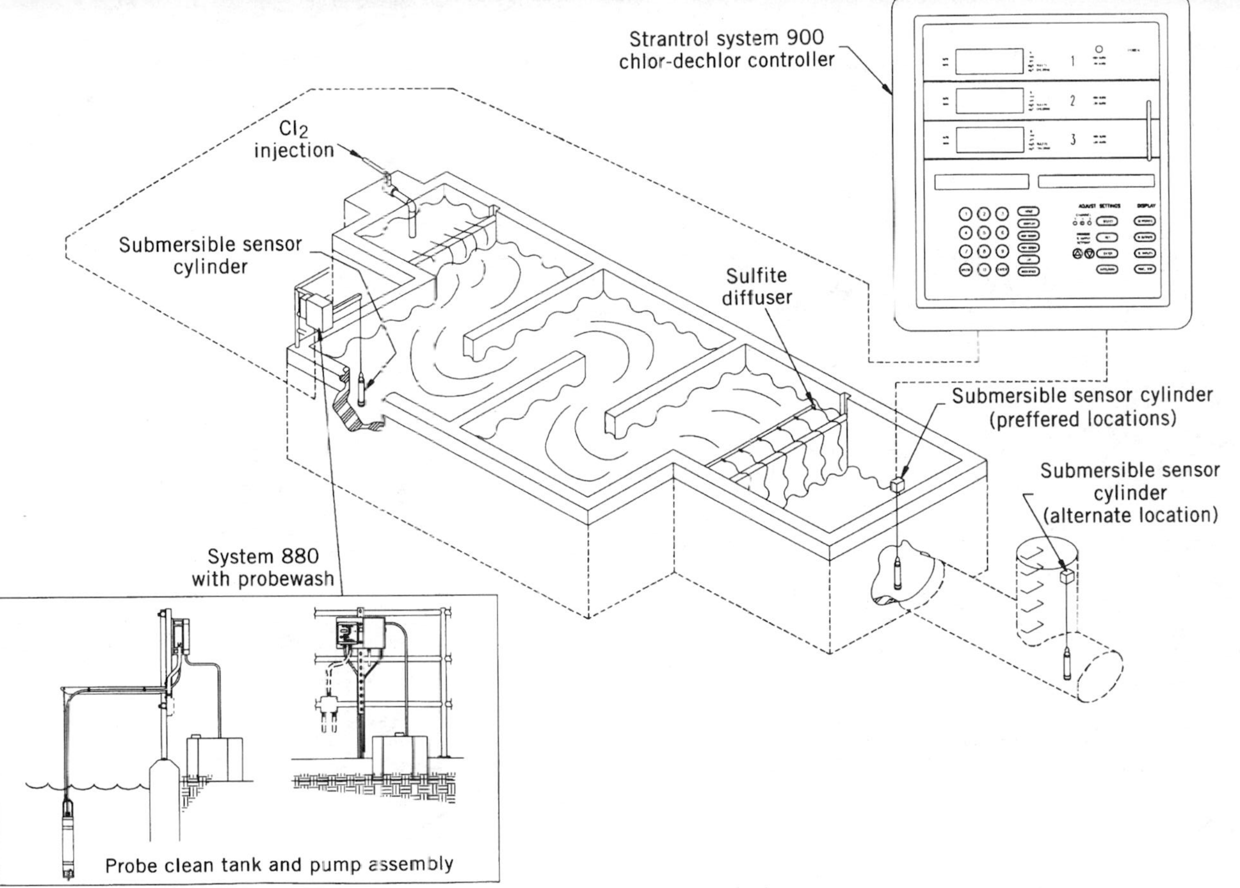

Fig. 9-96. Typical contact chamber with Strantrol dechloration and chlorination control.

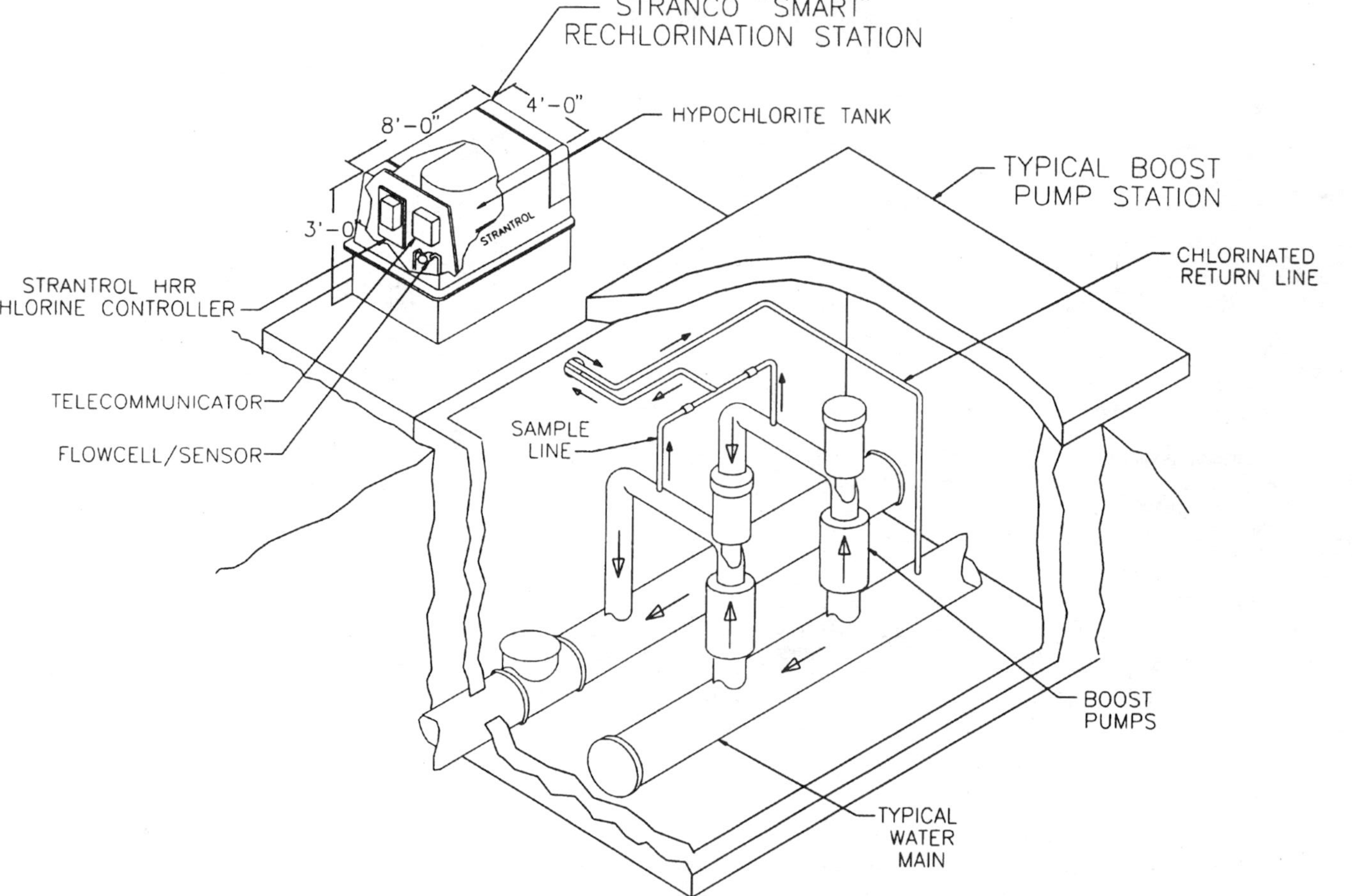

Fig. 9.97 Compact, low-cost Strantrol rechlorination station for water distribution systems.

reduction species and the reference electrode produces the measured voltage. Theoretically, no current needs to flow for the potential to exist. This causes the classic ORP response. Any changes of output that occurs as a result of influences other than the oxidant present causes errors. These situations are obviously intolerable. Therefore, the redox system must be designed to have electrodes of high purity and to remain stable in the presence of potentially interfering ionic species. This is why the Stranco system is unique—not just because of the 99.999 percent platinum reference electrode, but because the silver, silver chloride measuring electrode is immersed in a gel solution. (See Fig. 9-100.)

The Stranco HRR System electrodes are capable of operating in high oxidative environments, such as those found in high chlorine residual environments (15–20 mg/L). Previously the concept held by many was that operating in a significant chlorine residual environment would over time "poison" the ORP electrodes. This does not ever happen to the Stranco system; this is why it is so successful. This electrode cell system is shown in Fig. 9-101.

Moreover, this new technology is distinguished by its ability to remain stable in the presence of potentially interfering ionic species, and its ability to repeatedly differentiate differences as small as 0.1 mg/L.

Probe Contamination. In wastewater installations it is well known that there are many substances in treated sewage that can affect the electrochemical response of pure platinum electrodes. The surface of these electrodes can be treated to resist the chemical substance although this treatment affects their stability and sensitivity. Automated cleaning systems utilizing a small timer-actuated pump to periodically rinse the electrode with hydrochloric acid can quickly restore the surface of the electrodes.

Junction Potential. The liquid junction (salt bridge) can be thought of as the major artery of an ORP system. It is the pathway for ionic species to travel to and from the reference electrode. Some ion passage is required to transfer current, which allows a reading to be made, whereas other ions pass simply to satisfy concentration gradients. As with an artery, the effects of contamination or blockage are disastrous. If the current cannot pass, then the system simply will not function. If some ionic species can pass but others cannot, then a phenomenon known as charge separation occurs. This causes a potential to be set up across the liquid junction, resulting in instability. HRR instruments construct the junction from porous Teflon with large pores to facilitate free ionic movement. The Teflon is also inert, chemically resistant, and slippery enough to prevent many contaminants from adhering to it.

Electrolyte Loss. The electrolyte within the reference electrode conducts current and enters into necessary ionic changes with the reference element and the process. Owing to the large liquid junction pores in an HRR reference cell, the risk of losing electrolyte is high. For this reason a special gelling agent

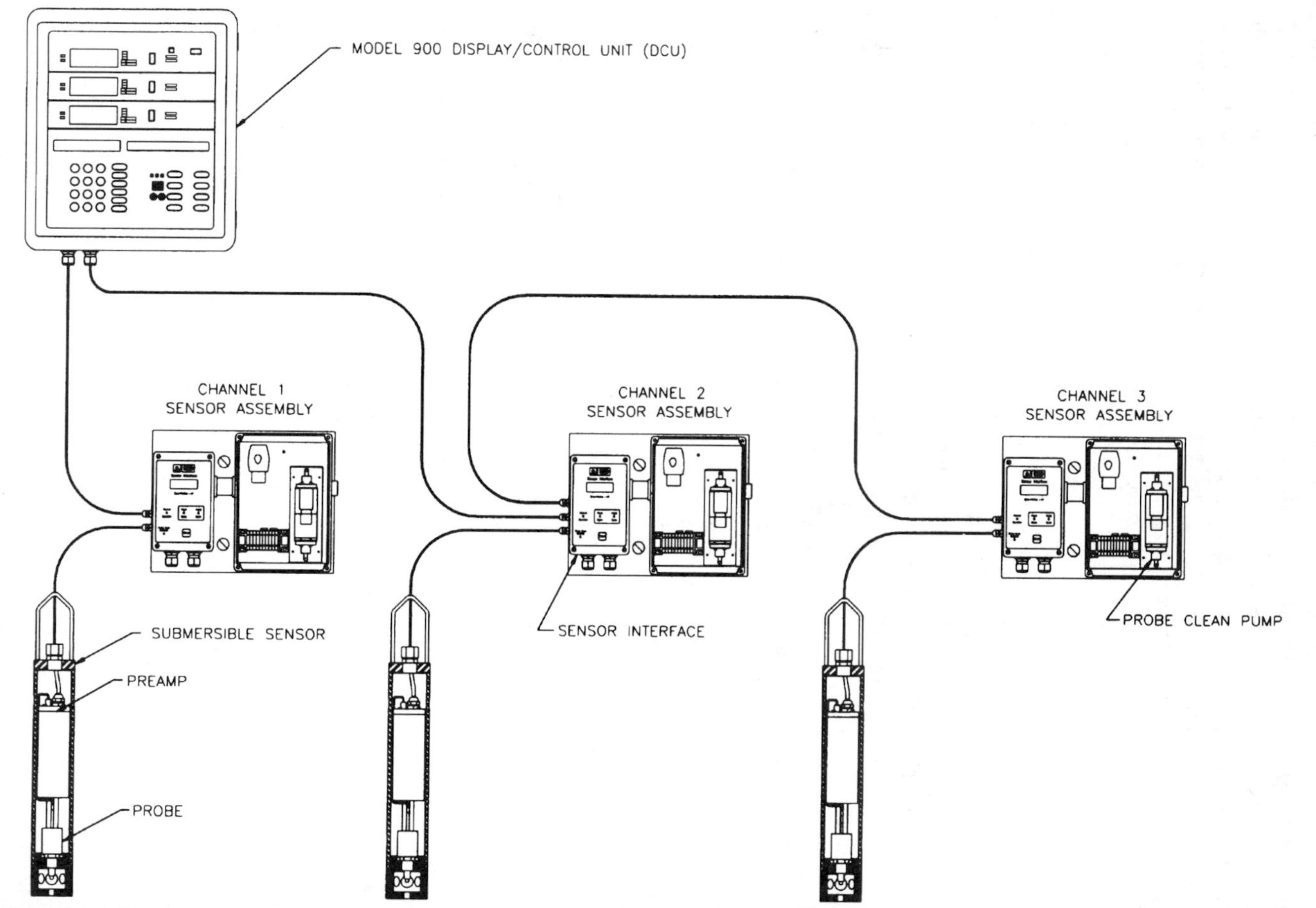
MODEL 900 DISPLAY/CONTROL UNIT (DCU)
CHANNEL 1
SENSOR ASSEMBLY
CHANNEL 2
SENSOR ASSEMBLY
CHANNEL 3
SENSOR ASSEMBLY
SUBMERSIBLE SENSOR
PREAMP
PROBE
SENSOR INTERFACE
PROBE CLEAN PUMP

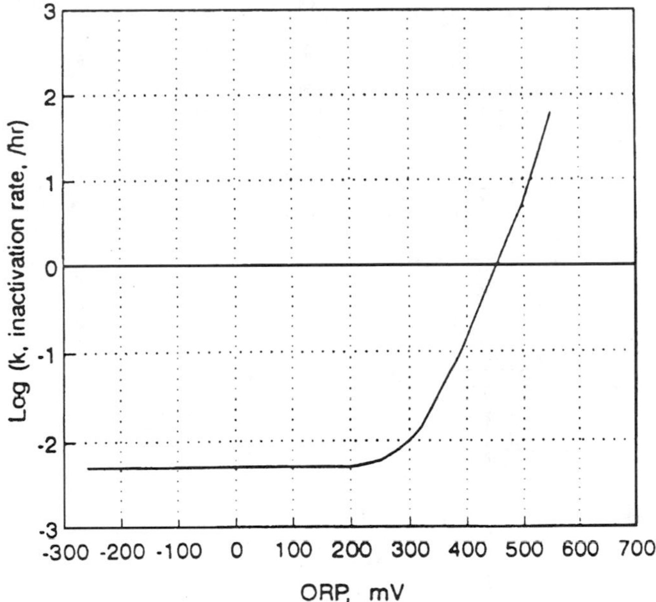

Fig. 9-99. Rate of poliovirus inactivation as a function of redox potential (Lund).

is used to thicken the solution. Standard industrial redox electrodes contain reference solutions that degrade rapidly in the presence of strong oxidants. For example, in high residual chlorine applications, reference solutions can fail within a matter of hours. This causes a breakdown in measurement and control accuracy. The HRR gel is impervious to chlorine or bromine oxidation, thus enabling long-term measurement and control stability.

Electronics. Another vital component of the HRR system is the electronics/electrode interface. For a high resolution device, the electronics need to be extremely high in impedance and stable in electrically noisy environments. The high impedance nature of this circuit is vital so that practically no current is drawn through the electrode. Current draw causes a rush of chemical changes within the reference cell, resulting in polarization. Additionally, too much current will cause the reduction of some ionic species in the water and mask the measurement of the chlorine residual.

Fig. 9-98. Strantrol 900 multiple control system (courtesy Stranco, Inc.). Unless otherwise specified, dimensions are in inches, tolerances on decimals/fractions. X/X—±1/16; .XX—±0.03; .XXX—±0.010; angles—±1.0'-0", surfaces 250$\surd$.

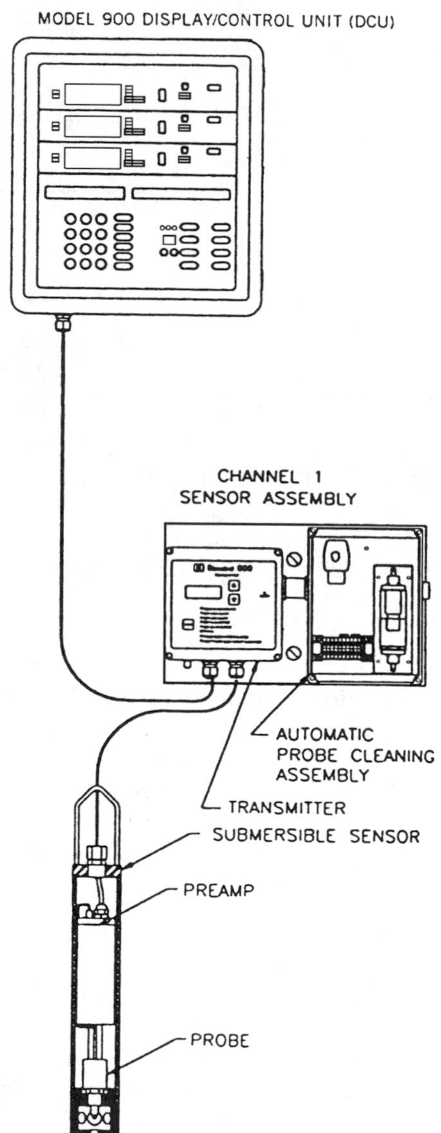

Fig. 9-100. The Strantrol silver chloride measuring electrode immersed in a gel solution (courtesy of Stranco Co.).

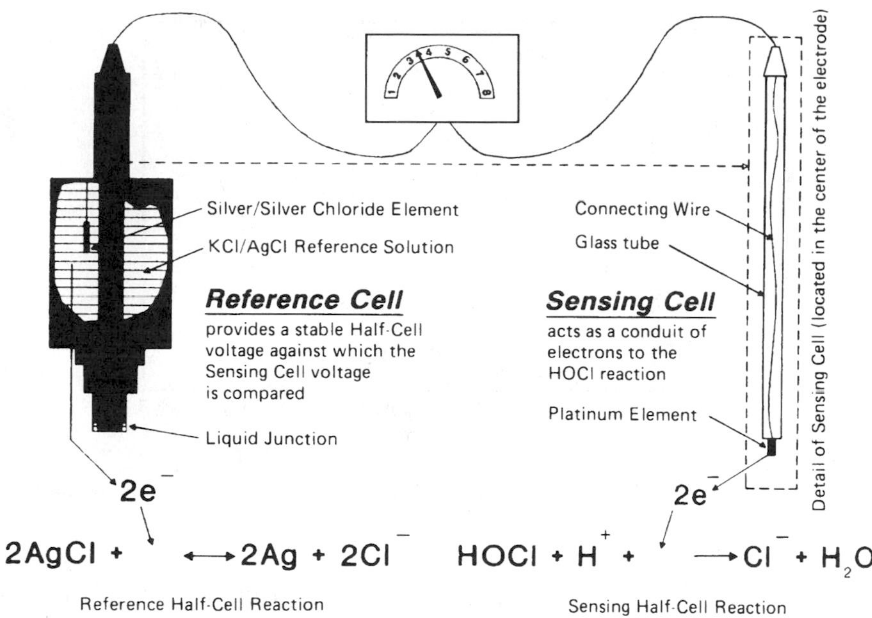

Fig. 9-101. Referencing and sensing cells ORP reactions (courtesy of Stranco).

Poise, Calibration, and Expectations. Every wastewater effluent will have a different poise. This ORP poise of the untreated water changes as the concentration of various electroactive contaminants changes, ultimately reaching an equilibrium. If this reading happens to be 200 mV, then the ORP contributed by the chlorine will be added to this number. Stranco has found that although poise varies from water to water, it affects mainly the zero point of the ORP–mg/L Cl_2 curve and not the span. Therefore, they do a single-point calibration (zero shift). This poise is usually determined after the system has been in operation for at least 24 hours and while normal operating conditions are being observed. When equilibration has occurred, the system is standardized with one of the titrimetric chlorine residual procedures. The expectations concern the weaknesses in the available chlorine residual measurements relative to establishing consistency necessary to meet the current NPDES regulations for wastewater dischargers. ORP measurements are much simpler to obtain, and it is hoped that the ORP cell will be easier to maintain than either the membrane or amperometric cells. It has been observed that there is a marked difference in the ORP mV curves between free chlorine residuals and chloramine residuals. Typically free chlorine at 1.0 mg/L shows approximately 750 mV versus 550 mV for a chloramine residual at 1.0 mg/L.

System Calibration. Figures 9-93a and 9-102 illustrate the logarithmic relationship between the ORP and mg/L of chlorine residual. Starting a system

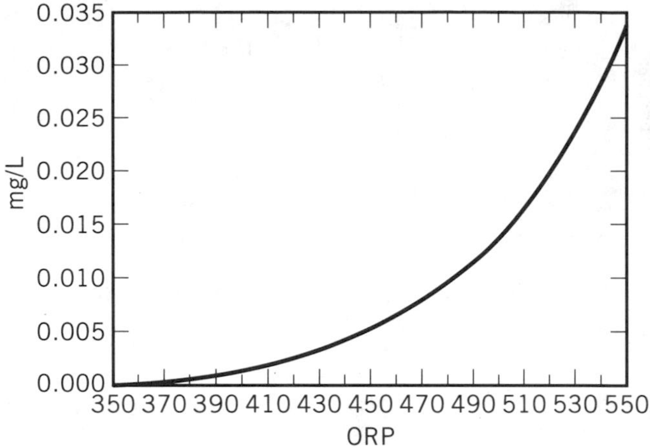

Fig. 9-102. ORP mV versus chlorine residual (Cl_2 residual 0.000 to 0.035 mg/L).

that covers a 1.0 mg/L range requires only a single-point calibration. This involves shifting the entire graph, either to the right or the left, depending upon the ORP poise of the water involved. For example, the theoretical ORP shown on Fig. 9-93a at a 0.5 mg/L residual is approximately 710 mV. This assumes that zero residual is 350 mV. Typically, however, when a system is started up, there is rarely an opportunity to observe a zero residual in the sample line. Therefore a residual test has to be made. If the measured residual is 0.5 mg/L, the unit is shifted so that it reads 710 mV and 0.5 mg/L. This works well in nearly all the cases.[99,100] Once the single calibration point is set correctly, the progression of the ORP readings will match the progression of the residuals. As an example, if 0.5 mg/L shows 710 mV, then 40 mV higher, at 750 mV, the chlorine residual should be 1.0 mg/L. But if at startup, the poise is such that 0.5 mg/L occurs at 600 mV ORP, then at 640 mV the chlorine residual should be reading 1.0 mg/L.

Figure 9-103 illustrates the difference in ORP readings between free residual chlorine and chloramines. This is an important illustration because the chlor–dechlor system residuals in wastewater treatment are predominantly chloramines.

INJECTOR SYSTEMS

General Description

The power developed by the injector allows the chlorine to flow from the supply containers through the chlorinator, which is the metering system, and then through the injector vacuum line to the injector inlet. At the injector, the chlorine dissolves in the injector water to form a mixture of hypochlorous

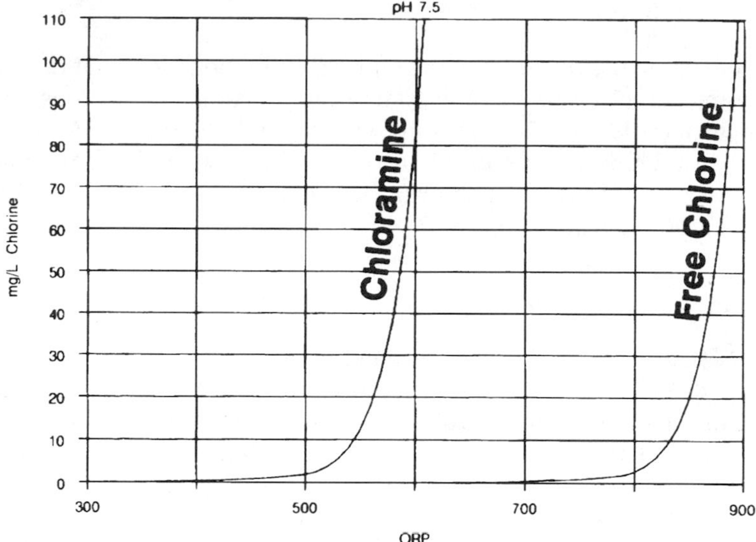

Fig. 9-103. ORP of free chlorine versus chloramine.

acid (HOCl) and molecular chlorine (Cl_2). This is the chlorine solution that flows in the solution lines to the diffuser at the point of application. This method of feeding chlorine gas is unique to water treatment processes (potable water, wastewater, and industrial waters). In the beginning, chlorinators metered and controlled directly into the process flow. This practice proved to be impractical over the long term. There were and are some exceptions. The difficulties with the direct feed system resulted from feed-rate sensitivity due to ambient temperature changes and low solubility of chlorine in water at atmospheric pressure. The vacuum system invented by C. F. Wallace overcame these difficulties and provided several advantages:

1. It is the easiest method of dissolving chlorine in water.
2. Chlorine is easily handled when in solution.
3. Since the injector creates a vacuum, it allows chlorine to be metered under a vacuum. This is the most accurate way of metering, since constant density is maintained and is not affected by ambient temperature changes.
4. Operating under a vacuum is safer than operating under pressure.
5. A metering system can be easily designed to stop automatically if the vacuum should fail.
6. Available vacuum can be used for automatic switchover of containers.
7. Injector vacuum level can be used to provide two alarm signals: (1) loss of chlorine supply; (2) injector malfunction or loss of water pressure.
8. Injector vacuum can be used to generate a variable chlorine feed rate as a function of chlorine residual (compound loop control).

System Description

The injector system is the heart of the entire chlorination facility. If this system is inoperable, no other part of the system can function. The various parts of this system include: (1) the operating water supply to the injector; (2) the injector; (3) the injector vacuum line from the chlorinator; (4) the injector discharge system, described as the chlorine solution line; and (5) the diffuser at the point of application.

Operating Water Supply

Water Quantity. Injectors are designed to develop about 25 inches of Hg vacuum at the injector inlet when the chlorinator is at maximum chlorine feed rate and the chlorine solution is discharging at atmospheric pressure (i.e., no back-pressure). The water flow through the injector must be sufficient to limit the chlorine solution strength to 3500 mg/L. Above this strength molecular chlorine appears in significant amounts, which breaks out of solution causing off-gassing at the point of application if open to the atmosphere, and gas binding in solution lines under low negative heads (see Table 4-1). Either condition is intolerable. Figure 9-104 shows the minimum amount of water required to limit the chlorine solution to a maximum strength of 3500 mg/L for various gas flow rates. A rule of thumb is: approx. 40 gal water/lb/day chlorine, more or less depending upon local conditions.

Water Pressure vs. Back-Pressure. A second factor is the pressure at the point of application of the chlorine. This is known as injector back-pressure. A higher back-pressure requires a higher injector inlet pressure and more operating water to make the injector function properly. The injector also has minimum operating water requirements. Chlorination equipment manufacturers use injector operating curves that specify how much water at which pressure is required for a given amount of chlorine to be applied against a given back-pressure. It is the designer's responsibility to make a hydraulic analysis of the chlorine solution line between the injector and the point of application to establish the amount of back-pressure to be expected. With this information and the maximum chlorine feed rate desired, the chlorinator manufacturer can then advise the necessary water supply inlet pressure required at the injector and the optimum injector operating water quantity. The back-pressure should never be allowed to drop below 2 psi.

In spite of manufacturers' injector curves it is unwise to operate any injector on less than 50 psi inlet pressure unless there are extenuating circumstances.

Flow Meters. Injectors are usually available in four sizes: 1-inch fixed-throat type and 2-, 3-, and 4-inch adjustable-throat types. The fixed-throat injector assembly utilizes interchangeable throats and tailways with various-size openings for different conditions of chlorine feed rate and hydraulic conditions

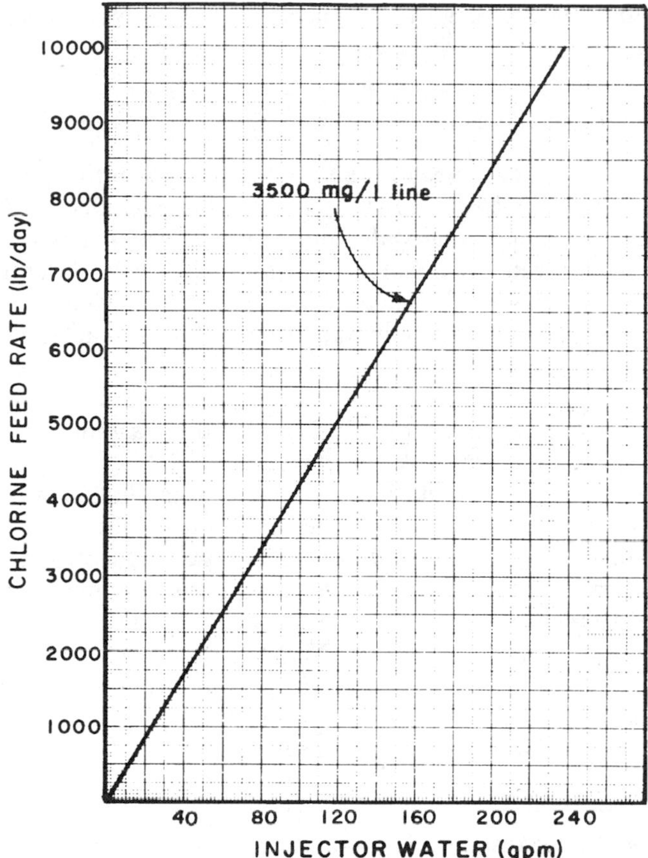

Fig. 9-104. This illustrates the amount of chlorine feed rate that limits the chlorine solution discharge concentration to 3500 mg/L.

(back-pressure). Therefore, the water flow through a fixed-throat injector is predictable. This is not the case with the adjustable-throat injector. For this reason, when adjustable-throat injectors are used, it is mandatory that the operator be provided with an indicating flow meter (for each injector) with a 3 to 1 operating range and a head loss at maximum flow not to exceed 5–6 ft.

Pressure-Regulating Valves. In cases where more than ample pressure is available to operate the chlorinator, there is no need to use pressure-regulating valves upstream from the injector except in cases of abnormally high pressure—that is, >150 psi.* The reason for not requiring pressure-reducing valves

*An exception occurs when booster pumps are used, or where cavitation of the injector tailway occurs.

is the hydraulic characteristics of an injector, which will consume all of the pressure available on the upstream side of the injector.

Hydraulic Gradient Analysis

Introduction. Since the injector system provides the motive power for any chlorination system, it is necessary to consult the manufacturer's injector curves to determine operating water pressure and water flow-rate requirements. The reader is cautioned that manufacturers may have different requirements for different modes of vacuum operation. As an example, Wallace & Tiernan recommends different minimum vacuum levels to achieve maximum chlorinator flow rates, as follows:

1. Pressure-to-the-machine type chlorinators require a minimum of 5 inches of Hg injector vacuum at the machine.
2. Remote vacuum type chlorinators require a minimum of 6 inches of Hg injector vacuum at the machine. These machines use a dual signal controller that programs the chlorine flow rate via the rotameter orifice.
3. The original compound loop control chlorinator system requires a minimum of 10 inches of Hg injector vacuum at the machine. This type of chlorination system controls the V-notch orifice by a water flow-rate transmitter, whereas the variable vacuum regulator is controlled from the residual analyzer transmitter (see Fig. 9-85). This system has been discontinued by W&T, but there are several hundred of them still in operation.

In all cases of injector operating requirements, it is always prudent to add a factor of safety by adding at least 3–5 psi to the calculated back-pressure. For system No. 3, add 6 psi.

Injectors can be installed in either the vertical or the horizontal position. The only precaution required is to change the position of the check valve inlet block on the 3- and 4-inch injectors so that the ball-check will operate as it does in the vertical position.

Typical Gradient. Figure 9-105 is a typical injector hydraulic gradient where $dF_1 = IOP$ = injector operating water pressure, and $dF_2 = BP$ = injector back-pressure, or hydraulic gradient immediately downstream from the injector (outlet of tailway). Injector water quantity Q is dependent upon three factors: (1) maximum chlorine gas feed rate; (2) size of injector; (3) back-pressure. However if the water pressure available to operate the injector is greater than required by the gradient analysis shown in Fig. 9-105, this water pressure will determine the amount of injector operating water for the case of the one-inch fixed-throat injectors. The throat size adjustment provided for in the 2-, 3-, and 4-inch injectors allows the operator to adjust the inlet water flow to conform to the hydraulic gradient analysis.

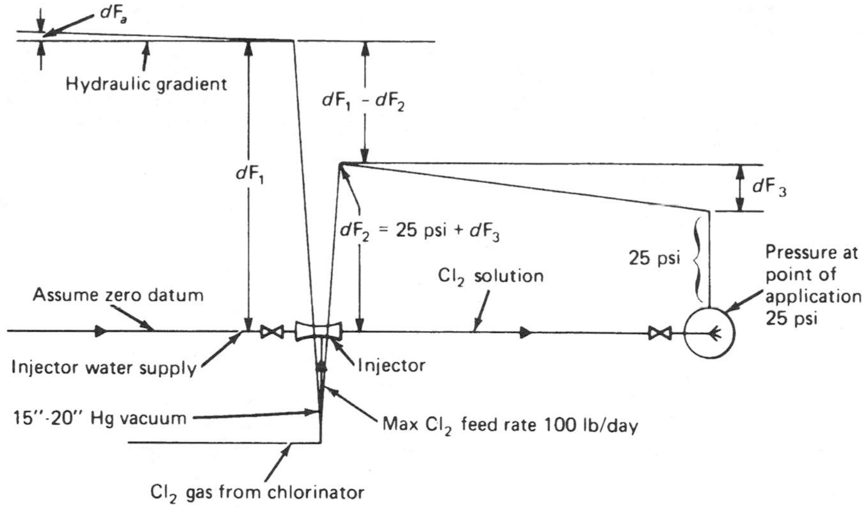

Fig. 9-105. Typical injector hydraulics, where dF_a = friction loss in injector water supply line to the injector; dF_1 = required operating pressure (consult manufacturer's curves); $dF_1 - dF_2$ = pressure drop across injector required for satisfactory operation; dF_3 = friction loss in chlorine solution line from injector to point of application. The chlorinator injector "back pressure" is dF_2. DF_1 is also dependent on the amount of chlorine to be injected as well as on the control mode, i.e., use of variable orifice positioner or differential regulator.

Example I. Figure 9-106a illustrates the hydraulic gradient of the injector system described as follows: Water is to be chlorinated at the outlet of an impounding reservoir. The location of the equipment will be such that the maximum static head at this point is 140 ft (60 psi). This then is the maximum back-pressure at any time. The peak water flow to be treated is estimated at two and one-half times the average daily consumption of 4 mgd = 10 mgd. Chlorine demand tests show that during summer months the demand is as high as 2.8 mg/L. Therefore, the chlorinator should be sized to feed at least 4 mg/L in order to provide an adequate free chlorine residual. The capacity of the chlorinator must be 4 mg/L × 8.33 lb/mg × 10 mgd = 330 lb/day. The selection would be a chlorinator with a maximum capacity of 400 lb/day—one of the breaking points on overall size of chlorinators. Consultation with the manufacturers' injector curves reveals that a fixed throat injector will be used, and that the most efficient size for feeding 350 lb/day maximum versus a 60 psi back-pressure will require an inlet pressure of 155 psi, which will produce a flow through the injector of Q = 20 gpm. Therefore, select a booster pump with 95 × 231 = 220 ft total net head lift and Q of 20 gpm plus a 25 percent (25 gpm) safety factor if the pump is to be centrifugal or 100 percent (40 gpm) if it is to be a turbine type. Additional allowance for friction loss in the solution line should be added to the pressure at the point of application. Assume that

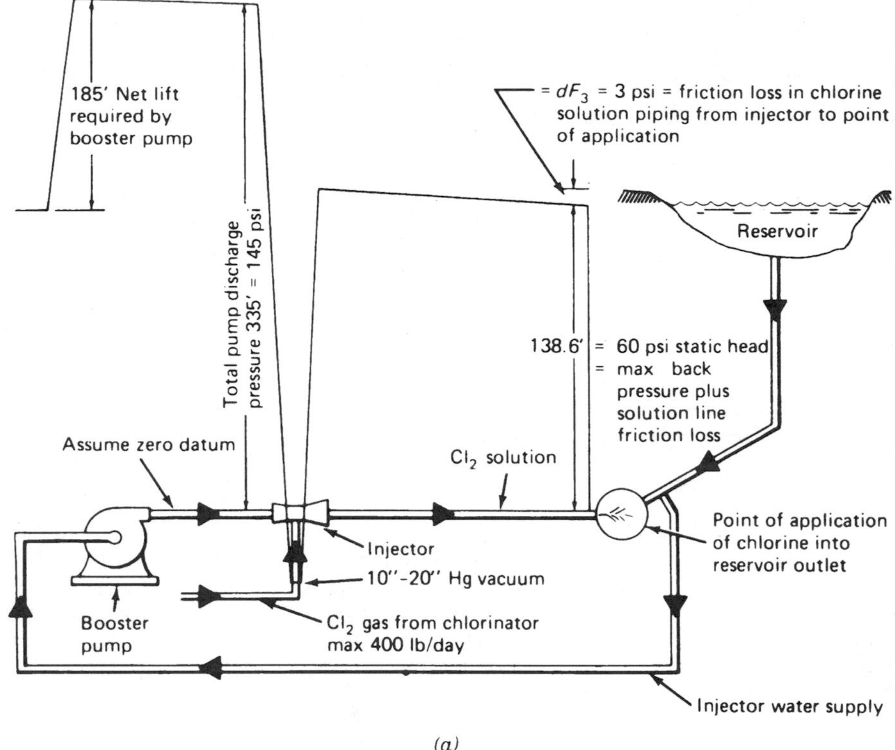

Fig. 9-106a. Hydraulic gradient injector system. Chlorine feed rate = 400 lb/day. Manufacturer's injector curves for this feed rate at 65 psi back pressure requires minimum injector water Q = 20 gpm. For the pump, the following amounts are practical in terms of safety factors: Q = 40 gpm for a turbine and Q = 25 gpm for a centrifugal pump.

the solution line has an equivalent length of 150 ft. A chlorine solution flow of 20 gpm* in a 1½-inch pipe will have a pressure drop of 4 ft per hundred. This will then raise the downstream back-pressure to approximately 63 psi. Since the pump chosen should have some factor of safety, we will select a pump to deliver 40 gpm at a total net head of 250 ft for a turbine and 25 gpm at a total net head of 250 ft for a centrifugal. Because of the high head, a turbine pump is preferred in this instance.

Example II. Installations in which the point of application is below the injector require that the chlorine solution line be sized so as to provide an artificial back-pressure of approximately 2 psi just downstream from the injector. This example involves a 2000 lb/day chlorinator using a 2-inch adjustable

*The flow through the injector is limited by the discharge pressure of the pump. In the case of the turbine pump, the excess flow recirculates in the bypass. (See Fig. 9-108.)

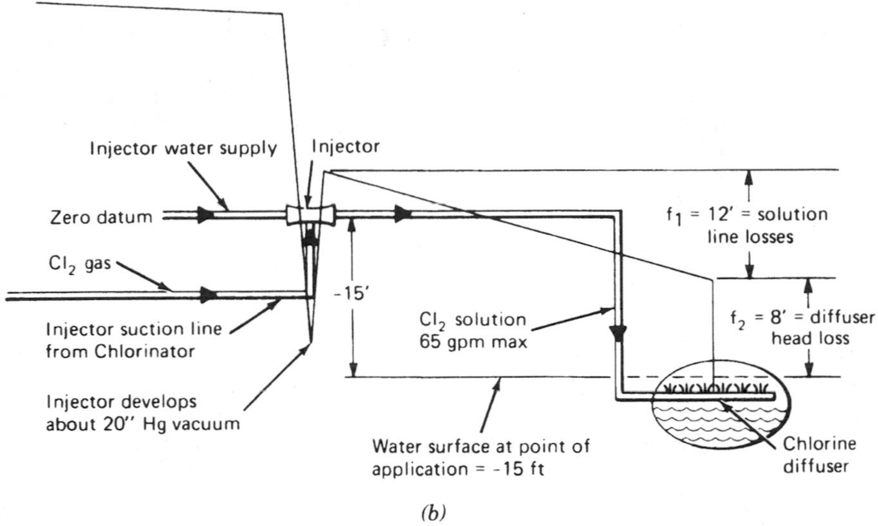

Fig. 9-106b. Hydraulic gradient of injector system to prevent negative head at the diffuser.

throat injector. The point of application is 25 ft below the injector, and the water surface is 15 ft below the injector. Therefore, the static back-pressure is −15 ft. The minimum amount of injector water will be 50 gpm. In negative head situations, there will be greater opportunity for developing chlorine fumes at the point of application owing to breakout of molecular chlorine under a negative pressure. Therefore, add 15 percent to the injector waterflow to decrease the chlorine concentration—try 65 gpm. The injector curves indicate a requirement of 25 psi with a back-pressure of 2 psi to meter 2000 lb chlorine per day.

Next choose a chlorine solution line small enough to produce 2 psi back pressure at the outlet of the injector. See Fig. 9-106b for the hydraulic gradient. Analysis of the solution line shows an equivalent length of 96 ft. Therefore, choose a 2-inch pipe size. At 65 gpm the friction loss in 96 ft of pipe is approximately 12 ft. Therefore, provide a diffuser with an 8 ft head loss. The total loss of 20 ft meets the requirement to provide a 2 psi back-pressure at the injector to prevent fuming at the diffuser. In all cases diffuser head losses of 8–10 ft are highly desirable because of the jet velocity attainable. This results in superior initial mixing. (See Chapter 8.)

Remote Injectors

Installation of injectors remote from the chlorinator location is more the rule than the exception. This is especially true in the wastewater application, which has already been described in Chapter 8. In potable water treatment there are two major advantages in having the injector adjacent to the diffuser: (1) it shortens the loop time in residual control application because gas velocities five times the chlorine solution velocities can be tolerated; and (2) head loss inherent in solution lines can be transferred and added to the diffuser loss to provide better mixing. See the "diffuser design" section in this chapter.

Vacuum Line Design

The design of a long vacuum line (>100 ft) should include the following features. The vacuum line (regardless of length) between the chlorinator and the injector should be sized so that the total pressure drop in this line is not more than 1½ inches of Hg. A shut-off valve should be provided at each end of this vacuum line. A vacuum gauge should be installed adjacent to, but downstream from, the shut-off valve at the injector end. The injector should always be installed in a horizontal position to achieve a low profile (better hydraulics) and to allow easy disassembly of the injector.

The optimum pipe size for vacuum lines between the metering equipment (chlorinators or sulfonators) and the injector is subject to a great deal of scrutiny. White has investigated the hydraulic characteristics of three remote injector systems varying in distances from 750 to 8740 ft. The most comprehensive study was of a chlorination system with three 2000 lb/day chlorinators remotely located from a single 3-inch injector connected by one 3-inch PVC vacuum line 8740 ft long.[25] Friction loss data were plotted showing the friction loss factor f as a function of Reynolds number, N_r (see Fig. 9 in the appendix). The response time (lag time) between the chlorinators and the injector was determined by 30-sec interval amperometric titrations of the chlorinated effluent immediately downstream from the chlorine diffuser, which was within 25 ft of the injector. As a check on these field observations, the Fischer and Porter engineering department set up a laboratory experiment using a 400 ft ¾-inch vacuum line to a remote injector and a variable chlorine feed rate of 150–500 lb/day. The results of this experiment when extrapolated agreed closely with the field results by White and Stone described above.[25]

Several important conclusions were drawn from these observations: (1) for any given system the lag time appears to be almost constant regardless of magnitude of change of the gas feed rate; (2) the higher the vacuum level in the vacuum line, the shorter the lag time because the lower density of the gas provides a higher velocity; (3) lag time is independent of total volume of the vacuum line; (4) friction factor f varies significantly with Reynolds number (see Fig. 9 in the appendix); and (5) when the total pressure drop is greater than about 1.5 inches of Hg, there is a noticeable decay in the vacuum level of the entire line (see Fig. 9-107). Referring to this figure, the total pressure drop was about 1.5 inches of Hg at flows less than 500 lb/day. Above this flow (i.e., 500–5500 lb/day), the pressure drop was practically constant. It varied from 2.88 to 3.06 inches of Hg, while the vacuum level varied from 5.25 to 22.25 inches of Hg. Over this range of vacuum level and at a constant temperature of

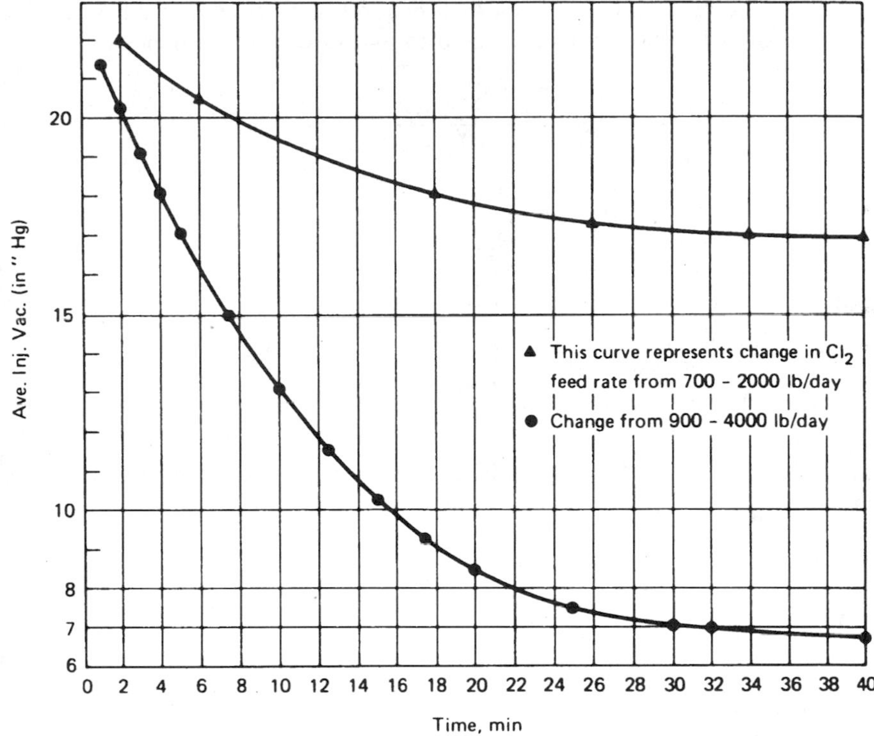

Fig. 9-107. Vacuum level decay in long vacuum lines due to excessive friction loss.

68°F the density of gas varied from 1.55 lb/ft^3 at the low vacuum to 0.048 lb/ft^3 at the high vacuum—this is a range of 32–1. From these values it is obvious that the flow conditions of this vacuum line are highly unstable. The instability of this system is a result of the high total pressure drop.

These observations bring up the question of the design procedure for an optimum size vacuum line for a given length and amount of gas flow. It would seem prudent to design for the severest conditions. Instead of designing for the minimum vacuum conditions allowable for the proper operation of chlorinators and sulfonators (about 8–10 in. Hg), the maximum injector vacuum level should be assumed (22–23 in. Hg). At this level the *total* pressure drop in the system, regardless of pipe length, should be limited to 1.50–1.75 inches of Hg. One such system designed to include these factors has been in operation long enough for operators to draw the following conclusion: it is a stable system with rapid response and therefore close to the optimum design. The maximum vacuum decay level from minimum to maximum feed rate is only about 3 inches of Hg (i.e., from 24 to 21 in. Hg, vacuum).

The design procedure is to first choose a line size. This particular case was for a dechlorination system—3300 ft from sulfonator to injector, 8000 lb/day maximum feed rate. Let us try a 4-inch Sch. 80 PVC pipe, $d = 3.826$ inches, using the following equation, corrected to pressure drop in inches of Hg:

$$\Delta P = \frac{11.89 \times L \times f \times W^2}{10^9 \times \rho \times d^5} \tag{9-8}$$

where

ΔP = total pressure drop, in. Hg
L = length of line in ft
f = friction factor from Fig. 9, appendix
W = lb/day Cl_2 or SO_2
ρ = density of gas, lb/ft^3*
d = inside pipe diameter, in.

To find f from Fig. 9 in the appendix, it is necessary to calculate the Reynolds number of the system:

$$N_r = \frac{6.32 \times \varpi}{\mu \times d} \tag{9-9}$$

where

ϖ = lb/hr gas flow
μ = viscosity of gas, cp†
d = inside pipe diameter, in.

So:

$$N_r = \frac{6.32 \times 8000}{0.0133 \times 3.826 \times 24}$$

$$N_r = 41{,}399$$

So from Fig. 9 in the appendix, $f = 0.027$; therefore:

$$\Delta P = \frac{11.89 \times 3300 \times 0.027 \times (8000)^2}{10^9 \times .05 \times (3.826)^5}$$

Also $\rho = 0.05$ lb/ft^3 at 22 in. Hg vac. and 68°F; so:

$$\Delta P = 1.65 \text{ in. Hg}$$

*For the gas density values, see Fig. 2 in Appendix I.
†For gas viscosity values see Fig. 3 in Appendix I.

At 6000 lb/day flow ΔP is about 0.16 inch of Hg. Now to calculate the lag time in this system it would be appropriate to calculate it for 4000 lb/day gas flow, but at 25 inches of Hg vacuum. Thus 4000 lb/day is 0.05 lb/sec, and the gas density at 25 inches of Hg vacuum and 68°F is 0.03 lb/ft³; so:

$$Q = \frac{0.05 \text{ lb/sec}}{0.03 \text{ lb/ft}^3} = 1.67 \text{ cfs}$$

$$V = \frac{1.67}{0.07986} = 20.91 \text{ ft/sec}$$

Therefore, the lag time for any change in gas feed rate is on the order of 3 min. We can show that the lag time would be nearly the same at 8000 lb/day, assuming a 22 inches of Hg vacuum level, 8000 lb/day = 0.09 lb/sec, and density = 0.048 lb/ft³:

$$Q = \frac{0.09 \text{ lb/sec}}{0.048 \text{ lb/ft}^3} = 1.88 \text{ cfs}$$

$$V = \frac{1.88}{0.07986} = 23.54 \text{ ft/sec}$$

So at double the gas flow the lag time decreased to about 2.35 min. Therefore, as the flow of gas changes, the density changes in a direction that provides an almost constant system lag time regardless of flow change.

It is to be noted that the physical properties of chlorine and sulfur dioxide gas under a vacuum are so similar that calculations for either are interchangeable; so, for simplicity, long vacuum lines can be designed for either gas based on the physical characteristics of chlorine.

Booster Pumps

Turbine vs. Centrifugal. Chlorinator or injector water booster pumps are as important as the injector or the chlorinator. If the booster pump is not large enough, the chlorinator will not operate. Therefore, never undersize or try to economize while attempting to select a pump. If anything, oversize. It is always a good idea to add 10 percent to the estimated pack-pressure, and be sure that the amount of chlorine to be fed is a little more than calculated.

Estimating the injector operating water flow depends on whether the pump to be used is a centrifugal or a turbine.

If the pump selected is a centrifugal, it will make a difference whether it is a unibuilt type or has an outboard bearing. If it is the latter, a higher wear factor for Q is allowable than if it is unibuilt. Picking a Q in excess of 15 percent of what the fixed throat injector will pass at the discharge pressure of the pump will overload the motor thrust bearing and cause undue wear.

A centrifugal pump with an outboard bearing can easily handle the injector Q plus 25 percent. With adjustable throat injectors this is not a problem because the injector Q is easily adjusted in the field.

A turbine pump, which is usually the one of choice for all fixed injector throat chlorinators, should be arranged with an adjustable pressure bypass assembly, as shown in Fig. 9-108. In these cases, the pump is selected to deliver $2Q$. In other words, if the injector requirement is 3 gpm, select a pump to deliver 6 gpm. The regulating bypass assembly allows the excess water to flow back into the suction. This is done by the needle valve in the bypass. As the pump wears over a period of time, and the pressure drops off, the bypass is closed off more to compensate for this wear.

With water that has a quantity of sand, it is more desirable to use a centrifugal pump because sand will ruin a turbine pump in a very short time.

Most of the chlorinator booster pump installations require a turbine pump because of high head and low Q requirements. For example, all chlorinators in the capacity range of 3 to 400 lb/day use a fixed-throat injector. The largest fixed-throat injector will pass about 22 gpm. Most injector water requirements are for less than 15 gpm.

Centrifugal pump selection usually begins when the injector operating water approaches 15 gpm.

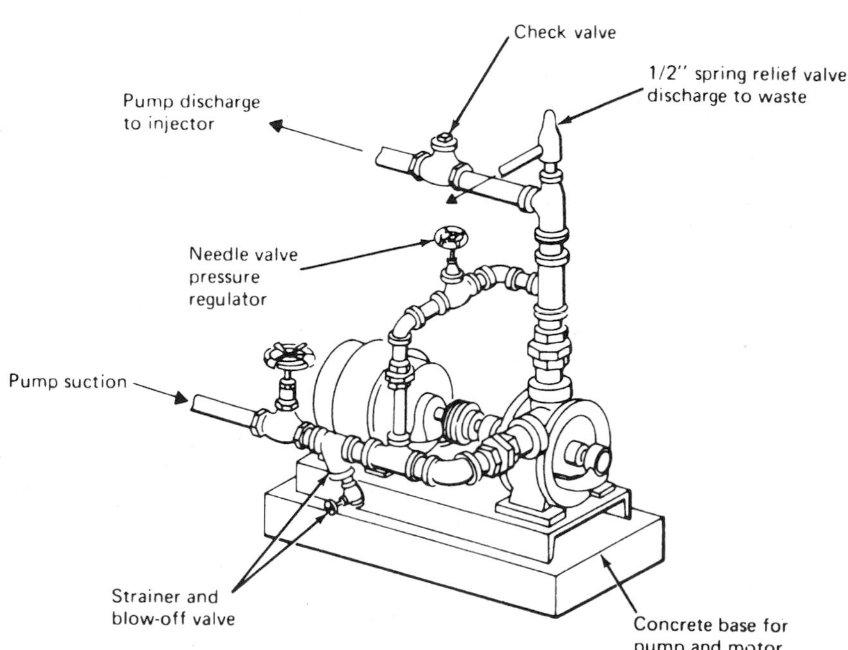

Fig. 9-108. Turbine type booster pump showing bypass piping (courtesy Wallace and Tiernan).

High Back-Pressure Conditions. There are situations of high static pressure that are beyond the ability of high-head multistage 3500 rpm turbine pumps and/or high-head centrifugal pumps. When conditions exist that are beyond the capabilities of these pumps, the chlorine solution has to be pumped. It is fair to say that the upper limit for a conventional turbine pump application would be a condition for a fixed-throat injector where the maximum chlorine feed rate is 400 lb/day and the maximum back-pressure 140 psi. When the chlorine feed rate and back-pressure combination exceeds 100 lb/day and 75 psi, respectively, pumping chlorine solution should be investigated. When a 2-inch injector is required, pumping chlorine solution should be investigated when the combination of chlorine feed rate and back-pressure is 1000 lb/day and 60 psi, respectively. For 3-inch and 4-inch injectors the chlorine solution should be pumped when the back-pressure reaches 15–20 psi.

When it becomes necessary to pump the chlorine solution, the pump system should be arranged as shown in Fig. 9-109. The two basic requirements for this system are: (1) the regulating valve should be adjusted so that there is always approximately 2–5 psi back-pressure on the injector (a negative head must be avoided); and (2) the pump should be sized to deliver the required injector water plus an additional 30–40 percent dilution water.

This system operates on the principle that the spring-loaded regulating valve creates a pressure drop when the pump is operating. The pressure on the downstream side of this valve becomes the injector "back-pressure." There-

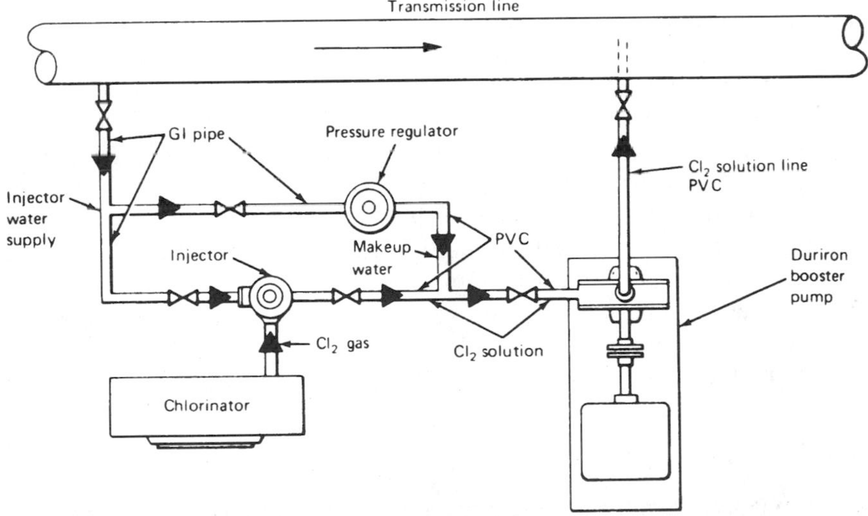

Fig. 9-109. Schematic arrangement showing booster pump downstream from injector. All pipe and valves downstream of injector and pressure regulator must be corrosion resistant, i.e., PVC, rubber-lined steel, Saran-lined steel, or fiber glass. Gauges should be installed to measure injector water supply pressure, reduced pressure to pump suction (2–5 psi), and pump discharge pressure.

fore the pressure drop across the regulating valve plus the friction loss in the system equals the TDH* requirement of the pump. If the chlorinator utilizes a fixed-throat injector, the largest-size throat should be used to ensure the minimum chlorine solution concentration.

Two types of pumps are available for this service. The type that has been in service for the longest time for pumping chlorine solution is the titanium pump line of the Duriron Co.[26] The City of San Francisco Water Dept. has reported excellent service by these pumps for more than 40 years. In recent years the Duriron Co. has made a line of reinforced Fiberglas pumps for this service.[27] The Fybroc Division of the Met Pro Corporation also has a line of Fiberglas pumps available for this service.[28]

Example. Assume a transmission line with a maximum static head of 150 psi and a chlorine requirement of 1000 lb/day. Using Wallace & Tiernan Injector Performance Curve 25.100.190.021 shows that the best a 2-inch injector can do with 150 psi initial water pressure is 80 psi back-pressure at 80 gpm. Therefore, the chlorine solution pump must be able to boost 80 gpm injector water plus about 25 gpm makeup water to 155 psi plus 10 psi friction loss, or 165 psi. So $165 - 80 = 85$ psi $\times$ $2.31 = 196$ ft TDH. This situation calls for a pump to deliver 105–110 gpm at a TDH of 205–210 ft.

Designer's Checklist for Injector System. The injector system and chlorine solution lines usually include all of the following components:

1. Injector water-pressure gauge.
2. Injector vacuum gauge for remote injector installations.
3. Injector vacuum line shut-off valve at remote injector location.
4. Chlorine solution pressure gauge located immediately downstream from the injector to indicate injector back-pressure (not required on fixed throat injector installations).
5. Injector water-pressure switch for low-water-pressure alarm.
6. Injector water flow meters for multiple chlorinator installations of chlorinators using 2-, 3-, and 4-inch injectors.
7. Chlorinator–sulfonator built-in vacuum switch and alarm for both high and low vacuum.
8. Compound back-pressure gauge for injector discharge.

Water Champ

This is probably one of the most important recent inventions for the application of chlorine gas to achieve disinfection of both potable water and wastewater. It is thoroughly described in Chapter 8.

*TDH = total dynamic head.

The following information pertains entirely to the application of chlorine gas in potable water systems. Since the publication of the third edition, a large number of Water Champ installations have been made for these systems.

Figure 9-110a illustrates how the Water Champ is installed in a water pipe or a transmission pipeline. Figure 9-110b illustrates another method of installing the Water Champ in pipelines.

High pH Waters. In two installations where the Water Champ was being used to disinfect the effluents of lime-treated surface waters, operating person-

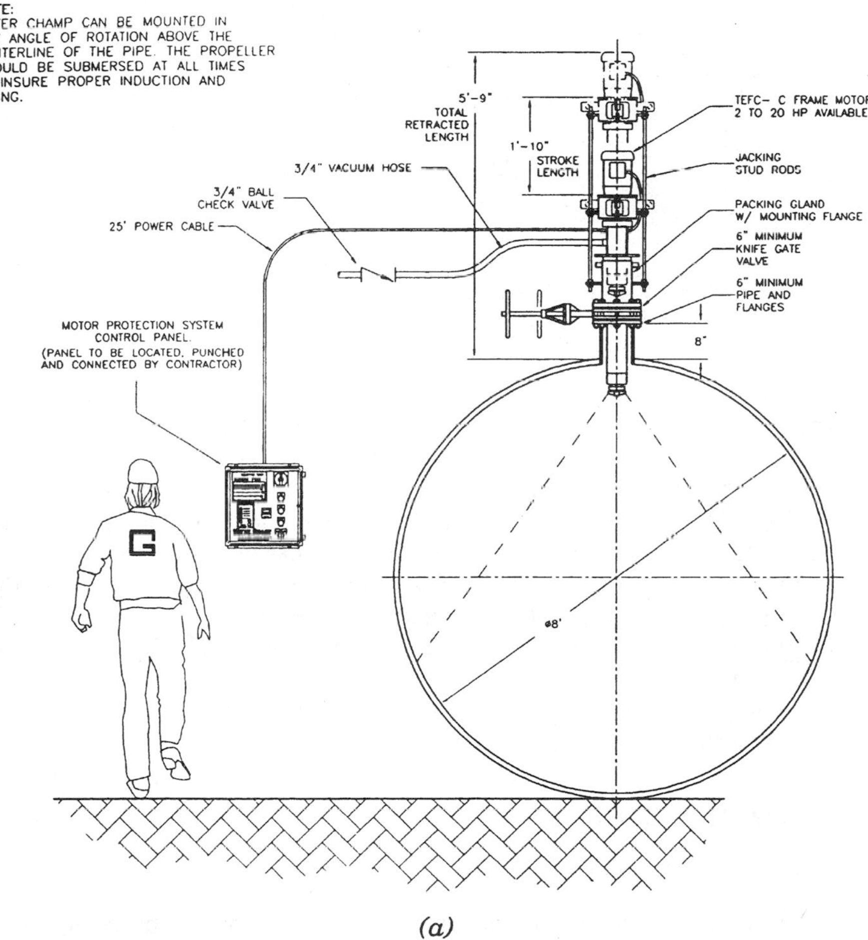

Fig. 9-110a. Typical Water Champ installation (courtesy of Gardiner Equipment Company). Note: Water Champ can be mounted in any angle of rotation above the centerline of the pipe. The propeller should be submersed at all times to ensure proper induction and mixing.

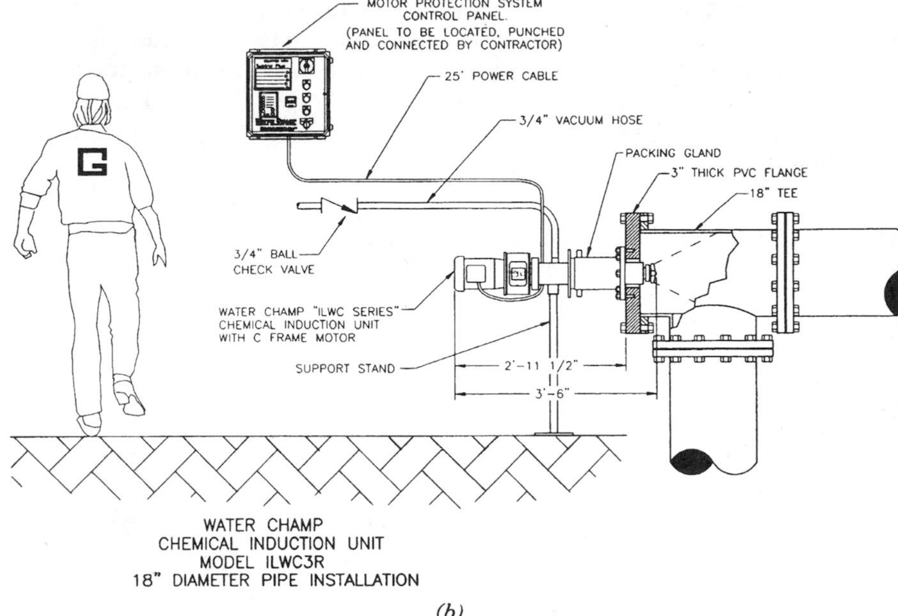

(b)

Fig. 9-110b. Water Champ chemical induction unit, model IL WC3R, 18″ diameter pipe installation (courtesy Gardiner Equipment Company).

nel called attention to the fact that after installing the Water Champ, they noticed a saving of up to 25–30 percent in chlorine consumption.

This proved that the Water Champ, which injects chlorine gas into the process water, simply utilizes chlorine at its maximum oxidizing potential of 1.36 mV, whereas in waters at pH 7.5 the HOCl oxidation potential is about 0.75 mV. This why in wastewater chlorination, it has been observed in several installations of small plants (10–15 mgd), where the W.C. is being used, that destruction of nitrite formation has been observed.

Without the W.C., the oxidation potential is so limited, using chloramines formed from the presence of ammonia N in the wastewater, that there is no possibility of destroying nitrite formation. This causes serious operating problems in these plants. (See Chapter 8.)

CHLORINATION STATIONS

Principal Considerations

The design and layout of any chlorination station must be based upon the following considerations:

1. Chlorine container selection
2. Chlorinator capacity

3. Points of application
4. Injector requirement
5. Methods of control
6. Alarms
7. Safety equipment
8. Chlorine residual testing facility
9. Chlorine storage inventory system
10. Housing requirements—space, ventilation, lighting, and heating
11. Reliability provisions

All of these items are addressed in detail elsewhere in this text.

Individual Deep-Well Station

This is probably the most widely used type of chlorinator station. The chlorine may be applied either down the well or at the well pump discharge. Figure 9-111 shows the application down the well. Whenever possible, this is the preferred method for a variety of reasons:

1. No injector booster pump is required.
2. Chlorine solution lines may be used simultaneously to feed a sequestering agent to prevent deposition of iron or manganese in the distribution system.
3. The chlorine solution line provides means for intermittent purging and cleaning of the aquifier with heavy doses of chlorine.
4. It provides longer contact time.

Special attention should be given to the point of application and the injector hydraulics. The chlorine solution line must terminate at least 10 ft below the pump bowls; the chlorine solution strength should be limited to about 100 mg/L; and the alkalinity of the water should be at least 100 mg/L.[29] These factors prevent chlorine corrosion of the pump and well casing.

The injector water line must be taken off the pump discharge line downstream from the check valve. Therefore, when the pump shuts down, the pipeline from the check valve drains back into the well casing. This creates a negative head on the upstream side of the injector, which soon equalizes to become the same as the negative head in the chlorine solution line. This equalizing of negative head on both sides of the injector allows the chlorinator to automatically shut down without any siphon action from the chlorine solution line. If the injector water line is taken off downstream from the pump discharge check valve, the chlorinator will continue to operate regardless of the well pump. Putting a solenoid valve in the injector water line at this point is useless because the chlorinator would continue to operate without injector water, owing to the siphon action of the negative head on the chlorine solution line.

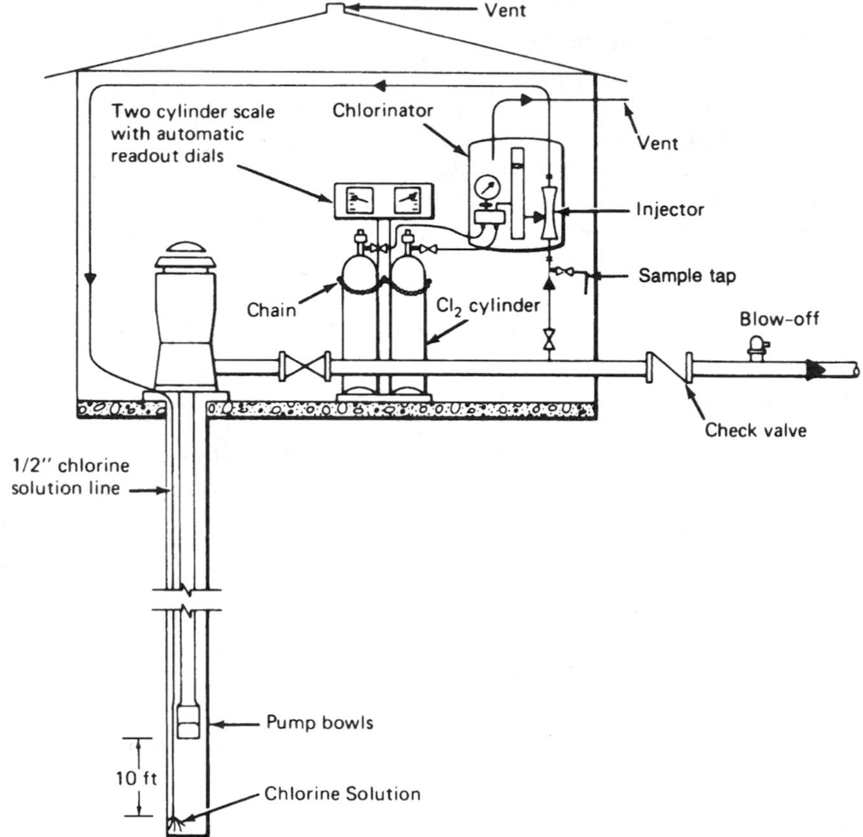

Fig. 9-111. Typical chlorinator installation with chlorine applied down the well.

When it is not convenient to put the solution line down the well, the alternative method is to utilize a chlorinator injector booster pump and apply the chlorine at the well discharge, as shown in Fig. 9-112. In this case the booster pump suction and the chlorine solution discharge must be downstream from the well pump check valve. Then, when the well pump shuts down, the injector pump also shuts down, and the pressure on both sides of the injector equalizes, thereby automatically stopping the chlorinator. This is known as semiautomatic operation.

Multiple Wells or Pump Stations

If there is more than one well station discharging into a common line, it is often preferable to use one chlorinator and one point of application in the common well discharge line. For such a situation the chlorinator control could

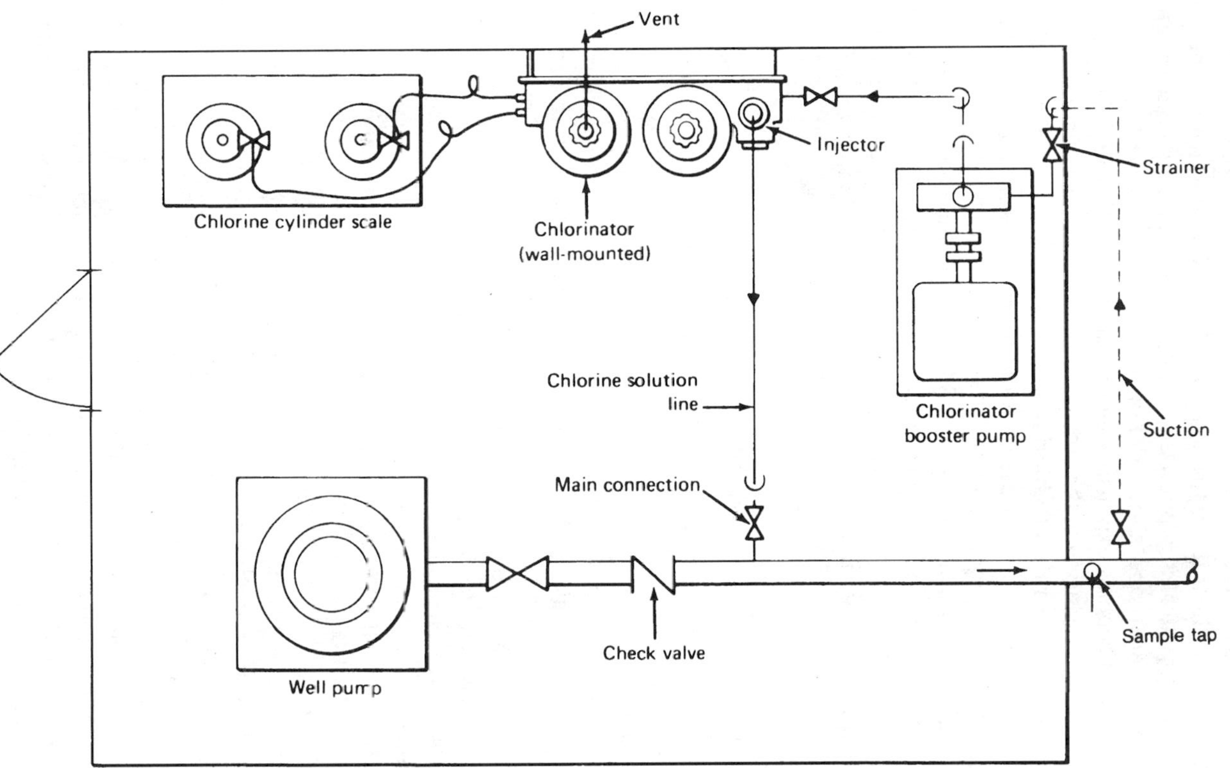

Fig. 9-112. Typical chlorinator installation with point of application in well pump discharge.

be either flow-proportional from a primary flow meter or step-rate from an adjustable-rate controller.

For purposes of illustration, the method of step-rate control will be used. This method is practical at a pumping station when the pump controls are at one location or near one another. For widely scattered well stations, the electrical wiring becomes expensive and cumbersome. Figure 9-113 shows a schematic of a pump station with three pumps of different capacities arranged to provide a separate rate of chlorination for the various combinations of pumps. The usefulness of step-rate control diminishes as the number of rates exceeds four or five, simply because the electrical interlocking gets overly complicated. However, if a primary metering device is not going to be provided at a multiple-pump station, the only method that can be used for chlorinator pacing is the step-rate type. With this in mind, the designer should select pumps of similar capacities, so that the control is limited to only four or five rates.

Relay Stations and Transmission Lines

Probably the only simple manual control chlorination stations are those that treat a constant flow of water that is being transferred from one reservoir system to another, with no user takeoffs between. These examples of constant flow are hydraulic rarities. There is usually some fluctuation in flow worth accounting for, particularly if the objective of chlorination is to provide a chlorine residual in the entire transmission line for purposes of maintaining the "C" factor or water quality control.

When it is necessary to maintain chlorine residuals in a distribution system, relay stations are often used. Because of problems of logistics, it is not necessarily convenient to install a primary metering device to pace the chlorinator; it is generally more desirable to rely entirely on straight residual control. The schematic for such a system is shown in Figs. 9-84 and 9-114. The key to a successful direct residual control system is to design a short loop time, 60–90 sec at average flow, and utilize the chlorinator injector pump to supply the sample to the residual analyzer. Both the chlorine solution diffuser and sample takeoff should incorporate the design principles described for Figs. 9-84 and 9-114.

This method is suitable if the flow changes in the distribution system reflect the usual diurnal domestic consumption. The chlorine residual controller can react fast enough to keep up with these fluctuations. Such a control cannot be used alone if the change in flow is abrupt, as reflected in a step-rate change caused by a nearby pumping station.

Reservoir Outlets

General Discussion. It is common engineering practice in distribution system design to use balancing reservoirs that float on the distribution system

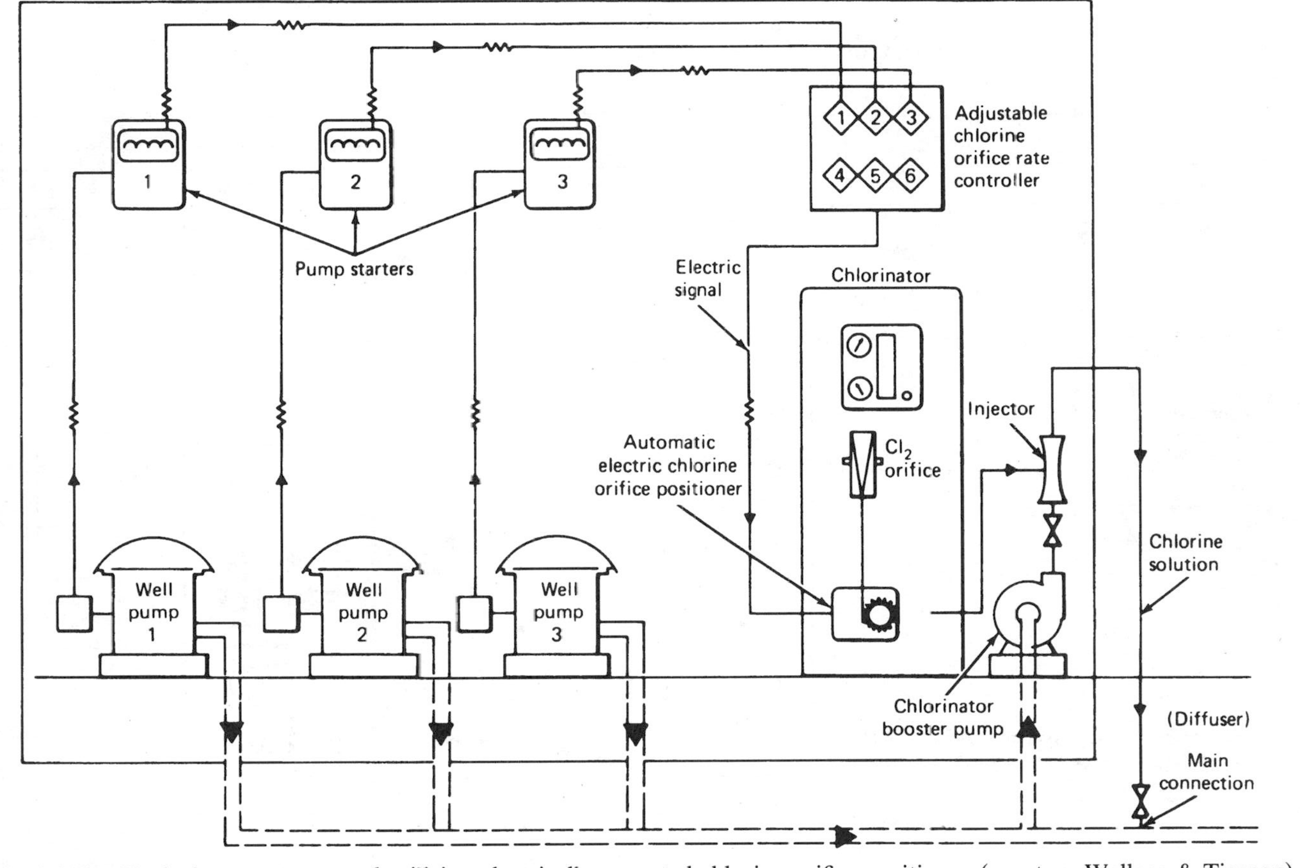

Fig. 9-113. Typical step-rate control utilizing electrically operated chlorine orifice positioner (courtesy Wallace & Tiernan).

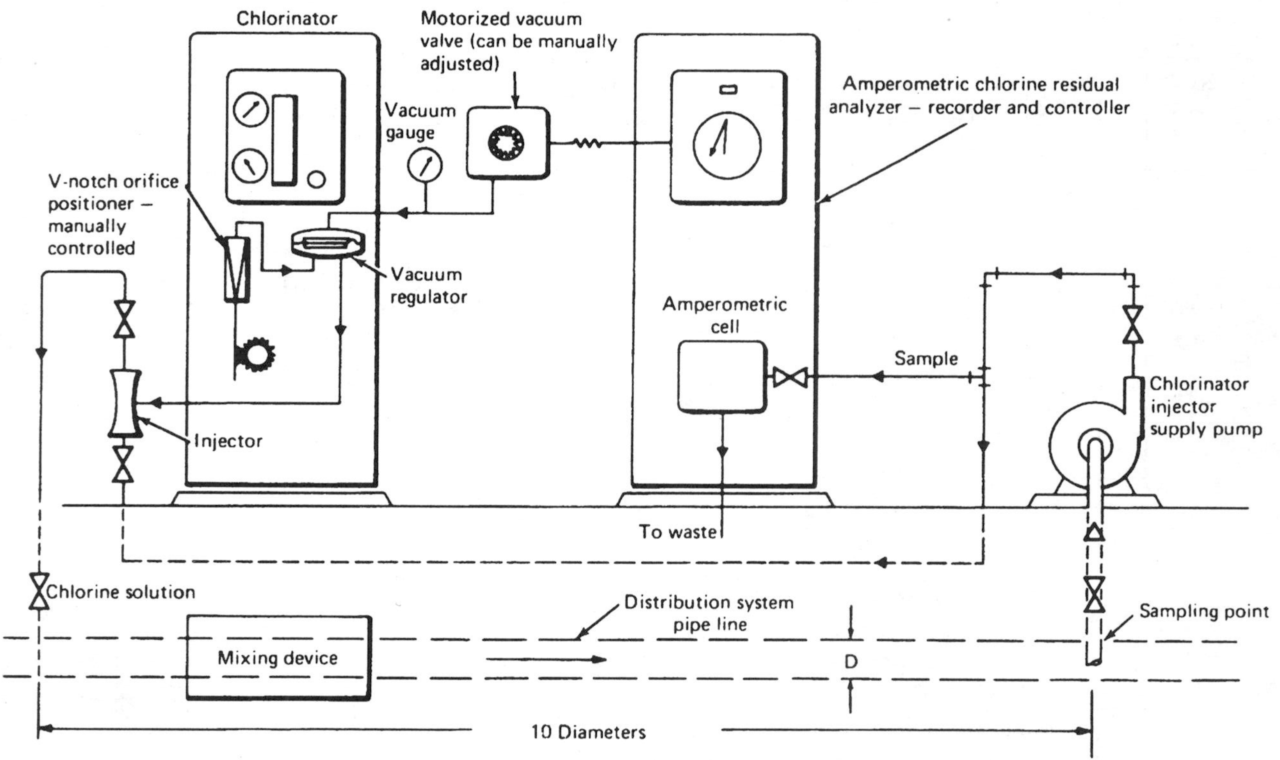

Fig. 9-114. Distribution system relay station with direct residual control.

hydraulic gradient. This is a convenient way to balance the flow in a large system. It also helps to solve the problem of pressure equalization by using reservoirs as pressure breaks in hilly terrain. The flow into and out of these reservoirs is almost always by a common pipe. Reversal of flow does occur in many of these systems where the outflow is a function of distribution system demand. During daytime hours the reservoir is supplying water for normal and peak demands, but during low-flow conditions in the early morning hours the system may be designed to fill the reservoir, thereby causing at least a once-daily flow reversal.

It is also common practice to chlorinate the outflow of a surface reservoir. This in itself is not an easy task. However, providing adequate and reliable chlorination for a reverse flow condition is even more challenging.

The usual method of chlorination control solely by flow pacing from a primary meter is a risky practice because of abnormally wide daily flow ranges experienced by these reservoir systems. Systems designed for flow reversal have much wider flow ranges because the flow has to reach zero before it can reverse. Therefore in every one of these systems the range is infinite (i.e., zero to any maximum value).

Chlorination control for reservoirs will be discussed below in two categories: (1) outflow only and (2) unpredictable reversal of flow. However, it is necessary first to discuss primary meters and reverse flow detection.

Primary Flow Meters. There is a wide variety of flow meters available, but none has the accuracy to provide the necessary chlorine dosage control over the flow ranges experienced at the majority of these systems. The designer or the operator should accept the flow meter as most accurate and acceptable for chlorinator control in the range of 4:1. Therefore, these chlorination systems should be supplemented with residual control. Retrofitting existing systems using existing flow meters must add residual control. The principal flow-meter accessory for chlorination control is an analog flow signal transmitter, preferably 4 to 20 mA or 10 to 50 mA. A pneumatic signal is acceptable but not preferred for a variety of reasons. The reasoning behind establishing an "accurate" range (4 to 1) for the flow meter is to allow this signal to blend properly with the residual signal. This is described below.

Flow-Reversal Detection. This is a difficult task in any hydraulic system because the basic requirement is the ability and assurance of being able to detect zero flow. Reliable devices such as the thermal probe system[30] are expensive. Laminar-type flow switches that flex with the waterflow are inherently insensitive to flows less than about 0.5 ft/sec. Other devices that operate on a turbine-type movement are more sensitive.

The need for a flow-reversal switching device in a chlorination system is a site condition requirement. The user may find it desirable to shut the chlorination system down during this period, or to be able to actuate valves to direct reverse flow to a compartmented reservoir.

Any chlorination system dealing with a reverse flow condition must have added protection of adjustable time delay relays. These relays will protect the chlorination control system from chatter due to the oscillating prism of water usually encountered when the flow makes the transition from forward to reverse. Appropriate switch gear can lock in the forward or reverse function long enough to allow the system to stabilize in either direction.

Reversal Detection by Chlorine Residual Analyzers. This type of reservoir chlorination system requires two chlorine residual analyzers arranged as shown in Fig. 9-115. In order to explain the operation of the analyzer control system the function of each principal element of the system will be described.

1. *Chlorine diffuser.* This is the jet injection type for a pressurized conduit. For pipes 36 inches in diameter or less, terminate the nozzle *D*/6 inches from the inside wall. For conduits larger than 36 inches, use an across-the-pipe diffuser perforated across the middle third of the diameter (see Fig. 9-83). The injection nozzle should be fitted with an orifice to provide an exit velocity of 22–26 ft/sec. Use the same velocity for the across-the-pipe diffuser. This diffuser design will eliminate the need for a mechanical mixer, provided that the sampling taps are approximately 10 diameters from the chlorine diffuser.
2. *Injector water supply.* Design piping and the injector pump to provide as much water as the injector will allow regardless of the chlorine flow rate. This will enhance mixing and help to maintain a reasonably constant loop time over a wide water flow range.
3. *Injector location.* The injector should be placed as close to the point of application as is practical.
4. *Sample taps.* The sample taps for each of the analyzers should terminate at the center of the pipe for diameters up to and including 36 inches. For larger pipes the sampling taps should be made the same as the chlorine diffusers. The sample piping for Analyzer No. 2 should be designed for about 3 gpm at 5 ft/sec to provide a scouring velocity and to minimize loop time. Diffuser-type taps should have sufficient holes to limit the entrance loss to 1 ft.
5. *Injector water pump.* This pump serves three functions: (1) it operates the injector; (2) it provides excess injector water to ensure good mixing and a more constant loop time; (3) it provides sample water to both analyzers.
6. *Analyzer No. 1.* This analyzer controls the chlorine residual in the outflow water to the distribution system. It is assumed that there will always be a chlorine residual in the sample line to this analyzer.
7. *Analyzer No. 2.* This analyzer operates on the biased sample principle. This arrangement is necessary in order that the analyzer may stay in calibration; that is, it must not be exposed to zero chlorine residual for any length of time. This is the analyzer that detects the flow reversal, as described below.

This system controls the chlorinator feed rate in the outflow mode by direct residual control, as described in "Control Strategies," earlier in this chapter. Assume that the residual control band is set between 0.4 and 0.6 mg/L. In the forward direction both analyzers will be receiving a sample in this band

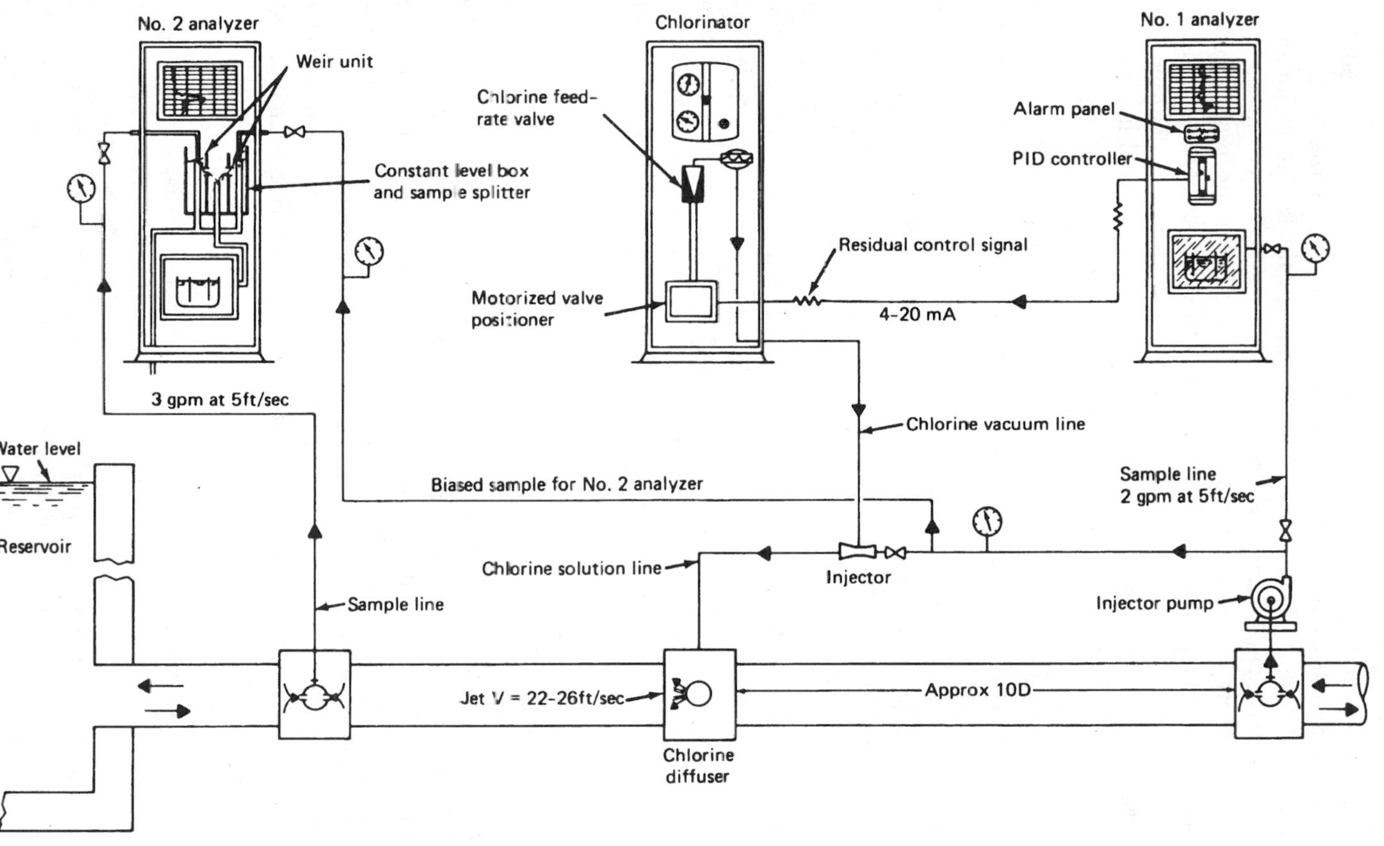

Fig. 9-115. Flow reversal detection and direct residual control using two chlorine residual analyzers (unpredictable reversal).

range, except that Analyzer No. 2 cell will receive a sample diluted 1:1 with unchlorinated reservoir water. Therefore this analyzer will actually be "seeing" a residual in the 0.2 to 0.3 range. This is sufficient to maintain its calibration equivalent to Analyzer No. 1.

When the reservoir flow begins to reverse to the inflow mode, nothing much will happen for several minutes. Then there will be evidence of residual decay in the distribution system, and Analyzer No. 1 will call for more chlorine; this will not be seen by Analyzer No. 1 but will appear as a residual increase on Analyzer No. 2. When this residual (which is diluted) reaches an arbitrary level of say 0.6 mg/L on Analyzer No. 2, an adjustable residual alarm switch is tripped actuating a time delay relay, which "freezes" the chlorine feed rate at that dosage level and locks out the residual control signal from Analyzer No. 1.

When the reversal begins to change to the outflow mode, the chlorinated portion of the reservoir inflow water will pass by Analyzer No. 2 without incident and begins to register on Analyzer No. 1. By field observation of this occurrence, the operator will be able to determine where to set the low-residual alarm switch on Analyzer No. 2. This probably will be about 0.1 mg/L. When this low-level alarm switch is tripped, it actuates a time delay relay, which restores the automatic chlorinator controls to Analyzer No. 1. The key to success of this system or any direct residual control system is a short loop time, 60 to 75 sec at average flow, and good mixing by short-circuiting the chlorine solution to the sample tap by utilizing the injector water pump as a sample pump.

Water Treatment Plants

General Discussion. A chlorination station for a conventional treatment plant should be designed for precise control, flexibility, reliability, and safety. It should also be provided with the necessary monitoring equipment and alarms for critical functions plus the required safety equipment for dealing with chlorine emissions. The principal considerations are: points of application, chlorinator capacity, control system, contact time, monitoring and control analyzers, injector system and chlorine solution lines, safety equipment, and standby equipment.

Points of Application. This is the first consideration because it prefaces the number of chlorinators required. In spite of the recent concern about the THM formation resulting from prechlorination, this remains one of the crucial points of application. This is described in Chapter 6. Normally this point is located to allow 2 or 3 min. of contact time ahead of the flash mixers. Although the plant may be designed to use the free residual process and carry a free residual through the entire plant to the clear well, an intermediate point of application should be provided just ahead of the filters. This point may be used intermittently on a manual basis for shock-type treatment to clean up any filter biofouling that may occur. In some plants this may occur often.

The other point of chlorination would be for final disinfection or postchlorination for distribution system residuals. In some plants that use the free residual process, this point of application is used for dechlorination to trim the residual entering the distribution system.

Chlorinator Capacity. Somewhere in the plant there will be a flow meter with a given range. The range of this meter should be the basis for the chlorinator capacity. For example, the engineers may be talking about a 10 mgd plant but will be providing a 20 mgd flowmeter. To simplify chlorinator selection, provide a pre- and postchlorinator with the capacity to treat 20 mgd—the full range of the flow meter. Then if one or the other is not in continuous use, it can be used for pre- and/or intermediate chlorination or as a standby. In any event, one chlorinator must be available for standby service.

The actual capacity of the equipment must be calculated on the basis of the 30-min. chlorine demand of the water to be treated. Chapter 6 delineates chlorine dosages for the various points of application.

Control System. If prechlorination is to be continuous, it should be flow-paced-controlled and monitored by a residual analyzer after 15–20 min. contact time. Prefilter chlorination should be manually controlled only. Postchlorination should be under a combination of flow-paced and residual control. Flow pacing for prechlorination should be from an influent meter, and postchlorination from an effluent meter. Flow pacing prechlorination by an effluent meter or postchlorination by an influent meter should never be attempted. The time lag and intermediate uses of plant water for backwashing, and so on, destroy the control function of flow pacing.

Mixing.* If the chlorine diffusers are designed in accordance with their recommendations of the "diffuser design" section of this chapter, there will be no need for additional mechanical mixers. (See Fig. 9-83.)

Contact Time. The only contact time that may be critical in a conventional treatment plant is postchlorination. Usually the clear well is used as a contact chamber. Otherwise the transmission line may be used to provide the necessary contact time before the water reaches the first consumer.

If the treatment plant is relying upon a free residual for disinfection, 10 min. contact time is sufficient. If chloramine treatment is used, the contact time should be increased to 50–60 min.

---

*See also diffusers as a mixing device in Chapter 8.

Monitoring and Control Analyzers. Chlorine residual analyzers should be used to monitor all prechlorination activities. Also analyzers should be used to monitor and control the residual in the water entering the distribution system. Large water systems will also find it useful to monitor chlorine residuals at strategic points or trouble spots in the distribution system.

Safety Equipment. All treatment plant chlorination stations should be equipped with the safety equipment described under that section.

Injector System and Chlorine Solution Lines. Experience has demonstrated that it is preferable to locate the injectors adjacent or near to the diffusers. The use of remote injectors eliminates the temptation to manifold the chlorine solution lines in the chlorinator room. Manifolding solution lines and using rotameters to split the chlorine flow downstream from the injector is excessively costly and reduces the chlorination system to an uncontrollable hazard resulting in an operator's nightmare. Locating the diffuser adjacent to the point of application requires that there be no more than one diffuser per injector and no more than one injector per diffuser. In some cases the vacuum lines to the various injectors may be manifolded at the chlorinators to achieve redundancy—but splitting the chlorine solution discharge from the injector should not be permitted.

CHLORINE RESIDUAL ANALYZERS

Historical Background

There are two methods by which the chlorine residual can be continuously analyzed: colorimetric and amperometric.

The colorimetric method, using orthotolidine (O-T) reagent, has been tried and discarded many times. Wallace & Tiernan Company began investigating the possibilities of automatic chlorine residual control using orthotolidine about 1927. Units were developed that could continuously record the O-T residual and at the same time adjust the chlorinator feed rate. The first of these was installed in 1929 at Little Falls, New Jersey; the second, in 1930 at Rahway, New Jersey.[31] A modification of these units was installed about 1930 in Los Angeles,[32] and operated for many years. One of these units is shown in Fig. 9-116. The latter units differed in that the O-T residual analyzer had a fixed color disk for a given O-T residual of 0.3 ppm. If the color in the analyzer did not match that of the disk, the error signal between the two drove the control valve on the "trimming chlorinator" until the two colors matched, and the error signal became zero. The residual record then was one of deviation. Some years later Caldwell[33] developed an O-T recorder based upon Harrington's work,[34] using neutral orthotolidine. The colorimetric recorders fell into disuse primarily because the photoelectric cells were not able to differentiate accurately enough between the relatively small changes in the

Fig. 9-116. Colorimetric O-T analyzer used for chlorine residual control by City of Los Angeles Department of Water and Power from 1930 to 1950s at McClay Highline Station.

color representing a residual change. Also, after it was realized how important free chlorine residual was, compared to total chlorine, another method had to be found, since it was not practically possible to do this with orthotolidine.[35] In addition, awkward mechanical operation helped make the system impractical.

Circa 1955 Fischer and Porter developed a sophisticated O-T analyzer, which survived for a few years until its amperometric analyzer was developed. Hach Chemical Co. also marketed an O-T analyzer about this same time, which was unable to survive competition from the amperometric analyzers. However, Hach did introduce the first successful DPD colorimetric analyzer in 1980. This unit is described below.

The search for a better method of chlorine residual determination resulted in the development by H. C. Marks of Wallace & Tiernan Company, about 1942,[36] of the amperometric titrator, in part inspired by Griffin's work in

1939–40 on the breakpoint phenomenon. The first amperometric chlorine residual recorder was placed on the market by Wallace & Tiernan Company about 1948. Among the first installations were those at Wyandotte, Michigan;[37] Allentown, Pennsylvania;[38] and Brantford, Ontario, Canada. Results from these early installations proved so dramatic that continuous residual recording became popular in the treatment of public water supplies. Operating experience from these installations demonstrated two important characteristics of the amperometric cell. These were predictable changes in cell output current due to changes in sample temperature and pH. Changes in pH were corrected by additions of pH 4 buffer solution, and temperature was corrected either by manual adjustment or automatically by the installation of a thermistor.[39] Most current models come equipped with a thermistor.

During this period Wallace & Tiernan Company made a polarographic cell for the City of Chicago Water Department that was arranged to measure combined (chloramine) chlorine only. It did not require chemicals but it used a large sample, one liter per minute. Twenty years later, when Chicago built the Central Water Filtration plant, analyzers were required to measure free chlorine only. The same type of analyzer was used except that the applied voltage across the electrodes was changed. This made the same cell specific for free chlorine.

Since the 1970s two notable contributions have been made. The first was the development by Morrow[40] for Fischer and Porter Co. of the Chlor-Trol analyzer, which was capable of measuring free chlorine in a sample containing significant chloramine concentrations. The second was the development of the selective membrane electrode by Johnson et al.[41] By this time many companies were actively engaged in developing instrumentation for continuous chlorine residual measurement. These participants included Foxboro, Orion, Uniloc (Rosemount), Delta Scientific, Enterra, Chlortect, and IBM. Other accomplishments have been modifications that have extended the use of chlorine residual monitoring in wastewater, and finally the step from monitoring to automatic residual control.

The step from residual recording to residual control took nearly 10 years to gain acceptance in the United States. Residual control was being used in England for the London water supply as early as 1950 and for controlling superchlorination and dechlorination stations of underground supplies where the only other treatment was micro-straining.

The use of continuous residual analyzers for wastewater chlorination began ca. 1960. The immediate purpose was to control the chlorination system. Prior to 1960 there was limited use of analyzers on highly polished tertiary effluents. The pathway to successful analyzer operation on wastewater effluents has been a rough one. One of the first successful installations on secondary treated effluents was made at the Napa Sanitary District, Napa, California, in 1961, where the unit installed was the same as the version used for potable water, except that a motorized filter was installed on the analyzer cell sample line.

When this unit was tried for operation on primary treated effluent, various problems were encountered, related to the quality and the characteristics of the wastewater. The high grease content in the primary effluent is particularly bothersome in that it coats the electrodes. The insulation effect of the grease causes a severe distortion of the calibration, resulting in erroneous readings. This became an unwarranted maintenance problem.

The amount of buffer solution required to lower the pH of most primary effluents is very high, and an appreciable excess of iodide is also necessary to make the reaction with combined residual chlorine proceed as fast as possible. This made the chemical cost of operation a most significant factor.

Experience has shown that chlorine residual analyzers are not reliable for use with primary effluents owing to the poor quality of the sample, and the inability to maintain a sample flow through the cell. Another factor discouraging to operating personnel is the time required to keep the electrode surfaces clean enough to maintain system calibration. Recalibrations for primary effluent samples need to be made at least twice a day.

For many years chlorine residual control was used solely as an adjunct to flow-paced control. Now there are many installations controlled solely by chlorine residual.

The advent of the programmable controller has stimulated interest in the use of residual control. This type of instrumentation allows the precise formulation of a residual control system to any site-specific case.

General Discussion and Definitions

Amperometric. Continuous analyzers that use two electrodes for measurement are usually categorized as amperometric. This was true when referring to the original Fischer and Porter and Wallace & Tiernan analyzers. However, it is no longer true with the array of analyzers currently available.

Electrode reactions are characterized by the transfer of electrons between the electrode and substances in the sample solution. The electrode reaction that initiates the electron transfer results in the flow of current between the electrodes. When the electrode reactions occur spontaneously upon short-circuiting of the two electrodes, the system is described as a galvanic cell. If, however, the reactions are forced to occur by the imposition of an external electromotive force, the cell is referred to as an electrolysis cell (i.e., with the generation of chlorine). In the above definitions the term "reactions" refers to the chemical changes of the solution that might occur. In the measurement of chlorine residuals the current flows between the electrodes primarily as a characteristic of the species involved. The dual electrode analyzers such as those made by Capital Controls, Fischer and Porter, and Wallace & Tiernan are galvanic cells—the amperometric type. These cells respond in a linear fashion only to the elemental halogens (bromine, chlorine, and iodine). They do respond to most chloramines, but not in a linear fashion.

Voltammetry. When the galvanic cell is subjected to an external applied voltage to provide the measurement of a particular chemical species in the presence of other species, the system is described as voltammetry. In this case the voltage applied provides qualitative information on the electroactive substance present, and the current measured provides quantitative information on the species selected by the applied voltage. This technique is used to isolate the free chlorine species that occurs in the presence of chloramines. This is a variation of the amperometric measurement technique. The term "amperometric" has come into use to indicate concentration measurement based upon the measurement of the current flowing between two electrodes at a constant potential at the indicator electrode.

Polarography. This is a broader term, used to describe analytical measurement of trace materials in solution, metallic ions, and so on.[42] This method utilizes the principles of voltammetry, that is, current–voltage curves, but with a polarographic cell. A polarographic cell is an electrolytic cell consisting of a nonpolarizable reference electrode, a readily polarizable electrode, and an electrolyte solution containing electro-oxidizable or electro-reducible material.

At this point the difference between a polarized and a depolarized electrode should be understood. An electrode is said to be polarized when it adopts the externally impressed potential (voltage) on it with little or no change in the rate of the electrode reaction (i.e., no change in current). Therefore, if only one electrode in the cell is polarized, its potential changes in the same amount as the change in applied voltage.

At the other extreme, an ideally depolarized electrode is one that retains a constant potential regardless of the magnitude of the current flowing between the electrodes. Therefore, the nonpolarizable electrode potential is not altered by the changes in the externally applied voltage.

When an increasing electromotive force (applied voltage) is impressed across the electrodes of a polarographic cell, and if the resulting current is plotted as a function of the applied voltage, a curve is obtained, as shown in Fig. 9-117. The extension of the wave curve along the current axis is directly related to the concentration of the material in the solution sample, and its inflection point (0.8 V) is at the applied voltage which is characteristic for this particular species of material. Examples of this type of cell for measuring species of chlorine residual are described below.

Membrane Cell. These probe-type chlorine residual monitors are descended from the work of Johnson et al.[41] They operate on the same principle as polarographic cells except that they use selective membranes—one for free chlorine and another for total chlorine residual. The exact materials for the membranes are always proprietary items. The cell design is such that it can monitor chlorine residuals without the need for chemical addition, which is required by the amperometric cells. The sensor consists of two noble metal

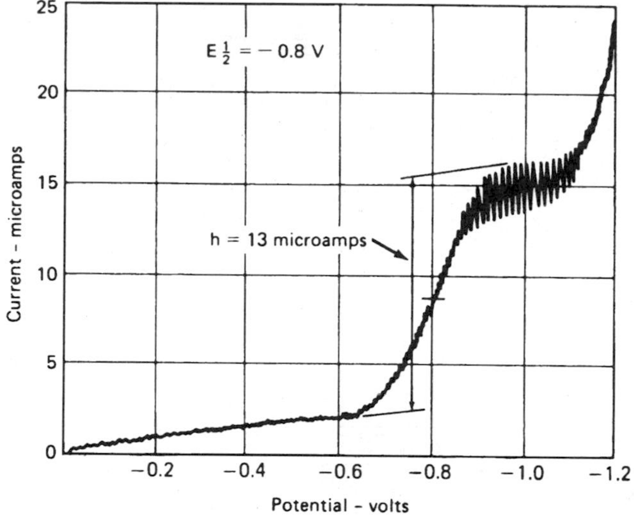

Fig. 9-117. Polarographic curve.

electrodes immersed in a common electrolyte. The electrodes are isolated from the monitored sample by the chlorine-permeable polymeric membrane. (See Fig. 9-118.) In essence the sensor is much like a battery, with the chlorine species generating the current flow.

Potentiometry. This method is used for measuring trace materials in solution by means of an electrode containing two sensing elements. The method is based upon the constant current technique of species separation, which measures the potential at each element. It is not widely used in the measurement of chlorine residuals.

The potentiometric method for chlorine residual is only applicable to the measurement of total residual chlorine because it measures the iodine species released from the potassium iodide electrolyte oxidized by total residual chlorine. A platinum (redox) sensing element develops a potential that depends upon the relative concentration of the iodide and iodine species in the sample. The second element, the iodide element, develops a potential that depends upon the iodide ion concentration in the sample. The electrode then in turn measures the difference between these two potentials developed at the two sensing elements. The net potential represents the iodine concentration directly related to the total residual chlorine as measured in the amperometric method.

Designer's Evaluation Checklist

The following limitations should be examined in selecting or evaluating an analyzer for site-specific conditions.

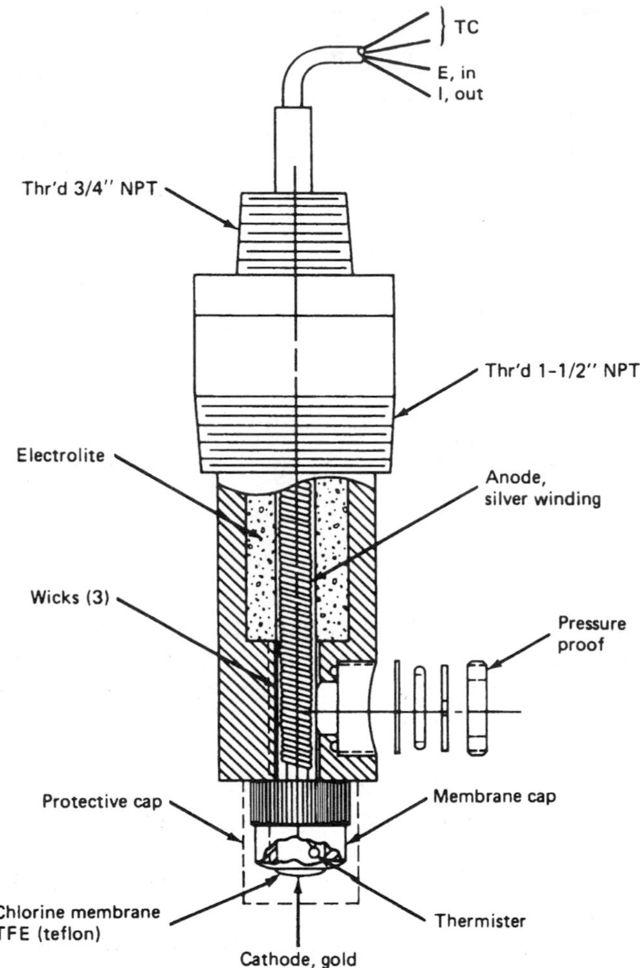

Fig. 9-118. Chlorine sensor polarographic membrane cell (courtesy Delta Analytical Div. Xertex Corp.)

Species Measurement. The most important function of any analyzer is its ability to measure any given species of chlorine residual without interference from other species present in the sample. For example, how dependable can a given analyzer be for measuring free chlorine residual in the presence of some chloramine residual? What proportion of chloramine residual will interfere with the accuracy of the free residual measurement? This is part of analyzer accuracy limitation.

Accuracy. This is the agreement between the amount of a component measured by the analytical method and the amount actually present. It is usually expressed as percent of full scale over a specified range.

Precision. This is referred to as reproducibility of measurement when repeated on a homogeneous sample under controlled conditions. Normally this should be expressed in percent of the full scale reading over a specified range.

Sensitivity. This is described as the least concentration of chlorine species detectable in any given sample.

Range. This is most important. Some analyzers are limited to a detectable range of only 0–2.0 mg/L. Analyzers capable of 0–10 or 0–20 mg/L are preferable to those with a limited range.

Response Time. This characteristic is of prime importance if the analyzer is to be used for residual control. The manufacturer usually expresses this time as that necessary to provide a percentage response to a step change in residual concentration after the sample has entered the analyzer.

For residual control purposes the 90 percent response time should not exceed 10 sec, and initial response should occur in less than 10 sec.

Temperature Compensation. All analyzers should be equipped with thermistors to compensate for temperature variations in the sample. These variations affect the response time and current output of the electrodes. Colorimetric analyzers experience major changes in response time due to changes in both sample and ambient temperatures.

Wallace & Tiernan Two-Electrode Cell

Note: The following is a descripton of analyzers made by W & T that have been superseded by the "New Era Micro 2000" series. The new analyzers are the three-electrode type, which provides a totally new and innovative approach to chlorine residual measurements. The information below, describing the W&T superseded analyzers, has been kept in this text for historical reasons.

General Discussion. This family of continuous analyzers was first to be marketed for measuring chlorine residuals. These analyzers can be divided into three categories: potable water, wastewater, and cooling water. The first two are amperometric. The third type (cooling water) is polarographic. They are described below.

Amperometric Cell Figure 9-119 illustrates a typical arrangement of the analyzer cell. The measuring electrode is copper, and the reference electrode is either platinum or gold. The electrodes are mounted coaxially so that the sample water flows continuously through the cell. The cell produces a small

(0–150 mA) DC current proportional to the amount of free halogen (chlorine, bromine, iodine, chlorine dioxide) in the sample. However, since it is capable of measuring an oxidation reaction, the cell will also produce a current due to the presence of chloramines in the sample. This response is not linear, but acts to interfere with free residual measurements. The limiting range of these cells is 0–20 mg/L, but they were available in ranges of: 0–1, 0–2, 0–5, 0–10, and 0–20 mg/L. The current response is linear over the entire range of 0–20 mg/L. Speed of response is 10 sec or less.

The basic principle of this cell is that a current will flow between the two electrodes proportional to the free halogen concentration in the flowing sample. To measure this current the electrodes are short-circuited.

A minor flow of current will always be present even in the absence of a free halogen; therefore the current-detecting device must be able to delete this current. This is shown as the zero adjustment in Fig. 9-119.

There are two ways of measuring the current produced by an amperometric cell. One way, illustrated above, measures the current flow directly. The other way is to pass the current through a known resistance and measure the potential drop. This is known as the IR drop method (see Capital Controls discussion, below).

Referring to Fig. 9-119, an external DC power source is used for comparing the current produced by the free halogen concentration in the cell. Current flows from *A* to *B*. As this current increases, the detecting device *D* indicates this and generates a signal that is amplified and drives a reversible motor, which positions a slider *R*. This movement increases the current from the external power source, which flows from *B* to *A*. As soon as the increase from the external source is sufficient to balance the potential between *A* and *B*, the motor stops, and the recording or indicating mechanism comes to rest (null position) at the new value of an increased current generated by the free halogen in the amperometric cell.

pH Effect. Current flow in amperometric cells is grossly affected by changes in pH. Figure 9-120 illustrates the magnitude of this effect upon a 1.0 mg/L free chlorine residual. This is called the pH coefficient. It is dealt with in either one of two ways by Wallace & Tiernan analyzers: addition of either carbon dioxide gas or liquid sodium acetate–acetic acid (pH 4 buffer) to the sample.[43] The cell current is most stable in the pH range of 4 to 4.5. The amount of buffer required depends upon the alkalinity of the sample. The analyzer is designed to meter either type of buffer additive.

Temperature Effect. All amperometric cells respond to changes in sample temperature. The change in cell current due to temperature change in the sample for a fixed residual is mathematically predictable, and therefore can be automatically compensated by the installation of a thermistor[39] in the circuitry as shown (temperature adjuster).

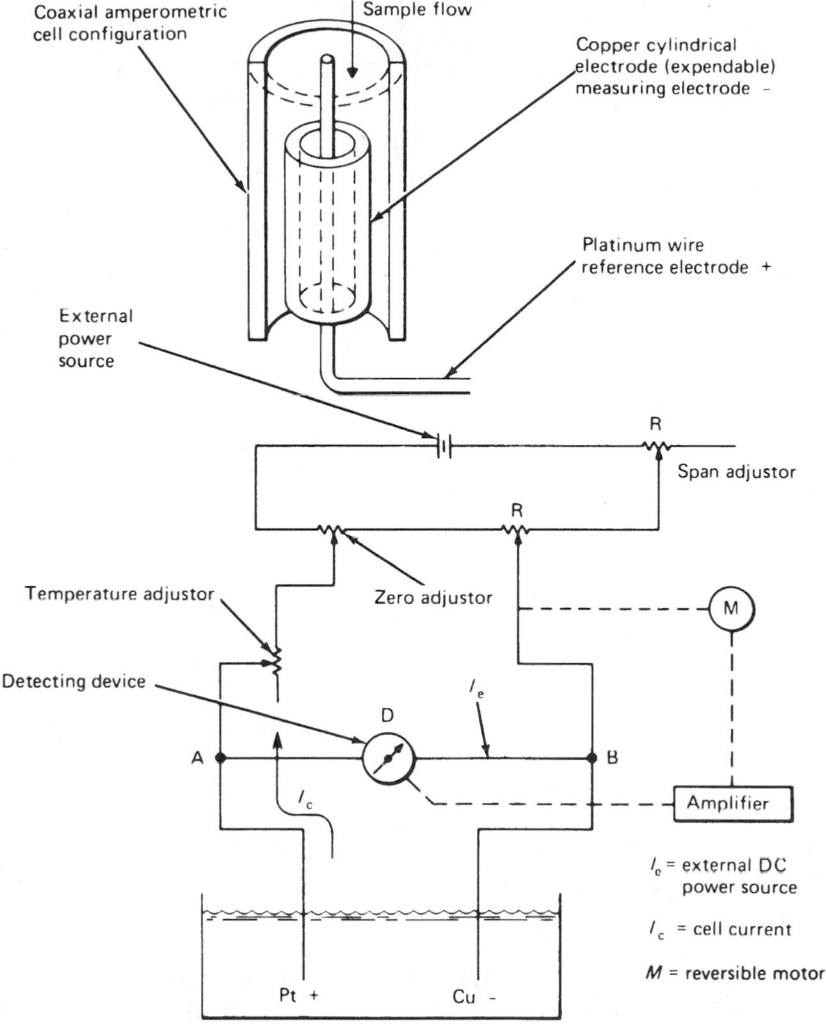

Fig. 9-119. Details of amperometric cell for chlorine residual analyzer (courtesy Wallace & Tiernan).

Diffusion Phenomenon. This condition was first recognized by Marks and Campbell in 1950.[44] When the cell current approaches zero or remains at zero in the absence of free halogen residual, a concentration of electrons begin to accumulate at the measuring electrode, thereby repelling the negative ions. This impedes the speed at which the free halogen can diffuse through the layers of electrons that accumulate under these conditions. Therefore, sample flushing action and electrode bombardment tend to alleviate this condition.

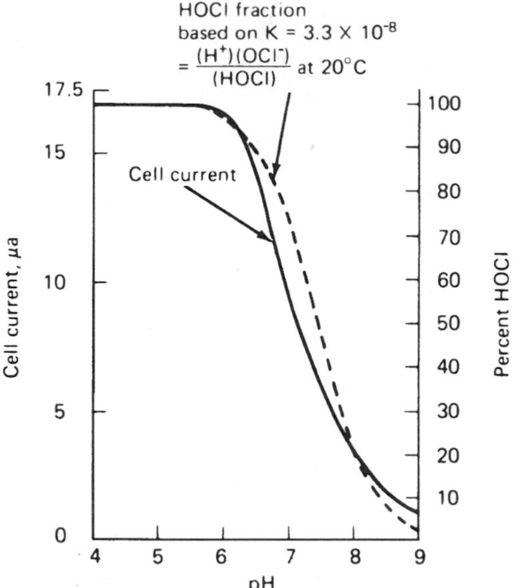

Fig. 9-120. Effect of pH on amperometric cell current. (All values are for 1.0 mg/L free chlorine residual.)

Electrode Fouling. The effect of this condition was discovered early in the development of these analyzers. To prevent the sample cell from being fouled by biological slime, organic debris, or chemical deposition (alumfloc, ferrifloc) provision is made to continuously recirculate a quantity of abrasive grit, which bombards both electrodes and the adjacent cell area. This keeps the electrodes clean and maintains the electrode surface at optimum molecular equilibrium.

Sample Flow Rate. Current flow in the cell will vary with the sample flow rate. It must be observed that in calibrating an analyzer with a sample containing free halogen, the zero check and adjustment is made by stopping the sample flow. This demonstrates the effect of sample flow rate upon cell current. The sample hydraulics must be such that the flow through the cell will be constant. Variable pressures in the sample source must be controlled by a suitable external pressure-regulating valve.

Zero Current Effect. It has been observed that there is no practical means of calibrating the amperometric cell when free halogen is absent from the sample.[45]* Consider the zero check described above, that is, completing the

*This problem has been eliminated by the advent of the Wallace & Tiernan three-electrode cell. (See below).

zero adjustment with the sample flow stopped while the sample contains a significant concentration of free halogen. When a sample not containing any free halogen is passed through the cell, the analyzer will always indicate a slightly negative reading—below zero.

Polarographic Cell. This cell uses a reference electrode and a measuring electrode as illustrated in Fig. 9-121. The measuring electrode is the only electrode that is in the sample. This cell is capable of measuring either free chlorine or free iodine, and it does not require any pH buffering or potassium iodide. The use of applied voltage to the electrodes allows the selection of either chlorine residual species. Figure 9-122 demonstrates that the free chlorine voltage current measurement curve obtains at +0.3 V applied voltage while combined chlorine obtains at −0.07 V (approx.). In this cell the reference electrode is silver–silver chloride, and the measuring electrode is platinum. The range of this analyzer is limited to 0–2 mg/L free chlorine. It is not affected by pH changes in the range of 5–8.5, nor is it affected by the presence

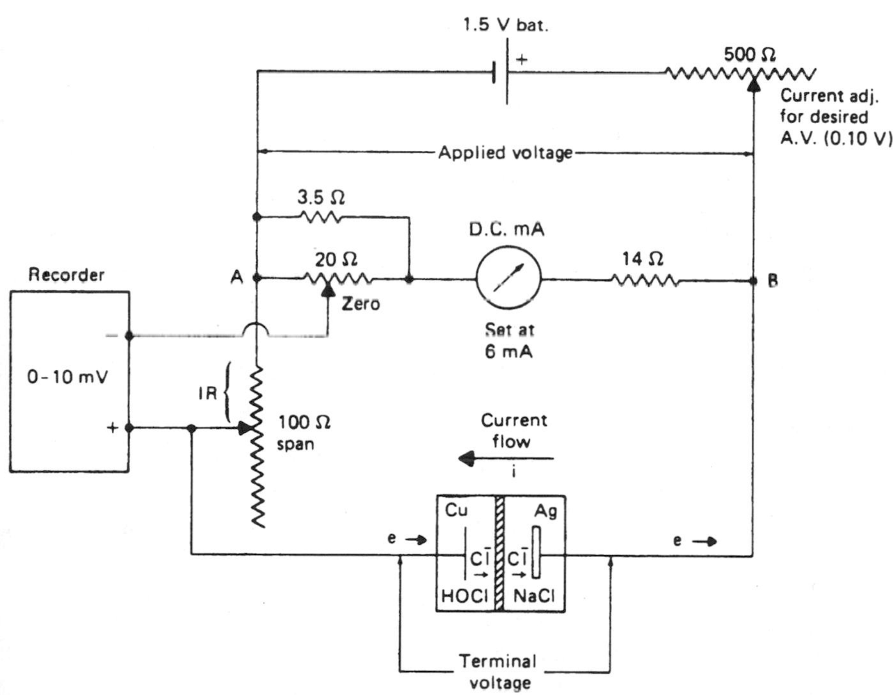

Fig. 9-121. Polarographic cell (courtesy Wallace & Tiernan).

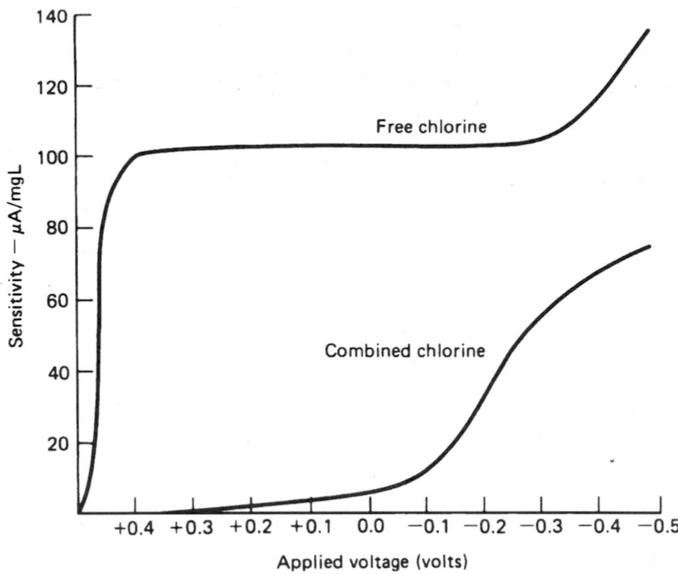

Fig. 9-122. Polarographic curve showing free chlorine versus chloramines.[58]

of combined chlorine residual. Since no chemicals are added to reduce pH, a true sample is analyzed. Changes in salinity and the presence of such cooling water treatment compounds as chromates, phosphates, and defoamers do not interfere with its accuracy. A thermistor compensates for sample temperature changes between 35°F and 120°F. Grit is replaced once a month, and electrolyte tablets (NaCl) are replaced once every 3 months. This unit was designed specifically for condenser cooling water systems.

Wallace & Tiernan Micro 2000 Three-Electrode Cell

Description. This analyzer can be used for measurements in potable water, wastewater, or cooling water. It analyzes a continuous sample of treated water amperometrically for free or total chlorine residual or for potassium permanganate residual. This analyzer is fully described and illustrated in W&T publication No. SB 50.505 UA.

Residuals can be measured in either freshwater or seawater. They are observed on an alphanumeric display, and the analyzer provides a proportional 4–20 mA analog output. Microprocessor-based electronics and a patented three-electrode measuring cell provide continuous on-line analysis of residual chlorine, sensitive to 0.001 mg/L, and the analyzer is capable of controlling residuals below 0.01 mg/L. The analyzer installation requires only piping of the sample line to the unit, provision for a drain, and a 120 V AC power supply. The operating range is 200 to 1, with eight operating ranges from 0–0.1 to 0–20 mg/L.

Figures 9-123 and 9-124 are schematic diagrams of the three-electrode cell and the Micro 2000 analyzer.

The sample flow system is composed of plug-in components on a swing-out panel. It has large ports throughout, grit bombardment of the measuring and counter electrodes, continuous metering of reagent chemicals and detergent into the sample, a self-cleaning main orifice, and a solenoid valve for automatic back-flushing cycles. A motor-driven impeller maintains a constant sample velocity at the measuring electrode. The electrode impeller and the grit bombardment are visible in the clear lucite measuring cell. The reagent pumps are the valveless peristaltic type with a brushless stepping motor.

The multifunction microprocessor-controlled electronics system is housed in a NEMA 4X enclosure. Two security-coded menus give access to two levels of operation: a scrollable, informative operator's menu with easy change of operating parameters and a supervisor's setup menu with instruction and error messages, to assure proper operation. On-line and off-line diagnostics are also available for servicing and calibration. The system provides separate, external high and low residual alarms. This analyzer can measure all species of total chlorine residuals, including organochloramines. In the absence of residuals (zero residuals), it maintains stability and calibration for long periods of time.

The Three-Electrode Cell: Important Features. It contains a spiral-wound platinum measuring electrode and a circular platinum counter electrode. The third electrode is a silver–silver chloride reference electrode that is surrounded by a potassium chloride gel in an upper electrolyte cavity. A replaceable porous reference junction is located behind the measuring electrode. It maintains a potential on the measuring electrode and makes possible the grit bombard-

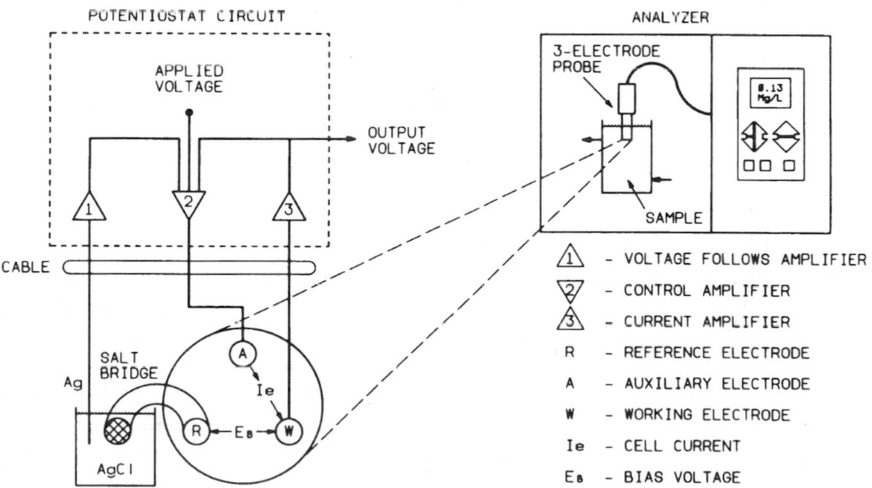

Fig. 9-123. Schematic of three-electrode cell (courtesy Wallace & Tiernan).

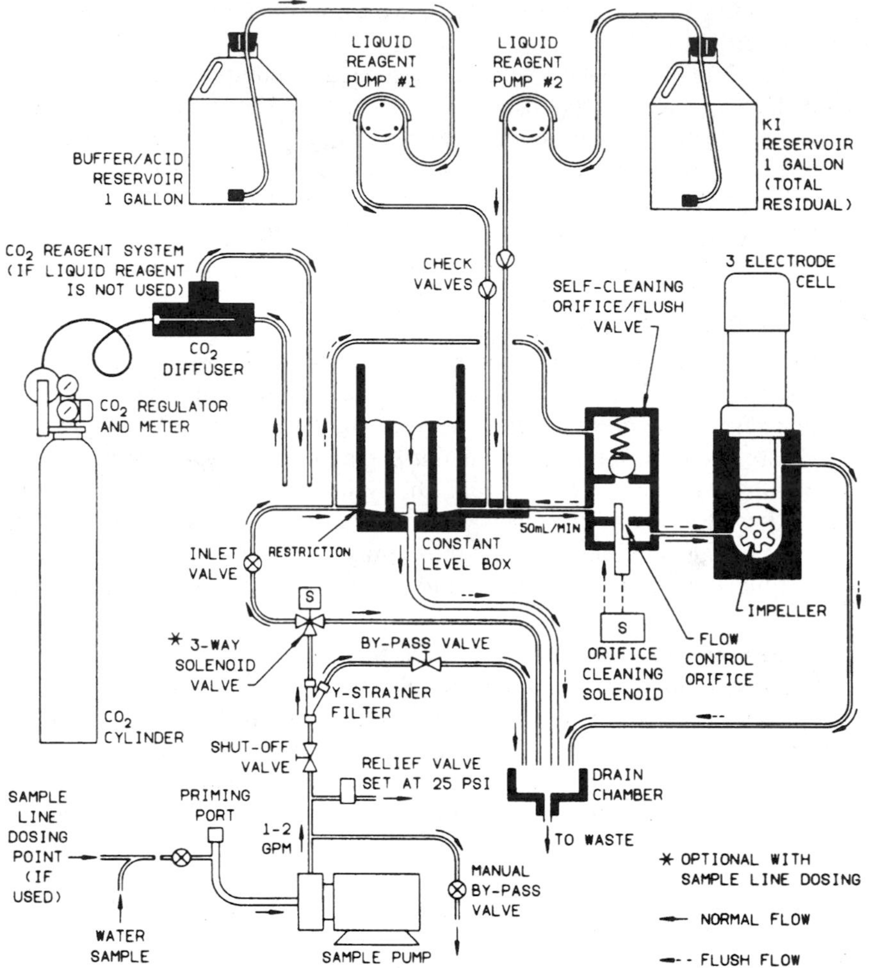

Fig. 9-124. Schematic of the Micro 2000 chlorine residual three-electrode cell analyzer (courtesy Wallace & Tiernan).

ment, to keep the electrode clean. A potentiostat circuit is also included, to eliminate background noise that might affect the stability of the reference electrode. This electrode cell configuration is a great improvement over the two-electrode amperometric technology[97,98] for the following reasons:

1. Zero calibration is eliminated because in the three-electrode system, current does not flow in the absence of a halogen residual.
2. Owing to the above, the accuracy and the sensitivity of residuals below 0.2 mg/L are in the one part per billion region.
3. Electrode fouling is completely eliminated.

4. The fixed potential on the measuring electrode yields greater stability plus zero cell current at zero residual.
5. The three-electrode system will maintain calibration for up to one year under favorable conditions.

Capital Controls Co.

Chlorine Residual Analyzer Series 1870E.[16] The technology of this analyzer has been independently tested, and it has proved to be a reliable analyzer for continuous and accurate measurement of free chlorine (HOCl 85 percent of total residual), combined chlorine, and oxidants such as chlorine dioxide, iodine, bromine, potassium, permanganate, or other oxidants that might be in the water being tested. (See Figs. 9-125a and 9-125b).

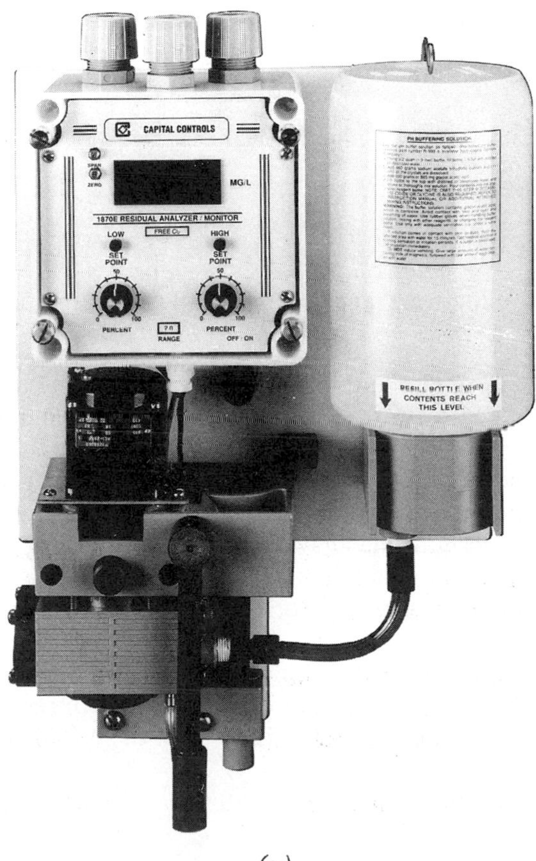

(a)

Fig. 9-125a. Capital Control chlorine residual analyzer, series 1870E (courtesy Capital Controls Co.).

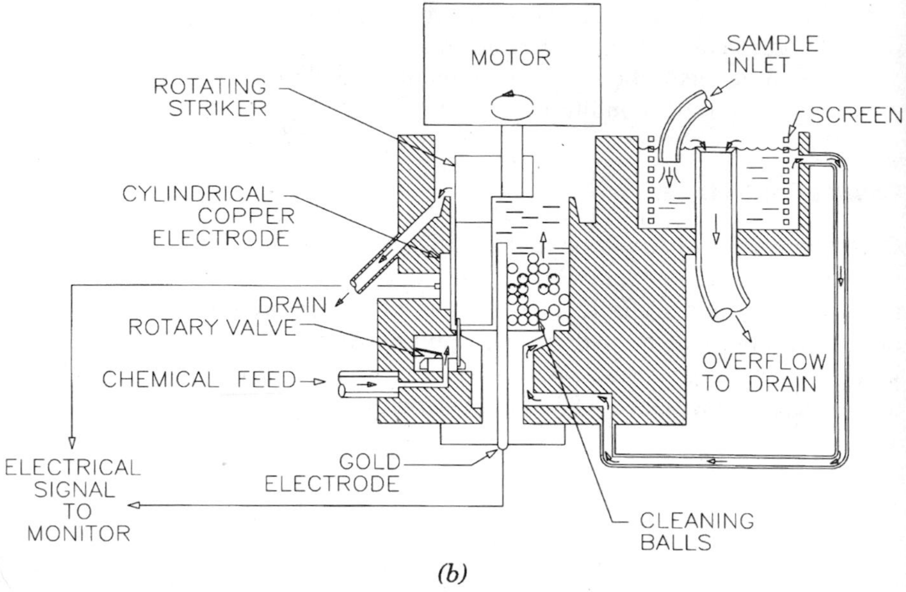

(b)

Fig. 9-125b. Capital Control chlorine residual analyzer, series 1870E flow diagram (courtesy Capital Controls Co.).

The following are the usual applications involved with the measurement of chlorine residuals:

1. Potable water: in-plant and finished water monitoring and control.
2. Wastewater: feed-forward dechlorination control and effluent monitoring.
3. Swimming pool disinfection: accurate control of disinfection to provide total health safety and the best comfort for the bathers.
4. Cooling water: monitoring and control of slime and algae control in the heat exchangers and system piping and throughout the tower.
5. Food and beverage: zero verification after dechlorination by carbon filtration.
6. Pharmaceuticals: again zero verification after dechlorination by carbon filtration. This also simplifies the validation procedures.
7. Power industry: effluent monitoring to meet the NPDES and MOE discharge limits.
8. Industrial wastewater: effluent control to meet local discharge requirements.

This amperometric-based chlorine residual analyzer, Series 1870E, illustrated in Fig. 9-125a, features a field-selectable monitoring range from 0–0.1

to 20 mg/L. These analyzers incorporate a constant, direct-drive electrode cleaning system, which eliminates signal drift and the need for frequent recalibration. Internal high and low setpoints are standard.

The water and wastewater sample to be monitored and the required reagent to be added are both fed to the analyzer by gravity. This eliminates the need for metering pumps. Extra-large gold and copper electrodes are used to provide maximum signal strength. Sample temperature variations are compensated within the measuring cell. This provides consistent residual values.

These Series 1870E analyzers are constructed of corrosion-resistant materials. Each unit is prepiped and prewired, requiring only field connection to the service points. All the various components and controls are accessible from the front of the unit to permit ease of observation of the solution level, sample flow, and the electrode cleaning system, and the required adjustments of the setpoints.

Principle of Operation—Series 1870E Analyzer. The sample flow is delivered to the constant head weir at about 500 ml/min. The excess overflows to the drain. (See the system diagram of Fig. 9-125b.) The sample then passes through the annular space between the two fixed electrodes in the "sensing" cell. As it passes through this cell, a small DC current is generated in direct linear proportion to the amount of residual present in the sample flow. This residual value is displayed on the digital indicator in mg/L.

The surfaces of both electrodes are kept clean by the action of the PVC spheres, which are continuously agitated by a motor-driven rotating striker. This constant cleaning eliminates signal drift and the necessity of frequent recalibrations, thereby providing an accurate residual measurement. In addition, a thermistor is provided that compensates for sample temperature variation.

The liquid reagent is stored in a single bottle and is fed from a constant head reservoir through a rotary valve. This particular configuration adds the precise amount of solution during each valve rotation. The reagent bottle supplies seven days of use before filling is necessary. An optional reagent feed system may be adapted for pH buffering in using CO_2 gas for certain water treatment applications.

Design Features. The following are the most important features of this relatively new model of chlorine residual analyzer:

1. *Accuracy:* the 2.0 percent accuracy of this unit is ideal for monitoring and control of potable water, wastewater, and industrial process water systems.
2. *Automatic cleaning system:* A continuous direct-drive cleaning system maintains a constant level of electrode cleanliness that is necessary to maintain the 2.0 percent accuracy.

3. *Safety provision:* The electronics enclosure is a NEMA 4X, which protects against corrosion due to the possible presence of chlorine vapor during some local maintenance operations.
4. *Gravity feed reagent:* This system provides the analyzer with a sample that has a pH of 4.5 to 4.8. This pH level ensures the stability and the strength of the generated residual value.
5. *Ease of use:* All components and controls are accessible from the front of the unit. This permits the operator ease of observation of solution level, sample flow, and electrode conditions and adjustment of setpoints.
6. *Mounting:* Easy mounting is accomplished through the mounting panel, where all the components are attached. Units can also be supplied in a floor or wall cabinet.
7. *Large cell:* The extra-large gold and copper electrodes provide maximum signal strength.
8. *High and low alarm setpoints:* These setpoints, which are required for the monitoring and control of residual values within a concentration band by using high and low alarms, are easily adjustable on the front panel. LED lights indicate that an alarm has occurred. A latching contact option is available to provide band control for high–low feed control systems.
9. *Ambient temperature:* 32°F to 120°F.
10. *Speed of response:* 4 sec from sample entry to display indication; 90 percent of full scale response within 1.5 to 2.0 min.
11. *Sample limitations:* Samples containing high concentrations of metal ions, or certain corrosion inhibitors, may affect the analyzer operation. Consult the factory for specific applications.

Capital Controls OXITRACE® MODEL 1871[117]

General Discussion. This relatively new analyzer represents one of the latest advances in residual analysis, Through microprocesser-based electronics, this unit offers precise PID control of the critical factors that involve the measurement of residual chlorine concentration. These include pH, temperature, and process water flow. (See Figs. 9-126a and 9-126b).

This is an amperometric-based instrument, designed to continuously analyze free (85% HOCI) chlorine or total chlorine plus chlorine dioxide, potassium permanganate, iodine, bromine, or other oxidants for potable water, wastewater, cooling water and other process water applications. All components and controls are accessible from the front to permit ease of setup and operation. The digital electronics system provides unequaled accuracy and reliability. The entire unit is constructed of corrosion-resistant materials. Each unit is prewired and prepiped, requiring only field connections to service points.

These analyzers use a direct-drive electrode cleaning system that eliminates signal drift. This minimizes the need for frequent recalibration. The extra-large gold and copper electrodes are used to maximize as much as possible

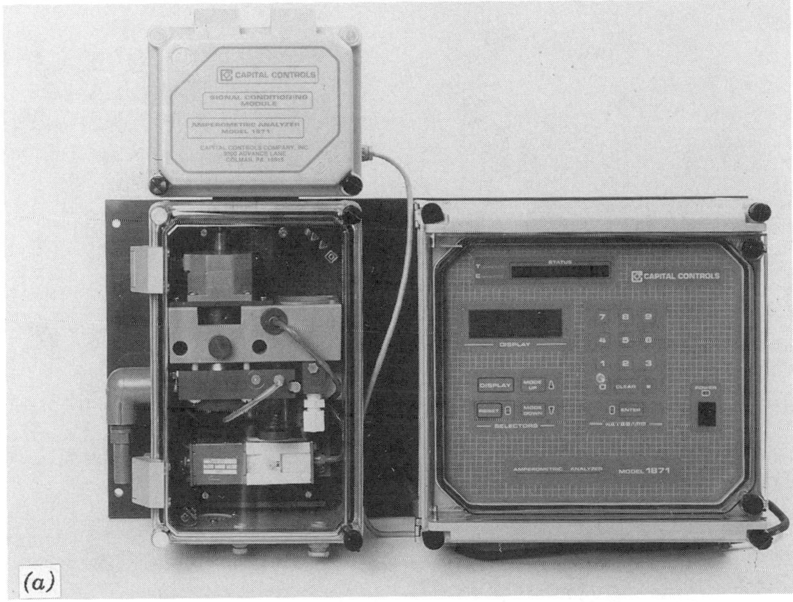

Fig. 9-126a. Capital Control on-track analyzer, model 1871 (courtesy Capital Controls Co.).

the signal strength. The ability to accurately measure low levels of chlorine residuals is enhanced by a cell pH monitoring and PID control system, plus an infrared flow detector and solid-state temperature compensation.

System Operation. (See Fig. 9-126b.) The sample flow is delivered to the sample filtering chamber at a rate between 400 and 700 ml/min., at a maximum pressure of 5 psig. Any excess overflows into a drain. The sample then passes through the annular space between the two fixed electrodes in the sensing cell. As the sample flow passes through the electrode cell, a small DC current is generated in direct linear proportion to the amount of residual present in the sample.

The surface of both electrodes are kept clean by the action of PVC spheres, which are dispersed by a motor-driven agitator. This continuous cleaning of the electrodes maintains an accurate residual measurement. In addition, a solid-state temperature device compensates for sample temperature variations. The residual value is displayed on the digital indicator in milligrams per liter (mg/L). A 4–20 mA DC residual output is displayed along with three auxiliary 4–20 mA DC outputs for pH, process flow, and temperature. Flow through the unit is monitored by a unique infrared flow detection system. The sample flow, in mL/min., may be accessed from the receiver.

A pH probe, and its associated PID function, controls the frequency of reagent feed through an electronic solenoid buffer pump. As the sample pH rises, the control system within this unit increases the reagent pumping

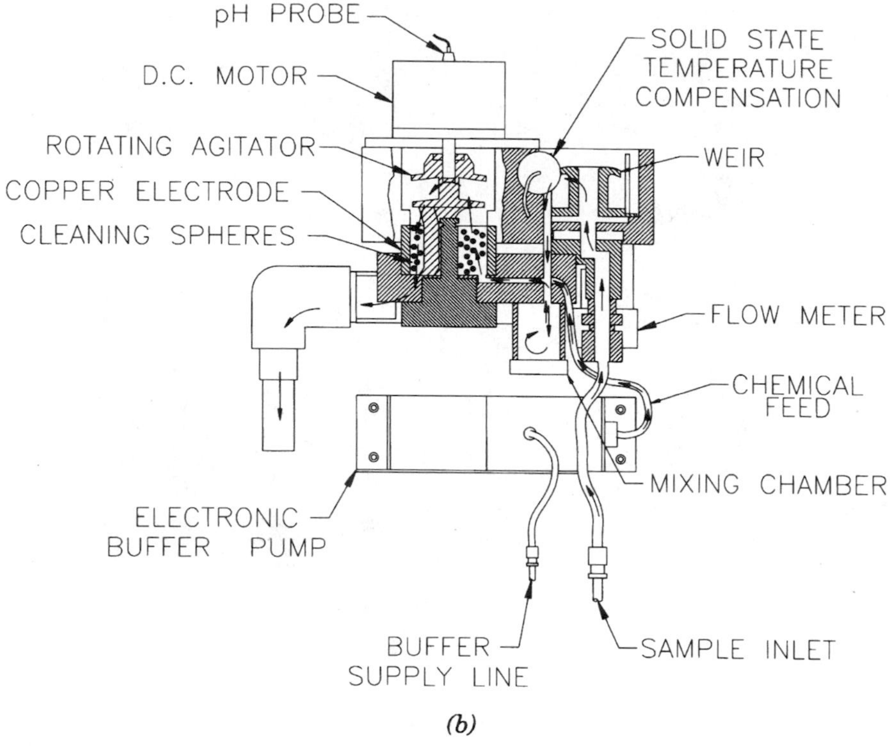

Fig. 9-126b. Capital Control on-track analyzer, OXITRACE© model 1871 flow diagram (courtesy Capital Controls Co.).

frequency to bring the pH back to the correct operating range. This range is between 4.3 and 4.5 pH. The one-gallon reagent bottle provides storage for 14 to 30 days' operation in most applications.

Applications. The applications for the OXITRACE® Model 1871 are the same as those for the Model Series 1870E outlined above.

Design Features. The important features of this new chlorine residual analyzer are outlined as follows:

1. *10 ppb detection:* Control of buffer addition through a unique infrared flow detector and PID control enables very accurate low (ppb) dechlorination with a resolution of 1 ppb.
2. *Status display:* A 16 digit alphanumeric display enables easy setup, indicates oxidant being measured, and announces alarm and fault conditions, along with setup parameters and self-diagnostics.
3. *Security code:* A four-digit security code prevents unauthorized access.

4. *Data logging:* Maximum, minimum, and average residual values are logged, and initial alarms are date/time-stamped at their first occurrence via an internal real-time clock.
5. *Accuracy:* The 1 percent accuracy of this unit is ideal for precise control of potable water, wastewater, and other process waters.
6. *Safety:* The electronics enclosure is NEMA to ensure protection of operating personnel.
7. *Automatic cleaning:* This is very important. A continuous direct-drive cleaning system maintains a constant level of electrode cleanliness. This maintains system accuracy.
8. *Ease of use:* All the operating parameters, such as relay operation and alarm setpoints, are entered from the key pad.
9. *Large cells:* The extra-large gold and copper electrodes will always be able to provide continuous maximum signal strength.
10. *Mounting:* Easy mounting is accomplished through the mounting panel where all the system components are attached.

ATI Residual Chlorine Monitor Series A 20[113]

General Description. The Series A 20 unit is based upon a polarographic membrane sensor, which measures residual chlorine directly without the need for sample pretreatment of any kind, except in some unusual case. The electronic package provides a real-time display of chlorine concentration, an external alarm, control outputs, and an isolated analog output. This unit assembly is illustrated by Fig. 9-127.

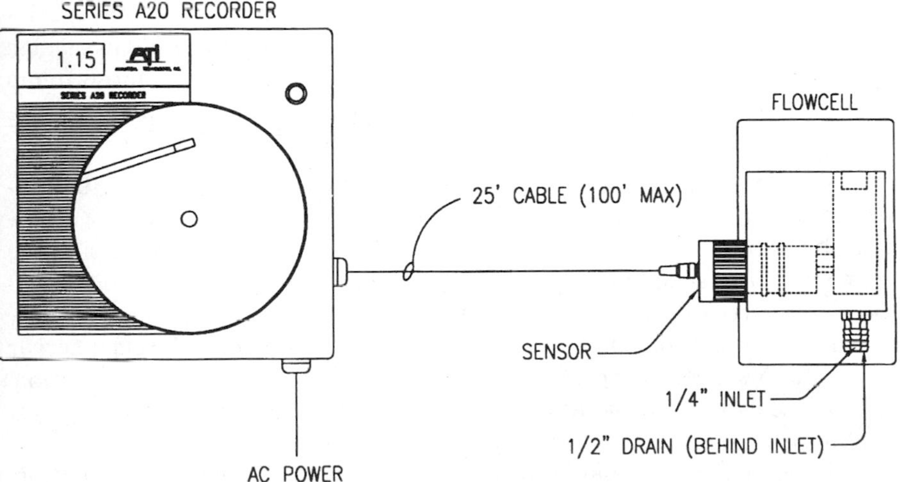

Fig. 9-127. ATI residual chlorine monitor, series A20 typical system diagram (courtesy Analytical Technology Inc.).

When this unit is being installed in a potable water or wastewater treatment facility, it should be housed in the available NEMA 4X fiberglass enclosure. This type of enclosure is immune to the usual chlorine rusting-type corrosion that occurs in these installations.

Five switches located on the front panel provide access to monitor programming functions, and allow operators to view information such as temperature, alarm setpoints, and an analog output value. An access code number is required in order to change any of the calibration or setup parameters, protecting the system from unauthorized tampering.

The Series A 20 monitors provide two selectable display ranges, 0 to 2 mg/L or 0 to 20 mg/L. The low-range display provides resolution to 0.001 mg/L and the high-range display provides 0.01 mg/L resolution. For special applications, a high-range system of 200 mg/L is available.

An isolated 4–20 mA output is provided for external recording or data logging. This output may be programmed for any required span using the front panel swithes, and may be changed to 20 mA and used with a shunt resistor to provide zero-based voltage outputs if required.

Residual Chlorine Sensors. The sensor consists of a pair of electrodes immersed in a conductive electrolyte and isolated from the sample by a chlorine-permeable membrane. The chlorine migrates through the membrane by simple diffusion and is reduced to chloride on the surface of the working electrode. This causes a flow of electrons through an external measuring circuit. This current flow is linearly proportional to chlorine concentration. The response of the sensor is also temperature-dependent, and an RTD in the sensor provides a temperature input to the electronics to allow for automatic temperature concentration.

Sensors have been developed for measuring the two most important species of chlorine residual. One is for free chlorine (HOCl), up to residuals as high as 200 mg/L, and the other is for chloramine residuals, such as those found in either chloraminated potable water supplies or wastewater effluents.

Flow Cell Assembly. The residual chlorine sensors provide the best sensitivity and stability when the sample flow is controlled. A constant head overflow assembly is provided to control both flow rate and pressure in the area of the membrane. The sample inlet flow and pressure from the system being monitored can vary widely without any effect upon the measurement. The only requirement is that the inlet flow rate be kept above 7 gal/hr. The clear acrylic flow cell allows operating personnel to easily observe any cell-fouling situation. A cover seals the sensor into the measuring chamber and protects against inlet overflows.

Flow-cell assemblies may have to be operated in a submerged condition. When this is the case, the submerged sensors require flow velocities of at least 0.5 ft/sec.

EIT[131]

Analyzer Models. There are two separate models available:

1. Model 8450 measures both HOCl and OCl^- ion. This analyzer provides superior free chlorine response and usually does not require chemical or buffer additives. A head flow cell assembly automatically regulates sample flow to ensure stable and continuous readings.
2. Model 8451-I has been engineered to accurately measure all combined (total) chlorine residuals regardless of the concentration variations of either ammonia N or organic N in the sample being measured for total chlorine residual. This unit is equipped with the EIT Infusor System. This system operates by converting any amount of free chlorine into combined chlorine. This is typical of any water with trace amounts or medium amounts of nitrogen compounds.

Monitors: Systems Description. These chlorine monitors utilize a direct measuring amperometric probe for chlorine residuals. These have been field-proved by thousands of installations. Each probe is hand-built and tested by EIT under stringent quality standards. Figure 9-128 illustrates the monitor for each of these systems.

The probes are constructed of PVC and consist of two primary components: a probe body and a cannister, as illustrated in Fig. 9-129. Each probe operates like that of a battery. The probe body's silver and platinum electrodes are immersed in an electrolyte contained within the probe cannister, and the cathode contacts the probe's polymeric membrane.

This membrane allows chlorine species to diffuse into the electrolyte near the cathode where chlorine is reduced to chloride ions. As the reduction of

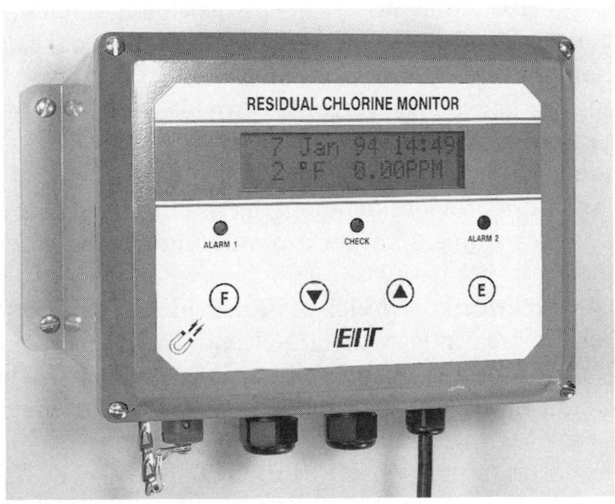

Fig. 9-128. Chlorine residual monitor (courtesy EIT).

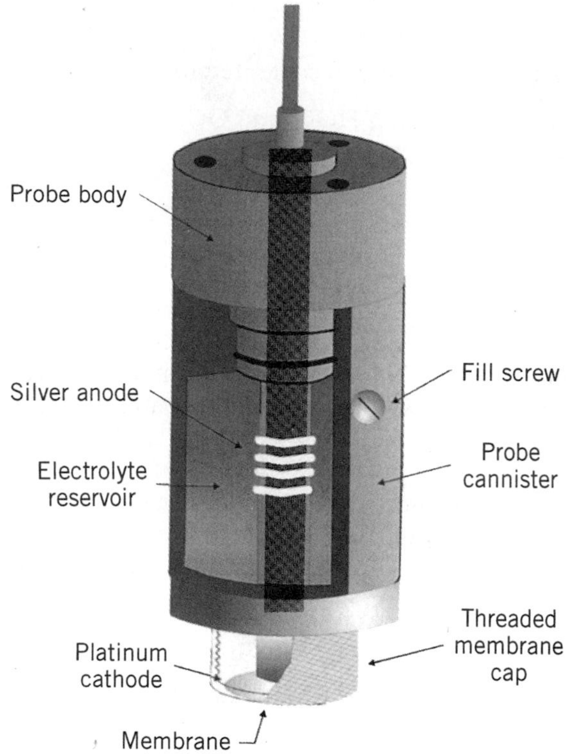

Fig. 9-129. EIT chlorine probe.

chlorine occurs at the cathode, the probe's silver anode reacts with the probe's electolyte, and by an oxidation process positively charged ions are produced—thus creating current flow that is proportional to the residual chlorine concentration in the sample.

The auto-ranging, backlit LCD display provides continuous indication of chlorine concentration, temperature, date, and time, while the front panel LEDs simultaneously indicate the alarm relay status. The optional digital communication feature provides data logging of output readings, and permits changing the system parameters from a remote monitoring site.

Clean Water Applications. Model 8450 provides a superior free chlorine response because it measures both HOCl and OCl^- and does not usually require chemical or buffer additives for proper operation. A head flow-cell assembly automatically regulates sample flow to ensure continuous stable readings.

Because it has a usable response to the OCl^- ion, the EIT probe can be used at higher pH values than it could if it only responded to HOCl. The output of the monitor will not be affected by moderate pH swings as long as the pH stays close to 7 at its highest swing. If the pH is above 7 but constant, the probe can be calibrated and used for the higher pH values.

For pH swings at pH 9 or above, EIT recommends the use of the optional CO_2 Sparger Buffering System.

Wastewater Applications. In these waters, there are usually enough compounds found that have sufficient concentrations of ammonia N for all concentrations of chlorine residuals to be combined chlorine species. The only exceptions to this are those plants that nitrify, or those that have oxidizing systems that cause the plants to automatically go in and out of nitrification. These are usually the smaller plants with capacities less than 20 mgd.

The EIT Model 8451 chlorine probe has been engineered to measure combined chlorine concentration with a high degree of accuracy, and specifically without the use of chemical additives.

Organic Content Fluctuations. Extensive research by EIT has resulted in the development of a single-probe, total chlorine monitoring system. This is the Model 1451-I, which includes the EIT Infuser System and produces accurate monitoring of any water sample containing both free and combined chlorine residuals.

Using a proprietary solution contained in a disposable reservoir, the Infusor System operates by converting the existing free chlorine into combined chlorine. The reservoirs require charging about every 10 weeks. *Note:* The EIT monitors can sit at zero almost indefinitely without substantial drift.[91]

Technical Specifications:

1. Repeatability and linearity: 0.5 percent, plus or minus.
2. Output range: 0–20 mg/L (selectable).
3. Displayed precision: 0.01, 0.1, 1.
4. Power: 110/220 VAC (switchable).
5. Monitor enclosures: NEMA 4X.
6. Monitor temperature limits: −4–122°F.
7. Probe temperature limits: 32–122°F.
8. Max. pressure at probe: 10 psi.
9. Max. flow rate in flow cell: 8–25 gph.

For details on the alarms see EIT publication DS-8450 (7/95) RA#46. For further information, call Customer Service, 800-634-4046.

Bailey–Fischer & Porter Co. CHLORTROL 5000™ Series 17B5000 Residual Chlorine Analyzer with Bare Electrode Cell

System Description.[56] This analyzer is the amperometric type with bare electrodes designed to provide continuous measurement of the chlorine residual content in the process water being used. The analyzer uses a flow-through measurement cell containing two dissimilar metal electrodes (see Fig. 9-130a). As the water sample flows past these electrodes, an electric current is generated that is directly proportional to the chlorine residual concentration.

One of the electrodes is rotated by an electric motor that imparts a swirling velocity to the water sample. This electrode rotation at a constant speed provides reproducible electrolytic conditions and makes the cell independent of sample flow variations. Inert plastic nonabrasive pellets in the cell keep the electrodes in a clean condition by scouring action. By proper reagent selection, either free chlorine (85 percent HOCl) or total chlorine residual (free chlorine plus combined chlorine) may be measured.

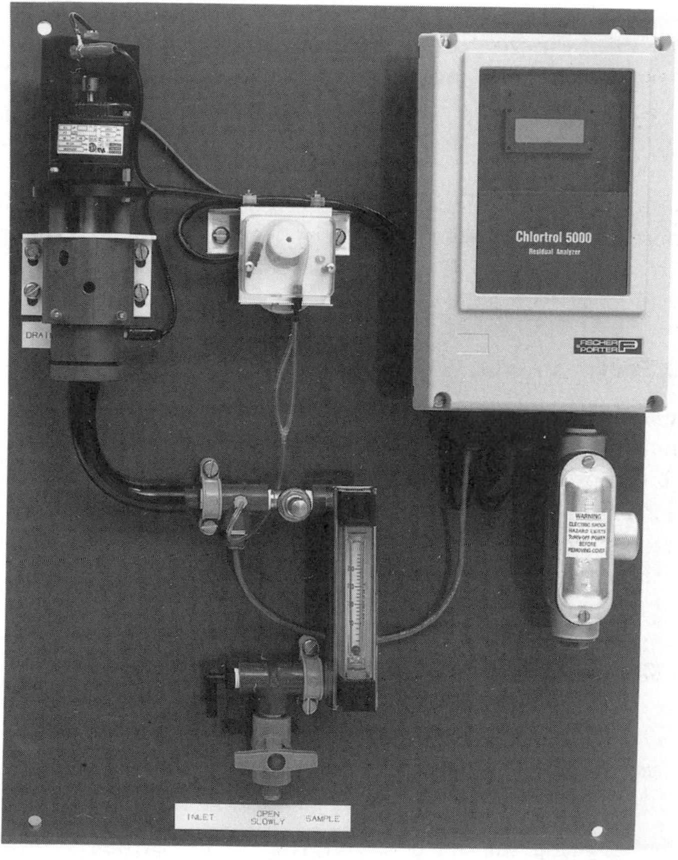

(a)

Fig. 9-130a. Chlortrol 5000™ bare electrode cell chlorine residual analyzer series 17B 5000 (courtesy Bailey–Fisher and Porter Co.).

A solid-state amplifier and signal conditioner convert the generated signal current to an isolated 4–20 mA DC output, suitable for use with standard electronic secondary instruments. The necessary zero and span adjustments are part of the circuitry, and the automatic temperature compensation of the cell output is included to eliminate errors due to changes in the sample water temperature. The operating range is field-selectable, and RFI immunity is built in.

The analyzer is also supplied with a digital indicator to eliminate the need for additional instrumentation at the point of measurement. The indicator is wall-mounted for operating and maintenance convenience. The reagent feed pump is a motor-driven peristaltic type, which pumps the necessary chemicals.

When acetic acid is used for pH control, an 8-gal-capacity reagent container will be provided when free chlorine is being measured. When total chlorine is being measured, two 8-gal containers will be furnished. These containers represent a 60-day supply in each case. When carbon dioxide is used for pH control, one less container is required.

For systems with the analyzer and associated instruments contained in the same floor-mounted cabinet, refer to Specification 17SB5000.

Design Features. The following design features of this chlorine residual analyzer should be helpful to operating personnel:

1. *Continuous monitoring:* This indicates and transmits residual chlorine levels. This information can be used with a controller to provide automatic control of the chlorinator or sulfonator. This frees the operating personnel for other duties by eliminating the need for frequent laboratory testing.
2. *Response time:* within 5 sec.
3. *Reliability:* The electrode surfaces are continuously cleaned by action of nonabrasive pellets. Automatic temperature compensation and RFI immunity are standard.
4. *Easy maintenance:* All components are easily accessible on the wall-mounted panel. The chemical supply lasts 60 days.
5. *Low installation and maintenance costs:* There is no need for expensive cabinets and instrumentation. The reagent cost is minimized by an automatic reagent feed system using either carbon dioxide or dilute acetic acid.
6. *Operating ranges:* 0–0.23, 0–0.5, 0–1.0, 0–2.0, 0–5.0, 0–10.0, and 0–20.0 mg/L, all field-selectable.
7. *Ambient temperature limits:* 33–122°F.
8. *System accuracy:* plus or minus 2 percent of span.
9. *Sensitivity:* The analyzer will recognize and respond to residual changes as low as 0.001 mg/L chlorine.
10. *Sample requirements:* The flow rate for flushing the "Y" strainer is 5–10 gpm, the flow rate to the measuring cell is 5–10 gpm, and the overall hydraulic pressure should be reasonably constant between 5 and 25 psig.

11. *Power requirements:* 110/120, 220/440 VAC, 50/60 Hz at 8 watts max.
12. *Output:* 4–20 mA DC into 600 ohms max. with built-in signal isolation.
13. *Interferences:* Turbidity and chemicals normally found in raw and treated waters do not affect cell operation, However, potassium permanganate and ozone do have an adverse effect.
14. *Shipping information:* When packed for shipping, the weight = 79 lb and the cubage = 12.8 cu ft.

Bailey–Fischer & Porter Anachlor II™ Series 17PC1000 Residual Chlorine Analyzer

System Description.[57] This series analyzer is an amperometric/polarographic device designed to provide continuous measurement of the concentration of residual chlorine in solution. The system includes an in situ nutating sensor assembly, an indicating transmitter, and a handrail mounting bracket/support post assembly. The sensor assembly contains electrodes and an electrolyte, which produce a low-level signal for the transmitter that is proportional to the residual concentration. (See Fig. 9-130b.)

A thermistor in the sensor automatically compensates for temperature changes in the process liquid. The sensor also contains a patented integralnutating device, which provides the necessary apparent sample velocity past the membrane, making the unit insensitive to process fluid velocities.

(b)

Fig. 9-130b. Bailey-Fisher and Porter Anachlor II residual chlorine analyzer.

The transmitter converts the signal produced by the sensor assembly to an isolated 4–20 mA DC signal over any of the field-selectable measurement ranges. An indicator shows the residual chlorine concentration. Output terminals are provided for applying the signal to compatible secondary instrumentation for recording/controlling/indicating the residual chlorine concentration of the process fluid.

Alternately the sensor may be used in a sampling mode by mounting it in a special sample tank which receives a continuous sample of the process fluid. See Technical Information Bulletin No. 17-15b for this and other application information.

Design Features:

1. Residuals are read directly (traceable to "Standard Methods") as a result of the patented chemistry, without any calculated or inferred values.
2. It is unaffected by variations in process flow down to zero velocity, or pH variations normally encountered in municipal or industrial potable water and wastewater systems.
3. This analyzer eliminates the requirements for the continuous feed of conditioning chemicals or reagents, and sample systems and pumps.
4. The system response is instantaneous.
5. Maintenance requirements are greatly reduced; recharging of the sensor with electrolyte and changing the membrane are required every two months, and typically area cleanup is accomplished in 10 min. In situ installations may require cleaning of the membrane with a squirt bottle and/or respanning at a frequency determined by the local operating conditions at each installation. In general, the higher the suspended solids are, in the process fluid, the more frequent the cleaning and the respanning will be.
6. Operating ranges are: 0–0.25, 0–0.5, 0–1.0, 0–2.0, 0–5.0, 0–10.0, and 0–20.0 mg/L.
7. Ambient temperature limits are: −20°F to 122°F
8. Power requirements are: 110/220, 220/440 V AC and 50/60 Hz at 8 watts max.
9. Output: 4–20 mA DC into 660 ohms max. with built-in signal isolation.
10. Enclosure classification: corrosion-resistant NEMA 4X (IP 66 per IEC529), watertight and dusttight for outdoor locations.
11. Case construction: a glass-filled polyester base with a polycarbonate cover. The base and cover are flame-retardant and rated UL 94V-0 and UL 94 V-1, respectively.
12. Mounting: field-mounted on a handrail mounting assembly.

EPCO: Chlortect® Chlorine Monitor[60]

This is one of two analyzers based upon the design of the "Flux Monitor" by Marinenko of the U.S. National Bureau of Standards.[61] The other is marketed by IBM (see below). This instrument was designed purposely to detect extremely low levels of chlorine residual in order to determine precisely the toxic effects of total chlorine residuals upon various species of fish and other aquatic life.

This instrument measures total chlorine residual only over a range of 2 mg/L. The least amount detectable is claimed to be 2μg/L (2 ppb). The digital display reads from zero to 1000 ppb. Internal calibration is selectable for 1, 10, 100, and 500 ppb.

Two reagents are required: potassium iodide and pH 4 buffer. These reagents and chlorinated sample are pumped by their separate peristaltic pumps. There are two cells in the analyzer; one is coulometric and the other is amperometric. The amperometric cell consists of a platinum measuring electrode (cathode) and a nonpolarizable reference electrode (Ag/AgCl) immersed in a salt solution. The coulometric cell consists of a platinum anode and a nonpolarizable reference electrode (Ag/AgCl). The operation of the analyzer utilizes three well-established electroanalytical techniques: coulometry, electrolysis, and amperometry. The coulometric cell provides the instrument with internal calibration by electrolytic on-site generation of a known amount of iodine.* This is accomplished by passing a fixed amount of current through the cell (see Fig. 9-131). The magnitude of the current depends upon the calibration range selected and is determined by Faraday's Law of Electrolysis.

The iodine generated by the stoichiometric reaction between the total chlorine residual in the sample and the potassium iodide added to the sample is measured by the amperometric cell as shown in Eq. (9-11):

$$2H^+ + OCl^- + 2I^- \xrightarrow{\text{pH 4}} I_2 + H_2O + Cl^- \tag{9-10}$$

$$I_2 + 2e^- \longrightarrow 2I^- \tag{9-11}$$

The current produced by the reduction of I_2 to I^- in the amperometric cell is directly proportional to the concentration of I_2, which in turn corresponds to the concentration of the total residual chlorine. This amount is displayed directly in ppb or ppm on the digital panel meter.

If any sample preparation is required owing the presence of undesirable foreign matter, the user must assume this responsibility.

IBM EC/250 Series Chlorine Analyzer[62,63]

This analyzer is based upon the same principles described above for the EPCO analyzer. These two analyzers have the same schematic arrangement of operation. There are some minor differences. For example, model 2A

*A chlorine-demand-free sample of deionized water must be available for the calibration procedure as shown in Fig. 9-131.

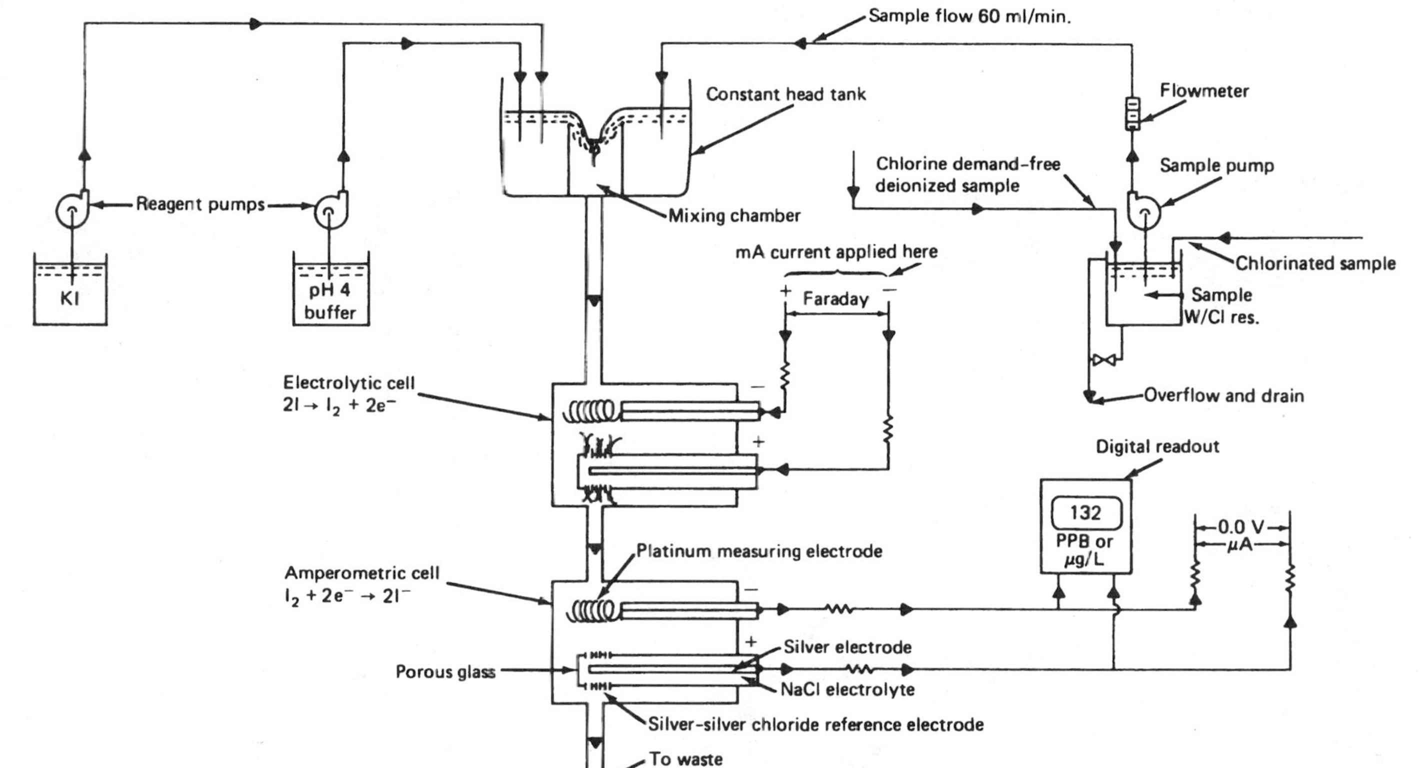

Fig. 9-131. High resolution amperometric cell with internal colometric calibrating circuit utilizing an electrolytic cell (courtesy of EPCO and IBM). Note: both cells are identical.

incorporates a heavy duty sample pump and a "self-cleaning" filter. Calibration ranges are 10, 100, and 500 ppb; detection limit = 2 ppb; linear range = 10 ppm; accuracy, ±10 percent for linear range; reproducibility, ±3 percent; reagent flow rate, 1 ml/min.; sample flow rate, 60 ml/min.; and response time, 90 percent in 3 min. Optional features are: automatic temperature compensation, automatic calibration, battery power adapter, high/low alarm, and 12 V DC power pack.

Orion Research

This manufacturer supplies two chlorine monitors for the continuous analysis of total chlorine residual—only. Model 1570 is for clean water[64] and Model 1770 is for "dirty water."[65] These analyzers use the potentiometric method for the determination of total chlorine residual. This was the only potentiometric analyzer commercially available as of 1992.

This method of analysis is based upon detection of the concentration of iodine in the sample using a special electrode. A pH buffer is added to the sample to lower the pH to between 3.0 and 4.0 KI is then added to the sample to convert all the residual chlorine to iodine. The electrode contains two sensing elements. A platinum electrode develops a potential that depends upon the relative concentration of iodine and iodide in solution:

$$E_1 = E_0 + (S/2)(\log [I_2]/[I^-]^2) \tag{9-12}$$

$$E_1 = E_0 + (S/2) \log [I_2] - S \log [I^-] \tag{9-13}$$

where

E_1 = potenetial developed by the platinum sensing element, mV
E_0 = a cell constant, mV
S = monovalent elcctrode slope (58 mV/decade at 20°C)
$[I_2]$ = iodine concentration, moles/L
$[I^-]$ = iodide concentration, moles/L

A second electrode, the iodide element, develops a potential that depends upon the iodide ion concentration in solution:

$$E_2 = E_0' - S \log [I^-] \tag{9-14}$$

where

E_2 = potential developed by the iodide sensing element, mV
E_0' = a cell constant, mV

The electrode thus measures the difference between potentials developed at the two sensing elements:

$$E = E_1 - E_2 = (E_0 - E_0') + (S/2) \log [I_2] \tag{9-15}$$
$$= E_0'' + (S/2) \log [I_2]$$

where

E_0'' = potential measured by the analyzer, mV

$E_0'' = E_0 - E_0'$ = a cell constant, mV

The net potential measured by the electrode is converted by analog electronics to read directly as residual chlorine in mg/L. The electrode is calibrated with a standard of known equivalent residual chlorine concentration.

The preceding equations demonstrate that the output of the electrode is proportional to the log of the iodine concentration, and thus proportional to the log of the total residual chlorine concentration. This enables the electrodes to measure the residual chlorine concentration over a four-decade range from 0.001 to 10 mg/L.

The only difference between the two models is in the quality of the sample water that is acceptable. Model 1770 is for so-called dirty water and is described here: The sample stream enters the inlet block and the flow is directed at high velocity parallel to the inlet screen; considerably less than 1 percent of this flow is taken through the screen into the monitor. The surface of the screen is subjected to considerable shear force, which keeps the screen surface clean. This design allows the analyzer to be used on unfiltered samples. The sample then flows to the flow cell block. The reagent is passed through a purification column to remove any background iodine that might have formed during storage, and then is delivered to the block by a separate reagent pump. Sample and reagent are combined within the flow cell block, where they are mixed and agitated by a stream of air. The reagent, which contains sufficient acid to give a resulting pH between 3.0 and 4.0 (depending upon the alkalinity of the sample), also contains iodide. The turbulent mixture is directed out of the flow cell block through a mixing loop. Under these conditions, any total residual chlorine reacts completely to form iodine; the resulting iodine concentration is equal to the sample residual chlorine concentration before reaction. The sample then returns to the flow cell block and is directed past a sensing electrode. The turbulence in the mixing loop and agitation against the electrode not only promote mixing but also help to minimize fouling of the electrode by sample debris.

Electrode response is affected by the sample temperature; so automatic temperature compensation circuitry is incorporated in the analyzer.

Specifications for the analyzer claim a sensitivity of .001 mg/L and an accuracy and a precision of ±10 percent and ±5 percent, respectively. Note that for precision and accuracy the specifications are stated as a percentage of the actual reading, whereas manufacturers of other residual chlorina analyzers state their specifications for precision and accuracy in terms of percentage of the full-scale reading. The analyzer response time is 2 min. to obtain a full response from a change in residual chlorine concentration in the sample. Minimum sample flow is 0.5 gpm, and minimum allowable water pressure is 20 psi. Maximum allowable is 100 psi.

The manufacturer recommends weekly checks for leaks, replenishment of reagent, and analyzer calibration monthly. The analyzer is equipped with a sample tap and a valve for calibration of grab samples. There are no operating data on how often this analyzer must be calibrated for wastewater practice.

Uniloc, Div. Rosemount Inc.*[66]

As of 1996 this unit no longer exists.

Hach Colorimetric[68,69]

The Hach company is the only supplier of continuous colorimetric analyzers in North America today (1997). These analyzers use the DPD colorimetric method of analysis. The 31100 model is used to analyze free chlorine over a range of 0–2 mg/L. The 61100 model covers a range of 0–5 mg/L. The 31300 model is designed to analyze total chlorine residual.

Figure 9-132 shows the basic mechanical operation and fluid path of sample and reagent through these analyzers. The reagent pump and sample pump pistons are synchronized precisely so that the reagent is added simultaneously with the sample. As the sample piston rises, drawing the sample in, the reagent piston discharges reagent into the sample stream. A mixing and delay coil provides time for the sample and the reagent to completely mix and the colorimetric reaction to progress to completion before the sample enters the colorimetric cell. After analysis (the other half of the pump cycle) the sample pump discharges while the reagent pump draws in the next portion. Valves (not shown in Fig. 9-132) open and close during the parts of the cycle when the pistons are stationary.

Figure 9-133 shows the optical design of the analyzers. The light beam passes through a lens before entering the cylindrical pump/sample cell. After passing through the cell, the light encounters a beam splitter, which directs part of the beam through a filter to a reference detector located at 90° to the light path. The other, main part of the light beam passes through a filter to a sample detector. The signals from the two detectors are fed to the electronics package for processing. The single-beam/dual-wavelength capability provides a correction for any changes in light source or sample cell condition, and

*Now called Rosemount Inc. They have taken over the Delta Scientific (Xertex) line of analyzers.

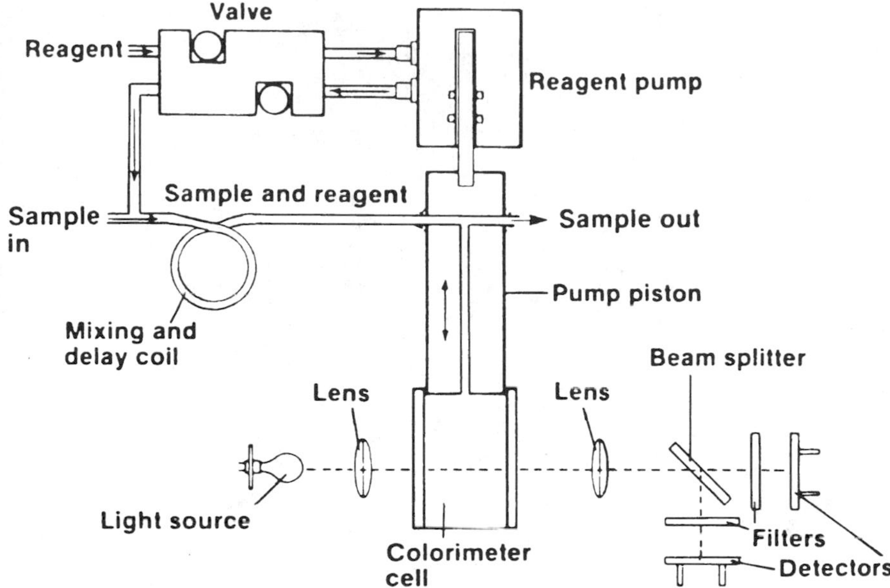

Fig. 9-132. Colorimetric analyzer showing fluid path and mechanical operation (courtesy Hach & Co.).

can compensate for moderate changes in sample turbidity and color by the reference wavelength used in the analyzer. Although these features increase the reliablity when the analyzer is used to determine residual chlorine concentrations in wastewaters that may contain suspended solids and other contaminants, sample pretreatment is still required for removal of debris and large solids that may foul or plug the analyzer.

A block diagram of the analyzer electrical system is shown in Fig. 9-134. Light, after passing through the sample, strikes the reference and sample silicon photocells. This produces a current, which goes to a log-ratio converter. The converter sends out a signal equal to the log of the ratio between the two currents. The log-ratio signal travels to an amplifier with gain and offset controls and then to a sample-and-hold circuit. A microswitch is activated when the sample piston is at the top of the cylinder/sample cell (allowing an unobstructed light path). This advises the sample-and-hold circuit to put the amplifier output into storage. The meter and other external readouts indicate this value, which is stored until the next cycle.

Free available chlorine DPD colorimetric analyzers are available with one of four factory-preset ranges. The widest range is 0–5.0 mg/L. Total residual chlorine DPD colorimetric analyzers are available with one of three factory-preset ranges. The widest range is 0–2.0 mg/L. The analyzers have no automatic temperature compensation. The effect of varying sample temperatures or ambient temperature can be significant.[52] Hach offers an optional sample

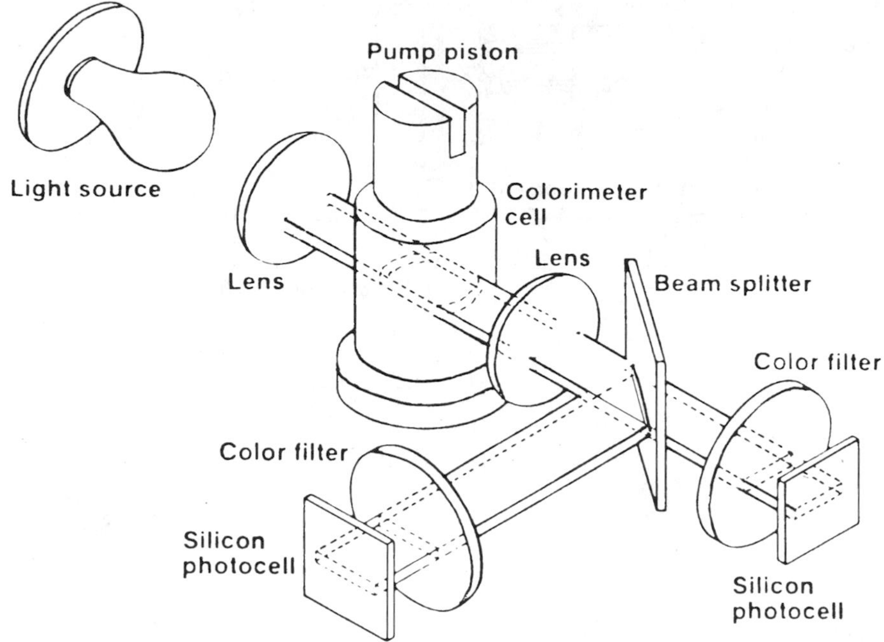

Fig. 9-133. DPD colorimetric analyzer, optical design (courtesy Hach & Co.).

heater that will maintain the sample temperature within ±1.5°C. This should be included as an integral part of these analyzers.

A major drawback of these analyzers is the response time to variations in sample concentrations. Unlike other analyzers, the concentration is determined by using discrete samples. One complete sample cycle requires time for entry of the sample into the analyzer, addition of reagent, color develop-

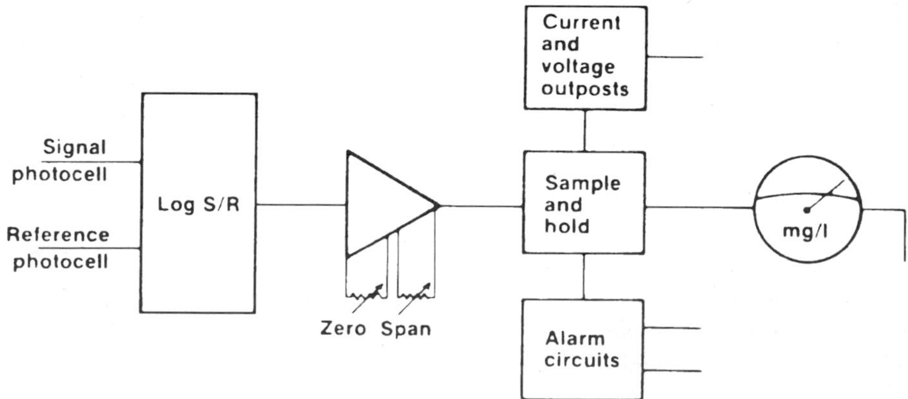

Fig. 9-134. DPD colorimetric analyzer, block diagram of electrical system (courtesy Hach & Co.).

ment, and expulsion of the sample from the analyzer. The supplier's specifications state the following response times:

Free available chlorine	Initial response in 30 sec, 97% in 2 min.
Total residual chlorine	Initial response in 5 min., 95% response in 7 min.

This response time is unacceptable for either residual control or the dechlorination process. These analyzers are relatively insensitive to drift, as demonstrated by the following supplier specification:

> A drift rate of less than 0.5 percent of full scale over 24 hours when measuring 0 mg/L residual chlorine with a constant sample temperature of 25°C and a constant ambient temperature of 30°C.

The sensitivity claimed by Hach is 0.02 mg/L, precision is ±0.5 percent of full scale at constant ambient and sample temperature, and accuracy is ±5 percent for the same conditions.

It is unlikely that these analyzers could be used for anything other than monitoring.

HOUSING

General

Many important design provisions for chlorination housing relate to the safe use of chlorine and the protection of those working with it. Consequently, many chlorine room design provisions are required elements of state standards.

Chlorinator and sulfonator rooms should be at or above ground level. Container storage should be planned so that it is separate from chlorinators and accessories. It is logical to locate the chlorination room near the point(s) of application to minimize the length of chlorine lines. Other general site considerations include a location that permits ease of access to facilitate container transport and handling, adequate drainage, and separation from other work areas.

Separation

Proper design standards require either a completely separate chlorination building or a room completely separate from the remainder of the building with access only through an outside door. There should be no apertures of any type from the chlorination room to other parts of a common building through which chlorine gas could enter other work areas.

Fire Hazard

The building should be designed and constructed to protect all elements of the chlorine system from fire hazards. If flammable materials are stored in the same building, a fire wall should separate the two areas. Fire-resistive construction is recommended. Water should be readily available for cooling cylinders in case of fire.

Space Requirements

Modern chlorination equipment is available in modules so that the chlorinators and accessory equipment can be arranged in a panel-like array. There should be about 4 ft between the front of a module and the nearest wall and about 2 ft on the sides and rear. Figure 9-18 illustrates space requirements for chlorinator-evaporator installations, and Fig. 9-23 illustrates space requirements for a ton container supply area.

The smallest area used for the installation of a chlorinator, weighing scales, and a spare cylinder* of chlorine should not be less than 6 ft by 6 ft.

There should be adequate room provided to allow ready access to all equipment for maintenance and repair. There should be sufficient clearance to allow safe handling of equipment containers. Absolute minimum clearance around and in back of equipment is 2 ft.

Some general minimum space guidelines are as follows:

1. Plants with one chlorinator feeding less than 200 lb/day should have at least 64 sq ft.
2. Plants using two chlorinators with a total feed rate of up to 400 lb/day should have at least 150 sq ft.
3. For each chlorinator-evaporator unit, 160 sq ft should be provided.

Ventilation

Adequate forced air ventilation is required for all chlorine equipment rooms. An exception to this would be small chlorinator installations (<100 lb/day) located in separate buildings if the windows and doors can provide the proper cross-circulation. For a small building, windows in opposite walls, a door with a louver near the floor, and a rotating-type vent in the ceiling usually provide the necessary cross-ventilation.

Factors to be considered in the design of a ventilation system are: air turnover rate, exhaust system type and location, intake location, and type, electrical controls, and temperature control.

A forced air system should be capable of providing one complete air change in 2–5 min. As chlorine gas is 2½ times heavier than air, it is logical to

*This space is for chlorination systems using a 150-lb cylinder.

provide air inlet openings for ventilation fans at or near floor level.* For small installations it is common to employ an exterior exhaust fan with the intake duct extending to the chlorine room floor. A wall-type exhaust fan is an acceptable alternative. The exhaust system should be completely separate from any other ventilation system. For larger installations a blower-type fan is needed. The use of free-moving, gravity-operated louvers may be advantageous in colder climates for conserving room heat when the blowers are not in operation; however, venting systems should not have covers. The fan discharge should be located so as to not contaminate the air supply of any other room or nearby habitations. It is mandatory that the ventilation discharge be located at a high-enough elevation to assure atmospheric dilution (e.g., at the roof of a single-story building).

Air inlets should be so located as to provide cross-ventilation. To prevent a fan from developing a vacuum in the room and thereby making it difficult to open the doors, louvers should be provided above the entrance door and opposite the fan suction. In some cases, it may be necessary to provide temperature control on the air supply so that the chlorination system is not adversely affected. A signal light indicating fan operation should be provided at each entrance when the fan can be controlled from more than one point.

Wind Socks

All installations should locate at least one wind sock on the chlorine supply structure. This is very valuable in the event of a leak. The sock should be as high as possible.

Doors

Exit doors from the chlorination room should be equipped with emergency hardware and open outward. Some design guides recommend two means of exit from each room or building in which chlorine is stored, handled, or used; however, this would not appear to be essential in most cases.

Inspection Window

A means should be provided to permit viewing the chlorinator and other equipment in the chlorination room without entering the room. A clear-glass, gastight window installed in an exterior door or an interior wall of the room is recommended. Door windows appear to be a logical provision even with a separate wall inspection window.

*Room ventilation air should always enter at floor level and exit at ceiling or roof level because vapor leaks will always follow the air circulation path.

Heating

The chlorinator room should be provided with a means for heating and controlling room air temperatures above 55°F. A minimum room temperature of 60°F has been recommended as a good practice. Ideally, the heating system should be able to reliably maintain a uniform moderate temperature throughout the chlorination room.

Hot water heating is generally preferred because of safety considerations and the uniformity of temperature that this method of heating provides, without the extremes that might be experienced with failure of a steam heating system. Electric heating is suitable, and forced air heating would be appropriate if an independent system is provided for the chlorination room or building. Central hot air heating is not acceptable because gas could escape through the heating system.

Chlorine vapor leaving a container will condense if the piping temperature is significantly lower than the temperature of the container. The design should provide a higher temperature in the chlorinator room than in the container room. This applies to systems using the gas phase from the containers. Elimination of unnecessary windows may aid in maintaining uniform building temperatures.

If container storage and chlorination equipment are in separate rooms, the temperature of the chlorine container should not be allowed to drop below 50°F if evaporators are not used

Drains

It is generally desirable to keep the plant floor drain system separate from that of the chlorinator. Drainage from a chlorinator drain relief valve may contain chlorine. Consequently, hose, plastic pipe, or tile drains are recommended. The discharge should be delivered to a point beyond a water-sealed trap or disposed of separately where there is ample dilution.

Scale pits are generally designed with floor drains having a water-sealed trap. In actual practice, most traps probably do not contain enough water to form a seal, and it would be preferable to provide a straight pipe drain outside to grade.

Vents

Chlorinators, sulfonators, external chlorine pressure-reducing valves, remote vacuum systems, and automatic switchover systems have vents to atmosphere. Since the advent of the Uniform Fire Code, these vents cannot necessarily be allowed to discharge directly to the outside air as has been practiced since the use of vacuum-operated chlorinators. This represents a span of over 70 years without any damaging results because chlorine vapor emissions (as opposed to liquid spills) are diluted quickly by the ambient air. Dealing with these vents to comply with the UFC is necessary.

These vent lines must be piped in such a manner that moisture is not allowed to accumulate in the piping. This means that some will have to be equipped with a condensate trap if the piping cannot be arranged to allow the moisture to drain from the vent line. Those vents that are required to "breathe" should be fitted with a wire screen at the discharge end to prevent the usual invasion of insects.

In all cases, manufacturers' instructions should be followed closely regarding piping requirements. It is acceptable to run the vent vertically (but no more than 25 ft) above the location of the unit that vents.

Evaporators are fitted with a water bath vapor vent, which can be manifolded together in a multiple evaporator installation and be discharged to atmosphere without traps. Evaporators are also fitted with a chlorine overpressure relief system vent. This system is fully described earlier in this chapter, under "Evaporator Pressure Relief System."

Electrical

Controls for fans and lights should operate automatically when the door is opened, and there should be provisions to activate them manually from outside the room. Switches for fans and lights should be outside the room at the entrance. A signal light indicating fan operation should be provided at each entrance when the fan can be controlled from more than one point.

Reliability Provisions. The need for continuous and dependable disinfection has been stressed. The chlorination system can fail for a number of reasons, and, therefore, the design of the system must include the necessary provisions to either prevent failures or allow immediate corrective action to be taken. Although assured reliability is essential, design provisions for this are often slighted.

Chlorine Supply

As a chemical feed process, one of the most frequent interruptions in treatment is caused by the exhaustion of the chlorine supply. Five features are essential to maintain continuous chlorine feed: (1) an adequate reserve supply of chlorine sufficient to meet normal needs and bridge delivery delays and other possible contingencies; (2) chlorine container scales; (3) a manifolded chlorine header system; (4) an automatic device for switching to a full chlorine container when the one in use becomes empty; and (5) an alarm system to alert operating personnel to imminent loss of chlorine supply. These five features are discussed elsewhere in this text. Without them it is not possible to assure uninterrupted chlorine feed even with full-time operator attendance and no equipment breakdowns.

The chlorine header system is needed both to provide a connected on-line chlorine supply adequate to assure uninterrupted flow of chlorine for whatever period that the system may be unattended and to allow switchover to a full cylinder without interruption of chlorine.

Power Failure

Power outage usually results in water supply failure, which in turn automatically shuts down the chlorination system. A range of special provisions can be employed to assure the reliability of the power and water supply, depending upon the particular situation. As discussed previously, these provisions may be in the form of a standby power source and pumps.

Standby Equipment

The design of the chlorine feed system should provide for continued operation in cases of equipment failure. Where both pre- and postchlorination are to be practiced, separate chlorination systems should be provided for each plus a standby system. If prechlorination is not to be continuously used, it may be possible to use this system as the standby system for disinfection. The units, piping, and accessories should be designed with this application in mind. If prechlorination must be carried out continuously, or if no prechlorination is to be done, a standby system, capable of replacing the postchlorination system during repairs, maintenance, or emergencies, should be provided. Standby equipment of sufficient capacity should be available to replace the largest unit during shutdowns. This includes standby pumps for the injector water supply.

In addition to standby equipment, the equipment manufacturer should be consulted regarding vulnerable components. These components should be a part of the plant's inventory of spare parts.

Water Supply

As mentioned above, during a power failure the injector water system will be shut down unless there is an alternate supply that does not require power, such as an elevated tank. Standby equipment to provide injector water in the event of a power failure would consist of an engine-driven injector supply pump. Every injector water supply system should have such a standby pumping unit. There is no way to operate the chlorination system without an adequate water supply.

Chlorine Residual Analyzers

Every system using an analyzer for chlorination control should be backed up by an effluent monitor analyzer that can be switched over to the control function in the event of control analyzer failure. Similarly, backup capability should be provided for analyzers controlling and monitoring the dechlorination process. Provisions should be made for standby sample pumps. Sample lines should be piped to facilitate flushing or purging to remove biological slimes.

ORP Backup

In all chlorination systems for either potable water systems or wastewater effluents, continuous ORP measurement of the oxidant level produced in these systems by chlorination is the best backup system an operator can have. Using a combination of chlorine residual analyzers and the Stranco High Resolution Redox System has proved to be the most effective means for reliability control of the disinfection process, whether it be for drinking water systems or wastewater effluents.

Operator Documentation of ORP Control

Records of the following items at the end of the chlorine contact chamber should be kept on a twice-daily basis, at peak flow and at low flow, if convenient to operating personnel:

1. ORP, mV.
2. Chlorine feed rate, lb/day.
3. Plant treated water effluent, mgd.
4. Chlorine residual, mg/L.

AMMONIATION FACILITY*

Useful Ammonia Compounds

Ammonia is commercially available in four forms: anhydrous ammonia, NH_3, commonly stored and transported as a liquid in pressure vessels; aqua ammonia, NH_4OH, most commonly a 15–25 percent solution of ammonia in deionized or softened water; white crystalline ammonium chloride, NH_4Cl, a substance very soluble in water; and ammonium sulfate, $(NH_4)_2SO_4$, a gray-green crystalline solid that is hygroscopic. The latter two must be dissolved in water to be useful in waterworks practice.

Source, Availability, and Uses of Ammonia

Ammonia as NH_3 does not appear free in nature. It occurs in compounds as the NH_4^+ ion. However, the primary source for the nitrogen atom is the

*As of 1997, the most popular method of adding ammonia N to produce chloramines in potable water or wastewater treatment is the use of a 15–20 percent ammonia hydroxide solution. The amount of NH_4OH solution to produce either a 3 to 1 or 6 to 1 ratio with chlorine allows for a much better mix than ammonia gas.

atmosphere. All ways of making ammonia consist of combining one atom of nitrogen with three atoms of hydrogen.

Ammonia is extremely important commercially. Its most notable uses are for refrigeration and fertilizers. Other uses occur in the explosives, petroleum refining, rubber, textiles, chemicals, pulp and paper, and metallurgical industries. It is a by-product of the destructive distillation of coal and coke, but most of the ammonia produced in the United States is made by the Haber process. This process produces ammonia by direct synthesis. Gaseous nitrogen and hydrogen are mixed in correct proportion and heated under pressure (up to 1000 atm.) at 400–600°C and passed over a catalyst. The pure nitrogen for the process is obtained from liquid air, whereas natural gas (methane, CH_3) is the usual hydrogen source. This is why oil companies are usually the important suppliers of ammonia. For example, Union Oil Co. is the main supplier on the West Coast of the United States, owing to its participation in the Alaska Pipeline. It has an ammonia manufacturing plant in Kenai, Alaska. From there ammonia is shipped to West Coast ports and exported to Far East countries.

About 30 billion tons of ammonia are produced annually in the United States. Its production is exceeded only by that of sulfuric acid and of oxygen.

Physical and Chemical Characteristics of Ammonia

Ammonia is a colorless gas with a very pungent, irritating odor. However, when it is released into the atmosphere, it creates a rather dense white fog, which is due to its immediate reaction with the moisture in the atmosphere. Ammonia is highly soluble in water.

At normal temperatures and pressures anhydrous ammonia is a gas. It is easily liquefied by pressurizing it in a container, and it is commonly stored and tranported as a liquid. When the liquid reverts to a gas, a great amount of heat is absorbed. This is why it is extensively used for refrigeration.

At atmospheric pressure liquid ammonia has a density of 42.6 lb/ft^3, approximately two-thirds that of water. The vapor pressure of ammonia at 70°F is 114 psi, compared to chlorine at 85 psi (see Fig. 13 in the appendix). The important properties of ammonia are shown in Table 9-14.

Characteristics that Affect Ammonia Use in Water Treatment

Weight. The molecular weight of ammonia (17.03) is about half that of chlorine (35.5). Therefore, a 2000 lb/day chlorinator can only feed 950 lb/day ammonia.

Ammonia gas is lighter than air; so leaking vapor will rise quickly. This feature eliminates any serious hazard from an ammonia leak.

Heat of Vaporization. Ammonia's heat of vaporization is nearly five times that of chlorine. Therefore, a 10,000 lb/day chlorine evaporator can vaporize only 2000 lb/day ammonia.

Table 9-14 Physical Properties of Ammonia[70]

Molecular symbol	NH_3
Molecular weight	17.031
Boiling point at 1 atmosphere*	−28°F (−33.3°C)*
Freezing point at 1 atmosphere	−107.9°F (−77.7°C)
Critical temperature	271.4°F (133.0°C)
Critical pressure	1657 psia (114.2 bars)
Latent heat at −28°F (−33.3°C) and 1 atmosphere	589.3 Btu/lb (13.71 × 10^5 J/kg)
Relative density of vapor compared to dry air at 32°F (0.0°C) and 1 atmosphere	0.5970
Vapor density at −28°F (−33.3°C) and 1 atmosphere	0.0555 lb/ft^3 (0.8899 kg/m^3)
Specific gravity of liquid at −28°F (−33.3°C) compared to water at 4°C	0.6819
Liquid density at −28°F (−33.3°C) and 1 atmosphere	42.57 lbs/ft^3 (681.9 kg/m^3)
Specific volume of vapor at 32°F (0.0°C) and 1 atmosphere	20.78 ft^3 (1.297 m^3/kg)
Heat of solution at 0% conc. by wt.	347.4 Btu/lb (8.081 × 10^5 J/kg)
Heat of solution at 28% conc. by wt.	214.9 Btu/lb (4.999 × 10^5 J/kg)

*atmosphere = 760 mm Hg = 1.01325 bars.

Solubility in Water. Ammonia's water solubility is almost 50 times that of chlorine. This simplifies the requirements for mixing at the point of application and reduces significantly the quantity of injector water required. This minimizes the burden when injector water must be softened to a hardness not exceeding the solubility of calcium carbonate (35 mg/L).

Reaction with Water. The use of 1 mg/L NH_3 will increase the alkalinity of water by 2.9 mg/L (as $CaCO_3$), whereas 1 mg/L chlorine reduces the alkalinity 1.4 mg/L.

Corrosivity. Dry ammonia, either liquid or gas, is not corrosive to any metals. Moist ammonia will not corrode iron or steel but will react (to corrode) with copper, brass, zinc, and most copper alloys. Never use galvanized pipe. Valves, flex connections, and other hardware used in chlorine supply systems contain pure copper and/or copper alloys—so they are not interchangeable with comparable ammonia supply hardware.

Water Softening Reaction. This is the one and only disadvantage of ammonia use in water treatment. Ammonia reacts with the hardness in water to produce a softening effect comparable to excess lime softening. This occurs at the interface between ammonia vapor and water: at the inlet to the injector throat and at the exit of the direct feed gas diffuser. At these two interfaces a scale will develop directly in proportion to the calcium and magnesium hardness. Stoppage occurs quickly due to this scale. This is why all injector water should be softened to the maximum solubility of calcium carbonate (35 mg/L). For the direct feed diffuser, the one supplier of this equipment solved the hardness problem with a diffuser design modification many years ago.

Physiological Effects

Persons having chronic respiratory disease or persons who have shown evidence of undue sensitivity to ammonia should not be employed where they will be exposed to ammonia.

Ammonia is not a cumulative metabolic poison; however, ammonia in the ambient air has an intense irritating effect upon the mucous membranes of the eyes, nose, throat, and lungs. High levels of ammonia can produce corrosive action on these tissues that can cause laryngeal and bronchial spasms, as well as edema, which will obstruct the breathing passages. Conscious people are protected by its pungent odor, but unconscious people are not. Table 9-15 lists the human physiological response to various concentrations of ammonia in air. It should be noted here that individuals differ in their sensitivity to ammonia. Some are highly reactive to low concentrations, whereas others show a significant tolerance to the irritative effects.

SUPPLY SYSTEM: ANHYDROUS AMMONIA

Cylinders

Two sizes, 100 lb and 150 lb, are usually available, but are not so common as they were. Cylinders are equipped with a dip tube so that when they are placed horizontally, liquid can be withdrawn.

800-lb Containers

These are still available but are not so popular as bulk storage tanks. They are similar to chlorine ton cylinders. They are equipped with dip tubes for

Table 9-15 Physiological Response to Ammonia[70]

	Concentration (ppm)
Least perceptible odor	5
Readily detectable odor	20–50
No discomfort or impairment of health for prolonged exposure	50–100
General discomfort and eye tearing; no lasting effect on short exposure	150–120
Severe irritation of eyes, ears, nose, and throat; no lasting effect on short exposure	400–700
Coughing, bronchial spasms	1,700
Dangerous, less than ½ hour exposure may be fatal	2,000–3,000
Serious edema, strangulation, asphyxia, rapidly fatal	5,000–10,000
Immediately fatal	10,000

vapor or liquid withdrawal and are the same physical size as chlorine ton containers. Therefore, a chlorine ton container lifting bar and trunnions will fit these cylinders.

Storage Tanks

General Consideration. Bulk storage tanks sized to fit the consumer's needs are the most popular means of on-site ammonia storage. For example, in the Southern California area the Union Oil Co., Chemical Division, has a large fleet of 8000-gal tank trucks used to service customer storage tanks on a routine basis. USS Agrichem (U.S. Steel Co.) and Chevron provide similar service in other areas.

Consumers in the water treatment industry are advised to purchase refrigeration-grade ammonia instead of commercial-grade because it is moisture-free. Commercial-grade ammonia purposely contains a small amount of moisture to reduce stress fractures (fatigue). This prolongs tanker life.

Design Factors. It is common practice in the industry for the supplier to build or furnish the consumer with a storage tank. Certain features should be a part of a proper storage tank. The shell should be designed for at least 250 lb working pressure.[70] There should be one liquid outlet with an excess flow check valve inside the tank and two vapor outlets, all with stainless steel trim shut-off angle valves (similar to those of a chlorine tank car). In the vicinity of these valves there should be a safety relief valve with a vent line that discharges 2–3 ft above "roof" height. It should end in a double 90-degree elbow to prevent rain from entering the vent, and it should have a moisture trap-leg adjacent to the valve discharge. This moisture trap will eliminate moisture interference with the relief valve.

Filling Density. The allowable filling density for uninsulated stationary storage tanks is 82 percent by volume and 56 percent by weight. Therefore, an integral part of the storage tank must be a device to measure the weight or the volume of ammonia at any time.

This is often overlooked by the user. Such an omission of scales can cause operating personnel serious problems, unless the supplier is aware of all operating requirements.

Inventory Control. The preferred method of determining the status of ammonia supply is by weight. Therefore, the contents of the tank should be weighed by either a set of industrial lever-type scales or a load cell. A load cell with a 4–20 mA signal for remote display is preferable to scales.

Withdrawal Rates. This is an important factor in selecting the size of a storage tank. Assume that the maximum withdrawal rate will never exceed 1000 lb/day, and assume that this much vapor will be needed at an ambient temperature of 50°F. How big will the storage tank have to be? The formula for withdrawal rates for ammonia is as follows:[2]

$$[\text{Room temp } (^\circ\text{F}) - \text{Liquid temp } (^\circ\text{F})] \times \text{withdrawal factor} = 1000 \text{ lb/day}$$

Liquid temperature is the temperature of the ammonia at minimum allowable inlet pressure to the ammoniator (10 psi). This pressure is required to actuate the remote vacuum regulator. At 10 psi the liquid temp is −5°F. Let the withdrawal factor be *W*:

$$[50 - (-5)]W = 1000 \text{ lb/day}$$

$$W = \frac{1000}{55} = 18.2$$

An 800-lb cylinder (192 gal) has a withdrawal factor of 3.2:[2]

$$\frac{W}{192} = \frac{18.2}{3.2}$$

$$= 1094 \text{ gal}$$

Therefore, an 1100-gal tank when full could provide a 1000 lb/day vapor withdrawal rate at 50°F ambient temperature.

Evaporators

Ammonia evaporators are identical to chlorine evaporators,* but the two must not be used interchangeably because this practice could produce an explosure mixture of chlorine and ammonia. The major difference between chlorine and ammonia evaporators is capacity. A 10,000 lb/day chlorine evaporator can only vaporize 2000 lb/day ammonia.

Evaporators for ammonia are being used in water treatment practice where the ammonia is supplied in 800-lb cylinders. Evaporators can and should be avoided by the use of custom-made storage tanks. All of the problems of evaporators and storage tanks, along with potential leaks, can be avoided by using liquid ammonium hydroxide at a strength of about 20 percent. This is probably the most popular method today (1977).

Leak Detectors

Continuous ammonia leak detector monitors similar to chlorine detectors are not available. However, American Gas and Chemical Co. of Northvale, New Jersey markets products for locating ammonia leaks at the source:

---

*Except that inlet and outlet valving have 316 SS trim.

- Its aerosol powder, ADP-219, can be sprayed onto areas of suspected leaks (pipe joints, fittings, valves, etc.) as a yellow powder coating that changes to a dark blue in the presence of a small leak. The coating can easily be removed with a damp cloth.
- Another aerosol spray, ADS-100, generates white smoke upon contact with ammonia fumes. This spray tends to neutralize the escaping ammonia, making it less hazardous, and procudes a visible fog that grows more intense as it approaches the leak.
- Its CG Tracer is a small, hand-held device sensitive to both ammonia and combustible gases. It can quickly locate an ammonia leak.

Potential Hazard of a Major Leak

There are two physical properties of anhydrous ammonia that reduce significantly the hazard of a major leak. These factors are graphically illustrated by the events of a major leak. A Chevron tanker truck overturned in Richmond, California, resulting in a 6-inch gash in the tank shell. The Chemtrec emergency response team arrived to see a dense white fog rising into the atmosphere above the overturned truck.[73] Firehoses were sprayed into the fog (ammonia plus atmospheric moisture), which disappeared immediately upon contact with the water. This is a result of the high solubility of ammonia in water. Ammonia is so much lighter than air that it rises quickly above and beyond people at ground level, and thus is greatly different from chlorine. After the fog was washed into the storm drains, water was sprayed into the gash on the tank. By this time the vaporization of the liquid due to the leak cooled the liquid sufficiently that the water formed an ice cover, which sealed off the liquid ammonia long enough for emergency workers to right the truck and move it to a safe place, where it was emptied without further incident. The high latent heat of vaporization of liquid ammonia is directly responsible for the rapid formation of the ice cover. This characteristic makes ammonia a universal refrigerant. From the above accident description it can be seen that ammonia is a relatively safe gas. Furthermore, the levels of concentration required to produce a dangerous environment are many times those for chlorine or sulfur dioxide.

Materials of Construction

The supply system under pressure should be Sch. 80 seamless steel pipe with 3000 lb forged steel fittings. Welded joints are preferred. Use bell reducer-type fittings. Bushings should never be used. Unions should be two-bolt flanged with a lead gasket joint. Valves should be steel with 316 stainless steel trim. All ammonia piping under a vacuum should be Sch. 80 PVC with solvent weld joints.

AMMONIATORS[75]

Direct Feed*

This type of ammoniator was available from only one manufacturer as of 1984 (Wallace & Tiernan). Metering capacities up to 1000 lb/day are available. All sizes come equipped with a special diffuser designed to prevent interference from scale formation and back-flooding of the equipment. The pressure at the point of application must not exceed 15 psi.

The manufacturer advised that there were over 200 direct-feed ammoniators in operation in 1984. Because of their stainless steel construction and the characteristics of ammonia there are no corrosion problems; hence they are rugged, low-maintenance devices. White has observed a pair of these units that have been in operation for over 60 years.

The direct-feed unit can be furnished for either manual control or automatic flow-paced control. The automatic version costs about three times as much as the manual unit. Operating experience with the automatic type has proved to be satisfactory.

Cautionary Instruction

Use of Anhydrous Ammonia. The use of ammonia in the gas form should be limited to cylinder deliveries. The use of storage tanks or railcars and tanker trucks is not recommended because of current interpretation of the Uniform Fire Code. (See above description of an ammonia tanker truck accident.)

The preference for capacity requirements exceeding the use of steel cylinders is to use an ammonia hydroxide solution with a strength of 15–20 percent ammonia N.

Solution Feed[76]

General Description. Solution-feed ammoniators are available from the top three chlorinator manufacturers. They are identical in design to chlorinators and sulfonators except for minor differences in materials of construction to conform to the chemical characteristics of ammonia vapor and aqueous solution. They are available up to capacities of 950 lb/day and are usually arranged for remote vaccum operation. Automatic switchover components are optional. The major design difference is in the sizing of the injectors, which is described below.

Control Strategies. Ammonia should be applied at a fixed ratio to chlorine for the best and most consistent results. Hence the ammoniator should be

*Owing to the popularity of using ammonia hydroxide solution, there is not much activity, if any, in the use of direct feed ammoniators (1997).

controlled by the same signals that control the chlorinator. The precise ratio of ammonia to chlorine can be achieved by rotameter feed-rate selection and use of the manual dosage adjustment on the ammoniator.

Injector System

Water Requirements. The major disadvantage of a solution-feed ammoniator is the softening reaction caused by the formation of ammonium hydroxide:

$$NH_3 + H_2O \rightarrow NH_4OH \tag{9-16}$$

The ammonium hydroxide precipitates the calcium and magnesium hardness down to the solubility limit of $CaCO_3$, which is 35 mg/L. Therefore, the injector water should be softened to this level of hardness to prevent the injector from plugging because of hardness scaling.* Owing to the high solubility of ammonia in water, much less water is required for solution-feed ammoniators than for chlorinators. For example, one manufacturer requires only 4.25 gpm with 50 psi injector water for 240 lb/day ammonia versus 6 psi back-pressure, and only 2.5 gpm versus 2 psi back-pressure.[73] Therefore, the manufacturer should be consulted early to determine injector water requirements, particularly if softening is to be involved.

The above figures are based upon the use of 1-inch fixed-throat injectors, which are limited in ammonia capacity to 240 lb/day: Multiple injectors are required to achieve a maximum capacity of 950 lb/day. In any case, duplicate injectors and duplicate diffusers are recommended for solution-free ammoniators.[77]

Vacuum Lines. Sizing of these lines is critical owing to the use of remote vacuum and remote injectors. The following analysis will serve as an example of the necessary calculations using these formulae:

Head loss in vacuum line:

$$\Delta P \text{ (in. Hg)} = \frac{11.89 \times L \times f \times W^2}{10^9 \times \rho \times d^5} \tag{9-17}$$

where

L = equivalent pipe length, fit

f = friction factor (see Figs. 10 and 11 in Appendix I)

*Capital Controls announced (1984) a special solution-feed ammonia injector with a flexible liner timed to flex by external water pressure on a programmed basis. The flexible liner is intended to shred the hardness scale buildup.

W = lb/day ammonia

ρ = density of NH_3, lb/ft^3

d = pipe diameter, inches

Reynolds number: $= N_R$

$$N_R = \frac{6.32 \times W}{\mu \times d \times 24} \tag{9-18}$$

where

W = lb/day ammonia

μ = viscosity of NH_3, centipoise

d = pipe diameter, inches

Use the Reynolds number to select the friction factor from Fig. 9 in the appendix. Calculate the density of ammonia at 20 inches of Hg vacuum and 50°F:

$$PV = NRT \tag{19-19}$$

where

P = 20 in. Hg or (30 − 20)/30 = 10/30 atm. = 0.33 atm.

T = 50°F or 283°K

R = 0.08205

Hence:

$$0.33V = (0.08205)(283)$$

$$V = 70.36 \text{ L/mole}$$

$$17.03 \text{ g/mole} \times \frac{1 \text{ mole}}{70.36} \times \frac{1 \text{ lb}}{454 \text{ gm}} \times \frac{28.3 \text{ L}}{\text{ft}^3} = 0.015 \text{ lb/ft}^3$$

Therefore:

$$P = 0.015 \text{ lb/ft}^3 \text{ at 20 in. Hg and 50°F}$$

Solve for Reynolds number using viscosity of NH_3 at 0.0095 centipoise at 50°F. Try 1.25-inch pipe and use max. capacity of the ammoniator (950 lb/day):

$$N_R = \frac{6.32 \times 950}{0.0095 \times 1.25 \times 24}$$

$$= 21{,}066$$

The value of f from Fig. 9 (appendix) is 0.03. Now solve for total head loss in 300 ft of pipe using Eq. (9-17):

$$\Delta P = \frac{11.89 \times 300 \times 0.03 \times (950)^2}{10^9 \times 0.015 \times (1.25)^5}$$

$$= 2.11 \text{ in. Hg}$$

Therefore a 1.25-inch-diameter PVC pipe is borderline; so use 100 ft of 1.5-inch and 200 ft of 1.25-inch.

Diffusers and Solution Lines

Mixing. The solution lines should be sized so as to not exceed 1.5 ft total head loss. The diffusers should be designed with holes that will create a head loss of 8–11 ft. This will translate to a G factor of 250–300, which is sufficient turbulence for a quick mix. The orifice jets must point upstream.

Back-pressure. The injector should be located as close to the diffuser as possible, which in nearly all instances calls for a remote injector. The centerline of the injector throat should be above the hydraulic gradient of the plant in order to keep the back-pressure as low as possible. If it is not practical to keep the back-pressure below 10 psi (to minimize injector water consumption), the ammonia solution should be pumped with a small turbine pump. Optimum back-pressure is about 2 psi.

Contact Time. The reaction of ammonia nitrogen with chlorine to form chloramines is practically instantaneous, so contact time is not a factor—rapid dispersion of the ammonia solution is the principal factor.

Flushing. Diffusers should not only be provided in duplicate; they should be arranged so that they can be flushed with acid or chlorine solution to dissolve any formation of carbonate scale.

ADVANTAGES OF SOLUTION-FEED ANHYDROUS AMMONIA (Gas form)

A great many chloramine systems in the United States use ammonium hydroxide solution. When faced with a decision to select aqua or anhydrous ammonia,

Engineering Science, consulting engineers, Pasadena, California, conducted a survey of both forms and concluded that anhydrous ammonia (the gas form) was more advantageous for the following reasons:[78]

- Much smaller space requirement for storage.
- Simplicity of metering and automatic control.
- Simplicity of operation and maintenance, as solution-feed ammoniators are virtually identical to chlorinators.
- Lower cost.

Now in 1997 for a variety of reasons, probably due to the UFC, the most popular system for forming chloramines uses ammonia hydroxide solution with metering pumps.

AQUEOUS AMMONIA (NH_4OH solution)

Source and Availability

Although there may be a single source of ammonia locally, there will be several distributors of ammonium hydroxide solution. The ammonia market—domestic and foreign—is so competitive that ammonia in either form is ubiquitous.

Local distributors have a fleet of trucks and will provide storage tanks upon request. Aqua ammonia costs about 14 percent more than anhydrous ammonia; however, storage tanks are less expensive than the pressurized tanks required for anhydrous ammonia.

Shipping can be made by 8000-gal tank cars, 4000-gal tank trucks, 375- and 750-gal drums, or 30-gal carboys.

Physical and Chemical Characteristics

Commercial strength aqua ammonia is approximately 20–30 percent. The density of Grade A, 29.4 percent solution, is 0.8974 at 60°F, compared to water at 1.

This strength of solution corrodes copper, aluminum alloys, and galvanized surfaces. When this solution is dispersed into the process water by the diffuser, a water-softening reaction will occur at the perforation in the diffuser. This will promote the deposit of calcium scale in proportion to the hardness of the water. Provisions should be made to allow the diffusers to be cleaned with acid or chlorine solution.

Potential Hazards

Ammonium hydroxide is a caustic solution and can be hazardous to persons in contact with the solution. Usual precautions required for caustic solutions should be observed.

If the solution is being applied at an installation using hypochlorite solution, steps must be taken to prevent aqua ammonia solution from being unloaded into the hypochlorite storage tank or vice versa. In such an event, an explosive mixture of nitrogen trichloride is likely to form with disastrous results. If not, the evolution of nitrogen trichloride is certain to cause a hazardous air pollution problem.

White noticed recently (1996) that there is an ammonia inhalant called AMOPLY by Johnson & Johnson for temporary relief of dizziness or fainting. It is intended for inhalation only.

SUPPLY SYSTEM FOR NH_4OH

Storage Tanks

The local supplier should be consulted on this matter. Make certain the supplier incorporates all of the necessary piping to facilitate unloading and that the tank has a proper sight glass. The tank and appurtenances should be made of mild steel. The words "AQUA AMMONIA—NH_4OH CAUSTIC" in large letters should be highly visible on each side of the tank.

METERING AND CONTROL SYSTEM FOR NH_4OH

Design Considerations

The most difficult part of the aqua ammonia facitity to design properly is the metering and control system. It is most exacting because ammonia N has to be applied at a constant ratio to chlorine. Depending upon local conditions, this ratio will be approximately 4 to 1 chlorine to ammonia. This means the feed rates will be 3 to 5 times lower than chlorine. Furthermore, because the aqua ammonia will be about 30 percent NH_3 or 300,000 mg/L, compared to 10,000 mg/L anhydrous ammonia solution in a solution-feed ammoniator, the designer is faced with the problem of precise control of small quantities of solution. For example, a water supply uses 50 lb/day chlorine, and the ammonia N requirement is 12.5 lb/day (4:1 ratio); then at 2.25 lb NH_3–N per gallon of 30 percent solution an aqua ammonia feed rate of only 0.23 gal/hr will be required. Obviously aqua ammonia is not suitable for small water supplies unless it is diluted with softened water.

Diaphragm Pumps

These are widely used in waterworks practice when they fit the requirements. If these pumps can fit the metering requirement, the control system should be patterned after those described in Chapter 2 for hypochlorite.

The designer should strive to eliminate pulsing activity of the diaphragm pumps in order to salvage the range and the accuracy of flow metering equipment required to control these pumps.

Diffusers and Solution Lines

Reasonable care should be used in the design of the solution lines in order to minimize transit time and thereby decrease system dead time. This will make the system more compatible with the chlorinator control system.

Owing to the fact there there are no real back-pressure (up to 125 psi) problems when using diaphragm pumps, the designer is free to put the required head loss into the diffuser (8–12 ft) to provide good mixing with the process water.

All diffusers should be supplied in duplicate so that they may be taken out of service for cleaning to remove scale deposits caused by hardness in the water.

RELIABILITY PROVISIONS FOR AMMONIA APPLICATION

This is not a factor in water treatment. Ammonia application interruptions for short periods will do little harm. Health effects of such an event are insignificant. Disappearance of ammonia in the treatment process may result in off-flavors at the consumer's tap. This may or may not result in consumer complaints.

REFERENCES

1. Anon., *Chlorine Institute Manual,* 4th ed. pp. 6, 10, 12, 13, and 14, New York, 1969.
2. Baker, R. J., "Maximum Withdrawal Rates from Chlorine, Sulfur Dioxide and Ammonia Cylinders," AWWA Publication, *OPflow,* p. 4, Apr. 1980.
3. Anon., "Vacuum Regulator Check Unit—3000 lb/da. Capacity," Wallace & Tiernan Div. Pennwalt Corp., Belleville, NJ, Book No. WBB50.177, Feb. 1982.
4. Whiteman, T. J., "Chlorine System Test," in-house report, East Bay Municipal Utility District WPCP, Oakland, CA, Feb. 14, 1980.
5. White, G. C., survey of five tank car and bulk storage air padding systems, Apr. 1982.
6. Anon., "Stop-Valve System for Chlorine Tank Car Padding," Pennwalt Corporation Dwg. No. T50W016, Tacoma, WA, July 9, 1968.
7. Beatty, A., private communication, Metropolitan Water District of Southern California, Los Angeles, CA, March 1982.
8. Albers, R. G., "Manufacturers' Data Reports," private communication issued by Pressed Steel Tank Co., Milwaukee, WI, Dec. 22, 1975.
9. "Chlorine Vaporizing Equipment," Chlorine Inst. Pamphlet No. 9, 2nd ed., New York, 1970.
10. "Technical Information for Handling Chlorine, Sulfur Dioxide, and Ammonia from Supply to Point of Application," Fischer and Porter Instr. Bull. 70-9001 Rev. 1, Publ. No. 22155, Warminster, PA, 1977.
11. "Evaporator Series 50.202," Wallace & Tiernan, Belleville, NJ, Rev. Jan. 1977.
12. Walker, T. B., personal communication, Wallace & Tiernan, June 1977.
13. Nagel, Wm., private communication, Fischer and Porter Co., July 1977.
14. Connell, Gerald F., "ADVANCE™ Single Point Gas Detector—Series 1610B," Bulletin No. 325.0001.0, Capital Controls, 3000 Advance Lane, P.O. Box 211, Colmar, PA, 18915, Sept. 1995.

14a. Connell, G. F., "ADVANCE™ Multipoint Gas Detector—Series 1620B", Bulletin No. 325.0005.0, Capital Controls, 3000 Advance Lane P.O. Box 211, Colmar, PA 18915, Oct. 1995.

15. Anon., "Draeger Safety Chloralarm," National Draeger Inc., Pittsburg, PA, Jan. 1983.

16. Anon., "Chlor-Guard Model 5152 Chlorine Gas Detector," Exidyne Inc., Exton, PA, May 1983.
17. Becker, J., private communication, Exidyne Inc., Exton, PA, June 1984.
18. Anon., "Chloralert Chlorine Detector Series 17CA 1000," Fischer and Porter Co., Mar. 1978.
19. Anon., "Series 50-125 Chlorine Detector Instruction Book," Wallace & Tiernan, Belleville, NJ, June 1980.
20. White, G. C., and Cariss, F., unpublished field data, East Bay Municipal Utility District, Oakland, CA, 1965.
21. Houston, R., "Friction in Pipes," *Product Engr.,* 191 (Aug. 1957).
22. Anon., "Wallace and Tiernan Series V-2000 V-Notch Chlorinators," Wallace & Tiernan, Belleville, NJ, Technical Data Sheet 25.055, Rev. Aug. 1989.
23. Finger, R. E., private communication, Renton, WA, Wastewater Treatment Plant, Metropolitan Seattle, WA, 1984.
24. Connell, G. F., private communication, Capital Controls Co., Colmar, PA, July 27, 1984.
25. White, G. C., and Stone, R. W., "Factors Affecting the Feed Rate Response Time in a Long Injector Vacuum Line for Chlorinators and Sulfonators," unpublished in-house report for Brown and Caldwell consulting engineers, Walnut Creek, CA, Mar. 1974.
26. Anon., Duriron Co. Bulletin No. P-10-101r, Dayton, OH, 1983.
27. Anon., Duriron Co. Bulletin No. P-17-101a, Dayton, OH, 1979.
28. Anon., MetPro Corp. Fybroc Div. Bulletin 15B1, Hatfield, PA, 1982.
29. Rossum, J. R., private communication, California Water Service Co., San Jose, CA, 1962.
30. Murphy, D., private communication, FCI Fluid Components, Inc. Bull. SF-3/79, San Marcos, CA, Jan. 1984.
31. Cutler, J. W., and Green, F. W., "Operating Experiences with a New Automatic Residual Control Recorder Controller," *J. AWWA,* **22,** 755 (1932).
32. Goudey, R. F., "Residual Chlorination on the Los Angeles System," *J. AWWA,* **28,** 1742 (1936).
33. Caldwell, D. H., "Automatic Chlorine Residual Indicator and Recorder," *J. AWWA,* **36,** 771 (1944).
34. Harrington, J. H., "Photo-cell Control of Water Chlorination," *J. AWWA,* **32,** 859 (1940).
35. Baker, R. J., and Griffin, A. E., "Development of Instrumentation in Chlorination," *J. AWWA,* **50,** 489 (1958).
36. Marks, H. C., Bannister, G. L., Glass, J. R., and Herrigel, E., "Amperometric Methods of Control of Water Chlorination," *Anal. Chem.,* **19,** 200 (1947).
37. Hazey, F. J., "Amperometric Chlorine Residual Recording," *J. AWWA,* **43,** 292 (1951).
38. Krum, H. J., "Residual Chlorine Recording Experiences," *Water Wks and Sew.,* **98,** 376 (Sept. 1951).
39. Clark, G. C., "Amperometric Techniques for Chlorine Residual," *Inst. and Automation* (Apr. 1954).
40. Morrow, J. J., U.S. Patent No. 3.4.3, 199 (Nov. 26, 1968).
41. Johnson, J. D., Edwards, J. W., and Keeslar, F., "Chlorine Residual Measurement Cell: The HOCl Membrane Electrode," *J. AWWA,* **70,** 341 (June 1978).
42. Kolthoff, I. M., and Lingane, J. J., *Polarography,* Vols. I and II, Interscience Publishers, New York, 1965.
43. Anon., "Residual Chlorine Analyzer for Water Treatment," Wallace & Tiernan, Belleville, NJ, Cat. No. 50.245, Rev. Jan. 1982.
44. Marks, H. C., and Campbell, G. A., "Dual Electrode Measuring Cell," in-house report, Wallace & Tiernan Inc., Belleville, NJ, 1950.
45. Huebner, W. B., private communication, Wallace & Tiernan, Belleville, NJ, Aug. 28, 1984.
46. Connell, Gerald F., "Advance Series 870 Chlorine Residual Analyzers," Capital Controls Co., Colmar, PA, Bulletin No. A1.1870.6, 1982.
47. Johnson, J. D., and Edwards, J. W., "An Amperometric Membrane Halogen Analyzer," *Proc. Div. Envir. Chem. ACS,* **14,** 169 (Apr. 1974).

48. Johnson, J. D., Edwards, J. W., and Keeslar, F., "Chlorine Residual Measurement Cell: The HOCl Membrane Electrode," *J. AWWA,* **70,** 341 (June 1978).
49. White, G. C., "Disinfection: Present and Future," *J. AWWA,* **66,** 689 (Dec. 1974).
50. Anon., "Directions for Series 8224/8324 and 8225/8325 Continuous Automatic Chlorine Monitor/Controller," Delta Scientific, National Sonics Div. Environtech Corp., Lindenhurst, NY, Mar. 1976.
51. Anon., "Polarographic Wet Chemistry Analysis," Series 8000 Analyzers, Delta Scientific Products National Sonics Div., Environtech Corp., Lindenhurst, NY, 1978.
52. Stanley, W., and Nossel, R., "Measurement of Residual Chlorine Compounds in Wastewater with Amperometric Membrane Electrodes," in R. L. Jolley et al. (Eds.), *Chemistry and Water Treatment of Water Chlorination,* Vol. 4, Book 1, p. 699, Ann Arbor Science (Butterworth Group) Ann Arbor, MI, 1983.
53. Anon., "Model 924 Chlorine Measurement System," Product Data Sheet 924, Delta Analytical (Xertex Corp), Hauppage, NY, 1983.
54. Anon., "Model 925 Chlorine Measurement System," Product Data Sheet 925, Delta Analytical (Xertex Corp), Hauppage, NY, 1983.
55. Obear, P. H., private communication, Xertex Corp. Huappage, NY, July 8, 1984.
56. Hayes, T. J., "CHLOR-TROL 5000™ Residual Chlorine Analyzer Series 17B5000," Fischer and Porter Co., Warminster, PA, Dec. 1996.
57. Hayes, T. J., "ANACHLOR II™ Residual Chlorine Analyzer Series 17PC1000" Fischer and Porter Co., Warminster, PA, Dec. 1996.
58. Morrow, J. J., and Roop, R. N., "Advances in Chlorine Residual Analysis," *J. AWWA,* **67,** 184 (Apr. 1975).
59. Anon., "CHLOR-TROL™ Free Residual Chlorine Analyzer, Series 17K1000," Fischer and Porter Co., Warminster, PA, June 1975.
60. Anon., "The Chlortect® Chlorine Monitor for the Accurate Measurement of Chlorine Residuals," U.S. Patent No. 3,966,413, EPCO, Danbury, CT, Oct. 1980.
61. Marinenko, G., Huggett, R. J., and Friend, D. G., "An Instrument with Internal Calibration for Monitoring Chlorine Residuals in Natural Waters," *J. Fisheries Research Board Canada,* **33,** 822 (Apr. 1976).
62. Anon., "EC/250 Chlorine Analyzer User's Manual," IBM Instruments Inc., Danbury, CT, 1982.
63. Kutt, J. C., and Vohra, S. K., "A Simple Approach to Chlorine Analysis," *Am. Lab.* (Dec. 1983).
64. Anon., "Instruction Manual, Model 1570 Chlorine Monitor," Orion Research Inc., Cambridge, MA, 1982.
65. Anon., "Instruction Manual, Model 1770 Chlorine Monitor," Orion Research Inc., Cambridge, MA, 1981.
66. Anon., "Instruction Manual, Model 853 Residual Chlorine Analyzer/Transmitter and Model 450 Sensor Assembly," Uniloc Div. of Rosemount Inc., Irvine, CA, Dec. 23, 1980.
67. Hoffman, F., private communication, Uniloc Div. of Rosemount Inc., Irvine, CA, May 13, 1981.
68. Anon., "Pump-Colorimeter Colorimetric Free and Total Chlorine," Models 31100, 31300, and 61100, Hach Company, Loveland, CO, Bulletin 1071, Feb. 1984.
69. Anon., "Free and Total Chlorine—Pump Colorimeter Colorimetric Models 31100, 31300, and 61100," Hach Company, Loveland, CO, Feb. 1984.
70. "Load Hugger 5000 lb. Capacity Bulletin," Lift All Products, Manheim, PA, (1974); and Liftex Slings, Inc. Bulletin A76-TD, Libertyville, IL, Dec. 1976.
71. Anon., "No. 53 MC Controller," product bulletin, Fischer and Porter Co., Warminster, PA, June 1982.
72. Hayes, T. H., private communication, Fischer and Porter Co., Warminster, PA, Aug. 15, 1984.
73. Sloat, R., private communication, Chief, Chemtrec Emergency Response Team, Dow Chemical Co., Pittsburgh, CA, Jan. 1984.

74. Anon., "Anhydrous Ammonia," CGA Pamphlet G-2, 6th ed., Compressed Gas Association Inc., New York, 1977.
75. Anon., "Direct Feed Ammoniator," Wallace & Tiernan–Pennwalt, Belleville, NJ, Catalog 60.215, 1980.
76. Anon., "V-Notch Ammoniator Series V-800 Remote Vacuum Arrangement," Wallace & Tiernan, Belleville, NJ Catalog 60.222, Aug. 1977.
77. Baker, R. J., and Rudolph, G. C., private communication, Wallace & Tiernan, Belleville, NJ, Mar. 1982.
78. Brentwood, R. W., Reichenberger, J. C., Suggs, D., and Clements, E. V., "Chloramine Disinfection for THM Control," paper presented at National Conference on Environmental Engineering, ASCE, Univ. of Southern California, June 26, 1984.
79. Goodwin, H. E., "Automated Chlorination System," *Pub. Wks.,* **93,** 74 (1962).
80. Stone, R. W., "Rancho Cordova Breakpoint Demonstration," report prepared by Sacramento Area Consultants, Sept. 1976.
81. Howerton, A. E., "Estimated Area Affected by a Chlorine Release," Chlorine Institute Report 71, Mar. 1969.
82. Horowitz, N. C., "Selecting Materials for Chlorine Gas Neutralization," *Chem. Engineering,* p. 105, Apr. 6, 1981.
83. Anon., "Emergency Chlorine Scrubbing Systems for Water Purification Plants," Ametek-Schutte and Koerting Div., Durham, N.C., Bulletin 7S/21B, 1977.
84. Dailey, L., private communication, Fischer and Porter Co., Warminster, PA, Jan. 29, 1985.
85. Hankins, R. A., private communication, Clorox Co. Tech Center, Pleasanton, CA, Jan. 31, 1985.
86. Roop, R. N., private communication, Fischer and Porter Co., Warminster, PA, Aug. 16, 1976.
87. Committee Report, "Chemical Hazards in Waterworks Plants," *J. AWWA,* **27,** 1225 (1935).
88. Chlorine Institute Inc., *Thermodynamic Properties of Chlorine in SI Units,* 72-Ref. 10.9.5, 2nd ed., Washington, DC, 1986.
89. Chlorine Institute Inc., *The Chlorine Manual,* Ref. 2.3.2, p. 8, 5th ed., Washington, DC, 1986.
90. Doyle, J., private communication, Chlorine Institute Inc., Washington, DC, 1984.
91. Becker, J., V. P., Sales, EIT, private communication, Exton, PA, Oct. 1990.
92. Doyle, J. H., personal communication, Chlorine Institute, letter dated Feb. 14, 1983.
93. White, G. C., personal investigation of nitrogen trichloride explosions at the chlor-alkali plant in Zipaquira, Colombia, S.A., Feb. 1981.
94. Powell, D., Videotape of 600-lb liquid chlorine leak, with commentary, Powell Fabrication and Mfg. Co., St. Louis, MI, 1985.
95. Hayes, T. J., private communication, Fischer and Porter Co., Warminster, PA, June 12, 1991.
96. Griffin, G., and Farrar, W., "Preventing Moisture Entry by Corking the Flex Line," Kirkwood WTP, Kirkwood, MO, AWWA *Opflow,* Feb. 1982.
97. Huebner, W. P., private communication, May 1991.
98. Stock, J. T., *Amperometric Titrations,* Interscience, New York, 1965.
99. Strand, R. L., "Optimization of Redox for Bromine and Chlorine Control," presented at the NACE Annual Conference and Corrosion Show, Cincinnati, OH, Mar. 11–15, 1991.
100. Strand, R. L., private communication, Nov. 13, 1991.
101. Kim, Y. H., Hensley, R., and Cooper, B., "Chlorine and Sulfur Dioxide Dosing Minimized through the Use of Redox Potential at Simi Valley, California: A Case Study," paper presented at the WEF Disinfection Specialty Conference, Portland OR, Mar. 17–20, 1996
102. Latimer, W. M., *Oxidation Potentials,* Prentice-Hall, New York, 1952.
103. Gurney, R. W., *Ionic Processes in Solution,* McGraw-Hill, New York, 1953.
104. Wareham, D. G., Hall, K. J., and Mavinic, D., S., "Real Time Control of Wastewater Treatment Systems Using ORP," *Water Science Technology,* **28,** No. 11–12, 273–282 (1993).
105. Kim, Y. H., "Evaluation of Redox Potential and Chlorine Residual as a Measure of Water Disinfection," paper presented at the 54th International Water Conference, Pittsburgh, PA, Oct. 11–13, 1993.
106. Wallace & Tiernan, Sales Service Department, "ORP in Wastewater Treatment," Nov. 1971.

107. Kim, Y. H., and Kiser, P., "Automatic Control of Dechlorination Process for Ion Exchange Systems," paper presented at the 55th International Water Conference, Pittsburgh, PA, Oct. 30–Nov. 2, 1994.
108. Victorin, K., Hellstrom, K., and Rylander, R., *Hyg. Camb.,* **70,** 313 (1972).
109. Lund, E., "Oxidative Inactivation of Poliovirus," Aarhuus Stiftsbogtrykker, Denmark, 1963.
110. Strand, R. L., and Kim, Y. H., "ORP as a Measure of Evaluating and Controlling Disinfection in Potable Water."
111. Carlson, S., Hasselbarth, U., and Mecke, P., *Archiv. fur Hyigene und Bakteriologie,* **152,** 306 (1968).
112. White, G. C., Personal experiences in the 1950s.
113. Becker, J. J., "Residual Chlorine Monitor," Analytical Technology Inc., 680 Hollow Road Box 879, Oaks, PA 19456, Customer Service 800-959-0299; 1995.
114. Midland Manufacturing Corp., "Angle Valves, Excess Flow Valves, Safety Relief Valves, and Valve Actuators on Tank Domes for Chlorine Road Tankers and Chlorine Storage Tanks," 7733 Gross Point Road, Skokie, IL 60077, 1993.
115. Bartel, M., "TGO Tetchnologies Inc.," 3450–C Regional Parkway, Santa Rosa, CA 95403-8247, Customer Service 800-543-6603; 1996.
116. White, G. C., Greenberg, A. E., and Kwan, Y. C., "Control of the Free Chlorine Residual Process by ORP Measurement," California Department of Health, June 1975.
117. Connell, G. F., "Microprocessor-Based Controller—Captrol® Series 1450," Capital Controls Co. Inc., 3000 Advance Lane, Colmar, PA 18915, Customer Service 215-997-9000; 1994.
118. Connell, G. F.; "Chlorine Residual Analyzer—OXITRACER Model 1871," Capital Controls Co. Inc., 3000 Advance Lane, P.O. Box 211, Colmar, PA 18915, 1996.
119. Becker, J., "Modular Chlorine Gas Detector," Analytical Technology Inc., 680 Hollow Road, Box 879, Oaks, PA 19456, 1994.
120. Anon., "Polytron Chlorine Cl_2L Sensing Head," National Draeger, Inc., 101 Technology Drive, P.O. Box 120, Pittsburgh, PA 15230, Customer Service 800-922-5518; Mar. 1993.
121. Cromer, R., "ATI—Series A15 Residual Chlorine Monitor," Analytical Technology Inc., 680 Hollow Road, Box 879, Oaks, PA 19456, Customer Service 800-959-0299; Aug. 1996.
122. Wilson, P., "LD Series Continuous Gas Monitoring System," Interscan Corporation, P.O. Box 2496, Chatsworth, CA 91313-2496, Customer Service 1-800-458-6153 (USA and Canada); 1964.
123. Schafer, G., "The V10K Chlorine Gas Feed System," Wallace & Tiernan Inc., 1901 W. Garden Road Vineland, NJ 08360, Customer Service 609-507-9000, Fax 609-507-4250; 1996.
124. Kinback, K., "Acutec 35 Gas Detection System," Publication No. SB 50.130UA, Oct. 1996, Wallace & Tiernan, 1901 W. Garden Road, Vineland NJ 08360, Customer Service 609-507-9000.
125. Cubellis, T., "Wallace & Tiernan PCU Process Control Unit," Publication TI 40.200UA, 1995.
126. Schafer, G., "Wallace & Tiernan SCU Signal Conditioning Unit," Publication TI 40.100UA, 1995, Customer Service 609-507-9000.
127. Schafer, G., "Wallace & Tiernan S10k Sonic Chlorinator," Publication No. TI25.200UA, 1996, 1901 W. Garden Road, Vineland, NJ 08360, Customer Service 609-507-9000.
128. Irey, R., "Guardian FX–1500/C12 Chlorine Gas Detector/Monitor," Foxcroft Equipment & Service Co. Inc., P.O. Box 39, 2101 Creek Road, GlenMoore, PA 19343, Customer Service 610-942-2888, Fax 610-942-2769.
129. Long, J., "Ecometrics Series 1710 Gas Detector," Ecometrics Inc., 130 W. Main St., Silverdale, PA 18962, Customer Service 215-453-9800; 1996.
130. Connell, G. F., "ADVANCE™ Electronic Dual-Platform Cylinder Scale—Series 1363B," Publication #320.0015.0, Apr. 1996, Capital Controls Co. Inc., 3000 Advance Lane Colmar, PA 18915, Customer Service 215-997-4000.
131. Eichelman, K., "On-line Measurement of Residual Chlorine Made Easy and Economical," EIT, 251 Welsh Pool Road Exton, PA 19341, Customer Service 1-800-634-4046; 1995.
132. Carsten, R. L., "Halogen Emergency Chlorine Valve Actuators," Halogen Valve Systems Inc., 1098 Redding Ave. Costa Mesa, CA 92626, Customer Service Fax 714-241-9709; 1995.

10
DECHLORINATION

HISTORICAL BACKGROUND

General Discussion

Before the days of the discovery of the breakpoint phenomenon (1939), potable water treatment plants practiced what was commonly called "superchlorination" to destroy tastes and odors caused by ordinary doses of chlorine. Now we know that this treatment was achieving free chlorine residuals, but they were far in excess of what was thought to be desirable at the consumer's tap (2–4 mg/L). The water was then dechlorinated to an acceptable distribution system residual (0.5–0.75 mg/L) with sulfur dioxide. This practice is continued today at a great many locations in the United States and Great Britain. Many water systems around the world use the so-called superchlor–dechlor method to compensate for limited contact time.

In some plants it has been found necessary to dechlorinate completely and rechlorinate to the desired residual as the water leaves the plant. This procedure removes all possibility of any off-flavors in the water due to the chlorination process.

Bottled water is largely supplied from specially treated ordinary tap water. Rarely is it ever "mountain spring" water. Thanks to knowledge gained from treatment plant experience, practically the raunchiest of waters can be made palatable from the superchlor–dechlor process.

In the making of bottled water, dechlorination is achieved by the use of activated carbon filters that are carefully monitored for zero residual. Free chlorine consumes the carbon, and combined chlorine destroys the adsorption power of carbon.

Military vessels employ the superchlor–dechlor method so that they can take on practically any raw water supply and still produce a palatable water. Researchers discovered this in the San Francisco Bay Area during World War II. Ships' systems were fitted with specially baffled contact tanks that provided 45 min. of detention, which ensured the destruction of amoebic cysts, as the chlorination system was designed to provide a 10 mg/L residual at the end of the contact time. A safety factor was built into the system by providing a carbon filter bypass, which allowed a continuous residual in a ship's distribution system at all times.

A great many industrial processes use zeolite softeners along with chlorination procedures. Residual chlorine will in time destroy the zeolite; so dechlorination must occur before the process water can enter the softeners. This is normally done with carbon filters.

Potable Water

The earliest reported use of dechlorination as a potable water treatment process was by Howard and Thompson[14] in 1926. Their report was based upon an exhaustive study on taste and odor control for the City of Toronto water supply. This study began in 1922. In the years following the discovery of the breakpoint phenomenon and the implementation of the free residual process, dechlorination became an important tool for the treatment plant operator. The free residual concept required that a prechlorination dose be sufficient to maintain an adequate residual throughout the entire plant. At some convenient place the plant effluent residual was trimmed by dechlorination to the appropriate level before entering the distribution system.[15] When the free residual is subject to decay by sunlight when passing through sedimentation basins and filters, the prechlorination dose can be adjusted automatically by a time clock to provide different dosage levels for nighttime and daytime operation. This reduces the operating range burden for the dechlorination system. Sulfur dioxide is the preferred chemical because the equipment and control modes are the same as for chlorination systems. In most cases the prechlorinator is flow-paced with automatic night–day dosage change, and the sulfonator operates by direct residual control without flow pacing.

Chlorination followed by dechlorination has been used in the British Isles for more than half a century. Its principal use is for disinfection of underground supplies where contact time is short. By using the chlorine residual–contact time concept, the chlorine residual can be elevated to the desired amount for the available contact time, which may be only 5–10 min. at best. The residual at the end of this time may be as high as a 3–5 mg/L, depending upon the local conditions. This residual is trimmed to the desired distribution system level by dechlorination.

This concept of the residual–contact time product has been used as an interim solution for *Giardia lamblia* destruction pending construction of filtration facilities at surface supplies contaminated by the intrusion of this pathogenic cyst. Cysts usually require a long free chlorine contact time to perform adequate disinfection. These cysts require at least 4 mg/L residual for one hour of contact time to be destroyed, or 8 mg/L for 30 min. Obviously either of these residuals would have to be trimmed by dechlorination before entering the distribution system.

Dechlorination has been found beneficial for waters that are burdened with high concentrations of ammonia N and organic N. When the free residual process is practiced in these waters, terminal dechlorination may be partial or complete. This depends upon the ratio of free chlorine to total combined chlorine residual just prior to the point of dechlorination. If at this point the free residual chlorine is 85 percent or more of the total, it is usually satisfactory to dechlorinate to a desired free chlorine residual because the amount of nuisance residuals—those made up of other than free chlorine–is so small that it will not be a factor in the palatability of the finished water or cause any problems in the distribution system, other things begin equal.

When the ratio of free to total residual drops below 85 percent, this usually indicates the presence of significant amounts of organic nitrogen in the raw water interfering with the free residual process. At this point the combined chlorine residual contains what appears to be dichloramine but may be some type of poly-N-chlor compounds as a result of the reaction of chlorine and organic nitrogen. This is an undesirable situation, as these nuisance residuals seem only to contribute to off-flavors in the water. Furthermore, it is at this point that nitrogen trichloride is most likely to form. Even in trace quantities, this compound gives off an obnoxious odor and burns the eyes. If the resulting combined residual is allowed to proceed into the distribution system, the remaining free chlorine continues to react with the combined residual to produce more nitrogen trichloride. Therefore, in such waters it is desirable to go to complete dechlorination. This removes all of the nuisance residuals and prevents further formation of either dichloramine or nitrogen trichloride, both of which produce off-flavors. This water, to which has been added a slight excess of sulfur dioxide, is then rechlorinated. Because dechlorination by sulfur dioxide will not remove any of the ammonia, there may be sufficient ammonia remaining after dechlorination to form a predominantly monochloramine residual upon rechlorination. This depends entirely on the amount of ammonia in the raw water that has not been destroyed by the free residual process. Usually postammoniation for such waters is required to produce the proper monochloramine residual. If this is done, the water leaving the treatment plant will not contain any nuisance residual fractions that will adversely affect the quality of water at the consumer's tap.

Other special applications requiring dechlorination are: ahead of demineralizers, boiler makeup water, certain food plant operations, and the beverage industry.[6] In these cases the dechlorination process is arranged to remove all remaining chlorine residual.

Special Situations

Dechlorination can be described as the practice of removing all or a specified fraction of the total chlorine residual. In potable water practice dechlorination is used to reduce the residual to a specified level at a point where the water enters the distribution system. In some cases where taste and odor control is a severe problem, control is achieved by complete dechlorination, followed by rechlorination. This removes the taste-producing nuisance residuals and prevents the formation of NCl_3 in the distribution system. Dechlorination of wastewater and power plant cooling water is required to eliminate chlorine residual toxicity, which is harmful to the aquatic life in the receiving waters.

The most practical method of dechlorination is by sulfur dioxide and/or aqueous solutions of sulfite compounds. Other methods used are granular activated carbon and hydrogen peroxide.

Ammonia has been described as a dechlorinating agent. This is a misnomer. The addition of ammonia converts the free chlorine to monochloramine. When this method of "dechlorination" is used, its sole purpose is to prevent the formation of NCl_3 in the distribution system. *NCl_3 will not form if HOCl is absent.* Nitrite (NO_2^-) has been used to dechlorinate the HOCl fraction of a total chlorine residual. This is a special case where the combined residual would interfere with the accuracy of an analyzer arranged to record only free chlorine residual. In this case the analyzer is arranged to measure total residual. At frequent intervals nitrite is automatically injected into the sample. The nitrite consumes the HOCl fraction, which is instantly reflected on the recorder chart. The dip on the chart is a record of the HOCl fraction which the operator needs to know in order to control the free residual process. This is a special case.

Wastewater

The widespread use of chlorine in wastewater discharges and cooling water treatment at steam power stations came under close scrutiny in the middle 1960s. Esvelt et al.[16] reported in 1971 on the toxicity of wastewater effluents being discharged into San Francisco Bay. This study covered a span of about three years and proved that chlorinated effluents were more toxic to aquatic life than unchlorinated effluents.

These studies also showed that a dechlorinated effluent is less toxic than either the chlorinated or the unchlorinated effluent. These studies used continuous-flow on-line bioassays with golden shiners (*Notemigonus chrysoleucas*). In these studies the fish were captive in tanks of various chlorine residuals for a minimum of 96 hours.

In 1973 Brungs[17] turned out a comprehensive report in which the toxicity of chlorine to aquatic life is quantified. This paper reviews and discusses the uses of chlorine and chlorine chemistry and emphasizes toxicity studies in the field and the laboratory. The literature review comprises more than 150 references. The following are the finite conclusions of Brungs' report:

1. In areas receiving wastes treated continuously with chlorine, total residual chlorine should not exceed 0.01 mg/L for the protection of more resistant organisms only, or exceed 0.002 mg/L for the protection of most aquatic organisms.
2. In areas receiving intermittently chlorinated wastes (power plants), total residual chlorine should not exceed 0.2 mg/L for a period of 2 hr/day for more resistant species of fish, or exceed 0.04 mg/L for a period of 2 hr/day for trout or salmon. If free chlorine persists, total residual

chlorine should not exceed 0.01 mg/L for a period of 30 min./day for areas with populations of trout and salmon.

To the chlorination–dechlorination practitioner, the above restrictions translate to a zero chlorine residual at all times. This is the interpretation of the California Fish and Game Commission as well as the California Water Quality Resources Board, which have jurisdiction over these matters. Zero residual is the consensus among other state regulatory boards; however, some exceptions do exist.

Therefore, wastewater dechlorination systems basically are designed to produce a zero chlorine residual in the effluent before it leaves the plant. There are always exceptions. In some cases the outfall may be long enough to consume up to 0.5 mg/L chlorine residual before reaching the receiving waters. In such a system a closed loop residual control system with a 0.5 mg/L set point can be used. The major problem with this kind of installation is the ability to produce a monitoring system to satisfy the regulatory agency. Most complaints by regulatory personnel in California are about noncompliance with maintaining a continuous zero residual in the effluents.

Fish That Thrive on Sewage[35]

The following was taken from an article by Reuters News Agency in London in 1975. White thinks it shows that maybe the British know something that the Americans do not. The article described what was happening at the Rye Meade WWTP, just outside London.

This sewage treatment plant serves a rural population of about 300,000. The area is dotted with lagoons that collect the treated wastewater effluent before it flows into the nearby River Lea. By the time the plant effluent has been discharged into the lagoons, it could be classified as a primary effluent with sedimentation. In the lagoons there is sufficient time for the effluent to undergo more sedimentation.

That some fish seemed to love settled sewage was the dream of area biologists, who realized that now any fear of anglers in the lagoons was thoroughly squashed. (These anglers believed all along that sewage killed fish in the rivers, period.) Furthermore, the biologists noticed that the wild fish in the treatment plant's lagoons, filled with the plant's effluent, grew bigger than those reared in the "source water" of the river. The very same thing happened when the biologists introduced carp into the lagoons—a fish that is a well-known traditional favorite in Europe and Southeast Asia. The fish seemed to have a ravenous appetite for the tiny worms and microscopic organisms that thrived on the "suspended soilds" in the plant effluent.

The wild fish population in these lagoons is mainly roach. Tony Dearsly, a fishing biologist, found that the one-year-olds were three times as fat as those bred in clean water. Then, when a lagoon was stocked with two-inch baby carp, they grew to ten inches in only two years according to biologist Reggie

Noble. He considered that to be an enormous growth rate. He also added that getting a marketable size like this in carp would require artificial feeding.

Reclaimed Water

Water conservation is receiving increased attention in the arid states. By its own nature water must always have a persisting residual to prevent biofouling in the distribution system. Therefore, these effluents do not require dechlorination unless they are discharged into an environment containing aquatic life.

California Wastewater Discharge Requirements

It is the intent of the California Water Code Section 13170.2(b) that the State Water Resources Control Board limit discharges of various pollutants to the waters of California and thereby prevent water quality degradation. One of these pollutants is chlorine. Chlorine is of concern because it is used extensively for disinfecting the effluents of wastewater treatment plants and is also known to be toxic to fish and other aquatic organisms in very low concentrations. For example, Table D of the current (1988) California Ocean Plan indicates the conservative estimate of chronic toxicity to be 0.01 mg/L. The 1988 Ocean Plan chlorine residual discharge limitations (Table B) are 0.01 mg/L daily maximum and 0.126 mg/L instantaneous maximum. Discharges to inland surface waters are governed by Basin Plans, which contain generalized narratives to regulate the discharge of toxic chemicals, and, with the exception of Regional Water Quality Board 2, San Francisco Bay Region, contain no numerical limits.

RWQCB-2 has greatly restricted the discharge of chlorine and adopted an effluent limitation of 0.00 mg/L in the Basin Plan. As there is no technology that can be used to control to this low a residual, all the plants in California had no other choice but to dechlorinate to a positive sulfite ion residual (a small excess of sulfur dioxide—1 mg/L).

For some unknown reasons other states either had no chlorine residual limitations, or they were much less stringent, that is, 0.5–1.0 mg/L. About 1979 the state of Washington gave Seattle's Renton Plant a 0.5 mg/L residual limitation; so they installed a residual control system that could in fact control their effluent residuals of approximately 1–3 mg/L to 0.05 mg/L with sulfur dioxide. This system had operated to everyone's satisfaction for at least two years when an 0.008 mg/L residual was imposed by the State authorities. The plant personnel under the direction of Richard E. Finger, process control supervisor, David Harrington, and Larry Paxton, instrument engineer, devised the "Renton Zero Residual Control System." It operated successfully for several years but was taken out of service when a two-million-dollar ocean outfall was completed. This outfall eliminated the need for dechlorination.

Wastewater Research

In the late 1970s the Sanitary Districts of Los Angeles County were commissioned by the EPA to make a comprehensive study of wastewater dechlorination. This investigation and study included a literature search and review, a pilot plant study, and a full-scale evaluation in the field. The field survey involved the canvassing of 55 operating plants in California by mail, telephone, and site visits to selected facilities. This project was reported in 1981 by Chen and Gan.[1] The important conclusions of this report are given below.

1. The sulfur dioxide process was the most cost effective of the three methods studied. To dechlorinate completely a 5 mg/L total chlorine residual the sulfur dioxide method total cost was 2 cents/1000 gal; the holding pond method cost was 4.5 cents, and carbon adsorption cost was 13.5 cents. These were the only methods studied.
2. The feed-forward method of control, with flow as the primary signal and residual as the secondary signal, was the most commonly used system for the sulfur dioxide method.
3. Adding an overdose of SO_2 was essential to achieve consistent dechlorination. This provided the sulfur dioxide method with generally good reliability.
4. The residual chlorine analyzer was the weakest link in the SO_2 system. Therefore it is the most important component. The inherent weakness of amperometric-type analyzers is their inability to maintain calibration stability for long-enough periods of time (~24 hours) and to remain in calibration at zero residual.
5. Depletion of dissolved oxygen or reduction in pH was not observed in the pilot plant studies at a sulfur dioxide to residual chlorine dosage ratio of 2 : 1.
6. No significant physical chemical degradation was found in the effluent after dechlorination by sulfur dioxide.
7. Bacterial regrowth in the 10-min. sampler after dechlorination was observed to be a one to two order of magnitude increase in total coliform density in all three processes.
8. The regrowth, which was predominantly in the total coliform group, seemed to originate from contamination by the existing microorganism communities in the dechlorinated effluent rather than from the reactivation of injured bacteria cells. Fecal streptococci in the effluent remained relatively unchanged after dechlorination.

Dechlorination Control by ORP (Redox)

Now (1997) that ORP methodology has been thoroughly established as the most accurate way to control the dechlorination process, it appears that most if not all of the problems described above can be eliminated by the use of the Stranco High Resolution Redox Control System. This control system and its capabilities are described later in this chapter.

SIGNIFICANCE OF CHLORINE SPECIES

Free Chlorine

In wastewater treatment this species never appeared in either prechlorination or postchlorination practices owing to the high levels of ammonia N in the sewage (10–15 mg/L). All the chlorine residuals were the combined species (mono- and dichloramine), except that the species that titrated as dichloramine is known to be organochloramine, which is always nongermicidal. This species is the result of the organic N in the wastewater. Although pure dichloramine is about 60 percent more germicidal than monochloramine, it rarely occurs in wastewater treatment. It will occur when the Cl_2 to NH_3-N ratio is 6 to 1. It may also occur when the pH is 7.0 or less.

Now that nitrogen removal is being practiced, we are seeing for the first time the free chlorine species. Although free chlorine is highly germicidal, it also is naturally highly reactive. Therefore, two side effects will occur that consume free chlorine, robbing it of a significant amount of disinfection power: (1) The most important of these side effects is its reaction with organic N: free chlorine (HOCl) reacts quickly with all the organic N present, whereas monochloramine reacts very slowly with organic N. This is why in many cases where total nitrification occurs (complete removal of NH_3-N), ammonium hydroxide has to be added to achieve the disinfection requirements. (2) If organic color is present in the effluent (San Jose), free chlorine is consumed by the natural bleaching effect. The loss of free chlorine will range from 10 to 15 mg/L.

Combined Chlorine

This includes those species that titrate as mono- and dichloramine by the forward amperometric titration procedure. The most important of them is monochloramine, for the reasons described above. Moreover, in wastewater disinfection it was found that where contact times approach 45–50 min., monochloramine is almost as germicidal as free chlorine. This came as quite a surprise to all the experts because all the free chlorine versus monochloramine germicidal efficiency studies were performed at contact times of 5–10 min.

Organic Chloramines

This species always occurs in wastewater chlorination. The amount that forms is a function of the concentration of organic N present and the detention time in the contact chamber. Since this species is not germicidal, its concentration should be monitored on a regular basis, as should too the organic N level before chlorination of the final effluent. The only acceptable means of evaluating the presence of this species is either by amperometric forward-titration or by the DPD-FAS titrimetric procedure.

It is believed that most of the organochloramines that appear in wastewater effluents are not toxic. However, laboratory tests on some specific organic N compounds have demonstrated toxicity.

Nitrogen Trichloride*

This species occurs only during the practice of breakpoint chlorination. It is merely a nuisance in wastewater practice. It is so volatile and unstable that it is difficult to quantify by analytical methods, but it is easily identified when it does occur. Anyone who comes in contact with this species will experience watering of the eyes immediately upon contact when it off-gases. If the person remains long enough in the off-gas area, difficulty in breathing will be noticeable. The only wastewater plant where it was evident is at the Irvine Ranch water reclamation plant in Southern California. While it is a powerful oxidant, it has no relevancy where wastewater disinfection is practiced except as an air pollution hazard.

DECHLORINATION CHEMICALS

Introduction

Although there have been many efforts to establish activated carbon as an acceptable dechlorination system, in the long run its expense usually defeats its acceptance. However, White decided to retain the description of this process in this 4th Edition. There are three currently active systems, two of which are about to replace the third system, originally known as the Renton system. This system was discontinued at Renton when a modified outfall was built that eliminated the need to dechlorinate. As of 1991 the modified and improved Renton system was still in operation at the Palo Alto, Calfiornia WWTP.

Activated Carbon

General Discussion. This chemical is packaged in two forms for water and wastewater use. The powdered form (PAC) is metered by weight and carried to the point of application as a water slurry. This is used as a pretreatment process to suppress taste and odor. It is often applied near the point of prechlorination. Operating experience indicates that only about 10 percent of the applied carbon reacts with the chlorine in a dechlorination reaction.[5] Granular activated carbon (GAC) is the other form of carbon used in water

*Sometimes known as trichloramine.

treatment processes. In this form it is used as a filter medium in conventional filter basins. It is used in vertical towers for the adsorption of organics and other undesirable compounds. PAC has not been used intentionally as a dechlorinating agent, but GAC has been used for many years.[6] It is the method of choice in the beverage industry. GAC has been used extensively as a filter medium in potable water treatment plants and has proved effective and reliable as a dechlorination agent.[7] It also produces a most palatable water.

GAC has not been successful as a dechlorinating agent in wastewater treatment. This is probably due to one or more of the following reasons: GAC may not have the ability to remove the organochloramines that form when significant concentrations of organic nitrogen are present; it may not be possible to design an effective carbon bed with the knowledge currently available; or it may be that the time required for the overall dechlorination reaction in wastewaters is much too long.

Though activated carbon can be a most effective method of dechlorination, its cost is almost prohibitive. This is one of the main reasons why sulfur dioxide–containing compounds are commonly used.

Chemistry of GAC Dechlorination

Definition of Carbon Mode of Action. The terms adsorption, saturation, and absorption are used frequently in describing the mode of action of GAC (granular activated carbon):

Adsorption is a phenomenon consisting of the adhesion in an extremely thin layer of the molecules of gases, dissolved substances, or liquids to the surfaces of solid bodies with which they are in contact.

Absorption is a process of soaking up like a sponge. This is in sharp contrast to adsorption.

Saturation is a volumetric expression. It is used in GAC reactions to explain what occurs when the attractive force of the carbon is depleted.

Free Chlorine. GAC reacts with free available chlorine as shown:

$$C^* + HOCl \rightarrow CO^* + H^+Cl^- \tag{10-1}$$

where C^* and CO^* indicate active carbon and a surface oxide on carbon, respectively.[8] If significant amounts of HOCl are allowed to react with the carbon, some of the oxygen attached to the surface may be emitted as CO or CO_2 gas. The stoichiometric reaction will be as follows:

$$C + 2Cl_2 + 2H_2O \rightarrow 4HCl + CO_2 \tag{10-2}$$

In this reaction 1.0 part chlorine will destroy 0.00845 part of carbon. As the ultimate reaction is chlorine conversion of the GAC to carbon dioxide, no regeneration of carbon is required—it must be physically replaced because it is destroyed.

Chloramines. Studies by Bauer and Snoeyink[9] demonstrated that pure chloramines can be dechlorinated by GAC as follows:

Monochloramine:

$$2NH_2Cl + H_2O + C^* \rightarrow NH_3 + HCl + CO^* \qquad (10\text{-}3)$$

Dichloramine:

$$2NHCl_2 + H_2O + C^* \rightarrow N_2 + 4HCl + CO^* \qquad (10\text{-}4)$$

The reaction time used in these observations was 20 hours. During these reactions the carbon apparently accumulates surface oxides, which partially oxidize the NH_2Cl and $NHCl_2$ nitrogen to N_2.

Chloramines are adsorbed by the GAC to the depletion of the attractive force, which is described above as saturation. Therefore, from the moment of saturation the chloramines will break through, and the carbon will have to be regenerated. Although both free chlorine and chloramines are adsorbed by the carbon, HOCl is adsorbed to the concentration reaction at equilibrium to form CO_2, and the chloramines are adsorbed to the point of depletion of the attractive force (saturation). Chloramines deplete the carbon, whereas HOCl destroys the carbon.[10]

GAC Filter Bed Design. In water treatment practice, water that passes through a GAC filter bed is considered completely dechlorinated unless depletion of the carbon occurs. Therefore the water so treated must be rechlorinated if a residual is required in the finished water.

A detailed study of the carbon–chlorine reaction[11] indicates the following relationship between flow rates, bed depths, concentration of influent and effluent chlorine, as well as granular carbon itself:

$$\log \frac{C_I}{C_E} = \frac{B \times \text{bed depth (ft)}}{\text{filtration rate (gpm/sq ft)}} = \frac{B}{V} \qquad (10\text{-}5)$$

in which C_I is the concentration of chlorine in influent (mg/L), C_E is the concentration of chlorine in effluent (mg/L), B is the efficiency constant for each carbon, and V is the flow rate (gpm/cu ft).

Hager and Flentje[7] report that a granular carbon medium in a 1 mgd filter at 2.5 gpm/sq ft and a 2.5 ft bed can process 700,000,000 gal of 4 mg/L free chlorine residual, and with 2 mg/L chlorine a similar bed would last approximately 6 years at a 1 mgd rate. The dechlorination reaction proceeds concurrently with the absorption of contaminants. Long-chain organic molecules, such as those of detergents, seem to reduce the dechlorination efficiency somewhat, but many common impurities and phenols have little effect upon the dechlorination reaction. A rise in temperature and a lowering of pH favor the reaction, and it has been found that chlorine–ammonia and other nitrogenous compounds of chlorine tend to react much more slowly than free chlorine.[7] Residual chlorine concentrations can be more easily maintained in the backwash water when combined chlorine residual, instead of free chlorine, is present.

Dechlorination by activated carbon for a conventional water treatment plant has certain limitations that should be recognized, and it should not be relied upon as the sole means of controlling the finished water residual. The exceptions are in the beverage industry, where the raw water is of superior quality, and the dechlorination must be complete.

In August 1977 the Cincinnati Water Works entered into a cooperative agreement with the USEPA to pursue a feasibility study of municipal water treatment using GAC adsorption and on-site carbon regeneration.[12] This project demonstrated that the GAC removed all of the free available chlorine and all but a trace of the combined chlorine. These removals permit the growth of bacteria within the carbon bed with potential for carryover into the distribution system. Therefore, a remedy for such a situation would be to construct a postchlorination facility followed by additional clear-well capacity to provide sufficient chlorine contact time for disinfection before the treated water enters the distribution system.

Sodium Thiosulfate—$Na_2S_2O_3$

This compound has always been used in the laboratory as a dechlorinating agent prior to routine coliform concentration determinations. It has never been used on a plant scale to control the dechlorinating process. It was tried many years ago in water treatment plants soon after the discovery of the breakpoint phenomenon (1939). However it was found lacking in several respects: (1) The reaction with residual chlorine was found to be stepwise, creating an unacceptable time factor. (2) Its reaction is stoichiometric only at pH 2. (3) Its chlorine removal factor varies significantly with pH. Its reaction with residual chlorine is as follows:

$$Na_2S_2O_3 + 4\ Cl_2 + 5\ H_2O \rightarrow 2\ NaHSO_4 + 8\ HCl \qquad (10\text{-}6)$$

Sodium thiosulfate has been and is being used in the pulp and paper industry and is currently being promoted for the use in wastewater treatment by Calabrian Corporation, a Houston, Texas-based firm, "as the most cost effective dechlorination chemical." While this may be true for the pulp and paper and chlor-alkali industries, it may not be suitable for wastewater treatment for the reasons described above.

SULFITE COMPOUNDS

General Discussion

Sulfite compounds are used in solution and are primarily for smaller installations where feed rates for sulfur dioxide (<100 lb/day) are not practical. Sulfites are also used on large installations where storage of sulfur dioxide might be considered a hazard. These solutions are applied with metering pumps. Feed-rate control and process monitoring of these systems need more attention and require more complex instrumentation than sulfur dioxide systems require.

There are four sulfur compounds to be considered as alternative chemicals to sulfur dioxide for dechlorination: sodium sulfite, sodium bisulfite, sodium metabisulfite, and sodium thiosulfate. Sodium sulfite is available only as a white powder or crystals. It is extremely difficult to handle in the dry form because it is hygroscopic; therefore, it is never used as a dechlorinating agent. Sodium thiosulfate is used in solution, but almost entirely as a laboratory chemical. It is not a satisfactory dechlorinating agent for treatment plant use because its reaction with chlorine is slow and not amenable to metering control situations. Therefore, only sodium bisulfite and sodium metabisulfite are practical alternative chemicals to sulfur dioxide.

Sodium Bisulfite ($NaHSO_3$)

This is a white powder or granular material, generally purchased as a solution in strengths up to 44 percent. It can be handled wet in stainless steel, PVC, or Fiberglas (tanks). It is usually metered as a dechlorinating agent by diaphragm-type pumps.

The reaction between bisulfite and chlorine residual is as follows:

$$NaHSO_3 + Cl_2 + H_2O \rightarrow NaHSO_4 + 2HCl \tag{10-7}$$

Each part of chlorine residual removed requires 1.46 parts of sodium bisulfite. The usual solution strength is 38 percent, with a specific gravity of 1.3. The manufacturer's data sheet[4] advises that at this specific gravity a 38 percent solution contains 3.5 lb of sulfite/gal. This calculates to 2.17 lb sulfur dioxide.

For each part chlorine removed, 1.38 parts of alkalinity as $CaCO_3$ will be consumed.

Sodium Metabisulfite ($Na_2S_2O_5$)

This is a cream-colored powder readily soluble in water and available in solutions of various strengths.

The reaction between metabisulfite and chlorine residual is as follows:

$$Na_2S_2O_5 + 2Cl_2 + 3H_2O \rightarrow 2NaHSO_4 + 4HCl \quad (10\text{-}8)$$

Therefore, each part of chlorine residual consumed requires 1.34 parts of sodium metabisulfite, and for each part of chlorine removed 1.38 parts of alkalinity as $CaCO_3$ will be consumed.

SULFUR DIOXIDE

Properties of Sulfur Dioxide

Physical Characteristics of SO_2. Sulfur dioxide is a colorless gas with a characteristic pungent odor. It may be cooled and compressed to a colorless liquid. When liquid sulfur dioxide in a closed container is in equilibrium with the SO_2 gas, the pressure within the container varies with the ambient temperature, which is similar to the characteristics of chlorine. The relationships for both sulfur dioxide and chlorine of vapor pressure versus temperature are shown in Fig. 10-6 in the appendix. Although sulfur dioxide is similar in many ways to chlorine, the most notable difference is in the vapor pressure. At 70°F, SO_2 vapor pressure is approximately 35 psi, while chlorine is approximately 90 psi. This is why the Sasco plastic blanket is so necessary in many climates. Keeping the cylinders warm and the pressure up eliminates a lot of operational problems.

The next most important difference is its solubility in water. The solubility of sulfur dioxide is 120 g/L, whereas that of chlorine is only 7 g/L, at 60°F.

The important characteristics of both sulfur dioxide and chlorine are shown in Table 10-1 (from the *International Critical Tables*).

Sulfur dioxide is not flammable or explosive in either the gaseous or liquid state. Like chlorine, *dry* sulfur dioxide is not corrosive to ordinary metals, but in the presence of any moisture it is extremely corrosive. Therefore, with one known exception the same materials of construction are used for handling both sulfur dioxide and chlorine. The exception is the use of 316 SS for sulfur

Table 10-1

Characteristic	SO_2	Cl_2
Molecular weight	64.06	70.91
Latent heat of vaporization at 32°F, Btu/lb	161.8	109.1
Liquid density at 60°F, lb/ft^3	87.2	88.8
Solubility in water at 60°F, g/l	120.0	7.0
Specific gravity of liquid 32°F (water = 1.0)	1.486	1.468
Vapor density at 32°F and 1 atm, lb/ft^3	0.1827	0.2006
Vapor density compared to dry air at 32°F and 1 atm, lb/ft^3	2.264	2.482
Specific volume of vapor at 32°F and 1 atm, ft^3/lb	5.47	4.98
Critical temperature, °F	314.8	291.2
Critical pressure, psia	1141.5	1118.4

dioxide diffusers and valve trim for valves in the SO_2 vapor and liquid times (monel trim is used for chlorine valves). The ability of 316 SS to withstand the corrosivity of SO_2 aqueous solution is helpful where PVC does not have the rigidity or structural strength needed. Although chlorinators and sulfonators may seem identical in construction, they should never be used interchangeably. If evaporators used for chlorine are to be used for SO_2, they must be thoroughly cleaned. *All of the corrosion products from handling chlorine must be removed from the container vessel before use with sulfur dioxide.*

Physiological Effects. Fatal accidents from sulfur dioxide vapor are rare because persons in the lethal concentration of SO_2 cannot breathe and are compelled to seek the open air.[2] Low concentrations cause a sensation of suffocation, coughing, sneezing, and tearing of the eyes. Following this low level exposure the victim will experience a rapid recovery from these symptoms after a few minutes in the open air free of SO_2. Liquid sulfur dioxide may cause severe injury to the skin and eyes due to the freezing action caused by the rapid evaporation of the liquid.

Physiological responses to various concentrations of SO_2 vapor are as follows[2]:

Effect	*SO_2 in Air, ppm*
Least detectable odor	3–5
Least amount causing immediate eye irritation	20
Least amount causing throat irritation	8–12
Least amount causing immediate coughing	20
Max. conc. allowable for prolonged exposure	10
Max. conc. allowable for ½ to 1 hr exposure	50–100
Dangerous for short exposure	400 500

First aid treatment for victims of SO_2 exposure is generally the same as for chlorine exposure. Remove all clothing; keep the patient warm with blankets; do not excite or exercise the patient. Inhalation of ammonium carbonate fumes relieves respiratory irritation, and coughing is relieved by drinking a mixture of 3–5 drops of chloroform in a full glass of water. If this is not available, have the patient sip 80 proof whiskey or vodka from a teaspoon or straw. Two ounces of either should be sufficient for complete relief from coughing. Depending upon the severity of the exposure, the patient should use copious amounts of water for flushing the eyes. This is particularly useful and necessary if liquid SO_2 has touched the victim's skin. The Compressed Gas Association should be consulted for the first aid supplies to be kept on hand. For skin burns, a 5 percent freshly made solution of tannic acid powder should be used to moisten sterile bandages placed over the skin burn area (8 level tablespoonfuls of tannic acid powder in an 8 oz glass of boiled water).

Operating Significance of SO_2 Characteristics

1. *Vapor Density.* The densities of SO_2 and chlorine gases are so nearly the same that a chlorine-indicating rotameter can be used to meter sulfur dioxide without appreciable consequences. To convert chlorine feed rate to sulfur dioxide multiply by 0.95. For example, a 1000 lb/day chlorine rotameter will become 950 lb/day sulfur dioxide.

Since the vapor density and the viscosity of sulfur dioxide are so close to those of chlorine, long vacuum line calculations can be sized using the same variables and formulas as for chlorine vacuum lines.

2. *Vapor Pressure.* At 70°F the vapor pressure of SO_2 is approximately 35 psi, while chlorine's is 90 psi. This low vapor pressure causes problems in the SO_2 supply system. One major effect is on the withdrawal rate. Assuming that 10 psi is the minimum operating pressure for a sulfonator and that the SO_2 ton container is at 70°F, the withdrawal rate will be:[3]

$$\text{(Cylinder pressure} - \text{Minimum operating pressure)} \times \text{Withdrawal factor} = \text{Maximum withdrawal rate, lb/day}$$

The withdrawal factor for a ton cylinder of SO_2 = 6.0; therefore:

$$(35 - 10) \times 6 = 150 \text{ lb/day}$$

This compares to a chlorine withdrawal rate of 440 lb/day, where the operating pressure at the chlorinator is 35 psi rather than 10 psi for SO_2.

The other major effect of the low vapor pressure is reliquefaction. This will be a chronic problem unless the pressure in the cylinder is raised artifically, or the SO_2 is transferred to an evaporator as a liquid. The ways to overcome the reliquefaction phenomenon are discussed later in this chapter (see discussion of the SASCO blanket).

3. *Solubility in Water.* The solubility of sulfur dioxide in water at 60°F is 120,000 mg/L as compared to 7000 mg/L for chlorine. This makes the handling of SO_2 solutions easier than the handling of chlorine. For example, negative pressures or excess turbulence at open channel diffusers must be avoided in the case of chlorine, but not so for sulfur dioxide. Consider the case of using a weir as a mixing device for a SO_2 solution. In the case of chlorine the diffuser should be placed downstream from the weir—in the zone of turbulence—to minimize off-gassing. For the SO_2 solution the diffuser is placed on the upstream side of the weir, and the full benefit of the weir as a mixing device is realized.

Reactions with Chlorine

The various reactions involved in the sulfur dioxide dechlorination process are shown below. The first reaction is the formation of sulfurous acid in the sulfonator injector:

$$SO_2 + H_2O \rightarrow H_2SO_3 \tag{10-9}$$

The sulfurous acid reacts with the various chlorine residual species as follows:

$$HOCl + H_2SO_3 \rightarrow HCl + H_2SO_4 \tag{10-10}$$

$$NH_2Cl + H_2SO_3 + H_2O \rightarrow NH_4Cl + H_2SO_4 \tag{10-11}$$

$$NHCl_2 + 2H_2SO_3 + 2H_2O \rightarrow NH_4Cl + HCl + 2H_2SO_4 \tag{10-12}$$

$$NCl_3 + 3H_2SO_3 + 3H_2O \rightarrow NH_4Cl + 2HCl + 3H_2SO_4 \tag{10-13}$$

These equations illustrate that all of the chlorine species can be dechlorinated with SO_2. The equations also illustrate that acids are produced that affect alkalinity and possibly pH. For each part of chlorine removed, 2.8 mg/L alkalinity as $CaCO_3$ is consumed.

The stoichiometric relationship requires 0.9 part SO_2 to remove 1.0 part of chlorine residual. In practice this ratio can be as high as 1.05 parts SO_2 to each part chlorine.

For practical purposes the reactions described by Eqs. (10-9) through (10-13) are complete in a matter of seconds (<10).

The great majority of potable waters have sufficient alkalinity buffering power (>40 mg/L) that there is no cause for concern about lowering the pH with SO_2 addition (2–3 mg/L max.). The same is true for wastewaters. The alkalinity of chlorinated wastewater effluents is usually in the range of 150–200 mg/L as $CaCO_3$. Dechlorination of a 12 mg/L chlorine residual will consume only 34 mg/L alkalinity as $CaCO_3$.

Reaction with Dissolved Oxygen

There has been some apprehension about the possibility that excess SO_2 might consume a significant amount of dissolved oxygen in the receiving waters downstream from a dechlorinated wastewater discharge. The reaction between the excess sulfite ion ($SO_3^=$) from dechlorination and the dissolved oxygen is:

$$SO_2 + H_2O + O_2 \rightarrow H_2SO_4 \tag{10-14}$$

This means it will require four parts of SO_2 to remove one part dissolved oxygen. Owing to the speed of reaction of Eq. (10-14) there would never be sufficient time available to complete the reaction at the optimum reactant concentrations. Plant-scale experience has confirmed this. The study by Chen and Gan[1] demonstrated no effect upon pH (7.2) due to an excess sulfite level of 2.6 mg/L in the plant effluent. Moreover the DO increased from 4.8 to 6.1 mg/L after dechlorination with sulfite. This was probably due to a high degree of turbulence caused by the mechanical mixers. There have not been any reports of sulfur compounds used for dechlorination having any effect upon dissolved oxygen consumption or pH change in the receiving waters or in the dechlorinated effluents.

SULFUR DIOXIDE FACILITY DESIGN

Factors Affecting System Efficiency

The principal factors affecting a successful operating installation are the same as for the chlorination system with one exception: monitoring for zero residual. The other principal factors are: the control system, adequate mixing, sufficient contact time, and trained and dedicated personnel. These factors are discussed in detail below.

Supply System

General Discussion. Sulfur dioxide is available in ton containers, tank trucks, or tank cars. Owing to the low vapor pressure of SO_2, special precautions must be taken when using ton containers to prevent reliquefication. Tank truck deliveries are by far the most popular means of delivery in California. This type of delivery requires the user to provide a storage tank. Some sulfur dioxide suppliers provide a storage tank as a ploy to have a continuing contract with the users. Tank cars are designed to be the same as those for chlorine; therefore, guidelines for chlorine tank car layouts can be used for SO_2 cars.

Ton Containers: Gas Withdrawal. Owing to its low vapor pressure, SO_2 gas reliquefies quite easily. This causes operating problems with the control equipment. This low vapor pressure is also a limiting factor in the movement of the sulfur dioxide from the supply system to the control equipment.

Withdrawing from the gas phase usually requires the application of heat to the cylinders. Gas-phase systems operate best at around 90–100°F, and it is acceptable practice to apply heat directly to SO_2 ton cylinders, provided that there is control to limit the heating to 100°F. The maximum gas withdrawal rate at 70°F from a ton container is about 180 lb/day without reliquefaction. At 100°F it is about twice as much.

The preferred method of applying heat to ton cylinders is to use a common ordinary electric blanket. This method has been used successfully in California for many years. It long ago superseded the nitrogen padding system. In addition to the bed blanket, there are two sources of blankets designed for use with sulfur dioxide ton cylinders. These blankets are called SASCO heaters by the Morgan Allen Co., Inc. of Houston, Texas,* and Metro-Quip Inc. of Arlington, Texas. The output temperature is 120°F, and the weight is 45 lb.

---

*Morgan Allen Co., Inc.: 6804 East Highway & South #327, Houston, TX 77083.

This temperature translates to a vapor pressure of 100 psi. Blankets for 150-lb cylinders are also available from these two sources.

Whenever heat is applied by these blankets for vapor withdrawal, a CPRV must be installed immediately downstream from the operating cylinder to reduce the pressure to 40–45 psi. This will eliminate the occurrence of reliquefaction, which is a common and annoying problem with sulfur dioxide.

Ton Containers: Liquid Withdrawal. When liquid withdrawal was used, nitrogen padding was usually suggested in earlier years. This is no longer necessary, since the electric blanket method has proved to be the most convenient one. The nitrogen system has only one advantage: it can be used to purge the piping system. However, since the advent of the remote vacuum system for both chlorinators and sulfonators, purging is not necessary because when these systems shut down, the entire piping system is under a vacuum.

Because evaporators are involved in liquid withdrawal, a pressure-reducing valve should be installed immediately downstream from the evaporator vapor outlet to prevent reliquefaction at the vacuum regulator check unit.

Storage Tanks. Owing to the widespread use of sulfur dioxide in industry and agriculture, a large proportion of SO_2 production is delivered by road tankers of 17–20 ton capacity. These suppliers have established the practice in many states of supplying stationary storage tanks for the users. These tanks do not conform to what might be described as "Chlorine Institute approved." They do not have excess flow valves, a safety relief valve, multiple vapor or liquid outlet valves, or any emergency device. Therefore, the user is advised to install a stationary tank equipped with a standard chlorine tank car dome assembly modified for SO_2 handling. This means replacing all monel trim valves including excess flow valves with 316 SS trim. This approach ensures greater safety and reliability. Moreover, the user can then utilize the chlorine tank car emergency kit. All piping and header systems, including expansion tanks, gauges, alarms, and so on should conform to the guidelines set forth in Chapter 9 for chlorine.

The vapor pressure of sulfur dioxide is so low (30 psi at 68°F) that it contributes to reliquefaction of the vapor. This event disrupts the metering and control equipment (sulfonators etc.) This occurs whether the containers are ton cylinders or storage tanks. The use of SO_2 storage tanks in California is practically universal. Most plants use evaporators in conjunction with these tanks.

Current practice is to use a separate room for the SO_2 tanks that is completely separated from the chlorination equipment. This serves two purposes: the room provides a way to isolate a leak, and it provides the means to increase the vapor pressure to a level that will prevent reliquefaction by keeping the room temperature to about 100°F. This will produce a vapor pressure of about 65 psi. Space heaters are commonly used for this purpose; however, the electric blanket approach is also acceptable.

For a completely stable withdrawal system that is fully compatible with the metering and control system, the vapor pressure should be reduced by a manually operated PRV located close to the discharge outlet, whether it be the outlet of the storage tank or the evaporator. The reduced pressure should be about 40 psi. This will prevent reliquefaction in the vapor piping and in the sulfonator, which is always a most troublesome situation in handling sulfur dioxide. The next most serious problem is the common occurrence of the presence of excessive moisture in the liquid SO_2 due to improper manufacturing procedures.

Tank Cars. Users of SO_2 in tank cars must provide a dry air padding system to raise the tank car pressure to at least 60 psi for the coldest ambient temperature. In warmer climes sunlight has been found to raise tank car pressures about 10–12 psi.

Flow of Sulfur Dioxide in Pipes. Owing to the physical and chemical similarity of chlorine and sulfur dioxide, the hydraulics of SO_2 vapor or liquid flow in pipes can be analyzed by using the same factors as those used for chlorine: viscosity, density, Reynolds number, and friction factor. This holds true for both pressure and vacuum calculations.

Materials of Construction. All of the piping, pipe fittings, pressure gauges, pressure switches, and expansion tanks shall be of the same material as for chlorine. The only exceptions are valves. Line valves and auxiliary valves must have 316 SS trim, rather than monel, which is used for chlorine.

Accessory Equipment. Sulfur dioxide ton containers are identical to chlorine containers except for the color—SO_2 tanks are usually blue, while chlorine tanks are silver-colored. Therefore, all of the accessory equipment such as weighing devices, trunnions, lifting bar, filters, and pressure reducing valves (with 316 SS trim) are the same as for chlorine. (See Chapter 9.)

Safety Equipment. The breathing apparatus and the container emergency kits are also the same as for chlorine.

Automatic Cylinder Switchover. Use the same system as for chlorine. (See Chapter 9.)

Remote Vacuum. This is the same as for chlorine.

Flexible Connections. Storage tanks eliminate the need for flexible connections. The flexible connections used for chlorine are still being used for SO_2 ton containers and tank cars without incident.

Sulfonators

Except for minor modifications in materials of construction, sulfonators are identical to chlorinators. (See Chapter 9.)

Evaporators

Evaporators for liquid sulfur dioxide are identical to those used for liquid chlorine. However, because of the difference in latent heat of vaporization, an 8000 lb/day chlorine evaporator will only vaporize about 5600 lb/day sulfur dioxide. This is based upon latent heats of vaporization of SO_2 at 164.5 and Cl_2 at 115.

Housing

The designer should use the same guidelines for housing SO_2 equipment as those used for chlorine. These design provisions are based upon safe operating conditions for personnel, which include but are not limited to space, ventilation, protection from fire, location of wind socks, door hardware, inspection windows, heating, electrical controls, drains, SO_2 vents, and evaporator vents. These considerations are described in detail in Chapter 9.

Separation

Chlorine gas and sulfur dioxide gas should never be allowed to come in contact with each other.* Therefore, it is imperative that the supply system piping should be so located and arranged that such an occurrence could never happen.

The most positive way to accomplish this is to have completely separate areas for the supply systems and house the sulfonators and evaporators in a separate room from the chlorination equipment.

Reliability

The dechlorination system can fail for the same reasons that cause failure of the chlorination system. Therefore, the reliability provisions described in Chapter 9 for chlorine are generally applicable to sulfonator systems.

Injector Systems

Water Requirements. Sulfonator manufacturers use injectors designed for the solubility of chlorine. Therefore, since sulfur dioxide is 17 times more soluble than chlorine, all SO_2 injectors are enormously oversized. This practice

*Although there are no known serious consequences from allowing chlorine and sulfur dioxide gases to come in contact with each other, the corrosive results could cause some operational problems.

wastes power and water, unless the water is recycled. As it would be a major project to redesign injectors for SO_2 installations, it would be easier and more practical to use the Water Champ instead (see Chapter 9). This approach eliminates the need for a water supply, a diffuser, and a mechanical mixer.

Materials of Construction. The SO_2 vacuum line and solution line require the exact same materials as for chlorine with one exception: the sulfur dioxide solution diffuser can be made of 316 SS.

Diffusers. The diffuser location is less critical for SO_2 solution than for chlorine because SO_2 gas is several times more soluble in water than chlorine. Therefore the danger of off-gassing due to turbulence at or adjacent to the diffuser is nonexistent with SO_2 application, whereas it is a serious problem with chlorine. For example, the SO_2 diffuser can be placed immediately adjacent to but upstream from a weir, and can utilize the turbulence on the downstream side for immediate mixing without fear of releasing SO_2 vapor due to the turbulence.

Mixing. The application of sulfur dioxide to wastewater differs considerably from the application of chlorine. The reaction is inorganic in nature. The $SO_3^{=}$ ion reacts with the chlorine residual to convert the active chlorine to chloride and the sulfite ion ($SO_3^{=}$) to sulfate ion ($SO_4^{=}$). This reaction takes precedence over any side reaction that might be encountered in wastewater. This reaction with chlorine occurs in a matter of seconds, probably 15–20 sec at the most. Mixing of the sulfur dioxide solution to achieve dechlorination is much less demanding than the mixing requirements for chlorine to achieve disinfection.

Mechanical mixers are not required if the structure is amenable to the required diffuser design. Examples for closed surcharged conduits and open channels for chlorine application are illustrated and described in Chapters 8 and 9. It is recommended that SO_2 solution diffusers be designed after the Egan concept of nozzle contraflow to produce a velocity gradient of 250–300.

Contact Time

Since the reaction time for the complete conversion of the total chlorine residual to chloride by the sulfite ion is on the order of seconds, there is no need for any contact chamber between the SO_2 diffuser and the receiving waters. However, common sense dictates that the structure downstream from the diffuser must be suitable for obtaining a dechlorinated sample about 5 ft downstream from the estimated mixing zone.

Sample Lines

The dechlorinated sample line must be kept free from biological slime growths. These lines carrying dechlorinated effluent will be subject to rapid development of organic debris and slimes throughout the transport system. Any such biofouling introduces errors in the analyzer system, whether it be used for control or for monitoring.

Two methods are capable of removing and/or controlling these growths: intermittent purging with either chlorine or caustic. Chlorine is the logical choice. It should be applied into the sample line collection pipe. (See page 1086 Fig. 10-5, the dechlor sample line purge system.) This should be made available for every chlorine residual analyzer.

All sample lines should be sized to provide line velocities up to 5 ft/sec. This acts to keep the line scoured and to decrease the dead time in the control loop.

The sampling point for coliform bacteria population measurement for proof of disinfection should be 5–10 ft upstream from the SO_2 diffuser. Otherwise effects of coliform regrowth after dechlorination will distort proof of disinfection.

Leak Detectors: SO_2

Introduction. There are two basic categories of sulfur dioxide detectors: ambient and source detectors. Ambient detectors are used for air pollution monitoring; they require greater sensitivity than source detectors. They are also described as toxic gas sensors.

Ambient So_2 detectors are usually made by manufacturers who also make NO_x detectors. An example of an ambient air SO_2 detector is the pulsed fluorescent analyzer by Thermo Electron Corporation, Waltham, MA.

Source detectors are used principally for monitoring industrial stack gases. The detectors for SO_2 leak detection at water and wastewater plants are required to detect concentrations of 5 ppm. Stack gas monitors must be able to detect in the 1 ppm range. Therefore, these monitors are in a different category from water and wastewater SO_2 leak detectors.

As of 1997, the most popular SO_2 leak detectors seem to be the ones described in the following paragraphs.

ATI (Analytical Technology Inc.)

This company produces a modular gas detector that can be used for either sulfur dioxide or chlorine vapors. Its Gas Sens systems consist of individual modules that can be located where desired. Each of these modules includes a digital display of gas concentration, isolated analog output, and four relay outputs (See Fig. 10-1.)

Receivers may be located up to 1000 ft from the sensor/transmitters, for remote location, or can provide local control functions such as valve shutoff while transmitting a 4–20 mA signal to remote displays or data loggers.

Sensor transmitters can be supplied with ATI's exclusive Auto–Test automatic sensor, that greatly reduces operator testing requirements. To meet the

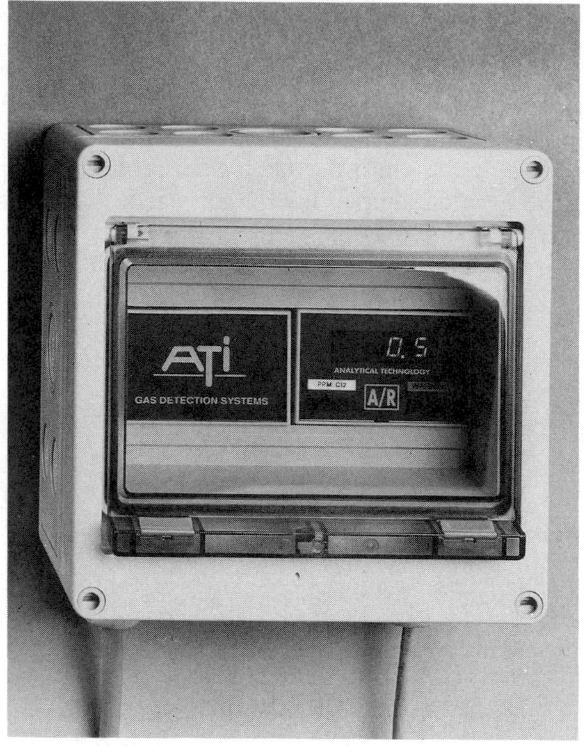

Fig. 10-1. ATI modular gas detector (courtesy Analytical Technologies Inc.).

needs of users around the globe, ATI has developed a compact universal power supply that will accept any AC or DC power supply input form 85 to 255 V.

To ensure that these detectors will remain operable when the power supply fails, ATI offers a separate battery-operated backup system. This system will operate a single point detector system for a minimum of 12 hours and an average of 24 hours.

The ranges of operation of these units are: 0–20 ppm for sulfur dioxide and 0–10 ppm for chlorine.

EIT

This company markets a line of toxic gas sensors that includes sensors for sulfur dioxide, chlorine, and hydrogen sulfide. These instruments are voltammetric sensors based upon proprietary electrochemical gas sensing electrodes. Each sensor contains a noble metal sensing electrode and a counter electrode immersed in a supporting acidic electrolyte. A diffusion membrane isolates the sensing electrode from the ambient air. This type of sensor is classified as a voltammetric diffusion limited electrochemical device. Figure 10-2 illustrates the geometry of the sensor.

The life of the sensor depends primarily upon the amount of lead used in the counter electrode and the loss of water from the sensor through evaporation. Life expectancy is on the order of two years, operating on low gas concentrations.

The standard range of this unit is 0-25 ppm SO_2. These detectors are factory-calibrated, and the alarm points are set at 5 ppm and 10 ppm.[19] At 5 ppm the DANGER lamp will begin to flash, and the SPDT relay will operate. At 10 ppm the ALARM lamp will light, the audible horn will sound, and the DPDT will operate. This is consistent with the physiological effects of SO_2 vapor; that is, the maximum concentration allowable for prolonged exposure is 10 ppm.[2]

Interscan Corporation. This company, founded in 1974, markets analyzer-detectors for toxic gas monitoring. These analyzers are equipped with a voltammetric sensor (U.S. Patent Number 4,017,373). (See Fig. 10-3). This sensor is an electrochemical gas detector operating under diffusion-controlled conditions. Model No. 1247 has an SO_2 operating range of 0–10 ppm.[18]

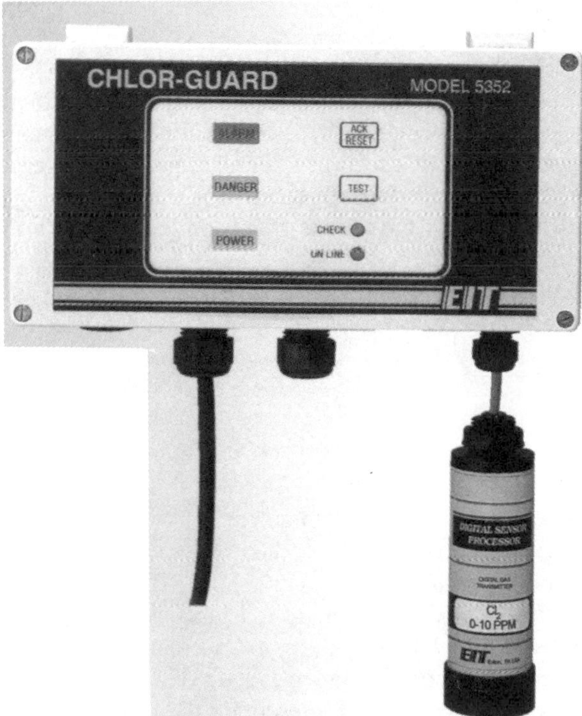

Fig. 10-2. Chlor-Guard single/dual channel toxic gas leak detectors (courtesy EIT).

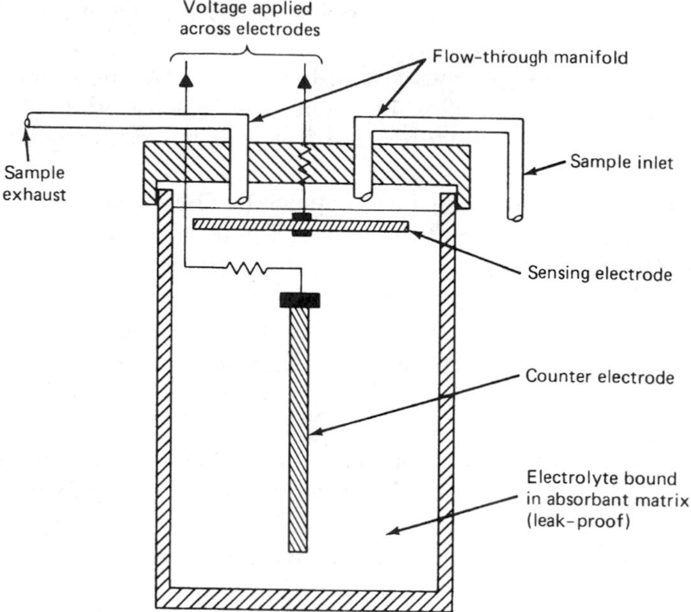

Fig. 10-3. Leak-proof electrochemical voltammetric sulfur dioxide sensor (courtesy of Interscan).

Gas molecules from the sample are adsorbed onto an electrocatalytic sensing electrode, after passing through a diffusion medium. These gas molecules are reacted electrochemically at an appropriate applied voltage to the sensing electrode. This reaction generates an electric current directly proportional to the gas concentration. This current is then converted to a voltage for local readout or a remote recorder.

The generated diffusion-limited current, I_{LIM}, is directly proportional to the gas concentration as follows:

$$I_{LIM} = \frac{nFADC}{L}$$

where

I_{LIM} = the limited diffusion current in amperes
F = the Faraday constant (96,500 coulombs)
A = the area of reaction interface, cm^2
n = the number of electrons/mole of reactant
L = the diffusion path length
C = the gas concentration, $moles/cm^3$
D = the gas diffusion constant, representing the product of the permeability and solubility coefficient of the gas in the diffusion medium

An external applied voltage maintains a constant potential on the sensing electrodes relative to a nonpolarizable reference counterelectrode. This electrode can sustain a current flow without suffering a change in potential; thus it acts as a reference electrode. This eliminates the need for a third (reference) electrode and a feedback circuit, as would be required for sensors using a polarizable air counterelectrode.

The sensor principal used by Interscan allows gas detectors to be made specific for each toxic gas. Sensor selection and design for specificity are accomplished by careful consideration of the following factors:

1. Selection of an appropriate sensing electrode potential within the voltammetric curve plateau for a given toxic gas.
2. Choice of an electrocatalyst for the sensing electrode.
3. The reaction kinetics of the interfering gas.
4. The solubility of the gas in the thin liquid film at the electrode.

Interscan claims an accuracy of 2 percent of full scale, limited only by the accuracy of the calibration standard.

Chemical reagents are not required. The sensor electrolyte (Fig. 10-3) is immobilized, similar to that in dry batteries. The bound electrolyte enables the analyzer to be operated in any position. The sensor has a leakproof reservoir. This provides longer sensor life and the ability to withstand pressurization, and eliminates the possibility of reference electrode contamination in a hostile environment.

Wallace & Tiernan Inc. Gas Detection System. This company's latest model chlorine gas detector (1997) is known as the Acutec 35 Gas Detection System. It utilizes the latest chlorine gas monitoring technology to provide a flexible and reliable self-testing detection system in a compact modular design. (See Fig. 10-4.)

Its sensors can detect the presence of chlorine, sulfur dioxide, and ammonia N in ambient air. It is ideal for detecting leaks from storage containers, process piping, or gas metering equipment in any type of water, wastewater, or industrial plant environment.

Its modular component design allows for a wide range of configurations that can be simply and easily installed, using standard enclosures. The receiver is available in a choice of either a detector or a monitor version, which has an LED display and a mA output signal. Both provide two selectable levels of alarm. A low-level warning alarm may simply indicate transient leak conditions, whereas a second alarm can initiate a full alert in the event of more serious leaks.

There is also an optional auto-test unit. This incorporates an integral gas generator that automatically tests the sensor daily, sounding alarms in the event of sensor failure. This added safety feature also reduces the costs incurred with manual testing procedures.

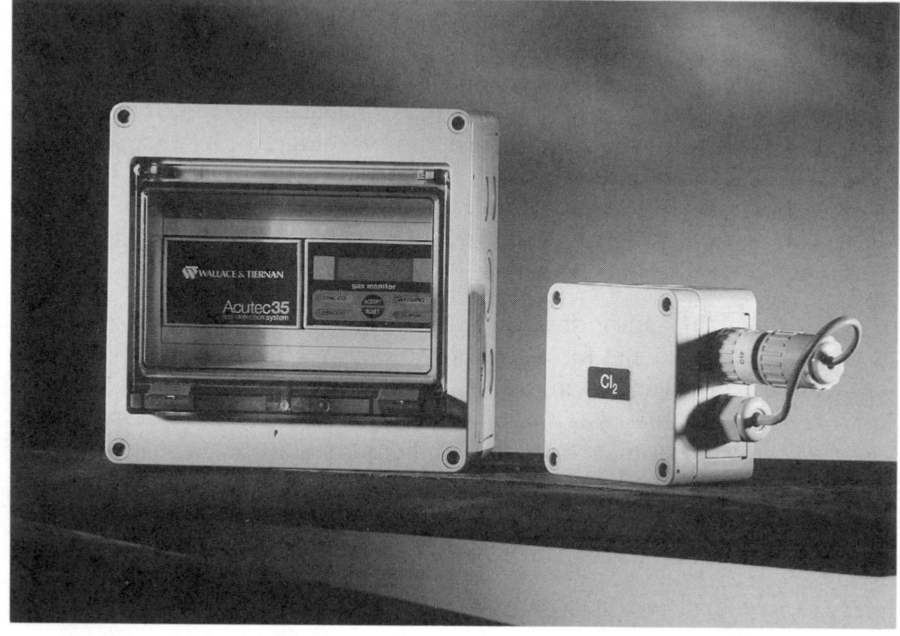

Fig. 10-4. Wallace & Tiernan Acutec 35 gas detection system.

Each point of detection requires the installation of one sensor/transmitter and one receiver module. The sensor/transmitter should be installed in the area to be monitored, and the receiver module can be located up to 1000 ft away.

A power supply module sufficient for use with up to two receivers is housed with the receivers in a NEMA 4X enclosure. All enclosures are provided with knockouts on all four sides to facilitate using the ½-inch FNPT conduit hubs that are provided. These NEMA 4X polystyrene enclosures are suitable for both indoor and outdoor installations. They are fitted with a tough, hinged polycarbonate window for clear visibility of all indicators and easy access to the control panel.

Included with the enclosure is a piezoelectric horn to provide audible local warning.

The Acutec 35 is equipped with a battery backup system. This consists of a sealed lead-acid battery mounted in its own enclosure. In the event of a power failure, it will maintain all receiver functions for a minimum of 12 hours on a single-point system, or 6 hours on a dual-point system. Battery charging is fully automatic and continuous through the power supply module.

Operating Ranges in ppm are: chlorine, 0–10; sulfur dioxide, 0–20; and ammonia 0–100.

Wallace & Tiernan operates in the United States, Canada, Australia, New Zealand, Mexico, South America, the United Kingdom, France, and Germany.

DECHLORINATION FACILITY DESIGN

Bisulfite Solution

General Discussion. Sodium bisulfite solution is the preferred chemical if for any reason sulfur dioxide is not being used. In large systems it is usually chosen because of potential hazards of SO_2 vapor. In small systems its use would be due to the difficulties that surround the metering of small quantities of SO_2 vapor—less than 10 lb/day.

Design Considerations

The components and their arrangement and selection are almost precisely the same as for hypochlorite facilities. These details are discussed in Chapter 2. In general there are three types of systems: (1) pumped, (2) gravity, and (3) eductor. Each of these systems provides the energy necessary to move the solution from the point of storage to the point of application.

In practically all cases the pumped system is preferred. The pumped system has two variations: (1) the positive-displacement-type metering pump system, which is controlled solely by the pump; and (2) the centrifugal-type pump system, which is controlled by a modulating control valve on the downstream side of the pump. These systems are discussed in Chapter 2.

Designer's Checklist

Positive-Displacement Pump System. The following are the vital components for this system:

1. Diaphragm pumps with back-pressure valves and accumulators on both the suction and discharge to dampen pulsation effect.
2. Magnetic flowmeter or rotameter to continuously monitor bisulfite feed rate with a remote recorder.
3. A flow pacing signal from treatment plant effluent flowmeter and a 3 or 4 to 1 ratio relay.
4. A monitoring system that will provide zero residual control.

These systems are discussed in this chapter.

Centrifugal Pump Systems. The components for this system are slightly different from those of the diaphragm pump system:

1. A centrifugal-type pump.
2. A modulating control valve that controls the pump discharge and meters the bisulfite solution.
3. A rotameter type flowmeter with an indicating transmitter and remote flow recorder to monitor the flow of bisulfite.

4. A flow pacing signal from the treatment plant effluent flowmeter and a 3 or 4 to 1 ratio delay.
5. A monitoring system with the same features as those for the diaphragm pump system.

Chemical Storage System. Regardless of the type of system, a storage vessel will be required. These tanks should be constructed of reinforced Fiberglas and should be equipped with a remote type of level sensing instrumentation, atmospheric vent, drain, pump suction, and sample tap for laboratory use in checking solution strength.

Most important is the piping configuration and safety valving used to load the tank. The objective is to prevent any other chemical from being loaded into this tank. Preferably this tank should be a discrete distance from the hypochlorite tank or other chemical tanks. Mixing bisulfite with hypochlorite produces a violent temperature reaction. Other mixtures may do the same or release a cloud of SO_2 vapor.

Design Calculation. The important mathematical relationship is: each gallon of 38 percent sodium bisulfite contains the equivalent of 2.17 lb sulfur dioxide.

ZERO RESIDUAL CONTROL

Analytical Procedures

Introduction. Because chlorine residual discharge requirements will most likely be based upon total residual chlorine, it is fair to say that the choice of analytical procedure will be left to the local operators.

It is important to realize that any chlorine residual in a wastewater effluent can be composed of free chlorine, monochloramine, dichloramine, and organochloramine. All of these species have an important relationship to the disinfection process but not to the dechlorination requirements, as the objective is usually zero residual control. Therefore, none of those procedures is discussed in this chapter. For any information on chlorine residual measurements, the reader is referred to Chapter 5.

However, where a chlorine residual analyzer is involved in wastewater treatment, it is calibrated to measure total chlorine residual; so the operator may want occasionally to check the overall accuracy of the analyzer readings. In these cases, the operator should use the forward-titration procedure, either the amperometric method or the DPD-FAS titrimetric method, as described in Chapter 5.

Sulfur Dioxide Residuals

Amperometric Back Titration. This is the most convenient method to use for the determination of sulfur dioxide residuals. The iodimetric method is described in Standard Methods, 19th edition, pages 4-131 and 4-132, with an equation to calculate the results. The following method is much easier than the iodometric method. This is a brief summary of the procedure:

Pour 200 ml of sample into the titrator cup. It is assumed that either (1) the chlorine concentration is less than 5 mg/L, or (2) there is no chlorine and possibly an excess of the sulfite ion in the sample from the application of sulfur dioxide or sodium bisulfite. Add 5 ml of PAO, and while stirring add 4 ml of pH 4.0 buffer solution and 1 ml of KI solution. Then titrate with 0.0282 *N* iodine solution. The end point is reached when the needle of the microammeter clearly deflects upward and does not quickly drop back. Record the amount of iodine solution used.

Calculation:

$$\text{mg/L } Cl_2 = (A - 5B) \times 200/C$$

where:

A = ml PAO
B = ml iodine solution
C = ml sample

When there is chlorine in the sample, less than 1 ml of I_2 is needed to titrate, and the above calculation will give a positive number. When no chlorine residual is present and there is an excess of the sulfite ion in the sample, caused by the addition of SO_2, the iodine titrant will react quantitatively with both the 5 ml of PAO and with the excess sulfite ion, which is also a reducing agent. Therefore, more than 1 ml of I_2 will be required to complete the titration procedure. For example, if $B = 2$ ml, then:

$$\text{mg/L } Cl_2 = (5 - 10) \times 200/200 = -5$$

This then is the amount of excess sulfite ion. As the mass ratio of sulfur dioxide to chlorine is about 1 to 1, a reasonable estimation of excess SO_2 concentration can be made.

The only suitable method for this in *Standard Methods,* 19th edition, is for "Sulfite," on pages 4-131 and 4-132.

DECHLORINATION CONTROL STRATEGIES

Introduction

Dechlorination practices in potable water are primarily concerned with obtaining zero chlorine residual, but unfortunatley chlorine analyzers are not capable of maintaining their calibration in a zero residual environment. This situation is due directly to the efforts of the federal Fish and Game Commission, which ran studies of fish trapped in tanks with heavily chlorinated water until they died. Then government regulators limited the total chlorine residuals in wastewater effluents discharging into rivers, lakes, seawater, and other fish habitats. These requirements varied from 0.02 to 0.2 mg/L.

Shortly thereafter, some graduate students doing ocean research at the Sanitary Engineering Laboratory of the University of California, Berkeley, found that there were no ill effects of sulfur dioxide residuals in San Francisco Bay water up to 10–15 mg/L. This was the answer for many plant operators wondering how to deal with the dechlorination problem—just add plenty of excess SO_2.

This entire idea of wastewater dechlorination has always been bothersome to White, who saw many years ago (1950s) how the French handled discharges into the river Seine. There were three large water treatment plants taking their supply from the river and several wastewater and industry discharges. The water plant operators had "fish cages" in the river that they used for clocking the speed of the fish swimming upstream. When the speed of the fish dropped below a certain value, the plant operators were so advised, and they in turn prepared the plant to handle what could be an industrial discharge, which required special attention to their pretreatment processes. Later, White was advised that the French and the Germans (on the Rhine) indicated that their discharges were easily handled by river water dilution.

Wastewater vs. Potable Water Dechlorination

Discussion. There are three major reasons why the strategies for dechlorination control must be different from those used for potable water: (1) there is no analytical system currently available to measure the concentration of the sulfite ion; (2) dechlorination must be complete (zero); and (3) chlorine residual analyzers are not capable of maintaining their calibration in a zero residual environment.

The following strategies are primarily for wastewater systems—zero control.

Previous Control Systems

In the years following institution of the dechlorination requirements, several methods were devised. These systems were all superseded in the early 1990s by the Stranco High Resolution Redox Control System and the Wallace & Tiernan Deox/2000[R] Dechlorination Analyzer. Both of these systems are described below, plus the Renton System.

The other three systems (described in the 3rd Edition) have been eliminated. They were too complicated and required constant operator attention.

Currently the big question is: should all wastewater effluents be required to dechlorinate to zero without any allowance for receiving water dilution?

The Renton System. (*Author's Note:* This system is being left in the text solely for historical purposes. It is no longer being used.) See page 1086.

Description. The concept of using a dechlorinated sample biased with an even split of the chlorinated sample was implemented by William Miks[20] at the Palo Alto Water Pollution Control plant in 1979. He diverted part of the chlorinated sample (used for chlorination control) to the freshwater dilution compartment of a Wallace & Tiernan analyzer. This sample was mixed 1 to 1 with the dechlorinated sample, which entered the other half of the sample dilution tank. This was convenient at Palo Alto because the chlorine control analyzer and the sulfonator control analyzer were located adjacent to each other. This method of "zero control" was still operating successfully in 1991. The sulfonator control analyzer was calibrated to provide about 0.3 mg/L negative residual. (See Figs. 10-5 and 10-6.)

A more sophisticated method using the above concept was developed by Kennedy/Jenks Engineers in 1980.[21] This method was presented at the WPCF annual conference in 1983.[22] The application of sulfur dioxide is controlled solely by residual measurement.

The Renton system was based upon the fundamental concept of the Kennedy/Jenks system. It operated almost flawlessly.[23] Flow pacing was not used. The control system provided for a slightly negative chlorine residual (−0.20 mg/L) to ensure reliability. This system is illustrated in Fig. 10-5. It was a retrofitted system, which utilized both Fischer and Porter and Wallace & Tiernan equipment. The success of this system was due in part to the short loop dead time—60 sec at 42 mgd average daily flow.

The dechlorinated effluent sample was biased with the chlorinated effluent sample to provide a sample that could be measured by a conventional residual analyzer. The chlorinated and dechlorinated sample flows were proportioned to a 1:1 ratio by using constant-head tanks. Unfortunately this proved to be the one major difficulty with this system. Owing to the numerous chlorine residual violations reported by the California Water Resources Surveillance group, a series of seminars was given for them by White in 1989. In the seminar for operating personnel, it was revealed that, for the most part, operators were trying to improve their ability to achieve this required even split of the two samples without much success. In the meantime the Renton, Washington plant had stopped dechlorinating. The Seattle Metro had built a long outfall that was so situated they received a waiver for the dechlorination requirement. This eliminated the search for any further refinements to the Renton system.

The chlorine residual in the biased sample of the dechlorinated effluent was calculated electronically from the measured residuals in the chlorinated and dechlorinated samples and the fixed mixing ratio of the two samples based upon the following equation:

$$C_D = 2C_M - C_C \tag{10-15}$$

where

C_D = chlorine residual of dechlorinated sample
C_C = chlorine residual of chlorinated sample at end of contact chamber

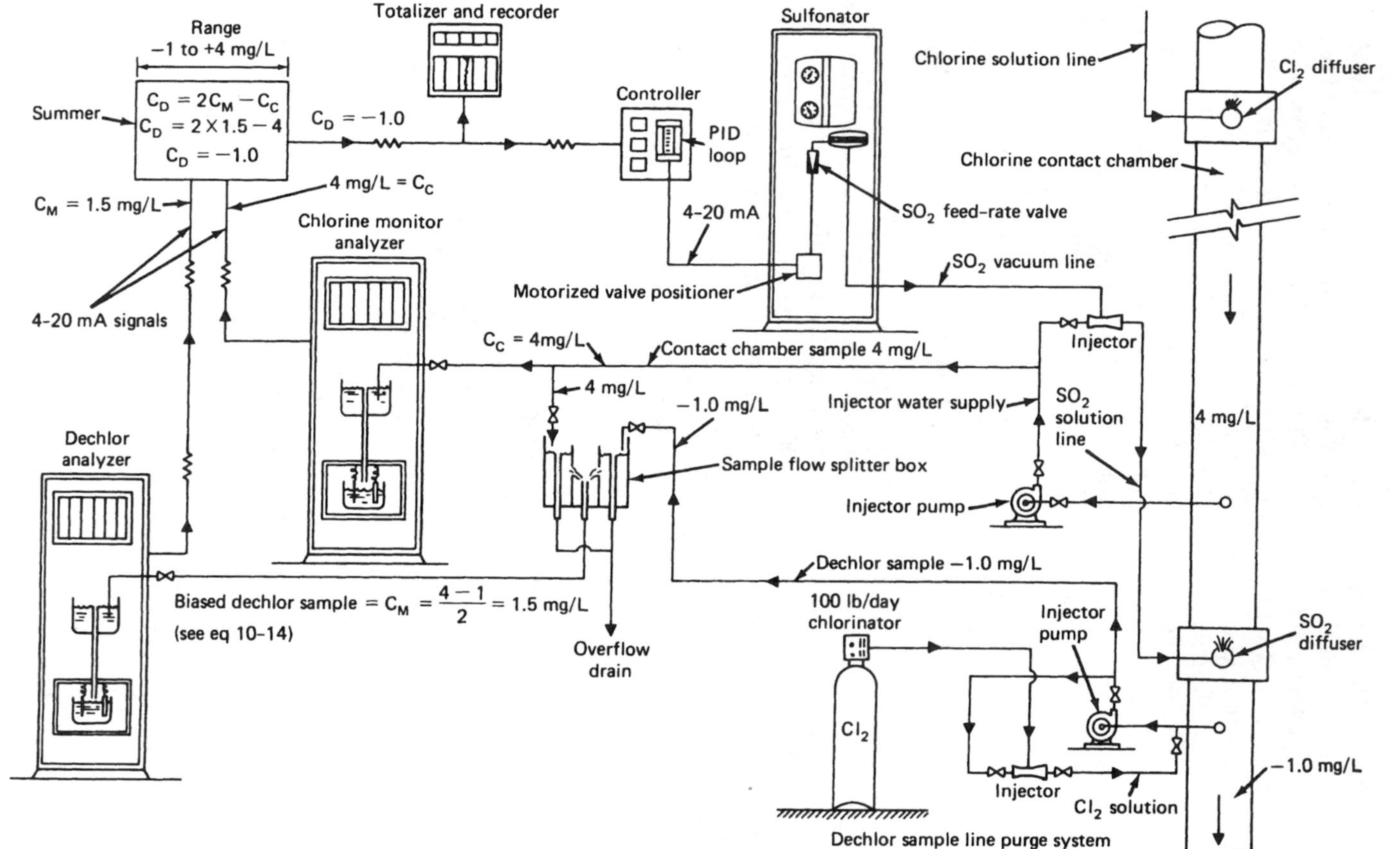

Fig. 10-5. Zero residual dechlorination system (after Renton, Washington WWTP[23]). Note: All of the control functions shown can be

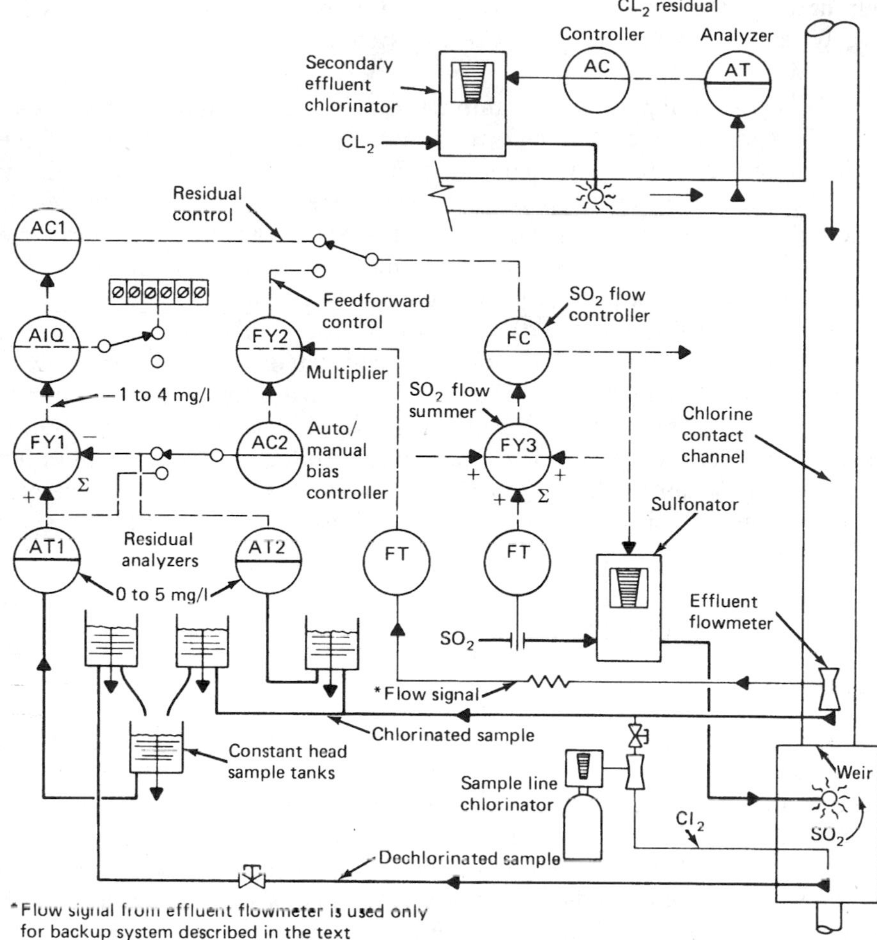

Fig. 10-6. Renton zero residual dechlorination system showing backup systems arrangement.[23]

C_M = chlorine residual of 1:1 mixture of chlorinated and dechlorinated sample

Each analyzer generated a signal proportional to a range of 0–5 mg/L. These two signals were sent to a "summer" (to add or subtract). The output of this summer went to a controller, which provided a PID loop to control the metering orifice on the sulfonator. The motor drive mechanism, which operated the SO_2 metering valve, was capable of changing the feed rate from zero to maximum in 15 sec. The deadband on the controller was equivalent to 0.1 mg/L residual. The speed of response of the SO_2 control valve in the sulfonator and the 60-sec loop dead time at 42-mgd average daily flow were largely responsible for the success of this system.

EXAMPLE. A hypothetical example shows how the system worked. Refer to Fig. 10-5. Assume the dechlorination system is not in operation and the chlorine residual exiting the contact basin is 4 mg/L, so $2C_M = 8$ or $C_M = 4$. Therefore, the signals going to the summer are 8 and 4, and the signal going to the controller is +4, calling for SO_2. The operator starts the sulfonator, and it overdoses to produce 1 mg/L of sulfite ion so that postresidual is −1 mg/L. From Eq. (10–15), $2C_M = +3$. Going to the summer are a +3 and a −4 (to be subtracted), which produce a −1 signal to the controller. The 4–20 mA output of the summer is calibrated so that a 4 mA signal represents a 1 mg/L chlorine residual signal, and a 20 mA signal represents 14 mg/L. This corresponds to the 0–5 mg/L range specified for the analyzers. The controller reduces the SO_2 feed rate to produce a zero residual in the dechlorinated sample; hence $C_M = (4 + 0)/2 = 2$ and $2C_m = 4$. Therefore, the two signals going to the summer are 4, and the output of the summer is a zero signal.

ZERO RESIDUAL CONTROL

The ORP Control System

Introduction. Since the publication of the 3rd Edition of this book, White was hired by the Stranco Corp. to evaluate the efficiency and reliability of its ORP control system as a wastewater disinfection process. After three years of investigating several installations of its High Resolution Redox Control Systems at WWTPs in California and Illinois, White concluded that the Stranco system was superior to all the currently available chlorination–dechlorination systems for wastewater effluents. However, he insists that the use of the ORP system still requires a chlorine residual analyzer to tell the operators if the chlorination equipment is operating properly.

As of 1997, it is well known that in measuring the total chlorine residual in wastewaters, there will always by part of that residual that is nongermicidal because of the presence of organic N compounds as well as nitrites, which detract from the germicidal efficiency of the applied chlorine.[31,32] In simpler terms, the measuring of chlorine residuals is not an accurate way to evaluate the germicidal efficiency of chlorine, particularly in wastewater.

As early as the 1950s, it was suggested that the measurement of redox potential (ORP) of chlorinated water could be used as an alternative way of assessing the disinfection efficiency of chlorination. This was largely a result of evidence produced regarding the wide variation of disinfection efficiency in the various chlorine species, as shown in Fig. 10-7.

As early as 1960, Ebba Lund,[30] a renowned scientist, proved conclusively that ORP measurement accurately predicts the rate of virus inactivation,

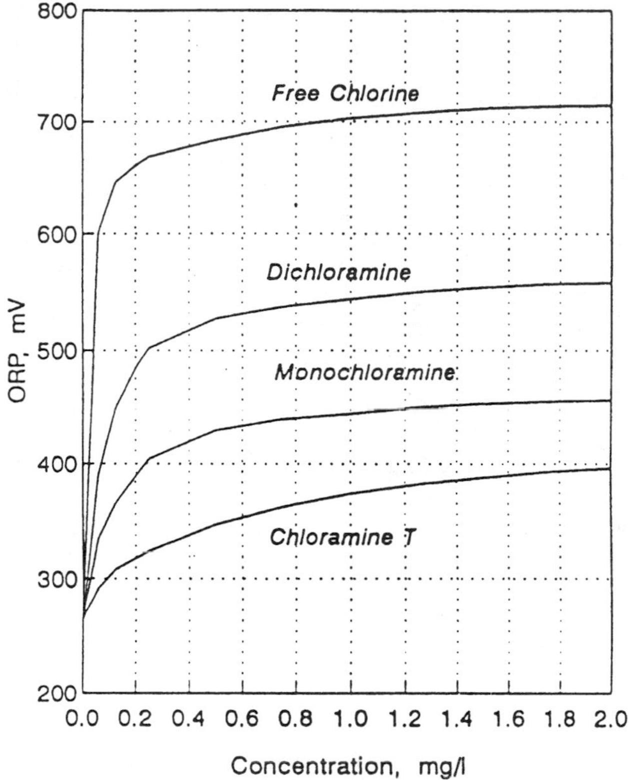

Fig. 10-7. Redox potential for various chlorine compounds (Victorin, et al.).

regardless of the oxidizers used in the experiments, as shown in Fig. 10-8. Now we know far more about the abilities of ORP measurements, as shown in Fig. 10-9, which illustrates redox potential as a function of free chlorine, chloramine, and sulfite. This clearly shows the dramatic change in the potential when the residual is near zero. This makes it quite easy to control extremely low concentrations of chlorine or sulfite, such as 0.1 mg/L.

The redox system differs from residual analyzers in several important aspects.[29,32] First, it measures the actual oxidant and reductant activity instead of residuals. Second, redox can measure the potential (voltage) of both sides of the chlorine and sulfite couple. This means that it can measure chlorine activity and sulfite activity, allowing the operator to lock in a slight residual of either chlorine or sulfite. Third, redox responses are logarithmic, meaning more precise dechlorination control where zero residual is the goal. Fourth, properly designed redox electrodes are current, and they can continue to operate properly, even after being partially coated with process contaminants.

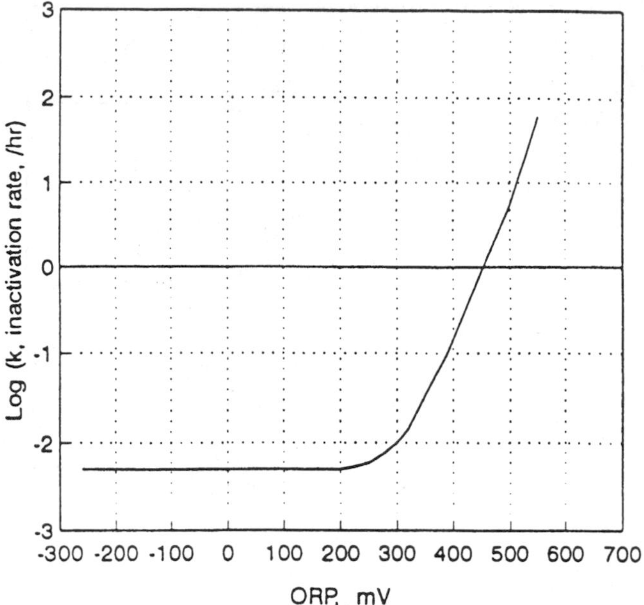

Fig. 10-8. Rate of poliovirus inactivation as a function of redox potential (Lund).

The following is a description of a typical WWTP in California using the Stranco Corp. High Resolution Redox System. There are about 300 of these chlor–dechlor systems operating in the USA today (1997).

Simi Valley Calfornia WWTP Installation. A description of the chlor–dechlor control system at this 12.5 mgd tertiary treatment facility, serving more than 100,000 residents and some 1100 local businesses, is appropriate, as it is one of the first installations of the Stranco High Resolution Redox systems in California.[29]

The plant effluent discharges into the Arroyo Simi Creek. A very short distance from the plant discharge, the creek discharges into the ocean in a very popular bathing area. Therefore, the NPDES discharge requirement is 2.2 MPN of total coliforms, and the chlorine residual must not exceed 0.1 mg/L total chlorine residual. As had been the usual case, the plant operators had to continuously overfeed the sulfur dioxide to meet the chlorine residual requirement. This was so largely because of the high degree of maintenance required using residual measurement by their existing chlorine residual analyzer.

In February of 1992 this plant had Stranco Corp. set up a pilot test system. A chlorine controller was installed to monitor the chlorine demand of the wastewater flow by suspending a Stranco ORP probe in the chlorine contact chamber about one minute downstream from the chlorine diffuser. The system

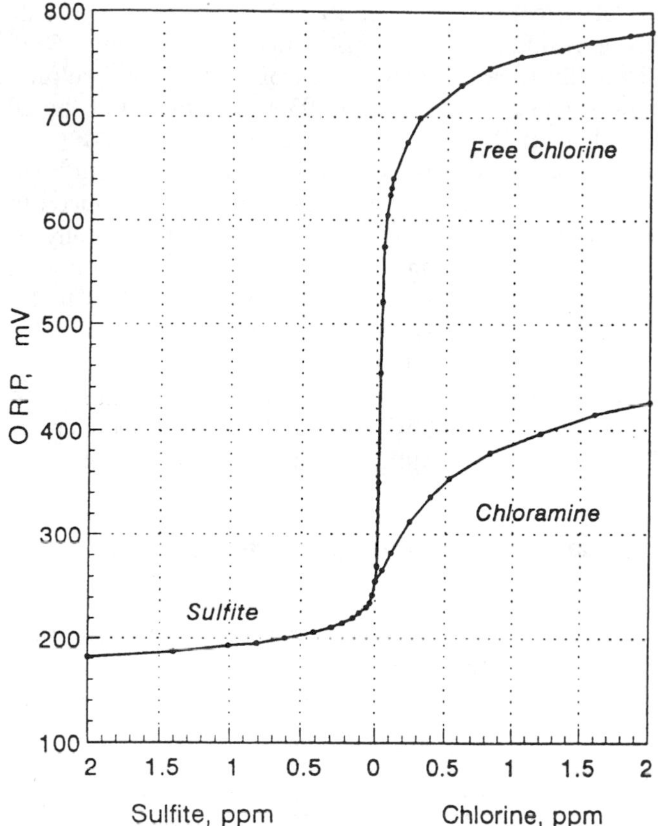

Fig. 10-9. ORP versus chlorine and sulfite concentrations (tap water, 21°C, pH 7.5).

controller converted the ORP signal to a 4–20 mA signal that was sent to a PID controller, which adjusted the chlorine feed rate.

A second ORP probe was located about 20 sec downstream from the sulfur dioxide diffuser. This probe sent a signal to another PID controller that modulated the sulfur dioxide feed rate through a controller in the sulfonator. After a month-long monitoring period, the units were placed in automatic control mode, with control setpoints set high enough to absolutely ensure that they would be in compliance. After a few weeks the setpoints were slowly lowered, reducing chlorine and sulfur dioxide feed rates, while the resulting effluent chlorine residuals and coliform counts were tracked. After this successful pilot study, the plant operators decided to change to the completely automatic control system (Stranco's HRR System) for the control of both chlorination and dechlorination systems.

Results from ORP Control. The chlorine controller currently maintains a 520 mV setpoint, which results in less than 2.2 MPN/100 ml of total coliforms.

The dechlorination controller is set to 200 mV, which meets the required effluent chlorine residual of less than 0.1 mg/L requirement. See Figs. 10-10 and 10-11. They illustrate the daily usage of chlorine and sulfur dioxide for the same months, August of 1991 and 1993, before and after the ORP control system was implemented.

The results of this new control system were quite amazing.[33] Not only was it far more accurate than their previous system, but it reduced the monthly chemical so much that the capital cost of the system was fully recovered in the first 6 months of operation. (See Fig. 10-12.) In addition to that, the operators are most fond of the system because it is easy to operate, and maintenance is truly minimal. All they have to do is clean the ORP chlorination probe once a month and the dechlorination probe once a week and no longer do they have to worry about making multiple daily changes in chemical feed rates.[33] As of late 1995, the ORP probes that were installed for the pilot project in 1991 were still operating without any replacement.

Wallace & Tiernan Deox/2000^R Dechlorination Analyzer

Introduction. This unit (Fig. 10-13) is a modification of the 1991 Deox 2000™ Iodine Bias Analyzer for Measuring and Control of Chlorine and Sulfur Dioxide Residuals.[25,26] This Deox/2000^R Dechlorination Analyzer 1995 unit makes direct measurements of SO_2 and Cl_2 residuals in one analyzer.

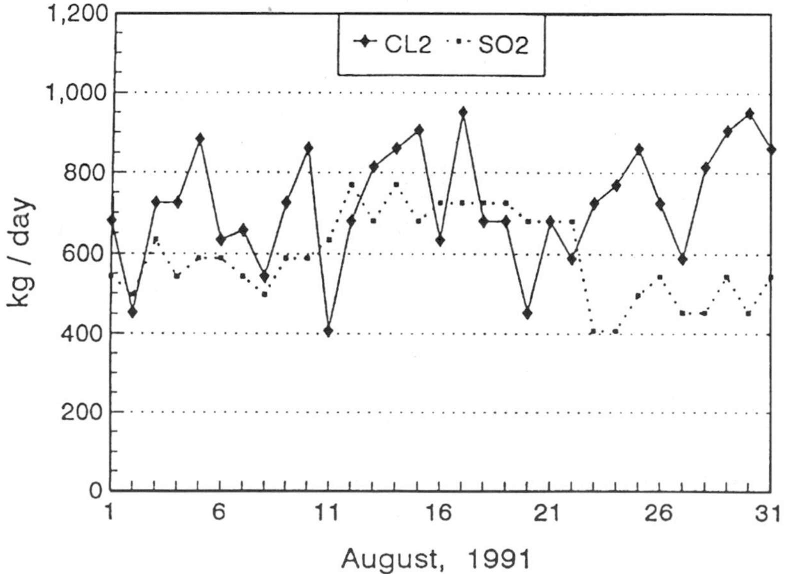

Fig. 10-10. Daily use of chlorine and sulfite (before ORP).

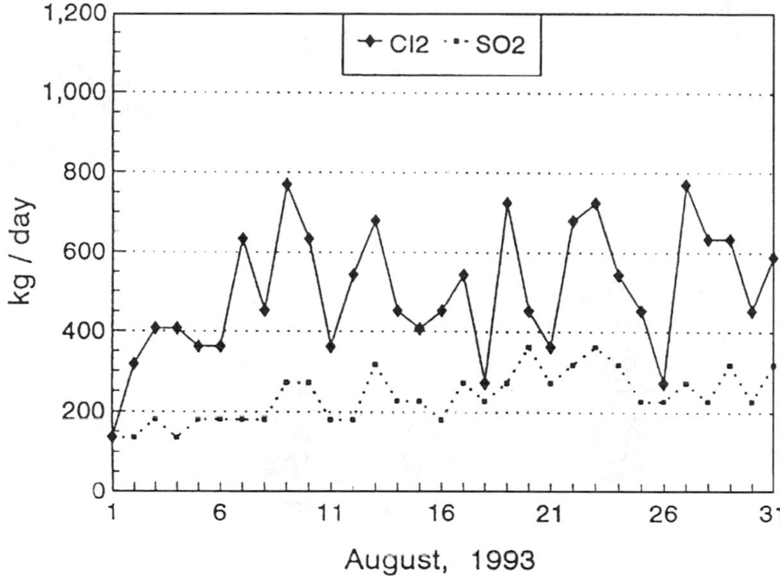

Fig. 10-11. Daily use of chlorine and sulfite (after ORP).

Until 1991, the preferred way to achieve zero chlorine residual was to use the Renton system (See Fig. 10-5 and 10.13.) Now in 1997 we have a unit that responds quantitatively to both chlorine and sulfur dioxide residuals (sulfite ion). This unit is the most effective on-line instrument for the accurate measurement of these residuals in wastewater plant effluents.

With its "center zero" residual analysis capability, the control to an SO_2 setpoint is achievable and ensures complete dechlorination at a level that supports compliance with discharge permits, while costly SO_2 overdosing is eliminated. See Fig. 10-13 of the Deox/2000^R analyzer.

Principle of Operation. The Deox/2000^R analyzer uses a unique process to measure the SO_2 (sulfite) residuals. It is based upon the principle of biasing the sample with an iodine source, which is unaffected by the sample flow quality. With a stable "iodine bias," the SO_2 or Cl_2 residuals are automatically determined from the measured amount of iodine that has reacted with the sample. (See Fig. 10-14.)

Three-Electrode Measuring Technology. By utilizing universally accepted and time-proven amperometric measurement technology, the Deox/2000^R Analyzer employs an innovative three-electrode measuring cell to provide direct measurement of residuals.

With its bare electrode design, the Deox/2000^R is not susceptible to the fouling that is typical in membrane type analyzers, caused by microorganisms,

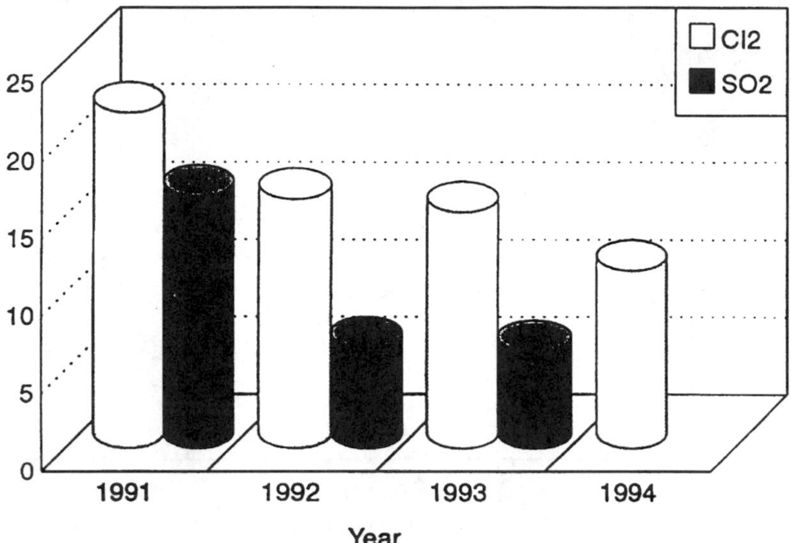

Fig. 10-12. Monthly usage of chlorination and dechlorination chemicals at the Simi Valley WWTP before and after the control by ORP.

grease, or turbidity. Because of the stability of the three-electrode measuring cell, daily maintenance for recalibration is virtually eliminated. In fact, the constant requirement for zero-adjustment, common in other analyzers, is eliminated entirely.

Display Details. The unit is housed in a NEMA 4X enclosure that provides a sunlight-readable alphanumeric display of residual levels, including type of residual, and alarm messages.

A six-button keypad provides access to the various displays, as well as setup, calibration, and diagnostic menus. Operator access can be restricted by two security codes for supervisory and operating personnel.

The unit can be easily reprogrammed for different operating ranges, alarm setpoints, and display information.

Arrangements. Because of their modular construction, the Deox/2000^R systems can be supplied in a variety of formats to suit any application.

Systems are available in free-standing cabinets incorporating circular chart records, or as wall- and panel-mounted units with the protection of watertight, corrosion-resistant NEMA 4X enclosures.

Technical Specifications.

Measuring ranges	5 discrete ranges adjustable from 0—0.5 mg/L SO_2, 0–0.5 mg/L Cl_2, up to 10 mg/L SO_2, 0–10 mg/L Cl_2

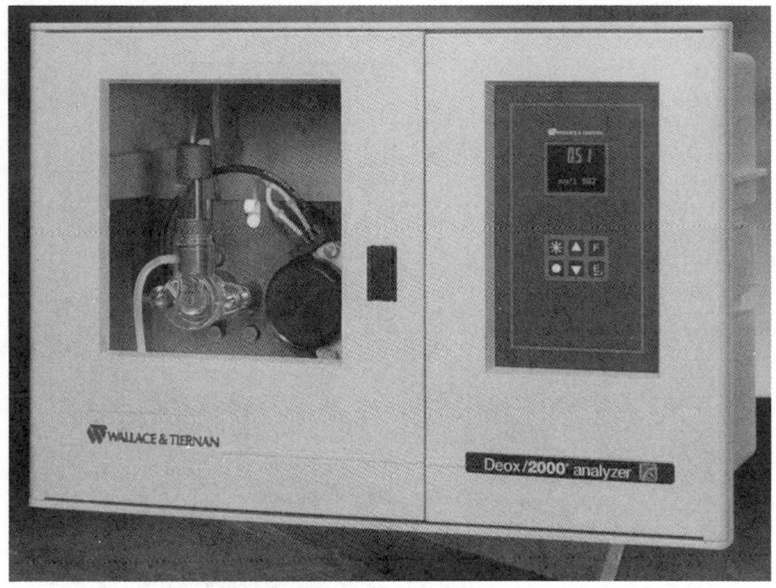

Fig. 10-13. Wallace & Tiernan Deox/2000[R] Dechlor Analyzer.

Accuracy	± 5% of full scale (over the full range from SO_2 to Cl_2 residual)
Power requirements	0.5 A at 115 VAC, 0.25 A at 230 VAC
Output	Isolated 4–20 mA proportional to residual
External alarms	Three relays user-configurable for high–low residual alarms, system alarms, and dosing control functions.
Dimensions	Wall-mounted 15 in. H, 22 in. W, 11 in. D Module-mounted 68¼ in. H, 27½ in. W, 16 in. D

See CN 150.530 UA & CN 150.531 UA for more dimension information.

See CN 150.535 UA & CN 150.536 for typical installation information.

For complete technical data and equipment specification, see publication TI 50. 515 UA.

Chemistry of Operation. This analyzer gives a continuous on-line determination of SO_2 residuals by metering a predetermined amount of iodine. Since iodine is the measured chemical in total chlorine residual analyzers, the change in iodine residual indicates either a chlorine or a sulfur dioxide residual, depending upon whether the change is positive or negative. Iodine reacts with

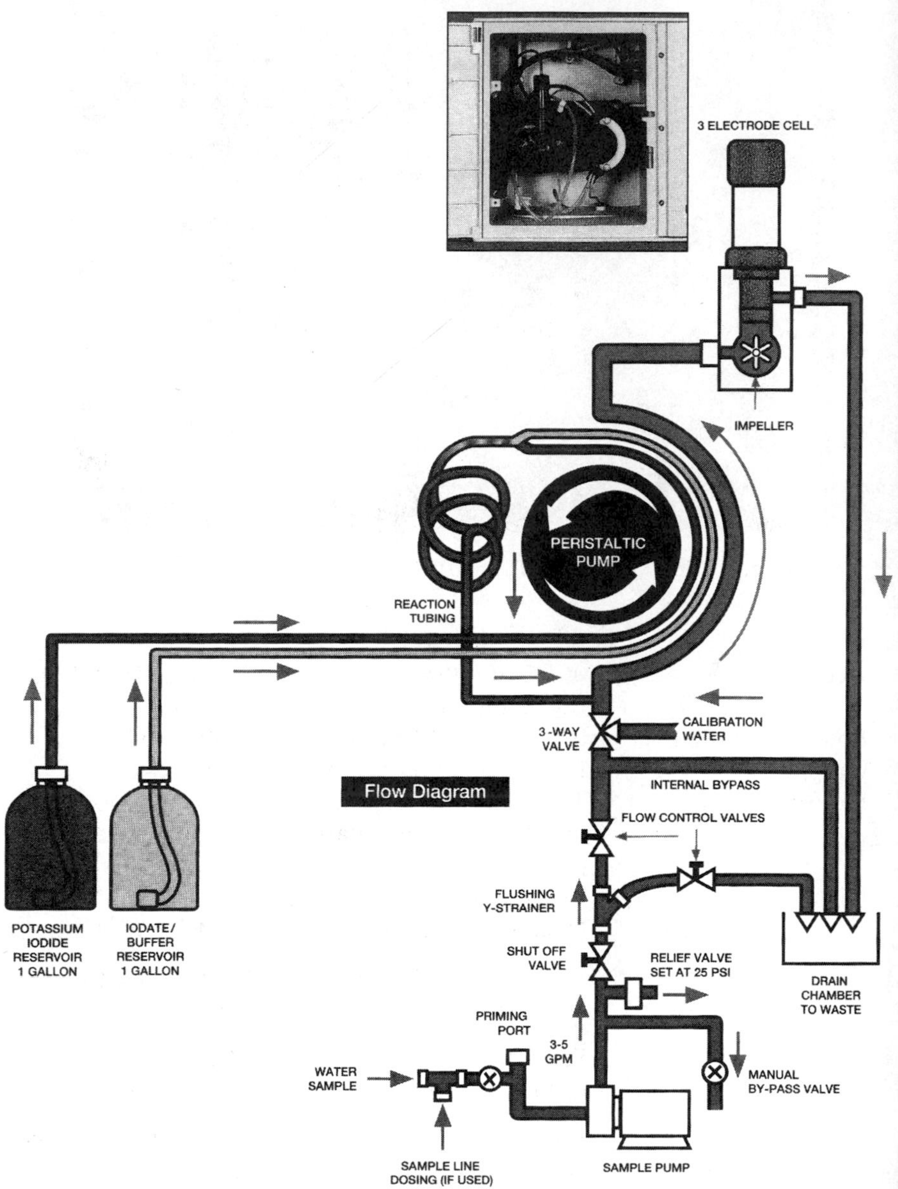

Fig. 10-14. Flow diagram for Wallace & Tiernan Deox/2000[R] Dechlor Analyzer.

sulfur dioxide in the same manner in which chlorine reacts with sulfur dioxide. Therefore, the amount of iodine consumed represents the amount of SO_2 in the water; alternatively, if there is excess iodine measured, then a chlorine residual is present, and can be accurately determined.[34]

Iodine is less likely to react with oxidative demand in the water, making determination of wastewater residuals more reliable than a hypochlorite biased measurement.

Since iodine is not stable when exposed to air and is very reactive with many materials, it is stored as potassium iodate (KIO_3) in order to maintain its stability. Then it is converted back to iodine just prior to use. The chemistry of the system is illustrated by the following equations:

$$KIO_3 + 5KI + 6HAc \rightarrow 3I_2 + 6KAc + 3H_2O$$

The reaction of SO_2 with iodine is:

$$H_2SO_3 + H_2O + I_2 \rightarrow H_2SO_4 + 2HI$$

The reaction of free chlorine with iodide is:

$$HOCl + 2KI \rightarrow KOH + KCl + I_2$$

A continuous sample is delivered to the analyzer. A flushing strainer divides the sample into two streams. The larger bypass stream continuously flushes the filter, and the smaller stream flows to the constant level chamber in the analyzer flow block. The constant level box inlet is designed for a 500 to 1000 mL/min. sample flow, but only about 50 mL/min. of this sample is used to make the analysis. The sample flows out through the cell block. Here a rotating impeller maintains a constant sample velocity and entrains grit to scour the electrode, which is the key to maintaining proper operation.

The three-electrode amperometric cell reduces a portion of the iodine to iodide at the working electrode. The resulting current (microamps) will be proportional to the residual in the cell. The working and the counter electrodes are made of the same material (platinum); so there is no current flowing at zero residual. This eliminates the need for a zero setting. The selectivity of the probe is set by biasing the working electrode with respect to a standard reference electrode, 0.2 V for iodine.

Due to the small sample flow through the unit, buildup of organic matter could cause a chlorine residual demand that would affect the system accuracy or completely block the sample flow. To minimize this possibility, the system periodically energizes the solenoid, which raises the orifice plunger. This removes all the unwanted material within the orifice. Directly above the orifice is a gas purge membrane used to remove any accumulated gas released from either the sample or the reagents because of the lowered pH. Flushing is not possible, as that would flush out the biasing agent and upset the equilibrium of the cell.

Due to reactivation of biological growth in the unit at low or negative chlorine residuals, and subsequent growth and demand in the sample lines, which would affect measurements, sample line dosing is a provided feature of the Deox 2000™. This allows for the periodic "shocking" of the sample line with a high residual to control growth and keep the lines clean.

Typically both wastewaters and condenser cooling tower waters usually must be dechlorinated. The reagent must be either sulfur dioxide gas or sodium bisulfite solution. Both of these chemicals produce sulfite as the dechlorinating agent. The chemistry is as follows:

$$\underset{\text{Chloramine}}{NH_2Cl} + \underset{\text{Sulfite}}{SO_3^=} + H_2O \rightarrow \underset{\text{Ammonium chloride}}{NH_4Cl} + \underset{\text{Sulfate}}{SO_4^=}$$

Sulfite reacts with all inorganic chlorine residuals to form sulfate and ammonium chloride. It requires 0.9 mg/L sulfur dioxide and 1.46 mg/L sodium bisulfite are needed to reduce each mg/L of chlorine residual.

Calibration. Simple automatic procedure. An analyzer is only as good as the method used to calibrate it. The Deox/2000™ uses a calibration method that is both simple and automatic. The analyzer display provides step-by-step instructions to prompt the operator during the procedure.

To calibrate the analyzer, the operator must only scroll to the iodine bias screen (in the main operating mode) and press the change key. This freezes the output signal and initiates automatic calibration of the iodine bias. After a stable iodine bias has been established in the cell (by the use of the demand-free water that has been fed with the calibration pump), the operator is prompted to take a sample for titration and scroll in the value as the iodine bias in mg/L. The analyzer will automatically calculate the process residual level (in mg/L, SO_2 or CL_2) from the actual measured residual and the iodine bias.

Process Control. Before this analyzer became available, it was not possible to accomplish control of chlorine residuals by dechlorination in wastewater discharges with satisfactory consistency or accuracy acceptable to the regulatory agencies in California and several other states. The various regulatory limits are usually 0.02 mg/L chlorine; so most dischargers went to at least a 2 mg/L excess of sulfur dioxide. With this analyzer, the bias stability is ±1.0 percent, and the overall analyzer is rated as better than ±5 percent of full scale.

Because both chlorine and sulfite can be accurately monitored with zero calibration by the presence of the third electrode, the bias analyzer can be used in a closed loop feedback control system for dechlorination by sulfur dioxide with an alarm to signal a chlorine residual violation. After calibration, the analyzer is returned to operation, and residuals are displayed on the alphanumeric display. The 4–20 mA analyzer output can be used to operate a recorder or drive a suitable controller in a feedback control system.

System Accuracy.[34] Under laboratory conditions, overall accuracy has been on the order of ±2 percent of full scale, that is, ±0.2 mg/L with a 5.0 mg/L bias or ±0.02 mg/L with a 0.5 mg/L bias. It is important to note that the establishment of the center point (zero mg/L of both chlorine and sulfur dioxide residual) is independent of analytical error. Overall analyzer accuracy will improve as residuals approach zero.

Bias Stability and Accuracy. One of the most critical factors in the overall performance of this analyzer is the bias accuracy. Over a two-week period, W & T reported an accuracy of 1.0 percent, with 100 percent stability reported during a two-month period. Stability in excess of two years is expected.[25] Both the sample feed-rate system and the iodate/KI solution exhibit negligible error over an eight-hour period.

Bailey–Fischer and Porter Z-CHLOR[R] Center Zero Dechlorination Analyzer Control System, Series 17SD4000

Introduction. Chlorine has long been used to treat municipal water and both municipal and industrial wastewater. It is a proven reliable and cost-effective disinfecting agent, for both potable water and wastewater.

In wastewater applications the chlorine is added to the treatment plant effluent just prior to the point of discharge, which is usually into a surface water stream or an ocean beach, to kill organisms that cause waterborne diseases. This also ensures safe conditions for recreational water uses, as well as for commercial operations such as fishing or clamming.

However, in recent years too many people have put environmental concerns ahead of public health issues. They worry about the effect of chlorine on aquatic life. This concern has resulted in increased requirements to monitor and control the amount of chlorine used, and, in many cases, to remove some or all of the residual chlorine in the wastewater plant effluent prior to being discharged. (See Figs. 10-15 and 10-16.)

System Description.[36] This analytical instrument system (Series 17SD4000) continuously measures the chlorine or sulfite ion residual in the wastewater treatment plant effluent flow, and controls the residual at a given preset value. This preset value may be a positive chlorine residual, a zero chlorine residual, or a positive sulfite ion residual value. The flow signal from the plant flow meter on the effluent line is combined with the residual signal in the controller.

The system is capable of measuring and controlling chlorine residuals down to zero mg/L of chlorine through a precisely controlled unique biasing system, which introduces a small constant quantity of oxidizing chlorine solution (a

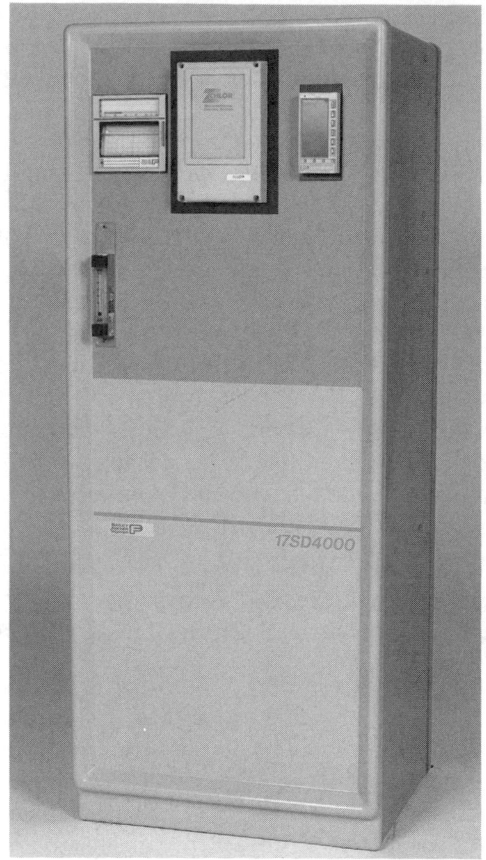

Fig. 10-15. Bailey–Fisher and Porter series 17SD 4000 Z-CHLOR center ZERO dechlorination analyzer with optional circular chart recorder (courtesy of Bailey–Fisher and Porter Co.).

dilute NaOCl solution) into the sample immediately prior to the analysis of residual chlorine. In this way, the chlorine residual analyzer (Series 17B5000) is continuously provided with a chlorinated sample.

The biased sample in the analyzer produces a signal equal to 50 percent of full scale on the controller display, or it records it on an optional recorder. The 50 percent point is equivalent to a zero chlorine or sulfite residual. If any chlorine is present in the sample, it is added to the bias value, and the analyzers will show a positive chlorine residual. Conversely, if sulfite is present in the sample, the signal will move below zero (a positive sulfite number) when the instantaneous reaction between the chlorine in the bias solution and sulfite ion reduces the chlorine in the analyzer cell.

The heart of the Z-CHLOR^{R} control system is the microprocessor-based controller, mounted on the cabinet face (see Fig. 10-15). It contains a specifically designed control logic keyed to the dechlorination process. It includes

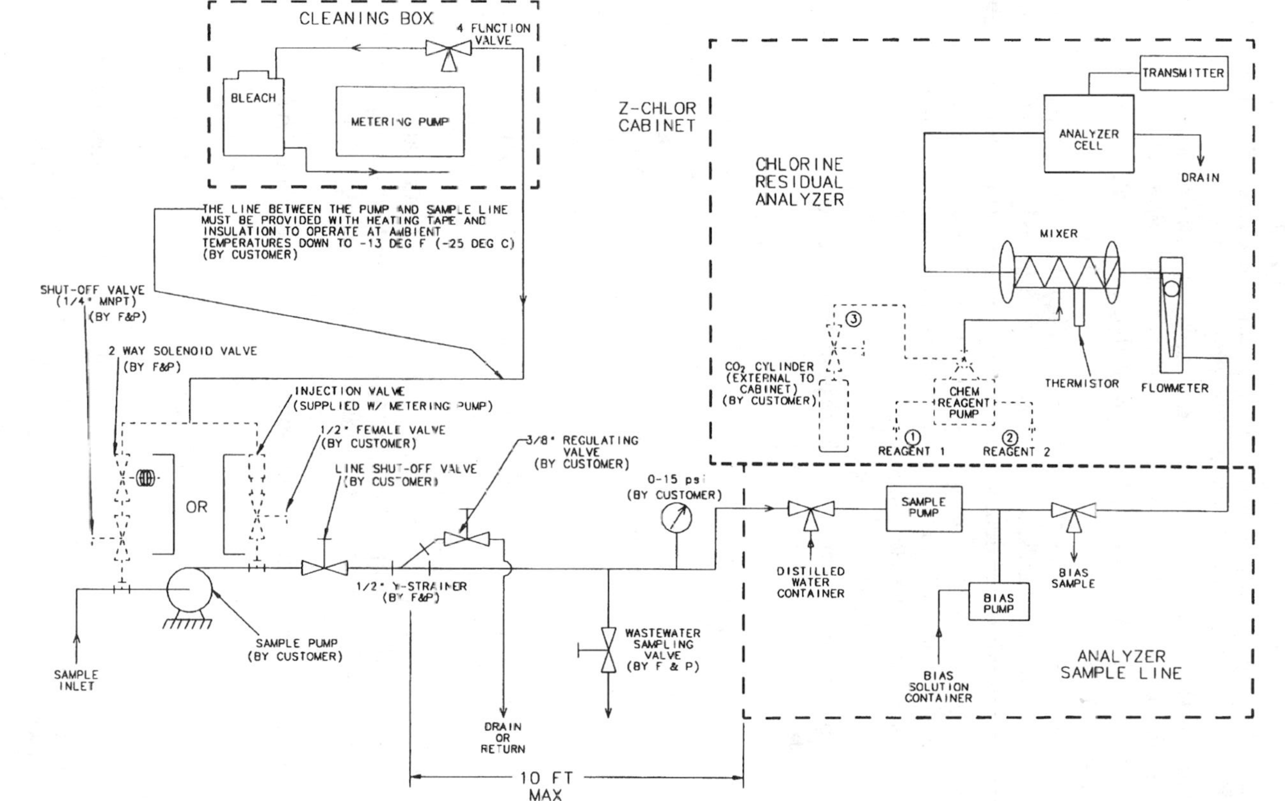

Fig. 10-16. Flow diagrams illustrating a variety of operations of the 17SD 4000 Z-CHLOR (courtesy Bailey–Fisher and Porter). Acetic acid or CO_2 for pH control of sample. One and two for total chlorine with acetic acid; two and three for total chlorine with CO_2.

a multistage alarm system, to alert operating personnel to residuals above or below the desired levels. An override system is activated should the residual levels reach the maximum permitted by the Plant Discharge Permit (PDP). Figure 10-16 shows the control system flow diagram.

The control system features a dot matrix display, which provides operating personnel with all the data necessary for proper operation. This is accomplished through the use of multiple displays that present all the data in an easily understood format.

The operator simply pushes a buttom to see digital and bar graph displays of chlorine or sulfite residuals; setpoint and output signal to the sulfonator; adjustable time trends of residual sulfite ion, and process water flows to show recent process performance; controller tuning parameters; various adjustable alarms; adjustable high and low output limiters; and various parameters to site-tune the sample in the cleaning cycle.

The Cleaning System Option. This system is available with an optional cleaning system as well as with an optional circular chart recorder mounted on the front of the cabinet (see Fig. 10-15). The cleaning system is necessary to keep the sampling line to the Z-CHLOR^R in a clean condition, free of slime and algae growth, which would otherwise reduce the level of chlorine residual reaching the analyzer.

It incorporates an external chemical pump, housed in a weatherproof enclosure for use in the feeding of sodium hypochlorite when the length of the sample line going to the analyzer exceeds 5 ft. A corrosion-resistant solenoid is included when the hypochlorite injection point is into the pump suction. The solenoid and chemical pump should be located at the beginning of the sample line. This option will help to maintain the inside of the sample line free from slime and algae buildup.

In operation, the chemical pump and solenoid are automatically activated for several minutes (field-adjustable) during every eight-hour period. A bottle of 5 percent sodium hypochlorite (bleach) is required for this cleaning system.

The controller provides all the operating controls for this cleaning system, including keeping the output to the sulfonator at a constant level during cleaning.

Design Features. This Z-CHLOR^R Series 17SD4000 system has several very important design features:

1. A series of displays on several dot matrix screens provide operating personnel with all the information they require. (See Fig. 10-15.)
2. It utilizes the time-proven, unique "center zero" feedback control concept.
3. It continuously monitors the chlorine or sulfite ion residual in a sample stream of potable water or wastewater being discharged from a treatment plant.

4. A state-of-the-art microprocessor-based controller handles the control of the sulfonator to maintain the residual at a preselected value.
5. The setpoint may be a positive chlorine residual value, zero chlorine residual value, or positive sulfite ion residual value to meet the plant discharge requirement.
6. An optional residual recorder provides a permanent record that the plant is meeting its effluent discharge requirements for chlorine residuals.
7. A multistep alarm system warns operating personnel, in steps of increasing severity, of abnormal conditions.
8. An override control system is triggered, should the chlorine residual reach the maximum permitted value.

Conclusions. Many refinements in monitoring and control technology have been made possible with the advent of today's microprocessor-based control system. In handling the specific tasks of chlorinating potable water and wastewater and removing excess chlorine from treated wastewater, their capabilities are being used to solve difficult control problems that are required to meet the present stringent requirements. These new solutions also provide operating personnel with all the information they need to achieve consistent disinfection process information.

Dechlorinated Sample Line Fouling

This problem, which has spoiled the reliability of many dechlorination systems, has been found to be site-specific. Although a rarity, there are some installations where biofouling is not a problem, whereas there are some others that require daily shock treatment with heavy doses of chlorine. (See Fig. 10-5.)

The best preventive maintenance tactic is to design the sample line piping to be small enough to provide a 5 ft/sec velocity—even if this requires a pump. This is considered to be a scouring velocity. In addition to the hydraulic characteristics, the sample line should be piped with the appropriate valves so that it can be shock-dosed with chlorine conveniently at any time. The dose, frequency of application, and sample line retention time all have to be determined by trial and error. This application is akin to the intermittent shock-dosing of condenser cooling water.

In wastewater treatment, the chlorine residuals are predominantly chloramines. The germicidal efficiency of this chlorine species may, in some cases, be insufficient to control the biofouling owing to lack of contact time. In these cases, flushing with caustic is the only solution. Therefore, in addition to the scouring velocity, consideration of a standby sample line might be very beneficial for greater system reliability. Preservation of the sample line's biological stability, such that dechlorination by the biomass in that line never occurs, is of the utmost importance.

Summary

Based upon years of operating experience, the task of choosing the appropriate control strategies is based upon answering the following questions:

1. Are the chlor–dechlor systems operating properly in accordance with their specifications?
2. Is the disinfection efficiency sufficient to meet discharge requirements?

The answer to question No. 1 can be satisfied by the two best analyzers on the market: the Wallace & Tiernan Deox/2000[R] Dechlorination (iodine bias) Analyzer and the Bailey–Fischer & Porter Z-CHLOR[R] Center Zero Dechlorination Analyzer Control System.

Owing to the existence of significant amounts (4–6 mg/L) of organic N compounds in wastewater effluents which titrate as dichloramines that are nongermicidal, all residual chlorine measurements will have inaccuracies due to the organochloramine compounds formed when organic N compounds are chlorinated.

The answer to question No. 2 is to eliminate this problem by determining the correct disinfection efficiency of the chlorine applied. The only way to achieve this is by using the Stranco High Resolution Redox Control System for the chlor–dechlor situation. This is a special ORP control system that effectively and correctly measures the disinfection efficiency of the chlorine applied.

Need for Chlorine Residual Analyzers. It should also be recognized that this redox measurement in no way affects the need for the chlorine residual analyzers. These analyzers are absolutely necessary for operating personnel to know the operating condition of all the chlor–dechlor equipment.

REFERENCES

1. Chen, C.-L., and Gan, H. B., "Wastewater Dechlorination State-of-the-Art Field Survey and Pilot Studies," Municipal Environ. Res. Lab., EPA-600/S2-81-169, Cincinnati, OH, Oct. 1981.
2. Anon., "Hazards of Sulfur Dioxide and Caustic Soda, A Committee Report," *J. AWWA,* **31,** 489 (Mar. 1939).
3. Baker, R. J., "Maximum Withdrawal Rates from Chlorine, Sulfur Dioxide and Ammonia Cylinders," *AWWA Op Flow,* p. 4, Apr. 1980.
4. Anon., "Sodium Bisulfite 38% Solution Data Sheet," Dupont Industrial Chemicals Dept., Wilmington, DE, Sept. 1975.
5. Vaughn, J. C., private communication, Southside Filter Plant, Chicago, IL, June 1968.
6. Baker, R. J., "Dechlorination ahead of Demineralizers and Similar Applications," paper presented at 23rd Int. Water Conf., Pittsburgh, PA, Oct. 24, 1962.
7. Hager, D. G., and Flentje, M. E., "Removal of Organic Contaminants by Granular Carbon Filtration," *J. AWWA,* **57,** 1440 (Nov. 1965).
8. Suidan, M. T., Snoeyink, V. L., and Schmitz, R. A., "Reduction of Aqueous HOCl with Activated Carbon," *J. Env. Engr. Div. ASCE,* **103,** 677 (Aug. 1977).

9. Bauer, R. C., and Snoeyink, V. L., "Reactions of Chloramines with Active Carbon," *J. WPCF,* **45,** 2290 (Nov. 1973).
10. Baker, R. J., private communication, Wallace & Tiernan, Div. of Pennwalt Corp., Belleville, NJ, Nov. 1980.
11. Magee, V., "The Application of Granular Activated Carbon for Dechlorination of Water Supplies," *Proc. Soc. Water Tr. and Exam.* (England), **5,** 17 (1956).
12. Miller, R., and Hartman, D. J., "Feasibility Study of Granular Activated Carbon Adsorption and On-site Regeneration," Municipal Envir. Res. Lab., EPA-600/S2-82-087, Cincinnati, OH, Nov. 1982.
13. Strockbine, W., "Superchlorination with Dechlorination by Ferrous Sulphate," *J. AWWA,* **32,** 1176 (1940).
14. Howard, N. J., and Thompson, R. E., "Chlorine Studies and Some Observations on Taste Producing Substances in Water and the Factors Involved in Treatment by the Super- and Dechlorination Method," *J. NEWWA,* **40,** 276 (1926).
15. White, G. C., "Chlorination and Dechlorination: A Scientific and Practical Approach," *J. AWWA,* **60,** 540 (May 1968).
16. Esvelt, L. A., Kaufman, W. J., and Selleck, R. E., "Toxicity Removal from Municipal Wastewater," SERL Report No. 71-7, Sanitary Engineering Research Lab., Univ. of California, Berkeley, Oct. 1971.
17. Brungs, W. A., "Effects of Residual Chlorine on Aquatic Life," *J. WPCF,* **45,** 2180 (Oct. 1973).
18. Anon., "Toxic Gas Monitoring Instrumentation," Interscan Corporation, Chatsworth, CA, 1984.
19. Anon., "TOX-ALARM Sulfur Dioxide Detector Model No. 5183 Instruction Manual," Exidyne Eastern Div., Exton, PA, Feb. 1983.
20. Miks, W. D., personal communication, Water Pollution Control Plant, Palo Alto, CA, 1979.
21. Calmer, J. C., private communication, Kennedy/Jenks Engrs., San Francisco, CA, 1980.
22. Adams, R. M., Calmer, J. C., and Nourse, J. C., "Dechlorination Process Control Using a Biased Dechlorinated Sample," paper presented at 55th Ann. Conf. Calif. Water Poll. Control Assoc., Palm Springs, CA, May 4–6, 1983.
23. Finger, R. E., Harrington, D., and Paxton, L. A., "Development of an On-line Zero Chlorine Residual Measurement and Control System," paper presented at Water Pollution Control Federation 56th Ann. Conf. Atlanta, GA, Oct. 4, 1983.
24. Anon., "Instruction Manual, Sulfur Dioxide Electrode Model No. 95-64," Orion Research Inc., Cambridge, MA, 1973.
25. Huebner, Wayne B., Stannard, James W., and Van Grouw, Albert, "Iodine Bias Analyzer for Monitoring and Control of Chlorine and Sulfur Dioxide Residuals," presented at the 63rd Annual WPCF Conf., Washington, DC, Oct. 1990.
26. Wallace & Tiernan, "DEOX/2000 Dechlorination Analyzer," Technical Data Sheet 50.510, Belleville, NJ, 1991.

11
OPERATION AND MAINTENANCE OF CHLORINATION AND DECHLORINATION EQUIPMENT

OPERATION: CHLORINATION EQUIPMENT

PHYSICAL FACILITIES: INVENTORY

The operator should have available for reference the design engineer's report delineating the design criteria. This information is vital to the overall operation of the system because it specifies the capabilities of the equipment. Of equal importance are the manufacturer's instruction books, one for each specific piece of equipment. It is imperative that operators take the time to familiarize themselves with the content of these books. As the manufacturers always provide supervision of installation, the operators should make it a point to have the manufacturers' representatives clarify any specific features of the equipment not adequately described in the instruction books.

If the chlorination system is part of a treatment plant complex, the following items should be available to the operator:

1. Operation and maintenance manual.
2. Contract plans and specifications.
3. A complete record of each piece of equipment including name of manufacturer, manufacturer's representative (if equipment not purchased directly from the manufacturer), identifying number (model), serial number, rated capacity, dates of purchase and date of installation.

SUPPLY SYSTEM PREPARATION

Introduction

The start-up of a new system or one that has been out of service for a considerable length of time is the most critical period of operation. Various factors contribute to this situation. One common cause is the failure of pipe fitters to clean up the pipe after threading and/or the use of improper thread lubricant. Another more serious situation is the accumulation of moisture in the system. This occurs in those instances where the system stands idle during construction for a long period of time with the supply system open to the atmosphere.

In one instance severe damage to a new system resulted from the accumulation of approximately 3 inches of water in the bottom of the evaporator chlorine container vessel. Therefore, certain precautions must be taken to ensure that the entire supply system is clean, dry, and gastight.

The recommended precautions are usually part of the contract specifications, which require the contractor to clean and dry the chlorine supply system and test it for leaks before any chlorine is allowed in the system. The procedure for cleaning, drying, and testing is outlined in the maintenance section of this chapter. The discussion on operation will assume that all of chlorine supply piping between the containers and the chlorinators and the evaporators are clean, dry, and gastight.

Small Systems

Installations that do not fall into the above category, such as those that involve less than 100 ft of pipe and do not utilize evaporators, are required only to thoroughly clean the pipe threads (before assembly) with trichlorethylene, or other suitable chlorinated solvent. Never use hydrocarbons or alcohols. The residual solvent may react adversely with the chlorine.

Following this, some suitable thread lubricant should be carefully applied to each joint. Teflon tape is satisfactory for pipes up to 1 inch in diameter. Other lubricants can be: John Crane Plastic Lead Seal No. 2; a mixture of linseed oil and graphite; a mixture of linseed oil and white lead; or a mixture of litharge and glycerine. The last must be mixed to a heavy, syruplike consistency from litharge powder and glycerine. As it hardens or "sets up" quite rapidly, it must not be made up too far in advance of its use. Although this mixture does harden in a short time, it should never be relied upon as a joint sealant. There is no substitute for good threads in a chlorine or sulfur dioxide piping system.

Obviously if the supply piping is assembled with welded, forged steel fittings, thread cleaning and lubrication are unnecessary.

START-UP (GAS SYSTEM)

Important Steps

The most critical period in the operation cycle occurs when a system is put into operation for the first time or after a prolonged shutdown. The following steps should be followed:

1. Start the injector water system, and make certain that the hydraulic conditions are satisfactory. This is done by checking the injector supply pressure gauge and the injector vacuum gauge. If the conditions are satisfactory, the vacuum gauge should show a reading above 10 in. Hg.

If the chlorinator is not equipped with a vacuum gauge, remove the tubing at the injector vacuum inlet and place a hand over the opening while the injector water is running. If the injector is performing properly, the suction will be felt instantly on the portion of the hand over the opening. This will be a strong suction requiring some effort to remove the hand. If the suction is feeble, the hydraulic conditions are not proper and should be investigated further.

When the injector system is operating properly, the chlorine gas may be turned on, but before this is done, the chlorine metering orifice should be in a partially open position, approximately 25 percent of maximum chlorine feed rate. To accomplish this, the metering valve control knob should be rotated to approximate this feed rate. If the chlorinator is arranged for some type of automatic operation, it must be disconnected from this mode of action and placed in manual control position.

2. Verify that all of the tubing, manifold, and auxiliary valve connections are correct and that all union joints are properly gasketed.

3. Check to see that all chlorine valves on the supply line to the chlorinator are closed.

4. "Crack" open the chlorine container valve; however, only open the container valve ¼ turn.

5. Then, with a bottle of ammonia, check all the joints between this valve and the next one downstream to make sure that there are no leaks. If there are none, fully open the next valve downstream and proceed as before. Continue until chlorine gas pressure is showing on the chlorinator gauge.

If there are no leaks, then the chlorinator is ready for further testing—that is, for rangeability, automatic control, and so on.

Procedure When a Leak Occurs

If there is a leak, the first step is to close the cylinder valve and relieve the entire supply system of chlorine pressure. **NEVER ATTEMPT TO REPAIR A LEAK BY TIGHTENING A PACKING GLAND, PIPE FITTING, OR ANYTHING ELSE WHEN GAS OR LIQUID PRESSURE EXISTS. THIS LEADS TO DISASTER.** If the cylinder valve is only cracked open, it can be closed quickly, as required for this procedure. Next open all other valves wide; increase the chlorine feed rate to maximum. This will reduce the pressure to zero very quickly. Repair any leaks that result from a poor fit at a gasketed joint. Doing this might require, in addition to a new gasket, refacing the cylinder valve seat with a flat file. When this is accomplished, the system is ready to be started again.

A gross leak, such as that caused by a missing gasket or a hole in a fitting, is another matter. Under these conditions, the operator should close the main cylinder valve and retreat to a safe area where the oxygen breathing apparatus is located, don this equipment, and return to perform the gas evacuation procedure through the chlorinator.

In small installments, where only one flexible connection is involved, the amount of gas that might leak out will be gone a few minutes after the cylinder valve is closed, and so the operator can return to correct the leak without the need of a gas mask in most cases.

Checking the Chlorine Supply System

On many occasions there can be an unforeseen obstruction in the supply system between the chlorine cylinders and the chlorinator. Therefore, after all the leaks have been corrected, verify that the chlorinator will reach its maximum capacity as specified. This is the most important criterion of any chlorinator installation. If it will not perform to capacity, the automatic mode of operation will be unreliable and inaccurate.

If the injector system is providing a stable 15 inches of Hg vacuum and the chlorine feed rate does not reach the maximum reading on the rotameter scale, then the difficulty is most likely in the supply system—such as a restriction due to a buildup of impurities from the chlorine gas. Location of the obstruction can be done as follows: If the obstruction is upstream from the chlorinator inlet pressure-reducing valve, the gas pressure on the chlorinator gauge will fall off rapidly when an attempt is made to reach maximum feed rate. If the obstruction is in the chlorinator inlet pressure-reducing valve, there will be no change in the chlorine pressure gauge at the chlorinator. To verify that the inlet CPRV is restricted, close the chlorine gas inlet valve to the chlorinator. If the gauge pressure does not fall off rapidly, then the obstruction is in the inlet pressure-reducing valve in the chlorinator.

Checking the Vacuum Relief System

It is easy to verify whether the fault is due to a weak spring in the vacuum relief device. While the chlorinator is operating, disconnect the vent line and plug the opening of the vacuum relief device. If the spring is weak, it is allowing air into the chlorinator. When the vent is plugged (with a thumb), the rotameter indicator will immediately jump to a higher position, showing an increase in chlorine feed rate. *Solution:* Replace the spring in the vacuum relief device.

Insufficient Vacuum: Checking the Injector System

If the chlorinator vacuum is not sufficient (less than 10 in. Hg), the difficulty is usually traceable to the injector system, which includes the injector operating water supply, the chlorine vacuum line, the chlorine solution line, and the diffuser system. The difficulty will consist of one or more of the following:

1. Vacuum leak in chlorinator or vacuum line to injector.
2. Insufficient water pressure.
3. Excessive friction loss in chlorine solution line caused by inadequate pipe size.
4. Too much head loss in diffuser due to insufficient holes, or holes too small.
5. Too much static head downstream from the injector.
6. A restriction in the injector throat and/or tailpiece.

A compound gauge with a protector diaphragm installed immediately downstream from the injector discharge will verify items 3, 4, and 5. The manufacturer's injector curves must be consulted to see if the back-pressure on the injector as indicated by the above gauge is within the limits for the existing operating conditions. This information together with a hydraulic analysis will identify the problem.

Injector Hydraulics; Effect of Air and Gas Binding

Another factor that contributes to excessive friction loss in the chlorine solution line is the phenomenon of air and gas binding. This problem is most likely to occur in solution lines longer than 100 ft that terminate in a closed conduit under some pressure.

There are two situations in which air and gas binding will reflect a vacuum of less than 10 inches of Hg on start-up under satisfactory hydraulic conditions:

1. In those installations where the injector is more than 300 ft from the chlorinator, the vacuum line between the chlorinator and the injector will become filled with air during start-up. If the line is long, this large amount of air reduces the injector vacuum, sometimes to less than 5 inches of Hg. The situation gradually corrects itself as chlorine gas displaces the air. The problem is caused by the fact that the air will not go into solution as does chlorine, and this greatly impairs the efficiency of the injector.
2. The same thing occurs in long solution lines. The air taken into the system via the vacuum relief system will reflect a deterioration of the injector vacuum. But when the chlorine is turned on and all the air is purged from the solution lines, the injector vacuum will return to normal.

Another phenomenon results from gas binding in the chlorine solution line. It should be explained, however, that the effect of air or gas binding in the solution line is to increase the back-pressure at the injector, thereby reducing its efficiency. At the injector there is a direct reaction by chlorine and water to produce a release of carbon dioxide in proportion to the alkalinity of the water. Other gases are also released, owing to the reaction between chlorine and the ammonia (NH_3) in the water. This is particularly pronounced in wastewater practice, where the effluent is used for injector water. These gases are forms of nitrogen resulting from the breakpoint reaction by chlorine. They have about the same solubility as carbon dioxide and so are expected to be released under these conditions. Copious quantities of such gases have been observed in various glass metering devices on chlorine solution lines. This phenomenon is characterized by a surging in the chlorine solution line. Elimination of this situation calls for patience on the part of the operator. The injector should be adjusted to give the maximum amount of operating water at the least back-pressure. This will provide maximum flushing capability at maximum injector efficiency. Each trial adjustment should be allowed to proceed for several hours before changing to another adjustment. The predominant symptom of this gas binding condition is a fluctuation on the vacuum gauge from 5 to 15 inches of Hg at a frequency of about once every 2 sec or less. At the same time, the chlorine feed rate may oscillate as much as 10 to 20 percent of scale, or even more, depending on how near to maximum the feed rate is when the fluctuation is greatest. During the experimental adjustment of the injector, try to keep the chlorine feed rate at about 50 percent of scale. The line will flush itself faster at this feed rate.

The best way of avoiding such a situation is to keep the chlorine solution lines as short as possible, locating the injectors remotely and keeping the solution lines at such an elevation as to reduce the injector back-pressure, and to add 15 to 30 percent more injector water pressure for any solution line in excess of 100 ft if the injector is larger than 2 inches. Injector systems utilizing the smaller sizes—that is, 1-inch and 2-inch—do not display the magnitude of surging noted in the 3-inch and 4-inch systems.

Vacuum Leaks

One other sure sign of trouble, either from injector hydraulics or a vacuum leak, is the behavior of the vacuum gauge when the chlorine feed rate is changed from minimum to maximum. If the vacuum at the lower feed rate is 20–22 inches of Hg and it deteriorates significantly (to 13–15 in. Hg) with increasing feed rates, this deterioration signifies trouble in the system and should be investigated.

The chlorinator itself can be a source of trouble if there is a vacuum or an air leak within the unit. If there is a sizable vacuum leak, it can be discovered by shutting off the injector water suddenly and applying ammonia on all the joints. This sudden removal of vacuum within the chlorinator momentarily puts it under a slight positive pressure, and chlorine will be expelled into the atmosphere. Very small vacuum leaks will not show up in this procedure.

The obvious points to check are the various tubing connections, which may have loosened in shipment. Next, the O-rings that seal the metering tube may be defective or may have been damaged during installation. Next, check the vacuum relief device, which sometimes arrives with a defective spring or a scored seat and so can let air into the control unit, thereby impairing the vacuum. This leak can be determined while the chlorinator is in operation, as follows: Put the chlorinator on manual control at about 50 percent feed rate, disconnect the vacuum relief vent line, and place your thumb over the opening. If there is a leak, the rotameter will "jump" to a higher reading.

START-UP (LIQUID SYSTEM)

Evaporator

General Discussion. The procedure for start-up on a system using liquid chlorine is generally similar to that using gas, but one big difference is in the role of the evaporator.

The evaporator is an extension of the chlorine system. Whatever happens in the container is reflected in the evaporator. The danger existing in a liquid system is the possibility of trapping chlorine liquid in a pipeline. If this occurs and there is a significant ambient temperature rise, the liquid chlorine will expand and rupture the pipe if the expansion of the liquid stresses the pipe beyond its ultimate tensile strength. **For this reason, the liquid line between the evaporator and the chlorine supply system should always remain open while the evaporator is operating.**

The first step preparatory to starting up a liquid system is to verify that the system is dry.* This can be done by heating the water in the evaporator and passing dry air (−40°F dew point) through the evaporator cylinder and all the chlorine lines between the containers and the chlorinators. This may take several hours. It is essential to do this because evaporators may sit idle on the job site for many months, accumulating moisture while the construction is being completed. This moisture collects in the bottom of the chlorine cylinder. If the moisture is not removed, the ferric chloride that is formed will pass through the chlorine control mechanism with disastrous results. Whenever this occurs, the entire chlorine supply system must be flushed with water and thoroughly dried with steam and hot, dry air. In addition, the chlorination equipment must be dismantled and cleaned. Care in cleaning all threaded pipe joints, to remove cutting oils and other foreign matter, and the use of Teflon tape instead of other joint lubricating or sealing compounds, will prevent the production of chlorine-induced impurities, which cause severe maintenance problems.

When the operator is convinced that the chlorine supply system is clean and dry, the next step is to start the evaporators. This is done by filling the water bath and adjusting the control devices so that the water level rides between the limits established by the manufacturer.

*See the "Maintenance" section in this chapter for further details on dealing with this problem.

Most evaporators are heated by electric immersion heaters; however, recirculated hot water or live steam is sometimes used to heat the water bath. Never energize the electric immersion heaters unless there is the proper amount of water in the evaporator. Usually this is provided for in the electric controls by an overriding circuit.

When the water bath temperature reaches 150°F, the chlorine pressure-reducing and shut-off valve will open, and the system is ready to operate; however, the operator must wait until the water bath reaches its maximum temperature and then adjusts the controls so that it operates between 170 and 180°F.

Now the system is ready to start. First, start the injector water system, proceeding exactly as in the start-up procedure, using gas; then follow the same procedure to check for chlorine leaks.

Always start a liquid system first with gas. If the system is started on liquid and there is a leak, many times more gas than the liquid used is released at the leak—one volume of liquid chlorine being equivalent to 456.8 volumes of gas. Furthermore, it takes that much longer to empty the system of liquid to repair the leak.

Once the system has been checked with gas for leaks, it is then ready to be switched to liquid and ready for control adjustments and calibration.

Cathodic Protection System. After the water bath is filled to the proper level add ¼ lb (max.) sodium sulfate (Na_2SO_4) to increase the conductivity of the water.[1] The instrument panel ammeter should be observed after the sodium sulfate has been dissolved in the water bath. When a new evaporator is placed into operation for the first time, the ammeter reading may be very low or zero. However, if any movement above zero is attained by turning the current knob clockwise, the cathodic protection system is satisfactory. A higher reading is not attained initially because of the insulating effect of the paint finish on the outside of the liquid chlorine container vessel. As this paint finish deteriorates with time, a higher ammeter reading will be observed.

If the ammeter continues to read zero, check the cathodic system electrical connections. If, after some months' operation, when the ammeter needle has moved into the midrange of scale, the needle decreases substantially, *check the anodes for their condition.* Whenever the ammeter reads in the high range (250 mA or more), turn the current knob counterclockwise to bring the needle near to the midpoint of the scale. A reading of 50–250 mA, or in the green area, is satisfactory—except at initial operation, when a low current flow is both normal and satisfactory. See the "maintenance" section of this chapter for routine inspection of the cathodic protection system.

Superheat. The proper operation of an evaporator depends upon the water bath temperature that will provide 20°F superheat to the gas being vaporized. This condition is dependent upon the vapor pressure of the liquid. For example, read the liquid vapor pressure on the evaporator instrument panel, and determine the corresponding liquid temperature from the chlorine vapor pressure curve. Then read the chlorine gas exiting the evaporator. This gas temperature should be kept 20°F higher than the corresponding liquid temperature obtained from the chlorine vapor pressure curve. When it is no longer possible to achieve 20°F of superheat, the chlorine container vessel in the evaporator must be cleaned, or the immersion heaters must be replaced.

Excessive Chlorine Pressure

Another condition that may be encountered in the start-up or routine operation of a liquid system is that of abnormally high pressure in ton containers or tank cars. In a tank car, high pressure is caused by air padding; but in ton containers, it results from exposure to heat, such as sunlight. Tank cars are insulated. The usual pressure range of ton containers is 85 to 100 psi; tank cars usually arrive with some air padding so as to show 125 psi. Whatever the incoming pressure, the evaporator efficiency will be greatly improved if the tank car or container pressure is reduced to 80 to 85 psi by operating from the gas phase until the pressure drops to this range.

The most significant case is that of the tank car, because they are not only often air-padded but are filled with liquid chlorine at very low temperatures. This requires more work for the evaporator.

Assume that the temperature of the liquid in a tank car is 60°F, and that the temperature of the evaporator water bath is 180°F. Figure 6 in the appendix shows that the vapor pressure of liquid chlorine at 60°F is approximately 70 psi without any air padding, and the thermal difference through the chlorine cylinder wall of the evaporator will be 120°F. With this temperature difference, the evaporator can transfer quantities of BTUs and thereby evaporate more chlorine. However, if the tank car happens to be air-padded to the maximum allowable pressure in accordance with *Chlorine Institute Pamphlet #54,* the chlorine pressure due to air padding can be 190 psi for liquid chlorine at 60°F. The pressure in the evaporator cylinder then becomes 190 psi, instead of 70 psi, owing to the air padding.

Again according to the vapor pressure curve, 190 psi vapor pressure is equivalent to 120°F. Therefore the evaporator "thinks" that the chlorine is at 120°F, so that there is now only a 60°F temperature differential between the water bath and the liquid chlorine, or half as much at the lower pressure; consequently, there will be only half as many Btu's transferred, and the evaporator can evaporate only half as much chlorine. If the evaporator is forced beyond this 50 percent capability, liquid chlorine will flash over into the gas phase and cause damage to the chlorinator equipment.

Important Tank Car Check

Tank cars should be delivered to chlorinator stations without air padding; but if one arrives padded, it is easy to remove the air padding simply by operating on the gas phase until the pressure drops to around 70 psi. Almost all chlorination equipment can operate to maximum capacity with 20 psi chlorine pressure at the chlorinator. This means that padding is required only when the tank car liquid temperature is in the 10°F range, and then only a small amount of padding is required.

In hot ton containers, it is not uncommon for the pressure to be as high as 180 psi (116°F) as a result of sitting in the sun on a hot day. The pressure can be quickly reduced by operating on the gas phase if the rate of withdrawal is somewhat in excess of 500 lb/day per cylinder. In warm climates, the "Arizona Blanket" system is used to keep a ton cylinder cool.*

TO STOP OR SECURE AN INSTALLATION

Four Steps

1. The first step is to shut off the chlorine supply system. If shutdown is to be for a short period, any auxiliary valve near the chlorinator may be used for this purpose. If it is to be for a long time or for major repairs, it is best to shut the system down at the main container valve. In this way all the chlorine in the system can be removed.
2. When the chlorine pressure gauge reaches zero, and if any disassembly of equipment is involved, remove the plastic plug usually located in the chlorine pressure-reducing valve assembly (in the chlorinator) to evacuate all the residual chlorine in the machine while the injector system is still running. If it is a small chlorinator, break the flexible connection at the cylinder. The injector vacuum will pull air through the flexible connection, thereby purging the entire system of chlorine. This also introduces moisture into the system from the atmosphere, but that is inevitable in the maintenance and operation of this equipment.
3. After the chlorine has been purged to the satisfaction of the operator, the injector system may be shut down, thus securing the entire installation.
4. At this point, reconnect either the flexible connection or the plastic plug, and then proceed with repairs as required.

TO SECURE AN EVAPORATOR

If the chlorination system is to be shut down temporarily for repairs or adjustments on any part downstream of the evaporator, it is necessary only to close the evaporator outlet (gas) valve (see Fig. 11-1, valve C).†

---

*By using the "Arizona Blanket" on the ton cylinder cooled by a small but continuous flow of water on the blanket, by a ⅓-inch pipe diffuser with numerous 1/32-inch holes the length of the container.

†There has only been one known case, as of 1997, where valve C was closed, and the power supply switch failed to close as the temperature went past the alarm shut-off point. This allowed the evaporator temperature to exceed some 400°F. This led to a meltdown of the chlorine container vessel in the evaporator, causing an enormous chlorine leak.

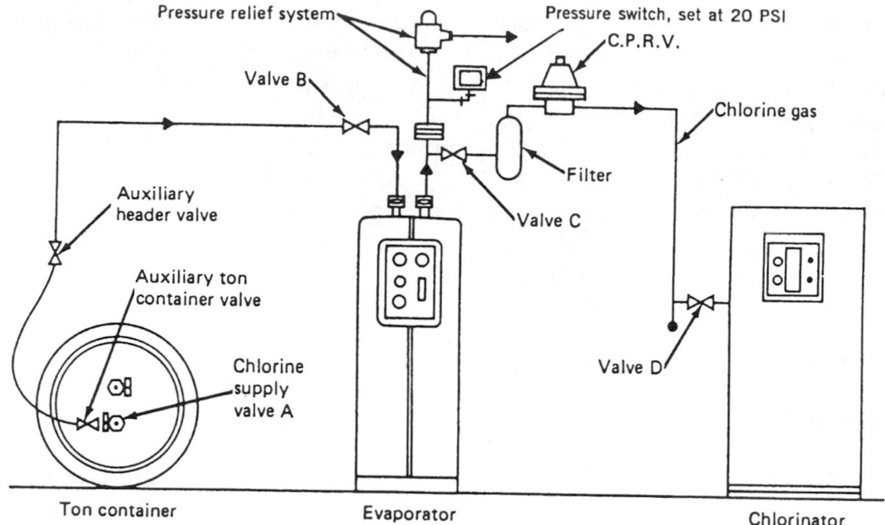

Fig. 11-1. Safest way to install an evaporator.

Always leave the liquid system intact. Think of the evaporator liquid, the liquid in the chlorine header piping, and the liquid in the storage system as one entity. Never separate one from the other by closing a valve.

In Fig. 11-1 valve *A* represents the shut-off of the storage system; valve *B*, the inlet to the evaporator; valve *C*, the outlet of the evaporator; and valve *D*, the inlet supply to the chlorinator. Valves *C* and *D* can be closed at any time and still leave the evaporator operating without any danger. In these instances, there will be a temperature rise in the liquid chlorine (because evaporation has ceased), and this tends to push the liquid back into the containers, thereby equalizing the pressure system.

Whenever it becomes necessary to close valve *B*, do not close valve *C* or *D* until the evaporator has been emptied of chlorine, as indicated on the evaporator pressure gauge.

If it becomes necessary to close valve *A*, allow the system to empty itself from valve *A* until the pressure at the evaporator reads zero before closing any other valve.

If there is any likelihood that valves *A* and *B* might be closed inadvertently without draining the liquid, then an expansion tank should be installed between these two points.

Figure 11-2 shows an installation with a long liquid line between the containers and the evaporators. This line has valves at either end that automatically close upon power failure, and is protected by an expansion tank. In a power failure, the automatic gas valve downstream from the evaporator also closes, trapping liquid in the evaporator—which heats up to the temperature of the water bath in the evaporator, as no gas is being evaporated.

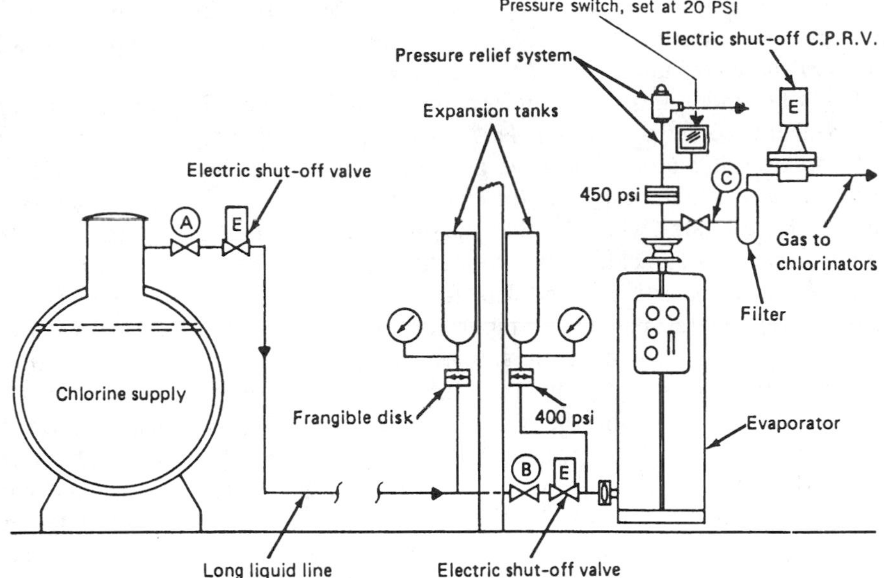

Fig. 11-2. Chlorine supply system with valves that close when power fails.

If the water bath temperature is as high as 180°F, the vapor pressure in the evaporator will rise to approximately 400 psi. This pressure will be exerted on the system between valve *B* and the chlorine pressure-reducing valve. This portion of the system may be able to accommodate this pressure. **Therefore, this portion of the system should be provided with an alarm system.** Connect the incoming liquid chlorine line downstream of the inlet shut-off valve to a frangible disk and expansion tank arrangement so that the disk ruptures at approximately 250 psi. The liquid will flow into the expansion tank connected to an alarm pressure switch set for any significant pressure—for example, 10 to 20 psi. The volume of the expansion tank will be sufficient to lower the pressure of this portion of the system to that of ambient temperature conditions.

CHLORINE SUPPLY SYSTEM

100- and 150-lb Cylinders

Installations using these small cylinders experience the most difficulty from the temperature-related properties of chlorine gas.

Never connect a cylinder that has stood in the hot sun for any length of time without allowing it to cool to the temperature of the chlorinator room. This usually takes overnight. If it is connected while hot, the gas between the cylinder and the chlorinator will cool rapidly at night, causing liquefaction of chlorine at the inlet pressure-reducing valve. Impurities in the chlorine gas precipitate out during this liquefaction process and quickly plug up the small ports in the chlorinator.

If a hot cylinder must be connected immediately to the chlorinator, it should be artificially cooled with water. This is accomplished by wrapping the cylinder with absorbent material, such as a blanket or burlap, and soaking with a continuous small stream of water. As a guideline to operators, particularly those in hot climates, when the chlorine pressure gauge reaches 120 psi, trouble can be expected. The chlorinator room should be sufficiently insulated, or the cylinder should be artificially cooled* to prevent the chlorine cylinder pressure from exceeding 120 psi. This temperature–pressure problem is accentuated even further in areas where the ambient temperature has a wide fluctuation from low to high in a 24-hour period. This situation exists in desert areas, such as the Far West and the Southwest. Insulation against this variation is the proper solution. This problem is also accentuated at those installations where the feed rate is well below the maximum draw-off rate for the chlorine cylinder. The maximum withdrawal rate for a 150-lb cylinder is 40 lb/day at 68°F. Higher rates will produce cooling of the liquid chlorine in the cylinder, thereby reducing the vapor pressure. At a multicylinder installation where the chlorine feed rate is in excess of 40 lb/day, an easy way to cool off a hot 150-lb cylinder is to connect this cylinder by itself to the chlorination equipment and withdraw the gas at a rate well in excess of 40 lb/day. As an example, withdrawing chlorine from a 100- or 150-lb cylinder at 100 lb/day will reduce the temperature and consequently the pressure by at least 20°F in 30 min. This temperature–pressure reduction will stabilize at the ambient temperature when other cylinders are connected so that the withdrawal rate is normal again.

Because of this temperature–pressure property of chlorine, the sun should never be allowed to shine on any chlorine cylinder discharging chlorine in the gas phase. Moreover, there is a potential danger of a serious leak by allowing cylinders to be exposed to direct sunlight. All chlorine cylinders are equipped with fusible plugs that melt at 158–165°F to prevent cylinder rupture in case of fire. Although the instances are rare, there have been cases of fusible plugs melting from the sun's heat.

Ton Containers

When these cylinders are used for gas withdrawal, they behave in the same way as do the 100- and 150-lb cylinders. The maximum rate of gas withdrawal is 400 lb/day at 68°F. These cylinders can be cooled by water as described above or by exceeding the vapor withdrawal rate. When multiple cylinders are used for gas withdrawal, it is customary to install a chlorine pressure-reducing valve in the chlorine header adjacent and close to the cylinders.

*This situation can be solved by using the "Arizona Blanket" to cool the cylinder.

Reducing the chlorine pressure to about 40 psi will eliminate liquefaction in the header and in the chlorinator. Reducing the pressure is like cooling the gas. The liquefaction will occur in the reducing valve; so to minimize fouling of this valve by chlorine impurities that drop out during liquefaction, this valve should always be preceded by a filter.

Ton containers have two outlet valves and six fusible plugs. One outlet valve is for gas withdrawal, and one for liquid withdrawal.

It is imperative to position these cylinders so that the two outlet valves line up in a vertical plane. In this position the top valve is for gas withdrawal, and the bottom valve is for liquid withdrawal. If these two outlet valves are not aligned vertically, serious difficulties will be encountered. A gas-phase withdrawal system may withdraw the liquid chlorine, which will severely damage the chlorination equipment. A liquid-withdrawal system will withdraw gas; this will immediately be reflected in a severe loss of chlorine supply.

In manifolding ton containers that are being used for liquid withdrawal, certain precautions must be taken. Never manifold a hot ton cylinder with others that are cooler. Cool the hot cylinder by the same procedure described for cooling the 100- and 150-lb cylinders. If this practice is not followed, the possibility arises of filling the partially empty, cool ton cylinders by one or more of the new hot cylinders. Filling a cylinder to liquid capacity could result in the eventual hydrostatic rupture of the filled cylinder(s). Such a situation would obviously cause a disastrous chlorine accident.* Therefore, always be sure that the new cylinder being connected is at the same temperature as the cylinders already connected before turning them into the system. This will be revealed by the vapor pressure of the cylinders. Operators are cautioned to connect the new cylinder(s) in such a way that they can verify and compare the vapor pressure between the new cylinders and those already in use. If there is a significant difference in these pressures, the new cylinders will have to be allowed to reach the temperature of the cylinders on the line by remaining in the storage area overnight before being connected. Otherwise, the new cylinders will have to be artificially cooled as described for 100- and 150-lb cylinders.

The best way to operate a ton container supply system is to use a group of cylinders until they are empty. When this occurs, secure the empty cylinders and activate a new group of full cylinders until empty and so on. It is best never to turn a new cylinder into a supply system of partly full cylinders. The maximum liquid withdrawal rate from a ton cylinder is 12,000 lb/day.

Tank Cars and Storage Tanks

The most important operating consideration for this type of supply system is proper instruction. A qualified representative from the chlorine supplier

*This is why the Chlorine Institute recommends against manifolding ton cylinders. However, such practice is common throughout the world.

should be engaged to demonstrate the use of safety equipment and to outline safety precautions and preventive maintenance procedures. The most critical time for these large systems is the original start-up, or start-up after a prolonged shutdown. Always start these systems on the gas phase first. A vapor leak is less lethal than a liquid leak. A tank car or storage tank installation capable of delivering either vapor or liquid should have two separate chlorine headers all the way to the evaporators, one for gas and one for liquid.

When starting liquid withdrawal, an operator must be careful to open the tank car (or storage tank) liquid valve very slowly and allow the piping system to fill up completely. Otherwise if the valve is opened quickly, the sudden flow of liquid chlorine will jam shut the emergency ball check valve that is in the liquid withdrawal tubes inside the tank dome assembly.

Chlorine Leak Detector

Unit Description. Most leak detectors for chlorine are of the amperometric type. However, there are others that respond to discoloration of impregnated strips or compounds, and some use a gas diffusion electrode.

The amperometric type utilizes dual platinum electrodes for a sensor. The sensor is continuously wetted with an electrolyte solution (high conductivity) that is in constant and continuous contact with the air sample of choice. The air sample is carried to the detector with either a built-in fan or a remote blower unit, depending upon the local conditions. Chlorine in the air sample is detected in seconds by the sensor. These detectors can detect 1 ppm or 3 ppm chlorine by volume in the air and can be altered in the field to alarm at 0.5 ppm or 3 ppm.

The detector has local indication by a red alarm light. Additional relay contacts are provided for external alarms and other equipment such as ventilating fans and emergency chlorine shut-off valves.

The electrolyte is a glycerine-based potassium iodide solution diluted with distilled water. The wetting action of this solution keeps the platinum electrodes of the sensor clean. When the air in contact with the sensor contains chlorine, it reacts with the potassium iodide by releasing free iodine, initiating an electric current across the platinum electrodes. Through an electronic circuit board, the detector activates the alarms.

The electrolyte tank is protected at all times from chlorine contamination by an activated carbon air filter. This tank is also equipped with a level indicator.

Preparation for Initial Operation.[5] Remove the sensor that contains the platinum electrodes. Be careful not to touch or disturb the electrodes. Body oil from the fingers or hands can insulate the electrodes. Clean the sensor with detergent (Tide) using a soft brush; then rinse the sensor with distilled water. Check to see that the glass orifice is clean and dry, and avoid air bubbles in the sensor system because they will block the flow of the electrolyte to the sensor. Reinstall the sensor.

Connect the drain tubing to provide for disposal of spent electrolyte into a suitable drain or container. The size of the electrolyte reservoir is usually about 5–6 quarts.

Now fill the reservoir with the electrolyte furnished with the unit. Follow the directions on the container. Do not leave any air space in the reservoir. Air space results in a faster drip rate, which increases unnecessary consumption of electrolyte.

Then check the electolyte flow to the sensor. The sensor and its electrodes must be completely wetted by the electrolyte. If not, use a cotton swab to spread the electrolyte around the sensor. Spraying the sensor with electrolyte or distilled water by means of a plastic wash bottle will also help wet the sensor.

The initial drip rate should be one drop every 5 sec to 1 min. After a few hours, depending upon the amount of air space in the reservoir, this rate should decrease to one drop every 2–5 min. The drip rate will vary with temperature and barometric changes.

The next step is to test the response of the detector to chlorine. First turn on the power. Next mix 3 or 4 drops of pH 4 buffer solution with a similar amount of household bleach (Clorox) in a small beaker. This mixture will generate a small amount of chlorine gas. Hold the beaker one inch below the suction inlet pipe. The detector should respond in seconds. If not, then consult the manufacturer's troubleshooting guide.

After this initial alarm—or any future alarm—flush the sensor with distilled water using the plastic wash bottle supplied with the detector.

Leak detectors should be checked on a routine basis at least once a week.

CHLORINE CONTROL AND METERING SYSTEM

Injector System

The control and metering system will not function without an adequate injector system. The injector system provides the power to pull gas from the containers through the chlorinator and then dissolves this gas into the injector water supply to provide a chlorine solution at the point of application. All injector systems should be operated with sufficient water to provide a solution concentration not to exceed 3500 mg/L of chlorine. This usually figures to be 40 gal/day/lb of chlorine.

The injector system should be evaluated as follows. Turn on the water supply, and allow the hydraulics of the system to stabilize. With the chlorine supply secured, the injector system should be indicating a vacuum in excess of 20 inches of Hg. Otherwise, something is wrong. Assuming the vacuum valve is operating properly, open the chlorine supply to the chlorinator and adjust the chlorine feed rate to about 25 percent of full scale. If the system is adequate, the admission of the chlorine to the injector may only lower the vacuum to about 15 inches of Hg or higher. As long as this vacuum does not deteriorate rapidly, it can be assumed that the injector system is in order. Let the system remain at this condition momentarily; then change the chlorine feed rate to 100 percent of full scale. If the injector system is adequate, the chlorine feed rate response will be instantaneous, and the injector vacuum gauge will not deteriorate much below 15 inches of Hg. In large installations where 100 percent of full scale is 2000 lb/day or more, and where the injector back pressure is greater than 3–5 psi, the injector vacuum might deteriorate to 10 inches of Hg or less upon feed-rate change from 25 to 100 percent of full scale on the rotameter feed-rate indicator. Chlorinators as a general rule will not operate properly at injector vacuum values much less than 10 inches of Hg. This figure should be checked with the manufacturer.

Booster Pumps

Pumps used to supply chlorinator injector operating water can be of either a centrifugal or a turbine type.

The centrifugal pump is used at installations where the total dynamic head is comparatively low (110–140 ft) and the volume is high (60–220 gpm). Centrifugal pumps are the pumps of choice for use at treatment plants, both potable water and wastewater. They are also used in instances where more than one injector is being operated from a single pump. The centrifugal pump has longer wear characteristics, requires less maintenance, and has no operating adjustments.

The turbine-type pump is an entirely different matter. It is used at installations where the chlorinator utilizes a 1-inch fixed-throat injector. The maximum capacity of these chlorinators is 500 lb/day, and the injector requirements vary from 1 to 25 gpm. The turbine pump is capable of operating these chlorinators over a wide range of injector back-pressures up to about 100 psi. These situations occur on a water transmission main or on the discharge of a well or a well field. A turbine pump is rarely ever used at a wastewater plant for chlorinator operation. By design the turbine is a positive-displacement type of pump, and therefore it must be equipped with an adjustable bypass assembly for regulating the discharge pressure.

This type of pump is selected to deliver almost twice the required amount of water passing through the injector, so that when it is new, the pump will bypass from the discharge line to the suction line through an adjustable needle valve. As the pump wears, the operator must throttle down on the needle valve in order to maintain proper pressure at the injector.

Upon start-up the needle valve is adjusted to give a discharge pressure that will just barely pull the maximum capacity of the chlorinator. This pressure is noted by the operator, who should then throttle the needle valve enough to raise the discharge pressure another 5 psi. This will provide an operating safety factor. As the pump wears with time, the operator will find it necessary to throttle the needle valve in order to maintain chlorinator performance.

Chlorinators

If the chlorinator feed-rate adjustment is manual, there are no critical adjustments to be made. The unit is adjusted manually to give the desired residual. No further adjustment or calibration is necessary. It is important, however, to choose a rotameter of the proper and most useful range. For example, if the desired feed rate is 15 lb/day, a 30 lb/day rotameter would be a wiser selection than a 100 lb/day size.

If the chlorinator is to be paced by a flow signal, the rotameter must be selected to fit the range of the flow meter. For example, suppose the meter range is 0–30 mgd and the desired dosage is 16 lb chlorine/mg; then the maximum chlorine feed rate will be 480 lb/day; so use a 500 lg/day rotameter.

When the chlorine feed rate is paced by a flow signal, the chlorinator must be adjusted for zero and span. These adjustments are carefully outlined in the manufacturer's instruction book. This adjustment is usually a combination of signal input and mechanical linkage, depending on the type of signal—pneumatic or electric.

The important point in attempting adjustments to the chlorinator control mechanism is that these adjustments can only be made properly when there is a flow of chlorine through the machine. Furthermore, a chlorinator is not designed for zero flow conditions while the injector is in operation. Thus, the zero adjustment is contingent upon a mechanical linkage. As long as the injector is operating and even though the chlorine metering orifice is adjusted to read zero on the rotameter scale, some chlorine will be passing through the chlorinator. This is easily verified. Shut off the inlet gas valve to the chlorinator, and watch the chlorine gas gauge pressure on the chlorinator instrument panel gradually creep toward zero.

The most important aid to chlorinator adjustment and calibration is the use of signal simulators to simulate input signals from flow and dosage measurement devices to the chlorinator. The signal simulators used should have the capability of providing input signals over the entire range of the chlorinator.

Special adjustments are required for chlorinators operating from a variable vacuum signal as the sole means of controlling the chlorine feed rate. This mode is commonly used for straight residual control. The special adjusting procedure is outlined in the manufacturer's instruction manual and has to do with the zero adjustment of the chlorine inlet reducing valve. The accuracy of this adjustment, as well as the overall accuracy of this method of control, is greatly improved by the use of an external pressure-reducing valve installed in the supply system upstream from the chlorinator inlet. This valve takes the burden off the pressure-reducing valve in the chlorinator. Without that additional valve, the chlorinator inlet pressure reducing valve would be the primary instrument of control in a variable vacuum system.

Chlorine Residual Analyzer

General Discussion. This unit has zero, span, and temperature adjustment. If a thermistor is used, it automatically compensates for temperature changes in the sample.

The zero adjustment is a simple one. The sample flow to the analyzer is shut off, or one lead to the cell is disconnected (depending on the manufacturer). Under these conditions the indicator or pen on the recorder should go to zero. If not, the potentiometer marked zero is rotated until the pen reads zero. The span adjustment depends upon the calibration.

The amperometric analyzer is generally calibrated by the use of an amperometric titrator. Other titrimeter methods can also be used such as starch–iodide or DPD–FAS. These methods are outlined in the 19th Edition of *Standard Methods.* In wastewater applications the back-titration method using the amperiometric titrator is preferred. The DPD–FAS and starch–iodide titrations are equally acceptable in wastewater practice. Colorimetric methods are not reliable for wastewater calibrations.

Advanced wastewater treatment plants that nitrify obtain free residual chlorine together with combined. In these cases it is desirable to differentiate the free from the combined; therefore, the forward-titration procedure must be used.

Whenever extensive calibration of the analyzer is required or desired, it is desirable to have a signal simulator available. This instrument provides checkpoints throughout the range of the analyzer cell output.

There is a significant difference between the operation of an analyzer for potable water treatment and the operation of one for wastewater.

Potable Water. There are many installations interested only in free chlorine. Measuring free chlorine can be done with or without chemicals. If the pH is constant, as it is in some instances, no buffer solution is required.* No KI is required because free chlorine is measured directly by the analyzer. In most of these cases there is not enough combined chlorine to cause any error in the free reading. If, however, the pH varies (which causes a variation in cell output), then pH 4 buffer solution or carbon dioxide must be added to the sample to stabilize the pH at 4.

Water systems that add ammonia to produce chloramines or those that treat water with a natural ammonia content are concerned with measuring total chlorine residual. In these instances it is imperative that the sample pH

*Constant pH is a rarity; moreover, maximum cell current occurs at pH 4, so it is prudent to add pH 4 buffer.

be lowered to 4 by buffer solution, and that potassium iodide be added to the buffer solution in sufficient quantity to convert the total residual to free iodine. The amount of KI to be added to the buffer solution depends upon the sample flow through the analyzer and the magnitude of the residual. The cell flow rates vary from one manufacturer to another (i.e., the Capital Controls cell flow is 100 ml/min, the Fischer and Porter cell flow is 150 ml/min, and the Wallace & Tiernan cell flow is 360 ml/min). The Wallace & Tiernan cell requires 15 g of KI/L of buffer per mg/L total chlorine residual.

Wastewater. All wastewater analyzers have (or should have) a sample inlet-line disk-type strainer (Cuno). The drain valve on this strainer should always be wide open for continuous flushing. The sample pump or sample supply should be able to deliver about 12 gpm sample at 25 psi at the inlet to the filter. Under these conditions there will be about 9–10 gpm discharging to waste through the filter, leaving sufficient capacity to supply the cell with an adequate continuous sample.

Chlorine residual analyzers for wastewater applications need many more calibration checks than those measuring residuals in potable waters. Furthermore, wastewaters generally have a significantly higher alkalinity than potable water; so pH control of the sample through the cell is critical. Similarly, the amount of KI added is also critical because of higher residuals required in wastewaters.

The sample pH must be kept at 4.5–5.0, preferably 4.5. Multicolored pH papers are satisfactory for this determination. Sufficient potassium iodide must be added to the pH 4 buffer solution to allow the chemical reaction between the KI and the chloramine residual to go to completion. The chlorine residual reacts with the potassium iodide to release free iodine in proportion to the total residual. The analyzer cell measures the free iodine. A rule of thumb for KI addition is 15 g/L of buffer solution per 150 ml/min cell flow of undiluted sample per mg/L of total chlorine residual. The Wallace & Tiernan wastewater analyzer is arranged for freshwater dilution of the sample. Dilution is usually on a 1:1 ratio.

The operator is advised to consult the manufacturer's instruction book to make sure of the proper calibration and operating procedures with respect to the buffer solution, KI addition, and so on. This is extremely important in the operation of a wastewater analyzer.

An easy way for the operator to determine whether or not sufficient KI is being added is to add about 1 ml of KI solution (from the amperometric titrator test solutions) to the sample head box or diversion box and observe the chlorine residual indicator on the analyzer. If it does not move, the KI addition is sufficient. If it deflects up scale, the KI addition is insufficient.

Owing to the importance of the addition of a buffer solution and KI, it is obvious that the operation of the reagent pumps is critical. Therefore, the operator must check the sample pH daily.

Once the analyzer has been put into operation, it is not necessary to respan the instrument each day. It is preferable to plot a graph of the indicated versus the titrated sample readings. The curves of each of these groups of readings should indicate a span adjustment either upward or downward. This procedure should be done at first on a weekly basis, inasmuch as most wastewater systems require respanning of the analyzer once a week.

Daily inspection of the sample system to the cell is necessary to assure a continuous sample. The noble metal electrode in the cell may be either gold or platinum, depending upon the manufacturer. This should be kept free of grease deposits by wiping it with a soft cloth. However, the sacrificial copper electrode should never be disturbed.

White has found that some wastewater residual control systems are operating without the benefit of KI addition to the buffer solution. This is proof that the amperometric cell is responsive to the combined chlorine residual. This procedure is considered unreliable because the cell response to chloramine is not linear, as it is with free iodine or free chlorine. However, these systems not using KI show remarkably consistent control, somewhat superior to that of some systems using KI for measuring total chlorine residual. This phenomenon requires further investigation.

Zero adjustment of analyzers used to monitor the dechlorinated sample will always expose a calibration problem that is best ignored. These analyzers are used primarily to monitor and control the sulfonation process. However, they are often arranged to monitor the dechlorinated effluent for 5 or 10 min. every hour. When the dechlorinated sample reaches the cell, the analyzer indicator or pen will drop below zero in spite of the fact that zero and span have already been adjusted for a sample containing chlorine residual. This phenomenon is probably due to the presence of excess SO_2 being applied. It does not signify a negative residual. It does, however, demonstrate that when a dechlorinated sample passes through the cell, it changes the molecular surface of the electrodes so that they are slightly different (electrochemically) from when the zero calibration check is made with a sample containing a chlorine residual. This phenomenon does not necessarily affect the span adjustment of the analyzer.

Amperometric Titrator

Operating personnel should practice the titration procedure initially two or three times a day, to establish confidence in the reliability of their results. The proper degree of confidence can come only after several titrations. Daily titrations after that are in order.

Before the start of any titrations (or when they are standing idle), it is helpful to keep the electrodes sensitized. This is done by immersing them in a sample jar containing tap water with about a 5 mg/L free chlorine residual. If the titrator is being used in the back-titration mode for wastewater, then the tap water should contain small concentrations of free iodine. This precaution provides continuous sensitivity of the electrodes, which results in greater repeatability and a higher degree of reliability.

In recent years there has been a demand of sorts to be able to read chlorine residuals in receiving waters as low as 0.005 mg/L. It has been shown that the sensitivity of the Wallace & Tiernan titrator can be increased from 0.1 mg/L to 0.001 mg/L by the following procedure:

1. Remove the ammeter from the titrator, and substitute a Rochester converter, which will deliver a milliamp output signal.
2. Carefully encase the sample jar in aluminum foil, and ground it.
3. Dilute the phenylarsene oxide four times. This modification can measure 0.001 mg/L chlorine residual.

Chlorine Flow Transmitters

The only satisfactory way to measure the flow of chlorine gas flowing through a chlorinator is by a differential transmitter. To accomplish this, an orifice is placed in the chlorinator gas discharge line. The orifice should be sized so that the vacuum differential will be 12 inches of water at maximum Cl_2 flow. The orifice is specific for a given size of rotameter. The transmitter consists of a spring-loaded diaphragm assembly, which must be mounted on top of the chlorinator in order to minimize access of moist chlorine gas to the diaphragm assembly. Since the relationship of the chlorine flow to the vacuum differential is a square root function, the diaphragm movement is converted to a linear function by a characterized spring. This makes the diaphragm movement directly proportional to gas flow. The diaphragm controls the armature in a differential transformer, which puts out an AC signal. This signal is converted to a 4–20 mA DC signal in a separate unit identified as a demodulator. Snap-in resistors are available to change the output signal to fit various categories of receivers.

It is first necessary to verify the differential across the orifice at 100 percent of chlorine flow. This can be done by connecting a water manometer across the tubing leads to the transmitter. This requires disconnecting the leads to the transmitter. The manometer should read between 10 and 14 inches of water differential. If not, check the size of the flow-measuring orifice with the manufacturer's instructions.

Assuming that the orifice is the correct size, adjust the zero setting of the demodulator to produce a milliampere current equal to the recorder (receiver) zero signal. This most likely will be 4 mA. Turn on the chlorinator, and operate it at 100 percent flow. Then adjust the coarse span and the fine span in the demodulator to obtain a milliampere signal equal to the receiver–recorder 100 percent signal. This will probably be 20 mA. Recheck zero and span, and adjust as required.

When not working on the electronic equipment inside, keep the demodulator cover closed at all times to prevent moisture and chlorine fumes from entering the enclosure.

The chlorine flow transmitting system is now ready to operate.

Alarms

Various types of alarm units may or may not require adjustment.

The chlorine high- and/or low-pressure supply alarms require only simple mechanical adjustments.

The high- and low-vacuum alarm switches on the chlorinator are usually preadjusted at the factory and need no further adjustment.

Other alarms maybe high- and low-limit switches on the chlorine orifice positioner that will require field adjustment. These alarms may also be installed in the chlorine flow recorders. Adjustments are made at the factory, but can be changed in the field.

Safety Equipment. See "Maintenance: Chlorination Equipment," later in this chapter.

OPERATION: DECHLORINATION EQUIPMENT

PHYSICAL FACILITIES

The operator should have available for reference the same material suggested for the chlorination equipment.

SUPPLY SYSTEM PREPARATION

The same precautions must be taken for the sulfur dioxide system as those described earlier in this chapter for the chlorination equipment.

The procedure for cleaning and drying the piping and the evaporators is also the same as it is for chlorine. (The procedure for cleaning and drying will be found in the "Maintenance" section of this chapter)

SULFUR DIOXIDE SUPPLY SYSTEM

General Discussion

The vapor characteristics of sulfur dioxide make it prone to reliquefaction—an undesirable situation when one is attempting to meter a gas. Unless the container is artificially padded with a propellant, the cylinder pressure at room temperature will be about 35 psi. (See Fig. 6 in the appendix for the sulfur dioxide vapor pressure curve.) If the cylinders are stored outdoors in a carport-type structure, the SO_2 gas will surely reliquefy in the header between the container and the sulfonator. The proper way to relieve this situation is by the application of heat on the cylinders. The best way to apply heat on ton containers is by using a SASCO Teflon-fabric electric blanket, by the Morgan Allen Co. of Houston, Texas (FAX 713-561-0002). Then, immediately downstream from the container header valve a pressure-reducing valve should be installed. An alternative method is to install an evaporator and artificially raise the vapor pressure of the cylinders by an air-pad system or a bottle(s) of nitrogen. The design considerations needed to provide an operable system have been described previously.

Sulfur dioxide is available in ton containers, tanker trucks, and tank cars. Tanker trucks are common throughout the United States. Suppliers using tanker trucks are usually able to furnish a storage tank for the user.

Ton Containers

If the gas phase of these containers is to be used, then heat must be applied to the containers. Sulfur dioxide gas-phase systems do not operate satisfactorily at ambient temperatures much below 85°F. They operate best at 100°F with an external reducing valve in the SO_2 header adjacent to the cylinders. The installation of an evaporator for these systems is the most practical solution.

Although manufacturers of sulfur dioxide may claim that gas withdrawal rates from ton containers can be as high as 500 lb/day, the operator is best served if the withdrawal rate is kept to 175–200 lb/day. The reliquefaction problem is kept to a minimum at this rate.

Stationary Storage Tanks

Owing to the fact that sulfur dioxide tankers are available and that the pressure of SO_2 vapor is so low (35 psi), the use of storage tanks is popular. The maximum delivery in road tankers is 20 tons. It takes about 2–3 hours to unload such a tanker. After the storage tank has received the tanker load, it should be air-padded to about 60 psi. This is a convenient operating pressure. At this figure, the gas leaving the evaporator should be reduced to a sulfonator inlet pressure of about 25 psi. The sulfur dioxide storage tank should be designed so that the suction line on the tanker's compressor can be connected to the gas phase of the storage tank. This speeds up the unloading operation by utilizing the SO_2 tank vapor pressure.

Safety Equipment

The safety equipment for sulfur dioxide operations is the same as that for chlorine.

SYSTEM START-UP

The operator is referred to the two "Start-up" sections under "Operation: Chlorination Equipment" in this chapter. The same guidelines apply.

Superheat

In the use of an SO_2 evaporator, which is a highly desirable piece of ancillary equipment, the same amount of superheat is required as with chlorine to assure satisfactory operation of this unit. Therefore, the operator should review the discussion of this topic in the "Evaporator" section under the heading "Chlorination Equipment: Operation," earlier in this chapter.

Sulfur Dioxide Supply Pressure

As compared to chlorine, the higher the SO_2 vapor pressure, the better it is for the operation of the facility. There is almost no likelihood that "high" SO_2 vapor pressure would ever be an operating problem. If such a situation were to occur, the same procedures as outlined for chlorine would be applicable.

TO STOP OR SECURE AN INSTALLATION

The operator should follow the same procedure as described under the heading "Chlorination Equipment: Operation." Identical guidelines apply for sulfur dioxide.

SULFUR DIOXIDE CONTROL AND METERING SYSTEM

Injector System

The only difference between the use of an injector system for sulfur dioxide and one for chlorine is that SO_2 gas is much more soluble in water than is chlorine. Therefore, if the operation of a sulfur dioxide injector system is fashioned in the same way as that for chlorine, the operator will be well on the side of safety. *The use of 3500 mg/L as the maximum strength of sulfurous* acid solution in the injector discharge will prevent breakout of molecular SO_2 at the point of application.

Booster Pumps

The same comments apply here as for the chlorination system.

Sulfonators

The same operating procedures are required for sulfonators as previously described for chlorinators. This includes adjustments and calibration. However, because a great many sulfonator control systems are based on feed-forward control, a slightly different set of operational conditions faces the operator. This difference concerns the function of the ratio station installed on the flow pacing signal, which is ahead of the flow-residual signal multiplier (see Chapter 9).

The ratio station is a precise dosage control instrument particularly valuable in "feed forward systems." Figure 11-3 provides the operator with dosage values for a given sulfonator orifice meter. The one shown is a 4000 lb/day meter used with a 200 mgd (maximum) plant flow meter. When the ratio station is set at unity (1.0), the SO_2 dosage will be 2.4 mg/L. At a ratio setting of 2.0, the dosage will be doubled, and so on.

EXAMPLE: Operating with a 1.0 ratio station setting, a 4000 lb/day SO_2 meter controlled from a 0–200 mgd flow meter, the SO_2 dosage would be

$$\frac{4000 \text{ lb/day}}{200 \text{ mgd} \times 8.34 \text{ lb/gm}} = 2.4 \text{ mg/L}$$

for any flow from 0 to 200 mgd.

Figure 11-4 provides the operator with the necessary information to set the system to dechlorinate a given chlorine residual to zero for any plant flow rate. The set of curves shown in Fig. 11-4 is for a 4000 lb/day SO_2 rotameter. The solid line represents a sulfur dioxide-to-chlorine ratio of 1:1 (by wt). The lower dotted line represents the stoichiometric ratio of 0.9:1. The upper dotted line is the probable upper limit of SO_2 to Cl_2 of 1.1:1.0.

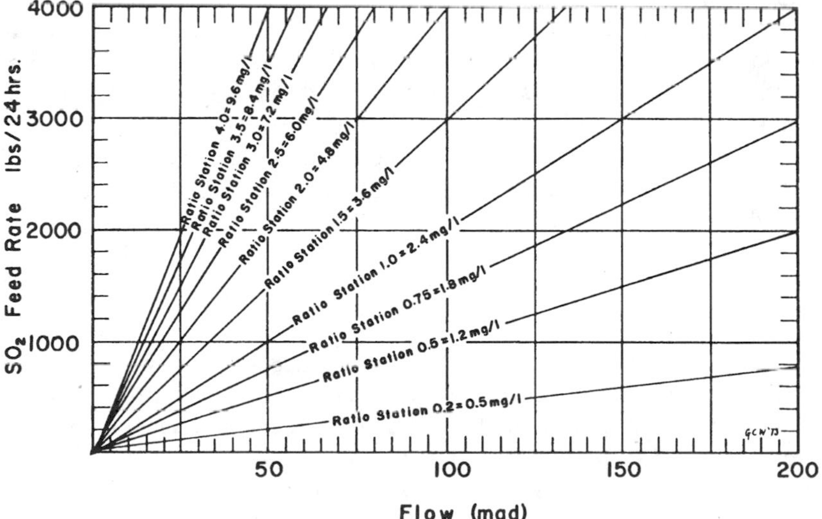

Fig. 11-3. Sulfur dioxide feed rate versus effluent flow for various ratio station settings.

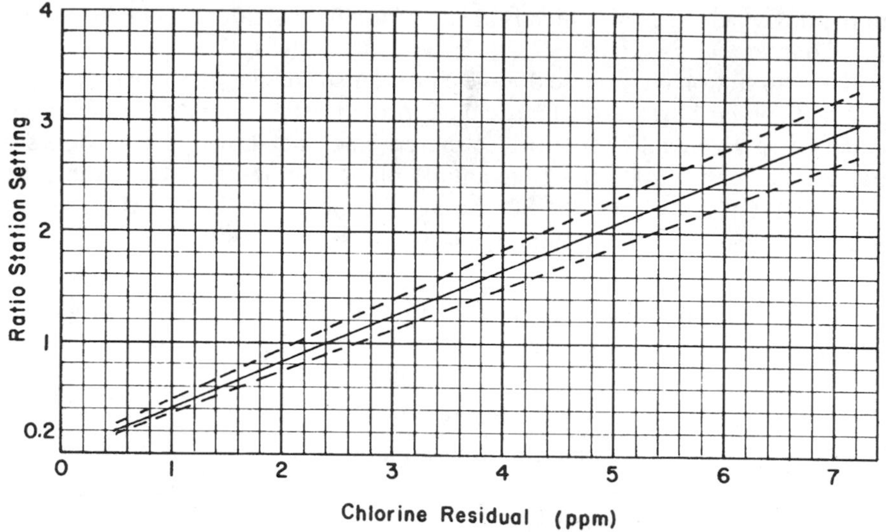

Fig. 11-4. Ratio station settings to achieve zero chlorine residual using a 4000 lb/day SO_2 rotameter. SO_2 to Cl_2 ratio: solid line 1:1, upper line 1.1:1, lower line 0.9:1.

Example: If the chlorine residual analyzer is showing 4 mg/L residual, the *proper setting* of the ratio station to feed enough SO_2 to produce a zero residual in the dechlorinated effluent will be about 1.65, using the solid line as a reference. If the selected ratio station setting to produce a zero residual falls above or below the dotted lines of Fig. 11-4 for a 4 mg/L chlorine residual (i.e., is greater than 1.8 or less than 1.5), *something is wrong*. Either the analyzer has drifted out of calibration, or the analytical technique of measuring the chlorine residual is in error. Therefore, the laboratory personnel should make further analyses of both chlorinated and dechlorinated effluent samples to verify the calibration of the analyzer. Figure 11-4 gives ratio station setting values for chlorine residuals regardless of plant flow between 0 and 200 mgd. The same approach would be used regardless of the effluent plant flow meter range.

Analyzers

Since about 1994, when Stranco Corp. introduced its patented Model 900 High Resolution Redox Control System for the chlor–dechlor process in the field of wastewater treatment, the role of chlorine residual analyzers has become limited, as ORP (redox) has proved highly successful owing to its performance at miniscule chlorine residuals with a minimum of maintenance. The only analyzer system that comes close to the Stranco system is the Wallace & Tiernan Deox-2000 unit. This unit uses an iodine bias system that limits the chlorine residual and prevents over-dechlorination. (See Chapters 8 and 9.)

MAINTENANCE: CHLORINATION EQUIPMENT

CHLORINE SUPPLY SYSTEM

General Discussion

The primary objective of maintenance of any chlorination facility is the prevention of leaks, primarily those occurring under pressure. Therefore, any joint in the supply system between the chlorine supply and the chlorine pressure-reducing valve in the chlorinator is a potential source of leaks. Beyond the inlet chlorine pressure-reducing valve, the chlorine gas is under a vacuum. The chlorine pressure gauge connection inside the chlorinator is also a source of leaks. The exception to this occurs with remote vacuum systems where the pressure in the supply system is converted to a vacuum adjacent to the chlorine cylinders. (See Chapter 9.) This eliminates chlorine under cylinder pressure in the piping between the cylinders and the chlorinators, or between the evaporators and the chlorinators. It does not eliminate the potential of leaks from flexible connections, header valves, cylinder valves, gasket failures, or fusible plugs on containers. These are the major sources of leaks. Serious leaks in the chlorine header piping proper are rare.

Chlorine gas leaks are insidious in nature. The usual method of detection (26° Baumé ammonia—not household grade) will not detect the very small leaks, the ones that cannot be detected by odor. These leaks can proceed unnoticed for weeks before reaching a critical stage. American Gas and Chemical Co. markets products for detecting small leaks of chlorine and ammonia. Its literature[2] describes a product (CDP-100) that is sprayed on critical joints. It reacts chemically with imperceptible chlorine leaks, causing a visible color change from white to yellow at the leak point. The product is easily removed with a damp cloth. The sensitivity of this detector is claimed to be 0.00001 Std. cc/sec.

Locating Leaks

The operator is advised to look for two different signs of a chlorine leak at a gas joint under pressure: discoloration and moisture.

Discoloration at Joint. Most chlorine gas header accessories are cadmium-plated over copper tubing and bronze or brass fittings. If a leak is in progress (even one undetectable by ammonia or by odor), the cadmium plating will disappear, and the base metal will take on a reddish color, signifying dezincified brass or corroded copper. Some green copper chloride scum will also appear around the edges of the corroded metal.

Moisture Formation. Small droplets of liquid may appear on the underside of a chlorine liquid or gas joint. This indicates the most insidious of all leaks, as it may go undetected on one-inch and ¾-inch steel header systems for a long period. Usually these lines are painted with a good grade of durable paint so that discoloration and products of corrosion may not be readily visible. It is best to paint these lines bright yellow. Then if a minute leak occurs, the first evidence will be a moist spot of brown rust, which is highly visible.

While these small leaks are in progress, corrosion is taking place at the threads of the joint and proceeds until a massive leak suddenly occurs. A continuous method of monitoring these leaks is by means of a chlorine gas leak detector, capable of detecting leaks as small as 0.5 ppm chlorine concentration in the air. These leak detectors monitor the air in either the chlorine container room or the chlorine equipment room, or in both, by drawing in a continuous sample of air.

Maintenance Steps to Prevent Leaks:

1. Each time a gasketed chlorine joint is broken, replace the gasket with a new one.
2. On a threaded pipe joint, wire-brush the threads and use Teflon tape for thread lubricant.
3. Replace all chlorine gas header valve packings at least once a year. Try to obtain Teflon-type packing where recommended. Failure of valve packing is one of the most common causes of chlorine valve leaks.

Other Sources of Leaks. Potential sources of gas and liquid leaks are the diaphragm protectors used on chlorine gas gauges and pressure switches. The silver diaphragms in these protectors fail in time owing to metal fatigue. The lives of these diaphragms vary widely, depending upon pressures carried and frequency of cycling, from seven to ten years.

When the silver diaphragm fails, the glycerine or fluorocarbon on the top side of the diaphragm diffuses back into the chlorine header system. The chlorine reacts with the glycerine to produce a large, gooey mass that resembles scrambled eggs and is difficult to remove. If this goes undetected, the chlorine will quickly corrode the bourdon tube in the gauge (or pressure switch), thereby causing a massive leak. These diaphragm failures are most likely to go undetected in large systems on the liquid chlorine phase, because the liquid chlorine carries away the small amount of glycerine into the evaporator, where the result may not be observed for some time.

In view of the overall consequences, the preventive maintenance program should consider the replacement of all gauges and pressure switches with diaphragm protectors once every five years.

Repairing Leaks

The most important advice to an operator is this: Never attempt to repair a leak on the chlorine supply system when it is under the chlorine container

pressure. Tightening packing glands, union joints, or flexible connections, etc., while the system is under pressure can only lead to disaster.

If the outlet valve of a container, tank car, or storage tank is leaking, the following steps should be taken in order: (1) Call the chlorine supplier to send expert help. (2) Don air or oxygen breathing apparatus (not a canister-type mask). (3) Ready the container emergency kit for action. (4) First try gingerly to tighten the valve packing; if this does not stop the leak, apply the emergency container kit and await the arrival of the chlorine packager representative.

Leaks at threaded joints of a flexible connection or at the union connection of an outlet valve require a thorough cleaning with a steel wire brush. These joints must be regasketed. Small and apparently insignificant leaks that can later result in devastating leaks often occur at the chlorine cylinder outlet valve. To prevent this situation, the operator should always carry a one-inch flat file to be able to reface any chlorine cylinder outlet valve. This allows the proper seating surface between the cylinder valve and the auxiliary cylinder valve. This maintenance procedure has proved to be well worth the effort.

Never reuse a gasket that has been in a joint after the joint has been disassembled.

The question of lead gaskets versus asbestos or paper composition gaskets continues to be controversial. The paper- or asbestos-type gaskets tend to cover up gross deficiencies in chlorine cylinder outlet valve maintenance. This gasket is favored by the chlorine cylinder packagers. On the other hand, once a lead gasket has passed the leak test of a joint, it will never result in a chlorine leak at that particular joint, provided that it is not disturbed by disassembly and reassembly. In contrast, a paper composition gasket can develop a serious leak by failure during operation. This type of gasket should be avoided because a gasket failure during operation can result in a much more serious leak than one by a lead gasket. Furthermore, the failure of a lead gasket has never proved to be the cause of a chlorine leak once that system was connected without leaks. A lead gasket joint will leak from the very beginning if it is not properly seated. This is the reason a lead gasket joint should be regasketed each time it is disassembled.

The use of yoke-type connections between flexible connections and cylinder valves or header valves minimizes the problem of leaks at these joints. Lead gaskets should be used at these joints. Tank car flexible connections use the two-bolt ammonia-type union with a replaceable lead gasket for attachment to the car outlet valve and the supply header. These gaskets must also be replaced after each disconnect.

Flexible Connections

The most vulnerable parts of a chlorine supply system are the flexible connections. These are usually made of 2000-lb annealed copper tubing, cadmium-plated. Whenever these connections are exposed to the moisture in the atmosphere during a cylinder change, corrosion sets in. Internal corrosion of these flexible connectors is a function of the number of times the tubing is exposed to the atmosphere. A positive way to determine the reliability of a flexible connection is to bend it slightly. If it "screeches," discard it immediately. This phenomenon signifies that internal corrosion is excessive, and the tubing is liable to rupture prematurely by crystallization. Flexible connections for 100-lb, 150-lb, and 1-ton cylinders should be discarded on a regular basis. For the smaller cylinders, discard after 10 cylinder changes, and for the ton containers once a year.

Flexible copper connections on tank cars have heavier walls than those on smaller containers, are subject to much less flexure, and can withstand more usage—up to 30 tank cars before replacement. These connectors are gradually being replaced by flexible metal hose connections that show an even longer safe life than the copper connectors. Whenever a tank car flexible connection is disconnected and not in use, it should be plugged at both ends with cork or wooden plugs and then stored in a warm place. This will practically eliminate internal corrosion. All unions on chlorine header systems should be the ammonia type (two-bolt) with a recessed gasketed joint. It is desirable to replace the gasket joints every five years, even if the joint has not been taken apart. **Ground joint unions should never be used.**

Safety Equipment

Safety equipment refers to breathing apparatus, chlorine container emergency kits, leak detectors, and automatic ventilating systems.

Chlorine container emergency kits for each size of container should be ready for instant use. Practice in using these devices should be a part of the maintenance schedule. The same applies to the emergency breathing apparatus. Special safety drills at prescribed intervals should be made available for all operating personnel on both the use and the handling of container kits and breathing apparatus. Demonstration of the chlorine leak detector response to chlorine should be a part of this drill. This will also familiarize operating personnel with the chlorine leak alarm system.

LEAK DETECTOR: AMPEROMETRIC TYPE

As there are several makes of leak detectors on the market, the operators should simply read the manufacturers' instruction literature. However, the following are some reminders to provide reliability of operation of the detector.

Weekly Checklist. Make the following checks:

1. Verify that there is electrolyte in the reservoir.
2. Check the drip rate from the sensor (2–10 min. per drop).
3. Check the detector response to chlorine (use test kit).

4. If the sensor is not clean and completely wet, flush it with distilled water.

Monthly Checklist. Make the following checks:

1. Verify the circuitry response to the test pushbutton.
2. Be sure the drop rate is not less than one drop every 10 min. or more than one drop every minute.
3. See that the sensor is clean and covered with electrolyte (no dry spots).
4. Check the reservoir level in the sight tube to see that it is not empty. Only fill the reservoir when required to. When filling is not required, electrolyte consumption will increase.
5. Check the drain container, and dispose of spent electrolyte if necessary.
6. Verify that the fan is operational and that the blade turns frequently.
7. Check for leaks around the grommet, O-rings, sight tube joints, and so on.
8. Check detector response to chlorine.

Annual Checklist. Check the following items:

1. Gasket behind the circuit board: Replace if ripped, cracked, or badly worn.
2. O-ring around the vent cap unit: Replace.
3. O-ring around the glass orifice: Replace.
4. Grommet (sensor mounting): Replace if leaking or badly worn.
5. Sight tube: Replace if yellowing or cracked.
6. Fan wiring, for loose or broken connections.
7. Drain tube connection, to see that it is not loose, cracked, or leaking.
8. Sensor, for loose or broken connections.
9. Mounting bolts: Replace if they are rusted or in any way corroded.
10. Carbon filter: Replace if carbon is wet.

CLEANING AND TESTING CHEMICAL PIPING

General Discussion

The chlorine supply system (including evaporator if used) should be cleaned after the passage of each 250 tons of chlorine, or before the start-up of a new system. The difficulty usually encountered with new systems occurs when new piping and equipment remains idle and open to the atmosphere for a significant length of time. Moisture collects in these systems during this construction period, and if they are not cleaned and dried before chlorine or sulfur dioxide is turned into the system, the resulting corrosion and corrosion products that are distributed throughout the facility can erupt into an unholy mess.

There are two methods for cleaning steel pipe used for chlorine or sulfur dioxide service. One is a chemical detergent method applied to an assembled system on-site. The other is the pickling system, where the pickling agent is either chlorine or sulfur dioxide, depending upon the service.

In either case, valves and gauges should be removed and/or isolated from the system. Reserve tanks can be cleaned along with the piping, but it is believed to be better practice to clean these two items separately.

Chemical Detergent Method

1. Degrease with caustic, trisodiumphosphate, sodium metasilicate, and a surfactant. Circulate this mixture of chemicals through the piping for 2 hours at 180–190°F.
2. Flush the system until the pH is neutral.
3. Descale with inhibited hydrochloric acid at 150–160°F for 2–4 hours.
4. Flush the system with citric flush.
5. Passivate the system with a passivating agent.
6. Blow it dry with nitrogen.

Pickling Method

Depending upon the condition of the system, it may be necessary to swab or flush the system to remove all vestiges of oil or grease deposited in the threading and coupling procedure during field fabrication. If this is necessary, use trichlorethylene or some other chlorinated solvent. If the piping is assembled with welded forged steel fittings, flush with a strong caustic solution. When solvent is used, it should be followed by a thorough flushing with cold water.

After the flushing procedure, the system should be completely assembled and pressurized with air to 300 psi and soap-tested, and all leaks should be repaired. During the pressure test, all equipment such as pressure gauges, pressure-reducing valves, pressure switches, rupture disks, chlorinators, and sulfonators should be isolated to prevent any possible overpressure damage.

A hydrostatic test may be substituted for air if it is more convenient.

The system should then be pickled in place by charging it with chlorine or sulfur dioxide (as the case may be) to normal operating pressure, and then checked for leaks using 26° Baumé ammonia solution. The pipe must be evacuated to repair any of the leaks. This is imperative if welding is required. After the leaks have been corrected, the system is again recharged with either chlorine or sulfur dioxide to operating pressure, and held for a 24-hour pickling period. At the end of the pickling period the gas is evacuated, and the system is flushed with fresh water at a minimum velocity of 2½ ft/sec for the time required to pass a quantity equal to its being fully flushed or until the water is clean. The system is then drained and flushed with steam, followed by purging with dry air at 140°F. This flushing cycle shall then continue until the air existing in the system is at zero relative humidity at 150 psi. Zero relative humidity occurs when the dew point is −40°F.

During the air flushing cycle it is sometimes desirable that all parts of the system be tapped sufficiently to remove any remaining bits of scale, rust, or other debris. The pickling process removes all of the mill scale and oxides of iron, which have a great tendency to cause maintenance problems in the entire system.

After the air drying cycle is completed, the equipment, valves, and other ancillary equipment removed or isolated are reassembled into the system. The system is then subjected to a 300 psi test with either dry air or nitrogen.

Both of the procedures for cleaning as described above should be supervised by a technician who has had extensive knowledge in these systems. It is this technician who will decide which pieces of the system have to be isolated, and which ones have to be removed.

TROUBLESHOOTING THE SUPPLY SYSTEM

The operator should become acquainted with the length of time it takes the chlorine supply system to empty itself at a given chlorine feed rate. With the chlorinator at a certain optimum feed rate, close the chlorine cylinder or header valves, and note how long it takes the pressure gauge on the chlorinator to reach zero. Any appreciable or noticeable sluggishness in the supply system to reach zero pressure condition signifies restrictions caused by impurities somewhere in the system between the cylinders and the chlorinator. This can occur in a single 150-lb cylinder hookup when there is stoppage, either in the auxiliary cylinder valve or in the 4-ft flexible connection between the cylinder and the chlorinator.

Sometimes the header system or flexible connections become so clogged that they cannot be emptied through the chlorinator. These cases require extreme caution. The operator must close off the outlet valve on each cylinder in operation, thereby securing the chlorine supply. Then the operator should don proper breathing apparatus and break one of the joints in the header system upstream of the stoppage point. It is best to break the joint nearest the auxiliary cylinder valve; this will allow the trapped gas in the header system to be released to the room ventilation system. The operator should then leave the area for enough time to allow the ventilation system to remove the gas.

This procedure places full dependence on the tight closing of the chlorine cylinder outlet valves. If one or more of the cylinder valves happen to be leaking, the header system should be reconnected as quickly as possible and immediate arrangements made to bypass the header system. (This is why large systems should have duplicate header systems.) It may not be possible to bypass the header system safely. In such a case, steps must be taken to utilize the appropriate Chlorine Institute Emergency Kit to seal off the leaking cylinders. Separate kits are available for 150-lb cylinders, ton containers, and tank cars. The tank car kit fits all sizes of tank cars.

When the cylinders have been secured and removed, then the header system can be "unplugged," as described above. The importance of keeping the header system in an operative condition at all times cannot be overemphasized.

Another situation that happens occasionally is a source of great dismay to operators, as well as being destructive to the chlorination equipment. This occurs in a gas system using ton cylinders when the feed tubes in the ton container develop a crack below the liquid level. These tubes are shown in Fig. 9-8. The crack usually develops where the tube is threaded into the cylinder outlet valve, and allows liquid instead of gas to enter the system. This is evidenced by icing or extreme cooling and the collection of moisture on the header system, and possible severe icing on the rotameter tube in the chlorinator. The liquid chlorine is trying to vaporize and, in so doing, is using up heat, thus causing the local cooling effect.

Liquid chlorine will quickly damage the plastic parts of the chlorinator. Upon recognizing the difficulty, the operator should immediately secure the cylinder and allow the header system to drain itself of liquid chlorine through the chlorinator; then the operator should reduce the chlorine feed rate to allow more vaporization to protect the chlorinator equipment. Next the operator should disconnect and rotate the defective cylinder 180 degrees. This will put the other feed tube in position. It is rare that both tubes will be defective in the same cylinder.

CHLORINE STORAGE TANKS

The most important maintenance here is on the valves in the dome of the tank. There should be on hand at all times a complete set of spare dome valves, including the safety valve. The safety valve should be replaced every two years, but the two liquid and two gas valves in the dome should be serviced annually and sent to a company specializing in valve repair. After cleaning and repair, the valves should be hydrostatically tested for 500 psi.

It is recommended that the storage tank be hydrostatically tested every three years. To simplify this procedure, it is desirable to have two extra 3- or 4-inch flanged connections at either end of the tank and on the crown. One of these is for filling the tank with water; the other is for the overflow. If special flange connections are not available, the safety valve can be removed and used as the water inlet. The two gas valves should be manifolded together for the overflow.

The procedure is to fill the tank with a fire hose as quickly as possible and to keep adding water until the overflow water becomes clear. When this is accomplished, either remove all the water with air padding or pump it out. Then go into the tank to inspect it, and remove any remaining scale or debris by scraping. Never enter a storage tank that has not been properly flushed out. Otherwise, remaining chlorine will combine with body perspiration to form strong hydrochloric acid, which can cause painful burning of the skin.

The hydrostatic test is made after the tank has been washed, cleaned, and inspected. It is then filled a second time for this test.

Prior to being filled with water, the tank must first be evacuated of all gas by having a vacuum pulled on it with the eductor especially provided for this purpose. After the tank is allowed to stay in a vacuum condition for three or four hours, it is ready to be filled with water. At this point, there is still plenty of residue in the tank that, if allowed to remain, would gas off for a long time.

EVAPORATORS

Liquid Chlorine Vessel

This piece of equipment is subject to filling with a gooey mass of sludge that accumulates from inherent impurities in liquid chlorine. The amount of sludge that accumulates in the buttom of the liquid chlorine vessel is primarily a function of the amount of chlorine passing through the evaporator. In general, evaporators should be inspected for sludge once a year or after passing 250 tons of chlorine, whichever comes first. In addition to this, the operator should keep a close watch on the evaporator superheat. If this cannot be maintained at 20°F, it signifies that the evaporator is losing capacity because of sludge buildup in the liquid container, or that there is an immersion heater failure.

Immersion Heaters

These heating elements are the heart of an evaporator. Scale deposits from hardness in the water-bath water can and do build up on these elements. These deposits can become so severe that annual replacement may be necessary. This phenomenon should be watched carefully during the first year of operation—by routine physical inspection every 90 days.

Cathodic Protection System*

The anodes should be inspected annually. Under normal conditions the anodes will be found nearly expended and should be replaced.

Before new anodes are installed, water should be drained from the system. Then the water-bath tank and chlorine vessel should be inspected for corrosion. If the cathodic protection system has been operated in the recommended range, little or no corrosion will be observed, and long service may be expected.

After the new anodes are in place, fill the water-bath with fresh water and add ¼ lb sodium sulfate to provide proper conductivity for the water.

*The MWD of Southern California has found that Beatty's Mixture is a fine substitute for cathodic protection. See Chapter 9 discussion of evaporators.

Cleaning the Liquid Chlorine Vessel

Cleaning the evaporator liquid container consists of dismantling and removing the chlorine vessel and flushing it with cold water until the flushing water is clean. The inside should be visually inspected for pitting. If the pitting is severe, the vessel should be replaced.

After all the flushing water has been removed, reassemble the evaporator, fill the water bath, and heat to 180°F. Then attach the aspirator so that a vacuum can be exerted on the inside of the vessel.[6] The vacuum should be about 25 inches of Hg and should continue for 24 hours with the water-bath at 180°F to remove all moisture from the inside of the chlorine vessel.

CHLORINE GAS FILTER

This unit should be inspected every six months. The condition of the filter element will give a clue to the condition of the header piping system and evaporator. The filter elements should be replaced at each inspection, and the sediment trap washed in cold water and dried before reassembly. Most chlorine impurities consist of ferric chloride, which dissolves readily in water. The complex chlorinated hydrocarbons and other gunk will not dissolve in water. If these accumulations appear in the filter or the external chlorine pressure-reducing valve, they can usually be removed by either trichloroethylene or isopropyl alcohol.

CHLORINE PRESSURE-REDUCING VALVE: EXTERNAL

This unit consists of a spring-loaded stem, actuated by a three-layer silver diaphragm assembly. The stem throttles the chlorine gas going through the seat assembly. This is the point of pressure drop, and this is where deposits of the chlorine impurities are most likely to occur. The filter placed upstream from the CPRV should remove most of these impurities. It is inevitable that some will deposit on the stem and seat assembly of the CPRV. Deposits that cannot be removed with a soft cloth can be removed with isopropyl alcohol, trichlorethylene, or Chlorothene-Nu. The latter is an inhibited 1,1,1-trichloroethane.

The spring opposing the silver diaphragm action will suffer fatigue and should be replaced every two or three years. Every five years the entire assembly should be dismantled and the diaphragm inspected for fatigue. The diaphragm as well as the lead gaskets on either side of it should be replaced at this time.

AUTOMATIC SWITCHOVER SYSTEM

Pressure Type

This system utilizes two chlorine pressure-reducing valves and is described in Chapter 9. For reliable operation these valves should be preceded by a gas filter to minimize valve maintenance. The area of vulnerability is the stem and seat assembly. These parts should be inspected and cleaned at least once every three months. Another area of vulnerability is the spring that is used to oppose the spring-loaded diaphragm. This spring operates entirely in a chlorine gas atmosphere and therefore suffers a form of corrosion fatigue. It should be replaced every two years. The spring that loads the diaphragm operates in ambient air and does not suffer from corrosion from normal operation. A ruptured diaphragm or a substantial gas leak in the vicinity of the valve will damage this spring.

Vacuum Type

This type of system utilizes two vacuum regulator units that contain a series of spring-loaded stem and seat assemblies. These units are designed to reduce cylinder pressure (85–100 psi) to about 60 inches of H_2O vacuum. This is a much greater pressure drop than for the pressure unit described above. The effect of this greater pressure drop is to precipitate more impurities on the seat and needle assembly. Therefore it is imperative to have a chlorine trap and filter assembly located upstream from the switchover unit. This assembly should be inspected on a regular basis. At the same time the seat and stem assemblies should be examined for deposits and cleaned according to manufacturer's instructions. All of the springs should be replaced every two years. They are subject to metal corrosion fatigue because they operate in a chlorine vapor atmopshere.

REMOTE VACUUM SYSTEM

This system is designed to reduce the chlorine or SO_2 pressure in the system, at the cylinder, to a vacuum as described above. The components include a vacuum regulator check unit and a standby pressure relief device. This system should also be protected by a trap and filter unit. It requires the same attention as the automatic vacuum switchover system.

CHLORINATORS

Modern chlorinators consist of a series of spring-loaded diaphragm units that form the basis for control of the chlorine gas through the installation, with certain vacuum values at various points. Therefore, it is essential that all the joints be vacuum-tight for proper operation.

Most of the maintenance problems occur from metal fatigue of the springs in each of the diaphragm assemblies and from improper stem and seat closure in these diaphragms caused by impurities in the chlorine gas. All springs should be replaced every two years; the stem and seat units should be inspected and cleaned annually.

The rotameter tube and float assembly should be removed and cleaned periodically at least every six months. The chlorine metering orifice should be dismantled for inspection after six months' operation because impurities deposited here will give a clue to the condition of the rest of the system.

Since the injector system is a vital part of the chlorinator, it should be inspected occasionally. By using a water pressure gauge on the water supply with an injector vacuum gauge on the chlorinator and a chlorine solution back-pressure gauge on the injector discharge, the difficulty with the injector system can be readily determined. If there is sufficient water pressure but the injector is not producing sufficient vacuum, only a couple of things can be wrong: the inlet to the injector throat may be partially plugged by ordinary debris or a tiny bit of gravel; or the discharge line to the point of application may have some stoppage, causing high back-pressure. These conditions are readily determined by inspection.

One other common condition unknown to novice operators is the deposition of iron and manganese on the injector throat. This problem is most prevalent on chlorinators with a fixed throat injector. Manganese deposits a slick black coating, which reduces the friction loss through the throat to such an extent that it prevents the formation of a vacuum at the throat, thereby making the chlorinator inoperative. Iron deposits will eventually reduce the size of the throat until it becomes so small that it will not pass sufficient water to produce a proper vacuum. This phenomenon is a result of the accumulation of sufficient amounts of iron and manganese in the injector water supply that become oxidized to their insoluble states at the point where the chlorine concentration is the highest—at the injector throat. Iron has a tendency also to slightly delayed reaction, and will sometimes leave deposits in the injector tailway, causing a restriction having the same net effect.

This situation is easily remedied by proper maintenance. The operator should keep on hand a spare injector throat and tailway and make an exchange at regular intervals consistent with proper operation. The throat and tailway are quickly cleaned by submerging the entire unit in muriatic acid. After a few minutes' soaking, remove it, and rinse it in clean water; it is then ready for reuse.

Injectors are equipped with a device to prevent back flow of water into the chlorinator. These are usually spring-loaded diaphragm assemblies. The springs operate in a moist chlorine atmosphere, and so they should be replaced routinely every two years.

BOOSTER PUMPS

Chlorinator booster pumps are as vital to the installation as is the power supply, the chlorine supply, or the chlorinator itself.

Two types of pumps are employed: turbine and centrifugal. The turbine is the more common on installations of 400 lb/day or less because the requirements are for high pressure and low volume. These units are furnished with a built-in adjustable bypass assembly that allows for wear on the pump impeller, which is critical for a turbine pump. The pump is usually designed to deliver twice the required amount of injector water.

When the pump is new, half of the pump discharge is bypassed to the suction. Adjusting the needle valve in the bypass will assure this function. As the pump wears, the discharge pressure falls off; then the operator must restore the proper pressure by readjusting the needle valve. When the needle valve is operating in an almost fully closed position, it is time to disassemble the pump and replace the impeller.

If sand is present, a turbine pump will wear out quickly. The sand must be removed, or the pump suction must be relocated to a point in the distribution system where sand is not present.

Centrifugal pumps are maintained in the usual way. However, they differ from turbine pumps in that they have no adjustable bypass assembly and no pressure-relief valve. Sand is not so critical in the operation of this type of pump. The best choice in a centrifugal pump is one that has an outboard bearing. The Unibuilt close-coupled type of pump will require a good bit of maintenance, as the inboard bearing at the motor is subject to frequent failure.

CONTROL DEVICES

Electric

Most control devices have a combination of plug-in transistorized components that give long and reliable service. However, the operator should be able to obtain replacements for any of these components in the event of failure. Devices with moving parts, such as potentiometers or safety devices dependent upon mercury switches, are subject to wear, and will require eventual replacement. The operator should have spares for these items also. Reversible motors used in these control devices, such as electric chlorine orifice positioners and motorized vacuum valves, are also subject to failure, even though they give long and reliable service. The operator should check the source and the availability of these motors. The manufacturers of these special-duty motors may run into production problems, resulting in delivery schedule delays. Most items that are components for electric control and alarm devices are off-the-shelf items and are readily available.

Pneumatic

Air control devices are subject more to maintenance problems caused by dirty or wet control air than they are to the wearing of moving parts. However, the moving parts, such as bellofram and other types of diaphragm assemblies, are subject to failure due to wear, and spares should be carried. Flapper valves and seat assemblies can be cleaned and reseated many times before replacement is required. The various springs required in a pneumatic system have a long life—as much as five or six years—because they operate in a normal atmospheric environment.

CHLORINE RESIDUAL ANALYZER

General Discussion

There are two principal categories of analyzers: potable water and wastewater. Both types are available to measure either free or total residual chlorine. The only analyzer available to measure free chlorine in wastewater (*in the presence of chloramine*) is the Bailey-Fischer and Porter unit with an impressed voltage applied across the electrodes.

The typical amperometric continuous analyzer consists of a cell assembly that accepts the sample from the sample piping and/or pumping system, plus electronic circuitry that interprets and records (or indicates) the intelligence from the cell. Wastewater analyzers differ from potable water analyzers in the way the sample is delivered to the cell, because wastewater analyzers are equipped with a special filter on the sample line.

Sample Line

The sample line to the analyzer is probably the most neglected component of a chlorine residual measuring system. It is subject to biofouling whether by potable water or by wastewater. This exerts a chlorine demand that gives a false reading at the analyzer. Usually the sample flow velocity is far below what would be considered a scouring velocity, and in wastewater systems, where grease and other organics are involved, the problem can be acute. Therefore, it is good practice to flush the sample line with caustic or chlorine on a routine basis. The frequency of flushing will have to be determined by experience and trial and error.

Some operators at wastewater plants have installed a point of chlorination into the dechlorinated sample line, which is chlorinated heavily for 10 min./day. This is probably the easiest way to keep this line from fouling-up the dechlor process.

Amperometric Cell

This unit consists of a two-electrode system, one platinum or gold and the other copper. The copper electrode loses metal ions in accordance with the amount of chlorine residual it is measuring, and so will have to be replaced in time.

The cell is equipped with some type of electrode bombardment system, which keeps the electrodes clean and in molecular balance. A common type is grit, which has to be replaced periodically in the cell.

If the cell is operating on potable water, little maintenance is required, depending upon how clean the water is. Some water, particularly raw or untreated, will require the installation of a manually operated filter on the sample line.

If the cell is operating on wastewater, it is imperative that the sample line be equipped with a motorized filter. This operation requires more maintenance on the sampling system because of the presence of suspended solids, fibrous materials, and grease, which are present in varying amounts, depending upon the type of wastewater treatment process.

A continuous and adequate flow of sample to the cell is the most important single factor in this type of operation. The maintenance and inspection program will vary widely from plant to plant.

Electrodes

The electrodes should be cleaned by flushing with fresh water. Do not disturb the sacrificial electrode. A typical amperometric analyzer consists of a copper sacrificial electrode and a noble metal (gold or platinum) permanent electrode.

The copper electrode needs to be replaced according to the amount of chlorine residual passing through the cell. Replacement periods might vary from six months to three years. A wastewater analyzer measuring residuals on the order of 5 mg/L will use up one copper electrode about every nine months. It is permissible to clean the platinum electrode with a dampened soft cloth and a mild abrasive.

If the output signal is erratic, and the cell refuses to remain in calibration, then both electrodes should be cleaned with the recommended abrasives.

It is imperative that maintenance personnel become familiar with the manufacturer's instructions before cleaning these electrodes, and for placing the analyzer back into operation.

Recorder or Indicator

This unit consists of plug-in transistorized components that are subject to the same maintenance problems as other electric control devices. The most common failures are in the potentiometers for the various adjustments, the plug-in amplifier, the chart motor, and the relay mechanism on the control circuits.

Inspections

The following is a suggested routine for daily, weekly, and quarterly maintenance and inspection for a wastewater analyzer:

Daily:

1. Check the sample flow as follows:
   a. Make certain that the pressure on the inlet line of the sample is adequate—about 25 psi just upstream of the motorized filter.

   b. See that the drain on the motorized filter is in the wide-open position, and that the filter is continuously flushing.
   c. Check the electrode bombardment system to see that the grit or other devices are in proper motion (swirling around the platinum electrode).
2. Check the buffer pumps. Count the number of drops per minute to assure the proper pH of the sample entering the cell.
3. Check to see that the sample flow through the cell is adequate.
4. With pH papers, check the pH of the sample as it leaves the cell. It should be between 4.5 and 5.0—never > 5.0.
5. Check the supply in buffer reservoirs.

Weekly:

1. Shut off the control signal (if any) to the chlorination equipment.
2. Flush the entire sample line up to and including the motorized filter with high-pressure plant water (50 to 75 psi).
3. Remove the platinum or gold electrode assembly, and wash it thoroughly.
4. Add grit to the cell upon restarting.
5. Recalibrate the cell (span adjustment and zero adjustment), using a back-titration method with an amperometric titrator. Be certain that the buffer pumps are in the off position during zero adjustment.
6. Do not give the reagent additive system any special maintenance unless the pumping system fails to perform properly.

Quarterly:

1. Clean the entire cell assembly with a soft, wet rag, to remove copper oxide from the copper electrode. This will result in erratic operation for about 12 hours. After this time, molecular equilibrium is reached, and system stability returns. This requires a second calibration with a cleaned copper electrode within 24 hours after starting.
2. Disassemble the outer case of the motorized filter strainer assembly; flush and clean it.

The filter is so important on a wastewater system that it is desirable to have available a complete unit with motor drive for replacement or cannibalization.

AMPEROMETRIC TITRATOR

The best maintenance of this unit is frequent use—at least once a day. When not in use, the electrodes should be allowed to soak in 1 to 2 mg/L free residual chlorine water. If the titrator is being used for wastewater analysis or for total chlorine residual only, then the electrodes should be allowed to soak in a 2 to 5 mg/L free iodine solution. This keeps the electrodes sensitized to either free chlorine or free iodine, as these are the only elements measured by the titrator—just the free halogens.

The platinum electrode or electrodes should be periodically cleaned with a mild abrasive household cleanser, applied with the fingers and then removed by washing in cold water. All electrode contact surfaces should be inspected and cleaned regularly.

If these steps are followed, the titrator will show sharp responses with clear-cut end points and repeatable results.

ELECTRICAL SWITCHGEAR

All electrical switchgear is normally housed in a room away from the chlorination equipment and chlorine supply. Such pieces of switchgear as the control and alarm devices should be in weatherproof enclosures so that any massive chlorine leak will not adversely affect the copper contact surfaces and exposed terminal strips. All switchgear manufactured by chlorinator manufacturers are protected by enclosures, so ancillary items manufactured by others should be similarly enclosed. In humid climates it is wise to place desiccators in the various types of switchgear housing if condensation is significant.

Good housekeeping is the best watchword throughout the chlorination facility for proper maintenance. The operator must not tolerate chlorine leaks or water leaks of any magnitude, and must not allow any equipment to become dirty either internally or externally. Any sign of corrosion should be immediately investigated. If these guidelines are followed, the equipment will perform properly and be free from the hazards of chlorine gas.

MAINTENANCE: DECHLORINATION EQUIPMENT

SULFUR DIOXIDE SUPPLY SYSTEM

General Discussion

The characteristics of sulfur dioxide are so similar to those of chlorine that almost everything said about maintenance for chlorine is applicable to the sulfur dioxide system. However, there are some exceptions.

The life of the diaphragm protectors on the vapor and liquid pressure gauges in an SO_2 system is much longer than for chlorine, because of the low vapor pressure of SO_2 (35 psi).

The corrosivity due to the entrance of moisture is almost the same as that for chlorine, but this action is different on different metals. For example, in a chlorine system, line valves and header valves use monel trim, but these are not interchangeable for sulfur dioxide. Valves for the latter should have 316 SS trim.

Moisture Problems

Sulfur dioxide producers claim that the moisture content does not exceed 40 ppm.[3] However, White[4] has found that this can escalate to more than 100 ppm when a packager transfers the SO_2 from the producer's tank car into the packager's tanker truck. However, certain packagers have a much better and less corrosive product than others who may be somewhat careless in their transfer of an otherwise acceptable product from a tank car to their tanker trucks. Experience in the San Francisco Bay Area indicates that some of the supplies of sulfur dioxide are more corrosive than chlorine by a factor of 10 to 1. This suggests lack of moisture control during the SO_2 transfer procedure. Attempts to quantify and monitor the moisture content by the user proved fruitless. However, a preventive maintenance procedure was worked out by one of the wastewater plant operators.[4] It is described below.

Preventive Maintenance

The following procedure requires an evaporator. In fact, it was the failure* of three SO_2 evaporators that inspired this procedure. It will boil off any excess moisture that may accumulate in the supply system.

Once every day or once every three days, depending upon the SO_2 usage, the evaporator under consideration should be shut down by closing the inlet liquid line. The sulfonator vacuum relief line should be plugged, and the sulfonator should be allowed to operate so as to create a vacuum within the SO_2 evaporator container vessel. The moisture that may collect in the supply system will always float upon the liquid SO_2 in the evaporator container vessel. Therefore, withdrawing all of the liquid SO_2 in the evaporator is the first step. Step 2 is the boiling off of the moisture in the container vessel by pulling a vacuum through the sulfonator while maintaining a temperature of 160–180°F in the SO_2 evaporator water-bath.

The amount of moisture that collects in the container will depend upon the moisture content in the SO_2 and the rate of SO_2 withdrawals. Therefore, each installation will require a special procedure for boil-off time and frequency of boil-off. In most cases moisture will be visible in the sulfonator rotameter tube, and when this moisture disappears, the boil-off procedure can be terminated. This procedure requires redundancy in the SO_2 evaporator equipment to allow intermittent evaporator shutdown periods.

Those installations not using evaporators will probably not experience corrosion as severe as those with evaporators. This problem is directly related to

*The failure, due to moisture in the supply, caused three tiny holes in the SO_2 evaporator container vessel. The holes all occurred about 12 inches above the bottom of the containers.

the increased corrosive activity of the heated SO_2 caused by the elevated temperature of the moisture in the evaporators.

However, those systems not using evaporators suffer a malaise of considerable importance. Owing to its low vapor pressure, SO_2 reliquefies readily in the supply system. This phenomenon interferes severely with the ability to meter the SO_2 with any degree of accuracy.

The best system for handling SO_2 dechlorination procedures uses a storage tank that can be pressurized to 75–100 psi and deliver the SO_2 to conventional vaporizing equipment. Boiling off the moisture described above should reduce the corrosive maintenance problem to within the limits currently experienced by the usual chlorination system.

The above moisture removal procedure eliminated corrosion failure of the evaporators at the plant where three sulfur dioxide evaporators failed in less than six months. The chlorine evaporators at this same plant have been in operation for over 20 years. This illustrates the magnitude of the problem and the vast difference between the packaging of chlorine versus sulfur dioxide.

The above procedure was devised after a careful examination was made of the lower section of the SO_2 container vessel in all three evaporators that failed. There was a series of pinholes distributed around the circumference of the vessel right at the SO_2 liquid–vapor interface.

LEAK DETECTION

One of the major maintenance differences between chlorine and sulfur dioxide is that of continuous leak detection equipment. Although the spot-check use of a solution of NH_4OH to locate leaks is the same for SO_2 as it is for chlorine, the continuous leak detectors are vastly different for the two. The procedure for detecting SO_2 in the air is much more involved, and therefore the equipment is more complex than a chlorine leak detector. The best advice is to adhere to the manufacturer's procedure for calibration and recommendations for maintenance.

SULFUR DIOXIDE CONTROL AND METERING SYSTEM

Everyting that has been said for chlorine under this section applies to sulfur dioxide. The only precaution to the operator is this: never interchange chlorination equipment, particularly evaporators, for sulfur dioxide duty. Although chlorination equipment and sulfonation equipment are, as presently manufactured, identical for all practical purposes, they should never be used alternately for chlorine and then sulfur dioxide. This is so because a violent heat of reaction occurs when sulfur dioxide comes in contact with corrosion products of chlorine in an evaporator or chlorinator, and vice versa. This reaction is most noticeable in an evaporator.

REFERENCES

1. Anon., "Installation, Operation and Maintenance Instructions," Book No. WAA 50.206, Wallace & Tiernan Div. of Pennwalt Corp., Belleville, NJ, 1979.
2. Anon., "Chlorine Leaks," American Gas and Chemical Co., Northvale, NJ, July 1983.
3. Anon., "Sulfur Dioxide Technical Handbook," 5th ed., Cities Service Co., Atlanta, 1979.
4. White, G. C., "Equipment Corrosion from Sulfur Dioxide," report to a client, Aug. 1980.
5. Anon., "Series 50–125 Chlorine Detector," Wallace & Tiernan Div. of Pennwalt Corp., Belleville, NJ, June 1980.
6. Anon., "Operation and Maintenance Instructions, Chlorine Evaporator," Book WAA 50.206, Wallace & Tiernan Div. of Pennwalt Corp., Belleville, NJ, 1979.

12
CHLORINE DIOXIDE

INTRODUCTION

The identification and the control of chlorine-resistant, disease-causing pathogens that have recently become apparent in the public water supply—especially the encysted parasite *Cryptosporidium parvum*—have become a top priority of potable-water-related research in the 1990s. The intense focus of, and the substantial resources devoted to, the scientific investigation of *Cryptosporidium* have attracted the attention of some of the finest scientists ever to work in the water-treatment field. Among their other accomplishments, these scientists have shown conclusively that chlorine dioxide is an effective disinfectant for controlling encysted parasites.[36,70]

Also during the 1990s, a wholly new method for generating high-purity chlorine dioxide—the *chlorine gas : solid chlorite* process—has become commercially available (CDG Technology, Inc., New York, New York and Bethlehem, Pennsylvania). Characterized by simple operation, flexible control, and reliable production of pure chlorine dioxide gas on demand, this process resolves essentially all of the problems associated historically with chlorine dioxide generation. Taken together with the *Cryptosporidium*-inactivation data, the *chlorine gas : solid chlorite* process may well be the most significant breakthrough in drinking water treatment since the advent of breakpoint chlorination.

Historical Background

Chlorine dioxide (ClO_2), a versatile, increasingly important industrial chemical, was discovered by Sir Humphrey Davy in 1814. Davy prepared the gas by pouring a strong solution of sulfuric acid on potassium chlorate. Subsequently, Millon replaced sulfuric acid with hydrochloric acid. This reaction has been practiced in more recent years for certain large-scale production situations using sodium chlorate instead of potassium chlorate:

$$2NaClO_3 + 4HCl \longrightarrow 2ClO_2 + Cl_2 + 2NaCl + 2H_2O \qquad (12\text{-}1)$$

Today, most chlorine dioxide is used for bleaching paper pulp because it produces a brighter, stronger fiber than does chlorine bleaching, and it does not form environmentally harmful chlorinated by-products (e.g., dioxins).

Virtually all chlorine dioxide used for pulp bleaching is generated on-site from sodium chlorate ($NaClO_3$) in tons/day quantities. These large-scale chlorine dioxide plants employ several similar processes to reduce sodium chlorate, for example, with sulfuric acid, methanol, or hydrogen peroxide.

Sodium chlorate–based methods are not used for chlorine dioxide generation in small quantities; rather, chlorine dioxide is made from sodium chlorite ($NaClO_2$). Sodium chlorite is costlier than sodium chlorate, but lends itself to smaller-scale chlorine dioxide production using inexpensive equipment.

Potable Water Treatment

The Mathieson Chemical Company first made sodium chlorite commercially available in the early 1940s; shortly thereafter, in January 1944, the first reported use of chlorine dioxide for potable water treatment in the United States was undertaken at Niagara Falls, New York.

Since the early days of chlorine dioxide use for potable water treatment, it has been used successfully to eliminate tastes and odors, particularly those caused by phenol pollution, as well as those associated with algae and decaying vegetation.[1] Chlorine dioxide has also been found to be superior to chlorine in the removal of iron and manganese.

In the early 1970s, it was discovered that chlorine dioxide does not promote formation of trihalomethanes (THMs) and is often effective in reducing THM precursors. Also, chlorine dioxide proved to be as effective a bactericide as chlorine, and far superior to chlorine as a viricide.

However, questions about the possible health effects of chlorine dioxide's principal inorganic by-products (especially chlorite ion, ClO_2^-), complicated analytical techniques, and awkward generation methods have until recently stalled chlorine dioxide's widespread adoption for water treatment.

The recent resurgence in interest in chlorine dioxide for drinking water is attributable to several factors:

- Chlorine dioxide does *not* produce chlorinated organic by-products—for example, trihalomethanes (THMs), which are the subject of increasingly strict governmental regulation.
- Chlorine dioxide has been proved effective for killing chlorine-resistant pathogens (e.g., *Giardia, Cryptosporidium*).
- Reliable analytical techniques for measuring chlorine dioxide and related oxychlorine species have been developed.
- Comprehensive toxicological studies on chlorine dioxide's principal by-product, chlorite ion (ClO_2^-), have allayed earlier concerns about possible adverse health effects.

Also, the practical difficulties once associated with the application of chlorine dioxide have been resolved by new chlorine dioxide generation methods. Notably, *chlorine gas : solid chlorite* process technology (also CDG Technology, Inc.) has afforded a reliable, user-friendly process for the on-site generation of high-purity chlorine dioxide—both in the laboratory and in the water treatment plant.[2,3]

Other Uses

Drinking-water treatment is the dominant chlorine dioxide disinfection application and is the principal focus of this chapter. However, chlorine dioxide's potent biocidal capabilities, together with advances in chlorine dioxide science and technology, have given rise to a number of important uses. These include wastewater treatment, industrial process-water treatment, mollusk control, food processing, food-equipment disinfection, oxidation of industrial wastes, and the gas sterilization of medical devices.

WATER TREATMENT PRACTICE

North America

The principal water-treatment use of chlorine dioxide in North America has historically been for the control of tastes and odors caused by phenolic compounds in the raw water supply. A Wallace & Tiernan survey in 1981 recorded chlorine dioxide use in North America for destruction of phenolic tastes and for iron and manganese removal. In another survey, Canadian installations used chlorine dioxide for the control of tastes and odors caused by industrial-phenolic pollution of the receiving waters.[4]

In 1972, the U.S. Environmental Protection Agency (USEPA) reported on the industrial pollution of the Lower Mississippi River in Louisiana.[5] The report disclosed the presence of chloroform in the New Orleans potable water supply; it was believed to have formed as a by-product of chlorination. This disclosure triggered further investigations confirming the formation of undesirable chloro-organics due to chlorination. A search to find alternatives to chlorination began in an attempt to reduce or eliminate the formation of trihalomethanes (THMs) caused by then-current chlorination practices.

Work by Miltner revealed that chlorine dioxide does not form THMs in drinking water; moreover, it apparently reduces THM-precursor concentration.[6] These findings spurred interest in the use of chlorine dioxide for potable water disinfection. Today, an estimated 700 to 900 U.S. drinking-water utilities use chlorine dioxide, largely for manganese oxidation, for taste and odor control, and to lower THM levels.[7] Many more are conducting pilot studies, and the use of chlorine dioxide, especially for protection of the public water supply against chlorine-resistant pathogens, is expected to burgeon over the coming years.

Europe

In the last two decades, two factors have changed the European approach to water treatment: one is the restriction of ammonia nitrogen content in finished water and the other, the desire to reduce THM formation. In December 1978 the Ministers of the European Economic Community (EEC) signed an agreement limiting the level of ammonia nitrogen (as NH_4) to 0.05 mg/L as a standard, with it never at any time to exceed 0.5 mg/L. It seems unlikely that the EEC, having further reduced the allowable ammonia N levels to 0.02 mg/L, will follow USEPA recommendations.

In Europe, where many surface water supplies historically have suffered from industrial pollution, particularly phenol spills, a 1977 survey reported several thousand chlorine dioxide installations at water treatment plants.[8] Chlorine dioxide's popularity resulted from its superior power to oxidize phenols and chlorophenols without imparting off-flavors and without adding a chlorinous taste, and from its superiority over chlorine for pretreatment to remove iron and manganese.

Prevention of THM formation was a key factor in using chlorine dioxide in many plants in France and other Western European countries, practically all of these installations being switchovers from chlorine.

The following train of processes is typical of a French water treatment plant:

- ClO_2 pretreatment at raw water pump station, to produce a residual of 0.5 to 2.0 mg/L.
- Flocculation and sedimentation.
- Filtration with sand and activated carbon.
- Ozonation (1–2 mg/L).
- Clearwell storage for 30 min.
- Breakpoint chlorination to remove ammonia N, if any.
- Another storage reservoir.
- Postchlorination of finished water for disinfection and distribution-system protection (0.3–0.5 mg/L residual).

A variation of this train of treatment might include chlorination and dechlorination, followed by chlorine dioxide for posttreatment.

Great Britain

The water supplies of Great Britain do not suffer from industrial pollution to the same extent as those in Western Europe; hence, British use of chlorine dioxide has been comparatively limited. However, chlorine dioxide has apparently been used successfully in the United Kingdom to control water quality problems in distribution systems.[9,10] For example, many plants turned to chlorine dioxide when the ammonia nitrogen content in the water escalated. This condition caused chlorine residuals far in excess of a plant's dechlorination equipment capacity. As chlorine dioxide does not react with ammonia nitrogen, it was possible to produce proper water quality with chlorine dioxide dosages of less than 0.5 mg/L, whereas chlorine dosages as high as 4.5 mg/L were required to produce sufficient free chlorine residuals to maintain proper water quality in the system.

Some chlorine dioxide installations in Britain have reportedly been used for the control of chlorophenolic tastes and other tastes and odors that are usually aggravated by chlorine; others have targeted color removal. Although the THM problem has increased interest in the use of ClO_2, there is no official move as yet to require abandonment of the free-residual process practiced in the United Kingdom almost universally since 1950.[11] Rather, the British regulatory authorities seem generally unconcerned with either THMs or ammonia N levels in potable water.

CHEMISTRY OF CHLORINE DIOXIDE

Chemical Properties

Chlorine dioxide—a relatively small, volatile, and highly energetic molecule—is a free radical. Chlorine dioxide is almost never used commercially as a gas at high concentrations because of its instability. For potable water and wastewater treatment processes, it has been produced and used until recently only in aqueous solutions.

Chlorine dioxide gas has an intense greenish yellow color with a distinctive odor similar to that of chlorine. The odor is evident at 0.3 ppm in air; at 45 ppm it is irritating.

A gas at ordinary temperatures, ClO_2 can be compressed to a liquid. It has a boiling point of 11°C and a melting point of −59°C.[12] Compressed, liquid chlorine dioxide is explosive at temperatures higher than −40°C.[13] Therefore, chlorine dioxide cannot be economically shipped and stored, but must be generated at the point of use. Concentrations of chlorine dioxide gas above about 10 percent in dry air (and somewhat higher in moist air) may give mild explosions or "puffs" on detonation.

One of the important physical properties of chlorine dioxide is its high solubility in water, particularly chilled water. The gas is soluble in water at room temperature. At 25°C it is about 23 times as concentrated in aqueous solution as in the gas phase at atmospheric pressure (760 mm Hg), with which it is in equilibrium. The water solubility of chlorine dioxide depends upon temperature and pressure: at 20°C and atmospheric pressure, the chlorine dioxide solubility in water is about 70 g/L. The solubility of chlorine is only about 7 g/L at the same conditions.[14] In waterworks practice, chlorine dioxide solution is made under partial pressure conditions (vacuum), with solutions having been known to reach concentrations of 40 g/L.

Chlorine dioxide exists in aqueous solution as a dissolved gas, but is extremely volatile and can be readily removed. For example, ClO_2 is easily expelled from solution in water by passing air through the solution. For this reason, the chlorine dioxide concentration in solution is not stable when the solution is in an open vessel. Under these circumstances, chlorine dioxide is lost to the atmosphere, and the solution strength deteriorates rapidly. Also, aqueous solutions of ClO_2 are subject to photolytic decomposition, the extent of which is a function of both time and the intensity of the ultraviolet component of the light source. However, aqueous solutions of ClO_2 will retain their strength for several months if properly stored in the dark.

Chemistry in Potable Water Treatment

Since publication of the benchmark treatise by Gordon et al.,[14] the chemistry of chlorine dioxide in water treatment has been increasingly well understood. Although chlorine dioxide does not belong to the family of "available chlorine" compounds—those chlorine compounds that hydrolyze to form hypochlorous acid—the oxidizing power of chlorine dioxide is often referred to as having an "available chlorine" content of 263 percent, calculated as follows:

> The chlorine in chlorine dioxide is 52.6 percent by weight. Since the chlorine atom undergoes five valence changes in the process of oxidation to the chloride ion:
>
> $$ClO_2 + 5e^- = Cl^- + 2O^{2-} \tag{12-2}$$
>
> the equivalent available chlorine content is 52.6 × 5 = 263 percent. In effect, this indicates that chlorine dioxide theoretically has 2.63 times the oxidizing power of chlorine.

This can be substantiated by the reactions that occur in the liberation of iodine from iodide in the acid starch–iodide analytical procedure:

Chlorine dioxide:

$$2ClO_2 + 10I^- + 8H^+ \longrightarrow 2Cl^- + 5I_2 + 4H_2O \tag{12-3}$$

Chlorine:

$$HOCl + 2I^- \longrightarrow OH^- + Cl^- + I_2 \tag{12-4}$$

In practice, however, chlorine dioxide is rarely reduced all the way to chloride ion, so its "263 percent equivalent available chlorine content" is not fully realized. Rather, chlorine dioxide's dramatic oxidizing power is more likely attributable to *chlorine dioxide : chlorite ion recycling,* where the chlorine dioxide is reduced to chlorite ion, which is oxidized back to chlorine dioxide, and so forth.

Selectivity as an Oxidant

Chlorine dioxide functions as a highly selective oxidant owing to unique, one-electron transfer mechanisms, wherein it attacks electron-rich centers in organic molecules and, in the process, is reduced to chlorite ion.

In contrast to chlorine, chlorine dioxide does not react with ammonia nitrogen or (to any significant extent) with primary amines, but it will oxidize nitrites to nitrates. It does not react by breaking carbon–carbon bonds, so mineralization of organics typically does not occur.[15] At neutral pH or above, sulfurous acid (H_2SO_3) reduces chlorine dioxide to chlorite ion. At low pH, the ClO_2 is reduced to chloride ion.

High-purity chlorine dioxide introduced into water as a gas produces fewer disinfection by-products (DBPs) than do solutions that contain chlorine and other oxychlorine species.[16]

In contrast to ozone, high-purity chlorine dioxide does not oxidize bromide ion to bromate ion, unless photolyzed.[17,18] Neither does it produce appreciable amounts of aldehydes, ketones, ketoacids, or other DBPs associated with ozonation of organic matter.[19]

ON-SITE GENERATION

General Discussion

The oft-stated goal of on-site ClO_2 generation is to achieve high generator "yield" or "efficiency." But both these terms are process-oriented; neither relates directly to the quality of the chlorine dioxide produced and, unfortunately, neither of these terms is used consistently or in accordance with any generally accepted definition. This has led to confusion both in the chlorine dioxide literature and marketplace.

Purity

The need for a simple, uniform, *product-oriented* standard for chlorine dioxide has given rise to the use of *Purity* as a precisely defined term-of-art; of necessity, it takes into account all of the following oxychlorine species:

Chlorine dioxide	ClO_2
Free available chlorine [FAC]	Cl_2; HOCl; OCl^-
Chlorite ion	ClO_2^-
Chlorate ion	ClO_3^-

The following, process-independent definition of Purity of ClO_2 *solutions* results:

$$ClO_2 \text{ Purity}_{(solution)} = \frac{[ClO_2]}{[ClO_2] + [FAC] + [ClO_2^-] + [ClO_3^-]} \quad (12\text{-}5)$$

For ClO_2 *gas,* neither the chlorite ion $[ClO_2^-]$ nor chlorate ion $[ClO_3^-]$ terms need be considered, since these ions don't exist in the gas phase. Similarly, the [HOCl] and $[OCl^-]$ terms drop out, and the FAC term is thus simplified to $[Cl_2]$. The following simple equation results:

$$ClO_2 \text{ Purity}_{(gas)} = \frac{[ClO_2]}{[ClO_2] + [Cl_2]} \quad (12\text{-}6)$$

Procurement Specifications

Consistent with the foregoing discussion of Purity as the appropriate standard for evaluating chlorine dioxide generator performance, the following language should be used to set forth the specifications for chlorine dioxide procurement:

> Vendor shall provide a means for the on-site production of chlorine dioxide (ClO_2), such that the chlorine dioxide produced is consistently of at least 99% *Purity*. For the purpose of this specification, *Purity* is defined as the weight ratio of chlorine dioxide concentration to the sum of the concentrations of chlorine dioxide (ClO_2), chlorite ion (ClO_2^-), chlorate ion (ClO_3^-), and FAC (FAC being Cl_2, HOCl and OCl^-), such that:

$$ClO_2 \text{ Purity} = \frac{[ClO_2]}{[ClO_2] + [FAC] + [ClO_2^-] + [ClO_3^-]} \times 100\% \quad (12\text{-}7)$$

> The chlorine dioxide production means shall be capable of maintaining the specified ClO_2 Purity across the specified turndown range [e.g., 10:1] without requiring recalibration between production-rate setpoints.

Commercial Methods

Chlorine dioxide for drinking water treatment has been generated most often from 25 percent aqueous sodium chlorite solution. In older methods, chlorine dioxide is produced by the reaction of sodium chlorite solution with acid (*acid:chlorite solution*). Other techniques employ the somewhat better reaction of sodium chlorite solution with a strong chlorine solution at low pH, sometimes with the "assistance" of acid addition (*chlorine solution:chlorite solution*). A variant of this process reacts chlorine *gas* with vaporized sodium chlorite solution at near-neutral pH (*chlorine gas:chlorite solution*).[20]

The state of the art is a process developed by Rosenblatt and co-workers that reacts chlorine *gas* with specially processed *solid* sodium chlorite to produce high-purity, chlorine-free chlorine dioxide *gas* (*chlorine gas : solid chlorite*).[21]

Acid : Chlorite Solution. Chlorine dioxide can be generated by acidification of sodium chlorite solution, and several stoichiometric reactions have been reported for such processes.[14]

When chlorine dioxide is generated from sodium chlorite by acid activation, the final composition is dependent upon several variables: sodium chlorite concentration, purity of sodium chlorite used, acid concentration, pH of reaction, and the presence of chloride ion.

Sulfuric Acid. The two reactions that are most widely accepted for sulfuric acid activation of sodium chlorite are:

$$4NaClO_2 + 2H_2SO_4 \longrightarrow 2Na_2SO_4 + 2ClO_2 + HCl + HClO_3 + H_2O \qquad (12\text{-}8)$$

and

$$10NaClO_2 + 5H_2SO_4 \longrightarrow 8ClO_2 + 5Na_2SO_4 + 2HCl + 4H_2O \qquad (12\text{-}9)$$

The first reaction is catalyzed by the chloride ion, which is also a product of the reaction. It has been reported not only that the reaction is accelerated by the chloride ion, but also that the stoichiometry is altered by the presence of chloride ion to that of the second reaction.[14]

Hydrochloric Acid. The stoichiometry expected for hydrochloric acid activation of sodium chlorite is:

$$5NaClO_2 + 4HCl \longrightarrow 4ClO_2 + 5NaCl + 2H_2O \qquad (12\text{-}10)$$

Research has shown HCl-activation of chlorite to achieve better results than those reported for sulfuric acid processes, with ClO_2 yields >77 percent.[22] Because of this relatively better yield, and because the presence of chloride ion is helpful, hydrochloric acid is generally the reagent of choice for sodium chlorite acidification processes.

When catalyzed by the presence of chloride ion, acid activation of sodium chlorite has a maximum possible yield of 80 percent of the quantity of ClO_2 that theoretically could be produced from reaction of the same amount of sodium chlorite with chlorine. Acid : chlorite solution processes can achieve production capacities of up to about 100 lb/day.[20] (Generators employing acid : chlorite solution processes are available from ProMinent Fluids Control, Pittsburgh, Pennsylvania.)

Chlorine Solution : Chlorite Solution. Chlorite ion (from dissolved sodium chlorite) will react in aqueous solution with chlorine or hypochlorous acid (HOCl) to form chlorine dioxide:

$$2ClO_2^- + HOCl + H^+ \longrightarrow 2ClO_{2(aq)} + Cl^- + H_2O \qquad (12\text{-}11)$$

Thus, two moles of chlorite ions theoretically will react with one mole of chlorine (i.e., as hypochlorous acid) to produce two moles of chlorine dioxide. If the chlorine and chlorite ion react stoichiometrically, the resulting pH is close to 7. In order to fully utilize sodium chlorite solution—the more expensive of the two ingredients—excess chlorine is often used, which lowers the pH and drives the reaction further toward completion.

Using a laboratory reactor to evaluate the chlorine solution : chlorite solution method, researchers[22] were able to achieve chlorine dioxide yields between 93 and 98 percent when chlorine was about 4 percent in excess of the chlorite-based stoichiometric requirement. The reaction time in the laboratory reactor was far in excess of that used in continuous commercial generators, and yields were better than might be found in a commercial production setting. Nonetheless, this method of generation has the potential of a significantly better yield than acid : chlorite solution methods. Some researchers have claimed maximum yield when the pH is on the order of 3.5–4.0.[12] Others have reported that even lower pH may be desirable.[23]

Because they are relatively slow, chlorine solution : chlorite solution processes are for all practical purposes generally limited to production capacities of up to about 250 lb/day.[20] (Generators employing chlorine solution : chlorite solution processes, some with "acid assistance," are manufactured by Drew Industrial Division of Ashland Chemical, Kansas City, Kansas; Wallace & Tiernan Inc., Vineland, New Jersey; Capital Controls, Inc., Colmar, Pennsylvania; and CIFEC, Inc., Paris, France.) (See Figs. 12-1 through 12-8.)

Chlorine Gas : Chlorite Solution. In a method that was for some time the industry standard, sodium chlorite solution is "vaporized" and reacted under a vacuum with molecular chlorine at near-neutral pH. This process uses undiluted reactants and is much more rapid than chlorine solution : chlorite solution methods. So long as they are well calibrated at least daily, properly maintained, and fed properly inventoried sodium chlorite solution, generators employing the chlorine gas : chlorite solution process can reliably supply chlorine dioxide solutions that approach 95 percent Purity. Benefiting from very fast reaction kinetics, the chlorine gas : chlorite solution process can produce many thousands of pounds of chlorine dioxide per day. (Generators employing the chlorine gas : chlorite solution process are manufactured by Vulcan Chemical Technologies, Inc.—formerly Rio Linda Chemical Company—Sacramento, California.) (See Fig. 12-7.)

Chlorine Gas : Solid Chlorite. Recently, a simple means for the generation of high-purity, chlorine-free chlorine dioxide *gas* has become commercially available.[2,3,24,25]

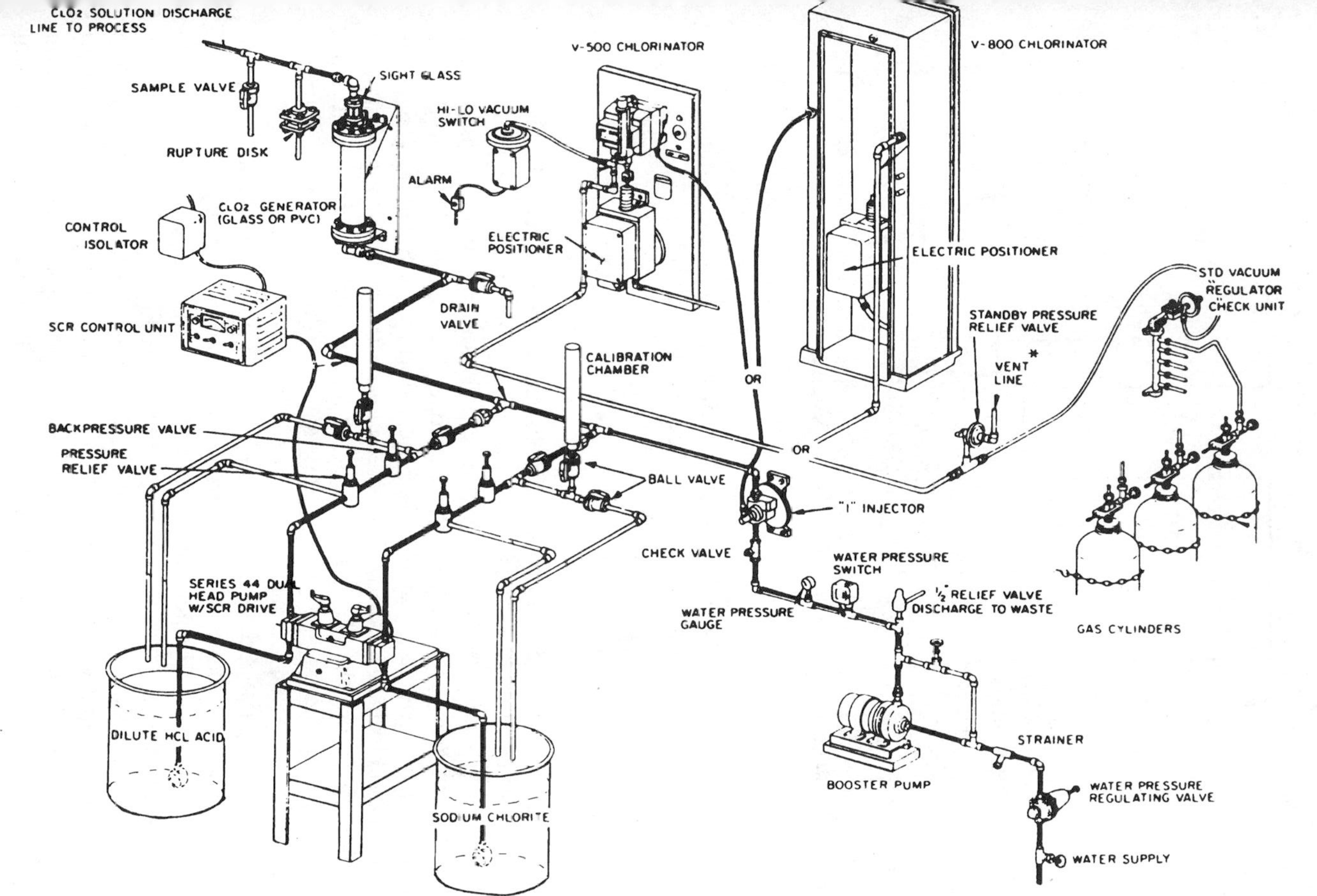

Fig. 12-1. Chlorine solution: chlorite solution chlorine dioxide generating system with supplemental acid feed (courtesy Wallace & Tiernan, Vineland, New Jersey).

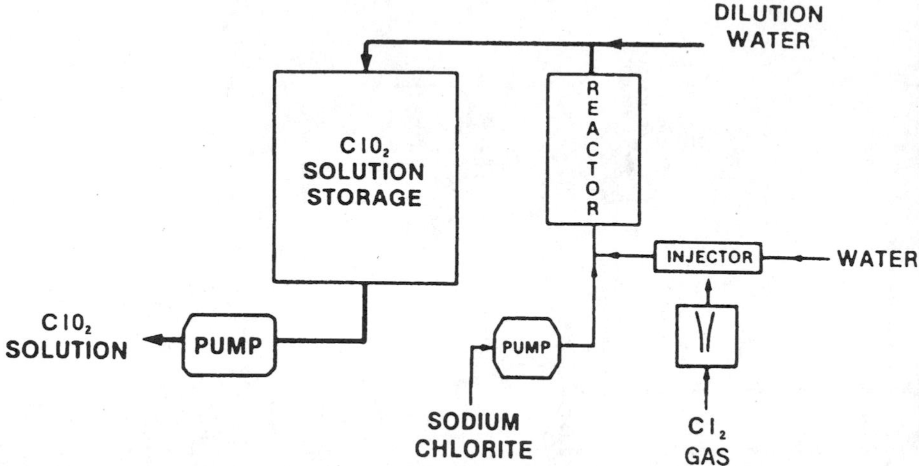

Fig. 12-2. Chlorine solution : chlorite solution chlorine dioxide generator with continuous batching (courtesy of Wallace & Tiernan, Vineland, New Jersey).

This method reacts dilute, humidified chlorine gas with specially processed solid sodium chlorite contained in a sealed reactor cartridge:

$$2\ NaClO_{2(solid)} + Cl_{2(gas)} \longrightarrow 2ClO_{2(gas)} + 2NaCl_{(solid)} \qquad (12\text{-}12)$$

The reaction is rapid and produces high-purity chlorine dioxide gas inherently free of chlorite and chlorate ions because these ions do not exist in the gas phase. Any unreacted sodium chlorite feedstock or impurities contained therein remain in the sealed reactor cartridge. Because the chlorine dioxide production rate is a function *solely* of the chlorine gas feedrate, generators that employ chlorine gas : solid chlorite technology are the only type of chlorine dioxide generators that are capable of effectively infinite turndown—that is, the chlorine dioxide production rate can be adjusted without requiring recalibration between settings.[20,27]

The chlorine gas : solid chlorite method, limited only by the available chlorine feedrate, can produce many thousands of pounds of high-purity chlorine dioxide gas per day. (Generators employing the chlorine gas : solid chlorite process are made by CDG Technology, Inc.) (See Fig. 12-8.)

It is essential to note that the chlorine gas : solid chlorite process described here is fundamentally different from superficially similar methods explored in the 1940s and 1950s.[26] In particular, the early attempts to use solid sodium chlorite in reaction with chlorine gas were inefficient at converting sodium chlorite to chlorine dioxide and were dangerous. These early processes tended to catch fire and, sometimes, explode; they were therefore abandoned. It is the modified chemistry of the specially processed solid sodium chlorite that

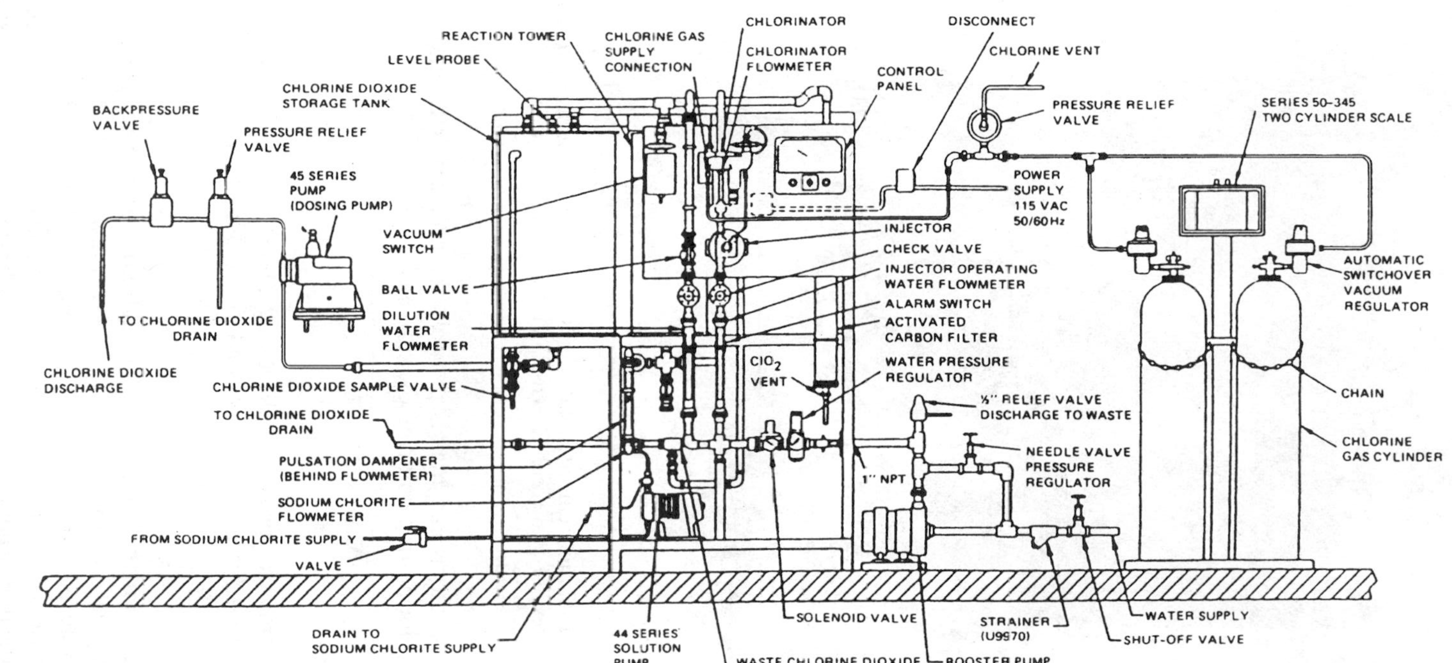

Fig. 12-3. Chlorine solution : chlorite solution chlorine dioxide generator. Schematic of a typical installation of a chlorine dioxide generator with continuous batching (courtesy of Wallace & Tiernan, Vineland, New Jersey).

Fig. 12-4. Chlorine solution : chlorite solution chlorine dioxide generator with continuous batching (courtesy of Wallace & Tiernan, Vineland, New Jersey).

resolved these problems and makes possible the efficient, safe, and reliable production of chlorine-free chlorine dioxide gas.

Other Methods

Besides the commercial processes discussed above, other chlorine dioxide generation methods have been promoted from time to time and warrant mention.

Electrolysis of Chlorite Solutions. Though not commercially available at this time, various methods for generation of chlorine dioxide by the electrolysis of sodium chlorite have been proposed. Griese and Rosenblatt[47] reported results of pilot work conducted at Evansville (Indiana) Water & Sewer Utility, using a generator that produced ~7g/L ClO_2 solution by the direct oxidation

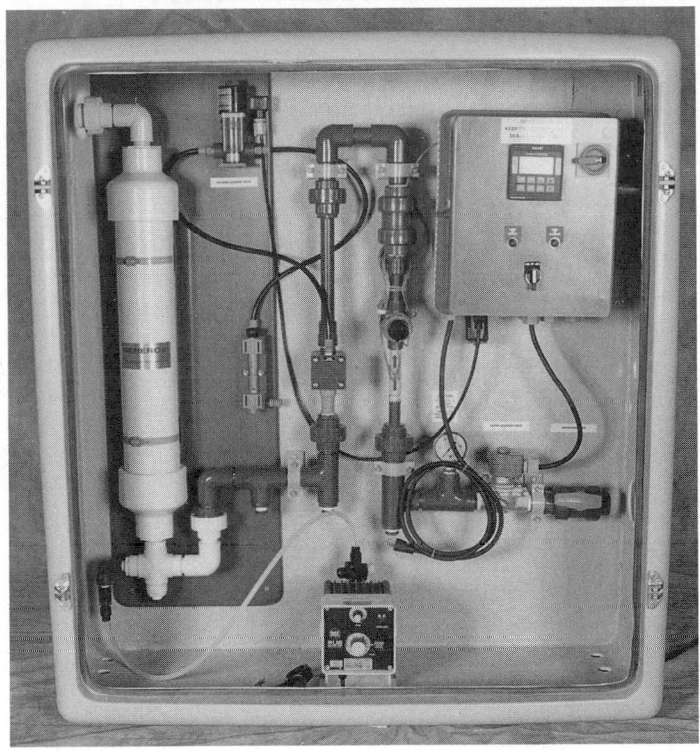

Fig. 12-5. Chlorine solution: chlorite solution chlorine dioxide generator (courtesy of Drew Industrial, Division of Ashland Chemical).

of ClO_2^- on a high-surface anode structure.[28,29] The overall net electrochemical reaction was:

$$NaClO_2 + H_2O \longrightarrow ClO_2 + NaOH + \frac{1}{2} H_2 \qquad (12\text{-}13)$$

The electrochemical generator produced unacceptably high levels of chlorate ion. Believed to be associated with one or more unanticipated and never fully understood reactions in the anode compartment, the chlorate ion levels were sufficiently high to cause the study to be terminated.

Recent work (1994) on electrochemical generation[30] oxidizes $NaClO_2$ in the anode compartment of a cation-exchange-membrane divided cell. Unreacted chlorite ion, chlorate ion, and other non-gas-phase species remain in the reaction liquor; dissolved ClO_2 gas is recovered from the liquor by being passed through a hydrophobic microporous membrane into recipient water running countercurrent to the stream of generator liquor. The ClO_2-containing recipient water constitutes the generated product. Water balance is achieved in continuous operation by maintaining a temperature differential between

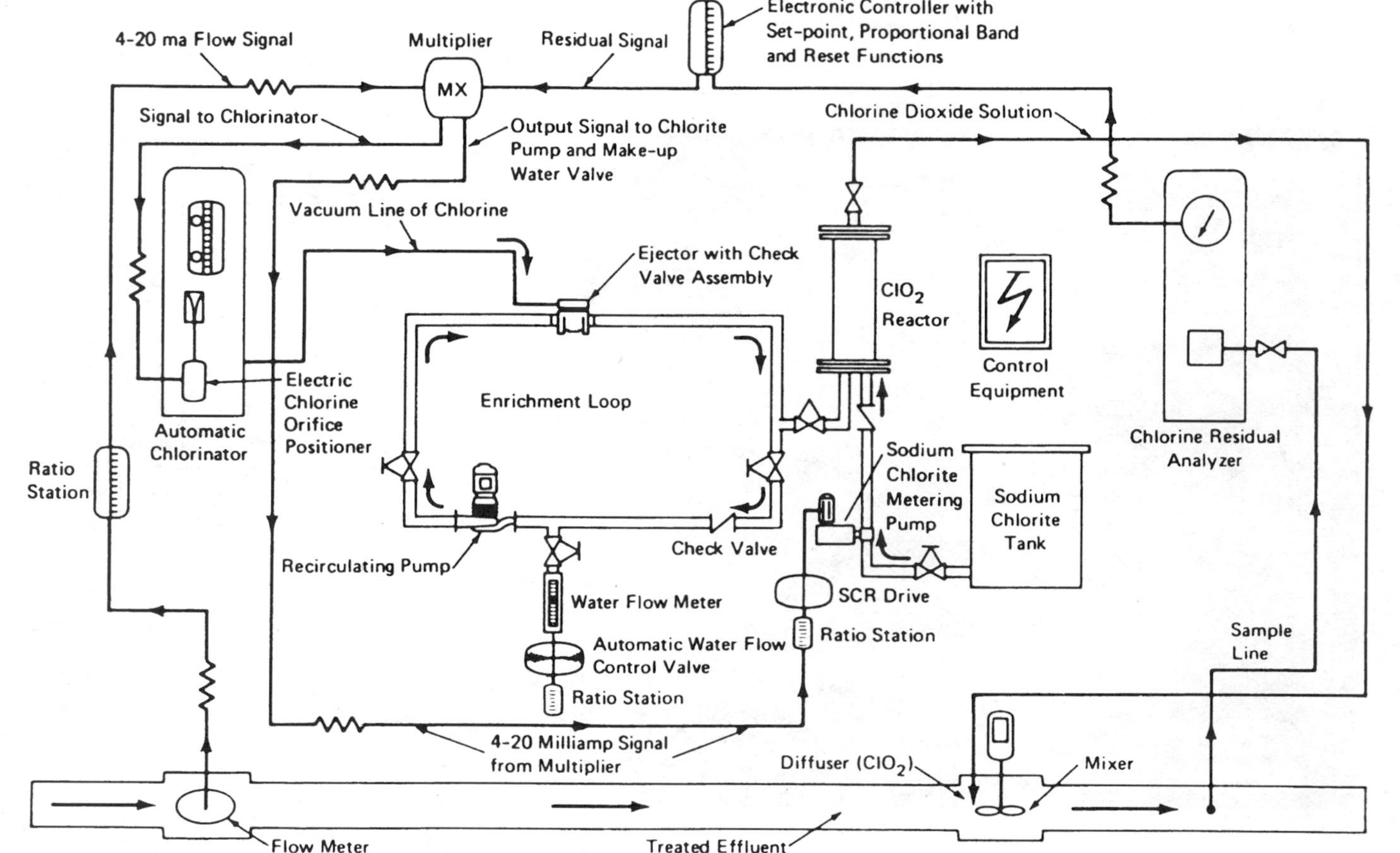

Fig. 12-6. Chlorine solution : chlorite solution chlorine dioxide generation system with automatic residual control (courtesy of CIFEC Paris, France).

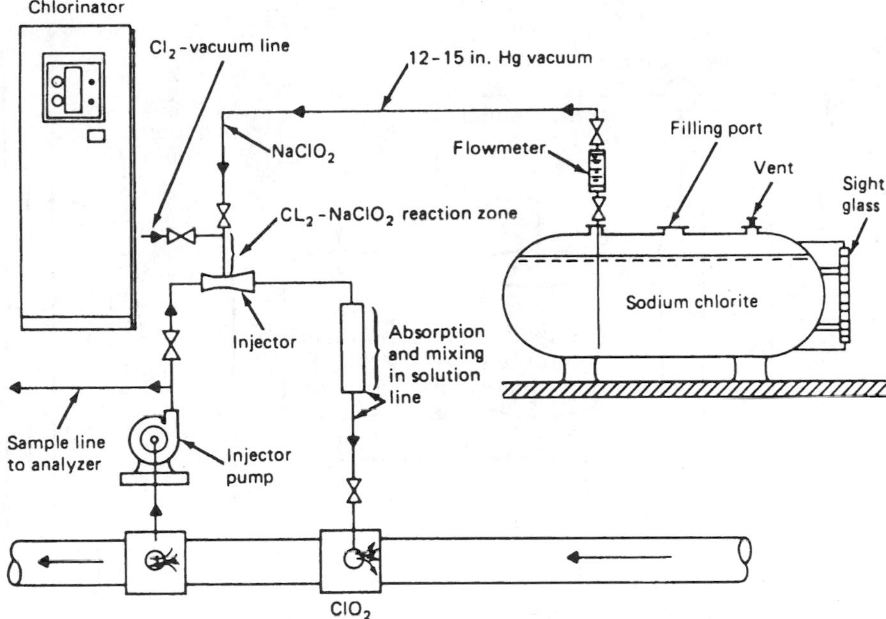

Fig. 12-7. Chlorine gas: chlorite solution vacuum-type chlorine dioxide generator system (courtesy of Vulcan Chemical Technologies).

the liquor and the receiving stream, so that water vapor is transported across the hydrophobic membrane, and also by transporting water across the cation exchange membrane. This process is claimed to be best carried out at neutral pH; therefore, special sodium chlorite solution formulations are required.

Chlorate Reduction. It is well understood that chlorate-based routes to chlorine dioxide generation, such as those used in pulp mills, are not satisfactory for most other, smaller-scale chlorine dioxide applications. Nonetheless, the prospect of being able to use a much-lower-cost feedstock to produce chlorine dioxide (i.e., sodium chlorate vs. sodium chlorite) is enticing. Especially for the unsuspecting newcomer to chlorine dioxide, some of the reasons for avoiding chlorate-based methods warrant review.

Attempts at small-scale, chlorate-based generation have combined concentrated acid such as hydrochloric acid with sodium chlorate using mechanical pumps and plastic mixers and reactors. Because it is inherently inefficient and slow, the associated reaction is driven by using excess acid. But the high acid concentration sets up a competing reaction that produces mostly chlorine:

$$NaClO_3 + 6HCl \longrightarrow 3Cl_2 + NaCl + 3H_2O \qquad (12\text{-}14)$$

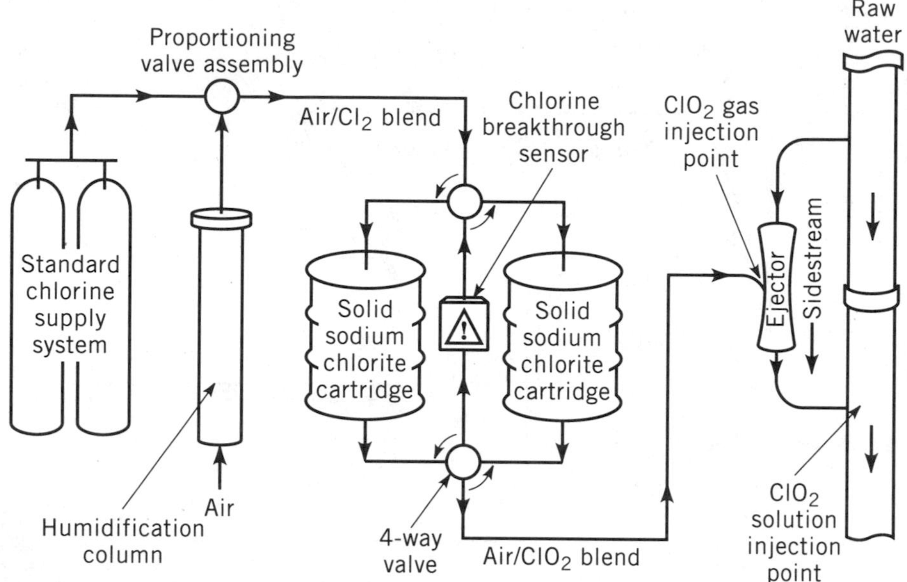

Fig. 12-8. Full-scale chlorine gas : solid chlorite generator. The full-scale chlorine gas : solid chlorite generator operates under vacuum created by means of an ejector situated on a raw-water sidestream. Chlorine from a standard vacuum chlorinator is pulled through the chlorine gas : solid chlorite system: first, the chlorine feedgas mixes with humidified, filtered air, then through a series of two or more reactor cartridges containing packed beds of specially processed solid sodium chlorite. In between the reactor cartridges, a sensor detects chlorine breakthrough, indicating that a reactor cartridge is spent (courtesy of CDG Technology, Inc., New York, New York and Bethlehem, Pennsylvania).

This chemistry is wholly unsatisfactory for most of the applications requiring small-scale generation, as it produces an effluent with a severely depressed pH that tends to contain high chlorine-impurity levels.[20] Processes for chemical reduction of chlorate ion—even some of the more recently developed processes that claim somewhat lower chlorine contaminant levels[98]—tend to be complicated, typically carry safety risks with consequences far more severe than those associated with chlorite-based methods, and are inherently incapable of reliably producing chlorine dioxide of the purity necessary for drinking water treatment. Although these risks and operational limitations are manageable in highly engineered, very large-scale installations such as those found in pulp mills, they are unsuitable for the environments in which water and wastewater treatment are practiced. Also, sodium chlorate is *not* an EPA approved or Federal Insecticide Fungicide and Rodenticide (FIFRA) registered precursor for chlorine dioxide production for drinking water treatment.

LABORATORY PREPARATION

General Discussion

In preparing chlorine dioxide stock solutions for testing in the laboratory, it is important to generate chlorine dioxide free from chlorine. Chlorine dioxide standard solutions should be prepared on the same day that samples are collected for analysis, and ClO_2 standards should always be shielded from light to prevent decomposition and the possible formation of ClO_3^-.

Chlorine dioxide samples should be analyzed within one hour of sampling and cannot be shipped overnight to a laboratory for analysis. Useful guidelines for sampling generator effluents are described in a 1994 AWWARF report.[31]

Bench-Scale Generation Methods

Potassium Persulfate: Chlorite Solution. A long-preferred procedure for laboratory preparation of chlorine dioxide stock solutions is the reaction of sodium chlorite solution with potassium persulfate.[32] The following procedure is suitable for the preparation of 300 mg/L chlorine dioxide solutions although solutions of up to 2 g/L can be prepared by increasing the reagent concentrations. The procedure uses 50 ml of 4 percent potassium persulfate, 25 ml of 16 percent sodium chlorite solution, and a generation apparatus with a three-bubble-tower arrangement. A nitrogen source is connected to the first tower. The second tower, filled with approximately 50 ml of water, is attached to the first tower. A scrubber containing sodium chlorite solution is deployed between the second and third bubble towers, and the outlet from the third tower is directed to a 500 ml glass reservoir containing 300 ml of water. The sodium chlorite and potassium persulfate solutions are poured into the first tower and swept with nitrogen so that the chlorine dioxide generated is stripped from solution and is passed through the system to the collection reservoir for 15 to 30 min. Care must be taken so that the chlorine dioxide is not stripped from the collection reservoir. After chlorine dioxide generation is complete (the collection reservoir liquid should have a green-yellow color), the chlorine dioxide concentration is measured by UV-VIS spectrophotometry.

Sulfuric Acid: Chlorite Solution. Another laboratory technique for preparation of ClO_2 solution is described in *Standard Methods.*[33] Used by Hood[34] and Roberts et al.,[22,35] this method utilizes the reaction of sulfuric acid with sodium chlorite solution. As with the persulfate method, nitrogen is used to transfer the chlorine dioxide gas from the reaction vessel to a final flask, which contains the resultant stock solution. (See Fig. 12-9)

Chlorine Gas : Solid Chlorite. A bench-scale counterpart to the commercial chlorine gas : solid chlorite generation method has become the ClO_2 generation

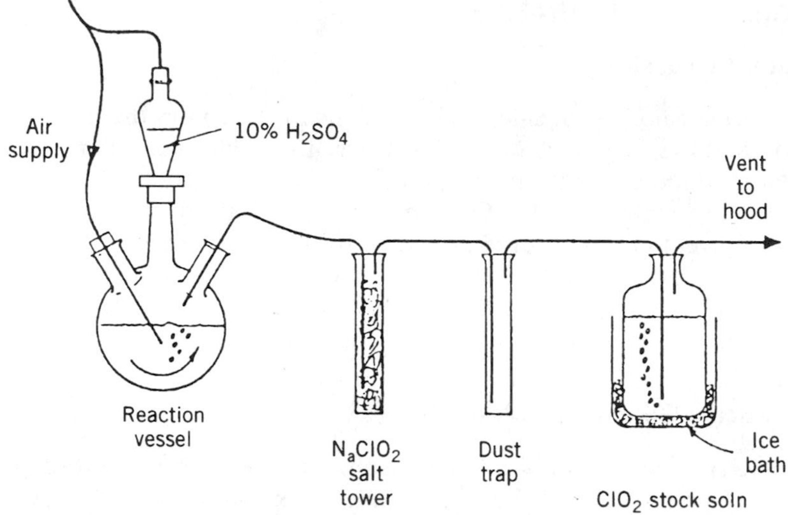

Fig. 12-9. Laboratory chlorine dioxide apparatus using acid activator of sodium chlorite. Approximately 500 mg/l ClO_2.

method of choice for leading researchers[36,37] because of the simplicity with which it can reproducibly generate high-purity, chlorine-free chlorine dioxide gas of known concentration. The process (Fig. 12-10) uses a preblended cylinder of 4 percent Cl_2 in an N_2 carrier (Air Products and Chemicals, Inc., Allentown, Pennsylvania), a humidification column, and a sealed reactor cartridge of specially processed solid sodium chlorite. The chlorine gas blend is introduced; it bubbles through water contained in the humidification column, and the dilute, humidified chlorine is introduced into the solid sodium chlorite reactor cartridge where it produces high-purity chlorine dioxide gas. The chlorine dioxide product gas, the flow rate of which is controlled by a rotameter, can be dissolved in water to produce a stock solution of up to about 7.5 g/L. The generation process is essentially instantaneous. Thc concentration of the chlorine dioxide product gas is a function solely of the chlorine gas feedrate. Because the reaction chemistry is stoichiometric according to Eq. (12-12), each mole of Cl_2 feedgas produces two moles of ClO_2 product gas. (Bench-scale chlorine gas: solid chlorite generators are available from CDG Technology, Inc.; see also Fig. 12-11.)

CHLORINE DIOXIDE MEASUREMENTS AND ANALYSES

Required Measurements

In the course of chlorine dioxide treatment of drinking water, chlorine dioxide, chlorite ion, and chlorate ion must be measured at several points, at various

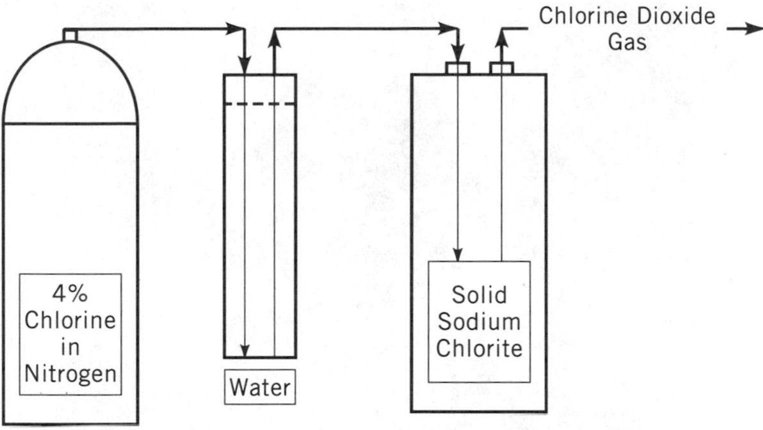

Fig. 12-10. Bench-scale chlorine gas:solid chlorite generator. The bench-scale embodiment of the chlorine gas:solid chlorite process uses a pre-blended cylinder of 4% Cl_2 in an N_2 carrier, a humidification column and a sealed reactor cartridge of specially-processed solid sodium chlorite. The chlorine gas blend is introduced under pressure (~15 psig); it bubbles through water contained in the humidification column and the dilute, humidified chlorine is introduced into the solid sodium chlorite reactor cartridge, where it reacts to produce high-purity chlorine dioxide gas in nitrogen (courtesy of CDG Technology, Inc., New York, New York and Bethlehem, Pennsylvania).

concentration levels, and in the presence of a variety of other chemical species. The necessary measurements must be made for various purposes, of which the following are particularly important.

Feedstock Analysis. For chlorine dioxide methods that use sodium chlorite solution, any impurities in this feedstock will carry through the generator and into the water being treated.[15] Commercial sodium chlorite solution is manufactured under tight production controls and is reliably of very high grade, but its specific composition is variable within a certain range. Also, if the material is allowed to age beyond its recommended shelf life, exposed to extreme temperatures, or otherwise mishandled, it may chemically degrade. This can result in additional, unwanted impurities being carried through the generator, and make it more difficult to maintain the tight chlorine:sodium chlorite solution ratios necessary to efficient generator performance. Therefore, it is essential that utilities institute proper handling protocols and analyze the composition of their sodium chlorite solution inventories at least monthly. Measured parameters should include specific gravity, bulk sodium chlorite concentration, and sodium chlorate concentration.

These feedstock-analysis procedures are not required with the chlorine gas:solid chlorite generation method, which uses factory-sealed cartridges containing specially processed solid sodium chlorite. Because the generator effluent contains *only* gas-phase constituents, even if the sodium chlorite in

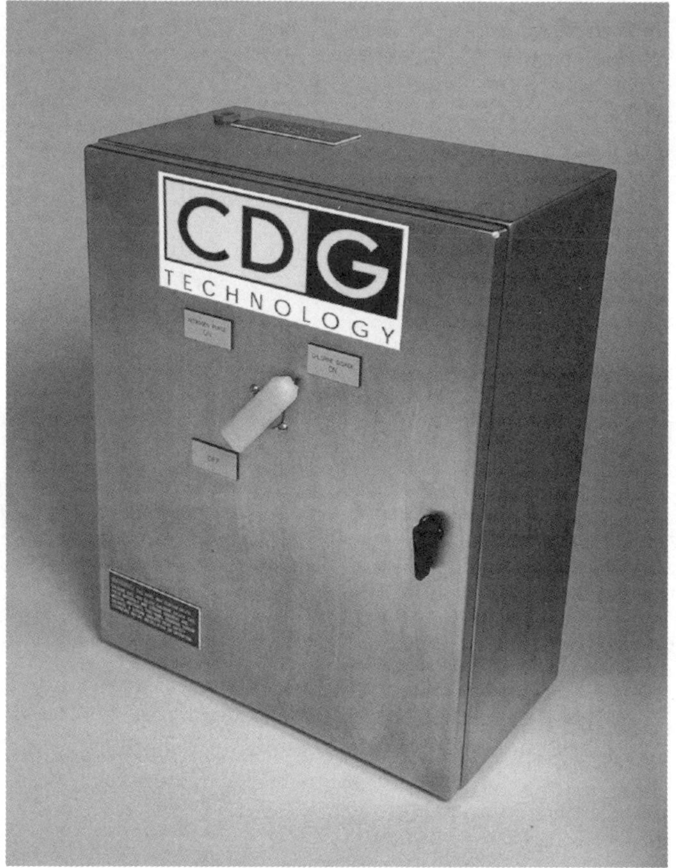

Fig. 12-11. Bench-scale chlorine gas : solid chlorite chlorine dioxide generator designed for laboratory preparation of stock solutions. Operation is by a single, 3-position 4-way hand valve (courtesy of CDG Technology, Inc., New York, New York and Bethlehem, Pennsylvania).

the cartridge deteriorates or contains impurities (e.g., chlorite or chlorate ions), the purity of the chlorine dioxide gas produced will be uncompromised.

Generator Performance. The USEPA's proposed Disinfectant/Disinfection By-Product (D/DBP) Rule (discussed further below, under "Regulatory Perspective") establishes performance criteria for chlorine dioxide generators. For this reason, and in order to know what ClO_2 dose is being applied, it is necessary to analyze the output of the generator. In choosing an analytical method for this purpose, it is essential that the various oxychlorine species that might be present can be detected and measured accurately at the concentration levels at which they would likely be present. For example, in the case of generators that chlorinate sodium chlorite solutions, possible species would include chlorine dioxide, chlorite ion, chlorate ion, and FAC (HOCl; OCl^-; Cl_2). In contrast, because neither chlorite ion nor chlorate ion exists in the gas phase, the only pertinent species in the effluent of a chlorine gas: solid chlorite generator are chlorine dioxide gas and chlorine gas.

Information Collection Rule (ICR) Compliance. Under the ICR (discussed below, under "Regulatory Perspective"), systems using chlorine dioxide are required to report measurements for chlorite and chlorate ions in the plant influent, before by-product removal treatment (e.g., with GAC, reduced iron, or sulfite ion), and at three points in the distribution system (beginning, middle, and end).[15] (See Tables 12-1 and 12-2.)

Operations. For systems using chlorine dioxide as a disinfectant, it is necessary to determine the chlorine dioxide residual in the water being treated in order to calculate disinfection credits. These measurements require a real-time method for measuring chlorine dioxide in the mg/L range, without interference from other species that may be in the treated water at the point(s) of measurement.

Workplace Monitoring. To comply with OSHA limits on allowable levels of ClO_2 exposure in the workplace, it is necessary to monitor the air where

Table 12-1 ICR Required Testing

Sampling Point	*Analyses*	*Frequency*
Plant influent	Chlorate ion	Quarterly
Prior to each ClO_2 application point	pH, alkalinity, tubidity, temperature, calcium, total hardness, TOC, UV_{254}, bromide ion	Monthly
Prior to 1st ClO_2 application point	Aldehydes, AOC/BDOC	Quarterly
Prior to application of Fe salts, sulfur reducing agents, or GAC	pH, residual ClO_2, chlorite ion, chlorate ion	Monthly
Before downstream chlorine or chloramine application	Aldehydes, AOC/BDOC	Quarterly
Distribution system entry	Residual ClO_2, chlorite ion, chlorate ion, bromate ion	Monthly
	Aldehydes, AOC/BDOC	Quarterly
Three distribution sampling points	Residual ClO_2, chlorite ion, chlorate ion, pH, and temperature	Monthly

Table 12-2 ICR Methods Specified by the USEPA for ClO_2, ClO_2^-, and ClO_3^-

Species	*Method*
Chlorine dioxide (ClO_2)	*Standard Methods* 4500-ClO_2 D (DPD Method)
	Standard Methods 4500-ClO_2 C (Amperometric Method I)
	Standard Methods 4500-ClO_2 E (Amperometric Method II)
Chlorite ion (ClO_2^-)	USEPA Method 300.0 (Ion Chromatography)
Chlorate ion (ClO_3^-)	USEPA Method 300.0 (Ion Chromatography)

chlorine dioxide is being generated and used. Various methods are available. Some (e.g., electrochemical sensors) can detect ClO_2 but also respond to Cl_2 without distinguishing between species (Analytic Technology, Inc., Oaks, Pennsylvania; Mine Safety Appliances, Pittsburgh, Pennsylvania). Other techniques, such as ion mobility spectroscopy (a time-of-flight technique), are highly sensitive and can selectively detect and quantify small amounts of airborne ClO_2, even in the presence of chlorine (Molecular Analytics, Inc., Baltimore, Maryland).

Analytical Methods[3,32]

A number of methods have been used or proposed for the measurement of chlorine dioxide. The analytical chemistry of chlorine dioxide, especially in solution, is complicated by its volatility, its sensitivity to light and instability over time, and interference from related redox species. Also, some of the techniques are labor-intensive, use expensive equipment, and require a high degree of technical skill. Different methods are required for the analyses of generator effluents and of distributed drinking water, and other water constituents can interfere with the measurement of target compounds.

DPD Reagent. DPD reagent, in wide use for chlorine measurement, is unsatisfactory for measurement of chlorine dioxide, chlorite ion, or chlorate ion because it is subject to severe interferences. Nonetheless, it is recommended by the USEPA for measurement of finished-water chlorine dioxide concentration and is used by some water treatment facilities.

Spectrophotometry. Chlorine dioxide can be measured spectrophotometrically either in solution or in the gas phase. The molar absorptivity of chlorine dioxide at its maximum absorbance wavelength (λ_{max} = 359 nm) can be used directly to calculate the concentration of "pure" chlorine dioxide, given the absorbance of the sample in a spectrophotometric cell of known pathlength.

Commercially available ultraviolet/visible (UV-VIS) spectrophotometers can be used to measure changes in light intensity through a sample cell. Typically transmittance or absorbance is measured. The transmittance of a sample is the amount of light passing through a sample as compared to the amount of light passing through a "blank" reference sample. The reference sample is normally the pure solvent in the absence of sample.

Actual concentrations of chlorine dioxide concentrations dissolved in water are calculated by using the Beer-Lambert law:

$$I = I_o \exp(-\varepsilon l C) \tag{12-15}$$

where:

I = intensity of the light transmitted through the sample
I_o = intensity of the light transmitted through the reference blank
ε = molar absorptivity of the species of interest (e.g., ClO_2)
l = pathlength of the spectrophotometer cell (typically 2.0 cm)
C = concentration of the species of interest (e.g., ClO_2)

Most spectrophotometers can reliably measure absorbance values over the 0.1 to 1.0 range. However, spectrophotometers are available that are reliable up to 2.0 absorbance units. When a 2 cm measuring cell is used, the maximum chlorine dioxide concentration that can be reliably measured is about 50 mg/L. When a 0.1 cm cell is used, about 1 g/L chlorine dioxide can be measured. The measurement of absorbance values above 2.0 usually is not reliable. Thus, for higher concentrations of chlorine dioxide, measurements should be made with shorter-pathlength cells. If different pathlength cells are not available, measurements can be made at a wavelength away from the wavelength of maximum absorbance (359 nm).

In solutions containing chlorine dioxide and chlorite ion, absorbance measurements are subject to interference from intermediate oxychlorine species. At wavelengths longer than 359 nm, such interference can become dominant, and the calculated chlorine dioxide concentration may differ from the true value by as much as a factor of two.[15] This technique is best used for strong chlorine dioxide solutions with relatively small chlorite ion concentration, for stock solutions prepared using demand-free water, or for gas-phase chlorine dioxide.[32]

Iodometry. Iodometric titration of chlorine dioxide, chlorine, chlorite ion, and, theoretically, chlorate ion is based on the oxidation of potassium iodide at various pH levels to distinguish these oxychlorine species from each other. These titrations can be performed manually using a starch indicator for endpoint analysis, as described in *Standard Methods* 4500-ClO_2 B, Iodometric Method (see page 4-54, 19th edition). However, iodometry is not readily practicable because it is subject to severe interferences, even with careful pH buffering and selective chemical masking. More accurate methods based on the same chemical principles have been developed—for example, the amperometric and FIA methods discussed below.

Amperometric Titration. Chlorine dioxide, free chlorine, and chlorite ion can be readily measured by amperometric titration at concentrations above 1 mg/L. Below this level, the accuracy of this method is highly dependent on the skill of the technician, and, although techniques exist for using amperometric titration to measure chlorate ion, such methods rely on the calculation of a small number by the difference between two large numbers. The results, especially for chlorate ion at levels less than 1 mg/L, are subject to large errors. There are two amperometric methods approved by the USEPA for ICR monitoring.[33]

Standard Method 4500-ClO_2 C, Amperometric Method I **(see page 4-55, 19th edition).** This titration is an extension of the amperometric procedure for measuring chlorine (*Standard Methods* 4500-Cl D). A dual platinum electrode assembly is used to measure parts per million (mg/L) concentrations of the residual oxychlorine species when chlorine dioxide is applied in water and wastewater treatment.

The amperometric titration is an electrochemical technique that measures the current (amperes) at a constant voltage. As titrant is added to the titration cell, the current will change, based on the reaction of phenylarsine oxide (PAO) with the oxyhalogen. When complete reaction is attained, a constant current is recorded, and the end point is reached.

$$C_6H_5AsO + Cl_2 + 2H_2O \longrightarrow C_6H_5AsO(OH)_2 + 2H^+ + 2Cl^- \qquad (12\text{-}16)$$

This method is a sequential titration procedure that measures the various oxyhalogen species. The sample pH is first raised to pH 12 to decompose the chlorine dioxide in the sample to chlorite and chlorate ions. Upon the lowering of the pH to 7, the sample is titrated for free chlorine. In a second sample, the pH again is first raised to pH 12 and then lowered to pH 7, and potassium iodide is added to the titration flask to react with monochloramine to produce iodine, which is then titrated. To a third sample buffered at pH 7, potassium iodide is added, and the solution is titrated for free chlorine, monochloramine, and one-fifth the chlorine dioxide ($ClO_2 \rightarrow ClO_2^-$, $1e^-$ or one fifth). To a fourth sample, potassium iodide is added, and the pH is lowered to pH 2, where free chlorine, monochloramine, chlorine dioxide, and chlorite ion ($ClO_2^- \rightarrow Cl^-$, $4e^-$ or four-fifths) react to give iodine, which is then titrated with phenylarsine oxide.

Standard Method 4500-ClO_2 E, Amperometric Method II **(see pages 4-57 and 4-58, 19th edition).** Like Amperometric Method I, this procedure uses successive titrations of combinations of chlorine species (as described for Amperometric Method I) whose concentrations can be calculated. Either phenylarsine oxide (PAO) or sodium thiosulfate (NTS) can be used as the titrant. The total mass of the oxidant species should be no greater than 15 mg. Sample volumes in the 200 to 300 ml range are preferred.

Free and combined chlorine and one-fifth of the chlorine dioxide are determined at pH 7. The pH is then lowered to pH 2 to determine the remaining four-fifths of the chlorine dioxide and all of the chlorite ion. Under highly acidic conditions, all the oxidized chlorine species, including chlorate ion, are determined.

Redox Flow Injection Analysis (FIA). FIA designates a group of iodometry-related analytical methods that can measure reliably and with great precision chlorine dioxide and chlorine at concentrations below 1 mg/L. (FIA can also be used to measure chlorite and chlorate ions, but with somewhat less precision.) However, monochloramine can interfere with the FIA redox method. Also, FIA requires a highly skilled technician and specialized, relatively expensive instrumentation.[15,39]

Ion Chromatography (IC). IC is the method of choice for measuring low levels of chlorite, chlorate, bromide, and bromate ions. The USEPA is requiring that IC be used for these measurements. Detection limits of 10–20 μg/L are possible for most laboratories; some have reported detection limits for these species of 1–2 μg/L. The IC method requires a skilled technician and expensive instrumentation. Ion chromatography is approved by USEPA for ICR monitoring of chlorite, chlorate, bromide, and bromate ions.

USEPA Method 300.0, Ion Chromatography.[32] This procedure is based on USEPA Method 300.0 using a borate eluent. Samples to be analyzed for chlorite and chlorate ions are purged with nitrogen to displace chlorine dioxide and further treated with ethylenediamine (EDA).

The sample (25–200 μL) is injected into the eluent at 1.0 ml/min. The sample passes through a metal-free column (MFC) to remove dissolved metals and a guard column before separation on the analytical column. An anion micromembrane suppressor (AMMS) is used with a weak sulfuric acid regenerant solution flowing at 10 ml/min (newer IC systems might use electrochemical suppression, which is equally effective). Detection is by conductivity. In addition, other anions are separated by this method including: F^-, Cl^-, NO_2^-, Br^-, NO_3^-, HPO_4^{2-}, and SO_4^{2-}.

This method may be used to measure chlorite and chlorate ions at 0.05 to 1.0 mg/L. Measurement at lower concentrations can be made by increasing the sample loop volume to 200 μL. USEPA reports the following method detection limits (MDLs): chlorite ion—3.4 μg/L; chlorate ion—9.4 μg/L.

Amaranth.[32] Amaranth is a colorimetric indicator that is selectively decolorized by mg/L concentrations of chlorine dioxide. At the concentrations normally encountered in drinking water, chlorite ion, chlorate ion, and monochloramine do not interfere with chlorine dioxide measurements by this method. Free chlorine will react with amaranth, but the reaction is slow as compared with chlorine dioxide. The chlorine dioxide–amaranth reaction is linear over the 0.25 to 5.5 mg/L range; the product of the reaction is relatively stable and does not require temperature control. Chlorine dioxide–amaranth measurements in the presence of free chlorine made within one minute of mixing the reagents can successfully avoid free-chlorine interference. Alternatively, the free chlorine can be effectively masked by use of reagents such as ethylenediamine. The amaranth method requires minimal technical skill and relatively inexpensive equipment.[15] Like CPR (below), the amaranth method can be readily used to determine residual ClO_2 levels in the treated water.

Chlorophenol Red (CPR). Chlorine dioxide can be measured by its reaction with CPR over the concentration range 0.2 to 2 mg/L, with a detection limit of 0.1 mg/L. The CPR method measures a decrease in color, owing to chlorine dioxide's bleaching action. Neither monochloramine nor chlorate ion interferes with CPR to any meaningful extent, nor does chlorite ion at concentrations below 1.5 mg/L do so, as long as the chlorite ion concentration does not exceed the chlorine dioxide concentration. In the latter case, an interfering intermediate-complex species may be formed. Free chlorine does interfere with CPR measurement of chlorine dioxide, but such interference can be successfully avoided by treatment of the sample with masking reagents that reduce free chlorine, such as oxalic acid. Use of the CPR method requires minimal technical skill and relatively inexpensive equipment.[40] CPR provides a convenient, easy way to determine residual ClO_2 levels in the treated water.

Lissamine Green B (LGB). Chlorine dioxide can be measured by using a spectrophotometric procedure based on ClO_2 decolorization of LGB dye. This procedure is useful for measurement at ClO_2 concentrations of 0.1–1.0 mg/L in treated water; the detection limit for the LGB method is about 0.1 mg/L. In contrast to some other methods, LGB is not subject to interference from chlorine—LGB reacts selectively with ClO_2 in the presence of free available chlorine (FAC), and is unreactive with combined forms of chlorine as well.[41]

DISINFECTION BY-PRODUCTS (DBPs)

General Discussion

Chlorate ion and chlorite ion are the main by-products of chlorine dioxide use. In the normal pH ranges found in potable water and wastewater treatment, chlorine dioxide is reduced principally to chlorite ion when it reacts with organic compounds.[42]

Chlorite Ion Occurrence

Chlorine dioxide's chlorite ion by-product derives from two principal sources:

1. *Generator-originated chlorite.* In generators that use sodium chlorite solution, some chlorite ion by-product can come from unreacted sodium chlorite feedstock that passes through the generator and into the water being treated.
2. *Reduction of ClO_2.* The major source of chlorite ion in the distribution system is the reduction of chlorine dioxide inherent in its desired action on target compounds.[43] This means that a significant percentage (up to ~70%) of the initial ClO_2 dosage ends up as chlorite ion.[44]

Chlorite Ion Minimization

The proposed D/DBP Rule maximum contaminant limit (MCL) on chlorite ion is 1 mg/L. Therefore, to comply with this regulation, ClO_2 dosages in potable water treatment should be limited to about 1.4 mg/L unless the excess chlorite ion is removed.

Avoidance. In ClO_2 production methods that use sodium chlorite solutions, generator-originated chlorite ion can be minimized by feedstock control, by frequent generator tuning, and by running *slightly* chlorine-rich. Alternatively, this source of chlorite ion can be completely eliminated by using a chlorine gas : solid chlorite generator, as any unreacted solid sodium chlorite feedstock remains in the reactor cartridge.

Removal. Chlorite ion formation is inherent in the use of chlorine dioxide. To expand the range of circumstances under which chlorine dioxide can be applied, especially in high-demand water, chlorite ion removal may be desirable.

Two chemical processes, described below, have been developed for chlorite ion removal; both use familiar water treatment chemicals.

Reduced Iron. The use of reduced iron (i.e., ferrous) to remove chlorite ion (by reduction to chloride ion) has been successfully accomplished under laboratory, pilot, and full-scale treatment conditions.[45,46] In the presence of reduced iron (Fe^{2+}), chlorite ion is reduced to chloride ion (Cl^-):

$$4Fe^{2+} + ClO_2^- + 10H_2O \longrightarrow 4Fe(OH)_{3(s)} + Cl^- + 8H^+ \quad (12\text{-}17)$$

The reaction is rapid over the pH 5 to 7 range, and is essentially complete in 15 sec or less for 1 mg/L chlorine dioxide, with no evidence of chlorate formation.[15,47] The reaction is still practicable, albeit somewhat slowed, at pH 7 to 10.[100]

Sulfite Ion. Chlorite ion removal by sulfite ion (by reduction to chloride ion) is efficient over the pH 5 to 6.5 range. Above about pH 7.0, the reaction begins to slow markedly, and is of essentially no utility at high pH.[15] Slootmaekers et al.[48] have made a comprehensive investigation of the use of sulfur dioxide/sulfite ion to remove the unwanted chlorite ion.

The stoichiometry of the sulfur compounds produces reactions according to the following equation:

$$2SO_3^{2-} + ClO_2^{-} \longrightarrow 2SO_4^{2-} + Cl^{-} \qquad (12\text{-}18)$$

This corresponds to two moles of sulfite ion consumed for every mole of chlorite ion removed. Acid solutions (in the presence of atmospheric air) in the pH range 5.5 to 6.0 favor the rapid removal of the chlorite ion and minimize the loss of the sulfite ion from the competing sulfite ion–oxygen reaction. The reaction of Eq. (12-18) is considered to be 95 percent effective; however, the reaction time is an important factor. It was reported that, with a tenfold excess of the sulfite ion and the chlorite ion residual in the 0.5 to 7.0 mg/L range, the complete removal of the chlorite ion occurred in less than one minute at pH 5.0 and below. At pH 6.5, not more than 15 min. was required. Based upon these studies,[48] it was concluded that the chlorite ion by-product could be reduced in the finished water to less than 0.1 mg/L. Table 12-3 shows the removal times for various concentrations of chlorite ion with sulfite ion from pH 5 to pH 8.5.

Chlorate Ion Occurrence

Chlorine dioxide's chlorate ion by-product derives from three principal sources:

Table 12-3 Prediction of Chlorite Ion Half-lives and Removal Times Using Sulfite Ion

pH	ClO_2^- (mg/L)	SO_3^{2-} (mg/L)	$t_{1/2}$ *(sec)*	*99% Removal*
5.0	0.5	5.0	2.9	0.3 min.
5.5	0.5	5.0	38.0	4.4 min.
5.5	1.0	10.0	9.5	1.1 min.
6.5	1.0	10.0	130.0	15.2 min.
6.5	1.0	20.0	32.6	3.8 min.
7.5	1.0	10.0	8,020.0	15.6 min.
7.5	1.0	100.0	80.2	9.4 min.
8.5	1.0	100.0	4×10^4	3.2 days

1. *Oxidation of ClO_2.* The reactions associated with both the chlorine gas: chlorite solution and the chlorine gas: solid chlorite methods (Eqs. 12-11 and 12-12) form an intermediate ($[Cl_2O_2]$) that is very important in chlorate ion by-product formation:[24,45]

$$Cl_2 + ClO_2^- \longrightarrow [Cl_2O_2] + Cl^- \tag{12-19}$$

Under conditions where both reactants are in high concentration, the intermediate is formed readily and produces mainly chlorine dioxide:

$$2[Cl_2O_2] \longrightarrow 2ClO_2 + Cl_2 \tag{12-20}$$

or

$$[Cl_2O_2] + ClO_2^- \longrightarrow 2ClO_2 + Cl^- \tag{12-21}$$

However, under conditions of low initial reactant concentrations or in the presence of excess chlorine (or hypochlorous acid), the intermediate forms chlorate ion:

$$[Cl_2O_2] + H_2O \longrightarrow ClO_3^- + Cl^- + 2H^+ \tag{12-22}$$

$$[Cl_2O_2] + HOCl \longrightarrow ClO_3^- + Cl_2 + H^+ \tag{12-23}$$

Thus, high concentrations of excess chlorite ion favor the production of chlorine dioxide. Chlorate ion production is favored in the presence of high concentrations of free chlorine, which can react with the $[Cl_2O_2]$ intermediate. The stoichiometry of the chlorate-forming reaction is:

$$ClO_2^- + HOCl \longrightarrow ClO_3^- + Cl^- + H^+ \tag{12-24}$$

$$ClO_2^- + Cl_2 + H_2O \longrightarrow ClO_3^- + 2Cl^- + 2H^+ \tag{12-25}$$

Besides the action by excess chlorine in the generator, chlorate ion can also be formed by oxidation of ClO_2 when other treatment chemicals, such as ozone, are fed simultaneously.

2. *Feedstock impurities.* Commercial solution chlorite feedstock is permitted to contain up to a certain percentage of sodium chlorate impurity (e.g., 3 percent of the total weight of a solution containing 25–34 percent sodium chlorite). And if the sodium chlorite is aged beyond its recommended shelf life or exposed to extreme temperatures, it may degrade such that its chlorate ion content increases. In ClO_2 production methods that use chlorite solutions, chlorate ion in the feedstock is passed through the generator into the water being treated.

3. *Photolytic decomposition of ClO_2.* Chlorine dioxide in solution can undergo a series of complex photochemical reactions to form chlorate ion; at least two intermediate species, [Cl_2O_2] and [Cl_2O_3], are thought to be involved, but the reaction mechanisms are not fully understood.[32]

Chlorate Ion Minimization

Removal. There are no extant, cost-effective, chemical removal methods for chlorate ion. For example, unlike chlorite ion, chlorate ion cannot be removed by sulfur compounds owing to its very long half-life at mg/L levels. Slootmaekers et al.[48] reported that the calculated half-life of the chlorate ion treated in solution with excess sulfur dioxide/sulfite ion exceeds many months!

Avoidance. Chlorate ion production is not inherent in chlorine dioxide use and can be substantially avoided. In ClO_2 production methods that react chlorine gas or chlorine solution with chlorite solutions, generator-contributed chlorate ion can be minimized through careful feedstock handling procedures and by frequent generator tuning to avoid excess chlorine. In chlorine gas : solid chlorite generators, any sodium chlorate impurity in the sodium chlorite feedstock remains in the reactor cartridge. Thus, there is no possibility of generator-originated chlorate ion with chlorine gas : solid chlorite-type generators. Beyond generator-related considerations, other steps necessary for chlorate ion avoidance are: (1) *never* mix chlorine dioxide with other oxidants (e.g., ozone), and (2) protect chlorine dioxide from exposure to light.

USES IN DRINKING WATER TREATMENT

General Discussion

Chlorine dioxide is a versatile water treatment chemical. As with other treatment modalities, each prospective application (e.g., taste, odor, and color control; manganese oxidation) may be subject to complex kinetic interrelationships in the series of reactions that occur in natural water. Therefore, to determine the appropriate ClO_2 application parameters, each water must be studied as a special case.

The measurement of oxidant demand—the difference between the added oxidant dose and the residual oxidant concentration after a prescribed contact time—is a necessary first step in applying chlorine dioxide or any other oxidant. Oxidant demand is highly dependent on the initial oxidant dose. Chlorine dioxide demand generally increases nonlinearly with time, and it must be characterized for a given contact time, under particular pH and temperature conditions. Oxidant consumption data generally cannot be extrapolated from one set of conditions to another; rather, studies should be performed corresponding to the range of conditions expected in the field, with samples taken at several time intervals so that proper demand curves can be constructed.

Disinfection

Germicidal Efficiency. Since the U.S. introduction of chlorine dioxide (1944) for the treatment of water supplies, many investigations have been made to determine its germicidal efficiency. The early investigations were carried out as a comparison to chlorine and were, for the most part, performed in sterile, chlorine-demand-free systems; chlorine dioxide was being compared to free chlorine in these instances. Subsequent investigations compared chlorine dioxide to chlorine in ammonia-laden wastewater. In 1947, Ridenour and Ingols[49] concluded that chlorine dioxide was at least as effective as chlorine, and, in contrast to chlorine, the bactericidal efficiency of chlorine dioxide was relatively unaffected by pH values between 6 and 10. Ridenour and Armbruster[50] found that less than 0.1 mg/L chlorine dioxide destroyed the common water pathogens *Eberthella typhosa, Shigella dysenteriae*, and *Salmonella paratyphi B* at temperatures between 5°C and 20°C at pH values above 7 with a 5-min. contact period. An increase in pH brought about an increase in germicidal efficiency.

Ridenour and Ingols[51] also reported that chlorine dioxide was clearly superior to chlorine in the destruction of spores on an equal residual basis. Using a 5-min. contact period, a 99.9 percent reduction of *Bacillus subtilus* in a chlorine-demand-free chlorine suspension required a 1.0 mg/L ClO_2 residual, as compared with a 3.5 mg/L free chlorine residual. They believed that in the case of the spores, the cell wall was penetrated by the use of the five valence changes in the oxidation of chlorine dioxide, but that vegetative cells do not utilize this phenomenon.

Malpas[52] concluded that the bactericidal efficiency of chlorine dioxide towards *E. coli, Salmonella typhosa*, and *Salmonella paratyphi* was as great as or greater than that of chlorine.

Hettche and Ehlbeck[53] investigated the action of chlorine dioxide on poliomyelitis virus and reported that it is more effective than either ozone or chlorine. This superior effectiveness may be explained by a chemical characteristic of chlorine dioxide reported by Ingols and Ridenour in 1948,[54] who found that chlorine dioxide (but not chlorine) reacted with peptone in amounts that followed the laws of adsorption. As viruses have a protein coat, it is likely that chlorine dioxide is adsorbed by the virus coat. This would cause higher local concentrations on the surface of the virus than would be expected from the measured residual and may account for the significant viricidal activity of chlorine dioxide. This theory appears to be compatible with the findings of Bernarde et al.,[55] who investigated the mechanism of kill by chlorine dioxide. This work indicates that ClO_2 does not react sufficiently with amino acids to alter their characteristic structures, thus eliminating the possibility of a reaction within the cell. They found, however, that in some unknown fashion ClO_2 abruptly inhibits protein synthesis; this action is probably the mechanism by which it destroys these vegetative organisms.

It was found that at pH 4.0, 6.45, and 8.42 the chlorine dioxide molecule remained unaltered and intact, and it was concluded that ClO_2 must therefore be the bactericidal compound. This is consistent with the fact that chlorine dioxide does not react with water to hydrolyze, as does chlorine.

Another important achievement of the work of Bernarde et al. was establishing the disinfecting ability of ClO_2 and its relative efficiency as a function of pH. Figure 12-12 illustrates the relative efficiency of both ClO_2 and chlorine with respect to pH. At pH 6.5, chlorine appears to be more efficient at the lower dosages. Both compounds are equally efficient at an initial dose of 0.75 mg/L. Increasing the pH to 8.5 results in a dramatic change in efficiency. As would be expected, the chlorine efficiency drops, for at this pH level only 8.72 percent of the residual is HOCl, whereas at pH 6.5 the residual is 89.2 percent HOCl. As the chlorine efficiency drops with the increase in pH, the chlorine dioxide efficiency increases. A 99+ percent destruction of *E. coli* in 15 sec with a 0.25 mg/L dose of ClO_2 is noted at pH 8.5, as compared with a 0.75 mg/L dose required for the same destruction by chlorine.

Longley et al.[57,58] compared the disinfection efficiencies of chlorine and chlorine dioxide in a sidestream from the effluent of a 90 million gallons per day (MGD) wastewater treatment plant; the flow rate of the sidestream through the contactor was about 0.9 MGD. This investigation showed that a 5.0 mg/L chlorine dioxide dose resulted in a 5 log reduction of fecal coliforms. This proved to be greater by more than one log reduction than using the same dose of chlorine at 30 min. contact time. Moreover, the inactivation of fecal coliforms by chlorine dioxide in 3 min. exceeded the inactivation by chlorine at 30 min. for the same dosage.

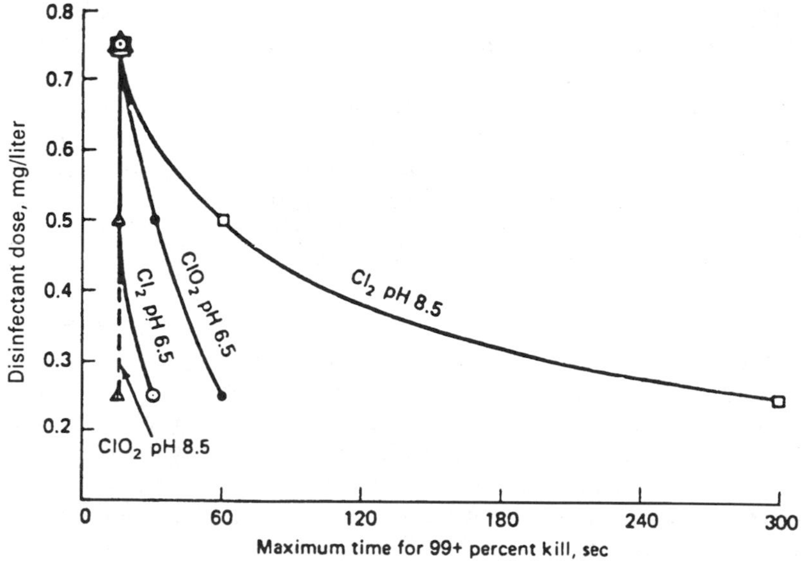

Fig. 12-12. Relative germicidal efficiency of chlorine and chlorine dioxide.[51] (Dosage in mg/L versus time required for a 99+ percent destruction of *E. coli* at the pH indicated).

Even more dramatic results were observed for the viricidal efficiency of chlorine dioxide relative to chlorine under similar test conditions. The observed chlorine dioxide coliphage inactivation of a 3 log reduction at 3 min. contact time exceeded the comparable chlorine inactivation by nearly a 2 log reduction.

A comprehensive investigation of chlorine dioxide as a disinfectant was undertaken by the Stanford group.[22,35,59–61] The investigation used the natural coliform populations in the effluents from three different wastewater treatment plants; all three plants produced conventional activated sludge effluents. Two plants filtered the secondary effluent. However, the focus of this investigation was on the nonfiltered secondary effluents.

All of the coliform studies were based on the natural population in the secondary effluent. The animal virus used was Poliovirus I, LCS strain grown in Buffalo Green Monkey kidney cells. Several disinfection experiments were run on different days. In order to compare the bactericidal efficiency of chlorine and chlorine dioxide and to determine the relationships of dosages and contact times to bactericidal efficiency, the logarithm of the organisms' survival ratio was calculated for each experiment. Results are shown in Fig. 12-13.

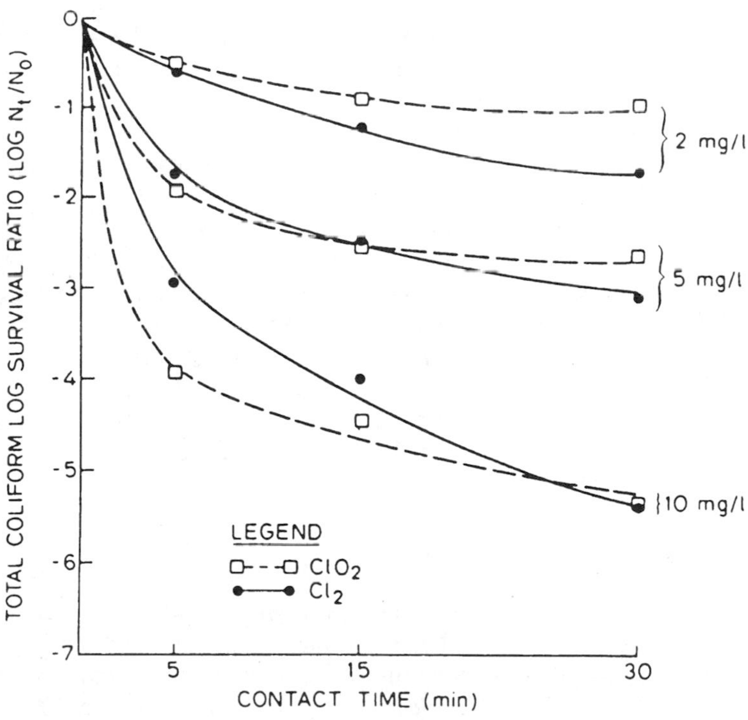

Fig. 12-13. Comparison of germicidal efficiency between chlorine dioxide and chlorine[23] showing coliform survival in a non-nitrified secondary affluent.

Chlorine dioxide demonstrated a more rapid coliform inactivation than chlorine at the shortest contact time (5 min.). The mean log survival ratio in one group of experiments was −4.40 for chlorine dioxide, compared to chlorine at −3.44. The dosage for each was 10 mg/L. It is clear that chlorine dioxide demonstrates a more rapid kill rate than chlorine at shorter contact times.

In the experiments with a nitrified and filtered effluent chlorine dioxide was shown to be superior to chlorine.[22] The advantage of chlorine dioxide was found to be due to filtration, not nitrification. The presence or absence of ammonia N has no effect on the germicidal efficiency of chlorine dioxide.

Viruses. Selected experiments were conducted to evaluate the viricidal effectiveness of chlorine dioxide as compared to chlorine (combined).[22] Palo Alto secondary effluent (nonnitrified) was dosed with 5 mg/L chlorine dioxide and chlorine. The sample used contained in situ coliphage and an inoculum of Poliovirus I. Samples were taken at several contact times and analyzed for coliphage, Poliovirus, and fecal coliforms. The results are shown in Fig. 12-14. These experiments demonstrate that chlorine dioxide is a far better viricide than combined chlorine. The reaction of combined chlorine with coliphage agrees with work by Snead and Olivieri.[62]

Encysted Parasites. Outbreaks of cryptosporidiosis and giardiasis on account of waterborne encysted parasites have been documented in the United States, Canada, and Great Britain.[63–66] At practical doses, chlorine, monochloramine, and UV irradiation have been shown wholly ineffective against *Giardia* cysts and *Cryptosporidium* oocysts.[67–69] *Cryptosporidium* oocysts are smaller, harder to detect, and more difficult to kill than *Giardia* cysts. Among alternative chemical disinfectants, chlorine dioxide and ozone are the only promising candidates for controlling encysted parasites. Using comprehensive animal-infectivity studies as a measure of oocyst viability, Finch and coworkers[36,99] have shown *quantitatively* that chlorine dioxide can achieve necessary levels of *Cryptosporidium* kill at practical dosage levels. This same research team has shown that sequential use of oxidants, such as ozone *followed* by chlorine dioxide,[70] results in remarkable, synergetic inactivation levels far greater than the simple addition of kill levels achieved with the application of these oxidants individually.

Control of Off-tastes and Odors

The first known use of chlorine dioxide in a water treatment process in the United States was at the No. 2 Niagara Falls, New York, Water Treatment Plant in January, 1944.[71–73] In September 1944, a chlorine dioxide generator was installed at the Western New York Water Company plant at Woodlawn, New York. Other plants that early adopted chlorine dioxide were those at Greenwood, South Carolina, and at Tonawanda, North Tonawanda, Lockport,

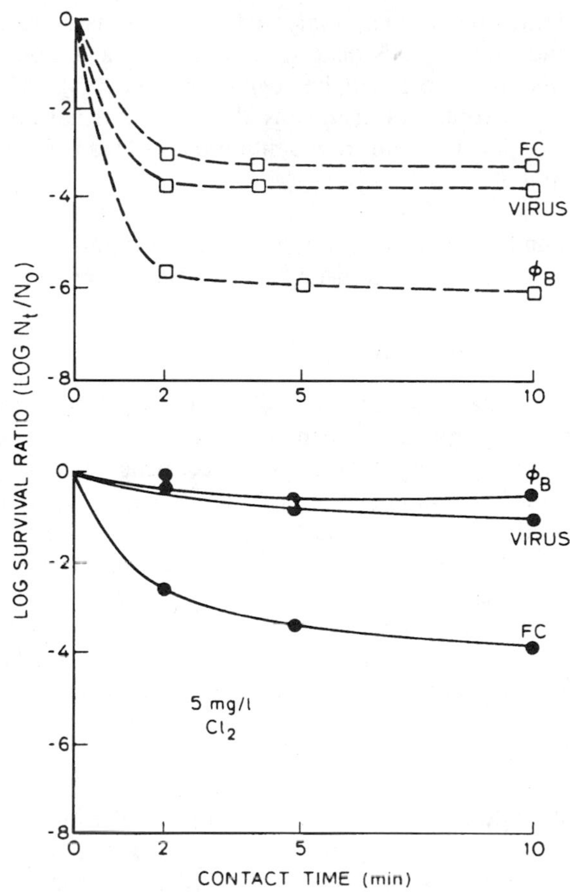

Fig. 12-14. Comparison of viricidal efficiency between chlorine dioxide and chlorine using an in situ coliphage and an inoculum of poliovirus I in a non-nitrified secondary effluent.[23] 5 mg/l ClO_2. FC = fecal coliform; ϕ_B = coliphage, *E. coli* B host; virus = polio virus.

and Port Colbourne, New York. All of these plants had experienced serious trouble from phenolic tastes and odors as a result of industrial waste pollution of the raw water. There were also varying degrees of taste and odors from algae and vegetation decomposition in the summer.

Chlorine dioxide seemed also to assist in reducing consumer complaints where it was necessary to carry free chlorine residuals in the distribution system. The success of this early work inspired further investigation into the properties of chlorine dioxide.

There are three principal reasons why chlorine dioxide is the chemical of choice in dealing with tastes and odors arising from phenolic compounds:

1. Chlorine dioxide reacts completely to destroy the taste-producing phenolic compounds many times faster than does free available chlorine.
2. In all cases studied so far, it has been reported that ClO_2 will always destroy any chlorophenol taste caused by prechlorination.
3. The efficiency of ClO_2 is improved, not impaired (as is HOCl), by a high-pH environment.

Chlorine dioxide also oxidizes some other off-taste and odor-causing compounds, such as mercaptans and disubstituted organic sulfides.

Algae and Decaying Vegetation

Chlorine dioxide is usually associated with correcting phenolic and chlorophenolic tastes and odors; however, in many cases, it has also been found to control successfully a variety of off-tastes occurring from both algae and decaying vegetation.[74–80] In some of these cases, it seems certain that actinomycetes are the offending organisms, and that resulting tastes can be controlled.

Chlorine dioxide has repeatedly proved successful in controlling musty tastes,[78] fishy tastes, and odors from *Anabena, Asterionella, Synura, Vorticella*, and others,[75,76] undoubtedly including actinomycetes. In nearly all these cases, the increased cost of using chlorine dioxide instead of chlorine has been offset to a certain extent by a reduction in the amount of activated carbon required.

Preoxidation

Preoxidation with chlorine dioxide has been highly effective for destruction of THM precursors and has been reported to improve coagulation and settling properties, resulting in substantially improved filter-run times.

Zebra Mussels

Chlorine dioxide has been proved effective in controlling mollusks, such as zebra mussels (*Dreissena polymorpha*), which can clog the intake pipes of water treatment plants.[47]

Manganese Removal

Manganese, a common ingredient of impounded water and of a great many well waters, can in minute quantities cause offensive black water in distribution systems. The difficulties encountered are well known: staining of clothes and plumbing fixtures, black water, and incrustation of mains, often giving rise to tremendous amounts of debris at the consumer's tap.

Some of the difficulty can be attributed to the action of free available chlorine, which reacts so slowly with manganese that 24 hours after treatment the manganese will become oxidized and slowly precipitate out of solution.

This reaction occurs in the distribution system. The overall reaction is as follows:

$$2ClO_2 + MnSO_4 + 4NaOH \rightarrow MnO_2 + 2NaClO_2 + Na_2SO_4 + 2H_2O \quad (12\text{-}26)$$

Recent research at Fort Collins, Colorado[81] has shown chlorine dioxide to be superior not only to chlorine but also to ozone and potassium permanganate for manganese oxidation.

Iron Removal

Chlorine dioxide can be used effectively for the oxidation of iron from the moderately soluble ferrous state to the ferric ion, resulting in the formation of ferric hydroxide as a heavy gelatinous brown floc, which can be removed by sedimentation followed by filtration. As with manganese, the optimum pH is always higher than 7, preferably 8 or 9, and must be determined by laboratory procedures. The rate at which chlorine dioxide reacts in the oxidation of iron is not so fast as with manganese. Chlorine is often the chemical of choice for iron removal, unless the water being treated has high THM formation potential, contains high levels of ammonia or other nitrogenous compounds that would exert a high chlorine demand, or contains phenols or other compounds that react with chlorine to produce off-tastes and odors.

Usually, the amounts of iron and manganese involved in the removal processes are so small that the consumption of alkalinity caused by the above reactions will be insignificant. Manganese removals are usually less than 1.0 mg/L, and those for iron less than 5.0 mg/L. However, if there is a significant reduction in alkalinity, it can be restored by the addition of commercially available chemicals, such as hydrated lime ($Ca(OH)_2$), soda ash (Na_2CO_3), or caustic (NaOH).

HEALTH AND SAFETY ISSUES

Fire and Explosion Hazard

Chlorine Dioxide. Chlorine dioxide has a significant vapor pressure, which increases as a function of dissolved chlorine dioxide and time. (See Table 12-4.) Care should be taken in handling concentrated aqueous solutions of ClO_2, especially in confinement, such as the headspace of a ClO_2 solution storage tank. Generally, vapor concentrations of chlorine dioxide in excess of 75–80 mm Hg are considered dangerous and should be avoided. This corresponds to the vapor pressure over a ~7.5 g/L ClO_2 solution at 25°C.

Sodium Chlorite. Sodium chlorite is a strong oxidizer. If sodium chlorite solution dries in contact with combustibles, it can ignite. For example, sodium

Table 12-4 ClO_2 Solubility vs. Vapor Pressure @ 25°C

ClO_2 (mg/L)	Vapor Pressure (mm Hg)	Atmospheres
100	1.0	0.001
500	5.1	0.007
1,000	10.3	0.013
2,000	20.5	0.027
4,000	41.0	0.054
6,000	61.6	0.081
8,000	82.1	0.108
10,000	102.6	0.135
15,000	153.9	0.202
20,000	205.2	0.270
25,000	256.5	0.337
30,000	307.8	0.405
40,000	410.4	0.540
50,000	513.0	0.675
55,000	564.3	0.742
60,000	615.6	0.810
65,000	666.9	0.877
70,000	718.2	0.945
75,000	769.5	1.012
80,000	820.7	1.080
85,000	872.0	1.147
90,000	923.3	1.215
95,000	974.6	1.282
100,000	1,025.9	1.350

chlorite solution spilled on wood can dry and then burst into flames, the wood serving as a fuel source. Similarly, contact with oil or grease, paper products, textiles, leather, and the like can cause dried sodium chlorite to combust.

Sodium chlorite is also reactive with a number of chemicals, such as acids and hypochlorites, contact with which can cause the uncontrolled release of chlorine dioxide gas. Pure, dry sodium chlorite is shock-sensitive. Also, dry sodium chlorite should be kept away from open flames, sparks, or excessive heat; it will decompose at temperatures above about 175°C (347°F). Commercial sodium chlorite formulations contain additives that reduce shock and heat sensitivity, and improve sodium chlorite's safety profile. So long as the manfuacturers' precautions are heeded, especially those concerning contamination with organics, commercial grades of sodium chlorite are relatively safe to handle.[83]

Toxicity

Chlorine Dioxide. ClO_2 is a mucosal irritant: acute exposure can cause eye, nose, throat, and lung irritation; prolonged exposure can cause bronchitis, reactive airways disease, and pulmonary edema. However, its health effects are not cumulative, and considerable evidence shows ClO_2 is unlikely to be carcinogenic.

Chlorite Ion. Animal studies have shown chlorite ion to be capable of inducing the sort of blood effects associated with oxidative stress, such as oxidative hemolytic anemia and methemoglobinemia.[84,85] Multigenerational rodent studies completed in 1996 dispelled earlier concerns raised about possible neonatal neurodevelopmental toxicity of chlorite ion.[38,86]

Chlorate Ion. Toxicological information on chlorate ion is limited, with only acute chlorate toxicity having been addressed.[87]

MYSTERIOUS TASTES AND ODORS ISSUE

A survey of 37 utilities using chlorine dioxide found that, although relatively rare, strange odors were sometimes associated with chlorine dioxide use.[88–90] Infrequent, seemingly random but nonetheless consistent customer complaints of unpleasant odors were reported when chlorine dioxide was applied to drinking water supplies. The strange aspects of the problem were not only the descriptions of the odors but their origin. The odors were variously described as kerosene- or cat urine–like. On investigation, these odors were determined to be associated with the presence of new carpeting in individual complainants' homes; the chlorine dioxide–associated odors were apparently being formed in the atmosphere around the tap rather than in the raw or finished water entering the distribution system.

On further investigation, it was learned that chlorine dioxide can be regenerated from chlorite ion, either directly or indirectly, by reactions with hypochlorous acid—even when ClO_2 dosage at the plant produces residuals at the consumer's tap of less than 0.20 mg/L. Apparently, the offensive odors are produced by reactions between regenerated ClO_2 vapor escaping from the tap and volatile organic compounds in the household air, such as those released by new carpeting. The investigation concluded that these odor occurrences can be prevented if either free residual chlorine is removed from the water at the entrance to the distribution system (e.g., by the use of chloramines instead of free chlorine for maintaining a distribution system), or chlorite ion is removed (e.g., by using sulfite ion or ferrous salts).

REGULATORY PERSPECTIVE

The central piece of legislation governing drinking water is the Safe Drinking Water Act (SDWA), under which there are various existing and proposed rules. One of these, the proposed Disinfectant/Disinfection By-Products (D/DBP) Rule, would set stricter standards for certain chemicals that are already regulated, such as total THMs, and would also set limits in allowable concentrations of other compounds for which there are no present standards, such as haloacetic acids (HAAs), bromate ion, and chlorite ion.[38]

Under the proposed D/DBP Rule, allowable levels of chlorite ion—chlorine dioxide's principal DBP—would be 1.0 mg/L, based on assessment of the risk associated with possible blood effects in the most sensitive subpopulations (*e.g.*, people genetically deficient in certain blood enzymes that protect against oxidative stress.)

As set forth in the text of the proposed D/DBP Rule, the USEPA has concluded that the presently available health-effects data are insufficient to establish safe chlorate ion exposure levels at this time. Further research is planned, and the USEPA has expressed its intention to set allowable chlorate ion limits in the future. In the meantime, chlorate ion levels should be kept as low as possible.

Related to the proposed D/DBP Rule, the Information Collection Rule (ICR), went into effect in early 1997. Under the ICR, large systems (*i.e.*, serving populations greater than 100,000) using chlorine dioxide are required to report measurements for chlorite and chlorate ions in the plant influent, before by-product removal treatment (e.g., with GAC, reduced iron, or sulfite ion), and at three points in the distribution system (beginning, middle, and end). Also, the ICR requires that bromate ion be measured by systems using chlorine dioxide because, at the time the ICR was drafted, it was unclear whether or not chlorine dioxide would behave like ozone and oxidize bromide ion to form bromate ion.

Chlorine dioxide vapor has a U.S. Occupational Health and Safety Administration (OSHA) permissible exposure limit (PEL) of 0.1 ppm, 8-hour time-weighted average (TWA), and a short-term exposure limit (STEL) of 0.3 ppm, 15-min. TWA.

OTHER CURRENT AND PROSPECTIVE APPLICATIONS

Wastewater Disinfection

Great interest periodically has centered upon the improvement of wastewater disinfection techniques in the United States.[91] The first critical study of chlorine dioxide for wastewater treatment was by Bernarde et al.[56] in 1965. This was a laboratory investigation using sterile solutions seeded with *E. coli*.

There is very little field experience with the application of chlorine dioxide as a disinfectant for either secondary or tertiary effluents. There are no known chlorine dioxide wastewater installations (other than pilot plants) in the United States at this time. Evaluation of chlorine dioxide as a wastewater disinfectant was undertaken by the Stanford group.[22] Their findings, which dealt with wastewater effluents and the native population of organisms, indicate that chlorine dioxide is only slightly superior to chlorine as a bactericide but is a much superior viricide. They concluded that chlorine dioxide's primary attraction for wastewater treatment is as a viricide. This is of considerable interest to health agencies where wastewater reclamation is considered.

The findings of the Stanford group were confirmed by a lengthy investigation in France. Covering a period of two to three years, this study examined the disinfection efficiency of various disinfectants on effluents of varying quality—primary, secondary, and filtered secondary.[92]

Combined Sewer Overflow (CSO)

An estimated 1,200 U.S. cities have combined sanitary and storm sewer systems. During heavy wet-weather events, the substantially increased water volume, comprising a mixture of runoff and raw sewage, may greatly exceed the sewer capacity. This excess water is discharged, generally into local waterways, and is often the cause for closings of nearby beaches, riverbank recreational areas, and so forth. There is increasing regulatory and political pressure to treat this sewage–runoff mixture prior to discharge. However, this material poses a number of difficult technical hurdles: it flows intermittently in huge volume at wildly (and quickly) variable rates, its pH is subject to great excursions, and it can contain high ammonia levels. Because of the sometimes very high pH and ammonia content, this water can exert a chlorine demand as high as 100 mg/L. And because of the high volumes and rapid flow rates, the available contact time may be very brief.

Because chlorine dioxide is unreactive with ammonia, is effective across a broad pH range, is more soluble in water than chlorine, has a front-end-weighted bacterial kill curve, and is a superior viricide relative to chlorine, it is an attractive candidate for CSO treatment.

Medical Device Sterilization

Chlorine dioxide gas is effective for the sterilization of packaged medical products.[93–95] This use of chlorine dioxide, developed by Rosenblatt and coworkers, was begun as an alternative to gas sterilization processes using ethylene oxide. It was also the application for which the chlorine gas : solid chlorite generation process was originally developed.

Food Processing

Chlorination of process water in canneries and frozen food packaging became an established practice about 1946 in the United States. It was so successful in potato dehydration plants that its use spread rapidly to fruit and vegetable canning operations.

Originally, the cannery water supply was a once-through system. However, after some 10 years of breakpoint chlorination, canneries attempted to conserve water, primarily to cut down on the cost of treating the wastewater. Thus a recycling system for food processing plants was evolved. It was soon discovered that the ammonia nitrogen concentration increased significantly in the recycled water, so that it became difficult to control the breakpoint process unless multiple points of application were used. It was found impractical to pursue the free residual process using breakpoint chlorination because of excessive chlorine consumption combined with the generation of intolerable amounts of nitrogen trichloride, which pervaded the working area.

In 1975, the Green Giant Co. decided to try chlorine dioxide.[96] Its Blue Earth plant chlorination system was retrofitted to generate chlorine dioxide. The prime objective was to determine the feasibility of water conservation through water recycling while maintaining acceptably low bacteria counts in the complete system. The experiment was a success. The chlorine dioxide treatment of the once-used water was highly effective in the control of bacteria growth and biofouling in pea and corn canneries. The persistent residuals and the lack of reaction with ammonia nitrogen made possible one-point application, rather than the multiple points of application necessary with rechlorination of used water. Chlorine dioxide residuals in the reuse water did not result in the generation of offensive odors in the plant, nor were there any off-favors produced in the product.

The success at this plant rapidly spread to other food processing plants. Since 1975 a majority of these processing plants have retrofitted their existing chlorination systems or purchased new chlorine dioxide systems. These plants include all types of food processing. The success of chlorine dioxide use in the recycling systems is the result of the chemical characteristics of chlorine dioxide: (1) it will not react with ammonia nitrogen, and (2) the residuals persist over a long period of time, so that a single point of application is sufficient. In most cases the chlorine dioxide dosages vary from 2 to 8 mg/L.

Chlorine dioxide has been proved effective for the disinfection of poultry chiller water, which is known for its high content of organic matter. Recently, the U.S. Food and Drug Administration (FDA) approved this use.[97]

SUMMARY

Advantages of Chlorine Dioxide

- Chlorine dioxide is an effective, fast-acting, broad-spectrum bactericide
- It is superior as a viricide to chlorine, which makes it a promising candidate for water reuse disinfection.
- It kills chlorine-resistant pathogens—for example, encysted parasites *Giardia* and *Cryptosporidium.*
- It does not react with ammonia nitrogen or primary amines.
- It does not react with oxidizable organic material to form trihalomethanes.
- It destroys THM precursors and enhances coagulation.
- It is excellent for the destruction of phenols, which cause taste and odor problems in potable water supplies.
- It has a long track record in the removal of iron and manganese. It is superior to chlorine, particularly when the iron and manganese occur in complexed compounds.

Disadvantages of Chlorine Dioxide

- The cost of chlorine dioxide, several times more expensive than chlorine, may make its use prohibitive for certain applications, especially in economically deprived (e.g., Third World) regions where even chlorination is not readily affordable.
- Chlorine dioxide cannot be transported as a compressed gas; it must be generated on-site.
- The chlorine dioxide prepared by some processes may contain significant amounts of free chlorine, which could defeat the objective of using ClO_2 to avoid the formation of THMs.

ACKNOWLEDGMENTS

The author acknowledges the valuable contributions to the revision of this chapter by Drs. Bernard Bubnis, Gilbert Gordon, Donald Gates and David Rosenblatt, as well as the information and suggestions provided by Drew Industrial Division of Ashland Chemical Company.

REFERENCES

1. Granstrom, M. L., and Lee, G. F., "Generation and Use of Chlorine Dioxide in Water Treatment," *J. AWWA,* **50,** 1453 (1958).
2. Benninger, R. W., Hoehn, R. C., and Rosenblatt, A. A., "Chlorine-Free Chlorine Dioxide Gas for Drinking Water Treatment at Roanoke County, Virginia," *Proceedings* of the American Water Works Association, Water Quality Technology Conference, Boston, MA, Nov. 17–21, 1996.
3. Hoehn, R. C., Rosenblatt, A. A., and Gates, D. J., "Considerations for Chlorine Dioxide Treatment of Drinking Water," *Proceedings* of the American Water Works Association, Water Quality Technology Conference, Boston, MA, Nov. 17–21, 1996.
4. Miller, G. W., et al., "An Assessment of Ozone and Chlorine Dioxide Technologies for Treatment of Municipal Water Supplies," Executive Summary, USEPA 60018-78-018, Cincinnati, OH, Oct. 1978.
5. "Industrial Pollution of the Lower Mississippi River in Louisiana, Based upon the Harris report" U.S. Environmental Protection Agency, Cincinnati, OH, Apr. 1972.*
6. Miltner, R. J., "The Effect of Chlorine Dioxide on Trihalomethanes in Drinking Water," master of science thesis, Univ. of Cincinnati, Cincinnati, OH, 1976.
7. Hoehn, R. C., "Chlorine Dioxide Use in Water Treatment: Key Issues," *Proceedings,* 2nd International Symposium, Chlorine Dioxide and Drinking Water Issues, Houston, TX, pp. 1–14, 1993 (held in 1992).
8. Symons, J. M., et al., "Ozone, Chlorine Dioxide, and Chloramines as Alternatives to Chlorine for Disinfection of Drinking Water," Environmental Protection Agency, Cincinnati, OH, Nov. 1977.
9. Kennett, C. A., "Experience with the Use of Chlorine Dioxide for Distribution Problems," Water Research Centre Conference, Medmenham Laboratory, Medmenham, Bucks, England, 1978.

*Formerly the U.S. Public Health Service

10. Dowling, L. T., "Chlorine Dioxide in Potable Water Treatment," *Water Treatment and Exam.,* **23,** 190 (1973).
11. McConnell, G., "Halo-organics in Water Supplies," *J. Inst. of Water Engrs. and Scientists,* **30,** 431 (Nov. 1976).
12. Sconce, J. S., *Chlorine: Its Manufacture, Properties and Use,* Reinhold, New York, 1962.
13. *Kirk-Othmer Encyclopedia of Chemical Technology,* 2nd ed. Vol. 5, Interscience, New York, 1964.
14. Gordon, G., Kieffer, R. G., and Rosenblatt, D. H., "The Chemistry of Chlorine Dioxide," in S. J. Lippard (Ed.), *Progress in Inorganic Chemistry,* Vol. XV, John Wiley & Sons, New York, 1972, pp. 202–286.
15. Gordon, G., and Bubnis, B., "Chlorine Dioxide Chemistry Issues," *Proceedings* of the Third International Symposium, Chlorine Dioxide: Drinking Water, Process Water, and Wastewater Issues, New Orleans, LA, Sept. 14–15, 1995.
16. Richardson, S. D., Thurston, A. D., Collette, T. W., Patterson, K. S., Lykins, B. W., Majetich, G., and Yong, Z., "Multispectral Identification of Chlorine Dioxide Disinfection Byproducts in Drinking Water," *Environ. Sci. Technol.,* **28**(4) (1994).
17. Gordon, G., and Bubnis, B., "Bromate Ion Formation in Water When Chlorine Dioxide Is Photolyzed in the Presence of Bromide Ion," *Proceedings* of the American Water Works Association, Water Quality Technology Conference, New Orleans, LA, Nov. 12–16, 1995.
18. Hoigne, J., and Bader, H., "Kinetics of Reactions of Chlorine Dioxide (OClO) in Water I. Rate Constants for Inorganic and Organic Compounds," *Water Res.,* **28**(1), 45–55 (1994).
19. Weinberg, H. S., Glaze, W. H., Krasner, S. W., and Sclimenti, M. J., "Formation and Removal of Aldehydes in Plants That Use Ozonation," *J. AWWA,* **85**(5), 2 (1993).
20. Pitochelli, A., "Chlorine Dioxide Generation Chemistry," *Proceedings* of the Third International Symposium, Chlorine Dioxide: Drinking Water, Process Water, and Wastewater Issues, New Orleans, LA, Sept. 14–15, 1995 (in press).
21. U.S. Patent No. 5,110,580, "Method and Apparatus for Chlorine Dioxide Manufacture," Rosenblatt, A. A., Rosenblatt, D. H., Feldman, D., Knapp, J. E., Battisti, D., and Morsi, B., May 5, 1992.
22. Roberts, P. V., Aieta, E. M., Berg, J. D., and Chow, B., "Chlorine Dioxide for Wastewater Disinfection: A Feasibility Evaluation," Technical Report No. 251, Civil Engineering Dept., Stanford Univ., Palo Alto, CA, Oct. 1980.
23. Jordan, R. W., Kosinski, A. J., and Baker, R. J., "Improved Method for Generating Chlorine Dioxide," TA 1053-C, Wallace & Tiernan, Belleville, NJ, Apr. 1980.
24. Gordon, G., and Rosenblatt, A. A., "Gaseous, Chlorine-Free Chlorine Dioxide for Drinking Water," *Proceedings* of the American Water Works Association, Water Quality Technology Conference, New Orleans, LA, Nov. 12–16, 1995.
25. Gordon, G., and Rosenblatt, A. A., "Gaseous, Chlorine-Free Chlorine Dioxide for Drinking Water," *Proceedings* of Chemical Oxidation: Technologies for the Nineties, Fifth International Symposium, Vanderbilt University, Nashville, TN, Feb. 15–17, 1995.
26. Woodway, E., "Chlorine Dioxide for Odor Control," *TAPPI,* **36**(5), 216–221 (May 1953).
27. Hoehn, R. C., and Rosenblatt, A. A., "Full-Scale, High-Performance Chlorine Dioxide Gas Generator," poster presented at the American Water Works Association, Engineering and Construction Conference, Denver, CO, Mar. 1996.
28. U.S. Patent No. 5,041,196, "Electrochemical Method for Producing Chlorine Dioxide Solutions," Cawlfield, D. W., and Kaczur, J. J., Aug. 20, 1991.
29. U.S. Patent No. 5,084,149, "Electrolytic Process Producing Chlorine Dioxide," Kaczur, J. J., and Cawlfield, D. W., Jan. 28, 1992.
30. International Patent Application No. PCT/CA94/00263, "Chlorine Dioxide Generation for Water Treatment," Cowley, G., Lipszatjn, M., and Ranger, G., published Nov. 24, 1994 (filed May 12, 1994).
31. Gallagher, D. L., Hoehn, R. C. and Dietrich, A. M., "Appendix A—Detailed Instructions for Amperometric Titration," in *Sources, Occurrence and Control of Chlorine Dioxide*

By-product Residuals in Drinking Water, American Water Works Association Research Foundation, Denver, CO, 1994.

32. Gates, D. J., *The Chlorine Dioxide Handbook,* American Water Works Association, Denver, CO, 1998.
33. *Standard Methods for the Examination of Water and Wastewater,* 19th edition, APHA, AWWA, WEF, Washington, DC, 1995.
34. Hood, N. J., "A Laboratory Evaluation of the DPD and Leuco Crystal Violet Methods for the Analysis of Residual Chlorine Dioxide in Water," master of science thesis, Virginia Polytechnic Institute, Blacksburg, VA, Oct. 1977.
35. Aieta, E. M., Berg, J. D., and Roberts, P. V., "Comparison of Chlorine Dioxide and Chlorine in the Disinfection of Wastewater," paper presented at the WPCF Annual Conference, Anaheim, CA, Oct. 1–6, 1978.
36. Finch, G., Liyanage, L. R. J., and Belosevic, M., "Effect of Chlorine Dioxide on *Giardia* and *Cryptosporidium,*" *Proceedings* of the Third International Symposium, Chlorine Dioxide: Drinking Water, Process Water, and Wastewater Issues, New Orleans, LA, Sept. 14–15, 1995.
37. LeChevallier, M. W., Arora, H., Battigelli, D., and Abbaszadegan, M., "Chlorine Dioxide for Control of Cryptosporidium and Disinfection By-Products" *Proceedings* of the American Water Works Association, Water Quality Technology Conference, Boston, MA, Nov. 17–21, 1996.
38. USEPA, "Proposed Disinfectants/Disinfection By-Products Rule," *Federal Register.* 59 FR, July 29, 1994.
39. Themelius, D., Wood, D., and Gordon, G., "Determination of Low Concentrations of Chlorite Ion and Chlorate Ion Using a Flow Injection System," *Anal. Chim. Acta* **225,** 437 (1989).
40. Sweetin, D. L., Sullivan, E., and Gordon, G., "The Use of Chlorophenol Red for the Selective Determination of Chlorine Dioxide in Drinking Water," *Talanta,* **43** (1996).
41. Chiswell, B., and O'Halloran, K. R., "Use of Lissamine Green B as a Spectrophotometric Reagent for the Determination of Low Levels of Chlorine Dioxide," *Analyst,* **116,** 657–661 (1991).
42. Masschelein, W. J., *Chlorine Dioxide Chemistry and Impact of Oxychlorine Compounds,* Ann Arbor Science, Ann Arbor, MI, 1979.
43. Gallagher, D. L., Hoehn, R. C., and Dietrich, A. M., *Sources, Occurrence, and Control of Chlorine Dioxide By-product Residuals in Drinking Water,* American Water Works Association Research Foundation, Denver, CO, 1994.
44. Werdehoff, K. S., and Singer, P. C., "Chlorine Dioxide Effects on THMFP, TOXFP, and the Formation of Inorganic By-products," *J. AWWA,* **79**(9), 107–113 (1987).
45. Knocke, W. R., and Iatrou, A., *Chlorite Ion Reduction by Ferrous Iron Addition,* American Water Works Association Research Foundation (AWWARF ISBN 0-89867-663-0), Denver, CO, 1993.
46. Griese, M. H., Kaczur, J. K., and Gordon, G., "Combining Methods for Reduction of Oxychlorine Compounds in Drinking Water," *J. AWWA,* **84,** 69–77 (1992).
47. Griese, M. H., and Rosenblatt, A. A., "Chlorine Dioxide's New Phase," *Proceedings* of the American Water Works Association, Annual Conference and Exposition, Toronto, Ontario, June 23–27, 1996.
48. Slootmaekers, B., Tachiyashiki, S., Wood, D., and Gordon, G., "The Removal of Chlorite Ion and Chlorate Ion from Drinking Water," Department of Chemistry, Miami University, Oxford, OH, 1989.
49. Ridenour, G. M., and Ingols, R. S., "Bactericidal Properties of Chlorine Dioxide," *J. AWWA,* **39,** 561 (1947).
50. Ridenour, G. M., and Armbruster, E. H., "Bactericidal Effects of Chlorine Dioxide," *J. AWWA,* **41,** 537 (1949).
51. Ridenour, G. M., Ingols, R. S., and Armbruster, E. H., "Sporicidal Properties of Chlorine Dioxide," *Water Wks. and Sew.,* **96,** 276 (1949).

52. Malpas, J. F. "Disinfection of Water Using Chlorine Dioxide," *Water Treat. Exam.*, **22,** 209 (1973).
53. Hettche, O., and Ehlbeck, H. W. S., "Epidemiology and Prophylaxes of Poliomyelitis in Respect of the Role of Water in Transfer," *Arch Hyg. Berlin,* **137,** 440 (1953).
54. Ingols, R. S., and Ridenour, G. M., "Chemical Properties of Chlorine Dioxide," *J. AWWA,* **40,** 1207 (1948).
55. Bernarde, M. A., Snow, W. B., Olivieri, V. P., and Davidson, B., "Kinetics and Mechanism of Bacterial Disinfection by Chlorine Dioxide," *Appl. Microbiol.,* **15,** 2,257 (1967).
56. Bernarde, M. A., Israel, B. M., Olivieri, V. P., and Granstrom, M. L., "Efficiency of Chlorine Dioxide as a Bactericide," *Appl. Microbiol.,* **13,** 776 (Sept. 1965).
57. Longley, K. E., Moore, B. E., and Sorber, C. A., "Relative Wastewater Disinfection Efficiencies of Chlorine and Chlorine Dioxide in a Gravity Flow Contactor," paper presented at the Annual WPCF Workshop, Houston, TX, Oct. 7, 1979.
58. Longley, K. E., Moore, B. E., and Sorber, C. A., "Comparison of Chlorine and Chlorine Dioxide as Wastewater Disinfectants," paper presented at the Annual Conference, Water Pollution Control Federation, Anaheim, CA, Oct. 3, 1978.
59. Aieta, E. M., Chow, B., and Roberts, P. V., "Chlorine Dioxide: Analytical Measurement and Pilot Plant Evaluation," paper presented at the National Symposium on Wastewater Disinfection, Cincinnati, OH, Sept. 18–20, 1978.
60. Berg, J. D., Aieta, E. M., and Roberts, P. V., "Effectiveness of Chlorine Dioxide as a Wastewater Disinfectant," paper presented at the National Symposium on Wastewater Disinfection, Cincinnati, OH, Sept. 18–20, 1978.
61. Aieta, E. M., and Roberts, P. V., "Disinfection with Chlorine and Chlorine Dioxide," *ASCE J. Environ. Engr.,* **109,** 783 (Aug. 1983).
62. Snead, M. C., and Olivieri, V. P., "Biological Evaluation of Methods for the Determination of Free Available Chlorine," Chap. 19, p. 401, in W. J. Cooper (Ed.), *Chemistry in Water Reuse,* Vol. 1, Ann Arbor Science, Ann Arbor, MI, 1981.
63. Gallaher, M. M., Herndon, J. L., Nims, L. J., Sterling, C. R., Grabowski, D. J., and Hull, H. F., "Cryptosporidiosis and Surface Water," *Am. J. Public Health,* **79**(1), 39–42 (1989).
64. Hayes, E. B., Matte, T. D., O'Brien, T. R., McKinley, T. W., Logsdon, G. S., Rose, J. B., Ungar, B. L. P., Word, D. M., Pinksky, P. F., Cummings, M. L., Wilson, M. A., Long, E. G., Hurwitz, E. S., and Juranek, D. D., "Large Community Outbreak of Cryptosporidiosis Due to Contamination of a Filtered Public Water Supply," *N. Eng. J. Med.* **320**(21), 1372–1376 (1989).
65. MacKenzie, W. R., Hoxie, N. J., Proctor, M. E., Gradus, M. S., Blair, K. A., Peterson, D. E., Kazmierczak, J. J., Addiss, D. G., Fox, K. R., Rose, J. B., and Davis, J. P., "A Massive Outbreak in Milwaukee of *Cryptosporidium* Infection Transmitted through the Public Water Supply," *N. Eng. J. Med.* **331,** 161–167 (1994).
66. Pett, B., Smith, F., Stendahl, D., and Welker, R., "Cryptosporidiosis Outbreak from an Operations Point of View: Kitchener–Waterloo, Ontario Spring 1993," *Proceedings* of the American Water Works Association, Water Quality Technology Conference, Miami, FL, Nov. 1993.
67. Korich, D. G., Mead, J. R., Madore, M. S., Sinclair, N. A., and Sterling, C. R., "Effects of Ozone, Chlorine Dioxide, Chlorine, and Monochloramine on *Cryptospridium parvum* Oocyst Viability," *Appl. Environ. Microbiol.* **56**(5), 1423–1428 (1990).
68. Lorenzo-Lorenzo, M. J., Ares-Mazas, M. E., Villacorta-Martinez de Maturana, I., and Duran-Oreiro, D., "Effect of Ultraviolet Disinfection of Drinking Water on the Viability of *Cryptosporidium parvum* Oocysts," *J. Parasitol.,* **79**(1), 67–70 (1993).
69. Ransome, M. E., Whitmore, T. N., and Carrington, E. G., "Effect of Disinfectants on the Viability of *Cryptosporidium parvum* Oocysts," *Water Supply,* **11,** 75–89 (1993).
70. Liyanage, R. J., Finch, G. R., and Belosevic, M., "Sequential Disinfection of *Cryptosporidium parvum* by Ozone and Chlorine Dioxide," presented at the International Ozone Association, Pan American Committee Meeting, Ottawa, Canada, Sept. 7–9, 1996.

71. Synan, J. F., MacMahon, J. D., and Vincent, G. P., "Chlorine Dioxide, a Development in the Treatment of Potable Water," *Water Wks. and Sew.*, **91,** 423 (1944).
72. Synan, J. F., MacMahon, J. D., and Vincent, G. P., "A Variety of Water Problems Solved by Chlorine Dioxide Treatment," *J. AWWA,* **37,** 869 (1945).
73. Aston, R. N., "Chlorine Dioxide Use in Plants on the Niagara River," *J. AWWA,* **39,** 687 (1947).
74. Mounsey, R. J., and Hagar, C., "Taste and Odor Control with Chlorine Dioxide," *J. AWWA,* **38,** 1051 (1946).
75. Coote, R., "Chlorine Dioxide Treatment at Valparaiso, Indiana," *Water and Sew. Wks.*, **97,** 13 (Jan. 1950).
76. Harlock, R., and Dowlin, R., "Chlorine and Chlorine Dioxide for Control of Algae Odors," *Water and Sew. Wks.,* **100,** 74 (Feb. 1953).
77. Ringer, W. C., and Campbell, S. J., "Use of Chlorine Dioxide for Algae Control at Philadelphia," *J. AWWA,* **47,** 740 (1955).
78. Malpas, J. F., "Use of Chlorine Dioxide in Water Treatment," *Eff. and Water Treatment J.* (England), p. 370 (July 1965).
79. Atkinson, J. W., *Water and Water Engr.* (England), **66,** 146 (1962).
80. Dietrich, A. M., Hoehn, R. C., Dufresne, L. C., Buffin, L. W., Rashash, D. M. C., and Parker, B. C., "Oxidation of Odorous Algal Metabolites by Permanganate, Chlorine and Chlorine Dioxide" (presented at the Fourth International Symposium on Off-flavors in the Aquatic Environment, Adelaide, South Australia, Oct. 2–7, 1994), *Water Sci. and Technol.,* **13**(11)(1995).
81. Gregory, D., and Carlson, M., "Oxidation of Dissolved Manganese in Natural Waters," *Proceedings* of the American Water Works Association, Annual Conference and Exposition, Toronto, Ontario, June 23–27, 1996.
82. Palin, A. T., "Chlorine Dioxide in Water Treatment," *J. Inst. Water Engrs.* (England), p. 61 (Feb. 1948).
83. Vulcan Chemicals (Div. of Vulcan Materials Company), "Materials Safety Data Sheet," Form 3239-640, Birmingham, AL, prepared July 1, 1993.
84. Heffernan, W. P., Guion, C., and Bull, R. J., "Oxidative Damage to the Erythrocyte Induced by Sodium Chlorite, *in vivo,*" *J. Environ. Pathol. Toxicol.,* No. 2, 1487–1499 (1979).
85. Heffernan, W. P., Guion, C., and Bull, R. J., "Oxidative Damage to the Erythrocyte Induced by Sodium Chlorite, *in vitro,*" *J. Environ. Pathol. Toxicol.,* No. 2, 1501–1510 (1979).
86. Gates, D. J., and Harrington, R. M., "Neuro-reproductive Toxicity Issues Concerning Chlorine Dioxide and the Chlorite Ion in Public Drinking Water Supplies," *Proceedings* of the American Water Works Association, Water Quality Technology Conference, New Orleans, LA, Nov. 12–16, 1995.
87. Bull, R. J., and Kopfler, F., *Health Effects of Disinfectants and Disinfection By-products,* AWWA Research Foundation (AWWARF ISBN 0-89867-566-9), Denver, CO, 1991.
88. Dietrich, A. M., Orr, M. P., Gallagher, D. L., and Hoehn, R. C., "Current Chlorine Dioxide Practices and Problems," presented at Annual AWWA Conference, Cincinnati, OH, June 1990.
89. Hoehn, R. C., Dietrich, A. M., Farmer, W. S., Lee, R. G., Aieta, E. M., Wood, D. W., and Gordon, G., "Household Odors Associated with the Use of Chlorine Dioxide during Drinking Water Treatment," *J. AWWA,* **90**(4) (1990).
90. Dietrich, A. M., and Hoehn, R. C., *Taste and Odor Problems Associated with Chlorine Dioxide,* AWWA Research Foundation, Denver, CO, 1991.
91. White, G. C., *Disinfection of Wastewater and Water for Reuse,* Van Nostrand Reinhold, New York, 1978.
92. Saunier, B. M., Michelon, B., and Jamody, M., "Essais de Disinfection des Eaux Usées Urbaines de Ville de Montpellier," report edited by Agence de Bassin Rhône Méditerranée Corse et Français Ministere L'Environnement, Sept. 1982.
93. Kowalski, J. B., Hollis, R. A., and Roman, C. A., "Sterilization of Overwrapped Foil Suture Packages with Gaseous Chlorine Dioxide," *Developments in Indust. Microbiol.* **29** (1988).

94. U.S. Patent No. 4,504,442, "The Use of Chlorine Dioxide Gas as a Chemosterilizing Agent," Rosenblatt, D. H., et al., issued Mar. 12, 1985.
95. U.S. Patent No. 4,681,739, "The Use of Chlorine Dioxide Gas as a Chemosterilizing Agent," Rosenblatt, D. H., et al., issued July 21, 1987.
96. Synan, J. F., and Malley, H. A., "Chlorine Dioxide: An Alternative to Chlorine for Disinfection in Water Systems," paper presented to the Engineering Panel of the Campden Food Preservation Research Assoc., Chipping Campden, Glos., England, Oct. 21, 1975.
97. Tsai, L.-S., Higby, R., and Schade, J., "Disinfection of Poultry Chiller Water with Chlorine Dioxide: Consumption and Byproduct Formation," *J. Agric. and Food Chem.,* **43** (1995).
98. Crump, B. R., Tenney, J., Gravitt, A., Isaac, T., and Ernst, W., "Design and Operation of a Small Scale Chlorine Dioxide Generator," *Proceedings* of Chemical Oxidation: Technologies for the Nineties, Seventh International Symposium, Vanderbilt University, Nashville, TN, Apr. 9–11, 1997 (in press).
99. Liyanage, L. R. J., Finch, G. R., and Belosevic, M., "Effect of Aqueous Chlorine and Oxychlorine Compounds on *Cryptosporidium parvum* Oocysts," *Environ. Sci. Technol.,* **31**(7), 1992–1994 (1997).
100. Hurst, G. H. and Knocke, W. R. "Evaluating ferrous iron for chlorite ion removal," *J. AWWA,* **89**(8), 98–105 (Aug. 1997).
101. Drew Industrial Division of Ashland Chemical Co., "Principles of Industrial Water Treatment" Eleventh Edition, 1994.

13
OZONE

POTABLE WATER TREATMENT

HISTORICAL BACKGROUND

Ozone (O_3) was first identified by Van Marum in 1785[90] in the vicinity of an electrical machine. The name "ozone" is derived from the Greek word *ozein,* "to smell." This gas, which is colorless at room temperature, owes its name to its characteristic odor, which is "pungent." Schonbein reported, in 1890, that the odor was due to a new substance, observed while he was performing some experiments on the electrolysis of sulfuric acid. Several years later, the substance was shown to be triatomic oxygen, O_3.

The first important use of ozone was in the disinfection of water. In 1886, the first experimental use of ozone in water was made by de Meritence.[2] As early as 1892, several experimental plants in European towns were in operation. The first full-scale plant was the work of a renowned innovator, Marius Paul Otto. He constructed an ozone generator, and had it installed to disinfect the water from the Vesubie River at the Bon Voyage plant in France, in about 1906. The second full-scale plant was at Rimiez in Nice, France, in 1909. This was the most important plant at that time in Western Europe.

Ozone thus has been used for over a century for water treatment on the European continent. More than a thousand municipal water treatment plants use ozone as part of their chemical treatment.[1] Most of them are in Western Europe, particularly France and Switzerland, but this usage is spreading to other countries. Most of these installations are primarily for taste and odor control and for color removal. They are, almost without exception, backed up by chlorination, in spite of the fact that ozone is an admirable disinfectant. This is due primarily to the fact that the superoxidation power of ozone breaks up a lot of organic matter in the water that supplies the existing microbial life in the water with a considerable amount of nutrients. This process needs to be stabilized with some protective chlorination to prevent microbiological instability in the water distribution system.

In July 1940,[90] the city of Whiting, Indiana installed an ozone generating system to eliminate and control the tastes and odors produced by the raw water chlorination process. This city has the longest operating experience with

O_3 of any American city. The ozone is used in a pretreatment process for the raw water, and chlorination is applied to the plant effluent.

In 1949, Philadelphia began operation of the world's largest O_3 plant for the removal of tastes, odors, and manganese from the grossly polluted Schuylkill River.[90] However, when this plant was expanded in 1959, the use of O_3 was discontinued because chlorine cost much less to achieve the same results.

In 1950,[90] part of the St. Maur supply for Paris, France, was put on O_3, principally because of complaints about the taste and odor of the water. During this time, Paris was ozonating about one-third of its supply, and it was reported that approximately 136 municipal water plants, serving 8 million people, were using ozonated water in France. England and Germany were also reported to have several municipal installations using ozone as the water disinfectant. Canada and Mexico were using O_3 to some extent, but in the United States there was very little activity (1950s).

In 1972 a modern ozonation facility was installed at Super-Rimiez, at Nice France, treating 24 mgd.[39] This water comes from the River Var, which originates close by in the Alps. The quality of this water is usually excellent and only deteriorates during flood periods following heavy rainfalls. The resulting pollution consists mainly of suspended materials, which are easily removed by sand filtration. Ozone is used to destroy coliform organisms and the grassy or earthy tastes that accompany periods of high runoff.[3]

The Nice plant is still in use. Enlargements made in 1922 and 1951 increased its capacity to about 25 mgd. As of 1976 the largest ozone installations are used for water treatment in the Paris area. Three separate plants treat water from the Oise, Marne, and Seine rivers. The combined capacity of these treatment plants is 360 mgd, and the ozone production (from air) is approximately 5 tons/day.

In Switzerland, ozone has been used for several decades for treating spring waters, groundwaters, and surface waters. Following a massive phenol discharge in 1957 at St. Gall, disinfection has been by ozone instead of chlorine. There are a great many (several hundred) installations using ozone for treatment of potable water and industrial processes. The largest has a capacity of 1750 lb/day. This facility is at the Lengg Lake plant in Zurich. The next largest systems are 650 lb/day in Geneva and St. Gall.[4]

The first large Russian filtration–ozonation plant was built in 1911 in Saint Petersburg with a total capacity of 12 mgd. Owing to difficulties in operation and maintenance this unit was shut down in 1922.[2] Subsequently, large ozonation installations have been included in water-treatment plants (in the former Soviet Union) for Moscow (317 mgd at 4 mg/L O_3), Kiev (106 mgd at 5 mg/L O_3), and Gorski (82 mgd at 2.0 mg/L O_3). Other plants using ozone as a disinfectant are those of Singapore; Lodz, Poland; and Chiba, Japan. None of these plants was designed for ozone usage capacities in excess of 5 mg/L, with the median design at 3.4 mg/L.[5]

CHEMICAL AND BIOLOGICAL PROPERTIES

Introduction

The ozone used in both potable water and wastewater treatment may be classified as both an oxidant and a germicidal compound. It thus has the same properties exhibited by aqueous chlorine; so there is tendency to view the two substances as competitors. However, the competition is neither exact nor universal, for the ways in which the functions are accomplished are somewhat different.[6] It should be emphasized that ozone and an aqueous chlorine solution can act in complementary fashion, each performing some tasks more usefully than the other. Therefore, there are three distinct facets of ozone: (1) as a bactericide; (2) as a viricide; and (3) as a powerful chemical oxidant.

Ozone as a Bactericide

In recent years there have been a number of laboratory and pilot-scale studies to evaluate the effectiveness of ozone in disinfecting wastewater. This work has also provided information on the mechanisms by which ozonation occurs.

The potent germicidal properties of ozone have been attributed to its high oxidation potential. Research studies indicate that disinfection by ozone is a direct result of bacterial cell wall disintegration, which is known as the lysis phenomenon. This mechanism of disinfection by ozone is indeed different from that by chlorine. Although the exact chemical action of chlorine is uncertain, it is generally agreed that the chlorine residual in an aqueous solution diffuses through the cell wall of the microorganisms and attacks the enzyme group, the destruction of which results in the death of the microorganism.

Ozone has long been recognized as an excellent disinfecting agent, but reliable quantitative studies of the fundamental germicidal activity of ozone are so few that our knowledge of its real potency is meager compared to what we know about chlorine and other disinfectants. This is partly because of the superior oxidizing power of ozone. It is most difficult to experimentally obtain the necessary time-dependent relations with extremely small ozone concentrations.[6]

Venosa[7] pointed out, in his comprehensive review of the literature dealing with the germicidal efficiency of ozone, that there exists much controversy, contradiction, confusion, and nonfactual subjective judgment on the use of ozone. One of the most serious failures by the various investigators has been their inability to distinguish between the concentration of ozone applied and the residual ozone necessary for effective disinfection. It must be recognized that the same principle for controlling chlorination, which is by the residual, should also be applied to the control of ozone.

In 1976, Morris[6] presented an excellent summary of what was then known about the germicidal efficiency of ozone. First he laid to rest the fallacy that ozone displays an "all-or-nothing" effect on bacterial kill, which was due to an interpretation of work done by Fetner and Ingols in 1956[8] that has been often quoted to substantiate this effect. Morris has emphasized that this so-called all-or-nothing effect is neither real nor significant.[6] The effect appears simply because of the inability or failure of investigators to space the dosage concentrations of ozone reagent close enough. Ozone is so strong a germicide that concentrations of only a few micrograms per liter are needed to measure germicidal action. The spacing of the concentrations used by Fetner and Ingols was about 0.1 mg/L or 100 μg/L, a large-enough gap to go from zero kill at a dosage just equal to demand, all the way to a very rapid kill at a concentration equal to demand plus 0.1 mg/L. This example is just one of many that confronts the researcher in attempting to evaluate the potency of disinfectants. To overcome this situation, Morris[9] developed the concept of the lethality coefficient for a given disinfectant. He treated all of the significant developed data on ozone, beginning with the work by Kessel et al. in 1943 to develop this lethality coefficient:

$$\Lambda = 4.6/Ct_{99} \tag{13-1}$$

where

Λ = specific lethality coefficient
C = residual concentration in mg/L
t_{99} = time in min. for 99 percent microorganism destruction (2-log destruction)

The weighted mean results of these evaluations are shown in Table 13-1. The values are considered by Morris to be valid only within a *factor of two.* The ranges of values do not warrant any greater confidence than this. Comparison of the values in this table with similar values obtained for chlorine

Table 13-1 Parameters For Disinfection by Ozone[6] (pH 7; 10–15°C)

Organism	Λ[a]	$C_{99:10}$[b]
Escherichia coli	500	0.001
Streptococcus faecalis	300	.0015
Polio virus	50	.01
Endamoeba histolytica	5	.1
Bacillus megatherium (spores)	15	.03
Mycobacterium tuberculosam	100	.005

[a] Λ = specific lethality coefficient = ln 100 ÷ Ct_{99}
[b] $C_{99:10}$ = concentration in mg/liter for 99 percent destruction or inactivation in 10 min.

(HOCl) is shown in Table 13-2. These values of the lethality coefficient are computed from the 1967 tabulation by Morris.[9] These tabulations by Morris clearly illustrate that ozone is a more powerful germicide against all classes of organisms listed, by factors of 10–100. The relative sensitivities of the various types of organisms are the same as for HOCl. As would be expected, bacteria are the most sensitive, and of the usual types of vegetative bacteria examined, all exhibit about the same sensitivity. Viruses are more resistant by about a factor of 10, but not enough forms have been tested with ozone to determine the range of resistances of the difficult viruses. Cysts and spores, as with aqueous chlorine, are about a factor of 10 times more resistant than viruses.

Not much experimentation has been done with the activity of ozone at various pH levels. The germicidal efficiency of ozone does not seem to be affected significantly within the pH range of 6 to 8.5. Even less is known about the effect of treated wastewater temperature on germicidal efficiency. One thing is certain, however: the higher the water temperature, the lower the efficiency of ozone mass transfer, which translates to lower germicidal efficiency.

For oxidizing compounds used as disinfectants, their superior oxidizing characteristics might provide them with high-lethality coefficients; however, these same characteristics might also cause them to have higher consumption rates if placed in an environment such as wastewater abounding in compounds that react rapidly with oxidizing agents. Therefore, the more reactive the compound, the fewer the "miles per gallon."

Ozone as a Viricide

Selna, Miele, and Baird[10] of the Sanitation Districts, Los Angeles County, made a comprehensive study of water reuse disinfection for unrestricted recreational purposes. According to the California Department of Public Health guidelines, in order to qualify for such use, a well-oxidized secondary effluent must be coagulated, settled, filtered, and disinfected to achieve a median total coliform MPN of 2.2/100 ml, or less. The required treatment is expensive from both a capital and an operational standpoint; therefore, the Sanitation Districts investigated less costly tertiary treatment alternatives to the required system during a two-year study at the Pomona research facility.

Table 13-2 Values of Λ at 5°C [(mg/liter)$^{-1}$ (min.)$^{-1}$]

Agent	*Enteric Bacteria*	*Amoebic cysts*	*Viruses*	*Spores*
O_3	500	0.5	<5	2
HOCl as Cl_2	20	0.05	1.0 up	0.05
OCl^- as Cl_2	0.2	0.0005	<0.02	<0.0005
NH_2Cl as Cl_2	0.1	0.02	0.005	0.001

One of the objectives of this study was to provide an effluent that would protect swimmers against viral illnesses. The pilot systems employed were from 25 to 100 gpm. Four treatment systems were evaluated, as follows:

1. Coagulation, sedimentation, filtration, and disinfection.
2. Coagulation, filtration, and disinfection.
3. Two-stage carbon adsorption and disinfection.
4. Nitrification, filtration, and disinfection.

Ozonation was tested as an alternative to chlorination in systems 1, 2, and 3. System 4 was an investigation of free residual chlorination with a two-hour contact time. All ozone contact times were designed for 18 min. Ozone dosages varied, with 10 mg/L for System 1, both 10 and 50 mg/L for System 2, and 6 mg/L for System 3 (carbon adsorption). Ammonia nitrogen concentrations in effluents 1, 2, and 3 were approximately 20 mg/L and suspended solids (predisinfection) about 1.5 mg/L. The total COD was on the order of 20–25 mg/L. Cumulative virus removal was best in System 1 (i.e., 5.5-log removal with an ozone dose of 10 mg/L). System 2 with an ozone dose of 50 mg/L was only about 5.4 logs removal. System 2 with a lowered ozone dose of 10 mg/L did almost as well in virus destruction as with 50 mg/L. System 3 using carbon adsorption and a 6 mg/L dose provided a 5.25-log removal. These results compared generally with those for chlorine (free or combined) of 4.6–5.25 logs removal, depending upon the system (see Chapter 8).

This investigation defines with confidence the viricidal capabilities of ozone under full-scale treatment plant conditions. It also reveals that the final ozonated effluent coliform concentrations did not routinely meet the required MPN standard of 2.2 MPN/100 ml. This continues to confirm the fact that ozone is a superior viricide but is not a reliable bactericide.

In general it is a poor disinfectant, owing to its short lived contact-time activity, which is a maximum of 5 to 6 min. When White was in Paris, France in the 1960s investigating ozone installations at the three large water treatment plants on the River Seine, the Trailagaz people operating the Choisy-le-Roi plant were very certain about this ozone contact time. This is the prime reason why ozone cannot under any circumstances be able to inactivate more than 3 logs of *Giardia* or *Cryptosporidium.* Much more research on this subject is called for.

Ozone as an Oxidant

Ozone has a wide array of attributes attractive to its use in potable water treatment, such as taste and odor control, color removal, and iron and manganese removal. These oxidizing powers are quite valuable in the polishing of low-quality supplies including water reuse situations.

Ozone oxidizes inorganic substances completely and rapidly (e.g., sulfides to sulfates, ferrous iron to ferric, manganous ion to manganese dioxide or permanganate, and nitrites to nitrate). Of even greater importance is ozone's capability of breaking down organic complexes of both iron and manganese, which usually defy the traditional procedures of iron and manganese removal from potable waters.

Oxidation of organic materials by ozone is rather selective and incomplete at the concentrations and pH values of aqueous ozonation. Unsaturated and aromatic compounds are oxidized and split at the classical double bonds, producing carboxylic acids and ketones as products.[11] Because of the high reactivity of ozone, oxidation of organic matter in the aqueous environment, whether it be potable water or wastewater, will consume ozone in varying amounts. Therefore, one of the most significant parameters for evaluating ozone is the determination of the immediate ozone demand. Oxidation of the (organic) material is usually incomplete. It is estimated that the reduction in TOC may be only 10–20 percent although decreases in COD and BOD are generally greater, ranging up to 50 percent COD reduction as with Montreal water.[12] There are also instances where COD has appeared to increase, resulting from conversion to more readily oxidized compounds.

Ozone exerts a powerful and effective bleaching action on the organic compounds that contribute to the color in wastewater and potable water.[36] The ability to attack these compounds that contribute to color, some of which are the humates and fulvates, makes ozone a fine wastewater polishing agent.

The ability of ozone to destroy taste-forming phenolic compounds is probably its most important contribution to the field of potable water treatment. Moreover, it appears to be capable of destroying other taste-forming compounds of unknown origin. There are two major mechanisms by which ozone may react with organic material.[13] The first of these is a direct additive attack, in which ozonides and ultimately peroxides are formed, together with a splitting of the organic molecule. The other mechanism is an accompaniment to the decomposition of ozone. This decomposition proceeds by way of the formation of the free radicals OH, HO_2, and HO_3, as described below. These free radicals, especially OH, are highly reactive against all sorts of organic material and may lead to autooxidation of a wide variety of organic matter, particularly substances present in wastewater effluents. The free radical autooxidation mechanism may well be involved in the disappearance of residual ozone after the initial rapid demand has been satisfied.[6]

Contact Time

One of the major problems in attempting to predict results for ozone dosages always involves the estimation of a suitable *Ct*. White always remembers what the chief operators had to say when he visited the three potable water treatment plants in Paris that were treating water from the River Rhine, particularly the Choisy-le-Roi plant, in the early 1970s. These operators insisited that the "*t*" of *Ct* should never exceed 5 min. They all claimed that after 5 min. the natural instability of ozone causes rapid disappearance of the ozone residual.

This is a pivotal reason for using the Stranco High Resolution Redox Control System, as it will reveal the precise amount of contact time at maximum ozone concentration = *Ct.* See Ref. 97, Chapter 13.

Color Removal[96]

In 1987 the City of Long Beach, California made an investigative study of color removal in its underground water supply of 50 mgd. This water is pumped from deep wells to a water treatment plant, which uses chemical coagulation with polymers to remove color bodies and multimedia filters to remove suspended particles plus chlorine for disinfection.

When ozone was used for color removal, they were able to reduce raw water color levels from 32–57 color units down to 1–4 color units with ozone dosages of 4–5 mg/L and residuals of 0.4–0.5 mg/L after about 5 min. of contact time. This case certainly proves the ability of ozone to destroy organic color in potable water quickly and thoroughly at low dosages. White's experience with chlorine for the same purpose in other waters always seemed to show the need for about 15 mg/L chlorine dosages.

Physical and Chemical Properties

Ozone (molecular weight 48) is an allotropic form of oxygen, a gas with a characteristic pungent odor to which it owes its name. It is produced commercially from dry air or oxygen and is formed by the corona discharge of high-voltage (4000–30,000 V) electricity. Ozone is also formed photochemically in the earth's atmosphere. It is one of the hazardous elements of smog, and its concentration in the atmosphere is an indicator of smog intensity. Ozone content in the atmosphere in excess of 0.25 ppm is generally considered injurious to human health. Ozone levels of 1.0 ppm in the atmosphere are extremely hazardous to health. It is colorless at room temperature, and it condenses to a dark blue liquid. It is generally encountered in dilute form in a mixture of oxygen and air. Liquid O_3 is very unstable and will readily explode. Concentrations of O_3 in air oxygen mixtures above 30 percent are easily exploded. Explosions may be caused by trace catalysts, organic materials, shocks, electric sparks, or sudden changes in temperature or pressure. Ozone absorbs light in the infrared, visible, and UV at certain wavelengths. It has an absorption maximum at 2537-A.

Ozone is more soluble in water than is oxygen, but, because of a much lower available partial pressure, it is difficult to obtain a concentration of more than a few milligrams per liter under normal conditions of temperature and pressure. The solubility of ozone in water is a complicated subject because it is dependent upon pressure and temperature. The instability of ozone presents a "one of a kind" problem in the disinfection of both drinking water and wastewater. An investigative study by J. W. Masschelein in 1962 made comparative solubilities in water of ozone, oxygen, nitrogen, carbon dioxide, chlorine, and chlorine dioxide. He found that chlorine gas was 7 times more soluble than ozone while chlorine dioxide gas was 94 times more soluble than ozone.

Ozone decomposes in water; this is probably due to its strong oxidizing ability rather than simple decomposition. Ozone is much more soluble in acetic acid, acetic anhydride, dichloroacetic acid, chloroform, and carbon tetrachloride than it is in water.

Ozone is reported to be naturally unstable and to decompose to ordinary oxygen slowly. Heat accelerates decomposition, and decomposition is instantaneous at temperatures of several hundred deg C. Moisture, silver, platinum, manganese dioxide, sodium hydroxide, soda lime, bromine, chlorine, and nitrogen pentoxide catalyze decomposition. Ozone is also decomposed photochemically. From a practical standpoint, decomposition is slow enough to permit the use of ozonized air or oxygen streams for water disinfection. Ozone weighs approximately 0.135 lb/ft^3 at one atm. It is a powerful oxidizing agent. Its oxidation potential is -2.07 V, referred to the hydrogen electrode at 25°C and at unity H-ion activity. Only fluorine has a more electrotnegative oxidation potential.[14] Ozone is extremely corrosive, so that materials of construction must be very carefully chosen. It attacks most metals except gold and platinum. Porcelain and glass do not react with ozone. PVC is used, but it is suspected that a reaction takes place, resulting in the loss of ozone.[15] It completely disintegrates rubber and attacks all plant life.

The solubility of ozone in water is a limiting factor that greatly affects the process of ozonation. At 20° Centigrade the solubility of ozone is only 570 mg/L.[17] Although ozone is more soluble than oxygen, chlorine is 12 times more soluble than ozone. In pure aqueous solution, ozone is thought to decompose as follows:[1]

$$O_3 + H_2O \rightarrow HO_3^+ + OH^- \tag{13-2}$$

$$HO_3^+ + OH^- \rightarrow 2HO_2 \tag{13-3}$$

$$O_3 + HO_2 \rightarrow HO + 2O_2 \tag{13-4}$$

$$HO + HO_2 \rightarrow H_2O + O_2 \tag{13-5}$$

The free radicals (HO_2 and HO) that form when ozone decomposes in aqueous solutions have great oxidizing power, and, in addition to disappearing rapidly (Eq. 13-4), may react with impurities (e.g., metal salts, organic matter, hydrogen, and hydroxide ions present in solution). These free radicals formed by the decomposition of ozone in water are apparently the principal reacting species. Ozone, while it exists, does not lose its oxidizing capacity in an aqueous solution. When ozone reacts with hydrogen peroxide, a powerful oxidant, the hydroxyl radical is formed. This is a totally different phenomenon. It is a very important reaction because when peroxone is formed by the addition of the correct amount of hydrogen peroxide, the stable OH radical is formed, which is more powerful than the OH ion that is formed in the chlorination "breakpoint" process (See Chapter 14.).*

Inorganic Reactions

Ozone reacts rapidly to oxidize ferrous and manganous ions into their insoluble (ferric and manganic) ions, resulting in either a floc that precipitates or a scum that clings to the water surface. Sulfides and sulfites are readily oxidized to sulfates, and nitrites to nitrates. The oxidation of iodides to iodine is the basis of the usual analytical determination of ozone. Bromides and chlorides are similarly oxidized to bromine (Br_2) bromate (BrO_3^-), and chlorine (Cl_2), respectively, and these reactions are slow and dependent upon the concentration of reactants.

The ammonium ion (NH_4^+) apparently is not attacked under the conditions normally found in wastewater treatment; so there is no waste of ozone oxidizing capacity or side reactions with the ammonia nitrogen in wastewater. However, at a 12:1 molar ratio of consumed ozone to ammonia, the ammonia will be completely oxidized to nitrate so long as the pH remains alkaline.[16]

Ozone reacts to oxidize ferrous and manganous compounds to form insoluble ferric and manganic compounds. When iron and manganese are present in complex organic compounds, as is often encountered in groundwaters, ozonation requires supplemental catalytic action by filtration or adsorption (GAC) to execute successful iron and manganese removal. When iron and manganese react with ozone, a brown scum will form on the surface of the treated water. As it is light and fluffy, it is difficult to remove. Therefore, ozone must be applied upstream from the coagulation process.

Ozone reacts effectively to destroy cyanides. When complex cyanides are present, ozone is more effective than chlorine. Inorganic cyanide reactions are as follows:

$$CN^- + O_3 \rightarrow CNO^- + O_2 \qquad (13\text{-}6)$$

$$2CNO^- + 3O_3 + H_2O \rightarrow N_2 + 2HCO_3 + 3O_2 \qquad (13\text{-}7)$$

4.5 mg/L ozone is required to destroy 1 mg/L cyanide as CN^-.[41]

Effect of Ozone Residuals on Chlorine and Chlorine Dioxide

French practice when using either chlorine or chlorine dioxide or both in the treatment train along with ozone is to apply the chlorine compounds where there are no ozone residuals. This is done because ozone reacts to destroy

*This is why White does not understand why the MWD of Southern California has not done an investigation of peroxone versus *Giardia* and *Crypto*.

both HOCl and ClO_2. If the chlorite ion is present, ozone converts it to chlorate.[49] Moreover, the hydroperoxides formed during ozonation will also act to dechlorinate both HOCl and ClO_2.

A recent study by Haag and Hoigné[42] has shown that ozone reacts with chlorine ($HOCl/OCl^-$) by the following reactions:

$$O_3 + OCl^- \rightarrow O_2 + [Cl - O - O^-] \rightarrow 2O_2 + Cl^- \ (77\%) \quad (13\text{-}8)$$

$$2O_3 + OCl^- \rightarrow 2O_2 + ClO_3^- \ (23\%) \quad (13\text{-}9)$$

That is, when a solution of hypochlorite is exhaustively ozonated, 77 percent of the chlorine is found as chloride ion, and 23 percent is found as the chlorate ion. Overall:

$$1.23O_3 + OCl^- \rightarrow 2O_2 + 0.77Cl^- + 0.23ClO_3^- \quad (13\text{-}10)$$

The above equations are written for the OCl^- ion instead of HOCl because the undissociated form does not react.

Ozone was found to react with monochloramine according to the following equation:

$$NH_2Cl + 3O_3 \rightarrow 2H^+ + NO_3^- + Cl^- + 3O_2 \quad (13\text{-}11)$$

Approximately 4 moles of ozone are lost per mole NH_2Cl. No chlorate is found in the reaction.

Ozonation of Seawater[42]

When seawater is ozonated, the following reactions occur:

$$O_3 + Br^- \rightarrow O_2 + OBr^- \quad (13\text{-}12)$$

$$O_3 + OBr^- \rightarrow [O_2 + BrOO^-] \rightarrow 2O_2 + Br^- \quad (13\text{-}13)$$

$$O_3 + 2Br^- + OH^- \rightarrow HOBr + BrO_3^- \quad (13\text{-}13a)$$

These reactions form a chain reaction that catalytically destroys ozone. As the concentration of Br^- in seawater is 65–75 mg/L, in the reaction with O_3 at concentrations for disinfection, the dominant mode of O_3 consumption is with Br^-. This is no disadvantage because an equivalent amount of HOBr is formed as O_3 is lost, and HOBr is an excellent disinfectant. These reactions will occur only if the contactor is designed for optimum plug-flow conditions with a transfer efficiency of at least 80–85 percent.

Organic Reactions

Ozone reacts readily with unsaturated organic compounds, adding all three oxygen atoms at a double or triple bond. The resulting compounds are called "ozonides." Decomposition of ozonides results in a rupture at the position of the double bond, causing the formation of aldehydes, ketones, and acids. Ozone readily destroys phenolic compounds* and is capable of bleaching the organic color found in some waters. These last two characteristics are responsible for the popularity of ozone in the treatment of low-quality surface waters in Western Europe.

Glaze et al.[18] reported on the ability of ozone to destroy humic acid, which is the precursor of THM (trihalomethane) formation.† Guirguis et al.[19] reported that ozone makes organic compounds more adsorbable by carbon. Prengle et al.[20] reported that, with time and proper dosage, ozone plus UV light can reduce malathion to carbon dioxide and water after forming three or four intermediate compounds such as alcohol, aldehydes, and oxalic acid after a one-hour contact time. Not only was the pesticide (malathion) destroyed, as indicated by gas chromatographic analysis, but also the total organic content of the water. Likewise Richard[21] revealed from his studies that ozone can degrade the pesticides parathion and marathion to phosphoric acid.

Toxicity of Ozone

During the International Ozone Institute meeting in Cincinnati, Ohio (1976), several speakers presented information on some of the known toxic properties of ozone. Falk and Moyer[22] reported on a review of scientific literature dealing with reactions of ozone with organic materials as they might occur under treatment conditions. All organic substances considered were taken from a list of compounds that had been found in drinking water by the EPA. Ozonolysis of some of these compounds forms a variety of hydroperoxides known to be mutagenic. Ozonolysis of pesticides can produce epoxides, some of which have been shown to be carcinogenic.

Hartemann, Block and Maugras from France[23] reported on the preliminary results of their studies that the toxicity induced by the ozonation of organic materials may be lower than that induced by chlorination, which was attributable to the chlorine residual.

Kinman et al.[24] found that ozonated wastewater was more toxic than unozonated wastewater.

Spanggord and McClurg,[25] of Stanford Research Institute (SRI), found that a very high concentration of ozone produces mutagenic compounds when reacted with ethanol.

---

*2.0 mg/L O_3 is required to destroy 1 mg/L phenols.[41]

†This conclusion has been modified: ozone only lowers the chlorine demand, so that there are lower chlorine dosages and hence lower THMs.

Simmon and Eckford,[26] also of SRI, found an increase in mutagenesis after ozonation of ethanol, benzidine, and nitrilotriacetic acid, but 26 other compounds studied were not found to be mutagenic after ozonation.

The public health significance of all these findings is not yet known. The French, who have been ozonating for many decades, do not express any concern, even though they are continuing their toxicity investigations. All of this plus personal investigations with health investigators in France has convinced White that our concerns over DBPs are greatly overexaggerated. As of 1997, the several years of cancer research work by Dr. Bruce Ames and his staff at the University of California, Berkeley, demonstrate quite conclusively that the DBPs caused by both chlorine and ozone in the treatment of potable water and wastewater are nothing more than harmless trivia. (See Chapter 6.)

Solubility of Ozone

Ozone is a gas; therefore, its solubility in a liquid is governed by Henry's law: the weight of any gas that dissolves in a given volume of a liquid, at constant temperature, is directly proportional to the pressure that the gas exerts above the liquid. In equation form:

$$Y = HX \tag{13-14}$$

where

Y = partial pressure of the gas above the liquid,mm Hg
X = concentration of the gas in the liquid at equilibrium with the gas above the liquid, moles gas/total moles of gas plus liquid
H = Henry's law constant (varies with temperature), mm Hg/mol fraction.

The terms in Eq. (13-14) are difficult to understand in practical terms. However, by converting them to units of concentration or mg/L, Henry's law is more easily understood. The terms then become:

Y = concentration of gas above the liquid in equilibrium with the gas dissolved in the liquid, mg/L
X = concentration of gas in the liquid in equilibrium with the gas above the liquid, mg/L

$$H = \frac{\text{mg gas/liter of gas}}{\text{mg gas/liter of liquid}} \tag{13-15}$$

Henry's law simply expresses the concentration of gas above the liquid that must exist in order for a given concentration of gas to be dissolved in the liquid. The lower the value of H, the more soluble the gas.

The solubilities of ozone and oxygen are compared in Table 13-3. Henry's constants were taken from the *International Critical Tables* and converted to units of concentration.[43]

Table 13-3 shows that ozone is about 13 times more soluble in water than oxygen at standard temperature and pressure ($H = 20.4$ for O_2 versus 1.56 for O_3). This means that only 1.56 mg/L ozone in air is required to maintain 1.0 mg/L ozone in water, whereas 20 mg/L oxygen in air is required to maintain 1.0 mg/L oxygen in water under equilibrium conditions at STP. The efficiency of production of ozone in air above about 1.0 weight percent (12.9 mg/L at STP) decreases substantially. Consequently, 8.3 mg/L is the maximum concentration that can be expected to dissolve in the water at that concentration in air, assuming 100 percent mass transfer efficiency and an ozone-demand-free water. Thus, even though ozone is more soluble than oxygen, when air is used as the carrier gas, less ozone will dissolve on an absolute basis owing to the lower concentration of ozone in the air (i.e., partial pressure). This emphasizes the necessity to achieve the maximum contactor efficiency (TE), owing to the difficulty of maintaining high partial pressures of ozone above the process liquid (potable water or wastewater).

Ozone Transfer Efficiency Concept

The transfer efficiency (TE) of a given gas–liquid contactor (mixing chamber) is an inherent property of the contactor and is a function of the gas* flow rate relative to the liquid flow rate. Transfer efficiency is defined as follows:

$$\mathrm{TE} = \frac{100(Y_1 - Y_2)}{Y_1} \tag{13-16}$$

where

Y_1 = mg O_3/L inlet carrier gas
Y_2 = mg O_3/L exhaust gas from contactor

TE is the fraction of ozone in the gas that has been transferred to the liquid expressed as percent.

Applied Ozone Dose. This is defined as follows:

$$D = Y_1 \left(\frac{Q_G}{Q_L}\right) \tag{13-17}$$

*Gas in this sense means the gas flow discharge from the ozone generator. It can be derived from prepared air or pure oxygen delivered to the generator.

Table 13-3 Solubility of Ozone and Oxygen in Water According to Henry's Law[a]

Temperature	Oxygen (air)			Ozone, 1.0 Weight %[b]		
	H, mg O_2/l air / mg O_2/l water	*Y*, mg O_2/l air	*X*, mg O_2/l water	*H*, mg O_3/l air / mg O_3/l liquid	*Y*, mg O_3/l air	*X*, mg O_3/l water
0	20.4	299	14.6	1.56	12.9	8.3
10	25.4	289	11.4	1.86	12.5	6.7
20	29.9	279	9.3	2.59	13.1	4.7
30	34.2	270	7.9	3.80	11.7	3.1

[a] Henry's constant *H* is a function of temperature, not concentration. Therefore the values of *H* for oxygen are the same whether the oxygen is 21 percent (i.e., in air) or 100 percent.
[b] 12 mg/l O_3 per liter carrier gas = 1.0 wt. percent at 20°C and atmospheric pressure, or lb O_3 per 100 lb carrier gas.

where

Q_G = gas flow rate, L/min.
Q_L = liquid flow rate, L/min.

The applied dose multiplied by the fraction transferred to the liquid is the *absorbed dose.*

Absorbed Ozone Dose. The transfer of ozone, also called the absorbed dose, is defined as follows:

$$
\begin{aligned}
T &= Y_1 \left(\frac{Q_G}{Q_L}\right)\left(\frac{Y_1 - Y_2}{Y_1}\right) \\
&= \frac{Q_G}{Q_L}(Y_1 - Y_2)
\end{aligned}
\tag{13-18}
$$

where T is the amount of ozone transferred to the liquid, mg/L. Equation (13-17) indicates the applied dose can be varied by changing either Y_1 (inlet carrier gas concentration) or the Q_G/Q_L ratio. Operating experience has shown that TE is more sensitive to shifts in the gas flow rate than ozone concentration in the carrier gas.[43] This is important, since an increase in the gas flow (Q_G) relative to the liquid flow may not result in a linear increase in absorbed ozone dose.

Application of Henry's law and the concept of ozone transfer efficiency is useful for designing and optimizing ozone contactors.

Control of Ozone System Efficiency

Factors Affecting System Efficiency. It is necessary to review these factors in order to understand an acceptable control strategy. First it is desirable to understand the proper terminology and to differentiate between applied and absorbed ozone dose:[44]

$$
\begin{aligned}
D &= \text{applied dose, mg/L to the liquid} \\
&= Y_1 Q_G / Q_L
\end{aligned}
\tag{13-19}
$$

where

Y_1 = ozone concentration in carrier gas, mg/L gas
Q_G = carrier gas flow rate L gas/min.
Q_L = liquid flow rate L liq./min. (this is process water flow through the contactor chamber)

When the ozone dose is increased by changing the Q_G/Q_L ratio, the transfer efficiency (TE) of the contactor decreases more rapidly than when the dose is increased by changing Y_1.[45,46] Therefore it is desirable to maintain a constant Q_G/Q_L ratio by flow pacing and to change dosage by changing Y_1.

Absorbed Ozone Dose (*T*). As already described, this is the arithmetical difference between the ozone concentration in the inlet carrier gas and the ozone concentration in the exhaust gas stream, multiplied by the Q_G/Q_L ratio:*

$$T = (Y_1 - Y_2)Q_G/Q_L \tag{13-20}$$

where

T = absorbed ozone dose, mg/L in the liquid†
Y_2 = concentration of ozone in the exhaust gas stream leaving the contactor, mg O_3/L gas.

Venosa and Meckes[36,45,46] have shown that disinfection efficiency of any given water source can be predicted if its demand properties and absorbed ozone dose are known.

Exhaust Gas Monitoring. Using this as an operational control strategy to achieve optimum disinfection efficiency has the following advantages: (1) true ozone is being measured, not total oxidant, which is what residual methods measure; (2) the reaction is instantaneous; (3) the measurement technology is convenient and reliable; (4) it is easily automated; (5) it is not subject to interferences; (6) it is not affected by sudden changes in effluent quality (ozone demand); and (7) it is useful over a wide range of secondary effluent quality.

The control arrangement, which is a variation of the compound-loop principle, is as follows:

1. Carrier gas flow to the contactor is flow-paced by a signal from the process flow meter.
2. The signal from the exhaust gas O_3 concentration monitor changes the generator power input, which changes the O_3 concentration in the inlet carrier gas.

This method of control was rigorously tested and was reported on by Venosa[45,74] in 1983 and 1985. The key to the success of this system depends upon the ability of the process flow metering device to maintain the carrier gas flow

*Liquid in process flow.
†Technically, T is not the absorbed dose but rather the transferred dose, for it consists of both consumed ozone and unconsumed residual.[45]

to the process flow constant over the range of process flow variation.* The process flow meter signal should be equipped with a 0.4–4.0 ratio relay and/or a microprocessor.

Ozone Residual Control

General Discussion. Experience to date (1997) indicates that residual control is doubtful as a control strategy for ozone application. The overriding factor contributing to this uncertainty is the rapid die-away phenomenon of an ozone residual. Analyzers cannot track fast enough to cope with the speed of residual decay. The other factor is the measurement of free ozone versus total oxidants. There is some doubt as to the ability of the galvanic analyzer to measure free ozone residuals. There is little doubt, however, about its ability to measure total oxidants (free ozone + ozonides + hyperperoxides). However, monitoring total oxidants by the galvanic analyzer is a possibility.

Investigations by White as a special consultant to Stranco from 1993 to 1997 convinced him that the Stranco High Resolution Redox Control System is the preferred method for measuring the levels of total oxidant in an ozone facility. This is fully described in Chapter 15.

Continuous Analyzers. There are three continuous monitors available for the measurement of ozone. One is the Bailey–Fischer and Porter Series 17T-2000 amperometric type, which is specific for free ozone. It can also be arranged to measure total residual oxidants. The others are the ATI Model A15/64 and the EIT Model 8422. These units are described in Chapter 15. Although of recent vintage (1994–97) they have been operating long enough to demonstrate their accuracy and reliability.

Another factor that is always a concern where oxidants are involved is the presence of substances that interfere with the analytical procedure. Copper ions poison the amperometric cell, and manganese ions interfere with some analytical procedures used in the calibration of the analyzers.

Residual control may never receive the consideration necessary to prove its practicality if the exhaust gas monitoring system continues to be as reliable as reported by the Venosa group. However, continuous residual recording for monitoring total oxidant residual may have practical value. It may be an important tool for increasing the reliability of the ozone process.

Now the Stranco ORP control system has proved to be the most reliable ozone monitoring method, and it has eliminated all the previous problems of compounds that adversely affect the amperometric cell.

*The transfer efficiency decreases when the gas-to-liquid flow ratio changes. This upsets the overall control system.

CURRENT PRACTICES: POTABLE WATER

United States

The first plant to use ozone in the United States is believed to be the one at Whiting, Indiana. This installation dates from 1940. Used primarily for the destruction of phenolic tastes and odors, it is still in use today. Ozone application is followed by chlorination to provide a persisting residual in the distribution system. As a result of applying the ozone ahead of chlorination, trihalomethanes were held to less than 1 μg/L.[27] Other installations were established at Strasburg, Pennsylvania, which began operating in 1973, and at Monroe and Bay City, Michigan, and at Saratoga, Wyoming, all of which began in 1978.

In Philadelphia, Pennsylvania, the Belmont Plant, with 36 mgd capacity, was commissioned in 1949 and operated successfully until 1959, when it was taken out of service. The ozone process was discontinued in favor of chlorination when the plant capacity was increased, and the contact basins provided contact exceeding 20 hours, making free residual chlorination much more economical than ozone.[2]

There are a variety of reasons why there has been limited use of ozone in the United States. This is mostly due to the availability of a superior quality of water for potable use in the United States as compared to the waters of Western Europe. The inadequacies of ozone—mainly, the interference by manganese and iron—are more pronounced in the United States. These factors, coupled with lower cost, greater flexibility, and better reliability of chlorination, contribute to ozone's limited use in the United States.

Now, however, in the 1990s, the effects of the increased pollution of both surface water and groundwater are causing a complete reevaluation of finished water quality. To improve this quality, the federal government enacted into law (1976) the Surface Water Treatment Rule.[51] This legislation established a variety of water quality requirements.[52] In addition to this rule the water industry is anticipating the Disinfection Byproducts Rule (DPB). As will be described later, there are many reasons for the use of ozone to improve both the quality and the reliability of the finished product. Currently (1997) ozone is being applied in all regions of the United States in response to the new regulations.[50] The most impressive such installation was put into service at the 150 mgd Hackensack Water Co. plant of Harrington Park, New Jersey, in 1983.[28] Since then, ozone installations have increased to 14 (1992), including the 600 mgd treatment plant for the City of Los Angeles Owens River supply. This facility was needed to meet the new turbidity requirements, which were satisfied by the use of preozonation because of its ability to improve coagulation and increase solids removal. The capacity of the ozonation system is 7900 lb/day. Posttreatment is by conventional chlorination. This plant is fully described in Chapter 15.

Current ozone installation projections called for 20 U.S. plants to have operating systems by the end of 1990, and 60 by 1997. Considering the progress of ozonation in comparison to total system capacity of 1000 lb/day in 1979, the units operating in June of 1990 had a total capacity of 23,000 lb/day, with the output expected to exceed 80,000 lb/day in 1997. This illustrates a tremendous surge of confidence in the idea that ozone is greatly needed to help solve our water quality problems.[45]

Canada

The province of Quebec has been the most active in the use of ozone. By 1978* there were 23 water treatment systems using ozone in Canada. All of these facilities but one, which is in Ontario, are in Quebec. This is undoubtedly a result of the close ties between Quebec and France. The first Canadian installation was in 1956 at the Ste-Therese Quebec filtration plant. During the next 13 years, ozonated water increased from 4 to 137 mgd (U.S. gallons). The largest plants include: City of Laval, 61 mgd (three systems); City of Quebec, 58 mgd; City of Chomeday, 47 mgd.

The newest and largest ozone facility in Canada is at the 300 mgd Charles J. Baillets WTP in Montreal, Quebec. The first phase was installed in 1978 with a capacity of 7500 lb/day.[30,31] The second phase increased the capacity to 9510 lb/day. This calculates to approximately a 4 mg/L dosage. The source water is from the St. Lawrence River, where the average raw water quality is 4 NTU turbidity, 3 mg/L TOC, and 350/100 ml total coliform. The ozone equipment consists of a 15 psi air feed system, refrigerant, and desiccant dryers operating at −71°C dew point. The contactor is an over/under baffle basin 25 ft deep with an 80 percent transfer efficiency. Theoretical detention time is 8 min. Tracer studies have yet to be made. The gas feed phase of ozone is measured optically, and the contactor residual is measured by an electrode system. Ozone from the newest installation is applied after filtration for color removal and disinfection. This plant also uses a postchlorination disinfection system to maintain a chlorine residual in the distribution system. Ozone was added to the 21 mgd Centrale de Traitement d'Eau plant in Quebec in 1976. The raw water is from the Rivieres-des-Prairies which has a turbidity of 4–8 NTU, 6.5 mg/L TOC, and 1000/100 ml total coliforms. This installation was used to control tastes and odors. Ozone is applied after filtration at a rate of 1.5 mg/L dose. The system capacity is 400 lb/day. It uses desiccant dryers for the air feed system to provide a −58°C dew point. The contactor is an over/under basin 18 ft deep and has a 4-min. theoretical contact time with liquid residual monitors. Operators have not provided ozone destruct equipment. The Sherbrooke plant, located in New Brunswick, Quebec, has a 500 lb/day ozonator where ozone is applied after microstraining treatment of a surface water. This is a 1974 installation. A WTP serving Roberval, Quebec since

*In the middle 1970s the U.S. government funded several surveys of potable water treatment practices in Canada, Great Britain, and Western Europe. The findings of these surveys on the use of ozone are contained in this text and are referenced accordingly.

1989 uses ozone for disinfection with the application point upstream from filtration. The ozone generator has a 600 lb/day capacity.[40]

Canadian plants report energy consumption for ozone treatment mostly in the range of 10 to 15 kWh/lb of O_3. Contact times vary from 2 to 20 min., depending upon the quality of the water and temperature. Nadeau and Pigeon[30] reported that it requires 12 kg/hr to produce a 0.4 mg/L ozone residual while treating 54 mgd (U.S. gallons). This calculates to a dosage of 1.4 mg/L ozone. Any water with such a relatively low oxidant demand must be of high quality.

The main use for ozone in Canada is to eliminate the seasonal taste and odor problems and to provide backup disinfection of surface waters.

France

By 1992 there were about 700 French water plants using ozone. Practically all of the plants are supplied with surface waters. The primary purposes cited for ozone use were taste and odor control, destruction of phenols, organics removal, viral inactivation, and bacterial destruction. Some of these plants use ozone for color removal and iron and manganese removal. Many plants report that ozone increases the efficiency of turbidity removal. Ozone dosages range from 0.15 mg/L to 10 mg/L. Contact time is usually 5–10 min. Average power consumption for 33 plants checked was 14 kWh/lb of ozone generated.

In 22 of 63 plants surveyed, ozone was the only oxidant used. Chlorine was being used for final disinfection in 26 plants, and chlorine dioxide was being used at 13 of 63 plants. Since the survey a great many chlorinator installations have been retrofitted to generate chlorine dioxide as a step to reduce THM formation. Before this conversion to chlorine dioxide, chlorine was being used for ammonia N removal. This required prechlorination dosages as high as 16 mg/L. However, when these plants convert to chlorine dioxide, it is necessary to revise the treatment process to remove ammonia N in order to meet the allowable limit required for drinking water. This is usually accomplished by biological nitrification.

The French water treatment process is similar to the conventional treatment train used in the United States. This consists of prechlorination followed by flash mixing, coagulation, sedimentation, filtration, ozonation, and terminal disinfection with either chlorine or chlorine dioxide. The only marked difference is the ozonation step. This allows lower dosages of the terminal disinfectant. Chlorine dosages are about 1.0 mg/L, and chlorine dioxide 0.6 mg/L in this step.

Switzerland

Switzerland has had over 40 years of experience in the use of ozone for the treatment of spring supplies, groundwater and surface waters. About 150 water plants use ozone. Many are small plants. Overall there is a variety of plant sizes.

Several are less than 1 mgd, with a gradation in sizes up to the largest, which is the Lengg plant in Zurich with a capacity of 60 mgd. Ozonator capacity is 260 lb/day (4.33 mg/L). The original unit processes were: prechlorination, rapid sand filtration (without chemical flocculation), slow sand filtration, and ozonation followed by postchlorination. The plant was to be expanded,[32] and chemical precipitation followed by sedimentation before rapid sand filters and GAC between ozonation and postchlorination were planned.

Ozone is used for the usual variety of purposes, such as bacterial and viral destruction, taste and odor removal, and organics removal. Ozone dosages range from 0.3 to 1.5 mg/L. For final or terminal disinfection it is estimated that 80 percent of the Swiss plants that use ozone also use chlorine dioxide as a final disinfectant. Some plants use GAC directly following ozonation. The average power consumption at five different plants was reported to be approximately 15 kWh/lb ozone for generation and contacting.

Germany

The approach to water treatment in Germany is substantially different from that in other European countries. For example, it is not uncommon to find fully chemically treated river water pumped back into the underground aquifers only to be pumped out of the ground as drinking water.[33] One of the major problems for Germany is the Rhine River, which is grossly polluted along its entire length. It is the source of water supply for many communities. German water specialists believe that the most serious problem with the Rhine is the high level of dissolved TOC, which ranges from 6 to 9 mg/L. Bernhardt[34] at Siegburg claims that if the TOC concentration exceeds 2 mg/L, it is not possible to maintain sufficient free residual to reduce the standard plate count below the German standard, which is 20 organisms per ml at 22°C. When such a situation occurs, additional treatment must be incorporated within the plant to provide a higher-quality water. Two parameters must be adhered to, that the treated water must have dissolved organics less than 2 mg/L TOC and an oxidant demand (chlorine demand) less than 0.5 mg/L. In other words if the treated water consumes more than 0.5 mg/L chlorine to produce a stable residual in the distribution system, then the water treatment process must be modified to reduce the chlorine demand.

Ozone usage in Germany is more varied than in any other country. The purposes of ozonation, dosage levels, methods of O_3 diffusion (contacting), and the assortment of equipment in use do not fall into any consistent pattern as in other countries. Power consumption also varies. It seems to be higher than in most plants in other countries. The lowest power consumption found was at Duisburg, 7 kWh/lb O_3.

Some 136 municipal water systems in western Germany are using ozone. The oldest installations date from about 1955 at the Dusseldorf water plants.

Of 31 answering the 1978 questionnaire, 24 used ozone for the reduction of organics, 13 for taste removal, 8 for viral inactivation, 7 for odor removal, and 5 for color removal. Ozone dosages ranged from 0.15 mg/L at the Diez-Lahn plant to 5.7 mg/L at the Osterode plant. Chlorine dioxide was used in 9 of the 31 plants for final disinfection, 10 used chlorine, 9 used ozone as the only treatment step, and 3 used ozone as the only oxidant. Ten plants were using GAC as an absorbent following the ozone step.

Owing to the conviction that the most urgent problem in treating water from the Rhine was too great a concentration of dissolved organics, Kühn et al.[35] developed a treatment train to deal with the problem. In this process the water was first allowed to filter through a berm into a small catchment area, which became the raw water supply. The first treatment step was ozonation. This was followed by chemical coagulation, flocculation, sedimentation, and GAC filtration. The final step was disinfection by either chlorine dioxide or chlorine. This complicated process produced a high-quality water with a chlorine demand less than 0.5 mg/L. The ozone step ahead of the GAC filters increased the TOC removal from 1 mg/L to 3 mg/L when the ozone dose was 3 mg/L.

While in pursuit of higher organic removal in potable water supplies, the German investigators were exploring better methods of treating wastewater discharges into the Rhine, hoping to decrease the dissolved organics at the source.

Austria

In Austria 42 plants were using ozone in 1978. A few plants were supplementing the ozone treatment with chlorine. The high quality of the water available throughout this country is reflected by the low ozone dosages used (0.06–1.2 mg/L). Some plants used ozone for color and organics removal. Power consumption for ozone generation ranged from 7 to 25 kWh/lb of O_3.

The Netherlands

Most ozone systems in the Netherlands were less than five years old in 1977. The Dutch philosophy exhibits a preference for physical and biological processes instead of chemical treatment. They rely heavily on large storage reservoirs that receive river water from infiltration systems. Their goal is to have several months' storage capacity. The flow sheet for a new plant at Amsterdam scheduled for completion in 1979 included pumping river water into a lake, mixing it one-to-one with groundwater consisting of 80 percent infiltration water, and coagulation with ferric chloride, followed by ozonation, powdered activated carbon, rapid sand filtration, slow sand filtration, and final chlorination.[34]

A new plant for Rotterdam was planned to take water from the Biesbosch reservoirs, which are equipped with destratification aerators to control the growth of plankton. After several months' detention in the reservoirs the water was to be coagulated and flocculated followed by sedimentation, ozonation, mixed media filtration, GAC, and final chlorination. Many plants were planning to switch final disinfection from chlorine to chlorine dioxide. Several undesirable experiences with biological regrowth problems when ozone was the last step resulted in its application well ahead of final chlorination.

Primary use of ozone in the Netherlands is for color removal and taste and odor control. Some plants use ozone for bacterial and viral destruction but rarely as a last step. Ozone dosages vary from 0.25 to 5 mg/L. Power consumption was reported to be about 8–9 kWh/lb O_3. Most plants practiced off-gas destruction. Of the seven plants reporting, four plants used chlorine for final disinfection. One plant used ozone as the last step, and in the other two ozone is followed by sand filtration as the last step.

Great Britain

The British rely heavily on physical and biological processes for water treatment and avoid chemical treatment whenever possible. The physical processes include microstraining and slow sand filters. They also look to open storage as a desirable treatment step. The Metropolitan Water Board of London is a strong proponent of huge storage reservoirs equipped with artificial destratification for plankton control.

There are 18 plants in Great Britain known to be using ozone, which is primarily for color removal. Most of them have been using ozone only a few years. The EPA survey team sent out 15 questionnaires to British plants using ozone; six responses were received.[29] Each of the six plants responding indicated that ozone was used only for color removal. One of the largest of these plants is the Watchgate treatment plant, which can supply up to 140 mgd to the city of Manchester. This plant, which uses ozone for color removal, began operating about 1975.[36] Disinfection is by chlorination. The chlorine residual (0.8 mg/L) is automatically controlled before entering the distribution system by the use of sulfur dioxide. Most of the plants using ozone have been in use for only a few years. Although the treatment processes vary from plant to plant, ozone is applied after a filtration step in each of the six plants responding. In four of the plants microstraining is the filtration method.

Ozone is also used for other purposes besides the main objective, which is color removal. These are taste and odor removal, iron and manganese removal, viral inactivation, and microorganisms destruction.

Chlorine is used as a final disinfectant in each of the six plants. In three of the six plants, chlorination follows ozonation, with no filtration step in between. Chlorine dioxide and ozone are not used together in any British plant. Chlorine dioxide is often used as a distinct unit process added stepwise (sometimes with chlorine) to ensure a better persisting residual.

Ozone dosages reported by four plants averaged 2.37 mg/L. Power consumption for ozone generation, air preparation, contacting, and off-gas treatment averaged 13.5 kWh/lb O_3 for the plants responding.

Other Installations

Some notable installations around the world are shown in Table 13-4. There are also installations at Singapore and at Chiba, Japan. Only one of the above illustrations has an ozone capacity greater than 5 mg/L, which is the rule of thumb for designers of these systems. This is probably the influence of the major supplier of ozone generators (Trailigaz of Paris), which has had long and successful experience with ozonation of potable water.*

THE OZONE–BIOLOGICAL ACTIVATED CARBON PROCESS

Historical Background

In the years following World War II the Germans began to study the use of activated carbon as a multipurpose tool in the processing of potable water. At first it was used as a dechlorinating agent. In those times it was standard practice to use chlorine for ammonia nitrogen removal in the heavily polluted waters of the Ruhr and Rhine rivers. This resulted in residuals much higher than desired in the finished waters; hence the necessity for dechlorination.

Another important factor in the German scheme of things is the philosophy of a drinking water supply. Germans are convinced that groundwater is the most important raw material for drinking water. In the overcrowded regions along the Rhine and the Ruhr there was no way to avoid the use of surface water as the raw material for drinking water supplies. Because of their successful experiences with groundwaters, treatment procedures were investigated that could produce a finished water with a 2 mg/L TOC or less and a chlorine demand of 0.5 mg/L or less.[35] In addition to the employment of water passage through the ground such as sand bank filtration or percolation systems, acti-

Table 13-4 Notable Ozone Installations: Water Treatment

Location	*O_3 lb/day*	*Capacity mg/l*
Moscow, Russia	10560	4.2
Kiev, Russia	4224	3.5
Manchester, U.K.	2640	2.5
Lodz, Poland	2112	3.3
Rotterdam, Netherlands	1200	4.8
Zurich, Switzerland	1560	2.8
Lake Constance, Germany	3000	2.1
Choisy-Le-Roi, Paris	6600	3.5
Neuilly-Sur-Marne, Paris	6400	4.8
Mery-Sur-Oise, Paris	3600	6.2

*Particularly because the water to all three WTPs in Paris prechlorinate the Seine River water with up to 16 mg/L of chlorine dioxide generated on-site.

vated carbon filters and ozone were thoroughly investigated as unit processes. Combinations of ozone and GAC were installed in water plants near Dusseldorf, West Germany, in the late 1950s, but the synergistic interaction of the two processes was not fully recognized until 10 years later.

Process Description

This is also known as the Mulheim Process.[37] This process produces a surface water by physiochemical procedures that is practically equivalent to groundwater. This scheme has been used at the Dohne plant in Mulheim, Germany, since April 1977. The treatment process is as follows: Ruhr River water is pumped from a side channel into a mixing chamber where chlorine used to be applied. Instead, ozone is applied here at an average dose of about 1 mg/L (depending upon turbidity). Coagulants are also added here. The water then passes to a pulsating type flocculating plant. The clarified water leaving the flocculating system passes through the ozonation contact basin, which has a mean retention time of 5 min. Ozone dosage here is 3 mg/L. The effluent from the ozone contactor goes to a double-layer group of pressure filters followed by an activated carbon filter. The water then flows into 15 injection wells and two filter basins. After the water is pumped from the injection wells to the distribution system, chlorine is added.

Significant Features

Preozonation leads to a transformation of the organic matter in the raw water. Ozone probably converts the long-chain, large-molecule organics, which are nonbiodegradable, to smaller, more biodegradable organics, and at the same time charges the water with dissolved oxygen. This effect of ozonation is the cornerstone of success of the Mulheim process. The water now has an abundance of oxygen together with biodegradable organic compounds, which together promote the growth of aerobic bacteria. The best place for this regrowth to take place is in a GAC filter. The organics there were both pore-adsorbed and surface-adsorbed on the granular activated carbon. The aerobic conditions within the carbon filters build up a biological system that achieves a high degree of nitrification. Bacterial regrowth after the water leaves the GAC filter is eliminated by passing the water through the ground or a slow sand filter.

Other Users. Users of this process, or modifications thereof, are: the Lengg plant, Zurich, Switzerland; the Bremen, Germany, plant on the Weser River; and La Chapelle at Rouen, France. The Rouen plant (14 mgd)[38] deserves special attention as a plant using a variation of the Mulheim process. The raw water is pumped from 100-ft wells located adjacent to the heavily polluted Seine River to an intake structure where it receives a preozonation dose of 0.7 mg/L and 3 min. of contact time.

In addition to the effect of preozonation in the Mulheim process, ozone at Rouen was found effective for iron and manganese removal. Preozonation is followed by direct sand filtration. The sand filters remove the oxidized iron and manganese plus some flocculated organics rendered insoluble by ozonation. Initial nitrification of the ammonia present takes place in the sand filter. There is no chemical addition other than ozone upstream from the sand filters. The granular activated carbon filters are located underneath the sand filters. The carbon bed is only 30 inches deep and has a theoretical contact time of 9 min. Both the sand and carbon filters are flat rectangular units rather than a columnar configuration. The carbon filters perform the removal of organics by adsorption, and ammonia removal is by the aerobic biological cultures found on the surface of the carbon. Postozonation for disinfection is 1.4 mg/L and is applied sequentially to two different diffuser chambers. The contact time is 12 min. in order to be certain of viral destruction. The dosage of ozone is consistent with French practice. The ozonators at the Rouen plant are automatically controlled by ozone residual monitors to provide an ozone residual of 0.4 mg/L at the end of the contact chamber. The finished water is stored in two underground reservoirs, each having a capacity of 350,000 gal. This stored water is treated with sufficient chlorine to provide a residual at the customer's tap of about 0.5 mg/L. The ammonia removal through the plant is about 86 percent. Therefore, because the final ammonia content is about 0.4 mg/L, the chlorine residual in the distribution system will be all combined residual. This will not promote the formation of THMs.*

At this time it is not possible to compare the overall performance of the Rouen plant with the Mulheim process because the organics concentrations are not reported in the same terms. The carbon filters operated 26 months at Rouen without any need to be regenerated. This was similar to the Mulheim experience. It is thought that the aerobic biological activity on the surface of the granular carbon does in some way effect a "regeneration" process.

The Canadian Experience

Introduction. In 1984 the city of Laval, Quebec, inaugurated the Ste. Rose 30 mgd water treatment plant, where it uses biological activated carbon (BAC) filtration at full-scale and 10 gpm pilot-scale investigations.[82–85]

This major step in advanced water treatment processes was inspired by concern about possible adverse health effects from disinfection by-products (DBPs) and compliance with stricter water quality regulations.[81] It is common knowledge that the elimination of total organic carbon (TOC) in drinking water would produce a water without any detectable DBPs. This is an impossibility with present treatment practices. However, a substantial reduction in dissolved organic carbon (DOC) prior to the application of chlorine usually will produce a high-quality water. Reduction of DOC by conventional treatment processes (coagulation, flocculation, and filtration) is limited. Water produced by these processes from surface waters can contain TOC levels

*There is not any European country, including Great Britain, that concerns itself with THMs.

varying from 1 to 20 mg/L.[88] As TOC contains THM precursors*[89] (of various molecular weights), removal of TOC becomes a primary goal in drinking water treatment processes. This procedure involves a lot of complex chemistry; so it is far more succinct to think in terms of reduction in chlorine demand.[84]

When DOC reduction was being achieved by GAC filtration, it turned out to be an expensive technology owing to periodic off-site regeneration of the saturated carbon.

BAC vs. GAC Filtration. It is important to realize the significant difference between these two carbons. GAC is a bituminous-based carbon (Calgon-F400). It accomplishes a removal of organics by adsorption, and requires backwashing and regeneration to restore its adsorption capacity. The BAC used in the Canadian studies is a new type of carbon identified as Picabiol,[88] which is a wood-based carbon. It is characterized by a superior macroporous structure that is favorable to biomass fixation. Microscopic examination of the carbon structure (after staining by acridine orange) shows a higher density of bacteria colonization of the carbon surface than with the bituminous carbon (Calgon-F400). Based upon the removal of assimilable organic carbon (AOC), preliminary data indicate that a shorter EBCT will be required when the Picabiol carbon is being used.

Assimilable Organic Carbon (AOC). These compounds contribute to biological instability in the distribution system, one of the consequences of ozonation. Ozone has the ability to break down the organics in such a way that they become appetizing nutrients for biological regrowth in the distribution system. The amount of AOC entering the distribution system is site-specific. Therefore, monitoring AOC is essential to plant performance evaluation. AOC compounds cannot be characterized chemically because they exist in large numbers at low concentrations, and their presence is irregular.

The Ste. Rose study[83] revealed that AOC can be completely removed if sufficient contact time is provided by the BAC filtration system. Therefore, it is reasonable to believe that increased AOC removal would bring a reduction in the chlorine demand. Furthermore, this would increase the stability of the chlorine residual in the distribution system. White points out that any kind of pretreatment process will always lower the chlorine demand.

Chlorine Demand. All the data in the study[84] strongly supported lowering the chlorine demand to limit THM formation at the chlorination step under plant operating conditions, rather than reducing the THM potential. (As of 1997 this situation has changed completely because the Ames cancer research group has demonstrated conclusively that THMs are trivia, and are in no way a public health risk. For further details see Chapter 6.) The study of chlorine

*"Precursors" are nothing more or less than "chlorine demand."

demand evolution through the various biological water treatment steps at the Ste. Rose plant demonstrated that chlorine demand can be reduced by coagulation, flocculation, and sedimentation by 55 percent for the short-term demand (4 hours) and 61 percent for the long-term demand (168 hours), and by BAC filtration by 25–60 percent for the short-term and 15–43 percent for the long-term.

The removal of the long-term chlorine demand requires an empty bed contact time (EBCT) of approximately 20 min. This removal by BAC filtration can be maintained even if low water temperatures (1–2°C) last for several months.

Measurement of Biodegradable Organic Carbon (BOC). A team of researchers from Gendron Lefevbre Consultants, the city of Laval, Quebec, Canada, and laboratory technicians from the Ste. Rose plant and the Engineering Department of Ecole Polytechnique, Montreal, made a comprehensive investigation of the bioassay-type techniques for the measurement of BOC.[87] These techniques are subject to bacterial and carbon contamination. In some cases they are difficult and costly to perform. The focus of this investigation was on a comparison of the three techniques that are commonly used:

1. *The van der Kooij AOC method:* This technique does not demonstrate a clear relationship between the removal of DOC and AOC in water treatment processes. The removal values obtained where the seasonal matrix varies with the seasons are unreliable for process control. In fact, this method will lead to an underevaluation of an adequate EBCT for removal of BOC and chlorine demand when one is defining design criteria for a BOC filtration system. However, using pure strains does bring some insight about the effect of treatment processes on specific groups of BOC compounds. This approach may prove useful in distribution system monitoring.
2. *The Servais-Billens BDOC method:* This is a simple and reliable method that requires two DOC measurements.[86] Although it is very sensitive to carbon contamination and has a limited sensitivity, it relates well to chlorine demand. The incubation period required for this method is probably too long for process control requirements.
3. *The Servais-Billens mortality flux method:* This procedure is labor-intensive and costly. With this method it is possible to determine both rapid and slow biodegradable fractions. This information is essential for the optimization and control of the biological process. It determines the amount of BOC that can be utilized with an economically acceptable contact time in the BAC filtration step. This method has shown that the rapidly biodegradable BOC is responsible for short-term chlorine demand (4 hours). An adapted inoculum should be used to provide representative results of the biomass present in the treatment process or in the distribution system.

After all of the effort required for this particular investigation, the question still remains: What method will provide reliable results for maintaining biological stability in the distribution system? It should be remembered that for regrowth prevention it is important to:

1. Limit the amount of nutrients available for the growth of suspended bacteria and fixed biomass.
2. Maintain a disinfectant residual to limit the growth of suspended bacteria.

Finally, the importance of BOC and chlorine demand associated with exported biomass from the BAC filters should be evaluated.

POTABLE WATER TREATMENT OBJECTIVES

Introduction

Legislative Issues. In the 1980s the utilities supplying drinking water realized the necessity to comply with the EPA MCLs (maximum contaminant levels). This became a part of the Safe Water Treatment Rule (SWTR).[51] It is fair to say that a large part of these legislative moves was a result of the discovery that the renowned free chlorine residual process (HOCl) generated trihalomethanes (THMs). These disinfection by-products (DBPs) are supposed to pose a health risk in the form of cancer.* The EPA set a THM MCL of 1.0 mg/L in the early 1980s and proposed a limit of 0.5 mg/L to become the rule in 1991—all of this in spite of the 1990 pronouncement, that drinking water is not a carcinogen, by the International Association for Cancer Research at its annual meeting in Lyon, France. In addition to this, the policymakers of the EPA ignore all the recent scientific evidence that all THMs are not a public health risk, and that only two or three DBPs might be a minor health risk.

Alternative Disinfectants. Further investigations revealed that the use of chloramines did not generate THMs. As converting from the free chlorine process to chloramines simply involved the addition of ammonia, there were a great many utilities that switched to the second-rate disinfection process—chloramination.

Following this shift in disinfection practices there were sudden, but probably related, waterborne outbreaks of gastroenteritis, which only recently became a reportable disease by the CDC in Atlanta, Georgia. It was determined that these outbreaks were caused by *Giardia lamblia* cysts and *Cryptosporidium parvum* oocysts. The immediate attention of the water industry was required, to examine and evaluate alternative disinfectants and better filtration practices. In this search ozone became a prime object for better disinfection practices.

*DBPs are misunderstood when they are considered as a health risk that causes cancer. The only exception is the formation of bromate and chlorate which affect the blood of dialysis patients.

Ozone had long been known as a most powerful oxidant and an excellent viricide, more lethal than chlorine.

Finished Water Quality

Introduction. After the above-described legislative moves, U.S. water producers soon understood that providing safe water to their consumers was only part of their responsibility. They soon realized that the water had to be aesthetically acceptable: no off-flavors, no unpleasant odors, and no unnatural color; otherwise, the consumer might think it was unsafe to drink. To accomplish all of these objectives, it has been the consensus of water quality investigators that some type of bench-scale or pilot-scale approach must be used—a primary reason for this conclusion being the fact that minor variations in raw water quality make every project site-specific.

Ozone has long been used for disinfection, T&O control, color removal, and iron and manganese removal. More recently it has been used to provide control of turbidity to meet the strict new EPA requirements and the oxidation of objectionable organics. The success of these applications depends upon the scope of the small-scale testing that is required.

Single- and Multiple-Step Systems. The pilot-scale system has to be flexible enough to determine the proper number of ozonation steps (pre-, intermediate, and/or post-). Then it must determine where disinfection credit is to be evaluated, plus the need for postoxidation to prevent regrowth of microorganisms in the distribution system.

When multiple points of application are to be used, the dosage and contact times have to be evaluated in order to provide the necessary hydraulic data for designing the prototype.[53]

Turbidity Control. In a conventional treatment plant for surface water, the initial treatment step is usually done for turbidity control. This depends largely upon the raw water quality. Therefore, the first objective is to determine whether preozonation dosages are to be limited to low levels.[48] With highly turbid waters, low ozone dosages reduce turbidity levels, whereas higher ozone dosages result in an increase in turbidity levels. If preozonation is limited to low levels, then two-stage ozonation in a conventional treatment plant system will be required. The low-level ozonation in the first step for turbidity control will also be able to remove iron and manganese if present. Some disinfection credit is usually achieved at this step. In the second stage of ozonation, the organics that are responsible for T&O, color, and DOC can be oxidized by using higher ozone dosages and longer contact times than those used in the first step.

With low-turbidity waters and most groundwaters, preozonation is not necessarily limited to low ozone dosages. Consequently, all oxidation steps can be achieved along with primary disinfection by using a single point of ozone application. Therefore, if filtration is called for, this is usually accomplished by simple direct filtration.[48]

Turbidity control by ozone is accomplished through its ability to destabilize the suspended particles by neutralizing the colloidal charges. The net effect is enhanced coagulation. This unique ability has resulted in such enormous savings in coagulation chemicals that the sizable capital cost of an ozone installation has been offset.

Disinfection By-products (DBPs). When strong oxidants such as ozone are used to treat either surface waters or groundwaters, by-products are always formed. In the absence of the bromide ion the usual by-products of ozonation are nonhalogenated low-molecular-weight acids, aldehydes, ketones, and alcohols. These oxidation products are easily biodegraded by soil and the microorganisms usually found in water. This degradation process of the DBPs is a short-cut, and is more reliable than attempting to solve the problem by so-called precursor removal.*

Assimilable Organic Carbon (AOC). The powerful characteristic of ozonation is viewed by some as a disadvantage and by others as an advantage. Ozonation produces high levels of AOC, and the entrance of AOC into the water distribution system must be avoided because it contributes to microorganism regrowth. This upsets the biological stability of the system. Therefore, the pilot-plant or the bench-scale study must evaluate biological filtration techniques to assure a significant amount of AOC reduction by mineralization to carbon dioxide and water.[48] If this is not successful, then the final step in the treatment train will have to be postdisinfection by either free chlorine or chloramination. This is imperative to prevent regrowth of the microorganisms in the distribution system. Only postchlorination has the ability to control and maintain biological stability in the distribution system. This facet of ozonation has been widely known for some, and observed for a long time.

If free chlorine is used in this step, the utility will only have THMs to worry about. Assuming that all of the upstream processes have been properly evaluated by either bench-scale or pilot-plant treatment, and that these processes have lowered the chlorine demand, it is reasonable to expect that the TTHMs will be less than 50 μg/L in bromide-ion-free water.

Chloramination. If the THM formation by free chlorine is in excess of the EPA MCL, then the best alternative is chloramination, which does not generate THMs, because it has a very low ORP reaction that results in a weak disinfectant. The only reported DBP resulting from chloramination following ozonation is acetonitrile,[48] a product of the reaction between chloramine and the DBP acetaldehyde, which is formed after ozonation.

---

*Precursor removal is nothing more or less than removal of chlorine demand.

A utility using chloramination as the posttreatment step must be warned that when the chloramine residual decays in the distribution system, both the remaining ammonia N from chloramination and the AOC from ozonation will still be available for microorganism regrowth.

Bromide Ion. Bromides are known to exist throughout the environment. Therefore, low concentrations of bromide ion can be found in most surface waters. Ozonation of these waters converts the bromide ion to an inorganic DBP, the bromate ion—BrO^-. This DBP is considered a health risk similar to the chlorate ion ClO^-, which is a DBP of chlorine dioxide treatment of water that is supposed to affect the blood supply of a dialysis patient.

The waters that suffer from seawater intrusion are the ones most concerned with the presence of the bromide ion. Seawater contains from 60 to 70 mg/L of the bromide ion, and some groundwaters have high levels of bromide ion due to the proximity of underground salt mines. Bromate in potable water supplies is likely to occur when there is seawater intrusion of groundwaters during a drought season. This occurs regularly along parts of the California coastline.

Monitoring the Distribution System

A great deal has been written about the need for monitoring the distribution system owing to new EPA regulations concerning sampling procedures, coliform counts, and the presence of HPC bacteria. Every utility with a sizable distribution system should establish a historical record of DO (dissolved oxygen) and chlorine residual throughout the distribution system. Before long this survey information will help to identify the trouble spots—if any. Therefore, such a survey is probably the most important one that has to be performed by every water utility. From this information the utility can evaluate the need for rechlorinating and/or increasing the chlorine residual entering the system. It will also identify the areas where the piping needs to be flushed and/or cleaned by physical means.

OZONATION OF WASTEWATER

USE OF OZONE IN WASTEWATER TREATMENT

General Discussion

The first reported investigation of ozone as a wastewater treatment process was by the U.S. Army ca. 1955.[54] The studies indicated that it was effective and economical despite the fact that ozone consumption might have to be as high as 100 mg/L. The laboratory results demonstrated that ozone could be used successfully for the sterilization of sewage seeded with *B. anthracis, B.*

subtilis, and influenza virus, and for inactivation of *Clostridium botulinum* toxin.

One of the first pilot-plant investigations of ozone for disinfecting sewage was carried out at the Eastern Sewage works in the London Borough of Redbridge, England. This work was reported in 1967.[55] The pilot plant received the equivalent of a secondary effluent and provided for microstraining (35 μ), prechlorination, ozonation, coagulation, and rapid sand filtration for a flow of about 35 gal/min. The ozone dose was kept between 20 and 25 mg/L to keep the color at or below 10° Hazen. When chlorine was used, the maximum dose used was 20 mg/L. With these dosages and on a microstrained secondary effluent containing 1.7×10^6 MPN total coliforms/100 ml, ozone produced a 4.5-log reduction down to 90/100 ml, whereas chlorine alone achieved a 6-log reduction down to 1 coliform/100 ml. (No data were available on contact time.) However, ozone performed well as a polishing agent. It achieved good color removal and was observed to break down the detergents.

In the United States, the Sanitation Districts of Los Angeles County made a comprehensive study on the use of ozone at their Pomona, California Water Renovation Plant. This study spanned a period of several years in the early 1970s.[10,56–58] Disinfection by ozone was compared to chlorine on the basis of both bactericidal and viricidal efficiency. Owing to the fact that treated wastewater from five activated sludge plants discharges into two rivers that contain undiluted secondary effluent during the swimming season, the disinfection requirement by the California State Department of Health specifies a seven-day median total coliform of 2.2 or less per 100 ml. Therefore this study examined this requirement as well as the EPA standard of 200/100 ml fecal coliforms and a variety of effluents. Using a well-oxidized secondary effluent, it required 50 mg/L dose of ozone to meet the 2.2/100 ml total coliform and only 10 mg/L dose to achieve 200/100 ml fecal coliforms. They observed the best kill at 3 min. of contact time, which is further evidence that ozone is a poor disinfectant for coliforms of every kind. This is so primarily because of the instability of even a small ozone residual over an extremely short contact time.

There is little or no current interest in wastewater disinfection in Europe or the British Isles. Therefore, there are no operating ozone installations in these areas for disinfection of wastewater effluents.

About 1974, the EPA, under the able and intelligent guidance of A. D. Venosa, Director, Wastewater Disinfection Research Program, organized several projects to examine the reliability and practicality of ozone as a wastewater disinfectant.[55] The most notable of these early investigations were carried out at Ford Southworth, Kentucky; Grandville, Michigan; Cleveland, Ohio; and New York City. The results were encouraging. The ozone dosage required to meet a fecal coliform standard of 200/100 ml MPN was on the order of 15–20 mg/L.

Venosa made some important observations resulting from these projects. He pointed out that the most serious practical problem was inadequate process control. When ozone was compared to chlorine (Grandville, Michigan[59]), chlorine produced the most uniform results because the dosage required to achieve consistent disinfection was easily and reliably controlled by continuous residual control equipment. This was not possible with ozone; therefore, manual control was the only means of dosage regulation.

Venosa further observed the excellent nonlinear correlation between the ozone concentration in the exhaust gas from the contactor and the fecal coliform concentration in the final effluent. This suggested that reliable disinfection could be achieved by the use of exhaust gas monitoring as a control strategy.[60]

Operating Installations

There are only a limited number of ozone installations at wastewater treatment plants. The following is a listing of installations in operative use in 1984 with the available detail data:

Location	*Plant Size, mgd*	*O_3 Capacity, lb/day*	*Max. Dose, mg/l*
Frankfort, KY	24	5	3
Southport Plant, Indianapolis, IN	125	8400	8
Brookings, SD	6	250	5
Vail, CO	—	—	2
Kennewick, WA	7.5	250	4
Cleveland Westerly AWTP, Northeast Ohio Regional Sewer Dist.	100	3900	4.7
Meander AWTP, Mahoning Co., OH	8	200	3

The system at Frankfort, Kentucky is apparently successful, having survived the usual startup problems. They reported experiencing difficulty with automatic control components.

The Indianapolis Southport system operates only in summer when disinfection is required.

The system at Brookings, SD experienced serious difficulties due to a poor contactor design.

The Kennewick, Washington system has operated with limited success due to high nitrite concentrations in the effluent, which consume a significant amount of ozone. There also has been a contactor efficiency problem. These difficulties combined with a capacity of only 4 mg/L conspire to limit the success of this installation.

The Cleveland Westerly AWTP ozone installation was originally designed as a disinfection system in 1974.[61,62] This was the result of a long-range economic evaluation by the design engineers. This perception changed in the intervening years during the construction period.* In 1984 the Westerly AWT plant was in the startup stage. The generators are by Emery Industries, Emerzone #EG-625. There are six units on line with one standby. The ozone production per generator is 556 lb/day at 1.5 weight percent ozone using air as the carrier gas. Activation of the plant occurred in the mid-1980s.

The facilities at the Westerly plant, designed by Engineering-Science, included conventional headworks, clarification and phosphorus removal with lime coagulation, recarbonation, preozonation, dual media sand filtration, granular activated carbon absorption, and disinfection by chlorination, followed by dechlorination. This process train flow using ozone ahead of the filters, which are followed by 17 ft GAC columns, was rigorously tested by pilot plant studies.[62] Results of the pilot study were similar to those reported by Trussell et al.[47,63] on preozonation of a tertiary effluent.

The Meander AWTP in Mahoning County, Ohio uses ozone for disinfection. The equipment was furnished by W. R. Grace Co., which sold its ozone line of equipment to Union Carbide Co. before the contract was completed. This system demonstrated remarkable disinfection efficiency in the initial stages of operation. When it was running on oxygen feed, fecal coliform counts of zero were obtained, and 10 or below when using air for carrier gas.[62] The current NPDES requirement is 200/100 ml MPN fecal coliforms. This is not considered to be strict enough to be called "disinfection of wastewater." The California requirements call for at least a 5-log reduction in total coliforms.

The Vail, Colorado installation has been touted as a most successful operating disinfection system,[73] owing to good design and construction features plus a good operator attitude toward the goals of the process. However, the success of this installation has been nullified by the NPDES requirements. The monthly average fecal coliform requirement is 6000/100 ml, and the weekly average is 12,000/100 ml (using a 1.5 mg/L ozone dose). This translates to approximately 24,000 and 48,000 total coliforms per 100 ml. How the EPA could consider that these numbers meet the disinfection requirement is a mystery, as the waters around Vail must be classified as recreational because the area is a fashionable ski resort. This means that such waters should not contain more than 200/100 ml fecal coliforms. However, the reliability of fecal coliforms as indicators of recreational waters has been questioned for the following reasons:

1. Pathogens, such as human enteric viruses, have been recovered from natural waters that were determined to be safe based upon low densities of fecal coliforms.[76,79]

*Ozone application was changed from disinfection to preozonation.

2. Fecal coliforms have been reported to be capable of multiplying in environmental waters under some conditions.[76]
3. Some fecal coliforms such as *Klebsiella pneumoniae* do not have a fecal source.[78]
4. Laboratory results show that fecal coliforms are less resistant than some pathogens (such as human enteric viruses) to chlorination and are less stable in natural waters.[80]

The Water Resources Research Center at the University of Hawaii has surveyed the quality of all the Hawaiian streams.[75] Because streams in Hawaii are all classified as recreational streams, they cannot contain more than 200/100 ml fecal coliforms. However, the above survey consistently recovered 10^2–10^4/100 ml of fecal coliforms and fecal streptococci. This includes those streams not known to be receiving contamination from any point source such as wastewater discharges.[80]

Three logical conclusions can be drawn from the results described above:

1. The assumption that wastewater is the primary source of fecal coliforms in streams is not valid.
2. The common practice of analyzing stream samples downstream from a wastewater discharge site for fecal coliforms and fecal streptococci, as a means of assessing the deleterious effects of this discharge, is inconclusive.
3. The hygienic quality of the streams in Hawaii cannot be properly evaluated by analyzing stream samples for fecal coliforms, as is mandated by law.

Preozonation: A Tertiary Process

General Discussion. Trussell et al.[47,63] reported in 1975 on a comprehensive study of an alternative tertiary treatment process that would produce an effluent to meet the strictest requirements of the California State Department of Health. The conventional process usually proposed by the State for similar situations requires coagulation–flocculation, followed by sedimentation, filtration, and disinfection by chlorine. The alternative process reported on here is the one suggested by the consultants directing this project.* This process consisted of preozonation of a secondary effluent followed by flocculation, coagulation, sedimentation, filtration, and disinfection with either ozone or chlorine. The investigation demonstrated the many benefits of preozonation followed by chlorine for final disinfection. Moreover, it proved to be less costly than the conventional method.

*James M. Montgomery Consulting Engineers, Pasadena, CA.

Table 13-5 Preozonation Summary of Results (Dose = 10 mg/L; Contact Time = 10 min)

Parameter	*Unit*	*Before O_3*	*After O_3*	*% Removal*
Turbidity	TU	5.5	2.8	50
Color	scu	46	14	70
SS	mg/L	12	5	60
COD	mg/L	87	44	50
Coliform	MPN/100 ml	3.8×10^5	1.9×10^3	99.5
Virus	PFU/1,000 gal	450,000	450	99.9

This alternative process was chosen because it was believed that it would produce an effluent to meet the following objectives:

1. Reduced incremental salt increase.
2. Greater virus removal efficiency.
3. Lower tertiary treatment costs.
4. Reduced organics.
5. Increased color and turbidity removal.
6. Disinfection to 2.2/100 ml coliforms.

The ozone pretreatment step at 10 mg/L and 10 min. of contact time consistently removed approximately 50 percent of the turbidity and suspended solids, and 70 percent of the color from the secondary effluent. COD was usually decreased, but not reproducibly. Coliform removals were usually 2–3 logs in the preozonation step; however, these removals were inconsistent.

Virus Destruction. The number of seeded viruses was generally reduced by 3 logs or more following pretreatment with 10 mg/L ozone. Results from preozonation are shown in Table 13-5. Virus concentrations are reported as PFU/1000 gal, based upon expected levels of enteric viruses in the secondary effluent and observed percent removals of seeded viruses tested in the ozonation process at the small-scale prototype plant. All the other data are median values.

One other significant achievement of preozonation was the effect on the alum dose. The results of coagulation, flocculation, and filtration are shown in Table 13-6. The alum dosages used were 10 and 20 mg/L. There was only

Table 13-6 Coagulation/Flocculation/Filtration Results (Alum Dose = 10–20 mg/L)

Parameter	*Unit*	*After O_3*	*After C/F/F*	*% Removal*
Turbidity	TU	2.8	1.2	60
Color	scu	14	11	20
SS	mg/L	5	3	40
COD	mg/L	44	35	20
Coliform	MPN/100 ml	1.9×10^3	9.8×10^2	50
Virus	PFU/1,000 gal	450	45	90

a slight improvement in results using 20 mg/L alum. Without preozonation, alum dosages of 90 mg/L were required to achieve comparable turbidities.[47]

Final Disinfection. Both clorine and ozone were compared as final disinfectants for bacterial destruction. Figure 13-1 shows this comparison. As shown, reozonation of the final effluent with ozone dosages as high as 10 mg/L failed to meet the 2.2/100 ml coliform objective. The chlorine residual in the contact basin was all "combined" because the NH_3–H content of this effluent was about 35 mg/L. In the full-scale plant the effluent will be nitrified, so free chlorine residuals will be formed. In the chlorination step for fecal disinfection a 2 mg/L dose removed approximately 2 logs of coliforms with combined residual after 30 min. of contact time. This dosage did not consistently achieve the coliform requirement of 2.2/100 ml. The suggested dosage for the full-scale plant is 4 mg/L, which resulted in virus removals greater than 2 logs. The final effluent was expected to have the following quality: turbidity, 0.5–1.5 TU; color, 5–10 SCU; suspended solids, 1–2 mg/L; coliform bacterial, 2.2 organisms/100 ml.[63] As described above, preozonation gave a 3–4 log removal of seeded viruses. Since ozone failed in the final disinfection process, reozonation for virus destruction was omitted from the investigation.[64] However, Fig.

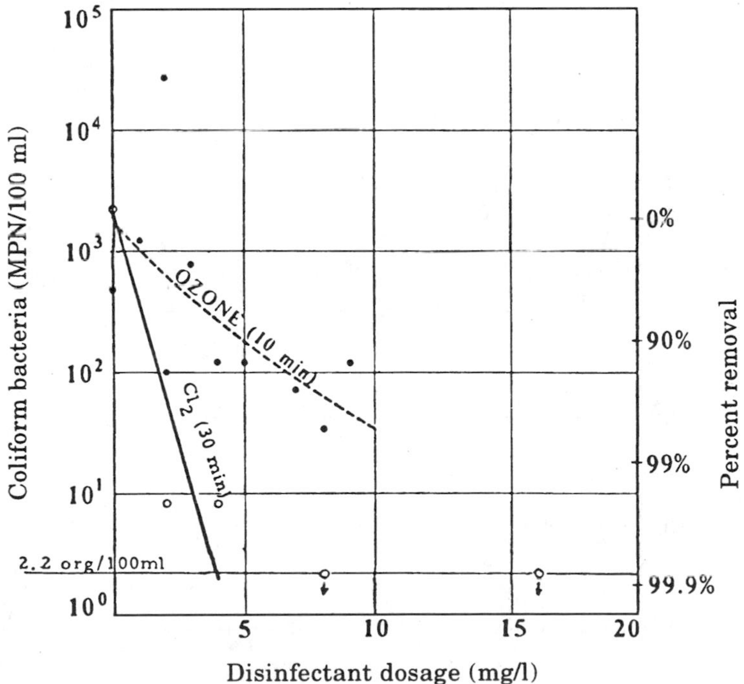

Fig. 13-1. Comparison of bacteria destruction: ozone versus chlorine.[47]

13-2 illustrates the shape of the curves comparing combined chlorine with ozone for seeded virus destruction with a one- or a two-run experiment. Therefore, the 5-log removal shown for ozone is not to be considered reliable.[64]

Projected Full-Scale Plant. Table 13-7 illustrates the probable effectiveness of ozone as a pretreatment process for wastewater reuse or tertiary effluents.

Control of Disinfection Efficiency

The reader is referred to the section "Control of Ozone System Efficiency" earlier in the chapter, and to Eqs. (13-16) through (13-20).

Venosa and Meckes[36,45,46] have shown that total and fecal coliform levels in a given effluent can be predicted if the demand properties of the effluent (TCOD, NO_2, TSS, and TOC) and the absorbed ozone dose are known.

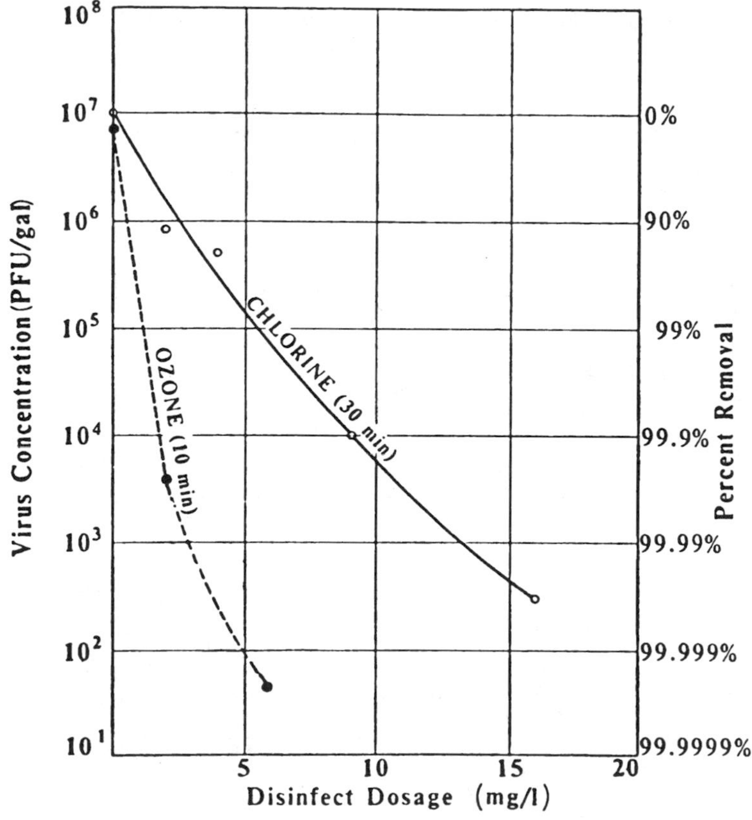

Fig. 13-2. Comparison of virus destruction: ozone versus chlorine.[47]

Table 13-7 Projected Quality Parameters of Full Scale Plant Effluent[47]

Parameter	Unit	Before Treatment	After Treatment	% Removal
Turbidity	TU	3–5	1.0	80
Color	SCU	20–40	5.10	75
SS	mg/L	10–15	1–3	75
COD	mg/L	40–60	10–20	60
Coliform[a]	MPN/100 ml	10^4–15^5	≤2.2	99.998
Virus[b]	PFU/1000 gal	450000	0.5–0.05	98.98

[a] The secondary coliform MPN is usually much higher than is shown in this table. The MPN for this investigation varied from 4.9×10^4 to 3.5×10^6 total coliforms.
[b] Seeded viruses.

Exhaust Gas Monitoring. Please see the earlier discussion of this topic under "Control of Ozone System Efficiency."

ORP Control

There is little doubt that, as of 1997, there is enough information from operating installations, particularly in Germany, to show that the use of ORP probes to measure the oxidation potential of ozone is a complete success.

Use of the Stranco High Resolution Redox System both measures the oxidant level on a continuous basis without any interference from pollution compounds and also produces a continuous and accurate record of the ozone demand.

Moreover, operating personnel soon develop complete satisfaction with the accuracy and the simplicity of the Stranco HRR system. This system has a successful operating background with ozone of at least three years since the early 1990s.

Residual Control

General Discussion Experience to date (1997) indicates that residual control is doubtful as a control strategy for ozone application. The overriding factor that contributes to this uncertainty is the rapid die-away phenomenon of an ozone residual. Analyzers cannot track fast enough to cope with the speed of residual decay. The other factor is the measurement of free ozone versus total oxidants. There is some doubt as to the ability of a galvanic analyzer to measure free ozone residuals. There is little doubt, however, about its ability to measure total oxidants (free ozone + ozonides + hyperperoxides). However, monitoring total oxidants by the galvanic analyzer is a possibility.

Continuous Analyzers. As of 1997 there are three continuous analyzers available for the measurement of ozone. One is the Bailey–Fischer and Porter

Series 17T-2000 amperometric analyzer, which is specific for free ozone. It can also be arranged to measure total residual oxidants. Then there is the ATI Model 15/64 dissolved ozone electronic monitor, plus the EIT Model 8422[98] microprocessor-based monitor, both with a 0 to 2.0 mg/L range. See Chapter 15 for further details concerning dissolved ozone analyzers.

The following comments are from the third edition of this text—but now, in 1997, after having had more than three years of success with ORP residual monitoring in chlorine residual analysis, it appears that the use of ORP control monitoring of ozone in solution will eliminate the following complaints about oxidant analysis:

Another factor always present where oxidants are involved is the presence of substances that interfere with the analytical procedure. Copper ions poison the amperometric cell, while manganese ions interfere with some analytical procedures used in the calibration of the analyzers.

Residual control may never receive the necessary consideration to prove its practicality if the exhaust gas monitoring system continues to be as reliable as reported by the Venosa group. However, continuous residual recording for monitoring total oxidant residual may have practical value. It may be an important tool to be used for increasing the reliability of the ozone process. (Therefore, be sure to read in Chapter 15 about the success that Germany is experiencing with the use of ORP control in the application of ozone to potable water treatment.[97]

Forecasting Coliform Concentrations in an Ozonated Effluent

Effluent Quality. In order for ozone to be effective in wastewater disinfection, the effluent must be of good quality. There is a significant difference between a filtered secondary effluent and a nonfiltered secondary. Other than the absorbed ozone dose, the total COD and nitrites are the most significant factors. In the case of filtration it is the reduction of COD and not suspended solids that has the greatest effect. The work by Venosa et al.[44] and Ghan et al.[56,57] confirm these observations.

Filtered Effluent. Venosa et al.[44] developed a mathematical model that can be used to predict either the total or fecal coliform concentrations in the ozonated effluent in accordance with the following equations:

$$\log_{10} \mathrm{TC} = 3.95 + 0.030\mathrm{TCOD} + 0.50\mathrm{NO_2^-\text{-}N} - 3.05 \log_{10} T \qquad (13\text{-}21)$$

$$\log_{10} \mathrm{FC} = 3.34 + 0.029\mathrm{TCOD} + 0.48\mathrm{NO_2^-\text{-}N} - 3.4 \log_{10} T \qquad (13\text{-}22)$$

where

TC = total coliforms/100 ml
FC = fecal coliforms/100 ml
T = absorbed ozone dose, mg/L

TCOD = total chemical oxygen demand, mg/L
NO_2^--N = nitrite nitrogen, mg/L

Unfiltered Secondary Effluent. Venosa et al. compared a poor-quality unfiltered effluent with the filtered effluent and developed the following equations to predict the final coliform concentrations in the ozonated effluent.

It is important to note that their analysis of the data indicated that additional factors were found to influence the correlation of effluent coliform densities.

$$\log_{10} TC_i = 0.96 \log_{10} TC_i - 0.89 + 0.012TCOD + 0.60\ NO_2^-N + 0.013TSS - 4.024 \log_{10} T - 0.57R^{1/2} \quad (13\text{-}23)$$

where
TSS = total suspended solids mg/L
TC_i = initial coliform concentration/100 ml
R = ozone residual mg/L

In this case of total coliforms the results were affected by both the initial coliform density and the ozone residual.

$$\log_{10} FC = 4.06 + 0.020TCOD = 0.37NO_2^-N + 0.012TSS - 3.94 \log_{10} T - 0.59R^{1/2} \quad (13\text{-}24)$$

where
TSS = total suspended solids mg/L
R = ozone residual mg/L

In Eq. (13-24) the final fecal coliforms were not affected by the initial fecal coliform concentration. However, both the final fecal and total coliforms were affected by the total suspended solids (TSS) and the ozone residual (R).

High-Level Disinfection with Ozone

Marlborough, Massachusetts Study. Owing to a current trend in certain coastal states to provide protection for shellfish growing areas and similar high quality receiving waters in other areas, standards have been imposed requiring high level disinfection of wastewater effluents. In 1977, Marlborough, Massachusetts built a pilot plant to receive secondary effluent from the Easterly Wastewater Plant. This pilot plant was arranged to ozonate either filtered or unfiltered secondary, filtered or unfiltered nitrified effluent, or a combination of these. The results of this study were reported by Stover and Jarnis in 1979.[65] They explored the efficiency of ozone to achieve final total coliforms of 70/100 ml, which is the level considered necessary to protect shellfish, and 2.2/100 ml. The latter level is required for water reuse situations and other sensitive areas in arid states where effluents flow in ephemeral streams where bathing can take place. (See Chapter 8 for situations where the 2.2 coliform level is mandatory in California.)

This study was aimed at determining the relationship between effluent quality, absorbed ozone, and total residual oxidants. The latter were measured by the back-titration method used to measure chlorine residuals with an amperometric end point.

Absorbed ozone concentrations of 2–35 mg/L were applied to the different effluents at contactor hydraulic reaction times of 1–10 min., yielding total effluent residual oxidant concentrations from 0.4 to 8.0 mg/L. The results showing the relationship of absorbed ozone dose and total residual oxidants to the final coliform count are presented in Figs. 13-3 and 13-4. The data for

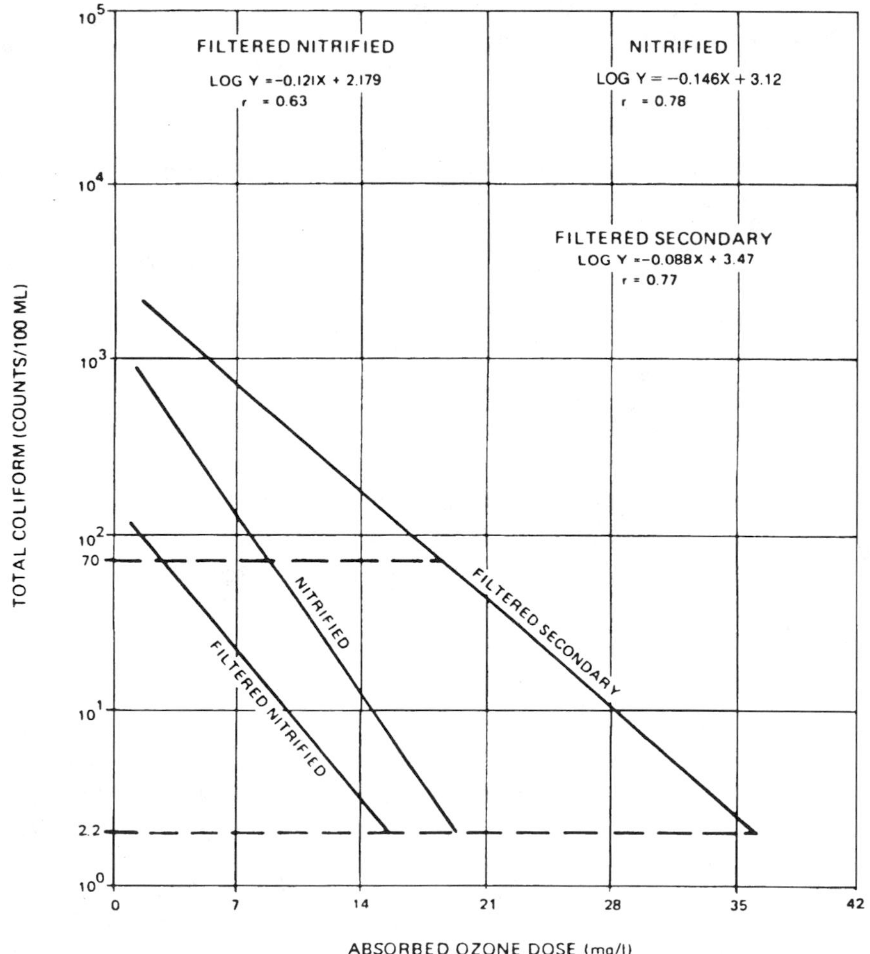

Fig. 13-3. Effluent total coliforms versus absorbed ozone dose[65] (courtesy *Journal WPCF*).

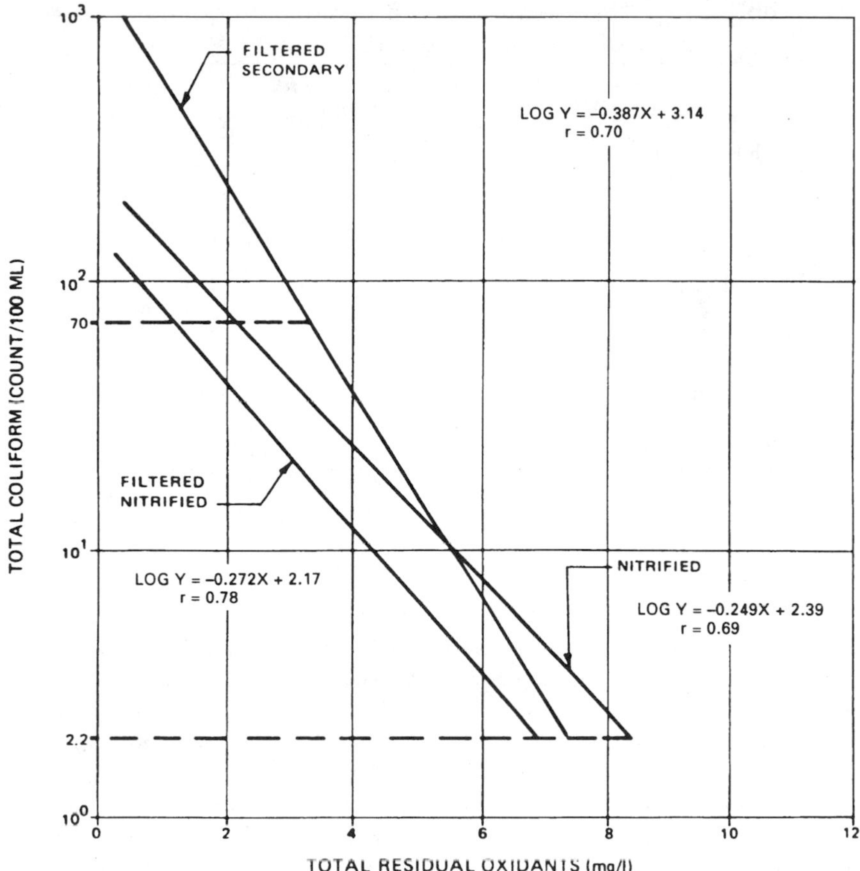

Fig. 13-4. Effluent total coliforms versus total residual oxidants[65] (courtesy *Journal WPCF*).

these plots are shown in Table 13-8. Fecal coliforms were never detected in the ozonated effluents at high-level disinfection.

The log total coliform reduction versus the log absorbed dose was plotted for all data points for both the nitrified effluents. Because the two nitrified effluents investigated were essentially of the same chemical water quality, with the exception of initial coliform concentration, the ozone requirement to achieve a certain log reduction would be expected to be the same for both effluents. The same plot was made for the combined secondary effluents. Then the required absorbed ozone dose was calculated from these plots to achieve a 70/100 ml total coliform. For the nitrified and secondary effluents the doses are 5.2 and 14.3 mg/L, respectively.

This study showed that the absorbed ozone dose gives good correlation with coliform reduction. It also demonstrated that the absorbed ozone dose is significantly dependent upon effluent quality.

Table 13-8 Absorbed Dose and Residual Requirements for Disinfection (Figs. 13-3 and 13-4)

Water Quality	*Absorbed Dose mg/L*	*Residual mg/L*
High-level disinfection (less than 2.2 total coliforms/100 ml)		
Filtered nitrified		
Nitrified	15	6.7
Filtered secondary and secondary	19	8.2
	36	7.2
70 Total Coliforms/100 ml		
Filtered nitrified	3	1.2
Nitrified	9	2.2
Filtered secondary and secondary	18	3.3

High-Level Disinfection with Ozone and UV

The EPA carried out a study to determine whether the combination of UV radiation and ozone would be sufficiently better than either disinfectant alone to justify the use of both techniques for wastewater disinfection.[66] This investigation revealed that the amount of applied ozone needed to achieve a fecal coliform limitation of 14/100 ml (approximately 60/100 ml total coliform) could be reduced as much as 80 percent if UV radiation either preceded or followed ozone addition. The ozone dose required to achieve the target coliform count alone was approximately 14 mg/L in all test runs. When UV was applied after ozone, the ozone dose could be reduced to about 3 mg/L. Simultaneous addition of UV and ozone in the same column resulted in a less than additive effect. Apparently UV reduces ozone to molecular oxygen. The sequential application of ozone first, followed by UV radiation exposure, was slightly better than UV radiation before ozone. All of this means, simply stated, that ozone is not a very good disinfectant (too short a contact time), and UV is only a good polishing agent to high quality effluents. The sequential addition of ozone and UV was found to be more cost-effective than either ozone or UV alone for plant sizes greater than 10 mgd. For smaller plants, UV alone appears to be a better choice than ozone.

This investigation was not intended to quantify the optimum UV dose. The purpose was to discover if UV did or did not have an effect upon ozonation. This is the reason why the UV dose is not mentioned.

OZONE FACILITY FOR WASTEWATER DISINFECTION

Application of ozone to a secondary effluent presents an entirely different chemical situation when compared to surface water supply treatment. In order for the ozone to be germicidally effective it first must overcome the initial immediate ozone demand (2–3 min.). Owing to the high reactivity of ozone, the demand can be extremely high. Moreover, this demand is variable and unpredictable over any 24-hour period. Therefore, ozone application must be ultimately controlled to reflect both process flow and ozone demand changes. Scientific advances in the field of control instrumentation now make this possible.

Disinfection Guidelines

There are several levels of disinfection for wastewater effluents. These levels are reported in one of the following ways:

1. MPN total coliforms surviving.
2. MPN fecal coliforms surviving.
3. Log reduction of initial coliform (total or fecal) concentration after disinfection.
4. Percent kill—99 percent = 2-log reduction, 99.999 percent = 5-log reduction, etc.

Current practice for disinfection guidelines is established either by the EPA requirement, which is 200/100 ml fecal coliforms MPN,* or by the state regulatory agency. Many states have disinfection requirements more stringent than the EPA 200/100 ml fecal coliforms. States with shellfish growing areas limit the total coliform concentration in the effluent to 70/100 ml. California uses several levels of total coliform concentration, depending upon the situation (240, 23, and 2.2). (See Chapter 8.) To convert fecal coliform concentration to total coliforms multiply by 4.

Therefore, to make a rational design of an ozone disinfection system it is necessary to know what the peak coliform concentration will be in the effluent before disinfection. Then, by knowing the disinfection requirement, the log reduction or the percent kill can be calculated arithmetically.

Ozone Demand and Absorption

This terminology is a carryover from chlorination practices. In chlorination it is easy to establish because it is the arithmetic difference between the chlorine dose and the chlorine residual after a specified contact time. This is difficult to determine for ozone because of the following factors: (1) there is no consensus on the optimum contact time; (2) ozone residual measurements are not reliable; and (3) ozone is so reactive that there is little if any true ozone residual after 3 to 5 min. in the contactor.

*This EPA fecal coliform concentration is equivalent to 1000/100 ml total coliforms.

Therefore, work by Venosa et al. on EPA-funded projects demonstrated that the best correlation for reliable disinfection by ozone is to continuously measure the ozone concentration in the exhaust gas. This was an important decision, because at that time (early 1980s) this was the only method available to measure the ozone *Ct*. By knowing this and the amount of ozone dosage, control is achieved by maintaining absorption in the contactor at a constant level for a given disinfection requirement. The amount of absorption is directly related to disinfection efficiency.[43,45,60] Now (1997) we have the practical use of ORP, which is easier to apply and extremely accurate.

Forecasting Ozonator Capacity

Secondary Effluents. Venosa et al.[60] studied the effect of various ozone dosages upon the relationship between coliform destruction (fecal and/or total) and absorbed ozone dose (Fig. 13-5). The same relationship was studied for ozone concentrations in the exhaust gas (Fig. 13-6). With these data, it is possible to estimate the required applied ozone dose for a given NPDES requirement.

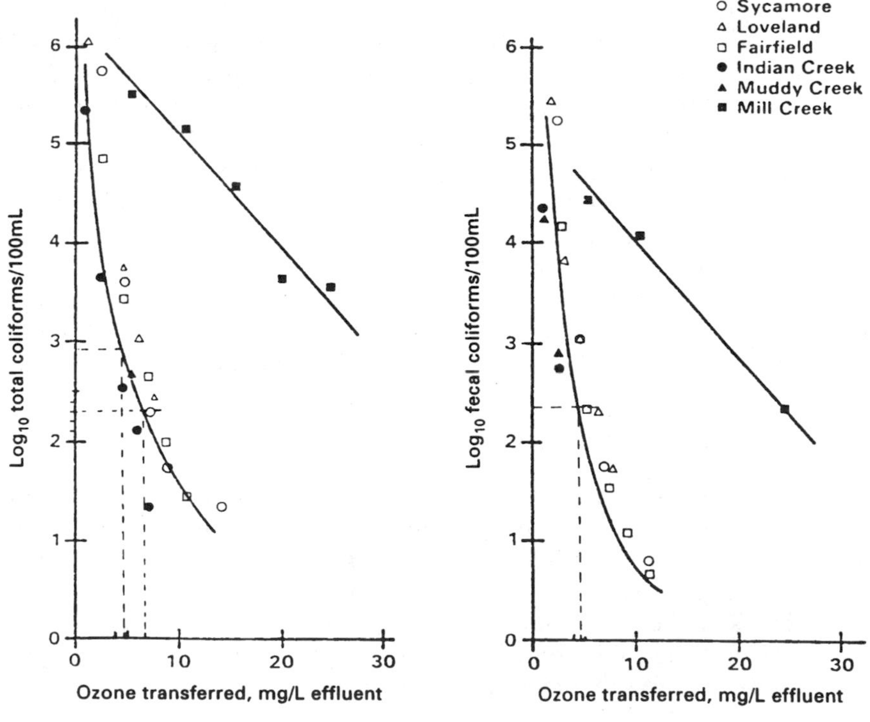

Fig. 13-5. Ozone absorbed (transferred) versus total coliforms and fecal coliforms[60] (courtesy *Journal WPCF*).

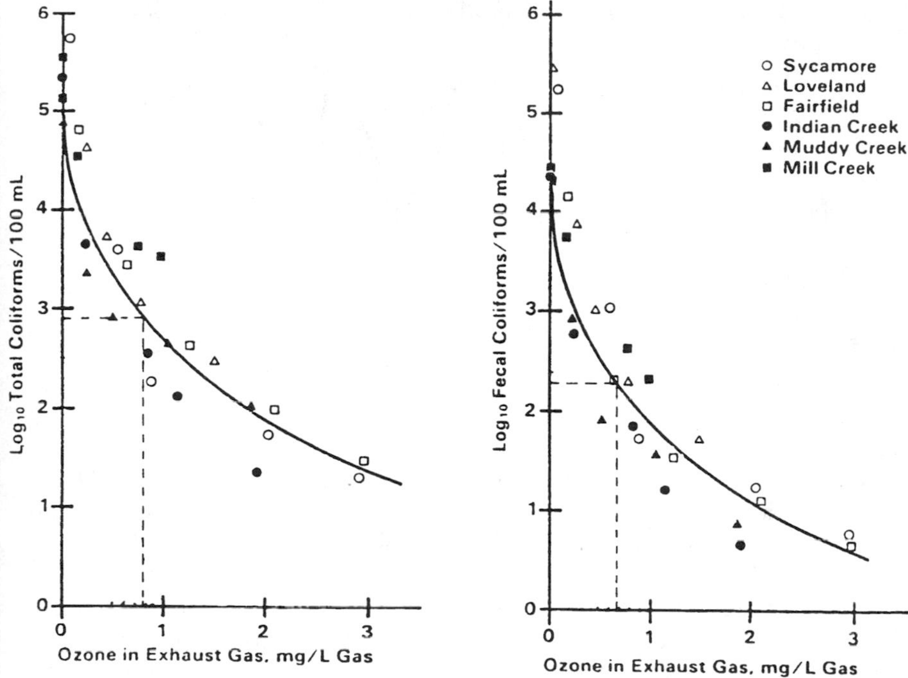

Fig. 13-6. Ozone in exhaust gas versus total coliforms and fecal coliforms[60] (courtesy *Journal WPCF*).

EXAMPLE. Assume NPDES requirement is 200/100 ml MPN fecal coliforms; $\log_{10} 200 = 2.2$. Examination of Fig. 13-5 shows that at log = 2.3 the absorbed ozone dose (ozone transferred) will be about 5 mg/L. A similar examination of Fig. 13-6 shows that ozone concentration in the exhaust gas will be about 0.7 mg/L.

The absorbed dose (T) equals the applied ozone dose (D) times the fraction that is transferred (see Eqs. 13-17 and 13-18):

$$T = Y_1 \left(\frac{Q_G}{Q_L}\right)\left(\frac{Y_1 - Y_2}{Y_1}\right)$$

where

Y_1 = concentration of ozone in carrier gas, mg/L
Q_G = carrier gas flow rate, L/min
Q_L = liquid flow rate, L/min
Y_2 = concentration of ozone in exhaust gas, mg/L

From Fig. 13-5, $T = 5$ mg/L and from Fig. 13-6, $Y_2 = 0.7$ mg/L. Solve for Y_1, in the above equation:

$$5 = Y_1 \left(\frac{Q_G}{Q_L}\right)\left(\frac{Y_1 - Y_2}{Y_1}\right)$$
$$= \frac{Q_G}{Q_L}(Y_1 - 0.7)$$

Assume $Q_G/Q_L = 0.5$; then:

$$Y_1 = \frac{5.35}{0.5} = 10.7 \text{ mg/L}$$

or, say, 11 mg/L.

To the above values for Y_1 a safety factor of 30 percent should be added to provide for unexpected levels of ozone demand or for disappointing contactor efficiency. Therefore the ozonator capacity should be $11 \times 1.3 = 14.3$, or, say, 15 mg/L = 125 lb/mg.

As a check on the above calculation, suppose the NPDES requirement is to be in equivalent total coliforms. This would be approximately four times the fecal coliforms or 800 total coliforms; $\log_{10} 800 = 2.9$. Inspection of Figs. 13-5 and 13-6 shows that the ozone absorbed and the ozone in the exhaust gas are practically identical to the amounts for fecal coliforms.

Preozonation: Tertiary Effluents. In the United States, pilot plant operation has been the basis for determination of ozonator capacity. The optimum ozone dosage is dependent upon water quality of each specific source. There are many variables to be considered affecting ozone consumption: color, turbidity, nitrates, COD, Fe, Mn, and a variety of organics.

Cost Studies

Several ozone feasibility studies have been made which compare ozone to chlorine specifically for wastewater disinfection.[67–72]

Table 13-9 Ozone Disinfection Cost

A. Ozone generated from air			
Plant size (mgd)	1	10	100
Capital cost	190,000	1,070,000	6,880,000
Disinfection cost ¢/1000 gal.	7.31	4.02	2.84
B. Ozone generated from oxygen			
Capital cost[a]	160,000	700,000	4,210,000
Disinfection cost ¢/1000 gal.	7.15	3.49	2.36

[a] The reduced cost here represents the elimination of drying atmosphere air for the process.

Table 13-10 Chlorination Disinfection Cost

Plant size	1	10	100
Capital cost	60,000	190,000	840,000
Disinfection cost ¢/1000 gal.	3.49	1.42	0.70
	Dechlorination with sulfur dioxide		
Plant size	1	10	1000
Capital cost	11,000	29,000	94,000
Disinfection cost ¢/1000 gal.	0.88	0.35	0.19

EPA Summary. Some of these studies and others were summarized in the EPA Task Force Report, March 1976.[71] These findings are shown in Table 13-9 for the ozonation process, with the chlorination/dechlorination process shown in Table 13-10.

The costs outlined above should be considered as tentative because the parameters of disinfection are not specific. The following cost comparisons are specific and therefore of more practical value for the designer.

Sacramento, California. In 1975, the Sacramento Regional County Sanitation District of California made an ozone feasibility study.[67] The disinfection requirement for the plant discharge is a total coliform MPN of 23/1200 ml. The PWWF is 240 mgd. The effluent from this plant is assumed to be a well-oxidized secondary effluent resulting from the activated sludge process. There is some cannery waste during the dry weather flow season of July to September.

Table 13-11 Cost Comparison of Ozonation and Chlorination/Dechlorination 240 mgd Design Capacity Current Cost Basis—ENR 2300 ($)

	Chlorine	*Ozone*	
Type of Cost	*10 mg/L Dosage*	*20 mg/L Dosage*	*40 mg/L Dosage*
Total capital cost	2,150,000	14,250,000	23,450,000
Average annual cost			
capital[a]	173,000	1,148,000	1,890,000
operating[b]	373,000	439,000	858,000
total	546,000	1,587,000	2,747,000
Average unit cost[c] (¢/1000 gal.)	1.2	3.5	6.0
Average annual local cost[d]	395,000	583,000	1,094,000

[a] Capital recovery at 7% for 30 years.
[b] Based on 125 mgd average annual flow.
[c] Based on total average annual cost.
[d] Sum of operating cost and local amortized capital cost share based on 87.5% Federal grant for capital facilities.

Table 13-12 Present Worth Analysis: Chlorination/ Dechlorination Versus Ozone

Cost Element	*Chlor./ Dechlor.*	*PCI Ozone*	*Union Carbide Ozone*
Total Capital	$990,000	1,380,000	1,570,000
Total O and M	2,600,000	4,620,000	4,870,000
Total Project	3,590,000	6,000,000	6,440,000
Annual O and M Cost	93,000	130,000	149,000

For this situation ozone equipment manufacturers recommended design capacity of 20 mg/L. However, the design engineers thought this would be much too low in view of recent operating experiences by the County Sanitation Districts of Los Angeles at their Pomona water reclamation plant. These studies indicated that a 40 mg/L dosage would be more appropriate to achieve the disinfection requirement of 23/100 ml MPN total coliforms. These results of this report are shown in Table 13-11. Power cost is based upon 0.9 ¢/kWh.

This analysis shows the capital cost of an ozonation system is between seven and eleven times as great as that of a chlorination/dechlorination system, and the annual operating cost for ozone is somewhere between 1.2 and 2.3 times as much as that of the chlorination/dechlorination method. On an average annual cost basis, the ozonation system is between three and five times as expensive as the chlorination/dechlorination system.[67]

Richmond, California. Another analysis for a less stringent disinfection requirement was made for the City of Richmond, California, discharging a well-oxidized activated sludge effluent into San Francisco Bay.[69] The effluent standards to be met were total coliform MPN of 70/100 ml with no more than 10 percent of these samples exceeding 230/100 ml.

Few reliable data are available to predict the ozone dosage required to achieve this particular disinfection objective. However, the following design parameters were selected based upon proposals from the ozone equipment suppliers:

Table 13-13 Annual Operating Cost

Operation of Maintenance	*Chlorination*	*Ozonation*
Disinfection 50 mgd flow	77,000[a]	91,000[b]
Deodorizing at headworks	16,000	1,000
Control of carbon column biofouling	16,500	5,000
Total O and M cost	109,500	97,000

[a] Includes cost of chemical, electrical power and maintenance.
[b] Includes cost of electric power and maintenance.

Table 13-14 Total Annual Cost of Disinfection

Itemized Costs	*Chlorination*	*Ozonation*
Capital cost	1,593,000	2,457,000
Amortized annual cost	136,000	205,000
O and M cost	109,500	97,000
Total annual cost	245,000	302,000

- Ozone dosage ADWF—15 mg/L.
- Ozone dosage PWWF—6 mg/L.
- Contact time at PWWF-10 min.

The system chosen, as recommended by the manufacturers, was based upon the use of high-purity oxygen recycled to the generators. Oxygen was to be supplied locally by the Airco plant at $50 or $60 per ton. Owing to certain cost differentials between the two ozone equipment suppliers, separate estimates are shown in Table 13-12. A separate calculation was made for a chlorination/dechlorination system, which is included in the tabulation. The analysis shown in Table 13-12 is based upon the present worth concept.

From the above two analyses, using the most optimistic dosage and contact times proposed by the ozone equipment manufacturers, disinfection is substantially more costly by ozone than it is by chlorination/dechlorination.

Cleveland, Ohio.* In contrast a report prepared in 1974 for the Cleveland Regional Sewer District[70] for the Westerly Advanced Wastewater Treatment Facility resulted in a recommendation for ozone rather than chlorination.† This was based upon an ozone disinfection dosage of 6 mg/L. Disinfection was not defined for MPN total coliforms. The comparative cost of the two systems is summarized in the tables.

It is clear from Tables 13-13 and 13-14 that the cost of disinfection by ozone is considerably greater than that of disinfection by chlorination. However, in this case, ozonation was selected on the basis of longer equipment life (20 to 40 years); the relatively inexpensive cost of deodorizing the screening and degritting facilities; and the control of the biological slimes in the carbon columns. That these additional factors might influence the choice of ozonation is at best tenuous. More operating data concerning these parameters must be developed.

---

*This is the Westerly Advanced Wastewater Treatment Plant owned by the Northeast Ohio Regional Sewer District and not the City of Cleveland.

†During the time between the report and the start of construction, this recommendation was changed. The role of ozonation was moved to preozonation, and chlorination was selected for disinfection.

Table 13-15 Present Worth Analysis of Chlorine Disinfection Versus Ozone[a]

Cost Items	*Chlorine*	*Ozone*
Capital cost	4,840,000	7,137,000
Present worth of interest	2,411,000	3,552,000
Present worth of annual costs	1,226,000	919,000
Net present worth	8,477,000	11,608,000

[a] A planning period of 20 yr. and an interest rate of 6.125 percent was used.

Other long-range economic evaluations have been made attempting to justify the use of an ozonation disinfection system instead of an existing chlorination process requiring the addition of a dechlorination facility. One such system involved the conversion of a primary treatment plant to provide secondary treatment. The Unox process was chosen for this conversion. Moreover, the existing chlorination system needed to be supplemented with dechlorination capability. In spite of all these factors, the long-range economic factors as presented by the engineers demonstrated that the chlorination/dechlorination process was the one of choice.[72]

Houston, Texas. Consultants for the city of Houston made a different type of disinfectant evaluation. They compared ozone, liquid–gas chlorine, and liquid hypochlorite for their new 400 mgd (peak flow) wastewater-treatment plant.[68] Then evaluations were prepared to determine capital cost and annual operation costs for each disinfection system. The present worth method was used to relate capital cost to annual cost over a 20-year design period. Parenthetically, it should be mentioned that the high-purity oxygen activated sludge treatment process (UNOX) was found to be the most cost-effective process.

The design dosage of 5 mg/L chlorine was assumed to provide a 1.0 mg/L residual at the end of 20 min. of contact time at peak flows. It was further assumed that a 5 mg/L ozone dose would be sufficient to meet or exceed the state standards for disinfection. This was also expected of chlorine, and no claim was made that either one would be more effective in eliminating total coliform organisms. In developing the costs of chlorine (in tank cars) versus ozone, the capital cost assessed chlorine at $4,640,000 for the contact chamber, and $1,000,000 for ozone. This calculates to an ozone contact time of 4.3 min. at peak flow.

Table 13-15 tabulates the Houston analysis; the ozone figures are based upon an arithmetical average of three equipment suppliers.*

*Houston chose commercial-grade sodium hypochlorite for disinfection for safety reasons. The only Houston installations using liquid chlorine are those supplied by 150-lb cylinders.

REFERENCES

1. McCarthy, J. J., and Smith, C. H., "A Review of Ozone and Its Application to Domestic Wastewater Treatment," *J. AWWA,* **66,** 718 (Dec. 1974).
2. Dyachov, A. V., "Recent Advances in Water Disinfection," paper presented at Ann. Conf. of Int. Water Poll. Research, Amsterdam, Netherlands, Sept. 1976.
3. Richards, W. N., and Shaw, B., "Developments in the Microbiology and Disinfection of Water Supplies," *J. Inst. Water Eng. Sci.,* **30,** 191 (June 1976).
4. Schalekamp, M., "Experience in Switzerland with Ozone, Particularly the Neutralization of Hygienically Undesirable Elements Present in Water," paper presented at the Annual AWWA Conf., Anaheim, CA, May 1977.
5. Stone, B. G., and Trussell, R., "Application of Ozone for Viral Disinfection," paper presented at the Annual Calif. Sect. Mtg. AWWA, San Diego, CA, Oct. 30, 1975.
6. Morris, J. C., "The Role of Ozone in Water Treatment," paper presented at the Annual Conference AWWA, New Orleans, LA, June 24, 1976.
7. Venosa, A. D., "Comparative Disinfection of Wastewater Effluent with Chlorine, Bromine Chloride, and Ozone," paper presented at the Forum on Disinfection with Ozone, Chicago, IL, June 2–4, 1976.
8. Fetner, R. H., and Ingols, R. S., "A Comparison of the Bactericidal Activity of Ozone and Chlorine against *Esch. coli* at 1°C," *J. Gen. Microbiol.,* **15,** 381 (1956).
9. Morris, J. C., "Aspects of the Quantitative Assessment of Germicidal Efficiency," in Chap. 1, J. D. Johnson (Ed.), *Disinfection, Water and Wastewater,* Ann Arbor Science, Ann Arbor, MI, 1975.
10. Selna, M. W., Miele, R. P., and Baird, R. B., "Disinfection for Water Reuse," paper presented at the Disinfection Seminar at the Annual AWWA Conf., Anaheim, CA, May 8, 1977.
11. Bailey, P. S., "Reactivity of Ozone with Various Organic Functional Groups Important to Water Purification," First Int. Symposium on Ozone for Water and Wastewater Treatment, Proc. IOI, Waterbury, CT, 1975.
12. Dellah, A., "Study of Ozone Reactions Involved in Water Treatment and the Present Chlorination Controversy," Proc. 2nd Int. Symposium on Ozone Technology, Montreal, PQ, Canada, May 11–14, 1975.
13. Hoigné, J., and Bader, H., "Identification and Kinetic Properties of the Oxidizing Decomposition Products of Ozone in Water and Its Impact on Water Purification," Proc. 2nd. Int. Symposium on Ozone Technology, Montreal, PQ, Canada, May 11–14, 1975.
14. Hann, V. A., and Manley, T. C., "Ozone," *Encyclopedia of Chemical Technology,* pp. 735–753, Interscience, New York, 1952.
15. O'Donovan, D. C., "Treatment With Ozone," *J. AWWA,* **57,** 1167 (1965).
16. Singer, P. C., and Zilli, W. B., "Ozonation of Ammonia in Wastewater," *Water Research,* **9,** 127 (1975).
17. Kinman, R. N., "Ozone in Water Disinfection," p. 123 in F. L. Evans, *Ozone in Water and Wastewater Treatment,* Ann Arbor Science, Ann Arbor, MI, 1972.
18. Glaze, W. H., Rawley, R., and Lin, S., "By-products of Organic Compounds in the Presence of Ozone and UV Light," paper presented at the IOI Meeting, Cincinnati, OH, Nov. 17–19, 1976.
19. Guirguis, W. A., Srivasta, P., Meister, T., Prober, R., and Hanna, Y., "Ozone Reactions with Organic Material in Sewage Non-sorbable by Activated Carbon," paper presented at the IOI Meeting, Cincinnati, OH, Nov. 17–19, 1976.
20. Prengle, H. W., Jr., Mauk, C. E., and Payne, J. E., "Ozone–UV Oxidation of Pesticides in Aqueous Solution," paper presented at the IOI Meeting, Cincinnati, OH, Nov. 17–19, 1976.
21. Richard, Y., "Organic Materials Produced Upon Ozonation of Water," paper presented at the IOI Meeting, Cincinnati, OH, Nov. 17–19, 1976.
22. Falk, H. L., and Moyer, J. E., "Ozone as a Disinfectant of Water," paper presented at the IOI Meeting, Cincinnati, OH, Nov. 17–19, 1976.
23. Hartemann, P., Block, J. C., and Maugras, M., "Biochemical Aspects of the Toxicity Involved by the Ozone Oxidation Products in Water," paper presented at the IOI Meeting, Cincinnati, OH, Nov. 17–19, 1976.

24. Kinman, R., Rickabaugh, J., Elia, V., McGinnis, K., Cody, T., Clark, S., and Christian, R., "Effect of Ozone on Hospital Wastewater Cytotoxicity," paper presented at the IOI Meeting, Cincinnati, OH, Nov. 17–19, 1976.
25. Spanggord, R. J., and McClurg, B. J., "Ozonation Methods and Ozone Chemistry for Selected Organic Compounds in Water," paper presented at the IOI Meeting in Cincinnati, OH, Nov. 17–19, 1976.
26. Simmon, V. F., and Eckford, S. L., "Methods for Evaluating the Mutagenic Activity of Ozonated Chemicals," paper presented at the IOI Meeting, Cincinnati, OH, Nov. 17–19, 1976.
27. White, G. C., "Other Methods of Disinfection," Chap. 2 in *The Quest for Pure Water,* Vol II, 2nd ed., AWWA, Denver, CO, 1981.
28. Hoven, D. L., Schwartz, B. J., and Weng, Cheng-Nan, "Ozone: an Economical Choice for Hackensack Water Company's Drinking Water," paper presented at Ann Conf. AWWA, St. Louis, MO, June 1983.
29. Miller, G. W., Rice, R. G., Robson, C. M., Scullin, R. L., Kuhn, W., Wolf, H., and Carswell, J. K., "An Assessment of Ozone and Chlorine Dioxide Technologies for Treatment of Municipal Water Supplies," *Executive Summary,* EPA 600/8-78-018, Cincinnati, OH, Oct. 1978.
30. Nadeau, M., and Pigeon, J. C., "The Evolution of Ozonation in the Province of Quebec, Canada," *Proceedings* First Int. Symposium on Ozone for Water and Wastewater Treatment, Washington, DC, Dec. 2–4, 1973, p. 157.
31. Dellah, A., "Ozonation at Charles-J. Des Baillets Water Treatment Plant, Montreal, Canada," *Proceedings* IOI Symposium, Montreal, Canada, May 11–14, 1975, p. 715.
32. Leduc, P., "Ozone Improves Taste, Odor and Color of Water," *Water and Sew. Wks.,* **125,** 49 (Dec. 1978).
33. Rook, J. J., "Developments in Europe," *J. AWWA,* **68,** 279 (June 1976).
34. Symons, J. M., "Trip Report for European Travel," EPA, Cincinnati, OH, Oct. 29, 1974.
35. Kühn, W., Sontheimer, H., and Kurz, R., "Use of Ozone and Chlorine in Water Works in the Federal Republic of Germany," *J. AWWA,* **70,** 326 (June 1978).
36. Anon., "Ozone Treatment Licks Color Problem," *Water and Sewage Works,* **122,** 52 (Apr. 1975).
37. Sontheimer, H., Heilker, E., Jekel, M. R., Nolte, H., and Vollmer, F. H., "The Mulheim Process," *J. AWWA,* **70,** 393 (July 1978).
38. Rice, R. G., Gomella, C., and Miller, G. W., "Rouen, France Water Treatment Plant," *Civil Engineering—ASCE,* **76** (May 1978).
39. Staff Report, "The Activated Carbon Dilemma," *Water and Sew. Wks.,* **125,** 34 (Dec. 1978).
40. Masschelein, J. W., "Overview of Ozone Installation in Europe," presented at the AWWA Annual Conference in Cincinnati, OH, June 17, 1990.
41. Anon., "Emerzone Systems Catalog," Emery Industries, Cincinnati, OH, Oct. 13, 1980.
42. Haag, W. R., and Hoigné, J., "Kinetics and Products of the Reactions of Ozone with Various Forms of Chlorine and Bromine in Water," paper presented at Ann. Conf. AWWA St. Louis, MO, June 1983.
43. Venosa, A. D., and Opatken, E. J., "Ozone Disinfection—State of the Art," paper presented at Pre-conf. Workshop, 52nd Ann.Conf. WPCF, Houston, TX, Oct. 7, 1979.
44. Venosa, A. D., Meckes, M. C., Opatken, E. J., and Evans, J. W., "Disinfection of Filtered and Unfiltered Secondary Effluents in Two Ozone Contactors," paper presented at 52nd Ann. WPCF Conf. Houston, TX, Oct. 7–11, 1979.
45. Venosa, A. D., "Effectiveness of Ozone as a Muncipal Wastewater Disinfectant," paper presented at Pre-conf. Workshop, 56th Ann. WPCF Conf., Atlanta, GA, Oct. 1, 1983.
46. Venosa, A. D., et al., "Comparative Efficiencies of Ozone Utilization and Microorganism Reduction in Different Ozone Contactors," in A. D. Venosa (Ed.), *Progress in Wastewater Disinfection Technology,* EPA-600/9-79-018 U.S. Environ. Prot. Agency, Cincinnati, OH, 1979.
47. Trussell, R., Nowak, T., Ismail, F., Jopling, W., and Cooper, R., "Ozone as a Pretreatment for Coagulation, Filtration and Disinfection," paper presented at the CWPCA Annual Conf., Los Angeles, CA, Apr. 1975.

48. Saunier, B. M., Selleck, R. E., and Trussell, R. R., "Preozonation as a Coagulant Aid in Drinking Water Treatment," *J. AWWA,* **75,** 239 (May 1983).
49. Lambert, M., personal communication, CIFEC, Paris, France, 1982.
50. Tate, Carol H., "Overview of Ozone Installations in North America," presented at the AWWA Annual Conference in Cincinnati, OH, June 1990.
51. National Primary Drinking Water Regulations, Final Rule, 40 CFR Pars 141 & 142, Part II EPA, *Federal Register,* June 29, 1990.
52. Hibler, Charles P., private communication, Aug. 1990.
53. Rice, R. G., "Ozone Process Design Considerations and Experiences," presented at the Ozone Seminar, AWWA Annual Conference in Cincinnati, OH, June 17, 1990.
54. Venosa, A. D., "Ozone as a Water and Wastewater Disinfectant: A Literature Review," in E. L. Evans H (Ed.), *Ozone in Water and Wastewater Treatment,* Ann Arbor Science, Ann Arbor, MI, 1972.
55. White, G. C., *Disinfection of Wastewater and Water for Reuse,* Van Nostrand Reinhold, New York, 1978.
56. Ghan, H. B., Chen, C. L., and Miele, R. P., "The Significance of Water Quality on Wastewater Disinfection with Ozone," paper presented at the Forum on Disinfection with Ozone, Chicago, IL, June 2–4, 1976.
57. Ghan, H. B., Chen, C. L., Miele, R. P., and Kugelman, I. J., "Wastewater Disinfection with Ozone," paper presented at the CWPCA Annual Conference, Los Angeles, CA, Apr. 1975.
58. Parkhurst, J. D., "Pomona Virus Study," Sanitation Districts of Los Angeles, Whittier, CA, Feb. 1977.
59. Ward, R. W., Giffin, R. D., DeGraeve, G. M., and Stone, R. A., "Disinfection Efficiency and Residual Toxicity of Several Wastewater Disinfectants," EPA Report, Grant No. S-802292, Cincinnati, OH, 1976.
60. Venosa, A. D., and Meckes, M. C., "Control of Ozone Disinfection by Exhaust Gas Monitoring," *J. WPCF,* **55,** 1163 (Sept. 1983).
61. "Feasibility Study of Ozone Disinfection of Wastewater Effluent for Westerly Advanced Wastewater Treatment Facility, Cleveland Sewer District," Engineering Science Ltd., Cleveland, OH, Mar. 1974.
62. Rownd, W. H., Mgr. Engineering-Science, Cleveland, OH, private communication, Mar. 13, 1984.
63. Trussell, R. R., et al., "Chino Basin Municipal Water District Prototype Plant Study," J. M. Montgomery Engineers, Pasadena, CA, Mar. 1975.
64. Trussell, R., private communication, J. M. Montgomery Engrs., Pasadena, CA, 1977.
65. Stover, E. L., and Jarnis, R. W., "Obtaining High Level Wastewater Disinfection with Ozone," *J. WPCF,* **53,** 1637 (Nov. 1981).
66. Venosa, A. D., Petrasek, A. C., Brown, D., Sparks, H. L., and Allen, D. M., "Disinfection of Secondary Effluents with Ozone and U.V.," *J. WPCF,* **56,** 137 (Feb. 1984).
67. Hoag, L. N., and Salo, J. E., "Ozonation Feasibility Study," Sacramento Regional Wastewater Management Program, Sacramento, CA, June 1975.
68. Matson, J. V., and Coneway, C. R., "Economics of Disinfection," paper presented at the IOI Forum on Ozone Disinfection, Chicago, IL, June 2–4, 1976.
69. Calmer, J., and Adams, R. M., "Evaluation of Disinfection by Ozone," in-house report, Kennedy Engineers, San Francisco, CA, 1977.
70. "Feasibility Study of Ozone Disinfection of Wastewater Effluent for Westerly Advanced Wastewater Treatment Facility, Cleveland Sewer District," Engineering Science Ltd., Cleveland, OH, Mar. 1974.
71. "Disinfection of Wastewater: EPA Task Force Report No. 430/9-75-012," Cincinnati, OH, March 1976.
72. Consoer, Townsend, and Assoc. Cons. Engrs. Chicago, IL, East Bay Mun. Util. Report, "Disinfection and Chlorine Residual Removal," 1974.
73. Rakness, Kerwin L., Stover, Enos L., and Krenek, David L., "Design, Start-up and Operation of an Ozone Disinfecting Unit," *J. WPCF,* **56,** 11, 1152 (Sept. 1984).

74. Venosa, Albert D., Rossman, Lewis A., and Sparks, Harold L., "Reliable Ozone Disinfection Using Off-gas Control," *J. WPCF,* **57,** 9, 929 (Sept. 1985).
75. Fujioka, Roger S., and Shizumura, Lyle K., "*Clostridium perfringens,* a Reliable Indicator of Stream Water Quality," *J. WPCF,* **57,** 10, 986 (Oct. 1985).
76. "Water Quality Standards" in "Public Health Regulations," Title 11, Chapter 4, Dept. of Health, State of Hawaii, 1982.
77. Dutka, B. J., "Coliforms Are an Inadequate Index for Water Quality," *J. Environ. Health,* **36,** 39 (1973).
78. Knittel, M. D., "Occurrence of *Klebsiella pneumonaie* in Surface Waters," *Appl. Microbiol.,* **29,** 595 (1975).
79. Berg, G., and Metcalf, T. G., "Indicators of Viruses in Water," in G. Berg (Ed.), *Indicators of Viruses in Water and Food,* Ann Arbor Science Publishers, Inc., Ann Arbor, MI, 1978.
80. Fujioka, R. S., "Stream Water Quality Assessment Based on Fecal Coliform and Fecal Streptococcus Analysis," Technical Memorandum Report No. 70, Water Resources Research Center, Univ. of Hawaii, 1983.
81. Richard, Y., "Biological Methods for the Treatment of groundwater," in H. Sontheimer and W. Kuhn (Eds.), "Oxidation Techniques in Drinking Water Treatment," EPA Report 57019-74-020. USEPA, Cincinnati, OH, 1979.
82. Prevost, M., Desjardins, R., Arcouette, N., Duchesne, D., and Coallier, J., "A Study of the Performance of Filtration and Biological Activated Carbon in Cold Waters," *Sciences et Techniques de L'Eau,* **23,** No. 1 (Feb. 1990).
83. Prevost, M., Duchesne, D., Coallier, J., Desjardins, R., and Lafrance, P., "Full-Scale Evaluation of Biological Activated Carbon Filtration for the Treatment of Drinking Water," presented at the Annual AWWA-WQTC Conference, Philadelphia, PA, Nov. 11–17, 1989.
84. Prevost, Michele, Desjardins, Raymond, Duchesne, Daniel, and Poirier, C., "Chlorine Demand Removal by Biological Activated Carbon Filtration in Cold Water," presented at the AWWA-WQTC Annual Conference, San Diego, CA, Nov. 1990.
85. Prevost, M., Lafrance, P., Desjardins, R., Huck, P. M., Anderson, W. D., Fedorak, P. M., and Bablon, G., "Biological Activated Carbon Filtration: The Canadian Experience," presented at the 5th World Filtration Congress, Acropolis, Nice, France, June 5–8, 1990.
86. Servais, P., Billen, G., Ventresque, C., and Bablon, G. B., "Microbial Activity in GAC Filters at the Choisy-le-Roi Treatment Plant," *J. AWWA,* p. 62 (Feb. 1991).
87. Prevost, M., Desjardins, R., Coallier, J., Duchesne, D., and Mailly, J., "Comparison of Biodegradable Organic Carbon Techniques for Process Control," presented at the Annual AWWA-NQTC Conference, San Diego, CA, Nov. 1990.
88. Montgomery, J. M., Consulting Engrs., Inc., *Water Treatment Principles and Design,* John Wiley & Sons, New York, 1985.
89. Rook, J. J., "Haloforms in Drinking Water,"*J. AWWA,* **68,** 168 (1976).
90. Kinman, R. N., "Ozone in Water Disinfection," in F. F. Evans, *Ozone in Water and Wastewater* Ann Arbor Science, Ann Arbor, MI, 1972.
91. Jacangelo, J. G., Patania, N. L., Reagan, K. M., Aieta, M. E., Krasner, S. W., and McGuire, M. J., "Ozonation: Assessing Its Role in the Formation and Control of Disinfection By-products," *J. AWWA,* p. 74 (Aug. 1989).
92. Ferguson, D. W., Gramith, J. T., and McGuire, M. J., "Applying Ozone for Organics Control and Disinfection: A Utility Perspective," *J. AWWA,* p. 32 (May 1991).
93. Gramith, J. T., Coffey, B. M., Krasner, S. W., Kuo, C.-Y., Means, E. G., "Demonstration-Scale Evaluation of Bromate Formation and Control, Due to Ozonation," Presented at Ann. AWWA Conf. in San Antonio, TX, June 10, 1993.
94. EPA, Proposed Rule of Primary Drinking Water Regulations for Disinfection By-products Caused by Ozonation, No. 811-Z-94-004, Vol. 59, No. 145, pp. 38735–38738, *Federal Register,* July 29, 1994.

95. Krasner, S. W., Glaze, W. H., Weinberg, H. S., Daniel, P. A., and Najm, I. N., "Formation and Control of Bromate During Ozonation of Waters Containing Bromide," *J. AWWA*, pp. 106–115 (Jan. 1995).
96. Joel, A. R. and Offner, H. G., "THM and Color Oxidation with Ozone," presented at Nat. Conf. on Wastewater, Edmonton, Alberta, Canada, by Capital Controls Co. 3000 Advance Lane, Colmar, PA 18915, 1987.
97. Green, D., "Control and Monitoring the Ozone Disinfection Process by the Stranco High Resolution Redox Control System at Dusseldorf Germany," The Stranco Corp. Bradley, IL., Customer Service 800-882-6466; 1995.
98. Bauer, J., "Microprocessor-Based Dissolved Ozone Monitor," EIT Model 8422, 251 Welsh Pool Road, Exton, PA 19341, Customer Service 800-634-4046; Nov. 1996.

14
OZONE, PEROXONE,* AND AO_xPs

INTRODUCTION

One of the most compelling reasons for the surge of interest in ozonation of potable water was the revelation (ca. 1974) that the renowned free residual chlorine process (HOCl) was responsible for generating trihalomethanes (THMs), which are purported to have undesirable health effects. Owing to this claim, these so-called disinfection byproducts (DBPs) were put under strict regulation by the EPA. Humic and fulvic acids have been identified as the precursors to the formation of the THMs. These acid compounds abound in the environment; they will always be with us. Moreover, because they are among nature's most important compounds, they are not easy to deal with.

The EPA set a maximum contaminant level (MCL) for THMs at 100 μg/L (ca. 1978) and proposed reducing the MCL to 50 μg/L by 1991. Investigations carried out since this ruling went into effect have shown that both chloramines and chlorine dioxide do not generate THMs. This then made these two chemicals the most likely alternative disinfectants.

The Metropolitan Water District of Southern California, which was formed in 1927[1] to transport Colorado River water 240 miles into the arid desert area of Southern California, was one of the first utilities to investigate alternative disinfectants to free chlorine. In 1981 it made a decision to switch from free chlorine to chloramination rather than invest in activated carbon to remove the THMs. This decision was governed by the high cost of the carbon treatment process. The entire system, which included five conventional WTPs, was converted to chloramines in 1984. In less than one year of operation MWD ran into trouble from this conversion to chloramines, with the users of kidney dialysis machines. The water used in these machines is tap water polished by activated carbon filtration to remove any trace of chlorine residual, which is harmful to the patient's blood supply. When free chlorine was being used, the dechlorination reaction by the activated carbon resulted in physical consumption of the carbon, so the operator always knew when to replace the carbon. However, dechlorinating a chloramine residual with activated carbon is quite different. Chloramines deplete the adsorption power of the carbon, and when this occurs, the operator has no forewarning until there is a chlorine

*Peroxone is a word that was coined by the Metropolitan Water District of Southern California during its extensive research projects using ozone and hydrogen peroxide in sequence with various ratios to each other.

breakthrough. In view of this serious situation, MWD had to shut down the chloramination procedure and revert to free residual chlorine until all the dialysis machines were equipped with dual carbon filters in series. By monitoring the discharge from the first carbon filter for chlorine residual, the operator knows when to replace the carbon in the primary filter. This retrofit of the dialysis water treatment systems took an entire year before MWD could go back to chloramination.

This episode generated more trouble for water utilities in California because the California Dialysis Council became acutely aware of the possible existence of harmful DBPs in other systems. This prompted a letter from the California Dialysis Council lawyer to the California Department of Health Services in Sacramento, California, asking how and when the DBPs of chlorine dioxide would be removed from those plants using chlorine dioxide. The DBPs in question were chlorite and chlorate ions, both harmful to dialysis patients. The Health Services group responded by shutting down all chlorine dioxide facilities that were treating drinking water. As of 1990, chlorine dioxide was still banned because there is not an acceptable method available for the removal of these two ions. As of 1996, chlorine dioxide is still banned in California owing to the extremely low concentration that is allowed for chlorate and chlorite ions. However, a study by Slootmaekers et al. at Miami University in Oxford, Ohio,[2] indicated some possible solutions using the sulfite ion. Nothing beneficial has been revealed on this project as of 1997.*

Also, White predicted trouble at the special seminar in 1981 when MWD announced going to chloramines to be able to meet the EPA THM limitations, because of the ammonia N added to the finished water to produce chloramines. In many cases utilities were using this chloraminated water and were situated in such a way that their water arrived with zero chlorine residual, but the ammonia N was still there. When this happens, the excess nitrogen in the water provides "caviar" for any and all microbiological life in the distribution system. To overcome this problem, these utilities had to rechlorinate the MWD water. This process is complicated and difficult to manage. Other utilities had to rechlorinate using the breakpoint chlorination process to overcome the use of chloramination by MWD.

The above difficulties are a good example of what can happen when the federal government gets involved in establishing nationwide regulations for drinking water quality. Recognizing this problem early on, the American Water Works Association Research Foundation reacted quickly to promote comprehensive investigations to meet these water quality challenges. The most outstanding of these efforts was the one by the Metropolitan Water District of Southern California, with headquarters in Los Angeles, California and ably assisted by James M. Montgomery Engineers of Pasadena, California.[3] These findings are incorporated in the following text.

*The California Health Services has relaxed their MCL for ClO_2 so that it is now (1997) available for potable water treatment in California.

UTILITIES' OBJECTIVES FOR FINISHED WATER

Water utilities are faced with the task of treating water with processes that combine to produce a water that not only meets the regulatory requirements but also meets the aesthetic requirements of the consumer. Therefore it has to taste good, smell good, and look good. Otherwise the public will become suspicious about whether or not it is safe to drink. MWD has concluded from its comprehensive investigation that ozone is one of the leading candidates that will allow utilities to meet these challenges in a cost-effective manner.

FEDERAL REGULATIONS

Safe Water Treatment Rule (SWTR)

This rule was issued on June 29, 1989, mandating specific filtration and disinfection requirements.[4] The focus of this rule is on turbidity and microbiological contaminants.

The microbiological contaminants specifically identified in this rule are *Giardia lamblia,* viruses, *Legionella,* heterotrophic plate count (HPC) bacteria, and coliform bacteria. However, an especially feared organism that is missing from this list is *Cryptosporidium,* which is highly resistant to chlorine but can probably be neutralized by a combination of peroxone and filtration.

Under the SWTR all public water systems using surface water sources will be required to achieve at least a 3-log removal or inactivation of *Giardia* cysts and at least a 4-log removal or inactivation of all enteric viruses by the treatment processes train. The unfortunate part of this rule is that the privately owned small supply systems in resort areas or vacation spots are not covered by it. Therefore, they become vulnerable to the possibility of a waterborne disease outbreak.

The weakness of these and other rules is that they allow 3- and 4-log removals, far below the California rule of 5 logs minimum.

The *Ct* Rule

The EPA has formulated a rule to establish the desired efficiency of the disinfection process. It is identified as the *Ct* rule and is part of the SWTR. It involves the product of the disinfectant residual C, in mg/L, and the exposure time t, in minutes, to this residual. The time t is identified as t_{10}, which represents the time that 90 percent of the treated water will be exposed to the disinfectant. The EPA expects that a utility will perform tracer studies or an equivalent demonstration. In the case of chlorine, the determination of t is easy if a chlorine residual analyzer is strategically located. Either shut off the chlorine and measure the time it takes for the appearance of the vertical movement downward of the pen on the chart, or reverse this process by increasing the dosage and timing the vertical movement upward. Measuring

t for ozone has been considerably more difficult because there was not an acceptable method to accurately measure a true ozone residual. As of 1992, the EPA was soon supposed to have a suitable method available. In 1997, Stranco found that its High Resolution Redox System, which it introduced to European ozone installations, finally provides an accurate method of measuring the oxidative power of an ozone residual.

All of the above rules will have to be changed in some way to provide at least a 5-log removal of all pathogenic organisms, including cysts and viruses. The reason for this is that two important things affect the way nature operates, whether we like it or not: (1) the more we develop antibiotics to deal with disease-causing organisms, the more vicious these organisms become by the simple process of mutation; and (2) our planet is approaching a population crisis in about 2025, when it will run out of water for 8.9 billion people, which is needed for agriculture as well as drinking water for both humans and animals (see Special Preface)

Disinfection By-products (DBPs)

The EPA released what would ultimately be known as the Disinfection and DBP rule (D/DBP) in 1992. The objective of this rule is to strike a balance between microbiological and chemical risks. The DBPs such as trihalomethanes (THMs) and haloacetic acids (HAAs) are to be regulated by MCLs for users of chlorine, with aldehydes and bromate regulated by MCLs for users of ozone. Chlorine residuals going into the distribution systems will probably be limited to some specific level for a given chlorine species. This move, as of 1997, does not appear to have had any benefits for the water utilities.

ADVANCED OXIDATION PROCESSES (AO$_x$Ps)

Description

AO$_x$Ps are identified as chemical processes that accelerate the decomposition of the ozone molecule to yield enough of the potent hydroxyl free radical to cause some specific improvement in the water treatment process train. These processes include peroxone, which uses hydrogen peroxide with ozone, ultraviolet radiation (UV) with ozone, elevated pH with ozone, and UV with hydrogen peroxide. Since the last process does not include ozone, it will not be mentioned further in this chapter.

Advantages

The available AO$_x$Ps have the potential to let water producers achieve definitive water quality goals at lower costs and greater reliability than conventional oxidation methods allow.

Chemistry

When the ozone molecule is used as the only oxidant in a treatment process for natural waters, its reactions are highly selective and relatively slow. However, when hydrogen peroxide is used with ozone, autocatalytic decomposition of the ozone molecule occurs.[5] This produces an intermediate oxidant, the potent hydroxyl free radical, as follows:

$$H_2O_2 + 2O_3 \rightarrow 2HO + 3O_2 \qquad (14\text{-}1)$$

This compound is less selective and more powerful than ozone. It is the same powerful intermediate that triggers the "breakpoint reaction" in chlorination practices. This free radical [HO] has also been observed to be solely responsible for the destruction of extremely objectionable phenolic tastes in a surface water. The water was treated by an "induced breakpoint" with chlorine and ammonia at a 20–1 ratio.

DISINFECTION BY-PRODUCTS (DBPs)

Occurrence

DBPs occur worldwide. A survey of 35 WTPs representing a broad range of raw water quality and treatment processes indicated that THMs and HAAs accounted for the majority of the DBPs detected. Chloraminated water was found to produce cyanogen chloride. It was also found that chlorination produced low levels (<18 mg/L) of the sum of formaldehyde and acetaldehyde, whereas ozonation produced levels up to 50 mg/L for both of these compounds.[6] All of this was observed without any reports of health risks.

Ozonation DBPs

Typical drinking water DBPs produced by ozone are: formaldehyde, acetaldehyde, acetone, and carboxylic acids. In ozonation of raw water containing bromide, the organic DBPs produced are: bromoform, dibromoacetic acid, bromopicrin, cyanogen bromide, and the inorganic bromate. When the peroxone process is used in high-level bromide waters, the formation of organic brominated DBPs is either eliminated or drastically reduced. However, the bromate production appears to be increased.[6]

Tan and Amy[24] found that the use of ozone in a water that contains bromide ions can lead to the formation of bromoform ($CHBr_3$) and bromate (BrO_3^-), depending upon pH and Br^- concentration. This demonstrates that aside from the DBPs formed by postchlorination, ozone itself, when used for DBP control, can actually form additional objectionable DBPs. This is not surprising because it is a rule of thumb that the stronger the oxidant, the more certain it is that the result will be more DBPs. This is why chloramines have little or no ability to form DBPs similar to ozone, bromine, chlorine, or chlorine dioxide. The ORP level of chloramines is insignificant compared to ozone. The reader is reminded that now (1997), cancer researchers who have been actively working on the causes of cancer for some 30 years have concluded that these THMs are trivia, owing mostly to their miniscule amounts. To make the amounts look serious, the EPA reports THMs in micrograms per liter, whereas everything else is in milligrams per liter. For instance, 50 μg/L equals 0.1 mg/L. Now try to picture what the "precurser" would be that removes 0.1 mg/L of chlorine demand.

Control of DBPs

Raw Water Quality. Currently the EPA is considering three alternatives for controlling the production of DBPs. A fourth method that should be actively pursued is to exercise stricter control of raw water quality. There are far too many uncontrolled and untreated discharges entering U.S. surface and ground waters. This is dramatically demonstrated by comparing the quality of the two source waters used by MWD of Southern California. The quality of the Colorado River water is much higher than that of the California State Project water. While the wastewater discharges in California are producing high-quality effluents, the runoffs from the large agricultural farmlands in the San Joaquin Valley discharge their pollutants into the Sacramento–San Joaquin Delta, which is the source of the California State Project water. This farmland irrigation runoff contributes significantly to the THM precursors found in the State Project water.

Removing DBP Precursors.* There are four strategies that might be used to remove DBP precursors (chlorine demand) as follows:

Membrane Technology. This is a site-specific possibility, but one burdened with a multitude of maintenance problems and reliability problems in addition to its high cost. This technology includes the following processes: microfiltration, ultrafiltration, reverse osmosis, nanofiltration, electrodialysis, and electrodialysis reversal.[7] The technology has been significantly improved over the past 25 years. The pilot plant for the Flagler Beach, Florida system is a good example of what could be expected from this technology. During a year-long operation, the THM concentration in the softened water was 350 μg/L (3.5 mg/L), while the membrane-treated water averaged 20 μ/L (0.20 mg/L).†

*This is simply lowering the chlorine demand by pre-treatment processes. Precursor is just a fancy word dreamed up by lab technicians.

†This type of situation is another good reason why the raw water quality criteria should include a limitation on the chlorine demand. White has suggested for years it should be about 1.5 mg/L after 20 min. of contact time. His prime example is the runoff water from the Andes Mountains in Chile, which travels several miles in a silt-laden river to a treatment plant that provides sedimentation and filtration. This treated water has a chlorine demand of only 0.20 mg/L after 20 min. of contact time.

Additionally, using the summed estimates of capital and O&M unit costs to construct a new 2.7 mgd plant for Flagler Beach, the membrane system would cost $1.33/1000 gal versus $1.34/1000 gal for the current softening process.[8]

GAC. The 1986 amendments to the Safe Drinking Water Act (SDWA) require the EPA to list the "best available technologies" (BATs) that are capable of meeting MCL regulations. GAC has been identified as the BAT for THM control, except for one serious reservation. In many waters that require frequent carbon regeneration cycles, it might not prove to be the BAT for THM control.[9] Moreover, GAC is cost-intensive.

Optimized Coagulation. Researchers have not found any significant DBP precursor (chlorine demand) destruction or removal by better coagulation and flocculation. In their search to see what could improve particle removal by improved coagulation and flocculation, various strategies were involved using ozone and peroxone. It was determined that preozonation followed by conventional treatment did not enhance the removal of TOC or THM precursors (chlorine demand). Where preoxidation by peroxone was used at a hydrogen peroxide-to-ozone ratio of 0.2, flocculation was improved at ozone dosages of both 1.0 and 2.5 mg/L. This meant greater particle removal, but there was no consistent THM precursor removal.[6]

Ozone. Research by Singer and Chang[10] has demonstrated that ozone alone has a negligible effect on the overall concentration of TOC in raw water. Therefore, it can be said that ozone will have no effect on the precursors (chlorine demand) in the raw water. In those reports where preozonation is said to have lowered the THMs, the reason is usually that the ozone has lowered the chlorine demand, which in turn lowers the required dosage. Some researchers have evidence the ozone can destroy precursors of chloroform and trichloroacetic acid, but the amounts and consistency of such destruction are not reliable enough to warrant using ozone as a DBP or a chlorine demand control strategy. This was found out many years ago in France, at the three water treatment plants using their raw water from the Seine River, which supplied drinking water to Paris consumers.

DBP Removal after Formation. This subject has received a significant amount of attention from researchers. The secret is to resort to a biological filtration system, where biodegradation of organic compounds will occur. Many water producers use preozonation followed by chlorination to control biological growths in the sedimentation system and/or filters. These plant effluents have concentration levels of formaldehyde comparable to those produced by the ozone contractor. However, when secondary disinfection is not applied until after filtration, a biofilm can develop on the filter media that is capable of biodegrading formaldehyde, AOC, and other ozonation by-products.[6] This is only part of the problem. When this strategy is used, adequate disinfection must be practiced in order to inactivate the microorganisms released from the filter media of the biodegradation step. If these organisms are not controlled by disinfection, regrowth of biological life in the distribution system will almost surely occur. Here again we are dealing with chlorine demand. This is another reason why the chlorine demand of a source water should be limited in some fashion—either that or the treated water. If this strategy is applied to the treated water, then it becomes a special investigation of what the treatment processes have to be.

Organics Removal

Ozone reacts selectively with organic compounds and usually cannot decompose them completely to carbon dioxide and water. Takahashi found that organic compounds such as carboxylic acids and aldehydes remain in solution after ozonation.[11]

However, if compounds were subjected to ozone and UV irradiation, a greater removal of TOC was observed than with ozone alone, and there was complete oxidation of low molecular weight organics to carbon dioxide and water.*

Taste and Odor Compounds

Owing to its superior oxidizing power, ozone is the oxidant of choice for use in the control of T&O in drinking water. Two of the compounds that are probably the worst offenders are MIB and geosmin. The MWD of Southern California made a two-year pilot study to evaluate both ozone and peroxone for their ability to remove these compounds. More than 100 pilot runs were made.[6] In each run the pilot plant influent was spiked with 100 mg/L of both MIB and geosmin. To achieve 90 percent removal by ozone alone required a dosage of greater than 4.0 mg/L, as compared with an ozone dose of 2.0 mg/L using peroxone. (Metropolitan was to confirm these findings at its demonstration plant, due to be on-line about 1993. See Ref. 26 and 27. Contact MWD for these documents.

Color Removal

Ozone has been used successfully since 1988 at the Myrtle Beach, South Carolina WTP to control the highly colored surface water caused by a variety of humic substances. Ozone reduces the influent color of 150–450 color units

*This oxidation phenomenon is called "mineralization."

to an average of 5 color units by an ozone dosage of up to 10 mg/L.[12] However, recoloration has occurred 1–2 hours downstream from the preozonation contactor. This phenomenon has been observed by other researchers when ozone dosages and contact times were insufficient, particularly if the color was derived from humic substances.[13] At Myrtle Beach this problem was solved by adding a postsedimentation step of applied ozone.

A 1990 study was made for the City of Long Beach, California to evaluate both ozone and peroxone for treating highly colored groundwater. It was found that using ozone alone was an affective method. Ozone dosages of 4 mg/L at an in-line ozone dissolving system and up to 6 mg/L by a conventional contacting system could reduce influent color levels of 50 color units to the target level of <10 color units. In contrast, peroxone was found to be less effective than ozone, especially with higher peroxide-to-ozone ratios. This finding strongly indicates that ozone, being a superoxidant, is far superior to the hydroxyl free radical for color removal.[6]

Oxidation of Volatile Organic Compounds (VOCs)

Ozone and ozone AO_xPs can be used to oxidize some selected VOCs. Several laboratory and pilot-scale studies were conducted in the years 1988–90 to investigate the ability of ozone and ozone AO_xPs to remove TCE, PCE, and carbon tetrachloride (CTC).

Glaze and Kang[14] conducted laboratory studies on the oxidation of TCE and PCE in a Southern California groundwater (LADWP, North Hollywood Wells). Various dosages of ozone and peroxone were used. The results demonstrated that direct ozonation of TCE and PCE in high-alkalinity groundwater was a relatively slow process, but when peroxone was used, the oxidation process was significantly accelerated, indicating that the hydroxyl free radical was responsible for the acceleration. It thus was concluded that not only is oxidation of TCE and PCE by peroxone a promising process, but because carbonates are known radical scavengers, water softening to reduce the carbonate and bicarbonate levels prior to oxidation can further increase the removal efficiency. The same Southern California groundwater was studied on a pilot scale by Aieta and co-workers to evaluate the function of the peroxide-to-ozone ratio and contact times versus the removal efficiency of TCE and PCE.[15]* It was shown that both TCE and PCE were effectively removed at ozone dosages from 1.0 to 9.0 mg/L utilizing a peroxide-to-ozone ratio of 0.5 (by wt). TCE required lower ozone dosages for the same percent removal when compared to PCE. The results further indicated that the removal efficiency was independent of contact time over the range studied, which was 2.7 to 25 min. It was the opinion of these researchers that the peroxone process was the BAT for VOC removal.

*This water was in the same well field at well No. 5, which was not the water used by Glaze and Kang.[14]

Another pilot-scale study was made to investigate another Southern California groundwater for VOC removal.[16] In addition to TCE and PCE, CTC was included. In this study the UV/ozone AO_xP was the primary process; however, peroxone was briefly tested. Because the best removal of CTC was only 80 percent (by wt) using UV/ozone and only 59 percent by peroxone, it was concluded that ozone AO_xPs were not sufficiently effective for removing CTC in order to meet the MCL requirements established by the EPA.

Other Organic Compounds

In addition to removing T&O compounds, color (humic substances), and VOCs, ozone and ozone AO_xPs can be used to oxidize a number of other organic compounds, such as pesticides, phenols, acetic and oxalic acids, nitrobenzene, and chlorobenzene compounds.

The pesticide atrazine can be effectively oxidized by ozone or peroxone.[17] In this study it was observed that chlorine, chloramines, or chlorine dioxide, as well as conventional treatment, provided little or no atrazine removal. Paillard and co-workers reported that the removal of common pesticides—atrazine and semazine—from drinking water by peroxone is much more efficient than removal by ozone alone.[18]

Ozone oxidizes nonhalogenated phenols, which are potential T&O problems, to tasteless compounds: glyoxal, oxalic acid, carbonic acid, and formic acid.[13] According to Rice, if high-enough ozone doses or longer contact times or both are used, the oxidation reaction may continue, and these compounds will eventually mineralize to carbon dioxide and water.[19]

DISINFECTION

Pilot Study: MWD of Southern California

This study examined the microbiocidal effectiveness of ozone and peroxone against *Escherichia coli,* two coliphages (MS2 and f2), *Giardia muris,* and HPC bacteria. With the exception of the HPC group, the microorganisms were seeded into the preoxidation disinfection contactors of the pilot plant and subjected to ozone dosages of 1.0, 2.0, and 4.0 mg/L. The peroxide-to-ozone ratio was 0.0, 0.05, 0.10, 0.20, and 3.0. Contact times were 6, 9, and 12 min.[20,21] The results of this study demonstrated that for MWD's source waters, peroxone at a peroxide-to-ozone ratio of 0.3 or less and ozone alone are comparable disinfectants at a 12-min. contact time. Ozone residual in the effluent was found to be the key to effective disinfection.

Bacterial Inactivation

E. coli inactivation greater than 5 logs was observed for all ozone dosages, peroxide-to-ozone ratios, and contact times that were evaluated. Not surprisingly, the naturally occurring HPC bacteria were found to be considerably more resistant than the other indicator groups. They were comparable in resistance to *Giardia muris* with a range of HPC inactivations from 2 to 3 logs.[20]

Viral Inactivation

For all the ozone dosages, peroxide-to-ozone ratios, and contact times that were studied, greater than 6.5 logs of both the MS2 and f2 coliphages were inactivated in both the State Project water and the Colorado River water.[21] Generally, when the ozone residual in the single stage contactor effluent was >0.10 mg/L, coliphage inactivation exceeded 6 logs.

Giardia Inactivation

In general only 2–3 logs of *Giardia muris* were inactivated under the conditions studied.[20] When the ozone residual was 0.10 mg/L leaving the single stage contactor, *G. muris* inactivation exceeded 0.5 log, but with an ozone residual of 4.0–5.0 mg/L, 2 logs inactivation was achieved. Owing to the lower ozone demand, greater levels of inactivation were observed for CRW than for SPW. The lower demand produced higher residuals.

ASSIMILABLE ORGANIC CARBON (AOC)

Occurrence of AOC

AOC is the fraction of dissolved organic carbon (DOC) that is degradable by bacteria. This is an important part of the oxidation process in potable water treatment practices. Ozone produces higher levels of AOC than any other oxidant. However, there is a problem to be faced with the production of AOC if, after optimum ozone oxidation conditions have been developed, it then is necessary to determine the use of the proper biological filtration techniques that will assure the required mineralization of the AOC* within the treatment plant. If mineralization does not occur, the AOC group becomes a potent source of nutrition for microorganisms. This causes significant regrowth of biological life in the distribution system, as was observed repeatedly during the early days of ozonated drinking water.[22]

If complete mineralization cannot be achieved within the plant, it will become necessary to add a post-disinfection step. If chloramination is used, it must be remembered that after the chlorine residual has disappeared in the reaches of the distribution system, the applied ammonia fraction in the chloramine remains as nutrition for the microorganisms. Therefore, the regrowth syndrome begins all over again. In instances of far-flung piping systems with

*Mineralization means to convert an organic compound to carbon dioxide and water.

dead-ends, careful monitoring for chlorine residual, DO, and biological activity must become a routine surveillance activity. The implications of AOC production by ozonation, which promotes bacteriological regrowth in the distribution system, include the inability to meet coliform standards and a degradation in the aesthetics of the finished water (i.e., off-flavors, odors, and color). This potential problem has led to a growing number of researchers conducting studies on the production and removal of AOC in drinking water.

Quantitative Evaluation of AOC

The pilot-scale study by MWD of Southern California has revealed some important definitive information.[6] MWD evaluated the AOC production of both ozone and peroxone on both the State Project Water and the Colorado River Water.

The background level of AOC in the ozone contactor effluent was 50–100 mg/L. This was increased by ozone application to 210–430 mg/L for the CRW and 380–480 mg/L for the SPW. As might be expected, peroxone yielded AOC levels that were 80–170 mg/L greater than those produced by ozone.

Most of the pilot-scale experiments were performed without the application of a secondary disinfectant following the ozone contactor. It was found that the AOC was biodegraded in the filters down to influent background levels or lower. This meant that the dual-media filters were biologically active. However, when chloramines were added to the contactor effluent during selected tests, AOC was only slightly reduced through the filters. This indicated that the biological population in the filters had been adversely affected by the addition of the disinfectant.

Postfiltration oxidation was also evaluated. When postoxidation was combined with preoxidation and conventional treatment, AOC levels produced in the preoxidation contactor were removed by the filters and then produced again by postfiltration oxidation but at slightly lower levels. Postoxidation alone also resulted in high concentrations of AOC.

Controlling AOC Levels by Filter Operation

The pilot-scale studies by MWD of Southern California demonstrated that oxidation by ozone and peroxone does lead to the production of very high levels of AOC. This becomes a burden upon any distribution system because the AOC factor stimulates biological regrowth. This upsets the system, causing a variety of tastes, odors, and depressing colors in the drinking water. These results, however, also demonstrate that AOC can be effectively removed through granular-media filters that are biologically active. This mode of filter operation is an accepted practice in Europe and Canada.

Other practices that are known to be effective in the removal of AOC are filters with GAC media followed by dual-media filters and rapid sand filtration.[23] It is also evident that postoxidation could cause serious regrowth problems in the distribution system.

Importance of Bench-, Pilot-, and Demonstration-Scale Studies

Because source waters come in a variety of quality levels, they will respond differently to ozone and ozone AO_xPs, particularly peroxone. Therefore it is imperative that water producers use either bench-, pilot-, or demonstration-scale studies to verify process efficiency and proper selection of the process treatment train. Furthermore, such testing will help to identify the proper selection of the contactor system, which is known to be one of the most important components of a successful ozone facility.

Bench-scale studies are used to determine the technical feasibility of a treatment strategy for a given water source. Bench-scale studies may utilize batch, semi-batch, or continuous-flow reactors, depending upon the bench area available.

Pilot-scale studies are generally designed for continuous flow, ranging in capacity from 1 to 25 gpm. These studies can determine both technical and economic feasibility data, as well as limited process refinements.

Demonstration-scale studies are sized large enough to permit scale-up of specific unit processes. This will provide data for estimating operating costs and materials evaluation. It can also be used to great advantage for operator training.

The operation of a demonstration-scale study will confirm process effectiveness, determine equipment reliability, address special issues such as air binding in the filtration system and potential for biological regrowth, and provide staff training. Additionally, it allows the evaluation of the unit processes at different times of the year so as to verify changes in the quality of the source water due to seasonal variations.

Demonstration-Scale Plant (5.5 mgd)

Introduction. The Metropolitan Water District of Southern California is the largest drinking water purveyor in North America. It owns and operates five filtration plants that by the year 1996 had a total combined capacity of approximately 2800 mgd.[25] The results from the two-year pilot-scale study for controlling T&O compounds, DBPs in low bromide water, and a variety of microorganisms demonstrated that the peroxone process was indeed successful. Therefore, MWD deemed it necessary to evaluate the peroxone technology at full scale before attempting to retrofit its filter plants.

Project Objectives. The oxidation demonstration plant objectives are as follows:

1. Verification of peroxone process effectiveness at plant scale.
2. Confirmation of process ability to control DBP, T&O removal, and disinfection efficiency using chloramines.
3. Determination of process refinement criteria at full-scale design.
4. Evaluation of both the process and the equipment by a variety of manufacturers.
5. Evaluation of a wide variety of equipment components that are impossible to evaluate at the pilot-scale level.
6. Evaluation of assimilable organic carbon (AOC) and DBP biodegradation in large-scale filters.

Plant Description. This 5.5 mgd demonstration plant is located at the MWD 500 mgd Weymouth Filtration Plant at La Verne, California. The source of water for MWD is either one or a mixture of both of the following: State project water (SPW) from the Sacramento and San Joaquin rivers in Northern California and Colorado River water.* The Colorado River water is transported some 240 miles to Lake Mathews through a 14–16-ft concrete-lined tunnel from Parker Dam on the California–Arizona border, near Needles, California. The SPW is piped from the Sacramento–San Joaquin Delta some 350 miles to Southern California, having been pumped about 3000 ft over the Tehachapi Mountains. All the filtration plants are downstream from Lake Mathews.

Figure 14-1 is a schematic layout of the demonstration plant, and Fig. 14-2 shows the plant layout. The flocculation and sedimentation basins to be used for the demonstration plant are part of a new 2.2 mgd washwater reclamation plant at the same location. The demonstration plant was designed by James M. Montgomery of Pasadena, California.

Equipment Evaluation. The following ozone and ozone-related equipment are to be evaluated as part of the objectives of this project:

1. Ozone generators—150 lb/day:
   (a) Trailigaz 60 Hz, standard horizontal tube.
   (b) Asea-Brown-Boveri, 40–1000 Hz, standard horizontal tube.
   (c) Capital Controls/Schmidding-Werke, 200–650 Hz, high-concentration oxygen feed.
2. Oxygen feed system: liquid carbonic, 6000-gal storage tank, GOX flow range 1–44 cfm.
3. Air feed system: capacity 130 cfm at 35 psi, outlet gas dewpoint − 100°F. Compressors are by SIHI; refrigerated dryers externally reactivated are by Zurn.

*The SPW project water contains a varying amount of bromide ion due to the seasonal variation of seawater intrusion in the Sacramento River.

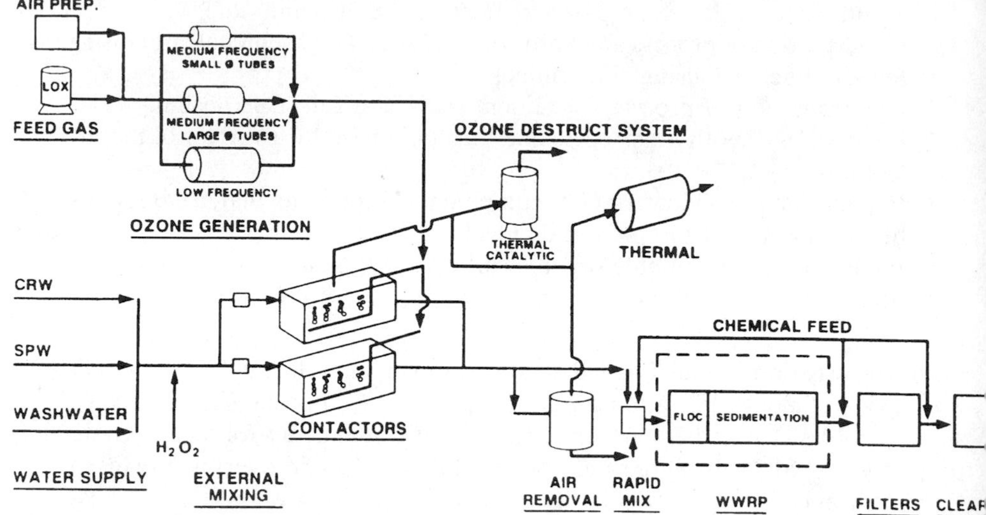

Fig. 14-1. Oxidation demonstration plant layout (courtesy of Metropolitan Water District of Southern California).

4. Ozone external mix system:
   (a) Upstream from ozone contactor No. 1—Koch Engineering Co. sidestream eductor.
   (b) Upstream from contactor No. 2—system A, sidestream eductor and system B, in-line injector, both by Komax.
5. Ozone contactor No. 1: rectangular, enclosed, flow-through with over–under baffles, cast-in-place concrete with 316 SS hardware; contact time 6–12 min. at 20-ft water depth with three countercurrent and three cocurrent ozone chambers. The diffusers consist of 80 rod-shaped fine bubble diffusers and 20 dome-shaped fine bubble diffusers. Ozone flow control is by 18 rotameters, and hydrogen peroxide feed is through three one-inch pipeline diffusers.
6. Ozone contactor No. 2: rectangular, enclosed, flow-through with serpentine baffling, cast-in-place concrete with 316 SS hardware; contact time 6–12 min. with 20-ft water depth. There are three ozone chambers, each with removable intermediate baffles. There are 40 rod-shaped fine bubble diffusers and 10 dome-shaped fine bubble diffusers. Ozone flow control is by nine rotameters and three one-inch pipeline diffusers for hydrogen peroxide feed.
7. Ozone destruct unit No. 1: thermal with a 40 kW heater, 150 lb/day ozone loading capacity, 299 cfm at a destruct temperature of 350°F; maximum ozone discharge residual 0.1 ppm; by Infilco Degremont Inc.
8. Ozone destruct units Nos. 2 and 3: thermal catalyst with 2.8 kW heaters using Carus Chemical Co., Carulite 200; operating temperature 15°F mini-

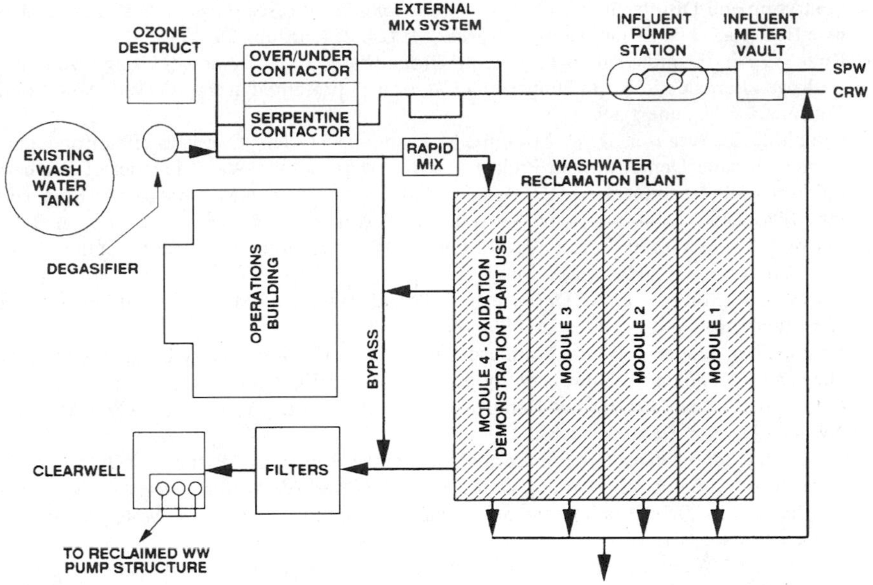

Fig. 14-2. Oxidation demonstration plant schematic (courtesy of Metropolitan Water District of Southern California).

mum above ambient; ozone loading capacity No. 2, 30 lb/day, 200 scfm max, and No. 3, 150 lb/day, 200 scfm max; 0.1 ppm ozone discharge residual; by Infilco Degremont Inc.

Plant Development Schedule. Experiments completed by the end in 1992 yielded process design criteria that allowed Metropolitan's Engineering Division to proceed with the design and construction of the ozone–peroxone facilities at each of the Metropolitan's five filtration plants. The full-scale facilities were expected to be on-line by late 1997, with this event heralding one of the most important processes in the development of advanced water treatment practices in the United States.[26,27]

REFERENCES

1. The State of California, "California Water Atlas," Governor's Office of Planning and Research, Sacramento, CA, 1978–79.
2. Slootmaekers, B., Tachiyashiki, S., Wood, D., and Gordon, G., "The Removal of Chlorite Ion and Chlorate Ion from Drinking Water," Dept. of Chemistry, Miami University, Oxford, OH, 1989.
3. McGuire, M. J., Ferguson, D. W., and Gramith, J. T., "Overview of Ozone Technology for Organics Control and Disinfection," presented by the Metropolitan Water District of Southern California at the AWWA Ann. Conf., Cincinnati, OH, June 1990.

4. "Filtration and Disinfection; Turbidity, *Giardia lamblia,* Viruses, *Legionella* and Heterotrophic Bacteria," Final Rule, *Federal Register* 54: 124: 27486, June 29, 1989.
5. Rulla, T. A., "Demonstration Project for Groundwater VOC Removal, Using Ozone and Hydrogen Peroxide in North Hollywood, California," presented at the AWWA Ann. Conf., Cincinnati, OH, June 1990.
6. Gramith, J. T., Ferguson, D. W., McGuire, M. J., and Tate, C. H., "Overview of Metropolitan's Ozone-Peroxide Demonstration Project," by the Metropolitan Water District of Southern California with Headquarters in Los Angeles, California and James M. Montgomery Engineers Inc. of Pasadena, California, presented at the AWWA Ann. Conf., Cincinnati, OH, June 1990.
7. Taylor, J. S., "Membrane Technology for Potable Water Treatment," presented at the AWWARF Conference, Oakland, CA, May 5–6, 1990.
8. Taylor, J. S., Mulford, L. A., Duranceau, S. J., and Barnet, W. M., "Cost and Performance of a Membrane Pilot Plant," *J. AWWA* (Nov. 1989).
9. Metropolitan Water District of Southern California, J. M. Montgomery Engineers Inc., and Michigan Technological University, "Optimization and Economic Evaluation of Granular Activated Carbon for Organics Removal," research report, prepared for AWWARF and AWWA, Denver, CO, June 1989.
10. Singer, P. C., and Chang, S., "Impact of Ozone on the Removal of Particles, TOC and THM Precursors," research report, prepared for AWWARF and AWWA, Denver, CO, June 1989.
11. Takahashi, N., "Ozonation of Several Organic Compounds Having Low Molecular Weight under Ultraviolet Irradiation," *Ozone Sci. & Eng.,* **12,** 1 (1990).
12. Dimitriou, M. A., and Ivanco, E., "The Myrtle Beach Ozone System and Water Treatment Plant," *Ozone in Water Treatment,* Vol. 1, Proceedings, Ninth Ozone World Congress, New York, 1989 L. J. Bollyky (Ed.), Intl. Ozone Assoc., Zurich, 1989.
13. Rice, R. G., and Nitzer, A., "Ozone for Drinking Water Treatment," *Handbook of Ozone Technology and Applications,* Vol. II, Butterworth Publ., Boston, MA, 1984.
14. Glaze, W. H., and Kang, J. W., "Advanced Oxidation Processes for Treating Groundwater Contaminated with TCE & PCE: Laboratory Studies," *J. AWWA* (May 1988).
15. Aieta, E. Marco, Reagan, Kevin M., Lang, John S., McReynolds, L., Kang, J.-W., and Glaze, Wm. H., "Advanced Oxidation Processes for Treating Groundwater Contaminated With TCE and OCE: Pilot Scale Evaluations," *J. AWWA* (May 1988).
16. City of Pasadena, California, Water and Power Department and James M. Montgomery Consulting Engineers Inc., "Advanced Oxidation Processes for the Control of Off-gas Emissions from VOC Stripping," research report, prepared for AWWARF and AWWA, Denver, CO, Oct. 1989.
17. Hulsey, R. A., Long, B. W., and Adams, C. W., "An Investigation of Oxidation Byproducts from Ozone Treatment of Atrazine," *Proceedings,* IOA Spring Conference on New Developments: Ozone in Water and Wastewater Treatment, Shreveport, LA, Mar. 1990.
18. Paillard, H., Valentis, G., Partington, J., and Taughe, N., "Removal of Atrazine and Semazine by O_3/H_2O_2 Oxidation in Potable Water Treatment," *Proceedings* IOA Spring Conference on New Developments: Ozone in Water and Wastewater Treatment, Shreveport, LA, Mar. 1990.
19. Rice, R. G., "Ozone for Treatment of Drinking Water," prepared for the Office of Drinking Water, USEPA under Contract No. 68-01-7289, with ICP Inc., Fairfax, VA, Mar. 1989.
20. Wolfe, R. L., Stewart, M. H., Scott, K. N., and McGuire, M. J., "Inactivation of *Giardia muris* and Indicator Organisms Seeded in Surface Water Supplies by Peroxone and Ozone," *Environ. Sci. Technol.,* **23,** 6, 744 (June 1989).
21. Wolfe, R. L., Stewart, M. H., Liang, S., and McGuire, M. J., "Disinfection of Model Indicator Organisms in a Drinking Water Pilot Plant by Using Peroxone," *Appl. Envir. Microbiol.,* **55,** 9, 2230 (Sept. 1989).
22. Rice, R. G., "Ozone Process Design Considerations and Experiences," presented at Ozone Seminar, AWWA Annual Conference, Cincinnati, OH, June 17, 1990.
23. American Water Words Research Foundation and Keurings Instituut voor Waterleiding Artikelen, "The Search for a Surrogate," a Cooperative Research Report, AWWA, Denver, CO, Oct. 1988.

24. Tan, L., and Amy, G. L., "Comparing Ozonation and Membrane Separation for Color Removal and Disinfection Byproduct Control," *J. AWWA,* pp. 74–79 (May 1991).
25. Gramith, J. T., and Ferguson, D. M., "Metropolitan's Oxidation Demonstration Project," unpublished report by the Metropolitan Water District of Southern California, Los Angeles, CA, June 1991.
26. Ferguson, D. W., McGuire, M. J., Koch, B., Wolfe, R. L., and Aieta, E. M., "Comparing PEROXONE and Ozone for Controlling Taste and Odor Compounds, Disinfection Byproducts, and Microorganisms" *J. AWWA, Research & Technology* pp, 181–191 April 1990.
27. Koch, B., Gramith, J. T., Dale, M. S., and Ferguson, D. W., "Control of 2-Methylisoborneol and Geosmin by Ozone and Peroxone: A Pilot Study. *Water Sci Tech* Vol 25, No. 2, pp 291–298, 1992. Printed in Great Britain.

15
OZONE FACILITY DESIGN

THE OZONATION PROCESS

Introduction

Ozone must be generated at the point of use because it is subject to rapid decomposition, reverting back to oxygen. Ozone can be generated from atmospheric air or from pure, commercially produced oxygen. The process may be achieved by electrical discharge or by photochemical action using ultraviolet light. The latter produces the ozone layer in the upper atmosphere. The most practical method for producing ozone is by electrical discharge.

Practices. In the early days of ozone treatment in Europe (1990s), it was the consensus that the ideal capacity of the system should be 5 mg/L. This was labeled "full ozonation." In France, for example, the rivers adjacent to Paris were all heavily polluted with untreated sewage and industrial waste. An ozone dose of 5 mg/L and a 4–5 min. residual of 0.4 mg/L seemed to provide adequate disinfection. This practice was followed partly because, owing to the limited solubility of ozone in "demand-free" water, ozonator capacities were arbitrarily sold with a maximum capacity of 5 mg/L.

Elements of a Conventional System. There are five basic components to the ozone process: (1) gas preparation to produce dry feed-gas, (2) electric power supply for the ozone generators, (3) the ozone generators, (4) the contractor where ozone is dissolved in the water, and (5) the exhaust gas destruct system. Each of these will be described below.

Theory of Generation

Ozone is produced from the oxygen in air or pure oxygen when a high-voltage alternating current is imposed across a discharge gap in the presence of either of these gases. Ozone is generated by electric discharge as follows.[1] The key to the process is in the field, between the electrodes, of stray electrons left over from previous discharge or from background radiation. These electrons become excited and accelerated within the high-energy field between the electrodes. The alternating current causes changing polarity, which in turn causes the negatively charged electrons to be attracted first to one electrode and then to the other, like bouncing balls. As the velocity of the electrons increases, they attain enough energy to split some O_2 molecules in two, which in turn combine with other O_2 molecules to form O_3. The visible effect of the incomplete oxygen molecule breakdown in the air gap between the highly charged electrodes is known as the *corona glow.*

Ozone production from either air or oxygen is generated in proportion to the energy applied to the ozone generator, as demonstrated by the following equation:[2]

$$3O_2 \leftrightarrow 2O_3 + 0.82 \text{ kWh/kg} \tag{15-1}$$

The ozone-generating reaction is initiated when free energetic electrons in the corona dissociate oxygen molecules:

$$e^- + O_2 \rightarrow 2O + e^- \tag{15-2}$$

This is followed by ozone formation from a three-body collision reaction:

$$O + O_2 + M \rightarrow O_3 + M \tag{15-3}$$

where M is any other molecule in the gas. At the same time, atomic oxygen and the available electrons react with ozone to form oxygen:[2]

$$O + O_3 \rightarrow 2O_2 \tag{15-4}$$

and

$$e^- + O_3 \rightarrow O_2 + O + e^- \tag{15-5}$$

Factors that must be considered in the design of electric discharge generators are: voltage, frequency, dielectric material property and thickness, discharge gap, and absolute pressure within the discharge gap.

Generator Yield

Under optimum conditions, ozone production depends upon the following relationships:[3]

$$V \sim pg \tag{15-6}$$

and

$$(Y/A) \sim f\varepsilon V^2/d \tag{15-7}$$

where

Y/A = ozone yield per unit area of electrode surface under optimum conditions
V = voltage across the discharge gap (peak V)
p = gas pressure in the discharge gap (psia)
g = width of discharge gap (in.)
g = frequency of applied voltage (Hz)
ε = dielectric constant
d = dielectric thickness (in.)

In studying the above relationships it is obvious that the ozone generator manufacturers are confronted with formidable problems. First of all, the basic method is inherently inefficient. Commercially available corona discharge generators produce ozone in the exiting gas in concentrations varying from 0.5 to 4.0 percent by weight of the carrier gas. Therefore, in making ozone from oxygen most of the oxygen passing through the generator is unchanged. So for economic reasons this oxygen must be used elsewhere in the overall scheme or be dried out and recycled through the generator.

The yield of a generator is related to the square of the voltage, yet high voltages increase the possibility of electrode failure caused by dielectric or electrode puncture. Higher voltages also result in higher pressures (the discharge gap being set), and this means higher operating temperatures. High operating temperatures increase the rate of ozone decomposition.

Dielectric thickness appears in the denominator of the yield relationship, which indicates that a thin dielectric is desirable. Thin dielectrics, however, are more susceptible to failure by puncture than thicker ones. It is evident that the problem is to attain maximum yield while simultaneously economizing on maintenance and replacement costs. The solutions to these problems are difficult to achieve. Some of the considerations in increasing ozonator efficiencies are as follows: High frequency is less damaging to dielectric surfaces than high voltage. Frequencies as high as 2000 Hz are now in use; emphasis is placed upon this parameter. Glass has been found to be practical dielectric material; it is cheap and readily available. Ingenious generator designs optimize heat removal. In addition, solid-state circuitry and acid-resistant materials reduce the power requirement cost and increase reliability.

So in a modern ozone generator with the latest manifestations of the ingenuity of the designers, only about 10 percent of the energy applied results in the production of ozone.

The largest portion of this energy loss is by the heat generated. The remainder of the loss is through light and sound. The decomposition of ozone back to oxygen is greatly accelerated with increasing temperature; so all high concentration ozonators must employ a method of heat removal.

Assuming that a clean, dry, oxygen-rich gas is being fed to an ozone generator and an efficient method of heat removal is available, production of ozone per unit area of electrode surface under optimum conditions is a function of:

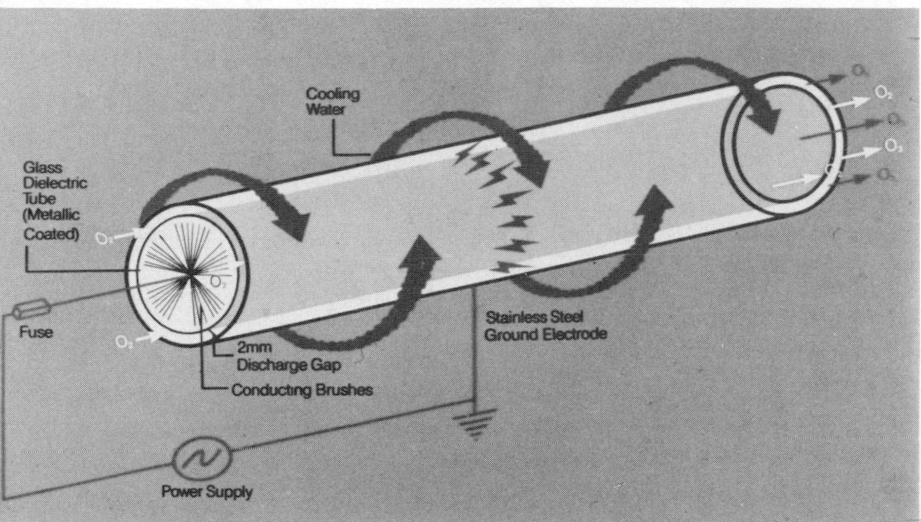

Fig. 15-1. Ozone generator tube unit schematic (courtesy Los Angeles Department of Water and Power).

1. Peak voltage across discharge gap.
2. Absolute gas pressure in gap.
3. Width of discharge gap.
4. Frequency of applied voltage.
5. Dielectric constant.
6. Thickness of the dielectric.

Therefore, to optimize the ozone yield, the following conditions should exist:

1. The combination of gas pressure and gap width should be arranged so the voltage may be kept relatively low for reasonable operating pressures.*
2. A thin material with a high dielectric constant should be used.
3. High frequency current (300–2000 Hz) should be used because high frequency increases the ozone yield and is less damaging to the dielectric than high voltage and prolongs the life of the equipment.
4. An efficient heat removal system is essential.

The most significant factor is the efficiency of heat removal. Although the voltage and/or the frequency can be continually increased to produce more and more ozone, the additional heat produced must be removed as efficiently as possible. Otherwise, if the temperature rises, the additional ozone produced will only decompose back to oxygen.

See Figs. 15-1 through 15-3 for generator illustrations.

*Lower voltage protects the dielectric and/or the electrode surfaces from high voltage failure.

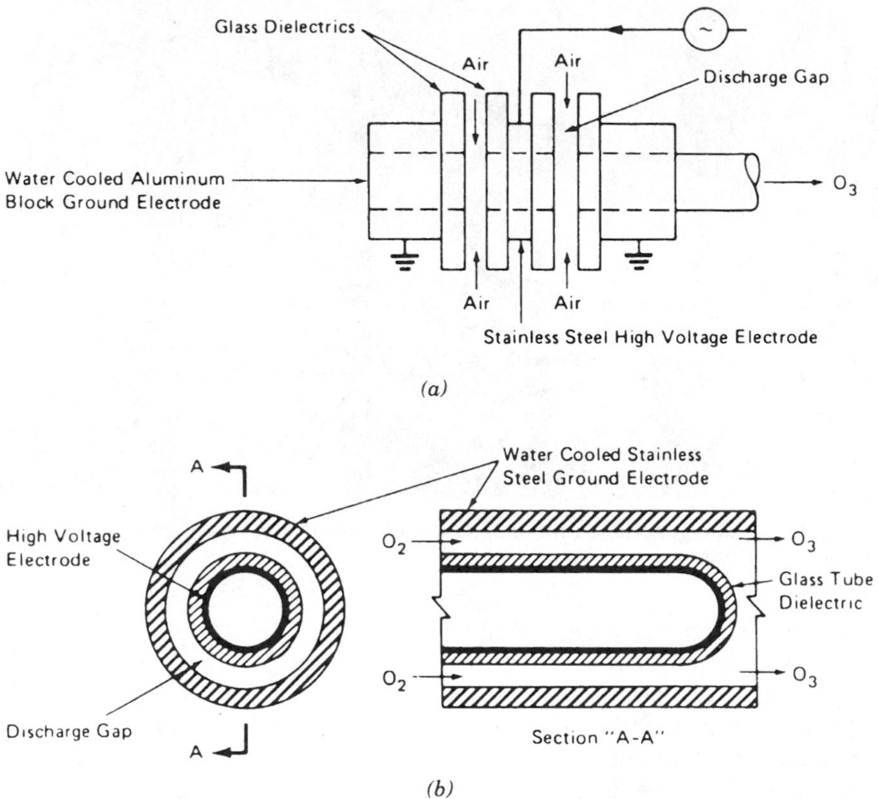

Fig. 15-2. Schematic illustration of commercial ozone generators. (*a*) Otto plate type generator unit; (*b*) tube type generator unit.

Types of Ozone Generators

Historical Note. In the 1970s the Lowther Plate was the newest species of ozone-generating equipment. It was marketed for several years in the United States by the W. R. Grace Co. in 1976 Union Carbide Co. purchased this system from Grace as a feature to supplement the Unox oxygen process. During these two ownerships, this generator was being marketed by a brilliant young engineer, the late Harvey Rosen, who was known in the trade as "Mr. Ozone." In 1992 the Union Carbide Ozone–Unox system was marketed by Lotepro Inc., a division of the German Linde Co.

Lowther Plate. The generator is air-cooled and can use either atmospheric air or pure oxygen. It is made up of a gastight "sandwich" consisting of an aluminum heat dissipator; a steel electrode coated with a ceramic dielectric; a silicone rubber spacer, which provides the precise amount of discharge gap; and a second ceramic-coated steel electrode with an air (or oxygen) inlet and an ozone outlet. The ozone exits through a second aluminum heat dissipator.

Fig. 15-3. Typical tube-type ozone generator (Trailigaz) (courtesy Compagnie Generale Des Eaux, Paris, France).

These sandwich-type units are pressed together in a frame and are manifolded for either air or oxygen inlet flow. Cooling is accomplished by a fan, which moves ambient air across the heat dissipators. This unit is illustrated in Fig. 15-4.

The manufacturers using this type of generator cite the following characteristics:[3]

1. It uses air cooling.
2. Thin dielectrics provide greater ozone yield, which in turn enhances heat removal.
3. The small discharge gap allows higher operating pressures.
4. Low operating voltages provide longer dielectric life.
5. High frequency operation allows higher ozone production without resorting to higher voltages.
6. Heat removal is more efficient.
7. It provides a high-yield efficiency that results in smaller space requirements for the equipment.
8. Power requirements are less than for generators based on older technology.

Unfortunately, this type of generator suffered failures and intensive maintenance problems after about three or four years of service. The ones now in

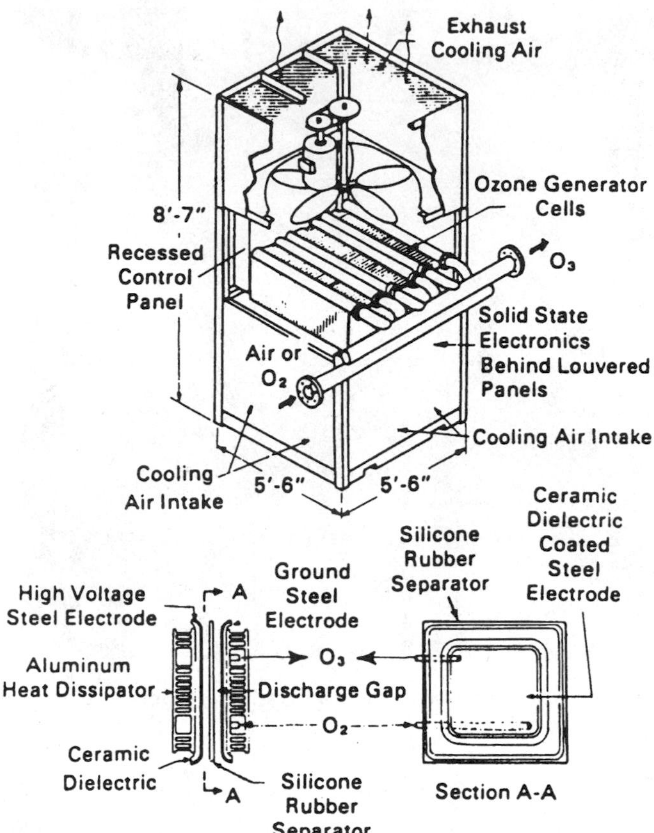

Fig. 15-4. Lowther air-cooled plate-type generator.

service in the United States are undergoing continuous field modifications. In the meantime the manufacturing rights have been purchased by Sumimoto of Japan. They have embarked on a redesign of the Lowther plate concept to eliminate the previous difficulties experienced by the users in the United States.[4]

Plate Type, Low Frequency. Ozone is produced as the air passes between the glass plates and is exposed to the corona discharge. Owing to the inherent low operating pressures of this cell, the older types of flat plate generators were limited to the application under negative pressure. Currently, air-cooled plate type units are available in the United States from Griffin Technics, Lodi, New Jersey. These units are limited to 100 lb/day capacity.

A German-made horizontal flat-plate, water-cooled generator, known as Hydrozon, will be licensed soon for manufacture in the United States. This unit is constructed so that the transformer, ozone generator, contactor, and ozone reaction chamber are all contained in a single housing.

Large-Diameter, Horizontal-Tube, Water-Cooled. Figures 15-5 and 15-6 illustrate the details of the tube assembly and the housing for the tubes. These tube types are based upon the Welsbach original element design: a dielectric glass tube closed at one end, internally metallized, and connected to the power supply by a stainless steel brush centered precisely inside a grounded metallic tube. Feed-gas flows through the gap between the two tubes. All the metallic parts should be type 316L stainless steel. Power is supplied to the inner metallic coating through an axial bus bar containing electrode brushes. Each corona tube can be fused individually to allow the generator to maintain continuous operation in the event of a single dielectric failure within the unit.

The larger ozone generators contain up to about 1000 high-voltage electrodes connected in parallel to the power supply. They operate under a pressure of 13–18 ft H_2O, which is required to discharge the ozone vapor through the diffusers in the contactor basins.

Cooling of the corona tube is done by passing either 58–68°F potable water or heat-exchanger-quality water along the outside of the outer chamber tube. Flow rates are 300 to 480 gal/lb of ozone generation capacity. This type of generator is usually designed to operate at a low frequency—50 Hz in Europe, 60 Hz in North America—and at medium frequencies of from 200 to 1000 Hz.

Fig. 15-5. Two Emerzone® model EG 250 tube-type ozone generators, capacity 250 lb/day (courtesy Emery Co., Cincinnati, Ohio).

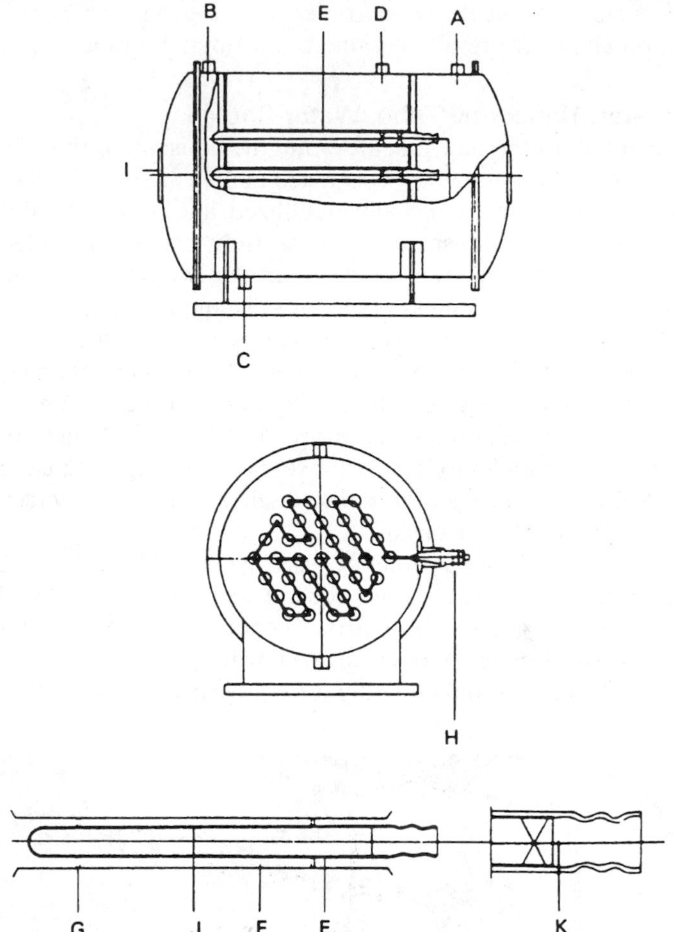

Fig. 15-6. Large-diameter, horizontal tube, water-cooled, voltage-controlled generator. A, air inlet; B, ozonized air outlet; C, coolant inlet; D, coolant outlet; E, dielectric tube; F, discharge zone; G, tube support; H, H.V. terminal; I, port; J, metallic coating; K, contact.

Currently these generators are marketed in the United States by Quantum Chemicals (Emery of Cincinnati, Ohio), Infilco Degremont, Trailegaz Ozone of America, ASEA-Brown-Boveri, Capital Controls, Hankin-Atlas, and Polymetrics/Welsbach.

Vertical Tube, Single-Fluid, Water-Cooled. This type of generator utilizes the cooling water as both the grounding electrode and the coolant (see Fig. 15-7). Three separate compartments are formed, for the feed-gas manifold, the ozone-rich product-gas manifold, and the cooling water. Dry air enters

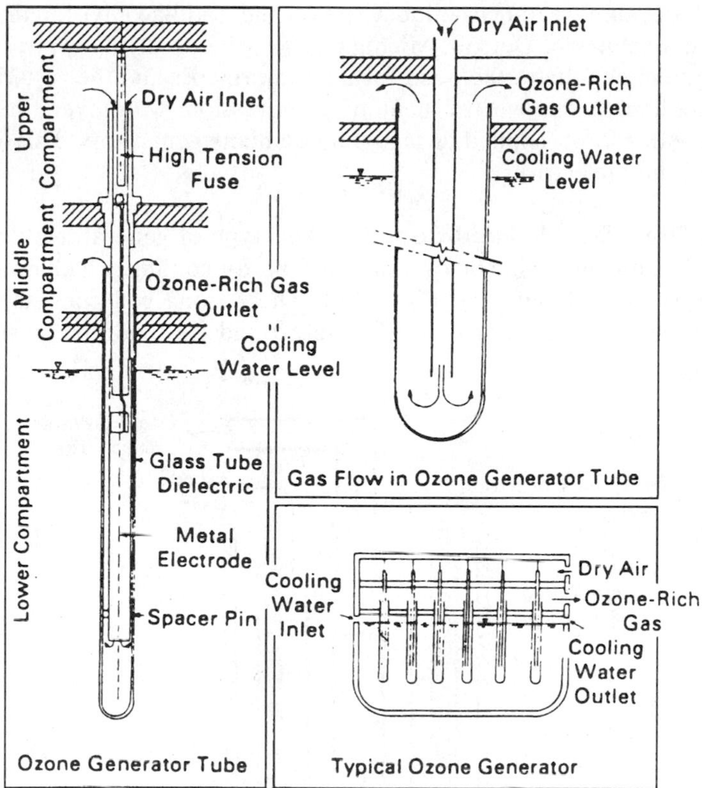

Fig. 15-7. Vertical-tube, single-fluid, water-cooled, voltage-controlled generator.

the top of the central metal tube, which also serves as the high voltage electrode. The ozone vapor travels downstream and emerges at the closed end of the dielectric tube and passes through the discharge gap. Each tube is almost entirely submerged in the cooling water of the lowest compartment.

Dielectric tubes are approximately one inch in diameter. This size is smaller than that of other tube-type generators. Cooling water requirements are low, about 230 gal/lb of ozone produced.[2] This system is unique because it operates under a negative head. Feed-gas is drawn into the entire system by the injector action of the aspirating turbine mixer in the contactor basin. The ozone generated in air has a concentration of 1.5–2.0 percent O_3, whereas in oxygen the ozone concentration is 3–5 percent. There are only two of these negative pressure systems currently operating in the United States: at Burke, Texas, and at Sturgeon Bay, Wisconsin. However, there are more than one hundred in Europe. The system was developed and formerly marketed under the trade name of Kerag, now a defunct company.

These systems are now made and marketed in the United States by Ozone Technology Inc., of Tyler, Texas. The largest U.S. installation of this system is a 640 lb/day unit treating cyanide wastes at the Cadillac Division of General Motors in a suburb of Detroit, Michigan.

This system eliminates the need for air compressors and the resulting need for air cooling to remove the heat of compression. Moreover, the system operates with a lower applied dosage than do many generators. This increases the life of the dielectrics.

Vertical-Tube, Double-Fluid-Cooled. This type of generator utilizes both water and a nonconducting oil or a similar fluid for cooling, which necessitates a rather complex design (see Fig. 15-8). The corona cell consists of three annular tubes: (1) the inner tube, (2) the ground electrode metal tube, and

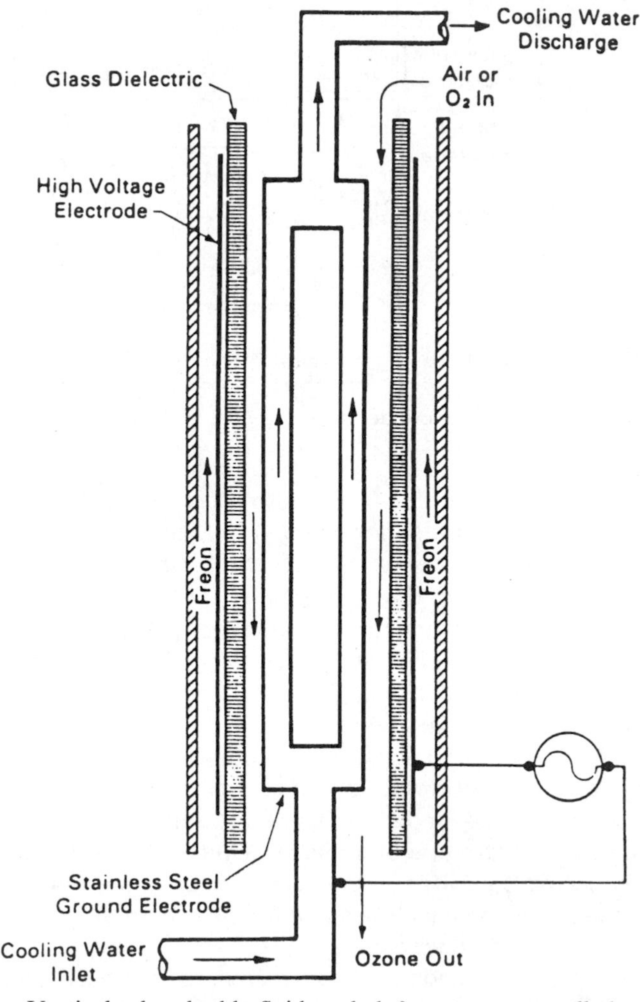

Fig. 15-8. Vertical-tube, double-fluid-cooled, frequency-controlled generator.

(3) the outer tube, a high voltage electrode that is a metal coating on the outer surface of the glass dielectric tube. The glass dielectric is encased in an outer tube that houses the cooling oil or the Freon, and the discharge gap is formed between the inner electrode and the glass electrode tube. This unit operates at high frequencies—1000 to 3000 Hz.

Cooling water flows through the inner electrode and the metal electrode, whereas the cooling oil flows along the outer metal electrode. The oil is cooled by an external oil-to-water heat exchanger for the final heat reduction step. Cooling water usage for this system is on the order of 300–480 gal/lb of ozone generation capacity.

This unit is manufacturered only by PCI Corporation, West Caldwell, New Jersey. There were ten units installed in 1992 in the 40 WTPs in the United States using ozone to treat potable water.

The high frequency generator usually operates at 10,000 V. Operating pressure is 20 psi, which is somewhat higher than that of conventional horizontal-tube generators. It is more complex than the horizontal-tube units and possibly costlier owing to the coding system, the cost of the dielectrics, and the close tolerances with the extremely narrow discharge gap of 0.0045 inch. However, the energy efficiency is reported to be higher than that of conventional horizontal-tube units because there is additional cooling, and fewer electrodes are required.

New Developments in Ozone Generation Technology. A generator called the Megos is a German development that uses a 0.35-inch horizontal tube, as compared to the more common horizontal-tube units that have tubes 2–3 inches in diameter. There are two of these at WTPs in Germany, at Moers and at Wuppertal. They are designed to operate on oxygen feed-gas. A schematic of this unit is shown in Fig. 15-9.

The glass dielectrics are open at both ends and are mounted in stainless steel tubes, like conventional dielectrics. An electrode, which is made up of four stainless steel rods, is positioned inside each dielectric. Oxygen flows in the annular space between the stainless steel tube and the glass dielectric, as well as in the hollow space inside the glass dielectric. Ozone is formed in both discharge gaps. This system is capable of producing ozone concentrations in oxygen as high as 10 percent by weight. This is far greater than the capability of any other design.

A high gas velocity through both gaps promotes efficient cooling, in addition to the water cooling system, as in conventional horizontal-tube-type generators. Cooling water requirements of the Megos generator are on the order of 180 gal/lb of ozone produced.

The Megos generators are based upon a relatively recent technology; so their use has been limited. There is none in use in the United States with the exception of some for test purposes. There are very few in Europe. The one at Moers has been operating since 1985. The maximum capacity of these

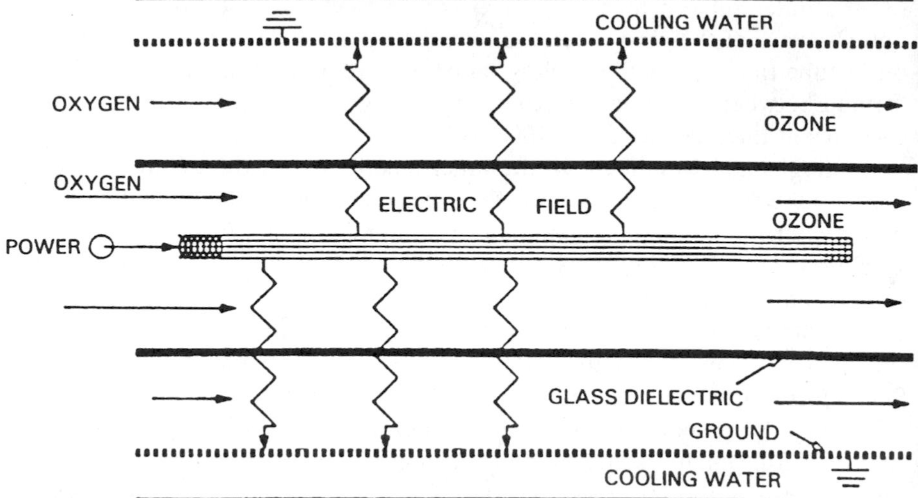

Fig. 15-9. Schematic diagram of Megos ozone generator (J. M. Montgomery Engrs., 1989).

compact units is limited to about 70 lb/day of ozone in concentrations of 6–9 percent from the oxygen feed-gas.

The advantages of this design are as follows: (1) reduced oxygen consumption, (2) smaller contactor basins, (3) reduction in amount of off-gas to be destroyed, (4) higher ozone TE, (5) smaller space requirements, and (6) lower capital costs. The power requirements are 5–6 kWh/lb of ozone generated.

The Megos unit is manufactured in Germany by Schmidding Werke, Koln, and is available in the United States from Capital Controls, Colmar, Pennsylvania.

In Japan, fine wire electrodes have been developed for a number of years. A schematic diagram of an experimental ozone generator element using fine wire electrodes is shown in Fig. 15-10.[2] The glass dielectric is similar to that of a conventional generator, but it has a stainless tube as an internal electrode fixed inside the glass tube and an external electrode along the outer surface of the glass tube. The discharge gap is filled with fine stainless steel wire. The corona discharge occurs between the wires and the outer electrodes as well as over the fine wire. According to the inventors, this type of generator may lead to improvements in generator efficiency, even at low frequencies.

Feed-Gas Quality

Important Considerations. The moisture content of the feed-gas to the ozone generator has a twofold influence on ozone production; a high level of moisture not only decreases the ozone production rate but also increases the

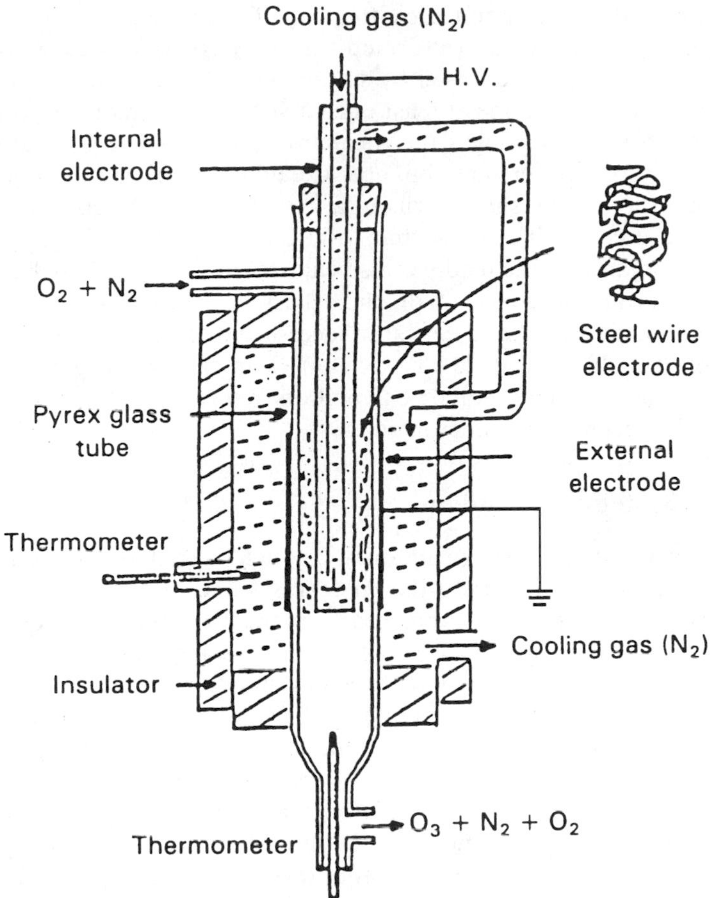

Fig. 15-10. Schematic diagram of fine wire, Japanese-type electrode, ozone generator (Okazaki et al., 1988).

contamination rate of the dielectrics. The contamination occurs whether the feed-gas is high purity oxygen or air. If oxygen is the feed-gas, hydrogen peroxide is formed in the presence of water vapor and creates deposits on the dielectrics. These deposits have to be removed by scrubbing with soapy water.

If air is the feed-gas, about one mole of nitrogen pentoxide (N_2O_5) will develop for every 100 moles of ozone formed.[5] The nitrogen pentoxide can decompose to form nitrogen dioxide (NO_2), which interferes directly with ozone production. When nitrogen pentoxide is in the presence of water vapor, nitric acid (HNO_3) will be formed. The nitric acid will cause severe corrosion within the generator and associated piping, and it can create a heat sink on the glass or ceramic dielectrics, which will increase the potential for dielectric breakage.

When the feed-gas dew point is less than −50°C, which is acceptable practice, about 3 to 5 g of nitric acid are formed for every 1000 g of ozone produced. Nitric acid can easily be removed by scrubbing with soapy water. This bit of maintenance should be done at least once a year. A high moisture content in the feed-gas not only damages the dielectrics, affecting the generator yield, but it escalates generator maintenance. Cleaning and repairing a generator require that the unit be shut down. In a survey of 37 U.S. plants, the trend was to achieve dew-points lower than −60°C.[6]

Dust and organics in the feed-gas also can create operating problems in the generator. The dust can collect on the dielectrics, decrease generator efficiency, increase dielectric stress, and cause unnecessary dielectric breakage. Therefore, filters should be installed up stream from the ozone generator. For maximum reliability there should be two filters in series: the first, a one-micron filter and the second, a 0.3-micron size.

Feed-Gas Systems Description

Feed-gas to an ozonator may be air, once-through oxygen, or recycled oxygen. The most common type in use is the air-feed system, accounting for about 95 percent of the installations. The remainder are the two types of oxygen systems. Once-through oxygen systems typically do not require further moisture removal, but filters to remove any particulates should be provided.

These feed-gas systems are classified by their operating pressures: negative, low, medium, and high. The most common are the low pressure systems, which operate at 10–20 psi. The medium and high pressure systems operate at 30–50 psi and 70–100 psi, respectively, and require pressure reduction upstream from the ozone generator.[7] It should be noted that this pressure regulator is used to control the upstream pressure. Negative pressure systems are designed to operate in conjunction with aspirating turbine contactors, where the injection device creates a vacuum on the feed-gas system, like an eductor or an injector. The negative pressure systems do not require the use of compressors. The system using the aspirating turbine mixer is a proprietary piece of equipment, used with the ozone generator originally marketed by Kerag. It is fully described in this text.

The decision to use either a high or a low pressure system is often based upon a qualitative evaluation of potential maintenance requirements in addition to the quantitative capital cost evaluation. Some of the considerations are as follows:

- The high pressure air pretreatment equipment usually has a higher maintenance requirement for the air compressors.
- The high pressure air pretreatment equipment has a lower maintenance requirement for the desiccant dryers.
- The high pressure air pretreatment system generally has a lower capital cost. At small and medium-sized installations, this lower cost may offset

the additional maintenance required for the compressors and the associated equipment, such as filters for the oil-type compressors. These installations should be investigated on the basis of potential maintenance requirements associated with the high and low pressure systems instead of evaluating the design on capital cost alone.

Air Preparation Systems

Low- and Medium-Pressure. Figure 15-11 shows an example of a low-pressure system, and Fig. 15-12 illustrates a nominal- or medium-pressure system. The pre-compressor filters are provided to protect the compressors

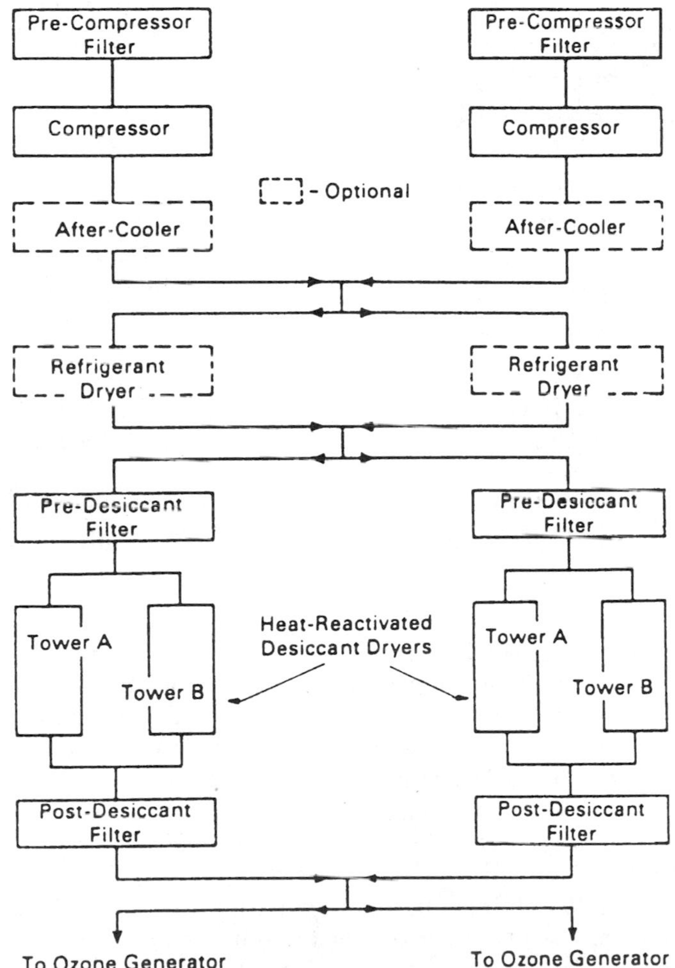

Fig. 15-11. Example low-pressure air feed-gas treatment schematic.

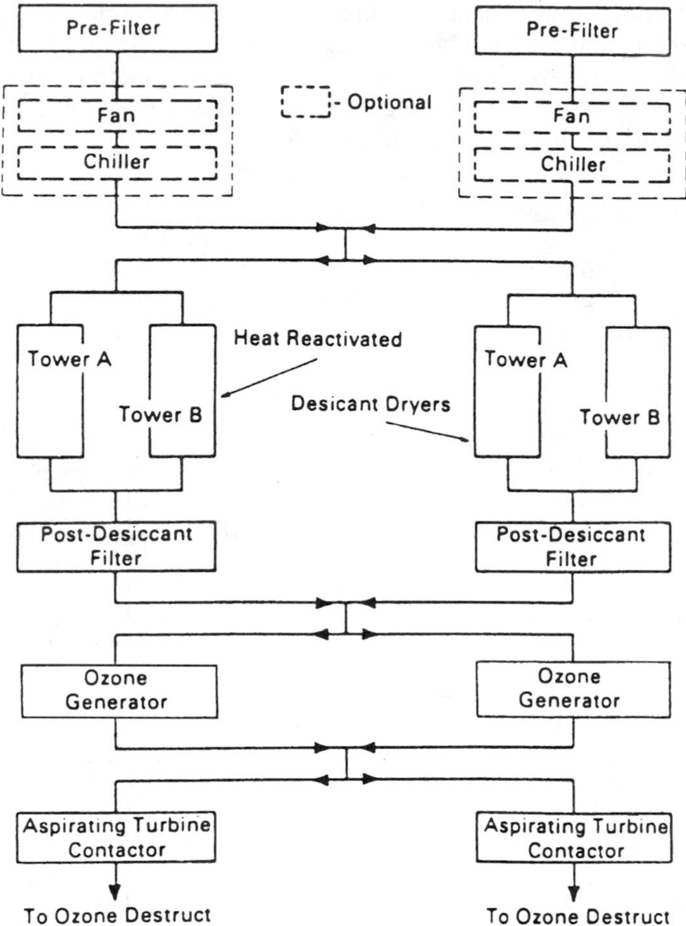

Fig. 15-12. Example nominal-pressure air feed-gas treatment schematic.

from any damage due to larger particulates. The compressors are typically positive-displacement, oil-less units. The positive-displacement feature is required to obtain a constant air flow at variable operating pressures. Variable pressures are often encountered owing to the variation in head losses in the downstream filters and diffusers in the contact basins. Oil-less compressors are used to eliminate oil contamination of the desiccant dryer medium and the ozonator dielectrics. Liquid-seal and rotary lobe compressors are used most frequently.

The after-cooler and refrigerant dryers are shown by dotted lines because they are optional. Usually either one or the other is furnished. The cooling mechanisms are used to remove moisture in the air flow at minimal operating expense.

The compressed, cooled gas is directed to a pre-desiccant filter, which is used to remove dust and dirt particles greater than 3–5 microns in diameter. This reduces plugging in the desiccant medium.

The most important component of the air treatment process is the desiccant dryer, which consists of two towers containing moisture-absorbing media. One tower operates while the other tower is regenerating. The low and medium pressure desiccant dryers use heat for reactivation of the desiccant.

The post-desiccant filters are installed to remove particulates smaller than 0.3–0.4 micron in diameter. Two-stage filtration at this point in the air flow system is preferred. The first stage removes particulates larger than 0.1 micron, and the second stage removes particulates less than 0.3–0.4 micron in diameter. These filters are not to be confused with the pre-compressor filters. The only difference between the low and medium pressure systems is the use of a pressure-reducing valve in the air feed line downstream from the post-desiccant filter and upstream from the ozone generator. This pressure-reducing valve is used to control the upstream pressure in the drying system, which is a critical requirement for reliable operation.

High-Pressure System. A schematic of this system is shown in Fig. 15-13. The pre-compressor filters and compressors are typically the same as in the low and medium pressure systems described above. The after-coolers that follow the compressors are essential, as they are used to remove the heat of compression owing to the higher discharge pressures involved. The pre-desiccant filters are used to remove particulates smaller than 3–5 microns in diameter when oil-less compressors are used. However, when oil-seal compressors are used, filtration must be provided to remove oil droplets smaller than 0.03 micron in diameter.

A major difference between the low- and high-pressure systems is the method of desiccant regeneration. High-pressure systems utilize pressure-swing heat-less regeneration. Adsorption of moisture occurs under high-pressure conditions, usually at 70–100 psi. When this pressure is reduced to atmospheric in the tower, desorption occurs. Purge air removes the moisture during the regeneration step, and purge air requirements are about 15–25 percent of the dry air flow. Regeneration cycles are much shorter than those for heat regeneration systems, about 1–5 min. The shorter cycle times for the heatless regeneration dryers translate into considerably smaller column sizes compared to the heat-regenerated dryers. The use of heat-less regeneration usually results in less stress on the desiccant, which leads to longer desiccant life expectancy, sometimes as long as 10 years.[2]

Negative-Pressure Systems. These systems are associated with the aspirating turbine contactor systems that create a slight negative pressure in the air feed system. Air is pulled through the system by a vacuum created by the aspirator (inductor) in the ozone contactor. The other elements of these systems are similar to those of the pressure systems described above.

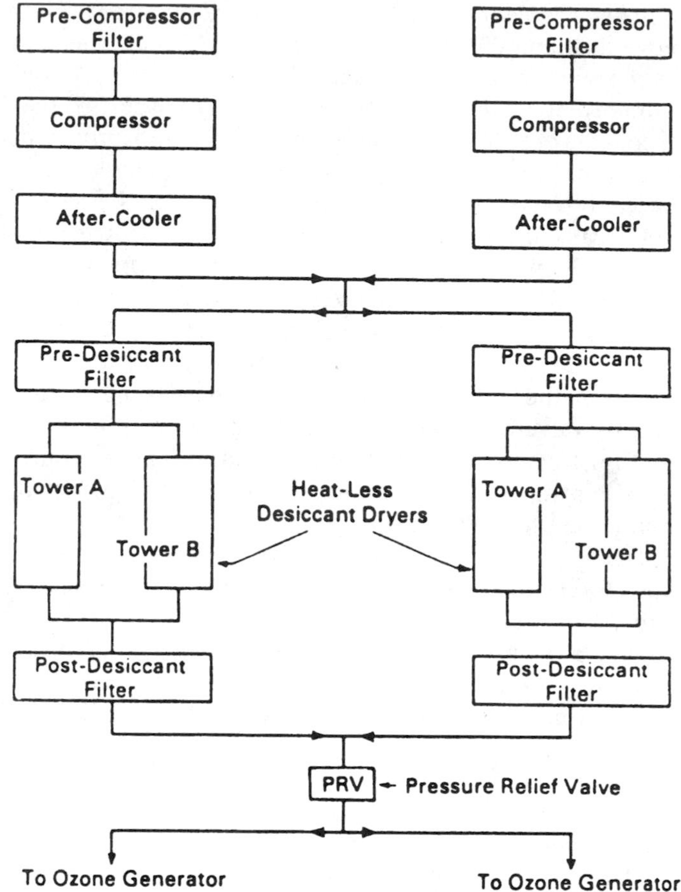

Fig. 15-13. Example high pressure air feed-gas treatment schematic.

The pre-desiccant filter is used to remove particulates larger than 5 microns in diameter, and the post-desiccant filter removes those greater than one micron. This system uses a heat-reactivated desiccant dryer, shown in Fig. 15-14.

The energy input and air supply to the turbine mixing and contacting unit can be controlled by a variable speed drive on the mixer to adjust its pumping action and by adjusting the water inlet orifice size to the turbine mixer.

Desiccant Dryers

Introduction. The air dryer is the most important component in the air feed system. Poor desiccant dryer performance reduces ozone production and causes damage to the internal parts of the generator. Although there are a great many dryers that are capable of producing a dew point of −40°C or

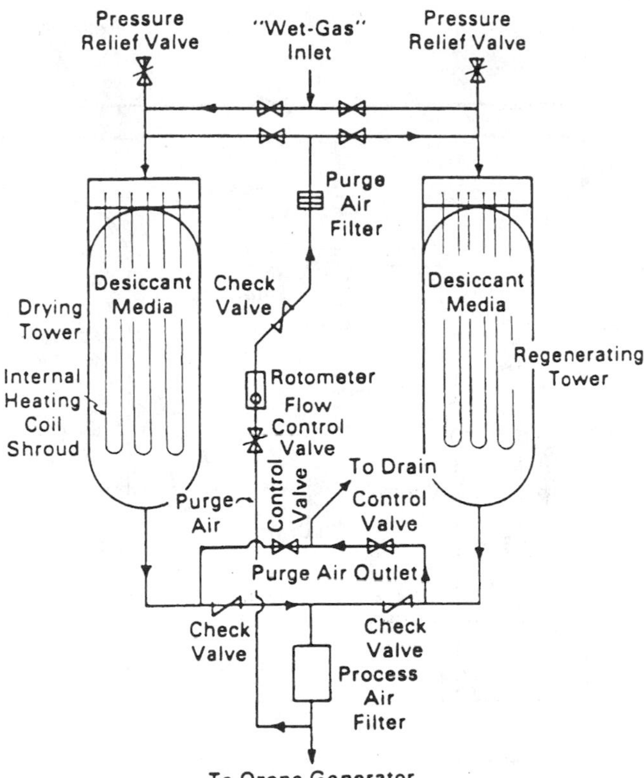

Fig. 15-14. Diagram of a heat-reactivated desiccant dryer with internal heating coils.

lower, the dew point for ozone generation must be at least −60°C. The dew point is a function of both the ambient temperature and atmospheric pressure. Since atmospheric pressure is a function of ground level, all dewpoint calculations must be based upon standard conditions of temperature and pressure.

There are two types of desiccant dryers, those with internal heaters and those with external heaters. These two types are described below.

Heat-Reactivated Desiccant Dryers. These dryers are used on low-, medium-, and negative-pressure systems. A schematic is shown in Fig. 15-15. Wet air is directed to the operating tower, where moisture in the air is adsorbed onto the desiccant. The desiccant is usually activated alumina and molecular sieves. After several hours of operation the desiccant becomes so laden with moisture that it is unable to maintain the targeted dew point, and when this occurs, it must be regenerated. The control system switches the wet air to the other tower via electrically operated stainless steel valves.

Regeneration requires heating the desiccant to a temperature between 90 and 260°C, followed by purging with dry air. A typical design heats to a

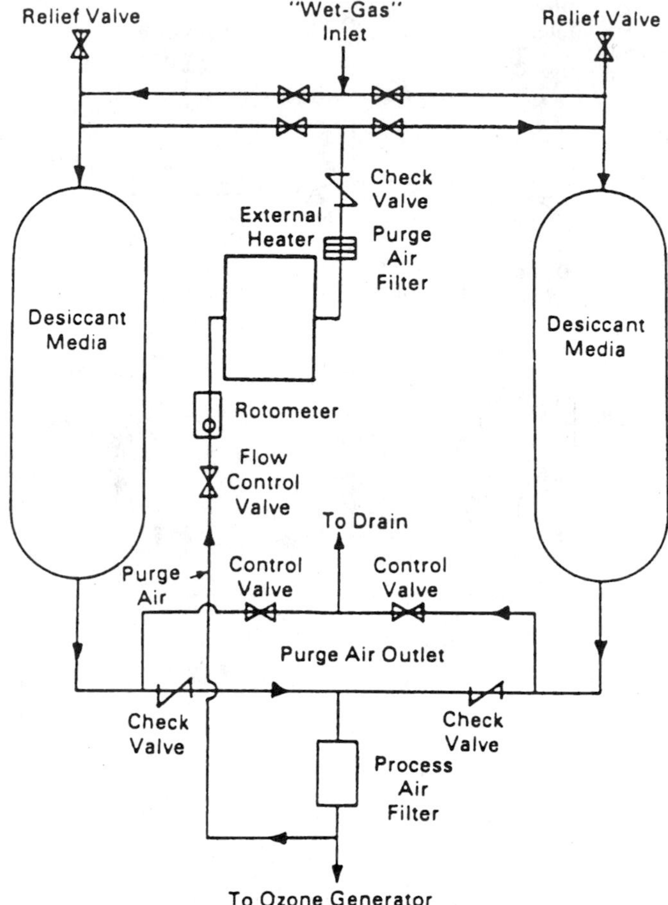

Fig. 15-15. Schematic of a heat-reactivated desiccant dryer with external heating equipment.

temperature between 120 and 170°C for 1–2 hours, followed by cooling for a minium of 6 hours. After heating is completed, the purge air continues to flow through the tower and cools the desiccant before the regenerated tower can be put back into service. It is of critical importance that the desiccant be completely cooled before going back into service, to prevent an unwanted "temperature spike." An aftercooler is sometimes used between the desiccant dryer and the dewpoint analyzer, fitted with an alarm to ensure that a temperature spike will not occur. Owing to the complexity of the air drying process, the dew point analyzer is one of the most important components of the air drying system.

The purge air is recycled dry air for regenerating each tower, which amounts to 15–20 percent of the dry air flow. In order to prevent equipment damage due to insufficient purge air flow, it is essential that this flow be automatically controlled and continuously monitored.

The purge air must be filtered twice to protect the equipment. It should be noted that in Fig. 15-16 purge air originates downstream from the initial process air filter, which removes particles larger than 0.3–0.4 micron in diameter. Then, after passing through the external heater but before returning to the dryers, the purge air is filtered for a second time. This eliminates any possibility that the purge air could cause plugging.

After months of operation, the desiccant will lose its moisture-absorbing capability. A life expectancy of three years is normal, but in practice it has been found to vary, ranging from one to five years.

The internal-type heating unit shown in Fig. 15-15 is used primarily in smaller installations because of capital cost considerations. One of the disadvantages of internal heating is the inability to achieve even heating; external heating is preferred because it provides a much better distribution of heat throughout the desiccant media. However, it increases the complexity of any installation because of the additional equipment required. Figure 15-16 illustrates a heat-reactivated desiccant dryer with external heating equipment.

The amount of desiccant used and the regeneration cycle time are the most important design considerations. The recommended minimum cycle time is 8 hours because of cooling considerations, but the cycle time must be longer than this because the desiccant loses its moisture-absorbing capability after a

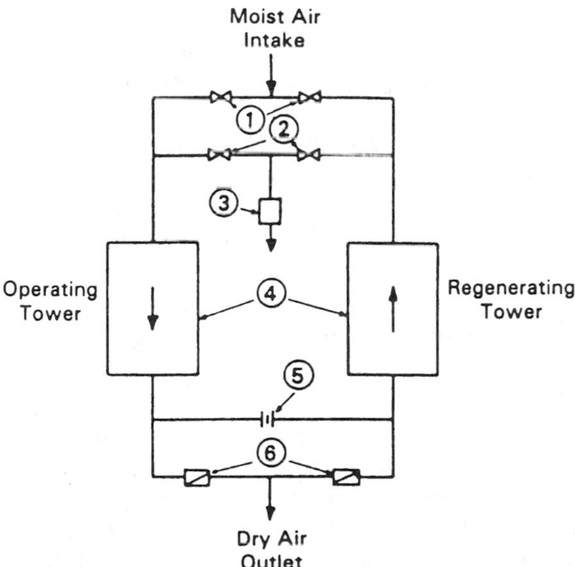

Fig. 15-16. Pressure swing (heatless) high-pressure desiccant dryer in purging mode. 1, Wet air inlet; 2, regenerated air outlet valve; 3, silencer; 4, adsorption desiccant; 5, orifice plate; 6, check valve.

few months of operation. At initial start-up, the cycle time should be longer than 8 hours to give the operator the flexibility to reduce the cycle time as the desiccant loses its moisture-absorbing capacity. When the minimum 8-hour cycle time is reached, the operator must replace the desiccant. Then the cycle time can be reset to the original design settings.

The regeneration cycle may be controlled on a time or a demand basis. The timed basis uses a preestablished time to initiate the regeneration cycle, whether or not the operating tower can achieve the desired dew point. The time cycle control is simply adjusted by changing the timer settings.

A regeneration cycle based upon demand is more expensive to install than one based upon time because of the controls involved, but the system has potential O&M cost savings. Using the demand system, the regeneration cycle is initiated when a preset maximum value for the dew point of the discharge air from the dryer is reached. The dew point temperature must be continuously monitored.

This method has definite advantages but is entirely dependent upon the reliability of the dew point monitor and the sensitivity of the control logic. Reliable process monitoring and control instrumentation should be provided as follows:

1. Feed-gas flow rate—monitors system loading.
2. Inlet temperature—monitors system loading.
3. Outlet temperature—monitors system performance.
4. Alarm system and shutdown capability.
5. Inlet and outlet pressure—monitors system operation.
6. Purge air flow rate—monitors system operation.
7. Discharge feed-gas dew point, in linemeter—monitors system performance.
8. Discharge feed-gas dewpoint recorder and system alarm, with shutdown capability.
9. Discharge feed-gas dew point and dew point cup measurement—measures dew point by manual methods to check on the reliability of the in-line meter.

Heat-Less Desiccant Dryers. These dryers operate at pressures ranging from 70 to 120 psi, and are also known as pressure-swing desiccant dryers because the units operate with varying pressures ranging from high to low. These dryers are considered to be acceptable alternatives for moisture removal in small to medium ozonation systems.

A schematic diagram is shown in Fig. 15-16. The principle of operation is adsorption of moisture by the desiccant under high pressure. After a period of 1–5 min. The two drying towers are switched, and the tower to be regenerated is reduced to atmospheric pressure and then purged with dry air from the operating tower. The moisture that has been adsorbed by the desiccant is evaporated into and carried away by the dry pure air, as air at the lower pressure has a greater capacity for holding moisture than air at the higher pressure. The purge air flow rate for these dryers is normally 15–25 percent of the process air flow.

These heat-less dryers have a history of reliable performance when properly sized and maintained. Like other dryers, they must be closely monitored and properly maintained for satisfactory operation.

Ozone Contacting

Introduction. This is probably the most important element in an ozonation installation. A combination of the contact chamber design and the selection of the diffuser device determines the transfer efficiency of the ozonation process. The object of the design is to maximize the transfer of the ozone vapor to the liquid phase in the contact chamber where it can react with the target contaminants.

Transfer Efficiency Concept. The transfer efficiency (TE) of a given gas–liquid contactor (mixing chamber) is an inherent property of the contactor and is a function of the gas* flow rate relative to the liquid flow rate. Transfer efficiency is defined as follows:

$$\mathrm{TE} = \frac{100\,(Y_1 - Y_2)}{Y_1} \tag{15-8}$$

where

Y_1 = mg O_3/L inlet carrier gas
Y_2 = mgO_3/L exhaust gas from contactor

TE is the fraction of ozone in the gas that has been transferred to the liquid expressed as percent.

Applied Ozone. This is defined as follows:

$$D = Y_1 \left(\frac{Q_G}{Q_L}\right) \tag{15-9}$$

where

Q_G = gas flow rate, L/min
Q_L = liquid flow rate, L/min

*Gas in this sense means the gas flow discharge from the ozone generator. It can be derived from prepared air or pure oxygen delivered to the generator.

The applied dose multiplied by the fraction transferred to the liquid is the absorbed dose.

Absorbed Ozone. The transfer of ozone, also called the absorbed dose, is defined as follows:

$$T = Y_1 \left(\frac{Q_G}{Q_L}\right) \left(\frac{Y_1 - Y_2}{Y_1}\right) \tag{15-10}$$

$$= \frac{Q_G}{Q_L} (Y_1 - Y_2)$$

where T is the amount of ozone transferred to the liquid in mg/L. Equation (15-9) indicates that the applied dose can be varied by changing either Y_1 (inlet carrier gas concentration) or the G_G/Q_L ratio. Operating experience has shown that TE (transfer efficiency) is more sensitive to shifts in the gas flow rate than ozone concentration in the carrier gas.[31] This is important, as an increase in the gas flow (Q_G) relative to the liquid flow may not result in a linear increase in absorbed ozone dose.

Application of Henry's law (see Chapter 13) and the concept of ozone transfer efficiency is useful for designing and optimizing ozone contactors.

Factors Affecting Ozone Transfer Efficiency (TE). The important physical factors affecting ozone TE are: pressure, temperature, ozone bubble size, contact basin flow characteristics, ozone concentration in the feed-gas, and ozone demand of the process water.

Design Considerations. The bubble diffuser is the most common type of ozone transfer device used in contacting basins around the world. A proprietary device that is also popular is the aspirating turbine mixer with the Kerag-design ozone generator. This text is based upon use of the bubble-type diffuser units.

Water quality as it affects ozone TE is so important that it can be concluded that the research work done on this subject using secondary and tertiary wastewater effluents is of great significance. These researchers all concluded that ozone transfer can be best described by the two-film theory.[8–11] In this theory the mass transfer of ozone per unit of time is a function of the two-film exchange area, the exchange potential, and a transfer coefficient. The exchange area for the bubble diffuser contactor is the surface area of the bubbles; the exchange potential is known as the driving force and is dependent upon the difference between the saturation ozone concentration and the residual ozone concentration. At this time, the above theory is not in use for design purposes, but only because design coefficients have not been documented.

Summary of Water Quality Factors Affecting Ozone Transfer Efficiency:

1. Ozone transfer efficiency (TE) will decrease as applied ozone dosage increases. A specified minimum design TE should be coupled with a specified ozone dose.
2. Ozone TE will increase as water quality deteriorates—this change increases the ozone demand. A specified minimum design TE should be coupled with a specified description of water quality. This could be related to the comparable chlorine demand of the water. (Sec Fig. 15-17).
3. Ozone TE will increase as the chemical quality of the water favors the presence of hydroxyl radicals (i.e., high pH or low alkalinity). A comparison of TEs for existing full-scale and pilot-scale results should be considered to give a strong indication of the relative quality of these waters.
4. Other factors that affect the ozone TE are the ratio of ozone gas to the process flow (G/L) and the ratio of ozone gas to contactor volume (G/V). Stover et al.[12] observed a decrease in the TE as the G/L ratio increased. A minor decrease in TE occurred as the G/L ratio increased from 0.2 to 0.5. The TE decrease was more pronounced as the G/L ratio

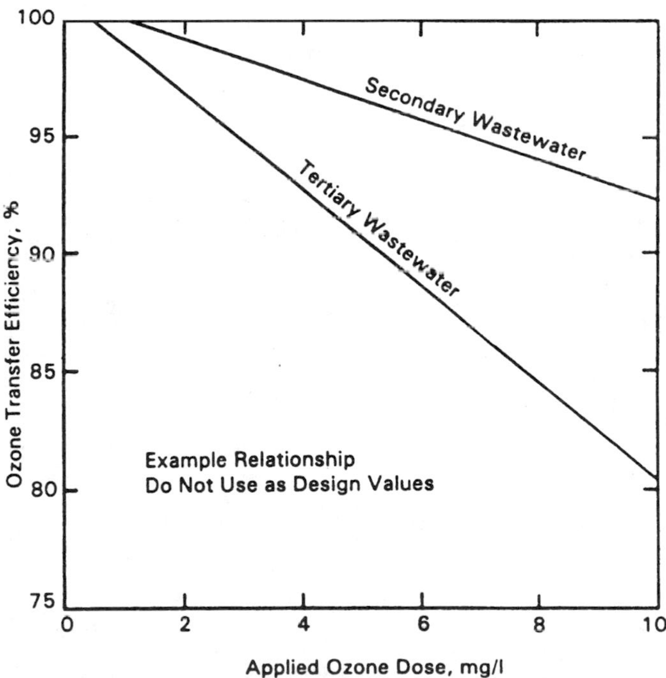

Fig. 15-17. Ozone transfer efficiency decreases as applied ozone dosage increases and as ozone demand of the water decreases.

approached 1.0. (See Fig. 15-18.) The units for the *G/L* ratio are cfm of gas to cfm of process water.

Grasso[13] and Given and Smith[14] reported a decrease in TE as the *G/V* ratio increased from 0.005 to 0.05. For best results, the *G/V* ratio should not exceed 0.03. The units for the *G/V* ratio are cfm of zone to cu ft of water volume in the contactor basin.

Summary of Contactor Basin Characteristics Affecting Ozone TE. These characteristics are most important for the design engineer because the hydraulic flow pattern and optimum diffuser submergence are crucial to the ozone TE efficiency in the contactor basin.

For plants located at sea level, the contactor depth for bubble diffusers should be from 16 to 20 ft. The air flow rate to the diffusers must be within the range recommended by the manufacturer. For a 60 sq in. porous disk diffuser, the flow rate should not exceed 2.0 cfm. Each of these disks should be secured with stainless steel holders and sealed with ozone-resistant materials such as Hypalon, Viton, or Teflon gaskets. These diffusers should be able to produce bubbles 2–3 mm in diameter.

The permeability of the porous media usually ranges from 12 to 20 cfm/sq fit per inch thickness at 2 inches of water column pressure. The disks are generally located from 6 to 12 inches from the bottom of the contactor basin.

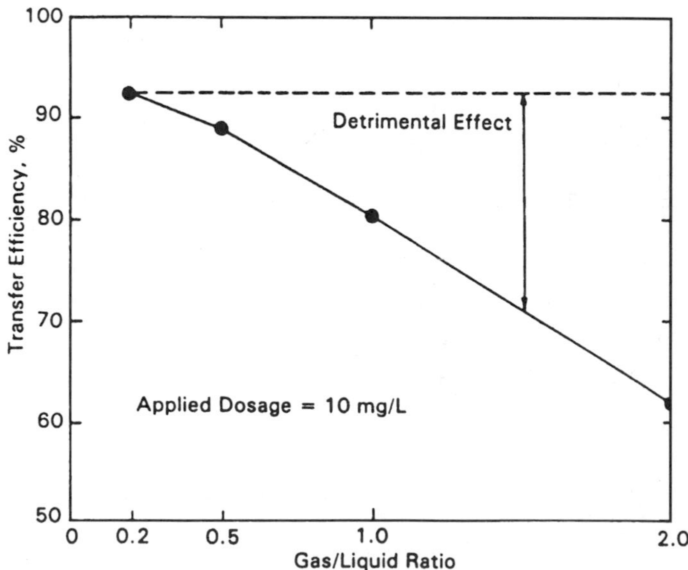

Fig. 15-18. An increase in the gas-to-liquid ratio causes a decrease in ozone transfer efficiency.

The diffuser flow pattern should oppose the water flow direction, and the location of the diffusers is also important. The initial point of ozone application will require the most diffusers. For a typical three-zone ozone application, the first stage will contain 50 percent of the diffusers, to take care of the immediate zone demand.

The flow through the contactor basin should approach plug-flow conditions. This minimizes or eliminates short-circuiting, which creates dead spots and reduces contact time. A conservative value for contact time at peak flow should be based upon the first appearance of a dye or other tracer. This eliminates all hydraulic variables that usually occur in any contact chamber. Use of this method for determining the optimum contact time has been most successful in achieving reliability of wastewater disinfection by chlorination in California, where many dischargers have to meet a 2.2/100 ml total coliform requirement on a daily basis.

Figure 15-19 illustrates a three-stage bubble diffuser contact basin, Fig. 15-20 illustrates the bubble diffuser contactor basin at the 600 mgd Los Angeles Aqueduct Filter Plant, and Fig. 15-21 illustrates an aspirating turbine-type mixer ozone contactor basin. This type of installation is unique because the turbine impeller creates a vacuum that draws the ozone gas directly into the process stream under a negative head. The design of this type is proprietary. The turbine reactor was evaluated by both Stover et al.[12] and Venosa et al.,[15] and in both studies the ozone transfer efficiency and disinfection performance were shown to be comparable to that of the bubble-type diffusers.

Multiple-stage ozone contactors, as shown in Fig. 15-19, reduce the effect of short-circuiting. For this reason the trend is definitely toward this type of design, but pilot studies are necessary to determine the use of a multiple-stage system.

There are two pressure control locations in the contactor basins. These components are used strictly to prevent structural damage to the basins. In Fig. 15-20 the pressure vacuum relief valve is usually set to provide for a 2-inch water column vacuum. If this vacuum decreases to a one-inch water column, the vacuum relief valve opens and relieves the pressure to atmosphere.

The ozone gas going to the destruct system is drawn from the contactor basin by a fan with an automatic pressure control valve. This fan is downstream from the demister and the destruct unit, with the discharge going to atmosphere. This part of the system controls the operation of the vacuum relief valves.

Ozone Destruction System

Introduction. These sytems are a vital part of an ozone facility because it is a physical impossibility to prevent off-gassing in the ozone contacting basins. In spite of this, many plants in Europe have operated for years without any destruct system. Off-gas was simply vented into the local atmosphere, as the

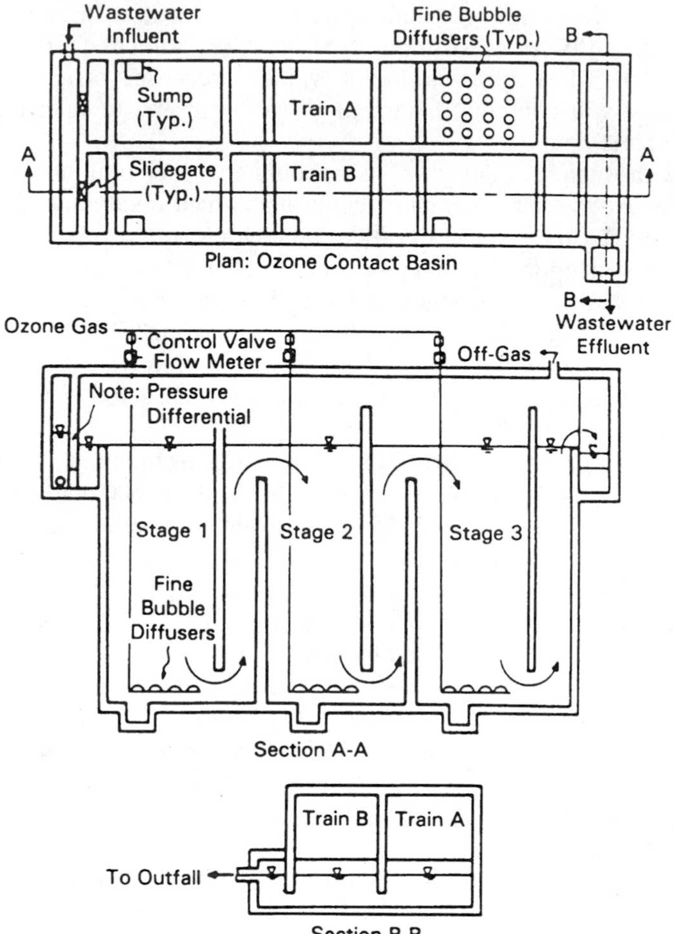

Fig. 15-19. Schematic of a 3-stage, bubble diffuser ozone contact basin.

only treatment was dilution with the ambient air. Ozone gas is extremely toxic. When it is ingested, the throat does not go into immediate spasms of closure as it does with chlorine gas. Consequently, a considerable amount of ozone enters the lungs, and it affects the lung tissue immediately—an extremely painful experience. So for safety reasons alone, units to destruct the off-gas must be a part of an ozone facility.

The minimum allowable ambient ozone concentration for an 8-hour working period is 0.1 ppm. This concentration is significantly less than the concentration in the off-gas, which is generally greater than 500 ppm by volume.

Current methods for the elimination of the off-gas from the ozone contactors are: thermal destruction, thermal/thermal catalytic destruction, and catalytic destruction.[16] Activated carbon has been used with some disastrous conse-

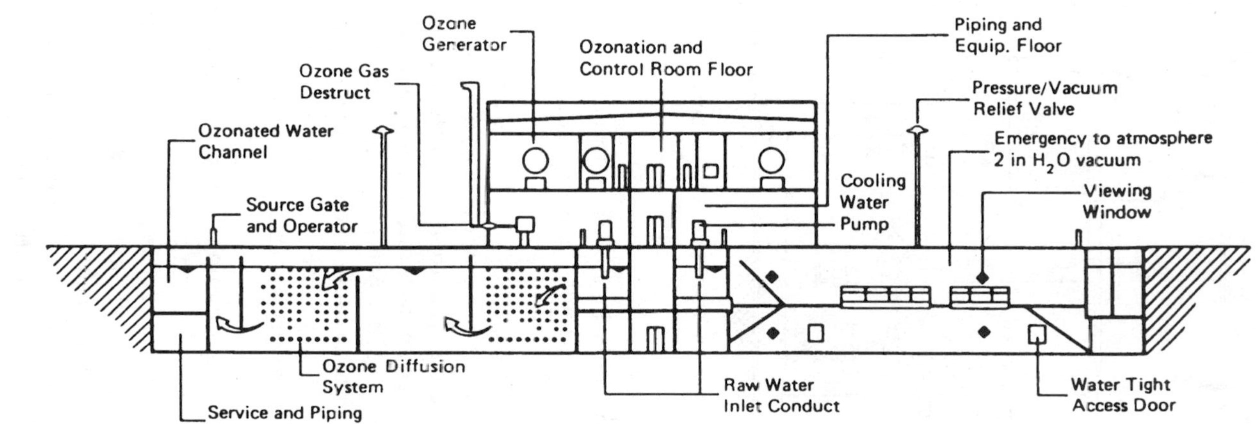

Fig. 15-20. Ozone contactor at the 600 mgd Los Angeles Aqueduct Filtration Plant, California (J. M. Montgomery, 1989).

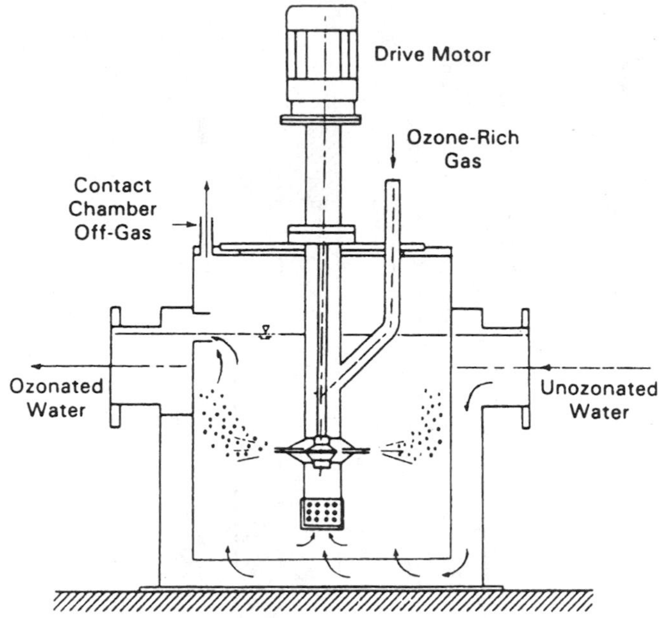

Fig. 15-21. Schematic of an aspirating turbine mixer ozone contractor.

quences. The ozone reacts with the carbon to form a powdery carbon that is highly explosive; therefore, it should never be used.

Thermal Destruction. This method is never used with oxygen feed-gas systems owing to the potential for uncontrollable fires caused by the presence of high oxygen concentrations. Thermal destruction of ozone involves the heating of the off-gas to a high temperature, which is maintained for a predetermined time. A temperature of 570 to 660°F, held for 3 sec, is required to achieve 99 percent ozone destruction. Because of the high temperature involved, heat recovery units are provided on all thermal destruct systems. The outlct gas temperatures for a heat recovery thermal destruct unit range from 160 to 230°F.[11,12]

A schematic of a thermal destruction unit with heat recovery is shown in Fig. 15-22. The off-gas passes through a pressure/vacuum relief valve and a demister before entering the heat exchanger. This valve is provided to protect the contactor basin from structural damage due to either excessive pressure or a vacuum buildup within the basin.

The stainless steel wire mesh demister is provided to reduce foam accumulation within the heat exchanger and heating elements. Different types of heat exchangers can be used, including cross-flow, shell and tube, and plate types. A fan is provided to move gas through the system when the contactor is operated under a slightly negative pressure, which is usually the case. The

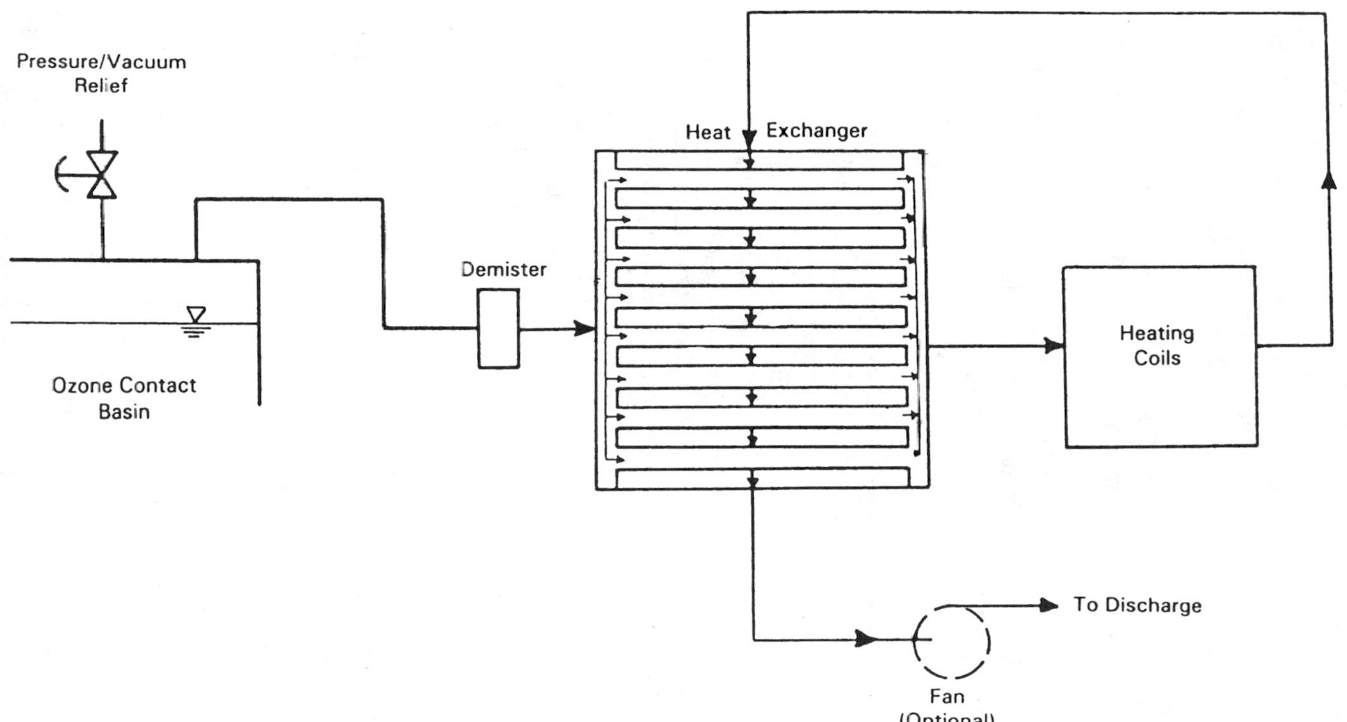

Fig. 15-22. Diagrammatic example of an ozone thermal destruct unit with a heat exchanger.

heat exchanger allows the heat from the gas stream exiting the heating coils to preheat the gas stream about to enter the heating coils. The head loss for thermal systems with heat exchangers can be as high as 3 ft of the water column. Every installation should be provided with a standby unit in order to achieve adequate reliability for the ozonation facility.

Instrumentation for Monitoring System Performance. The following are required for a proper level of monitoring:

1. Inlet and outlet gas temperature—monitors system performance.
2. Inlet gas flow-rate—monitors system loading.
3. Inlet ozone concentration—monitors system loading.
4. Outlet ozone concentration using meter—monitors system performance.

Thermal Catalytic and Catalytic Destruction.[11] A schematic of a metal catalyst ozone destruct unit is shown in Fig. 15-23. Because it is important to keep any foam away from the metal catalyst, the off-gas passes through the demister before entering the unit. A flow rate per volume of catalyst of 20 scfm/cu ft has been successfully used, and flows as high as 50 scfm/cu ft have been reported.[11] Because all catalytic systems are of a proprietary nature, the size of the metal catalyst chamber varies with each manufacturer. Here, again, a fan is shown as an optimal item, depending upon whether or not the contact basin is to be operated under pressure or vacuum conditions.

Metal catalysts are now being used most often because they are more active than metal oxides. Metals such as finely divided platinum or palladium can operate at temperatures as low as 85°F.[17] Their use promotes lower operating costs. The required temperature rise is much lower than for the thermal destruct systems. Furthermore, the life expectancy of the catalyst is about five years.[16] By comparison, a metal oxide such as aluminum oxide that contains palladium operates at a temperature range between 120 and 160°F. The heating elements of these catalysts should be within easy reach for cleaning and other maintenance requirements.

In addition to the monitoring requirements for the thermal catalysts described above, the gas pressure differential across the catalyst should be monitored.[11]

Automatic Control Systems

General Discussion. The decision to automate the control of the ozonation process presents some perplexing problems; so very few ozonation facilities are automated. In general, the purpose of automation has been to reduce energy consumption rather than to improve oxidation and disinfection performance. Ozone overdoses will not affect appreciably the finished water quality, except perhaps in the instance of microflocculation and turbidity control.

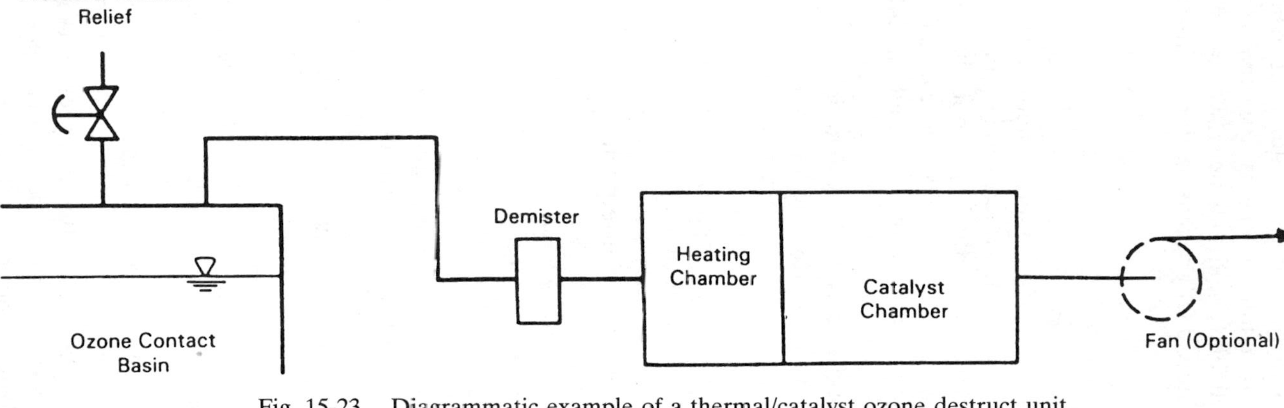

Fig. 15-23. Diagrammatic example of a thermal/catalyst ozone destruct unit.

The primary consideration in evaluating process control and automation needs is the potential payback of the capital investment. Operating flexibility should be provided to minimize energy consumption. This approach also improves process reliability and enhances system maintenance. Each case should be evaluated individually because water quality control is almost always site-specific.

Control Strategies

Ozone Dosage Control. The simplest method of control is to manually change the applied ozone dosage rate to correspond to changes in the water flow rate. Unfortunately this method does not allow the system to respond on its own when the water flow rate changes. Manual operation requires the operator to adjust the generator power setting to maintain the desired ozone feed rate.

The simplest method of automatic control is to vary the power supply to the ozone generator in proportion to the water flow rate. This is readily accomplished by the use of a microprocessor. The unit receives such information as water flow rate, feed-gas flow rate, and ozone concentration in the feed-gas. It then calculates the applied ozone dosage and compares this to the setpoint value. The output signal from the microprocessor will increase or decrease the ozone generator power supply in order to have the calculated applied ozone dosage equal to the setpoint. This will cause some changes in the ozone concentration because the feed-gas flow rate will remain the same. However, if there happen to be large variations in the ozone demand, the air supply rate will have to be adjusted for the difference. In order to avoid undesirable excursions from the setpoint, the microprocessor should incorporate a P.I.D. control loop.

The microprocessor method has already been established as a reliable way to achieve automatic process control, provided that the proper equipment has been selected. This method can give adequate control over a wide range of water flow rates if the water flow meter has an accuracy commensurate with the system range of water flow rates.

Ozone Residual Control. This type of control strategy has yet to be properly explored. It has been used for more than **half a century** 45 years in the chlorination of drinking water, but in spite of its success with chlorination, there is an overriding factor that contributes to the uncertainty of using residual control as a practical strategy for ozone: the rapid die-away phenomenon of the ozone residual. The concern is that analyzers may not be able to track it fast enough to cope with the speed of residual decay. There is also doubt about whether the bare electrode analyzers can measure dissolved ozone residuals consistently and reliably. However, there is little doubt about their ability to measure total oxidants (free ozone + ozonides + hyperperoxides). There may turn out to be a different scenario if future research ever finds a consistent relationship between total oxidants and dissolved ozone. (See below for ORP control of ozone residual.)

Off-Gas Concentration Control Method. This method has been rigorously tested and was reported by Venosa (and co-workers)[18,35] in 1983 and 1985. The key to the success of this system depends upon the ability of the process flow metering device to keep the carrier gas flow to the process flow constant over the range of process flow variation. The process flow metering device must be equipped with a microprocessor.*

Using this control strategy to achieve optimum operating efficiency has the following advantages:

1. True ozone is being measured, not total oxidant, which is measured by the bare electrode analyzers.
2. The reaction is instantaneous.
3. The measurement technology is convenient and reliable.
4. It is easily automated.
5. It is not subject to interferences.
6. It is not affected by sudden changes in water quality (ozone demand).
7. It is useful over a wide range of water quality.

The control arrangement, which is a variation of the compound-loop principle, is as follows:

1. Carrier gas flow to the contactor is flow-paced by a signal from the process flow meter.
2. The signal from the exhaust gas ozone concentration monitor changes the generator power input, thereby changing the ozone concentration in the inlet carrier gas.

This method of control was in use (in 1992) at three plants† in Deusseldorf, Germany.[2] These plants treat riverbank-filtered water with ozone, primarily to oxidize iron and manganese. Owing to the copious formation of ferric and manganic hydroxides and oxides, it is impossible to rely upon the measurement of dissolved ozone residuals for control. In addition, iron and manganese levels in the raw water vary, as do other ozone-demanding components (organics). This in itself makes applied ozone control very difficult.

---

*The transfer efficiency decreases when the gas-to-liquid flow ratio changes. This upsets the overall control system.

†These three plants began using ORP control by the Stranco HRR System in 1995, with complete success. This system eliminated the problem caused by iron and manganese compounds in the process water.

Ozone Feed-Rate Monitor

Photometer Calibration. This instrument performs the same function as a rotameter in a chlorinator—it provides the operator with the ozone dosage feed rate. It operates on the phenomenon of ultraviolet (UV) light absorption by the ozone molecule. When UV light passes through an air sample with a given pathlength and a given ozone concentration, the measured light intensity will be markedly less than when there is no ozone. The intensity of the light passing through the monitor depends upon the length of the light path, ozone concentration in the air sample, and the wavelength of the light source. This phenomenon of UV absorbance is expressed by the Beer–Lambert law, as follows:

$$I = I_o\, 10(-xLC) \tag{15-11}$$

where

I_o = UV intensity at reference ozone concentration.
I = UV intensity at ozone sample concentration.
x = molar extinction coefficient in 1 $mol^{-1}cm^{-1}$
L = pathlength of light source, cm
C = ozone concentration, Mg/L

Equation (15-11) is used as a standard measure of ozone concentration in air or oxygen in the range of 1 to 300 g/cm^3.

The sample flow rate has no effect upon the ozone concentration readings. Also the light intensity is not affected by other substances so long as those substances do not change during adjacent reference and sample cycles. Ambient temperature and pressure variations can be determined and corrected for by using the ideal gas laws and by following the manufacturer's recommended procedures for calibration. All available photometers should have an electronic system that provides automatic ambient temperature and pressure compensation.

In order for a photometer to be considered accurate and reliable, it must incorporate the following requirements:

1. An air sample with a known ozone concentration.
2. An air sample identical to the first in its light-absorbing properties, except for an unknown ozone concentration.
3. A UV light source of known wavelength and fixed intensity.
4. A stable detector.
5. A known and fixed light pathlength.
6. The value of x, which is the UV absorption coefficient at the wavelength being used.

With good instrumentation, an absolute accuracy of about plus or minus 2 percent can be attained.[30] There is no interference from other species that may possibly be found in the output of ozone generators, such as nitrogen oxides, hydrogen peroxide, and nitric acid.

The photometer just described can also be used to measure the off-gas ozone concentration. Using these two photometers in consort with each other provides the best means for automatic control of an ozone generator.[18]

These photometers are available in the United States from Daisibi and PCI, and in Switzerland from Segrist.

MEASUREMENT OF OZONE RESIDUALS IN AQUEOUS SOLUTIONS

Units for Oxidant Residuals after Ozonation

The conversion to moles, equivalents, or normality from units of mg/L (as ozone, chlorine species, or chlorine dioxide) can be confusing. Gordon et al.[19] recommended that all oxidizing agents and their by-products be reported in molar units (M) and if necessary in mg/L of the oxidizing agent. Furthermore, they recommend that oxidizing equivalents per mole of oxidant be reported to minimize additional confusion. For example, when ClO_2 is reduced to ClO_2^-, this corresponds to 1 eq/mol, and when ClO_2 is reduced to Cl^-, this corresponds to 5 eq/mol.

A summary of molecular weights and oxidizing equivalents for the various chlorine species, oxychlorine species, and ozone is given in Table 15-1.

***Standard Methods,* 19th Edition—1996.** It is most important to realize that the only method listed in this highly informative document is the Indigo

Table 15-1 Equivalent Weights for Calculating Concentrations Based on Mass

Species	*Molecular Weight g/mol*	*Electrons Transferred Number*	*Equivalent Weight g/eq*
Chlorine	70.906	2	35.453
Monochloramine	51.476	2	25.738
Dichloramine	85.921	4	21.480
Trichloramine	120.366	6	20.061
Chlorine dioxide	67.452	1	67.452
Chlorine dioxide	67.452	5	13.490
Chlorite ion	67.452	4	16.863
Chlorate ion	83.451	6	13.909
Ozone	47.998	2	23.999
Ozone	47.998	6	8.000

Colorimetric Method (p. 4–104). This method requires the use of a photometer, or a filter colorimeter for use at 600 plus or minus 5 nm.

The minimum detectable ozone residual in an aqueous solution is determined by the spectrophotometer procedure using thermostated cells and a high quality photometer. The low range procedure will measure ozone down to 2 μg/L. For the visual method the detection limit is 10 μg/L.

The ORP Method of O_3 Residual Control[38]

After years of optimistic searching, White has found proof that the Redox method pioneered by Stranco Inc. is the system of choice. This information that provided the proof came from an installation in Dusseldorf, Germany.

The company responsible for the treatment processes used in the abstraction of the water and the protection of the distribution system is Stadtweke Dusseldorf AG. The chief engineer who was largely responsible for organizing and operating the various treatment processes is Jurgen Schubert, who used Redox as the principal parameter in work on the overall effectiveness of treatment improvements, distribution system stability, buffer storage renewal, and overall cleanliness. Since 1982 Mr. Schubert has measured the Redox of the water flowing in and out of the buffer storage system from distribution.

Across the city at the top of a hilly area are two storage vessels, each with a capacity of 15,852,000 gal. These fill up at night, and water flows out during the day. The storage time in these vessels can be 12 or more hours per day.

Near the end of 1995, Mr. Schubert installed the Strantrol HRR Control System to improve the accuracy and reliability of the overall processes. (This can be seen from the data in Fig. 15-23a.) When the Redox span is from 600

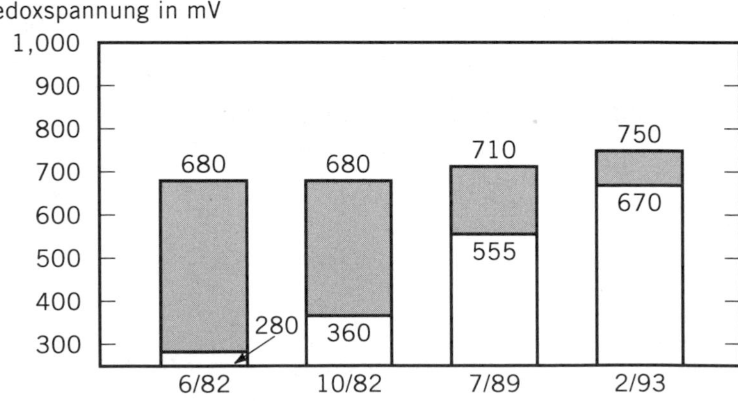

Fig. 15-23a. A German WWTP illustration of disinfection improvement after ORP control was accomplished beginning in July 1989.

to 700 mV between raw water and the finished water, this is indicative of superb treatment. By 1997, Mr. Schubert had installed three new Strantrol control systems, which are used for the following purposes:

1. To monitor ozone depletion at about 880 mV, a sensor cleaning system because iron and manganese are present at this stage.
2. To monitor the correct level of ClO_2 injection at the entrance to the distribution/storage system, resulting in a steady redox value of 700 mV.
3. To take Stranco HRR Redox measurements at the inlet and outlet of the buffer storage vessels—currently about 665 mV, which indicates an extremely low loss of oxidative protection throughout the system.

As redox changes become more discrete, the stability and reliability of the Strantrol HRR Redox control system is very important to the confidence in the information produced

The Dusseldorf, Germany, Potable Water Treatment Process.

For over 35 years, this process has been used for treating water from the River Rhine, including the water from this river that is known as "river bank filtration." The following are the main features of the system at the water treatment plant:

1. Water entering a specially designed absorption tank is dosed with air/ozone.
2. Next, the ozonated water enters a covered intermediate basin for contact and ozone removal.
3. Then the water enters a multilayered activated carbon reaction vessel, where optimum steady flow rates have been established.
4. Then the water is treated with minute quantities of phosphate and silicate to neutralize carbonic acid formation.
5. Addition of chlorine dioxide (60 μg/L) is done to maintain a small but effective oxidizing residual, to maintain a clean and protected distribution system.
6. The large buffer storage at the end of distribution enables more even flow rates and enhanced treatment qualities at the production plants.

The Redox (ORP) Process.[38] The basic things to know about this process are poise and demand. These are the two unique properties of redox that no other oxidant has. The poise measures the changes in water quality that occur when a source water is being treated, and the difference between the poise millivolts and the demand millivolts establishes the water quality. The only reason why redox can accomplish this feat is that ORP measurements are not afflicted with any errors due to any objectionable compounds in the water. The poise measurement is made prior to the addition of ozone or chlorine dioxide. This measurement is always considered a measurement of pollution or possibly antioxidants. The other measurement is made after the ozone or chlorine dioxide is applied and after the desired contact time has arrived.

Both of these measurements are recorded in millivolts (mV). The difference between the two is a reading of the ozone demand or the chlorine dioxide demand; therefore, operating personnel have a perfect way to realize how well their process water is behaving under any given treatment process that is being used.

Figure 15-23a illustrates how the water treatment distribution and quality have been constantly improving over the years. The data show the variation in the oxidative state of the water at the inlet and the outlet of the buffer storage vessels. Now, in 1997, the variations are quite small (30 to 40 mV less than the value leaving the treatment works). This clearly gives an overall evaluation of the correctness of water collection, treatment, distribution cleanliness, very low treated water demand, and buffer storage cleanliness.

Owing to greatly improved water quality, now the ozone dosage of this well water is down to 1 mg/L, and the activated carbon usage is up to 2.5 to 3.5 mg/L water treated.

Other Field Studies by Stranco. There are several of these going on in other parts of Germany where groundwater (leachate) runoff contaminates underground waters. Redox is a very important control method because it minimizes the amounts of ozone residual that is needed prior to the biological treatment stage. These ozone amounts are as follows:

1. Runoff water (leachate) −100 mV.
2. After oxygen is applied, +50 mV.
3. After ozone is applied and residual is 0.4 mg/L, then +100 mV is needed.

For example, after the addition of oxygen, 50 mV of ozone is needed for preparation of the leachate. Sometimes the leachate background Redox can change. Therefore, the amount of oxygen addition after stage 2 above will also change, as follows:

1. Runoff water (leachate) −140 mV.
2. After oxygen is applied, +10 mV.
3. After ozone is applied, about +50 mV excess of Redox will be needed.

In Germany, other ORP users have told Stranco that the Strantrol control system is superior to all other methods, or procedures.

Cooling Tower Installations. In these installations there is a primary need to ensure ozone residuals as low as parts per billion where open type towers are used because off-gassing at higher residuals becomes a health hazard near the tower area. In Germany, this is a health and safety hazard.

At a German chemical factory using open cooling towers, it was necessary to control the biological growths appearing on the towers, which affected the proper operation of the cooling systems. Therefore, ozone was chosen as the control application. In this case the ozone was made from liquid oxygen. A Strantrol HRR Monitoring system was used, which was measuring 0.01, 0.02, and 0.04 mg/L, with completely accurate repeatability, against the Indigo Colorimetric Method to meet the health and safety standards.

A Special Note for the Reader

In view of the foregoing information, all the methods described next in this text, after the "General Discussion" and down to but not including "On-line Analyzers for Measuring Ozone in Solution," are in fact obsolete; but White is leaving them in for historical reasons.

General Discussion

The problem encountered in measuring ozone residuals in aqueous solution is strictly a consequence of the nature of ozone: it is a powerful oxidant suffering from continuous self-destruction, off-gassing from solution, and instantaneous reaction with many organic and inorganic contaminants usually found in water. Knowing the ozone residual after a given contact time is crucial to the successful performance and reliability of an ozone facility. Adherence to the Safe Water Treatment Rule (SWTR) and the *Ct* rule requires that all disinfectants rely upon a measurable residual at a specified contact time in the finished water that is going to the consumer. Furthermore, demonstration projects have concluded that the ultimate success of an ozone facility depends largely upon the ability of operating personnel to accurately measure the dissolved ozone residual.

As a result of the many legislative issues formulated in the United States during the mid-1980s, the AWWA Research Foundation funded and organized projects to evaluate the various techniques of residual measurements for all disinfectants.[19–21] The following discussion is based largely upon the work of three researchers mentioned above and on recent field experience of operators and equipment manufacturers.

Ozonation of water not only produces an ozone residual but also produces other oxidants. These "other" oxidants, broadly classified as ozonides and hydroperoxides, are thought to be nongermicidal, so they are considered to be interferences; they interfere with the quantitative analysis of a true ozone residual. Thus, some methods of ozone analysis are not specific for ozone because their results include these other oxidants. There is a parallel case in the measurement of chlorine species. Whenever water or wastewater contains any organic N, all of the chlorine species will react with these compounds to form organochloramines. This chlorine species is not only nongermicidal; it is only detectable by forward-titration using either the amperometric or the

DPD-FAS titrimetric method because it appears in the dichloramine fraction. Therefore, all total chlorine residuals will be false when organic N is present. Owing to this often-encountered false chlorine residual, the users of ozone believe it is entirely possible to achieve reliable disinfection by adjusting to "false" ozone residuals that contain ozone plus the other oxidants. Continuous analyzers that measure ozone plus the total oxidants have been available for many years. These analyzers are classified as the bare electrode type; the membrane type is designed to be ozone-specific.

Most of the ozone residual methods are modifications of chlorine residual methods that determine total oxidants in solution, and no one method has been selected as the referee method. Therein lies the difficulty with ozone residual determination.

Although the 19th edition of *Standard Methods* describes a method for ozone gas stripped from solution in water, researchers question the accuracy and reliability of the iodometric procedure. A summary of the task force investigators' findings is described below.[19]

Iodometric Method

Owing to several problems with the stripping of ozone from solution, the consensus is to abandon this approach. In a detailed comparison of eight analytical methods for determining residual ozone, Grumwell et al.[22] and Gordon et al.[23] concluded that there is no iodometric method that can be recommended for the determination of ozone in aqueous solution because of the unreliability of the method and the difficulty of comparing results obtained when minor modifications were made in the iodometric method itself.

Arsenic III Back-Titration

In this method, inorganic arsenic III reacts with ozone at adjusted pH values between 6.5 and 7.0. The excess arsenic III species is back-titrated with a standard iodine solution to a starch end point. The primary advantages of this method are: minimal interferences, good precision in the hands of an experienced technician, and good overall accuracy. This method continued to be recommended.[19]

Indigo Trisulfonate

Characteristics. This method is the one recommended over all others for measuring ozone residuals in aqueous solution.[19] This colorimetric method is quantitative, selective, and simple, and it replaces methods based upon the measurements of total oxidants. It is applicable to lakewater, river infiltrate, groundwater containing manganese, extremely hard groundwater, and biologically treated wastewater.[24]

Principle. The decrease in absorbance is linear with increasing concentration. The proportionality constant at 600 nm is 0.42 + 0.01 cm/mg/L (Ae = 20,000/M × cm), compared to ultraviolet absorption of pure ozone of e = 2950/M × cm at 258 nm.[25,26]

Interferences. Hydrogen peroxide and organic peroxides decolorize the indigo reagent very slowly. If measurement is made less than 6 hours after the reagent is added, there will be no interference. Organic peroxides may react more swiftly. Fe III does not interfere, but Mn II, if oxidized by ozone, forms compounds that decolorize the indigo reagent. This interference can be corrected by making a measurement relative to a blank in which the ozone has been destroyed selectively. Without the corrective step, 0.1 mg/L of ozonated manganese gives a response of about 0.08 mg/L of apparent ozone. Chlorine also interferes, but it can be masked by the addition of malonic acid. When bromide ion (Br^-) is present, ozone oxidizes it to bromine (Br_2), which interferes at a ratio of 1 mole HOBr to 0.04 mole ozone.

Residual Measurement. This is accomplished by the spectrophotometric procedure using a high-quality photometer. The lower limit of detection is listed as 2 μg/L ozone.[24]

Calculations—Spectrophotometer Procedure:

$$\text{Ozone, mg/L} = \frac{100 \times \Delta A}{f \times b \times V} \tag{15-12}$$

where

ΔA = difference in absorbance between sample and blank
b = pathlength of cell, cm
V = volume of sample, cm (normally is 90 ml)
f = 0.42

Amperometry

Introduction. Amperometric analyzers have been in use to measure chlorine residuals since the early 1950s. These analyzers are sometimes referred to as the bare electrode type. In the late 1970s Stanley and Johnson[27] reported on the membrane electrode amperometric analyzers. The following is a brief discussion of these different types.

Bare Electrode Analyzer. These units require agitation of the sample at the electrode surface to establish a diffusion layer. The electric current that is measured across the indicating and reference electrodes is directly proportional to the concentration of the dissolved oxidant. This type of analyzer has good sensitivity and is applicable as a continuous nonselective monitor for ozone. It is often used to measure total oxidants when measuring dissolved ozone. When chlorine, chlorine dioxide, bromine, and iodine are present, then these oxidants are interferences. The nature and the magnitude of these interferences have yet to be determined.

The bare electrode units suffer from loss of sensitivity and calibration due to an accumulation of impurities that tend to form on the electrode surfaces. These occurrences are widely variable because they are site-specific. In addition to the problem of impurities, the electrode response is adversely influenced by numerous surface-active agents, halogens, and oxygen.

Amperometric Membrane Sensors. The use of gas-permeable membranes instead of bare electrodes is viewed as an improvement in the amperometric method for ozone analyzers, because of the increased selectivity and the elimination of electrode fouling.[27–29] These membrane electrodes exhibit less than 2 percent interference, in terms of electric current response, from bromine, hypobromous acid, hypochlorous acid, nitrogen chloride, chlorine dioxide, and hydrogen peroxide.[27,28]

In addition to the above, the use of positive voltage potentials and polymeric membranes that are selectively permeable to gases has improved the possibility of selective measurement of ozone, which would be a further improvement over the bare electrode configuration. The task force group[19] strongly recommended further research and development for this approach, using different applied voltages across the membrane electrode and the reference electrode.

Ultraviolet Measurements

Ultraviolet absorption measurements can also be used for residual aqueous ozone at 258–260 nm. While the molar absorptivity for gaseous ozone has a generally agreed-on value of $3000 + 301\ cm^{-1}\ mol^{-1}$, uncertainty about ozone's aqueous molar absorptivity is critical to the future use of UV methods. Therefore, the task force has strongly recommended further research on this problem. For example, if the molar absorptivity for ozone is unambiguous, then UV absorption is, in principle, an absolute method for measuring ozone concentration because it does not depend upon calibration or standardization against other analytical methods.

Summary

1. The UV spectrophotometer method is the only one that is recommended for the determination of ozone concentration in the gas phase.
2. Accurate determination of ozone in the aqueous phase is complicated by its rapid decomposition and reactivity with other species present, plus the formation of ozone DBPs.

3. There is no iodometric-based chemistry that is acceptable for determining ozone residuals in water.
4. Indigo trisulfonate and arsenic III are two direct oxidation methods that are acceptable. The indigo method is now a standard in Germany and Switzerland. It is described in the 17th edition of *Standard Methods.*
5. Amperometry techniques continue to improve and keep alive the hopes for a practical and reliable method for automatic ozone residual control. The inability to accomplish this type of control is the weakest part of an ozone facility.

On-line Analyzers for Measuring Ozone in Solution

Introduction. Measurement of ozone residuals in aqueous solution has been described by Gordon et al. as amperometry.[19] There are two different types of amperometry: the bare-electrode method and the permeable gas membrane method. The bare-electrode method has good sensitivity and is reliable for the halogens, where it has established its accuracy and reliability. However, when a powerful oxidant such as ozone in solution produces other oxidants that interfere with the bare-electrode method,* the consensus is that the bare electrode system measures total oxidants and is not able to be selective for ozone.

This is where the permeable gas membrane system differs from the bare-electrode system. The manufacturers of these systems (EIT and ATI) have developed the use of proprietary membranes and electrolytes that are selective for ozone. These polymeric membranes show less than 2 percent interference, in terms of current response, from the presence of bromine, hypobromous acid, hypochlorous acid, nitrogen trichloride, and chlorine dioxide. Manufacturers of the bare-electrode technology have not attempted to refine the method to better cope with the selective measurement of ozone.

ATI-Series A15/64.[39]

Dissolved Ozone Monitor. It is an advanced electrochemical system for continuously monitoring dissolved ozone concentration in water.

It uses a direct sensing ozone probe to measure dissolved ozone specifically, without interference from other sample components, such as residual chlorine. It is engineered to perform reliably with a minimum of maintenance. It is ideal for potable water treatment, or virtually any ozonated water stream process. This instrument is illustrated in Figure 15-24 and Figure 15-25.

Electronic Monitor. This is housed in a NEMA 4X fiberglass enclosure to prevent the possibility of any corrosion from an ozone leak. There are five

*The oxidation–reduction potential (ORP) of ozone is 2.07 volts (E_0), as compared to 1.36 for chlorine.

Fig. 15-24. Series A15/64 dissolved ozone monitor (courtesy ATI).

switches on the front panel that provide access to monitor programming functions, and allow operators to view information such as sample temperature, alarm setpoints, and analog output value. It is also equipped with an access code number to avoid tampering.

The DO_3 monitor may be programmed for either low range or high range measurement. In the low range, the monitor displays ozone over a range of 0 to 2.000 ppm, with a resolution of 0.001 ppm (1 ppb). In the high range, the display is 0 to 20.00 with a resolution of 0.01 ppm.

For special applications, a high range system of 0 to 200 ppm is available. An isolated 4 to 20 mA output is provided for external recording or data logging. This output may be programmed for any required span using the front panel switches, and may be changed to 0 to 20 mA and used with a shunt resistor to provide zero-based voltage output if required.

All variables such as alarms or control setpoints, deadband, and time delay are programmable from the front panel keys.

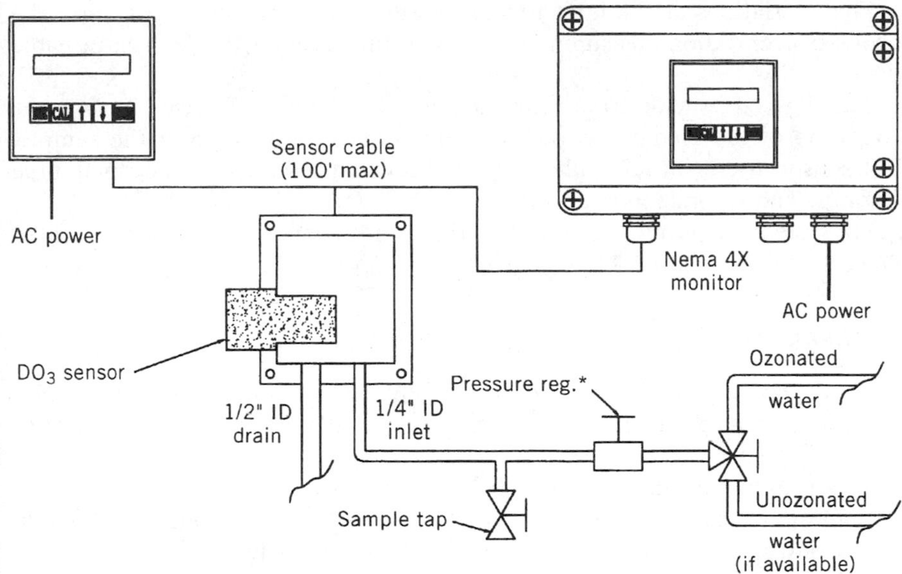

Fig. 15-25. Dissolved ozone monitor flow-cell installation diagram (courtesy ATI). *Pressure regulator recommended if inlet pressure varies > 10 psi.

Dissolved Sensing Element. The A15/64 sensing element is a polargraphic membrane sensor that is specific to ozone. In operation, ozone diffuses through a membrane at the tip of the sensor and is reduced to oxygen. This reaction causes an electric current to flow between the sensing and counter electrodes, with the current being directly proportional to the dissolved ozone concentration. In essence, the sensor operates like a battery, with the ozone causing the current to flow. Because the sensor consumes ozone in the measurement, sample flow is required for the measurement. A sample velocity of at least 0.2 ft/sec is recommended to ensure accurate data.

As the permeable membrane isolates the sensing electrodes from the solution being monitored, the dissolved ozone sensor exhibits excellent zero stability. Zero drift is generally less than 5 ppb over several months of operation. The response to ozone is fast, reaching 90 percent in less than one minute, and the polymeric membrane is not degraded by the normal constituents in water. Should coatings become deposited on the membrane, they can be quickly and easily removed by wiping with a soft cloth. Sensors normally last five years or more, and they are designed to make normal membrane changes simple and inexpensive.

Flowcell Assembly. Maximum stability from a DO_3 sensor is obtained when the sample flow and pressure are stable. The A15/64 is provided with a constant head overflow chamber that automatically regulates the sample flow past the sensor.

Flow variations at the inlet to the overflow cell have no effect upon the ozone concentration measurement. This feature ensures the highest possible data validity.

The flowcell is made from clear cast acrylic, allowing the user to view the condition of the membrane without removing the sensor from the sample. The sensor inserts into the side of the flowcell and is sealed in place by double O-rings. The recommended flow rate is 15 gal/hr.

For systems requiring a flow system that is not open to atmosphere, a sealed flowcell with ⅛-inch FNPT fittings for the inlet and the outlet is available.

Additional Details:

1. Display: 16-character alphanumeric liquid crystal display with back light.
2. Measurement range: programmable 0 to 20.00 ppm standard or 0 to 2.00 ppm.
3. Sensitivity: 0.001 ppm above 0.005 ppm.
4. Output range: programmable for any range from 0–1 ppm to 0–20 ppm.
5. Control relays: two SPDT, 5 A, 230 V AC resistive, with programmable setpoints.
6. Power: 110 V AC, 50/60 Hz, 5 V A max.

EIT[40]–Model 84[22]. Figure 15-26 illustrates this microprocessor-based dissolved ozone monitor. It provides a continuous highly accurate ozone measure-

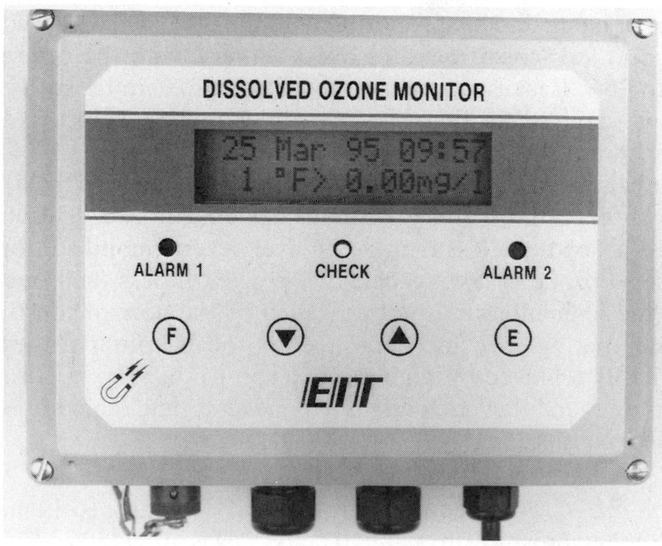

Fig. 15-26. Model 84 microprocessor-based ozone monitor (courtesy EIT).

ments for many diverse ozone process users. The monitor is sealed in a NEMA 4X enclosure that never has to be opened once the wiring has been completed.

Ozone concentration readings are made by a direct-measuring, low-maintenance sensor that does not require any chemical treatment of the sample. This sensor is not subject to any interference from other dissolved ions, including residual chlorine.

Parameter changes and calibration procedures are performed by applying a small magnet to the front panel icons. Menu prompt software is included as a guide to system setup and operating details, such as:

1. Auto-ranging, back-lighted display.
2. System damping and password security.
3. Optional RS232/RS485 communications for data logging.

Request data sheet 84[22].

Calibration. The indigo trisulfonate method is recommended over all others. A second choice would be the arsenic III method. The indigo method is a colorimetric procedure that is quantitative, selective, and simple to perform, but the operator must have a high-quality spectrophotometer. The lower limit of detection is listed as 0.002 mg/L.[24] (See also above discussion of the method.)

Current Practices. The European Economic Community (Great Britain, France, Germany, Belgium, Netherlands, and Switzerland) has accomplished very little in developing analyzers for the measurement of ozone in solution. Its operators seem to want to rely primarily upon dosage control. In spite of this apparent lack of interest, there is what appears to be a very reliable Swiss-made analyzer. One of these analyzers has been in operation for several years at the Chas. J. Baillets WTP. It was made by Polymetron S.A. of Switzerland and is identified as the Ozonmat Model 8880. Plant personnel have praised the reliability and the accuracy of this unit.[32] However, a lack of response from the manufacturer to the author's requests accounts for the lack of any information on the Ozonmat Model 8880 in this text.

Manufacturers in the United States have been far more interested in producing an ozone analyzer to measure the residual in the contact chamber. A very scientifically advanced unit is the one made by EIT. Several of these are in use at the 600 mgd WTP built by the Los Angeles Department of Water and Power to treat the water from the Owens River aqueduct. These units have been in operation for some time measuring ozone residual in solution in the range of 0.01 to 0.05 mg/L.[33] In the start-up phase of operation they experienced trouble with the sample pressure variation, which caused wide swings in calibration accuracy. This problem was corrected by the installation of a constant head cell provided by the manufacturer, EIT. The only major maintenance activity is replacement of the membranes every three months. There is a Rosemount unit at this plant that is suffering from the same sample pressure compensation problem.

LOS ANGELES AQUEDUCT FILTRATION PLANT*

Introduction

This plant is probably the largest water treatment plant in the world at 600 mgd capacity. It is easily the most technically advanced facility of its kind in North America. The source water is from the snow pack on the eastern slope of the Sierra Nevada Mountains west of but close by Mono Lake. This water flows through the 338-mile-long Los Angeles Aqueduct to Sylmar, California, where the filtration plant is located. The plant was built by the Los Angeles Department of Water and Power, and was dedicated April 2, 1987, although water had started flowing through this plant in November 1986.[34]

The plant was designed by the joint venture of Brown & Caldwell and Camp Dresser & Mckee, in Pasadena, California. The primary contractor for plant construction was the M. A. Mortenson Company. The oxygen/ozone system was supplied under a separate contract with Air Products and Chemicals, Inc. (APCI). The Brown Boveri Company (BBC) supplied the ozone generators to APCI.

In spite of the high quality of the source water, this plant became imperative in the mid-1970s in part because of tightened federal and state turbidity requirements. Another major factor was dwindling of the settling and storage capacity along the aqueduct route caused by a continued decrease in the snow pack and rainfall. These events made it necessary to provide a treatment plant that would reduce the turbidity and bring the flow into compliance with the new water quality standards.

Plant Processes

These processes are illustrated in Fig. 15-27. They include the following: screening, ozonation, direct filtration, chemical addition, rapid mixing, coagulation, flocculation, filtration, and postchlorination. The simplified oxygen/ozone process is illustrated by Fig. 15-28.

Ozonation Facility

General Description. The ozonation facility shown in Fig. 15-29 generates oxygen for the ozone generator feed-gas and is composed of five key components:

*The author has chosen to describe this facility in as much detail as possible with text and illustrations because it is an outstanding example of modern technology in the pretreatment of drinking water supplies.

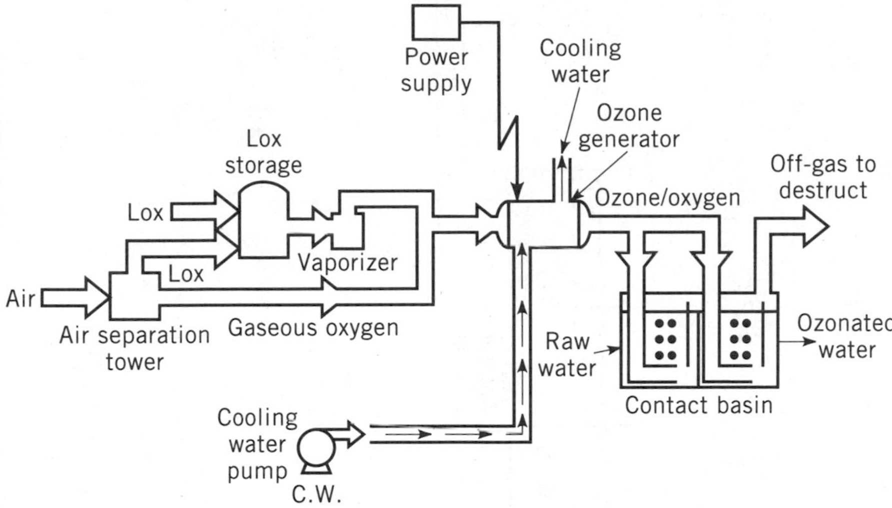

Fig. 15-27. Simplified oxygen/ozone process (courtesy Los Angeles Department of Water and Power).

1. A 50-ton per day cryogenic oxygen plant, built by Air Products & Chemicals Inc. (APCI).
2. Four on-line ozone generators, plus one standby and provision for expansion to a sixth unit. This provides an ozone capacity of 7900 lb/day, which is equivalent to about 8 mg/L for a 600 mgd flow.
3. Ozone gas diffusion equipment for dissolving the ozone into the raw water supply. (See Fig. 15-30.)
4. Ozone off-gas destruct units to convert the ozone back to oxygen before it is vented to the outside atmosphere.
5. Computerized control equipment to match ozone production to treatment plant flow and ozone demand and to optimize production efficiency.

A two-story ozone building, shown in Figure 15-31, and an outdoor oxygen generation facility were built near the plant's inlet structure. The ozone generators, power supply units, and control room are located on the second floor. The first floor houses the vent gas blowers, ozone destruct units, process piping, and electrical switchgear equipment. Ozone is injected into the raw water as it flows through four contact basins constructed beneath the ozone building.

Ozone is made up of three oxygen atoms that are produced in a tube-type water-cooled generator, as shown in Figure 15-32. There are 866 of these dielectric tubes per generator. As described elsewhere, the oxygen feed-gas flows through a uniformly charged air space with electrical energy splitting the bond of a portion of the diatomic oxygen molecules to form "active"

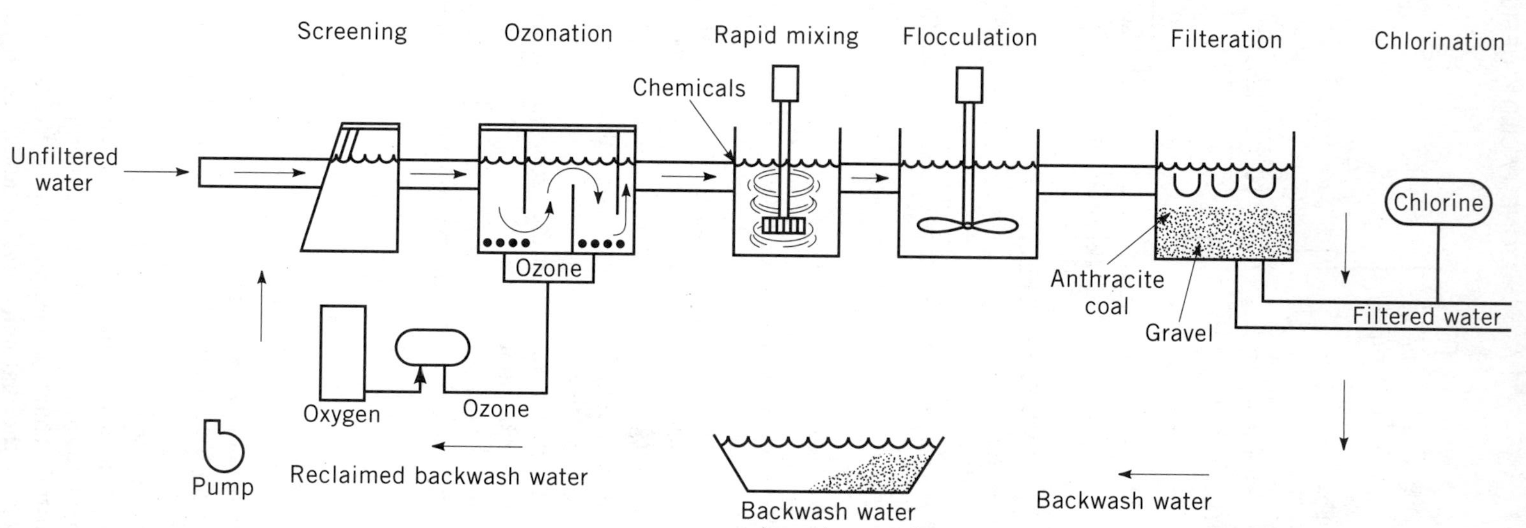

Fig. 15-28. Treatment plant processes (courtesy Los Angeles Department of Water and Power).

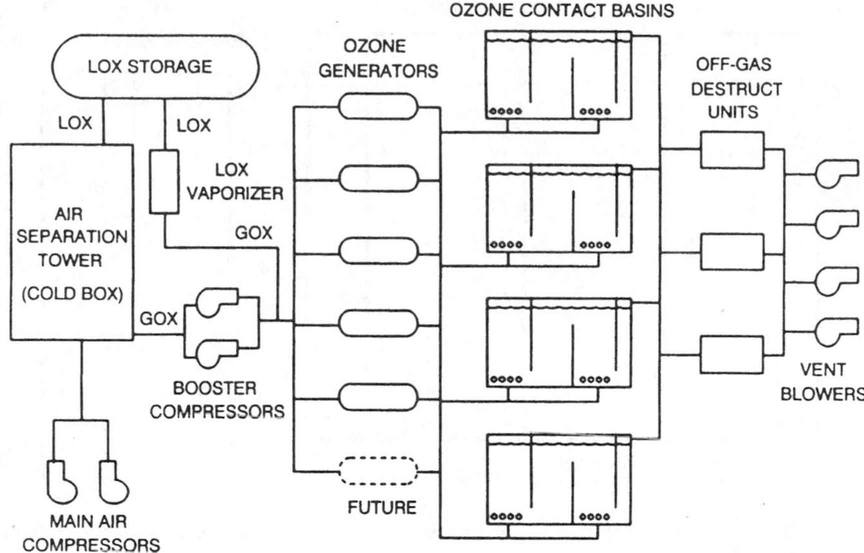

Fig. 15-29. Ozonation system schematic (courtesy Los Angeles Department of Water and Power).

elemental oxygen atoms. These active atoms combine with oxygen molecules to form triatomic oxygen. As the need for ozone changes, the number of ozone generators will automatically adjust from one to four units, according to the ozone demand and the rate of raw water flow. This provides a 20 to 100 percent range of system capacity. In addition to this, the oxygen feed-gas flow rate and power to the ozone generators will adjust automatically to maintain specific operating conditions. Although the system capacity calculates to 8 mg/L, the applied dosage currently ranges from 1.0 to 1.5 mg/L to provide about a 0.2 mg/L or higher ozone residual in the contact basin effluent. This depends upon the variations in the raw water quality, which varies on a seasonal basis.

The goal of primary disinfection by ozone is to achieve a one-log reduction of pathogens in the raw water effluent from the contact basins. These basins have a water depth that varies from 17 to 20 ft. They are equipped with a vast array of fine bubble diffusers, made of ceramic fused alumina rod-type material. All ozone and oxygen that is not absorbed in the water is released into the headspace above the water line in the contact basins as off-gas. A vent blower system produces a 2-inch water column vacuum in the head space to capture and deliver the off-gas to the ozone thermal catalytic destruct unit before it is safely discharged to the atmosphere as oxygen and less than 0.1 ppm ozone by volume.

Control System. The operation of the ozone generation and the oxygen plant is controlled and monitored by a dedicated programmable controller.

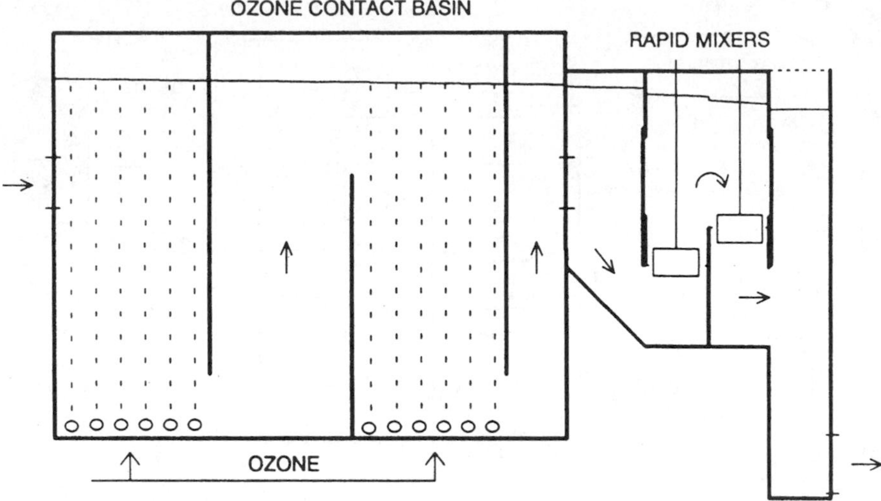

Fig. 15-30. Ozone contact basin, sectional view (courtesy Los Angeles Department of Water and Power).

This controller receives "information" on water flow, applied ozone dosage, ozone residual in the contact chamber, and purity and flow of the oxygen feed-gas. From these data, the controller calculates and adjusts the oxygen and ozone production rates to achieve optimum efficiency. A remote terminal unit (RTU) interfaces with the ozone programmable controller. It relays the status of all critical process and safety parameters to the central computer control system located in the plant's administration building. Air monitors continuously sample the environment within and around the ozone building for excessive levels, above ambient conditions, of both ozone and oxygen concentrations. If the ozone concentration exceeds a preset level or oxygen concentration exceeds standard atmospheric levels, the ozonation system will automatically shut down. During such an emergency, audio and visual alarms will alert the control room operator to make an announcement over the public address system and activate a standby blower system to supplement the normal ventilation system.

In addition to ozone production safety and off-gas emission control systems, the ozonation system is equipped with a safety valve on the oxygen feed-gas line that runs from the oxygen plant to the ozone building. This valve provides a gastight shut-off of oxygen flowing into the ozone building in case of an emergency, such as a seismic disturbance. The ozone leak potential after automatic shutdown is limited to only that ozone contained under low pressure in the piping between the ozone generator and the contact basins.

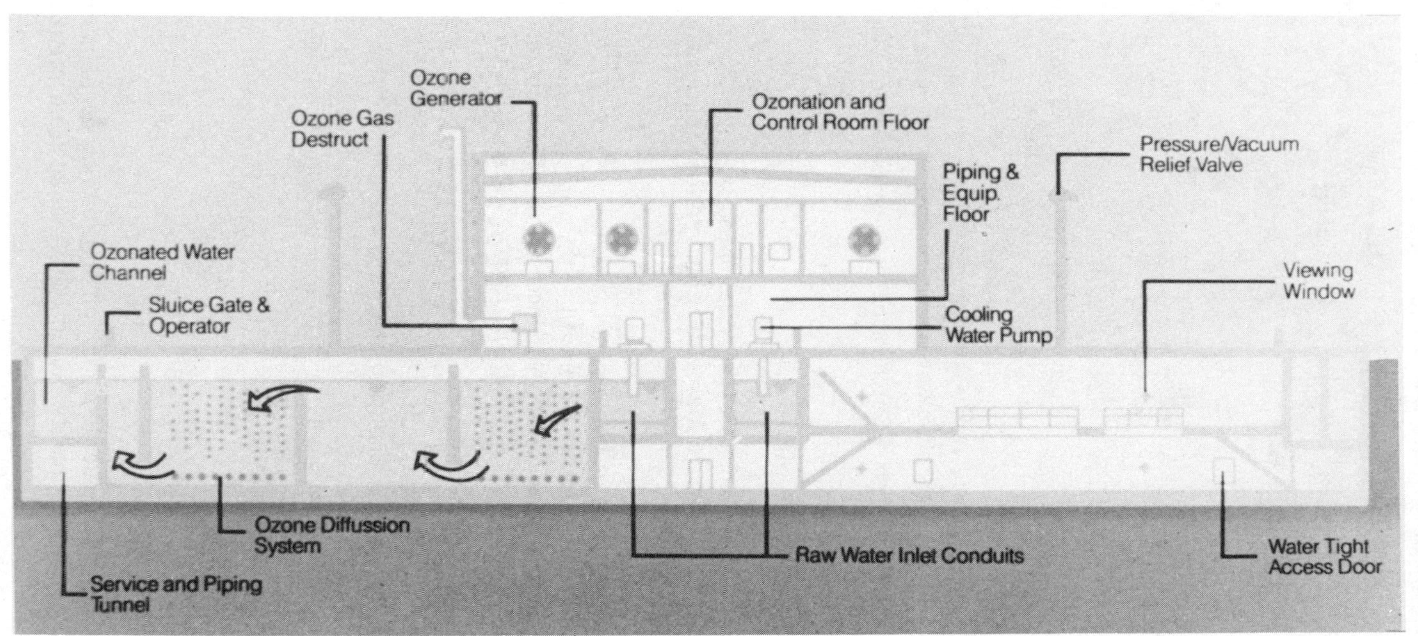

Fig. 15-31. Ozone building, sectional view (courtesy Los Angeles Department of Water and Power).

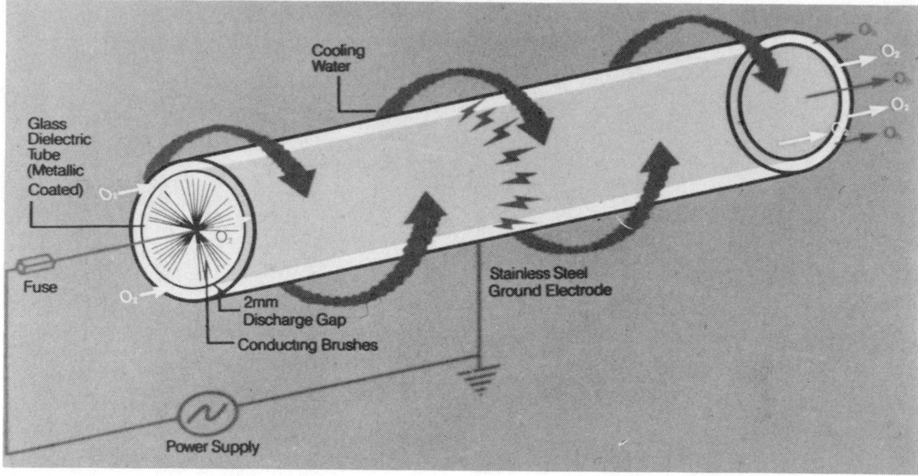

Fig. 15-32. Ozone generator tube unit schematic (courtesy Los Angeles Department of Water and Power).

Advantages of Preozonation

The application of ozone ahead of all other processes has, by this plant, demonstrated the following advantages over conventional treatment alternatives:*

1. It allows filtration rates to be increased by 50 percent (from 9 to 13.5 gpm/sq ft). This reduces by one-third the number of filters required.
2. It decreases the time necessary for flocculation by 50 percent (from 20 to 10 min.). This reduces by half the number of flocculation compartments required.
3. It increases filter run times between backwash cycles, thus reducing the size of backwash facilities required.
4. It reduces chemical coagulant demand by 33 percent.
5. It reduces chlorine demand by 50 percent. This in turn reduces THMs. The result is an annual saving of 1600 tons of chlorine per year that is normally used for in-aqueduct treatment.
6. It reduces filter backwash waste sludge in proportion to the reduction in coagulant usage.

Water Quality Benefits

In addition to its providing microflocculation and primary disinfection,† several other important water quality benefits are gained from preozonation.

*These advantages are a direct result of the enormous oxidation power of ozone.
†Ozone is a powerful viricide.

These include reductions in the levels of color, taste, odors, and the particulates remaining after filtration, which are detected by a particle counter rather than a turbidimeter.

Costs

The total project cost was $146 million. The total estimated capital savings from the use of preozonation as a pretreatment process was approximately $2 million. The average specific energy required to generate ozone is 7 kWh/lb of ozone. The oxygen and ozone generation facilities account for about 68 percent of the electrical power consumed at the filtration plant. It is estimated that this type of ozone application costs about $4 per million gallons of water treated with 1.0 mg/L of ozone. The elimination of in-aqueduct chlorination by the use of preozonation at the plant results in an annual saving of $0.2 million.

OZONIA®—THE CLEAN TECHNOLOGY

General Discussion

OZONIA®–North America, is part of the worldwide Ozonia organization, which also includes Ozonia International in Paris and Ozonia AG in Zurich. They are now the global leaders (1997) in the generation and application of ozone, as follows:[36]

1. Providing systems for both municipal and industrial customers, with units as large as 4000 lb/day of ozone.
2. With capabilities in both air and oxygen as well as oxygen/ozone recycle.
3. Possessing long-term experience inherited through Asea Brown Boveri, the pioneers in ozone generation and application.
4. Featuring an extensive research and development program.
5. Based in the United States, with engineering, support services, spare parts and manufacturing, all done in North America.
6. Supported by the expertise of Degremont, S.A. and Air Liquide.
7. Having municipal and industrial installations throughout the United States and Canada, as well as internationally. Thousands of pounds of ozone are produced every day by Ozonia's technology—safely, efficiently, and reliably. See Fig. 15-33.

Historical Background

Ozonia was established by Degremont, S. A. and Air Liquide to provide advanced ozone technology worldwide. It was founded to unite the expertise of Degremont in the application of ozone with the ozone technology purchased from Asea Brown Boveri, and the gas handling expertise of Air Liquide. This uniting of individuals, possessing a wealth of experience in ozone, enabled

Fig. 15-33. Ozone generator systems (courtesy Ozonia North America).

them to focus exclusively on providing the highest quality, most efficient, most reliable, and most innovative ozone system available in North America.

They supply a system of ozone generation equipment that includes all the major components and services necessary for a complete installation. They believe from their own experience that single source responsibility allows them to fully integrate all aspects of the process into a reliable and efficient ozone installation.

Ozone Gas

This is a naturally occurring triatomic form of oxygen (O_3) which, under atmospheric temperature and pressure, is an unstable gas that decomposes rapidly into molecular oxygen. It is a very powerful oxidant with a redox potential of 2.07, compared to chlorine gas at 1.36.

The considerable oxidizing power of ozone and its molecular oxygen by-products makes it highly desirable for the treatment of potable water.

Ozone has several significant advantages over the oxidants as follows:

1. It can be generated on-site.
2. It rapidly decomposes to oxygen, leaving no traces.
3. Reactions in potable water do not produce any toxic compounds that may be injurious to human health.
4. It reacts swiftly and effectively on many species of viruses.
5. In the pretreatment process it breaks down organic compounds caused by pollution of surface waters, allowing them to be easily removed.

How Ozone Works

It acts by direct or indirect oxidation, by ozonolysis, and by catalysis. The three major pathways occur as follows:

1. Direct oxidation reactions of ozone, resulting from the action of an atom of oxygen, are typical first-order high redox potential reactions.
2. In the indirect oxidation reactions of ozone, the ozone molecule decomposes to form free radicals (OH), which quickly react to oxidize organic and inorganic compounds.
3. Ozone may also act by ozonolysis, by fixing the complete reaction of double-linked atoms, producing two simple molecules with differing properties and molecular characteristics.

Water Treatment Applications

Based upon the history of ozone usage in the twentieth century, the various applications in water treatment include the following:

1. Elimination of taste and odor.
2. Color removal: It oxidizes both organic and inorganic compounds that discolor drinking water.
3. Potable water disinfection: It inactivates pathogenic viruses very quickly, along with some protozoa. However, its disinfection efficiency is restricted to its limited contact time, because when ozone is injected into water it decomposes rapidly into molecular oxygen. One of the most experienced users of ozone, Compagnie Genaral Des Eaux, Paris, claimed that the maximum possible contact time for O_3 in a water solution could never be greater than five minutes.[36]
4. Elimination of iron and manganese: It transforms ferrous and manganous salts into ferric and manganic hydroxide, which are removed by sedimentation and filtration.
5. TOC reduction, with ozone acting together with biologically activated carbon.
6. Reduction of organics: It partially oxidizes organic matter and can cause a 10 percent reduction in organic levels. It also acts in making significant improvement in water quality by reducing organic levels in filtration or clarification by initiating changes in the solubility of the colloidal organics.
7. Oxidation of trace organics and hydrocarbons: It oxidizes phenols, benzenes, and other aromatic hydrocarbons.

Ozone in Industry. The use of ozone in industry is primarily for color removal or just plain bleaching. This would include the pulp and paper industry, textiles, and kaolin clay bleaching. It is also used in cooling water treatment to prevent a buildup of algae slime on the towers and to prevent microbiological corrosion in the heat exchangers.

Ozonia O_3 Generators

Introduction. Ozonia pioneered in the industry with large-capacity, medium-frequency ozone generators. These generators are designed to provide the most efficient and reliable ozone generation commercially available. This system provides the highest level of power efficiency and the most advanced electrical systems with the highest quality materials of construction. A major factor in this achievement was the extensive research and development by Ozonia in the areas of ozone dielectrics, materials of construction, the ozone generation reaction, and power supply technology, together with close attention to construction techniques and cooling water needs.

The Ozone System. It is composed of five separate steps, as follows:

1. Feed gas preparation.
2. Ozone generation.
3. Cooling.
4. Ozone contacting.
5. Vent gas collection.

As illustrated in Fig. 15-34, the stages are separate subsystems, integrated into a single process by a centralized controller. The first step provides a contaminant-free, high purity feed gas. The second step generates ozone. The third step removes the excess heat produced by the ozone generation process. The fourth step contacts the ozonated gas stream with the material to be treated. In the fifth and last step, the residual ozone is removed from the vent gas before this gas is sent to the atmosphere or recycled.

These systems are complete from gas preparation to ozone contacting and ozone destruct equipment. Each system includes basic instrumentation, interlocks, and controls that can be completely automated in semiautomatic or fully automated PLC/PC based designs.

Ozone Generation. Ozone can be produced from oxygen in the atmosphere or from pure oxygen. This can be achieved by several methods; however, the silent electrical discharge process is the most common method.

Ozone is produced when a gaseous dry oxygen or air gas stream is subjected to a high voltage/high density electrical current, which provides the energy to drive the reaction. The field acts between two electrodes separated by a dielectric, forming a gap across which the energy discharge occurs. Ozone is formed by splitting molecular O_2 into two atoms, which recombine with the other oxygen molecules to produce ozone (O_3).

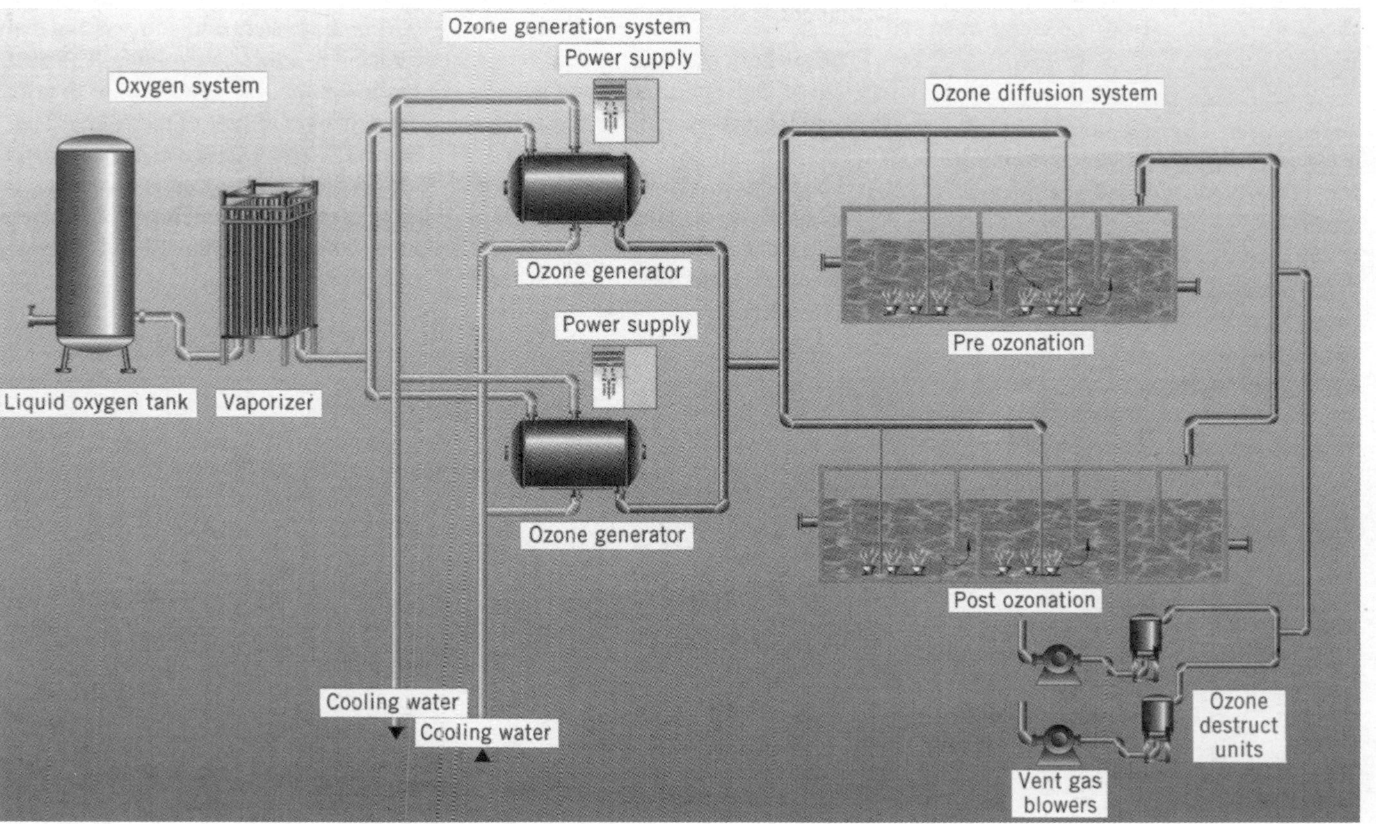

Fig. 15-34. Ozone generation system including oxygen supply with pre- and post-ozonation diffusion systems and vent gas blowers.

The principal elements of the ozone generating cell are: (1) an electrical source providing energy, (2) a discharge gap containing the oxygen-rich feed gas, (3) a dielectric material to prevent short-circuiting, and (4) a heat removal mechanism to dissipate waste heat, which is a by-product of the exothermic reactions that produce ozone.

Feed Gas—Air or Oxygen. To ensure efficient and reliable ozone generation, atmospheric air presents so many problems that have to be eliminated that it does not seem worthwhile. The air preparation process required to convert atmospheric air into feed gas to the ozone generator is as follows:

1. The air must be absolutely free of dust, oil, and any sort of particulates.
2. It must be completely dry—no water vapor.
3. It must be very close to ambient temperature.
4. It must be at a pressure suitable for the ozone generator.

When the feed gas is ambient air, the ozone generator performance is optimized at a concentration of 2 to 4 percent by weight. However, when the feed gas is oxygen, the optimum concentration is about 10 to 12 percent by weight. Therefore, it is easy to understand why oxygen from a cryogenic plant is the method of choice for generating ozone.

Ozone may be produced from oxygen stored on-site as liquid oxygen (LOX) or from oxygen production systems, which separate and condense oxygen and remove impurities and nonessential gases. The oxygen production system may be cryogenic air separation, a pressure swing adsorption (PSA) process, or a vacuum swing adsorption (VSA) process.

Cooling Water. The cooling water flow required for an ozone generator changes with the ozone production and cooling water temperature requirements. The generator is most efficient at cooling water temperatures of 700°F, or less, using glass type dielectrics. However, Ozonia's AT technology is not so sensitive to water temperature fluctuations. Typically, a maximum cooling water temperature rise of 100°F is used for the design.

As with any cooling water application, the water must be properly treated to minimize corrosion, scaling, and microbiological fouling of the water-side of the ozone generator.

External Gas Recovery. Oxygen is recovered from the contactor, stripped of residual ozone, dried, and purified. It is then recompressed and fed back into the ozone generator.

This process has been pioneered by Ozonia in a number of industrial applications. The technology allows the integrated oxygen/ozone system to reduce operating costs by recycling the majority of the oxygen.

Services. Ozonia manufactures its equipment at the Lodi, New Jersey corporate headquarters (FAX 201-778-2131). Sales, design, and engineering support are available from both the Lodi office and the Richmond, Virginia corporate office (FAX 804-756-0519).

Ozonia Advanced Technology (OZAT)

The Ozat Compact O_3 Generation Units.[37] Ozonia has developed and standardized a patented and proprietary new dielectric constructed of composite materials, which are more efficient, more stress-resistant, and operate at lower voltages than glass dielectrics. The last feature accounts for significantly less breakage during operation and maintenance when compared to glass. Moreover, this new advanced technology, with its modular construction, results in a very compact unit capable of producing high ozone concentrations for the economic production of ozone.

This revolutionary new technology allows for an increase in ozone concentrations typically from 2 to 4 percent in atmospheric air and 6 to 15 percent in oxygen. It has lowered operating costs 30 to 40 percent by using less power and oxygen feed gas. This new Ozonia AT system can be retrofitted into existing Ozonia systems.

These AT systems have an additional feature in that there is minimal impact from high cooling water temperatures, ensuring stable ozone production throughout the operating year and eliminating the need for costly chilled water systems.

The voltage of these new dielectric tubes is significantly lower than the voltage of conventional tubes, making them much more reliable. Since the new tubes are constructed of nonglass materials, they are high in endurance and shock-resistant, thereby reducing maintenance costs.

Key Improvements of OZAT. These are as follows:

1. Ozone concentrations in excess of 10 percent can be produced from oxygen feed gas.
2. The ozone yield for a given electrode area has been dramatically improved.
3. The specific energy consumption of the Advanced Technology has been reduced by up to 60 percent for equal ozone concentrations when compared with the conventional Ozonia technology, which is reknowned for low consumption.
4. Most important are the high-endurance, shock-resistant Advaned Technology (AT) dielectrics, with optimized properties and geometry.

OZAT O_3 Generators. These generators have been specifically designed and manufactured with the smaller applications in mind. With this new Ozonia product line, a completely new approach to ozone generation has been taken, which makes the very same technology, as used for the large installations, available to small consumers for the very economic production of ozone.

This OZAT system is produced in six unit sizes that range from 0.2 lb/day to a maximum of 162 lb/day of ozone. As the product is intended for all general applications, special attention has been given to design features ensuring that the product is absolutely reliable and easy to operate, requiring minimum maintenance and supervision.

These ozone generators are complete, fully factory-assembled units that can be easily integrated into all types of systems, old or new, with the minimum amount of installation time and space requirement. The complete apparatus is contained in a very compact enclosure, and depending upon the model in question, is either for wall, tabletop, or floor-standing installation.

Ancillary Equipment. There are a few pieces of this kind of equipment required to form a functional process, described as follows:

1. Feed gas equipment. Most state of the art ozone generators use oxygen as the feed gas. The preferred method is to use liquid oxygen. This can be purchased from gas companies, that is, delivered in special containers at regular intervals. This oxygen is of very high purity, and it is easy to maintain.
2. The contacting system, one of the most important accessories needed in the application of ozone. From past experience it is almost mandatory to use the porous diffuser method for getting the best mix of the ozone into the process flow.
3. Vent ozone destructors.
4. Water chiller/cooler unit.

Applications. Besides the usual applications for these small capacity units being used for the disinfection of potable water and wastewater, there are other important uses for these OZAT units, as follows:

1. Wafer etching in the semiconductor industry.
2. Disinfection of water used in the beverage industry.
3. Disinfection of air conditioning systems.
4. Disinfection of washing water in hospitals for patients, nurses, and doctors.
5. Advanced oxidation processes (AOP).

Contact. If your plant or project fits into the size range of the OZAT generators, and you want the latest in ozone generation, call Ozonia at 800-439-4040, or write to 2924 Emerywood Parkway, Richmond, Virginia, VA 23294 (FAX: 804-756-0519).

REFERENCES

1. Ogden, M., "Ozonation Today," *Ind. Water Eng.,* **7,** 36 (June 1970).
2. Rice, R. G., "Ozone for Treatment of Drinking Water," prepared for USEPA, Washington, DC, Contract No. 68-01-7289, 1991.
3. Rosen, H. M., "Use of Ozone and Oxygen in Advanced Wastewater Treatment," *J. WPCF,* **45,** 521 (Dec. 1973).
4. Rice, R. G., private communication, Dec. 1990.
5. Masschelein, W. J., "The Direct Action of Dispersed Ozonized Gas Bubbles," in *The Ozone Manual for Water and Wastewater Treatment,* John Wiley & Sons, New York, 1982.
6. Tate, C. H., "Overview of Ozone Installations in North America," presented at the AWWA Annual Conference in Cincinnati, OH, June 1990.
7. Rakness, K. L., Renner, R. C., and Hegg, B. A., "Ozone System Design for Water and Wastewater," in D. W. Smith and G. W. Finch (Eds.), *The Role of Ozone in Water and Wastewater Treatment,* Proceedings of the Second International Conference, Kitchener, Ontario, Canada, TekTran International, Ltd., pp. 135–152, 1987.
8. Hill, A. G., and Spencer, H. T., "Mass Transfer in a Gas Sparged Ozone Reactor," First International Symposium on Ozone for Water and Wastewater Treatment, International Ozone Institute, 1975.
9. Opatken, E. J., "Economic Evaluation of Ozone Contactors," in *Progress in Wastewater Disinfection Technology,* Proceedings of the National Symposium, Cincinnati, OH, EPA-600/9-79-018, NTIS No. PB-299338, USEPA, 1979.
10. Rouston, M., et al., "Mass Transfer of Ozone to Water: A Fundamental Study," *Ozone Sci.,* **2,** 337–344 (1981).
11. Design Manual: *Municipal Wastewater Disinfection,* Water Engineering Research Laboratory, Cincinnati, OH 45268, EPA/625/1-86/021, 1986.
12. Stover, E. L., et al., "High Level Ozone Disinfection of Municipal Wastewater Effluents," EPA Grant No. R804946, 1980.
13. Grasso, N., "The Effect of Gas Flow Rate to Static Liquid Volume Ratio on Disinfection in a Diffuser Bubble Ozone Contactor," master's thesis, Purdue University, May 1979.
14. Given, P. W., and Smith, D. W., "Pilot Studies on Ozone Disinfection and Transfer in Wastewater," in *Municipal Wastewater Disinfection,* Proceedings of Second National Symposium, Orlando, FL, EPA-600/9-83-009, NTIS No. PB83-263848, USEPA, Cincinnati, OH, 1983.
15. Venosa, A. D., et al., "Comparative Efficiencies of Ozone Utilization and Microorganism Reduction in Different Ozone Contactors," in *Progress in Wastewater Technology,* Proceedings of the National Symposium, Cincinnati, OH, EPA 60/9-79-018, NTIS No. PB-299338, USEPA, 1979.
16. Horst, M., "Removal of the Residual Ozone in the Air after the Application of Ozone," in *Ozone Manual for Water and Wastewater Treatment,* John Wiley & Sons, New York, 1982.
17. Orgler, K., "Methods and Operating Costs of Ozone Destruction in Off-Gas," in *Ozone Manual for Water and Wastewater Treatment,* John Wiley & Sons, New York, 1982.
18. Venosa, A. D., "Effectiveness of Ozone as a Municipal Wastewater Disinfectant," paper presented at preconference workshop, 56th Ann. Conf. WPCF, Atlanta, GA, Oct. 1, 1983.
19. Gordon, G., Cooper, W. J., Rice, R. G., and Pacey, G. E., "Methods of Measuring Disinfectant Residuals," *J. AWWA,* p. 94 (Sept. 1988).
20. Gordon, G., et al., "Disinfectant Residual Measurement Methods," AWWARF, Denver, CO, 1987.
21. Gordon, G., et al., "A Survey of the Current Status of Residual Disinfectant Measurement Methods for All Chlorine Species and Ozone," AWWARF, Denver, CO, 1987.
22. Grumwell, J., et al., "A Detailed Comparison of Analytical Methods for Residual Ozone Measurements," *Ozone Sci. Eng.,* **5,** 203 (1983).
23. Gordon, G., Rakness, K., Vornehm, David, and Wood, Delmer, "Limitations of the Iodometric Determination of Ozone," *J. AWWA,* p. 72 (June 1989).

24. "4500—O_3 Ozone Residual," *Standard Methods,* 17th Edition, pp. 4–162.
25. Hoigné, J., and Bader, H., "Bestimmung von Ozone and Chlordioxid in Wasser mit der Indigo Methode," *Vom Wasser,* **55,** 261 (1980).
26. Bader, H., and Hoigné, J., "Determination of Ozone in Water by the Indigo Method," *Water Research,* **15,** 449 (1981).
27. Stanley, J. W., and Johnson, J. D., "Amperometric Membrane Electrode for Measurement of Ozone in Water," *Anal. Chem.,* **51,** 2144 (1979).
28. Stanley, J. W., and Johnson, J. D., "Analysis of Ozone in Aqueous Solution," in R. G. Rice and A. Netzer (Eds.), *Handbook of Ozone Technology and Application,* Vol. I, Ann Arbor Science, Ann Arbor, MI, 1982.
29. Masschelin, W. J., "Continuous Amperometric Residual Ozone Analysis in the Tailfer Plant, Brussels, Belgium," in W. J. Masschelin (Ed.), *Ozonation Manual for Water and Wastewater Treatment,* John Wiley & Sons, New York, 1982.
30. "Ozone Concentration Measurement in a Process Gas by U.V. Absorption," International Ozone Association Standardization Committee—Europe 002/87 (F), c/o CIBE, 764 Chausee de Waterloo, B-1180 Brussels, Belgium.
31. Venosa, A. D., and Opatken, E. J., "Ozone Disinfection—State of the Art," paper presented at preconference workshop, 52nd Ann. Conf. WPCF, Houston, TX, Oct. 7, 1979.
32. White, G. C., personal inspection of the ozone and chlorination facilities at the Chas. J. Baillets WTP, Montreal, Quebec, Nov. 17, 1990.
33. White, G. C., private communication with instrumentation engineer, George Pence, at the Los Angeles Department of Water and Power 100 mgd Owens River aqueduct WTP, Oct. 23, 1990.
34. Stolarik, G., "Ozonation: Pretreatment to Filtration at the Los Angeles Aqueduct Filtration Plant," Los Angeles Department of Water and Power, Los Angeles, CA, June 1988.
35. Venosa, A. D., Rossman, L. A., and Sparks, H. L., "Reliable Ozone Disinfection Using Off-gas Control," *J. WPCF,* pp. 929–934 (Sept. 1985).
36. Dimitriou, M. A., "Ozone Technology and Equipment Design," Ozonia North America, 2924 Emerywood Parkway, Richmond, VA 23294, 1996.
37. Dimitriou, M. A., "The OZAT Compact Ozone Generation Units," Ozonia North America, 2924 Emerywood Parkway, Richmond, VA 23294, 1996.
38. Green, D., "Control and Monitoring the Ozone Disinfection Process by the Stranco High Resolution Redox Control System at Dusseldorf Germany," The Stranco Corp., Bradley IL, Customer Service 800-882-6466; 1995.
39. Cromer, R. B., Dissolved Ozone Monitor Model A15/64," ATI 680 Hollow Road, Box 879, Oaks, PA 19456, Customer Service 800-959-0299; July 1996.
40. Bauer, J., "Microprocessor-Based Dissolved Ozone Monitor Model 8422," EIT 252 Welsh Pool Road, Exton, PA 19341, Customer Service 800-634-4046; Nov. 1996.

16
BROMINE, BROMINE CHLORIDE, AND IODINE

BROMINE (Br_2)

Introduction

All bromine species used in water and wastewater treatment revert to bromides after being consumed in the oxidation process. This in itself is not an issue. However, when a potable water treatment plant chlorinates water containing bromides, they are oxidized to hypobromous acid and bromamines if any ammonia nitrogen is present. These compounds react with natural precursors in the water to form yet another series of trihalomethanes, which are considered to be carcinogenic. Therefore, the use of bromine as an alternative or a supplement to chlorination is not considered practical or acceptable from an environmental point of view. The consensus is that there are more than enough bromides occurring naturally in the environment to be dealt with, so that adding more bromides to the waterways only compounds the problem.

The following text on bromine is an attempt to describe what is known about its properties, characteristics, occurrence, water chemistry, and how it is used.

Occurrence

Bromine was discovered in seawater, in 1826, by Antoine J. Balard. It derives its name from its offensive odor: The Greek word *bromos* means stench. It does not occur in nature as a free element. It exists primarily in the bromide form, and is found widely distributed in relatively small proportions. Bromides available for extracting bromine occur in the ocean, salt lakes, brines or salt deposits left after these waters evaporated during earlier geological periods, and from the mineral bromyrite. The bromide content of seawater is about 70 mg/L. The total bromine content in the earth's crust is estimated at 10^{15} to 10^{16} tons.[1]

The Dead Sea in Israel is one of the richest sources of bromine in the world, containing nearly 0.4 percent at the surface and up to 0.6 percent at deeper levels. In Western Europe the most significant source is the salt deposits at Strassfurt, Germany. Principal sources in the United States are the brine wells in Arkansas (Arkansas produces about three-fourths of the national bromine output), Ohio, Michigan, and West Virginia, with a bromide content ranging from 0.2 to 0.4 percent. Those in Michigan underlie a large area of the Great Lakes region and occur in various sandstone strata at depths of 700–8000 ft. Here the bromide contents vary from 0.05 to 0.3 percent, and are generally higher in the deeper levels.

Bromine Production

The recovery of bromine from seawater was first achieved on a commercial scale in 1924 by the Ethyl Corporation.[2] This process involved treatment of the seawater with chlorine and analine. The first successful bromine plant was put into operation at Kure Beach, North Carolina, 1933, and was capable of extracting 3000 tons of bromine per year. In this plant, the process consisted of adjusting the pH of seawater to 3.5 with sulfuric acid, followed by the application of chlorine. The bromine, liberated by the chlorine, was removed as a dilute bromine gas with a current of air and absorbed in a sodium carbonate solution, from which it was recovered by acidification and stripping with stream. The critical part of this type of bromine extraction is control of the pH at 3.5.

Oxidation of bromide to bromine can be accomplished either chemically or electrochemically. The electrochemical methods are no longer significant for commercial production. Chemical oxidation can be effected by either chlorine compounds or oxygen-containing compounds such as manganese dioxide, bromate, or chlorate.

The extraction of bromine from bromide compounds requires four steps: (1) oxidation of bromide to elemental bromine (Br_2); (2) separation of the bromine from solution; (3) condensation and isolation of the bromine vapor; (4) purification. Current bromine production methods are based on the modified Kubierschky steaming-out process and the H. H. Dow blowing-out process.

Kubierschky Process. In this process the raw brine is preheated to about 90°C, treated with chlorine in a packed tower, and then placed in a steaming-out tower into which steam and additional chlorine are injected. The outgoing brine is neutralized with caustic and used to preheat the raw brine. From the top of the steaming-out tower, the halogen and steam vapor passes into a condenser and then into a gravity separator. Vent gases from the separator return to the chlorination system, the upper water layer containing Br_2 and Cl_2 is returned to the steaming-out tower, and the lower layer containing crude bromine passes on to a stripping tower. From the stripping column, bromine is purified in a fractionating column, which produces a 99.8 percent pure liquid bromine as the final product.

The H. H. Dow Process. This process utilizes air instead of steam for the "blowing-out" step in the extraction of bromine. It is a more economical extracting agent than steam, especially when the bromine source is as dilute as in seawater. In the process, the halogens are absorbed from the air in a sodium carbonate solution, or by sulfur dioxide reduction:[1]

$$Br_2\ (Cl_2) + SO_2 + 2H_2O \rightarrow 2HBr\ (2HCl) + H_2SO_4 \qquad (16\text{-}1)$$

Bromine can then be separated by chlorinating the mixed acids in the blowing-out tower. The theoretical yield is 2.2 tons of bromine per ton of chlorine.[3]

From 1973 to 1975, the estimated total annual bromine production in the United States was about 220,000 tons.[4]

In the years following World War I, the demand for bromine was for pharmaceutical bromides, the organic chemical industry, and photography. However, the biggest boon to the bromine industry was the discovery of tetraethyl lead as an antiknock ingredient in gasoline to accommodate the powerful high-compression automobile engines. But this ingredient posed a serious problem: deposits of lead in the engine. It was found that a mixture of ethylene dibromide and ethylene dichloride added to the tetraethyl lead was an excellent scavenger, which prevented lead deposition in the engine. These lead halides were sufficiently volatile to be expelled in the engine exhausts. It is estimated that about 70 percent of the 1973–75 production of bromine was used to make ethylene dibromide for gasoline. However, air pollution controls later led to a ban on the use of tetraethyl lead in gasoline, thereby wiping out the major market for bromine production. This might be a boon to the water pollution control industry. This will be discussed later in this chapter. Bromine at a lower price becomes a most interesting disinfectant, particularly for water reuse situations.

Physical and Chemical Properties

Bromine is a dark brownish red, heavy, mobile liquid. It gives off, even at ordinary temperatures, a heavy, brownish red vapor with a sharp, penetrating, suffocating odor. The vapor is extremely irritating to the mucous membranes of the eyes, nasal passages, and throat, and is extremely corrosive to most metals. Liquid bromine is likewise corrosive and destructive to organic tissues. In contact with the skin, it produces painful burns, which are slow to heal.

Bromine (Br_2; atomic number, 35; molecular weight, 159.832; specific gravity, 3.12) weighs 26.0 lb/gal, and has a boiling point of 58.78°C. Of the metals used to handle bromine, lead is the most versatile.[3] Bromine reacts with lead to form a dense superficial coating of lead bromide, which, if not distrubed, prevents further attack. This is similar to the reaction of chlorine and silver. Tantalum is completely resistant to bromine, wet or dry, at temperatures up to 300°F.

Nickel and its alloy, Monel, resist dry bromine and are especially useful as a material for shipping containers. Other nickel alloys, including the hastelloys, are less suitable. Iron, steel, cast iron, stainless steel, and copper are attacked by bromines, either wet or dry. Silver withstands dry bromine.

Bromine handled in lead, nickel, or Monel containers should be dry (less than 0.003 percent moisture)[3] and should be protected from ordinary air, from which it can readily absorb enough moisture to make it severely corrosive to these materials.

Bromine is three times as soluble as chlorine (i.e., 3.13 g/100 ml water at 30°C). This is an important characteristic when one is considering the physical aspects of applying bromine to a process stream. Dispersion and diffusion is made "easier," and diffuser design to prevent off-gassing is less of a problem than with chlorine.

Chemistry of Bromine in Water and Wastewater

Bromine is unique in being the only nonmetallic element that is liquid at ordinary temperatures. It reacts with ammonia compounds in solution to form bromamines and displays the breakpoint phenomenon similarly to chlorine.

Bromine in water hydrolyzes:

$$Br_2 + H_2O \leftrightharpoons HOBr + H^+ + Br^- \qquad (16\text{-}2)$$

for which reaction the equilibrium constant is 5.8×10^{-9}.

Depending upon the pH, the proportion of dissociation of hypobromous acid (HOBr) and hypobromite is:

$$\frac{[OBr^-][H^+]}{[HOBr]} = K = 2 \times 10^{-9} \qquad (16\text{-}3)$$

Like chlorine, bromine reacts with ammonia forming bromamine. Both Galal-Gorchev and Morris[5] and Johnson and Overby[6] have identified and studied the rate reactions of the compounds NH_2Br, $NHBr_2$, and NBr_3 using the ultraviolet absorption spectrophotometry technique. They reported rapid formation of all bromamine species; however, once the bromamines have formed, a series of decomposition reactions take place. The major chemical difference between the bromamine species and the chloramine species is that the formation of the bromamine species is reversible from monobromamine through dibromamine to tribromamine and back again by fast reactions, simply by changing the pH of the solution.

Bromine, like chlorine, displays a breakpoint, and it is the decomposition of the dibromamine that is the basis for this reaction. Tribromamine is the major species of combined residual bromine present beyond the breakpoint. In the pH range of 7 to 8 it decomposes in accordance with the following equation:

$$2NBr_3 + 3H_2O \rightarrow N_2 + 3HOBr + 3Br^- + 3H^+ \qquad (16\text{-}4)$$

La Pointe, Inman, and Johnson[7] have shown that the breakpoint occurs when the bromine-to-ammonia nitrogen molar ratio is 1.5. This is precisely the stoichiometric amount of bromine required to oxidize all of the ammonia to nitrogen gas.

For wastewater disinfection, it is of considerable practical significance that the predominant species of bromine compounds is dibromamine over a pH range of 7 to 8.5. This is so because dibromamine has a germicidal efficiency almost equal to that of free chlorine. Dibromamine is very active and usually displays a rapid decomposition, reverting to the bromide ion. At this point the bromine residual is extinguished. This feature of bromine reactions does not always occur in wastewater application, in spite of the presence of excess ammonia N. To date this aberration is thought to be due to the formation of stable organic bromamines resulting from the presence of organic N.

Free bromine residuals (hypobromous acid, HOBr), which would occur in highly nitrified effluents, do not decompose nearly as rapidly as the bromamines. Their persistence increases with decrease in halogen demand of the water being treated.

Reactions with Chlorine

The reactions with chlorine and the bromide ion are the only ones of particular interest. It is important to realize that free chlorine (HOCl) has the ability to oxidize bromide ions to form hypobromous acid (HOBr) in the pH range of 7 to 9. This phenomenon was the basis of a patent issued to Marks and Strandkov.[8] It was put to use in the application of chlorine to recirculated condenser cooling water at gasoline refineries in the 1950s.* Bromide salts were added to the cooling water in the atmospheric tower basin. This was followed by intermittent chlorination. The only chemical consumption was that of chlorine. The chlorine added converted the bromides to hypobromous acid. When the bromine residuals disappeared, they reverted to the bromide ion. Thus the intermittent chlorination kept repeating the process to control algae in the towers and tower basins and slime in the condenser tubes.

However, chloramine residuals will not oxidize the bromide ion at pH 4, but free chlorine will. This is the basis of a free chlorine residual analyzer developed by Fischer and Porter in 1968.[9] This analyzer does not suffer from any significant chloramine interference when high concentrations of combined residual are present with free chlorine. A bromide salt is added along with

*It was thought at the time that combined chlorine could oxidize bromide ions, which it cannot do. This was discovered during the investigations of THM formation by chlorination.

pH 4 buffer to the sample. The free chlorine oxidizes the bromide ion to free bromine,* which is measured by the analyzer cell.

Other applications of this phenomenon are described below.

USE OF BROMINE IN WATER PROCESSES

Potable Water

The use of free bromine (Br_2) in potable water is probably nonexistent.

The only known use in municipal potable water treatment was at Irvington, California about 1938. It was discontinued after a reasonable trial period because it did not solve the distribution system problem of water quality degradation. The bromine applied reacted so quickly and completely with the zoological slimes on the walls of pipes that it was impossible to obtain a residual downstream from the point of application. It also imparted a high-intensity medicinal taste to the water.[2]

As early as 1955, significant efforts were being made to produce solid or dry granular disinfectants using the best attributes of bromine. U.S. Patent 781,730 was issued to the Diversey Corporation of Chicago, Illinois for the invention of a stable dry product composed of hypochlorite and alkali metal bromide. This product was claimed to have extraordinary disinfectant properties when placed in aqueous solution due to the formation of the hypochlorite–hyprobromite mixture.

In 1967 and 1969 patents were issued to Jack F. Mills et al. of the Dow Chemical Co.[10,11] The invention described in these patents represents a process for treating water with elemental bromine obtained from the polybromide form of an anion exchange resin. An effective method for the preparation of this resin is to pass an essentially saturated solution of bromine in aqueous sodium bromide slowly up through a bed of quaternary ammonium anion exchange resin. The resulting polybromide resin in wet form contains about 48 percent bromine. Development of the polybromide resin system as a practical means for disinfection has been concentrated on units capable of treating small quantities of water—potable water for household use and swimming pools.[12] The Everpure Co. of Chicago has developed disposable cartridges containing bromine-impregnated resin to feed predetermined amounts of bromine into water for disinfection.[13] The polybromide resin is sealed permanently into the cartridge to prevent its escape into the water system. Polybromide resins with bromine loadings of 25 percent have a very low acute oral toxicity. Direct contact with undiluted 25 percent resin is only moderately irritating to the skin but is capable of producing uncomfortable irritation upon direct contact with the eyes.

The disposable cartridge-type brominator has been installed and operated aboard offshore oil well drilling rigs, at some remote land stations, and on ocean-going vessels that use seawater distilling systems as a source for their potable water supply.

---

*Free bromine is Br_2.

Wastewater

The use of bromine in wastewater or water reuse situations is unknown in the United States or Canada as of 1997. The only use that would be acceptable would be in a wastewater discharging into seawater. In the early 1980s Dow Cemical Co. made a test project of bromine for disinfection of the East Bay Municipal Utility District WWTP effluent that discharges into the San Francisco Bay on the left-hand side of the highway entrance to the San Francisco Bay Bridge en route to San Francisco. The project failed because of the high oxidative power of the bromine. In other words the bromine demand was "out of site." The required dosages were cost-prohibitive.

However, when a wastewater plant is near enough to a seawater supply, such as the California coast, if it is convenient, the seawater can be used to operate the chlorinator injectors. The 60–80 mg/L of bromides in the seawater will be converted to bromamines in the chlorine solution. As the oxidative power of bromamines is equivalent to that of free chlorine (HOCl), the total effect by a comparative test at the Santa Cruz, California WWTP was a disinfection efficiency increase of about 15 percent.

Cooling Water

About 1983 the electric power industry began an investigation using bromine in condenser cooling water treatment. The objective was to determine whether or not dechlorination could be avoided by using bromine, owing to the expected rapid decay of bromine residuals. The system adopted by some of the steam generating plants amounted to pumping a bromide salt solution into the chlorine solution discharge of existing conventional chlorination equipment. The chlorine oxidizes the bromide ion in the salt solution to free bromine, which goes into solution as hypobromous acid, the active ingredient. Whether or not the resulting bromine residuals will decay fast enough to meet the NPDES discharge requirements is unknown at this time.

Swimming Pools

The most widespread use of bromine as a disinfectant began in Illinois during World War II. The Illinois State Department of Health, under the direction of C. W. Klassen, began an investigation of the use of liquid bromine in the elemental form as a substitute for chlorine when the latter became scarce during the war yeras. After a trial run in a few swimming pools for a couple of years, it was concluded by bacteriological evidence that it was doing an efficient job of disinfection. Permission to use elemental bromine was granted to a number of additional pools in 1947, and before the end of that year there were about 25 outdoor and 30 indoor pools where bromine was applied instead of the difficult-to-obtain chlorine. However, in order to avoid the hazards of handling liquid bromine, bromine in stick form was introduced about 1958.

The stick is a compound of both bromine and chlorine known as bromo-chloro-dimethyl hydantoin (Dihalo). The use of this chemical was documented by Brown et al.[14] It is thought that this compound, which hydrolyzes to form hypobromous acid, has an additional capability in that it also releases some HOCl by hydrolysis, which reacts with the reduced bromides to form more hypobromous acid. This form of bromine can produce satisfactory results for indoor and very small outdoor swimming pools. The high cost of the bromine sticks makes its use on large outdoor pools decidedly uneconomical.

Br_2 Facility Design

Current practice is limited to the use of bromine in stick form or resin impregnation systems described above for swimming pool and potable water treatment. Neither of these methods is practical for wastewater or water reuse situations. The hazards of handling and the difficulties in metering molecular bromine (liquid or vapor) make it impractical to use. These difficulties stem from the unique characteristic of bromine that allows it to be in the liquid form at room temperature and pressure. This is one of the major reasons why BrCl is being pursued as a practical means for the bromination of wastewater. BrCl has a vapor pressure of 30 psi at room temperature (20°C). This simplifies the packaging and handling problems.

Owing to the undesirable handling characteristics of molecular bromine, there can be no valid recommendations for a facility to handle molecular bromine.

BROMIDES: ON-SITE GENERATION OF Br_2

System Description

The on-site oxidation of bromide salts in an aqueous solution to form free bromine (Br_2) is an old concept. In 1948, a patent was issued to Marks and Strandkov of Wallace & Tiernan Co. that involved on-site production of either iodine or bromine by the oxidation of the respective salts by either free or combined chlorine at pH 7–8.[8] Chemically the iodine release was predictable and quantitative, but the bromine release was not quantitative or predictable at this pH. The iodine system failed because control was difficult. However, since then there has been a renewed interest in the on-site generation of bromine from bromide for disinfection of reclaimed water. A patent was issued in 1973 to A. Derreumaux (France) for such a system.[15] This system is different from the one proposed by Marks in that the bromide salt solution is oxidized by a chlorine solution at a carefully controlled pH of 1–2. It conforms to all the established kinetics of bromine chemistry. Free bromine is produced in the Derreumaux process by injecting a bromide salt solution of known concentration into the injector discharge of a conventional chlorinator. The reaction between the chlorine and bromide ion to convert the bromide to hypobromous acid takes place in a reactor where the pH for the reaction is kept below 2. The reaction proceeds as follows:

$$Cl_2 + 2NaBr \rightarrow 2NaCl + Br_2 \tag{16-5}$$

Thus 2.90 lb of pure sodium bromide (NaBr) will react with 1.0 lb chlorine to produce 2.25 lb bromine (Br_2).

The patented Derreumaux system is illustrated in Fig. 16-1. It consists of conventional chlorination equipment, a reaction vessel, and a metering pump for the addition of the bromide salt. The success of this system is its ability to consistently produce an almost pure solution of hypobromous acid at a controlled pH of less than 2. This ensures maximum quantitative conversion of the bromide ion in accordance with Eq. (16-5). In cases where the chlorinator injector water is plant effluent (i.e., high NH_3 concentration), there will not be any chlorine lost in side reactions with ammonia nitrogen at this low pH and with the reaction time involved. After the bromide is oxidized to free bromine, it will remain as Br_2. Therefore, at this controlled low pH in the reactor zone, the free bromine will not be wasted on side reactions with the ammonia nitrogen present.

The most important feature of this process is the sequential addition of chlorine followed by bromine. In practice the chlorine is added first to satisfy the immediate halogen demand (3–5 min. contact time) with a conventional

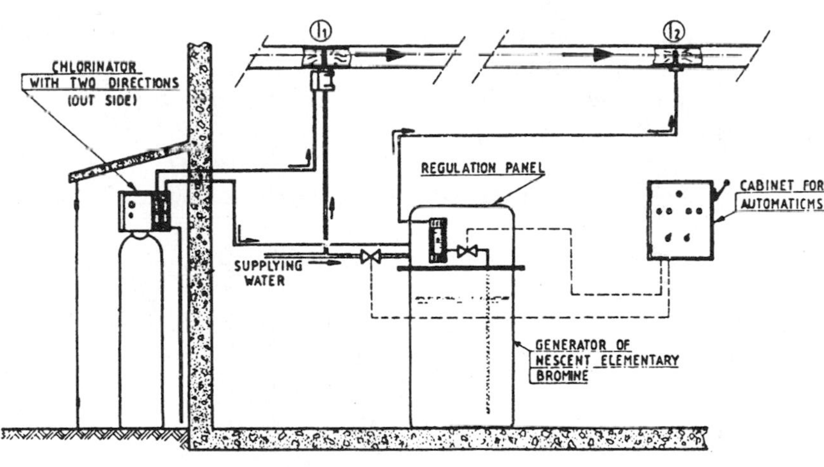

Fig. 16-1. The Derreumaux bromine generating system (courtesy CIFEC Co.). 1, First point of injection, chlorine water; 2, second point of injection, chlorine and nescent bromine.

compound loop residual control system. Then bromine is added on a flow-proportional basis relying on manual control of the bromine dosage.

This procedure serves three purposes: (1) it reduces significantly the amount of bromine required; (2) it improves the disinfection efficiency;[16] (3) it solves the problem of trying to cope with bromine demand changes normally found in wastewater effluents. Field experience has shown that owing to the rapid die-away of bromamine residuals, it is not practical to control the dosage by residual control.[17] In a comparative situation it was shown that chlorine is far more able than bromine to meet the diurnal variations in halogen demand, particularly those caused by the usual unpredictable and uncontrollable plant upsets. This is to be expected because of the much greater reactivity of bromine with the various constituents in wastewater.

This method has been used in France for both wastewater and reclaimed water treatment.

The presumed advantages of bromine in wastewater are considered to be: (1) the presence of ammonia N in wastewater converts the bromine applied to the dibromamine, which is nearly as germicidal as free chlorine; (2) dibromamine, the dominant species, is so highly reactive with seawater constitutents that the residual should die away rapidly;* (3) owing to these two characteristics, contact times may be reduced from current chlorination practices of 45–60 min. to perhaps 20–30 min., thus reducing debromination requirements if necessary.

However, the process must be cost-effective. Most of the bromide salts marketed in the United States and Canada are very high-grade and are used primarily in photographic emulsions; so their cost is relatively high—in excess of 70¢/lb for available bromine.

Dow Chemical produces a low-grade calcium bromide solution (specific gravity of 1.7–1.72 g/ml); (2) an 80 percent calcium bromide flake. The bromide ion content in the solution is about 42–43 percent and in the solid form is about 64 percent. Assuming the following reaction:

$$Cl_2 + CaBr_2 \rightarrow Br_2 + CaCl_2 \tag{16-6}$$

it would take approximately 2.82 lb of pure calcium bromide plus 1 lb of chlorine to make 2.25 lb of bromine.

Tank-truck lots of the solution cost 28.5¢/lb in 3500 lb lots FOB Ludington, Michigan (1977 prices).

Current U.S. Practices

The only known installations are those used by the electric power industry. Several power plants are experimenting with this method of on-site generation

*Owing to the presence of significant concentrations of organic *N* (3–10 mg/L) in wastewaters, organic bromamines will form as their chlorine counterparts always do. This species persists as a measurable residual for extended periods of time (several hours).

of bromine with the hope that they can escape the need to dechlorinate the cooling water discharge. These discharges usually contain chlorine residuals in the range of 0.5–2.0 mg/L total residual chlorine. The hope is based upon the rapid die-away phenomenon of bromine residuals. The success or failure will depend upon site-specific conditions, which will vary depending upon the water quality at each location. The two most significant water quality parameters in addition to halogen demand are: (1) ammonia N and (2) organic N concentration. If sufficient ammonia N is present to convert all of the residual to bromamine, and if there is insufficient organic N (and contact time) to form organobromamines, the residuals should die away in a few minutes.

From the power industry's point of view, it is much more appealing to retrofit an existing chlorination sytsem to on-site generation of bromine than it is to install a dechlorination facility. From an environmental point of view, it would be more appropriate to relax the intermittent chlorine residual discharge requirements than to force a situation where an alternative will contribute to adding more bromides to the environment.

Comparison with Other Methods

This system of on-site generation of free bromine has the following advantages and disadvantages over the application of molecular bromine and bromine chloride:

Advantages:

1. It can be adapted to conventional chlorination equipment and controls.
2. Conventional metering equipment (diaphragm pumps) can be used for feeding the bromine salt solution.
3. It is readily adaptable to conventional compound loop control utilizing existing chlorine residual analyzers.
4. The equipment specific for this process has been proved satisfactory by sufficient field experience.
5. It eliminates hazards of handling liquid molecular bromine or bromine chloride.
6. It eliminates equipment problems of corrosion and the chemical problem of dissociation encountered in the use of bromine chloride and molecular bromine.

Disadvantages:

1. The major disadvantage of this process is chemical cost. The least expensive high-grade bromide salt costs about $0.70/lb in 500-lb lots (1976). The low-grade $CaBr_2$ may be the answer.
2. The availability of bromine compounds is limited. However, the mandatory elimination of leaded gasoline for automobile engines in the United States eliminated the production of bromine compounds for use as lead scavengers in gasoline. (In 1973, 70 percent of the total U.S. bromine production was for this purpose.)

3. The reaction of bromamines in wastewater effluents with organics and the subsequent formation of undesirable bromoorganics is a potential hazard.
4. The end product of bromination—bromide ion—can be a potential hazard to the environment.
5. There is insufficient field experience for proper evaluation.

BROMINE CHLORIDE, BrCl

Physical and Chemical Properties

Bromine chloride is classified as an interhalogen compound because it is formed from two different halogens. These compounds resemble the halogens themselves in their physical and chemical properties except where differences in electronegativity are noted. Bromine chloride at equilibrium is a fuming dark-red liquid below 5°C. It can be withdrawn as a liquid from storage vessels equipped with dip tubes under its own pressure (30 psig 25°C). Liquid BrCl can be vaporized and metered as a vapor in equipment similar to that used for chlorine.

Bromine chloride is an extremely corrosive compound in the presence of low concentrations of moisture. Although it is less corrosive than bromine,[18] great care must be exercised in the selection of materials for metering equipment. Like chlorine and bromine, it may be stored in steel containers—assuming that the BrCl or Br_2 is packaged in an environment when the air is dry (i.e., a dew point not higher than −40°F).

When in contact with skin and other tissues, liquid BrCl, like Br_2, causes severe burns. Low concentrations of the vapor are extremely irritating to the eyes and respiratory tract.

Bromine chloride exists in equilibrium with bromine and chlorine in both gas and liquid phases as follows:[6]

$$2BrCl \leftrightharpoons Br_2 + Cl_2 \tag{16-7}$$

There is little information on the equilibrium in the liquid state. A study by Mills[19] and an earlier investigation by Cole and Elverum[20] indicate less dissociation in the liquid (20 percent) than in the vapor state. The equilibrium constant for the vapor phase dissociation of BrCl is close to 0.34, which corresponds to a degree of dissociation of 40.3 percent at 25°C.[1,20]

The density of BrCl is 2.34 g/cc at 20°C.

Figure 16-2 compares the vapor pressure–temperature curves for bromine chloride, bromine, and chlorine.

The solubility of BrCl in water is 8.5 g/100 cc at 20°C. This is 2.5 times the solubility of bromine and 11 times that of chlorine.[1]

Bromine chloride forms a yellow crystalline hydrate, $BrCl \cdot 7.34\ H_2O$ at 18°C and 1 atm. This compares to the formation of chlorine hydrate ($Cl_2 \cdot 8H_2O$), which forms at 9.6°C. This is a significant characteristic because the formation of these hydrates causes operational problems in metering equipment.

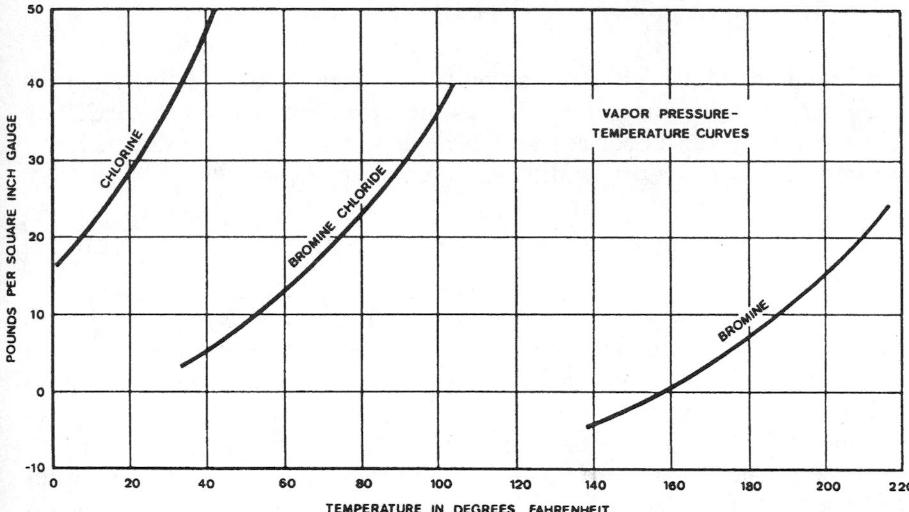

Fig. 16-2. Vapor pressure versus temperature curves from bromine, bromine chloride, and chlorine (courtesy Capital Controls Co.).

Preparation of Bromine Chloride

Bromine chloride is prepared by adding an equivalent amount of chlorine (as a gas or a liquid) to bromine until the mixture has increased in weight by 44.3 percent:

$$Br_2 + Cl_2 \leftrightharpoons 2BrCl \qquad (16\text{-}8)$$

Bromine chloride may also be prepared by the reaction of bromine in an aqueous hydrochloric acid solution. In the laboratory it can be prepared by oxidizing a bromide salt in a solution containing hydrochloric acid. This produces the following reaction:

$$KBrO_3 + 2KBr + 6HCl \rightarrow 3BrCl + 3KCl + 3H_2O \qquad (16\text{-}9)$$

Chemistry of Bromine Chloride in Water

Bromine chloride vapor appears to hydrolyze exclusively to hypobromous acid and hydrochloric acid:

$$BrCl + H_2O \leftrightharpoons HOBr + HCl \qquad (16\text{-}10)$$

whereas bromine vapor (or liquid) hydrolyzes to hypobromous acid and hydrogen bromide:

$$Br_2 + H_2O \leftrightharpoons HOBr + HBr \tag{16-11}$$

The formation of HBr represents a significant loss in the disinfecting potential of the expensive bromine molecules. In the hydrolysis reaction of BrCl, any HBr formed by the dissociation of elemental bromine is presumed to be oxidized quickly to HOBr by the HOCl remaining in solution:

$$HBr + HOCl \rightarrow HOBr + HCl \tag{16-12}$$

However, in the case of wastewaters, the ammonia nitrogen present would immediately convert any HOCl in solution to chloramines in near-neutral pH environments. Chloramines cannot oxidize any HBr formed by hypobromous acid (HOBr) at these pH levels.[21] Therefore, the problem of dissociation of bromine chloride is significant in the presence of ammonium ions.

It is of practical significance that the hydrolysis constant for BrCl in water is 2.94×10^{-5} at 0°C, compared with the same constant for chlorine which is 1.45×10^{-4} at 0°C. It is paradoxical that BrCl is several times more soluble in water than chlorine, yet it hydrolyzes 10 times more slowly. The hydrolysis constant of molecular bromine is 0.7×10^{-9} at 0°C, which is significantly different from that of bromine chloride.

Bromine chloride combines with ammonia in the same manner as molecular bromine to form bromamines. At the usual pH levels encountered in wastewaters (7–8.5), the dominant species will be dibromamine.[24] Typical reactions are as follows:

$$NH_3 + HOBr \leftrightharpoons NH_2Br + H_2O \tag{16-13}$$

$$NH_2Br + HOBr \leftrightharpoons NHBr_2 + H_2O \tag{16-14}$$

$$NHBr_2 + HOBr \leftrightharpoons NBr_3 + H_2O \tag{16-15}$$

Bromine chloride is presumed to have a higher speed of reactivity than bromine. In wastewater, the formation of bromamines is probably much faster than the formation of chloramines. Of greater practical significance is the rapid die-away of the bromamine residuals. The half-life of bromamine residuals in secondary effluents is about 10 min. If organobromamines are formed, the half-life is much longer. In highly nitrified effluents the predominant species would be HOBr. This species of bromine residual persists for a significant length of time in a low-halogen-demand environment.*

*An example of this is the practice of power plant condenser cooling water chlorination, where seawater is used. If the ammonia nitrogen content is less than 0.2 mg/L and the chlorine dose is 1–2 mg/L, the chlorine applied immediately converts (stoichiometrically) an equivalent amount of the bromide ions present into HOBr. The resulting bromine residual persists in the condenser discharge plume nearly as long as would comparable chlorine residuals.

Bromine Chloride Facility Design

Current Practice. The first attempt at plant-scale use of BrCl was at Grandville, Michigan[22] in 1974 and 1975. This was an EPA-sponsored project intended to evaluate several wastewater disinfectants. It was not a strict exercise in disinfection because the coliform reduction ratio was too low. The MPN of total coliforms before disinfection ranged between 8×10^4/100 ml and 3×10^6/100 ml. The disinfection requirement was to achieve 10^3/100 ml MPN.* According to the model developed by Collins et al.,[23] the required chlorine residual contact time for the destruction of the lower range of coliform (8×10^4) is 7 and for the higher range (3×10^6) is 23.5. Therefore, if mixing at the point of application is adequate, and if 30 min. of contact time is available, assuming 80–90 percent plug-flow conditions in the contact chamber, the predicted chlorine residual requirement would range from 0.23 to 0.8 mg/L to meet the 1000/100 ml total coliform disinfection requirements. For comparable quality effluents this level of chlorine residual is considered very low.

The chlorine dosage at the beginning of this project was 2.9 mg/L, resulting in a 2.0 mg/L residual at the end of 30 min. This demonstrates a chlorine demand of 0.9 mg/L in 30 min, which appears to be unreasonably low for such an effluent.

At the middle of the project the chlorine dosage was lowered to 2.3 mg/L, which produced a 1.0 g/L residual at the end of 30 min. This residual compares to the mathematical model prediction of 0.8 mg/L for a total coliform concentration before chlorination (y_0) of 3×10^6/100 ml. This is the high range of y_0 for the above project.

The BrCl dosage was lowered from 3.6 to 3.0 mg/L early in the project, and about the middle of the project it was lowered to 2.0 mg/L, where it remained for the rest of the project. Because of the rapid die-away of the BrCl residual, this process could only be controlled by dosage, whereas the chlorination process was controlled by residual.

The performance of BrCl was disappointing. Chlorine succeeded in meeting the requirements on the Grandville project 90 percent of the time, but BrCl met the requirements only 80 percent of the time, largely because of equipment failures. The reasons are described later in this chapter.

BrCl System Evaluation. The Potomac Electric Power Company studied and compared two parallel systems at the Morgantown Steam Electric Station.[25] This is an 1100-MWe, fossil-fuel, two-unit generating facility using low-salinity (7000 mg/L) estuarine water for once-through condenser cooling. One system used chlorine, and the other used bromine chloride. Both disinfectants were applied at levels of 0.5 mg/L or less. Two properties of bromine chloride were examined: decay rate of bromine residuals and biofouling control effectiveness. The BrCl meeting and control unit is shown in Fig. 16-3.

*Good disinfection is usually considered as a 5-log reduction of initial total coliform content.

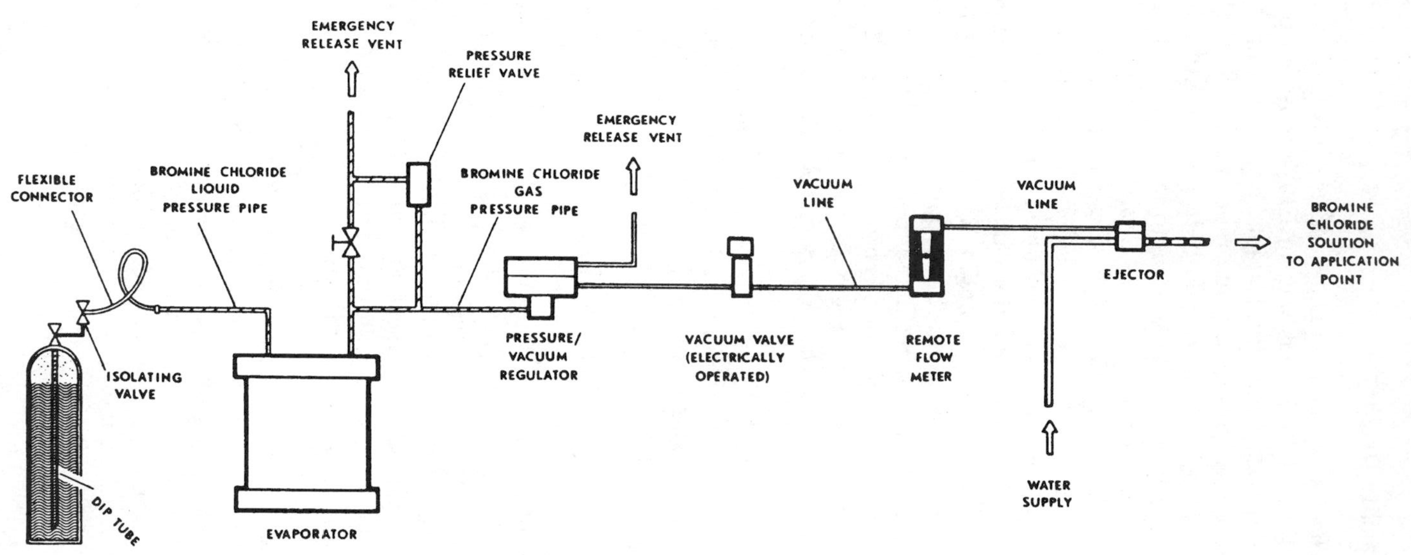

Fig. 16-3. Bromine chloride metering and control equipment (courtesy Capital Controls Co.).

On an equal weight basis, the continuous low-level application of BrCl resulted in equally good control of biofouling in the condensers. On an equimolar basis, therefore, BrCl would be nearly twice as effective but would cost about 2½ times as much as chlorine at current market prices.

The flaw in this investigation stems from the fact that the chlorine applied converted the bromide ion* to HOBr, which reacted with the ammonia N present (0.25–0.50 mg/L) to produce bromamine. This would explain the nearly equal effectiveness of the two halogens on an equal weight basis.

The BRCl and chlorine residuals dissipated in three decay phases. A very rapid residual loss occurred in the first phase, which lasted less than 1 min. Quasi first-order decay kinetics characterized the second phase, which lasted about 10 min. In the third phase the decay was relatively slow, more than 45 min. The BrCl-induced oxidants dissipated faster than chlorine-induced oxidants. The effluent of the chlorobrominated cooling water contained only two-thirds to one-half the amount of oxidant present in the chlorinated cooling water, even though the BrCl and chlorine were applied at equimolar concentrations.[25]

The report on this parallel study of BrCl and chlorine recommended the consideration of BrCl as an alternative to chlorine for biofouling control. This recommendation was based upon the ability of BrCl to control biofouling at low-level concentrations (less than 1 mg/L) and its relatively rapid residual dissipation compared to chlorine. Whether or not the use of BrCl will eliminate the need for debromination to meet NPDES discharge requirements has not yet been determined.

Metering and Control Equipment. A limited field experience quickly revealed that a new species of equipment would have to be designed to solve the complex problems of construction materials that could stand up to the corrosivity of bromine chloride.

At the Grandville, Michigan project[22] the equipment chosen was a modified Wallace & Tiernan pressure gas-feed chlorinator fitted with an injector. This model (No. 20-055) was chosen because it was constructed of materials thought to be able to withstand bromine chloride. Many problems concerning corrosion and dissociation of the vapor were encountered, requiring various modifications to the equipment.

At about the same time the Public Service Electric and Gas Co. of New Jersey entered into a project at one of its power stations to evaluate the difference between chlorine and bromine chloride using conventional chlorination equipment for both gases on a side-by-side installation. It took only a few hours' operation to observe that conventional chlorination equipment is not suitable for bromine chloride.[26] Most of the plastic and rubber materials were found to be entirely unsuitable because of permeation and/or direct

*There were sufficient bromide ions in the estuarine water to effect this conversion.

chemical attack, leading to nearly complete failure of this material. A successful experimental model was constructed on the direct gas-fed approach using slightly modified current production components such as a stainless steel indicating rotameter, chlorine line valves, and chlorine pressure-reducing valves. Kynar pipe and valves were used adjacent to the injector where moist BrCl vapor might be encountered. The "suck-back" phenomenon was dealt with by using a modified chlorine check valve adjacent to but not part of the injector assembly.

Figure 16-3 illustrates a system subsequently developed by Capitol Controls, Colmar, Pennsylvania. This system is essentially the same concept as the Public Service Co. experimental model described by Cole.[26] Figure 16-3 represents the system provided by Capital Controls Co. for the Potomac Electric Power Co. investigation.

Injectors. The injectors for use with BrCl may require modifications. The one used on the project described by Cole[26] was a 3-inch Wallace & Tiernan Model A452. This unit was about 10 years old and did not require any modification. The one used on the Grandville project did require some modification, but that used by Capital Controls on the Potomac project was a standard chlorine injector.

Automatic Controls. The system shown in Fig. 16-3 is readily adaptable to automatic control by simply automating the control valve.

Conventional amperometric analyzers may be calibrated to record bromine residuals, but because of the evanescent characteristics of bromine residuals, automatic residual control might not be practical. Field experience is too limited to permit a judgment on this facet.

Solution Lines and Diffusers. The material and the design for this part of the system can be the same as used for chlorine.

Storage and Handling. As in all liquid–vapor exchange chemical systems, the problem of reliquefaction is of importance. This problem increases with decreasing vapor pressure equilibrium of the liquefied gas. Sulfur dioxide reliquefies much more easily than chlorine because of its low vapor pressure (i.e., 35 psi at 68°F). This compares with a vapor pressure of 30 psi at 68°F for BrCl. In the case of BrCl it is more significant because dissociation occurs during reliquefaction, and all of the chemical advantages of bromine chloride are lost.

Reliquefaction can be prevented by either raising the temperature of the liquid–vapor container (bromine cylinder) or by external pressure padding with compressed air or nitrogen.

Padding the container to about 50 psi and withdrawing the liquid to an evaporator, which is similar to the practice of handling sulfur dioxide, is a preferred procedure. This requires that the bromine chloride containers be fitted with dip tubes and outlet valves similar to chlorine or sulfur dioxide ton containers. One connection would be for pressurization of the vapor and the other for liquid withdrawal. If this were the case, then the container storage and handling system would be the same as for chlorine or SO_2.

Evaporators. It is generally conceded that the most practical way to handle bromine chloride is to withdraw the liquid and vaporize it under controlled conditions.

Evaporators used for both chlorine and sulfur dioxide can be used without modification for bromine chloride. The piping arrangement and accessories would be the same as for chlorine. The downstream reducing valve would be set at 25 psi and the containers pressurized to 50 psi. The evaporator should be large enough to provide 30°F superheat.

Materials of Construction. Although bromine chloride is much less corrosive than bromine,[18] great care must be taken in the selection of materials throughout the design of this facility. The material for handling the liquid and/or the vapor under pressure is the same as for chlorine. Therefore, Sch. 80 seamless steel pipe can be used between the containers and the evaporators, and between the evaporators and the metering system.[26]

Bromine chloride vapor or liquid cannot be handled in PVC or ABS plastics, even at very low pressures, as can chlorine and sulfur dioxide. The preferred piping material for moist BrCl vapor is Kynar, a vinylidenefluoride resin. Similarly, Kynar valves should be used when required with this piping material.

Safety Equipment. Chlorine leak detectors can be readily adapted to detect BrCl in the atmosphere.

Chlorine container kits can be used on BrCl containers if the BrCl is packaged in containers of similar design.

Comparison with Other Methods

Advantages:

1. Use of BrCl eliminates the hazards of handling Br_2.
2. It is presumed to have greater germicidal efficiency than Br_2 and chloramines.
3. At the 1976 price of 20¢/lb it is cheaper than on-site generated bromine.
4. Contact time should be less than for chlorine; so contact chambers would be smaller.

Disadvantages:

1. Equipment for metering and injection of BrCl is still in the experimental design stage.
2. There is limited field experience.

3. If it is used alone, residual control will be difficult owing to rapid die-away of bromine residuals.
4. It is subject to reliquefaction, which causes dissociation of the vapor. This defeats the purpose of using bromine chloride.

Costs. The following is the estimated cost for wastewater disinfection using bromine chloride as determined by the 1976 EPA Task Force Report:[27]

Plant size (mgd)	1	10	100
Capital cost ($)	47,000	129,000	414,000
Disinfection cost (¢/1000 gal.)	4.52	3.04	2.65

GERMICIDAL EFFICIENCY (Br_2 AND BrCl)

A review of the literature from 1945 to 1976 revealed a considerable lack of agreement on the germicidal efficiency of bromine compounds; that is, free bromine (Br_2), hypobromous acid (HOBr), monobromamine (NH_2Br), and dibromamine (NHB_2). The major difficulty in sorting out the literature was to determine which species was in fact being investigated. This has to do with bromine chemistry. The publication edited by Johnson[28] is the most informative on the germicidal chemistry of bromine. The most informative investigations of the germicidal efficiency of bromine are those by Wyss and Stockton,[29] Marks and Strandkov,[30] Johannesson,[31] McKee et al.,[32] Koski & Stuart,[33] Schaffer and Mills,[34] Sollo et al.,[35] and Johnson and Sun.[24]

Although this work was done in the laboratory under ideal conditions, the following pertinent conclusions can be made with respect to wastewater disinfection: In a wastewater effluent containing an ammonia nitrogen content of 10–30 mg/L and a pH from 7 to 8, the dominant species of bromine will be dibromamine.[24] Dibromamine appears to be almost as germicidal as hypobromous acid (HOBr) but not as germicidal as free bromine (Br_2), which only exists below pH 7. Therefore, the germicidal efficiency of dibromamine is almost comparable to that of free chlorine (HOCl), as it has been demonstrated that hypobromous acid and hypochlorous acid have about the same germicidal efficiency.[6,24,28] This is significant in wastewater treatment because the application of bromine to wastewater will result in the formation of dibromamine, which is *theoretically* many times more germicidal than chloramines. The difficulty in resolving these comparisons is the total lack of field experience. As described above, the Grandville, Michigan work[22] revealed disappointment in the germicidal efficiency of bromine. This is probably due to the fact that bromine is such a powerful oxidant that much of it is lost in side reactions with organic matter in the sewage.

Koski and Stuart,[33] using the AOAC (Association of Official Agricultural Chemists) method for evaluating disinfectants, reported that liquid bromine at 1.0 ppm is as effective as the 0.6 ppm of available chlorine control, with

E. coli as the test organism, but that 2.0 ppm liquid bromine is necessary to provide activity equivalent to the 0.6 ppm available chlorine control when *S. faecalis* is the test organism. McKee et al.[32] found that bromine in settled sewage at 15-min. disinfectant contact times was less effective than either chlorine or iodine. It is to be presumed that this phenomenon is a result of bromine becoming irretrievably bound up with the organic matter in sewage, or forming bromamines that decompose rapidly into components that do not have any disinfecting power. The interesting point here is that the residual bromine, which is a combination of HOBr and bromamine, appears to be as colicidal as chlorine, but the difficulty is that not much of the bromine becomes a stable residual. This is based on milliequivalents per liter of the halogen. On a dosage basis, however, chlorine is far superior to either bromine or iodine because much of the bromine or iodine is utilized in satisfying the halogen demand. Consequently, they are the least effective of the halogens where sewage is to be disinfected.[32] Eight mg/L of chlorine will achieve disinfection comparable to 45 mg/L of bromine or iodine. This is why the literature on this subject stresses the necessity of first satisfying the halogen demand of the sewage with the old reliable and less expensive chlorine. Kott in 1969[36] and subsequent investigations has demonstrated the superior germicidal efficiency of a combination of chlorine and bromine. First, chlorine is applied to satisfy the immediate halogen demand, followed by the application of bromine.

This corresponds to the work of Sollo et al.[16,35] and to the field experiences of CIFEC.[37] The latter reports that a 3 mg/L dose of chlorine in a wastewater effluent at Bormes, France, containing 90 mg/L ammonia nitrogen followed by an 8 mg/L dose of bromine gave a 100 percent reduction of 10^6 coliforms with only 10 min. of contact.

According to Mills, bromine chloride is superior in germicidal efficiency to molecular bromine. This notion is derived from the fact that the chemical reactivity of BrCl is much faster than that of molecular bromine. So far there is too little information to substantiate this supposition.

TOXICITY OF BROMINE RESIDUALS

The acute toxicity tests with chlorobrominated wastewater effluent reported by Ward et al.[22] on the Grandville, Michigan project indicate that bromine residuals are not so lethal to fish as those of chlorine compounds. This supports the finding of Mills,[39] who concluded that chlorobromination produced a less toxic effluent because bromamines are less stable than chloramines and thus do not persist as long as they do. This would be true regardless of how the active bromine compounds were applied. The active ingredient would be mostly dibromamine at the pH normally encountered in wastewater treatment. However, if organobromamines are formed because of the presence of certain organic N compounds, the total bromine residual may persist for hours in a wastewater effluent.

BROMO-ORGANIC COMPOUNDS

There is continuing concern over the use of halogens as disinfectants of potable water, wastewater, and water for reuse, due to some so-called potentially harmful disinfection by-products (DPBs).

In addition to the chloro-organic compounds, a variety of bromo-organic compounds were also detected. These included bromodichloromethane, dibromochloromethane, and bromoform. The sources of these compounds, generally referred to as precursors, are uncertain, but they appear to be distributed by nature as a natural phenomenon. The precursors of these compounds have been found in soil and in the chlorophyll of blue-green algae. It has been suggested that the occurrence of halogenated bromine compounds is due to fallout from automobile exhausts. It was postulated that because bromine was used as a lead scavenger in leaded gasoline, these bromine compounds were deposited as a residue on paved roads and therefore reached water supplies by natural runoff. This possibility is probably overemphasized because bromides are found almost everywhere in nature. However, the elimination of leaded gasoline certainly must have curtailed this source as a precursor.

An example of how the ubiquitous bromide ion in the environment can affect a potable water supply is found in the Contra Costa County Water District experience.[38] This water utility serves 160,000 people in the cities of Concord, Clayton, Port Costa, Walnut Creek, Pleasant Hill, and Martinez, California. The principal source of water supply is from the Sacramento–San Joaquin delta. During two back-to-back drought years, salt water intrusion from the tidal effect in San Francisco Bay increased the bromide concentration in the water going to the treatment plant. Chlorination of this water converted the bromides to bromine species. These bromine compounds reacted with the chlorine demand in the water to form trihalomethanes in excess of the EPA maximum contaminant level of 0.10 mg/L. (In the 1990s, any sane person could hardly get worried about 0.10 mg/L of one of nature's compounds in drinking water, let alone wastewater!) Nevertheless, this forced the water district to take corrective measures. After studying several treatment changes in an effort to reduce the chlorine demand ("THM precursors"), none of which was entirely successful, the district had to switch its chlorination practice to chloramines. Owing to the lower oxidative power of chloramines, this brought the TTHM levels well below the EPA trivia maximum contaminant level (MCl) for THMs.*

It is important to recognize that all of the bromine species are better oxidizing agents than their analogous chlorine species. For example, the oxidation of cellulose by hypobromous acid is much faster than by hypochlorous acid,

*This problem of seawater intrusion is cyclical owing to the drought years. Adding ammonia *N* to any water supply for no good reason, such as THM trivia, can produce far more water quality degradation than 0.1. or 0.90 mg/L of THMs, simply because the added ammonia *N* will upset the microbiological life in the distribution system when the chlorine residual disappears.

and the oxidation of glucose to gluconic acid by hypobromite is 1300 times faster than by hypochlorite. It is not surprising that bromine is a lesser halogenating agent than chlorine, simply because it is a more potent oxidizing agent.

There are many organic reducing agents in wastewaters. These include organic alcohols, aldehydes, amines, and mercaptans. They are completely oxidized by bromine chloride, resulting in the formation of inorganic chloride and bromide salts as the major byproducts.

The carbon–bromine bond is less stable than the carbon–chlorine bond, and there is a possibility that in addition to metabolism by hydroxylation, well-established for chlorobiphenyls, a reductive debromination may be a degradation pathway of brominated aromatics. In general, compounds that are more readily brominated are also more susceptible to either hydrolytic or photochemical degradation. This reduces the incidence of bromo-organic compound formation. Moreover, the chemical bond strengths show that bromine bonds are weaker than chlorine bonds. Thus bromine compounds are generally more unstable than those of chlorine and are therefore easier to chemically degrade to innocuous inorganic compounds. The fact that bromine residuals die away after a few minutes' contact time in potable water does not mean that a similar die-away pattern will occur in highly polluted water or wastewater effluents. There is evidence that certain species of organic N compounds react with the bromine species to form organobromamines that have a very slow decay rate similar to that of organochloramines. Therefore, much more investigative work would have to be done to enable us to predict with reasonable accuracy the characteristics of bromine residual die-away in various situations.

MEASUREMENT OF BROMINE RESIDUALS

General Discussion

Regardless of the method of bromine application, whether it is generated on site or is injected as molecular bromine, or is an aqueous solution of either bromine or bromine chloride, the resulting residuals respond the same to all analytical procedures.

At the Grandville, Michigan project[22] the bromine residuals were determined by a spectrophotometric procedure developed by Mills of Dow Chemical Co.[39] Other researchers used the DPD–FAS titrimetric method,[35] and Johnson and coworkers[6,24,28] used a thiosulfate–iodine amperometric titration that measured the oxidation of iodide to iodine at pH 7 by the bromine residuals. Larson and Sollo[40] used the bromcresol purple colorimetric procedure.

There are no known analytical methods available to an operator or a laboratory technician capable of distinguishing between the three bromine species: hypobromous acid (HOBr), monobromamine (NH_2Br), and dibromamine ($NHBr_2$). Therefore, the only measurement available is total bromine residual.

When chlorine and bromine are used sequentially, as in water reclamation, there may be a mixture of chloramine and bromamine owing to the presence of ammonia N in the wastewater. Total residual, measured as either bromine or chlorine, is the preferred choice because most instances would involve the use of a continuous residual analyzer. This instrument can only be calibrated for total residual in the event of a mixture of chlorine and bromine residuals. This can be done by using either the amperometric forward-titration method or the DPD–FAS titrimetric method.

Differentiating between total chlorine and total bromine in the same sample can be done by Palin's DPD method[41] described below.

Amperometric Method

The amperometric method is well suited to this measurement. As the amperometric cell has a linear sensitivity to all the free halogens, and free and combined bromine compounds will release iodine quantitatively at pH 4, the procedure for total available bromine is determined on the titrator by using the procedure given for total residual chlorine—using buffer solution at pH 4 and potassium iodide solution. Then, titrating for bromine, the pipette reading in milliliters must be multiplied by 2.25 to express the answer in ppm of bromine.

DPD Differentiation Method[41]

The following procedure assumes free chlorine, combined chlorine, and free and combined bromine residuals. Hereafter the term "residual bromine" is taken to mean free bromine plus bromamines.

The reagents required are: (1) standard ferrous ammonium sulfate (FAS) solution (1 ml = 0.100 mg available chlorine); (2) DPD No. 1 powder (a combined buffer–indicator reagent); (3) potassium iodide crystals.

Free Chlorine. To a 100 ml sample add 0.5 g DPD No. 1 powder, mix rapidly to dissolve, and titrate immediately with the FAS solution. This is reading *A*. (If free chlorine is not present, this step may be omitted.)

Combined Available Chlorine. Add to the previous sample several crystals (approximately 1.0 g) of potassium iodide, mix to dissolve, and, after it stands for 2 min., continue titration with FAS solution. This is reading *C*.

Residual Bromine. To a *second* 100 ml sample add 2 ml of 10 percent wt/v glycine solution and mix.* Then add approximately 0.5 g DPD No. 1 powder, mix, and titrate with the standard FAS solution. This is reading Br.

---

*If it has already been determined that free chlorine is absent, this step may be omitted.

Calculations. For a 100 ml sample, 1 ml FAS solution = 1 mg/L available chlorine.

Residual bromine = Br
Free available chlorine = A − Br
Combined chlorine = $C - A$

If it is desired to report the residual bromine in terms of bromine, multiply the Br reading result by 2.25. (This is the ratio of the molecular weight of bromine to chlorine.)

EFFECT OF SEAWATER CHLORINATION

Chlorination of seawater and estuarine saline waters reacts to convert the excess bromides into the bromine species. If ammonia N is not present, the species will be hypobromous acid. If ammonia N is present, mono- and dibromamine will be present. Therefore all of the remaining residual after chlorination will be bromine residual. This chemistry is explained in Chapter 4. Measuring these residuals is usually carried out for total residual and reported as chlorine residuals.

IODINE

Physical and Chemical Characteristics

Iodine, a nonmetallic element with an atomic weight of 126.92—the heaviest of the halogen group—was discovered by Curtois in 1811. It is the only halogen that is solid at room temperature, and it can change spontaneously into the vapor state without first passing through a liquid phase. It is isolated as a shining blackish-gray crystalline solid of specific gravity 4.93 with a peculiar chlorine-like odor. It melts at 113.6°C to a black mobile liquid and boils at 184°C under atmospheric pressure to produce the characteristic violet-colored vapor.[42]

Occurrence and Production

Iodine is always found combined, as in the iodides. It is prepared from kelp and from crude Chile saltpeter. This saltpeter ($NaNO_3$) contains about 0.2 percent sodium iodate ($NaIO_3$). Historically, the United States has depended upon imports for almost all of its iodine needs. Originally from Chile, iodine is produced as a coproduct from natural nitrate production. In recent times, however, Japan has become the dominant factor in iodine world trade. For example, the 1974–76 U.S. consumption of iodine was on the order of 6–7 million lb; and during 1975, U.S. imports totaled 5.3 million lb, of which 93 percent was from Japan and the remainder from Chile.[43] Chile's production appears to have stabilized at a maximum of 4.4 million lb; this is from natural sources.

The small U.S. output in 1975 came entirely from the Dow Chemical Co. in Midland, Michigan, which recovered iodine and several other chemical products from a mixed-salts type of natural well brine. There are some reportedly high-grade brines in northern Oklahoma, which may be exploited by several companies. In the fall of 1975, Houston Chemical Co. broke ground for a 2 million lb/yr iodine plant at Woodward, Oklahoma; nevertheless, most of the iodine comes from South America.

Uses

Iodine and its compounds are used as catalysts in the chemical industry (production of synthetic rubber), and in food products, pharmaceutical preparations, stabilizers (as nylon precursors), antiseptics, medicine (treatment of cretinism and goiter), inks and colorants, and industrial and household disinfectants.

Iodine was first used in medical practice for treatment of goiter by Prout in 1816.[44] The first specific reference to the use of iodine in wounds was in 1839. Iodine was officially recognized by the *Pharmacopoeia* of the United States in 1830 as tincture of iodine. The first official United States tincture was a 5 percent solution of iodine in diluted alcohol.

Use in Water Treatment

The first known field use for iodine in water treatment was in World War I by Vergnoux,[45] who reported rapid sterilization of water for troops. Investigations on the use of iodine as an alternative method of disinfecting water supplies for United States troops were made in the years following World War I. Hitchens (1922)[46] recommended the use of 5 ml of a 7 percent tincture of iodine for a Lyster bag (about 140 L or 37 gal), and stated that, even with raw Potomac River water, such a treatment rendered the water safe for drinking in 30 min. This was followed by Dunham's[46] recommendation (1930) to increase the dosage to 10 ml of a 7 percent tincture of iodine per Lyster bag, or 2 to 3 drops per canteen. Pond and Willard[47] verified in 1937 that 2 drops/L of a 7 percent tincture of iodine is sufficient to render any water innocuous within 15 min., and that 3 drops/L did not give any better results.

During World War II, a series of studies at Harvard University by Chang, Morris, et al.[48,49] led to the development of globaline tablets for disinfecting small or individual supplies for the U.S. Army. Each tablet contained 20 mg tetraglycine hydroperiodide, of which 40 percent is I_2 and 20 percent iodide, combined with 85 mg acid pyrophosphate. One tablet imparts 8 mg I_2 to a liter of water. Chang and Morris[46] list the following advantages of the globaline tablet over the tincture: (1) Iodide is present in half the amount of I_2. (2) There is a definite dosage. (3) I_2 loss is insignificant. (4) The presence of an acid salt counteracts high pH. (5) The treated water is more palatable (less taste and odor). (6) The tablets are convenient for field use. The globaline tablets superseded the Halazone tablets for troop use because one globaline tablet per quart or liter destroys the cysts of *Entamoeba histolytica* in 10 min., where six Halazone tablets are required to accomplish the same kill.[46] Likewise, the human enteric pathogens (bacterial, viral, and protozoan) are satisfactorily destroyed in the same time[48–50] under ordinary conditions. If the water is very cold, deeply colored, or highly turbid, the dosage should be increased to two tablets per liter and the contact time extended to 20 min.

Iodination of water supplies has been limited largely to emergency treatment by the military. The use of iodine as a disinfectant for water has been recognized for a long time, but has never generated enough interest to displace the popular use of chlorine. As compared to chlorine, the very high cost of iodine and its possible physiologic effects are the main reasons for its limited use. The military did not give unqualified approval to the use of iodine as a disinfectant for water until a six-month study was made at a naval installation in the Marshall Islands in 1949–50. The use of iodine in the drinking water revealed no untoward effects from its ingestion by a large group of service men,[51] whose average intake was 12 mg/day for 16 weeks and 19.2 mg/day for 10 weeks.

In 1965, Black and co-workers[52,53] completed a demonstration project, with a grant from the U.S. Public Health Service, to study the effects of prolonged use of iodine at three correctional institutions in Florida. Iodination of these water supplies was carried out over a period of 19 months, serving approximately seven hundred persons under carefully planned chemical, medical, and bacteriological controls. Iodine proved to be a satisfactory disinfectant and in doses up to 1.0 ppm did not produce any discernible color, taste, or odor in the water. There was no evidence from the medical investigation that this level of iodine has any adverse effect upon the general health or thyroid function.

Montezuma's Revenge

The most popular application is in tablet form or tincture of iodine solution for emergency treatment of small supplies on a batch basis such as water supplies for troops on bivouac, or travelers in foreign countries where "traveler's diarrhea" is a common occurrence. This type of emergency treatment has proved to be reliable and effective.

The U.S. Public Health Service lists the iodine dosage for emergency water treatment as: 5 drops of a 2 percent tincture for 30 min. *per quart* (or one liter) for clear water and 10 drops for turbid water. The potential for unpleasant tastes with iodine is about the same as with chlorine. Iodine does not react with nitrogenous compounds as chlorine does; so the resulting taste in polluted waters will be different.

Chemistry of Iodination

There are two methods of dosing a water supply with iodine. The most practical method is the use of iodine crystals. The discussion that follows will be limited to this method. The other possible method is the use of the "iodine bank" scheme. This procedure requires the presence of an excess of iodide. Elemental iodine is released from the iodide quantitatively by the application of an oxidant such as chlorine. This is the procedure used in swimming pools and has no apparent practical application in water supply treatment. This method utilizes the only advantage of iodine over chlorine—that is, that iodine will not react with nitrogenous compounds to form iodomines as chlorine does to form the objectionable chloramines.

Therefore, the following chemistry of weak solutions of elemental iodine is based on the use of iodine crystals without the presence of added iodine. The work by Chang[50] (1958) is the basis of this chemistry.

Four different substances must be considered in the reactions of elemental iodine with water:

Elemental iodine—I_2
Hypoiodus acid—HIO
Periodide or tri-iodide ion—I_3^-
Iodate ion—IO_3^-

First is the hydrolysis of I_2 to form hypoiodus acid:

$$I_2 + H_2O \leftrightharpoons HIO + H^+ + I^- \tag{16-16}$$

$$\frac{[HIO][H]^+[I]^-}{[I_2]} = K_h \tag{16-17}$$

$$K_h = 3 \times 10^{-13} \text{ at } 25°C$$

Figure 16-4 shows the I_2 hydrolysis data for concentrations of iodine in the 0.1–5 mg/L range at 18°C and pH 6.0–8.0, the probable range of use for small water supplies. Table 16-1 shows the effect of pH on the hydrolysis of iodine with a total iodine residual of 0.5 ppm.

Second is the possible dissociation of hypoiodous acid with variation in pH to form hypoiodite ion. It is of interest to compare this phenomenon with that of hypochlorous acid in the case of chlorine, and how it affects the germicidal efficiency of that compound.

$$HIO \leftrightharpoons H^+ + IO^- \tag{16-18}$$

$$\frac{[H^+][IO^-]}{[HIO]} = K_a = 4.5 \times 10^{-13} \tag{16-19}$$

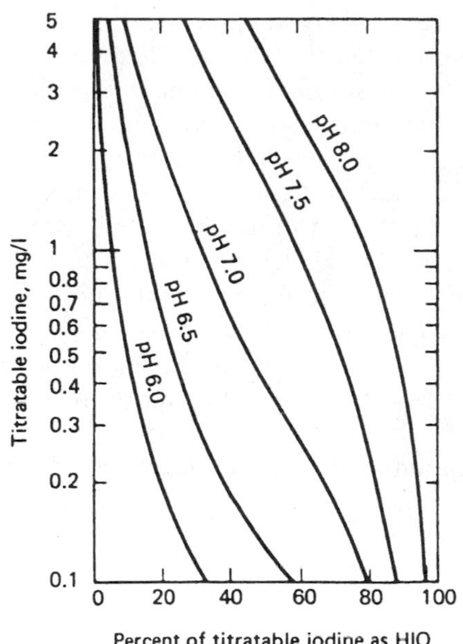

Fig. 16-4. Percent of titrable iodine as I_2 and HIO in water at 18°C and pH as indicated.[50]

$$\frac{[\mathrm{HIO}]}{[\mathrm{IO}^-]} = \frac{[\mathrm{H}^+]}{K_a} \tag{16-20}$$

From Eq. (16-20) it is possible to calculate the ratio of undissociated acid to hypoiodite ion at any pH. The dissociation constant of hypoiodous acid reveals that is is only slightly stronger as an acid than pure water. The formation of hypoiodite ion (IO^-) is insignificant, as can be seen from Table 16-1.

Table 16-1

	Content of Residual Percent		
pH	**I_2**	**HIO**	**IO^-**
5	99	1	0
6	90	10	0
7	52	48	0
8	12	88	0.005

Compare the formation of hypoiodite ion with that of the hypochlorite ion for chlorine in Table 16-2.

Both elemental iodine and hypoiodous acid are effective germicides. At pH 8, the effective chlorine germicide HOCl is present at 21.5 percent, whereas the ineffective hypochlorite ion is present at 78.5 percent. With iodine, however, the ineffective hypoiodite ion is present only at 0.005 percent with elemental iodine at 12 percent and hypoiodous acid at 88 percent. Therefore, the formation of the ineffective hypoiodite ion can be ignored.

Third is the possibility of the formation of the bactericidally ineffective tri-iodine ion I_3^-. Chang reports that when using either iodine crystals, iodine tablets, or tincture of iodine, the formation of tri-iodine ion can be ignored. The amount of iodine formed from hydrolysis of I_2 at the 1 to 2 ppm level and at pH values of 6.5 to 8.0 would be so small that the tri-iodide ion I_3^- formed in the water should not be much more than 1 percent of the total titrable iodide.

Fourth is the formation of the iodate ion, which has no apparent cysticidal or viricidal activity. Iodate formation in solutions of elemental iodine (without added iodide) proceeds as follows:

$$I_2 + H_2O \leftrightarrows HIO + H^+ + I^- \tag{16-21}$$

$$3HIO + 3(OH)^- \leftrightarrows (IO_3)^- + 2I^- + 3H_2O \tag{16-22}$$

These equations show that alkaline pH values shift the reactions to the right, and acidic pH values shift them to the left. Iodate formation is not likely to be a problem interfering with the disinfection efficiency of iodine until the pH rises above 8.0 and the contact time runs over 30 min. Since most of the disinfection action of iodination is accomplished in the first 15 to 20 min., the small losses of titrable iodine to iodate need not be a problem in treating waters at pH levels below 8.4.

Germicidal Efficiency

It is the consensus of many investigators[46,47,52,54–58] that elemental iodine I_2 and hypoiodous acid HIO are the two most powerful disinfecting agents among

Table 16-2

	Content of Residual Percent at 20°C		
pH	**Cl_2**	**HOCl**	**OCl^-**
5	0.0	99.74	0.26
6	0.0	97.45	2.55
7	0.0	79.29	20.71
8	0.0	27.69	72.31

the titrable iodine species. The relative amounts of titrable iodine existing as I_2 and HIO in an aqueous solution of elemental iodine depend primarily on the strength and pH of the solution and, to a much lesser extent, upon the temperature. The comparative disinfection efficiency of I_2 and HIO varies considerably, depending on the type of organism. Wyss and Strandkov[54] reported that against *B. metiens* spores, I_2 is three to six times greater than HIO, but Chang[46] reported that in the destruction of cysts of *E. histolytica,* I_2 is two to three times more effective than HIO. On the other hand, Clark et al.[56] found that hypoiodous acid destroys viruses at a rate considerably faster than that achieved by I_2; HIO has also been claimed to be three to four times more active than I_2 in destroying *E. coli.*[46] The difference in effectiveness of hypoiodous acid and elemental iodine on the type of organism suggests a different mode of action, depending on whether or not the organism is protected by a membrane such as a spore or a cyst. The superiority of I_2 over HIO in the destruction of spores and cysts is apparently due to the greater penetrating power of I_2. In the case of vegetative bacteria, which have a physiologically active cell membrane, the oxidizing power of the compound becomes the important factor. This is supported by the fact that the normal redox potential (E_0 at 25°C) of HIO is 990 mV, and that of I_2 is 536 mV.[46] In virus destruction, which probably proceeds as a function of the oxidizing power of the compound, this can be supported further when comparing the viricidal efficiency of hypoiodous acid and that of hypochlorous acid. HOCl has a normal redox potential of 1490 mV; its viricidal efficiency is five times greater than that of HIO,[56] whereas that of HIO is at least 40 times greater than that of I_2.[46] These differences support the belief that virus destruction is a result of the reaction of the disinfectant with the protein shell and not with the nucleic acid core; if the latter were involved, the protein shell would act as a protective shield, and the destructive process would not be much different from that of spores and cysts.

In order to relate these differences in the germicidal efficiency of both I_2 and HIO, Chang[46] prepared Fig. 16-5, which shows the relation to titrable iodine and contact time for the 99.9 percent destruction of cysts, a virus, and bacteria by I_2 and HIO at 18°C. In practice, these parameters allow a good margin of safety. Satisfactory bactericidal results can be obtained with 0.1 to 0.2 ppm in less than 4 min. Virus destruction by I_2 lies beyond the practical limit of 1 to 2 ppm, but for HIO it is well within these limits. The I_2 cysticidal curve lies close to the upper limit of practical range: 1.0 ppm in 30 min to 2 ppm in 13 min.

Compared with free chlorine residuals, iodine is definitely inferior. Free chlorine is four times more cysticidal than HIO, and it is two hundred times more viricidal and two times more cysticidal than I_2.

The 1960 report by McKee et al.[32] demonstrates that iodine is an effective disinfectant in wastewater; however, wasted side reactions due to the halogen demand of wastewater and its heavier molecular weight *requires 45* mg/L *iodine* to achieve the same level of disinfection as *8* mg/L *of chlorine.*

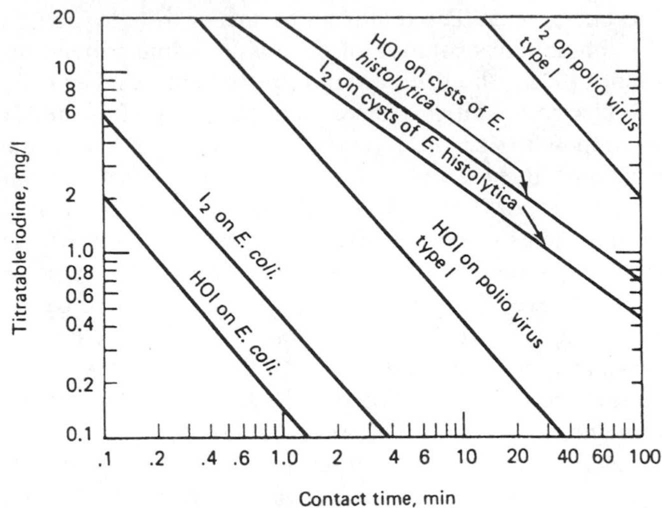

Fig. 16-5. Time versus concentration relationship in the destruction (99.0) of cysts, viruses and bacteria by I_2 and HIO at 18°C.[46]

Limitations of Iodination

Iodination can be used where chlorination or more complete treatment of water either is not possible or is impractical. Such use suggests that this process would probably be limited to remote rural areas or undeveloped countries where little or no adequate supervision and expertise are available. Further, it is the consensus that on a prolonged treatment basis, because of possible undesirable physiological effects, dosages of 1–2 mg/L with occasional dosages of 4–5 mg/L are the practical limits. This would require an investigative procedure to determine the desirability of iodination.

Comparison with Chlorination

The cost of iodination is an important factor against its use. Iodine crystals were available at approximately $2.50 per lb FOB New York from the Chilean Nitrate Sales Corporation in 150-lb containers, compared with chlorine at 15¢ per lb. However, in a remote location, transport of the iodine crystals and the installation of an electrically driven pump are about the only considerations for applying iodine. The apparatus for a hypochlorite installation is not too much different from this.

Free chlorine (HOCl) is *five times more viricidal* and *four times* more *cysticidal than HIO,* and is two hundred times more viricidal and two times more cysticidal than I_2. This obvious superiority of chlorination in disinfecting power, together with its low cost and pharmacological inertness, makes chlorination the method of choice over iodination in water treatment. The emphasis here is on free rather than combined chlorine.

Iodination Facility

The application of iodine for the treatment of water is best accomplished by the use of a saturator. Iodine is the least soluble halogen in water, and its solubility is significantly dependent on temperature,[52] as shown in Fig. 16-6. For the usual water supply, the strength of the iodine solution will be in the range of about 200 to 300 ppm. At these concentrations, 316 SS is satisfactory for piping and metering pumps. Not all plastics are suitable: nearly all clear plastic tubing will discolor rather quickly, and some will disintegrate in a matter of months. Figure 16-7 illustrates an adequate arrangement for the application of iodine into a water supply. The system consists of a saturator for dissolving the iodine crystals. The temperature of the water will determine the strength of the solution to be injected by the metering pumps. As the accuracy of dosage is critical, it is strongly recommended that only metering pumps be used instead of injection devices that produce a vacuum by means of water pressure. The latter devices cannot provide the necessary dependable and accurate feed rates, compared to a positive displacement metering pump. Furthermore, the metering pumps should be paced from a flow-measuring device through either an automatic stroke positioner or a variable speed SCR drive if the water flow is variable.

Wastewater. Iodine has never been tried on a plant-scale basis as a wastewater disinfectant, but has been used only in the laboratory for purposes of investigating its germicidal efficiency. The cost and nonavailability of iodine militate heavily against its use as a disinfectant for wastewater.

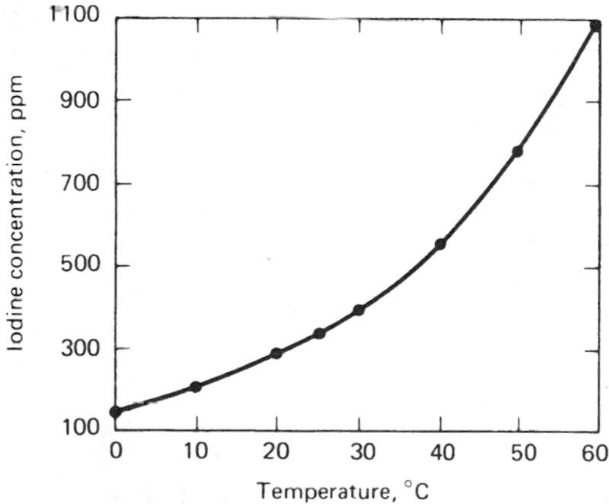

Fig. 16-6. Solubility of iodine in water (concentration as a function of temperature).

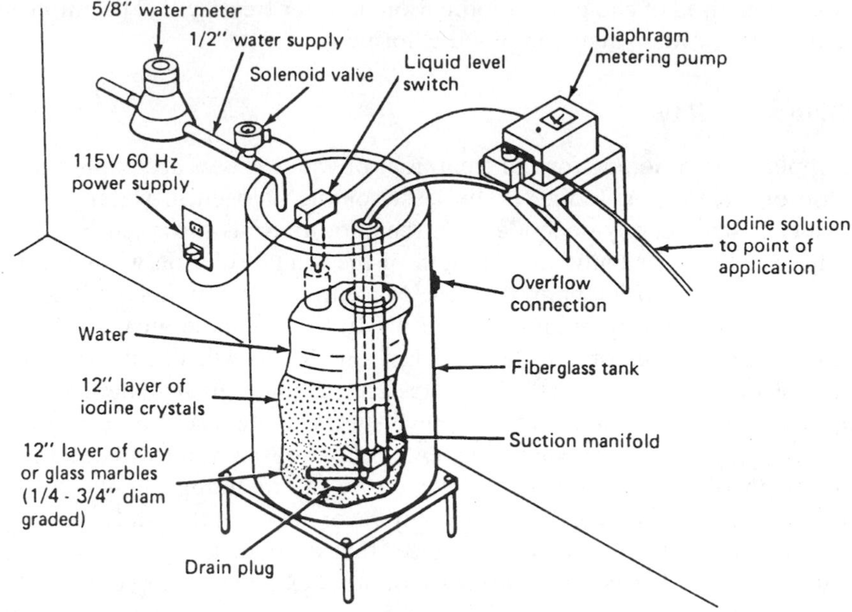

Fig. 16-7. Iodination installation.

Cost. The 1976 price for crude iodine crystals in 100-lb drums was $2.60/lb, and for USP granular iodine was $4.00–$5.00/lb in 100-lb drums. This compared to chlorine at 10–15¢/lb and bromine chloride at 20¢/lb.

Determination of Iodine Residuals

The Standard Methods Committee approved the methods of measuring both iodine and iodide in 1993; see pages 4-70 to 4-73 in the 19th Edition of *Standard Methods.* These measurements of iodine residuals are by the leuco crystal violet method, and the amperometric titration method. The amperometric titration method is the more practical method of the two. Moreover, it is more precise because the electrodes can be sensitized in a few minutes with a dilute solution of iodine. This gives a sharp end point.

The DPD–FAS titrimetric method has been found to be equally satisfactory for precise measurements.

SUMMARY

Bromine

This is considered the most reactive oxidant of all the halogens, except fluorine which is not considered here. For this reason it has some desirable chemical characteristics. The hydrolysis of bromine in an aqueous solution is almost identical to that of chlorine. Bromine added to water reacts to form hypobromous acid, and in the presence of ammonia nitrogen in the quantities usually found in wastewater bromamines are formed. *The germicidal efficiency of these bromamines in a pH environment similar to that of wastewater is practically equal to that of free chlorine.*

Bromine is highly reactive; therefore, its halogen demand is greatly distorted beyond its usefulness. So any system in wastewater treatment attempting to exploit the features of bromine should be arranged to use chlorine to minimize the "bromine demand."

Bromine as Br_2 is a difficult and hazardous chemical to handle because it is in liquid form at room temperature. This led to the development of bromine chloride, which has the same vapor pressure characteristics as sulfur dioxide. BrCl is much more desirable than elemental bromine from the standpoint of chemical handling facilities.

Bromine Chloride

Bromine chloride has been investigated thoroughly as a practical substitute for elemental bromine and liquid–gas chlorine installations. Those investigations have shown conclusively that a new species of equipment will be required for the metering and control of BrCl. The factors of increased corrosivity of BrCl over chlorine and the delicate problem of BrCl dissociation have determined the necessity of fundamental design changes.

On-site generation of bromine from bromide salts using chlorine as the oxidizer has interesting possibilities. Injecting a bromide salt into a chlorinator solution discharge line (similar to the generation of chlorine dioxide) causes the formation of elemental bromine, which hydrolyzes immediately to hypobromous acid. This concept can be used to great advantage at existing chlorination installations for the following reasons. The existing chlorination system can be arranged to produce a 2–5 min. controlled chlorine residual of 0.3–0.5 mg/L (instead of the usual 5–7 mg/L). This sequesters the halogen demand. Following this is the formation of elemental bromine from the combination of the chlorine solution discharge with the bromide salt injection from the second chlorinator application. This becomes the second point of "chlorine" application downstream from the original point of chlorination.

This system has many desirable features: it can utilize existing chlorination facilities; it eliminates the hazardous problems of handling bromine; and it can utilize the superior germicidal properties of bromine without sacrificing other design considerations. Moreover, the bromamine residual die-away phenomenon may be so rapid that debromination by SO_2 may not be required, and the contact time for these "hot" bromine residuals can be as short as 5 min. and not longer than 15 min. This would bring contact chamber requirements to the lowest recommended for ozone, which are the lowest for all disinfectants.

Iodine

Iodine has been used as a temporary expedient for the disinfection of small water supplies and as an alternative to chlorine in the treatment of swimming pools.

It has never been tried even on a pilot-plant experimental basis for the disinfection of wastewater. The cost and the reliability of iodine supply make it an impossible choice as a disinfectant for wastewaters or waters for reuse.

REFERENCES

1. Mills, J. F., "Interhalogens and Halogen Mixtures as Disinfectants," Chapter 6 in J. D. Johnson (Ed.), *Disinfection: Water and Wastewater,* Ann Arbor Science, Ann Arbor, MI, 1975.
2. Riley, J. P., and Skirrow, G., *Chemical Oceanography,* Vol. I, p. 36, Academic Press, New York, 1965.
4. "Bromine, Its Properties and Uses," Michigan Chem. Corp., Chicago, IL, 1958.
4. "Bromine Outlook Tied to Clean Air Rules," *Chem. and Engr. News,* p. 11 (Feb. 25, 1974).
5. Galal-Gorchev, Hend, and Morris, J. C., "Formation and Stability of Bromamide and Nitrogen Tribromide in Aqueous Solution," *Inorganic Chem.,* **4,** 899 (1965).
6. Johnson, D. J., and Overby, R., "Bromine and Bromamine Disinfection Chemistry," *ASCE J. San. Eng. Div.,* p. 617 (Oct. 1971).
7. La Pointe, T. F. Inman, G., and Johnson, J. D., "Kinetics of Tribromamine Decomposition," Chap. 15 in J. D. Johnson (Ed.), *Disinfection: Water and Wastewater,* Ann Arbor Science, Ann Arbor, MI, 1975.
8. U.S. Patent, 2,443,429, "Procedure for Disinfecting Aqueous Liquid," Marks and Strandkov, and Wallace & Tiernan Co., Inc., Belleville, NJ, June 15, 1948.
9. Morrow, J. J., U.S. Patent No, 3,4.3,199 to Fischer and Porter Co., Warminster, PA, Nov. 26, 1968.
10. Mills, J. F., assignor to Dow Chemical Co. "Control of Microorganisms With Polyhalide Resins," U.S. Patent 3,462,363, Aug. 19, 1969.
11. Mills, J. F., Goodenough, R. D., and Nekervis, W. F., assignors to Dow Chemical Co., "Process for Treating Water With Bromine," U.S. Patent No. 3,316,173, Apr. 25, 1967.
12. Goodenough, R. D., Mills, J. F., and Place J., "Anion Exchange Resin (Polybromide Form) as a Source of Active Bromine for Water Disinfection," *Env. Sci. Tech.,* **3,** 354 (Sept. 1969).
13. Regunathan, P., and Brejcha, R. J., "A New Technology in Potable Water Disinfection for Offshore Rigs," paper presented at the Ann. Mtg. Soc. of Petroleum Engrs. of AIME, Dallas, TX, Sept. 28–Oct. 1, 1975.
14. Brown, J. R., McLean, D. M., and Nixon, M. C., "Bromine Disinfection of a Large Swimming Pool," *Can. J. Public Health,* **55,** 251 (June 1964).
15. Derreumaux, A., "Process and Apparatus for Disinfecting or Sterilizing by the Combination of Chlorine and Other Halogens," French patent No. 2,171,890 (1973).
16. Sollo, F. W., Mueller, H. F., and Larson, T. E., "Prechlorination Enhanced Disinfection by Bromine," paper presented at the 7th Int. Conf. on Water Poll. Res., Paris, France, Sept. 9–12, 1974.
17. Venosa, A. D., "Disinfection: State of the Art, Alternatives to Chlorination for Wastewater," paper presented at a Disinfection Seminar, Dept. of Ecology, Univ. of Washington, Seattle, WA, May 26–27, 1976.
18. Mills, J. F., and Oakes, B. D., "Bromine Chloride Less Corrosive than Bromine," *Chem. Eng.,* pp. 102–106 (Aug. 1973).
19. Mills, J. F., "Disinfection of Sewage by Chlorobromination," paper presented at Ann. Chem. Soc. Meeting, Dallas, TX, Apr. 1973.

20. Cole, L. G., and Elverum, G. W., "Thermodynamic Properties of the Diatomic Interhalogens from Spectroscopic Data," *J. Chem. Phys.,* **20,** 1543 (1952).
21. Jolles, Z. E., *Bromine and Its Compounds,* Chap. 1, Ernest Benn Ltd., London, England, 1966.
22. Ward, R. W., Giffin, R. D., De Graeve, G. M., and Stone, R. A., "Disinfection Efficiency and Residual Toxicity of Several Wastewater Disinfectants," Vol. I, Grandville, Mich., EPA Pre-publication Report, Project No. S-802292 (1976).
23. Collins, H. F., Selleck, R. E., and White, G. C., "Problems in Obtaining Adequate Sewage Disinfection," *ASCE J. Env. Eng. Div.,* **97,** 549 (Oct. 1971).
24. Johnson, J. D., and Sun, W., "Bromine Disinfection of Wastewater," Chap. 9 in J. D. Johnson (Ed.), *Disinfection: Water and Wastewater,* Ann Arbor Science, Ann Arbor, MI, 1975.
25. Bongers, L. H., O'Connor, T. P., and Burton, D. T., "Bromine chloride—An Alternative to Chlorine for Fouling Control in Condenser Cooling Systems," EPA Report No. 600/7-77-053, National Technical Information Service, Springfield, VA, May 1977.
26. Cole, S. A., "Experimental Equipment for Feeding Bromine Chloride," paper presented at the Int. Water Conf., Pittsburgh, PA, Oct. 31, 1974.
27. Task Force Report, "Disinfection of Wastewater, EPA-430/9-75-012," U.S. Env. Prot. Agency, Cincinnati, OH, Mar. 1976.
28. Johnson, J. D. (Ed.), *Disinfection: Water and Wastewater,* Ann Arbor Science, Ann Arbor, MI, 1975.
29. Wyss, O., and Stockton, R. J., "The Germicidal Action of Bromine," *Arch. Biochem.,* **12,** 267 (1947).
30. Marks, H. C., and Strandkov, F. B., "Halogens and Their Mode of Action," *Ann. NY Acad. Sci.,* **53,** 163 (1950).
31. Johannesson, J. K., "Anomalous Bactericidal Action of Bromamine," *Nature,* **181,** 1799 (1958).
32. McKee, J. E., Brokaw, C. J., and McLaughlin, R. T., "Chemical and Colicidal Effects of Halogen in Sewage," *J. WPCF,* **32,** 795 (1960).
33. Koski, T. A., Stuart, L. S., and Ortenzio, L. F., "Comparison of Chlorine, Bromine, and Iodine as Disinfectants for Swimming Pool Water," *Appl. Micro.,* p. 276 (Mar. 1966).
34. Schaffer, R. B., and Mills, J. F., *Proceedings of the National Symposium on Quality Standards for Natural Waters,* p. 158, University of Press, Ann Arbor, MI, 1966.
35. Sollo, F. W., Mueller, H. F., Larson, T. E., and Johnson, J. D., "Bromine Disinfection of Wastewater Effluent," Chap. 8 in J. D. Johnson (Ed.), *Disinfection: Water & Wastewater,* Ann Arbor Science, Ann Arbor, MI, 1975.
36. Kott, Y., "Effect of Halogens on Algae—III Field Experiment," *Water Research,* p. 265 (1969).
37. Derreumaux, A., private communication, CIFEC, Neuilly-sur-Seine, France, 1976.
38. Lange, A. L. and Kawczynski, Eliz., "Controlling Organics: The Contra Costa County Water District Experience," *J. AWWA,* **70,** 653 (Nov. 1978).
39. Mills, J. F., "A Spectrophotometric Method for Determining Microquantities of Various Halogen Species," Dow Chemical Co., Midland, MI, 1971.
40. Larson, T., and Sollo, F. W., "Determination of Free Bromine in Water," Annual Progress Report, U.S. Army Medical R & D Command, Contract No. DA-49-193-MD-2909 (1967).
41. Palin, A. T., "Analytical Control of Water Disinfection With Special Reference to Differential DPD Methods for Chlorine, Chlorine Dioxide, Bromine, Iodine and Ozone," *J. Inst. Water Eng.,* **28,** 139 (1974).
42. Anon., *Handbook of Chemistry and Physics,* 44th ed., pp. 1732–1744, Chemical Rubber Co., Cleveland, OH, 1962.
43. "Mineral Industry Surveys," U.S. Dept. of Interior Bureau of Mines, Washington, DC, 1975.
44. Lawrence, C. A., and Block, S. S., *Disinfection, Sterilization and Preservation,* Lea and Febiger, Philadelphia, PA, 1968.
45. Vergnoux, "Examinen rapide et sterilization des eaux pour les troupes en campagne," *L'Union Pharmaceutique* (France), pp. 194–201 (1915).
46. Chang, S. L., "Iodination of Water," U.S. Dept. H.E.W., Taft San. Engr. Center, Cincinnati, OH, 1966.

47. Pond, M. A., and Willard, W. R., "Emergency Iodine Sterilization for Small Samples of Drinking Water," *J. AWWA,* **29,** 1995 (1937).
48. Chang, S. L., and Morris, J. C., "Elemental Iodine as a Disinfectant for Drinking Water," *Ind. and Eng. Chem.,* **45,** 1009 (1953).
49. Morris, J. C., Chang, S. L., Fair, G. M., and Conant, G. H., Jr., "Disinfection of Drinking Water under Field Conditions," *Ind. and Eng. Chem.,* **45,** 1013 (1953).
50. Chang, S. L., "The Use of Iodine as a Water Disinfectant," *J. Am. Pharm. Assoc. Sc. Ed.,* **47,** 417 (1958).
51. Morgan, D. P., and Karpen, R. J., "Test of Chronic Toxicity of Iodine as Related to the Purification of Water," *U.S. Armed Forces Med. J.,* **4,** 725 (1953).
52. Black, A. P., Kinman, R. N., Thomas, W. C., Jr., Freund, G., and Bird, E. D., "Use of Iodine for Disinfection," *J. AWWA,* **57,** 1401 (1965).
53. Freund, G., Thomas, W. C., Jr., Bird, E. D., Kinman, R. N., and Black, A. P., "Effect of Iodinated Water Supplies on Thyroid Function," *J. Clin. Endocr.,* **26,** 619 (1966).
54. Wyss, O., and Strandkov, F. B., "The Germicidal Action of Iodine," *Arch. Biochem.,* **6,** 261 (1945).
55. Chambers, C. W., Kabler, P. W., Malaney, G., and Bryant, A., "Iodine as a Bactericide," *Soap and San. Chem.,* **28,** 149 (1952).
56. Clark, N. A., Berg, G., Kabler, P. W., and Chang, S. L., "Human Enteric Viruses in Water: Source, Survival and Removability," *Int. Conf. Water Poll. Res.,* London, Pergamon Press, 1962.
57. Gershenfeld, L., and Whitlin, B., "Free Halogens—A Comparative Study of Their Efficiencies as Bactericidal Agents," *Am. J. Pharm.,* **121,** 95 (1949).
58. Marks, H. C., and Strandkov, F. B., "Halogens and Their Mode of Action," *Ann. NY Acad. Sci.,* **53,** 163 (1950).

17
ULTRAVIOLET RADIATION AND AO_xPS

INTRODUCTION

Historical Background

The text of this chapter originally was written to describe the use of UV as a disinfectant for the effluent of a secondary or tertiary wastewater treatment plant discharging into the ocean surf or a surface water stream providing a 100 to 1 dilution. For this kind of situation, various states involved as well as the EPA were satisfied that using a fecal coliform requirement of 200 MPN/100 mL would be acceptable disinfection. This is the equivalent of 1000 MPN/100 mL of total coliforms. Several states, including California, will not accept fecal coliform requirements for a variety of reasons, primarily because they are unreliable, plus the fact that the Health Services agency in California always required a 5-log removal of total coliforms for all waters discharging into surface or bay waters.[34] Ocean surf waters used for bathing had a variety of requirements, depending upon the location. The most lenient requirement was 1000/100 mL of total coliforms.

It is important that the reader be reminded that as of 1997, the arid western states are predicted to "run out of water" in 2015, while the planet runs out in 2025, all due to overpopulation. California is going to lead the pack, owing to its continuing increase in population. This is why there are some 20 water reuse projects that are either under design or are under construction in California.

These projects are focusing on water reuse systems that convert properly treated wastewater that will be used for drinking water. Up until now (1997), reuse water has been primarily used for agricultural irrigation, golf course watering, and other municipal green areas. UV will probably be the final polishing agent. The reader is referred to the Denver, Colorado reuse project that is described in Chapter 9.

Sampling

In trying to achieve valid information concerning the disinfection of wastewater of any kind, there is the problem of sampling. Grab sampling can and does produce enormous errors. To avoid these errors, all water for sampling should be pumped from the effluent flow channel, as in the sampling used for operating a chlorine residual analyzer.

In the early 1960s California was chlorinating several primary effluents to achieve acceptable disinfection. Because of this action by the State Health Department, Wallace & Tiernan Co. had developed a new chlorine residual analyzer designed to measure chlorine residuals in primary effluents. White happened to be installing one of the new W&T analyzers capable of measuring chlorine residuals in primary effluents at the same time that city and state inspectors were checking the disinfection efficiency of the primary effluent at the San Francisco North Point WWTP.

The city took a grab sample from the effluent discharge channel, while White gave them a pumped sample from the sample line to the analyzer, which was only a few feet downstream from where the grab sample was taken. The grab sample revealed a total coliform count of 250/100 mL, whereas the pumped sample had greater than 39,000/100 mL total coliform. This was a real shocker, which eventually led the state authorities to declare that the disinfection of primary effluents was not an acceptable possibility because of evidence of natural break-up of the settleable solids when subjected to a mixed device.

This means that any grab sample should always be subjected to a motorized mixer that will break up any solids that might be present in the grab sample.

Current Situation (1997)

In the years since the third edition of this book was published, there have been many changes in the U.S. society that have had serious and negative effects on potable water and wastewater systems. Particularly ominous is the water shortage crisis that is predicted for 2015 in California and for 2025 for the earth's water supply. Hence water reclamation projects are on a very steep rise, particularly in California where the population is still increasing. All of this has resulted in a report by the National Water Research Institute* titled "UV Disinfection Guidelines for Wastewater Reclamation in California & UV Disinfection Research Priorities.[28,33]

The guidelines presented in the report are intended to provide guidance to regulatory staffs of the State of California Regional Water Quality Control Boards and the Department of Health Services when reviewing applications for the use of UV disinfection systems in wastewater and water reuse applications. Prepared by an independent panel of experts, it is the most important document to date for people in this business. Every agency involved in the

*This is of particular interest to the arid regions of the 48 states, such as Colorado, Arizona, New Mexico, and Texas, in addition to California. Nevada is definitely an arid state, but it is blessed with the Hoover Dam on the Colorado River. Since California has always suffered from water shortages due to a rising population and severe droughts, it is the most active state in water reuse programs.

supply and treatment of potable water or wastewater should have a copy of this report.

The guidelines that follow, which are from the report, have no binding regulatory effect unless they are declared by the California Department of Health Services as official regulations. However, there is no doubt that these regulations will soon apply to all states because disinfection is the most serious public health risk issue facing consumers. The reader should be interested in the comparison between the California Health Services regulations established in 1978.[34]

Guidelines for UV Disinfection Systems[28]

This report is for UV as an alternative disinfectant, which is required to meet the California Wastewater Reclamation Criteria Title 22, Division 4, Chapter 3, of the California Code of Regulations[28] for uses where an oxidized, coagulated, filtered, disinfected effluent is required. The guidelines developed in this document were developed on the basis of the demonstrated "equivalency" of UV disinfection to chlorine disinfection as it is used in wastewater reclamation plants. Equivalency of UV disinfection to a conventional process used in wastewater reclamation and reuse must be demonstrated by the following criteria:

1. Filtered effluent turbidity equal to or less than 2 NTU, met with the same statistical frequency as required for chlorine disinfection.
2. Total coliform count equal to or less than 2.2/100 mL, met with the same statistical frequency as required for disinfection with chlorine.
3. Virus inactivation efficiency equivalent to that achieved with chlorine disinfection, 4 logs of inactivation (99.99 percent reduction), based upon poliovirus.[29,35,38]

Furthermore, these criteria are based upon currently available UV technology using low pressure mercury vapor UV lamps, with flow parallel to the lamps in nonpressurized channels. Owing to limited experience with the most restrictive wastewater reclamation applications (2.2 or less total coliform/ 100 mL), the guidelines should be considered as interim, subject to revision as experience is gained with the demonstrated field technology, with variations of the demonstrated technology, and with newly evolving technology.

Nonconforming UV Systems

UV disinfection systems that do not conform to the requirements set forth in this report may be acceptable if it can be demonstrated, to the satisfaction of the California Department of Health Services, that they provide a degree of treatment and reliability at least equal to systems that have been shown to be acceptable to the regulatory agencies.

Determination of equivalency may require studies directed at the inactivation of viruses or other microorganisms by the proposed UV system. The determination of equivalency using a standard predetermined protocol would be required in the following (and possibly other) instances:

1. When UV disinfection systems with different lamp orientations are proposed.
2. When a UV disinfection system is proposed with a different UV dose from that called for in the guidelines.
3. When medium and high energy UV lamps are proposed.
4. When a UV disinfection system is proposed for a wastewater reclamation application where the transmittance of the filtered effluent is below 55 percent.

ESSENTIAL UV SYSTEM ELEMENTS

UV Dose.

The UV dose is defined as the product of the average UV intensity, expressed in milliwatts per square centimeter (mW/cm^2), and the average exposure time of the fluid to be treated, expressed in seconds (s). So the UV dose is expressed in units of milliwatts seconds per square centimeter ($10^3 \mu W\text{-sec/cm}^2$).

A UV disinfection system for the most restrictive reuse applications specified in the Wastewater Reclamation Criteria (2.2 total coliform/100 mL) should be designed to deliver, under worst operating conditions, a minimum UV design dose of 140 $\mu W\text{-sec/cm}^2$ at maximum week flow and 100 $\mu W\text{-sec/cm}^2$ at peak flow (maximum day). The minimum required design dose must be based upon the following conditions:

1. UV lamp output = 70 percent of normal (new) lamp output.
2. Transmittance through quartz sleeves = 70 percent.
3n Minimum allowable wastewater transmittance = 55 percent. (If continuous transmittance data have been collected for a minimum period of six months, including wet weather periods, a higher transmittance valuc may be allowed.)
4. UV dose calculation method = point source summation.
5. UV dose to be achieved with a minimum of three UV banks in series.

Rationale.

Based upon pilot testing of UV disinfection with filtered secondary effluent, it has been found that a UV dose varying from 100 to 120 $\mu W\text{-sec/cm}^2$ will achieve 4 logs ($\log(N/N_0) = -4$) of inactivation of poliovirus.[29,30,32,33]

To be equivalent to Wastewater Reclamation Criteria with chlorine disinfection, 4 logs of virus inactivation are required. The reported UV doses in the cited studies were calculated for a horizontal lamp configuration, using a proprietary series dose estimation model. When the UV dose is recomputed using the point source summation (PSS) method for computing the UV intensity, the corresponding UV dose values are approximately 120 to 140 μW-sec/cm^2.

Based upon the test results cited above, the operational average series, the dose should be equal to or greater than 140 μW-sec/cm^2. The recommended dose estimation procedure is based upon restrictive worst-case conditions. This design procedure provides a common basis for all UV manufacturers on which to base their equipment requirements.

In wastewater reclamation facilities where a constant flow is treated, the design UV dose corresponds to the operating UV dose (140 μW-sec/cm^2). In wastewater reclamation facilities where a variable flow is treated, the probability that minimum fluid transmission, peak flow rate, lamp coating at its design limit, and lamps at the end of their useful life will all occur simultaneously is highly unlikely. Should all of the design conditions occur simultaneously, the standby UV bank or channel (see below) would be put into operation to maintain the operational average UV dose.

To standardize the computation of the average UV radiation (intensity) within the UV reactor, the EPA point source summation method (PSS) is to be used. In the new EPA UV disinfection model (UVDIS Version 3.1), currently under review, the UV intensity computation is based upon the PSS method.

Reactor Design.

UV disinfection contact chambers should be designed with inlet, channel approach, and outlet conditions that promote plug flow within the system. Inlets should be designed to allow for equal flow distribution among the UV channels. Channel approach conditions should allow sufficient distance to develop a uniform flow field upstream from the first bank of UV lamps in a system. The outlet condition should be such that the hydraulic behavior within the last bank of lamps is not affected by the outlet fluid level control device. Good transverse (cross-sectional fluid-level control should also be promoted within the system. To ensure proper inlet and outlet flow conditions, the following criteria are suggested:

1. Unobstructed approach channel length before the first bank = 2 × (channel water depth) or 4 ft minimum.
2. Unobstructed downstream channel length following last bank of UV lamps before fluid level control device = 2 × (channel water depth) or 4 ft minimum.
3. Spacing between UV banks = minimum spacing required for maintenance and access.
4. To avoid the growth of algae containing biofilms, the upstream and downstream portions of the UV reactor channel and the sections of the channel between the UV banks must be covered (lighttight).

Rationale.

Based upon current theory, "turbulent plug flow" is considered to be conducive to good performance within a UV reactor. A properly designed inlet structure and approach channel will help to ensure that plug-flow conditions are imposed upon the first bank of lamps in a UV disinfection channel. Tests performed upon horizontal lamp arrays indicate that the lamps are effective in promoting plug-flow conditions. Therefore, hydraulic conditions are forced into a "plug-flow-like" condition by passing through UV lamp modules.

A properly designed outlet structure will ensure that outlet conditions do not affect fluid behavior within the last UV array because the lamps themselves virtually guarantee plug-flow behavior. Therefore, favorable hydraulic conditions may be ensured in the entire reactor, if inlet and outlet structures are designed properly.

Reliability Design for UV Systems.

As the requirements for the reclamation and reuse of wastewater effluent are necessarily stringent, special attention must be devoted to the reliability of any proposed UV system.

The design of the system, such as the number and the configuration of UV banks, plus the provision of standby power, are of critical importance. Another condition affecting installations in California is the continuous seismic activity. This is an important design consideration.

Number of Banks. Key elements of the UV design include a minimum of three banks in series and the provision for UV system redundancy. The UV channel system should be designed to deliver the minimum design dose under the worst operating conditions, with a minimum of three UV banks in series.

System Redundancy. Standby capacity, which is the basis of redundance, can be provided as follows:

1. Provide either one standby UV bank per channel, or one or more standby channels, depending upon the size of the installation. Where a standby UV bank is used, a minimum of four UV banks must be installed in each disinfection channel.
2. Provide automatic alternation of the standby bank or standby channel.

Rationale. Based upon current experience with the most reactive wastewater reclamation criteria, it appears that a minimum of three banks is necessary to minimize short-circuiting and inadequate disinfection that could occur with fewer banks.

For example, with a single bank of lamps in service, an individual lamp failure may result in organisms passing through the bank without receiving an adequate dose of UV.

Also it must be considered that one bank at a time must be removed for cleaning and maintenance. This means that there is a need for at least four bank locations to be provided for each channel. Any event, such as multiple lamp failure or a decrease in filter effluent quality, that reduces the effectiveness of an individual bank of lamps can be offset by activating a standby bank.

Backup Power. The reliability of backup power and power supply should be provided as required in the Wastewater Reclamation Criteria. In addition, to ensure a continuous supply of power to all facilities, should one of the power supply lines fail, a looped power distribution system should be provided.

Another feature that could enhance the system reliability at minimal cost increase would be to divide the disinfection system components of the same type (i.e., banks) between two or more power distribution panel boards or switchboards.

Rationale. As the UV disinfection system cannot operate without electrical power, backup power is essential to ensure continuous disinfection, unless the reclamation plant has alternative reliability provisions or capabilities as outlined in the Wastewater Reclamation Criteria.

Using multiple panel boards of switchboards would allow part of the system to remain on-line, even if one of the power distribution panel boards should fail.

Electrical Safety Design. All UV systems must be provided with approved ground fault interrupt circuitry.

Rationale. This type of electrical circuitry is required to minimize the hazard to plant personnel in the event of a lamp break or other event that creates direct electrical contact with water, such as the system's continuous fluid flow channels.

Seismic Design. The UV system should be designed in accordance with the seismic design requirements of section 2312 of the Uniform Building Code for special occupancy structures in the appropriate seismic zone. These same seismic design standards shall apply to structures in which UV replacement lamps are stored on-site.

Rationale. Seismic design considerations are particularly important for UV systems because of the fragile components (especially lamps and quartz sleeves) used in these systems.

The seismic safety design of the UV disinfection system should be at least equivalent to the design of the reclamation facility prior to disinfection. This provision will ensure that whenever the plant is capable of producing effluent, the UV system will provide adequate disinfection at all times.

Monitoring and Alarms

Continuous Monitoring. The ability to monitor operating parameters continuously is of fundamental importance in the operation of any UV disinfection system. The alarm system design is critical in maintaining the performance of the UV system.

Wastewater:

1. Flow rate
2. Fluid transmittance
3. Turbidity
4. Liquid level in UV disinfection channels

UV Disinfection System:

1. Status of each UV bank, On/Off
2. Status of each UV lamp, On/Off
3. UV intensity measured by at least one probe per bank
4. Lamp age in hours

Rationale. The parameters identified above are needed to determine the UV dose and to monitor the operation of the UV system. Flow rate and transmittance are needed to determine the average UV dose. Turbidity monitoring can be used to turn on an additional UV bank, or a UV reactor in response to deteriorating filter effluent quality.

The depth of water in the UV channels must be controlled carefully to prevent the depth of water above the highest UV lamps from exceeding the predetermined design maximum value. This would prevent organisms from being exposed to a lower-than-average UV intensity and would prevent lamps from being out of the flow and losing the effect of their UV light owing to low water levels. The status of each bank and UV lamp is needed to provide on-line monitoring of the operation of the UV system and to control the alarm system. The UV intensity and lamp age are used to determine the average UV intensity.

Continuous Monitoring of UV Dose. The average UV dose delivered to each UV reactor is to be monitored continuously. The average UV intensity within the reactor is to be computed by using the PSS method. Input data for the computation of UV intensity include: (1) fluid transmittance, (2) the degree of lamp fouling, and (3) average age of UV lamps.

The average UV dose is then computed by multiplying the exposure time, determined from the flow rate, by the average UV intensity. The UV dose should be increased automatically to compensate for high turbidity readings or low UV dose, or both, by activating an additional bank or channel, depending upon the method used to provide redundancy (see above).

The alarm system should be activated if the average UV dose falls below 140 μW-sec/cm^2 for more than 3 min.

Rationale. Continuous determination of UV dose is technologically feasible and is consistent with the current requirement for continuous chlorine residual monitoring. The use of the average UV intensity to calculate the UV dose is a reasonable approach advocated by the EPA and the industry. The average UV intensity accounts for the UV light that an organism would be subjected to if it were traveling through the reactor under turbulent flow conditions.

Alarms. Both high-priority and low-priority alarms are required to operate a UV disinfection system effectively. Major alarms, if left unattended, may compromise the performance of the UV system. Minor alarms, if left unattended, will not compromise the UV system performance. The requirements for major and minor alarms are as follows:

High-Priority (Major) Alarms:

- Adjacent lamp failure—two or more adjacent lamps have failed.
- Multiple lamp failure—more than 5 percent of the lamps in a bank have failed.
- Low-low UV intensity—the lamp bank probe intensity drops below a field-adjustable setpoint.
- Low-low UV transmittance—the influent wastewater UV transmittance drops below a field-adjustable setpoint.
- High-high turbidity—the influent wastewater turbidity exceeds a field-adjustable setpoint.

Rationale. The UV dosage should be increased by activating the redundant channel or lamp banks in response to high-priority alarms when the UV disinfection performance is being threatened.

Low-Priority (Minor) Alarms:

- Individual lamp failure—location of the lamp is indicated by the bank module and lamp sequence.
- Low UV intensity—the lamp bank probe intensity drops below a field-adjustable setpoint.
- Low UV transmittance—the influent wastewater UV transmittance drops below a field-adjustable setpoint.
- High turbidity—the influent wastewater turbidity exceeds a field-adjustable setpoint.
- High or low water levels in the disinfection channel

Rationale. Low-priority alarm indicates that maintenance is required. For example, low or high water levels in the UV channel should cause a low-priority alarm, requiring the operator to investigate the problem. The operator may activate an additional UV channel during investigation or repair.

UV Alarm Records. The UV system should be designed to record automatically all high- and low-priority alarm conditions.

Field Testing before Startup

Hydraulic Residence Time Testing. In addition to field testing of the UV electrical system components, the following hydraulic tests should be conducted to verify the physical performance of the UV disinfection system.

Tracer studies of the UV reactors should be conducted to define residence time distributions. These tests should be conducted at the minimum anticipated flow and at 50, 75, and 100 percent of the peak design flow.

Tests should be conducted with all of the UV banks and lamps in place. In large open channel systems, where complete mixing of the tracer before entering the UV bank cannot be achieved, the tracer should be injected at nine points, uniformly distributed over the cross-sectional plane, perpendicular to the fluid flow.

Hydraulic residence time testing may be waived for small treatment plants on the approval of the Department of Health Services. Approval to waive on-site tracer testing should be based upon objectively substantiated tracer testing on an equivalent installation of a particular UV disinfection unit.

Rationale. The numerical relationships between the average exposure time and flow rates for a reactor can only be verified by a tracer study.

Water Level Verification. The downstream water level control system should be capable of maintaining the water level within a prescribed range.

Rationale. Verification of the variation of depth of flow in the UV channel is of critical importance in computing the average UV dose.

Compliance Monitoring

Importance of Sampling. Compliance monitoring for UV disinfection systems is without a doubt the most important phase of the entire operation. In order to take into account the well-known "clumping phenomenon," White insists that "grab samples" be extracted from the UV channel via a pump just as a chlorine residual analyzer receives its sample. The sample should flow to the tap at about a 5 ft/sec velocity to break up any clumping solids and at the same time keep the piping clean. Over the years White has found, in practically every case, that there was always a higher coliform count in the analyzer's sample than in a grab sample.

Grab Samples. Compliance monitoring for UV disinfection systems includes grab samples, a variety of continuous on-line measurements, and continuous monitoring of the UV dose.

Routine monitoring based upon representative grab samples should include the following:

1. Total coliform bacteria, daily.
2. Suspended solids, daily.

The samples for both coliform bacteria and suspended solids shall be collected daily at a time when wastewater characteristics are most demanding on the treatment facilities and disinfection facilities. Suspended solids may already be collected in connection with the effluent filtration system.

Rationale. The required sampling program for compliance is consistent with the existing program set forth in the Wastewater Reclamation Criteria. The results of the suspended solids testing should ultimately be correlated to the corresponding turbidity readings. Furthermore, White claims that grab samples do not account for the clumping phenomenon. He points out the enormous differences that he has found on-site between the total coliform count of a sample taken from the sample pumped out of a channel going to a chlorine residual analyzer versus a grab sample out of the open channel. On secondary effluents the difference is very significant; all one has to do is try it.

Continuous Measurement. Routine UV disinfection system monitoring, based upon continuous on-line measurement, should include:

Wastewater:

1. Flow rate
2. Fluid transmittance
3. Turbidity
4. Liquid level in the UV disinfection channels

The use of fluid transmittance measurements based upon grab samples may be allowed for small treatment plants by the approval of the Department of Health Services.

UV Disinfection System:

1. UV intensity
2. Lamp age in hours

Rationale. Submission of data on the above parameters will serve to demonstrate that the UV disinfection system was operational on a continuous basis.

UV Dose. The operational average dose must at all times be at least 140 μW-sec/cm^2, based upon continuous monitoring. The average dose is the product of the average UV intensity and the average exposure time of the fluid. The average UV intensity is determined by using the point source summation (PSS) method.

The average exposure time shall be determined from the residence time distribution curves as determined by field testing before start-up.

Rationale. Continuous disinfection in conventional wastewater reclamation treatment processes is indicated by the presence of a specified disinfectant residual because coliform bacteria are measured infrequently. As this concept of chlorine residual is not applicable to UV, a different means of assuring that the minimum average UV dose is maintained is required.

The continuous on-line monitoring of the average UV dose, in conjunction with the other continuous parameter monitoring data, is to serve as a substitute for the measurement of residual in chlorine disinfection.

Engineering Report*

For existing and proposed wastewater reclamation facilities, for which an engineering report has not been submitted, a complete engineering report must be submitted prior to implementation of disinfection with a UV system.†

For existing wastewater reclamation facilities for which an engineering report acceptable to the regulatory agencies has been submitted, and for which UV disinfection is proposed, the following types of reports may be required:

1. A complete, updated engineering report may be required if, since submission of the original engineering report, changes or modifications have occurred in the production of reclaimed water, such as raw or treated water quality, treatment processes, plant reliability features, or monitoring, or operation and maintenance procedures, reclaimed water transmission, including the distribution system, or water reuse areas (e.g., type of reuse, use area controls, or use area design). The submission of a complete, updated engineering report in lieu of an abbreviated report in which only the UV disinfection system is addressed will be at the discretion of the regulatory agencies.
2. An abbreviated engineering report in which only the UV disinfection system and attendant treatment and reliability features is addressed is acceptable only if proposed modifications solely involve the disinfection processes, such as replacing or enhancing existing disinfection facilities with UV disinfection facilities.

---

*For complete details required by this report the reader is referred to the National Water Research Institute report of September 1993. See Ref. 28.

†According to the manufacturer. See Ref. 41.

Historical Background of the UV Process

Ultraviolet radiation from mercury arcs has been used since 1909 for the sterilization of water.[1] A number of cities in Europe have had installations on water supplies with capacities up to 400–500 gpm. However, the costs of these installations greatly exceeded those using chlorination systems, which explains in part the limited use of the UV method on public water supplies.

In the past, before the adoption of chlorine, the major use of UV radiation was for the disinfection of potable water. Potable water is never free of bacteria; there is always a low level of bacteria colonies in any potable water supply, which can be substantiated by the heterotrophic plate count (HPC). Therefore, it is reasonable to expect that many processes using potable water want or require close control for complete elimination of all bacteria in the system. Some of these applications include continuous UV exposure to recycled distilled water in the pharmaceutical and cosmetic industries and UV radiation for kidney dialysis water, as well as many other applications involving specialty products that require an almost sterile water supply. The beverage market is a big one for UV systems, which are used to sterilize makeup water for frozen juice concentrates and final rinse water for bottles and containers, and as a final treatment for bottled water. They are also used almost exclusively for cottage cheese washwater. Fish hatcheries have used UV to disinfect the water to avoid the toxic effect of chlorine.

Since 1975, interest in UV as an alternate disinfectant for wastewater in the United States has gained momentum, particularly after the publication of proceedings of a national symposium on wastewater disinfection in 1979. The application of UV disinfection of secondary wastewater effluents has increased in both the United States and Canada. The principal forces behind the use of UV as an alternative to chlorination include concern about the possible harmful effects of the DBPs caused by chlorination on aquatic life in the receiving waters, as well as the risk factor in handling chlorine. Now (1996) aquatic life has long been protected by dechlorination, so that is no longer a valid point; and DBPs have been shown to be trivia. However, the Uniform Fire Code has done the most damage to disinfection of potable water and the safety of plant operators. We can never recover from that harm.

Evidence of the widespread use of UV in recent years has been substantiated by the number of manufacturers engaged in this process: Aquafine Corp. (Sanitron Purifiers), California Ultraviolet Co., Capital Controls, Fischer & Porter Co., Hanovia Ltd., Katadyn Systems Inc., Pure Water Systems Inc., Steri-Tronics, Steroline, Ultra Dynamics Corp., and Ultraviolet Purification Systems, Inc.

MAJOR UV EQUIPMENT SUPPLIERS (1996)

Now (1997) many of the above names have either been changed or replaced by new companies. Currently one of the largest and most active company seems to be Trojan Technologies Inc.[31] with headquarters in Ontario, Canada, followed by Ultra Guard,[36] of Burnaby, BC Canada. The latter company is a descendant of Asea Brown Boveri of Switzerland, and its claim for special attention is its high efficiency, high intensity UV lamps, and proprietary flow control technology, plus the ability to handle effluent flows up to 10 billion gallons per day. This company is represented by Service Systems International, Ltd., White Rock, BC (phone 800-488-4544). There are other worldwide companies with both large and small systems, as follows.

Infilco Degremont Inc.[41]: Aquaray 40 VLS Vertical Lamp System

Introduction. Since its introduction in 1986, this system (VLS), according to the manufacturer, remains the most advanced UV disinfection and data management system available. Moreover, this company has regional sales offices in many areas outside the United States, including Canada, South America, France, Italy, Africa, India, Spain, Switzerland, China, Japan, England, Belgium, Turkey, Portugal, the Philippines, Thailand, Singapore, India, Indonesia, Mexico, and Germany.

This UV system uses energy-efficient low-pressure lamps in the vertical position, which produces efficient flow pacing with low head loss. This low-head-loss lamp array promotes highly desirable semiturbulent plug flow to ensure that all water passing through the lamp array receives maximum exposure to the germicidal wavelength of the UV light. (See Fig. 17-1.)

Equipment maintenance is simple, and regular maintenance can be automated. Its features include full computer control and monitoring, along with individual lamp management. The system is easily expanded to meet future needs, and its most important feature is that it can meet the most stringent water reclamation applications.

Vertical Lamp System Details. This system employs the very latest control and monitoring techniques,* which result in the most cost-efficient life cycle operation of any UV system. Its power consumption is lower than that of other UV systems due to the use of high efficiency lamps along with an enhanced electronic power management system.

All microorganisms contain proteins and nucleic acid as their main constituents. The germicidal wavelength of UV light (254 nm) disrupts these constituents and destroys the ability of the microorganisms to reproduce. Therefore, the microbes have to be exposed to a lethal dose of UV energy.

UV dosage is a function of average nominal UV intensity multiplied by contact time. The average nominal UV intensity is conservatively calculated

*This means UV Transmission monitor with 253.7 nm wave length microprocessor for automatically monitoring the UV transmissivity.

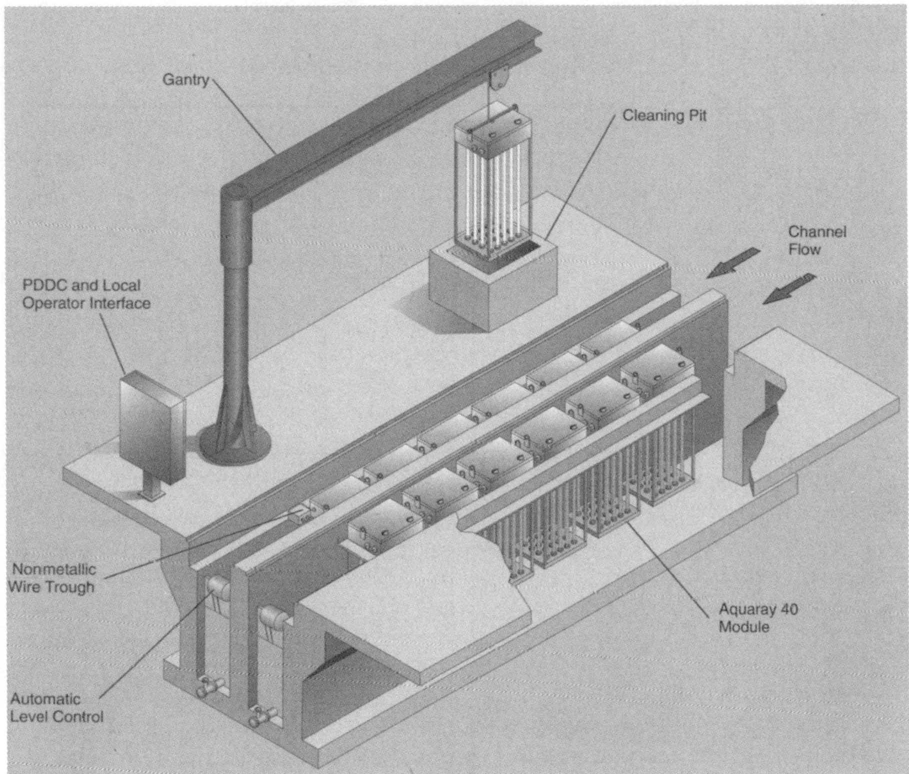

Fig. 17-1. Typical Aquaray 40 VLS wastewater system layout (courtesy of Infilco Degremonte Inc.).

by using the EPA UV intensity requirement, which establishes an industry standard for each lamp array.

The Aquaray 40 VLS uses an advanced, uniform, staggered array, with the flow direction perpendicular to the lamps (Fig. 17-2). This staggered lamp array has proved to dramatically enhance system performance. It also minimizes the effects of failed lamps in the array because it is impossible for an organism to avoid the UV radiation field. Even with a failed lamp, the organisms will still encounter repeated areas of maximum intensity of UV energy.

The heart of the system is the lamp module. This unit contains forty 58-inch arc length high-performance UV lamps arranged in a uniform staggered array of five rows of eight lamps each. The lamps are housed inside precision quartz jackets with test-tube ends that are supported in neoprene holders. The open top end terminates inside the top enclosure, and is sealed within the enclosure by using a neoprene compression gland to ensure a waterproof seal.

The module itself is made of 304L and 316L stainless steel, with a NEMA 4X top enclosure, four support legs, and a bottom support pan. The module

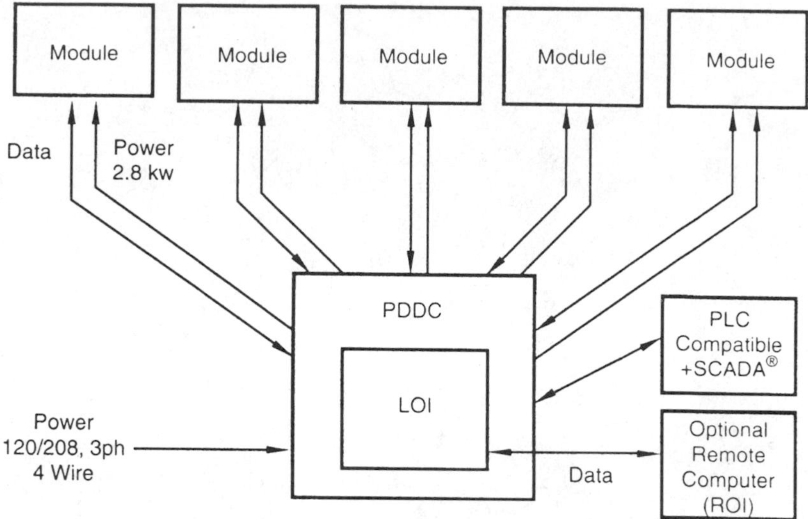

Fig. 17-2. Schematic diagram of the Aquaray 40 VLS system (courtesy of Infilco Degremonte Inc.).

houses electronic ballasts and lamp control assemblies, plus the data control assembly. Each of these plug-in components can be easily accessed through the open lid of the module without removal of the module from the channel. The module bottom pan includes orifices adjacent to each lamp, which allows air to be precisely introduced into the lamp array for effective and fully automatic lamp precleaning and cleaning.

Lamps are easily and safely accessed for service with no need to remove or raise equipment from the process water flow. Raising the module cover causes the lamps to be switched off automatically.

Level Control. Since any UV disinfection system has to react to a wide variety of process flows, from minimum to wet weather, two effective methods are available to achieve water level control:

1. Automatic flap-gate level controllers. They provide reliable level control, and are used when space is at a premium.
2. Extended straight edge weir designs. With the vertical lamp system, the acceptable peak crest can be twice that of the horizontal system.

Flow Pacing. In response to the treatment plant flow meter, the Aquaray control system adjusts the number of modules/rows of lamps in service. Since each row of lamps within the module can be independently controlled, the switching increments can be better optimized, and can ensure that the minimum number of lamps necessary are used for treatment. The Aquaray system is designed to minimize energy consumption to 70 watts per lamp, and extend the useful lamp life, by very accurately controlling the number of lamps in service.

Maintenance. Routine maintenance on the Aquaray systems is safe, simple, and quick. The operator opens the appropriate module. A safety interlock automatically switches off the lamps. The operator can then check the particular lamp, and it is a simple matter to unplug and remove the failed lamp from its quartz jacket. A new lamp is then put into place and plugged in.

With the lid secured and latched, the UV lamps are automatically switched back on. Typically this entire procedure takes less than 2 min. When one is working on the Aquaray systems, it is not necessary to lift individual modules out of the channel; there is no waiting for high temperature lamps to cool down; and there is no complicated dissassembly or assembly required of worn-out wipers or other parts, as seems to be necessary with other UV systems.

The high performance Aquaray electronic ballasts can be easily accessed and changed without removing the module from the channel. When the module enclosure lid is opened, the lamps turn off automatically. The ballasts are mounted around the inside of the module enclosure; so it is a simple matter to unplug a ballast and lift it out of the enclosure. A replacement can then be installed and secured. Once the lid is closed and relatched, the lamps will automatically switch back on. Typically this procedure takes just 5 min.

Quartz Jacket Fouling. In any UV system, the primary fouling problem encountered is the deposition of minerals on the outside of the lamp's protective quartz jacket. Under the influence of heat from the UV lamp itself, calcium bicarbonate in the process water forms a precipitate, which deposits itself as a scale on the lamp's protective quartz jacket.

With the passage of time, this occurrence reduces system efficiency. This phenomenon is particularly evident in medium-pressure systems, where UV lamp surfaces reach 600°C to 900°C. In these cases fouling can occur in a few hours of operation.

The Aquaray 40 VLS module has a patented integral "air-scrub" feature, which serves the dual purpose of assisting during chemical cleaning and functioning as a precleaning method. Precleaned air is automatically introduced to the lamp array through the bottom support pan for about 15 to 30 min., once or twice a day. The effect of this air is to vigorously agitate the process water. By regularly subjecting the quartz screens to this air scrubbing, the mineral deposition rate is retarded considerably so that the period between cleaning operations is greatly extended.

This air-scrub feature is built in to every Aquaray 40 VLS module. When used to enhance the chemical cleaning cycle, the air-scrub is connected to either an air supply header or a small on-site compressor.

System Cleaning. The Aquaray 40 VLS System offers a fully automatic cleaning system that significantly reduces maintenance. The air-scrub technology is simple and cost-effective. There is no need for complicated wiper systems, which also demand frequent maintenance or replacement after as few as 1500 cycles.

Ecometrics, Inc.[40]

Ecometrics, Inc., of Silverdale, Pennsylvania, is a fairly recent newcomer to the UV disinfection market. Its main focus is primarily upon the smaller municipal treatment facilities and the industrial water treatment market. Many of these systems are doing strictly what UV has long been used for: polishing high quality potable water to kill and control the microbiological life systems that interfere with the processing of pharmacy products, cosmetics, breweries, chemical plants, canneries, hospitals, kidney dialysis machines, reverse osmosis, and fish hatcheries.

Ecometrics makes both horizontal and vertical in-channel systems for both municipal and industrial applications. The horizontal units are available in peak flow ranges from 25,000 gpd with one module up to 4 mgd with 20 modules. (See Fig. 17-3.) The vertical units are available in peak flow ranges from 0.375 mgd and one module up to 15 mgd and 16 modules. Then, under the heading of "liquid purifier specifications," there is system capacity for clear wastewater of 2 gpm with one 17-inch lamp up to 2000 gpm with one hundred twenty 64-inch lamps. For "clear fresh water" it is 3 to 2600 gpm with the same number of lamps, and for "high purity water" it is 4 gpm to 3200 gpm also with the same number of lamps.

This equipment is suitable for either indoor or outdoor installations. The company also offers UV systems for closed pipe applications that range from 2 to 2000 gpm capacity. (See Fig. 17-4).

Capital Controls Company Inc.

Probably the most active company worldwide is Capital Controls Company of Colmar, Pennsylvania.[39] Their systems are described below. This company makes two modular open channel OXYFREE® UV disinfection systems that cover all the various sizes and effluent qualities of the wastewater to be treated. The Series 8300 (see Fig. 17-5) is limited to effluent flows up to 1 mgd. The series 8200 (see Fig. 17-6) is for any system larger than 1 mgd, with no current limit on the upper size.

All their systems have been bioassay-proved to meet EPA guidelines, by a full-scale test by an independent laboratory. Wastewater disinfection is achieved by UV lamps suspended horizontally on modules and submerged within the process wastewater parallel to the liquid flow. Several modules are arranged adjacent to one another to form a bank of lamps. (See Fig. 17-7) Each lamp bank may have as many as 240 lamps or as few as 4 lamps,

Fig. 17-3. Horizontal open channel UV system (courtesy Econometrics Inc.).

depending upon the volume and the quality of the wastewater being treated. Multiple lamp banks may be arranged in series and/or in parallel to achieve the desired level of disinfection, as well as control and access for maintenance. The single lamp module is shown in Fig. 17-8.

Each lamp bank incorporates an optional UV intensity sensor to monitor the level of UV intensity entering the process water that is being treated. A local control panel houses lamp power supplies, provides for control and monitoring of each lamp bank, and supplies power to the power distribution box. From this box the power is distributed to each lamp module and its respective lamps through watertight cables and connectors.

Large capacity systems can be controlled from a remote location by an interface to the plant SCADA system. Alternatively, all functions can be contained locally at the power supply panel. Lamp banks can be switched on and off, based upon process flow. Historical data such as hour run on lamps, UV intensity, and lamp status can be recorded and displayed on the operator

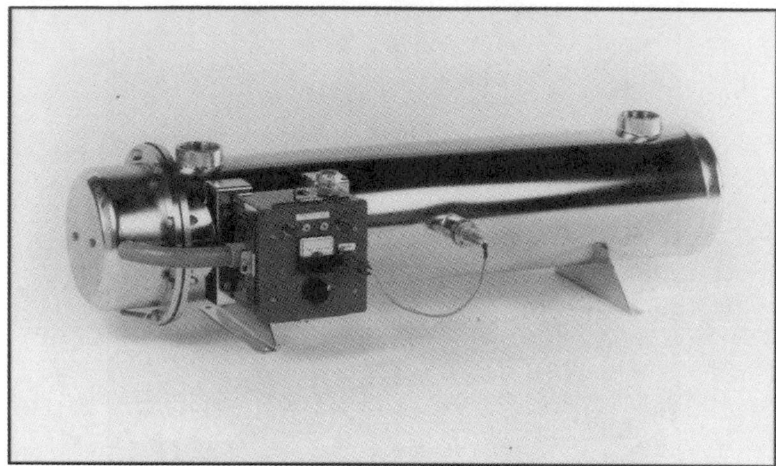

Fig. 17-4. UV system for closed pipe applications (courtesy Econometrics Inc.).

interface. Liquid level within the channel is controlled by an automatic, counter-balanced gate or a fixed weir.

The lamps are nonpropietary, industry standard G64T5L, low pressure mecury vapor, horizontal germicidal lamps. They are configured for proper flow distribution. The lamp module frames are lightweight 316 stainless steel that allows ease of maintenance and reliable operation.

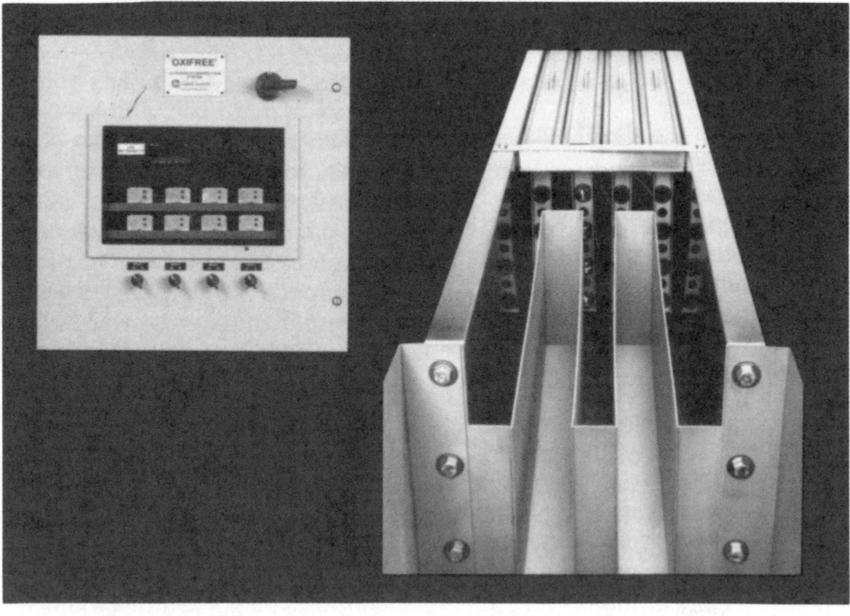

Fig. 17-5. OXIFREE® Series 8300 packaged open channel UV disinfection system (courtesy of Capital Controls).

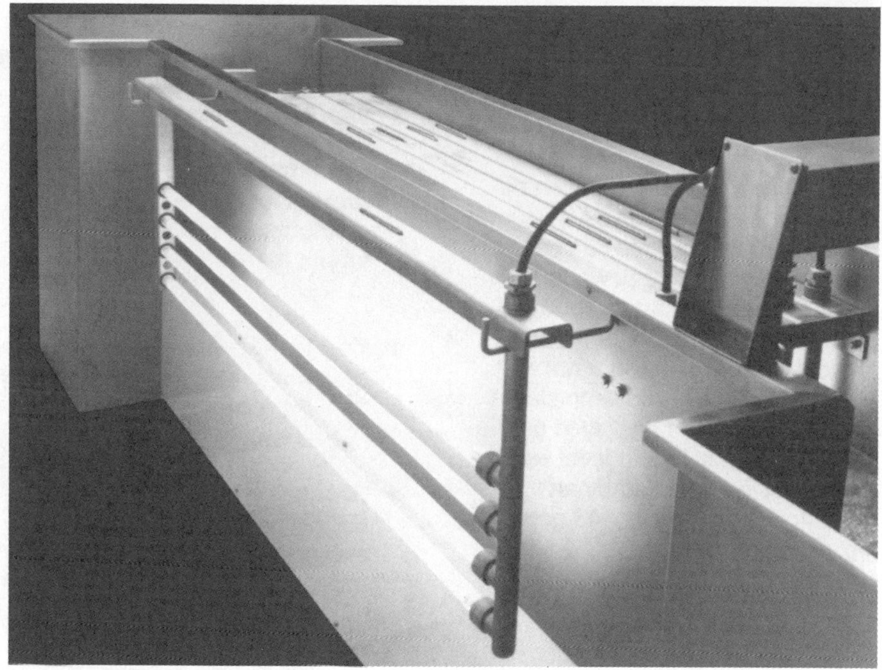

Fig. 17-6. OXIFREE® Series 8200 modular open channel UV disinfection system (courtesy of Capital Controls).

Trojan Technologies Inc.

System Description. This describes the Series UV 3000™ Modular System. This unit incorporates major technological and microprocessing advances in this industry, which have resulted in new standards of system efficiency and cost-effective UV treatment of wastewater effluents.

Gravity flow brings the wastewater to the UV system. The lamp modules are installed in an open channel outdoors, and the system is completely weatherproof. It has a compact system control center, which incorporates menu-driven screens, for operator monitoring and control. This control center can be positioned locally or at a remote location for optimum access to all control/monitoring information. See Fig. 17-9, and Fig. 17-10.

The UV modules are available for a variety of system capacities. There are three for flows under one mgd, one at 1.5 mgd, and another at 3 mgd, as well as one for 50 mgd and one for 500 mgd. These modules, which hold the lamps and outer quartz sleeves in position, are completely immersible because they are contained in a 316 stainless steel frame installed in the channel. No mounting is necessary. A "light lock" on the UV module frame prevents any UV energy from escaping outside the channel. These modules are powered through a weatherproof, industrial-quality cable with a molded PVC connector.

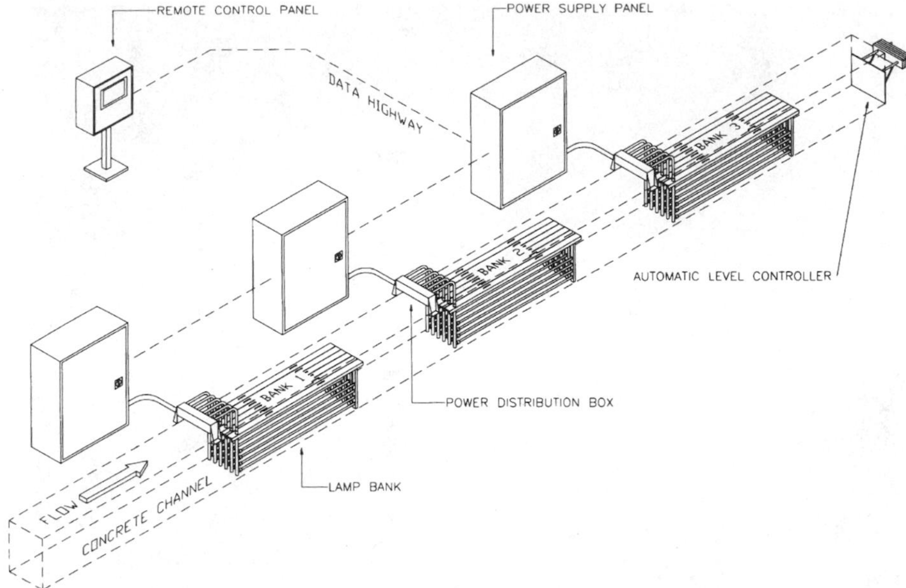

Fig. 17-7. Series 8200 multiple bank UV system with distributed control (courtesy of Capital Controls).

Depending upon local requirements, the instant-start lamps have a 30- or 58-inch effective arc length and will be assembled in modules with two, four, eight, or sixteen lamps each. Each lamp within the module will be separately powered. This permits local or remote status monitoring for each lamp. Both pins on these instant-start lamps are at one end, and are separated by a dielectric barrier that prevents arcing. Expected lamp life is from 10,000 to 15,000 hours, depending upon site-specific conditions, such as on/off lamp cycles, quality of effluent, and frequency of cleaning.

The quartz sleeves, with a nominal wall thickness of 1.5 mm, are sturdy enough for use in a wastewater plant, yet will allow transmission of 89 percent of the output of the UV lamps. The lamps are sealed inside the quartz sleeves by multiple seals, which effectively maintain a watertight barrier around the internal wiring while individually isolating each lamp. These quartz sleeves are designed to last the life of the system, unless accidentally broken or fractured.

At the heart of the Trojan System UV 3000™ is the electronic ballast. This ballast weighs less than 30 percent of an electromagnetic ballast and generates significantly less waste heat. With these improvements in ballast design, Trojan engineers were able to mount the ballast within an enclosure on the module frame. As normal convection cooling is adequate, the Trojan UV system does not need any mechanical cooling.

Instrumentation and Automation. **The Power Distribution Center (PDC)** enclosure is constructed of heavy gauge 304 stainless steel, and is mounted

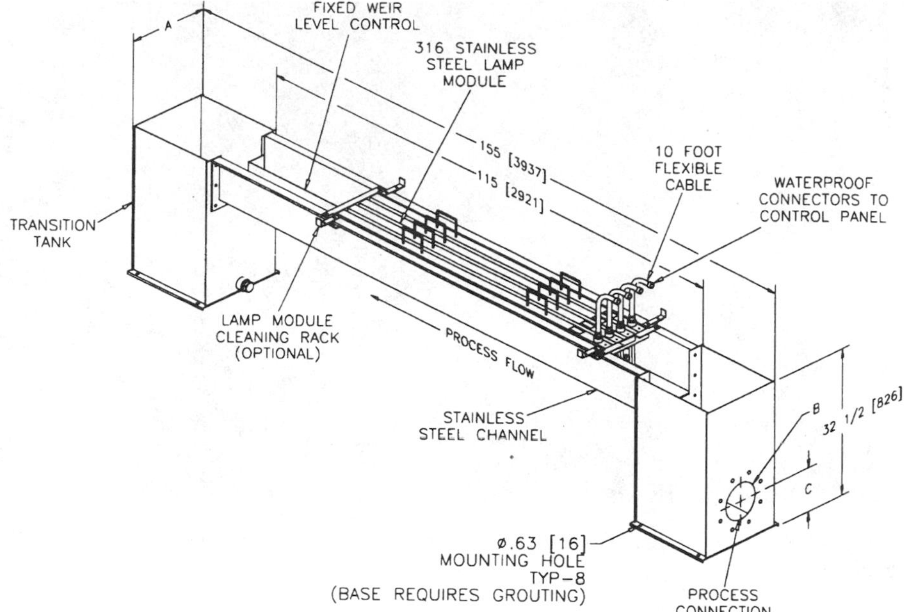

Fig. 17-8. OXIFREE® Series 8200 modular open channel UV disinfection system (courtesy of Capital Controls). Dimensions = inches [millimeters].

above the channel. See Fig. 17-11 The PDC consists of a service entrance and a bus bar power distribution system. The service entrance provides service power to the bus bar. It can be configured to match site power requirements and can be mounted on either end of the PDC enclosure. Power is relayed from a bus bar system to individual UV modules through stainless steel receptacles. All UV modules are individually ground-fault- and overload-protected for safety. Receptacles are not fully mated with a UV module.

The **System Control Center (SCC)** provides monitoring and control of all UV functions. The Trojan On-Line GUI System™ (Graphical User Interface) software, in addition to being a PC-based, "real-time" data collector and controller, is an innovative program that logs and displays historical UV data for process control and refinement. An optional modem hook-up, the Trojan **On-Line RMI System**™ (Remote Monitoring Interface) linking Trojan's factory service to the equipment, is available for providing instantaneous access for review of UV system issues. System data such as UV intensity, individual lamp status, lamp operating hours, UV transmission, UV dose applied, and other parameters are all readily available to the operator. Setpoints are field-adjustable and password-protected to prevent unauthorized changes. Setpoints can be readily adjusted to reflect varying field conditions and specific operating conditions.

For example, the Trojan System UV 3000™ can be flow-paced to turn off or shut down banks of UV lamps in response to changes in the process flow

Fig. 17-9. UV 3000 open channel disinfection system (courtesy of Trojan Technologies, Inc.).

rate. Systems with multiple UV banks and channels can be automatically controlled by the SCC to maintain equal wear on the UV lamps. See Fig. 17-12 for the System Control Center.

Each installation is provided with a UV detection system. It consists of a submersible, stainless steel UV detection probe. One is provided for each UV bank. It transmits a precise measurement of the UV intensity to the Power Distribution Center via a waterproof cable. The UV probe monitors short-term variations in the UV-absorbing qualities of the effluent, any fouling of the quartz sleeves, and any long-term decrease in UV lamp output.

Alarm System. The Trojan System UV3000™ differentiates between major alarm conditions, where disinfection could be compromised, and minor alarm conditions, where routine maintenance is required. The alarm system will even identify the exact location of a failed lamp by its unique bank address.

Many other alarm conditions can be programmed into the software to create a tailored alarm environment. All alarm histories are stored to provide a record for plant maintenance personnel.

To ensure optimum disinfection, the Trojan-designed Automatic Level Controller (ALC) keeps the level of effluent in the channel containing the UV system within one inch of optimum, irrespective of flow rate.

Channel Level Control. Another very important part of the overall system is the Trojan UVT2537 microprocessor-based instrument for automatically monitoring the UV transmissivity of wastewater at the 254 nm wavelength.

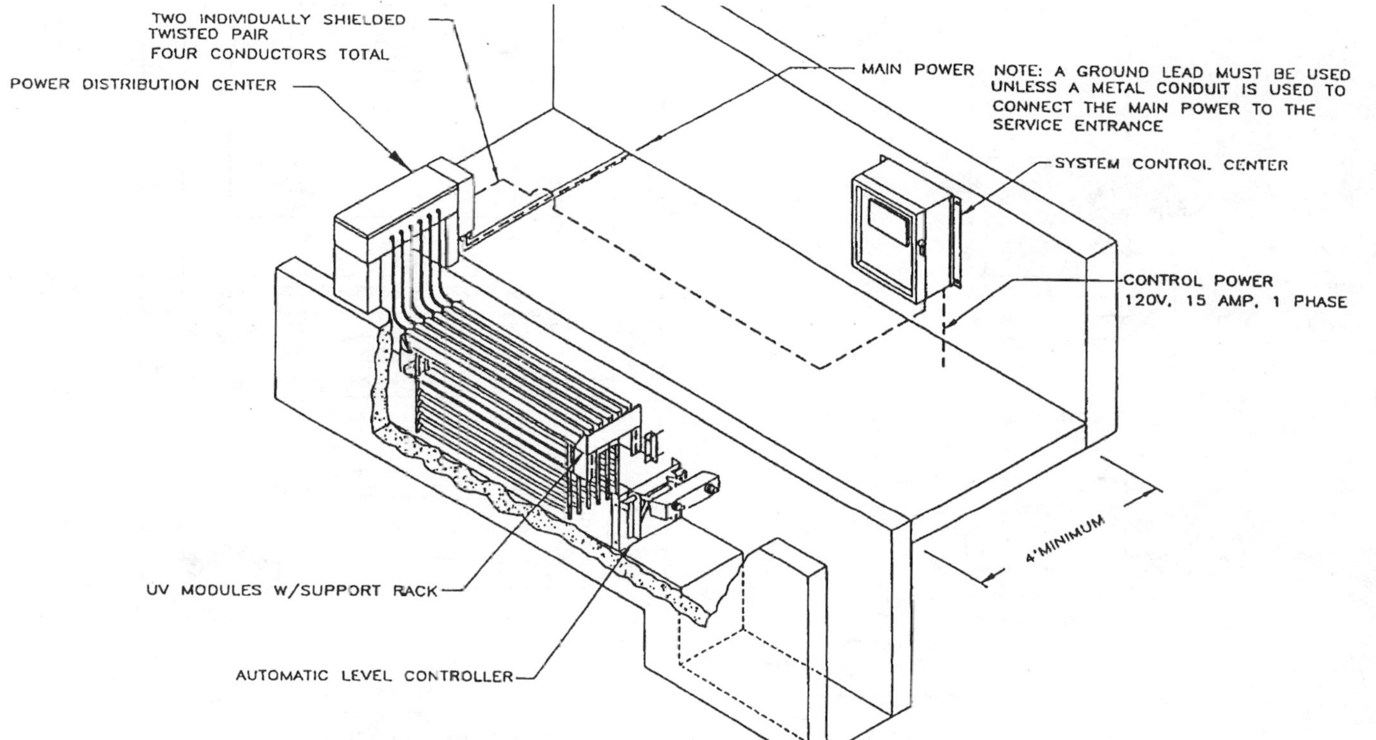

Fig. 17-10. UV 3000 disinfection system installation diagram (courtesy of Trojan Technologies, Inc.). Unless otherwise specified, dimensions are in inches; tolerances as per Trojan Technologies; engineering reference standards.

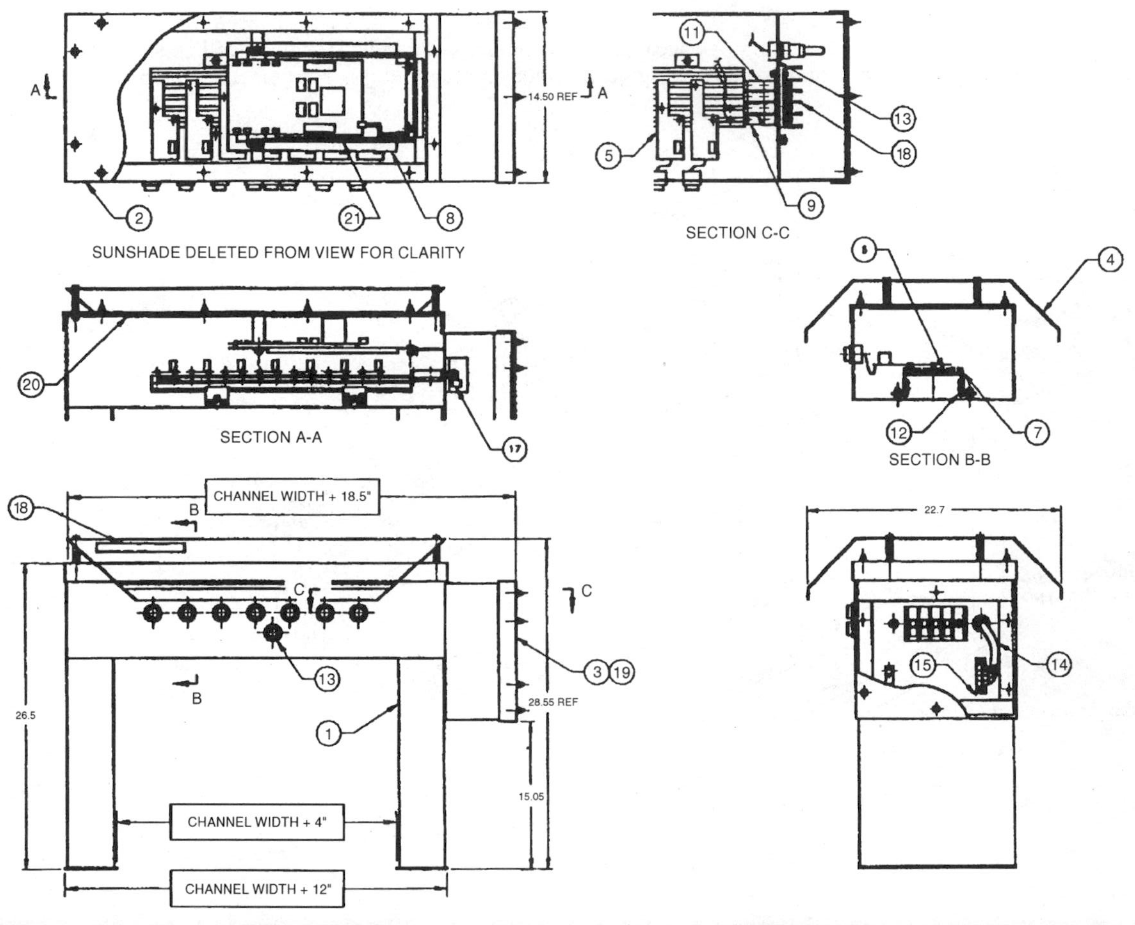

14.50 REF
SUNSHADE DELETED FROM VIEW FOR CLARITY
SECTION C-C
SECTION A-A
SECTION B-B
CHANNEL WIDTH + 18.5"
22.7
28.55 REF
26.5
15.05
CHANNEL WIDTH + 4"
CHANNEL WIDTH + 12"

In-line calibration checks ensure that continuous, accurate readings are available for process control. The LCD display screen provides a visual trending of the UV transmission over a 60-min. or a 24-hour period.

Cleaning Systems. The Trojan System UV 3000™ has been designed to facilitate easy cleaning of the UV modules. Individual modules can be removed from service and cleaned manually or in a custom-made cleaning tank. The custom-made mobile cleaning tank is made of 304 stainless steel, and comes equipped with an air compressor for cleaning solution agitation, as well as with a support rack for servicing individual components within the UV modules. In large installations, cleaning is accomplished by an overhead crane, which relocates large banks of UV modules to a cleaning basin. The rugged stainless steel module frames and support racks permit transport of the modules, either individually or as a bank, without any threat of breakage.

Maintenance. With no moving parts and very little wiring and electrical connections, a Trojan system requires very little maintenance. Moreover, the system continually monitors itself and incorporates alarms that will alert operators of any malfunction.

Regular cleaning of the quartz sleeves helps to maintain efficiency of operation. Suspended matter, such as algae, plastics, and so on, can easily be hosed off with water while the system remains in operation.

Lamps will need changing after approximately 13,000 hours (18 months) of continuous operation, but with flow-pacing, may last up to three years. Replacing lamps is an easy task: no tools are required, and it takes less than two hours to replace a 50-lamp system.

Customizing the System. Prospects for new installations of a UV disinfection system should let the Trojan engineers customize the system that is just right for your needs. They will need to know the following:

- average and peak flowrates of process water;
- type of process water: primary, secondary, or filtered;
- secondary (tertiary);
- influent conditions, such as turbidity (NTU), total suspended solids (TSS);

Fig. 17-11. Power distribution center UV 3000 disinfection system (courtesy of the Trojan Technologies, Inc.). Notes: Material--304 SST, PDC rating NEMA 4. 1, PDC weldment; 2, PDC busway cover; 3, PDC service entrance cover (STD); 4, PDC sunshade; 5, OES relay board assembly; 6, copper bus bar; 7, insulation extrusion; 8, communication controller assembly; 9, feed-thru rod; 10, feed-thru rod bushing; 11, insulating shrink sleeve; 12, busway mounting lug; 13, probe receptacle; 14, communication harnass; 15, terminal strip; 16, feed-thru rod insulator; 17, power terminal; 18, label--12.5" logo; 19, label--10" logo; 20, gasket--PDC covers; 21, interconnect cable.

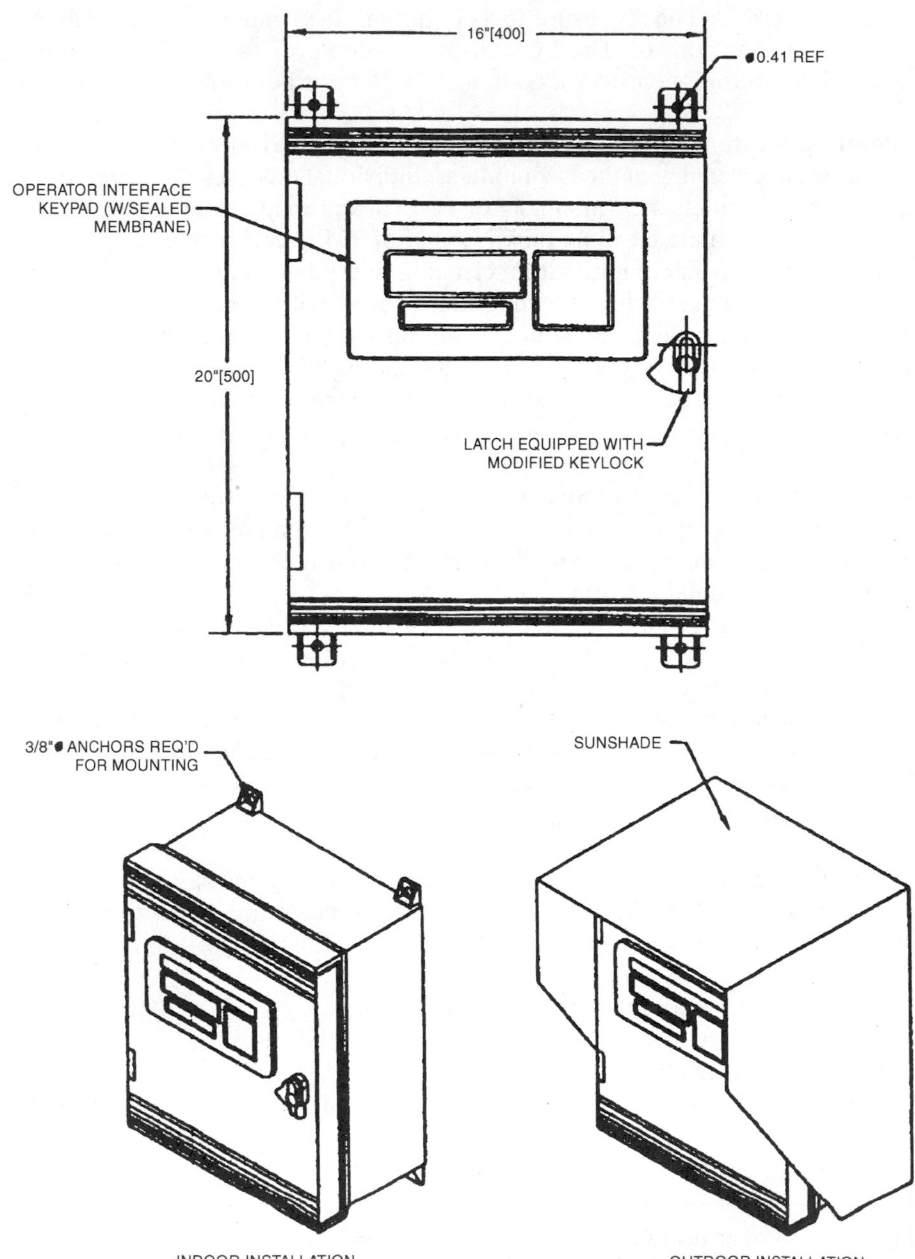

Fig. 17-12. Type AFW-UV3000 (SCC) disinfection system, installation details (courtesy of Trojan Technologies, Inc.). Notes: (1) Panel is supplied with protective hood for outdoor installations; (2) panel size is approximately 20" × 16" × 9"; (3) panel is NEMA 4 rated by Canadian Standards Association; (4) mounting by others; (5) electrical service required--12V 60Hz 15 amps.

- disinfection standard to be achieved.

System Description II This is a description of System UV4000™. Recent advances in UV technology have resulted in the development of this system to overcome the operational limitations of low intensity lamp systems for larger scale applications and for the treatment of lower quality wastewater. This unit is illustrated in Figure 17-13.

System UV4000™ is a reliable alternative to conventional disinfection methods. Minimal space requirements and reduced labor costs ensure cost-effective operation. See Figure 17-14 for a typical operating configuration. Key features of the UV4000™ include:

- use of fully automated, self-cleaning technology;
- use of high lamps, with >90 percent reduction in lamp requirements;
- UV dosing based upon continuous effluent monitoring and variable lamp output;
- multiple UV banks in series,
- open-channel, gravity flow configuration (see Fig. 17-15);
- fully automated self-cleaning technology.

Fig. 17-13. UV 4000™ disinfection system (courtesy of Trojan Technologies, Inc.).

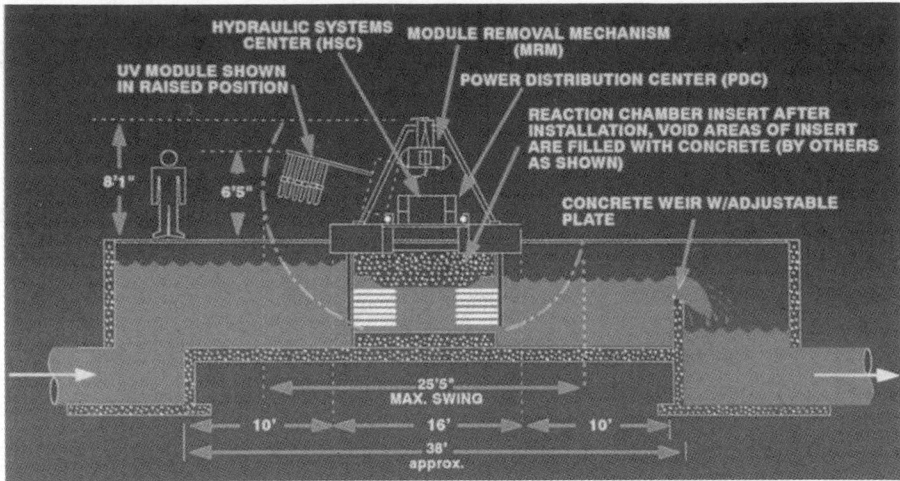

Fig. 17-14. Typical system configuration of UV 4000 (courtesy of Trojan Technologies, Inc.).

High-Intensity Lamps. The use of high-intensity lamps extends the range of UV disinfection to poorer quality effluents such as primary effluents, combined sewer overflows, and stormwater.

The high-intensity UV lamps are positioned within the UV reactor in an array that provides controlled water layer geometry at all flows. The controlled geometry of the UV reactor eliminates the potential for short-circuiting of flow at high velocities or low transmittances.

Fig. 17-15. System UV 4000™ in operation, treating 19 MGD of wastewater (courtesy of Trojan Technologies, Inc.).

Cleaning of the Quartz Sleeves. These sleeves are indeed important because they protect the UV lamps. Cleaning is accomplished by using a dramatically new technique developed by Trojan's scientists and engineers. A combination of mechanical and chemical cleaning is used to aggressively remove the deposits. The complete cleaning cycle takes place with the UV modules in their normal operating position, without interrupting the disinfection process.

Cleaning cycles are activated by timer or UV light sensors, and are programmed to conduct sequential cleaning of all modules within each operative bank. Even large systems can be cleaned in minutes, without the need for a system shutdown or bypass.

Automated Cleaning Cycle. The fully automated cleaning cycle is programmed for each installation and is set to operate as frequently as once per hour, depending upon the rate of fouling. Continuous operation of the UV system with clean sleeves will enhance overall system performance by eliminating the energy that is wasted due to system fouling. Under no-flow conditions, lamps are turned off automatically to ensure that no energy is wasted.

Customer Service Trojan Technologies main office: 3020 Gore Road London, Ontario, Canada N5V 4T7. Tel: 519-457-3400 Fax: 519-457-3030.

Safe Water Solutions Inc.

Introduction. This Clean Water System[42] using UV as the primary disinfectant was developed in Brown Deer, Wisconsin by G. Michael Furst, Jr., P.E. the product manager. This is a very special type of UV system, because it was developed solely for the inactivation of the oocysts of *Cryptosporidium* and *Giardia*. However, if this device can achieve a proper level of disinfection of these two organisms, it will undoubtedly be able to disinfect all the other pathogenic organisms that are found in potable water supplies.

It is not surprising that the inventors of this system are from Wisconsin, since the largest U.S. drinking water system to be struck by *Crypto* was the City of Milwaukee, where more than 400,000 residents suffered terrible consequences, such as high fevers, vomiting, diarrhea, abdominal cramps, and severe headaches.

The innovative technology for the inactivation of these oocysts is used without any chemical addition or filtrate disposal. The marketing people simply state that the basic technology is "enhanced UV" application. This reminds us that UV still will not kill or inactivate any organism or particle that can be seen with the naked eye.

This system, which the owners call a "device," was tested by Clancy Environmental Consultants[43] of St. Albans, Vermont, and after that infectivity studies were conducted by the University of Arizona, College of Agriculture, Department of Veterinary Science[44] in Tucson, Arizona.

Additional information for this work was obtained from researchers in Scotland.[45] This work was done at the University of Strathclyde, Scottish Parasite Diagnostic Laboratory, Stobhill NHS Trust, Springburn, Glasgow, Scotland G21-3UW. The report on all this work was received by the Clancy group in 1995. The information reported by this group from Scotland carries great weight because both England and Scotland have been dealing with *Cryptosporidium* in some of their large potable water distribution systems since 1988–89, and several hundred thousand water consumers have been poisoned by this oocyst. White strongly suggests that the SWS systems be tested for both pre- and postchlorination procedures. Ordinarily prechlorination has been found by overseas health organizations to make oocyst removal by filtration much more effective, and postchlorination could only improve disinfection without any worries from DBPs.

System Description Model CID-2S. This system is designed to be able to inactivate both *Cryptosporidium* and *Giardia* oocysts using an entirely different inactivation system from those used for other pathogenic organisms.

The SWS Model CID (Cryptosporidium Inactivation Device) has two treatment chambers. Each chamber is divided into two discrete cells by a specially designed filter screen, which has nominal 2-micron openings. There are three sets of low pressure UV lamps on either side of the screen. These lamps generate a 253.7 nm wavelength, which is within the germicidal wavelength band of 250 to 270 nm.

The lamps are enclosed in Teflon-coated quartz sleeves to minimize fouling. Each of the sleeves has two lamps in it, one on duty and one on standby. Three of these mercury vapor UV lamps are in operation in each cell at all times. The lamp operation status is continuously monitored and is displayed on the control panel. (See Figures 17-16, and 17-17.)

Water flow into the CID system is continuous. Water flows through the screen in chamber one, where it is irradiated by three UV lamps. Then the flowing water departs from chamber one and goes into chamber two through another identical screen. After a preset dosage of 4000 μW-sec/cm^2 irradiation

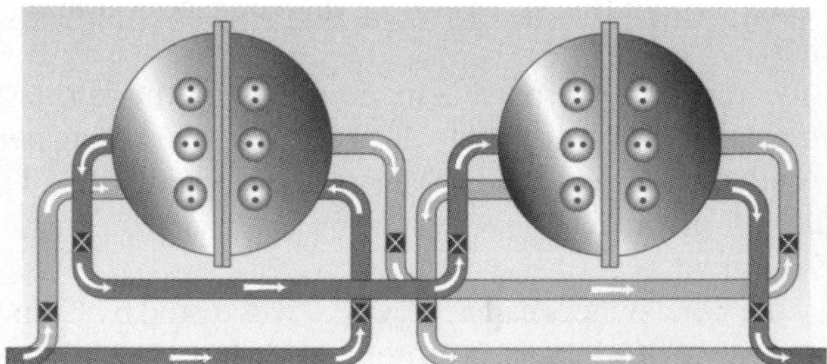

Fig. 17-16. Flow diagram of model CID-25 Cryptosporidium inactivation device (courtesy of Safe Water Solutions, Inc.).

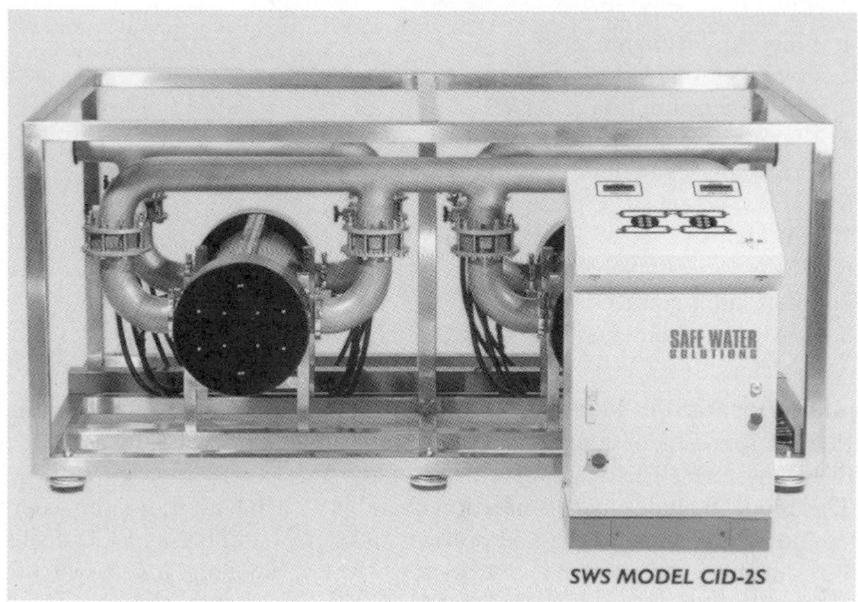

Fig. 17-17. Model CID-25 Cryptosporidium inactivation device (courtesy of Safe Water Solutions, Inc.).

per chamber, a series of eight motorized valves are automatically operated, which reverse the flow from chamber two to chamber one. Any microorganisms trapped on the screen in chamber one are flushed off of that screen and onto the screen in chamber two, where they all receive a second irradiation dosage of 4000 mW-sec/cm^2. After this dosage is achieved, the valves reverse again, sending the inactivated microorganisms out of the machine. All organisms in the water flow receive two doses of irradiation of 4000 μW-sec/cm^2 each. Therefore, with each pass through the system, each microorganism will have absorbed 8000 μW-sec/cm^2 of irradiation by UV light. Infectivity tests indicate that this system removes greater than 3 logs of the oocycts.

Key Features:

1. Flow rates up to 880 gpm.
2. Unique in-line system.
3. Delivery of a double dose of UV irradiation independent of flow rate.
4. Adjustable dosage control.
5. High quality, low maintenance components.
6. Low power consumption: 2.2 kWh for the CID-25.
7. Integral standby UV lamps.
8. Teflon-coated quartz sleeves to minimize fouling.

Technical Data CID-2S:

1. Flow rate: 440 gpm.
2. Overall frame size: 7 ft 8 in. L × 4 ft 1 in. W × 3 ft 6 in. H.
3. Power consumption: 2.2 kWh.
4. Materials of construction: 316L stainless steel.
5. Treatment chambers: 2.
6. Surface screens: 2.
7. UV lamps: 12 × 85W LP mercury vapor.
6. Valves: 8 motor-operated.
7. Inlet/outlet connection: 4 in.
8. Power requirements: 240 V AC, 60 Hz.

System Description Model SWS DH-5. This is an advanced sensor design with a microprocessor control to overcome problems traditionally associated with UV systems. This system is illustrated by Figs. 17-18 and 17-19.

This model includes five disinfection chambers, each with four lamp assemblies and a UV detector. The microprocessor control unit has an LCD display that is integrally mounted in a control panel where the lamp drives, switching contactors, and safety circuits are located. The entire assembly is conveniently skid-mounted with a 316L stainless steel frame to inhibit any corrosion.

Fig. 17-18. SWS Model DH-5 microprocessor control with skid mounting (courtesy of Safe Water Solutions, Inc.).

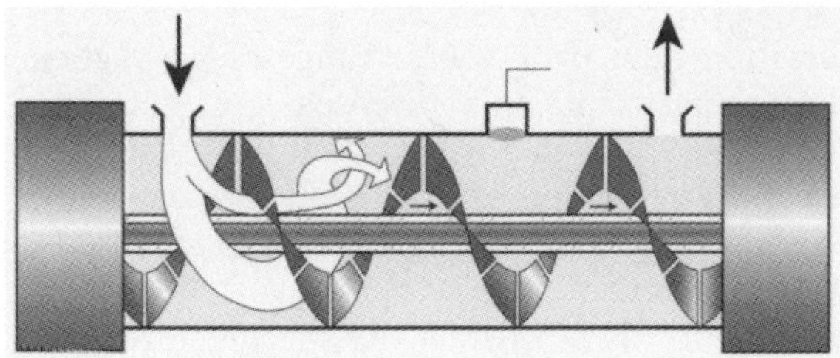

Fig. 17-19. SWS Model DH-5 flow diagram (courtesy of Safe Water Solutions, Inc.).

Operation. Process water enters the system after passing through a flow meter. Detectors mounted on the outer wall of the irradiation chambers measure the intensity of the 253.7 nm signal received as the process flow passes through the chamber.

From the flow rate and intensity signals, the microprocessor computes and displays the actual dose, water translucency, and flow rate. The flow within each disinfection chamber (Fig. 17-21) is directed by fixed stator blades on the inner wall of the chamber. This induces a double helical flow pattern that prevents "shadowing." This ensures that the flowing process water receives multiple doses of the germicidal UV wavelength.

Clumps and slimes are drawn through the gaps between the stator blades and broken down by the shearing action of the process flow. They are then irradiated as they pass through the chamber. The Teflon-coated quartz sleeve, housing the UV lamps, has a nonstick surface, which minimizes any fouling problems. The vigorous turbulence in the chamber creates a natural scouring action.

Maintenance. The microprocessor monitors and controls the key components and actuates the appropriate alarms that are on the LCD display panel. This system is designed for ease of maintenance by operating personnel. The design enables components to be removed or replaced quickly. If preferred, disinfection chambers can be completely removed for both servicing and maintenance. This can be accomplished easily by one person. Prebuilt lamp assemblies and printed circuit boards reduce servicing time to a minimum.

Flexibility. The microprocessor and the modular design of the Model DH-5 offer flexibility in both application and operation. Higher flow rates can be accommodated with multiple installations. The system is rugged and reliable, and the skid-mounted design can easily be used by mobile forces or for disaster relief. This SWS system can be readily configured for remote monitoring via an RS232 port.

Key Features:

1. High inactivation rates over a wide range of pathogenic organisms, viruses, and acute gastrointestinal diseases.
2. Precisely controlled flow pattern that delivers multiple doses of UV to ensure proper log-reduction of pathogenic organisms.
3. Distinctive double helical flow action that enhances organism inactivation in highly turbid water.
4. Continuous monitoring of dose at the specific germicidal wavelength.
5. Design that eliminates short-circuiting, overcomes shadowing, and removes shear clumps and slimes.
6. Anti-fouling design, self-cleaning, and no moving parts.
7. Teflon-coated quartz sleeves to resist fouling and reduce maintenance.
8. Low power consumption
9. High quality, low maintenance components.

Technical Data DH-5:

1. Treatment chambers: 5.
2. Sensor: UV 253.7 nm specific-, electronically stable.
3. UV lamps: 4 × 85W LP mercury per chamber.

Ultra Guard™: UV Disinfection of Wastewater

Introduction. This UV system is manufactured by Ultra Guard, in Barnaby, British Columbia, Canada. The company specializes in wastewater disinfection, with claims that its treatment module can eliminate such bacteria pathogens as viruses, spores, and other disease causing entities, plus the fact that the modular system has not been troubled by murkiness in the treated sewage effluent. (See Fig. 17-20.)

The main thrust of the Ultra Guard system is its high efficiency, high intensity lamps. These lamps are primarily for wastewater systems, not potable water systems. Low intensity lamps are simply not able to deal with the amount of suspended solids that can be penetrated by these high intensity lamps. Ultra Guard claims a 4.7-log reduction of microorganisms. Its field tests indicate the systems are capable of treating up to at least 600,000 gal per day per single lamp system. The power consumption is about 10 kWh per 250,000 gal of treated effluent.

For a two-lamp system, the power consumption would be doubled. Actual capacities depend upon the effluent quality. Fortunately, the Ultra Guard system has a power-saving feature that is continuously regulated by the suspended solids demands and the flow rate of the effluent. This in turn ensures optimum efficiency at all times.

System Features. Ultra Guard uses in-stream flow sensors to detect any variation in flow. The sensor data are fed to each individual UV lamp control-

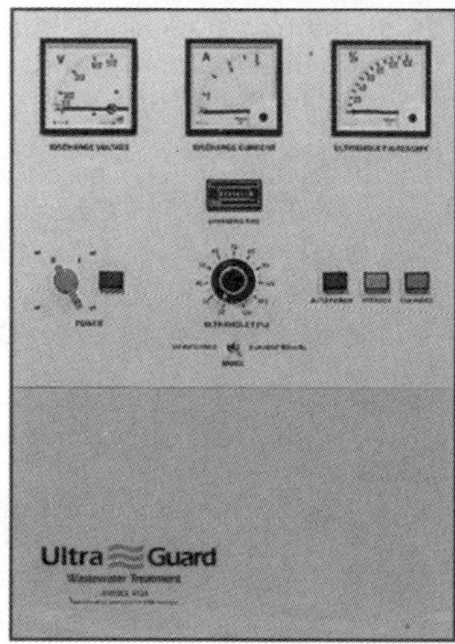

Fig. 17-20. The control panel that provides either automatic or manual adjustment of the intensity of each UV lamp (courtesy of Ultra Guard).

ler. The output of these individual lamps is based upon setpoints that are established for the specific effluent conditions at each installation.

The effluent flow channel is equipped with a patented diffuser screen. This screen intercepts the incoming flow of treated effluent pouring into the UV channel, mixes the layers, and directs the flow toward the lamp module. This allows the Ultra Guard system to set lamp intensity based upon a consistent predictable effluent flow rate—anywhere from 100 to 800 gpm.

A flow-pacing device determines the effluent flow volume headed for the disinfection modules. This provides continuous readings of the effluent transmission quality. These readings are fed to a program resident in the system control scheme where they are analyzed to select proper power settings for UV lamp intensity.

In multichannel installations, powered flow control gates are provided to balance the flow control system. This controls the number of modules in operation, or the removal of a channel for maintenance or lamp replacement. The gates are an integral component of the diffuser mechanism, operating like a second screen, which needs to travel only a short distance to become engaged.

This gauge operation can be remotely controlled, and the remote system can also take over other control and monitoring features provided by panel-mounted instrumentation.

Disinfection. The basis of identifying disinfection efficiency is based upon the usual UV lamp germicidal range of 250–280 nm. This system has been tested in effluent flows from 100 to 800 gpm and with suspended solids as high as 100 mg/L with UV transmission from 18 to 60 percent and log cycle reductions of pathogenic organisms as high as 4.8.

Distributor. The system is distributed by Service Systems International Ltd., Suite 203, 12840 16th Avenue, White Rock, BC, Canada, V4A IN6 (Customer Service 800-488-4544; Fax 604-541-1699).

Bailey–Fisher and Porter: Series 70UV-6000 Open Channel Disinfection System

This particular system is the largest UV disinfection system in the world, capable of treating 265 mgd and requiring 11,520 lamps. The system is installed in Calgary, Canada. Energy efficiency and ease of operation were critical parts of the basic design. No special trade skills are required for operating personnel. (See Figures 17-21 & 17-22) and Specification No. E97-902; (see Fig. 17-23).

No hard wiring is needed to change critical electrical or electronic components. All lamp modules use industrial quick-disconnect connectors, and all ballasts simply "plug in."

All the Bailey–Fischer and Porter UV systems are designed for water flow parallel to the UV lamps. This results in significantly less fouling and entrapment of debris within the lamp modules.

When lamp modules are being removed, all of these systems continue to operate. The lightweight eight-lamp rack can be handled by one person. The weight of the rack is 44 lb.

Design Features

Modular Design. The 70UV-6000 Series design uses a rugged modular concept engineered to meet the needs of even the largest plants. It is installed into fabricated concrete channels, and it incorporates numerous features including energy-efficient electronic ballast technology. It can be configured to suit virtually any field situation, and can be furnished with the simplest of manual controls or the most sophisticated distributed control systems.

Enhanced Operator Safety. It employs five mA Ground Fault Interrupters (GFIs) to optimize the safety of operating and maintenance personnel. It also provides lamp and alarm monitoring with self-diagnostic circuitry.

The system comes with full manual backup control capabilities, as well as optional enhanced automatic controls, which are easily interfaced with plantwide control systems that will expand monitoring capabilities, including remote alarming.

Energy Efficiency. The system uses solid-state electronic ballasts to reduce operating costs and maximize lamp life, plus built-in automatic control features that improve operation and prolong lamp life. This reduces operating costs.

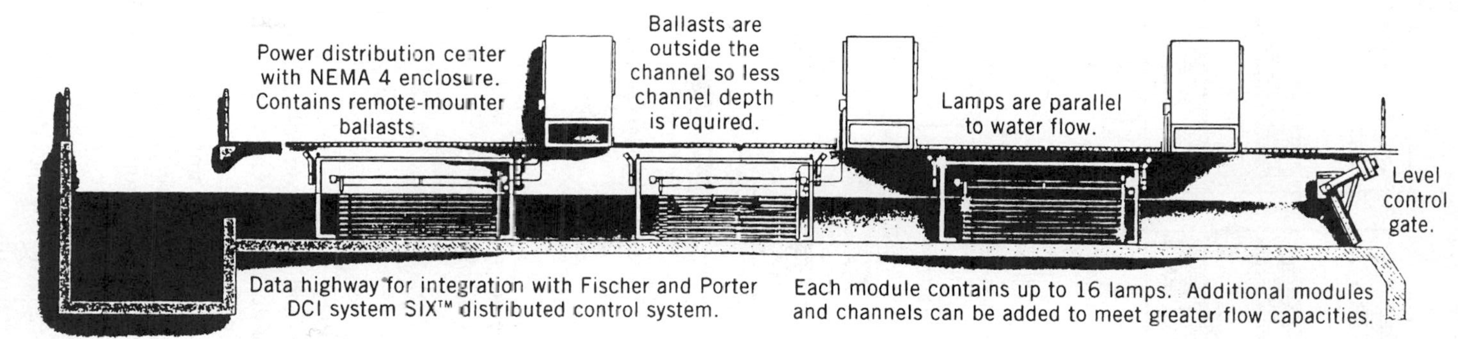
Power distribution center
with NEMA 4 enclosure.
Contains remote-mounter
ballasts.
Ballasts are
outside the
channel so less
channel depth
is required.
Lamps are parallel
to water flow.
Level
control
gate.
Data highway for integration with Fischer and Porter
DCI system SIX™ distributed control system.
Each module contains up to 16 lamps. Additional modules
and channels can be added to meet greater flow capacities.

Fig. 17-21. UV disinfection system Series 70 UV-6000, data highway and ballasts for integration with DCI System Six control system (courtesy of Bailey–Fischer and Porter).

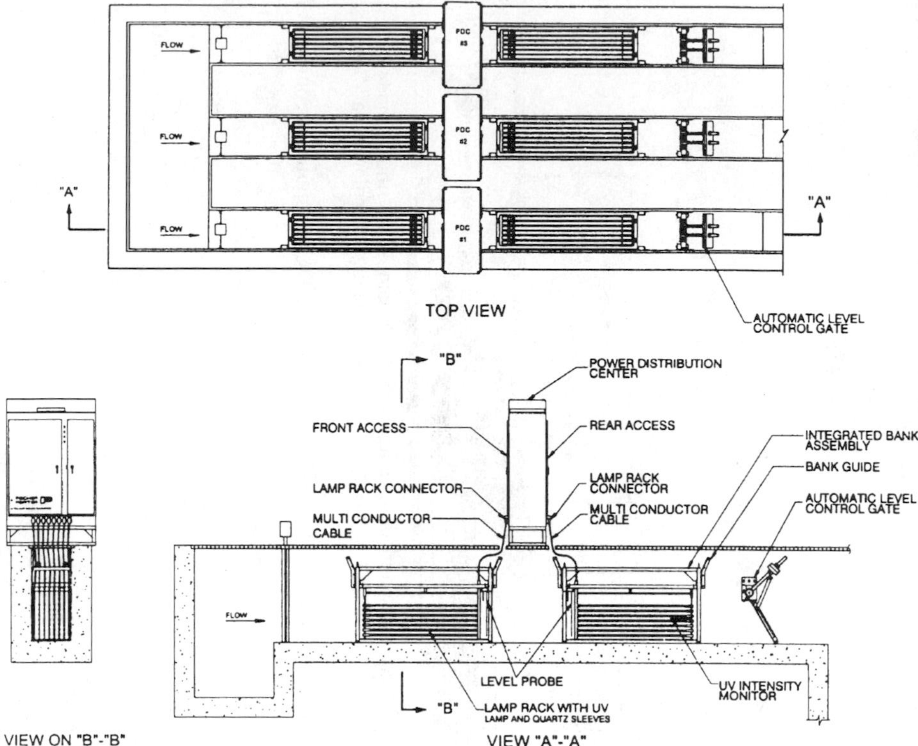

Fig. 17-22. Layout diagrams for Series 70 UV 6000 for UV disinfection system (courtesy of Bailey–Fischer and Porter).

Start-up System. This includes a channel level control gate or weir, plus preassembled quick-disconnect cables.

Engineering Details

UV Transmittance Analyzer. The new Bailey–Fischer and Porter 17UVT100 analyzer provides additional control and efficiency in the operation of the family of all UV disinfection systems offered by Bailey–Fischer and Porter for wastewater treatment systems. (See Fig. 17-24.)

Typically UV system control is based upon an algorithm using the lowest transmittance of UV light expected for any given treated wastewater-effluent. However, transmittance can vary significantly; and the 17UVT100 allows customers to take advantage of this characteristic.

The transmittance analyzer uses an absorption photometer to continuously measure the transmittance of 254 nm UV radiation in the treatment plant effluent. This information is then fed to the UV control system, and the level of light

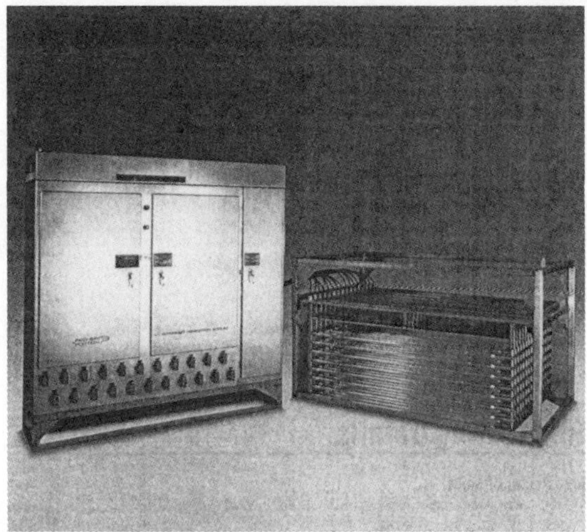

Fig. 17-23. Series 70 UV 6000 specification E97-902 UV open channel disinfection system (courtesy of Bailey–Fischer and Porter).

applied is either increased or decreased to match the UV output required to perform the disinfection requirements as the needs change in the effluent.

This ensures that the disinfection permit is met in the most efficient manner. As a result of this enhanced control system, operating costs can be dramatically reduced.

Input Power Choices:

1. 120 V AC, 60 Hz, single-phase.
2. 208/120 V AC, 60 Hz, three-phase.
3. 220/240 V AC, 50 or 60 Hz single-phase.
4. 380/415 V AC, 50 Hz, three-phase.
5. 460/480 V AC, 60 Hz, three-phase.
6. 600 V AC, 60 Hz, three-phase.
7. Other conveniently available single- or three-phase power supplies.

Lamp Power:

1. Demand is 70 watts per lamp, including ballast.
2. Power factor is 0.97.
3. Total harmonic distortion is less than 7 percent (nominal).

Monitors, Alarms, and Indicators:

1. Lamp status monitor with one indicator lamp per UV lamp.
2. Ground fault trip monitor with contact closure for remote alarm.

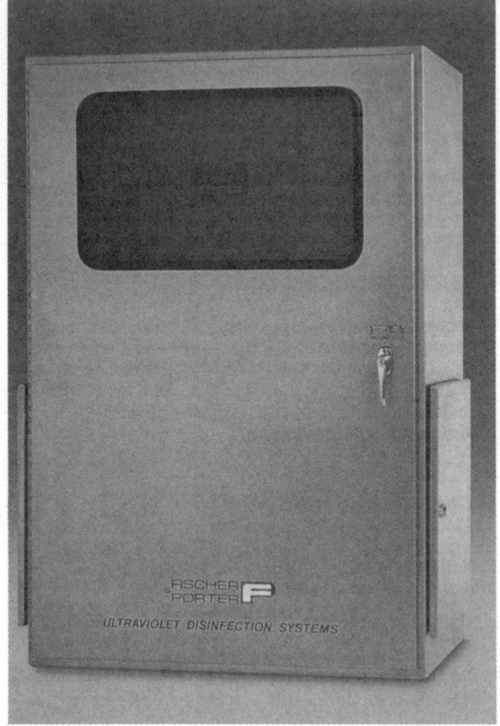

Fig. 17-24. UV Transmittance analyzer Series 17 UVT-100 and system enclosure (courtesy of Bailey–Fischer and Porter).

3. Elapsed time counter with digital LCD readout.
4. Remote annunciation of common alarm (optional).

Acceptable Ambient Conditions:

1. −34°F to 122°F, or −30°C to 50°C.
2. Relative humidity 5 to 95 percent.

Telemetry. Optional digital telemetry is available for the transmission of system status and alarms to and from remote locations.

Power Distribution Center:

1. A central station for all system control monitoring and alarm functions.
2. NEMA 4X free-standing cabinet.
3. Thermostatically controlled heater and ventilation fan.

Lamp Rack Controller. This includes one to eight two-lamp ballasts with lamp status, GFI trip, and power-on indicators with manual or automatic switch.

Lamp Rack Assembly. This assembly supports 4 to 16 slimline instant-start germicidal lamps in quartz sleeves. A quick-disconnect cable on each rack makes their removal fast and easy. The lightweight eight-lamp rack weighs only 44 lb.

UV Intensity Monitor:

1. Fiber optic sensor without submerged electrical connections.
2. Analog display with dual adjustable alarm setpoints for low warning and low-low intensity alarm.
3. 4–20 mA DC output.

System Options

Enhanced UV Control System. Such a system is available that will further simplify system operation, as well as optimize performance, and will minimize the cost of operation and maintenance. Three options are available: the Bailey DCI SYSTEM SIX™, the Bailey INFI 90, or one's choice of PLC. Each of these options offers full system monitoring and alarm control.

The Bailey control systems optimize UV system performance by employing CRT-based operator interface; full alarm and status monitoring displays; flow indication; lamp and bank operating time; channel and bank selection algorithm; flow pacing; power failure detection and alarm; on-line detection and alarm.

Integrated Bank Assembly. A stainless steel supporting rack for assemblies within the channel permits easy removal of an entire bank of lamps for cleaning, removal of an individual lamp for cleaning, or removal of an individual lamp for servicing. The integrated bank assembly is composed of source rack assemblies.

Photometer. This instrument enables operators to monitor treated effluent quality by measuring the UV transmittance in the wastewater. This permits system operation to enhance disinfection performance.

Portable Cleaning Tank. A portable cleaning tank with an air agitation system facilitates the cleaning of three to eight racks simultaneously.

Wide Selection of Complementary Flow Meters. These options may be added for measuring wastewater flow rates, and to provide input for flow pacing of the system.

Customer Service. Contact David C. Frost Associates, Inc. at 2652 San Antonio Drive, Walnut Creek, CA 94598 (Phone 510-947-6733; Fax 510-947-6784).

OPERATING AND TECHNICAL ASPECTS

Previous Experiences with UV Systems

Historically, the use of UV as an alternative to chlorine does not have a good record. The use of UV on board cruise vessels has proved to be unreliable, and system inefficiency has led to serious gastrointestinal outbreaks in these vessels. The major complaint against UV for potable water systems is lack of proof of system reliability (absence of residual measurement) and lack of a persisting disinfecting residual.

In 1975 the EPA explored the use of UV for small public water supplies[2] by a grant to the Vermont State Department of Health. This was a demonstration grant designed to compare ozone and UV with chlorine as practical methods for disinfecting small water supplies. This report was definitely unfavorable to UV as a practical alternative to chlorine. The specific complaints were that both UV and ozone units are many times more costly than chlorination units, that they incur higher operation and maintenance costs, and that they are less reliable. Most important, though, was that neither ozone nor UV could provide a residual disinfectant to protect the water in the distribution system. For this project the capital cost for the UV system was $1,995 versus $550 for a hypochlorinator, and the annual operation and maintenance cost for a 20,000 GPD system and 2 mg/L chlorine dosage was $141/yr for chlorine and $458/yr for UV. It is evident that public health agencies are not enthusiastic about the use of UV on public water supplies as an alternative to chlorine.

The Current Approach to UV Disinfection

As the reader may expect, the thought of using UV as a primary disinfectant for potable water is almost beyond imagination. The Public Health Service had been recommending this method of disinfection for many years to "polish" potable water in hospitals, including doctor's surgical rooms, plus recirculated air to provide a bacteria-free environment. Then, all of a sudden, the potential use of UV was directed to wastewater effluents, by the present EPA (formerly part of the Public Health Service) under the able directon of A. D. Venosa, an accomplished sanitary engineer.

Venosa planned and funded four investigations of UV as a wastewater disinfectant based upon the EPA discharge requirement of 200/100 ml MPN fecal coliforms. These projects included the plant-scale investigations of St. Michaels, Maryland,[3] Waldwick, New Jersey,[4] and Port Richmond, New York,[5,6] and the bioassay-dose measurement studies by Johnson et al.[7,8]

The big mistake here was the selection by EPA of fecal coliforms instead of total coliforms as the basis for disinfection. This was contrary to the 30-year effort of the Sanitary Division of the California State Department of Health, which used total coliforms to force the City of Los Angeles to treat their 150 mgd wastewater discharge into the truly great Santa Monica bathing beach.*

This effort had proved to the California State Department of Health that the only way to properly evaluate disinfection was by the total coliform count. Fecal coliforms are too unreliable a measure, and in the long run are meaningless. As an example, White was called in as a consultant by an industry discharging its waste into the Ohio River, that could not meet the fecal coliform discharge requirements. There was a sudden large increase in the fecal coliforms that could not be accounted for, which caused problems that could not be solved. They could not properly determine the dechlorination needed because the effluent detention time was too small (15 min.) for the amount of chlorine required. Using total coliforms always requires 40 to 60 min. of detention time at peak daily flows, whereas 15 min. for fecals is standard practice. The reader is reminded that the equivalent of a fecal coliform limit of 200 MPN/100 ml in total coliforms is 1000 MPN/100 ml.

This chapter is concerned with water reclamation for water reuse, and is based upon total coliform standards as developed by California Health Services. Therefore, the reader must remember that the UV applications (descended from the third edition) described below are all based upon fecal coliforms, which are unacceptable for water reuse and water reclamation. Furthermore, we are talking only about the treatment of wastewater (sewage) effluents.

UV Disinfection Studies by the EPA in 1975

The results of the studies by the EPA under the direction of A. D. Venosa generated a genuine interest in the acceptability of UV as an alternative to chlorine under certain circumstances, as described below. The EPA had been duly impressed with the performance of UV as a wastewater disinfectant. By 1984 about 100 systems had been funded, and of these about 50 were being installed, and approximately 15 were in operation. It should be pointed out that effluent quality is all-important because suspended solids concentration is critical, as they shield the organism from the rays. It is the consensus that the suspended solids concentration should not exceed 20 mg/L for reliable operation, and 10 mg/L is preferred.[10]

*Since White grew up in that area, and was a beach goer, he remembers this situation and its various developments. The final result was the Hyperion WWTP, built in 1950. The big mistake here was made by the engineers who designed the main Los Angeles sewage outfall to discharge into the best bathing in the whole of North America. The Point Loma sewage outfall in San Diego is another similar case.

There are still many questions to be answered before standards of design, acceptability, and reliability can be formulated to the satisfaction of the EPA[33,34] and the state health agencies.

Physics of Radiation

Radiation results from the emission of energy, which can be converted into thermal, chemical, or mechanical energy. Radiations are usually divided into two main groups—corpuscular and electromagnetic wave radiations. UV is an example of electromagnetic radiations, and consequently has no matter associated with it. It is similar in this respect to X-rays and gamma rays. It differs from corpuscular radiations exemplified by electrons, protons, neutrons, and certain other subatomic particles. Electromagnetic radiations are believed to be made up of waves that comprise the wide range of radiations from electric waves, radio waves, infrared or heat rays, and visible light to UV light rays, roentgen rays, gamma rays, and secondary cosmic rays. The entire group of such radiations is known as the electromagnetic spectrum. This group of rays travel at the speed of light; however, they differ widely in wavelength from 10^{11} cm for electric waves down to 10^{-11} for secondary cosmic rays.[11]

It has been well established that bacteria, viruses, and fungi can be destroyed by UV whether suspended in air, suspended in liquid, or deposited on surfaces.[37] The killing or inactivation of microorganisms as a result of exposure to UV radiation is expressed as the product of the intensity, I, of UV energy and the time of exposure, t. This leads to the postulation that the same rate of kill can be obtained using either high-intensity UV energy for a short time or low-intensity UV for a proportionally longer time, so that the product $I \times t$ is equal for both cases.

Germicidal Lamps

UV radiation is generated by special sources of mercury vapor known commonly as germicidal lamps. The principle of all germicidal lamps is the same: that of electron flow between electrodes through ionized mercury vapor. The arc in a fluorescent lamp operates on the same basic principle and produces the same type of UV energy. The difference between the two is that the bulb of the fluorescent lamp is coated with a *phosphor* compound that converts UV to visible light. The germicidal lamp is made of special glass that is not coated, so it transmits UV rays generated by the arc. The low-pressure mercury arc lamps emit their maximum energy output at a wavelength of 253.7 nm.

Two types of lamps are used in these commercial UV systems. They are described as high intensity and low intensity mercury arc lamps. Some systems prefer the low intensity lamps, claiming that these lamps require longer "contact time" in the process flow channels, thus increasing the mercury arc detention time and thus improving the disinfection process.

Organism Killing Action by UV

We always have to keep reminding ourselves that nature has the last word regarding disinfection; UV cannot inactivate any organism that can be seen with the human eye, such as *Giardia* and *Cryptosporidium* oocysts.

The radiation energy of UV rays can be used to destroy microorganisms. In order to kill, the electromagnetic waves of ultraviolet irradiation must actually strike the organism. In this process some of the radiation energy is absorbed by the organism and other constituents in the medium surrounding the organism. There is sufficient evidence to conclude that if sufficient dosages of UV energy reach the organisms, water or wastewater can be disinfected to any degree required. The germicidal effect of ultraviolet energy is thought to be associated with its absorption by various organic molecular components essential to the functioning of cells. Energy dissipation by excitation, causing disruptions of unsaturated bonds, particularly of the purine and pyridimine components of nucleoproteins, appears to produce a progressive lethal biochemical change.[12] The UV treatment does not alter water chemically; nothing is being added except energy, which produces heat, resulting in a temperature rise in the treated water.

The question comes up quite often—how does UV kill? It seems as if the emitted rays of light energy "stab" the microorganism because there is no chemical reaction according to many knowledgeable electrochemical participants in this field. If so, says White, then there cannot be any DBPs. That would be great news for everyone.

The UV rays penetrate the cell walls of microorganisms. In accordance with the laws of photochemical action, the energy that kills a bacterium is that absorbed by the organism. The only radiant energy effective in killing bacteria is that which reaches the bacteria; therefore, the water must be free of any particles that would act as a shield. This is one of the main disadvantages of the UV process; other disadvantages are the lack of a field test to readily establish the efficiency of the process, and its inability to provide any residual disinfecting power.

The manufacturers of germicidal lamps caution in their literature that UV radiation will only kill invisible organisms. In other words UV will not kill organisms visible to the naked eye. (See above.)

Degree of Inactivation

The degree of inactivation by ultraviolet radiation is directly related to the UV dose applied to the water or wastewater. Dose is described as the product of the rate at which the energy is emitted (intensity) and the time the organism is exposed to the energy.

$$D = I \times t \tag{17-1}$$

where

D = dose, microwatt-seconds per cm^2
I = irradiation, microwatts per cm^2
t = time, seconds

Therefore, the principal design factors for any UV system for disinfection are the intensity of radiant energy that is able to reach the organism and the time of exposure.

Germicidal Efficiency

It has been determined that the germicidal effect of ultraviolet rays is maximal at wavelengths between 250 and 265 nm. There is an abrupt decrease at 290–300 nm.[13,14] As a consequence, low-pressure mercury arc lamps have served as the ideal source of germicidal energy because their maximum energy output is at a wavelength of 253.7 nm.

Various investigations have shown a wide range of sensitivity of different microorganisms to ultraviolet energy.[37] In 1959, Kawabata and Harada[15] reported the following contact times required to achieve a 99.9 percent kill (3-log reduction)* at a fixed UV intensity for the following organisms:

E. coli	60 sec
Shigella	47 sec
S. typhosa	49 sec
Streptococcus faecalis	165 sec
B. subtilis	240 sec
B. subtilis spores	369 sec

Ultraviolet radiation has also been shown to be effective in the inactivation of viruses. In 1965, Huff et al.[12] reported satisfactory results, which included studies of several strains of polio, Echo 7, and Coxsackie 9 viruses. The intensities varied from 7000 to 11,000 μW-sec/cm^2.

Hill et al. reported in 1971[16] that the infectivity loss of Polio 1 virus exceeded 3 logs after 15.7 sec of UV exposure in continuous flowing seawater. Based upon these findings the use of UV should be highly effective in the disinfection of flowing seawater for use in artificial shellfish purification systems.

Owing to the long exposure times (47–369 sec) listed in the above investigations, the data are misleading compared to what was known in 1984. The work done at St. Michaels, Maryland[3] used a 130-sec exposure and a correspondingly low UV intensity to achieve the desired germicidal efficiency. Current designs are for high intensity and lower exposure times—6 to 10 sec.

---

*Standard practice in California has always been to design wastewater disinfection facilities to accomplish a 5-log kill, or 99.999 percent removal.

There is no doubt that the germicidal efficiency of UV is predictable for a given species of organism on the basis of the UV intensity–exposure time product. In practice it will be necessary to prove and evaluate a given installation to confirm the design parameters.

The Sanitation Districts of Los Angeles County, in 1977, prepared a report for the California State Water Resources Control Board and the EPA, a UV virus study performed at their special wastewater research treatment plant in Pomona, California.[35] There is no reference to the ability of ultraviolet radiation to destroy cysts. It is unlikely that UV would have any effect on this type of organism, as it is the consensus that UV radiation will not kill any organism which can be seen with the naked eye. This is well known phenomenon.

The most discouraging fact of the apparent germicidal efficiency of UV is that the best we can expect is 99.9 percent or a 3-log kill of viruses and bacteriophage, and zero if the organism can be seen with the naked eye.

This is all the more reason why sampling for proof of disinfection should never be based upon "grab samples." All sampling should originate with a pump in the open channel that will send a sample to the tap at 5 ft/sec. This will keep the piping system free of slime, and the pump will break up any solids that could have been instrumental in blocking the UV path, thereby damaging the disinfection efficiency.

UV Dose vs. Bacterial Kill*

This relationship is characterized by a mathematical model that assumes second-order kinetics when the coliform concentrations are in the range where disinfection usually takes place, as follows:

$$\frac{dN}{dt} - kN^2I \qquad (17\text{-}2)$$

Integrated, this becomes:

$$\frac{1}{N} - \frac{1}{N_0} = kIt \qquad (17\text{-}3)$$

where

N = coliform concentration, MPN/100 ml, at time t
N_0 = influent coliform concentration, MPN/100 ml
k = rate constant
I = the average ultraviolet intensity in the exposure chamber
t = exposure time

*The problem with this study is the use of fecal coliforms rather than total coliforms.

The influent coliform concentration is usually so much greater than the final concentration that the term $1/N_0$ becomes negligible, and Eq. (17-3) can be simplified to:

$$\frac{1}{N} = kIt \tag{17-4}$$

On the basis of 350 samplings conducted throughout a one-year pilot program, Scheible and Bassell[4] were able to show a quite favorable correlation between UV dose and coliform kill. Their data are illustrated by Fig. 17-25.

The mathematical model represented by Fig. 17-25 is:

$$\text{Effluent fecal coliform} = (1.26 \times 10^{13})(\text{UV dose})^{-2.27} \tag{17-5}$$

To compute the UV dose required to achieve a fecal coliform concentration at time t, Eq. (17-5) must be rearranged as follows:

$$(\text{UV dose})^{-2.27} = \frac{\text{Eff. FC}}{1.23 \times 10^{13}}$$

$$(\text{UV dose})^{+2.27} = \frac{1.23 \times 10^{10}}{\text{Eff. FC}}$$

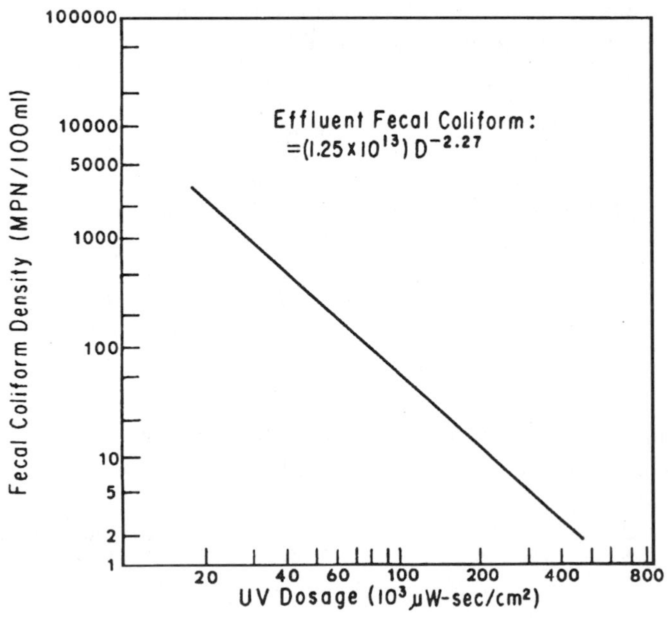

Fig. 17-25. UV dose versus fecal coliform destruction.[4]

$$[(\text{UV dose})^{+2.27}]^{0.4405} = \frac{(1.23 \times 10^{13})^{0.4405}}{(\text{Eff. FC})^{0.4405}}$$

$$\text{UV dose} = \frac{5.836 \times 10^5}{(\text{Eff. FC})^{0.4405}}$$

EXAMPLE. Compute UV dose to achieve 200/100 ml MPN fecal coliform using Eq. (17-6):

$$\text{UV dose} = \frac{5.836 \times 10^5}{(200)^{0.4405}}$$

$$= 56560\ \mu\ \text{W-sec/cm}^2$$

The dose used by Scheible and Bassell[4] was 60,000 μW-sec/cm^2. The above example is site-specific because it is based upon an equation that is site-specific. Equation (17-6) was derived from a 10-month field study at the North Bergen County Wastewater Treatment Plant, Waldwick, New Jersey.[7] Nevertheless it is important because it demonstrates that a correlation between UV dose and organism destruction does exist. A more reliable and practical approach to dose versus bacterial kill is described below under the heading "UV Facility Design."

Process Variables that Affect Disinfection Efficiency

UV Absorption. This may also be described as UV demand. There are various constituents in wastewater that affect the rate of absorption of the ultraviolet energy as it passes through the wastewater in the disinfection chamber. UV energy is also absorbed within the reactor as it passes through quartz sleeves and Teflon, and by the shadowing effect of the reactor walls. During the St. Michaels study[3] the absorption coefficient was found to vary from 0.052 to 0.722, a factor of 14, depending upon the effluent. The absorption coefficient was calculated for each observation from the equation

$$I/I_0 = e^{-K_a L} \qquad (17\text{-}7)$$

where

I = radiation intensity at the bottom of disinfection chamber
I_0 = initial radiation intensity
K_a = absorption coefficient
L = distance in cm between points of measurement of I and I_0
e = natural log base

At the North Bergen County Water Pollution Control Plant a different approach was used.[4] The investigators sought to find the average intensity of radiation based upon lamp design and configuration, which also allowed for the effect of UV absorption as shown in the following equation:

$$I/I_0 = \frac{r}{r + L} e^{K_a L} \tag{17-8}$$

where

r = radius of the lamp
L = distance from lamp surface

The average absorbance coefficient (K_a sec^{-1}) was found to be 0.39, and the range over a 30-day period was from 0.22 to 0.55.

Each of these methods illustrates the process variables due to effluent quality and reactor design.

An entirely different approach was used by Scheible[5,6] at the Port Richmond Wastewater Treatment Plant, New York City, investigation.

Effluent Quality. There are many factors that influence the efficiency of UV radiation energy, all of which affect the absorption coefficient (i.e., affect the transmittance efficiency). The St. Michaels study indicated that the amount of COD might be a better indicator than turbidity. The turbidity ranged from 1.5 to 17 JTU and the COD from 15 to 65 mg/L. When the turbidity was 10 JTU and the COD 65 mg/L, the absorption coefficient was at the highest level, 0.722; but at a turbidity of 17 JTU and a COD of 35 mg/L, the absorption coefficient was only 0.356. This confuses the issue of turbidity as an effluent quality parameter.

Petrasek et al.[17] observed that both organic nitrogen and ammonia nitrogen interfered with the transmittance of UV radiation.

Johnson et al.[7,9] proved conclusively that filtration of a secondary effluent improved the disinfection efficiency significantly. Their study demonstrated that one of the chief factors limiting ideal UV radiation efficiency in practice was directly related to the protection of bacterial colonies inside sewage particles that represent suspended solids in the wastewater. This confirms the belief of Venosa,[10] who summarized the experience gained from the comprehensive EPA research program that 20 mg/L suspended solids was the upper limit of practicality for UV as a disinfectant, and 10 mg/L SS was highly preferable. When compared to chlorine, UV has much less penetrating power. Collins et al. found that suspended solids concentration as high as 100 mg/L did not interfere with the disinfection efficiency of chlorine.[20]

Photoreactivation

Until Kelner's report[18] in 1949, it was generally believed that when bacterial cells were exposed to UV light, the DNA of these cells was permanently destroyed. Kelner reported partial reactivation of bacterial cells when exposed to visible light after absorbing UV light in lethal doses.

The damage incurred by UV absorption is the dimerization of thymine, one of the components of DNA. If this dimerization is not too extensive, it can be reversed in some microorganisms by exposing the irradiated cells to visible light in the range of 330–500 nm.[5,11]

Scheible made a comprehensive analysis of photoreactivation effects at the North Bergen County plant scale investigation.[5] This was carried out for both total and fecal coliforms. In order to overcome the effects of photoreactivation based upon 200/100 MPN fecal coliform, the UV dose would have to approximate 200×10^3 μW-sec/cm^2 as compared to 70×10^3 μW-sec/cm^2 for the nonphotoreactivated samples. This would represent a 460 percent increase in the system sizing and would have a direct effect upon the capital cost, which represents approximately 43 percent of the annual costs.

Toxicity of Radiation

Radiation technology has been used to destroy substances known to be responsible for foul tastes in potable water. Beta radiation is effective in the destruction of geosmin and phenols at a penetration distance of 10 cm; however, gamma radiation has been discovered to produce mutagenesis in the surviving viruses during potable-water and water-reuse treatment. Some of the side effects of UV on plants and animals include retardation of cell division, increase in the rate of mutation, erythema, chromosomal aberration, and changes in cellular viscosity.[11]

UV FACILITY DESIGN

Introduction

The text under this heading must be separated into two principal categories: potable water and wastewater. This is the result of a long-term interest by the U.S. Public Health Service in the disinfection of water. The use of UV for wastewater disinfection to meet the EPA discharge standards is of fairly recent origin (ca. 1978), and it is too soon to establish design standards. The "EPA Design Manual for UV Disinfection of Wastewater" was published in 1986, and it has been superseded by a 1995 Water Environment Research Foundation report titled "Comparison of UV Irradiation to Chlorination: Guidance for Achieving Optimal UV Performance."*

*The most important report ever published on the use of UV radiation to disinfect water for reuse is the one by the National Water Research Institute in Fountain Valley, California. It is listed as "UV Disinfection Guidelines for Wastewater Reclamation in California and UV Disinfection Research Needs Identification." See Ref. 28 this Chapter.

There is such a great difference between the end uses of potable water and treated sewage effluents that there is no way to compare them. Now we are talking solely about treating sewage effluents for reuse, which may be used for potable water via groundwater recharge. White has decided to leave the "potable water" section in this edition simply for historical purposes.

Potable Water

The following are excerpts from a policy statement on the use of the ultraviolet process by the U.S. Public Health Service, "Criteria for the Acceptability of an Ultra-Violet Disinfecting Unit."[19]

1. Ultraviolet radiation at a level of 253.7 nm must be applied at a minimum dosage of 16,000 microwatt-seconds per square centimeter at all points throughout the water disinfection chamber.
2. Maximum water depth in the chamber, measured from the tube surface to the chamber wall, shall not exceed three inches.
3. The ultraviolet tubes shall be:
   a. Jacketed so that a proper operating tube temperature of about 105°F is maintained, and
   b. the jacket shall be of quartz or high silica glass with similar optical characteristics.
4. A time delay mechanism shall be provided to permit a two-minute tube warm-up period before water flows from the unit. One manufacturer recommends a 5-minute warm-up period.
5. The unit shall be designed to permit frequent mechanical cleaning of the water contact surface of the jacket without disassembly of the unit.
6. An automatic flow control valve, accurate within the expected pressure range, shall be installed to restrict flow to the maximum design flow of the treatment unit.
7. An accurately calibrated ultraviolet intensity meter, properly filtered to restrict its sensitivity to the disinfection spectrum, shall be installed in the wall of the disinfection chamber at the point of greatest water depth from the tube or tubes.
8. A flow diversion valve or automatic shut-off valve shall be installed which will permit flow into the potable water system only when at least the minimum ultra-violet dosage is applied. When power is not being supplied to the unit, the valve should be in a closed (fail-safe) position so as to prevent the flow of water into the potable system.
9. An automatic audible alarm shall be installed to warn of malfunction or impending shutdown if considered necessary by the Control or Regulatory agency.
10. The materials of construction shall not impact toxic materials into the water, either as a result of the presence of toxic constituents in the materials of construction or as a result of physical or chemical changes resulting from exposure to ultra-violet energy.

11. The unit shall be designed to protect the operator against electrical shock or excessive radiation.

As with any potable water treatment process, due consideration must be given to the reliability, economy, and competent operation of the disinfection process and related equipment, including:

1. Installation of the unit in a protected enclosure not subject to extremes of temperature which could cause malfunction.
2. Provision of a spare ultraviolet tube and other necessary equipment to effect proper repair by qualified personnel properly instructed in the operation and maintenance of equipment.
3. Frequent inspection of the unit and keeping a record of all operations, including maintenance problems.

Of all the above items listed by the U.S. Public Health Service bulletin, probably the most important are: UV intensity meter and the fail-safe control.

One other item worthy of mention is personnel exposure to ultraviolet radiation. Workers in the area of ultraviolet radiation can suffer severe burns if not properly protected. Protection of the eyes is extremely important. Tolerance limits for exposure to ultraviolet have been set by the American Medical Association. Prolonged exposure can cause permanent damage to the eyesight.

The equipment manufacturers of ultraviolet purifiers recommend the following maintenance procedures:

1. Wipe the quartz jacket of the lamp at least once a month.
2. Let the ultraviolet lamps warm up for at least 5 minutes before allowing the use of treated water.
3. Lamps should be replaced when the intensity meter indicates less than 70 percent of the rated lamp intensity; in refrigerated or very cold water, replace at 50 percent of initial rate of intensity.
4. Sterilize the entire system, including the purifier and treated water system, prior to the use of the ultraviolet process.

Ultraviolet radiation has proved successful as a disinfectant under certain conditions. It can be used as such for small water supplies under the various requirements of the U.S. Public Health Service. It is far more expensive than chlorination, and so will find its application only under special circumstances that would not warrant the use of chlorine.

Wastewater

It is abundantly clear from the data collected by the EPA-funded research projects, ably reported by Scheible et al.[5,6] and Qualls and Johnson,[8] that the manufacturers must take steps to provide the consumer with guaranteed "capacity" and "range" of operation for a prescribed accuracy. This would be akin to data sheets published by pump manufacturers, which relate pump capacity to total dynamic head, horsepower required, and calculated efficiencies.

Critical parameters for a UV reactor completely dependent upon manufacturer's design, type of lamp selection, and configuration are: *average intensity,* the *dispersion coefficient,* and *hydraulic head loss.*

Water Reclamation and Reuse. According to the National Water Research Council Report,[28] these facilities will have to meet the same disinfection requirements as those for potable water using chlorine, namely 2.2 MPN/100 ml total coliforms. To achieve this goal, the overall treatment system will probably have to be a filtered secondary effluent equipped with a special UV contact channel that can receive two consecutive doses of UV. The filtered effluent turbidity will have to be less than 2 NTU, at the same statistical frequency as required with chlorination, and virus inactivation will have to be 99.99 percent reduction (4-logs). This will be based upon poliovirus, the same as required for chlorine.

UV Intensity. The principal factors for the optimum design of a UV system are described in the following text. The rate of disinfection is directly related to the average intensity of the UV. This is the rate at which the radiation energy is delivered to the wastewater by the UV source. This energy transmittance is dependent upon lamp selection (quartz-Teflon, etc.) and lamp spacing. There is no practical way to measure the true intensity of a UV system in the field by the consumer. This must be done by the manufacturer of the UV reactor. Scheible et al.[7] have developed a mathematical model that calculates the intensity at any point within the UV reactor. This is used to estimate the average intensity emitted by any specific unit. The mathematical model described below is based upon the point source summation method (PSS) described by Jacob and Dranoff[21] and applied to UV disinfection systems by Qualls and Johnson.[8] The mathematical techniques rely upon the basic physical properties of the UV lamps, configuration of multilamp chambers, and assumed properties of the wastewater (absorbance coefficient).

Scheible[6] describes the UV intensity at any given distance from a point source of energy as:

$$I = \left(\frac{S}{4\pi R^2}\right) e^{-aR} \tag{17-9}$$

where

I = intensity at a given point, watts/cm^2
S = UV energy output at source, watts
a = absorbance coefficient, cm^{-1} to the base e
R = distance from energy source to the point of receiver location

Figure 17-26 illustrates a tubular germicidal lamp, which is treated as a series of point sources of energy. The UV intensity at a specific point is the sum of the intensities from the individual point sources. Scheible et al.[7] developed the following equation which represents UV intensity at a specific point (see Fig. 17-26) in the reactor:

$$I(r, z) = \sum_{n=1}^{n=N} \frac{S/N}{4\pi(r^2 + z^2)} \exp[-a(r^2 + z^2)^{1/2}] \tag{17-10}$$

where

$z = L\left(\frac{n-1}{N-1}\right)$
N = number of point sources into which line source is divided
L = lamp arc length
r, z = coordinates of the point receiver
a = absorbance coefficient of the wastewater, cm^{-1} to the base e
$R = \sqrt{r^2 + z^2}$

Equation (17-10) derives from an assumption that the point receiver located at (r, z) is spherical and infinitely small. Therefore, energy radiating from any point source element of the lamp will strike the receiver normal to its surface.

To calculate the intensity at point (r, z), Eq. (17-10) must be solved N times and all the solutions summed. This mathematical exercise can be programmed into a computer or a calculator. Then the average intensity of UV in a system can be estimated by averaging the computed intensities at a representative number of points in a cross-sectional plane of the UV reactor.

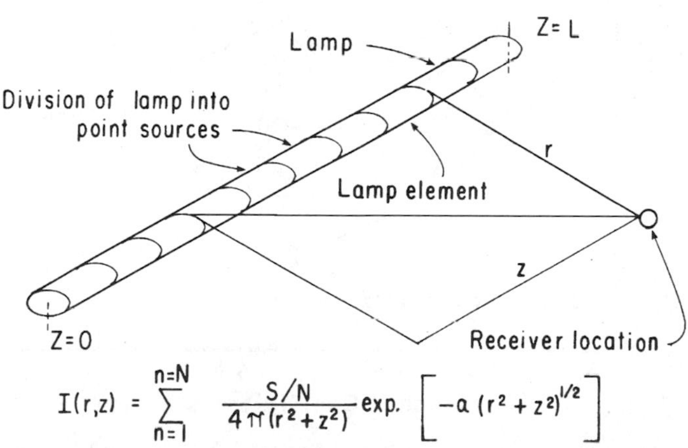

Fig. 17-26. Lamp geometry for point intensity approximation.[7]

Equation (17-10) is capable of accounting for the absorption of energy as it passes through the wastewater, quartz sleeves, Teflon, air, neighboring lamps, and reactor shadowing, for any lamp configuration where lamps are parallel to each other, any lamp rating output, any number of lamps per battery, and any variation in energy output.

Figure 17-27 illustrates the calculated values of average intensity for the two quartz systems installed at the Port Richmond WPCP, New York City.[6] The loss of efficiency displayed by the closely spaced unit is attributed to the shadowing effect of neighboring lamps. If lamp spacing allows UV energy to collide with a neighboring lamp, it will be absorbed by that lamp.

Equation (17-10) demonstrates that the variables that directly affect the reactor intensity are: (1) UV source output, watts; (2) a, absorption coefficient of wastewater (varies with quality); and (3) the distance R the energy will be required to travel through the wastewater. The value of R will be a function of the lamp spacing, which is governed by the absorbance characteristics of the wastewater to be treated.

Hydraulic Characteristics. The efficiency of any UV reactor is intensely dependent upon the dispersion index (E_x), which is related directly to exposure

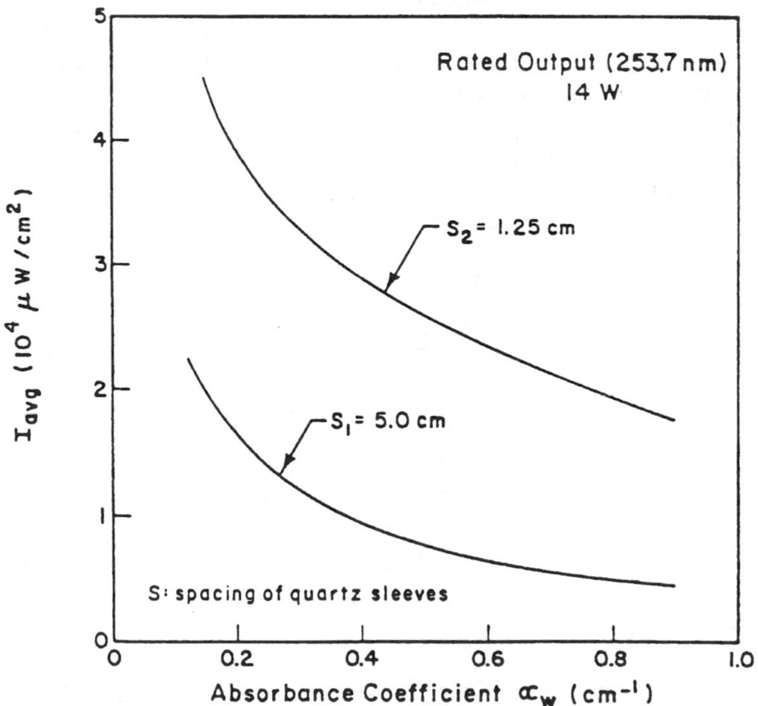

Fig. 17-27. Computed intensity for two different UV reactors at Point Richmond WPCP.[6]

time and lamp configuration (spacing). E_x can be determined in the same fashion as the dispersion index for a chlorine contact chamber—with a dye study (see Chapter 8). At ideal plug flow E_x is zero.

Assuming a one-dimensional forward (advective–dispersive) flow, Scheible[6] developed a mathematical expression that incorporates the dispersion index:

$$\frac{L}{L_0} = \exp\left[\left(\frac{ux}{2E_x}\right)\left\{1 - \left(1 + \frac{4kE_x}{u^2}\right)^{1/2}\right\}\right] \tag{17-11}$$

where

$$\frac{L}{L_0} = \exp\left[\left(\frac{ux}{2E_x}\right)\left\{1 - \left(1 + \frac{4kE_x}{u^2}\right)^{1/2}\right\}\right]$$

E_x = dispersion coefficient, cm^2/sec
L = concentration of organisms after UV exposure, MPN/100 ml
L_0 = concentration of organisms before UV exposure, MPN/100 ml
u = fluid velocity, cm/sec
x = forward distance traveled during exposure, cm
k = rate constant, sec^{-1}

Figure 17-28 illustrates a series of solutions to Eq. (17-11), which demon-

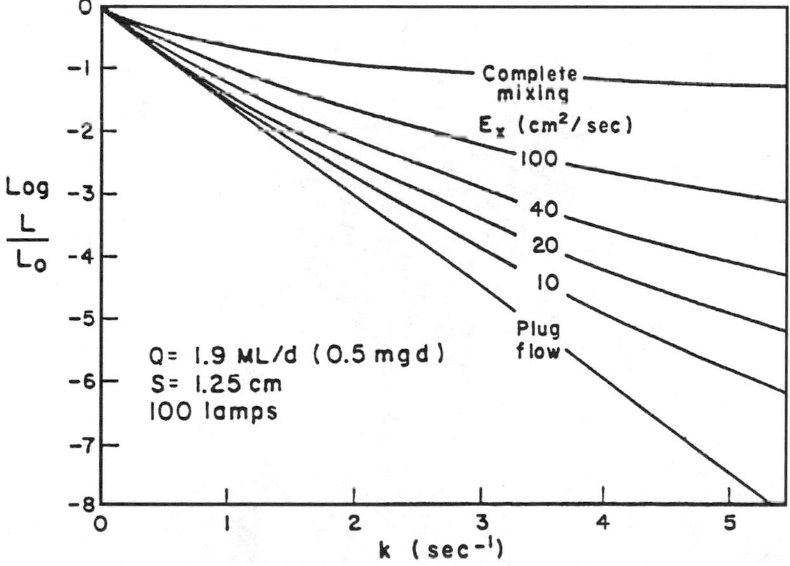

Fig. 17-28. Sensitivity of disinfection efficiency to the dispersion coefficient.[6]

strate the sensitivity of UV system disinfection efficiency to the dispersion coefficient.

Figure 17-29 illustrates the modern concept of advective–dispersive flow utilizing a lamp configuration to achieve the best possible dispersion coefficient at highest average intensity and minimum head loss through the reactor (6–8 inches). The longer the length x can be, the better the efficiency.

Exposure Time. For a conventional modern system, exposure time is in the range of 10 to 30 sec. Exposure time is a function of lamp spacing. Closely spaced lamps have a high UV intensity. Teflon-coated lamps require a longer exposure time because these lamps have a lower intensity.

Figure 17-30 illustrates the concept of turbulence through a typical UV reactor. The design of the reactor should strive for maximum turbulence at minimum head loss. Turbulence is stimulated by high velocities, closer lamp spacing, staggered lamp spacing, perpendicular flow to lamps, and longer length.

Banks of lamps in series are preferred to banks in parallel.

Effluent Quality. The single most damaging variable to UV efficiency is the concentration of suspended solids. Experience indicates that this parame-

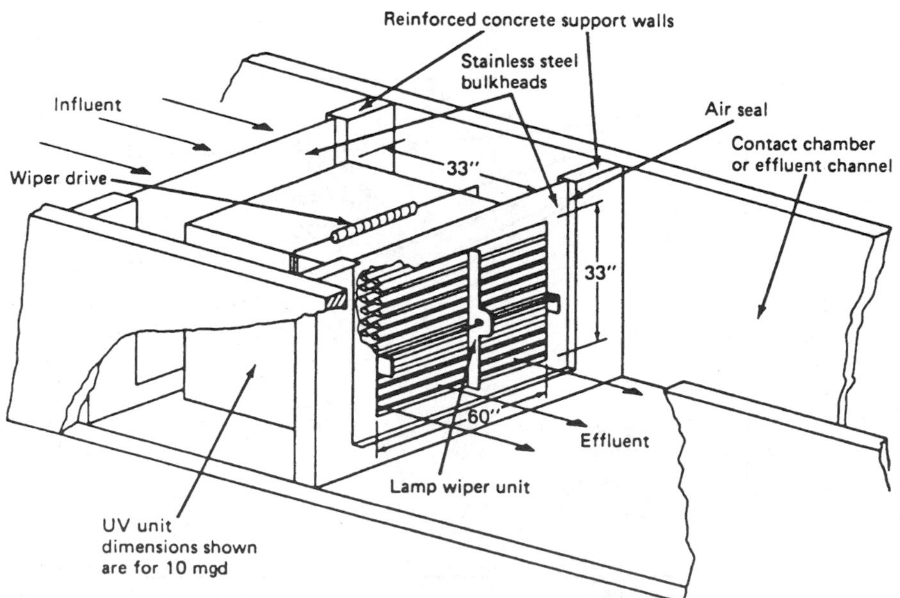

Fig. 17-29. Typical lamp configuration for a wastewater UV reactor where $x = 33$ inches.

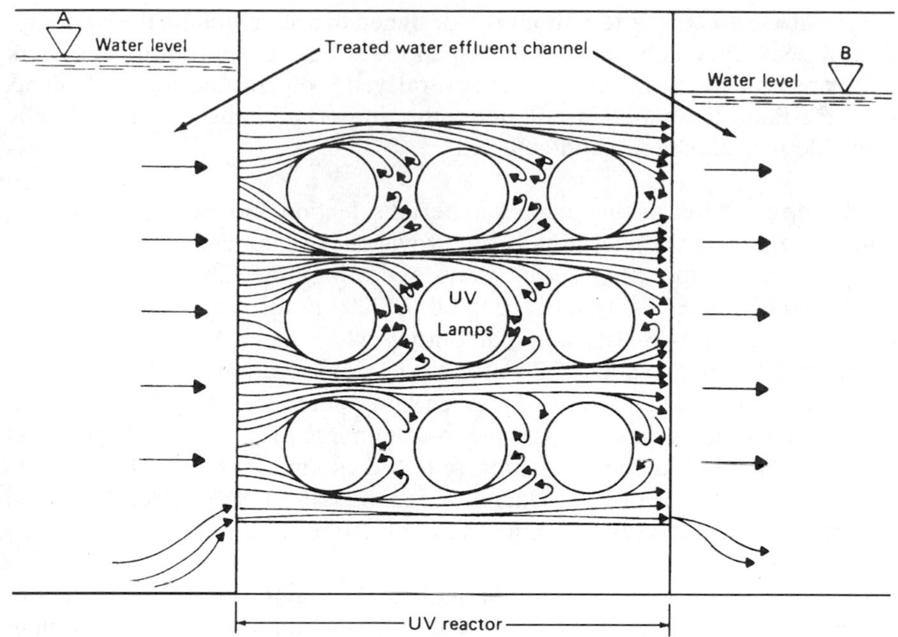

Fig. 17-30. Flow pattern showing turbulance through a typical UV reactor with perpendicular flow. Note: Water level A–B = head loss through reactor.

ter should be limited to 10–15 mg/L.* Scheible has modified Eq. (17-11) to account for this variable as follows:[22]

$$\frac{L}{L_0} = \exp\left[\left(\frac{ux}{2E_x}\right)\left\{1 - \sqrt{1 + \left(\frac{4kE_x}{u^2}\right)}\right\}\right] + C(\mathrm{SS})^2 \qquad (17\text{-}12)$$

where SS is suspended solids, mg/L and

$C = 0.25$ for fecal coliforms
$C = 0.9$ for total coliform
$k = 0.0000145\ I^{1.3}$

where I = computed intensity.

This mathematical model was developed during the calibration of the Point Richmond UV reactors. It is now a predictive tool sensitive to all the process variables.

*For the development of UV systems to meet the requirements for water reclamation and water reuse effluents, the parameter for TSS (total suspended solids) will have to be on the order of 2–3 mg/L.

A total wastewater system should be designed to accommodate the combination of peak daily flow, maximum suspended solids concentration, and peak daily absorbance coefficient. After several years plus a multitude of plants since the Point Richmond installation, other important items are undoubtedly available to equipment manufacturers.

UV Lamps. After having made a material selection and an arc length commitment, the next consideration is the provision for lamp aging. Figure 17-31 illustrates the lamp output decay versus operating time. The serviceable life for germicidal lamps is usually accepted as 8000 hours. The lamp chosen in Fig. 17-31 has a 14.3-watt nominal output when new. At 8000 hours the output deteriorates to about 5.8 watts; so the deterioration is 5.8/14.3 = 0.41. Therefore, the original output of a given lamp deteriorates by a factor of 59 percent. This means that to satisfy the loss of energy over the life (8000 hours) of any lamp, the number of lamps required at time zero will have to be increased by a factor of 2.44. Ozone emission is another consideration. Use of lamps that produce excessive amounts of ozone should be avoided.

Provisions for Cleaning. Biofouling of the UV lamp surface in direct contact with the wastewater is a serious and never-ending maintenance problem with UV reactors. This is best studied by seeking operating experience from an existing installation. There are four categories of cleaning techniques: mechanical wiper, ultrasonics, high-pressure wash, and chemical cleaning with either acid, caustic, or detergent. This problem must be dealt with regardless of cost.

Control System. The UV system must be equipped with controls capable of switching banks of lamps on or off to achieve the necessary dose proportional to flow. It is also desirable to have a slave lamp operating to check wastewater quality absorbance. Other features should include the ability to monitor lamp output with time, in-place intensity monitors, power meters, lamp operation indicators, elapsed time monitors, lamp temperature indicators, ballast panel temperature indicators, and appropriate alarms to assist the operators.

Maintenance. The reactor should be equipped with a system drain and have the ability to isolate modules. There should also be complete redundancy so that the system efficiency will not be impaired because of either routine or emergency maintenance. Lamps and ballasts must be accessible. An inventory of lamps, quartz sheaths, and ballasts should be maintained. Records should be kept of lamp use, lamp life, and replacement cycles for both lamps and ballasts.

Safety Considerations. Operators should be instructed about the dangers of UV. They should be provided with proper goggles and clothing to protect

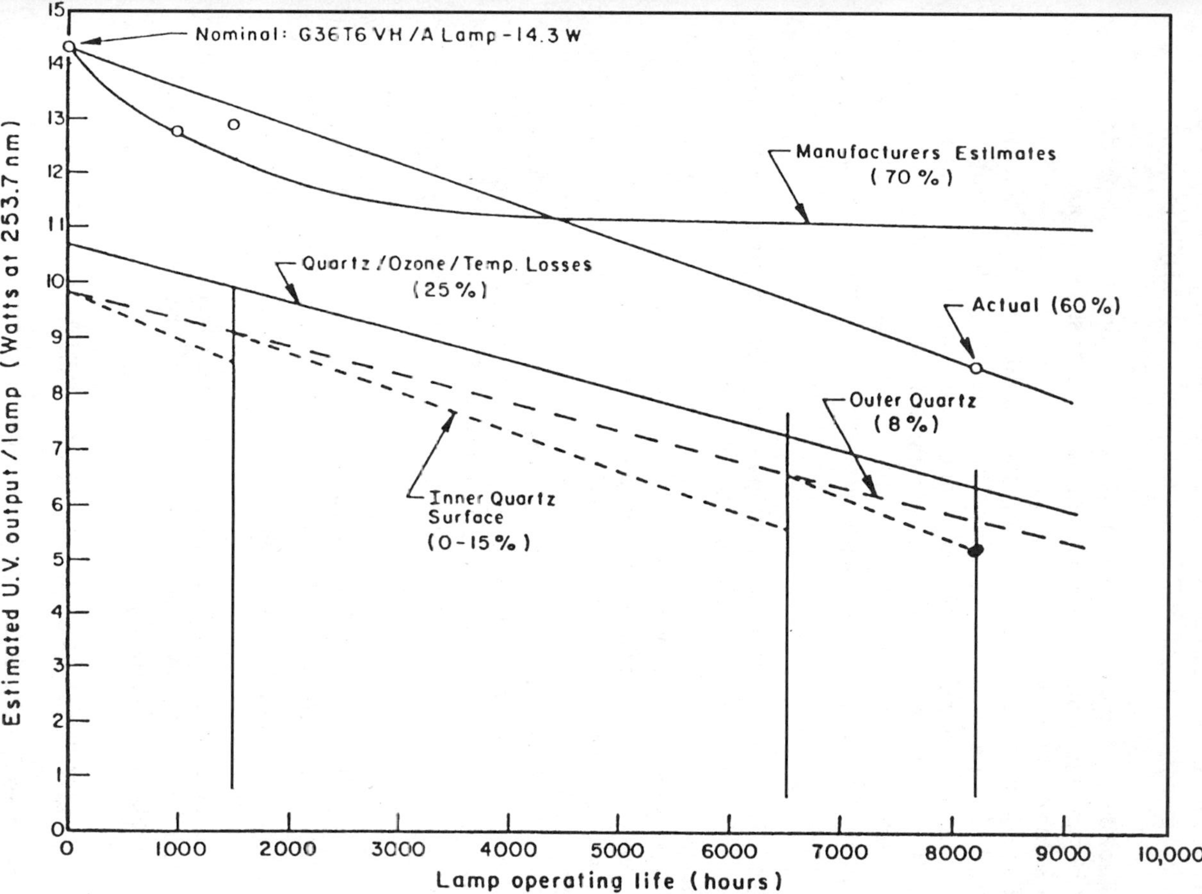

Fig. 17-31. Estimated UV output per lamp versus operating life.[6]

themselves from UV radiation. The proximity of the water to electricity is such that extra precautions *must* be taken. Safety equipment such as automatic power shut-off for hazardous circumstances should be provided. Proper storage and means of disposal of UV supplies should be available in every system.

Reliability. The performance of a UV reactor is completely dependent upon the power supply and the functioning of the lamps. Therefore a redundant but separate power supply is necessary.

Since lamp failures require equipment shutdown for repairs, provisions should be made to activate a redundant module while lamp replacement or other maintenance is in progress. There is at this time (1997) insufficient operating experience to conclude how much reactor redundancy should be provided. If a UV system is to be compared to a chlorine disinfection system, complete redundancy should be provided, that is, complete duplication of the UV reactor.

Cost. Scheible and Bassell[4] developed costs over a range of wastewater flows of 1–100 mgd. The following is an example for a 10 mgd plant that requires a 100-kW UV reactor system. The costs are based upon: a peak to average power requirement ratio of 2.1; 50 percent replication at peak; 1.25-cm spacing; average absorption coefficient of 0.5 cm^{-1}; an NPDES requirement of 200/100 ml MPN fecal coliforms (max. 30 day mean); and EPA index = 330.

For 100 kW, total estimated capital costs were $720,000. This included UV reactor, cleaning system, power supply and switch gear, instrumentation and controls, contact chamber structure, housing and ancillary equipment plus installation and start-up costs, and engineering contingencies:

Labor	10,400
Materials (lamps, ballasts, etc.)	43,000
Power—416,000 kWh at $0.04545/kWh	18,907
Total O & M Costs	$72,307

Assuming 20 years at 6 percent, capital recovery factor = 0.087, then amortized cost of capital investment = $60,800, which brings total annual cost to $134,000 or 3.6 cents/1000 gal.

IMPORTANT FIELD AND LABORATORY STUDIES

Deleterious Effects of Photoreactivation

Field and laboratory studies of a secondary effluent treated by UV radiation were conducted at a 1.0 mgd plant in Hyrum, Utah by George D. Harris et al.[23] They observed that two different reactor types demonstrated different photoreactivation potentials. The study compared the performances of the quartz-sleeve and the polytetrafluoroethylene (PFTE)-tube reactors. These are the most commonly used UV disinfection reactor designs because of the high transmittance of short-wave UV radiation through quartz and PFTE. Typically fused quartz sleeves and virgin PFTE tubes have UV transmittances of 90–95 percent and 70–85 percent, respectively.

Another objective of the study was to correlate commonly measured wastewater quality parameters with reactor efficiency and UV absorbance using field and laboratory data collected over a nine-month period. The third objective was to determine, if possible, the potential for photoreactivation of indicator bacteria in the disinfected wastewater. This is of particular concern in treatment plants where these effluents containing UV-inactivated bacteria are being discharged into lakes or rivers. Once into these surface waters, significant fractions of bacteria may be reactivated by sunlight. There is increasing evidence of photoreactivation of UV-inactivated bacteria in secondary effluents.[4,9,23]

The inactivation of bacteria by UV radiation results primarily from the adsorption of radiation by the deoxyribonucleic acid (DNA) of the microorganisms, followed by dimerization of thymine bases in DNA. These thymine dimers distort the double helix conformation of DNA and may block replication, effectively inactivating the bacteria.

The germicidal effect of UV radiation is greatest in the wavelength range of 190 to 300 nm. Exposure of UV-damaged cells to higher wavelengths of light, primarily in the visible range, often may repair much of the damage to the DNA. This repair is a result of the pyrimidine monomerization process, which is called "photoreactivation." It occurs primarily at wavelengths between 300 and 500 nm, with the most effective wavelengths dependent upon the particular organism.[24] Reactivation of the UV-inactivated bacteria can also occur in the dark. This repair by darkness is believed to be an excision repair mechanism, whereby thymine dimers are excised by a multienzymatic mechanism. The subsequent synthesis of a new component of the DNA then occurs from information on the complementary strand.[27]

This study concluded that the efficiency of disinfection was significantly decreased by photoreactivation caused by sunlight. It was also found that disinfection efficiency was consistently lower in the quartz-sleeve reactors. The total coliform increase in the quartz-sleeve reactor was 2.0 logs, whereas the PFTE tube reactor increase was only 0.5 log. For fecal coliforms the increase was 1.5 logs for the quartz-sleeve and 0.5 log for the PFTE tube.

As for the relative effect of other water quality parameters, there was a highly significant correlation between SS and fecal coliform survival at 5000 μW-sec/cm^2 and absorbance at 254 nm. Turbidity was also found to be a good surrogate of disinfection efficiency and UV absorbance. Other commonly measured water quality parameters such as BOD and TOC were found to be significantly correlated with fecal coliform survival and UV absorbance.[23] This study demonstrated the consistent advantage of the PFTE tube reactor compared to the quartz-sleeve reactor.

Evaluation of Reactor Hydraulics

The importance of hydraulics in a UV disinfection unit cannot be overestimated. Ideally the reactor unit must be able to provide 100 percent plug-flow conditions. This means that the molecules of fluid flowing through the reactor must all have the same residence time. Furthermore, for ideal conditions, the motion of flow should be turbulent radially from the direction of flow.[25] Achieving these two objectives allows each molecule of water to receive the same overall average intensity of radiation in the nonuniform intensity field that exists in a UV reactor. The tradeoff here is that some axial dispersion will occur, producing a nonideal flow pattern. Finally, maximum use must be made of the entire reactor volume that contains the UV lamps. Dead spaces must be minimized as much as possible.

The comprehensive study by Kreft, Scheible, and Venosa[25] concluded that PFTE-tube reactor units provide for near plug-flow hydraulic conditions. Two different types of quartz-sleeve units also were evaluated for their hydraulic characteristics. Both units exhibited significant short-circuiting, which reduces the effective contact time with the UV radiation. This short-circuiting also reduces energy efficiency because a portion of the UV light is consumed in the "dead spaces."

Reactor Unit Cleaning Requirements

Cleaning was found to be imperative for all units tested.[25] This is necessary to preserve the UV transmittance properties. The tube units have been marketed as a nonfouling material; however, observations and measurements during these and other studies have demonstrated that secondary wastewater effluents do cause tube fouling, and in some cases the fouling is severe. Frequent visual monitoring of these tubes in operating units demonstrated that there is a need for both physical and chemical cleaning of these tubes. A high pressure nozzle-type cleaning device is available for scouring fouled surfaces. Swabbing the tubes when practical is also recommended.

Quartz tubes will foul either biologically or chemically, or in both ways. The fouling rate depends upon the water quality and whether the lamps are on or off. Accessory cleaning equipment, either ultrasonic or mechanical wipers, has been provided with most quartz units. Wipers appear to have a greater potential for keeping quartz surfaces clean than do the ultrasonics, but neither device will preclude the need for chemical cleaning. Quartz units should be provided with some sort of chemical cleaning, preferably a recirculation system.[25]

Operating Installations: A 1984 Study

Description of the Study. This study was conducted in the United States and Canada to determine the extent to which UV is used in municipal wastewater treatment.[26] The study identified 36 facilities that were designed with UV disinfection, 31 that were under construction, 52 that were in operation, and 8 that had been in use but were abandoned. Design flow capacities were approximately 1.0 mgd with one exception, the 50 mgd plant for Madison, Wisconsin. One of the purposes of this study was to select six operational plants that could be visited for detailed inspections. The sites selected consisted of three sets of two facilities chosen to represent each of the three most commonly used brands. Of the 52 plants identified, the most common forms of secondary treatment were the oxidation ditch (31 percent) and activated sludge (25 percent). None of the six selected used filtration.

The study group concluded that until a comprehensive guide is published to assist engineers in making informed decisions instead of depending upon manufacturers, they can follow the recommendations described below.

Performance. To adequately evaluate the performance of a UV unit the following data should be made available:

1. Coliforms entering the unit (Y_0).
2. Coliforms exiting the unit (Y).
3. TSS entering the unit.
4. Absorbance of wastewater (or percent transmittance).
5. Flow rate at time of sampling.
6. Number of lamps illuminated at time of sampling.
7. Pumping rate. All samples required to confirm that the UV-treated-effluent quality meets the discharge requirements should be pumped out of the effluent channel exactly like a sample going to a chlorine residual analyzer. This will confirm or deny any effect of the clumping phenomenon. The pumping rate of the sample should be about 5 ft/sec. This will keep the piping clean and free of biological slime. This is most important.

Maintenance. This is the most important aspect of operating these units and should consist of the following primary tasks:

1. Routine chemical cleaning of the system.
2. Periodic replacement of lamps, ballasts, and quartz or nonreactive polymer surfaces.
3. Periodic repair or replacement of electrical components, such as meters, relays, and indicator lights.
4. An annual cleaning, overhaul, and repair of the entire system.

Pilot Testing. There is no consensus on the dose required for disinfection; therefore, pilot testing is the only reliable way to establish design criteria. On-site testing also will provide an opportunity to predict cleaning requirements.

In 1992 Chen and Kuo reported on the pilot testing of UV disinfection using a filtered secondary effluent.[29] It was found that a UV dose varying from 100 to 120 mW-sec/cm^2 will achieve 99.99 percent (4 logs) of inactivation of poliovirus. To be equivalent to wastewater reclamation criteria using chlorine as the disinfectant, 4 logs of virus inactivation are required. Similarly, in 1992 CH_2M-Hill[30] performed a disinfection pilot study of the rapid infiltration/ extraction (RIX) demonstration project for the Santa Ana, California, Watershed Project Authority. This project included the cities of San Bernardino and Colton. These cities are located in the most arid area of the entire state of California.)

Adequate Space Requirements. Sufficient space must be provided around the UV unit to accommodate maintenance activities. All UV units must be designed so that all lamps can be easily changed by one person without the use of special tools or ladders. The ballasts should be arranged so that the operator can remove or replace any ballast without having to remove others. Ballasts should not be gang-mounted such that access to one is hindered by one or more others, as this is not only inconvenient, wasting time and effort, but it can lead to overheating and ventilation problems.

The engineer also must provide floor space for any separate UV system components, in addition to the UV reactor. Power supply cabinets and cleaning equipment must be included in the space layout.

Adequate Ventilation. The engineer should require the UV manufacturer to provide assurance of adequate ventilation of the power supply cabinets to prevent failure from overheating. That assurance should cite the critical temperature at which the first component would fail (ballasts, for example). Failure to achieve this level of ventilation has been a frequent cause of electrical problems at the operating installations.[26]

Miscellaneous Requirements. The design engineers should be sure to require the manufacturers to equip each reactor unit with combinations of UV lamps and ballasts that are matched and intended for combination use. This will improve a lamp's output and extend its useful life.

The equipment design should include a proper drain system with a bottom cleanout port to allow removal of deposits or accumulations inherent in the operation of these systems.

Modular construction should be specified when multiple units are required. This will allow uninterrupted service when maintenance is performed.

The electrical power consumption of the UV system should be metered separately for process documentation. Also, operators should be able to determine quickly if each individual lamp is illuminated.

The system should be designed so that power consumption will vary with the process flow (i.e., low flow—decreased power consumption).

Finally, the engineer should insist upon sufficient staffing for data collection and monitoring, in order to evaluate performance and reliability.

All samples for evaluating performance should be pumped out of the effluent channel at about 0.5 gpm and 5 ft/sec to keep the piping clean and free from biological slime. This will confirm the validity of the NTU performance.

UV Disinfection Research Priorities[28]

During the meeting of a panel of the National Water Research Institute it became obvious that a number of important issues related to UV disinfection must be resolved before use of this alternate disinfectant could be considered routine for water reclamation and reuse. Of the issues presented and that were discussed by the panel, the following list represented those that were the most important.

Standardized Determination of UV Dose

Problem. The UV dose is defined as the average UV intensity in the UV reactor multiplied by the average exposure time. Unfortunately, no universally accepted method is available that can be used to determine the average UV intensity and the average exposure time in operating full-scale installations. The average intensity must be known under a variety of operating conditions, including the failure of one or more lamps. The average exposure time depends upon hydraulic characteristics of the UV reactor, which include the effects of axial turbulence and the upstream and downstream flow conditions.

Objectives:

1. To develop a standardized procedure that can be used to determine the UV intensity within a UV disinfection reactor subject to varying water quality and for a variety of reactor configurations.
2. To develop a standardized procedure that can be used to determine the average exposure time for a variety of reactor configurations.

Suggested Approach. With respect to UV intensity, the approach should be to develop a mathematical model that has been verified with bioassays and chemical actinometry. The UV intensity model must be capable of predicting average intensity within the UV reactor with respect to varying wastewater quality along with one or more lamps out of service. Before development of any new models, the adequacy of the new EPA model should be evaluated over a wide range of operating conditions.

A procedure must be developed and tested to determine the average exposure time in both parallel and perpendicular UV flow reactors. To the extent possible, the effects of axial turbulence as well as channel geometry, including inlet and outlet conditions, must be evaluated.

UV Disinfection Testing Protocol

Problem. Because UV disinfection represents an attractive business opportunity, a number of equipment manufacturers have entered the field. Moreover, not all of the UV disinfection systems offered for sale have been tested under field conditions; so there is no way that the claims of manufacturers can be evaluated. Further, because a myriad of analytical techniques have been used, generalizations and extrapolations from these efforts are impossible.

Objective. Researchers are to develop a standardized protocol that manufacturers should follow in testing their systems prior to making marketing claims.

Suggested Approach. The protocol to be developed should address the items identified in this guidance document.[28]

Effect of Reliability Issues on UV Disinfection Performance in Wastewater Reclamation

Problem. Reliability issues are of critical importance in the use of UV disinfection in applications where a restrictive effluent bacterial quality standard must be met. Important operational issues include lamp outage, varying influent quality, and plant upsets. For example, as UV lamps age, they will begin to fail. When failure occurs, it may not be possible to change the lamp or lamps immediately. If it is not possible to change the lamps immediately, what effect will the failure of one or more lamps have on the overall performance of the UV disinfection system? Another important operational issue is: What effect will the varying influent quality or plant upsets have on the overall performance of the UV system?

Objective. Researchers are to define what effect reliability issues will have on the overall performance of the UV disinfection system, with special emphasis on UV systems designed to meet the restrictive California Wastewater Reclamation Criteria[34] (2.2 coliforms/100 ml or less).

Suggested Approach. Using a number of well-defined UV installations, collect field data on the effect of lamp failure. Both lamp and power failure can be simulated by selectively turning off lamps. A wide range of lamp failure modes should be tested, including the failure of adjacent lamps. The data obtained to evaluate the effect of lamp failure can also be used to verify the dose model to be developed as described in Research Topic 1. The effect of varying water quality and plant upsets can best be studied by selecting several plants with different treatment processes and monitoring more closely the effect of these conditions on the performance of the UV disinfection system. Such studies should be conducted over a period of a year, or during periods of the year that are known to cause operational problems, such as periods of stormwater infiltration.

Assessment of Importance of Pretreatment on UV Disinfection Performance

Problem. It has been observed that, for a given wastewater treatment plant, the distribution of particle sizes in secondary effluent is dependent upon the nature of the pretreatment process and the design of the secondary settling facilities. Typically the volume distribution of particle sizes in secondary effluent is bimodal, with larger particle sizes controlling the effectiveness of UV disinfection. It has also been observed that the effluent distribution of particle sizes for a given plant does not change significantly with increasing suspended solids. What does change is the total number of particles. To meet the most restrictive California Wastewater Reclamation Criteria consistently, effluent filtration has been required to reduce the number of particles in the effluent because UV cannot kill or inactivate anything that can be seen with the human eye.

Objectives:

1. To assess the effects of various pretreatment processes and the design of the secondary setting facilities on the performance of UV disinfection with and without effluent filtration.
2. To assess how the various pretreatment processes affect the removal of the larger pathogenic organisms such as the cysts of *Giardia* and *Cryptosporidium.*
3. To quantify whether changes in treatment plant design and operation can bring about changes in the performance of the UV disinfection system.

Suggested Approach

The first phase of the assessment should be to conduct a comprehensive review of the literature to identify the effects of pretreatment that have been documented.

The second phase of the assessment is to conduct a comprehensive study of existing UV systems, paying special attention to the findings from the first phase of the assessment. Based upon these findings from the literature review and the field studies, the effect of changes in plant operation on UV disinfection performance should be quantified.

Development of a Suitable Biological Indicator to Assess UV Disinfection Performance

Problem. At the present time there is no suitable microorganism that can be used as an indicator of the performance of UV disinfection. A variety of organisms have been suggested for use as potential indicator organisms on a routine basis to assess the performance of UV disinfection. Male-specific coliphages, *Clostridium perfringens,* and heterotrophic organisms as measured by the heterotrophic plate count have been suggested.

Objective. To identify and test one or more organisms that may be suitable for use as an indicator organism to assess the performance of UV disinfection.

Suggested Approach. One or more full-scale UV systems should be selected and monitored over a one-year period to determine the removal of male-specific coliphages, *Clostridium perfringens,* heterotrophic organisms measured by the heterotrophic plate count, and any other appropriate organisms. Reduction of these organisms before and after UV treatment should be determined. Various water quality characteristics such as percent transmittance at 254 nm, suspended solids, particle size distribution, and turbidity should be monitored to assess their influence.

If a reliable indicator organism can be identified, the next step would be to quantify its relationship to the inactivation of adenovirus, reovirus, retrovirus, and rotavirus (see research topic photoreactivation following UV disinfection. If a pilot-scale system is available, then seeded work with reoviruses and rotaviruses should also be conducted to determine if naturally occurring strains of these viruses are more resistant than laboratory strains.

Effects of UV Disinfection on Protozoa and Protozoan Cysts

Problem. The effectiveness of UV disinfection in the inactivation of protozoa, and especially the cysts of protozoa under field conditions, is not well documented in the literature. This research topic is a high priority for all modes of disinfection. The principal problem with the identification of protozoan cysts is to determine cyst viability. At present it appears that the only reliable method is to use animals. This is an expensive and tedious task. Although the importance of this research topic is not in question, it represents a long-term effort.

Objective. Researchers are to define the effectiveness of UV disinfection in the inactivation of Giardia, Cryptosporidium, and their respective cysts.

Suggested Approach. Use *Giardia, Cryptosporidium,* and their cysts as found in treated effluents to assess the performance of UV disinfection. The survival and the viability of protozoan cysts should be related to UV dose.

Identification of By-products Formed in UV Disinfection and Their Potential Health Effects

Problem. At the UV dosages now used in low energy applications, the formation of UV disinfection by-products (DBPs) appears to be minimal.

However, as medium and high energy UV lamps become available, it is anticipated that the formation of these DBPs may increase. The principal concern with the DBPs is with public health. A secondary concern is the effects of any by-products on the aesthetic quality of the water. For this reason, the by-products formed during UV disinfection should be identified, and their health effects evaluated. (White suggests that to accomplish this, get Bruce Ames and his group at U.C. Berkeley to investigate this issue.)

Objectives:

1. To identify the UV DBPs formed at various UV intensities.
2. To assess the health effects of the by-products that may be of health concern, but have not been studied.
3. To determine whether any of the by-products formed during UV disinfection can be used as a surrogate chemical measure to assure the performance of the process in a manner comparable to the use of chlorine residual in chlorine disinfection.

Suggested Approach. Using the new medium and high energy UV lamps, identify the types of compounds that are formed. Look for these compounds with the low energy lamps. Depending upon the energy spectrum of the newer medium and high energy lamps, it may not be possible to correlate the results to the low energy lamps. The health effects of UV DBPs that are unknown should be assessed, on the basis of known information. Further health effects studies should be proposed as required.

Determine if any of the DBPs formed during UV disinfection, which can be identified reliably, can be correlated to the efficiency of the UV disinfection system.

Characterization of Organic Precursors in the Formation of DBPs

Problem. Total organic carbon (TOC) is currently used as a surrogate for measuring DPB precursors. Although it is well established that DBPs originate from organic compounds, the use of TOC as a surrogate measure of DBP precursors is inadequate. Only 10 percent of the compounds that comprise TOC in wastewater have been identified. Furthermore, the organic compounds in wastewater are oxidized to a different extent by the various physical and chemical methods used for the treatment of wastewater. To develop effective control strategies to minimize the production and release of DBPs, more of the compounds that comprise the TOC must be identified. Any classification scheme based upon chemical characterization will improve the understanding of DBP production, and will lead to more effective control strategies.

Objectives:

1. To provide further identification and characterization of UV DBP precursor materials.
2. To identify UV DBP precursor control technologies.
3. To determine if UV disinfection accelerates or exacerbates the production of DBPs.

Suggested Approach. Using new liquid chromatograph techniques, such as preparative liquid chromatography, empore disks, etc., determine the feasibility of fractionating the TOC into chemical classes. Such a classification scheme, based upon how the solutes are recovered, can be used to characterize the TOC more fully.

Photoreactivation Following UV Disinfection Problem

Although a considerable amount of information is available on the photoreactivation of microorganisms following UV disinfection, there is little consensus on how the results reported in the literature relate to the actual health risks in full-scale operating systems. It has been suggested that photoreactivation is not a significant problem because UV-injured organisms die off faster in natural waters. Information from operating UV systems is needed to assess the effects of UV dose upon the photoreactivation of pathogenic bacteria and other potential indicator organisms.

Objectives:

1. To determine the effects of UV dose upon the photoreactivation of pathogenic bacteria and other potential indicator organisms.
2. To determine the survival and fate of partially damaged organisms when discharged to streams and other bodies of water exposed to sunlight.
3. To assess whether particular design and/or operating features can be utilized to minimize photoreactivation.

Suggested Approach. The objectives would be accomplished by laboratory studies on the bacterial pathogens and other indicator organisms. Field studies on naturally occurring organisms would also be necessary. Specifically:

1. Evaluate the influence of water quality parameters, UV dose, design and operating features, exposure of organisms to light before UV treatment, etc., on the degree of photoreactivation

2. Compare photoreactivation of laboratory strains of bacteria added to wastewater versus naturally occurring strains
3. Determine if photoreactivation occurs in selected waterborne bacterial pathogens (*Salmonella, Shigella,* etc.), and in bacterial indicators that could be used for monitoring UV performance such as male-specific coliphages, *Clostridium perfringens,* and heterotrophic organisms.
4. If photoreactivation of pathogens occurs, determine if their virulence is altered—such as whether the genes that control the photo repair could also influence the virulence genes.
5. Determine the survival and the fate of UV-injured microorganisms after discharge into streams of natural waters.

Effects of UV Disinfection on Selected Viruses

Problem. The effects of UV disinfection on adenovirus, double-stranded RNA viruses (reovirus), retrovirus, and rotavirus are not well understood, especially under field conditions. Reoviruses and rotaviruses are the enteric viruses most resistant to UV irradiation, yet no data exist on the removal of rotaviruses by pilot plants or in operating systems. Only one study exists on the removal of reoviruses, and in that study it was concluded that reoviruses were not inactivated completely by UV.

Objective. To determine the comparative removal of reovirus, retrovirus, and rotavirus by UV disinfection.

Suggested Approach. A one-year study would be necessary because the concentrations of reoviruses and rotaviruses varies greatly over the year, with the highest concentrations in the winter. Various water quality characteristics, such as percent UV light transmittance, suspended solids, and turbidity, should be monitored to assess their influence. If a pilot-scale system is available, then seeded work with reoviruses and rotaviruses should also be conducted to determine if naturally occurring strains of these viruses are more resistant than laboratory strains.

UV ADVANCED OXIDATION PROCESS

Purus UV Process

Introduction. Purus Inc. is a Silicon Valley firm located in San Jose, California. It was founded in 1989 to develop an advanced ultraviolet radiation process for the removal of volatile organic chemicals (VOCs) from water and air. One of the great environmental challenges facing the United States and other industrial countries is the task of cleaning up groundwater resources and soils contaminated with VOCs. Recognizing these hazards, the USEPA has established regulations for most VOCs. These MCLs include trichloroethylene (TCE), tetrachloroethylene (PCE), carbon tetrachloride, Freon 113, acetone, and benzene. Cleaning up VOC-contaminated sites is difficult, expensive, and usually long-term. Furthermore, most current methods are less than ideal, in that the contaminant is merely transferred to another medium, such as an activated carbon adsorbent or the atmosphere. This is an important reason why there is a definite need for an inexpensive process that can destroy these contaminants on-site and convert them to harmless by-products without the necessity of transporting or handling the contaminants.

The Purus technology provides a process that meets all these requirements. This process is based upon VOC destruction via photolytic oxidation, using a greatly improved and scientifically advanced ultraviolet light source.

Current VOC Treatment Technologies

Almost all VOC removal from groundwaters is based upon well extraction followed by treatment, commonly referred to as pump-and-treat remediation. Affected soil layers also must be treated to prevent future groundwater contamination. Soil treatments include volatilization, aeration (soil venting), vacuum extraction, soil washing, and thermal catalytic destruction. These treatment options for cleaning up VOC-contaminated groundwater can be divided into five groups, as follows:

1. Air stripping/vacuum extraction with emission controls.
2. Liquid phase GAC adsorption
3. Biological treatment.
4. Membrane technology.
5. AO_xPs (includes UV-based systems).

The Purus technology is an AO_xP. It is most attractive because it converts the VOCs into innocuous by-products such as carbon dioxide, water, and an inorganic compound. This process is usually referred to as mineralization. The equation below illustrates the complete mineralization of TCE by the Purus AO_xP:

Purus method, advanced UV photolysis:

$$\underset{\text{TCE}}{Cl_2C{=}CClH} + UV \longrightarrow 2CO_2 + 3Cl^- \quad (17\text{-}13)$$

There are several arrangements of AO_xPs that are used for contaminant destruction:

1. Ozone alone.
2. Peroxone = $O_3 + H_2O_2$.
3. Ozone + UV.
4. Peroxone + UV.
5. H_2O_2 + UV.
6. H_2O_2 + metal catalyst + UV.
7. Fenton's reagent.
8. Sunlight + metal catalyst.

The primary reaction mechanism of these conventional AO_xPs involves two steps. The first step is the formation of a hydroxyl radical, a powerful oxidizer. The second step is the reaction of the hydroxyl radical with the organic contaminant. This sets off continuous reaction propagations until complete mineralization of the contaminant occurs. Equation (17-14) shows the two-step reaction of a conventional AO_xP:

Step 1:

$$\left.\begin{array}{l} H_2O_2 + O_3 \\ H_2O_2 + UV \\ H_2O_2 + O_3 + UV \\ H_2O_2 + M^+ + UV \end{array}\right] \rightarrow *OH$$

Step 2:

$$\underset{\text{TCE}}{Cl_2C{=}CClH} + *OH \longrightarrow 2CO_2 + 3Cl^- \qquad (17\text{-}14)$$

The destruct mechanism depends entirely upon the reaction rate between the [OH] radical and the organic substrate. The reaction is favored if the substrate has electron-dense molecular bonds, such as double bonds, triple bonds, or aromatic configurations. Organic molecules with saturated bonds are not good targets for [OH] radical attack. Table 17-1 (from Purus Inc.) shows the rate constants for the [OH] radical, which illustrate this point.

Purus Process Description. This can be characterized as a secondary destruct mechanism via direct photolysis, which occurs when sufficient UV energy is absorbed by the organic contaminant. The absorbed energy transforms the atomic electrons into higher energy states, causing molecular bonds to break. If an oxygen source such as moisture or air is present when the bond

Table 17-1 Rate Constants for the Hydroxyl Radical

Compound	*Rate of Reaction* ($K \times 10^{-9}$ M^{-1} S^{-1})
(Unsaturated organics)	
Benzene	7.8
Vinyl chloride	7.1
Trichloroethylene (TCE)	4.0
(Saturated organics)	
Dichloromethane	0.058
Chloroform	0.005
Carbon tetrachloride	Nonreactive
Trichloroethane (TCA)	Very slow
Dichloroethane (DCA)	Very slow

Ref: Burton, et al., *J. Phys. Chem. CEF. Data,* 17:2:513 (1988).

breakage occurs, complete mineralization of the organic molecule to carbon dioxide is possible, as shown in Eq. (17-13).

Advanced oxidation process (AOP):

$$\underset{\text{TCE}}{Cl_2C{=}CHCl} \xrightarrow{\text{AOP}} 2CO_2 + 3Cl^- \tag{17-15}$$

The Purus process is built upon the "flashlamp" technology that was originally developed for the laser industry; the Purus system represents a new generation of the xenon UV flashlamp. A flashlamp is an arc lamp that operates in the pulsed mode; it is capable of converting stored electrical energy into intense outbursts of radiant energy. The Purus light is generated by a high temperature xenon plasma contained within a chamber of UV-transmissive quartz. The plasma is produced by a pulse discharge of electrical energy across two electrodes contained in the quartz chamber. This discharge quickly heats the xenon gas to an extremely high temperature, more than 13,000 K. The optical properties of the plasma approach those of an ideal black-body radiator, which has a peak emission wavelength defined by its characteristic temperature. The greater the energy poured into the xenon plasma, the hotter it will become, the lower the wavelength will be, and the higher the energy will be at its maximum emission point. Therefore, the Purus flashlamp technology produces an exponentially increasing number of photons as the temperature of the xenon plasma increases—**this is Wien's displacement law.*** By comparison, the physics of the mercury lamp is vastly different.

*This law states: The number of photons emitted by a black-body is proportional to the 4th power of the temperature.

The Purus process for recycled water is illustrated by Eq. (17-15).

Purus Technology vs. the Conventional Mercury Lamp Technology. Almost all the radiation emitted by the low pressure mercury lamp is resonance radiation at 253.7 and 185.0 nm. At these two wavelengths, mercury lamps are very efficient. However, the mercury lamp is handicapped because mercury sources have relatively low intensity. This is a result of the reversible process by which a photon is emitted from an excited atom; the mercury atoms will begin to reabsorb radiation at the two characteristic lines as the energy is increased. If the power into a mercury lamp is increased, the efficiency of producing UV decreases significantly and simultaneously with the rate of increase in power input, whereas with the Purus process significant amounts of energy can be discharged into the xenon plasma before practical limitations are reached. Furthermore, the efficiency of the Purus technology allows the system to be selective for the particular contaminant involved. The peak power to the lamp determines the UV content of the spectral emission, and the frequency of the pulse determines the average power. The mineralization characteristics of various organic contaminants are a function of both the peak power (UV spectra) and the average power.

One of the more important features of the Purus technology is the ability to shift the maximum UV spectra output to more closely match the absorption characteristics of a specific organic compound. This is done by careful control of the pulse discharge used to heat the plasma. The average power, single pulse power, pulse shape, quartz envelope temperature, and average current have all been examined by the designers of the Purus process. Therefore, the present Purus system employs customized electronics and microprocessor control to regulate these parameters.

As of 1992, Purus is developing organic sensors that will allow closed loop control of the UV power input, depending upon the influent level and desired effluent level because the input level of a toxic stream will vary over a period of time. This ability to optimize the power input requirement for a given contaminant level can produce significant treatment cost savings, protection from unexpected surges of high concentration pockets, and enhancement of system reliability.

The Purus aqueous destruction rate of contaminants is typically described in the form of a rate graph, expressed as contaminant concentration versus exposure time in seconds as illustrated by Eq. (17-16):

$$k = -\mathrm{Ln}\frac{A_i}{A_0}/t \tag{17-16}$$

where

k = rate constant

A_0 = initial contaminant concentration

A_i = contaminant concentration at time t
t = exposure time in seconds
Ln = logarithm

Table 17-2, from Purus Inc., lists comparative destruction rates by the Purus technology versus published rates for the mercury-based systems. The Purus rates have been normalized to 1.000, with other rates expressed as the relative ratio. For example, the Purus rate for benzene destruction is 42 times faster than the published rate for mercury lamps. Here again the reader is reminded that the Purus pulse plasma flashlamp can produce the fastest known photolysis destruction rates because it yields such a high intensity of UV radiation in the 180 to 240 nm wavelength range.

THM Photolysis. The Purus system has achieved the fastest THM photolysis rates of any published UV-based data. It has been found that absorption of the halide molecules is continuous in nature. The absorption wavelength maximum shifts in the longer wavelengths as the electronegativity of the halogen substituent is lowered from chlorine to bromine, or as the number of bonded halogens increases. For example, the maximum wavelength of $CHBr_3$ is longer than that for $CHCl_3$. Likewise, the maximum wavelength for $CHCl_3$ is longer than that for CH_3Cl. The spectra of the higher halogen-substituted compounds also show a broadening of the absorption region because of splitting from the halogen–halogen interaction.

Numerous variables have been examined—parameters associated with UV lamp operation and construction, as well as parameters critical to the reaction chamber design. In addition, various catalysts, including TiO_2, have been studied. As a result of this work, a commercially competitive AO_xP has evolved, with features and benefits superior to those of the conventional AO_xPs previously available. Ten patent applications are in various stages of submission to the U.S. Patent Office as of 1997.

Experiments conducted by Purus have shown that photolysis rates have yielded two increasing trends:

Table 17-2 Relative Destruction Rate Constant Comparison

Source	*Chemical*	*"K" Rate Ratio Purus/Others*
Purus/Symons et al.	Benzene	42:1
Purus/Symons et al.	Carbon tetrachloride	860:1
Purus/Ultrox	I,I-DCA	9:1
Purus/Ultrox	I,I,I-TCA	21:1
Purus/Peroxidation Systems	TCE	4:1

$$CCl_4 \rightarrow CHCl_3 \rightarrow CH_2Cl_2 \tag{17-17}$$

and

$$CHI_3 \rightarrow CHBr_3 \rightarrow CHCl_3 \tag{17-18}$$

The photochemistry of alkyl halides, RXn, where X is chlorine or bromine, all involve the loss of the weakest bound halogen ion. The reaction is described by the following equation:

$$RX + UV \rightarrow R^* + X^- \tag{17-19}$$

where RX is the organic compound (example, $CHCl_3$), R is the organic portion of the molecule, X is the halogen portion of the molecule, UV is the ultraviolet light energy, R^* is the organic radical after bond breakage, and X^- is the halide ion (Cl^-) resulting from bond breakage. *Note.* In the presence of an oxygen source, such as air or water, the organic radical will be completely oxidized to carbon dioxide.

The solvent cage around this alkyl halide is important in the process. Thus in the aqueous phase, continuous reaction propagation will continue to occur until the organic carbon is completely mineralized to inorganic carbon dioxde, with the halogen remaining in solution as a free halide salt. Direct photolysis of the THMs can be achieved only with UV photons in the alkyl halide absorption region. Therefore, the 254 nm emission line of a conventional mercury lamp will not yield a sufficient photolysis rate for chloroform.

Key Applications of Purus Photolysis. This system can fulfill many environmental applications, such as groundwater remediation, air stripping emission control, and soil venting. The following compounds can be effectively destroyed by the Purus process:

1. Double-bonded organics: TCE, PCE, DCE.
2. Saturated organics: chloroform, TCA, methylene chloride, carbon tetrachloride, and DCA.
3. Petroleum compounds: benzene, toluene, ethyl benzene, xylene, MTBE.
4. Pesticides: EDB, DBCP.
5. Metal chelants: cyanides, EDTA.
6. Other organics: PCBs, explosives, dioxyn, acetone, IPA.

Summary. All of the information described above was developed by Purus with its own equipment. The majority of the data was taken on an HP .3850 gas chromatograph with either an ECD or an FID detector. Some of the samples were run on a Varian Saturn GC/MS utilizing an ion tap detector. Purus periodically sends out duplicate samples to local EPA-certified labs for independent confirmation of the data.

REFERENCES

1. Nagy, R., "Research Report BL-R-6-1059-3023-1, Water Sterilization by UV Radiation," Westinghouse Electric Co., Lamp Division, Bloomfield, NJ, Feb. 1955.
2. Witherell, L. E., Solomon, R. L., and Stone, K. M., "Ozone and Ultraviolet Radiation Disinfection for Small Community Water Systems," Municipal Env. Res. Lab, U.S. EPA report No. EPA 600/2-79-060, Cincinnati, OH, July 1979.
3. Roeber, J. A., and Hoot, F. M., "Ultraviolet Disinfection of Activated Sludge Effluent Discharging to Shellfish Waters," EPA 600/2-75-060, Municipal Environmental Research Lab., Cincinnati, OH, Dec. 1975.
4. Scheible, O. K., and Bassell, C. D., "Ultraviolet Disinfection of a Secondary Wastewater Treatment Plant," EPA Municipal Env. Res. Lab report EPA-600/S2-81-152, Cincinnati, OH, Sept. 1981.
5. Scheible, O. K., Forndran, A., and Leo, W. M., "Pilot Investigation of Ultraviolet Wastewater Disinfection at the New York City Port Richmond Plant," paper presented at 2nd Nat. Symposium of Municipal Wastewater Disinfection, in Symposium Proceedings, EPA/600-9-83-009, Cincinnati, OH, July 1983.
6. Scheible, O. K., "Design and Operation of UV Systems," presented at Ann. Conf. WPCF Atlanta, GA, Oct. 1983.
7. Johnson, J. D., Qualls, R. G., Aldrich, K. H., and Flynn, M. P., "UV Disinfection of Secondary Effluent: Dose Measurement and Filtration Effects," 2nd Ann. Seminar, Wastewater Disinfection, sponsored by EPA, Orlando, FL, Jan. 26–28, 1982.
8. Qualls, R. G., and Johnson, J. D., "Bioassay and Dose Measurement in UV Disinfection," *Applied and Environmental Microbiol.,* p. 872 (Mar. 1983).
9. Qualls, R. G., Flynn, M. P., and Johnson, J. D., *"The Role of Suspended Particles in Ultraviolet Disinfection,"* Univ. of North Carolina, Chapel Hill, NC, 1983.
10. Venosa, A. D., Director, Wastewater Disinfection Research Program, private communication, May 1984.
11. Lawrence, C. A., and Block, S. S., *Disinfection, Sterilization and Preservation,* Lea and Febiger, Philadelphia, PA, 1968.
12. Hufff, C. B., Smith, B. S., Boring W. D., and Clark, N. A., "Study of Ultraviolet Disinfection of Water and Factors in Treatment Efficiency," *Public Health Reports,* **80,** 695 (Aug. 1965).
13. Luckiesh, M., and Holladay, L. L., "Disinfection of Water by Means of Germicidal Lamps," *General Electric Review,* p. 45 (Apr. 1944).
14. Loofbourow, J. R., "The Effects of Ultraviolet Radiation on Cells," *Growth,* **12,** 77 (1948).
15. Kawabata, T., and Harada, T., "The Disinfection of Water by the Germicidal Lamp," *J. Illumination Soc.,* **36,** 89 (1959).
16. Hill, Jr., W. F., Akin, E. W., Benton, W. H., and Hambley, F. E., "Viral Disinfection of Estuarine Water by UV," *J. San. Engr. Div., ASCE,* p. 601 (OCt. 1971).
17. Petrasek, Jr., A. C., Andrews, D. C., and Wolf, H. W., "Ultraviolet Disinfection of Wastewater Effluents," paper presented at the Disinfection Seminar, AWWA Ann. Conf., Anaheim, CA, May 8, 1977.
18. Kelner, A., "Effect of Visible Light on the Recovery of *Streptomyces griseus* conidia from Ultraviolet Irradiation Injury," *Proc. Nat. Acad. of Sci. U.S.,* **35,** 73 (1949).
19. Anon., "Policy Statement on the Use of Ultra-violet Process for Disinfection of Water," Dept. of H.E.W., Div. of Env. Engr. and Food Prot., Washington, DC, Apr. 1, 1967.
20. Selleck, R. E., private communication, Univ. of California, Berkeley, CA, June 1984.
21. Jacob, S. M., and Dranoff, J. S., "Light Intensity Profiles in a Perfectly Mixed Photoreactor," *J. AIChE,* **16,** 359 (May 1970).
22. Scheible, O. K., "UV Disinfection," paper presented at the Disinfection of Water and Wastewater Seminar, University of Wisconsin Ext. Div., Milwaukee, WI, May 16–18, 1984.
23. Harris, George, D., Adams, Dean, Sorenson, Darwin L., and Dupont, R. Ryan, "The Influence of Reactivation and Water Quality on UV Disinfection of Secondary Municipal Wastewater," *J. WPCF,* **59,** 8, 781–789 (Aug. 1987).

24. Jagger, J. H., *Introduction to Research in Photobiology,* Prentice-Hall Inc., Englewood Cliffs, NJ, 1967.
25. Kreft, P., Scheible, O. K., and Venosa, A. D., "Hydraulic Studies and Cleaning Evaluation of UV Disinfection Units," *J. WPCF,* **58,** 12, 1129–1137 (Dec. 1986).
26. White, S. C., Jernigan, E. B., and Venosa, A. D., "A Study of Operational UV Disinfection Equipment at Secondary Treatment Plants," *J. WPCF,* **58,** 3, 181–192 (Mar. 1986).
27. Watkin, E. M., "Dark Repair Mutations Induced by *Escherichia coli* by UV Light," in F. H. Sobek (Ed.), *Repair from Genetic Radiation Damage,* Pergamon Press, London, 1963.
28. "UV Disinfection Guide Lines for Wastewater Reclamation in California & UV Disinfection Research Priorities," prepared by the National Water Research Institute, Fountain Valley, CA for the State of California Department of Health Services, Sacramento, CA, Sept. 1993.
29. Chen, C. L., and Kuo, J. F., "UV Inactivation of Bacteria and Viruses in Tertiary Effluents," presentation to the DHS UV Disinfection Committee on Research at the County Sanitation Districts of Los Angeles County, 1992.
30. CH_2M-Hill, "Disinfection Pilot Study—Rapid Infiltration/Extraction (RIX) Demonstration Project," prepared for the Santa Ana Watershed Project Authority, City of San Bernardino and the City of Colton, CA, 1992.
31. Trojan Technologies Inc., "The Basic Design, Operational and Water Quality Conditions and Constraints by UV Systems Intended to Produce a Pathogen Free Effluent," prepared for the California Dept. of Health Services, UV Disinfection Committee, 1992.
32. Snider, K. E., Darby, J. L., and Tchobanoglous, Geo., "Evaluation of Ultraviolet Disinfection for Wastewater Reuse Applications in California," Dept. of Civil Engineering, University of California, Davis, 1991.
33. U.S. Environment Protection Agency, "Design Manual: Municipal Wastewater Disinfection," EPA/625/1-86/021, Office of Research and Development, Water Engineering Research Laboratory, Center for Environmental research Information, Cincinnati, OH, 1986.
34. California Code of Regulations, "Title 22, Division 4, Wastewater Reclamation Criteria, Environmental Health," Chapter 3: "Reclamation Criteria," State of California, Department of Health Services, Sanitary Engineering Section, pp. 1557–1610, 1978.
35. Sanitation Districts of Los Angeles County, "Pomona Virus Study, Final Report," prepared for California State Water Resources Control Board and the U.S. Environmental Protection Agency, 1977.
36. UV Systems Technology Inc. (Ultra Guard), 2800 Ingleton Ave., Burnaby, BC, Canada V5C 6G7, Phone 800-801-5656; 1996.
37. Wilson, B. P., Roessler, E., Van Dellen, M., Abbaszadegan, and Gerba, C. P., "Coliphage MS-2 as a UV Water Disinfection Efficacy Test Surrogate for Bacterial and Viral Pathogens," poster portion of AWWA Water Quality Technology Conf., 1992.
38. Havelaar, A. H., Nieustad, T. J., Meullemans, F., and Van Olphen, M. F., "Specific RNA Bacteriophages as Model Viruses in UV Disinfection of Wastewater," *Water Science Technology,* **24,** No. 2 (1991).
39. Capital Controls, "Series 8200 OXYFREE® Ultraviolet Disinfection System," 3000 Advance Lane, Colmar, PA 18915 (A Severn-Trent Company, England), 1996.
40. Ecometrics Inc., "UV Open Channel Disinfection Technology," 130 W. Main Street, Silverdale, PA 18962, 1995.
41. Bowen, M. W., Aquaray® 40 VLS UV Disinfection," Infilco Degremont Inc. P.O. Box 71390, Richmond, VA 23255-1390, Aug. 1996.
42. Furst, G. Michael, Jr., P. E., "*Cryptosporidium* & *Giardia* Inactivation Device," Safe Water Solutions L.L.C., 9333 No. 49th St., Brown Deer, WI, 53223, Customer Service 414-365-2377; 1996.
43. Clancy, J., "Clancy Environmental Consultants Inc.," St. Albans, VT.
44. Sterling, C. R., et al., "Infectivity Studies Showing Greater than 3-Log Removal of Oocysts," University of Arizona, College of Agriculture, Department of Veterinary Science, Tucson, AZ.
45. Campbell, A. T., Robertson, L. J., Snowball, M. R., and Smith, H. V., "Inactivation of Oocysts of *Cryptosporidium parvum* by UV Irradiation," *Water Research* Vol. **29,** No. 11, pp. 2583–2586 (1995), Elsevier Science Ltd., Great Britain.

Appendix I

FRICTION LOSS FACTORS AND PIPE DATA

The nomograph in Fig. 16 allows a direct determination of the equivalent length of pipe for all appurtenances from the K factor. A list of K factors is shown in Table 1.

The nomograph is by S. Labella* and was derived as follows:

$$Hf = K\frac{V2}{2G} \text{ Darcy Equation} \tag{1}$$

Table 1 Values of K for Determining "Equivalent Length of Pipe" from the Nomograph on Page 1506

*Resistance of Valves and Fittings**

Type of Resistance	*K*	*Type of Resistance*	*K*
90° E11 std.	0.9	Outlet loss:	
90° E11 flanged	0.25	pipe to atmosphere	1.0
Ditto-long radius	0.2	main connection type	
		chlorine diffuser	1.0
Tee – std:	1.8	Valves in open position:	
	1.8	angle valve	5
	0.6	globe valve	10
45° E11 screwed	0.4	gate valve	0.2
Long radius flanged	0.2	plug valve	0.77
Sudden enlargement:		stem-type chlorine	
		gas valve	10
d/D – ¼	0.92	diaphragm valve	2.3
d/D – ½	0.56	ball valve	0.8
d/D – ¾	0.19	ordinary entrance	0.5
Sudden contraction:			
d/D – ¼	0.42		
d/D – ½	0.33		
d/D – ¾	0.19		

*Values given are based on consensus of Crane Company catalog #52; Bulletin 252, Univ. of Wisconsin; Giesecke & Badgett; "Hydraulics" by Daugherty; and Tentative Standards Hydraulic Institute.

*Reprinted from p. R-238 and R-239, 1967 WATER AND SEWAGE WORKS.

Solving the Hazen Williams equation for *Hf*

$$Hf = \left[\frac{VL^{0.54}(4)^{0.64}}{131.8(d)^{0.63}}\right]^{1.85} \quad (2)$$

Substitute $V = Q/A$, where Q = discharge in gpm into equations (1) and (2) and equating (1) to (2), solve for L, which becomes:

$$L = \frac{K(Q^{0.15})d^{0.87}}{0.81}$$

Example: Find the equivalent length of 6″ $C = 100$, in pipe for a valve with $K = 1.0$ and an average flow of 300 gpm.

Solution: Connect K and the diameter (6); find the point on the turning line. Connect the turning line point to the discharge (300). Read the answer, *14 ft,* on the length axis.

Example: 10 gpm through 1¼″ rubber lined valve

Solution: From Table, $C_v = 24$

$P = (Q/C_v)^2 = (10/24)^2 = 0.174$ psi

0.174 psi × 2.31 = 0.4 ft H_2O head loss

Table 2 Pipe Specifications

Nominal Diameter	*Inside Diameter*	A (*in.*^{2})	*Volume (ft*^{3}*/ft) Area (ft*^{2}*)*
Schedule 80 Steel for Gas and Liquid Chlorine Supply Lines			
¾	0.742	0.4324	0.00300
1	0.957	0.7193	0.00499
1¼	1.278	1.2813	0.00891
1½	1.500	1.767	0.01225
2	1.939	2.953	0.02050
Schedule 80 PVC for Chlorine Solution Piping and Injector Vacuum Lines			
½	0.546	0.2341	0.00163
¾	0.742	0.4324	0.00300
1	0.957	0.7193	0.00499
1¼	1.278	1.2813	0.00891
1½	1.500	1.767	0.01225
2	1.939	2.953	0.02050
2½	2.323	4.298	0.02942
3	2.900	6.605	0.04587
4	3.826	11.50	0.07986
6	5.761	26.07	0.1810
8	7.625	45.66	0.3171

Table 3 Friction Loss in PVC Pipe Schedule 80 (for Chlorine Solution Lines)

Velocity measured in ft/sec, Loss in feet of water head per 100 ft of pipe

Gal.	½″		¾″		1″		1¼″		1½″		2″		2½″		3″		3½″		4″		*Gal.*
min.	*Vel.*	*Loss*	*Vel.*	*Loss*	*Vel.*	*Loss*	*Vel.*	*Loss*	*Vel.*	*Loss*	*Vel.*	*Loss*	*Vel.*	*Loss*	*Vel.*	*Loss*	*Vel.*	*Loss*	*Vel.*	*Loss*	*min.*
2	2.74	6.72	1.48	1.51																	2
4	5.48	24.2	2.97	5.45	1.79	1.54	1.00	.39	.73	.177											4
6	8.23	51.2	4.45	11.5	2.68	3.34	1.50	.82	1.09	.375	.65	.107									6
8	11.0	86.9	5.94	19.6	3.57	5.69	2.00	1.39	1.45	.64	.87	.183	.61	.077							8
10	13.7	132.0	7.42	29.6	4.46	8.60	2.50	2.10	1.82	.96	1.09	.276	.76	.115	.485	.039					10
12			8.91	41.5	5.36	12.0	3.00	2.94	2.18	1.35	1.30	.387	.91	.161	.572	.055					12
15			11.1	62.7	6.7	22.9	3.76	4.45	2.72	2.04	1.63	.585	1.14	.243	.727	.083	.54	.035			15
18			13.4	87.9	8.03	25.5	4.50	6.25	3.27	2.86	1.96	.818	1.36	.340	.873	.116	.65	.056			18
20	5″ PIPE		14.8	107	8.92	30.9	5.00	7.57	3.63	3.47	2.17	.996	1.51	.414	.97	.140	.72	.068	.56	.037	20
25					11.2	58.8	6.25	11.4	4.55	5.25	2.71	1.51	1.9	.625	1.21	.212	.90	.103	.695	.055	25
30	.53	.025			13.4	65.3	7.50	16.0	5.45	7.38	3.26	2.11	2.27	.874	1.44	.297	1.08	.145	.84	.077	30
35	.62	.034			15.6	86.9	8.75	21.3	6.38	9.78	3.80	2.81	2.65	1.16	1.70	.396	1.26	.192	.973	.103	35
40	.71	.043			17.9	111	10.0	27.3	7.26	12.5	4.35	3.59	3.03	1.49	1.94	.507	1.44	.246	1.12	.132	40
45	.795	.054	6″ PIPE				11.2	33.9	8.26	15.6	4.89	4.46	3.41	1.86	2.18	.629	1.63	.306	1.25	.164	45
50	.88	.065	.62	.027			12.5	41.3	9.08	18.9	5.43	5.41	3.79	2.25	2.42	.766	1.80	.372	1.40	.199	50
55	.973	.078	.676	0.32			13.7	49.2	10.00	32.0	5.98	6.44	4.16	2.68	2.67	.912	1.99	.443	1.53	.237	55
60	1.06	.091	.74	.039			15.0	57.8	10.9	26.5	6.52	7.61	4.54	3.16	2.92	1.07	2.17	.522	1.67	.279	60
65	1.15	.106	.80	.044			16.1	67.0	11.8	30.7	7.06	8.84	4.92	3.66	3.14	1.25	2.35	.604	1.81	.323	65
70	1.23	.121	.86	.051			17.5	77.1	12.7	35.3	7.61	10.1	5.30	4.20	3.39	1.43	2.53	.691	1.95	.371	70
75	1.33	.138	.923	.057			18.8	87.4	13.6	40.1	8.15	11.5	5.68	4.79	3.64	1.62	2.70	.787	2.08	.421	75
80	1.41	.155	.98	.065			20.0	98.2	14.5	45.2	8.69	12.9	6.05	5.36	3.88	1.83	2.89	.888	2.23	.475	80
85	1.50	.174	1.04	.072			21.2	110	15.4	50.3	9.03	14.5	6.43	6.02	4.10	2.04	3.05	.992	2.34	.531	85
90	1.59	.193	1.11	.080			22.5	122	16.3	55.9	9.78	16.1	6.81	6.53	4.33	2.27	3.25	1.10	2.51	.592	90
95	1.67	.213	1.20	.089					17.2	62.0	10.3	17.8	7.19	7.38	4.57	2.51	3.42	1.21	2.64	.652	95

100	1.76	.234	1.23	.098					18.2	68.2	10.9	19.6	7.57	8.13	4.85	2.76	3.67	1.34	2.79	.719	100
110	1.95	.279	1.36	.117					20.0	81.3	12.0	23.4	8.33	9.68	5.33	3.29	3.97	1.60	3.07	.855	110
120	2.11	.329	1.48	.137	8″ PIPE				21.8	95.4	13.0	27.4	9.08	11.4	5.80	3.87	4.33	1.88	3.35	1.00	120
130	2.3	.381	1.60	.159					23.6	111	14.1	31.8	9.84	13.2	6.30	4.48	4.69	2.18	3.63	1.16	130
140	2.47	.437	1.72	.182	.98	.047			25.4	127	15.2	36.5	10.6	15.1	6.80	5.12	5.05	2.50	3.91	1.33	140
150	2.65	.496	1.85	.207	1.05	.054					16.3	41.5	11.3	17.2	7.27	5.87	5.41	2.84	4.19	1.52	150
160	2.82	.559	1.97	.234	1.12	.059					17.4	46.7	12.1	19.4	7.75	6.58	5.78	3.20	4.47	1.71	160
170	3.0	.626	2.08	.261	1.19	.067					18.5	52.2	12.9	21.7	8.20	7.37	6.14	3.58	4.75	1.91	170
180	3.16	.696	2.22	.290	1.26	.074					19.6	58.3	13.6	24.1	8.60	8.18	6.50	3.97	5.02	2.12	180
190	3.36	.769	2.34	.321	1.33	.082					20.6	64.4	14.4	26.6	9.20	9.05	6.85	4.39	5.30	2.35	190
200	3.52	.846	2.46	.353	1.41	.090					21.7	70.5	15.1	29.3	9.70	9.96	7.22	4.84	5.58	2.58	200
220	3.88	1.01	2.71	.421	1.55	.108					23.9	84.1	16.7	34.9	10.6	11.9	7.94	5.78	6.14	3.08	220
2	2.74	6.72	1.48	1.51																	2
4	5.48	24.2	2.97	5.45	1.79	1.54	1.00	.39	.73	.177											4
6	8.23	51.2	4.45	11.5	2.68	3.34	1.50	.82	1.09	.375	.65	.107									6
8	11.0	86.9	5.94	19.6	3.57	5.69	2.00	1.39	1.45	.64	.87	.183	.61	.077							8
10	13.7	132.0	7.42	29.6	4.46	8 60	2.50	2.10	1.82	.96	1.09	.276	.76	.115	.485	.039					10
240	4.23	1.18	2.96	.484	1.69	.126					26.1	98.7	18.2	41.0	11.6	13.9	8.66	6.77	6.70	3.62	240
260	4.58	1.37	3.20	.573	1.83	.147					28.3	115	19.7	47.5	12.6	16.2	9.38	7.85	7.26	4.19	260
280	4.94	1.57	3.45	.658	1.97	.168							21.2	54.5	13.5	18.6	10.1	9.02	7.82	4.79	280
300	5.29	1.79	3.69	.747	2.11	.191							22.7	62.0	14.4	21.1	10.8	10.2	8.38	5.45	300
320	5.64	2.01	3.94	.841	2.24	.215							24.2	69.9	15.5	23.7	11.5	11.5	8.94	6.16	320
340	5.99	2.26	4.19	.940	2.39	.240							25.8	78.2	16.3	26.6	12.3	12.9	9.50	6.91	340
360	6.35	2.51	4.43	1.05	2.64	.261							27.2	86.9	17.4	29.5	13.0	14.3	10.0	7.66	360
380	6.70	2.77	4.68	1.16	2.68	.295							28.8	96.1	18.6	32.6	13.7	15.8	10.6	8.46	380
400	7.05	3.05	4.93	1.27	2.81	.325							30.3	106	19.4	35.9	14.4	17.4	11.2	9.31	400
450	7.95	3.79	5.54	1.53	3.16	.404									21.8	44.6	16.2	21.6	12.5	11.6	450
500	8.82	4.61	6.16	1.92	3.51	.493									23.2	54.1	18.1	26.3	14.0	14.1	500
550	9.70	5.50	6.77	2.29	3.86	587									26.5	64.9	19.9	31.4	15.3	16.8	550

Table 4 Head-Loss Calculations for Saunders-Type Diaphragm Valves

Liquid flow formula:

$$C_v = Q\sqrt{\frac{G}{\Delta P}} \qquad (1)$$

Sizing formula from Fluid Controls Institute, Standard FCI 62-1:

$$Q = C_v\sqrt{\frac{\Delta P}{G}} \qquad (2)$$

$$\Delta P = G\left(\frac{Q}{C_v}\right)^2 \qquad (3)$$

Where: C_v = valve sizing constant
Q = flow in gallons per minute (U.S. gpm)
G = specific gravity of liquid (water = 1.0)
P = pressure drop ($P_1 - P_2$) in psi

Valve Sizing Constant C_v

Valve Size	*Rubber Lined*	*Plastic Lined*	*Solid Plastic*
¾″	9.0	7.0	10.0
1″	14	19.5	16.5
1¼″	24	19.5	19.2
1½″	39	40	30
2″	57	60	54
2½″	101	84	—
3″	180	140	110
4″	350	310	—
6″	890	580	—
8″	1670	1250	—

CONTACT CHAMBER EVALUATION METHODS

Dispersion Index "d." An example of the computation of the Chemical Engineering Dispersion Index Number d where C = fluorometer units (dye concentration), and t = residence time (min.).*

*Courtesy Endel Sepp California State Dept. of Health.

t	C	$t \times C$	$t^2 \times C$
40	0	—	—
45	10	450	20,250
50	49	2450	122,500
55	78	4290	235,950
60	72	4320	259,200
65	61	3965	257,725
70	50	3500	245,000
75	37	2775	208,125
80	21	1680	134,400
85	15	1275	108,375
90	9	810	72,900
95	5	475	45,125
100	4	400	40,000
105	2	210	22,050
Sums	413	26,600	1,771,600

$$T = V/Q = 72.3 \text{ min.}$$

$$t_g = \frac{\Sigma tC}{\Sigma C} = \frac{26,600}{413} = 64.4 \text{ min.}$$

$$t_g^2 = 4148.2$$

$$\sigma_t^2 = \frac{\Sigma t^2 C}{\Sigma C} - t_g^2 = \frac{1,771,600}{413} - 4148.2$$

$$= 4289.6 - 4148.2 = 141.4$$

$$\sigma^2 = \sigma_t^2 / t_g^2 = \frac{141.4}{4148.2} = 0.034$$

$$d = \sigma^2/2 = \frac{0.034}{2} = 0.017$$

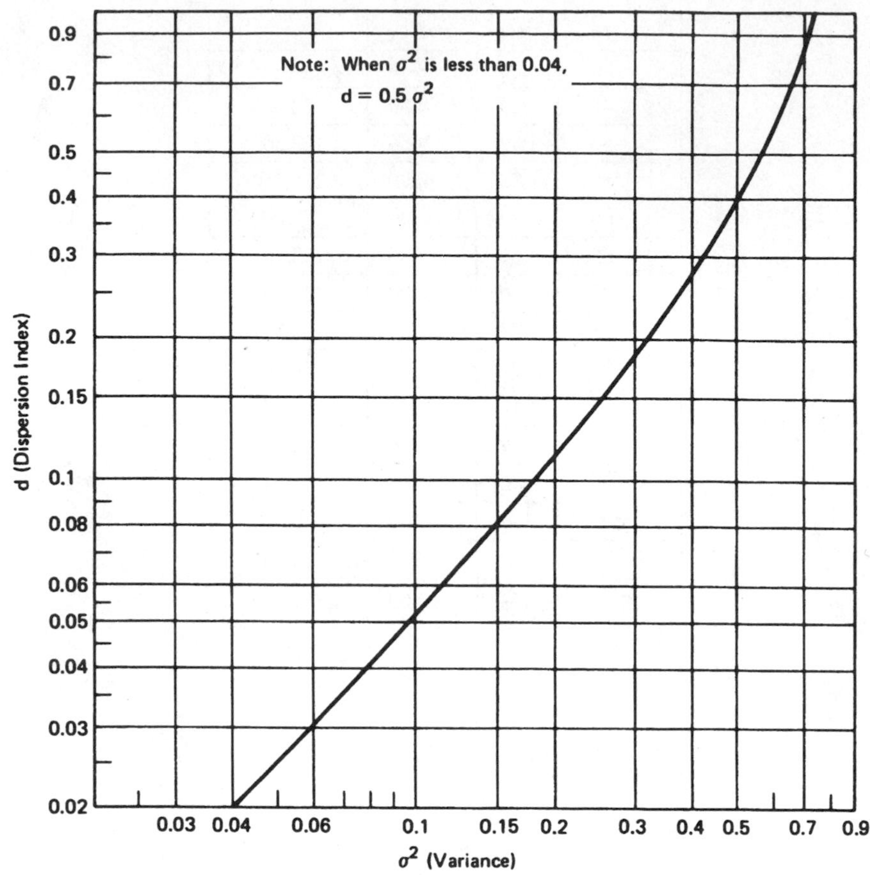

σ^2 (variance) versus "d" the chemical engineering index

t_g = ave. contact time = time to centroid of curve

Calculator Solution. The following is the keystroke sequence for an HP 21 calculator to determine the value of d for any given value of σ^2 based on the formula:

$$\sigma^2 = 2d - 2d^2(1 - e^{-1/d})$$

First assume $d = \frac{1}{2}\sigma^2$, then by trial and error find d by selecting values of d on either side of $\frac{1}{2}\sigma^2$, as follows:

DSP, 3; d = approximately $0.5\sigma^2$

ENTER	ENTER	ENTER
$1/x$	CHS	e^x
1	$x \leftrightarrows y$	−
$x \rightleftarrows y$	ENTER	2
[hatched key]	y^x	2
×	×	$x \rightleftarrows y$
2	×	$x \rightleftarrows y$
−	Read σ^2 value on the register	

Repeat this process until the value of d is within 0.005 for a given value of σ^2.

Morrill Index. An example of the Morrill Index computation, where

t = residence time (min.)
C = fluorometer units (dye concentration)

t	C	Cumulative C	$\frac{C}{\Sigma C}$ (percent)
0	0	—	—
20	10	10	6
25	15	25	14
30	25	50	28
35	30	80	44
40	60	140	78
45	30	170	94
50	10	180[a]	100

[a] $\Sigma C = 180$

The values of $C/\Sigma C$ are plotted against time on the probability scale.

The *Morrill Index* is the ratio of the time required for the passage of 90 percent of the dye to the time of passage of 10 percent of the dye.

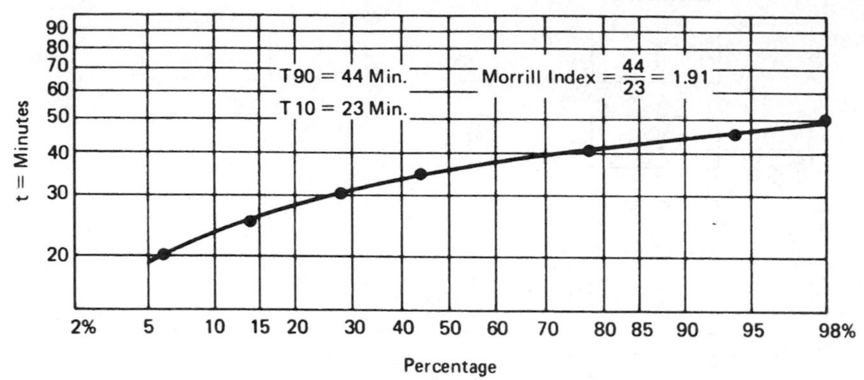

Morrill index.

CONVERSION FACTORS

U.S. (English System)

1 ppm	= 8.34 lbs/mg
1 gpm	= 1440 GPD
1 gpm	= 0.0022 cfs
1 gal	= 0.832 Imperial gal
1 gal	= 0.1337 ft^3
1 gal	= 8.34 lbs
1 ft^3	= 7.48 gal
1 mgd	=1.547 cfs
1 mgd	= 694.44 gpm
1 cfs	= 0.646 mgd
1 cfs	= 448.8 gpm
1 day	= 1440 min
1 day	= 86,400 sec
1 hp	= 550 ft-lb per sec
1 psi	= 2.04 in. Hg at 60°F
1 psi	= 2.31 ft H_2O
1 acre	= 43,560 ft^2 or SQFT

U.S. to Metric

1 in.	= 2.54 cm
1 ft	= 0.3048 meter
1 in.^{3}	= 16.387 cm^3
1 ft^3	= 0.0283 cu meters
1 oz	= 0.0295 liters
1 gal	= 3.7854 liters
1 gal	= 0.003785 cu meters
1 lb (mass)	= 0.45359 kilograms
1 oz	= 28.349 grams
1 watt hour	= 3600 joules
1 hp	= 746 W
1 lb (force)	= 4.448 Newtons
1 gpm	= 0.063 liter per sec
1 cfs	= 2446.6 cu meters per day
1 cfs	= 0.02832 cu meters per sec
1 mgd	= 3785.4 cu meters per day
1 psi	= 6.893 Kilo-Pascals
1 ppm (by wt.) × S.G.	= milligrams per liter

METRIC TO U.S. (ENGLISH SYSTEM)

To convert from metric to U.S. take the reciprocal of the metric values. For example:

To convert from cubic meters to gallons multiply cubic meters by 1/0.003785.

$$\therefore 1 \text{ cu meter} \times \frac{1}{0.003785} = 264.2 \text{ gal}$$

Temperature

Fahrenheit = 1.8°C + 32

Centigrade = 5⁄9(°F − 32)

Kelvin = °C + 273.15

Kelvin = 5⁄9(°F + 459.67)

Other Calculator Solutions. The following are the keystroke sequences for an HP 21 calculator for two often used equations in this text.

$$\Delta P \text{ (in. Hg)} = \frac{L \times 5.83 \times f \times W^2 \times 2.04}{10^9 \times \rho \times d^5} \qquad *$$

*Equation (3-1) is shown in this text.

L [Enter ↑] 5.83 [×] f [×] W [Enter ↑] [×] [×] 2.04 [×]

[EEX] 9 [÷] ρ [÷] d [Enter ↑] 5 [▨] [y^x] [÷] ----→ ΔP

and

$$y/y_0 = [1 + 0.23\ ct]^{-23} \quad *$$

Solving for ct:

y_0 [Enter ↑] y [÷] 3 [$1/x$] [▨] [y^x] 1 [−] 0.23 [÷] ----→ ct

CONTROL SYSTEMS DEFINITIONS

Common terminology used in describing automatic control logic and functions is as follows:

Algorithm. This is a sequence of calculations performed to obtain a given result.

Example. A proportional controller uses an algorithm. The controller output is $m = K_cE$, where K_c = proportional gain and E = error.

Analog. This is a representation of numerical values by means of physical variables such as voltage, current, resistance, rotation, etc.

Example. The chlorine residual span of a continuous analyzer is 0–5 mg/l. This is represented as a 4–20 mA transmitter output signal where 4 mA = 0 mg/l and 20 mA = 5 mg/l.

Proportional Band. This describes the type of control action caused by a change in the process variable. The ability of a controller to produce a small or large output change (i.e., valve movement) for a small process variable change is described as a narrow or a wide proportional band adjustment. A wide band results from low gain and a narrow band from high gain.†

The gain, or proportional band, is adjustable at the controller, but must be such that there will always be some finite error to produce a correction.

Reset Action (Integral). Proportional controllers used in chlorine residual control loops should be provided with reset action. Reset action combined with proportional band is known as PI—for proportional plus integral—

*Equation (2-11) is in this text.
†Gain is the numerical reciprocal of proportional band.

because the reset action corresponds to a time integration or summation of the error.

Reset action in a controller responds to both the duration and the magnitude of the error signal. Its output feeds into the proportional unit. This output increases in proportion to the error signal and the length of time the error signal exists. Whether the error signal is positive or negative, reset action will act upon the proportional controller until the error signal is zero. It supplements proportional action so as to wipe out the error signal. The error signal cannot be eliminated by proportional band alone, as described above.

Derivative Action. This mode of control action is often added to PI controllers. When this is done, it becomes a PID controller. This mode of control has been called "rate action." It corresponds to the derivative of the error signal. Mathematician Klaf defines a derivative as: "the instantaneous rate of change of a function."

Derivative action supplements the corrective action of proportional control by augmenting the signal by an amount proportional to the rate of change of the error signal. This is why it is sometimes referred to as "rate action."

Summary. A PID control loop contains the following three important modes of control action. Each action can be adjusted to site-specific conditions or characteristics.

1. Proportional action with adjustable gain to obtain stability.
2. Reset action to compensate for load changes which wipes out the resulting error signal.
3. Derivative action (rate control) which speeds up control action when rapid load changes occur.

WATER VISCOSITY

Water viscosity is expressed in English units as absolute viscosity = lb-sec/ft^2 as follows:

$$2.735 \times 10^{-5} \text{ at } 50°\text{F}$$
$$2.05 \times 10^{-5} \text{ at } 70°\text{F}$$

The symbol for absolute viscosity is μ.

Water viscosity is also expressed as kinematic viscosity in English units as ft^2/sec as follows:

$$1.41 \times 10^{-5} \text{ at } 50°\text{F}$$
$$1.06 \times 10^{-5} \text{ at } 70°\text{F}$$

The symbol for absolute viscosity is ν, where

$$\nu = \frac{\mu}{62.4}$$

Calculator (HP25) solution for velocity gradient *G*, where:

$$G = \sqrt{\frac{3 \times 550}{2.735 \times 10^{-5} \times 77.37}}$$

$$\mu = 2.735 \times 10^{-5} \text{ lb-sec/ft}^2$$

$$\text{hp} = 3$$

$$V = 77.37 \text{ ft}^3$$

Solution:

3 [ENTER] 550 [X] 2.735 [ENTER] 5 (CHS) $g \cdot 10^x$
[×] [÷] 77.37 [÷] $f\sqrt{x} = 883 \text{ sec}^{-1}$

THE EGAN JET DIFFUSER CONCEPT

The design of this type of jet mixing is based upon the notion that sufficient mixing of chlorine solution occurs when the velocity gradient G approximates 500.

The following are the calculations for the Egan diffuser retrofit described in Chapter 8 (see Figs. 8-16 and 8-17).

Velocity Gradient *G*:

$$G = \left(\frac{550 \times P(\text{hp})}{\mu(\text{lb} - \text{sec/ft}^2) \times V(\text{ft}^3)}\right)^{1/2} \tag{8.11}$$

where

P = the power being dissipated in the mixing zone

$$= \frac{Qph}{60 \times 550 \times \text{eff}}$$

with:

Q = total injector discharge flow, gal/min
p = water density, lb/gal
h = head loss through diffuser, ft
60 = sec/min
550 = ft-lb/sec/hp
eff = system efficiency

Continuing with Eq. (8-11):

u = absolute fluid viscosity, lb-sec/ft^2
V = process water volume in mixing zone, ft^3. (In this instance it is the distance between baffles B and C = 3.34 ft.)

Velocity Gradient of Jet System. Assume 3-¾″ holes, where Q = 67 and flow per hole is approx. 22 gpm. This creates a diffuser head loss of approx. 4.03 ft. Therefore:

$$P = \frac{67 \times 4.03}{3960 \times 1.0} = 0.068 \text{ hp}$$

$$G = \left(\frac{550 \times 0.068}{2.5 \times 10^{-5} \times 41.8}\right)^{1/2} = 189 \text{ sec}^{-1}$$

Velocity Gradient of Baffles. Assume baffle A provides desired water surface level only—does not provide any mixing energy.

The mixing zone in this particular case is the volume contained between baffles B and C. Effluent flow is 10 mgd = 6944 gpm. This calculates to a velocity = 1.23 ft/sec, so $V^2/2\,gh$ = 0.023 ft.

Assume efficiency of baffle energy generation to be 50 percent because the jet stream is centered in the upper half of the 4-ft RCP. Said another way, the cone of influence of baffle mixing extends to about half of the volume between baffles B and C. So

$$P = \frac{6944 \text{ gpm} \times 8.34 \text{ lb/ft}^3 \times 0.023 \text{ ft}}{60 \text{ sec} \times 550 \text{ ft-lb/min} \times 0.5}$$

$$= 0.08 \text{ hp}$$

and therefore

$$G = \left(\frac{550 \times 0.08}{2.5 \times 10^{-5} \times 41.8}\right)^{1/2}$$

$$= 205 \text{ sec}^{-1}$$

Velocity Gradient of Contraflow Effect. In order to perceive the mixing effect of contraflow, i.e., the diffuse jet opposing the process water flow, we must bear in mind that mixing results from turbulence, and turbulence is the manifestation of hydraulic energy loss. There are no proven mathematical relationships available to quantify this effect. However, Egan has offered the following approach by using what he calls "free-wheeling" assumptions.*

Referring to Fig. 8-17, the premise is that the diffuser jet stream and the opening created by baffle *B* act as orifices imparting jet stream energy in opposite directions. Some of the energy created will be cancelled out due to dissipation by the opposing velocity head energy. The remaining balance will be available for mixing energy. If the diffuser jet stream were directed downstream instead of upstream, this differential energy would not be generated for mixing.

Egan defines these opposing energies as follows (see Fig. 17): assume the effect of the vertical component (V_B) is negligible. At baffle C, V_B = 2.45 ft/sec and V = 16.1 ft/sec.

Converting these values to velocity heads:

$$h_B = 0.068 \text{ ft}$$
$$h_J = 4.03$$

Convert velocity heads to energy assuming time t = one second for flash mixing duration.

Pounds of water imparted by diffusion jet:

$$67 \text{ gpm} \times \frac{1 \text{ min}}{60 \text{ sec}} \times 8.34 \text{ lb/gal} = 9.3 \text{ lb/sec}$$

Pounds of water imparted by baffle:

$$6944 \text{ gpm} \times \frac{1 \text{ min}}{60 \text{ sec}} \times 8.34 \text{ lb/gal} = 965.2 \text{ lb/sec}$$

Determine energy of both streams:

$$\text{energy} = \text{velocity head} \times \text{lb water/sec.}$$

Diffuser jet energy:

$$E_J = 4.03 \text{ ft} \times 9.31 \text{ lb/sec} = 37.52 \text{ ft-lb/sec}$$

Baffle energy:

$$E_B = 0.068 \text{ ft} \times 965.2 \text{ lb/sec} = 65.63 \text{ ft-lb/sec}$$

*See Ref. 96, Chapter 8.

Net energy available for mixing:

$$E_M = E_B + E_J$$
$$= 65.63 \text{ ft-lb/sec} - 37.52 \text{ ft-lb/sec}$$
$$= 28.11 \text{ ft-lb/sec}$$

Convert to hp:

$$\frac{28.11}{550} = 0.051 \text{ hp}$$

$$G_C = \sqrt{\frac{550 \times 0.051}{2/5 \times 10^{-5} \times 20.9^*}} = 232 \text{ sec}^{-1}$$

In this case of any diffuser acting as an obstruction to flow in a pipe or channel, will dissipate energy. This energy loss is used for mixing and can be calculated as additional velocity gradient.

DEBYE—HUCKEL EQUATION CONSTANTS†

$$f_x = \text{antilog} \frac{AZ_x^2(I)^{1/2}}{1 + Ba_x^0(I)^{1/2}}$$

Temp. °*C*	*Unit Volume of Solvent*	
	A	*B*
0	0.4918	0.3248
5	0.4952	0.3256
10	0.4989	0.3264
15	0.5028	0.3273
20	0.5070	0.3282
25	0.5115	0.3291
30	0.5161	0.3301
35	0.5211	0.3312
40	0.5262	0.3323
45	0.5317	0.3334

In Chapter 4, Eq. (4-9a), the ionic radii for H^+ = 9.0 and for OCl^- – 4.0.

*Use restricted flash mixing zone as volume.

†*Source:* Electric Power Research Inst., *Dechlorination Technology Manual,* EPRI, CS-3748, Palo Alto, CA, Nov. 1984.

f_x = activity coefficient of the ionic species x
Z_x = ionic charge of the ionic species x
A = a constant which depends upon absolute temperature and dielectric constant of solvent
B = a constant which depends upon the absolute temperature and dielectric constant of the solvent
a_x^0 = approximate effective radius of ionic species x in aqueous solution (Å)
I = ionic strength of the solution (mol/I)
Å = angstroms

The values for unit weight of solvent (molality scale) can be obtained by multiplying the corresponding value for unit volume by the square root of the density of water at the approximate temperature.

GLOSSARY

A	angstrom = 1×10^{-10} meters = 1×10^{-7} millimeter
ACC	automatic control center
AOC	assimable organic carbon
AO_xP	advanced oxidation processes
AS	analog subsystem
AWT	advanced wastewater treatment
AWWA	American Water Works Association
BAC	biological activated carbon
BOD	biological oxygen demand
B-P	breakpoint
Btu	British thermal unit
CA	cellulose acetate
CAA	chloroacetic acid
CAC	combined available chlorine
CCCS	central computer control system
COD	chemical oxygen demand
CPRV	chlorine pressure-reducing valve
CRC	combined residual chlorine
CRW	Colorado River water
CS	control system
Ct	chlorine residual × contact time
CTC	central terminal control
DBP	disinfection byproducts
DE	diatomaceous earth
DIW	deionized water
DNA	deoxyribonucleic acid
DO	dissolved oxygen
DOC	dissolved organic carbon

DPD	diethyl-*p*-phenylenediamine
DPP	differential pulse polarography
DTs	delirium tremens
EBCT	empty bed contact time
ED	electrodialysis
EDP	ethyl dibromide
EPA	environmental protection agency (also USEPA)
EPDM	ethylene propylene diene monomer
FAC	free available chlorine
FACTS	free available chlorine test—syringaldazine
FAS	ferrous ammonium sulfate
FIA	flow injection analysis
FRC	free residual chlorine
FRP	reinforced fiberglass
GAC	granular activated carbon
HAA	haloacetic acid
HFF	hollow-fine fiber
HPC	heterotrophic plate count
IDLH	immediate danger to life and health
IX	anion exchange
JTU	Jackson turbidity units
LSL	lower sensitivity limits
μA	microamps
mA	milliamps
micron	0.001 millimeter
μg	microgram = 0.001 milligram
mg	milligram = 0.001 gram
MCL	maximum contaminant level
MF	microfiltration
MFI	modified fouling index
MGD	million gallons per day
MPN	most probable number
MWC	minimum water column
MWD	Metropolitan Water District
MWD	maximum water depth
MWD	minimum water depth
MLVSS	mixed liquor volatile suspended solids
mV	millivolts
NTU	nephelometric turbidity units
nm	nanometer = 1×10^{-9} meter
nano-	one billionth part of
NF	nanofiltration
NMW	nominal molecular weight
NPDES	national pollutant discharge elimination system
ORP	oxidation reduction potential

OT	orthotolidine
OTA	orthotolidine arsenite
PAC	powdered activated carbon
PAO	phenylarsine oxide
PC	programmable controller
PCB	polychlorinated biphenyl
PCS	process control software
PDWF	peak dry weather flow
PFU	plaque-forming units
PICS	process instrument control system
PSS	point summation sources
PVC	polyvinyl chloride
RAS	return activated sludge
RCP	reinforced concrete pipe
RMPP	risk management protection plan
RO	reverse osmosis
RTU	remote terminal unit
SDH	succinic dehydrogenase
SDI	silt density index
SHMP	sodium hexametaphosphate
SLS	sodium laurel sulfate
SOC	synthetic organic compound
SPC	standard plate count
SPW	state water project (California)
SS	suspended solids
SW	spiral wound
SWDA	safe water drinking act
SWTR	safe water treatment rule
TCA	trichloroacetic acid
TCE	trichlorethylene
TDD	triple distilled deionized
TDH	total dehydrogenase
TDH	total dynamic head
TDS	total dissolved solids
TFC	thin-film composites
THM	trihalomethane
THMFP	trihalomethane formation potential
TOC	total organic carbon
TRC	total residual chlorine
TTCE	tetrachloroethylene
TTHM	total trihalomethane
UF	ultrafiltration
VOC	volatile organic carbon
WPCF	water pollution control federation
WPCP	water pollution control plant

WTP	water treatment plant
WWTP	wastewater treatment plant

Cubic Equation. The following cubic equation was used to generate the data shown in Tables 4-1 and 4-4 (Chapter 4). These data illustrate the molecular chlorine, hypochlorous acid, and hypochlorite ion equilibria under various conditions of concentration, temperature, and pH.

$$\left\{-\left(\frac{1}{K_D}\right)^2 (H^+)^3 - K_A \left(\frac{1}{K_D}\right)^2 (H^+)^2 + 4\left(\frac{1}{K_D}\right) K_3([\mathrm{H}^+] + K_A)^2\right\} [\mathrm{HOCL}]^3$$

$$\left\{-4\left(\frac{1}{K_D}\right) K_3 T[\mathrm{H}^+]([\mathrm{H}^+] + K_A)\right\} [\mathrm{HOCL}]^2$$

$$\left\{+\left(\frac{1}{K_D}\right) T[\mathrm{H}^+]^2 + K_A + [\mathrm{H}^+] + \left(\frac{1}{K_D}\right) K_3 T^2[\mathrm{H}^+]^2\right\} [\mathrm{HOCL}] - T[\mathrm{H}^+] = 0$$

where:

T = total halogen added

K_D = hydrolysis of Cl_2 at 25°C = 3.944×10^{-4}

K_A = acid dissociation of HOCL = 2.904×10^{-8}

K_3 = formation of trichloride ion = 0.191

Table 5 Oxidation-Reduction Potentials (E_0) at 25°C for Titrable Species of Halogens and Ozone

Reaction	*Potential in Volts* (E_0)
$O_3 + 2H^+ + 2e = O_2 + H_2O$	2.07
$HOCl + H^+ + 2e = Cl^- + H_2O$	1.49
$Cl_2 + 2e = 2Cl^-$	1.36
$HOBr + H^+ = 2e = Br^- + H_2O$	1.33
$O_3 + H_2O + 2e = O_2 + 2OH^-$	1.24
$ClO_2 + e = ClO_2^-$	1.15
$Br_2 + 2e = 2Br$	1.07
$HOI + H_+ + 2e = I^- + H_2O$	0.99
$ClO_2\ (aq) + e + ClO_2^-$	0.95
$OCl^- + H_2O + 2e = Cl^- + 2\ OH^-$	0.90
$OBr^- + H_2O + 2e = Br^- + OH^-$	0.70
$I_2 + 2e = 2I^-$	0.54
$I_3 + 2e = 3I^-$	0.53
$OI^- + H_2O + 2e = I^- + 2\ OH^-$	0.49

Table 5a Comparative Oxidation Potentials—mVs

1. Fluorine	3.0
2. Hydroxyl radical	2.8
3. Ozone	2.1
4. Hydrogen peroxide	1.8
5. Potassium permanganate	1.7
6. Hypochlorous acid	1.5
7. Chlorine dioxide	1.5
8. Chlorine	1.4
9. Oxygen	1.2
10. Hydroxyl ion	0.4

**Chemical Engineering,* September 1994.

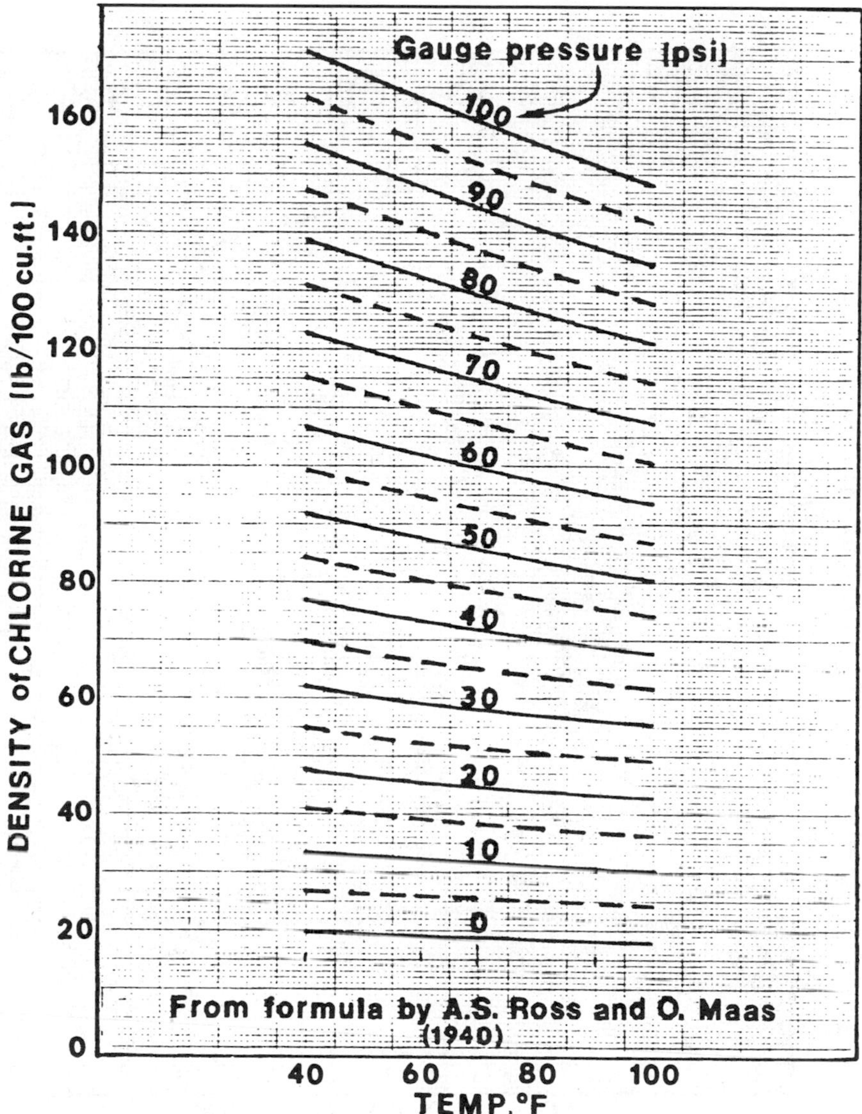

Fig. 1. Density of chlorine gas under pressure. Calculated from formula by A. S. Ross and O. Mass, "The Density of Gaseous Chlorine," *Can. J. Res.* **18B,** 55–65 (1940).

Chlorine Gas Density

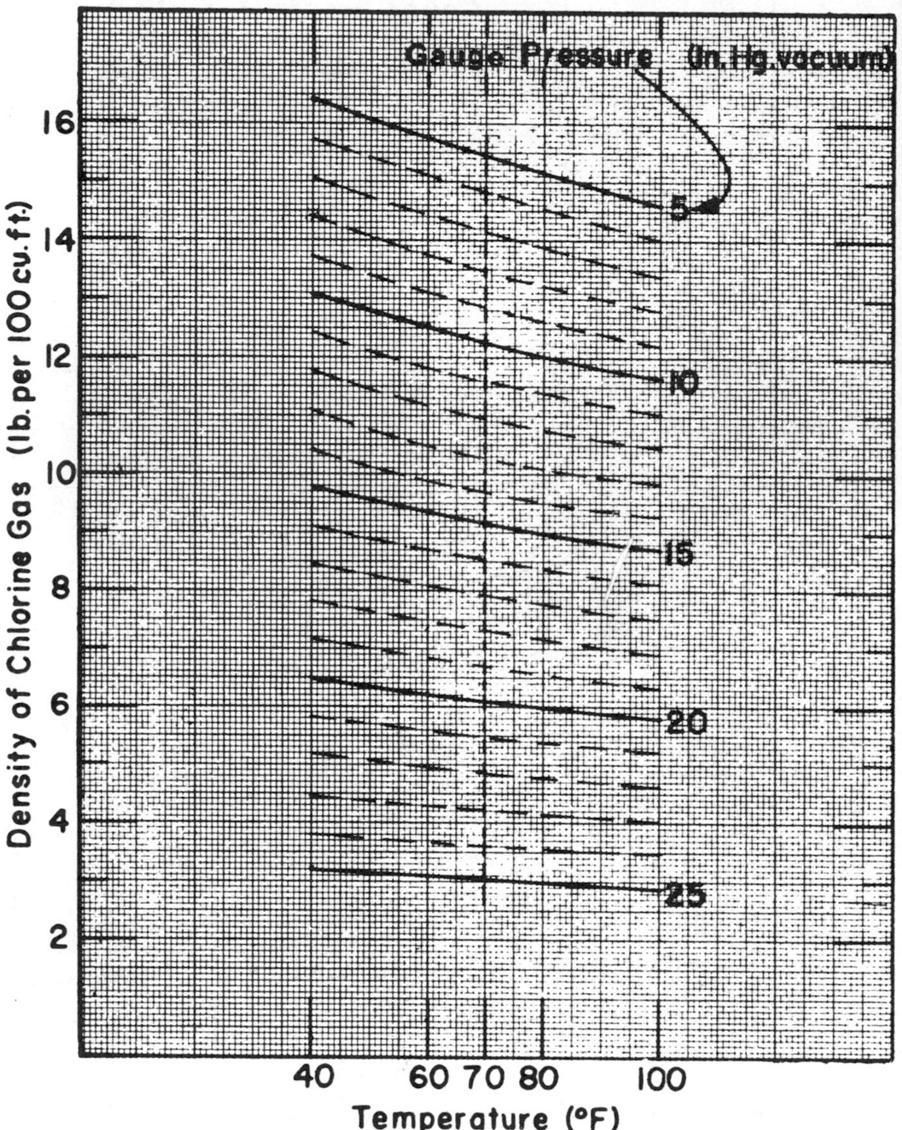

Fig. 2. Density of chlorine gas under vacuum (from Ross and Mass).

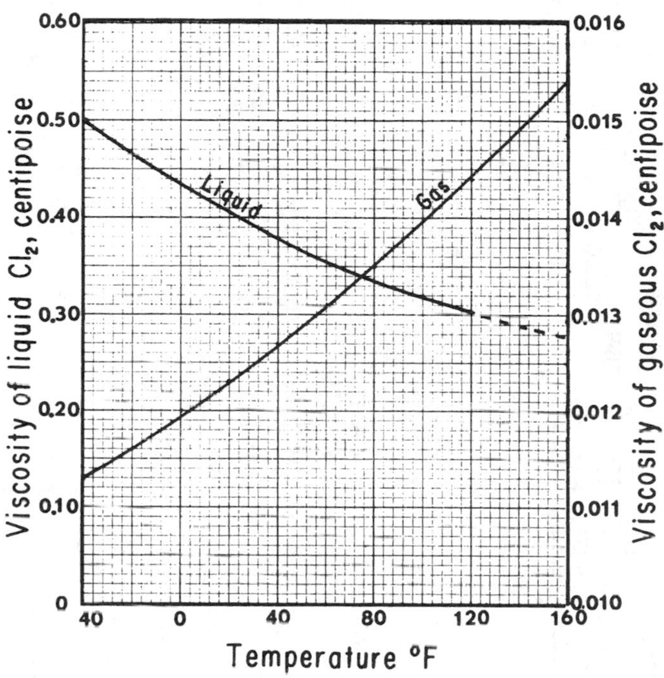

Fig. 3. Viscosity of chlorine liquid and gas (from Steacie and Johnson).

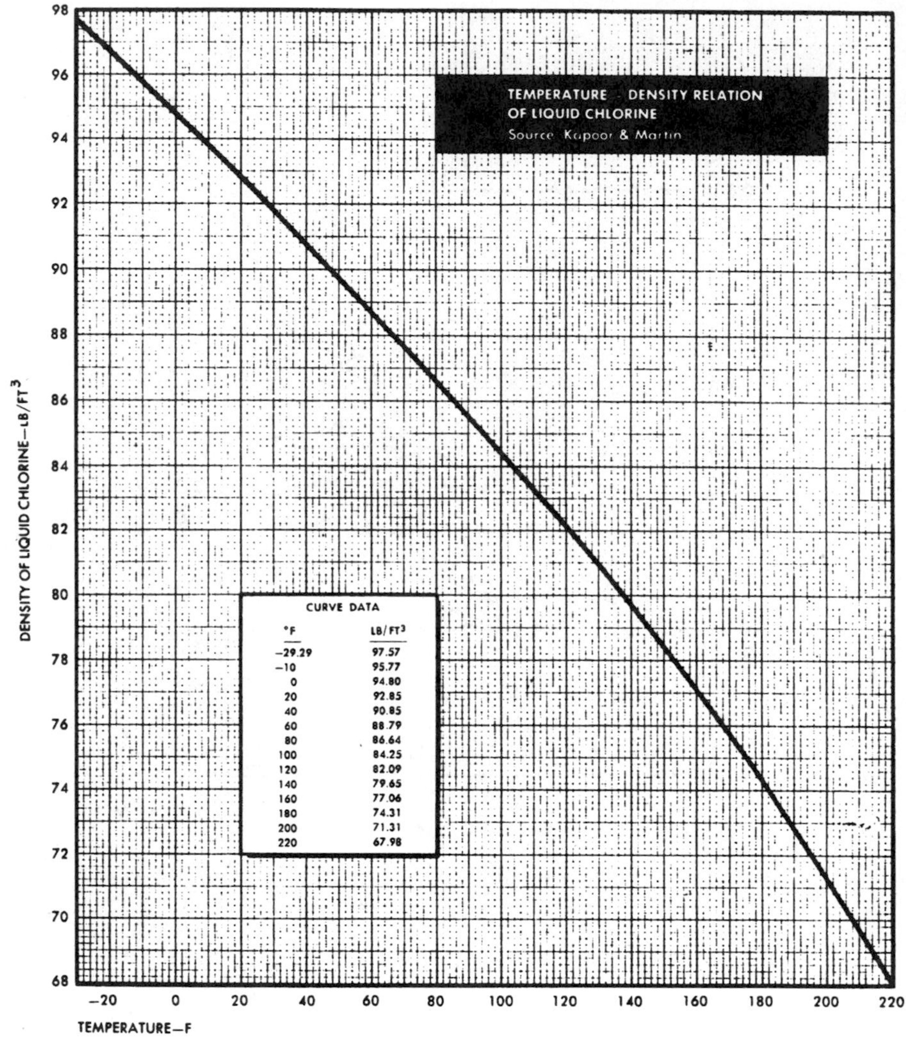

Fig. 4. Temperature–density relation of liquid chlorine (from Kapoor and Martin).

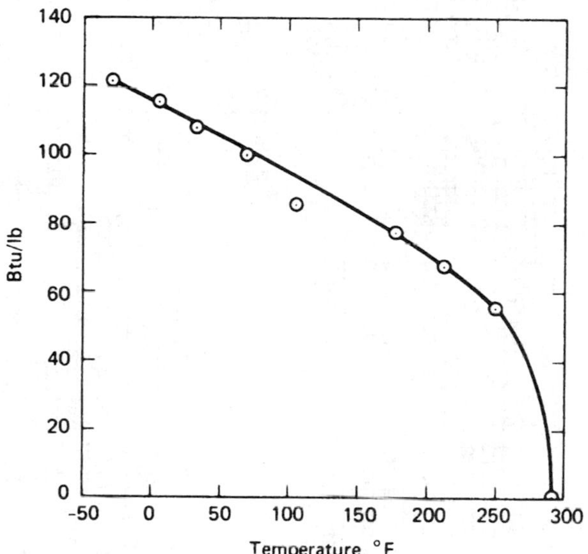

Fig. 5. Latent heat of vaporization of liquid chlorine (from Pellaton).

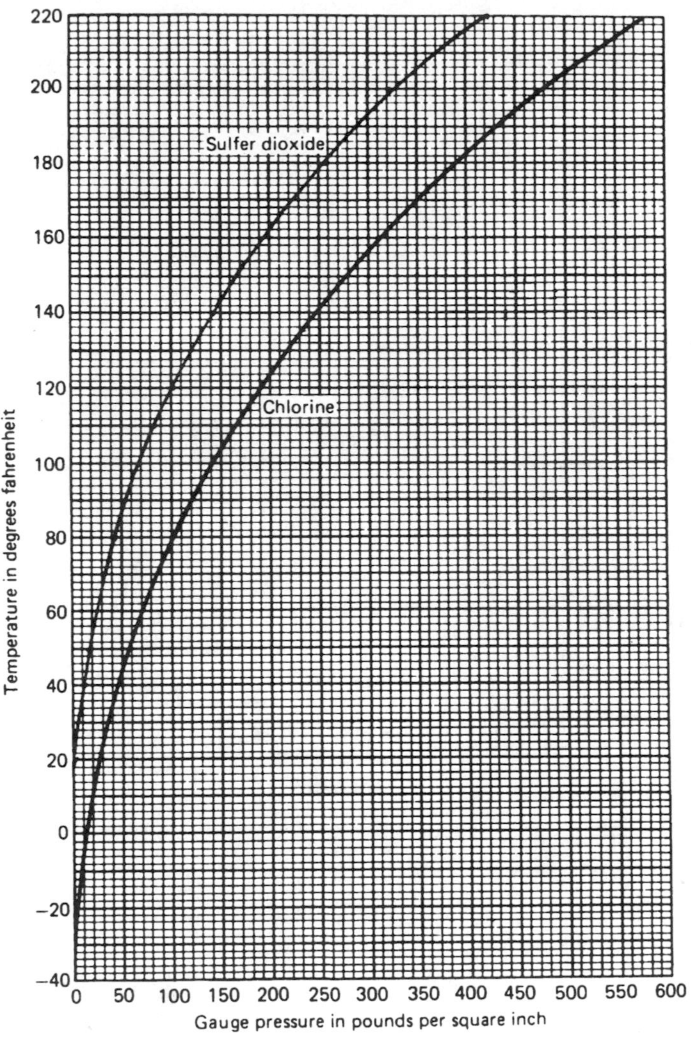

Fig. 6. Vapor pressure versus temperature of liquid chlorine and sulfur dioxide.

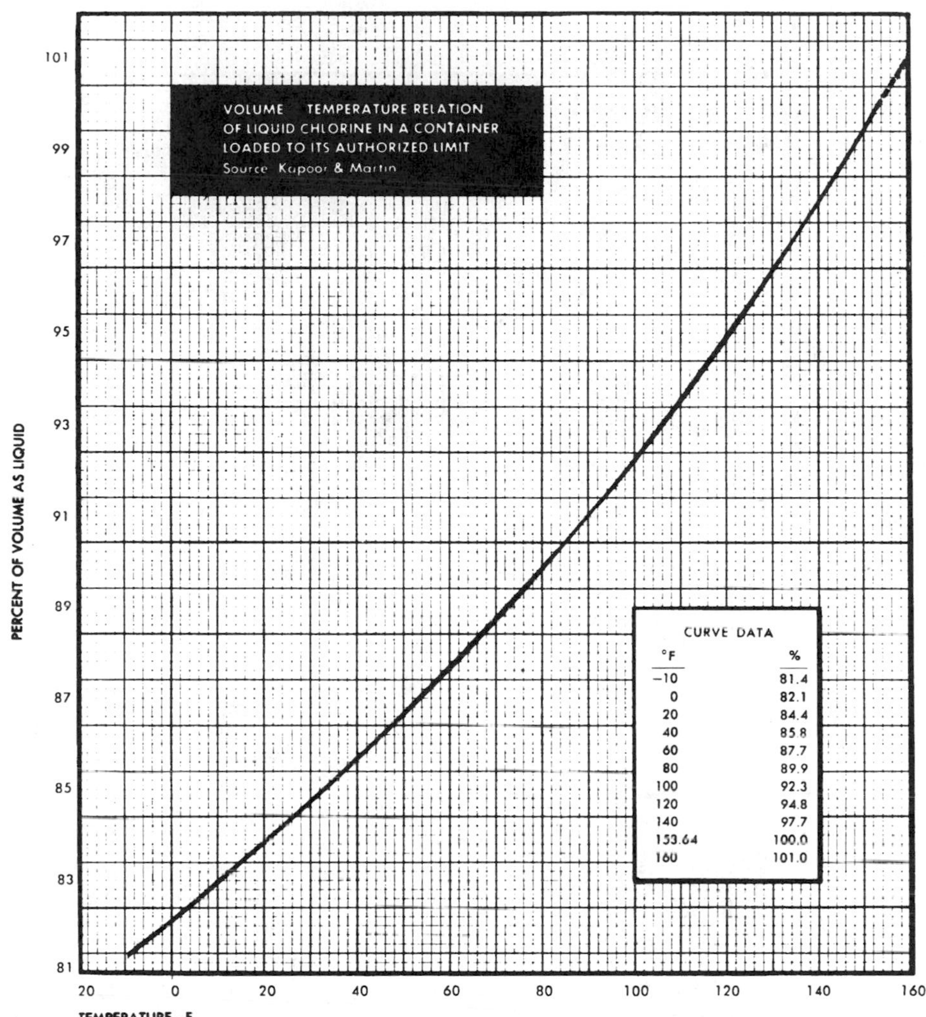

Fig. 7. Volume–temperature relation of liquid chlorine in container loaded to its authorized limit (from Kapoor and Martin).

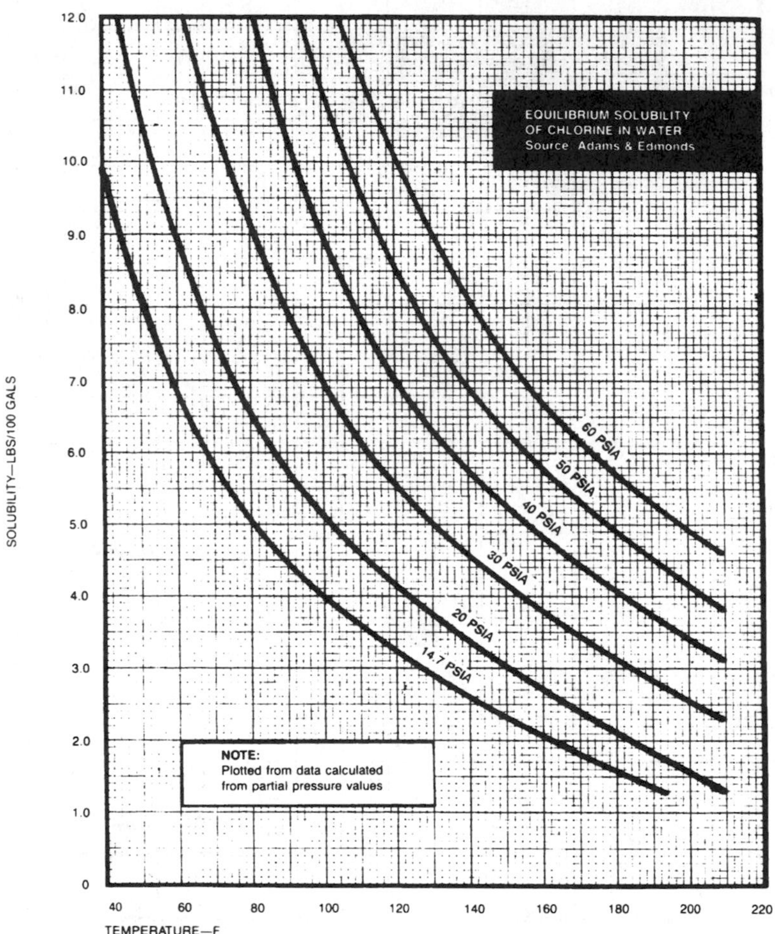

Fig. 8. Solubility of chlorine in water (Adams and Edmonds).

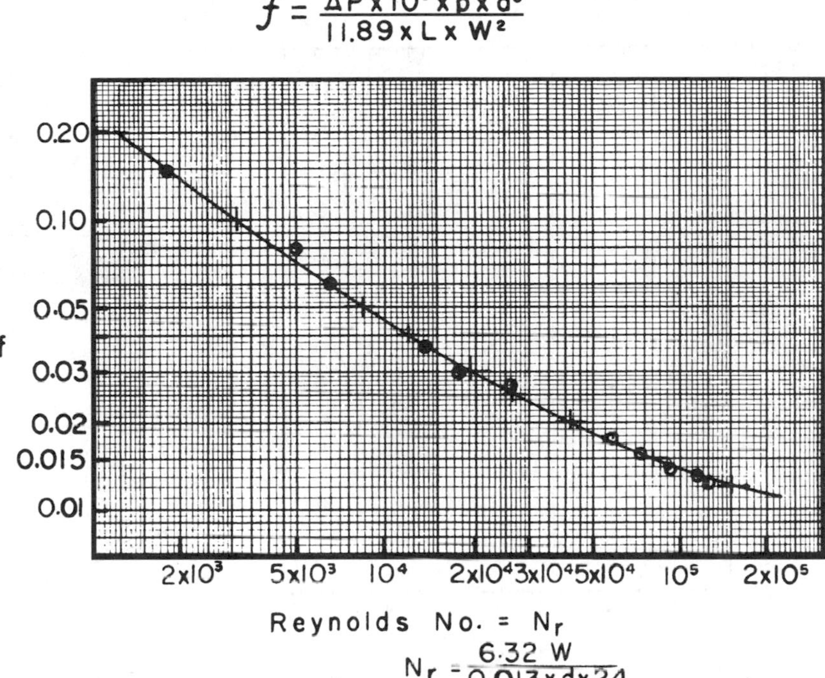

Fig. 9. Friction factor f as a function of Reynolds number, N, for PVC injector vacuum lines—for either chlorine or sulfur dioxide gas flow.

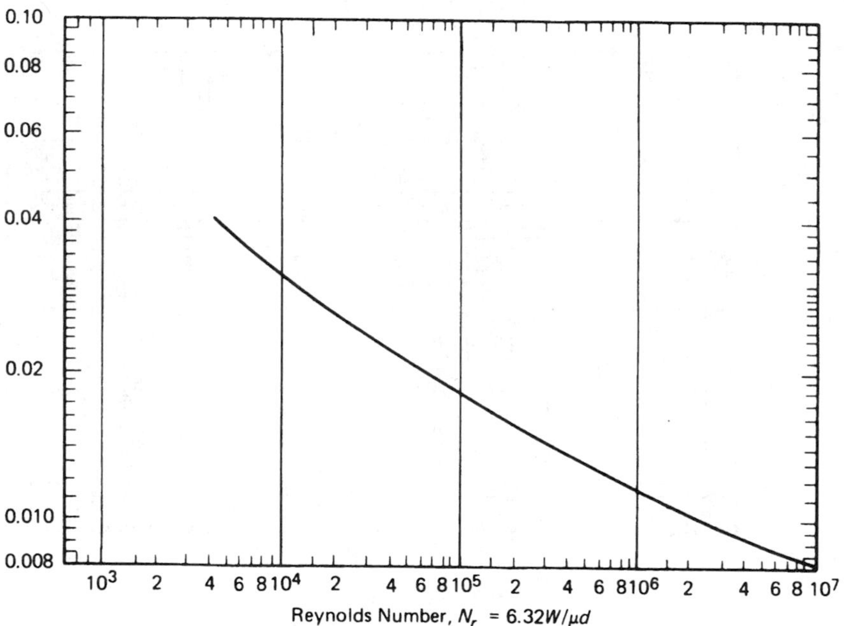

Fig. 10. Darcy–Weisbach friction factor *f* for fluid flow in PVC pipe (courtesy Alden Hydraulic Laboratory, Worcester Polytechnic Institute).

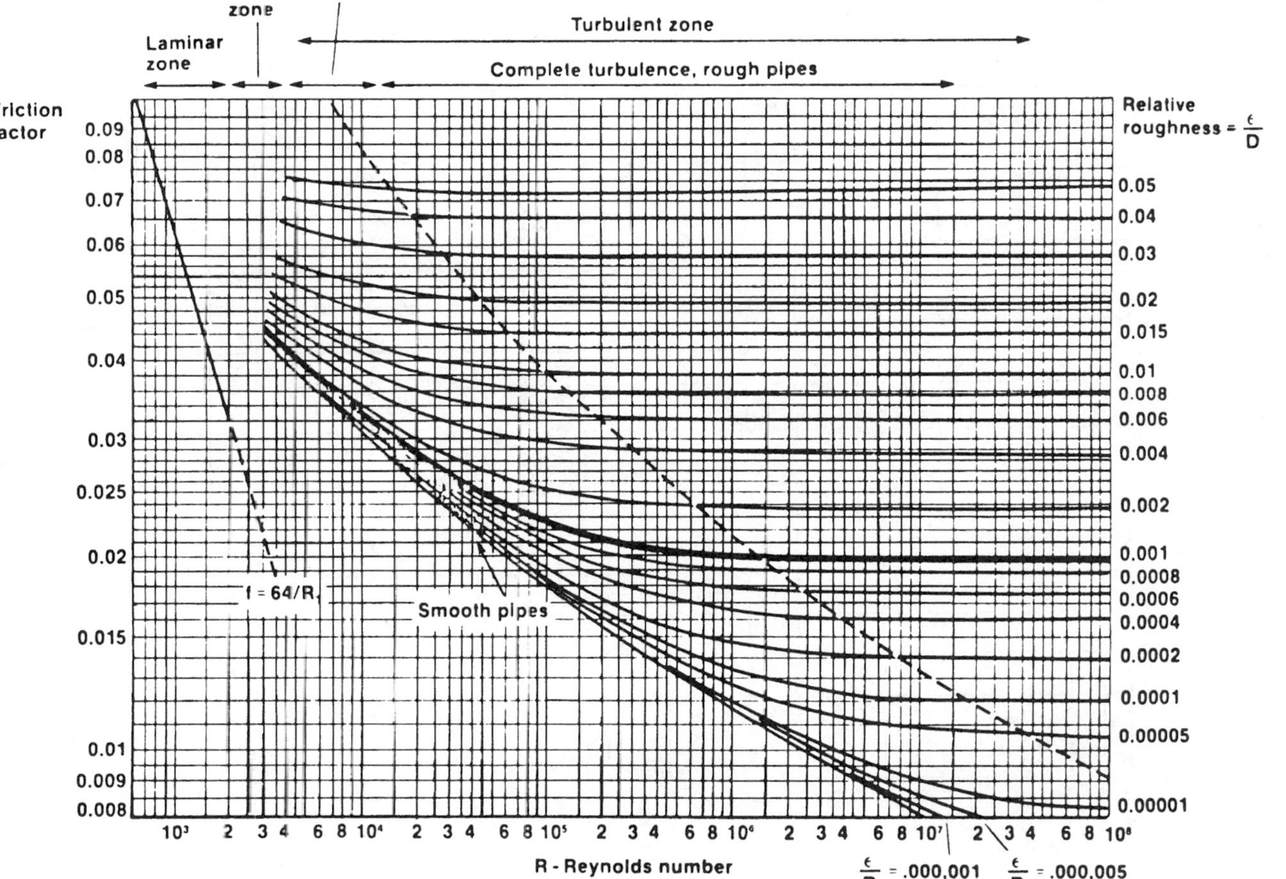

Fig. 11. Friction factor (f) as a function of Reynolds number and relative roughness (from L. F. Moody, "Friction Factors for Pipe Flow," *Trans. Am. Soc. Mech. Engrs.*, 671 (1944).

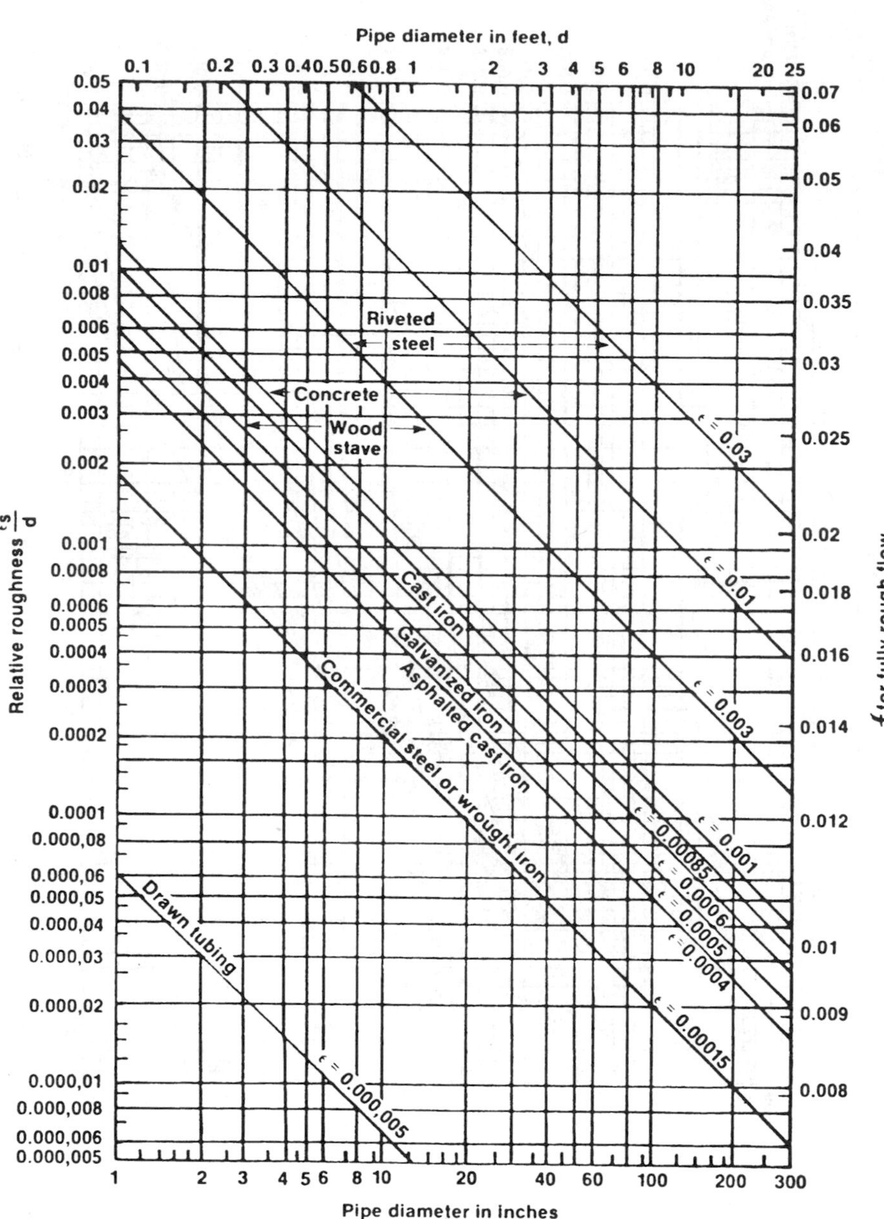

Fig. 12. Relative roughness *Es/d* as a function of hydraulic diameter of pipe (after Moody—see Fig. 11).

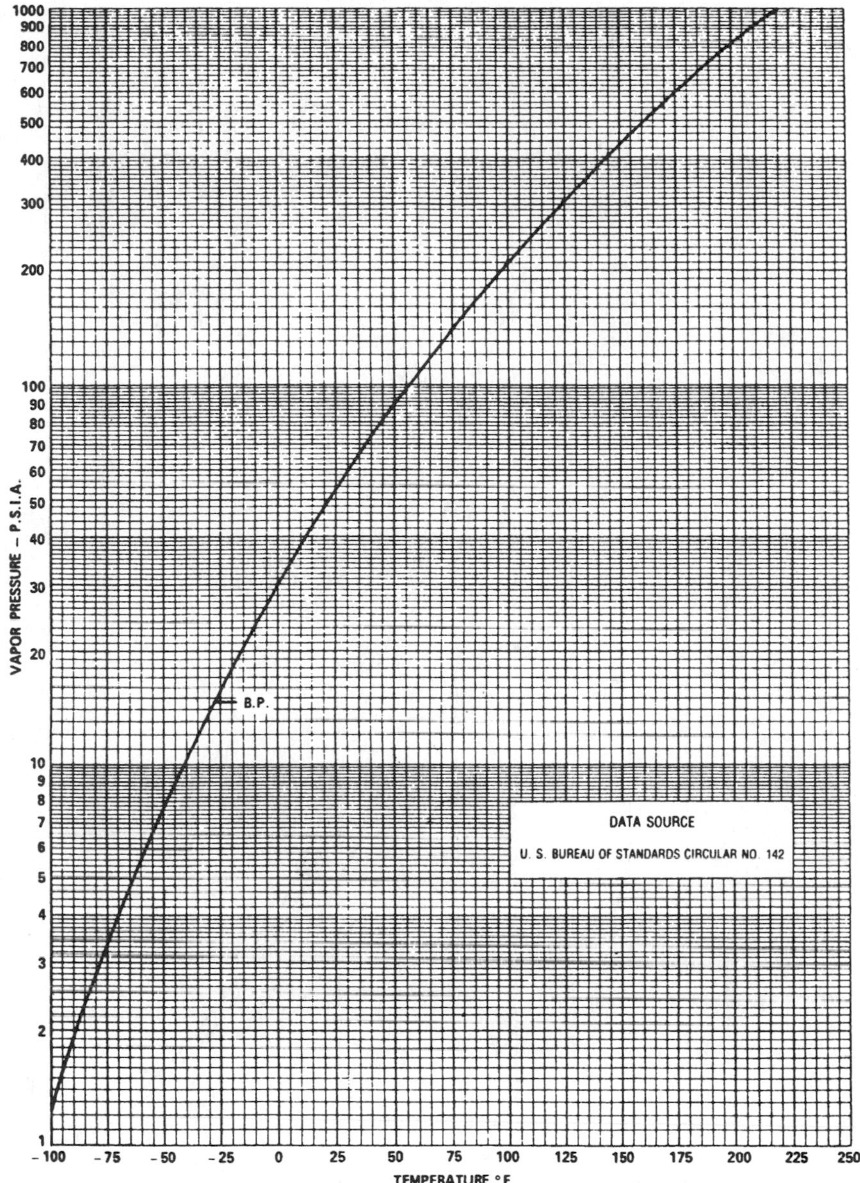

Fig. 13. Anhydrous ammonia vapor pressure versus temperature °F (U.S. Bureau of Standards Circular No. 142).

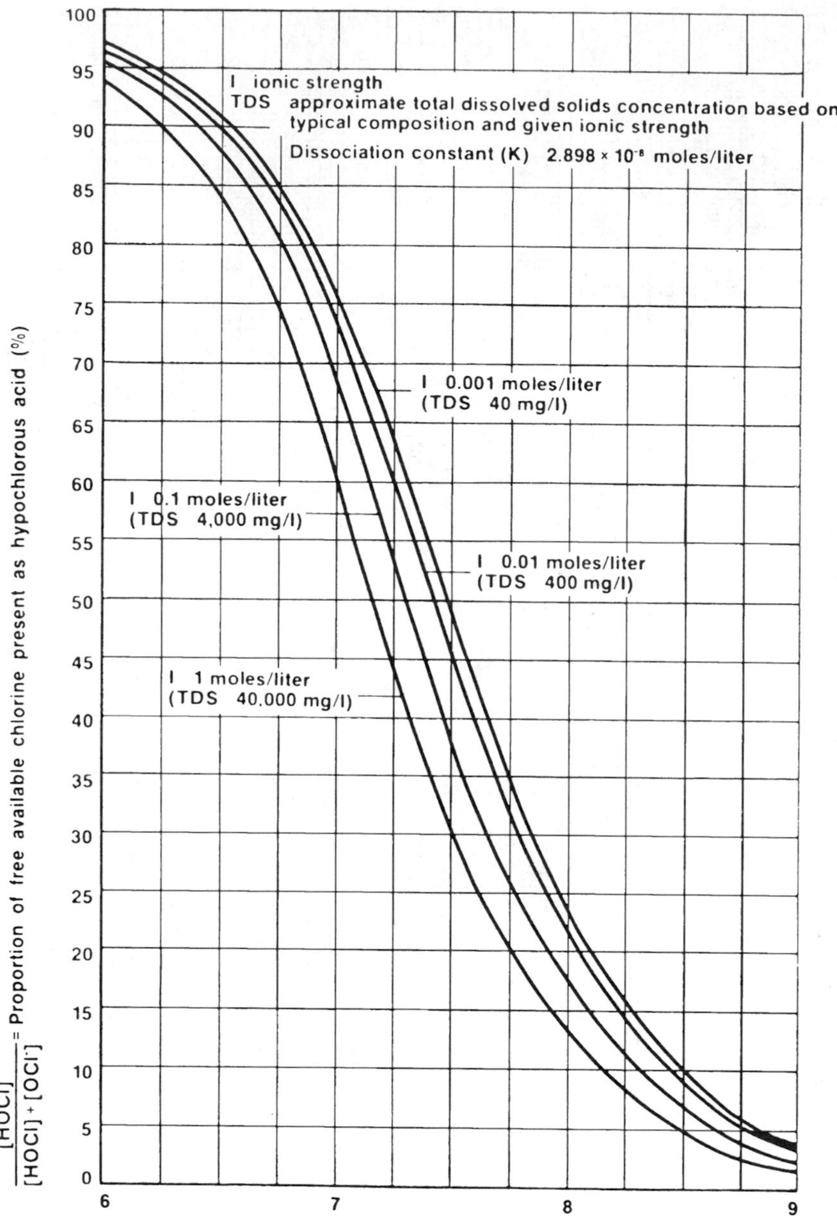

Fig. 14. Effect of pH and ionic strength on the dissociation of HOCl at 25°C.

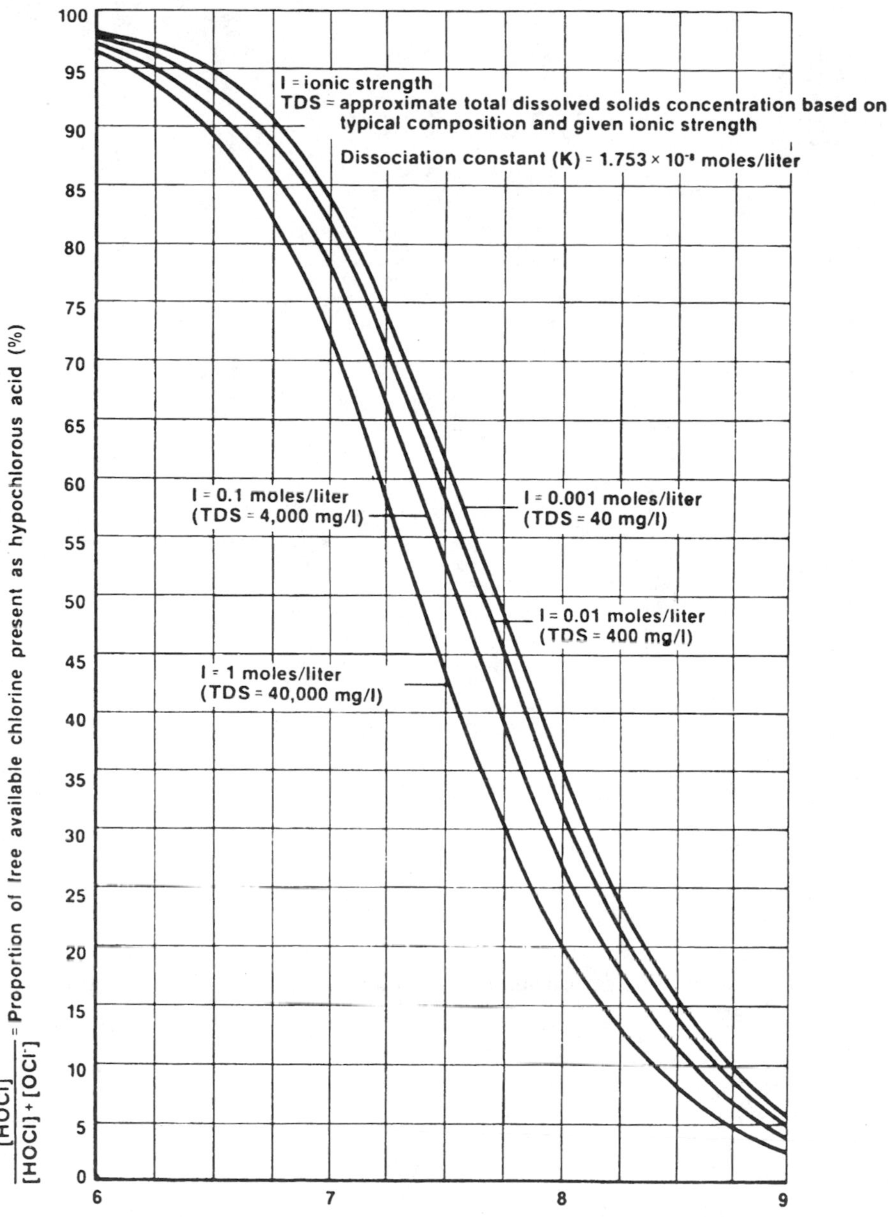

Fig. 15. Effect of pH and ionic strength on the dissociation of HOCl at 5°C.

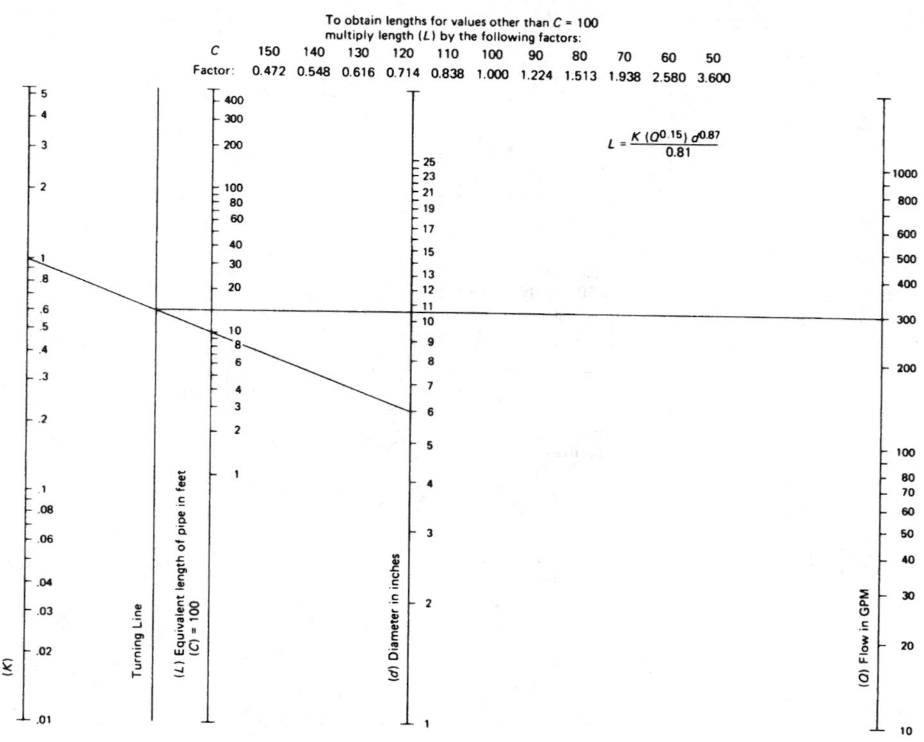

Fig. 16. Nomograph for calculating the equivalent length of pipe for fittings.

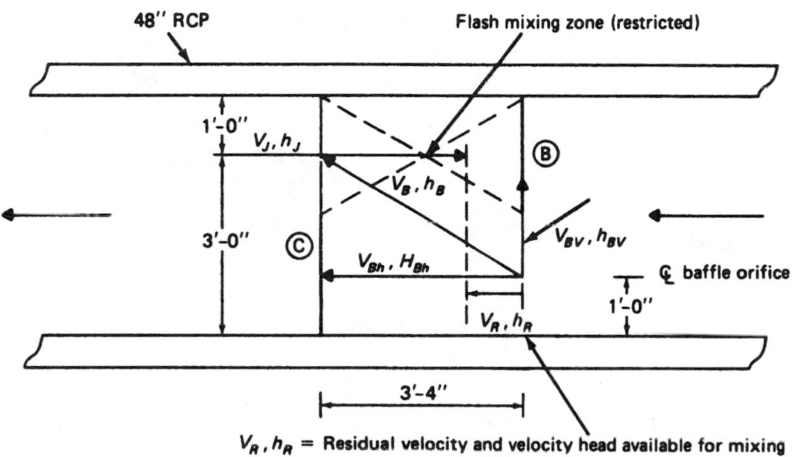

Fig. 17. Free body diagram of contraflow effect in flash mixing zone (Ref. 96, Chpt. 8).

Appendix II
SAN FRANCISCO WATER DEPT. ANNUAL WATER QUALITY REPORT 1991

PRIMARY STANDARDS—Mandatory Health-Related Standards				
			TREATED 5FWD SYSTEM WATER(1)	
PARAMETER	*UNIT*	*MAXIMUM CONTAMINANT LEVEL*	*RANGE*	*AVERAGE*
CLARITY				
Turbidity	NTU	1.0	0.1–1.0	0.3
MICROBIOLOGICAL	% Tests			
Total Coliform Bacteria	Positive	10	none	0.0
ORGANIC CHEMICALS				
Atrazine	mg/l	0.003	ND	ND
Bentazon	mg/l	0.018	ND	ND
Benzene	mg/l	0.001	ND	ND
Carbofuran	mg/l	0.018	W	
Carbon tetrachloride	mg/l	0.0005	ND	ND
Chlordane	mg/l	0.0001	W	
2,4-D	mg/l	0.1	ND	ND
1,2-Dibromo-3-chloropropane	mg/l	0.0002	ND	ND
1,4-Dichlorobenzene	mg/l	0.005	ND	ND
1,1-Dichloroethane	mg/l	0.005	ND	ND
1,2-Dichloroethane	mg/l	0.0005	ND	ND
1,1-Dichloroethylene	mg/l	0.006	ND	ND
Cis-1,2-Dichloroethylene	mg/l	0.006	ND	ND
Trans-1,2-Dichloroethylene	mg/l	0.01	ND	ND
1,2-Dichloropropane	mg/l	0.005	ND	ND
1,3-Dichloropropene	mg/l	0.0005	ND	ND
Di(2-ethylhexyl)phthalate	mg/l	0.004	W	
Endrin	mg/l	0.0002	ND	ND
Ethylbenzene	mg/l	0.680	ND	ND
Ethylene dibromide	mg/l	0.00002	ND	ND
Glyphosate	mg/l	0.7	W	
Heptachlor	mg/l	0.00001	W	
Heptachlor epoxide	mg/l	0.00001	W	
Lindane	mg/l	0.004	ND	ND
Methoxychlor	mg/l	0.1	ND	ND
Molinate	mg/l	0.02	ND	ND
Monochlorobenzene	mg/l	0.030	ND	ND
Simazine	mg/l	0.01	ND	ND
1,1,2,2-Tetrachloroethane	mg/l	0.001	ND	ND
Tetrachloroethylene	mg/l	0.005	ND	ND

PRIMARY STANDARDS—Mandatory Health-Related Standards				
			TREATED 5FWD SYSTEM WATER(1)	
PARAMETER	*UNIT*	*MAXIMUM CONTAMINANT LEVEL*	*RANGE*	*AVERAGE*
Thiobencarb	mg/l	0.07	ND	ND
Total trihalomethanes (TTHM)	mg/l	0.10	0.030–0.175	0.082
Toxaphene	mg/l	0.005	ND	ND
2,4,5-TP (Silvex)	mg/l	0.01	ND	ND
1,1,1-Trichloroethane	mg/l	0.200	ND	ND
1,1,2-Trichloroethane	mg/l	0.032	ND	ND
Trichloroethylene	mg/l	0.005	ND	ND
Trichlorofluoromethane (Freon II)	mg/l	0.15	ND	ND
1,1,2-Trichloro-1,2,2-trifluorethane (Freon 113)	mg/l	1.2	ND	ND
Vinyl chloride	mg/l	0.0005	ND	ND
Xylenes	mg/l	1.750	ND	ND
INORGANIC CHEMICALS				
Aluminum	mg/l	1.0	0.07–0.14	0.1
Arsenic	mg/l	0.05	<0.005	<0.005
Barium	mg/l	1.0	0.01–0.04	0.02
Cadmium	mg/l	0.01	<0.0005	<0.0005
Chromium	mg/l	0.05	<0.005	<0.005
Fluoride (2)	mg/l	1.4–2.4	0.7–1.1	0.9
Lead (3)	mg/l	0.05	<0.005	<0.005
Mercury	mg/l	0.002	<0.0002	<0.0002
Nitrate (as NO3)	mg/l	45	0.1–0.5	0.09
Selenium	mg/l	0.01	<0.005	<0.005
Silver	mg/l	0.05	<0.001	<0.001
RADIONUCLIDES (5)				
Gross alpha (including radium 226 and 228)	pCi/L	15	ND-0.7	0.1
Gross beta	pCi/L	50	ND-3.0	1.0
Strontium-90	pCi/L	8	ND	ND
Tritium	pCi/L	20.000	ND	ND
SECONDARY STANDARDS–Aesthetic Standards				
Chloride	mg/l	250	13–19	16
Color	units	15	0	0
Copper	mg/l	100	<0.01–0.1	0.01
Foaming agents (MBAS)	mg/l	0.5	ND	ND
Iron	mg/l	0.3	0.02–0.15	0.07
Maganese	mg/l	0.05	<0.01–0.02	0.01
Odor-threshold	units	3	<1	<1
Specific conductance	umho/cm	900	132–254	172
Sulfate	mg/l	250	7–22	12
Total dissolved solids (TDS)	mg/l	500	79–152	103
Zinc	mg/l	5.0	0.01–0.09	0.06
ADDITIONAL CONSTITUENTS ANALYZED				
Alkalinity (CaCO3)	mg/l	no standard	32–71	47
Asbestos (4.6)	MFL	no standard	ND	ND
Calcium	mg/l	no standard	10–34	18

PRIMARY STANDARDS—Mandatory Health-Related Standards

PARAMETER	*UNIT*	*MAXIMUM CONTAMINANT LEVEL*	*TREATED 5FWD SYSTEM WATER(1)*	
			RANGE	*AVERAGE*
Chlorine (Free)	mg/l	no standard	0.1–1.1	0.4
Hardness (CaCO3)	mg/l	no standard	36–90	54
pH	units	no standard	7.9–9.1	8.7
Potassium	mg/l	no standard	0.1–1.6	0.6
Sodium	mg/l	no standard	11–15	13

(1) System-wide data, average over time and location.
(2) Fluoride standard depends on temparature.
(3) The average lead in our source water is <0.001 mg/l, however, the amount of lead found at consumer's faucets varies according to the type and age of the plumbing materials used in their homes.
(4) Proposed Maximum Contaminant Level of 7.1 MFL exceeding 10 μm in Length (long fibres).
(5) Based on values measured during the calendar year 1988.
(6) Based on values measured during the calendar year 1987.

ND = Not detected
mg/l = miligrams per liter (part per million)
μm = micron
WD = Waiver from State DHS
pCi/L = pico Curies per liter
umhos/cm = micromho per centimeter
MFL = million fibers per liter
< = less than

If you have additional questions or concerns regarding the quality of your water, please call the Water Quality Division at 415-872-5942.

INDEX